S0-CBI-355

Organic Electronic Spectral Data
Volume XXV 1983

Organic Electronic Spectral Data

Volume XXV 1983

JOHN P. PHILLIPS, DALLAS BATES
HENRY FEUER & B.S. THYAGARAJAN
EDITORS

CONTRIBUTORS

Dallas Bates
H. Feuer
L.D. Freedman
C.M. Martini
F.C. Nachod
J.P. Phillips

AN INTERSCIENCE ® PUBLICATION
JOHN WILEY & SONS
New York • Chichester • Brisbane • Toronto • Singapore

An Interscience ® Publication

Library of Congress Catalog Card Number: 60-16428

ISBN 0-471-51505-1

Printed in the United States of America

10 9 8 7 6 5 4 3 2 1

INTRODUCTION TO THE SERIES

In 1956 a cooperative effort to abstract and publish in formula order all the ultraviolet-visible spectra of organic compounds presented in the journal literature was organized through the enterprise and leadership of M.J. Kamlet and H.E. Ungnade. Organic Electronic Spectral Data was incorporated in 1957 to create a formal structure for the venture, and coverage of the literature from 1946 onward was then carried out by chemists with special interests in spectrophotometry through a page by page search of the major chemical journals. After the first two volumes (covering the literature from 1946 through 1955) were produced, a regular schedule of one volume for each subsequent period of two years was introduced. In 1966 an annual schedule was inaugurated.

Altogether, more than fifty chemists have searched a group of journals totalling more than a hundred titles during the course of this sustained project. Additions and subtractions from both the lists of contributors and of journals have occurred from time to time, and it is estimated that the effort to cover all the literature containing spectra may not be more than 95% successful. However, the total collection is by far the largest ever assembled, amounting to well over a half million spectra in the twenty-five volumes so far.

Volume XXVI is in preparation.

PREFACE

Processing of the data provided by the contributors to Volume XXV as to the last several volumes was performed at the University of Louisville.

John P. Phillips
Dallas Bates
Henry Feuer
B.S. Thyagarajan

ORGANIZATION AND USE OF THE DATA

The data in this volume were abstracted from the journals listed in the reference section at the end. Although a few exceptions were made, the data generally had to satisfy the following requirements: the compound had to be pure enough for satisfactory elemental analysis and for a definite empirical formula; solvent and phase had to be given; and sufficient data to calculate molar absorptivities had to be available. Later it was decided to include spectra even if solvent was not mentioned. Experience has shown that the most proable single solvent in such circumstances is ethanol.

All entries in the compilation are organized according to the molecular formula index system used by Chemical Abstracts. Most of the compound names have been made to conform with the Chemical Abstracts system of nomenclature. For the benefit of those who prefer less rigorous names for common compounds the Merck Index has been generally used for those compounds with spectra in it.

Solvent or phase appears in the second column of the data lists, often abbreviated according to standard practice; there is a key to less obvious abbreviations on the next page. Anion and cation are used in this column if the spectra are run in relatively basic or acidic conditions respectively but exact specifications cannot be ascertained.

The numerical data in the third column present wavelength values in nanometers (millimicrons) for all maxima, shoulders and inflections, with the logarithms of the corresponding molar absorptivities in parentheses. Shoulders and inflections are marked with a letter s. In spectra with considerable fine structure in the bands a main maximum is listed and labelled with a letter f. Numerical values are given to the nearest nanometer for wavelength and nearest 0.01 unit of the logarithm of the molar absorptivity. Spectra that change with time or other common conditions are labelled "anom." or "changing", and temperatures are indicated if unusual.

The reference column contains the code number of the journal, the initial page number of the paper, and in the last two digits the year (1983). A letter is added for journals with more than one volume of section in a year. The complete list of all articles and authors thereof appears in the References at the end of the book.

Several journals that were abstracted for previous volumes in this series have been omitted, usually for lack of useful data, and several new ones have been added. Most Russian journals have been abstracted in the form of the English translation editions.

ABBREVIATIONS

s	shoulder or inflection
f	fine structure
n.s.g.	no solvent given in original reference
$C_6H_{11}Me$	methylcyclohexane
C_6H_{12}	cyclohexane
DMF	dimethylformamide
DMSO	dimethylsulfoxide
THF	tetrahydrofuran

Other solvent abbreviations generally follow the practice of Chemical Abstracts.

Underlined data were estimated from graphs.

JOURNALS ABSTRACTED

Journal	No.	Journal	No.
Acta Chem. Scand.	1	Tetrahedron Letters	88
Indian J. Chem.	2	Angew. Chem., Intl. Ed.	89
Anal. Chem.	3	Polyhedron	90
J. Heterocyclic Chem.	4	J. Applied Chem. U.S.S.R.	93
Ann. Chem. Liebigs	5	Chem. Pharm. Bull. Japan	94
Ann. chim. (Rome)	7	J. Pharm. Soc. Japan	95
Chemica Scripta	11	The Analyst	96
Australian J. Chem.	12	Z. Chemie	97
Steroids	13	J. Agr. Food Chem.	98
Bull. Chem. Soc. Japan	18	Theor. Exptl. Chem.	99
Bull. Polish Acad. Sci.	19	J. Natural Products	100
Bull. soc. chim. Belges	20	J. Organometallic Chem.	101
Bull. soc. chim. France	22	Phytochemistry	102
Can. J. Chem.	23	Khim. Geterosikl. Soedin.	103
Chem. Ber.	24	Zhur. Organ. Khim.	104
Chem. and Ind. (London)	25	Khim. Prirodn. Soedin.	105
Chimia	27	Die Pharmazie	106
Compt. rend.	28	Synthetic Comm.	107
Doklady Akad. Nauk S.S.S.R.	30	Israel J. Chem.	108
Experientia	31	Doklady Phys. Chem.	109
Gazz. chim. ital.	32	Russian J. Phys. Chem.	110
Helv. Chim. Acta	33	European J. Med. Chem.	111
J. Chem. Eng. Data	34	Spectroscopy Letters	112
J. Am. Chem. Soc.	35	Macromolecules	116
J. Pharm. Sci.	36	Org. Preps. and Procedures	117
J. Chem. Soc., Perkin Trans. II	39B	Synthesis	118
J. Chem. Soc., Perkin Trans. I	39C	S. African J. Chem.	119
Nippon Kagaku Kaishi	40	Pakistan J. Sci. Ind. Research	120
J. Chim. Phys.	41	J. Macromol. Sci.	121
J. Indian Chem. Soc.	42	Ukrain. Khim. Zhur.	124
J. Org. Chem.	44	Inorg. Chem.	125
J. Phys. Chem.	46	Makromol. Chem.	126
J. Polymer Sci., Polymer Chem. Ed.	47	Croatica Chem. Acta	128
J. prakt. Chem.	48	Bioorg. Chem.	130
Monatsh. Chem.	49	J. Mol. Structure	131
Rec. trav. chim.	54	J. Appl. Spectroscopy S.S.S.R.	135
Polish J. Chem.	56	Carbohydrate Research	136
Spectrochim. Acta	59	Finnish Chem. Letters	137
J. Chem. Soc., Faraday Trans. I	60	Chemistry Letters	138
Ber. Bunsen Gesell. Phys. Chem.	61	P and S and Related Elements	139
Z. phys. Chem.	62	J. Anal. Chem. S.S.S.R.	140
Z. Naturforsch.	64	Heterocycles	142
Zhur. Obshchei Khim.	65	Arzneimittel. Forsch.	145
Biochemistry	69	Photochem. Photobiol.	149
Izvest. Akad. Nauk S.S.S.R.	70	J. Chem. Research	150
Coll. Czech. Chem. Comm.	73	J. Photochem.	151
Mikrochim. Acta	74	Nouveau J. Chim.	152
J. Chem. Soc., Chem. Comm.	77	Photobiochem. Photobiophys.	156
Tetrahedron	78	Organometallics	157
Revue Roumaine Chim.	80	J. Antibiotics	158
Arch. Pharm.	83	J. Carbohydrate Chem.	159
Talanta	86	Anal. Letters	160
J. Med. Chem.	87	Il Farmaco	161
		Merck Index, Tenth Edition	162

Organic Electronic Spectral Data
Volume XXV 1983

Compound	Solvent	$\lambda_{max}(\log \epsilon)$	Ref.
CBr_4Hg_4			
Mercury, tetrabromo-μ_4-methanetetrayl-tetra-	DMSO	257(1.26)	101-0217-83P
$CClF_2NO_2$			
Methane, chlorodifluoronitro-	gas	195(--),279(--)	151-0009-83A
CCl_2FNO_2			
Methane, dichlorofluoronitro-	gas	198(--),276(--)	151-0009-83A
CCl_3NO_2			
Methane, trichloronitro-	gas	202(--),272(--)	151-0009-83A
CCl_4Hg_4			
Mercury, tetrachloro-μ_4-methanetetrayl-tetra-	DMSO	252(1.32)	101-0217-83P
CF_2Se			
Carbonoselenoic difluoride	gas	236(--),434f(--)	23-1743-83
CF_3NO_2			
Methane, trifluoronitro- (cross section given)	gas	277($5x10^{-19}cm^2$)	151-0009-83A
$CHBr_3Hg_3$			
Mercury, tribromo-μ_3-methylidynetri-	DMSO	251(1.28)	101-0217-83P
$CHCl_3Hg_3$			
Mercury, trichloro-μ_3-methylidynetri-	DMSO	249(1.15)	101-0217-83P
$CHCl_3O_2S$			
Methanesulfinic acid, trichloro-, tetraethylammonium salt	H_2O	238(3.35)	70-1016-83
	MeCN	248(--),273(--)	70-1016-83
$CHHg_3I_3$			
Mercury, triiodo-μ_3-methylidynetri-	DMSO	256(1.64)	101-0217-83P
$CH_2Br_2Hg_2$			
Mercury, dibromo-μ-methylenedi-	MeOH	211s(1.13)	101-0217-83P
$CH_2Cl_2Hg_2$			
Mercury, dichloro-μ-methylenedi-	MeOH	216(1.20)	101-0217-83P
$CH_2Hg_2I_2$			
Mercury, diiodo-μ-methylenedi-	MeOH	c.199(--),227s(1.26), 241(1.30)	101-0217-83P
	DMSO	252(1.42)	101-0217-83P
CH_2O			
Formaldehyde	EtOH	270(0.78)	110-0378-83
CH_3BrHg			
Mercury, bromomethyl-	MeOH	200(c.0.85)	101-0217-83P
CH_3ClHg			
Mercury, chloromethyl-	MeOH	c.198(c.0.53)	101-0217-83P
CH_3HgI			
Mercury, iodomethyl-	MeOH	c.199(c.1.00),227(0.40)	101-0217-83P

Compound	Solvent	$\lambda_{max}(\log \epsilon)$	Ref.
CH_3N_5O			
5H-Tetrazol-5-one, 1-amino-1,2-dihydro-	MeOH	220(3.31)	35-0902-83
CHg_4I_4			
Mercury, tetraiodo-μ4-methanetetrayl-	DMSO	255(1.72),280s(1.40)	101-0217-83P
$C_2H_2F_3NO_4S$			
Methane, trifluoro[(nitromethyl)sulfonyl]-, cesium salt	H_2O	295(4.11)	70-1016-83
C_2H_3			
Vinyl radical	gas	423.2f(1.48)	23-0993-83
C_2H_3HgN			
Mercury, (cyano-C)methyl-	MeOH	200(0.32)	101-0217-83P
C_2H_3HgNS			
Mercury, methyl(thiocyanato-S)-	MeOH	204(1.32)	101-0217-83P
$C_2H_4N_4$			
1,2,4,5-Tetrazine, 1,4-dihydro-	dioxan	298(1.78)	24-2261-83
$(C_2H_4O)_n$			
Polyvinyl alcohol, reacted with 2,6-dichlorobenzaldehyde	dioxan	274(2.73),281(2.71)	116-0291-83
C_2H_4OS			
Ethanethioic acid, cesium salt	EtOH	249(4.02)	64-1585-83B
Ethanethioic acid, potassium salt	EtOH	249(3.90)	64-1585-83B
Ethanethioic acid, rubidium salt	EtOH	250(3.69)	64-1585-83B
Ethanethioic acid, sodium salt	EtOH	249(3.47)	64-1585-83B
C_2H_4S			
Thiirane	C_6H_{12}	243(1.5),259(1.5)	46-4585-83
$C_2H_6N_2$			
Ethanimidamide	n.s.g.	224(3.60)	162-0006-83
C_2H_6S			
Methane, thiobis-, compd. with iodine	gas	285(4.13),457(2.85)	23-1933-83
	heptane	300(4.45),441(3.26)	23-1933-83

Compound	Solvent	λ_{max}(log ε)	Ref.
$C_3H_2FN_3O_2$			
1,2,4-Triazine-3,5(2H,4H)-dione, 6-fluoro-	0.05M HCl	262.5(3.62)	73-2676-83
	0.05M NaOH	289.1(3.59)	73-2676-83
$C_3H_2Hg_2N_2$			
Mercury, bis(cyano-C)-μ-methylenedi-	MeOH	201(1.08)	101-0217-83P
$C_3H_2Hg_2N_2S_2$			
Mercury, μ-methylenebis(thiocyanato-S)di-	MeOH	c.202(c.1.11),230(1.48)	101-0218-83P
$C_3H_2N_4O_4$			
1,2,4-Triazine-3,5(2H,4H)-dione, 6-nitro-	0.05M HCl	281(3.66)	73-2676-83
	0.05M NaOH	263(3.21),356(3.75)	73-2676-83
$C_3H_2O_2S$			
1,3-Oxathiol-2-one	MeOH	215(3.61),225(3.61)	152-0269-83
$C_3H_2O_3$			
1,3-Dioxol-2-one	gas	200s(3.0),230s(2.3)	151-0001-83A
C_3H_3NOSSe			
2-Selenazolidinone, 4-thioxo-	n.s.g.	260(4.18),315(4.26), 440(--)	103-0273-83
$C_3H_3N_3O_2$			
1H-Imidazole, 2-nitro-	pH 13	374(4.11)	162-0132-83
$C_3H_3N_3O_2S$			
1H-Imidazole-4-thiol, 5-nitro-	H_2O	217(4.47),372(4.14)	103-0650-83
$C_3H_3N_3O_3$			
Furazancarboxylic acid, amino-	MeOH	209(3.42),287(3.33)	142-2351-83
$C_3H_3N_3O_5S$			
1H-Imidazole-4-sulfonic acid, 5-nitro-, monosodium salt	H_2O	220(3.79),293(3.93)	103-0650-83
$C_3H_4N_2O_3S$			
1H-Imidazole, sulfur trioxide adduct	H_2O	207(3.58)	104-1935-83
1H-Imidazole-4-sulfonic acid, barium salt(2:1)	H_2O	204(4.14)	104-1935-83
$C_3H_4N_2S_2$			
1,2,3-Thiadiazolium, 4-mercapto-3-methyl-, hydroxide, inner salt	EtOH	256(3.48),404(3.98)	94-1746-83
1,2,3-Thiadiazolium, 5-mercapto-3-methyl-, hydroxide, inner salt	EtOH	266(3.82),414(3.85)	94-1746-83
$C_3H_4N_4OS$			
1,2,4-Triazin-3(2H)-one, 6-amino-4,5-dihydro-5-thioxo-	pH 1	243.5(3.83),303(3.92), 352(3.82)	44-1271-83
	pH 11	245.5(3.83),329(4.03)	44-1271-83
	MeOH	245(3.82),304(3.89), 357(3.79)	44-1271-83
$C_3H_4N_4S_2$			
1,2,4-Triazine-3,5(2H,4H)-dione, 6-amino-	pH 1	255s(3.87),299.5(4.36), 329s(4.10)	44-1271-83
	pH 11	281(4.18),325(4.00)	44-1271-83

Compound	Solvent	λ_{max}(log ε)	Ref.
1,2,4-Triazine-3,5(2H,4H)-dithione, 6-amino- (cont.)	MeOH	255s(3.83),300.5(4.35), 333.5s(4.09)	44-1271-83
$C_3H_4N_6O$			
1,3,4-Oxadiazole-2-methanamine, 5-azido-	MeOH	238(4.01)	35-0902-83
C_3H_4O			
Acrolein (changing)	H_2O	210(4.1),315(1.7)	112-0601-83
$C_3H_4O_3$			
Ethylene carbonate	gas	190(1.8)(end abs.)	151-0001-83A
2-Propenal, 2,3-dihydroxy-	pH 3.8	268(4.21)	18-0467-83
	pH 8	293(4.39)	18-0467-83
	pH 13.5	323(4.37)	18-0467-83
C_3H_5BrO			
2-Propanone, 1-bromo-	hexane	300(2.05)	39-1053-83B
C_3H_5ClO			
2-Propanone, 1-chloro-	hexane	291(1.56)	39-1053-83B
C_3H_5FO			
2-Propanone, 1-fluoro-	hexane	280(1.10)	39-1053-83B
C_3H_5IO			
2-Propanone, 1-iodo-	hexane	306s(2.50)	39-1053-83B
C_3H_5NO			
2-Propenal, 3-amino-	MeOH	272(4.50)	131-0001-83K
(E-syn-E)-	C_6H_{12}	252(--)	131-0001-83K
	dioxan	261(4.38)	131-0001-83K
(Z-syn-Z)-	C_6H_{12}	287(4.06)	131-0001-83K
	dioxan	295(--)	131-0001-83K
$C_3H_5N_3OS$			
Ethylenethiourea, nitroso-	H_2O	220(4.00),261(4.00), 310s(3.53),406.5(1.95), 425(1.82)	94-3678-83
	CH_2Cl_2	250(4.09),308(3.48), 414(1.85),432(1.78)	94-3678-83
$C_3H_5N_5O$			
1,2,4-Triazin-3(2H)-one, 5,6-diamino-	pH 1	240s(3.83),315(3.49)	44-1271-83
	pH 11	227.5(4.01),247s(3.79), 301(3.48)	44-1271-83
	MeOH	225.5(4.11),249.5s(3.78), 307.5(3.45)	44-1271-83
$C_3H_6N_2O_2$			
Ethenamine, N-methyl-N-nitro-	H_2O	272(3.96)	104-1270-83
Ethenamine, N-methyl-2-nitro-	H_2O	236(3.51),344(4.28)	104-1270-83
$C_3H_6N_2S$			
2-Imidazolidinethione	H_2O	233.0(4.18)	94-3678-83
C_3H_6O			
Oxirane, methyl-, (S)-(-)-	gas	157(3.8),160.0(3.8), 174.0(3.3)	35-1738-83

Compound	Solvent	λ_{max}(log ϵ)	Ref.
C_3H_6S			
Thietane	C_6H_{12}	215(3.33),275(1.52)	101-0C73-83L
C_3H_7Cl			
Propane, 1-chloro-	gas	184.9(1.72)	151-0103-83C
C_3H_7I			
Propane, 1-iodo-	C_6H_{12}	256(2.70)	44-1732-83
$C_3H_8N_2O_3$			
1-Propanol, 2-(hydroxynitrosoamino)-,	pH 2	224(4.14)	158-0916-83
calcium salt (nitrosofungin)	pH 12	245(4.25)	158-0916-83
C_3H_8S			
Ethane, (methylthio)-, compd. with	gas	289(4.16),450(2.92)	23-1933-83
iodine	heptane	301(4.46),438(3.28)	23-1933-83

Compound	Solvent	λ_{max}(log ε)	Ref.
$C_4F_{12}N_4S_4$			
1,3,5,7,2,4,6,8-Tetrathiatetrazocine, 2,4,6,8-tetrakis(trifluoromethyl)-	gas	190(3.30),220(3.14), 332(2.46)	24-1257-83
$C_4HCl_2NO_2$			
1H-Pyrrole-2,5-dione, 3,4-dichloro-	C_6H_{12}	268(3.9),330s(1.4)	18-1362-83
	EtOH	236(4.1),287(2.8)	24-2591-83
$C_4HHg_3N_3$			
Mercury, tris(cyano-C)-μ_3-methylidyne-tri-	MeOH	202(1.40)	101-0217-83P
$C_4HHgN_3S_3$			
Mercury, μ_3-methylidynetris(thiocyanato-S)tri-	MeOH	202(1.00),234(0.90)	101-0217-83P
	DMSO	253(1.43)	101-0217-83P
$C_4H_2N_4O_2$			
1,2,4-Triazine-6-carbonitrile, 2,3,4,5-tetrahydro-3,5-dioxo-	0.05M HCl	204(3.91),280(4.20)	73-2676-83
	0.05M NaOH	242(3.99),313(4.02)	73-2676-83
$C_4H_2N_6$			
1H-Imidazole-4-carbonitrile, 5-azido-	n.s.g.	213(3.97),258(4.05)	103-1231-83
$C_4H_3ClN_2O_2$			
1H-Pyrrole-2,5-dione, 3-amino-4-chloro-	EtOH	230(4.1),355(3.3)	24-2591-83
$C_4H_3FN_2O_2$			
2,4(1H,3H)-Pyrimidinedione, 5-fluoro-	pH 1	266(3.85)	162-0599-83
$C_4H_4ClNO_4S$			
Thiophene, 3-chloro-2,5-dihydro-4-nitro-, 1,1-dioxide	n.s.g.	258(3.74)	104-1568-83
$C_4H_4FN_3O$			
Cytosine, 5-fluoro-	pH 1	285(3.95)	162-0590-83
$C_4H_4N_2$			
Pyrimidine	EtOH	245(3.52),460(2.66)	35-0040-83
$C_4H_4N_4O_4S$			
4H-[1,2,5]Oxadiazolo[3,4-c][1,2,6]-thiadiazin-7(6H)-one, 4-methyl-, 5,5-dioxide, sodium salt	MeOH	210(3.45),262(3.42)	142-2351-83
$C_4H_4O_3$			
Succinic anhydride	gas	220(1.40)	23-2790-83
$C_4H_5NO_2$			
2H-Pyrrol-2-one, 1,5-dihydro-5-hydroxy-	$CHCl_3$	237(3.01)	102-1278-83
Succinimide	MeCN	224(2.00),230(1.97), 245(1.86)	23-1890-83
silver salt	MeCN-base	223(3.65)	23-1890-83
silver salt after reduction	MeCN-base	227(4.04)	23-1890-83
$C_4H_5NO_2S$			
Carbo(isothiocyanatidic) acid, ethyl ester	dioxan	257(3.25)	24-1297-83
C_4H_5NTe			
Isotellurazole, 3-methyl-	MeCN	218(3.73),235s(3.61),	118-0824-83

Compound	Solvent	$\lambda_{max}(\log \epsilon)$	Ref.
Isotellurazole, 3-methyl- (cont.)		306(3.72)	118-0824-83
$C_4H_5N_2$			
Pyrazinium radical	acid	565(1.45)	125-0655-83
$C_4H_5N_3O_2$			
1H-Imidazole-4-carboxamide, 5-hydroxy-	H_2O	236(3.75),275(4.13)	4-0875-83
	M HCl	241(4.00),277s(--)	4-0875-83
	M NaOH	232(3.57),277(4.12)	4-0875-83
1H-Pyrrole-2,5-dione, 3,4-diamino-	H_2O	223(4.1),405(3.2)	24-2591-83
	dioxan	223(4.15),400(3.42)	24-2591-83
1,2,4-Triazine-3,5(2H,4H)-dione, 6-methyl-	pH 1	261(3.91)	162-0130-83
	pH 13	246(3.68)	162-0130-83
$C_4H_5N_3O_3$			
Furazancarboxylic acid, (methylamino)-, monosodium salt	MeOH	214(3.42),303(3.18)	142-2351-83
1H-Imidazole, 4-methoxy-5-nitro-	H_2O	225(3.75),387(4.11)	103-0650-83
$C_4H_5N_5$			
3H-Pyrazolo[1,5-d]tetrazole, 3-methyl-	CH_2Cl_2	283(3.49)	4-1629-83
$C_4H_5N_5O_3S$			
3H-[1,2,5]Oxadiazolo[3,4-c][1,2,6]-thiadiazin-7-amine, N-methyl-, 5,5-dioxide	H_2O	210(4.03),240s(3.74), 260s(3.66),272s(3.58), 334(3.40)	142-2351-83
$C_4H_6N_2S_2$			
1,2,3-Thiadiazolium, 4-mercapto-3,5-dimethyl-, hydroxide, inner salt	EtOH	254(4.27),418(3.46)	94-1746-83
$C_4H_6N_4OS$			
1,2,4-Triazin-3(2H)-one, 6-amino-5-(methylthio)-	pH 1	235(4.02),287(3.85), 348.5s(3.49)	44-1271-83
	pH 11	237(4.01),272.5(3.64), 349(3.58)	44-1271-83
	MeOH	237(4.07),284(3.86), 350(3.46)	44-1271-83
$C_4H_6N_4O_2$			
1H-Imidazol-4-amine, N-methyl-5-nitro-	H_2O	209(4.06),382(4.18)	103-0650-83
$C_4H_6O_2S$			
2-Propenoic acid, 3-(methylthio)-	EtOH	284(4.05)	102-1619-83
$C_4H_7ClN_2$			
1,6-Diazabicyclo[3.1.0]hexane, 6-chloro-	heptane	219(3.10)	70-1754-83
$C_4H_7ClN_2O_2$			
Ethenamine, N-(2-chloroethyl)-N-nitro-	H_2O	273(3.87)	104-1270-83
$C_4H_7N_2O_3P$			
Phosphonic acid, (1H-imidazol-1-ylmethyl)-	EtOH	204(1.46),243(1.46), 279(0.9)	65-1555-83
$C_4H_7N_3O$			
4(1H)-Pyrimidinone, 2-amino-5,6-dihydro-	pH 12	210(3.84),238(3.85)	142-1769-83
	MeOH	225(3.92),251(3.71)	142-1769-83

Compound	Solvent	$\lambda_{max}(\log \epsilon)$	Ref.
$C_4H_7N_3O_3$			
Acetic acid, [[imino(methylamino)methyl]amino]oxo-	EtOH	end absorption	142-1067-83
$C_4H_7N_5O$			
1H-Imidazole-4-carboxamide, 5-hydrazino-, monohydrochloride	n.s.g.	268(4.00)	103-1235-83
$C_4H_7N_5S$			
1,2,4-Triazine-5,6-diamine, 3-(methylthio)-	pH 1	257.5(4.35),320s(3.45)	44-1271-83
	pH 11	250(4.15),320(3.68)	44-1271-83
	MeOH	250(4.17),324(3.70)	44-1271-83
$C_4H_8BrNO_2S$			
1,2-Thiazetidine, 2-bromo-3,3-dimethyl-, 1,1-dioxide	CH_2Cl_2	296(2.38)	44-0537-83
$C_4H_8N_2O_2$			
Cyclopropanamine, N-methyl-N-nitro-	H_2O	244(3.81)	104-1463-83
$C_4H_8N_2O_3$			
Ethenamine, N-(methoxymethyl)-N-nitro-	H_2O	273(3.74)	104-1270-83
$C_4H_8N_4$			
1,2,4,5-Tetrazine, 1,4-dihydro-1,4-dimethyl-	dioxan	237(3.82),315(1.84)	24-2261-83
1,2,4,5-Tetrazine, 1,4-dihydro-3,6-dimethyl-	EtOH	300(2.03)	24-2261-83
$C_4H_8N_4O_2$			
2-Butenediamine, 2,3-diamino-, (E)-	H_2O	310(3.83)	24-2591-83
$C_4H_8N_4S_2$			
1,2,4,5-Tetrazine, 1,4-dihydro-3,6-bis(methylthio)-	dioxan	220(4.20),300s(2.26)	24-2261-83
C_4H_8O			
Oxirane, 2,3-dimethyl-, (-)-(S,S)-	gas	147(4.2)	35-1738-83
C_4H_8OS			
2-Propanone, 1-(methylthio)-	hexane	302(2.33)	39-1053-83B
$C_4H_8O_2$			
2-Propanone, 1-methoxy-	hexane	283(1.23)	39-1053-83B
$C_4H_8O_2S$			
Sulfoxonium, dimethyl-, 2-oxoethylide	hexane	248(--)	99-0023-83
	EtOH	250(3.51)	99-0023-83
C_4H_8S			
Thiirane, 2,3-dimethyl-, (+)-(R,R)-	gas	159(4.0),172(3.9), 198f(3.5),210(3.6), 250(--)	46-4585-83
$C_4H_8S_2$			
Ethane(dithioic) acid, ethyl ester	H_2O	208(1.2),308(1.8)	23-1440-83
C_4H_8Se			
Ethene, (ethylseleno)-	heptane	218s(3.53),242(3.89), 271s(3.28)	104-1562-83
	EtOH	217s(3.53),240(3.85),	104-1562-83

Compound	Solvent	$\lambda_{max}(\log \epsilon)$	Ref.
Ethene, (ethylseleno)- (cont.)		270s(3.30)	104-1562-83
$C_4H_{10}GeS$			
1,3-Thiagermetane, 3,3-dimethyl-	C_6H_{12}	213(3.60),260(2.32)	101-0C73-83L
$C_4H_{10}S$			
Ethane, 1,1'-thiobis-, compd. with iodine	gas	290(4.23),452(3.02)	23-1933-83
	heptane	303(4.42),437(3.29)	23-1933-83
$C_4H_{10}Se$			
Ethane, 1,1'-selenobis-	MeOH	250(1.70)	101-0031-83M
$C_4H_{10}Te$			
Ethane, 1,1'-tellurobis-	MeOH	235(3.84)	101-0031-83M
$C_4H_{11}ClSi$			
Silane, (chloromethyl)trimethyl-	isooctane	203(2.45),224(0.56), 262(0.51)	35-5665-83
$C_4H_{12}Bi_2$			
Dibismuthine, tetramethyl-	pentane	220(4.57),264(3.86)	157-1859-83
$C_4H_{12}NO_2$			
Tetramethylammonium superoxide dimer	MeCN	257(4.00)(changing)	125-2577-83
	MeCN	289(4.06)	125-2577-83
C_4OS_5			
[1,2]Dithiolo[4,3-c]-1,2-dithiol-3(6H)-one, 6-thioxo-	MeCN	233(4.19),250(3.87), 289(3.82),347(3.24), 457s(3.88),476(3.91)	88-3577-83
$C_4O_2S_4$			
[1,2]Dithiolo[4,3-c]-1,2-dithiole-3,6-dione	MeCN	267(3.62),399(3.73), 414(3.74)	88-3577-83
C_4S_6			
[1,2]Dithiolo[4,3-c]-1,2-dithiole-3,6-dithione	MeCN	240(4.26),257(4.27), 308(3.84),344s(3.53), 495s(3.97),524(4.05)	88-3577-83

Compound	Solvent	$\lambda_{max}(\log \epsilon)$	Ref.
$C_5Cl_3F_3N_2$			
Pyrimidine, 2,4,5-trichloro-6-(trifluoromethyl)-	MeCN	226(4.04),281(3.65)	4-0219-83
$C_5Cl_4F_3N_2O_2P$			
Phosphorodichloridic acid, 4,5-dichloro-6-(trifluoromethyl)-2-pyrimidinyl ester	MeCN	222(4.00),279(3.71)	4-0219-83
$C_5HCl_3F_3N_2O_2P$			
Phosphorodichloridic acid, 4-chloro-6-(trifluoromethyl)-2-pyrimidinyl ester	MeCN	262(3.74)	4-0219-83
$C_5H_2ClF_3N_2O_2$			
2,4(1H,3H)-Pyrimidinedione, 5-chloro-6-(trifluoromethyl)-	MeOH	278(3.83),329(2.69)	4-0219-83
$C_5H_3ClF_3N_3$			
2-Pyrimidinamine, 4-chloro-6-(trifluoromethyl)-	MeOH	234(4.20),306(3.57)	4-0219-83
4-Pyrimidinamine, 2-chloro-6-(trifluoromethyl)-	MeOH	270(4.10),285(3.63)	4-0219-83
$C_5H_3ClF_3N_3O$			
2(1H)-Pyrimidinone, 4-amino-5-chloro-6-(trifluoromethyl)-, monohydrochloride	MeOH	240(4.04),323(3.54)	4-0219-83
$C_5H_3ClN_4$			
1H-Purine, 6-chloro-	pH 7.2	265(3.96)	162-0304-83
	pH 13	274(3.94)	162-0304-83
$C_5H_3ClN_4O$			
4-Pyrimidinecarbonitrile, 2-amino-6-chloro-1,4-dihydro-4-oxo-	EtOH	234(4.00),300(4.13)	4-0041-83
$C_5H_3Cl_2N_3O$			
5-Pyrimidinecarboxaldehyde, 2-amino-4,6-dichloro-	EtOH	239(3.97),283(4.13)	4-0041-83
$C_5H_3F_3N_2OS$			
4(1H)-Pyrimidinone, 2,3-dihydro-2-thioxo-6-(trifluoromethyl)-	MeOH	212.5(4.45),270(4.51)	4-0219-83
$C_5H_3F_3N_2O_2$			
2,4(1H,3H)-Pyrimidinedione, 6-(trifluoromethyl)-	MeOH	260(3.88),307(2.59)	4-0219-83
$C_5H_3N_3O_2$			
2-Furancarboxaldehyde, 5-azido-	C_6H_{12}	311(4.20)	12-0963-83
$C_5H_4BrNO_3$			
Furan, 2-(bromomethyl)-5-nitro-	MeOH	203(4.04),304(4.05)	73-2682-83
$C_5H_4Br_2N_2O$			
2(1H)-Pyrazinone, 3,5-dibromo-1-methyl-	EtOH	345(3.71)	4-0919-83
$C_5H_4ClN_3O_2$			
5-Pyrimidinecarboxaldehyde, 2-amino-	EtOH	244(3.90),313(4.27)	4-0041-83

Compound	Solvent	λ_{max}(log ε)	Ref.
4-chloro-1,6-dihydro-6-oxo- (cont.)			4-0041-83
$C_5H_4Cl_2N_2O$			
2(1H)-Pyrazinone, 3,5-dichloro-1-methyl-	EtOH	344(3.80)	4-0919-83
$C_5H_4Cl_2N_4O$			
5-Pyrimidinecarboxaldehyde, 2-amino-4,6-dichloro-, oxime	EtOH	268(4.23),306s(3.69)	4-0041-83
$C_5H_4F_3N_3O$			
2(1H)-Pyrimidinone, 4-amino-6-(trifluoromethyl)-, monohydrochloride	MeOH	230(4.04),307.5(3.62)	4-0219-83
4(1H)-Pyrimidinone, 2-amino-6-(trifluoromethyl)-	MeOH	220(3.92),295(3.95)	4-0219-83
$C_5H_4N_4O$			
1H-Pyrazolo[3,4-d]pyrimidin-4-ol (allopurinol)	pH 1	250(3.88)	162-0043-83
	pH 13	257(3.86)	162-0043-83
	MeOH	252(3.86)	162-0043-83
$C_5H_4N_4O_2$			
7H-Pyrazolo[4,3-d]pyrimidin-7-one, 2,6-dihydro-3-hydroxy-	MeOH	230(3.97),288(3.60)	94-1228-83
$C_5H_4N_4O_3$			
1H-Pyrazole-4-carbonitrile, 4,5-dihydro-3-(nitromethyl)-5-oxo-, compd. with hydrazine (1:2)	EtOH	256(3.88)	48-0041-83
$C_5H_4N_4S$			
6-Purinethiol	H_2O	229(3.88),322(4.35)	36-0372-83
C_5H_4OS			
2H-Thiopyran-2-one	C_6H_{12}	328(3.46),334(3.46), 349s(3.32)	139-0059-83B
$C_5H_4O_2$			
2H-Pyran-2-one	EtOH	216(3.36),289(3.69)	139-0059-83B
$C_5H_5FN_2O_2$			
1H-Pyrazole-3-carboxylic acid, 4-fluoro-, methyl ester	MeOH	218(3.88),242(3.65)	104-0403-83
$C_5H_5NO_4$			
3-Isoxazoleacetic acid, 4,5-dihydro-5-oxo-	EtOH	256(3.55)	4-1597-83
$C_5H_5N_3O_2$			
1H-Imidazole, 4-ethenyl-5-nitro-	MeOH	223(4.21),322(3.96)	4-0629-83
$C_5H_5N_3O_3$			
Formamide, N-(4-amino-2,5-dihydro-2,5-dioxo-1H-pyrrol-3-yl)-	EtOH	227(4.17),363(3.43)	24-2591-83
2(1H)-Pyridinone, 5-amino-6-nitro-	EtOH	227(4.30),438(3.96)	103-0303-83
$C_5H_5N_5$			
Adenine	pH 1	261(4.14)	94-3454-83
	H_2O	259(4.15)	94-3454-83
	pH 7.0	207(4.37),260.5(4.13)	162-0023-83
	pH 13	267(4.12)	94-3454-83

Compound	Solvent	$\lambda_{max}(\log \epsilon)$	Ref.
$C_5H_5N_5O$			
1H-Imidazo[4,5-d]-1,2,3-triazine, 4-methoxy-	n.s.g.	203(4.37),253(3.78)	103-1231-83
C_5H_5P			
Phosphorin	EtOH	213(4.28),246(3.93)	24-0445-83
$C_5H_6AgNO_2$			
2,6-Piperidinedione, silver salt	MeCN-base	231(3.95)	23-1890-83
after reduction	MeCN-base	237(4.00)	23-1890-83
$C_5H_6ClN_3O$			
2(1H)-Pyrazinone, 3-amino-5-chloro-1-methyl-	EtOH	326(3.91)	4-0919-83
$C_5H_6N_2$			
2-Propenenitrile, 3-(1-aziridinyl)-, trans	heptane	251(--)	70-2322-83
	MeOH	253(4.22)	70-2322-83
2-Pyridinamine	heptane	230(4.2),290(3.63)	70-2437-83
	PrOH	296(3.64)	70-2437-83
	MeCN	293(3.68)	70-2437-83
$C_5H_6N_2S$			
2-Thiophenecarboxaldehyde, hydrazone	EtOH	305(3.99),340(3.93)	34-0132-83
$C_5H_6N_4O_2S_2$			
Thiourea, (4-methyl-5-nitro-2-thiazolyl)-	$CHCl_3$	272(3.71),362(4.23)	74-0075-83B
$C_5H_6N_6$			
1H-Imidazo[4,5-d]-1,2,3-triazin-4-amine, N-methyl-	n.s.g.	218(4.13),264(3.96), 303(3.75)	103-1231-83
1H-Purine-2,6-diamine	pH 1.9	241(3.98),282(4.00)	162-0431-83
C_5H_6O			
Acetaldehyde, cyclopropylidene-	hexane	220(3.96)	78-3307-83
C_5H_7Cl			
Cyclopentene, 1-chloro-	EtOH	200(3.66)	35-6907-83
C_5H_7I			
Cyclobutane, (iodomethylene)-	EtOH	248(2.62)	35-6907-83
C_5H_7Li			
Lithium, pentadienyl-	THF	358(3.73)	157-0021-83
$C_5H_7NO_2$			
2,6-Piperidinedione (glutarimide)	MeCN	233s(2.33)	23-1890-83
silver salt	MeCN-base	231(3.95)	23-1890-83
$C_5H_7NO_3$			
2-Propenoic acid, 2-(formylamino)-, methyl ester	MeOH	246.2(3.75)	118-0539-83
C_5H_7NTe			
Isotellurazole, 3,5-dimethyl-	MeCN	221(3.83),239s(3.65), 304(3.72)	118-0824-83
Isotellurazole, 3-ethyl-	MeCN	217(3.76),236(3.63), 305(3.73)	118-0824-83

Compound	Solvent	λ_{max}(log ϵ)	Ref.
$C_5H_7N_2$			
Pyrazinium, methyl-, radical	acid	590(1.45)	125-0655-83
$C_5H_7N_3OS$			
1,2,4-Triazin-5(2H)-one, 2-methyl-3-(methylthio)-	EtOH	236(4.39)	44-4585-83
1,2,4-Triazin-5(4H)-one, 4-methyl-3-(methylthio)-	EtOH	230(3.76),296(3.88)	44-4585-83
$C_5H_7N_3O_2$			
1H-Imidazole-4-carboxamide, 5-hydroxy-N-methyl-	H_2O	238(3.62),276(4.09)	4-0875-83
	M HCl	244(3.98),280s(--)	4-0875-83
	M NaOH	276(4.06)	4-0875-83
1H-Imidazole-4-carboxamide, 5-hydroxy-1-methyl-	H_2O	239(3.66),278(4.11)	4-0875-83
	M HCl	240(3.81),278(3.86)	4-0875-83
	M NaOH	240s(--),277(4.12)	4-0875-83
1H-Imidazole-4-carboxamide, 5-methoxy-	H_2O	256(4.12)	4-0875-83
	M HCl	241(4.02)	4-0875-83
	M NaOH	268(4.13)	4-0875-83
1H-Imidazole-5-carboxamide, 4-hydroxy-1-methyl-	H_2O	237(3.60),277(4.07)	4-0875-83
	M HCl	244(3.89),275s(--)	4-0875-83
	M NaOH	231(3.40),287(4.09)	4-0875-83
$C_5H_7N_3O_3$			
1H-Imidazole, 5-ethoxy-4-nitro-	H_2O	230(3.62),390(4.10)	103-0650-83
1H-Imidazole-4-methanol, 1-methyl-5-nitro-	MeOH	231(3.51),304(3.90)	4-0629-83
1H-Pyrazole-4-carboxamide, 3-amino-5-(nitromethyl)-	H_2O	226s(3.92),256(3.79)	48-0041-83
$C_5H_7N_5O_3S$			
4H-[1,2,5]Oxadiazolo[3,4-c][1,2,6]-thiadiazin-7-amine, N,4-dimethyl-, 5,5-dioxide	H_2O	209(3.99),273(3.69), 300s(3.56)	142-2351-83
C_5H_8			
1,4-Pentadiene	gas	178(4.23)	54-0302-83
$C_5H_8N_2O$			
1H-Imidazole-4-ethanol	pH 2+	250s(2.28)	35-2382-83
2-Propenenitrile, 3-(methoxymethyl-amino)-	heptane	253(3.92)	70-2322-83
	MeOH	261(4.13)	70-2322-83
3(4H)-Pyridinone, 5,6-dihydro-, oxime, trifluoroacetate	MeOH	257(4.11)	44-3574-83
$C_5H_8N_4O_2$			
1H-Imidazol-4-amine, N,N-dimethyl-5-nitro-	H_2O	211(4.05),398(4.06)	103-0650-83
$C_5H_8N_4S$			
5H-Tetrazole-5-thione, 1,4-dihydro-1-methyl-4-(2-propenyl)-	hexane	205(3.76),249(4.09)	24-3427-83
$C_5H_8N_4S_2$			
1,2,4-Triazin-6-amine, 3,5-bis(methyl-thio)-	pH 1	269(4.24),308(3.93)	44-1271-83
	pH 11	262(4.25),338(3.53)	44-1271-83
	MeOH	264(4.29),350(3.54)	44-1271-83
$C_5H_8N_6O_2$			
1H-Imidazole-4-carboxamide, 5-[2-(ami-nocarbonyl)hydrazino]-	n.s.g.	227(3.68),265(4.10)	103-1235-83

Compound	Solvent	λ_{max}(log ϵ)	Ref.
$C_5H_8O_2$			
2-Butenoic acid, 2-methyl-, (Z)-	H_2O	217(3.71)	162-0094-83
$C_5H_9ClN_2$			
1,6-Diazabicyclo[3.1.0]hexane, 6-chloro-5-methyl-	heptane	217(3.11)	70-1754-83
C_5H_9NOS			
2-Thietanone, 3-amino-4,4-dimethyl-, monohydrobromide, (R)-	EtOH	216(3.36),236(3.30)	39-2259-83C
monohydrochloride	EtOH	213(3.53),237(3.26)	39-2259-83C
$C_5H_9N_3O$			
4(1H)-Pyrimidinone, 2-amino-5,6-dihydro-1-methyl-	pH 12	207(4.13),238(4.08)	142-1769-83
	MeOH	205(4.15),240(4.14)	142-1769-83
$C_5H_9S_2$			
4H-1,3-Dithiinium, 5,6-dihydro-2-methyl-, tetrafluoroborate	CH_2Cl_2	228(--),287(--), 310(--)(decomposes)	1-0687-83
$C_5H_{10}N_2O_2S$			
Ethanimidothioic acid, N-[[(methylamino)carbonyl]oxy]-, methyl ester (methomyl)	H_2O	233(3.94)	98-0625-83
$C_5H_{10}N_2O_3$			
Ethenamine, N-(ethoxymethyl)-N-nitro-	H_2O	278(3.90)	104-1270-83
$C_5H_{10}N_2S_2$			
2H-1,3,5-Thiadiazine-2-thione, tetrahydro-3,5-dimethyl- (dazomet)	C_6H_{12}	242(3.85),289(4.00)	162-0409-83
$C_5H_{10}N_4S_2$			
1,2,4,5-Tetrazine, 1,2-dihydro-1-methyl-3,6-bis(methylthio)-	dioxan	226(4.22),303(2.27)	24-2261-83
$C_5H_{10}N_4Se$			
1,2,3,4-Selenatriazol-5-amine, N,N-diethyl-	CCl_4	278(3.77)	1-0585-83
$C_5H_{10}O$			
2-Pentanone	hexane	279(1.20)	39-1053-83B
$C_5H_{10}OS$			
2-Propanone, 1-(ethylthio)-	hexane	301(2.35)	39-1053-83B
$C_5H_{10}S$			
Propanethial, 2,2-dimethyl-	MeCN	508(1.20)	35-1683-83
$C_5H_{11}NO$			
2-Propanone, 1-(dimethylamino)-	hexane	291(1.45)	39-1053-83B
$C_5H_{11}NO_2Si$			
Silane, trimethyl(2-nitroethenyl)-	EtOH	235(3.08)	44-3189-83
$C_5H_{11}N_3Si$			
Silacyclopentane, 1-azido-1-methyl-	C_6H_{12}	268(2.18)	152-0645-83
$C_5H_{11}S_5Si$			
Silane, trimethyl-1,2,3,4,5-pentathiepan-6-yl-	hexane	233(3.75),281(3.43)	101-0133-83L

Compound	Solvent	λ_{max}(log ϵ)	Ref.
$C_5Hg_4N_4$			
Mercury, tetrakis(cyano-C)-μ_4-methane-tetrayltetra-	MeOH	202(1.51)	101-0217-83P
$C_5Hg_4N_4S_4$			
Mercury, μ_4-methanetetrayltetra-kis(thiocyanato-S)tetra-	DMSO	255(1.48),260s(1.45)	101-0217-83P

Compound	Solvent	λ_{max}(log ϵ)	Ref.
$C_6H_2D_2O_2$			
2-Propynoic acid, 2-propynyl-1,1-d ester	EtOH	207s(3.51)	33-2322-83
$C_6H_2N_4O_5$			
Benzofurazan, 4,6-dinitro-	MeOH	300(3.85),338(3.65), 461(4.48)	12-1227-83
$C_6H_2N_4S_4$			
1,2,7,9-Tetrathia-3,6,8,10-tetraaza-cyclohept[e]indene	CH_2Cl_2	342(4.0),366(4.0), 582(4.0),625(4.0)	23-1562-83
$C_6H_3Cl_2F_3N_2S$			
Pyrimidine, 4,5-dichloro-2-(methyl-thio)-6-(trifluoromethyl)-	MeCN	238(3.59),270(3.46), 320(3.39)	4-0219-83
$C_6H_3N_3O_6$			
Benzene, 1,2,4-trinitro-	MeCN-buffer pH 8.5	234(4.12)	138-1445-83
Benzene, 1,3,5-trinitro-	EtOH	222(3.74)	12-1227-83
	EtOH-NaOH	250(3.38),429(3.74), 501(3.60)	12-1227-83
	DMSO-NH_3	442(4.39),520(4.29)	74-0301-83C
	DMSO-$BuNH_2$	440(4.16),520(3.98)	74-0301-83C
	DMSO-Et_2NH	440(4.13),520(4.03)	74-0301-83C
	DMSO-Et_3N	436(3.62),516(3.55)	74-0301-83C
	DMSO-$PhNH_2$	466(1.63),518(--)	74-0301-83C
$C_6H_3N_3O_7$			
Picric acid, potassium salt	pH 7-8	352(4.11)	77-0995-83
	dioxan	349(4.16)	35-4337-83
18C6 crown complex	toluene	360(4.19)	35-4337-83
	dioxan	360(4.18)	35-4337-83
Picric acid, sodium salt	dioxan	347(4.16)	35-4337-83
18C6 crown complex	toluene	360(4.16)	35-4337-83
	dioxan	360(4.20)	35-4337-83
$C_6H_4BrNO_2$			
2-Pyridinecarboxylic acid, 5-bromo-	MeOH	233(4.03),273(3.68)	106-0591-83
$C_6H_4ClF_3N_2OS$			
4(1H)-Pyrimidinone, 5-chloro-2-(meth-ylthio)-6-(trifluoromethyl)-	MeOH	237(3.77),306(3.99), 315(3.98)	4-0219-83
$C_6H_4ClNO_2S$			
Benzenesulfenyl chloride, 4-nitro-	pentane	231(--),302(4.01), 390s(--)	65-0668-83
	CH_2Cl_2	316(4.03),390s(--)	65-0668-83
$C_6H_4Cl_2S$			
Benzenesulfenyl chloride, 4-chloro-	pentane	247(3.95),395(2.61)	65-0668-83
	CH_2Cl_2	248(4.01),395(2.78)	65-0668-83
$C_6H_4INO_2$			
2-Pyridinecarboxylic acid, 5-iodo-	MeOH	246(3.95),280(3.79)	106-0591-83
$C_6H_4N_2$			
3-Pyridinecarbonitrile	EtOH	217(4.00),259(3.35), 265(3.37),271(3.24)	35-0040-83
4-Pyridinecarbonitrile	EtOH	211(3.95),220(3.88), 257(3.46)	35-0040-83

Compound	Solvent	λ_{max}(log ε)	Ref.
$C_6H_4N_2O_4$			
2-Pyridinecarboxylic acid, 5-nitro-	MeOH	245(3.92),280(3.77)	106-0591-83
$C_6H_4N_2S_2$			
1,3,2,4-Benzodithiadiazine-3-S^{IV}	CH_2Cl_2	283(4.3),291(4.3), 371(3.0),617(2.7)	77-0073-83
$C_6H_4N_3O_6$			
1,3-Cyclohexadiene, 1,3,5-trinitro-, ion(1-), tetramethylammonium salt	DMSO	478(4.38),585(4.04)	39-1197-83B
$C_6H_4N_4O_6$			
4-Benzofurazanol, 1,4-dihydro-5,7-di- nitro-, monopotassium salt	EtOH	248(3.98),298(3.89), 337(3.63),460(4.46)	12-1227-83
$C_6H_4N_6O_2$			
1-Triazene-1-carbonitrile, 3-(2-nitro- 3-pyridinyl)-, potassium salt	H_2O	200(4.20),318(4.18)	103-0303-83
$C_6H_4N_6O_4$			
Benzenamine, 2-azido-3,5-dinitro-	EtOH	237(4.22),301(4.02), 386(3.82)	12-1227-83
$C_6H_4N_6O_4S_2$			
1H-Imidazole, 4,4'-dithiobis[5-nitro-	H_2O	357(4.03)	103-0650-83
$C_6H_4O_2$			
1,4-Benzenediol	EtOH	294(3.51)	98-0734-83
$C_6H_5BrN_4$			
1H-Pyrrolo[2,3-d]pyrimidin-4-amine, 5-bromo-	pH 1	232(4.22),282(3.94)	12-0165-83
	pH 11	225(4.17),279(3.95)	12-0165-83
C_6H_5BrO			
2,5-Cyclohexadien-1-one, 4-bromo-	H_2O	c.240(c.4.0)(unstable)	44-0759-83
$C_6H_5ClN_4$			
1H-Pyrrolo[2,3-d]pyrimidin-2-amine, 4-chloro-	MeOH	232(4.46),253(3.58), 317(3.74)	5-0137-83
C_6H_5ClO			
Phenol, 4-chloro-	39.2% H_2SO_4	278.5(3.15)	54-0299-83
	90.4% H_2SO_4	273.0(2.77)	54-0299-83
C_6H_5ClS			
Benzenesulfenyl chloride	pentane	236(3.90),394(2.54)	65-0668-83
	CH_2Cl_2	239(3.89),394(2.45)	65-0668-83
Benzenethiol, 4-chloro-	pentane	240(4.00),287(2.72)	65-0668-83
	CH_2Cl_2	244(4.04),285(2.74)	65-0668-83
$C_6H_5F_3N_2OS$			
4-Pyrimidinol, 2-(methylthio)-6-(tri- fluoromethyl)-	MeOH	226(3.80),296(3.93)	4-0219-83
C_6H_5NO			
Benzene, nitroso-	MeCN	282(4.00)	44-2897-83
4-Pyridinecarboxaldehyde	EtOH	258(3.85)	140-0215-83
$C_6H_5NO_2$			
Benzene, nitro-	MeCN	262(4.00)	44-2897-83

Compound	Solvent	λ_{max}(log ε)	Ref.
$C_6H_5NO_2S$			
Benzenethiol, 4-nitro-	pentane	220(3.84),301(4.05)	65-0668-83
	CH_2Cl_2	315(4.07)	65-0668-83
$C_6H_5NO_3$			
Acetaldehyde, (3-amino-5-oxo-2(5H)-furanylidene)- (basidalin)	pH 2	251(4.22),308(3.98)	158-0448-83
	pH 12	260(4.31)	158-0448-83
	MeOH	220(3.89),277(4.20)	158-0448-83
2-Pyridinecarboxylic acid, 5-hydroxy-	MeOH	243(3.92),279(3.77), 327(2.95)	106-0591-83
$C_6H_5N_2O_4$			
Pyrazinium, 2,3-dicarboxy-, radical	acid	680(1.63)	125-0655-83
$C_6H_5N_3$			
Benzene, azido-	MeCN	250(4.00)	44-2897-83
1H-Pyrrolo[2,3-d]pyrimidine	MeOH	222(4.38),270(3.60)	5-1576-83
$C_6H_5N_3O_2$			
Ethanone, 1-(5-azido-2-furanyl)-	C_6H_{12}	302(4.20)	12-0963-83
$C_6H_5N_3O_4$			
Formamide, N,N'-(2,5-dihydro-2,5-dioxo-1H-pyrrole-3,4-diyl)bis-	EtOH	217(4.15),363(3.43)	24-2591-83
$C_6H_5N_3S$			
4H-Pyrrolo[2,3-d]pyrimidine-4-thione, 1,7-dihydro-	MeOH	269(3.59),326(4.37)	5-1576-83
$C_6H_5N_5O_6$			
4-Benzofurazanamine, 1,4-dihydro-5,7-dinitro-, 3-oxide, mono-ammonium salt	MeOH	265.5(4.18),304.5(4.18), 355(3.98),459.5(4.52)	12-0297-83
$C_6H_6BrN_5$			
1H-Purin-6-amine, 8-bromo-N-methyl-	pH 1	206(4.28),268(4.26)	94-3454-83
	H_2O	215.5(4.33),275(4.26)	94-3454-83
	pH 13	219.5(4.28),275(4.26)	94-3454-83
C_6H_6NS			
Phenylthio, 4-amino- (radical)	C_6H_{12}	547(4.04)	151-0157-83A
	MeOH	594(4.18)	151-0157-83A
	MeCN	585(4.28)	151-0157-83A
$C_6H_6N_2O_2$			
Benzenamine, 2-nitro-	n.s.g.	270(3.8),350(3.8)	65-1072-83
1:1 $GaCl_3$ complex	n.s.g.	285(3.7),510(1.54)	65-1072-83
1:2 $GaCl_3$ complex	n.s.g.	290(3.8)	65-1072-83
2-Pyridinecarboxylic acid, 5-amino-	MeOH	274(4.15)	106-0591-83
$C_6H_6N_2O_3$			
2-Pyridinecarboxylic acid, 5-(hydroxy-amino)-	MeOH	277(4.01)	106-0591-83
$C_6H_6N_3O_4$			
1,4-Cyclohexadien-1-amine, 2,4-dini-tro-, ion(1-), tetramethylammonium salt	DMSO	274(3.90),347(4.20), 567(4.26)	39-1197-83B
$C_6H_6N_4O_2$			
3H-Pyrazole-4-carbonitrile, 3,3-di-	MeCN	257(4.08),342s(2.46)	78-1247-83

Compound	Solvent	λ_{max}(log ϵ)	Ref.
methyl-5-nitro- (cont.)			78-1247-83
7H-Pyrazolo[4,3-d]pyrimidin-7-one, 2,6-dihydro-3-hydroxy-2-methyl-	MeOH	227(4.13),231s(4.12), 298(3.73)	94-1228-83
$C_6H_6N_4S$			
1H-Purine, 6-(methylthio)-	EtOH	217(4.0),226s(3.9), 282(4.13),288(4.13)	36-0372-83
C_6H_6O			
Phenol	hexane	213(3.75),273(3.23)	99-0034-83
	39.2% H_2SO_4	269.0(3.15)	54-0299-83
	86.9% H_2SO_4	263.5(2.87)	54-0299-83
	H_2SO_4	211(3.78),270(3.18)	99-0612-83
$C_6H_6O_3$			
2,4-Hexadienoic acid, 6-oxo-	EtOH	267(4.52)	111-0447-83
C_6H_6S			
Benzenethiol	pentane	231(4.07),276(2.73)	65-0668-83
	hexane	236(3.92),279(2.78)	99-0034-83
	CH_2Cl_2	238(4.08),276(2.75)	65-0668-83
	H_2SO_4	236(3.92),279(2.78)	99-0612-83
$C_6H_7BO_2$			
Boronic acid, phenyl-	H_2O	220(3.93),268(2.73)	32-0515-83
$C_6H_7BrO_2$			
2,4-Hexadienoic acid, 4-bromo-, (E,E)-	EtOH	269(4.32)	111-0457-83
$C_6H_7ClN_2O$			
Pyrazine, 2-chloro-3,5-dimethyl-, 4-oxide	EtOH	229(3.83),269(3.90), 298(3.89),304(2.64), 310(2.63)	4-0311-83
$C_6H_7ClN_2O_2$			
2,3-Pyrazinedione, 5-chloro-1,4-dihydro-1,4-dimethyl-	EtOH	318(3.72)	4-0919-83
2(1H)-Pyrazinone, 5-chloro-3-methoxy-1-methyl-	EtOH	315(3.85)	4-0919-83
$C_6H_7Cl_3O$			
3-Hexen-2-one, 6,6,6-trichloro-	EtOH	216(4.20),280(3.11)	104-1561-83
C_6H_7N			
Aniline	hexane	235(3.97),287(3.30)	99-0034-83
$C_6H_7NO_2$			
Butanenitrile, 2-acetyl-3-oxo-	EtOH	232s(3.45),278(4.03)	4-0645-83
Ethanone, 1-(5-methyl-4-isoxazolyl)-	EtOH	232(3.80),275(3.07)	4-0645-83
$C_6H_7NO_3$			
2(5H)-Furanone, 5-(2-aminoethylidene)-4-hydroxy-, hydrobromide	EtOH	241(4.10),306(3.87)	39-2983-83C
	HCl	251(4.15)	39-2983-83C
$C_6H_7N_3$			
1H-Imidazole-1-carbonitrile, 2,5-dimethyl-	EtOH	232(4.01)	4-1277-83
$C_6H_7N_3OS$			
4H-Thiazolo[2,3-c][1,2,4]triazin-4-one, 6,7-dihydro-7-methyl-	EtOH	211(4.05),296(3.73)	44-4585-83

Compound	Solvent	$\lambda_{max}(\log \epsilon)$	Ref.
7H-Thiazolo[3,2-b][1,2,4]triazin-7-one, 2,3-dihydro-2-methyl-	EtOH	233(4.28)	44-4585-83
$C_6H_7N_3O_3$			
Acetamide, N-(4-amino-2,5-dihydro-2,5-dioxo-1H-pyrrol-3-yl)-	EtOH	227(4.21),380(3.40)	24-2591-83
3-Pyridinamine, 6-methoxy-2-nitro-	MeOH	229(4.30),431(3.89)	103-0303-83
5-Pyrimidineacetic acid, 2-amino-4,5-dihydro-4-oxo-	50% MeOH	289(3.81)	73-0292-83
	+ HCl	221(4.02),261(3.92)	73-0292-83
	+ NaOH	231(3.98),281(3.86)	73-0292-83
$C_6H_7N_3O_4$			
1H-Pyrazole-3-carboxylic acid, 5-hydroxy-4-nitroso-, ethyl ester	MeOH	264(3.94),374(3.46)	94-1228-83
$C_6H_7N_4O_2$			
Pyrazinium, 2,3-dicarbamoyl-, radical	acid	440(1.60)	125-0655-83
$C_6H_7N_5$			
1H-Purin-6-amine, N-methyl-	pH 1	266(4.20)	94-3454-83
	H_2O	266(4.21)	94-3454-83
	pH 7	266(4.20)	94-4270-83
	pH 13	272(4.21)	94-3454-83
	pH 13	273(4.19)	94-4270-83
	EtOH	266(4.20)	94-4270-83
1H-Purin-6-amine, 1-methyl-, hydriodide	pH 1	258(4.07)	104-2094-83
	pH 14	270(4.20)	104-2094-83
3H-Purin-6-amine, 3-methyl-, hydriodide	pH 1	274(4.27)	104-2094-83
	pH 7	274(4.17)	104-2094-83
	pH 14	273(4.17)	104-2094-83
9H-Purin-6-amine, 9-methyl-	pH 1	261(4.17)	104-2094-83
	pH 14	261(4.21)	104-2094-83
Pyrazine, 3-azido-2,5-dimethyl-	EtOH	275(3.90)	4-1277-83
$C_6H_7N_5O$			
1H-Imidazo[4,5-d]-1,2,3-triazine, 4-ethoxy-	n.s.g.	202(4.37),243(3.79)	103-1231-83
9H-Purin-6-amine, 8-hydroxy-N-methyl-	pH 1	274(4.15)	94-3454-83
	H_2O	272(4.22)	94-3454-83
	pH 13	281(4.25)	94-3454-83
C_6H_8			
Cyclopropane, 2-propenylidene-	hexane	226.5(4.48)	24-0882-83
1-Penten-3-yne, 2-methyl-	EtOH	224(4.08)	104-1621-83
$C_6H_8Br_2N_2$			
2,3-Diazabicyclo[2.2.2]oct-2-ene, 1,4-dibromo-	benzene	366(2.11)	35-7108-83
$C_6H_8Br_2N_2O$			
2,3-Diazabicyclo[2.2.2]oct-2-ene, 1,4-dibromo-, N-oxide	benzene	232(3.83)	35-7108-83
$C_6H_8Br_2N_2O_2$			
2,3-Diazabicyclo[2.2.2]oct-2-ene, 1,4-dibromo-, 2,3-dioxide	MeOH	272(3.85)	35-7108-83
$C_6H_8ClN_3O_4S_2$			
1,3-Benzenedisulfonamide, 4-amino-6-chloro-	EtOH	224(4.62),266(4.27),313(3.59)	162-0289-83

Compound	Solvent	λ_{max}(log ε)	Ref.
C_6H_8IN			
1H-Pyrrole, 1-(2-iodoethyl)-	MeOH	225(3.77),260s(2.90)	23-0454-83
C_6H_8N			
Benzenaminium ion	H_2SO_4	203(3.88),254(2.30)	99-0612-83
$C_6H_8N_2O$			
Pyrazinemethanol, 6-methyl-	EtOH	271(3.89),275(3.90), 310s(2.66)	4-0311-83
$C_6H_8N_2OS$			
Isothiazolo[4,5-c]pyridin-3(2H)-one, 4,5,6,7-tetrahydro-, monohydrobromide	MeOH	263(3.68)	87-0895-83
Isothiazolo[5,4-c]pyridin-3(2H)-one, 4,5,6,7-tetrahydro-, monohydrochloride	MeOH	263(3.68)	87-0895-83
$C_6H_8N_2O_3$			
3-Isoxazoleacetamide, 2,5-dihydro-2-methyl-5-oxo-	EtOH	266(4.06)	4-1597-83
3-Isoxazoleacetamide, 5-methoxy-	EtOH	224(3.86)	4-1597-83
1H-Pyrrole, 2-methoxy-1-methyl-5-nitro-	MeOH	280(3.8),359(3.6)	44-0162-83
1H-Pyrrole, 3-methoxy-1-methyl-4-nitro-	MeOH	297(4.0)	44-0162-83
$C_6H_8N_2O_5$			
2-Propenoic acid, 3-[[(methyl-aci-nitro)acetyl]amino]- (enteromycin)	MeOH	230(3.95),275(4.11), 298(4.20)	162-0518-83
$C_6H_8N_4$			
4H-Imidazole-4-carbonitrile, 5-amino-4-ethyl-	EtOH-pH 7	270(3.71)	44-0003-83
	EtOH-pH 13	283(3.88)	44-0003-83
$C_6H_8N_4OS$			
1H-Pyrimido[5,4-b][1,4]thiazin-4(6H)-one, 2-amino-7,8-dihydro-	pH 1	277s(3.80),304(4.00)	87-0559-83
	pH 6.8	223(4.33),266(3.75), 302(3.91)	87-0559-83
	pH 13	270(3.83),289(3.87)	87-0559-83
$C_6H_8N_4O_2S$			
1H-Pyrimido[5,4-b][1,4]thiazin-4(6H)-one, 2-amino-7,8-dihydro-, 5-oxide	pH 1	224(4.23),259(4.09)	87-0559-83
	pH 6.8	224(4.53),260(4.09)	87-0559-83
	pH 13	260(4.00)	87-0559-83
$C_6H_8N_4O_2S_2$			
Thiourea, N-methyl-N'-(4-methyl-5-nitro-2-thiazolyl)-	$CHCl_3$	258(3.85),364.5(4.20)	74-0075-83B
$C_6H_8N_4O_3$			
5-Pyrimidineacetic acid, 2,6-diamino-1,4-dihydro-4-oxo-	50% MeOH	278(4.13)	73-0292-83
	+ HCl	270(4.24)	73-0292-83
	+ NaOH	226(3.96),271(4.02)	73-0292-83
C_6H_8O			
2-Propanone, 1-cyclopropylidene-	hexane	219(4.08)	78-3307-83
$C_6H_8O_2$			
2,4-Hexadienoic acid	EtOH	256(4.42)	111-0447-83

Compound	Solvent	λ_{max}(log ε)	Ref.
$C_6H_8O_3$			
4H-1,3-Dioxin-6-carboxaldehyde, 2-methyl-, (R)-	MeOH	250(2.88)	136-0286-83G
2,5-Furandione, dihydro-3,4-dimethyl-, cis	n.s.g.	230(1.4)	23-2790-83
trans	n.s.g.	225(1.4)	23-2790-83
2H-Pyran-2-one, 5,6-dihydro-4-methoxy-	n.s.g.	232(4.08)	158-83-150
C_6H_9Br			
1-Hexyne, 1-bromo-	C_6H_{12}	219(2.67)	44-1732-83
	MeOH	214(2.69)	44-1732-83
C_6H_9Cl			
Cyclohexene, 1-chloro-	EtOH	200(3.64)	35-6907-83
C_6H_9I			
Cyclopentane, (iodomethylene)-	EtOH	251(2.61)	35-6907-83
1-Hexyne, 1-iodo-	C_6H_{12}	210s(2.64),254(2.54)	44-1732-83
	MeOH	207s(2.81),239(2.60)	44-1732-83
	CH_2Cl_2	251(2.61)	44-1732-83
C_6H_9NO			
2(1H)-Pyridinone, 3,4-dihydro-6-methyl-	MeOH	249(3.31)	5-0220-83
$C_6H_9NO_2$			
3-Penten-2-one, 4-hydroxy-3-(iminomethyl)-	EtOH	250.5(4.14),280(4.15)	4-0649-83
$C_6H_9NO_2S$			
Acetamide, N-(tetrahydro-2-oxo-3-thienyl)- (citiolone)	n.s.g.	238(3.64)	162-0330-83
$C_6H_9NO_3$			
2(5H)-Furanone, 5-(2-aminoethyl)-4-hydroxy-	EtOH	249(4.16)	39-2983-83C
	HCl	223(3.98)	39-2983-83C
$C_6H_9NO_5$			
2-Butenoic acid, 3-methoxy-4-nitro-, methyl ester, (E)-	EtOH	231(4.09)	33-1475-83
C_6H_9NTe			
Isotellurazole, 3-propyl-	MeCN	217(3.77),237(3.65), 305(3.73)	118-0824-83
$C_6H_9N_2$			
Pyrazinyl, dihydro-2,5-dimethyl-	acid	570(1.48)	125-0655-83
Pyrazinyl, dihydro-2,6-dimethyl-	acid	570(1.45)	125-0655-83
$C_6H_9N_3O_2$			
Acetamide, N-(4,5-dihydro-1-methyl-4-oxo-1H-imidazol-2-yl)-	EtOH	223(3.44),255(3.60)	103-0481-83
1H-Imidazole-4-carboxamide, 5-ethoxy-	H_2O	257(4.18)	4-0875-83
	M HCl	241(4.07)	4-0875-83
	M NaOH	268(4.19)	4-0875-83
1H-Imidazole-4-carboxamide, N-ethyl-5-hydroxy-	H_2O	237(3.68),275(4.15)	4-0875-83
	M HCl	243(4.05),280s(--)	4-0875-83
	M NaOH	276(4.04)	4-0875-83
1H-Imidazole-4-carboxamide, 1-ethyl-5-hydroxy-	H_2O	238(3.66),278(4.09)	4-0875-83
	M HCl	239(3.79),278(3.87)	4-0875-83
	M NaOH	277(4.20)	4-0875-83

Compound	Solvent	λ_{max}(log ϵ)	Ref.
1H-Imidazole-4-carboxamide, 5-methoxy-1-methyl-	H_2O	245(3.96)	4-0875-83
	M HCl	227(4.00)	4-0875-83
	M NaOH	244(3.96)	4-0875-83
1H-Imidazole-5-carboxamide, 1-ethyl-4-hydroxy-	H_2O	235s(--),280(3.92)	4-0875-83
	M HCl	244(3.87),278s(--)	4-0875-83
	M NaOH	233(3.36),287(4.09)	4-0875-83
1H-Imidazole-5-carboxamide, 4-methoxy-1-methyl-	H_2O	258(4.09)	4-0875-83
	M HCl	243(3.97)	4-0875-83
	M NaOH	285(4.08)	4-0875-83
5-Pyrimidinemethanol, 4-amino-2-methoxy- (bacimethrin)	H_2O	227(3.88),271(3.86)	162-0135-83
	pH 1	229(3.92),261(3.98)	162-0135-83
C_6H_{10}			
1-Hexyne, homopolymer	hexane	285(3.20)	18-2798-83
$C_6H_{10}N_2$			
1H-Pyrazole, 1,3,5-trimethyl-	MeOH	208s(3.62),220(3.70)	24-1520-83
$C_6H_{10}N_2O_3$			
3-Penten-2-one, 4-(methylamino)-3-nitro-	EtOH	302(4.01),352(3.65)	40-0088-83
Pyrazinol, 2,3-dihydro-3,3-dimethyl-, 1,4-dioxide	EtOH	357(4.16)	103-1128-83
$C_6H_{10}N_4O_2$			
1H-Imidazole-4-carboxylic acid, 5-hydrazino-, ethyl ester, dihydrochloride	n.s.g.	266(4.01)	103-1235-83
$C_6H_{10}N_4O_2S$			
1H-Imidazole-4-carboxylic acid, 2,3-dihydro-5-hydrazino-2-thioxo-, ethyl ester, dihydrochloride	n.s.g.	294(4.12)	103-1235-83
$C_6H_{10}N_6O$			
1H-Imidazole-4-carboxamide, 5-(3,3-dimethyl-1-triazenyl)- (dacarbazine)	pH 1	223(3.88)	162-0405-83
	pH 7	237(4.05)	162-0405-83
$C_6H_{10}O$			
3-Penten-2-one, 4-methyl-	alkane	231(4.06)	78-3307-83
$C_6H_{10}OS_2$			
Ethenone, bis(ethylthio)-	MeOH	288(3.78)	18-0171-83
$C_6H_{10}O_2$			
3-Penten-2-one, 4-methoxy-	EtOH	257(4.11)	39-2353-83C
$C_6H_{11}NO$			
3-Buten-2-one, 3-methyl-4-(methylamino)-	MeOH	296(4.33)	131-0001-83K
(E-syn-E)-	C_6H_{12}	279(--)	131-0001-83K
	dioxan	283(4.38)	131-0001-83K
(Z-syn-Z)-	C_6H_{12}	310(4.08)	131-0001-83K
	dioxan	283(--)	131-0001-83K
$C_6H_{11}NOS$			
2-Thietanone, 3-amino-3,4,4-trimethyl-, hydrochloride, (3RS)-	EtOH	221(3.34),235(3.36), 280(2.42)	39-2259-83C
$C_6H_{11}N_3O$			
4(1H)-Pyrimidinone, 2-amino-5,6-di-	pH 12	207(4.02),236(3.98)	142-1769-83

Compound	Solvent	λmax(log ε)	Ref.
hydro-1,5-dimethyl- (cont.)	MeOH	205(4.06),238(4.02)	142-1769-83
4(1H)-Pyrimidinone, 2-amino-1-ethyl-	pH 12	208(4.09),238(4.04)	142-1769-83
5,6-dihydro-	MeOH	206(4.12),240(4.11)	142-1769-83
4(1H)-Pyrimidinone, 5,6-dihydro-2-(di-	pH 12	232(4.13)	142-1769-83
methylamino)-	MeOH	236(4.16)	142-1769-83
4(1H)-Pyrimidinone, 5,6-dihydro-	pH 12	219(4.02),236(3.87)	142-1769-83
1-methyl-2-(methylamino)-	MeOH	220(4.05),238(4.03)	142-1769-83
$C_6H_{11}N_3O_3$			
Acetic acid, [[imino(methylamino)meth-yl]amino]oxo-, ethyl ester	EtOH	230s(2.75)	142-1067-83
$C_6H_{12}N_2S_2$			
Ethanedithioamide, tetramethyl-	EtOH	259s(4.3),273(4.3)	49-0101-83
$C_6H_{12}N_4$			
1,2,4,5-Tetrazine, 1,4-dihydro-1,3,4,6-tetramethyl-	EtOH	235(3.83),310(2.08)	24-2261-83
$C_6H_{12}N_4S_2$			
1,2,4,5-Tetrazine, 1,4-dihydro-1,4-dimethyl-3,6-bis(methylthio)-	dioxan	230(4.24),312(2.26)	24-2261-83
$C_6H_{12}OSi$			
2-Propenal, 3-(trimethylsilyl)-	C_6H_{12}	217(3.89),340(1.49), 354(1.49),371(1.34), 390(0.9)	157-0332-83
Silane, trimethyl(1-oxo-2-propenyl)-	C_6H_{12}	213(3.94),434(1.98)	78-0949-83
$C_6H_{12}O_2Si$			
2-Propanone, 1-oxo-1-(trimethylsilyl)-	C_6H_{12}	285(1.60),296(1.61), 535(2.00)	78-0949-83
$(C_6H_{12}Si)_n$			
Silane, trimethyl-1-propynyl-, polymer	C_6H_{12}	273(2.08)	35-7473-83
$C_6H_{13}N_3O_2$			
1-Triazenecarboxylic acid, 3,3-dimeth-yl-, 1-methylethyl ester	EtOH	264(4.21)	33-1416-83
$C_6H_{14}N_2O_2$			
Diazene, bis(1-methylethoxy)-	heptane	224.2(3.88)	44-3728-83
$C_6H_{14}N_4O$			
1-Triazene-1-carboxamide, 3,3-dimeth-yl-N-(1-methylethyl)-	EtOH	265(4.18)	33-1416-83
$C_6H_{18}Ge_2$			
Digermane, hexamethyl-	C_6H_{12}	202(--)	101-0149-83G

Compound	Solvent	λ_{max}(log ε)	Ref.
$C_7H_2Br_2O_4$			
1,3-Benzodioxole-4,7-dione, 5,6-dibromo-	$CHCl_3$	325(4.15),512(2.52)	64-0392-83B
$C_7H_2Cl_4O_4$			
Cyclopentadiene-1,2-dicarboxylic acid, tetrachloro-	hexane	328(3.30)	18-0481-83
$C_7H_2N_6O_9$			
2H-Benzimidazol-2-one, 1,3-dihydro-4,5,6,7-tetranitro-	73% H_2SO_4	385(3.8)	104-0376-83
	89.3% H_2SO_4	368(3.8)	104-0376-83
	94.0% H_2SO_4	360(3.8)	104-0376-83
	97.5% H_2SO_4	350s(3.7)	104-0376-83
$C_7H_3BrO_4$			
1,3-Benzodioxole-4,7-dione, 5-bromo-	$CHCl_3$	307(4.06),500(2.84)	64-0392-83B
$C_7H_4N_2OSSe$			
4H-Pyrido[3,2-e]-1,3-selenazin-4-one, 2,3-dihydro-2-thioxo-	MeOH	226(4.13),265(4.46)	73-3567-83
$C_7H_4N_2OSe_2$			
4H-Pyrido[3,2-e]-1,3-selenazin-4-one, 2,3-dihydro-2-selenoxo-	MeOH	228(4.16)	73-3567-83
$C_7H_4O_4$			
1,3-Benzodioxole-4,7-dione	$CHCl_3$	262(4.10),472(3.07)	64-0392-83B
1:1 compd. with 1,3-benzodioxol-4-ol	$CHCl_3$	262(4.16),478(3.17)	64-0392-83B
$C_7H_4O_6$			
4H-Pyran-2,6-dicarboxylic acid, 4-oxo- (chelidonic acid)	H_2O	270(4.06)	162-0285-83
$C_7H_5BF_3NOS$			
Boron, trifluoro(thieno[2,3-b]pyridine-7-oxide-O)-, (T-4)-	EtOH	237(4.55),285(3.78), 312(3.61)	4-0213-83
$C_7H_5BrClNO_2$			
Furo[3,2-c]pyridin-6(2H)-one, 7-bromo-4-chloro-3,5-dihydro-	EtOH	298(3.75)	120-0007-83
$C_7H_5ClN_2$			
3H-Diazirine, 3-chloro-3-phenyl-	isooctane	325s(--),339(2.26) 350s(--),356(2.30)	35-6513-83
C_7H_5ClOS			
Benzenecarbothioic acid, 2-chloro-, cesium salt	EtOH	242(3.73),252s(3.72)	64-1585-83B
potassium salt	EtOH	232s(4.02),255(3.92)	64-1585-83B
rubidium salt	EtOH	247s(4.17)	64-1585-83B
sodium salt	EtOH	255s(3.72)	64-1585-83B
Benzenecarbothioic acid, 3-chloro-, potassium salt	EtOH	289(3.95),296s(3.93)	64-1585-83B
rubidium salt	EtOH	288(3.83),299s(3.79)	64-1585-83B
sodium salt	EtOH	288(4.14),297s(4.11)	64-1585-83B
	EtOH	288(3.56),295s(3.53)	64-1585-83B
Benzenecarbothioic acid, 4-chloro-, cesium salt	EtOH	241(4.02),289s(3.72), 299(3.75)	64-1585-83B
lithium salt	EtOH	239(3.81),290s(3.50), 299(3.53)	64-1585-83B
potassium salt	EtOH	290s(3.78),298(3.80)	64-1585-83B
rubidium salt	EtOH	291s(3.73),300(3.76)	64-1585-83B

Compound	Solvent	λ_{max}(log ε)	Ref.
Benzenecarbothioic acid, 4-chloro-, sodium salt	EtOH	288(3.72),297(3.76)	64-1585-83B
$C_7H_5Cl_2$			
Methyl, (2,4-dichlorophenyl)-	C_6H_{12}	330(3.70)	46-1960-83
Methyl, (2,6-dichlorophenyl)-	C_6H_{12}	336.5(3.70)	46-1960-83
$C_7H_5FN_2$			
3H-Diazirine, 3-fluoro-3-phenyl-	isooctane	348s(--),366(2.47), 382s(--),386(2.46)	35-6513-83
C_7H_5NOS			
Thieno[2,3-b]pyridine, 7-oxide	EtOH	235(4.55),281(3.97), 310(3.69),320(3.69)	4-0213-83
Thieno[3,2-c]pyridine, 5-oxide	EtOH	253(4.46),281(4.03)	4-0213-83
$C_7H_5NO_2$			
Benzonitrile, 2,3-dihydroxy-	MeOH	211(4.37),232(3.95), 318(3.68)	64-0866-83B
$C_7H_5NO_2S_2$			
2,6-Pyridinedicarbothioic acid	CH_2Cl_2	265(3.89)	139-0361-83B
compd. with pyridine	isoPrOH	254(3.84),382s(--)	139-0361-83B
$C_7H_5NO_3$			
Benzaldehyde, 4-nitro-	EtOH	265(4.00)	140-0215-83
1,2-Benzisoxazol-3(2H)-one, 6-hydroxy-	MeOH	219(4.27),252(3.84), 260s(--),280s(--), 283(3.81),290(3.78)	158-0445-83
	90%MeOH-HCl	206(4.33),252(3.92), 260s(--),280s(--), 283(3.80),290(3.80)	158-0445-83
	+ NaOH	220s(--),237(4.26), 264(3.64),273s(--), 296(3.75)	158-0445-83
$C_7H_5NO_3S$			
Benzenecarbothioic acid, 4-nitro-, cesium salt	EtOH	259(4.00),338(3.65)	64-1585-83B
lithium salt	EtOH	262(3.92),354(3.50)	64-1585-83B
potassium salt	EtOH	259(4.11),340(3.74)	64-1585-83B
rubidium salt	EtOH	259(4.11),340(3.74)	64-1585-83B
sodium salt	EtOH	260(4.04)	64-1585-83B
$C_7H_5NO_4$			
2-Propenal, 3-(5-nitro-2-furanyl)-	EtOH	345(4.32)	140-0215-83
$C_7H_5N_3OS$			
Pyrido[2,3-d]pyrimidin-4(1H)-one, 2,3-dihydro-2-thioxo-	MeOH	254(3.01)	73-3315-83
4H-Pyrido[3,2-e]-1,3-thiazin-4-one, 2-amino-	MeOH	222(5.48)	73-3315-83
$C_7H_5N_3O_2$			
[1,2,3]Triazolo[1,5-a]pyridine-3-carboxylic acid	EtOH	237(3.78),286(4.07), 300s(--)	150-1341-83M
$C_7H_6BrNO_2S$			
1,2-Benzisothiazole, 2-bromo-2,3-dihydro-, 1,1-dioxide	EtOH	255(3.20),298(2.90)	44-0537-83

Compound	Solvent	λ_{max}(log ϵ)	Ref.
C_7H_6Cl			
Methyl, (2-chlorophenyl)-	C_6H_{12}	325(3.91)	46-1960-83
Methyl, (3-chlorophenyl)-	C_6H_{12}	324(3.85)	46-1960-83
Methyl, (4-chlorophenyl)-	C_6H_{12}	317(3.86)	46-1960-83
$C_7H_6ClNO_2$			
Furo[3,2-c]pyridin-6(2H)-one, 4-chloro-3,5-dihydro-	EtOH	289(3.73)	120-0007-83
$C_7H_6ClNO_2S$			
1,2-Benzisothiazole, 2-chloro-2,3-dihydro-, 1,1-dioxide	EtOH	266(3.02)	44-0537-83
C_7H_6ClNS			
Benzenecarbothioamide, 4-chloro-	EtOH	304(3.80),380(2.44)	80-0555-83
	CCl_4	306(3.88),415(2.44)	80-0555-83
$C_7H_6ClN_3OS$			
3-Pyridinecarboxamide, N-(aminothioxomethyl)-2-chloro-	MeOH	222(5.13)	73-3315-83
$C_7H_6ClN_3O_2$			
Methanehydrazonoyl chloride, 1-nitro-N-phenyl-	MeOH	239(4.13),385(4.22)	104-0783-83
$C_7H_6Cl_2N_2O_4$			
2(1H)-Pyridinone, 6-chloro-5-(2-chloroethyl)-4-hydroxy-3-nitro-	EtOH	347(3.69)	120-0007-83
$C_7H_6N_2$			
Benzene, (diazomethyl)-	n.s.g.	277(4.49)	44-4407-83
1H-Benzimidazole, SO_3 adduct	H_2O	242(3.66),267(3.77), 273(3.76)	104-1935-83
$C_7H_6N_2O_2S$			
5H-Thiazolo[3,2-a]pyrimidine-3,5(2H)-dione, 6-methyl-	MeOH	235(3.74),285(3.55)	56-1027-83
$C_7H_6N_2O_6$			
Phenol, 2-methoxy-4,6-dinitro-	pH 13	385(4.2),425s(4.0)	104-0689-83
	NaOH	385(4.21)	104-1845-83
$C_7H_6N_4O$			
Pyrido[3,4-d]-1,2,3-triazin-4(3H)-one, 3-methyl-	EtOH	286(4.00),304s(--), 317(3.70)	95-1129-83
$C_7H_6N_4O_5$			
Benzoic acid, 3,5-dinitro-, hydrazide	H_2O	208(4.31),235(4.26)	140-0091-83
$C_7H_6N_4O_6$			
Benzofurazan, 1,4-dihydro-4-methoxy-5,7-dinitro-, potassium salt	EtOH	260(3.58),301(3.84), 340(3.63),460(4.51)	12-1227-83
$C_7H_6N_4O_7$			
Benzofurazan, 1,4-dihydro-4-methoxy-5,7-dinitro-, 3-oxide, potassium salt	MeOH	268(4.01),308(4.02), 355(3.64),462(4.49)	12-0297-83
$C_7H_6N_8O_2$			
Pyrimido[5,4-e]tetrazolo[1,5-b][1,2,4]-triazine-7,9(6H,8H)-dione, 6,8-dimethyl-	EtOH	238(4.37),310(3.56)	44-1628-83

Compound	Solvent	$\lambda_{max}(\log \epsilon)$	Ref.
C_7H_6OS			
Benzenecarbothioic acid, cesium salt	EtOH	282(3.76),293(3.75)	64-1585-83B
lithium salt	EtOH	282(3.65),286s(3.64)	64-1585-83B
potassium salt	EtOH	283(3.75),295(3.74)	64-1585-83B
rubidium salt	EtOH	282(3.63),296(3.62)	64-1585-83B
sodium salt	EtOH	282(3.65),295(3.63)	64-1585-83B
$C_7H_6O_2$			
Benzaldehyde, 4-hydroxy-	pH 1	217(4.04),279(4.17)	98-0780-83
	H_2O	217(4.04),279(4.17)	98-0780-83
	pH 13	233(3.88),324(4.42)	98-0780-83
$C_7H_6O_3$			
Benzoic acid, 4-hydroxy-	EtOH	250(4.1)	35-7396-83
	EtOH-HCl	252(4.2)	35-7396-83
	EtOH-NaOH	273(4.3)	35-7396-83
2,4,6-Cycloheptatrien-1-one, 2,7-dihydroxy-	MeOH	245(4.56),324(3.78), 355s(--),365(3.85)	158-83-64
	EtOH	244(4.60),324(3.83), 365(3.87),375(3.94)	158-83-64
	EtOH-base	258(4.47),336(4.10), 412(4.00)	158-83-64
1,4-Dioxocin-6-carboxaldehyde	hexane	250s(3.65)	89-0571-83S
$C_7H_6O_4$			
1,4-Dioxocin-6-carboxylic acid	MeCN	255s(3.65)	89-0571-83S
C_7H_7			
Cycloheptatrienylium hexafluoroantimonate	MeCN	226(4.13),275(3.56)	44-0596-83
Methyl, phenyl-	C_6H_{12}	316(3.94)	46-1960-83
$C_7H_7BrN_2O$			
Benzoic acid, 2-bromo-, hydrazide	pH 1	270(2.74)	140-0091-83
	H_2O	270(2.74)	140-0091-83
Benzoic acid, 4-bromo-, hydrazide	pH 1	248(4.10)	140-0091-83
	H_2O	247(4.11)	140-0091-83
C_7H_7BrO			
Benzene, 1-bromo-4-methoxy-	39.2% H_2SO_4	278.5(3.12)	54-0299-83
	89.1% H_2SO_4	273.0(2.90)	54-0299-83
$C_7H_7ClN_2O$			
Benzoic acid, 2-chloro-, hydrazide	pH 1	260(4.09)	140-0091-83
	H_2O	258(4.08)	140-0091-83
Benzoic acid, 3-chloro-, hydrazide	pH 1	233(3.88),286(2.78)	140-0091-83
	H_2O	230(3.84),270(2.84)	140-0091-83
Benzoic acid, 4-chloro-, hydrazide	pH 1	242(4.16)	140-0091-83
	H_2O	243(4.15)	140-0091-83
1,2-Diazabicyclo[5.2.0]nona-2,4-dien-9-one, 8-chloro-, cis	CH_2Cl_2	298(3.92)	5-2141-83
C_7H_7ClOS			
Benzenesulfenyl chloride, 4-methoxy-	pentane	278(3.62),395(2.16)	65-0668-83
	CH_2Cl_2	283(3.62),395(2.18)	65-0668-83
C_7H_7ClS			
Benzenesulfenyl chloride, 4-methyl-	pentane	242(3.89),397(2.45)	65-0668-83
	CH_2Cl_2	245(3.91),396(2.45)	65-0668-83

Compound	Solvent	λ_{max}(log ε)	Ref.
$C_7H_7Cl_2NO_2$			
2(1H)-Pyridinone, 6-chloro-5-(2-chloro-ethyl)-4-hydroxy-	EtOH	275(3.72)	120-0007-83
C_7H_7FO			
Benzene, 1-fluoro-4-methoxy-	39.2% H_2SO_4	276.0(3.31)	54-0299-83
	86.9% H_2SO_4	269.5(3.20)	54-0299-83
$C_7H_7FO_6$			
1-Butene-1,2,4-tricarboxylic acid, 1-fluoro-, cis	EtOH	222(4.13)	118-1010-83
trans	EtOH	217(4.00)	118-1010-83
C_7H_7FS			
Benzene, 1-fluoro-4-(methylthio)-	decane	252(3.81)	65-0449-83
$C_7H_7F_3N_2OS$			
Pyrimidine, 4-methoxy-2-(methylthio)-6-(trifluoromethyl)-	MeOH	228(3.73),251(4.13), 284(3.62)	4-0219-83
C_7H_7NO			
Formamide, N-phenyl-	C_6H_{12}	240(4.10)	162-0605-83
	H_2O	240(4.05)	162-0605-83
	EtOH	243(4.14)	162-0605-83
	MeCN	243(4.11)	23-1890-83
silver salt	MeCN-base	209(3.95),242(4.02)	23-1890-83
2-Furanacetonitrile, 5-methyl-	$CHCl_3$	219(2.94)	13-0339-83A
$C_7H_7NO_2$			
1,2-Benzisoxazol-4(5H)-one, 6,7-di-hydro-	EtOH	229.5(3.95),270(2.83)	4-0645-83
Benzoic acid, 2-amino-	EtOH	337(3.70)	98-0734-83
	EtOH	220(4.4),245(3.8), 338(3.7)	145-0621-83
Cyclohexanecarbonitrile, 2,6-dioxo-	EtOH	259.5(4.22)	4-0645-83
1H-Pyrrole-2,5-dicarboxaldehyde, 1-methyl-	MeOH	298(4.28)	54-0347-83
$C_7H_7NO_3$			
Phenol, 2-methoxy-4-nitroso-	pH 13	392.5(4.41)	104-0689-83
Phenol, 4-methyl-2-nitro-	neutral	325(3.98)	126-2361-83
	anion	417(4.33)	126-2361-83
2-Pyridinecarboxylic acid, 5-methoxy-	MeOH	246(3.99),281(3.73)	106-0591-83
$C_7H_7NO_4$			
Phenol, 2-methoxy-4-nitro-	pH 13	432(4.27)	104-0689-83
	NaOH	430(4.26)	104-1845-83
C_7H_7NS			
Benzenecarbothioamide	EtOH	244(3.97),296(3.85), 370(2.40)	80-0555-83
	CCl_4	239(3.94),298(3.81), 412(2.40)	80-0555-83
$C_7H_7N_3$			
1H-Indazol-5-amine	H_2O	317(3.73)	35-6223-83
	MeOH	320(3.52)	35-6223-83
	MeCN	326(3.62)	35-6223-83
7H-Pyrrolo[2,3-d]pyrimidine, 7-methyl-	MeOH	271(3.58),293s(3.34)	5-1576-83

Compound	Solvent	$\lambda_{max}(\log \epsilon)$	Ref.
$C_7H_7N_3O$			
[1,2,3]Triazolo[1,5-a]pyridine, 5-methoxy-	EtOH	260(3.94),267(3.94)	150-1341-83M
$C_7H_7N_3OS_3$			
Thiazolo[5,4-d]pyrimidin-7(4H)-one, 5,6-dihydro-6-methyl-2-(methylthio)-5-thioxo-	MeOH	215(4.21),305(3.95), 340(4.19)	2-0243-83
$C_7H_7N_3O_2$			
4-Pyrimidinecarbonitrile, 1,2,3,6-tetrahydro-1,3-dimethyl-2,6-dioxo-	MeCN	289(3.89)	35-0963-83
$C_7H_7N_3O_3$			
Benzoic acid, 2-nitro-, hydrazide	pH 1	266(3.78)	140-0091-83
	H_2O	266(3.78)	140-0091-83
Benzoic acid, 4-nitro-, hydrazide	pH 1	266(4.02)	140-0091-83
	H_2O	266(4.02)	140-0091-83
Isoxazolo[4,5-c]pyridine-4,6(5H,7H)-dione, 5-amino-3-methyl-	MeOH	208(4.12),289(4.01)	142-1315-83
$C_7H_7N_3S$			
Pyrimidine, 1,4-dihydro-4-(2-thiazolyl)-	EtOH	240(3.74),294(3.31)	54-0364-83
$C_7H_7N_5O_2$			
7H-Oxazolo[3,2-a]pyrimidin-7-one, 2-(azidomethyl)-2,3-dihydro-	EtOH	227(3.85),258(3.77)	128-0125-83
Pyrimido[5,4-e]-1,2,4-triazine-5,7(6H,8H)-dione, 6,8-dimethyl-	EtOH	238(4.27),275(3.20), 340(3.62)	162-0583-83
C_7H_8			
Toluene	H_2SO_4	208(3.90),260(2.30)	99-0612-83
	n.s.g.	263(3.78)	99-0034-83
C_7H_8ClNO			
1H-Pyrrole-2-carboxaldehyde, 1-(2-chloroethyl)-	MeOH	207(3.38),258(3.87), 288(4.16)	23-1697-83
$C_7H_8Cl_2N_2O$			
2(1H)-Pyrazinone, 3,5-dichloro-6-ethyl-1-methyl-	EtOH	349(3.90)	4-0919-83
$C_7H_8IN_5O_2$			
2,4(1H,3H)-Pyrimidinedione, 1-(2-azido-3-iodopropyl)-	EtOH	264(4.01)	128-0125-83
$C_7H_8N_2$			
3H-Pyrazole, 5-ethynyl-3,3-dimethyl-	MeCN	260(3.43),350(2.30)	88-1775-83
Pyrrolo[1,2-a]pyrazine, 3,4-dihydro-	EtOH	266s(--),294(4.02)	104-0393-83
$C_7H_8N_2O$			
Benzenecarboximidamide, N-hydroxy-	EtOH	260(3.77)	32-0845-83
Benzoic acid hydrazide	pH 1	231(4.04),270(3.08)	140-0091-83
	H_2O	227(4.02)	140-0091-83
Ethanone, 1-(4-methyl-5-pyrimidinyl)-	EtOH	224(3.87),299(3.25)	4-0649-83
$C_7H_8N_2O_2$			
Benzoic acid, 2-hydroxy-, hydrazide	pH 1	239(3.94),303(3.57)	140-0091-83
	H_2O	240(3.88),298(3.53)	140-0091-83
Urocanic acid, 2-methyl-	pH 7.5	281(4.21)	78-3523-83

Compound	Solvent	λ_{max}(log ϵ)	Ref.
$C_7H_8N_2O_3$			
7H-Oxazolo[3,2-a]pyrimidin-7-one, 2,3-dihydro-2-(hydroxymethyl)-	EtOH	227(3.92),259(3.85)	128-0125-83
2,4(1H,3H)-Pyrimidinedione, 1-(oxiranylmethyl)-	EtOH	267(3.80)	128-0125-83
$C_7H_8N_4O$			
Imidazo[4,5-d][1,3]diazepin-8(3H)-one, 6,7-dihydro-5-methyl-, monohydrochloride	MeOH	228(4.30),297(3.59), 334s(--)	87-1478-83
4H-Pyrazolo[3,4-d]pyrimidin-4-one, 1,5-dihydro-1,6-dimethyl-	EtOH	209.5(4.40),253(3.85)	11-0171-83B
3-Pyridinecarbonitrile, 2,4-diamino-6-methoxy-	EtOH	227(4.49),295(3.88)	49-0973-83
3-Pyridinecarbonitrile, 4,6-diamino-2-methoxy-	EtOH	224(4.42),270(4.14), 280s(4.13)	49-0973-83
7H-Pyrrolo[2,3-d]pyrimidin-2-amine, 4-methoxy-	MeOH	222(4.41),257(3.86), 287(3.85)	5-0137-83
$C_7H_8N_4O_7$			
2(1H)-Pyridinone, 5,6-dihydro-1-methyl-3,5-dinitro-6-(nitromethyl)-, compd. with guanidine	MeOH	321(4.07),472(4.53)	138-0715-83
$C_7H_8N_4S$			
3H-Purine, 3-methyl-6-(methylthio)-	pH 1	235(3.92),276(3.65), 316(4.41)	94-3149-83
	pH 7	237(4.07),312(4.31)	94-3149-83
C_7H_8O			
Benzene, methoxy-	39.2% H_2SO_4	268.5(3.13)	54-0299-83
	86.9% H_2SO_4	264.0(2.92)	54-0299-83
Phenol, 4-methyl-	EtOH	280(3.20)	98-0734-83
	39.2% H_2SO_4	275.5(3.20)	54-0299-83
	86.9% H_2SO_4	271.5(2.86)	54-0299-83
C_7H_8OS			
Benzenethiol, 4-methoxy-	pentane	234(3.87),295(3.05)	65-0668-83
	CH_2Cl_2	240(3.90),293(3.08)	65-0668-83
$C_7H_8O_2S$			
Benzene, (methylsulfonyl)-	H_2SO_4	258(2.78),264(3.00), 271(2.90)	99-0612-83
$C_7H_8O_3$			
2-Furanacetic acid, 5-methyl-	$CHCl_3$	218(2.69)	13-0339-83A
$C_7H_8O_6$			
1-Butene-1,2,4-tricarboxylic acid, cis	EtOH	210(4.00)	118-1010-83
trans	EtOH	218(4.01)	118-1010-83
C_7H_8S			
Benzene, (methylthio)-	decane	214s(3.99),254(4.00)	65-0449-83
Benzenethiol, 4-methyl-	pentane	235(3.90),283(2.76)	65-0668-83
	CH_2Cl_2	239(3.91),282(2.77)	65-0668-83
$C_7H_9Br_2NO$			
2,4-Pentadienal, 2,4-dibromo-5-(dimethylamino)-	MeCN	388(4.56)	33-1427-83

Compound	Solvent	$\lambda_{max}(\log \epsilon)$	Ref.
$C_7H_9ClN_2O_2$			
2(1H)-Pyrazinone, 5-chloro-3-ethoxy-1-methyl-	EtOH	316(3.79)	4-0919-83
C_7H_9N			
Benzenamine, N-methyl-	hexane	240(4.03),293(3.28)	99-0034-83
	n.s.g.	284(3.17)	32-0069-83
C_7H_9NO			
Ethanone, 1-(4-methyl-1H-pyrrol-3-yl)-	EtOH	249(3.92),280s(3.40)	94-0925-83
4-Penten-2-ynal, 5-(dimethylamino)-	MeCN	346(4.18)	33-1427-83
$C_7H_9NO_2$			
2-Cyclohexen-1-one, 3-hydroxy-2-(iminomethyl)-	EtOH	247(4.15),294(4.09)	4-0649-83
5(4H)-Oxazolone, 2-methyl-4-(1-methylethylidene)-	EtOH	213(3.48),268(3.67)	39-2259-83C
$C_7H_9NO_4$			
3-Isoxazoleacetic acid, 2,5-dihydro-2-methyl-5-oxo-, methyl ester	EtOH	267(4.05)	4-1597-83
3-Isoxazoleacetic acid, 5-methoxy-, methyl ester	EtOH	224(3.86)	4-1597-83
$C_7H_9NO_5$			
2-Butenedioic acid, 2-(formylamino)-, dimethyl ester, (E)-	MeOH	260.5(4.04)	118-0539-83
(Z)-	MeOH	278(4.18)	118-0539-83
$C_7H_9N_3$			
Cyclohexene, 1-azido-6-methylene-	EtOH	228(3.96),254(3.82)	97-0096-83
$C_7H_9N_3O$			
1H-1,2-Diazepine-1-carboxamide, 5-methyl-	MeOH	219(4.08),370(2.40)	5-1361-83
Ethanone, 1-(2-amino-4-methyl-5-pyrimidinyl)-	EtOH	271.5(4.32)	4-0649-83
$C_7H_9N_3OS$			
4H-Thiazolo[2,3-c][1,2,4]triazin-4-one, 6,7-dihydro-3,7-dimethyl-	EtOH	212(4.05),293(3.85)	44-4585-83
4H-Thiazolo[2,3-c][1,2,4]triazin-4-one, 6,7-dihydro-6,7-dimethyl-, cis	EtOH	212(4.07),296(3.81)	44-4585-83
trans	EtOH	214(4.18),299(3.89)	44-4585-83
4H-Thiazolo[2,3-c][1,2,4]triazin-4-one, 6,7-dihydro-7,7-dimethyl-	EtOH	212(4.02),297(3.73)	44-4585-83
7H-Thiazolo[3,2-b][1,2,4]triazin-7-one, 2,3-dihydro-2,2-dimethyl-	EtOH	234(4.37)	44-4585-83
7H-Thiazolo[3,2-b][1,2,4]triazin-7-one, 2,3-dihydro-2,3-dimethyl-, cis	EtOH	235(4.36)	44-4585-83
trans	EtOH	236(4.35)	44-4585-83
$C_7H_9N_3O_3$			
2,4-Hexadienoic acid, 6-[(aminocarbonyl)hydrazono]-	5% Na_2CO_3	308(4.69)	111-0447-83
3-Pyridinamine, 6-ethoxy-2-nitro-	H_2O	225(4.16),425(3.71)	103-0303-83
$C_7H_9N_3O_4$			
1H-Pyrazole-3-carboxylic acid, 5-hydroxy-1-methyl-4-nitroso-, ethyl ester	MeOH	266(3.95),371(4.46)	94-1228-83

Compound	Solvent	$\lambda_{max}(\log \epsilon)$	Ref.
$C_7H_9N_3O_6$			
2(1H)-Pyridinone, 5,6-dihydro-6-methoxy-1-methyl-3,5-dinitro-, sodium salt	MeOH	449(4.49)	138-0715-83
$C_7H_9N_5$			
Imidazo[1,5-a]-1,3,5-triazin-4-amine, 8-ethyl-	EtOH-pH 1	274(3.85),309(3.74)	44-0003-83
	EtOH-pH 7	271(3.94),322(3.61)	44-0003-83
	EtOH-pH 13	286(3.95),307(3.92)	44-0003-83
3H-Purin-6-amine, N,3-dimethyl-	pH 1	282(4.27)	94-4270-83
	pH 7	285(4.16)	94-4270-83
	pH 13	286(4.15)	94-4270-83
	EtOH	291(4.13)	94-4270-83
monohydriodide	pH 1	224(4.35),282(4.27)	94-4270-83
	pH 7	224(4.39),285(4.17)	94-4270-83
	pH 13	223.5(4.40),286(4.16)	94-4270-83
	EtOH	290(4.13)	94-4270-83
6H-Purin-6-imine, 1,9-dihydro-1,9-dimethyl-, monohydriodide	pH 7	261(4.12)	104-2094-83
	pH 14	261(4.00)	104-2094-83
6H-Purin-6-imine, 3,7-dihydro-3,7-dimethyl-, monohydriodide	pH 7	276(4.23)	104-2094-83
perchlorate	pH 1	223(4.06),277(4.22)	94-3149-83
	pH 7	223(4.06),277(4.22)	94-3149-83
	pH 13	282(4.17)	94-3149-83
	EtOH	225(3.98),279(4.17)	94-3149-83
$C_7H_9N_5O$			
1H-Purin-6-amine, N-methoxy-N-methyl-	pH 1	283.5(4.23)	94-3149-83
	pH 7	215(4.20),278.5(4.16)	94-3149-83
	pH 13	280(4.14)	94-3149-83
	EtOH	278(4.16)	94-3149-83
1H-Purin-6-amine, N-methoxy-1-methyl-	pH 1	228(3.79),278.5(4.00)	94-3149-83
	pH 7	273.5(4.11)	94-3149-83
	pH 13	273(4.19)	94-3149-83
	EtOH	275(4.08)	94-3149-83
3H-Purin-6-amine, N-methoxy-3-methyl-	pH 1	223(4.01),283(4.27)	94-3149-83
	pH 7	297(4.18)	94-3149-83
	pH 13	287(4.07)	94-3149-83
	EtOH	295(4.16)	94-3149-83
$C_7H_9N_5O_2$			
2,4(3H,5H)-Pyrimidinedione, 5-diazo-6-(dimethylamino)-3-methyl-	MeCN	214(4.41),281(4.21)	35-4809-83
$C_7H_9N_5O_3$			
2,4(1H,3H)-Pyrimidinedione, 1-(2-azido-3-hydroxypropyl)-	EtOH	264(4.12)	128-0125-83
2,4(1H,3H)-Pyrimidinedione, 1-(3-azido-2-hydroxypropyl)-	EtOH	266(3.95)	128-0125-83
$C_7H_9N_5O_4$			
Acetic acid, 2-[3-(aminocarbonyl)-5-(hydroxyamino)-2-oxo-2H-pyrrol-4-yl]hydrazide	DMF	274(4.21),426(3.83)	48-0041-83
C_7H_{10}			
Bicyclo[4.1.0]hept-2-ene	pentane	200s(3.8)	35-0514-83
$C_7H_{10}BrNO$			
2,4-Pentadienal, 2-bromo-5-(dimethylamino)-, (Z,E)-	MeCN	379(4.77)	33-1427-83

Compound	Solvent	$\lambda_{max}(\log \epsilon)$	Ref.
$C_7H_{10}Br_2$			
Cyclohexane, (dibromomethylene)-	pentane	207(4.15),215(4.00), 219s(3.92)	35-6907-83
	MeOH	208(4.08),214(4.00), 220(3.90)	35-6907-83
$C_7H_{10}Cl_2$			
Cyclohexane, (dichloromethylene)-	pentane	200s(4.04),204(4.08), 208s(4.04)	35-6907-83
	MeOH	207(3.92)	35-6907-83
$C_7H_{10}N_2$			
2,3-Diazatricyclo[3.2.2.0^{5,7}]non-2-ene, endo	benzene	368s(2.06),379(2.33)	35-7102-83
2-Propenenitrile, 3-(2,2-dimethyl-1-aziridinyl)-, cis	heptane	258(4.15)	70-2322-83
	MeOH	262(4.07)	70-2322-83
$C_7H_{10}N_2OS$			
2,4-Diazabicyclo[4.1.0]hept-2-en-5-one, 4-methyl-3-(methylthio)-	EtOH	213(4.05)	39-2645-83C
2H-Isothiazolo[4,5-d]azepin-3(4H)-ol, 5,6,7,8-tetrahydro-, dihydrobromide	MeOH	264(3.52)	87-0895-83
$C_7H_{10}N_2O_3$			
2-Furanmethanamine, N,N-dimethyl-5-nitro-	MeOH	214(3.90),227(3.71), 312(4.13)	73-2682-83
$C_7H_{10}N_4O$			
Pyrazolo[3,4-d][1,3]diazepin-4-ol, 1,4,5,6-tetrahydro-1-methyl-	pH 1	244(3.84)	88-4789-83
	pH 11	274(4.06)	88-4789-83
	MeOH	275(4.01)	88-4789-83
2,3,6,7-Tetraazatricyclo[5.2.2.0^{5,9}]-undeca-2,6-diene, 2-oxide, endo	EtOH	230(3.83),322(2.62)	35-7102-83
$C_7H_{10}N_4OS$			
1H-Pyrimido[5,4-b][1,4]thiazin-4(6H)-one, 2-amino-7,8-dihydro-6-methyl-	pH 1	277s(3.80),305(3.99)	87-0559-83
	pH 6.8	223(4.33),267(3.77), 302(3.93)	87-0559-83
	pH 13	270(3.81),290(3.86)	87-0559-83
$C_7H_{10}N_4O_2S$			
1H-Pyrimido[5,4-b][1,4]thiazin-4(6H)-one, 2-amino-7,8-dihydro-6-methyl-, 5-oxide	pH 1	225(4.27),259(4.10)	87-0559-83
	pH 6.8	225(4.52),261(4.06)	87-0559-83
	pH 13	261(4.01)	87-0559-83
$C_7H_{10}N_4O_2S_2$			
Thiourea, N,N-dimethyl-N'-(4-methyl-5-nitro-2-thiazolyl)-	$CHCl_3$	300(3.78),362(4.20)	74-0075-83B
$C_7H_{10}N_5$			
7H-Purinium, 6-amino-7,9-dimethyl-, perchlorate	pH 7	270(4.12)	104-2094-83
$C_7H_{10}O$			
2-Butanone, 3-cyclopropylidene-	hexane	229(4.15)	78-3307-83
2-Cyclohepten-1-one	EtOH	225(4.09),328(1.46)	44-1670-83
1-Cyclopentene-1-carboxaldehyde, 5-methyl-	EtOH	237.5(4.01)	39-1901-83C
$C_7H_{10}O_2$			
2,4-Hexadienoic acid, methyl ester	EtOH	262(4.30)	70-1897-83

Compound	Solvent	λ_{max}(log ε)	Ref.
$C_7H_{10}O_3$			
2-Hexenoic acid, 4,5-epoxy-, methyl ester, (E)-	EtOH	220(4.14)	111-0447-83
$C_7H_{10}S$			
Bicyclo[2.2.1]heptane-2-thione	benzene	497(0.78)	44-0214-83
$C_7H_{11}BrNO_2S_3$			
Methanaminium, N-(3a-bromotetrahydro-thieno[3,4-d]-1,3-dithiol-2-ylidene)-N-methyl-, bromide, S,S-dioxide	H_2O	254(4.23)	103-1289-83
perchlorate	H_2O	253(4.26)	103-1289-83
$C_7H_{11}I$			
Cycloheptene, 1-iodo-	EtOH	263(2.54)	35-6907-83
Cyclohexane, (iodomethylene)-	EtOH	251(2.62)	35-6907-83
$C_7H_{11}NO$			
1-Azabicyclo[2.2.2]octan-3-one, conjugate acid	H_2O	283(1.16)	35-0265-83
1H-Pyrrole-2-methanol, α,1-dimethyl-	MeOH	222(3.87)	44-2488-83
$C_7H_{11}NO_2$			
3H-Pyrrol-3-one, dihydro-5-methoxy-4,4-dimethyl-	$CHCl_3$	263(2.56)	49-0983-83
$C_7H_{11}NO_2S$			
Acetamide, N-(2,2-dimethyl-4-oxo-3-thietanyl)-, (R)-	EtOH	237(3.25)	39-2259-83C
$C_7H_{11}NO_3$			
2-Butenoic acid, 2-(acetylamino)-3-methyl-	EtOH	225(3.84)	39-2259-83C
$(C_7H_{11}NO_3)_n$			
L-Alanine, N-(2-methyl-1-oxo-2-propenyl)-, homopolymer	H_2O	190(3.72)	116-1564-83
$C_7H_{11}NO_3S$			
Propanoic acid, 3-(dimethylamino)-2-oxo-3-thioxo-, ethyl ester	MeCN	255(4.32),299(3.62), 366(3.27)	5-1694-83
1H-Pyrrole-1-ethanol, methanesulfonate	MeOH	220(3.64)	23-0454-83
$C_7H_{11}N_3O_2$			
1H-Imidazole-5-carboxamide, 1-ethyl-4-methoxy-	H_2O	258(4.15)	4-0875-83
	M HCl	244(4.03)	4-0875-83
	M NaOH	258(4.15)	4-0875-83
1H-Imidazolium, 4-(aminocarbonyl)-3-ethyl-5-hydroxy-1-methyl-, hydroxide, inner salt	H_2O	236(3.67),279(4.12)	4-0875-83
	M HCl	241(3.70),280(3.99)	4-0875-83
	M NaOH	239(3.57),279(4.12)	4-0875-83
$C_7H_{11}N_5$			
1H-Purin-6-amine, 2,3-dihydro-1,3-dimethyl-	pH 1	225(4.00),317.5(3.81)	94-3521-83
	pH 7	319(3.81)	94-3521-83
	pH 13	235s(3.91),303(3.79)	94-3521-83
	EtOH	240s(3.83),327(3.92)	94-3521-83
monohydriodide	pH 1	317(3.81)	94-3521-83
	pH 7	319(3.81)	94-3521-83
	pH 13	303(3.78)	94-3521-83
	EtOH	323(3.88)	94-3521-83

Compound	Solvent	λ_{max}(log ε)	Ref.
$C_7H_{11}N_5O_2$			
2,4,5(3H)-Pyrimidinetrione, 6-(dimethylamino)-3-methyl-, 5-hydrazone	MeCN	297(4.09)	35-4809-83
$C_7H_{11}N_5O_3$			
1H-Imidazole-4-carboxylic acid, 5-[2-(aminocarbonyl)hydrazino]-, ethyl ester	n.s.g.	223(3.55),267(4.14)	103-1235-83
$C_7H_{12}BrI$			
Cyclohexane, (bromoiodomethyl)-	EtOH	220(3.14),276(2.60)	44-2084-83
$C_7H_{12}Br_2$			
Cyclohexane, (dibromomethyl)-	EtOH	224(2.57)	44-2084-83
$C_7H_{12}I_2$			
Cyclohexane, (diiodomethyl)-	EtOH	253(2.87),292(3.10)	44-2084-83
	CH_2Cl_2	248(2.89),291(3.12)	44-2084-83
$C_7H_{12}N_2O$			
Ethanone, 1-(4,5-dihydro-5,5-dimethyl-1H-pyrazol-3-yl)-	C_6H_{12}	276(4.18)	44-3189-83
Furazan, butylmethyl-	EtOH	218(4.02)	94-2269-83
3-Pyridinecarboxamide, 1,4,5,6-tetrahydro-1-methyl-	EtOH	300(4.45)	150-2326-83M
$C_7H_{12}N_2OS_2$			
Propanedithioamide, N,N,N',N'-tetramethyl-2-oxo-	MeCN	267(4.29),302(3.99), 374(2.42)	5-1116-83 +5-1694-83
$C_7H_{12}N_2O_2$			
1H-Pyrazole-3-carboxylic acid, 4,5-dihydro-5,5-dimethyl-, methyl ester	C_6H_{12}	287(4.18)	44-3189-83
$C_7H_{12}N_2O_2S$			
Propanamide, 3-(dimethylamino)-N,N-dimethyl-2-oxo-3-thioxo-	MeCN	252(4.05),296(3.62), 366s(2.18)	5-1694-83
$C_7H_{12}N_2O_3$			
Propanediamide, N,N,N',N'-tetramethyl-2-oxo-	EtOH	212(3.98),255(3.81), 347(2.01)	5-1694-83
$C_7H_{12}N_4O_2S$			
1H-Imidazole-4-carboxylic acid, 5-hydrazino-2-(methylthio)-, ethyl ester, hydrochloride	n.s.g.	281(4.36)	103-1235-83
$C_7H_{12}N_4O_3$			
1H-Imidazole-4-carboxamide, 5-amino-2-[(2-hydroxyethoxy)methyl]-	EtOH	243(3.89),273(4.08)	4-1169-83
$C_7H_{12}O$			
Ethanone, 1-(2,2-dimethylcyclopropyl)-	C_6H_{12}	281(2.43)	88-3025-83
3-Hexen-2-one, 5-methyl-	MeOH	221(4.09),306(1.48)	44-0584-83
4-Hexen-2-one, 5-methyl-	MeOH	283(1.99)	44-0584-83
$C_7H_{12}O_2$			
2-Butenoic acid, 2,3-dimethyl-, methyl ester	MeOH	223(3.91)	24-0264-83

Compound	Solvent	$\lambda_{max}(\log \epsilon)$	Ref.
$C_7H_{12}O_2S$			
Sulfonium, dimethyl-, 1-acetyl-2-oxo-propylide	hexane	244(4.02),268(3.94)	99-0023-83
	EtOH	233(3.9),271(4.05)	99-0023-83
$C_7H_{12}O_3S$			
Sulfoxonium, dimethyl-, 1-acetyl-2-oxopropylide	EtOH	216(--),262(4.03)	99-0023-83
	ether	220(3.85),257(3.97)	99-0023-83
$C_7H_{12}O_4S$			
Sulfonium, dimethyl-, 2-methoxy-1-(methoxycarbonyl)-2-oxoethylide	n.s.g.	230(4.11),240(4.11)	121-0433-83B
$C_7H_{12}S_4$			
1,4,7,10-Tetrathiaspiro[5.5]undecane	n.s.g.	208.5(3.44),230.8(3.05)	4-0033-83
$C_7H_{13}I$			
2-Pentene, 3-iodo-2,4-dimethyl-	EtOH	256(2.53)	35-6907-83
$C_7H_{13}NO$			
3-Buten-2-one, 4-(dimethylamino)-3-methyl-	MeOH	304(4.42)	131-0001-83K
(E-syn-E)-	C_6H_{12}	287(4.37)	131-0001-83K
3-Buten-2-one, 4-(ethylamino)-3-methyl-	MeOH	298(4.49)	131-0001-83K
(E-syn-E)-	C_6H_{12}	277(--)	131-0001-83K
	dioxan	283(4.39)	131-0001-83K
(Z-syn-Z)-	C_6H_{12}	312(4.08)	131-0001-83K
	dioxan	283(--)	131-0001-83K
1-Penten-3-one, 2-methyl-1-(methyl-amino)-	MeOH	297(4.40)	131-0001-83K
(E-syn-E)-	C_6H_{12}	277(--)	131-0001-83K
	dioxan	285(4.29)	131-0001-83K
(Z-syn-Z)-	C_6H_{12}	310(4.08)	131-0001-83K
	dioxan	285(--)	131-0001-83K
$C_7H_{13}NO_3$			
L-Alanine, N-(2-methyl-1-oxopropyl)-	H_2O	187(3.90)	116-1564-83
$C_7H_{13}N_3S$			
2,4-Thiazolediamine, N,N,N',N'-tetra-methyl-, compd. with 2,2'-(2,5-cyclohexadiene-1,4-diylidene)-bis[propanedinitrile](1:1)	MeCN	275(4.59),392(4.48), 662(4.28),742(4.37), 762(4.30),820s(4.49), 845(4.65)	88-3563-83
$C_7H_{14}NO$			
4H-1,2-Oxazinium, 2-ethyl-5,6-dihydro-3-methyl-, tetrafluoroborate	MeCN	213(3.66)	5-0897-83
2H-Pyrrolium, 1-ethoxy-3,4-dihydro-5-methyl-, tetrafluoroborate	MeCN	201(3.68)	5-0897-83
$C_7H_{14}N_2O_2$			
1-Buten-2-amine, N,3,3-trimethyl-N-nitro-	H_2O	238(3.69)	104-1270-83
1-Buten-2-amine, N,3,3-trimethyl-1-nitro-	H_2O	237(3.32),343(3.20)	104-1270-83
$C_7H_{14}OSi$			
2-Butenal, 3-(trimethylsilyl)-	C_6H_{12}	235(4.20),335(1.66), 347(1.72),362(1.75), 377(1.63),395(1.23)	157-0332-83

Compound	Solvent	$\lambda_{max}(\log \epsilon)$	Ref.
3-Buten-2-one, 4-(trimethylsilyl)-	C_6H_{12}	220(4.14),338(1.68), 350(1.63),365(1.43)	157-0332-83
2-Propenal, 2-methyl-3-(trimethyl-silyl)-	C_6H_{12}	232(3.80),335(1.49), 348(1.46),363(1.30), 380(0.85)	157-0332-83
$C_7H_{16}S_2Si$			
Silane, 1,3-dithian-2-yltrimethyl-	hexane	241(2.43)	101-0143-83L

Compound	Solvent	$\lambda_{max}(\log \epsilon)$	Ref.
$C_8Cl_4N_2$			
Propanedinitrile, (2,3,4,5-tetra-chloro-2,4-cyclopentadien-1-ylidene)-	C_6H_{12}	306(4.37),320(4.40), 541(2.14)	5-0154-83
C_8Cl_6O			
1(2H)-Pentalenone, 2,2,3,4,5,6-hexa-chloro-	hexane	245(4.17),274(4.21), 285s(4.11),318(3.94), 331(4.06),346(3.97), 415(2.97)	18-0481-83
1(5H)-Pentalenone, 2,3,4,5,5,6-hexa-chloro-	hexane	260(4.32),270(4.34)	18-0481-83
C_8Cl_8			
Pentalene, 1,1,2,3,4,5,5,6-octachloro-1,5-dihydro-	hexane	262(4.01),325(3.38)	18-0481-83
Pentalene, 1,1,2,3,4,5,6,6a-octa-chloro-1,6a-dihydro-	hexane	260(3.99),352(3.67)	18-0481-83
C_8Cl_8O			
1(3aH)-Pentalenone, 2,3,3a,6a,?,?,?,?-octachlorodihydro-	hexane	248(3.90)	18-0481-83
$C_8Cl_{10}O$			
1(2H)-Pentalenone, 2,2,3,3,3a,6a-?,?,?,?-decachlorotetrahydro-	hexane	230(3.98)	18-0481-83
$C_8HCl_5O_2$			
1(6aH)-Pentalenone, 2,4,5,6,6a-penta-chloro-3-hydroxy-	hexane	220(3.62),276(4.56), 357(4.23)	18-0481-83
C_8HCl_7			
Benzene, pentachloro(1,2-dichloro-ethenyl)-, (E)-	EtOH	216(4.81),240(4.08), 293(2.77),300(2.81)	1-0823-83
(Z)-	EtOH	216(4.81),242(4.11), 295(2.82),304(2.80)	1-0823-83
Benzene, pentachloro(2,2-dichloro-ethenyl)-	EtOH	217(4.70),240(4.15), 290(2.12),300(2.58)	1-0823-83
C_8HN_5			
1-Propene-1,1,2,3,3-pentacarbonitrile, cesium salt	MeOH	200f(--),250(--), 400(--),405(--)	46-4641-83
$C_8H_2Cl_6$			
Benzene, pentachloro(1-chloroethenyl)-	EtOH	216(4.80),239(3.99), 292(2.70),297(2.73)	1-0823-83
Benzene, pentachloro(2-chloroethenyl)-, (E)-	EtOH	227(4.55),244(4.30), 310(2.60),325(2.13)	1-0823-83
(Z)-	EtOH	213(4.65),240(4.06), 290(2.47),298(2.46)	1-0823-83
$C_8H_3ClN_2S$			
Thieno[2,3-b]pyridine-6-carbonitrile, 4-chloro-	EtOH	248(4.58),289(3.98), 321(3.74)	4-0213-83
$C_8H_3Cl_5$			
Benzene, pentachloroethenyl-	EtOH	221(4.64),240(4.24), 297(2.50)	1-0823-83
$C_8H_3N_3O_2$			
5H-1-Pyrindine-5,7(6H)-dione, 6-diazo-	MeOH	207(4.37),229(4.35), 248s(4.18),309(4.00)	30-0289-83

Compound	Solvent	λ_{max}(log ε)	Ref.
5H-2-Pyrindine-5,7(6H)-dione, 6-diazo-	MeOH	208s(4.33),219(4.40), 248(4.31),287(3.87)	30-0289-83
$C_8H_4AgNO_2$			
1H-Isoindole-1,3(2H)-dione, silver salt	MeCN-base	230(4.19),236(4.19), 226[sic](3.96)	23-1890-83
C_8H_4ClNOS			
8H-Thieno[3,2-c]azepin-8-one, 7-chloro-	EtOH	255(4.2),290(4.0), 350(3.8)	138-0145-83
$C_8H_4ClNO_4S$			
4H,5H-Pyrano[3,4-e][1,3]oxazine-4,5-dione, 7-chloro-2-(methylthio)-	$CHCl_3$	282(4.05),344(4.21)	24-3725-83
$C_8H_4F_4HgO_3$			
Mercury, (5-fluoro-2-hydroxyphenyl)-[trifluoroacetato-O]-	DMSO	280(3.45)	69-1342-83
$C_8H_4N_2$			
1,2-Benzenedicarbonitrile	EtOH	237(4.02),276(3.05), 282(3.22),291(3.23)	35-0040-83
1,3-Benzenedicarbonitrile	EtOH	206(4.66),272(2.56), 280(2.69),288(2.73)	35-0040-83
1,4-Benzenedicarbonitrile	EtOH	235(4.35),248(4.35), 278(3.25),287(2.73)	35-0040-83
$C_8H_4N_2O_4$			
1H-1,3-Dioxolo[4,5-f]indazole-4,8-dione	acetone	247(3.11),317(3.53)	64-0392-83B
$C_8H_4N_2S$			
Thieno[2,3-b]pyridine-6-carbonitrile	EtOH	243(4.46),286(3.81)	4-0213-83
Thieno[3,2-b]pyridine-5-carbonitrile	EtOH	244(4.68)	4-0213-83
Thieno[3,2-c]pyridine-4-carbonitrile	EtOH	236(4.51)	4-0213-83
$C_8H_4N_4O_6$			
Furazan, (2,4-dinitrophenyl)-, 2-oxide	10% MeOH	245(4.24)	32-0811-83
	+ NaOH	402(3.89)	32-0811-83
$C_8H_4O_3$			
1,3-Isobenzofurandione	EtOH	224(3.90),276(3.05), 281(3.02)	73-0112-83
$C_8H_5BrF_3NO$			
Acetamide, N-(2-bromophenyl)-2,2,2-trifluoro-	EtOH	203(4.40)	78-3767-83
$C_8H_5BrN_2OS$			
1,3,4-Oxadiazole-2(3H)-thione, 5-(3-bromophenyl)-	EtOH	255(4.12),305(4.17)	142-2211-83
1,3,4-Oxadiazole-2(3H)-thione, 5-(4-bromophenyl)-	EtOH	240(4.13),305(4.18)	142-2211-83
$C_8H_5Br_2NO_3$			
Ethanone, 2,2-dibromo-1-(2-nitrophenyl)-	EtOH	211.0(4.07),262.0(3.77)	23-0400-83
$C_8H_5ClN_2OS$			
1,3,4-Oxadiazole-2(3H)-thione, 5-(3-chlorophenyl)-	EtOH	257(4.11),310(4.17)	142-2211-83

Compound	Solvent	λ_{max}(log ϵ)	Ref.
1,3,4-Oxadiazole-2(3H)-thione, 5-(4-chlorophenyl)-	EtOH	248(4.50),340(4.16)	142-2211-83
C_8H_5NOS			
2-Benzothiazolecarboxaldehyde	EtOH	285(3.45)	140-0215-83
$C_8H_5NO_2$			
1H-Isoindole-1,3(2H)-dione	EtOH	229s(4.25),238s(4.08), 290(3.27),298(3.25)	73-0112-83
$C_8H_5NO_3$			
Compd. from p-coumaric acid and $NaNO_2$	pH 1	229(4.13),317(4.09)	98-0780-83
	H_2O	229(4.11),261(3.52)	98-0780-83
	pH 13	268(4.16),291(4.11)	98-0780-83
2-Octene-4,6-diynoic acid, 8-amino-8-oxo-, (E)-	EtOH	224(4.58),274(4.22), 291(4.36),309.5(4.28)	162-0433-83
1H-Pyrano[4,3-c]pyridine-1,4(3H)-dione	MeOH	263(3.55),281(3.63), 365(3.15)	103-0882-83
6H-Pyrano[3,4-b]pyridine-5,8-dione	MeOH	246(3.59),344(3.19)	103-0882-83
$C_8H_5N_3O_3S$			
1,3,4-Oxadiazole-2(3H)-thione, 5-(3-nitrophenyl)-	EtOH	224(4.40),330(4.10)	142-2211-83
1,3,4-Oxadiazole-2(3H)-thione, 5-(4-nitrophenyl)-	EtOH	257(4.35),340(4.16)	142-2211-83
	pH 1	342(--)	142-2211-83
	pH 12.7	386(--)	142-2211-83
$C_8H_5N_3O_4$			
Furazan, (2-nitrophenyl)-, 3-oxide	10% MeOH	258(4.01)	32-0811-83
	+ NaOH	303(4.05)	32-0811-83
Furazan, (4-nitrophenyl)-, 3-oxide	10% MeOH	255(3.96),309(4.12)	32-0811-83
	+ NaOH	385(4.25)	32-0811-83
$C_8H_5N_5S$			
10H-2,3,6,9-Tetrazaphenothiazine	EtOH	273(4.14),278(4.03)	4-1047-83
C_8H_6BrN			
1H-Indole, 5-bromo-	EtOH	226(4.67),281(3.77), 290(3.59)	128-0157-83
	EtOH-NaOH	228(--),282(--), 284(--)	128-0157-83
	C_6H_{12}	224(--),267(--), 273(--),286(--)	128-0157-83
	MeOH	224(--),277(--), 287(--)	128-0157-83
$C_8H_6BrNO_3$			
Ethanone, 2-bromo-1-(2-nitrophenyl)-	EtOH	213.0(4.00),261.5(3.78)	23-0400-83
$C_8H_6Br_3NO_3$			
2,5-Cyclohexadien-1-one, 3,4,5-tribromo-2,6-dimethyl-4-nitro-	C_6H_{12}	267(4.14)	12-0839-83
$C_8H_6Br_4O$			
2,5-Cyclohexadien-1-one, 3,4,4,5-tetrabromo-2,6-dimethyl-	$CHCl_3$	269(4.19)	12-0839-83
$C_8H_6ClNS_2$			
2H-1,4-Benzothiazine-3(4H)-thione, 6-chloro-	MeOH	280(4.39)	87-0845-83

Compound	Solvent	$\lambda_{max}(\log \epsilon)$	Ref.
2H-1,4-Benzothiazine-3(4H)-thione, 7-chloro-	MeOH	280(4.38)	87-0845-83
$C_8H_6ClN_3$			
1-Phthalazinamine, 4-chloro-	MeOH	212(4.68),233(4.00), 260(3.63),316(3.88)	103-0666-83
	MeCN	212(4.56),230s(3.85), 256(3.48),310(3.75)	103-0666-83
sulfate	MeOH	213(4.64),265(3.76), 295(3.85)	103-0666-83
	MeCN	220(4.60),244s(4.18), 267(3.95),294(3.70)	103-0666-83
$C_8H_6ClN_5$			
1H-Tetrazol-1-amine, 5-chloro-N-(phenylmethylene)-	MeOH	216(3.98),285(4.31)	35-0902-83
$C_8H_6Cl_2$			
Benzene, (2,2-dichloroethenyl)-	n.s.g.	260(4.20)	1-0823-83
$C_8H_6F_4O$			
2,5-Cyclohexadien-1-one, 2,3,5,6-tetrafluoro-4,4-dimethyl-	heptane	229(4.02),269(2.7)	104-0884-83
$C_8H_6N_2OS$			
1,3,4-Oxadiazole-2(3H)-thione, 5-phenyl-	EtOH	250(4.05),295(4.32)	142-2211-83
Thieno[2,3-b]pyridine-6-carboxamide	EtOH	241(4.38),286(4.14)	4-0213-83
Thieno[3,2-b]pyridine-5-carboxamide	EtOH	241(4.72)	4-0213-83
$C_8H_6N_2O_2$			
Furoxan, 3-phenyl-	10% MeOH	231(4.20),282(3.94)	32-0811-83
	+ NaOH	328(4.16)	32-0811-83
$C_8H_6N_2O_2S$			
Thieno[2,3-b]pyridine, 2-methyl-5-nitro-	EtOH	220(4.04),264(4.36), 292(3.87)	103-1163-83
$C_8H_6N_2O_3$			
Phenol, 4-furazanyl-, N-oxide	pH 1	239(4.20),286(3.98), 295s(--)	98-0780-83
	H_2O	238(4.19),286(3.97), 295s(--)	98-0780-83
	pH 13	248(4.12),352(4.19)	98-0780-83
$C_8H_6N_2S_2$			
Thieno[2,3-b]pyridine-6-carbothioamide	EtOH	244(4.26),300s(4.12), 322(4.20)	4-0213-83
Thieno[3,2-b]pyridine-5-carbothioamide	EtOH	251(4.59),322(4.09)	4-0213-83
$C_8H_6N_4$			
1H-Benzotriazole, 1-(isocyanomethyl)-	EtOH	253(3.82),266(3.76), 283(3.62)	94-0723-83
1H-Indole, 4-azido-	EtOH	223(4.81),292(4.33)	44-5130-83
1H-Indole, 5-azido-	EtOH	244(4.38),300(3.60), 311(3.43)	44-5130-83
C_8H_6O			
Bicyclo[4.2.0]octa-1,3,5-trien-7-one	3-Mepentane at 77°K	250(4.04),290(3.51)	151-0131-83B

Compound	Solvent	$\lambda_{max}(\log \epsilon)$	Ref.
Methanone, (4-methylene-2,5-cyclohexadien-1-ylidene)-	3-Mepentane at 77°K	400(3.51)	151-0131-83B
$C_8H_6O_4$			
1,2-Benzenedicarboxylic acid	EtOH	225s(3.89),278s(3.105), 282(3.105),294s(2.87)	73-0112-83
1,3-Benzodioxole-4,5-dione, 6-methyl-	MeCN	214(3.82),348(3.08), 570(2.19)	64-0752-83B
1,4-Cyclohexadiene-1-carboxylic acid, 3,6-dioxo-, methyl ester	hexane	241.5(4.12),300(2.60)	131-0055-83N
$C_8H_6S_4$			
1,3-Dithiole, 2,2'-(1,2-ethanediylidene)bis-	$CHCl_3$	387(4.31),406(4.33)	88-3469-83
$C_8H_7BrO_4$			
Benzaldehyde, 3-bromo-2,5-dihydroxy-4-methoxy-	MeOH	275(2.60),372(3.05)	78-0667-83
	MeOH-NaOH	216(3.48),314(3.95), 411(2.57)	78-0667-83
Benzeneacetic acid, 5-bromo-2,3-dihydroxy-	EtOH	208(4.44),290(3.44)	104-0458-83
C_8H_7Cl			
Benzene, (2-chloroethenyl)-, trans	n.s.g.	254(4.22)	1-0823-83
$C_8H_7ClN_2O_2S$			
2H-1,2,4-Benzothiadiazine, 7-chloro-3-methyl-, 1,1-dioxide	MeOH	268(4.05)	162-0435-83
$C_8H_7ClN_4O_3$			
7H-Pyrrolo[2,3-d]pyrimidine, 2-chloro-4-methoxy-7-methyl-6-nitro-	MeOH	240(4.00),342(4.10)	5-0137-83
C_8H_7ClO			
Ethanone, 1-(3-chlorophenyl)-	MeOH	240(4.00)	44-4808-83
Ethanone, 1-(4-chlorophenyl)-	MeOH	250(4.21)	44-4808-83
$C_8H_7ClO_3$			
Benzaldehyde, 5-chloro-2-hydroxy-3-methoxy-	EtOH	227(4.16),268(3.85), 354(3.41)	104-0458-83
$C_8H_7ClO_4$			
Benzeneacetic acid, 5-chloro-2,3-dihydroxy-	EtOH	210(4.40),290(3.45)	104-0458-83
$C_8H_7Cl_2NO_3$			
2,5-Cyclohexadiene-1-acetamide, 3,5-dichloro-1-hydroxy-4-oxo-	MeOH	245(3.95)	31-1091-83
C_8H_7FO			
Ethanone, 1-(4-fluorophenyl)-	MeOH	243(4.07)	44-4808-83
$C_8H_7FO_2$			
Benzeneacetic acid, 4-fluoro-	pH 1.2	264(2.8),270(2.8)	41-0559-83
	pH 3.95	264(2.9),270(2.8)	41-0559-83
	pH 10	264(2.9),270(2.9)	41-0559-83
$C_8H_7HgN_3O_2$			
Mercury, (acetato-O)(4-azidophenyl)-	DMSO	380(4.51)	69-1342-83

Compound	Solvent	$\lambda_{max}(\log \epsilon)$	Ref.
C_8H_7N			
1H-Indole	EtOH	225(4.53),272(3.78), 288(3.65)	128-0157-83
	C_6H_{12}	217(--),268(--), 289(--)	128-0157-83
	MeOH	224(--),273(--), 288(--)	128-0157-83
	EtOH-NaOH	226(--),273(--), 289(--)	128-0157-83
C_8H_7NO			
2-Propenal, 3-(4-pyridinyl)-	EtOH	340(4.43)	140-0215-83
C_8H_7NOS			
Benzoxazole, 2-(methylthio)-	$CHCl_3$	280(4.13),288(4.12)	94-1733-83
2(3H)-Benzoxazolethione, 3-methyl-	$CHCl_3$	309(4.48)	94-1733-83
Thieno[2,3-b]pyridine, 4-methyl-, 7-oxide	EtOH	236(4.54),285(3.89), 312(3.70),322(3.72)	4-0213-83
C_8H_7NOSe			
1,2-Benzisoselenazole, 3-methyl-, 2-oxide	EtOH	238(4.29),296(4.10)	78-0831-83
$C_8H_7NO_2$			
Acetamide, N-(4-oxo-2,5-cyclohexadien-1-ylidene)-	hexane	263(4.52),376(2.20)	162-0009-83
Benzene, 1-ethenyl-4-nitro-	M KOH	311(4.11)	35-0265-83
Benzene, (2-nitroethenyl)-	n.s.g.	309(4.22)	22-0339-83
2-Cyclopenten-1-one, 2-ethenyl-5-hydroxy-5-isocyano-, (+)-	MeOH	236(4.24)	158-83-98
2-Pyridinecarboxaldehyde, 6-acetyl-	n.s.g.	231(3.82),272(3.58)	64-0516-83B
Spiro[6-oxabicyclo[3.1.0]hex-3-en-2,2'-oxirane], 4-isocyano-3'-methyl- (isonitrin A)	MeOH	230.4(4.14)	158-83-98
$C_8H_7NO_2S_2$			
2,4-Thiazolidinedione, 5-(3-methyl-2-thienyl)-	$CHCl_3$	210(2.49),238(--), 282(2.41),407(3.17)	13-0339-83A
$C_8H_7NO_3$			
Benzeneacetaldehyde, 4-hydroxy-α-(hydroxyimino)-	pH 1	225(4.00),299(4.06)	98-0780-83
	H_2O	228(4.04),304(4.09)	98-0780-83
	pH 13	256(3.90),284(3.95), 360(4.31)	98-0780-83
Ethanone, 2-hydroxy-1-(2-nitrosophenyl)-	EtOH	286(3.68),310(3.67)	23-0400-83
Ethanone, 1-(3-nitrophenyl)-	MeOH	227(4.35)	44-4808-83
Ethanone, 1-(4-nitrophenyl)-	MeOH	263(4.14)	44-4808-83
Oxirane, (2-nitrophenyl)-	EtOH	206.5(4.14),261.5(3.81)	23-0400-83
$C_8H_7NO_6$			
Benzoic acid, 4-hydroxy-3-methoxy-5-nitro-	NaOH	278(4.32),430(3.63)	104-1845-83
$C_8H_7NS_2$			
2H-1,4-Benzothiazine-3(4H)-thione	MeOH	275(4.32)	87-0845-83
$C_8H_7N_2$			
Quinoxalinyl, dihydro-	acid	357(4.15)	125-0655-83

Compound	Solvent	$\lambda_{max}(\log \epsilon)$	Ref.
$C_8H_7N_3OS$			
Pyrido[2,3-d]pyrimidin-4(1H)-one, 2,3-dihydro-1-methyl-2-thioxo-	MeOH	293(5.19)	73-3315-83
4H-Pyrido[3,2-e]-1,3-thiazin-4-one, 2-(methylamino)-	MeOH	228(4.08)	73-3315-83
$C_8H_7N_3O_2$			
1,2,4-Triazolidine-3,5-dione, 4-phenyl-, 1,4-dimethoxybenzene complex	CH_2Cl_2	457(c.2),545s(--), 580s(--)	44-1708-83
1,2,4-Triazolo[4,3-a]pyridinium, 1-acetyl-2,3-dihydro-3-oxo-, hydroxide, inner salt	THF	374(3.79)	18-2969-83
1,2,4-Triazolo[4,3-a]pyridin-3(2H)-one, 2-acetyl-	THF	351(3.57)	18-2969-83
$C_8H_7N_3O_5$			
Benzenecarboximidic acid, 3,5-dinitro-, methyl ester	MeOH-NaOMe	378(2.9),480(2.8)	18-1206-83
$C_8H_7N_3S$			
Thieno[2,3-b]pyridine-6-carboximidamide	EtOH	245(4.41),291(4.19)	4-1717-83
$C_8H_7N_4O_7$			
Benzofurazan, 4,?-dihydro-4,4-dimethoxy-5,7-dinitro-, ion(1-)-	MeOH	256(3.65),302.5(3.93), 338(3.65),453.5(4.50)	12-0297-83
$C_8H_7N_5O$			
1H-Pyrrolo[2,3-d]pyrimidine-5-carbonitrile, 2-amino-4-methoxy-, monohydrochloride	MeOH	225(4.27),292(3.68)	88-0495-83
5H-Tetrazol-5-one, 1,2-dihydro-1-[(phenylmethylene)amino]-	MeOH	215(3.98),266s(--), 301(4.10)	35-0902-83
$C_8H_7N_7O_2$			
Pyrimido[5,4-e]tetrazolo[1,5-b]pyridazine-7,9(6H,8H)-dione, 6,8-dimethyl-	EtOH	235(4.45),295(3.79)	44-1628-83
Pyrimido[5,4-e]-1,2,4-triazolo[4,3-b]-[1,2,4]triazine-7,9(6H,8H)-dione, 6,8-dimethyl-	EtOH	232(4.25),340(3.63)	44-1628-83
C_8H_8AsN			
1H-1,3-Benzazarsole, 1-methyl-	MeOH	235(4.47),258s(3.72), 297s(3.46),319(3.77), 325s(3.76)	101-0257-83R
C_8H_8ClNS			
Benzeneethanethioamide, 4-chloro-	heptane	268(4.03),375(2.30)	80-0555-83
	EtOH	273(4.04),325(2.00)	80-0555-83
$C_8H_8ClN_3O$			
7H-Pyrrolo[2,3-d]pyrimidine, 2-chloro-4-methoxy-7-methyl-	MeOH	224(4.33),268(3.84)	5-0137-83
$C_8H_8ClN_3OS$			
3-Pyridinecarboxamide, 2-chloro-N-[(methylamino)thioxomethyl]-	MeOH	235(5.09)	73-3315-83
$C_8H_8Cl_2O$			
Benzene, 1-(dichloromethyl)-3-methoxy-	EtOH	225(3.94),282(3.57)	142-0481-83

Compound	Solvent	λ_{max}(log ε)	Ref.
$C_8H_8N_2$			
1H-Indazole, 1-methyl-	pH 7	254(3.53),292(3.72)	138-0805-83
2H-Indazole, 2-methyl-	pH 7	275(3.80),295(3.78)	138-0805-83
$C_8H_8N_2O_2S$			
2-Pyridinecarbothioic acid, 6-(amino-carbonyl)-, S-methyl ester	n.s.g.	275(3.81)	64-0516-83B
$C_8H_8N_2O_3$			
2-Pyridinecarboxylic acid, 5-(acetyl-amino)-	MeOH	256(4.15),287(3.99)	106-0591-83
$C_8H_8N_2O_3S$			
2-Propenal, 3-[(5-methyl-2-thienyl)-amino]-2-nitro-	EtOH	224(3.92),280(3.77), 387(4.20)	103-1163-83
$C_8H_8N_2S$			
2(1H)-Quinazolinethione, 3,4-dihydro-	MeOH	277(4.17)	83-0379-83
$C_8H_8N_2S_2$			
2H-Imidazole-2-thione, 1,3-dihydro-1-(2-thienylmethyl)-	0.5M HCl	237(4.13),260(4.05)	142-2019-83
	0.5M NaOH	237(4.24)	142-2019-83
	MeOH	235(4.17),265(4.11)	142-2019-83
$C_8H_8N_4O$			
Formamide, N-(1H-benzotriazol-1-yl-methyl)-	EtOH	254(3.80),260(3.78), 281(3.63)	94-0723-83
$C_8H_8N_4O_2$			
Pyrido[3,4-d]-1,2,3-triazin-4(3H)-one, 6-methoxy-3-methyl-	EtOH	211(4.38),294s(--), 303(4.11),340s(--)	95-1129-83
Pyrido[3,4-d]-1,2,3-triazin-4(3H)-one, 8-methoxy-3-methyl-	EtOH	217(4.22),233(4.10), 294(3.86),304s(--), 318(3.65)	95-1129-83
$C_8H_8N_4O_6$			
Acetamide, N-(2-amino-6-hydroxy-3,5-dinitrophenyl)-	EtOH	221s(3.91),239s(3.65), 290s(3.90),326(4.14), 400(3.92)	12-1227-83
C_8H_8O			
Acetophenone	hexane	238(4.10)	99-0034-83
	MeOH	241(4.09)	44-4808-83
photooxidant product	EtOH	278(3.00),320(1.70)	98-0734-83
C_8H_8OS			
Benzenecarbothioic acid, 2-methyl-, cesium salt	EtOH	259(3.74)	64-1585-83B
potassium salt	EtOH	260(3.66)	64-1585-83B
rubidium salt	EtOH	256(3.81)	64-1585-83B
sodium salt	EtOH	258(3.69)	64-1585-83B
Benzenecarbothioic acid, 3-methyl-, cesium salt	EtOH	285(3.81),294s(3.75)	64-1585-83B
potassium salt	EtOH	291(3.63),296s(3.57)	64-1585-83B
rubidium salt	EtOH	286(3.78),294s(3.74)	64-1585-83B
sodium salt	EtOH	284(3.71),294s(3.65)	64-1585-83B
Benzenecarbothioic acid, 4-methyl-, cesium salt	EtOH	242(4.06),283(3.93), 291s(3.91)	64-1585-83B
lithium salt	EtOH	241(3.93),284(3.74), 286s(3.73)	64-1585-83B
potassium salt	EtOH	283(3.84),293(3.84)	64-1585-83B

Compound	Solvent	λ_{max}(log ε)	Ref.
Benzenecarbothioic acid, 4-methyl-, rubidium salt	EtOH	283(3.85),294(3.84)	64-1585-83B
sodium salt	EtOH	285(3.81),299(3.80)	64-1585-83B
Ethanone, 1-(4-mercaptophenyl)-	hexane	223(3.90),277(4.20)	99-0034-83
$C_8H_8O_2$			
Ethanone, 1-(4-hydroxyphenyl)-	hexane	258(4.10)	99-0034-83
Methanediol, 2,4,6-cycloheptatrien-1-ylidene-, dilithium salt	THF-hexane at -25°C	654(3.15)	24-1777-83
3-Oxatetracyclo[3.3.1.0^{2,4}.0^{6,8}]nonan-9-one	benzene	284(1.40)	44-1718-83
$C_8H_8O_2S$			
Benzenecarbothioic acid, 4-methoxy-, cesium salt	EtOH	257(4.07),292(3.99)	64-1585-83B
lithium salt	EtOH	256(3.88),293(3.81)	64-1585-83B
potassium salt	EtOH	257(4.09),291(4.02)	64-1585-83B
rubidium salt	EtOH	293(3.92)	64-1585-83B
sodium salt	EtOH	258(3.97),294(3.90)	64-1585-83B
$C_8H_8O_3$			
2,5-Cyclohexadiene-1,4-dione, 2-methoxy-6-methyl-	MeOH	265(4.27),365(3.09)	158-83-100
Ethanone, 1-(2,4-dihydroxyphenyl)-	EtOH	277(4.44),315(4.18)	23-0378-83
3-Furancarboxaldehyde, 2-(2-oxo-propyl)-	MeCN	201(4.24),217s(3.91), 263(3.75)	24-3366-83
4H-Furo[3,2-c]pyran-4-one, 6,7-di-hydro-6-methyl-	MeCN	211(3.93),243(3.67)	24-3366-83
$C_8H_8O_3S$			
2-Furanpropanoic acid, 5-methyl-α-thioxo-	$CHCl_3$	220(2.41),322(2.82)	13-0339-83A
$C_8H_8O_4$			
Benzoic acid, 4-hydroxy-3-methoxy- (vanillic acid)	NaOH	298(4.18)	104-1845-83
2,5-Cyclohexadiene-1,4-dione, 2,6-di-methoxy-	MeOH	288(4.10)	100-0248-83
	$CHCl_3$	289(4.20),379(2.67)	94-3865-83
1,4-Dioxocin-6-carboxylic acid, methyl ester	MeCN	255s(3.35)	89-0571-83S
Ethanone, 1-(2,3,4-trihydroxyphenyl)-	MeOH	237(3.93),296(4.10)	162-0620-83
2H-Pyran-2,4(3H)-dione, 3-acetyl-6-methyl-	EtOH	224(4.06),308(4.10)	103-0481-83
$C_8H_8O_5$			
Benzaldehyde, 2,3,4-trihydroxy-6-(hy-droxymethyl)- (fomecin A)	EtOH	241(4.03),304(4.18)	162-0603-83
C_8H_9			
Methyl, (2-methylphenyl)-	C_6H_{12}	321.5(4.01)	46-1960-83
Methyl, (3-methylphenyl)-	C_6H_{12}	321(3.83)	46-1960-83
Methyl, (4-methylphenyl)-	C_6H_{12}	320(3.87)	46-1960-83
$C_8H_9ClN_2O$			
1,2-Diazabicyclo[5.2.0]nona-2,4-dien-9-one, 8-chloro-8-methyl-, trans	CH_2Cl_2	303(3.96)	5-2141-83
$C_8H_9ClN_2O_3S$			
2-Propenoic acid, 3-[(chloroacetyl)-amino]-2-cyano-3-(methylthio)-, methyl ester	EtOH	295(3.80),322(2.68)	142-1745-83

Compound	Solvent	$\lambda_{max}(\log \epsilon)$	Ref.
5-Pyrimidinecarboxylic acid, 2-(chloromethyl)-1,4-dihydro-6-(methylthio)-4-oxo-, methyl ester	EtOH	244(2.83),295(4.41)	142-1745-83
$C_8H_9ClN_4O_2$			
Formamide, N-[4-chloro-6-(dimethylamino)-5-formyl-2-pyrimidinyl]-	EtOH	253(4.51),272s(4.26), 325(3.83)	4-0041-83
$C_8H_9ClO_2$			
Phenol, 4-chloro-2-methoxy-6-methyl-	EtOH	208(4.56),237(3.75), 286(3.32)	104-0458-83
C_8H_9ClS			
Benzene, 1-chloro-4-(ethylthio)-	decane	224s(3.88),264(4.08)	65-0449-83
$C_8H_9Cl_2N$			
Benzenemethanamine, 3,5-dichloro-α-methyl-, (S)-	C_6H_{12}	264(2.23),270(2.32), 272s(2.30),277(2.20), 280(2.18)	35-1578-83
	MeOH	263(2.23),269(2.32), 276(2.20),278s(2.15)	35-1578-83
	10% HCl	266(2.46),273(2.61), 281(2.53)	35-1578-83
$C_8H_9Cl_3N_4$			
Imidazo[1,2-c]pyrimidium, 5-amino-4-chloro-7-(2-chloroethyl)-8,9-dihydro-, chloride	EtOH	213(4.21),280(3.92), 342(3.99)	103-1120-83
$C_8H_9Cl_3N_4O_2$			
4-Pyrimidinamine, 6-chloro-N,N-bis(2-chloroethyl)-5-nitro-	EtOH	208(4.11),253(4.07), 348(3.85)	103-1120-83
C_8H_9FS			
Benzene, 1-(ethylthio)-4-fluoro-	decane	218s(3.70),256(3.72)	65-0449-83
$C_8H_9IO_2$			
Benzene, 2-iodo-1,4-dimethoxy-	MeOH	238(4.54),288(4.62)	64-0866-83B
C_8H_9N			
2,4-Cyclopentadiene-1-carbonitrile, 1,2-dimethyl-	hexane	201(2.48),253(3.58)	24-0299-83
2,4-Cyclopentadiene-1-carbonitrile, 1,3-dimethyl-	hexane	202(3.01),250(3.32)	24-0299-83
2-Cyclopentene-1-carbonitrile, 1-methyl-4-methylene-	hexane	227(4.00)	24-0299-83
3-Cyclopentene-1-carbonitrile, 1-methyl-2-methylene-	hexane	222(4.27)	24-0299-83
Pyridine, 2-cyclopropyl-	MeOH	268(3.97)	48-0517-83
Pyridine, 4-cyclopropyl-	MeOH	258(3.92)	48-0517-83
C_8H_9NO			
Acetamide, N-phenyl- (acetanilide)	MeOH	200(4.53),242(4.11)	116-1564-83
Ethanone, 1-(4-aminophenyl)-	hexane	224(4.02),283(4.29)	99-0034-83
3,5-Hexadiyn-2-one, 6-(dimethylamino)-	MeCN	215(4.24),330(4.02)	33-1631-83
C_8H_9NOS			
Benzenecarbothioamide, 4-methoxy-	EtOH	306(4.04),375(2.65)	80-0555-83
	CCl_4	316(4.19),407(2.64)	80-0555-83

Compound	Solvent	$\lambda_{max}(\log \epsilon)$	Ref.
$C_8H_9NO_2$			
Acetamide, N-(4-hydroxyphenyl)-	EtOH	250(4.14)	162-0007-83
$C_8H_9NO_3$			
Benzene, 1,2-dimethoxy-4-nitroso-	MeOH	207(4.03),214s(--), 246.5(3.79),331.5s(--), 348(4.0)	44-5041-83
6-Oxabicyclo[3.1.0]hex-3-ene-2-methanol, 2-hydroxy-4-isocyano-α-methyl- (isonitrin B)	MeOH	217(3.91)	158-83-98
1(2H)-Pyridinepropanoic acid, 2-oxo-	EtOH	227(3.84),302(3.76)	94-2552-83
$C_8H_9NO_3S$			
Ethanone, 2-(methylsulfonyl)-1-(2-pyridinyl)-	EtOH	236(3.41),271(3.16)	103-0518-83
Ethanone, 2-(methylsulfonyl)-1-(3-pyridinyl)-	EtOH	234(3.71),268(3.31), 330(2.90)	103-0518-83
Ethanone, 2-(methylsulfonyl)-1-(4-pyridinyl)-	EtOH	224(3.83),284(3.20)	103-0518-83
$C_8H_9NO_4$			
1H-Azepine-4-carboxylic acid, 6,7-dihydro-5-methoxy-7-oxo-	MeOH	218(4.27),263(3.83)	44-5041-83
1,2-Ethanediol, 1-(2-nitrophenyl)-	EtOH	206.0(3.44),259.0(3.72)	23-0400-83
2-Furanpropanoic acid, α-(hydroxyimino)-5-methyl-	$CHCl_3$	214(2.64)	13-0339-83A
1-Propanone, 2-methyl-1-(5-nitro-2-furanyl)-	MeOH	222(3.90),303(4.30)	12-0963-83
C_8H_9NS			
Benzeneethanethioamide	heptane	268(4.03),375(1.60)	80-0555-83
	EtOH	273(4.04),338(2.00)	80-0555-83
Cyclohexene, 1-isothiocyanato-6-methylene-	EtOH	220(3.86),265(3.44)	97-0096-83
$C_8H_9N_3$			
2-Pyrimidineacetonitrile, 4,6-dimethyl-	MeCN	247(3.60)	44-0575-83
$C_8H_9N_3O$			
2H-Isoxazolo[2,3-a]pyrimidin-2-imine, 5,7-dimethyl-, monohydrochloride	MeCN	274(4.24),315(3.48)	44-0575-83
3H-Pyrazole-4-carbonitrile, 5-acetyl-3,3-dimethyl-	MeCN	263(3.59)	78-1247-83
2-Pyrimidineacetonitrile, 4,6-dimethyl-, 1-oxide	H_2O	258(3.88),290(3.30)	44-0575-83
	MeCN	270(4.02),315(3.45)	44-0575-83
5(6H)-Quinazolinone, 2-amino-7,8-dihydro-	EtOH	274(4.05)	4-0649-83
$C_8H_9N_3O_2$			
Benzene, 4-azido-1,2-dimethoxy-	MeOH	212(4.76),258.5(4.58), 287s(--)	44-5041-83
1-Propanone, 1-(5-azido-2-furanyl)-2-methyl-	C_6H_{12}	303(4.20)	12-0963-83
3H-Pyrazole-4-carboxylic acid, 5-cyano-3,3-dimethyl-, methyl ester	MeCN	254(3.84)	78-1247-83
3H-Pyrazole-5-carboxylic acid, 4-cyano-3,3-dimethyl-, methyl ester	MeCN	227(3.83)	78-1247-83
1H-Pyrazolo[1,2-a][1,2,4]triazol-4-ium, 2,3-dihydro-5,6,7-trimethyl-1,3-dioxo-, hydroxide, inner salt	MeCN	233(4.139),231.5(4.136), 247s(4.049),256s(3.79), 288.5(2.615)	142-0023-83

Compound	Solvent	$\lambda_{max}(\log \epsilon)$	Ref.
1H-2-Pyrindine-5,7(2H,6H)-dione, 6-diazotetrahydro-, monohydrobromide	EtOH	217(4.05),244(4.08)	12-1221-83
$C_8H_9N_3O_3$			
Isoxazolo[4,5-c]pyridin-4(5H)-one, 5-amino-6-methoxy-3-methyl-	MeOH	213(4.21),258(3.84), 287.5(4.02)	142-1315-83
$C_8H_9N_3O_5$			
Glycine, N-[(1,2,3,4-tetrahydro-6-methyl-2,4-dioxo-5-pyrimidinyl)-carbonyl]-	H_2O	269.5(3.98)	94-0135-83
$C_8H_9N_3S$			
7H-Pyrrolo[2,3-d]pyrimidine, 7-methyl-4-(methylthio)-	MeOH	248(3.68),294(4.09)	5-1576-83
C_8H_9O			
Ethylium, 1-hydroxy-1-phenyl-	H_2SO_4	295(4.31)	99-0612-83
Methyl, (2-methoxyphenyl)-	C_6H_{12}	332(3.56)	46-1960-83
Methyl, (3-methoxyphenyl)-	C_6H_{12}	321(3.28)	46-1960-83
Methyl, (4-methoxyphenyl)-	C_6H_{12}	321(3.46)	46-1960-83
C_8H_{10}			
Bicyclo[4.2.0]octa-2,8-diene	hexane	240(4.11),271.5s(3.55)	24-0894-83
Cyclohexene, 1-ethynyl-	EtOH	224(4.07)	70-0569-83
1,3,5-Hexatriene, 2-ethenyl-, cis	hexane	219(4.56),225.5(4.52), 241s(4.02),258s(3.92)	24-0894-83
$C_8H_{10}BrCl_2NPt$			
Platin um , (3-bromobenzenamine)di-chloro(η^2-ethene)-, trans	n.s.g.	241(3.88),299(3.23)	131-0309-83H
Platinum, (4-bromobenzenamine)di-chloro(η^2-ethenyl)-, trans	n.s.g.	252(3.99),301(3.33)	131-0309-83H
$C_8H_{10}BrN$			
Benzenemethanamine, 4-bromo-α-methyl-, (S)-	C_6H_{12}	261(2.43),268(2.49), 276(2.34)	35-1578-83
	MeOH	260(2.40),267(2.43), 276(2.26)	35-1578-83
hydrochloride	MeOH	258(2.32),264(2.36), 269s(2.20),274(2.08)	35-1578-83
	10% HCl	264(2.38),268s(2.26), 273s(2.08)	35-1578-83
$C_8H_{10}BrNO_2$			
1H-2-Pyrindine-5,7(2H,6H)-dione, 6-bromotetrahydro-, trifluoroacetate	H_2O	272(4.39)	12-1221-83
$C_8H_{10}BrN_5O_2$			
1,2-Propanediol, 3-(6-amino-8-bromo-9H-purin-9-yl)-, (±)-	pH 2 and 12	267(4.28)	73-1910-83
$C_8H_{10}ClN$			
Benzenemethanamine, 4-chloro-α-methyl-, (S)-	C_6H_{12}	261(2.45),268(2.53), 276(2.43)	35-1578-83
	MeOH	261(2.40),268(2.49), 276(2.38)	35-1578-83
hydrochloride, (R)-	MeOH	258(2.30),265(2.34), 269s(2.20),274(2.15)	35-1578-83
	10% HCl	263(2.36),268s(2.23), 272(2.11)	35-1578-83

Compound	Solvent	$\lambda_{max}(\log \epsilon)$	Ref.
$C_8H_{10}Cl_2FNPt$			
Platinum, dichloro(η^2-ethene)(3-fluoro-benzenamino)-, trans	n.s.g.	248(3.72),264s(3.67), 296(3.18)	131-0309-83H
$C_8H_{10}Cl_2INPt$			
Platinum, dichloro(η^2-ethene)(3-iodo-benzenamino)-, trans	n.s.g.	241(4.05),255s(3.87), 299(3.26)	131-0309-83H
Platinum, dichloro(η^2-ethene)(4-iodo-benzenamino)-, trans	n.s.g.	247(3.75),261(3.83), 302(3.15)	131-0309-83H
$C_8H_{10}Cl_2N_2O_2Pt$			
Platinum, dichloro(η^2-ethene)(3-nitro-benzenamine-N^1)-, trans	n.s.g.	250(4.02),292(3.40)	131-0309-83H
Platinum, dichloro(η^2-ethene)(4-nitro-benzenamine-N^1)-, trans	n.s.g.	240(3.93),288(3.97), 345(4.06)	131-0309-83H
$C_8H_{10}Cl_2N_4$			
Pteridine, 4-chloro-8-(2-chloroethyl)-5,6,7,8-tetrahydro-	EtOH	216(4.38),289(3.94), 316(4.03)	103-1120-83
$C_8H_{10}Cl_3NPt$			
Platinum, dichloro(3-chlorobenzen-amine)(η^2-ethene)-, trans	n.s.g.	249(3.80),300(3.23)	131-0309-83H
$C_8H_{10}N_2$			
Pyrrolo[1,2-a]pyrazine, 3,4-dihydro-6-methyl-	EtOH	254(3.58),306(4.07)	104-0393-83
1H-Pyrrolo[2,3-b]pyridine, 2,3-di-hydro-4-methyl-	heptane	240(3.9),303(3.66)	70-2437-83
	H_2O	305(--)	70-2437-83
	PrOH	308(3.66)	70-2437-83
	HOAc	317(--)	70-2437-83
	MeCN	305(--)	70-2437-83
$C_8H_{10}N_2O$			
Acetic acid, 2-phenylhydrazide	hexane	231.5(3.90),282.5(3.00)	162-0015-83
Benzoic acid, 2-methyl-, hydrazide	pH 1	240(3.96),302(3.57)	140-0091-83
	H_2O	302(3.56)	140-0091-83
Benzoic acid, 3-methyl-, hydrazide	pH 1	236(3.98),280(2.95)	140-0091-83
	H_2O	234(3.94)	140-0091-83
Benzoic acid, 4-methyl-, hydrazide	pH 1	228(3.95),322(2.84)	140-0091-83
	H_2O	226(3.95)	140-0091-83
1,2-Diazabicyclo[5.2.0]nona-2,4-dien-9-one, 5-methyl-	MeOH	250(3.62),298(3.98)	5-2141-83
Ethanone, 1-(2,4-dimethyl-5-pyrimi-dinyl)-	EtOH	232.5(4.00)	4-0649-83
$C_8H_{10}N_2O_2$			
Acetamide, 2-cyano-2-(tetrahydro-2H-pyran-2-ylidene)-	n.s.g.	255(4.19)	103-1050-83
Benzoic acid, 2-methoxy-, hydrazide	pH 1	240(3.91),302(3.52)	140-0091-83
	H_2O	294(3.46)	140-0091-83
Benzoic acid, 4-methoxy-, hydrazide	pH 1	258(4.16)	140-0091-83
	H_2O	253(4.14)	140-0091-83
Carbamic acid, (4-cyano-1,3-butadien-yl)-, ethyl ester	MeOH	303(4.43)	18-0175-83
Pyrazinemethanol, 6-methyl-, acetate	EtOH	223s(2.68),269(3.88), 274(3.88),295s(2.66)	4-0311-83
Pyrazinol, 3,5-dimethyl-, acetate	EtOH	273(3.42),294s(3.13)	4-0311-83
$C_8H_{10}N_2O_2S_2$			
5-Pyrimidinecarboxylic acid, 1,4-di-	EtOH	252(4.00),279(3.43),	142-1745-83

Compound	Solvent	λ max(log ε)	Ref.
hydro-2-methyl-6-(methylthio)-4-thioxo-, methyl ester (cont.)		305(3.64),350(3.86)	142-1745-83
$C_8H_{10}N_2O_3$			
4(1H)-Pyrimidinone, 2-methoxy-1-(oxiranylmethyl)-	EtOH	233(3.90),255(3.81)	128-0125-83
$C_8H_{10}N_2O_3S$			
Acetic acid, [(1,4-dihydro-5-methyl-4-oxo-2-pyrimidinyl)thio]-, methyl ester	MeOH	288(3.95)	56-1027-83
Isothiazolo[4,5-c]pyridine-5(4H)-carboxylic acid, 2,3,6,7-tetrahydro-3-oxo-, methyl ester	MeOH	214(3.57),263(3.79)	87-0895-83
Isothiazolo[5,4-c]pyridine-6(2H)-carboxylic acid, 3,4,5,6-tetrahydro-3-oxo-, methyl ester	MeOH	216(3.61),264(3.82)	87-0895-83
$C_8H_{10}N_2O_5S$			
7H-Oxazolo[3,2-a]pyrimidin-7-one, 2,3-dihydro-2-[[(methylsulfonyl)oxy]-methyl]-	EtOH	227(3.70),257(3.62)	128-0125-83
$C_8H_{10}N_4O$			
Imidazo[4,5-d][1,3]diazepin-8(1H)-one, 5-ethyl-4,7-dihydro-, hydrochloride	MeOH	229(4.30),296(3.61), 330s(--)	87-1478-83
4H-Pyrazolo[3,4-d]pyrimidin-4-one, 1,5-dihydro-1,5,6-trimethyl-	EtOH	209.5(4.32),255(3.75)	11-0171-83B
7H-Pyrrolo[2,3-d]pyrimidin-2-amine, 4-methoxy-7-methyl-	MeOH	227(4.48),263(3.95), 286(3.94)	5-0137-83
$C_8H_{10}N_4OS$			
Ethanol, 2-[(2-amino-1H-pyrrolo-[2,3-d]pyrimidin-4-yl)thio]-	MeOH	231(4.48),318(3.98)	5-0137-83
$C_8H_{10}N_4O_5$			
2H-Pyrrole-3-carboxylic acid, 4-(2-acetylhydrazino)-5-(hydroxyamino)-2-oxo-, methyl ester	DMF	274(4.17),377(4.00)	48-0041-83
$C_8H_{10}N_6$			
1H-Imidazole, 2-(1H-imidazol-2-ylazo)-4,5-dimethyl-	MeOH	467(4.40)	56-0547-83
2-Pyrimidinamine, 4,6-dimethyl-N-1H-1,2,4-triazol-3-yl-	MeOH	249(4.33)	118-0222-83
$C_8H_{10}O$			
Benzene, 1-methoxy-4-methyl-	39.2% H_2SO_4	275.5(3.19)	54-0299-83
	86.9% H_2SO_4	269.0(4.00)	54-0299-83
Cyclopentanone, 2-cyclopropylidene-	alkane	240(4.00)	78-3307-83
$C_8H_{10}O_2$			
Bicyclo[2.1.0]pent-2-ene-5-carboxylic acid, 5-methyl-, methyl ester, endo-exo	hexane at 0°	261(2.65)	24-0299-83
exo,endo	hexane at 0°	230(2.57)	24-0299-83
2,4-Cyclopentadiene-1-carboxylic acid, 1-methyl-, methyl ester	C_6H_{12}	253(3.32)	24-0299-83
2,5(1H,3H)-Pentalenedione, tetrahydro-, trans	MeCN	289s(1.75),297(1.85), 308(1.85),318(1.57)	35-0145-83
2-Propanone, 1-(3-methyl-2-furanyl)-	MeCN	217(3.91),275(2.27)	24-3366-83

Compound	Solvent	λ_{max}(log ε)	Ref.
$C_8H_{10}O_3$			
Bicyclo[2.2.1]heptane-2-carboxylic acid, 6-oxo-	MeCN	292(1.65)	44-1862-83
2-Cyclopentene-1-acetic acid, 4-oxo-, methyl ester	MeOH	217(4.02),313(1.52)	35-1839-83
3-Furancarboxaldehyde, 2-(2-hydroxy-propyl)-	MeCN	202(4.21),217s(3.90), 263(3.80)	24-3366-83
$C_8H_{10}O_4$			
2-Propenoic acid, 3-(2-methyl-4H-1,3-dioxin-6-yl)-, [R-(E)]-	MeOH	270(4.26)	136-0286-83G
[S-(E)]-	MeOH	270(4.19)	136-0286-83G
$C_8H_{10}O_5$			
3-Furancarboxylic acid, tetrahydro-2-methyl-4,5-dioxo-, ethyl ester	MeOH	257(4.03),295s(3.23)	12-0311-83
$C_8H_{10}S$			
Benzene, (ethylthio)-	decane	214s(3.95),256(3.93)	65-0449-83
$C_8H_{11}BrO_2$			
2,4-Hexadienoic acid, 6-bromo-, ethyl ester	EtOH	259(4.26)	111-0447-83
$C_8H_{11}ClN_4O_4$			
Ethanol, 2,2'-[(6-chloro-5-nitro-4-pyrimidinyl)imino]bis-	EtOH	206(4.21),255(4.13)	103-1120-83
$C_8H_{11}ClO$			
Bicyclo[2.2.1]heptan-2-one, 6-(chloro-methyl)-, endo	MeCN	292(1.58)	44-1862-83
$C_8H_{11}Cl_2NPt$			
Platinum, (benzenamine)dichloro(η^2-ethene)-	n.s.g.	254(3.80),304(3.21)	131-0309-83H
$C_8H_{11}Cl_4NO$			
3-Butenamide, 2,3,4,4-tetrachloro-N-(1,1-dimethylethyl)-	heptane	212(3.96),230s(3.85)	5-1496-83
$C_8H_{11}N$			
Benzenamine, N-ethyl-	n.s.g.	286(3.17)	32-0069-83
Benzenemethanamine, α-methyl-, (S)-	C_6H_{12}	258(2.28),264(2.18), 268(2.00)	35-1578-83
	MeOH	257(2.26),264(2.11), 267(1.93)	35-1578-83
hydrochloride	MeOH	257(2.32),260(2.20), 263(2.26),267(2.08)	35-1578-83
	10% HCl	256(2.34),260(2.26), 262(2.26),267(2.08)	35-1578-83
$C_8H_{11}NO_2$			
1H-2-Pyrindine-5,7(2H,6H)-dione, tetrahydro-	EtOH	265(4.45)	12-1221-83
1H-Pyrrole-2,5-dione, 3,4-diethyl-(xeronimide)	MeOH	220(4.08)	78-1893-83
$C_8H_{11}NO_2S$			
Pyridine, 2-[(1-methylethyl)sulfonyl]-	EtOH	213(3.72),258(3.51)	131-0277-83H

Compound	Solvent	λ_{max}(log ϵ)	Ref.
$C_8H_{11}NO_4$			
4-Oxa-1-azabicyclo[3.2.0]heptan-7-one, 2-hydroxy-3-(2-methoxyethylidene)-	EtOH	243(3.76)	39-2513-83C
2-Oxazoleacetic acid, 2,5-dihydro-5-(2-hydroxyethylidene)-, methyl ester, [R-(Z)]-	EtOH	271(4.08)	39-2513-83C
2-Propenoic acid, 2,2'-iminobis-, dimethyl ester	CCl_4	282(3.73),301(3.74)	4-0001-83
$C_8H_{11}NO_5$			
2-Butenedioic acid, 2-(acetylamino)-, dimethyl ester, (E)-	MeOH	264(3.90)	118-0539-83
(Z)-	MeOH	275(4.23)	118-0539-83
$C_8H_{11}N_3$			
1H-Imidazole-1-carbonitrile, 2,5-diethyl-	EtOH	232(4.06)	4-1277-83
$C_8H_{11}N_3O$			
1H-1,2-Diazepine-1-carboxamide, N,5-dimethyl-	MeOH	371(2.45)	5-2141-83
$C_8H_{11}N_3O_3$			
5-Pyrimidineacetic acid, 2-amino-dihydro-4-oxo-, ethyl ester	50% MeOH	292(3.89)	73-0292-83
	+ HCl	221(4.06),261(3.95)	73-0292-83
	+ NaOH	231(3.96),281(3.89)	73-0292-83
$C_8H_{11}N_3O_4$			
5-Pyrimidinecarboxamide, 1,2,3,4-tetrahydro-N-(2-hydroxyethyl)-6-methyl-2,4-dioxo-	H_2O	268.3(4.05)	94-0135-83
$C_8H_{11}N_3O_6$			
6-Azauridine	H_2O	262(3.79)	162-0131-83
1H-Imidazole, 1-α-D-arabinofuranosyl-2-nitro-	EtOH	225(3.58),320(3.92)	87-0020-83
β-	EtOH	225(3.58),318(3.91)	87-0020-83
$C_8H_{11}N_3O_6S$			
6-Azauracil, 5-(α-D-ribofuranosylthio)-	pH 1	303(3.86)	80-0733-83
	pH 13	303(3.68)	80-0733-83
anomer	pH 1	303(3.70)	80-0733-83
	pH 13	303(3.82)	80-0733-83
$C_8H_{11}N_4$			
Imidazo[1,2,3-ij]pteridinium, 5,6,7,8,9,10-hexahydro-, chloride, hydrochloride [sic]	EtOH	211(4.08),286(3.76), 343(3.78)	103-1120-83
$C_8H_{11}N_4S$			
4H-Pyrimido[4,5-d][1,3]thiazinium, 2-amino-6,7-dimethyl-, perchlorate	pH 9.18	225(4.09),268(3.95), 328(4.36)	44-2476-83
$C_8H_{11}N_5$			
Adenine, N^6,1,9-trimethyl-, perchlorate	pH 1 and 7	263(4.14)	94-4270-83
	pH 13	263(4.17),270s(4.11), 295s(3.53)	94-4270-83
	EtOH	263(4.13)	94-4270-83
Adenine, N^6,3,9-trimethyl-, hydriodide	pH 1	221(4.37),281(4.20)	94-4270-83
	pH 7	221(4.36),280(4.20)	94-4270-83
	EtOH	282(4.19)	94-4270-83

Compound	Solvent	λ_{max}(log ε)	Ref.
Pyrazine, 3-azido-2,5-diethyl-	EtOH	273(3.93)	4-1277-83
$C_8H_{11}N_5O$			
Adenine, N[6]-methoxy-1,9-dimethyl-, perchlorate	pH 1	231(3.83),283(3.99)	94-3149-83
	pH 7	271(4.16),312s(3.36)	94-3149-83
	pH 13	271(4.15),312s(3.34)	94-3149-83
	EtOH	273(4.08)	94-3149-83
Adenine, N[6]-methoxy-3,7-dimethyl-, perchlorate	pH 1	276(4.15)	94-3149-83
	pH 7	290(4.08)	94-3149-83
	EtOH	287(4.04)	94-3149-83
9H-Purin-6-amine, N-methoxy-N,9-dimethyl-	pH 1	275.5(4.22)	94-3149-83
	pH 7 and 13	276(4.25)	94-3149-83
	EtOH	276(4.25)	94-3149-83
hydriodide	pH 1	220(4.36),276(4.20)	94-3149-83
	pH 7	220(4.41),276(4.24)	94-3149-83
	pH 13	276(4.22)	94-3149-83
	EtOH	219(4.50),277(4.22)	94-3149-83
$C_8H_{11}N_5O_2$			
3H-Pyrazolo[1,5-d]tetrazole-7-carboxylic acid, 3,6-dimethyl-, ethyl ester	CH_2Cl_2	288(3.87),333s(3.18)	4-1629-83
$C_8H_{11}N_5O_3$			
Guanine, 8-[(2-hydroxyethoxy)methyl]-	pH 1	251(4.01),272(3.92)	4-1169-83
	pH 5	249(4.04),273(3.91)	4-1169-83
	pH 11	278(3.95)	4-1169-83
8H-Purin-8-one, 6-amino-9-(2,3-dihydroxypropyl)-7,9-dihydro-, (±)-	pH 3	270(4.04),282(4.04)	73-1910-83
	pH 7	271(4.13)	73-1910-83
	pH 12	281(4.20)	73-1910-83
4(1H)-Pyrimidinone, 1-(2-azido-3-hydroxypropyl)-2-methoxy-	EtOH	228(4.05),251(3.97)	128-0125-83
$C_8H_{11}N_5O_5S$			
2,4(1H,3H)-Pyrimidinedione, 1-[2-azido-3-[(methylsulfonyl)oxy]propyl]-	EtOH	263(4.00)	128-0125-83
2,4(1H,3H)-Pyrimidinedione, 1-[3-azido-2-[(methylsulfonyl)oxy]propyl]-	EtOH	264(3.97)	128-0125-83
$C_8H_{11}O_5P$			
Phosphonic acid, (2,5-dihydroxyphenyl)-, dimethyl ester	EtOH	224.5(3.76),313(3.71)	70-1532-83
C_8H_{12}			
Cyclohexane, 1,2-bis(methylene)-	isooctane	220(4.00)	54-0302-83
Cyclohexene, ethenyl-	hexane	219s(4.21),227s(4.34), 230.9(4.37),238s(4.19)	131-0291-83N
	MeOH	219s(4.23),226s(4.35), 230.4(4.38),238s(4.23)	131-0291-83N
$C_8H_{12}AgNO_2$			
2,5-Pyrrolidinedione, 3,3,4,4-tetramethyl-, silver(1+) salt	MeCN-base	227(3.60)	23-1890-83
after reduction	MeCN-base	230(3.76)	23-1890-83
$C_8H_{12}Br_2N_4$			
Imidazo[1,2,3-ij]pteridinium, 5,6,7,8,9,10-hexahydro-, bromide, hydrobromide	EtOH	208(4.22),283(3.87), 345(3.90)	103-1120-83
$C_8H_{12}Br_2N_4O$			
1H,10H-Imidazo[1,2,3-ij]pteridinium,	EtOH	208(4.24),284(3.89),	103-1120-83

Compound	Solvent	$\lambda_{max}(\log \epsilon)$	Ref.
5,6,7,8-tetrahydro-4-hydroxy-, bromide, hydrobromide (cont.)		345(3.92)	103-1120-83
$C_8H_{12}Cl_2N_4$			
Imidazo[1,2,3-ij]pteridinium, 4-chloro-5,6,7,8,9,10-hexahydro-, chloride	EtOH	206(4.26),289(3.91), 345(3.95)	103-1120-83
Imidazo[1,2-c]pyrimidin-8-amine, 7-chloro-1-(2-chloroethyl)-1,2,3,8a-tetrahydro-, monohydrochloride	EtOH	208(4.19),274(3.93), 342(3.91)	103-1120-83
$C_8H_{12}CoN_4O_4$			
Cobalt, [N,N'-bis(1-methyl-2-nitroethylidene)-1,2-ethanediaminato(2-)-N^1,N^2,O^1,O^2]-	DMSO	311(4.18),440(3.34), 600(2.60)	40-0088-83
$C_8H_{12}CuN_4O_4$			
Copper, [N,N'-bis(1-methyl-2-nitroethylidene)-1,2-ethanediaminato(2-)-N^1,N^2,O^1,O^2]-	DMSO	338(4.28),590(2.50)	40-0088-83
$C_8H_{12}N_2O$			
1H-1,2-Diazepine, 1-acetyl-6,7-dihydro-5-methyl-	MeOH	211(3.88),245(3.78), 305(3.76)	5-2141-83
$C_8H_{12}N_2O_2S$			
Sulfoxonium, dimethyl-, (1,6-dihydro-1-methyl-6-oxo-2-pyrimidinyl)methylide	EtOH	269(3.85),325(4.31)	39-2645-83C
$C_8H_{12}N_2O_5S$			
4(1H)-Pyrimidinone, 2,3-dihydro-6-(D-arabino-tetrahydroxybutyl)-2-thioxo-	pH 1	210(4.65),270(4.54)	136-0333-83M
	pH 7	255(4.32),305(4.28)	136-0333-83M
	pH 11	260(4.32),310(4.26)	136-0333-83M
$C_8H_{12}N_2O_6S$			
2,4(1H,3H)-Pyrimidinedione, 1-[3-hydroxy-2-[(methylsulfonyl)oxy]propyl]-	EtOH	263(4.02)	128-0125-83
$C_8H_{12}N_2S$			
2(1H)-Pyrazinethione, 3,6-diethyl-	EtOH	242(3.41),281(3.86)	142-0797-83
$C_8H_{12}N_4$			
4H-Imidazole-4-carbonitrile, 5-amino-4-butyl-	EtOH-pH 7	260(3.74)	44-0003-83
	EtOH-pH 13	276(4.06)	44-0003-83
$C_8H_{12}N_4NiO_4$			
Nickel, [N,N'-bis(1-methyl-2-nitroethylidene)-1,2-ethanediaminato(2-)-N^1,N^2,O^1,O^2]-	DMSO	280(3.95),449(3.90), 540(3.43)	40-0088-83
$C_8H_{12}N_4O_5$			
1,3,5-Triazin-2(1H)-one, 4-amino-1-β-D-ribofuranosyl- (azacitidine)	H_2O	241(3.94)	162-0128-83
	pH 2	249(3.49)	162-0128-83
	pH 12	223(4.38)	162-0128-83
$C_8H_{12}N_4S_2$			
1,3,4-Thiadiazole, 2,3-dihydro-2,5-dimethyl-2-[(5-methyl-1,3,4-thiadiazol-2-yl)methyl]-	EtOH	252s(3.22)	4-0073-83

Compound	Solvent	$\lambda_{max}(\log \epsilon)$	Ref.
$C_8H_{12}N_5O$			
1H-Purinium, 6-(methoxyamino)-1,3-dimethyl-, iodide	pH 1	223(4.31),274(4.16)	94-3149-83
perchlorate	pH 1	275(4.17)	94-3149-83
7H-Purinium, 6-(methoxyamino)-7,9-dimethyl-, iodide	pH 1	226(4.28),283(3.97)	94-3149-83
	pH 7	226(4.28),283(3.96)	94-3149-83
	EtOH	219(4.32),291(3.93)	94-3149-83
perchlorate	pH 1	232(3.85),283(3.98)	94-3149-83
	pH 7	232(3.85),283(3.97)	94-3149-83
	EtOH	235(3.90),291(3.91)	94-3149-83
$C_8H_{12}N_8O_2$			
Pyridazino[4,5-d]pyridazine-1,4,5,8-tetrone, 2,3,6,7-tetrahydro-2,6-dimethyl-, 4,8-dihydrazone	MeOH	265(4.06),284(4.01), 445(3.57)	161-0842-83
$C_8H_{12}O_2$			
Bicyclo[2.2.1]heptan-2-one, 6-(hydroxymethyl)-, endo	MeCN	293(1.56)	44-1862-83
2,4-Hexadienoic acid, ethyl ester	EtOH	259(4.42)	111-0447-83
$C_8H_{12}O_3$			
Cyclopentaneacetic acid, 3-oxo-, methyl ester, (-)-	EtOH	285(1.40)	35-1839-83
2,4-Hexadienoic acid, 6-hydroxy-, ethyl ester	EtOH	256(4.28)	111-0447-83
2-Hexenoic acid, 6-oxo-, ethyl ester, (E)-	C_6H_{12}	217(3.95),290(1.58)	22-0175-83
(Z)-	C_6H_{12}	215(4.06),290(1.52)	22-0175-83
$C_8H_{12}O_3S_2$			
3(2H)-Thiophenone, 4-(butylthio)-, 1,1-dioxide	EtOH	203(4.00),307(3.76)	104-1447-83
	EtOH-KOH	270(3.91),330(3.56), 422(3.38)	104-1447-83
$C_8H_{12}O_4$			
4-Hexenoic acid, 5-methoxy-3-oxo-, methyl ester	EtOH	260(4.14)	39-2353-83C
D-threo-Hex-2-enonic acid, 2,3-dideoxy-4,6-O-ethylidene-, (E)-(R)-	MeOH	216(3.82)	136-0286-83G
(S)-	MeOH	220(3.53)	136-0286-83G
2-Pentenoic acid, 2-methoxy-4-oxo-, ethyl ester, (E)-	MeOH	252(3.76)	33-0606-83
	hexane	246(--)	33-0606-83
	CH_2Cl_2	250(--)	33-0606-83
(Z)-	MeOH	262(3.97),270s(3.95)	33-0606-83
	hexane	254(--)	33-0606-83
	CH_2Cl_2	257(--)	33-0606-83
3-Pentenoic acid, 4-methoxy-2-oxo-, ethyl ester, (E)-	MeOH	288(4.02)	33-0606-83
(Z)-	MeOH	265(3.81)	33-0606-83
$C_8H_{12}S_2$			
1,3-Cyclobutanedithione, 2,2,4,4-tetramethyl-	hexane	227(4.33),298(2.61), 500(1.35)	44-4482-83
2-Thietanethione, 3,3-dimethyl-4-(1-methylethylidene)-	C_6H_{12}	244(3.58),340(3.65), 460(1.08)	44-4482-83 +78-2719-83
$C_8H_{12}S_3$			
2,4(3H,5H)-Thiophenedithione, 3,3,5,5-tetramethyl-	C_6H_{12}	268(4.21),310(4.23), 480(1.60)	44-4482-83

Compound	Solvent	λ_{max}(log ε)	Ref.
$C_8H_{13}ClN_4O_2$			
Ethanol, 2,2'-[(5-amino-6-chloro-4-pyrimidinyl)imino]bis-	EtOH	214(4.14),275(3.80), 316(4.00)	103-1120-83
$C_8H_{13}ClO_2$			
2-Hexenoic acid, 6-chloro-3-methyl-, methyl ester, (E)-	EtOH	218(4.41)	23-1053-83
(Z)-	EtOH	217(4.20)	23-1053-83
$C_8H_{13}I$			
Cyclooctene, 1-iodo-	EtOH	259(2.58)	35-6907-83
$C_8H_{13}NO$			
2H-Indole, 3,3a,4,5,6,7-hexahydro-, 1-oxide	MeCN	244(3.99)	5-0897-83
3(2H)-Pyridinone, 1,6-dihydro-1,2,5-trimethyl-, hydrochloride	EtOH	331(4.14)	70-1655-83
3(2H)-Pyridinone, 1,6-dihydro-1,5,6-trimethyl-, hydrochloride	EtOH	231(4.27)	70-1655-83
1H-Pyrrole-2-methanol, α-ethyl-1-methyl-	MeOH	222(3.94)	44-2488-83
$C_8H_{13}NO_2$			
2H-Pyran-2-one, 3-[(dimethylamino)-methylene]tetrahydro-	n.s.g.	302(4.30)	103-1050-83
2,5-Pyrrolidinedione, 3,3,4,4-tetra-methyl-	MeCN	232(2.18),251(2.11)	23-1890-83
$C_8H_{13}NS$			
2-Cyclohexene-1-thione, 3-amino-5,5-dimethyl-	pH 1	310(4.25),350(3.82)	78-3405-83
	H_2O	352(4.57)	78-3405-83
	pH 13	352(4.22)	78-3405-83
$C_8H_{13}N_2S$			
Methanaminium, N-(4,6-dimethyl-2H-1,3-thiazin-2-ylidene)-N-methyl-, perchlorate	MeCN	224(4.26),254(3.75), 352(3.44)	118-0827-83
$C_8H_{13}N_3O$			
1H-1,2-Diazepine-7-acetamide, 6,7-di-hydro-5-methyl-	MeOH	239(3.59),300(3.65)	5-2141-83
$C_8H_{13}N_3O_2$			
1H-Imidazole-4-carboxamide, 5-ethoxy-1-ethyl-	H_2O	246(3.94)	4-0875-83
	M HCl	228(3.96)	4-0875-83
	M NaOH	245(3.94)	4-0875-83
1H-Imidazolium, 4-(aminocarbonyl)-1,3-diethyl-5-hydroxy-, hydroxide, inner salt	H_2O	239(3.52),279(4.06)	4-0875-83
	M HCl	240(3.61),280(3.92)	4-0875-83
	M NaOH	280(4.06)	4-0875-83
$C_8H_{13}N_3O_3$			
1H-Imidazole-4-carboxylic acid, 2,5-dihydro-2,2,5,5-tetramethyl-1-nitroso-	EtOH	235(3.96)	70-2501-83
$C_8H_{13}N_3O_4$			
1H-Imidazole-4-carboxylic acid, 2,5-dihydro-2,2,5,5-tetramethyl-1-nitroso-, 3-oxide	EtOH	245(4.08)	70-2501-83

Compound	Solvent	λ_{max}(log ε)	Ref.
$C_8H_{13}N_3O_5$			
4(1H)-Pyrimidinone, 2-amino-6-(1,2,3,4-tetrahydroxybutyl)-, [1S-(1R*,2S*-3S*)]-	pH 1	216(3.75),260(3.79)	136-0333-83M
	pH 7	255(3.83),277(3.71)	136-0333-83M
	pH 11	220(3.84),270(3.76)	136-0333-83M
$C_8H_{13}N_5O$			
1H-Purin-6-amine, 2,3-dihydro-N-methoxy-1,3-dimethyl-	pH 1	241(4.00),327(3.83)	94-3521-83
	pH 7	286(4.00)	94-3521-83
	pH 13	260s(3.87),291(4.04)	94-3521-83
	EtOH	289(4.04)	94-3521-83
$C_8H_{13}N_5O_2$			
1H-Imidazole-4-carboximidamide, N-methoxy-N'-methyl-5-(formylmethamino)-	pH 7	219(4.11)	94-3521-83
	EtOH	220(4.12)	94-3521-83
$C_8H_{13}N_5O_3S$			
1H-Imidazole-4-carboxylic acid, 5-[2-(aminocarbonyl)hydrazino]-2-(methylthio)-, ethyl ester	n.s.g.	215(4.07),240(3.96), 285(4.32)	103-1235-83
$C_8H_{13}N_5O_4$			
Acetamide, N-(2,6-diamino-1,4-dihydro-4-oxo-5-pyrimidinyl)-2-(2-hydroxyethoxy)-	pH 5	267(4.07)	4-1169-83
	pH 11	267(4.09)	4-1169-83
$C_8H_{14}NO$			
1-Azoniabicyclo[2.2.2]octane, 1-methyl-3-oxo-, iodide	H_2O	281(1.03)	35-0265-83
$C_8H_{14}N_2O_2$			
Piperidine, 1-(1-methyl-2-nitroethenyl)-	EtOH	248(3.38),361(4.30)	40-0088-83
Pyrazine, 2,5-dihydro-2,2,5,5-tetramethyl-, 1,4-dioxide	EtOH	234(4.22)	103-1128-83
$C_8H_{14}N_2O_2S$			
3-Pyridinesulfonamide, 1,4-dihydro-N,N,1-trimethyl-	CH_2Cl_2	315(3.5)	64-0873-83B
3-Pyridinesulfonamide, 1,6-dihydro-N,N,1-trimethyl-	CH_2Cl_2	342(3.4)	64-0873-83B
$C_8H_{14}N_2O_3$			
Pyrazinol, 2,3-dihydro-3,3,5,6-tetramethyl-, 1,4-dioxide	EtOH	345(4.14)	103-1128-83
$C_8H_{14}N_4O$			
1H-Pyrazole-4-carboxamide, 5-amino-1-methyl-N-propyl-	EtOH	229.5(3.63),254.5(3.64)	11-0171-83B
5H-Tetrazol-5-one, 1-(1,1-dimethylethyl)-1,4-dihydro-4-(2-propenyl)-	hexane	227(3.66)	24-3427-83
$C_8H_{14}N_4O_2$			
Ethanol, 2,2'-[(5-amino-4-pyrimidinyl)-imino]bis-, monohydrochloride	EtOH	214(4.01),323(3.92)	103-1120-83
$C_8H_{14}N_4O_4$			
1,2-Ethanediamine, N,N'-bis(1-methyl-2-nitroethenyl)-	CH_2Cl_2	335(4.18),360(4.18)	40-0088-83
$C_8H_{14}N_4O_4S_2$			
1,2,4,5-Tetrazine, 3,6-bis[(ethylsul-	MeOH	219(4.14),278(3.30),	104-0446-83

Compound	Solvent	λ_{max}(log ε)	Ref.
fonyl)methyl]- (cont.)		531(2.64)	104-0446-83
$C_8H_{14}N_4S$			
5H-Tetrazole-5-thione, 1-(1,1-dimeth-ylethyl)-1,4-dihydro-4-(2-propenyl)-	hexane	206(3.76),253(3.97)	24-3427-83
$C_8H_{14}S$			
Cyclobutanethione, 2,2,4,4-tetra-methyl-	hexane	215(3.70),230(3.85), 500(1.08)	44-4482-83
2-Thiabicyclo[3.2.2]nonane	EtOH	214(3.17)	104-0265-83
2-Thiabicyclo[3.3.1]nonane	EtOH	214(3.23)	104-0265-83
2-Thiabicyclo[2.2.2]octane, 3-methyl-	EtOH	216(3.21),240(1.88)	104-0265-83
6-Thiabicyclo[3.2.1]octane, 7-methyl-, exo	EtOH	216(3.22),238(1.76)	104-0265-83
$C_8H_{14}S_2$			
2(3H)-Thiophenethione, dihydro-3,3,5,5-tetramethyl-	C_6H_{12}	312(4.20),470(1.34)	44-4482-83
$C_8H_{15}LiSi$			
Lithium, [1-(trimethylsilyl)-2,4-penta-dienyl]-, (E)-	THF	370(3.53)	157-0021-83
$C_8H_{15}NO$			
3-Buten-2-one, 3-methyl-4-[(1-methyl-ethyl)amino]-	MeOH	298(4.47)	131-0001-83K
(E-syn-E)-	C_6H_{12}	278(--)	131-0001-83K
	dioxan	283(4.42)	131-0001-83K
(Z-syn-Z)-	C_6H_{12}	311(4.11)	131-0001-83K
	dioxan	286(--)	131-0001-83K
1-Hexen-3-one, 2-methyl-1-(methyl-amino)-	MeOH	299(4.42)	131-0001-83K
(E-syn-E)-	C_6H_{12}	277(--)	131-0001-83K
	dioxan	284(4.33)	131-0001-83K
(Z-syn-Z)-	C_6H_{12}	312(4.07)	131-0001-83K
	dioxan	313(--)	131-0001-83K
1,2-Oxazepine, 2-ethyl-2,5,6,7-tetra-hydro-3-methyl-	MeCN	228(3.12)	5-0897-83
1-Penten-3-one, 1-(dimethylamino)-2-methyl-	MeOH	304(4.30)	131-0001-83K
(E-syn-E)-	C_6H_{12}	288(4.32)	131-0001-83K
1-Penten-3-one, 1-(ethylamino)-2-methyl-	MeOH	297(4.43)	131-0001-83K
(E-syn-E)-	C_6H_{12}	275(--)	131-0001-83K
	dioxan	284(4.29)	131-0001-83K
(Z-syn-Z)-	C_6H_{12}	314(4.08)	131-0001-83K
	dioxan	284(--)	131-0001-83K
$C_8H_{15}NOS$			
Butanethioamide, N,N,3,3-tetramethyl-2-oxo-	MeOH	246(3.97),269(3.90), 308s(3.19),321s(2.97), 360s(2.37)	5-1116-83
$C_8H_{15}NS_2$			
Butanethioamide, N,N,3,3-tetramethyl-2-thioxo-	isooctane	265(4.11),343(3.32), 401(3.28),541(2.18)	5-1116-83
	MeOH	529(--)	5-1116-83
$C_8H_{15}N_3OS$			
4-Morpholinecarbothioamide, N-[(di-methylamino)methylene]-	MeCN	224s(3.90),275(4.33), 305(3.93)	97-0019-83

Compound	Solvent	λ_{max}(log ϵ)	Ref.
$C_8H_{15}N_3O_2$			
Diazenecarboxylic acid, 1-pyrrolidinyl-, 1-methylethyl ester	EtOH	268(4.22)	33-1416-83
$C_8H_{15}N_3O_3$			
Diazenecarboxylic acid, [[(1-methylethyl)amino]carbonyl]-, 1-methylethyl ester	EtOH	267(3.23),415(1.60)	33-1416-83
$C_8H_{15}S_2$			
4H-1,3-Dithiin-4-ium, 2-butyl-5,6-dihydro-, tetrafluoroborate	CH_2Cl_2	312(3.53)	1-0687-83
$C_8H_{16}NO$			
1,2-Oxazepinium, 2-ethyl-4,5,6,7-tetrahydro-3-methyl-, tetrafluoroborate	MeCN	198(3.68)	5-0897-83
Pyridinium, 1-ethoxy-2,3,4,5-tetrahydro-6-methyl-, tetrafluoroborate	MeCN	213(3.68)	5-0897-83
$C_8H_{16}N_4O_4S_2$			
1,2,4,5-Tetrazine, 3,6-bis[(ethylsulfonyl)methyl]-1,2-dihydro-	MeOH	222(3.58),323(2.08)	104-0446-83
4H-1,2,4-Triazol-4-amine, 3,5-bis[(ethylsulfonyl)methyl]-	MeOH	207(4.07)	104-0446-83
$C_8H_{16}O$			
Butane, 3-(ethenyloxy)-2,2-dimethyl-, (S)-	gas	196(4.08),207s(--), 214s(--),220s(--)	150-0236-83S
	heptane	199(4.05)	150-0236-83S
	MeOH	198(4.06)	150-0236-83S
$C_8H_{16}Si$			
Silane, trimethyl(2-methylene-3-butenyl)-	hexane	231.0(4.07)	78-0883-83
Silane, trimethyl-2,4-pentadienyl-, (E)-	THF	236(4.42)	157-0021-83
$C_8H_{17}BrN_2Si$			
3H-Pyrazole, 3-bromo-4,5-dihydro-5,5-dimethyl-3-(trimethylsilyl)-	EtOH	246(3.68)	44-3189-83
$C_8H_{17}N_3O_2$			
1-Triazenecarboxylic acid, 3,3-diethyl-, 1-methylethyl ester	EtOH	267.5(4.22)	33-1416-83
$C_8H_{17}N_3Si$			
Silacyclopentane, 1-azido-1-(1,1-dimethylethyl)-	C_6H_{12}	214(2.69)	152-0645-83
$C_8H_{18}N_2O_2$			
Diazene, bis(1,1-dimethylethoxy)-	heptane	224.6(3.85)	44-3728-83
	65% MeCN	226.6(3.83)	44-3728-83
$C_8H_{19}NO_2Si_2$			
Silane, (2-nitroethenylidene)bis[trimethyl-	EtOH	230(3.00)	44-3189-83

Compound	Solvent	λmax(log ε)	Ref.
C_9BrF_7O			
1H-Inden-1-one, 2-bromo-2,3,3,4,5,6,7-heptafluoro-2,3-dihydro-	heptane	250s(3.98),252(4.00), 286(3.15),292(3.14)	104-1880-83
C_9BrF_9			
1H-Indene, 1-bromo-1,2,2,3,3,4,5,6,7-nonafluoro-2,3-dihydro-	heptane	271(3.04)	104-1880-83
1H-Indene, 2-bromo-1,1,2,3,3,4,5,6,7-nonafluoro-2,3-dihydro-	heptane	266(2.95)	104-1880-83
C_9ClF_9			
1H-Indene, 1-chloro-1,2,2,3,3,4,5,6,7-nonafluoro-2,3-dihydro-	heptane	268(2.93)	104-1880-83
1H-Indene, 5-chloro-1,1,2,2,3,3,4,6,7-nonafluoro-2,3-dihydro-	heptane	271s(3.02),278(3.06)	104-1880-83
$C_9Cl_2F_8$			
1H-Indene, 1,2-dichloro-1,2,3,3,4,5-6,7-octafluoro-2,3-dihydro-	heptane	270(3.02)	104-1880-83
1H-Indene, 5,6-dichloro-1,1,2,2,3,3-4,7-octafluoro-2,3-dihydro-	heptane	279(3.25),288(3.32)	104-1880-83
$C_9Cl_3F_7$			
1H-Indene, 1,1,2-trichloro-2,3,3,4,5-6,7-heptafluoro-2,3-dihydro-	heptane	267s(2.97),271(3.01)	104-1880-83
$C_9Cl_4F_6$			
1H-Indene, 1,1,3,3-tetrachloro-2,2,4,5,6,7-hexafluoro-2,3-dihydro-	heptane	272s(2.93),276(2.96)	104-1880-83
$C_9Cl_5F_5$			
1H-Indene, 1,1,3,3,5-pentachloro-2,2,4,6,7-pentafluoro-2,3-dihydro-	heptane	278(3.11),286(3.15)	104-1880-83
$C_9F_{15}NS_5$			
1H-Pyrrole, 1,2,3,4,5-pentakis[(trifluoromethyl)thio]-	hexane	223(4.00),255(3.89)	24-3325-83
C_9HClF_8			
1H-Indene, 3-chloro-1,1,2,2,4,5,6,7-octafluoro-2,3-dihydro-	heptane	269(2.96)	104-1880-83
$C_9HCl_3F_6$			
1H-Indene, 1,1,3-trichloro-2,2,4,5,6,7-hexafluoro-2,3-dihydro-	heptane	270s(2.87),274(2.89)	104-1880-83
C_9HF_9			
1H-Indene, 1,1,2,2,3,4,5,6,7-nonafluoro-2,3-dihydro-	heptane	263s(2.89),268(2.93)	104-1880-83
$C_9H_3Cl_5O_2$			
1(6aH)-Pentalenone, 2,4,5,6,6a-pentachloro-3-methoxy-	hexane	220(3.70),277(4.44), 355(4.13)	18-0481-83
$C_9H_4N_2O$			
Propanedinitrile, (4-oxo-2,5-cyclohexadien-1-ylidene)-	CH_2Cl_2	323(3.23)	44-0129-83
$C_9H_4O_3$			
1H-Indene-1,2,3-trione	CH_2Cl_2	606(1.30)	5-1694-83

Compound	Solvent	$\lambda_{max}(\log \epsilon)$	Ref.
$C_9H_5BrO_2S$			
4H-1-Benzothiopyran-4-one, 2-bromo-3-hydroxy-	MeOH	264(4.40),370(4.08)	83-0921-83
$C_9H_5ClO_2$			
2H-1-Benzopyran-2-one, 5-chloro-	MeOH	283(4.06),315(3.58)	44-2578-83
$C_9H_5F_3N_2O_2$			
Benzoic acid, 4-(1-diazo-2,2,2-trifluoroethyl)-	EtOH	284(4.34),450(1.83)	5-1510-83
Benzoic acid, 4-[3-(trifluoromethyl)-3H-diazirin-3-yl]-	EtOH	233(4.23),276(3.04), 284s(2.95),348(2.59)	5-1510-83
$C_9H_5NO_4S$			
4H-1-Benzothiopyran-4-one, 3-hydroxy-2-nitro-	MeOH	272(4.16),424(3.60)	83-0921-83
$C_9H_6Br_2N_2$			
8-Quinolinamine, 5,7-dibromo-	acid	275.5(4.54),434(--)	74-0095-83C
	base	255.5(4.48),350(--)	74-0095-83C
$C_9H_6ClNO_5$			
4H,5H-Pyrano[3,4-e]-1,3-oxazine-4,5-dione, 7-chloro-2-ethoxy-	$CHCl_3$	278(3.87),284(3.85), 327(3.76)	24-3725-83
$C_9H_6Cl_2N_2$			
8-Quinolinamine, 5,7-dichloro-	acid	272(4.50),424(--)	74-0095-83C
	base	255(4.44),344(--)	74-0095-83C
$C_9H_6I_2N_2$			
8-Quinolinamine, 5,7-diiodo-	acid	282(4.37),450(--)	74-0095-83C
	base	262(4.50),355(--)	74-0095-83C
$C_9H_6N_2S$			
Thieno[2,3-b]pyridine-6-carbonitrile, 4-methyl-	EtOH	245(4.55),287(4.00)	4-0213-83
$C_9H_6N_2S_2$			
Propanedinitrile, [(methylthio)-2-thienylmethylene]-	EtOH	222(4.11),348(3.81)	142-1793-83
$C_9H_6N_4S$			
4H-Pyridazino[4,5-b]pyrido[3,2-e]-[1,4]thiazine	EtOH	254(4.21),283(4.16)	4-1047-83
C_9H_6OS			
2H-1-Benzothiopyran-2-one	hexane	338(3.40)	139-0059-83B
4H-1-Benzothiopyran-4-one	EtOH	287(3.45),336(4.03)	139-0059-83B
$C_9H_6O_2$			
2H-1-Benzopyran-2-one (coumarin)	EtOH	238(4.10),314(3.88)	139-0059-83B
	CH_2Cl_2	275(4.1),310(4.0)	35-3344-83
	+ BF_3	305(4.2)	35-3344-83
Isocoumarin	EtOH	228(4.47),239(4.22), 253(3.86),261(3.87), 318(3.58)	102-1489-83
$C_9H_6O_2S$			
2H-1-Benzothiopyran-2-one, 4-hydroxy-	EtOH	232(3.92),320(3.52)	139-0059-83B
	dioxan	302(--),332(--)	139-0059-83B

Compound	Solvent	$\lambda_{max}(\log \epsilon)$	Ref.
$C_9H_6O_3$			
2H-1-Benzopyran-2-one, 7-hydroxy-	pH <5	324(4.0)	18-0006-83
	pH >11	366(4.1)	18-0006-83
$C_9H_7BrN_4$			
1H-Imidazole, 2-[(4-bromophenyl)azo]-	MeOH	240(3.97),364(4.50), 372(4.49),443s(3.36)	56-0971-83
	MeOH-HCl	238(3.76),362(4.46), 374(4.43),420s(3.33)	56-0971-83
	MeOH-NaOH	248(3.87),272s(3.51), 400(4.41),420s(4.33)	56-0971-83
	50% MeOH	236(3.84),362(4.42), 374(4.41),428s(3.40)	56-0971-83
	MeCN	239(3.93),365(4.45), 374(4.44),436s(3.44)	56-0971-83
	1,2-$C_2H_4Cl_2$	242(3.92),367(4.41), 378(4.40),448(3.30)	56-0971-83
C_9H_7BrO			
1H-Inden-1-one, 5-bromo-2,3-dihydro-	EtOH	257(4.36),287(3.80), 294(3.79)	22-0096-83
1H-Inden-1-one, 6-bromo-2,3-dihydro-	EtOH	244(4.25),302(3.77)	22-0096-83
$C_9H_7BrO_4$			
1,3-Benzodioxole-4,5-dione, 6-bromo-7-ethyl-	MeCN	213(3.49),288(3.02), 580(2.83)	64-0752-83B
$C_9H_7ClN_2O$			
4H-Pyrido[1,2-a]pyrimidin-4-one, 3-chloro-2-methyl-	EtOH	248(4.08),256(4.05), 345(4.06)	142-1083-83
$C_9H_7ClN_2O_2$			
4H-Pyrido[1,2-a]pyrimidin-4-one, 3-chloro-9-hydroxy-2-methyl-	EtOH	245(3.97),260(3.98), 270(3.98),320(3.84), 350(4.10),367(4.23)	142-1083-83
$C_9H_7ClN_2O_5$			
4H,5H-Pyrano[3,4-e]-1,3-oxazine-4,5-dione, 7-chloro-2-[(2-hydroxyethyl)amino]-	$CHCl_3$	243(3.75),292(4.02), 339(4.14)	24-3725-83
$C_9H_7Cl_2NO_3$			
Furan, 2-(4,4-dichloro-3-methyl-1,3-butadienyl)-5-nitro-, (E)-	MeOH	215(3.88),280(4.23), 390(4.21)	73-1062-83
Furan, 2-(4,4-dichloro-3-methyl-1-cyclobuten-1-yl)-5-nitro-	MeOH	235(4.18),255(4.19), 320(3.95)	73-1062-83
$C_9H_7F_3O$			
Ethanone, 1-[3-(trifluoromethyl)-phenyl]-	MeOH	235(4.02)	44-4808-83
$C_9H_7IO_4$			
1,2-Benziodoxol-3(1H)-one, 1-acetoxy-	DMF	276(3.40)	35-0681-83
C_9H_7N			
Cyclopent[b]azepine	n.s.g.	250(4.7),350f(3.4), 600f(2.5)	77-1246-83B
Isoquinoline, perchlorate	MeCN	336(3.51)	35-1204-83
Quinoline, perchlorate	MeCN	312(3.82)	35-1204-83

Compound	Solvent	$\lambda_{max}(\log \epsilon)$	Ref.
C_9H_7NO			
1H-Indole-3-carboxaldehyde	EtOH	219(4.46),245(4.12), 260(4.04),298(4.10)	128-0157-83
	EtOH-NaOH	224(--),267(--), 328(--)	128-0157-83
	C_6H_{12}	212(--),236(--), 282(--)	128-0157-83
	MeOH	221(--),242(--), 262(--),298(--)	128-0157-83
Isoxazole, 3-phenyl-	C_6H_{12}	238(4.23)	44-3189-83
Pyridine, 2-(2-furanyl)-	pH 2	332(4.27)	23-0334-83
	EtOH	270(4.12),303(4.16)	23-0334-83
Pyridine, 2-(3-furanyl)-	pH 2	245(3.86),307(3.99)	23-0334-83
	EtOH	227s(--),255(3.77), 285(3.74)	23-0334-83
Pyridine, 3-(2-furanyl)-	pH 2	236(4.00),290(4.29)	23-0334-83
	EtOH	236(3.72),283(4.21)	23-0334-83
Pyridine, 3-(3-furanyl)-	pH 2	243(4.03)	23-0334-83
	EtOH	225s(--),257(3.86)	23-0334-83
Pyridine, 4-(2-furanyl)-	pH 2	237(3.84),332(4.41)	23-0334-83
	EtOH	235(3.94),294(4.37)	23-0334-83
Pyridine, 4-(3-furanyl)-	pH 2	239(3.98),250s(--), 304(3.98)	23-0334-83
	EtOH	223s(--),270(4.01)	23-0334-83
8-Quinolinol	pH 1.6	310f(3.4),355(3.4)	35-5695-83
	H_2O	280(3.5),301(3.5)	35-5695-83
	pH 10	325(3.6),370(3.6)	35-5695-83
$C_9H_7NO_2$			
1H-Indole-2-carboxylic acid	EtOH	226(4.49),294(4.23)	128-0157-83
	EtOH-NaOH	231(--),293(--)	128-0157-83
	C_6H_{12}	219(--),294(--)	128-0157-83
	MeOH	226(--),290(--)	128-0157-83
1H-Indole-3-carboxylic acid	MeOH	280(4.01),286(3.98)	64-0108-83B
$C_9H_7NO_2S$			
4H-1-Benzothiopyran-4-one, 2-amino-3-hydroxy-	MeOH	241(4.39),280(3.72), 368(4.06),370(4.06)	83-0921-83
$C_9H_7NO_3$			
Spiro[oxirane-2,8'(7'H)-[5H]pyrano-[4,3-b]pyridin]-5'-one	MeOH	207(4.09),225(3.99), 272(3.48)	103-0886-83
$C_9H_7N_3$			
1H-Benzimidazole, 1-(isocyanomethyl)-	EtOH	245(3.87),274(3.63), 280(3.65)	94-0723-83
$C_9H_7N_3O_3$			
2-Pyridinecarboxylic acid, 3-(diazo-acetyl)-, methyl ester	MeOH	248s(4.01),297(3.98)	103-0882-83
3-Pyridinecarboxylic acid, 2-(diazo-acetyl)-, methyl ester	MeOH	207(4.08),229s(3.95), 308(3.93)	30-0289-83
	MeOH	308(3.93)	103-0882-83
3-Pyridinecarboxylic acid, 4-(diazo-acetyl)-, methyl ester	MeOH	208(4.11),219s(4.08), 248(3.90),295(3.96), 302s(3.94)	30-0289-83
	MeOH	248(3.90),295(3.96), 302s(3.94)	103-0882-83
4-Pyridinecarboxylic acid, 3-(diazo-acetyl)-, methyl ester	MeOH	208(4.26),219s(4.04), 248s(3.85),292(4.00)	30-0289-83
	MeOH	248s(3.85),292(4.00)	103-0882-83

Compound	Solvent	λ_{max}(log ϵ)	Ref.
$C_9H_7N_5O_2$			
1H-Imidazole, 2-[(2-nitrophenyl)azo]-	MeOH	c.200(4.35),370(4.64), 436s(3.57)	56-0971-83
	MeOH-H_2SO_4	c.195(4.59),220(4.42), 256(4.29),352(4.72), 414s(3.35)	56-0971-83
	MeOH-KOH	228(4.47),404(4.74), 450s(4.51)	56-0971-83
	1,2-$C_2H_4Cl_2$	210(4.22),372(4.66), 438(3.58)	56-0971-83
1H-Imidazole, 2-[(3-nitrophenyl)azo]-	MeOH	197(4.53),366(4.54), 420s(3.85)	56-0971-83
	MeOH-H_2SO_4	195(4.72),223(4.64), 255(4.00),355(4.41), 410s(3.60)	56-0971-83
	1,2-$C_2H_4Cl_2$	213(4.22),374(4.48), 430s(3.82)	56-0971-83
1H-Imidazole, 2-[(4-nitrophenyl)azo]-	MeOH	216(3.93),276(3.84), 385(4.44),460s(3.40)	56-0971-83
	MeOH-HCl	210(3.92),267(3.62), 358(4.41)	56-0971-83
	MeOH-NaOH	215(3.88),262(3.75), 458(4.40)	56-0971-83
	20% MeOH	218(3.89),274(3.78), 383(4.40)	56-0971-83
	MeCN	250(4.41),388(4.41), 478s(3.40)	56-0971-83
	1,2-$C_2H_4Cl_2$	226(3.85),278(3.81), 390(4.36),475s(3.17)	56-0971-83
$C_9H_7N_7O_2S$			
1H-Purine, 6-[(1-methyl-4-nitro-1H-imidazol-5-yl)thio]- (azathioprine)	pH 1	280(4.24)	162-0130-83
	pH 13	285(4.19)	162-0130-83
	MeOH	276(4.26)	162-0130-83
$C_9H_8Br_4O_2$			
Benzene, 1,2,4-tribromo-3-(bromomethyl)-5,6-dimethoxy-	$CHCl_3$	297(4.10)	94-1754-83
$C_9H_8ClNO_3$			
Furo[3,2-c]pyridin-6-ol, 4-chloro-2,3-dihydro-, acetate	EtOH	295(3.76)	120-0007-83
C_9H_8ClNS			
2-Propenethioamide, 3-(2-chlorophenyl)-	EtOH	337(3.81)	80-0555-83
2-Propenethioamide, 3-(4-chlorophenyl)-	EtOH	337(3.91),430(2.44)	80-0555-83
$C_9H_8ClN_3$			
1-Phthalazinamine, 4-chloro-N-methyl-	MeOH	212(4.72),233(4.00), 270(3.70),324(3.93)	103-0666-83
	MeCN	210(4.68),235s(3.95), 260(3.63),325(3.88)	103-0666-83
sulfate(2:1)	MeOH	214(4.54),266(4.00), 309(3.83)	103-0666-83
	MeCN	220(4.62),247s(4.14), 268(4.14),307(3.78)	103-0666-83
1(2H)-Phthalazinimine, 4-chloro-2-methyl-	MeOH	211(4.58),242s(3.85), 274(3.98),322(3.72)	103-0666-83
	MeCN	210(4.65),235s(3.82), 273(3.99),330(3.71)	103-0666-83

Compound	Solvent	λ_{max}(log ε)	Ref.
1(2H)-Phthalazinimine, 4-chloro-2-methyl-, sulfate(2:1)	MeOH	212(4.54),245(4.13), 267(3.83),293(3.80)	103-0666-83
	MeCN	222(4.56),243(4.15), 264(3.85),288(3.76)	103-0666-83
$C_9H_8ClN_5O$			
1,3,5-Triazin-2(1H)-one, 4,6-diamino-1-(4-chlorophenyl)-, hydrochloride	M HCl	233(4.37)	94-2473-83
	H_2O	222(4.30)	94-2473-83
$C_9H_8Cl_3N_3O_5$			
2-Propanone, 1,1,1-trichloro-3-(2,6-dimethoxy-5-nitro-4-pyrimidinyl)-	CCl_4	233(4.33),310(4.24)	103-0435-83
$C_9H_8N_2$			
Cyclohept[b]pyrrol-2-amine	n.s.g.	268(4.58),277.5(4.53), 298(3.51),370(3.67)	28-0661-83B
8-Quinolinamine	acid	260(--),395(2.99)	74-0095-83C
	base	247.5(--),332.5(3.44)	74-0095-83C
$C_9H_8N_2O$			
Pyridine, 2-(5-methyl-3-isoxazolyl)-	MeOH	232(4.03),266(3.80), 270(3.79),277s(3.64)	142-0501-83
Pyridine, 2-(3-methyl-5-isoxazolyl)-	MeOH	251(4.17),282(4.13), 290s(4.02)	142-0501-83
Pyridine, 3-(3-methyl-5-isoxazolyl)-	MeOH	257(4.17),277s(4.00)	142-0501-83
Pyridine, 3-(5-methyl-3-isoxazolyl)-	MeOH	232(4.10),266(3.58)	142-0501-83
Pyridine, 4-(3-methyl-5-isoxazolyl)-	MeOH	262(4.24)	142-0501-83
Pyridine, 4-(5-methyl-3-isoxazolyl)-	MeOH	232(4.12),265s(3.52), 275s(3.37)	142-0501-83
2(1H)-Quinoxalinone, 3-methyl-	EtOH	340(3.95)	18-0326-83
$C_9H_8N_2OS$			
1,3,4-Oxadiazole-2(3H)-thione, 5-(3-methylphenyl)-	EtOH	247(4.17),295(4.46)	142-2211-83
1,3,4-Oxadiazole-2(3H)-thione, 5-(4-methylphenyl)-	EtOH	240(4.26),300(4.27)	142-2211-83
1,2,3-Thiadiazolium, 4-hydroxy-3-methyl-5-phenyl-, hydroxide, inner salt	EtOH	256(3.97),395(4.04)	94-1746-83
$C_9H_8N_2OSSe$			
4H-Pyrido[3,4-e]-1,3-selenazin-4-one, 2,3-dihydro-5,7-dimethyl-2-thioxo-	MeOH	220(3.30),305(4.10)	73-3567-83
$C_9H_8N_2OS_2$			
1,2,3-Thiadiazolium, 4-mercapto-3-(4-methoxyphenyl)-, hydroxide, inner salt	EtOH	252(3.77),343(3.77)	94-1746-83
$C_9H_8N_2OSe_2$			
4H-Pyrido[3,4-e]-1,3-selenazin-4-one, 2,3-dihydro-5,7-dimethyl-2-selenoxo-	MeOH	217(4.01),365(4.10)	73-3567-83
$C_9H_8N_2O_2$			
1,3(2H,4H)-Isoquinolinedione, 1-oxime	EtOH	220(4.07),259(3.95), 293(3.18)	39-1813-83C
2H-Pyrano[3,2-c]pyridine-8-carbonitrile, 3,4,6,7-tetrahydro-7-oxo-	n.s.g.	225(4.46),260(3.51), 267s(3.46),320(3.83)	103-1050-83
$C_9H_8N_2O_2S$			
1H-Imidazole-5-carboxylic acid, 1-(2-thienylmethyl)-	MeOH	233(4.24)	142-2019-83

Compound	Solvent	$\lambda_{max}(\log \epsilon)$	Ref.
1,3,4-Oxadiazole-2(3H)-thione, 5-(4-methoxyphenyl)-	EtOH	245(3.90),296(3.98)	142-2211-83
$C_9H_8N_2O_2S_2$			
1H-Imidazole-4-carboxylic acid, 2,3-dihydro-3-(2-thienylmethyl)-2-thioxo-	0.5M HCl	257s(3.98),289(4.05)	142-2019-83
	0.5M NaOH	235s(4.18),287(4.11)	142-2019-83
	MeOH	236(4.11),263(3.98), 290(4.01)	142-2019-83
$C_9H_8N_2O_2S_3$			
Benzenesulfenamide, N-1,3-dithiolan-2-ylidene-2-nitro-	1,2-$C_2H_4Cl_2$	306s(3.8),375(3.59), 408s(3.4)	118-0050-83
$C_9H_8N_2O_4$			
Compd. from ferulic acid and $NaNO_2$	pH 1	240(4.13),307(3.93)	98-0780-83
	H_2O	241(4.12),309(3.92)	98-0780-83
	pH 13	254(4.13),365(4.16)	98-0780-83
Phenol, 4-furazanyl-2-methoxy-, N-oxide	pH 1	346(4.01)	98-0780-83
	H_2O	351(4.00)	98-0780-83
	pH 13	283(3.74),376(4.15)	98-0780-83
$C_9H_8N_2S_2$			
1,2,3-Thiadiazolium, 4-mercapto-3-methyl-5-phenyl-, hydroxide, inner salt	EtOH	298(4.10),459(3.49)	94-1746-83
$C_9H_8N_4$			
1H-Imidazole, 2-(phenylazo)-	MeOH	230(3.90),356(4.38), 366(4.35),426s(3.25)	56-0971-83
	MeOH-HCl	229(3.74),342(4.34), 353(4.38),366(4.32)	56-0971-83
	MeOH-NaOH	233(3.82),268(3.43), 387(4.32),424s(4.06)	56-0971-83
	20% MeOH	230(3.81),358(4.32), 366s(4.31),392s(4.00)	56-0971-83
	MeCN	238(4.02),256(3.81), 344s(4.27),358(4.37), 372(4.35),434s(3.26)	56-0971-83
	1,2-$C_2H_4Cl_2$	238(3.53),352s(4.29), 362(4.33),372(4.32), 440s(3.18)	56-0971-83
$C_9H_8N_4O$			
1H-1,2,3-Triazole-4-carboxaldehyde, 5-amino-1-phenyl-	EtOH	230(4.15),281(3.95)	95-0594-83
1H-1,2,3-Triazole-4-carboxaldehyde, 5-(phenylamino)-	EtOH	246(4.18),340(3.91)	95-0594-83
$C_9H_8N_4O_2$			
1H-Imidazol-4-amine, 5-nitro-N-phenyl-	H_2O	242(4.05),392(4.15)	103-0650-83
$C_9H_8N_4O_4$			
Benzenepropanoic acid, 4-azido-2-nitro-	MeOH	245(4.25)	64-0049-83C
$C_9H_8N_4O_5$			
1H,3H,8H-Pteridine-6-carboxylic acid, 3,8-dimethyl-2,4,7-trioxo-	H_2O	275s(--),287(4.27), 430(3.91)	35-3304-83
$C_9H_8N_4S$			
3-Pyridinamine, 2-(pyrazinylthio)-	MeOH	235(4.10),300(3.96), 348(4.11)	4-1047-83

Compound	Solvent	$\lambda_{max}(\log \epsilon)$	Ref.
$C_9H_8N_6$			
3H-1,2,4-Triazolo[1,5-d]tetrazole, 3-methyl-6-phenyl-	CH_2Cl_2	244(4.23),266(4.18)	4-1629-83
$C_9H_8N_6O_2$			
Pyrimido[5,4-e]-1,2,4-triazolo[4,3-b]-pyridazine-7,9(6H,8H)-dione, 6,8-dimethyl-	EtOH	233(4.25),315(3.66)	44-1628-83
$C_9H_8N_6O_3$			
1,3,5-Triazin-2(1H)-one, 4,6-diamino-1-(4-nitrophenyl)-, hydrochloride	M HCl	232(4.13)	94-2473-83
C_9H_8O			
1H-Inden-1-one, 2,3-dihydro-	MeOH	243(4.08),287(3.31), 294(3.34)	94-3113-83
2-Propenal, 3-phenyl-, trans	MeOH	281(4.30)	94-1544-83
$C_9H_8O_3$			
4(5H)-Benzofuranone, 6,7-dihydro-5-(hydroxymethylene)-	EtOH	216.5(3.63),270s(3.29), 301(3.46)	111-0113-83
1(3H)-Isobenzofuranone, 7-methoxy-	MeOH	234(3.97),296(3.75)	2-1257-83
2-Propenoic acid, 3-(2-hydroxyphenyl)-	MeOH	215(5.22),273.5(5.22), 323.5(4.94)	94-3865-83
2-Propenoic acid, 3-(4-hydroxyphenyl)-	EtOH	223(4.16),286(4.28)	162-0366-83
$C_9H_8O_4$			
1,3-Benzodioxole-4,5-dione, 6,7-dimethyl-	MeCN	209(3.47),272(3.08), 575(2.90)	64-0752-83B
C_9H_9			
Bicyclo[5.2.0]nonatrienylium hexafluoroantimonate	MeCN	230(4.28),295(3.54)	44-0596-83
$C_9H_9BrN_2O_6$			
6,2'-O-Cyclouridine, 5-bromo-	pH 1	266.5(4.13)	44-2719-83
	H_2O	266(4.12)	44-2719-83
	pH 13	265(3.99)	44-2719-83
6,3'-O-Cyclouridine, 5-bromo-	0.05M HCl	274(--)	44-2719-83
	H_2O	274(--)	44-2719-83
	0.05M NaOH	273.5(--)	44-2719-83
6,5'-O-Cyclouridine, 5-bromo-	pH 1	278(4.05)	44-2719-83
	H_2O	278(4.06)	44-2719-83
	pH 13	277(3.92)	44-2719-83
Uracil, 6,2'-anhydro-5-bromo-6-hydroxy-1-β-D-lyxofuranosyl-	pH 1	266.5(4.12)	44-2719-83
	H_2O	266(4.12)	44-2719-83
	pH 13	265.5(3.98)	44-2719-83
$C_9H_9BrO_4$			
Benzeneacetic acid, 5-bromo-2,3-dihydroxy-, methyl ester	EtOH	208(4.48),290(3.46)	104-0458-83
$C_9H_9Br_3O_3$			
Benzenemethanol, 2,3,6-tribromo-4,5-dimethoxy-	$CHCl_3$	283(3.97),293(3.87)	94-1754-83
$C_9H_9ClN_2$			
2H-Indol-2-imine, 3-chloro-1,3-dihydro-1-methyl-, monohydrochloride	n.s.g.	216(4.16),264(3.87)	103-1175-83

Compound	Solvent	$\lambda_{max}(\log \epsilon)$	Ref.
$C_9H_9ClN_2O_6$			
6,2'-O-Cyclouridine, 5-chloro-	pH 1	264(4.14)	44-2719-83
	H_2O	264(4.14)	44-2719-83
	pH 13	264(4.03)	44-2719-83
6,5'-O-Cyclouridine, 5-chloro-	pH 1	276.5(4.06)	44-2719-83
	H_2O	276.5(4.06)	44-2719-83
	pH 13	276.5(3.93)	44-2719-83
$C_9H_9ClO_2$			
3-Benzofuranol, 5-chloro-2,3-dihydro-3-methyl-	EtOH	229(3.84),286(3.30)	39-1649-83C
$C_9H_9ClO_3$			
Benzaldehyde, 5-chloro-2,3-dimethoxy-	EtOH	226(4.17),264(3.72), 332(3.32)	104-0458-83
$C_9H_9ClO_4$			
Benzeneacetic acid, 5-chloro-2,3-dihydroxy-, methyl ester	EtOH	209(4.40),290(3.45)	104-0458-83
$C_9H_9F_3O$			
2,5-Cyclohexadien-1-one, 2,3,5-trifluoro-4,4,6-trimethyl-	heptane	228(4.17),257(2.94)	104-0884-83
$C_9H_9IN_2O_6$			
6,2'-O-Cyclouridine, 5-iodo-	pH 1	265(4.17)	44-2719-83
	H_2O	266(4.17)	44-2719-83
	pH 13	263.5(4.07)	44-2719-83
6,3'-O-Cyclouridine, 5-iodo-	pH 1	273.5(4.01)	44-2719-83
	H_2O	274(4.00)	44-2719-83
	pH 13	270.5(3.92)	44-2719-83
6,5'-O-Cyclouridine, 5-iodo-	pH 1	281(3.96)	44-2719-83
	H_2O	281(3.96)	44-2719-83
	pH 13	276(3.86)	44-2719-83
C_9H_9N			
1H-Indole, 3-methyl-	EtOH	225(4.59),275s(3.81), 282(3.83),290s(3.76)	128-0157-83
	EtOH-NaOH	230(--),276s(--), 282(--),292s(--)	128-0157-83
	C_6H_{12}	222(--),272s(--), 280(--),292(--)	128-0157-83
	MeOH	223(--),278s(--), 283(--),291s(--)	128-0157-83
C_9H_9NO			
1H-Indole, 6-methoxy-	EtOH	274(3.54),292(3.66), 301(3.52)	151-0171-83C
2-Propenamide, 3-phenyl-, (E)-	EtOH	217(4.22),222(4.17), 274(4.22)	88-4381-83
C_9H_9NOSe			
1,2-Benzisoselenazole, 3-ethyl-, 2-oxide	EtOH	238(4.36),295(4.18)	78-0831-83
$C_9H_9NO_2$			
Benzonitrile, 2,5-dimethoxy-	MeOH	213(4.35),235(4.06), 267(3.93),315(3.75)	64-0866-83B
1,3-Butanedione, 1-(2-pyridinyl)-	MeOH	235(3.81),245s(3.70), 280s(3.74),312(4.08)	142-0501-83
1,3-Butanedione, 1-(3-pyridinyl)-	MeOH	233(3.96),307(4.21)	142-0501-83

Compound	Solvent	$\lambda_{max}(\log \epsilon)$	Ref.
1,3-Butanedione, 1-(4-pyridinyl)-	MeOH	221(3.91),310(4.08)	142-0501-83
2,4-Pentadiynal, 5-morpholino-	MeCN	213(4.27),340(4.08)	33-1631-83
Propanoic acid, 3-(3-isocyano-2-cyclo-penten-1-ylidene)-	MeOH	218(3.58),273(3.96)	158-83-99
	ether	270(4.13)	158-83-99
$C_9H_9NO_2S_2$			
2,6-Pyridinedicarbothioic acid, O,S-dimethyl ester	n.s.g.	280(4.07),413(2.15)	64-0516-83B
$C_9H_9NO_3$			
Formamide, N-(2-acetyl-4-hydroxy-phenyl)-	MeOH	235(4.83),265(4.30), 350(3.96)	24-1309-83
$C_9H_9NO_5$			
Acetaldehyde, (2-acetoxy-7-oxo-4-oxa-1-azabicyclo[3.2.0]hept-3-ylidene)-	EtOH	254(4.16)	39-2513-83C
Benzoic acid, 4-(hydroxymethyl)-3-nitro-, methyl ester	$CHCl_3$	244(3.87),260s(3.81), 300s(3.23)	136-0023-83M
6-Oxa-1-azabicyclo[5.2.0]nona-2,4-diene-2-carboxylic acid, 3-methoxy-9-oxo-, lithium salt	H_2O	297(3.84)	39-0115-83C
C_9H_9NS			
2-Propenethioamide, 3-phenyl-	EtOH	330(3.82),420(2.49)	80-0555-83
$C_9H_9N_3$			
Pyrimidine, 1,4-dihydro-4-(2-pyri-dinyl)-	EtOH	261(3.70),290s(3.23)	54-0364-83
Pyrimidine, 1,4-dihydro-4-(4-pyri-dinyl)-	EtOH	255(3.57),303(3.19)	54-0364-83
$C_9H_9N_3O$			
Formamide, N-(1H-benzimidazol-1-yl-methyl)-	EtOH	245(3.83),273(3.65), 280(3.62)	94-0723-83
3,5-Isoxazolediamine, 4-phenyl-	MeCN	265(3.96)	44-0575-83
$C_9H_9N_3OS$			
Pyrido[2,3-d]pyrimidin-4(1H)-one, 1-ethyl-2,3-dihydro-2-thioxo-	MeOH	313(3.14)	73-3315-83
4H-Pyrido[3,2-e]-1,3-thiazin-4-one, 2-(ethylamino)-	MeOH	221(5.54),315(3.39)	73-3315-83
4H-Pyrido[3,4-e]-1,3-thiazin-4-one, 2-amino-5,7-dimethyl-	MeOH	239(4.42)	73-3426-83
$C_9H_9N_3O_2$			
1,2,3-Triazolo[1,5-a]pyridine-3-carb-oxylic acid, ethyl ester	EtOH	245(3.75),286(4.17), 300s(--)	150-1341-83M
$C_9H_9N_3O_2S$			
1H-Imidazole-5-carboxamide, N-hydroxy-1-(2-thienylmethyl)-	MeOH	233(4.22)	142-2019-83
$C_9H_9N_3O_3$			
Benzoic acid, 4-azido-2-methoxy-, methyl ester	MeOH	224(4.22),268(4.21), 302(3.92)	44-5041-83
1,2,4-Triazolo[4,3-a]pyridinium, 1-(ethoxycarbonyl)-2,3-dihydro-3-oxo-, hydroxide, inner salt	THF	378(3.74)	18-2969-83
$C_9H_9N_5$			
Benzenamine, 4-(1H-imidazol-2-ylazo)-	MeOH	201(3.84),263(3.48),	56-0971-83

Compound	Solvent	$\lambda_{max}(\log \epsilon)$	Ref.
(cont.)		310(3.16),406?(4.09), 432(4.09)	56-0971-83
1,3,5-Triazine-2,4-diamine, 6-phenyl-	EtOH	249(4.40)	162-0155-83
$C_9H_9N_5O$			
1,3,5-Triazin-2(1H)-one, 4,6-diamino-1-phenyl-, hydrochloride	M HCl	241(4.16)	94-2473-83
	H_2O	232(4.18)	94-2473-83
1H-1,2,3-Triazole-4-carboxaldehyde, 5-amino-1-phenyl-, oxime	EtOH	270(4.16)	95-0594-83
1H-1,2,3-Triazole-4-carboxaldehyde, 5-(phenylamino)-, oxime	EtOH	271(4.24),300s(4.01)	95-0594-83
$C_9H_9N_5O_2S$			
Benzenesulfonamide, 4-(1H-imidazol-2-ylazo)-	MeOH	250(4.24),364(4.25)	56-0971-83
$C_9H_9N_7O_2$			
Pyrimido[5,4-e]-1,2,4-triazolo[4,3-b]-[1,2,4]triazine-7,9(6H,8H)-dione, 3,6,8-trimethyl-	EtOH	232(4.37),347(3.83)	44-1628-83
$C_9H_9S_2$			
1,3-Dithiol-1-ium, 4,5-dihydro-2-phenyl-, tetrafluoroborate	CH_2Cl_2	235(4.28),348(3.58)	1-0687-83
$C_9H_{10}AsNS$			
1H-1,3-Benzazarsole, 1-methyl-2-(methylthio)-	MeOH	255(4.30),292s(3.54), 331(3.43)	101-0257-83R
$C_9H_{10}BrClO_2$			
Benzene, 5-bromo-1-(chloromethyl)-2,3-dimethoxy-	EtOH	212(4.56),226s(--), 291(3.34)	104-0458-83
Benzene, 1-(bromomethyl)-5-chloro-2,3-dimethoxy-	EtOH	216(4.48),297(3.33)	104-0458-83
$C_9H_{10}Br_2O_2$			
Benzene, 5-bromo-1-(bromomethyl)-2,3-dimethoxy-	EtOH	214(4.40),297(3.30)	104-0458-83
$C_9H_{10}ClF_3N_2O_2$			
Pyrimidine, 5-chloro-2,4-diethoxy-6-(trifluoromethyl)-	MeOH	223(3.99),281(3.80)	4-0219-83
$C_9H_{10}ClNO_4S$			
Carbamic acid, (chlorosulfonyl)-2,4,6-cycloheptatrien-1-yl-, methyl ester	MeCN	217(4.6),277(3.7)	24-3516-83
$C_9H_{10}ClN_3OS$			
3-Pyridinecarboxamide, 2-chloro-N-[(ethylamino)thioxomethyl]-	MeOH	236(4.81)	73-3315-83
$C_9H_{10}Cl_2O_2$			
Benzene, 5-chloro-1-(chloromethyl)-2,3-dimethoxy-	EtOH	211(4.57),229(3.98), 290(3.33)	104-0458-83
$C_9H_{10}FN_5O_4$			
Uridine, 3'-azido-2',3'-dideoxy-5-fluoro-	pH 2	270(3.85)	87-1691-83
	pH 12	270(3.84)	87-1691-83
	MeOH	270(3.89)	87-1691-83

Compound	Solvent	$\lambda_{max}(\log \epsilon)$	Ref.
$C_9H_{10}F_3N$			
Benzenemethanamine, α-methyl-4-(trifluoromethyl)-, (S)-	C_6H_{12}	258(2.49),263(2.46), 268(2.30)	35-1578-83
	MeOH	263(2.53),268(2.38)	35-1578-83
hydrochloride	MeOH	257(2.62),262(2.74), 268(2.68)	35-1578-83
	10% HCl	257(2.70),262(2.80), 268(2.72)	35-1578-83
$C_9H_{10}NPS$			
1H-1,3-Benzazaphosphole, 1-methyl-2-(methylthio)-	MeOH	250(3.37),319(2.86)	101-0257-83R
$C_9H_{10}N_2$			
Benzonitrile, 4-(1-aminoethyl)-, (S)-	C_6H_{12}	267(2.77),269(2.71), 273(2.68),274(2.65), 279(2.58)	35-1578-83
1H-Indole-3-methanamine	MeOH	279(3.76),287(3.68)	64-0108-83B
$C_9H_{10}N_2O$			
Isoquinoline, 1,2,3,4-tetrahydro-2-nitroso-	MeOH	223(3.98),347(1.98)	4-0121-83
5(6H)-Quinazolinone, 7,8-dihydro-2-methyl-	EtOH	219s(3.76),233(3.92), 257s(3.64),266s(3.49), 285(2.75)	4-0649-83
$C_9H_{10}N_2OS$			
Formamide, N-[(methylphenylamino)-thioxomethyl]-	MeCN	225(4.24),281(4.24), 339s(2.26)	97-0019-83
$C_9H_{10}N_2O_2$			
Benzenamine, N-(1-methyl-2-nitroethenyl)-	EtOH	234(3.93),354(4.27)	40-0088-83
Benzenecarboximidamide, N-acetoxy-	EtOH	248(3.94)	32-0845-83
Pyrano[2,3-c]pyrazol-6(2H)-one, 2,3,4-trimethyl-	MeOH	225(3.34),305(4.11)	20-0451-83
$C_9H_{10}N_2O_2S_2$			
Aureothricin	n.s.g.	248(4.79),312(3.59), 388(4.04)	162-0127-83
$C_9H_{10}N_2O_3$			
Carbamic acid, [(phenylamino)carbonyl]-, methyl ester	EtOH	240(4.19)	22-0073-83
2-Pyridinecarboxylic acid, 5-[(1-oxopropyl)amino]-	MeOH	257(4.13),285(4.01)	106-0591-83
$C_9H_{10}N_2O_4$			
Glycine, 2-nitrobenzyl-	MeOH	206(4.3),256(3.9)	5-0231-83
$C_9H_{10}N_2O_5$			
4(1H)-Pyridazinone, 6,2'-anhydro-6-hydroxy-1-β-D-arabinofuranosyl-	pH 1	249s(3.87),254(3.88), 268(3.89)	44-3765-83
	pH 11	251s(3.80),273(3.90)	44-3765-83
	MeOH	249s(3.80),254(3.82), 274(3.84)	44-3765-83
4(1H)-Pyridazinone, 6,5'-anhydro-6-hydroxy-1-β-D-ribofuranosyl-	pH 1	268(4.19)	44-3765-83
	pH 11	265.5(4.15)	44-3765-83
	MeOH	267(4.21)	44-3765-83
Uracil, 6,2'-anhydro-6-hydroxy-1-(5-deoxy-β-D-lyxofuranosyl)-	pH 1 and H_2O	251(4.21)	44-2719-83
	pH 13	254(4.06)	44-2719-83

Compound	Solvent	λ_{max}(log ε)	Ref.
$C_9H_{10}N_2O_6$			
Uracil, 6,2'-anhydro-6-hydroxy-1-β-D-lyxofuranosyl-	pH 1	252(4.23)	44-2719-83
	H_2O	251(4.22)	44-2719-83
	pH 13	254(4.09)	44-2719-83
$C_9H_{10}N_4O_3$			
Pyrimido[4,5-c]pyridazine-5,7(6H,8H)-dione, 3-methoxy-6,8-dimethyl-	EtOH	228(4.55),347(3.17)	44-1628-83
Pyrimido[4,5-c]pyridazine-3,5,7(2H,5H-7H)-trione, 2,6,8-trimethyl-	EtOH	233(4.25),380(3.08)	44-1628-83
$C_9H_{10}N_4O_3S$			
7-Pteridinesulfenic acid, 1,2,3,4-tetrahydro-1,3,6-trimethyl-2,4-dioxo-	pH 2	224(4.27),262(3.75), 360(4.10)	88-5047-83
	pH 9	226(4.19),294(3.16), 416(4.10)	88-5047-83
$C_9H_{10}N_4O_4$			
2-Furancarboxamide, 5-(4-amino-2-oxo-1(2H)-pyrimidinyl)-4,5-dihydro-4-hydroxy-, (2R-trans)-	H_2O	227(4.14),263(3.98)	65-1712-83
$C_9H_{10}N_4O_5S$			
7-Pteridinesulfonic acid, 1,2,3,4-tetrahydro-2,4-dioxo-	pH 7	242(4.24),345(3.96)	88-5047-83
$C_9H_{10}O$			
Ethanone, 1-(4-methylphenyl)-	hexane	245(4.24)	99-0034-83
	MeOH	252(4.17)	44-4808-83
$C_9H_{10}O_2$			
3-Benzofuranol, 2,3-dihydro-3-methyl-	EtOH	212(3.76),275(3.25)	39-1649-83C
Bicyclo[3.2.0]hept-2-ene-6-acetaldehyde, 7-oxo-, (1α,5α,6α)-	hexane	304(1.36)	44-1718-83
2,4,6-Cycloheptatrien-1-one, 2-hydroxy-3,7-dimethyl-	MeOH	250(4.5),325(3.9), 370(3.9)	88-3879-83
	MeOH-acid	252(4.6),310(3.8), 350(3.8)	88-3879-83
	MeOH-base	255(4.5),335(4.0), 410(4.0)	88-3879-83
Ethanone, 1-(4-methoxyphenyl)-	MeOH	271(4.21)	44-4808-83
3-Oxatetracyclo[3.3.2.0^{2,4}.0^{6,8}]-decan-9-one, (1α,2α,4α,5α,6α,8α)-	benzene	281(1.67)	44-1718-83
(1α,2α,4α,5α,6β,8β)-	benzene	289(1.80)	44-1718-83
(1α,2β,4β,5α,6α,8α)-	benzene	296(1.65)	44-1718-83
$C_9H_{10}O_3$			
1-Propanone, 3-hydroxy-1-(4-hydroxyphenyl)-	EtOH	220(4.15),273s(--), 279(4.32)	102-0749-83
$C_9H_{10}O_4$			
Ethanone, 1-(2,6-dihydroxy-4-methoxyphenyl)-	EtOH	285(4.2)	83-0971-83
$C_9H_{10}O_5$			
6H-Furo[2,3-b]pyran-2,6(3H)-dione, 3a,7a-dihydro-4-methoxy-3-methyl-	EtOH	234(4.09)	94-0925-83
$C_9H_{11}BrO_2$			
Benzene, 5-bromo-1,2-dimethoxy-3-methyl-	EtOH	207(4.48),283(3.27)	104-0458-83

Compound	Solvent	λ_{max}(log ε)	Ref.
Benzene, 1-(bromomethyl)-2,3-dimethoxy-	MeOH	219(4.10),286(3.24)	36-0792-83
$C_9H_{11}BrO_3$			
Benzenemethanol, 5-bromo-2,3-dimethoxy-	EtOH	208(4.52),222s(--), 283(3.27)	104-0458-83
$C_9H_{11}ClN_2O_6$			
3(2H)-Pyridazinone, 6-chloro-5-hydroxy-2-β-D-ribofuranosyl-	MeOH	274(3.70),303s(3.47)	4-0369-83
	pH 1	254(3.67),284s(3.46)	4-0369-83
	pH 11	274(3.81),302s(3.45)	4-0369-83
$C_9H_{11}ClO_2$			
Benzene, 5-chloro-1,2-dimethoxy-3-methyl-	EtOH	208(4.55),224(3.89), 280(3.13)	104-0458-83
$C_9H_{11}ClO_3$			
Benzenemethanol, 5-chloro-2,3-dimethoxy-	EtOH	207(4.50),224(3.85), 282(3.22)	104-0458-83
$C_9H_{11}ClS$			
Benzene, 1-chloro-4-[(1-methylethyl)-thio]-	decane	226(3.97),265(3.90)	65-0449-83
Benzene, 1-chloro-4-(propylthio)-	decane	225s(3.85),264(4.04)	65-0449-83
$C_9H_{11}DO_3$			
3-Oxabicyclo[3.2.1]octane-2,4-dione-1-d, 8,8-dimethyl-, (1R)-	dioxan	230f(2.2)	39-0305-83C
$C_9H_{11}D_5N_2O_3S$			
3-Pyridine-2,6-d_2-sulfonamide, 1,6-dihydro-6-(methoxy-d_3)-N,N,1-trimethyl-	CH_2Cl_2	306(3.6)	64-0873-83B
$C_9H_{11}FN_2O_5$			
Uridine, 2'-deoxy-5-fluoro-	pH 7.2	268(3.88)	162-0589-83
	pH 14	270(3.81)	162-0589-83
$C_9H_{11}FN_2O_6$			
2,4(1H,3H)-Pyrimidinedione, 1-β-D-arabinofuranosyl-5-fluoro-	pH 7	269.5(3.91)	48-0387-83
	pH 12	270.0(--)	48-0387-83
$C_9H_{11}FN_6O_4$			
2H-1,2,3-Triazolo[4,5-d]pyrimidin-7-amine, 2-β-D-ribofuranosyl-5-fluoro-	pH 1	247(3.74),292(4.00)	87-1483-83
	pH 7	247(3.75),292(4.01)	87-1483-83
	pH 13	243(3.66),297(3.97)	87-1483-83
3H-1,2,3-Triazolo[4,5-d]pyrimidin-7-amine, 3-β-D-ribofuranosyl-5-fluoro-	pH 1	270(4.12)	87-1483-83
	pH 7	270(4.12)	87-1483-83
	pH 13	271.5(4.04)	87-1483-83
$C_9H_{11}FS$			
Benzene, 1-fluoro-4-[(1-methylethyl)-thio]-	decane	216s(3.87),258(3.61)	65-0449-83
$C_9H_{11}F_3N_2O_2$			
Pyrimidine, 2,4-diethoxy-6-(trifluoromethyl)-	MeCN	212(3.88),264(3.77)	4-0219-83
$C_9H_{11}IN_2O_5$			
2,4(1H,3H)-Pyrimidinedione, 1-(5-deoxy-β-D-lyxofuranosyl)-5-iodo-	pH 1	287(3.88)	44-2719-83
	H_2O	287.5(3.89)	44-2719-83
	pH 13	280(3.78)	44-2719-83

Compound	Solvent	$\lambda_{max}(\log \epsilon)$	Ref.
$C_9H_{11}IN_2O_6$			
2,4(1H,3H)-Pyrimidinedione, 5-iodo-1-β-D-lyxofuranosyl-	pH 1	287(3.89)	44-2719-83
	H_2O	287(3.89)	44-2719-83
	pH 13	280(3.79)	44-2719-83
$C_9H_{11}NO$			
Benzaldehyde, 4-(dimethylamino)-	H_2O	355(4.53)	2-1217-83
	H_2O	240(3.8),353(4.5)	140-0681-83
	0.2M HCl	245(4.2),350(3.5)	140-0681-83
	MeOH	342(4.50)	2-1217-83
	EtOH	340(4.53)	2-1217-83
	acetone	340(4.53)	2-1217-83
	dioxan	332(4.52)	2-1217-83
	MeCN	340(4.53)	2-1217-83
	DMF	340(4.53)	2-1217-83
	DMSO	342(4.53)	2-1217-83
	$CHCl_3$	340(4.54)	2-1217-83
Bicyclo[2.2.1]heptane-2-acetonitrile, 6-oxo-, endo	MeCN	290(1.60)	44-1862-83
Ethanone, 1-[4-(methylamino)phenyl]-	hexane	229(3.83),307(4.35)	99-0034-83
$C_9H_{11}NO_2$			
1,2-Benzisoxazol-4(5H)-one, 6,7-dihydro-6,6-dimethyl-	EtOH	231(3.90),272(3.77)	4-0645-83
Cyclohexanecarbonitrile, 4,4-dimethyl-2,6-dioxo-	EtOH	233.5(3.95),268(4.25)	4-0645-83
2-Heptenenitrile, 6,6-dimethyl-4,5-dioxo- (12:88 E:Z)	MeOH	239(3.80),310(2.48)	12-0963-83
$C_9H_{11}NO_3$			
2-Pyridinecarboxylic acid, 5-propoxy-	MeOH	244(4.03),282(3.76)	106-0591-83
Tyrosine, photooxidation product	H_2O	275(3.00)	98-0734-83
$C_9H_{11}NO_4$			
1H-Azepine-4-carboxylic acid, 6,7-dihydro-5-methoxy-7-oxo-, methyl ester	MeOH	220(3.97),267(3.57)	44-5041-83
1-Propanone, 2,2-dimethyl-1-(5-nitro-2-furanyl)-	MeOH	302(4.20)	12-0963-83
$C_9H_{11}NO_5$			
6-Oxa-1-azabicyclo[5.2.0]non-2-ene-2-carboxylic acid, 3-methoxy-9-oxo-, lithium salt	H_2O	251(3.74)	39-0115-83C
$C_9H_{11}NO_{10}S_2$			
1-Azabicyclo[3.2.0]hept-2-ene-2-carboxylic acid, 7-oxo-3-sulfo-1-[1-(sulfooxy)ethyl]- (SF-2103A)	pH 7.2	230(3.74),266(3.76)	158-83-57
$C_9H_{11}NS$			
Benzeneethanethioamide, 2-methyl-	heptane	268(4.06),373(2.00)	80-0555-83
	EtOH	268(4.06),330(2.00)	80-0555-83
$C_9H_{11}N_3$			
Ethenamine, N-methyl-2-(phenylazo)-	EtOH	339(4.41)	65-1902-83
	EtOH-H_2SO_4	379(4.48)	65-1902-83
	50%EtOH-H_2SO_4	379(4.51)	65-1902-83
$C_9H_{11}N_3O$			
Benzaldehyde, 4-(3,3-dimethyl-1-triazenyl)-	MeOH	202(3.87),229(3.93), 342(4.42)	83-0271-83

Compound	Solvent	λ max(log ε)	Ref.
$C_9H_{11}N_3O_2$			
1,2-Diazabicyclo[5.2.0]nona-3,5-diene-2-carboxamide, 5-methyl-9-oxo-	MeOH	276(3.84)	5-1361-83
1-Propanone, 1-(5-azido-2-furanyl)-2,2-dimethyl-	$CHCl_3$	311(4.18)	12-0963-83
$C_9H_{11}N_3O_3$			
Isoxazolo[4,5-c]pyridin-4(5H)-one, 6-methoxy-3-methyl-5-(methylamino)-	MeOH	211(4.14),257(3.84), 286(4.02)	142-1315-83
Morpholine, 4-(3-nitro-4-pyridinyl)-	H_2O	225(4.16),400(3.38)	103-0303-83
$C_9H_{11}N_3O_4$			
1H-Pyridazin-4-imine, 6,2'-anhydro-6-hydroxy-1-β-D-arabinofuranosyl-, hydrochloride	pH 1	257s(3.83),262(3.88), 282(3.97)	44-3765-83
	pH 11	260s(3.79),280(3.89)	44-3765-83
	MeOH	257s(3.83),263(3.84), 283(3.93)	44-3765-83
$C_9H_{11}N_3O_5$			
D-Alanine, N-[(1,2,3,4-tetrahydro-6-methyl-2,4-dioxo-5-pyrimidinyl)-carbonyl]-	H_2O	269(4.00)	94-0135-83
L-	H_2O	269(4.03)	94-0135-83
Glycine, N-[(1,2,3,4-tetrahydro-6-methyl-2,4-dioxo-5-pyrimidinyl)-carbonyl]-, methyl ester	H_2O	270(3.97)	94-0135-83
D-erythro-Hex-1-enitol, 1,5-anhydro-2,3-dideoxy-3-(2-nitro-1H-imidazol-1-yl)-	EtOH	226(3.49),316(3.78)	87-0020-83
1H-Imidazole, 1-(2,3-dideoxy-α-D-erythro-hex-2-enopyranosyl)-2-nitro-	EtOH	216(3.64),315(3.80)	87-0020-83
β-	EtOH	214(3.54),314(3.66)	87-0020-83
$C_9H_{11}N_3O_6$			
2(1H)-Pyrimidinone, 5,6-dihydro-1-methyl-3,5-dinitro-6-(2-oxopropyl)-, compd. with guanidine(1:1)	MeOH	234(3.90),327(4.04), 476(4.48)	138-0715-83
D-Serine, N-[(1,2,3,4-tetrahydro-6-methyl-2,4-dioxo-5-pyrimidinyl)-carbonyl]-	H_2O	271(3.97)	94-0135-83
L-	H_2O	271(3.98)	94-0135-83
$C_9H_{11}N_5O_3$			
Biopterin	0.08M HCl	247(4.04)	162-0174-83
$C_9H_{11}N_5O_4$			
Eritadenine	H_2O	261(4.16)	162-0530-83
Uridine, 3'-azido-2',3'-dideoxy-	pH 1	262(3.98)	87-0891-83
	pH 2	260(4.03)	87-0544-83
	pH 12	260(3.93)	87-0544-83
	pH 13	262(3.83)	87-0544-83
$C_9H_{11}O$			
Ethylium, 1-hydroxy-1-(4-methylphenyl)-	H_2SO_4	312(4.39)	99-0612-83
$C_9H_{11}OS$			
Ethylium, 1-hydroxy-1-[4-(methylthio)phenyl]-	H_2SO_4	258(3.57),418(4.29)	99-0612-83
$C_9H_{11}O_2$			
Ethylium, 1-hydroxy-1-(4-methoxyphenyl)-	H_2SO_4	233(3.38),350(4.32)	99-0612-83

Compound	Solvent	λ_{max}(log ϵ)	Ref.
$C_9H_{11}O_3S$			
Ethylium, 1-hydroxy-1-[4-(methylsulfonyl)phenyl]-	H_2SO_4	276(4.25)	99-0612-83
C_9H_{12}			
1,3,5-Cycloheptatriene, 7,7-dimethyl-	C_6H_{12}	203(4.03),267(3.40)	24-1848-83
$C_9H_{12}ClN$			
Benzeneethanamine, 4-chloro-α-methyl-, (S)-	C_6H_{12}	262(2.49),269(2.63), 277(2.62)	35-1578-83
	MeOH	261(2.45),268(2.57), 276(2.51)	35-1578-83
hydrochloride	MeOH	261(2.34),268(2.45), 276(2.36)	35-1578-83
$C_9H_{12}ClNO_2S$			
Thiophene, 2-(1-chloro-2,2-dimethylpropyl)-5-nitro-	MeOH	316(3.95)	44-4202-83
$C_9H_{12}ClN_3O_4$			
3(2H)-Pyridazinone, 5-amino-6-chloro-2-(2-deoxy-β-D-erythro-pentofuranosyl)-	pH 1	291s(4.14),299(4.16), 313s(3.95)	4-0369-83
	pH 11	291s(4.14),299(4.16), 313s(3.95)	4-0369-83
	MeOH	292s(4.13),301(4.16), 315s(3.94)	4-0369-83
$C_9H_{12}ClN_3O_5$			
3(2H)-Pyridazinone, 5-amino-6-chloro-2-β-D-ribofuranosyl-	pH 1	292s(4.08),300(4.10), 312s(3.90)	4-0369-83
	pH 11	291s(4.09),300(4.12), 312s(3.94)	4-0369-83
	MeOH	292s(4.06),301(4.09), 315s(3.88)	4-0369-83
$C_9H_{12}DNO_3S$			
Acetamide, N-acetyl-N-(2,2-dimethyl-4-oxo-3-thietanyl-3-d)-, (±)-	EtOH	215(3.52)	39-2259-83C
$C_9H_{12}FN_3O_4$			
Uridine, 3'-amino-2',3'-dideoxy-5-fluoro-	pH 1	268(3.91)	87-0891-83
	pH 2	270(4.10)	87-1691-83
	pH 12	270(4.05)	87-1691-83
	pH 13	269(3.78)	87-0891-83
	MeOH	271(4.12)	87-1691-83
$C_9H_{12}NO_5PS$			
Phosphorothioic acid, O,O-dimethyl O-(3-methyl-4-nitrophenyl) ester	n.s.g.	269.5(3.83)	162-0573-83
$C_9H_{12}NO_6P$			
L-Tyrosine, dihydrogen phosphate	0.05M HCl	265(2.85)	118-0030-83
$C_9H_{12}N_2$			
2(1H)-Isoquinolinamine, 3,4-dihydro-, monohydrochloride	MeOH	264(2.50),272(2.50)	4-0121-83
2-Pyridinemethanamine, N,N-dimethyl-α-methylene-	heptane	218(3.98),242(3.78), 275(3.53)	103-0518-83
3-Pyridinemethanamine, N,N-dimethyl-α-methylene-	heptane	214(4.37),268(3.52)	103-0518-83
4-Pyridinemethanamine, N,N-dimethyl-	heptane	214(4.27),275(3.51)	103-0518-83

Compound	Solvent	λ_{max}(log ε)	Ref.
α-methylene- (cont.)			103-0518-83
$C_9H_{12}N_2O$			
4(1H)-Cinnolinone, 5,6,7,8-tetrahydro-3-methyl-	EtOH	278(4.09)	142-0061-83
1,2-Diazabicyclo[5.2.0]nona-2,4-dien-9-one, 5,8-dimethyl-, cis	MeOH	250(3.64),299(4.00)	5-2141-83
4H-Pyrido[1,2-a]pyrimidin-4-one,	EtOH	227(3.84),278(3.70)	39-0369-83C
6,7,8,9-tetrahydro-6-methyl-	EtOH	227(3.83),279(3.70)	39-1413-83B
(R)-	H_2O	227(3.83),276(3.73)	39-1413-83B
	M HCl	231(3.91),259(3.59)	39-1413-83B
	dioxan	228(3.79),284(3.67)	39-1413-83B
1H-Pyrrolo[2,3-b]pyridine, 2,3-dihy-	heptane	308(3.87)	70-2437-83
dro-6-methoxy-4-methyl-	PrOH	310(3.89)	70-2437-83
	MeCN	310(--)	70-2437-83
$C_9H_{12}N_2O_2$			
2,4-Hexadienamide, N-(4,5-dihydro-2-oxazolyl)-, monohydriodide, (E,E)-	EtOH	270(4.51)	111-0441-83
$C_9H_{12}N_2O_2S_2$			
5-Pyrimidinecarboxylic acid, 2-methyl-4,6-bis(methylthio)-, methyl ester	EtOH	270(3.62),296(4.87)	142-1745-83
$C_9H_{12}N_2O_3S$			
Acetic acid, [(1,4-dihydro-5-methyl-4-oxo-2-pyrimidinyl)thio]-, ethyl ester	MeOH	287(3.69)	56-1027-83
6H-Isothiazolo[4,5-d]azepine-6-carboxylic acid, 2,3,4,5,7,8-hexahydro-3-oxo-, methyl ester	MeOH	266(3.71)	87-0895-83
$C_9H_{12}N_2O_4$			
Morpholine, 4-[(5-nitro-2-furanyl)-methyl]-	MeOH	215(3.95),230(3.77), 313(4.19)	73-2682-83
$C_9H_{12}N_2O_4S$			
6aH-Thieno[2,3-c]pyrazole-5,6a-dicarboxylic acid, 3,3a,4,5-tetrahydro-, dimethyl ester	MeOH	263(2.56),326(2.11)	70-1894-83
$C_9H_{12}N_2O_6$			
3(2H)-Pyridazinone, 5-hydroxy-2-α-D-	pH 1	248(3.63),273s(3.52)	44-3765-83
arabinofuranosyl-	pH 11	267(3.80),291s(3.58)	44-3765-83
	MeOH	254(3.62),272s(3.58)	44-3765-83
2,4(1H,3H)-Pyrimidinedione, 1-β-D-	pH 7	262.0(4.02)	48-0387-83
arabinofuranosyl-	pH 12	262.5(--)	48-0387-83
$C_9H_{12}N_3O_2$			
1H-Pyrazolo[1,2-a][1,2,4]triazol-4-ium, 2,3-dihydro-2,5,6,7-tetramethyl-1,5-dioxo-, tetrafluoroborate	MeCN	233.5(4.234),235(4.233), 239s(4.219),250s(3.97), 307(3.459)	142-0023-83
$C_9H_{12}N_4OS$			
7H-Pyrrolo[2,3-d]pyrimidin-4-amine, 7-(methoxymethyl)-2-(methylthio)-	MeOH	234(4.42),280(4.18)	5-0876-83
$C_9H_{12}N_4O_2$			
Ethanol, 2-[(8-amino[1,2,4]triazolo[1,5-a]pyridin-2-yl)methoxy]-	EtOH	212(4.37),276(3.05)	4-1169-83

Compound	Solvent	λ$_{max}$(log ε)	Ref.
$C_9H_{12}N_4O_2S$			
Thiophene, 2-(1-azido-2,2-dimethyl-propyl)-4-nitro-	EtOH	228(4.08),272(3.88)	44-4202-83
Thiophene, 2-(1-azido-2,2-dimethyl-propyl)-5-nitro-	EtOH	324(3.92)	44-4202-83
$C_9H_{12}N_4O_3$			
Imidazo[4,5-d][1,3]diazepin-8(3H)-one, 6,7-dihydro-3-[(2-hydroxyethoxy)-methyl]-	MeOH	227(4.34),251(3.78), 299(3.47),349(3.57)	87-1478-83
4H-Pyrrolo[2,3-d]pyrimidin-4-one, 2-amino-1,7-dihydro-7-[(2-hydroxy-ethoxy)methyl]-	pH 1	258(4.01)	5-0137-83
	pH 7	258(4.07),280s(--)	5-0137-83
	pH 13	259(4.06)	5-0137-83
	MeOH	259(4.10),280s(3.89)	5-0137-83
$C_9H_{12}N_4O_7$			
2(1H)-Pyrimidinone, 4-amino-1-(2-O-nitro-β-D-arabinofuranosyl)-	pH 1	275(4.11)	87-0280-83
	H_2O	231(3.92),268(3.93)	87-0280-83
	pH 13	230(3.92),268(3.97)	87-0280-83
	EtOH	239(3.90),269(3.92)	87-0280-83
hydrochloride	pH 1	276(4.12)	87-0280-83
	H_2O	236(3.86),270(3.98)	87-0280-83
	EtOH	273(3.94)	87-0280-83
$C_9H_{12}N_6$			
2-Pyrimidinamine, 4,6-dimethyl-N-(5-methyl-1H-1,2,4-triazol-3-yl)-	MeOH	248(4.30)	118-0222-83
$C_9H_{12}N_6O_3$			
Cytidine, 3'-azido-2',3'-dideoxy-	pH 1	279(4.13)	87-0891-83
	pH 2	278(4.13)	87-0544-83
	pH 12	268(4.04)	87-0544-83
	pH 13	271(3.93)	87-0891-83
$C_9H_{12}N_6O_4$			
3(2H)-Pyridazinone, 5-amino-2-(2-azido-2-deoxy-β-D-ribofuranosyl)-	pH 1	274(3.88),300s(3.64)	44-3765-83
	pH 11	274(3.87),301s(3.64)	44-3765-83
	MeOH	278(4.03),302s(3.82)	44-3765-83
$C_9H_{12}N_6O_5$			
5H-1,2,3-Triazolo[4,5-d]pyrimidin-5-one, 7-amino-2,3-dihydro-2-β-D-ribofuranosyl-	pH 1	291(3.99)	87-1483-83
	pH 7	262(4.02),287(4.04)	87-1483-83
	pH 13	259(3.71),317(3.91)	87-1483-83
$C_9H_{12}O$			
Benzenemethanol, α-ethyl-	MeOH	250(2.24),260(2.06)	162-0547-83
Cyclohexanone, 2-cyclopropylidene-	alkane	238(3.89)	78-3307-83
2-Cyclohexen-1-one, 2-ethenyl-3-methyl-	hexane	222(3.79),252.5(3.77), 325(1.71)	23-0936-83
3-Cyclohexen-1-one, 2-ethylidene-3-methyl-, (E)-	hexane	224(3.84),270(3.63), 316s(2.34),330s(2.15), 342s(2.06)	23-0936-83
(Z)-	hexane	223(3.96),269(3.65), 324s(1.95)	23-0936-83
Cyclopentanone, 2-(cyclopropylmethyl-ene)-, (E)-	MeOH	267(4.17)	23-0288-83
Furan, 3-methyl-2-(2-methyl-2-prop-enyl)-	MeCN	220(3.67),279(2.18)	24-3366-83
2-Propynal, 3-(1-methylcyclopentyl)-	C_6H_{12}	225(3.94),233(3.91)	33-2760-83
Spiro[2.5]octan-4-one, 8-methylene-	hexane	217(3.24),284(1.77)	23-0078-83

Compound	Solvent	λ_{max}(log ε)	Ref.
Spiro[2.5]oct-7-en-4-one, 8-methyl-	hexane	222(3.06),232s(2.97), 275(1.89)	23-0078-83
	EtOH	218(3.05),235(2.76), 285s(1.90)	23-0078-83
$C_9H_{12}OS_2$			
Carbonodithioic acid, S-methyl S-(6-methylene-1-cyclohexen-1-yl) ester	EtOH	225(3.91),247(3.98)	97-0096-83
$C_9H_{12}O_2$			
2-Cyclohexen-1-one, 2-(2-propenyloxy)-	EtOH	260(3.80)	44-4241-83
2,4-Cyclopentadiene-1-carboxylic acid, 1,2-dimethyl-, methyl ester	hexane	258(3.47)	24-0299-83
2,4-Cyclopentadiene-1-carboxylic acid, 1,3-dimethyl-, methyl ester	hexane	253(3.15)	24-0299-83
2-Cyclopentene-1-carboxylic acid, 1-methyl-4-methylene-, methyl ester	hexane	233(4.15)	24-0299-83
3-Cyclopentene-1-carboxylic acid, 1-methyl-2-methylene-, methyl ester	hexane	228(4.52)	24-0299-83
2-Cyclopenten-1-one, 3-acetyl-4,4-dimethyl-	MeOH	243(4.16)	39-1465-83C
2-Nonen-4-ynoic acid, (E)-	$CHCl_3$	251(3.74)	13-0339-83A
$C_9H_{12}O_2Se$			
7-Selenabicyclo[2.2.1]heptane-2,3-dione, 1,5,5-trimethyl-	C_6H_{12}	304(3.39),435(2.62)	39-0333-83C
	EtOH	311(3.47),423(2.81)	39-0333-83C
$C_9H_{12}O_3$			
2,5-Cyclohexadien-1-one, 4-ethoxy-4-methoxy-	EtOH	217.5(4.15)	39-1255-83C
2H-Pyran-2-one, 5-ethyl-4-methoxy-6-methyl- (macomelin)	EtOH	223s(--),286(3.82)	94-3781-83
$C_9H_{12}O_4$			
2-Propenoic acid, 3-(2-methyl-4H-1,3-dioxin-6-yl)-, methyl ester, [S-(E)]-	MeOH	270(4.19)	136-0286-83G
2H-Pyran-2-one, 5-(1-hydroxyethyl)-4-methoxy-6-methyl-	EtOH	284(3.82)	94-3781-83
2H-Pyran-2-one, 5-(2-hydroxyethyl)-4-methoxy-6-methyl-	EtOH	285(3.82)	94-3781-83
$C_9H_{12}O_5$			
6H-Furo[2,3-b]pyran-6-one, 2,3,3a,7a-tetrahydro-2-hydroxy-4-methoxy-3-methyl- (astepyrone)	EtOH	235(4.06)	94-0925-83
2H-Pyran-2-one, 5-(1,2-dihydroxyethyl)-4-methoxy-6-methyl-, (S)-	EtOH	284(3.81)	94-3781-83
$C_9H_{13}ClSSi$			
Silane, [(4-chlorophenyl)thio]trimethyl-	heptane	230(4.1),247(4.1)	65-2021-83
$C_9H_{13}Cl_2NOPt$			
Platinum, dichloro(η^2-ethene)(4-methoxybenzenamine-N)-, trans	n.s.g.	241(3.68),273(3.95), 325(3.28)	131-0309-83H
$C_9H_{13}Cl_2NPt$			
Platinum, dichloro(η^2-ethene)(3-methylbenzenamine)-, trans	n.s.g.	248(3.82),257(3.82), 303(3.29)	131-0309-83H
Platinum, dichloro(η^2-ethene)(4-methylbenzenamine)-, trans	n.s.g.	247(3.73),263(3.84), 304(3.28)	131-0309-83H

Compound	Solvent	λ_{max}(log ε)	Ref.
$C_9H_{13}IN_4O_3$			
L-Histidine, N-β-alanyl-5-iodo- (monoiodocarnosine)	H_2O	188(4.18)	94-3302-83
$C_9H_{13}N$			
Benzenamine, N-(1-methylethyl)-	n.s.g.	288(3.06)	32-0069-83
Benzenamine, N-propyl-	n.s.g.	285(3.47)	32-0069-83
Benzenamine, N,N,4-trimethyl-	76% H_2SO_4	251(2.33),257(2.36)	39-1191-83B
Benzeneethanamine, α-methyl-, (S)-	C_6H_{12}	253(2.23),259(2.32), 265(2.20),268(2.18)	35-1578-83
	MeOH	259(2.30),261(2.26), 264(2.20),268(2.15)	35-1578-83
hydrochloride	MeOH	258(2.23),261s(2.08), 264(2.15),267(1.90)	35-1578-83
Benzenemethanamine, α,4-dimethyl-, (S)-	C_6H_{12}	265(2.54),273(2.51)	35-1578-83
	MeOH	264(2.53),267(2.49), 273(2.49)	35-1578-83
hydrochloride	MeOH	262(2.40),267(2.26), 271(2.23)	35-1578-83
	10% HCl	262(2.42),266s(2.28), 271(2.23)	35-1578-83
$C_9H_{13}NO$			
Benzenemethanol, α-methyl-4-(N-methyl-amino)- (dication)	H_2SO_4	276(4.10)	99-0612-83
1H-Pyrrole-2-carboxaldehyde, 4-ethyl-3,5-dimethyl-	MeOH	312(3.95)	78-1893-83
$C_9H_{13}NO_2$			
2-Cyclohexen-1-one, 3-hydroxy-2-(imi-nomethyl)-5,5-dimethyl-	EtOH	248(4.16),293(4.15)	4-0649-83
$C_9H_{13}NO_3S$			
Acetamide, N-acetyl-N-(2,2-dimethyl-4-oxo-3-thietanyl)-, (R)-	EtOH	214(3.43)	39-2259-83C
2-Thiophenemethanol, α-(1,1-dimethyl-ethyl)-5-nitro-	MeOH	212(3.72),321(4.03)	44-4202-83
$C_9H_{13}NO_5S$			
1-Azetidineacetic acid, α-(1-methyl-ethylidene)-2-oxo-4-sulfino-, α-methyl ester, (R)-	EtOH	223(2.60)	39-2241-83C
4-Thia-1-azabicyclo[3.2.0]heptane-2-carboxylic acid, 3,3-dimethyl-7-oxo-, methyl ester, 4,4-dioxide	EtOH	213(2.42),225s(2.00)	39-2241-83C
$C_9H_{13}N_2O_2$			
Benzenaminium, N,N,4-trimethyl-2-nitro-	76% H_2SO_4	273(3.81)	39-1191-83B
Benzenaminium, N,N,4-trimethyl-3-nitro-	76% H_2SO_4	256(3.76)	39-1191-83B
$C_9H_{13}N_3O$			
Benzoic acid, 4-(dimethylamino)-, hydrazide	pH 1	225(4.03),325(2.70)	140-0091-83
	H_2O	221(3.85),317(4.08)	140-0091-83
$C_9H_{13}N_3O_3$			
Morpholine, 4-(1-methyl-4-nitro-1H-pyrrol-2-yl)-	MeOH	280(3.8),348(3.6)	44-0162-83
Morpholine, 4-(1-methyl-4-nitro-1H-pyrrol-3-yl)-	MeOH	308(3.9)	44-0162-83

Compound	Solvent	λ_{max}(log ε)	Ref.
1H-Pyrazole-4-carboxylic acid, 5-acetylamino-1-methyl-, ethyl ester	EtOH	227.5(4.02)	11-0171-83B
$C_9H_{13}N_3O_3S$			
5-Pyrimidinepentanoic acid, 6-amino-1,2,3,4-tetrahydro-4-oxo-2-thioxo-	50% MeOH	242(4.24),265(4.18), 298(4.18)	73-0292-83
	+ HCl	244(3.93),285(4.33)	73-0292-83
	+ NaOH	242(4.24),265(4.07), 298(4.19)	73-0292-83
$C_9H_{13}N_3O_4$			
3(2H)-Pyridazinone, 5-amino-2-(2-deoxy-β-D-erythro-pentofuranosyl)-	pH 1	274(3.85),297s(3.61)	44-3765-83
	pH 11	274(3.82),295s(3.61)	44-3765-83
	MeOH	275(3.79),302s(3.52)	44-3765-83
5-Pyrimidinecarboxamide, 1,2,3,4-tetrahydro-N-(2-hydroxy-1-methylethyl)-6-methyl-2,4-dioxo-, (R)-	H_2O	267.5(4.01)	94-0135-83
(S)-	H_2O	267.5(4.03)	94-0135-83
5-Pyrimidinepentanoic acid, 6-amino-1,2,3,4-tetrahydro-2,4-dioxo-	50% MeOH	276(4.27)	73-0292-83
	+ HCl	226.5(3.80),274(4.32)	73-0292-83
	+ NaOH	277.5(4.25)	73-0292-83
Uridine, 3'-amino-2',3'-dideoxy-	pH 1	262(3.93)	87-0891-83
	pH 2	260(3.96)	87-0544-83
	pH 12	260(3.87)	87-0544-83
	pH 13	263(3.86)	87-0891-83
$C_9H_{13}N_3O_5$			
2(1H)-Pyrazinone, 3-amino-1-β-D-ribofuranosyl-	pH 3	235(3.86),318(3.92), 328s(3.91),345s(3.72)	87-0283-83
	pH 7	211(3.89),243(3.89), 320(3.93)	87-0283-83
3(2H)-Pyridazinone, 5-amino-2-β-D-arabinofuranosyl-	pH 1	274(3.92),302s(3.58)	44-3765-83
	pH 11	275(3.86),300s(3.52)	44-3765-83
	MeOH	277(3.92),303s(3.60)	44-3765-83
4(1H)-Pyridazinone, 6-amino-1-β-D-arabinofuranosyl-	pH 1	253(3.64),278s(3.42)	44-3765-83
	pH 11	270(3.82)	44-3765-83
	MeOH	278(3.83)	44-3765-83
5-Pyrimidinecarboxamide, 1,2,3,4-tetrahydro-N-[2-hydroxy-1-(hydroxymethyl)ethyl]-6-methyl-2,4-dioxo-	H_2O	269(4.07)	94-0135-83
2(1H)-Pyrimidinone, 4-amino-1-β-D-arabinofuranosyl-	pH 2	212.5(4.01),281(4.12)	162-0401-83
	pH 7	271.0(3.97)	48-0387-83
	pH 12	272.5(3.97)	162-0401-83
$C_9H_{13}N_3O_6$			
Bredinin	M HCl	243(3.81),279(4.09)	4-0875-83
	H_2O	244(3.81),278(4.16)	4-0875-83
	M NaOH	240s(--),276(4.21)	4-0875-83
2,3-Pyrazinedione, 1-amino-1,4-dihydro-4-β-D-ribofuranosyl-	pH 7	219(3.95),240s(3.71), 310(3.83),323s(3.79), 338s(3.52)	87-0283-83
$C_9H_{13}N_5$			
Imidazo[1,5-a]-1,3,5-triazin-4-amine, 8-butyl- (spectra in ethanol)	pH 1	274(4.10),310(3.91)	44-0003-83
	pH 7	265(3.65),322(4.19)	44-0003-83
	pH 13	285(3.90),310(3.85)	44-0003-83
$C_9H_{13}N_5O_4$			
Guanine, 9-[(1,3-dihydroxy-2-propoxy)-methyl]-	MeOH	254(4.11),270s(3.96)	87-0759-83

Compound	Solvent	$\lambda_{max}(\log \epsilon)$	Ref.
$C_9H_{13}N_7O_4$			
2H-1,2,3-Triazolo[4,5-d]pyrimidine-5,7-diamine, 2-β-D-ribofuranosyl-	pH 1	264(4.12),286(4.11)	87-1483-83
	pH 7	261(3.76),314(3.91)	87-1483-83
	pH 13	261(3.76),313(3.93)	87-1483-83
3H-1,2,3-Triazolo[4,5-d]pyrimidine-5,7-diamine, 3-β-D-ribofuranosyl-	pH 1	257(4.06),283(3.91)	87-1483-83
	pH 7	225.5(4.35),256s(--), 285(4.03)	87-1483-83
	pH 13	225.5(4.34),257s(--), 286(4.03)	87-1483-83
C_9H_{14}			
1,3,4-Hexatriene, 3-ethyl-5-methyl-	EtOH	221(4.15)	70-0569-83
$C_9H_{14}Br_2N_2O_3$			
Pyrazine, 5,6-bis(bromomethyl)-2,3-dihydro-3-methoxy-2,2-dimethyl-, 1,4-dioxide	EtOH	270(4.12),368(4.07)	103-1128-83
$C_9H_{14}ClN_2O_3P$			
Phosphonic acid, ethenyl-, 2-chloroethyl 2-(1H-imidazol-1-yl)ethyl ester, monohydrochloride	EtOH	274(0.76),280(0.79)	65-1555-83
$C_9H_{14}ClN_5O_5$			
Adenine, N^6-ethyl-N^6-methoxy-3-methyl-, perchlorate	pH 1	227(3.98),292(4.33)	94-3521-83
	pH 7	226(4.07),293(4.28)	94-3521-83
	pH 13	226(4.10),295(4.23)	94-3521-83
	EtOH	227(4.06),295(4.30)	94-3521-83
$C_9H_{14}CuN_4O_4$			
Copper, [N,N'-bis(1-methyl-2-nitroethylidene)-1,3-propanediaminato(2-)-N^1,N^3,O^1,O^3]-	DMSO	339(4.24),620(2.48)	40-0088-83
$C_9H_{14}N_2O_2$			
3H-Pyrazole-4-carboxylic acid, 3,3,5-trimethyl-, ethyl ester	MeCN	260(3.70)	78-1247-83
$C_9H_{14}N_2O_3$			
4,6(3H,5H)-Pyrimidinedione, 5,5-diethyl-2-(1-methylethoxy)-	anion	272(4.01)	111-0521-83
Pyrimidinedione, 5,5-diethyl-4-(1-methylethoxy)-	anion	244(3.85)	111-0521-83
$C_9H_{14}N_2O_5S$			
4(1H)-Pyrimidinone, 2-(methylthio)-6-(tetrahydroxybutyl)-, [1R-(1R*,2S*,3R*)]-	pH 1	240(3.96),282(3.91)	136-0333-83M
	pH 7	236(4.03),282(3.96)	136-0333-83M
	pH 11	245(3.96),276(3.81)	136-0333-83M
4-Thia-1-azabicyclo[3.2.0]heptane-2-carboxylic acid, 6-amino-3,3-dimethyl-7-oxo-, methyl ester, 4,4-dioxide	EtOH	218(3.36),223s(3.32), 274(2.78)	39-2241-83C
$C_9H_{14}N_2O_6S$			
4(1H)-Pyrimidinone, 1-[3-hydroxy-2-[(methylsulfonyl)oxy]propyl]-2-methoxy-	EtOH	233(3.90),255(3.81)	128-0125-83
$C_9H_{14}N_2O_8S_2$			
2,4(1H,3H)-Pyrimidinedione, 1-[2,3-	EtOH	262(3.91)	128-0125-83

Compound	Solvent	λ_{max}(log ε)	Ref.
bis[(methylsulfonyl)oxy]propyl]-			128-0125-83
$C_9H_{14}N_4O_2S_2$			
Thiourea, N,N-diethyl-N'-(4-methyl-5-nitro-2-thiazolyl)-	$CHCl_3$	299(3.75),363(4.07)	74-0075-83B
palladium chelate	$CHCl_3$	261(4.48),303s(4.20), 338(4.19),415(4.63)	74-0075-83B
$C_9H_{14}N_4O_3$			
Cytidine, 3'-amino-2',3'-dideoxy-	pH 2	277(4.07)	87-0544-83
	pH 12	268(3.89)	87-0544-83
Imidazo[4,5-d][1,3]diazepin-8-ol, 3,6,7,8-tetrahydro-3-[(2-hydroxyethoxy)methyl]-, (±)-	MeOH	283(3.91)	87-1478-83
5-Pyrimidinepentanoic acid, 2,6-diamino-1,4-dihydro-4-oxo-	50% MeOH	241(3.73),278.5(4.18)	73-0292-83
	+ HCl	274(4.27)	73-0292-83
	+ NaOH	243(3.78),272(4.06)	73-0292-83
$C_9H_{14}N_4O_4$			
1,2,4-Triazine-3,6-dicarboxylic acid, 4-(dimethylamino)-4,5-dihydro-, dimethyl ester	CH_2Cl_2	346(3.62)	83-0472-83
$C_9H_{14}N_4O_5$			
1H-Imidazole-4-carboxamide, 5-(D-ribosylamino)-	H_2O	231(3.68),270(4.10)	136-0001-83B
2(1H)-Pyrazinone, 4-amino-3,4-dihydro-3-imino-1-β-D-ribofuranosyl-, monohydrochloride	pH 3	220(3.96),240s(3.79), 313(3.91),325s(3.71)	87-0283-83
	pH 11	247(3.95),322(3.83), 335s(3.81),350s(3.60)	87-0283-83
$C_9H_{14}N_5O$			
1H-Purinium, 6,7-dihydro-6-(methoxyimino)-1,7,9-trimethyl-, iodide	pH 1	226(4.31),289(3.94)	94-3149-83
	pH 7	226(4.31),289(3.94)	94-3149-83
	EtOH	294(3.90)	94-3149-83
perchlorate	pH 1	229(3.85),289(3.95)	94-3149-83
	pH 7	229(3.85),289(3.94)	94-3149-83
	EtOH	233(3.87),294(3.90)	94-3149-83
1H-Purinium, 1-ethyl-6-(methoxyamino)-3-methyl-, iodide	pH 1	223(4.32),275(4.16)	94-3521-83
	EtOH	277(4.15),340(3.15)	94-3521-83
$C_9H_{14}O$			
Acetaldehyde, (2-methylcyclohexylidene)-, (E,2S)-(-)-	C_6H_{12}	232(4.25),322(1.57), 333(1.67),347(1.71), 373(1.60),380(1.23)	35-3252-83
	C_6H_{12}	232(4.25)	35-3264-83
(Z,2S)-(+)-	C_6H_{12}	231(4.12),322(1.42), 334(1.49),348(1.52), 363(1.40),381(1.04)	35-3252-83
	C_6H_{12}	231(4.12)	35-3264-83
Acetaldehyde, (3-methylcyclohexylidene)-, (E,3R)-(-)-	C_6H_{12}	231(4.22),322(1.62), 333(1.71),347(1.73), 363(1.61),381(1.23)	35-3252-83
	C_6H_{12}	231(4.22)	35-3264-83
(Z,3R)-(-)-	C_6H_{12}	232(4.25),322(1.61), 347(1.76),363(1.62), 381(1.23)	35-3252-83
	C_6H_{12}	232(4.25)	35-3264-83
Acetaldehyde, (4-methylcyclohexylidene)-, (aR)-(-)-	C_6H_{12}	230(4.29),321(1.61), 333(1.71),348(1.73), 362(1.62),381(1.26)	35-3252-83

Compound	Solvent	$\lambda_{max}(\log \epsilon)$	Ref.
Furan, 3-methyl-2-(2-methylpropyl)-	MeCN	220(3.78)	24-3366-83
Norcamphor	MeCN	291(1.40)	44-1862-83
$C_9H_{14}OS$			
Bicyclo[2.2.1]heptane-2-thione, 3,3-dimethyl-, S-oxide	$CHCl_3$	273(3.76),315s(2.78)	44-0214-83
$C_9H_{14}O_4$			
Pentanoic acid, 4,4-dimethyl-2,3-dioxo-, ethyl ester	MeCN	290s(1.93),400(1.72)	5-1694-83
$C_9H_{14}O_5$			
D-threo-Hex-2-enonic acid, 2,3-dideoxy-4,6-O-ethylidene-, methyl ester, (R)-	MeOH	220(3.81)	136-0286-83G
(S)-	MeOH	208(4.43)	136-0286-83G
$C_9H_{14}S$			
Bicyclo[2.2.1]heptane-2-thione, 3,3-dimethyl- (thiocamphenilone)	benzene	494(1.15)	44-0214-83
$C_9H_{15}BrNO_2S_3$			
Ethanaminium, N-(3a-bromotetrahydrothieno[3,4-d]-1,3-dithiol-2-ylidene)-N-ethyl-, bromide, S,S-dioxide	H_2O	254(4.14)	103-1289-83
perchlorate	H_2O	255(4.14)	103-1289-83
$C_9H_{15}Cl_2N_2O_3P$			
Phosphonic acid, [2-(1H-imidazol-1-yl)-ethyl]-, bis(2-chloroethyl) ester	EtOH	206(3.79),273(1.49), 280(1.51)	65-1555-83
$C_9H_{15}NO$			
1H-Pyrrole-2-methanol, 1-methyl-α-(1-methylethyl)-	MeOH	222(3.98)	44-2488-83
Quinoline, 2,3,4,4a,5,6,7,8-octahydro-, 1-oxide	MeCN	252(3.87)	5-0897-83
$C_9H_{15}N_3O$			
1H-1,2-Diazepine-7-acetamide, 6,7-dihydro-α,5-dimethyl-, (R*,R*)-	MeOH	240(3.58),301(3.65)	5-2141-83
$C_9H_{15}N_3O_2S$			
Ergothioneine (dihydrate)	H_2O	258(4.20)	162-0529-83
$C_9H_{15}N_3O_3$			
1H-Imidazole-4-carboxylic acid, 2,5-dihydro-2,2,5,5-tetramethyl-1-nitroso-, methyl ester	EtOH	204(3.55),235(3.87)	70-2501-83
$C_9H_{15}N_3O_4$			
1H-Imidazole-4-carboxylic acid, 2,5-dihydro-2,2,5,5-tetramethyl-1-nitroso-, methyl ester, 3-oxide	EtOH	233(3.89),274(4.05)	70-2501-83
$C_9H_{15}N_5O_2$			
5-Pyrimidinepentanoic acid, 2,4,6-triamino-	50% MeOH	276(4.06)	73-0292-83
	+ HCl	218(4.29),283(4.23)	73-0292-83
	+ NaOH	276(4.05)	73-0292-83
$C_9H_{15}S_3$			
Cyclopropenylium, tris(ethylthio)-, tetrafluoroborate	MeOH	272(3.49)	18-0171-83

Compound	Solvent	$\lambda_{max}(\log \epsilon)$	Ref.
$C_9H_{16}Br_2N_2O_2$			
1H-Imidazole, 4-(dibromomethyl)-2,5-dihydro-1-methoxy-2,2,5,5-tetramethyl-, 3-oxide	EtOH	202(3.78),268(3.94)	70-2123-83
$C_9H_{16}Cl_2N_2O_2$			
1H-Imidazole, 4-(dichloromethyl)-2,5-dihydro-1-methoxy-2,2,5,5-tetramethyl-, 3-oxide	EtOH	203(3.60),256(4.03)	70-2123-83
$C_9H_{16}N$			
Pyrrolidinium, 1-(3-methyl-2-butenylidene)-, perchlorate	MeCN	268(4.08)	35-4033-83
$C_9H_{16}NO$			
Cyclopent[c][1,2]oxazinium, 1-ethyl-3,4,4a,5,6,7-hexahydro-, tetrafluoroborate	MeCN	220(3.71)	5-0897-83
$C_9H_{16}N_2$			
3H-Cyclooctapyrazole, 3a,4,4,5,5,6,6-7,7,8,8,9,9,9a-tetradecahydro-	hexane	328(2.67)	39-0983-83B
2-Propenenitrile, 3-[bis(1-methylethyl)amino]-, cis	heptane	257(4.19)	70-2322-83
	MeOH	269(4.4)	70-2322-83
trans	heptane	264(4.22)	70-2322-83
	MeOH	269(4.39)	70-2322-83
$C_9H_{16}N_2O$			
2-Furanamine, 5-ethyl-4-(ethylimino)-4,5-dihydro-5-methyl-	n.s.g.	273(4.30)	77-1455-83
$C_9H_{16}N_2O_3$			
Pyrazine, 2,3-dihydro-3-methoxy-2,2,5,6-tetramethyl-, 1,4-dioxide	EtOH	348(4.21)	103-1128-83
$C_9H_{16}N_2O_3S$			
3-Pyridinesulfonamide, 1,6-dihydro-6-methoxy-N,N,1-trimethyl-	CH_2Cl_2	306(3.6)	64-0873-83B
$C_9H_{16}N_2S$			
2-Pyrimidinethione, 4,6-diethyl-1,2,3,4-tetrahydro-4-methyl-	MeOH	261(4.23)	104-2148-83
$C_9H_{16}N_4O$			
5H-Tetrazol-5-one, 1-(3-butenyl)-4-(1,1-dimethylethyl)-1,4-dihydro-	hexane	228(3.67)	24-3427-83
5H-Tetrazol-5-one, 1-(1,1-dimethylethyl)-1,4-dihydro-4-(2-methyl-2-propenyl)-	hexane	225(3.66)	24-3427-83
$C_9H_{16}N_4O_4$			
1,3-Propanediamine, N,N'-bis(1-methyl-2-nitroethenyl)-	CH_2Cl_2	338(4.18),358(4.18)	40-0088-83
$C_9H_{16}OS_2$			
1,2-Dithiolane, 3-methoxy-4,4-dimethyl-5-(1-methylethylidene)-	C_6H_{12}	250(3.63),345(2.48)	44-4482-83 +78-2719-83
$C_9H_{16}OSi$			
2-Cyclohexen-1-one, 3-(trimethylsilyl)-	C_6H_{12}	230(4.16),340(1.76), 353(1.81),368(1.67)	157-0332-83

Compound	Solvent	λ_{max}(log ε)	Ref.
Silane, (1,1-dimethylethyl)dimethyl-(1-oxo-2-propynyl)-	C_6H_{12}	219(3.67),432(2.17)	78-0949-83
$C_9H_{16}O_5S$			
Sulfoxonium, dimethyl-, 1-(ethoxy-carbonyl)-2-ethoxy-2-oxoethylide	EtOH	227(4.11)	99-0023-83
	hexane	218(--)	99-0023-83
$C_9H_{17}BrN_2O_2$			
Piperidine, 4-bromo-2,2,6,6-tetra-methyl-4-nitro-	EtOH	285(2.60)	70-0752-83
hydrobromide	EtOH	294(2.15)	70-0752-83
$C_9H_{17}ClN_2O_2$			
Piperidine, 4-chloro-2,2,6,6-tetra-methyl-4-nitro-	EtOH	282(1.95)	70-0752-83
hydrochloride	EtOH	285(1.52)	70-0752-83
$C_9H_{17}IN_2O_2$			
Piperidine, 4-iodo-2,2,6,6 - tetra-methyl-4-nitro-	EtOH	308(3.00)	70-0752-83
$C_9H_{17}NO$			
3-Buten-2-one, 4-(diethylamino)-3-methyl-	MeOH	304(4.46)	131-0001-83K
(E-syn-E)-	C_6H_{12}	287(4.37)	131-0001-83K
3-Buten-2-one, 4-[(1,1-dimethylethyl)-3-methyl-	MeOH	299(4.37)	131-0001-83K
(E-syn-E)-	C_6H_{12}	279(--)	131-0001-83K
	dioxan	286(4.22)	131-0001-83K
(Z-syn-Z)-	C_6H_{12}	308(4.05)	131-0001-83K
	dioxan	286(--)	131-0001-83K
3,5-Heptadien-2-ol, 7-(dimethylamino)-	MeCN	219(2.34)	33-1427-83
1-Hepten-3-one, 2-methyl-1-(methyl-amino)-	MeOH	298(4.42)	131-0001-83K
(E-syn-E)-	C_6H_{12}	277(--)	131-0001-83K
	dioxan	285(4.31)	131-0001-83K
(Z-syn-Z)-	C_6H_{12}	312(4.09)	131-0001-83K
	dioxan	285(--)	131-0001-83K
1-Hexen-3-one, 1-(dimethylamino)-2-methyl-	MeOH	304(4.58)	131-0001-83K
	dioxan	288(4.25)	131-0001-83K
1-Hexen-3-one, 2-ethyl-1-(methyl-amino)-	MeOH	299(4.34)	131-0001-83K
(E-syn-E)-	C_6H_{12}	278(--)	131-0001-83K
	dioxan	286(--)	131-0001-83K
(Z-syn-Z)-	C_6H_{12}	313(4.03)	131-0001-83K
	dioxan	320(--)	131-0001-83K
$C_9H_{17}N_3O_2$			
Diazenecarboxylic acid, 1-piperidino-, 1-methylethyl ester	EtOH	270(4.26)	33-1416-83
2-Propenamide, 3-amino-N-[[(1,1-di-methylethyl)amino]carbonyl]-2-methyl-, (E)-	H_2O	293(4.23)	35-0956-83
(Z)-	H_2O	301(4.15)	35-0956-83
$C_9H_{17}N_3O_3$			
2-Butenamide, N-[2-(hydroxynitroso-amino)-3-methylbutyl)- (dopastin)	pH 2	215(4.32)	162-0498-83
1H-Imidazole-4-carboxaldehyde, 2,5-di-hydro-1-methoxy-2,2,5,5-tetramethyl-, oxime, 3-oxide	EtOH	203(3.85),242(3.85), 295(4.16)	70-2123-83

Compound	Solvent	$\lambda_{max}(\log \epsilon)$	Ref.
1H-Imidazole-4-carboxamide, 2,5-dihydro-1-methoxy-2,2,5,5-tetramethyl-, 3-oxide	EtOH	262(4.01)	70-2123-83
$C_9H_{17}N_3O_4$			
Piperidine, 2,2,6,6-tetramethyl-4,4-dinitro-	EtOH	282(2.18)	70-0752-83
hydrochloride	EtOH	282(1.90)	70-0752-83
$C_9H_{18}GeOSi$			
Silane, trimethyl[3-oxo-3-(trimethylgermyl)-1-propynyl]-	n.s.g.	222(4.04)	77-0239-83
	C_6H_{12}	222(4.04)	78-3073-83
$C_9H_{18}N_2O_2$			
1H-Imidazole, 2,5-dihydro-1-methoxy-2,2,4,5,5-pentamethyl-, 3-oxide	EtOH	232(3.98)	70-2123-83
1-Penten-1-amine, 1-(1,1-dimethylethyl)-N-nitro-	H_2O	231(3.53)	104-1270-83
Piperidine, 2,2,6,6-tetramethyl-4-nitro-	EtOH	280(1.95)	70-0752-83
$C_9H_{18}N_6$			
1,3,5-Triazine-2,4,6-triamine, N,N,N',N',N",N"-hexamethyl-	EtOH	226(4.69)	162-0048-83
$C_9H_{18}OSiSn$			
Silane, trimethyl[3-oxo-3-(trimethylstannyl)-1-propynyl]-	n.s.g.	225(3.93)	77-0239-83
$C_9H_{18}OSi_2$			
Silane, (3-oxo-1-propyne-1,3-diyl)-bis[trimethyl-	n.s.g.	224(3.93)	77-0239-83
	C_6H_{12}	222(4.04)	78-3073-83
$C_9H_{22}N_3O_2P$			
Phosphorus(1+), (2-methoxy-2-oxoethyl)-tris(N-methylmethanaminato)-, hydroxide, inner salt	hexane	232(4.12)	99-0023-83
	EtOH	234(4.20)	99-0023-83

Compound	Solvent	λ_{max}(log ϵ)	Ref.
$C_{10}Cl_6N_2$			
Cyclobuta[1,2-b:4,3-b']dipyridine, hexachloro-	C_6H_{12}	259(4.72),268(4.71), 379(4.35)	77-0425-83
$C_{10}Cl_6N_4$			
Dipyrido[3,2-c:2',3'-e]pyridazine, hexachloro-	C_6H_{12}	244.5(4.40),271.5(4.67), 291(4.42),309.5(4.09), 321.5(4.01),334(3.61), 350.5(3.72),367.5(3.73)	77-0425-83
$C_{10}Cl_{10}O$			
Cyclobuta[1,2:3,4]dicyclopenten-1(3aH)-one, 2,3,3a,3b,4,4,5,6,6a,6b-decachloro-3b,4,6a,6b-tetrahydro-, (3aα,3bα,6aα,6bα)-	hexane	234(3.98)	98-0621-83
Cyclobuta[1,2:3,4]dicyclopenten-1(3aH)-one, 2,3,3a,3b,4,5,6,6,6a,6b-decachloro-3b,6,6a,6b-tetrahydro-, (3aα,3bα,6aα,6bα)-	hexane	235(4.13)	98-0621-83
4,7-Methano-1H-inden-1-one, 2,3,3a,4-5,6,7,7a,8,8-decachloro-3a,4,7,7a-tetrahydro-, (3aα,4α,7α,7aα)-	hexane	243(3.90),268(3.43)	98-0621-83
$C_{10}F_6N_2$			
Cyclobuta[1,2-b:4,3-b']dipyridine, hexafluoro-	C_6H_{12}	231(4.37),229(4.38), 322(3.67),338(3.91)	77-0425-83
$C_{10}F_6N_4$			
Dipyrido[3,2-c:2',3'-e]pyridazine, hexafluoro-	C_6H_{12}	243(3.40),314(1.95), 326(1.95)	77-0425-83
$C_{10}H_2Br_2O_2S_2$			
Benzo[1,2-b:4,5-b']dithiophene-2,6-dione, 3,7-dibromo-	CH_2Cl_2	242(4.37),374(4.07), 402s(4.34),421(4.56), 442(4.59),488s(2.88)	138-0905-83
$C_{10}H_2Cl_2O_2S_2$			
Benzo[1,2-b:4,5-b']dithiophene-2,6-dione, 3,7-dichloro-	CH_2Cl_2	243(4.38),374s(4.16), 391s(4.35),411(4.56), 431(4.58),478s(2.83)	138-0905-83
$C_{10}H_2N_4O_2$			
Propanedinitrile, 2,2'-(3,4-dihydroxy-3-cyclobutene-1,2-diylidene)bis-, dipotassium salt	H_2O	380(4.53)	88-1567-83
$C_{10}H_5Br_2N_3$			
1H-Pyrido[3,2,1-ij][1,2,4]benzotriazin-4-ium, 8,10-dibromo-, hydroxide, inner salt	EtOH	266(--),273(--), 320s(--),337(--), 351(--),420(--), 438(--),480(--), 513(--),551(--)	39-0349-83C
$C_{10}H_5Cl_3O_2S$			
Benzo[b]cyclobuta[d]thiophene, 1,2a,7b-trichloro-2a,7b-dihydro-, 3,3-dioxide, cis	EtOH	221(3.83),258(2.4), 264(2.63),269(2.75), 287(2.69)	44-1725-83
$C_{10}H_5Cl_3S$			
Benzo[b]cyclobuta[c]thiophene, 1 (and 2)2a,7b-trichloro-2a,7b-dihydro-, cis	$CHCl_3$	291(3.23),298(3.22)	44-1725-83

Compound	Solvent	λ_{max}(log ε)	Ref.
$C_{10}H_5Cl_5O_2$			
1(6aH)-Pentalenone, 2,4,5,6,6a-penta-chloro-3-ethoxy-	hexane	220(3.65),276(4.50), 357(4.19)	18-0481-83
$C_{10}H_5NO_4$			
2H,5H-Pyrano[4,3-b]pyran-3-carbo-nitrile, 7-methyl-2,5-dioxo-	EtOH	369(4.15)	5-0695-83
	toluene	378(--)	5-0695-83
	DMSO	372(--)	5-0695-83
$C_{10}H_5N_3O_2$			
1H-Naphtho[2,3-d]triazole-4,9-dione	EtOH	246(4.73),270(4.31), 330(3.59)	87-0714-83
$C_{10}H_6BrNO_3$			
5,8-Quinolinedione, 6-bromo-7-methoxy-	$CHCl_3$	260(3.97),277(3.87), 383(2.98)	142-1031-83
$C_{10}H_6BrN_3$			
1H-Pyrido[3,2,1-ij][1,2,4]benzotria-zin-4-ium, 8-bromo-, hydroxide, inner salt	EtOH	260(4.30),265(4.27), 318s(3.64),332(3.81), 347(4.08),414(3.51), 429(3.56),480(3.42), 510(3.40),552(3.04)	39-0349-83C
1H-Pyrido[3,2,1-ij][1,2,4]benzotria-zin-4-ium, 10-bromo-, hydroxide, inner salt	EtOH	273(4.36),278(4.37), 317s(3.38),332(3.79), 345(4.03),406(3.44), 426(3.52),475(3.35), 506(3.31),546(2.92)	39-0349-83C
$C_{10}H_6Br_2Cl_2S$			
Benzo[b]thiophene, 3-bromo-2-(2-bromo-1,2-dichloroethyl)-	EtOH	233(4.27),278(4.04), 298(3.72),309(3.68)	44-1725-83
$C_{10}H_6Br_2N_2$			
2,2'-Bipyridine, 4,4'-dibromo-	EtOH	214(4.64),238s(4.01), 246(4.06),253s(4.04), 275s(4.12),282(4.15), 290s(4.01)	44-0283-83
$C_{10}H_6Br_2N_2O$			
Ethanone, 2,2-dibromo-1-(2-quinoxa-linyl)-	EtOH	241(4.28),257(4.32), 317(3.86)	106-0829-83
$C_{10}H_6Br_2N_2O_2$			
2,2'-Bipyridine, 4,4'-dibromo-, 1,1'-dioxide	EtOH	223(4.40),254(4.24), 277(4.38)	44-0283-83
$C_{10}H_6Br_3NO$			
1H-Indole-3-carboxaldehyde, 2,5,6-tri-bromo-1-methyl-	MeOH	222(4.38),255(4.19), 277(3.92),309(3.92)	88-0481-83
$C_{10}H_6ClNOSSe$			
2-Selenazolidinone, 5-[(2-chlorophen-yl)methylene]-4-thioxo-	n.s.g.	269(4.10),337(4.14), 419(4.12)	103-0273-83
2-Selenazolidinone, 5-[(4-chlorophen-yl)methylene]-4-thioxo-	n.s.g.	275(4.12),330(4.06), 420(4.16)	103-0273-83
$C_{10}H_6ClN_3O_2$			
2,2'-Bipyridine, 4-chloro-4'-nitro-	EtOH	239(4.18),274s(3.81), 280(3.84),308s(3.49)	44-0283-83

Compound	Solvent	λ_{max}(log ϵ)	Ref.
$C_{10}H_6Cl_2N_2$			
2,2'-Bipyridine, 4,4'-dichloro-	EtOH	243(4.08),251s(4.04), 273s(4.11),280(4.14), 291s(3.97)	44-0283-83
$C_{10}H_6Cl_2N_2O$			
2(1H)-Pyrazinone, 3,5-dichloro-1-phenyl-	EtOH	346(3.95)	4-0919-83
$C_{10}H_6Cl_2N_2O_2$			
2,2'-Bipyridine, 4,4'-dichloro-, 1,1'-dioxide	EtOH	223(4.38),250(4.17), 277(4.32)	44-0283-83
$C_{10}H_6Cl_4S$			
Benzo[b]cyclobuta[d]thiophene, 1,2,2a,7b-tetrachloro-1,2,2a,7b-tetrahydro-, trans	n.s.g.	224(3.91),242(3.78), 296(2.81)	44-1725-83
$C_{10}H_6CrN_2O_5Se$			
Chromium, pentacarbonyl(5,6-dihydro-4H-cyclopenta-1,2,3-selenadiazole-N^2)-, (OC-6-22)-	hexane	240(4.53),455(3.72)	24-0230-83
$C_{10}H_6N_2O_2$			
2H-1-Benzopyran-2-one, 4-(diazomethyl)-	EtOH	244.5(4.23),251(4.25), 283(3.93),320.5(4.20)	94-3014-83
$C_{10}H_6N_2O_3$			
2H-1-Benzopyran-2-one, 4-(diazomethyl)-7-hydroxy-	EtOH	242(4.15),320(4.48)	94-3014-83
$C_{10}H_6N_2O_3SSe$			
2-Selenazolidinone, 5-[(3-nitrophenyl)methylene]-4-thioxo-	n.s.g.	265(4.12),340(4.01), 432(4.14)	103-0273-83
2-Selenazolidinone, 5-[(4-nitrophenyl)methylene]-5-thioxo-	n.s.g.	280(4.18),358(4.06), 434(4.20)	103-0273-83
$C_{10}H_6N_2S$			
Cyclohepta[4,5]pyrrolo[2,3-d]thiazole	C_6H_{12}	252(3.96),296(4.80), 312(4.10),343(3.80), 359(3.86),525(2.90), 546(2.85),563(2.84), 588(2.61),608(2.58)	18-1247-83
$C_{10}H_6N_4$			
Pyridazino[4,5-g]phthalazine	DMSO	305(3.71),317(3.77), 331(3.76),338s(3.65)	142-1279-83
$C_{10}H_6N_4O$			
Pyrido[3,2-f][1,2,3]benzotriazin-1(2H)-one	EtOH	212.5(4.26),227s(4.18), 248(4.15),313(4.59), 324(3.61),343(3.60)	95-0601-83
$C_{10}H_6N_4O_2$			
Pyridazino[4,5-g]phthalazine-1,4-dione, 2,3-dihydro-	MeOH	232(4.49),263(3.72), 291s(3.65),333(3.69)	142-1279-83
$C_{10}H_6N_4O_4$			
2,2'-Bipyridine, 4,4'-dinitro-	EtOH	214s(4.24),228(4.29), 304(3.70)	44-0283-83

Compound	Solvent	λ_{max}(log ε)	Ref.
$C_{10}H_6N_4O_5$			
2,2'-Bipyridine, 4,4'-dinitro-, 1-oxide	EtOH	217(4.40),267(4.16), 323(4.07)	44-0283-83
$C_{10}H_6N_4O_6$			
2,2'-Bipyridine, 4,4'-dinitro-, 1,1'-dioxide	EtOH	232s(4.17),328(4.30)	44-0283-83
$C_{10}H_6OS_3$			
Dithieno[2,3-b:2',3'-e]thiepin-4-one, 4,8-dihydro-	MeOH	263(4.06),281(3.95), 313(3.98)	4-1085-83
Dithieno[2,3-b:3',2'-e]thiepin-4-one, 4,8-dihydro-	MeOH	259(4.25),345(3.52)	4-1085-83
Dithieno[3,2-b:3',2'-e]thiepin-4-one, 4,9-dihydro-	MeOH	229(4.08),307(3.98)	4-1085-83
Dithieno[3,2-b:2',3'-e]thiepin-9-one, 4,9-dihydro-	MeOH	251(3.64),325(4.11)	4-1085-83
$C_{10}H_7BrClNO_2$			
1H-Indole-2,3-dione, 5-bromo-1-(2-chloroethyl)-	EtOH	215(4.29),253(4.18), 257(4.16),301(3.18), 436(2.48)	103-0286-83
$C_{10}H_7BrN_2O$			
Ethanone, 2-bromo-1-(2-quinoxalinyl)-	EtOH	253(4.36),310(3.66)	106-0829-83
$C_{10}H_7BrN_2OS$			
4(5H)-Thiazolone, 2-amino-5-[(4-bromophenyl)methylene]-	n.s.g.	244(3.79),290(3.96), 335(4.26)	103-0745-83
$C_{10}H_7BrN_2O_2S$			
4(1H)-Pyrimidinone, 5-bromo-2,3-dihydro-6-(2-hydroxyphenyl)-2-thioxo-	MeOH	215(4.44),280(4.44), 310s(4.18)	94-3728-83
$C_{10}H_7Br_3S$			
Benzo[b]thiophene, 3-bromo-2-(2,2-dibromoethyl)-	EtOH	268(3.92),275(3.94), 294(3.63),304(3.66)	44-1725-83
$C_{10}H_7ClN_2O$			
Pyrimidine, 2-chloro-4-phenyl-, 1-oxide	EtOH	288(4.26),326(4.21)	103-0091-83
Pyrimidine, 2-chloro-5-phenyl-, 1-oxide	EtOH	261(4.37),333(3.53)	103-0091-83
$C_{10}H_7ClN_2OS$			
4(5H)-Thiazolone, 2-amino-5-[(4-chlorophenyl)methylene]-	n.s.g.	241(4.12),290(4.22), 335(4.44)	103-0745-83
$C_{10}H_7ClN_2O_2S$			
4(1H)-Pyrimidinone, 5-chloro-2,3-dihydro-6-(2-hydroxyphenyl)-2-thioxo-	MeOH	215(4.59),280(4.60), 310s(4.28)	94-3728-83
$C_{10}H_7ClN_2O_4$			
1H-Indole-2,3-dione, 1-(2-chloroethyl)-5-nitro-	EtOH	209(4.26),326(4.06)	103-0286-83
$C_{10}H_7ClO_4$			
2(3H)-Benzofuranone, 7-acetoxy-5-chloro-	EtOH	208(4.11),220(3.82), 283(3.03)	104-0458-83

Compound	Solvent	λmax(log ε)	Ref.
$C_{10}H_7Cl_2NO$			
8-Quinolinol, 5,7-dichloro-2-methyl-	hexane	253(4.68),320(3.47)	74-0371-83A
	benzene	318(3.50)	74-0371-83A
	$CHCl_3$	253.5(4.65),317(3.51)	74-0371-83A
	isoamyl alcohol-$HClO_4$	260(4.63)	74-0371-83A
$C_{10}H_7Cl_2NO_2$			
1H-Indole-2,3-dione, 5-chloro-1-(2-chloroethyl)-	EtOH	213(4.26),251(4.15), 256(4.12),300(3.15), 435(2.41)	103-0286-83
$C_{10}H_7FN_2OS$			
4(5H)-Thiazolone, 2-amino-5-[(4-fluorophenyl)methylene]-	n.s.g.	230(4.17),285(4.16), 331(4.35)	103-0745-83
$C_{10}H_7NO$			
4-Quinolinecarboxaldehyde	EtOH	295(3.78)	140-0215-83
$C_{10}H_7NOSSe$			
2-Selenazolidinone, 5-(phenylmethylene)-4-thioxo-	n.s.g.	269(4.07),331(4.05), 416(4.17)	103-0273-83
$C_{10}H_7NO_2$			
1H-Indene-1,3(2H)-dione, 2-(aminomethylene)-	EtOH	235(4.43),248s(4.22), 281s(4.21),290(4.44), 313s(4.28),323(4.34)	4-0649-83
$C_{10}H_7N_3$			
Cyclohepta[b]pyrrole-3-carbonitrile, 2-amino-	EtOH	242s(4.11),250(4.12), 284(4.53),291(4.53), 320s(3.95),366(3.86), 423(3.54),433(3.54)	18-3703-83
1H-Pyrido[3,2,1-ij][1,2,4]benzotriazin-4-ium, hydroxide, inner salt	EtOH	262(4.32),266s(4.30), 310s(3.41),325(3.76), 338(3.94),396(3.46), 415(3.52),470(3.42), 502(3.40),542(3.09)	39-0349-83C
	EtOH-HCl	259(4.34),265(4.30), 285(3.36),300(3.54), 313(3.64),350(3.20), 366(3.27),429(3.33)	39-0349-83C
	dioxan	265s(4.29),270(4.30), 320s(3.40),335(3.83), 350(4.12),414(3.52), 430(3.57),486(3.44), 516(3.44),557(3.12)	39-0349-83C
	$CHCl_3$	266(4.29),271s(4.28), 318s(3.39),333(3.76), 346.5(4.00),409(3.47), 425(3.55),483(3.45), 514(3.47),552(3.17)	39-0349-83C
$C_{10}H_7N_3OS$			
Imidazo[2,1-b][1,3,5]benzothiadiazepin-5(6H)-one	MeOH	206(4.46),234(4.31), 282s(3.31)	4-1287-83
$C_{10}H_7N_3O_2$			
2,2'-Bipyridine, 4-nitro-	EtOH	236(4.13),277(3.86), 314(3.45)	44-0283-83

Compound	Solvent	λ_{max}(log ϵ)	Ref.
$C_{10}H_7N_3O_3$			
2,2'-Bipyridine, 4-nitro-, 1-oxide	EtOH	227(4.09),285(4.17), 330(4.00)	44-0283-83
2,2'-Bipyridine, 4-nitro-, 1'-oxide	EtOH	238(4.52),281(3.94), 320(3.58)	44-0283-83
$C_{10}H_7N_3O_3S$			
4(5H)-Thiazolone, 2-amino-5-[(4-nitrophenyl)methylene]-	n.s.g.	224(4.05),293(4.12), 354(4.29)	103-0611-83
$C_{10}H_7N_3O_4$			
2,2'-Bipyridine, 4-nitro-, 1,1'-dioxide	EtOH	216(4.21),256(3.84), 331(4.05)	44-0283-83
$C_{10}H_7N_3S$			
Cyclohepta[4,5]pyrrolo[2,3-d]thiazol-2-amine	MeOH	216(4.63),237(4.45), 298(5.09),327(4.98), 361(4.21),378(4.10), 493(3.85)	18-1247-83
$C_{10}H_7N_3S_2$			
Imidazo[2,1-b][1,3,5]benzothiadiazepine-5(6H)-thione	MeOH	218(4.66),241(4.34), 278(4.06),294s(3.26), 304s(3.20)	4-1287-83
$C_{10}H_7N_5O$			
4H-Imidazo[4,5-d]-1,2,3-triazin-4-one, 3,5-dihydro-3-phenyl-	n.s.g.	224(3.94),295(4.40)	103-1235-83
$C_{10}H_8$			
Naphthalene	EtOH	221(5.06),276(3.74), 298(2.51),312(2.41)	35-6096-83
Tricyclo[4.2.2.0^{1,6}]deca-2,4,7,9-tetraene	EtOH	204(3.69),271(3.29), 279(3.29)	88-3361-83
$C_{10}H_8BrNO_2$			
7-Isoquinolinol, 8-bromo-6-methoxy-, perchlorate	MeOH	241(4.43),258(4.24), 318(3.54)	83-0520-83
$C_{10}H_8BrN_3$			
2-Pyrimidinamine, 5-bromo-4-phenyl-	MeOH	245(4.35),328(3.73)	103-0091-83
$C_{10}H_8BrN_3O$			
Ethanone, 2-bromo-1-(2-quinoxalinyl)-, oxime	EtOH	254(4.23),333(3.69)	106-0829-83
2-Pyrimidinamine, 5-bromo-4-phenyl-, 1-oxide	EtOH	247(4.22),274s(3.75), 368(4.09)	103-0091-83
$C_{10}H_8BrN_3S$			
Imidazo[1,2-c]quinazoline-5(6H)-thione, 9-bromo-2,3-dihydro-	EtOH	249(4.21),283s(4.66), 292(4.68)	97-0215-83
$C_{10}H_8ClNO_2$			
1H-Indole-2,3-dione, 1-(2-chloroethyl)-	EtOH	213(4.18),244(4.34), 251(4.27),301(3.45), 420(2.70)	103-0286-83
$C_{10}H_8ClNO_5$			
4H,5H-Pyrano[3,4-e]-1,3-oxazine-4,5-dione, 7-chloro-2-propoxy-	$CHCl_3$	278(3.92),284(3.92), 330(3.87)	24-3725-83

Compound	Solvent	$\lambda_{max}(\log \epsilon)$	Ref.
$C_{10}H_8ClN_3O$			
2-Pyrimidinamine, 5-chloro-4-phenyl-, 1-oxide	EtOH	247(4.09),269s(4.00), 370(3.98)	103-0091-83
$C_{10}H_8Cl_4O_3$			
1,6-Dioxadispiro[2.1.4.3]dodec-10-en-12-one, 4,4,10,11-tetrachloro-	EtOH	262(4.00)	23-0545-83
$C_{10}H_8N_2O$			
Benzonitrile, 4-(4,5-dihydro-2-oxazolyl)-	C_6H_{12}	259.5(4.09),292s(3.48)	88-2279-83
2,2'-Bipyridine, 1-oxide	EtOH	238(4.35),267(4.06), 284s(3.97)	44-0283-83
Ethanone, 1-(5-quinoxalinyl)-	MeOH	214(4.11),238(3.83), 315s(3.45),322(3.46)	18-3358-83
Pyridine, 3,3'-oxybis-	EtOH	216(4.30),268(4.07), 274(4.02)	4-1411-83
$C_{10}H_8N_2OS$			
4(5H)-Thiazolone, 2-amino-5-(phenylmethylene)-	n.s.g.	238(4.00),288(4.12), 330(4.34)	103-0611-83
5H-Thiazolo[2,3-b]quinazolin-3(2H)-one	MeOH	222(4.21),296(3.98)	83-0569-83
$C_{10}H_8N_2O_2$			
2,2'-Bipyridine, 1,1'-dioxide	EtOH	221(4.31),238(4.18), 267(4.18)	44-0283-83
[2,3'-Bipyridine]-3,4'-diol	MeOH	293(4.57),349(3.53)	102-2847-83
	MeOH-HCl	302(4.14),393(3.49)	102-2847-83
	MeOH-NaOH	297(3.92),307(3.87), 408(3.49)	102-2847-83
[2,2'-Bi-1H-pyrrole]-5,5'-dicarboxaldehyde	MeOH	250(3.76),362(4.40), 377.7(4.40)	35-6429-83
2-Furancarboxaldehyde, (2-furanylmethylene)hydrazone	EtOH	237(3.42),327(4.55), 333(4.57),336(4.57), 341(4.41)	34-0132-83
$C_{10}H_8N_2O_5$			
2-Furancarboxylic acid, 5-(3,4-dihydro-2,4-dioxo-1(2H)-pyrimidinyl)-, methyl ester	MeOH	265(4.08),295(4.06)	65-1712-83
$C_{10}H_8N_2S$			
Acetonitrile, (1H-indol-3-ylthio)-	MeOH	270(3.89),277(3.89), 287(3.82)	87-0230-83
$C_{10}H_8N_2S_2$			
2-Thiophenecarboxaldehyde, (2-thienylmethylene)hydrazone	EtOH	270(4.13),342(4.58)	34-0132-83
$C_{10}H_8N_4$			
4H-Imidazole-4-carbonitrile, 5-amino-4-phenyl-	EtOH-pH 7	268(3.72)	44-0003-83
	EtOH-pH 13	288(3.85)	44-0003-83
$C_{10}H_8N_4O$			
3H-Pyrazol-3-one, 4-diazo-2,4-dihydro-2-methyl-5-phenyl-	EtOH	249(3.93),273s(3.77), 340(3.58)	44-1069-83
3H-Pyrazol-3-one, 4-diazo-2,4-dihydro-5-methyl-2-phenyl-	EtOH	247(4.48),280s(3.69), 322(3.25)	44-1069-83

Compound	Solvent	$\lambda_{max}(\log \epsilon)$	Ref.
$C_{10}H_8N_6O$			
1H-Imidazo[1,2-b][1,2,4]triazole-5,6-dione, 6-(phenylhydrazone)	EtOH	285s(3.50),325(4.09)	4-0639-83
$C_{10}H_8O$			
3-Butyn-2-one, 4-phenyl-	EtOH	215(4.08),274(4.18)	65-2058-83
1-Naphthalenol	MeOH	213(4.66),234(4.57), 276(3.72),292(3.80), 311(3.72),324(3.61)	48-1002-83
2-Naphthalenol	MeOH	210(4.58),226(4.97), 264(3.85),276(3.90), 285(3.76),326(3.55), 330(3.54)	48-1002-83
	MeOH	254(3.45),263(3.59), 273(3.66),285(3.52), 318(3.24),330(3.31)	160-1403-83
	MeOH-base	273(3.71),282(3.83), 293(3.72),345(3.46)	160-1403-83
$C_{10}H_8OS$			
2H-1-Benzothiopyran-2-one, 4-methyl-	EtOH	298(3.85),342(3.46)	139-0059-83B
	C_6H_{12}	297(--),337(--)	139-0059-83B
4H-1-Benzothiopyran-4-one, 2-methyl-	EtOH	247(4.38),335(4.06)	139-0059-83B
$C_{10}H_8O_2$			
Benzo[1,2:4,5]dicyclobutene-3,6-dione, 1,2,3,4,5,6-hexahydro-	C_6H_{12}	261(4.22),270(4.24), 346(2.21),464(1.45)	88-1727-83
2H-1-Benzopyran-2-one, 4-methyl-	EtOH	274(3.91),313(3.72)	139-0059-83B
	C_6H_{12}	279(--),314(--)	139-0059-83B
4H-1-Benzopyran-4-one, 6-methyl-	MeOH	230(4.14),245(4.06), 260s(3.85),310(3.84)	118-0310-83
4H-1-Benzopyran-4-one, 8-methyl-	MeOH	230(4.0),249(4.4), 265(3.81),281(4.61), 305(3.81)	118-0310-83
Ethanone, 1-(2-benzofuranyl)-	MeOH	225(3.96),230(3.92), 292(4.41)	94-3728-83
$C_{10}H_8O_2S$			
4H-1-Benzothiopyran-4-one, 3-methoxy-	MeOH	258(4.40),369(4.03)	83-0921-83
$C_{10}H_8O_2S_3$			
3-Thiophenecarboxylic acid, 2-[(3-thienylthio)methyl]-	MeOH	247(4.08)	4-1085-83
$C_{10}H_8O_2Te$			
4H-1-Benzo[b]tellurin-4-one, 7-methoxy-	CH_2Cl_2	370(3.82)	35-0883-83
$C_{10}H_8O_3$			
1H-2-Benzopyran-5-carboxaldehyde, 3,4-dihydro-1-oxo-	n.s.g.	223(4.30),290(3.13)	162-0531-83
1H-2-Benzopyran-1-one, 8-hydroxy-3-methyl-	EtOH	228(4.26),234.5(4.28), 256(4.01),271s(3.77), 341(3.77)	102-1489-83
2H-1-Benzopyran-2-one, 7-hydroxy-	pH <5	321(4.0)	18-0006-83
4-methyl-	pH >11	230(4.1),361(4.1)	18-0006-83
with β-cyclodextrin	H_2O	322(4.19)	95-0193-83
2H-1-Benzopyran-2-one, 5-methoxy-	MeOH	295(4.01),243(3.70)	44-2578-83
$C_{10}H_8O_4$			
1,3-Benzodioxole-4,7-dione, 5-(2-propenyl)-	$CHCl_3$	275(4.09),495(2.93)	64-0392-83B

Compound	Solvent	λ_{max}(log ϵ)	Ref.
4,7-Benzofurandione, 6-methoxy-2-methyl- (acamelin)	EtOH	219(4.32),263(3.97),307(4.03),417(3.91)	44-0127-83
2H-1-Benzopyran-2-one, 5-hydroxy-6-methoxy- (floribin)	MeOH	224(4.19),290(3.68),343(4.01)	106-0072-83
	MeOH-NaOAc	300(--),385(--)	106-0072-83
2H-1-Benzopyran-2-one, 7-hydroxy-6-methoxy-	MeOH	229(4.14),252(3.96),260s(3.94),297(3.98),344(4.13)	100-0222-83
$C_{10}H_8O_4S$			
4H-1-Benzothiopyran-4-one, 3-methoxy-, 1,1-dioxide	MeOH	266(4.22),302(3.77),304(3.77)	83-0921-83
1-Naphthalenesulfonic acid, 2-hydroxy-	MeOH	210(4.41),230(4.91),271(3.24),280(3.33),290(3.24),337(3.17),348(3.11)	48-1002-83
1-Naphthalenesulfonic acid, 4-hydroxy-	MeOH	222(4.44),234(4.41),291(3.77),305(3.79),325(3.62)	48-1002-83
2-Naphthalenesulfonic acid, 4-hydroxy-	MeOH	220(4.27),238(4.06),312(3.37),330(3.43),350(3.35)	48-1002-83
$C_{10}H_8O_5S$			
2-Naphthalenesulfonic acid, 4,6-dihydroxy-	MeOH	221(4.44),229(4.48),244(4.57),299(3.68),327(3.22),341(3.04)	48-1002-83
2-Naphthalenesulfonic acid, 4,7-dihydroxy-	MeOH	228(4.37),243(4.27),286(3.49),293(3.49),306(3.20),384(2.97)	48-1002-83
2-Naphthalenesulfonic acid, 6,7-dihydroxy-	MeOH	212(4.07),236(4.55),292(3.65),318(3.37),333(3.42)	48-1002-83
$C_{10}H_8O_7S_2$			
1,6-Naphthalenedisulfonic acid, 8-hydroxy-	MeOH	218(4.40),247(4.32),316(3.48),325(3.54),337(3.48)	48-1002-83
2,7-Naphthalenedisulfonic acid, 4-hydroxy-	MeOH	222(4.30),248(4.27),305(3.27),323(3.31),337(3.27)	48-1002-83
$C_{10}H_8O_{10}S_3$			
1,3,6-Naphthalenetrisulfonic acid, 7-hydroxy-	MeOH	226(4.28),250(4.49),302(3.93),312(3.95),370(3.65)	48-1002-83
$C_{10}H_8S_2$			
2H-1-Benzothiopyran-2-thione, 4-methyl-	EtOH	302(4.24),420(4.05)	139-0059-83B
4H-1-Benzothiopyran-4-thione, 2-methyl-	EtOH	293(3.88),412(4.32)	139-0059-83B
$C_{10}H_9BrN_2O_2$			
[2,2'-Bi-1H-pyrrole]-5-carboxaldehyde, 3'-bromo-4-methoxy-	MeOH	247(3.57),357(3.83)	44-2314-83
[2,2'-Bi-1H-pyrrole]-5-carboxaldehyde, 5'-bromo-4-methoxy-	MeOH	256(3.67),364(3.86)	44-2314-83
$C_{10}H_9BrN_2O_3$			
Ethanone, 1-(5-bromo-2,3-dimethoxy-	EtOH	211s(--),225(4.27),	104-0458-83

Compound	Solvent	$\lambda_{max}(\log \epsilon)$	Ref.
phenyl)-2-diazo- (cont.)		258s(--),305(3.85)	104-0458-83
$C_{10}H_9BrN_2S_2$			
2,4(LH,3H)-Quinazolinedithione, 6-bromo-3-ethyl-	EtOH	258(4.09),291(4.33), 310(4.37),333(4.06), 388(3.69)	97-0215-83
$C_{10}H_9BrN_4$			
1H-Imidazole, 2-[(4-bromophenyl)azo]-1-methyl-	MeOH	242(3.85),368(4.24), 380(4.20),440s(3.41)	56-0971-83
	MeOH-HCl	242(3.78),369(4.44), 380(4.42),445s(3.31)	56-0971-83
	MeOH-NaOH	242(3.89),366(4.35), 380(4.33),434s(3.48)	56-0971-83
	20% MeOH	237(3.81),369?(4.35), 376(4.35),420s(3.75)	56-0971-83
	MeCN	244(4.49),351s(4.34), 368(4.43),378(4.42), 446s(3.49)	56-0971-83
	1,2-$C_2H_4Cl_2$	238(3.87),369(4.45), 382(4.43),434s(3.56)	56-0971-83
$C_{10}H_9BrO$			
1(2H)-Naphthalenone, 6-bromo-3,4-dihydro-	EtOH	250(3.88),305(3.13)	22-0096-83
$C_{10}H_9ClN_2O$			
1,8-Naphthyridin-4(1H)-one, 3-chloro-2,7-dimethyl-	EtOH	252(4.38),282(3.36), 292(3.43),330(3.97)	142-1083-83
4H-Pyrido[1,2-a]pyrimidin-4-one, 3-chloro-2,6-dimethyl-	EtOH	255(4.07),261(4.06), 363(4.02)	142-1083-83
$C_{10}H_9ClN_2OS$			
Benzamide, 4-chloro-N-(4,5-dihydro-2-thiazolyl)-	C_6H_{12}	202(4.3),280(4.5), 295(4.4)	41-0603-83
$C_{10}H_9ClN_2O_3$			
Ethanone, 1-(5-chloro-2,3-dimethoxyphenyl)-2-diazo-	EtOH	209(4.41),226(4.46), 263(4.02),301(4.22)	104-0458-83
$C_{10}H_9ClN_2O_4$			
4H,5H-Pyrano[3,4-e]-1,3-oxazine-4,5-dione, 7-chloro-2-(propylamino)-	$CHCl_3$	298(4.08),339(4.18)	24-3725-83
$C_{10}H_9ClN_2O_5$			
4H,5H-Pyrano[3,4-e]-1,3-oxazine-4,5-dione, 7-chloro-2-[(3-hydroxypropyl)amino]-	$CHCl_3$	244(3.84),297(4.15), 342(4.25)	24-3725-83
$C_{10}H_9ClO$			
Benzofuran, 6-chloro-2-ethenyl-2,3-dihydro-	EtOH	219(3.83),283(3.42), 287(3.40)	78-0169-83
$C_{10}H_9ClOS$			
2,4,6-Cycloheptatrien-1-one, 2-[(2-chloro-2-propenyl)thio]-	MeOH	243(4.25),290(4.22), 349(3.54),364(3.30)	142-1709-83
$C_{10}H_9ClO_3$			
2-Propenoic acid, 3-(2-chloro-6-methoxyphenyl)-, (Z)-	MeOH	260(3.58)	44-2578-83

Compound	Solvent	λ_{max}(log ε)	Ref.
$C_{10}H_9Cl_2NO_3$			
Furan, 2-(4,4-dichloro-3-ethyl-1-cyclo-buten-1-yl)-5-nitro-	MeOH	235(4.17),255(4.19), 320(3.98)	73-1062-83
Furan, 2-[3-(dichloromethylene)-1-pen-tenyl]-5-nitro-, (E)-	MeOH	215(3.98),280(4.25), 390(4.23)	73-1062-83
$C_{10}H_9Cl_3N_2O_3$			
Carbamic acid, [(phenylamino)carbo-nyl]-, 2,2,2-trichloroethyl ester	EtOH	246(4.20)	22-0073-83
$C_{10}H_9Cl_3O_3$			
1,6-Dioxadispiro[2.1.4.3]dodec-10-en-12-one, 4,4,10-trichloro-	EtOH	244(4.06)	23-0545-83
$C_{10}H_9N$			
1H-Indole, 1-ethenyl-	EtOH	212(4.78),253(4.85), 294(4.39),300(4.40)	104-0189-83
1-Naphthalenamine	MeOH	213(4.76),243(4.41), 321(3.74),332(3.72)	48-1002-83
2-Naphthalenamine	MeOH	214(4.44),238(4.73), 272(3.75),282(3.83), 292(3.70),342(3.36)	48-1002-83
$C_{10}H_9NO$			
Ethanone, 1-(1H-indol-3-yl)-	EtOH	217(4.53),240(--), 252s(4.13),297(4.11)	128-0157-83
	+ NaOH	226(--),266(--), 332(--)	128-0157-83
	C_6H_{12}	212(--),234(--), 282(--)	128-0157-83
	MeOH	219(--),240(--), 252s(--),297(--)	128-0157-83
1-Naphthalenol, 2-amino-	MeOH	207(4.01),238(4.25), 300(3.33),338(3.37)	48-1002-83
2-Propenenitrile, 2-methoxy-3-phenyl-, (E)-	EtOH	208(3.90),218(3.89), 275(4.07)	78-1551-83
(Z)-	EtOH	205(4.00),219(4.01), 281(4.45)	78-1551-83
Quinoline, 2-methoxy-, perchlorate	MeCN	312(3.92)	35-1204-83
$C_{10}H_9NOS_2$			
2(3H)-Benzothiazolone, 3-(thiiranyl-methyl)-	EtOH	284(--),292(3.29)	103-1170-83
$C_{10}H_9NO_2$			
2H-1-Benzopyran-2-one, 7-amino-4-methyl-	EtOH	355(4.26)	139-0059-83B
with β-cyclodextrin	H_2O	345(4.19)	95-0193-83
5(4H)-Oxazolone, 2-methyl-4-phenyl-	C_6H_{12}	277(4.24)	44-0695-83
1H-Pyrano[3,4-c]pyridin-1-one (gentianine)	n.s.g.	220(4.38)	162-0627-83
$C_{10}H_9NO_3$			
1H-Indole-2-carboxylic acid, 5-meth-oxy-	EtOH	223(4.39),296(4.17), 325s(3.60)	128-0157-83
	EtOH-NaOH	230(--),292(--)	128-0157-83
	C_6H_{12}	220(--),297(--), 331s(--)	128-0157-83
	MeOH	225(--),292(--), 325s(--)	128-0157-83

Compound	Solvent	$\lambda_{max}(\log \epsilon)$	Ref.
$C_{10}H_9NO_3S$			
1-Naphthalenesulfonic acid, 2-amino-	MeOH	219(4.04),247(4.55), 276(3.56),285(3.61), 292(3.31),349(3.30)	48-1002-83
1-Naphthalenesulfonic acid, 4-amino-	MeOH	216(4.61),248(4.39), 333(3.80)	48-1002-83
1-Naphthalenesulfonic acid, 5-amino-	MeOH	213(4.65),249(4.35), 335(3.78)	48-1002-83
1-Naphthalenesulfonic acid, 6-amino-	MeOH	217(4.13),247(4.39), 278(3.46),286(3.49), 292(3.42),348(3.12)	48-1002-83
1-Naphthalenesulfonic acid, 7-amino-	MeOH	218(4.28),244(4.65), 289(3.81),299(3.77), 350(3.54)	48-1002-83
1-Naphthalenesulfonic acid, 8-amino-	MeOH	214(4.73),253(4.54), 342(3.89)	48-1002-83
2-Naphthalenesulfonic acid, 1-amino-	MeOH	217(4.60),251(4.31), 336(3.74)	48-1002-83
2-Naphthalenesulfonic acid, 4-amino-	MeOH	216(4.62),247(4.31), 332(3.71)	48-1002-83
2-Naphthalenesulfonic acid, 5-amino-	MeOH	214(4.65),250(4.38), 336(3.70)	48-1002-83
2-Naphthalenesulfonic acid, 6-amino-	MeOH	217(4.06),245(4.35), 295(3.61),346(3.07)	48-1002-83
2-Naphthalenesulfonic acid, 7-amino-	MeOH	217(4.25),246(4.70), 279(3.65),285(3.69), 292(3.54),347(3.17)	48-1002-83
2-Naphthalenesulfonic acid, 8-amino-	MeOH	215(4.69),249(4.43), 336(3.78)	48-1002-83
$C_{10}H_9NO_4$			
2H-1-Benzopyran-3-carboxamide, 5,6,7,8-tetrahydro-2,5-dioxo-	EtOH	313(3.72)	5-0695-83
	toluene	317(--)	5-0695-83
	DMSO	321(--)	5-0695-83
$C_{10}H_9NO_4S$			
1-Naphthalenesulfonic acid, 3-amino-4-hydroxy-	MeOH	216(4.33),249(4.51), 307(3.82),340(3.60)	48-1002-83
1-Naphthalenesulfonic acid, 4-amino-5-hydroxy-	MeOH	217(4.30),233(4.63), 322(4.06),336(4.13), 345(4.15)	48-1002-83
2-Naphthalenesulfonic acid, 3-amino-4-hydroxy-	MeOH	215(3.97),249(4.10), 307(3.12),341(3.16)	48-1002-83
2-Naphthalenesulfonic acid, 6-amino-4-hydroxy-	MeOH	220(4.42),252(4.63), 306(3.95),346(3.33)	48-1002-83
2-Naphthalenesulfonic acid, 7-amino-4-hydroxy-	MeOH	228(4.39),253(4.30), 306(3.51),348(3.21)	48-1002-83
$C_{10}H_9NO_5S$			
2-Naphthalenesulfonic acid, 5-amino-4,6-dihydroxy-	MeOH	221(4.17),253(4.44), 330(3.30),360(3.21)	48-1002-83
2-Naphthalenesulfonic acid, 8-amino-4,7-dihydroxy-	MeOH	227(4.53),243(4.71), 322(3.73),370(3.71)	48-1002-83
$C_{10}H_9NO_6S_2$			
1,5-Naphthalenedisulfonic acid, 2-amino-	MeOH	217(3.97),249(4.42), 278(3.53),288(3.57), 298(3.40),359(3.37)	48-1002-83
1,5-Naphthalenedisulfonic acid, 4-amino-	MeOH	219(4.43),258(4.37), 300(3.18),348(3.66)	48-1002-83

Compound	Solvent	λ_{max}(log ε)	Ref.
1,6-Naphthalenedisulfonic acid, 8-amino-	MeOH	223(4.53),255(4.21), 346(3.65)	48-1002-83
2,6-Naphthalenedisulfonic acid, 3-amino-	MeOH	221(4.19),252(4.55), 285(3.54),288(3.54), 293(3.49),353(3.18)	48-1002-83
2,6-Naphthalenedisulfonic acid, 4-amino-	MeOH	221(4.52),251(4.35), 340(3.65)	48-1002-83
2,7-Naphthalenedisulfonic acid, 3-amino-	MeOH	222(4.45),251(4.78), 285(3.91),292(4.00), 301(3.97),354(3.56)	48-1002-83
2,7-Naphthalenedisulfonic acid, 4-amino-	MeOH	221(4.51),255(4.23), 343(3.62)	48-1002-83
$C_{10}H_9NO_7S_2$			
1,3-Naphthalenedisulfonic acid, 4-amino-5-hydroxy-	MeOH	217(3.90),237(4.33), 332(3.48),359(3.57), 370(3.53)	48-1002-83
1,5-Naphthalenedisulfonic acid, 3-amino-4-hydroxy-	MeOH	226(4.36),262(4.43), 316(3.62),330(3.68), 380(3.41)	48-1002-83
1,6-Naphthalenedisulfonic acid, 7-amino-8-hydroxy-	MeOH	227(4.32),262(4.40), 316(3.57),330(3.65), 380(3.33)	48-1002-83
2,7-Naphthalenedisulfonic acid, 3-amino-4-hydroxy-	MeOH	220(4.51),262(4.44), 328(3.60),379(3.49)	48-1002-83
2,7-Naphthalenedisulfonic acid, 4-amino-5-hydroxy-	MeOH	217(3.98),238(4.47), 331(3.89),349(3.95), 365(3.74)	48-1002-83
$C_{10}H_9NO_8S_2$			
2,7-Naphthalenedisulfonic acid, 3-amino-4,5-dihydroxy-	MeOH	210(3.75),247(3.87), 330(3.14),348(3.08)	48-1002-83
$C_{10}H_9NO_9S_3$			
1,3,6-Naphthalenetrisulfonic acid, 7-amino-	MeOH	225(3.97),257(4.22), 302(3.67),365(3.47)	48-1002-83
1,3,6-Naphthalenetrisulfonic acid, 8-amino-	MeOH	223(4.51),257(4.21), 355(3.59)	48-1002-83
1,4,6-Naphthalenetrisulfonic acid, 3-amino-	MeOH	222(3.68),255(3.99), 290(3.12),365(3.47)	48-1002-83
1,4,6-Naphthalenetrisulfonic acid, 8-amino-	MeOH	223(4.54),255(4.24), 354(3.66)	48-1002-83
$C_{10}H_9NSSe$			
2(3H)-Thiazoleselone, 3-methyl-4-phenyl-	EtOH	346(4.25)	56-0875-83
$C_{10}H_9NS_2$			
2(3H)-Thiazolethione, 3-methyl-4-phenyl-	EtOH	322(4.17)	56-0875-83
$C_{10}H_9NS_3$			
2(3H)-Benzothiazolethione, 3-(thiiranylmethyl)-	EtOH	328(4.39)	103-1170-83
$C_{10}H_9NTe$			
Isotellurazole, 3-methyl-5-phenyl-	MeCN	294(4.03),306s(3.92), 331s(3.56)	118-0824-83
$C_{10}H_9N_3$			
1H-Benzimidazole, 1-(isocyanomethyl)-	EtOH	243(4.89),249(4.85),	94-0723-83

Compound	Solvent	λ_{max}(log ε)	Ref.
2-methyl- (cont.)		274(4.65),281(4.72)	94-0723-83
$C_{10}H_9N_3O$			
Formamide, N-(1-amino-3-isoquinolinyl)-	EtOH	220(4.52),249(4.29), 310(4.11),340s(3.68)	39-1137-83C
4H-Indol-4-one, 5-diazo-1-ethenyl-1,5,6,7-tetrahydro-	MeOH	241(4.28),269s(3.79), 326(3.95)	23-1697-83
2-Pyrimidinamine, 5-phenyl-, 1-oxide	EtOH	240(4.27),255(4.29), 356(3.81)	103-0091-83
$C_{10}H_9N_3S$			
Thieno[2,3-b]pyridine, 6-(4,5-dihydro-1H-imidazol-2-yl)-	EtOH	241(4.41),292(4.16)	4-1717-83
$C_{10}H_9N_5$			
3H-Pyrazolo[1,5-d]tetrazole, 3-methyl-7-phenyl-	CH_2Cl_2	249(4.09),258s(4.02), 267s(3.97),319(3.81)	4-1629-83
$C_{10}H_9N_5O_2$			
1H-Imidazole, 1-methyl-2-[(4-nitrophenyl)azo]-	MeOH	224(4.10),274(4.00), 362(4.26),384(4.32), 460s(3.53)	56-0971-83
	MeOH-HCl	208s(3.88),270(3.64), 365(4.41),440s(3.13)	56-0971-83
	MeOH-NaOH	218(3.87),278(3.79), 390(4.31),450s(3.73)	56-0971-83
	50% MeOH	214(3.96),281(3.82), 390(4.36),460s(3.61)	56-0971-83
	MeCN	249(4.41),360s(4.18), 388(4.27),456s(3.60)	56-0971-83
	1,2-$C_2H_4Cl_2$	230(3.90),278(3.83), 394(4.40),470s(3.50)	56-0971-83
1H-Imidazole, 2-[(5-methyl-2-nitrophenyl)azo]-	MeOH	198(4.67),234(4.33), 366(4.62),452s(3.48)	56-0971-83
	MeOH-H_2SO_4	195(4.63),224(4.54), 354(4.15),420s(3.00)	56-0971-83
	MeOH-KOH	c.205(4.41),410(4.54), 480s(4.00)	56-0971-83
	1,2-$C_2H_4Cl_2$	212(4.29),375(4.51), 434s(3.71)	56-0971-83
1H-Imidazole, 2-[(2-nitro-4-methylphenyl)azo]-	MeOH	c.198(4.60),220(4.28), 370(4.63),440s(3.74)	56-0971-83
	MeOH-H_2SO_4	c.195(4.56),220(4.32), 440s(3.63)	56-0971-83
	MeOH-KOH	c.205(4.41),406(4.58), 430s(4.11)	56-0971-83
	1,2-$C_2H_4Cl_2$	213(4.34),378(4.62), 460s(3.20)	56-0971-83
$C_{10}H_{10}BrCl_3N_2O_4S$			
Ceph-3-em-4-carboxylic acid, 7-amino-3-(bromomethyl)-, 2,2,2-trichloroethyl ester, 1-oxide, hydrobromide, (1S,6R,7R)-	MeOH	283(3.92)	78-0461-83
$C_{10}H_{10}BrNO_2$			
Benzeneacetonitrile, 5-bromo-2,3-dimethoxy-	EtOH	209(4.48),231(3.90), 285(3.31)	104-0458-83
$C_{10}H_{10}BrN_3O$			
Tambjamine B	MeOH	255(3.66),397(4.30)	44-2314-83

Compound	Solvent	λ_{max}(log ε)	Ref.
$C_{10}H_{10}ClNOS_2$			
2-Propanol, 1-(2-benzothiazolylthio)-3-chloro-	EtOH	282(3.98),292(3.94),302(3.84)	103-1170-83
$C_{10}H_{10}ClNO_2$			
Benzene, 1-(3-chloro-1-butenyl)-4-nitro-, (E)-	EtOH	221(4.07),304(4.17)	12-0527-83
Benzeneacetonitrile, 5-chloro-2,3-dimethoxy-	EtOH	208(4.50),233(3.86),287(3.34)	104-0458-83
$C_{10}H_{10}ClNO_2S$			
2(3H)-Benzothiazolone, 3-(3-chloro-2-hydroxypropyl)-	EtOH	282(--),290(3.84)	103-1170-83
$C_{10}H_{10}ClNO_4S$			
4H,5H-Pyrano[3,4-e][1,3]oxazine-4,5-dione, 7-chloro-2-(propylthio)-[sic]	$CHCl_3$	284(4.03),346(4.20)	24-3725-83
$C_{10}H_{10}ClNO_5S$			
2H-Pyran-5-carbothioic acid, 4-chloro-6-[(ethoxycarbonyl)amino]-2-oxo-, S-methyl ester	$CHCl_3$	310(4.24)	24-3725-83
2H-Pyran-3-carboxylic acid, 6-chloro-4-[[(methylthio)carbonyl]amino]-2-oxo-, ethyl ester	$CHCl_3$	244(3.94),345(4.21)	24-3725-83
$C_{10}H_{10}ClN_3$			
Methanamine, N-(4-chloro-2-methyl-1(2H)-phthalazinylidene)-	MeOH	212(4.54),228s(4.22),273(3.94),325(3.81)	103-0666-83
	MeCN	212(4.54),228(4.15),271(3.98),335(3.73)	103-0666-83
sulfate(2:1)	MeOH	222(4.48),247s(4.08),272(3.88),305(3.83)	103-0666-83
	MeCN	222(4.51),246s(4.08),272(3.90),307(3.85)	103-0666-83
1-Phthalazinamine, 4-chloro-N,N-dimethyl-	MeOH	217(4.67),260s(3.60),330(3.85)	103-0666-83
	MeCN	214(4.56),262s(3.43),322(3.66)	103-0666-83
sulfate(2:1)	MeOH	220(4.54),247(4.00),277(3.95),322(3.88)	103-0666-83
	MeCN	217(4.52),246(3.88),276(3.93),321(3.82)	103-0666-83
$C_{10}H_{10}Cl_4N_2O_4S$			
Ceph-3-em-4-carboxylic acid, 7-amino-3-(chloromethyl)-, 2,2,2-trichloroethyl ester, 1-oxide, (1S,6R,7R)-	MeOH	277(3.84)	78-0461-83
$C_{10}H_{10}Co$			
Cobaltocenium, hexafluorophosphate	H_2O	264(4.52),300s(3.08),409(2.30)	101-0093-83K
$C_{10}H_{10}CuN_6O_2$			
Copper, [[3,3'-(1,2-ethanediyl)diimino]bis[3-imino-2-formylpropanenitrilato]](2-)]-	DMF	575(2.16)	12-2413-83
$C_{10}H_{10}F_3N_5O_4$			
Thymidine, 3'-azido-3'-deoxy-α,α,α-trifluoro-	pH 13	260(3.86)	87-1691-83
	MeOH	262(4.01)	87-1691-83

Compound	Solvent	λ max (log ε)	Ref.
$C_{10}H_{10}N$			
Isoquinolinium, 2-methyl-, perchlorate	MeCN	334(3.66)	35-1204-83
Quinolinium, 1-methyl-, perchlorate	MeCN	315(3.86)	35-1204-83
$C_{10}H_{10}NOS_2$			
2H-[1,3]Thiazino[2,3-b]benzothiazol-ium, 3,4-dihydro-3-hydroxy-, chloride	EtOH	315(4.05)	103-1173-83
$C_{10}H_{10}N_2$			
Methanamine, N-(1H-indol-3-ylmethyl-ene)-	MeOH	214(4.18),242(4.02), 260(3.95),296(4.01)	104-1367-83
1,2-Naphthalenediamine	MeOH	216(4.19),249(4.13), 326(3.44)	48-1002-83
1,4-Naphthalenediamine	MeOH	212(4.55),253(4.13), 347(3.69)	48-1002-83
$C_{10}H_{10}N_2O$			
Ethanone, 1-(1H-indol-3-yl)-	n.s.g.	226(4.28),257(4.08), 283(3.97)	103-0184-83
1H-Indole-1-carboxaldehyde, 3-methyl-, oxime (spectrum of class)	EtOH	250(4.3),295-310(3.88-4.11)	2-0927-83
Methanamine, N-(1H-indol-3-ylmethyl-ene)-, N-oxide	EtOH	225(4.35),253(3.99), 277(3.92),322(4.27)	104-1367-83
1-Naphthalenol, 5,6-diamino-	MeOH	210(4.33),226(4.50), 270(3.57),317(3.57), 341(2.90)	48-1002-83
1-Naphthalenol, 7,8-diamino-	MeOH	210(4.41),230(4.29), 270(3.65),318(3.60), 342(3.54)	48-1002-83
$C_{10}H_{10}N_2OSe$			
2H-1,3-Benzoselenazin-2-imine, 4,8-di-methyl-, 3-oxide	EtOH	258(4.44),390(4.23)	78-0831-83
$C_{10}H_{10}N_2O_2$			
[2,2'-Bi-1H-pyrrole]-5-carboxaldehyde, 4-methoxy-	MeOH	252(3.56),364(3.79)	44-2314-83
1(2H)-Isoquinolinone, 2-amino-3-meth-oxy-	MeOH	223(4.45),250s(4.02), 287(3.90),341(3.69)	142-1315-83
$C_{10}H_{10}N_2O_2S_2$			
1H-Imidazole-4-carboxylic acid, 2,3-dihydro-3-(2-thienylmethyl)-2-thioxo-, methyl ester	0.5M NaOH	236(4.14),307(4.16)	142-2019-83
	MeOH	235(4.10),262(3.97), 298(4.02)	142-2019-83
1H-Imidazole-5-carboxylic acid, 2-(methylthio)-1-(2-thienylmethyl)-	0.5M HCl	233(4.14),259(4.01)	142-2019-83
	0.5M NaOH	235(4.13),263(4.01)	142-2019-83
	MeOH	236(4.09),270(4.09)	142-2019-83
$C_{10}H_{10}N_2O_3$			
Acetic acid, [(2-acetylphenyl)amino]-imino]-	EtOH	207(3.89),242s(3.60), 295s(3.43),306(3.46)	158-83-69
$C_{10}H_{10}N_2O_3S$			
1-Naphthalenesulfonic acid, 3,4-di-amino-	MeOH	222(4.20),253(4.24), 320(3.31),368(3.20)	48-1002-83
1-Naphthalenesulfonic acid, 5,6-di-amino-	MeOH	218(4.02),256(4.08), 327(3.11),370(3.04)	48-1002-83
1-Naphthalenesulfonic acid, 5,8-di-amino-	MeOH	216(4.66),260(4.30), 368(3.68)	48-1002-83

Compound	Solvent	λ_{max}(log ε)	Ref.
2-Naphthalenesulfonic acid, 1,4-di-amino-	MeOH	218(4.33),253(4.42), 358(3.55)	48-1002-83
2-Naphthalenesulfonic acid, 5,6-di-amino-	MeOH	219(4.16),250(4.20), 337(3.30),368(3.20)	48-1002-83
2-Naphthalenesulfonic acid, 5,8-di-amino-	MeOH	216(4.48),262(4.04), 363(3.56)	48-1002-83
	MeOH	217(4.55),263(4.10), 365(3.62)	48-1002-83
2-Naphthalenesulfonic acid, 7,8-di-amino-	MeOH	218(4.16),255(4.24), 330(3.50),370(3.40)	48-1002-83
$C_{10}H_{10}N_2O_4S$			
1-Naphthalenesulfonic acid, 5,8-di-amino-4-hydroxy-	MeOH	210(4.21),238(4.39), 265(4.00),322(3.67), 350(3.60)	48-1002-83
2-Naphthalenesulfonic acid, 3,6-di-amino-4-hydroxy-	MeOH	220(3.91),252(4.17), 308(3.55),350(3.07)	48-1002-83
2-Naphthalenesulfonic acid, 3,7-di-amino-4-hydroxy-	MeOH	211(4.10),248(4.17), 308(3.77),349(3.61)	48-1002-83
2-Naphthalenesulfonic acid, 3,8-di-amino-4-hydroxy-	MeOH	210(4.13),243(4.42), 278(3.82),322(3.49), 350(3.53)	48-1002-83
2-Naphthalenesulfonic acid, 5,6-di-amino-4-hydroxy-	MeOH	215(4.32),243(4.40), 265(4.00),325(3.50), 342(3.53)	48-1002-83
2-Naphthalenesulfonic acid, 5,8-di-amino-4-hydroxy-	MeOH	210(4.06),236(4.38), 265(3.70),323(3.56), 342(3.50)	48-1002-83
2-Naphthalenesulfonic acid, 7,8-di-amino-4-hydroxy-	MeOH	210(4.31),236(4.37), 278(4.08),322(3.85), 347(3.71)	48-1002-83
$C_{10}H_{10}N_2O_6$			
6H-Furo[2',3':4,5]oxazolo[3,2-a]pyrim-idine-2-carboxylic acid, 2,3,3a,9a-tetrahydro-3-hydroxy-6-oxo-, methyl ester, [2S-(2α,3β,3aβ,9aβ)]-	MeOH	225(4.07),250(3.99), 275s(--)	65-1712-83
$C_{10}H_{10}N_2O_6S_2$			
1,3-Naphthalenedisulfonic acid, 5,6-diamino-	MeOH	219(4.40),265(4.50), 330(3.49),386(3.36)	48-1002-83
1,5-Naphthalenedisulfonic acid, 3,4-diamino-	MeOH	223(4.10),267(4.40), 330(3.50),380(3.51)	48-1002-83
2,6-Naphthalenedisulfonic acid, 1,4-diamino-	MeOH	223(4.28),265(4.33), 337(3.50),381(3.04)	48-1002-83
2,6-Naphthalenedisulfonic acid, 3,4-diamino-	MeOH	222(4.46),266(4.63), 330(3.58),385(3.41)	48-1002-83
2,7-Naphthalenedisulfonic acid, 1,4-diamino-	MeOH	218(4.20),262(4.27), 335(3.45),380(3.40)	48-1002-83
2,7-Naphthalenedisulfonic acid, 3,4-diamino-	MeOH	218(4.36),265(4.46), 334(3.55),384(3.20)	48-1002-83
$C_{10}H_{10}N_2O_7S_2$			
1,3-Naphthalenedisulfonic acid, 5,8-diamino-4-hydroxy-	MeOH	210(4.14),247(4.32), 265(3.87),326(3.45), 348(3.59)	48-1002-83
2,7-Naphthalenedisulfonic acid, 3,5-diamino-4-hydroxy-	MeOH	213(4.42),244(4.48), 278(4.26),320(3.47), 348(3.40)	48-1002-83

Compound	Solvent	λ max(log ε)	Ref.
$C_{10}H_{10}N_2O_9S_3$			
1,3,6-Naphthalenetrisulfonic acid, 7,8-diamino-	MeOH	222(4.00),260(3.74), 352(3.08)	48-1002-83
1,4,6-Naphthalenetrisulfonic acid, 5,8-diamino-	MeOH	228(4.45),258(4.53), 310(3.63),355(3.54)	48-1002-83
$C_{10}H_{10}N_2S$			
5H-Thiazolo[2,3-b]quinazoline, 2,3-dihydro-	MeOH	225(4.18),298(4.01)	83-0569-83
$C_{10}H_{10}N_2S_2$			
1,2,3-Thiadiazolium, 3-ethyl-4-mercapto-5-phenyl-, hydroxide, inner salt	EtOH	300(4.06),459(3.41)	94-1746-83
1,2,3-Thiadiazolium, 4-mercapto-5-methyl-3-(phenylmethyl)-, hydroxide, inner salt	EtOH	257(4.00),430(3.22)	94-1746-83
$C_{10}H_{10}N_4$			
1H-Imidazole, 1-methyl-2-(phenylazo)-	MeOH	232(3.84),360(4.28), 372(4.24),430s(3.28)	56-0971-83
	MeOH-HCl	230(3.73),357(4.28), 372s(4.22)	56-0971-83
	MeOH-NaOH	228(3.74),360(4.16), 368(4.15),420s(3.37)	56-0971-83
	20% MeOH	228(3.73),360(4.22), 368s(4.21),418s(3.30)	56-0971-83
	MeCN	239(4.18),250(4.13), 352s(4.26),361(4.30), 386?(4.15),412s(3.85)	56-0971-83
	1,2-$C_2H_4Cl_2$	232(3.88),352s(3.48), 363(4.34),374(4.32), 440s(3.32)	56-0971-83
1H-Pyrrole-2-carboxaldehyde, (1H-pyrrol-2-ylmethylene)hydrazone	EtOH	255(3.68),356(4.60)	34-0132-83
$C_{10}H_{10}N_4O$			
1H-Imidazole, 2-[(4-methoxyphenyl)-azo]-	MeOH	248(3.88),368(4.40), 380(4.41),406s(4.10), 440s(3.54)	56-0971-83
	MeOH-HCl	252(3.76),388(4.43), 394(4.44),422s(4.20)	56-0971-83
	MeOH-NaOH	216(3.82),240(3.82), 282(3.51),374(4.32), 385(4.35),412s(4.20)	56-0971-83
	20% MeOH	246(3.82),372(4.37), 379(4.38),410s(4.11)	56-0971-83
	MeCN	250(4.47),370(4.47), 382(4.48),410s(4.17), 450s(3.48)	56-0971-83
	1,2-$C_2H_4Cl_2$	250(3.95),372(3.44), 383(4.45),410s(4.10)	56-0971-83
1H-Imidazole-4-carboxamide, 5-amino-N-phenyl-	n.s.g.	247(3.98),285(4.23)	103-1235-83
2,4(1H,3H)-Pyrimidinedione, 4-(phenylhydrazone)	MeOH	238(4.04),277(3.85)	4-1037-83
$C_{10}H_{10}N_4OS$			
4(3H)-Pyrimidinone, 2,6-diamino-5-(phenylthio)-	pH 6.8	246(4.21),267(4.16)	87-0559-83

Compound	Solvent	λ_{max}(log ε)	Ref.
$C_{10}H_{10}N_4O_2$			
1,3,4(2H)-Isoquinolinetrione, 2-amino-1,2,3,4-tetrahydro-, 4-(methylhydrazone)	MeOH	224(4.29),261(4.14), 310(3.59),376(4.31)	142-1315-83
1H-Pyrazol-5-amine, 3-(nitromethyl)-1-phenyl-	EtOH	240(4.04),413(2.05)	48-0041-83
$C_{10}H_{10}N_4O_4$			
Benzenebutanoic acid, 4-azido-2-nitro-	MeOH	245(4.26)	64-0049-83C
$C_{10}H_{10}N_4O_4S$			
Benzenesulfonamide, 4-amino-N-(1,2,3,4-tetrahydro-2,4-dioxo-5-pyrimidinyl)-	pH 7	268(4.32)	161-0352-83
$C_{10}H_{10}N_6O_2$			
Pyrimido[5,4-e]-1,2,4-triazolo[4,3-b]-pyridazine-7,9(6H,8H)-dione, 3,6,8-trimethyl-	EtOH	233(4.57),325(3.85)	44-1628-83
$C_{10}H_{10}N_6O_3S$			
Benzenesulfonamide, N-(aminocarbonyl)-4-(1H-imidazol-2-ylazo)-	MeOH	230(4.03),370?(4.32), 376(4.32),440s(3.84)	56-0971-83
$C_{10}H_{10}O$			
Benzofuran, 2-ethenyl-2,3-dihydro-	EtOH	216(3.69),278(3.41), 286(3.33)	78-0169-83
1H-Inden-1-one, 2,3-dihydro-2-methyl-	MeOH	245(4.11),286(3.34), 293(3.41)	94-3113-83
1H-Inden-1-one, 2,3-dihydro-5-methyl-	MeOH	252(4.01),284(4.39), 293(3.44)	94-3113-83
1H-Inden-1-one, 2,3-dihydro-7-methyl-	MeOH	249(4.12),289(3.37), 297(3.40)	94-3113-83
$(C_{10}H_{10}O)_n$			
2-Propen-1-one, 2-methyl-1-phenyl-, polymer	dioxan	245(3.9),320(2.3)	151-0073-83C
$C_{10}H_{10}O_2$			
4H-1-Benzopyran-4-one, 2,3-dihydro-6-methyl-	MeOH	225(4.15),255(3.99), 325(3.55)	118-0310-83
4H-1-Benzopyran-4-one, 2,3-dihydro-8-methyl-	MeOH	257(4.08),330(3.71)	118-0310-83
Cyclopropanecarboxylic acid, 2-phenyl-, cis	n.s.g.	219(3.88),261(2.30)	78-2965-83
trans?	n.s.g.	222(4.04),267(2.58)	78-2965-83
Spiro[7-oxabicyclo[2.2.1]heptane-2,2'-oxirane], 3,5,6-tris(methylene)-	EtOH	220s(3.51),245(3.54)	88-3603-83
Tetracyclo[3.3.0.0^{2,8}.0^{4,6}]octane-3,7-dione, 1,5-dimethyl-	EtOH	313(1.82)	89-0661-83S
Tricyclo[6.2.0.0^{3,6}]deca-1,3(6),7-triene-2,7-diol	EtOH	221s(3.96),276s(3.06), 282(3.12)	88-1727-83
Tricyclo[6.2.0.0^{3,6}]dec-1(8)-ene-2,7-dione	C_6H_{12}	245s(4.05),250(4.10), 257s(3.96),389(1.88)	88-1727-83
$C_{10}H_{10}O_3$			
1H-2-Benzopyran-1-one, 3,4-dihydro-8-hydroxy-3-methyl- (mellein)	EtOH	246(3.81),314(3.60)	39-2185-83C
4H-1-Benzopyran-4-one, 2,3-dihydro-5-hydroxy-2-methyl-	EtOH	270(3.97),347(3.49)	39-2185-83C
2-Furanmethanol, α-2-furanyl-α-methyl-	EtOH	218(4.00)	103-0474-83

Compound	Solvent	λ_{max}(log ε)	Ref.
1(3H)-Isobenzofuranone, 3-ethyl-7-hydroxy- (isoochracein)	EtOH	214(4.26),234(3.85), 300(3.66)	39-2185-83C
$C_{10}H_{10}O_4$			
1(3H)-Isobenzofuranone, 6,7-dimethoxy- (meconine)	MeOH	307(3.66)	2-1257-83
1(3H)-Isobenzofuranone, 3-ethyl-4,7-dihydroxy-	EtOH	237(3.78),327(3.62)	39-2185-83C
2-Propenoic acid, 3-(3,4-dihydroxyphenyl)-, methyl ester	EtOH	219(3.96),237(3.83), 245(3.85),301s(3.96), 332(4.10)	102-1619-83
$C_{10}H_{10}O_6$			
1,3-Cyclohexadiene-1-carboxylic acid, 5-[(1-carboxyethenyl)oxy]-6-hydroxy- (isochorismic acid)	H_2O	280(4.09)	35-3346-83
1,5-Cyclohexadiene-1-carboxylic acid, 3-[(1-carboxyethenyl)oxy]-4-hydroxy-, hydrate (chorismic acid)	H_2O	275(3.42)	162-0315-83
barium salt	H_2O	272(3.43)(unstable)	162-0315-83
4-Epichorismic acid	pH 7	267(3.42)	35-6264-83
$C_{10}H_{10}Rh$			
Rhodocenium hexafluorophosphate	H_2O	232(4.79),275s(3.20)	101-0093-83K
$C_{10}H_{10}S_2$			
1,3-Dithiolane, 2-(2,4,6-cycloheptatrien-1-ylidene)-	C_6H_{12}	239(3.9),262(3.8), 325(3.8)	24-1154-83
	MeCN	245(4.2),255(4.0), 266(3.9),322(3.5)	24-1154-83
$C_{10}H_{11}$			
Azulenylium, 1,2,3,?-tetrahydro-, (OC-6-11)-, hexafluoroantimonate	MeCN	223(4.25),304(3.71)	44-0596-83
$C_{10}H_{11}BrO_4$			
Benzeneacetic acid, 5-bromo-2,3-dimethoxy-	EtOH	210(4.23),226s(--), 282(3.01)	104-0458-83
$C_{10}H_{11}Br_3O_3$			
Benzene, 1,2,4-tribromo-5,6-dimethoxy-3-(methoxymethyl)-	$CHCl_3$	283(4.00),293(3.94)	94-1754-83
$C_{10}H_{11}ClN_2OS$			
Benzamide, 2-chloro-N-[(ethylamino)-thioxomethyl]-	C_6H_{12}	236(4.15),285(3.89)	41-0603-83
Benzamide, 4-chloro-N-[(ethylamino)-thioxomethyl]-	C_6H_{12}	246(4.35)	41-0603-83
$C_{10}H_{11}ClN_2O_2$			
Acetamide, 2-chloro-N-[2-(hydroxyimino)-2-phenylethyl]-, (E)-	EtOH	246(4.0)	5-2038-83
$C_{10}H_{11}ClN_2O_3$			
Carbamic acid, [(phenylamino)carbonyl]-, 2-chloroethyl ester	EtOH	242(4.21)	22-0073-83
1,2-Diazabicyclo[5.2.0]nona-3,5-diene-2-carboxylic acid, 8-chloro-9-oxo-, ethyl ester, cis	CH_2Cl_2	281(4.03)	5-1361-83

Compound	Solvent	λ_{max}(log ε)	Ref.
$C_{10}H_{11}ClO_4$			
Benzeneacetic acid, 5-chloro-2,3-dimethoxy-	EtOH	208(4.38),226(3.80), 283(2.94)	104-0458-83
$C_{10}H_{11}FN_2O_8$			
β-D-Glucopyranuronic acid, 1-deoxy-1-(5-fluoro-3,4-dihydro-2,4-dioxo-1(2H)-pyrimidinyl)-	H_2O	266(3.90)	65-1483-83
β-D-Glucopyranuronic acid, 1-deoxy-1-(5-fluoro-3,6-dihydro-2,6-dioxo-1(2H)-pyrimidinyl)-	H_2O	272(3.81)	65-1483-83
$C_{10}H_{11}F_3N_2O_6$			
2,4(1H,3H)-Pyrimidinedione, 1-β-D-arabinofuranosyl-5-(trifluoromethyl)-	pH 1	263(4.09)	87-0598-83
	pH 12	263(3.93)	87-0598-83
	MeOH	263(4.11)	87-0598-83
$C_{10}H_{11}NO$			
Abikoviromycin	pH 1	236(3.99),341(4.05)	162-0001-83
	EtOH	218(3.83),244(3.99), 289(3.94)	162-0001-83
Acetamide, N-(1-phenylethenyl)-	C_6H_{12}	248(3.80)	44-0695-83
1(2H)-Isoquinolinone, 3,4-dihydro-3-methyl-, (-)-	EtOH	231(3.91),280s(3.0)	103-0528-83
2(1H)-Quinolinone, 3,4-dihydro-3-methyl-, (+)-	EtOH	250(4.19),285s(3.16)	103-0528-83
$(C_{10}H_{11}NO)_n$			
Propenamide, 2-methyl-N-phenyl-, polymer	MeOH	200(4.30),242(4.11)	116-1564-83
$C_{10}H_{11}NOS$			
Benzeneethanethioamide, N,N-dimethyl-α-oxo-	MeOH	255(4.31),401s(2.43)	5-1116-83
2-Propenethioamide, 3-(4-methoxyphenyl)-	EtOH	342(4.12),425(2.63)	80-0555-83
$C_{10}H_{11}NO_2$			
Benzonitrile, 2-methoxy-4,6-dimethyl-, N-oxide	MeOH	219(4.34),267(4.33)	39-1569-83B
Benzonitrile, 4-methoxy-2,6-dimethyl-, N-oxide	MeOH	222(4.34),260s(4.05), 267(4.06),305(3.60)	4-1609-83 +39-1569-83B
Gentianine, dihydro-	n.s.g.	270(3.4)	162-0627-83
3,5-Hexadiyn-2-one, 6-morpholino-	MeCN	214(4.31),326(4.04)	33-1631-83
$C_{10}H_{11}NO_2S$			
2-Thietanone, 3-[(2-furanylmethylene)amino]-4,4-dimethyl-, (R)-	EtOH	220(3.87),274(4.21)	39-2259-83C
$C_{10}H_{11}NO_2S_2$			
2H-1,4-Benzothiazine-3(4H)-thione, 6,7-dimethoxy-	MeOH	285(4.46)	87-0845-83
$C_{10}H_{11}NO_3$			
3-Buten-2-ol, 4-(4-nitrophenyl)-, (E)-	EtOH	222(4.04),308(4.14)	12-0527-83
2,4-Pentadiynoic acid, 5-morpholino-, methyl ester	MeCN	218(4.36),314(3.98)	33-1631-83
3H-Pyrrolizine-7-carboxylic acid, 1,2-dihydro-3-oxo-, ethyl ester	EtOH	223(4.30),290s(--)	5-0521-83
3H-Pyrrolizine-7-carboxylic acid, 5,6-dihydro-3-oxo-, ethyl ester	EtOH	272(3.95),350(3.90)	5-0521-83

Compound	Solvent	λ_{max}(log ε)	Ref.
$C_{10}H_{11}NO_4$			
Pyridinium, 2-methoxy-1-(methoxycarbonyl)-2-oxoethylide	H_2O	366(3.05)	65-2073-83
	EtOH	398(3.18)	65-2073-83
	85% EtOH	390(3.17)	65-2073-83
	$CHCl_3$	429(3.38)	65-2073-83
$C_{10}H_{11}NO_4S$			
Thieno[2,3-c]pyridin-3(2H)-one, 5-ethyl-4-hydroxy-7-methyl-, 1,1-dioxide	EtOH	225(4.15),348(3.68)	70-1320-83
	EtOH-KOH	230(4.08),264(3.76), 330(3.37),380(3.66)	70-1320-83
$C_{10}H_{11}NO_5$			
Benzoic acid, 2-hydroxy-4,6-dimethyl-3-nitro-, methyl ester	EtOH	244(3.94),313(3.51)	33-1475-83
$C_{10}H_{11}NS_2$			
Benzeneethanethioamide, N,N-dimethyl-α-thioxo-	MeOH	233(3.69),272(4.13), 322(4.08),409(2.39), 584(2.27),623s(2.13)	5-1116-83
	isooctane	537(--)	5-1116-83
$C_{10}H_{11}N_2$			
1(4H)-Quinoxalinyl, 2,3-dimethyl-	acid	370(4.04)	125-0655-83
$C_{10}H_{11}N_3$			
8-Cycloheptapyrazolamine, N,3-dimethyl-	MeOH	233(4.21),255(4.33), 309(3.87),366(3.58), 437(3.67)	142-0009-83
Propanedinitrile, [5-(dimethylamino)-2,4-pentadienylidene]-	$CHCl_3$	478(4.91)	70-0780-83
	EtOH	482(--)	70-0780-83
$C_{10}H_{11}N_3O$			
Formamide, N-[(2-methyl-1H-benzimidazol-1-yl)methyl]-	EtOH	245(4.85),250(4.83), 274(4.68),281(4.73)	94-0723-83
Methanamine, 1-[3-methoxy-5-(1H-pyrrol-2-yl)-2H-pyrrol-2-ylidene]- (tambjamine A)	MeOH	255(3.66),397(4.30)	44-2314-83
4H-Pyrido[1,2-a]pyrimidine-3-carbonitrile, 6,7,8,9-tetrahydro-6-methyl-4-oxo-	EtOH	231(3.67),307(3.90)	39-0369-83C
$C_{10}H_{11}N_3O_2$			
2-Pyrimidineacetonitrile, α-acetoxy-4,6-dimethyl-	MeCN	244(3.59)	44-0575-83
$C_{10}H_{11}N_3O_6$			
Acetamide, N-[2-(2,6-dinitrophenoxy)-ethyl]-	DMSO	305(3.29)	18-2173-83
Acetamide, N-(2,6-dinitrophenyl)-N-(2-hydroxyethyl)-	DMSO	320(3.62)	18-2173-83
Ethanol, 2-[(2,6-dinitrophenyl)amino]-, acetate	DMSO	428(3.78)	18-2173-83
$C_{10}H_{11}N_4S$			
5H-Pyrimido[4,5-d]thiazolo[3,2-a]pyrimidinium, 2,3-dimethyl-, perchlorate	pH 6.86	202(4.32),210s(--), 230(3.95),283(3.34), 382(4.35)	44-2476-83
	3M HCl	220(3.94),241(3.76), 337(4.16)	44-2476-83

Compound	Solvent	λ_{max}(log ε)	Ref.
$C_{10}H_{11}N_5$			
Benzenamine, 4-[(1-methyl-1H-imidazol-2-yl)azo]-, (E)-	MeOH	250(3.94),310(3.59), 408(4.05),440(4.02)	56-0971-83
$C_{10}H_{11}N_5O$			
1H-Imidazole-4-carboxamide, 5-hydrazino-N-phenyl-, monohydrochloride	n.s.g.	271(4.41)	103-1235-83
1,3,5-Triazin-2(1H)-one, 4,6-diamino-1-(4-methylphenyl)-, hydrochloride	M HCl	233(4.31)	94-2473-83
	H_2O	232(4.10)	94-2473-83
$C_{10}H_{11}N_5O_2$			
1,3,5-Triazin-2(1H)-one, 4,6-diamino-1-(4-methoxyphenyl)-, hydrochloride	M HCl	232(4.41)	94-2473-83
	H_2O	230(4.39)	94-2473-83
$C_{10}H_{11}N_5O_3S$			
Benzenesulfonamide, 4-amino-N-(6-amino-1,4-dihydro-4-oxo-5-pyrimidinyl)-	pH 7	264(4.31)	161-0352-83
$C_{10}H_{11}N_5O_4S$			
Benzenesulfonamide, 4-amino-N-(6-amino-1,2,3,4-tetrahydro-2,4-dioxo-5-pyrimidinyl)-	pH 7	267(4.44)	161-0352-83
$C_{10}H_{11}N_5O_7$			
6H-Purin-6-one, 1,9-dihydro-9-(2-O-nitro-β-D-arabinofuranosyl)-	pH 1	248(4.02)	88-3183-83
	pH 13	252(4.08)	88-3183-83
$C_{10}H_{11}N_7O_2S$			
Benzenesulfonamide, N-(aminoiminomethyl)-4-(1H-imidazol-2-ylazo)-	MeOH	234(3.94),370(4.43), 374(4.42),440s(3.33)	56-0971-83
$C_{10}H_{11}S_2$			
4H-1,3-Dithiin-1-ium, 5,6-dihydro-2-phenyl-, tetrafluoroborate	CH_2Cl_2	236(3.66),266(3.41), 335(4.04)	1-0687-83
1,3-Dithiol-1-ium, 4,5-dihydro-2-(4-methylphenyl)-, tetrafluoroborate	CH_2Cl_2	235(3.68),267(3.42), 373(4.32)	1-0687-83
1,3-Dithiol-1-ium, 4,5-dihydro-2-(phenylmethyl)-, tetrafluoroborate	CH_2Cl_2	307(4.29)	1-0687-83
$C_{10}H_{12}BrNO_4$			
Benzeneacetamide, 5-bromo-N-hydroxy-2,3-dimethoxy-	EtOH	208(4.54),225s(--), 284(3.25)	104-0458-83
$C_{10}H_{12}ClNO_2$			
2-Pyridinecarboxylic acid, 5-butyl-4-chloro-	MeOH	269(3.54),276(3.48)	106-0591-83
$C_{10}H_{12}ClNO_4$			
Benzeneacetamide, 5-chloro-N-hydroxy-2,3-dimethoxy-	EtOH	209(4.30),225s(--), 283(3.26)	104-0458-83
$C_{10}H_{12}Cl_2N_2O$			
2(1H)-Pyrazinone, 3,5-dichloro-1-cyclohexyl-	EtOH	342(3.92)	4-0919-83
$C_{10}H_{12}FN_3O_7$			
β-D-Glucopyranuronamide, 1-deoxy-1-(5-fluoro-3,4-dihydro-2,4-dioxo-1(2H)-pyrimidinyl)-	H_2O	265(3.85)	65-1483-83
β-D-Glucopyranuronamide, 1-deoxy-1-(5-fluoro-3,6-dihydro-2,6-dioxo-	H_2O	272(3.81)	65-1483-83

Compound	Solvent	$\lambda_{max}(\log \epsilon)$	Ref.
1(2H)-pyrimidinyl)- (cont.)			65-1483-83
$C_{10}H_{12}F_3N$			
Benzeneethanamine, α-methyl-4-(tri-fluoromethyl)-, (S)-	C_6H_{12}	258(2.49),264(2.42), 269(2.20)	35-1578-83
	MeOH	258(2.52),263(2.48), 269(2.32)	35-1578-83
hydrochloride	MeOH	258(2.54),263(2.60), 269(2.49)	35-1578-83
$C_{10}H_{12}F_3N_3O_4$			
Thymidine, 3'-amino-3'-deoxy-α,α,α-trifluoro-	pH 2	261(3.98)	87-1691-83
	pH 12	260(3.82)	87-1691-83
	MeOH	262(3.97)	87-1691-83
$C_{10}H_{12}NO_2$			
Pyridinium, 3-(methoxycarbonyl)-1-(2-propenyl)-, perchlorate	MeOH	266(3.61)	35-1204-83
$C_{10}H_{12}N_2$			
1H-Indole-3-methanamine, N-methyl-	MeOH	279(3.75),287(3.68)	64-0108-83B
Pyrimidine, 1,4,5,6-tetrahydro-2-phenyl-	EtOH	203(4.38),230(4.16), 275(3.26)	39-1223-83C
$C_{10}H_{12}N_2O$			
Ethanone, 1-(1,2,3,4-tetrahydro-5-quinoxalinyl)-	MeOH	215(4.34),262(4.06), 419(3.72)	18-3358-83
$C_{10}H_{12}N_2OS$			
Acetamide, N-[4-(2-amino-2-thioxo-ethyl)phenyl]-	EtOH	267(4.05),346(4.20)	80-0555-83
Benzamide, N-[(ethylamino)thioxo-methyl]-	C_6H_{12}	239(4.33),284(3.93)	41-0603-83
$C_{10}H_{12}N_2OS_2$			
4-Thiazolidinone, 5-[5-(dimethylami-no)-2,4-pentadienylidene]-2-thioxo-	EtOH	418(3.62),540(4.75)	70-0780-83
$C_{10}H_{12}N_2O_2$			
Benzenecarboximidamide, N-acetoxy-3-methyl-	EtOH	245(3.93)	32-0845-83
Benzenecarboximidamide, N-acetoxy-4-methyl-	EtOH	225(4.09),246(4.01)	32-0845-83
Benzenemethanamine, N-(1-methyl-2-nitroethenyl)-	EtOH	236(3.56),348(4.30)	40-0088-83
1,3-Butanedione, 1-phenyl-, dioxime	EtOH	213(4.1),250(4.0)	56-0129-83
Pyrido[1,2-a]pyrimidine-5-carboxalde-hyde, 1,6,7,8-tetrahydro-6-methyl-4-oxo-	EtOH	224(4.02),250s(3.44), 340(4.31)	39-0369-83C
$C_{10}H_{12}N_2O_2S$			
Benzenesulfonamide, 2-(2-butenylidene-amino)-, (E,E)-	EtOH	210(4.47),252(4.15), 317(3.58)	104-1402-83
(Z,E)-	EtOH	209(3.96),251(3.51), 315(3.00)	104-1402-83
$C_{10}H_{12}N_2O_3$			
1,3-Butanedione, 1-(2-hydroxyphenyl)-, dioxime	MeOH	219(4.16),259(3.92), 307(3.54)	56-0129-83
Carbamic acid, [(phenylamino)carbo-nyl]-, ethyl ester	MeOH at 50°	241(4.18)	22-0073-83
	EtOH	242(4.21)	22-0073-83

Compound	Solvent	λ_{max}(log ε)	Ref.
4H-Pyrido[1,2-a]pyrimidine-3-carboxylic acid, 6,7,8,9-tetrahydro-6-methyl-4-oxo-	EtOH	230(3.74),300(3.90)	39-0369-83C
	EtOH	207(3.74),230(3.74), 303(3.92)	39-1413-83B
$C_{10}H_{12}N_2O_4$			
Benzenebutanoic acid, 4-amino-2-nitro-	MeOH	270(4.15)	64-0049-83C
Glycine, N-[(2-nitrophenyl)methyl]-, methyl ester, monohydrochloride	MeOH	257(3.7)	5-0231-83
2-Pyridinecarboxylic acid, 5-butyl-4-nitro-	MeOH	289(3.43)	106-0591-83
$C_{10}H_{12}N_2O_7S$			
Morpholine, 4-[[2-(5-nitro-2-furanyl)-2-oxoethyl]sulfonyl]-	MeOH	225(4.0),305(4.13)	73-1062-83
$C_{10}H_{12}N_2SSe$			
6-Selena-6a-thia-1,2-diazapentalene, 3,4-dimethyl-1-phenyl-	C_6H_{12}	204(4.27),254(4.43), 281s(4.12),514(4.08)	39-0777-83C
$C_{10}H_{12}N_4$			
1,4-Phthalazinediamine, N,N'-dimethyl-	MeOH	204(4.54),222(4.40), 344(3.70)	103-0666-83
	MeCN	222(4.36),348(3.72)	103-0666-83
$C_{10}H_{12}N_4O$			
4H-Pyrazolo[3,4-d]pyrimidin-4-one, 1,5-dihydro-1,6-dimethyl-5-(2-propenyl)-	EtOH	212(4.45),255.5(3.86)	11-0171-83B
$C_{10}H_{12}N_4O_4$			
5-Pyrimidinepentanoic acid, 2-(cyanoamino)-1,4-dihydro-6-hydroxy-4-oxo-	50% MeOH	222(4.178),274(4.229)	73-0304-83
	+ HCl	221(4.003),254(4.228)	73-0304-83
	+ NaOH	226(4.213),278(4.186)	73-0304-83
$C_{10}H_{12}N_4O_4S$			
6H-Purine-6-thione, 9-β-D-arabinofuranosyl-1,9-dihydro-	MeOH	323(4.4)	36-0372-83
$C_{10}H_{12}N_5O_6P$			
Adenosine cyclic 3',5'-(hydrogen phosphate)	pH 2	256(4.16)	162-0387-83
	pH 7	258(4.17)	162-0387-83
$C_{10}H_{12}N_5O_7P$			
Guanosine cyclic 3',5'-(hydrogen phosphate), calcium salt	pH 1	256.5(4.06)	162-0388-83
	pH 7	254(4.11)	162-0388-83
	pH 12	262(4.09)	162-0388-83
$C_{10}H_{12}N_6O_2S$			
Benzenesulfonamide, 4-amino-N-(2,4-diamino-5-pyrimidinyl)-	pH 7	267(4.42)	161-0352-83
Benzenesulfonamide, 4-amino-N-(4,6-diamino-5-pyrimidinyl)-	pH 7	267(4.35)	161-0352-83
$C_{10}H_{12}N_6O_3S$			
Benzenesulfonamide, 4-amino-N-(2,6-diamino-1,4-dihydro-4-oxo-5-pyrimidinyl)-	pH 7	269(4.34)	161-0352-83
$C_{10}H_{12}N_6O_4$			
Urea, N-(1,2,3,4,7,8-hexahydro-3,8-dimethyl-2,4,7-trioxo-6-pteridinyl)-N-	neutral	292(4.07),340(4.07)	35-3304-83
	anion	305(4.14),361(4.11)	35-3304-83

Compound	Solvent	λmax(log ε)	Ref.
methyl- (cont.)			35-3304-83
$C_{10}H_{12}N_6O_6$			
9H-Purin-6-amine, 9-(2-O-nitro-β-D-arabinofuranosyl)-	pH 1	256(4.16)	88-3183-83
	pH 13	258(4.15)	88-3183-83
$C_{10}H_{12}N_8O_4$			
Acetic acid, azido-, 2-[(2-amino-1,6-dihydro-6-oxo-9H-purin-9-yl)methoxy]ethyl ester	pH 7	253(3.93),270s(--)	87-0602-83
$C_{10}H_{12}O$			
1(2H)-Azulenone, 3,4,5,8-tetrahydro-	MeOH	236(3.98)	23-1226-83
Benzene, 1-methoxy-4-(1-propenyl)-	EtOH	253.5(4.27)	162-0094-83
Spiro[bicyclo[2.2.1]heptan-2,2'-oxiran], 5,6-bis(methylene)-, endo	isooctane	237s(3.94),247(3.98), 255s(3.78)	88-3603-83
exo	isooctane	237s(3.92),244(3.95), 253s(3.77)	88-3603-83
$C_{10}H_{12}O_2$			
Acetic acid, (2-cyclopenten-1-ylidene)-, 2-propenyl ester	EtOH	270(4.22)	33-1876-83
3-Benzofuranol, 2,3-dihydro-3,7-dimethyl-	EtOH	215(3.79),277(3.30)	39-1649-83C
Cyclopropa[cd]pentalene-2,3(1H,2aH)-dione, tetrahydro-4a,4b-dimethyl-	EtOH	291(2.01)	89-0661-83S
Ethanone, 1-(2-methoxy-6-methylphenyl)-	MeOH	220(4.85),242(3.63), 280(3.56)	100-0235-83
1H-Indene-1,5(6H)-dione, 2,3,7,7a-tetrahydro-7a-methyl-	EtOH	240(4.05)	12-0117-83
3-Oxatetracyclo[3.3.3.0^{2,4}.0^{6,8}]undecan-9-one, (1α,2α,4α,5α,6β,8β)-	benzene	285(1.62)	44-1718-83
$C_{10}H_{12}O_3$			
4H-1,3-Benzodioxin-6-ol, 2,4-dimethyl-	heptane	220(3.65),227(3.60), 295(3.47)	103-0954-83
3-Benzofuranol, 2,3-dihydro-5-methoxy-3-methyl-	EtOH	227(3.71),293(3.43)	39-1649-83C
$C_{10}H_{12}O_4$			
2,5-Cyclohexadiene-1,4-dione, 2,3-dimethoxy-5,6-dimethyl- (aurantiogliocladin)	n.s.g.	275(4.58),407(3.08)	162-0126-83
$C_{10}H_{12}O_4S_2$			
1,3-Dithiole-4,5-dicarboxylic acid, 2-(1-methylethylidene)-, dimethyl ester	isoPrOH	246(4.15),340(3.04), 410(3.36)	104-1925-83
$C_{10}H_{12}O_5$			
Benzoic acid, 3,4,5-trimethoxy-	EtOH	256(3.83)	151-0171-83C
$C_{10}H_{12}S$			
Benzene, [1-[(methylthio)methyl]ethenyl]-	MeOH	239(3.85)	44-0835-83
Thiophene, 3-(1-cyclohexen-1-yl)-	MeCN	221(3.74),240(3.91), 244(3.90)	4-0729-83
$C_{10}H_{12}S_4$			
Tetramethyltetrathiafulvalene bromide	MeCN	400(--),420s(--), 462(4.16),530(3.52),	104-1092-83

Compound	Solvent	$\lambda_{max}(\log \epsilon)$	Ref.
(cont.)		650(3.81)	104-1092-83
Tetramethyltetrathiafulvalene chloride	MeCN	400s(--),420s(--), 462(4.16),530(3.53), 658(3.82)	104-1092-83
perchlorate	MeCN	400s(--),420s(--), 462(4.16),530(3.53), 658(3.82)	104-1092-83
$C_{10}H_{13}BrN_2O_4$			
2,4(1H,3H)-Pyrimidinedione, 5-bromo-1-[2-hydroxy-4-(hydroxymethyl)cyclopentyl]-, (1α,2β,4α)-(±)-	pH 1	211(3.99),284(4.00)	87-0156-83
	pH 7	210(3.99),284(4.00)	87-0156-83
	pH 13	280(3.88)	87-0156-83
2,4(1H,3H)-Pyrimidinedione, 5-bromo-1-[3-hydroxy-4-(hydroxymethyl)cyclopentyl]-, (1α,3β,4α)-(±)-	pH 1	212(3.97),284(4.00)	87-0156-83
	pH 7	212(3.98),284(4.00)	87-0156-83
	pH 13	280(3.87)	87-0156-83
$C_{10}H_{13}BrN_2O_5$			
2,4(1H,3H)-Pyrimidinedione, 5-bromo-1-[2,3-dihydroxy-4-(hydroxymethyl)-cyclopentyl]-, (1α,2β,3β,4α)-(±)-	pH 1	212(3.98),284(4.01)	87-0156-83
	pH 7	211(3.98),283(4.00)	87-0156-83
	pH 13	280(3.89)	87-0156-83
$C_{10}H_{13}BrO_3$			
2(5H)-Furanone, 5-(2-bromoethylidene)-4-(1,1-dimethylethoxy)-	EtOH	267(4.28)	39-2983-83C
$C_{10}H_{13}ClN_2O_3S$			
Benzenesulfonamide, 4-chloro-N-[(propylamino)carbonyl]-	pH 2	232.5(4.22)	162-0309-83
$C_{10}H_{13}ClN_2O_5$			
3(2H)-Pyridazinone, 6-chloro-2-(2-deoxy-β-D-erythro-pentofuranosyl)-5-methoxy-	pH 1	250(3.89),280(3.65)	4-0369-83
	pH 11	250(3.89),280(3.66)	4-0369-83
	MeOH	250(3.84),286(3.68)	4-0369-83
$C_{10}H_{13}ClN_2O_6$			
3(2H)-Pyridazinone, 6-chloro-5-methoxy-2-β-D-ribofuranosyl-	pH 1	250(3.69),280s(3.44)	4-0369-83
	pH 11	250(3.72),277s(3.58)	4-0369-83
	MeOH	251(3.59),287(3.41)	4-0369-83
$C_{10}H_{13}ClN_6O_6S$			
Adenosine, 2-chloro-, 5'-sulfamate	H_2O	264(4.15)	158-83-106
	0.05M HCl	264(4.12)	158-83-106
	0.05M NaOH	263.5(4.15)	158-83-106
$C_{10}H_{13}ClS$			
Benzene, 1-chloro-4-[(1,1-dimethylethyl)thio]-	decane	229(4.23),273(3.51)	65-0449-83
$C_{10}H_{13}DO_2$			
Geranial-1-d, 9-oxo-	EtOH	237(4.38)	23-1053-83
Geranial-1-d, 10-oxo-	EtOH	234(4.42)	23-1053-83
Neral-1-d, 9-oxo-	EtOH	236(4.35)	23-1053-83
Neral-1-d, 10-oxo-	EtOH	229(4.32),235s(4.29)	23-1053-83
$C_{10}H_{13}FS$			
Benzene, 1-[(1,1-dimethylethyl)thio]-4-fluoro-	decane	230(4.03),262(3.18)	65-0449-83
$C_{10}H_{13}F_3O$			
3-Decyn-2-one, 1,1,1-trifluoro-	C_6H_{12}	230(3.93),237s(3.88), 312(1.26)	44-1925-83

Compound	Solvent	λ_{max}(log ε)	Ref.
$C_{10}H_{13}IN_2O_4$			
2,4(1H,3H)-Pyrimidinedione, 1-[2-hydroxy-4-(hydroxymethyl)cyclopentyl]-5-iodo-, (1α,2β,4α)-(±)-	pH 1	217(4.04),293(3.94)	87-0156-83
	pH 7	217(4.04),294(3.94)	87-0156-83
	pH 13	284(3.81)	87-0156-83
2,4(1H,3H)-Pyrimidinedione, 1-[3-hydroxy-4-(hydroxymethyl)cyclopentyl]-5-iodo-, (1α,3β,4α)-(±)-	pH 1	217(4.04),292(3.95)	87-0156-83
	pH 7	217(4.04),293(3.94)	87-0156-83
	pH 13	283(3.81)	87-0156-83
$C_{10}H_{13}IN_2O_5$			
2,4(1H,3H)-Pyrimidinedione, 1-[2,3-dihydroxy-4-(hydroxymethyl)cyclopentyl]-5-iodo-, (1α,2β,3β,4α)-(±)-	pH 1	217(4.03),293(3.94)	87-0156-83
	pH 7	217(4.03),292(3.93)	87-0156-83
	pH 13	284(3.80)	87-0156-83
$C_{10}H_{13}N$			
2,4,6-Octatrienenitrile, 3,7-dimethyl-	EtOH	315(4.42)	35-4033-83
$C_{10}H_{13}NO$			
2,4,6-Cycloheptatriene-1-carboxamide, N,N-dimethyl-, lithium salt	1:1 THF-hexane at -20°	656(3.60)	24-1777-83
	2:1 THF-hexane at -50°	656(3.7),690(3.7)	24-1777-83
	at 25°	655(3.6)	24-1777-83
$C_{10}H_{13}NO_2$			
1,3(2H,5H)-Isoquinolinedione, 6,7,8,8a-tetrahydro-8a-methyl-	EtOH	233(4.12)	2-1103-83
2,4-Pentadiynoic acid, 5-(diethylamino)-, methyl ester	MeCN	213(4.58),318(4.05)	33-1631-83
$C_{10}H_{13}NO_3$			
Acetic acid, cyano(tetrahydro-2H-pyran-2-ylidene)-, ethyl ester	n.s.g.	263(4.26)	103-1050-83
2-Pyridinecarboxylic acid, 5-butoxy-	MeOH	244(4.02),281(3.74)	106-0591-83
3H-Pyrrolizine-7-carboxylic acid, 1,2,5,6-tetrahydro-3-oxo-, ethyl ester	EtOH	280(4.34)	5-0521-83
$C_{10}H_{13}NO_4$			
3H-Azepine-5-carboxylic acid, 2,4-dimethoxy-, methyl ester	MeOH	222(4.36),267(4.02)	44-5041-83
$C_{10}H_{13}N_3O$			
5(6H)-Quinazolinone, 2-amino-7,8-dihydro-7,7-dimethyl-	EtOH	276(4.29)	4-0649-83
$C_{10}H_{13}N_3O_2$			
1,2-Diazabicyclo[5.2.0]nona-3,5-diene-2-carboxamide, 5,8-dimethyl-9-oxo-, cis	MeOH	276(3.79)	5-1393-83
4,7-Methano-3H-indazole, 3a,4,7,7a-tetrahydro-3,3-dimethyl-7a-nitro-, (3aα,4β,7β,7aα)-	MeCN	247(3.08),335(2.30)	78-1247-83
Pyridine, 3-nitro-4-piperidino-	H_2O	250(4.27),390(3.48)	103-0303-83
4H-Pyrido[1,2-a]pyrimidine-3-carboxamide, 1,6,7,8-tetrahydro-1-methyl-4-oxo-, (R)-	M HCl	207(3.82),232(3.88), 280(3.83)	39-1413-83B
	EtOH	206(3.82),231(3.83), 303(3.94)	39-1413-83B
1H-Pyrrole-1-butanoic acid, α-diazo-, ethyl ester	MeOH	218(3.92),262(4.00)	23-0454-83

Compound	Solvent	λ_{max}(log ε)	Ref.
$C_{10}H_{13}N_3O_3$			
Isoxazolo[4,5-c]pyridin-4(5H)-one, 5-(dimethylamino)-6-methoxy-3-methyl-	MeOH	202(4.18),256(3.91), 287(4.04)	142-1315-83
$C_{10}H_{13}N_3O_5$			
D-Alanine, N-[(1,2,3,4-tetrahydro-6-methyl-2,4-dioxo-5-pyrimidinyl)-carbonyl]-, methyl ester	H_2O	268.5(3.98)	94-0135-83
L-	H_2O	268.5(4.03)	94-0135-83
$C_{10}H_{13}N_3O_5S$			
D-Cysteine, S-methyl-N-[(1,2,3,4-tetrahydro-6-methyl-2,4-dioxo-5-pyrimidinyl)carbonyl]-	H_2O	271.5(4.02)	94-0135-83
L-	H_2O	270.5(4.07)	94-0135-83
$C_{10}H_{13}N_3O_6$			
D-Serine, N-[(1,2,3,4-tetrahydro-6-methyl-2,4-dioxo-5-pyrimidinyl)-carbonyl]-, methyl ester	H_2O	271(4.04)	94-0135-83
L-	H_2O	270.5(4.03)	94-0135-83
$C_{10}H_{13}N_3S$			
Thiourea, [(dimethylamino)methylene]-phenyl-	MeCN	297(4.47)	97-0019-83
$C_{10}H_{13}N_3Si$			
Silacyclopentane, 1-azido-1-phenyl-	C_6H_{12}	264(2.60)	152-0645-83
$C_{10}H_{13}N_5O$			
9H-Purin-6-amine, 9-[2-(2-propenyl-oxy)ethyl]-	pH 1	262(4.04)	111-0555-83
	pH 13	263(4.01)	111-0555-83
$C_{10}H_{13}N_5O_3$			
Adenosine, 3'-deoxy-	MeOH	260(4.16)	102-2509-83
	EtOH	259(4.14)	118-0304-83
	EtOH	260(4.16)	162-0360-83
$C_{10}H_{13}N_5O_4$			
Adenosine	n.s.g.	260(4.18)	162-0023-83
$C_{10}H_{13}N_5O_4S$			
6H-Purine-6-thione, 2-amino-1,9-dihydro-9-β-D-arabinofuranosyl-	MeOH	335(4.36),344(4.42)	36-0372-83
$C_{10}H_{13}N_5O_5$			
Guanosine	pH 5.5	188.3(4.43),252.5(4.14)	162-0657-83
$C_{10}H_{13}N_7O_2S$			
Benzenesulfonamide, 4-amino-N-(2,4,6-triamino-5-pyrimidinyl)-	pH 7	267(4.41)	161-0352-83
$C_{10}H_{13}O$			
Ethylium, 1-(4-ethylphenyl)-1-hydroxy-	H_2SO_4	313(4.42)	99-0612-83
$C_{10}H_{14}$			
Bicyclo[4.1.0]hept-2-ene, 7,7-dimethyl-4-methylene-, (-)-	EtOH	232(3.72),254(4.07)	104-0582-83

Compound	Solvent	λ_{max}(log ϵ)	Ref.
$C_{10}H_{14}Br_2O$			
Bicyclo[2.2.1]heptan-2-one, 3,3-di-bromo-1,7,7-trimethyl-, (1R)-	C_6H_{12}	323(1.88)	162-0437-83
$C_{10}H_{14}ClN_3O_5$			
3(2H)-Pyridazinone, 6-chloro-5-(meth-ylamino)-2-β-D-ribofuranosyl-	pH 1	232(3.67),239s(3.58), 301(3.86),314s(3.70)	4-0369-83
	pH 11	232(3.72),239s(3.65), 290s(3.81),301(3.88), 314s(3.72)	4-0369-83
	MeOH	232(3.69),239s(3.58), 290s(3.79),301(3.85), 314s(3.67)	4-0369-83
$C_{10}H_{14}D_2O_2$			
Geraniol-1,1-d_2, 9-oxo-	EtOH	237(4.09)	23-1053-83
Geraniol-1,1-d_2, 10-oxo-	EtOH	229(4.20)	23-1053-83
Nerol-1,1-d_2, 9-oxo-	EtOH	237(4.08)	23-1053-83
Nerol-1,1-d_2, 10-oxo-	EtOH	229(4.29)	23-1053-83
$C_{10}H_{14}D_2O_3$			
2,6-Octadienal-8,8-d_2, 8-hydroxy-2-(hydroxymethyl)-6-methyl- (isomer mixture)	EtOH	234(4.01)	23-1053-83
$C_{10}H_{14}N$			
Pyridinium, 1-(3-methyl-2-butenyl)-, perchlorate	MeOH	259(3.63)	35-1204-83
$C_{10}H_{14}N_2O$			
Cinnolinium, 5,6,7,8-tetrahydro-4-hy-droxy-2,3-dimethyl-, hydroxide, inner salt	EtOH	265(3.49),313(3.79)	142-0061-83
4(1H)-Cinnolinone, 5,6,7,8-tetra-hydro-1,3-dimethyl-	EtOH	280(4.02)	142-0061-83
4H-Pyrido[1,2-a]pyrimidin-4-one, 1,6,7,8-tetrahydro-1,6-dimethyl-	MeCN	344(3.19)	39-1413-83B
4H-Pyrido[1,2-a]pyrimidin-4-one, 6,7,8,9-tetrahydro-3,6-dimethyl-	EtOH	231(3.93),278(3.98)	39-0369-83C
$C_{10}H_{14}N_2O_2$			
Benzenamine, N,N,4,5-tetramethyl-2-nitro- (cation)	65.8% H_2SO_4	291(3.89)	39-1191-83B
2-Pyridinecarboxylic acid, 4-amino-5-butyl-	MeOH	223(4.40),274(4.04)	106-0591-83
$C_{10}H_{14}N_2O_2S$			
3-Thietanamine, N,N-dimethyl-3-(2-pyridinyl)-, 1,1-dioxide, di-hydrochloride	EtOH	254(3.80),258(3.81)	103-0518-83
3-Thietanamine, N,N-dimethyl-3-(3-pyridinyl)-, 1,1-dioxide, di-hydrochloride	EtOH	255(3.50),260(3.51)	103-0518-83
3-Thietanamine, N,N-dimethyl-3-(4-pyridinyl)-, 1,1-dioxide, di-hydrochloride	EtOH	259(3.64)	103-0518-83
$C_{10}H_{14}N_2O_3$			
Piperidine, 1-[(5-nitro-2-furanyl)-methyl]-	MeOH	213(3.91),229(3.72), 311(4.15)	73-2682-83
2-Pyridinecarboxylic acid, 5-butyl-	MeOH	226(4.35),282(3.99)	106-0591-83

Compound	Solvent	λ_{max}(log ϵ)	Ref.
4-(hydroxyamino)- (cont.)			106-0591-83
$C_{10}H_{14}N_2O_3S$			
Acetic acid, [(1,4-dihydro-5-methyl-4-oxo-2-pyrimidinyl)thio]-, propyl ester	MeOH	288(3.98)	56-1027-83
Sulfoxonium, dimethyl-, 1-(1,6-dihydro-1-methyl-6-oxo-2-pyrimidinyl)-2-oxopropylide	EtOH	250(3.83),325s(--)	39-2645-83C
$C_{10}H_{14}N_2O_4$			
Benzenepropanoic acid, α-hydrazino-3,4-dihydroxy-α-methyl-, (S)-	MeOH	282.5(3.47)	162-0249-83
$C_{10}H_{14}N_2O_4S$			
1H-Imidazole-1-acetic acid, 5-(ethoxycarbonyl)-4-mercapto-, ethyl ester	EtOH	270(3.74),322(3.87)	2-0030-83
5-Pyrimidinepentanoic acid, dihydro-5-methyl-4,6-dioxo-2-thioxo-	50%MeOH-HCl	238(4.20),286(3.90)	73-0137-83
	+ NaOH	253(3.97),305(4.02)	73-0137-83
$C_{10}H_{14}N_2O_5$			
1(2H)-Pyrimidinebutanoic acid, 3,4-dihydro-β-hydroxy-2,4-dioxo-, ethyl ester	EtOH	265(3.95)	128-0125-83
5-Pyrimidinepentanoic acid, hexahydro-5-methyl-2,4,6-trioxo-	50%MeOH-HCl	213.5(3.57),237s(3.31)	73-0137-83
	+ NaOH	262.5(4.00)	73-0137-83
$C_{10}H_{14}N_2O_5S$			
Pseudouridine, 1'-methyl-2-thio-, (±)-	pH 1	215(4.04),275(4.16), 290s(4.13)	18-2680-83
	pH 13	221(4.10),262(4.09), 295s(3.82)	18-2680-83
	MeOH	215(4.04),276(4.16), 290(4.12)	18-2680-83
Pseudouridine, 2'-methyl-2-thio-, (±)-	pH 1	215(4.01),275(4.04)	18-2680-83
	pH 13	220(4.24),263(4.16)	18-2680-83
	MeOH	215(3.98),276(4.06)	18-2680-83
Pseudouridine, 4'-methyl-2-thio-, (±)-	pH 1	214(4.12),283(4.19)	18-2680-83
	pH 13	223(4.10),264(4.05), 286(3.93)	18-2680-83
	MeOH	215(4.04),276(4.19), 291(4.15)	18-2680-83
Pseudouridine, 5'-methyl-2-thio-, (±)-	pH 1	213(3.78),280(3.67), 295(3.75)	18-2680-83
	pH 13	221(4.06),264(4.01), 285(3.89)	18-2680-83
	MeOH	215(3.82),277(3.92), 291(3.88)	18-2680-83
$C_{10}H_{14}N_2O_6$			
Pseudouridine, 1'-methyl-, (±)-	pH 1	263(--)	18-2680-83
	pH 13	265s(--),289(--)	18-2680-83
	MeOH	263(--)	18-2680-83
Pseudouridine, 2'-methyl-, (±)-	pH 1	263(3.94)	18-2680-83
	pH 13	285(3.90)	18-2680-83
	MeOH	264(3.81)	18-2680-83
Pseudouridine, 4'-methyl-, (±)-	pH 1	264(3.88)	18-2680-83
	pH 13	286(3.88)	18-2680-83
	MeOH	264(3.89)	18-2680-83
Pseudouridine, 5'-methyl-, (±)-	pH 1	263(3.91)	18-2680-83
	pH 13	287(3.93)	18-2680-83

Compound	Solvent	λ_{max}(log ε)	Ref.
Pseudouridine, 5'-methyl-, (±)-	MeOH	264(3.75)	18-2680-83
3(2H)-Pyridazinone, 5-methoxy-2-β-D-ribofuranosyl-	pH 1	247(3.75),270s(3.57)	4-0369-83
	pH 11	247(3.71),270(3.53)	4-0369-83
	MeOH	247(3.67),278(3.51)	4-0369-83
Uracil, 1-(3-C-methyl-β-D-ribofuranosyl)-	pH 1-7	262(4.02)	136-0075-83M
	pH 13	262(3.89)	136-0075-83M
Uracil, 3-(3-C-methyl-β-D-ribofuranosyl)-	pH 1-7	264(3.92)	136-0075-83M
	pH 13	292(4.06)	136-0075-83M
Uracil, 5-[(β-D-ribofuranosyl)methyl]-	M HCl	266(3.85)	18-2700-83
	M NaOH	285(3.85)	18-2700-83
	MeOH	265(3.85)	18-2700-83
$C_{10}H_{14}N_2O_6S$			
Pseudouridine, 1'-(hydroxymethyl)-2-thio-, (±)-	pH 1	213(4.14),274(4.13), 297s(4.05)	18-2680-83
	pH 13	265(4.10),284(4.02)	18-2680-83
	MeOH	213(4.09),275(4.10), 299s(4.01)	18-2680-83
Pseudouridine, 2'-(hydroxymethyl)-2-thio-, (±)-	pH 1	215(3.97),276(4.04)	18-2680-83
	pH 13	220(4.07),264(3.96)	18-2680-83
	MeOH	215(3.96),277(4.07)	18-2680-83
Pseudouridine, 4'-(hydroxymethyl)-2-thio-, (±)-	pH 1	214(4.13),273(4.13), 292s(4.08)	18-2680-83
	pH 13	265(4.13),285(4.11)	18-2680-83
	MeOH	213(4.15),275(4.20), 296s(4.12)	18-2680-83
$C_{10}H_{14}N_2O_7$			
Pseudouridine, 2'-(hydroxymethyl)-, (±)-	pH 1	266(3.71)	18-2680-83
	pH 13	285(3.58)	18-2680-83
	MeOH	267(3.76)	18-2680-83
Pseudouridine, 4'-(hydroxymethyl)-, (±)-	pH 1	264(3.78)	18-2680-83
	pH 13	287(3.83)	18-2680-83
	MeOH	264(3.82)	18-2680-83
3(2H)-Pyridazinone, 5-hydroxy-6-methoxy-2-β-D-ribofuranosyl-, monohydrochloride	pH 1	243(3.88),280(3.53)	4-0369-83
	pH 11	266(4.17)	4-0369-83
	MeOH	247(3.83),272s(3.65)	4-0369-83
$C_{10}H_{14}N_2S$			
Benzenecarboximidamide, N,N-dimethyl-N'-(methylthio)-	hexane	314(3.34)	77-0390-83
$C_{10}H_{14}N_4$			
5,5'-Bi-3H-pyrazole, 3,3,3',3'-tetramethyl-	MeCN	262(3.96),271(3.87), 349(2.63)	88-1775-83
$C_{10}H_{14}N_4O$			
4H-Pyrazolo[3,4-d]pyrimidin-4-one, 1,5-dihydro-1,6-dimethyl-5-(1-methylethyl)-	EtOH	210(4.15),257.7(3.90)	11-0171-83B
4H-Pyrazolo[3,4-d]pyrimidin-4-one, 1,5-dihydro-1,6-dimethyl-5-propyl-	EtOH	209.5(4.12),256(3.51)	11-0171-83B
$C_{10}H_{14}N_4O_3$			
Ethanol, 2-[(2-amino-4-methoxy-7H-pyrrolo[2,3-d'pyrimidin-7-yl)-methoxy]-	MeOH	259(4.01),285(3.87)	5-0137-83
$C_{10}H_{14}N_4O_5$			
2(1H)-Pyridinone, 5,6-dihydro-1-methyl-3,5-dinitro-6-pyrrolidino-,	MeOH	314(4.03),448(4.03)	138-0715-83

Compound	Solvent	$\lambda_{max}(\log \epsilon)$	Ref.
compd. with pyrrolidine (cont.)			138-0715-83
$C_{10}H_{14}N_4O_6$			
β-D-Glucopyranuronamide, 1-(4-amino-2-oxo-1(2H)-pyrimidinyl)-1-deoxy-	H_2O	270(3.85)	105-0580-83
$C_{10}H_{14}N_5O_7P$			
5'-Adenylic acid	pH 7.0	259(4.19)	162-0024-83
$C_{10}H_{14}N_5O_8P$			
5'-Guanylic acid, barium salt	pH 2	256(4.09)	162-0658-83
	pH 12	260(4.08)	162-0658-83
$C_{10}H_{14}N_6$			
1H-Imidazole, 4,5-diethyl-2-(1H-imidazol-2-ylazo)-	MeOH	469(4.44)	56-0547-83
$C_{10}H_{14}N_6O_4$			
Glycine, 2-[(2-amino-1,6-dihydro-6-oxo-9H-purin-9-yl)methoxy]ethyl ester, monohydrochloride	pH 1	253(3.93)	87-0602-83
$C_{10}H_{14}N_6O_4S$			
2H-1,2,3-Triazolo[4,5-d]pyrimidin-7-amine, 5-(methylthio)-2-β-D-ribofuranosyl-	pH 1	230(4.09),245s(--), 284(4.13),305(4.10)	87-1483-83
	pH 7	224(4.18),247(4.13), 279(3.99),308(3.98)	87-1483-83
	pH 13	222(4.16),246(4.19), 277(4.02),308(4.02)	87-1483-83
3H-1,2,3-Triazolo[4,5-d]pyrimidin-7-amine, 5-(methylthio)-3-β-D-ribofuranosyl-	pH 1	241(4.14),277(4.21)	87-1483-83
	pH 7	247(4.30),288(4.13)	87-1483-83
	pH 13	247(4.31),288(4.13)	87-1483-83
$C_{10}H_{14}N_6O_6S$			
2H-1,2,3-Triazolo[4,5-d]pyrimidin-7-amine, 5-(methylsulfonyl)-2-β-D-ribofuranosyl-	pH 1	227(4.14),255(3.54), 301(3.98)	87-1483-83
	pH 7	226(4.15),255(3.56), 300(3.99)	87-1483-83
	pH 13	315(3.86),334s(--)	87-1483-83
3H-1,2,3-Triazolo[4,5-d]pyrimidin-7-amine, 5-(methylsulfonyl)-3-β-D-ribofuranosyl-	pH 1	268s(--),263s(--), 288(3.99)	87-1483-83
	pH 7	258s(--),263s(--), 288(4.00)	87-1483-83
	pH 13	255s(--),290(4.03)	87-1483-83
$C_{10}H_{14}O$			
4(1H)-Azulenone, 2,3,3a,7,8,8a-hexahydro-	EtOH	227(4.10),329(1.51)	44-1670-83
Bicyclo[3.1.1]hept-2-en-6-one, 2,7,7-trimethyl- (chrysanthenone)	n.s.g.	290(2.08)	162-0320-83
Cyclohexanone, 2-(cyclopropylmethylene)-, (E)-	MeOH	265(4.07)	23-0288-83
Furan, 3-methyl-2-(2-methyl-2-butenyl)-	MeCN	219(4.03),270(2.37)	24-3366-83
4H-Inden-4-one, 1,2,3,5,6,7-hexahydro-1-methyl-	n.s.g.	251(4.1055)	2-1154-83
2,4,6-Octatrienal, 3,7-dimethyl-, (E)-	EtOH	335(4.45)	35-4033-83
Phenol, 2-methyl-5-(1-methylethyl)- (carvacrol)	EtOH	277.5(3.262)	162-0262-83
$C_{10}H_{14}O_2$			
Acetic acid, (3,5-dimethyl-2-cyclo-	EtOH	274(4.35)	111-0447-83

Compound	Solvent	λ_{max}(log ϵ)	Ref.
hexen-1-ylidene)- (cont.)			111-0447-83
Benzeneethanol, 2-methoxy-6-methyl-	MeOH	220(4.61),275(3.13), 280(3.13)	100-0235-83
Benzenemethanol, 3-hydroxy-α,α,4-tri-methyl-	MeOH	207(4.56),221(4.49), 277(3.97)	102-2080-83
2-Cyclohexen-1-one, 2-(3-butenyloxy)-	EtOH	260(3.61)	44-4241-83
2-Cyclohexen-1-one, 2-[(2-methyl-2-propenyl)oxy]-	EtOH	260(3.69)	44-4241-83
2-Cyclopenten-1-one, 3-methoxy-2-meth-yl-5-(2-propenyl)-	EtOH	253(4.04)	39-1885-83C
$C_{10}H_{14}O_3$			
3(2H)-Benzofuranone, 4,5,6,7-tetra-hydro-2-methoxy-2-methyl-	EtOH	280(4.01)	142-0061-83
1H-2-Benzopyran-1-one, 3,4,4a,5,6,7-hexahydro-8-hydroxy-3-methyl- (ramulosin)	EtOH	265(4.01)	39-2185-83C
2(5H)-Furanone, 4-(1,1-dimethyleth-oxy)-5-ethylidene-, (Z)-	EtOH	259(4.24)	39-2983-83C
2(5H)-Furanone, 3-(1,1-dimethylethyl)-5-ethylidene-4-hydroxy-	EtOH	249(4.12),313(3.97)	39-2983-83C
	HCl	269(4.17)	39-2983-83C
$C_{10}H_{14}O_4$			
Cyclohexanecarboxylic acid, 1,2,2-tri-methyl-4,5-dioxo-, (±)-	EtOH	268(3.81)	39-1465-83C
	aq KOH	312(3.72)	39-1465-83C
2-Cyclohexene-1,4-dione, 2,3-dimeth-oxy-5,6-dimethyl-, trans-(+)-	n.s.g.	289(4.39)	162-0127-83
$C_{10}H_{14}S$			
Benzene, [(1,1-dimethylethyl)thio]-	decane	218(4.07),268(3.18)	65-0449-83
Benzene, [(2-methylpropyl)thio]-	decane	215s(3.91),256(3.91)	65-0449-83
2-Thiabicyclo[3.3.1]nona-3,7-diene, 4,8-dimethyl-	EtOH	229(3.55),263(2.78)	104-0289-83
2-Thiabicyclo[3.3.1]non-7-ene, 8-meth-yl-4-methylene-	EtOH	244(2.82)	104-0283-83
$C_{10}H_{15}BrO$			
d-Camphor, 3β-bromo-	C_6H_{12}	312(1.95)	162-0194-83
$C_{10}H_{15}ClN_2O$			
2(1H)-Pyrazinone, 5-chloro-3,6-bis(1-methylethyl)-	EtOH	225(3.91),307(3.82)	4-0951-83
$C_{10}H_{15}ClO$			
d-Camphor, 3-chloro-, endo	C_6H_{12}	305(1.72)	162-0299-83
d-Camphor, 3-chloro-, exo	C_6H_{12}	306(1.75)	162-0299-83
$C_{10}H_{15}Cl_2NO_3$			
2-Butenoic acid, 4,4-dichloro-3-mor-pholino-, ethyl ester, (E)-	heptane	200(3.56),293(4.09)	5-0013-83
$C_{10}H_{15}Cl_2N$ Pt			
Platinum, dichloro(3,4-dimethylbenz-enamine)(η^2-ethene)-	n.s.g.	244(3.89),264(3.97), 301(3.48)	131-0309-83H
$C_{10}H_{15}N$			
Benzenamine, N-butyl-	n.s.g.	285(3.17)	32-0069-83
Benzenamine, N-(1,1-dimethylethyl)-	n.s.g.	233(3.82)	32-0069-83
Benzenamine, N-(1-methylpropyl)-	n.s.g.	286(3.14)	32-0069-83
Benzenamine, N-(2-methylpropyl)-	n.s.g.	292(3.20)	32-0069-83
Benzenamine, N,N-diethyl-	isooctane	259(4.22),303(3.37)	162-0452-83

Compound	Solvent	$\lambda_{max}(\log \epsilon)$	Ref.
Benzenamine, N,N,3,4-tetramethyl-	65.8% H_2SO_4	260(2.57),278(2.45)	39-1191-83B
Benzenemethanamine, α,3,5-trimethyl-, (S)-	C_6H_{12}	263s(2.28),265(2.36), 267s(2.32),272(2.28), 274s(2.15)	35-1578-83
	MeOH	262s(2.32),265(2.38), 267(2.36),271(2.30), 274(2.23)	35-1578-83
hydrochloride	MeOH	262s(2.48),265(2.56), 267(2.58),270s(2.51), 275(2.57)	35-1578-83
	10% HCl	262(2.52),265s(2.59), 267(2.61),270s(2.54), 274(2.58)	35-1578-83
$C_{10}H_{15}NO$			
3,5,7-Octatrien-2-one, 8-(dimethyl-amino)-	EtOH	438(4.63)	70-0780-83
	$CHCl_3$	420(--)	70-0780-83
d-Pseudoephedrine, hydrochloride	EtOH	208(3.92),251(2.21), 257(2.30),264(2.21)	162-0520-83
$C_{10}H_{15}NOS$			
2H-1,4-Benzothiazin-8(5H)-one, 3,4,6,7-tetrahydro-6,6-dimethyl-	EtOH	290(4.39)	142-2391-83
$C_{10}H_{15}NO_2$			
Pentanenitrile, 4-methyl-2-(2-methyl-1-oxopropyl)-3-oxo-	EtOH	280.5(4.06)	4-0645-83
1-Propanone, 2-methyl-1-[5-(1-methyl-ethyl)-4-isoxazolyl]-	EtOH	238(3.87)	4-0645-83
$C_{10}H_{15}NO_3$			
2-Pyrrolidinone, 4-acetoxy-1-methyl-5-(1-methylethylidene)-	EtOH	231(4.10)	88-4755-83
$C_{10}H_{15}NO_3S$			
1,2-Benzoxathiin-4-amine, 5,6,7,8-tetrahydro-N,N-dimethyl-, 2,2-dioxide	EtOH	257(3.84),290s(3.72)	4-1549-83
$C_{10}H_{15}NO_5S$			
1-Azetidineacetic acid, α-(1-methyl-ethylidene)-2-(methylsulfonyl)-4-oxo-, methyl ester, (R)-	EtOH	211(2.57),222s(2.40)	39-2241-83C
$C_{10}H_{15}NO_7$			
2H-Pyrrol-2-one, 5-(β-D-glucopyrano-syloxy)-1,5-dihydro-	MeOH	198(3.95),234(3.17)	102-1278-83
$C_{10}H_{15}N_2O$			
4H-Pyrido[1,2-a]pyrimidinium, 6,7,8,9-tetrahydro-1,6-dimethyl-4-oxo-, (R)-	H_2O	201(3.60),234(3.98), 265(3.61)	39-1413-83B
$C_{10}H_{15}N_2OS$			
Morpholinium, 4-(4,6-dimethyl-2H-1,3-thiazin-2-ylidene)-, perchlorate	MeCN	225(4.21),255(4.01), 360(3.58)	118-0827-83
$C_{10}H_{15}N_3$			
1H-Imidazole-1-carbonitrile, 2,5-bis(1-methylethyl)-	EtOH	232(4.08)	4-1277-83
1H-Imidazole-1-carbonitrile, 2,5-dipropyl-	EtOH	232(4.01)	4-1277-83

Compound	Solvent	λ_{max}(log ε)	Ref.
$C_{10}H_{15}N_3O_2$			
Hydrazinecarboxamide, 2-(3-acetyl-4,4-dimethyl-2-cyclopenten-1-ylidene)-	MeOH	309(4.46)	39-1465-83C
Piperidine, 1-(1-methyl-4-nitro-1H-pyrrol-2-yl)-	MeOH	281(3.9),358(3.6)	44-0162-83
Piperidine, 1-(1-methyl-4-nitro-1H-pyrrol-3-yl)-	MeOH	310(3.9)	44-0162-83
$C_{10}H_{15}N_3O_3$			
5-Pyrimidineacetic acid, 2-amino-1,4-dihydro-4-oxo-, butyl ester	50% MeOH	292(3.90)	73-0292-83
	+ HCl	221(4.06),261(3.94)	73-0292-83
	+ NaOH	233(3.97),281(3.90)	73-0292-83
$C_{10}H_{15}N_3O_4$			
2,4(1H,3H)-Pyrimidinedione, 5-amino-1-[2-hydroxy-4-(hydroxymethyl)cyclopentyl]-, (1α,2β,4α)-(±)-	pH 1	271(3.92)	87-0156-83
	pH 7	225s(3.81),297(3.82)	87-0156-83
	pH 13	233s(--),292(3.76)	87-0156-83
2,4(1H,3H)-Pyrimidinedione, 5-amino-1-[3-hydroxy-4-(hydroxymethyl)cyclopentyl]-, (1α,3β,4α)-(±)-, sulfate	pH 1	270(3.97)	87-0156-83
	pH 7	226(3.85),298(3.88)	87-0156-83
	pH 13	233s(--),291(3.81)	87-0156-83
5-Pyrimidinepentanoic acid, 2-amino-1,4-dihydro-6-hydroxy-4-oxo-, methyl ester	50% MeOH	271(4.088)	73-0304-83
	+ HCl	267(4.08)	73-0304-83
	+ NaOH	271(4.07)	73-0304-83
$C_{10}H_{15}N_3O_4S$			
5-Pyrimidinecarboxamide, 1,2,3,4-tetrahydro-N-[1-(hydroxymethyl)-2-(methylthio)ethyl]-6-methyl-2,4-dioxo-	H_2O	269(3.98)	94-0135-83
$C_{10}H_{15}N_3O_5$			
Cytidine, 3'-C-methyl-	pH 1	280(4.10)	136-0075-83M
	pH 7-13	271(3.94)	136-0075-83M
Isocytosine, 5-[(β-D-ribofuranosyl)-methyl]-, hydrochloride	MeOH	224(4.06),263(3.89)	18-2700-83
	M HCl	223(4.09),264(3.98)	18-2700-83
	M NaOH	232(4.05),281(3.96)	18-2700-83
Pseudoisocytidine, 1'-methyl-, hydrochloride	pH 1	221(3.94),262(3.81)	18-2680-83
	pH 13	230(4.04),278(3.92)	18-2680-83
	MeOH	222(3.98),262(3.80),300s(3.01)	18-2680-83
Pseudoisocytidine, 2'-methyl-, hydrochloride	pH 1	222(3.76),264(3.63)	18-2680-83
	pH 13	233(3.96),280(3.83)	18-2680-83
	MeOH	244(4.03),265(3.84)	18-2680-83
Pseudoisocytidine, 4'-methyl-, hydrochloride	pH 1	220(4.17),262(4.04)	18-2680-83
	pH 13	238(4.07),280(3.66)	18-2680-83
	MeOH	223(4.09),263(3.91),290(3.74)	18-2680-83
Pseudoisocytidine, 5'-methyl-, hydrochloride	MeOH	223(4.02),265(3.85),290(3.59)	18-2680-83
3(2H)-Pyridazinone, 6-(methylamino)-2-β-D-ribofuranosyl-	pH 1	232(4.50),340(3.52)	4-0369-83
	MeOH	234(4.51),347(3.54)	4-0369-83
2,4(1H,3H)-Pyrimidinedione, 5-amino-1-[2,3-dihydroxy-4-(hydroxymethyl)-cyclopentyl]-, (1α,2β,3β,4α)-(±)-	pH 1	269(3.96)	87-0156-83
	pH 7	225s(3.84),297(3.87)	87-0156-83
	pH 13	233s(--),291(3.80)	87-0156-83
2(1H)-Pyrimidinone, 4-amino-1-(5-deoxy-β-D-ribohexopyranosyl)-	H_2O	230s(3.77),271(3.93)	136-0099-83J
$C_{10}H_{15}N_3O_6$			
Pseudoisocytidine, 2'-(hydroxymethyl)-, hydrochloride, (±)-	pH 13	232(3.94),280(3.81)	18-2680-83
	MeOH	224(4.10),265(3.89)	18-2680-83

Compound	Solvent	λ max(log ε)	Ref.
Pseudoisocytidine, 4'-(hydroxymethyl)-,	pH 1	222(3.98),263(3.88)	18-2680-83
hydrochloride, (±)-	pH 13	233(4.04),278(3.93)	18-2680-83
	MeOH	224(3.97),263(3.90)	18-2680-83
1H-Pyrazole-5-carboxamide, 4-hydroxy-	H_2O	223(3.94),266(3.71)	18-2700-83
3-[(β-D-ribofuranosyl)methyl]-	pH 13	235(3.64),311(3.74)	18-2700-83
$C_{10}H_{15}N_3O_8$			
2,4(1H,3H)-Pyrimidinedione, 6-amino-5-β-D-glucopyranosyloxy- (convicine)	pH 7	245(3.43),271(4.16)	162-0358-83
$C_{10}H_{15}N_5$			
3H,10bH-Imidazo[1,2,3-ij]pteridin-10-amine, 5,6-dihydro-N,N-dimethyl-, chloride	EtOH	227(4.06),258(4.32), 332(4.01)	103-1120-83
Pyrazine, 3-azido-2,5-bis(1-methylethyl)-	EtOH	275(3.92)	4-1277-83
Pyrazine, 3-azido-2,5-dipropyl-	EtOH	275(3.94)	4-1277-83
$C_{10}H_{15}N_5O_3$			
1,2-Propanediol, 3-[2-(6-amino-1H-	pH 1	262(4.17)	111-0555-83
purin-9-yl)ethoxy]-	pH 13	263(4.17)	111-0555-83
$C_{10}H_{15}N_5O_{10}P_2$			
Adenosine diphosphate	pH 7	259(4.19)	162-0023-83
$C_{10}H_{15}N_7OS$			
Guanidine, N-cyano-N'-methyl-N"-[2-[[(5-methyl-1H-imidazol-4-yl)methyl]thio]ethyl]-N'-nitroso-	EtOH	385s(--),397(2.30), 414s(--)	23-1771-83
$C_{10}H_{16}$			
1,3-Cycloheptadiene, 1,5,5-trimethyl-	EtOH	252(3.88)	105-0422-83
1,3-Cycloheptadiene, 2,6,6-trimethyl-	EtOH	254(3.72)	105-0422-83
1,3-Cyclohexadiene, 5,5,6,6-tetramethyl-	C_6H_{12}	258(3.6)	24-1897-83
Cyclohexane, 1-methyl-2-(2-propenylidene)-, [S-(E)]-	C_6H_{12}	233(4.37),239(4.41), 247(4.23)	35-3252-83 +35-3264-83
[S-(Z)]-	C_6H_{12}	231(4.42),238(4.46), 246(4.28)	35-3252-83
Cyclohexane, 1-methyl-3-(2-propenylidene)-, [R-(E)]-	C_6H_{12}	231(4.44),238(4.48), 246.5(4.30)	35-3252-83
[R-(Z)]-	C_6H_{12}	231(4.46),238(4.50), 246(4.32)	35-3252-83 +35-3264-83
Cyclohexane, 1-methyl-4-(2-propenylidene)-, (R)-	C_6H_{12}	231(4.34),237.5(4.39), 246(4.21)	35-3252-83
3,4,6-Nonatriene, 3-methyl-	EtOH	228(4.05)	70-0569-83
2-Nonyne, 4-methylene-(mixture with isomer)	EtOH	224(4.04)	104-1621-83
4-Nonyne, 6-methylene-	EtOH	225(4.01)	104-1621-83
$C_{10}H_{16}GeOS$			
Germane, [(4-methoxyphenyl)thio]trimethyl-	heptane	234(4.25),244s(--), 283(3.23),291(3.18)	65-2021-83
$C_{10}H_{16}NO_2$			
Pyrrolidinium, 2-carboxy-1-(3-methyl-2-butenylidene)-, perchlorate	CD_3CN	275(4.11)	35-4033-83
$C_{10}H_{16}N_2O$			
Bicyclo[2.2.1]heptane-2,3-dione,	heptane	293(4.00),377(2.50)	78-0733-83
1,7,7-trimethyl-, 3-hydrazone, (+)-	CF_3CH_2OH	292(3.83),364(2.45)	78-0733-83

Compound	Solvent	$\lambda_{max}(\log \epsilon)$	Ref.
Bicyclo[2.2.1]heptane-2,3-dione, 1,7,7-trimethyl-, 3-hydrazone, (-)-	heptane	248s(3.80),263(3.87), 340s(1.67)	78-0733-83
	CF_3CH_2OH	244s(3.82),284(4.11), 330s(3.65)	78-0733-83
2(1H)-Pyrazinone, 3,6-bis(1-methylethyl)-	EtOH	225(2.91),320(2.87)	4-0951-83
2(1H)-Pyrazinone, 3,6-dipropyl-	EtOH	227(3.67),323(3.68)	4-0951-83
$C_{10}H_{16}N_2O_2$			
2(1H)-Pyrazinone, 3,6-bis(1-methylethyl)-, 4-oxide	EtOH	225(4.27),230(4.26), 276(3.91),330(3.90)	4-0951-83
$C_{10}H_{16}N_2O_3$			
2-Cyclohexen-1-one, 3-(dimethylamino)-5,5-dimethyl-2-nitro-	H_2O	284(4.15),355(3.61)	78-3405-83
	pH 1	285(4.15),355(3.63)	78-3405-83
	pH 13	262(4.16),337(3.42)	78-3405-83
2-Quinoxalinol, 2,3,5,6,7,8-hexahydro-3,3-dimethyl-, 1,4-dioxide	EtOH	350(4.19)	103-1128-83
$C_{10}H_{16}N_2S$			
2(1H)-Pyrazinethione, 3,6-bis(1-methylethyl)-	EtOH	247(2.74),283(3.21)	142-0797-83
2(1H)-Pyrazinethione, 3,6-dipropyl-	EtOH	248(2.63),283(3.09)	142-0797-83
$C_{10}H_{16}N_4O_2$			
1H-1,2-Diazepine-3-acetamide, 1-(aminocarbonyl)-2,3-dihydro-α,5-dimethyl-, (R*,R*)-	MeOH	287(3.95)	5-2141-83
1H-Pyrazole-4-carboxamide, 5-acetamido-1-methyl-N-propyl-	EtOH	222(4.00)	11-0171-83B
$C_{10}H_{16}N_4O_5$			
2(1H)-Pyridinone, 6-(diethylamino)-5,6-dihydro-1-methyl-3,5-dinitro-, compd. with triethylamine	MeOH	314(4.11),449(4.10)	138-0715-83
$C_{10}H_{16}N_5O_{13}P_2$			
Adenosine triphosphate	pH 7.0	259(4.19)	162-0023-83
$C_{10}H_{16}O$			
Bicyclo[2.2.1]heptan-2-one, 1,3,3-trimethyl- (fenchone)	heptane	289(1.27)	46-3034-83
2-Cyclohexen-1-one, 3-methyl-6-(1-methylethyl)- (piperitone)	EtOH	233(4.14),318(1.75)	102-2227-83
2-Cyclopenten-1-one, 2,3-dimethyl-4-(1-methylethyl)-	EtOH	237(4.15),300(1.75)	5-0761-83
4,6-Octadien-3-one, 2,6-dimethyl-	EtOH	280(4.30)	5-0761-83
4-Octyn-2-ol, 2-methyl-3-methylene-	EtOH	222(4.03)	104-1621-83
Spiro[2.5]octan-4-one, 6,6-dimethyl-	hexane	285(1.56)	23-0936-83
$C_{10}H_{16}OS$			
Thiocamphor S-oxide	$CHCl_3$	270(3.71),310s(2.70)	44-0214-83
Thiofenchone S-oxide	$CHCl_3$	270(3.71),310s(2.74)	44-0214-83
$C_{10}H_{16}OSSi$			
Silane, [(4-methoxyphenyl)thio]trimethyl-	heptane	238(4.20),282.5(3.15), 289(3.10)	65-2021-83
$C_{10}H_{16}OSSn$			
Stannane, [(4-methoxyphenyl)thio]trimethyl-	heptane	240(4.28),252s(--), 286(3.29),294(3.20)	65-2021-83

Compound	Solvent	$\lambda_{max}(\log \epsilon)$	Ref.
$C_{10}H_{16}OS_2$			
Methanethione, (tetrahydro-3,3,5,5-tetramethyl-4H-thiopyran-4-ylidene)-, S-oxide	isooctane	237(3.64),568(1.04)	77-0797-83
$C_{10}H_{16}O_2$			
Cicrotoic acid	n.s.g.	219(4.08)	162-0323-83
Cyclohexanone, 2-[(1-methylethoxy)-methylene]-	EtOH	277(4.23)	44-4272-83
1-Cyclohexene-1-carboxaldehyde, 2-(1-methylethoxy)-	EtOH	276(4.19)	44-4272-83
2-Cyclohexen-1-one, 4-(hydroxymethyl)-4,5,5-trimethyl-	EtOH	230(3.89)	39-1465-83C
$C_{10}H_{16}O_2S_2$			
Methanethione, (tetrahydro-3,3,5,5-tetramethyl-4H-thiopyran-4-ylidene)-, S,S-dioxide	isooctane	237(3.56),565(1.08)	77-0797-83
$C_{10}H_{16}O_3S_2$			
Methanethione, (tetrahydro-3,3,5,5-tetramethyl-4H-thiopyran-4-ylidene)-, S,S,thiono-trioxide	MeCN	249(2.53)	77-0797-83
3(2H)-Thiophenone, 4-(hexylthio)-, 1,1-dioxide	EtOH	203(4.03),307(3.73)	104-1447-83
	EtOH-KOH	238(3.99),325(3.79), 427(3.54)	104-1447-83
$C_{10}H_{16}O_5$			
2-Propenoic acid, 3-methoxy-3-(tetrahydro-5-methoxy-4-methyl-3-furanyl)-	EtOH	233(4.06)	94-0925-83
$C_{10}H_{16}O_8P_2$			
Phosphonic acid, (2,5-dihydroxy-1,4-phenylene)bis-, tetramethyl ester	MeOH	220.5(4.04),337(3.81)	70-1532-83
$C_{10}H_{16}S$			
2-Hexen-4-yne, 2-(butylthio)-, (Z)-	MeCN	210(3.70),275(4.06)	4-0729-83
Thiocamphor	benzene	498(1.11)	44-0214-83
Thiofenchone	benzene	487(1.95)	44-0214-83
	heptane	240(4.00),486(1.09)	46-3034-83
	$CHBr_3$	484(1.71)	46-3034-83
$C_{10}H_{16}SSi$			
Silane, trimethyl[(4-methylphenyl)-thio]-	heptane	230(4.1),250(4.0)	65-2021-83
$C_{10}H_{16}S_2$			
Methanethione, (tetrahydro-3,3,5,5-tetramethyl-4H-thiopyran-4-ylidene)-	isooctane	238(3.60),573(1.08)	77-0797-83
$C_{10}H_{16}Se$			
Selenofenchone, (-)-	heptane	224(3.58),271(4.01), 625(1.62)	46-3034-83
	dioxan	219(3.61),275(4.00), 607(1.63)	46-3034-83
	$CHBr_3$	576(1.89)	46-3034-83
$C_{10}H_{17}NO_2$			
4-Hepten-3-one, 5-hydroxy-4-(iminomethyl)-2,6-dimethyl-	EtOH	259(4.12),280(4.16)	4-0649-83

Compound	Solvent	λmax(log ε)	Ref.
2,4-Hexadienoic acid, 6-(dimethylamino)-, ethyl ester, (E,E)-	EtOH	255(4.47)	70-1444-83
$C_{10}H_{17}NO_2S$			
Pyrrolidine, 1-(5,6-dihydro-2-methyl-2H-thiopyran-3-yl)-, S,S-dioxide	EtOH	263(4.1)	39-2735-83C
$C_{10}H_{17}NO_3S$			
Morpholine, 4-(5,6-dihydro-2-methyl-2H-thiopyran-3-yl)-, S,S-dioxide	EtOH	225(3.7)	39-2735-83C
$C_{10}H_{17}N_2O_2$			
Piperidinyloxy, 4-isocyanato-2,2,6,6-tetramethyl-	MeCN	241(3.30),465(1.00)	70-1906-83
$C_{10}H_{17}N_3OS$			
4-Morpholinecarbothioamide, N-(1-pyrrolidinylmethylene)-	MeCN	279(4.13),308(3.98)	97-0019-83
$C_{10}H_{17}N_3O_2S$			
4-Morpholinecarbothioamide, N-(4-morpholinylmethylene)-	MeCN	276(4.47),306(4.04)	97-0019-83
$C_{10}H_{17}N_3O_3$			
Hydrazinecarboxamide, 2-(1-formyl-3,3-dimethyl-5-oxohexylidene)-	EtOH	273(4.26)	39-1465-83C
$C_{10}H_{17}N_3S$			
1-Pyrrolidinecarbothioamide, N-(1-pyrrolidinylmethylene)-	MeCN	229(3.82),272(4.39), 302s(4.00)	97-0019-83
$C_{10}H_{18}ClO_3P$			
Phosphonic acid, (2-chloro-3-methyl-1,3-pentadienyl)-, diethyl ester	C_6H_{12}?	255(c.2.5)	65-1054-83
$C_{10}H_{18}NO$			
3H-2,1-Benzoxazinium, 1-ethyl-4,4a,5,6,7,8-hexahydro-, tetrafluoroborate	MeCN	218(3.70)	5-0897-83
2H-Indolium, 1-ethoxy-3,3a,4,5,6,7-hexahydro-, tetrafluoroborate	MeCN	208(3.72)	5-0897-83
$C_{10}H_{18}NO_2$			
Methanaminium, N-[(6-ethoxy-3,4-dihydro-2H-pyran-5-yl)methylene]-N-methyl-, tetrafluoroborate	n.s.g.	312(4.48)	103-1050-83
$C_{10}H_{18}N_2$			
Bicyclo[2.2.1]heptan-2-one, 1,7,7-trimethyl-, hydrazone	heptane	210(3.90),228(3.63)	78-0733-83
	CF_3CH_2OH	203(3.57),228(4.00)	78-0733-83
$C_{10}H_{18}N_2O_3S$			
3-Pyridinesulfonamide, 6-ethoxy-1,6-dihydro-N,N,1-trimethyl-	CH_2Cl_2	304(3.6)	64-0873-83B
$C_{10}H_{18}N_2O_6$			
Methyl α-D-kijanoside	CF_3CH_2OH	199(3.70)	39-1497-83C
$C_{10}H_{18}N_4O$			
5H-Tetrazol-5-one, 1-(1,1-dimethylethyl)-1,4-dihydro-4-(4-pentenyl)-	hexane	227(3.66)	24-3427-83

Compound	Solvent	λ_{max}(log ε)	Ref.
$C_{10}H_{18}N_4O_4$			
1,4-Butanediamine, N,N'-bis(1-methyl-2-nitroethenyl)-	CH_2Cl_2	338(4.18),358(4.17)	40-0088-83
$C_{10}H_{18}O$			
Geraniol	n.s.g.	190-195(4.26)	162-0629-83
$C_{10}H_{18}OS_2$			
Ethenone, bis[(1,1-dimethylethyl)-thio]-	MeOH	251(3.83)	18-0171-83
$C_{10}H_{18}OSi$			
Silane, (1,1-dimethylethyl)dimethyl-(1-oxo-2-butynyl)-	C_6H_{12}	227(3.87),420(2.23)	78-0949-83
$C_{10}H_{18}O_3$			
2(3H)-Furanone, dihydro-5-(1-hydroxy-hexyl)-	EtOH	212(3.11)	158-83-61
$C_{10}H_{18}S$			
2-Thiabicyclo[2.2.2]octane, 1,3,3-tri-methyl-	EtOH	215(3.04)	104-0283-83
6-Thiabicyclo[3.2.1]octane, 4,7,7-tri-methyl-, endo	EtOH	213(3.08),240(3.06)	104-0283-83
$C_{10}H_{18}S_2$			
Propane, 2,2'-[1,2-ethynediylbis-(thio)]bis[2-methyl-	CCl_4	329(3.88),400(3.52), 520(2.11)	54-0083-83
$C_{10}H_{18}Si_3$			
1,2,5-Trisilacyclohepta-3,6-diyne, 1,1,2,2,5,5-hexamethyl-	n.s.g.	243s(3.64),251(3.70), 260(2.41)	35-3359-83
$C_{10}H_{19}BrOSi$			
Silane, (2-bromo-1-oxo-2-butenyl)(1,1-dimethylethyl)dimethyl-	C_6H_{12}	251(3.85),398(2.11), 416(2.10)	78-0949-83
$C_{10}H_{19}N$			
2-Propen-1-amine, 1-ethyl-N-(2,2-di-methylpropylidene)-	hexane	247(1.97)	116-1679-83
$C_{10}H_{19}NO$			
4-Heptanone, 3-[(dimethylamino)meth-ylene]-	MeOH	303(4.36)	131-0001-83K
(E-syn-E)-	C_6H_{12}	287(4.26)	131-0001-83K
4-Heptanone, 3-[(ethylamino)methyl-ene]-	MeOH	299(4.40)	131-0001-83K
(E-syn-E)-	C_6H_{12}	277(--)	131-0001-83K
	dioxan	286(4.33)	131-0001-83K
(Z-syn-Z)-	C_6H_{12}	313(4.05)	131-0001-83K
	dioxan	312(--)	131-0001-83K
3-Pentanone, 2-[(diethylamino)meth-ylene]-	MeOH	305(4.49)	131-0001-83K
(E-syn-E)-	C_6H_{12}	288(4.33)	131-0001-83K
$C_{10}H_{19}NOS$			
Butanethioamide, N,N-diethyl-3,3-di-methyl-2-oxo-	MeOH	276(3.91),309s(3.31), 322s(3.04),364(2.24)	5-1116-83
Pentanethioamide, N-butyl-3-methyl-2-oxo-	MeOH	248(3.90),266(3.92), 366(2.19)	5-1116-83

Compound	Solvent	$\lambda_{max}(\log \epsilon)$	Ref.
$C_{10}H_{19}NO_2$			
L-Alanine, N-(2,2-dimethylpropylidene)-, ethyl ester	hexane	c.220(c.2.0)	116-1679-83
1-Propanol, 2-[(2,2-dimethylpropylidene)amino]-, acetate, (S)-	hexane	c.230(c.2.0)	116-1679-83
$C_{10}H_{19}N_3O_2$			
1H-Imidazole-4-carboxaldehyde, 2,5-dihydro-1-methoxy-2,2,5,5-tetramethyl-, O-methyloxime	EtOH	236(4.16)	70-2123-83
$C_{10}H_{19}N_3O_3S$			
4(5H)-Thiazolone, 2-[[(diethylamino)-methyl]amino]-5,5-bis(hydroxymethyl)-	EtOH	226(4.24),250(4.09)	103-1076-83
$C_{10}H_{19}N_3O_4$			
1H-Imidazole, 4-(dimethoxymethyl)-2,5-dihydro-2,2,5,5-tetramethyl-1-nitroso-, 3-oxide	EtOH	240(4.22)	70-2501-83
$C_{10}H_{20}N_2O_2$			
1H-Imidazole, 4-ethyl-2,5-dihydro-1-methoxy-2,2,5,5-tetramethyl-, 3-oxide	EtOH	236(4.00)	70-2123-83
1-Penten-1-amine, 1-(1,1-dimethylethyl)-N-methyl-N-nitro-	H_2O	236(3.54)	104-1270-83
$C_{10}H_{21}N_3O_4$			
1,2-Triazanedicarboxylic acid, 3,3-dimethyl-, bis(1-methylethyl) ester	EtOH	210(2.46)(end abs.)	33-1416-83
$C_{10}H_{22}GeS$			
1,3-Thiagermetane, 3,3-dibromo-	C_6H_{12}	217(3.59),261(2.30)	101-0C73-83L
$C_{10}H_{22}N_2O_2$			
Hyponitrous acid, bis(1,1-dimethylpropyl) ester	isooctane	225.1(3.85)	44-3728-83
$C_{10}H_{24}S_3Si_2$			
Silane, 1,2,5-trithiepane-3,7-diylbis[trimethyl-	hexane	223(3.10),250(2.57)	101-0133-83L
$C_{10}H_{24}S_4Si_2$			
Silane, 1,2,5,6-tetrathiocane-3,8-diylbis[trimethyl-	hexane	220(3.11),250(2.78)	101-0133-83L
$C_{10}H_{30}Cl_4N_7O_2RuS$			
Pentammine[S-[β-(4-pyridinyl)ethyl]-cysteine]ruthenium(III) tetrachloride	pH 1	254(0.85),260(0.80),268(0.74)	87-0737-83

Compound	Solvent	$\lambda_{max}(\log \epsilon)$	Ref.
$C_{11}H_4ClNO_4S_2$			
4H,5H-Pyrano[3,4-e][1,3]oxazine-4,5-dione, 7-chloro-2-(2-thienylthio)-	$CHCl_3$	283(4.14),346(4.26)	24-3725-83
$C_{11}H_5ClOS_2$			
4H-Thieno[2,3-b][1]benzothiopyran-4-one, 6-chloro-	MeOH	263(4.44),288s(3.97), 307s(3.67),366(3.86)	73-2970-83
$C_{11}H_5Cl_2NO_2S$			
9H-[1]Benzothiopyrano[2,3-d]oxazol-9-one, 2-(dichloromethyl)-	MeOH	294(3.79),296(3.80), 350(3.93),352(3.93)	83-0921-83
$C_{11}H_5N_7O_4$			
5H-Pyrido[3',2':4,5][1,2,3]triazolo-[1,2-a]benzotriazol-6-ium, 2,10-dinitro-, hydroxide, inner salt	EtOH	235(3.82),286(3.49), 335(4.38),425(3.80)	20-0067-83
$C_{11}H_5N_7O_{10}$			
2-Pyridinamine, 3,5-dinitro-N-(2,4,6-trinitrophenyl)-	neutral	380(4.280)	103-0511-83
	acid	280(4.270)	103-0511-83
$C_{11}H_6BrN_5O_6$			
2-Pyridinamine, 5-bromo-N-(2,4,6-trinitrophenyl)-	neutral	385(4.060)	103-0511-83
	acid	364(4.130)	103-0511-83
$C_{11}H_6ClN_5O_6$			
2-Pyridinamine, 5-chloro-N-(2,4,6-trinitrophenyl)-	neutral	380(4.120)	103-0511-83
	acid	361(4.070)	103-0511-83
$C_{11}H_6IN_5O_6$			
2-Pyridinamine, 5-iodo-N-(2,4,6-trinitrophenyl)-	neutral	367(4.233)	103-0511-83
	acid	356(4.086)	103-0511-83
$C_{11}H_6N_2O$			
Propanedinitrile, [(4-formylphenyl)-methylene]-	EtOH	220(4.26)	140-0215-83
$C_{11}H_6N_4$			
1,3,5-Heptatriene-1,1,7-tricarbonitrile, 7-carbonimidoyl-	EtOH	550(4.95)	70-0780-83
	$CHCl_3$	560(--)	70-0780-83
$C_{11}H_6N_6O_2$			
5H-Pyrido[3',2':4,5][1,2,3]triazolo-[1,2-a]benzotriazol-6-ium, 10-nitro-, hydroxide, inner salt	EtOH	234(4.47),283(4.04), 318(3.91),342(3.70), 401(4.45)	20-0067-83
$C_{11}H_6N_6O_8$			
2-Pyridinamine, N-(2,4-dinitrophenyl)-3,5-dinitro-	neutral	385(4.380)	103-0511-83
	acid	335(4.150)	103-0511-83
2-Pyridinamine, 3-nitro-N-(2,4,6-trinitrophenyl)-	neutral	385(3.970)	103-0511-83
	acid	342(3.941)	103-0511-83
2-Pyridinamine, 5-nitro-N-(2,4,6-trinitrophenyl)-	neutral	390(4.302)	103-0511-83
	acid	312(4.107)	103-0511-83
$C_{11}H_6O_5$			
Naphtho[2,3-d]-1,3-dioxole-4,9-dione, 6-hydroxy-	DMSO	478(2.84)	64-0392-83B
Naphtho[2,3-d]-1,3-dioxole-5,8-dione, 6-hydroxy-	EtOH	210(3.93),270(4.20), 308(3.84)	107-0691-83

Compound	Solvent	$\lambda_{max}(\log \epsilon)$	Ref.
$C_{11}H_7Br$			
1H-Cyclobuta[de]naphthalene, 1-bromo-	EtOH	225(4.83),284(3.67), 307(2.86),320(2.76)	35-6096-83
$C_{11}H_7BrO_3S$			
4H-1-Benzothiopyran-4-one, 3-acetoxy-2-bromo-	MeOH	264(4.17),345(3.98)	83-0921-83
$C_{11}H_7ClN_4O_2$			
7H-Pyrazolo[4,3-d]pyrimidin-7-one, 2-(4-chlorophenyl)-2,4-dihydro-3-hydroxy-	MeOH	251(4.35),304(3.81)	94-1228-83
$C_{11}H_7ClO_2S_2$			
Benzoic acid, 5-chloro-2-(2-thienylthio)-	MeOH	257(4.18),323(3.59)	73-2970-83
$C_{11}H_7Cl_5O_2$			
1(6aH)-Pentalenone, 2,4,5,6,6a-pentachloro-3-(1-methylethoxy)-	hexane	220(3.68),277(4.27), 355(4.16)	18-0481-83
$C_{11}H_7F_3O$			
2-Naphthalenol, 1-(trifluoromethyl)-	MeOH	223(4.84),266(3.60), 276(3.65),287(3.52), 319(3.39),331(3.44)	23-0368-83
$C_{11}H_7IN_2$			
3H-Benz[e]indazole, 9-iodo-	EtOH	207(4.39),247(4.61), 283(3.95),334(3.40)	35-6096-83
$C_{11}H_7NO$			
Furo[2,3-g]quinoline	EtOH	243(4.78),321(4.12), 332(4.17)	44-0774-83
Furo[3,2-g]quinoline	EtOH	244(4.77),321(4.09), 333(4.13)	44-0774-83
$C_{11}H_7NOS_2$			
4H-Indeno[1,2-c]isothiazol-4-one, 3-(methylthio)-	EtOH	235(4.14),270(4.52), 305(4.20),340(3.85)	95-1243-83
$C_{11}H_7NO_2S$			
9H-[1]Benzothiopyrano[2,3-d]oxazol-9-one, 2-methyl-	MeOH	262(4.45),350(3.90), 380(3.43)	83-0921-83
$C_{11}H_7NO_3$			
1H-Cyclobuta[de]naphthalen-1-ol, nitrate	EtOH	224(4.84),272(3.67), 302(2.71),312(2.53)	35-6096-83
$C_{11}H_7NO_3S$			
[1]Benzothiopyrano[3,2-b]-1,4-oxazine-3,10(2H,4H)-dione	MeOH	260(4.37),262(4.37), 352(4.11),354(4.11)	83-0921-83
$C_{11}H_7NO_4$			
7H-Furo[2,3-f][1]benzopyran, 2-nitro-	EtOH	222(4.396),227(4.367), 255(3.847),300(3.900), 377(3.994)	111-0015-83
$C_{11}H_7NO_5S$			
4H-1-Benzothiopyran-4-one, 3-acetoxy-2-nitro-	MeOH	234(4.39),258(4.30), 391(3.71)	83-0921-83

Compound	Solvent	λ_{max}(log ε)	Ref.
$C_{11}H_7N_3$			
2-Propenenitrile, 3-(2-quinoxalinyl)-, (E)-	EtOH	208(4.20),258(4.48), 339(4.01)	12-0963-83
$C_{11}H_7N_3S$			
Propanedinitrile, (2H-1,4-benzothiazin-3(4H)-ylidene)-	MeOH	279(4.33),313(4.16), 339(4.19)	87-0845-83
Quinoxaline, 2-(4-thiazolyl)-	EtOH	241(3.87),266(3.75), 341(3.63),354s(3.55)	106-0829-83
$C_{11}H_7N_5$			
5H-Pyrido[3',2':4,5][1,2,3]triazolo-[1,2-a]benzotriazol-6-ium, hydroxide, inner salt	EtOH	231(4.54),267(3.74), 280(3.86),288(3.90), 364(4.47)	20-0067-83
$C_{11}H_7N_5O_6$			
2-Pyridinamine, 3,5-dinitro-N-(4-nitrophenyl)-	neutral	366(4.280)	103-0511-83
	acid	301(4.250)	103-0511-83
2-Pyridinamine, N-(2,4,6-trinitrophenyl)-	neutral	380(4.110)	103-0511-83
	acid	357(4.064)	103-0511-83
$C_{11}H_7N_7O_5$			
1H-Pyrazole-4-carbonitrile, 2,3-dihydro-5-[nitro[(4-nitrophenyl)azo]-methyl]-3-oxo-	MeCN	396(4.29)	48-0041-83
$C_{11}H_8$			
1H-Cyclobuta[de]naphthalene	EtOH	224(4.84),272(3.67), 312(2.53)	35-6104-83
$C_{11}H_8BrP$			
Phosphorin, 3-(4-bromophenyl)-	EtOH	212(4.45),244(4.58), 265(4.27),307(3.14)	24-0445-83 +24-1756-83
$C_{11}H_8Br_2N_2O_3$			
Acetic acid, bromo(1-bromo-3,4-dihydro-3-oxo-2(1H)-quinoxalinylidene)-, methyl ester	EtOH	234.5(4.33),292.5(3.88), 349.0(3.92)	94-3009-83
$C_{11}H_8ClNO_2S$			
3-Buten-2-one, 1-chloro-4-(2-thioxo-3(2H)-benzoxazolyl)-	EtOH	315(5.16)	105-0478-83
$C_{11}H_8ClNO_3S$			
Acetamide, 2-chloro-N-(3-hydroxy-4-oxo-4H-1-benzothiopyran-2-yl)-	$CHCl_3$	312(3.70),374(4.28)	83-0921-83
$C_{11}H_8ClN_3OS$			
Acetamide, 2-(6-chloro-2H-1,4-benzothiazin-3(4H)-ylidene)-2-cyano-	MeOH-$CHCl_3$	282(4.47),319(4.23)	87-0845-83
Acetamide, 2-(7-chloro-2H-1,4-benzothiazin-3(4H)-ylidene)-2-cyano-	MeOH-$CHCl_3$	282(4.38)	87-0845-83
$C_{11}H_8Cl_2N_2$			
Pyrazine, 2,6-dichloro-3-methyl-5-phenyl-	EtOH	243(3.70),255(3.81), 309(3.75)	4-0311-83
$C_{11}H_8Cl_2N_2O$			
Pyrazine, 2,6-dichloro-3-methyl-5-phenyl-, 4-oxide	EtOH	214(4.22),242(4.29), 262s(4.13),282s(3.91), 318(3.59)	4-0311-83

Compound	Solvent	λ_{max}(log ε)	Ref.
$C_{11}H_8Cl_2N_2O_3$			
Acetic acid, chloro(1-chloro-3,4-dihydro-3-oxo-2(1H)-quinoxalinylidene)-, methyl ester	EtOH	233(4.34),290(3.88),355(3.89)	94-3009-83
$C_{11}H_8CrN_2O_5Se$			
Chromium, pentacarbonyl(4,5,6,7-tetrahydro-1,2,3-benzoselenadiazole-N^2)-, (OC-6-22)-	hexane	237(4.63),450(3.81)	24-0230-83
	MeOH	240(4.54),412(3.79)	24-0230-83
$C_{11}H_8F_3NOS$			
2-Propanone, 1,1,1-trifluoro-3-(3-methyl-2(3H)-benzothiazolylidene)-	EtOH	214(3.66),246(3.18),356(3.80),367(3.84)	39-0795-83C
$C_{11}H_8N$			
Azepino[2,1,7-cd]pyrrolizinium perchlorate	CF_3COOH	259(4.14),296s(3.97),324(4.01),435(3.28)	24-1174-83
$C_{11}H_8N_2$			
3H-Benz[e]indazole	EtOH	230(4.52),242(4.43),287(3.94),328(3.36)	35-6096-83
Pyrazino[1,2-a]indole	EtOH	224s(--),253(4.68),307(3.44),320(3.62),335(3.54),366s(--),388s(--),405(3.48),426(3.39),453s(--)	118-1037-83
Pyridazino[1,6-a]indole	EtOH	226(4.25),255s(--),262(4.67),278s(--),289s(--),314(3.40),325(3.48),342(3.47),385s(--),399(3.15),422(3.18),448s(--)	118-1037-83
Pyrido[1,2-a]benzimidazole	EtOH	206(4.84),242(4.82),246(4.84),258(4.26),268(4.18),300(3.76),312(3.78),330(3.76),342(3.94),360(3.79),380(3.48)	103-1102-83
9H-Pyrido[3,4-b]indole	MeOH	233(4.61),248s(4.37),281s(4.09),288(4.29),336(3.67),349(3.68)	18-1450-83
	EtOH	234(4.90),250s(4.86),282(4.52),289(4.60),339(4.20),352(4.24)	78-0033-83
Pyrimido[1,2-a]indole	EtOH	223(4.21),259s(--),264(4.82),285s(--),310(3.51),322(3.57),338(3.47),433(3.10)	118-1037-83
Pyrimido[1,6-a]indole	EtOH	224(4.10),250s(--),257(4.61),270s(--),280s(--),293s(--),297(3.70),305(3.78),311s(--),319(3.81),357s((--),376(3.07),395(3.07),420s(--)	118-1037-83
$C_{11}H_8N_2O_2$			
2H-1-Benzopyran-2-one, 4-(diazomethyl)-7-methyl-	EtOH	245(4.11),252.5(4.08),289(3.97),320(4.21)	94-3014-83
5-Quinoxalinecarboxaldehyde, 8-acetyl-	MeOH	218(4.28),239(4.05),313(3.72),322s(3.70)	18-3358-83

Compound	Solvent	λ_{max}(log ε)	Ref.
$C_{11}H_8N_2O_3$			
2,4(1H,3H)-Pyrimidinedione, 1-benzoyl-	MeOH	260(4.15)	65-1294-83
$C_{11}H_8N_2O_3S_2$			
4-Thiazolecarboxylic acid, 4,5-dihydro-2-(6-hydroxy-2-benzothiazolyl)-	H_2O	268(3.88),327(4.27)	162-0586-83
$C_{11}H_8N_4$			
1,1-Cyclobutanedicarbonitrile, 3-(dicyanomethylene)-2,2-dimethyl-	EtOH	224(3.96),275(4.00)	44-1500-83
$C_{11}H_8N_4O_2$			
7H-Pyrazolo[4,3-d]pyrimidin-7-one, 2,6-dihydro-3-hydroxy-2-phenyl-	MeOH	248(4.28),304(3.77)	94-1228-83
Pyridazino[4,5-g]phthalazine-1,4-dione, 2,3-dihydro-2-methyl-	MeOH	234(4.70),263(3.61), 274(3.63),291(3.53), 340(3.68)	142-1279-83
2(1H)-Quinoxalinone, 3,4-dihydro-3-(1,3,4-oxadiazol-2-ylmethylene)-	EtOH	265.3(3.93),290.0(3.84), 364.5(4.8),384.5(4.24), 407(4.06)	142-1917-83
	70% DMSO-EtOH	350(3.88),363(3.93), 383.5(3.87),406(3.66)	142-1917-83
	70% TFA-EtOH	398.5(3.87),419.5(4.02), 445.0(3.88)	142-1917-83
$C_{11}H_8N_4O_3$			
Benzo[g]pteridine-2,4(1H,3H)-dione, 8-hydroxy-7-methyl-	pH 3	384-7(4.4)	65-2351-83
	MeOH	230(4.7),255(4.5), 385(4.4)	65-2351-83
	HOAc	392(4.3)	65-2351-83
	pyridine	392-5(4.3)	65-2351-83
$C_{11}H_8N_4O_4$			
2-Pyridinamine, N-(2,4-dinitrophenyl)-	neutral	365(4.255)	103-0511-83
	acid	360(4.220)	103-0511-83
2-Pyridinamine, 3,5-dinitro-N-phenyl-	neutral	345(4.127)	103-0511-83
	acid	306(4.114)	103-0511-83
$C_{11}H_8N_4S$			
Quinoxaline, 2-(2-amino-4-thiazolyl)-	EtOH	250(4.12),277s(3.84), 310s(3.32),379(3.55)	106-0829-83
$C_{11}H_8N_6O_3$			
4H-Pyrazolo[3,4-d]-1,2,3-triazin-4-one, 1,7-dihydro-5-(nitromethyl)-7-phenyl-	EtOH	228(4.36),305(3.93)	48-0041-83
$C_{11}H_8O_2S_2$			
1H-Indene-2-carbodithioic acid, 2,3-dihydro-1,3-dioxo-, methyl ester	EtOH	233(4.50),303(4.08), 380(4.49)	95-1243-83
$C_{11}H_8O_3$			
4H-1-Benzopyran-4-one, 2-acetyl-	EtOH	203(4.19),235s(4.10), 242(4.13),263s(3.71), 306(3.76)	33-0586-83
1,4-Naphthalenedione, 5-hydroxy-2-methyl- (plumbagin)	EtOH	211(4.39),266(3.99), 404(3.52)	102-1245-83
$C_{11}H_8O_3S$			
Benzo[b]thiophen-3(2H)-one, 2-(1,3-dioxolan-2-ylidene)-	EtOH	209(4.42),248(4.36), 256.5(4.36),284(4.15),	104-1530-83

Compound	Solvent	λ_{max}(log ε)	Ref.
(cont.)		300(4.20)	104-1530-83
4H-1-Benzothiopyran-4-one, 3-acetoxy-	MeOH	256(4.37),348(4.12)	83-0921-83
$C_{11}H_8O_4$			
2H-1-Benzopyran-2-one, 6-acetyl-7-hydroxy-	EtOH	207(4.05),227(4.15),256(4.47),309(4.02),341(4.06)	44-2709-83
2H-1-Benzopyran-2-one, 8-acetyl-7-hydroxy-	EtOH	209(4.23),233(3.91),242(3.92),267(3.93),317(4.06),344s(3.93)	44-2709-83
1,4-Naphthalenedione, 3,5-dihydroxy-2-methyl- (droserone)	EtOH	228(4.62),278(4.52),400(3.66)	102-1245-83
$C_{11}H_9Br$			
Bicyclo[4.4.1]undeca-1,3,5,7,9-pentaene, 3-bromo-	C_6H_{12}	265(4.79),305(3.82),367(2.38),375(2.36),383(2.30),392s(2.14),400s(1.97),406(1.88)	35-3375-83
$C_{11}H_9BrN_2OS_2$			
5H-Thiazolo[2,3-b]quinazoline-5-thione, 7-bromo-2,3-dihydro-2-methoxy-	EtOH	300(4.07),309(4.07),352(3.86),368(4.02),388(3.93)	97-0215-83
$C_{11}H_9BrN_2O_2S$			
4(5H)-Thiazolone, 5-[(4-bromophenyl)-methylene]-2-[(hydroxymethyl)amino]-	n.s.g.	243(3.99),290(4.12),335(4.37)	103-0745-83
5H-Thiazolo[2,3-b]quinazolin-5-one, 7-bromo-2,3-dihydro-2-methoxy-	EtOH	281(3.92),323(3.48),333(3.43)	97-0215-83
$C_{11}H_9BrN_2O_3$			
Acetic acid, (1-bromo-3,4-dihydro-3-oxo-2(1H)-quinoxalinylidene)-, methyl ester	EtOH	230(4.35),286(3.86),350(3.81)	94-3009-83
$C_{11}H_9ClN_2$			
Pyrazine, 2-chloro-3-methyl-5-phenyl-	EtOH	253(4.12),290(3.94),311(4.04)	4-0311-83
Pyrazine, 2-chloro-5-methyl-3-phenyl-	EtOH	235(3.80),252s(3.73),290(3.79),304(3.70)	4-0311-83
$C_{11}H_9ClN_2O$			
Pyrazine, 2-chloro-3-methyl-5-phenyl-, 1-oxide	EtOH	231s(3.99),264(4.45),327(3.71)	4-0311-83
Pyrazine, 2-chloro-5-methyl-3-phenyl-, 1-oxide	EtOH	232(4.22),254(4.22),316(3.59)	4-0311-83
Pyrazine, 3-chloro-2-methyl-6-phenyl-, 1-oxide	EtOH	253(4.29),290s(3.78),348s(3.45)	4-0311-83
Pyrazine, 3-chloro-6-methyl-2-phenyl-, 1-oxide	EtOH	232(4.33),248s(4.24),280s(3.95),313(3.59)	4-0311-83
$C_{11}H_9ClN_2O_2$			
Pyrazine, 2-chloro-3-methyl-5-phenyl-, 1,4-dioxide	EtOH	210(4.19),266(4.45),316(4.29)	4-0311-83
Pyrazine, 2-chloro-5-methyl-3-phenyl-, 1,4-dioxide	EtOH	242(4.22),259(4.05),311(4.26)	4-0311-83
$C_{11}H_9ClN_2O_2S$			
4(5H)-Thiazolone, 5-[(4-chlorophenyl)-methylene]-2-[(hydroxymethyl)-amino]-, (Z)-	n.s.g.	241(4.15),291(4.25),334(4.52)	103-0745-83

Compound	Solvent	λ_{max}(log ε)	Ref.
$C_{11}H_9ClN_2O_3$			
Acetic acid, (1-chloro-3,4-dihydro-3-oxo-2(1H)-quinoxalinylidene)-, methyl ester	EtOH	230(4.37),285.5(3.86), 350(3.83)	94-3009-83
$C_{11}H_9Cl_4NO$			
3-Butenamide, 2,3,4,4-tetrachloro-N-(4-methylphenyl)-	heptane	197(4.26),237(4.07)	5-1496-83
$C_{11}H_9FN_2O_2S$			
4(5H)-Thiazolone, 5-[(4-fluorophenyl)-methylene]-2-[(hydroxymethyl)amino]-	n.s.g.	238(3.84),285(3.99), 330(4.19)	103-0745-83
$C_{11}H_9NOS_2$			
1H-Indene-2-carbodithioic acid, 3-amino-1-oxo-	EtOH	214(4.24),236(4.46), 262(4.12),310(4.26), 368(4.53)	95-1243-83
$C_{11}H_9NO_2S$			
3-Buten-2-one, 4-(2-thioxo-3(2H)-benzoxazolyl)-	EtOH	310(4.45)	105-0478-83
$C_{11}H_9NO_2SSe$			
2-Selenazolidinone, 5-[(3-methoxyphen-yl)methylene]-4-thioxo-	n.s.g.	243(4.01),273(3.86), 344(4.17),432(4.23)	103-0273-83
2-Selenazolidinone, 5-[(4-methoxyphen-yl)methylene]-4-thioxo-	n.s.g.	240(3.91),282(3.73), 354(4.14),434(4.26)	103-0273-83
$C_{11}H_9NO_3S$			
1H-Pyrrole-2-carboxaldehyde, 1-(phen-ylsulfonyl)-	MeOH	224(4.01),238(3.97), 275(3.94),283s(3.94)	44-3214-83
$C_{11}H_9N_2O_3$			
Pyridinium, 1-[2-(5-nitro-2-furanyl)-ethenyl]-, bromide, (E)-	MeOH	227(3.37),362(3.63)	73-1891-83
(Z)-	MeOH	225(3.27),261(3.01), 340(3.05)	73-1891-83
iodide	MeOH	225(3.27),261(3.01), 337(3.05)	73-1891-83
perchlorate	MeOH	226(3.20),263(2.86), 341(2.97)	73-1891-83
picrate	MeOH	223(3.53),351(3.48)	73-1891-83
$C_{11}H_9N_3$			
1H-Imidazole-1-carbonitrile, 2-methyl-4-phenyl-	EtOH	252(3.77)	4-1277-83
1H-Imidazole-1-carbonitrile, 2-methyl-5-phenyl-	EtOH	217(4.40),258(4.26)	4-1277-83
1H-Imidazole-1-carbonitrile, 4-methyl-2-phenyl-	EtOH	273(4.15)	4-1277-83
1H-Imidazole-1-carbonitrile, 5-methyl-2-phenyl-	EtOH	274(4.27)	4-1277-83
1H-Imidazole-1-carbonitrile, 5-methyl-4-phenyl-	EtOH	248(3.90)	4-1277-83
1H-Pyrido[3,2,1-ij][1,2,4]benzotria-zin-4-ium, 2-methyl-, hydroxide, inner salt	EtOH	262(4.17),265(4.20), 310s(3.19),324(3.56), 337(3.77),369(3.26), 414(3.35),468(3.29), 502(3.27),534(3.00)	39-0349-83C
1H-Pyrido[3,2,1-ij][1,2,4]benzotria-zin-4-ium, 7-methyl-, hydroxide,	EtOH	261(4.49),314s(3.53), 330(3.92),343(4.11),	39-0349-83C

Compound	Solvent	$\lambda_{max}(\log \epsilon)$	Ref.
inner salt (cont.)		422(3.79),460(3.76), 484(3.74),520(3.45)	39-0349-83C
1H-Pyrido[3,2,1-ij][1,2,4]benzotria- zin-4-ium, 8-methyl-, hydroxide, inner salt	EtOH	265(4.45),271(4.45), 314s(3.58),330(3.91), 344(4.11),406(3.56), 422(3.61),487(3.54), 520(3.52),560(3.20)	39-0349-83C
1H-Pyrido[3,2,1-ij][1,2,4]benzotria- zin-4-ium, 10-methyl-, hydroxide, inner salt	EtOH	268(4.49),273(4.49), 314s(3.52),328(3.87), 342(4.09),405(3.56), 421(3.61),487(3.52), 518(3.46),560(3.11)	39-0349-83C
1H-Pyrido[2,3-b]indol-2-amine	MeOH	231(4.10),260(3.66), 337(3.76)	18-1450-83
9H-Pyrido[3,4-b]indol-1-amine	MeOH	240(4.70),278(3.90), 288(4.00),299(3.47), 326(3.73),337(3.85), 350(3.78)	18-1450-83
9H-Pyrido[3,4-b]indol-3-amine	MeOH	236(4.53),248s(4.49), 292s(4.00),299(4.10)	18-1450-83
$C_{11}H_9N_3OS$			
Acetamide, 2-(2H-1,4-benzothiazin- 3(4H)-ylidene)-2-cyano-	MeOH	276(4.40)	87-0845-83
$C_{11}H_9N_3O_2$			
Formamide, N,N'-1,3-isoquinoline- diylbis-	EtOH	240(4.58),277s(3.85), 290(4.06),301(4.08), 347(3.79)	39-1137-83C
Pyridine, 2-(phenyl-ONN-azoxy)-, 1-oxide	n.s.g.	268(4.29),330(3.89)	142-1751-83
Pyridine, 3-(phenyl-NNO-azoxy)-, 1-oxide	n.s.g.	278(4.34),331(4.17)	142-1751-83
Pyridine, 3-(phenyl-ONN-azoxy)-, 1-oxide	n.s.g.	278(4.34),308(4.20)	142-1751-83
Pyridine, 4-(phenyl-NNO-azoxy)-, 1-oxide	n.s.g.	232(4.12),349(4.38)	142-1751-83
Pyridine, 4-(phenyl-ONN-azoxy)-, 1-oxide	n.s.g.	289(3.91),367(4.38)	142-1751-83
$C_{11}H_9N_3O_3$			
2,2'-Bipyridine, 4-methoxy-4'-nitro-	EtOH	217(4.56),246s(4.19), 313(3.57)	44-0283-83
$C_{11}H_9N_3O_4$			
2,2'-Bipyridine, 4-methoxy-4'-nitro-, 1-oxide	EtOH	237(4.21),282(3.71), 349s(2.88)	44-0283-83
$C_{11}H_9N_3O_5$			
2,2'-Bipyridine, 4-methoxy-4'-nitro-, 1,1'-dioxide	EtOH	267(4.15),329(4.12)	44-0283-83
$C_{11}H_9N_3S$			
10H-Pyrazino[2,3-b][1,4]benzothiazine, 2-methyl-	EtOH	246(4.50),290(3.27), 322(3.56)	87-0564-83
$C_{11}H_9N_3S_2$			
Imidazo[2,1-b][1,3,5]benzothiadiaze- pine, 5-(methylthio)-	MeOH	206(4.56),232(4.18), 244(4.17),262s(3.08)	4-1287-83

Compound	Solvent	$\lambda_{max}(\log \epsilon)$	Ref.
$C_{11}H_9N_5$			
Imidazo[1,5-a]-1,3,5-triazine, 4-amino-8-phenyl-	pH 1	258s(4.00),285(4.07), 333(3.93)	44-0003-83
	pH 7	284(4.11),295s(4.07), 340(3.92)	44-0003-83
	pH 13	328(4.23)	44-0003-83
Pyrazine, 2-azido-3-methyl-5-phenyl-	EtOH	221(4.06),240(4.32)	4-1277-83
Pyrazine, 2-azido-3-methyl-6-phenyl-	EtOH	232s(3.91),268s(3.87), 307(4.11)	4-1277-83
Pyrazine, 2-azido-5-methyl-3-phenyl-	EtOH	222(4.64),240(4.48), 263s(3.70)	4-1277-83
Pyrazine, 2-azido-6-methyl-3-phenyl-	EtOH	235(3.68),261s(3.87), 310(4.30),317(4.31), 336s(3.98)	4-1277-83
Pyrazine, 2-azido-6-methyl-5-phenyl-	EtOH	242(4.21),271(4.06), 305s(3.67)	4-1277-83
$C_{11}H_9N_5O$			
4H-Pyrazolo[3,4-d]pyridazin-4-one, 3-amino-2,5-dihydro-2-phenyl-	EtOH	240(4.47),269s(3.83), 305(3.71)	48-0041-83
$C_{11}H_9N_5O_2$			
Pyrido[4,3-g]pteridine-2,4(3H,10H)-dione, 3,10-dimethyl- (oxidized)	n.s.g.	317(3.75),442(3.83)	138-0775-83
reduced	n.s.g.	359(3.92),470(3.30)	138-0775-83
$C_{11}H_9P$			
Phosphorin, 3-phenyl-	EtOH	208(4.38),234(4.53), 262(4.16),302(2.69)	24-0445-83
Phosphorin, 4-phenyl-	EtOH	216(4.15),284(4.18)	24-0445-83
$C_{11}H_{10}BrCl_3N_2O_5S$			
5-Thia-1-azabicyclo[4.2.0]oct-2-ene-2-carboxylic acid, 3-(bromomethyl)-7-(formylamino)-8-oxo-, 2,2,2-trichloroethyl ester, 5-oxide, [5S-(5α,6β,7α)]-	EtOH	283(4.00)	78-0337-83
$C_{11}H_{10}BrNO_2$			
Isoquinoline, 8-bromo-6,7-dimethoxy-	MeOH	238(4.67),296(3.57)	83-0520-83
$C_{11}H_{10}BrN_5O_2$			
Pyrrolo[2,3-c]azepin-8(1H)-one, 4-(2-amino-1,5-dihydro-5-oxo-4H-imidazol-4-ylidene)-2-bromo-4,5,6,7-tetrahydro- (hymenialdisine)	MeOH	227s(4.05),270(4.07), 346(4.20)	94-2321-83
$C_{11}H_{10}Br_2$			
1,3,5-Cycloheptatriene, 1-(2,2-dibromoethenyl)-6-ethenyl-	C_6H_{12}	255(4.41),332(3.88)	35-3375-83
$C_{11}H_{10}Br_2N_2O$			
Ethanone, 1-(5,7-dibromo-2,3-dihydro-1H-cycloheptapyrazin-9-yl)-	MeOH	233(4.13),270(4.07), 386(3.75),461(3.80)	18-3358-83
$C_{11}H_{10}ClNO_2$			
Ethanone, 2-chloro-1-(5-methoxy-1H-indol-3-yl)-	MeOH	220(4.30),252(4.04), 275(3.95),302(3.91)	100-0732-83
1H-Indole-2,3-dione, 1-(2-chloroethyl)-5-methyl-	EtOH	216(4.14),249(4.33), 255(4.26),303(3.41), 435(2.71)	103-0286-83

Compound	Solvent	λ_{max}(log ε)	Ref.
1H-Indole-2,3-dione, 1-(2-chloroethyl)-7-methyl-	EtOH	218(4.17),248(4.30), 253(4.26),312(3.56), 427(2.78)	103-0286-83
$C_{11}H_{10}ClNO_5$			
4H,5H-Pyrano[3,4-e][1,3]oxazine-4,5-dione, 2-butoxy-7-chloro-	$CHCl_3$	277(4.08),285(4.07), 390(4.01)	24-3725-83
$C_{11}H_{10}ClN_3O_3S$			
5H-Pyrimido[4,5-b][1,4]thiazine-6-carboxylic acid, 7-acetyl-4-chloro-, ethyl ester	EtOH	224(4.13),254(4.14), 295(3.27),365(3.59)	103-0618-83
$C_{11}H_{10}Cl_3NO_2$			
2-Butenamide, 2,4,4-trichloro-3-hydroxy-N-(4-methylphenyl)-	heptane	195(4.36),220(3.38), 290(4.22)	5-1496-83
$C_{11}H_{10}N_2$			
Benzonitrile, 4-(3,4-dihydro-2H-pyrrol-5-yl)-	C_6H_{12}	257.5(4.31),296s(3.00)	88-2279-83
Pyrazine, 2-methyl-6-phenyl-	EtOH	247(4.03),290(4.03)	4-0311-83
3-Quinolinecarbonitrile, 1,2-dihydro-1-methyl-	80% isoPrOH	248(4.42),295(3.75), 416(3.45)	44-2053-83
3-Quinolinecarbonitrile, 1,4-dihydro-1-methyl-	80% isoPrOH	233(4.04),241(3.94), 334(4.08)	44-2053-83
$C_{11}H_{10}N_2O$			
2,2'-Bipyridine, 4-methoxy-	EtOH	218(4.41),242(4.02), 277(4.13)	44-0283-83
Pyrazine, 2-methyl-6-phenyl-, 1-oxide	EtOH	249(4.35),282(3.94)	4-0311-83
Pyrazine, 2-methyl-6-phenyl-, 4-oxide	EtOH	229s(4.04),262(4.46), 321(3.78)	4-0311-83
Pyrazinemethanol, 6-phenyl-	EtOH	251(3.88),288(3.83), 362(3.20)	4-0311-83
2(1H)-Pyrazinone, 5-methyl-3-phenyl-	EtOH	227(3.62),252(3.65), 355(3.76)	4-0311-83
$C_{11}H_{10}N_2O_2$			
2,2'-Bipyridine, 4-methoxy-, 1-oxide	EtOH	237(4.03),272(3.90), 320(3.04)	44-0283-83
Pyrazine, 2-methyl-6-phenyl-, 1,4-dioxide	EtOH	260(4.35),276.5s(4.14), 315.5(4.15)	4-0311-83
$C_{11}H_{10}N_2O_2S$			
4(5H)-Thiazolone, 2-amino-5-[(4-methoxyphenyl)methylene]-	n.s.g.	246(4.09),296(3.93), 350(4.39)	103-0611-83
5H-Thiazolo[2,3-b]quinazolin-5-one, 2,3-dihydro-2-methoxy-	EtOH	273(4.20),281s(4.15), 313(3.55),323(3.47)	97-0215-83
$C_{11}H_{10}N_2O_2S_5$			
Benzenesulfenamide, N-[4,5-bis(methylthio)-1,3-dithiol-2-ylidene]-2-nitro-	1,2-$C_2H_4Cl_2$	312(4.14),420(3.77)	118-0050-83
$C_{11}H_{10}N_2O_3$			
Acetic acid, (3,4-dihydro-3-oxo-2(1H)-quinoxalinylidene)-, methyl ester	$CHCl_3$	262(3.92),285(3.79), 359(4.15),377(4.20), 396(3.99)	142-1917-83
	DMSO	288(3.53),359(4.26), 377(4.34),396(4.09)	142-1917-83
	TFA	331(3.74),381(3.50)	142-1917-83

Compound	Solvent	λ_{max}(log ϵ)	Ref.
2,2'-Bipyridine, 4-methoxy-, 1,1'-di-oxide	EtOH	217(4.34),237(4.15), 269(4.25),319s(3.32)	44-0283-83
5H-Cyclopentapyrazine, 2-(2-furanyl)-6,7-dihydro-, 1,4-dioxide	EtOH	275(4.29),298(4.53), 318s(4.28),333s(4.23)	103-1104-83
2-Furanmethanamine, 5-nitro-N-phenyl-	MeOH	204(4.35),242(4.15), 314(4.60)	73-2682-83
$C_{11}H_{10}N_2O_4$			
Benzenamine, 2-[(5-nitro-2-furanyl)-methoxy]-	MeOH	211(3.20),238(1.26), 315(2.72)	73-2682-83
Phenol, 2-[[(5-nitro-2-furanyl)-methyl]amino]-	MeOH	208(4.61),242(4.14), 312(4.33)	73-2682-83
Phenol, 3-[[(5-nitro-2-furanyl)-methyl]amino]-	MeOH	207(4.57),243(4.14), 318(4.14)	73-2682-83
Phenol, 4-[[(5-nitro-2-furanyl)-methyl]amino]-	MeOH	204(4.29),242(4.07), 317(4.13)	73-2682-83
$C_{11}H_{10}N_2O_4S$			
Thieno[2,3-b]pyridine-2-carboxylic acid, 3-methyl-5-nitro-, ethyl ester	EtOH	225(3.85),272(4.54), 306(3.80)	103-1163-83
$C_{11}H_{10}N_2O_6$			
Pyrano[2,3-c]pyrazole-3,4-dicarboxylic acid, 1,6-dihydro-1-methyl-6-oxo-, dimethyl ester	MeOH	226(3.43),243s(3.54), 322(3.97)	24-1525-83
$C_{11}H_{10}N_2S$			
4(3H)-Pyrimidinethione, 3-(phenyl-methyl)-	EtOH	295(4.05),350s(--)	39-2645-83C
$C_{11}H_{10}N_4$			
4H-Imidazole-4-carbonitrile, 5-amino-4-(phenylmethyl)-	EtOH-pH 7	274(3.39)	44-0003-83
	EtOH-pH 13	286(3.77)	44-0003-83
Pyridine, 2-(3-phenyl-1-triazenyl)-	toluene	342(4.1)	80-0065-83
[1,2,4]Triazino[4,3-b]indazole, 3-ethyl-	EtOH	271(4.43),284(4.20), 338(3.53),357(3.59)	44-2330-83
$C_{11}H_{10}N_4O_2$			
1H-Pyrazolium, 1-[(1,3-dihydro-1,3-di-oxo-2H-isoindol-2-yl)amino]-4,5-di-hydro-, hydroxide, inner salt	EtOH	219(4.53),233s(4.30), 240s(4.24),264(4.07)	104-2317-83
1H-1,2,3-Triazole-4-carboxaldehyde, 1-acetyl-5-(phenylamino)-	EtOH	247(4.32),337(4.00)	95-0594-83
$C_{11}H_{10}N_4O_4$			
Hydrazinecarboxylic acid, 2-(2-quinox-alinylmethylene)-, N,N'-dioxide (carbadox)	H_2O	236(4.04),251(4.04), 303(4.56),366(4.21), 373(4.21)	162-0245-83
4-Pyrazolecarboxylic acid, 5-amino-3-(nitromethyl)-1-phenyl-	EtOH	232(4.35),255s(3.97)	48-0041-83
$C_{11}H_{10}N_6$			
6H-Imidazo[1,2-b][1,2,4]triazol-6-one, 5-methyl-, phenylhydrazone	EtOH	274(3.14),334(3.71)	4-0639-83
1H-Purine-6,8-diamine, N^8-phenyl-	pH 1	300(4.29)	94-3454-83
	H_2O	292(4.38)	94-3454-83
	pH 13	300(4.44)	94-3454-83
$C_{11}H_{10}N_6O_3$			
Carbamic acid, [(5-azido-1,3,4-oxadia-zol-2-yl)methyl]-, phenylmethyl ester	MeOH	237(4.03)	35-0902-83

Compound	Solvent	$\lambda_{max}(\log \epsilon)$	Ref.
$C_{11}H_{10}N_6O_4$			
Pyrimido[5,4-g]pteridine-2,4,6,8-tetrone, 1,3,7,9-tetrahydro-3,7,10-trimethyl-	neutral	279(4.19),406(4.37)	35-3304-83
	anion	241(4.67),288(3.98), 423(4.58)	35-3304-83
$C_{11}H_{10}N_8O_5$			
1H-Pyrrole-4-carboxamide, 3-amino-5-[nitro[(4-nitrophenyl)azo]phenyl]-	HOAc	368(3.78)	48-0041-83
$C_{11}H_{10}O$			
3a(4H)-Azulenol, 4-methylene-	hexane-ether	265(4.06),392(3.61)	88-4261-83
Naphthalene, 2-methoxy-	MeOH and MeOH-base	261(3.6),271(3.6), 281(3.5),313(3.18), 328(3.28)	160-1403-83
$C_{11}H_{10}O_2$			
4H-1-Benzopyran-4-one, 2-ethyl-	EtOH	223(4.32),241s(3.96), 262(3.85),295(3.87), 299s(3.86)	33-0586-83
Pentacyclo[5.4.0.0^{2,6}.0^{3,10}.0^{5,9}]undecane-8,11-dione	MeCN	301(1.49)	44-1862-83
$C_{11}H_{10}O_2S$			
2H-1-Benzothiopyran-2-one, 7-methoxy-4-methyl-	EtOH	330(3.79),343(3.79)	139-0059-83B
	C_6H_{12}	333(--),342(--)	139-0059-83B
4H-1-Benzothiopyran-4-one, 6-methoxy-2-methyl-	EtOH	260(4.38),357(3.94)	139-0059-83B
4H-1-Benzothiopyran-4-one, 7-methoxy-2-methyl-	EtOH	268(4.42),328(4.05)	139-0059-83B
$C_{11}H_{10}O_2S_2$			
6H-Cyclohepta[c]furan-6-one, 5,7-bis(methylthio)-	CH_2Cl_2	300(4.2),310(4.16), 352(3.7),376(3.7), 474(3.2)	83-0730-83
$C_{11}H_{10}O_2S_3$			
2-Thiophenecarboxylic acid, 3-[(2-thienylthio)methyl]-, methyl ester	MeOH	253(4.11)	4-1085-83
3-Thiophenecarboxylic acid, 2-[(2-thienylthio)methyl]-, methyl ester	MeOH	245(4.15)	4-1085-83
3-Thiophenecarboxylic acid, 2-[(3-thienylthio)methyl]-, methyl ester	MeOH	246(4.04)	4-1085-83
$C_{11}H_{10}O_3$			
2H-1-Benzopyran-2-one, 7-methoxy-4-methyl-	EtOH	310(3.97),322(4.00)	139-0059-83B
	C_6H_{12}	310(--),319(--)	139-0059-83B
with β-cyclodextrin	H_2O	322(4.16)	95-0193-83
3,6-Dioxabicyclo[3.1.0]hexan-2-one, 4-methyl-1-phenyl-, (1α,4α,5α)-	EtOH	230(3.30),254(2.74)	54-0515-83
2(5H)-Furanone, 4-hydroxy-5-methyl-3-phenyl-	EtOH	260(4.13),278(4.10)	54-0515-83
Furo[2,3-f]-1,3-benzodioxole, 6-ethenyl-6,7-dihydro-	EtOH	238(3.39),300(3.63), 310(3.57)	78-0169-83
$C_{11}H_{10}O_3S$			
Sulfonium, dimethyl-, 2,4-dioxo-2H-1-benzopyran-3(4H)-ylide	MeOH	280(4.02)	39-0841-83B
	EtOH	212(4.47),225(4.32), 244(3.84),280(4.03)	104-0283-83
$C_{11}H_{10}O_4$			
Benzenepropanoic acid, α,β-dioxo-,	CH_2Cl_2	261(3.30),402(1.43)	5-1694-83

Compound	Solvent	λ_{max}(log ε)	Ref.
ethyl ester (cont.)			5-1694-83
1H-2-Benzopyran-5-carboxaldehyde, 3,4-dihydro-8-hydroxy-3-methyl-1-oxo-	EtOH	276(4.15),314(3.58)	39-2185-83C
2H-1-Benzopyran-2-one, 6,7-dimethoxy-	MeOH	220(4.34),225(4.3), 250(3.95),272(3.93), 295(3.95),310(3.95), 340(4.05)	120-0367-83
$C_{11}H_{10}O_4S_2$			
1,3-Benzodioxole-4,5-dione, 6-(1,3-dithian-2-yl)-	MeCN	210(4.14),252(3.64), 435(3.10)	64-0752-83B
$C_{11}H_{10}O_5$			
Benzoic acid, 2,3-diformyl-6-methoxy-5-methyl-	n.s.g.	214(4.27),271(3.84), 304(3.51)	162-0634-83
1H-2-Benzopyran-5-carboxylic acid, 3,4-dihydro-8-hydroxy-3-methyl-1-oxo-	EtOH	226(4.16),244(3.74), 314(3.23)	39-2185-83C
$C_{11}H_{10}S_2$			
4H-1-Benzothiopyran-4-thione, 2,6-dimethyl-	3-Mepentane	410(4.30)	23-0894-83
$C_{11}H_{11}$			
Tricyclo[7.2.0.0^{3,6}]undeca-2,6,8-trien-1-ylium hexafluoroantimonate	MeCN	233(4.29),308(3.58)	44-0596-83
$C_{11}H_{11}BrN_2O$			
Ethanone, 1-(5-bromo-2,3-dihydro-1H-cycloheptapyrazin-9-yl)-	MeOH	245(4.06),252(4.06), 270s(4.04),380(3.64), 449(3.82)	18-3358-83
1,2,4-Oxadiazole, 3-(4-bromo-2,6-dimethylphenyl)-5-methyl-	MeOH	270s(2.78),278s(2.75)	4-1609-83
$C_{11}H_{11}BrN_2O_2$			
5-Quinoxalinecarboxaldehyde, 8-acetyl-6-bromo 1,2,3,4-tetrahydro-	MeOH	232(4.24),280(3.81), 331(3.98),493(3.74), 515s(3.70)	18-3358-83
Tryptophan, 2-bromo-	H_2O	218(4.48),279(3.84), 288(3.75)	88-5555-83
$C_{11}H_{11}BrO$			
5H-Benzocyclohepten-5-one, 1-bromo-6,7,8,9-tetrahydro-	EtOH	249(3.85),299(3.18)	22-0096-83
5H-Benzocyclohepten-5-one, 3-bromo-6,7,8,9-tetrahydro-	EtOH	248(3.78),290(3.06)	22-0096-83
$C_{11}H_{11}ClN_2O_4$			
4H,5H-Pyrano[3,4-e][1,3]oxazine-4,5-dione, 7-chloro-2-(diethylamino)-	$CHCl_3$	302(4.08),342(4.25)	24-3725-83
$C_{11}H_{11}Cl_3N_2O_5S$			
5-Thia-1-azabicyclo[4.2.0]oct-2-ene-2-carboxylic acid, 7-(formylamino)-3-methyl-8-oxo-, 2,2,2-trichloroethyl ester, 5-oxide, [5S-(5α,6β,7α)]-	EtOH	269(3.89)	78-0337-83
$C_{11}H_{11}F_5N_2O_5$			
Uridine, 2'-deoxy-5-(trifluoroethyl)-	pH 2	263(4.08)	87-0598-83
	pH 12	262(3.94)	87-0598-83
	MeOH	263(4.09)	87-0598-83

Compound	Solvent	$\lambda_{max}(\log \epsilon)$	Ref.
$C_{11}H_{11}N$			
1H-Indole, 5-ethenyl-1-methyl-	EtOH	208(4.6),249(4.7), 278(4.0)	104-0388-83
1H-Indole, 5-ethenyl-2-methyl-	EtOH	208(4.05),252(4.7)	104-0388-83
1H-Indole, 5-ethenyl-3-methyl-	EtOH	208(4.03),250(4.72)	104-0388-83
$C_{11}H_{11}NO$			
2-Propenenitrile, 2-ethoxy-3-phenyl-, (E)-	EtOH	219(3.91),277(4.12)	78-1551-83
(Z)-	EtOH	205(3.95),219(3.97), 281(4.39)	78-1551-83
$C_{11}H_{11}NOS$			
2H-Pyrrol-2-one, 1,5-dihydro-3,4-dimethyl-5-(2-thienylmethylene)-, (E)-	MeOH	342(4.13)	54-0114-83
(Z)-	MeOH	352(4.37)	54-0114-83
Quinoline, 4-methoxy-3-(methylthio)-	EtOH	332(3.12)	49-0281-83
4(1H)-Quinolinone, 1-methyl-3-(methylthio)-	EtOH	338(3.95),378(3.34)	49-0281-83
$C_{11}H_{11}NO_2$			
Ethanone, 1-(4,5-dihydro-3-phenyl-5-isoxazolyl)-	C_6H_{12}	275(4.10)	44-3189-83
1H-Indole-5-carboxaldehyde, 1-ethenyl-4,5,6,7-tetrahydro-4-oxo-	MeOH	239(4.16),274(4.10), 322s(3.70)	23-1697-83
1H-Indole-3-carboxylic acid, 2-methyl-, methyl ester	EtOH	217(4.58),230(4.27), 256(4.00),285(4.06)	12-1419-83
Naphthalene, 1,2-dihydro-3-(nitromethyl)-	MeOH	265(4.04)	22-0061-83
2H-Pyrrol-2-one, 5-(2-furanylmethylene)-1,5-dihydro-3,4-dimethyl-, (E)-	MeOH	357(4.35)	54-0114-83
(Z)-	MeOH	350(4.55)	54-0114-83
$C_{11}H_{11}NO_2S$			
Benzenepropanethioamide, N,N-dimethyl-α,β-dioxo-	MeCN	227(3.69),266(3.87), 318s(3.39),405(2.78)	5-1694-83
2(3H)-Benzoxazolethione, 3-(tetrahydro-2-furanyl)-	$CHCl_3$	304(4.47)	94-1733-83
$C_{11}H_{11}NO_3$			
Benzenepropanamide, N,N-dimethyl-α,β-dioxo-	MeCN	268(4.73),405(2.01)	5-1694-83
$C_{11}H_{11}NO_4S$			
2-Naphthalenesulfonic acid, 4-hydroxy-7-(methylamino)-	MeOH	228(4.33),257(4.36), 270(4.28),292(3.75), 306(3.59),348(3.23)	48-1002-83
$C_{11}H_{11}NO_6$			
Benzenebutanoic acid, 2-(formylamino)-α,5-dihydroxy-γ-oxo-, calcium salt	H_2O	233(4.4),344(3.6)	5-0226-83
$C_{11}H_{11}NSSe$			
2(3H)-Thiazoleselone, 3-ethyl-4-phenyl-	EtOH	348(4.33)	56-0875-83
$C_{11}H_{11}NS_2$			
2(3H)-Thiazolethione, 3-ethyl-4-phenyl-	EtOH	325(4.24)	56-0875-83
$C_{11}H_{11}N_3$			
6,7,8-Triazatricyclo[3.3.0.0^{2,4}]oct-6-ene, 8-phenyl-	EtOH	223s(3.8),242(3.6), 274(3.2),350(2.9)	88-5727-83

Compound	Solvent	λ_{max}(log ε)	Ref.
$C_{11}H_{11}N_3O$			
4H-Isoxazolo[5',4':3,4]cyclohepta-[1,2-b]pyrazine, 5,6-dihydro-3-methyl-	MeOH	241s(3.67),260(3.71), 302(4.03)	142-1117-83
2-Pyrimidinamine, 4-methyl-6-phenyl-, 3-oxide	EtOH	276(3.79),346(4.04)	103-0091-83
$C_{11}H_{11}N_3OS$			
1,2,4-Triazin-5(2H)-one, 2-methyl-3-(methylthio)-6-phenyl-	EtOH	210(3.93),245(4.01), 325(4.11)	44-4585-83
1,2,4-Triazin-5(4H)-one, 4-methyl-3-(methylthio)-6-phenyl-	EtOH	211(4.16),250(3.85), 334(4.15)	44-4585-83
$C_{11}H_{11}N_3O_2$			
Ethanone, 1-[3-(1H-benzotriazol-1-yl)-3-methyloxiranyl]-	EtOH	257(4.9),293(4.62)	39-0581-83C
Ethanone, 1-[3-(2H-benzotriazol-2-yl)-3-methyloxiranyl]-	EtOH	255(4.79),284(4.56)	39-0581-83C
1H-Imidazole-4-carboxamide, 5-hydroxy-N-(phenylmethyl)-	H_2O	236s(--),278(4.21)	4-0875-83
	M HCl	246(4.14),280s(--)	4-0875-83
	M NaOH	279(3.99)	4-0875-83
1H-Imidazole-4-carboxamide, 5-hydroxy-1-(phenylmethyl)-	H_2O	241(3.76),279(4.18)	4-0875-83
	M HCl	241(3.83),280(4.05)	4-0875-83
	M NaOH	278(4.21)	4-0875-83
1H-Imidazole-4-carboxamide, 5-(phenylmethoxy)-	H_2O	256(4.11)	4-0875-83
	M HCl	237(3.98)	4-0875-83
	M NaOH	268(4.12)	4-0875-83
1H-Imidazole-5-carboxamide, 4-hydroxy-1-(phenylmethyl)-	H_2O	238(3.72),279(4.06)	4-0875-83
	M HCl	244(3.88),277(3.63)	4-0875-83
	M NaOH	287(4.10)	4-0875-83
3H-Pyrazole, 3,3-dimethyl-5-nitro-4-phenyl-	MeCN	232(3.53),296(3.18)	78-1247-83
4H-Pyrido[1,2-a]pyrimidine-3-carbonitrile, 9-formyl-1,6,7,8-tetrahydro-6-methyl-4-oxo-	EtOH	267(3.51),320s(3.96), 358(4.38)	39-0369-83C
$C_{11}H_{11}N_3O_3$			
1H-Imidazole-4-methanol, 5-nitro-1-(phenylmethyl)-	MeOH	231(3.57),302(3.88)	4-0629-83
$C_{11}H_{11}N_5$			
1H-Pyrazol-5-amine, 4-(1H-benzimidazol-2-yl)-1-methyl-	MeOH	235(4.04),263(3.82), 272(3.83),302(4.31), 312(4.28)	94-2114-83
$C_{11}H_{11}N_5O$			
Acetamide, N-[4-(1H-imidazol-2-ylazo)phenyl]-, (E)-	MeOH	202(3.87),246(3.64), 373(4.15),386?(4.15), 412s(3.85)	56-0971-83
$C_{11}H_{11}N_5O_2$			
Pyrrolo[2,3-c]azepin-8(1H)-one, 4-(2-amino-1,5-dihydro-5-oxo-4H-imidazol-4-ylidene)-4,5,6,7-tetrahydro-, (Z)-	MeOH	228(4.06),265(4.02), 344(4.28)	94-2321-83
1H-1,2,3-Triazole-4-carboxaldehyde, 1-acetyl-5-(phenylamino)-, 4-oxime	EtOH	330(3.92)	95-0594-83
1H-1,2,3-Triazole-4-carboxaldehyde, 5-amino-1-phenyl-, O-acetyloxime	EtOH	288(4.15)	95-0594-83
1H-1,2,3-Triazole-4-carboxaldehyde, 5-(phenylamino)-, O-acetyloxime	EtOH	330(4.01)	95-0594-83
L-Tryptophan, 6-azido-	pH 1	210(4.30),244(4.34),	44-5130-83

Compound	Solvent	$\lambda_{max}(\log \epsilon)$	Ref.
L-Tryptophan, 6-azido- (cont.)		285(3.99)	44-5130-83
	pH 7	210(4.28),242(4.35), 287(4.00)	44-5130-83
	pH 13	220(4.12),245(4.29), 289(3.92)	44-5130-83
	EtOH	214(4.31),247(4.34), 289(3.99)	44-5130-83
$C_{11}H_{11}N_5O_3$			
1H-Pyrazole-4-carboxamide, 5-amino-3-(nitromethyl)-1-phenyl-	EtOH	233(4.36),253s(4.07)	48-0041-83
$C_{11}H_{11}N_5O_3S$			
4H-[1,2,5]Oxadiazolo[3,4-c][1,2,6]-thiadiazin-7-amine, 4-methyl-N-(phenylmethyl)-, 5,5-dioxide	EtOH-H_2O	217(3.93),280(3.67)	142-2351-83
$C_{11}H_{11}S_2$			
1,3-Dithiol-1-ium, 4,5-dihydro-2-(2-phenylethenyl)-, tetrafluoroborate	CH_2Cl_2	244(3.87),436(4.40)	1-0687-83
$C_{11}H_{12}$			
1,3,5-Cycloheptatriene, 1,6-diethenyl-	3-Mepentane	236(4.65),242.5(4.73), 325(3.79)	35-6211-83
$C_{11}H_{12}BrNO_3$			
Benzamide, N-acetoxy-4-bromo-2,6-dimethyl-	MeOH	271(2.85),274(2.85), 278s(2.82)	32-0011-83
2(1H)-Isoquinolinecarboxaldehyde, 8-bromo-3,4-dihydro-7-hydroxy-6-methoxy-	MeOH	289(3.50)	83-0520-83
$C_{11}H_{12}FeO_4$			
Iron, tricarbonyl[(1,2,3,4-η)(5,5-dimethyl-1,3-cyclohexadien-2-ol)]-	hexane	217(4.1),279s(3.4)	24-0243-83
$C_{11}H_{12}NS$			
Thieno[2,3-c]pyridinium, 6-methyl-7-(2-propenyl)-, iodide	MeOH	220(4.37),239(4.32), 268(3.77),275s(3.72), 317(3.86),327(3.97)	83-0831-83
$C_{11}H_{12}N_2$			
1,2-Diazabicyclo[3.1.0]hex-2-ene, 6-methyl-3-phenyl-, exo	EtOH	255(4.08)	35-0933-83
1H-Pyrrolo[1,2,3-de]quinoxaline, 2,3-dihydro-5-methyl-	MeOH	226(4.44),282(3.89)	18-3358-83
$C_{11}H_{12}N_2O$			
1H-1,3-Benzodiazepine, 7-methoxy-2-methyl-	EtOH	220(4.40),262(4.31), 270s(--)	142-2173-83
monohydriodide	EtOH	220(4.32),242(4.24), 250(4.28),263(4.45)	142-2173-83
Ethanone, 1-(2,3-dihydro-1H-cycloheptapyrazin-9-yl)-	MeOH	254(4.25),303(3.61), 386(3.77),458(3.97)	18-3358-83
Ethanone, 1-(1-methyl-1H-indol-3-yl)-, oxime	n.s.g.	230(4.28),259(4.11), 290(3.96)	103-0184-83
1,2,4-Oxadiazole, 3-(2,6-dimethylphenyl)-5-methyl-	MeOH	271s(2.59),278s(2.56)	4-1609-83
1-Propanone, 1-(1H-indol-3-yl)-, oxime	n.s.g.	225(4.25),258(4.11), 284(3.97)	103-0184-83

Compound	Solvent	$\lambda_{max}(\log \epsilon)$	Ref.
2H-Pyrrol-2-one, 1,5-dihydro-3,4-dimethyl-5-(1H-pyrrol-2-ylmethylene)-	MeOH	378(4.38)	54-0347-83
$C_{11}H_{12}N_2OS$			
Acetamide, N-[4-(3-amino-3-thioxo-1-propenyl)phenyl]-	EtOH	345(4.15)	80-0555-83
$C_{11}H_{12}N_2O_2$			
5H-Cycloheptapyrazin-5-one, 6-acetyl-1,2,3,4-tetrahydro-	MeOH	246(4.22),282(4.17), 308(4.17),378(3.97), 468(3.86)	142-1117-83
1H-Indene-2-carboxamide, N-(aminocarbonyl)-2,3-dihydro-	EtOH	247(2.44),260(2.63), 267(2.83),273(3.12)	120-0364-83
1(2H)-Isoquinolinone, 3-methoxy-2-(methylamino)-	MeOH	224s(4.39),228(4.40), 280s(3.93),286(3.94), 342(3.64)	142-1315-83
4(5H)-Oxazolone, 2-(ethylamino)-5-phenyl-	EtOH	221(4.42)	162-0574-83
5-Quinoxalinecarboxaldehyde, 8-acetyl-1,2,3,4-tetrahydro-	MeOH	214(4.21),240(4.26), 277(4.00),328(4.07), 490(3.74)	18-3358-83
Tryptophan	EtOH	281(3.80)	98-0734-83
$C_{11}H_{12}N_2O_2S$			
1H-Imidazole-5-carboxylic acid, 1-(2-thienylmethyl)-, ethyl ester	MeOH	235(4.26)	142-2019-83
$C_{11}H_{12}N_2O_2S_2$			
1H-Imidazole-4-carboxylic acid, 2,3-dihydro-3-(2-thienylmethyl)-2-thioxo-, ethyl ester	0.5M HCl	236(4.09),292(4.04)	142-2019-83
	0.5M NaOH	232(4.16),308(4.15)	142-2019-83
	MeOH	202(4.05),235(4.07), 262(3.95),298(4.00)	142-2019-83
1,2,3-Thiadiazolium, 4-ethoxy-5-mercapto-3-(4-methoxyphenyl)-, hydroxide, inner salt	EtOH	245(3.87),303(3.18)	94-1746-83
$C_{11}H_{12}N_2O_3$			
3-Penten-2-one, 3-nitro-4-(phenylamino)-	EtOH	312(4.10),360s(3.76)	40-0088-83
4H-Pyrido[1,2-a]pyrimidine-3,9-dicarboxaldehyde, 1,6,7,8-tetrahydro-6-methyl-4-oxo-	EtOH	227(4.09),245s(3.60), 321s(3.84),373(4.25)	39-0369-83C
$C_{11}H_{12}N_2O_3S$			
Cyclopentanone, 2-[[2-oxo-2-(2-thienyl)ethylidene]amino]-, 1-oxime, N-oxide	EtOH	266s(3.88),312(4.11)	103-1104-83
Morpholine, 4-[(3-nitrophenyl)thioxomethyl]-	heptane	250(4.16),290(4.00), 398s(--)	80-0555-83
	EtOH	252(4.18),281(4.08)	80-0555-83
Morpholine, 4-[(4-nitrophenyl)thioxomethyl]-	heptane	259(4.04),320(3.66), 400s(--)	80-0555-83
	EtOH	264(4.24),307(3.86)	80-0555-83
2-Propenal, 2-nitro-3-[(4,5,6,7-tetrahydrobenzo[b]thien-2-yl)amino]-	EtOH	228(3.96),278(3.91), 398(4.08)	103-1163-83
$C_{11}H_{12}N_2O_4$			
Cyclopentanone, 2-[[2-(2-furanyl)-2-oxoethylidene]amino]-, 1-oxime, N-oxide	EtOH	333(4.35)	103-1104-83
4H-Pyrido[1,2-a]pyrimidine-3-carboxylic	EtOH	222(4.09),271(3.36),	39-0369-83C

Compound	Solvent	λ_{max}(log ϵ)	Ref.
acid, 9-formyl-1,6,7,8-tetrahydro-6-methyl-4-oxo-		317s(3.89),357(4.20)	39-0369-83C
$C_{11}H_{12}N_2O_4S$			
2-Naphthalenesulfonic acid, 3-amino-4-hydroxy-7-(methylamino)-	MeOH	212(4.43),248(4.60), 307(3.90),351(3.60)	48-1002-83
$C_{11}H_{12}N_2O_5$			
Pyrano[2,3-c]pyrazole-3,4-dicarboxylic acid, 1,6-dihydro-1-methyl-, dimethyl ester	MeOH	210(4.26),339(4.08), 303(3.86)	24-1525-83
$C_{11}H_{12}N_2O_5S$			
2-Thiophenecarboxylic acid, 3-methyl-5-[(2-nitro-3-oxo-1-propenyl)amino]-, ethyl ester	EtOH	224(4.01),268(3.91), 391(4.20)	103-1163-83
$C_{11}H_{12}N_2O_6$			
2,4(1H,3H)-Pyrimidinedione, 1-β-D-arabinofuranosyl-5-ethynyl-	pH 7	229(4.02),288(4.06)	87-0661-83
$C_{11}H_{12}N_2O_8$			
Furo[3,4-d]pyrimidine-2,4,7(3H)-trione, 1,5-dihydro-1-β-D-ribofuranosyl-	H_2O	285(3.94)	94-3074-83
	pH 13	265(3.91)	94-3074-83
	EtOH	285(3.90)	94-3074-83
$C_{11}H_{12}N_4O$			
1H-Imidazole, 2-[(4-ethoxyphenyl)azo]-	MeOH	250(3.93),376(4.45), 380(4.46),405s(4.19)	56-0971-83
1H-Imidazole, 2-[(4-methoxyphenyl)-azo]-1-methyl-	MeOH	250(3.89),374?(4.35), 385(4.35),418s(4.00), 450s(3.58)	56-0971-83
	MeOH-HCl	256(3.88),386(4.37), 400(4.42),425s(4.21)	56-0971-83
	MeOH-NaOH	250(3.87),380(4.39), 385?(4.39),416s(4.09)	56-0971-83
	20% MeOH	250(3.86),374(4.39), 385(4.40),420s(4.15)	56-0971-83
	MeCN	250(3.86),370(4.40), 382(4.39),408s(4.08), 436s(3.70)	56-0971-83
	1,2-$C_2H_4Cl_2$	252(3.78),375(4.25), 385(4.24),420s(3.88), 456s(3.40)	56-0971-83
1H-Imidazole-4-carboxamide, 5-amino-N-(phenylmethyl)-	n.s.g.	241(4.10),267(4.19)	103-1235-83
2,4(1H,3H)-Pyrimidinedione, 1-methyl-, 4-(phenylhydrazone)	MeOH	238(4.23),284(4.10)	4-1037-83
2,4(1H,3H)-Pyrimidinedione, 3-methyl-, 4-(phenylhydrazone)	MeOH	239s(3.81),260(3.97), 280s(3.94),349(3.57)	4-1037-83
$C_{11}H_{12}N_4OS$			
Phenol, 4,5-dimethyl-2-[(5-methyl-1,3,4-thiadiazol-2-yl)azo]-	pH 4.2	450(1.8)	73-0471-83
	pH 11.3	530(2.0)	73-0471-83
copper complex	pH 7-9	610(4.03)	73-0471-83
nickel complex	pH 7-9	625(4.30)	73-0471-83
$C_{11}H_{12}N_4O_2$			
Furazancarboxamide, 4-(methylamino)-N-(phenylmethyl)-	EtOH	232(3.91),250(3.51)	142-2351-83

Compound	Solvent	$\lambda_{max}(\log \epsilon)$	Ref.
$C_{11}H_{12}N_4O_4S$			
1H-Pyrazole, 3-[(2,4-dinitrophenyl)-thio]-4,5-dihydro-5,5-dimethyl-	EtOH	265(4.32)	44-3189-83
$C_{11}H_{12}N_6$			
4,6-Pyrimidinediamine, N-methyl-5-(phenylazo)-	pH 1	245(4.30),386(4.23)	94-3454-83
	H_2O	248(4.28),388(4.25)	94-3454-83
	pH 13	248(4.28),388(4.25)	94-3454-83
$C_{11}H_{12}N_6O_3$			
1H-Imidazo[4,5-g]pteridine-6,8(4H,7H)-dione, 2-methoxy-1,4,7-trimethyl-	n.s.g.	430(4.23)	35-3304-83
	cation	390(4.21)	35-3304-83
1H-Imidazo[4,5-g]pteridine-2,4,7-tri-one, 3,6,8,9-tetrahydro-3,6,8,9-tetramethyl-	DMF	290(4.31),466(3.97)	35-3304-83
1H-Pyrazole-4-carboxylic acid, 5-ami-no-3-(nitromethyl)-1-phenyl-, hydrazide	EtOH	234(4.32),251s(4.04)	48-0041-83
$C_{11}H_{12}N_6O_4$			
Pyrimido[5,4-g]pteridine-2,4,6,8-tetrone, 1,3,5,7,9,10-hexahydro-3,7,10-trimethyl-	neutral	292(4.01)	35-3304-83
	cation	292(3.28)	35-3304-83
	anion	292(--)	35-3304-83
	dianion	292(4.23)	35-3304-83
$C_{11}H_{12}O$			
Benzofuran, 2,3-dihydro-2-(1-methyl-ethenyl)-	EtOH	216(3.72),226(3.69), 279(3.40),287(3.39)	78-0169-83
Bicyclo[4.3.1]deca-2,4,8-trien-7-one, 9-methyl-	EtOH	227(4.38),253s(3.56), 331(2.68)	24-2914-83
1H-Cyclopenta[a]pentalen-1-one, 3a,3b,6,6a,7,7a-hexahydro-, (3aα,3bα,6aα,7aα)-	EtOH	213(4.14)	88-2781-83
1H-Inden-1-one, 2,3-dihydro-2,5-di-methyl-	MeOH	254(4.17),286(3.48), 293(3.44)	94-3113-83
1H-Inden-1-one, 2,3-dihydro-2,6-di-methyl-	MeOH	247(4.05),295(3.45), 303(3.47)	94-3113-83
1H-Inden-1-one, 2,3-dihydro-2,7-di-methyl-	MeOH	249(4.09),291(3.33), 298(3.34)	94-3113-83
Pentacyclo[5.4.0.0^{2,6}.0^{3,10}.0^{5,9}]un-decan-8-one	MeCN	299(1.26)	44-1862-83
$C_{11}H_{12}O_2$			
5H-Benzocyclohepten-5-one, 1,2,3,4-tetrahydro-6-hydroxy-	C_6H_{12}	240s(4.38),245(4.39), 310s(3.71),324(3.80), 336s(3.76),349(3.71), 360(3.65),369(3.64)	39-0285-83C
5H-Benzocyclohepten-5-one, 6,7,8,9-tetrahydro-3-hydroxy-	EtOH	232(4.03),261(3.45), 328(2.97)	78-0581-83
	EtOH-NaOH	249(3.98),290(3.56), 392(3.08)	78-0581-83
7H-Benzocyclohepten-7-one, 1,2,3,4-tetrahydro-6-hydroxy-	C_6H_{12}	237(4.46),308s(3.08), 323(3.88),343s(3.78), 357(3.82),375(3.72)	39-0285-83C
Benzofuran, 2-ethenyl-2,3-dihydro-4-methoxy-	EtOH	225(3.69),230(3.89), 275(3.13),281(3.13)	78-0169-83
Benzofuran, 2-ethenyl-2,3-dihydro-5-methoxy-	EtOH	218(3.68),229(3.78), 300(3.59)	78-0169-83
Benzofuran, 2-ethenyl-2,3-dihydro-6-methoxy-	EtOH	225(3.69),288(3.61), 292(3.57)	78-0169-83

Compound	Solvent	λ_{max}(log ε)	Ref.
2H-1-Benzopyran-3(4H)-one, 2,2-dimethyl-	EtOH	212(4.04),219s(3.96), 280(3.34)	39-1431-83C
2,5-Dioxabicyclo[4.1.0]heptane, 7-phenyl-, endo	C_6H_{12}	207(3.86),220(3.70)	4-0251-83
1-Naphthalenol, 5,6-dihydro-4-methoxy-	EtOH	217(4.28),270(3.86), 280s(3.64),323(3.64)	94-2662-83
Spiro[4.6]undeca-8,10-diene-6,7-dione	$CHCl_3$	263s(3.26),310(3.52), 328s(3.47),450(2.42)	39-0285-83C
$C_{11}H_{12}O_3$			
1H-2-Benzopyran-1-one, 3,4-dihydro-8-hydroxy-3,5-dimethyl-	EtOH	247(3.64),323(3.43)	39-2185-83C
Bicyclo[3.1.0]hex-3-en-2-one, 1,5-diacetyl-4-methyl-	MeOH	245(3.56)	32-0427-83
Ethanone, 1,1'-(4-methoxy-1,3-phenylene)bis-	EtOH	273(4.00),268(3.81), 310(3.13)[sic]	102-1515-83
1(2H)-Naphthalenone, 3,4-dihydro-4,8-dihydroxy-3-methyl-	EtOH	217(4.52),259(4.09), 335(3.76)	102-1245-83
1(2H)-Naphthalenone, 3,4-dihydro-8-hydroxy-5-methoxy-	EtOH	236(4.15),266(3.92), 370(3.53)	94-2662-83
2,4,6-Nonatrien-8-ynoic acid, 3-hydroxy-7-methyl-, methyl ester, (Z,E,Z)-	ether	223s(3.65),231(3.68), 320(4.46)	44-2379-83
$C_{11}H_{12}O_4$			
Benzenepentanoic acid, 2-hydroxy-δ-oxo-	EtOH	252(3.99),324(3.60)	2-0331-83
Bicyclo[2.2.1]hepta-2,5-diene-2,3-dicarboxylic acid, dimethyl ester	MeOH	240(3.41)	24-1520-83
2,4,6-Cycloheptatrien-1-one, 3-(2-acetoxyethyl)-2-hydroxy-	EtOH	236(4.28),322(3.70), 338.8(3.63),354(3.66), 370(3.61),396s(2.99)	39-0285-83C
$C_{11}H_{12}O_5$			
Elenolide	EtOH	225(4.29),317(1.75)	162-0512-83
$C_{11}H_{12}S_2$			
1,3-Dithiane, 2-(2,4,6-cycloheptatrien-1-ylidene)-	C_6H_{12}	212(4.1),242(3.9), 337(4.1)	24-1154-83
$C_{11}H_{13}$			
1H-Benzocycloheptenylium, 2,3,4,?-tetrahydro-, hexafluoroantimonate	MeCN	224(4.25),297(3.71)	44-0596-83
$C_{11}H_{13}BrN_2O_3$			
1,2-Diazabicyclo[5.2.0]nona-3,5-diene-2-carboxylic acid, 8-bromo-5-methyl-9-oxo-, ethyl ester, cis	MeOH	273(3.83)	5-1361-83
$C_{11}H_{13}BrO$			
Benzene, [[(4-bromo-2-butenyl)oxy]-methyl]-	EtOH	220(4.21),264(3.18), 271(3.28),277(3.19)	78-0169-83
$C_{11}H_{13}BrO_4$			
Benzeneacetic acid, 5-bromo-2,3-dimethoxy-, methyl ester	EtOH	209(4.46),287(3.28)	104-0458-83
$C_{11}H_{13}Br_3O_3$			
Benzene, 1,2,4-tribromo-3-(ethoxymethyl)-5,6-dimethoxy-	$CHCl_3$	287(3.57),293(3.51)	94-1754-83

Compound	Solvent	λ_{max}(log ε)	Ref.
$C_{11}H_{13}ClN_2$			
2H-Indol-2-imine, 3-(2-chloroethyl)-1,3-dihydro-1-methyl-, monohydrochloride	n.s.g.	216(4.16),264(3.87)	103-1175-83
$C_{11}H_{13}ClN_2O_2$			
Acetamide, 2-chloro-N-[2-(hydroxyimino)-2-(4-methylphenyl)ethyl]-, (E)-	EtOH	252(4.1)	5-2038-83
$C_{11}H_{13}ClN_2O_3$			
Acetamide, 2-chloro-N-[2-(hydroxyimino)-2-(4-nethoxyphenyl)ethyl]-, (E)-	EtOH	263(4.2)	5-2038-83
1,2-Diazabicyclo[5.2.0]nona-3,5-diene-2-carboxylic acid, 8-chloro-5-methyl-9-oxo-, ethyl ester, cis	CH_2Cl_2	277(3.86)	5-1361-83
$C_{11}H_{13}ClOS$			
1-Propanethione, 1-(4-chlorophenyl)-2,2-dimethyl-, S-oxide	$CHCl_3$	263(3.82),310s(3.34)	44-0214-83
$C_{11}H_{13}ClO_3$			
2-Cyclopenten-1-one, 4,5-diacetyl-4-(chloromethyl)-3-methyl-	MeOH	215(3.86),243(3.88), 248(3.87),292(3.81), 310(3.73)	32-0427-83
$C_{11}H_{13}ClO_4$			
Benzeneacetic acid, 5-chloro-α,2-dihydroxy-α-methyl-, ethyl ester	EtOH	229(3.96),284(3.32), 288(3.30)	39-1649-83C
Benzeneacetic acid, 5-chloro-2,3-dimethoxy-, methyl ester	EtOH	208(4.19),283(3.24)	104-0458-83
$C_{11}H_{13}ClS$			
1-Propanethione, 1-(4-chlorophenyl)-2,2-dimethyl-	benzene	560(1.90)	44-0214-83
$C_{11}H_{13}Cl_3N_2O_3S_2$			
5-Thia-1-azabicyclo[4.2.0]oct-2-ene-2-carboxylic acid, 7-amino-3-[(methylthio)methyl]-8-oxo-, 2,2,2-trichloroethyl ester, monohydrochloride, (6R-trans)-	MeOH	272.5(3.82)	78-0337-83
$C_{11}H_{13}FOS$			
1-Propanethione, 1-(4-fluorophenyl)-2,2-dimethyl-, S-oxide	$CHCl_3$	263(3.81),310s(3.31)	44-0214-83
$C_{11}H_{13}FS$			
1-Propanethione, 1-(4-fluorophenyl)-2,2-dimethyl-	benzene	263(3.40),560(1.78)	44-0214-83
$C_{11}H_{13}IN_4O_3$			
7H-Pyrrolo[2,3-d]pyrimidin-4-amine, 7-(5-deoxy-β-D-ribofuranosyl)-5-iodo-	pH 1	238.5(4.29),286(3.92)	12-0165-83
	pH 12	232s(4.12),280(3.95)	12-0165-83
$C_{11}H_{13}N$			
1H-Indole, 5-ethyl-1-methyl-	EtOH	206(4.52),273(4.01), 288(3.96),298(3.80)	104-0388-83

Compound	Solvent	$\lambda_{max}(\log \epsilon)$	Ref.
1H-Indole, 5-ethyl-2-methyl-	EtOH	225(4.37),275(3.8), 285(3.8),295(3.6)	104-0388-83
1H-Indole, 5-ethyl-3-methyl-	EtOH	225(4.7),277(3.8), 285(3.8),298(3.7)	104-0388-83
$C_{11}H_{13}NO$			
1H-2-Benzazepin-1-one, 2,3,4,5-tetrahydro-3-methyl-, (R)-	EtOH	220(4.13),226(4.15), 272(3.26)	103-0528-83
2H-1-Benzazepin-2-one, 1,3,4,5-tetrahydro-3-methyl-, (R)-	EtOH	239(4.12),268(3.25), 275s(3.0)	103-0528-83
3-Butenamide, 3-methyl-4-phenyl-	n.s.g.	249(4.1)	162-0149-83
Cyclopent[b]indol-8b(1H)-ol, 2,3,3a,4-tetrahydro-, cis	EtOH	205(4.40),245(3.90), 300(3.40)	78-3767-83
2-Propenal, 3-[4-(dimethylamino)-phenyl]-	H_2O	400(4.56)	2-1217-83
	MeOH	390(4.58)	2-1217-83
	EtOH	388(4.58)	2-1217-83
	acetone	385(4.58)	2-1217-83
	dioxan	388(4.59)	2-1217-83
	MeCN	385(4.58)	2-1217-83
	DMF	385(4.59)	2-1217-83
	DMSO	387(4.59)	2-1217-83
	$CHCl_3$	388(4.59)	2-1217-83
$C_{11}H_{13}NOS$			
Benzoxazole, 2-[(1,1-dimethylethyl)-thio]-	$CHCl_3$	281(4.03),288(4.05)	94-1733-83
2(3H)-Benzoxazolethione, 3-(1,1-dimethylethyl)-	$CHCl_3$	308(4.45)	94-1733-83
Morpholine, 4-(phenylthioxomethyl)-	heptane	249(3.65),291(3.68), 400(2.30)	80-0555-83
	EtOH	235(4.00),287(4.05), 366(2.60)	80-0555-83
$C_{11}H_{13}NO_2$			
2-Azabicyclo[4.3.1]deca-4,7-diene-3,9-dione, 5,7-dimethyl-	EtOH	220(4.15),232(4.08), 335(2.23),350(2.18)	39-2541-83C
4(5H)-Benzofuranone, 5-[(dimethyl-amino)methylene]-6,7-dihydro-	EtOH	250(3.93),355(4.27)	111-0113-83
6-Isoquinolinol, 3,4-dihydro-7-methoxy-1-methyl-	pH 1	242.5(4.22),300(3.96), 346(3.94)	94-2583-83
	pH 13	241.5(4.04),302s(3.95), 325.5(4.18)	94-2583-83
	EtOH	226(4.13),269s(4.03), 273.5(4.05),311(3.70), 401(4.36)	94-2583-83
5(4H)-Oxazolone, 4-ethyl-2-methyl-4-phenyl-	C_6H_{12}	257(3.30)	44-0695-83
$C_{11}H_{13}NO_2S$			
Morpholine, 4-[(2-hydroxyphenyl)thioxomethyl]-	heptane	245(3.90),296(4.02), 389(2.84)	80-0555-83
	EtOH	240(3.86),284(4.15), 362(2.60)	80-0555-83
2-Thietanone, 3-[(2-furanylmethylene)-amino]-3,4,4-trimethyl-	EtOH	223(3.95),273(4.04)	39-2259-83C
$C_{11}H_{13}NO_3$			
Benzamide, N-acetoxy-2,6-dimethyl-	MeOH	267(2.88),274(2.88)	32-0011-83
$C_{11}H_{13}NO_3S$			
2-Butenamide, 2-(methylsulfonyl)-N-	EtOH	253(4.02)	104-1314-83

Compound	Solvent	λ$_{max}$(log ε)	Ref.
phenyl- (cont.)			104-1314-83
2-Butenamide, 3-(methylsulfonyl)-N-phenyl-	EtOH	220(4.21),293(3.99)	104-1314-83
2-Butenamide, 4-(methylsulfonyl)-N-phenyl-, (E)-	EtOH	281(4.13)	104-1314-83
3-Butenamide, 3-(methylsulfonyl)-N-phenyl-	EtOH	245(4.16)	104-1314-83
2-Propenamide, 2-methyl-3-(methylsulfonyl)-N-phenyl-	EtOH	220(4.11),270(3.85)	104-1314-83
2-Propenamide, 2-[(methylsulfonyl)-methyl]-N-phenyl-	EtOH	266(3.87)	104-1314-83
$C_{11}H_{13}NO_4$			
2-Pyridinecarboxylic acid, 6-(1,3-dioxan-2-yl)-, methyl ester	n.s.g.	218(3.89),265(3.60)	64-0516-83B
$C_{11}H_{13}NO_4S$			
2-Azetidinone, 4-[[[(4-methylphenyl)-sulfonyl]oxy]methyl]-	EtOH	221s(3.86),231(3.94), 263(2.91),274(2.86)	39-0649-83C
$C_{11}H_{13}N_2O_4P$			
Phosphonic acid, [3-(4-methylphenyl)-1,2,4-oxadiazol-5-yl]-, dimethyl ester	MeOH	271s(3.81),287s(3.75)	44-4232-83
$C_{11}H_{13}N_2S$			
Thiazolium, 4-amino-2,3-dimethyl-5-phenyl-, perchlorate	MeCN	225s(3.82),259(3.75), 324(3.87)	118-0582-83
$C_{11}H_{13}N_3OS$			
4H-Pyrido[3,4-e]-1,3-thiazin-4-one, 2-(ethylamino)-5,7-dimethyl-	MeOH	246(4.11)	73-3426-83
$C_{11}H_{13}N_3O_4$			
7H-Pyrrolo[2,3-d]pyrimidine, 7-β-D-arabinofuranosyl-	MeOH	270(3.57)	5-1576-83
$C_{11}H_{13}N_3O_4S$			
4H-Pyrrolo[3,2-d]pyrimidine-4-thione, 1,7-dihydro-7-β-D-ribofuranosyl-	pH 1	263(3.69),336(4.25)	44-0780-83
	pH 7	263(3.80),328(4.36)	44-0780-83
	pH 13	250s(3.84),313(4.09)	44-0780-83
$C_{11}H_{13}N_3O_5$			
4H-Pyrido[3,2-d]pyrimidin-4-one, 1,5-dihydro-7-β-D-ribofuranosyl-	pH 1	238(4.43),262s(3.98)	44-0780-83
	pH 7	231(4.46),261(3.98)	44-0780-83
	pH 13	228(4.43),267(3.96)	44-0780-83
$C_{11}H_{13}N_3S$			
1H-Pyrazole-1-carbothioamide, 4,5-dihydro-N-methyl-3-phenyl-	EtOH	325(4.44)	4-1359-83
$C_{11}H_{13}N_5$			
Benzenamine, 4-(1H-imidazol-2-ylazo)-N,N-dimethyl-	MeOH	472(4.52)	56-0547-83
$C_{11}H_{13}N_5O$			
1H-Imidazole-4-carboxamide, 5-hydrazino-N-(phenylmethyl)-, monohydrochloride	n.s.g.	274(4.45)	103-1235-83

Compound	Solvent	λ_{max}(log ε)	Ref.
$C_{11}H_{13}N_5O_3$			
1,2-Diazabicyclo[5.2.0]nona-3,5-diene-2-carboxylic acid, 8-azido-5-methyl-9-oxo-, ethyl ester, cis	MeOH	274(3.81)	5-1361-83
Neplanocin A	H_2O	262(4.16)	33-1915-83
$C_{11}H_{13}N_5O_5$			
Uridine, 3'-azido-2',3'-dideoxy-, 5'-acetate	MeOH	260(3.96)	87-0544-83
$C_{11}H_{13}S_2$			
4H-1,3-Dithiin-1-ium, 5,6-dihydro-2-(4-methylphenyl)-, tetrafluoroborate	CH_2Cl_2	242(3.34),276(2.94), 366(4.11)	1-0687-83
4H-1,3-Dithiin-1-ium, 5,6-dihydro-2-(phenylmethyl)-, tetrafluoroborate	CH_2Cl_2	309(4.19),334(3.89)	1-0687-83
$C_{11}H_{14}BrNO$			
Propanamide, N-(2-bromophenyl)-2,2-dimethyl-	EtOH	206(4.30),239(3.80)	78-3767-83
$C_{11}H_{14}BrNO_2$			
Carbamic acid, (2-bromophenyl)-, 1,1-dimethylethyl ester	EtOH	208(4.50),233(4.10)	78-3767-83
$C_{11}H_{14}ClNO_2$			
Benzene, 1-(1-chloro-2,2-dimethylpropyl)-3-nitro-	MeOH	259(3.99)	12-0081-83
$C_{11}H_{14}ClN_5$			
1,3,5-Triazine-2,4-diamine, 1-(4-chlorophenyl)-1,6-dihydro-6,6-dimethyl- (cycloguanil)	H_2O	241(4.11)	162-0390-83
monohydrochloride	H_2O	241(4.12)	162-0390-83
$C_{11}H_{14}ClN_5O_2$			
Acetamide, N-[2-[(6-chloro-7H-purin-7-yl)methoxy]ethyl]-N-methyl-	H_2O	272(4.05)	83-0146-83
Acetamide, N-[2-[(6-chloro-9H-purin-9-yl)methoxy]ethyl]-N-methyl-	H_2O	264(3.94)	83-0146-83
$C_{11}H_{14}NS$			
Thieno[2,3-c]pyridinium, 4,8-dihydro-6-methyl-5-(2-propenyl)-, perchlorate	MeOH	235(3.51),265(3.14), 323(4.20)	83-0831-83
$C_{11}H_{14}N_2$			
2H-Azirin-3-amine, N,2,2-trimethyl-N-phenyl-	hexane	254(4.24),274s(3.32), 283s(3.16),291s(2.85)	33-0262-83
	CH_2Cl_2	256(4.23),282s(3.26), 290s(3.00)	33-0262-83
1,5-Ethano-2H-1,5-benzodiazepine, 3,4-dihydro-	n.s.g.	210(3.70),230(2.22)	103-0553-83
$C_{11}H_{14}N_2O$			
Ethanone, 1-(1,2,3,4-tetrahydro-1-methyl-5-quinoxalinyl)-	MeOH	212(4.31),268(3.93), 418(3.67)	18-3358-83
$C_{11}H_{14}N_2OS$			
Benzamide, N-[(ethylamino)thioxomethyl]-2-methyl-	C_6H_{12}	234(4.26),282(4.07)	41-0603-83

Compound	Solvent	$\lambda_{max}(\log \epsilon)$	Ref.
Benzamide, N-[(ethylamino)thioxomethyl]-4-methyl-	C_6H_{12}	244(4.37),283(4.00)	41-0603-83
$C_{11}H_{14}N_2O_2$			
4H-Pyrido[1,2-a]pyrimidine-9-carboxaldehyde, 1,6,7,8-tetrahydro-3,6-dimethyl-4-oxo-	EtOH	227(3.98),341(4.31)	39-0369-83C
$C_{11}H_{14}N_2O_2S$			
Benzenesulfonic acid, (2-methyl-2-propenylidene)hydrazide	heptane	227(4.11),276(4.25), 418(2.02)	44-3994-83
$C_{11}H_{14}N_2O_3$			
Carbamic acid, [(phenylamino)carbonyl]-, 1-methylethyl ester	EtOH	246(4.20)	22-0073-83
1,2-Diazabicyclo[5.2.0]nona-3,5-diene-5-carboxylic acid, 5-methyl-9-oxo-, ethyl ester	MeOH	212(3.65),272(3.90)	5-1361-83
1,2-Diazabicyclo[5.2.0]nona-3,5-diene-2-carboxylic acid, 8-methyl-9-oxo-, ethyl ester, cis	CH_2Cl_2	278(3.92)	5-1361-83
4H-Pyrido[1,2-a]pyrimidine-3-carboxylic acid, 6,7,8,9-tetrahydro-1,6-dimethyl-4-oxo-	MeCN	365(3.47)	39-1413-83B
4H-Pyrido[1,2-a]pyrimidine-3-carboxylic acid, 6,7,8,9-tetrahydro-6-methyl-4-oxo-, methyl ester	EtOH	207(3.76),231(3.76), 305(3.91)	39-1413-83B
$C_{11}H_{14}N_2O_4$			
L-Alanine, N-[[3-hydroxy-5-(hydroxymethyl)-2-methyl-4-pyridinyl]methylene]-	MeOH	251(4.01),285s(3.63), 336(3.55),421(3.37)	35-0803-83
$C_{11}H_{14}N_2O_4S$			
L-Cysteine, N-[[3-hydroxy-5-(hydroxymethyl)-2-methyl-4-pyridinyl]methylene]-	MeOH	250s(3.53),272s(3.52), 294(3.74),335(3.00), 405(2.86)	35-0803-83
$C_{11}H_{14}N_2O_5$			
1H-Pyrazole-4-acetic acid, α-ethenyl-5-hydroxy-3-(methoxycarbonyl)-1-methyl-, methyl ester	MeOH	206(3.75),231(3.82), 305(3.18)	24-1525-83
1H-Pyrazole-4-acetic acid, α-ethylidene-5-hydroxy-3-(methoxycarbonyl)-1-methyl-, methyl ester, (E)-	MeOH	210(4.20),228s(4.13), 315(3.23)	24-1525-83
1H-Pyrazole-3-carboxylic acid, 5-[(4-methoxy-4-oxo-2-butenyl)oxy]-1-methyl-, methyl ester	MeOH	208(4.09),220s(4.03), 253s(3.43)	24-1525-83
2,4(1H,3H)-Pyrimidinedione, 1-(2-deoxy-α-D-erythro-pentofuranosyl)-5-ethenyl- (partial spectra)	pH 6	301(4.16)	44-1854-83
	pH 13	298(4.15)	44-1854-83
L-Serine, N-[[3-hydroxy-5-(hydroxymethyl)-2-methyl-4-pyridinyl]-methylene]-	MeOH	253(3.98),285s(3.55), 336(3.57),420(3.35)	35-0803-83
$C_{11}H_{14}N_2S$			
1-Propanamine, 3-(1H-indol-3-ylthio)-	MeOH	273(3.79),279(3.82), 287(3.76)	87-0230-83
2-Propanamine, 1-(2H-indol-3-ylthio)-	MeOH	272(3.79),279(3.82), 288(3.78)	87-0230-83

Compound	Solvent	$\lambda_{max}(\log \epsilon)$	Ref.
$C_{11}H_{14}N_4O$			
[1,2,3]Triazolo[1,5-a]pyridine-3-carboxamide, N,N-diethyl-	EtOH	245(3.89),277s(--), 287(4.1),305s(--)	150-1341-83M
$C_{11}H_{14}N_4O_2S$			
Propanoic acid, 2,2-dimethyl-, (1H-purin-6-ylthio)methyl ester	MeOH	215(4.07),223s(4.0), 278(4.14),284(4.12)	36-0372-83
$C_{11}H_{14}N_4O_2S_2$			
2,4(1H,3H)-Pteridinedione, 7-(ethyldithio)-1,3,6-trimethyl-	MeOH	210(4.29),254(4.03), 357(4.24)	88-5047-83
$C_{11}H_{14}N_4O_3$			
7H-Pyrrolo[2,3-d]pyrimidin-4-amine, 7-(2-deoxy-β-D-erythro-pentofuranosyl)-	MeOH	270(4.09)	5-0876-83
	MeOH	227(4.38),271(4.13)	35-4059-83
$C_{11}H_{14}N_4O_4$			
3-Deazaguanosine, 2'-deoxy-	pH 1	284(4.04),310(3.82)	87-0286-83
	pH 7	271(4.02),301(3.92)	87-0286-83
	pH 12	274(4.02),296s(3.93)	87-0286-83
1H-Imidazole-4-carboxamide, 5-(cyanomethyl)-1-(2-deoxy-β-D-ribofuranosyl)-	MeOH	234(3.94)	87-0286-83
Imidazo[4,5-c]pyridin-4(5H)-one, 6-amino-1-(2-deoxy-α-D-ribofuranosyl)-	pH 1	284(4.04),312(3.83)	87-0286-83
	pH 7	270(4.03),301(3.92)	87-0286-83
	pH 12	273(4.03),299s(3.93)	87-0286-83
Imidazo[4,5-c]pyridin-4(5H)-one, 6-amino-3-(2-deoxy-α-D-ribofuranosyl)-	pH 1	277(--),314(--)	87-0286-83
	pH 7	258(--),314(--)	87-0286-83
	pH 12	257(--),313(--)	87-0286-83
β-	pH 1	278(4.04),317(3.76)	87-0286-83
	pH 7	259(3.79),316(3.86)	87-0286-83
	pH 12	259(3.79),316(3.85)	87-0286-83
9H-Purine, 6-methyl-9-β-D-ribofuranosyl-	pH 1.0	264.5(3.86)	94-3104-83
	pH 5.4	261(3.90)	94-3104-83
4H-Pyrrolo[2,3-d]pyrimidin-4-one, 2-amino-7-(2-deoxy-β-D-erythro-pentofuranosyl)-3,7-dihydro-	MeOH	217(4.29),257(4.11), 281s(3.86)	44-3119-83
$C_{11}H_{14}N_4O_5$			
7-Deazaguanosine	pH 1	258(4.07)	5-0137-83
	pH 7	258(4.10),280s(--)	5-0137-83
	pH 13	260(4.10)	5-0137-83
9H-Purine-6-methanol, 9-β-D-ribofuranosyl-	MeOH	263(3.93)	94-3104-83
2,4(1H,3H)-Pyrimidinedione, 5-hydroxy-1-methyl-6-[2-(3-methyl-2,5-dioxo-4-imidazolidinyl)ethyl]-	pH 4	288(3.92)	4-0753-83
	pH 14	310(3.91)(unstable)	4-0753-83
3H,5H-Pyrrolo[3,2-d]pyrimidin-4-one, 2-amino-7-β-D-ribofuranosyl-	pH 1	234(4.25),271(4.17)	44-0780-83
	pH 7	230(4.35),269(3.95)	44-0780-83
	pH 13	223(4.37),257(3.82), 285(3.84)	44-0780-83
$C_{11}H_{14}N_6O_3$			
1H-Imidazo[4,5-g]pteridine-2,6,8-(3H,4H,7H)-trione, 5,9-dihydro-1,3,4,7-tetramethyl-	n.s.g.	276(4.22)	35-3304-83
	anion	276(4.11)	35-3304-83
$C_{11}H_{14}O$			
1H-Benzocyclohepten-1-one, 2,3,4,5,6,9-hexahydro-	MeOH	246(3.89)	23-1226-83

Compound	Solvent	$\lambda_{max}(\log \epsilon)$	Ref.
Bicyclo[4.3.1]deca-2,4-dien-7-one, 9-methyl-, exo	C_6H_{12}	244(3.78),253(3.98), 263(3.96),287s(2.83), 294(2.90),303(2.95), 313(2.89),324(2.58)	24-2914-83
Dispiro[2.1.2.3]dec-8-en-4-one, 8-methyl-	hexane	221(3.17),230s(3.06), 280(1.88)	23-0078-83
Spiro[2.5]oct-5-en-4-one, 5-ethenyl-6-methyl-	hexane	254(3.75),314s(1.88)	23-0936-83
Spiro[2.5]oct-6-en-4-one, 5-ethylidene-6-methyl-, (E)-	hexane	220(3.48),262(3.28), 318s(2.05)	23-0936-83
(Z)-	hexane	222(4.01),277(3.74), 323s(2.26)	23-0936-83
Tricyclo[3.3.1.1^{3,7}]decanone, 4-methylene-	heptane	295(2.04)	35-0079-83
$C_{11}H_{14}OS$			
1-Propanethione, 2,2-dimethyl-1-phenyl-, S-oxide	$CHCl_3$	265(3.84),308s(3.31)	44-0214-83
$C_{11}H_{14}O_2$			
Acetic acid, (2-cyclopenten-1-ylidene)-, 3-butenyl ester	EtOH	269(4.25)	33-1876-83
1,3-Dioxane, 2-(3-methyl-1,3-hexadien-5-ynyl)-	hexane	259(4.23)	33-1148-83
$C_{11}H_{14}O_3$			
Benzeneacetic acid, α,α-dimethyl-3-methoxy-	MeOH	222(3.05),273(2.39), 281(2.31)	142-0451-83
Benzoic acid, 4-hydroxy-2,3-dimethyl-, ethyl ester	EtOH	259(4.04)	39-0667-83C
2-Cyclopenten-1-one, 3-acetoxy-2-methyl-5-(2-propenyl)-	EtOH	234(4.08)	39-1885-83C
2-Cyclopenten-1-one, 3-acetoxy-4(and 5)-methyl-5(and 4)-(2-propenyl)- (1:1 mixture)	EtOH	234(4.12),287(2.43)	39-1885-83C
2-Cyclopenten-1-one, 3-acetoxy-5-(2-methyl-2-propenyl)-	EtOH	239.5(4.10)	39-1885-83C
6-Nonen-8-ynoic acid, 7-methyl-3-oxo-, methyl ester	ether	225(3.84),250(3.08)	44-2379-83
Tricyclo[3.2.1.0^{3,6}]octan-2-one, 6-acetoxy-7-methyl-, anti	EtOH	222(3.18),294(1.95)	39-1893-83C
syn	EtOH	222(3.20),293(1.93)	39-1893-83C
$C_{11}H_{14}O_4$			
Benzeneacetic acid, α,2-dihydroxy-α-methyl-, ethyl ester	EtOH	213(3.76),275(3.29), 279(3.28)	39-1649-83C
3,6(2H,3aH)-Benzofurandione, 4,5-dihydro-2-methoxy-2,7-dimethyl-	EtOH	242(3.75),258(3.89), 368(4.11)	88-2331-83
3(2H)-Benzofuranone, 2-acetoxy-4,5,6,7-tetrahydro-2-methyl-	EtOH	272(3.96)	142-0061-83
[2,3'-Bifuran]-5,5'(2H,2'H)-dione, 3',4'-dihydro-2,2',2'-trimethyl-	MeOH	207.5(3.91)	5-1760-83
Ethanone, 1,1',1",1"'-cyclopropanediylidenetetrakis-	MeOH	208(3.51),300(2.40)	32-0427-83
Ethanone, 1-(2-hydroxy-4,6-dimethoxy-3-methylphenyl)-	EtOH	288(4.2)	83-0971-83
$C_{11}H_{14}O_5$			
Benzeneacetic acid, α,2,4-trihydroxy-α-methyl-, ethyl ester	EtOH	220(3.97),280(3.51), 283(3.48)	39-1649-83C

Compound	Solvent	$\lambda_{max}(\log \epsilon)$	Ref.
$C_{11}H_{14}O_6$			
Cyclopenta[c]pyran-4-carboxylic acid, 1,4a,5,7a-tetrahydro-1,7-dihydroxy-7-(hydroxymethyl)-, methyl ester	MeOH	242(3.92)	100-0532-83
$C_{11}H_{14}S$			
1-Propanethione, 2,2-dimethyl-1-phenyl-	benzene	314(3.28),560(1.85)	44-0214-83
$C_{11}H_{15}$			
Cycloheptatrienylium, 1,3-diethyl-, perchlorate	10% HCl	237(4.57),289(3.75)	78-4011-83
$C_{11}H_{15}BrN_2$			
1H-1,5-Benzodiazepine, 1-(2-bromoethyl)-2,3,4,5-tetrahydro-	n.s.g.	210(3.98),230(3.38), 250(3.81),290(3.13), 300(2.83)	103-0553-83
$C_{11}H_{15}BrO$			
2(3H)-Naphthalenone, 1-bromo-4,4a,5,6,7,8-hexahydro-4a-methyl-, (S)-	EtOH	259(4.05)	107-1013-83
$C_{11}H_{15}ClN_2O_7$			
3(2H)-Pyridazinone, 4-chloro-5,6-dimethoxy-2-β-D-ribofuranosyl-	pH 1	245(3.74),294(3.57)	4-0369-83
	pH 11	245(3.81),292(3.65)	4-0369-83
	MeOH	245(3.68),300(3.52)	4-0369-83
3(2H)-Pyridazinone, 5-chloro-4,6-dimethoxy-2-β-D-ribofuranosyl-	pH 1	248(3.65),296(3.63)	4-0369-83
	pH 11	248(3.65),296(3.61)	4-0369-83
	MeOH	250(3.59),300(3.59)	4-0369-83
$C_{11}H_{15}ClO_4$			
Ethanone, 1,1'-[3-(chloromethyl)-2,3-dihydro-2-hydroxy-2,5-dimethyl-3,4-furandiyl]bis-	MeOH	268(4.09)	32-0427-83
$C_{11}H_{15}NOS$			
Propanethioamide, N-(2-hydroxyphenyl)-N,2-dimethyl-	EtOH	278(4.18)	33-0262-83
$C_{11}H_{15}NO_2$			
Propanamide, N-(2-hydroxyphenyl)-N,2-dimethyl-	EtOH	277(3.50),280s(3.48)	33-0262-83
2-Pyridinecarboxylic acid, 5-butyl-4-methyl-	MeOH	230(3.92),266(3.70)	106-0591-83
$C_{11}H_{15}NO_3$			
Benzenemethanol, α-(1,1-dimethylethyl)-3-nitro-	EtOH	268(3.89)	12-0081-83
8-Isoquinolinol, 1,2,3,4-tetrahydro-6,7-dimethoxy- (anhalamine)	EtOH	274(2.90)	162-0095-83
Phenol, 2-(1,1-dimethylethyl)-6-methyl-4-nitro-	neutral	333(3.97)	126-2361-83
	anion	447(4.47)	126-2361-83
2-Pyridinecarboxylic acid, 5-butyl-4-methoxy-	MeOH	242(3.84)	106-0591-83
1(2H)-Pyridinehexanoic acid, 2-oxo-	EtOH	229(3.80),303(3.69)	94-2552-83
1(2H)-Pyridinepropanoic acid, 2-oxo-4-propyl-	EtOH	231(3.62),298(3.66)	94-2552-83
1H-Pyrrole-2-carboxylic acid, 3-hydroxy-5-methyl-4-(2-propenyl)-, ethyl ester	MeCN	273(4.18)	33-1902-83

Compound	Solvent	$\lambda_{max}(\log \epsilon)$	Ref.
$C_{11}H_{15}NO_4$			
Ribopyranosylamine, N-phenyl-	EtOH	236(4.16),283(3.30)	136-0001-83B
$C_{11}H_{15}NO_6$			
1-Azetidinepropanoic acid, 2-acetoxy-α-methyl-β,4-dioxo-, ethyl ester	MeCN	208(5.01),262(3.59)	32-0855-83
1-Propanone, 1-(2-acetoxy-2,5-dihydro-5-nitro-2-furanyl)-2,2-dimethyl-	EtOH	215(3.44),295(2.37)	12-0963-83
$C_{11}H_{15}N_2O_3$			
4H-Pyrido[1,2-a]pyrimidinium, 3-carboxy-6,7,8,9-tetrahydro-1,6-dimethyl-4-oxo-, (R)-	H_2O	204(3.85),238(3.88), 282(3.76)	39-1413-83B
$C_{11}H_{15}N_3$			
2(1H)-Pyridinone cyclohexylidenehydrazone	EtOH	271(4.34),303s(3.72)	25-0202-83
	EtOH-HCl	260(4.20),332(4.00)	25-0202-83
$C_{11}H_{15}N_3O_2$			
1,2-Diazabicyclo[5.2.0]nona-3,5-diene-2-carboxamide, N,5,8-trimethyl-9-oxo-, cis	MeOH	278(3.91)	5-2141-83
4H-Pyrido[1,2-a]pyrimidine-3-carboxamide, 1,6,7,8-tetrahydro-1,6-dimethyl-4-oxo-, (R)-	EtOH	205(3.90),225s(3.81), 260(4.40),361(3.49)	39-1413-83B
$C_{11}H_{15}N_3O_3$			
1,2-Diazabicyclo[5.2.0]nona-3,5-diene-2-carboxylic acid, 8-amino-5-methyl-9-oxo-, ethyl ester, cis-(±)-	MeOH	278(3.89)	5-1374-83
$C_{11}H_{15}N_3O_5S$			
D-Cysteine, S-methyl-N-[(1,2,3,4-tetrahydro-6-methyl-2,4-dioxo-4-pyrimidinyl)carbonyl]-, methyl ester	H_2O	271(3.99)	94-0135-83
L-	H_2O	271(4.02)	94-0135-83
D-Methionine, N-[(1,2,3,4-tetrahydro-6-methyl-2,4-dioxo-5-pyrimidinyl)-carbonyl]-	H_2O	270(4.00)	94-0135-83
L-	H_2O	270(4.05)	94-0135-83
$C_{11}H_{15}N_3O_6$			
5-Pyrimidinepentanoic acid, 2-[(carboxymethyl)amino]-1,4-dihydro-6-hydroxy-4-oxo-	50% MeOH	245s(3.794),275(4.118)	73-0304-83
	+ HCl	233(3.828),274(4.03)	73-0304-83
	+ NaOH	243(3.726),275(4.083)	73-0304-83
$C_{11}H_{15}N_3Si$			
Silacyclohexane, 1-azido-1-phenyl-	C_6H_{12}	264(2.94)	152-0645-83
$C_{11}H_{15}N_5$			
Methanimidamide, N'-[1-(dicyanomethylene)-3-(dimethylamino)-2-propenyl]-N,N-dimethyl-	EtOH	329s(4.2),374(4.7)	5-1107-83
$C_{11}H_{15}N_5O_2$			
1H-Imidazo[2,1-f]purine-2,4(3H,6H)-dione, 7-ethyl-7,8-dihydro-1,3-dimethyl-	EtOH	299(4.40)	118-0577-83
9H-Purine-9-ethanol, 6-amino-2-[(2-propenyloxy)methyl]-	pH 1	262(4.11)	111-0555-83
	pH 13	263(4.14)	111-0555-83

Compound	Solvent	λ_{max}(log ϵ)	Ref.
$C_{11}H_{15}N_5O_5$			
Adenosine, 1-methoxy-	pH 1	258(4.12)	94-3149-83
	pH 7	258(4.11)	94-3149-83
	EtOH	251s(4.03),256.5(4.11), 264.5(4.03),289(3.56)	94-3149-83
$C_{11}H_{15}OS$			
Sulfonium, dimethyl(1-methyl-2-oxo-2-phenylethyl)-, tetrafluoroborate	n.s.g.	250(4.09),280(3.38)	116-0864-83
Sulfonium, ethylmethyl(2-oxo-2-phenyl-ethyl)-, tetrafluoroborate	n.s.g.	248(4.03),290(3.60)	116-0864-83
$C_{11}H_{16}$			
Bicyclo[4.1.1]octa-2,4-diene, 2,7,7-trimethyl-, (1R)-(+)-	EtOH	283(3.55),294(3.50), 305(3.19)	35-6123-83
Bicyclo[4.1.1]octa-2,4-diene, 3,7,7-trimethyl-, (1S)-(+)-	EtOH	283(3.40),294(3.35), 305(3.00)	35-6123-83
1H-Cyclopentacyclooctene, 2,4,5,6,7,8-hexahydro-	EtOH	249(4.09)	33-2519-83
1H-Indene, 2,4,5,6-tetrahydro-6,6-di-methyl-	EtOH	243(4.15)	33-2519-83
$C_{11}H_{16}ClN$			
Benzeneethanamine, 4-chloro-N,N,α-tri-methyl-, (S)-	C_6H_{12}	269(2.70),271s(2.68), 277(2.62),280s(2.43)	35-1578-83
	MeOH	262(2.54),268(2.62), 277(2.53)	35-1578-83
hydrochloride	MeOH	261(2.34),268(2.45), 276(2.36)	35-1578-83
$C_{11}H_{16}ClN_3O_5$			
3(2H)-Pyridazinone, 6-chloro-4-(di-methylamino)-2-β-D-ribofuranosyl-	pH 1	241(3.75),250s(3.59), 320(4.03)	4-0369-83
	pH 11	242(3.60),320(4.02)	4-0369-83
	MeOH	242(3.67),250s(3.53), 280s(3.99),319(4.08), 334s(3.90)	4-0369-83
3(2H)-Pyridazinone, 6-chloro-5-(di-methylamino)-2-β-D-ribofuranosyl-	pH 1	242(4.16),301(3.73)	4-0369-83
	pH 11	242(4.12),303(3.68)	4-0369-83
	MeOH	239(4.15),301(3.76)	4-0369-83
$C_{11}H_{16}ClN_5O_2$			
1H-Purine-2,6-dione, 8-[[1-(chloro-methyl)propyl]amino]-3,7-dihydro-1,3-dimethyl-	EtOH	296(4.28)	118-0577-83
$C_{11}H_{16}FN_3O_3$			
1(2H)-Pyrimidinecarboxamide, 5-fluoro-N-hexyl-3,4-dihydro-2,4-dioxo-	$CHCl_3$	258(4.06)	162-0256-83
$C_{11}H_{16}I_2$			
Tricyclo[3.3.1.1^{3,7}]decane, 1-iodo-3-(iodomethyl)-	hexane	327(4.11)	99-0206-83
$C_{11}H_{16}N_2O$			
1H-1,5-Benzodiazepine-1-ethanol, 2,3,4,5-tetrahydro-	n.s.g.	210(4.00),230(3.38), 250(3.81),290(3.13), 300(2.81)	103-0553-83
1-Propanone, 2-methyl-1-[4-(1-methyl-ethyl)-5-pyrimidinyl]-	EtOH	219(3.75),249s(3.37), 282(2.87)	4-0649-83

Compound	Solvent	$\lambda_{max}(\log \epsilon)$	Ref.
$C_{11}H_{16}N_2O_2$			
Phenol, 4-(dimethylamino)-3-methyl-, methylcarbamate	EtOH	248.5(4.82)	162-0064-83
$C_{11}H_{16}N_2O_3$			
1H-1,2-Diazepine-1-carboxylic acid, 2-acetyl-2,3-dihydro-5-methyl-, ethyl ester	MeOH	270(3.95)	5-2141-83
$C_{11}H_{16}N_2O_3S$			
Acetic acid, [(1,4-dihydro-5-methyl-4-oxo-2-pyrimidinyl)thio]-, butyl ester	MeOH	288(3.97)	56-1027-83
Acetic acid, [(1,4-dihydro-5-methyl-4-oxo-2-pyrimidinyl)thio]-, isobutyl ester	MeOH	288(4.03)	56-1027-83
$C_{11}H_{16}N_2O_4$			
2,4(1H,3H)-Pyrimidinedione, 1-[2-hydroxy-4-(hydroxymethyl)cyclopentyl]-5-methyl-, (1α,2α,4α)-(±)-	pH 1	211(3.92),272(4.03)	4-0655-83
	pH 7	211(3.92),273(4.03)	4-0655-83
	pH 13	271(3.90)	4-0655-83
$C_{11}H_{16}N_2O_4S$			
1H-Imidazole-1-acetic acid, 3-(ethoxycarbonyl)-4-(methylthio)-, ethyl ester	EtOH	286(4.10)	2-0030-83
$C_{11}H_{16}N_2O_5$			
5-Pyrimidinepentanoic acid, 5-ethylhexahydro-2,4,6-trioxo-	50%MeOH-HCl	208s(4.02)	73-0137-83
	+ NaOH	241.5(4.03)	73-0137-83
$C_{11}H_{16}N_2O_5S$			
Pseudouridine, 1',4'-dimethyl-2-thio-, (±)-	pH 1	216(4.11),277(4.22), 285(4.21)	18-2680-83
	pH 13	222(4.20),262(4.17), 290(3.97)	18-2680-83
	MeOH	216(4.13),277(4.28), 290(4.24)	18-2680-83
Pseudouridine, 5',5'-dimethyl-2-thio-, (±)-	pH 1	214(3.88),274(3.94), 289(3.95)	18-2680-83
	pH 13	222(3.90),263(3.83), 284(3.70)	18-2680-83
	MeOH	214(3.76),276(3.82), 292(3.78)	18-2680-83
$C_{11}H_{16}N_2O_6$			
Pseudouridine, 1',4'-dimethyl-, (±)-	pH 13	287(3.68)	18-2680-83
	MeOH	265(3.79)	18-2680-83
3(2H)-Pyridazinone, 2-(2-deoxy-β-D-erythro-pentofuranosyl)-5,6-dimethoxy-	pH 1	239(4.02),279(3.68)	4-0369-83
	pH 11	239(4.02),280(3.70)	4-0369-83
	MeOH	240s(3.99),285(3.53)	4-0369-83
2,4(1H,3H)-Pyrimidinedione, 5-[3,4-dihydroxy-5-(1-hydroxy-1-methylethyl)-tetrahydrofuran-2-yl]-, (2R*,3R*-4S*,5S*)-	pH 1	264(3.92)	18-2680-83
	pH 13	218(3.92),287(4.04)	18-2680-83
	MeOH	212(3.90),263(3.82)	18-2680-83
$C_{11}H_{16}N_2O_6S$			
Uridine, 6-(ethylthio)-	H_2O	283(4.17)	94-1222-83
$C_{11}H_{16}N_2O_7$			
3(2H)-Pyridazinone, 4,6-dimethoxy-2-β-	pH 1	282(3.73)	4-0369-83

Compound	Solvent	λ_{max}(log ε)	Ref.
D-ribofuranosyl- (cont.)	pH 11	283(3.71)	4-0369-83
	MeOH	286(3.72)	4-0369-83
3(2H)-Pyridazinone, 5,6-dimethoxy-2-β-D-ribofuranosyl-	pH 1	239(3.82),280(3.47)	4-0369-83
	pH 11	240(3.88),280(3.60)	4-0369-83
	MeOH	240(3.74),285(3.45)	4-0369-83
$C_{11}H_{16}N_2O_7S$			
Uridine, 6-[(2-hydroxyethyl)thio]-	MeOH	279.5(4.12)	94-1222-83
$C_{11}H_{16}N_3O_2$			
4H-Pyrido[1,2-a]pyrimidinium, 3-(aminocarbonyl)-6,7,8,9-tetrahydro-1,6-dimethyl-4-oxo-, (R)-	H_2O	208(3.90),235(3.93), 286(3.82)	39-1413-83B
$C_{11}H_{16}N_4$			
1H-Pyrazole, 4,4'-methylenebis[3,5-dimethyl-	MeOH	268(4.09)	32-0427-83
2,3,7,8-Tetraazatetracyclo[7.2.1.0^{4,12}.0^{6,10}]trideca-2,7-diene, 1,6-dimethyl-	EtOH	335(2.61)	39-2545-83C
2,3,7,8-Tetraazatetracyclo[7.3.1.0^{4,12}.0^{6,10}]trideca-1,6-diene, 4,9-dimethyl-	EtOH	268(3.57)	39-2541-83C
$C_{11}H_{16}N_4O$			
4H-Pyrazolo[3,4-d]pyrimidin-4-one, 5-butyl-1,5-dihydro-1,6-dimethyl-	EtOH	208(4.70),256(3.96)	11-0171-83B
$C_{11}H_{16}N_4O_5$			
1H-Purine-6-methanol, 6,9-dihydro-9-β-D-ribofuranosyl-, (R)-	pH 7	293(3.62)	94-3104-83
(S)-	pH 7	297(3.68)	94-3104-83
1,2,4-Triazine-3,6-dicarboxylic acid, 4,5-dihydro-4-morpholino-, dimethyl ester	CH_2Cl_2	345(3.67)	83-0472-83
$C_{11}H_{16}N_6O_2$			
Acetamide, N-[2-[(6-amino-9H-purin-9-yl)methoxy]ethyl]-N-methyl-	H_2O	261(4.25)	83-0146-83
$C_{11}H_{16}N_6O_4$			
L-Alanine, 2-[(2-amino-1,6-dihydro-6-oxo-9H-purin-9-yl)methoxy]ethyl ester, monohydrochloride	pH 1	254(4.04),273s(--)	87-0602-83
β-Alanine, 2-[(2-amino-1,6-dihydro-6-oxo-9H-purin-9-yl)methoxy]ethyl ester, monohydrochloride	pH 1	254(4.04),273s(--)	87-0602-83
$C_{11}H_{16}N_6O_5$			
3H-1,2,3-Triazolo[4,5-d]pyrimidin-7-amine, 5-ethoxy-3-β-D-ribofuranosyl-	pH 1	270(4.00)	87-1483-83
	pH 7	273(3.88)	87-1483-83
	pH 13	275(3.89)	87-1483-83
$C_{11}H_{16}O$			
Bicyclo[4.3.1]deca-2,4-dien-7-ol, 9-methyl-, (7-endo,9-exo)	EtOH	239s(3.79),245(3.93), 254(3.97),264(3.74)	24-2914-83
exo,exo	EtOH	238s(3.79),244(3.95), 253(3.99),263(3.77)	24-2914-83
Cyclohexanol, 1-ethenyl-2-(2-propenylidene)-	C_6H_{12}	237(4.34)	44-4272-83

Compound	Solvent	$\lambda_{max}(\log \epsilon)$	Ref.
Cyclohexanone, 2-(cyclopropylmethyl-ene)-3-methyl-, cis	MeOH	265(4.06)	23-0288-83
trans	MeOH	263(4.07)	23-0288-83
2-Cyclohexen-1-one, 2-ethenyl-3,5,5-trimethyl-	EtOH	220(3.95),259(3.93)	23-0936-83
3-Cyclohexen-1-one, 2-ethylidene-3,3,5-trimethyl-, (E)-	hexane	224(3.94),269(3.76), 318(2.21),330(2.17), 344(2.04),354(1.72)	23-0936-83
	EtOH	224(3.90),277(3.58)	23-0936-83
(Z)-	hexane	221.5(4.08),268(3.75), 322(2.11)	23-0936-83
	EtOH	223(4.00),279(3.58)	23-0936-83
4H-Inden-4-one, 1,2,3,5,6,7-hexahydro-6,6-dimethyl-	EtOH	250(3.76)	23-0936-83
1(2H)-Naphthalenone, 3,4,5,6,7,8-hexa-hydro-5-methyl-	EtOH	246.5(4.033)	2-0331-83
Spiro[2.5]oct-7-en-4-one, 6,6,8-tri-methyl-	hexane	217(3.34),232s(3.03), 284(1.63)	23-0078-83
$C_{11}H_{16}O_2$			
Acetic acid, (3,5,5-trimethyl-2-cyclo-hexen-1-ylidene)-	EtOH	274(4.35)	111-0447-83
5,7-Octadienal, 3,3,7-trimethyl-4-oxo-, (E)-	pentane	340(1.85)	33-1638-83
$C_{11}H_{16}O_3$			
3-Cyclohexene-1-propanoic acid, β,4-dimethyl-2-oxo-	EtOH	231(4.53)	64-0497-83B
$C_{11}H_{16}O_4$			
Cyclohexanepentanoic acid, δ,2-dioxo-	EtOH	291(4.017)	2-0331-83
Cyclopentanepentanoic acid, δ, 2-dioxo-, methyl ester	n.s.g.	286(3.5798)	2-1154-83
3-Furancarboxylic acid, 2-(2-hydroxy-propyl)-, 1-methylethyl ester	MeCN	210s(3.83),243(3.84)	24-3366-83
$C_{11}H_{17}ClN_2O$			
Pyrazine, 3-chloro-5-(1-methylethyl)-2-(2-methylpropyl)-, 4-oxide	EtOH	232(4.39),271(4.06), 301(3.52),312s(3.51)	94-0020-83
Pyrazinemethanol, 6-chloro-α,α-di-methyl-5-(2-methylpropyl)-	EtOH	231(4.27),271(3.96), 299s(3.48),311s(3.41)	94-0020-83
$C_{11}H_{17}N$			
Benzenamine, N-pentyl-	n.s.g.	286(3.20)	32-0069-83
Benzeneethanamine, N,N,α-trimethyl-, (S)-	C_6H_{12}	265(2.48),268(2.32), 272(2.28)	35-1578-83
	MeOH	262(2.38),264(2.32), 268(2.23)	35-1578-83
hydrochloride	MeOH	258(2.28),264(2.18), 268(1.90)	35-1578-83
Cyclohexanamine, N-(1-methyl-2-butyn-ylidene)-	C_6H_{12}	218(3.80),259s(2.51)	44-1925-83
Pyridine, 2-(1,2,2-trimethylpropyl)-, (R)-	heptane	252s(3.36),257(3.45), 261(3.44),268s(3.29)	39-0399-83B
	MeOH	253s(3.42),257(3.52), 262(3.51),269(3.35)	39-0399-83B
	3-Mepentane at -180°	252s(3.42),257(3.51), 262(3.51),269(3.46)	39-0399-83B
$C_{11}H_{17}NOS$			
2H-1,4-Benzothiazin-8(5H)-one,	EtOH	228(3.91),333(4.13)	142-2391-83

Compound	Solvent	λ_{max}(log ε)	Ref.
3,4,6,7-tetrahydro-3,6,6-trimethyl-			142-2391-83
$C_{11}H_{17}NO_2$			
Benzeneethanamine, 2-ethoxy-3-methoxy-, hydrochloride	MeOH	216s(4.06),272(3.46)	83-0845-83
2,4-Hexadienoic acid, 6-piperidino-, hydrochloride, (E,E)-	EtOH	247(4.48)	111-0447-83
$C_{11}H_{17}NO_2Si$			
3-Pyridinecarboxylic acid, 2-(trimethylsilyl)ethyl ester	MeOH	222(3.76),256(3.45), 262s(3.49),268(3.36)	5-1374-83
$C_{11}H_{17}NO_4$			
2-Butenedioic acid, 2-piperidino-, dimethyl ester	MeOH	288(3.07)	2-0290-83
$C_{11}H_{17}N_2O_2$			
1(2H)-Pyridinyloxy, 3,6-dihydro-4-(isocyanatomethyl)-2,2,6,6-tetramethyl-	MeCN	234(3.48),446(0.88)	70-1659-83
$C_{11}H_{17}N_3O$			
Hydrazinecarboxamide, 2-(1,2,3,5,6,7-hexahydro-6-methyl-4H-inden-4-ylidene)-	n.s.g.	269(4.4771)	2-1154-83
1-Propanone, 1-[2-amino-4-(1-methylethyl)-5-pyrimidinyl]-2-methyl-	EtOH	270.5(4.24)	4-0649-83
$C_{11}H_{17}N_3O_4$			
2,4(1H,3H)-Pyrimidinedione, 1-[2-hydroxy-4-(hydroxymethyl)cyclopentyl]-5-(methylamino)-, (1α,2β,4α)-(±)-	pH 1	272(4.00)	87-0156-83
	pH 7	234(3.88),302(3.86)	87-0156-83
	pH 13	235s(--),294(3.79)	87-0156-83
2,4(1H,3H)-Pyrimidinedione, 1-[3-hydroxy-4-(hydroxymethyl)cyclopentyl]-5-(methylamino)-, (1α,3β,4α)-(±)-	pH 1	270(3.97)	87-0156-83
	pH 7	236(3.83),303(3.81)	87-0156-83
	pH 13	235s(--),294(3.76)	87-0156-83
5-Pyrimidinepentanoic acid, 2-(ethylamino)-1,4-dihydro-6-hydroxy-4-oxo-	MeOH	241(3.830),275(4.141)	73-0304-83
	MeOH-HCl	240s(3.871),268(4.116)	73-0304-83
	MeOH-NaOH	241(3.826),275(4.121)	73-0304-83
$C_{11}H_{17}N_3O_4S$			
5-Pyrimidinecarboxamide, 1,2,3,4-tetrahydro-N-[1-(hydroxymethyl)-3-(methylthio)propyl]-6-methyl-2,4-dioxo-, (R)-	H_2O	268.5(4.02)	94-0135-83
(S)-	H_2O	268.5(3.93)	94-0135-83
$C_{11}H_{17}N_3O_5$			
Pseudoisocytidine, 1',4'-dimethyl-, hydrochloride, (±)-	MeOH	223(4.27),263(4.10), 292s(3.52)	18-2680-83
	pH 1	222(3.98),263(3.84)	18-2680-83
	pH 13	231(3.96),278(3.84)	18-2680-83
Pseudoisocytidine, 5',5'-dimethyl-, hydrochloride, (±)-	MeOH	223(4.02),265(3.86), 290(3.59)	18-2680-83
	pH 1	221(3.95),262(3.84)	18-2680-83
	pH 13	233(3.94),276(3.84)	18-2680-83
3(2H)-Pyridazinone, 4-(dimethylamino)-2-β-D-ribofuranosyl-	pH 1	317(3.99)	4-0369-83
	pH 11	317(4.01)	4-0369-83
	MeOH	317(4.05)	4-0369-83
3(2H)-Pyridazinone, 6-(dimethylamino)-2-β-D-ribofuranosyl-	pH 1	245(4.43),353(3.47)	4-0369-83
	pH 11	245(4.40),353(3.41)	4-0369-83
	MeOH	245(4.41),353(3.49)	4-0369-83

Compound	Solvent	λ_{max}(log ε)	Ref.
2,4(1H,3H)-Pyrimidinedione, 1-[2,3-di-hydroxy-4-(hydroxymethyl)cyclopent-yl]-5-(methylamino)-, (1α,2β,3β,4α)-	pH 1	271(3.96)	87-0156-83
	pH 7	235(3.82),303(3.80)	87-0156-83
	pH 13	235s(--),294(3.78)	87-0156-83
$C_{11}H_{17}N_5O_3$			
1H-Purine-2,6-dione, 3,7-dihydro-8-[[1-(hydroxymethyl)propyl]-amino]-1,3-dimethyl-	EtOH	297(4.37)	118-0577-83
$C_{11}H_{17}N_5O_4$			
1,2-Propanediol, 3-[3-(6-amino-9H-purin-9-yl)-2-hydroxypropoxy]-	pH 1	262(4.10)	111-0555-83
	pH 13	263(4.11)	111-0555-83
1(2H)-Pyrimidineacetic acid, α-[3-[(aminoiminomethyl)amino]propyl]-3,4-dihydro-5-methyl-2,4-dioxo-, (S)-	H_2O	273(3.91)	35-6989-83
$C_{11}H_{18}$			
4-Decyne, 6-methylene-	EtOH	224(4.05)	104-1621-83
2,3-Heptadiene, 4-ethenyl-2,6-di-methyl-	EtOH	227(4.00)	70-0569-83
$C_{11}H_{18}NO_3$			
Pyridinium, 4-(carboxymethyl)-1,2,3,6-tetrahydro-2,2,6,6-tetramethyl-1-oxo-, chloride	H_2O	237s(3.26),417(2.04), 472s(1.93)	70-1659-83
nitrate	H_2O	247s(3.23),395(2.11), 485s(1.89)	70-1659-83
perchlorate	H_2O	237s(3.32),417(2.04), 470s(1.94)	70-1659-83
	MeCN	227s(3.32),420(2.08), 471s(1.99)	70-1659-83
1(2H)-Pyridinyloxy, 4-(carboxymethyl)-3,6-dihydro-2,2,6,6-tetramethyl-	EtOH	231(3.43),434(0.87)	70-1659-83
$C_{11}H_{18}N_2O$			
Pyrazine, 3-methoxy-2,5-bis(1-methyl-ethyl)-	EtOH	216(3.93),281(3.79), 294(3.85)	4-0951-83
$C_{11}H_{18}N_2O_3$			
2(1H)-Pyrazinone, 1-hydroxy-6-(1-hy-droxy-1-methylethyl)-3-(2-methyl-propyl)-	EtOH	235.5(3.89),339(3.66)	94-0020-83
2,4,6(1H,3H,5H)-Pyrimidinetrione, 5,5-diethyl-1-(1-methylethyl)-	anion	244(3.88)	111-0521-83
$C_{11}H_{18}N_2O_6$			
Carbamic acid, [3-[(2-deoxy-β-D-ery-thro-pentofuranosyl)amino]-2-methyl-1-oxo-2-propenyl]-, methyl ester	H_2O	290(4.10)	35-0956-83
$C_{11}H_{18}N_6O_2$			
Hydrazinecarboxamide, 2-[1-[3-[(amino-carbonyl)hydrazono]-5,5-dimethyl-1-cyclopenten-1-yl]ethylidene]-	MeOH	314(4.48)	39-1465-83C
$C_{11}H_{18}O$			
2-Cyclohexen-1-one, 2-ethyl-3,5,5-tri-methyl-	EtOH	244(3.96)	23-0078-83
$C_{11}H_{18}O_2$			
2-Cyclohexen-1-one, 2-(2-ethoxyethyl)-	EtOH	244(4.04),310(1.81)	23-0078-83

Compound	Solvent	λ_{max}(log ϵ)	Ref.
3-methyl- (cont.)	EtOH	245(4.02),310(1.84)	23-0078-83
$C_{11}H_{18}O_4$			
Furan, 2-(2,2-dimethoxypropyl)-3-(methoxymethyl)-	MeCN	215(3.89)	24-3366-83
$C_{11}H_{18}O_5S$			
D-xylo-Hexofuranos-5-ulose, 3-O-methyl-6-S-methyl-1,2-O-(1-methylethylidene)-6-thio-	EtOH	214(2.71),247(2.44), 300(2.39)	159-0139-83
$C_{11}H_{19}NO$			
3-Buten-2-one, 4-(cyclohexylamino)-3-methyl-	MeOH	299(4.48)	131-0001-83K
(E-syn-E)-	C_6H_{12}	277(--)	131-0001-83K
	dioxan	284(4.44)	131-0001-83K
(Z-syn-Z)-	C_6H_{12}	316(3.99)	131-0001-83K
	dioxan	315(--)	131-0001-83K
$C_{11}H_{19}N_2O_2$			
Piperidinyloxy, 4-(isocyanatomethyl)-2,2,6,6-tetramethyl-	MeCN	242(3.46),463(1.02)	70-1659-83
$C_{11}H_{19}N_2O_3$			
1H-Pyrrol-1-yloxy, 2,5-dihydro-3-[[(2-hydroxyethyl)amino]carbonyl]-2,2,5,5-tetramethyl-	MeOH	274(3)	28-0043-83B
$C_{11}H_{19}N_3OS$			
4-Morpholinecarbothioamide, N-(piperidinylmethylene)-	MeCN	276(4.51),302(4.09)	97-0019-83
$C_{11}H_{19}N_3O_3S$			
2H-Thiazolo[3,2-a]-1,3,5-triazin-6(7H)-one, 3-(1,1-dimethylethyl)-3,4-dihydro-7,7-bis(hydroxymethyl)-	EtOH	219(4.12)	103-0961-83
$C_{11}H_{19}N_3O_5$			
2-Propenamide, 3-[(2-deoxy-β-D-erythro-pentofuranosyl)amino]-2-methyl-N-[(methylamino)carbonyl]-, (E)-	H_2O	289(4.26)	35-0956-83
$C_{11}H_{19}N_7O$			
9H-Purine-9-ethanol, 6-amino-α-[[(3-aminopropyl)amino]methyl]-, (±)-	pH 2,7,12	262(4.12)	73-1910-83
$C_{11}H_{19}N_7O_2$			
1,2-Propanediol, 3-[6-amino-8-[(3-aminopropyl)amino]-9H-purin-9-yl]-, (±)-	pH 2	280(4.12)	73-1910-83
	pH 7 and 12	280(4.19)	73-1910-83
8H-Purin-8-one, 6-amino-9-[2-[(3-aminopropyl)amino]-3-hydroxypropyl]-7,9-dihydro-	pH 2	270(--),281(4.01)	73-1910-83
	pH 7	273(4.11)	73-1910-83
	pH 12	283(4.17)	73-1910-83
$C_{11}H_{20}Ge_2$			
Digermane, pentamethylphenyl-	C_6H_{12}	228(4.05)	101-0149-83G
$C_{11}H_{20}NO$			
2,1-Benzoxazepinium, 1-ethyl-3,4,5,5a-6,7,8,9-octahydro-, tetrafluoroborate	MeCN	220(3.75)	5-0897-83

Compound	Solvent	λ_{max}(log ε)	Ref.
Quinolinium, 1-ethoxy-2,3,4,4a,5,6,7,8-octahydro-, tetrafluoroborate	MeCN	198(3.72)	5-0897-83
$C_{11}H_{20}NO_3$			
Piperidinium, 4-(carboxymethyl)-2,2,6,6-tetramethyl-1-oxo-, chloride	H_2O	230(3.28),464(1.32), 477(1.32),504s(1.18)	70-1659-83
nitrate	H_2O	244s(3.31),466(1.34), 479(1.35),503s(1.18)	70-1659-83
1-Piperidinyloxy, 4-(carboxymethyl)-2,2,6,6-tetramethyl-	EtOH	243(3.56),451(1.04)	70-1659-83
$C_{11}H_{20}NS$			
Methanaminium, N-[5,5-dimethyl-3-(methylthio)-2-cyclohexen-1-ylidene]-N-methyl-, iodide	H_2O	225(4.23),330(4.45)	78-3405-83
	pH 1	225(4.24),265(4.24), 330(4.45)	78-3405-83
$C_{11}H_{20}N_2$			
1H-Imidazole, 2,4-bis(1-methylpropyl)-	EtOH	212(3.96)	4-1277-83
1H-Imidazole, 2,4-bis(2-methylpropyl)-	EtOH	213(4.04)	4-1277-83
1H-Imidazole, 2,4-dibutyl-	EtOH	213(4.11)	4-1277-83
2-Propenenitrile, 3-[bis(1,1-dimethylethyl)amino]-, (Z)-	heptane	208(3.91),266(3.66), 300s(--)	70-2322-83
	MeOH	212(3.63),270(4.06)	70-2322-83
$C_{11}H_{20}N_2O$			
3-Furanamine, N,N,2-triethyl-2,5-dihydro-5-imino-2-methyl-	n.s.g.	280-284(4.34-4.48) (class spectrum)	77-1455-83
$C_{11}H_{20}N_2O_3S$			
3-Pyridinesulfonamide, 1,6-dihydro-N,N,1-trimethyl-6-(1-methylethoxy)-	CH_2Cl_2	304(3.7)	64-0873-83B
$C_{11}H_{20}O_4Si$			
Pentanedioic acid, 2-methylene-4-(trimethylsilyl)-, dimethyl ester	C_6H_{12}	215(2.60)	44-3189-83
$C_{11}H_{20}O_5S$			
α-D-Glucofuranose, 3-O-methyl-6-S-methyl-1,2-O-(1-methylethylidene)-6-thio-	EtOH	213(2.53),223s(--)	159-0139-83
$C_{11}H_{21}INO$			
1-Piperidinyloxy, 4-(2-iodoethyl)-2,2,6,6-tetramethyl-	hexane	246(3.70),490(1.23)	70-1757-83
$C_{11}H_{21}NO$			
3-Buten-2-one, 4-[bis(1-methylethyl)-amino]-3-methyl-	MeOH	306(4.44)	131-0001-83K
(E-syn-E)-	C_6H_{12}	288(4.42)	131-0001-83K
4-Heptanone, 3-[[(1-methylethyl)amino]methylene]-	MeOH	299(4.22)	131-0001-83K
(E-syn-E)-	C_6H_{12}	281(--)	131-0001-83K
(Z-syn-Z)-	C_6H_{12}	314(3.93)	131-0001-83K
$C_{11}H_{24}Si_2$			
Silane, 1,3-pentadiene-1,5-diylbis-[trimethyl-, (E,E)-	THF	249(4.42)	157-0021-83
cesium salt	THF	396(3.60)	157-0021-83
lithium salt	THF	386(3.79)	157-0021-83
potassium salt	THF	393(3.62)	157-0021-83

Compound	Solvent	$\lambda_{max}(\log \epsilon)$	Ref.
$C_{12}F_{18}S_6$			
Benzene, hexakis[(trifluoromethyl)-thio]-	hexane	305(3.48),350(3.78)	104-2313-83
$C_{12}F_{18}Se_6$			
Benzene, hexakis[(trifluoromethyl)-seleno]-	hexane	330(3.23),412(3.66)	104-2313-83
$C_{12}H_4Br_2N_2O_4$			
Acenaphthylene, 1,2-dibromo-5,6-di-nitro-	THF	231(4.46),287(4.15), 346(4.15),452(2.96)	44-2949-83
$C_{12}H_4Cl_2N_6O_9$			
Diazene, bis(2-chloro-3,5-dinitro-phenyl)-, 1-oxide	EtOH	254(4.16),300s(3.73)	12-1227-83
$C_{12}H_4N_4$			
Propanedinitrile, 2,2'-(2,5-cyclohexa-diene-1,4-diylidene)bis-	H_2O	392(4.8)	88-1567-83
	MeCN	390(4.80)	104-0742-83
$C_{12}H_5N_5O_9S$			
10H-Phenothiazine, 1,3,7,9-tetra-nitro-, 5-oxide	DMF	480(4.74)	80-0381-83
$C_{12}H_5N_5O_{10}S$			
10H-Phenothiazine, 1,3,7,9-tetra-nitro-, 5,5-dioxide	DMF	480(4.74)	80-0381-83
$C_{12}H_6$			
Benzocyclooctene, 5,6,9,10-tetrade-hydro-	pentane	228(4.38),233(4.37), 238s(4.43),244(4.67), 249(4.61),257(4.86), 327(2.76),332(2.77), 341(2.88),345(2.91), 361(2.95),366(2.88)	78-0427-83
$C_{12}H_6ClN_3OS$			
2H-Pyrido[3,4-b][1,4]benzothiazine-4-carbonitrile, 7-chloro-3,5-di-hydro-3-oxo-	MeOH	253(4.56)	87-0845-83
2H-Pyrido[3,4-b][1,4]benzothiazine-4-carbonitrile, 8-chloro-3,5-di-hydro-3-oxo-	MeOH	256(4.60)	87-0845-83
$C_{12}H_6ClN_3O_2S$			
2H-Pyrido[3,4-b][1,4]benzothiazine-4-carbonitrile, 7-chloro-3,5-di-hydro-3-oxo-, 10,10-dioxide	MeOH	258(4.59)	87-0845-83
$C_{12}H_6Cl_6O_2S$			
Benzo[b]thiophene-3-carboxylic acid, 4,5,6,7-tetrachloro-2-(dichloro-methyl)-, ethyl ester	heptane	194(4.30),218(4.39), 240(4.53),282s(4.09), 290(4.13),316(3.72), 328(3.71)	5-0013-83
$C_{12}H_6N_2$			
1,5-Naphthalenedicarbonitrile	EtOH	228(5.01),288(4.15), 300(4.29),309(4.16), 312(4.16),328(3.97)	35-0040-83

Compound	Solvent	λ_{max}(log ε)	Ref.
$C_{12}H_6N_2O_2$			
1,7-Phenanthroline-5,6-dione	MeOH	230(4.51),283(3.97)	35-4431-83
	tert-BuOH	252(4.49),285s(3.77)	35-4431-83
	CH_2Cl_2	255(4.51),286s(3.43)	35-4431-83
	base	230(--),287(3.96)	35-4431-83
1,10-Phenanthroline-5,6-dione	MeOH	251(4.33),292s(3.83)	35-4431-83
	tert-BuOH	255(4.51),295s(3.63)	35-4431-83
	CH_2Cl_2	257(4.50),295s(3.40)	35-4431-83
	base	295(3.95)	35-4431-83
4,7-Phenanthroline-5,6-dione	MeOH	230(4.20),282s(3.97)	35-4431-83
	tert-BuOH-MeCN	250(4.46),285s(4.02)	35-4431-83
	CH_2Cl_2	253(4.49),285s(3.79)	35-4431-83
	base	227(4.40),282(4.02)	35-4431-83
$C_{12}H_6N_4O_7S$			
10H-Phenothiazine, 1,3,7-trinitro-, 5-oxide	DMF	500(4.73)	80-0381-83
$C_{12}H_6N_4O_8S$			
10H-Phenothiazine, 1,3,7-trinitro-, 5,5-dioxide	DMF	505(4.73)	80-0381-83
$C_{12}H_6O_2S$			
Naphtho[2,3-b]thiophene-4,9-dione	EtOH	248(4.475),252(4.49), 270s(4.08),280(4.13), 325(3.73)	73-0112-83
$C_{12}H_6O_3$			
Naphtho[2,3-c]furan-1,3-dione	EtOH	237(4.75),272(3.755), 280(3.745),293s(3.51), 320(3.12),336(3.22)	73-0112-83
1H,3H-Naphtho[1,8-cd]pyran-1,3-dione	EtOH	228(4.42),328(3.90), 339(3.89)	73-0112-83
$C_{12}H_7Br_3O_3$			
1(4H)-Naphthalenone, 5-acetoxy-2,4,4-tribromo-	hexane	261(4.04),313(3.41)	44-5359-83
$C_{12}H_7ClN_2$			
1,8-Phenanthroline, 4-chloro-	EtOH	213s(4.24),237(4.53), 246s(4.52),285s(3.94), 297(3.94),334(3.22), 350(3.31)	33-0620-83
$C_{12}H_7ClN_2O$			
2-Phenazinol, 3-chloro-	MeOH	260(4.36),265(3.46)	64-0866-83B
2-Phenazinol, 4-chloro-	MeOH	262(4.00),364(3.00)	64-0866-83B
$C_{12}H_7ClOS_2$			
Thieno[2,3-b][1]benzothiepin-4(5H)-one, 7-chloro-	MeOH	258(4.33),313(3.95)	73-2970-83
$C_{12}H_7Cl_2NO_2$			
Benzene, 2,4-dichloro-1-(4-nitrosophenoxy)-	n.s.g.	758(1.40)	98-0227-83
$C_{12}H_7Cl_7O_2S$			
3-Butenoic acid, 4,4-dichloro-3-[(pentachlorophenyl)thio]-, ethyl ester	heptane	195s(4.36),218(4.77), 243s(4.33),272s(4.02), 302(3.92)	5-0013-83

Compound	Solvent	λ_{max}(log ε)	Ref.
$C_{12}H_7NOS$			
2H-Benzo[b]cyclobuta[d]thiopyran-1-carbonitrile, 1,3-dihydro-3-oxo-	MeOH	211(4.43),222(4.39), 237(4.37),265(3.69), 304(3.93),343(3.59)	142-1275-83
$C_{12}H_7NO_4$			
5H-[1]Benzopyrano[2,3-b]pyridin-5-one, 6,8-dihydroxy-	EtOH	209(4.36),225(4.32), 256(4.12),283(3.73), 305(3.97),355s(3.73), 390s(3.17)	39-0219-83C
$C_{12}H_7N_2$			
Azepino[2,1,7-cd]pyrrolizinium, 1-cyano-, perchlorate	EtOH	207(4.13),253(4.23), 282s(4.02),295s(3.96), 313s(2.89),446(3.45)	24-1174-83
$C_{12}H_7N_3OS$			
2H-Pyrido[3,4-b][1,4]benzothiazine-4-carbonitrile, 3,5-dihydro-3-oxo-	MeOH	252(4.63)	87-0845-83
$C_{12}H_7N_3O_2S$			
2-Thiazolecarboxylic acid, 4-(2-quinoxalinyl)-	EtOH	246(4.04),267s(3.74), 292s(3.44),342(3.55)	106-0829-83
potassium salt	H_2O	246(3.72),269(3.56), 344(3.39),354s(3.35)	106-0829-83
$C_{12}H_7N_9O_{12}$			
4-Benzofurazanamine, N-(1,4-dihydro-5,7-dinitro-4-benzofurazanyl)-1,4-dihydro-5,7-dinitro-, N,3-dioxide, dipotassium salt	MeOH	266(4.32),304(4.32), 352(3.92),460(4.78)	12-0297-83
$C_{12}H_8$			
1H-Cyclobuta[de]naphthalene, 1-methylene-	EtOH	221(4.80),257(4.10), 311(3.08),323(3.14)	35-6104-83
$C_{12}H_8Br_2S_2$			
Disulfide, bis(2-bromophenyl)	heptane	240(4.24),265s(3.62)	65-0519-83
	MeOH	235(4.31),270s(3.57)	65-0519-83
Disulfide, bis(3-bromophenyl)	heptane	240(4.25),270s(3.65)	65-0519-83
	MeOH	230s(4.39),270s(3.63)	65-0519-83
Disulfide, bis(4-bromophenyl)	heptane	251(4.39),277s(3.90), 302s(3.51)	65-0519-83
	MeOH	248(4.34),270s(3.90), 305s(3.34)	65-0519-83
	MeCN	248(4.36),270s(3.89), 295s(3.52)	65-0519-83
$C_{12}H_8Br_2Se_2$			
Diselenide, bis(2-bromophenyl)	heptane	246(4.32),275s(3.41), 311s(2.92)	65-0519-83
	MeOH	244(4.39),275s(3.42), 311s(2.84)	65-0519-83
Diselenide, bis(3-bromophenyl)	heptane	245(4.25),275s(3.61), 331(2.82)	65-0519-83
Diselenide, bis(4-bromophenyl)	heptane	258(4.40),276s(4.15), 340(3.20)	65-0519-83
	MeOH	253(4.25),265s(4.06), 310s(3.23)	65-0519-83
	MeCN	253(4.42),270s(4.09), 305s(3.28)	65-0519-83

Compound	Solvent	λ max(log ε)	Ref.
$C_{12}H_8ClF$			
1,1'-Biphenyl, 3-chloro-4'-fluoro-	hexane	248(4.18)	64-0621-83B
$C_{12}H_8ClN$			
9H-Carbazole, 3-chloro-	MeOH	215(3.31),223(3.30), 229(3.32),236(3.34), 246(3.09),260(3.05), 291s(2.78),297(2.94), 319s(2.13),330(2.22), 343(2.18)	100-0852-83
$C_{12}H_8ClNO_3$			
5H-Furo[2,3-f]indole-2-carboxylic acid, 8-chloro-, methyl ester	CH_2Cl_2	306(4.16),316(4.26)	4-1059-83
$C_{12}H_8ClN_4$			
Benzenediazonium, 4-[(4-chlorophenyl)-azo]-, tetrafluoroborate	H_2O	274(3.91),363(4.44), 470s(3.11)	104-1493-83
$C_{12}H_8Cl_2$			
1,1'-Biphenyl, 3,3'-dichloro-	hexane	245(4.21)	64-0621-83B
$C_{12}H_8Cl_2S_2$			
Disulfide, bis(2-chlorophenyl)	heptane	246(4.31),265(3.57)	65-0519-83
	MeOH	242(4.33),265s(3.70)	65-0519-83
Disulfide, bis(4-chlorophenyl)	heptane	250(4.34),277s(3.68), 300s(3.42)	65-0519-83
	MeOH	246(4.39),270s(3.79), 295s(3.43)	65-0519-83
	MeCN	245(4.38),270s(3.88)	65-0519-83
$C_{12}H_8Cl_2Se_2$			
Diselenide, bis(2-chlorophenyl)	heptane	244(4.25),265s(3.56), 305s(3.00)	65-0519-83
	MeOH	241(4.31),275s(3.59), 312s(2.94)	65-0519-83
Diselenide, bis(3-chlorophenyl)	heptane	242(4.22),275s(3.54), 330(2.98)	65-0519-83
Diselenide, bis(4-chlorophenyl)	heptane	256(4.38),271s(4.01), 340(3.13)	65-0519-83
	MeOH	250(4.37),275s(3.73), 330(3.11)	65-0519-83
	MeCN	250(4.40),275s(3.77), 332(3.13)	65-0519-83
$C_{12}H_8Cl_2Te_2$			
Ditelluride, bis(4-chlorophenyl)	heptane	258(4.38),276s(4.24), 407(2.95)	65-0519-83
$C_{12}H_8CsN$			
Cesium, 9H-indeno[2,1-b]pyridin-9-yl-	$(MeOCH_2)_2$	361(4.04),484(3.18), 506(3.19)	104-1592-83
$C_{12}H_8FNOS$			
10H-Phenothiazin-3-ol, 7-fluoro-	benzene	207(4.42),255(4.54), 307(3.64)	4-0803-83
$C_{12}H_8FNO_2S$			
10H-Phenothiazin-3-ol, 7-fluoro-, 5-oxide	EtOH	239(4.56),274(4.09), 299(3.91),342(3.78)	4-0803-83

Compound	Solvent	λ_{max}(log ε)	Ref.
$C_{12}H_8F_2$			
1,1'-Biphenyl, 3,3'-difluoro-	hexane	242(4.14)	64-0621-83B
1,1'-Biphenyl, 3,4'-difluoro-	hexane	245(4.16)	64-0621-83B
$C_{12}H_8F_2S_2$			
Disulfide, bis(4-fluorophenyl)	heptane	235(4.28),270s(3.57), 305s(3.17)	65-0519-83
$C_{12}H_8F_2Se_2$			
Diselenide, bis(4-fluorophenyl)	heptane	245(4.07),273s(3.65), 342(2.87)	65-0519-83
	MeOH	238(4.13),275s(3.52), 332(2.98)	65-0519-83
	MeCN	236(4.27),270s(3.64), 333(3.01)	65-0519-83
$C_{12}H_8F_2Te_2$			
Ditelluride, bis(4-fluorophenyl)	heptane	248(4.35),264s(4.16), 409(2.90)	65-0519-83
$C_{12}H_8I_2S_2$			
Disulfide, bis(4-iodophenyl)	heptane	255(4.48),282s(4.02), 305s(3.65)	65-0519-83
$C_{12}H_8I_2Se_2$			
Diselenide, bis(4-iodophenyl)	heptane	262(4.52),281s(4.15), 364(3.31)	65-0519-83
	MeOH	245(4.26),260s(4.20), 310s(3.34)	65-0519-83
	MeCN	257(4.23),310s(3.06)	65-0519-83
$C_{12}H_8KN$			
Potassium, 9H-indeno[2,1-b]pyridin-9-yl-	$(MeOCH_2)_2$	361(4.04),484(3.26), 504(3.27)	104-1592-83
$C_{12}H_8NNa$			
Sodium, 9H-indeno[2,1-b]pyridin-9-yl-	$(MeOCH_2)_2$	361(4.04),483(3.16), 506(3.16)	104-1592-83
$C_{12}H_8N_2$			
Bullvalenedicarbonitrile	MeOH	245(3.80),255s(3.75)	24-3800-83
2,7-Diazaphenanthrene	EtOH	234(4.75),239(4.80), 267(4.04),277(4.18), 288(4.20),312s(3.00), 325(3.38),340(3.67), 357(3.73)	4-1107-83
Phenazine	acid	255(4.87),276s(3.30), 364s(4.11),382(4.36), 426(3.40),450s(3.20)	125-0655-83
$C_{12}H_8N_2O$			
1,8-Phenanthrolin-4(1H)-one	EtOH	210(4.35),229s(4.38), 241(4.44),247s(4.39), 317(3.97),342(3.83)	33-0620-83
$C_{12}H_8N_2OS$			
Pyridine, 2-[5-(2-thienyl)-3-isoxazolyl]-	MeOH	234(4.11),237s(4.11), 247(4.08),270s(4.15), 285(4.28),291(4.29)	142-0501-83
Pyridine, 3-[5-(2-thienyl)-3-isoxazolyl]-	MeOH	237(4.12),265s(4.09), 291(4.27)	142-0501-83

Compound	Solvent	$\lambda_{max}(\log \epsilon)$	Ref.
Pyridine, 4-[3-(2-thienyl)-5-isoxazolyl]-	MeOH	268(4.52)	142-0501-83
Pyridine, 4-[5-(2-thienyl)-3-isoxazolyl]-	MeOH	234(4.21),237s(4.20), 246s(4.14),267s(4.14), 285(4.23),295(4.23)	142-0501-83
Thiazolo[3,2-a]pyrimidin-4-ium, 3-hydroxy-2-phenyl-, hydroxide, inner salt	MeCN	510(4.10)	124-0857-83
	benzene	550(--)	124-0857-83
	MeOH	475(--)	124-0857-83
	EtOH	481(--)	124-0857-83
	dioxan	542(--)	124-0857-83
	HOAc	449(--)	124-0857-83
	CH_2Cl_2	526(--)	124-0857-83
	$CHCl_3$	526(--)	124-0857-83
	CCl_4	558(--)	124-0857-83
$C_{12}H_8N_2O_2$			
1,7-Phenanthroline-5,6-diol, monohydrochloride	M HCl	295(4.36),357(3.78)	35-4431-83
1,10-Phenanthroline-5,6-diol, monohydrochloride	0.05M HCl	294(4.51)	35-4431-83
4,7-Phenanthroline-5,6-diol, monohydrochloride	pH 1	258(4.27),278(4.16), 356(3.93)	35-4431-83
1H,5H-Pyrrolo[2,3-f]indole-3,7-dicarboxaldehyde	EtOH	204.5(4.13),206s(3.95), 222(4.39),229.8(4.39), 279(4.45),300s(4.01), 345(4.22)	103-0871-83
3H,6H-Pyrrolo[3,2-e]indole-1,8-dicarboxaldehyde	EtOH	216(4.57),274(4.55), 325(3.95)	103-0871-83
$C_{12}H_8N_2O_2S$			
2H-1-Benzopyran-2-one, 3-(2-amino-4-thiazolyl)-	MeCN	359(4.18)	48-0551-83
$C_{12}H_8N_2O_4$			
2H-1-Benzopyran-2-one, 7-acetoxy-4-(diazomethyl)-	EtOH	254(4.32),319(4.34)	94-3014-83
$C_{12}H_8N_2O_4S_2$			
Disulfide, bis(2-nitrophenyl)	heptane	241(3.97),260s(3.70), 353(3.48)	65-0519-83
Disulfide, bis(3-nitrophenyl)	heptane	239(4.15),306(3.12)	65-0519-83
Disulfide, bis(4-nitrophenyl)	heptane	227s(3.95),305(4.20)	65-0519-83
	MeOH	315(4.38)	65-0519-83
	MeCN	314(4.35)	65-0519-83
	DMF	320(4.49)	65-0519-83
	CH_2Cl_2	318(4.40)	65-0519-83
	$CHCl_3$	314(4.38)	65-0519-83
	CCl_4	310(4.36)	65-0519-83
$C_{12}H_8N_2O_4Se_2$			
Diselenide, bis(2-nitrophenyl)	heptane	236(3.99),270s(3.72), 370(3.60)	65-0519-83
Diselenide, bis(3-nitrophenyl)	heptane	246(4.43),319(3.29)	65-0519-83
Diselenide, bis(4-nitrophenyl)	heptane	222s(4.22),308(4.30)	65-0519-83
$C_{12}H_8N_2S$			
Benzothiazole, 2-(2-pyridinyl)-	CH_2Cl_2	250(3.81),310(4.30), 325s(4.16)	4-1481-83
$C_{12}H_8N_4$			
Dibenzo[b,e]-1,3a,6,6a-tetraazapenta-	EtOH	234(4.54),271(3.67),	20-0067-83

Compound	Solvent	$\lambda_{max}(\log \epsilon)$	Ref.
lene- (cont.)		280(3.92),343(4.51), 356(4.60)	20-0067-83
$C_{12}H_8N_4O$			
Benzenediazonium, 4-[(4-hydroxyphenyl)-azo]-, anion	H_2O	284(3.73),344(3.71), 589(4.68)	104-1493-83
$C_{12}H_8N_4OS$			
2H-Pyrido[3,4-b][1,4]benzothiazine-4-carbonitrile, 8-amino-3,5-dihydro-3-oxo-, monohydrochloride	MeOH	259(4.57)	87-0845-83
2-Thiazolecarboxamide, 4-(2-quinoxalinyl)-	EtOH	246(3.93),266s(3.76), 293s(3.45),342(3.45), 355s(3.36)	106-0829-83
$C_{12}H_8N_5O_2$			
Benzenediazonium, [(4-nitrophenyl)-azo]-, tetrafluoroborate	H_2O	284s(3.76),334(4.52), 473(2.91)	104-1493-83
$C_{12}H_8O$			
Cyclobuta[b]naphthalen-1(2H)-one	EtOH	243(4.60),292s(3.90), 303(4.02)	39-2659-83C
Dibenzofuran	ether	228s(4.34),241(4.04), 245(4.05),249(4.31), 276s(4.11),280(4.25), 286(4.26),290s(4.06), 295(3.97),297(3.98), 301(3.61)	33-1672-83
$C_{12}H_8OS_2$			
Thieno[2,3-c][2]benzothiepin-4(9H)-one	MeOH	234s(4.00),261(4.23), 354(3.63)	73-0623-83
$C_{12}H_8O_3$			
2H,8H-Benzo[1,2-b:3,4-b']dipyran-2-one	EtOH	217(4.412),287s(--), 294(4.064),329(4.064)	111-0009-83
2H,8H-Benzo[1,2-b:5,4-b']dipyran-2-one	EtOH	223(4.304),267(4.209), 347(3.957)	111-0009-83
7H-Furo[2,3-f][1]benzopyran-2-carboxaldehyde	EtOH	223(4.210),227(4.224), 290(4.202),338(4.125)	111-0015-83
7H-Furo[3,2-g][1]benzopyran-2-carboxaldehyde	EtOH	240(4.324),260s(--), 312(4.001),355(4.168)	111-0015-83
$C_{12}H_8O_4$			
2,3-Naphthalenedicarboxylic acid	EtOH	243(4.75),270(3.75), 280(3.74),326s(3.13), 334.5(3.18),343s(3.03)	73-0112-83
$C_{12}H_8S_4$			
1,3-Benzodithiole, 2-(1,3-dithiol-2-ylideneethylidene)-	$CHCl_3$	380(4.19),400(4.14)	88-3469-83
$C_{12}H_9BrClNO_2$			
1H-Indole-2,3-dione, 5-bromo-1-(3-chloro-2-butenyl)-	EtOH	215(4.38),254(4.62), 305(3.54),439(2.42)	103-0286-83
$C_{12}H_9BrN_2O_3$			
Isoxazole, 5-[2-(4-bromophenyl)ethenyl]-3-methyl-4-nitro-	MeOH	245(4.08),268(4.15), 357(4.40)	4-0105-83

Compound	Solvent	λmax(log ε)	Ref.
$C_{12}H_9BrN_2O_4$			
Phenol, 4-bromo-2-[2-(3-methyl-4-nitro-5-isoxazolyl)ethenyl]-	EtOH	227(4.27),263s(4.08), 268(4.10),387(4.26)	142-0263-83
$C_{12}H_9ClN_2O$			
Phenol, 4-[(4-chlorophenyl)azo]-	C_6H_{12}	241(4.17),356(4.51), 426s(4.24)	59-0729-83
	EtOH	245(4.06),360(4.47), 428s(3.32)	59-0729-83
	ether	245(4.08),356(4.50), 430(3.13)	59-0729-83
	DMF	365(4.47),430s(3.37)	59-0729-83
	$CHCl_3$	357(4.46),428s(3.23)	59-0729-83
$C_{12}H_9ClN_2O_2$			
3-Pyridinecarboxamide, 1-(4-chloro-	H_2O	273(4.18)	54-0331-83
phenyl)-1,4-dihydro-4-oxo-	M HCl	271(4.06)	54-0331-83
3-Pyridinecarboxamide, 1-(4-chloro-	H_2O	258(4.13),304(3.69)	54-0331-83
phenyl)-1,6-dihydro-6-oxo-	M HCl	258(4.13),304(3.70)	54-0331-83
$C_{12}H_9ClN_2O_4$			
1H-Indole-2,3-dione, 1-(3-chloro-2-butenyl)-5-nitro-	EtOH	207(4.49),255(4.20), 336(4.26)	103-0286-83
$C_{12}H_9ClN_4O$			
Benzenamine, 4-[(4-chlorophenyl)azo]-N-nitroso-, sodium salt	H_2O	235(4.11),360(4.30), 440s(3.20)	104-1493-83
$C_{12}H_9ClN_4O_3S$			
Diazene, [4-[(4-chlorophenyl)azo]-phenyl](sulfinooxy)-, sodium salt	H_2O	351(4.42),440s(3.12)	104-1493-83
$C_{12}H_9ClOS_2$			
Ethanone, 1-[5-chloro-2-(2-thienyl-thio)phenyl]-	MeOH	233(4.48),260(4.07), 339(3.63)	73-2970-83
$C_{12}H_9ClO_2S_2$			
Benzeneacetic acid, 5-chloro-2-(2-thienylthio)-	MeOH	251(4.20),285s(3.71)	73-2970-83
$C_{12}H_9Cl_2NO_2$			
1H-Indole-2,3-dione, 5-chloro-1-(3-chloro-2-butenyl)-	EtOH	212.5(4.48),251(4.38), 301.5(3.49),438(2.41)	103-0286-83
$C_{12}H_9F$			
1,1'-Biphenyl, 3-fluoro-	hexane	245(4.17)	64-0621-83B
$C_{12}H_9FN_2O_2$			
3-Pyridinecarboxamide, 1-(4-fluoro-	H_2O	270(4.22)	54-0331-83
phenyl)-1,4-dihydro-4-oxo-	M HCl	268(4.09)	54-0331-83
3-Pyridinecarboxamide, 1-(4-fluoro-	H_2O	259(4.15),304(3.70)	54-0331-83
phenyl)-1,6-dihydro-6-oxo-	M HCl	259(4.14),303(3.69)	54-0331-83
$C_{12}H_9F_3O$			
Naphthalene, 2-methoxy-1-(trifluoro-methyl)-	MeOH	224(4.74),270(3.58), 279(3.65),291(3.54), 319(3.39),333(3.43)	23-0368-83
2-Naphthalenol, 5-methyl-1-(trifluoro-methyl)-	MeOH	222(4.67),281(3.67), 293(3.61),322(3.39), 335(3.44)	23-0368-83

Compound	Solvent	λ_{max}(log ε)	Ref.
$C_{12}H_9N$			
5H-Benzocycloheptene-5-carbonitrile	EtOH	274.5(3.89)	18-3449-83
7H-Benzocycloheptene-7-carbonitrile	EtOH	227(--)	18-3449-83
Bicyclo[4.4.1]undeca-1,3,5,7,9-pentaene-2-carbonitrile	dioxan	265(4.69),326(3.89), 362s(2.86),370s(2.84), 381s(2.76),395s(2.52)	35-3375-83
1,6-Methano[10]annulene, 3-cyano-	dioxan	263(4.83),311(3.91), 362s(2.56),370s(2.62), 378(2.67),388(2.70), 398(2.70),408(2.55)	35-3375-83
$C_{12}H_9NO$			
4H-Furo[3,2-b]pyrrole, 2-phenyl-	MeOH	246(4.02),324(4.49)	73-0772-83
$C_{12}H_9NOS$			
Benzothiopyrano[4,3-b]pyrrol-4-one, 2,3-dihydro-	MeOH	236(4.61),264s(3.87), 348(3.98)	83-0897-83
$C_{12}H_9NO_2S$			
Benzeneacetic acid, α-(2-thiazolylmethylene)-, (E)-	MeOH	238(3.81),245s(3.78), 283s(3.91),308(4.08)	4-0005-83
(Z)-	MeOH	228(3.98),316(4.38)	4-0005-83
1,3-Propanedione, 1-(2-pyridinyl)-3-(2-thienyl)-	MeOH	224(3.96),272(3.91), 354(4.37)	142-0501-83
1,3-Propanedione, 1-(3-pyridinyl)-3-(2-thienyl)-	MeOH	224(3.98),269(3.90), 352(4.40)	142-0501-83
1,3-Propanedione, 1-(4-pyridinyl)-3-(2-thienyl)-	MeOH	217(4.12),282(3.98), 350(4.36)	142-0501-83
$C_{12}H_9NO_4$			
7H-Furo[3,2-g][1]benzopyran, 7-methyl-2-nitro-	EtOH	240(4.414),260(4.409), 312(4.05),357(4.26)	111-0015-83
$C_{12}H_9NO_6$			
Benzoic acid, 2-[(5-nitro-2-furanyl)-methoxy]-	MeOH	210(3.66),233(2.42), 306(2.60)	73-2682-83
$C_{12}H_9NS$			
Azuleno[2,1-d]thiazole, 2-methyl-	hexane	222(3.98),296(4.81), 301(4.79),335(3.47), 367(3.82),385(3.42), 621(2.59)	142-1263-83
10H-Phenothiazine, radical cation	MeCN	438(3.64),516(3.84), 735(3.01),815(2.97)	152-0105-83
$C_{12}H_9N_2$			
Phenazinyl, dihydro-	acid	252s(4.71),255(4.73), 308(3.18),370(3.78), 432(3.90),443(4.03), 586s(3.95),636(3.11)	125-0655-83
$C_{12}H_9N_3$			
Benzo[f]quinoxalin-6-amine	CH_2Cl_2	250(4.31),292(4.29), 327(3.93),390(3.86)	83-0889-83
2-Propenenitrile, 3-(3-methyl-2-quinoxalinyl)-, (E)-	EtOH	210(4.28),265(4.40), 337(4.06)	12-0963-83
$C_{12}H_9N_3O$			
4,7-Phenanthrolin-5-ol, 6-amino-, hydrochloride	pH 1	255(4.32),275(4.20), 353(3.97)	35-4431-83

Compound	Solvent	$\lambda_{max}(\log \epsilon)$	Ref.
$C_{12}H_9N_3OS$			
Acetamide, N-cyclohepta[4,5]pyrrolo-[2,3-d]thiazol-2-yl-	MeOH	217(3.78),244(3.64), 303(4.36),325(4.16), 352(3.60),368(3.63)	18-1247-83
5-Pyrimidinecarbonitrile, 1,4-dihydro-6-(methylthio)-4-oxo-2-phenyl-	EtOH	254(3.55),300(3.32), 318(4.32)	142-1745-83
5H-1,3,4-Thiadiazolo[3,2-a]pyrimidin-5-one, 7-methyl-2-phenyl-	EtOH	256(2.49),342(4.40)	2-0815-83
$C_{12}H_9N_3OS_3$			
Thiazolo[5,4-d]pyrimidin-7(4H)-one, 5,6-dihydro-2-(methylthio)-6-phenyl-5-thioxo-	MeOH	215(4.42),306(4.12), 342(4.27)	2-0243-83
$C_{12}H_9N_3O_2$			
1,3-Benzenediol, 4-(2H-benzotriazol-2-yl)-	$CHCl_3$	251(3.90),343(4.37)	49-0937-83
Diazene, (4-nitrophenyl)phenyl	n.s.g.	340(4.3),465(2.6)	151-0249-83C
1H-Naphtho[2,3-d]triazole-4,9-dione, 6,7-dimethyl-	EtOH	255(4.85),283(4.30), 338(3.61)	87-0714-83
$C_{12}H_9N_3O_2S$			
2(3H)-Thiazolone, 4-(2-oxo-2H-1-benzopyran-3-yl)-, 2-hydrazone	MeCN	362(4.05)	48-0551-83
$C_{12}H_9N_3O_3$			
Phenol, 4-[(4-nitrophenyl)azo]-	MeOH	256(3.9),382(5.35)	80-0903-83
	$C_2H_4Cl_2$	373(4.37)	80-0903-83
$C_{12}H_9N_3S$			
Quinoxaline, 2-methyl-3-(4-thiazolyl)-	EtOH	242(3.92),266s(3.54), 338(3.44)	106-0829-83
Quinoxaline, 2-(2-methyl-4-thiazolyl)-	EtOH	244(4.10),270s(3.80), 348(3.75),360s(3.69)	106-0829-83
Thionine	pH 7	288(4.58),598(4.74)	151-0233-83C
	pH 14	270(4.36),518(4.27)	151-0233-83C
$C_{12}H_9N_4$			
Benzenediazonium, 4-(phenylazo)-, tetrafluoroborate	H_2O	275(3.72),360(4.38), 475(3.04)	104-1493-83
	n.s.g.	270(3.8),350(4.4)	104-0322-83
$C_{12}H_9N_4O$			
Benzenediazonium, 4-[(4-hydroxyphenyl)azo]-, tetrafluoroborate	H_2O	225(4.11),266(4.11), 282(4.10),425(4.45)	104-1493-83
$C_{12}H_9N_5OS$			
2-Thiazolecarboxylic acid, 4-(2-quinoxalinyl)-, hydrazide	EtOH	247(4.00),266s(3.87), 293s(3.63),342(3.59), 353s(3.50)	106-0829-83
$C_{12}H_9N_5O_2$			
2,4(1H,3H)-Pteridinedione, 3-amino-7-phenyl-	pH 4.0	224(4.29),250s(4.03), 276s(3.82),348(4.34)	5-0852-83
	pH 10.0	238(4.27),267(4.34), 300s(3.85),373(4.10)	5-0852-83
$C_{12}H_9N_5O_3$			
Benzenamine, 4-[(4-nitrophenyl)azo]-N-nitroso-, sodium salt	H_2O	250s(4.04),378(4.40), 290s(3.93)	104-1493-83

Compound	Solvent	λ_{max}(log ε)	Ref.
4H-Pyrazolo[3,4-d]pyrimidin-4-one, 1,5-dihydro-3-(nitromethyl)-1-phenyl-	EtOH	232(4.46),272(4.04), 311s(2.18)	48-0041-83
$C_{12}H_9N_5O_5S$			
Diazene, [4-[(4-nitrophenyl)azo]-phenyl](sulfinooxy)-, sodium salt	H_2O	353(4.51),440s(3.26)	104-1493-83
$C_{12}H_9N_5O_6$			
Benzenamine, 4-(1,4-dihydro-5,7-di-nitro-4-benzofurazanyl)-, N-oxide	MeOH	248(4.22),295(3.95), 364(3.82),473(4.39)	88-1555-83
$C_{12}H_{10}$			
1H-Cyclobuta[de]naphthalene, 1-methyl-	EtOH	226(4.83),272(3.67), 277(3.64),282(3.65), 302(2.76),311(2.58), 316(2.42),322(1.96)	35-6104-83
2,5-Etheno[4.2.2]propella-3,7,9-triene	EtOH	245s(2.3)	88-3361-83
$C_{12}H_{10}BrN_3O$			
4-Pyridine carboxylic acid 2-(4-bromo-phenyl)hydrazide, hydrochloride	acid	220(4.3),277(3.9)	56-1371-83
	pH 10	275s(3.6)	56-1371-83
$C_{12}H_{10}BrN_3O_4$			
Benzoic acid, 4-[(2-amino-5-bromo-1,4,5,6-tetrahydro-4,6-dioxo-5-pyrimidinyl)methyl]-	50%MeOH-HCl	234(4.34)	73-0299-83
	+ NaOH	238(4.31)	73-0299-83
$C_{12}H_{10}BrP$			
Phosphorin, 3-(4-bromophenyl)-4-methyl-	EtOH	212(4.38),244(4.37), 262s(4.14),304(2.85)	24-1756-83
$C_{12}H_{10}ClN$			
[1,1'-Biphenyl]-4-amine, 4'-chloro-	pH 1	254(4.34)	162-0064-83
$C_{12}H_{10}ClNO_2$			
[1]Benzopyrano[4,3-b]pyrrol-4(1H)-one, 8-chloro-2,3-dihydro-1-methyl-	MeOH	217(4.59),232s(4.29), 246s(4.04),286(3.53), 320(3.90),350(3.86)	5-0165-83
1H-Indole-2,3-dione, 1-(3-chloro-2-butenyl)-	EtOH	213.5(4.28),247(4.32), 303(3.33),428(2.67)	103-0286-83
1,4-Naphthalenedione, 2-chloro-3-(dimethylamino)-	EtOH	498(--)	150-0168-83S
	dioxan	482(3.49)	150-0168-83S
$C_{12}H_{10}ClNO_3$			
4(1H)-Pyridinone, 1-(4-chlorophenyl)-5-hydroxy-2-(hydroxymethyl)-	MeOH	224s(4.40),290(4.27)	18-1879-83
$C_{12}H_{10}ClN_2O$			
Pyridinium, 3-(aminocarbonyl)-1-(4-chlorophenyl)-, chloride	H_2O	277(3.91)	54-0331-83
$C_{12}H_{10}ClN_3$			
Benzenamine, 4-[(4-chlorophenyl)azo]-	MeOH	392(4.19)	80-0903-83
$C_{12}H_{10}ClN_3O_4$			
Benzoic acid, 4-[(2-amino-5-chloro-1,4,5,6-tetrahydro-4,6-dioxo-5-pyrimidinyl)methyl]-	50%MeOH-HCl	236(4.31)	73-0299-83
	+ NaOH	234(4.33),270s(3.77)	73-0299-83
1H-Pyrazole-3-carboxylic acid, 1-(4-chlorophenyl)-5-hydroxy-4-nitroso-, ethyl ester	MeOH	265(4.42),390(3.23)	94-1228-83

Compound	Solvent	λ_{max}(log ε)	Ref.
$C_{12}H_{10}ClP$			
Phosphorin, 3-(4-chlorophenyl)-4-methyl-	EtOH	210(4.36),237(4.22),256s(4.09),304(2.95)	24-1756-83
$C_{12}H_{10}Cl_2N_2O_2$			
2(1H)-Pyrazinone, 3,5-dichloro-6-(3-methoxyphenyl)-1-methyl-	EtOH	348(3.95)	4-0919-83
$C_{12}H_{10}Cl_2O$			
1,4-Methano-6H-benzocyclohepten-6-one, 7,8-dichloro-1,2,3,4-tetrahydro-	hexane	234(4.11),249(4.00),245(3.95),260(4.00),316(3.65),326(3.64),344(3.46),357(3.21),365(2.70)	88-4747-83
$C_{12}H_{10}Cl_2O_2$			
1H-Inden-1-one, 7-(dichloroacetyl)-7,7a-dihydro-7a-methyl-, trans	EtOH	241(3.84),338(3.82)	39-2399-83C
$C_{12}H_{10}CrN_2O_5Se$			
Chromium, pentacarbonyl(5,6,7,8-tetrahydro-4H-cyclohepta-1,2,3-selenadiazole-N^2)-, (OC-6-22)-	hexane	242(4.61),448(3.80)	24-0230-83
$C_{12}H_{10}FN_2O$			
Pyridinium, 1-(4-fluorophenyl)-3-(aminocarbonyl)-, chloride	H_2O	274(3.89)	54-0331-83
$C_{12}H_{10}FP$			
Phosphorin, 3-(4-fluorophenyl)-4-methyl-	EtOH	208(4.34),235(4.26),260(4.10),274s(3.76),304(2.95)	24-1756-83
$C_{12}H_{10}F_3NO$			
1H-Indole, 1,2-dimethyl-3-(trifluoroacetyl)-	EtOH	210(4.53),253(4.18),268(3.96),323(4.10)	39-2425-83C
$C_{12}H_{10}N_2$			
Benzenamine, N-(2-pyridinylmethylene)-, (E)-	$C_6H_{11}Me$	228(4.03),235(4.05),257(3.97),277(4.00)	33-1961-83
Benzenamine, N-(3-pyridinylmethylene)-, (E)-	$C_6H_{11}Me$	228s(--),236(4.16),254(4.10),280s(--),322(3.78)	33-1961-83
Benzenamine, N-(4-pyridinylmethylene)-, (E)-	$C_6H_{11}Me$	226(4.15),232s(--),260(4.05),325(3.68)	33-1961-83
Diazene, diphenyl-	gas	302(4.31)	135-1183-83
	C_6H_{12}	229(4.17),319(4.41),446(2.72)	59-0729-83
	EtOH	230(4.14),318(4.40),444(2.67)	59-0729-83
	DMF	322(4.30)	135-1183-83
	50% EtOH	434(2.81)	59-0729-83
	+ HCl	434(2.78)	59-0729-83
	ether	230(4.16),317(4.45),447(2.68)	59-0729-83
	50% DMF	436(2.83)	59-0729-83
	$CHCl_3$	320(4.42),442(2.79)	59-0729-83
Diazene, diphenyl-, (E)-	benzene	323(4.3),445(2.6)	151-0249-83C
	MeCN	320(4.36)	44-2897-83
1H-Indole-3-acetonitrile, 4-ethenyl-	MeOH	217(4.36),295(3.91)	78-3695-83

Compound	Solvent	λmax(log ε)	Ref.
Phenazine, dihydro-	acid	243(4.26),273(4.04), 303s(3.85)	125-0655-83
2-Pyridinamine, N-(phenylmethylene)-, (E)-	$C_6H_{11}Me$	261(4.11)	33-1961-83
3-Pyridinamine, N-(phenylmethylene)-, (E)-	$C_6H_{11}Me$	268(4.14)	33-1961-83
Pyrido[1,2-a]benzimidazole, 2-methyl-	EtOH	210(3.92),240(4.26), 247(4.33),261(3.82), 270(3.72),302(3.28), 314(3.34),337(3.30), 346(3.32),364(3.32), 384(3.04)	103-1102-83
Pyrido[1,2-a]benzimidazole, 4-methyl-	EtOH	214(4.16),240(4.70), 245(4.66),258(4.32), 266(4.18),298(3.74), 310(3.78),338(3.88), 352(3.86),372(3.62)	103-1102-83
9H-Pyrido[3,4-b]indole, 1-methyl-	MeOH	234(4.63),238s(4.62), 248(4.43),281s(4.07), 287(4.29),333(3.75), 347(3.76)	18-1450-83
	EtOH	234(4.89),238(4.90), 249s(4.75),281(4.20), 288(4.25),335(3.67), 348(3.69)	78-0033-83
	EtOH-HCl	236(4.73),239(4.93), 253s(4.60),280(4.35), 300(4.28),360(3.68)	78-0033-83
	CH_2Cl_2	236(4.85),238(4.92), 250s(4.70),285(4.21), 290(4.28),328(3.67), 342(3.68)	78-0033-83
	CH_2Cl_2-HCl	238(4.90),280(4.35), 298(4.12),332(3.62), 360(3.54)	78-0033-83
Semibullvalene-3,7-dicarbonitrile, 1,5-dimethyl-	dioxan	213(4.06),252(3.81)	24-3751-83
$C_{12}H_{10}N_2O$			
Diazene, diphenyl-, 1-oxide, (E)-	MeCN	321(4.15)	44-2897-83
Diazene, diphenyl-, 1-oxide, (Z)-	MeCN	239(--),327(3.59)	44-2897-83
Ethanone, 1-phenyl-2-(4-pyrimidinyl)-	CCl_4	227(4.45),282s(--)	103-0435-83
Ethanone, 2-(2-pyridinyl)-1-(3-pyridinyl)-	$C_6H_{11}Me$	238(3.7),300s(4.1), 345(4.3)	61-0391-83
	MeOH	230(3.9),300f(3.9), 340(4.2),403(3.8)	61-0391-83
Phenol, 2-(phenylazo)-	MeCN	327(4.26)	44-2897-83
Phenol, 4-(phenylazo)-	gas	316(4.32)	135-1183-83
	C_6H_{12}	244(3.78),348(4.37), 428s(3.23)	59-0729-83
	EtOH	238(3.82),351(4.39), 430s(3.31)	59-0729-83
	50% EtOH	420s(3.24)	59-0729-83
	+ HCl	420s(3.24)	59-0729-83
	ether	240(4.00),348(4.43), 435(3.18)	59-0729-83
	DMF	358(4.39),435(3.35)	59-0729-83
	DMF	356(4.35)	135-1183-83
	50% DMF	426s(3.29)	59-0729-83
	$CHCl_3$	347(4.32),428s(3.33)	59-0729-83

Compound	Solvent	$\lambda_{max}(\log \epsilon)$	Ref.
Phenol, 4-(phenylazo)-, trans	EtOH	235(3.0),330(3.4), 430s(--)	151-0255-83B
Pyrimido[2,1-a]isoindol-4(6H)-one, 2-methyl-	EtOH	206(4.34),245(3.95), 297(3.93)	124-0755-83
1H,5H-Pyrrolo[2,3-f]indole, 1-acetyl-	EtOH	206.5(4.28),217(4.22), 268(4.35),307.7(4.02), 322(3.97),334(3.99)	103-0871-83
3H,6H-Pyrrolo[3,2-e]indole, 3-acetyl-	EtOH	206(4.54),217s(4.49), 260(4.39),268(4.38), 317.5(4.27)	103-0871-83
$C_{12}H_{10}N_2OS$			
4H,6H-[1,3]Thiazino[2,3-b]quinazolin-4-one, 2-methyl-	MeOH	228(4.25),311(4.09)	83-0379-83
$C_{12}H_{10}N_2O_2$			
5H-Cycloheptapyrazin-5-one, 6-acetyl-2(and 3)-methyl-	MeOH	238(4.16),260s(3.94), 348(3.75)	142-1117-83
Phenol, 2-[(4-hydroxyphenyl)azo]-	C_6H_{12}	261(4.08),390(4.35)	59-0729-83
	EtOH	243(4.01),350(4.30), 478(3.64)	59-0729-83
	50% EtOH	350(4.32),475(3.55)	59-0729-83
	ether	250(4.16),368(4.38)	59-0729-83
	DMF	350(3.98),506(4.34)	59-0729-83
	50% DMF	350(4.30),478(3.30)	59-0729-83
	$CHCl_3$	384(4.46)	59-0729-83
3-Pyridinecarboxamide, 1,4-dihydro-4-oxo-1-phenyl-	H_2O	271(4.31)	54-0331-83
	M HCl	267(4.20)	54-0331-83
3-Pyridinecarboxamide, 1,6-dihydro-6-oxo-1-phenyl-	H_2O	259(4.16),303(3.72)	54-0331-83
	M HCl	259(4.15),302(3.73)	54-0331-83
$C_{12}H_{10}N_2O_2S$			
Pyridine, 3-[(4-methylphenyl)thio]-2-nitro-	EtOH	222(4.40),272(4.02), 360(3.65)	103-0303-83
$C_{12}H_{10}N_2O_3$			
3-Pyridinecarboxamide, 1,4-dihydro-1-(4-hydroxyphenyl)-4-oxo-	H_2O	275(4.17)	54-0331-83
	M HCl	282(3.97)	54-0331-83
3-Pyridinecarboxamide, 1,6-dihydro-1-(4-hydroxyphenyl)-6-oxo-	H_2O	259(4.14),305(3.64)	54-0331-83
	M HCl	259(4.14),305(3.67)	54-0331-83
$C_{12}H_{10}N_2O_3S$			
5H-Thiazolo[2,3-b]quinazoline-2-acetic acid, 2,3-dihydro-3-oxo-	MeOH	219(4.13),293(4.15)	83-0569-83
$C_{12}H_{10}N_2O_4$			
Phenol, 2-[2-(3-methyl-4-nitro-5-isoxazolyl)ethenyl]-	MeOH	215(4.22),252s(3.95), 273(4.08),386(4.31)	142-0263-83
$C_{12}H_{10}N_2O_5$			
Benzoic acid, 2-[[(5-nitro-2-furanyl)-methyl]amino]-	MeOH	220(7.63!),250(2.42), 322(2.98)	73-2682-83
Benzoic acid, 4-[[(5-nitro-2-furanyl)-methyl]amino]-	MeOH	208s(1.50),211(2.41), 292(4.90)	73-2682-83
Oxepino[4,5-d]pyridazine-1,4-dicarboxylic acid, dimethyl ester	CH_2Cl_2	254(4.19),306(3.55)	24-0097-83
$C_{12}H_{10}N_4$			
2,3-Phenazinediamine	heptanol	280(4.5),435(4.2)	140-0666-83
Pyrazolo[5,1-c][1,2,4]triazine, 8-methyl-7-phenyl-	EtOH	267(4.52),329(3.42), 390(3.11)	44-2330-83

Compound	Solvent	λ_{max}(log ε)	Ref.
2-Pyridinecarboxaldehyde, (2-pyridinylmethylene)hydrazone	EtOH	268(4.20),276(4.25), 286(4.27),298(4.16), 308(3.83)	34-0132-83
$C_{12}H_{10}N_4O$			
Benzenamine, N-nitroso-4-(phenylazo)-, sodium salt	H_2O	238(4.10),354(4.29), 435s(3.65)	104-1493-83
Pyrazolo[1,5-a]pyrimidin-7(1H)-one, 1-methyl-5-(4-pyridinyl)-	n.s.g.	220(4.34),285(3.94), 375(3.72)	56-1219-83
$C_{12}H_{10}N_4OS_2$			
Thiazolo[5,4-d]pyrimidin-7(4H)-one, 5,6-dihydro-6-methyl-2-(phenylamino)-5-thioxo-	MeOH	224s(--),312(4.35), 347(4.38)	2-0243-83
$C_{12}H_{10}N_4O_2$			
Benzenamine, 4-[(4-nitrophenyl)azo]-	gas	353(4.49)	135-1183-83
	MeOH	277(3.98),444(4.36)	80-0903-83
	DMF	458(4.45)	135-1183-83
	$C_2H_4Cl_2$	273(3.95),426(4.31)	80-0903-83
Phenol, 4-[[4-(nitrosoamino)phenyl]-azo]-, dianion	H_2O	250(4.22),320(4.02), 410s(4.21),450(4.50)	104-1493-83
7H-Pyrazolo[4,3-d]pyrimidin-7-one, 2,4-dihydro-3-hydroxy-2-(3-methylphenyl)-	MeOH	250(4.26),310(3.77)	94-1228-83
7H-Pyrazolo[4,3-d]pyrimidin-7-one, 2,4-dihydro-3-hydroxy-2-(4-methylphenyl)-	MeOH	249(4.33),305(3.86)	94-1228-83
Pyridazino[4,5-g]phthalazine-1,4-dione, 2,3-dihydro-2,3-dimethyl-	MeOH	232(4.69),238(4.67), 270(3.59),334(3.62)	142-1279-83
Pyridazino[4,5-g]phthalazin-1(2H)-one, 4-methoxy-2-methyl-	MeOH	234(4.49),254(3.70), 274(3.74),291(3.64), 338(3.81)	142-1279-83
2(1H)-Quinoxalinone, 3,4-dihydro-3-[(5-methyl-1,3,4-oxadiazol-2-yl)methylene]-	EtOH	265(3.90),290(3.82), 364.5(4.19),384.5(4.26), 407.0(4.08)	142-1917-83
	70% DMSO-EtOH	350(3.88),363(3.99), 383.5(3.87),406.0(3.66)	142-1917-83
	70% TFA-EtOH	400(4.08),422(4.14), 445(3.86)	142-1917-83
$C_{12}H_{10}N_4O_3S$			
Diazene, [4-(phenylazo)phenyl](sulfinooxy)-, sodium salt	H_2O	348(4.34),455s(3.00)	104-1493-83
$C_{12}H_{10}N_4O_6$			
1H-Pyrazole-3-carboxylic acid, 5-hydroxy-1-(4-nitrophenyl)-4-nitroso-, ethyl ester	MeOH	222(4.16),286(4.19), 312(4.29)	94-1228-83
$C_{12}H_{10}N_4S$			
2-Thiazolamine, 4-(3-methyl-2-quinoxalinyl)-	EtOH	246(3.95),367(3.20)	106-0829-83
$C_{12}H_{10}O$			
7bH-Cyclopent[cd]inden-5-ol, 7b-methyl-	EtOH	293(4.48),326s(3.60), 467(2.54)	77-0317-83
Ethanone, 1-(2-naphthalenyl)-	MeOH	333(3.20),339(3.22)	126-1143-83
	dioxan	327.5(3.20),341(3.21)	126-1143-83

Compound	Solvent	$\lambda_{max}(\log \epsilon)$	Ref.
$C_{12}H_{10}O_3$			
2H-1-Benzopyran-2-one, 7-hydroxy-8-(2-propenyl)-	EtOH	252(4.81),260(3.62),339(4.21)	44-2709-83
$C_{12}H_{10}O_4$			
2H-1-Benzopyran-2-one, 4-(acetoxy-methyl)-	EtOH	271(4.02),311.5(3.78)	94-3014-83
2H-1-Benzopyran-2-one, 8-acetyl-7-hydroxy-4-methyl-	EtOH	211(4.38),240s(--),268(3.97),273(3.96),312(4.02),334(3.98)	44-2709-83
1-Benzoxepin-2-carboxaldehyde, 2,5-di-hydro-7-(hydroxymethyl)-5-oxo-	EtOH	202(4.09),219(4.08),257.5(3.86),332(3.15)	39-1718-83C
2,5-Furandione, 4-benzoyldihydro-5-methyl-	EtOH	271(4.07),345(3.11)	12-0311-83
1,4-Naphthalenedione, 2,5-dihydroxy-3,8-dimethyl-	n.s.g.	208(4.62),236(4.59),284(4.52),433(3.91)	88-1333-83
	base	220(4.60),238(4.58),270(4.59),394(3.91),495(3.89)	88-1333-83
$C_{12}H_{10}O_5$			
2H-1-Benzopyran-2-one, 4-(acetoxy-methyl)-7-hydroxy-	EtOH	325(4.16)	94-3014-83
$C_{12}H_{10}O_7$			
2-Furancarboxylic acid, 5,5'-(1-hy-droxyethylidene)bis-	EtOH	245(4.10)	103-0474-83
$C_{12}H_{10}S_2$			
Disulfide, diphenyl	heptane	244(4.27),276s(3.59),289s(3.33)	65-0519-83
	MeOH	240(4.25),260s(3.62)	65-0519-83
	MeCN	239(4.37),260s(3.81)	65-0519-83
$C_{12}H_{10}Se_2$			
Diselenide, diphenyl	heptane	246(4.28),263s(4.11),337(3.04)	65-0519-83
	MeOH	238(4.32),275s(3.45),331(3.03)	65-0519-83
	MeCN	239(4.36),270s(3.64),327(2.99)	65-0519-83
$C_{12}H_{10}Te_2$			
Ditelluride, diphenyl	heptane	248(4.33),272s(4.06),406(2.91)	65-0519-83
$C_{12}H_{11}BrO$			
1(2H)-Acenaphthylenone, 8-bromo-2a,3,4,5-tetrahydro-	EtOH	261(3.56),309(3.94)	22-0096-83
5(1H)-Acenaphthylenone, 6-bromo-2,2a,3,4-tetrahydro-	EtOH	255(4.27),262(4.32),270(4.18),315(3.50)	22-0096-83
5(1H)-Acenaphthylenone, 7-bromo-2,2a,3,4-tetrahydro-	EtOH	254(4.02),313(3.37)	22-0096-83
$C_{12}H_{11}ClN_2O_3S$			
1H-Pyrido[2,3-b][1,4]thiazine-2-carb-oxylic acid, 3-acetyl-6-chloro-, ethyl ester	EtOH	231(4.14),247(4.14),282s(3.35),370(3.38)	103-0618-83
$C_{12}H_{11}ClN_2O_4$			
4H,5H-Pyrano[3,4-e][1,3]oxazine-	$CHCl_3$	304(4.12),344(4.29)	24-3725-83

Compound	Solvent	λ_{max}(log ε)	Ref.
4,5-dione, 7-chloro-2-piperidino-			24-3725-83
$C_{12}H_{11}ClN_4O_3$			
4H-Imidazole-2-carboxylic acid, 5-[2-(4-chlorophenyl)hydrazino]-4-oxo-, ethyl ester	EtOH	205(4.11),227(4.02), 298(4.08)	103-0571-83
$C_{12}H_{11}ClN_{10}S_2$			
4,6-Pyrimidinediamine, 5,5'-[(5-chloro-2,3-pyrazinediyl)bis(thio)]bis-	MeOH	251(3.95),323(3.78)	4-1047-83
$C_{12}H_{11}ClO_6$			
Benzeneacetic acid, 2,3-diacetoxy-5-chloro-	EtOH	207(4.33),275(2.78)	104-0458-83
$C_{12}H_{11}Cl_3N_2O_4S$			
5-Thia-1-azabicyclo[4.2.0]oct-2-ene-2-carboxylic acid, 3-ethenyl-7-(formylamino)-8-oxo-, 2,2,2-trichloroethyl ester, (6R-trans)-	EtOH	297(4.14)	78-0461-83
$C_{12}H_{11}N$			
Benzenamine, N-phenyl-	EtOH	285(4.3)	145-0621-83
cation	MeCN	680(4.18)	70-0251-83
radical	MeCN	770(3.59)	70-0251-83
1H-Pyrrole, 1-ethenyl-2-phenyl-	C_6H_{12}	246(4.2),272(4.2)	104-1151-83
$C_{12}H_{11}NO$			
3-Buten-2-one, 4-(1H-indol-4-yl)-	MeOH	252(4.15),366(4.08)	5-2135-83
6-Oxa-7-azatricyclo[3.3.0.0^{2,4}]oct-7-ene, 8-phenyl- (1α,2β,4β,5α)-	EtOH	265(4.1)	88-5727-83
$C_{12}H_{11}NO_2$			
[1]Benzopyrano[4,3-b]pyrrol-4(1H)-one, 2,3-dihydro-1-methyl-	MeOH	234(4.27),246s(4.12), 305(3.80),340(4.00)	5-0165-83
1,4-Naphthalenedione, 2-(dimethylamino)-	C_6H_{12}	444(3.53)	150-0168-83S
	EtOH	471(--)	150-0168-83S
1H-Pyrrole-2-carboxylic acid, 1-phenyl-, methyl ester	MeCN	239s(4.05),277(4.46)	24-0563-83
1H-Pyrrole-3-carboxylic acid, 1-phenyl-, methyl ester	MeCN	241s(2.95),260(4.34)	24-0563-83
$C_{12}H_{11}NO_2S$			
1-Penten-3-one, 1-(2-thioxo-3(2H)-benzoxazolyl)-	EtOH	310(4.29)	105-0478-83
$C_{12}H_{11}NO_3$			
2H-1-Benzopyran-3-carbonitrile, 5,6,7,8-tetrahydro-7,7-dimethyl-2,5-dioxo-	n.s.g.	373(4.24)	5-0695-83
	toluene	323(--)	5-0695-83
	DMSO	322(--)	5-0695-83
4(1H)-Pyridinone, 5-hydroxy-2-(hydroxymethyl)-1-phenyl-	MeOH	224s(4.33),287(4.25)	18-1879-83
$C_{12}H_{11}NO_3S$			
1H-Pyrrole, 2-acetyl-1-(phenylsulfonyl)-	MeOH	224(4.02),238(4.07), 268s(3.95),274.5(3.98)	44-3214-83
1H-Pyrrole, 3-acetyl-1-(phenylsulfonyl)-	MeOH	220(4.10),238(4.28)	44-3214-83
Thieno[2,3-b]pyridine-6-propanoic acid, β-oxo-, ethyl ester	EtOH	235(4.24),249(4.29), 291(4.23)	4-1717-83

Compound	Solvent	λ_{max}(log ϵ)	Ref.
$C_{12}H_{11}NO_4$			
1,3-Dioxolo[4,5-h]quinolin-8(9H)-one, 6-methoxy-9-methyl-	n.s.g.	226(4.50)	162-0264-83
1H-Indole-4,6-dicarboxylic acid, dimethyl ester	MeOH	234(4.40),266s(2.93), 292s(3.26),337(4.02)	44-2488-83
$C_{12}H_{11}NO_4S$			
1H-Pyrrole-2-carboxylic acid, 1-(phenylsulfonyl)-, methyl ester	MeOH	229(4.18),259(3.97)	44-3214-83
$C_{12}H_{11}NO_5$			
2H-1,4-Benzoxazine-5-carboxylic acid, 3,4-dihydro-7-methoxy-2-methylene-3-oxo-, methyl ester	MeOH	222(4.41),267(3.54), 300(3.54),351(3.98)	158-0200-83
$C_{12}H_{11}N_2O$			
Pyridinium, 3-(aminocarbonyl)-1-phenyl-, chloride	H_2O	272(3.88)	54-0331-83
$C_{12}H_{11}N_2O_2$			
Pyridinium, 3-(aminocarbonyl)-1-(4-hydroxyphenyl)-, chloride	H_2O	246(6.64),318(4.81)	54-0331-83
$C_{12}H_{11}N_2O_3$			
Pyridinium, 2-methyl-1-[2-(5-nitro-2-furanyl)ethenyl]-, bromide, (E)-	MeOH	227(3.20),268(2.91), 348(3.26)	73-1891-83
(Z)-	MeOH	224(3.27),268(2.92), 333(3.06)	73-1891-83
Pyridinium, 3-methyl-1-[2-(5-nitro-2-furanyl)ethenyl]-, bromide, (E)-	MeOH	223(3.32),267(3.00), 361(3.35)	73-1891-83
(Z)-	MeOH	223(3.25),268(3.00), 334(3.05)	73-1891-83
iodide	MeOH	223(3.50),356(3.29)	73-1891-83
picrate	MeOH	216(3.46),360(3.72)	73-1891-83
perchlorate	MeOH	223(3.25),356(3.30)	73-1891-83
Pyridinium, 4-methyl-1-[2-(5-nitro-2-furanyl)ethenyl]-, bromide, (E)-	MeOH	223(3.22),360(3.41)	73-1891-83
(Z)-	MeOH	223(3.25),263(2.93), 351(3.04)	73-1891-83
$C_{12}H_{11}N_3$			
Benzenamine, 4-(phenylazo)-	gas	337(4.51)	135-1183-83
	DMF	463(4.45)	135-1183-83
	MeOH	248(3.94),390(4.35)	80-0903-83
	$C_2H_4Cl_2$	382(4.33)	80-0903-83
4-Pyridinecarboxaldehyde, phenylhydrazone	hexane	362(4.49)	140-0215-83
	EtOH	365(--)	140-0215-83
5H-Pyrido[4,3-b]indol-3-amine, 1-methyl-	MeOH	241(4.61),263(4.87), 303(4.00),315(4.00)	18-1450-83
9H-Pyrido[3,4-b]indol-3-amine, 1-methyl-	MeOH	240(4.56),291s(4.00), 297(4.09)	18-1450-83
$C_{12}H_{11}N_3O$			
Phenol, 4-[(4-aminophenyl)azo]-	MeOH	251(4.02),390(4.39), 418(4.47),450(4.47)	80-0903-83
4H-Pyrido[1,2-a]pyrimidine-7-carbonitrile, 3-ethyl-2-methyl-4-oxo-	EtOH	232(4.30),350(4.01)	106-0218-83
$C_{12}H_{11}N_3OS$			
4H-Thiazolo[2,3-c][1,2,4]triazin-4-one, 6,7-dihydro-7-methyl-3-phenyl-	EtOH	207(4.24),244(4.00), 325(4.24)	44-4585-83

Compound	Solvent	$\lambda_{max}(\log \epsilon)$	Ref.
7H-Thiazolo[3,2-b][1,2,4]triazin-7-one, 2,3-dihydro-2-methyl-6-phenyl-	EtOH	202(4.27),234(4.31), 306(4.14)	44-4585-83
$C_{12}H_{11}N_3O_2$			
1H-Imidazole, 1-methyl-4-nitro-5-(2-phenylethenyl)-, (E)-	MeOH	267(4.34),358(4.08)	4-0629-83
1H-Imidazole, 1-methyl-5-nitro-4-(2-phenylethenyl)-, (E)-	MeOH	272(4.39),376(4.23)	4-0629-83
$C_{12}H_{11}N_3O_3S$			
Benzenesulfonic acid, 4-[(4-amino-phenyl)azo]-, sodium salt	H_2O	389(4.03)	80-0903-83
	MeOH	262(4.04),401(4.39)	80-0903-83
$C_{12}H_{11}N_3O_4$			
1H-Pyrazole-3-carboxylic acid, 5-hy-droxy-4-nitroso-1-phenyl-, ethyl ester	MeOH	263(4.33),390(3.29)	94-1228-83
$C_{12}H_{11}N_3S$			
Thiourea, N-phenyl-N'-2-pyridinyl-	H_2O	247(4.30),260(4.28), 291(4.36),308(4.23)	65-1315-83
	$CHCl_3$	265(4.30),298(4.43)	65-1315-83
$C_{12}H_{11}N_3S_2$			
5H-Thiazolo[3,4-b][1,2,4]triazol-4-ium, 6-methyl-3-(methylthio)-1-phenyl-, hydroxide, inner salt	MeCN	257(4.27),400(4.21)	103-0749-83
$C_{12}H_{11}N_5$			
Imidazo[1,5-a]-1,3,5-triazin-4-amine, 8-(phenylmethyl)- (spectra in EtOH)	pH 1	270(3.92),302(3.75)	44-0003-83
	pH 7	271(3.83),328(3.77)	44-0003-83
	pH 13	269(3.92),302(3.95)	44-0003-83
3-Pyridinecarbonitrile, 4,6-diamino-2-(phenylamino)-	EtOH	226(4.47),260(4.44), 301(4.27)	49-0973-83
$C_{12}H_{11}N_5O_2$			
2,4(1H,3H)-Pyrimidinedione, 5-diazo-3-methyl-6-(methylphenylamino)-	MeCN	204(4.42),222s(--), 284(4.20)	35-4809-83
$C_{12}H_{11}N_5O_4S_2$			
Thiourea, N,N-dimethyl-N'-[5-nitro-4-(4-nitrophenyl)-2-thiazolyl]-	$CHCl_3$	268(4.35),292s(4.31), 363(3.99)	74-0075-83B
palladium chelate	$CHCl_3$	345s(4.30),430(4.45)	74-0075-83B
$C_{12}H_{11}OP$			
Phosphorin, 3-(4-methoxyphenyl)-	EtOH	212(4.70),247(4.62), 282(4.45)	24-0445-83
	EtOH	212(4.70),247(4.62), 282(4.45),311s(2.60)	24-1756-83
$C_{12}H_{11}P$			
Phosphorin, 3-(4-methylphenyl)-	EtOH	240(4.51),274(4.17)	24-0445-83 +24-1756-83
Phosphorin, 4-methyl-3-phenyl-	EtOH	216(4.25),251(4.42), 320(3.00)	24-1756-83
$C_{12}H_{12}$			
Bicyclo[5.4.1]dodeca-2,5,7,9,11-penta-ene	3-Mepentane	245(4.60),322.5(3.56)	35-6211-83
Bicyclo[2.2.2]oct-2-ene, 5,6,7,8-tetrakis(methylene)-	isooctane	221(3.97),227.5(4.01), 236(3.99),252(3.96),	33-1134-83

Compound	Solvent	λ_{max}(log ε)	Ref.
(cont.)		260(3.95),269s(3.73)	33-1134-83
Bicyclo[4.4.1]undeca-1,3,5,7,9-penta-ene, 2-methyl-	C_6H_{12}	259(4.78),309(3.81), 357s(2.21),366(2.25), 374(2.26),382(2.25), 391(2.24),396(2.24), 399s(2.17),407(2.16)	35-3375-83
Bicyclo[4.4.1]undeca-1,3,5,7,9-penta-ene, 3-methyl-	C_6H_{12}	260(4.82),299(3.79), 357(2.30),364(2.33), 372(2.33),380(2.27), 403(2.00)	35-3375-83
$C_{12}H_{12}BrN_3O_4$			
2,4(1H,3H)-Pyrimidinedione, 5-(4-bromo-2,5-dihydro-1-methyl-2,5-dioxo-1H-pyrrol-3-yl)-1,3,6-trimethyl-	MeOH	247(4.06),268(3.90), 328(3.16)	89-0120-83S
$C_{12}H_{12}BrN_4$			
5H-Pyrido[1,2-a]pyrimido[4,5-d]pyrimidinium, 8-bromo-2,3-dimethyl-, perchlorate	pH 6.86	213(4.38),260(3.64), 302(4.06),410(4.34)	44-2476-83
$C_{12}H_{12}Br_4$			
2,6-Methano-1H-cycloprop[a]azulene, 1,1,3,4(and 1,1,4,5)-tetrabromo-1a,1b,2,3,6,6a,7,7a-octahydro-	EtOH	218(3.86)	104-1787-83
$C_{12}H_{12}ClN$			
Pyridine, 3-chloro-1,4-dihydro-1-(phenylmethyl)-	EtOH	233(3.56),308(3.31)	150-2326-83M
$C_{12}H_{12}ClNO_2S_2$			
2-Propanol, 1-(2-benzothiazolylthio)-3-chloro-, acetate	EtOH	280(4.01),290(3.94), 301(3.80)	103-1173-83
$C_{12}H_{12}ClNO_3$			
2H-Pyrrol-2-one, 3-chloro-1,5-dihydro-4-hydroxy-5-methoxy-1-(4-methylphenyl)-	MeOH	202(4.16),227(4.08), 290(3.64)	5-1504-83
$C_{12}H_{12}ClN_2S$			
1,2,3-Thiadiazolium, 5-(2-chloro-1-methylethenyl)-4-methyl-2-phenyl-, perchlorate	MeOH	208(4.15),245s(3.63), 372(4.08)	39-0777-83C
$C_{12}H_{12}N$			
Isoquinolinium, 2-(2-propenyl)-, perchlorate	MeCN	336(3.63)	35-1204-83
Quinolinium, 1-(2-propenyl)-, perchlorate	MeCN	317(3.87)	35-1204-83
$C_{12}H_{12}NO_2S_2$			
2H-[1,3]Thiazino[2,3-b]benzothiazolium, 3-acetoxy-3,4-dihydro-, chloride	EtOH	310(4.05)	103-1173-83
$C_{12}H_{12}N_2$			
Pyrazolo[1,5-a]quinoline, 4,5-dihydro-4-methyl-	EtOH	261(4.12)	88-4679-83
3H-Pyrido[3,4-b]indole, 4,9-dihydro-1-methyl-	EtOH	235(4.20),238s(4.19), 315(4.18)	4-0267-83
	EtOH-HCl	244(4.03),345(4.33)	4-0267-83

Compound	Solvent	$\lambda_{max}(\log \epsilon)$	Ref.
$C_{12}H_{12}N_2O$			
1H-1,2-Diazepine, 1-benzoyl-4,5-dihydro-	MeOH	268(4.04)	5-1361-83
Pyrazine, 2-methoxy-3-methyl-5-phenyl-	EtOH	212(3.96),255(4.06), 313(3.97)	4-0311-83
Pyrazine, 2-methoxy-5-methyl-3-phenyl-	EtOH	221(3.90),245(4.01), 318(4.12)	4-0311-83
3H-Pyrazol-3-one, 2,4-dihydro-2-phenyl-5-(1-propenyl)-	EtOH	260(4.27)	142-1801-83
2H-Pyrrol-2-one, 1,5-dihydro-3,4-dimethyl-5-(2-pyridinylmethylene)-, (E)-	MeOH	315(4.31)	54-0114-83
(Z)-	MeOH	338(4.37)	54-0114-83
2H-Pyrrol-2-one, 1,5-dihydro-3,4-dimethyl-5-(4-pyridinylmethylene)-, (E)-	MeOH	304(4.08)	54-0114-83
(Z)-	MeOH	320(4.30)	54-0114-83
$C_{12}H_{12}N_2OSSe$			
2-Selenazolidinone, 5-[[4-(dimethylamino)phenyl]methylene]-4-thioxo-	n.s.g.	241(3.86),310(3.69), 382(3.21),504(4.64)	103-0273-83
$C_{12}H_{12}N_2O_2$			
Acetamide, N-[2-(1H-indol-3-yl)-2-oxoethyl]-	MeOH	240(4.06),259(3.91), 297(4.02)	158-0913-83
1H-1,2-Benzodiazepine-4-carboxylic acid, ethyl ester	EtOH	211(4.13),258(4.22), 303s(3.45)	138-0463-83
[1]Benzopyrano[3,4-c]pyrazol-1(4H)-one, 2,3-dihydro-4,4-dimethyl-	EtOH	214(4.24),255(3.72), 273(3.78),294(3.79), 302(3.78)	39-1431-83C
2,2'-Bipyridine, 4,4'-dimethoxy-	EtOH	212(4.59),256(3.98), 269s(2.96)	44-0283-83
Bullvalenedicarboxaldehyde dioxime	MeOH	219s(4.27),245(4.34)	24-3800-83
Cyclohepta[b]pyrrole-3-carboxylic acid, 2-amino-, ethyl ester	EtOH	247(4.13),287s(4.55), 293(4.58),320s(3.89), 365(3.85),424s(3.64), 432(3.66)	18-3703-83
1-Naphthalenamine, N,N-dimethyl-4-nitro-	pH 1	240s(--),334(4.60)	104-0520-83
	MeOH	260(3.99),406(4.00)	104-0520-83
	acetone	415(4.01)	104-0520-83
	DMSO	270(3.92),435(4.00)	104-0520-83
1-Naphthalenamine, N,N-dimethyl-5-nitro-	pH 1	328(4.63)	104-0520-83
	MeOH	268(4.38),375(3.82)	104-0520-83
	acetone	380(3.98)	104-0520-83
	DMSO	275(4.05),395(3.48)	104-0520-83
1,6-Phenazinedione, 2,3,4,7,8,9-hexahydro-	MeOH	230(3.56),315(3.66)	24-1309-83
1H-Pyrrole-2-carboxaldehyde, 5-[(1,5-dihydro-3,4-dimethyl-5-oxo-2H-pyrrol-2-ylidene)methyl]-, (Z)-	MeOH	375(4.65),395s(4.48)	54-0347-83
1H-Pyrrole-2-carboxaldehyde, 1,1'-(1,2-ethanediyl)bis-	EtOH	260s(4.25),287(4.41)	104-0393-83
3H-Pyrrol-3-one, 4-acetyl-2,4-dihydro-5-methyl-2-phenyl-	DMF	262(4.61)	90-0309-83
$C_{12}H_{12}N_2O_2S$			
5H-Cyclopentapyrazine, 6,7-dihydro-2-methyl-3-(2-thienyl)-, 1,4-dioxide	EtOH	244(4.25),268(4.15), 298(4.39)	103-1104-83
Methanone, (4,5,6,7-tetrahydro-1H-benzimidazol-2-yl)-2-thienyl-	EtOH	282(4.92),357(4.24)	103-1104-83

Compound	Solvent	λ_{max}(log ε)	Ref.
2-Propenoic acid, 3-amino-3-thieno-[2,3-b]pyridin-6-yl-, ethyl ester	EtOH	244(4.28),335(4.28)	4-1717-83
Quinoxaline, 5,6,7,8-tetrahydro-2-(2-thienyl)-, 1,4-dioxide	EtOH	272(4.18),304(4.57), 338s(4.17)	103-1104-83
$C_{12}H_{12}N_2O_3$			
Acetamide, N-[1-hydroxy-2-(1H-indol-3-yl)-2-oxoethyl]-	MeOH	242(4.03),262(3.91), 302(4.05)	158-0913-83
2,2'-Bipyridine, 4,4'-dimethoxy-, 1-oxide	EtOH	223(4.42),246(4.18), 272(4.06),321(3.18)	44-0283-83
5H-Cycloheptapyrazine, 2-(2-furanyl)-4,7-dihydro-3-methyl-, 1,4-dioxide	EtOH	242(4.11),270(4.27), 291(4.48)	103-1104-83
2-Furanmethanamine, N-(2-methyl-phenyl)-5-nitro-	MeOH	207(4.43),240(2.10), 314(4.01)	73-2682-83
2-Furanmethanamine, N-(3-methyl-phenyl)-5-nitro-	MeOH	205(4.56),245(4.20), 313(4.16)	73-2682-83
2-Furanmethanamine, N-(4-methyl-phenyl)-5-nitro-	MeOH	206(4.43),244(4.16), 315(4.11)	73-2682-83
Methanone, 2-furanyl(4,5,6,7-tetra-hydro-1-hydroxy-1H-benzimidazol-2-yl)-	EtOH	290(4.95),351(4.24)	103-1104-83
4H-Pyrido[1,2-a]pyrimidine-9-carbox-ylic acid, 3-ethyl-2-methyl-4-oxo-	EtOH	259(4.12),265s(4.10), 335(3.97)	106-0218-83
Quinoxaline, 2-(2-furanyl)-5,6,7,8-tetrahydro-, 1,4-dioxide	EtOH	270(4.29),295(4.59), 320s(4.30),328s(4.25)	103-1104-83
$C_{12}H_{12}N_2O_3S$			
2-Benzofurancarboxamide, N-(amino-thioxomethyl)-3-ethoxy-	MeOH	220(4.37),233(4.37), 311(4.64)	94-3728-83
2-Thietanone, 4,4-dimethyl-3-[[(4-ni-trophenyl)methylene]amino]-, (R)-	EtOH	215(4.23),281(4.32)	39-2259-83C
$C_{12}H_{12}N_2O_4$			
2,2'-Bipyridine, 4,4'-dimethoxy-, 1,1'-dioxide	EtOH	213(4.42),246s(4.12), 272(4.32),320s(3.50)	44-0283-83
2-Furanmethanamine, N-(4-methoxyphen-yl)-5-nitro-	MeOH	204(4.36),242(4.18), 316(4.17)	73-2682-83
$C_{12}H_{12}N_2O_5$			
Oxepino[4,5-d]pyridazine-1,4-dicarb-oxylic acid, 2,4a-dihydro-, dimethyl ester	CH_2Cl_2	279(4.19),395(3.74)	24-0097-83
$C_{12}H_{12}N_2S$			
4(1H)-Pyridazinethione, 3,5-dimethyl-1-phenyl-	C_6H_{12}	203(4.18),251(3.69), 391(4.54)	39-0777-83C
$C_{12}H_{12}N_2S_2$			
4(3H)-Pyrimidinethione, 2-(methyl-thio)-3-(phenylmethyl)-	EtOH	238(4.01),295(3.63), 360(3.57)	39-2645-83C
2-Thiophenecarboxaldehyde, 5-methyl-, [(5-methyl-2-thienyl)methylene]hy-drazone	EtOH	270(4.11),352(4.68)	34-0132-83
$C_{12}H_{12}N_3S_2$			
5H-Thiazolo[3,4-b][1,2,4]triazol-4-ium, 6-methyl-3-(methylthio)-1-phenyl-, perchlorate	HOAc	253(3.99),375(4.18)	103-0749-83
$C_{12}H_{12}N_4OS$			
3H-Pyrimido[5,4-b][1,4]thiazin-4(6H)-	pH 1	280(3.93),303(4.03)	87-0559-83

Compound	Solvent	$\lambda_{max}(\log \epsilon)$	Ref.
one, 2-amino-7,8-dihydro-6-phenyl- (cont.)	pH 6.8	222(4.43),277s(3.86), 300(3.95)	87-0559-83
	pH 13	287(3.95)	87-0559-83
$C_{12}H_{12}N_4O_2$			
Benzoic acid, 4-(1H-imidazol-2-yl-azo)-, ethyl ester	MeOH	234(3.82),368(4.36), 382(4.33),438s(3.29)	56-0971-83
3H-Pyrazolium, 1-[(1,3-dihydro-1,3-di-oxo-2H-isoindol-2-yl)amino]-4,5-di-hydro-3-methyl-, hydroxide, inner salt	EtOH	219(4.51),233s(4.31), 240s(4.25),265(4.11)	104-2317-83
3H-Pyrazolium, 1-[(1,3-dihydro-1,3-di-oxo-2H-isoindol-2-yl)amino]-4,5-di-hydro-5-methyl-, hydroxide, inner salt	EtOH	219(4.49),233s(4.28), 240s(4.21),265(4.03)	104-2317-83
$C_{12}H_{12}N_4O_2S$			
3H-Pyrimido[5,4-b][1,4]thiazin-4(6H)-	pH 1	225(4.41),260(4.06)	87-0559-83
one, 2-amino-7,8-dihydro-6-phenyl-,	pH 6.8	226(4.57),261(4.05)	87-0559-83
5-oxide	pH 13	261(4.00)	87-0559-83
$C_{12}H_{12}N_4O_2S_2$			
Thiourea, N,N-dimethyl-N'-[4-(4-nitro-phenyl)-2-thiazolyl]-	$CHCl_3$	267(4.29),290s(4.16), 348(4.17)	74-0075-83B
palladium chelate	$CHCl_3$	290(4.58),334(4.44), 365s(4.27)	74-0075-83B
$C_{12}H_{12}N_4O_3$			
1H-Imidazole-1-acetamide, 2-nitro-N-(phenylmethyl)-	EtOH	313(3.88)	162-0154-83
Urea, N-methyl-N'-(6-phenylpyrazinyl)-, N,N'-dioxide	pH 5.0	242(4.29),272(4.51), 302(4.18),353(3.84)	5-0852-83
	pH 13.0	258(4.30),290(4.52), 396(3.91)	5-0852-83
$C_{12}H_{12}N_4O_4$			
1H-Pyrazole-4-carboxylic acid, 5-amino-3-(nitromethyl)-1-phenyl-, methyl ester	EtOH	233(4.37),253s(4.03)	48-0041-83
$C_{12}H_{12}N_6$			
1H-Purine-6,8-diamine, N^6-methyl-	pH 1	238(4.14),307(4.31)	94-3454-83
N^8-phenyl-	H_2O	223(4.24),241s(4.13), 295(4.38)	94-3454-83
	pH 13	305(4.41)	94-3454-83
$C_{12}H_{12}N_6O_4$			
Pyrimido[5,4-g]pteridine-2,4,6,8-(1H,3H,7H,10H)-tetrone, 1,3,7,10-tetramethyl-	n.s.g.	285(3.98),417(3.95)	35-3304-83
$C_{12}H_{12}N_6S$			
Ethenetricarbonitrile, [2,4-bis(di-methylamino)-5-thiazolyl]-	CH_2Cl_2	230(4.18),310(3.76), 350(3.73),492(4.50)	88-3563-83
$C_{12}H_{12}O$			
Bicyclo[2.2.2]octanone, 5,6,7,8-	EtOH	251(4.06),302(2.65)	33-1134-83
tetrakis(methylene)-	dioxan	251(4.06),305(2.64)	33-1134-83 +88-1983-83
Bicyclo[4.4.1]undeca-1,3,5,7,9-penta-ene, 2-methoxy-	C_6H_{12}	255(4.63),265s(4.54), 318(3.77),403(2.93),	35-3375-83

Compound	Solvent	$\lambda_{max}(\log \epsilon)$	Ref.
(cont.)		410s(2.90)	35-3375-83
Bicyclo[4.4.1]undeca-1,3,5,7,9-pentaene, 3-methoxy-	dioxan	264(4.77),301(3.80), 371s(2.82),378(2.89), 387(2.91),396(2.85), 406(2.59)	35-3375-83
7bH-Cyclobut[cd]azulen-7b-ol, 1,1a-dihydro-1-methyl-	C_6H_{12}	239(3.90),302(3.47)	88-4261-83
2-Cyclohexen-1-one, 2-phenyl-	MeCN	217(4.04),255(3.72), 328(1.72)	44-1654-83
1H-Indene-5-carboxaldehyde, 3,7-dimethyl-	C_6H_{12}	208(4.15),236(4.61), 261(3.92),277(3.21)	35-4056-83
Naphthalene, 2-ethoxy-	MeOH and MeOH-base	261(3.6),271(3.6), 281(3.5),313(3.18), 328(3.28)	160-1403-83
Tricyclo[5.3.1.0^{4,11}]undeca-2,5,7-trien-8-one	EtOH	236(3.98)	24-2903-83
$C_{12}H_{12}O_2$			
Benzocyclooctene-5,6-dione, 7,8,9,10-tetrahydro-	$CHCl_3$	260(4.08),298(3.36), 368(1.98)	39-1983-83C
5-Benzofurancarboxaldehyde, 2,3-dihydro-2-(1-methylethenyl)-	EtOH	229(4.21),284(4.25), 293(4.26)	78-0169-83
2H-Cyclopent[cd]inden-2-one, 1,2a,7a-7b-tetrahydro-2a-hydroxy-7b-methyl-	EtOH	307(3.78),337s(3.45)	39-2399-83C
Ethanone, 1-(2-ethenyl-2,3-dihydro-5-benzofuranyl)-	EtOH	226(4.05),278(4.13), 295(4.05)	78-0169-83
4a,8a-Methanonaphthalen-1(4H)-one, 4-methoxy-	CH_2Cl_2	266(3.40)	118-0463-83
$C_{12}H_{12}O_3$			
2H-1-Benzopyran-2-one, 7-ethoxy-4-methyl- (with β-cyclodextrin)	H_2O	322(4.20)	95-0193-83
1,4-Naphthalenedione, 2,3-dihydro-2-hydroxy-2,3-dimethyl-	MeOH	223(4.54),252(4.15), 302(3.23),330s(2.42)	44-3301-83
$C_{12}H_{12}O_3S$			
2-Propenethioic acid, 3-(2-oxo-2-phenylethoxy)-, S-methyl ester	MeOH	247(4.17),276(4.41)	88-5563-83
$C_{12}H_{12}O_4$			
Cyclopropa[cd]pentalene-2,3-dicarboxylic acid, 2a,2b,4a,4b-tetrahydro-4a,4b-dimethyl-	dioxan	220(3.945),257(3.685)	24-3751-83
4H,5H-Pyrano[4,3-b]pyran-4,5-dione, 2,3-dihydro-2-methyl-7-(1-propenyl)-, [S-(E)]- (deoxyradicinin)	EtOH	222(4.21),269(3.72), 281s(3.62),343(4.25)	158-83-84
$C_{12}H_{12}O_4Se$			
Spiro[1,3-benzoxaselenole-2,1'-cyclohexane]-2',4,6'(5H)-trione, 6,7-dihydro-	$CHCl_3$	295(3.30),398(2.17)	39-0333-83C
$C_{12}H_{12}O_5$			
1,3-Benzodioxole-5-acetic acid, 6-acetyl-, methyl ester	EtOH	230(4.08),273(3.63), 306(3.64)	107-0691-83
1H-2-Benzopyran-5-carboxylic acid, 3,4-dihydro-8-hydroxy-3-methyl-1-oxo-, methyl ester, (R)-	EtOH	226(4.40),250(3.93), 307(3.52)	39-2185-83C
2-Furancarboxylic acid, 5-[1-hydroxy-1-(5-methyl-2-furanyl)ethyl]-	EtOH	249(3.4)	103-0474-83

Compound	Solvent	λ_{max}(log ϵ)	Ref.
$C_{12}H_{12}O_6S$			
2H-1-Benzopyran-2-one, 7-methoxy-4-[[(methylsulfonyl)oxy]methyl]-	EtOH	322.5(4.11)	94-3014-83
$C_{12}H_{13}$			
1,4-Methano-1H-benzocycloheptenylium, 2,3,4,?-tetrahydro-, hexafluoroantimonate	MeCN	232(--),295(--)	44-0596-83
tetrafluoroborate	MeCN	237(4.50),280(3.53), 299(3.59)	78-4011-83
$C_{12}H_{13}As$			
1H-Arsole, 2,5-dimethyl-1-phenyl-	C_6H_{12}	207(4.52),263(4.56)	101-0335-83I
$C_{12}H_{13}BrN_2O$			
1,2,4-Oxadiazole, 3-(4-bromo-2,6-dimethylphenyl)-5-ethyl-	MeOH	270s(2.83),274s(2.85)	4-1609-83
$C_{12}H_{13}BrN_2O_2$			
[2,2'-Bi-1H-pyrrole]-5-carboxaldehyde, 3'-bromo-4-methoxy-1,1'-dimethyl-	MeOH	311(4.32)	44-2314-83
[2,2'-Bi-1H-pyrrole]-5-carboxaldehyde, 5'-bromo-4-methoxy-1,1'-dimethyl-	MeOH	327(4.30)	44-2314-83
$C_{12}H_{13}BrN_2O_2S_2$			
2,4(1H,3H)-Quinazolinedithione, 6-bromo-3-(2,2-dimethoxyethyl)-	EtOH	260(4.20),292(4.45), 312(4.49),333s(4.10), 389(3.72)	97-0215-83
$C_{12}H_{13}BrN_6S_2Zn$			
Zinc, [[2,2'-[1-(4-bromophenyl)-1,2-ethanediylidene]bis[N-methylhydrazinecarbothioamidato]](2-)-N^2,$N^{2'}$,S,S']-, (T-4)-	EtOH	292(4.42),455(4.16)	161-0143-83
$C_{12}H_{13}Br_3N_2$			
1H-Indole-3-methanamine, 2,5,6-tribromo-N,N,1-trimethyl-	MeOH	232(4.67),298(4.00), 308(3.96)	88-0481-83
$C_{12}H_{13}Br_3N_2O$			
1H-Indole-3-methanamine, 2,5,6-tribromo-N,N,1-trimethyl-, N-oxide	MeOH	230(4.59),294(4.07), 306(4.03)	88-0481-83
$C_{12}H_{13}ClN_6S_2Zn$			
Zinc, [[2,2'-[1-(4-chlorophenyl)-1,2-ethanediylidene]bis[N-methylhydrazinecarbothioamidato]](2-)-N^2,$N^{2'}$,S,S']-, (T-4)-	EtOH	287(4.32),450(4.07)	161-0143-83
$C_{12}H_{13}Cl_2NO_3$			
Benzamide, N-acetoxy-3,5-dichloro-2,4,6-trimethyl-	MeOH	277(3.72),285(2.70)	32-0011-83
$C_{12}H_{13}Cl_3N_2O_4S$			
5-Thia-1-azabicyclo[4.2.0]oct-2-ene-2-carboxylic acid, 3-ethyl-7-(formylamino)-8-oxo-, 2,2,2-trichloroethyl ester, (6R-trans)-	EtOH	266(3.77)	78-0461-83
$C_{12}H_{13}Cl_3N_2O_4S_2$			
5-Thia-1-azabicyclo[4.2.0]oct-2-ene-	EtOH	270.5(3.88)	78-0337-83

Compound	Solvent	$\lambda_{max}(\log \epsilon)$	Ref.
2-carboxylic acid, 7-(formylamino)-3-[(methylthio)methyl]-8-oxo-, 2,2,2-trichloroethyl ester, (6R-trans)-			78-0337-83
$C_{12}H_{13}Cl_3N_2O_5S$			
5-Thia-1-azabicyclo[4.2.0]oct-2-ene-2-carboxylic acid, 3-ethyl-7-(formylamino)-8-oxo-, 2,2,2-trichloroethyl ester, 5-oxide, [5S-(5α,6β-7α)]-	EtOH	271(3.89)	78-0461-83
$C_{12}H_{13}Cl_3N_2O_5S_2$			
5-Thia-1-azabicyclo[4.2.0]oct-2-ene-2-carboxylic acid, 7-(formylamino)-3-[(methylthio)methyl]-8-oxo-, 2,2,2-trichloroethyl ester, 5-oxide, [5S-(5α,6β,7α)]-	EtOH	275(3.93)	78-0337-83
$C_{12}H_{13}F_7O$			
5-Dodecyn-4-one, 1,1,1,2,2,3,3-heptafluoro-	C_6H_{12}	234(4.03),242s(3.97), 317(1.61)	44-1925-83
$C_{12}H_{13}N$			
1H-Indole, 5-ethenyl-2,3-dimethyl-	EtOH	206(4.0),255(4.6), 282(3.8)	104-0388-83
$C_{12}H_{13}NO$			
Pyrido[2,1-a]isoindol-6(2H)-one, 1,3,4,10b-tetrahydro-	$CHCl_3$	247(3.77),270s(3.48)	78-2691-83
$C_{12}H_{13}NOS$			
2H-Pyrrol-2-one, 1,5-dihydro-1,3,4-trimethyl-5-(2-thienylmethylene)-, (E)-	MeOH	336(4.08)	54-0114-83
(Z)-	MeOH	337(4.23)	54-0114-83
2-Thietanone, 4,4-dimethyl-3-[(phenylmethylene)amino]-, (R)-	EtOH	212(4.27),253(4.15)	39-2259-83C
$C_{12}H_{13}NO_2$			
Benzene, 1-(4-methyl-1,3-pentadienyl)-4-nitro-, (E)-	MeOH	252(4.20),360(4.34)	12-0527-83
2H-1-Benzopyran-2-one, 7-(dimethylamino)-4-methyl- (with β-cyclodextrin)	H_2O	373(4.23)	95-0193-83
2H-Pyrrol-2-one, 5-(2-furanylmethylene)-1,5-dihydro-1,3,4-trimethyl-, (E)-	MeOH	356(4.28)	54-0114-83
(Z)-	MeOH	349(4.40)	54-0114-83
$C_{12}H_{13}NO_2S$			
1-Benzothiepin-4-carboxamide, 2,3,4,5-tetrahydro-4-methyl-5-oxo-	EtOH	244(3.99)	118-0727-83
2(3H)-Benzoxazolethione, 3-(tetrahydro-2H-pyran-2-yl)-	$CHCl_3$	303(4.46)	94-1733-83
$C_{12}H_{13}NO_3$			
1-Penten-3-one, 4-methyl-1-(4-nitrophenyl)-, (E)-	EtOH	218(4.29),297(4.25)	12-0527-83
$C_{12}H_{13}NO_3S$			
2-Butenamide, 4-(ethenylsulfonyl)-N-	EtOH	221(4.27),281(4.12)	104-1314-83

Compound	Solvent	$\lambda_{max}(\log \epsilon)$	Ref.
phenyl-, (E)- (cont.)			104-1314-83
$C_{12}H_{13}NO_4$			
3-Buten-2-ol, 4-(4-nitrophenyl)-, acetate, (E)-	EtOH	219(4.08),296(4.03)	12-0527-83
$C_{12}H_{13}NO_8$			
1H-Pyrrole-2,3,4,5-tetracarboxylic acid, tetramethyl ester	EtOH	226(3.85),268(3.81)	4-0001-83
$C_{12}H_{13}N_2S$			
5H-Pyrrolo[2,1-b]thiazolium, 3-amino-6,7-dihydro-2-phenyl-, perchlorate	MeCN	227s(3.93),264(3.93), 327.5(4.13)	118-0582-83
$C_{12}H_{13}N_3$			
1H-Imidazol-5-amine, 1-methyl-4-(2-phenylethenyl)-, monohydrochloride, (E)-	MeOH	231(4.04),327(4.38)	4-0629-83
Propanedinitrile, [7-(dimethylamino)-2,4,6-hexatrienylidene]-	$CHCl_3$	550(4.94)	70-0780-83
	EtOH	580(--)	70-0780-83
[1,2,4]Triazolo[1,5-a]pyridine, 5,6,7,8-tetrahydro-2-phenyl-	EtOH	245(4.22)	4-1657-83
1,2,4-Triazolo[4,3-a]pyridine, 5,6,7,8-tetrahydro-3-phenyl-	EtOH	239(4.07)	4-1657-83
$C_{12}H_{13}N_3O$			
Ethanone, 1-(5-methyl-3-isoxazolyl)-, phenylhydrazone, (Z)-	MeOH	335(4.26)	4-0931-83
4H-Isoxazolo[5',4':3,4]cyclohepta[1,2-b]pyrazine, 5,6-dihydro-3,8(and 9)-dimethyl-	MeOH	245s(3.55),260(3.61), 307(3.91)	142-1117-83
$C_{12}H_{13}N_3OS$			
Isothiazolo[5,1-e][1,2,3]thiadiazole-7-S^{IV}, 1,6-dihydro-6-hydroxy-3,4-dimethyl-1-phenyl-	C_6H_{12}	200(4.28),237(3.93), 278(3.98),436(4.22)	39-0777-83C
4H-Pyrido[3,2-e]-1,3-thiazin-4-one, 2-piperidino-	MeOH	256(4.73)	73-3315-83
$C_{12}H_{13}N_3O_2$			
1H-Imidazole-5-carboxamide, 4-methoxy-1-(phenylmethyl)-	H_2O	258(4.06)	4-0875-83
	M HCl	245(3.93)	4-0875-83
	M NaOH	258(4.06)	4-0875-83
1H-Imidazolium, 4-(aminocarbonyl)-5-hydroxy-1-methyl-3-(phenylmethyl)-, hydroxide, inner salt	H_2O	242(3.64),281(4.07)	4-0875-83
	M HCl	242(3.67),282(3.96)	4-0875-83
	M NaOH	230(3.60),285(4.08)	4-0875-83
4H-Pyrido[1,2-a]pyrimidine-7-carboxamide, 3-ethyl-2-methyl-4-oxo-	EtOH	233(4.33),347(4.07)	106-0218-83
4H-Pyrido[3.2-a]pyrimidine-8-carboxamide, 3-ethyl-2-methyl-4-oxo-	EtOH	243(3.99),260(4.09), 353(3.96)	106-0218-83
$C_{12}H_{13}N_3O_4$			
Carbamic acid, [2-(5-nitro-1H-indol-3-yl)ethyl]-, methyl ester	EtOH	260s(4.16),274(4.26), 329(3.92)	94-1856-83
$C_{12}H_{13}N_3O_4S$			
5H-Pyrimido[4,5-b][1,4]thiazine-6-carboxylic acid, 7-acetyl-4-methoxy-, ethyl ester	EtOH	257(3.83),327(3.54), 480(3.15)	103-0618-83

Compound	Solvent	λ_{max}(log ε)	Ref.
$C_{12}H_{13}N_3O_5S$			
Carbamic acid, [[5-(methylsulfonyl)-1,3,4-oxadiazol-2-yl]methyl]-, phenylmethyl ester	MeOH	274s(3.53)	35-0902-83
$C_{12}H_{13}N_3S$			
Thieno[2,3-b]pyridine, 6-(4,5,6,7-tetrahydro-1H-1,3-diazepin-2-yl)-	EtOH	247(4.50),291(4.29)	4-1717-83
$C_{12}H_{13}N_4$			
5H-Pyrido[1,2-a]pyrimido[4,5-d]pyrimidinium, 2,3-dimethyl-, perchlorate	pH 6.86	202(4.45),220s(--), 251(3.73),288(4.02), 397(4.33)	44-2476-83
conjugate acid	pH 1	200(4.48),217s(--), 244(3.87),266(4.02), 343(4.23)	44-2476-83
$C_{12}H_{13}N_5O$			
Acetamide, N-[4-[(1-methyl-1H-imidazol-2-yl)azo]phenyl]-, (E)-	MeOH	202(3.91),248(3.67), 378(4.04),388(4.05), 413s(3.80)	56-0971-83
$C_{12}H_{13}N_5O_2$			
2,4,5(3H)-Pyrimidinetrione, 3-methyl-6-(methylphenylamino)-, 5-hydrazone	MeCN	299(4.18)	35-4809-83
[1,2,4]Triazolo[5,1-c][1,2,4]triazine, 4,6-dihydro-4,4-dimethoxy-7-phenyl-	EtOH	233(4.29),260(4.16)	44-2330-83
mixture with 4,8-dihydro isomer	EtOH	234(4.21),258(4.16)	44-2330-83
$C_{12}H_{13}N_5O_3$			
7H-Pyrrolo[2,3-d]pyrimidine-5-carbonitrile, 4-amino-7-(2-deoxy-β-D-erythro-pentofuranosyl)-	pH 1	231(4.25),273(4.09)	35-4059-83
	pH 13	231(4.08),278(4.18)	35-4059-83
	MeOH	229(4.10),278(4.20)	35-4059-83
$C_{12}H_{13}N_7O_2S_3Zn$			
Zinc, [[2,2'-[1-(4-nitrophenyl)-1,2-ethanediylidene]bis[N-methylhydrazinecarbothioamidato]](2-)-$N^2,N^{2'}$,S,S']-, (T-4)-	EtOH	368(4.47),475(4.24)	161-0143-83
$C_{12}H_{13}S_2$			
4H-1,3-Dithiin-1-ium, 2-(2-phenylethenyl)-, tetrafluoroborate	CH_2Cl_2	303(4.43),431(3.49)	1-0687-83
$C_{12}H_{14}$			
Azulene, 3,3a-dihydro-1,4-dimethyl-	hexane	211s(--),228(4.41), 234(4.38),312(3.73)	35-4056-83
Bicyclo[4.3.1]deca-2,4,7-triene, 7-methyl-9-methylene-	C_6H_{12}	224(4.41),230(4.45), 241s(4.16),249s(4.06)	24-2914-83
Dicyclopropa[cd,gh]pentalene, octahydro-2c,2d-dimethyl-1,2-bis(methylene)-	C_6H_{12}	228(4.246)	88-5063-83
$C_{12}H_{14}BrNOS$			
Morpholine, 4-[2-(4-bromophenyl)-1-thioxoethyl]-	heptane	282(4.17)	80-0555-83
	EtOH	281(4.21)	80-0555-83
$C_{12}H_{14}BrNO_3$			
2(1H)-Isoquinolinecarboxaldehyde, 8-bromo-3,4-dihydro-6,7-dimethoxy-	MeOH	286(3.27)	83-0520-83

Compound	Solvent	λ_{max}(log ε)	Ref.
$C_{12}H_{14}ClNOS$			
Morpholine, 4-[2-(4-chlorophenyl)-1-thioxoethyl]-	heptane	282(4.21)	80-0555-83
	EtOH	281(4.20)	80-0555-83
$C_{12}H_{14}ClNO_2$			
Benzene, 1-(3-chloro-4-methyl-1-pentenyl)-4-nitro-, (E)-	EtOH	217(4.04),286(4.04)	12-0527-83
$C_{12}H_{14}ClNS$			
Piperidine, 1-[(4-chlorophenyl)thioxomethyl]-	heptane	255(3.94),288(3.86), 401(2.60)	80-0555-83
	EtOH	248(4.11),282(4.05), 365(2.60)	80-0555-83
$C_{12}H_{14}Co$			
Cobaltocinium, 1,1'-dimethyl-, hexafluorophosphate	H_2O	267(4.53),308s(3.08), 416(2.41)	101-0093-83K
$C_{12}H_{14}CuN_8$			
Copper, [7,12-diimino-1,4,8,11-tetraazacyclotetradeca-4,14-diene-6,13-dicarbonitrilato(2-)-N^1,N^4,N^8,N^{11}]-	EtOH-DMF	298(4.02)	12-2413-83
$C_{12}H_{14}F_2N_2O_7S_2$			
Pyridinium, 1-methyl-, 4,4'-oxybis-, bis(fluorosulfonate)	pH 5.1	255(4.31)	64-1034-83B
$C_{12}H_{14}N_2$			
1,2-Diazabicyclo[3.1.0]hex-2-ene, 6,6-dimethyl-3-phenyl-	EtOH	256(4.15)	35-0933-83
Pyrimidine, 1,2-dihydro-1,4-dimethyl-6-phenyl-	EtOH	355(3.59)	39-1799-83C
Pyrimidine, 1,2-dihydro-4,6-dimethyl-1-phenyl-	EtOH	343(3.86)	39-1799-83C
Pyrimidine, 1,4-dihydro-4,6-dimethyl-1-phenyl-	EtOH	258(3.77)	39-1799-83C
Pyrimidine, 1,6-dihydro-4,6-dimethyl-1-phenyl-	EtOH	311(3.72)	39-1799-83C
1H-Pyrrolo[1,2,3-de]quinoxaline, 2,3-dihydro-1,5-dimethyl-	MeOH	223(4.32),284(3.84)	18-3358-83
$C_{12}H_{14}N_2O$			
1-Butanone, 1-(1H-indol-3-yl)-, oxime, (E)-	n.s.g.	226(4.34),275(3.92)	103-0184-83
1H-Cyclohepta[b]pyrazine, 5-acetyl-2,3-dihydro-1-methyl-	MeOH	259(4.18),379(3.75), 466(3.93)	18-3358-83
1H-1,2-Diazepine, 1-benzoyl-4,5,6,7-tetrahydro-	MeOH	256(3.06)	5-1361-83
1,2,4-Oxadiazole, 5-ethyl-3-(2,6-dimethylphenyl)-	MeOH	268s(2.83),274s(2.85)	4-1609-83
1-Propanone, 1-(1-methyl-1H-indol-3-yl)-, oxime	n.s.g.	229(4.29),260(4.16), 290(3.97)	103-0184-83
Pyridinium, 3,3'-oxybis[1-methyl-, diiodide	H_2O	226(4.53),272(3.93)	4-1411-83
2H-Pyrrol-2-one, 1,5-dihydro-3,4-dimethyl-5-[(1-methyl-1H-pyrrol-2-yl)methylene]-	MeOH	381(4.49)	54-0347-83
$C_{12}H_{14}N_2O_2$			
[1]Benzopyrano[3,4-c]pyrazol-1(4H)-one, 2,3,3a,9b-tetrahydro-4,4-di-	EtOH	209(4.05),220s(3.94), 277(3.32),284(3.32)	39-1431-83C

Compound	Solvent	λ_{max}(log ε)	Ref.
methyl-, (±)- (cont.)			39-1431-83C
[2,2'-Bi-1H-pyrrole]-5-carboxaldehyde, 4-methoxy-1,1'-dimethyl-	MeOH	330(4.20)	44-2314-83
5H-Cycloheptapyrazin-5-one, 6-acetyl-1,2,3,4-tetrahydro-2(and 3)-methyl-	MeOH	247(4.29),281(4.25), 307(4.24),378(4.04), 468(3.86)	142-1117-83
1(2H)-Isoquinolinone, 2-(dimethyl-amino)-3-methoxy-	MeOH	205(4.38),229(4.37), 277(4.04),284s(4.03), 343(3.64)	142-1315-83
2-Naphthalenecarboxamide, N-(amino-carbonyl)-1,2,3,4-tetrahydro-	EtOH	209(3.9),278(3.3)	120-0364-83
1,2,4-Oxadiazole, 3-(2-methoxy-4,6-dimethylphenyl)-5-methyl-	MeOH	282(3.32)	4-1609-83
1,2,4-Oxadiazole, 3-(4-methoxy-2,6-dimethylphenyl)-5-methyl-	MeOH	280(3.08)	4-1609-83
5-Quinoxalinecarboxaldehyde, 8-acetyl-1,2,3,4-tetrahydro-4-methyl-	MeOH	245(4.30),322(3.62), 467(3.66)	18-3358-83
$C_{12}H_{14}N_2O_2S$			
Piperidine, 1-[(3-nitrophenyl)thioxo-methyl]-	heptane	254(4.17),290(3.95), 397(2.70)	80-0555-83
	EtOH	256(4.20),280(4.12)	80-0555-83
Piperidine, 1-[(4-nitrophenyl)thioxo-methyl]-	heptane	259(4.22),326(3.80), 391(3.14)	80-0555-83
	EtOH	266(4.27),314(4.07), 375(3.39)	80-0555-83
$C_{12}H_{14}N_2O_2S_2$			
1H-Imidazole-5-carboxylic acid, 2-(methylthio)-1-(2-thienyl-methyl)-, ethyl ester	0.5M HCl	235(4.13),263(4.05)	142-2019-83
	0.5M NaOH	225(4.22),231(4.21), 264(4.11)	142-2019-83
	MeOH	201(3.98),236(4.08), 272(4.15)	142-2019-83
$C_{12}H_{14}N_2O_3$			
Pentaleno[?,1-d]pyrimidine-2,4,8(3H)-trione, 1,4b,5,6,7,7a-hexahydro-1,3-dimethyl-	MeCN	249(3.27),313(3.82)	35-0963-83
3-Penten-2-one, 3-nitro-4-[(phenyl-methyl)amino]-	EtOH	306(4.06),350(3.85)	40-0088-83
$C_{12}H_{14}N_2O_3S$			
Morpholine, 4-[2-(3-nitrophenyl)-1-thioxoethyl]-	heptane	280(4.28)	80-0555-83
	EtOH	275(4.19)	80-0555-83
$C_{12}H_{14}N_2O_5$			
1,2-Hydrazinedicarboxylic acid, 1-(2-oxo-2-phenylethyl)-, dimethyl ester	EtOH	245(4.05),283(2.97)	104-0110-83
Uridine, 2'-deoxy-5-(1-propynyl)-	pH 7	231(4.02),291(4.01)	87-0661-83
$C_{12}H_{14}N_2O_6$			
L-Aspartic acid, N-[[3-hydroxy-5-(hy-droxymethyl)-2-methyl-4-pyridinyl)-methylene]-	MeOH	250s(3.99),290s(3.60), 335(3.51),422(3.44)	35-0803-83
2,4(1H,3H)-Pyrimidinedione, 1-β-D-arabinofuranosyl-5-(1-propynyl)-	pH 7	230(4.03),290(4.04)	87-0661-83
Sarubicin A	EtOH	212(4.25),263(4.13), 272s(4.08),294s(3.78), 469(3.22)	158-0976-83
Uridine, 2'-deoxy-5-(3-hydroxy-1-pro-pynyl)-	pH 6	234(4.10),288(4.04)	44-1854-83
	pH 13	234(4.10),287(3.98)	44-1854-83

Compound	Solvent	λ_{max}(log ε)	Ref.
$C_{12}H_{14}N_2O_7$			
2-Butenedioic acid, 2-[5-hydroxy-3-(methoxycarbonyl)-1-methyl-1H-pyrazol-4-yl]-, dimethyl ester, (E)-	MeOH	224(4.07),316(3.34)	24-1525-83
$C_{12}H_{14}N_3O_5PS$			
Phosphonothioic acid, [3-(4-nitrophenyl)-1,2,4-oxadiazol-5-yl]-, O,O-diethyl ester	MeOH	270s(4.22),276s(4.20)	44-4232-83
$C_{12}H_{14}N_3O_6P$			
Phosphonic acid, [3-(4-nitrophenyl)-1,2,4-oxadiazol-5-yl]-, diethyl ester	MeOH	271s(3.885),286s(3.84)	44-4232-83
$C_{12}H_{14}N_4$			
1H-Pyrrole-2-carboxaldehyde, 1-methyl-, [(1-methyl-1H-pyrrol-2-yl)methylene]hydrazone	EtOH	260(3.87),353(4.64)	34-0132-83
$C_{12}H_{14}N_4O$			
2,4(1H,3H)-Pyrimidinedione, 1,3-dimethyl-, 4-(phenylhydrazone)	MeOH	234s(4.06),263(4.24), 284s(4.15),355(3.80)	4-1037-83
$C_{12}H_{14}N_4O_2$			
1,3,4(2H)-Isoquinolinetrione, 2-(dimethylamino)-, 4-methylhydrazone	MeOH	222(4.36),257(4.21), 373(4.28)	142-1315-83
4H-Pyrido[1,2-a]pyrimidine-7-carboxylic acid, 3-ethyl-2-methyl-4-oxo-, hydrazide	EtOH	235(4.32),348(4.05)	106-0218-83
$C_{12}H_{14}N_4O_2S_2$			
4(1H)-Pyrimidinone, 2-(methylthio)-1-[2-[2-(methylthio)-6-oxo-1(6H)-pyrimidinyl]ethyl]-	EtOH	248(4.17),285(4.06), 298s(--),310s(--)	39-2645-83C
4(3H)-Pyrimidinone, 3,3'-(1,2-ethanediyl)bis[2-(methylthio)-	EtOH	227(3.12),290(5.1), 313s(--)	39-2645-83C
$C_{12}H_{14}N_4O_3S$			
2,4(1H,3H)-Pyrimidinedione, 1-methyl-3-[2-[2-(methylthio)-6-oxo-1(6H)-pyrimidinyl]ethyl]-	EtOH	250(4.16),275s(--)	39-2645-83C
$C_{12}H_{14}N_4O_6$			
2,4(1H,3H)-Pyrimidinedione, 6,6'-(1,2-ethanediyl)bis[5-hydroxy-1-methyl-	pH 2	288(4.18)	4-0753-83
	pH 14	245s(4.02),320(4.18)	4-0753-83
$C_{12}H_{14}N_4O_7$			
7H-Pyrrolo[2,3-d]pyrimidine-5-carboxylic acid, 2-amino-3,4-dihydro-4-oxo-7-β-D-ribofuranosyl- (cadeguomycin)	pH 1 and H_2O	232(4.30),272(3.84), 298(3.88)	142-0027-83
	H_2O	233(3.9),272(3.5), 299(3.6)	88-3647-83
	base	225s(3.9),268(3.6), 282s(3.5)	88-3647-83
Spiro[6H-pyrano[3,2-d]pyrimidine-6,4'(1'H)-pyrimidine]-2,2',4,6'-(1H,3H,3'H,5'H)-tetrone, 7,8-dihydro-5',5'-dihydroxy-1,3'-dimethyl-	H_2O	286(3.88)	4-0753-83
$C_{12}H_{14}N_6$			
Methanimidamide, N'-[2-cyano-1-(di-	EtOH	260(3.8),372(4.1)	5-1107-83

Compound	Solvent	λ_{max}(log ε)	Ref.
cyanomethylene)-3-(dimethylamino)-2-propenyl]-N,N-dimethyl- (cont.)			5-1107-83
4,6-Pyrimidinediamine, N,N'-dimethyl-5-(phenylazo)-	pH 1	257(4.33),390(4.21)	94-3454-83
	H_2O	251(4.30),396(4.22)	94-3454-83
	pH 13	250(4.31),396(4.21)	94-3454-83
4,6-Pyrimidinediamine, N-ethyl-5-(phenylazo)-	pH 1	248(4.28),375(4.22)	94-3454-83
	H_2O	249(4.27),380(4.23)	94-3454-83
	pH 13	249(4.27),390(4.24)	94-3454-83
$C_{12}H_{14}N_6OS_2Zn$			
Zinc, [[2,2'-[1-(4-hydroxyphenyl)-1,2-ethanediylidene]bis[N-methylhydrazinecarbothioamidato]](2-)-N^2,$N^{2'}$,S,S']-, (T-4)-	EtOH	286(4.23),454(4.00)	161-0143-83
$C_{12}H_{14}N_6O_3S_3Zn$			
Zinc, [2,2'-[1-(4-sulfophenyl)-1,2-ethanediylidene]bis[N-methylhydrazinecarbothioamidato]](2-)-N^2,$N^{2'}$,S,S']-	EtOH	293(4.33),456(4.00)	161-0143-83
$C_{12}H_{14}N_6O_5$			
6-Pteridinecarboxamide, 1,2,3,4,7,8-hexahydro-N,3,8-trimethyl-N-[(methylamino)carbonyl]-2,4,7-trioxo-	neutral	330(3.71)	35-3304-83
	anion	351(3.73)	35-3304-83
$C_{12}H_{16}N_6S_2Zn$			
Zinc, [2,2'-(1-phenyl-1,2-ethanediylidene)bis[N-methylhydrazinecarbothioamidato]](2-)-N^2,$N^{2'}$,S,S']-, (T-4)-	EtOH	288(4.19),450(4.00)	161-0143-83
$C_{12}H_{14}O$			
Benzofuran, 2-ethenyl-2,3-dihydro-5,7-dimethyl-	EtOH	216(3.71),272(3.40), 284(3.38)	78-0169-83
Bicyclo[2.2.2]octan-2-ol, 5,6,7,8-tetrakis(methylene)-, (±)-	EtOH	233(4.00),239(3.99), 254(3.93)	33-1134-83
	dioxan	233(4.06),256(3.96)	33-1134-83
1H-Cyclopenta[a]pentalen-1-one, 3a,3b,4,6a,7,7a-hexahydro-6-methyl-, (3aα,3bα,6aα,7aα)-	EtOH	220(3.95)	88-2781-83
1H-Inden-1-one, 2,3-dihydro-2,4,7-trimethyl-	MeOH	254(4.04),303(3.40)	94-3113-83
1H-Inden-1-one, 2,3-dihydro-2,5,6-trimethyl-	MeOH	254(4.17),297(3.31)	94-3113-83
1H-Inden-1-one, 2,3-dihydro-2,5,7-trimethyl-	MeOH	258(4.14),287(3.29), 298(3.31)	94-3113-83
1H-Inden-1-one, 2,3-dihydro-2,6,7-trimethyl-	MeOH	252(4.41),303(3.86)	94-3113-83
Pentacyclo[5.4.0.$0^{2,6}$.$0^{3,10}$.$0^{5,9}$]undecan-8-one, 11-methyl-	MeCN	295(1.28)	44-1862-83
1-Penten-3-one, 4-methyl-1-phenyl-, (E)-	EtOH	223(4.02),289(4.35)	12-0527-83
3-Penten-1-one, 4-methyl-1-phenyl-	C_6H_{12}	238(3.73)	44-0695-83
$C_{12}H_{14}O_3$			
Benzaldehyde, 2-hydroxy-4-[(3-methyl-2-butenyl)oxy]-	MeOH	255(3.29),292(3.51)	2-0276-83
2H-1-Benzopyran-3(4H)-one, 7-methoxy-2,2-dimethyl-	EtOH	212(4.18),226(3.86), 282(3.32),288(3.30)	39-1431-83C
2-Furanmethanol, α,5-dimethyl-α-(5-methyl-2-furanyl)-	EtOH	226(3.95)	103-0474-83

Compound	Solvent	λmax(log ε)	Ref.
$C_{12}H_{14}O_4$			
1H-2-Benzopyran-1-one, 3,4-dihydro-8-hydroxy-6-methoxy-3,5-dimethyl-	EtOH	266(4.10),310(3.78)	39-2185-83C
Bicyclo[1.1.0]butane-2,4-dipropenoic acid, dimethyl ester, (E,E)-endo,endo	MeOH	210(4.18)	24-0681-83
Bicyclo[2.2.2]octa-2,5-diene-2,3-dicarboxylic acid, dimethyl ester	hexane	217(3.73)	24-0681-83
	EtOH	218.5(3.72)	24-0681-83
$C_{12}H_{14}O_4S$			
Sulfonium, methylphenyl-, 2-methoxy-1-(methoxycarbonyl)-2-oxoethylide	EtOH	222(4.33)	121-0433-83B
$C_{12}H_{14}O_5$			
4-Heptenoic acid, 7-(2,3,5-trioxocyclopentyl)-	MeOH-HCl	278(4.04)	87-0786-83
2-Propenoic acid, 3-(3,4,5-trimethoxyphenyl)-	EtOH	297(4.17)	151-0171-83C
$C_{12}H_{14}O_6$			
1H-2-Benzopyran-1-one, 3,4-dihydro-4,8-dihydroxy-6,7-dimethoxy-3-methyl-, (3R-cis)-	EtOH	219(4.38),228s(4.34), 273(4.12),307(3.63)	94-0925-83
1(3H)-Isobenzofuranone, 5-hydroxy-4-(hydroxymethyl)-3,7-dimethoxy-6-methyl-	MeOH	220(4.4),264(4.0), 303s(3.3)	158-83-101
	MeOH-KOH	218(4.2),241(4.0), 309(4.3)	158-83-101
1,1(2H)-Pentalenedicarboxylic acid, hexahydro-3,4-dioxo-, dimethyl ester	MeOH	300(2.04)	5-0112-83
	MeCN	300(2.12)	5-0112-83
	H_2SO_4	315(3.17)(changing)	5-0112-83
$C_{12}H_{14}Rh$			
Rhodocinium, 1,1'-dimethyl-, hexafluorophosphate	H_2O	238(4.67),275(3.44)	101-0093-83K
$C_{12}H_{14}S_4$			
1,3-Dithiole, 2,2'-(1,2-ethanediylidene)bis[4,5-dimethyl-	$CHCl_3$	394(4.45),415(4.47)	88-3469-83
$C_{12}H_{15}BrS$			
Thiophene, 3-bromo-4-(1-cyclohexen-1-yl)-2,5-dimethyl-	MeCN	190(4.27),241(3.87)	4-0729-83
$C_{12}H_{15}ClN_2$			
2H-Indol-2-imine, 3-(3-chloropropyl)-1,3-dihydro-1-methyl-, monohydrochloride	n.s.g.	216(4.16),264(3.87)	103-1175-83
$C_{12}H_{15}ClO_4$			
2-Butanone, 4-(5-chloro-2,3-dimethoxyphenyl)-4-hydroxy-	EtOH	208(4.59),221(3.94), 283(3.25)	104-0458-83
$C_{12}H_{15}N$			
Cyclobutanecarbonitrile, 3-(dicyclopropylmethylene)-	hexane	214(4.05)	44-1500-83
1H-Indole, 5-ethyl-2,3-dimethyl-	EtOH	207(4.4),230(4.6), 286(4.0),296(3.9)	104-0388-83
1-Propen-1-amine, N-[1-(2-methylphenyl)ethylidene]-, (E,E and E,Z)-	EtOH	216(4.21),266(4.07)	44-5348-83
(Z,E and Z,Z)-	EtOH	211(4.14),240(4.06)	44-5348-83

Compound	Solvent	λ_{max}(log ε)	Ref.
$C_{12}H_{15}NO$			
3H-Indole, 2-ethoxy-2,3-dimethyl-	EtOH	253(3.88)	94-2986-83
1H-Pyrrole, 2,5-dihydro-2-[(2-methoxyphenyl)methyl]-	MeOH	215(4.20),270(3.28), 277(3.25)	87-0469-83
1H-Pyrrole, 2,5-dihydro-2-[(3-methoxyphenyl)methyl]-	MeOH	220s(3.92),272(3.31), 279(3.28)	87-0469-83
$C_{12}H_{15}NOS$			
Butanethioamide, 3,3-dimethyl-2-oxo-N-phenyl-	MeOH	251(3.81),263(3.72), 300(4.11),315s(4.04), 390s(1.39)	5-1116-83
Morpholine, 4-(2-phenyl-1-thioxoethyl)-	heptane	281(4.17)	80-0555-83
	EtOH	280(4.20)	80-0555-83
Piperidine, 1-[(2-hydroxyphenyl)-thioxomethyl]-	heptane	244(3.90),296(4.06), 388(2.87)	80-0555-83
	EtOH	239(3.89),283(4.20), 362(2.70)	80-0555-83
Piperidine, 1-[(4-hydroxyphenyl)-thioxomethyl]-	heptane	282(--),291(4.18)	80-0555-83
	EtOH	239(3.92),282(4.23), 369(2.82)	80-0555-83
$C_{12}H_{15}NO_2$			
Acetic acid, cyano(3,5,5-trimethyl-2-cyclohexen-1-ylidene)-	EtOH	302.5(4.33)	111-0447-83
2-Azetidinone, 4-ethoxy-3-methyl-1-phenyl-, (±)-	MeOH	244(4.20)	5-1393-83
Isoquinoline, 5-ethoxy-3,4-dihydro-6-methoxy-, hydrochloride	MeOH	225s(4.15),276(3.98), 325(3.89)	83-0845-83
$C_{12}H_{15}NO_2S$			
Morpholine, 4-[(2-methoxyphenyl)thioxomethyl]-	heptane	249(3.80),289(4.10), 389(2.80)	80-0555-83
	EtOH	244(3.90),286(4.20), 369(2.60)	80-0555-83
Morpholine, 4-[(4-methoxyphenyl)thioxomethyl]-	heptane	255(4.03),282(4.12), 403(2.70)	80-0555-83
	EtOH	238(3.93),281(4.20), 369(2.84)	80-0555-83
$C_{12}H_{15}NO_3$			
Benzamide, N-acetoxy-2,4,6-trimethyl-	MeOH	268(2.57),270(2.58), 276(2.53)	32-0011-83
1-Penten-3-ol, 4-methyl-1-(4-nitrophenyl)-, (E)-	EtOH	220(4.00),298(4.07)	12-0527-83
$C_{12}H_{15}NO_4$			
Benzamide, N-acetoxy-4-methoxy-2,6-dimethyl-	MeOH	270(3.05),280(2.95)	32-0011-83
Flavipucine	EtOH	330(3.73)	162-0587-83
$C_{12}H_{15}NS$			
Piperidine, 1-(phenylthioxomethyl)-	heptane	252(3.91),293(3.97), 400(2.48)	80-0555-83
	EtOH	242(4.01),285(4.07), 363(2.47)	80-0555-83
$C_{12}H_{15}N_2$			
Quinolizinium, 2-[(1-methylethyl)-amino]-, bromide	MeOH	238(4.40),311(4.20), 354(3.83)	4-0407-83

Compound	Solvent	λ_{max}(log ε)	Ref.
$C_{12}H_{15}N_3OS$			
4H-Pyrido[3,4-e][1,3]thiazin-4-one, 5,7-dimethyl-2-(propylamino)-	MeOH	245(4.14)	73-3426-83
$C_{12}H_{15}N_3O_2$			
Cyclopentanecarbonitrile, 2-(1,2,3,4-tetrahydro-1,3-dimethyl-2,4-dioxo-5-pyrimidinyl)-	MeCN	269(3.91)	35-0963-83
$C_{12}H_{15}N_3O_4$			
5-Pyrimidinepentanoic acid, 2-amino-1,4,5,6-tetrahydro-4,6-dioxo-5-(2-propynyl)-	50%MeOH-HCl	216(3.96)	73-0137-83
	+ NaOH	228(4.20),263(3.98)	73-0137-83
$C_{12}H_{15}N_3O_4S$			
5H-Pyrrolo[3,2-d]pyrimidine, 4-(methylthio)-7-β-D-ribofuranosyl-	pH 1	245s(3.79),312(4.37)	44-0780-83
	pH 7	247(4.00),294(4.19)	44-0780-83
	pH 13	252(4.23),296(4.16)	44-0780-83
7H-Pyrrolo[2,3-d]pyrimidine, 7-α-D-arabinofuranosyl-4-(methylthio)-	MeOH	249(3.79),292(4.11)	5-1576-83
β-	MeOH	249(3.83),292(4.10)	5-1576-83
$C_{12}H_{15}N_3O_5$			
Imidazo[1,2-c]pyrimidin-5(6H)-one, 2-methyl-6-β-D-ribofuranosyl-	pH 1.4	290(4.22)	56-0779-83
	pH 7	276(4.18)	56-0779-83
Imidazo[1,2-c]pyrimidin-5(6H)-one, 3-methyl-6-β-D-ribofuranosyl-	pH 1.4	294(4.24)	56-0779-83
	pH 7	278(4.23)	56-0779-83
$C_{12}H_{15}N_3O_6$			
2(1H)-Pyridinone, 5,6-dihydro-1-methyl-3,5-dinitro-6-(2-oxocyclohexyl)-, compd. with diethylamine	MeOH	233(3.97),328(4.10), 479(4.40)	138-0715-83
$C_{12}H_{15}N_5O_4$			
7H-Pyrrolo[2,3-d]pyrimidine-5-carboxamide, 4-amino-7-(2-deoxy-β-D-erythro-pentofuranosyl)-	pH 1	230(4.16),274(4.10)	35-4059-83
	pH 13	278(4.17)	35-4059-83
	MeOH	231(4.01),279(4.17)	35-4059-83
$C_{12}H_{15}N_5O_5$			
β-D-arabino-Heptofuranuronic acid, 1-(6-amino-9H-purin-9-yl)-1,5,6-trideoxy-	MeOH	259(4.15)	87-1530-83
$C_{12}H_{15}N_7OS$			
Guanidine, N-cyano-N'-[2-[(5-methyl-1H-imidazol-4-yl)methylthio]ethyl]-N'-nitroso-N"-(2-propynyl)-	EtOH	390s(--),402(2.19), 420s(--)	23-1771-83
$C_{12}H_{16}$			
4,6-Methano-1H-indene, 2,4,5,6-tetrahydro-5,5-dimethyl-, (4R)-	EtOH	256(3.96)	33-2519-83
Propadiene, cyclopropyl-, dimer	hexane	267(4.14)	44-1500-83
$C_{12}H_{16}BrNO_2$			
Isoquinoline, 8-bromo-1,2,3,4-tetrahydro-5,6-dimethoxy-2-methyl-, hydrochloride	MeOH	275(3.71),310(3.40)	83-0801-83
$C_{12}H_{16}ClNO$			
Cyclohexanol, 1-(2-aminophenyl)-2-chloro-	EtOH	204(4.40),239(3.80), 292(3.30)	78-3767-83

Compound	Solvent	$\lambda_{max}(\log \epsilon)$	Ref.
$C_{12}H_{16}FN_2O_8P$			
Uridine, 2'-deoxy-5-fluoro-5'-(O-1,3,2-dioxaphosphorinan-2-yl)-, P-oxide	EtOH	268(3.92)	87-1153-83
$C_{12}H_{16}INO_4$			
1H-Pyrrole-3-acetic acid, 2-(ethoxycarbonyl)-5-iodo-4-methyl-, ethyl ester	$CHCl_3$	281(4.14)	78-1849-83
$C_{12}H_{16}NO$			
2H-Isoindol-2-yloxy, 1,3-dihydro-1,1,3,3-tetramethyl-	EtOH	229(3.50),250s(3.25) 268s(2.95),417(0.60)	12-0397-83
$C_{12}H_{16}NO_2$			
Pyridinium, 3-(methoxycarbonyl)-1-(3-methyl-2-butenyl)-, perchlorate	MeOH	264(3.69)	35-1204-83
$C_{12}H_{16}N_2$			
1H-Indole-3-ethanamine, α-ethyl- (etryptamine)	n.s.g.	220.5(4.31),281(3.78), 289.5(3.72)	162-0562-83
Pyrrolo[2,3-b]indole, 1,2,3,3a,8,8a-hexahydro-1,3a-dimethyl-	EtOH	214(3.93),244(4.00), 299(3.57)	1-0803-83
	EtOH-HCl	210(3.86),238(3.99), 193(3.53)[sic]	1-0803-83
$C_{12}H_{16}N_2O$			
1H-Indol-5-ol, 3-[2-(dimethylamino)-ethyl]- (bufotenine)	n.s.g.	220(4.0),265(3.7)	162-0204-83
$C_{12}H_{16}N_2OS$			
Morpholine, 4-[[(methylthio)imino]-phenylmethyl]-, (E)-	hexane	310(3.46)	77-0390-83
$C_{12}H_{16}N_2O_2$			
2,4,6-Cycloheptatrien-1-one, 2-acetyl-7-[[2-(methylamino)ethyl]amino]-	MeOH	247(4.17),349(3.80), 421(3.96)	18-3358-83
$C_{12}H_{16}N_2O_3$			
1H-Cyclopentapyrimidine-2,4,7(3H)-trione, 5,6-dihydro-1,3,5,6,6-pentamethyl-	MeCN	247(3.12),311(3.83)	35-0963-83
1,2-Diazabicyclo[5.2.0]nona-3,5-diene-2-carboxylic acid, 5,8-dimethyl-9-oxo-, ethyl ester, cis	MeOH	273(3.86)	5-1393-83
1,2-Diazabicyclo[5.2.0]nona-3,5-diene-2-carboxylic acid, 8,8-dimethyl-9-oxo-, ethyl ester	MeOH	277(3.95)	5-1361-83
4H-Pyrido[1,2-a]pyrimidine-2-carboxylic acid, 6,7,8,9-tetrahydro-6-methyl-4-oxo-, ethyl ester	EtOH	230(3.81),301(3.96)	39-0369-83C
(R)-	EtOH	207(3.76),231(3.75), 304(3.91)	39-1413-83B
$C_{12}H_{16}N_2O_4$			
Isoquinoline, 1,2,3,4-tetrahydro-5,6-dimethoxy-2-methyl-8-nitro-	MeOH	214s(4.24),302(3.88), 340(3.90)	83-0801-83
$C_{12}H_{16}N_2O_5$			
5-Pyrimidinepentanoic acid, hexahydro-2,4,6-trioxo-5-(2-propenyl)-	50%MeOH-HCl	220(3.51)	73-0137-83
	+ NaOH	242(3.82)	73-0137-83

Compound	Solvent	λ_{max}(log ε)	Ref.
L-Threonine, N-[[3-hydroxy-5-(hydroxymethyl)-2-methyl-4-pyridinyl]methylene]-	MeOH	255(4.00),288s(3.63), 336(3.58),420(3.33)	35-0803-83
$C_{12}H_{16}N_2O_6$			
2,4(1H,3H)-Pyrimidinedione, 1-(2,4-diacetoxybutyl)-	EtOH	264(3.98)	128-0125-83
$C_{12}H_{16}N_4O_3S$			
2,4(1H,3H)-Pteridinedione, 1,3,6-trimethyl-7-[(1-methylethyl)sulfinyl]-	MeOH	245(4.16),353(4.00)	88-5047-83
7H-Pyrrolo[2,3-d]pyrimidin-4-amine, 7-(2-deoxy-β-D-erythro-pentofuranosyl)-2-(methylthio)-	MeOH	234(4.41),281(4.18)	5-0876-83
$C_{12}H_{16}N_4O_4$			
7H-Pyrrolo[2,3-d]pyrimidin-2-amine, 7-(2-deoxy-β-D-erythro-pentofuranosyl)-4-methoxy-	MeOH	259(4.02),285(3.87)	5-0137-83
	MeOH	225(4.40),259(3.98), 285(3.88)	44-3119-83
$C_{12}H_{16}N_4O_5$			
7H-Pyrrolo[2,3-d]pyrimidin-2-amine, 7-α-D-arabino-furanosyl-4-methoxy-	MeOH	225(4.40),260(3.98), 285(3.88)	136-0029-83G
β-	MeOH	218(4.39),260(3.99), 285(3.88)	136-0029-83G
$C_{12}H_{16}O$			
Bicyclo[4.3.1]deca-2,4,8-trien-7-ol, 7,9-dimethyl-, endo	EtOH	247s(3.70),253(3.71)	24-2914-83
4,7-Methano-1H-inden-1-one, 3a,4,5,6-7,7a-hexahydro-3a,7a-dimethyl-	EtOH	234(3.79)	33-0627-83
2(4aH)-Naphthalenone, 5,6,7,8-tetrahydro-4,4a-dimethyl-	EtOH	244(4.22)	24-3884-83
1-Penten-3-ol, 4-methyl-1-phenyl-, (E)-	MeOH	254(4.19),284(2.08), 292(1.95)	12-0527-83
2-Propyn-1-one, 1-(2,3-dimethylbicyclo[2.2.1]hept-2-yl)-, 2-exo-3-endo)-(±)-	EtOH	211(3.66),218s(3.54)	33-0627-83
Spiro[4.5]deca-6,9-dien-8-one, 6,10-dimethyl-	EtOH	246.5(4.29)	24-3884-83
Spiro[bicyclo[2.2.1]heptane-2,1'-[3]-cyclopenten]-2'-one, 3-methyl-, (1α,2α,3β,4α)-(±)-	EtOH	220(3.94)	33-0627-83
Tricyclo[5.3.1.0^{4,11}]undec-9-en-8-one, 10-methyl-	EtOH	240(4.11)	24-2903-83
$C_{12}H_{16}OS$			
1-Propanethione, 1-(4-methoxyphenyl)-2,2-dimethyl-	benzene	310(3.97),560(2.41)	44-0214-83
$C_{12}H_{16}O_2$			
Cyclohexanone, 2-(1-oxo-2,4-hexadienyl)-, (E,E)-	MeOH	242(3.57),296(4.14), 308(4.15),354(4.31)	78-4243-83
	MeOH-NaOH	248(3.65),295(4.13), 305(4.12),353(4.27)	78-4243-83
1(4H)-Naphthalenone, 4a,5,8,8a-tetrahydro-4-hydroxy-2,3-dimethyl-, (4α,4aβ,8aβ)-	$CHCl_3$	317(1.66)	35-5354-83
Spiro[4.5]dec-7-ene-1,6-dione, 4,8-dimethyl-	EtOH	233(4.14)	64-0497-83B

Compound	Solvent	$\lambda_{max}(\log \epsilon)$	Ref.
Tricyclo[7.3.0.0^{3,7}]dodecane-5,11-dione, (1S,3R,7R,9S)-	MeCN	288s(1.61),295(1.68), 305(1.63),315(1.36)	35-0145-83
$C_{12}H_{16}O_2S$			
1-Propanethione, 1-(4-methoxyphenyl)-2,2-dimethyl-, S-oxide	$CHCl_3$	263(3.79),306s(3.40)	44-0214-83
$C_{12}H_{16}O_3$			
Benzenepropanoic acid, 2-methoxy- ,4-dimethyl-	EtOH	274(3.80),280(3.78)	64-0497-83B
Benzoic acid, 4-hydroxy-2,3,5-trimethyl-, ethyl ester (6-^{14}C)	EtOH	267(3.95)	39-0667-83C
2-Cyclopenten-1-one, 3-acetoxy-5-(2-butenyl)-2-methyl-	EtOH	236(4.10)	39-1885-83C
2-Cyclopenten-1-one, 3-acetoxy-4(and 5)-methyl-5(and 4)-(2-methyl-2-propenyl)-	EtOH	233(4.11)	39-1885-83C
Spiro[1,3-dioxolane-2,1'-[1H]inden]-5'(4'H)-one, 2',3',3'a,7'a-tetrahydro-7'a-methyl-	EtOH	240(4.13)	12-0117-83
Spiro[1,3-dioxolane-2,5'-[5H]inden]-1'(4'H)-one, 3'a,6',7',7'a-tetrahydro-7'a-methyl-	EtOH	225(4.04)	12-0117-83
$C_{12}H_{16}O_4$			
Benzeneacetic acid, α,2-dihydroxy-α,3-dimethyl-, ethyl ester	EtOH	218(3.95),280(3.38)	39-1649-83C
Benzoic acid, 2,4-dihydroxy-3,5,6-trimethyl-, ethyl ester (2-^{14}C)	EtOH	218(4.43),270(4.15)	39-0667-83C
1(3H)-Isobenzofuranone, 3-butylidene-4,5,6,7-tetrahydro-6,7-dihydroxy-, (Z)- (ligustilidiol)	n.s.g.	270(3.92)	88-4675-83
Oxiranebutanoic acid, 3-(5-oxo-1,3-pentadienyl)-, methyl ester, (-)-(2S,3S,1E,3E)-	EtOH	276(4.54)	35-3661-83
$C_{12}H_{16}O_5$			
Benzeneacetic acid, α,2-dihydroxy-5-methoxy-α-methyl-, ethyl ester	EtOH	225(3.79),295(3.49)	39-1649-83C
$C_{12}H_{16}O_6$			
2-Butenedioic acid, 2,3-diacetyl-, diethyl ester	MeOH	216(3.92),250(3.44), 276(3.25)	32-0489-83
1-Cyclopentene-1,2-dicarboxylic acid, 3-hydroxy-3-methyl-5-oxo-, diethyl ester	MeOH	212.5(3.93)	32-0489-83
$C_{12}H_{16}O_7$			
Arbutin	EtOH	285(3.42)	102-0223-83
$C_{12}H_{16}S_2$			
Dispiro[4.1.4.1]dodecane-6,12-dithione	$CHCl_3$	263(3.35),298(2.62), 492(1.08)	44-4482-83
2-Thiaspiro[3.4]octane-1-thione, 3-cyclopentylidene-	benzene	270(4.11),350(3.98), 410s(--)	44-4482-83 +78-2719-83
$C_{12}H_{16}S_3$			
12-Thiadispiro[4.1.4.2]tridecane-6,13-dithione	$CHCl_3$	260(4.14),311(4.19), 480(1.51)	44-4482-83

Compound	Solvent	$\lambda_{max}(\log \epsilon)$	Ref.
$C_{12}H_{17}Br_2N_4O_3P$			
Phosphonic acid, (1,2-dibromoethyl)-, bis[2-(1H-imidazol-1-yl)ethyl] ester, dihydrochloride	EtOH	255(2.57)	65-1555-83
$C_{12}H_{17}ClN_2$			
Spiro[cyclohexane-1,2'-[2H-cyclopentapyrimidine], 4'-chloro-1',5',6',7'-tetrahydro-, monophosphorodichloridate	MeOH	365(3.73)	103-0095-83
$C_{12}H_{17}ClO$			
Benzene, 1-(3-chloro-1-methylpropyl)-2-methoxy-4-methyl-	EtOH	275(3.37),281(3.36)	64-0497-83B
$C_{12}H_{17}ClOSi$			
Ethanone, 2-chloro-1-[2-[(trimethylsilyl)methyl]phenyl]-	hexane	248s(3.98),253(3.99), 304(3.51)	78-1109-83
$C_{12}H_{17}FN_3O_7P$			
Uridine, 2'-deoxy-5-fluoro-3'-(O-tetrahydro-2H-1,3,2-oxazaphosphorin-2-yl)-, P-oxide	EtOH	268(3.90)	87-1153-83
Uridine, 2'-deoxy-5-fluoro-5'-(O-tetrahydro-2H-1,3,2-oxazaphosphorin-2-yl)-, P-oxide	EtOH	269(3.89)	87-1153-83
$C_{12}H_{17}N$			
Ethanamine, N-(1-phenylbutylidene)-, (E)-	heptane	218(3.63),240(3.93)	104-0110-83
Ethenamine, N,N-diethyl-2-phenyl-	heptane	224(4.00),280(3.45)	104-0110-83
$C_{12}H_{17}NO$			
2-Azatricyclo[4.4.0.0^{2,8}]dec-9-en-7-one, 6,8,10-trimethyl-	MeOH	215(3.66),319(2.36)	35-3273-83
Cyclopentanone, 2-[5-(dimethylamino)-2,4-pentadienylidene]-	EtOH	455(4.67)	70-0780-83
	$CHCl_3$	435(--)	70-0780-83
3,5,7,9-Decatetraen-2-one, 10-(dimethylamino)-	EtOH	468(4.56)	70-0780-83
	$CHCl_3$	450(4.68)	70-0780-83
$C_{12}H_{17}NO_2$			
Isoquinoline, 1,2,3,4-tetrahydro-5,6-dimethoxy-2-methyl-	MeOH	227(3.35),233s(3.28), 280(3.10)	83-0801-83
Pentanamide, N-hydroxy-N-(phenylmethyl)-	EtOH	209(4.08)	34-0433-83
$C_{12}H_{17}NO_3$			
8-Isoquinolinol, 1,2,3,4-tetrahydro-6,7-dimethoxy-1-methyl- (anhalonidine)	EtOH	270(2.81)	162-0095-83
2-Pyridinecarboxylic acid, 5-butyl-4-ethoxy-	MeOH	242(3.85)	106-0591-83
1H-Pyrrole-2-carboxylic acid, 4-(3-butenyl)-3-hydroxy-5-methyl-, ethyl ester	MeCN	274(4.23)	33-1902-83
$C_{12}H_{17}NO_3S$			
Pyrrolidine, 1-(5,6,7,8-tetrahydro-1,2-benzoxathiin-4-yl)-, S,S-dioxide	EtOH	251(3.87),297(3.80)	4-1549-83

Compound	Solvent	λ_{max}(log ϵ)	Ref.
$C_{12}H_{17}NO_4$			
6-Oxa-1-azabicyclo[5.2.0]non-2-ene-2-carboxylic acid, 9-oxo-, 1,1-dimethylethyl ester	EtOH	242(3.97)	44-1841-83
Propanedioic acid, [5-(dimethylamino)-2,4-pentadienylidene]-, dimethyl ester	EtOH	460(4.76)	70-0780-83
1H-Pyrrole-3-acetic acid, 2-(ethoxy-carbonyl)-4-methyl-, ethyl ester	$CHCl_3$	272(4.07)	78-1849-83
$C_{12}H_{17}NO_4S$			
Morpholine, 4-(5,6,7,8-tetrahydro-1,2-benzoxathiin-4-yl)-, S,S-dioxide	EtOH	263(3.89),288s(3.73)	4-1549-83
$C_{12}H_{17}NO_4S_2$			
2-Thietanone, 3-amino-4,4-dimethyl-, p-toluenesulfonate, (3R)-	EtOH	228(4.13),239(3.43)	39-2259-83C
$C_{12}H_{17}NO_6$			
2H-1,4-Oxazin-2-one, 5,5-bis(acetoxy-methyl)-3-ethyl-5,6-dihydro-	isoPrOH	320(2.02)	44-2989-83
$C_{12}H_{17}N_3O$			
2,4-Cyclohexadien-1-one, 6-(3-azido-propyl)-2,4,6-trimethyl-	MeOH	318(3.60)	35-3273-83
4H-1,2,3-Triazolo[4,5,1-ij]quinolin-7-one, 5,6,6a,7,9a,9b-hexahydro-6a,8,9a-trimethyl-	MeOH	226(4.15),258(3.74)	35-3273-83
isomer m. 79-80°	MeOH	271(3.65)	35-3273-83
4H-1,2,3-Triazolo[4,5,1-ij]quinolin-9-one, 5,6,6a,9,9a,9b-hexahydro-6a,8,9a-trimethyl-	MeOH	204(4.38),236(4.05), 280(3.35),350(2.68)	44-2432-83
$C_{12}H_{17}N_3O_2$			
1H-Cyclopentapyrimidine-2,4(3H,5H)-dione, 6,7-dihydro-7-imino-1,3,5,6,6-pentamethyl-	MeCN	306(3.88)	35-0963-83
5-Pyrimidinepropanenitrile, 1,2,3,4-tetrahydro-α,α,β,1,3-pentamethyl-2,4-dioxo-	MeCN	271(3.90)	35-0963-83
$C_{12}H_{17}N_3O_4$			
Agaritine	H_2O	237.5(4.08),280(3.15)	162-0028-83
5-Pyrimidinepentanoic acid, 2-amino-1,4,5,6-tetrahydro-4,6-dioxo-5-(2-propenyl)-	50%MeOH-HCl	217(4.03)	73-0137-83
	+ NaOH	226(4.21),262(3.00)	73-0137-83
5-Pyrimidinepentanoic acid, 1,4-di-hydro-6-hydroxy-4-oxo-2-(2-propen-ylamino)-	50% MeOH	244(3.811),275(4.138)	73-0304-83
	+ HCl	240s(3.853),269(4.120)	73-0304-83
	+ NaOH	244(3.798),275(4.130)	73-0304-83
$C_{12}H_{17}N_3O_5S$			
D-Methionine, N-[(1,2,3,4-tetrahydro-6-methyl-2,4-dioxo-5-pyrimidinyl)-carbonyl]-, methyl ester	H_2O	269(4.00)	94-0135-83
L-	H_2O	269(4.06)	94-0135-83
$C_{12}H_{17}N_3O_7S$			
Glycine, N-[(1,2,3,4-tetrahydro-4-oxo-1-β-D-ribofuranosyl-2-thioxo-5-pyrimidinyl)methyl]-	pH 2	272(4.16)	18-5398-83
	pH 12	272(4.14)	18-5398-83

Compound	Solvent	$\lambda_{max}(\log \epsilon)$	Ref.
$C_{12}H_{17}N_3O_8$			
Glycine, N-[(1,2,3,4-tetrahydro-2,4-dioxo-1-β-D-ribofuranosyl-5-pyrimidinyl)methyl]-	pH 2	267(3.96)	88-5395-83
	pH 12	265(3.75)	88-5395-83
$C_{12}H_{17}N_4O_3P$			
Phosphonic acid, ethenyl-, bis[2-(1H-imidazol-1-yl)ethyl] ester, dihydrochloride	EtOH	207(3.96),247(2.24), 279(1.60)	65-1555-83
$C_{12}H_{17}N_5O_3$			
9H-Purin-6-amine, 9-[(2,2-dimethyl-1,3-dioxolan-4-yl)methyl]-8-methoxy-	MeOH	261(4.08)	73-1910-83
$C_{12}H_{17}N_5O_4$			
9H-Purin-6-amine, 9-(5,6-dideoxy-β-D-arabino-heptofuranosyl)-	H_2O	259(4.15)	87-1530-83
$C_{12}H_{17}N_5O_5$			
Adenosine, N-(2-hydroxyethyl)-	EtOH	213.5(4.19),267(4.20)	102-2509-83
Adenosine, N^6-methoxy-1-methyl-	pH 1	233(3.86),282(3.97)	94-3149-83
	pH 7	270(4.17)	94-3149-83
	pH 13	270(4.16)	94-3149-83
	EtOH	270(4.15),321(3.30)	94-3149-83
Adenosine, 1-methoxy-N^6-methyl-	pH 1	262(4.16)	94-3149-83
	pH 7	262(4.15)	94-3149-83
	EtOH	260.5(4.15),268s(4.06)	94-3149-83
$C_{12}H_{18}$			
Azulene, 2,3,3a,4,5,6-hexahydro-1,4-dimethyl- (clavukerin A)	n.s.g.	238(4.05),245(4.14), 253(4.04)	12-0211-83
	MeOH	244(4.32)	94-2160-83
Benzene, hexamethyl-, complex with diethyl 1,2-dicyanoafumarate	CH_2Cl_2	450(3.43)	35-1276-83
complex with fumaronitrile	Pr_2O	308(5.15)	35-1276-83
Bicyclo[6.2.2]dodeca-1,8-diene	C_6H_{12}	202(3.82),216s(3.74)	44-0551-83
Bicyclo[6.2.2]dodeca-1(11),8-diene	C_6H_{12}	205(3.58),232s(2.92)	44-0551-83
$C_{12}H_{18}BrN_4O_3P$			
Phosphonic acid, (2-bromoethyl)-, bis[2-(1H-imidazol-1-yl)ethyl] ester, dihydrochloride	EtOH	269(0.95)	65-1555-83
$C_{12}H_{18}Cl_2N_2$			
Pyrazine, 2,5-dichloro-3-(1-methylpropyl)-6-(2-methylpropyl)-	EtOH	221(4.13),285s(3.92), 297(3.99)	94-0020-83
$C_{12}H_{18}N$			
Pyridinium, 1-methyl-2-(4-methyl-3-pentenyl)-, perchlorate	MeOH	268(3.86)	35-1204-83
$C_{12}H_{18}N_2$			
Methanediamine, 1-(2,4,6-cyclohepta-trien-1-ylidene)-N,N,N',N'-tetramethyl-	ether	335(3.4),430(2.45)	24-1777-83
	THF	338(--),492s(--)	24-1154-83
2-Propenenitrile, 3-(3,6-dihydro-2,2,6,6-tetramethyl-1(2H)-pyridinyl)-, cis	heptane	266(4.23)	70-2322-83
	MeOH	272(4.23)	70-2322-83
trans	heptane	264(4.10)	70-2322-83
	MeOH	268(6.36!)	70-2322-83

Compound	Solvent	λ_{max}(log ε)	Ref.
$C_{12}H_{18}N_2O$			
1-Propanone, 2-methyl-1-[2-methyl-1-[2-methyl-4-(1-methylethyl)-5-pyrimidinyl]-	EtOH	228.5(3.88),252s(3.56), 282(2.78)	4-0649-83
2-Propenenitrile, 3-(2,2,6,6-tetramethyl-4-oxo-1-piperidinyl)-, (Z)-	heptane	270(4.0)	70-2322-83
	MeOH	274(4.27)	70-2322-83
Spiro[cyclohexane-1,8'(5'H)-imidazo-[1,2-a]pyridin]-2-ol, 6',7'-dihydro- (nitrabirine)	EtOH	212(3.84)	105-0202-83
$C_{12}H_{18}N_2O_2$			
8-Isoquinolinamine, 1,2,3,4-tetrahydro-5,6-dimethoxy-2-methyl-	MeOH	207(4.43),235(3.74), 290(3.30)	83-0801-83
$C_{12}H_{18}N_2O_4S$			
5-Pyrimidinepentanoic acid, hexahydro-4,6-dioxo-2-propyl-2-thioxo-	50%MeOH-HCl	237(3.99),286.5(4.35)	73-0137-83
	+ NaOH	253.5(3.91),306(4.40)	73-0137-83
$C_{12}H_{18}N_2O_5$			
5-Pyrimidinepentanoic acid, hexahydro-2,4,6-trioxo-5-propyl-	50%MeOH-HCl	210(3.90),240(3.02)	73-0137-83
	+ NaOH	240(4.02)	73-0137-83
$C_{12}H_{18}N_4O_4$			
Imidazo[4,5-d][1,3]diazepin-8-ol, 3-(2-deoxy-α-D-erythro-pentofuranosyl)-3,6,7,8-tetrahydro-5-methyl-	pH 7	280(3.93)	87-1478-83
(8R)-	pH 7	281(3.96)	87-1478-83
1,2,4-Triazine-3,6-dicarboxylic acid, 4,5-dihydro-4-piperidino-, dimethyl ester	CH_2Cl_2	350(3.67)	83-0472-83
$C_{12}H_{18}N_4O_5$			
1H-Imidazole-4-carboxamide, 5-[[2,3-O-(1-methylethylidene)-β-D-ribofuranosyl]amino]-	EtOH	235(3.69),272(4.05)	136-0001-83B
$C_{12}H_{18}N_4O_6$			
1,2-Ethanediamine, N,N'-bis(3-nitro-2-oxo-3-penten-4-yl)-	CH_2Cl_2	314(4.18),355(3.85)	40-0088-83
$C_{12}H_{18}N_6$			
1H-Imidazole, 2-(1H-imidazol-2-ylazo)-4,5-dipropyl-	MeOH	471(4.44)	56-0547-83
$C_{12}H_{18}O$			
Cyclopentanone, 3,3-dimethyl-2-(3-methyl-1,3-butadienyl)-, (E)-	pentane	229s(4.31),233(4.34), 240s(4.18),296(2.52)	33-2236-83
1(2H)-Naphthalenone, 3,4,5,6,7,8-hexahydro-5,5-dimethyl-	MeOH	245(4.10)	39-0751-83C
1(4H)-Naphthalenone, 4a,5,6,7,8,8a-hexahydro-4a,5-dimethyl-	EtOH	231(3.76)	39-0161-83C
$C_{12}H_{18}OSi$			
Ethanone, 1-[2-[(trimethylsilyl)-methyl]phenyl]-	isoPrOH	248(3.89),300(3.35)	78-1109-83
$C_{12}H_{18}O_2$			
1,2-Benzenediol, 3-(1,1-dimethylethyl)-5-ethyl-	hexane	283(3.17)	70-0550-83
Benzenepropanol, 2-methoxy-α,4-dimethyl-	EtOH	275(2.36),281(3.33)	64-0497-83B

Compound	Solvent	$\lambda_{max}(\log \epsilon)$	Ref.
2-Cyclohexen-1-one, 3-[(1-methyl-4-pentenyl)oxy]-	EtOH	251(3.93)	142-1005-83
$C_{12}H_{18}O_3$			
1-Cyclopentene-1-heptanoic acid, 5-oxo-	EtOH	229(4.04)	39-0319-83C
$C_{12}H_{18}O_4$			
Cyclohexanecarboxylic acid, 2-(2-methoxy-2-oxoethylidene)-1-methyl-, methyl ester	EtOH	234(4.12)	2-1103-83
1-Cyclopentene-1-heptanoic acid, 2-hydroxy-3-oxo-	EtOH	262(4.03)	39-0319-83C
Pentanoic acid, 5-[(6-oxo-2,4-hexadienyl)oxy]-, methyl ester, (E,E)-	n.s.g.	267(4.26)	44-4413-83
$C_{12}H_{19}ClN_2$			
Pyrazine, 2,5-dibutyl-3-chloro-	EtOH	213(3.93),280(3.81), 301s(3.48)	4-1277-83
$C_{12}H_{19}ClN_2O$			
Pyrazine, 3-chloro-5-(1-methylpropyl)-2-(2-methylpropyl)-, 1-oxide	EtOH	307(4.40),271(4.06), 302(3.56),312s(3.51)	94-0020-83
Pyrazinemethanol, 6-chloro-α-ethyl-α-methyl-5-(1-methylpropyl)-	EtOH	232.5(4.22),272(3.86), 301s(3.41)	94-0020-83
$C_{12}H_{19}Cl_2NPt$			
Platinum, dichloro[4-(1,1-dimethylethyl)benzenamine](η^2-ethene)-	n.s.g.	243(3.68),262(3.75), 302(3.20)	131-0309-83H
$C_{12}H_{19}NO$			
Morpholine, 4-(4,4-dimethyl-1,5-cyclohexadien-1-yl)-	C_6H_{12}	218(3.94),283(3.28)	33-0735-83
Morpholine, 4-(1-ethenyl-4-methyl-1,3-pentadienyl)-	C_6H_{12}	259(3.74),303(3.69)	33-0735-83
$C_{12}H_{19}NO_2$			
2,4-Hexadiyn-1-amine, 6,6-diethoxy-N,N-dimethyl-	hexane	217(3.37)	33-1427-83
Pentanenitrile, 2-(2,2-dimethyl-1-oxopropyl)-4,4-dimethyl-3-oxo-	EtOH	285(4.04)	4-0645-83
1-Propanone, 1-[5-(1,1-dimethylethyl)-4-isoxazolyl]-2,2-dimethyl-	EtOH	227(3.72),280(2.60)	4-0645-83
$C_{12}H_{19}NO_2Si$			
Glycine, N-phenyl-, (trimethylsilyl)-methyl ester	C_6H_{12}	242(4.26),294(3.52)	65-0814-83
$C_{12}H_{19}NO_3S$			
1,2-Benzoxathiin-4-amine, N,N-diethyl-5,6,7,8-tetrahydro-, 2,2-dioxide	EtOH	259(3.82),293s(3.73)	4-1549-83
$C_{12}H_{19}N_3$			
1H-Imidazole-1-carbonitrile, 2,5-bis(1-methylpropyl)-	EtOH	233(3.96)	4-1277-83
1H-Imidazole-1-carbonitrile, 2,5-bis(2-methylpropyl)-	EtOH	234(4.16)	4-1277-83
1H-Imidazole-1-carbonitrile, 2,5-dibutyl-	EtOH	233(3.93)	4-1277-83
$C_{12}H_{19}N_3O$			
Hydrazinecarboxamide, 2-(3,4,5,6,7,8-	EtOH	268(4.48)	2-0331-83

Compound	Solvent	$\lambda_{max}(\log \epsilon)$	Ref.
hexahydro-5-methyl-1(2H)-naphthalen-dene)- (cont.)			2-0331-83
$C_{12}H_{19}N_3O_4$			
2,4(1H,3H)-Pyrimidinedione, 5-(dimeth-ylamino)-1-[3-hydroxy-4-(hydroxy-methyl)cyclopentyl]-, (1α,3β,4α)-(±)-	pH 1	270(3.98)	87-0156-83
	pH 7	230(3.88),294(3.82)	87-0156-83
	pH 13	233s(--),284(3.76)	87-0156-83
5-Pyrimidinepentanoic acid, 1,4-di-hydro-6-hydroxy-4-oxo-2-(propyl-amino)-	50% MeOH	242(3.824),275(4.126)	73-0304-83
	+ HCl	240s(3.864),269(4.102)	73-0304-83
	+ NaOH	242(3.814),275(4.116)	73-0304-83
5-Pyrimidinepropanoic acid, 3,4-di-hydro-6-hydroxy-4-oxo-2-(propyl-amino)-, ethyl ester	50% MeOH	273(4.22)	73-0304-83
	+ HCl	272(4.118)	73-0304-83
	+ NaOH	276(4.116)	73-0304-83
$C_{12}H_{19}N_5$			
Pyrazine, 3-azido-2,5-bis(1-methyl-propyl)-	EtOH	277(3.92)	4-1277-83
Pyrazine, 3-azido-2,5-bis(2-methyl-propyl)-	EtOH	282(3.93)	4-1277-83
Pyrazine, 3-azido-2,5-dibutyl-	EtOH	212(3.88),279(3.82)	4-1277-83
$C_{12}H_{19}N_5O_3$			
1H-Purine-2,6-dione, 3,7-dihydro-8-[[1-(hydroxymethyl)propyl]-amino]-1,3,7-trimethyl-	EtOH	295(4.42)	118-0577-83
$C_{12}H_{20}$			
4-Undecyne, 6-methylene-	EtOH	225(4.07)	104-1621-83
5-Undecyne, 7-methylene-	EtOH	224(4.08)	104-1621-83
$C_{12}H_{20}Bi_2$			
Dibismuthine, tetrakis(1-methyl-ethenyl)-	pentane	230(4.18),270(3.79)	157-1859-83
$C_{12}H_{20}N_2O$			
Cyclohexanone, 2-[(4-methyl-1-pipera-zinyl)methylene]-, (E)-	EtOH	330.5(4.34)	4-0839-83
$C_{12}H_{20}N_2O_2$			
Aspergillic acid	pH 8	235(4.02),328(3.93)	162-0122-83
$C_{12}H_{20}N_2O_2SSe$			
4,6(1H,5H)-Pyrimidinedione, 5-ethyl-dihydro-5-[3-[(1-methylethyl)sel-eno]propyl]-2-thioxo-	MeOH	236.4(3.87),286.3(4.29)	101-0171-83S
$C_{12}H_{20}N_2O_2STe$			
4,6(1H,5H)-Pyrimidinedione, 5-ethyl-dihydro-5-[3-[(1-methylethyl)tell-uro]propyl]-2-thioxo-	MeOH	234.9(4.21),287.2(4.47)	101-0171-83S
$C_{12}H_{20}N_2O_3$			
Pyrazinemethanol, α-ethyl-6-hydroxy-α-methyl-5-(2-methylpropyl)-, 1-oxide	EtOH	235(4.16),333(3.84)	94-0020-83
$C_{12}H_{20}N_2O_3Se$			
2,4,6(1H,3H,5H)-Pyrimidinetrione, 5-ethyl-5-[3-[(1-methylethyl)-seleno]propyl]-	MeOH	212.2(3.31),279(2.02)	101-0171-83S

Compound	Solvent	λ_{max}(log ε)	Ref.
$C_{12}H_{20}N_2O_3Te$			
2,4,6(1H,3H,5H)-Pyrimidinetrione, 5-ethyl-5-[3-[(1-methylethyl)-telluro]propyl]-	MeOH	212.2(4.06),230.6(3.57)	101-0171-83S
$C_{12}H_{20}N_2S$			
2(1H)-Pyrazinethione, 3,6-bis(2-methylpropyl)-	EtOH	246(3.38),283(3.78)	142-0797-83
$C_{12}H_{20}O$			
Acetaldehyde, [4-(1,1-dimethylethyl)-cyclohexylidene]-, (aR)-(-)-	C_6H_{12}	231(4.19),321(1.64), 333(1.67),347(1.68), 363(1.57),381(1.20)	35-3252-83
	C_6H_{12}	231(4.19)	35-3264-83
Acetaldehyde, [5-methyl-2-(1-methylethyl)cyclohexylidene]-, [2S-(1E,2α,5β)]-	C_6H_{12}	236(4.10)	35-3264-83
$C_{12}H_{20}OS$			
Cyclohexanone, 3-(ethylthio)-2-methyl-5-(1-methylethenyl)-, (5R)-	C_6H_{12}	296(1.40)	49-0195-83
	MeOH	242(2.69),293s(1.60)	49-0195-83
(5S)-	MeOH	239s(3.75),290(1.95)	49-0195-83
$C_{12}H_{20}O_2$			
2,4-Decadienoic acid, ethyl ester	EtOH	265(4.35)	70-1897-83
$C_{12}H_{20}O_3$			
Cyclohexanecarboxylic acid, 5-(1,1-dimethylethyl)-1-methyl-2-oxo-, axial	aq dioxan-KOH	284(1.66)	44-4497-83
equatorial	aq dioxan-KOH	286(1.63)	44-4497-83
$C_{12}H_{20}O_3S_2$			
3(2H)-Thiophenone, 4-(octylthio)-, 1,1-dioxide	EtOH	203(3.88),307(3.71)	104-1447-83
	EtOH-KOH	273(3.79),328(3.68), 425(3.51)	104-1447-83
$C_{12}H_{20}O_4$			
1,4-Cyclohexadiene, 3,6-diethoxy-3,6-dimethoxy-	EtOH	207.5(3.21)	39-1255-83C
$C_{12}H_{20}O_5S$			
α-D-xylo-Hexofuranos-5-ulose, 6-deoxy-6-(dimethylsulfonio)-3-O-methyl-1,2-O-(1-methylethylidene)-	EtOH	261(3.91)	159-0139-83
$C_{12}H_{20}S$			
Bicyclo[2.2.1]heptane-2-thione, 1,3,3,7,7-pentamethyl-	benzene	490(0.78)	44-0214-83
$C_{12}H_{20}Sb_2$			
Distibine, tetrakis(1-methylethenyl)-	C_6H_{12}	208(4.54),290(3.99)	157-1573-83
$C_{12}H_{21}Bi$			
Bismuthine, tris(2-methyl-1-propenyl)-	C_6H_{12}	270s(3.57)	157-1859-83
$C_{12}H_{21}FO_8S_2$			
α-D-xylo-Hexofuranos-5-ulose, 6-deoxy-6-(dimethylsulfonio)-3-O-methyl-1,2-O-(1-methylethylidene)-, fluorosulfonate	EtOH	210(2.53)	159-0139-83

Compound	Solvent	$\lambda_{max}(\log \epsilon)$	Ref.
$C_{12}H_{21}N_2O$			
Methanaminium, N-[[3-[(dimethylamino)-methylene]-2-methoxy-1-cyclopenten-1-yl]methylene]-N-methyl-, perchlorate	MeOH	457(4.97)	104-1854-83
$C_{12}H_{21}N_2O_3$			
1H-Pyrrol-1-yloxy, 2,5-dihydro-3-[[(2-hydroxypropyl)amino]carbonyl]-2,2,5,5-tetramethyl-	MeOH	278(3)	28-0043-83B
$C_{12}H_{21}Sb$			
Stibine, tris(1-methyl-1-propenyl)-, (E,E,E)-	C_6H_{12}	248(3.71)	157-1573-83
Stibine, tris(2-methyl-1-propenyl)-	C_6H_{12}	<220(--)	157-1573-83
$C_{12}H_{22}N_2O_2$			
Hyponitrous acid, dicyclohexyl ester	heptane	226.0(3.86)	44-3728-83
$C_{12}H_{22}N_2O_7$			
β-D-xylo-Hexopyranoside, 3-hydroxypropyl 2,3,4,6-tetradeoxy-4-[(methoxycarbonyl)amino]-3-C-methyl-3-nitro-	CF_3CH_2OH	197(3.60)	39-1497-83C
$C_{12}H_{22}N_4O_4$			
1,6-Hexanediamine, N,N'-bis(1-methyl-2-nitroethenyl)-	CH_2Cl_2	338(4.18),358(4.18)	40-0088-83
$C_{12}H_{22}N_4O_4S_2$			
1,2,4,5-Tetrazine, 3,6-bis[(butylsulfonyl)methyl]-	MeOH	220(4.13),278(3.27), 531(2.64)	104-0446-83
$C_{12}H_{22}OS$			
Cyclohexanone, 3-(ethylthio)-2-methyl-5-(1-methylethyl)-, (R,S,S)-	C_6H_{12}	296(1.42)	49-0195-83
	MeOH	220(3.03),238(2.81), 290s(1.60)	49-0195-83
(S,S,S)-	MeOH	240(2.54),290(1.88)	49-0195-83
$C_{12}H_{22}O_2$			
2,4-Hexadiene, 1,6-dimethoxy-2,3,4,5-tetramethyl- (E,Z)-	hexane	220(3.61),240(3.18), 260(2.53)(end abs.)	24-0264-83
$C_{12}H_{23}NO$			
4-Heptanone, 3-[(diethylamino)methylene]-	MeOH	303(4.40)	131-0001-83K
(E-syn-E)-	C_6H_{12}	287(4.32)	131-0001-83K
4-Heptanone, 3-[[(1,1-dimethylethyl)-amino]methylene]-	MeOH	301(4.32)	131-0001-83K
(E-syn-E)-	C_6H_{12}	281(--)	131-0001-83K
(Z-syn-Z)-	C_6H_{12}	314(3.99)	131-0001-83K
1-Hepten-3-one, 1-[(1,1-dimethylethyl)amino]-2-methyl-	MeOH	299(4.36)	131-0001-83K
(E-syn-E)-	C_6H_{12}	278(--)	131-0001-83K
(Z-syn-Z)-	C_6H_{12}	315(4.06)	131-0001-83K
$C_{12}H_{23}NOS$			
Butanethioamide, 3,3-dimethyl-N,N-bis(1-methylethyl)-2-oxo-	MeOH	248(3.83),278(3.85), 309s(3.28),322s(3.06), 369(2.20)	5-1116-83

Compound	Solvent	λ_{max}(log ε)	Ref.
$C_{12}H_{23}O_4P$			
Phosphonium, diethylmethyl-, 2-ethoxy-1-(ethoxycarbonyl)-2-oxoethylide	hexane	220s(--),234(4.05)	99-0023-83
	EtOH	210(--),240(--)	99-0023-83
$C_{12}H_{24}N_4O_4S_2$			
1,2,4,5-Tetrazine, 3,6-bis[(butylsulfonyl)methyl]-1,2-dihydro-	MeOH	226(3.00),323(2.18)	104-0446-83
4H-1,2,4-Triazol-4-amine, 3,5-bis-[(butylsulfonyl)methyl]-	MeOH	208(4.09)	104-0446-83
$C_{12}H_{24}O_2Te$			
Pentanoic acid, 5-(hexyltelluro)-, methyl ester	MeOH	233.0(3.68)	101-0031-83M
$C_{12}H_{24}Si$			
Silane, (1-butyl-2,4-pentadienyl)-trimethyl-	THF	238(3.63)	157-0021-83
Silane, (3-isobutyl-1,4-pentadienyl)-trimethyl-	THF	194(3.78)	157-0021-83
Silane, (3-sec-butyl-1,4-pentadienyl)-trimethyl-	THF	196(3.95)	157-0021-83
Silane, (3-tert-butyl-1,4-pentadienyl)trimethyl-	THF	194(3.30)	157-0021-83
$C_{12}H_{24}Si_4$			
1,2,5,6-Tetrasilacycloocta-3,7-diyne, 1,1,2,2,5,5,6,6-octamethyl-	n.s.g.	226.5(4.21),236s(4.03)	35-3359-83
$C_{12}H_{25}N_2O_2$			
1-Piperidinyloxy, 4-[(2-hydroxypropyl)-amino]-2,2,6,6-tetramethyl-	MeOH	278(3)	28-0043-83B
$C_{12}H_{26}N_2$			
Diazene, bis(1,2,2-trimethylpropyl)-	dodecane	367(1.28)	24-1787-83
$C_{12}H_{26}N_2Si$			
1,3-Diaza-2-silacyclopent-4-ene, 1,3-bis(1,1-dimethylethyl)-2,2-dimethyl-	hexane	232(4.21),255(3.98), 392s(3.25),330s(2.8)	157-0903-83
$C_{12}H_{26}N_3O_3P$			
Phosphorus(1+), [1-(ethoxycarbonyl)-2-oxopropyl]tris(N-methylmethan-aminato)-, hydroxide, inner salt	hexane	215(3.86),250(3.98)	99-0023-83
$C_{12}H_{28}Sn$			
Stannane, tributyl-	isooctane	217(2.73),248s(0.64)	35-5665-83
$C_{12}H_{36}OSi_6$			
Oxahexasilacycloheptane, dodecamethyl-	C_6H_{12}	215.5(4.18),242.5(3.72)	157-0903-83
$C_{12}H_{36}O_2Si_6$			
1,3-Dioxa-2,4,5,6,7,8-hexasilacyclooctane, 2,2,4,4,5,5,6,6,7,7,8,8-dodecamethyl-	C_6H_{12}	217.3(4.20),237.2(3.89)	157-0903-83
$C_{12}H_{36}O_3Si_6$			
1,3,5-Trioxa-2,4,6,7,8,9-hexasilacyclononane, 2,2,4,4,6,6,7,7,8,8,9,9-dodecamethyl-	C_6H_{12}	209(4.13),229.5(3.82)	157-0903-83

Compound	Solvent	$\lambda_{max}(\log \epsilon)$	Ref.
$C_{12}H_{36}O_4Si_6$			
1,3,5,7-Tetraoxa-2,4,6,8,9,10-hexa-silacyclodecane, 2,2,4,4,6,6,8,8-9,9,10,10-dodecamethyl-	C_6H_{12}	220(3.96)	157-0903-83
$C_{12}N_6$			
1-Cyclopropene-1,2-diacetonitrile, α,α'-dicyano-3-(dicyanomethyl-ene)-, ion(2-)	H_2O	315(4.52)	88-1567-83

Compound	Solvent	λ_{max}(log ϵ)	Ref.
$C_{13}H_5NO_4$			
2H,5H-Pyrano[3,2-c][1]benzopyran-3-carbonitrile, 2,5-dioxo-	EtOH	379(4.19)	5-0695-83
	toluene	379(--)	5-0695-83
	DMSO	376(--)	5-0695-83
$C_{13}H_6BrCl_2NO$			
Benzoxazole, 2-(4-bromophenyl)-4,6-dichloro-	n.s.g.	311(4.52)	18-1514-83
$C_{13}H_6Br_2ClNO$			
Benzoxazole, 4,6-dibromo-2-(4-chlorophenyl)-	n.s.g.	316(4.51)	18-1514-83
$C_{13}H_6Br_3NO$			
Benzoxazole, 4,6-dibromo-2-(4-bromophenyl)-	n.s.g.	315(4.58)	18-1514-83
$C_{13}H_6Cl_2N_2OS_2$			
Benzenesulfenamide, N-(5,7-dichloro-6-oxo-2(6H)-benzothiazolylidene)-, (E)-	$CHCl_3$	584(4.42)	150-0480-83M +150-0028-83S
$C_{13}H_6Cl_3NO$			
Benzoxazole, 4,6-dichloro-2-(4-chlorophenyl)-	n.s.g.	310(4.50)	18-1514-83
$C_{13}H_6Cl_4O_3$			
Spiro[3,5-cyclohexadiene-1,2'-[3,6]-methano[2H]cyclopenta[b]furan]-2,7'-dione, 3,4,5,6-tetrachloro-3',3'a,6',6'a-tetrahydro-	MeCN	217(4.07),246(3.86), 314(3.45),348(3.34), 380(3.24)	142-1017-83
Spiro[3,5-cyclohexadiene-1,4'-[3]oxatricyclo[4.2.1.0^{2,5}]non[7]ene]-2,9'-dione, 3,4,5,6-tetrachloro-	MeCN	210(4.19),245(3.77), 335(3.41),380s(3.18)	142-1017-83
isomer	MeCN	201(4.09),221(4.01), 243s(3.78),330(3.49), 390(3.24)	142-1017-83
$C_{13}H_6N_4$			
Propanedinitrile, 2,2'-(2-methyl-2,5-cyclohexadiene-1,4-diylidene)bis-	MeCN	390(4.67)	104-0742-83
$C_{13}H_7Br_2NO$			
Benzoxazole, 4,6-dibromo-2-phenyl-	n.s.g.	310(4.46)	18-1514-83
$C_{13}H_7ClN_2OS_2$			
Benzenesulfenamide, N-(5-chloro-6-oxo-2(6H)-benzothiazolylidene)-, (E)-	$CHCl_3$	565(4.32)	150-0480-83M
Benzenesulfenamide, N-(7-chloro-6-oxo-2(6H)-benzothiazolylidene)-, (E)-	$CHCl_3$	568(4.38)	150-0480-83M
$C_{13}H_7Cl_2NO$			
Benzoxazole, 4,6-dichloro-2-phenyl-	n.s.g.	305(4.41)	18-1514-83
$C_{13}H_7F_6NO_2S$			
2,4-Pentanedione, 1,1,1,5,5,5-hexafluoro-3-(3-methyl-2(3H)-benzothiazolylidene)-	EtOH	232(3.72),292(3.92), 365(3.77)	39-0795-83C
$C_{13}H_7NO_2$			
Indeno[1,2-b]pyran-2-carbonitrile,	MeOH	207(4.37),246(4.01),	83-0264-83

Compound	Solvent	λ_{max}(log ϵ)	Ref.
4,5-dihydro-4-oxo- (cont.)		315(4.24)	83-0264-83
$C_{13}H_7NO_5$			
2H,5H-Pyrano[3,2-c][1]benzopyran-3-carboxamide, 2,5-dioxo-	EtOH	426(4.08)	5-0695-83
	toluene	373(--)	5-0695-83
	DMSO	369(--)	5-0695-83
$C_{13}H_7N_3O$			
1-Phenazinecarbonitrile, 2-hydroxy-	MeOH	265(4.62),291(4.06), 345(3.58),456(3.62)	64-0866-83B
$C_{13}H_7N_3O_5$			
Benzoxazole, 2-(3,5-dinitrophenyl)-	C_6H_{12}	317(4.13)	41-0595-83
$C_{13}H_8BrCs$			
Cesium, (2-bromo-9H-fluoren-9-yl)-	$(MeOCH_2)_2$	376(4.17),458(3.26), 488(3.32),525(3.14)	104-1592-83
$C_{13}H_8BrNO$			
Benzoxazole, 2-(4-bromophenyl)-	C_6H_{12}	306(4.48)	41-0595-83
$C_{13}H_8BrN_3OS$			
Pyrido[2,3-d]pyrimidin-4(1H)-one, 1-(4-bromophenyl)-2,3-dihydro-2-thioxo-	MeOH	281(3.51)	73-3315-83
$C_{13}H_8Br_2O$			
Methanone, bis(4-bromophenyl)-	MeOH	266(4.44)	12-0409-83
$C_{13}H_8Cl_2OS$			
Methanethione, bis(4-chlorophenyl)-, S-oxide	$CHCl_3$	269(4.19),336(4.16)	44-0214-83
$C_{13}H_8Cl_2S$			
Methanethione, bis(4-chlorophenyl)-	benzene	328(4.02),605(2.03)	44-0214-83
$C_{13}H_8Cl_4O_3$			
3,10-Methanobenzo[b]cyclopenta[e][1,4]-dioxepin-11-ol, 5,6,7,8-tetrachloro-3,8a,10,10a-tetrahydro-, (R)-	$CHCl_3$	293s(2.94),299(2.98)	142-0197-83
(S)-	$CHCl_3$	293s(3.13),298(3.20)	142-0197-83
1,4-Methanodibenzo[b,e][1,4]dioxin-11-ol, 6,7,8,9-tetrachloro-1,4,4a,10a-tetrahydro-	$CHCl_3$	292s(3.22),299(3.31)	142-0197-83
stereoisomer	$CHCl_3$	293(3.12),299(3.17)	142-0197-83
Spiro[3,5-cyclohexadiene-1,4'-[3]oxa-tricyclo[4.2.1.0^{2,5}]non[7]en]-2-one, 3,4,5,6-tetrachloro-9'-hydroxy-	MeCN	218(4.22),355(3.51)	142-0197-83
$C_{13}H_8F_4$			
1,1'-Biphenyl, 3-fluoro-4'-(trifluoro-methyl)-	hexane	250(4.16)	64-0621-83B
$C_{13}H_8N_2OS_2$			
Benzenesulfenamide, N-(6-oxo-2(6H)-benzothiazolylidene)-, (E)-	$CHCl_3$	554(4.34)	150-0480-83M +150-0028-83S
$C_{13}H_8N_2O_4$			
5H-[1]Benzopyrano[4,3-b]pyridine-3-carboxamide, 1,2-dihydro-2,5-dioxo-	EtOH	353(4.15)	5-0695-83
	toluene	360(--)	5-0695-83
	DMSO	358(--)	5-0695-83

Compound	Solvent	λ_{max}(log ε)	Ref.
2H-Pyrano[3,2-c]quinoline-3-carboxamide, 5,6-dihydro-2,5-dioxo-	EtOH	389(4.05)	5-0695-83
	toluene	407(--)	5-0695-83
	DMSO	400(--)	5-0695-83
$C_{13}H_8N_2S$			
Azuleno[2,1-d]thiazole-4-carbonitrile, 2-methyl-	MeOH	208(4.42),226s(4.16), 256s(4.00),308(4.81), 320(4.84),360s(3.82), 374(3.94),388s(3.63), 548(2.66),590s(2.64)	142-1263-83
Thiazolo[3,2-a]perimidine	MeOH	210(4.42),239(4.39), 338(4.21)	83-0728-83
$C_{13}H_8N_2S_2$			
Propanedinitrile, [benzo[b]thien-2-yl-(methylthio)methylene]-	EtOH	224(4.77),270(4.40), 346(4.16)	142-1793-83
$C_{13}H_8N_4$			
1,3,5,7-Nonatetraene-1,1,9-tricarbonitrile, 9-carbonimidoyl-	EtOH-acetone	650(4.71)	70-0780-83
$C_{13}H_8N_4O_2$			
Indeno[1,2-b]pyran-4(5H)-one, 2-(1H-1-tetrazol-5-yl)-	MeOH	205(4.01),212(3.99), 252(3.91),260(3.89), 290(3.91),310(4.00)	83-0264-83
$C_{13}H_8N_4O_3$			
1H-Pyrazolo[3,4-b]pyridine-3-carboxaldehyde, 5-nitro-1-phenyl-	DMF	278(4.17),317(3.88)	103-1163-83
$C_{13}H_8N_4O_3S$			
Pyrido[2,3-d]pyrimidin-4(1H)-one, 2,3-dihydro-1-(3-nitrophenyl)-2-thioxo-	MeOH	274(3.42)	73-3315-83
Pyrido[2,3-d]pyrimidin-4(1H)-one, 2,3-dihydro-1-(4-nitrophenyl)-2-thioxo-	MeOH	277(3.46)	73-3315-83
4H-Pyrido[3,2-e]-1,3-thiazin-4-one, 2-[(3-nitrophenyl)amino]-	MeOH	230(3.46)	73-3315-83
4H-Pyrido[3,2-e]-1,3-thiazin-4-one, 2-[(4-nitrophenyl)amino]-	MeOH	230(4.18)	73-3315-83
$C_{13}H_8N_4O_8S$			
Thieno[2,3-b]pyridinium, 7-hydroxy-, picrate	EtOH	236(4.81),326s(4.26), 358(4.39)	4-0213-83
$C_{13}H_8O$			
9H-Fluoren-9-one	EtOH	249(4.75),257(4.94), 276s(3.43),296(3.52), 307s(3.29),313s(3.24), 322s(3.12),328s(3.07), 380(2.42)	73-0112-83
$C_{13}H_8OS$			
9H-Xanthene-9-thione	CH_2Cl_2	578(1.40)	54-0083-83
$C_{13}H_8O_2$			
Benzo[f]chromone	MeOH	233(4.32),261(4.28), 270s(4.21),305(4.81), 340s(3.68)	118-0310-83

Compound	Solvent	λ_{max}(log ε)	Ref.
4H-Dibenzo[b,d]pyran-6-one	EtOH	225s(4.43),253(4.02), 260(4.04),271(4.02), 287(3.57),298(3.63), 316(3.62)	73-0112-83
	EtOH	316(3.82),326(3.60)	139-0059-83B
2H-Naphtho[1,2-b]pyran-2-one	EtOH	349(3.88),364(3.89)	139-0059-83B
2H-Naphtho[2,3-b]pyran-2-one	C_6H_{12}	323(4.32),360(3.20)	139-0059-83B
9H-Xanthen-9-one	EtOH	231s(4.60),239(4.63), 261(4.10),288s(3.64), 337(3.83),346s(3.72)	73-0112-83
$C_{13}H_8O_3$			
Indeno[1,2-b]pyran-3-carboxaldehyde, 4,5-dihydro-4-oxo-	dioxan	212(4.26),253(4.03), 261(4.03),283(4.08), 305(4.01)	83-0264-83
$C_{13}H_8O_6$			
Lamellicolic anhydride	EtOH	250(4.18),292(3.76), 352(4.00),368s(3.89)	78-2283-83
	EtOH-NaOH	251s(4.18),314(4.36), 372(4.04)	78-2283-83
$C_{13}H_8S_2$			
9H-Thioxanthene-9-thione	CH_2Cl_2	605(1.83)	54-0083-83
$C_{13}H_9BrClN_3OS$			
3-Pyridinecarboxamide, N-[[(4-bromo-phenyl)amino]thioxomethyl]-2-chloro-	MeOH	260(4.13)	73-3315-83
$C_{13}H_9BrN_2O$			
11H-Pyrido[2,1-b]quinazolin-11-one, 2-(bromomethyl)-	MeCN	369.2(4.16)	64-0248-83B
11H-Pyrido[2,1-b]quinazolin-11-one, 7-(bromomethyl)-	MeOH	325(4.100),349(4.190)	64-0248-83B
$C_{13}H_9BrO_2$			
Propenal, 3-(4-bromo-1-hydroxy-2-naphthalenyl)-	benzene	315(3.67),325(3.68), 338(3.63),355(3.60)	103-0146-83
	EtOH	290(3.82),310(3.67), 338(3.60),355(3.56), 385(3.18)	103-0146-83
	dioxan	328(3.67),338(3.68), 355(3.52),375(3.15)	103-0146-83
	acetone	322(3.62),338(3.62), 355(3.58),375(3.42)	103-0146-83
	MeCN	335(3.64),355(3.60), 365(3.36)	103-0146-83
	DMSO	310(3.87),390(3.26), 485(3.79)	103-0146-83
	CCl_4	320(3.63),332(3.63), 340(3.58),357(3.56)	103-0146-83
$C_{13}H_9ClN_4O_2$			
2,4(1H,3H)-Pteridinedione, 6-chloro-3-methyl-7-phenyl-	pH 4.0	227(4.32),244s(4.16), 285s(3.64),355(4.23)	5-0852-83
	pH 10.0	223(4.30),269s(4.22), 278(4.32),386(4.00)	5-0852-83
	MeOH	227(4.30),244s(4.11), 280s(3.72),355(4.20)	5-0852-83
2,4(1H,3H)-Pteridinedione, 7-chloro-3-methyl-6-phenyl-	pH 4.0	246(4.22),268(4.17), 348(4.09)	5-0852-83

Compound	Solvent	λmax(log ε)	Ref.
2,4(1H,3H)-Pteridinedione, 7-chloro-3-methyl-6-phenyl- (cont.)	pH 12.0	254(4.25),290(4.30), 378(4.00)	5-0852-83
	MeOH	248(4.20),270(4.18), 349(4.06)	5-0852-83
$C_{13}H_9ClN_4O_3$			
2,4(1H,3H)-Pteridinedione, 6-chloro-3-methyl-7-phenyl-	pH 2.0	251(4.44),360(3.95), 370s(3.91)	5-0852-83
	pH 8.0	269(4.28),282(4.30), 402(3.85)	5-0852-83
	MeOH	259(4.32),364(3.73), 374s(3.71)	5-0852-83
$C_{13}H_9ClN_4O_3S$			
3-Pyridinecarboxamide, 2-chloro-N-[[(3-nitrophenyl)amino]thioxomethyl]-	MeOH	244(4.11)	73-3315-83
$C_{13}H_9ClO_3$			
2H-Pyrano[3,2-d]benzoxepin-2-one, 3-chloro-5,6-dihydro-	EtOH	215.5(4.20),247(3.78), 356(4.07)	4-0539-83
$C_{13}H_9ClS$			
Methanethione, (4-chlorophenyl)phenyl-	benzene	322(3.95),603(1.94)	49-0214-83
$C_{13}H_9Cl_2NO_2$			
Benzamide, 2-chloro-N-(2-chlorophenyl)-N-hydroxy-	n.s.g.	260(4.11)	42-0686-83
Benzamide, 2-chloro-N-(3-chlorophenyl)-N-hydroxy-	n.s.g.	265(4.00)	42-0686-83
$C_{13}H_9FO_2S$			
Benzoic acid, 2-[(2-fluorophenyl)-thio]-	MeOH	253(3.97),275s(3.70), 313(3.66)	73-1187-83
$C_{13}H_9N$			
Acridine	EtOH	250(5.03),280(3.17), 324s(3.56),331s(3.63), 339s(3.87),348s(3.96), 356(4.03),380s(3.50)	73-0112-83
Benzo[f]quinoline	EtOH	266(4.06),316(3.18), 331(3.41),347(3.54)	162-0158-83
Cyclobut[a]azulene-8-carbonitrile, 2a,8a-dihydro-	C_6H_{12}	230(4.13),270s(3.80), 347(4.11),361(4.19), 380(3.98),436(2.64), 466(2.67),504(2.61), 546(2.43),598(2.08), 658(1.37)	35-6718-83
1-Heptalenecarbonitrile	C_6H_{12}	267(4.34),362(3.69)	35-6718-83
$C_{13}H_9NO$			
1,2-Benzisoxazole, 3-phenyl-	n.s.g.	287(3.85)	18-1514-83
Benzoxazole, 2-phenyl-	C_6H_{12}	260s(--),270s(--), 280s(--),287(--), 292(--),299(4.45), 305(--),314(--)	41-0595-83
	n.s.g.	299(4.33)	18-1514-83
$C_{13}H_9NOS$			
Azuleno[2,1-d]thiazole-4-carboxaldehyde, 2-methyl-	MeOH	220(4.34),235(4.27), 264(4.13),322s(4.59),	142-1263-83

Compound	Solvent	$\lambda_{max}(\log \epsilon)$	Ref.
(cont.)		334(4.70),379(3.83), 399(3.87),413(2.84), 548(3.66)	142-1263-83
Isoxazole, 3-phenyl-5-(2-thienyl)-	MeOH	246(4.18),292(4.28)	142-0501-83
Isoxazole, 5-phenyl-3-(2-thienyl)-	MeOH	272(4.45)	142-0501-83
$C_{13}H_9NOSe$			
1,2-Benzisoselenazole, 3-phenyl-, 2-oxide	EtOH	232(4.42),309(4.03)	78-0831-83
$C_{13}H_9NO_2$			
Benzo[b]cyclohept[e][1,4]oxazin-6(11H)-one	MeOH	205(4.40),228(4.34), 260(4.20),270s(4.15), 292(4.10),310s(4.04), 323s(3.91),400(3.88)	18-2756-83
	MeOH-HCl	205(4.35),228(4.40), 281(4.28),418(3.93)	18-2756-83
	MeOH-NaOH	260(4.20),290(4.10), 418(3.88)	18-2756-83
Benzo[b]cyclohept[e][1,4]benzoxazin-10(11H)-one	MeOH	205(4.24),227(4.28), 259(4.21),270(4.21), 284(4.05),305s(3.72), 320s(3.59),415s(3.62), 483(3.88)	18-2756-83
	MeOH-HCl	205(4.15),227(4.29), 260s(4.14),271(4.20), 284(4.19),320s(3.59), 420s(3.68),480(3.83)	18-2756-83
	MeOH-NaOH	213(4.41),259(4.17), 270(4.18),284s(4.05), 305s(3.72),320s(3.59), 415s(3.60),483(3.83)	18-2756-83
10H-Phenoxazine-1-carboxaldehyde	MeOH	207(4.28),227(4.43), 275(3.88),310(3.48), 435(3.74)	18-2756-83
$C_{13}H_9NO_3$			
4H-Furo[3,2-b]pyrrole-5-carboxylic acid, 2-phenyl-	MeOH	233(3.84),337(4.80)	73-0772-83
Indeno[1,2-b]pyran-2-carboxaldehyde, 4,5-dihydro-4-oxo-, 2-oxime	MeOH	203(4.28),220(4.23), 255(4.18),263(4.17), 295(4.25),307(4.27)	83-0264-83
$C_{13}H_9NO_4$			
5H-[1]Benzopyrano[2,3-b]pyridin-5-one, 6-hydroxy-8-methoxy-	EtOH	209(4.38),225(4.31), 256(4.15),283s(3.83), 303(4.04),355s(3.70), 390s(3.13)	39-0219-83C
Naphtho[2,1-b]furan, 7-methoxy-2-nitro-	C_6H_{12}	320(4.13),342(4.12), 378(4.16),397(4.03)	41-0363-83
	EtOH	335s(3.85),385(4.09)	41-0363-83
	DMF	350s(3.78),392(4.11)	41-0363-83
	DMSO	350s(3.78),392(4.11)	41-0363-83
$C_{13}H_9NO_6$			
2-Naphthalenecarboxylic acid, 5,6-dihydro-7-nitro-5,6-dioxo-, ethyl ester	EtOH	207(4.42),245(3.32), 265(4.30),420s(3.04)	103-0876-83
$C_{13}H_9N_2S_3$			
Thiazolo[3,2-b]isothiazolium, 7-cyano-	$HClO_4$	249(4.38),267s(3.99),	49-0999-83

Compound	Solvent	$\lambda_{max}(\log \epsilon)$	Ref.
6-(methylthio)-3-phenyl-, perchlorate		356(4.19)	49-0999-83
$C_{13}H_9N_3$			
Pyrido[2,3-d]pyridazine, 5-phenyl-	EtOH	232s(4.08),281(3.78)	95-0631-83
$C_{13}H_9N_3O$			
Pyrido[2,3-d]pyridazine, 5-phenyl-, 7-oxide	EtOH	253(4.13),295(4.06), 365(3.34)	95-0631-83
$C_{13}H_9N_3OS$			
5H-Pyrido[3,4-b][1,4]benzothiazine-4-carbonitrile, 2,3-dihydro-2-methyl-3-oxo-	MeOH	253(4.64),300(4.02)	87-0845-83
5H-Pyrido[3,4-b][1,4]benzothiazine-4-carbonitrile, 2,3-dihydro-5-methyl-3-oxo-	MeOH	251(4.53)	87-0845-83
1H-Pyrido[2,3-d]pyrimidin-4-one, 2,3-dihydro-1-phenyl-2-thioxo-	MeOH	270(3.38)	73-3315-83
4H-Pyrido[3,2-e]-1,3-thiazin-4-one, 2-(phenylamino)-	MeOH	237(4.31)	73-3315-83
$C_{13}H_9N_3OSe$			
4H-Pyrido[3,2-e]-1,3-selenazin-4-one, 2-(phenylamino)-	MeOH	211(4.23),273(4.46)	73-3567-83
$C_{13}H_9N_3O_2$			
1,2,4-Triazolo[4,3-a]pyridin-3(2H)-one, 2-benzoyl-	THF	358(3.63)	18-2969-83
$C_{13}H_9N_3O_2S$			
2-Thiazolecarboxylic acid, 4-(3-methyl-2-quinoxalinyl)-	EtOH	240(3.97),264s(3.57), 338(3.48)	106-0829-83
$C_{13}H_9N_3O_3$			
Benzo[1,2-b:3,4-b']dipyrrole-8-carboxaldehyde, 1,6-dihydro-3-(2-nitroethenyl)-	EtOH	238(4.29),285(4.16), 280s(4.13),311(3.98)	104-0385-83
3-Quinolinecarboxylic acid, 2-(diazoacetyl)-, methyl ester	MeOH	208(4.11),243(4.27), 288(3.67),294s(3.66), 313s(3.55)	103-0886-83
$C_{13}H_9N_5O_3$			
Imidazo[1,2-a]pyridine-2,3-dione, 3-[(4-nitrophenyl)hydrazone]	EtOH	264(4.76),410(3.72)	4-0639-83
1H-Pyrazolo[4,3-b]pyridine-3-carboxamide, 6-nitro-1-phenyl-	DMF	307(4.01),393(4.08)	103-1163-83
$C_{13}H_{10}$			
1H-Cyclobuta[de]naphthalene-, 1-ethylidene-	EtOH	218(4.87),258(4.22), 309(3.21),321(2.97)	35-6104-83
$C_{13}H_{10}BrClN_2$			
Benzenecarbohydrazonoyl chloride, N-(4-bromophenyl)-	EtOH	240(4.13),304(4.15), 338(4.30)	70-1439-83
$C_{13}H_{10}BrO_3P$			
10H-Phenoxaphosphine, 1-bromo-10-hydroxy-8-methyl-, 10-oxide	MeOH	205(4.23),222(4.26), 246s(3.83),282(3.23), 292(3.41),301(3.51)	4-1601-83
10H-Phenoxaphosphine, 7-bromo-10-hydroxy-2-methyl-, 10-oxide	MeOH	206s(4.22),225(4.34), 290(3.36),295(3.44)	4-1601-83

Compound	Solvent	λ_{max}(log ε)	Ref.
$C_{13}H_{10}ClNO_2$			
Benzamide, 3-chloro-N-(2-hydroxy-phenyl)-	EtOH	294.5(3.94)	39-0141-83C
$C_{13}H_{10}ClNO_3$			
2H-Furo[2,3-h]-1-benzopyran-2-one, 3-chloro-4-(dimethylamino)-	EtOH	217.5(4.34),243s(4.39), 247.5(4.40),295s(3.96), 328(4.09)	111-0113-83
$C_{13}H_{10}ClN_3OS$			
Benzamide, N-[[(6-chloro-2-pyridinyl)-amino]thioxomethyl]-	MeOH	250(4.28)	73-3315-83
$C_{13}H_{10}ClN_3OSe$			
3-Pyridinecarboxamide, 2-chloro-N-[(phenylamino)selenoxomethyl]-	MeOH	215(4.16),250(4.08)	73-3567-83
$C_{13}H_{10}ClO_3P$			
10H-Phenoxaphosphine, 1-chloro-10-hy-droxy-8-methyl-, 10-oxide	MeOH	219(4.34),245(3.97), 279s(3.25),290s(3.44), 300(3.60)	4-1601-83
10H-Phenoxaphosphine, 7-chloro-10-hy-droxy-2-methyl-, 10-oxide	MeOH	222(4.56),246s(4.07), 288s(3.54),298(3.67)	4-1601-83
$C_{13}H_{10}CsN$			
Cesium, (3-methyl-9H-indeno[2,1-c]-pyridin-9-yl)-	$(MeOCH_2)_2$	359.5(4.10),497(3.28), 530(3.15)	104-1592-83
$C_{13}H_{10}F_3NO$			
2-Propanone, 1,1,1-trifluoro-3-(1-meth-yl-2(1H)-quinolinylidene)-	EtOH	222(4.40),298(4.13), 410(4.41),428(4.32)	39-0795-83C
$C_{13}H_{10}N_2$			
1H-Benzimidazole, 2-phenyl-	C_6H_{12}	302(4.25)	59-0609-83
	H_2O	298(4.37)	59-0609-83
	MeOH	301(4.39)	59-0609-83
	EtOH	302(4.36)	59-0609-83
	MeCN	301(4.28)	59-0609-83
$C_{13}H_{10}N_2OS$			
Azuleno[2,1-d]thiazole-4-carboxalde-hyde, 2-methyl-, oxime	MeOH	234(4.35),256(4.16), 319(4.65),387(3.81), 404(3.76),420(3.58), 524(2.50),630(2.54), 680s(2.45)	142-1263-83
Phenol, 5-amino-2-(2-benzothiazolyl)-	EtOH	357(4.57)	4-1517-83
Thiazolo[3,2-a]pyrimidin-4-ium, 3-hy-droxy-7-methyl-2-phenyl-, hydrox-ide, inner salt	MeCN	488(4.02)	124-0857-83
5H-Thiazolo[3,2-a]pyrimidin-5-one, 7-methyl-3-phenyl-	EtOH	265(2.42),325(5.80)	2-0815-83
$C_{13}H_{10}N_2OS_2$			
2-Propenamide, 3-benzo[b]thien-2-yl-2-cyano-3-(methylthio)-	EtOH	228(3.74),262(4.35), 268(4.35),324(4.24)	142-1793-83
$C_{13}H_{10}N_2O_2$			
1H-Benz[de]isoquinoline-1,3(2H)-dione, 6-(methylamino)-	MeOH	440(4.04)	104-2273-83
Phenazine, 2-methoxy-, 5-oxide	MeOH	270(4.96),378(4.11)	64-0866-83B
Phenol, 5-amino-2-(2-benzoxazolyl)-	EtOH	337(4.58)	4-1517-83

Compound	Solvent	λ_{max}(log ϵ)	Ref.
$C_{13}H_{10}N_2O_2$			
3H-Phenoxazin-3-one, 2-(methylamino)-	MeOH	205(4.56),238(4.57), 270s(4.26),420(4.40), 435(4.41)	18-2756-83
	MeOH-HCl	238(4.57),465(4.02), 520(3.90)	18-2756-83
$C_{13}H_{10}N_2O_3$			
Benzamide, N-(4-nitrophenyl)-	EtOH	232(4.10),323(4.26)	104-2104-83
Benzoic acid, 4-[(4-hydroxyphenyl)-azo]-	C_6H_{12}	257(4.10),368(4.49), 440s(3.15)	59-0729-83
	EtOH	243(4.09),363(4.48), 438s(3.22)	59-0729-83
	50% EtOH	430s(3.37)	59-0729-83
	ether	254(4.13),364(4.51), 434(3.09)	59-0729-83
	DMF	370(4.47),436s(3.24)	59-0729-83
	50% DMF	430s(3.39)	59-0729-83
	$CHCl_3$	365(4.45),440s(3.23)	59-0729-83
Benzo[b][1,8]naphthyridin-5(10H)-one, 6,8-dihydroxy-10-methyl-	MeOH	202s(4.33),216s(4.38), 226s(4.42),235(4.46), 271(4.72),328(3.97), 382(3.86)	39-0219-83C
Phenazine, 2-methoxy-, 5,10-dioxide	MeOH	250(4.10),286(4.66), 403(3.87),482(3.93)	64-0866-83B
Pyrano[2,3-c]pyrazole-3,6-dione, 1,2-dihydro-4-methyl-1-phenyl-	MeOH	250(4.30),315(4.18)	20-0451-83
5H-Pyrrolo[2,1-b][3]benzazepin-11-one, 6,11-dihydro-9-nitro-	EtOH	208(4.18),246(4.25), 340(4.06)	44-3220-83
$C_{13}H_{10}N_2O_3S$			
Acetic acid, (3-oxo-5H-thiazolo[2,3-b]quinazolin-2(3H)-ylidene)-, methyl ester	MeOH	213(4.37),362(4.26)	83-0569-83
$C_{13}H_{10}N_4OS$			
2-Thiazolecarboxamide, N-methyl-4-(2-quinoxalinyl)-	EtOH	248(3.93),266s(3.79), 293s(3.49),342(3.49), 354s(3.44)	106-0829-83
2-Thiazolecarboxamide, 4-(3-methyl-2-quinoxalinyl)-	EtOH	243(4.17),266s(3.71), 289s(3.50),338(3.56)	106-0829-83
$C_{13}H_{10}N_4O_2$			
2,4(1H,3H)-Pteridinedione, 3-methyl-7-phenyl-	pH 5.0	223(4.37),272s(3.85), 349(4.32)	5-0852-83
	pH 11.0	233(4.34),272(4.28), 373(4.07)	5-0852-83
$C_{13}H_{10}N_4O_3$			
2,4,6(3H)-Pteridinetrione, 3-methyl-7-phenyl-	pH 3.0	259(4.08),293(3.83), 397(4.11)	5-0852-83
	pH 8.6	225(4.25),268(4.17), 405(4.09)	5-0852-83
	pH 13.0	225(4.23),275(4.25), 420(4.00)	5-0852-83
2,4,7(3H)-Pteridinetrione, 3-methyl-6-phenyl-	pH 1.0	232(4.11),282(4.03), 349(4.28)	5-0852-83
	pH 6.0	234s(4.15),288(4.06), 348(4.29)	5-0852-83
	pH 13.0	230(4.61),272(4.03), 291(4.05),358(4.30)	5-0852-83

Compound	Solvent	$\lambda_{max}(\log \epsilon)$	Ref.
Pyrazolo[1,5-d][1,2,4]triazine-2-carboxylic acid, 6,7-dihydro-7-oxo-4-phenyl-, methyl ester	EtOH	238(4.20),282(4.11)	44-1069-83
$C_{13}H_{10}N_4O_4$			
2,4,6(3H)-Pteridinetrione, 1,5-dihydro-3-methyl-7-phenyl-, 8-oxide	pH 2.0	232(4.30),257(4.26), 320s(3.60),390(4.02)	5-0852-83
	pH 6.5	229(4.35),267(4.23), 320s(3.52),410(4.01)	5-0852-83
	pH 12.0	233(4.33),281(4.34), 325s(3.46),446(4.01)	5-0852-83
$C_{13}H_{10}N_4O_4S_2$			
2-Naphthalenesulfonic acid, 6-amino-4-hydroxy-3-(2-thiazolylazo)-	H_2O	513(4.05)	7-0155-83
	HCOOH-DMF	503(3.90)	7-0155-83
$C_{13}H_{10}N_4O_7S_3$			
2,7-Naphthalenedisulfonic acid, 5-amino-4-hydroxy-3-(2-thiazolylazo)-	H_2O	554(4.15)	7-0155-83
	HCOOH-DMF	580(3.70)	7-0155-83
$C_{13}H_{10}N_6$			
1H-Benzotriazole, 1-(2H-benzotriazol-2-ylmethyl)-	EtOH	279(4.23),286(4.19)	142-1787-83
1H-Benzotriazole, 1,1'-methylenebis-	EtOH	254(4.54),282(3.86)	142-1787-83
2H-Benzotriazole, 2,2'-methylenebis-	EtOH	282(4.37),289(4.37)	142-1787-83
7H-Pyrrolo[2,3-d]pyrimidine, 7,7'-methylenebis-	MeOH	270(3.89)	5-1576-83
$C_{13}H_{10}O$			
8a,4a-Propenonaphthalen-11-one	MeCN	216(4.20),270(3.32)	35-2800-83
$C_{13}H_{10}OS$			
6H-Dibenzo[b,d]thiopyran-6-ol	MeOH	238(4.26),251(4.25), 271(4.01),313(3.42)	12-0795-83
$C_{13}H_{10}O_2$			
Benzo[f]chromanone	MeOH	240(4.31),262(3.89), 315(3.81),353(3.62)	118-0310-83
4H-Benz[3,4]cyclohepta[1,2-b]furan-4-one, 1,2-dihydro-	EtOH	240(4.34),247.5(4.33), 252(4.33),280(4.69), 292.5(4.23),320(3.70), 336.5(3.74),370(3.66), 380s(3.63)	39-0285-83C
Methanone, (4-hydroxyphenyl)phenyl-	EtOH	228(4.0),295(4.2)	65-0243-83
2-Propenal, 3-(4-hydroxy-1-naphthalenyl)-	EtOH	375(3.89),468(3.04)	104-1864-83
	acetone	363(4.24)	104-1864-83
	DMSO	375(3.79),485(3.08)	104-1864-83
	CCl_4	357(3.91)	104-1864-83
$C_{13}H_{10}O_2S$			
1,3-Propanedione, 1-phenyl-3-(2-thienyl)-	MeOH	258(3.98),355(4.40)	142-0501-83
$C_{13}H_{10}O_3$			
2H,8H-Benzo[1,2-b:3,4-b']dipyran-2-one, 8-methyl-	EtOH	217(4.415),287s(--), 294(4.089),330(4.113)	111-0009-83
2H,8H-Benzo[1,2-b:5,4-b']dipyran-2-one, 8-methyl-	EtOH	225(4.364),267(4.291), 348(4.088)	111-0009-83
7H-Furo[3,2-g][1]benzopyran, 2-acetyl-7-methyl-	EtOH	239(4.35),262(4.156), 308(3.978),350(4.206)	111-0015-83

Compound	Solvent	λ_{max}(log ε)	Ref.
7H-Furo[2,3-f][1]benzopyran-2-carboxaldehyde, 7-methyl-	EtOH	223s(--),227(4.253), 290(4.226),341(4.166)	111-0015-83
7H-Furo[3,2-g][1]benzopyran-2-carboxaldehyde, 7-methyl-	EtOH	240(4.409),260s(--), 312(4.050),357(4.260)	111-0015-83
Indeno[1,2-b]pyran-4(5H)-one, 2-(hydroxymethyl)-	MeOH	203(4.40),289(4.27), 299(4.29)	83-0264-83
$C_{13}H_{10}O_5$			
1-Benzoxepin-3-carboxylic acid, 7-formyl-2,5-dihydro-5-oxo-, methyl ester	$CHCl_3$	253.5(4.41),270s(4.28), 340(3.32)	39-1719-83C
$C_{13}H_{10}S$			
Azuleno[1,2-b]thiophene, 2-methyl-	C_6H_{12}	284(4.47),311(4.69), 322(4.73),350(3.42), 366(3.66),385(3.80), 405(3.90),627(2.62), 663(2.57),686(2.59), 736(2.24),763(2.28)	138-1721-83
Methanethione, diphenyl-	CH_2Cl_2	590(2.00)	54-0083-83
Thiophene, 2-phenyl-5-(1-propynyl)-	MeOH	310(4.22)	100-0646-83
$C_{13}H_{11}BrN_2$			
Benzaldehyde, (4-bromophenyl)hydrazone	EtOH	240(4.12),316(4.24), 350(4.43)	70-1439-83
Benzaldehyde, 4-bromo-, phenylhydrazone	EtOH	244(4.20),305(4.00), 357(4.42)	70-1439-83
$C_{13}H_{11}BrN_2S$			
Benzenecarbothioamide, 2-amino-5-bromo-N-phenyl-	EtOH	318(4.00),372(3.43)	97-0215-83
$C_{13}H_{11}Cl$			
1,1'-Biphenyl, 3-chloro-6-methyl-	n.s.g.	235(3.46)	39-0859-83B
1,1'-Biphenyl, 4-chloro-2-methyl-	n.s.g.	240(3.26)	39-0859-83B
$C_{13}H_{11}ClN_2$			
Benzaldehyde, (4-chlorophenyl)hydrazone	isooctane	234(4.24),294(4.06), 328(4.39)	70-0485-83
Benzaldehyde, 4-chloro-, phenylhydrazone	EtOH	244(4.18),304(4.01), 356(4.39)	70-1439-83
Benzenecarbohydrazonoyl chloride, N-phenyl-	EtOH	234(4.22),295(4.03), 333(4.33)	70-1439-83
$C_{13}H_{11}ClN_2O_2$			
Pyrazinol, 6-chloro-3-methyl-5-phenyl-, acetate	EtOH	241(4.01),250(4.01), 303(3.99)	4-0311-83
Pyrazinol, 6-chloro-5-methyl-3-phenyl-, acetate	EtOH	251(4.03),307(3.95)	4-0311-83
$C_{13}H_{11}ClN_2O_3S$			
2-Propenoic acid, 3-[(4-chlorobenzoyl)amino]-2-cyano-3-(methylthio)-, methyl ester	EtOH	250(3.76),283(3.92), 327(3.79)	142-1745-83
5-Pyrimidinecarboxylic acid, 2-(4-chlorophenyl)-1,4-dihydro-6-(methylthio)-4-oxo-, methyl ester	EtOH	218(4.42),257(4.23), 310(4.51)	142-1745-83
$C_{13}H_{11}ClN_2O_4$			
1H-Indole-2,3-dione, 1-(3-chloro-2-butenyl)-7-methyl-5-nitro-	EtOH	214(4.42),260(3.68), 335(4.18)	103-0286-83

Compound	Solvent	$\lambda_{max}(\log \epsilon)$	Ref.
$C_{13}H_{11}ClN_4O$			
4H-Pyrazolo[3,4-d]pyrimidin-4-one, 5-(2-chlorophenyl)-1,5-dihydro-1,6-dimethyl-	EtOH	211.5(4.42),255(3.85)	11-0171-83B
4H-Pyrazolo[3,4-d]pyrimidin-4-one, 5-(3-chlorophenyl)-1,5-dihydro-1,6-dimethyl-	EtOH	213(4.35),256(3.88)	11-0171-83B
$C_{13}H_{11}ClO_3$			
2H-1-Benzopyran-2-one, 8-(2-chloro-2-propenyl)-7-hydroxy-4-methyl-	EtOH	248(3.62),258(3.65), 326(4.19)	44-2709-83
$C_{13}H_{11}F_3O$			
Naphthalene, 2-methoxy-5-methyl-1-(trifluoromethyl)-	MeOH	224(4.67),236(4.48), 286(3.69),297(3.65), 323(3.42),336(3.46)	23-0368-83
$C_{13}H_{11}N$			
Pyridine, 4-(2-phenylethenyl)-, rheniumtricarbonyl complex, cis	MeCN	330(4.42)	35-7241-83
trans	MeCN	330(4.69)	35-7241-83
$C_{13}H_{11}NO$			
9H-Carbazol-3-ol, 6-methyl- (glycozolinine)	EtOH	224(3.92),254(4.02), 269(3.96),298(4.04)	102-1064-83
4H-Furo[3,2-b]pyrrole, 2-(4-methylphenyl)-	MeOH	205(3.97),320(4.52)	73-0772-83
4H-Furo[3,2-b]pyrrole, 4-methyl-2-phenyl-	MeOH	240(4.18),312(4.64)	73-0772-83
Methanone, (4-aminophenyl)phenyl-	MeOH	242(4.23),335(4.32)	12-0409-83
$C_{13}H_{11}NO_2$			
1,2-Benzisoxazol-4(5H)-one, 6,7-dihydro-6-phenyl-	EtOH	216s(4.07),230s(3.98)	4-0645-83
Cyclohexanecarbonitrile, 2,6-dioxo-4-phenyl-	EtOH	231(3.96),272(4.30)	4-0645-83
$C_{13}H_{11}NO_3$			
Furo[2,3-b]quinoline, 4,8-dimethoxy-	n.s.g.	238(4.76),332(3.88), 370(3.89)	162-0567-83
$C_{13}H_{11}NO_4$			
7H-Furo[2,3-f][1]benzopyran, 7,7-dimethyl-2-nitro-	EtOH	223(4.42),228s(--), 255(3.804),301(4.034)	111-0015-83
7H-Furo[3,2-g][1]benzopyran, 7,7-dimethyl-2-nitro-	EtOH	226(4.321),250s(--)	111-0015-83
1H-Pyrrole-3-carboxylic acid, 4,5-dihydro-4,5-dioxo-2-phenyl-, ethyl ester	dioxan	275(4.0),398(3.7)	94-0356-83
$C_{13}H_{11}NO_4S$			
Acetamide, N-(3-acetoxy-4-oxo-4H-1-benzothiopyran-2-yl)-	MeOH	236(4.33),274(4.31), 380(4.21)	83-0921-83
$C_{13}H_{11}NS$			
10H-Phenothiazine, 10-methyl-, radical cation	MeCN	441(3.56),512(3.96), 760(3.08),843(3.08)	152-0105-83
2-Propenethioamide, 3-(2-naphthalenyl)-, cis	benzene	298(4.18),308(4.20), 400s(2.90)	151-0183-83C
	EtOH	258(4.35),264(4.34), 292(4.22),304(4.20)	151-0183-83C

Compound	Solvent	$\lambda_{max}(\log \epsilon)$	Ref.
2-Propenethioamide, 3-(2-naphthalenyl)-, trans	benzene	307(4.26),317(4.29), 348(4.18),360(4.15), 450s(2.45)	151-0183-83C
	EtOH	254s(4.24),269(4.35), 278(4.36),305(4.33), 313(4.35),342(4.19), 358(4.15)	151-0183-83C
$C_{13}H_{11}N_2$			
Phenazinium , N-methyl-	1.2M $HClO_4$	268(4.96),370s(4.18), 387(4.46),434(3.51), 464s(3.30)	125-0655-83
radical	1.2M $HClO_4$	256(4.72),290(3.34), 310(3.63),370(3.90), 434s(4.00),445(4.04), 460s(3.90),594s(3.00), 644(3.15)	125-0655-83
$C_{13}H_{11}N_2S$			
Thiazolo[3,2-a]pyridinium, 3-amino-2-phenyl-, perchlorate	MeCN	226s(4.27),264(3.90), 353(4.14)	118-0582-83
Thiazolo[3,4-a]pyrimidin-5-ium, 6-methyl-8-phenyl-, perchlorate	MeCN	243(3.31),267(2.94), 382(2.76)	103-0037-83
$C_{13}H_{11}N_3$			
3,6-Acridinediamine	pH 7.0	440(4.61)	156-0177-83B
1H-Benzimidazole, 1-methyl-2-(2-pyridinyl)-	MeOH	240s(4.04),300(4.28)	4-1481-83
1H-Benzimidazole, 2-(2-pyridinylmethyl)-	MeOH	225(3.97),255(3.97), 260s(3.94),325(3.83)	4-1481-83
$C_{13}H_{11}N_3O$			
Ethanone, 1-(3-diazo-5-methyl-2-phenyl-3H-pyrrol-4-yl)-	EtOH	244(4.08)	142-0255-83
Phenol, 5-amino-2-(1H-benzimidazol-2-yl)-	EtOH	338(4.53)	4-1517-83
Pyrazolo[1,5-a]pyrimidin-7(1H)-one, 1-methyl-5-phenyl-	n.s.g.	241(4.49),280(3.78), 319(3.59)	56-1219-83
2-Pyrimidineacetonitrile, 4-methyl-6-phenyl-, 1-oxide	MeCN	250(4.52),279(3.95), 343(3.55)	44-0575-83
4H-Pyrrolo[3,4-c]pyridazin-4-one, 1,8-dihydro-5-methyl-7-phenyl-	EtOH	217(4.20),222(4.15), 297(4.28)	142-0255-83
$C_{13}H_{11}N_3O_2$			
Benzaldehyde, (4-nitrophenyl)hydrazone	isooctane	293(3.87),365(4.63)	104-2104-83
	EtOH	300(3.79),406(4.46)	104-2104-83
	dioxan	298(3.47),394(4.59)	104-2104-83
	MeCN	299(3.87),402(4.50)	104-2104-83
	CCl_4	376(--),395(--)	104-2104-83
Benzaldehyde, 4-nitro-, phenylhydrazone	hexane	394(4.42)	140-0215-83
	EtOH	422(--)	140-0215-83
Benzoic acid, 4-[(4-aminophenyl)azo]-	H_2O	390(4.23)	80-0903-83
	MeOH	268(4.11),411(4.41)	80-0903-83
4H-Furo[3,2-b]pyrrole-5-carboxylic acid, 2-phenyl-, hydrazide	MeOH	279(3.24),333(3.47)	73-1878-83
Pyrimido[4,5-b]quinoline-2,4(3H,10H)-dione, 3,10-dimethyl-	pH 5.0	390(4.04)	12-1873-83
$C_{13}H_{11}N_3O_5$			
Benzenamine, N-(2-methoxyphenyl)-2,4-dinitro-	benzene	360(4.18)	44-1613-83

Compound	Solvent	$\lambda_{max}(\log \epsilon)$	Ref.
$C_{13}H_{11}N_3S$			
Quinoxaline, 2-(2,4-dimethyl-5-thiazolyl)-	EtOH	256s(3.70),275(3.74), 285s(3.73),355(3.64)	106-0829-83
$C_{13}H_{11}N_4$			
Benzenediazonium, 4-[(4-methylphenyl)-azo]-, tetrafluoroborate	H_2O	278(4.07),372(4.40), 470s(3.25)	104-1493-83
$C_{13}H_{11}N_4O$			
Benzenediazonium, 4-[(4-methoxyphenyl)-azo]-, tetrafluoroborate	H_2O	222(4.12),268(4.09), 277(4.08),405(4.43)	104-1493-83
$C_{13}H_{11}N_5$			
Imidazo[1,2-a]pyrimidine, 2-methyl-3-(phenylazo)-	EtOH	263(4.30),360(3.86), 420(3.78)	4-0639-83
$C_{13}H_{11}N_5OS$			
Benzamide, N-[(1H-purin-6-ylthio)-methyl]-	MeOH	225(4.19),282(4.13), 288(4.12)	36-0372-83
5H-[1,3,5]Thiadiazepino[5,6,7-gh]-purin-7-ol, 6,7-dihydro-7-phenyl-	MeOH	225(4.25),287(4.22), 296(4.29)	36-0372-83
2-Thiazolecarboxylic acid, 4-[3-methyl-2-quinoxalinyl)-, hydrazide	EtOH	243(4.02),267s(3.70), 297s(3.57),335(3.55)	106-0829-83
$C_{13}H_{11}N_5O_4$			
1H-Pyrazole-3-carboxamide, 4-[(2-nitro-3-oxo-1-propenyl)amino]-1-phenyl-	DMF	308(4.08),381(4.29)	103-1163-83
$C_{13}H_{11}N_5O_5$			
2-Propenal, 2-nitro-3-[[3-(nitromethyl)-1-phenyl-1H-pyrazol-5-yl]amino]-	EtOH	224(4.02),262(3.88)	103-1163-83
$C_{13}H_{11}N_5O_6$			
Benzenamine, 4-(1,4-dihydro-5,7-dinitro-4-benzofurazanyl)-N-methyl-, N-oxide	MeOH	256(4.28),294(3.98), 364(3.83),472(4.41)	88-1555-83
$C_{13}H_{12}$			
Benzene, 1,1'-methylenebis-, cation	H_2SO_4	440(4.50)	44-3458-83
	FSO_3H-SbF_5	440(4.58)	44-3458-83
anion	Li-ether	407(4.49)	44-3458-83
	Li-DMF	448(4.57)	44-3458-83
	Li-THF	418(--),443(--)	44-3458-83
Benzene, 1,5-heptadien-3-ynyl-, (E,E)-	C_6H_{12}	305s(--),311(3.57), 338s(--)	44-5379-83
(E,Z)-	C_6H_{12}	311(4.48)	44-5379-83
(Z,E)-	C_6H_{12}	232(4.53),310(4.64), 328(4.49)	44-5379-83
$C_{13}H_{12}BrNO_6$			
3,5-Isoxazoledicarboxylic acid, 4-(4-bromophenyl)-4,5-dihydro-, dimethyl ester, 2-oxide	MeOH	263(4.02)	40-1678-83
$C_{13}H_{12}ClNO_2$			
1H-Indole-2,3-dione, 1-(3-chloro-2-butenyl)-5-methyl-	EtOH	217(4.31),251(4.51), 304(3.88),510(2.78)	103-0286-83
1H-Indole-2,3-dione, 1-(3-chloro-2-butenyl)-7-methyl-	EtOH	219(4.31),248(4.44), 315(3.65),436(2.75)	103-0286-83

Compound	Solvent	λ_{max}(log ε)	Ref.
$C_{13}H_{12}ClNO_3$			
2H-Furo[2,3-h]-1-benzopyran-2-one, 3-chloro-5,6-dihydro-4-(dimethylamino)-	EtOH	236(3.94),266s(3.74), 356(4.15)	111-0113-83
1H-Indole-2,3-dione, 1-(3-chloro-2-butenyl)-5-methoxy-	EtOH	216(4.44),257(4.48), 309(3.32),510(2.78)	103-0286-83
$C_{13}H_{12}ClNO_6$			
3,5-Isoxazoledicarboxylic acid, 4-(3-chlorophenyl)-4,5-dihydro-, dimethyl ester, 2-oxide	MeOH	264(3.94),269(3.93)	40-1678-83
3,5-Isoxazoledicarboxylic acid, 4-(4-chlorophenyl)-4,5-dihydro-, dimethyl ester, 2-oxide	MeOH	264(4.02)	40-1678-83
$C_{13}H_{12}ClN_2$			
Azepino[2,1,7-cd]pyrrolizinium, 2-chloro-8-(dimethylamino)-, perchlorate	CF_3COOH	287(4.21),304(4.26), 354(3.98),390(3.31), 431(3.05)	24-1174-83
$C_{13}H_{12}ClN_2S$			
1,2,3-Benzothiadiazolium, 7-(chloromethylene)-4,5,6,7-tetrahydro-2-phenyl-, perchlorate	MeOH	207(4.12),240s(3.75), 397(4.21)	39-0777-83C
$C_{13}H_{12}FNO_6$			
3,5-Isoxazoledicarboxylic acid, 4-(3-fluorophenyl)-4,5-dihydro-, dimethyl ester, 2-oxide	MeOH	262(3.97),267(3.96)	40-1678-83
3,5-Isoxazoledicarboxylic acid, 4-(4-fluorophenyl)-4,5-dihydro-, dimethyl ester, 2-oxide	MeOH	265(4.00)	40-1678-83
$C_{13}H_{12}F_3NO_4$			
2-Butenoic acid, 4-hydroxy-4-[2-(trifluoroacetyl)amino]phenyl]-, methyl ester	EtOH	208(4.40),245(4.00)	78-3767-83
$C_{13}H_{12}N_2$			
Benzaldehyde, phenylhydrazone	isooctane	236(4.20),295(4.06), 338(4.37)	70-0485-83
	EtOH	237(4.15),304(4.08), 347(4.38)	70-1439-83 +104-2104-83
	MeCN	302(4.05),342(4.35)	104-2104-83
	CCl_4	307(--),345(--)	104-2104-83
1H-Indole, 1-methyl-3-(1H-pyrrol-2-yl)-	EtOH	212(4.19),245(4.30), 320(3.72)	103-0289-83
1H-Indole, 3-(3-methyl-1H-pyrrol-2-yl)-	EtOH	212(4.24),240(4.26), 300(3.62)	103-0289-83
Phenazine, 5,10-dihydro-5-methyl-	1.2M $HClO_4$	245(4.08),268(4.23), 300s(3.88)	125-0655-83
3-Pyridinecarbonitrile, 1,6-dihydro-1-(phenylmethyl)-	MeOH	250(3.59),348(3.49)	32-0569-83
$C_{13}H_{12}N_2O$			
Benzamide, N-(2-aminophenyl)-	H_2O	290(3.6)	93-0783-83
1H-1,2-Diazepine, 1-benzoyl-5-methyl-	MeOH	225(4.18)	5-1361-83
Ethanone, 1-(4-methyl-2-phenyl-5-pyrimidinyl)-	EtOH	282.5(4.36)	4-0649-83
1H-Indeno[1,2-d]pyridazin-1-one, 2,9-dihydro-9,9-dimethyl-	MeOH	239(4.43),246(4.45), 290(3.96)	2-0230-83

Compound	Solvent	λ_{max}(log ε)	Ref.
1H-Indole-3-acetonitrile, 4-(2-methoxyethenyl)-, (Z)-	MeOH	220(4.36),299(4.07), 316(3.88)	78-3695-83
Methanone, bis(4-aminophenyl)-	MeOH	236(4.20),340(4.45)	12-0409-83
Phenol, 4-[(4-methylphenyl)azo]-	C_6H_{12}	242(4.13),349(4.47), 422s(3.12)	59-0729-83
	EtOH	243(4.11),353(4.51), 424s(3.19)	59-0729-83
	50% EtOH-HCl	420s(3.42)	59-0729-83
	ether	242(4.19),350(4.52), 427(3.02)	59-0729-83
	DMF	360(4.47),426s(3.19)	59-0729-83
	$CHCl_3$	350(4.46),424s(3.18)	59-0729-83
9H-Pyrido[3,4-b]indole, 7-methoxy-1-methyl-	EtOH	241(4.96),261s(4.86), 301(4.66),325(4.20), 338(4.32)	78-0033-83
	EtOH	243(4.71),303(4.22), 327(3.78),339(3.74)	105-0376-83
Pyrimido[2,1-a]isoindol-4(1H)-one, 1,2-dimethyl-	EtOH	204(4.27),233(4.47), 280(3.95),292(3.66), 356(3.81),370(3.88), 411(3.47)	124-0755-83
Pyrimido[2,1-a]isoindol-4(6H)-one, 2,3-dimethyl-	EtOH	205(4.35),245(3.87), 303(3.98)	124-0755-83
5H-Pyrrolo[2,1-b][3]benzazepin-11-one, 9-amino-6,11-dihydro-	EtOH	206(4.20),238(4.27), 325(4.18)	44-3220-83
Pyrrolo[1,2-a]pyrimidin-2(6H)-one, 7,8-dihydro-7-phenyl-	EtOH	207(4.07),241(4.17)	4-0393-83
9H-Pyrrolo[3,2-f]quinolin-9-one, 3,6-dihydro-2,7-dimethyl-	EtOH	212(4.54),243(4.30), 252(4.30),284(4.15), 347(4.08)	103-0401-83
$C_{13}H_{12}N_2O_2$			
5H-Cyclopentapyrazine, 6,7-dihydro-2-phenyl-, 1,4-dioxide	EtOH	267(4.41),315(4.28)	103-1104-83
Phenol, 4-[(4-methoxyphenyl)azo]-	C_6H_{12}	245(4.26),352(4.53), 426s(3.37)	59-0729-83
	EtOH	245(4.27),355(4.58), 426s(3.47)	59-0729-83
	ether	245(4.27),352(4.59), 426(3.32)	59-0729-83
	DMF	361(4.59),428s(3.48)	59-0729-83
	$CHCl_3$	356(4.53),426s(3.48)	59-0729-83
Pyrazinemethanol, 6-phenyl-, acetate	EtOH	251(3.90),281(3.91), 360(3.25)	4-0311-83
Pyrazinol, 3-methyl-5-phenyl-, acetate	EtOH	232(3.71),246(3.70), 288(3.82),302(3.80)	4-0311-83
Pyrazinol, 5-methyl-3-phenyl-, acetate	EtOH	210(3.92),248(4.00), 285(3.90),305(3.94)	4-0311-83
Pyrazolo[1,5-a]quinoline-3-carboxylic acid, 4,5-dihydro-4-methyl-	EtOH	267(4.20)	88-4679-83
Pyrazolo[1,5-a]quinoline-3-carboxylic acid, 4,5-dihydro-5-methyl-	EtOH	268(4.18)	88-4679-83
3-Pyridinecarboxamide, 1,4-dihydro-1-(4-methylphenyl)-4-oxo-	H_2O	273(4.24)	54-0331-83
	M HCl	272(4.08)	54-0331-83
3-Pyridinecarboxamide, 1,6-dihydro-1-(4-methylphenyl)-6-oxo-	H_2O	258(4.21),304(3.76)	54-0331-83
	M HCl	258(4.20),303(3.76)	54-0331-83
3-Pyridinecarboxamide, 1,6-dihydro-6-oxo-1-(phenylmethyl)-	H_2O	260(4.13),298(3.73)	54-0331-83
	M HCl	260(4.11),296(3.73)	54-0331-83
$C_{13}H_{12}N_2O_2S_3$			
Benzenesulfenamide, 2-nitro-N-(4,5,6,7-	1,2-$C_2H_4Cl_2$	315(4.07),423(3.73)	118-0050-83

Compound	Solvent	$\lambda_{max}(\log \epsilon)$	Ref.
tetrahydro-1,3-benzodithiol-2-ylidene)- (cont.)			118-0050-83
Benzenesulfenamide, 4-nitro-N-(4,5,6,7-tetrahydro-1,3-benzodithiol-2-ylidene)-	1,2-$C_2H_4Cl_2$	303(3.95),395(4.17)	118-0050-83
$C_{13}H_{12}N_2O_3$			
Isoxazole, 3-methyl-5-[2-(3-methylphenyl)ethenyl]-4-nitro-	MeOH	250s(3.98),268(4.02), 360(4.19)	4-0105-83
3-Pyridinecarboxamide, 1,4-dihydro-1-(4-methoxyphenyl)-4-oxo-	H_2O	276(4.30)	54-0331-83
	M HCl	282(4.14)	54-0331-83
3-Pyridinecarboxamide, 1,6-dihydro-1-(4-methoxyphenyl)-6-oxo-	H_2O	258(4.10),305(3.66)	54-0331-83
	M HCl	258(4.10),305(3.68)	54-0331-83
$C_{13}H_{12}N_2O_3S$			
2-Propenoic acid, 3-(benzoylamino)-2-cyano-3-(methylthio)-, methyl ester	EtOH	242(4.42),278(4.44), 328(4.33)	142-1745-83
$C_{13}H_{12}N_2O_4$			
1H-1,2-Benzodiazepine-3,4-dicarboxylic acid, dimethyl ester	EtOH	204(4.21),257(4.16), 285s(4.02)	138-0463-83
2-Butenedioic acid, 2-[(2-cyanophenyl)amino]-, dimethyl ester, (E)-	MeOH	218(4.42),330(4.21)	142-0489-83
$C_{13}H_{12}N_2O_5$			
Phenol, 2-methoxy-4-[2-(3-methyl-4-nitro-5-isoxazolyl)ethenyl]-, (E)-	MeOH	230(4.06),265s(4.00), 285(4.06),405(4.36)	142-0263-83
Phenol, 4-methoxy-2-[2-(3-methyl-4-nitro-5-isoxazolyl)ethenyl]-, (E)-	MeOH	245s(4.02),273(4.00), 345(4.04),415(4.15)	142-0263-83
Phenol, 5-methoxy-2-[2-(3-methyl-4-nitro-5-isoxazolyl)ethenyl]-, (E)-	MeOH	220s(4.14),287(4.05), 412(4.41)	142-0263-83
$C_{13}H_{12}N_2O_6$			
3-Oxa-1-azabicyclo[3.2.0]heptane-2-carboxylic acid, 7-oxo-, (4-nitrophenyl)methyl ester, cis	EtOH	214(4.41),265(4.54)	39-0649-83C
$C_{13}H_{12}N_2O_7$			
3,5-Isoxazoledicarboxylic acid, 4,5-dihydro-4-(4-nitrosophenyl)-, dimethyl ester, 2-oxide	MeOH	264(4.24)	40-1678-83
$C_{13}H_{12}N_2O_8$			
3,5-Isoxazoledicarboxylic acid, 4,5-dihydro-4-(2-nitrophenyl)-, dimethyl ester, 2-oxide, cis	MeOH	262(4.08)	40-1678-83
3,5-Isoxazoledicarboxylic acid, 4,5-dihydro-4-(3-nitrophenyl)-, dimethyl ester, 2-oxide, cis	MeOH	262(4.30)	40-1678-83
3,5-Isoxazoledicarboxylic acid, 4,5-dihydro-4-(4-nitrophenyl)-, dimethyl ester, 2-oxide, cis	MeOH	266(4.24)	40-1678-83
$C_{13}H_{12}N_2S$			
Acetonitrile, [[1-(2-propenyl)-1H-indol-3-yl]thio]-	MeOH	280(3.90),292s(3.80)	87-0230-83
$C_{13}H_{12}N_2SSe$			
[1,2]Thiaselenolo[4,5,1-hi][1,2,3]-benzothiadiazole-3S^{IV}, 2,6,7,8-	C_6H_{12}	204(4.32),250(4.49), 279s(4.12),300s(4.10)	39-0777-83C

Compound	Solvent	λ_{max}(log ε)	Ref.
tetrahydro-2-phenyl- (cont.)		525(4.07)	39-0777-83C
$C_{13}H_{12}N_4O$			
Benzenamine, 4-[(4-methylphenyl)azo]-N-nitroso-, sodium salt	H_2O	235(4.26),350(4.34), 440s(3.90)	104-1493-83
3H-Pyrazolo[3,4-c]pyridazine, 4-methoxy-3a-methyl-3-phenyl-	EtOH	253(4.31),291(4.17), 296(4.16),390(3.79)	44-2330-83
4H-Pyrazolo[3,4-d]pyrimidin-4-one, 1,5-dihydro-1,6-dimethyl-5-phenyl-	EtOH	210.5(4.60),256(3.90)	11-0171-83B
Pyrazolo[1,5-a]pyrimidin-7(4H)-one, 2-amino-5-methyl-3-phenyl-	MeCN	244(4.32),265(4.30)	39-0011-83C
Pyrazolo[5,1-c][1,2,4]triazine, 4-methoxy-8-methyl-7-phenyl-	EtOH	264(4.53),298(3.54), 360(3.48)	44-2330-83
$C_{13}H_{12}N_4O_2$			
Benzenamine, 4-[(4-methoxyphenyl)-azo]-N-nitroso-, sodium salt	H_2O	236(4.19),360(4.33), 435s(3.85)	104-1493-83
Imidazo[1,5-a]pyrazine-6,8(5H,7H)-dione, 1-[(phenylmethyl)amino]-	EtOH	326(4.07)	2-0030-83
$C_{13}H_{12}N_4O_4S$			
Diazene, [4-[(4-methoxyphenyl)azo]-phenyl](sulfinooxy)-, sodium salt	H_2O	360(4.41)	104-1493-83
$C_{13}H_{12}N_4O_5$			
2H-Pyrrole-3-carboxylic acid, 4-(2-benzoylhydrazino)-5-(hydroxyimino)-2-oxo-, methyl ester	DMF	392(4.10)	48-0041-83
$C_{13}H_{12}N_4S$			
Dithizone	hexane	450(4.4),615(4.5)	97-0056-83
$C_{13}H_{12}N_6O$			
3-Pyridinecarbonitrile, 2,4-diamino-6-methoxy-5-(phenylazo)-	EtOH	374(4.52)	49-0973-83
3-Pyridinecarbonitrile, 4,6-diamino-2-methoxy-5-(phenylazo)-	EtOH	328(4.38),384(4.34)	49-0973-83
$C_{13}H_{12}N_6S$			
Carbonothioic dihydrazide, bis(2-pyridinylmethylene)-	pH 1.8	360(3.6)	86-0555-83
	pH 4-9	327(3.6)	86-0555-83
$C_{13}H_{12}O$			
7bH-Cyclopent[cd]indene, 5-methoxy-7b-methyl-	EtOH	464(2.61)	77-0317-83
$C_{13}H_{12}OS$			
3H-Benzo[b]cyclobuta[d]thiopyran-3-one, 1,2-dihydro-1,1-dimethyl-	MeOH	212(4.41),230(4.42), 264(3.74),305(3.96), 340(3.63)	142-1275-83
$C_{13}H_{12}OS_2$			
7H-Benzocyclohepten-7-one, 6,8-bis-(methylthio)-	CH_2Cl_2	322(4.6),353(4.1)	83-0730-83
$C_{13}H_{12}O_2$			
Bicyclo[4.4.1]undeca-1,3,5,7,9-pentaene-2-carboxylic acid, methyl ester	C_6H_{12}	267(4.59),328(3.91), 377(2.91),383s(2.88), 398s(2.68),406s(2.25)	35-3375-83
Naphtho[1,2-b]furan-3-ol, 2,3-dihydro-3-methyl-	EtOH	211(4.47),237(4.46), 299(3.56),315(3.45)	39-1649-83C

Compound	Solvent	$\lambda_{max}(\log \epsilon)$	Ref.
$C_{13}H_{12}O_3$			
6H-Dibenzo[b,d]pyran-6-one, 7,8,9,10-tetrahydro-3-hydroxy-	MeOH	321(4.18)	78-1265-83
Ethanone, 1-(2-hydroxy-7-methoxy-1-naphthalenyl)-	EtOH	264(3.36),328(3.66)	104-0183-83
Ethanone, 1-(3-hydroxy-6-methoxy-2-naphthalenyl)-	EtOH	270(3.87),324(4.05)	104-0183-83
Euparin	EtOH	263(4.54),358(3.77)	162-0563-83
$C_{13}H_{12}O_4$			
2H-1-Benzopyran-2-one, 4-(acetoxymethyl)-7-methyl-	EtOH	279.5(4.08),314(3.98)	94-3014-83
2-Cyclopenten-1-one, 4-(benzoyloxy)-2-(hydroxymethyl)-, (R)-	EtOH	210s(3.52),229(3.80)	39-2441-83C
1,4-Naphthalenedione, 2,7-dimethoxy-5-methyl-	EtOH	263(4.26),291(4.00), 343(3.23),384(3.11)	78-2283-83
1H,3H-Pyrano[4,3-b][1]benzopyran-10-one, 4,10-dihydro-3-hydroxy-3-methyl-	EtOH	226(4.44),265(3.89), 297(3.93)	77-0335-83
$C_{13}H_{12}O_4S_2$			
1,3-Dithiole-4,5-dicarboxylic acid, 2-phenyl-, dimethyl ester	MeOH	230s(3.98),300s(3.62), 350s(3.41)	104-0750-83
$C_{13}H_{12}O_5$			
2H-1-Benzopyran-2-one, 4-(acetoxymethyl)-7-methoxy-	EtOH	321(4.14)	94-3014-83
1-Benzoxepin-3-carboxylic acid, 2,5-dihydro-7-(hydroxymethyl)-5-oxo-, methyl ester	ether	205(4.23),220(4.15), 250(3.74),279(3.64), 348(3.08)	39-1718-83C
$C_{13}H_{13}BrNOS$			
Thiazolium, 3-[2-(4-bromophenyl)-2-oxoethyl]-2,4-dimethyl-, perchlorate	EtOH	264(4.20)	65-2424-83
$C_{13}H_{13}BrO$			
2H-Benz[cd]azulen-2-one, 3-bromo-1,6,7,8,9,9a-hexahydro-	EtOH	257(4.20),297(3.63)	22-0096-83
2H-Benz[cd]azulen-2-one, 5-bromo-1,6,7,8,9,9a-hexahydro-	EtOH	264(4.43),300(3.82)	22-0096-83
$C_{13}H_{13}ClNOS$			
Thiazolium, 3-[2-(4-chlorophenyl)-2-oxoethyl]-2,4-dimethyl-, perchlorate	EtOH	260(4.19)	65-2424-83
$C_{13}H_{13}ClN_2$			
Pyrimidine, 2-(chlorophenylmethyl)-4,6-dimethyl-	MeCN	248(3.63)	44-0575-83
$C_{13}H_{13}ClN_2OS$			
Morpholine, 4-[4-(4-chlorophenyl)-5-isothiazolyl]-	EtOH	243(4.34),291(3.97)	97-0020-83
$C_{13}H_{13}ClN_2O_3S$			
1H-Pyrido[2,3-b][1,4]thiazine-2-carboxylic acid, 6-chloro-3-(1-oxopropyl)-, ethyl ester	EtOH	256(4.17),312(3.4)	103-0618-83
$C_{13}H_{13}ClO_6$			
Benzeneacetic acid, 2,3-diacetoxy-5-chloro-, methyl ester	EtOH	206(4.20),217(4.06), 274(2.77)	104-0458-83

Compound	Solvent	$\lambda_{max}(\log \epsilon)$	Ref.
$C_{13}H_{13}Cl_4NO$			
3-Butenamide, 2,3,4,4-tetrachloro-N-(2,4,6-trimethylphenyl)-	heptane	198(4.61),230s(4.14)	5-1496-83
$C_{13}H_{13}FO_3$			
Benzoic acid, 3-fluoro-4-(2-methyl-3-oxo-1-propenyl)-, ethyl ester, (E)-	MeCN	278(4.29)	87-1282-83
$C_{13}H_{13}IN_2O_6$			
1-Azetidineacetic acid, α-hydroxy-2-(iodomethyl)-4-oxo-, (4-nitrophenyl)methyl ester	EtOH	214(4.44),265(4.55)	39-0649-83C
$C_{13}H_{13}N$			
[1,1'-Biphenyl]-2-amine, N-methyl-	EtOH	360(3.6)	103-0970-83
Pyridine, 2,6-dimethyl-4-phenyl-	neutral	308(3.38)	39-0045-83B
	protonated	287(3.13)	39-0045-83B
1H-Pyrrole, 1-methyl-2-(2-phenylethenyl)-	MeOH	329(4.41)	44-2488-83
$C_{13}H_{13}NO$			
5H-Benzocycloheptene-5-carbonitrile, 6,7-dihydro-6-methoxy-, trans	EtOH	252(4.14)	18-3449-83
2H-Pyrrol-2-one, 1,5-dihydro-3,4-dimethyl-5-(phenylmethylene)-, (E)-	MeOH	312(4.20)	54-0114-83
(Z)-	MeOH	323(4.30)	54-0114-83
2H-Pyrrol-2-one, 5-ethylidene-1,5-dihydro-4-methyl-3-phenyl-	MeOH	228(3.77),292(4.40)	4-0687-83
$C_{13}H_{13}NOS$			
Benzo[b]thiophen-3(2H)-one, 2-[3-(dimethylamino)-2-propenylidene]-	hexane	355(4.2),470(4.5), 490(4.6)	30-0331-83
$C_{13}H_{13}NO_2$			
2-Cyclohexen-1-one, 3-hydroxy-2-(iminomethyl)-5-phenyl-	EtOH	248.5(4.17),294(4.19)	4-0649-83
1-Naphthalenecarboxylic acid, 5-(dimethylamino)-	MeOH	243(4.13)	64-0049-83C
5(2H)-Oxazolone, 2-methyl-4-phenyl-2-(2-propenyl)-	C_6H_{12}	264(4.12)	44-0695-83
$C_{13}H_{13}NO_2S$			
1-Hexen-3-one, 1-(2-thioxo-3(2H)-benzoxazolyl)-	EtOH	310(4.74)	105-0478-83
$C_{13}H_{13}NO_3$			
4(1H)-Pyridinone, 5-hydroxy-2-(hydroxymethyl)-1-(4-methylphenyl)-	MeOH	244(4.43),288(4.30)	18-1879-83
$C_{13}H_{13}NO_4$			
1H-Indole-4,6-dicarboxylic acid, 1-methyl-, dimethyl ester	MeOH	237(4.41),331(3.90)	44-2488-83
3-Oxa-1-azabicyclo[3.2.0]heptane-2-carboxylic acid, 7-oxo-, phenylmethyl ester, cis	EtOH	219(3.34),229s(2.98), 253(2.63),259(2.63), 264(2.54),269(2.46)	39-0649-83C
4(1H)-Pyridinone, 5-hydroxy-2-(hydroxymethyl)-1-(3-methoxyphenyl)-	MeOH	222(4.46),288(4.29)	18-1879-83
4(1H)-Pyridinone, 5-hydroxy-2-(hydroxymethyl)-1-(4-methoxyphenyl)-	MeOH	244(4.43),288(4.30)	18-1879-83

Compound	Solvent	$\lambda_{max}(\log \epsilon)$	Ref.
$C_{13}H_{13}NO_7$			
3,5-Isoxazoledicarboxylic acid, 4,5-dihydro-4-(3-hydroxyphenyl)-, dimethyl ester, 2-oxide, cis	MeOH	266(4.05)	40-1678-83
3,5-Isoxazoledicarboxylic acid, 4,5-dihydro-4-(4-hydroxyphenyl)-, dimethyl ester, 2-oxide, cis	MeOH	268(4.02)	40-1678-83
$C_{13}H_{13}N_2O$			
Pyridinium, 3-(aminocarbonyl)-1-(4-methylphenyl)-, chloride	H_2O	281(3.76)	54-0331-83
Pyridinium, 4-(aminocarbonyl)-1-(phenylmethyl)-, bromide	EtOH	252(3.31),370(3.04)	150-2326-83M
Pyrimido[2,1-a]isoindolium, 4,6-dihydro-1,2-dimethyl-4-oxo-, methyl sulfate	EtOH	204(4.43),257(4.07), 300(3.92)	124-0755-83
$C_{13}H_{13}N_2O_2$			
Pyridinium, 3-(aminocarbonyl)-1-(4-methoxyphenyl)-, chloride	H_2O	245(3.85),317(3.79)	54-0331-83
$C_{13}H_{13}N_2O_3S$			
Thiazolium, 2,4-dimethyl-3-[2-(4-nitrophenyl)-2-oxoethyl]-, perchlorate	EtOH	261(4.26)	65-2424-83
$C_{13}H_{13}N_3$			
Benzenamine, 4-[(4-methylphenyl)azo]-	MeOH	389(4.37)	80-0903-83
	$C_2H_4Cl_2$	382(4.40)	80-0903-83
$C_{13}H_{13}N_3O$			
Benzamide, 4-amino-N-(2-aminophenyl)-	H_2O	292(4.3)	93-0783-83
$C_{13}H_{13}N_3OS$			
2H-Isothiazolo[4,5,1-hi][1,2,3]benzothiadiazole-3-S^{IV}, 4,6,7,8-tetrahydro-4-hydroxy-2-phenyl-	C_6H_{12}	205(4.39),221s(4.17), 278(3.98),431(4.33), 452(4.31)	39-0777-83C
4H-Thiazolo[2,3-c][1,2,4]triazin-4-one, 6,7-dihydro-6,7-dimethyl-3-phenyl-	EtOH	205(4.17),245(3.90), 327(4.14)	44-4585-83
trans	EtOH	206(4.25),245(4.02), 326(4.25)	44-4585-83
7H-Thiazolo[3,2-b][1,2,4]triazin-7-one, 2,3-dihydro-2,3-dimethyl-6-phenyl-, cis	EtOH	204(4.13),237(4.24), 307(4.08)	44-4585-83
trans	EtOH	206(4.06),240(4.19), 310(4.01)	44-4585-83
$C_{13}H_{13}N_3O_2S_3$			
4a,10a-(Iminomethaniminomethano)-10H-1,4-dithiino[2,3-b][1,4]benzothiazine-11,13-dione, 2,3-dihydro-12-methyl-	MeCN	213(4.65),305(3.15)	138-1093-83
$C_{13}H_{13}N_3O_3$			
Acetamide, N,N'-(4-phenyl-3,5-isoxazolediyl)bis-	MeCN	238(4.03)	44-0575-83
$C_{13}H_{13}N_3O_3S$			
Acetamide, 2-cyano-2-(6,7-dimethoxy-2H-1,4-benzothiazin-3(4H)-ylidene)-	MeOH	286(4.45),360(4.20)	87-0845-83

Compound	Solvent	λ_{max}(log ϵ)	Ref.
$C_{13}H_{13}N_3O_3S_2$			
4a,10a-(Iminomethaniminomethano)-10H-1,4-oxathiino[3,2-b][1,4]benzothiazine-11,13-dione, 2,3-dihydro-12-methyl-	MeCN	218(4.49),298(3.36)	138-1093-83
$C_{13}H_{13}N_3O_4$			
1H-Pyrazole-3-carboxylic acid, 5-hydroxy-1-(2-methylphenyl)-4-nitroso-, ethyl ester	MeOH	256(4.07),370(3.47)	94-1228-83
1H-Pyrazole-3-carboxylic acid, 5-hydroxy-1-(3-methylphenyl)-4-nitroso-, ethyl ester	MeOH	254(4.40),393(3.23)	94-1228-83
1H-Pyrazole-3-carboxylic acid, 5-hydroxy-1-(4-methylphenyl)-4-nitroso-, ethyl ester	MeOH	264(4.33),390(3.25)	94-1228-83
$C_{13}H_{13}N_3O_6$			
1,3,5-Triazine-2,4(1H,3H)-dione, 6-(6,7-dihydro-6,7-dioxo-1,3-benzodioxol-5-yl)dihydro-	MeCN	214(4.11),232(4.05), 445(3.06)	64-0752-83B
$C_{13}H_{13}N_3O_6S$			
Cephacetrile (sodium)	0.1N $NaHCO_3$	260(3.97)	162-0274-83
$C_{13}H_{13}N_4O_2$			
Imidazo[1,5-a]pyrazinium, 1-amino-5,6,7,8-tetrahydro-6,8-dioxo-2-(phenylmethyl)-, bromide	H_2O	251(3.68),296(4.03)	2-0030-83
$C_{13}H_{13}N_5O$			
3H-Purin-6-amine, 3-methyl-N-(phenylmethoxy)-	pH 1	287(4.28)	94-3521-83
	pH 7	295(4.21)	94-3521-83
	pH 13	288(4.10)	94-3521-83
	EtOH	294(4.19)	94-3521-83
$C_{13}H_{13}N_5O_3$			
1H-1,2,3-Triazole-4-carboxaldehyde, 1-acetyl-5-(phenylamino)-, 4-(O-acetyloxime)	EtOH	330(3.98)	95-0594-83
$C_{13}H_{13}N_5O_4$			
Adenine, N^6-(1,2-O-isopropylidene-α-D-glycero-pent-3-enofuranurononoyl)-	MeOH	211(4.08),257(4.15), 293(3.78)	65-1480-83
Pyrimido[1,2-a]purin-10(3H)-one, 3-(2-deoxy-β-D-erythro-pentofuranosyl)-	H_2O	214(4.38),253(4.22), 300(3.54),320(3.58), 350(3.51)	18-1799-83
$C_{13}H_{13}N_5O_5$			
Pyrimido[1,2-a]purin-10(3H)-one, 3-β-D-ribofuranosyl-	pH 4.0	220(4.3),255(4.1), 330(3.6)	18-1799-83
$C_{13}H_{13}N_7$			
Carbonimidic dihydrazide, bis(2-pyridinylmethylene)-	pH 1.0	335(3.6)	86-0555-83
	pH 4.5	310(3.6)	86-0555-83
	pH 10.6	348(3.6)	86-0555-83
	pH 15	414(3.6)	86-0555-83
$C_{13}H_{13}N_7S_3Zn$			
Zinc, [[2,2'-[1-(4-cyanophenyl)-1,2-ethanediylidene]bis[N-methylhydra-	EtOH	335(4.51),461(4.17)	161-0143-83

Compound	Solvent	$\lambda_{max}(\log \epsilon)$	Ref.
zinecarbothioamidato]](2-)- $N^2,N^{2'}$,S,S']-, (T-4)- (cont.)			161-0143-83
$C_{13}H_{13}OP$			
Phosphorin, 3-(4-methoxyphenyl)- 4-methyl-	EtOH	210(3.81),250(3.64), 284s(3.32),312s(2.85)	24-1756-83
$C_{13}H_{13}P$			
Phosphorin, 4-methyl-3-(4-methyl- phenyl)-	EtOH	211(4.34),239(4.25), 256s(4.10),274s(3.64), 310(2.78)	24-1756-83
$C_{13}H_{14}$			
Benzene, 1,3,5-heptatrienyl-, (E,Z,E)-	C_6H_{12}	304(4.54),317(4.61), 327(4.46)	44-5379-83
(E,Z,Z)-	C_6H_{12}	307s(--),318(4.63), 331s(--)	44-5379-83
(Z,Z,E)-	C_6H_{12}	297(4.60)	44-5379-83
Cyclohexene, 6-(2-butynylidene)-1-(1- propynyl)-	EtOH	272(3.93)	44-4272-83
$C_{13}H_{14}ClNOS$			
Morpholine, 4-[3-(2-chlorophenyl)- 1-thioxo-2-propenyl]-	heptane	247(4.07),282(4.21)	80-0555-83
	EtOH	234(4.13),286(4.30), 300(4.23)	80-0555-83
Morpholine, 4-[3-(4-chlorophenyl)- 1-thioxo-2-propenyl]-	heptane	250(4.10),307(4.30)	80-0555-83
	EtOH	227(4.10),295(4.34), 395s(--)	80-0555-83
$C_{13}H_{14}ClN_3O$			
Morpholine, 4-[4-(4-chlorophenyl)-1H- pyrazol-3-yl]-	MeCN	231(4.01),252(3.89), 264s(3.94),274(4.01)	48-0689-83
$C_{13}H_{14}DNO_3S$			
Carbamic acid, (2,2-dimethyl-4-oxo- 3-thietanyl-3-d)-, phenylmethyl ester, (±)-	EtOH	215(3.89),235(3.26)	39-2259-83C
$C_{13}H_{14}F_3NO_2$			
Oxazole, 2,5-dihydro-5-methoxy-2,2- dimethyl-4-phenyl-5-(trifluoro- methyl)-	EtOH	244(4.17)	33-0262-83
$C_{13}H_{14}INO_4$			
1-Azetidineacetic acid, α-hydroxy- 2-(iodomethyl)-4-oxo-, phenyl- methyl ester	EtOH	219(3.36),230s(2.95), 253(2.63),259(2.59), 264(2.54),269(2.43)	39-0649-83C
$C_{13}H_{14}NOS$			
Thiazolium, 2,4-dimethyl-3-(2-oxo- 2-phenylethyl)-, perchlorate	EtOH	252(4.31)	65-2424-83
$C_{13}H_{14}N_2$			
[1,1'-Biphenyl]-2,4'-diamine, N^2- methyl-	EtOH	250(3.98),310(2.66)	103-0970-83
Cyclobutanecarbonitrile, 3-(cyano- methylene)-2,2-dicyclopropyl-	hexane	213(4.07)	44-1500-83
1,1-Cyclobutanedicarbonitrile, 3-(di- cyclopropylmethylene)-	EtOH	216(4.08)	44-1500-83
1,2-Cyclobutanedicarbonitrile, 3-(di- cyclopropylmethylene)-, cis	hexane	215(4.05)	44-1500-83

Compound	Solvent	λ_{max}(log ϵ)	Ref.
1,2-Cyclobutanedicarbonitrile, 3-(dicyclopropylmethylene)-, trans	hexane	217(3.87)	44-1500-83
Pyrazolo[1,5-a]quinoline, 4,5-dihydro-4,5-dimethyl-	EtOH	263(4.13)	88-4679-83
Pyrazolo[1,5-a]quinoline, 4,5-dihydro-5,5-dimethyl-	EtOH	263(4.17)	88-4679-83
3-Pyridinecarbonitrile, 1,4,5,6-tetrahydro-1-(phenylmethyl)-	EtOH	281(4.18)	150-2326-83M
$C_{13}H_{14}N_2O$			
4,7-Methano-3H-pyrrolo[4,3,2-fg][2,3]-benzoxazocine, 1,4,5,7-tetrahydro-5-methyl-, (±)-	MeOH	222(4.57),287(3.95)	78-3695-83
3H-Pyrazol-3-one, 2,4-dihydro-5-(1-methyl-1-propenyl)-2-phenyl-	EtOH	258(4.36)	142-1801-83
3H-Pyrazol-3-one, 2,4-dihydro-5-(2-methyl-1-propenyl)-2-phenyl-	EtOH	266(4.30)	142-1801-83
3H-Pyrido[3,4-b]indole, 4,9-dihydro-6-methoxy-1-methyl-	EtOH	235(4.20),239s(4.19), 316(4.18)	4-0267-83
3H-Pyrido[3,4-b]indole, 4,9-dihydro-7-methoxy-1-methyl- (harmaline)	MeOH	218(4.27),260(3.90), 376(4.02)	162-0665-83
2H-Pyrrol-2-one, 1,5-dihydro-1,3,4-trimethyl-5-(2-pyridinyl)-, (E)-	MeOH	271(4.00)	54-0114-83
(Z)-	MeOH	307(4.08)	54-0114-83
2H-Pyrrol-2-one, 1,5-dihydro-1,3,4-trimethyl-5-(4-pyridinyl)-, (E)-	MeOH	284(4.00)	54-0114-83
(Z)-	MeOH	294(4.11)	54-0114-83
$C_{13}H_{14}N_2OS$			
Morpholine, 4-(4-phenyl-5-isothiazolyl)-	EtOH	240(4.16),282(3.93)	97-0020-83
$C_{13}H_{14}N_2O_2$			
2H-1,2-Benzodiazepine-2-carboxylic acid, 1,3-dihydro-3-methylene-, ethyl ester	EtOH	224(4.07),241(4.09), 289(4.02),299s(--)	94-3684-83
1H-Pyrrole-2-carbonitrile, 5-[(1,5-dihydro-3,4-dimethyl-5-oxo-2H-pyrrol-2-ylidene)methyl]-1-methyl-, (Z)-	MeOH	378(4.67),396s(4.53)	54-0347-83
Spiro[cyclopropane-1,4'(1'H)-isoquinoline]-1',3'(2'H)-dione, 2'-(dimethylamino)-	MeOH	213(4.46),245(3.99), 292(3.43)	142-1315-83
$C_{13}H_{14}N_2O_2S$			
2,5-Pyrrolidinedione, 1-[2-(6,7-dihydrothieno[3,2-c]pyridin-4-yl)-ethyl]-	MeOH	223(4.19),228(4.18), 258(3.64)	13-0493-83B
Quinoxaline, 5,6,7,8-tetrahydro-2-methyl-3-(2-thienyl)-, 1,4-dioxide	EtOH	244(4.24),266(4.10), 297(4.35)	103-1104-83
$C_{13}H_{14}N_2O_3$			
5H-Cyclopentapyrazine, 6,7-dihydro-2-methyl-3-(5-methyl-2-furanyl)-, 1,4-dioxide	EtOH	244(4.05),270(4.12), 298(4.48),338s(4.05)	103-1104-83
Quinoxaline, 2-(2-furanyl)-5,6,7,8-tetrahydro-3-methyl-, 1,4-dioxide	EtOH	241(4.05),268(4.11), 292(4.36)	103-1104-83
$C_{13}H_{14}N_2O_3S$			
Morpholine, 4-[3-(3-nitrophenyl)-1-thioxo-2-propenyl]-	heptane	280(4.40)	80-0555-83
	EtOH	272(4.40)	80-0555-83

Compound	Solvent	λ_{max}(log ε)	Ref.
Morpholine, 4-[3-(4-nitrophenyl)-1-thioxo-2-propenyl]-	EtOH	280(4.34),322s(--)	80-0555-83
$C_{13}H_{14}N_2O_4$			
DL-Tryptophan, N-acetyl-6-hydroxy-	EtOH	235(4.34),267s(3.57), 276(3.59),295(3.62)	102-1026-83
	EtOH-NaOH	235(4.45),268s(3.70), 278s(3.63),313(3.67)	102-1026-83
$C_{13}H_{14}N_2O_4S_2$			
Gliotoxin	n.s.g.	270(3.65)	162-0635-83
$C_{13}H_{14}N_2O_7$			
6H-Furo[2',3':4,5]oxazolo[3,2-b]pyridazin-6-one, 3-acetoxy-2-(acetoxymethyl)-2,3,3a,9a-tetrahydro-, monohydrobromide, [2R-(2α,3β,3aβ-9aβ)]]-	pH 1	254(3.87),274(3.92)	44-3765-83
	pH 11	254(3.85),275(3.91)	44-3765-83
	MeOH	253(3.97),272(3.97)	44-3765-83
$C_{13}H_{14}N_4$			
Benzimidazole, 2-[1-cyano-2-(dimethylamino)ethenyl]-1-methyl-, perchlorate	EtOH	320(4.51)	124-0297-83
$C_{13}H_{14}N_4O$			
Benzamide, 2,4-diamino-N-(4-aminophenyl)-, cation	acid	<u>292(4.3)</u>	93-0783-83
dication	acid	<u>300(3.7)</u>	93-0783-83
trication	acid	<u>240(4.1)</u>	93-0783-83
$C_{13}H_{14}N_4O_2$			
Acetamide, N-(1-acetyl-5-amino-4-phenyl-1H-pyrazol-3-yl)-	MeCN	245(4.25)	39-0011-83C
3H-Pyrazolium, 1-[(1,3-dihydro-1,3-dioxo-2H-isoindol-2-yl)amino]-4,5-dihydro-3,3-dimethyl-, hydroxide, inner salt	EtOH	219(4.51),233s(4.29), 240s(4.23),266(4.09)	104-2317-83
3H-Pyrazolium, 1-[(1,3-dihydro-1,3-dioxo-2H-isoindol-2-yl)amino]-4,5-dihydro-5,5-dimethyl-, hydroxide, inner salt	EtOH	219(4.51),233s(4.29), 240s(4.23),266(4.09)	104-2317-83
$C_{13}H_{14}N_4O_3$			
4H-Imidazole-2-carboxylic acid, 5-[2-(2-methylphenyl)hydrazino]-4-oxo-, ethyl ester	EtOH	206(4.33),222(4.14), 299(4.16)	103-0571-83
4H-Imidazole-2-carboxylic acid, 5-[2-(3-methylphenyl)hydrazino]-4-oxo-, ethyl ester	EtOH	206(4.27),229(4.05), 296(4.08)	103-0571-83
4H-Imidazole-2-carboxylic acid, 5-[2-(4-methylphenyl)hydrazino]-4-oxo-, ethyl ester	EtOH	205(4.21),226(4.08), 297(4.04)	103-0571-83
1H-Imidazolium, 4-amino-5-(aminocarbonyl)-1-(carboxymethyl)-3-(phenylmethyl)-, hydroxide, inner salt	H_2O	268(3.88)	2-0030-83
$C_{13}H_{14}N_4O_4$			
1H-Purine-2,6-dione, 7-(5,6-dihydro-6-methyl-5-oxo-2H-pyran-2-yl)-3,7-dihydro-1,3-dimethyl-, (2S-cis)-	$CHCl_3$	278(3.95)	136-0301-83A

Compound	Solvent	$\lambda_{max}(\log \epsilon)$	Ref.
$C_{13}H_{14}N_4O_5$			
1H-Purine-2,6-dione, 7-(2,3-anhydro-6-deoxy-β-L-lyxo-hexopyranos-4-ulos-1-yl)-3,7-dihydro-1,3-dimethyl-	MeOH	279(4.04)	136-0301-83A
$C_{13}H_{14}O$			
Naphthalene, 2-propoxy- (class spectrum)	MeOH and MeOH-base	261(3.6),271(3.6), 281(3.5),313(3.18), 328(3.28)	160-1403-83
8a,4a-Propenonaphthalen-11-one, 1,2,3,4-tetrahydro-	MeCN	222(4.04),268(3.36)	35-2800-83
$C_{13}H_{14}OS_2$			
Spiro[1,3-dithiolan-2,5'(1'aH)-[1,2,4]-ethanylylidene[1'H]cyclobuta[cd]-pentalen]-7'-one, hexahydro-	MeCN	252(2.47),303(1.48)	44-1862-83
$C_{13}H_{14}OS_3$			
6H-Cyclohepta[c]thiophen-6-one, 1,3-dimethyl-5,7-bis(methylthio)-	CH_2Cl_2	326(4.4),401(3.95)	83-0730-83
$C_{13}H_{14}O_2$			
2H-Cyclopent[cd]inden-2-one, 1,2a,7a,7b-tetrahydro-2a-methoxy-7b-methyl-	EtOH	240(3.43),310(3.75), 342s(3.29)	39-2399-83C
1H-Cyclopropa[a]naphthalene-1-carboxylic acid, 1a,2,3,7b-tetrahydro-1-methyl-, (1α,1aα,7bα)-	n.s.g.	227(3.88),268(2.67)	78-2965-83
Ethanone, 1-[2,3-dihydro-2-(1-methylethenyl)-5-benzofuranyl]-	EtOH	226(4.06),282(4.08), 287(4.08)	78-0169-83
4a,8a-Propanonaphthalene-9,10-dione, 1,4,5,8-tetrahydro-	isooctane	465(1.30)	35-2800-83
$C_{13}H_{14}O_3$			
4,7-Benzofurandione, 2,3-dihydro-2,2-dimethyl-6-(2-propenyl)-	EtOH	272(4.06),426(3.07)	12-1263-83
Ethanone, 1-[2,3-dihydro-4-hydroxy-2-(1-methylethenyl)-5-benzofuranyl]-	EtOH	214(4.26),235(3.93), 241(3.92),289(4.17)	78-0169-83
Ethanone, 1-[2,3-dihydro-6-hydroxy-2-(1-methylethenyl)-5-benzofuranyl]-	EtOH	214(4.22),237(4.05), 280(4.07),327(3.89)	78-0169-83
Ethanone, 1-[2-(hydroxymethyl)-2-methyl-2H-1-benzopyran-6-yl]- (artemisenol)	EtOH	250(3.91),280(3.74), 330(3.02)	102-2587-83
2-Naphthalenecarboxaldehyde, 5,6-dihydro-1,4-dimethoxy-	EtOH	250(4.47),300(3.73), 350(3.57)	94-2662-83
Spiro[1,3-dioxolan-2,5'(1'aH)-[1,2,4]-ethanylylidene[1H]cyclobuta[cd]-pentalen]-7'-one, hexahydro-	MeCN	295(1.28)	44-1862-83
$C_{13}H_{14}O_4$			
2H-1-Benzopyran-3-carboxylic acid, 7-methoxy-2,2-dimethyl-	EtOH	210s(4.08),222(4.20), 247(4.23),298s(3.94), 310(3.97),337(4.04)	39-0197-83C
Ethanone, 1-[2,3-dihydro-4,6-dihydroxy-2-(1-methylethenyl)-5-benzofuranyl]-	EtOH	213(4.14),222(4.08), 292(4.19),321(3.53)	78-0169-83
Ethanone, 1-[2,3-dihydro-4,6-dihydroxy-2-(1-methylethenyl)-7-benzofuranyl]-	EtOH	216(4.01),228(4.02), 288(4.18),331(3.42)	78-0169-83

Compound	Solvent	$\lambda_{max}(\log \epsilon)$	Ref.
$C_{13}H_{14}O_4S_2$			
1,3-Benzodioxole-4,5-dione, 6-(1,3-dithian-2-yl)-7-ethyl-	MeCN	214(4.21),273(3.83), 565(3.81)	64-0752-83B
Butanedioic acid, [(phenylthioxomethyl)thio]-, dimethyl ester	MeCN	226s(3.84),305(4.13), 494(2.09)	104-0750-83
$C_{13}H_{14}O_9$			
Norbergenin	EtOH	222(4.27),289(3.93)	102-1053-83
	EtOH-NaOAc	212(--),245s(--), 338(--)	102-1053-83
	EtOH-$AlCl_3$	223(--),240s(--), 325(--)	102-1053-83
$C_{13}H_{15}$			
1H-Cyclopent[f]azulenylium, hexahydro-, hexafluoroantimonate	MeCN	242(4.19),327(3.74)	44-0596-83
$C_{13}H_{15}BrN_2O$			
1,2,4-Oxadiazole, 3-(4-bromo-2,6-dimethylphenyl)-5-(1-methylethyl)-	MeOH	270s(2.84),278s(2.79)	4-1609-83
1,2,4-Oxadiazole, 3-(4-bromo-2,6-dimethylphenyl)-5-propyl-	MeOH	270s(2.88),278s(2.81)	4-1609-83
$C_{13}H_{15}BrN_2O_6$			
1H-Pyrazole-4-acetic acid, 5-acetoxy-α-(2-bromoethylidene)-3-(methoxycarbonyl)-1-methyl-, methyl ester, (E)-	MeOH	217(4.28)	24-1525-83
$C_{13}H_{15}BrS$			
1H-Indene-1-thione, 4-bromo-2,3-dihydro-2,2,3,3-tetramethyl-	3-Mepentane	335(3.94)	23-0894-83
$C_{13}H_{15}ClN_2$			
Pyrimidine, 1-(4-chlorophenyl)-1,4-dihydro-4,4,6-trimethyl-	EtOH	265(3.82)	39-1799-83C
$C_{13}H_{15}ClN_2O_5$			
1,2-Diazabicyclo[5.2.0]nona-3,5-diene-2,6-dicarboxylic acid, 8-chloro-9-oxo-, diethyl ester	CH_2Cl_2	321(4.24)	5-1361-83
$C_{13}H_{15}Cl_3N_2O$			
2-Butenamide, 2,4,4-trichloro-3-(dimethylamino)-N-(4-methylphenyl)-	heptane	198(4.04),225(3.77), 335(3.80)	5-1496-83
3-Butenamide, 2,4,4-trichloro-3-(dimethylamino)-N-(4-methylphenyl)-	heptane	200(4.50),240(4.26)	5-1496-83
$C_{13}H_{15}FO_2$			
Benzoic acid, 3-fluoro-4-(2-methyl-1-propenyl)-, ethyl ester	EtOH	270(4.21)	87-1282-83
$C_{13}H_{15}N$			
Bicyclo[4.4.1]undeca-1,3,5,7,9-pentaen-2-amine, N,N-dimethyl-	C_6H_{12}	251(4.62),271s(4.46), 351(4.05),405s(3.34)	35-3375-83
Bicyclo[4.4.1]undeca-1,3,5,7,9-pentaen-3-amine, N,N-dimethyl-	dioxan	246(4.33),282(4.58), 342(4.03),408(3.44)	35-3375-83
$C_{13}H_{15}NO$			
4aH-Carbazol-4a-ol, 2,3,4,4a-tetrahydro-9-methyl-	EtOH	230(4.28),281s(3.52), 289(3.54)	39-0141-83C

Compound	Solvent	$\lambda_{max}(\log \epsilon)$	Ref.
2-Propenenitrile, 2-(1,1-dimethylethoxy)-3-phenyl-	EtOH	206(3.92),219(3.93), 278(4.18)	78-1551-83
1H-Pyrrole-2-methanol, 1-methyl-α-(phenylmethyl)-	MeOH	237(3.77)	44-2488-83
2H-Pyrrol-2-one, 5-ethyl-1,5-dihydro-4-methyl-3-phenyl-	MeOH	232(4.10)	4-0687-83
$C_{13}H_{15}NOS$			
Morpholine, 4-(3-phenyl-1-thioxo-2-propenyl)-	heptane	244(3.96),287(4.27), 297(4.23)	80-0555-83
	EtOH	228(4.11),294s(4.32), 385s(--)	80-0555-83
$C_{13}H_{15}NOS_2$			
Cyclohepta[c]pyrrol-6(2H)-one, 1,3-dimethyl-5,7-bis(methylthio)-	CH_2Cl_2	326(4.5),417(4.2), 438(4.2)	83-0730-83
$C_{13}H_{15}NO_2$			
4(3H)-Benzofuranone, 6,7-dihydro-5-(1-pyrrolidinylmethylene)-	EtOH	252(3.93),356(4.29)	111-0113-83
Pyridinium, 4,4-dimethyl-2,6-dioxocyclohexylide	H_2O	342(3.08)	65-2073-83
	EtOH	367(3.24)	65-2073-83
	85% EtOH	359(3.24)	65-2073-83
	$CHCl_3$	408(3.45)	65-2073-83
1H-Pyrrole, 2-[(2,3-dimethoxyphenyl)-methyl]-	MeOH	215s(4.19),269(3.10)	83-0773-83
1H-Pyrrole, 3-[(2,3-dimethoxyphenyl)-methyl]-	MeOH	215s(4.18),270(3.22)	83-0773-83
$C_{13}H_{15}NO_3$			
4(5H)-Benzofuranone, 6,7-dihydro-5-(4-morpholinylmethylene)-	EtOH	247.5(3.89),353(4.28)	111-0113-83
2-Naphthalenecarboxamide, 1,2,3,4-tetrahydro-6-methoxy-2-methyl-1-oxo-	EtOH	224(4.03),279(4.17)	118-0727-83
$C_{13}H_{15}NO_3S$			
Carbamic acid, (2,2-dimethyl-4-oxo-3-thietanyl)-, phenylmethyl ester, (R)-	EtOH	213(3.93),237s(3.28)	39-2259-83C
$C_{13}H_{15}NO_4$			
1-Benzoxepin-4-carboxamide, 2,3,4,5-tetrahydro-8-methoxy-4-methyl-5-oxo-	EtOH	274(4.01)	118-0727-83
1-Cyanovinyl camphanate, (-)-	dioxan	212(3.74)(end abs.)	33-1865-83
$C_{13}H_{15}N_2OS$			
Thiazolium, 4-(acetylamino)-2,3-dimethyl-5-phenyl-, perchlorate	MeCN	279.5(4.08)	118-0582-83
$C_{13}H_{15}N_2S$			
Thiazolo[3,2-a]pyridinium, 3-amino-5,6,7,8-tetrahydro-2-phenyl-, perchlorate	MeCN	225.5s(3.99),264.5(3.95), 326.5(4.16)	118-0582-83
$C_{13}H_{15}N_3$			
3,5-Pyridinedicarbonitrile, 1-ethenyl-1,4-dihydro-2,4,4,6-tetramethyl-	EtOH	216(4.36),260(3.78), 333(3.75)	73-0617-83
$C_{13}H_{15}N_3O$			
Morpholine, 4-(4-phenyl-1H-pyrazol-3-yl)-	MeCN	222s(4.04),252(3.92), 276(3.93)	48-0689-83

Compound	Solvent	$\lambda_{max}(\log \epsilon)$	Ref.
$C_{13}H_{15}N_3OS$			
Isothiazolo[5,1-e][1,2,3]thiadiazole-7-S^{IV}, 1,6-dihydro-6-methoxy-3,4-dimethyl-1-phenyl-	C_6H_{12}	202(4.31),241(4.06), 273(4.02),432(4.32)	39-0777-83C
Pyrido[2,3-d]pyrimidin-4(1H)-one, 1-cyclohexyl-2,3-dihydro-2-thioxo-	MeOH	296(3.20)	73-3315-83
Thiazolo[3,2-a]-1,3,5-triazin-6(7H)-one, 7-ethyl-3,4-dihydro-3-phenyl-	EtOH	236(4.21)	103-0961-83
$C_{13}H_{15}N_3O_2$			
1(4H)-Pyridineacetic acid, 3,5-dicyano-2,4,4,6-tetramethyl-	EtOH	219(4.46),340(3.85)	73-0617-83
4H-Pyrido[1,2-a]pyrimidine-7-carboxamide, 3-ethyl-N,2-dimethyl-4-oxo-	EtOH	233(4.33),347(4.08)	106-0218-83
$C_{13}H_{15}N_3O_2S$			
4H-Pyrido[3,4-e]-1,3-thiazin-4-one, 5,7-dimethyl-2-morpholino-	MeOH	270(3.98)	73-3426-83
2H-Thiazolo[3,2-a]-1,3,5-triazin-6(7H)-one, 3,4-dihydro-7-(hydroxymethyl)-7-methyl-3-phenyl-	EtOH	235(4.14)	103-0961-83
$C_{13}H_{15}N_3O_4S$			
5H-Pyrimido[4,5-b][1,4]thiazine-6-carboxylic acid, 4-methoxy-7-(1-oxopropyl)-, ethyl ester	EtOH	256(4.11),326(3.81), 476(3.85)	103-0618-83
$C_{13}H_{15}N_3O_7$			
Hex-1-enitol, 1,5-anhydro-2,3-dideoxy-3-(2-nitro-1H-imidazol-1-yl)-, 4,6-diacetate	EtOH	226(3.58),312(3.81)	87-0020-83
1H-Imidazole, 1-(4,6-di-O-acetyl-2,3-dideoxy-α-D-erythro-hex-2-enopyranosyl)-2-nitro-	EtOH	220(3.73),320(3.84)	87-0020-83
β-	EtOH	220(3.66),318(3.81)	87-0020-83
$C_{13}H_{15}N_4$			
5H-Pyrido[1,2-a]pyrimido[4,5-d]pyrimidinium, 2,3,9-trimethyl-, perchlorate	pH 6.86	204(4.47),230(4.24), 288(3.92),391(4.36)	44-2476-83
conjugate acid	pH 1	204(4.46),230(4.24), 267(3.89),340(4.25)	44-2476-83
5H-Pyrido[1,2-a]pyrimido[4,5-d]pyrimidinium, 2,3,10-trimethyl-, perchlorate	pH 6.86	203(4.40),223(4.12), 253(3.91),287(4.01), 400(4.35)	44-2476-83
conjugate acid	pH 1	202(4.47),227s(--), 247(3.82),264(3.99), 345(4.21)	44-2476-83
$C_{13}H_{15}N_5O_4$			
3H-Imidazo[2,1-i]purine, 7-methyl-	pH 1.4	278(4.08)	56-0779-83
3-β-D-ribofuranosyl-	pH 7	261(3.65),269(3.76), 279(3.76),305(3.46)	56-0779-83
3H-Imidazo[2,1-i]purine, 8-methyl-	pH 1.4	276(4.08)	56-0779-83
3-β-D-ribofuranosyl-	pH 7	261(3.77),268(3.83), 278(3.79),296(3.48)	56-0779-83
$C_{13}H_{15}N_5O_4S$			
2,4-Thiazolediamine, 5-(2,4-dinitrophenyl)-N,N,N',N'-tetramethyl-	CH_2Cl_2	233(4.30),563(4.19)	88-3563-83

Compound	Solvent	$\lambda_{max}(\log \epsilon)$	Ref.
$C_{13}H_{15}O_2P$			
Phosphorin, 1,1-dihydro-1,1-dimethoxy-3-phenyl-	hexane	245(4.5),350(4.2)	24-0445-83
$C_{13}H_{16}BrNS$			
Piperidine, 1-[2-(4-bromophenyl)-1-thioxoethyl]-	heptane	281(4.16)	80-0555-83
	EtOH	280(4.19)	80-0555-83
$C_{13}H_{16}BrN_3$			
[1,2]Diazepino[5,4-b]indole, 1,2,3,4-5,6-hexahydro-9-bromo-2,3-dimethyl-	EtOH	233(4.04),286s(3.31), 294(3.36),303(3.30)	39-1937-83C
$C_{13}H_{16}ClNO_2$			
Benzene, 1-(3-chloro-4,4-dimethyl-1-pentenyl)-4-nitro-, (E)-	EtOH	221(4.06),304(4.15)	12-0527-83
$C_{13}H_{16}ClNS$			
Piperidine, 1-[2-(4-chlorophenyl)-1-thioxoethyl]-	heptane	281(4.20)	80-0555-83
	EtOH	280(4.20)	80-0555-83
$C_{13}H_{16}ClN_3$			
[1,2]Diazepino[5,4-b]indole, 9-chloro-1,2,3,4,5,6-hexahydro-2,3-dimethyl-	EtOH	233(4.26),285s(3.55), 293(3.59),302(3.55)	39-1937-83C
$C_{13}H_{16}ClN_3OS$			
3-Pyridinecarboxamide, 2-chloro-N-[(cyclohexylamino)thioxomethyl]-	MeOH	240(5.03)	73-3315-83
$C_{13}H_{16}ClN_3O_5S_2$			
Ambuside	n.s.g.	343(4.52)	162-0057-83
$C_{13}H_{16}IN_3$			
3,5-Pyridinedicarbonitrile, 1,4-dihydro-1-(2-iodoethyl)-2,4,4,6-tetramethyl-	EtOH	221(4.41),337(3.80)	73-0617-83
$C_{13}H_{16}N_2$			
Pyrimidine, 1,2-dihydro-4,6-dimethyl-1-(4-methylphenyl)-	EtOH	343(3.71)	39-1799-83C
Pyrimidine, 1,4-dihydro-4,6-dimethyl-1-(4-methylphenyl)-	EtOH	255(3.78)	39-1799-83C
Pyrimidine, 1,6-dihydro-4,6-dimethyl-1-(4-methylphenyl)-	EtOH	310(3.67)	39-1799-83C
Pyrimidine, 1,4-dihydro-4,4,6-trimethyl-1-phenyl-	EtOH	257(3.82)	39-1799-83C
$C_{13}H_{16}N_2O$			
1-Butanone, 1-(1H-indol-3-yl)-3-methyl-, oxime	n.s.g.	222(4.29),259(4.08), 284(3.96)	103-0184-83
1,2,4-Oxadiazole, 3-(2,6-dimethylphenyl)-5-(1-methylethyl)-	MeOH	268(2.84),274(2.86)	4-1609-83
1,2,4-Oxadiazole, 3-(2,6-dimethylphenyl)-5-propyl-	MeOH	269(2.81),274(2.82)	4-1609-83
9-Oxaergoline, 2,3-dihydro-	EtOH	244(3.74),292(3.32)	87-0522-83
	EtOH-HCl	263(2.88),272(2.81)	87-0522-83
1-Pentanone, 1-(1H-indol-3-yl)-, oxime	n.s.g.	224(4.30),258(4.13), 284(3.98)	103-0184-83
3-Pyridinecarboxamide, 1,4,5,6-tetrahydro-1-(phenylmethyl)-	EtOH	302(4.17)	150-2326-83M
1H-Pyrido[3,4-b]indole, 2,3,4,9-tetrahydro-6-methoxy-1-methyl-	EtOH	225(4.34),280(3.86)	162-0026-83
	pH 1	220(4.40),273(3.86)	162-0026-83

Compound	Solvent	λ_{max}(log ϵ)	Ref.
1H-Pyrido[3,4-b]indole-1-methanol, 2,3,4,9-tetrahydro-2-methyl-	EtOH	226(4.08),284(3.58), 291(3.57)	100-0310-83
Pyrimidine, 1,2-dihydro-1-(4-methoxyphenyl)-4,6-dimethyl-	EtOH	343(3.68)	39-1799-83C
Pyrimidine, 1,4-dihydro-1-(4-methoxyphenyl)-4,6-dimethyl-	EtOH	250(3.60)	39-1799-83C
Pyrimidine, 1,6-dihydro-1-(4-methoxyphenyl)-4,6-dimethyl-	EtOH	302(3.57)	39-1799-83C
$C_{13}H_{16}N_2OS$			
5H-Pyrrolo[1,2-a]thieno[3',2':3,4]pyrido[1,2-c]pyrimidin-7(8H)-one, 3b,4,9,9a,11,12-hexahydro-, cis-(±)-	MeOH	233(3.91)	13-0493-83B
$C_{13}H_{16}N_2O_2$			
Ethanone, 2-(dimethylamino)-1-(5-methoxy-1H-indol-3-yl)-	EtOH	220(4.27),253(4.27), 274(3.97),305(3.84)	100-0732-83
1,2,4-Oxadiazole, 5-ethyl-3-(2-methoxy-4,6-dimethylphenyl)-	MeOH	283(3.23)	4-1609-83
1,2,4-Oxadiazole, 5-ethyl-3-(4-methoxy-2,6-dimethylphenyl)-	MeOH	280(3.15)	4-1609-83
2-Propanol, 2-[(3-ethenylphenyl)azo]-, acetate	CH_2Cl_2	400(2.23)	126-2285-83
2-Pyrrolidinone, 1-methyl-3-[2-(methylamino)benzoyl]-	MeOH	236(4.39),271(3.83)	83-0897-83
$C_{13}H_{16}N_2O_2S$			
Acetamide, N-[4-(4-morpholinylthioxomethyl)phenyl]-	EtOH	219(4.14),273(4.38), 370(2.90)	80-0555-83
$C_{13}H_{16}N_2O_3S$			
1-Pyrrolidinepropanamide, 2,5-dioxo-N-[2-(2-thienyl)ethyl]-	MeOH	232(3.96)	13-0493-83B
$C_{13}H_{16}N_2O_4$			
6,9-Methano-2H,6H-pyrimido[2,1-b][1,3]-oxazocin-2-one, 7-acetoxy-7,8,9,10-tetrahydro-3-methyl-, (6α,7β,9α)-	pH 1	239(3.92)	4-0655-83
	pH 7	248(4.04)	4-0655-83
	EtOH	245(4.09)	4-0655-83
6,9-Methano-2H,6H-pyrimido[2,1-b][1,3]-oxazocin-2-one, 8-acetoxy-7,8,9,10-tetrahydro-3-methyl-, (6α,8β,9α)-	pH 1	238(3.92)	4-0655-83
	pH 7	247(4.03)	4-0655-83
	EtOH	243(4.08)	4-0655-83
4H-[1,3]Oxazino[3,2-b][1,2]diazepine-6(10aH)-carboxylic acid, 2,9-dimethyl-4-oxo-, ethyl ester	MeOH	216(3.93),254(4.03)	5-1393-83
4H-Pyrido[1,2-a]pyrimidine-3-carboxylic acid, 9-formyl-1,6,7,8-tetrahydro-6-methyl-4-oxo-, ethyl ester	EtOH	223(4.20),265(3.49), 315s(3.86),358(4.34)	39-0369-83C
$C_{13}H_{16}N_2O_4S$			
Glycine, N-[N-[(phenylmethoxy)carbonyl]thioglycyl]-, methyl ester	$CHCl_3$	266(4.10)	78-1075-83
$C_{13}H_{16}N_2O_5$			
3a,6a(1H,6H)-Cyclopentapyrazoledicarboxylic acid, 4-methyl-6-oxo-, diethyl ester	MeOH	230(3.97),297s(3.30)	32-0489-83
4,5'-Cyclopseudouridine, 2',3'-O-isopropylidene-2'-methyl-	MeOH	293(3.49)	18-2680-83
1,2-Diazabicyclo[5.2.0]nona-3,5-diene-2-carboxylic acid, 8-acetoxy-5-methyl-9-oxo-, ethyl ester, cis	MeOH	274(3.85)	5-1361-83

Compound	Solvent	$\lambda_{max}(\log \epsilon)$	Ref.
2-Hexanone, 5-methyl-5-nitro-4-(4-nitrophenyl)-	EtOH	268(4.03)	12-0527-83
1,2-Hydrazinedicarboxylic acid, 1-(1-methyl-2-oxo-2-phenylethyl)-, dimethyl ester	EtOH	248(4.05),280(3.34), 312s(2.03)	104-0110-83
Uridine, 5-(1-butynyl)-2'-deoxy-	pH 6	232(4.03),293(4.03)	44-1854-83
	pH 13	230(4.11),287(3.93)	44-1854-83
$C_{13}H_{16}N_2O_5S$			
1-Azabicyclo[3.2.0]hept-2-ene-2-carboxylic acid, 3-[[2-(acetylamino)-ethenyl]thio]-6-(1-hydroxyethyl)-7-oxo-, (E)-	H_2O	231(4.14),308(4.08)	87-0271-83
(Z)-	H_2O	226(4.05),306(3.97)	87-0271-83
$C_{13}H_{16}N_2O_6$			
L-Glutamic acid, N-[[3-hydroxy-5-(hydroxymethyl)-2-methyl-4-pyridinyl]-methylene]-	MeOH	254(4.00),290s(3.65), 336(3.54),420(3.37)	35-0803-83
2,4(1H,3H)-Pyrimidinedione, 1-β-D-arabinofuranosyl-5-(1-butynyl)-	pH 7	232(4.07),292(4.03)	87-0661-83
Uridine, 2'-deoxy-5-(4-hydroxy-1-butynyl)-	pH 6	233(4.08),293(4.08)	44-1854-83
	pH 13	228s(4.15),288(3.98)	44-1854-83
Uridine, 2'-deoxy-5-(3-methoxy-1-propynyl)-	pH 7	230(4.03),288(4.08)	87-0661-83
$C_{13}H_{16}N_2O_6S$			
1-Azabicyclo[3.2.0]hept-2-ene-2-carboxylic acid, 3-[[2-(acetylamino)ethenyl]sulfinyl]-6-(1-hydroxyethyl)-7-oxo-, sodium salt, (R)-	H_2O	238(4.08),291(3.92)	87-0271-83
Epithienamycin B sulfoxide	H_2O	244(4.14),287(4.06)	158-83-107
$C_{13}H_{16}N_2O_8S_2$			
1-Azabicyclo[3.2.0]hept-2-ene-2-carboxylic acid, 3-[[2-(acetylamino)ethenyl]thio]-7-oxo-6-[1-(sulfooxy)ethyl]-, [5R-[3(E),5α,6β(S*)]-, disodium salt	H_2O	228(4.12),307(3.91)	87-0271-83
	H_2O	227(4.17),307(4.19)	78-2551-83
$C_{13}H_{16}N_2O_9S$			
1-Azabicyclo[3.2.0]hept-2-ene-2-carboxylic acid, 3-[[2-(acetylamino)ethenyl]sulfinyl]-7-oxo-6-[1-(sulfooxy)ethyl]- (MM-4550)	H_2O	240(4.13),287(4.08)	78-2551-83
disodium salt	H_2O	240(4.16),292(4.01)	87-0271-83
$C_{13}H_{16}N_4O_4$			
1H-Purine-2,6-dione, 7-(5,6-dihydro-5-hydroxy-6-methyl-2H-pyran-2-yl)-3,7-dihydro-1,3-dimethyl-, [2S-(2α,5β,6α)]-	MeOH	275(4.04)	136-0301-83A
$C_{13}H_{16}N_4O_5$			
1H-Purine-2,6-dione, 7-(2,3-anhydro-6-deoxy-β-L-talo-pyranosyl)-3,7-dihydro-1,3-dimethyl-	MeOH	275(3.90)	136-0301-83A
4H-Pyrazolo[3,4-d]pyrimidin-4-one, 1,5-dihydro-1-[2,3-O-(1-methylethylidene)-β-D-ribofuranosyl]-	MeOH	250(4.80)	87-1601-83

Compound	Solvent	$\lambda_{max}(\log \epsilon)$	Ref.
$C_{13}H_{16}N_6OS_2Zn$			
Zinc, [[2,2'-[1-(4-methoxyphenyl)-1,2-ethanediylidene]bis[N-methylhydrazinecarbothioamidato]](2-)-N^2,$N^{2'}$-S,S']-, (T-4)-	EtOH	288(4.27),452(3.98)	161-0143-83
$C_{13}H_{16}N_6O_2S$			
2,4-Thiazolediamine, N,N,N',N'-tetramethyl-5-[(4-nitrophenyl)azo]-	$CHCl_3$	328(4.02),435(4.09), 545(4.52)	88-3563-83
$C_{13}H_{16}N_6O_2S_3Zn$			
Zinc, [[2,2'-[1-[4-(methylsulfonyl)-phenyl]-1,2-ethanediylidene]bis-[N-methylhydrazinecarbothioamid-ato]](2-)]-, (T-4)-	EtOH	328(4.33),460(4.01)	161-0143-83
$C_{13}H_{16}N_6S_2Zn$			
Zinc, [[2,2'-[1-(4-methylphenyl)-1,2-ethanediylidene]bis[N-methylhydrazinecarbothioamidato]](2-)-N^2,$N^{2'}$-S,S']-, (T-4)-	EtOH	288(4.36),450(4.11)	161-0143-83
$C_{13}H_{16}O$			
Benzene, 1-(1-cyclohexen-1-yl)-4-methoxy-	EtOH	255(4.17)	12-0361-83
$C_{13}H_{16}O_2$			
5H-Benzocycloheptene-1,2(5H)-dione, 6,7,8,9-tetrahydro-3,4-dimethyl-	hexane	213(4.51),417(3.22)	24-3884-83
Cyclohexanone, 2-(4-methoxyphenyl)-	EtOH	224(4.00),275(3.23), 281(3.17)	12-0361-83
$C_{13}H_{16}O_3$			
2H-1-Benzopyran-3-methanol, 7-methoxy-2,2-dimethyl-	EtOH	212s(3.85),224(4.05), 282(3.80),306(3.70), 314s(3.66)	39-0197-83C
Naphthalene, 1,2-dihydro-8-methoxy-5-(methoxymethoxy)-	EtOH	222(4.21),226s(3.90), 270(3.90),316(3.54)	94-2662-83
$C_{13}H_{16}O_4$			
4H-1,3-Benzodioxin-4-one, 2-(1,1-dimethylethoxy)-2-methyl-	pH 7.4	301(3.40)	1-0351-83
1,3-Benzodioxole-4,5-dione, 6,7-dipropyl-	MeCN	208(3.37),275(1.83), 562(1.26)	64-0752-83B
4,7-Benzofurandione, 2,3-dihydro-6-(2-hydroxypropyl)-2,2-dimethyl-	EtOH	278(4.02),426(2.95)	12-1263-83
2-Propenoic acid, 3-(2-hydroxy-3-methoxyphenyl)-2-methyl-, ethyl ester	MeOH	228.5(4.46),278(4.43), 318.5s(3.91)	2-0352-83
$C_{13}H_{16}O_5$			
Ethanone, 1-(3,4-dihydro-3,4,7-trihydroxy-2,2-dimethyl-2H-1-benzopyran-6-yl)-	EtOH	290(3.98),320(3.95)	39-0883-83C
$C_{13}H_{16}O_6$			
1H-Indene-1,1-dicarboxylic acid, octahydro-3,4-dioxo-, dimethyl ester	MeOH	279(3.97)	5-0112-83
	MeOH-base	315(4.20)	5-0112-83
	MeCN	278(3.94)	5-0112-83
	H_2SO_4	312(4.29)	5-0112-83
1H-Indene-4,4-dicarboxylic acid, octahydro-1,7-dioxo-, dimethyl ester	MeOH	281(3.95)	5-0112-83
	MeOH-base	312(4.21)	5-0112-83

Compound	Solvent	λ_{max}(log ε)	Ref.
(cont.)	MeCN	278(3.90)	5-0112-83
	H_2SO_4	311(4.22)	5-0112-83
$C_{13}H_{16}S$			
1H-Indene-1-thione, 2,3-dihydro-2,2,3,3-tetramethyl-	3-Mepentane	230f(--),308(--),322(4.11),540(--),565(--)	23-0894-83
$C_{13}H_{17}BrS$			
Thiophene, 3-bromo-2,5-dimethyl-4-(2-methyl-1-cyclohexen-1-yl)-	MeCN	193(4.45),241(3.87)	4-0729-83
$C_{13}H_{17}Cl$			
Benzene, (3-chloro-4,4-dimethyl-1-pentenyl)-, (E)-	EtOH	250(3.70)	12-0527-83
$C_{13}H_{17}N$			
5,8-Methanoquinoline, 5,6,7,8-tetrahydro-8,9,9-trimethyl-, (5S)-	heptane	185(4.6),215(3.5),270(3.6),277(3.6)	77-0287-83
$C_{13}H_{17}NO$			
Benzenebutanenitrile, 2-methoxy-γ,4-dimethyl-	EtOH	281(3.30)	64-0497-83B
Cyclohexanone, 2-(methylphenylamino)-	MeOH	251(4.04),294.5(3.27)	39-0573-83C
Pentacyclo[5.4.0.0^{2,6}.0^{3,10}.0^{5,9}]undecan-8-one, 11-(dimethylamino)-, syn	MeCN	219(3.68),280s(1.64)	44-1862-83
1H-Pyrrole, 2,5-dihydro-2-[1-(4-methoxyphenyl)ethyl]-	MeOH	224(3.96),275(3.14),282(3.08)	87-0469-83
$C_{13}H_{17}NOS$			
Butanethioamide, N,3,3-trimethyl-2-oxo-N-phenyl-	MeOH	260(4.12),278(4.07),365(2.28)	5-1116-83
Morpholine, N-[2-(4-methylphenyl)-1-thioxoethyl]-	heptane	282(4.17)	80-0555-83
	EtOH	279(4.23),345(2.38)	80-0555-83
Piperidine, N-[(2-methoxyphenyl)thioxomethyl]-	heptane	249(3.20),288(3.22),394(2.32)	80-0555-83
	EtOH	244(3.52),284(3.76),363(2.47)	80-0555-83
Piperidine, N-[(4-methoxyphenyl)thioxomethyl]-	heptane	256(4.09),283(4.13),401(2.77)	80-0555-83
	EtOH	243(3.99),279(4.22),365(2.84)	80-0555-83
$C_{13}H_{17}NO_2$			
4(5H)-Benzofuranone, 5-[(diethylamino)methylene]-6,7-dihydro-	EtOH	249.5(3.93),365(4.30)	111-0113-83
3H-Indole-3-ethanol, 2-ethoxy-3-methyl-	EtOH	254(3.85)	94-2986-83
2H-Pyrrole, 2-[(2,3-dimethoxyphenyl)methyl]-3,4-dihydro-	MeOH	220(3.88),268(2.93)	83-0773-83
$C_{13}H_{17}NO_3$			
8-Isoquinolinecarboxaldehyde, 1,2,3,4-tetrahydro-5,6-dimethoxy-2-methyl-	MeOH	229(4.15),277(3.79),317(3.55)	83-0801-83
1-Penten-3-ol, 4,4-dimethyl-1-(4-nitrophenyl)-, (E)-	EtOH	220(4.05),302(4.11)	12-0527-83
$C_{13}H_{17}NO_5$			
Cyclopropanecarbonitrile, 2-(1,2-O-isopropylidene-3-O-methyl-α-D-xylo-	EtOH	217(2.60)	159-0139-83

Compound	Solvent	λ_{max}(log ε)	Ref.
furanuronoyl)- (cont.)			159-0139-83
$C_{13}H_{17}NO_6$			
1H-Pyrrole-2,5-dicarboxylic acid, 3-(2-ethoxy-2-oxoethyl)-4-methyl-, 2-ethyl ester	$CHCl_3$	280(4.28),290s(4.21)	78-1849-83
$C_{13}H_{17}NS$			
Piperidine, 1-(2-phenyl-1-thioxo-ethyl)-	heptane	281(4.19),378(2.60)	80-0555-83
	EtOH	279(4.21)	80-0555-83
$C_{13}H_{17}N_2O$			
1H-Imidazol-1-yloxy, 2,5-dihydro-2,2,5,5-tetramethyl-4-phenyl-	EtOH	234(4.32),284(3.86)	70-2501-83
$C_{13}H_{17}N_2O_4P$			
Phosphonic acid, [3-(4-methylphenyl)-1,2,4-oxadiazol-5-yl]-, diethyl ester	MeOH	272s(3.87),287s(3.835)	44-4232-83
$C_{13}H_{17}N_3$			
1H-Benzimidazole, 2-(1-methyl-2-piperidinyl)-	MeOH	245(3.74),275(3.77), 280(3.74)	4-1481-83
[1,2]Diazepino[5,4-b]indole, 1,2,3,4-5,6-hexahydro-2,3-dimethyl-	EtOH	226(4.24),275s(3.55), 284(3.60),292(3.56)	39-1937-83C
$C_{13}H_{17}N_3O$			
3,5-Pyridinedicarbonitrile, 1,4-dihydro-1-(2-hydroxyethyl)-2,4,4,6-tetramethyl-	EtOH	222(4.57),340(3.81)	73-0617-83
$C_{13}H_{17}N_3O_2$			
2H-Azirine-2-carboxamide, 3-(dimethylamino)-N,N-dimethyl-2-phenoxy-	EtOH	263s(2.86),269(2.92), 275(2.79)	33-0262-83
Hexanenitrile, 2-[(1,2,3,4-tetrahydro-1,3-dimethyl-2,4-dioxo-5-pyrimidinyl)methylene]-, (E)-	MeCN	258(3.79),303(4.02)	35-0963-83
(Z)-	MeCN	259(3.92),313(4.05)	35-0963-83
4H-Pyrido[1,2-a]pyrimidine-3-carboxaldehyde, 9-[(dimethylamino)methylene]-6,7,8,9-tetrahydro-6-methyl-4-oxo-	EtOH	262(3.83),308(3.54), 417(4.69)	39-0369-83C
$C_{13}H_{17}N_3O_3$			
Acetic acid, (dimethylamino)oxo-, 2-acetyl-2-methyl-1-phenylhydrazide	hexane	248.5(4.20)	162-0482-83
$C_{13}H_{17}N_3O_4$			
Benzoic acid, 4-[3-(2-ethoxy-2-oxo-ethyl)-1-triazenyl]-, ethyl ester	hexane	221(4.00),289(4.29)	23-1549-83
	EtOH	223(4.20),310(4.28)	23-1549-83
1-Propanone, 1-[2-[(1,1-dimethylethyl)-NNO-azoxy]-4-nitrophenyl]-	EtOH	201(3.1),239(2.8), 313(2.8)	103-0593-83
$C_{13}H_{17}N_3O_9$			
Propanedioic acid, (1,2,3,4-tetrahydro-1-methyl-3,5-dinitro-6-oxo-2-pyridinyl)-, diethyl ester, compd. with diethylamine	MeOH	310(3.06),449(3.25)	138-0715-83
deuterated analog	MeOH	310(3.09),450(3.30)	138-0715-83
$C_{13}H_{17}N_5$			
Benzenamine, N,N-diethyl-4-(1H-	MeOH	477(4.57)	56-0547-83

Compound	Solvent	$\lambda_{max}(\log \epsilon)$	Ref.
imidazol-2-ylazo)- (cont.)			56-0547-83
$C_{13}H_{17}N_5O_3$			
1H-Imidazo[1,2-f]purine-2,4(3H,6H)-dione, 8-acetyl-7-ethyl-7,8-dihydro-1,3-dimethyl-	EtOH	292(4.27)	118-0577-83
$C_{13}H_{17}N_5O_6$			
Kanagamicin	pH 1	231(4.15),272(3.80), 297(3.84)	142-0027-83
	H_2O	233(4.20),272(3.80), 298(3.84)	142-0027-83
$C_{13}H_{17}OS$			
Thiophenium, tetrahydro-1-(1-methyl-2-oxo-2-phenylethyl)-, tetrafluoroborate	n.s.g.	250(4.11),280(3.47)	116-0864-83
Thiophenium, tetrahydro-1-[2-(4-methylphenyl)-2-oxoethyl]-, tetrafluoroborate	n.s.g.	260(4.16),300(3.58)	116-0864-83
2H-Thiopyranium, tetrahydro-1-(2-oxo-2-phenylethyl)-, tetrafluoroborate	n.s.g.	248(4.04),290(3.65)	116-0864-83
$C_{13}H_{17}O_2S$			
Thiophenium, tetrahydro-1-[2-(4-methoxyphenyl)-2-oxoethyl]-, tetrafluoroborate	n.s.g.	222(3.96),287(4.26)	116-0864-83
$C_{13}H_{18}BrNO_2$			
Isoquinoline, 8-bromo-5-ethoxy-1,2,3,4-tetrahydro-6-methoxy-2-methyl-	MeOH	226s(4.15),286(3.72)	83-0845-83
Isoquinoline, 8-bromo-1,2,3,4-tetrahydro-5,6-dimethoxy-2,3-dimethyl-	MeOH	223s(4.05),285(3.31)	83-0801-83
$C_{13}H_{18}ClNO$			
Oxazolidine, 2-(4-chlorophenyl)-3-methyl-4-(1-methylethyl)-, (2S-cis)	EtOH	216(4.15),254(3.66), 281s(2.72),291(2.51)	142-0607-83
$C_{13}H_{18}NO_2$			
Isoquinolinium, 3,4-dihydro-5,6-dimethoxy-2,3-dimethyl-, iodide	MeOH	217(4.70),330(4.51)	83-0801-83
Isoquinolinium, 5-ethoxy-3,4-dihydro-6-methoxy-2-methyl-, iodide	MeOH	216(4.52),331(4.32)	83-0845-83
$C_{13}H_{18}N_2OS$			
Morpholine, 4-[[4-(dimethylamino)phenyl]thioxomethyl]-	heptane	258(3.96),298(4.28), 408(3.20)	80-0555-83
	EtOH	253(3.99),293(4.21), 374(3.65)	80-0555-83
$C_{13}H_{18}N_2O_2$			
1-Propanone, 1-[2-[(1,1-dimethylethyl)-NNO-azoxy]phenyl]-	EtOH	207(3.4),243(4.0), 318(3.0)	103-0593-83
$C_{13}H_{18}N_2O_2S$			
Benzenesulfonic acid, 4-methyl-, (2-methylcyclopentylidene)hydrazide	EtOH	206(4.05),229(4.08), 275(2.92)	39-1901-83C
$C_{13}H_{18}N_2O_3$			
1,2-Diazabicyclo[5.2.0]nona-3,5-diene-2-carboxylic acid, 5,8,8-trimethyl-	MeOH	275(3.82)	5-1361-83

Compound	Solvent	λ_{max}(log ε)	Ref.
9-oxo-, ethyl ester (cont.)			5-1361-83
4H-Pyrido[1,2-a]pyrimidine-3-acetic acid, 6,7,8,9-tetrahydro-6-methyl-4-oxo-, ethyl ester	EtOH	229(3.70),280(3.75)	39-0369-83C
$C_{13}H_{18}N_2O_4$			
Isoquinoline, 5-ethoxy-1,2,3,4-tetrahydro-6-methoxy-2-methyl-8-nitro-	MeOH	216s(4.25),238s(3.96), 297(3.77),344(3.75)	83-0845-83
Isoquinoline, 1,2,3,4-tetrahydro-5,6-dimethoxy-2,3-dimethyl-8-nitro-	MeOH	212s(4.44),238s(4.04), 300(3.71),340(3.71)	83-0801-83
L-Valine, N-[[3-hydroxy-5-(hydroxymethyl)-2-methyl-4-pyridinyl]methylene]-	MeOH	255(4.00),290(3.65), 337(3.55),420(3.44)	35-0803-83
$C_{13}H_{18}N_2O_4S$			
1-Azabicyclo[3.2.0]hept-2-ene-2-carboxylic acid, 6-(1-hydroxyethyl)-7-oxo-3-(3-pyrrolidinylthio)-, (R)-	H_2O	297(3.92)	158-1034-83
L-Methionine, N-[[3-hydroxy-5-(hydroxymethyl)-2-methyl-4-pyridinyl]methylene]-	MeOH	253(3.22),290s(--), 336(3.99),423(3.22)	35-0803-83
$C_{13}H_{18}N_2O_5$			
2,5'-Anhydro-5,6-dihydrouridine, 2',3'-O-isopropylidene-5-methyl-, (5R,5S)-	n.s.g.	253(3.99)	33-0687-83
2,4(1H,3H)-Pyrimidinedione, 1-[3-acetoxy-4-(hydroxymethyl)cyclopentyl]-5-methyl-, (1α,3β,4α)-(±)-	pH 1 and 7	272(4.01)	4-0655-83
$C_{13}H_{18}N_2O_5S$			
2-Thiouracil, 5-[(2R*,3R*,4R*,5R*)-5-[(R*)-1-hydroxyethyl]-3,4-isopropylidenedioxytetrahydrofuran-2-yl]-	pH 13	223(4.21),264(4.18), 284(4.06)	18-2680-83
	MeOH	213(4.10),274(4.16), 286(4.12)	18-2680-83
2-Thiouracil, 5-[(hydroxymethyl)-3,4-isopropylidenedioxy-2-methyltetrahydrofuran-2-yl]-, (2R*,3R*,4R*,5R*)	pH 13	223(4.07),261(4.07), 292(3.83)	18-2680-83
	MeOH	214(3.98),275(4.11), 290(4.08)	18-2680-83
(2R*,3S*,4S*,5R*)-	pH 13	223(3.98),264(3.94), 289(3.83)	18-2680-83
	MeOH	215(3.98),276(4.16), 290(4.12)	18-2680-83
(2S*,3R*,4R*,5R*)-	pH 13	221(4.14),262(4.08), 298(3.91)	18-2680-83
	MeOH	215(4.10),275(4.20), 297(4.17)	18-2680-83
$C_{13}H_{18}N_2O_6$			
Uracil, 5-[(2R*,3R*,4R*,5R*)-5-[(R*)-(1-hydroxyethyl)-3,4-isopropylidenedioxytetrahydrofuran-2-yl]-	pH 13	285(3.87)	18-2680-83
	MeOH	262(3.92)	18-2680-83
Uracil, 5-[5-(hydroxymethyl)-3,4-isopropylidenedioxy-2-methyltetrahydrofuran-2-yl]-, (2R*,3R*,4R*,5R*)-	pH 13	288(4.03)	18-2680-83
	MeOH	263(3.93)	18-2680-83
(2R*,3S*,4S*,5R*)-	pH 13	285(3.88)	18-2680-83
	MeOH	266(3.86)	18-2680-83
(2S*,3R*,4R*,5R*)-	pH 13	289(3.89)	18-2680-83
	MeOH	265(3.87)	18-2680-83

Compound	Solvent	λ_{max}(log ε)	Ref.
$C_{13}H_{18}N_2O_6S$			
2-Thiouracil, 5-[(2R*,3S*,4S*)-5,5-bis(hydroxymethyl)-3,4-isopropylidenedioxytetrahydrofuran-2-yl]-	pH 13	213(4.05),265(3.95), 292(3.86)	18-2680-83
Uridine, 2'-deoxy-5-[1-[(2-hydroxyethyl)thio]ethenyl]-	pH 6	254(4.08)	69-1696-83
$C_{13}H_{18}N_2O_7$			
Uracil, 5-[(2R*,3S*,4S*)-5,5-bis(hydroxymethyl)-3,4-isopropylidenedioxytetrahydrofuran-2-yl]-	pH 13	284(3.67)	18-2680-83
	MeOH	263(3.69)	18-2680-83
$C_{13}H_{18}N_2O_7S$			
2-Thiouracil, 5-[(2R*,3R*,4R*,5R*)-2,5-bis(hydroxymethyl)-3,4-isopropylidenedioxytetrahydrofuran-2-yl]-	pH 13	223(4.11),264(3.93), 289(3.85)	18-2680-83
	MeOH	214(3.94),276(4.06), 293(4.00)	18-2680-83
$C_{13}H_{18}N_2O_8S_2$			
1-Azabicyclo[3.2.0]hept-2-ene-2-carboxylic acid, 3-[[2-(acetylamino)-ethyl]thio]-7-oxo-6-[1-(sulfooxy)-ethyl]-, disodium salt	H_2O	298(3.93)	78-2551-83
$C_{13}H_{18}N_4O$			
Pyrido[2,3-b]pyrazin-3(4H)-one, 4-[3-(dimethylamino)propyl]-2-methyl-	n.s.g.	222(4.4),327(4.0)	161-0330-83
Pyrido[3,4-b]pyrazin-2(1H)-one, 1-[3-(dimethylamino)propyl]-3-methyl-	n.s.g.	231(4.4),305(3.8)	161-0330-83
$C_{13}H_{18}O$			
1H-Benzocyclohepten-1-one, 2,3,4,5,6,9-hexahydro-6,6-dimethyl-	MeOH	248(3.83)	23-1226-83
2H-Benzocyclohepten-2-one, 4a,5,6,7,-8,9-hexahydro-4,4a-dimethyl-	EtOH	244(4.22)	24-3884-83
2-Cyclohexen-1-one, 3-(3-ethenyl-2,2-dimethylcyclopropyl)-, trans	MeOH	256(4.17)	23-1226-83
1,3-Methanonaphthalen-5(1H)-one, 2,3,4,6,7,8-hexahydro-2,2-dimethyl-	EtOH	262(4.06)	44-3750-83
1(2H)-Naphthalenone, 3,4,5,6,7,8-hexahydro-7-(2-propenyl)-, (+)-	EtOH	246(4.04)	44-3750-83
2(3H)-Naphthalenone, 4,6,7,8-tetrahydro-1,3,8-trimethyl-	EtOH	295(3.92)	78-3929-83
8a,4a-Propenonaphthalen-11-one, 1,2,3,4,5,6,7,8-octahydro-	MeCN	231(3.81)	35-2800-83
Spiro[5.5]undeca-1,4-dien-3-one, 1,5-dimethyl-	EtOH	247(4.25)	24-3884-83
$C_{13}H_{18}O_2$			
Cyclobutanecarboxylic acid, 3-(dicyclopropylmethylene)-, methyl ester	hexane	214(4.01)	44-1500-83
2-Naphthalenecarboxylic acid, 1,2,3-5,6,7-hexahydro-, ethyl ester	EtOH	241(4.15)	44-4272-83
2-Propenoic acid, 3-(2-ethenyl-1-cyclohexen-1-yl)-, ethyl ester, (E)-	EtOH	219(4.21),299(4.36), 305(4.23)	44-4272-83
$C_{13}H_{18}O_2S$			
2(5H)-Thiophenone, 3,5-diethyl-4-hydroxy-5-(2-methyl-1,3-butadienyl)-, (E)-(+)- (thiotetromycin)	EtOH	238(4.48),300(3.67)	158-0109-83
	EtOH	238(4.48),300(3.67)	158-1781-83

Compound	Solvent	λ_{max}(log ε)	Ref.
$C_{13}H_{18}O_3$			
3-Buten-2-one, 4-(3,7,7-trimethyl-2,8-dioxabicyclo[3.2.1]oct-3-en-1-yl)-, (E)-	pentane	226(c.4.30)	33-1061-83
2-Cyclopenten-1-one, 3-acetoxy-4(and 5)-methyl-5(and 4)-(2-methylbut-2-enyl)-, (1:1 mixture)	EtOH	243(4.22)	39-1885-83C
3aH-Indene-3a-carboxylic acid, 1,2,3-4,5,6-hexahydro-4,4-dimethyl-6-oxo-, methyl ester	EtOH	240(3.98)	44-4642-83
2-Naphthalenecarboxylic acid, 1,2,3,4-5,6,7,8-octahydro-5-methyl-1-oxo-, methyl ester (higher R_F isomer)	EtOH	247(4.049)	2-0331-83
isomer	EtOH	245.5(4.061)	2-0331-83
$C_{13}H_{18}O_4$			
4-Heptenoic acid, 7-(3-hydroxy-5-oxo-1-cyclopenten-1-yl)-, methyl ester, (Z)-(±)-	MeOH	221(3.95)	87-0786-83
$C_{13}H_{18}O_6$			
Benzeneacetic acid, α,2-dihydroxy-4,5-dimethoxy-α-methyl-, ethyl ester	EtOH	225(3.84),291(3.67)	39-1649-83C
Cyclopropanecarboxaldehyde, 2-(1,2-O-isopropylidene-3-O-methyl-α-D-xylo-furanuronoyl)-	EtOH	223(2.33),280(1.94)	159-0139-83
$C_{13}H_{19}$			
Cycloheptatrienylium, 1,3-bis(1-methylethyl)-, perchlorate	10% HCl	273(4.56),290(3.81)	78-4011-83
$C_{13}H_{19}BrO_3$			
1-Cyclopentene-1-heptanoic acid, 4-bromo-5-oxo-, methyl ester	EtOH	237(3.71)	104-0261-83
$C_{13}H_{19}ClN_2$			
Spiro[cyclohexane-1,2'(1'H)-quinazoline], 4'-chloro-5',6',7',8'-tetrahydro-, monohydrochloride	MeOH	367(3.80)	103-0095-83
monophosphorodichloridate	MeOH	365(3.70)	103-0095-83
$C_{13}H_{19}FO_3$			
3-Buten-2-one, 4-(4-fluoro-5-hydroxy-2,2,6-trimethyl-7-oxabicyclo[4.1.0]-hept-1-yl)-, [1α(E),4α,5β,6α]-	EtOH	228(4.09)	33-1061-83
$C_{13}H_{19}NO$			
Cyclohexanone, 2-[5-(dimethylamino)-2,4-pentadienylidene]-	EtOH	452(4.65)	70-0780-83
	$CHCl_3$	430(--)	70-0780-83
1H-Isoindole, 2,3-dihydro-2-methoxy-1,1,3,3-tetramethyl-	EtOH	215(3.43),257(2.75), 263(2.95),270(3.00)	12-1957-83
Oxazolidine, 3-methyl-4-(1-methylethyl)-2-phenyl-	EtOH	203(4.23),246(3.50), 278s(2.45),287s(2.34)	142-0607-83
$C_{13}H_{19}NO_2$			
Isoquinoline, 5-ethoxy-1,2,3,4-tetrahydro-6-methoxy-2-methyl-	MeOH	221s(3.99),278(3.35)	83-0845-83
Isoquinoline, 1,2,3,4-tetrahydro-5,6-dimethoxy-2,3-dimethyl-	MeOH	222s(3.92),277.5(3.12)	83-0801-83
Pyrrolidine, 2-[(2,3-dimethoxyphenyl)-methyl]-	MeOH	218s(3.68),272(3.15)	83-0773-83

Compound	Solvent	λ_{max}(log ε)	Ref.
$C_{13}H_{19}NO_3$			
Benzaldehyde, 4-[2-(dimethylamino)ethoxy]-3-ethoxy-	MeOH	230(4.26),274(4.10), 308(3.99)	73-1173-83
8-Isoquinolinemethanol, 1,2,3,4-tetrahydro-5,6-dimethoxy-2-methyl-	MeOH	226s(3.83),282(3.04)	83-0801-83
Retronecine, 9(and 7)-O-(3-methyl-2-butenoyl)-	EtOH	220(4.04)	39-1819-83C
$C_{13}H_{19}NO_3S$			
Piperidine, 1-(5,6,7,8-tetrahydro-1,2-benzoxathiin-4-yl)-, S,S-dioxide	EtOH	263(3.83),290s(3.70)	4-1549-83
$C_{13}H_{19}NO_4$			
2-Azabicyclo[2.2.2]oct-5-ene-6-carboxylic acid, 7-(2-methyl-1,3-dioxolan-2-yl)-, methyl ester	EtOH	308(4.06)	44-4262-83
1H-Pyrrole-3-acetic acid, 2-(ethoxycarbonyl)-4,5-dimethyl-, ethyl ester	$CHCl_3$	283(4.20)	78-1849-83
Ribofuranosylamine, N-(2-phenylethyl)-	EtOH	257(4.30)	136-0001-83B
$C_{13}H_{19}NO_4S_2$			
2-Thietanone, 3-amino-3,4,4-trimethyl-, p-toluenesulfonate	EtOH	229(3.76),232s(3.45)	39-2259-83C
$C_{13}H_{19}N_2O_3$			
4H-Pyrido[1,2-a]pyrimidinium, 3-(ethoxycarbonyl)-6,7,8,9-tetrahydro-1,6-dimethyl-4-oxo-	H_2O	204(3.86),238(3.88), 287(3.82)	39-1413-83B
$C_{13}H_{19}N_2O_8P$			
Thymidine, 5'-O-1,3,2-dioxaphosphorinan-2-yl-, P-oxide	EtOH	267(3.93)	87-1153-83
$C_{13}H_{19}N_3O_2$			
5-Pyrimidinepropanenitrile, α-butyl-1,2,3,4-tetrahydro-1,3-dimethyl-2,4-dioxo-	MeCN	270(3.93)	35-0963-83
$C_{13}H_{19}N_3O_5$			
Isocytosine, 5-[(2R*,3R*,4R*,5R*)-5-[(R*)-(1-hydroxyethyl)-3,4-isopropylidenedioxytetrahydrofuran-2-yl]-	pH 13	232(3.96),276(3.84)	18-2680-83
	MeOH	226(3.86),290(3.86)	18-2680-83
Isocytosine, 5-[(2R*,3R*,4R*,5R*)-5-hydroxymethyl-3,4-isopropylidenedioxy-2-methyltetrahydrofuran-2-yl]-	pH 13	231(3.89),278(3.77)	18-2680-83
	MeOH	224(4.02),290(3.93)	18-2680-83
(2R*,3S*,4S*,5R*)-	pH 13	233(4.01),276(3.90)	18-2680-83
	MeOH	225(4.14),289(4.05)	18-2680-83
5-Pyrimidinepentanoic acid, 3,4-dihydro-6-hydroxy-2-morpholino-4-oxo-	50% MeOH	250s(3.868),275(4.072)	73-0304-83
	+ HCl	240(3.964),283(3.998)	73-0304-83
	+ NaOH	250s(3.872),274(4.052)	73-0304-83
$C_{13}H_{19}N_3O_6$			
Isocytosine, 5-[(2R*,3S*,4S*)-5,5-bis(hydroxymethyl)-3,4-isopropylidenedioxytetrahydrofuran-2-yl]-	pH 13	233(4.01),277(3.89)	18-2680-83
	MeOH	227(3.78),290(3.76)	18-2680-83
$C_{13}H_{19}N_5O_4$			
1H-Purine-2,6-dione, 8-[[(1-acetoxymethyl)propyl]amino]-3,7-dihydro-1,3-dimethyl-	EtOH	296(4.33)	118-0577-83

Compound	Solvent	$\lambda_{max}(\log \epsilon)$	Ref.
1H-Pyrazolo[3,4-d]pyrimidin-4-amine, N-propyl-1-β-D-ribofuranosyl-	pH 11	265s(--),282(--)	87-1601-83
	MeOH	265s(--),283(4.10)	87-1601-83
$C_{13}H_{20}N_2O$			
1,4,6,8-Nonatetraen-3-one, 1,9-bis(dimethylamino)-	EtOH	256(4.09),450(4.74)	70-1882-83
	$CHCl_3$	430(--)	70-1882-83
$C_{13}H_{20}N_2O_3S$			
Etozolin	MeOH	243(4.0),283(4.32)	162-0561-83
Piperazine, 1-methyl-4-(5,6,7,8-tetrahydro-1,2-benzoxathiin-4-yl)-, S,S-dioxide	EtOH	263(3.93),285s(3.79)	4-1549-83
$C_{13}H_{20}N_2O_4$			
Propanedioic acid, amino[2-(1H-pyrazol-1-yl)ethyl]-, diethyl ester	MeOH	222(3.62),303(3.39)	23-0454-83
$C_{13}H_{20}N_2O_5$			
5-Pyrimidinepentanoic acid, 5-butyldihydro-2,4,6-trioxo-	50%MeOH-HCl	211.5(3.95)	73-0137-83
	+ NaOH	241(3.97)	73-0137-83
$C_{13}H_{20}N_2O_6S$			
Uridine, 6-(butylthio)-	MeOH	279(4.11)	94-1222-83
$C_{13}H_{20}N_4O_4$			
Imidazo[4,5-d][1,3]diazepin-8-ol, 3-(2-deoxy-β-D-erythro-pentofuranosyl)-5-ethyl-3,6,7,8-tetrahydro-	pH 7	281(3.97)	87-1478-83
1,2,4-Triazine-3,6-dicarboxylic acid, 4-(hexahydro-1H-azepin-1-yl)-4,5-dihydro-	CH_2Cl_2	351(3.67)	83-0472-83
$C_{13}H_{20}O$			
2-Buten-1-one, 1-(2,6,6-trimethyl-2-cyclohexen-1-yl)-, (E)-	EtOH	225(4.16),333(1.90)	54-0307-83
2-Buten-1-one, 1-(2,6,6-trimethyl-1-cyclohexen-1-yl)- (β-damascone)	EtOH	224(4.21),274s(3.13), 335(1.90)	54-0307-83
$C_{13}H_{20}OS$			
2(3H)-Naphthalenone, 4,4a,5,6,7,8-hexahydro-4a,5-dimethyl-1-(methylthio)-, cis-(±)-	EtOH	245(3.97)	39-0161-83C
$C_{13}H_{20}OS_2$			
2,3-Dithiaspiro[4.4]nonane, 1-cyclopentylidene-4-methoxy-	CCl_4	263(3.70),320(2.40)	44-4482-83
$C_{13}H_{20}O_2$			
1,2-Benzenediol, 3-(1,1-dimethylethyl)-5-(1-methylethyl)-	hexane	279(3.15)	70-0550-83
2(3H)-Naphthalenone, 4,4a,5,6,7,8-hexahydro-5-hydroxy-1,3,8-trimethyl-	EtOH	248(4.00)	78-3929-83
2(3H)-Naphthalenone, 4,4a,5,6,7,8-hexahydro-1-methoxy-4a,5-dimethyl-, cis-(±)-	EtOH	254(3.99)	39-0161-83C
$C_{13}H_{20}O_3$			
1-Cyclopentene-1-heptanoic acid, 5-oxo-, methyl ester	EtOH	228.5(3.91)	104-0261-83

Compound	Solvent	λ_{max}(log ϵ)	Ref.
$C_{13}H_{20}O_3S_4$			
2-Hexenoic acid, 4,5-di-1,3-dithiolan-2-yl-3-methoxy-	EtOH	238(4.01)	94-0925-83
$C_{13}H_{20}O_4$			
1-Cyclopentene-1-heptanoic acid, 4-hydroxy-5-oxo-, methyl ester	EtOH	230(3.93)	39-0319-83C
$C_{13}H_{20}O_6S$			
α-D-xylo-Hex-5-enofuranose, 3-O-methyl-6-S-methyl-1,2-O-(1-methylethylidene)-6-thio-, acetate, (Z)-	EtOH	228(3.77),246(3.78)	159-0139-83
$C_{13}H_{21}N$			
Piperidine, 1-(4,4-dimethyl-1,5-cyclohexadien-1-yl)-	C_6H_{12}	223(3.93),286(3.24)	33-0735-83
$C_{13}H_{21}NO$			
Morpholine, 4-[4-methyl-1-(1-methylethenyl)-1,3-pentadienyl]-, (Z)-	C_6H_{12}	265(3.77)	33-0735-83
Morpholine, 4-(2,4,4-trimethyl-1,5-cyclohexadien-1-yl)-	C_6H_{12}	238(3.66),270s(--)	33-0735-83
$C_{13}H_{21}NO_2$			
2-Cyclohexen-1-one, 3-methyl-2-[2-(4-morpholinyl)ethyl]-	EtOH	243(3.97)	23-0078-83
2,4-Hexadienoic acid, 6-piperidino-, ethyl ester, (E,E)-	EtOH	253(4.49)	111-0447-83
$C_{13}H_{21}NO_5$			
1H-Pyrrole-2,4-dicarboxylic acid, 2,3-dihydro-3-hydroxy-1,3,5-trimethyl-, diethyl ester	EtOH	271(4.16)	94-1474-83
$C_{13}H_{21}NO_5S_2$			
Butanoic acid, 2-amino-3-mercapto-2,3-dimethyl-, p-toluenesulfonate	EtOH	225(3.70),257(2.23), 262(2.23)	39-2259-83C
$C_{13}H_{21}N_2O_4$			
1H-Pyrrol-1-yloxy, 3-[[(2-ethoxy-2-oxoethyl)amino]carbonyl]-2,5-dihydro-2,2,5,5-tetramethyl-	MeOH	269(3)	28-0043-83B
$C_{13}H_{21}N_3O$			
1-Propanone, 1-[2-amino-4-(1,1-dimethylethyl)-5-pyrimidinyl]-2,2-dimethyl-	EtOH	224(3.92),262(3.84), 294s(3.65)	4-0649-83
$C_{13}H_{21}N_3O_2$			
3-Heptenoic acid, 2,4-dimethyl-7-(4-methyl-1H-1,2,3-triazol-1-yl)-, methyl ester, (E)-	MeOH	207(3.72)	44-2432-83
$C_{13}H_{21}N_3O_4$			
5-Pyrimidinepentanoic acid, 2-(butylamino)-1,4-dihydro-6-hydroxy-4-oxo-	50% MeOH	241(3.817),276(4.134)	73-0304-83
	+ HCl	240s(3.909),269(4.114)	73-0304-83
	+ NaOH	241(3.813),276(4.120)	73-0304-83
5-Pyrimidinepentanoic acid, 2-(ethylamino)-1,4-dihydro-6-hydroxy-4-oxo-, ethyl ester	50% MeOH	270(4.212)	73-0304-83
	+ HCl	240s(3.864),268(4.086)	73-0304-83
	+ NaOH	242(3.818),275(4.122)	73-0304-83

Compound	Solvent	λ_{max}(log ε)	Ref.
$C_{13}H_{21}N_7O_4$			
Adenosine, 8-[(3-aminopropyl)amino]-	pH 2	280(4.04)	73-1910-83
	pH 12	280(4.09)	73-1910-83
$C_{13}H_{22}$			
Cyclododecene, 3-methylene-, (E)-	EtOH	231(4.11)	33-2519-83
Cyclohexane, 1-(1,1-dimethylethyl)-4-(2-propenylidene)-, (R)-	C_6H_{12}	231(4.46),238(4.50), 246(4.32)	35-3252-83 +35-3264-83
Cyclohexane, 4-methyl-1-(1-methylethyl)-2-(2-propenylidene)-, [1S-(1α,2E,4β)]-	C_6H_{12}	233(4.41),239(4.45), 248(4.27)	35-3264-83
2,3-Decadiene, 4-ethenyl-2-methyl-	EtOH	225(4.16)	70-0569-83
$C_{13}H_{22}N_2O$			
Pyrazine, 3-methoxy-2,5-bis(2-methylpropyl)-	EtOH	222(3.74),283(3.74), 298(3.77)	4-0951-83
$C_{13}H_{22}N_4O_4$			
Morpholine, 4,4'-(2,3-dihydro-1-methyl-4-nitro-1H-pyrrole-2,3-diyl)bis-, trans	MeOH	230(3.7),375(4.3)	44-0162-83
$C_{13}H_{22}O$			
Acetaldehyde, [4-(1,1-dimethylethyl)-2-methylcyclohexylidene]-, (E,2R,4R)-	C_6H_{12}	232(4.25)	35-3264-83
(E,2S,4R)-(+)-	C_6H_{12}	233(4.18),332(1.94), 346(1.85),362(1.66), 380(1.26)	35-3252-83
(Z,2S,4R)-(+)-	C_6H_{12}	232(4.26),322(1.59), 333(1.70),347(1.75), 363(1.65),382(1.30)	35-3252-83
$C_{13}H_{22}O_2S_2$			
Thiophene, 2,3-dihydro-3-[(octylthio)-methylene]-, 1,1-dioxide	EtOH	213(3.78),305(4.44)	104-1447-83
$C_{13}H_{22}O_3S_2$			
3(2H)-Thiophenone, 5-methyl-4-(octylthio)-, 1,1-dioxide	EtOH	258(3.52),307(2.71)	104-1447-83
	EtOH-KOH	255(3.71),300(2.97)	104-1447-83
$C_{13}H_{23}NO$			
3(2H)-Pyridinone, 1-(1,1-dimethylethyl)-1,6-dihydro-5-methyl-4-(1-methylethyl)-, hydrochloride	n.s.g.	243(4.11)	70-2119-83
4(1H)-Quinolinone, octahydro-1-(1-methylpropyl)-, trans	isooctane	211(3.09),230s(2.26), 296(1.62)	103-0425-83
$C_{13}H_{23}NO_2$			
2,4-Heptadienoic acid, 7-(diethylamino)-, ethyl ester, (E,E)-	EtOH	262(4.33)	70-1444-83
$C_{13}H_{23}NO_3$			
Cyclohexanone, 2-[[bis(2-methoxyethyl)amino]methylene]-, (E)-	EtOH	331(4.29)	4-0839-83
$C_{13}H_{23}NO_3Si_2$			
6H-[1,2,5]Oxadisilolo[3,4-d]azepine-6-carboxylic acid, 1,3,3a,8a-tetrahydro-1,1,3,3-tetramethyl-, ethyl ester, trans	MeOH	238(4.37)	18-0175-83

Compound	Solvent	$\lambda_{max}(\log \epsilon)$	Ref.
$C_{13}H_{23}N_2O$			
Methanaminium, N-[[3-[(dimethylamino)-methylene]-2-methoxy-1-cyclohexen-1-yl]methylene]-N-methyl-, perchlorate	MeOH	453(4.97)	104-1854-83
$C_{13}H_{23}N_2O_3$			
1H-Pyrrol-1-yloxy, 2,5-dihydro-3-[[(2-hydroxy-2-methylpropyl)amino]carbonyl]-2,2,5,5-tetramethyl-	MeOH	276(3)	28-0043-83B
$C_{13}H_{24}O_2S$			
Cyclohexanone, 3-(ethylthio)-5-(1-methoxy-1-methylethyl)-2-methyl-, (R,S,S)-	MeOH	218(3.03),237(2.90), 288s(1.60)	49-0195-83
(S,S,S)-	MeOH	237(2.99),292(1.85)	49-0195-83
$C_{13}H_{24}O_4$			
2(3H)-Furanone, dihydro-4-(hydroxymethyl)-3-(1-hydroxy-6-methylheptyl)-	EtOH	215(2.08)	158-83-166
$C_{13}H_{25}NO_4Si_2$			
1H-Azepine-1-carboxylic acid, 4,5-dihydro-4,5-bis(hydroxydimethylsilyl)-, ethyl ester, trans	MeOH	240(4.09)	18-0175-83
$C_{13}H_{25}O_3P$			
Phosphonic acid, (2-methyl-1,3-butadien-1-yl)-, bis(1,1-dimethylethyl) ester, (E)-	C_6H_{12}?	236.5(4.35)	65-1054-83
$C_{13}H_{25}O_4P$			
Phosphonium, triethyl-, 2-ethoxy-1-(ethoxycarbonyl)-2-oxoethylide	hexane	222s(--),235(4.05)	99-0023-83
	EtOH	210(--),241(--)	99-0023-83
$C_{13}H_{25}O_5P$			
Phosphorus(1+), ethoxy[2-ethoxy-1-(ethoxycarbonyl)-2-oxoethyl]-diethyl-, hydroxide, inner salt, (T-4)-	hexane	228(4.05)	99-0023-83
$C_{13}H_{25}O_7P$			
Phosphorus(1+), triethoxy[2-ethoxy-1-(ethoxycarbonyl)-2-oxoethyl]-, hydroxide, inner salt, (T-4)-	hexane	216s(4.07)	99-0023-83
	EtOH	205(3.84),228(3.96)	99-0023-83
$C_{13}H_{26}O_2Te$			
Octanoic acid, 8-(butyltelluro)-, methyl ester	MeOH	233.0(3.71)	101-0031-83M
Pentanoic acid, 5-(heptyltelluro)-, methyl ester	MeOH	232.0(3.68)	101-0031-83M
$C_{13}H_{28}N_2O_4P$			
1-Piperidinyloxy, 4-[(diethoxyphosphinyl)amino]-2,2,6,6-tetramethyl-	EtOH	242(3.32),452(1.03)	70-1906-83

Compound	Solvent	λ_{max}(log ε)	Ref.
$C_{14}H_5Cl_2D_7N_2O$			
3-Pyridinecarbonitrile-2,6-d_2, 1-[(2,6-dichlorophenyl)methyl-d_2]-1,6-dihydro-6-(methoxy-d_3)-	CH_2Cl_2	298(3.8),313(3.8)	64-0873-83B
$C_{14}H_6Cl_6$			
Benzene, 1,1'-(1,2-dichloro-1,2-ethenediyl)bis[2,4-dichloro-, (E)-	heptane	236(4.29)	18-3009-83
(Z)-	heptane	266(3.81)	18-3009-83
$C_{14}H_6O_8$			
Ellagic acid	EtOH	255(4.60),366(3.93)	162-0512-83
$C_{14}H_6S_4$			
Anthra[1,9-cd:4,10-c'd']bis[1,2]dithiole	$CHCl_3$	288(4.36),437(3.71), 520(3.93),548(3.98)	88-2401-83
$C_{14}H_7BrCl_2S$			
Dibenzo[b,f]thiepin, 10-bromo-2,8-dichloro-	MeOH	226(4.56),266(4.31), 295(3.85)	73-1187-83
$C_{14}H_7ClN_2O_6S$			
4H,5H-Pyrano[3,4-e][1,3]oxazine-4,5-dione, 7-chloro-2-[[(2-nitrophenyl)-methyl]thio]-	$CHCl_3$	272(4.16),344(4.16)	24-3725-83
$C_{14}H_7ClO_2$			
9,10-Anthracenedione, 2-chloro-	polyethylene	257.5(4.62),277(4.29), 323?(3.25?)	135-1304-83
$C_{14}H_7F_5O_4$			
2(5H)-Furanone, 4-hydroxy-5-methyl-3-[1-oxo-3-(pentafluorophenyl)-2-propenyl]-, (E)-	MeOH	200(4.04),237(4.14), 279(4.14),333(4.17)	83-0115-83
$C_{14}H_7NO_3$			
5H-Furo[3',2':6,7][1]benzopyrano[3,4-c]pyridin-5-one	EtOH	240(4.44),280(3.92), 325(3.83)	39-0225-83C
$C_{14}H_8BrClOS$			
Dibenzo[b,f]thiepin-10(11H)-one, 11-bromo-2-chloro-	MeOH	231(4.35),260s(3.97), 327(3.61)	73-0906-83
$C_{14}H_8BrNO_5S$			
2-Anthracenesulfonic acid, 1-amino-4-bromo-9,10-dihydro-9,10-dioxo-	DMF	490(3.82)	104-0382-83
$C_{14}H_8ClN$			
Cyclopent[4,5]azepino[2,1,7-cd]pyrrolizine, 2-chloro-	EtOH	212(4.21),244(4.02), 325(3.36),350(4.13), 440(3.14)	24-1174-83
trifluoroacetate	CF_3COOH	293(4.39),305(4.41), 341(3.87),352(4.08), 383(3.25),403(3.43), 425(3.16)	24-1174-83
$C_{14}H_8ClNO_4S_2$			
4H,5H-Pyrano[3,4-e][1,3]oxazine-4,5-dione, 7-[(4-chlorophenyl)thio]-2-(methylthio)-	$CHCl_3$	248(4.10),326(3.91), 380(4.14)	24-3725-83

Compound	Solvent	$\lambda_{max}(\log \epsilon)$	Ref.
$C_{14}H_8Cl_4$			
Benzene, 1,1'-(1,2-dichloro-1,2-ethenediyl)bis[2-chloro-, (E)-	heptane	251(3.71)	18-3009-83
(Z)-	heptane	265(3.71)	18-3009-83
$C_{14}H_8Cl_6O_2$			
Spiro[2H-cyclohepta[b]furan-2,1'-[3,5]-cyclohexadien]-2'-one, 3,3,3',4',5'-6'-hexachloro-3,3a,4,8a-tetrahydro-, (2'α,3'αa,8'aα)-	EtOH	229(4.22),337(3.48)	78-1281-83
(2'α,3'aβ,8'aβ)-	EtOH	232(4.13),348(3.36)	78-1281-83
$C_{14}H_8CsN$			
Cesium, (2-cyano-9H-fluoren-9-yl)-	$(MeOCH_2)_2$	357(4.12),519(3.42)	104-1592-83
Cesium, (9-cyano-9H-fluoren-9-yl)-	$(MeOCH_2)_2$	416(3.35),439(3.35)	104-1592-83
$C_{14}H_8FNO_3S$			
Dibenzo[b,f]thiepin-10(11H)-one, 3-fluoro-8-nitro-	MeOH	225(4.26),255s(4.10), 335(4.04)	73-0144-83
$C_{14}H_8LiN$			
Lithium, (2-cyano-9H-fluoren-9-yl)-	$(MeOCH_2)_2$	360(4.08),379(4.00), 499(3.30),529(3.43), 569(3.32)	104-1592-83
Lithium, (9-cyano-9H-fluoren-9-yl)-	$(MeOCH_2)_2$	406(3.32),428(3.31)	104-1592-83
$C_{14}H_8N_2O$			
6H-Indolo[3,2,1-de][1,5]naphthyridin-6-one (canthin-6-one)	EtOH	252(4.08),261(4.05), 271(4.00),294(3.85), 364(4.02),381(3.97)	94-3198-83
$C_{14}H_8N_2O_2$			
Canthine-2,6-dione	MeOH	227(4.44),247(4.34), 253(4.31),291(3.99), 300(4.01),400s(3.91), 421(4.11),445(4.14)	100-0222-83
Canthin-6-one, 10-hydroxy-	MeOH	228s(4.19),269s(4.22), 275(4.25),304s(4.04), 312(4.07),352(4.17)	100-0222-83
	MeOH-NaOMe	238s(4.20),283s(4.14), 297(4.19),320s(4.11), 356(4.04),428(3.93)	100-0222-83
$C_{14}H_8N_2O_2S$			
2-Thiazoleacetonitrile, 4-(2-oxo-2H-1-benzopyran-3-yl)-	MeCN	334(4.23)	48-0551-83
$C_{14}H_8N_2O_3S$			
4H-Indeno[1,2-c]isothiazole-3-acetic acid, α-cyano-4-oxo-, methyl ester	EtOH	220(--),237(--), 261(--),271(--), 293(--),320(--), 354(--),410(--)	95-1243-83
$C_{14}H_8N_2O_7S$			
4H,5H-Pyrano[3,4-e][1,3]oxazine-4,5-dione, 2-(methylthio)-7-(4-nitrophenoxy)-	$CHCl_3$	273(4.32),337(4.14)	24-3725-83
$C_{14}H_8N_4O_5$			
Bicyclo[4.2.0]octa-1,3,5-triene-7,8-dione, mono(2,4-dinitrophenylhydrazone)	CH_2Cl_2	236(3.96),256s(3.86), 382(4.08)	18-0965-83

Compound	Solvent	λ_{max}(log ε)	Ref.
$C_{14}H_8N_6O_{12}$			
Benzene, 1,1'-(1,2-ethanediyl)bis- [2,4,6-trinitro-	DMSO-base	380(4.11),620(4.62)	150-0314-83S
$C_{14}H_8N_8S_4$			
2-Thiazolamine, N,N'-[3-(2-thiazolyl)- 2H-thiazolo[3,2-a]-1,3,5-triazine- 2,4(3H)-diylidene]bis-	dioxan	262(3.73),315(3.75), 375(3.64)	48-0463-83
$C_{14}H_8O_2$			
1,4-Anthracenedione	EtOH	233(4.82),251(4.24), 288(4.12),298(4.15), 410(3.63)	73-0112-83
9,10-Anthracenedione	EtOH	252(4.65),272(4.16), 324(3.69)	73-0112-83
	96% H_2SO_4	272(4.66),311(4.08), 411(3.99)	23-1965-83
	polyethylene	252(4.60),273.5(4.29), 323(3.60)	135-1304-83
$C_{14}H_8O_3$			
9,10-Anthracenedione, 1-hydroxy-	96% H_2SO_4	216(4.15),261(4.43), 305(4.00),458(3.97)	23-1965-83
9,10-Anthracenedione, 2-hydroxy-	96% H_2SO_4	248(4.17),289(4.41), 321(4.01),415(3.79), 484(3.89)	23-1965-83
$C_{14}H_8O_4$			
9,10-Anthracenedione, 1,2-dihydroxy-	MeOH	248(4.18),265s(4.08), 275s(3.95)	94-2353-83
	96% H_2SO_4	274(4.49),286(4.44), 323(3.84),498(4.08)	23-1965-83
9,10-Anthracenedione, 1,4-dihydroxy-	96% H_2SO_4	253(4.36),278(4.50), 330(3.83),478(3.86), 511(4.23),548(4.41)	23-1965-83
9,10-Anthracenedione, 1,8-dihydroxy-	n.s.g.	250(4.60),430(4.35)	162-0406-83
9,10-Anthracenedione, 2,3-dihydroxy-	96% H_2SO_4	206(4.18),246(4.01), 301(4.51),425(3.86)	23-1965-83
$C_{14}H_8O_4Ru_2$			
Ruthenium, tetracarbonylbis[μ-(η:η^5- 2,4-cyclopentadien-1-ylidene)]di-	THF	243(3.99),273(4.04), 329(3.85),388s(3.18)	35-1676-83
$C_{14}H_8O_7$			
2H,8H-Benzo[1,2-b:3,4-b']dipyran-10- carboxaldehyde, 4-hydroxy-5-methoxy- 2,8-dioxo-	EtOH	228(4.26),260(3.89), 280(3.22),325(4.55)	78-2277-83
$C_{14}H_8Se_4$			
1,3-Benzdiselenole, 2-(1,3-benzodi- selenol-2-ylidene)-	MeCN-THF	221(4.45),286(4.21), 320s(3.70),470(2.20)	77-0295-83
$C_{14}H_9BrN_2O$			
Bicyclo[4.2.0]octa-1,3,5-triene-7,8- dione, mono[(4-bromophenyl)hydra- zone], (E)-	CH_2Cl_2	258(4.26),342(4.10), 407(4.12)	18-0965-83
1(2H)-Phthalazinone, 2-(4-bromo- phenyl)	CH_2Cl_2	236(4.42),270(3.89), 303(3.97)	18-0965-83
$C_{14}H_9BrN_2S_2$			
2,4(1H,3H)-Quinazolinedithione,	EtOH	293(4.12),308(4.14),	97-0215-83

Compound	Solvent	λ_{max}(log ε)	Ref.
6-bromo-3-phenyl- (cont.)		360(3.90),377(3.94), 395(3.82)	97-0215-83
$C_{14}H_9BrO_2$			
2-Biphenylenol, 3-bromo-, acetate	EtOH	245(4.60),254(4.97), 345(3.47),365(3.67)	18-1192-83
$C_{14}H_9BrO_5$			
2H,5H-Pyrano[3,2-c][1]benzopyran-2,5-dione, 4-(bromomethyl)-8-methoxy-	MeCN	278.2(4.003),356.0(4.30)	64-0248-83B
$C_{14}H_9ClN_2O_4S$			
Spiro[benzoxazole-2(3H),5'-[4H,5H]-pyrano[3,4-e][1,3]oxazin]-4'-one, 7'-chloro-2'-(methylthio)-	$CHCl_3$	279(4.06),338(4.18), 352(4.20)	24-3725-83
$C_{14}H_9Cl_2NO_3$			
Furan, 2-(4,4-dichloro-3-phenyl-1-cyclobuten-1-yl)-5-nitro-	MeOH	208(4.15),232(4.17), 310(4.52),320s(4.44)	73-1062-83
$C_{14}H_9Cl_2N_3O$			
Pyrido[2,3-d]pyrimidin-7(1H)-one, 6-(2,6-dichlorophenyl)-2-methyl-	MeOH	304(4.17),314(4.22), 326s(--)	87-0403-83
	MeOH-HCl	218(3.69),308(4.19), 300s(--)	87-0403-83
	MeOH-KOH	270s(--),330(4.17)	87-0403-83
$C_{14}H_9Cl_5O$			
Benzenemethanol, 4-chloro-α-(4-chlorophenyl)-α-(trichloromethyl)-	EtOH	226(4.43),258(2.82), 266(2.85),276(2.60)	162-0448-83
$C_{14}H_9F_3N_2OS$			
Thiazolo[3,2-a]pyrimidin-4-ium, 3-hydroxy-5-methyl-2-phenyl-7-(trifluoromethyl)-, hydroxide, inner salt	MeCN	580(4.01)	124-0857-83
	benzene	618(--)	124-0857-83
	MeOH	545(--)	124-0857-83
	EtOH	556(--)	124-0857-83
	dioxan	614(--)	124-0857-83
	CH_2Cl_2	598(--)	124-0857-83
	$CHCl_3$	600(--)	124-0857-83
	CCl_4	626(--)	124-0857-83
	quinoline	613(--)	124-0857-83
$C_{14}H_9N$			
Indolizino[5,4,3-aij]quinoline	EtOH	240(4.7),280(4.0), 374f(4.22)	33-2135-83
$C_{14}H_9NO$			
9-Acridinecarboxaldehyde	EtOH	365(3.91)	140-0215-83
$C_{14}H_9NO_3$			
4H-3,1-Benzoxazin-4-one, 7-hydroxy-2-phenyl- (diantalexin)	MeOH	261(4.65),292(4.15), 302(4.13),320s(--)	88-0051-83
Pyridinium, 1-(4-hydroxy-2-oxo-2H-1-benzopyran-3-yl)-, hydroxide, inner salt	EtOH	240(4.03),275(4.09), 358(3.58)	104-0283-83
	50% EtOH	340(3.44)	104-0283-83
	$CHCl_3$	387(3.73)	104-0283-83
$C_{14}H_9N_3O$			
1-Phenazinecarbonitrile, 2-methoxy-	MeOH	212(3.97),266(4.56), 290(4.31),348(3.64), 456(3.69)	64-0866-83B

Compound	Solvent	$\lambda_{max}(\log \epsilon)$	Ref.
$C_{14}H_9N_3OS$			
Furo[2',3':4,5]pyrrolo[1,2-d]-1,2,4-triazine-8(7H)-thione, 2-phenyl-	MeOH	335(3.50),400(3.01)	73-1878-83
$C_{14}H_9N_3O_2$			
Furo[2',3':4,5]pyrrolo[1,2-d]-1,2,4-triazin-8(7H)-one, 2-phenyl-	MeOH	279(3.23),333(3.64)	73-1878-83
$C_{14}H_9N_3O_3$			
Furo[2',3':4,5]pyrrolo[1,2-d]-1,2,4-triazine-5,8-dione, 6,7-dihydro-2-phenyl-	MeOH	270(2.88),342(3.58)	73-1878-83
1,3,4-Oxadiazole, 2-(4-nitrophenyl)-5-phenyl-	toluene	317(4.28)	103-0022-83
$C_{14}H_9N_3Se$			
12H-Quinoxalino[2,3-b][1,4]benzo-selenazine	EtOH	222(4.52),254(4.92), 426(4.11)	78-0831-83
$C_{14}H_{10}$			
Anthracene	MeOH	221(4.06),252(5.34), 374(3.93)	97-0341-83
	EtOH	252(5.31),294(2.84), 310(3.085),324(3.46), 340(3.74),357(3.91), 376.5(3.89)	73-0112-83
Phenanthrene	EtOH	245s(4.73),251(4.84), 273.5(4.19),281(4.09), 293.5(4.18),309(2.28), 315.5(2.365),323(2.45), 330(2.52),338.5(2.51), 346(2.49),356.5(2.19), 370s(1.85),376(2.17)	73-0112-83
$C_{14}H_{10}AgNO_2$			
Silver, (N-benzoylbenzamidato-O,O')-	MeCN-Et_4N^+-BF_4^-	241(4.45)	23-1890-83
$C_{14}H_{10}BNS_2$			
Dithieno[2,3-c:2',3'-e][1,2]azaborine, 4,5-dihydro-5-phenyl-	C_6H_{12}	197(4.63),232(4.36), 247(4.38),269(4.12), 321(4.25),334(4.19)	11-0055-83B
Dithieno[2,3-c:3',2'-e][1,2]azaborine, 4,5-dihydro-5-phenyl-	C_6H_{12}	193(4.68),226(4.11), 246(4.44),259(4.27), 295(3.91),321(4.12)	11-0055-83B
Dithieno[2,3-c:3',4'-e][1,2]azaborine, 4,5-dihydro-5-phenyl-	C_6H_{12}	205(4.62),234(4.32), 282(4.00),309(4.35)	11-0055-83B
$C_{14}H_{10}BrClN_2O_3S$			
Acetamide, 2-bromo-N-[3-chloro-4-[(phenylsulfonyl)imino]-2,5-cyclo-hexadien-1-ylidene]-	EtOH	265(3.95)	44-3146-83
$C_{14}H_{10}BrN$			
Acetonitrile, bicyclo[5.4.1]dodeca-2,5,7,9,11-pentaen-4-ylidenebromo-	EtOH	394(4.02)	88-2151-83
1H-Indole, 2-(2-bromophenyl)-	EtOH	237s(4.16),306(4.22)	39-2417-83C
$C_{14}H_{10}BrNO_5S$			
Ethanone, 1-(4-bromophenyl)-2-[(4-nitrophenyl)sulfonyl]-	EtOH	214(4.24),265(4.45)	42-0137-83
	neutral	258(4.39)	42-0137-83

Compound	Solvent	λ_{max}(log ϵ)	Ref.
(cont.)	anion	358(3.86)	42-0137-83
Ethanone, 2-[(4-bromophenyl)sulfonyl]-1-(4-nitrophenyl)-	neutral	238(4.24)	42-0137-83
	anion	322(3.90)	42-0137-83
$C_{14}H_{10}Br_2$			
Benzene, 1,1'-(1,2-ethenediyl)bis[4-bromo-	heptane	318(4.60)	94-0373-83
$C_{14}H_{10}Br_2O_3S$			
Ethanone, 1-(4-bromophenyl)-2-[(4-bromophenyl)sulfonyl]-	EtOH	267(4.43)	42-0137-83
	neutral	268(4.29)	42-0137-83
	anion	315(4.15)	42-0137-83
$C_{14}H_{10}Br_2O_4$			
2,4-Pentadienoic acid, 5-(4,5-dibromo-2-furanyl)-3-(2-furanyl)-, methyl ester, (E,E)-	MeOH	282(4.33),326(4.40)	73-3559-83
(E,Z)-	MeOH	271(4.27),313(4.31)	73-3559-83
$C_{14}H_{10}Br_3NO$			
Acetamide, N-(5,8,10-tribromo-7-benzocyclooctenyl)-	EtOH	220(4.3)	18-1192-83
$C_{14}H_{10}ClNO_6$			
2(5H)-Furanone, 3-[3-(4-chloro-3-nitrophenyl)-1-oxo-2-propenyl]-4-hydroxy-5-methyl-, (E)-	MeOH	201(4.22),221(4.27), 295(4.16),348(4.20)	83-0115-83
$C_{14}H_{10}ClN_3$			
1-Phthalazinamine, 4-chloro-N-phenyl-	MeOH	213(4.72),248(4.16), 288(3.81),344(4.02)	103-0666-83
	MeCN	213(4.72),248(4.20), 272s(3.90),285s(3.84), 338(4.06)	103-0666-83
sulfate (2:1)	MeOH	218(4.58),248s(4.08), 277(4.00),319(3.88)	103-0666-83
	MeCN	221(4.67),248s(4.15), 277(4.11),312(3.90)	103-0666-83
$C_{14}H_{10}Cl_2$			
Benzene, 1,1'-(1,2-ethenediyl)bis[4-chloro-	heptane	317(4.54)	94-0373-83
$C_{14}H_{10}Cl_2N_2O_2$			
1,2-Diazabicyclo[5.2.0]nona-3,5-dien-9-one, 2-benzoyl-8,8-dichloro-	CH_2Cl_2	290(3.93)	5-1361-83
$C_{14}H_{10}Cl_2N_4$			
Pyrido[2,3-d]pyrimidin-7-amine, 6-(2,6-dichlorophenyl)-2-methyl-	pH 1	331(4.19)	87-0403-83
	MeOH-pH 7	225(4.72),329(4.17)	87-0403-83
$C_{14}H_{10}Cl_2OS$			
Ethanone, 1-[5-chloro-2-[(4-chlorophenyl)thio]phenyl]-	MeOH	230(4.38),263(3.99), 285s(3.83),345(3.63)	73-1173-83
$C_{14}H_{10}Cl_2O_4$			
5,10-Benzocyclooctenedione, 6,7-dichloro-1,4-dimethoxy-	$CHCl_3$	300(3.09)	39-1983-83C
2(5H)-Furanone, 3-[3-(3,4-dichlorophenyl)-1-oxo-2-propenyl]-4-hydroxy-5-methyl-	MeOH	203(4.35),237(4.21), 355(4.30)	83-0115-83

Compound	Solvent	$\lambda_{max}(\log \epsilon)$	Ref.
$C_{14}H_{10}FNO_2S$			
10H-Phenothiazin-3-ol, 7-fluoro-, acetate	EtOH	205(4.42),243(4.41), 279(4.26)	4-0803-83
$C_{14}H_{10}FNO_3S$			
10H-Phenothiazin-3-ol, 7-fluoro-, acetate, 5-oxide	EtOH	207(4.40),253(4.58), 267(4.43),361(3.61)	4-0803-83
$C_{14}H_{10}N_2O$			
Bicyclo[4.2.0]octa-1,3,5-triene-7,8-dione, mono(phenylhydrazone), (E)-	CH_2Cl_2	228(4.10),249(4.09), 290(3.79),332(4.01), 408(3.91)	18-0965-83
(Z)-	CH_2Cl_2	252(4.01),258(4.02), 293(3.88),334(4.20), 432(4.05)	18-0965-83
Methanone, 1H-indol-3-yl-2-pyridinyl-	EtOH	256.2(3.49?),267.5(3.47?), 327.5(3.44?)	35-0907-83
1,3,4-Oxadiazole, 2,5-diphenyl-	toluene	284(4.42)	103-0022-83
Pyridine, 2-(3-phenyl-5-isoxazolyl)-	MeOH	238(4.29),250s(4.25), 280(4.26)	142-0501-83
Pyridine, 2-(5-phenyl-3-isoxazolyl)-	MeOH	238(4.24),272(4.40)	142-0501-83
Pyridine, 3-(3-phenyl-5-isoxazolyl)-	MeOH	244(4.31),261(4.31)	142-0501-83
Pyridine, 3-(5-phenyl-3-isoxazolyl)-	MeOH	240(4.28),273(4.35)	142-0501-83
Pyridine, 4-(3-phenyl-5-isoxazolyl)-	MeOH	245s(4.25),263(4.35)	142-0501-83
Pyridine, 4-(5-phenyl-3-isoxazolyl)-	MeOH	240(4.32),266(4.28)	142-0501-83
5H-Pyrrolo[2,1-b][3]benzazepine-9-carbonitrile, 6,11-dihydro-11-oxo-	EtOH	202(4.54),223(4.62), 250(4.20),337(4.14)	44-3220-83
$C_{14}H_{10}N_2O_4$			
Benzaldehyde, 4-nitroso-, dimer	EtOH	340(4.08)	44-0136-83
Benzene, 1,1'-(1,2-dinitro-1,2-ethenediyl)bis-	n.s.g.	315(4.13)	22-0339-83
3H-Phenoxazine-8-carboxylic acid, 2-(methylamino)-3-oxo- (texazone)	MeOH	247(4.34),257(4.37), 426(4.20),436(4.21), 485s(3.70)	158-0688-83
$C_{14}H_{10}N_2O_4S_2$			
4H,5H-Pyrano[3,4-e][1,3]oxazine-4,5-dione, 7-[(4-aminophenyl)thio]-2-(methylthio)-	$CHCl_3$	246(4.19),292(4.39), 346(4.00)	24-3725-83
$C_{14}H_{10}N_2O_7S$			
Ethanone, 1-(4-nitrophenyl)-2-[(4-nitrophenyl)sulfonyl]-	EtOH	258(4.19)	42-0137-83
	neutral	260(4.33)	42-0137-83
	anion	370(3.97)	42-0137-83
$C_{14}H_{10}N_2S$			
2-Propenenitrile, 3-(1,2-dihydro-1-phenyl-2-thioxo-3-pyridinyl)-	EtOH	213(4.26),242(4.42), 324(4.01),421(3.62)	4-1651-83
$C_{14}H_{10}N_3OS_3$			
Bisthiazolo[3,4-a:5',4'-e]pyrimidin-9-ium, 1,2-dihydro-8-(methylthio)-2-oxo-6-phenyl-, perchlorate	DMF	300(4.47),313(4.48), 450(3.92)	103-0754-83
$C_{14}H_{10}N_4OS_2$			
4-Pyridinecarboxamide, N-(1,4-dihydro-2,4-dithioxo-3(2H)-quinazolinyl)-	EtOH	253(4.11),284(4.30), 306(4.16),375(3.77)	49-0915-83
$C_{14}H_{10}N_4O_2$			
Cyclohexaneacetic acid, 2,5-bis(di-	MeCN	245(4.26)	44-3852-83

Compound	Solvent	λ max (log ε)	Ref.
cyanomethylene)- (cont.)			44-3852-83
$C_{14}H_{10}N_4O_3$			
Benzenamine, 2-[5-(4-nitrophenyl)-1,3,4-oxadiazol-2-yl]-	MeOH	258(4.12),282(4.19), 353(4.11)	118-0842-83
5H-1,3,4-Benzotriazepin-5-one, 1,4-dihydro-2-(4-nitrophenyl)-	MeOH	260(4.18),295(4.13)	118-0842-83
$C_{14}H_{10}N_4O_4$			
Benzo[1,2-b:3,4-b']dipyrrole, 1,6-dihydro-3,8-bis(2-nitroethenyl)-	EtOH	205(4.53),222(4.46), 280(4.14),438(4.51)	104-0385-83
$C_{14}H_{10}O$			
9(10H)-Anthracenone	96% H_2SO_4	212(4.39),248(3.81), 284(4.02),352(4.38)	23-1965-83
$C_{14}H_{10}OS$			
1H-Naphtho[2,1-b]thiopyran-1-one, 3-methyl-	EtOH	340(3.92),356(3.72)	139-0059-83B
3H-Naphtho[2,1-b]thiopyran-3-one, 1-methyl-	C_6H_{12}	333(4.08),380(3.36)	139-0059-83B
$C_{14}H_{10}O_2$			
Benzil	EtOH	260(4.34)	162-0153-83
$C_{14}H_{10}O_3$			
2,4,6-Cycloheptatrien-1-one, 4-[3-(2-furanyl)-1-oxo-2-propenyl]-2-hydroxy-, (E)-	MeOH	250(4.41),347(4.33)	18-3099-83
	MeOH-NaOH	264(4.19),345(4.50), 445(3.77)	18-3099-83
7bH-Cyclopent[cd]indene-1,2-dicarboxylic acid, 7b-methyl-	EtOH	204(3.30),222(3.08), 258(3.93),312.5(4.58), 347(3.85),480(3.49)	39-0083-83C
7H-Furo[3,2-g][1]benzopyran-7-one, 9-acetyl-2-methyl-	EtOH	236(4.40),240s(4.38), 270(4.19),295s(3.99), 349(3.83)	44-2709-83
$C_{14}H_{10}O_5$			
9H-Xanthen-9-one, 1,7-dihydroxy-3-methoxy-	MeOH	260(4.35),275(4.30), 315(4.10),410(3.70)	162-0628-83
$C_{14}H_{10}O_6$			
2H,8H-Benzo[1,2-b:3,4-b']dipyran-2,8-dione, 4-hydroxy-5-methoxy-10-methyl-	EtOH	225(4.37),252(3.68), 280(4.00),320(3.72)	78-2277-83
1H,3H-Naphtho[1,8-cd]pyran-1,3-dione, 4,9-dihydroxy-6-methoxy-7-methyl-	$CHCl_3$	254(4.00),285s(3.61), 290(3.62),350(3.95), 367(3.83)	78-2283-83
$C_{14}H_{10}O_7$			
1,3-Benzodioxole-4,7-dione, compd. with 1,3-benzodioxol-4-ol	$CHCl_3$	262(4.16),478(3.17)	64-0392-83B
$C_{14}H_{11}BrN_2O_2$			
1,2-Diazabicyclo[5.2.0]nona-3,5-dien-9-one, 2-benzoyl-8-bromo-, cis	MeOH	290(3.98)	5-1361-83
$C_{14}H_{11}BrO_3S$			
Ethanone, 1-(4-bromophenyl)-2-(phenylsulfonyl)-	EtOH	217(4.05),267(4.25)	42-0137-83
	neutral	267(4.35)	42-0137-83
	anion	306(4.22)	42-0137-83
Ethanone, 2-[(4-bromophenyl)sulfonyl]-1-phenyl-	EtOH	220(4.38),254(4.24), 282(4.39)	42-0137-83

Compound	Solvent	λ_{max}(log ε)	Ref.
(cont.)	neutral	265(4.37)	42-0137-83
	anion	304(4.11)	42-0137-83
$C_{14}H_{11}BrO_4$			
2,4-Pentadienoic acid, 5-(5-bromo-2-furanyl)-3-(2-furanyl)-, methyl ester, (E,E)-	MeOH	278(4.42),319(4.49)	73-3559-83
(E,Z)-	MeOH	265(4.38),306(4.24)	73-3559-83
$C_{14}H_{11}Cl$			
Benzene, 1-chloro-3-(2-phenylethenyl)-, (E)-	EtOH	221(4.23),228(4.27), 233(4.26),297(4.50), 307(4.30)	39-2215-83C
$C_{14}H_{11}ClN_2O_2$			
1,2-Diazabicyclo[5.2.0]nona-3,5-dien-9-one, 2-benzoyl-8-chloro-, cis	MeOH	290(3.98)	5-1361-83
Pyrano[2,3-c]pyrazol-4(1H)-one, 5-chloro-3,6-dimethyl-1-phenyl-	EtOH	246(4.40)	44-4078-83
$C_{14}H_{11}ClN_2O_3$			
Carbamic acid, [(phenylamino)carbonyl]-, 3-chlorophenyl ester	EtOH	241(4.22)	22-0073-83
Carbamic acid, [(phenylamino)carbonyl]-, 4-chlorophenyl ester	EtOH	241(4.25)	22-0073-83
$C_{14}H_{11}ClO_4$			
2H-1-Benzopyran-2-one, 8-acetyl-6-(2-chloro-2-propenyl)-7-hydroxy-	EtOH	228s(4.03),234s(4.02), 247(4.00),271(4.00), 277s(3.96),319(4.04), 346(4.00),353s(4.00)	44-2709-83
2H-1-Benzopyran-2-one, 8-acetyl-7-[(2-chloro-2-propenyl)oxy]-	EtOH	304s(4.09),321(4.17)	44-2709-83
2(5H)-Furanone, 3-[3-(4-chlorophenyl)-1-oxo-2-propenyl]-4-hydroxy-5-methyl-, (E)-	MeOH	201(4.09),234(4.11), 356(4.28)	83-0115-83
$C_{14}H_{11}Cs$			
Cesium, (2-methyl-9H-fluoren-9-yl)-	$(MeOCH_2)_2$	362(4.11)	104-1592-83
Cesium, (9-methyl-9H-fluoren-9-yl)-	$(MeOCH_2)_2$	377.5(4.10),474(3.02), 506(3.11),539(2.95)	104-1592-83
$C_{14}H_{11}FO_4$			
2(5H)-Furanone, 3-[3-(4-fluorophenyl)-1-oxo-2-propenyl]-4-hydroxy-5-methyl-, (E)-	MeOH	202(4.14),217(4.10), 233(4.09),359(4.29)	83-0115-83
$C_{14}H_{11}Li$			
Lithium, (2-methyl-9H-fluoren-9-yl)-	$(MeOCH_2)_2$	370(4.08)	104-1592-83
Lithium, (9-methyl-9H-fluoren-9-yl)-	$(MeOCH_2)_2$	385.5(4.01),482(3.92), 516(3.05),556(2.93)	104-1592-83
$C_{14}H_{11}N$			
Acetonitrile, bicyclo[5.4.1]dodeca-2,5,7,9,11-pentaen-4-ylidene-	EtOH	385(3.93)	88-2151-83
$C_{14}H_{11}NO$			
Benzoxazole, 2-(4-methylphenyl)-	C_6H_{12}	302(4.50)	41-0595-83
$C_{14}H_{11}NOS$			
Benzoxazole, 2-[(phenylmethyl)thio]-	$CHCl_3$	282(4.21),289(4.21)	94-1733-83

Compound	Solvent	λ max(log ε)	Ref.
2(3H)-Benzoxazolethione, 3-(phenyl-methyl)-	$CHCl_3$	309(4.50)	94-1733-83
$C_{14}H_{11}NO_2$			
Benzene, 1,1'-(1-nitro-1,2-ethene-diyl)bis-	n.s.g.	314(4.13)	22-0339-83
Benzoxazole, 2-(4-methoxyphenyl)-	C_6H_{12}	307(4.53)	41-0595-83
1,3-Propanedione, 1-phenyl-3-(2-pyri-dinyl)-	MeOH	237(3.96),342(4.31)	142-0501-83
1,3-Propanedione, 1-phenyl-3-(3-pyri-dinyl)-	MeOH	230(3.98),244(3.98), 342(4.36)	142-0501-83
1,3-Propanedione, 1-phenyl-3-(4-pyri-dinyl)-	MeOH	223s(4.06),290s(3.95), 338(4.28)	142-0501-83
$C_{14}H_{11}NO_2S_2$			
2-Propenoic acid, 3-benzo[b]thien-2-yl-2-cyano-3-(methylthio)-, methyl ester	EtOH	228(3.51),262(4.32), 268(4.32),328(4.13)	142-1793-83
$C_{14}H_{11}NO_3$			
4H-Furo[3,2-b]pyrrole-5-carboxylic acid, 2-(4-methylphenyl)-	MeOH	295(4.31),321(4.50)	73-0772-83
4H-Furo[3,2-b]pyrrole-5-carboxylic acid, 4-methyl-2-phenyl-	MeOH	207(4.14),335(4.68)	73-0772-83
1H,3H-Naphtho[1,8-cd]pyran-1,3-dione, 6-(dimethylamino)-	o-xylene	410(3.98)	104-2273-83
	MeOH	420(4.08)	104-2273-83
	isoAmOH	420(4.04)	104-2273-83
	DMSO	430(4.08)	104-2273-83
Spiro[oxirane-2,4'(3'H)-[1H]pyrano-[4,3-b]quinolin]-1'-one, 10'-methyl-	MeOH	220(4.38),245(4.91), 294(3.81)	103-0886-83
$C_{14}H_{11}NO_4$			
5H-[1]Benzopyrano[2,3-b]pyridin-5-one, 6,8-dimethoxy-	MeOH	209(4.53),219(4.52), 226s(4.49),286s(4.09), 303(4.25),324s(4.11)	39-0219-83C
$C_{14}H_{11}NO_5S$			
Ethanone, 1-(4-nitrophenyl)-2-(phenyl-sulfonyl)-	EtOH	220(4.01),267(4.18)	42-0137-83
	neutral	265(4.27)	42-0137-83
	anion	358(3.91)	42-0137-83
$C_{14}H_{11}NO_6$			
2,4-Pentadienoic acid, 3-(2-furanyl)-5-(5-nitro-2-furanyl)-, methyl ester, (E,E)-	MeOH	276(4.41),360(4.40)	73-3559-83
(E,Z)-	MeOH	276(4.42),320(4.35)	73-3559-83
$C_{14}H_{11}NS$			
9(10H)-Acridinethione, 10-methyl-	3-Mepentane	488(4.32)	23-0894-83
$C_{14}H_{11}NS_3$			
Ethane(dithioic) acid, (diphenylami-no)thioxo-, cyclohexylamine salt	CH_2Cl_2	283s(4.00),356s(3.63)	97-0247-83
dimethylamine salt	CH_2Cl_2	262s(4.08),288s(4.06), 368(3.75)	97-0247-83
dipropylamine salt	CH_2Cl_2	267(4.10),292(4.07), 369(3.81)	97-0247-83
morpholine salt	CH_2Cl_2	265(4.12),298(4.04), 369(3.80)	97-0247-83
piperidine salt	CH_2Cl_2	271(4.16),292s(4.08), 369(3.83)	97-0247-83

Compound	Solvent	λ_{max}(log ε)	Ref.
pyrrolidine salt (cont.)	CH_2Cl_2	266(4.11),292(4.07), 369(3.83)	97-0247-83
$C_{14}H_{11}N_3O$			
Benzenamine, 2-(5-phenyl-1,3,4-oxadiazol-2-yl)-	MeOH	258(4.06),281(4.19), 352(4.09)	118-0842-83
5H-1,3,4-Benzotriazepin-5-one, 3,4-dihydro-2-phenyl-	MeOH	280(4.14)	118-0842-83
$C_{14}H_{11}N_3OS$			
Pyrido[2,3-d]pyrimidin-4(1H)-one, 2,3-dihydro-1-(2-methylphenyl)-2-thioxo-	MeOH	280(3.40)	73-3315-83
Pyrido[2,3-d]pyrimidin-4(1H)-one, 2,3-dihydro-1-(4-methylphenyl)-2-thioxo-	MeOH	294(3.47)	73-3315-83
Pyrido[2,3-d]pyrimidin-4(1H)-one, 2,3-dihydro-1-(phenylmethyl)-2-thioxo-	MeOH	293(3.40)	73-3315-83
Pyrido[3,2-e]-1,3-thiazin-4-one, 2-(methylphenylamino)-	MeOH	254(4.03)	73-3315-83
Pyrido[3,2-e]-1,3-thiazin-4-one, 2-[(2-methylphenyl)amino]-	MeOH	230(3.33)	73-3315-83
Pyrido[3,2-e]-1,3-thiazin-4-one, 2-[(4-methylphenyl)amino]-	MeOH	248(3.70)	73-3315-83
Pyrido[3,2-e]-1,3-thiazin-4-one, 2-[(phenylmethyl)amino]-	MeOH	235(3.40)	73-3315-83
$C_{14}H_{11}N_3OSe$			
Pyrido[3,2-e]-1,3-selenazin-4-one, 2-[(2-methylphenyl)amino]-	MeOH	213(4.40),328(4.21)	73-3567-83
Pyrido[3,2-e]-1,3-selenazin-4-one, 2-[(4-methylphenyl)amino]-	MeOH	213(4.46),327(4.16)	73-3567-83
$C_{14}H_{11}N_3O_2$			
1,2,4-Triazolo[4,3-a]pyridinium, 2,3-dihydro-3-oxo-1-(phenylacetyl)-, hydroxide, inner salt	THF	232(3.91),258(3.49), 384(3.88)	18-2969-83
1,2,4-Triazolo[4,3-a]pyridin-3(2H)-one, 2-(phenylacetyl)-	THF	257(3.63),266(3.61), 350(3.60)	18-2969-83
$C_{14}H_{11}N_3O_2S$			
Pyrido[2,3-d]pyrimidin-4(1H)-one, 2,3-dihydro-1-(4-methoxyphenyl)-2-thioxo-	MeOH	280(3.41)	73-3315-83
Pyrido[3,2-e]-1,3-thiazin-4-one, 2-[(4-methoxyphenyl)amino]-	MeOH	246(3.83)	73-3315-83
2-Thiazolecarboxylic acid, 4-(2-quinoxalinyl)-, ethyl ester	dioxan	246(3.97),263(3.83), 297(3.58),338(3.57), 351s(3.49)	106-0829-83
$C_{14}H_{11}N_3O_2Se$			
4H-Pyrido[3,2-e]-1,3-selenazin-4-one, 2-[(4-methoxyphenyl)amino]-	MeOH	214(4.29),330(4.35)	73-3567-83
$C_{14}H_{11}N_3O_3$			
Benzo[1,2-b:3,4-b']dipyrrole-8-carboxaldehyde, 1,6-dihydro-3-(2-nitro-1-propenyl)-	EtOH	208(4.36),272(4.10), 315(3.86),419(4.04)	103-0488-83
Benzoic acid, 4-nitro-, (phenylmethylene)hydrazide	EtOH	270(4.15),308(4.17)	104-2104-83
	MeCN	273(4.25),302(4.24)	104-2104-83
1,2,4-Triazolo[4,3-a]pyridine-2(3H)-carboxylic acid, 3-oxo-, phenylmethyl ester	THF	254(3.54),264(3.54), 276(3.34),347(3.56)	18-2969-83

Compound	Solvent	λ_{max}(log ϵ)	Ref.
1,2,4-Triazolo[4,3-a]pyridinium, 2,3-dihydro-3-oxo-1-[(phenylmethoxy)-carbonyl]-, hydroxide, inner salt	THF	224(3.68),254(3.63), 314(3.52),377(3.79)	18-2969-83
$C_{14}H_{11}N_3O_3S$			
2H-Pyrido[3,4-b][1,4]benzothiazine-4-carbonitrile, 3,5-dihydro-7,8-dimethoxy-3-oxo-	MeOH	251(4.67),310(3.76)	87-0845-83
$C_{14}H_{11}N_3O_3S_3$			
Acetic acid, [[6,7-dihydro-2-(methylthio)-7-oxo-6-phenylthiazolo[5,4-d]-pyrimidin-5-yl]thio]-	MeOH	217(4.44),285(3.98), 306(4.23),342(4.07)	2-0243-83
$C_{14}H_{11}N_3O_5$			
Carbamic acid, [(phenylamino)carbonyl]-, 4-nitrophenyl ester	dioxan	250(4.18)	22-0073-83
$C_{14}H_{11}N_3S$			
2-Benzothiazolecarboxaldehyde, phenylhydrazone	hexane	393(3.57)	140-0215-83
	EtOH	393(--)	140-0215-83
$C_{14}H_{11}N_5O$			
Furo[2',3':4,5]pyrrolo[1,2-d]-1,2,4-triazin-8(7H)-one, 2-phenyl-, hydrazone	MeOH	354(3.59)	73-1878-83
$C_{14}H_{11}N_5O_2$			
1,2-Diazabicyclo[5.2.0]nona-3,5-dien-9-one, 8-azido-2-benzoyl-, cis	CH_2Cl_2	288(3.51)	5-1361-83
Imidazo[1,2-a]pyridine, 2-methyl-3-[(4-nitrophenyl)azo]-	EtOH	264(4.56),420(4.12)	4-0639-83
$C_{14}H_{11}N_5O_8$			
Benzenemethanamine, α-methyl-4-nitro-N-(2,4,6-trinitrophenyl)-	$CHCl_3$	340(4.18),405(3.75)	104-1122-83
$C_{14}H_{11}N_7O_2$			
Pyrimido[5,4-e]-1,2,4-triazolo[4,3-b]-[1,2,4]triazine-7,9(6H,8H)-dione, 6,8-dimethyl-3-phenyl-	EtOH	227(4.67),268(4.54), 360(3.99)	44-1628-83
$C_{14}H_{12}$			
Anthracene, 9,10-dihydro-	MeOH	264(3.02),271(3.02)	97-0341-83
cation	H_2SO_4	430(4.50)	44-3458-83
anion	Li-THF	400(4.63)	44-3458-83
Benzene, 1,1'-(1,2-ethenediyl)-	hexane	229(4.18),308(4.43)	94-0373-83
Benzene, 1,1'-ethenylidenebis-	n.s.g.	205(4.6),229s(4.2), 250(4.1)	62-0950-83
1H-Cyclobuta[de]naphthalene, 1-(1-methylethylidene)-	EtOH	222(4.91),262(4.25), 309(3.27)	35-6104-83
$C_{14}H_{12}BrClN_2O_3S$			
Acetamide, 2-bromo-N-[3-chloro-4-(phenylsulfonylamino)phenyl]-	EtOH	265(4.38)	44-3146-83
$C_{14}H_{12}BrNO_2$			
Benzamide, 4-bromo-N-hydroxy-N-(phenylmethyl)-	EtOH	205(4.48),238(4.14)	34-0433-83

Compound	Solvent	λ_{max}(log ϵ)	Ref.
$C_{14}H_{12}Br_2O_4$			
1,2-Benzenediol, 6-bromo-3-[(3-bromo-4-hydroxyphenyl)methyl]-4-(hydroxymethyl)- (avrainvilleol)	MeOH	288(3.64)	102-0743-83
$C_{14}H_{12}ClN$			
Aziridine, 1-chloro-2,2-diphenyl-, (R)-(-)-	isooctane	217s(4.16),253(3.08), 259(3.07),265(3.02)	32-0799-83
$C_{14}H_{12}ClNO_2$			
Benzamide, 2-chloro-N-hydroxy-N-(phenylmethyl)-	EtOH	205(4.48)	34-0433-83
$C_{14}H_{12}ClN_3OS$			
3-Pyridinecarboxamide, 2-chloro-N-[[(2-methylphenyl)amino]-thioxomethyl]-	MeOH	230(4.16)	73-3315-83
3-Pyridinecarboxamide, 2-chloro-N-[[(4-methylphenyl)amino]-thioxomethyl]-	MeOH	223(3.14)	73-3315-83
3-Pyridinecarboxamide, 2-chloro-N-[[(phenylmethyl)amino]-thioxomethyl]-	MeOH	245(4.18)	73-3315-83
$C_{14}H_{12}ClN_3OSe$			
3-Pyridinecarboxamide, 2-chloro-N-[[(2-methylphenyl)amino]-selenoxomethyl]-	MeOH	213(4.28),250(4.11)	73-3567-83
3-Pyridinecarboxamide, 2-chloro-N-[[(4-methylphenyl)amino]-selenoxomethyl]-	MeOH	212(4.12),250(4.01)	73-3567-83
$C_{14}H_{12}ClN_3O_2S$			
3-Pyridinecarboxamide, 2-chloro-N-[[(4-methoxyphenyl)amino]-thioxomethyl]-	MeOH	240(4.03)	73-3315-83
$C_{14}H_{12}ClN_3O_2Se$			
3-Pyridinecarboxamide, 2-chloro-N-[[(4-methoxyphenyl)amino]-selenoxomethyl]-	MeOH	220(4.03),247(4.34)	73-3567-83
$C_{14}H_{12}FNO_2$			
Benzamide, 4-fluoro-N-hydroxy-N-(phenylmethyl)-	EtOH	207(4.25)	34-0433-83
$C_{14}H_{12}F_3NO_2$			
Acetamide, 2,2,2-trifluoro-N-[2-(6-oxo-1-cyclohexen-1-yl)phenyl]-	EtOH	207(4.40)	78-3767-83
$C_{14}H_{12}F_3NO_6$			
3,5-Isoxazoledicarboxylic acid, 4,5-dihydro-4-[4-(trifluoromethyl)phenyl]-, dimethyl ester, 2-oxide, cis	MeOH	262(4.05),268(4.05)	40-1678-83
$C_{14}H_{12}F_3N_3O$			
Acetamide, 2,2,2-trifluoro-N-[1-methyl-4-(2-phenylethenyl)-1H-imidazol-5-yl]-	MeOH	226(4.11),233(4.06), 306(4.40)	4-0629-83

Compound	Solvent	λ_{max}(log ε)	Ref.
$C_{14}H_{12}N$			
Benzo[a]quinolizinium, 2-methyl-, perchlorate	EtOH	322(3.72),337(4.05), 353(4.16)	77-1037-83
$C_{14}H_{12}NO_3S$			
1,3-Oxathiol-1-ium, 2-(dimethylamino)-5-(2-oxo-2H-1-benzopyran-3-yl)-, perchlorate	MeCN	354(4.27)	48-0551-83
$C_{14}H_{12}N_2$			
Benzaldehyde, (phenylmethylene)hydrazone	n.s.g.	210(4.3),230s(4.0), 300(4.3)	131-0239-83N
Benzenamine, 2-(1H-indol-2-yl)-	EtOH	224(4.49),246s(4.17), 286s(3.97),295(4.02), 318(4.05)	39-2417-83C
	EtOH-acid	216s(4.37),234s(4.25), 302(4.17)	39-2417-83C
2,4,6-Cycloheptatrien-1-one, 2,4,6-cycloheptatrien-1-ylidenehydrazone	EtOH	245(4.10),285s(2.74), 388(4.22)	24-1963-83
monotetrafluoroborate	EtOH	250s(4.22),267s(3.98), 290(3.71),415(4.40)	24-1963-83
Dicyclohepta[c,e]pyridazine, 5,8(and 1,8)-dihydro-	EtOH	240s(3.91),315(3.90), 350s(3.69)	24-1963-83
1,6:8,13-Diimino[14]annulene, syn	MeOH	305(5.02),340s(4.09), 376(3.92),532(2.85)	35-6982-83
3,8-Phenanthroline, 4,7-dimethyl-	EtOH	218(4.48),241(4.78), 276(4.11),287(4.28), 326(3.48),341(3.74), 358(3.78)	4-1107-83
$C_{14}H_{12}N_2O$			
Benzenamine, 4-(2-benzoxazolyl)-N-methyl-	C_6H_{12}	326(4.42)	41-0595-83
Benzo[4,5]cyclohepta[1,2-c]pyrazol-4(1H)-one, 1,3-dimethyl-	EtOH	255(4.45),306s(3.92), 318(4.02)	142-1581-83
Benzoic acid, (phenylmethylene)hydrazide	EtOH	219(4.18),296(4.41)	104-2104-83
	dioxan	222(4.18),295(4.37)	104-2104-83
	MeCN	219(4.18),294(4.37)	104-2104-83
	isooctane	218(--),291(--)	104-2104-83
Ethanone, 1-[4-(phenylazo)phenyl]-	dioxan	327.5(4.37),448(2.85)	126-1143-83
Phenazine, 3-methoxy-1-methyl-	MeOH	258(4.92),354(4.00), 390(3.82)	64-0866-83B
Phenazine, 3-methoxy-2-methyl-	MeOH	257(4.86),369(4.08)	64-0866-83B
2(10H)-Phenazinone, 3,10-dimethyl-	MeOH	232(4.18),287(4.33), 358(3.71),513(3.68)	64-0866-83B
2(10H)-Phenazinone, 4,10-dimethyl-	MeOH	229(4.38),282(4.35), 355(3.81),505(3.71)	64-0866-83B
1,8-Phenanthroline, 4-ethoxy-	EtOH	209(4.35),217s(4.28), 247(4.60),289(4.04), 300(4.10),330(3.26), 346(3.36)	33-0620-83
9H-Pyrido[3,4-b]indole, 1-ethenyl-4-methoxy-	EtOH	228(4.15),246(4.15), 270(3.98),358(3.55)	94-3198-83
5H-Pyrrolo[2,1-b][3]benzazepine-3-carbonitrile, 6,11-dihydro-11-hydroxy-	EtOH	209(4.30),261(4.18)	44-3220-83
5(6H)-Quinazolinone, 7,8-dihydro-2-phenyl-	EtOH	290(4.39)	4-0649-83
$C_{14}H_{12}N_2OS$			
2-Propenamide, 3-(1,2-dihydro-1-phen-	EtOH	211(4.35),239(4.33),	4-1651-83

Compound	Solvent	$\lambda_{max}(\log \epsilon)$	Ref.
yl-2-thioxo-3-pyridinyl)-, (E)-		325(4.06),410(3.65)	4-1651-83
Thiazolo[3,2-a]pyrimidin-4-ium, 3-hydroxy-5,7-dimethyl-2-phenyl-, hydroxide, inner salt	MeCN	496(4.03)	124-0857-83
$C_{14}H_{12}N_2O_2$			
Benzenemethanamine, α-(nitrophenylmethylene)-, (Z)-	EtOH	362(4.26)	22-0339-83
1H-Benz[de]isoquinoline-1,3(2H)-dione, 6-(dimethylamino)-	MeOH	420(4.00)	104-2273-83
1H-Benz[de]isoquinoline-1,3(2H)-dione, 2-methyl-6-(methylamino)-	MeOH	440(4.20)	104-2273-83
Phenazine, 3-methoxy-1-methyl-, 5-oxide	MeOH	223(4.03),274(4.80), 376(3.95),438(3.79)	64-0866-83B
2-Propenoic acid, 3-[3-(cyanomethyl)-1H-indol-4-yl]-, methyl ester, (E)-	MeOH	207(4.46),233(4.21), 350(3.92)	78-3695-83
Pyrano[2,3-c]pyrazol-4(1H)-one, 3,6-dimethyl-1-phenyl-	EtOH	244(4.36)	44-4078-83
Pyrano[2,3-c]pyrazol-6(1H)-one, 3,4-dimethyl-1-phenyl-	MeOH	260(4.26),320(4.04)	20-0451-83
1H,5H-Pyrrolo[2,3-f]indole, 1,5-diacetyl-	EtOH	204(3.8),256(4.27), 286(3.95),293(3.98), 298.5(4.07),321(3.91), 334(4.08)	103-0871-83
3H,6H-Pyrrolo[3,2-e]indole, 3,6-diacetyl-	EtOH	205.8(4.38),218s(4.06), 223(4.01),270(4.43), 298.5(4.36),312(4.20), 333(3.81)	103-0871-83
$C_{14}H_{12}N_2O_2S$			
4H-Indeno[1,2-c]isothiazol-4-one, 3-morpholino-	EtOH	261(4.67),310(4.17), 370(3.75)	95-1243-83
$C_{14}H_{12}N_2O_3$			
Benzo[b][1,8]naphthyridin-5(10H)-one, 6,8-dimethoxy-	MeOH	201(4.26),221(4.39), 233(4.38),267(4.77), 290s(3.90),314(3.83), 367(3.96)	39-0219-83C
Benzo[b][1,8]naphthyridin-5(10H)-one, 6-hydroxy-8-methoxy-10-methyl-	MeOH	203s(3.98),228s(4.21), 235(4.26),272(4.60), 328(3.79),385(3.71)	39-0219-83C
Carbamic acid, [(phenylamino)carbonyl]-, phenyl ester	EtOH	242(4.26)	22-0073-83
1H-Furo[2,3-c]pyrazole-5-carboxylic acid, 1-methyl-3-phenyl-, methyl ester	EtOH	221(3.92),252(4.09), 302(4.40)	44-1069-83
1H-Furo[2,3-c]pyrazole-5-carboxylic acid, 3-methyl-1-phenyl-, methyl ester	EtOH	256(4.37),263s(4.36), 279s(4.19)	44-1069-83
Phenazine, 2-methoxy-3-methyl-, 5,10-dioxide	MeOH	227(3.92),292(4.64), 410(3.81),481(3.85)	64-0866-83B
$C_{14}H_{12}N_2O_4$			
Benzamide, N-hydroxy-2-nitro-N-(phenylmethyl)-	EtOH	207(4.43),257(3.86)	34-0433-83
Benzamide, N-hydroxy-3-nitro-N-(phenylmethyl)-	EtOH	209(4.42),253(4.06)	34-0433-83
$C_{14}H_{12}N_2O_5$			
1-Azabicyclo[3.2.0]hept-2-ene-2-carboxylic acid, 7-oxo-, (4-nitrophen-	MeCN	270(4.11)	158-83-81

Compound	Solvent	λ max(log ε)	Ref.
yl)methyl ester (cont.)			158-83-81
(R)-	EtOH	267(4.17)	23-2257-83
(S)-	EtOH	267(4.19)	23-2257-83
(±)-	EtOH	267(4.20)	23-2257-83
$C_{14}H_{12}N_4$			
1,2,4,5-Tetrazine, 1,4-dihydro-3,6-diphenyl-	dioxan	246(4.46),283s(3.65), 357s(2.55)	24-2261-83
$C_{14}H_{12}N_4OS$			
2-Thiazolecarboxamide, N,N-dimethyl-4-(2-quinoxalinyl)-	EtOH	246(3.92),273(3.87), 295s(3.79),345(3.53), 357s(3.51)	106-0829-83
2-Thiazolecarboxamide, N-ethyl-4-(2-quinoxalinyl)-	EtOH	245(3.86),267s(3.69), 293s(3.44),343(3.34), 355s(3.26)	106-0829-83
2-Thiazolecarboxamide, N-methyl-4-(3-methyl-2-quinoxalinyl)-	EtOH	242(4.06),266s(3.74), 292s(3.56),337(3.57)	106-0829-83
$C_{14}H_{12}N_4O_2$			
Acetic acid, [(8-formyl-1,6-dihydrobenzo[1,2-b:3,4-b']dipyrrol-3-yl)methylene]hydrazide	EtOH	213(4.26),224(4.28), 273(4.02),314(4.20)	104-0385-83
3-Pyridinecarbonitrile, 1-[[(5-cyano-1(2H)-pyridinyl)methoxy]methyl]-1,6-dihydro-6-oxo-	CH_2Cl_2	252(4.2),308s(3.7), 313(3.7),334s(3.7)	64-0878-83B
$C_{14}H_{12}N_4O_3$			
Pyrazolo[1,5-d][1,2,4]triazine-2-carboxylic acid, 6,7-dihydro-6-methyl-7-oxo-4-phenyl-, methyl ester	EtOH	236(4.16),255s(2.96), 290(4.09)	44-1069-83
Pyrazolo[1,5-d][1,2,4]triazine-3-carboxylic acid, 6,7-dihydro-4-methyl-7-oxo-6-phenyl-, methyl ester	EtOH	288(4.16)	44-1069-83
Pyrazolo[1,5-d][1,2,4]triazine-3-carboxylic acid, 6,7-dihydro-6-methyl-7-oxo-4-phenyl-, methyl ester	EtOH	242(4.05),259s(4.00), 299(4.01)	44-1069-83
$C_{14}H_{12}N_4O_4$			
Benzoic acid, 2-amino-, 2-(4-nitrobenzoyl)hydrazide	MeOH	239(4.14),250(4.13), 258(4.16),270(4.14)	118-0842-83
$C_{14}H_{12}N_4O_6$			
Benzenemethanamine, α-methyl-N-(2,4,6-trinitrophenyl)-	$CHCl_3$	338(4.19),410(3.79)	104-1122-83
$C_{14}H_{12}N_4S$			
Propanedinitrile, [2-[(dimethylamino)methylene]-2H-1,4-benzothiazin-3(4H)-ylidene]-	MeOH	220(--),237s(--), 274(4.33),313(--), 339(--)	87-0845-83
$C_{14}H_{12}O$			
Ethanone, 1-[1,1'-biphenyl]-4-yl-	hexane	275(4.37)	99-0034-83
Naphtho[2,1-b]furan, 2-ethenyl-1,2-dihydro-	EtOH	211(4.40),232(4.65), 271(3.62),281(3.70), 293(3.57)	78-0169-83
3H-Naphtho[2,1-b]pyran, 3-methyl-	EtOH	242(4.61)	39-0883-83C
$C_{14}H_{12}OS$			
6H-Dibenzo[b,d]thiopyran-6-ol, 6-methyl-	MeOH	255(4.17),323(3.23)	12-0795-83

Compound	Solvent	$\lambda_{max}(\log \epsilon)$	Ref.
Ethanone, 1-[4-(phenylthio)phenyl]-	hexane	237(3.94),255(3.85), 208(3.51),301(4.27)	99-0034-83
Methanethione, (4-methoxyphenyl)-phenyl-	benzene	361(3.98),594(2.19)	44-0214-83
$C_{14}H_{12}OS_2$			
Ethanone, 1-(4,9-dihydrothieno[2,3-c]-[2]benzothiepin-2-yl)-	MeOH	217(4.15),272(3.65), 356(4.23)	73-0623-83
$C_{14}H_{12}O_2$			
Ethanone, 2-hydroxy-1,2-diphenyl-	EtOH	247(4.16)	162-0156-83
Ethanone, 1-(4-phenoxyphenyl)-	hexane	267(4.26)	99-0034-83 +124-1211-83
Phenol, 2,2'-(1,2-ethenediyl)bis-	EtOH	278(4.13),321(4.16)	39-1083-83C
2-Propenal, 3-(1-hydroxy-4-methyl-2-naphthalenyl)-	acetone	333(3.89),353(3.75), 377(3.59),490(3.30)	103-0146-83
Tetracyclo[4.2.2.2^{2,5}.0^{1,6}]dodeca-3,7,9,11-tetraene-3-carboxylic acid, methyl ester	EtOH	274(3.12)	88-3361-83
$C_{14}H_{12}O_3$			
7H-Furo[2,3-f][1]benzopyran-2-carbox-aldehyde, 7,7-dimethyl-	EtOH	222s(--),227(4.185), 290(4.155),343(4.097)	111-0015-83
7H-Furo[3,2-g][1]benzopyran-2-carbox-aldehyde, 7,7-dimethyl-	EtOH	240(4.372),260(4.194), 312(4.025),358(4.244)	111-0015-83
Seselin	EtOH	217(4.452),285s(--), 294(4.103),331(4.116)	111-0009-83
Xanthyletin	EtOH	225(4.369),265(4.289), 348(4.103)	111-0009-83
$C_{14}H_{12}O_3S$			
Ethanone, 1-phenyl-2-(phenylsulfonyl)-	EtOH	212(4.18),253(4.14)	42-0137-83
	neutral	256(4.21)	42-0137-83
	anion	302(4.15)	42-0137-83
$C_{14}H_{12}O_4$			
1,3-Benzenediol, 2-[2-(3,5-dihydroxy-phenyl)ethenyl]-, (E)- (gnetol)	EtOH	219(4.45),309(4.44), 318(4.43),337(4.2)	2-0101-83
2H-1-Benzopyran-2-one, 6-acetyl-7-hy-droxy-8-(2-propenyl)-	EtOH	228(4.16),263(4.54), 308s(4.06),326(4.08), 347(4.06)	44-2709-83
2H-1-Benzopyran-2-one, 8-acetyl-7-hy-droxy-6-(2-propenyl)-	EtOH	213(4.35),255(3.99), 272(4.00),324(4.03), 347(4.00),357s(3.98)	44-2709-83
2H-1-Benzopyran-2-one, 6-acetyl-7-(2-propenyloxy)-	EtOH	217(4.26),253(4.35), 307(4.10),328(4.15)	44-2709-83
Ethanone, 1-(2,4-dihydroxyphenyl)-2-(3-hydroxyphenyl)-	MeOH	212(4.36),230s(3.99), 278(4.16),315(3.91)	100-0852-83
Fulvoplumierin	EtOH	272(3.85),365(4.53)	162-0611-83
2(5H)-Furanone, 4-hydroxy-5-methyl-3-(1-oxo-3-phenyl-2-propenyl)-, (E)-	MeOH	203(4.14),232(4.08), 356(4.32)	83-0115-83
7H-Furo[3,2-g][1]benzopyran-2-carbox-ylic acid, ethyl ester	EtOH	237(4.337),258(4.321), 334(4.112)	111-0015-83
$C_{14}H_{12}O_5$			
1H-Naphtho[2,3-c]pyran-1-one, 3,4-di-hydro-7,9,10-trihydroxy-3-methyl-, (R)- (7-de-O-methylsemivioxanthin)	n.s.g.	268(4.58),380(3.94)	119-0082-83
$C_{14}H_{12}O_6$			
2H-1-Benzopyran-4-acetic acid, 7-acet-	EtOH	276.5(4.07),311.5(4.03)	94-3014-83

Compound	Solvent	λ_{max}(log ϵ)	Ref.
oxy-2-oxo-, methyl ester (cont.)			94-3014-83
$C_{14}H_{13}BrN_2O_2S$			
Methanone, (4-bromophenyl)(2-morpholino-5-thiazolyl)-	MeCN	259(4.00),352(4.36)	97-0019-83
$C_{14}H_{13}Cl$			
1,1'-Biphenyl, 2'-chloro-2,5-dimethyl-	n.s.g.	272(2.92)	39-0859-83B
1,1'-Biphenyl, 3'-chloro-2,5-dimethyl-	n.s.g.	275(3.43)	39-0859-83B
1,1'-Biphenyl, 4'-chloro-2,5-dimethyl-	n.s.g.	244(3.28)	39-0859-83B
$C_{14}H_{13}ClN_2OS$			
Methanone, (4-chlorophenyl)(2-pyrrolidino-5-thiazolyl)-	MeCN	250(4.02),356(4.45)	97-0019-83
$C_{14}H_{13}ClN_4$			
4H-Pyrrolo[2,3-d]pyrimidin-4-imine, 3-(4-chlorophenyl)-3,7-dihydro-2,5-dimethyl-	EtOH	227(4.31),273(3.83)	5-2066-83
$C_{14}H_{13}ClN_4OS$			
Acetamide, 2-(7-chloro-2H-1,4-benzothiazin-3(4H)-ylidene)-2-cyano-N-(dimethylamino)-	MeOH	235(--),365(4.49)	87-0845-83
$C_{14}H_{13}Cl_3O_2$			
Benzene, [(2,3,5-trichloro-4,4-dimethoxy-2-cyclopenten-1-ylidene)methyl]-	$CHCl_3$	266(4.12),344(4.23)	104-0469-83
$C_{14}H_{13}FN_2O$			
3-Pyridinecarbonitrile, 1-[(4-fluorophenyl)methyl]-1,6-dihydro-6-methoxy-	CH_2Cl_2	240(3.7),313(3.8), 360(3.4)	64-0873-83B
$C_{14}H_{13}NO$			
Ethanone, 1-[4-(phenylamino)phenyl]-	hexane	224(3.90),236(3.90), 292(3.52),321(4.42)	99-0034-83
	isooctane	224(3.90),236(3.90), 321(4.42)	124-1211-83
4H-Furo[3,2-b]pyrrole, 4-methyl-2-(4-methylphenyl)-	MeOH	205(4.22),320(4.46)	73-0772-83
Pyridine, 1-acetyl-1,4-dihydro-4-(phenylmethylene)-	MeOH	236(3.73),349(4.46)	24-3192-83
$C_{14}H_{13}NO_2$			
Benzamide, N-hydroxy-N-(phenylmethyl)-	EtOH	205(4.35)	34-0433-83
Flindersine	MeOH	235(4.42),333(4.00), 350(4.10),365(3.93)	162-0588-83
Methanone, (2-methoxy-6-methyl-3-pyridinyl)phenyl-	EtOH	250(4.3),292(4.1)	104-1190-83
	15% H_2SO_4	260(4.1),300(4.2)	104-1190-83
	95% H_2SO_4	310(4.2),355(4.1)	104-1190-83
$C_{14}H_{13}NO_2S$			
Benzene, 1-[[(4-methylphenyl)thio]methyl]-3-nitro-	EtOH	252(3.83)	12-0081-83
$C_{14}H_{13}NO_3S$			
7H-Indol-7-one, 1,4,5,6-tetrahydro-1-(phenylsulfonyl)-	MeOH	223(3.95),260s(4.07), 267(4.08),273(4.08)	44-3214-83
1H-Pyrrole, 2-(cyclopropylcarbonyl)-1-(phenylsulfonyl)-	MeOH	224(4.10),237(4.09), 274(4.06)	44-3214-83

Compound	Solvent	λ_{max}(log ϵ)	Ref.
1H-Pyrrole, 3-(cyclopropylcarbonyl)-1-(phenylsulfonyl)-	MeOH	239(4.29)	44-3214-83
$C_{14}H_{13}NO_4$			
Phenol, 2-[(2-hydroxy-5-methylphenyl)-methyl]-4-nitro-	neutral	327(3.96)	126-2361-83
	anion	404(4.24)	126-2361-83
	dianion	420(4.30)	126-2361-83
4(1H)-Pyridinone, 1-(4-acetylphenyl)-5-hydroxy-2-(hydroxymethyl)-	MeOH	228(4.31),292(4.27)	18-1879-83
1H-Pyrrole-3-carboxylic acid, 4,5-di-hydro-1-methyl-4,5-dioxo-2-phenyl-, ethyl ester	dioxan	280s(3.8),410(3.6)	94-0356-83
3,4-Quinolinedicarboxylic acid, 2-methyl-, dimethyl ester	EtOH	225(4.31),245(4.45), 295(3.65)	12-1419-83
$C_{14}H_{13}NO_4S$			
Benzene, 1-[[(4-methylphenyl)sulfonyl]methyl]-3-nitro-	EtOH	258(3.49)	12-0081-83
$C_{14}H_{13}NO_5$			
Furo[3,4-c]quinoline-3a(3H)-carboxylic acid, 1,4,5,9b-tetrahydro-5-methyl-1,4-dioxo-, methyl ester	EtOH	258(3.92)	12-1419-83
3,4-Quinolinedicarboxylic acid, 1,2-di-hydro-1-methyl-2-oxo-, dimethyl ester	EtOH	236(3.80),292(3.88), 355(3.72)	12-1419-83
3,4-Quinolinedicarboxylic acid, 2-methyl-, dimethyl ester, 1-oxide	EtOH	249(4.51),338(3.92)	12-1419-83
$C_{14}H_{13}NO_5S$			
1H-Pyrrole-2-butanoic acid, γ-oxo-1-(phenylsulfonyl)-	MeOH	237(4.11),268s(4.02), 274(4.03)	44-3214-83
1H-Pyrrole-3-butanoic acid, γ-oxo-1-(phenylsulfonyl)-	MeOH	237(4.33)	44-3214-83
$C_{14}H_{13}NO_6$			
Benzoic acid, 2-[(5-nitro-2-furanyl)-methoxy]-, ethyl ester	MeOH	211(2.92),232(1.92), 308(2.41)	73-2682-83
1H-Indole-4,6,7-tricarboxylic acid, trimethyl ester	MeOH	222(4.21),253(4.40), 334(3.97)	44-2488-83
$C_{14}H_{13}N_3$			
Acetaldehyde, (phenylimino)-, phenyl-hydrazone	EtOH	375(4.52)	65-1902-83
	EtOH	240(4.2),300(3.7), 375(4.5)	99-0539-83
tautomer	hexane	246(4.2),283(3.5), 406(4.3)	99-0539-83
2-Propenal, 3-(4-pyridinyl)-, phenyl-hydrazone	hexane	366(4.54)	140-0215-83
	EtOH	390(--)	140-0215-83
$C_{14}H_{13}N_3O$			
Ethanone, 1-[4-[(4-aminophenyl)azo]-phenyl]-	MeOH	273(4.12),420(4.34)	80-0903-83
	$C_2H_4Cl_2$	404(4.37)	80-0903-83
2H-Isoxazolo[2,3-a]pyrimidin-2-imine, 5,7-dimethyl-3-phenyl-	MeCN	273(4.01),321(3.52)	44-0575-83
5(6H)-Quinazolinone, 2-amino-7,8-di-hydro-7-phenyl-	EtOH	276.5(4.28)	4-0649-83
$C_{14}H_{13}N_3O_2$			
4H-Furo[3,2-b]pyrrole-5-carboxylic acid, 2-(4-methylphenyl)-, hydrazide	MeOH	211(3.11),342(3.77)	73-1878-83

Compound	Solvent	λ_{max}(log ε)	Ref.
2,4(1H,3H)-Pyrimidinedione, 3-methyl-5-(1-methyl-1H-indol-3-yl)-	MeCN	231(4.48),305(3.98)	94-3496-83
2,4(1H,3H)-Pyrimidinedione, 3-methyl-6-(1-methyl-1H-indol-3-yl)-	MeCN	222.5(4.48),262(3.98), 320(4.32)	94-3496-83
$C_{14}H_{13}N_3O_2S_2$			
Diazene, (4-nitrophenyl)(4,5,6,7-tetrahydro-1,3-benzodithiol-2-ylidene)methyl]-	MeCN	545(4.40)	118-0840-83
$C_{14}H_{13}N_3O_3$			
Benzamide, N-[(2,3-dihydro-7-oxo-7H-oxazolo[3,2-a]pyrimidin-2-yl)methyl]-	EtOH	227(4.22),255s(3.87)	128-0125-83
$C_{14}H_{13}N_3O_4$			
Benzenemethanamine, N-(2,4-dinitrophenyl)-α-methyl-	$CHCl_3$	345(4.28),390(3.82)	104-1122-83
$C_{14}H_{13}N_3O_5$			
Ethanone, 2-(2,6-dimethoxy-5-nitro-4-pyrimidinyl)-1-phenyl-	CCl_4	246(4.24),338(4.31)	103-0435-83
$C_{14}H_{13}N_3O_6$			
2(1H)-Pyridinone, 5,6-dihydro-1-methyl-3,5-dinitro-6-(2-oxo-2-phenylethyl)-, compd. with diethylamine	MeOH	246(4.24),326(3.97), 478(4.42)	138-0715-83
$C_{14}H_{13}N_5$			
1H-Tetrazol-5-amine, N-phenyl-1-(phenylmethyl)-	n.s.g.	208(4.17),246(4.23)	78-2599-83
1H-Tetrazol-5-amine, 1-phenyl-N-(phenylmethyl)-	n.s.g.	209(4.15),248(3.83)	78-2599-83
$C_{14}H_{13}N_5O_6$			
Benzenamine, 4-(1,4-dihydro-5,7-dinitro-4-benzofurazanyl)-N,N-dimethyl-, N-oxide	MeOH	262(4.30),294(3.97), 364(3.79),474(4.36)	88-1555-83
$C_{14}H_{13}O$			
Ethylium, 1-(4-biphenylyl)-1-hydroxy-	H_2SO_4	397(4.36)	99-0612-83
$C_{14}H_{13}OS$			
Ethylium, 1-hydroxy-1-[4-(phenylthio)-phenyl]-	H_2SO_4	250(3.92),279(3.75), 402(4.54)	99-0612-83
$C_{14}H_{13}O_2$			
Ethylium, 1-hydroxy-1-(4-phenoxyphenyl)-	H_2SO_4	226(3.89),237(3.78), 345(4.47)	99-0612-83
$C_{14}H_{13}O_3P$			
10H-Phenoxaphosphine, 10-hydroxy-1,7-dimethyl-, 10-oxide	MeOH	218(4.32),245(3.77), 264s(3.09),275(3.32), 295(3.52)	4-1601-83
$C_{14}H_{13}O_3S$			
Ethylium, 1-hydroxy-1-[4-(phenylsulfonyl)phenyl]-	H_2SO_4	286(4.19)	99-0612-83
$C_{14}H_{14}$			
Anthracene, 1,2,3,4-tetrahydro-	MeOH	227s(4.95),230(5.02), 254s(3.26),267(3.56),	97-0341-83

Compound	Solvent	λ max (log ε)	Ref
Anthracene, 1,2,3,4-tetrahydro- (cont.)		324(2.88)	97-0341-83
Anthracene, 1,4,9,10-tetrahydro-	MeOH	227s(1.66),230s(1.72), 254s(2.28),267(2.65)	97-0341-83
Benzene, (1-phenylethyl)-, cation	FSO_3H-SbF_5	422(4.57)	44-3458-83
anion	Cs-$C_6H_{11}NH_2$	477(4.49)	44-3458-83
Bicyclo[2.2.2]octane, hexakis(methylene)-	isooctane	247(4.27),268s(3.81)	33-0019-83
1,3,5-Cycloheptatriene, 7-methyl-7-phenyl-	hexane	254(3.49),258(3.50), 264(3.48)	24-1848-83
$C_{14}H_{14}ClNO$			
1H-Cyclopent[cd]indene-2-carbonitrile, 2-chloro-2,2a,7a,7b-tetrahydro-2a-methoxy-7b-methyl-, exo	EtOH	305(3.84)	39-2399-83C
$C_{14}H_{14}ClNO_2$			
1,4-Naphthalenedione, 2-chloro-3-(diethylamino)-	dioxan	491(3.49)	150-0168-83S
	EtOH	498(--)	150-0169-83S
$C_{14}H_{14}ClNO_3S$			
1-Butanone, 1-[1-(phenylsulfonyl)-2-pyrrolyl]-4-chloro-	MeOH	224(4.07),237(4.11), 268s(4.01),274(4.05)	44-3214-83
1-Butanone, 1-[1-(phenylsulfonyl)-3-pyrrolyl]-4-chloro-	MeOH	219(4.19),238(4.37)	44-3214-83
$C_{14}H_{14}ClNO_3Se$			
4-Oxa-1-azabicyclo[3.2.0]heptan-7-one, 2-[(4-chlorophenyl)seleno]-3-(2-methoxyethylidene)-, (Z)-	EtOH	233(4.21)	39-2513-83C
2-Oxazoleethaneselenoic acid, 2,5-dihydro-5-(2-methoxyethylidene)-, Se-(4-chlorophenyl) ester, [R-(Z)]-	EtOH	229(4.21),268(4.08)	39-2513-83C
$C_{14}H_{14}F_3NO_2$			
4aH-Carbazol-4a-ol, 1,2,3,4,9,9a-hexahydro-9-(trifluoroacetyl)-, cis	EtOH	208(4.40)	78-3767-83
$C_{14}H_{14}N_2$			
Benzaldehyde, methylphenylhydrazone	isooctane	234(4.14),292(4.11), 326(4.29)	70-0485-83
Benzenamine, 4,4'-(1,2-ethenediyl)bis-	MeCN	340(4.46)	94-0373-83
6,7-Diazatetradeca-3,5,7,9,11-pentaene-1,13-diyne, 3,12-dimethyl-	THF	273(3.46),308s(3.79), 326s(4.30),338(4.38), 353(4.39),370(4.21)	18-1467-83
Hydrazinedium, 1,2-bis(2,4,6-cycloheptatrien-1-ylidene)-, bis(tetrafluoroborate)	EtOH	250s(4.15),267s(3.89), 292(3.59),428(4.40)	24-1963-83
1H-Indole, 1-methyl-3-(3-methyl-1H-pyrrol-2-yl)-	EtOH	212(4.27),242(4.33), 305(3.65)	103-0289-83
3,8-Phenanthrolinium, 3,8-dimethyl-, diiodide	H_2O	245(4.83),269(4.11), 284(4.02),368(3.72), 388(3.79)	4-1107-83
$C_{14}H_{14}N_2O$			
Acetamide, N-[4-(2-pyridinylmethyl)-phenyl]-	EtOH	245(4.44),385(3.59)	103-0970-83
Benzaldehyde, 4-methoxy-, phenylhydrazone	EtOH	246(4.11),305(4.22), 348(4.40)	70-1439-83
Benzo[4,5]cyclohepta[1,2-c]pyrazol-4(1H)-one, 9,10-dihydro-1,3-dimethyl-	EtOH	277(4.09)	142-1581-83

Compound	Solvent	λ_{max}(log ε)	Ref.
Diazene, bis(2-methylphenyl)-, 1-oxide	EtOH-MeOH	305(4.0)	151-0149-83A
1H-Pyrazino[3,2,1-jk]carbazol-2(3H)-one, 7,8,9,10-tetrahydro-	n.s.g.	220(4.31)	103-1312-83
9H-Pyrido[3,4-b]indole, 1-ethyl-4-methoxy-	EtOH	238(4.81),244(4.84), 259(4.07),288(4.32), 336(4.00),350(4.07)	94-3198-83
Pyrimido[2,1-a]isoindol-4(1H)-one, 1,2,3-trimethyl-	EtOH	238(4.40),282(4.05), 294(3.76),362(3.82), 374(3.91),414(3.49)	124-0755-83
5H-Pyrrolo[2,1-c][1,4]benzodiazepin-5-one, 2-ethylidene-1,2,3,11a-tetrahydro- (prothracarcin)	MeOH	218(4.32),239s(4.04), 255s(3.85),316(3.60)	158-83-102
Pyrrolo[1,2-a]pyrimidin-2-one, 2,6,7,8-tetrahydro-4-methyl-7-phenyl-	EtOH	242(4.16)	4-0393-83
	EtOH-HCl	206(4.38),240(4.23)	4-0393-83
Pyrrolo[1,2-a]pyrimidin-4-one, 4,6,7,8-tetrahydro-2-methyl-7-phenyl-	EtOH	204(4.22),225(3.85), 268(3.68)	4-0393-83
9H-Pyrrolo[2,3-f]quinolin-9-one, 1,6-dihydro-2,3,7-trimethyl-	EtOH	217(4.75),280(4.77), 343(4.04)	103-0401-83
9H-Pyrrolo[3,2-f]quinolin-9-one, 3,8-dihydro-1,2,7-trimethyl-	EtOH	217(4.54),245(4.24), 254(4.27),289(4.04), 348(4.06)	103-0401-83
6H-Pyrrolo[3,2-g]quinolin-6-one, 1,5-dihydro-2,3,8-trimethyl-	EtOH	200(4.24),237(4.62), 267(4.17),350(4.20)	103-0401-83
$C_{14}H_{14}N_2O_2$			
5H-Cyclopentapyrazine, 6,7-dihydro-2-methyl-3-phenyl-, 1,4-dioxide	EtOH	243(4.30),253s(4.24), 307(4.31)	103-1104-83
Methanone, phenyl(4,5,6,7-tetrahydro-1-hydroxy-1H-benzimidazol-2-yl)-	EtOH	258(4.86),328(4.10)	103-1104-83
Pyrano[2,3-c]pyrazol-4(1H)-one, 5,6-dihydro-3,5-dimethyl-1-phenyl-	EtOH	256(4.29)	118-0844-83
Pyrano[2,3-c]pyrazol-4(1H)-one, 5,6-dihydro-3,6-dimethyl-1-phenyl-	EtOH	256(4.31)	118-0844-83
Pyrazolo[1,5-a]quinoline-3-carboxylic acid, 4,5-dihydro-4,5-dimethyl-	EtOH	268(4.16)	88-4679-83
Pyrazolo[1,5-a]quinoline-3-carboxylic acid, 4,5-dihydro-5,5-dimethyl-	EtOH	268(4.21)	88-4679-83
Quinoxaline, 5,6,7,8-tetrahydro-2-phenyl-, 1,4-dioxide	EtOH	264(4.34),318(4.21)	103-1104-83
$C_{14}H_{14}N_2O_2S$			
Methanone, (2-morpholino-5-thiazolyl)-phenyl-	MeCN	245(3.91),347(4.36)	97-0019-83
$C_{14}H_{14}N_2O_3$			
1,3-Butanedione, 1-(2,5-dihydro-3-methyl-5-oxo-1-phenyl-1H-pyrazol-4-yl)-	EtOH	262(4.12),318(4.04)	44-4078-83
5,6-Diazaspiro[2.4]hept-4-ene-1-carboxylic acid, 4-methyl-7-oxo-6-phenyl-, methyl ester	EtOH	232(4.24),262(3.54)	44-1069-83
5,6-Diazaspiro[2.4]hept-4-ene-1-carboxylic acid, 6-methyl-7-oxo-4-phenyl-, methyl ester, cis	EtOH	223(3.63),278(3.90)	44-1069-83
trans	EtOH	222(3.89),301(3.98)	44-1069-83
2H-Pyrimidin-2-one, 4-hydroxy-6-methyl-3-[1-(phenylhydrazono)ethyl]-	EtOH	300(3.97),350(3.92)	44-4078-83
$C_{14}H_{14}N_2O_3S$			
1H-Imidazole-5-carboxylic acid, 4-mer-	EtOH	239(4.10),282(3.68),	2-0030-83

Compound	Solvent	λ_{max}(log ε)	Ref.
capto-1-(2-oxo-2-phenylethyl)-, ethyl ester (cont.)		321(3.68)	2-0030-83
Morpholine, 4-(5-nitro-3-phenyl-2-thienyl)-	MeCN	230(4.19),288(3.91), 439(4.18)	48-0168-83
$C_{14}H_{14}N_2O_4$			
1H-Imidazole-5-carboxylic acid, 4,5-dihydro-4-oxo-1-(2-oxo-2-phenyl-ethyl)-, ethyl ester	EtOH	243(4.4)	2-0030-83
6,7-Phthalazinedicarboxylic acid, diethyl ester	MeOH	222(4.75),306(3.31), 317(3.31)	142-1279-83
3-Pyridinecarboxylic acid, 2-[(3,5-di-methoxyphenyl)amino]-	MeOH	208(4.54),229s(4.24), 289(4.42),340(3.75)	39-0219-83C
$C_{14}H_{14}N_2O_4S$			
Ethanone, 1-(2-pyridinyl)-, O-[(4-methylphenyl)sulfonyl]oxime, N-oxide, (Z)-	EtOH	269.5(4.09)	94-2269-83
2H-[1,2,5]Oxadiazolo[2,3-a]pyridine, 3-methyl-2-[[(4-methylphenyl)sul-fonyl]oxy]-	EtOH	283(4.05)	94-2269-83
$C_{14}H_{14}N_2O_5$			
Benzoic acid, 2-[[(5-nitro-2-furanyl)-methyl]amino]-, ethyl ester	MeOH	222(9.44),252(3.72), 323(4.84)	73-2682-83
Benzoic acid, 4-[[(5-nitro-2-furanyl)-methyl]amino]-, ethyl ester	MeOH	208s(1.66),222(2.27), 294(5.26)	73-2682-83
1H-Cyclopenta[d]pyridazine-4,7-dicarb-oxylic acid, 3-acetyl-2-methyl-, dimethyl ester	CH_2Cl_2	308(4.3),364(3.5)	24-0097-83
Oxepino[4,5-d]pyridazine-1,4-dicarbox-ylic acid, 6,8-dimethyl-, dimethyl ester	CH_2Cl_2	262(4.35),330(3.38)	24-0097-83
2,4(1H,3H)-Pyrimidinedione, 1-acetyl-5-(2,5-dimethoxyphenyl)-	pH 1	260(4.00),292s(--)	87-1028-83
	H_2O	260(3.99),292s(--)	87-1028-83
	0.3M NaOH	292(4.12)	87-1028-83
$C_{14}H_{14}N_2O_6$			
Phenol, 2,6-dimethoxy-4-[2-(3-methyl-4-nitro-5-isoxazolyl)ethenyl]-, (E)-	MeOH	232(4.18),258s(3.93), 392(4.02),410(4.37)	142-0263-83
$C_{14}H_{14}N_4$			
Acetaldehyde, [(4-aminophenyl)imino]-, phenylhydrazone	EtOH	394(4.59)	65-1902-83
conjugate acid	EtOH	347(4.34),488(4.40)	65-1902-83
	EtOH-H_2SO_4	429(4.67)	65-1902-83
Pyrazolo[1,5-a]pyrimidin-2-amine, 5,7-dimethyl-3-phenyl-	MeCN	249(4.19),278(4.17)	39-0011-83C
4H-Pyrrolo[2,3-d]pyrimidin-4-imine, 3,7-dihydro-2,5-dimethyl-3-phenyl-	EtOH	233(4.20),277(3.75)	5-2066-83
$C_{14}H_{14}N_4O$			
4H-Pyrazolo[3,4-d]pyrimidin-4-one, 1,5-dihydro-1,6-dimethyl-5-(2-methyl-phenyl)-	EtOH	212(4.64),256(3.93)	11-0171-83B
4H-Pyrazolo[3,4-d]pyrimidin-4-one, 1,5-dihydro-1,6-dimethyl-5-(phenyl-methyl)-	EtOH	211(4.57),261.5(3.87)	11-0171-83B
$C_{14}H_{14}N_4OS$			
Acetamide, 2-(2H-1,4-benzothiazin-	MeOH	219(--),240s(--),	87-0845-83

Compound	Solvent	$\lambda_{max}(\log \epsilon)$	Ref.
ylidene)-2-cyano-N-[(dimethyl-amino)methylene]- (cont.)		365(4.20)	87-0845-83
Benzoic acid, 2-amino-, 2-[(2-amino-phenyl)thioxomethyl]hydrazide	EtOH	358(4.13)	49-0915-83
$C_{14}H_{14}N_4O_2$			
Benzenamine, N,N-dimethyl-4-[(4-nitro-phenyl)azo]-	MeOH	282(4.11),481(4.44)	80-0903-83
	$C_2H_4Cl_2$	482(4.36)	80-0903-83
cis	gas	272(3.2),410(3.0)	151-0245-83B
trans	gas	270(3.2),390(3.5)	151-0245-83B
4H-Pyrazolo[3,4-d]pyrimidin-4-one, 1,5-dihydro-5-(2-methoxyphenyl)-1,6-dimethyl-	EtOH	212(4.62)	11-0171-83B
$C_{14}H_{14}N_4O_6$			
3-Furancarboxaldehyde, 2-(2-hydroxy-propyl)-, 2,4-dinitrophenylhydrazone	MeCN	202(4.21),217s(3.90), 263(3.80)	24-3366-83
$C_{14}H_{14}N_5$			
Benzenediazonium, 4-[[4-(dimethyl-amino)phenyl]azo]-, (E)-, tetra-fluoroborate	H_2O	265(3.71),298(3.45), 320(3.48),610(4.35)	104-1493-83
	pH 1	515(4.7)	104-1493-83
	pH 3	510(4.6),610s(3.7)	104-1493-83
(also other pH values)	pH 13	480(4.2)	104-1493-83
$C_{14}H_{14}N_6O$			
3-Pyridinecarbonitrile, 4,6-diamino-2-ethoxy-5-(phenylazo)-	EtOH	243(4.35),384(4.34)	49-0973-83
$C_{14}H_{14}N_6O_8$			
1,8-Naphthalenediamine, N,N,N',N'-tetramethyl-2,4,5,7-tetranitro-	MeOH	355(3.58),458(3.80)	104-0520-83
$C_{14}H_{14}N_8O_4S_3$			
Cefazolin	pH 6.4	272(4.12)	162-0269-83
$C_{14}H_{14}O$			
Chamavioline	n.s.g.	214(0.930),228(1.12), 240(1.06),311(1.53), 380(0.56),519(0.46)	83-0908-83
$C_{14}H_{14}OS_2$			
Bicyclo[4.4.1]undeca-3,6,8,10-tetraen-2-one, 5-[bis(methylthio)methylene]-	CH_2Cl_2	298(3.88),402(4.22)	118-0463-83
4a,8a-Methanonaphthalen-1(2H)-one, 2-[bis(methylthio)methylene]-	CH_2Cl_2	300(3.94),383(3.82)	118-0463-83
$C_{14}H_{14}O_2$			
[Bi-2,4-cycloheptadien-1-yl]-7,7'-di-one	MeOH	241(3.69)	18-0175-83
$C_{14}H_{14}O_2S_2$			
Disulfide, bis(4-methoxyphenyl)	heptane	245(4.30),272s(3.98), 316s(3.44)	65-0519-83
	MeOH	245(4.26),265s(3.99), 300s(3.61)	65-0519-83
	MeCN	245(4.29),267s(4.11)	65-0519-83
$C_{14}H_{14}O_2Se_2$			
Diselenide, bis(4-methoxyphenyl)	heptane	236(4.40),262(4.22), 288s(3.25)	65-0519-83

Compound	Solvent	$\lambda_{max}(\log \epsilon)$	Ref.
Diselenide, bis(4-methoxyphenyl) (cont.)	MeOH	240s(4.25),265s(4.11), 333s(3.25)	65-0519-83
	MeCN	240s(4.25),270s(4.12), 325s(3.48)	65-0519-83
$C_{14}H_{14}O_2Te_2$			
Ditelluride, bis(4-methoxyphenyl)	heptane	256(4.29),273s(4.17), 408(3.04)	65-0519-83
$C_{14}H_{14}O_3$			
6H-Dibenzo[b,d]pyran-6-one, 7,8,9,10-tetrahydro-3-hydroxy-4-methyl-	MeOH	247(3.71),256(3.70), 321(4.20)	78-1265-83
Ethanone, 1-(2,6-dimethoxy-1-naphthalenyl)-	EtOH	260(3.37),314(3.62), 326(3.79)	104-0183-83
Ethanone, 1-(3,6-dimethoxy-2-naphthalenyl)-	EtOH	268(3.59),312(3.79)	104-0183-83
1H-Indene-2-carboxylic acid, 3-formyl-1,1-dimethyl-, methyl ester	MeOH	228(4.11),232(4.08), 290(4.11)	2-0230-83
Kawain	MeOH	210(4.38),245(4.44), 282(2.81)	162-0760-83
1H-Naphtho[2,1-b]pyran-1,2-diol, 2,3-dihydro-3-methyl-	EtOH	250(4.60),271(4.56)	39-0883-83C
2H-Pyran-2-one, 5,6-dihydro-4-methoxy-6-(2-phenylethenyl)-, cis (cis-kawain)	MeOH	238(4.33)	64-0658-83B
$C_{14}H_{14}O_4$			
Tricyclo[4.2.2.0^{2,5}]deca-3,6,9-triene-7,8-dicarboxylic acid, dimethyl ester	ether	246(3.33)	88-3361-83
$C_{14}H_{14}O_6$			
1,4-Naphthalenedione, 5-hydroxy-6-(1-hydroxyethyl)-2,7-dimethoxy-	MeOH	221(4.14),231s(4.07), 258s(3.88),263(3.89), 310(3.72)	158-83-38
$C_{14}H_{14}S_2$			
Disulfide, bis(2-methylphenyl)	heptane	242(4.23),262s(3.64)	65-0519-83
	MeOH	239(4.17),270s(3.46)	65-0519-83
Disulfide, bis(3-methylphenyl)	heptane	243(4.21),263s(3.64)	65-0519-83
	MeOH	242(4.24),263s(3.63)	65-0519-83
Disulfide, bis(4-methylphenyl)	heptane	245(4.35),272s(3.98), 316s(3.44)	65-0519-83
	MeOH	241(4.33),260s(3.80), 300s(3.29)	65-0519-83
	MeCN	240(4.12),265s(3.52)	65-0519-83
$C_{14}H_{14}Se_2$			
Diselenide, bis(2-methylphenyl)	heptane	240(4.23),270s(3.84), 330(2.91)	65-0519-83
Diselenide, bis(3-methylphenyl)	heptane	243(4.16),270s(3.59), 334(2.97)	65-0519-83
	MeOH	239(4.55),270s(3.97), 331(3.13)	65-0519-83
Diselenide, bis(4-methylphenyl)	heptane	250(4.32),265s(4.10), 341(3.13)	65-0519-83
	MeOH	242(4.25),275s(3.62), 333(3.12)	65-0519-83
	MeCN	243(4.28),270s(3.82), 330(3.10)	65-0519-83

Compound	Solvent	$\lambda_{max}(\log \epsilon)$	Ref.
$C_{14}H_{14}Te_2$			
Ditelluride, bis(4-methylphenyl)	heptane	255(4.35),275s(4.12), 409(2.90)	65-0519-83
$C_{14}H_{15}BrN_2O$			
9-Oxaergoline, 2-bromo-6-methyl-, oxalate	EtOH	224(4.57),276s(3.95), 281(3.96),290s(3.87)	87-0522-83
$C_{14}H_{15}BrN_2O_3$			
Pyrrolo[2,3-b]indole-1(2H)-carboxylic acid, 8-acetyl-6-bromo-3,3a,8,8a-tetrahydro-, methyl ester	EtOH	216(4.47),248(4.07), 284(3.44),291(3.43)	94-1806-83
$C_{14}H_{15}ClN_2O_4$			
Tryptophan, 5-chloro-N-(methoxy-carbonyl)-, methyl ester	EtOH	227.5(4.56),283(3.73), 290(3.75),300(3.64)	94-1856-83
$C_{14}H_{15}Cl_2NOPt$			
Platinum, dichloro(η^2-ethenyl)(4-phen-oxybenzenamine-N)-, trans	n.s.g.	241(3.99),266(3.97), 300s(3.46)	131-0309-83H
$C_{14}H_{15}N$			
[1,1'-Biphenyl]-2-amine, N-ethyl-	EtOH	367(3.6)	103-0970-83
4H,8H-Cyclopenta[4,5]pyrrolo[3,2,1-ij]quinoline, 5,6,9,10-tetrahydro-	MeOH	233(4.38),285(3.86)	39-0505-83C
Pyridine, 2-[(4-ethylphenyl)methyl]-	EtOH	260(3.66)	103-0970-83
$C_{14}H_{15}NO$			
6H-Cyclohept[cd]indol-6-one, 1,3,4,5-tetrahydro-1,2-dimethyl-	EtOH	249(4.16),370(3.76)	39-2425-83C
1-Naphthalenecarbonitrile, 3,4-di-hydro-3-(3-hydroxypropyl)-	EtOH	224(4.58),231(3.55), 275(3.26)	44-3269-83
2H-Pyrrol-2-one, 1,5-dihydro-1,3,4-trimethyl-5-(phenylmethylene)-, (E)-	MeOH	276(4.18)	54-0114-83
(Z)-	MeOH	304(4.23)	54-0114-83
$C_{14}H_{15}NOS$			
2H-[1]Benzothieno[3,2-b]pyran-2-amine, N,N,3-trimethyl-	hexane	348(4.6),457s(4.1), 485(4.2)	30-0331-83
$C_{14}H_{15}NO_2$			
7-Azabicyclo[4.1.0]hept-2-ene-3-carb-oxylic acid, 7-phenyl-, methyl ester	MeCN	234(4.33),271(3.83)	24-0563-83
Flindersine, dihydro-	MeOH	225(4.44),272(3.94), 283(3.96),312(3.93)	162-0588-83
1,4-Naphthalenedione, 2-(diethylamino)-	C_6H_{12}	451(3.51)	150-0168-83S
	EtOH	472(--)	150-0168-83S
2-Propenoic acid, 3-(1,3-dimethyl-1H-indol-2-yl)-, methyl ester, (E)-	MeOH	221(4.15),259(3.60), 347(3.81)	39-0501-83C
2H-Pyrrol-2-one, 1,5-dihydro-5-[(4-methoxyphenyl)methylene]-3,4-di-methyl-, (E)-	MeOH	266(3.34),333(4.20)	54-0114-83
(Z)-	MeOH	238(3.62),345(4.38)	54-0114-83
$C_{14}H_{15}NO_2S$			
1-Hepten-3-one, 1-(2-thioxo-3(2H)-benzoxazolyl)-	EtOH	310(4.79)	105-0478-83
1-Hexen-3-one, 5-methyl-1-(2-thioxo-3(2H)-benzoxazolyl)-	EtOH	310(5.01)	105-0478-83

Compound	Solvent	λ_{max}(log ε)	Ref.
$C_{14}H_{15}NO_2S_2$			
1H-Isoindole-1,3(2H)-dione, 2-[3-(1,3-dithiolan-2-yl)propyl]-	MeOH	220(4.68),233(4.23), 241(4.09),293(3.34)	78-2691-83
$C_{14}H_{15}NO_4$			
Pyrrolo[1,2-b]isoquinoline-3,10(2H,5H)-dione, 1,10a-dihydro-7,8-dimethoxy-, (S)-(+)-	MeOH	234(4.32),278(4.09) 3.5(3.93)[sic]	44-4222-83
$C_{14}H_{15}NO_6$			
1H-Indole-2,3-dicarboxylic acid, 1-acetyl-2,3-dihydro-2-hydroxy-, dimethyl ester	EtOH	248(4.11),277(3.58), 286(3.51)	12-1419-83
3,5-Isoxazoledicarboxylic acid, 4,5-dihydro-4-(2-methylphenyl)-, dimethyl ester, 2-oxide	MeOH	262(4.01)	40-1678-83
3,5-Isoxazoledicarboxylic acid, 4,5-dihydro-4-(3-methylphenyl)-, dimethyl ester, 2-oxide	MeOH	270(4.03)	40-1678-83
3,5-Isoxazoledicarboxylic acid, 4,5-dihydro-4-(4-methylphenyl)-, dimethyl ester, 2-oxide	MeOH	264(4.02)	40-1678-83
$C_{14}H_{15}NO_7$			
3,5-Isoxazoledicarboxylic acid, 4,5-dihydro-4-(2-methoxyphenyl)-, dimethyl ester, 2-oxide	MeOH	268(4.06)	40-1678-83
3,5-Isoxazoledicarboxylic acid, 4,5-dihydro-4-(3-methoxyphenyl)-, dimethyl ester, 2-oxide	MeOH	268(4.04)	40-1678-83
$C_{14}H_{15}N_2O$			
Pyrimido[2,1-a]isoindolium, 4,6-dihydro-1,2,3-trimethyl-4-oxo-, salt with 4-methylbenzenesulfonic acid	EtOH	210(4.19),258(4.07), 305(3.93)	124-0755-83
$C_{14}H_{15}N_3$			
Benzenamine, 2-methyl-4-[(2-methylphenyl)azo]-	50% EtOH-HCl	326(4.28),490(3.40)	162-0062-83
$C_{14}H_{15}N_3O$			
Acetamide, N-[1-methyl-4-(2-phenylethenyl)-1H-imidazol-5-yl]-, (E)-	MeOH	226(4.14),233(4.10), 300(4.42),308(4.43)	4-0629-83
$C_{14}H_{15}N_3OS$			
2H-Isothiazolo[4,5,1-hi][1,2,3]benzothiadiazole-3-S^{IV}, 4,6,7,8-tetrahydro-4-methoxy-2-methyl-	C_6H_{12}	202(4.41),227s(4.03), 280(4.02),434(4.34), 453(4.34)	39-0777-83C
Pyrido[2,3-e]-1,3-thiazin-4-one, 2-(cyclohexylamino)-	MeOH	326(5.43)	73-3315-83
$C_{14}H_{15}N_3O_2$			
2H-1-Benzopyran-2-one, 4-(diazomethyl)-7-(diethylamino)-	EtOH	258.5(4.26),333(4.43), 395(4.28)	94-3014-83
Pyrimido[1,2-a]indole, 1,2,3,4-tetrahydro-3-methyl-10-(2-nitroethenyl)-	MeOH	476(4.32)	103-0507-83
Pyrimido[4,5-b]quinoline-2,4(1H,3H)-dione, 5,10-dihydro-1,3,10-trimethyl-	pH 5.0	318(4.15)	12-1873-83

Compound	Solvent	λ_{max}(log ϵ)	Ref.
$C_{14}H_{15}N_3O_2S_3$			
4-Thiazolecarboxylic acid, 2-(methylthio)-5-[[(phenylamino)thioxomethyl]amino]-, ethyl ester	MeOH	215(4.48),282(4.21), 334(4.19)	2-0243-83
$C_{14}H_{15}N_3O_3S$			
Methyl Orange (sodium salt)	pH 7	378(3.91),467(4.41)	18-2535-83
	pH 10	465(5.3)	60-0155-83
	+ $CaCl_2$	460(4.9)	60-0155-83
$C_{14}H_{15}N_3O_4$			
Benzoic acid, 4-[(2-amino-1,4-dihydro-6-hydroxy-4-oxo-5-pyrimidinyl)methyl]-, ethyl ester	MeOH	207(4.47),245(4.30), 267(4.18)	73-0299-83
Pyrrolo[1,2-a]pyrazine-1,4-dione, hexahydro-2-[(2-nitrophenyl)-methyl]-, (S)-	MeOH	256(4.6)	5-0231-83
$C_{14}H_{15}N_3O_5$			
1,2-Diazabicyclo[5.2.0]nona-3,5-diene-2-carboxylic acid, 8-(2,5-dioxo-1-pyrrolidinyl)-9-oxo-, ethyl ester, cis	MeOH	278(3.90)	5-1374-83
Pyrrolo[2,3-b]indole-1(2H)-carboxylic acid, 8-acetyl-3,3a,8,8a-tetrahydro-5-nitro-	EtOH	226(4.03),324.5(4.09)	94-1856-83
$C_{14}H_{15}N_3O_5S$			
Uridine, 5'-S-2-pyridinyl-5'-thio-	MeOH	251(4.22)	39-1315-83C
$C_{14}H_{15}N_3O_6$			
Tryptophan, N-(methoxycarbonyl)-5-nitro-, methyl ester	EtOH	257(4.16),271(4.24), 326(3.91)	94-1856-83
$C_{14}H_{15}N_5$			
1H-Pyrazolo[3,4-d]pyrimidin-4-amine, 1,6-dimethyl-N-(2-methylphenyl)-	EtOH	206(5.53),283(4.12)	11-0171-83B
$C_{14}H_{15}N_5O$			
Benzenamine, N,N-dimethyl-4-[[4-(nitrosoamino)phenyl]azo]-, sodium salt	H_2O	245(4.11),320(4.02), 410s(4.08),460(4.18)	104-1493-83
3H-Purin-6-amine, N,3-dimethyl-N-(phenylmethoxy)-, monohydrochloride	pH 1	225s(4.00),294(4.27)	94-3521-83
	pH 7	223s(4.08),296(4.24)	94-3521-83
	pH 13	296.5(4.22)	94-3521-83
	EtOH	225(4.10),296.5(4.25)	94-3521-83
$C_{14}H_{15}N_5O_3S$			
Benzenamine, N,N-dimethyl-4-[[4-[(sulfinooxy)azo]phenyl]azo]-, sodium salt	H_2O	280(3.54),505(4.30)	104-1493-83
$C_{14}H_{15}N_5O_7$			
2-Butenal, 4-methoxy-2-[(4-oxo-2-azetidinyl)oxy]-, 1-(2,4-dinitrophenylhydrazone)	EtOH	256(4.11),285s(3.86), 369(4.39)	39-2513-83C
$C_{14}H_{16}$			
Azulene, 7-ethyl-1,4-dimethyl-(chamazulene)	n.s.g.	370(3.7)	162-0283-83
Benzene, [2,2-dimethyl-1-(1-propynyl)-cyclopropyl]-	C_6H_{12}	206.5(3.81)	12-0581-83

Compound	Solvent	$\lambda_{max}(\log \epsilon)$	Ref.
$C_{14}H_{16}Br_2ClNO_4$			
L-Phenylalanine, 3,5-dibromo-4-chloro-N-[(1,1-dimethylethoxy)carbonyl]-	EtOH	281(2.98)	33-0960-83
$C_{14}H_{16}ClNS$			
Piperidine, 1-[3-(2-chlorophenyl)-1-thioxo-2-propenyl]-	heptane	251(4.12),279(4.22)	80-0555-83
	EtOH	254(4.15),283(4.30)	80-0555-83
Piperidine, 1-[3-(4-chlorophenyl)-1-thioxo-2-propenyl]-	heptane	254(4.06),299(4.29), 309(4.20)	80-0555-83
	EtOH	228(4.11),288(4.37), 300(4.33),385s(--)	80-0555-83
$C_{14}H_{16}Cl_4N_2$			
Pyrrolidine, 1,1'-[(2,3,4,5-tetrahydro-2,4-cyclopentadien-1-ylidene)methylene]bis-	DMF	270(3.51),346(3.86)	5-0154-83
$C_{14}H_{16}Cl_4N_2O_2$			
Morpholine, 4,4'-[(2,3,4,5-tetrachloro-2,4-cyclopentadien-1-ylidene)methylene]bis-	DMF	275(4.02),352(4.23)	5-0154-83
$C_{14}H_{16}F_3NO$			
Cyclohexanone, 2-[phenyl(2,2,2-trifluoroethyl)amino]-	MeOH	243(4.01),283.5(3.27)	39-0573-83C
$C_{14}H_{16}FeO_4$			
Iron, tricarbonyl[(1,2,3,8a-η)-4,4a,5,6,7,8-hexahydro-4a-methyl-2-naphthalenol]-	hexane	229(4.2),284s(3.5)	24-0243-83
$C_{14}H_{16}N$			
Isoquinolinium, 2-(3-methyl-2-butenyl)-, perchlorate	MeCN	336(3.62)	35-1204-83
Quinolinium, 1-(3-methyl-2-butenyl)-, perchlorate	MeCN	313(3.89)	35-1204-83
$C_{14}H_{16}NOS$			
Thiazolium, 2,4-dimethyl-3-[2-(4-methylphenyl)-2-oxoethyl]-, perchlorate	EtOH	260(4.33)	65-2424-83
$C_{14}H_{16}NO_2S$			
Thiazolium, 3-[2-(4-methoxyphenyl)-2-oxoethyl]-2,4-dimethyl-, perchlorate	EtOH	286(4.29)	65-2424-83
$C_{14}H_{16}N_2O$			
9-Oxaergoline, 6-methyl-, monohydrochloride, (±)-	EtOH-HCl	224(4.53),276(3.79), 281(3.82),291(3.78)	87-0522-83
$C_{14}H_{16}N_2OS$			
Morpholine, 4-[4-(4-methylphenyl)-5-isothiazolyl]-	EtOH	243(4.27),289(3.97)	97-0020-83
$C_{14}H_{16}N_2OSSe$			
2-Selenazolidinone, 5-[[4-(diethylamino)phenyl]methylene]-4-thioxo-	n.s.g.	240(3.90),306(3.88), 353(4.26),518(4.41)	103-0273-83
$C_{14}H_{16}N_2O_2$			
2H-1,2-Benzodiazepine-2-carboxylic acid, 1,3-dihydro-7-methyl-3-methylene-, ethyl ester	EtOH	225(4.00),242(4.00), 289(3.93),298s(--)	94-3684-83

Compound	Solvent	λ_{max}(log ε)	Ref.
Butanamide, N-(2,3-dimethyl-1H-indol-5-yl)-3-oxo-	EtOH	245(4.45),300(3.85)	103-0401-83
Butanamide, N-(2,3-dimethyl-1H-indol-6-yl)-3-oxo-	EtOH	218(4.29),247(4.37), 296(4.11)	103-0401-83
Indolo[4,3-ef][2,1]benzisoxazole, 4,6,6a,7,9,9a-hexahydro-9-methoxy-7-methyl-, (6aα,9α,9aβ)-(±)-	MeOH	221(4.55),279(3.86), 289(3.75)	78-3695-83
(6aα,9β,9aα)-(±)-	MeOH	222(4.50),280(3.82), 291(3.74)	78-3695-83
2-Pentenamide, 2-[1-(methylimino)ethyl]-4-oxo-N-phenyl-	MeOH	239(4.2),322(4.18)	83-1000-83
2-Propenoic acid, 3-(1H-indol-3-ylmethylamino)-, ethyl ester	EtOH	220(4.35),289(4.38)	104-1367-83
2H-Pyrido[3,4-b]indole-2-carboxylic acid, 1,3,4,9-tetrahydro-, ethyl ester	EtOH	200(4.30),223(4.60), 277(3.90),289(3.80)	78-3767-83
Spiro[cyclopropan-1,4'(1'H)-isoquinoline]-1',3'(2'H)-dione, 2'-(dimethylamino)-2-methyl-	MeOH	214(4.47),245(3.99), 293(3.48)	142-1315-83
$C_{14}H_{16}N_2O_2S$			
Morpholine, 4-[4-(4-methoxyphenyl)-5-isothiazolyl]-	EtOH	248(4.21),288(3.99)	97-0020-83
Piperidine, 1-[3-(3-nitrophenyl)-1-thioxo-2-propenyl]-	heptane	255(4.34),266(4.34), 280(4.31)	80-0555-83
	EtOH	270(4.44)	80-0555-83
$C_{14}H_{16}N_2O_2S_2$			
1,4-Benzenediethanethioamide, N,N,N',N'-tetramethyl-α,α'-dioxo-	MeCN	249s(4.45),268(4.62), 303s(3.80),400(3.65), 421s(3.48)	5-1116-83
$C_{14}H_{16}N_2O_3$			
2H-1,2-Benzodiazepine-2-carboxylic acid, 1,3-dihydro-7-methoxy-3-methylene-, ethyl ester	EtOH	235(3.97),242(3.97), 287(3.91),296s(--)	94-3684-83
Pyrido[1,2-a]pyrimidine-7-carboxylic acid, 3-ethyl-2-methyl-4-oxo-, ethyl ester	EtOH	236(4.37),350(4.10)	106-0218-83
Pyrido[1,2-a]pyrimidine-8-carboxylic acid, 3-ethyl-2-methyl-4-oxo-, ethyl ester	EtOH	264(4.07),270(4.06), 348s(3.85),372(3.90)	106-0218-83
Pyrido[1,2-a]pyrimidine-9-carboxylic acid, 3-ethyl-2-methyl-4-oxo-, ethyl ester	EtOH	247(4.19),333(4.05)	106-0218-83
Quinoxaline, 5,6,7,8-tetrahydro-2-methyl-3-(5-methyl-2-furanyl)-, 1,4-dioxide	EtOH	242(4.06),268(4.06), 298(4.52),338s(4.06)	103-1104-83
$C_{14}H_{16}N_2O_3S$			
1H-Indene-2-carbothioamide, N-(2-hydroxyethyl)-3-[(2-hydroxyethyl)-amino]-1-oxo-	EtOH	231(4.40),270(4.33), 301(4.45),329(4.22), 342(4.28)	95-1243-83
$C_{14}H_{16}N_2O_4$			
1H-Pyrazole-3,4-dicarboxylic acid, 4,5-dihydro-1-phenyl-, 3-ethyl 4-methyl ester	EtOH	236(3.96),297(3.53), 350(4.23)	44-3189-83
$C_{14}H_{16}N_2O_5$			
Oxepino[4,5-d]pyridazine-1,4-dicarbox-	CH_2Cl_2	284(4.18),402(3.79)	24-0097-83

Compound	Solvent	λ_{max}(log ε)	Ref.
ylic acid, 2,4a-dihydro-6,8-dimethyl-, dimethyl ester (cont.)			24-0097-83
2,4(1H,3H)-Pyrimidinedione, 5-[(2,3,4-trimethoxyphenyl)methyl]-	pH 1	266(3.94)	87-0667-83
	pH 13	280(3.83)	87-0667-83
$C_{14}H_{16}N_2O_6$			
[3,3'-Bi-3H-pyrrole]-5,5'-dicarboxylic acid, 4,4'-dihydroxy-, diethyl ester	MeCN	305(4.02)	33-1902-83
$C_{14}H_{16}N_2S$			
2-Propenenitrile, 3-(1-cyclohexyl-1,2-dihydro-2-thioxo-3-pyridinyl)-	EtOH	213(3.64),242(3.57), 325(3.42),408(3.23)	4-1651-83
$C_{14}H_{16}N_4$			
Benzenamine, 4-[(4-aminophenyl)azo]-N,N-dimethyl-	MeOH	256(4.03),322(2.78), 370(4.34)	80-0903-83
	$C_2H_4Cl_2$	420(4.48)	80-0903-83
1(2H)-Phthalazinone, (1,3-dimethyl-2-butenylidene)hydrazone	MeOH	208(4.43),240(4.95), 289(4.30),357(4.18)	162-0202-83
$C_{14}H_{16}N_4O_2$			
3H-Pyrazoline, 1-[(1,3-dihydro-1,3-dioxo-2H-isoindol-2-yl)amino]-4,5-dihydro-3,3,5-trimethyl-, hydroxide, inner salt	EtOH	219(4.49),233s(4.27), 240s(4.21),267(4.05)	104-2317-83
3H-Pyrazoline, 1-[(1,3-dihydro-1,3-dioxo-2H-isoindol-2-yl)amino]-4,5-dihydro-3,5,5-trimethyl-, hydroxide, inner salt	EtOH	219(4.50),233s(4.28), 241s(4.23),268(4.07)	104-2317-83
$C_{14}H_{16}N_4O_2S$			
1,2,4-Triazine-4(5H)-acetic acid, α-(aminothioxomethyl)-6-phenyl-, ethyl ester	EtOH	300(4.46)	2-0030-83
$C_{14}H_{16}N_4O_3S$			
Cytidine, 2'-deoxy-5'-S-2-pyridinyl-5'-thio-	pH 1	251(3.95),281(4.13), 310(3.89)	39-1315-83C
$C_{14}H_{16}N_4O_4$			
L-Histidine, N-[[3-hydroxy-5-(hydroxymethyl)-2-methyl-4-pyridinyl]methylene]-	MeOH	253(3.99),285s(3.70), 335(3.38),412(3.15)	35-0803-83
1H-Imidazole-4-carboxamide, 5-amino-2-[[(2-benzoyloxy)ethoxy]methyl]-	EtOH	230(4.28),261(4.15)	4-1169-83
$C_{14}H_{16}N_4O_4S$			
Cytidine, 5'-S-2-pyridinyl-5'-thio-	pH 1	251(3.93),281(4.12), 310(3.89)	39-1315-83C
$C_{14}H_{16}N_4O_5S$			
2-Propenoic acid, 3-[(1,2,3,4-tetrahydro-1,3,6-trimethyl-2,4-dioxo-7-pteridinyl)sulfinyl]-, ethyl ester, (E)-	CH_2Cl_2	245(4.31),353(4.02)	88-5047-83
$C_{14}H_{16}N_4O_8$			
1H-Pyrano[3,2-d:5,6-d']dipyrimidine-2,4,6,8(3H,4aH,7H,9H)-tetrone, 10a-(acetoxymethyl)-10,10a-dihydro-4a-hydroxy-1,9-dimethyl-	H_2O	283.5(3.85)	4-0753-83

Compound	Solvent	λ_{max}(log ε)	Ref.
$C_{14}H_{16}N_5O$			
1H-Purinium, 1,3-dimethyl-6-[(phenylmethoxy)amino]-, chloride	pH 1	274.5(4.20)	94-3521-83
Pyridinium, 3-[[4-(acetylamino)-2-aminophenyl]azo]-1-methyl-, iodide	H_2O	467(4.582)	7-0055-83
	DMSO-p-tolueneSO$_3$H	492(4.384)	7-0055-83
$C_{14}H_{16}N_6S$			
Benzonitrile, 4-[[2,4-bis(dimethylamino)-5-thiazolyl]azo]-	$CHCl_3$	320(4.13),495(4.64)	88-3563-83
$C_{14}H_{16}O$			
Benzene, 2-(2,4-cyclohexadien-1-yl)-1-methoxy-3-methyl-	MeOH	208(4.06),275(3.62), 292(3.62),310(3.57)	100-0235-83
Naphthalene, 2-butoxy- (class spectrum)	MeOH and MeOH-base	261(3.6),271(3.6), 281(3.5),313(3.18), 328(3.28)	160-1403-83
$C_{14}H_{16}O_2$			
1H-Cyclopropa[a]naphthalene-1-carboxylic acid, 1a,2,3,7b-tetrahydro-1-methyl-, methyl ester, (1α,1aα,7bα)-	n.s.g.	224(4.00),267(2.68)	78-2965-83
1,4-Hexadien-3-one, 2-(4-methoxyphenyl)-5-methyl-	n.s.g.	280(4.37)	12-0311-83
1H-Indene-2-carboxylic acid, 1,1,3-trimethyl-, methyl ester	MeOH	228(4.15),234(4.08), 288(4.24)	2-0230-83
Naphthalene, 1-ethyl-2,7-dimethoxy-	EtOH	298(3.28),308(3.19), 326(3.12)	104-0183-83
Naphthalene, 2-ethyl-3,6-dimethoxy-	EtOH	276(3.27),310(3.31), 324(3.49)	104-0183-83
$C_{14}H_{16}O_3$			
2H-1-Benzopyran-4-carboxylic acid, 2,2-dimethyl-, ethyl ester	EtOH	225(4.00),278(3.49), 317(3.23)	39-1431-83C
2H-1-Benzopyran-2-one, 7-butoxy-	$CHCl_3$	324(4.22)	44-1872-83
syn H-T dimer	$CHCl_3$	249(3.60),280(3.51)	44-1872-83
Naphthalene, 1,3,6-trimethoxy-8-methyl-	EtOH	237(4.52),249s(4.42), 269(3.45),281(3.52), 297(3.54)	78-2283-83
$C_{14}H_{16}O_3S$			
Bicyclo[2.2.1]heptan-2-one, 6-[(phenylsulfonyl)methyl]-	EtOH	254(2.75),259(2.93), 265(3.07),272(2.99)	104-0835-83
Cyclopentanone, 3-[1-(phenylsulfonyl)-2-propenyl]-	EtOH	259(2.90),265.5(3.04), 272.5(2.97)	104-0835-83
Cyclopentanone, 3-[3-(phenylsulfonyl)-1-propenyl]-, (E)-	EtOH	253(2.67),258(2.86), 264.5(3.02),271.5(2.96)	104-0835-83
Cyclopentanone, 3-[3-(phenylsulfonyl)-2-propenyl]-, (E)-	EtOH	260.5(2.83),267(2.93), 274(2.85)	104-0835-83
$C_{14}H_{16}O_4$			
2H-1-Benzopyran-4-carboxylic acid, 3-hydroxy-2,2-dimethyl-, ethyl ester	EtOH	214(4.00),242(3.98), 250s(3.93),285(3.66), 292s(3.61)	39-1431-83C
Carbonic acid, 2,2-dimethyl-2H-1-benzopyran-3-yl ethyl ester	EtOH	224(4.34),267(3.79), 276s(3.72),304(3.56), 312s(3.30)	39-1431-83C
2-Naphthalenecarboxaldehyde, 5,6-dihydro-4-methoxy-1-(methoxymethoxy)-	EtOH	250(4.40),298(3.66), 349(3.49)	94-2662-83
Semibullvalene-3,7-dicarboxylic acid, 1,5-dimethyl-, dimethyl ester	dioxan	220(4.034),263(3.757)	24-3751-83

Compound	Solvent	λ_{max}(log ε)	Ref.
$C_{14}H_{16}O_4S$			
1,4-Naphthalenedione, 2,3-dihydro-2-hydroxy-3-[(2-hydroxyethyl)thio]-2,3-dimethyl-, trans-(±)-	MeOH	225(4.54),254(4.21), 297(3.35),337s(2.83)	44-3301-83
$C_{14}H_{16}O_5$			
2H,5H-Pyrano[3,2-c]benzopyran-2,5-dione, 8-methoxy-4-methyl-	MeOH	277.6(4.058),355.3(4.377)	64-0248-83B
$C_{14}H_{16}O_7$			
1,4,7,10,13-Benzopentaoxacyclopentadecin-3,11(2H,12H)-dione, 5,6,8,9-tetrahydro-	MeOH	270(3.19)	49-0359-83
$C_{14}H_{17}$			
Cycloheptatrienylium, bicyclo[2.2.1]-hept-1-yl-, perchlorate	10% HCl	221(4.55),309(3.94)	78-4011-83
Cycloheptatrienylium, bicyclo[2.2.1]-hept-2-yl-, perchlorate	10% HCl	229(4.45),307(3.82)	78-4011-83
$C_{14}H_{17}BrN_2O_6$			
2,4(1H,3H)-Pyrimidinedione, 1-[3-acetoxy-4-(acetoxymethyl)cyclopentyl]-5-bromo-, (1α,3β,4α)-(±)-	pH 1	212(4.00),284(4.02)	87-0156-83
	pH 7	211(3.98),282(4.00)	87-0156-83
	pH 13	280(3.88)	87-0156-83
2,4(1H,3H)-Pyrimidinedione, 1-[2-acetoxy-4-(acetoxymethyl)cyclopentyl]-5-bromo-, (1α,2β,4α)-(±)-	pH 1	211(3.97),283(4.00)	87-0156-83
	pH 7	212(4.00),283(4.00)	87-0156-83
	pH 13	280(3.86)	87-0156-83
$C_{14}H_{17}ClN_2O_6$			
2(1H)-Pyrimidinone, 4-chloro-1-(3,5-di-O-acetyl-2-deoxy-β-D-ribofuranosyl)-5-methyl-	$CHCl_3$	318(3.76)	32-0863-83
$C_{14}H_{17}Cl_3N_2O$			
2-Butenamide, 2,4,4-trichloro-3-[(1-methylethyl)amino]-N-(4-methylphenyl)-, (Z)-	heptane	193(4.18),222s(3.66), 308(4.19)	5-1496-83
$C_{14}H_{17}F_3N_2O_2$			
4-Oxazolamine, 2,5-dihydro-5-methoxy-N,2,2-trimethyl-N-phenyl-5-(trifluoromethyl)-	EtOH	237(3.84)	33-0262-83
$C_{14}H_{17}IN_2O_6$			
2,4(1H,3H)-Pyrimidinedione, 1-[3-acetoxy-4-(acetoxymethyl)cyclopentyl]-5-iodo-, (1α,3β,4α)-(±)-	pH 1	218(4.01),293(3.91)	87-0156-83
	pH 7	217(4.01),292(3.90)	87-0156-83
	pH 13	284(3.76)	87-0156-83
$C_{14}H_{17}N$			
4-Azuleneethanamine, N,N-dimethyl-	EtOH	240(4.36),265s(4.35), 270s(4.49),275s(4.60), 278(4.62),283(4.61), 298(3.70),317s(3.34), 328(3.51),342(3.64), 354(3.05),530s(2.48), 550s(2.56),568(2.61), 587s(2.58),615(2.55), 642s(2.29),675(2.12)	18-2311-83
hydrochloride	EtOH	240(4.41),270s(4.55), 275s(4.66),278(4.68), 283(4.67),298(3.78),	18-2311-83

Compound	Solvent	λ$_{max}$(log ε)	Ref.
(cont.)		318s(3.48),329(3.62) 342(3.72),354(3.24), 531s(2.51),553s(2.59), 571(2.65),588s(2.61), 617(2.57),644s(2.33), 677(2.14)	18-2311-83
Bicyclo[3.2.1]oct-3-en-2-amine, 3-phenyl-, exo-, hydrochloride	EtOH	244(4.05)	87-0055-83
$C_{14}H_{17}NO$			
Acetamide, N-(4-methyl-1-phenyl-3-pentenylidene)-	C_6H_{12}	244(3.69)	44-0695-83
Azocino[2,1-a]isoindol-5(7H)-one, 8,9,10,11,12,12a-hexahydro-	$CHCl_3$	251(3.66),270s(3.46), 280(3.29)	78-2691-83
1H-Cyclohept[cd]indol-6-ol, 3,4,5,6-tetrahydro-1,2-dimethyl-	EtOH	232(4.57),296(3.92)	39-2425-83C
2-Cyclohexen-1-one, 5,5-dimethyl-3-(phenylamino)-	EtOH	201(3.95),229(3.78), 312(4.38)	104-2027-83
Ethanone, 1-[1,4,5,6-tetrahydro-1-(phenylmethyl)-3-pyridinyl]-	EtOH	318(4.49)	150-2326-83M
$C_{14}H_{17}NO_2$			
4(5H)-Benzofuranone, 6,7-dihydro-5-(piperidinomethylene)-	EtOH	252.5(3.96),358(4.36)	111-0113-83
Cyclohexanone, 2-[[(4-methoxyphenyl)-amino]methylene]-	CH_2Cl_2	366(4.30)	24-0152-83
2,4,6-Heptatrien-1-one, 7-(dimethyl-amino)-1-(5-methyl-2-furanyl)-	EtOH	500(4.73)	70-0780-83
	$CHCl_3$	470(--)	70-0780-83
5(4H)-Oxazolone, 4,4-dimethyl-2-(2,4,6-trimethylphenyl)-	EtOH	235(3.87),265s(2.92)	39-0979-83C
3-Pyridinecarboxylic acid, 1,4,5,6-tetrahydro-1-(phenylmethyl)-, methyl ester	EtOH	298(4.38)	150-2326-83M
$C_{14}H_{17}NO_2S$			
Morpholine, 4-[3-(2-methoxyphenyl)-1-thioxo-2-propenyl]-	heptane	238(4.28),302(4.30), 321(4.30)	80-0555-83
	EtOH	237(4.23),294(4.30), 326(4.27)	80-0555-83
Morpholine, 4-[3-(3-methoxyphenyl)-1-thioxo-2-propenyl]-	heptane	230(4.05),300(4.22), 312(4.20)	80-0555-83
	EtOH	230(4.10),304(4.26), 315(4.27)	80-0555-83
$C_{14}H_{17}NO_3$			
5H-Benzocycloheptene-6-carboxamide, 6,7,8,9-tetrahydro-2-methoxy-6-methyl-5-oxo-	EtOH	225(3.99),278(4.03)	118-0727-83
$C_{14}H_{17}NO_4$			
1,4-Dioxaspiro[4.5]decane, 2-(2-nitro-phenyl)-	EtOH	206.5(4.10),212(3.74)	23-0400-83
2H-Pyrrol-2-one, 1,5-dihydro-4,5,5-trimethoxy-1-(4-methylphenyl)-	heptane	198(4.28),228s(3.96), 267(3.43)	5-1504-83
$C_{14}H_{17}NO_7$			
Dhurrin	n.s.g.	228(4.11)	162-0427-83
$C_{14}H_{17}NO_7S$			
1-Azetidineacetic acid, α-hydroxy-2-[[[(4-methylphenyl)sulfonyl]oxy]-	EtOH	220s(3.99),230(4.05), 263(3.00),274(2.93)	39-0649-83C

Compound	Solvent	λ_{max}(log ε)	Ref.
methyl]-4-oxo-, methyl ester (cont.)			39-0649-83C
$C_{14}H_{17}NO_8$			
7-Azabicyclo[2.2.1]hept-2-ene-1,2,3,4-tetracarboxylic acid, tetramethyl ester	CCl_4	254(3.27)	4-0001-83
1H-Pyrrole-2,3,4,5-tetracarboxylic acid, 3,4-diethyl 2,5-dimethyl ester	EtOH	224(3.87),267(3.80)	4-0001-83
$C_{14}H_{17}NS$			
Piperidine, 1-(3-phenyl-1-thioxo-2-propenyl)-	heptane	244(4.04),284(4.31), 295(4.27)	80-0555-83
	EtOH	222(4.13),288(4.32), 385s(--)	80-0555-83
$C_{14}H_{17}N_3O$			
Morpholine, 4-[4-(4-methylphenyl)-1H-pyrazol-3-yl]-	MeCN	226(3.99),253s(3.96), 266(3.90)	48-0689-83
$C_{14}H_{17}N_3O_2$			
Morpholine, 4-[4-(4-methoxyphenyl)-1H-pyrazol-3-yl]-	MeCN	230(4.07),235(4.12), 265(4.15)	48-0689-83
1,8-Naphthalenediamine, N,N,N',N'-tetramethyl-4-nitro-	pH 1	240s(--),338(3.68)	104-0520-83
	MeOH	280(4.04),357(3.69), 470(4.07)	104-0520-83
	acetone	360(3.73),470(4.08)	104-0520-83
	HOAc	327(3.70),492(2.97)	104-0520-83
	DMSO	288(4.08),360(3.68), 490(4.15)	104-0520-83
1(4H)-Pyridinecarboxylic acid, 3,5-dicyano-2,4,4,6-tetramethyl-, ethyl ester	EtOH	212(4.32),255(4.13), 310(3.45)	73-0608-83
4H-Pyrido[1,2-a]pyrimidine-7-carboxamide, 3-ethyl-N,N,2-trimethyl-4-oxo-, monohydrochloride	EtOH	234(4.28),344(4.08)	106-0218-83
$C_{14}H_{17}N_3O_3S$			
2H-Thiazolo[3,2-a]-1,3,5-triazin-6(7H)-one, 3,4-dihydro-7,7-bis(hydroxymethyl)-3-(phenylmethyl)-	EtOH	210(4.21)	103-0961-83
$C_{14}H_{17}N_4$			
Pyridinium, 3-[[4-(dimethylamino)-phenyl]azo]-1-methyl-, iodide	H_2O	499(4.537)	7-0055-83
	DMSO-p-tolu-eneSO_3H	497(4.525)	7-0055-83
$C_{14}H_{17}N_4O$			
Pyridinium, 3-[[4-[(2-hydroxyethyl)-amino]phenyl]azo]-1-methyl-, iodide	H_2O	473(4.369)	7-0055-83
	DMSO-p-tolu-eneSO_3H	487(4.413)	7-0055-83
$C_{14}H_{17}N_5O$			
1,2,4-Triazolo[5,1-c][1,2,4]triazine, 4-ethoxy-3,4-dihydro-3,3-dimethyl-1-phenyl-	EtOH	244(4.36),320(3.24)	44-2330-83
$C_{14}H_{17}N_5O_3$			
4H-Cyclopenta-1,3-dioxole-6-methanol, 4-(6-amino-9H-purin-9-yl)-3a,6a-dihydro-2,2-dimethyl-, (3aα,4α,6aα)-(±)-	MeOH	262(4.16)	33-1915-83

Compound	Solvent	λ_{max}(log ϵ)	Ref.
$C_{14}H_{17}N_5O_4$			
4,13-Epoxy-1,3-dioxolo[6,7][1,3]diazocino[1,2-e]purin-9(6H)-one, 3a,4,11,12,13,13a-hexahydro-2,2,11-trimethyl-	MeOH	264(4.24)	138-1017-83
$C_{14}H_{17}N_5O_5$			
Adenosine, 3'-deoxy-, 2',5'-diacetate	EtOH	258(4.13)	118-0304-83
$C_{14}H_{17}N_7OS_2Zn$			
Zinc, [N-[4-[bis[[(methylamino)thioxomethyl]hydrazono]ethyl]phenyl]acetamidato(2-)]-, (T-4)-	EtOH	303(4.54),337(4.47), 460(4.21)	161-0143-83
$C_{14}H_{17}O_7P$			
α-D-Xylofuranose, 1,2-O-(1-methylethylidene)-, cyclic phenyl phosphate, (R)-	EtOH	208.5(3.34)	23-1387-83
(S)-	EtOH	210(3.42)	23-1387-83
$C_{14}H_{18}$			
Anthracene, 1,2,3,4,5,6,7,8-octahydro-	MeOH	206(4.64),276(3.20), 286(3.30)	97-0341-83
Benzene, [2,2-dimethyl-1-(1-propenyl)-cyclopropyl]-, 9:1 E:Z	C_6H_{12}	211(4.05)	12-0581-83
$C_{14}H_{18}BrN_3O$			
Tambjamine D	MeOH	257(3.79),401(4.36)	44-2314-83
$C_{14}H_{18}ClN_3$			
[1,2]Diazepino[5,4-b]indole, 9-chloro-2-ethyl-1,2,3,4,5,6-hexahydro-3-methyl-	EtOH	235(4.23),286s(3.57), 293(3.65),303(3.55)	39-1937-83C
$C_{14}H_{18}FN_2O_9P$			
Uridine, 2'-deoxy-5'-O-1,3,2-dioxaphosphorinan-2-yl)-5-fluoro-, 3-acetate, P-oxide	EtOH	268(3.94)	87-1153-83
$C_{14}H_{18}N_2$			
Pyrimidine, 1,4-dihydro-4,4,6-trimethyl-1-(3-methylphenyl)-	EtOH	257(3.71)	39-1799-83C
Pyrimidine, 1,4-dihydro-4,4,6-trimethyl-1-(4-methylphenyl)-	EtOH	255(3.86)	39-1799-83C
Pyrimidine, 1,4-dihydro-4,4,6-trimethyl-1-(phenylmethyl)-	EtOH	254(3.49)	39-1799-83C
$C_{14}H_{18}N_2O$			
2H-Azirin-3-amine, N,2,2-trimethyl-N-[2-(2-propenyloxy)phenyl]-	EtOH	244(3.89),278(3.54)	33-0262-83
1-Butanone, 3-methyl-1-(1-methyl-1H-indol-3-yl)-, oxime	n.s.g.	227(4.31),259(4.09), 290(3.96)	103-0184-83
9-Oxaergoline, 2,3-dihydro-6-methyl-(3α)-(±)-	EtOH	211(4.47),216(3.78), 293(3.34)	87-0522-83
	EtOH-HCl	263(2.87),271(2.81)	87-0522-83
Pyridinium, 3,3'-oxybis[1-ethyl-, diiodide	H_2O	225(4.55),271(3.93)	4-1411-83
Pyrimidine, 1,4-dihydro-1-(4-methoxyphenyl)-4,4,6-trimethyl-	EtOH	251(3.91)	39-1799-83C

Compound	Solvent	$\lambda_{max}(\log \epsilon)$	Ref.
$C_{14}H_{18}N_2OS$			
Acetamide, N-[4-(1-piperidinylthioxo-methyl)phenyl]-	EtOH	245(4.41),271(4.43), 364(3.11)	80-0555-83
$C_{14}H_{18}N_2O_2$			
1,2,4-Oxadiazole, 3-(2-methoxy-4,6-di-methylphenyl)-5-(1-methylethyl)-	MeOH	288(3.26)	4-1609-83
1,2,4-Oxadiazole, 3-(4-methoxy-2,6-di-methylphenyl)-5-(1-methylethyl)-	MeOH	280(3.11)	4-1609-83
1,2,4-Oxadiazole, 3-(2-methoxy-4,6-di-methylphenyl)-5-propyl-	MeOH	283(3.23)	4-1609-83
1,2,4-Oxadiazole, 3-(4-methoxy-2,6-di-methylphenyl)-5-propyl-	MeOH	280(3.20)	4-1609-83
$C_{14}H_{18}N_2O_2S$			
Diazene, (cyclohexylidenemethyl)[(4-methylphenyl)sulfonyl]-	heptane	228(4.11),284(4.34), 421(2.03)	44-3994-83
$C_{14}H_{18}N_2O_4$			
4H-Pyrido[1,2-a]pyrimidine-3-acetic acid, 9-formyl-1,6,7,8-tetrahydro-6-methyl-4-oxo-, ethyl ester	EtOH	226(3.97),248s(3.41), 342(4.36)	39-0369-83C
$C_{14}H_{18}N_2O_4S$			
2H-[1,2,5]Oxadiazolo[2,3-a]pyridine, 4,5,6,7-tetrahydro-3-methyl-2-[[(4-methylphenyl)sulfonyl]oxy]-	EtOH	263.2(2.72)	94-2269-83
L-Phenylalanine, N-(1-thio-L-α-aspar-tyl)-, 1-methyl ester	EtOH	267(3.98)	78-4121-83
$C_{14}H_{18}N_2O_5$			
1,2-Hydrazinedicarboxylic acid, 1-(1,1-dimethyl-2-oxo-2-phenylethyl)-, dimethyl ester	EtOH	245(3.97),280(3.00), 312(2.04)	104-0110-83
Uridine, 2'-deoxy-5-(1-pentynyl)-	pH 6	230(4.09),292(4.03)	44-1854-83
	pH 13	230(4.17),287(3.94)	44-1854-83
$C_{14}H_{18}N_2O_5S$			
Asparenomycin C, dihydro-, sodium salt	H_2O	228(4.33),282(3.94), 326s(--)	158-83-126
1-Azabicyclo[3.2.0]hept-2-ene-2-carb-oxylic acid, 3-[[2-(acetylamino)eth-enyl]thio]-6-(1-hydroxy-1-methyleth-yl)-7-oxo-, sodium salt	H_2O	230(4.14),310(4.19)	39-0403-83C
	H_2O	230(4.19),310(4.17)	78-0075-83
(Z)-	H_2O	232(4.20),308(4.16)	87-0271-83
$C_{14}H_{18}N_2O_6$			
Uridine, 2'-deoxy-5-(5-hydroxy-1-pent-ynyl)-	pH 6	232(4.05),292(4.05)	44-1854-83
	pH 13	232s(4.12),287(3.95)	44-1854-83
$C_{14}H_{18}N_2O_6S$			
1-Azabicyclo[3.2.0]hept-2-ene-2-carb-oxylic acid, 3-[[2-(acetylamino)eth-enyl]sulfinyl]-6-(1-hydroxy-1-methyl-ethyl)-7-oxo-, (R)-, sodium salt	H_2O	244(4.17),290(4.11)	39-0403-83C
	H_2O	241(4.16),293(4.00)	87-0271-83
(S)-	H_2O	252(4.16),287(4.08)	39-0403-83C
	H_2O	237(4.19),290(4.00)	87-0271-83
$C_{14}H_{18}N_2O_7$			
L-Threonine, N-[N-(2,3-dihydroxybenz-oyl)-L-alanyl]-	EtOH and EtOH-HCl	213(4.31),249.5(3.94), 315(3.51)	158-1396-83

Compound	Solvent	λ_{max}(log ε)	Ref.
L-Threonine, N-[N-(2,3-dihydroxybenzoyl)-L-alanyl]- (cont.)	EtOH-NaOH	224(4.16),238s(4.11), 262s(3.90),328(3.59)	158-1396-83
$C_{14}H_{18}N_2O_8S_2$			
1-Azabicyclo[3.2.0]hept-2-ene-2-carboxylic acid, 3-[[2-(acetylamino)ethenyl]thio]-6-[1-methyl-1-(sulfooxy)ethyl]-7-oxo-, disodium salt, (Z)-	H_2O	232(4.11),307(4.11)	87-0271-83
$C_{14}H_{18}N_2O_9S_2$			
1-Azabicyclo[3.2.0]hept-2-ene-2-carboxylic acid, 3-[[2-(acetylamino)ethenyl)sulfinyl]-6-[1-methyl-1-(sulfooxy)ethyl]-7-oxo-, (R)-, disodium salt	H_2O	241(4.21),291(4.05)	87-0271-83
(S)-	H_2O	237(4.12),292(3.96)	87-0271-83
$C_{14}H_{18}N_4OS_2$			
Hydrazinecarbothioamide, 2-[3-(4-morpholinyl)-2-phenyl-3-thioxo-1-propenyl]-	MeCN	282(4.49),406(2.70)	48-0689-83
$C_{14}H_{18}N_6$			
4,6-Pyrimidinediamine, N,N'-diethyl-5-(phenylazo)-	pH 1	257(4.34)	94-3454-83
	H_2O	218(4.30),254(4.29)	94-3454-83
	pH 13	254(4.29)	94-3454-83
$C_{14}H_{18}N_6O_3$			
4,13-Epoxy-1,3-dioxolo[6,7][1,3]diazocino[1,2-e]purin-9-amine, 3a,4,11,12-13,13a-hexahydro-2,2,11-trimethyl-	MeOH	273(4.31)	138-1017-83
1H-1,2,4-Triazol-5-amine, 3-(hexahydro-1H-azepin-1-yl)-1-methyl-N-[(5-nitro-2-furanyl)methylene]-	$CHCl_3$	340(4.35),496(4.04)	24-1547-83
$C_{14}H_{18}N_8$			
1H-Benzotriazole, 1-[[5-(1-piperidinylmethyl)-1H-tetrazol-1-yl]methyl]-	EtOH	232(3.94),256(3.56), 282(3.52)	94-0723-83
$C_{14}H_{18}OS$			
Spiro[6H-cyclohepta[b]thiophene-6,1'-cyclohexan]-4(5H)-one, 7,8-dihydro-	n.s.g.	256(3.85)	42-0303-83
$C_{14}H_{18}OSi$			
7bH-Cyclobut[cd]azulen-7b-ol, 1,1a-dihydro-1-(trimethylsilyl)-, (1α,1aα,7bα)-	C_6H_{12}	225s(3.44),249(3.58), 303(3.10)	88-4261-83
$C_{14}H_{18}O_2$			
1H-Benz[e]indene-3,7(2H,3aH)-dione, 4,5,8,9,9a,9b-hexahydro-3a-methyl-, [3aS-(3aα,9aα,9bα)]-	EtOH	238(4.11)	39-2353-83C
Spiro[cyclopropan-1,5'-[5H]inden]-3'(2'H)-one, 1',4'-dihydro-4'-hydroxy-2',4',6'-trimethyl-	MeOH-NaOH	215(3.95),321(4.00) (unstable)	88-4117-83
Spiro[5.5]undeca-7,10-diene-2,9-dione, 1,5,5-trimethyl-, (±)-	EtOH	241(4.23)	94-2308-83
$C_{14}H_{18}O_3$			
Benzaldehyde, 4-(methoxymethoxy)-3-(3-methyl-2-butenyl)-	MeOH	250(3.41),295(3.35)	2-1061-83

Compound	Solvent	λ_{max}(log ε)	Ref.
2-Cyclopenten-1-one, 3-acetoxy-5-(1-cyclopentenylmethyl)-2-methyl-	EtOH	237(4.13)	39-1885-83C
2-Cyclopenten-1-one, 3-acetoxy-5(and 4)-[(5-methyl-1-cyclopenten-1-yl)-methyl]-	EtOH	236(4.04)	39-1901-83C
as-Indacene-8a(1H)-carboxylic acid, 2,3,3a,4,6,7,8,8b-octahydro-4-oxo-, methyl ester	EtOH	237(3.98)	44-4642-83
Sorbicillin, 2',3'-dihydro-	MeOH	238s(3.90),285(4.15), 330(3.76)	78-4243-83
	MeOH-NaOH	256(3.78),351(4.53)	78-4243-83
$C_{14}H_{18}O_4$			
4H-1,3-Benzodioxin-4-one, 2-(2,2-dimethylpropoxy)-2-methyl-	pH 7.4	298(3.74)	1-0351-83
2H-1-Benzopyran-4-carboxylic acid, 3,4-dihydro-3-hydroxy-2,2-dimethyl-, ethyl ester	EtOH	211(3.85),218s(3.86), 225s(3.76),277(3.30), 284(3.28)	39-1431-83C
2H-Pyran-5-carboxylic acid, 3,4-dihydro-2-methyl-4-oxo-6-(1,3-pentadienyl)-, ethyl ester, (E,E)-	EtOH	334(4.34)	118-0948-83
$C_{14}H_{18}O_5$			
1-Butanone, 1-(5-acetyl-2-hydroxy-4-methoxyphenyl)-3-hydroxy-3-methyl-	EtOH	252(4.44),279(4.03), 320(3.66)	102-1512-83
1,3-Cyclohexanedione, 2-(4-acetyltetrahydro-2,2-dimethyl-5-oxo-3-furanyl)-	MeOH	262(4.13)	104-2027-83
$C_{14}H_{18}O_6$			
β-D-Allopyranoside, 4-ethenylphenyl (glycoside C)	MeOH	257(4.12),290s(3.33), 301s(3.09)	95-0679-83
Benzoic acid, 2-(2-ethoxy-1,1-dimethyl-2-oxoethoxy)-4-methoxy-	EtOH	214(4.32),253(4.03), 283(3.41)	39-1431-83C
Bicyclo[5.3.0]decane-8,8-dicarboxylic acid, 2,10-dioxo-, dimethyl ester	MeOH	288(3.98)	5-0112-83
	+tert-BuOK	310(4.19)	5-0112-83
	MeCN	287(3.96)	5-0112-83
	H_2SO_4	311(4.19)	5-0112-83
1,7-Dioxacyclotetradec-3-ene-2,8,11,12-tetrone, 6,14-dimethyl-	EtOH	428(1.34)	88-0287-83
1,1(2H)-Naphthalenedicarboxylic acid, octahydro-4,5-dioxo-, dimethyl ester	MeOH	293(3.94)	5-0112-83
	+tert-BuOK	303(3.96)	5-0112-83
	MeCN	293(3.94)	5-0112-83
	H_2SO_4	307(4.15)	5-0112-83
$C_{14}H_{18}Si$			
1,6-Methano[10]annulene, 3-(trimethylsilyl)-	C_6H_{12}	261(3.83),299(3.82), 356s(2.34),369s(2.32), 378s(2.28),386s(2.23), 391s(2.22),402(2.11)	35-3375-83
$C_{14}H_{19}NO$			
Cyclohexanone, 2-[methyl(4-methylphenyl)amino]-	MeOH	238(4.05),252(4.08), 300(3.49)	39-0573-83C
Cyclopentanone, 2-[7-(dimethylamino)-2,4,6-heptatrien-1-ylidene]-	$CHCl_3$	473(4.81)	70-0780-83
	EtOH	481(--)	70-0780-83
2,4,6,8-Nonatetraenal, 9-piperidino-	MeOH	477(4.79)	35-0646-83
Pyrrolo[1,2-a]quinolin-2-ol, 1,2,3-3a,4,5-hexahydro-3,3-dimethyl-	MeCN	257(3.75)	35-1204-83

Compound	Solvent	λ_{max}(log ϵ)	Ref.
$C_{14}H_{19}NOS$			
Benzeneethanethioamide, 4-(1,1-dimethylethyl)-N,N-dimethyl-α-oxo-	MeOH	261(4.46),374(3.14)	5-1116-83
Propanethioamide, N,2-dimethyl-N-[2-(2-propenyloxy)phenyl]-	EtOH	277(4.23),281s(4.21)	33-0262-83
$C_{14}H_{19}NO_2$			
3H-Indole-3-ethanol, 2-ethoxy-3-ethyl-	EtOH	256(3.62)	94-2986-83
2,4-Pentanedione, 3-[7-(dimethylamino)-2,4,6-heptatrien-1-ylidene]-	$CHCl_3$	510(4.49)	70-0780-83
	EtOH	545(--)	70-0780-83
Propanamide, N,2-dimethyl-N-[2-(2-propenyloxy)phenyl]-	EtOH	275(3.46),280(3.35)	33-0262-83
Pyrrolo[1,2-b]isoquinoline, 1,2-dimethoxy-5,7,8,9,9a,10-hexahydro-	MeOH	225s(3.80),273(3.34)	83-0773-83
$C_{14}H_{19}NO_3$			
1H-2-Benzopyran-1-one, 3-(1-amino-3-methylbutyl)-3,4-dihydro-8-hydroxy-, hydrochloride	MeOH	246(3.80),314(3.61)	87-1370-83
Ethanone, 1-(1,2,3,4-tetrahydro-5,6-dimethoxy-2-methyl-8-isoquinolinyl)-	MeOH	225.5(4.30),268(3.91), 306(3.56)	83-0801-83
8-Isoquinolinecarboxaldehyde, 1,2,3,4-tetrahydro-5,6-dimethoxy-2,3-dimethyl-	MeOH	229(4.27),276(3.98), 313(3.69)	83-0801-83
Pyrrolo[1,2-b]isoquinolin-10-ol, 5,7,8,9,9a,10-hexahydro-2,3-dimethoxy-	MeOH	208(4.24),230(3.63), 285(3.32)	83-0773-83
$C_{14}H_{19}NO_4$			
Anisomycin	n.s.g.	224(4.03),277(3.26), 283(3.20)	162-0098-83
Propanedioic acid, [7-(dimethylamino)-2,4,6-heptatrienylidene]-, dimethyl ester	$CHCl_3$	480(4.71)	70-0780-83
3,4-Pyrrolidinediol, 2-[(2-methoxyphenyl)methyl]-, 3-acetate, (2α,3α,4β)-(±)-	MeOH	221(3.87),271(3.35), 277(3.33)	87-0469-83
3,4-Pyrrolidinediol, 2-[(3-methoxyphenyl)methyl]-, 3-acetate, (2α,3α,4β)-(±)-	MeOH	216(3.88),272(3.33)	87-0469-83
β-D-Ribofuranosylamine, 2,3-O-(1-methylethylidene)-N-phenyl-	EtOH	236(4.28),282(3.51)	136-0001-83B
$C_{14}H_{19}NO_7$			
2-Oxa-6-azabicyclo[3.2.1]oct-6-ene-1,7-dicarboxylic acid, 8-hydroxy-3-oxo-, 7-(1,1-dimethylethyl) 1-ethyl ester, syn-(±)-	EtOH	225(2.75),250(2.18)	77-0508-83
$C_{14}H_{19}NS$			
Piperidine, 1-[2-(4-methylphenyl)-1-thioxoethyl]-	heptane	281(4.20)	80-0555-83
	EtOH	279(4.21)	80-0555-83
$C_{14}H_{19}NS_2$			
Benzeneethanethioamide, 4-(1,1-dimethylethyl)-N,N-dimethyl-α-thioxo-	MeOH	240(3.89),273(4.16), 336(4.25),402s(3.38), 580(2.37),618s(2.18)	5-1116-83
	isooctane	593(--),628s(--)	5-1116-83
$C_{14}H_{19}N_3$			
[1,2]Diazepino[5,4-b]indole, 2-ethyl-	EtOH	226(4.22),278s(3.59),	39-1937-83C

Compound	Solvent	λ_{max}(log ε)	Ref.
1,2,3,4,5,6-hexahydro-3-methyl-		285(3.61),293(3.56)	39-1937-83C
[1,2]Diazepino[5,4-b]indole, 1,2,3,4-5,6-hexahydro-2,3,9-trimethyl-	EtOH	229(4.19),281s(3.61), 288(3.62),299(3.55)	39-1937-83C
$C_{14}H_{19}N_3O$			
2(1H)-Quinoxalinone, 1-[3-(dimethyl-amino)propyl]-3-methyl-	n.s.g.	230(4.3),279(3.8), 338(3.8)	161-0330-83
Tambjamine C	MeOH	258(3.76),405(4.36)	44-2314-83
$C_{14}H_{19}N_3S$			
Ethanimidamide, N'-[2-(1H-indol-3-yl-thio)ethyl]-N,N-dimethyl-, mono-(cyclohexylsulfamate)	MeOH	272(3.79),279(3.81), 287(3.77)	87-0230-83
$C_{14}H_{19}N_5O_5$			
β-D-arabino-Heptofuranuronic acid, 1-(6-amino-9H-purin-9-yl)-1,5,6-trideoxy-, ethyl ester	MeOH	259(4.20)	87-1530-83
$C_{14}H_{19}N_5O_6$			
Guanosine, N-(2-methyl-1-oxopropyl)-	MeOH	255(4.23),259(4.23), 280(4.10)	33-2069-83
$C_{14}H_{19}N_5S$			
1H-1,2,4-Triazol-3-amine, 5-(hexahy-dro-1H-azepin-1-yl)-1-methyl-N-(2-thienylmethylene)-	$CHCl_3$	344(4.11)	24-1547-83
1H-1,2,4-Triazol-5-amine, 3-(hexahy-dro-1H-azepin-1-yl)-1-methyl-N-(2-thienylmethylene)-	$CHCl_3$	320(4.42),408(4.36)	24-1547-83
$C_{14}H_{19}N_7O_6$			
1H-Imidazolium, 2,4-bis(dimethylamino)-1-methyl-3-(2,4,6-trinitro-2,4-cy-clohexadien-1-yl)-, hydroxide, inner salt	MeCN	452(4.32),562(4.02)	88-3563-83
$C_{14}H_{20}Cl_6$			
Bicyclo[2.2.2]octane, 2,3,5,6,7,8-hexakis(chloromethyl)-	MeCN	215(1.60)(end abs.)	33-0019-83
$C_{14}H_{20}N_2O$			
1H-Imidazole, 2,5-dihydro-1-methoxy-2,2,5,5-tetramethyl-4-phenyl-	EtOH	205(4.23),239(4.00)	70-2123-83
$C_{14}H_{20}N_2O_4$			
L-Leucine, N-[[3-hydroxy-5-(hydroxy-methyl)-2-methyl-4-pyridinyl]meth-ylene]-	MeOH	255(4.01),290s(3.63), 388(3.59),422(3.38)	35-0803-83
$C_{14}H_{20}N_2O_5$			
Propanedioic acid, (formylamino)-[2-(1H-pyrrol-1-yl)ethyl]-, diethyl ester	MeOH	222(3.74)	23-0454-83
$C_{14}H_{20}N_2O_5S$			
1-Azabicyclo[3.2.0]hept-2-ene-2-carb-oxylic acid, 3-[[2-(acetylamino)-ethyl]thio]-6-(1-hydroxy-1-methyl-ethyl)-7-oxo-, sodium salt	H_2O	298(3.97)	39-0403-83C
	H_2O	298(3.94)	158-0943-83

Compound	Solvent	λ_{max}(log ϵ)	Ref.
2-Thiouracil, 5-[2,5-dimethyl-5-(hydroxymethyl)-3,4-isopropylidenedioxytetrahydrofuran-2-yl]-, (2R*,3R*,4S*,5R*)-	pH 13	223(4.17),261(4.15), 292(3.94)	18-2680-83
	MeOH	215(4.10),276(4.21), 290(4.17)	18-2680-83
2-Thiouracil, 5-[5-(1-hydroxy-1-methylethyl)-3,4-isopropylidenedioxytetrahydrofuran-2-yl]-, (2R*,3R*,4S*,5S*)-	pH 13	222(4.17),264(4.08), 285(3.98)	18-2680-83
	MeOH	213(4.05),276(4.14), 290(4.11)	18-2680-83
2-Thiouracil, 6-(1,2:3,4-di-O-isopropylidene-D-arabino-tetritol-1-yl)-	pH 1	285(4.02),295(3.96)	136-0333-83M
	pH 7	255(4.13),315(4.11)	136-0333-83M
	pH 11	255(4.12),310(4.10)	136-0333-83M
$C_{14}H_{20}N_2O_6$			
Uracil, 5-[2,5-dimethyl-5-(hydroxymethyl)-3,4-isopropylidenedioxytetrahydrofuran-2-yl]-, (2R*,3R*,4S*,5R*)-	pH 13	217(4.00),287(3.88)	18-2680-83
	MeOH	264(3.82)	18-2680-83
Uracil, 5-[5-(1-hydroxy-1-methylethyl)-3,4-isopropylidenedioxytetrahydrofuran-2-yl]-, (2R*,3R*,4S*,5S*)-	pH 13	286(3.73)	18-2680-83
	MeOH	263(3.59)	18-2680-83
$C_{14}H_{20}N_2O_6S$			
Carpetimycin C	H_2O	285(4.00)	158-0943-83
$C_{14}H_{20}N_2O_7S$			
Uridine, 6-mercapto-5'-O-(methoxymethyl)-2',3'-O-(1-methylethylidene)-	MeOH	238(3.92),316(4.37)	94-1222-83
$C_{14}H_{20}N_2O_8$			
Propanedioic acid, 2,2'-[1,2-ethanediylbis(iminomethylidyne)]bis-	MeCN	220(4.41),272(4.48), 291(4.47)	97-0064-83
$C_{14}H_{20}N_2O_8S$			
2,4(1H,3H)-Pyrimidinedione, 5-[2-C-methyl-2,3-O-(1-methylethylidene)-5-O-(methylsulfonyl)-β-DL-ribofuranosyl]-	MeOH	261(3.85)	18-2680-83
$C_{14}H_{20}N_2O_9S_2$			
1-Azabicyclo[3.2.0]hept-2-ene-2-carboxylic acid, 3-[[2-(acetylamino)ethyl]sulfinyl]-6-[1-methyl-1-(sulfooxy)ethyl]-7-oxo-, disodium salt (carpetimycin D)	H_2O	285(4.10)	158-0943-83
$C_{14}H_{20}N_2S$			
Piperidine, 1-[[4-(dimethylamino)phenyl]thioxomethyl]-	heptane	263(--),294(4.29), 405(3.07)	80-0555-83
	EtOH	263(--),291(4.26), 373(3.55)	80-0555-83
$C_{14}H_{20}N_2S_2$			
3-Pentenenitrile, 3,3'-[1,2-ethanediylbis(thio)]bis[4-methyl-	EtOH	207(4.32),254(3.92)	39-1223-83C
$C_{14}H_{20}N_4O_8S$			
1H-Purine-2,6-dione, 7-[6-deoxy-2-O-(methylsulfonyl)-β-L-galactopyranosyl]-3,7-dihydro-1,3-dimethyl-	MeOH	276(4.04)	136-0301-83A

Compound	Solvent	λ_{max}(log ε)	Ref.
$C_{14}H_{20}N_8O_2$			
1H-Imidazole-2,4-diamine, N,N,N',N',1-pentamethyl-5-[3-(4-nitrophenyl)-1-triazenyl]-	pyridine	532(4.15)	88-3563-83
	acetone	514(--)	88-3563-83
$C_{14}H_{20}O$			
2-Cyclohexen-1-one, 3-(3-ethenyl-2,2-dimethylcyclopropyl)-2-methyl-, cis	MeOH	257(4.04)	23-1226-83
trans	MeOH	257(4.09)	23-1226-83
2-Cyclohexen-1-one, 3-(4,5-heptadienyl)-4-methyl-	hexane	227(4.04)	44-2584-83
2-Cyclohexen-1-one, 2-methyl-3-[1-(1-methylethenyl)-2-butenyl]-	MeOH	245(4.11)	23-1226-83
Cyclopenta[1,4]cyclobuta[1,2]benzen-5(6H)-one, 4-ethylideneoctahydro-8-methyl-, cis?	hexane	297(2.04)	44-2584-83
trans?	hexane	299(1.77),308(1.76)	44-2584-83
3a,7-Methano-3aH-cyclopentacycloocten-2(3H)-one, 4,5,6,7,8,9-hexahydro-1,7-dimethyl-, (±)-	$CHCl_3$	397(4.48)	44-2318-83
Spiro[4.5]dec-6-en-8-one, 1-methyl-4-(1-methylethenyl)-	MeOH	230(3.78)	44-0670-83
$C_{14}H_{20}O_2$			
Bicyclo[3.2.0]heptane-6-carboxaldehyde, 3,3,6-trimethyl-7-(1-methylethenyl)-2-oxo-	pentane	295(1.79)	33-1638-83
isomer 18b	pentane	265(2.43)	33-1638-83
Bicyclo[3.2.0]heptan-2-one, 6-acetyl-3,3-dimethyl-7-(1-methylethenyl)-	EtOH	265s(2.34)	33-1638-83
1,3-Butadiene, 1-(1,2:3,4-diepoxy-2,6,6-trimethyl-1-cyclohexyl)-3-methyl-	EtOH	232(3.99)	33-1638-83
3-Buten-2-one, 4-(3,4,4-trimethyl-2-methylene-7-oxabicyclo[4.2.0]hept-3-yl)-	pentane	220(4.17)	33-1638-83
2,7,9-Decatrienal, 2,5,5,9-tetramethyl-6-oxo-, (E,E)-	pentane	330(1.95)	33-1638-83
2,7-Dioxabicyclo[2.2.1]heptane, 3-ethylidene-6,6-dimethyl-1-(3-methyl-1,3-butadienyl)-, (E,E)-	pentane	228(4.42)	33-1638-83
3,8-Dioxatricyclo[5.1.0.0^{2,4}]octane, 1,6,6-trimethyl-7-(3-methyl-1,3-butadienyl)-, (E)-	EtOH	235(4.36)	33-1061-83
3-Hepten-2-one, 6-methyl-6-(3-methyl-2H-pyran-6-yl)-, (E)-	pentane	217(4.18),281(3.65)	33-1638-83
(Z)-	pentane	225(4.18),280(3.75)	33-1638-83
1(4H)-Naphthalenone, 4a,5,8,8a-tetrahydro-4-hydroxy-2,3,4a,8a-tetramethyl-, (4α,4aα,8aα)-	$CHCl_3$	324(1.82),345s(1.63)	35-5354-83
(4α,4aβ,8aβ)-	$CHCl_3$	324(1.81),345s(1.66)	35-5354-83
1(4H)-Naphthalenone, 4a,5,8,8a-tetrahydro-4-hydroxy-2,3,6,7-tetramethyl-	$CHCl_3$	318(1.72)	35-5354-83
7-Oxabicyclo[4.1.0]heptan-2-one, 1,5,5-trimethyl-6-(3-methyl-1,3-butadienyl)-, (E)-	pentane	238(4.36),290(2.37), 300s(2.35),305s(2.33)	33-2236-83
7-Oxabicyclo[4.1.0]heptan-2-one, 3,4,4-trimethyl-3-(3-methyl-1,3-butadienyl)-	pentane	232(4.30),311(2.78), 319(2.75)	33-1638-83
7-Oxabicyclo[4.1.0]heptan-2-one, 1,5,5-trimethyl-6-(3-methyl-2-	pentane	300(1.58)	33-2236-83

Compound	Solvent	$\lambda_{max}(\log \epsilon)$	Ref.
cyclobuten-1-yl)- (cont.)			33-2236-83
4-Oxatricyclo[6.1.0.0^{3,5}]nonan-2-one, 7,7,8-trimethyl-9-(1-methylethenyl)-	pentane	292(1.98)	33-1638-83
2-Propanone, 1-[dihydro-4,4-dimethyl-5-(3-methyl-1,3-butadienyl)-2(3H)-furanylidene]-, (E,E)-	EtOH	227(4.21),267(4.25)	33-1638-83
(Z,E)-	EtOH	227(4.31),264(4.08)	33-1638-83
Spiro[5.5]undeca-1,4-dien-3-one, 10-hydroxy-7,7,11-trimethyl-, cis	EtOH	248(4.15)	94-2308-83
3,8,10-Undecatriene-2,7-dione, 6,6,10-trimethyl-, (E,E)-	pentane	366(2.00)	33-1061-83
(E,Z)-	MeCN	334(2.15)	33-1638-83
$C_{14}H_{20}O_2S$			
2(5H)-Thiophenone, 3,5-diethyl-4-methoxy-5-(2-methyl-1,3-butadienyl)-, (E)-(+)-	EtOH	238(4.22)	158-1781-83
3(2H)-Thiophenone, 2,4-diethyl-5-methoxy-2-(2-methyl-1,3-butadienyl)-, (E)-	EtOH	235(4.21),313(3.79)	158-1781-83
$C_{14}H_{20}O_3$			
4aH-Benzocycloheptene-4a-carboxylic acid, 2,3,4,5,6,7,8,9-octahydro-4-methyl-2-oxo-, methyl ester	EtOH	241(3.98)	44-4642-83
Bicyclo[4.3.1]dec-7-ene-1-carboxylic acid, 7,9-dimethyl-10-oxo-, methyl ester	EtOH	222(3.12),255(2.16), 261(2.16),273(2.06)	44-4642-83
2-Naphthalenecarboxylic acid, 1,2,3,4,5,6,7,8-octahydro-2,5-dimethyl-1-oxo-, methyl ester, high R_F isomer	EtOH	246.5(4.101)	2-0331-83
low R_F isomer	EtOH	246(4.081)	2-0331-83
4a(2H)-Naphthalenecarboxylic acid, 1,3,4,5,6,7-hexahydro-5,5-dimethyl-7-oxo-, methyl ester	EtOH	240(3.97)	44-4642-83
1-Propanone, 1,1'-(2,5-furandiyl)bis-[2,2-dimethyl-	EtOH	220(3.94),288(4.38)	12-0963-83
12-Tetradecene-8,10-diyne-1,4,5-triol, [R*,R*-(E)]-	MeOH	212(4.57),226(3.70), 239(3.85),251(3.99), 266(4.14),282(4.05)	102-0592-83
$C_{14}H_{20}O_4$			
Cyclohexanecarboxaldehyde, 6-(1,3-heptadienyl)-3,4-dihydroxy-2-oxo-	dioxan	232(4.52),290(3.74)	162-0608-83
1,4-Dioxaspiro[2.4]hept-5-en-7-one, 5-(1,1-dimethylethyl)-2-(2,2-dimethyl-1-oxopropyl)-, (Z)-	MeOH	275(3.80)	4-1389-83
3-Hepten-2-one, 6-methyl-6-(1-methyl-2,3,5-trioxabicyclo[2.2.2]oct-7-en-4-yl)-, (E)-	pentane	219(4.11)	33-1638-83
2(1H)-Naphthalenone, 6,7,8,8a-tetrahydro-8-[(2-methoxyethoxy)methoxy]-, trans-(±)-	MeOH	289(4.16)	88-3687-83
$C_{14}H_{20}O_5$			
1,4,7,10,13-Benzopentaoxacyclopentadecin, 2,3,5,6,8,9,11,12-octahydro-	MeOH	223(3.83),275(3.36)	44-2029-83
1-Cyclopentene-1-heptanoic acid, 4-acetoxy-5-oxo-	EtOH	230(3.95)	39-0319-83C

Compound	Solvent	λ_{max}(log ϵ)	Ref.
$C_{14}H_{20}O_6$			
6-Oxabicyclo[3.2.0]hept-2-ene-1,2-dicarboxylic acid, 3-methoxy-5-methyl-, diethyl ester	MeOH	250(3.83)	32-0489-83
$C_{14}H_{20}S$			
Thiophene, 3-[5-methyl-2-(1-methylethyl)-1-cyclohexen-1-yl]-, (R)-	MeCN	207(4.09),240(3.76)	4-0729-83
$C_{14}H_{20}S_2$			
Dispiro[5.1.5.1]tetradecane-7,14-dithione	$CHCl_3$	244(4.20),310(3.04), 504(1.57)	44-4482-83
2-Thiaspiro[3.5]nonane-1-thione, 3-cyclohexylidene-	$CHCl_3$	248(3.68),353(3.85), 463(1.62)	44-4482-83
	$CHCl_3$	246(3.68),353(3.85), 463(1.62)	78-2719-83
$C_{14}H_{20}S_3$			
14-Thiadispiro[5.1.5.2]pentadecane-7,15-dithione	$CHCl_3$	260(4.12),310(4.15), 497(1.40)	44-4482-83
$C_{14}H_{21}NO_2$			
Isoquinoline, 8-ethyl-1,2,3,4-tetrahydro-5,6-dimethoxy-2-methyl-, hydrochloride	MeOH	226s(3.60),280(3.37)	83-0801-83
Oxazolidine, 2-(4-methoxyphenyl)-3-methyl-4-(1-methylethyl)-, (2S-cis)-	EtOH	224(4.15),271(3.74), 278(3.66),294s(3.23)	142-0607-83
$C_{14}H_{21}NO_3$			
8-Isoquinolinemethanol, 1,2,3,4-tetrahydro-5,6-dimethoxy-2,3-dimethyl-	MeOH	222s(3.97),278(3.67)	83-0801-83
$C_{14}H_{21}NO_4$			
2,4-Cyclohexadien-1-one, 5-amino-6-hydroxy-3-methoxy-2,6-dimethyl-4-(2-methyl-1-oxobutyl)-, (R*,R*)-	EtOH	214s(4.05),239(3.93), 316(4.07),399(3.67)	88-2527-83
$C_{14}H_{21}NO_5$			
3,9-Dioxabicyclo[4.2.1]nonan-4-one, 5-(dimethylaminomethylene)-7,8-isopropylidenedioxy-1-methyl-, (1R*,6R*,7S*,8S*)-	MeOH	297(4.30)	18-2680-83
3,9-Dioxabicyclo[4.2.1]nonan-4-one, 5-(dimethylaminomethylene)-7,8-isopropylidenedioxy-2-methyl-, (1R*,2R*,6R*,7R*,8R*)-	MeOH	298(4.22)	18-2680-83
3,9-Dioxabicyclo[4.2.1]nonan-4-one, 5-(dimethylaminomethylene)-7,8-isopropylidenedioxy-7-methyl-, (1R*,6S*,7R*,8R*)-	MeOH	300(4.21)	18-2680-83
$C_{14}H_{21}NO_7$			
D-glycero-D-galacto-Heptonic acid, 2-deoxy-2-[(phenylmethyl)amino]-	H_2O	203(3.89),255(2.42)	136-0158-83C
D-glycero-D-gulo-Heptonic acid, 2-deoxy-2-[(phenylmethyl)amino]-	H_2O	204(3.93),255(2.40)	136-0158-83C
D-glycero-D-ido-Heptonic acid, 2-deoxy-2-[(phenylmethyl)amino]-	H_2O	203(3.94),255(2.51)	136-0158-83C
D-glycero-D-talo-Heptonic acid, 2-deoxy-2-[(phenylmethyl)amino]-	H_2O	203(3.97),255(2.38)	136-0158-83C

Compound	Solvent	$\lambda_{max}(\log \epsilon)$	Ref.
D-glycero-L-gluco-Heptonic acid, 2-deoxy-2-[(phenylmethyl)amino]-	H_2O	205(3.92),261(3.08)	136-0158-83C
D-glycero-L-manno-Heptonic acid, 2-deoxy-2-[(phenylmethyl)amino]-	H_2O	204(3.93),255(2.51)	136-0158-83C
$C_{14}H_{21}N_3O_4$			
5-Pyrimidinepentanoic acid, 1,4-dihydro-6-hydroxy-4-oxo-2-(2-propenylamino)-, ethyl ester	50% MeOH	242s(3.734),272(4.180)	73-0304-83
	+ HCl	237(3.879),270(4.084)	73-0304-83
	+ NaOH	242(3.830),275(4.120)	73-0304-83
$C_{14}H_{21}N_3O_5$			
Isocytosine, 5-[2,5-dimethyl-5-(hydroxymethyl)-3,4-isopropylidenedioxytetrahydrofuran-2-yl]-, (2R*,3R*,4S*,5R*)-	pH 13	231(3.94),278(3.84)	18-2680-83
	MeOH	225(4.00),292(3.91)	18-2680-83
Isocytosine, 5-[5-(1-hydroxy-1-methylethyl)-3,4-isopropylidenedioxytetrahydrofuran-2-yl]-, (2R*,3R*,4S*,5S*)-	pH 13	233(3.87),276(3.77)	18-2680-83
	MeOH	227(3.73),290(3.73)	18-2680-83
Isocytosine, 6-(1,2:3,4-di-O-isopropylidene-D-arabino-tetritol-1-yl)-	pH 1	220(4.03),263(3.87)	136-0333-83M
	pH 7	235(4.01),280(3.87)	136-0333-83M
	pH 11	225(4.01),280(3.88)	136-0333-83M
5-Pyrimidinepentanoic acid, 1,4-dihydro-6-hydroxy-2-(4-morpholinyl)-4-oxo-, methyl ester	50%MeOH-HCl	240(4.066),283(4.092)	73-0304-83
	+ NaOH	250s(3.928),275(4.094)	73-0304-83
$C_{14}H_{21}N_3O_5S_2$			
Uridine, 5'-thio-5'-(diethylcarbamodithioate)	H_2O	260(4.24)	39-1315-83C
$C_{14}H_{21}N_3O_6$			
5,5(4H)-Pyrimidinedipentanoic acid, 2-amino-1,6-dihydro-4,6-dioxo-	50%MeOH-HCl	207(4.32)	73-0137-83
	+ NaOH	224(4.37),263(4.00)	73-0137-83
$C_{14}H_{22}N$			
Pyrrolidinium, 1-(3,7-dimethyl-2,4,6-octatrienylidene)-, (E,E)-, perchlorate	$CHCl_3$	406(4.51)	35-4033-83
$C_{14}H_{22}NO_2$			
Pyrrolidinium, 1-[3-(2-carboxycyclopentyl)-2-butenylidene]-, perchlorate, (Z)-	MeCN	276(4.08)	35-4033-83
sodium salt	MeCN	297(4.04)	35-4033-83
$C_{14}H_{22}N_2O_3Si$			
1,2-Diazabicyclo[5.2.0]nona-2,4-diene-4-carboxylic acid, 8-methyl-9-oxo-, (2-trimethylsilylethyl) ester, cis	MeOH	307(3.73)	5-1393-83
$C_{14}H_{22}N_2O_5S$			
Pseudouridine, 1'-pentyl-2-thio-, (±)-	MeOH	215(4.04),277(4.16), 291(4.12)	18-2680-83
	pH 1	213s(3.86),273(3.93), 295s(3.79)	18-2680-83
	pH 13	220s(3.97),262(3.97), 297s(3.66)	18-2680-83
Pseudouridine, 4'-pentyl-2-thio-, (±)-	MeOH	215(4.12),276(4.27), 291(4.23)	18-2680-83
	pH 1	214(4.16),283(4.22)	18-2680-83
	pH 13	223(4.14),264(4.08), 286(3.97)	18-2680-83

Compound	Solvent	λ_{max}(log ε)	Ref.
Pseudouridine, 5'-pentyl-2-thio-, (±)-	MeOH	214(3.64),276(3.71), 291(3.67)	18-2680-83
	pH 1	215(3.85),276(3.92), 290(3.92)	18-2680-83
	pH 13	222(3.72),264(3.64), 285(3.53)	18-2680-83
$C_{14}H_{22}N_2O_6$			
Pseudouridine, 1'-pentyl-, (±)-	MeOH	265(3.78)	18-2680-83
	pH 1	265(3.82)	18-2680-83
	pH 13	266s(3.51),289(3.69)	18-2680-83
Pseudouridine, 4'-pentyl-	MeOH	264(3.98)	18-2680-83
	pH 1	264(3.96)	18-2680-83
	pH 13	287(3.92)	18-2680-83
Pseudouridine, 5'-pentyl-	MeOH	212(3.93),265(3.84)	18-2680-83
	pH 1	264(3.55)	18-2680-83
	pH 13	217(4.04),286(3.86)	18-2680-83
$C_{14}H_{22}N_3O_2$			
2-Aza-1-azoniabicyclo[5.2.0]nona-3,5,9-triene, 9-(dimethylamino)-2-(ethoxycarbonyl)-8,8-dimethyl-, chloride	MeOH	266(3.73)	5-1361-83
$C_{14}H_{22}N_4$			
2-Pentenenitrile, 3,3'-(1,2-ethanediyldiimino)bis[4-methyl-	EtOH	262(4.48)	39-1223-83C
$C_{14}H_{22}N_4O_5$			
5-Pyrimidinepentanoic acid, 2-[(ethoxyiminomethyl)amino]-1,4-dihydro-6-hydroxy-4-oxo-, ethyl ester	50% MeOH	221s(3.988),285(4.136)	73-0304-83
	+ HCl	253(4.102)	73-0304-83
	+ NaOH	231(4.128),268(4.128), 280s(4.102)	73-0304-83
$C_{14}H_{22}O$			
2(3H)-Naphthalenone, 4,4a,5,6,7,8-hexahydro-5-methyl-8-(1-methylethyl)-	EtOH	241(4.19)	33-1835-83
$C_{14}H_{22}OSi$			
Silane, trimethyl[(9-methylbicyclo[4.3.1]deca-2,4,7-trien-7-yl)oxy]-, exo	C_6H_{12}	242(3.76),249(3.76)	24-2914-83
$C_{14}H_{22}O_2$			
Cyclohexanone, 3-hydroxy-2,4,4-trimethyl-3-(3-methyl-1,3-butadienyl)-, [2α,3α,3(E)]-	MeCN	228s(4.39),232(4.41), 287s(1.81)	33-2236-83
[2α,3α,3(Z)]-	pentane	292s(1.68),300s(1.60), 314s(1.30)	33-2236-83
[2α,3β,3(E)]-	MeCN	233(4.33)	33-2236-83
[2α,3β,3(Z)]-	pentane	275s(1.85)	33-2236-83
1,5-Cyclooctanedione, 2,2,6-trimethyl-7-(1-methylethenyl)-	pentane	304(1.62)	33-2236-83
isomer	pentane	285(1.86)	33-2236-83
Ethanone, 1-[1-hydroxy-3,3-dimethyl-2-(3-methyl-1,3-butadienyl)-, [1α,2α(E)]-	pentane	227(4.35),233(4.36)	33-2236-83
isomer	pentane	229(4.33),234(4.34)	33-2236-83
2-Heptanone, 3-hydroxy-6-methyl-6-[3-(1-methylethenyl)-1-cyclopropen-1-yl]-	pentane	227s(2.51)	33-2236-83

Compound	Solvent	λ_{max}(log ε)	Ref.
3-Oxepanone, 2,6,6-trimethyl-7-(3-methyl-1,3-butadienyl)-	pentane	229(4.40),302(1.43)	33-2236-83
isomer	pentane	230(4.38),286(2.19)	33-2236-83
Spiro[5.5]undec-1-en-3-one, 10-hydroxy-7,7,11-trimethyl-	MeCN	232(4.01)	94-2308-83
8,10-Undecadiene-3,7-dione, 6,6,10-trimethyl-, (E)-	pentane	264(4.33),341s(2.15)	33-2236-83
$C_{14}H_{22}O_3$			
Cyclopentanone, 2-(1-hydroxy-2-methyl-2-propenyl)-5,5-dimethyl-3-(2-oxopropyl)-	EtOH	293(1.83)	33-1638-83
$C_{14}H_{22}O_4$			
Acetic acid, (2-ethoxycarbonyl-2-methylcyclohexan-1-ylidene)-, ethyl ester	EtOH	233(4.11)	2-1103-83
3-Cyclohexene-1-carboxylic acid, 4-ethoxy-1,6,6-trimethyl-2-oxo-, ethyl ester, (±)-	EtOH	256(3.76)	39-1465-83C
$C_{14}H_{22}O_7$			
D-arabino-3-Heptulosonic acid, 2-deoxy-4,5:6,7-bis-O-(1-methylethylidene)-, methyl ester	MeOH or H_2O	240(3.34)	136-0333-83M
$C_{14}H_{22}S_3$			
Bicyclo[2.2.1]hept-2-ene, 2-[[1,2-bis(methylthio)ethenyl]thio]-1,7,7-trimethyl-	hexane	251(2.80),312(2.88), 380(1.99),488(0.85)	54-0103-83
$C_{14}H_{23}ClOPdS_2$			
Palladium, [2,4-bis[1,1-dimethyl-2-(methylthio)ethyl]-3-furanyl-C,S,S']chloro-, (SP-4-4)-	MeOH	228(4.32),244(4.18), 300(3.34)	101-0C19-83F
$C_{14}H_{23}N$			
Cyclohexanamine, N-(1,4,4-trimethyl-2-pentynylidene)-	C_6H_{12}	220(4.04),270(2.45)	44-1925-83
$C_{14}H_{23}NO$			
2,6,8-Decatrienamide, N-(2-methylpropyl)- (affinin)	EtOH	228.5(4.53)	162-0027-83
$C_{14}H_{23}NO_2Si$			
Glycine, N-(2,4-dimethylphenyl)-, (trimethylsilyl)methyl ester	C_6H_{12}	242(4.20),293(3.48)	65-0814-83
$C_{14}H_{23}N_3O_4$			
5-Pyrimidinepentanoic acid, 1,4-dihydro-6-hydroxy-4-oxo-2-(propylamino)-, ethyl ester	50% MeOH	270(4.23)	73-0304-83
	+ HCl	238(3.845),269(4.102)	73-0304-83
	+ NaOH	242(3.804),265(4.126)	73-0304-83
$C_{14}H_{23}N_3O_5$			
Pseudoisocytidine, 1'-pentyl-, hydrochloride, (±)-	MeOH	210s(4.09),225(3.90), 261(3.64),300s(4.07)	18-2680-83
	pH 1	210(4.09),225s(3.94), 262(3.68)	18-2680-83
	pH 13	230s(3.85),277(3.68)	18-2680-83
Pseudoisocytidine, 4'-pentyl-, hydrochloride, (±)-	MeOH	225(3.88),266(3.69), 290(3.48)	18-2680-83
	pH 1	220(4.03),262(3.91)	18-2680-83

Compound	Solvent	$\lambda_{max}(\log \epsilon)$	Ref.
(cont.)	pH 13	231(4.10),276(3.94)	18-2680-83
Pseudoisocytidine, 5'-pentyl-, (±)-	MeOH	224(3.98),266(3.79), 290(3.62)	18-2680-83
	pH 1	220(4.07),262(3.96)	18-2680-83
	pH 13	233(3.87),276(3.77)	18-2680-83
$C_{14}H_{23}N_5O_4$			
5,5(4H)-Pyrimidinedipentanoic acid, 2,6-diamino-4-imino-	50% MeOH	242(4.39),283(4.01)	73-0292-83
	+ HCl	242(4.39),282(4.07)	73-0292-83
	+ NaOH	234(4.29),286(3.89)	73-0292-83
$C_{14}H_{24}$			
1,3-Cyclohexadiene, 1,4-bis(1,1-dimethylethyl)-	EtOH	260(3.65)	44-4186-83
1,3-Cyclohexadiene, 1,5-bis(1,1-dimethylethyl)-	C_6H_{12}	265(3.77)	44-4190-83
1,3-Cyclohexadiene, 2,5-bis(1,1-dimethylethyl)-	C_6H_{12}	260(3.78)	44-4186-83
1,3-Cyclohexadiene, 2,6-bis(1,1-dimethylethyl)-	C_6H_{12}	254(4.04)	44-4190-83
Cyclohexane, 4-(1,1-dimethylethyl)-2-methyl-1-(2-propenylidene)-, (E,2R,4R)-(-)	C_6H_{12}	232(4.38),239(4.42), 247(4.26)	35-3264-83
(E,2S,4R)-(+)-	C_6H_{12}	232(4.39),239(4.42), 246(4.24)	35-3252-83
(Z,2S,4R)-(+)-	C_6H_{12}	232(4.39),239(4.43), 246(4.27)	35-3264-83
6-Tridecyne, 8-methylene-	EtOH	224(4.11)	104-1621-83
$C_{14}H_{24}N_2$			
Pyrrolo[3,2-b]pyrrole, 3,6-bis(1,1-dimethylethyl)-3,3a,6,6a-tetrahydro-	C_6H_{12}	233(3.20),290(3.02), 318s(2.18)	138-0743-83
$C_{14}H_{24}N_2S$			
1,3,4-Thiadiazole, 2,5-dihydro-2,2-dimethyl-5-(2,2,6,6-tetramethylcyclohexylidene)-	isooctane	220(4.3),253(4.2), 345(2.6)	24-0066-83
$C_{14}H_{24}N_4$			
1,2,4,5-Tetrazine, 3,6-dicyclohexyl-1,4-dihydro-	dioxan	289(2.01)	24-2261-83
$C_{14}H_{24}N_8S_2$			
Methanaminium, N,N'-[azinobis[4-(dimethylamino)-5,2-thiazolediylidene]]bis[N-methyl-, bis(tetrafluoroborate)	MeCN	369(4.52),539(2.22)	88-3563-83
$C_{14}H_{24}O$			
2-Cyclohexen-1-one, 2,4,4-trimethyl-3-(3-methylbutyl)-	pentane	240(4.18),327(1.57)	33-2236-83
2-Cyclotridecen-1-one, 2-methyl-, (E)-	isooctane	228.5(4.13),326.5(1.62)	152-0399-83
(Z)-	isooctane	233.5(3.75),322.5(1.70)	152-0399-83
$C_{14}H_{24}OS$			
Cyclohexanone, 3-[(1,1-dimethylethyl)-thio]-2-methyl-5-(1-methylethenyl)-, [2S-(2α,3α,5β)]-	C_6H_{12}	296(1.51)	49-0195-83
	MeOH	244(2.52),289(1.48)	49-0195-83
$C_{14}H_{24}O_2$			
Cyclopentanemethanol, 1-hydroxy-α,3,3-	pentane	235(4.13)	33-2236-83

Compound	Solvent	$\lambda_{max}(\log \epsilon)$	Ref.
trimethyl-2-(3-methyl-1,3-butadienyl)-, (E)- (cont.)			33-2236-83
isomer	pentane	235(4.26)	33-2236-83
7-Oxabicyclo[4.1.0]heptan-2-one, 3,4,4-trimethyl-3-(3-methylbutyl)-, (1α,3β,6α)-	pentane	307(1.60)	33-1638-83
3-Oxepanol, 2,6,6-trimethyl-7-(3-methyl-1,3-butadienyl)-, (E)-	pentane	230(4.36)	33-2236-83
isomer	pentane	230(4.39)	33-2236-83
$C_{14}H_{24}O_4$			
1,4-Cyclohexadiene, 3,6-dimethoxy-3,6-bis(1-methylethoxy)-	EtOH	207.5(3.48)	39-1255-83C
1,4-Cyclohexadiene, 3,6-dimethoxy-3,6-dipropoxy-	EtOH	207.5(3.48)	39-1255-83C
$C_{14}H_{25}NO$			
2,4-Decadienamide, N-(1-methylpropyl)-	EtOH	202(3.93)	70-1897-83
1-Hepten-3-one, 1-(cyclohexylamino)-	MeOH	300(4.45)	131-0001-83K
(E-syn-E)-	C_6H_{12}	278(--)	131-0001-83K
	dioxan	288(--)	131-0001-83K
(Z-syn-Z)-	C_6H_{12}	315(4.12)	131-0001-83K
	dioxan	313(--)	131-0001-83K
$C_{14}H_{25}N_3O_5$			
2-Propenamide, 3-[(2-deoxy-β-D-erythro-pentofuranosyl)-N-[[(1,1-dimethylethyl)amino]carbonyl]-2-methyl-, (E)-	H_2O	291(4.38)	35-0956-83
$C_{14}H_{26}N_2O_8$			
β-D-xylo-Hexopyranoside, 2-hydroxy-1-(2-hydroxyethyl)propyl 2,3,4,6-tetradeoxy-4-[(methoxycarbonyl)-amino]-3-C-methyl-3-nitro-	CF_3CH_2OH	199(3.62)	39-1497-83C
$C_{14}H_{26}N_4O_{11}P_2$			
Citicoline	pH 1	280(4.11)	162-0329-83
$C_{14}H_{26}OS$			
Cyclohexanone, 3-[(1,1-dimethylethyl)-thio]-2-methyl-5-(1-methylethyl)-, [2S-(2α,3α,5β)]-	C_6H_{12}	298(1.32)	49-0195-83
	MeOH	245(2.45),290(1.48)	49-0195-83
$C_{14}H_{26}O_2$			
Cyclohexanone, 3-hydroxy-2,4,4-trimethyl-3-(3-methylbutyl)-	pentane	289(1.36)	33-2236-83
isomer 14B	pentane	290(1.30)	33-2236-83
3,7-Undecanedione, 6,6,10-trimethyl-	pentane	286(1.75)	33-2236-83
$C_{14}H_{26}O_6$			
Bicyclo[2.2.2]octane-2,3,5,6,7,8-hexamethanol, (2-endo,3-exo,5-endo-6-exo,7-syn,8-anti)	EtOH	215(2.18)(end abs.)	33-0019-83
$C_{14}H_{27}NO$			
4-Heptanone, 3-[[bis(1-methylethyl)-amino]methylene]-	MeOH	304(4.41)	131-0001-83K
(E-syn-E)-	C_6H_{12}	287(4.34)	131-0001-83K

Compound	Solvent	$\lambda_{max}(\log \epsilon)$	Ref.
$C_{14}H_{28}NO_4P$			
Phosphorus(1+), [2-ethoxy-1-(ethoxy-carbonyl)-2-oxoethyl]diethyl(1-propanaminato)-, hydroxide, inner salt	hexane	218s(--),236(4.02)	99-0023-83
$C_{14}H_{28}N_3O_4PS_3$			
Ethanimidothioic acid, N-[[[[[(5,5-dimethyl-1,3,2-dioxaphosphorinan-2-yl)-(1,1-dimethylethyl)amino]thio]methylamino]carbonyl]oxy]-, methyl ester, P-sulfide	H_2O	233(4.01)	98-0625-83
$C_{14}H_{30}OSn$			
Stannane, acetyltributyl-	ether	346s(--),359s(--), 375(2.3),391(2.3)	101-0175-83K
$C_{14}H_{30}Si_4$			
Disilane, 1,1,2,2-tetramethyl-1,2-bis[(trimethylsilyl)ethynyl]-	n.s.g.	223(4.03),230(4.06)	35-3359-83
$C_{14}H_{30}Si_5$			
1,2,3,6,7-Pentasilacyclonona-4,8-diyne, 1,1,2,2,3,3,6,6,7,7-decamethyl-	n.s.g.	242s(3.97),250(4.13)	35-3359-83
$C_{14}H_{31}LiSi_3$			
Lithium, [1,3,5-tris(trimethylsilyl)-2,4-pentadienyl]-	THF	423(3.67)	157-0021-83
$C_{14}H_{32}Si_3$			
Silane, 1,4-pentadiene-1,3,5-tris[trimethyl-, (E,E)-	THF	219(4.00)	157-0021-83

Compound	Solvent	$\lambda_{max}(\log \epsilon)$	Ref.
$C_{15}H_7Cl_2NS$			
Dibenzo[b,f]thiepin-10-carbonitrile, 2,8-dichloro-	MeOH	230(4.56),252(4.31), 268(4.33),302(3.99), 351s(2.72)	73-1187-83
$C_{15}H_8ClNS$			
Dibenzo[b,f]thiepin-10-carbonitrile, 2-chloro-	MeOH	217.5(4.50),228(4.50), 267.5(4.32),304(4.07)	73-1187-83
$C_{15}H_8FNS$			
Dibenzo[b,f]thiepin-2-carbonitrile, 7-fluoro-	MeOH	230(4.50),272.5(4.43), 303s(3.66)	73-0144-83
$C_{15}H_8N_2O_2$			
Indolo[1,2-b][2,6]naphthyridine-5,11-dione	EtOH	229(4.58),244s(4.57), 252(4.59),266s(4.29), 276s(4.18),316s(4.16), 324(4.17),332s(4.16), 404(4.45)	39-2413-83C
Indolo[1,2-b][2,6]naphthyridine-5,12-dione	EtOH	234(4.36),277(4.11), 310s(3.81),380(3.96)	44-2690-83
Indolo[1,2-b][2,7]naphthyridine-6,12-dione	EtOH	233(4.52),244(4.57), 269(4.12),290s(4.05), 300s(4.12),309(4.17), 316s(4.13),332s(3.90), 405(4.03)	39-2413-83C
5H-Pyrido[2,3-b]carbazole-5,11(10H)-dione	EtOH	214(4.39),230(4.22), 270(4.25),394(3.66)	39-2409-83C
5H-Pyrido[3,2-b]carbazole-5,11(6H)-dione	EtOH-NaOH	272(4.23),327(4.16), 445(3.75)	39-2409-83C
5H-Pyrido[3,4-b]carbazole-5,11(10H)-dione	EtOH-NaOH	270(4.30),303(4.37), 450(3.72)	39-2409-83C
$C_{15}H_8N_2O_3$			
Pyridinium, 4-cyano-, 2,4-dioxo-2H-1-benzopyran-3(4H)-ylide	EtOH	207(4.60),222(4.41), 274(4.21),430(3.86)	104-0283-83
	50% EtOH	418(3.73)	104-0283-83
	$CHCl_3$	468(3.98)	104-0283-83
$C_{15}H_8N_4$			
1,3-Butadiene-1,1,2-tricarbonitrile, 4-(1H-indol-1-yl)-	THF	210(4.50),256(3.97), 292(3.77),454(4.51)	103-0765-83
$C_{15}H_8N_4O_2$			
1,4-Cyclohexadiene-1-acetic acid, 3,6-bis(dicyanomethylene)-, methyl ester	MeCN	404(4.78)	44-3852-83
1,4-Cyclohexadiene-1-propanoic acid, 3,6-bis(dicyanomethylene)-	MeCN	397(4.72)	44-3852-83
$C_{15}H_8O_6$			
Rhein	MeOH	229(4.57),258(4.30), 435(4.05)	162-1179-83
$C_{15}H_9BrO_8$			
[5,5'-Bi-1,3-benzodioxole]-4,7,7'(4'H)-trione, 6-bromo-4'-methoxy-	$CHCl_3$	298(4.11),350(2.91), 490(2.77)	64-0392-83B
$C_{15}H_9Br_2N$			
4H-Cyclopenta[def]phenanthren-1-amine, 2,8-dibromo-	C_6H_{12}	238(4.60),254(4.40), 290(4.53),319(3.93), 366(3.70),383(3.72)	18-1259-83

Compound	Solvent	$\lambda_{max}(\log \epsilon)$	Ref.
4H-Cyclopenta[def]phenanthren-2-amine, 1,3-dibromo-	C_6H_{12}	269(4.86),291(4.27), 313(4.14),363(3.65), 380(3.70)	18-1259-83
4H-Cyclopenta[def]phenanthren-3-amine, 2,8-dibromo-	C_6H_{12}	228(4.49),257(4.70), 292(4.05),330(4.13), 348(3.95),365(3.89)	18-1259-83
4H-Cyclopenta[def]phenanthren-8-amine, 3,9-dibromo-	C_6H_{12}	231(4.52),257(4.65), 325(4.12),370(3.49)	18-1259-83
$C_{15}H_9ClN_2O_2$			
Benzoyl chloride, 4-(5-phenyl-1,3,4-oxadiazol-2-yl)-	toluene	312(4.39)	103-0022-83
$C_{15}H_9ClO_3$			
4H-1-Benzopyran-4-one, 2-(4-chlorophenyl)-3-hydroxy-	MeOH	242s(4.51),247(4.52), 308(4.36),346(4.49)	118-0835-83
$C_{15}H_9FOTe$			
4H-1-Benzotellurin-4-one, 7-fluoro-2-phenyl-	CH_2Cl_2	385(3.89)	35-0883-83
$C_{15}H_9F_2N_3O_5S$			
1,3,4-Oxadiazole, 2-[4-(difluoromethyl)sulfonyl]phenyl]-5-(4-nitrophenyl)-	toluene	310(4.49)	103-0022-83
$C_{15}H_9F_6NO_2$			
2,4-Pentanedione, 1,1,1,5,5,5-hexafluoro-3-(1-methyl-2(1H)-quinolinylidene)-	EtOH	239(4.25),307(4.03), 406(3.53)	39-0795-83C
$C_{15}H_9IO_3$			
Iodonium, (4-hydroxy-2-oxo-2H-1-benzopyran-3-yl)phenyl-, hydroxide, inner salt	EtOH	236(4.25),243(4.19), 298(3.99)	104-0319-83
$C_{15}H_9NO_3$			
6H-Anthra[1,9-cd]isoxazol-6-one, 5-methoxy-	EtOH	244(4.99),249(5.1), 445(4.16),467(4.16)	103-1057-83
5H-Furo[3',2':6,7][1]benzopyrano[3,4-c]pyridin-5-one, 7-methyl-	EtOH	240(4.39),285(4.01), 310s(3.92),330s(3.82)	39-0225-83C
$C_{15}H_9NS$			
Dibenzo[b,f]thiepin-10-carbonitrile	MeOH	211.5(4.39),227(4.40), 265(4.27),303.5(4.03)	73-1187-83
$C_{15}H_9N_5O$			
Furo[2',3':4,5]pyrrolo[1,2-d]-1,2,4-triazolo[3,4-f]-1,2,4-triazine, 9-phenyl-	MeOH	237(3.37),344(3.66)	73-1878-83
$C_{15}H_9N_5O_6$			
Benzeneacetonitrile, 2-[(2,4,6-trinitrophenyl)methylene]amino]-	$CHCl_3$	353(3.47)	44-2468-83
$C_{15}H_{10}BrF_3N_2O_3S$			
Acetamide, 2-bromo-N-[4-[(phenylsulfonyl)amino]-3-(trifluoromethyl)-2,5-cyclohexadien-1-ylidene]-	EtOH	260(4.09)	44-3146-83

Compound	Solvent	$\lambda_{max}(\log \epsilon)$	Ref.
$C_{15}H_{10}BrN$			
4H-Cyclopenta[def]phenanthren-1-amine, 2-bromo-	C_6H_{12}	236(4.62),282(4.45), 310(3.93),356(3.68), 373(3.74)	18-1259-83
4H-Cyclopenta[def]phenanthren-2-amine, 1-bromo-	C_6H_{12}	243(4.38),268(4.84), 289(4.24),309(3.98), 364(3.49),373(3.49), 380(3.51)	18-1259-83
4H-Cyclopenta[def]phenanthren-3-amine, 8-bromo-	C_6H_{12}	252(4.64),290(4.01), 329(4.12),363(3.67)	18-1259-83
4H-Cyclopenta[def]phenanthren-8-amine, 9-bromo-	C_6H_{12}	229(4.49),254(4.66), 323(4.05),364(3.45)	18-1259-83
$C_{15}H_{10}ClNSSe$			
2(3H)-Thiazoleselone, 3-(4-chlorophenyl)-4-phenyl-	EtOH	360(4.27)	56-0875-83
$C_{15}H_{10}ClNS_2$			
2(3H)-Thiazolethione, 3-(4-chlorophenyl)-4-phenyl-	EtOH	333(4.19)	56-0875-83
$C_{15}H_{10}ClN_3O_3$			
2H-1,4-Benzodiazepin-2-one, 5-(2-chlorophenyl)-1,3-dihydro-7-nitro- (clonazepam)	isoPrOH-MeOH	248(4.16),310(4.07)	162-0340-83
$C_{15}H_{10}Cl_2N_2S$			
2H-1,4-Benzodiazepine-2-thione, 7-chloro-5-(2-chlorophenyl)-1,3-dihydro-	MeOH	303(4.40)	73-0123-83
$C_{15}H_{10}Cl_2O_5$			
Spiro[benzofuran-2(3H),1'-[3,5]cyclohexadiene]-2',3-dione, 5,5'-dichloro-4',6-dihydroxy-4,6'-dimethyl-	EtOH	233(4.39),280(4.30), 322(4.04)	39-0413-83C
$C_{15}H_{10}Cl_4O$			
2-Propanone, 1,3-bis(2,4-dichlorophenyl)-	C_6H_{12}	295s(2.38)	46-1960-83
2-Propanone, 1,3-bis(2,6-dichlorophenyl)-	C_6H_{12}	295s(2.01)	46-1960-83
$C_{15}H_{10}FOTe$			
1,2-Oxatellurol-1-ium, 5-(3-fluorophenyl)-3-phenyl-	CH_2Cl_2	436(4.40)	35-0883-83
$C_{15}H_{10}F_2N_2O_3S$			
1,3,4-Oxadiazole, 2-[4-[(difluoromethyl)sulfonyl]phenyl]-5-phenyl-	toluene	300(4.38)	103-0022-83
$C_{15}H_{10}N_2$			
1H-Phenanthro[9,10-d]imidazole	hexane	245(--),252(--), 280(--),300(--)	39-1641-83B
	H_2O	245s(4.6),252(4.7), 272(4.0),298(3.8), 332(3.0),348(3.0)	39-1641-83B
1H-Pyrrolo[2,3-c]phenanthridine	EtOH	215(4.69),261(4.79), 337(4.22),352(4.19), 370(4.20)	103-1135-83
1H-Pyrrolo[3,2-i]phenanthridine	EtOH	208(4.56),220(4.55), 240(4.82),254(4.69),	103-1135-83

Compound	Solvent	$\lambda_{max}(\log \epsilon)$	Ref.
(cont.)		275(4.54),345(4.08)	103-1135-83
$C_{15}H_{10}N_2O_2$			
6H-Indolo[3,2,1-de][1,5]naphthyridin-6-one, 5-methoxy-	EtOH	252(4.08),261(4.05), 271(4.00),294(3.85), 364(4.02),381(3.97)	94-3198-83
6H-Indolo[3,2,1-de][1,5]naphthyridin-6-one, 10-methoxy-	MeOH	212(4.20),232s(3.81), 266s(3.95),274(4.06), 310(3.69),352(3.83)	100-0222-83
$C_{15}H_{10}N_2O_2S$			
4H-1-Benzothiopyran-4-one, 3-hydroxy-2-(phenylazo)-	MeOH	372(4.19),374(4.19), 458(3.92),460(3.92)	83-0921-83
$C_{15}H_{10}N_2O_2S_3$			
Benzenesulfenamide, 2-nitro-N-(4-phenyl-1,3-dithiol-2-ylidene)-	1,2-$C_2H_4Cl_2$	316(3.94),418(3.47)	118-0050-83
$C_{15}H_{10}N_2O_3$			
3H-Indolo[3,2,1-de][1,5]naphthyridine-2,6-dione, 3-methoxy-	MeOH	227(4.24),232s(4.24), 248(4.22),253s(4.21), 292(4.01),302(4.01), 325(3.81),403s(3.97), 420(4.13)	100-0222-83
$C_{15}H_{10}N_2O_4$			
3H-Indolo[3,2,1-de][1,5]naphthyridine-2,6-dione, 10-hydroxy-3-methoxy-	MeOH	220(4.32),248s(4.25), 272(4.09),288(4.05), 365(4.15),382(4.17), 400(4.18),423(4.25), 448(4.24)	100-0222-83
$C_{15}H_{10}N_4$			
Propanedinitrile, 2,2'-[2-(1-methylethyl)-2,5-cyclohexadiene-1,4-diylidene]bis-	MeCN	245(2.83),340(3.20), 400(2.60)	104-0742-83
1,2,4-Triazino[4,3-b]indazole, 3-phenyl-	EtOH	248(3.97),293(4.53), 362(4.54)	44-2330-83
$C_{15}H_{10}N_4O_4$			
Benzeneacetonitrile, 2-[[(2,4-dinitrophenyl)methylene]amino]-	$CHCl_3$	350(4.00)	44-2468-83
1,3,5-Triazabicyclo[3.1.0]hexane-2,4-dione, 6-(4-nitrophenyl)-3-phenyl-	$CHCl_3$	260(4.1),330(2.6)	103-1345-83
1,2,4-Triazolidinium, 1-[(4-nitrophenyl)methylene]-3,5-dioxo-4-phenyl-, hydroxide, inner salt	$CHCl_3$	260(4.0),330(3.4), 430(3.7)	103-1345-83
$C_{15}H_{10}N_4O_6$			
Pyrido[1,2-a]pyrimidin-4-one, 2-methyl-3-(2,4-dinitrophenoxy)-	EtOH	244(4.29),280(4.07), 346(4.19)	142-1083-83
$C_{15}H_{10}N_6$			
Tetrazolo[1',5':1,2]quino[3,4-b]quinoxaline, 9,14-dihydro-	MeCN	230(4.51),270(4.63), 360(4.04)	103-0222-83
$C_{15}H_{10}OS$			
4H-1-Benzothiopyran-4-one, 2-phenyl-	EtOH	272(4.36),345(4.04)	139-0059-83B
$C_{15}H_{10}OSe$			
4H-1-Benzoselenin-4-one, 2-phenyl-	CH_2Cl_2	357(3.75)	35-0883-83

Compound	Solvent	$\lambda_{max}(\log \epsilon)$	Ref.
$C_{15}H_{10}O_2$			
9-Anthracenecarboxylic acid	MeOH	361.5(3.86)	151-0131-83C
	EtOH	361.5(3.89)	151-0131-83C
	ether	361.0(3.88)	151-0131-83C
with sodium dodecyl sulfate	n.s.g.	364.5(3.92)	151-0131-83C
$C_{15}H_{10}O_3$			
9,10-Anthracenedione, 1-hydroxy-2-methyl-	MeOH	248(4.18),265s(4.08), 275s(3.95)	94-2353-83
	MeOH	226(4.24),247(4.40), 254(4.42),279(4.02), 325(3.43),408(3.74)	102-0737-83
Benz[a]azulene-1,4-dione, 10-methoxy-	MeCN	265(4.3),315(4.3), 472(3.8),690(3.2)	24-2408-83
Benzo[f]coumarin, 3-acetyl-	benzene	387(4.14)	151-0325-83A
4H-1-Benzopyran-4-one, 3-hydroxy-2-phenyl-	C_6H_{12}	340(4.23),352(4.13)	151-0067-83A
	toluene	345(4.22),357s(--)	151-0067-83A
	MeOH	238(4.31),243s(4.31), 306(4.08),344(4.23)	118-0835-83
	MeOH	345(4.19),354s(--)	151-0067-83A
	EtOH	344(4.21),352s(--)	151-0067-83A
	ether	342(4.24),355s(--)	151-0067-83A
	pH 0	344(4.23)	151-0067-83A
	pH 3	342(4.23)	151-0067-83A
	pH 7	342(4.23)	151-0067-83A
	pH 10	402(4.07)	151-0067-83A
	pH 13	402(4.20)	151-0067-83A
	H_o -0.31	345(4.24)	151-0067-83A
	H_o -1.72	359.5(4.26)	151-0067-83A
	H_o -3.38	373(4.40)	151-0067-83A
	H_o -5.80	378(4.40)	151-0067-83A
$C_{15}H_{10}O_4$			
9,10-Anthracenedione, 1,3-dihydroxy-2-methyl- (rubiadin)	EtOH	246(4.39),280(4.52), 415(3.87)	162-1193-83
9,10-Anthracenedione, 1,8-dihydroxy-3-methyl- (chrysophanic acid)	n.s.g.	226(4.61),256(4.45), 278(4.15),288(4.15), 436(4.07)	162-0321-83
4H-1-Benzopyran-4-one, 5,7-dihydroxy-2-phenyl- (chrysin)	n.s.g.	270(4.40),329(3.90)	162-0321-83
4H-1-Benzopyran-4-one, 7-hydroxy-2-(4-hydroxyphenyl)-	MeOH	256(4.11),315(4.08), 331(4.15)	161-0067-83
4H-1-Benzopyran-4-one, 7-hydroxy-3-(4-hydroxyphenyl)- (diadzein)	n.s.g.	250(4.44)	162-0405-83
$C_{15}H_{10}O_5$			
9,10-Anthracenedione, 1,3,4-trihydroxy-2-methyl-	MeOH	284(4.63),304s(4.22), 321(3.82),424(3.83)	94-2353-83
9,10-Anthracenedione, 1,3,8-trihydroxy-6-methyl- (emodin)	EtOH	254(4.19),292(4.14), 450(3.84)	94-4543-83
	EtOH	222(4.55),252(4.26), 265(4.27),289(4.34), 437(4.10)	162-0514-83
4H-1-Benzopyran-4-one, 5,7-dihydroxy-2-(4-hydroxyphenyl)- (apigenin)	EtOH	269(4.27),340(4.32)	162-0107-83
relative absorbance given	MeOH	267(0.92),296s(0.68), 336(1.00)	102-2107-83
	MeOH-$AlCl_3$	276(0.81),301(0.78), 348(1.00),384(0.85)	102-2107-83
	MeOH-$AlCl_3$-HCl	276(0.88),299(0.85), 340(1.00),381(0.83)	102-2107-83

Compound	Solvent	λ_{max}(log ϵ)	Ref.
4H-1-Benzopyran-4-one, 5,7-dihydroxy-3-(4-hydroxyphenyl)- (genistein)	n.s.g.	262.5(2.14)	162-0626-83
4H-1-Benzopyran-4-one, 5,6,7-trihydroxy-2-phenyl- (baicalein)	EtOH	276(4.42),324(4.18)	162-0136-83
$C_{15}H_{10}O_6$			
9,10-Anthracenedione, 1,3,4,5-tetrahydroxy-2-methyl-	MeOH	256(4.46),274(4.21), 293s(4.08),350(3.80), 464s(4.05),490(3.98), 525(3.86)	102-2583-83
	MeOH-base	225(--),290(--), 350s(--),490s(--), 515(--),551(--)	102-2583-83
4H-1-Benzopyran-4-one, 2-(3,4-dihydroxyphenyl)-5,7-dihydroxy- (relative absorbance given)	MeOH	242s(0.76),253(0.86), 267(0.77),291s(0.54), 348(1.00)	102-2107-83
	MeOH-AlCl3	272(0.81),300s(0.32), 330s(0.21),423(1.00)	102-2107-83
	MeOH-AlCl3-HCl	264s(0.95),276(1.00), 296(0.66),358(1.00), 385(0.98)	102-2107-83
4H-1-Benzopyran-4-one, 3,5,7-trihydroxy-2-(2-hydroxyphenyl)-	EtOH	264.0(4.265),375(4.005)	162-0408-83
4H-1-Benzopyran-4-one, 5,6,7-trihydroxy-2-(4-hydroxyphenyl)-	EtOH	286(4.22),339(4.26)	162-1210-83
(relative absorbance given)	MeOH	284(0.84),336(1.00)	102-2107-83
	MeOH-AlCl3	262(0.12),292s(0.64), 304(0.86),375(1.00)	102-2107-83
	MeOH-AlCl3-HCl	262s(0.16),292s(0.68), 302(0.84),364(1.00)	102-2107-83
4H-1-Benzopyran-4-one, 5,7,8-trihydroxy-2-(4-hydroxyphenyl)-	EtOH	286s(4.32),306(4.38), 335s(4.15)	18-3773-83
	EtOH-NaOAc	304(4.30),375(4.05)	18-3773-83
	EtOH-AlCl3	289(4.24),317(4.39), 343s(4.27),412(3.75)	18-3773-83
(relative absorbance given)	MeOH	280(0.95),305(1.00), 330s(0.88),364s(0.56)	102-2107-83
	MeOH-AlCl3	250s(0.61),299(0.80), 326(1.00),360s(0.56), 432(0.38)	102-2107-83
	MeOH-AlCl3-HCl	262s(0.64),286(0.83), 316(1.00),347(0.92), 404(0.51)	102-2107-83
$C_{15}H_{10}O_7$			
4H-1-Benzopyran-4-one, 2-(3,4-dihydroxyphenyl)-3,5,7-trihydroxy-	EtOH	258(2.75),375(2.75)	162-1160-83
4H-1-Benzopyran-4-one, 2-(3,4-dihydroxyphenyl)-5,6,7-trihydroxy- (relative absorbance given)	MeOH	248s(0.82),283(0.77), 346(1.00)	102-2107-83
	MeOH-AlCl3	248s(0.99),270(0.78), 302(0.63),420(1.00)	102-2107-83
	MeOH-AlCl3-HCl	258(0.76),296(0.76), 372(1.00)	102-2107-83
4H-1-Benzopyran-4-one, 2-(3,4-dihydroxyphenyl)-5,7,8-trihydroxy-	EtOH	284(4.23),334(4.23)	18-3773-83
	EtOH-NaOAc	303s(4.12),327(4.14), 386(4.10)	18-3773-83
	EtOH-AlCl3	287(4.23),305s(4.17), 358(4.21),402s(3.98)	18-3773-83
4H-1-Benzopyran-4-one, 5,7-dihydroxy-2-(3,4,5-trihydroxyphenyl)- (relative absorbance given)	MeOH	260(0.81),269(0.80), 302s(0.50),354(1.00)	102-2107-83
	MeOH-AlCl3	272(0.75),310(0.30),	102-2107-83

Compound	Solvent	λ_{max}(log ε)	Ref.
(cont.)		424(1.00)	102-2107-83
	MeOH-$AlCl_3$-HCl	277(0.94),306(0.52), 364s(0.89),393(1.00)	102-2107-83
Hypolactin (relative absorbance given)	MeOH	256(0.82),280(1.00), 302s(0.82),340(0.93)	102-2107-83
	MeOH-$AlCl_3$	274s(0.90),296(1.00), 322s(0.75),342s(0.67), 396(0.95)	102-2107-83
	MeOH-$AlCl_3$-HCl	210(0.77),286(0.90), 306(0.80),324s(0.77), 358(1.00),430s(0.58)	102-2107-83
$C_{15}H_{10}O_8$			
4H-1-Benzopyran-4-one, 2-(3,4-dihydroxyphenyl)-3,5,6,7-tetrahydroxy-	EtOH	259(4.23),361(4.34)	162-1160-83
Carbonic acid, 4,9-dihydroxy-7-methyl-1,3-dioxo-1H,3H-naphtho[1,8-cd]pyran-6-yl methyl ester	EtOH	226(3.62),247(3.69), 282(3.08),313(3.23), 341(3.28),363(3.34), 395(3.04)	78-2283-83
	EtOH-NaOH	251s(3.90),314(4.04), 372(3.78)	78-2283-83
$C_{15}H_{10}O_{10}S$			
Benzenesulfonic acid, 2,3-dihydroxy-5-(3,5,7-trihydroxy-4-oxo-4H-1-benzopyran-2-yl)-	H_2O	256(4.37),365(4.29)	104-1490-83
4H-1-Benzopyran-8-sulfonic acid, 2-(3,4-dihydroxyphenyl)-3,5,7-trihydroxy-4-oxo-	H_2O	257(4.26),372(4.36)	104-1490-83
	MeOH	256(4.37),272(4.38)	104-1490-83
$C_{15}H_{10}O_{13}S_2$			
4H-1-Benzopyran-8-sulfonic acid, 2-(3,4-dihydroxy-5-sulfophenyl)-3,5,7-trihydroxy-4-oxo-, disodium salt	H_2O	262(4.72),366(4.34)	104-1490-83
$C_{15}H_{10}S$			
5H-Dibenzo[a,d]cycloheptene-5-thione	3-Mepentane	403(3.78),596(2.31)	23-0894-83
$C_{15}H_{11}$			
Cyclohept[a]azulenylium, tetrafluoroborate	MeCN	251(4.10),285s(3.95), 333(4.76),356s(3.90), 380(2.80),444s(3.15), 476(3.84),506(4.43), 556(3.12),611(2.93), 673(2.71)	88-0069-83
$C_{15}H_{11}Br$			
1H-Indene, 2-bromo-1-phenyl-	hexane	273(4.07),291(3.58)	104-1438-83
$C_{15}H_{11}ClO_2S$			
Benzenecarbothioic acid, S-[2-(4-chlorophenyl)-2-oxoethyl] ester	MeOH	252(4.22),340(2.89)	44-0835-83
$C_{15}H_{11}ClO_3S$			
4H-1,3-Benzodioxin-4-one, 2-[(4-chlorophenyl)thio]-2-methyl-	pH 7.4	304(3.30)	1-0351-83
$C_{15}H_{11}ClO_4$			
4H-1,3-Benzodioxin-4-one, 2-(4-chlorophenyl)-2-methyl-	pH 7.4	300(3.58)	1-0351-83

Compound	Solvent	$\lambda_{max}(\log \epsilon)$	Ref.
$C_{15}H_{11}ClS$			
Benzo[b]thiophene, 5-chloro-2-methyl-3-phenyl-	EtOH	239(4.60),262(3.98), 298(3.62),308(3.60)	44-1275-83
$C_{15}H_{11}Cl_2N_3O$			
Pyrido[2,3-d]pyrimidin-7(8H)-one, 6-(2,6-dichlorophenyl)-2,8-dimethyl-	MeOH	306(--),315(4.20), 326s(--)	87-0403-83
	MeOH-5M HCl	221(4.26),301(--), 310(4.17),320s(--)	87-0403-83
$C_{15}H_{11}Cl_2N_5O$			
Urea, [6-(2,6-dichlorophenyl)-2-methylpyrido[2,3-d]pyrimidin-7-yl]-	MeOH	226(4.72),319(4.20), 328(4.19)	87-0403-83
$C_{15}H_{11}CsO_2$			
9H-Fluorene-9-carboxylic acid, methyl ester, ion(1-)-, cesium	$(MeOCH_2)_2$	403(3.79),420(3.75)	104-1592-83
$C_{15}H_{11}FN_2$			
Quinazoline, 2-(4-fluorophenyl)-6-methyl-	EtOH	263(4.59),286s(4.11), 330(3.53),342s(3.39)	88-4351-83
$C_{15}H_{11}LiO_2$			
9H-Fluorene-9-carboxylic acid, methyl ester, ion(1-)-, lithium	$(MeOCH_2)_2$	385(3.75),409(3.68)	104-1592-83
$C_{15}H_{11}N$			
6H-Isoindolo[2,1-a]indole [sic]	EtOH	224(4.18),249s(3.91), 257(3.94),312s(4.14), 324(4.20),337(4.00)	39-2417-83C
$C_{15}H_{11}NO$			
2-Propenenitrile, 2-phenoxy-3-phenyl-, (E)-	EtOH	205(4.17),216(4.14), 284(4.23)	78-1551-83
(Z)-	EtOH	206(4.24),216(4.25), 282(4.42)	78-1551-83
$C_{15}H_{11}NOS$			
[1]Benzothieno[3,2-b]azet-2(1H)-one, 2a,7b-dihydro-7b-phenyl-	EtOH	250(3.96),290(2.98), 300(2.90)	44-1275-83
$C_{15}H_{11}NO_3$			
3H-Benz[e]indole-4,5-dione, 1-acetyl-2-methyl-	EtOH	220(4.31),265(4.28), 380s(3.72)	103-1086-83
[1]Benzopyrano[3,4-b]cyclopenta[d]-pyridine-6,8-dione, 7,9,10,11-tetrahydro-	dioxan	240(4.07),330(4.45)	4-0775-83
$C_{15}H_{11}NO_4$			
2(5H)-Furanone, 3-[3-(4-cyanophenyl)-1-oxo-2-propenyl]-4-hydroxy-5-methyl-, (E)-	MeOH	201(4.15),229(4.16), 299(4.26),312(4.26), 351(4.32)	83-0115-83
Indeno[1,2-b]pyran-2-carboxaldehyde, 4,5-dihydro-4-oxo-, 2-(O-acetyloxime)	MeOH	200(4.34),213(4.30), 218(4.29),253(4.20), 262(4.18),277(4.13), 291(4.17),311(4.16)	83-0264-83
$C_{15}H_{11}NO_4S$			
Benzenecarbothioic acid, S-[2-(4-nitrophenyl)-2-oxoethyl] ester	MeOH	261(4.13),342(2.76)	44-0835-83

Compound	Solvent	$\lambda_{max}(\log \epsilon)$	Ref.
$C_{15}H_{11}NO_4S_2$			
4H,5H-Pyrano[3,4-e][1,3]oxazine-4,5-dione, 7-[(4-methylphenyl)thio]-2-(methylthio)-	$CHCl_3$	246(4.19),297(4.40), 348(4.02)	24-3725-83
$C_{15}H_{11}NO_5S$			
Propanedioic acid, (4-oxo-4H-indeno-[1,2-c]isothiazol-3-yl)-, dimethyl ester	EtOH	263(4.55),269(4.53), 306(3.65),354(3.99), 444(3.89)	95-1243-83
$C_{15}H_{11}NO_6$			
1,2-Naphthalenedione, 4-(1-acetyl-2-oxopropyl)-3-nitro-	EtOH	258(3.87),350s(3.20)	103-1086-83
$C_{15}H_{11}NS$			
Dibenzo[b,e]thiepin-11-carbonitrile, 6,11-dihydro-	MeOH	263(4.01)	73-1898-83
$C_{15}H_{11}NSSe$			
2(3H)-Thiazoleselone, 3,4-diphenyl-	EtOH	360(4.30)	56-0875-83
$C_{15}H_{11}NS_2$			
2(3H)-Thiazolethione, 3,4-diphenyl-	EtOH	333(4.13)	56-0875-83
$C_{15}H_{11}N_3OS$			
Furo[2',3':4,5]pyrrolo[1,2-d]-1,2,4-triazine-8(7H)-thione, 2-(4-methylphenyl)-	MeOH	313(3.07),400(3.05)	73-1878-83
Furo[2',3':4,5]pyrrolo[1,2-d]-1,2,4-triazine-8(7H)-thione, 5-methyl-2-phenyl-	MeOH	335(3.16),400(3.01)	73-1878-83
$C_{15}H_{11}N_3OS_2$			
Benzamide, N-(1,4-dihydro-2,4-dithioxo-3(2H)-quinazolinyl)-	EtOH	253(4.20),285(4.50), 308(4.47),382(3.81)	49-0915-83
$C_{15}H_{11}N_3O_2$			
Benzeneacetonitrile, 2-[[(4-nitrophenyl)methylene]amino]-	$CHCl_3$	333(4.00)	44-2468-83
Furo[2',3':4,5]pyrrolo[1,2-d]-1,2,4-triazin-8(7H)-one, 2-(4-methylphenyl)-	MeOH	227(3.27),335(3.60)	73-1878-83
$C_{15}H_{11}N_3O_2S_3$			
4H-Bisthiazolo[3,2-a:4',5'-e]pyrimidinium, 8-hydroxy-5-methyl-2-(methylthio)-4-oxo-7-phenyl-, hydroxide, inner salt	MeCN	218(4.30),276(4.22), 316(4.29),374(3.97), 438(3.85)	2-0243-83
	benzene	460(--)	2-0243-83
	CH_2Cl_2	450(--)	2-0243-83
$C_{15}H_{11}N_3O_3$			
Furo[2',3':4,5]pyrrolo[1,2-d]-1,2,4-triazine-5,8-dione, 6,7-dihydro-2-(4-methylphenyl)-	MeOH	270(2.72),342(3.50)	73-1878-83
2H-Indol-2-one, 1,3-dihydro-4-[[(4-nitrophenyl)imino]-	$CHCl_3$	346(3.94)	44-2468-83
2H-Pyrano[3,2-c]quinoline-3-carbonitrile, 8-(dimethylamino)-5,6-dihydro-2,5-dioxo-	EtOH	462(4.34)	5-0695-83
	toluene	453(--)	5-0695-83
	DMSO	467(--)	5-0695-83

Compound	Solvent	λ_{max}(log ε)	Ref.
$C_{15}H_{11}N_3O_4$			
1,3,4-Oxadiazole, 2-(4-methoxyphenyl)-5-(4-nitrophenyl)-	toluene	337(4.27)	103-0022-83
$C_{15}H_{11}O_2$			
9H-Fluorene-9-carboxylic acid, methyl ester, ion(1-), cesium	$(MeOCH_2)_2$	403(3.79),420(3.75)	104-1592-83
lithium	$(MeOCH_2)_2$	385(3.75),409(3.68)	104-1592-83
$C_{15}H_{11}O_5$			
Pelargonidin (chloride)	EtOH-HCl	530(4.51)	162-1014-83
$C_{15}H_{12}$			
Anthracene, 9-methyl-	C_6H_{12}	366.5(4.00)	151-0061-83C
	benzene	369.5(3.96)	151-0061-83C
	MeOH	366.0(3,98)	151-0061-83C
	ether	366.0(3.99)	151-0061-83C
$C_{15}H_{12}BrF_3N_2O_3S$			
Acetamide, 2-bromo-N-[4-[(phenylsulfonyl)amino]-3-(trifluoromethyl)-phenyl]-	EtOH	270(4.24)	44-3146-83
$C_{15}H_{12}BrN_3O$			
1,2,4-Triazin-3(2H)-one, 6-(4-bromophenyl)-4,5-dihydro-2-phenyl-	MeOH	246(4.38),308(4.24)	104-1533-83
$C_{15}H_{12}BrN_3OS$			
4H-Pyrido[3,4-e]-1,3-thiazin-4-one, 2-[(4-bromophenyl)amino]-5,7-dimethyl-	MeOH	260(4.40)	73-3426-83
$C_{15}H_{12}BrN_3OSe$			
4H-Pyrido[3,4-e]-1,3-selenazin-4-one, 2-[(4-bromophenyl)amino]-5,7-dimethyl-	MeOH	245(3.34),270(3.28)	73-3567-83
$C_{15}H_{12}Br_2$			
Benzene, 1,1'-(2,3-dibromo-1-propenylidene)bis-	hexane	240(4.00),270(4.01)	104-1438-83
$C_{15}H_{12}Br_2O_4$			
3,5-Cyclohexadiene-1,2-dione, 6-bromo-3-[(3-bromo-4-hydroxyphenyl)methyl]-4-(methoxymethyl)-	MeOH	287(3.60)	102-0743-83
$C_{15}H_{12}ClNO_2S$			
1H-Indole, 3-chloro-2-methyl-1-(phenylsulfonyl)-	EtOH	216(4.4),257(4.15), 290s(3.54)	39-2417-83C
$C_{15}H_{12}ClNO_3$			
2H-Furo[2,3-h]-1-benzopyran-2-one, 3-chloro-4-pyrrolidino-	EtOH	242(4.43),328(4.09)	111-0113-83
$C_{15}H_{12}ClNO_4$			
2H-Furo[2,3-h]-1-benzopyran-2-one, 3-chloro-4-morpholino-	EtOH	217s(4.19),243s(4.21), 248(4.22),295s(3.90), 325(3.99)	111-0113-83
$C_{15}H_{12}ClNO_5S$			
2H-Pyran-5-carbothioic acid, 4-chloro-	$CHCl_3$	312(4.15)	24-3725-83

Compound	Solvent	λ_{max}(log ε)	Ref.
6-[(ethoxycarbonyl)amino]-2-oxo-			24-3725-83
$C_{15}H_{12}ClN_3$			
Benzenamine, N-(4-chloro-2-methyl-1(2H)-phthalazinylidene)-, (E)-	MeOH	211(4.69),238(4.23), 275(4.26),338(3.90)	103-0666-83
	MeCN	212(4.68),242s(4.18), 275(4.26),340(3.89)	103-0666-83
sulfate	MeOH	225(4.52),277(3.94), 295(3.88),332(3.88)	103-0666-83
	MeCN	226(4.51),276(4.00), 328(3.88)	103-0666-83
1-Phthalazinamine, 4-chloro-N-methyl-N-phenyl-	MeOH	215(4.63),253(3.94), 285(3.81),340(3.81)	103-0666-83
	MeCN	215(4.68),253(4.02), 285s(3.85),333(3.85)	103-0666-83
sulfate	MeOH	222(4.54),282(3.97), 328(3.92)	103-0666-83
	MeCN	222(4.63),272(4.02), 326(3.95)	103-0666-83
$C_{15}H_{12}ClN_3O$			
1H-Pyrrolo[3,2-c]pyridine-7-carboxamide, 6-chloro-1-(phenylmethyl)-	EtOH	229(4.96),240(4.12)	103-0052-83
$C_{15}H_{12}Cl_2N_4$			
Pyrido[2,3-d]pyrimidin-7-amine, 6-(2,6-dichlorophenyl)-N,2-dimethyl-	MeOH	225(4.59),274(3.81), 337(4.16)	87-0403-83
	MeOH-HCl	355(4.18)	87-0403-83
$C_{15}H_{12}Cl_2N_4O_2S$			
Methanesulfonamide, N-[6-(2,6-dichlorophenyl)-2-methylpyrido[2,3-d]pyrimidin-7-yl]-	MeOH	217(4.64),322(4.19), 335(4.22),349(4.05)	87-0403-83
	MeOH-HCl	206(4.60),261(3.79), 303(4.14)	87-0403-83
	MeOH-KOH	279(3.87),345(4.24)	87-0403-83
$C_{15}H_{12}Cl_2O$			
2-Propanone, 1,3-bis(2-chlorophenyl)-	C_6H_{12}	295s(2.20)	46-1960-83
2-Propanone, 1,3-bis(3-chlorophenyl)-	C_6H_{12}	296(2.37)	46-1960-83
2-Propanone, 1,3-bis(4-chlorophenyl)-	C_6H_{12}	295(2.39)	46-1960-83
$C_{15}H_{12}F_2O_3$			
Propanoic acid, 2-[(2',4'-difluoro-[1,1'-biphenyl]-4-yl)oxy]- ($E^{1\%}$ given)	EtOH	209(725),252(6.13)	145-0198-83
$C_{15}H_{12}F_9O_6Rh$			
Rhodium, tris(1,1,1-trifluoro-2,4-pentanedionato-O,O)-, trans	C_6H_{12}	265(3.9),325(3.9)	35-6038-83
	benzene	327(--)	35-6038-83
	EtOH	323(--)	35-6038-83
$C_{15}H_{12}FeO_4$			
Iron, tricarbonyl[η^4-5,6,7,8-tetrakis(methylene)bicyclo[2.2.2]octanone]-	EtOH	285s(3.46),315s(3.28)	33-1134-83
$C_{15}H_{12}N_2$			
Benzenamine, N-4H-quinolizin-4-ylidene-	EtOH	254(4.03),395(4.02)	4-0407-83
	EtOH-HBr	262(3.99),384(4.13)	4-0407-83
	EtOH-NaOH	253(4.05),285(3.92), 417(4.08)	4-0407-83

Compound	Solvent	λ_{max}(log ϵ)	Ref.
1H-Pyrazole, 3,5-diphenyl-	hexane	252(3.9)	151-0245-83A
	pH 7	253(4.0)	151-0245-83A
	MeOH	254(4.1)	151-0245-83A
	MeCN	254(4.1)	151-0245-83A
$C_{15}H_{12}N_2O_2$			
Benzo[1,2-b:3,4-b']dipyrrole-8-carboxaldehyde, 1,6-dihydro-3-(3-oxo-1-butenyl)-	EtOH	201(4.30),217(4.64), 252s(3.92),268s(4.02), 294(4.23)	103-0488-83
1,3,4-Oxadiazole, 2-(4-methoxyphenyl)-5-phenyl-	toluene	298(4.46)	103-0022-83
1,4-Phthalazinedione, 2,3-dihydro-2-methyl-3-phenyl-	EtOH	216(4.65),237(4.10), 305(3.70)	104-2260-83
4H-Pyrido[1,2-a]pyrimidin-4-one, 2-methyl-3-phenoxy-	EtOH	245(4.18),342(4.10)	142-1083-83
$C_{15}H_{12}N_2O_2S$			
Acetamide, N-[4-(2-benzothiazolyl)-3-hydroxyphenyl)-	EtOH	344(4.52)	4-1517-83
$C_{15}H_{12}N_2O_2S_2$			
2-Thiazolecarbothioamide, N,N-dimethyl-4-(2-oxo-2H-1-benzopyran-3-yl)-	MeCN	346(4.31)	48-0551-83
$C_{15}H_{12}N_2O_3$			
Acetamide, N-[4-(2-benzoxazolyl)-3-hydroxyphenyl]-	EtOH	330(4.56)	4-1517-83
Acetic acid, (3,4-dihydro-3-oxonaphtho[1,8-ef]-1,4-diazepin-2(1H)-ylidene)-, methyl ester	MeOH	229(4.61),324s(--), 332(4.43),404(3.54)	18-2338-83
Hydrofuramide	n.s.g.	215(4.16),259(4.18)	162-0695-83
1,5-Naphthyridin-8(8aH)-one, 4-hydroxy-6-methyl-7-phenoxy-	EtOH	265(4.06),346(4.14), 362(4.22)	142-1083-83
1,7-Phenanthroline-5,6-dione, acetone adduct	EtOH	229(4.44),286(4.08)	35-4431-83
1,10-Phenanthroline-5,6-dione, acetone adduct	EtOH	228(4.36),296(4.00)	35-4431-83
4,7-Phenanthroline-5,6-dione, acetone adduct	EtOH	230(4.40),284(4.03)	35-4431-83
Pyrano[2,3-c]pyrazol-4(1H)-one, 5-acetyl-3-methyl-1-phenyl-	EtOH	312(4.10)	44-4078-83
$C_{15}H_{12}N_2O_6$			
1,3-Dioxolane, 4-(2-nitrophenyl)-2-(4-nitrophenyl)-, cis	EtOH	203(4.33),261(4.20)	23-0400-83
trans	EtOH	205(4.30),261.5(4.22)	23-0400-83
$C_{15}H_{12}N_2S_2$			
1,2,3-Thiadiazolium, 4-mercapto-5-phenyl-3-(phenylmethyl)-, hydroxide, inner salt	EtOH	303(4.07),470(3.44)	94-1746-83
$C_{15}H_{12}N_4$			
8-Quinolinamine, 5-(phenylazo)-	aq EtOH	202(--),250(--), 425(4.21)	74-0011-83C
	acid	202(--),262(--), 522(4.59)	74-0011-83C
$C_{15}H_{12}N_4O_2$			
1-Triazenium, 1-(1,3-dihydro-1,3-dioxo-2H-isoindol-2-yl)-2-methyl-3-	EtOH	218(4.57),239s(4.26), 264(4.10)	104-2260-83

Compound	Solvent	$\lambda_{max}(\log \epsilon)$	Ref.
phenyl-, hydroxide, inner salt, (E)-			104-2260-83
(Z)-	EtOH	221(4.54),331(4.20)	104-2260-83
$C_{15}H_{12}N_4O_4$			
2,4(1H,3H)-Pteridinedione, 6-acetoxy-3-methyl-7-phenyl-	MeOH	222(4.30),243(4.10), 274s(3.86),352(4.23)	5-0852-83
Pyrrolo[3,2-b]pyridine-3-carboxamide, 1-(4-methoxyphenyl)-6-nitro-	EtOH	248(4.31),270(4.20), 319(3.83),339(3.76)	103-1163-83
$C_{15}H_{12}N_4O_5$			
2,4(1H,3H)-Pteridinedione, 6-acetoxy-3-methyl-7-phenyl-, 8-oxide	MeOH	251(4.32),294s(3.91), 316s(3.78),357(3.83)	5-0852-83
	MeCN	257(4.35),314(3.84), 352(3.87)	5-0852-83
Pyrazolo[1,5-d][1,2,4]triazine-2,3-dicarboxylic acid, 6,7-dihydro-7-oxo-4-phenyl-, dimethyl ester	EtOH	232(4.08),282(4.04)	44-1069-83
$C_{15}H_{12}N_6O_2$			
Pyrimido[5,4-e]-1,2,4-triazolo[4,3-b]-pyridazine-7,9(6H,8H)-dione, 6,8-dimethyl-3-phenyl-	EtOH	228(4.13),268(4.16), 335(3.30)	44-1628-83
$C_{15}H_{12}N_6O_3S$			
2(3H)-Benzothiazolone, [1-[(2-hydroxy-4-nitrophenyl)azo]ethylidene]hydrazone	EtOH	500(4.51)	65-1033-83
	EtOH-base	595(3.98)	65-1033-83
	acetone	500(4.44),630(3.91)	65-1033-83
$C_{15}H_{12}O$			
9-Anthracenemethanol	C_6H_{12}	365.5(3.99)	151-0131-83C
	benzene	367.5(3.96)	151-0131-83C
	MeOH	364.0(3.96)	151-0131-83C
	ether	364.0(3.96)	151-0131-83C
in cetyltrimethylammonium bromide	n.s.g.	367.0(3.93)	151-0131-83C
in sodium dodecyl sulfate	n.s.g.	366.5(3.93)	151-0131-83C
9(8aH)-Anthracenone, 8a-methyl-	EtOH	222(4.48),254(4.19), 260(4.16),275(4.04), 285(3.96),315s(3.74), 329(3.86),342(3.86), 407(3.62),428s(3.48)	35-3234-83
$C_{15}H_{12}OS$			
6H-Dibenzo[b,d]thiopyran, 6-ethenyl-, 5-oxide	MeOH	256(3.84),271(3.85), 296(3.66)	12-0795-83
6H-Dibenzo[b,d]thiopyran, 6-ethylidene-, 5-oxide, (E)-	MeOH	221(4.45),239(4.40), 307(3.69)	12-0795-83
(Z)-	MeOH	225(4.47),242(4.37), 310(3.76)	12-0795-83
$C_{15}H_{12}O_2$			
Benzeneacetic acid, α-(phenylmethylene)-, cis	EtOH	223(4.51),280(4.29)	162-1050-83
trans	EtOH	222(4.16),289(4.35)	162-1050-83
$C_{15}H_{12}O_2S$			
Benzenecarbothioic acid, S-(2-oxo-2-phenylethyl) ester	MeOH	243(4.07),335(2.66)	44-0835-83
$C_{15}H_{12}O_3$			
Naphtho[1,2-b]furan-4,5-dione, 2,3-dihydro-3,3-dimethyl-2-methylene-	MeOH	260(4.33),267s(4.28), 311(3.74),440(3.37)	102-0737-83

Compound	Solvent	λ max (log ε)	Ref.
2-Propenal, 3-(4-acetoxy-1-naphthalenyl)-	CCl_4	334(4.19)	104-1864-83
	DMSO	349(4.06)	104-1864-83
$C_{15}H_{12}O_4$			
Benzaldehyde, 2-(3-formylphenoxy)-3-methoxy-	MeOH	251(4.08),312(3.56)	83-0624-83
Benzaldehyde, 2-(4-formylphenoxy)-3-methoxy-	MeOH	269(4.19)	83-0624-83
Benzaldehyde, 3-(3-formylphenoxy)-4-methoxy-	MeOH	226(4.46),271(4.15)	83-0624-83
4H-1,3-Benzodioxin-4-one, 2-methyl-2-phenoxy-	pH 7.4	300(3.26)	1-0351-83
4H-1-Benzopyran-4-one, 2,3-dihydro-	MeOH	289(4.59)	94-1544-83
5,7-dihydroxy-2-phenyl- (pinocembrin)	MeOH	246(3.71),323(4.46)	102-0573-83
	MeOH-HCl	229(4.17),290(4.46)	102-0573-83
	MeOH-NaOH	246(3.87),324(4.41)	102-0573-83
7H-Furo[3,2-g][1]benzopyran-7-one, 9-acetyl-2,5-dimethyl-	EtOH	238(4.39),240s(4.39), 260(4.18),287s(4.03), 346(3.88)	44-2709-83
1H-Phenalen-1-one, 3-hydroxy-4,9-dimethoxy-	CH_2Cl_2	227s(4.18),237(4.24), 248(4.33),363(4.25), 398s(3.80),409s(3.72), 435s(3.36)	44-2115-83
$C_{15}H_{12}O_4S_2$			
Dibenzo[b,f]thiepin, 2-(methylsulfonyl)-, 5,5-dioxide	MeOH	229.5(4.54),267s(3.82), 302(3.96)	73-0906-83
$C_{15}H_{12}O_5$			
2H,8H-Benzo[1,2-b:3,4-b']dipyran-3-carboxylic acid, 2-oxo-, ethyl ester	EtOH	229(4.410),293s(--), 301(4.108),355(4.286)	111-0015-83
2H,8H-Benzo[1,2-b:5,4-b']dipyran-3-carboxylic acid, 2-oxo-, ethyl ester	EtOH	229(4.436),282(4.286), 314s(--),379(4.181)	111-0015-83
2H-1-Benzopyran-2-one, 8-(2,5-dihydro-4-methyl-2-oxo-3-furanyl)-7-methoxy- (microminutin)	EtOH	268(3.59),321(4.23)	44-0268-83
$C_{15}H_{12}O_6$			
4H-1-Benzopyran-4-one, 2-(3,4-dihydroxyphenyl)-2,3-dihydro-5,7-dihydroxy- (eriodictyol)	EtOH	290(2.54),326(2.16)	162-0530-83
9H-Xanthen-9-one, 2,5-dihydroxy-1,6-dimethoxy- (same spectrum with $AlCl_3$ or NaOAc-boric acid added)	EtOH	250(3.2),315(2.95), 355(2.903)	102-0233-83
$C_{15}H_{12}O_7$			
1H-Phenalen-1-one, 3,4,7,8,9-pentahydroxy-2-methoxy-6-methyl-	EtOH	314(4.08),399(3.97)	18-3661-83
$C_{15}H_{12}S$			
5H-Dibenz[a,d]cycloheptene-5-thione, 10,11-dihydro-	3-Mepentane	316(4.15),608(2.43)	23-0894-83
$C_{15}H_{13}BrN_2O_2$			
1-Propanone, 1-[4-bromo-2-(phenyl-ONN-azoxy)phenyl]-	EtOH	254(4.33),316(3.32)	103-0559-83
$C_{15}H_{13}BrN_2O_4S$			
Acetamide, 2-bromo-N-(3-methoxy-	EtOH	260(4.02),280(3.90)	44-3146-83

Compound	Solvent	λ_{max}(log ϵ)	Ref.
4-(phenylsulfonylimino)-3,5-cyclo-hexadien-1-ylidene]- (cont.)			44-3146-83
$C_{15}H_{13}BrO_3S$			
Ethanone, 1-(4-bromophenyl)-2-[(4-methylphenyl)sulfonyl]-	EtOH	221(4.25),267(4.27)	42-0137-83
	neutral	268(4.29)	42-0137-83
	anion	306(4.12)	42-0137-83
Ethanone, 2-[(4-bromophenyl)sulfonyl]-1-(4-methylphenyl)-	EtOH	215(4.26),240(4.36), 263(4.32)	42-0137-83
	neutral	260(4.27)	42-0137-83
	anion	310(4.18)	42-0137-83
$C_{15}H_{13}BrO_4S$			
Ethanone, 2-[(4-bromophenyl)sulfonyl]-1-(4-methoxyphenyl)-	EtOH	233(4.35),292(4.30)	42-0137-83
	neutral	294(4.24)	42-0137-83
	anion	308(4.20)	42-0137-83
$C_{15}H_{13}ClN_2O_2$			
1,2-Diazabicyclo[5.2.0]nona-3,5-dien-9-one, 2-benzoyl-8-chloro-8-methyl-, cis	CH_2Cl_2	289(3.97)	5-1361-83
Pyrano[2,3-c]pyrazol-4(1H)-one, 5-chloro-3-ethyl-6-methyl-1-phenyl-	EtOH	246(4.40)	44-4078-83
$C_{15}H_{13}ClN_2O_3$			
Carbamic acid, methyl[(phenylamino)-carbonyl]-, 3-chlorophenyl ester	EtOH	242.5(4.23)	22-0073-83
Carbamic acid, methyl[(phenylamino)-carbonyl]-, 4-chlorophenyl ester	EtOH	241(4.30)	22-0073-83
$C_{15}H_{13}ClO_2$			
3(2H)-Benzofuranone, 2-[(4-chlorophen-yl)methylene]-4,5,6,7-tetrahydro-	EtOH	235(4.03),316(4.32)	4-0543-83
$C_{15}H_{13}ClO_4$			
2H-1-Benzopyran-2-one, 8-acetyl-6-(2-chloro 2 propenyl) 7 hydroxy 4 methyl-	EtOH	226s(4.11),245s(3.95), 272(4.04),276s(4.03), 311(4.00),348(4.01)	44-2709-83
2H-1-Benzopyran-2-one, 8-acetyl-7-[(2-chloro-2-propenyl)oxy]-4-methyl-	EtOH	297(4.04),318(4.18)	44-2709-83
$C_{15}H_{13}ClO_5$			
Benzocyclooctene-5,6-dione, 9-chloro-2,3,4-trimethoxy-	$CHCl_3$	300s(2.81)	39-1983-83C
$C_{15}H_{13}Cl_3N_4O_4$			
Benzo[g]pteridine-2,4-dione, 2,3,4,10-tetrahydro-1,3,10-trimethyl-, salt with trichloromethanesulfonic acid	M HCl	262(4.44),372(4.17)	78-3359-83
$C_{15}H_{13}NO$			
8H-Indeno[6,5,4-cd]indol-8-one, 4,6,6a,7-tetrahydro-9-methyl-	MeOH	212(4.33),262(4.23), 359(3.99)	39-1545-83C
$C_{15}H_{13}NOS$			
Thieno[3,2-c]pyridine, 4-[(4-methoxy-phenyl)methyl]-	MeOH	223s(4.64),270(3.96), 286s(3.81),295(3.51)	83-0244-83
$C_{15}H_{13}NO_2$			
9H-Carbazol-3-ol, 6-methyl-, acetate	EtOH	230(4.62),239(4.14), 266(4.26),299(4.20), 332(3.48)	102-1064-83

Compound	Solvent	λ_{max}(log ε)	Ref
2H-Indol-2-one, 1,3-dihydro-3-(4-hydroxy-2-methylphenyl)-, (±)-	MeOH	252(4.12),282s(3.39), 288(3.32)	5-1744-83
11H-Pyrrolo[2,1-b][3]benzazepin-11-one, 2-acetyl-5,6-dihydro-	EtOH	212(4.31),247(4.30), 321(4.16)	44-3220-83
Spiro[2,5-cyclohexadiene-1,4'(1'H)-quinoline]-2',4a(3'H)-dione, 2-methyl-, (±)-	MeOH	212(4.89),244(4.69)	5-1744-83
$C_{15}H_{13}NO_2S$			
Azuleno[2,1-d]thiazole-4-carboxylic acid, 2-methyl-, ethyl ester	hexane	212(4.32),254(3.97), 260(3.97),300s(4.45), 311(4.73),324(4.77), 363s(3.73),382(3.89), 402(3.64),585(3.54)	142-1263-83
1H-Indole, 2-methyl-1-(phenylsulfonyl)-	EtOH	212(4.33),252(4.10), 284s(3.25)	39-2417-83C
1H-Indole, 1-(methylsulfonyl)-2-phenyl-	EtOH	216(4.35),272(4.25)	39-2417-83C
2-Propenoic acid, 3-(1,2-dihydro-1-phenyl-2-thioxo-3-pyridinyl)-, methyl ester	EtOH	234(4.18),324(4.00), 400(3.61)	4-1651-83
2-Thiopheneacetonitrile, α-[(2,5-dimethoxyphenyl)methylene]-, (E)-	EtOH	245(3.85),330(4.11), 382(4.16)	94-2023-83
(Z)-	EtOH	244(4.03),305(3.84), 360(3.83)	94-2023-83
$C_{15}H_{13}NO_3$			
4H-Furo[3,2-b]pyrrole-5-carboxylic acid, 4-methyl-2-(4-methylphenyl)-	MeOH	207(4.14),335(4.68)	73-0772-83
1H-Indeno[1,2-b]pyridine-2-carboxylic acid, 1-ethyl-4,5-dihydro-4-oxo-	MeOH	238(4.27),303(4.32), 313(4.37)	83-0264-83
$C_{15}H_{13}NO_5$			
9(10H)-Acridinone, 1,3,5-trihydroxy-4-methoxy-10-methyl-	MeOH	209(4.15),223(4.16), 265(4.59),285s(4.25), 319(4.08),334s(4.00), 412(3.55)	94-0901-83
$C_{15}H_{13}NO_5S$			
Ethanone, 2-[(4-methylphenyl)sulfonyl]-1-(4-nitrophenyl)-	EtOH	227(4.29),284(4.30)	42-0137-83
	neutral	264(4.20)	42-0137-83
	anion	360(3.88)	42-0137-83
$C_{15}H_{13}NO_6$			
2,4-Pentadienoic acid, 3-(2-furanyl)-5-(5-nitro-2-furanyl)-, ethyl ester, (E)-	MeOH	318(4.21),365(4.15)	73-3559-83
(Z)-	MeOH	312(4.35),356(4.18)	73-3559-83
$C_{15}H_{13}NO_6S$			
Ethanone, 1-(4-methoxyphenyl)-2-[(4-nitrophenyl)sulfonyl]-	EtOH	226(4.15),252(4.38), 288(4.26)	42-0137-83
	neutral	265(4.27)	42-0137-83
	anion	365(3.76)	42-0137-83
$C_{15}H_{13}NS_3$			
Ethane(dithioic) acid, (diphenylamino)thioxo-, methyl ester	CH_2Cl_2	262(4.10),289(4.10), 343s(3.85),434s(3.01)	97-0247-83
$C_{15}H_{13}N_3$			
Pyrimidine, dihydro-6-phenyl-4-(2-pyridinyl)-	EtOH	239(4.21),286(3.49), 330s(2.94)	54-0364-83

Compound	Solvent	$\lambda_{max}(\log \epsilon)$	Ref.
$C_{15}H_{13}N_3O$			
Benzenamine, 2-[5-(4-methylphenyl)-1,3,4-oxadiazol-2-yl]-	MeOH	258(4.14),282(4.21), 295(4.15),353(4.12)	118-0842-83
5H-1,3,4-Benzotriazepin-5-one, 1,4-dihydro-2-(4-methylphenyl)-	MeOH	245(4.19),252(4.17), 275(4.11)	118-0842-83
Phenol, 2-(2H-benzotriazol-2-yl)-4-(1-methylethenyl)-	$CHCl_3$	241(4.26),270(4.19), 303(4.23),346(4.12)	49-0937-83
1,2,4-Triazin-3(2H)-one, 4,5-dihydro-2,6-diphenyl-	MeOH	244(4.34),302(4.18)	104-1533-83
$C_{15}H_{13}N_3OS$			
4H-Pyrido[3,4-e]-1,3-thiazin-4-one, 5,7-dimethyl-2-(phenylamino)-	MeOH	264(4.13)	73-3426-83
$C_{15}H_{13}N_3OSe$			
4H-Pyrido[3,4-e]-1,3-selenazin-4-one, 5,7-dimethyl-2-(phenylamino)-	MeOH	213(3.03),269(4.96)	73-3567-83
$C_{15}H_{13}N_3O_2$			
Acetamide, N-[4-(1H-benzimidazol-2-yl)-3-hydroxyphenyl]-	EtOH	327(4.56)	4-1517-83
Acetamide, N-(5-methyl-7-phenyl-2H-isoxazolo[2,3-a]pyrimidin-2-ylidene)-	MeCN	257(4.13),272(4.15), 328(4.47)	44-0575-83
perchlorate	MeCN	255(4.24),326(4.64)	44-0575-83
Benzenamine, 2-[5-(4-methoxyphenyl)-1,3,4-oxadiazol-2-yl]-	MeOH	258(4.05),285(4.08), 352(4.04)	118-0842-83
5H-1,3,4-Benzotriazepin-5-one, 1,4-dihydro-2-(4-methoxyphenyl)-	MeOH	244(4.15),255(4.17), 290(4.18)	118-0842-83
Diazene, 1-[2-(4-methylphenyl)ethenyl]-2-(4-nitrophenyl)-	EtOH	403(4.64)	65-1902-83
conjugate acid	$EtOH-H_2SO_4$	465(4.64)	65-1902-83
Ethanone, 1-[3-(2H-benzotriazol-2-yl)-2-hydroxy-5-methylphenyl]-	CH_2Cl_2	252(4.10),295(4.08), 346(4.09)	47-1263-83
1H-Indole-2-methanamine, 5-nitro-	MeOH	270(4.4),330(3.92)	103-1302-83
Phenol, 2-(2H-benzotriazol-2-yl)-4-methyl-, acetate	CH_2Cl_2	300(4.28)	47-1263-83
$C_{15}H_{13}N_3O_2S$			
2-Thiazolecarboxylic acid, 4-(3-methyl-2-quinoxalinyl)-, ethyl ester	EtOH	243(4.14),264(3.85), 300(3.67),335(3.69)	106-0829-83
$C_{15}H_{13}N_3O_4$			
Benzenecarboximidamide, 2-methyl-N'-[(4-nitrobenzoyl)oxy]-, (Z)-	EtOH	258(4.08),296s(--)	32-0845-83
Benzenecarboximidamide, 3-methyl-N'-[(4-nitrobenzoyl)oxy]-	EtOH	248(4.28)	32-0845-83
Benzenecarboximidamide, 4-methyl-N'-[(4-nitrobenzoyl)oxy]-	EtOH	253(4.29)	32-0845-83
Benzoic acid, 4-[[2-amino-1,4,5,6-tetrahydro-4,6-dioxo-5-(2-propynyl)-5-pyrimidinyl]methyl]-	50%MeOH-HCl	230s(4.27)	73-0299-83
	+ NaOH	225(4.50),265(4.01)	73-0299-83
4H-Furo[3,2-b]pyrrole-5-carboxylic acid, 2-phenyl-, 2-(methoxycarbonyl)hydrazide	MeOH	234(2.81),339(3.69)	73-1878-83
$C_{15}H_{13}N_3O_5$			
Carbamic acid, methyl[(phenylamino)carbonyl]-, 4-nitrophenyl ester	EtOH	243(4.23)	22-0073-83
$C_{15}H_{13}N_5O$			
Furo[2',3':4,5]pyrrolo[1,2-d]-1,2,4-	MeOH	349(3.56)	73-1878-83

Compound	Solvent	$\lambda_{max}(\log \epsilon)$	Ref.
triazin-8(7H)-one, 2-(4-methylphenyl)-, hydrazone (cont.)			73-1878-83
Furo[2',3':4,5]pyrrolo[1,2-d]-1,2,4-triazin-8(7H)-one, 5-methyl-2-phenyl-, hydrazone	MeOH	347(3.43)	73-1878-83
Imidazo[2,1-f]purin-2(8H)-one, 1,8-dimethyl-7-phenyl-	MeOH	262(4.23),328(4.24)	105-0029-83
	MeOH-acid	253(4.26),312(4.13)	105-0029-83
	MeOH-base	262(4.21),328(4.22)	105-0029-83
$C_{15}H_{13}N_5OS$			
1H-Imidazo[2,1-f]purin-2(3H)-one, 4,8-dihydro-1,8-dimethyl-7-phenyl-4-thioxo-	MeOH	212(4.33),251(4.26),348(4.43)	105-0029-83
	MeOH-acid	212(4.36),251(4.29)	105-0029-83
	MeOH-base	280(4.39),332(4.35)	105-0029-83
$C_{15}H_{13}N_5O_2$			
1,2-Diazabicyclo[5.2.0]nona-3,5-dien-9-one, 8-azido-2-benzoyl-5-methyl-, cis	MeOH	284(3.86)	5-1361-83
$C_{15}H_{14}BrNOS$			
Ethanone, 1-[1-(4-bromophenyl)-1,2-dihydro-4,6-dimethyl-2-thioxo-3-pyridinyl]-	EtOH	255(3.85),290(3.85),379(3.83)	118-1025-83
$C_{15}H_{14}Br_2O_2$			
Benzene, 4-(bromomethyl)-2-[4-(bromomethyl)phenoxy]-1-methoxy-	MeOH	245(4.28)	83-0445-83
$C_{15}H_{14}ClNOS$			
Ethanone, 1-[1-(4-chlorophenyl)-1,2-dihydro-4,6-dimethyl-2-thioxo-3-pyridinyl]-	EtOH	255(3.81),290(3.83),377(3.82)	118-1025-83
$C_{15}H_{14}ClNOS_2$			
Ethanone, 1-[1-(4-chlorophenyl)-1,2-dihydro-6-methyl-4-(methylthio)-2-thioxo-3-pyridinyl]-	EtOH	285(4.50),385(3.92)	118-1025-83
$C_{15}H_{14}ClNO_2S$			
Ethanone, 1-[1-(4-chlorophenyl)-1,2-dihydro-4-methoxy-6-methyl-2-thioxo-3-pyridinyl]-	EtOH	236(4.41),292(3.92),357(3.83)	118-1025-83
$C_{15}H_{14}ClNO_3$			
2H-Furo[2,3-h]-1-benzopyran-2-one, 3-chloro-4-(diethylamino)-	EtOH	218s(4.26),243s(4.35),247.5(4.36),303s(4.01),326(4.04)	111-0113-83
2H-Furo[2,3-h]-1-benzopyran-2-one, 3-chloro-5,6-dihydro-4-pyrrolidino-	EtOH	235s(3.94),267s(3.79),354(4.12)	111-0113-83
2H-Pyrano[3,2-d][1]benzoxepin-2-one, 3-chloro-4-(dimethylamino)-5,6-dihydro-	EtOH	213.5(4.15),233s(3.81),264(4.08),335(4.06)	4-0539-83
$C_{15}H_{14}ClNO_4$			
2H-Furo[2,3-h]-1-benzopyran-2-one, 3-chloro-5,6-dihydro-4-morpholino-	EtOH	236(3.85),266(3.62),359(4.04)	111-0113-83
$C_{15}H_{14}ClN_3O_4S$			
5-Thia-1-azabicyclo[4.2.0]oct-2-ene-2-carboxylic acid, 7-[(aminophenyl-	pH 7	265(3.83)	162-0267-83

Compound	Solvent	$\lambda_{max}(\log \epsilon)$	Ref.
acetyl)amino]-3-chloro-8-oxo- (cefachlor) (cont.)			162-0267-83
$C_{15}H_{14}Cl_2O$			
Spiro[7H-cyclobut[a]indene-7,1'-cyclopentan]-1(2H)-one, 2,2-dichloro-2a,7a-dihydro-	MeOH	238(3.09),245s(3.08), 250s(3.04),255s(2.97), 257s(2.92),261s(2.94), 269s(3.07),276(3.12), 279s(3.10),310s(1.98), 322(1.99)	39-0285-83C
$C_{15}H_{14}CsN$			
Cesium, [2-(dimethylamino)-9H-fluoren-9-yl]-	$(MeOCH_2)_2$	384(4.09)	104-1592-83
$C_{15}H_{14}FNOS$			
10H-Phenothiazine, 3-fluoro-7-(1-methylethoxy)-	benzene	204(4.32),256(4.47), 306(3.55)	4-0803-83
$C_{15}H_{14}FNOS_2$			
Ethanone, 1-[1-(4-fluorophenyl)-1,2-dihydro-6-methyl-4-(methylthio)-2-thioxo-3-pyridinyl]-	EtOH	284(4.49),384(3.95)	118-1025-83
$C_{15}H_{14}FNO_2S$			
Ethanone, 1-[1-(4-fluorophenyl)-1,2-dihydro-4-methoxy-6-methyl-2-thioxo-3-pyridinyl]-	EtOH	235(4.32),295(3.86), 357(3.80)	118-1025-83
10H-Phenothiazine, 3-fluoro-7-(1-methylethoxy)-, 5-oxide	benzene	216(4.36),234(4.45), 273(4.09),301(3.93)	4-0803-83
$C_{15}H_{14}FeO_4$			
Iron, tricarbonyl[η^4-5,6,7,8-tetrakis(methylene)bicyclo[2.2.2]octan-2-ol]-	EtOH	285s(3.45)	33-1134-83
$C_{15}H_{14}LiN$			
Lithium, [2-(dimethylamino)-9H-fluoren-9-yl]-	$(MeOCH_2)_2$	374(4.15)	104-1592-83
$C_{15}H_{14}NS$			
Thieno[3,2-c]pyridinium, 5-methyl-4-(phenylmethyl)-, iodide	MeOH	237(4.54),286(3.56), 310(3.60)	83-0244-83
$C_{15}H_{14}N_2$			
1H-Pyrazole, 4,5-dihydro-1,3-diphenyl-	C_6H_{12}	239(4.04),357(4.29)	39-1679-83B
$C_{15}H_{14}N_2O$			
Benzenamine, 4-(2-benzoxazolyl)-N,N-dimethyl-	C_6H_{12}	332(4.66)	41-0595-83
Indolo[4,3-fg]isoquinolin-9(4H)-one, 6,6a,7,8-tetrahydro-8-methyl-	MeOH	211(4.25),256(4.18), 360(3.90)	39-1545-83C
5(6H)-Quinazolinone, 7,8-dihydro-2-methyl-7-phenyl-	EtOH	238(4.07),255s(3.88), 294(3.61)	4-0649-83
$C_{15}H_{14}N_2OS$			
Indeno[1,2-c]isothiazol-4-one, 3-piperidino-	EtOH	227(4.14),262(4.74), 305(4.18),380(3.82)	95-1243-83
$C_{15}H_{14}N_2O_2$			
Benzenemethanamine, N-methyl-α-(nitro-	EtOH	366(4.18)	22-0339-83

Compound	Solvent	λ_{max}(log ε)	Ref.
phenylmethylene)-, (Z)- (cont.)			22-0339-83
1H-Benzimidazole, 5,6-dimethoxy-2-phenyl-	EtOH	206(4.51),326(4.41)	94-2910-83
1H-Benz[de]isoquinoline-1,3(2H)-dione, 6-(dimethylamino)-2-methyl-	MeOH	420(4.04)	104-2273-83
Benzo[a]pyrrolo[3,4-f]quinolizine-1,12-dione, 2,3,5,6,10b,11-hexahydro-	MeOH	242(4.26),304(4.26)	135-0303-83
1,2-Diazabicyclo[5.2.0]nona-3,5-dien-9-one, 2-benzoyl-5-methyl-	MeOH	224(3.98),285(3.94)	5-1393-83
1,2-Diazabicyclo[5.2.0]nona-3,5-dien-9-one, 2-benzoyl-7-methyl-	MeOH	285(3.94)	5-1393-83
1-Propanone, 1-[2-(phenyl-ONN-azoxy)-phenyl]-	EtOH	249(4.19),323(3.34)	103-0559-83
	EtOH	253(4.1),335(5.02)	103-0593-83
Pyrano[2,3-c]pyrazol-4(1H)-one, 3-ethyl-6-methyl-1-phenyl-	EtOH	244(4.41)	44-4078-83
Pyrazolo[1,5-a]quinoline-3-carboxylic acid, 4,5-dihydro-5-(1-propenyl)-, (E:Z 1:1)	EtOH	267(4.17)	88-4679-83
9H-Pyrido[3,4-b]indole, 1-ethenyl-4,9-dimethoxy-	EtOH	234(4.09),256(4.16), 272(3.89),360(3.46)	94-3198-83
	EtOH	234(4.41),244(4.36), 270(4.03),350(3.60)	94-3198-83
1H-Pyrrolo[2,3-d]carbazole-2,5(3H,4H)-dione, 3a,7-dihydro-3a-methyl-, (3aR*,11bS*)-(±)-	MeOH	237(4.11),297(3.91), 342(4.17)	5-1744-83
$C_{15}H_{14}N_2O_2Se$			
2-Propenamide, 3-[1,2-dihydro-1-(4-methoxyphenyl)-2-seleno-3-pyridinyl]-	EtOH	219(4.48),242(4.32), 351(4.08),429(3.62)	4-1651-83
$C_{15}H_{14}N_2O_3$			
Benzo[b][1,8]naphthyridin-5(10H)-one, 6,8-dimethoxy-10-methyl-	MeOH	201(4.33),227(4.42), 236(4.45),269(4.81), 292s(3.92),318(4.03), 370(4.02)	39-0219-83C
Carbamic acid, methyl[(phenylamino)-carbonyl]-, phenyl ester	EtOH	243(4.23)	22-0073-83
1H-1,2-Diazepine-4-carboxylic acid, 1-benzoyl-, ethyl ester	MeOH	230(4.15),270s(3.93), 393(3.00)	5-1374-83
1H-1,2-Diazepine-6-carboxylic acid, 1-benzoyl-, ethyl ester	MeOH	231(4.12),292(3.78)	5-1374-83
Phenol, 4-(5,6-dimethoxy-1H-benzimid-azol-2-yl)-	EtOH	208(4.53),325(4.41)	94-2910-83
Pyridinium, 4-benzoyl-1-[(ethoxycarbonyl)amino]-, hydroxide, inner salt	MeOH	330(3.69)	5-1374-83
$C_{15}H_{14}N_2O_3S$			
Acetamide, N-[3-methyl-4-[(phenyl-sulfonyl)imino]-2,5-cyclohexa-dien-1-ylidene]-	EtOH	265(4.11),300(4.20)	44-3146-83
$C_{15}H_{14}N_2O_4$			
Carbamic acid, [(phenylamino)carbonyl]-, 4-methoxyphenyl ester	EtOH	240(4.30)	22-0073-83
$C_{15}H_{14}N_2O_7$			
Uridine, 2'-deoxy-5-(3,6-dioxo-1,4-cyclohexadien-1-yl)-	pH 4.4	250(4.15),275s(--), 385(3.26)	87-1028-83

Compound	Solvent	λ_{max}(log ε)	Ref.
$C_{15}H_{14}N_4$			
1,2,4,5-Tetrazine, 1,4-dihydro-1-methyl-3,6-diphenyl-	hexane	236(4.38),290(3.66), 360(2.76)	24-2261-83
	EtOH	236(4.36),275s(3.60), 355s(2.67)	24-2261-83
	dioxan	238(4.35),291(3.59), 360s(2.73)	24-2261-83
$C_{15}H_{14}N_4O$			
Ethanone, 1-(1,5-diphenylformazanyl)-	EtOH	262(3.82),305(4.20), 445(4.35)	104-2104-83
	MeCN	263(3.83),296(4.18), 440(4.32)	104-2104-83
	CCl_4	298(4.16),442(4.31)	104-2104-83
	isooctane	266(--),292(--), 440(--)	104-2104-83
3-Pyridinecarbonitrile, 1-[3-(3-cyano-1(4H)-pyridinyl)propyl]-1,6-dihydro-6-oxo-	CH_2Cl_2	250(4.1),320(3.7), 348(3.7)	64-0878-83B
3-Pyridinecarbonitrile, 1-[3-(5-cyano-1(2H)-pyridinyl)propyl]-1,6-dihydro-6-oxo-	CH_2Cl_2	247(4.2),319(3.7), 334(3.7),352s(3.7)	64-0878-83B
$C_{15}H_{14}N_4OS$			
Pyrido[2,3-d]pyrimidin-4(1H)-one, 1-[4-(dimethylamino)phenyl]-2,3-dihydro-2-thioxo-	MeOH	280(3.29)	73-3315-83
4H-Pyrido[3,2-e]-1,3-thiazin-4-one, 2-[[4-(dimethylamino)phenyl]amino]-	MeOH	224(3.16)	73-3315-83
$C_{15}H_{14}N_4OSe$			
4H-Pyrido[3,2-e]-1,3-selenazin-4-one, 2-[[4-(dimethylamino)phenyl]amino]-	MeOH	213(3.19)	73-3567-83
$C_{15}H_{14}N_4O_5$			
1H-Pyrrole-3-carboxamide, 1-(4-methoxyphenyl)-4-[(2-nitro-3-oxo-1-propenyl)amino]-	EtOH	240(4.10),302(4.16), 388(4.17)	103-1163-83
$C_{15}H_{14}N_4S$			
Propanedinitrile, [2-[(dimethylamino)-methylene]-4-methyl-2H-1,4-benzothiazin-3(4H)-ylidene]-	MeOH	224(4.27),276(4.29), 348(4.21)	87-0845-83
$C_{15}H_{14}N_6S_2$			
7H-Pyrrolo[2,3-d]pyrimidine, 7,7'-methylenebis[4-(methylthio)-	MeOH	249(4.05),294(4.48)	5-1576-83
$C_{15}H_{14}O$			
9(2H)-Anthracenone, 1,9a-dihydro-9a-methyl-	EtOH	252(4.58),260(4.59), 290(4.08),297(4.09), 307s(3.99),363s(3.53), 373(3.57),383s(3.50)	35-3234-83
9(4H)-Anthracenone, 1,9a-dihydro-9a-methyl-	EtOH	237(4.50),260s(3.84), 272s(3.67),282s(3.47), 333(3.40)	35-3234-83
Bicyclo[3.3.1]nona-3,6-dien-2-one, 6-phenyl-	EtOH	237(4.25)	2-0619-83
Ethanone, 1-[4-(phenylmethyl)phenyl]-	hexane	249(4.19)	99-0034-83
	dioxan	249(4.16)	124-1211-83
2-Propanone, 1,3-diphenyl-	C_6H_{12}	294(2.39)	46-1960-83

Compound	Solvent	λ_{max}(log ϵ)	Ref.
$C_{15}H_{14}O_2$			
Benzenemethanol, α-(methoxyphenyl-methylene)-, (Z)-	EtOH at 6°	301(3.69)	89-0405-83 +89-0551-83S
Methanone, (4-hydroxy-3,5-dimethyl-phenyl)phenyl-	EtOH	240(4.07),304(4.08)	24-0970-83
Phenol, 3-methoxy-5-(2-phenylethen-yl)-, (E)-	MeOH	211(4.39),228s(4.27), 234s(4.25),300(4.46), 308(4.46)	100-0852-83
Spiro[5H-benzocycloheptene-5,1'-cyclo-pentane]-6,7-dione	MeOH	225(3.78),243(3.94), 250s(3.93),255s(3.79), 257s(3.66),260s(3.57), 265s(3.33),277s(3.11), 316(3.11)	39-0285-83C
$C_{15}H_{14}O_2S$			
Methanethione, bis(4-methoxyphenyl)-	benzene	354(4.39),591(2.45)	44-0214-83
$C_{15}H_{14}O_3$			
2H-1-Benzopyran-7-ol, 3,4-dihydro-3-(3-hydroxyphenyl)-	MeOH	222s(4.27),276s(3.69), 282(3.75),290s(3.54)	100-0852-83
2H-1-Benzopyran-7-ol, 3,4-dihydro-3-(4-hydroxyphenyl)-, (S)-(-)-equol	pH 1	221(4.25),280(3.66)	100-0852-83
	pH 13	240(4.35),294(3.83)	100-0852-83
	MeOH	203(4.97),225s(4.45), 280(3.71),283(3.71), 289s(3.57)	100-0852-83
Furan, 2,2'-(2-furanylmethylene)bis-[5-methyl-	EtOH	226(4.38)	103-0478-83
Lapachol	n.s.g.	251.5(4.38),278(4.28), 331(3.43)	162-0772-83
Naphtho[2,3-b]furan-4,9-dione, 2,3-dihydro-2,3,3-trimethyl- (α-dunni-one)	MeOH	248s(4.13),253(4.17), 288(3.91),335(3.27), 350(2.97)	102-0737-83
2,5-Phenanthrenediol, 9,10-dihydro-7-methoxy-	MeOH	274s(4.21),281(4.28), 295s(4.14)	102-1011-83
$C_{15}H_{14}O_3S$			
Ethanone, 1-(4-methylphenyl)-2-(phen-ylsulfonyl)-	EtOH	213(4.33),266(4.27)	42-0137-83
	neutral	267(4.29)	42-0137-83
	anion	303(4.22)	42-0137-83
Ethanone, 2-[(4-methylphenyl)sulfon-yl]-1-phenyl-	EtOH	220(4.12),267(4.15)	42-0137-83
	neutral	252(4.15)	42-0137-83
	anion	300(4.13)	42-0137-83
$C_{15}H_{14}O_3S_2$			
1,4-Dithiaspiro[4.4]non-6-ene-6-carb-oxaldehyde, 8-(benzoyloxy)-, (R)-	EtOH	232(4.33),274(3.34), 281(3.32)	39-2441-83C
$C_{15}H_{14}O_4$			
7H-Benzocyclohepten-7-one, 5-(2-acet-oxyethyl)-6-hydroxy-	EtOH	241(4.45),244(4.45), 268(4.35),279(4.66), 315s(3.70),370(3.61), 390s(3.53)	39-0285-83C
Benzoic acid, 2-hydroxy-6-[2-(4-hy-droxyphenyl)ethyl]-	pH 12	238(4.25),293(3.72)	35-4480-83
2H-1-Benzopyran-2-one, 7-methoxy-8-(3-methyl-2-oxo-3-butenyl)-	EtOH	246(3.69),256(3.67), 322(4.17)	102-0792-83
Dibenzofuran, 1,2,3-trimethoxy-	EtOH	226(4.56),258(4.15), 290(4.32),298(4.28)	39-2267-83C
Dibenzofuran, 1,2,4-trimethoxy-	EtOH	220s(4.44),231(4.50), 259(4.13),278s(4.03), 283(4.16),317(3.63)	39-2267-83C

Compound	Solvent	λ_{max}(log ε)	Ref.
Dibenzofuran, 1,3,4-trimethoxy-	EtOH	220(4.59),234(4.58), 265(4.23),283(4.30), 299(4.05),310(4.21)	39-2267-83C
Dibenzofuran, 2,3,4-trimethoxy-	EtOH	233(4.45),256(4.04), 290(4.17),297s(4.09), 316(3.67)	39-2267-83C
6H-Dibenzo[b,d]pyran-6-one, 3-acetoxy-7,8,9,10-tetrahydro-	MeOH	272(4.02),282(4.02), 310(4.05)	78-1265-83
6H-Dibenzo[b,d]pyran-6-one, 4-acetyl-7,8,9,10-tetrahydro-3-hydroxy-	MeOH	208(4.41),271(4.00), 275(4.00),305(4.00), 336(3.87)	78-1265-83
Ethanone, 1-(2,4-dihydroxyphenyl)-2-(3-methoxyphenyl)-	pH 13	253s(3.92),280s(3.65), 334(4.27)	100-0852-83
	MeOH	213(4.42),238s(3.90), 280(4.20),317(3.97)	100-0852-83
Ethanone, 1-(2,4-dihydroxyphenyl)-2-(4-methoxyphenyl)-	MeOH	213(4.39),217s(4.38), 229s(4.21),239s(3.90), 278(4.21),318(3.97)	100-0852-83
Indeno[1,2-b]pyran-4(5H)-one, 2-(ethoxyhydroxymethyl)-	MeOH	200(4.35),220(4.08), 289(4.25),299(4.27)	83-0264-83
1,4-Naphthalenedione, 2-(1,1-dimethyl-2-oxopropyl)-3-hydroxy- (streptocarpone)	MeOH	246s(4.14),252(4.18), 275(4.19),329(3.36), 460(3.09)	102-0737-83
Naphtho[1,2-b]furan-4,5-dione, 2,3-dihydro-6-hydroxy-2,3,3-trimethyl-	MeOH	242s(4.35),260(4.40), 294(3.89),412(3.84)	102-0737-83
Naphtho[1,2-b]furan-4,5-dione, 2,3-dihydro-7-hydroxy-2,3,3-trimethyl-	MeOH	270(4.52),277(4.55), 306(3.83),495(3.31)	102-0737-83
2,4-Pentadienoic acid, 3,5-di-2-furanyl-, ethyl ester, (E,E)-	MeOH	278(4.31),321(4.39)	73-3559-83
(E,Z)-	MeOH	265(4.25),312(4.50)	73-3559-83
2,4-Pentadienoic acid, 3-(2-furanyl)-5-(5-methyl-2-furanyl)-, methyl ester, (E,E)-	MeOH	278(4.33),357(4.38)	73-3559-83
(E,Z)-	MeOH	278(4.33),315(4.34)	73-3559-83
Peucedanin	MeOH	255(4.40),295(4.05), 340(3.70)	162-1033-83
Xanthoxyletin	EtOH	277(4.28),269(4.32), 322(3.99),347(4.05)	162-1446-83
Yangonin	EtOH	360(4.33)	162-1451-83
$C_{15}H_{14}O_4S$			
Ethanone, 1-(4-methoxyphenyl)-2-(phenylsulfonyl)-	EtOH	220(4.16),280(4.33)	42-0137-83
	neutral	298(4.20)	42-0137-83
	anion	305(4.15)	42-0137-83
$C_{15}H_{14}O_5$			
1,3-Benzenediol, 5-[2-(3,5-dihydroxyphenyl)ethenyl]-2-methoxy-, cis	EtOH	225(4.22),294(3.67)	102-2819-83
	EtOH-NaOEt	240(--),305(--)	102-2819-83
trans	EtOH	305(3.59),315(3.56)	102-2819-83
	EtOH-NaOEt	310(--),320(--)	102-2819-83
2H-1-Benzopyran-2-one, 8-acetyl-7-hydroxy-4-methyl-6-(2-oxopropyl)-	EtOH	213(4.37),246s(3.93), 273(4.01),277s(4.01), 311(3.97),348(3.98)	44-2709-83
2H-1-Benzopyran-2-one, 5,7-dimethoxy-8-(3-oxo-1-butenyl)-	MeOH	214(4.34),248(3.96), 320(4.45)	94-3330-83
2-Dibenzofuranol, 1,3,4-trimethoxy- (α-pyrufuran)	EtOH	224(4.44),259(4.07), 287(4.23),315s(3.46)	39-2267-83C
3-Dibenzofuranol, 1,2,4-trimethoxy- (β-pyrufuran)	EtOH	217(4.51),227(4.49), 261(4.05),289(4.24), 301s(4.05)	39-2267-83C
	EtOH	220(4.42),229(4.42),	39-2267-83C

Compound	Solvent	$\lambda_{max}(\log \epsilon)$	Ref.
(cont.)		261(4.02),289(4.18), 300s(4.05)	39-2267-83C
4-Dibenzofuranol, 1,2,3-trimethoxy-	EtOH	220(4.50),231(4.50), 262(4.11),285(4.22), 304(3.75)	39-2267-83C
1,4-Dioxaspiro[4.4]non-6-ene-6-carboxaldehyde, 8-(benzoyloxy)-, (R)-	EtOH	215s(3.79),232(4.00), 275(2.74)	39-2441-83C
$\Delta^{1,2}$-Furanoeremophilane-3,6,9-trione, 10α-hydroxy-	EtOH	304.5(3.95)	94-3544-83
2(5H)-Furanone, 4-hydroxy-3-[3-(4-methoxyphenyl)-1-oxo-2-propenyl]-5-methyl-, (E)-	MeOH	203(4.16),239(4.10), 401(4.39)	83-0115-83
Methanone, bis(2-hydroxy-4-methoxyphenyl)-	n.s.g.	284(4.12),340(4.12)	162-0157-83
Methysticin	EtOH	226(4.40),267(4.14), 306(3.93)	162-0879-83
$C_{15}H_{14}O_6$			
2H-1-Benzopyran-3,5,7-triol, 2-(3,5-dihydroxyphenyl)-3,4-dihydro-	MeOH	214(4.47),225(4.46), 281(3.76),394(3.16)	102-0565-83
Neoliacine	EtOH	205(3.60)	77-1107-83
$C_{15}H_{14}S$			
Benzene, [(2-phenyl-2-propenyl)thio]-	MeOH	243(3.97)	44-0835-83
Methanethione, bis(4-methylphenyl)-	benzene	329(4.36),600(2.35)	44-0214-83
$C_{15}H_{15}$			
Cycloheptatrienylium, (2,3-dimethylphenyl)-, tetrafluoroborate	n.s.g.	372(3.75)	137-0104-83
Cycloheptatrienylium, (2,4-dimethylphenyl)-, tetrafluoroborate	n.s.g.	389(3.88)	137-0104-83
Cycloheptatrienylium, (2,5-dimethylphenyl)-, tetrafluoroborate	n.s.g.	361(3.76)	137-0104-83
Cycloheptatrienylium, (2,6-dimethylphenyl)-, tetrafluoroborate	n.s.g.	279s(3.67)	137-0104-83
Cycloheptatrienylium, (3,4-dimethylphenyl)-, tetrafluoroborate	n.s.g.	397(4.17)	137-0104-83
Cycloheptatrienylium, (3,5-dimethylphenyl)-, tetrafluoroborate	n.s.g.	379(4.10)	137-0104-83
Cycloheptatrienylium, (4-ethylphenyl)-, tetrafluoroborate	10% HCl	277(4.01),396(4.24)	78-4011-83
$C_{15}H_{15}AsN$			
3H-1,3-Bezazarsole, 3-ethyl-2-phenyl-, tetrafluoroborate	MeOH	245(3.97),320(3.85) (anom.)	101-0257-83R
$C_{15}H_{15}BrN_2O_3S$			
Acetamide, 2-bromo-N-[3-methyl-4-[(phenylsulfonyl)amino]phenyl]-	EtOH	265(4.20)	44-3146-83
$C_{15}H_{15}BrN_2O_4S$			
Acetamide, 2-bromo-N-[3-methoxy-4-[(phenylsulfonyl)amino]phenyl]-	EtOH	265(4.13),295(4.08)	44-3146-83
$C_{15}H_{15}BrO_6$			
3-Benzofurancarboxylic acid, 6-bromo-4,5-dihydroxy-2-methyl-7-(2-oxopropyl)-, ethyl ester	EtOH	215(4.59),250(4.14)	103-0945-83
$C_{15}H_{15}ClN_2O$			
4(1H)-Cinnolinone, 3-[(4-chlorophen-	EtOH	277(4.02)	4-0543-83

Compound	Solvent	$\lambda_{max}(\log \epsilon)$	Ref.
yl)methyl]-5,6,7,8-tetrahydro- (cont.)			4-0543-83
$C_{15}H_{15}ClN_4OSe$			
3-Pyridinecarboxamide, 2-chloro-N-[[[4-(dimethylamino)phenyl]amino]-selenoxomethyl]-	MeOH	210(4.33),265(5.43)	73-3567-83
$C_{15}H_{15}ClO_2$			
Spiro[7H-cyclobut[a]indene-7,1'-cyclo-pentan]-1(2H)-one, 2-chloro-2a,7a-dihydro-7a-hydroxy-, (2α,2aα,7aα)-	EtOH	249(2.60),255(2.65), 264(2.73),268s(2.72), 274(2.72),286s(2.25), 308(2.44)	39-0285-83C
$C_{15}H_{15}ClO_6$			
3-Benzofurancarboxylic acid, 6-chloro-4,5-dihydroxy-2-methyl-7-(2-oxo-propyl)-, ethyl ester	EtOH	205(4.28),250(4.10)	103-0945-83
$C_{15}H_{15}DS$			
Benzene, [(1-methyl-2-phenylethyl-2-d)thio]-	EtOH	210(4.15),260(3.70)	22-0180-83
$C_{15}H_{15}DSe$			
Benzene, [(1-methyl-2-phenylethyl-2-d)seleno]-	EtOH	210(4.26),220(4.20), 240(3.49),270(3.40)	22-0180-83
$C_{15}H_{15}FN_2O$			
3-Pyridinecarbonitrile, 6-ethoxy-1-[(4-fluorophenyl)methyl]-1,6-dihydro-	CH_2Cl_2	240(3.7),313(3.8), 356s(3.3)	64-0873-83B
$C_{15}H_{15}F_5N_2O_7$			
Uridine, 2'-deoxy-5-(pentafluoro-ethyl)-, 3',5'-diacetate	MeOH	263(4.05)	87-0598-83
$C_{15}H_{15}NO$			
Ethanone, 1-[2'-(methylamino)[1,1'-bi-phenyl]-4-yl]-	EtOH	245(4.3)	103-0970-83
Ethanone, 1-[4-(methylphenylamino)-phenyl]-	hexane	228(3.92),241(3.86), 286(3.56),317(4.40)	99-0034-83
2-Propanone, 1-[4-(phenylamino)phen-yl]-	dioxan	228(3.92),241(3.86), 276(3.56),317(4.40)	124-1211-83
$C_{15}H_{15}NOS$			
Ethanone, 1-(1,2-dihydro-4,6-dimethyl-1-phenyl-2-thioxo-3-pyridinyl)-	EtOH	284(4.41),384(3.85)	118-1025-83
$C_{15}H_{15}NOS_2$			
Ethanone, 1-[1,2-dihydro-6-methyl-4-(methylthio)-1-phenyl-2-thioxo-3-pyridinyl]-	EtOH	283(4.41),384(3.85)	118-1025-83
$C_{15}H_{15}NO_2$			
Apo-β-erythroidine	EtOH	240(4.39),345(3.54)	162-0109-83
Benzoic acid, 2-[(2,3-dimethylphenyl)-amino]-	EtOH-HCl	220(4.5),281(3.9), 352(3.8)	145-0621-83
Benzoic acid, 2-[(2,6-dimethylphenyl)-amino]-	EtOH-HCl	220(4.5),260(3.9), 345(3.8)	145-0621-83
Benzoic acid, 2-[(3,4-dimethylphenyl)-amino]-	EtOH-HCl	286(4.2),355(3.9)	145-0621-83
Cyclopenta[b]pyrrole-4-carboxylic acid, 1,4,5,6-tetrahydro-, phenyl-	MeOH	217(3.95),241(3.76), 266(2.66),268(2.49)	23-1697-83

Compound	Solvent	λ_{max}(log ϵ)	Ref.
methyl ester (cont.)			23-1697-83
Isoapo-β-erythroidine	EtOH	253(4.23),288(4.03), 379(3.81)	162-0738-83
$C_{15}H_{15}NO_2S$			
Ethanone, 1-(1,2-dihydro-4-methoxy-6-methyl-1-phenyl-2-thioxo-3-pyridinyl)-	EtOH	235(4.35),295(3.87), 355(3.81)	118-1025-83
$C_{15}H_{15}NO_3$			
Benzamide, N-hydroxy-2-methoxy-N-(phenylmethyl)-	EtOH	207(4.42),281(3.45)	34-0433-83
1H-Indole-3-acetic acid, 2-methyl-1-(3-oxo-1-butenyl)-	MeOH	222(4.18),274(3.70), 339(3.90)	5-1744-83
$C_{15}H_{15}NO_4$			
Phenol, 2-[(2-hydroxy-5-nitrophenyl)-methyl]-4,6-dimethyl-	neutral	326(3.95)	126-2361-83
	anion	404(4.24)	126-2361-83
	dianion	417(4.28)	126-2361-83
Phenol, 4-[(2-hydroxy-5-nitrophenyl)-methyl]-2,6-dimethyl-	neutral	328(4.00)	126-2361-83
	anion	416(4.30)	126-2361-83
	dianion	419(4.30)	126-2361-83
1H-Pyrrole-3-carboxylic acid, 1-ethyl-4,5-dihydro-4,5-dioxo-2-phenyl-, ethyl ester	dioxan	282(3.7),405(3.6)	94-0356-83
$C_{15}H_{15}NO_5S$			
1H-Pyrrole-2-butanoic acid, γ-oxo-1-(phenylsulfonyl)-, methyl ester	MeOH	237(4.10),269s(4.01), 274(4.03)	44-3214-83
1H-Pyrrole-3-butanoic acid, γ-oxo-1-(phenylsulfonyl)-, methyl ester	MeOH	238(4.29)	44-3214-83
$C_{15}H_{15}NO_6$			
1H-Indole-4,5,6-tricarboxylic acid, 1-methyl-, trimethyl ester	MeOH	258(4.52),316s(3.73), 332(3.80)	44-2488-83
$C_{15}H_{15}N_2S$			
Thiazolo[3,4-a]pyrimidin-5-ium, 2,4,6-trimethyl-8-phenyl-, perchlorate	MeCN	246(3.31),373(2.82)	103-0037-83
$C_{15}H_{15}N_3$			
Benzenamine, 4-methyl-N-[2-(phenylazo)ethenyl]-	EtOH	348(4.36),376(4.55)	65-1902-83
conjugate acid	EtOH-H_2SO_4	348(4.36),442(4.53)	65-1902-83
	50% EtOH-47% H_2SO_4	442(4.71)	65-1902-83
2-Propenenitrile, 3-[3-(1,1-dimethylethyl)-2-quinoxalinyl]-, (E)-	MeOH	239(4.23),256(4.30), 325(3.99)	12-0963-83
(Z)-	MeOH	240(4.30),256(4.38), 325(3.99)	12-0963-83
$C_{15}H_{15}N_3O$			
Benzenamine, 2-methoxy-N-[2-(phenylazo)ethenyl]-	EtOH	375(4.50)	65-1902-83
conjugate acid	EtOH-H_2SO_4	450(4.58)	65-1902-83
1H-Pyrrolo[3,2-c]pyrimidine-7-carboxamide, 2,3-dihydro-1-(phenylmethyl)-	EtOH	208(4.42),275(4.41)	103-0052-83
$C_{15}H_{15}N_3O_2$			
Benzoic acid, 2-amino-, 2-(4-methylbenzoyl)hydrazide	MeOH	240(4.12),245(4.13), 260(4.17),270(4.13)	118-0842-83

Compound	Solvent	λ_{max}(log ϵ)	Ref.
2,4(1H,3H)-Pyrimidinedione, 1,3-dimethyl-5-(1-methyl-1H-indol-3-yl)-	MeCN	231(4.48),310(3.99)	94-3496-83
2,4(1H,3H)-Pyrimidinedione, 1,3-dimethyl-6-(1-methyl-1H-indol-3-yl)-	MeCN	221(4.48),275(4.04), 305(4.04)	94-3496-83
$C_{15}H_{15}N_3O_2S_2$			
Diazene, (4-nitrophenyl)[1-(4,5,6,7-tetrahydro-1,3-benzodithiol-2-ylidene)ethyl]-	MeCN	557(4.38)	118-0840-83
$C_{15}H_{15}N_3O_3$			
Benzoic acid, 2-amino-, 2-(4-methoxybenzoyl)hydrazide	MeOH	242(4.11),250(4.16), 258(4.13),270(4.13)	118-0842-83
$C_{15}H_{15}N_3O_4$			
Benzoic acid, 4-[[2-amino-1,4,5,6-tetrahydro-4,6-dioxo-5-(2-propenyl)-5-pyrimidinyl]methyl]-	50%MeOH-HCl	230s(4.09)	73-0299-83
	+ NaOH	217(4.60),264(3.82)	73-0299-83
$C_{15}H_{15}N_5O$			
1H-Tetrazol-5-amine, N-(4-methoxyphenyl)-1-(phenylmethyl)-	n.s.g.	207(4.28),248(4.24), 293s(3.38)	78-2599-83
1H-Tetrazol-5-amine, 1-(4-methoxyphenyl)-N-(phenylmethyl)-	n.s.g.	206(4.23),232(4.08), 245s(4.02)	78-2599-83
$C_{15}H_{15}O$			
Ethylium, 1-hydroxy-1-[4-(phenylmethyl)phenyl]-	H_2SO_4	222(4.20),313(4.40)	99-0612-83
$C_{15}H_{16}$			
Naphthalene, 1,4-dimethyl-6-(1-methylethenyl)-	EtOH	218(4.21),250(4.59), 283(3.82),293(3.88), 304(3.75)	2-0824-83
$C_{15}H_{16}BrNO_6$			
3-Benzofurancarboxylic acid, 6-bromo-4,5-dihydroxy-7-[2-(hydroxyimino)propyl]-2-methyl-, ethyl ester	EtOH	217(4.62),250(4.15), 300s(3.60)	103-0945-83
$C_{15}H_{16}ClNO_3$			
2H-Furo[2,3-h]-1-benzopyran-2-one, 3-chloro-4-(diethylamino)-5,6-dihydro-	EtOH	237s(3.94),270(3.68), 357(4.18)	111-0113-83
$C_{15}H_{16}ClNO_6$			
3-Benzofurancarboxylic acid, 6-chloro-4,5-dihydroxy-7-[2-(hydroxyimino)-propyl]-2-methyl-, ethyl ester	EtOH	220(4.48),254(4.14), 300s(3.66)	103-0945-83
$C_{15}H_{16}Cl_2N_2O_2$			
3-Pyridinecarboxamide, 1-[(2,6-dichlorophenyl)methyl]-1,6-dihydro-6-methoxy-N-methyl-	CH_2Cl_2	256(3.9),315(3.8)	64-0873-83B
$C_{15}H_{16}Cl_3NO_3$			
1H-Pyrrole-1-carboxylic acid, 2,5-dihydro-2-[(2-methoxyphenyl)methyl]-, 2,2,2-trichloroethyl ester	MeOH	222(3.91),272(3.41), 279(3.38)	87-0469-83
1H-Pyrrole-1-carboxylic acid, 2,5-dihydro-2-[(3-methoxyphenyl)methyl]-, 2,2,2-trichloroethyl ester	MeOH	223s(4.38),272(3.34), 279(3.32)	87-0469-83

Compound	Solvent	λ_{max}(log ϵ)	Ref.
$C_{15}H_{16}Cl_3NO_4$			
6-Oxa-3-azabicyclo[3.1.0]hexane-3-carboxylic acid, 2-[(2-methoxyphenyl)methyl]-, 2,2,2-trichloroethyl ester	MeOH	233(3.92),272(3.20), 279(3.24),292(2.98)	87-0469-83
6-Oxa-3-azabicyclo[3.1.0]hexane-3-carboxylic acid, 2-[(3-methoxyphenyl)methyl]-, 2,2,2-trichloroethyl ester	MeOH	230(3.92),273(3.16), 280(3.18),290s(3.96)	87-0469-83
$C_{15}H_{16}N_2$			
Ethanone, 1-phenyl-, methylphenylhydrazone	MeCN	253(4.28),296(3.40), 352(3.60)	70-0485-83
1H-Indole, 3-[3-(1-methylethyl)-1H-pyrrol-2-yl]-	EtOH	213(4.27),240(4.30), 300(3.63)	103-0289-83
Indolo[4,3-fg]isoquinoline, 4,6,6a,7-8,9-hexahydro-8-methyl-	MeOH	243(4.37),266(4.37), 311(4.03)	39-1545-83C
2(1H)-Isoquinolinamine, 3,4-dihydro-3-phenyl-, monohydrochloride	MeOH	252(2.64),264(2.74), 272(2.60)	4-0121-83
$C_{15}H_{16}N_2O$			
Benzaldehyde, 4-ethoxy-, phenylhydrazone	EtOH	245(4.10),304(4.21), 349(4.39)	70-1439-83
[1,4]Diazepino[3,2,1-jk]carbazol-5(4H)-one, 6,7,9,10,11,12-hexahydro-	n.s.g.	315(3.94)	103-1312-83
4H-Indeno[1,2-b]pyridin-4-one, 1,5-dihydro-1-methyl-2-[(methylamino)-methyl]-	MeOH	204(4.28),226(3.96), 294(4.33)	83-0264-83
1H-Pyrazino[3,2,1-jk]carbazol-2(3H)-one, 7,8,9,10-tetrahydro-9-methyl-	n.s.g.	250(4.13)	103-1312-83
1H-Pyrido[3,4-b]indole, 2-acetyl-1-ethylidene-2,3,4,9-tetrahydro-	MeOH	228(4.38),304(4.30), 310(4.30)	4-0183-83
4H-Pyrido[1,2-a]pyrimidin-4-one, 6,7,8,9-tetrahydro-2-methyl-7-phenyl-	EtOH	205(4.20)	4-0393-83
4H-Pyrido[1,2-a]pyrimidin-4-one, 6,7,8,9-tetrahydro-6-methyl-3-phenyl-	EtOH	223(3.72),245(3.73), 302(4.00)	39-0369-83C
$C_{15}H_{16}N_2O_2$			
Pyrano[2,3-c]pyrazol-4(1H)-one, 5,6-dihydro-3,6,6-trimethyl-1-phenyl-	EtOH	260(4.33)	118-0844-83
Pyrano[4,3-c]pyrazol-3(2H)-one, 6,7-dihydro-4,6,6-trimethyl-2-phenyl-	EtOH	264(4.29)	142-1801-83
1H-Pyrazole-4-carboxylic acid, 5-(1-methylethenyl)-1-phenyl-, ethyl ester	EtOH	244(4.09)	118-0566-83
1H-Pyrazole-4-carboxylic acid, 1-phenyl-5-(1-propenyl)-, ethyl ester	EtOH	257(4.15)	118-0566-83
3H-Pyrazol-3-one, 4-acetyl-2,4-dihydro-5-(2-methyl-1-propenyl)-2-phenyl-	EtOH	259(4.34)	142-1801-83
3-Pyridinecarboxamide, 1,4-dihydro-4-oxo-1-(2,4,6-trimethylphenyl)-	H_2O	262(4.20)	54-0331-83
	M HCl	250(3.99)	54-0331-83
1H-Pyrrolo[2,3-d]carbazole-2,5(3H,4H)-dione, 3a,6,6a,7-tetrahydro-3a-methyl-, (3aS*R*,6aS*R*,11bS*R*)-	MeOH	245(3.78),299(3.48)	5-1744-83
Quinoxaline, 5,6,7,8-tetrahydro-2-methyl-3-phenyl-, 1,4-dioxide	EtOH	242(4.28),253s(4.17), 307(4.24)	103-1104-83
$C_{15}H_{16}N_2O_3$			
1,3-Butanedione, 1-(3-ethyl-2,5-dihydro-5-oxo-1-phenyl-1H-pyrazol-4-yl)-	EtOH	260(4.10),318(4.02)	44-4078-83

Compound	Solvent	$\lambda_{max}(\log \epsilon)$	Ref.
Indolo[4,3-ef][2,1]benzisoxazole-9-carboxylic acid, 4,6,6a,7,9,9a-hexahydro-7-methyl-, methyl ester, (6aα,9α,9aα)-(±)-	MeOH	222(4.49),280(3.84), 291(3.77)	78-3695-83
4,7-Methano-3H-pyrrolo[4,3,2-fg][2,3]-benzoxazocine-11-carboxylic acid, 1,4,5,7-tetrahydro-5-methyl-, methyl ester, (4α,7α,11R*)-(±)-	MeOH	220(3.57),287(2.97)	78-3695-83
2H-Pyran-2-one, 4-hydroxy-6-methyl-3-[1-(phenylhydrazono)propyl]-	EtOH	308(4.02),340(3.86)	44-4078-83
$C_{15}H_{16}N_2O_3S$			
Acetamide, N-[3-methyl-4-[(phenylsulfonyl)amino]phenyl]-	EtOH	255(4.28)	44-3146-83
1H-Imidazole-5-carboxylic acid, 4-(methylthio)-1-(2-oxo-2-phenylethyl)-, ethyl ester	EtOH	241(4.21),286(4.10)	2-0030-83
5H-Pyrrolo[2,1-b][3]benzazepine-2-sulfonamide, 6,11-dihydro-N,N-dimethyl-11-oxo-	EtOH	209(4.44),262(4.03), 309(4.16)	44-3220-83
$C_{15}H_{16}N_2O_5$			
5-Pyrimidinepentanoic acid, hexahydro-2,4,6-trioxo-5-phenyl-	50%MeOH-HCl	215(3.77),235.5(3.45)	73-0137-83
	+ NaOH	240(3.97)	73-0137-83
$C_{15}H_{16}N_2O_5S$			
2-Thiopseudouridine, 4'-phenyl-, (±)-	pH 1	276(4.13),289(4.14)	18-2680-83
	pH 13	216(4.29),264(4.14), 289(4.14)	18-2680-83
	MeOH	276(4.22),290(4.17)	18-2680-83
2-Thiopseudouridine, 5'-phenyl-, (±)-	pH 1	275(3.84),290(3.81)	18-2680-83
	pH 13	264(3.81),285(3.70)	18-2680-83
	MeOH	212(3.64),277(3.56), 290(3.51)	18-2680-83
Uridine, 5'-deoxy-5'-(phenylthio)-	pH 1	254(4.14)	39-1315-83C
	MeOH	256(4.15)	39-1315-83C
$C_{15}H_{16}N_2O_6$			
Pseudouridine, 4'-phenyl-, (±)-	pH 1	263(3.94)	18-2680-83
	pH 13	288(3.71)	18-2680-83
	MeOH	264(3.73)	18-2680-83
Pseudouridine, 5'-phenyl-, (±)-	pH 1	264(3.86)	18-2680-83
	pH 13	285(3.97)	18-2680-83
	MeOH	265(3.87)	18-2680-83
$C_{15}H_{16}N_2O_6S$			
Uridine, 6-(phenylthio)-	MeOH	282(4.04)	94-1222-83
$C_{15}H_{16}N_2S$			
Azuleno[2,1-d]thiazole-4-methanamine, N,N,2-trimethyl-	MeOH	222(4.03),257(3.97), 300(4.83),306s(4.79), 352(3.69),368(3.83), 608(2.56),650s(2.50), 720s(2.12)	142-1263-83
$C_{15}H_{16}N_4$			
4H-Pyrrolo[2,3-d]pyrimidin-4-imine, 3,7-dihydro-2,5-dimethyl-3-(3-methylphenyl)-	EtOH	234(4.14),275(3.81)	5-2066-83
4H-Pyrrolo[2,3-d]pyrimidin-4-imine, 3,7-dihydro-2,5-dimethyl-3-(4-	EtOH	234(4.39),277(3.90)	5-2066-83

Compound	Solvent	λ_{max}(log ϵ)	Ref
methylphenyl)- (cont.)			5-2066-83
$C_{15}H_{16}N_4O_7$			
Pyridinium, 1-methyl-, 2,4,6-trinitro-1-(2-oxopropyl)cyclohexadienide, cation	MeCN	259(4.08)	103-0773-83
anion	MeCN	461(4.00),581(3.96)	103-0773-83
$C_{15}H_{16}N_6O$			
3-Pyridinecarbonitrile, 4,6-diamino-5-(phenylazo)-2-propoxy-	EtOH	238(4.37),384(4.33)	49-0973-83
$C_{15}H_{16}N_6O_2S$			
Adenosine, 2'-deoxy-5'-S-(2-pyridinyl)-5'-thio-	pH 1	251(4.16),310(3.79)	39-1315-83C
$C_{15}H_{16}N_6O_4S$			
Guanosine, 5'-S-(2-pyridinyl)-5'-thio-	pH 1	251(4.16),310(3.79)	39-1315-83C
$C_{15}H_{16}O$			
Bicyclo[3.3.1]non-1-en-3-one, 2-phenyl-	MeCN	242(c.3.66),350(c.2.48)	44-1654-83
$C_{15}H_{16}O_2$			
1,4-Naphthalenedione, 2,5-dimethyl-8-(1-methylethyl)-	EtOH	250(4.19),268(3.92), 360(3.54)	22-0112-83
Naphtho[2,3-b]furan-4(6H)-one, 8a,9-dihydro-3,5,8a-trimethyl-, (S)-	EtOH	214(3.85),259(3.38), 286(3.36)	102-0187-83
4(1H)-Phenanthrenone, 2,3,9,10-tetrahydro-7-methoxy-	EtOH	246(4.17),290(3.69), 310(3.61)	78-4221-83
$C_{15}H_{16}O_2S_2$			
4a,8a-Methanonaphthalen-1(2H)-one, 2-[bis(methylthio)methylene]-4-methoxy-	CH_2Cl_2	262(4.00),398(3.94)	118-0463-83
$C_{15}H_{16}O_3$			
2H-1-Benzopyran-2-one, 7-methoxy-8-(3-methyl-2-butenyl)- (osthol)	n.s.g.	258(3.63),322(3.90)	162-0989-83
2-Cyclohexene-1-carboxylic acid, 2-benzoyl-1-methyl-	EtOH	268(3.78),281(3.72)	150-1301-83M
Cyclopenta[g]-2-benzopyran-1(6H)-one, 7,8-dihydro-9-hydroxy-5,7,7-trimethyl-	EtOH	231(4.41),239s(4.20), 260(4.02),273(3.74), 347(3.60)	102-1489-83
1H-Fluorene-1-carboxylic acid, 2,3,4,4a,9,9a-hexahydro-1-methyl-9-oxo-, cis	EtOH	245(4.17)	150-1301-83M
Naphtho[1',2':1,3]cyclopropa[1,2-b]-furan-2(3H)-one, 3a,3b,4,5-tetrahydro-7-methoxy-3b-methyl-	EtOH	284(3.24),291(3.23)	39-0985-83C
Yomogin, (-)-	EtOH	212(4.11),238(4.11), 263s(--)	94-3397-83
$C_{15}H_{16}O_3S_2$			
3(2H)-Thiophenone, 4-(butylthio)-2-(phenylmethylene)-, 1,1-dioxide	EtOH	205(4.15),290(4.28), 320(4.27),370(3.95)	104-1447-83
$C_{15}H_{16}O_4$			
Benzene, 1,2,3-trimethoxy-4-phenoxy-	EtOH	212(4.33),264s(3.22), 270(3.33),277(3.34)	39-2267-83C

Compound	Solvent	$\lambda_{max}(\log \epsilon)$	Ref.
Benzene, 1,2,3-trimethoxy-5-phenoxy-	EtOH	216(4.45),238s(3.98), 265(3.33),272(3.35), 278(3.30)	39-2267-83C
Benzene, 1,2,4-trimethoxy-5-phenoxy-	EtOH	211(4.26),273s(3.50), 280s(3.62),292(3.77)	39-2267-83C
Benzene, 1,2,5-trimethoxy-3-phenoxy-	EtOH	228(4.55),265s(3.77), 272(3.92),278(3.98)	39-2267-83C
Benzenemethanol, 3-[4-(hydroxymethyl)-phenoxy]-4-methoxy-	MeOH	223(4.24),275(3.51)	83-0445-83
2-Naphthaleneacetic acid, α,1-dihydroxy-α-methyl-, ethyl ester	EtOH	214(4.54),238(4.47), 299(3.64),311(3.56), 327(3.51)	39-1649-83C
Pentacyclo[5.4.0.0^{2,9}.0^{3,5}.0^{6,8}]undec-10-ene-10,11-dicarboxylic acid, dimethyl ester	MeCN	235(3.70)	88-2147-83
Tetracyclo[5.4.0.0^{4,11}.0^{8,10}]undeca-2,5-diene-2,3-dicarboxylic acid, dimethyl ester	MeCN	236(3.69)	88-2147-83
Yangonin, dihydro-	MeOH	228(3.71),274(3.82)	162-1451-83
$C_{15}H_{16}O_5$			
9,10-Anthracenedione, 1,2,3,4,4a,9a-hexahydro-1,2,8-trihydroxy-6-methyl-(trichodermaol)	MeOH	234.9(4.46),277.9(3.81), 346.7(3.80)	138-0923-83
2H,4H-Benzo[1,2-b:4,3-c']dipyran-2,6(8H)-dione, 9,10-dihydro-5-hydroxy-4,8,8-trimethyl- (fuscin)	n.s.g.	352(4.92)	23-2285-83
1H-2-Benzopyran-3,7-dione, 6,8-dihydroxy-1-methyl-5-(3-methyl-2-butenyl)- (secofuscin)	n.s.g.	349(4.92)	23-2285-83
1-Cyclohexene-1-carboxylic acid, 4-hydroxy-2-[2-(4-hydroxyphenyl)ethyl]-6-oxo-, (S)-	pH 2	222(4.08),240(3.98), 280s(--)	35-4480-83
	pH 7	220(4.00),257(4.02)	35-4480-83
final spectrum	pH 12	238(4.25),293(3.72)	35-4480-83
Guillonein	EtOH	224(4.065)	102-0987-83
Lactucin	n.s.g.	257(4.15)	162-0770-83
Methysticin, dihydro-	MeOH	232(4.18),288(3.56)	162-0879-83
Vernolepin	MeOH	220(4.29)(end abs.)	100-0161-83
$C_{15}H_{16}O_6$			
1,3-Dioxolane-2-propanal, α-(benzoyloxy)-2-(2-oxoethyl)-, (R)-	EtOH	210(3.57),228(4.02)	39-2441-83C
$C_{15}H_{16}O_9$			
2H-1-Benzopyran-2-one, 6-β-D-glucopyranosyl-5,7-dihydroxy- (daurosside D)	EtOH	225(4.28),253s(3.99), 262(3.99),333(4.22)	105-0413-83
$C_{15}H_{17}ClN_4O$			
5-Pyrimidinecarboxaldehyde, 2-amino-4-chloro-6-[(4-phenylbutyl)amino]-	EtOH	335(4.16)	4-0041-83
$C_{15}H_{17}ClN_4S$			
1H-Pyrazino[2,3-b][1,4]benzothiazine-1-propanamine, 8-chloro-N,N-dimethyl-, dihydrochloride	EtOH-HCl	246s(--),252(4.54), 258s(--)	87-0564-83
10H-Pyrazino[2,3-b]benzothiazine-10-propanamine, 8-chloro-N,N-dimethyl-, monohydrochloride	EtOH-HCl	250(4.53)	87-0564-83

Compound	Solvent	λ_{max}(log ε)	Ref.
$C_{15}H_{17}Cl_3N_2O_2$			
2-Butenamide, 2,4,4-trichloro-N-(4-methylphenyl)-3-morpholino-, (E)-	heptane	196(4.50),225(3.99), 331(3.93)	5-1496-83
4-Morpholinepropanamide, α-chloro-β-(dichloromethylene)-N-(4-methylphenyl)-	heptane	195(4.50),239(4.13)	5-1496-83
$C_{15}H_{17}NO$			
Benzenemethanol, 4-(methylphenylamino)-α-methyl- (dication)	H_2SO_4	279(4.21)	99-0612-83
$C_{15}H_{17}NO_2$			
7-Azabicyclo[4.1.0]hept-2-ene-3-carboxylic acid, 7-(phenylmethyl)-, methyl ester	MeCN	257(3.66)	24-0563-83
2-Cyclohexene-1-carboxylic acid, 4-[(phenylmethylene)amino]-, methyl ester	MeCN	247(4.16),278(3.24), 287(3.00)	24-0563-83
5(2H)-Oxazolone, 2-(1,1-dimethyl-2-propenyl)-2-methyl-4-phenyl-	C_6H_{12}	260(4.00)	44-0695-83
5(2H)-Oxazolone, 2-methyl-2-(3-methyl-2-butenyl)-4-phenyl-	C_6H_{12}	264(4.09)	44-0695-83
5(4H)-Oxazolone, 2-methyl-4-(3-methyl-2-butenyl)-4-phenyl-	C_6H_{12}	256(3.15)	44-0695-83
5(4H)-Oxazolone, 4-methyl-4-(3-methyl-2-butenyl)-2-phenyl-	C_6H_{12}	243(4.23)	44-0695-83
1H-Pyrrole-2-carboxaldehyde, 3,4-dimethyl-1-[(phenylmethoxy)methyl]-	$CHCl_3$	285s(4.08),303(4.14)	49-0753-83
2H-Pyrrol-2-one, 1,5-dihydro-5-[(4-methoxyphenyl)methylene]-1,3,4-trimethyl-, (E)-	MeOH	265(3.45),327(4.04)	54-0114-83
(Z)-	MeOH	230(3.20),331(4.26)	54-0114-83
$C_{15}H_{17}NO_2S_2$			
1H-Isoindole-1,3(2H)-dione, 2-[3-(1,3-dithian-2-yl)propyl]-	MeOH	241(4.07),293(3.34)	78-2691-83
$C_{15}H_{17}NO_3$			
4H-[1,4]Oxazecino[5,4-a]isoindole-2,13(1H,5H)-dione, 6,7,8,8a-tetrahydro-	$CHCl_3$	243(3.75),270s(3.29), 280(3.16)	78-2691-83
$C_{15}H_{17}NO_3S$			
1,2-Benzoxathiin-4-amine, 5,6,7,8-tetrahydro-N-methyl-N-phenyl-, 2,2-dioxide	EtOH	250s(3.72),280s(3.80), 302(3.87)	4-1549-83
$C_{15}H_{17}NO_4$			
2,6-Piperidinedione,4-[2-(2-hydroxy-3,5-dimethylphenyl)-2-oxoethyl]-	n.s.g.	262(4.04),345(3.66)	162-0022-83
$C_{15}H_{17}NO_6$			
3,5-Isoxazoledicarboxylic acid, 4-(2,4-dimethylphenyl)-4,5-dihydro-, dimethyl ester, 2-oxide, cis	MeOH	263(4.06)	40-1678-83
α-D-Xylofuranose, 5-C-(carboxyphenylamino)-1,2-O-(1-methylethylidene)-, intramol. 5,3-ester	MeOH	236(4.26)	65-1917-83
$C_{15}H_{17}NO_7$			
Carbonic acid, ethyl 7-methoxy-2,2-di-	EtOH	212s(4.11),217(4.30),	39-0197-83C

Compound	Solvent	$\lambda_{max}(\log \epsilon)$	Ref.
methyl-6-nitro-2H-1-benzopyran-4-yl ester (cont.)		223s(4.28),263(4.20), 296s(3.76),360(3.72)	39-0197-83C
$C_{15}H_{17}NO_8$			
3,5-Isoxazoledicarboxylic acid, 4-(2,3-dimethoxyphenyl)-4,5-dihydro-, dimethyl ester, 2-oxide, cis	MeOH	266(4.03)	40-1678-83
3,5-Isoxazoledicarboxylic acid, 4-(2,4-dimethoxyphenyl)-4,5-dihydro-, dimethyl ester, 2-oxide, cis	MeOH	270(4.05)	40-1678-83
$C_{15}H_{17}NS$			
Thieno[3,2-c]pyridine, 4,5,6,7-tetrahydro-5-methyl-4-(phenylmethyl)-	MeOH	238(3.87),268.5(2.58)	83-0244-83
$C_{15}H_{17}N_2$			
Pyridinium, 1-methyl-2-[3-(1-methyl-2(1H)-pyridinylidene)-1-propenyl]-, perchlorate	MeOH	558(5.04)	104-2114-83
	CH_2Cl_2	563(--)	104-2114-83
Pyridinium, 1-methyl-4-[3-(1-methyl-4(1H)-pyridinylidene)-1-propenyl]-, perchlorate	MeOH	604(5.21)	104-2114-83
	CH_2Cl_2	612(--)	104-2114-83
$C_{15}H_{17}N_2O$			
Pyridinium, 3-(aminocarbonyl)-1-(2,4-6-trimethylphenyl)-, chloride	H_2O	264(3.67)	54-0331-83
$C_{15}H_{17}N_3$			
Benzaldehyde, 4-(dimethylamino)-, phenylhydrazone	benzene	360(4.5)	140-0094-83
$C_{15}H_{17}N_3O_4$			
Benzoic acid, 4-[(2-amino-1,4-dihydro-6-hydroxy-4-oxo-5-pyrimidinyl)methyl]-, propyl ester	MeOH	244(4.39),262s(4.28)	73-0299-83
Benzoic acid, 4-[(2-amino-1,4,5,6-tetrahydro-4,6-dioxo-5-propyl-5-pyrimidinyl)methyl]-	50%MeOH-HCl	230s(4.18),260s(3.57)	73-0299-83
	+ NaOH	225(4.49),265(4.06)	73-0299-83
$C_{15}H_{17}N_3O_4S$			
Thymidine, 5'-S-2-pyridinyl-5'-thio-	H_2O	247(3.89),270(3.83)	39-1315-83C
	MeOH	249(3.99)	39-1315-83C
$C_{15}H_{17}N_3O_4S_2$			
Carbamic acid, [4-(phenylmethyl)-1,2,4-dithiazolidine-3,5-diylidene]bis-, diethyl ester	CH_2Cl_2	268(4.31),c.300(3.61)	24-1297-83
$C_{15}H_{17}N_3O_5$			
1,2-Diazabicyclo[5.2.0]nona-3,5-diene-2-carboxylic acid, 8-(2,5-dioxo-1-pyrrolidinyl)-5-methyl-9-oxo-, ethyl ester, cis-(±)-	MeOH	275(3.89)	5-1374-83
Pseudoisocytidine, 4'-phenyl-, hydrochloride, (±)-	pH 1	263(3.90)	18-2680-83
	pH 13	231(4.15),276(3.99)	18-2680-83
	MeOH	217(4.25),262(3.98)	18-2680-83
Pseudoisocytidine, 5'-phenyl-, hydrochloride, (±)-	pH 1	263(4.01)	18-2680-83
	pH 13	233(4.12),277(4.00)	18-2680-83
	MeOH	225(4.10),264(3.92), 290(3.60)	18-2680-83
Uridine, 2'-deoxy-5-(phenylamino)-	pH 7	253(4.09),321(3.56)	33-0534-83

Compound	Solvent	λ max (log ε)	Ref.
$C_{15}H_{17}N_3O_5S_2$			
1,2,4-Triazin-3(2H)-one, 4,5-dihydro-6-[(phenylmethyl)thio]-2-β-D-ribofuranosyl-5-thioxo-	pH 1	324(3.93)	80-0989-83
	pH 13	336(3.96)	80-0989-83
	EtOH	324(4.02)	80-0989-83
$C_{15}H_{17}N_3O_6$			
Uridine, 5-(phenylamino)-	pH 7	252(4.09),322(3.58)	33-0534-83
$C_{15}H_{17}N_3O_7$			
4-Pyrimidinecarbonitrile, 3-[5-O-acetyl-2,3-O-(1-methylethylidene)-β-D-ribofuranosyl]-1,2,3,4-tetrahydro-2,6-dioxo-	EtOH	281(3.93)	35-0963-83
$C_{15}H_{17}N_5O$			
5-Pyrimidinecarbonitrile, 2-amino-1,6-dihydro-6-oxo-4-[(4-phenylbutyl)-amino]-	EtOH	228(4.65),273(4.12)	4-0041-83
$C_{15}H_{17}N_5O_4S$			
Benzenesulfonic acid, 4-[4-[(2-amino-5-cyano-1,6-dihydro-6-oxo-4-pyrimidinyl)amino]butyl]-	EtOH	228(4.71),273(4.16)	4-0041-83
$C_{15}H_{17}N_5O_5$			
7H-Dipyrimido[1,6-a:4',5'-d]pyrimidin-7-one, 5,8-dihydro-2-methyl-8-β-D-ribofuranosyl-	MeOH	235(3.93),250s(3.78), 275s(3.54),356(4.31)	77-0183-83
$C_{15}H_{17}N_5S$			
Imidazo[2,1-b][1,3,5]benzothiadiazepine, 5-(4-methyl-1-piperazinyl)-	MeOH	210(4.41),222s(4.30), 258(4.13),284(3.85)	4-1287-83
$C_{15}H_{18}BrClO$			
2H-Oxocin, 8-(1-bromopropylidene)-3-chloro-3,4,7,8-tetrahydro-2-(2-penten-4-ynyl)- (3Z-venustin)	EtOH	214s(4.20),222(4.23), 231s(4.08)	138-0779-83
$C_{15}H_{18}BrClO_2$			
4,9-Dioxabicyclo[6.1.0]nonane, 3-(1-bromopropylidene)-6-chloro-5-(2-penten-4-ynyl)-	EtOH	212(4.26),220s(4.23), 231(4.02)	138-0779-83
$C_{15}H_{18}ClN_5O$			
5-Pyrimidinecarboxaldehyde, 2-amino-4-chloro-6-[(4-phenylbutyl)amino]-, oxime	EtOH	226(4.44),280(4.11), 309(4.20),319s(4.10)	4-0041-83
$C_{15}H_{18}ClP$			
Phosphorin, 1-chloro-1-(1,1-dimethylethyl)-1,1-dihydro-3-phenyl-	hexane	242(2.9),350(3.76)	24-0445-83
$C_{15}H_{18}Cl_2FeSi$			
Ferrocene, 1,1'-[(dichlorosilylene)-methylene]-2',3',4',5'-tetramethyl-	isooctane	325s(--),430(2.08)	35-0181-83
$C_{15}H_{18}Cl_2N_2O_3S$			
3-Pyridinesulfonamide, 1-[(2,6-dichlorophenyl)methyl]-1,6-dihydro-6-methoxy-N,N-dimethyl-	CH_2Cl_2	305(3.8)	64-0873-83B

Compound	Solvent	$\lambda_{max}(\log \epsilon)$	Ref.
$C_{15}H_{18}FP$			
Phosphorin, 1-(1,1-dimethylethyl)-1-fluoro-1,1-dihydro-3-phenyl-	hexane	235(4.2),248s(4.1),363(3.76)	24-0445-83
$C_{15}H_{18}FeO_3$			
Iron, dicarbonyl[η^6-4-(2,6,6-trimethyl-1,3-cyclohexadien-1-yl)-3-buten-2-one]-	EtOH	258(3.99),321(3.57)	88-1611-83
$C_{15}H_{18}N_2O$			
2-Butenamide, N-[2-(1H-indol-3-yl)-ethyl]-3-methyl-	EtOH	223(4.07),275(3.49),283(3.50),291(3.42)	100-0310-83
9-Oxaergoline, 1,6-dimethyl-, (±)-, oxalate	EtOH-HCl	228(4.50),283s(3.77),290(3.80)	87-0522-83
3-Penten-2-one, 4-[[2-(1H-indol-3-yl)-ethyl]amino]-	EtOH	221(4.56),283s(4.07),290s(4.14),312(4.30)	4-0267-83
2-Propanone, 1-(2,3,4,9-tetrahydro-1-methyl-1H-pyrido[3,4-b]indol-1-yl)-, monohydrochloride	EtOH	221(4.56),275s(3.86),288.5(3.74)	4-0267-83
$C_{15}H_{18}N_2O_2$			
2-Butenoic acid, 3-[(2-methyl-1H-indol-5-yl)amino]-, ethyl ester	EtOH	219(4.39),294(4.51)	103-0401-83
1H-Pyrido[3,4-b]indol-1-ol, 2-acetyl-1-ethyl-2,3,4,9-tetrahydro-	MeOH	235(4.18),310(4.27)	4-0183-83
	MeOH-HCl	235(4.18),310(4.27)	4-0183-83
1H-Pyrrole-2-carboxaldehyde, 5-[(3-ethyl-1,4-dihydro-4-methyl-5-oxo-2H-pyrrol-2-ylidene)methyl]-3,4-dimethyl-, (E)-	EtOH	277(4.19),315(4.11),375(4.14)	49-0753-83
1H-Pyrrolo[2,3-d]carbazol-2(3H)-one, 3a,4,5,6,6a,7-hexahydro-5-hydroxy-3a-methyl-	MeOH	246(3.83),297(3.43)	5-1744-83
3:1 cis-trans-(±)-	MeOH	245(3.71),297(3.30)	5-1744-83
$C_{15}H_{18}N_2O_2S$			
Acetamide, N-[4-(3-morpholino-3-thioxo-1-propenyl)phenyl]-	EtOH	238(4.10),294(4.35)	80-0555-83
$C_{15}H_{18}N_2O_3$			
2-Cyclohexen-1-one, 2-nitro-5,5-dimethyl-3-[(phenylmethyl)amino]-	pH 1	268(4.32),335(3.92)	78-3405-83
	H_2O	268(4.29),335(3.92)	78-3405-83
	pH 13	269(4.26),340(3.85)	78-3405-83
3-Pyridinecarboxylic acid, 1,4,5,6-tetrahydro-5-methyl-4-oxo-1-(phenylamino)-, ethyl ester	EtOH	237(4.20),306(4.19)	118-0566-83
3-Pyridinecarboxylic acid, 1,4,5,6-tetrahydro-6-methyl-4-oxo-1-(phenylamino)-, ethyl ester	EtOH	238(4.22),307(4.18)	118-0566-83
2,4(3H,5H)-Pyrimidinedione, 5-ethyl-6-(1-methylethoxy)-5-phenyl-	anion	244(3.85)	111-0521-83
4,6(1H,5H)-Pyrimidinedione, 5-ethyl-2-(1-methylethoxy)-5-phenyl-	anion	274(4.05)	111-0521-83
2,4,6(1H,3H,5H)-Pyrimidinetrione, 5-ethyl-1-(1-methylethyl)-5-phenyl-	anion	247(3.94)	111-0521-83
SEN-215	MeOH	240(4.02),270s(--),344(3.52)	158-83-176
$C_{15}H_{18}N_2O_4$			
L-Tryptophan, N-acetyl-6-methoxy-, methyl ester	EtOH	225(4.29),275(3.51),294(3.54)	102-1026-83

Compound	Solvent	$\lambda_{max}(\log \epsilon)$	Ref.
$C_{15}H_{18}N_2O_6$			
3,5-Isoxazoledicarboxylic acid, 4-[4-(dimethylamino)phenyl]-4,5-dihydro-, dimethyl ester, 2-oxide, cis	MeOH	261(4.36)	40-1678-83
$C_{15}H_{18}N_2O_9$			
2,4(1H,3H)-Pyrimidinedione, 1-(2,3,5-tri-O-acetyl-β-D-arabinofuranosyl)-, (as 5-iodo derivative)	MeOH	283(3.95)	87-0598-83
$C_{15}H_{18}N_2S$			
2-Butenethioamide, N-[2-(1H-indol-3-yl)ethyl]-3-methyl-	EtOH	224(4.05),274(3.47), 283(3.50),292(3.43)	100-0310-83
$C_{15}H_{18}N_3O_8PS_2$			
1,2,4-Triazin-3(2H)-one, 4,5-dihydro-6-[(phenylmethyl)thio]-2-(5-O-phosphono-β-D-ribofuranosyl)-5-thioxo-	pH 1	324(4.04)	80-0989-83
	H_2O	324(4.04)	80-0989-83
	pH 13	336(3.99)	80-0989-83
$C_{15}H_{18}N_4$			
Pyrazolo[1,5-a]pyrimidin-2-amine, 6,7-dihydro-5,7,7-trimethyl-3-phenyl-	MeCN	270(4.12)	39-0011-83C
$C_{15}H_{18}N_4O_2$			
5-Pyrimidinecarboxaldehyde, 2-amino-1,6-dihydro-6-oxo-4-[(4-phenylbutyl)amino]-, monohydrochloride	EtOH	224s(4.09),260(3.46), 309(4.40)	4-0041-83
$C_{15}H_{18}N_4O_3$			
5-Pyrimidinecarboxamide, 1,2,3,4-tetrahydro-2,4-dioxo-6-[(4-phenylbutyl)amino]-	EtOH	264(4.36)	4-0041-83
$C_{15}H_{18}N_4O_4$			
L-Histidine, N-[[3-hydroxy-5-(hydroxymethyl)-2-methyl-4-pyridinyl]methylene]-1-methyl-	MeOH	252s(4.04),290s(--), 336(3.58),423(3.30)	35-0803-83
9H-Purine, 9-(3a,6a-dihydro-6-(hydroxymethyl)-2,2-dimethyl-4H-cyclopenta-1,3-dioxol-4-yl)-6-methoxy-, (±)-	MeOH	249(3.85)	33-1915-83
$C_{15}H_{18}N_4O_{10}$			
Acetamide, N-[1-(3,5-di-O-acetyl-2-O-nitro-β-D-arabinofuranosyl)-1,2-dihydro-2-oxo-4-pyrimidinyl]-	pH 1	246(4.09),300(3.99)	87-0280-83
	EtOH	249(4.18),297(3.99)	87-0280-83
$C_{15}H_{18}N_6O_3$			
Carbamic acid, [1-(5-azido-1,3,4-oxadiazol-2-yl)-2-phenylethyl]-, 1,1-dimethylethyl ester, (S)-	MeOH	239(4.19)	35-0902-83
$C_{15}H_{18}O$			
Naphthalene, 2-(pentyloxy)- (class spectrum)	MeOH and MeOH-base	261(3.6),271(3.6), 281(3.5),313(3.18), 328(3.28)	160-1403-83
2-Naphthalenemethanol, α,α,5,8-tetramethyl-	EtOH	230(4.78),277s(3.73), 286(3.79)	2-0824-83
Naphtho[2,3-b]furan, 4,4a,8a,9-tetrahydro-3,5,8a-trimethyl-, (4aS-trans)-	EtOH	217(3.86),263(3.62)	102-0187-83

Compound	Solvent	$\lambda_{max}(\log \epsilon)$	Ref.
$C_{15}H_{18}OS$			
Bicyclo[2.2.1]heptan-2-one, 6-[[(phenylmethyl)thio]methyl]-, endo	MeCN	289(1.94)	44-1862-83
$C_{15}H_{18}O_2$			
Azuleno[4,5-b]furan-2(3H)-one, 3a,4,5-6,6a,7,9a,9b-octahydro-9-methyl-3,6-bis(methylene)-	n.s.g.	206(4.19)	102-1993-83
Cyclodeca[b]furan-4(7H)-one, 8,11-dihydro-3,6,10-trimethyl-, (Z,Z)-	EtOH	252(3.49)	102-1207-83
1H-Cyclopropa[a]naphthalene-1-carboxylic acid, 1a,2,3,7b-tetrahydro-1-(1-methylethyl)-	n.s.g.	224(3.93),268(2.71)	78-2965-83
1,4-Naphthalenedione, 2,3-dihydro-2,5-dimethyl-8-(1-methylethyl)-	EtOH	229(4.42),255(3.77), 315(3.32)	22-0112-83
2-Phenanthrenol, 1,2,3,4,9,10-hexahydro-7-methoxy-	EtOH	271.5(4.21)	39-0903-83C
$C_{15}H_{18}O_3$			
Acetic acid, cyclohexylidenephenoxy-, methyl ester	MeOH	223(4.26),270(3.30)	44-3408-83
Achalensolide	MeOH	241(4.00)	44-4038-83
Ambrosin	EtOH	217(4.13),324(1.56)	162-0056-83
1,3-Benzodioxol-5(7aH)-one, 7a-(3-methyl-2-butenyl)-6-(2-propenyl)-	EtOH	212(4.05),247(3.83), 309(3.53)	94-2879-83
2H-1-Benzopyran-2-one, 7-(hexyloxy)-	$CHCl_3$	324(4.27)	44-1872-83
H-T dimer, syn	$CHCl_3$	247(3.83),280(3.61)	44-1872-83
1H-Cyclopropa[a]naphthalene-1-carboxylic acid, 1a,2,3,7b-tetrahydro-6-methoxy-1,7-dimethyl-, (1aα,7bα)-	n.s.g.	209(4.30),280(3.31)	78-2965-83
11αH-Eudesma-1,4-dien-8β,13-olide, 3-oxo-	EtOH	240(4.05),263s(--)	94-3397-83
11βH-	EtOH	240(4.08),263s(--)	94-3397-83
3-Furancarboxylic acid, 5-(2,6-dimethyl-1,5,7-octatrienyl)-, (E,E)-	n.s.g.	212(4.59),234(4.75), 257(4.73),262(4.72)	12-0371-83
Hispanolide	EtOH	213(3.79),254(3.49)	102-1985-83
Illicinole	EtOH	210(4.20),240(3.81), 302(3.79)	94-2879-83
2,6,8-Nonatrien-4-ynal, 9-(1,3-dioxan-2-yl)-2,7-dimethyl-	hexane	235(3.88),328(4.41), 348(4.31)	33-1148-83
Spiro[benzofuran-3(2H),1'-cyclohexane]-2-carboxylic acid, methyl ester	MeOH	214(3.73),277(3.36), 284(3.31)	44-3408-83
	ether	279(3.56),285(3.53)	44-3408-83
Xanthatatin	n.s.g.	213(3.86),275(4.36)	162-1444-83
$C_{15}H_{18}O_4$			
2H-1-Benzopyran-3-carboxylic acid, 7-methoxy-2,2-dimethyl-, ethyl ester	EtOH	208(4.04),223(4.11), 249(4.20),253s(4.20), 297s(3.90),306(3.95), 345(4.10)	39-0197-83C
Helenalin	n.s.g.	223(3.28)	162-0667-83
2-Naphthaleneacetic acid, 1,2,3,4,4a,7-hexahydro-α,4a,8-trimethyl-1,7-dioxo- (3,6-dioxoeudesma-1,4-dienoic acid)	EtOH	248.5(4.07)	94-3397-83
Naphtho[2,3-b]furan-4,6-dione, 4a,5,7-8,8a,9-hexahydro-8a-hydroxy-3,4a,5-trimethyl-, (4aα,5α,8aα)-(±)-	EtOH	269(3.49)	94-3544-83
Parthenin	n.s.g.	215(4.18),340(1.34)	162-1011-83

Compound	Solvent	λ max(log ε)	Ref.
$C_{15}H_{18}O_4S$			
Bicyclo[2.2.1]heptan-2-one, 6-[[[(4-methylphenyl)sulfonyl]oxy]methyl]-, endo	MeCN	294(1.56)	44-1862-83
$C_{15}H_{18}O_5$			
2H-1-Benzopyran-3-carboxylic acid, 3,4-dihydro-7-methoxy-2,2-dimethyl-4-oxo-, ethyl ester	EtOH	214(4.23),234(4.00), 237s(4.00),277(4.11), 316(3.80)	39-0197-83C
2H-1-Benzopyran-4-carboxylic acid, 3-hydroxy-7-methoxy-2,2-dimethyl-, ethyl ester	EtOH	214(4.34),248(4.18), 297(3.71),305(3.73), 316s(3.59)	39-1431-83C
Carbonic acid, ethyl 5-methoxy-2-(3-methyl-1-oxo-2-butenyl)phenyl ester	EtOH	208(3.98),230(4.08), 284(4.18)	39-0197-83C
Ethanone, 1-(2,3-dihydro-8-hydroxy-5,5-dimethyl-5H-1,4-dioxino[2,3-c]-[1]benzopyran-9-yl)-, cis	EtOH	284(3.90),320(3.74)	39-0883-83C
$C_{15}H_{18}O_6$			
1,4-Dioxaspiro[4.5]decane-7,8,9-triol, 7-benzoate, [7R-(7α,8β,9β)]-	EtOH	208s(3.46),230(4.00), 273(2.90)	39-2441-83C
$C_{15}H_{19}$			
1H-Dibenzo[a,d]cycloheptenylium, 2,3,4,6,7,8,9,?-octahydro-, hexafluoroantimonate	MeCN	247(4.14),314(3.77)	44-0596-83
$C_{15}H_{19}Br$			
2,4-Pentadecadiene-6,9-diyne, 1-bromo-, (E,E)-	EtOH	275(4.44)	35-3656-83
$C_{15}H_{19}BrO_2$			
Bicyclo[8.3.2]pentadeca-10(15),11,13-trien-14-one, 12-bromo-15-hydroxy-	MeOH	251(4.04),269s(3.98), 286s(3.65),368(3.71), 380s(3.69),414s(3.24)	88-3879-83
$C_{15}H_{19}Br_2ClO_3$			
5,11,14-Trioxatricyclo[10.2.1.0^{4,6}]-pentadec-8-ene, 2-chloro-13-(1,2-dibromoethenyl)-10-methyl- (poitediene)	EtOH	214(3.82)	88-4649-83
$C_{15}H_{19}ClN_2O_2$			
2H-Pyrrol-2-one, 3-chloro-1,5-dihydro-5-methoxy-4-[(1-methylethyl)amino]-1-(4-methylphenyl)-	heptane	193(4.14),238(4.05), 285(3.62)	5-1504-83
$C_{15}H_{19}ClO$			
2H-Oxocin, 3-chloro-3,4-dihydro-2-(2-penten-4-ynyl)-8-propyl-, (Z)-	EtOH	214s(3.99),221(4.03), 231s(3.90),267(3.69)	138-0779-83
$C_{15}H_{19}FN_2O_3$			
1-Piperazinecarboxylic acid, 4-[2-(4-fluorophenyl)-2-oxoethyl]-, ethyl ester	MeOH	244(3.92)	73-2977-83
$C_{15}H_{19}N$			
1-Azuleneethanamine, N,N,4-trimethyl-, hydrochloride	EtOH	241(4.42),271s(4.47), 277s(4.62),281(4.67), 286(4.66),290s(4.57), 301(3.98),318s(3.45),	18-2311-83

Compound	Solvent	$\lambda_{max}(\log \epsilon)$	Ref.
(cont.)		332(3.60),341(3.68), 353(3.40),550(2.70), 584s(2.66),617s(2.45), 643(2.28),680(1.72)	18-2311-83
2,4,6,8,10-Dodecapentaenenitrile, 3,7,11-trimethyl-, trans	EtOH	372(4.59)	35-4033-83
4H-Pyrrolo[3,2,1-ij]quinoline, 5,6-dihydro-1-methyl-2-propyl-	MeOH	210s(4.05),232(4.40), 287(3.85)	39-0501-83C
$C_{15}H_{19}NOS$			
1-Propanol, 2-[[2-phenyl-1-(2-thienyl)ethyl]amino]-, hydrochloride	EtOH	207(4.01),265s(2.36), 268s(2.11)	94-0031-83
2-Propene-1-thione, 1-(4-methoxyphenyl)-3-piperidino-	heptane	282(4.12),301(4.28), 312(4.27)	80-0555-83
	EtOH	230(4.15),304(4.34), 316(4.35)	80-0555-83
$C_{15}H_{19}NO_2$			
Cyclopentanecarboxylic acid, 2-(6-cyano-1,5-dimethyl-1,3,5-hexatrienyl)-	EtOH	315(4.38)	35-4033-83
2H-Pyrrol-2-one, 5-ethoxy-5-ethyl-1,5-dihydro-4-methyl-3-phenyl-	MeOH	225(4.09)	4-0687-83
$C_{15}H_{19}NO_5$			
2-Pyrrolidinecarboxylic acid, N-(3,4-dimethoxybenzoyl)-, methyl ester	MeOH	210(4.23),254(3.71), 285(3.53)	83-0773-83
$C_{15}H_{19}NO_6$			
1,3-Dioxolane, 2-[4-(2,3-dimethoxyphenyl)-3-nitro-3-butenyl]-	MeOH	226s(4.06),314(3.85)	83-0773-83
Propanedioic acid, methyl[(3-nitrophenyl)methyl]-, diethyl ester	EtOH	262(3.89)	12-0081-83
$C_{15}H_{19}N_3O_2$			
1(4H)-Pyridineacetic acid, 3,5-dicyano-2,4,4,6-tetramethyl-, ethyl ester	EtOH	217(4.63),332(3.99)	73-0608-83
2H-Pyrido[3,4-b]indole-2-carboxamide, 1-ethyl-1,3,4,9-tetrahydro-1-hydroxy-N-methyl-	MeOH	234(4.19),310(4.28)	4-0183-83
$C_{15}H_{19}N_3O_2S$			
Carbamic acid, [3-(1-methylethyl)-4-[(1-methylethyl)imino]-1,3-thiazetidin-2-ylidene]-, phenyl ester	dioxan	255(4.28)	24-1297-83
$C_{15}H_{19}N_3O_8$			
2(1H)-Pyrimidinone, 4-amino-1-(2,3,5-tri-O-acetyl-β-D-arabinofuranosyl)-	pH 1	240(3.96),307(4.13)	87-0280-83
	pH 13	276(3.96),303(3.97)	87-0280-83
	EtOH	247(4.19),299(3.90)	87-0280-83
$C_{15}H_{19}N_3S$			
Ethanamine, 2-(1H-indol-3-ylthio)-N-(1-methyl-2-pyrrolidinylidene)-	MeOH	273(3.78),279(3.81), 287(3.77)	87-0230-83
$C_{15}H_{19}N_4$			
Pyridinium, 3-[[4-(ethylamino)-2-methylphenyl)azo]-1-methyl-, iodide	H_2O	490(4.473)	7-0055-83
	DMSO-p-tolueneSO_3H	499(4.579)	7-0055-83

Compound	Solvent	λ_{max}(log ε)	Ref.
$C_{15}H_{19}N_5$			
[1,2,4]Triazino[4,3-b]indazole, 3,4-dihydro-3,3-dimethyl-4-pyrrolidino-	EtOH	251(3.91),362(3.97)	44-2330-83
$C_{15}H_{19}N_5O_2$			
5-Pyrimidinecarboxamide, 2-amino-1,4-dihydro-4-oxo-6-[(4-phenylbutyl)-amino]-	EtOH	230(4.62),273.5(4.20)	4-0041-83
1H-1,2,4-Triazol-3-amine, N-[(4-methoxyphenyl)methylene]-1-methyl-5-morpholino-	$CHCl_3$	242(3.84),325(4.34)	24-1547-83
1H-1,2,4-Triazol-5-amine, N-[(4-methoxyphenyl)methylene]-1-methyl-3-morpholino-	$CHCl_3$	242(3.87),312(4.17), 360(4.10)	24-1547-83
$C_{15}H_{19}N_5O_5$			
1H-Purine-2,6-dione, 7-[2-(acetylamino)-2,3,6-trideoxy-β-L-threo-hexopyranos-4-ulos-1-yl]-3,7-dihydro-1,3-dimethyl-	H_2O	275(3.85)	136-0301-83A
$C_{15}H_{19}N_5O_5S$			
Benzenesulfonic acid, 4-[4-[[2-amino-5-(aminocarbonyl)-1,6-dihydro-6-oxo-4-pyrimidinyl]amino]butyl]-	EtOH	224.5(4.67),273(4.15)	4-0041-83
$C_{15}H_{19}N_5O_9$			
1,2-Hydrazinedicarboxylic acid, 1,1'-[2-oxo-2-(2-pyridinyl)ethylidene]-bis-, tetramethyl ester	EtOH	222(3.98),283(3.65)	104-0110-83
$C_{15}H_{19}N_7$			
1H-Benzimidazole, 1-[[5-(1-piperidinylmethyl)-1H-tetrazol-1-yl]methyl]-	EtOH	244(3.34),274(3.01), 281(3.04)	94-0723-83
$C_{15}H_{20}$			
Benzene, 1-methyl-2-(1-methyl-1,6-heptadienyl)-	heptane	240(3.98)	44-0090-83
$C_{15}H_{20}BrClO$			
2H-Pyran, 3-bromo-8-chloro-2-ethyl-2,3,4,7,8,9-hexahydro-9-(2-penten-4-ynyl)- (srilankenyne)	MeOH	227(4.13)	44-0395-83
Spiro[5.5]undec-3-en-2-one, 8-bromo-9-chloro-5,5,9-trimethyl-1-methylene-	n.s.g.	233(3.80)	18-3824-83
$C_{15}H_{20}BrN_6O_3P$			
Phosphonic acid, [1-bromo-2-(1H-imidazol-1-yl)ethyl]-, bis[2-(1H-imidazol-1-yl)ethyl] ester, monohydrobromide dihydrochloride	EtOH	269(2.70)	65-1555-83
$C_{15}H_{20}Br_2N_2O_4$			
Carbamic acid, [3-[4-[2-(acetylamino)-ethyl]-2,6-dibromophenoxy]propyl]-, methyl ester	MeOH	254(4.32),309(4.18)	78-0667-83
$C_{15}H_{20}ClN_3O$			
2H-Pyrrol-2-one, 3-chloro-4,5-bis(dimethylamino)-1,5-dihydro-1-(4-methylphenyl)-	heptane	194(4.10),244(3.94), 288(3.85)	5-1504-83

Compound	Solvent	λ max(log ε)	Ref.
$C_{15}H_{20}Cl_2O$			
2,5-Cyclohexadien-1-one, 4-(dichloromethylene)-2,6-bis(1,1-dimethylethyl)-	isooctane	320(4.45)	73-2376-83
$C_{15}H_{20}N_2O$			
1H-Pyrido[3,4-b]indole-1-methanol, 2,3,4,9-tetrahydro-α,α,1-trimethyl-	EtOH	225(4.56),274s(3.86), 280(3.87),289(2.78)	4-0267-83
2H-Pyrrol-2-one, 5-[(3,4-dimethyl-1H-pyrrol-2-yl)methylene]-4-ethyl-1,5-dihydro-1,3-dimethyl-, (Z)-	$CHCl_3$	264(4.02),385(4.08)	49-0753-83
2H-Pyrrol-2-one, 4-ethyl-1,5-dihydro-3-methyl-5-[(1,3,4-trimethyl-1H-pyrrol-2-yl)methylene]-, (Z)-	$CHCl_3$	263(4.08),378(4.05)	49-1107-83
$C_{15}H_{20}N_2OS$			
Acetamide, N-[4-[2-(1-piperidinyl)-2-thioxoethyl]phenyl]-	EtOH	234(4.19),306(4.40), 315(4.41)	80-0555-83
$C_{15}H_{20}N_2O_2Si$			
1H-Pyrazole-3-carboxylic acid, 1-phenyl-4-(trimethylsilyl)-, ethyl ester	EtOH	244(3.88)	44-3189-83
$C_{15}H_{20}N_2O_3$			
Pifoxime	n.s.g.	210(4.36),258(4.28)	162-1070-83
$C_{15}H_{20}N_2O_4$			
Benzene, 1-[4-methyl-1-(1-methyl-1-nitroethyl)-2-pentenyl]-4-nitro-, (E)-	EtOH	271(4.03)	12-0527-83
2H-Pyrido[1,2-a]pyrimidine-3-carboxylic acid, 9-(ethoxymethylene)-6,7,8,9-tetrahydro-6-methyl-4-oxo-, ethyl ester	EtOH	273(4.02),350(4.38)	39-0369-83C
β-D-Ribofuranosylamine, N-[2-(1H-indol-3-yl)ethyl]-	EtOH	221(4.59),283(3.89), 290(3.81)	136-0001-83B
$C_{15}H_{20}N_2O_4S$			
L-Alanine, N-[N-[(phenylmethoxy)carbonyl]thio-L-alanyl]-, methyl ester	$CHCl_3$	270(4.04)	78-1075-83
$C_{15}H_{20}N_2O_5$			
3-Heptanone, 2,6-dimethyl-6-nitro-5-(4-nitrophenyl)-	EtOH	219(3.94),270(4.03)	12-0527-83
Uridine, 2'-deoxy-5-(3,3-dimethyl-1-butynyl)-	pH 6	232(4.09),292(4.05)	44-1854-83
	pH 13	229(4.16),287(3.96)	44-1854-83
Uridine, 2'-deoxy-5-(1-hexynyl)-	pH 6	233(4.06),293(4.07)	44-1854-83
	pH 13	232(4.11),287(3.97)	44-1854-83
$C_{15}H_{20}N_2O_6$			
2,4(1H,3H)-Pyrimidinedione, 1-β-D-arabinofuranosyl)-5-(3,3-dimethyl-1-butynyl)-	pH 7	231(4.06),292(4.05)	87-0661-83
$C_{15}H_{20}N_2O_6S$			
Pyrimidine, 1,4,5,6-tetrahydro-1-methyl-2-[2-(2-thienyl)ethenyl]-, tartrate	H_2O	312(4.27)	162-1148-83
$C_{15}H_{20}N_3O_3P$			
Sydnone imine, N-(4-methyl-1,3,2-	MeCN	340(c.3.78)	65-1263-83

Compound	Solvent	$\lambda_{max}(\log \epsilon)$	Ref.
dioxaphosphorinan-2-yl)-3-(1-methyl-2-phenylethyl)- (cont.)			65-1263-83
$C_{15}H_{20}N_4O$			
1(4H)-Pyridineacetamide, 3,5-dicyano-N,N,2,4,4,6-hexamethyl-	EtOH	218(4.51),339(3.78)	73-0617-83
$C_{15}H_{20}N_4O_2$			
Benzo[1,2-c:5,4-c']dipyrazole, 1,7-di-acetyl-1,3a,4,7,8,8a-hexahydro-3,5,8a-trimethyl-, cis	EtOH	242(4.87)	39-2545-83C
$C_{15}H_{20}N_6O_4$			
1H-Imidazole-4-carboxamide, 1-[3-(3,4-dihydro-3,5-dimethyl-2,4-dioxo-1(2H)-pyrimidinyl)propyl]-5-(form-ylamino)-N-methyl-	H_2O	250(4.07),267s(4.01)	89-0623-83
3,4-Propano-7bH-imidazo[4',5':3,4]-cyclobuta[1,2-d]pyrimidine-7b-carb-oxamide, 3a-(formylamino)-3a,3b,5,6-7,7a-hexahydro-N,6,7a-trimethyl-5,7-dioxo-	H_2O	222(3.87)	89-0623-83
$C_{15}H_{20}O$			
Benzene, [2-(methoxymethyl)-2-methyl-1-(1-propenyl)cyclopropyl]-	EtOH	202.5(4.07)	12-0581-83
2,4,6,8,10-Dodecapentaenal, 3,7,11-trimethyl-	EtOH	378(4.61)	35-4033-83
Furan, 2-(2,6-dimethyl-1,5,7-octa-trienyl)-4-methyl-, (E,E)-	n.s.g.	204(4.00),229(4.33),265(3.93),273(3.80),283(3.56)	12-0371-83
Furan, 2-(2,6-dimethyl-2,5,7-octa-trienyl)-4-methyl-, (Z,E)-	n.s.g.	210(3.70),232(4.07),275(2.60)	12-0371-83
Phenanthrene, 1,2,3,4,4aα,9,10,10aα-octahydro-7-methoxy-	EtOH	279(3.34),287.5(3.31)	39-0903-83C
Solanoscone, 2,3-dehydro-	EtOH	250(3.53)	102-1819-83
$C_{15}H_{20}OS$			
Spiro[6H-cyclohepta[b]thiophene-6,1'-cyclohexan]-4(5H)-one, 7,8-dihydro-2-methyl-	n.s.g.	257(4.0)	42-0303-83
$C_{15}H_{20}O_2$			
Alantolactone	EtOH	212(3.98)	162-0032-83
Alloalantalactone	MeOH	212(3.98)	2-0286-83
Azuleno[5,6-c]furan-4-ol, 4,4a,5,6,7-7a,8,9-octahydro-6,6-dimethyl-8-methylene-, [4S-(4α,4aα,7aα)]-	EtOH	none above 210 nm	88-4631-83
Bicyclo[8.3.2]pentadeca-10(15),11,13-trien-14-one, 15-hydroxy-	MeOH	262(4.3),382(3.9)	88-3879-83
Cyclodeca[b]furan-4(5H)-one, 6,7,8,11-tetrahydro-3,6,10-trimethyl-, (E)-	EtOH	211(4.03),262(3.00)	102-1207-83
Methanone, 3-furanyl[4-methyl-2-(2-methylpropyl)-2-cyclopenten-1-yl]-(isomyomontanone)	EtOH	215(4.34),252(3.95)	88-1749-83
$C_{15}H_{20}O_2S$			
Benzene, [(1-methyl-1,3-octadienyl)-sulfonyl]-, (E,E)-	MeOH	260(4.37)	88-4315-83

Compound	Solvent	$\lambda_{max}(\log \epsilon)$	Ref.
$C_{15}H_{20}O_3$			
9aH-Benz[e]indene-9a-carboxylic acid, 1,2,3,3a,4,6,7,8,9,9b-decahydro-4-oxo-, methyl ester	EtOH	239(3.98)	44-4642-83
9bH-Benz[e]indene-9b-carboxylic acid, 1,2,3,5,5a,6,7,8,9,9a-decahydro-5-oxo-, methyl ester	EtOH	239(3.99)	44-4642-83
Ethanone, 1-[2,4-dimethoxy-3-(3-methyl-2-butenyl)phenyl]-	MeOH	268(4.39),302(3.93)	2-0274-83
Ethanone, 1-[2,4-dimethoxy-5-(3-methyl-2-butenyl)phenyl]-	MeOH	267(4.10),310(3.97)	2-0274-83
3-Furancarboxylic acid, 5-(2,6-dimethyl-5,7-octadienyl)-, (E)-	n.s.g.	212(3.85),231(4.09), 253(3.46),262(3.40)	12-0371-83
(Z)-	n.s.g.	212(4.03),235(4.17), 263(3.46),275(3.22)	12-0371-83
Illudin M	EtOH	228(4.14),318(3.56)	162-0714-83
Laureacetal C	EtOH	217(3.73)	138-0029-83
Naphtho[2,3-b]furan-2,6(3H,4H)-dione, 3a,7,8,8a,9,9a-hexahydro-3,5,8a-trimethyl-	EtOH	246(4.06)	94-3397-83
isomer	EtOH	247(4.20)	94-3397-83
Naphtho[2,3-b]furan-2(3H)-one, decahydro-4a-hydroxy-8a-methyl-3,5-bis(methylene)-	EtOH	215(3.9)	102-2767-83
Naphtho[2,3-b]furan-4(4aH)-one, 5,6,7-8,8a,9-hexahydro-8a-hydroxy-3,4a,5-trimethyl-, (4aα,5α,8aα)-(±)-	EtOH	266(3.57)	94-3544-83
2,4,6-Octatrienal, 7-(2-carboxycyclopentyl)-3-methyl-	EtOH	335(4.38)	35-4033-83
Phomenone, 13-deoxy-	EtOH	245(4.34)	102-2082-83
Terrecyclic acid A	EtOH	236(3.80)	158-83-86
$C_{15}H_{20}O_4$			
Barrelin	EtOH	214(2.63)	102-0777-83
Benghalensin A	EtOH	242(4.15)	44-5318-83
Benghalensin B	EtOH	241(3.95)	44-5318-83
2,5(3H,4H)-Benzofurandione, tetrahydro-6-methyl-3-methylene-6-(4-oxo-pentyl)-, [3aR-(3aα,6α,7aα)]-	EtOH	220(4.3)	39-2705-83C
2H-1-Benzopyran-3-carboxylic acid, 3,4-dihydro-7-methoxy-2,2-dimethyl-, ethyl ester, (±)-	EtOH	211(4.28),223s(3.99), 283(3.60),289(3.56)	39-0197-83C
1,2-Cyclobutanedicarboxylic acid, 3-(dicyclopropylmethylene)-, dimethyl ester, cis	hexane	210(3.84)	44-1500-83
trans	hexane	212(3.92)	44-1500-83
Ethanone, 1-[2-hydroxy-4-(methoxymethyl)-3-(3-methyl-2-butenyl)phenyl]-	MeOH	235(3.28),320(3.95)	2-1061-83
Illudin S	EtOH	233(4.12),319(3.56)	162-0714-83
Ivaxillin, 11(13)-dehydro-	MeOH	214(4.25)	102-2773-83
Naphtho[1,2-b]furan-2,8(3H,8H)-dione, 3a,5,5a,6,7,9b-hexahydro-6-hydroxy-3,5a,9-trimethyl-	MeOH	209s(3.50),243(4.02)	39-0355-83C
Naphtho[1,2-b]furan-2(3H)-one, decahydro-9a-hydroperoxy-5a-methyl-3,9-bis(methylene)-	MeOH	214(4.17)	39-0355-83C
Naphtho[2,3-b]furan-2(3H)-one, 3a,4-4a,5,8,8a,9,9a-octahydro-5-hydroperoxy-5,8a-dimethyl-3-methylene-(5-deoxy-5-hydroperoxytelekin)	EtOH	210(4)	102-2767-83
(5-deoxy-5-hydroperoxy-5-epitelekin)	EtOH	212(4.3)	102-2767-83

Compound	Solvent	λ_{max}(log ϵ)	Ref.
2,4-Pentadienoic acid, 5-(1-hydroxy-2,6,6-trimethyl-4-oxo-2-cyclohexen-1-yl)-3-methyl- (abscisic acid)	MeOH	252(4.40)	162-0002-83
2H-Pyran-5-carboxylic acid, 3,4-dihydro-2,2-dimethyl-4-oxo-6-(1,3-pentadienyl)-, ethyl ester, (E,E)-	EtOH	334(4.37)	118-0948-83
$C_{15}H_{20}O_4Se$			
β-D-Ribofuranoside, methyl 2,3-O-(1-methylethylidene)-5-Se-phenyl-5-seleno-	EtOH	243(3.54)	78-0759-83
$C_{15}H_{20}O_5$			
Ethanone, 1-(3,4-dihydro-7-hydroxy-3,4-dimethoxy-2,2-dimethyl-2H-1-benzopyran-6-yl)-	EtOH	287(3.92),320(3.96)	39-0883-83C
2,4-Pentadienoic acid, 5-[1-hydroxy-2-(hydroxymethyl)-6,6-dimethyl-4-oxo-2-cyclohexen-1-yl]-3-methyl-, [S-(Z,E)]-	MeOH	256(3.24)	102-1277-83
Phaseoline	EtOH	245(4.00)	158-83-168
$C_{15}H_{20}O_6$			
Benzeneacetic acid, 2-(2-ethoxy-1,1-dimethyl-2-oxoethoxy)-4-methoxy-	EtOH	210(4.20),228(4.04), 280(3.51),285(3.30)	39-1431-83C
1H-Benzocycloheptene-1,1-dicarboxylic acid, decahydro-4,5-dioxo-, dimethyl ester	MeOH	288(3.92)	5-0112-83
	+ t-BuOK	306(3.95)	5-0112-83
	MeCN	288(3.92)	5-0112-83
	H_2SO_4	310(4.11)	5-0112-83
Vomitoxin	EtOH	218(3.65)	162-1440-83
$C_{15}H_{20}O_7$			
Nivalenol	MeOH	218(3.80)	162-0955-83
$C_{15}H_{21}ClN_3O_3P$			
1,3,2-Dioxaphosphorinane, 2-chloro-2-[[2,3-dihydro-3-(1-methyl-2-phenylethyl)-1,2,3-oxadiazol-5-yl]imino]-2,2-dihydro-4-methyl-	MeCN	300(3.80)	65-1263-83
$C_{15}H_{21}ClO_2$			
Laurencial	EtOH	227(4.03)	138-0299-83
$C_{15}H_{21}NO$			
Cyclohexanone, 2-[5-(dimethylamino)-2,4-pentadienylidene]-6-ethylidene-	EtOH	480(4.46)	70-0780-83
	$CHCl_3$	460(--)	70-0780-83
1H-Isoindole, 2,3-dihydro-1,1,3,3-tetramethyl-2-(2-propenyloxy)-	EtOH	214(3.41),257(2.85), 263(3.00),270(3.00)	12-1957-83
2,4,6,8-Nonatetraenal, 5-methyl-9-piperidino-, (all-E)-	MeOH	475(4.87)	35-0646-83
Pyrrolo[1,2-a]quinoline, 1,2,3,3a,4,5-hexahydro-2-methoxy-3,3-dimethyl-	MeCN	263(3.75)	35-1204-83
$C_{15}H_{21}NO_2$			
4(5H)-Benzofuranone, 5-[[bis(1-methylethyl)amino]methylene]-6,7-dihydro-	EtOH	251.5(3.97),356(4.30)	111-0113-83
3-Cyclohexene-1-carbonitrile, 6-(3,3-dimethyl-1,2-dioxobutyl)-3,4-dimethyl-	MeOH	297(2.90)	12-0963-83
1H-Pyrrolo[1,2-c][1,3]oxazin-1-one, 4,6-bis(1,1-dimethylethyl)-	EtOH	223(3.97),228(3.96), 271(3.97)	138-0743-83

Compound	Solvent	λ max(log ε)	Ref.
$C_{15}H_{21}NO_3$			
2H-Benzo[b]quinolizin-2-ol, 1,3,4,6-11,11a-hexahydro-8,9-dimethoxy-	n.s.g.	234(4.16),286(3.81)	12-0149-83
3-Heptanone, 2,6-dimethyl-6-nitro-5-phenyl-	MeOH	259(1.70),265(1.60)	12-0527-83
2-Pyrrolidinone, 5-(2,4-dihydroxy-6-pentylphenyl)-	EtOH	284.5(3.42)	150-1156-83M
2-Pyrrolidinone, 5-(2,6-dihydroxy-4-pentylphenyl)-	EtOH	283.5(3.19)	150-1156-83M
$C_{15}H_{21}NO_4$			
3,4-Pyrrolidinediol, 2-[1-(4-methoxy-phenyl)ethyl]-, 3β-acetate	MeOH	223(4.02),275(3.35), 282(3.27)	87-0469-83
β-D-Ribofuranosylamine, 2,3-O-(1-methylethylidene)-N-(phenylmethyl)-	EtOH	262(3.94)	136-0001-83B
$C_{15}H_{21}NO_7$			
Sesbanimide	n.s.g.	<220 (end absorption)	35-3739-83
$C_{15}H_{21}NSSi$			
Pyrrolidine, 2-[2-(trimethylsilyl)-benzo[b]thien-4-yl]-	n.s.g.	237(4.11),268(3.71), 276(3.73),298(3.38), 308(3.40)	44-3428-83
$C_{15}H_{21}N_2O_5$			
1(2H)-Pyridinyloxy, 4-[2-[(2,5-dioxo-1-pyrrolidinyl)oxy]-2-oxoethyl]-3,6-dihydro-2,2,6,6-tetramethyl-	MeCN	221s(3.48),445(0.87)	70-1659-83
$C_{15}H_{21}N_2O_9P$			
Thymidine, 5'-O-1,3,2-dioxaphosphorin-an-2-yl-, 3'-acetate, P-oxide	EtOH	268(3.97)	87-1153-83
$C_{15}H_{21}N_3O$			
Pyrazolidinium, 5-hexyl-3-oxo-1-(2-pyridinylmethylene)-, hydroxide, inner salt	toluene	364(4.35)	48-0205-83
	MeCN	354(4.38)	48-0205-83
Pyrazolidinium, 5-hexyl-3-oxo-1-(4-pyridinylmethylene)-, hydroxide, inner salt	toluene	357(4.38)	48-0205-83
	MeCN	355(4.41)	48-0205-83
$C_{15}H_{21}N_3O_2$			
Cyclooctanecarbonitrile, 2-(1,2,3,4-tetrahydro-1,3-dimethyl-2,4-dioxo-5-pyrimidinyl)-	MeCN	260(3.91)	35-0963-83
$C_{15}H_{21}N_3O_4$			
1-Triazenium, 2,3-bis[(1-methylethoxy)-carbonyl]-1-(4-methylphenyl)-, hydroxide, inner salt	EtOH	238(3.79),363(4.30)	33-1599-83
$C_{15}H_{21}N_3O_4S$			
1-Carbapen-2-em-3-carboxylic acid, 6-[(R)-1-hydroxyethyl]-2-[(S)-1-acetimidoylpyrrolidin-3-ylthio]-, (5R,6S)- (RS-533)	H_2O	298(4.02)	158-1034-83
$C_{15}H_{21}N_6O_3P$			
Phosphonic acid, [2-(1H-imidazol-1-yl)-ethyl]-, bis[2-(1H-imidazol-1-yl)-ethyl] ester, dihydrochloride	EtOH	280(1.47)	65-1555-83

Compound	Solvent	$\lambda_{max}(\log \epsilon)$	Ref.
monohydrobromide dihydrochloride	EtOH	280(0.72)	65-1555-83
$C_{15}H_{22}$			
Cyclohexene, 4-(3-ethenyl-5-methyl-3,4-hexadienyl)-	EtOH	224(4.27)	70-0569-83
Spiro[4.5]dec-6-ene, 1-methyl-8-methylene-4-(1-methylethenyl)-, [1R-(1α,4β,5β)]-	MeOH	237(4.33)	44-0670-83
$C_{15}H_{22}Br_2O$			
Spiro[5.5]undec-7-en-1-ol, 3-bromo-9-(bromomethylene)-1,5,5-trimethyl-	EtOH	246(4.59)	88-0847-83
$C_{15}H_{22}NO_2$			
Pyrrolidinium, 2-carboxy-1-(3,7-dimethyl-2,4,6-octatrienylidene)-, [S-(E,E,?)]-, perchlorate	$CHCl_3$	415(4.48)	35-4033-83
$C_{15}H_{22}NO_4$			
Pyrrolidinium, 2-carboxy-1-[3-(2-carboxycyclopentyl)-2-butenylidene]-, perchlorate	$CHCl_3$	282(4.00)	35-4033-83
$C_{15}H_{22}N_2O$			
Pyrrolo[1,2-c]pyrimidin-1(2H)-one, 4,6-bis(1,1-dimethylethyl)-	EtOH	277(4.04)	138-0743-83
$C_{15}H_{22}N_2O_5$			
1,3,6-Oxadiazepine, 4,7-dimethoxy-2-[1-methylene-3-[(tetrahydro-2H-pyran-2-yl)oxy]propyl]-	EtOH	220(4.23),354(3.48)	77-0399-83
$C_{15}H_{22}N_2O_5S$			
4(1H)-Pyrimidinone, 2-(methylthio)-6-(2,2,2',2'-tetramethyl[4,4'-bi-1,3-dioxolan]-5-yl)-, [4S-[4α(S*),5β]]-	pH 1	240(3.97),280(3.95)	136-0333-83M
	pH 7	255(4.01),280(3.52)	136-0333-83M
	pH 11	240(4.01),280(3.81)	136-0333-83M
$C_{15}H_{22}O$			
Bicyclo[5.3.0]dec-5-en-10-one, 2,6-dimethyl-9-(1-methylethylidene)-, (1R*,2R*,7R*)-	EtOH	265(4.08)	88-0947-83
(1R*,2R*,7S*)-	EtOH	256(3.96)	88-0947-83
Bicyclo[3.3.1]non-3-en-2-one, 5-methyl-1-(2-methylenebutyl)-, (±)-	MeOH	231(3.87)	44-2318-83
Cyclohexanol, 3-(cyclohexylideneethylidene)-4-methylene-, (Z)-(±)-	MeOH	260(4.02)	88-4257-83
5H-Cyclopropa[a]naphthalen-5-one, 1,1a,2,3,6,7,7a,7b-octahydro-1,1,7,7a-tetramethyl-, [1aR-(1aα,7α,7aα,7bα)]-	EtOH	243(3.85)	102-2753-83
Furan, 2-(2,6-dimethylocta-5,7-dienyl)-4-methyl-, (E)-	n.s.g.	206s(3.75),229(4.30), 272(3.68)	12-0371-83
2-Hepten-4-one, 2-methyl-6-(4-methylene-2-cyclohexen-1-yl)- (curlone)	MeOH	235.5(4.32)	102-0596-83
Tricyclo[5.3.0.0^{2,6}]decan-3-one, 1,7-dimethyl-4-(1-methylethylidene)-, (1R*,2R*,6R*,7S*)-	EtOH	247(3.99)	88-0947-83
(1R*,2R*,6S*,7S*)-	EtOH	265(4.09)	88-0947-83
(1R*,2S*,6R*,7S*)-	EtOH	262(3.85)	88-0947-83
(1R*,2S*,6S*,7S*)-	EtOH	244(3.78)	88-0947-83

Compound	Solvent	$\lambda_{max}(\log \epsilon)$	Ref.
$C_{15}H_{22}O_2$			
1,3-Cyclohexanediol, 5-(cyclohexylideneethylidene)-4-methylene-, (1α,3β,5Z)-(±)-	ether	261(4.27)	88-4257-83
2,4,6,10-Dodecatetraenoic acid, 3,7,11-trimethyl-, (E,E,E)-	hexane	307(4.58)	33-0494-83
3-Furanmethanol, α-[(4-methyl-2-(2-methylpropyl)-1-cyclopenten-1-yl]-	EtOH	209(4.04),271(2.95)	88-1749-83
Helminthosporal	EtOH	266(4.04)	162-0669-83
Isogermacrone epoxide	EtOH	251(3.88)	78-3397-83
2-Naphthaleneacetaldehyde, decahydro-2-hydroxy-4a-methyl-α,8-bis(methylene)- (7-hydroxycostal)	n.s.g.	215(3.73)	77-0353-83
1,2-Naphthalenedicarboxaldehyde, 1,4,4a,5,6,7,8,8a-octahydro-5,5,8a-trimethyl- (polygodial)	MeOH	230(3.85)	44-1866-83
2(3H)-Naphthalenone, 4,4a,5,6,7,8-hexahydro-6-hydroxy-4a,5-dimethyl-3-(1-methylethenyl)- (petasol)	EtOH	236(3.97)	102-1619-83
isopetasol	EtOH	247(3.95),280(3.80)	102-1619-83
	EtOH	243(3.89),278(3.51)	102-1619-83
2,4-Pentadienoic acid, 3-methyl-5-(2,6,6-trimethyl-2-cyclohexen-1-yl)-, (E,E)-	EtOH	261(4.39)	33-1148-83
$C_{15}H_{22}O_3$			
4aH-Benzocycloheptene-4a-carboxylic acid, 2,3,4,5,6,7,8,9-octahydro-4,4-dimethyl-2-oxo-, methyl ester	EtOH	243(3.96)	44-4642-83
Bicyclo[4.3.1]dec-7-ene-1-carboxylic acid, 7,9,9-trimethyl-10-oxo-, methyl ester	EtOH	201(3.61),284(1.94)	44-4642-83
Carabrol	MeOH	213(3.92)	102-2773-83
2-Hexenoic acid, 5-(3-methyl-2H-pyran-6-yl)-2,5-dimethyl-, methyl ester	pentane	215(4.07),282(3.56)	33-1638-83
2-Naphthaleneacetic acid, decahydro-8,8a-dimethyl-α-methylene-4-oxo- (fluorensic acid)	EtOH	203(3.95)	39-0161-83C
$C_{15}H_{22}O_4$			
1-Butanone, 1-(4-hydroxy-2,6-dimethoxy-3,5-dimethylphenyl)-2-methyl-, (±)-	EtOH	217s(4.12),262(3.60)	88-2531-83
	EtOH-NaOH	331(3.94)	88-2531-83
Cyclodeca[b]furan-2(3H)-one, 3a,4,5,8,9,11a-hexahydro-4,9-dihydroxy-3,6,10-trimethyl-	EtOH	218(3.31)	102-0197-83
Isogermacrone diepoxide	EtOH	254(3.62)	78-3397-83
$C_{15}H_{22}O_4$(cont.)			
Mukaadial	EtOH	210(3.63),235s(3.30)	138-0979-83
$C_{15}H_{22}O_5$			
Naphtho[1,2-b]furan-2(3H)-one, 3a,4,5,5a,6,7,8,9b-octahydro-8-hydroperoxy-6-hydroxy-3,5a,9-trimethyl-	MeOH	212(3.68)	39-0355-83C
$C_{15}H_{23}NO$			
2-Aza-10-tert-butyl-6,8-dimethyltricyclo[4.4.0.$0^{2,8}$]dec-9-en-7-one	MeOH	217(3.78),317(2.48)	35-3273-83
$C_{15}H_{23}NOSi_2$			
Isoxazole, 3-phenyl-4,5-bis(trimethyl-	C_6H_{12}	215(3.87)	44-3189-83

Compound	Solvent	λ_{max}(log ϵ)	Ref.
silyl)- (cont.)			44-3189-83
$C_{15}H_{23}NO_2$			
1,3-Benzenediol, 5-pentyl-2-(2-pyrrolidinyl)-	EtOH	283(3.18)	150-1156-83M
1,3-Benzenediol, 5-pentyl-4-(2-pyrrolidinyl)-	EtOH	283.5(3.22)	150-1156-83M
2(3H)-Naphthalenone, 4,4a,5,6,7,8-hexahydro-4a-methyl-1-morpholino-	EtOH	244(4.04)	107-1013-83
2(3H)-Naphthalenone, 4,4a,5,6,7,8-hexahydro-4a-methyl-8-morpholino-	EtOH	238.5(3.95)	107-1013-83
$C_{15}H_{23}NO_3$			
2,5-Cyclohexadien-1-one, 2,6-bis(1,1-dimethylethyl)-4-methyl-4-nitro-	C_6H_{12}	234(4.04)	12-2339-83
1(2H)-Pyridinepropanoic acid, 4-heptyl-2-oxo-	EtOH	231(3.80),298(3.83)	94-2552-83
$C_{15}H_{23}NO_4$			
1,3-Dioxolane-2-propanamine, α-[(2,3-dimethoxyphenyl)methyl]-	MeOH	218s(3.93),278(3.09)	83-0773-83
$C_{15}H_{23}NO_5$			
3,9-Dioxabicyclo[4.2.1]nonan-4-one, 7,8-isopropylidenedioxy-2,2-dimethyl-5-(dimethylaminomethylene)-, (1S*,6R*,7R*,8S*)-	MeOH	300(4.20)	18-2680-83
$C_{15}H_{23}N_3$			
1-Buten-1-amine, N-(1-methylethyl)-3-[(1-methylethyl)imino]-4-(2-pyridinyl)-, monohydrobromide	MeOH	262(3.62),268(3.58), 317(4.58)	4-0407-83
$C_{15}H_{23}N_3O$			
2,4-Cyclohexadien-1-one, 6-(3-azidopropyl)-2,6-dimethyl-4-(1,1-dimethylethyl)-	MeOH	315(3.62)	35-3273-83
7H-1,2,3-Triazolo[4,5,1-ij]quinolin-7-one, 9a-(1,1-dimethylethyl)-4,5,6,6a,9a,9b-hexahydro-	MeOH	228(4.12)	35-3273-83
$C_{15}H_{23}N_3O_4$			
1,2-Triazanedicarboxylic acid, 3-(4-methylphenyl)-, bis(1-methylethyl) ester	EtOH	234(4.09),286(3.18)	33-1599-83
$C_{15}H_{23}N_3O_6$			
Glycine, N-[5-(5-ethylhexahydro-2,4,6-trioxo-5-pyrimidinyl)-1-oxopentyl]-, ethyl ester	MeOH	212.5(3.87)	73-0137-83
5-Pyrimidinepentanoic acid, 2-[(2-ethoxy-2-oxoethyl)amino]-1,4-dihydro-6-hydroxy-4-oxo-, ethyl ester	50% MeOH	233(3.88),275(4.14)	73-0304-83
	+ HCl	232(3.842),275(4.111)	73-0304-83
	+ NaOH	243(3.836),274(4.160)	73-0304-83
$C_{15}H_{23}N_3O_7$			
2-Cyclohexen-1-one, 2,6-bis(1,1-dimethylethyl)-4-methyl-4,5,6-trinitro-, (4α,5α,6α)-	$CHCl_3$	242(3.99)	12-2339-83
$C_{15}H_{24}$			
1H-Cyclopentacyclododecene, 2,4,5,6,7-	EtOH	246(4.23)	33-2519-83

Compound	Solvent	$\lambda_{max}(\log \epsilon)$	Ref.
8,9,10,11,12-decahydro- (cont.)			33-2519-83
α-Farnesene, (E,E)-	EtOH	233(4.43)	162-0567-83
α-Farnesene, (Z,E)-	EtOH	238(4.05)	162-0567-83
β-Farnesene, (E)-	hexane	224(4.15)	162-0568-83
β-Farnesene, (Z)-	hexane	224(4.24)	162-0568-83
Germacrene D	hexane	259(3.52)	94-1743-83
δ-Selinene, (+)-	MeOH	241(4.23),247(4.25)	94-1991-83
Striatene	n.s.g.	232s(--),238(4.39), 245s(--)	18-1125-83
$C_{15}H_{24}N_2O$			
2(3H)-Naphthalenone, 4,4a,5,6,7,8-hexahydro-4a-methyl-1-piperazino-	EtOH	242(4.04)	107-1013-83
2(3H)-Naphthalenone, 4,4a,5,6,7,8-hexahydro-4a-methyl-8-piperazino-	EtOH	238.5(3.98)	107-1013-83
$C_{15}H_{24}N_2O_6$			
2-Cyclohexen-1-one, 2,6-bis(1,1-dimethylethyl)-6-hydroxy-4-methyl-4,5-dinitro-, (4α,5α,6α)-	$CHCl_3$	240(3.68)	12-2339-83
$C_{15}H_{24}N_3O_2$			
2-Aza-1-azoniabicyclo[5.2.0]nona-3,5,9-triene, 9-(dimethylamino)-2-(ethoxycarbonyl)-5,8,8-trimethyl-, chloride	MeOH	265(3.80)	5-1361-83
$C_{15}H_{24}N_4$			
2-Pentenenitrile, 3,3'-(1,3-propanediyldiimino)bis[4-methyl-	EtOH	262(4.48)	39-1223-83C
$C_{15}H_{24}N_4O$			
4H-Pyrazolo[3,4-d]pyrimidin-4-one, 1,5-dihydro-1,6-dimethyl-5-octyl-	EtOH	210(4.38),256(3.82)	11-0171-83B
$C_{15}H_{24}N_8O_3$			
L-Lysine, N^2-[3-(6-amino-9H-purin-9-yl)-DL-alanyl]-, methyl ester, dihydrobromide	pH 1	259(4.18)	103-0328-83
	pH 7	268(4.12)	103-0328-83
$C_{15}H_{24}O$			
Bisabolene, α-oxo-	$CHCl_3$	230(4.1)	100-0424-83
Cadina-4,10(15)-dien-3-one	EtOH	238(3.98)	33-1835-83
1,5,7,11-Dodecatetraen-3-ol, 3,7,11-trimethyl-, [S-(E,E)]- (isofokienol)	EtOH	231(4.21)	102-2235-83
Drim-8-en-7-one	EtOH	249.5(4.14)	105-0139-83
Fokienol	EtOH	225(4.29)	102-2235-83
Germacra-4(15),5(E),10(14)-trien-1β-ol	EtOH	239(4.17)	94-1743-83
1-Naphthalenol, 1,2,3,7,8,8a-hexahydro-4,8a-dimethyl-6-(1-methylethyl)-	hexane	240(4.30),246(4.32), 253s(4.15)	44-4410-83
7-Oxabicyclo[4.1.0]heptane, 1,2,3-trimethyl-2-(3-methyl-2,4-pentadienyl)-, α-	C_6H_{12}	238(4.26)	18-1125-83
β-	C_6H_{12}	237(4.29)	18-1125-83
11-Oxabicyclo[8.1.0]undec-5-ene, 1-methyl-7-methylene-4-(1-methylethyl)-	hexane	234(4.10)	94-1743-83
$C_{15}H_{24}OS_2$			
14-Thiadispiro[5.1.5.2]pentadecane-7-thione, 15-methoxy-	CCl_4	264(4.04),510(2.47)	44-4482-83

Compound	Solvent	λ_{max}(log ϵ)	Ref.
$C_{15}H_{24}O_2$			
Bicyclo[8.1.0]undecane-2,6-dione, 1,5,8,8-tetramethyl-, [1S-(1R*,5S*,10R*)]-	EtOH	208(3.54),288(1.86)	102-1507-83
2(1H)-Naphthalenone, octahydro-1,4a-dimethyl-7-(2-methyloxiranyl)-	MeOH	280(1.53)	23-1111-83
$C_{15}H_{24}O_3$			
1(4H)-Naphthalenone, 4a,5,6,7,8,8a-hexahydro-8a-hydroxy-4-(hydroxymethyl)-3,4a,8,8-tetramethyl-, [4S-(4α,4aα,8aβ)]- (uvidin E)	MeOH	244.5(3.95)	39-2739-83C
2(1H)-Naphthalenone, octahydro-5,6-dihydroxy-5,8a-dimethyl-3-(1-methylethylidene)-, [4aR-(4aα,5α,6α,8aβ)]-	EtOH	258(4.03)	100-0671-83
$C_{15}H_{24}O_3S$			
1,4-Dioxaspiro[4.5]decane, 8-[(tert-butylsulfinyl)ethylidene]-7-(methylene)-	EtOH	235(4.06)	78-1123-83
1,4-Dioxaspiro[4.5]dec-7-ene, 7-[(butylsulfinyl)methyl]-8-ethenyl-	EtOH	246(4.17)	78-1123-83
$C_{15}H_{24}O_3Si$			
Butanoic acid, 2-(2-methylphenoxy)-, (trimethylsilyl)methyl ester	heptane	217(3.88),272(3.16)	65-1039-83
Butanoic acid, 2-(4-methylphenoxy)-, (trimethylsilyl)methyl ester	heptane	224(3.95),280(3.20)	65-1039-83
$C_{15}H_{24}O_{11}$			
Scabrosidol	MeOH	218(3.8)	100-0614-83
$C_{15}H_{25}NOSi_2$			
Isoxazole, 4,5-dihydro-3-phenyl-5,5-bis(trimethylsilyl)-	C_6H_{12}	290(3.91)	44-3189-83
$C_{15}H_{25}NO_2$			
2-Cyclohexen-1-one, 3,5,5-trimethyl-2-(2-morpholinoethyl)-	EtOH	245.5(3.97)	23-0078-83
$C_{15}H_{25}NO_2Si$			
Glycine, N-phenyl-, (triethylsilyl)-methyl ester	C_6H_{12}	242(4.22),296(3.53)	65-0814-83
$C_{15}H_{25}N_3O_4$			
5-Pyrimidinepentanoic acid, 2-(butylamino)-1,4-dihydro-6-hydroxy-4-oxo-, ethyl ester	50% MeOH	270(4.226)	73-0304-83
	+ HCl	238(3.909),269(4.108)	73-0304-83
	+ NaOH	242(3.820),275(4.132)	73-0304-83
$C_{15}H_{26}N_4$			
1,2,4,5-Tetrazine, 3,6-dicyclohexyl-1,4-dihydro-1-methyl-	dioxan	229(3.68),296(2.04)	24-2261-83
$C_{15}H_{26}N_4O_2$			
Piperidine, 1,1'-(2,3-dihydro-1-methyl-4-nitro-1H-pyrrole-2,3-diyl)bis-, trans	MeOH	230(3.7),376(4.3)	44-0162-83
$C_{15}H_{26}O$			
2,6,10-Dodecatrien-1-ol, 3,7,11-trimethyl-, trans-trans (farnesol)	n.s.g.	194(4.46)	162-0568-83

Compound	Solvent	λ_{max}(log ϵ)	Ref.
$C_{15}H_{26}O_2$			
6,11-Dodecadiene-2,3-diol, 2,6-dimethyl-10-methylene-, (E)-(+)-	EtOH	225(4.45)	102-1245-83
1,9,11-Dodecatriene-3,7-diol, 3,7,11-trimethyl- (hydroxyfokienol)	EtOH	226(4.37)	102-2235-83
2(1H)-Naphthalenone, octahydro-6-(1-methylethyl)-1,4a-dimethyl-	MeOH	282(1.45)	23-1111-83
Spiro[4.5]decan-7-one, 2-(1-hydroxy-1-methylethyl)-6,10-dimethyl-, [5S-[5α(S*),6β,10α]]-	EtOH	290(1.62)	94-1991-83
$C_{15}H_{26}O_3$			
6-Nonen-3-one, 2,6-dimethyl-9-(2-methyl-1,3-dioxolan-2-yl)-	EtOH	295(2.10)	22-0175-83
1,2-Propanediol, 3-(1,3,5-dodecatrienyloxy)-, (E,E,E)-	n.s.g.	265(4.20)	54-0465-83
(Z,E,E)-	n.s.g.	267(4.23)	54-0465-83
$C_{15}H_{27}OS_5$			
Sulfonium, [2-(ethylthio)-2-oxo-1-[1,2,3-tris(ethylthio)-2-cyclopropen-1-yl]ethyl]dimethyl-, perchlorate	MeCN	244(3.92),282(3.85), 306(4.00)	18-0171-83
tetrafluoroborate	MeCN	243(4.23),281s(4.15), 306(4.34)	18-0171-83
$C_{15}H_{27}O_2S_4$			
Sulfonium, [2-ethoxy-2-oxo-1-[1,2,3-tris(ethylthio)-2-cyclopropen-1-yl]ethyl]dimethyl-, perchlorate	MeCN	270(4.15),280(4.08)	18-0171-83
tetrafluoroborate	MeCN	270(4.52),281s(4.45)	18-0171-83
$C_{15}H_{27}S_3$			
Cyclopropenylium, tris[(1,1-dimethylethyl)thio]-, tetrafluoroborate	MeOH	272(3.55)	18-0171-83
$C_{15}H_{28}O_2S$			
Cyclohexanone, 3-[(1,1-dimethylethyl)thio]-5-(1-methoxy-1-methylethyl)-2-methyl-, [2S-(2α,3α,5β)]-	C_6H_{12}	295(1.30)	49-0195-83
	MeOH	245(2.46),291(1.60)	49-0195-83
$C_{15}H_{28}O_3$			
1,2-Propanediol, 3-(1,3-dodecadienyloxy)-, (E,E)-	n.s.g.	237(4.26)	54-0465-83
(Z,E)-	n.s.g.	239(4.15)	54-0465-83
$C_{15}H_{28}O_4Se$			
Propanedioic acid, ethyl[3-[(1-methylethyl)seleno]propyl]-, diethyl ester	MeOH	218.7(3.99),255(2.07)	101-0171-83S
$C_{15}H_{28}O_4Te$			
Propanedioic acid, ethyl[3-[(1-methylethyl)telluro]propyl]-, diethyl ester	MeOH	232.8(3.81),280(2.29), 355(1.84)	101-0171-83S
	MeOH	233.2(3.70),270(2.48), 355.0(1.92)	101-0171-83S
$C_{15}H_{29}NO_2Si_2$			
1H-Azepine-1-carboxylic acid, 4,5-dihydro-4,5-bis(trimethylsilyl)-, ethyl ester, trans	MeOH	240(4.16)	18-0175-83

Compound	Solvent	$\lambda_{max}(\log \epsilon)$	Ref.
$C_{15}H_{29}O_6PS$			
Phosphorus(1+), (ethanethiolato)[2-ethoxy-1-(ethoxycarbonyl)-2-oxo-ethyl]bis(2-propanolato)-, hydroxide, inner salt, (T-4)-	hexane	215(3.48),253(3.60)	99-0023-83
	EtOH	212(--),257(--)	99-0023-83
$C_{15}H_{30}$			
3-Hexene, 3-(1,1-dimethylethyl)-2,2,4,5,5-pentamethyl-	n.s.g.	234(3.70)	88-4821-83
$C_{15}H_{30}NO_4P$			
Phosphorus(1+), [2-ethoxy-1-(ethoxy-carbonyl)-2-oxoethyl]diethyl(N-ethylethanaminato)-, hydroxide, inner salt, (T-4)-	hexane	228s(--),235(4.02)	99-0023-83
	EtOH	212(--),241(--)	99-0023-83
$C_{15}H_{30}NO_6P$			
Phosphorus(1+), diethoxy[2-ethoxy-1-(ethoxycarbonyl)-2-oxoethyl](N-ethylethanaminato)-, hydroxide, inner salt, (T-4)-	hexane	217s(4.07)	99-0023-83
	EtOH	210(3.90),231(3.95)	99-0023-83
Phosphorus(1+), diethoxy[2-ethoxy-1-(ethoxycarbonyl)-2-oxoethyl](2-methyl-2-propanaminato)-, hydroxide, inner salt, (T-4)-	hexane	208(3.95),231(4.10)	99-0023-83
	EtOH	201(--),234(--)	99-0023-83
$C_{15}H_{30}O_2Te$			
Octanoic acid, 8-(hexyltelluro)-, methyl ester	MeOH	233.0(3.66)	101-0031-83M
$C_{15}H_{32}N_4O_4$			
Istamycin A_1	H_2O	end absorption	158-83-40
$C_{15}H_{35}CoN_5$			
Cobalt(2+), (4,8,11-trimethyl-1,4,8,11-tetraazacyclotetradecan-1-ethan-amine-N^{α},N^1,N^4,N^8,N^{11})-, (SP-5-54)-, diperchlorate	H_2O	464(1.46),516(1.40), 581(1.32),660(1.11)	33-2086-83
$C_{15}H_{35}CuN_5$			
Copper(2+), (4,8,11-trimethyl-1,4,8,11-tetraazacyclotetradecan-1-ethan-amine-N^{α},N^1,N^4,N^8,N^{11})-, diperchlorate	H_2O	684(1.31)	33-2086-83
$C_{15}H_{35}N_5Ni$			
Nickel(2+), (4,8,11-trimethyl-1,4,8,11-tetraazacyclotetradecan-1-ethan-amine-N^{α},N^1,N^4,N^8,N^{11})-, diperchlorate	H_2O	379(2.06),588(1.66)	33-2086-83

Compound	Solvent	λ_{max}(log ε)	Ref.
$C_{16}Cl_{12}$			
Pentalene, hexachloro-, dimer	CH_2Cl_2	250(3.57),290s(3.86), 300(3.96),390-460(2.6)	18-0481-83
$C_{16}H_2N_8$			
1-Cyclobutene-1,2-diacetonitrile, α,α'-dicyano-3,4-bis(dicyanomethylene)-, (dianion)	H_2O	400(5.07)	88-1567-83
$C_{16}H_4Cl_2N_2S$			
Biphenyleno[2,3-c]thiophene-4,9-dicarbonitrile, 1,3-dichloro-	EtOH	277s(4.45),281(4.56), 307(4.49),319(4.56), 362s(3.76),380(3.99), 401(4.15),426(4.05)	39-1443-83C
$C_{16}H_5N_5O_6$			
Propanedinitrile, (2,4,7-trinitro-9H-fluoren-9-ylidene)-	CH_2Cl_2	365(4.38)	162-0450-83
$C_{16}H_7ClN_2S$			
Dibenzo[b,f]thiepin-2(or 8),10-dicarbonitrile, 8(or 2)-chloro-	MeOH	232(4.60),277(4.34), 303s(3.92)	73-1187-83
$C_{16}H_8Cl_2O_6$			
2-Anthracenecarboxylic acid, 5,8-dichloro-9,10-dihydro-1,4-dihydroxy-9,10-dioxo-, methyl ester	dioxan	350(3.47),471(3.88)	104-0533-83
$C_{16}H_8N_2O_4$			
5H-Benzo[b]carbazole-6,11-dione, 8-nitro-	EtOH	212(4.56),253(4.45), 299(4.29),404(3.64)	39-2409-83C
[2,2'-Bi-2H-isoindole]-1,1',3,3'-tetrone	EtOH	240(4.25),296(3.68), 303(3.67)	104-2270-83
Indolo[1,2-b]isoquinoline-6,12-dione, 8-nitro-	EtOH	228(4.20),248(4.23), 283(3.73),324s(4.18), 330(3.76),346(3.66), 415(3.80)	39-2413-83C
$C_{16}H_8N_2O_4S$			
9H-[1]Benzothiopyrano[2,3-d]oxazol-9-one, 2-(4-nitrophenyl)-	dioxan	324(4.43),368(4.04)	83-0921-83
$C_{16}H_8O_4$			
2(3H)-Benzofuranone, 3-(2-oxo-3(2H)-benzofuranylidene)-, (E)-	CH_2Cl_2	227(4.15),252(4.10), 254(4.10),395(4.12), 433(4.13)	78-2147-83
$C_{16}H_8O_6$			
Medicagol	EtOH	245(4.29),270(3.91), 297(3.85),310(4.03), 348(4.46)	162-0824-83
$C_{16}H_9BrCl_2N_4$			
4H-[1,2,4]Triazolo[4,3-a][1,4]benzodiazepine, 1-bromo-8-chloro-6-(2-chlorophenyl)-	MeOH	222(4.58),250s(4.06)	73-2395-83
$C_{16}H_9Cl$			
Azuleno[2,1,8-ija]azulene, 5-chloro-	MeOH	240(4.15),284(5.13), 310(4.36),372(3.90), 388s(3.85),394(5.04),	88-0781-83

Compound	Solvent	$\lambda_{max}(\log \epsilon)$	Ref.
(cont.)		460s(3.25),490(3.73), 590(2.60)	88-0781-83
$C_{16}H_9ClN_2O_3$			
1H-Naphth[2,3-g]indole-2,6,11(3H)-trione, 5-amino-4-chloro-	dioxan	294(4.63),513(4.05)	104-1892-83
$C_{16}H_9ClN_2O_4S_2$			
Thiocyanic acid, 2-[(7-chloro-4,5-dioxo-4H,5H-pyrano[3,4-e][1,3]oxazin-2-yl)thio]-1-phenylethyl ester	$CHCl_3$	280(4.10),344(4.17)	24-3725-83
$C_{16}H_9ClO_6$			
2-Anthraceneacetic acid, 3-chloro-9,10-dihydro-1,4-dihydroxy-9,10-dioxo-	dioxan	291(3.85),471(3.85)	104-1900-83
$C_{16}H_9NO_2$			
Fluoranthene, 1-nitro-	MeCN	260(4.75),290(4.67), 392(4.20)	1-0833-83
Fluoranthene, 3-nitro-	MeCN	257(4.50),277(4.48), 383(4.30)	1-0833-83
Fluoranthene, 7-nitro-	MeCN	239(4.84),314(4.43), 364(4.06)	1-0833-83
Fluoranthene, 8-nitro-	MeCN	234(4.84),306(4.52), 326(4.47),370(4.33)	1-0833-83
$C_{16}H_9NO_2S$			
Benzo[e]phenothiazine-6,11-dione	DMF	315(4.31),700(3.01)	88-3567-83
9H-[1]Benzothiopyrano[2,3-d]oxazol-9-one, 2-phenyl-	$CHCl_3$	316(4.31),350(3.97), 364(3.96)	83-0921-83
$C_{16}H_9NO_3$			
2-Pyrenol, 1-nitro-	THF	219(5.34),257.5(5.28), 266(5.48),309.5(5.59), 349(5.30),474(4.49)	138-0347-83
$C_{16}H_9NO_4$			
1H-Naphth[2,3-g]indole-2,6,11(3H)-trione, 5-hydroxy-	dioxan	279(4.08),474(3.99)	104-1892-83
$C_{16}H_9N_5$			
1,1,2,2-Cyclobutanetetracarbonitrile, 3-(1H-indol-1-yl)-	THF	211(4.34),260(3.86), 291(3.59)	103-0765-83
$C_{16}H_{10}ClFN_4O_3$			
3H-1,4-Benzodiazepine-2-carboxaldehyde, 7-chloro-5-(2-fluorophenyl)-α-nitro-, oxime	isoPrOH	243s(4.48),275(4.14), 335(3.72)	4-0551-83
$C_{16}H_{10}Cl_2N_2$			
Pyrazine, 2,6-dichloro-3,5-diphenyl-	EtOH	235s(3.58),257(3.75), 322(3.62)	4-0311-83
$C_{16}H_{10}Cl_2N_4O$			
1H-[1,2,4]Triazolo[4,3-a][1,4]benzodiazepin-1-one, 8-chloro-6-(2-chlorophenyl)-2,4-dihydro-	MeOH	237s(4.27),307(3.14)	73-2395-83
$C_{16}H_{10}Cl_2O_3$			
2,4,6-Cycloheptatrien-1-one, 4-[3-(2,4-	MeOH	254(4.43),320(4.34),	18-3099-83

Compound	Solvent	λ_{max}(log ε)	Ref.
dichlorophenyl)-1-oxo-2-propenyl]-		380s(3.82)	18-3099-83
2-hydroxy-, (E)- (cont.)	MeOH-NaOH	240s(4.25),283(4.29), 445(3.63)	18-3099-83
$C_{16}H_{10}Cl_2O_7$			
Erdin, (+)-	n.s.g.	284(4.32)	162-0525-83
$C_{16}H_{10}Cl_3N_3O_4$			
2,4(1H,3H)-Pyrimidinedione, 5-[4-chloro-1-(3,5-dichlorophenyl)-2,5-dihydro-2,5-dioxo-1H-pyrrol-3-yl]-1,3-dimethyl-	MeOH	236(3.95),272(3.74), 362(3.31)	89-0120-83S
$C_{16}H_{10}Cl_4O_4$			
Benzoic acid, 2,3,4,5-tetrachloro-6-(4-hydroxy-3,5-dimethylbenzoyl)-	EtOH	209(4.74),308(4.21)	24-0970-83
$C_{16}H_{10}Cl_4O_5$			
11H-Dibenzo[b,e][1,4]dioxepin-11-one, 2,4,7,9-tetrachloro-3-hydroxy-8-methoxy-1,6-dimethyl- (diploicin)	n.s.g.	270(3.79)	162-0488-83
$C_{16}H_{10}N_2$			
Benzonitrile, 4,4'-(1,2-ethenediyl)bis-	MeCN	326(4.65)	94-0373-83
$C_{16}H_{10}N_2O$			
4-Isoquinolinecarbonitrile, 1,2-dihydro-1-oxo-3-phenyl-	EtOH	215(4.56),233s(4.29), 244s(4.26),310(4.20), 325s(4.11),340s(3.81)	39-1813-83C
$C_{16}H_{10}N_2O_2$			
5H-Benzo[b]carbazole-6,11-dione, 8-amino-	EtOH	215(4.40),235(4.30), 258(4.37),291(4.53), 380(4.09),502(3.15)	39-2409-83C
[1]Benzopyrano[4,3-c]pyrazol-4(1H)-one, 1-phenyl-	EtOH	260(4.06),270(4.07), 297(3.82),307(3.77)	88-0381-83
Indolo[1,2-b]-2,6-naphthyridine-5,11-dione, 12-methyl- (9λ,8ε)	EtOH	230(4.33),250s(4.38), 256(4.42),274s(4.12), 281(4.04),314s(4.04), 326(4.02),343s(4.14), 403(?)	39-2413-83C
5H-Pyrido[3,4-b]carbazole-5,11-dione, 10-methyl-	EtOH	250(--),280(--), 390(--)	39-2409-83C
$C_{16}H_{10}N_2O_3$			
1H-Naphth[2,3-g]indole-2,6,11(3H)-trione, 5-amino-	dioxan	250(4.61),515(4.03)	104-1892-83
Pyrido[3,4-b]carbazole-5,11(10H)-dione, 7-methoxy-	EtOH	269s(4.20),279(4.25), 324s(3.28),442(3.68)	44-2690-83
$C_{16}H_{10}N_2O_4$			
6H-[2]Benzoxepino[4,3-b]indol-11(12H)-one, 9-nitro-	EtOH	250(4.13),262s(4.01), 343(3.98)	39-2409-83C
$C_{16}H_{10}N_2O_6$			
2,3-Pyridinedicarboxylic acid, 3-methyl 2-(5-oxo-5H-pyrano[4,3-b]pyridin-8-yl) ester	MeOH	208s(3.99),266s(3.60), 270(3.61),310(3.22)	103-0886-83
$C_{16}H_{10}N_2S$			
Benzeneacetonitrile, α-(2-benzothiazo-	MeOH	225s(3.73),227s(4.21),	4-0005-83

Compound	Solvent	λ_{max}(log ε)	Ref.
lylmethylene)-, (Z)- (cont.)		280s(3.73),354(4.48)	4-0005-83
$C_{16}H_{10}N_4O_5S_3$			
4H-Bisthiazolo[3,2-a:4',5'-e]pyrimidinium, 8-hydroxy-5-methyl-2-(methylthio)-7-(4-nitrobenzoyl)-4-oxo-, hydroxide, inner salt	MeCN	217(4.42),260(4.22), 300(4.33),380(4.21)	2-0243-83
$C_{16}H_{10}N_4S_2$			
Pyrimidine, 4,4'-[2,2'-bithiophene]-3,3'-diylbis-	$CHCl_3$	240(4.35),275(4.58)	24-0479-83
$C_{16}H_{10}O$			
Cyclobuta[l]phenanthren-1(2H)-one	3-Mepentane at 77°K	245(--),265s(--), 320s(--),360(--), 365(--)	151-0131-83B
1-Fluoranthenol	MeOH	222(4.81),245(4.79), 270(4.20),275(4.40), 281(4.37),286(4.64), 310(3.82),325(4.08), 368(4.08),380(4.06)	44-2360-83
2-Fluoranthenol	MeOH	238(4.61),256(4.33), 270(4.29),281(4.26), 292(4.28),312(4.55), 327(3.75),347(3.78), 365(3.83)	44-2360-83
3-Fluoranthenol	MeOH	222(4.55),240(4.58), 267(3.92),284(4.16), 296(4.42),306(3.68), 321(3.72),347(3.75), 361(3.84)	44-2360-83
7-Fluoranthenol	MeOH	242(4.44),264(4.26), 270(4.23),284(3.70), 292(3.64),308(3.52), 323(3.70),370(3.94)	44-2360-83
8-Fluoranthenol	MeOH	238(4.56),274(4.02), 286(4.36),297(4.62), 310(3.59),324(3.68), 342(3.68),352(3.64), 359(3.78)	44-2360-83
Methanone, (10-methylene-9(10H)-phenanthrenylidene)-	3-Mepentane at 77°K	247(4.26),262(4.20)	151-0131-83B
$C_{16}H_{10}O_3$			
9,10-Anthracenedione, 2-acetyl-	dioxan	326(3.73)	126-1143-83
Naphth[2,3-b]oxirene-2,7-dione, 1a,7a-dihydro-1a-phenyl-	benzene	308(3.48),348s(2.55)	44-4968-83
	MeCN	231(4.45),262(3.68), 305(3.30),340s(2.45)	44-4968-83
	EPA	304(3.48),336s(2.60)	44-4968-83
$C_{16}H_{10}O_4$			
1,4-Anthracenedione, 9-acetoxy-	EtOH	410(3.72)	99-0420-83
photoinduced form	EtOH	468(4.00)	99-0420-83
9,10-Anthracenedione, 1-acetoxy-	EtOH	330(3.77)	99-0420-83
5H-Benzofuro[6,5-c][2]benzopyran-5-one, 7-methoxy-	MeOH-DMSO	323(3.89)	78-1265-83
Benzoic acid, 2-(1-oxo-1H-2-benzopyran-3-yl)-	EtOH	232(4.452),283(4.12), 291(4.082),323(3.911)	2-1108-83
$C_{16}H_{10}O_5$			
9,10-Anthracenedione, 1-acetoxy-5-hydroxy-	EtOH	340(3.77)	99-0420-83

Compound	Solvent	λ_{max}(log ϵ)	Ref.
9,10-Anthracenedione, 1-acetoxy-8-hydroxy-	EtOH	340(3.72)	99-0420-83
$C_{16}H_{10}O_8$			
Kermesic acid	n.s.g.	276(4.52),312(4.12), 498(3.96)	162-0760-83
$C_{16}H_{10}S$			
Anthra[2,3-b]thiophene	dioxan	276(4.48)	44-4419-83
$C_{16}H_{10}S_4$			
1,3-Benzodithiole, 2,2'-(1,2-ethanediylidene)bis-	MeCN	374(4.81),392(4.94)	77-0004-83
$C_{16}H_{11}BrO_3$			
2,4,6-Cycloheptatrien-1-one, 4-[3-(4-bromophenyl)-1-oxo-2-propenyl]-2-hydroxy-, (E)-	MeOH	252(4.39),322(4.39), 385s(3.77)	18-3099-83
	MeOH-NaOH	324(4.35),440(3.73)	18-3099-83
$C_{16}H_{11}BrO_6$			
4H-1-Benzopyran-4-one, 6-bromo-5,7,8-trihydroxy-2-(4-methoxyphenyl)-	EtOH	304(4.47)	18-3773-83
	EtOH-NaOAc	306(4.39)	18-3773-83
	EtOH-AlCl$_3$	306s(4.37),325(4.47)	18-3773-83
$C_{16}H_{11}ClN_2O$			
Pyrazine, 2-chloro-3,5-diphenyl-, 1-oxide	EtOH	268.5(4.16),340(3.24)	4-0311-83
2(1H)-Quinoxalinone, 3-[2-(4-chlorophenyl)ethenyl]-	EtOH	390(4.64)	18-0326-83
$C_{16}H_{11}ClN_2O_3$			
Benzoyl chloride, 4-[5-(4-methoxyphenyl)-1,3,4-oxadiazol-2-yl]-	toluene	330(4.36)	103-0022-83
$C_{16}H_{11}ClN_2O_4$			
9,10-Anthracenedione, 2-acetoxy-1,4-diamino-	dioxan	256(4.59),494(3.89)	104-1892-83
$C_{16}H_{11}ClN_4$			
Benzo[1,2-b:4,3-b']dipyrrole, 1-[(4-chlorophenyl)azo]-3,6-dihydro-	EtOH	207.5(4.49),270(4.03), 322(3.95),242(4.53)	103-0871-83
$C_{16}H_{11}ClO_3$			
2,4,6-Cycloheptatrien-1-one, 4-[3-(4-chlorophenyl)-1-oxo-2-propenyl]-2-hydroxy-, (E)-	MeOH	254(4.42),322(4.39), 380s(3.86)	18-3099-83
	MeOH-NaOH	284(4.28),316(4.31), 440(3.66)	18-3099-83
$C_{16}H_{11}Cl_2NO_3$			
2,4,6-Cycloheptatrien-1-one, 4-[3-(2,4-dichlorophenyl)-1-(hydroxyimino)-2-propenyl]-2-hydroxy-, (?,E)-	MeOH	238(4.49),293(4.47), 363s(3.83)	18-3099-83
	MeOH-NaOH	248(4.48),331(4.45), 406(4.13)	18-3099-83
$C_{16}H_{11}Cl_6NO_4$			
Spiro[2,4-cyclohexadiene-1,2'-[2H]-furo[3,2-b]azepine]-4'(3'H)-carboxylic acid, 2,3,3',3',4,5-hexachloro-3'a,8'a-dihydro-6-oxo-, ethyl ester, (2'α,3'aα,8'aα)-	EtOH	260(4.10),340(3.45)	78-1281-83
(2'α,3'aβ,8'aβ)-	EtOH	263(3.95),350(3.28)	78-1281-83

Compound	Solvent	λ_{max}(log ε)	Ref.
$C_{16}H_{11}FOS$			
Ethanone, 1-(7-fluorodibenzo[b,f]thiepin-2-yl)-	MeOH	241(4.39),255s(4.34), 276(4.37),305s(3.71)	73-0144-83
$C_{16}H_{11}IN_2O$			
2(1H)-Pyrazinone, 6-iodo-3,5-diphenyl-	EtOH	231(3.88),263(3.83), 332(3.83)	4-0311-83
$C_{16}H_{11}N$			
Indolo[1,2-b]isoquinoline	EtOH	229(4.37),238s(--), 244(4.53),256s(--), 268s(--),281s(--), 284(4.69),301(4.52), 325s(--),340s(--), 357(3.96),372s(--), 396s(--)	118-1037-83
Indolo[1,2-a]quinoline	EtOH	223(4.48),241(4.36), 251s(--),254s(--), 265s(--),275(4.67), 294s(--),318s(--), 333(3.66),355s(--), 373(3.95),392(3.99), 415(3.76)	118-1037-83
$C_{16}H_{11}NO$			
1(4H)-Naphthalenone, 4-(phenylimino)-	n.s.g.	330(3.71)	39-1759-83C
$C_{16}H_{11}NO_2$			
9-Anthraceneacetonitrile, 9,10-dihydro-9-hydroxy-10-oxo-	EtOH	235s(3.96),275(4.21)	70-2379-83
Benzenepropanenitrile, α-benzoyl-β-oxo-	EtOH	233(4.03),327(4.09)	4-0645-83
Methanone, phenyl(5-phenyl-4-isoxazolyl)-	EtOH	252(4.18),284(4.04)	4-0645-83
1(4H)-Naphthalenone, 4-(phenylimino)-, N-oxide	n.s.g.	390(4.2)	39-1759-83C
Spiro[2H-indene-2,2'-[2H]indole]-1,3'-dione, 1',3-dihydro-	EtOH	237(4.38),250s(4.34), 330(3.65),390(3.54)	39-2413-83C
$C_{16}H_{11}NO_2S$			
Benzeneacetic acid, α-(2-benzothiazolylmethylene)-, (E)-	MeOH	207(4.51),225(4.40), 247s(3.83),255s(3.77), 315(4.27)	4-0005-83
(Z)-	MeOH	213(4.39),262(3.84), 329(4.48)	4-0005-83
$C_{16}H_{11}NO_3$			
6H-Anthra[1,9-cd]isoxazol-6-one, 5-ethoxy-	EtOH	244(5.00),249(5.10), 445(4.12),466(4.12)	103-1057-83
Furo[2,3-c]acridin-6(1H)-one, 5-hydroxy-11-methyl- (furofoline)	MeOH	226(4.27),245(4.43), 266s(4.62),274(4.76), 312s(4.01),323(4.02), 405(3.62)	39-1681-83C
2H-Pyrano[3',2':5,6]benzofuro[3,2-c]pyridin-2-one, 4,11-dimethyl-	EtOH	250(4.39),328(4.14)	39-0225-83C
7H-Pyrrolo[3,2,1-de]phenanthridin-7-one, 10-hydroxy-9-methoxy- (pratorimine)	MeOH	226(4.18),235(4.21), 248(4.37),255(4.34), 285(3.87),295(4.12), 337(3.26),348(3.59), 362(3.51)	102-2305-83
	MeOH-NaOAc	226(4.60),236(4.51), 250(4.79),256(4.71),	102-2305-83

Compound	Solvent	λ_{max}(log ϵ)	Ref.
(cont.)		285(4.11),295(4.43), 337(3.55),348(3.66), 365(3.59)	102-2305-83
	MeOH-NaOMe	222(--),236(--), 245s(--),274(--), 292(--),303(--), 395(--)	102-2305-83
$C_{16}H_{11}NO_3S$			
4H-Furo[3,4-b]indole, 4-(phenylsulfonyl)-	EtOH	245s(4.13),256s(4.08), 274(3.71),293(3.69)	88-5435-83
$C_{16}H_{11}NO_4$			
Acetamide, N-(9,10-dihydro-8-hydroxy-9,10-dioxo-1-anthracenyl)-	benzene	442(3.86)	104-0745-83
9,10-Anthracenedione, 1-acetoxy-8-amino-	benzene	472(3.72)	104-0745-83
$C_{16}H_{11}NO_5$			
2-Anthraceneacetamide, 9,10-dihydro-1,4-dihydroxy-9,10-dioxo-	dioxan	283(3.97),482(3.97)	104-0533-83
2,4,6-Cycloheptatrien-1-one, 2-hydroxy-4-[3-(3-nitrophenyl)-1-oxo-2-propenyl]-, (E)-	MeOH	259(4.40),300(4.25), 386(3.68)	18-3099-83
	MeOH-NaOH	270(4.26),345s(3.90), 444(3.49)	18-3099-83
2,4,6-Cycloheptatrien-1-one, 2-hydroxy-4-[3-(4-nitrophenyl)-1-oxo-2-propenyl]-, (E)-	MeOH	255(4.35),318(4.51), 400s(3.73)	18-3099-83
	MeOH-NaOH	271s(4.27),319(4.43), 448(4.68)	18-3099-83
3H-Phenoxazin-3-one, 1,7-diacetyl-4-hydroxy-	$CHCl_3$	390s(--),415(3.95), 440(3.95),470s(--)	44-3649-83
$C_{16}H_{11}NO_7$			
3H-Phenoxazine-1,7-dicarboxylic acid, 4-hydroxy-3-oxo-, dimethyl ester	$CHCl_3$	390s(--),426(3.9), 445(3.9),470s(--)	44-3649-83
3H-Phenoxazine-1,9-dicarboxylic acid, 2-hydroxy-3-oxo-, dimethyl ester	$CHCl_3$	398s(--),424(4.10), 446(4.10),477s(--)	44-3649-83
$C_{16}H_{11}NS_2$			
2H-1,3-Thiazine-2-thione, 4,6-diphenyl-	MeCN	240(4.03),316(4.44), 475(3.31)	118-0827-83
$C_{16}H_{11}N_3$			
Cyclohept[d]imidazo[1,2-a]imidazole, 2-phenyl-	EtOH	232(4.50),245s(4.42), 334(4.70),421(3.62), 428s(3.60)	18-3703-83
1H-Imidazole-1-carbonitrile, 2,4-diphenyl-	EtOH	242(4.48)	4-1277-83
1H-Imidazole-1-carbonitrile, 2,5-diphenyl-	EtOH	293(4.30)	4-1277-83
1H-Imidazole-1-carbonitrile, 4,5-diphenyl-	EtOH	228(4.34),272(4.03)	4-1277-83
4-Isoquinolinecarbonitrile, 1-amino-3-phenyl-	EtOH	217(4.53),257(4.52), 330(4.16)	39-1813-83C
1H-Pyrido[3,2,1-ij][1,2,4]benzotriazin-4-ium, 2-phenyl-, hydroxide, inner salt	EtOH	246(4.29),264(4.34), 284s(4.13),342s(3.71), 355(3.46),406(3.56), 425(3.13),480(3.38), 515(3.34)	39-0349-83C

Compound	Solvent	$\lambda_{max}(\log \epsilon)$	Ref.
$C_{16}H_{11}N_3O_2S$			
Anthra[1,2-c][1,2,5]thiadiazole-6,11-dione, 4-(dimethylamino)-	toluene	535(4.07)	135-0283-83
	MeOH	548(--)	135-0283-83
	HOAc	548(3.96)	135-0283-83
	dioxan	530(4.10)	135-0283-83
	$CHCl_3$	550(4.06)	135-0283-83
$C_{16}H_{11}N_3O_2Se$			
Anthra[1,2-c][1,2,5]selenadiazole-6,11-dione, 4-(dimethylamino)-	toluene	570(4.05)	135-0283-83
	MeOH	580(--)	135-0283-83
	HOAc	582(4.04)	135-0283-83
	dioxan	565(4.00)	135-0283-83
	$CHCl_3$	585(4.07)	135-0283-83
$C_{16}H_{11}N_3O_3$			
Anthra[1,2-c][1,2,5]oxadiazole-6,11-dione, 4-(dimethylamino)-	toluene	535(4.10)	135-0283-83
	MeOH	550(--)	135-0283-83
	HOAc	550(4.16)	135-0283-83
	dioxan	538(4.14)	135-0283-83
	$CHCl_3$	545(4.16)	135-0283-83
2(1H)-Quinoxalinone, 3-[2-(4-nitrophenyl)ethenyl]-	EtOH	396(4.35)	18-0326-83
$C_{16}H_{11}N_3O_4$			
Phthalazino[2,3-b]phthalazine-5,14-dione, 7,12-dihydro-9-nitro-	EtOH	237(4.14),244(4.09), 265(4.01),303(3.9)	104-2270-83
$C_{16}H_{11}N_5$			
Pyrazine, 2-azido-3,5-diphenyl-	EtOH	238s(4.17),278s(4.03), 343(4.61)	4-1277-83
Pyrazine, 2-azido-3,6-diphenyl-	EtOH	238s(4.17),280s(4.05), 343(4.16)	4-1277-83
$C_{16}H_{11}N_5O$			
Furo[2',3':4,5]pyrrolo[1,2-d]-1,2,4-triazolo[3,4-f][1,2,4]triazine, 3-methyl-9-phenyl-	MeOH	237(3.39),347(3.75)	73-1878-83
Furo[2',3':4,5]pyrrolo[1,2-d]-1,2,4-triazolo[3,4-f][1,2,4]triazine, 9-(4-methylphenyl)-	MeOH	328(3.94),342(3.36)	73-1878-83
$C_{16}H_{11}N_5O_2$			
1H,5H-Pyrrolo[2,3-f]indole, 3-[(4-nitrophenyl)azo]-	EtOH	205(4.46),239(4.37), 294(4.25),328(4.31), 250[sic](4.39)	103-0871-83
3H,6H-Pyrrolo[3,2-e]indole, 1-[(4-nitrophenyl)azo]-	EtOH	208(4.45),268(3.96), 339(3.96),204.5(3.96), 253(4.54)[sic]	103-0871-83
$C_{16}H_{11}N_7O_3$			
2H-1,2,3-Triazole, 4-nitro-2-phenyl-5-[(2-phenyl-2H-1,2,3-triazol-4-yl)oxy]-	EtOAc	273.7(4.43),311.8(4.04), 355.9(3.60)	1-0097-83
$C_{16}H_{12}$			
9,10-Ethenoanthracene, 9,10-dihydro-	hexane-isooctane	272(3.37),280(3.60)	35-3226-83
	MeCN	254(2.99)(not a maximum)	35-3226-83
3,4:6,7-Dibenzotricyclo[3.3.0.0^{2,8}]-octa-3,6-diene-	hexane-$isoPr_2O$	274(3.36),282(3.23)	35-3226-83

Compound	Solvent	λ_{max}(log ϵ)	Ref.
Fluoranthene, 1,10b-dihydro-	MeOH	259s(4.51),268(4.58), 280s(4.41),312(3.62), 325(3.54)	44-2360-83
Pyrene, 4,5-dihydro-	MeOH	259(4.61),282(4.18), 288(4.13),299(4.09)	54-0014-83
$C_{16}H_{12}AsCl$			
1H-Arsole, 1-chloro-2,5-diphenyl-	CH_2Cl_2	246(4.23),402(4.07)	101-0269-83H
$C_{16}H_{12}BrNO_4$			
3H-Benz[c]indole-1-carboxylic acid, 8-bromo-4,5-dihydro-2-methyl-4,5-dioxo-, ethyl ester	EtOH	208(4.39),235(4.24), 281(4.39),350(3.92)	103-1086-83
$C_{16}H_{12}ClFN_4O$			
3H-1,4-Benzodiazepine-2-carboximidamide, 7-chloro-5-(2-fluorophenyl)-N-hydroxy-	isoPrOH	224(4.50),310(4.09)	4-0551-83
$C_{16}H_{12}ClFOS$			
Ethanone, 1-(11-chloro-7-fluoro-10,11-dihydrodibenzo[b,f]thiepin-2-yl)-	MeOH	239(4.15),302(4.09)	73-0144-83
$C_{16}H_{12}ClNO_3$			
2,4,6-Cycloheptatrien-1-one, 4-[3-(4-chlorophenyl)-1-(hydroxyimino)-2-propenyl]-2-hydroxy-, (?,E)-	MeOH	245(4.38),293(4.48), 380(3.74)	18-3099-83
$C_{16}H_{12}ClN_3O$			
Bicyclo[4.4.1]undeca-4,6,8,10-tetraene-2,3-dione, 2-[(2-chloro-3-pyridinyl)hydrazone]	CH_2Cl_2	438(3.86)	138-0653-83
Oxazolo[2,3-b][1,3,4]benzotriazepine, 8-chloro-2,3-dihydro-6-phenyl-	MeOH	227(4.55),263(4.28), 300s(3.73)	106-0081-83
$C_{16}H_{12}Cl_2$			
1,3,9,11-Cyclotridecatetraene-5,7-diyne, 13-(dichloromethylene)-4,9-dimethyl-	CH_2Cl_2	269s(4.23),287(4.35), 355(3.91)	39-2997-83C
$C_{16}H_{12}Cl_2N_4$			
3-Pyridinecarbonitrile, 1-[(2,6-dichlorophenyl)methyl]-1,6-dihydro-6-(1H-imidazol-1-yl)-	CH_2Cl_2	308(3.8),358s(3.0)	64-0878-83B
$C_{16}H_{12}Cl_2N_4O$			
Acetamide, N-[6-(2,6-dichlorophenyl)-2-methylpyrido[2,3-d]pyrimidin-7-yl]-	pH 1	266(3.95),297(4.10) (changing)	87-0403-83
	MeOH-pH 7	320(4.09)	87-0403-83
	MeOH	213(4.62),320(4.12)	87-0403-83
	MeOH-HCl	224(4.49),267(3.94), 299(4.12)	87-0403-83
$C_{16}H_{12}Cl_2N_4O_2$			
Carbamic acid, [6-(2,6-dichlorophenyl)-2-methylpyrido[2,3-d]pyrimidin-7-yl]-, methyl ester	pH 1	218(4.56),265(3.88), 301(4.12)	87-0403-83
	MeOH-pH 7	225(4.68),318(4.11), 324(4.13),350s(2.98)	87-0403-83
	MeOH-NaOH	290(3.84),358(4.22)	87-0403-83

Compound	Solvent	λ_{max}(log ε)	Ref.
$C_{16}H_{12}N_2$			
Isoindolo[2,1-a]quinoxaline, 9-methyl-	EtOH	205s(4.51),225(4.56), 236(4.42),258(4.64), 278(4.43),290(4.28), 310(4.08),323(4.18), 356s(4.12),373(4.21), 405(3.97),425(3.88)	104-0586-83
Pyridazine, 3,6-diphenyl-	MeOH	230(3.79),280(4.45)	35-0933-83
$C_{16}H_{12}N_2O$			
Pyrazine, 2,6-diphenyl-, 4-oxide	EtOH	270(4.63),340(3.80)	4-0311-83
2(1H)-Quinoxalinone, 3-(2-phenyl-ethenyl)-	EtOH	390(4.19)	18-0326-83
$C_{16}H_{12}N_2O_2$			
Phthalazino[2,3-b]phthalazine-5,12(7H,14H)-dione	EtOH	236(4.05),307(3.8)	104-2270-83
2(1H)-Quinoxalinone, 3-[2-(2-hydroxy-phenyl)ethenyl]-	EtOH	400(4.23)	18-0326-83
2(1H)-Quinoxalinone, 3-[2-(4-hydroxy-phenyl)ethenyl]-	EtOH	406(4.57)	18-0326-83
$C_{16}H_{12}N_2O_3$			
Benzoic acid, 4-(5-phenyl-1,3,4-oxa-diazol-2-yl)-, methyl ester	toluene	300(4.38)	103-0022-83
$C_{16}H_{12}N_2O_4$			
3H-Phenoxazin-3-one, 1,7-diacetyl-4-amino-	$CHCl_3$	428(4.18),448(4.18)	44-3649-83
$C_{16}H_{12}N_2O_4S$			
Benzenesulfonic acid, 3-[(4-hydroxy-1-naphthalenyl)azo]-, sodium salt	pH 7.5	475(4.45)	94-0162-83
Benzenesulfonic acid, 4-[(4-hydroxy-1-naphthalenyl)azo]-, sodium salt	pH 7.5	475(4.43)	94-0162-83
$C_{16}H_{12}N_2O_5$			
2,4,6-Cycloheptatrien-1-one, 2-hydr-oxy-4-(1-hydroxyimino)-3-(3-nitro-phenyl)-2-propenyl]-, (?,E)-	MeOH	249(4.52),272(4.55), 377(3.79)	18-3099-83
	MeOH-NaOH	227s(4.13),273(4.26), 403(3.55)	18-3099-83
$C_{16}H_{12}N_2O_6$			
3H-Phenoxazine-1,7-dicarboxylic acid, 4-amino-3-oxo-, dimethyl ester	$CHCl_3$	416(4.21),444(4.21)	44-3649-83
$C_{16}H_{12}N_2O_7S_2$			
1-Naphthalenesulfonic acid, 4-hydroxy-3-[(4-sulfophenyl)azo]-, disodium salt	pH 7.5	495(4.35)	94-0162-83
$C_{16}H_{12}N_2S$			
2H-1,3-Thiazin-2-imine, 4-phenyl-6-(phenylthio)-, monoperchlorate	MeCN	250s(4.23),325(4.48), 396(4.32)	118-0827-83
$C_{16}H_{12}N_4O_8$			
Benzeneacetic acid, 2-[[(2,4,6-tri-nitrophenyl)methylene]amino]-, methyl ester	$CHCl_3$	365(3.38)	44-2468-83

Compound	Solvent	λ_{max}(log ε)	Ref.
$C_{16}H_{12}N_6$			
6H-Imidazo[1,2-b][1,2,4]triazol-6-one, 5-phenyl-, phenylhydrazone	EtOH	275s(3.55),350(4.27)	4-0936-83
$C_{16}H_{12}O_2$			
9,10-Anthracenedione, 2-ethyl-	polyethylene	258(4.63),276(4.28), 323(3.60)	135-1304-83
9-Anthracenol, acetate	EtOH	208(4.30),253(5.22), 346(3.78),383(3.93)	2-1191-83
Benz[a]azulene-1,4-dione, 2,3-dimethyl-	MeCN	250(4.2),297(4.3), 361(3.8),429(3.5), 580(2.8),618(2.9), 665(2.7)	24-2408-83
17,18-Dioxatricyclo[12.2.1.1^{6,9}]octadeca-2,4,6,8,10,12,14,16-octaene, (E,E,Z,Z)-	hexane	278(4.86),331(3.65), 530(2.43)	89-0480-83S
2,3-Fluoranthenediol, 2,3-dihydro-, trans	MeOH	226(4.38),230(4.40), 249(4.30),259(4.46), 282(3.70),288(3.71), 313(3.67)	44-2360-83
$C_{16}H_{12}O_2S$			
4H-1-Benzothiopyran-4-one, 6-methoxy-2-phenyl-	EtOH	282(4.41),362(3.93)	139-0059-83B
18,19-Dioxa-12-thiatricyclo[13.2.1-1^{6,9}]nonadeca-2,4,6,8,10,13,15,17-octaene, (E,Z,Z,E)-	hexane	220(3.85),241(3.90), 321(4.33),331(4.36), 394(3.22),413(3.34), 436(3.38)	89-0480-83S
(Z,Z,E,Z)-	hexane	243(4.18),331(4.63), 316.5(4.56),390(3.56), 415(3.75),439(3.82)	89-0480-83S
$C_{16}H_{12}O_2Te$			
4H-1-Benzotellurin-4-one, 7-methoxy-2-phenyl-	CH_2Cl_2	388(3.72)	35-0883-83
$C_{16}H_{12}O_3$			
9,10-Anthracenedione, 2-hydroxy-1,3-dimethyl-	EtOH	242(4.36),276(4.53), 375(3.61),515(2.51)	24-0970-83
4H-1-Benzopyran-4-one, 3-hydroxy-2-(4-methylphenyl)-	MeOH	242(4.29),247(4.29), 310(4.08),350(4.27)	118-0835-83
2,4,6-Cycloheptatrien-1-one, 2-hydroxy-4-(1-oxo-3-phenyl-2-propenyl)-	MeOH	252(4.43),318(4.37), 380s(3.82)	18-3099-83
	MeOH-NaOH	265s(4.23),318(4.36), 444(3.80)	18-3099-83
1H-Indene-1,3(2H)-dione, 2-(4-methoxyphenyl)-	pH 12	300(4.5),345(4.0), 480(3.2)	151-0061-83C
12,18,19-Trioxatricyclo[13.2.1.1^{6,9}]-nonadeca-2,4,6,8,10,13,15,17-octaene, (Z,Z,E,Z)-	hexane	236(4.35),242(4.35), 287s(4.54),298(4.64), 309(4.67),384(3.89), 401(4.05),426(4.06)	89-0480-83S
$C_{16}H_{12}O_4$			
4H-1-Benzopyran-4-one, 3-hydroxy-2-(4-methoxyphenyl)-	MeOH	233(3.90),252(3.85), 318s(3.71),356(4.01)	118-0835-83
4H-1-Benzopyran-4-one, 3-hydroxy-8-methoxy-2-phenyl-	MeOH	240(4.20),253(4.32), 315(3.99),350(4.12)	118-0835-83
4H-1-Benzopyran-4-one, 7-hydroxy-2-(4-methoxyphenyl)-	MeOH	253(4.27),324(3.70)	161-0067-83
4H-1-Benzopyran-4-one, 7-hydroxy-3-(4-methoxyphenyl)-	EtOH	250(4.44),300(4.05)	162-0606-83

Compound	Solvent	λ_{max}(log ε)	Ref
2,4,6-Cycloheptatrien-1-one, 2-hydroxy-4-[3-(2-hydroxyphenyl)-1-oxo-2-propenyl]-, (E)-	MeOH	250(4.40),320(4.16), 370(4.20)	18-3099-83
	MeOH-NaOH	259(4.34),335(4.21), 444(4.17)	18-3099-83
[2.2]Metaparacyclophanequinone	$CHCl_3$	251(4.35)	35-6650-83
$C_{16}H_{12}O_4S_4$			
1,3-Dithiole-4,5-dicarboxylic acid, 2-(1,3-benzodithiol-2-ylideneethylidene)-, dimethyl ester	$CHCl_3$	374(4.56),393(4.53)	88-3469-83
$C_{16}H_{12}O_5$			
9,10-Anthracenedione, 1,6-dihydroxy-8-methoxy-3-methyl- (questin)	EtOH	251(4.20),286(4.24), 441(2.88)	94-4543-83
2H-1-Benzopyran-2-one, 7-hydroxy-4-(3-hydroxy-4-methoxyphenyl)- (volubolin)	MeOH	224(4.19),284(4.40), 340(3.98)	102-2625-83
4H-1-Benzopyran-4-one, 3,7-dihydroxy-2-(4-methoxyphenyl)-	MeOH	336(4.17)	161-0067-83
4H-1-Benzopyran-4-one, 5,7-dihydroxy-2-(4-hydroxyphenyl)-8-methyl- (relative absorbance given)	MeOH	274(1.00),300s(0.77), 328(0.87)	102-2107-83
	MeOH-$AlCl_3$	262s(0.60),282(0.96), 296s(0.83),310(1.00), 346(1.00),392(0.56)	102-2107-83
	MeOH-$AlCl_3$-HCl	260s(0.55),282(0.92), 307(1.00),342(0.95), 392(0.35)	102-2107-83
4H-1-Benzopyran-4-one, 5,7-dihydroxy-2-(4-methoxyphenyl)- (acacetin) (relative absorbance given)	MeOH	269(0.98),303s(0.84), 327(1.00)	102-2107-83
	MeOH-$AlCl_3$	259s(0.52),277(0.85), 292s(0.78),302(0.87), 344(1.00),382(0.71)	102-2107-83
	MeOH-$AlCl_3$-HCl	260s(0.63),279(0.92), 294s(0.90),300(0.43), 338(1.00),379(0.64)	102-2107-83
4H-1-Benzopyran-4-one, 3-hydroxy-2-(4-hydroxyphenyl)-7-methoxy-	EtOH	352(4.35)	142-0039-83
4H-1-Benzopyran-4-one, 5-hydroxy-2-(4-hydroxyphenyl)-7-methoxy- (genkwanin) (relative absorbance given)	MeOH	269(0.86),336(1.00)	102-2107-83
	MeOH-$AlCl_3$	277(0.87),294s(0.60), 303(0.69),348(1.00), 384(0.85)	102-2107-83
	MeOH-$AlCl_3$-HCl	276(0.89),294s(0.74), 301(0.79),343(1.00), 382(0.75)	102-2107-83
$C_{16}H_{12}O_6$			
4H-1-Benzopyran-4-one, 5,6-dihydroxy-2-(4-hydroxyphenyl)-7-methoxy-(sorbifolin)(relative absorbance given)	MeOH	285(0.83),335(1.00)	102-2107-83
	MeOH-$AlCl_3$	266s(0.26),292s(0.66), 303(0.78),367(1.00)	102-2107-83
	MeOH-$AlCl_3$-HCl	266s(0.29),290s(0.69), 302(0.82),362(1.00)	102-2107-83
4H-1-Benzopyran-4-one, 5,7-dihydroxy-2-(3-hydroxy-4-methoxyphenyl)-(diosmetin)(relative absorbance given)	n.s.g.	253(4.28),268(4.25), 345(4.32)	162-0481-83
	MeOH	240s(0.70),252(0.78), 267(0.80),291s(0.49), 344(1.00)	102-2107-83
	MeOH-$AlCl_3$	267(0.87),273(0.83), 296(0.60),362(0.95), 390(1.00)	102-2107-83
	MeOH-$AlCl_3$-HCl	264s(0.84),276(0.85), 295(0.66),351(1.00), 383(0.92)	102-2107-83

Compound	Solvent	λ_{max}(log ε)	Ref.
4H-1-Benzopyran-4-one, 5,7-dihydroxy-2-(4-hydroxy-3-methoxyphenyl)- (relative absorbance given)	MeOH	241(0.85),249s(0.82), 269(0.81),347(1.00)	102-2107-83
	MeOH-$AlCl_3$	262(0.85),274(0.85), 296(0.53),366s(0.91), 390(1.00)	102-2107-83
	MeOH-$AlCl_3$-HCl	259(0.95),276(0.96), 294(0.67),353(1.00), 386(0.94)	102-2107-83
4H-1-Benzopyran-4-one, 5,7-dihydroxy-2-(4-hydroxyphenyl)-3-methoxy- (relative absorbance given)	MeOH	269(1.00),298(0.68), 322s(0.71),350(0.82)	102-2107-83
	MeOH-$AlCl_3$	277(1.00),304(0.70), 350(0.74),399(0.67)	102-2107-83
	MeOH-$AlCl_3$-HCl	278(1.00),304(0.69), 348(0.75),398(0.64)	102-2107-83
4H-1-Benzopyran-4-one, 5,7-dihydroxy-2-(4-hydroxyphenyl)-6-methoxy- (hispidulin)	EtOH	217(4.58),277(4.30), 339(4.44)	105-0763-83
relative absorbance given	MeOH	275(0.65),336(1.00)	102-2107-83
	MeOH-$AlCl_3$	262s(0.35),282s(0.55), 294s(0.60),303(0.64), 362(1.00)	102-2107-83
	MeOH-$AlCl_3$-HCl	260s(0.31),287s(0.65), 303(0.74),356(1.00)	102-2107-83
4H-1-Benzopyran-4-one, 5,7-dihydroxy-2-(4-hydroxyphenyl)-8-methoxy- (relative absorbance given)	MeOH	274(1.00),300(0.82), 324(0.92),350s(0.72)	102-2107-83
	MeOH-$AlCl_3$	282(0.86),310(0.87), 349(1.00),398(0.54)	102-2107-83
	MeOH-$AlCl_3$-HCl	282(0.89),309(0.98), 344(1.00),397(0.49)	102-2107-83
4H-1-Benzopyran-4-one, 5,7-dihydroxy-3-(3-hydroxy-4-methoxyphenyl)-	EtOH	263(4.53)	162-1108-83
4H-1-Benzopyran-4-one, 5,8-dihydroxy-2-(4-hydroxyphenyl)-7-methoxy-	EtOH	279(4.32),309(4.33), 326s(4.27),367s(3.96)	18-3773-83
	EtOH-NaOAc	278(4.30),309(4.26), 326s(4.21),391(4.09)	18-3773-83
	EtOH-$AlCl_3$	285(4.25),319(4.37), 345s(4.31),406(3.87)	18-3773-83
relative absorbance given	MeOH	280(0.96),306(1.00), 330s(0.77),366s(0.38)	102-2107-83
	MeOH-$AlCl_3$	285(0.82),321(1.00), 351(0.95),424(0.30)	102-2107-83
	MeOH-$AlCl_3$-HCl	285(0.80),320(1.00), 348(0.89),424(0.27)	102-2107-83
4H-1-Benzopyran-4-one, 5,6,7-trihydroxy-2-(4-methoxyphenyl)- (relative absorbance given)	MeOH	286(0.86),335(1.00)	102-2107-83
	MeOH-$AlCl_3$	276s(0.50),302(1.00), 380(0.90)	102-2107-83
	MeOH-$AlCl_3$-HCl	262s(0.28),292s(0.86), 302(0.94),358(1.00)	102-2107-83
4H-1-Benzopyran-4-one, 5,7,8-trihydroxy-2-(4-methoxyphenyl)-	EtOH	287s(4.37),304(4.41), 367s(3.83)	18-3773-83
	EtOH-NaOAc	297(4.39)	18-3773-83
	EtOH-$AlCl_3$	293s(4.27),316(4.41), 345s(4.22),402(3.88)	18-3773-83
1,4-Naphthalenedione, 2-(3,6-dihydro-5-methyl-6-oxo-2H-pyran-2-yl)-5,8-dihydroxy-	MeOH	275(3.50),330(2.74), 487(3.44),514(3.46), 557(3.23)	105-0532-83
$C_{16}H_{12}O_7$			
4H-1-Benzopyran-4-one, 2-(3,4-dihydroxy-5-methoxyphenyl)-5,7-dihydroxy- (relative absorbance given)	MeOH	250s(0.79),268(0.69), 300s(0.48),350(1.00)	102-2107-83
	MeOH-$AlCl_3$	270(0.71),310s(0.26),	102-2107-83

Compound	Solvent	$\lambda_{max}(\log \epsilon)$	Ref.
(cont.)		344s(0.18),430(1.00)	102-2107-83
	MeOH-AlCl$_3$-HCl	275(0.88),303(0.51), 363(0.95),386(1.00)	102-2107-83
4H-1-Benzopyran-4-one, 2-(3,4-dihydroxyphenyl)-3,5-dihydroxy-7-methoxy- (rhamnetin)	EtOH	256(4.40),371(4.41)	162-1178-83
4H-1-Benzopyran-4-one, 2-(3,4-dihydroxyphenyl)-5,6-dihydroxy-7-methoxy- (pedalitin)(relative absorbance given)	MeOH	256s(0.69),284(0.65), 345(1.00)	102-2107-83
	MeOH-AlCl$_3$	246s(0.56),274(0.55), 302(0.61),343s(0.32), 425(1.00)	102-2107-83
	MeOH-AlCl$_3$-HCl	260(0.47),296(0.70), 372(1.00)	102-2107-83
4H-1-Benzopyran-4-one, 2-(3,4-dihydroxyphenyl)-5,7-dihydroxy-3-methoxy- (relative absorbance given)	MeOH	256(1.00),268s(0.87), 292s(0.55),359(0.86)	102-2107-83
	MeOH-AlCl$_3$	277(0.93),303s(0.37), 339(0.32),440(1.00)	102-2107-83
	MeOH-AlCl$_3$-HCl	268(1.00),277s(0.94), 303(0.57),360(0.75), 402(0.85)	102-2107-83
4H-1-Benzopyran-4-one, 2-(3,4-dihydroxyphenyl)-5,7-dihydroxy-6-methoxy- (nepetin) (relative absorbance given)	MeOH	257(0.90),272(0.90), 348(1.00)	102-2107-83
	MeOH-AlCl$_3$	274(0.78),306s(0.39), 338(0.23),428(1.00)	102-2107-83
	MeOH-AlCl$_3$-HCl	262(0.78),280(0.82), 298(0.63),367(1.00)	102-2107-83
4H-1-Benzopyran-4-one, 2-(3,4-dihydroxyphenyl)-5,7-dihydroxy-8-methoxy- (onopordin) (relative absorbance given)	MeOH	258(0.88),274(1.00), 296s(0.64),349(0.95)	102-2107-83
	MeOH-AlCl$_3$	277(0.82),306s(0.31), 336s(0.14),433(1.00)	102-2107-83
	MeOH-AlCl$_3$-HCl	264s(0.82),280(1.00), 301(0.71),358(0.98), 400(0.66)	102-2107-83
4H-1-Benzopyran-4-one, 2-(3,4-dihydroxyphenyl)-5,8-dihydroxy-7-methoxy-	EtOH	278(4.27),301(4.15), 345(4.23)	18-3773-83
	EtOH-NaOAc	275(4.26),299s(4.10), 347(4.13),390s(4.05)	18-3773-83
	EtOH-AlCl$_3$	294(4.25),310(4.13), 358(4.21),417(3.98)	18-3773-83
4H-1-Benzopyran-4-one, 5,6,7-trihydroxy-2-(3-hydroxy-4-methoxyphenyl)- (relative absorbance given)	MeOH	242s(0.80),284(0.78), 342(1.00)	102-2107-83
	MeOH-AlCl$_3$	255s(0.49),296(0.64), 315s(0.57),380(1.00)	102-2107-83
	MeOH-AlCl$_3$-HCl	257s(0.71),294(0.69), 365(1.00)	102-2107-83
4H-1-Benzopyran-4-one, 5,6,7-trihydroxy-2-(4-hydroxy-3-methoxyphenyl)- (nodifloretin) (relative absorbance given)	MeOH	276(0.90),344(1.00)	102-2107-83
	MeOH-AlCl$_3$	262s(--),302(1.00), 390(1.00)	102-2107-83
	MeOH-AlCl$_3$-HCl	255s(--),290(1.00), 370(1.00)	102-2107-83
$C_{16}H_{12}O_8$			
4H-1-Benzopyran-4-one, 5,7-dihydroxy-3-methoxy-2-(3,4,5-trihydroxyphenyl)- (relative absorbance given)	MeOH	254(0.94),268s(0.88), 305s(0.67),361(1.00)	102-2107-83
	MeOH-AlCl$_3$	271(1.00),312(0.28), 428(1.00)	102-2107-83
	MeOH-AlCl$_3$-HCl	273(1.00),308(0.43), 360(0.70),405(0.84)	102-2107-83
4H-1-Benzopyran-4-one, 5,8-dihydroxy-7-methoxy-2-(3,4,5-trihydroxyphenyl)-	EtOH	278(4.22),312(4.06), 350(4.22)	18-3773-83
	EtOH-NaOAc	267(4.22),310(3.93),	18-3773-83

Compound	Solvent	λ_{max}(log ϵ)	Ref.
(cont.)		353s(4.03),417(4.14)	18-3773-83
	EtOH-AlCl$_3$	283(4.25),324(4.03), 365(4.16),433(4.14)	18-3773-83
4H-1-Benzopyran-4-one, 2-(3,4-dihydroxyphenyl)-5,6,7-trihydroxy-3-methoxy-	MeOH	258(--),270s(--), 350(--)	102-2107-83
	MeOH-AlCl$_3$	274(--),305s(--), 426(--)	102-2107-83
	MeOH-AlCl$_3$-HCl	262(--),276s(--), 381(--)	102-2107-83
4H-1-Benzopyran-4-one, 2-(3,4-dihydroxyphenyl)-5,6,7-trihydroxy-8-methoxy-	EtOH	284(4.26),353(4.24)	40-0161-83
	EtOH-NaOAc	266s(4.22),330s(4.09), 390(4.23)	40-0161-83
	EtOH-AlCl$_3$	258s(4.13),300(4.24), 385(4.40)	40-0161-83
	EtOH-AlCl$_3$-HCl	258(4.14),302(4.31), 378(4.43)	40-0161-83
$C_{16}H_{13}As$			
1H-Arsole, 2,5-diphenyl-	C_6H_{12}	225(4.22),368(4.04)	101-0335-83I
$C_{16}H_{13}Br$			
Benzene, 1,1'-(2-bromo-1,3-butadienylidene)bis-	hexane	287(4.03)	104-1438-83
Benzene, 1,1'-(2-bromo-3-methylene-1-propene-1,3-diyl)bis-	hexane	256(4.40),272(4.42)	104-1438-83
1H-Indene, 2-bromo-1-methyl-3-phenyl-	hexane	269(4.05)	104-1438-83
1H-Indene, 2-bromo-3-methyl-1-phenyl-	hexane	266(4.23)	104-1438-83
$C_{16}H_{13}BrN_2O_2$			
1H-Indole, 2-(bromomethyl)-1-methyl-5-nitro-3-phenyl-	dioxan	285(4.38),330(3.95)	103-1302-83
$C_{16}H_{13}Br_2NO_2$			
2-Propenamide, 2,3-dibromo-N-hydroxy-3-phenyl-N-(phenylmethyl)-	EtOH	208(4.35),244(4.03)	34-0433-83
$C_{16}H_{13}Cl$			
Benzene, 1,1'-(2-chloro-1,3-butadienylidene)bis-	hexane	280(4.12)	104-1438-83
Benzene, 1,1'-(2-chloro-3-methylene-1-propene-1,3-diyl)bis-	hexane	252(4.00),273(4.08), 282(4.07)	104-1438-83
1H-Indene, 2-chloro-1-methyl-3-phenyl-	hexane	268(3.97)	104-1438-83
1H-Indene, 2-chloro-3-methyl-1-phenyl-	hexane	263(3.95)	104-1438-83
$C_{16}H_{13}ClIN_3O$			
2H-1,3,4-Benzotriazepin-2-one, 7-chloro-1,3-dihydro-3-(2-iodoethyl)-5-phenyl-	dioxan	254s(4.11),310(3.73)	106-0081-83
$C_{16}H_{13}ClN_2OS$			
Acetamide, 2-chloro-N-[4-(6-methyl-2-benzothiazolyl)phenyl]-	MeOH	327(4.54)	49-0599-83
$C_{16}H_{13}ClOS$			
Benzenecarbothioic acid, S-[2-(4-chlorophenyl)-2-propenyl] ester	MeOH	245(4.02)	44-0835-83
$C_{16}H_{13}ClOTe$			
1,2-Oxatellurol-1-ium, 5-(3-methylphenyl)-3-phenyl-, chloride	CH_2Cl_2	422(4.36)	35-0883-83

Compound	Solvent	λ_{max}(log ϵ)	Ref.
$C_{16}H_{13}ClO_2Se$			
1,2-Oxaselenol-1-ium, 5-(4-methoxy-phenyl)-3-phenyl-, chloride	CH_2Cl_2	378(4.40)	35-0883-83
$C_{16}H_{13}ClO_2Te$			
1,2-Oxatellurol-1-ium, 5-(4-methoxy-phenyl)-3-phenyl-, chloride	n.s.g.	435(4.48)	35-0875-83
$C_{16}H_{13}ClO_6$			
1H,3H-Naphtho[1,8-cd]pyran-1,3-dione, 5-chloro-4,6,9-trimethoxy-7-methyl-	EtOH	256(4.11),323s(3.58), 338(3.66),368s(3.58), 386s(3.49)	78-2283-83
	EtOH-NaOH	248(4.26),303(3.60), 313(3.60),345(3.45)	78-2283-83
$C_{16}H_{13}Cl_2NO_2$			
2,4,6-Cycloheptatrien-1-one, 4-[1-amino-3-(2,4-dichlorophenyl)-2-propenyl]-, (E)-	MeOH	260(4.45),335(3.99), 373(3.74),394(3.77)	18-3099-83
$C_{16}H_{13}Cl_2N_5S$			
Thiourea, N-[6-(2,6-dichlorophenyl)-2-methylpyrido[2,3-d]pyrimidin-7-yl]-N'-methyl-	MeOH	340(4.32)	87-0403-83
$C_{16}H_{13}F$			
Benzene, 1,1'-(2-fluoro-1,3-butadien-ylidene)bis-	hexane	285(4.47)	104-1438-83
$C_{16}H_{13}FO_2S$			
Ethanone, 1-(7-fluoro-10,11-dihydro-11-hydroxydibenzo[b,f]thiepin-2-yl)-	MeOH	238(4.09),252s(3.91), 303(4.08)	73-0144-83
$C_{16}H_{13}IN_2OS$			
Acetamide, 2-iodo-N-[4-(6-methyl-2-benzothiazolyl)phenyl]-	MeOH	330(4.58)	49-0599-83
$C_{16}H_{13}NO$			
1H-Indole-3-carboxaldehyde, 1-(phenyl-methyl)-	EtOH	212(4.18),247(3.88), 300(3.90)	104-1367-83
2-Propenenitrile, 3-phenyl-2-(phenyl-methoxy)-, (E)-	EtOH	208(4.15),278(4.19)	78-1551-83
(Z)-	EtOH	206(4.23),282(4.44)	78-1551-83
$C_{16}H_{13}NOSSe$			
2(3H)-Thiazoleselone, 3-(4-methoxy-phenyl)-4-phenyl-	EtOH	358(4.32)	56-0875-83
$C_{16}H_{13}NOS_2$			
2(3H)-Thiazolethione, 3-(4-methoxy-phenyl)-4-phenyl-	EtOH	330(4.22)	56-0875-83
$C_{16}H_{13}NO_2S$			
2(3H)-Benzoxazolethione, 3-(3,4-di-hydro-1H-2-benzopyran-1-yl)-	$CHCl_3$	302(4.46)	94-1733-83
$C_{16}H_{13}NO_3$			
3-Buten-1-one, 1-(4-nitrophenyl)-4-phenyl-, (E)-	EtOH	255(4.11)	35-0933-83
2,4,6-Cycloheptatrien-1-one, 2-hydroxy-4-[1-(hydroxyimino)-3-phenyl-2-	MeOH	246(4.43),290(4.44), 372(3.79)	18-3099-83

Compound	Solvent	$\lambda_{max}(\log \epsilon)$	Ref.
propenyl]-, lower melting isomer			18-3099-83
isomer	MeOH	245(4.45),288(4.51), 372(3.78)	18-3099-83
$C_{16}H_{13}NO_3S$			
Benzenecarbothioic acid, S-[2-(4-nitrophenyl)-2-propenyl] ester	MeOH	298(3.59)	44-0835-83
1-Naphthalenesulfonic acid, 8-(phenylamino)-	MeOH	219(4.40),270(3.97), 348(3.47),375(3.54)	48-1002-83
ammonium salt (average absorbances)	n.s.g.	220(4.65),270(4.27), 355s(3.78),375(3.81)	35-6236-83
$C_{16}H_{13}NO_4$			
3H-Benz[e]indole-1-carboxylic acid, 4,5-dihydro-2-methyl-4,5-dioxo-, ethyl ester	EtOH	210(4.42),265(4.40), 360s(3.75)	103-1086-83
2H-1-Benzopyran-3-carboxamide, 5,6,7,8-tetrahydro-2,5-dioxo-7-phenyl-	toluene	318(3.94)	5-0695-83
	EtOH	312(--)	5-0695-83
	DMSO	320(--)	5-0695-83
$C_{16}H_{13}NO_4S$			
2-Naphthalenesulfonic acid, 4-hydroxy-6-(phenylamino)-	MeOH	225(4.51),282(4.40), 330(4.27)	48-1002-83
2-Naphthalenesulfonic acid, 4-hydroxy-7-(phenylamino)-	MeOH	230(4.54),290(4.36), 321(4.16),356(3.33)	48-1002-83
$C_{16}H_{13}NO_5S$			
4H,5H-Pyrano[3,4-e][1,3]oxazine-4,5-dione, 7-(1-methylethoxy)-2-(phenylthio)-	$CHCl_3$	280(4.26),326(3.89)	24-3725-83
4H,5H-Pyrano[3,4-e][1,3]oxazine-4,5-dione, 2-(phenylthio)-7-propoxy-	$CHCl_3$	287(4.32),328(3.76)	24-3725-83
$C_{16}H_{13}NSSe$			
2(3H)-Thiazoleselone, 3-(4-methylphenyl)-4-phenyl-	EtOH	358(4.29)	56-0875-83
$C_{16}H_{13}NS_2$			
2(3H)-Thiazolethione, 3-(4-methylphenyl)-4-phenyl-	EtOH	333(4.21)	56-0875-83
$C_{16}H_{13}N_3$			
4-Quinolinecarboxaldehyde, phenylhydrazone	hexane	365(4.40)	140-0215-83
	EtOH	390(--)	140-0215-83
$C_{16}H_{13}N_3OS$			
Furo[2',3':4,5]pyrrolo[1,2-d]-1,2,4-triazine-8(7H)-thione, 5-methyl-2-(4-methylphenyl)-	MeOH	312(3.16),400(3.13)	73-1878-83
$C_{16}H_{13}N_3OS_2$			
Benzamide, N-(1,4-dihydro-2,4-dithioxo-3(2H)-quinazolinyl)-4-methyl-	EtOH	253(4.26),284(4.42), 306(4.33),385(3.69)	49-0915-83
$C_{16}H_{13}N_3O_2$			
Furo[2',3':4,5]pyrrolo[1,2-d]-1,2,4-triazin-8(7H)-one, 5-methyl-2-(4-methylphenyl)-	MeOH	205(2.90),320(3.52)	73-1878-83
Phthalazino[2,3-b]phthalazine-5,14-dione, 9-amino-7,12-dihydro-	EtOH	237(4.33),244(4.29), 303(3.96)	104-2270-83

Compound	Solvent	λ_{max}(log ϵ)	Ref.
$C_{16}H_{13}N_3O_3$			
Nimetazepam	MeOH	259(4.20),308(3.98)	162-0940-83
$C_{16}H_{13}N_3O_4$			
9aH-2a,4,9b-Triazacyclopenta[cd]phena-lene-1,2-dicarboxylic acid, dimethyl ester	EtOH	223(4.47),245(4.41), 249(4.42),271(4.20), 279s(4.15),335(4.03)	39-0349-83C
$C_{16}H_{13}N_3O_4S$			
2-Naphthalenesulfonic acid, 6-amino-4-hydroxy-5-(phenylazo)-	pH 7.00	482(4.24)	33-2002-83
$C_{16}H_{13}N_3O_6$			
Benzeneacetic acid, 2-[[(2,4-dinitro-phenyl)methylene]amino]-, methyl ester	$CHCl_3$	350(4.15)	44-2468-83
$C_{16}H_{13}N_3O_7$			
Phenol, 2-[bis[(5-nitro-2-furanyl)-methyl]amino]-	MeOH	211(5.21),235(2.97), 288(1.48)	73-2682-83
$C_{16}H_{13}N_5$			
1H-Pyrazol-5-amine, 4-(1H-benzimida-zol-2-yl)-1-phenyl-	MeOH	240(3.73),263(3.44), 272(3.37),303(3.79), 313(3.77)	94-2114-83
$C_{16}H_{13}N_5O_2$			
Pyrimido[4,5-b]quinoline-2,4(3H,10H)-dione, 5-(1H-imidazol-1-yl)-3,10-dimethyl-	MeOH	221(4.61),268(4.64), 320(4.07),412(4.09), 433s(4.02)	83-0476-83
Quino[8,7-g]pteridine-9,11(7H,10H)-dione, 2,4,7-trimethyl-	MeOH	317(4.16),437(3.95)	18-1694-83
$ZrCl_4$ salt	MeOH	328(4.44),434(4.14)	18-1694-83
$C_{16}H_{13}N_5O_4$			
Benzoic acid, [1-[(4-nitrophenyl)azo]-2-oxopropylidene]hydrazide	EtOH	278(4.20),370(3.86)	104-2104-83
	dioxan	278(4.18),380(4.11)	104-2104-83
	MeCN	277(4.16),370(4.12)	104-2104-83
	CCl_4	278(4.24),383(3.24)	104-2104-83
	isooctane	275(--),376(--)	104-2104-83
$C_{16}H_{13}OP$			
Cyclohepta[c]phosphole, 1-methoxy-3-phenyl-	hexane	273s(4.00),291(4.06), 303(4.03),331(4.06), 386(3.15),398s(3.14), 726s(2.32),768(2.35), 860s(2.10)	89-0879-83
$C_{16}H_{13}OTe$			
1-Benzotellurinium, 7-methoxy-2-phen-yl-, perchlorate	CH_2Cl_2	285(4.47),386(3.94)	35-0883-83
1,2-Oxatellurol-1-ium, 5-(3-methyl-phenyl)-3-phenyl-, chloride	CH_2Cl_2	422(4.36)	35-0883-83
$C_{16}H_{13}O_2$			
1-Benzopyrylium, 2-(2-hydroxyphenyl)-4-methyl-, perchlorate	MeCN	424(4.39)	103-0243-83
$C_{16}H_{13}O_2Se$			
1,2-Oxaselenol-1-ium, 5-(4-methoxy-phenyl)-3-phenyl-, chloride	CH_2Cl_2	378(4.40)	35-0883-83

Compound	Solvent	$\lambda_{max}(\log \epsilon)$	Ref.
$C_{16}H_{13}O_2Te$			
1,2-Oxatellurol-1-ium, 5-(4-methoxy-phenyl)-3-phenyl-, chloride	n.s.g.	435(4.48)	35-0875-83
$C_{16}H_{13}P$			
Cyclohepta[b]phosphole, 2-(phenyl-methyl)-	hexane	246(4.24),310(4.58), 322(4.61),364(3.51), 380(3.72),400(3.61), 588s(2.25),640(2.46), 684(2.43),700(2.43), 756s(2.07),788s(1.97)	89-0057-83 +89-0075-83S
$C_{16}H_{14}$			
Benzo[1,2:3,4]dicycloheptene, 12a,12b-dihydro-, trans	isooctane	233(4.10),284(3.78), 366(3.80)	24-1963-83
Cycloprop[a]indene, 1,1a,6,6a-tetra-hydro-1a-phenyl-	MeOH	209(4.12),218s(4.05), 228s(3.97),269s(3.24), 276s(3.04)	44-2202-83
Cycloprop[a]indene, 1,1a,6,6a-tetra-hydro-6-phenyl-, (1aα,6α,6aα)-	MeOH	214(4.09),268s(3.18), 276(3.15)	44-2202-83
Cycloprop[a]indene, 1,1a,6,6a-tetra-hydro-6a-phenyl-	MeOH	204(4.58),230s(4.04)	44-2202-83
1H-Indene, 2-(phenylmethyl)-	MeOH	206(4.42),257(4.17)	44-2202-83
$C_{16}H_{14}Br_2$			
Benzene, 1,1'-(2,3-dibromo-1-butenyli-dene)bis-	hexane	240(4.07),267(4.05)	104-1438-83
Biphenylene, 2,6-dibromo-3,7-diethyl-	EtOH	210.2(3.91),252(4.61), 261(5.00),372(3.82)	18-1192-83
$C_{16}H_{14}Br_4$			
Benzocyclooctene, 2,5,7,10-tetrabromo-3,8-diethyl-	EtOH	230(4.62)	18-1192-83
$C_{16}H_{14}ClNO_2$			
2,4,6-Cycloheptatrien-1-one, 4-[1-amino-3-(4-chlorophenyl)-2-prop-enyl]-2-hydroxy-	MeOH	248(4.44),335(3.90), 373(3.83),400s(3.50)	18-3099-83
$C_{16}H_{14}ClNO_2S_3$			
Morpholine, 4-[2-[5-chloro-2-(2-thien-ylthio)phenyl]-2-oxo-1-thioxoethyl]-	MeOH	237(4.50),271(4.37), 358(3.54)	73-2970-83
$C_{16}H_{14}ClNO_3$			
2H-Furo[2,3-h]-1-benzopyran-2-one, 3-chloro-4-piperidino-	EtOH	217.5s(4.28),243s(4.36), 247.5(4.37),295s(3.95), 328(4.05)	111-0113-83
$C_{16}H_{14}ClN_3O$			
3H-1,3,4-Benzotriazepine, 7-chloro-2-methoxy-3-methyl-5-phenyl-	MeOH	227(4.47),273(4.07), 303s(3.56),350(2.78)	106-0081-83
$C_{16}H_{14}Cl_2$			
3,5,8,10-Tridecatetraene-1,12-diyne, 7-(dichloromethylene)-3,11-dimethyl-, (E,E,Z,Z)-	CH_2Cl_2	247s(4.00),256(4.18), 265(4.19),299s(4.33), 317(4.48),331(4.39)	39-2997-83C
$C_{16}H_{14}Cl_2N_6$			
Guanidine, N-[6-(2,6-dichlorophenyl)-2-methylpyrido[2,3-d]pyrimidin-7-yl]-N'-methyl-	MeOH	293(3.99),360(4.38)	87-0403-83

Compound	Solvent	λ_{max}(log ε)	Ref.
$C_{16}H_{14}Cl_3IN_2O_5S_2$			
5-Thia-1-azabicyclo[4.2.0]oct-2-ene-2-carboxylic acid, 3-(iodomethyl)-8-oxo-7-[(2-thienylacetyl)amino]-, 2,2,2-trichloroethyl ester, 5-oxide, [5S-(5α,6β,7α)]-	EtOH	227.5(4.10),294(3.97)	78-0461-83
$C_{16}H_{14}Cl_4N_2O_5S_2$			
5-Thia-1-azabicyclo[4.2.0]oct-2-ene-2-carboxylic acid, 3-(chloromethyl)-8-oxo-7-[(2-thienylacetyl)amino]-, 2,2,2-trichloroethyl ester, 5-oxide, [5S-(5α,6β,7α)]-	EtOH	234(4.00),276(3.89)	78-0461-83
$C_{16}H_{14}FNO_2S_2$			
Dibenzo[b,f]thiepin-2-sulfonamide, 7-fluoro-N,N-dimethyl-	MeOH	270(4.46),300s(3.70)	73-0144-83
$C_{16}H_{14}FNO_3S_2$			
Dibenzo[b,f]thiepin-2-sulfonamide, 7-fluoro-10,11-dihydro-N,N-dimethyl-11-oxo-	MeOH	242(4.33),280(4.08), 328(3.74)	73-0144-83
$C_{16}H_{14}F_3NO_2$			
Acetic acid, trifluoro-, 3,4-dihydro-1,2-dimethyl-1H-cyclohept[cd]indol-6-yl ester	EtOH	244(4.39),336(3.77)	39-2425-83C
$C_{16}H_{14}N_2$			
1,2-Diazabicyclo[3.1.0]hex-2-ene, 3,6-diphenyl-, cis	MeOH	257(4.06)	35-0933-83
Pyridazine, 1,4-dihydro-3,6-diphenyl-	MeOH	245(4.26)	35-0933-83
Quinoxaline, 6,7-dimethyl-2-phenyl-	C_6H_{12}	210(3.53),216s(3.49), 228s(3.29),264(3.55), 269s(3.54),342(3.14), 359s(2.99)	4-1739-83
$C_{16}H_{14}N_2O$			
Benzenamine, N-(1H-indol-3-ylmethylene)-4-methyl-, N-oxide	EtOH	220(4.52),270(4.05), 290(3.94),360(3.82)	104-1367-83
Benzenamine, N-[(1-methyl-1H-indol-3-yl)methylene]-, N-oxide	EtOH	240(4.48),266(4.05), 365(4.51)	104-1367-83
Benzenamine, N-[(2-methyl-1H-indol-3-yl)methylene]-, N-oxide	EtOH	205(3.48),230(3.55), 278(3.04),370(3.48)	104-1367-83
Ethanone, 1-(1H-indol-3-yl)-2-phenyl-, oxime, (E)-	n.s.g.	225(4.32),261(4.00), 287(3.96)	103-0184-83
$C_{16}H_{14}N_2O_2$			
1H-Indole, 1,2-dimethyl-5-nitro-3-phenyl-	MeOH	276(4.31),322(3.87)	103-1302-83
3H-Indole, 3,3-dimethyl-2-(4-nitrophenyl)-	$C_6H_{11}Me$	345(4.305)	151-0251-83A
3H-Indole, 3,3-dimethyl-5-nitro-2-phenyl-	$C_6H_{11}Me$	333(4.386)	151-0251-83A
1,8-Naphthyridin-4(1H)-one, 2,7-dimethyl-3-phenoxy-	EtOH	248(4.51),276(3.74), 288(3.67),332(4.09)	142-1083-83
1,4-Phthalazinedione, 2-ethyl-2,3-dihydro-3-phenyl-	EtOH	216(4.66),237(4.12), 305(3.70)	104-2260-83
1(2H)-Phthalazinone, 4-(4-hydroxy-3,5-dimethylphenyl)-	EtOH	209(4.70),305(3.97)	24-0970-83

Compound	Solvent	λ_{max}(log ε)	Ref.
4H-Pyrido[1,2-a]pyrimidin-4-one, 2,6-dimethyl-3-phenoxy-	EtOH	255(4.13),360(4.04)	142-1083-83
$C_{16}H_{14}N_2O_3$			
Benzo[1,2-b:4,3-b']dipyrrole, 1,3,6-triacetyl-3,6-dihydro-	EtOH	206(4.45),232(4.40), 253(4.52),289.8(4.26), 314(4.30)	103-0871-83
Benzo[1,2-b:4,5-b']dipyrrole, 1,3,5-triacetyl-1,5-dihydro-	EtOH	204(--),243.5(--), 252(--),277(--), 287s(--),322(--), 335(--)	103-0871-83
1H-Indole-2-methanol, 1-methyl-5-nitro-3-phenyl-	dioxan	226s(4.3),282(4.32), 336(3.9)	103-1302-83
1,3,4-Oxadiazole, 2,5-bis(4-methoxyphenyl)-	toluene	300(4.51)	103-0022-83
Pyrano[2,3-c]pyrazol-4(1H)-one, 5-acetyl-3-ethyl-1-phenyl-	EtOH	312(4.10)	44-4078-83
$C_{16}H_{14}N_2O_3S$			
1-Naphthalenesulfonic acid, 5-amino-8-(phenylamino)-	MeOH	215(4.44),245(4.09), 277(4.05),380(3.59)	48-1002-83
1H-Thieno[2',3':4,5]pyrrolo[1,2,3-de]-quinoxaline-9-carboxylic acid, 2,3-dihydro-8-methyl-2-oxo-, ethyl ester	dioxan	225(4.48),400(4.34), 420(4.26)	103-0959-83
$C_{16}H_{14}N_2O_4$			
Benzeneacetic acid, 2-[[(4-nitrophenyl)methylene]amino]-, methyl ester	$CHCl_3$	343(4.01)	44-2468-83
Carbamic acid, [(phenylamino)carbonyl]-, 3-acetylphenyl ester	EtOH	240.5(4.44)	22-0073-83
	dioxan	249(4.20)	22-0073-83
Carbamic acid, [(phenylamino)carbonyl]-, 4-acetylphenyl ester	EtOH	247(4.50)	22-0073-83
	dioxan	249(4.41)	22-0073-83
1,2-Diazabicyclo[5.2.0]nona-3,5-diene-4-carboxylic acid, 2-benzoyl-8-methyl-9-oxo-, cis	MeOH	310(3.80)	5-1393-83
$C_{16}H_{14}N_2O_4S$			
2-Naphthalenesulfonic acid, 3-amino-4-hydroxy-6-(phenylamino)-	MeOH	210(4.19),230(4.14), 298(4.27),322(3.94)	48-1002-83
2-Naphthalenesulfonic acid, 3-amino-4-hydroxy-7-(phenylamino)-	MeOH	212(4.38),232(4.27), 271(4.24),313(4.18)	48-1002-83
2-Naphthalenesulfonic acid, 5-amino-4-hydroxy-6-(phenylamino)-	MeOH	210(4.38),232(4.35), 256(4.20),329(4.00)	48-1002-83
$C_{16}H_{14}N_2O_5$			
Oxepino[4,5-c]pyridine-3,4-dicarboxylic acid, 1-cyano-6,8-dimethyl-, dimethyl ester	CH_2Cl_2	274(4.21),348(3.24)	24-0097-83
2H-2-Pyrindine-3,4-dicarboxylic acid, 5-acetyl-1-cyano-6-methyl-, dimethyl ester	CH_2Cl_2	256(4.24),332(4.12), 345s(4.08),405(3.79)	24-0097-83
$C_{16}H_{14}N_2O_5S$			
2-Thiophenecarboxylic acid, 3-[(2-nitro-3-oxo-1-propenyl)amino]-5-phenyl-, ethyl ester	EtOH	245(3.98),314(4.30), 380(4.27)	103-1163-83
$C_{16}H_{14}N_2O_7$			
2-Butenedioic acid, 2-[[(1,3-dihydro-1,3-dioxo-2H-isoindol-2-yl)acetyl]-amino]-, dimethyl ester, (E)-	EtOH	264(4.01)	118-0539-83

Compound	Solvent	λ_{max}(log ε)	Ref.
(Z)- (cont.)	EtOH	275(4.29)	118-0539-83
6-Oxa-1-azabicyclo[5.2.0]nona-2,4-diene-2-carboxylic acid, 3-methoxy-9-oxo-, (4-nitrophenyl)methyl ester	EtOH	260(4.08),313(4.06)	39-0115-83C
$C_{16}H_{14}N_4O_2$			
1-Triazenium, 1-(1,3-dihydro-1,3-dioxo-2H-isoindol-2-yl)-2-ethyl-3-phenyl-, hydroxide, inner salt, (E)-	EtOH	217(4.57),239s(4.25), 265(4.08)	104-2260-83
(Z)- (contains some isomer)	EtOH	222(4.56),333(4.08)	104-2260-83
$C_{16}H_{14}N_4O_2S$			
Morpholine, 4-[[4-(2-quinoxalinyl)-2-thiazolyl]carbonyl]-	EtOH	247(3.99),268s(3.81), 297s(3.57),343(3.50), 357s(3.46)	106-0829-83
$C_{16}H_{14}N_4O_3$			
Benzenamine, N,N-dimethyl-4-[5-(4-nitrophenyl)-1,3,4-oxadiazol-2-yl]-	toluene	307(4.49),390(4.09)	103-0022-83
	EtOH	310(4.48),385(4.3)	103-0022-83
$C_{16}H_{14}N_4O_4$			
Benzo[1,2-b:3,4-b']dipyrrole, 1,6-dihydro-3,8-bis(2-nitro-1-propenyl)-	EtOH	204(4.22),233(4.15), 284(3.87),432(4.17)	103-0488-83
$C_{16}H_{14}N_4O_4S_2$			
1,2,4,5-Tetrazine, 3,6-bis[(phenylsulfonyl)methyl]-	MeOH	219(4.35),261(3.54), 267(3.60),274(3.57), 531(2.58)	104-0446-83
$C_{16}H_{14}N_4O_5$			
Ethanol, 2-[(8-nitro-1,2,4-triazolo-[1,5-a]pyridin-2-yl)methoxy]-, benzoate	EtOH	230(4.50),325(3.73)	4-1169-83
Pyrazolo[1,5-d][1,2,4]triazine-2,3-dicarboxylic acid, 6,7-dihydro-4-methyl-7-oxo-6-phenyl-, dimethyl ester	EtOH	292(4.03)	44-1069-83
Pyrazolo[1,5-d][1,2,4]triazine-2,3-dicarboxylic acid, 6,7-dihydro-6-methyl-7-oxo-4-phenyl-, dimethyl ester	EtOH	232(4.11),288(4.08)	44-1069-83
$C_{16}H_{14}N_4O_6$			
Pyrrolidine, 1-(2',4',6'-trinitro-[1,1'-biphenyl]-2-yl)-	$CHCl_3$	270(4.5),504(3.91)	44-4649-83
$C_{16}H_{14}N_4O_7$			
Morpholine, 4-(2',4',6'-trinitro-[1,1'-biphenyl]-2-yl)-	$CHCl_3$	440(3.71)	44-4649-83
$C_{16}H_{14}N_6O_3S$			
2(3H)-Benzothiazolone, [1-[(2-hydroxy-4-nitrophenyl)azo]propylidene]hydrazone	EtOH	505(4.44)	65-1033-83
	EtOH-base	600(4.23)	65-1033-83
	acetone	505(4.40),625(3.98)	65-1033-83
$C_{16}H_{14}O$			
2-Propen-1-one, 2-(4-methylphenyl)-1-phenyl-	MeOH	251(4.27)	12-0311-83
$C_{16}H_{14}OS$			
6H-Dibenzo[b,d]pyran, 6-(1-methylethylidene)-, 5-oxide	MeOH	227(4.50),313(3.84)	12-0795-83

Compound	Solvent	λ_{max}(log ε)	Ref.
$C_{16}H_{14}O_2$			
9,10-Anthracenedione, 1,9a-dihydro-1,9a-dimethyl-, trans	MeOH	231(3.99),247(4.37),325s(3.84),338(3.89)	44-2412-83
9,10-Anthracenedione, 1,9a-dihydro-6,9a-dimethyl-	MeOH	233(4.00),249(4.38),325s(3.85),340(3.90)	44-2412-83
Ethanone, 1,1'-[1,1'-biphenyl]-4,4'-diylbis-	hexane	287(3.86)	99-0034-83
2,4,10,12-Tetradecatetraene-6,8-diyne-dial, 5,10-dimethyl-, (E,E,Z,Z)-	THF	245s(4.12),254(4.24),293(4.38),339(4.41),362(4.40),387(4.31)	18-1467-83
$C_{16}H_{14}O_2S$			
Ethanone, 1,1'-(thiodi-4,1-phenylene)-bis-	hexane	239(4.20),290(4.23),319(4.08)	99-0034-83
$C_{16}H_{14}O_3$			
Ethanone, 1,1'-(oxydi-4,1-phenylene)-bis-	hexane	270(4.43)	99-0034-83
$C_{16}H_{14}O_3$(cont.)			
Ketoprofen	MeOH	255(4.33)	162-0762-83
$C_{16}H_{14}O_4$			
4H-1,3-Benzodioxin-4-one, 2-methyl-2-(phenylmethoxy)-	pH 7.4	300(3.43)	1-0351-83
3(2H)-Benzofuranone, 2-(1,3-benzodiox-ol-5-ylmethylene)-4,5,6,7-tetra-hydro-	EtOH	244(3.90),266(3.98),376(4.26)	4-0543-83
5H-Benzofuro[6,5-c][2]benzopyran-5-one, 1,2,3,4-tetrahydro-7-methoxy-	MeOH	210(4.47),248(4.38),263(4.22),299(4.05)	78-1265-83
Benzoic acid, 2-(4-hydroxy-3,5-dimethylbenzoyl)-	EtOH	296(4.17)	24-0970-83
4H-1-Benzopyran-4-one, 2,3-dihydro-7-hydroxy-3-[(4-hydroxyphenyl)-methyl]-	EtOH	278(4.06),310(3.86),372(3.42)	142-0039-83
2aH-Cyclopent[cd]indene-1,2-dicarbox-ylic acid, 2a-methyl-, dimethyl ester	EtOH	205(4.42),252(3.95),270(3.92),283(3.88),311(3.84)	39-0083-83C
7bH-Cyclopent[cd]indene-1,2-dicarbox-ylic acid, 7b-methyl-, dimethyl ester	EtOH	217(3.99),262s(3.82),305(4.60),336s(3.84),471(3.25)	39-0083-83C
Furopinnarin, demethyl-	MeOH	224.5(4.43),246(4.11),251.5(4.13),273.5(4.31),314.5(4.09)	94-2712-83
Imperatorin	n.s.g.	250(4.24),265(4.00),302(3.95)	162-0716-83
Medicarpin	EtOH	287(3.90)	149-0323-83B
1H-Phenalen-1-one, 3,4,9-trimethoxy-	hexane	250(4.35),323s(3.54),362(4.16),392s(3.78),413s(3.52)	44-2115-83
Tetracyclo[4.2.2.2^{2,5}.0^{1,6}]dodeca-3,7,9,11-tetraene-3,4-dicarboxylic acid, dimethyl ester	hexane	273(3.00)	88-3361-83
	EtOH	277.5(2.93)	88-3361-83
Tricyclo[9.2.2.1^{4,8}]hexadeca-4,6,8(16),13,15-pentaene-12,14-dione, 6,16-dihydroxy-	THF	245(4.00),315(3.50),355(2.92),490(2.83)	35-6650-83
$C_{16}H_{14}O_5$			
9,10-Anthracenedione, 1-acetoxy-1,4,4a,9a-tetrahydro-8-hydroxy-	EtOH	230(4.28),348(3.68)	44-3252-83

Compound	Solvent	λ max (log ε)	Ref.
6H-Dibenzo[b,d]pyran-6-one, 3,8,9-trimethoxy-	$CHCl_3$	252.5(4.25),259(4.31), 312(3.93),337.5(3.70)	39-2031-83C
Isoflavan, 2',7-dihydroxy-4',5'-(methylenedioxy)-, [³H]-(3R)-	EtOH	289(3.84),298(--)	102-1591-83
$C_{16}H_{14}O_6$			
3(2H)-Benzofuranone, 2,6-dihydroxy-2-[(4-hydroxyphenyl)methyl]-4-methoxy-, (-)- (carpusin)	MeOH	290(4.41),324s(--)	102-0794-83
	MeOH-NaOAc	317(--)	102-0794-83
Homoeriodictyol	EtOH	290(2.26),328(2.33)	162-0685-83
1H,3H-Naphtho[1,8-cd]pyran-1,3-dione, 4,6,9-trimethoxy-7-methyl-	EtOH	227(3.77),252(4.11), 261s(4.08),346(3.76), 363(3.69),380(3.58)	78-2283-83
2-Pentenal, 5-(1,4-dihydro-5,8-dihydroxy-1,4-dioxo-2-naphthalenyl)-5-hydroxy-2-methyl-	MeOH	278(3.91),330(3.02), 486(3.78),517(3.81), 555(3.59)	105-0532-83
$C_{16}H_{14}O_7$			
2-Pentenoic acid, 5-(1,4-dihydro-5,8-dihydroxy-1,4-dioxo-2-naphthalenyl)-5-hydroxy-2-methyl-	MeOH	276(3.85),330(3.00), 486(3.70),514(3.80), 553(3.58)	105-0532-83
$C_{16}H_{14}O_8S_4$			
1,3-Dithiole-4,5-dicarboxylic acid, 2,2'-(1,2-ethanediylidene)bis-, tetramethyl ester	$CHCl_3$	374(4.45),389(4.42)	88-3469-83
$C_{16}H_{14}S$			
10,9-(Epithiomethano)anthracene, 9,10-dihydro-12-methyl-	MeCN	225(3.91),278(3.43)	78-1487-83
$C_{16}H_{15}Br$			
Benzene, 1,1'-(2-bromo-1-butenylidene)bis-	hexane	246(4.03)	104-1438-83
$C_{16}H_{15}ClN_2O_3$			
4H-1,3-Benzodioxin-6-ol, 7-[(2-chlorophenyl)azo]-2,4-dimethyl-	EtOH	362(4.23),458(3.71)	103-0954-83
$C_{16}H_{15}ClN_4O_3$			
Carbamic acid, [5-amino-3-(4-chlorophenyl)-2H-pyrido[4,3-b][1,4]oxazin-7-yl]-, ethyl ester	pH 1	298(3.86),307s(3.84), 367s(4.33),376(4.42) (changing)	87-1614-83
$C_{16}H_{15}ClN_6S_2$			
2-Pyridinamine, 3,3'-[(5-chloro-2,3-pyrazinediyl)bis(thio)]bis[6-methyl-	MeOH	260(4.17),310(3.77)	4-1047-83
$C_{16}H_{15}ClO_4$			
1,4-Naphthalenedione, 2-(1-chloro-4-methyl-3-pentenyl)-5,8-dihydroxy-	MeOH	274(3.93),330(2.96), 489(3.73),517(3.73), 556(3.50)	105-0532-83
$C_{16}H_{15}ClS$			
Benzene, 1-chloro-4-[1-[[(phenylmethyl)thio]methyl]ethenyl]-	MeOH	245(4.02)	44-0835-83
$C_{16}H_{15}Cl_3N_2O_4S$			
5-Thia-1-azabicyclo[4.2.0]oct-2-ene-2-carboxylic acid, 7-amino-8-oxo-	EtOH	262s(2.94),268s(3.88), 273.5(3.98)	78-0461-83

Compound	Solvent	λ_{max}(log ε)	Ref.
3-(phenylmethyl)-, 2,2,2-trichloroethyl ester, 5-oxide, monohydrochloride, [5S-(5α,6β,7α)]- (cont.)			78-0461-83
$C_{16}H_{15}NO$			
Acetamide, N-(1,2-diphenylethylidene)-	C_6H_{12}	248(3.52)	44-0695-83
Pyridine, 1,4-dihydro-1-(1-oxo-2-butenyl)-4-(phenylmethylene)-	MeOH	374(4.40)	24-3192-83
$C_{16}H_{15}NO_2$			
Benzo[a]cyclopenta[f]quinolizine-1,12-dione, 2,3,5,6,10b,11-hexahydro-, (R)-	MeOH	258(4.32),291(4.09)	135-0303-83
4(5H)-Benzofuranone, 6,7-dihydro-5-[(methylphenylamino)methylene]-	EtOH	262s(3.93),362(4.31)	111-0113-83
2,4,6-Cycloheptatrien-1-one, 4-(1-amino-3-phenyl-2-propenyl)-2-hydroxy-	MeOH	245(4.49),335(3.96), 395(3.57)	18-3099-83
8H-Indeno[6,5,4-cd]indol-8-one, 4-acetyl-4,5,5a,6,6a,7-hexahydro-	MeOH	236(4.50),294(4.42)	39-1545-83C
1-Oxa-3-azaspiro[4.5]deca-3,6,9-trien-8-one, 2,2-dimethyl-4-phenyl-	EtOH	245(3.98)	33-2252-83
$C_{16}H_{15}NO_2S$			
Benzene, 1-nitro-4-[1-[[(phenylmethyl)thio]methyl]ethenyl]-	MeOH	298(3.59)	44-0835-83
Thieno[2,3-c]pyridine, 7-[(3,4-dimethoxyphenyl)methyl]-	MeOH	226(4.51),298(3.96), 309(3.96)	83-0912-83
Thieno[3,2-c]pyridine, 4-[(3,4-dimethoxyphenyl)methyl]-	MeOH	222(4.64),268(3.95), 276s(3.95),295s(3.60)	83-0244-83
2-Thietanone, 3-[[(2-hydroxy-1-naphthalenyl)methylene]amino]-4,4-dimethyl-, (R)-	EtOH	229(4.73),305(4.04), 317(4.08),352(3.90)	39-2259-83C
$C_{16}H_{15}NO_3$			
1H-Benzo[a]furo[3,4-f]quinolizine-1,12(3H)-dione, 5,6,10b,11-tetrahydro-3-methyl-	MeOH	238(4.19),292(4.21)	135-0303-83
Bicyclo[2.2.1]heptan-2-one, 6-(N-phthalimidomethyl)-, endo	MeCN	292(3.27)	44-1862-83
4H-Furo[3,2-b]pyrrole-5-carboxylic acid, 2-(4-methylphenyl)-, ethyl ester	MeOH	233(3.84),342(4.79)	73-0772-83
4H-Furo[3,2-b]pyrrole-5-carboxylic acid, 4-methyl-2-phenyl-, ethyl ester	MeOH	279(4.24),333(4.74)	73-0772-83
Lycorin-2-one	$CHCl_3$	245(3.49),292(3.47)	102-2193-83
1H,3H-Naphtho[1,8-cd]pyran-1,3-dione, 6-(diethylamino)-	MeOH	420(3.90)	104-2273-83
$C_{16}H_{15}NO_3S$			
2-Butenamide, N-phenyl-2-(phenylsulfonyl)-, (Z)-	EtOH	231(4.27),258(4.03)	104-1314-83
2-Butenamide, N-phenyl-3-(phenylsulfonyl)-	EtOH	230(4.33),300(4.03)	104-1314-83
2-Butenamide, N-phenyl-4-(phenylsulfonyl)-, (E)-	EtOH	220(4.40),281(4.11)	104-1314-83
3-Butenamide, N-phenyl-3-(phenylsulfonyl)	EtOH	220(4.08),246(4.15)	104-1314-83
2-Propenamide, 2-methyl-N-phenyl-3-(phenylsulfonyl)-, (E)-	EtOH	220(4.30),270(3.90)	104-1314-83

Compound	Solvent	$\lambda_{max}(\log \epsilon)$	Ref.
2-Propenamide, 2-(methylsulfonyl)-N,3-diphenyl-	EtOH	219(4.16),250(4.34),270(4.24)	104-1314-83
2-Propenamide, N-phenyl-2-[(phenylsulfonyl)methyl]-	EtOH	218(4.30),266(3.90)	104-1314-83
$C_{16}H_{15}NO_4$			
3,4-Pyridinedicarboxylic acid, 5-methyl-6-phenyl-, dimethyl ester	EtOH	254.5(3.98),289(4.03)	33-0262-83
$C_{16}H_{15}NO_5$			
9(10H)-Acridinone, 1,3-dihydroxy-5,6-dimethoxy-10-methyl- (grandisine II)	MeOH	220(4.19),269(4.80),296(4.20),332(4.16),390(3.78)	102-1493-83
	MeOH-NaOMe	224(--),265(--),285s(--),365(--),400s(--)	102-1493-83
	MeOH-$AlCl_3$	228(--),263s(--),275(--),310s(--),362(--),425(--)	102-1493-83
9(10H)-Acridinone, 1,5-dihydroxy-3,6-dimethoxy-10-methyl- (grandisine I)	MeOH	216(4.24),256s(4.76),265(4.86),282s(4.37),328(4.04),382(3.78)	102-1493-83
	MeOH-NaOMe	215(--),256s(--),265(--),282s(--),328(--),382(--)	102-1493-83
	MeOH-$AlCl_3$	224(--),255(--),275(--),300s(--),346(--),416(--)	102-1493-83
9(10H)-Acridinone, 1,6-dihydroxy-3,5-dimethoxy-10-methyl- (citpressine I)	MeOH	220(4.09),265s(4.64),271(4.65),294s(4.06),332(3.96),384(3.58)	94-0895-83
	MeOH-$AlCl_3$	220(--),273(--),308s(--),352(--),420(--)	94-0895-83
6-Oxa-1-azabicyclo[5.2.0]nona-2,4-diene-2-carboxylic acid, 3-methoxy-9-oxo-, phenylmethyl ester	EtOH	235s(3.72),314(3.93)	39-0115-83C
$C_{16}H_{15}NS_3$			
Ethane(dithioic) acid, (diphenylamino)thioxo-, ethyl ester	CH_2Cl_2	262(4.08),290(4.09),342s(3.85),439s(2.99)	97-0247-83
$C_{16}H_{15}N_3O$			
Benzenamine, N,N-dimethyl-4-(5-phenyl-1,3,4-oxadiazol-2-yl)-	toluene	335(4.56)	103-0022-83
	dioxan	340(4.3)	103-0022-83
	MeCN	348(4.3)	103-0022-83
	DMSO	338(4.3)	103-0022-83
$C_{16}H_{15}N_3OS$			
Pyrido[3,4-e]-1,3-thiazin-4-one, 5,7-dimethyl-2-(methylphenylamino)-	MeOH	250(4.11)	73-3426-83
Pyrido[3,4-e]-1,3-thiazin-4-one, 5,7-dimethyl-2-[(4-methylphenyl)amino]-	MeOH	280(4.00)	73-3426-83
Pyrido[3,4-e]-1,3-thiazin-4-one, 5,7-dimethyl-2-[(phenylmethyl)amino]-	MeOH	247(4.16)	73-3426-83
4-Thiazolidinone, 2-(diphenylaminomethylimino)-	EtOH	226(4.39),240(4.37)	103-1076-83
$C_{16}H_{15}N_3OSe$			
4H-Pyrido[3,4-e]-1,3-selenazin-4-one,	MeOH	243(4.16),288(4.00)	73-3567-83

Compound	Solvent	$\lambda_{max}(\log \epsilon)$	Ref.
5,7-dimethyl-2-[(4-methylphenyl)-amino]- (cont.)			73-3567-83
$C_{16}H_{15}N_3O_2$			
1H-Indole-2-methanamine, 1-methyl-5-nitro-3-phenyl-	MeOH	266s(4.35),280(4.45), 336(3.95)	103-1302-83
$C_{16}H_{15}N_3O_2S$			
4H-Pyrido[3,4-e]-1,3-thiazin-4-one, 2-[(4-methoxyphenyl)amino]-5,7-dimethyl-	MeOH	263(4.05)	73-3426-83
$C_{16}H_{15}N_3O_2Se$			
4H-Pyrido[3,4-e]-1,3-selenazin-4-one, 2-[(4-methoxyphenyl)amino]-5,7-dimethyl-	MeOH	213(3.01),260(4.91)	73-3567-83
$C_{16}H_{15}N_3O_3$			
4,7-Phenanthrolin-5-ol, 6-(3-hydroxy-4-morpholinyl)-, hydrochloride	pH 1	258(4.28),385(3.59)	35-4431-83
2-Propenoic acid, 2-azido-3-[5-(4-methylphenyl)-2-furanyl]-, ethyl ester	MeOH	268(3.92),378(4.57)	73-0772-83
$C_{16}H_{15}N_3O_4$			
4H-Furo[3,2-b]pyrrole-5-carboxylic acid, 2-(4-methylphenyl)-, 2-(methoxycarbonyl)hydrazide	MeOH	313(3.07),400(3.05)	73-1878-83
$C_{16}H_{15}N_3O_5$			
4H-1,3-Benzodioxin-6-ol, 2,4-dimethyl-7-[(2-nitrophenyl)azo]-	EtOH	264(3.98),356(4.14), 517(3.85)	103-0954-83
4H-1,3-Benzodioxin-6-ol, 2,4-dimethyl-7-[(4-nitrophenyl)azo]-	EtOH	271(3.79),355(4.34), 494(3.92)	103-0954-83
$C_{16}H_{15}N_5O$			
Furo[2',3':4,5]pyrrolo[1,2-d]-1,2,4-triazin-8(7H)-one, 5-methyl-2-(4-methylphenyl)-, hydrazone	MeOH	352(3.43)	73-1878-83
$C_{16}H_{15}N_5OS$			
1H-Imidazo[2,1-f]purin-2(8H)-one, 1,8-dimethyl-4-(methylthio)-7-phenyl-	MeOH	256(4.45),319(4.36)	105-0029-83
	MeOH-acid	249(4.42),332(4.37)	105-0029-83
	MeOH-base	256(4.47),319(4.35)	105-0029-83
$C_{16}H_{15}N_5O_5$			
Carbamic acid, [5-amino-3-(4-nitrophenyl)-2H-pyrido[4,3-b]-1,4-oxazin-7-yl]-, ethyl ester	pH 1	254(4.20),310s(3.75), 398(4.36)(changing)	87-1614-83
$C_{16}H_{16}$			
Benzene, 1,1'-(1,2-ethenediyl)bis[4-methyl-	heptane	315(4.48)	94-0373-83
Bicyclo[5.4.1]dodeca-2,5,7,9,11-pentaene, 3,5-diethenyl-	3-Mepentane	277.5(4.86),286(4.87), 367(3.72)	35-6211-83
Cyclopenta[ef]heptalene, 2a,10b-dihydro-2a,10b-dimethyl-	hexane	234(4.22),273(4.16), 283(4.19),438(3.31)	89-0282-83S
Cyclopenta[ef]heptalene, 6,10b-dihydro-6,10b-dimethyl-	hexane	247(4.09),255(4.11), 276(4.17),420(3.38)	89-0282-83S
Cyclopenta[ef]heptalene, 6a,10a-dihydro-6a,10b-dimethyl-	hexane	232(3.93),290(3.92), 300(3.87),475(3.31)	89-0282-83S

Compound	Solvent	$\lambda_{max}(\log \epsilon)$	Ref.
$C_{16}H_{16}BrNO_2S$			
Methanone, (4-bromophenyl)[4-methyl-5-(4-morpholinyl)-2-thienyl]-	MeCN	259(4.10),371(4.20)	48-0168-83
$C_{16}H_{16}BrN_3$			
6H-Pyrido[2',1':3,4][1,4]diazepino-[1,2-a]benzimidazol-5-ium, 7,8-dihydro-2-methyl-, bromide	EtOH	344(4.23)	4-0029-83
$C_{16}H_{16}ClNO_2$			
1,4-Naphthalenedione, 2-chloro-3-[2-(diethylamino)ethenyl]-	C_6H_{12}	547(4.11)	150-0168-83S
	EtOH	589(--)	150-0168-83S
$C_{16}H_{16}ClNO_3$			
2H-Furo[2,3-h]-1-benzopyran-2-one, 3-chloro-5,6-dihydro-4-piperidino-	EtOH	234s(4.23),265s(4.06), 349(4.43)	111-0113-83
$C_{16}H_{16}Cl_2N_2O$			
3-Pyridinecarbonitrile, 1-[(2,6-dichlorophenyl)methyl]-1,6-dihydro-6-(1-methylethoxy)-	CH_2Cl_2	314(3.8),356s(3.2)	64-0873-83B
3-Pyridinecarbonitrile, 1-[(2,6-dichlorophenyl)methyl]-1,6-dihydro-6-propoxy-	CH_2Cl_2	238(3.7),312(3.8), 355s(3.3)	64-0873-83B
$C_{16}H_{16}Cl_4O_4$			
Dicyclopropa[a,c]naphthalene, 1,1,2,2-tetrachloro-1,1a,1b,2,2a,6b-hexahydro-2a,3,6,6b-tetramethoxy-, (1aα,1bβ-2aβ,6bα)-	MeOH	313.5(3.11),325s(2.97)	39-1983-83C
$C_{16}H_{16}F_3NO$			
Ethanone, 2,2,2-trifluoro-1-(3,4,4a,9-tetrahydro-4a,9-dimethyl-2H-carbazol-1-yl)-	EtOH	202(4.14),240(3.82), 387(4.29)	39-0795-83C
$C_{16}H_{16}NOS$			
Thieno[3,2-c]pyridinium, 4-[(methoxyphenyl)methyl]-5-methyl-, iodide	MeOH	238(4.74),285(3.81), 311(3.69)	83-0244-83
$C_{16}H_{16}N_2$			
Benzaldehyde, 4-methyl-, [(4-methylphenyl)methylene]hydrazone, compd. with ethenetetracarbonitrile (1:1)	CH_2Cl_2	500(1.23)	59-0933-83
complex with DDQ	CH_2Cl_2	565(1.83)	59-0933-83
2,4,6-Nonatrien-8-ynal, 7-methyl-, (3-methyl-2-penten-4-ynylidene)hydrazone , (?,?,E,Z,Z,E)-	THF	233(4.13),277(3.93), 289(3.95),342s(4.66), 359(4.81),377(4.84), 395(4.70)	18-1467-83
$C_{16}H_{16}N_2O$			
Indolo[4,3-fg]isoquinolin-9(4H)-one, 6,6a,7,8-tetrahydro-8,10-dimethyl-	MeOH	210(4.35),244(4.23), 353(3.94)	39-1545-83C
5(6H)-Quinazolinone, 7,8-dihydro-7,7-dimethyl-2-phenyl-	EtOH	292(4.40)	4-0649-83
$C_{16}H_{16}N_2O_2$			
Benzaldehyde, 2-methoxy-, [(2-methoxyphenyl)methylene]hydrazone, DDQ complex	CH_2Cl_2	570(1.48)	59-0933-83

Compound	Solvent	$\lambda_{max}(\log \epsilon)$	Ref.
Benzaldehyde, 4-methoxy-, [(4-methoxyphenyl)methylene]hydrazone, compd. with ethenetetracarbonitrile (1:1)	CH_2Cl_2	580(1.32)	59-0933-83
complex with DDQ	CH_2Cl_2	675(1.92)	59-0933-83
Benzenamine, N,N-dimethyl-4-[2-(4-nitrophenyl)ethenyl]-	toluene	300(4.08),432(4.36)	103-0029-83
Benzenemethanamine, N-ethyl-α-(nitrophenylmethylene)-, (Z)-	EtOH	367(4.28)	22-0339-83
1,2-Diazabicyclo[5.2.0]nona-3,5-dien-9-one, 2-benzoyl-5,8-dimethyl-	MeOH	220(3.99),287(3.94)	5-1393-83
1,2-Diazabicyclo[5.2.0]nona-3,5-dien-9-one, 2-benzoyl-7,8-dimethyl-	MeOH	286(4.01)	5-1393-83
Indolo[4,3-fg]isoquinolin-9(4H)-one, 4-acetyl-5,5a,6,6a,7,8-hexahydro-	MeOH	256(4.54),296(4.24)	39-1545-83C
4H-Pyrido[1,2-a]pyrimidine-9-carboxaldehyde, 1,6,7,8-tetrahydro-6-methyl-4-oxo-3-phenyl-	EtOH	236(4.07),266s(3.85), 358(4.42)	39-0369-83C
Rugulovasine A	EtOH	224(4.37),277(3.70), 288(3.78),295(3.77)	162-1194-83
Rugulovasine B	EtOH	227(4.16),278(3.68), 288(3.73),295(3.72)	162-1194-83
$C_{16}H_{16}N_2O_3$			
4(1H)-Cinnolinone, 3-(1,3-benzodioxol-5-ylmethyl)-5,6,7,8-tetrahydro-	EtOH	281(4.14)	4-0543-83
1,2-Diazabicyclo[5.2.0]nona-3,5-diene-2-carboxylic acid, 9-oxo-8-phenyl-, ethyl ester, cis	CH_2Cl_2	281(3.98)	5-1361-83
$C_{16}H_{16}N_2O_4$			
3-Butenoic acid, 3-methyl-2-[[[5-oxo-2-(phenylmethyl)-4(5H)-oxazolylidene]methyl]amino]-	MeCN	317(4.30)	142-1001-83
1H-Perimidine-2-acetic acid, 2,3-dihydro-2-(methoxycarbonyl)-, methyl ester	MeOH	232(4.62),322s(--), 330(4.03),343(4.04)	18-2338-83
$C_{16}H_{16}N_2O_5$			
D-gluco-Oct-2-enononitrile, 2-cyano-2,3-dideoxy-6,8-O-(phenylmethylene)-, (S)-	MeOH	250(3.95)	87-0030-83
$C_{16}H_{16}N_2O_7$			
6-Oxa-1-azabicyclo[5.2.0]non-2-ene-2-carboxylic acid, 3-methoxy-9-oxo-, (4-nitrophenyl)methyl ester	EtOH	268(4.26)	39-0115-83C
$C_{16}H_{16}N_3$			
6H-Pyrido[2',1':3,4][1,4]diazepino-[1,2-a]benzimidazol-5-ium, 7,8-dihydro-2-methyl-, bromide	EtOH	344(4.23)	4-0029-83
$C_{16}H_{16}N_4$			
Pyrazolo[5,1-c][1,2,4]benzotriazine, 6,7,8,9-tetrahydro-3-methyl-2-phenyl-	EtOH	268(4.60),306(3.53), 341(3.51),376(3.46)	44-2330-83
1,2,4,5-Tetrazine, 1,2-dihydro-1,2-dimethyl-3,6-diphenyl-	dioxan	255(4.34),330(4.12)	24-2261-83
1,2,4,5-Tetrazine, 1,4-dihydro-1,4-dimethyl-3,6-diphenyl-	dioxan	226(4.30),293(3.62)	24-2261-83

Compound	Solvent	λ_{max}(log ε)	Ref.
$C_{16}H_{16}N_4O$			
2H-Imidazo[4,5-b]quinoxalin-2-one, 1,3,3a,4,9,9a-hexahydro-4-methyl-1-phenyl-, cis	EtOH	214(4.62),242(4.25), 297(3.54)	103-1333-83
3-Pyridinecarbonitrile, 1-[4-(5-cyano-1(2H)-pyridinyl)butyl]-1,6-dihydro-6-oxo-	CH_2Cl_2	248(4.2),319(3.7), 334(3.7),353s(3.7)	64-0878-83B
5-Pyrimidinecarbonitrile, 4-[2-(dimethylamino)ethenyl]-1,6-dihydro-6-oxo-1-(phenylmethyl)-	EtOH	214(4.20),287(4.37), 376(4.31)	103-0657-83
5-Pyrimidinecarbonitrile, 6-[2-(dimethylamino)ethenyl]-1,4-dihydro-4-oxo-1-(phenylmethyl)-	EtOH	212(4.16),238(4.19), 290(4.23),378(4.37)	103-0657-83
$C_{16}H_{16}N_4OS$			
2-Thiazolecarboxamide, N-butyl-4-(2-quinoxalinyl)-	EtOH	247(3.96),267s(3.82), 293s(3.56),342(3.51), 354s(3.44)	106-0829-83
$C_{16}H_{16}N_4O_2$			
Benzeneethanimidamide, N-hydroxy-2-[(hydroxyamino)iminomethyl]-α-(phenylmethylene)-	EtOH	230(4.12),321(3.85)	39-1813-83C
3-Formazancarboxylic acid, 1,5-diphenyl-, ethyl ester	EtOH	255(4.50),290(4.59), 440(4.74)	104-0585-83
$C_{16}H_{16}N_4O_2S$			
Carbamic acid, (5-amino-3-phenyl-2H-pyrido[4,3-b]-1,4-thiazin-7-yl)-, ethyl ester	pH 1	258(4.43),393(4.28)	87-1614-83
$C_{16}H_{16}N_4O_3$			
Carbamic acid, (5-amino-3-methyl-2H-pyrido[4,3-b][1,4]oxazin-7-yl)-, ethyl ester	pH 1	297(3.85),303s(3.83), 362s(4.34),372(4.37) (changing)	87-1614-83
Ethanol, 2-[(6-amino[1,2,4]triazolo-[1,5-a]pyridin-2-yl)methoxy]-, benzoate	EtOH	239(4.58),257(3.72), 280(3.29)	4-1169-83
Ethanol, 2-[(8-amino[1,2,4]triazolo-[1,5-a]pyridin-2-yl)methoxy]-, benzoate	EtOH	216(4.38),277(4.04)	4-1169-83
$C_{16}H_{16}N_4O_4$			
1H-Pyrazolo[3,4-b]pyrazine, 3-phenyl-1-β-D-ribofuranosyl-	MeOH	235(4.43),285(3.99)	161-0024-83
4-Pyridinecarboxylic acid, [[5-(acetoxymethyl)-3-hydroxy-2-methyl-4-pyridinyl]methylene]hydrazide, (E)-	EtOH	215(4.20),295(4.13), 340(3.77)	87-0298-83
(Z)-	EtOH	215(4.30),296(4.21), 340(3.92)	87-0298-83
$C_{16}H_{16}N_4O_4S_2$			
1,2,4,5-Tetrazine, 1,2-dihydro-3,6-bis[(phenylsulfonyl)methyl]-	MeOH	219(4.35),259(3.11), 266(3.26),273(3.18), 313(2.15)	104-0446-83
4H-1,2,4-Triazol-4-amine, 3,5-bis[(phenylsulfonyl)methyl]-	MeOH	220(4.27),258(3.34), 266(3.41),273(3.31)	104-0446-83
$C_{16}H_{16}N_4O_8S$			
Cefuroxime	pH 6	274(4.25)	162-0272-83
sodium salt	H_2O	274(4.24)	162-0272-83

Compound	Solvent	$\lambda_{max}(\log \epsilon)$	Ref.
$C_{16}H_{16}N_4S$			
2H-Imidazo[4,5-b]quinoxaline-2-thione, 1,3,3a,4,9,9a-hexahydro-4-methyl-1-phenyl-	EtOH	219(4.64),242(4.35), 297(3.70)	103-1333-83
$C_{16}H_{16}N_6O_5Se$			
Adenosine, 5'-Se-(2-nitrophenyl)-5'-seleno-	EtOH	257(4.66)	78-0759-83
$C_{16}H_{16}N_6O_6Se$			
Adenosine, 5'-deoxy-5'-[(2-nitrophenyl)seleninyl]-	EtOH	262(4.32)	78-0759-83
$C_{16}H_{16}O$			
Bicyclo[3.3.1]nona-3,6-dien-2-one, 1-methyl-6-phenyl-	EtOH	237(4.25)	2-0619-83
2-Butanone, 1,3-diphenyl- (with cetyltrimethylammonium chloride)	H_2O	260(2.95),290(2.56)	35-1309-83
(with sodium dodecyl sulfate)	H_2O	260(2.55),290(2.39)	35-1309-83
$C_{16}H_{16}O_2$			
Benzene, 1,1'-(1,2-dimethoxy-1,2-ethenediyl)bis-, (Z)-	EtOH	295(4.01)	89-0551-83S
Benzene, 1,1'-(1,2-ethenediyl)bis[4-methoxy-	heptane	324(4.55)	94-0373-83
Benzoic acid, 4-(1-methyl-2,4,6-cycloheptatrien-1-yl)-, methyl ester	hexane	195(4.79),236(4.26), 274(3.61),283(3.48)	24-1848-83
[1,1'-Biphenyl]-4,4'-dimethanol, α,α'-dimethyl- (dication)	H_2SO_4	366(4.65)	99-0612-83
1-Cyclohexene-1-carboxaldehyde, 2-[(4-methoxyphenyl)ethynyl]-	C_6H_{12}	247(4.34),259(4.20), 320(4.38)	44-4272-83
2,4,6,8,10,12,14-Hexadecaheptaenedial, all-trans (verpacrocein)	acetone	404s(5.04),423(5.22), 448(5.21)	64-0492-83C
$C_{16}H_{16}O_2S$			
Benzene, [3-[(phenylmethyl)sulfonyl]-1-propenyl]-, trans	EtOH	206(4.51),217s(--), 255(4.31),283s(--), 293(3.15)	44-4022-83
Benzenemethanol, 4,4'-thiobis[α-methyl- (dication)	H_2SO_4	370(4.19),434(4.48)	99-0612-83
Ethanone, 1-(4-methoxyphenyl)-2-[(phenylmethyl)thio]-	MeOH	274(4.16),338(2.78)	44-0835-83
$C_{16}H_{16}O_2S_2$			
Benzoic acid, 2-[[4-(methylthio)-phenyl]thio]-, ethyl ester	MeOH	268(4.23),285s(4.12), 320(3.80)	73-0906-83
$C_{16}H_{16}O_3$			
Benzenemethanol, 4,4'-oxybis[α-methyl- (dication)	H_2SO_4	331(4.29),360(4.63)	99-0612-83
3(2H)-Benzofuranone, 4,5,6,7-tetrahydro-2-[(4-methoxyphenyl)methylene]-	EtOH	242(4.04),369(4.31)	4-0543-83
2H-1-Benzopyran-7-ol, 3,4-dihydro-3-[(4-hydroxyphenyl)methyl]-	EtOH	281(3.64)	142-0039-83
Furan, 2,2',2"-methylidynetris[5-methyl-	EtOH	226(4.32)	103-0360-83 +103-0478-83
2,4-Pentadienal, 5-(6-hydroxy-2-methyl-2H-1-benzopyran-2-yl)-2-methyl-, (E,E)-	EtOH	217(4.60),274(4.59)	39-0039-83C
4a(2H)-Phenanthrenecarboxylic acid, 3,4,9,10-tetrahydro-2-oxo-, methyl ester	EtOH	234(4.04)	39-2519-83C

Compound	Solvent	$\lambda_{max}(\log \epsilon)$	Ref.
Phenol, 4-(3,4-dihydro-7-methoxy-2H-1-benzopyran-3-yl)-, (±)-	MeOH	224(4.29),279(3.71),283(3.71),289s(3.61)	100-0852-83
2-Propenoic acid, 3-(3-hydroxy-2-naphthalenyl)-2-methyl-, ethyl ester, (E)-	MeOH	224(4.49),263.5(4.59),306(4.22)	2-0352-83
$C_{16}H_{16}O_3S$			
Ethanone, 1-(4-methylphenyl)-2-[(4-methylphenyl)sulfonyl]-	EtOH	219(4.28),265(4.31)	42-0137-83
	neutral	264(4.21)	42-0137-83
	anion	300(4.12)	42-0137-83
$C_{16}H_{16}O_3S_2$			
6,10-Dithiaspiro[4.5]dec-1-ene-1-carboxaldehyde, 3-(benzoyloxy)-, (R)-	EtOH	207s(3.89),233(4.22),272(3.18),297s(3.63)	39-2441-83C
$C_{16}H_{16}O_4$			
Benzoic acid, 4-methoxy-, 2-(4-hydroxyphenyl)ethyl ester	MeOH	204.5(4.35),208(4.35),224.5(4.04),256.5(4.33)	94-2712-83
6H-Dibenzo[b,d]pyran-6-one, 3-acetoxy-7,8,9,10-tetrahydro-4-methyl-	MeOH	277(4.07),309(3.96)	78-1265-83
6H-Dibenzo[b,d]pyran-6-one, 7,8,9,10-tetrahydro-4-hydroxy-3-(2-propenyloxy)-	MeOH	260(4.09),313(4.14)	78-1265-83
Ethanone, 1,1'-(2,7-dimethoxy-1,8-naphthalenediyl)bis-	EtOH	318(3.62),332(3.62)	104-0183-83
1,4-Naphthalenedione, 5,8-dihydroxy-2-(4-methyl-3-pentenyl)-	MeOH	275(3.94),330(2.97),483(3.80),511(3.82),547(3.57)	105-0532-83
1,4-Naphthalenedione, 2-(1,1-dimethyl-2-oxopropyl)-3-methoxy-	MeOH	245(4.06),250(4.05),275(3.85),327(3.27)	102-0737-83
Pentacyclo[$6.4.0.0^{2,10}.0^{3,6}.0^{7,9}$]dodeca-4,11-diene-11,12-dicarboxylic acid, dimethyl ester	MeCN	235(3.70)	88-2147-83
2,4-Pentadienoic acid, 3-(2-furanyl)-5-(5-methyl-2-furanyl)-, ethyl ester, (E,E)-	MeOH	320(4.25),366(4.20)	73-3559-83
(E,Z)-	MeOH	312(4.36),356(4.26)	73-3559-83
1,5-Phenanthrenediol, 9,10-dihydro-2,7-dimethoxy- (eulophiol)	MeOH	214(4.59),281.7(4.31),308.4(4.05)	102-0747-83
2H-Pyran-5-carboxylic acid, 3,4-dihydro-4-oxo-6-(2-phenylethenyl)-, ethyl ester	EtOH	340(4.36)	118-0948-83
$C_{16}H_{16}O_4S$			
Benzenemethanol, 4,4'-sulfonylbis[α-methyl- (dication)	H_2SO_4	294(4.46)	99-0612-83
Ethanone, 1-(4-methoxyphenyl)-2-[(4-methylphenyl)sulfonyl]-	EtOH	227(4.31),292(4.17)	42-0137-83
	neutral	292(4.24)	42-0137-83
	anion	302(4.22)	42-0137-83
$C_{16}H_{16}O_5$			
9(4H)-Anthracenone, 4-acetoxy-1,4a,9a,10-tetrahydro-5,10-dihydroxy-, (4α,4aβ,9aβ,10α)-	EtOH	224(4.04),257(3.78),313(3.32)	44-3252-83
5H-Benzofuro[6,5-c][2]benzopyran-5-one, 1,2,3,4,9,10-hexahydro-9-hydroxy-7-methoxy-, (±)-	MeOH	207(4.71),256(3.74),325(4.20)	78-1265-83
Dibenzofuran, 1,2,3,4-tetramethoxy-	EtOH	220s(4.56),229(4.60),261(4.21),285(4.34),297s(3.92),309s(3.75)	39-2267-83C
	EtOH	218s(4.52),227(4.56),260(4.17),284(4.30),	39-2267-83C

Compound	Solvent	$\lambda_{max}(\log \epsilon)$	Ref.
(cont.)		294s(3.89),307s(3.71)	39-2267-83C
1,4-Naphthalenedione, 5,8-dihydroxy-2-(1-hydroxy-4-methyl-3-pentenyl)- (shikonin)	MeOH	276(3.88),330(3.05), 484(3.78),513(3.82), 551(3.60)	105-0532-83
1,4-Naphthalenedione, 5,8-dihydroxy-2-(tetrahydro-5,5-dimethyl-2-furanyl)-	MeOH	275(3.81),330(2.94), 487(3.40),513(3.74), 551(3.51)	105-0532-83
$C_{16}H_{16}O_6$			
1,4-Naphthalenedione, 2-(1,5-dihydroxy-4-methyl-3-pentenyl)-5,8-dihydroxy-	MeOH	275(3.88),330(2.97), 484(3.78),514(3.81), 552(3.58)	105-0532-83
$C_{16}H_{16}O_7$			
9,10-Anthracenedione, 1,2,3,4-tetrahydro-1,2,3,5-tetrahydroxy-7-methoxy-2-methyl- (dactylariol)	EtOH	220(4.65),238(3.97), 269(4.19),284s(3.99), 420(3.74)	23-0372-83
9,10-Anthracenedione, 1,2,3,4-tetrahydro-2,3,5,8-tetrahydroxy-7-methoxy-2-methyl-	n.s.g.	230(4.51),306(3.94), 478(3.86),505(3.92), 542(3.73)	158-83-50
4aH-Xanthene-4a-carboxylic acid, 2,3,4,9-tetrahydro-1,4,8-trihydroxy-6-methyl-9-oxo-, methyl ester	EtOH	250(3.70),261(3.87), 328(3.58)	39-1365-83C
isomer (β-diversonolic ester)	EtOH	228(4.23),239(4.24), 250(4.03),259(4.02), 326(3.40)	39-1365-83C
$C_{16}H_{16}S$			
Benzene, [1-[[(phenylmethyl)thio]-methyl]ethenyl]-	MeOH	240(4.01)	44-0835-83
$C_{16}H_{17}$			
Cycloheptatrienylium, [4-(1-methylethyl)phenyl]-, tetrafluoroborate	10% HCl	275.5(4.06),394(4.26)	78-4011-83
Cycloheptatrienylium, (2,4,6-trimethylphenyl)-, tetrafluoroborate	n.s.g.	279s(3.81)	137-0104-83
$C_{16}H_{17}ClN_2O$			
Cinnolinium, 3-[(4-chlorophenyl)methyl]-5,6,7,8-tetrahydro-4-hydroxy-2-methyl-, hydroxide, inner salt	EtOH	265(3.53),320(3.78)	4-0543-83
Ethanone, 2-(4-chlorophenyl)-1-(4,5,6-7-tetrahydro-1-methyl-1H-indazol-3-yl)-	EtOH	217(4.05),246(4.15)	4-0543-83
Tetrazepam	EtOH	227(4.45)	162-1322-83
$C_{16}H_{17}ClN_2O_5$			
Pyrrolo[2,3-b]indole-1,2(2H)-dicarboxylic acid, 8-acetyl-5-chloro-3,3a,8-8a-tetrahydro-, dimethyl ester, (2α,3aβ,8aβ)-(±)-	EtOH	252(4.19),286(3.30), 294s(3.26)	94-1856-83
DL-Tryptophan, 1-acetyl-5-chloro-N-(methoxycarbonyl)-, methyl ester	EtOH	243.5(4.36),267(3.93), 296(3.81),305.5(3.86)	94-1856-83
$C_{16}H_{17}CsSi$			
Cesium, [9-(trimethylsilyl)-9H-fluoren-9-yl]-	$(MeOCH_2)_2$	357(4.09),460(3.19)	104-1592-83
$C_{16}H_{17}DO_3S$			
Benzeneethan-β-d-ol, α-methyl-, 4-methylbenzenesulfonate, (R*,S*)-	EtOH	212(4.07),227(4.06)	22-0180-83

Compound	Solvent	$\lambda_{max}(\log \epsilon)$	Ref.
$C_{16}H_{17}FN_2O$			
3-Pyridinecarbonitrile, 1-[(4-fluoro-phenyl)methyl]-1,6-dihydro-6-propoxy-	CH_2Cl_2	313(3.8),356s(3.3)	64-0873-83B
$C_{16}H_{17}F_3N_2O_9$			
2,4(1H,3H)-Pyrimidinedione, 1-(2,3,5-tri-O-acetyl-β-D-arabinofuranosyl)-5-(trifluoromethyl)-	pH 2	260(4.02)	87-0598-83
	pH 13	260(3.87)	87-0598-83
	MeOH	260(4.08)	87-0598-83
$C_{16}H_{17}LiSi$			
Lithium, [9-(trimethylsilyl)-9H-fluoren-9-yl]-	$(MeOCH_2)_2$	362.5(4.09),442(3.16), 467(3.26),498(3.09)	104-1592-83
$C_{16}H_{17}NO$			
1(4H)-Acridinone, 4a,9,9a,10-tetra-hydro-2,9a-dimethyl-4-methylene-, cis	MeOH	255(4.11),272(4.11)	35-3273-83
2-Aza-6,8,10-trimethyl-3,4-benzotri-cyclo[4.4.0.0^{2,8}]dec-8-en-7-one	MeOH	209(4.06),229(4.09), 313(3.20)	35-3273-83
2-Aza-6,8,10-trimethyl-3,4-benzotri-cyclo[4.4.0.0^{2,8}]dec-9-en-7-one	MeOH	212(4.09),319(2.61)	35-3273-83
1-Naphthalenol, 4-[3-[(1-methylethyl)-imino]-1-propenyl]-, (E,?)-	toluene	337(3.91),460(3.59)	104-1864-83
	DMSO	350(3.86),515(4.17)	104-1864-83
	+ NaOMe	370(3.92),440(4.18)	104-1864-83
Oxaziridine, 2-(1-methylethyl)-1,1-di-phenyl-, (S)-	isooctane	215s(4.31),253(2.78), 269(2.79),265(2.69), 272(2.38)	32-0799-83
Pyridine, 1,4-dihydro-1-(2-methyl-1-oxopropyl)-4-(phenylmethylene)-	MeOH	234(3.76),349(4.47)	24-3192-83
Pyridine, 1,4-dihydro-1-(1-oxobutyl)-4-(phenylmethylene)-	MeOH	238(3.67),350(4.40)	24-3192-83
$C_{16}H_{17}NOS$			
Cyclopentanone, 2-[(3-ethyl-2(3H)-benzothiazolylidene)ethylidene]-	MeOH	464(4.73)	104-1854-83
	benzene	425(--),440(--)	104-1854-83
	DMF	444(--)	104-1854-83
Ethanone, 1-[1,2-dihydro-4,6-dimethyl-1-(2-methylphenyl)-2-thioxo-3-pyri-dinyl]-	EtOH	297(3.80),374(3.82)	118-1025-83
Ethanone, 1-[1,2-dihydro-4,6-dimethyl-1-(phenylmethyl)-2-thioxo-3-pyri-dinyl]-	EtOH	295(3.73),373(3.77)	118-1025-83
$C_{16}H_{17}NOS_2$			
Ethanone, 1-[1,2-dihydro-6-methyl-1-(2-methylphenyl)-4-(methylthio)-2-thi-oxo-3-pyridinyl]-	EtOH	282(4.49),382(3.97)	118-1025-83
Ethanone, 1-[1,2-dihydro-6-methyl-4-(methylthio)-1-(phenylmethyl)-2-thioxo-3-pyridinyl]-	EtOH	283(4.39),382(3.85)	118-1025-83
$C_{16}H_{17}NO_2$			
1-Butanone, 1-(2-hydroxyphenyl)-4-(phenylamino)-	MeOH	257(4.24),298(3.5), 330s(3.43),448(2.9)	5-0165-83
$C_{16}H_{17}NO_2S$			
Ethanone, 1-[1,2-dihydro-4-methoxy-6-methyl-1-(2-methylphenyl)-2-thioxo-3-pyridinyl]-	EtOH	235(4.37),296(3.87), 355(3.80)	118-1025-83
Ethanone, 1-[1,2-dihydro-1-(4-methoxy-	EtOH	264(3.92),285(3.87),	118-1025-83

Compound	Solvent	$\lambda_{max}(\log \epsilon)$	Ref.
phenyl)-4,6-dimethyl-2-thioxo-3-pyridinyl]- (cont.)		374(3.85)	118-1025-83
Ethanone, 1-[5-(4-morpholinyl)-4-phenyl-2-thienyl]-	MeCN	229(4.16),267(4.20), 296s(3.73),359(4.10)	48-0168-83
$C_{16}H_{17}NO_2S_2$			
Ethanone, 1-[1,2-dihydro-1-(2-methoxyphenyl)-6-methyl-4-(methylthio)-2-thioxo-3-pyridinyl]-	EtOH	284(4.49),383(3.93)	118-1025-83
Ethanone, 1-[1,2-dihydro-1-(4-methoxyphenyl)-6-methyl-4-(methylthio)-2-thioxo-3-pyridinyl]-	EtOH	218(4.25),284(4.44), 384(3.86)	118-1025-83
Sulfilimine, S-(4-ethenylphenyl)-S-methyl-N-[(4-methylphenyl)sulfonyl]-	EtOH	261(4.35)	121-0017-83A
$C_{16}H_{17}NO_3$			
4aH-Carbazole-4a-acetic acid, 1,2,9,9a-tetrahydro-4-methyl-2-oxo-, methyl ester, cis	MeOH	232(4.42),318(3.31)	5-1744-83
1H-Indole-3-acetic acid, 2-methyl-1-(3-oxo-1-butenyl)-, methyl ester	MeOH	216(4.41),276(4.08), 337(4.10)	5-1744-83
$C_{16}H_{17}NO_3S$			
Ethanone, 1-[1,2-dihydro-4-methoxy-1-(2-methoxyphenyl)-6-methyl-2-thioxo-3-pyridinyl]-	EtOH	233(4.40),284(3.98), 355(3.83)	118-1025-83
Ethanone, 1-[1,2-dihydro-4-methoxy-1-(4-methoxyphenyl)-6-methyl-2-thioxo-3-pyridinyl]-	EtOH	234(4.43),290(3.91), 355(3.82)	118-1025-83
$C_{16}H_{17}NO_4$			
7-Azabicyclo[4.1.0]hept-4-ene-1,4-dicarboxylic acid, 7-phenyl-, dimethyl ester	MeCN	233(4.34),272(3.85)	24-0563-83
1H-Azepine-2,5-dicarboxylic acid, 2,3-dihydro-1-phenyl-, dimethyl ester	MeCN	240(3.71),341(3.69)	24-0563-83
2-Butenedioic acid, 2-(1,3-dimethyl-1H-indol-2-yl)-, dimethyl ester, (E)-	McOH	225(4.34),295(3.98)	39-0501-83C
(Z)-	MeOH	226(4.34),260(3.90), 277(3.78),340(3.98)	39-0501-83C
2(5H)-Furanone, 3-[3-[4-(dimethylamino)phenyl]-1-oxo-2-propenyl]-4-hydroxy-5-methyl-, (E)-	MeOH	203(4.39),238(4.05), 323(3.99),396(4.16), 503(4.42)	83-0115-83
Lunine	EtOH	222(4.31),247(4.60), 314(4.02),325(4.01)	162-0801-83
1H-Pyrrole-3-carboxylic acid, 4,5-dihydro-1-(1-methylethyl)-4,5-dioxo-2-phenyl-, ethyl ester	dioxan	282(3.7),405(3.6)	94-0356-83
1H-Pyrrole-3,4-dicarboxylic acid, 2,2-dimethyl-5-phenyl-, dimethyl ester	EtOH	231(3.87)	33-0262-83
$C_{16}H_{17}NO_5$			
6-Oxa-1-azabicyclo[5.2.0]non-2-ene-2-carboxylic acid, 3-methoxy-9-oxo-, phenylmethyl ester	EtOH	265(3.92)	39-0115-83C
2(1H)-Pyridinone, 1-methyl-3-(2,4,6-trimethoxybenzoyl)-	EtOH	205(4.57),270s(3.92), 282s(3.63),290s(3.68), 344(3.97)	39-0219-83C
$C_{16}H_{17}N_3$			
Acetaldehyde, [(4-methylphenyl)imino]-,	EtOH	364(4.54)	65-1902-83

Compound	Solvent	λ_{max}(log ε)	Ref.
methylphenylhydrazone (cont.)			65-1902-83
conjugate acid	EtOH-H_2SO_4	331(4.41),420(3.90)	65-1902-83
Benzo[f]quinoxalin-6-amine, N,N-diethyl-	EtOH	218(4.47),240(4.42), 277(4.18),315s(--), 375(3.86)	83-0283-83
1H-Indazol-1-amine, N-ethyl-N-methyl-3-phenyl-	n.s.g.	216(4.38),247(3.91), 272.6(3.70),311.5(3.99)	77-1344-83
$C_{16}H_{17}N_3O$			
2H-Pyrrole-3,3-dicarbonitrile, 4-ethyl-3,4-dihydro-2,2-dimethyl-5-phenyl-	EtOH	249(4.17)	33-0262-83
5H-[1,2,3]Triazolo[4,5,1-de]acridin-5-one, 2a,5a,6,11a-tetrahydro-2a,4,5a-trimethyl-	MeOH	206(4.58),226(4.55), 274(3.94),332(4.05)	35-3273-83
$C_{16}H_{17}N_3O_2S$			
2H-1-Benzopyran-2-one, 3-(2-amino-4-thiazolyl)-7-(diethylamino)-	MeCN	410(4.38)	48-0551-83
$C_{16}H_{17}N_3O_4$			
Anthramycin	MeCN	235(4.26),333(4.50)	162-0100-83
2-Propenoic acid, 3-(benzoylamino)-2-cyano-3-morpholino-, methyl ester	EtOH	255(3.56),322(3.66)	142-1745-83
$C_{16}H_{17}N_3O_4S$			
Cephalexin	n.s.g.	260(3.89)	162-0275-83
$C_{16}H_{17}N_3O_7$			
Pyrrolo[2,3-b]indole-1,2(2H)-dicarboxylic acid, 8-acetyl-3,3a,8,8a-tetrahydro-5-nitro-, dimethyl ester, (2α,3aβ,8aβ)-(±)-	EtOH	226(4.01),323(4.08)	94-1856-83
DL-Tryptophan, 1-acetyl-N-(methoxycarbonyl)-5-nitro-, methyl ester	MeCN	229.5(3.97),258s(4.34), 268(4.38),302(3.94)	94-1856-83
$C_{16}H_{17}N_5O_3Se$			
Adenosine, 5'-Se-phenyl-5'-seleno-	EtOH	259(4.26)	78-0759-83
$C_{16}H_{17}N_5O_4S$			
Benzenamine, 4-[(9-β-D-ribofuranosyl-9H-purin-6-yl)thio]-	EtOH	260(4.30),281(4.35)	130-0045-83
$C_{16}H_{17}N_5O_4Se$			
Adenosine, 5'-deoxy-5'-(phenylselenyl)-	EtOH	257(4.25)	78-0759-83
$C_{16}H_{17}O$			
Benzenemethanol, α-methyl-4-(2-phenylethyl)- (cation)	H_2SO_4	222(4.15),314(4.40)	99-0612-83
$C_{16}H_{18}ClN_3OS$			
4(5H)-Thiazolone, 5-[(4-chlorophenyl)-methylene]-2-[(piperidinomethyl)-amino]-, (E)-	n.s.g.	231(4.06),293(4.12), 345(4.31)	103-0745-83
$C_{16}H_{18}Cl_3NO_3$			
1H-Pyrrole-1-carboxylic acid, 2,5-dihydro-2-[1-(4-methoxyphenyl)ethyl]-, 2,2,2-trichloroethyl ester	MeOH	224(4.1),274(3.2), 281(3.15)	87-0469-83

Compound	Solvent	$\lambda_{max}(\log \epsilon)$	Ref.
$C_{16}H_{18}Cl_3NO_4$			
6-Oxa-3-azabicyclo[3.1.0]hexane-3-carboxylic acid, 2-[1-(4-methoxyphenyl)-ethyl]-, 2,2,2-trichloroethyl ester	MeOH	224(4.07),275(3.22), 282(3.16)	87-0469-83
$C_{16}H_{18}F_3N_3$			
Benzenamine, N-methyl-2-[4,5,6,7-tetrahydro-7-methyl-3-(trifluoromethyl)-1H-indazol-7-yl]-	EtOH	246(4.01),294(3.46)	39-0795-83C
$C_{16}H_{18}FeO_4$			
Iron, tricarbonyl[η^4-4-(2,6,6-trimethyl-1,3-cyclohexadien-1-yl)-3-buten-2-one]-	n.s.g.	283s(3.81)	88-1611-83
$C_{16}H_{18}NO_5P$			
α-D-Xylofuranose, 3-C-(cyanomethyl)-3-deoxy-3-(hydroxyphenylphosphinyl)-1,2-O-(1-methylethylidene)-, intramol. 3,5-ester, (R)-	EtOH	207(3.76),218(4.07), 224(3.91)	159-0019-83
(S)-	EtOH	207.5(3.73),218.5(4.00), 223.5(3.90)	159-0019-83
$C_{16}H_{18}N_2$			
Agroclavine	n.s.g.	225(4.47),284(3.88), 293(3.81)	162-0028-83
1H-Indole, 1-methyl-3-[3-(1-methylethyl)-1H-pyrrol-2-yl]-	EtOH	213(4.29),240(4.33), 300(3.66)	103-0289-83
Indolo[4,3-fg]isoquinoline, 4,6,6a,7-8,9-hexahydro-8,10-dimethyl-	MeOH	237(4.31),309(4.00)	39-1545-83C
$C_{16}H_{18}N_2O$			
Elymoclavine	n.s.g.	227(4.31),283(3.84), 293(3.76)	162-0512-83
Indolo[4,3-fg]isoquinolin-9(4H)-one, 6,6a,7,8,10,10a-hexahydro-8,10-dimethyl-, (6aR*,10R*,10aR*)-	MeOH	228(4.46),284(3.83)	39-1545-83C
(6aR*,10S*,10aS*)-	MeOH	226(4.48),284(3.88)	39-1545-83C
$C_{16}H_{18}N_2O_2$			
4(1H)-Cinnolinone, 5,6,7,8-tetrahydro-3-[(4-methoxyphenyl)methyl]-	EtOH	275(4.03)	4-0543-83
Pyrano[4,3-c]pyrazol-4(1H)-one, 4-ethyl-6,7-dihydro-6,6-dimethyl-2-phenyl-	EtOH	275(4.29)	142-1801-83
Pyrano[4,3-c]pyrazol-4(1H)-one, 6,7-dihydro-6-methyl-1-phenyl-3-propyl-	EtOH	256(4.35)	118-0844-83
1H-Pyrazole-4-carboxylic acid, 5-(1-methyl-1-propenyl)-1-phenyl-, ethyl ester, (E)-	EtOH	246(4.15)	118-0566-83
1H-Pyrazole-4-carboxylic acid, 5-(2-methyl-1-propenyl)-1-phenyl-, ethyl ester	EtOH	251(4.15)	118-0566-83
3H-Pyrazol-3-one, 2,4-dihydro-5-(2-methyl-1-propenyl)-4-(1-oxopropyl)-2-phenyl-	EtOH	259(4.31)	142-1801-83
3-Pyridinecarboxamide, 1,6-dihydro-6-oxo-1-[(2,4,6-trimethylphenyl)-methyl]-	H_2O	261(4.12),300(3.74)	54-0331-83
	M HCl	260(4.10),296(3.74)	54-0331-83
4H-Pyrido[1,2-a]pyrimidin-4-one, 6,7,8,9-tetrahydro-2,6-dimethyl-3-phenoxy-	EtOH	235(3.76),277(3.88)	142-1083-83

Compound	Solvent	$\lambda_{max}(\log \epsilon)$	Ref.
$C_{16}H_{18}N_2O_2S$			
Benzenesulfonamide, N-(3,4-dihydro-2(1H)-isoquinolinyl)-4-methyl-	MeOH	255s(3.08),263(3.06), 272(2.97)	4-0121-83
$C_{16}H_{18}N_2O_3$			
Pilosine	EtOH	210(4.10)	162-0747-83
$C_{16}H_{18}N_2O_4$			
Diazene, bis(3,4-dimethoxyphenyl)-	MeOH	208(4.02),251.5(3.86), 372(4.08),383(4.08)	44-5041-83
$C_{16}H_{18}N_2O_5S$			
3(2H)-Pyridazinone, 6-[(phenylmethyl)-thio]-2-β-D-ribofuranosyl-	pH 1	247(4.40),329(3.37)	4-0369-83
	pH 11	247(4.39),326(3.36)	4-0369-83
	MeOH	246(4.40),329(3.38)	4-0369-83
$C_{16}H_{18}N_4$			
4H-Pyrrolo[2,3-d]pyrimidin-4-imine, 3-(2,6-dimethylphenyl)-3,7-dihydro-2,5-dimethyl-	EtOH	233(4.01),273(3.66)	5-2066-83
4H-Pyrrolo[2,3-d]pyrimidin-4-imine, 3-(4-ethylphenyl)-3,7-dihydro-2,5-dimethyl-	EtOH	233(4.26),276(3.84)	5-2066-83
$C_{16}H_{18}N_4O_2$			
4H,7H-Benz[g]imidazo[1,2,3-ij]pteridine-4,6(5H)-dione, 7-ethyl-1,2-dihydro-9,10-dimethyl-	H_2O	242(4.26),290s(3.59), 316(3.59)	35-6679-83
protonated	H_2O	222(4.36),246(3.97), 284(4.03),304s(3.86)	35-6679-83
diprotonated	H_2O	251(4.05),326(3.87)	35-6679-83
anion	H_2O	323(3.48)	35-6679-83
4H-Benz[g]imidazo[1,2,3-ij]pteridin-3-ium, 7-ethyl-1,2,5,6-tetrahydro-9,10-dimethyl-4,6-dioxo-, perchlorate	H_2O	215(4.40),259(4.58), 348(4.05),492(3.93)	35-6679-83
anion	H_2O	330(4.03),358(4.03), 486(3.93)	35-6679-83
$C_{16}H_{18}N_4O_3$			
Ethanol, 2-[ethyl[4-[(4-nitrophenyl)-azo]phenyl]amino]-	gas	378(4.49)	135-1185-83
	DMF	506(4.52)	135-1185-83
$C_{16}H_{18}N_4O_7$			
9H-Purine, 9-(2,3,5-tri-O-acetyl-β-D-ribofuranosyl)-	MeOH	263(3.92)	94-3104-83
Pyridinium, 1-ethyl-, 1-acetonyl-2,4,6-trinitrocyclohexadienate, cation	MeCN	261(4.29)	103-0773-83
anion	MeCN	457(4.03),581(3.99)	103-0773-83
$C_{16}H_{18}N_4O_7S$			
Inosine, 6-thio-, 2',3',5'-triacetate	MeOH	236(4.13),309(4.37)	36-0372-83
$C_{16}H_{18}N_4O_8$			
4H-Pyrazolo[3,4-d]pyrimidin-4-one, 1,5-dihydro-1-(2,3,5-tri-O-acetyl-β-D-ribofuranosyl)-	MeOH	251(3.82)	87-1601-83
	pH 1	251(--)	87-1601-83
	pH 11	270(--)	87-1601-83
$C_{16}H_{18}O$			
1-Acenaphthylenol, 1-ethyl-1,2-dihydro-2,2-dimethyl-	hexane	268s(--),280(3.79), 288(3.85),300(3.66),	104-0520-83

Compound	Solvent	λ_{max}(log ϵ)	Ref.
(cont.)		306(3.46),315(2.99), 320(2.79)	104-0520-83
$C_{16}H_{18}O_2$			
1H-Benz[e]inden-1-one, 2,3,4,5-tetra-hydro-7-methoxy-3,3-dimethyl-	EtOH	246(4.35),290(3.93)	78-4221-83
1-Cyclohexene-1-carboxaldehyde, 2-[2-(4-methoxyphenyl)ethenyl]-, (E)-	C_6H_{12}	256(4.04),336(4.30)	44-4272-83
$C_{16}H_{18}O_3$			
Camphononic acid benzylidene deriv.	EtOH	297(4.34)	39-1465-83C
2-Pentenal, 5-(6-hydroxy-2-methyl-2H-1-benzopyran-2-yl)-2-methyl-, (E)-(elaeagin)	EtOH	227(4.31)	39-0039-83C
$C_{16}H_{18}O_3S_2$			
6,10-Dithiaspiro[4.5]dec-1-ene-1-meth-anol, 3-(benzoyloxy)-, (R)-	EtOH	207(3.94),233(4.18), 272s(3.60),279s(2.86)	39-2441-83C
$C_{16}H_{18}O_4$			
2-Naphthaleneacetic acid, 1,4,4a,7-tetrahydro-α,4a,8-trimethyl-1,7-dioxo-, methyl ester, [S-(R*,R*)]-	EtOH	253(4.02),280(3.92)	94-3397-83
2-Naphthaleneacetic acid, 1,4,4a,7-tetrahydro-α,4a,8-trimethyl-3,7-dioxo-, methyl ester, [S-(R*,R*)]-	EtOH	252(3.84),310(4.21)	94-3397-83
4a(4H)-Phenanthrenecarboxylic acid, 1,4b,5,6,7,8,10,10a-octahydro-1,4-dioxo-, methyl ester	EtOH	205(3.89),226(4.00)	32-0757-83
$C_{16}H_{18}O_5$			
Benzene, 1,2,3,4-tetramethoxy-5-phen-oxy-	EtOH	210(4.36),263(3.14), 271(3.30),277(3.36)	39-2267-83C
D-erythro-Hex-2-enitol, 1,5-anhydro-2,3-dideoxy-1-C-phenyl-, diacetate, (R)-	EtOH	257(2.74)	23-0533-83
1-Naphthalenecarboxylic acid, 2,5,7-trimethoxy-4-methyl-, methyl ester	EtOH	236(4.23),245s(4.30), 249(4.32),307(3.79)	78-2283-83
$C_{16}H_{18}O_6$			
1,4-Naphthalenedione, 6-(1-ethoxyeth-yl)-5-hydroxy-2,7-dimethoxy-	MeOH	221(4.08),234s(4.01), 259s(3.78),264(3.79), 310(3.64)	158-83-38
$C_{16}H_{18}O_9$			
4H-1-Benzopyran-4-one, 6-β-D-gluco-pyranosyl-5,7-dihydroxy-2-methyl-	MeOH	234(4.35),255(3.70), 280(3.62),325s(3.20)	102-2591-83
	MeOH-NaOMe	234(--),255s(--), 278(--),305s(--), 327(--),396(--)	102-2591-83
	MeOH-AlCl$_3$	242(--),262(--), 295(--),335(--), 395(--)	102-2591-83
$C_{16}H_{19}$			
Dicyclopent[e,g]azulenylium, 1,2,3,4-5,7,8,9,10,?-decahydro-, hexa-fluoroantimonate	MeCN	257(4.65),308(3.66)	44-0596-83
$C_{16}H_{19}BrN_2O$			
9-Oxaergoline, 2-bromo-6-propyl-,	EtOH	224(4.56),276s(3.94),	87-0522-83

Compound	Solvent	$\lambda_{max}(\log \epsilon)$	Ref.
monohydrochloride, (±)- (cont.)		281(3.96),290s(3.87)	87-0522-83
$C_{16}H_{19}BrO_3$			
2-Naphthaleneacetic acid, 6-bromo-3,4,4a,7-tetrahydro-α,4a,8-trimethyl-7-oxo-, methyl ester, [S-(R*,R*)]-	EtOH	238(4.11),267(3.87), 317(4.06)	94-3397-83
$C_{16}H_{19}ClN_2O_3$			
2H-Pyrrol-2-one, 3-chloro-1,5-dihydro-5-methoxy-1-(4-methylphenyl)-4-morpholino-	heptane	195(4.22),240(4.01), 301(3.75)	5-1504-83
$C_{16}H_{19}ClO_3$			
Bucloxic acid	EtOH	255(4.19)	162-0201-83
$C_{16}H_{19}F_6N_5O$			
Hydrazinium 1,4,5,10-tetrahydro-5,10,10-trimethyl-3,4-bis(trifluoromethyl)pyrazolo[4,3-c][1]-benzazepin-4-olate	EtOH	273(4.41),279(4.36)	39-0795-83C
$C_{16}H_{19}N$			
Benzenamine, 4-(1,1-dimethylethyl)-N-phenyl-, cation	MeCN	690(4.28)	70-0251-83
radical	MeCN	760(3.58)	70-0251-83
$C_{16}H_{19}NO$			
Butanamide, N-[2-(1-azulenyl)ethyl]-	EtOH	237(4.12),259s(4.13), 264s(4.30),268s(4.44), 274s(4.57),278(4.64), 282s(4.39),287s(4.44), 297s(3.64),320s(3.22), 328s(3.39),334s(3.42), 343(3.58),358(3.27), 516(2.02),534s(2.16), 558s(2.30),580s(2.38), 597(2.36),618s(2.37), 652(2.35),683s(3.05), 722(1.90)	18-2059-83
$C_{16}H_{19}NO_2$			
2,4,6,8-Nonatetraen-1-one, 9-(dimethylamino)-1-(5-methyl-2-furanyl)-	$CHCl_3$	505(4.70)	70-0780-83
	EtOH	520(--)	70-0780-83
$C_{16}H_{19}NO_2S_2$			
1H-Isoindole-1,3(2H)-dione, 2-[5-(1,3-dithiolan-2-yl)pentyl]-	$CHCl_3$	244(4.16),295(3.36)	78-2691-83
Spiro[azocino[2,1-a]isoindole-12(7H)-2'-[1,3]dithiolan]-5(12aH)-one, 8,9,10,11-tetrahydro-12a-hydroxy-	MeOH	240s(3.70)	78-2691-83
Thiophene, 2-[2,2-dimethyl-1-[(4-methylphenyl)thio]propyl]-4-nitro-	EtOH	223(4.23)	44-4202-83
Thiophene, 2-[2,2-dimethyl-1-[(4-methylphenyl)thio]propyl]-5-nitro-	EtOH	220(4.04),245(3.89), 326(3.86)	44-4202-83
$C_{16}H_{19}NO_3$			
α-Erythroidine, hydrochloride	EtOH	224(4.55)	162-0531-83
1H-Pyrrole-2-carboxylic acid, 3-(2-hydroxyethyl)-4,5-dimethyl-, phenylmethyl ester	$CHCl_3$	284(4.24)	78-1849-83

Compound	Solvent	λ_{max}(log ε)	Ref.
$C_{16}H_{19}NO_4$			
2H-1-Benzopyran-2-one, 4-(acetoxymethyl)-7-(diethylamino)-	EtOH	246(4.08),378(4.30)	94-3014-83
$C_{16}H_{19}NO_4S_2$			
Thiophene, 2-[2,2-dimethyl-1-[(4-methylphenyl)sulfonyl]propyl]-5-nitro-	MeOH	221(4.20),298(3.57)	44-4202-83
$C_{16}H_{19}N_2O$			
Pyridinium, 3-(aminocarbonyl)-1-[(2,4,6-trimethylphenyl)methyl]-	H_2O	265(3.64)	54-0331-83
$C_{16}H_{19}N_3$			
Benzenamine, N,N-diethyl-4-(phenylazo)-	gas	367(4.47)	135-1183-83
	DMF	425(4.45)	135-1183-83
$C_{16}H_{19}N_3O_2$			
Propanedinitrile, [2,2-dimethyl-1-(3-nitrophenyl)propyl]ethyl-	EtOH	259(3.90)	12-0081-83
2-Pyrimidineacetamide, α-ethoxy-4,6-dimethyl-α-phenyl-	MeCN	245(3.58)	44-0575-83
$C_{16}H_{19}N_3O_3$			
1H-Pyrrolo[1,2-a][1,3]diazepine-5-carboxylic acid, 3-acetyl-9-cyano-4,5-dihydro-7,8-dimethyl-, ethyl ester	EtOH	354(3.97)	4-0081-83
Pyrrolo[1,2-a]pyrimidine-4-carboxylic acid, 3-acetyl-8-cyano-1,4-dihydro-4,6,7-trimethyl-, ethyl ester	EtOH	366(4.05)	4-0081-83
$C_{16}H_{19}N_3O_4$			
Benzoic acid, 4-[(2-amino-1,4-dihydro-6-hydroxy-4-oxo-5-pyrimidinyl)methyl]-, butyl ester	MeOH	241(4.29),265(4.24)	73-0299-83
5-Pyrimidinepentanoic acid, 1,4-dihydro-6-hydroxy-4-oxo-2-[(phenylmethyl)amino]-	50% MeOH	275(4.006)	73-0304-83
	+ HCl	228(3.799),270(3.97)	73-0304-83
	+ NaOH	275(4.023)	73-0304-83
$C_{16}H_{19}N_3O_4S$			
3(2H)-Pyridazinone, 5-amino-2-[2-S-(phenylmethyl)-2-thio-β-D-ribofuranosyl]-	pH 1	276(3.82),300s(3.62)	44-3765-83
	pH 11	275(3.65),301s(3.59)	44-3765-83
	MeOH	276(3.85),302s(3.61)	44-3765-83
$C_{16}H_{19}N_3O_5$			
Uridine, 2'-deoxy-5-[(4-methylphenyl)amino]-	pH 7	255(4.14),320(3.58)	33-0534-83
$C_{16}H_{19}N_3O_5S$			
Cefroxadine	pH 1	267(3.79)	162-0271-83
$C_{16}H_{19}N_3O_5S_2$			
1,2,4-Triazin-3(2H)-one, 5-(methylthio)-6-[(phenylmethyl)thio]-2-β-D-ribofuranosyl-	pH 1	302(3.85)	80-0989-83
	pH 13	302(3.57)	80-0989-83
	EtOH	303(3.55)	80-0989-83
$C_{16}H_{19}N_3O_6$			
Uridine, 5-[(4-methylphenyl)amino]-	pH 7	254(4.14),320(3.59)	33-0534-83
$C_{16}H_{19}N_5O_7S$			
Guanosine, 6-thio-, 2',3',5'-triacetate	MeOH	251(3.92),317(4.13)	36-0372-83

Compound	Solvent	λ_{max}(log ε)	Ref.
$C_{16}H_{20}Br_2O_3$			
2-Naphthaleneacetic acid, 1,6-dibromo-1,2,3,4,4a,7-hexahydro-α,4a,8-trimethyl-7-oxo-, methyl ester, [1R-[1α,2α(S*),4aα]]-	EtOH	207(4.08),262(4.11)	94-3397-83
$C_{16}H_{20}Cl_4N_2$			
Piperidine, 1,1'-[(2,3,4,5-tetrachloro-2,4-cyclopentadien-1-ylidene)methylene]bis-	DMF	274(3.85),348(4.22)	5-0154-83
$C_{16}H_{20}CuN_8O_2$			
Copper, [[4,4'-(1,2-ethanediyldiimino)-bis[6-(dimethylamino)-5-pyrimidinecarboxaldehydato]](2-)-N^4,$N^{4'}$,O^5,$O^{5'}$]-	$CHCl_3$	342(4.30)	12-2413-83
$C_{16}H_{20}Ge_2$			
Dibenzo[c,e][1,2]digermanin, 5,6-dihydro-5,5,6,6-tetramethyl-	n.s.g.	237(4.40),270s(3.58)	35-7469-83
$C_{16}H_{20}NO_2$			
Ethanaminium, N-(3-ethoxy-4-oxo-1(4H)-naphthalenylidene)-N-ethyl-, tetrafluoroborate	MeCN	253(4.16),300(4.17), 400s(3.35)	88-3567-83
Ethanaminium, N-(4-ethoxy-1-oxo-2(1H)-naphthalenylidene)-N-ethyl-, tetrafluoroborate	$CHCl_3$	265(4.35),305(3.92), 465(3.67)	88-3567-83
$C_{16}H_{20}N_2$			
Indolo[4,3-fg]isoquinoline, 4,6,6a,7-8,9,10,10a-octahydro-8,10-dimethyl-, (6aR*,10R*,10aS*)-	MeOH	228(4.47),285(3.85)	39-1545-83C
(6aR*,10S*,10aR*)-	MeOH	226(4.20),284(3.52), 294(3.46)	39-1545-83C
$C_{16}H_{20}N_2O$			
Chanoclavine	n.s.g.	225(4.44),284(3.82), 293(3.72)	162-0284-83
9-Oxaergoline, 6-propyl-, monohydrochloride	EtOH-HCl	222(4.52),275s(3.78), 280(3.81),290(3.77)	87-0522-83
$C_{16}H_{20}N_2O_2$			
2-Butenoic acid, 3-[(2,3-dimethyl-1H-indol-5-yl)amino]-, ethyl ester	EtOH	227(4.36),294(4.50)	103-0401-83
2-Butenoic acid, 3-[(2,3-dimethyl-1H-indol-6-yl)amino]-, ethyl ester	EtOH	222(4.42),294(4.39)	103-0401-83
2-Pentanone, 4-[[2-(1H-indol-3-yl)-ethyl]imino]-1-methoxy-	EtOH	221(4.57),284s(4.06), 291(4.11),315(4.37)	4-0267-83
2-Pentanone, 4-[[2-(1H-indol-3-yl)-ethyl]imino]-5-methoxy-	EtOH	221(4.56),291(4.22), 311(4.25)	4-0267-83
2-Pentanone, 4-[[2-(5-methoxy-1H-indol-3-yl)ethyl]imino]-	EtOH	222(4.56),283s(4.07), 312(4.30)	4-0267-83
2-Pentynoic acid, 4-methyl-4-[[(methylphenylamino)methylene]amino]-, ethyl ester, (E)-	EtOH	260(4.27)	33-0262-83
2-Propanone, 1-[2,3,4,9-tetrahydro-1-(methoxymethyl)-1H-pyrido[3,4-b]indol-1-yl-, monohydrochloride	EtOH	221(4.45),272(3.95), 294s(3.70),307s(3.59)	4-0267-83
2-Propanone, 1-(2,3,4,9-tetrahydro-6-methoxy-1-methyl-1H-pyrido[3,4-b]-indol-1-yl)-, monohydrochloride	EtOH	222(4.56),276(3.86), 289(3.74)	4-0267-83

Compound	Solvent	$\lambda_{max}(\log \epsilon)$	Ref.
1H-Pyrrole-2-carboxaldehyde, 5-[(3-ethyl-1,5-dihydro-1,4-dimethyl-5-oxo-2H-pyrrol-2-ylidene)methyl]-3,4-dimethyl-, (E)-	$CHCl_3$	268(4.25),305(3.99), 373(4.20)	49-0753-83
(Z)-	$CHCl_3$	268(4.26),307(3.99), 374(4.21)	49-0753-83
1H-Pyrrole-2-carboxaldehyde, 5-[(3-ethyl-1,5-dihydro-4-methyl-5-oxo-2H-pyrrol-2-ylidene)methyl]-1,3,4-trimethyl-, (Z)-	$CHCl_3$	268(4.11),296(4.08), 365(4.28)	108-0187-83
2H-Pyrrole-3-carboxylic acid, 2,2-dimethyl-5-(methylphenylamino)-, ethyl ester	EtOH	249(4.18),319(3.33)	33-0262-83
	EtOH	218s(4.11),248.5(4.18), 319(3.33)	33-0262-83
$C_{16}H_{20}N_2O_3$			
3-Pyridinecarboxylic acid, 1,4,5,6-tetrahydro-5,6-dimethyl-4-oxo-1-(phenylamino)-, ethyl ester	EtOH	238(4.24),308(4.18)	118-0566-83
3-Pyridinecarboxylic acid, 1,4,5,6-tetrahydro-6,6-dimethyl-4-oxo-1-(phenylamino)-, ethyl ester	EtOH	237(4.22),306(4.20)	118-0566-83
$C_{16}H_{20}N_2O_3Te$			
2,4,6(1H,3H,5H)-Pyrimidinetrione, 5-ethyl-5-[4-(phenyltelluro)butyl]-	MeOH	222.9(3.65),269.4(3.65), 326.2(2.80)	101-0171-83S
$C_{16}H_{20}N_3O_8PS_2$			
1,2,4-Triazin-3(2H)-one, 5-(methylthio)-6-[(phenylmethyl)thio]-2-(5-O-phosphono-β-D-ribofuranosyl)-, monoammonium salt	pH 1	302(3.92)	80-0989-83
	H_2O	303(3.87)	80-0989-83
	pH 13	302(3.91)	80-0989-83
$C_{16}H_{20}N_4O_3$			
2,4,5(3H)-Pyrimidinetrione, 6-[[2-(dimethylamino)-4,5-dimethylphenyl]methylamino]-3-methyl-	MeCN	229(4.39),245s(--), 281(4.11),342(3.85)	35-4809-83
$C_{16}H_{20}N_4O_4$			
4(1H)-Azulenone, octahydro-, 2,4-dinitrophenylhydrazone	EtOH	235(4.08),379(4.15)	44-1661-83
$C_{16}H_{20}N_4O_6$			
Cyclohexanecarboxylic acid, 4-[(2,4-dinitrophenyl)hydrazono]-1,2,2-trimethyl-, (±)-	EtOH	355(4.24)	39-1465-83C
$C_{16}H_{20}N_4O_9$			
1,2-Hydrazinedicarboxylic acid, 1,1'-(2-oxo-2-phenylethylidene)bis-, tetramethyl ester	EtOH	250(4.04),280(3.12)	104-0110-83
$C_{16}H_{20}N_5O_5$			
5H-Dipyrimido[1,6-a:4',5'-d]pyrimidinium, 7,8-dihydro-2,3-dimethyl-7-oxo-8-β-D-ribofuranosyl-, perchlorate	pH 1	223(4.31),347(4.50)	44-2481-83
	pH 6.86	220(4.25),235(4.15), 250(3.94),382(4.53)	44-2481-83
$C_{16}H_{20}N_6$			
[1,2,4]Triazolo[5,1-c][1,2,4]triazine, 3,4-dihydro-3,3-dimethyl-7-phenyl-4-pyrrolidino-	EtOH	246(4.36),319(3.35)	44-2330-83

Compound	Solvent	$\lambda_{max}(\log \epsilon)$	Ref.
$C_{16}H_{20}N_6O_2$			
2,4(3H,5H)-Pyrimidinedione, 5-diazo-6-[[2-(dimethylamino)-4,5-dimethylphenyl]methylamino]-3-methyl-	MeCN	205s(--),225(4.52), 255s(--),281(4.24), 350s(--)	35-4809-83
1H-1,2,4-Triazol-3-amine, 5-(hexahydro-1H-azepin-1-yl)-1-methyl-N-[(2-nitrophenyl)methylene]-	$CHCl_3$	356(3.73)	24-1547-83
1H-1,2,4-Triazol-5-amine, 3-(hexahydro-1H-azepin-1-yl)-1-methyl-N-[(2-nitrophenyl)methylene]-	$CHCl_3$	314(3.85),448(3.73)	24-1547-83
$C_{16}H_{20}N_6O_6$			
β-D-arabino-Heptofuranuronamide, 1-(6-amino-9H-purin-9-yl)-1,5,6-trideoxy-, 2,3-diacetate	MeOH	259(4.17)	87-1530-83
$C_{16}H_{20}N_8NiO_2$			
Nickel, [[4,4'-(1,2-ethanediyldiimino)-bis[6-(dimethylamino)-5-pyrimidinecarboxaldehydato](2-)-N^4,$N^{4'}$,O^5,$O^{5'}$]-	$CHCl_3$	282(4.43),370(4.16), 432(3.53)	12-2413-83
$C_{16}H_{20}OS$			
Cyclohexanone, 2-methyl-5-(1-methylethenyl)-3-(phenylthio)-, (S,S,R)-	MeOH	217(3.97),258(3.73)	49-0195-83
epimer	MeOH	260(3.65)	49-0195-83
$C_{16}H_{20}OSe$			
4(1H)-Azulenone, octahydro-5-(phenylseleno)-	EtOH	236(3.63),315(2.56)	44-1670-83
$C_{16}H_{20}O_2$			
Arnebinol	EtOH	213.2(4.42),294.2(3.83)	88-2407-83
Cyclopentanone, 3-(hydroxymethyl)-2,2,3-trimethyl-5-(phenylmethylene)-	EtOH	297(4.35)	39-1465-83C
1H-Cyclopropa[a]naphthalene-1-carboxylic acid, 1a,2,3,7b-tetrahydro-1-(1-methylethyl)-, methyl ester, (1α,1aα,7bα)-	n.s.g.	225(3.82),268(2.60)	78-2965-83
$C_{16}H_{20}O_3$			
Azuleno[4,5-b]furan-4-one, 4,6a,7,8,9-9a-hexahydro-7-methoxy-3,6,9-trimethyl-	EtOH	216(4.13),239(3.71), 318(3.57),330s(3.49)	102-1207-83
1H-Benz[e]indene-3,7(2H,3aH)-dione, 6-acetyl-4,5,8,9,9a,9b-hexahydro-3a-methyl-, [3aS-(3aα,9aα,9bβ)]-	EtOH	241(3.88)	39-2353-83C
Dibenzo[b,d]pyran-1,7-dione, 1,2,3,4-7,8,9,10-octahydro-3,3,6-trimethyl-	EtOH	202(3.81),231(3.92), 245(3.82),312(3.56)	104-2027-83
3-Furancarboxylic acid, 5-(2,6-dimethyl-1,5,7-octatrienyl)-, methyl ester, (E,E)-	n.s.g.	210(4.12),233(4.28), 257(4.23),262(4.23)	12-0371-83
(E,Z)-	n.s.g.	210(4.38),236(4.51), 255(4.49),262(4.48)	12-0371-83
4a(4H)-Phenanthrenecarboxylic acid, 1,4b,5,6,7,8,10,10a-octahydro-1-oxo-, methyl ester, (4aα,4bα,10aβ)-	EtOH	219(3.89)	32-0187-83
1(2H)-Phenanthrenone, 3,4,4a,9,10,10a-hexahydro-5-methoxy-4a-methyl-, cis	MeOH	274(3.15)	39-0751-83C
isomer?	MeOH	279(3.16)	39-0751-83C
2-Propenoic acid, 3-cyclohexyl-2-phenoxy-, methyl ester, (E)-	MeOH	219.7(4.09),270(3.31)	44-3408-83

Compound	Solvent	λ_{max}(log ϵ)	Ref.
2-Propenoic acid, 3-cyclohexyl-2-phenoxy-, methyl ester, (Z)-	MeOH	217(4.31),270(3.42)	44-3408-83
$C_{16}H_{20}O_4$			
2-Naphthaleneacetic acid, 1,2,3,4,4a,7-hexahydro-α,4a,8-trimethyl-1,7-dioxo-, methyl ester	EtOH	249(4.11)	94-3397-83
2-Naphthaleneacetic acid, 1,2,3,4,4a,7-hexahydro-α,4a,8-trimethyl-3,7-dioxo-, methyl ester	EtOH	245.5(4.10)	94-3397-83
2-Naphthaleneacetic acid, 3,4,4a,5,6,7-hexahydro-α,4a,8-trimethyl-3,7-dioxo-, methyl ester	EtOH	308(4.35)	94-3397-83
4a(4H)-Phenanthrenecarboxylic acid, 1,4b,5,6,7,8,10,10a-octahydro-4-hydroxy-1-oxo-, methyl ester	EtOH	230(3.62)	32-0183-83
$C_{16}H_{20}O_4S_2$			
1,5-Dithiaspiro[5.5]undecane-8.9,10-triol, 8-benzoate, [8R-(8α,9β,10β)]-	EtOH	209s(3.56),229(3.81), 272s(2.83),280s(2.71)	39-2441-83C
$C_{16}H_{20}O_4Se$			
Spiro[1,3-benzoxaselenole-2,1'-cyclohexane]-2',4,6'(5H)-trione, 6,7-dihydro-4',4',6,6-tetramethyl-	$CHCl_3$	300(3.59),402(2.15)	39-0333-83C
$C_{16}H_{20}O_5$			
Curvularin	EtOH	223(4.05),272(3.80), 304.5(3.71)	162-0382-83
$C_{16}H_{20}O_5S_2$			
1,4-Naphthalenedione, 2-[(2,3-dihydroxy-4-mercaptobutyl)thio]-2,3-dihydro-3-hydroxy-2,3-dimethyl-	MeOH	225(4.54),254(4.18), 297(3.23),337s(2.74)	44-3301-83
$C_{16}H_{20}O_6$			
9(2H)-Anthracenone, 1,3,4,4a,9a,10-hexahydro-2,3,8,10-tetrahydroxy-6-methoxy-3-methyl-	EtOH	281(4.55),315(4.17)	23-0378-83
$C_{16}H_{20}O_8$			
1,4,7,10,13,16-Benzohexaoxacyclooctadecin-3,14(2H,15H)-dione, 5,6,8,9-11,12-hexahydro-	MeOH	272(3.22)	49-0359-83
$C_{16}H_{20}O_9$			
Gentiopicrin	MeOH	270(3.96)	162-0628-83
$C_{16}H_{20}Si_2$			
9,10-Disilaphenanthrene, 9,10-dihydro-9,9,10,10-tetramethyl-	n.s.g.	239(4.45),280(3.49)	35-7469-83
$C_{16}H_{21}BrO_3$			
2-Naphthaleneacetic acid, 6-bromo-1,2,3,4,4a,7-hexahydro-α,4a,8-trimethyl-7-oxo-, methyl ester	EtOH	206(4.00),254(4.10)	94-3397-83
$C_{16}H_{21}Br_3O_3$			
2-Naphthaleneacetic acid, 1,6,6-tribromo-1,2,3,4,4a,5,6,7-octahydro-α,4a,8-trimethyl-7-oxo-, methyl ester	EtOH	274(4.02)	94-3397-83

Compound	Solvent	$\lambda_{max}(\log \epsilon)$	Ref.
$C_{16}H_{21}N$			
1-Azuleneethanamine, N,N-diethyl-	EtOH	238(4.27),263s(4.44), 268s(4.46),273s(4.71), 277(4.76),281s(4.73), 287s(4.58),297s(3.96), 315s(3.34),322s(3.43), 334(3.59),344(3.74), 359(3.48),537s(2.35), 560s(2.47),581s(2.54), 601(2.59),622s(2.54), 655(2.51),687s(2.23), 724(2.09)	18-2059-83
hydrochloride	EtOH	238(4.36),267s(4.57), 273s(4.69),277(4.74), 282s(4.71),288s(4.55), 298s(4.03),328s(3.90), 331(3.90),344(3.97), 359(3.84),547s(2.38), 563s(2.46),589(2.53), 610s(2.49),636(2.46), 705(2.04)	18-2059-83
4-Azuleneethanamine, N,N,6,8-tetra-methyl-	EtOH	244(4.43),271s(4.31), 275s(4.45),281s(4.61), 285(4.67),290(4.67), 294s(4.61),305s(3.93), 322s(3.45),328s(3.51), 334(3.59),344s(3.64), 349(3.71),359s(3.06), 547(2.71),564s(2.69), 581s(2.65),637s(2.20)	18-2311-83
hydrochloride	EtOH	244(4.48),270s(4.32), 275s(4.50),281s(4.67), 285(4.73),294s(6.67), 305s(4.00),328s(3.62), 334(3.70),343(3.74), 348(3.80),359s(2.73), 547(2.78),565s(2.76), 580s(2.73),635s(2.31)	18-2311-83
6-Azuleneethanamine, N,N,4,8-tetra-methyl-	EtOH	244(4.51),275s(4.53), 281(4.68),284(4.74), 289(4.73),294s(4.66), 304s(4.05),322s(3.50), 328s(3.57),333(3.64), 343s(3.69),348(3.75), 548(2.69),562s(2.67), 584s(2.62),638s(2.18)	18-2311-83
hydrochloride	EtOH	245(4.43),270s(4.26), 275s(4.46),281s(4.63), 285(4.69),294s(4.61), 306s(3.90),327s(3.69), 333(3.75),342s(3.78), 348(3.84),520s(2.53), 555(2.64),575s(2.61), 592s(2.56),650s(2.09)	18-2311-83
2-Azulenemethanamine, N,N,4,6,8-penta-methyl-	EtOH	246(5.40),287(5.72), 293(5.72),335(4.55), 351(4.68),537(2.65), 560s(2.62),616s(2.20)	18-2311-83
hydrochloride	EtOH	246(4.53),287(4.79), 293(4.79),335(4.00), 351(4.04),536(2.72),	18-2311-83

Compound	Solvent	$\lambda_{max}(\log \epsilon)$	Ref.
(cont.)		560s(2.70),615s(2.33), 624s(2.27)	18-2311-83
Bicyclo[3.2.1]oct-3-en-2-amine, N,N-dimethyl-3-phenyl-, endo, (Z)-2-butenedioate	EtOH	236(4.12)	87-0055-83
$C_{16}H_{21}NO_2$			
6H-Cyclopenta[5,6]naphtho[2,1-c]isoxazol-6-one, 3b,4,5,5a,7,8,8a,8b,9,10-decahydro-3,5a-dimethyl-, [3bS-(3bα,5aβ,8aα,8bβ)]-	EtOH	205(3.38),227(3.70)	39-2353-83C
$C_{16}H_{21}NO_3$			
3-Buten-2-one, 4-(1,2,3,4-tetrahydro-5,6-dimethoxy-2-methyl-8-isoquinolinyl)-	MeOH	222s(4.08),234s(4.02), 307(4.06)	83-0801-83
Ethanone, 1-(1,2,3,5,10,10a-hexahydro-8,9-dimethoxypyrrolo[1,2-b]isoquinolin-6-yl)-	MeOH	225(4.24),267(4.83), 306(3.49)	83-0773-83
$C_{16}H_{21}NO_3S$			
1,2-Benzoxathiin-4-amine, 3,4,5,6,7,8-hexahydro-N,N-dimethyl-3-phenyl-, 2,2-dioxide, cis	EtOH	230s(3.50),309(3.25)	4-0839-83
trans	EtOH	230s(3.44)	4-0839-83
$C_{16}H_{21}NO_4$			
Propanedioic acid, [9-(dimethylamino)-2,4,6,8-nonatetraenylidene]-, dimethyl ester	$CHCl_3$	512(4.64)	70-0780-83
	EtOH	520(--)	70-0780-83
$C_{16}H_{21}NO_8$			
7-Azabicyclo[2.2.1]hept-2-ene-1,2,3,4-tetracarboxylic acid, 2,3-diethyl 1,4-dimethyl ester	CCl_4	262(3.47)	4-0001-83
1H-Pyrrole-2,3,4,5-tetracarboxylic acid, 2,5-dimethyl 3,4-bis(1-methylethyl) ester	EtOH	225(4.15),267(4.13)	4-0001-83
1H-Pyrrole-2,3,4,5-tetracarboxylic acid, 2,5-dimethyl 3,4-dipropyl ester	EtOH	224(3.89),267(3.81)	4-0001-83
$C_{16}H_{21}N_3O_3S$			
Carbamic acid, [1-[5-(methylthio)-1,3,4-oxadiazol-2-yl]-2-phenylethyl]-, 1,1-dimethylethyl ester, (S)-	MeOH	236s(3.86)	35-0902-83
$C_{16}H_{21}N_3O_4$			
3H-1,2,4-Triazole-1,2-dicarboxylic acid, 1,2-dihydro-3,3-dimethyl-5-phenyl-, diethyl ester	EtOH	237(4.12)	33-0262-83
$C_{16}H_{21}N_3O_6$			
1,2-Diazabicyclo[5.2.0]nona-3,5-diene-2-carboxylic acid, 8-[(4-methoxy-1,4-dioxobutyl)amino]-5-methoxy-9-oxo-, ethyl ester, cis-(±)-	MeOH	275(4.00)	5-1374-83
1-Triazenium, 2,3-bis[(1-methylethoxy)carbonyl]-1-[4-(methoxycarbonyl)phenyl]-, hydroxide, inner salt	EtOH	231(4.05),259(3.95), 352(4.11),361s(4.07)	33-1599-83

Compound	Solvent	λ_{max}(log ε)	Ref.
$C_{16}H_{21}N_4$			
Pyridinium, 3-[[4-(diethylamino)phenyl]azo]-1-methyl-, iodide	H_2O	511(4.604)	7-0055-83
	DMSO-p-tolueneSO_3H	503(4.634)	7-0055-83
$C_{16}H_{21}N_4O$			
Pyridinium, 3-[[4-[ethyl(2-hydroxyethyl)amino]phenyl]azo]-1-methyl-, iodide	H_2O	500(4.593)	7-0055-83
	DMSO-p-tolueneSO_3H	502(4.586)	7-0055-83
$C_{16}H_{21}N_5O_2$			
Benzoic acid, 4-(1H-imidazol-2-ylazo)-, 2-(diethylamino)ethyl ester	MeOH	234(3.90),368(4.43), 378(4.42),450s(3.28)	56-0971-83
$C_{16}H_{21}N_5O_2S$			
Benzoic acid, 4-[[2,4-bis(dimethylamino)-5-thiazolyl]azo]-, ethyl ester	$CHCl_3$	319(4.11),495(4.58)	88-3563-83
$C_{16}H_{21}N_7$			
1H-Benzimidazole, 2-methyl-1-[[5-(1-piperidinylmethyl)-1H-tetrazol-1-yl]methyl]-	EtOH	244(3.93),250(3.88), 275(3.64),281(3.70)	94-0723-83
$C_{16}H_{21}OP$			
Phosphorin, 1-(1,1-dimethylethyl)-1,1-dihydro-1-methoxy-3-phenyl-	hexane	245(4.0),349(3.76)	24-0445-83
$C_{16}H_{22}$			
5H-Indene, 2,3,4,5,5,6,7-heptamethyl-	MeCN	220(4.12),248(4.34), 253(4.29),265(3.75), 420(3.03)	44-0309-83
$C_{16}H_{22}Br_2O_3$			
2-Naphthaleneacetic acid, 1,6-dibromo-1,2,3,4,4a,5,6,7-octahydro-α,4a,8-trimethyl-7-oxo-, methyl ester, [1R-[1α,2α(S*),4aα,6β)]-	EtOH	263(4.07)	94-3397-83
$C_{16}H_{22}Ge_2$			
Digermane, 1,1,2,2-tetramethyl-1,2-diphenyl-	C_6H_{12}	233(4.30)	101-0149-83G
$C_{16}H_{22}N_2O$			
Indeno[5,4-e]indazol-6(1H)-one, 3b,4-5,5a,7,8,8a,8b,9,10-decahydro-3,5a-dimethyl-, [3bS-(3bα,5aβ,8aα,8bβ)]-	EtOH	204(3.55),226(3.63)	39-2353-83C
9-Oxaergoline, 2,3-dihydro-6-propyl-, (3β)-(±)-	EtOH	211(4.49),245(3.79), 293(3.36)	87-0522-83
	EtOH-HCl	265(2.85),272(2.83)	87-0522-83
$C_{16}H_{22}N_2O_4$			
Benzene, 1-[4,4-dimethyl-1-(1-methyl-1-nitroethyl)-2-pentenyl]-4-nitro-, (E)-	EtOH	270(4.05)	12-0527-83
$C_{16}H_{22}N_2O_4S$			
Glycine, N-[N-[(1,1-dimethylethoxy)-carbonyl]thioglycyl]-, phenylmethyl ester	n.s.g.	266(4.11)	78-1075-83

Compound	Solvent	λ_{max}(log ϵ)	Ref.
$C_{16}H_{22}N_2O_5$			
3-Heptanone, 2,2,6-trimethyl-6-nitro-5-(4-nitrophenyl)-	EtOH	269(4.19)	12-0527-83
Uridine, 2'-deoxy-5-(1-heptynyl)-	pH 6	234(4.04),293(4.04)	44-1854-83
	pH 13	233(4.12),290(3.96)	44-1854-83
$C_{16}H_{22}N_4$			
Benzo[1,2-b:4,3-b']dipyrrole-1,8-dimethanamine, 3,6-dihydro-N,N,N',N'-tetramethyl-	EtOH	205s(4.01),223(4.44), 240s(4.07),278s(3.83), 289(4.03),303(4.07)	103-0871-83
Benzo[1,2-b:4,5-b']dipyrrole-3,7-dimethanamine, 1,5-dihydro-N,N,N',N'-tetramethyl-	EtOH	202.5(4.19),205s(4.57), 220(4.87),238s(4.65), 305(4.35),330(4.17), 340(4.17)	103-0871-83
$C_{16}H_{22}N_4O_3$			
2,4(1H,3H)-Pyrimidinedione, 6-[[2-(dimethylamino)-4,5-dimethylphenyl)-methylamino]-5-hydroxy-3-methyl-	MeCN	318(4.10)(changing)	35-4809-83
$C_{16}H_{22}N_6O_2$			
2,4,5(3H)-Pyrimidinetrione, 6-[[2-(dimethylamino)-4,5-dimethylphenyl]-methylamino]-3-methyl-, 5-hydrazone	MeCN	232(4.42),260s(--), 300(4.20)	35-4809-83
$C_{16}H_{22}OS$			
Cyclohexanone, 2-methyl-5-(1-methylethyl)-3-(phenylthio)-, [2S-(2α,3α,5β)]-	MeOH	216(3.96),258(3.76) (anom.)	49-0195-83
Spiro[6H-cyclohepta[b]thiophene-6,1'-cyclohexan]-4(5H)-one, 2-ethyl-7,8-dihydro-	n.s.g.	258(3.9)	42-0303-83
$C_{16}H_{22}O_2$			
Bicyclo[8.3.2]pentadeca-10(15),11,13-trien-14-one, 15-methoxy-	MeOH	251(4.17),318(3.55), 350s(3.51)	88-3879-83
Cyclodeca[b]furan, 4,7,8,11-tetrahydro-8-methoxy-3,6,10-trimethyl-, [S-(E,E)]-	EtOH	208(4.03)	102-1207-83
1,4-Ethanonaphthalene-2,5(1H,3H)-dione, 4,6,7,8-tetrahydro-8,8,9,9-tetramethyl-	C_6H_{12}	246(3.97),254(3.98), 298(2.97)	33-0735-83
1-Naphthalenemethanol, 5,6,7,8-tetrahydro-2-methoxy-4-methyl-6-(1-methylethenyl)-, (R)-	EtOH	284(3.29),292(3.31)	78-2647-83
$C_{16}H_{22}O_3$			
2-Naphthaleneacetic acid, 1,2,3,4,4a,7-hexahydro-α,4a,8-trimethyl-7-oxo-, methyl ester	EtOH	240(4.06)	94-3397-83
4a(2H)-Phenanthrenecarboxylic acid, 1,3,4,4b,5,6,7,8,8a,9-decahydro-9-oxo-, methyl ester	EtOH	238(4.00)	44-4642-83
$C_{16}H_{22}O_4$			
Cyclohexanone, 2-acetyl-3-(4,4-dimethyl-2,6-dioxocyclohexyl)-	EtOH	266(4.13)	104-2027-83
2-Naphthaleneacetic acid, 1,2,3,4,4a-5,6,7-octahydro-α,4a,8-trimethyl-3,7-dioxo-, methyl ester, [2R-[2α(S*),4aα]]-	EtOH	244(4.17)	94-3397-83

Compound	Solvent	λ_{max}(log ε)	Ref.
Naphtho[2,3-b]furan-2(3H)-one, decahydro-8a-methyl-4a-(methyldioxy)-3,5-bis(methylene)-	EtOH	210(4)	102-2767-83
epimer	EtOH	210(4.2)	102-2767-83
Propanoic acid, 2-[5-methoxy-2-(2-propenyl)phenoxy]-2-methyl-, ethyl ester	EtOH	212(4.20),219(4.04), 272(2.99),279(2.99)	39-1431-83C
$C_{16}H_{22}O_8$			
β-D-Glucopyranoside, 4-(3-hydroxy-1-propenyl)-2-methoxyphenyl (coniferin)	MeOH	258.5(4.26),292s(3.75)	102-0553-83
$C_{16}H_{22}O_{10}$			
Swertiamarin	MeOH	238(3.93)	162-1295-83
$C_{16}H_{22}O_{11}$			
Monotropein, monohydrate	n.s.g.	235(3.98)	162-0895-83
$C_{16}H_{23}BrO_2S$			
Thiophene, 3-bromo-2,5-dimethyl-3-[5-methyl-2-(1-methylethyl)-1-cyclohexen-1-yl]-, 1,1-dioxide, (+)-	MeCN	220(3.78),300(2.20)	4-0729-83
$C_{16}H_{23}BrS$			
Thiophene, 3-bromo-2,5-dimethyl-4-[5-methyl-2-(1-methylethyl)-1-cyclohexen-1-yl]-	MeCN	193(4.39),244(3.87)	4-0729-83
$C_{16}H_{23}ClN_3O_3P$			
1,3,2-Dioxaphosphorinane, 2-chloro-2-[[2,3-dihydro-3-(1-methyl-2-phenylethyl)-1,2,3-oxadiazol-5-yl]imino]-2,2-dihydro-5,5-dimethyl-	MeCN	300(3.81)	65-1263-83
$C_{16}H_{23}IN_3O_3P$			
1,3,2-Dioxaphosphorinane, 2-iodo-2-[[2,3-dihydro-3-(1-methyl-2-phenylethyl)-1,2,3-oxadiazol-5-yl]imino]-2,2-dihydro-5,5-dimethyl-	EtOH	306(3.79)	65-1263-83
$C_{16}H_{23}NO$			
Bicyclo[8.3.2]pentadeca-10(15),11,13-trien-14-one, 15-(methylamino)-	MeOH	257.5(4.01),287(3.82), 376s(3.77),412(3.82)	88-3879-83
1H-Isoindole, 2-(2-butenyloxy)-2,3-dihydro-1,1,3,3-tetramethyl-, (E)-	EtOH	215(3.41),257(2.89), 263(3.00),270(3.00)	12-1957-83
(Z)-	EtOH	215(3.43),256(2.89), 264(3.03),270(3.05)	12-1957-83
1H-Isoindole, 2,3-dihydro-1,1,3,3-tetramethyl-2-[(1-methyl-2-propenyl)oxy]-	EtOH	215(3.47),256(2.86), 263(2.97),270(2.99)	12-1957-83
1H-Isoindole, 2,3-dihydro-1,1,3,3-tetramethyl-2-[(2-methyl-2-propenyl)oxy]-	EtOH	215(3.33),256(2.82), 263(2.99),269(3.04)	12-1957-83
$C_{16}H_{23}NO_3$			
3-Buten-2-ol, 4-(1,2,3,4-tetrahydro-5,6-dimethoxy-2-methyl-8-isoquinolinyl)-	MeOH	260(3.99),300(3.38)	83-0801-83
3-Heptanone, 2,2,6-trimethyl-6-nitro-5-phenyl-	EtOH	256(1.70),263(1.70), 267(1.70)	12-0527-83

Compound	Solvent	$\lambda_{max}(\log \epsilon)$	Ref.
2-Piperidinone, 6-(2,4-dihydroxy-6-pentylphenyl)-	EtOH	284(3.49)	150-1156-83M
2-Piperidinone, 6-(2,6-dihydroxy-4-pentylphenyl)-	EtOH	282.5(3.15)	150-1156-83M
$C_{16}H_{23}NO_4$			
4-Morpholinecarboxaldehyde, 3-(2,4-dihydroxy-6-pentylphenyl)-	EtOH	285(3.36)	150-1156-83M
4-Morpholinecarboxaldehyde, 3-(2,6-dihydroxy-4-pentylphenyl)-	EtOH	285(3.36)	150-1156-83M
β-D-Ribofuranosylamine, 2,3-O-(1-methylethylidene)-N-(2-phenylethyl)-	EtOH	257(4.30)	136-0001-83B
$C_{16}H_{23}NO_7$			
Acetic acid, (2-nitrophenoxy)-, 2-(2-butoxyethoxy)ethyl ester	MeOH	250(3.52),315(3.32)	49-0359-83
$C_{16}H_{23}NO_8$			
Bakankosin	EtOH	235(4.2)	162-0137-83
$C_{16}H_{23}N_2O_{10}P$			
Phosphoric acid, diethyl 3a,4,6,7,8,10-11,11a-octahydro-10-hydroxy-2,2-dimethyl-4,11-epoxy-1,3-dioxolo[4,5-e]pyrimido[1,6-a]azepin-9-yl ester, [3aR-(3aα,4β,10α,11β,11aα)]-	10% EtOH	275(3.96)	4-0753-83
$C_{16}H_{23}N_3O_5$			
1-Triazenium, 1-(4-ethoxyphenyl)-2,3-bis[(1-methylethoxy)carbonyl]-, hydroxide, inner salt	EtOH	231(3.79),248s(3.77), 382(4.36)	33-1599-83
$C_{16}H_{23}N_3O_5S$			
Piperidine, 1-[[2-(5-nitro-2-furanyl)-1-(1-piperidinyl)ethenyl]sulfonyl]-	MeOH	230(4.06),306(4.03), 405(3.95)	73-1062-83
$C_{16}H_{23}N_3O_6$			
1,2-Triazanedicarboxylic acid, 3-[4-(methoxycarbonyl)phenyl]-, bis(1-methylethyl) ester	EtOH	216(3.99),280(4.26)	33-1599-83
$C_{16}H_{23}N_5O_4$			
1H-Pyrazolo[3,4-d]pyrimidin-4-amine, N-cyclohexyl-1-β-D-ribofuranosyl-	MeOH	265(4.90),284(5.07)	87-1601-83
$C_{16}H_{23}N_5O_8S$			
1,2-Hydrazinedicarboxylic acid, 1,1'-[(dimethylamino)-3-thienylethenylidene]bis-, tetramethyl ester	EtOH	245(4.03),300(3.76)	104-0110-83
$C_{16}H_{24}$			
Cyclobutane, tetrakis(1-methylethylidene)-	C_6H_{12}	260s(4.176),272(4.279), 283s(4.23),307s(3.857)	77-1058-83
Naphthalene, 7-(1,1-dimethylethyl)-1,2,3,4-tetrahydro-1,1-dimethyl-	MeOH	270(2.76),278(2.73)	2-0215-83
$C_{16}H_{24}N_2O$			
Cyclohexanone, 2,6-bis[3-(dimethylamino)-2-propenylidene]-	EtOH	267(4.31),277(4.23), 305(4.17),383(4.44), 469(4.82)	70-1882-83

Compound	Solvent	λ_{max}(log ε)	Ref.
$C_{16}H_{24}N_2O_2$			
Ethanone, 2-[1-(3-aminopropyl)-2-pyrrolidinyl]-1-(2-hydroxy-6-methylphenyl)-	MeOH	225(3.93),255(3.51), 315(3.18)	100-0235-83
1H,7H-Pyrazolo[1,2-a]pyrazole-1,7-dione, 2,6-bis(1,1-dimethylethyl)-3,5-dimethyl-	dioxan	232(4.11),255s(3.68), 370(3.90)	151-0171-83A
$C_{16}H_{24}N_2O_4$			
1,4-Benzenedipropanoic acid, β,β'-diamino-, diethyl ester, dihydrochloride	H_2O	257(2.51),262(2.56), 268(2.44)	4-1107-83
$C_{16}H_{24}N_2O_4$(cont.)			
2,8-Diazabicyclo[3.2.1]octa-3,6-diene-2,8-dicarboxylic acid, 4,7-bis(1-methylethyl)-, dimethyl ester	C_6H_{12}	221s(3.80),248(3.70)	88-2275-83
Pyrrolo[3,2-b]pyrrole-1,4-dicarboxylic acid, 3a,6a-dihydro-3,6-bis(1-methylethyl)-, dimethyl ester, cis	C_6H_{12}	226(4.37)	88-2275-83
$C_{16}H_{24}N_3O_2$			
1-Piperidinyloxy, 2,2,6,6-tetramethyl-4-[[(phenylamino)carbonyl]amino]-	EtOH	241(4.39),268s(3.38), 274s(3.31),452(1.03)	70-1906-83
$C_{16}H_{24}N_4O_3$			
Arphamenine A, hydrochloride	H_2O	257(2.26)	158-1576-83
$C_{16}H_{24}N_4O_4$			
Arphamenine B, hydrochloride	H_2O	275(3.02)	158-1576-83
Arphenamine B	H_2O	275(3.0)	158-1572-83
$C_{16}H_{24}O$			
Benzenemethanol, 2,3,4,5,6-pentamethyl-α-(1-methyl-1-propenyl)-, (E)-	96% H_2SO_4	260(4.18),313(3.70), 351(3.66),483(3.97)	44-0309-83
Cedrene, 9-acetyl-	C_6H_{12}	247(3.82)	39-1373-83C
2-Cyclohexen-1-one, 3-(7,7-dimethyl-4,5-octadienyl)-	hexane	227(4.10)	44-2584-83
Cyclopenta[1,4]cyclobuta[1,2]benzen-5(6H)-one, 4-(2,2-dimethylpropylidene)octahydro-, cis?	hexane	297(2.10)	44-2584-83
trans?	hexane	301(1.97)	44-2584-83
$C_{16}H_{24}O_2$			
2,4,6,10-Dodecatetraenoic acid, 3,7,11-trimethyl-, (E,E,Z)-, methyl ester	hexane	297s(4.46),308(4.56), 318s(4.49)	39-3005-83C
(Z,E,E)-	hexane	313(4.42),319s(4.40)	39-3005-83C
(Z,Z,E)-	hexane	312(4.37)	39-3005-83C
Farnesylic acid, 4,5-didehydro-, methyl ester	hexane	307(4.56)	33-0494-83
1-Naphthalenol, 1,2,3,7,8,8a-hexahydro-4,8a-dimethyl-6-(1-methylethyl)-, formate, (1R-cis)-	hexane	240(4.30),246(4.33), 254s(4.17)	44-4410-83
$C_{16}H_{24}O_3$			
4-Eudesmenoic acid, 3-oxo-, methyl ester	EtOH	247(4.02)	94-3397-83
Oxiranebutanoic acid, 3-(1-nonen-3-ynyl)-, methyl ester, cis	EtOH	229(4.15)	35-3656-83

Compound	Solvent	λ_{max}(log ε)	Ref.
$C_{16}H_{24}O_4$			
Benzeneheptanoic acid, 3,5-dimethoxy-, methyl ester	EtOH	221.5(3.91),273(3.18), 282(3.18)	39-2211-83C
Brefeldin A	EtOH	215(4.05)	162-0189-83
8,10-Undecadiene-3,7-dione, 4-acetoxy-6,6,10-trimethyl-, (E)-	EtOH	270(4.27)	33-1638-83
$C_{16}H_{24}O_5$			
Acetic acid, phenoxy-, 2-(2-butoxyethoxy)ethyl ester	MeOH	270(3.17),275(3.09)	49-0359-83
1,4,7,10,13-Benzopentaoxacyclopentadecin, 5,6,8,9,11,12,14,15-octahydro-5,15-dimethyl-, (5R,15R)-	MeOH	221(3.83),275(3.28)	44-2029-83
lithium perchlorate complex	MeOH	217(3.85),275(3.28)	44-2029-83
potassium perchlorate complex	MeOH	219(3.85),275(3.28)	44-2029-83
rubidium perchlorate complex	MeOH	220(3.84),275(3.28)	44-2029-83
sodium perchlorate complex	MeOH	217(3.88),273(3.26), 277s(--)	44-2029-83
1,4,7,10,13-Benzopentaoxacyclopentadecin, 5,6,8,9,11,12,14,15-octahydro-6,14-dimethyl-, (6S,14S)-	MeOH	223(3.85),276(3.36)	44-2029-83
cesium perchlorate complex	MeOH	223(3.85),275(3.36)	44-2029-83
potassium perchlorate complex	MeOH	221(3.87),274(3.36)	44-2029-83
sodium perchlorate complex	MeOH	220(3.85),273(3.30)	44-2029-83
1,4,7,10,13-Benzopentaoxacyclopentadecin, 5,6,8,9,11,12,14,15-octahydro-8,12-dimethyl-, (6S,12S)-	MeOH	223(3.84),277(3.38)	44-2029-83
	CH_2Cl_2	227(3.92),277(3.46)	44-2029-83
lithium perchlorate complex	MeOH	223(3.72),275(3.26)	44-2029-83
potassium perchlorate complex	MeOH	222(3.73),274(3.49)	44-2029-83
2-Cyclopenten-1-one, 5-acetoxy-2-(7-acetoxyheptyl)-	EtOH	232(3.96)	39-0319-83C
$C_{16}H_{24}O_{10}$			
8-Epiloganic acid	EtOH	235(4.05)	95-0508-83
$C_{16}H_{24}S$			
Cyclohexene, 1-[4-(butylthio)-3-penten-1-ynyl]-2-methyl-, (Z)-	MeCN	215(3.85),280(3.92)	4-0729-83
$C_{16}H_{25}BrO_4$			
2,3,13,14-Pentadecanetetrone, 8-(bromomethyl)-	EtOH	262(2.73),420(1.48)	5-0181-83
$C_{16}H_{25}NO_2$			
1H-Azepine-1-carboxylic acid, 2,5-bis(1,1-dimethylethyl)-, methyl ester	C_6H_{12}	208(4.15),245s(3.58)	88-2275-83
1H-Azepine-1-carboxylic acid, 3,6-bis(1,1-dimethylethyl)-, methyl ester	C_6H_{12}	212(4.29),233s(3.66), 285s(2.59)	88-2275-83
1,3-Benzenediol, 5-pentyl-2-(2-piperidinyl)-	EtOH	283(3.20)	150-1156-83M
1,3-Benzenediol, 5-pentyl-4-(2-piperidinyl)-	EtOH	283.5(3.22)	150-1156-83M
Isoquinoline, 8-butyl-1,2,3,4-tetrahydro-5,6-dimethoxy-2-methyl-	MeOH	220s(3.86),280(3.05)	83-0801-83
Nonanamide, N-hydroxy-N-(phenylmethyl)-	EtOH	209(4.10)	34-0433-83
$C_{16}H_{25}NO_3$			
8-Isoquinolinepropanol, 1,2,3,4-tetrahydro-5,6-dimethoxy-α,2-dimethyl-	MeOH	226s(3.89),282(3.09)	83-0801-83

Compound	Solvent	λ_{max}(log ε)	Ref.
$C_{16}H_{25}NO_3$			
1(2H)-Pyridineundecanoic acid, 2-oxo-	EtOH	229(3.75),303(3.65)	94-2552-83
$C_{16}H_{25}N_3O_5$			
1,2-Triazanedicarboxylic acid, 3-(4-ethoxyphenyl)-, bis(1-methylethyl) ester	EtOH	235(4.13),295(3.45)	33-1599-83
$C_{16}H_{25}N_7O_8$			
Gougerotin	pH 1	275(4.13)	162-0651-83
	H_2O	235(3.97),267(3.97)	162-0651-83
	pH 13	267(3.99)	162-0651-83
$C_{16}H_{26}$			
1H-Cyclopentacyclododecene, 2,4,5,6,7-8,9,10,11,12-decahydro-2-methyl-	EtOH	247(4.21)	33-2519-83
$C_{16}H_{26}BrNO_4S_2Si$			
4,5-Dithia-1-azabicyclo[4.2.0]oct-2-ene-2-carboxylic acid, 3-(bromomethyl)-7-[1-[[(1,1-dimethylethyl)dimethylsilyl]oxy]ethyl]-8-oxo-, methyl ester, [6α,7α(R*)]-	$CHCl_3$	282(3.87),336(3.49)	88-3283-83
$C_{16}H_{26}N_2O$			
Acetamide, N-[4-(octylamino)phenyl]-	MeOH	269(4.23)	5-0802-83
2(3H)-Naphthalenone, 4,4a,5,6,7,8-hexahydro-4a-methyl-1-(4-methyl-1-piperazinyl)-, (S)-	EtOH	242(3.85)	107-1013-83
2(3H)-Naphthalenone, 4,4a,5,6,7,8-hexahydro-4a-methyl-8-(4-methyl-1-piperazinyl)-, (S)-	EtOH	237(3.95)	107-1013-83
$C_{16}H_{26}N_2O_2$			
Pyrrolo[3,2-b]pyrrole-1(3aH)-carboxylic acid, 3,6-bis(1,1-dimethylethyl)-6,6a-dihydro-, methyl ester	C_6H_{12}	228(3.94),287s(2.36)	138-0743-83
$C_{16}H_{26}N_4$			
2-Heptenenitrile, 3,3'-(1,2-ethanediyldiimino)bis-	n.s.g.	263(4.53)	39-1223-83C
2-Hexenenitrile, 3,3'-(1,2-ethanediyldiimino)bis-	n.s.g.	263(4.54)	39-1223-83C
	n.s.g.	263(4.58)	39-1223-83
$C_{16}H_{26}O$			
1-Penten-3-one, 4,4-dimethyl-1-(2,6,6-trimethyl-1-cyclohexen-1-yl)-, (E)-	EtOH	218(3.80),298(4.01)	54-0302-83
1-Penten-3-one, 4,4-dimethyl-1-(2,6,6-trimethyl-2-cyclohexen-1-yl)-	EtOH	233(4.14),318(2.02)	54-0302-83
$C_{16}H_{27}NOSi$			
Methanamine, 1-(2,4,6-cycloheptatrien-1-ylidene)-N,N-dimethyl-1-[(triethylsilyl)oxy]-	THF	306(3.9),437(<2)	24-1777-83
$C_{16}H_{27}NO_4S_2Si$			
4,5-Dithia-1-azabicyclo[4.2.0]oct-2-ene-2-carboxylic acid, 7-[1-[[(1,1-dimethylethyl)dimethylsilyl]oxy]ethyl]-3-methyl-8-oxo-, methyl ester, [6R-[6α,7α(R*)]]-	EtOH	233(3.68),277(3.80), 326(3.47)	88-1631-83

Compound	Solvent	$\lambda_{max}(\log \epsilon)$	Ref.
$C_{16}H_{27}O_3P$			
Phosphorin, 1,4-bis(1,1-dimethylethyl)-1,1-dihydro-1-methoxy-2-(methoxycarbonyl)-	EtOH	216(4.30),244s(3.38),385(4.00)	88-5051-83
$C_{16}H_{28}Bi_2$			
Dibismuthine, tetrakis(2-methyl-1-propenyl)-	pentane	265s(4.20)	157-1859-83
$C_{16}H_{28}N_4$			
1,2,4,5-Tetrazine, 3,6-dicyclohexyl-1,2-dihydro-1,2-dimethyl-	hexane	270(3.48)	24-2261-83
	EtOH	288(3.42)	24-2261-83
	dioxan	274(3.44)	24-2261-83
1,2,4,5-Tetrazine, 3,6-dicyclohexyl-1,4-dihydro-1,4-dimethyl-	hexane	244(3.74),305s(2.14)	24-2261-83
	EtOH	242(3.77),305(2.06)	24-2261-83
	dioxan	243(3.78),304(2.05)	24-2261-83
$C_{16}H_{28}N_4O_2S$			
Compd. III, m. 122-5°	EtOH	221(4.30)	103-1076-83
Spiro[2H-1,3-oxazine-5(6H),7'(6'H)-[2H]thiazolo[3,2-a][1,3,5]triazin]-6'-one, 3,3'-bis(1,1-dimethylethyl)-3,3',4,4'-tetrahydro-	EtOH	217(4.05)	103-0961-83
4(5H)-Thiazolone, 5-(hydroxymethyl)-5-(1-piperidinylmethyl)-2-[(1-piperidinylmethyl)amino]-	EtOH	221(4.29),250(3.91)	103-1076-83
$C_{16}H_{28}N_4O_7$			
L-Lysine, N^6-[[[3-[(2-deoxy-β-D-erythro-pentofuranosyl)amino]-2-methyl-1-oxo-2-propenyl]amino] c arbonyl]-, (E)-	MeOH	286(4.86)	35-6989-83
$C_{16}H_{28}N_6O_7$			
L-Arginine, N^6-[[[3-[(2-deoxy-β-D-erythro-pentofuranosyl)amino]-2-methyl-1-oxo-2-propenyl]amino]carbonyl]-, (E)-	H_2O	290(4.25)	35-6989-83
$C_{16}H_{28}O_4$			
1,4-Cyclohexadiene, 3,6-dibutoxy-3,6-dimethoxy-	EtOH	205.5(3.30)	39-1255-83C
$C_{16}H_{28}O_7$			
β-D-Glucopyranoside, 8-hydroxy-3,7-dimethyl-2,6-octadienyl	EtOH	212(3.59)	95-0508-83
$C_{16}H_{28}P_2$			
Phosphine, 1,3,5,7-octatetraene-1,8-diylbis[diethyl-, (E,E,Z,Z)-	hexane	329(4.59)	88-1955-83
$C_{16}H_{28}Sb_2$			
Distibine, tetrakis(2-methyl-1-propenyl)-	C_6H_{12}	207(4.36),294(3.21),312(3.21)	157-1573-83
$C_{16}H_{29}NO$			
2-Cyclohexen-1-one, 5,5-dimethyl-3-(octylamino)-	EtOH	204(3.60),293(4.49)	104-2027-83
$C_{16}H_{30}N_8$			
4H-Imidazolium, 4-[2,5-bis(dimethylamino)-3-methyl-4H-imidazolium-4-	acetone	466(4.31)	88-3563-83

Compound	Solvent	λ_{max}(log ε)	Ref.
ylidene]-2,5-bis(dimethylamino)-1-methyl-, bis(tetrafluoroborate) (cont.)			88-3563-83
$C_{16}H_{32}O_2Te$			
Propanoic acid, 3-(undecyltelluro)-, ethyl ester	MeOH	233.0(3.59)	101-0031-83M
$C_{16}H_{35}O_2PS_2$			
Phosphorodithioic acid, bis(2-ethylhexyl) ester, barium salt	hexane	227(4.0),255(3.5) (anom.)	30-0316-83
$C_{16}H_{36}Si_2$			
Disilene, tetrakis(1,1-dimethylethyl)-	C_6H_{12}	305(3.72),433(3.45)	157-1464-83
$C_{16}H_{38}N_2Si_2$			
1,2-Ethenediamine, N,N'-bis(1,1-dimethylethyl)-N,N'-bis(trimethylsilyl)-, (E)-	hexane	240(3.95),327(2.55) (decomposes)	24-0136-83
$C_{16}H_{40}Si_4$			
Cyclotetrasilane, octaethyl-	isooctane	211s(4.36),228s(4.15), 304s(2.26)	157-1792-83

Compound	Solvent	λ_{max}(log ε)	Ref.
$C_{17}H_6Cl_2N_2O_4$			
Anthra[1,2-b]furan-3-carbonitrile, 2-amino-7,10-dichloro-5-hydroxy-6,11-dioxo-	dioxan	327(3.84),469(3.89)	104-0533-83
$C_{17}H_7ClN_2O_4$			
Anthra[1,2-b]furan-3-carbonitrile, 2-amino-4-chloro-6,11-dihydro-5-hydroxy-6,11-dioxo-	dioxan	311(4.20),464(3.96)	104-1900-83
$C_{17}H_8ClNO_4S$			
4H,5H-Pyrano[3,4-e][1,3]oxazine-4,5-dione, 7-chloro-2-(1-naphthalenylthio)-	$CHCl_3$	288(4.13),346(4.14)	24-3725-83
$C_{17}H_8N_2O_4$			
Anthra[1,2-b]furan-3-carbonitrile, 2-amino-6,11-dihydro-5-hydroxy-6,11-dioxo-	dioxan	306(4.19),470(4.01)	104-0533-83
$C_{17}H_9NO_4$			
Oxoanolobine	EtOH	217(4.24),249(4.43), 274(4.35),324s(3.84), 370(3.65),442(3.76)	100-0761-83
$C_{17}H_9NO_5$			
Tuberosinone	EtOH	238(4.57),277(4.12), 286(4.15),318(4.05), 330(4.16),372(3.92), 390(3.94),476(4.22)	100-0761-83
$C_{17}H_9N_3O$			
Isoindolo[2,1-b]isoquinoline-12-carbonitrile, 5,7-dihydro-5-imino-7-oxo-	EtOH	220(4.56),237(4.49), 257s(4.30),286(3.77), 318(3.95),335s(3.91), 360s(3.83),392(3.74), 416s(3.51)	39-1813-83C
$C_{17}H_{10}N_2O_4$			
5H-Benzo[b]carbazole-6,11-dione, 5-methyl-8-nitro-	EtOH	209(4.48),252s(4.45), 257(4.47),270s(4.40), 291s(4.29),337s(3.85), 394s(3.78),438s(3.56)	39-2409-83C
$C_{17}H_{10}O$			
1-Pyrenecarboxaldehyde	hexane	363(4.32),389(4.37)	149-0141-83B
	benzene	363(4.26),396(4.25)	149-0141-83B
	MeOH	361(4.24),392(4.12)	149-0141-83B
	EtOH	361(4.23),392(4.14)	149-0141-83B
	CCl_4	368(4.26),396(4.28)	149-0141-83B
flash photolysis triplet	benzene	443(4.32)	149-0141-83B
$C_{17}H_{10}O_4$			
Spiro[furan-2(3H),1'(3'H)-isobenzofuran]-3,3'-dione, 4-phenyl- (fluorescamine)	ether	235(4.41),276(3.60), 284(3.61),306(3.58)	162-0594-83
$C_{17}H_{10}O_5$			
5H-Furo[4',3',2':1,10]phenanthro[3,4-d]-1,3-dioxol-5-one, 8-methoxy- (aristololide)	EtOH	233(4.46),254(4.26), 279(4.18),289s(4.13), 322(3.87),340s(3.70), 396(3.78)	142-0771-83

Compound	Solvent	$\lambda_{max}(\log \epsilon)$	Ref.
Aristololide (cont.)	EtOH-KOH	234(4.39),253(4.35), 278(4.12),289s(4.03), 329(3.84),340s(3.77), 397(3.67)	142-0771-83
$C_{17}H_{10}O_7$			
5H-Phenanthro[4,5-bcd]pyran-5,9,10-tri-one, 1,8-dihydroxy-2-methoxy-7-methyl- (biruloquinone)	$CHCl_3$	290(4.03),307(4.17), 320(4.12),440(3.33), 568(3.42)	105-0270-83
	MeOH	234(--),268(--), 323(--),403(--), 565(--)	105-0270-83
$C_{17}H_{11}BrO_2Te$			
[1,2]Oxatellurolo[2,3-b][1,2]oxa-tellurole-7-Te^{IV}, 3-bromo-2,5-diphenyl-	CH_2Cl_2	426(4.28)	44-5149-83
$C_{17}H_{11}Br_2NO$			
Acetamide, N-(2,8-dibromo-4H-cyclo-penta[def]phenanthren-1-yl)-	C_6H_{12}	235(4.36),263(4.75), 301(4.12),313(4.13), 341(3.50),358(3.50)	18-1259-83
Acetamide, N-(2,8-dibromo-4H-cyclo-penta[def]phenanthren-3-yl)-	C_6H_{12}	230(4.42),264(4.80), 311(4.15),339(3.37), 357(3.37)	18-1259-83
$C_{17}H_{11}Br_3O_2Te$			
[1,2]Oxatellurolo[2,3-b][1,2]oxa-tellurole-7-Te^{IV}, 3,7,7-tribromo-7,7-dihydro-2,5-diphenyl-	CH_2Cl_2	340(4.28),555(4.57)	44-5149-83
$C_{17}H_{11}ClN_2O$			
Pyrimido[2,1-a]isoindol-4(6H)-one, 2-(4-chlorophenyl)-	EtOH	203(4.53),266(4.54), 310s(3.85)	124-0755-83
$C_{17}H_{11}ClN_2O_2$			
[1]Benzopyrano[2,3-c]pyrazol-4(1H)-one, 6-chloro-3-methyl-1-phenyl-	EtOH	242(4.40),264(4.16)	118-0214-83
[1]Benzopyrano[4,3-c]pyrazol-4(1H)-one, 8-chloro-3-methyl-1-phenyl-	EtOH	262(4.07),273(4.07), 310(3.84)	88-0381-83
[1]Benzopyrano[4,3-c]pyrazol-4(2H)-one, 8-chloro-3-methyl-2-phenyl-	EtOH	260s(4.16),298(3.86), 309(3.83)	88-0381-83
1H-Naphth[2,3-g]indole-6,11-dione, 5-amino-4-chloro-2-methyl-	dioxan	278(4.47),451(3.77), 550(3.98)	104-1892-83
$C_{17}H_{11}ClO_2Te$			
[1,2]Oxatelluro[2,3-b][1,2]oxatellu-role-7-Te^{IV}, 3-chloro-2,5-diphenyl-	CH_2Cl_2	427(4.505)	44-5149-83
$C_{17}H_{11}Cl_2FN_2O_2$			
1H-Pyrazole-3-acetic acid, 5-chloro-4-(4-chlorophenyl)-1-(4-fluoro-phenyl)-	MeOH	252(3.3)	83-0608-83
$C_{17}H_{11}Cl_2NO_2$			
[1]Benzopyrano[4,3-b]pyrrol-4(1H)-one, 1-(3,4-dichlorophenyl)-2,3-dihydro-	MeOH	266(4.00),313(3.97), 346(4.14)	5-0165-83
$C_{17}H_{11}Cl_3O_2Te$			
[1,2]Oxatellurolo[2,3-b][1,2]oxatellu-role-7-Te^{IV}, 3,7,7-trichloro-7,7-di-	CH_2Cl_2	315(4.25),548(4.58)	44-5149-83

Compound	Solvent	$\lambda_{max}(\log \epsilon)$	Ref.
hydro-2,5-diphenyl- (cont.)			44-5149-83
$C_{17}H_{11}Cs$			
Cesium, 1H-benzo[a]fluoren-1-yl-	$(MeOCH_2)_2$	436(3.91)	104-1592-83
$C_{17}H_{11}Li$			
Lithium, 1H-benzo[a]fluoren-1-yl-	$(MeOCH_2)_2$	465(3.93)	104-1592-83
$C_{17}H_{11}N$			
10-Phenanthrenecarbonitrile, 1-ethenyl-	MeOH	252(4.74),308(4.12), 320(4.11)	39-1015-83B
$C_{17}H_{11}NO$			
Benzoxazole, 2-(2-naphthalenyl)-	C_6H_{12}	314(4.46)	41-0595-83
Naphth[2,3-d]oxazole, 2-phenyl-	C_6H_{12}	327(3.90)	41-0595-83
$C_{17}H_{11}NO_2$			
[1]Benzopyrano[4,3-b]pyrrol-4(1H)-one, 1-phenyl-	MeOH	230(4.47),276s(3.84), 286(3.95),308(4.11), 320(4.10)	5-0165-83
$C_{17}H_{11}NO_2S$			
4H-[1]Benzothieno[3,2-b]pyran-4-one, 2-amino-3-phenyl-	MeOH	205(4.46),235(4.42), 255(4.40),330(4.08)	103-1167-83
9H-1-Benzothiopyrano[2,3-d]oxazol-9-one, 2-(4-methylphenyl)-	$CHCl_3$	354(4.22),370(4.22)	83-0921-83
$C_{17}H_{11}NO_3$			
5H-Benzo[b]carbazole-6,11-dione, 8-methoxy- (6λ,8ε)	EtOH	212(4.26),221(--), 272s(--),276(--), 304s(--),355(--)	39-2409-83C
$C_{17}H_{11}NO_3S$			
Furo[2,3-g]quinoline, 9-(phenylsulfonyl)-	EtOH	237(4.86),327(4.32)	44-0774-83
2-Propen-1-one, 3-(5-nitrobenzo[b]thien-2-yl)-1-phenyl-	MeOH	250(4.03),335(4.28)	4-0129-83
$C_{17}H_{11}NO_4$			
4,5-Dioxodehydroasimilobine	EtOH	246(4.70),292s(4.14), 305(4.26),318(4.28), 459(4.23)	100-0761-83
$C_{17}H_{11}NO_7$			
Aristolochic acid	EtOH	250(4.43),318(4.08), 390(3.81)	162-0114-83
$C_{17}H_{11}N_2O_2S$			
Acridinium, 10-(5-formyl-2,3-dihydro-2-oxo-4-thiazolyl)-, chloride	EtOH	250(5.10),366(4.30)	103-0226-83
$C_{17}H_{11}N_3OS$			
2-Naphthalenol, 1-(2-benzothiazolyl-	acid	445(4.51)	140-0272-83
	neutral	490(4.58)	140-0272-83
	base	560(4.68)	140-0272-83
Pyrido[2,3-d]pyrimidin-4(1H)-one, 2,3-dihydro-1-(2-naphthalenyl)-2-thioxo-	MeOH	292(3.38)	73-3315-83
4H-Pyrido[3,2-e]-1,3-thiazin-4-one, 2-(2-naphthalenylamino)-	MeOH	225(3.12)	73-3315-83

Compound	Solvent	$\lambda_{max}(\log \epsilon)$	Ref.
$C_{17}H_{11}N_3O_4$			
[1]Benzopyrano[2,3-c]pyrazol-4(1H)-one, 3-methyl-1-(4-nitrophenyl)-	EtOH	218(4.56),290(4.35)	118-0214-83
$C_{17}H_{11}N_3S$			
Quinoxaline, 2-(2-phenyl-4-thiazolyl)-	EtOH	258(4.10),301(3.80), 349(3.57),361s(3.49)	106-0829-83
$C_{17}H_{11}N_5O_3S$			
Imidazo[2,1-b]thiazole-5,6-dione, 3-phenyl-, 6-[(4-nitrophenyl)hydrazone]	EtOH	380(4.23)	4-0639-83
$C_{17}H_{12}$			
Dicyclopenta[ef,kl]heptalene, 1-methyl-	MeOH	283(5.23),296s(4.58), 310(4.46),322s(3.83), 354s(3.71),370(3.97), 376s(3.79),386(3.93), 391(4.17),432s(2.90), 460s(3.23),492(3.78)	88-0781-83
Pyrene, 4-methyl-	C_6H_{12}	232(4.61),240(4.86), 252s(4.07),262(4.35), 274(4.56),306(3.99), 320(4.38),335(4.57)	54-0014-83
$C_{17}H_{12}BrNO$			
Acetamide, N-(1-bromo-4H-cyclopenta-[def]phenanthren-2-yl)-	C_6H_{12}	241(4.17),273(4.80), 309(3.95),330(3.05), 346(3.24),364(3.33)	18-1259-83
Acetamide, N-(2-bromo-4H-cyclopenta-[def]phenanthren-1-yl)-	C_6H_{12}	218(4.44),262(4.71), 297(4.08),308(4.10), 337(3.16),354(3.07)	18-1259-83
Acetamide, N-(8-bromo-4H-cyclopenta-[def]phenanthren-3-yl)-	C_6H_{12}	222(4.26),259(4.56), 311(4.16),355(3.10)	18-1259-83
Acetamide, N-(9-bromo-4H-cyclopenta-[def]phenanthren-8-yl)-	C_6H_{12}	238(4.48),257(4.69), 302(4.09),313(4.09), 335(3.23),353(3.15)	18-1259-83
$C_{17}H_{12}BrN_3O_3$			
Bicyclo[4.4.1]undeca-3,6,8,10-tetraene-2,5-dione, mono[(2-bromo-4-nitrophenyl)hydrazone]	CH_2Cl_2	419(4.66)	33-2369-83
	EtOH-KOH	624(4.44)	33-2369-83
$C_{17}H_{12}Br_2N_2O$			
Bicyclo[4.4.1]undeca-3,6,8,10-tetraene-2,5-dione, mono[(2,5-dibromophenyl)-hydrazone], (±)-	CH_2Cl_2	400(4.40)	33-2369-83
	EtOH-KOH	554(4.71)	33-2369-83
$C_{17}H_{12}Br_2O_2Te$			
[1,2]Oxatellurolo[2,3-b][1,2]oxatellurole-7-Te^{IV}, 7,7-dibromo-7,7-dihydro-2,5-diphenyl-	CH_2Cl_2	330(4.146),547(4.505)	44-5149-83
$C_{17}H_{12}ClFN_4O_3$			
3H-1,4-Benzodiazepine-2-carboxaldehyde, 7-chloro-5-(2-fluorophenyl)-α-nitro-, O-methyloxime	isoPrOH	215s(4.50),247(4.48), 280s(4.19),325s(3.70), 347(3.74)	4-0551-83
$C_{17}H_{12}ClNO_2$			
[1]Benzopyrano[4,3-b]pyrrol-4(1H)-one, 8-chloro-2,3-dihydro-1-phenyl-	MeOH	220(4.50),256(4.01), 320s(3.91),350(4.00)	5-0165-83

Compound	Solvent	$\lambda_{max}(\log \epsilon)$	Ref.
$C_{17}H_{12}ClN_3O_3$			
Bicyclo[4.4.1]undeca-3,6,8,10-tetraene-2,5-dione, mono[(4-chloro-2-nitrophenyl)hydrazone]	CH_2Cl_2	450(4.19)	138-0653-83
$C_{17}H_{12}ClN_3S$			
5-Thiazolecarbonitrile, 2-amino-4-[4-chloro-2-(phenylmethyl)phenyl]-	EtOH	215s(4.40),247(4.04), 303(3.92)	87-0100-83
$C_{17}H_{12}Cl_2N_4$			
4H-[1,2,4]Triazolo[4,3-a][1,4]benzodiazepine, 8-chloro-6-(2-chlorophenyl)-1-methyl-	MeOH	275s(3.93),290(3.20)	73-0123-83
$C_{17}H_{12}Cl_2O_2Te$			
[1,2]Oxatellurolo[2,3-b][1,2]oxatellurole-7-Te^{IV}, 7,7-dichloro-7,7-dihydro-2,5-diphenyl-	CH_2Cl_2	315(4.28),544(4.50)	44-5149-83
$C_{17}H_{12}Cl_2O_7$			
Spiro[benzofuran-2(3H),1'-[2,5]cyclohexadiene]-2'-carboxylic acid, 5,7-dichloro-4-hydroxy-6'-methoxy-6-methyl-3,4'-dioxo-, methyl ester (geodin)	n.s.g.	284(4.28)	162-0525-83
$C_{17}H_{12}Cl_3N_3O_4$			
2,4(1H,3H)-Pyrimidinedione, 5-[4-chloro-1-(3,5-dichlorophenyl)-2,5-dihydro-2,5-dioxo-1H-pyrrol-3-yl]-1,3,6-trimethyl-	MeOH	265(4.06),304(3.37)	89-0120-83S
$C_{17}H_{12}F_3N_3O_4S$			
2-Naphthalenesulfonic acid, 6-amino-4-hydroxy-3-[[3-(trifluoromethyl)phenyl]azo]-	pH 7.00	508(4.07)	33-2002-83
2-Naphthalenesulfonic acid, 6-amino-4-hydroxy-5-[[3-(trifluoromethyl)phenyl]azo]-	pH 7.00	490(4.15)	33-2002-83
$C_{17}H_{12}N_2$			
Cyclohepta[4,5]pyrrolo[1,2-a]imidazole, 2-phenyl-	EtOH	247(4.28),260s(4.23), 320(4.61),340s(4.45), 384(3.57),404(3.63), 429(3.58),470(2.70), 505(2.76),545(2.74), 587s(2.61),650s(2.28)	18-3703-83
Propanedinitrile, (bicyclo[5.4.1]dodeca-2,5,7,9,11-pentaen-4-ylideneethylidene)-	EtOH	493(4.43)	88-2151-83
$C_{17}H_{12}N_2O$			
Benzeneacetonitrile, 2-cyano-α-[(4-methoxyphenyl)methylene]-	EtOH	216(4.27),240s(4.09), 260s(3.68),336(4.39)	39-1813-83C
4-Isoquinolinecarbonitrile, 1-methoxy-3-phenyl-	EtOH	217(4.49),256(4.46), 308(4.13),325s(4.02), 337s(3.78)	39-1813-83C
Methanone, phenyl(4-phenyl-5-pyrimidinyl)-	EtOH	258(4.25),277s(4.08)	4-0649-83
Pyrimido[2,1-a]isoindol-4(6H)-one, 2-phenyl-	EtOH	203(4.42),261(4.51), 310s(3.80)	124-0755-83

Compound	Solvent	λmax(log ε)	Ref.
$C_{17}H_{12}N_2OS$			
4H-Indeno[1,2-c]isothiazol-4-one, 3-[(phenylmethyl)amino]-	EtOH	261(4.62),292(4.23), 320(4.28),330(3.82), 344(3.93)	95-1243-83
$C_{17}H_{12}N_2O_2$			
5H-Benzo[b]carbazole-6,11-dione, 8-amino-5-methyl-	EtOH	213(4.23),228s(4.08), 278s(4.18),290(4.28), 310s(4.04),374(3.79), 500(2.90)	39-2409-83C
[1]Benzopyrano[2,3-c]pyrazol-4(1H)-one, 3-methyl-1-phenyl-	EtOH	242(4.45),264(4.23)	118-0214-83
[1]Benzopyrano[4,3-c]pyrazol-4(1H)-one, 3-methyl-1-phenyl-	EtOH	261(4.09),272(4.09), 298(3.87),308(3.82)	88-0381-83
[1]Benzopyrano[4,3-c]pyrazol-4(2H)-one, 3-methyl-2-phenyl-	EtOH	282s(4.03),299(3.88)	88-0381-83
4-Isoquinolinecarbonitrile, 1,2-dihydro-3-(4-methoxyphenyl)-1-oxo-	EtOH	216(4.51),235s(4.13), 254(4.05),271s(3.88), 317(4.18)	39-1813-83C
Telazoline	EtOH	242s(4.52),251(4.53), 283(4.47),317s(3.93), 470(4.08)	100-0761-83
$C_{17}H_{12}N_2O_4$			
6H-[2]Benzoxepino[4,3-b]indol-11(12H)-one, 12-methyl-9-nitro-	EtOH	225(4.30),246(4.32), 337(4.20),406(3.70)	39-2409-83C
1,4-Naphthalenedione, 2-(methylamino)-	toluene	447(3.46)	150-0168-83S
3-(4-nitrophenyl)-	EtOH	463(--)	150-0168-83S
Spiro[2H-indene-2,2'-[2H]indole]-1,3'-(1'H,3H)-dione, 1'-methyl-6-nitro-	$CHCl_3$	260s(3.96),280s(3.73), 407(3.53)	39-2413-83C
$C_{17}H_{12}N_4$			
Propanedinitrile, (4-formylbenzylidene)-, phenylhydrazone	hexane	440(4.38)	140-0215-83
	EtOH	461(--)	140-0215-83
$C_{17}H_{12}N_4OS$			
4H-Thiazolo[3,2-a]-1,3,5-triazin-4-one, 2,3-dihydro-3-phenyl-2-(phenylimino)-	dioxan	234s(4.11),265(5.00), 322s(3.90)	48-0463-83
$C_{17}H_{12}N_4OS_2$			
Thiazolo[5,4-d]pyrimidin-7(4H)-one, 5,6-dihydro-6-phenyl-2-(phenylamino)-5-thioxo-	MeOH	224s(--),312(4.31), 356(4.35)	2-0243-83
$C_{17}H_{12}N_4O_2S_2$			
5H-Thiazolo[3,4-b][1,2,4]triazol-4-ium, 3-(methylthio)-6-(4-nitrophenyl)-1-phenyl-, hydroxide, inner salt	MeCN	257(4.32),298(4.42), 426(4.11)	103-0749-83
$C_{17}H_{12}N_4O_4S_2$			
2-Naphthalenesulfonic acid, 6-amino-	H_2O	544(4.30)	7-0155-83
3-(2-benzothiazolylazo)-4-hydroxy-	HCOOH-DMF	589(4.22)	7-0155-83
$C_{17}H_{12}O_2S$			
Benzeneacetic acid, α-(benzo[b]thien-2-ylmethylene)-, (E)-	MeOH	215(4.38),234(4.27), 258(3.91),265s(3.85), 323(4.48)	4-0005-83
$C_{17}H_{12}O_2Te$			
[1,2]Oxatellurolo[2,3-b][1,2]oxatellurole-7-Te^{IV}, 2,5-diphenyl-	CH_2Cl_2	424(4.61)	44-5149-83

Compound	Solvent	$\lambda_{max}(\log \epsilon)$	Ref.
$C_{17}H_{12}O_3$			
1,4-Naphthalenedione, 2-(4-methoxy-phenyl)-	C_6H_{12}	402(3.62)	150-0168-83S
	EtOH	411(--)	150-0168-83S
$C_{17}H_{12}O_4$			
Benzoic acid, 2-[(1-oxo-1H-2-benzo-pyran-3-yl)methyl]-	EtOH	230(3.69),240(3.457), 257(3.184),265(3.219), 274(3.165),326(3.845)	2-1108-83
1H-Indene-1,3-dione, 2,3-dihydro-2-(3,6-dimethyl-2,5-dioxocyclo-hexadien-1-yl)-	EtOH	234(4.26),254(4.36), 296(3.71),298(3.66)	39-1753-83C
$C_{17}H_{12}O_5$			
9,10-Anthracenedione, 1-acetoxy-2-methoxy-	EtOH?	374(3.80)	99-0420-83
photoinduced form	EtOH?	526(4.04)	99-0420-83
9,10-Anthracenedione, 1-acetoxy-4-methoxy-	EtOH?	402(3.78)	99-0420-83
photoinduced form	EtOH?	510(3.98)	99-0420-83
2,4,6-Cycloheptatrien-1-one, 4-[3-(1,3-benzodioxol-5-yl)-1-oxo-2-propenyl]-2-hydroxy-	MeOH	255(4.33),335(4.30)	18-3099-83
	MeOH-NaOH	270(4.27),350(4.39), 430s(3.79)	18-3099-83
$C_{17}H_{12}O_6$			
Aflatoxin B_1	EtOH	223(4.41),265(4.13), 362(4.34)	162-0027-83
8H-1,3-Dioxolo[4,5-g][1]benzopyran-8-one, 7-(4-hydroxyphenyl)-9-methoxy- (irisolone)	n.s.g.	270(4.62),330s(3.97)	162-0735-83
$C_{17}H_{12}O_7$			
Aflatoxin G_1	EtOH	243(4.06),257(4.00), 264(4.00),362(4.21)	162-0027-83
Aflatoxin M_1	EtOH	226(4.36),265(4.06), 357(4.28)	162-0028-83
$C_{17}H_{13}AsCl_2$			
1H-Arsole, 1-(dichloromethyl)-2,5-di-phenyl-	C_6H_{12}	213(4.23),244(4.19), 380(4.14)	101-0335-83I
$C_{17}H_{13}BrN_2O$			
Bicyclo[4.4.1]undeca-3,6,8,10-tetraene-2,5-dione, mono[(2-bromophenyl)hydra-zone], (±)-	CH_2Cl_2	406(4.39)	33-2369-83
	EtOH-KOH	538(4.58)	33-2369-83
$C_{17}H_{13}ClN_2O$			
Bicyclo[4.4.1]undeca-4,6,8,10-tetraene-2,3-dione, 2-[(4-chlorophenyl)hydra-zone]	CH_2Cl_2	468(4.16)	138-0653-83
$C_{17}H_{13}ClN_2O_2$			
Lonazolac	pH 13	281(4.38)	162-0797-83
	MeOH	281(4.39)	162-0797-83
$C_{17}H_{13}ClN_4$			
Alprazolam	EtOH	222(4.60)	162-0047-83
$C_{17}H_{13}F_3O_3$			
Acetic acid, trifluoro-, 2-methoxy-1,2-diphenylethenyl ester, (Z)-	EtOH	285(4.00)	89-0551-83S

Compound	Solvent	$\lambda_{max}(\log \epsilon)$	Ref.
$C_{17}H_{13}N$			
Benzeneacetonitrile, 2-ethenyl-α-(phenylmethylene)-, trans	MeOH	250(4.03),272(4.04), 295(4.15)	39-1015-83B
90% cis	MeOH	218(4.17),224(4.18), 249(4.12),275(4.15)	39-1015-83B
2-Butenenitrile, 4-phenyl-2-(phenylmethylene)-	MeCN	336(4.36)	118-0917-83
Cycloprop[a]indene-1a(1H)-carbonitrile, 6,6a-dihydro-1-phenyl-, exo	MeOH	229(4.06),259(3.63), 266(3.56),276(3.36)	39-1015-83B
$C_{17}H_{13}NOS$			
Benzeneacetamide, α-(benzo[b]thien-2-ylmethylene)-, (E)-	MeOH	215(4.45),234(4.29), 258(3.90),267s(3.84), 321(4.49)	4-0005-83
(Z)-	MeOH	218(4.34),232s(4.15), 265s(3.88),273(3.91), 325s(4.51),330(4.52), 345s(4.35)	4-0005-83
[1]Benzopyrano[4,3-b]pyrrole-4(1H)-thione, 2,3-dihydro-1-phenyl-	MeOH	232(4.17),250s(4.01), 290(4.02),408(4.28)	5-0165-83
[1]Benzothiopyrano[4,3-b]pyrrol-4(1H)-one, 2,3-dihydro-1-phenyl-	MeOH	240(4.31),266s(3.85), 368(3.86)	83-0897-83
$C_{17}H_{13}NO_2$			
[1]Benzopyrano[4,3-b]pyrrol-4(1H)-one, 2,3-dihydro-1-phenyl- (3λ,4ε)	MeOH	228s(4.05),306s(2.53), 340(4.02),?(4.05)	5-0165-83
Spiro[2H-indene-2,2'-[2H]indole]-1,3'(1'H,3H)-dione, 1'-methyl-	EtOH	230(4.36),241(4.39), 251s(4.34),299(4.66), 407(4.42)	39-2413-83C
$C_{17}H_{13}NO_3$			
6H-Anthra[1,9-cd]isoxazol-6-one, 5-(1-methylethoxy)-	EtOH	244(4.99),249(5.06), 446(4.12),465(4.12)	103-1057-83
6H-Anthra[1,9-cd]isoxazol-6-one, 5-propoxy-	EtOH	244(4.99),249(5.10), 445(4.14),466(4.13)	103-1057-83
Pratosine	MeOH	228(4.18),237(4.20), 248(4.39),254(4.12), 305s(3.99),335(3.88), 350(3.72)	102-2305-83
7-Quinolinecarboxylic acid, 6-hydroxy-5-phenyl-, methyl ester	EtOH	238(4.59),293(3.68), 370(3.62)	95-0631-83
4H-Quinolizin-4-one, 1-acetyl-2-hydroxy-3-phenyl-	MeOH	230(4.19),285(4.20), 386(4.02)	49-0349-83
Telitoxine	EtOH	233(4.29),243(4.30), 277(4.14),288(4.14), 298(4.10),307s(3.70), 322(3.55),350(3.41), 367(3.55)	100-0761-83
$C_{17}H_{13}NO_3S$			
Pyrrole, 2-benzoyl-1-(phenylsulfonyl)-	MeOH	248(4.19),288(4.08)	44-3214-83
Pyrrole, 3-benzoyl-1-(phenylsulfonyl)-	MeOH	248(4.28)	44-3214-83
$C_{17}H_{13}NO_4$			
3H,7H-Pyrano[2',3':7,8][1]benzopyrano-[2,3-b]pyridin-7-one, 6-hydroxy-3,3-dimethyl-	EtOH	236(4.16),279(4.26), 282(4.26),331(3.93)	39-0219-83C
$C_{17}H_{13}NO_5$			
2-Anthraceneacetamide, 9,10-dihydro-1,4-dihydroxy-N-methyl-9,10-dioxo-	dioxan	283(3.82),486(3.85)	104-0533-83

Compound	Solvent	λ_{max}(log ε)	Ref.
$C_{17}H_{13}NO_5$			
2,4,6-Cycloheptatrien-1-one, 4-[3-(1,3-benzodioxol-5-yl)-1-(hydroxyimino)-2-propenyl]-2-hydroxy-, lower melting	MeOH	245(4.53),333(4.50), 402(3.96)	18-3099-83
higher melting	MeOH	243(4.53),330(4.47), 402(3.90)	18-3099-83
$C_{17}H_{13}NO_5S$			
2-Naphthalenesulfonic acid, 7-(benzoylamino)-4-hydroxy-	MeOH	225(4.48),270(4.32), 310(3.82),348(3.45)	48-1002-83
$C_{17}H_{13}NO_6$			
2,4,6-Cycloheptatrien-1-one, 2-hydroxy-4-[3-(4-methoxy-3-nitrophenyl)-1-oxo-2-propenyl]-, (E)-	MeOH	252(4.26),326(4.30), 390s(3.70)	18-3099-83
	MeOH-NaOH	278(4.28),330(4.44), 440(3.74)	18-3099-83
$C_{17}H_{13}NO_8S_2$			
1,7-Naphthalenedisulfonic acid, 4-(benzoylamino)-5-hydroxy-	MeOH	206(4.35),245(4.52), 263(4.09),332(4.03), 346(4.09),355(4.00)	48-1002-83
2,7-Naphthalenedisulfonic acid, 4-(benzoylamino)-5-hydroxy-	MeOH	207(4.26),243(4.55), 326(3.90),339(3.92), 351(3.90)	48-1002-83
$C_{17}H_{13}NS$			
Dibenzo[b,f]cyclopropa[d]thiocin-11b(7H)-carbonitrile, 1,1a-dihydro-	MeOH	253(3.95),262s(3.89)	73-1898-83
Dibenzo[b,e]thiepin-11-carbonitrile, 11-ethenyl-6,11-dihydro-	MeOH	266(3.92)	73-1898-83
6,11-Ethanodibenzo[b,e]thiepin-11(6H)-carbonitrile	MeOH	262.5(3.92)	73-1898-83
$C_{17}H_{13}NS_2$			
[1]Benzothiopyrano[4,3-b]pyrrole-4(1H)-thione, 2,3-dihydro-1-phenyl-	MeOH	238(4.23),266(4.18), 396(4.25)	83-0897-83
$C_{17}H_{13}N_3$			
Cyclohept[d]imidazo[1,2-a]imidazole, 2-(4-methylphenyl)-	EtOH	235(4.42),255(4.31), 338(4.60),423(3.56), 428s(3.54)	18-3703-83
$C_{17}H_{13}N_3O$			
Methanone, (2-amino-4-phenyl-5-pyrimidinyl)phenyl-	EtOH	251(4.37),285s(4.09)	4-0649-83
$C_{17}H_{13}N_3O_3$			
Bicyclo[4.4.1]undeca-3,6,8,10-tetraene-2,5-dione, mono[(2-nitrophenyl)hydrazone]	CH_2Cl_2	441(4.24)	138-0653-83
$C_{17}H_{13}N_3O_3S$			
3-Thiophenecarbonitrile, 4-amino-2-(dimethylamino)-5-[(2-oxo-2H-1-benzopyran-3-yl)carbonyl]-	MeCN	348(4.26)	48-0551-83
$C_{17}H_{13}N_3O_8$			
Benzoic acid, 2-[bis[(5-nitro-2-furanyl)methyl]amino]-	MeOH	209s(1.09),220(3.66), 294(7.57)[sic]	73-2682-83
Benzoic acid, 4-[bis[(5-nitro-2-furanyl)methyl]amino]-	MeOH	220(4.03),294(8.19)[sic]	73-2682-83

Compound	Solvent	λ_{max}(log ε)	Ref.
$C_{17}H_{13}N_3S_2$			
5H-Thiazolo[3,4-b][1,2,4]triazol-4-ium, 3-(methylthio)-1,6-diphenyl-, hydroxide, inner salt	MeCN	283(4.43),416(4.24)	103-0749-83
$C_{17}H_{13}N_4O_2S_2$			
5H-Thiazolo[3,4-b][1,2,4]triazol-4-ium, 3-(methylthio)-6-(4-nitrophenyl)-1-phenyl-, perchlorate	HOAc	265(4.09),295(4.38), 420(4.09)	103-0749-83
$C_{17}H_{13}N_5O$			
Furo[2',3':4,5]pyrrolo[1,2-d]-1,2,4-triazolo[3,4-f][1,2,4]triazine, 3,6-dimethyl-9-phenyl-	MeOH	237(3.39),344(3.74)	73-1878-83
Furo[2',3':4,5]pyrrolo[1,2-d]-1,2,4-triazolo[3,4-f][1,2,4]triazine, 3-methyl-9-(4-methylphenyl)-	MeOH	238(2.96),347(3.34)	73-1878-83
$C_{17}H_{13}O$			
Pyrylium, 2,6-diphenyl-, perchlorate	MeCN	215(4.25),275(4.25), 400(4.44)	22-0115-83
$C_{17}H_{13}O_2$			
Cyclohept[a]azulenylium, 11-(methoxycarbonyl)-, tetrafluoroborate	MeCN	255(4.35),285s(4.19), 297s(4.32),332.5(4.87), 348s(4.57),460s(3.77), 490(4.38),560(3.21), 612(2.96),676(2.88)	88-0069-83
$C_{17}H_{14}$			
Dibenzo[a,e]cyclooctene, 5-methyl-	hexane-isooctane	236(4.45)	35-3226-83
3,4:6,7-Dibenzotricyclo[3.3.0.0^{2,8}]-octa-3,6-diene, 1-methyl-	hexane-isooctane	254(3.26),274(3.40), 282(3.27)	35-3226-83
3,4:6,7-Dibenzotricyclo[3.3.0.0^{2,8}]-octa-3,6-diene, 2-methyl-	hexane-isooctane	274(3.40),282(3.27)	35-3226-83
3,4:6,7-Dibenzotricyclo[3.3.0.0^{2,8}]-octa-3,6-diene, 5-methyl-	hexane-isooctane	274(3.39),282(3.29)	35-3226-83
Pyrene, 4,5-dihydro-4-methyl-	C_6H_{12}	258(4.57),280s(4.05), 286(4.01),298(3.99)	54-0014-83
$C_{17}H_{14}AsCl$			
1H-Arsole, 1-(chloromethyl)-2,5-diphenyl-	C_6H_{12}	238(4.15),375(4.19)	101-0335-83I
$C_{17}H_{14}BrFe$			
Methylium, (4-bromophenyl)ferrocenyl-	50% H_2SO_4	225s(--),255(4.09), 288(4.10),335(4.21)	32-0721-83
$C_{17}H_{14}BrNO_2$			
1,3-Dioxolo[4,5-g]isoquinoline, 5-[(2-bromophenyl)methyl]-7,8-dihydro-	EtOH	215s(3.46),230s(3.36), 318(2.91)	12-1061-83
$C_{17}H_{14}ClFN_4O$			
3H-1,4-Benzodiazepine-2-carboximidamide, 7-chloro-5-(2-fluorophenyl)-N-hydroxy-N'-methyl-	isoPrOH	226(4.51),319(3.97)	4-0551-83
$C_{17}H_{14}ClN_3O$			
Benzenepropanenitrile, 4-chloro-α-[[4-(dimethylamino)phenyl]imino]-β-oxo-	EtOH	490(4.5)	64-0930-83B

Compound	Solvent	$\lambda_{max}(\log \epsilon)$	Ref.
$C_{17}H_{14}Cl_2N_4O$			
Acetamide, N-[6-(2,6-dichlorophenyl)-2,8-dimethylpyrido[2,3-d]pyrimidin-7(8H)-ylidene]-	pH 1	283(3.87),347(4.26) (changing)	87-0403-83
	MeOH-pH 7	221(4.63),284(3.76), 341(4.23)	87-0403-83
	MeOH-NaOH	284(3.74),341(4.22), 350(4.22)	87-0403-83
Acetamide, N-[6-(2,6-dichlorophenyl)-2-methylpyrido[2,3-d]pyrimidin-7-yl]-N-methyl-	MeOH-pH 1	295(4.12)	87-0403-83
	MeOH-pH 7	214(4.69),320(4.04)	87-0403-83
	MeOH-NaOH	320(4.04)	87-0403-83
Acetic acid, 2-[7-chloro-5-(2-chlorophenyl)-3H-1,4-benzodiazepin-2-yl]-hydrazide	MeOH	280(4.21),315(3.68), 340(3.35)	73-0123-83
$C_{17}H_{14}Cl_2O_5$			
Spiro[benzofuran-2(3H),1'-cyclohexa-2',5'-diene]-3,4'-dione, 5,5'-dichloro-2',6-dimethoxy-4,6'-dimethyl-	$CHCl_3$	278(4.25),334(4.07)	39-0413-83C
Spiro[benzofuran-2(3H),1'-cyclohexa-3',5'-diene]-2',3-dione, 5,5'-dichloro-4',6-dimethoxy-4,6'-dimethyl-	$CHCl_3$	278(4.50),332(4.84)	39-0413-83C
$C_{17}H_{14}Cl_2O_7$			
Benzoic acid, 2-(3,5-dichloro-2,6-dihydroxy-4-methylbenzoyl)-5-hydroxy-3-methoxy-, methyl ester, (dihydrogeodin)	EtOH	283(3.82)	94-4543-83
$C_{17}H_{14}FFe$			
Methylium, ferrocenyl(4-fluorophenyl)-	50% H_2SO_4	255(4.07),280s(--), 334(4.15)	32-0721-83
$C_{17}H_{14}F_6N_4O_6$			
Riboflavin, $\alpha^7,\alpha^7,\alpha^7,\alpha^8,\alpha^8,\alpha^8$-hexafluoro-	n.s.g.	427(4.01)	64-0701-83C
$C_{17}H_{14}FeNO_2$			
Methylium, ferrocenyl(4-nitrophenyl)-	50% H_2SO_4	251(4.12),278(4.15), 325(4.20)	32-0721-83
$C_{17}H_{14}NS$			
1,3-Thiazin-1-ium, 2-methyl-4,6-diphenyl-, perchlorate	MeCN	257(4.09),283(3.97), 389(4.07)	118-0827-83
$C_{17}H_{14}NS_2$			
1,3-Thiazin-1-ium, 2-(methylthio)-4,6-diphenyl-, perchlorate	MeCN	266(4.28),372(4.57), 433(4.34)	118-0827-83
$C_{17}H_{14}N_2$			
Ellipticine	n.s.g.	239(4.23),277(4.61), 286(4.76),294(4.74), 332(3.65),382(3.61), 400(3.53)	162-0512-83
Flavopereirine	EtOH	230(4.40),238(4.43), 248(4.39),294(4.14), 351(4.25),390(4.14)	162-0587-83
perchlorate	EtOH-HCl	238(4.57),294(4.22), 350(4.31),389(4.21)	162-0587-83
1H,7H-Indolo[2',3':4,5]pyrrolo[3,2,1-ij]quinoline, 2,3-dihydro-	dioxan	213(4.22),240(4.24), 264(4.66),327(4.33)	103-0959-83

Compound	Solvent	λ_{max}(log ϵ)	Ref.
Olivacine	EtOH	224(4.39),238(4.33), 276(4.70),287(4.85), 292(4.83),314(3.66), 329(3.80),375(3.66)	162-0981-83
$C_{17}H_{14}N_2O$			
Azobenzene, 2-propanoyl-	EtOH	228(3.05),322(4.30), 445(2.73)	103-0559-83
4-Isoquinolinecarbonitrile, 3,4-dihydro-1-methoxy-3-phenyl-	EtOH	211(4.42),247(3.85), 262s(3.64)	39-1813-83C
Pyrazine, 2-methoxy-3,5-diphenyl-	EtOH	234(4.06),264(3.98), 335(3.87)	4-0311-83
2(1H)-Pyrimidinone, 6-methyl-1,4-diphenyl-	MeOH	209(4.35),234(4.28), 315(3.91)	39-1773-83C
2(1H)-Quinoxalinone, 3-[2-(4-methylphenyl)ethenyl]-	EtOH	394(4.47)	18-0326-83
$C_{17}H_{14}N_2OS_4$			
2-Propanol, 1,3-bis(2-benzothiazolylthio)-	EtOH	282(4.37),292(4.32), 302(4.25)	103-1170-83
$C_{17}H_{14}N_2O_3S$			
2-Thiophenecarboxylic acid, 3-[(5-oxobicyclo[4.4.1]undeca-3,6,8,10-tetraen-2-ylidene)hydrazino]-, methyl ester, (±)-	CH_2Cl_2	320(3.89),414(4.46)	33-2369-83
	EtOH-KOH	421(3.86),567(4.57)	33-2369-83
$C_{17}H_{14}N_2O_4$			
Acetamide, N-[3-acetoxy-4-(2-benzoxazolyl)phenyl]-	EtOH	316(4.51)	4-1517-83
Benzoic acid, 4-[5-(4-methoxyphenyl)-1,3,4-oxadiazol-2-yl]-, methyl ester	toluene	310(4.52)	103-0022-83
$C_{17}H_{14}N_2O_5S$			
Calmagite	pH 10.1	610(4.31)	162-0236-83
2-Naphthalenesulfonic acid, 7-[(4-aminobenzoyl)amino]-4-hydroxy-	MeOH	224(4.49),258(4.19), 301(4.38)	48-1002-83
$C_{17}H_{14}N_2O_8S_2$			
2,7-Naphthalenedisulfonic acid, 3-amino-5-(benzoylamino)-4-hydroxy-	MeOH	210(4.30),230(4.32), 242(4.31),269(4.28), 355(3.47)	48-1002-83
$C_{17}H_{14}N_3S_2$			
5H-Thiazolo[3,4-b][1,2,4]triazol-4-ium, 3-(methylthio)-1,6-diphenyl-, perchlorate	HOAc	280(4.38),398(4.15)	103-0749-83
$C_{17}H_{14}N_4$			
2,2'-Bi-1H-pyrrole, 5-[[5-(1H-pyrrol-2-yl)-2H-pyrrol-2-ylidene]methyl]-	MeOH	283(3.99),326(3.95), 337(3.99),410(3.57), 435(3.61),643(5.08)	5-0894-83
$C_{17}H_{14}N_4OS$			
Propanedinitrile, [[(4,6-dimethyl-2-pyrimidinyl)thio]phenylacetyl]-	MeOH	254(4.06),320(3.76)	103-0492-83
$C_{17}H_{14}N_6$			
6H-Imidazo[1,2-b][1,2,4]triazol-6-one, 5-(4-methylphenyl)-, phenylhydrazone	EtOH	270(3.55),350(4.04)	4-0639-83
6H-Imidazo[1,2-b][1,2,4]triazol-6-one,	EtOH	285s(3.41),360(4.44)	4-0639-83

Compound	Solvent	λ_{max}(log ϵ)	Ref.
5-phenyl-, (4-methylphenyl)hydrazone			4-0639-83
Tetrazolo[1',5':1,2]quino[3,4-b]quinoxaline, 9,14-dihydro-7,12-dimethyl-	MeCN	235(4.58),280(4.68), 390(4.08)	103-0222-83
$C_{17}H_{14}O$			
1,4-Pentadien-3-one, 1,5-diphenyl-, cis-cis	n.s.g.	287(4.04)	162-0435-83
cis-trans	n.s.g.	295(4.30)	162-0435-83
trans-trans	n.s.g.	330(4.54)	162-0435-83
$C_{17}H_{14}OS_3$			
Ethane(dithioic) acid, (methylthio)-9H-xanthen-9-ylidene-, methyl ester	CCl_4	332(2.92),428(2.22), 525(1.20)	54-0091-83
$C_{17}H_{14}O_2$			
9-Anthracenepropanoic acid	MeOH	366.5(4.00)	151-0131-83C
	EtOH	367.0(3.97)	151-0131-83C
	ether	366.5(4.03)	151-0131-83C
in cetyltrimethylammonium bromide	n.s.g.	371.0(3.94)	151-0131-83C
in sodium dodecyl sulfate	n.s.g.	369.0(3.94)	151-0131-83C
2-Cyclopenten-1-one, 4-hydroxy-3,4-diphenyl-	MeOH	285(4.36)	24-1309-83
5,10-Methanocyclopentacycloundecene-2-carboxylic acid, methyl ester	EtOH	350(4.6),425(3.6), 672(2.95)	88-2151-83
$C_{17}H_{14}O_2S$			
Propanoic acid, 2-(2-phenanthrenylthio)-, (S)-	MeCN	208s(--),220(4.29), 251(4.53),264(4.58), 274(4.55),294(4.21), 308s(--)	56-0767-83
$C_{17}H_{14}O_3$			
Benz[a]azulene-1,4-dione, 10-methoxy-2,3-dimethyl-	MeCN	262(4.4),283(4.5), 306(4.4),462(3.8), 687(3.2)	24-2408-83
2,4,6-Cycloheptatrien-1-one, 2-hydroxy-4-[3-(4-methylphenyl)-1-oxo-2-propenyl]-, (E)-	MeOH	253(4.33),328(4.27), 384s(3.87)	18-3099-83
	MeOH-NaOH	266(4.14),326(4.27), 440(3.57)	18-3099-83
1,3,9,11-Cyclotridecatetraene-5,7-diyne-1-carboxylic acid, 4,9-dimethyl-13-oxo-, methyl ester, (Z,E,Z,Z)-	ether	246s(4.03),260s(4.28), 277(4.47),390(3.08)	44-2379-83
$C_{17}H_{14}O_3Te$			
4H-1-Benzotellurin-4-one, 5,7-dimethoxy-2-phenyl-	CH_2Cl_2	381(3.81)	35-0883-83
$C_{17}H_{14}O_4$			
4H-1-Benzopyran-4-one, 2,3-dihydro-3-[(4-hydroxyphenyl)methylene]-7-methoxy- (bonducellin)	EtOH	318(4.16),358(4.21)	102-2835-83
	EtOH-NaOMe	270(3.69),312(3.95), 394(4.23)	102-2835-83
	EtOH-NaOAc	315(4.14),385(4.22)	102-2835-83
	EtOH-$AlCl_3$	318(4.28),358(4.36)	102-2835-83
	+ HCl	318(4.28),358(4.36)	102-2835-83
	EtOH-NaOAc-boric acid	318(4.36),358(4.43)	102-2835-83
4H-1-Benzopyran-4-one, 5,7-dimethoxy-2-phenyl-	EtOH	262(4.42),307(4.16)	102-0625-83
4H-1-Benzopyran-4-one, 3-hydroxy-8-methoxy-2-(4-methylphenyl)-	MeOH	242(4.23),256(4.28), 318(4.07),350(4.21)	118-0835-83

Compound	Solvent	$\lambda_{max}(\log \epsilon)$	Ref.
2,4,6-Cycloheptatrien-1-one, 2-hydroxy-3-[3-(4-methoxyphenyl)-1-oxo-2-propenyl]-	MeOH	250(4.45),348(4.38)	18-3099-83
	MeOH-NaOH	267(4.23),347(4.52), 440(3.79)	18-3099-83
2,4-Hexadienoic acid, 6-(1,3-dihydro-1,3-dioxo-2H-inden-2-ylidene)-, ethyl ester, (E,E)-	EtOH	212(4.26),241(4.20), 248s(--),265s(--), 272(4.18),313(4.81)	111-0447-83
1H-Indene-1,3(2H)-dione, 2-(4,4-dimethyl-2,6-dioxocyclohexylidene)-	MeCN	229(4.41),258(4.38), 323(3.42)	32-0507-83
$C_{17}H_{14}O_4P$			
Pyrylium, 2,6-diphenyl-4-phosphono-, perchlorate	MeCN	220(4.42),280(4.53), 310(4.07),410(4.51)	22-0115-83
$C_{17}H_{14}O_5$			
9,10-Anthracenedione, 2-(ethoxymethyl)-1,3-dihydroxy-	MeOH	240(3.08),244(4.08), 278(4.04),406(3.45)	94-2353-83
2H-1-Benzopyran-2-one, 4-(4-hydroxyphenyl)-5,7-dimethoxy-	MeOH	256(4.04),324(4.22)	102-1657-83
	MeOH-NaOMe	256(--),368(--)	102-1657-83
4H-1-Benzopyran-4-one, 2-(3,4-dimethoxyphenyl)-3-hydroxy-	MeOH	248(4.59),325(4.30), 363(4.66)	118-0835-83
4H-1-Benzopyran-4-one, 5-hydroxy-3,7-dimethoxy-2-phenyl-	EtOH	267(4.45),325(4.06)	102-0625-83
	$EtOH-AlCl_3$-HCl	250(4.24),280(4.42), 329(4.116),398(3.89)	102-0625-83
4H-1-Benzopyran-4-one, 5-hydroxy-2-(4-hydroxyphenyl)-7-methoxy-6-methyl- (relative absorbance given)	MeOH	274(0.85),332(1.00)	102-2107-83
	$MeOH-AlCl_3$	263s(0.48),288(0.84), 301(0.87),351(1.00), 390s(0.45)	102-2107-83
4H-1-Benzopyran-4-one, 5-hydroxy-7-methoxy-2-(4-methoxyphenyl)- (relative absorbance given)	MeOH	270(0.93),298s(0.70), 329(1.00)	102-2107-83
	$MeOH-AlCl_3$	262s(0.53),280(0.89), 294s(0.77),303(0.83), 344(1.00),382(0.64)	102-2107-83
	$MeOH-AlCl_3$-HCl	264s(0.54),278(0.85), 294s(0.71),304(0.80), 342(1.00),384(0.69)	102-2107-83
2,4,6-Cycloheptatrien-1-one, 2-hydroxy-4-[3-(3-hydroxy-4-methoxyphenyl)-1-oxo-2-propenyl]-, (E)-	MeOH	258(4.35),350(4.41)	18-3099-83
	MeOH-NaOH	268(4.39),345(4.43), 426(4.28)	18-3099-83
$C_{17}H_{14}O_6$			
Aflatoxin B_2	EtOH	265(4.07),363(4.37)	162-0027-83
9,10-Anthracenedione, 4,5-dihydroxy-1,2-dimethoxy-7-methyl-	MeOH	230(4.48),257(4.32), 295(4.01),440(4.02)	102-2583-83
4H-1-Benzopyran-4-one, 5,6-dihydroxy-7-methoxy-2-(4-methoxyphenyl)- (relative absorbance given)	MeOH	286(0.89),331(1.00)	102-2107-83
	$MeOH-AlCl_3$	265s(0.29),292s(0.73), 303(0.86),362(1.00)	102-2107-83
	$MeOH-AlCl_3$-HCl	264s(0.32),292s(0.78), 302(0.89),358(1.00)	102-2107-83
4H-1-Benzopyran-4-one, 5,7-dihydroxy-2-(3,4-dimethoxyphenyl)- (relative absorbance given)	MeOH	240(0.92),248s(0.87), 269(0.87),291s(0.58), 340(1.00)	102-2107-83
	$MeOH-AlCl_3$	261(0.91),276(0.92), 295(0.59),359(1.00), 387(0.97)	102-2107-83
	$MeOH-AlCl_3$-HCl	259(0.77),279(0.90), 293s(0.69),348(1.00), 381s(0.72)	102-2107-83
4H-1-Benzopyran-4-one, 5,7-dihydroxy-2-(4-hydroxyphenyl)-3-methoxy-6-methyl- (relative absorbance given)	MeOH	271(1.00),338(0.96)	102-2107-83
	$MeOH-AlCl_3$	260s(0.84),280(0.96), 298s(0.84),306s(0.84), 363(1.00),403s(0.76)	102-2107-83

Compound	Solvent	λ_{max}(log ϵ)	Ref.
(cont.)	MeOH-$AlCl_3$-HCl	260s(0.72),282(0.96), 306s(0.82),358(1.00), 404s(0.65)	102-2107-83
4H-1-Benzopyran-4-one, 5,7-dihydroxy-3-methoxy-2-(4-methoxyphenyl)- (relative absorbance given)	MeOH	268(1.00),298s(0.65), 320s(0.74),347(0.82)	102-2107-83
	MeOH-$AlCl_3$	278(1.00),296s(0.65), 304(0.72),350(0.92), 400(0.77)	102-2107-83
	MeOH-$AlCl_3$-HCl	278(1.00),297(0.71), 304(0.72),345(0.90), 399(0.70)	102-2107-83
4H-1-Benzopyran-4-one, 5,7-dihydroxy-6-methoxy-2-(4-methoxyphenyl)- (pectolinarigenin)(relative absorbance given)	MeOH	276(0.78),330(1.00)	102-2107-83
	MeOH-$AlCl_3$	263s(0.41),280s(0.66), 303(0.77),352(1.00)	102-2107-83
	MeOH-$AlCl_3$-HCl	262s(0.42),293s(0.85), 301(0.86),350(1.00)	102-2107-83
4H-1-Benzopyran-4-one, 5,7-dihydroxy-8-methoxy-2-(4-methoxyphenyl)- (relative absorbance given)	MeOH	275(1.00),299(0.85), 322(0.83),358s(0.51)	102-2107-83
	MeOH-$AlCl_3$	284(0.84),310(0.97), 346(1.00),398(0.43)	102-2107-83
	MeOH-$AlCl_3$-HCl	285(0.88),309(1.00), 342(0.94),398(0.40)	102-2107-83
4H-1-Benzopyran-4-one, 5,8-dihydroxy-7-methoxy-2-(4-methoxyphenyl)-	EtOH	282(4.34),308(4.39), 325s(4.28),363s(3.85)	18-3773-83
	EtOH-NaOAc	282(4.34),308(4.40), 325s(4.29),363s(3.84)	18-3773-83
	EtOH-$AlCl_3$	288(4.25),318(4.43), 343s(4.28),407(3.77)	18-3773-83
4H-1-Benzopyran-4-one, 5-hydroxy-2-(3-hydroxy-4-methoxyphenyl)-7-methoxy- (pilloin)(relative absorbance given)	MeOH	244(0.84),252(0.92), 268(0.84),293s(0.62), 344(1.00)	102-2107-83
	MeOH-$AlCl_3$	266(0.96),275(1.00), 296(0.70),360(0.93), 386(0.93)	102-2107-83
	MeOH-$AlCl_3$-HCl	262(0.96),276(1.00), 296s(0.76),355(0.96), 384(0.88)	102-2107-83
4H-1-Benzopyran-4-one, 5-hydroxy-2-(4-hydroxyphenyl)-6,7-dimethoxy- (cirsimaritin)(relative absorbance given)	MeOH	277(0.73),331(1.00)	102-2107-83
	MeOH-$AlCl_3$	264s(0.39),280s(0.65), 296s(0.76),303(0.77), 356(1.00)	102-2107-83
	MeOH-$AlCl_3$-HCl	263s(0.40),295s(0.85), 303(0.86),353(1.00)	102-2107-83
C H O			
Aflatoxin G_2	EtOH	265(3.99),363(4.32)	162-0027-83
Aflatoxin M_2	EtOH	221(4.30),264(4.04), 357(4.32)	162-0028-83
4H-1-Benzopyran-4-one, 5,7-dihydroxy-2-(4-hydroxy-3,5-dimethoxyphenyl)- (tricin)(relative absorbance given)	MeOH	244s(0.87),268(0.69), 298s(0.49),350(1.00)	102-2107-83
	MeOH-$AlCl_3$	256s(0.75),277(0.82), 307(0.82),365s(0.90), 393(1.00)	102-2107-83
	MeOH-$AlCl_3$-HCl	257(0.79),278(0.90), 306(0.54),362(1.00), 390(0.98)	102-2107-83
4H-1-Benzopyran-4-one, 5,6-dihydroxy-2-(3-hydroxy-4-methoxyphenyl)-7-methoxy-	MeOH	246s(0.69),284(0.75), 341(1.00)	102-2107-83
	MeOH-$AlCl_3$	261(0.59),297(0.77), 373(1.00)	102-2107-83

Compound	Solvent	$\lambda_{max}(\log \epsilon)$	Ref.
(cont.)	MeOH-AlCl$_3$-HCl	259(0.61),296(0.81), 367(1.00)	102-2107-83
4H-1-Benzopyran-4-one, 5,7-dihydroxy-2-(3-hydroxy-4-methoxyphenyl)-3-methoxy- (relative absorbance given)	MeOH	253(1.00),266s(0.87), 293s(0.53),353(0.91)	102-2107-83
	MeOH-AlCl$_3$	268(1.00),278s(0.95), 302(0.59),366(0.69), 406(0.77)	102-2107-83
	MeOH-AlCl$_3$-HCl	268(1.00),277s(0.98), 300(0.57),360(0.74), 402(0.74)	102-2107-83
4H-1-Benzopyran-4-one, 5,7-dihydroxy-2-(3-hydroxy-4-methoxyphenyl)-6-methoxy- (relative absorbance given)	MeOH	244s(0.71),254s(0.70), 274(0.70),342(1.00)	102-2107-83
	MeOH-AlCl$_3$	262(0.66),284(0.70), 298s(0.69),370(1.00)	102-2107-83
	MeOH-AlCl$_3$-HCl	260(0.66),286(0.77), 296s(0.76),364(1.00)	102-2107-83
4H-1-Benzopyran-4-one, 5,7-dihydroxy-2-(4-hydroxy-3-methoxyphenyl)-3-methoxy-	EtOH	256(4.24),270(4.21), 360(4.24)	102-2881-83
	EtOH-NaOEt	277(4.30),334(3.98), 416(4.43)	102-2881-83
	EtOH-NaOAc	279(4.32),321(4.03), 389(4.14)	102-2881-83
	EtOH-NaOAc-boric acid	257(4.21),270(4.24), 360(4.30)	102-2881-83
	EtOH-AlCl$_3$	268(4.25),277(4.22), 301(3.85),367(4.12), 406(4.17)	102-2881-83
	EtOH-AlCl$_3$-HCl	268(4.16),278(4.18), 299(3.87),357(4.11), 402(4.08)	102-2881-83
relative absorbance given	MeOH	256(1.00),268s(0.84), 292s(0.46),356(0.88)	102-2107-83
	MeOH-AlCl$_3$	267(1.00),275s(0.95), 300s(0.44),354(0.72), 404(0.77)	102-2107-83
	MeOH-AlCl$_3$-HCl	264(1.00),276(0.98), 298s(0.50),359(0.75), 402(0.74)	102-2107-83
4H-1-Benzopyran-4-one, 5,7-dihydroxy-2-(4-hydroxy-3-methoxyphenyl)-6-methoxy- (jaceosidin) (relative absorbance given)	MeOH	245s(0.71),275(0.67), 344(1.00)	102-2107-83
	MeOH-AlCl$_3$	261(0.68),284(0.68), 298s(0.62),370(1.00)	102-2107-83
	MeOH-AlCl$_3$-HCl	258(0.69),286(0.73), 298s(0.69),366(1.00)	102-2107-83
4H-1-Benzopyran-4-one, 5,7-dihydroxy-2-(4-hydroxy-3-methoxyphenyl)-8-methoxy- (relative absorbance given)	MeOH	253(0.89),273(1.00), 294s(0.65),337(0.85)	102-2107-83
	MeOH-AlCl$_3$	262s(0.84),282(1.00), 304s(0.67),362(1.00), 401(0.72)	102-2107-83
	MeOH-AlCl$_3$-HCl	262s(0.82),284(0.98), 304s(0.71),356(1.00), 400(0.61)	102-2107-83
4H-1-Benzopyran-4-one, 5,8-dihydroxy-2-(3-hydroxy-4-methoxyphenyl)-7-methoxy-	EtOH	279(4.34),298s(4.21), 339(4.24)	18-3773-83
	EtOH-NaOAc	279(4.33),298s(4.22), 337(4.25)	18-3773-83
	EtOH-AlCl$_3$	286(4.29),307(4.23), 354(4.28),416(3.87)	18-3773-83
4H-1-Benzopyran-4-one, 5,8-dihydroxy-2-(4-hydroxy-3-methoxyphenyl)-7-methoxy-	EtOH	255(4.13),279(4.26), 340(4.24)	18-3773-83
	EtOH-NaOAc	277(4.24),341(4.14),	18-3773-83

Compound	Solvent	$\lambda_{max}(\log \epsilon)$	Ref.
(cont.)		410(4.12)	18-3773-83
	EtOH-AlCl$_3$	260s(4.08),286(4.21), 328s(4.19),356(4.28), 412(3.93)	18-3773-83
4H-1-Benzopyran-4-one, 5,7-dihydroxy-2-(4-hydroxyphenyl)-3,6-dimethoxy- (relative absorbance given)	MeOH	271(0.84),342(1.00)	102-2107-83
	MeOH-AlCl$_3$	268s(0.62),280(0.79), 298s(0.63),305s(0.62), 367(1.00),404s(0.78)	102-2107-83
	MeOH-AlCl$_3$-HCl	262s(0.63),282(0.86), 305s(0.68),364(1.00), 404s(0.67)	102-2107-83
4H-1-Benzopyran-4-one, 5,7-dihydroxy-2-(4-hydroxyphenyl)-3,8-dimethoxy- (relative absorbance given)	EtOH	274(1.00),305s(0.75), 327(0.82),359(0.75)	102-2107-83
	EtOH-AlCl$_3$-HCl	284(1.00),313(0.86), 351(0.95),415(0.65)	102-2107-83
4H-1-Benzopyran-4-one, 5,7-dihydroxy-2-(4-hydroxyphenyl)-6,8-dimethoxy- (relative absorbance given)	MeOH	282(0.90),296s(0.87), 332(1.00)	102-2107-83
	MeOH-AlCl$_3$	266(0.46),290s(0.71), 310(0.92),360(1.00), 412s(0.36)	102-2107-83
	MeOH-AlCl$_3$-HCl	266s(0.44),290s(0.77), 310(0.98),356(1.00), 412s(0.27)	102-2107-83
4H-1-Benzopyran-4-one, 2-(3,4-dihydroxyphenyl)-5-hydroxy-3,7-dimethoxy- (relative absorbance given)	MeOH	258(1.00),270s(0.77), 296s(0.37),357(0.90)	102-2107-83
	MeOH-AlCl$_3$	276(1.00),302s(0.28), 320s(0.20),440(0.96)	102-2107-83
	MeOH-AlCl$_3$-HCl	272(1.00),278s(0.96), 302s(0.36),368(0.60), 402(0.76)	102-2107-83
4H-1-Benzopyran-4-one, 2-(3,4-dihydroxyphenyl)-5-hydroxy-6,7-dimethoxy- (relative absorbance given)	MeOH	243(0.80),256(0.79), 274(0.79),346(1.00)	102-2107-83
	MeOH-AlCl$_3$	275(0.70),302s(0.63), 338(0.52),428(1.00)	102-2107-83
	MeOH-AlCl$_3$-HCl	262s(0.80),286(0.85), 297s(0.84),372(1.00)	102-2107-83
4H-1-Benzopyran-4-one, 2-(3,4-dimethoxyphenyl)-5,7,8-trihydroxy-	EtOH	286(4.30),317(4.28)	18-3773-83
	EtOH-NaOAc	298s(4.28),317(4.31)	18-3773-83
	EtOH-AlCl$_3$	292(4.24),327(4.28), 347s(4.26),405s(3.77)	18-3773-83
$C_{17}H_{14}O_8$			
4H-1-Benzopyran-4-one, 5,7-dihydroxy-3-methoxy-6-methyl-2-(3,4,5-trihydroxyphenyl)- (alluandiol) (relative absorbance given)	MeOH	257(0.91),266s(0.82), 300s(0.48),360(1.00)	102-2107-83
	MeOH-AlCl$_3$	272(1.00),322(0.42), 374s(0.63),430(0.96)	102-2107-83
	MeOH-AlCl$_3$-HCl	262s(0.87),277(0.90), 313(0.62),369(1.00), 414s(0.60)	102-2107-83
4H-1-Benzopyran-4-one, 2-(3,4-dihydroxyphenyl)-5,6-dihydroxy-7,8-dimethoxy-	EtOH	258(4.12),291(4.25), 402(4.42)	40-0161-83
	EtOH-NaOAc	273(4.17),408(4.24)	40-0161-83
	EtOH-AlCl$_3$	263(4.09),305(4.27), 374(4.37)	40-0161-83
	EtOH-AlCl$_3$-HCl	262(4.14),304(4.30), 370(4.44)	40-0161-83
4H-1-Benzopyran-4-one, 2-(3,5-dihydroxy-4-methoxyphenyl)-5,7-dihydroxy-3-methoxy- (relative absorbance given)	MeOH	267(1.00),294s(0.49), 350(0.74)	102-2107-83
	MeOH-AlCl$_3$	275(1.00),304(0.48), 349(0.64),399(0.57)	102-2107-83

Compound	Solvent	$\lambda_{max}(\log \epsilon)$	Ref.
(cont.)	MeOH-AlCl$_3$-HCl	275(1.00),302(0.45), 346(0.65),398(0.50)	102-2107-83
4H-1-Benzopyran-4-one, 2-(3,5-dihydroxy-4-methoxyphenyl)-5,7-dihydroxy-8-methoxy- (relative absorbance)	MeOH	275(1.00),327(0.70)	102-2107-83
	MeOH-AlCl$_3$	285(--),310(--), 350(--)	102-2107-83
	MeOH-AlCl$_3$-HCl	285(--),305(--), 345(--)	102-2107-83
4H-1-Benzopyran-4-one, 2-(3,4-dihydroxyphenyl)-5,6-dihydroxy-3,7-dimethoxy- (relative absorbance given)	MeOH	259(0.84),279(0.82), 348(1.00)	102-2107-83
	MeOH-AlCl$_3$	281(0.83),300s(0.65), 344(0.52),436(1.00)	102-2107-83
	MeOH-AlCl$_3$-HCl	267(0.85),295(0.86), 379(1.00)	102-2107-83
4H-1-Benzopyran-4-one, 2-(3,4-dihydroxyphenyl)-5,7-dihydroxy-3,6-dimethoxy- (axillarin) (relative absorbance given)	MeOH	260(0.97),272s(0.88), 354(1.00)	102-2107-83
	MeOH-AlCl$_3$	278(0.92),306s(0.49), 342s(0.39),440(1.00)	102-2107-83
	MeOH-AlCl$_3$-HCl	269(0.98),282s(0.93), 302s(0.71),372(1.00), 410s(0.84)	102-2107-83
4H-1-Benzopyran-4-one, 2-(3,4-dihydroxyphenyl)-5,7-dihydroxy-3,8-dimethoxy- (relative absorbance given)	MeOH	261(1.00),273(0.99), 298s(0.60),336s(0.73), 363(0.84)	102-2107-83
	MeOH-AlCl$_3$	281(1.00),312s(0.39), 342s(0.35),451(0.92)	102-2107-83
	MeOH-AlCl$_3$-HCl	272s(0.99),282(1.00), 306s(0.62),361(0.78), 418(0.39)	102-2107-83
4H-1-Benzopyran-4-one, 2-(3,4-dihydroxyphenyl)-5,7-dihydroxy-6,8-dimethoxy- (relative absorbance)	MeOH	256s(0.74),280(0.85), 346(1.00)	102-2107-83
	MeOH-AlCl$_3$	275(--),340(--), 430(--)	102-2107-83
	MeOH-AlCl$_3$-HCl	260(--),300(--), 370(--)	102-2107-83
4H-1-Benzopyran-4-one, 5,6,7-trihydroxy-2-(4-hydroxy-3-methoxyphenyl)-3-methoxy- (relative absorbance given)	MeOH	240s(0.74),257s(0.66), 280(0.69),349(1.00)	102-2107-83
	MeOH-AlCl$_3$	260s(0.58),302(0.66), 312s(0.61),396(1.00)	102-2107-83
	MeOH-AlCl$_3$-HCl	262(0.63),296(0.70), 376(1.00)	102-2107-83
Carbonic acid, 4,9-dimethoxy-7-methyl-1,3-dioxo-1H,3H-naphtho[1,8-cd]-pyran-6-yl methyl ester	EtOH	257(4.18),323s(3.36), 340(3.58),370(3.57), 384(3.52)	78-2283-83
	EtOH-NaOH	249(4.30),304(3.46), 312(3.45),345s(3.11)	78-2283-83
$C_{17}H_{14}S_4$			
Ethane(dithioic) acid, (methylthio)-9H-thioxanthen-9-ylidene-, methyl ester	CCl$_4$	323(3.10),412(2.43), 510(1.38)	54-0091-83
11bH-Thiopyrano[4,3,2-kl]thioxanthene, 2,3-bis(methylthio)-	CCl$_4$	293(3.18),358(2.85)	54-0091-83
$C_{17}H_{15}As$			
1H-Arsole, 1-methyl-2,5-diphenyl-	C_6H_{12}	218(4.16),367(4.27)	101-0335-83I
$C_{17}H_{15}Br$			
Benzene, 1,1'-(2-bromo-1,3-pentadienylidene)bis-	hexane	237(4.06),285(4.09)	104-1438-83
1H-Indene, 2-bromo-1-ethyl-3-phenyl-	hexane	271(3.95)	104-1438-83

Compound	Solvent	$\lambda_{max}(\log \epsilon)$	Ref.
$C_{17}H_{15}BrFeO$			
Ferrocene, [(4-bromophenyl)hydroxymethyl]-	EtOH	443(2.05)	32-0721-83
$C_{17}H_{15}BrO_2$			
Furan, 2,2'-[(4-bromophenyl)methylene]-bis[5-methyl-	EtOH	225(4.31)	103-0478-83
$C_{17}H_{15}Cl$			
Benzene, 1,1'-(2-chloro-1,3-pentadienylidene)bis-, (Z)-	n.s.g.	290(4.04)	104-1438-83
$C_{17}H_{15}ClN_2O_2S_2$			
2-Propanol, 1-(2-benzothiazolylthio)-3-chloro-, phenylcarbamate	EtOH	290(4.00),297(4.07), 300(3.90)	103-1173-83
$C_{17}H_{15}ClO_2$			
Furan, 2,2'-[(4-chlorophenyl)methylene]bis[5-methyl-	EtOH	224(4.37)	103-0478-83
$C_{17}H_{15}ClO_7$			
Benzoic acid, 2-(3-chloro-2,6-dihydroxy-4-methylbenzoyl)-5-hydroxy-3-methoxy-, methyl ester	EtOH	282(3.58)	94-4543-83
$C_{17}H_{15}Cl_2N_5$			
Methanimidamide, N'-[6-(2,6-dichlorophenyl)-2-methylpyrido[2,3-d]pyrimidin-7-yl]-N,N-dimethyl-	MeOH	240(--),260s(--), 289(3.96),365(4.44)	87-0403-83
$C_{17}H_{15}Cl_2N_5O$			
Urea, N-[6-(2,6-dichlorophenyl)-2-methylpyrido[2,3-d]pyrimidin-7-yl[N'-ethyl-, (hemihydrate)	MeOH	224(4.70),320(4.19), 330(4.19)	87-0403-83
$C_{17}H_{15}Cl_2N_5S$			
Carbamimidothioic acid, N-[6-(2,6-dichlorophenyl)-2-methylpyrido[2,3-d]-pyrimidin-7-yl]-N'-methyl-, methyl ester	MeOH	289(3.99),360(4.50), 374(4.51)	87-0403-83
$C_{17}H_{15}Cl_3N_2O_5S$			
5-Thia-1-azabicyclo[4.2.0]oct-2-ene-2-carboxylic acid, 7-(formylamino)-8-oxo-3-(phenylmethyl)-, 2,2,2-trichloroethyl ester, 5-oxide, [5S-(5α,6β,7α)]-	EtOH	262s(3.92),268s(3.97), 272.5(3.99)	78-0461-83
$C_{17}H_{15}FFeO$			
Ferrocene, [(4-fluorophenyl)hydroxymethyl]-	EtOH	440(2.03)	32-0721-83
$C_{17}H_{15}FeNO_3$			
Ferrocene, [hydroxy(4-nitrophenyl)-methyl]-	EtOH	446(2.10)	32-0721-83
$C_{17}H_{15}NO$			
1H-Indene-2-carboxamide, N-(4-methylphenyl)-	dioxan	303(4.3)	104-1363-83
Isoquinoline, 3-ethoxy-1-phenyl-	EtOH	279(3.81),348(3.93)	44-1275-83

Compound	Solvent	λ_{max}(log ϵ)	Ref.
2H-Pyrrol-2-one, 1,5-dihydro-4-methyl-3,5-diphenyl-	MeOH	230(4.17)	4-0687-83
$C_{17}H_{15}NOS$			
[1]Benzothieno[3,2-b]azete, 2-ethoxy-2a,7b-dihydro-7b-phenyl-	EtOH	251(3.93),291(3.15), 300(3.08)	44-1275-83
$C_{17}H_{15}NO_2$			
Benzeneacetonitrile, 4-methoxy-α-[(4-methoxyphenyl)methylene]-	MeOH	205(4.05),233(4.16), 342(4.40)	83-0271-83
1H-Indene-1,3(2H)-dione, 2-[4-(dimethylamino)phenyl]-	C_6H_{12}	225(4.3),250(4.2), 300(3.3)	151-0061-83C
	pH 12	297(4.5),345(4.1)	151-0061-83C
Methanone, [4,5-dihydro-1-phenyl-1H-pyrrol-3-yl)(2-hydroxyphenyl)-	MeOH	220s(4.16),250s(3.87), 358(4.45)	5-0165-83
5(4H)-Oxazolone, 2-methyl-4-phenyl-4-(phenylmethyl)-	C_6H_{12}	256(3.20)	44-0695-83
2-Pyrrolidinone, 3-[(2-hydroxyphenyl)-methylene]-1-phenyl-	MeOH	308s(4.41),338(4.27)	5-0165-83
2H-Pyrrol-2-one, 1,5-dihydro-5-hydroxy-4-methyl-3,5-diphenyl-	MeOH	260(3.80)	4-0687-83
$C_{17}H_{15}NO_3$			
2,4,6-Cycloheptatrien-1-one, 2-hydroxy-4-[1-(hydroxyimino)-3-(4-methylphenyl)-2-propenyl]-, (?,E)-	MeOH	231(4.43),302(4.45), 370(3.70)	18-3099-83
	MeOH-NaOH	230(4.21),301(4.28), 402(3.51)	18-3099-83
1H-Furo[3,4-e]indole-1,3(3aH)-dione, 4,5,6,8b-tetrahydro-6-methyl-4-phenyl-	MeOH	229(3.77)	44-2488-83
1H-Indole-3-carboxylic acid, 2-(4-methoxyphenyl)-, methyl ester	EtOH	219(4.63),248(4.45), 305(4.30)	12-1419-83
Noroliveroline	EtOH	233s(4.09),245s(3.99), 265s(4.01),272(4.07), 281s(3.97),315(3.51)	100-0761-83
$C_{17}H_{15}NO_3S$			
1-Naphthalenesulfonic acid, 8-[(methylphenyl)amino]-	dioxan	255(4.20),340s(--), 365(3.32)	35-6236-83
$C_{17}H_{15}NO_4$			
3H-Benz[e]indole-1-carboxylic acid, 4,5-dihydro-2,3-dimethyl-4,5-dioxo-, ethyl ester	EtOH	210(4.41),266(4.37), 380(3.80)	103-1086-83
2,4,6-Cycloheptatrien-1-one, 4-[1-amino-3-(1,3-benzodioxol-5-yl)-2-propenyl]-2-hydroxy-, (E)-	MeOH	245(4.41),317(4.12), 395(3.72)	18-3099-83
	MeOH-HCl	237(4.46),317(4.17), 355s(3.82)	18-3099-83
2,4,6-Cycloheptatrien-1-one, 2-hydroxy-4-[1-(hydroxyimino)-3-(4-methoxyphenyl)-2-propenyl]-, (?,E)-	MeOH	233(4.44),317(4.50), 378(3.83)	18-3099-83
Furan, 2,2'-[(3-nitrophenyl)methylene]-bis[5-methyl-	EtOH	231(4.24)	103-0478-83
Furan, 2,2'-[(4-nitrophenyl)methylene]-bis[5-methyl-	EtOH	223(4.26)	103-0478-83
Norcularicine, (+)-	MeOH	206(4.57),224s(4.23), 287(3.79)	100-0881-83
$C_{17}H_{15}NO_5$			
2,3-Furandicarboxylic acid, 5-[3-(phenylimino)-1-propenyl]-, dimethyl ester, (E,E)-	EtOH	208(4.15),227(4.16), 325(4.31),353(4.22)	77-1216-83

Compound	Solvent	$\lambda_{max}(\log \epsilon)$	Ref.
6-Oxa-2-azaspiro[4.5]deca-3,7-diene-4-carboxylic acid, 1,9-dioxo-3-phenyl-, ethyl ester	EtOH	250(4.15),297(3.89)	94-0356-83
$C_{17}H_{15}NO_6$			
3,5-Isoxazoledicarboxylic acid, 4,5-dihydro-4-(1-naphthalenyl)-, dimethyl ester, 2-oxide	MeOH	273(4.20)	40-1678-83
3,5-Isoxazoledicarboxylic acid, 4,5-dihydro-4-(2-naphthalenyl)-, dimethyl ester, 2-oxide	MeOH	268(4.20)	40-1678-83
$C_{17}H_{15}NO_9S_3$			
2,7-Naphthalenedisulfonic acid, 4-hydroxy-5-[[(4-methylphenyl)sulfonyl]-amino]-	MeOH	206(4.32),226(4.68), 240(4.42),257(4.30), 360(3.86)	48-1002-83
$C_{17}H_{15}N_3O$			
Propanedinitrile, [3-(diethylamino)-4-oxo-1(4H)-naphthalenylidene]-	$CHCl_3$	255(4.42),320(4.11), 390(3.87),540(4.13)	88-3567-83
$C_{17}H_{15}N_3OS$			
1H-Pyrrole-2,5-dione, 3-(methylthio)-4-phenyl-, 2-(phenylhydrazone)	DMF	417(4.46)	48-0041-83
$C_{17}H_{15}N_3O_2S_3$			
6H-Thiazolo[3,4-a]pyrimidin-2(3H)-one, 6-(3-ethyl-4-oxo-2-thioxo-5-thiazolidinylidene)-3,4-dihydro-8-phenyl-	DMF	455(4.59)	103-0752-83
$C_{17}H_{15}N_3O_3$			
Formamide, N-[(1-methyl-5-nitro-3-phenyl-1H-indol-2-yl)methyl]-	MeOH	260(4.2),270(4.22), 336(3.92)	103-1302-83
$C_{17}H_{15}N_3O_5S$			
2-Naphthalenesulfonic acid, 3-amino-7-[(4-aminobenzoyl)amino]-4-hydroxy-	MeOH	210(4.59),232(4.55), 313(4.57),373(4.03)	48-1002-83
$C_{17}H_{15}N_5O_7$			
Benzeneacetamide, N,N-dimethyl-2-[[(2,4,6-trinitrophenyl)methylene]amino]-	$CHCl_3$	376(3.40)	44-2468-83
$C_{17}H_{15}O_2$			
1-Benzopyrylium, 5-methoxy-4-methyl-2-phenyl-, perchlorate	MeCN	419(4.44)	103-0243-83
1-Benzopyrylium, 6-methoxy-4-methyl-2-phenyl-, perchlorate	MeCN	369(4.34),413(4.22)	103-0243-83
1-Benzopyrylium, 7-methoxy-4-methyl-2-phenyl-, perchlorate	MeCN	418(4.58)	103-0243-83
1-Benzopyrylium, 8-methoxy-4-methyl-2-phenyl-, perchlorate	MeCN	381(4.39)	103-0243-83
1-Benzopyrylium, 2-(2-methoxyphenyl)-4-methyl-, perchlorate	MeCN	359(4.12),425(4.32)	103-0243-83
1-Benzopyrylium, 2-(3-methoxyphenyl)-4-methyl-, perchlorate	MeCN	374(4.39)	103-0243-83
$C_{17}H_{16}$			
Benzene, 1-ethenyl-2-(1-methyl-2-phenylethenyl)-, cis	MeOH	248(4.19)	39-1015-83B

Compound	Solvent	$\lambda_{max}(\log \epsilon)$	Ref.
Tricyclo[9.4.1.1^{3,9}]heptadeca-2,4,7,9,11,13,15-heptaene	3-Mepentane	288(4.90),370(3.59)	35-6211-83
$C_{17}H_{16}BN_5O_6$			
Formazan, 5-(diacetoxyboryl)-3-nitro-1,5-diphenyl-	hexane	508(4.26)	65-2061-83
	EtOH	240(4.0),335(3.9), 510(4.27)	65-2061-83
$C_{17}H_{16}BrNO_2$			
Ethanone, 1,1'-[4-(4-bromophenyl)-2,6-dimethyl-3,5-pyridinediyl]bis-	EtOH	208(4.45),228s(4.28), 244s(4.25)	103-0415-83
$C_{17}H_{16}ClFN_4O$			
1H-1,4-Benzodiazepine-2-carboximidamide, 7-chloro-5-(2-fluorophenyl)-2,3-dihydro-N-hydroxy-N'-methyl-	isoPrOH	223s(4.51),266s(3.96), 368(3.57)	4-0551-83
$C_{17}H_{16}ClNOS$			
2-Propen-1-one, 1-(4-chlorophenyl)-3-[1,2-dihydro-1-(1-methylethyl)-2-thioxo-3-pyridinyl]-, (E)-	EtOH	276(4.33),341(3.94), 425(3.71)	4-1651-83
$C_{17}H_{16}ClNO_3$			
2H-Pyrano[3,2-d][1]benzoxepin-2-one, 3-chloro-5,6-dihydro-4-pyrrolidino-	EtOH	211.5(4.22),235s(3.88), 268(4.20),337(4.09)	4-0539-83
$C_{17}H_{16}Cl_2$			
Benzene, 1,2-dichloro-4-(3-methyl-3-phenyl-1-butenyl)-, (E)-	n.s.g.	220.5(4.41),259(4.36), 264.5(4.34),270(4.26), 297(3.29),308(3.10)	12-0565-83
$C_{17}H_{16}Cl_2N_4O$			
3-Pyridinecarboxamide, 1-[(2,6-dichlorophenyl)methyl]-1,4-dihydro-4-(1H-imidazol-1-yl)-N-methyl-	CH_2Cl_2	308(3.7),365(3.4)	64-0878-83B
$C_{17}H_{16}Cl_4O_5$			
3H-Benzo[a]dicyclopropa[c,e]cyclohepten-3-one, 1,1,2,2-tetrachloro-1,1a,1b,2,2a,7b-hexahydro-2a,4,5,6-tetramethoxy-, (1aα,1bβ,7aβ,7bα)-	MeOH	281(3.49)	39-1983-83C
$C_{17}H_{16}FeO$			
Ferrocene, (hydroxyphenylmethyl)-	EtOH	330s(--),440(2.03)	32-0721-83
	50% H_2SO_4	230s(--),253(4.05), 283(4.00),328(4.06), 405(3.55)	32-0721-83
$C_{17}H_{16}N_2$			
1H-1,4-Diazepine, 2,3-dihydro-2,3-diphenyl-, cis	MeCN	306(3.826)	5-1207-83
perchlorate	MeCN	333(4.12)	5-1207-83
1H-1,4-Diazepine, 2,3-dihydro-2,3-diphenyl-, trans	MeCN	306(3.83)	5-1207-83
perchlorate	MeCN	334(4.12)	5-1207-83
1H-Pyrrolo[1,2,3-de]quinoxaline, 2,3-dihydro-5-methyl-6-phenyl-	EtOH	215(4.41),235(4.54), 305(4.19)	103-1340-83
Spiro[6H-cyclohepta[b]quinoxaline-6,1'-cyclopentane]	MeOH	234(4.28),346(3.92)	39-0285-83C

Compound	Solvent	$\lambda_{max}(\log \epsilon)$	Ref.
$C_{17}H_{16}N_2O$			
6-Cinnolinol, 5,7,8-trimethyl-3-phenyl-	EtOH	273(4.64)	39-1753-83C
1,2-Diazabicyclo[3.1.0]hex-2-ene, 3-(4-methoxyphenyl)-6-phenyl-, cis	EtOH	281(4.31)	35-0933-83
Ethanone, 1-(1-methyl-1H-indol-3-yl)-2-phenyl-, oxime, (E)-	n.s.g.	226(4.31),277(3.98)	103-0184-83
$C_{17}H_{16}N_2O_2$			
20,21-Dinoraspidospermidine-4,10-dione, 2,3-didehydro-, (±)-	EtOH	239(4.16),297(3.90), 344(4.23)	78-3719-83
2-Naphthalenecarbonitrile, 3-[2-(diethylamino)ethenyl]-1,4-dihydro-1,4-dioxo-	C_6H_{12}	544(4.16)	150-0168-83S
	EtOH	568(--)	150-0168-83S
$C_{17}H_{16}N_2O_2S$			
Carbamic acid, [[(diphenylmethylene)-amino]thioxomethyl]-, ethyl ester	CH_2Cl_2	276(4.55),c.355(3.40)	24-1297-83
Carbamic acid, (3,4-diphenyl-1,3-thiazetidin-2-ylidene)-, ethyl ester	dioxan	233(4.13),265(4.33), c.322(2.9)	24-1297-83
3H-Thieno[2,3-d]-1,3-diazepine-3-carboxylic acid, 2,5-dimethyl-, phenylmethyl ester	EtOH	227(4.12),255s(--), 300(3.48)	94-3684-83
$C_{17}H_{16}N_2O_3$			
Butanamide, N-[4-(2-benzoxazolyl)-3-hydroxyphenyl]-	EtOH	330(4.56)	4-1517-83
$C_{17}H_{16}N_2O_3S$			
Benzenesulfonamide, N-(3-acetyl-1H-indol-2-yl)-4-methyl-	$CHCl_3$	247(4.30),275(4.20), 322(3.98)	103-0868-83
Thieno[2',3':4,5]pyrrolo[1,2,3-ef]-[1,5]benzodiazepine-10-carboxylic acid, 1,2,3,4-tetrahydro-9-methyl-3-oxo-, ethyl ester	dioxan	200(4.15),225(4.53), 250(4.34),310(4.06)	103-0959-83
$C_{17}H_{16}N_2O_4$			
Carbamic acid, methyl[(phenylamino)-carbonyl]-, 3-acetylphenyl ester	EtOH	240(4.44)	22-0073-83
Carbamic acid, methyl[(phenylamino)-carbonyl]-, 4-acetylphenyl ester	EtOH	248.5(4.48)	22-0073-83
1,2-Diazabicyclo[5.2.0]nona-3,5-dien-9-one, 8-acetoxy-2-benzoyl-5-methyl-, cis	MeOH	288(3.91)	5-1361-83
$C_{17}H_{16}N_2O_5$			
3-Pyridinecarboxamide, N-[2-[(2-methyl-4-oxo-4H-1,3-benzodioxin-2-yl)oxy]-ethyl]-	pH 7.4	300(3.38)	1-0351-83
Uridine, 2'-deoxy-5-(phenylethynyl)-	pH 6	264(4.12),279(4.09), 307(4.21)	44-1854-83
	pH 13	266(4.10),282(4.14), 305(4.22)	44-1854-83
$C_{17}H_{16}N_2O_9S_3$			
2,7-Naphthalenedisulfonic acid, 3-amino-4-hydroxy-5-[[(4-methylphenyl)-sulfonyl]amino]-	MeOH	210(4.19),228(4.18), 243(4.15),267(4.05), 357(3.10)	48-1002-83
$C_{17}H_{16}N_4O$			
3-Buten-2-one, 4-[(7-methyl-3-phenyl-pyrazolo[1,5-a]pyrimidin-2-yl)amino]-	MeCN	277(4.34),333(4.34)	39-0011-83C

Compound	Solvent	$\lambda_{max}(\log \epsilon)$	Ref.
$C_{17}H_{16}N_4OS$			
Piperidine, 1-[[4-(2-quinoxalinyl)-2-thiazolyl]carbonyl]-	EtOH	247(4.04),267s(3.86),298s(3.63),343(3.57),357s(3.50)	106-0829-83
$C_{17}H_{16}N_4OS_2$			
Propanenitrile, 2-(2,3-dihydro-1,3-dimethyl-1H-benzimidazol-2-ylidene)-3-(3-ethyl-4-oxo-2-thioxothiazolidin-5-ylidene)-	benzene	481(4.79)	124-0297-83
	EtOH	471(4.88)	124-0297-83
	$CHCl_3$	482(4.82)	124-0297-83
$C_{17}H_{16}N_4O_2$			
1-Triazenium, 1-(1,3-dihydro-1,3-dioxo-2H-isoindol-2-yl)-3-phenyl-2-propyl-, hydroxide, inner salt, (E)-	EtOH	218(4.56),238s(4.24),268(4.06)	104-2260-83
(Z)-	EtOH	223(4.55),332(4.20)	104-2260-83
$C_{17}H_{16}N_4O_3$			
2H-Pyrrolo[2,3-b]quinoxalin-2-one, 1,3,3a,4,9,9a-hexahydro-4-methyl-3-(4-nitrophenyl)-	EtOH	222(4.57),265(4.10),303(3.89)	103-0901-83
$C_{17}H_{16}N_4O_5$			
Benzeneacetamide, 2-[[(2,4-dinitrophenyl)methylene]amino]-N,N-dimethyl-	$CHCl_3$	363(4.12)	44-2468-83
$C_{17}H_{16}N_4O_6$			
Piperidine, 1-(2',4',6'-trinitro[1,1'-biphenyl]-2-yl)-	$CHCl_3$	476(3.72)	44-4649-83
$C_{17}H_{16}O$			
1H-Cyclopenta[a]pentalen-1-one, 3a,3b,6,6a,7,7a-hexahydro-6-phenyl-	EtOH	219(4.17),247(3.96)	88-2781-83
Pentacyclo[5.4.0.0^{2,6}.0^{3,10}.0^{5,9}]undecan-8-one, 11-phenyl-, anti	MeCN	258(2.36),297(1.34)	44-1862-83
$C_{17}H_{16}O_2$			
3-Buten-1-one, 1-(4-methoxyphenyl)-4-phenyl-, (E)-	EtOH	270(4.36)	35-0933-83
Ethanone, 1,1'-(methylenedi-4,1-phenylene)bis-	hexane	254(4.52)	99-0034-83
Furan, 2,2'-(phenylmethylene)bis[5-phenyl-	EtOH	230(4.14)	103-0478-83
$C_{17}H_{16}O_3$			
Benzenemethanol, α-(methoxyphenylmethylene)-, acetate, (Z)-	EtOH	282(4.05)	89-0551-83S
5H-Benzofuro[6,5-c][2]benzopyran-5-one, 1,2,3,4-tetrahydro-7,9-dimethyl-	MeOH	209(4.44),249(4.44),297(4.02)	78-1265-83
5H-Benzofuro[6,5-c][2]benzopyran-5-one, 1,2,3,4-tetrahydro-7,10-dimethyl-	MeOH	211(4.50),249(4.40),263(4.10),301(4.10)	78-1265-83
1,3,9-Cyclotridecatriene-5,7-diyne-1-carboxylic acid, 4,9-dimethyl-13-oxo-, methyl ester, (Z,Z,Z)-	ether	262(3.43),275(3.48),358(3.02)	44-2379-83
3,9-Phenanthrenedione, 4,4a-dihydro-6-methoxy-1,4a-dimethyl-, (S)-	EtOH	289(4.16),301s(4.07),331(3.77)	23-2461-83
$C_{17}H_{16}O_4$			
6H-Benzofuro[3,2-c][1]benzopyran, 6a,11a-dihydro-3,9-dimethoxy-	EtOH	282(3.69),287(3.61),309(3.45)	102-2031-83

Compound	Solvent	λ_{max}(log ε)	Ref.
Benzoic acid, 2-(4-hydroxy-3,5-dimethylbenzoyl)-, methyl ester	EtOH	299(4.17)	24-0970-83
Ethanone, 1-(3,4-dihydro-3,7-dihydroxy-2-phenyl-2H-1-benzopyran-6-yl)-	MeOH	279(4.71),325(4.23)	2-1116-83
Ethanone, 1-[2,4-dihydroxy-5-[(3-phenyloxiranyl)methyl]phenyl]-	MeOH	279(4.71),325(4.27)	2-1116-83
2H-Naphtho[1,2-b]pyran-5-carboxylic acid, 6-hydroxy-2,2-dimethyl-, methyl ester	MeOH	238(3.85),247(3.86), 265(3.86),273(3.92), 282(3.93),392(3.32)	94-2353-83
$C_{17}H_{16}O_4S$			
Sulfonium, diphenyl-, 2-methoxy-1-(methoxycarbonyl)-2-oxoethylide	EtOH	218(4.4),250s(3.9)	121-0433-83B
$C_{17}H_{16}O_5$			
4H-1-Benzopyran-4-one, 2,3-dihydro-5,7-dihydroxy-3-[(4-hydroxyphenyl)-methyl]-6-methyl-	EtOH	295(3.99)	142-0039-83
2,4-Hexadienoic acid, (7-methoxy-2-oxo-2H-1-benzopyran-4-yl)-, methyl ester	EtOH	262(4.47),321.5(4.16)	94-3014-83
$C_{17}H_{16}O_6$			
Benzaldehyde, 2-(5-formyl-2-methoxyphenoxy)-3,4-dimethoxy-	MeOH	222(4.47),276(4.34)	83-0624-83
Benzaldehyde, 3-(5-formyl-2-methoxyphenoxy)-4,5-dimethoxy-	MeOH	226(4.50),274(4.32)	83-0624-83
Biacangelicol, 1-	EtOH	222(4.40),241(4.14), 249(4.14),271(4.21), 313(4.05)	105-0610-83
2-Propen-1-one, 1-(3,4-dimethoxyphenyl)-3-(2,4,6-trihydroxyphenyl)-	MeOH	378(4.28)	150-2601-83M
$C_{17}H_{16}O_7$			
Benzoic acid, 2-(2,6-dihydroxy-4-methylbenzoyl)-5-hydroxy-3-methoxy-, methyl ester (sulochrin)	EtOH	283(4.13)	94-4543-83
$C_{17}H_{16}S_3$			
1H-2-Benzothiopyran, 3,4-bis(methylthio)-1-phenyl-	CCl_4	269s(2.43),344(2.23)	54-0091-83
2-Propene(dithioic) acid, 2-(methylthio)-3,3-diphenyl-, methyl ester	CCl_4	317(2.96),420(2.45), 525(1.15)	54-0091-83
$C_{17}H_{17}BN_4O_4$			
Boron, bis(acetato-O)[[1,1'-methylenebis[2-phenyldiazenato]](1-)-N^2,N^2]-	hexane	505(4.31)	65-2061-83
	EtOH	508(4.28)	65-2061-83
$C_{17}H_{17}BrN_2O_3S$			
Propanamide, 2-bromo-2-methyl-N-[3-methyl-4-[(phenylsulfonyl)imino]-2,5-cyclohexadien-1-ylidene]-	EtOH	275(4.18)	44-3146-83
$C_{17}H_{17}Cl$			
Benzene, 1-chloro-3-(2,2-dimethyl-3-phenylcyclopropyl)-, trans	n.s.g.	212(4.28),220(4.23), 259(3.08),264.5(3.08), 272(3.01)	12-0565-83
Benzene, 1-chloro-4-(2,2-dimethyl-3-phenylcyclopropyl)-, trans	n.s.g.	210.5(4.12),220.5(4.08), 229.5(4.10),258(3.82), 290(2.80),301(2.51)	12-0565-83
Benzene, 1-chloro-3-(3-methyl-3-phenyl-1-butenyl)-, (E)-	n.s.g.	220(4.32),252.5(4.30), 291(3.17),300.5(3.01)	12-0565-83

Compound	Solvent	$\lambda_{max}(\log \epsilon)$	Ref.
Benzene, 1-chloro-4-(3-methyl-3-phenyl-1-butenyl)-, (E)-	n.s.g.	212.5(4.41),257(4.49), 290(3.48),300(3.22)	12-0565-83
$C_{17}H_{17}ClN_2O$			
4H-3,1-Benzoxazin-2-amine, 6-chloro-N-ethyl-4-methyl-4-phenyl- (etifoxine)	EtOH	273(4.33)	162-0558-83
$C_{17}H_{17}ClN_2O_2$			
Acetic acid, chloro[[2-(phenylmethyl)-phenyl]hydrazono]-, ethyl ester	EtOH	230(4.03),292(3.98), 319(4.25)	4-0225-83
Acetic acid, chloro[[4-(phenylmethyl)-phenyl]hydrazono]-, ethyl ester	EtOH	230(4.08),295(4.11), 321(4.33)	4-0225-83
Acetic acid, chloro[phenyl(phenyl-methyl)hydrazono]-, ethyl ester	EtOH	225(4.08),290(3.92), 313(3.54)	4-0225-83
$C_{17}H_{17}ClN_2O_6S$			
1,2-Diazabicyclo[5.2.0]nona-3,5-diene-2-carboxylic acid, 8-[[(4-chloro-phenyl)sulfonyl]oxy]-5-methyl-9-oxo-, ethyl ester, cis	MeOH	271(3.95)	5-1361-83
$C_{17}H_{17}Cs$			
Cesium, [9-(1,1-dimethylethyl)-9H-fluoren-9-yl]-	$(MeOCH_2)_2$	377(4.06),471(2.97), 532(2.84),498(3.04)	104-1592-83
$C_{17}H_{17}IO_5$			
Spiro[1,3-dioxolane-2,10'-[10H-3,10a]-methano[1H]indeno[2,1-c]oxepin-1'-one, 3',4',5',5'a-tetrahydro-4'-iodo-8'-methoxy-, (3'aα,4'β,5'aα-10'aα)-	EtOH	220(3.97),255s(3.97), 281(3.48),288.5(3.42)	44-1643-83
$C_{17}H_{17}Li$			
Lithium, [9-(1,1-dimethylethyl)-9H-fluoren-9-yl]-	$(MeOCH_2)_2$	386(4.12),476(3.01), 507(3.12),544(2.93)	104-1592-83
$C_{17}H_{17}NOS_2$			
Piperidine, 4-thieno[2,3-c][2]benzo-thiepin-4(9H)-ylidene-, S-oxide	MeOH	271(4.11)	73-0623-83
$C_{17}H_{17}NO_2$			
Benzenepropanenitrile, β-ethyl-4-hy-droxy-α-(4-hydroxyphenyl)-	MeOH	208(4.10),224(4.39), 276(3.55),285(3.41)	83-0271-83
2,4,6-Cycloheptatrien-1-one, 4-[1-ami-no-3-(4-methylphenyl)-2-propenyl]-2-hydroxy-	MeOH	247(4.45),335(3.99), 394(3.77)	18-3099-83
Cyclopenta[b]pyrrole-4-carboxylic acid, 1-ethenyl-1,4,5,6-tetrahydro-, phenylmethyl ester	MeOH	217(4.14),248(4.16)	23-1697-83
1H-Dibenzo[a,f]quinolizine-1,13(2H)-dione, 3,4,6,7,11b,12-hexahydro-	MeOH	265(4.16),306(4.23)	135-0303-83
Ethanone, 1,1'-(2,6-dimethyl-4-phenyl-3,5-pyridinediyl)bis-	EtOH	206(4.37),226s(4.46), 240s(4.18),280s(3.86)	103-0415-83
Ethanone, 1,1'-[(methylimino)di-4,1-phenylene]bis-	hexane	236(4.20),296(4.13), 350(4.37)	99-0034-83
8H-Indeno[6,5,4-cd]indol-8-one, 4-ace-tyl-9-methyl-4,5,5a,6,6a,7-hexa-hydro-	MeOH	261(4.44),293(4.43)	39-1545-83C
1-Oxa-3-azaspiro[4.5]deca-3,6,9-trien-8-one, 2,2,6-trimethyl-4-phenyl-	EtOH	248(4.26)	33-2252-83

Compound	Solvent	λ_{max}(log ε)	Ref.
$C_{17}H_{17}NO_2S$			
Benzo[b]thiophene-2-carboxylic acid, 3-amino-2,3-dihydro-3-phenyl-, ethyl ester	EtOH	249(3.76),278(3.27), 288(3.23),296(3.05)	44-1275-83
1H-Thieno[2',3':4,5]pyrrolo[3,2,1-ij]-quinoline-9-carboxylic acid, 2,3-dihydro-8-methyl-, ethyl ester	dioxan	220(4.42),240(4.56), 270(3.90),315(4.03)	103-0959-83
2-Thietanone, 3-[[(2-hydroxy-1-naphthalenyl)methylene]amino]-3,4,4-trimethyl-, (±)-	EtOH	228(4.78),315(4.04), 350(3.83)	39-2259-83C
$C_{17}H_{17}NO_3$			
Anaxagoreine	EtOH	213(--),273(4.12), 311(3.48)	100-0761-83
1H,13H-Benzo[a]pyrano[4,3-f]quinolizine-1,13-dione, 3,4,6,7,11b,12-hexahydro-3-methyl-	MeOH	248(4.16),302(4.22)	135-0303-83
2,4,6-Cycloheptatrien-1-one, 4-[1-amino-3-(4-methoxyphenyl)-2-propenyl]-2-hydroxy-	MeOH	247(4.44),334(4.02), 372s(3.74),394(3.76)	18-3099-83
4H-Furo[3,2-b]pyrrole-5-carboxylic acid, 4-methyl-2-(4-methylphenyl)-, ethyl ester	MeOH	233(3.84),340(4.79)	73-0772-83
$C_{17}H_{17}NO_4$			
Benzoic acid, 2,4,6-trimethyl-, (3-nitrophenyl)methyl ester	EtOH	250(4.03)	12-0081-83
2H-1-Benzopyran-3(4H)-one, 4,4-dimethoxy-2-phenyl-, oxime	n.s.g.	231(3.65),275(3.46)	88-2209-83
$C_{17}H_{17}NO_4S$			
Benzenemethanol, α-[1-[(4-methylphenyl)sulfinyl]-2-propenyl]-2-nitro-	n.s.g.	221s(4.32)	12-1049-83
Benzenemethanol, α-[1-[(4-methylphenyl)sulfinyl]-2-propenyl]-4-nitro-	n.s.g.	269(4.19)	12-1049-83
Benzenemethanol, α-[3-[(4-methylphenyl)sulfinyl]-2-propenyl]-2-nitro-, (E)-	n.s.g.	229s(4.25)	12-1049-83
Benzenemethanol, α-[3-[(4-methylphenyl)sulfinyl]-2-propenyl]-4-nitro-, (E)-	n.s.g.	230(4.21),267(4.03)	12-1049-83
$C_{17}H_{17}NO_5$			
9(10H)-Acridinone, 1-hydroxy-3,5,6-trimethoxy-10-methyl- (citpressine II)	MeOH	220(4.14),270(4.79), 300s(4.15),333(4.12), 382(3.74)	94-0895-83
	MeOH-$AlCl_3$	230(--),265s(--), 277(--),353(--), 428(--)	94-0895-83
$C_{17}H_{17}NO_6$			
9(10H)-Acridinone, 1,5-dihydroxy-2,3,4-trimethoxy-10-methyl- (citbrasine)	MeOH	209(4.18),225(4.08), 262s(4.51),272(4.58), 325(3.96),426(3.59)	94-0901-83
$C_{17}H_{17}NS_2$			
Piperidine, 4-thieno[2,3-c][2]benzothiepin-4(9H)-ylidene-	MeOH	283(3.81)	73-0623-83
$C_{17}H_{17}NS_3$			
Ethane(dithioic) acid, (diphenylamino)-thioxo-, propyl ester	CH_2Cl_2	262(4.11),290(4.12), 369s(3.85),433s(3.02)	97-0247-83

Compound	Solvent	$\lambda_{max}(\log \epsilon)$	Ref.
$C_{17}H_{17}N_3$			
Benzo[f]quinoxaline, 6-piperidino-	EtOH	242(4.45),280(4.16), 318(3.86),376(3.86)	83-0283-83
$C_{17}H_{17}N_3OS$			
2H-Thiazolo[3,2-a]-1,3,5-triazin-6(7H)-one, 7-ethyl-3,4-dihydro-3-(2-naphthalenyl)-	EtOH	215(4.89),241(4.98)	103-0961-83
$C_{17}H_{17}N_3O_2$			
Benzenamine, 4-[5-(4-methoxyphenyl)-1,3,4-oxadiazol-2-yl]-N,N-dimethyl-	toluene	335(4.58)	103-0022-83
Butanamide, N-[4-(1H-benzimidazol-2-yl)-3-phydroxyphenyl]-	EtOH	327(4.56)	4-1517-83
$C_{17}H_{17}N_3O_2S$			
Thiazolo[3,2-a]pyrimidin-4-ium, 3-hydroxy-5-methyl-7-morpholino-2-phenyl-, hydroxide, inner salt	MeCN	440(4.17)	124-0857-83
	benzene	469(--)	124-0857-83
	MeOH	416(--)	124-0857-83
	EtOH	419(--)	124-0857-83
	$HOCH_2CH_2OH$	410(--)	124-0857-83
	HOAc	390(--)	124-0857-83
	CH_2Cl_2	449(--)	124-0857-83
	$CHCl_3$	450(--)	124-0857-83
	CCl_4	477(--)	124-0857-83
	quinoline	460(--)	124-0857-83
$C_{17}H_{13}N_3O_3$			
Benzeneacetamide, N,N-dimethyl-2-[[(4-nitrophenyl)methylene]amino]-	$CHCl_3$	346(4.02)	44-2468-83
$C_{17}H_{13}N_3O_3S_3$			
4H-Bisthiazolo[3,2-a:4',5'-e]pyrimidine-4,8(7H)-dione, 5a-ethoxy-5,5a-dihydro-5-methyl-2-(methylthio)-7-phenyl-	MeOH	220(4.43),276(3.87), 285(3.88),325(4.26)	2-0243-83
$C_{17}H_{17}N_5O_2$			
Propanenitrile, 3-[ethyl[4-[(4-nitrophenyl)azo]phenyl]amino]-	gas	360(4.49)	135-1185-83
	DMF	485(4.52)	135-1185-83
$C_{17}H_{17}N_5O_6$			
Guanosine, N-benzoyl-	MeOH	236(4.21),258(4.16), 264(4.16),294(4.17)	33-2069-83
$C_{17}H_{17}O_4$			
Phenanlenylium, 1,3,4,9-tetramethoxy-, tetrafluoroborate	MeCN	220(4.38),236(4.34), 267s(4.04),374(4.29), 399(4.30),423(4.38)	44-2115-83
$C_{17}H_{17}O_4P$			
Methanone, [4-[(4-methyl-1,3,2-dioxaphosphorinan-2-yl)oxy]phenyl]phenyl-	EtOH	279(4.2)	65-0243-83
$C_{17}H_{18}$			
Benzene, 1,1'-(3,3-dimethyl-1,2-cyclopropanediyl)bis-, trans	n.s.g.	212.5(4.21),223(4.27), 252.5(3.34),265(3.20), 275(2.76),284.5(2.28), 293(2.10)	12-0565-83
Biphenylene, 2,3-dihydro-1,2,3,4-tetramethyl-3-methylene- (same spectrum in	CF_3COOH	262(3.79),272(3.74), 282(3.72),338(3.60),	104-0899-83

Compound	Solvent	λ_{max}(log ϵ)	Ref.
96% H_2SO_4) (cont.)		386(3.96),578(3.04)	104-0899-83
	hexane	230(4.28),238(4.30), 269(4.31),275(4.30), 315(3.76),330s(--)	104-0899-83
$C_{17}H_{18}BrN$			
Benzenemethanamine, 2-[2-(4-bromophenyl)ethenyl]-N,N-dimethyl-, hydrochloride, (E)-	HOAc	295(4.42)	39-1503-83B
$C_{17}H_{18}BrNO_2$			
Ethanone, 1,1'-[4-(4-bromophenyl)-1,4-dihydro-2,6-dimethyl-3,5-pyridinediyl]bis-	EtOH	204(4.25),219s(4.16), 255(4.22),384(3.85)	103-0415-83
$C_{17}H_{18}BrN_3OS$			
4H-Cyclopentapyrimidine-4-thione, 1-(4-bromophenyl)-1,5,6,7-tetrahydro-2-morpholino-	MeOH	225(4.25),344(4.48)	73-3573-83
$C_{17}H_{18}ClNO_3$			
2H-Pyrano[3,2-d][1]benzoxepin-2-one, 3-chloro-4-(diethylamino)-5,6-dihydro-	EtOH	214(4.19),239s(3.82), 264.5(3.94),340(4.11)	4-0539-83
$C_{17}H_{18}ClN_3O$			
11H-Dibenzo[b,e][1,4]diazepin-11-one, 7-chloro-10-[2-(dimethylamino)ethyl]-5,10-dihydro- (clobenzepam)	EtOH	230(4.52)	162-0336-83
$C_{17}H_{18}ClN_3OS$			
4H-Cyclopentapyrimidine-4-thione, 1-(4-chlorophenyl)-1,5,6,7-tetrahydro-2-morpholino-	MeOH	221(4.25),345(4.49)	73-3575-83
$C_{17}H_{18}Cl_2N_2O$			
3-Pyridinecarbonitrile, 1-[(2,6-dichlorophenyl)methyl]-6-(1,1-dimethylethoxy)-1,6-dihydro-	CH_2Cl_2	314(3.8),356s(3.2)	64-0873-83B
$C_{17}H_{18}INO_4$			
1H-Pyrrole-2-carboxylic acid, 3-(2-acetoxyethyl)-5-iodo-4-methyl-, phenylmethyl ester	$CHCl_3$	280(4.23)	78-1849-83
$C_{17}H_{18}NO_2S$			
Thieno[2,3-c]pyridinium, 7-[(3,4-dimethoxyphenyl)methyl]-6-methyl-, iodide	MeOH	242(4.37),273(3.88), 321(3.86)	83-0912-83
perchlorate	MeOH	237(4.18),270(3.71), 287(3.44),316(3.57)	83-0912-83
Thieno[3,2-c]pyridinium, 4-[(3,4-dimethoxyphenyl)methyl]-5-methyl-, iodide	MeOH	235(4.69),284(3.85), 325(3.51)	83-0244-83
$C_{17}H_{18}NO_4$			
Propylium, 1,1-bis(4-methoxyphenyl)-2-nitro-	H_2SO_4-CF_3COOH	398(4.39)	104-0610-83
$C_{17}H_{18}N_2$			
2H-Azirine-3-amine, 2-ethyl-N-methyl-	EtOH	254(4.29),277s(3.52),	33-0262-83

Compound	Solvent	$\lambda_{max}(\log \epsilon)$	Ref.
N,2-diphenyl- (cont.)		281s(3.33),289s(2.86)	33-0262-83
	CH_2Cl_2	257.5(4.33),277s(3.69), 285s(3.28),291s(2.91)	33-0262-83
Benzeneethanimidamide, α-ethylidene-N-methyl-N-phenyl-	EtOH	248(4.16),283.5s(3.41), 292s(3.10)	33-0262-83
Methanimidamide, N-methyl-N-phenyl-N'-(1-phenyl-1-propenyl)-	EtOH	261(3.96)	33-0262-83
$C_{17}H_{18}N_2O$			
20,21-Dinoraspidospermidin-4-one, 2,3-didehydro-, (±)-	EtOH	240(4.10),295(3.88), 345(4.15)	78-3719-83
$C_{17}H_{18}N_2OS$			
Azuleno[2,1-d]thiazole, 2-methyl-4-(4-morpholinylmethyl)-	MeOH	222(4.04),255(3.96), 300(4.83),352(3.68), 368(3.81),380s(3.24), 610(2.58),654(2.55), 720(2.22)	142-1263-83
Thiazolo[3,2-a]pyrimidin-4-ium, 7-(1,1-dimethylethyl)-3-hydroxy-5-methyl-2-phenyl-, hydroxide, inner salt	MeCN	500(4.02)	124-0857-83
$C_{17}H_{18}N_2O_2$			
20,21-Dinoraspidospermidine-4,10-dione, (±)-	MeOH	242(3.87),297(3.45)	78-3719-83
Furo[2,3-g]indolo[2,3-a]quinolizin-2(1H)-one, 3a,4,6,7,12,12b,13,13a-octahydro-	EtOH	226(4.65),276(3.91), 282(3.95),291(3.87)	77-1120-83
epimer	EtOH	227(4.55),277(3.86), 283(3.88),291(3.79)	77-1120-83
Indolo[4,3-fg]isoquinolin-9(4H)-one, 4-acetyl-5,5a,6,6a,7,8-hexahydro-8-methyl-	MeOH	255(4.51),297(4.17)	39-1545-83C
Indolo[4,3-fg]isoquinolin-9(4H)-one, 4-acetyl-5,5a,6,6a,7,8-hexahydro-10-methyl-	MeOH	255(4.49),292(4.24)	39-1545-83C
Indolo[4,3-fg]isoquinolin-9(4H)-one, 4-acetyl-5,5a,6,8,10,10a-hexahydro-10-methyl-	MeOH	214(4.39),261(4.32)	39-1545-83C
1H-Pyrazole-4-carboxylic acid, 5-(1,3-pentadienyl)-1-phenyl-, ethyl ester, (E,E)-	EtOH	296(4.25)	118-0566-83
$C_{17}H_{18}N_2O_3$			
Cinnolinium, 3-(1,3-benzodioxol-5-ylmethyl)-5,6,7,8-tetrahydro-4-hydroxy-2-methyl-, hydroxide, inner salt	EtOH	274(3.62),290(3.66), 319(3.71)	4-0543-83
4(1H)-Cinnolinone, 3-(1,3-benzodioxol-5-ylmethyl)-5,6,7,8-tetrahydro-1-methyl-	EtOH	285(4.21)	4-0543-83
Ethanone, 2-(1,3-benzodioxol-5-yl)-1-(4,5,6,7-tetrahydro-1-methyl-1H-indazol-3-yl)-	EtOH	243(4.06),283(3.78)	4-0543-83
1H-Pyrrole-3-carboxylic acid, 1-acetyl-3-cyano-2,3-dihydro-2,5-dimethyl-4-phenyl-, methyl ester	MeOH	265(4.26)	22-0195-83
1H-Pyrrolo[2,3-d]carbazole-2,5(3H,4H)-dione, 7-acetyl-3a,6,6a,7-tetrahydro-3a-methyl-, (3aα,6aβ,11bR*)-(±)-	MeOH	251(4.10),280s(3.36), 290s(3.29)	5-1744-83

Compound	Solvent	$\lambda_{max}(\log \epsilon)$	Ref.
$C_{17}H_{18}N_2O_4$			
1H-Benz[de]isoquinoline-1,3(2H)-dione, 4,7-dimethoxy-2,6-dimethyl-9-(methylamino)-	$CHCl_3$	270(4.70),302(4.18), 329s(4.15),251(4.28), 426(4.34)	78-2283-83
3-Butenoic acid, 3-methyl-2-[[[5-oxo-2-(phenylmethyl)-4(5H)-oxazolylidene]methyl]amino]-, methyl ester	MeCN	317(4.30)	142-1001-83
Ethanone, 1,1'-[1,4-dihydro-2,6-dimethyl-4-(4-nitrophenyl)-3,5-pyridinediyl]bis-	EtOH	205(4.25),217s(4.11), 257(4.34),387(3.76)	103-0415-83
L-Phenylalanine, N-[[3-hydroxy-5-(hydroxymethyl)-2-methyl-4-pyridinyl]methylene]-	MeOH	255(4.00),272s(3.88), 290s(3.54),335(3.60), 425(3.32)	35-0803-83
$C_{17}H_{18}N_2O_5$			
2-Propenoic acid, 2-[[(1,3-dihydro-1,3-dioxo-2H-isoindol-2-yl)acetyl]amino]-, 1,1-dimethylethyl ester	MeOH	240.9(4.09)	118-0539-83
1H-Pyrrole-3-propanoic acid, 2-carboxy-5-[(3-ethenyl-1,5-dihydro-4-methyl-5-oxo-2H-pyrrol-2-ylidene)methyl]-4-methyl-	2M NaOH	242(4.31),421(4.40)	5-0585-83
1H-Pyrrole-3-propanoic acid, 2-carboxy-5-[(4-ethenyl-1,5-dihydro-3-methyl-5-oxo-2H-pyrrol-2-ylidene)methyl]-4-methyl-	2M NaOH	231(4.31),265(4.16)	5-0585-83
L-Tyrosine, N-[[3-hydroxy-5-(hydroxymethyl)-2-methyl-4-pyridinyl]methylene]-	MeOH	250s(4.00),270s(3.90), 285s(3.78),335(3.57), 425(3.42)	35-0803-83
$C_{17}H_{18}N_2O_5S$			
1-Azabicyclo[3.2.0]hept-2-ene-2-carboxylic acid, 6-ethyl-3-(methylthio)-7-oxo-, (4-nitrophenyl)methyl ester	THF	269(3.97),320(3.98)	158-0407-83
$C_{17}H_{18}N_2O_6$			
L-Tyrosine, 3-hydroxy-N-[[3-hydroxy-5-(hydroxymethyl)-2-methyl-4-pyridinyl]methylene]-	MeOH	255s(3.93),283(3.94), 330s(3.22),406(3.25)	35-0803-83
$C_{17}H_{18}N_4O_2S$			
Carbamic acid, (5-amino-2-methyl-3-phenyl-2H-pyrido[4,3-b]-1,4-thiazin-7-yl)-, ethyl ester	pH 1	258(4.43),391(4.30)	87-1614-83
$C_{17}H_{18}N_4O_3$			
Carbamic acid, (5-amino-2-methyl-3-phenyl-2H-pyrido[4,3-b]-1,4-oxazin-7-yl)-, ethyl ester	pH 1	299(3.85),306s(3.76), 358s(4.34),371(4.38)	87-1614-83
$C_{17}H_{18}N_4O_3S$			
1H-Benzimidazol-2-amine, 6-[(hydroxyimino)phenylmethyl]-1-[(1-methylethyl)sulfonyl]-, (E)- (enviroxime)	MeOH	218(4.66),290(4.43)	162-0519-83
(Z)- (zinviroxime)	MeOH	254(4.32),285(4.12)	162-0519-83
Carbamic acid, [5-amino-3-(3-methoxyphenyl)-2H-pyrido[4,3-b]-1,4-thiazin-7-yl]-, ethyl ester	pH 1	256(4.41),304s(3.92), 395(4.28)	87-1614-83
Carbamic acid, [5-amino-3-(4-methoxyphenyl)-2H-pyrido[4,3-b]-1,4-thiazin-7-yl]-, ethyl ester	pH 1	250(4.36),291(4.29), 399(4.37)	87-1614-83

Compound	Solvent	λ_{max}(log ϵ)	Ref.
$C_{17}H_{18}N_4O_4$			
Benzoic acid, 4-methyl-, 2-[(7,8-dihydro-8-oxoimidazo[4,5-d][1,3]diazepin-3(6H)-yl)methoxy]ethyl ester	MeOH	232(4.52),350(3.58)	87-1478-83
Carbamic acid, [5-amino-2-(3-methoxyphenyl)-2H-pyrido[4,3-b]-1,4-oxazin-7-yl]-, ethyl ester	pH 1	300s(3.84),305(3.84), 361s(4.34),374(4.40)	87-1614-83
$C_{17}H_{18}N_4O_5$			
1H-Pyrazolo[3,4-b]pyrazine, 1-β-D-glucopyranosyl-3-phenyl-	MeOH	240(4.37),285(3.94)	161-0024-83
$C_{17}H_{18}N_4O_5S$			
4H-Thiazolo[5',4':3,4]pyrrolo[1,2-a]-indole-6,9-dione, 8-amino-10-[[(aminocarbonyl)oxy]methyl]-2-ethoxy-3a,10b-dihydro-7-methyl-, (3aR-cis)-	MeOH	255(4.15),309(3.92), 345(3.59)	44-5026-83
$C_{17}H_{18}N_6O_6$			
Adenosine, N-[(4-nitrophenyl)methyl]-	EtOH	268(4.32)	130-0045-83
$C_{17}H_{18}O$			
Benzene, 1,1'-(4-methoxy-1-butene-1,4-diyl)bis-, (E)-	MeOH	252(4.29)	35-0933-83
1-Butanone, 3-methyl-1,3-diphenyl-	n.s.g.	211(3.78),238(3.90), 278.5(2.89),287.5(2.76)	12-0097-83 12-0565-83
Oxirane, 2-(1-methyl-1-phenylethyl)-3-phenyl-, trans	n.s.g.	217.5(4.05),253(2.64), 258(2.69),261(2.68), 264(2.64),267.5(2.56)	12-0097-83
3-Pentanone, 2,4-diphenyl-, (±)-	benzene	295(2.56),304(2.53), 320(1.97)	35-1309-83
meso-	benzene	297(2.49),305(2.44), 320(2.05)	35-1309-83
2-Propanone, 1,3-bis(2-methylphenyl)-	C_6H_{12}	296(2.38)	46-1960-83
2-Propanone, 1,3-bis(3-methylphenyl)-	C_6H_{12}	292(2.40)	46-1960-83
2-Propanone, 1,3-bis(4-methylphenyl)-	C_6H_{12}	295(2.49)	46-1960-83
$C_{17}H_{18}OS$			
Benzene, 1-methoxy-4-[1-[[(phenylmethyl)thio]methyl]ethenyl]-	MeOH	260(3.91)	44-0835-83
$C_{17}H_{18}O_2$			
Benzene, 1,1'-[(2-methyl-2-butene-1,4-diyl)bis(oxy)]bis-	EtOH	221(4.30),267(3.75), 271(3.82),278(3.76)	78-0169-83
Benzenemethanol, 4,4'-methylenebis[α-methyl-, dication	H_2SO_4	320(4.59)	99-0612-83
9(1H)-Phenanthrenone, 2,3,4,4a-tetrahydro-6-methoxy-4a-methyl-1-methylene-, (S)-	EtOH	238(3.62),310(3.63)	23-2461-83
9(3H)-Phenanthrenone, 4,4a-dihydro-6-methoxy-1,4a-dimethyl-, (S)-	EtOH	233(3.87),253(3.74), 292(3.96),321(3.83)	23-2461-83
$C_{17}H_{18}O_3$			
1H-Benz[e]inden-1-one, 2,3-dihydro-5,7-dimethoxy-3,3-dimethyl-	EtOH	221(4.58),250(4.78), 310(4.16),342(4.02), 355(4.04)	78-4221-83
2-Buten-1-one, 1-(4,6-dimethoxy-1-naphthalenyl)-3-methyl-	EtOH	256(4.48),318(4.00), 345(4.01)	78-4221-83
6H-Dibenzo[b,d]pyran-6-one, 7,8,9,10-tetrahydro-3-hydroxy-4-methyl-2-(2-propenyl)-	MeOH	207(4.65),327(4.18)	78-1265-83

Compound	Solvent	λ_{max}(log ε)	Ref.
$C_{17}H_{18}O_3$			
6H-Dibenzo[b,d]pyran-6-one, 7,8,9,10-tetrahydro-4-methyl-3-(2-propenyloxy)-	MeOH	204(4.77),244(3.77), 256(3.73),320(4.21)	78-1265-83
Furan, 2,2',2"-ethylidenetris[5-methyl-	EtOH	229(4.31)	103-0474-83
6-Nonen-8-ynoic acid, 7-methyl-2-(3-methyl-2-penten-4-ynylidene)-3-oxo-, methyl ester	ether	235(2.48)	44-2379-83
1H-Phenalen-1-one, 2,3-dihydro-5,7-dimethoxy-3,3-dimethyl-	EtOH	224(4.35),254(4.52), 324(3.96),346(3.98)	78-4221-83
9(1H)-Phenanthrenone, 2,3,4,4a-tetrahydro-10-hydroxy-6-methoxy-4a-methyl-1-methylene-, (R)-	EtOH	222(4.10),277(4.01)	23-2461-83
9(3H)-Phenanthrenone, 4,4a-dihydro-10-hydroxy-6-methoxy-1,4a-dimethyl-, (R)-	EtOH	279(4.17),326(4.11)	23-2461-83
2-Propanone, 1,3-bis(2-methoxyphenyl)-	C_6H_{12}	300s(2.46)	46-1960-83
2-Propanone, 1,3-bis(3-methoxyphenyl)-	C_6H_{12}	300s(2.66)	46-1960-83
2-Propanone, 1,3-bis(4-methoxyphenyl)-	C_6H_{12}	300s(2.76)	46-1960-83
$C_{17}H_{18}O_4$			
3(2H)-Benzofuranone, 2-[(3,4-dimethoxyphenyl)methylene]-4,5,6,7-tetrahydro-	EtOH	260(3.94),372(4.34)	4-0543-83
2H-1-Benzopyran-5-ol, 3,4-dihydro-2-(4-hydroxyphenyl)-7-methoxy-8-methyl-	EtOH	275(3.51),281(3.45)	142-0039-83
2H-1-Benzopyran-7-ol, 3,4-dihydro-2-(3-hydroxy-4-methoxyphenyl)-8-methyl-, (S)-	MeOH	226s(4.21),283(3.73), 288s(3.71)	94-2146-83
2H-1-Benzopyran-7-ol, 3,4-dihydro-2-(4-hydroxy-3-methoxyphenyl)-8-methyl-, (S)-	EtOH	281(3.62),285(3.59)	142-0039-83
2H-1-Benzopyran-7-ol, 3,4-dihydro-3-[(4-hydroxyphenyl)methyl]-8-methoxy-	EtOH	279(3.54)	142-0039-83
6H-Dibenzo[b,d]pyran-6-one, 7,8,9,10-tetrahydro-3-hydroxy-4-methoxy-2-(2-propenyl)-	MeOH	207(4.75),325(4.18)	78-1265-83
6H-Dibenzo[b,d]pyran-6-one, 7,8,9,10-tetrahydro-4-methoxy-3-(2-propenyloxy)-	MeOH	205(4.72),246(3.80), 255(3.81),315(4.18)	78-1265-83
Ethanone, 2-(4-ethoxyphenyl)-1-(2-hydroxy-4-methoxyphenyl)-	MeOH	213(4.38),217(4.38), 228s(4.27),238s(3.94), 276(4.21),317(3.92)	100-0852-83
2H-Pyran-5-carboxylic acid, 3,4-dihydro-2-methyl-4-oxo-6-(2-phenylethenyl)-, ethyl ester, (E)-	EtOH	340(4.43)	118-0948-83
$C_{17}H_{18}O_4S$			
1H-Indene-1,5(6H)-dione, 2,3,7,7a-tetrahydro-7a-methyl-4-[(phenylsulfonyl)methyl]-, (S)-	EtOH	219(4.09),250(3.99)	39-2337-83C
$C_{17}H_{18}O_5$			
Benzaldehyde, 3,4-dimethoxy-5-(2-methoxy-4-methylphenoxy)-	MeOH	281(4.02)	83-0624-83
Benzaldehyde, 3-(2,3-dimethoxy-5-methylphenoxy)-4-methoxy-	MeOH	227(4.36),273(4.07)	83-0624-83
2H-1-Benzopyran-5,7-diol, 2-(3,4-dimethoxyphenyl)-3,4-dihydro-, (S)-	MeOH	278(3.03)	150-2601-83M
Spiro[1,3-dioxolane-2,9'-[9H]fluorene]-8'a(4'bH)-carboxylic acid, 5',8'-di-	EtOH	213(3.79),221(3.79), 226(3.78),283.5(3.39),	44-1643-83

Compound	Solvent	λ_{max}(log ϵ)	Ref
hydro-2'-methoxy-, cis (cont.)		290(3.33)	44-1643-83
Spiro[1,3-dioxolane-2,10'-[10H-3,10a]-methano[1H]indeno[2,1-c]oxepin]-1'-one, 3',4',5',5'a-tetrahydro-8'-methoxy-,(3'α,5'aα,10'aα)-	EtOH	220(3.81),282(3.39), 288.5(3.35)	44-1643-83
$C_{17}H_{18}O_6$			
Javunicine	MeOH	480s(3.78),507.3(3.83), 543(3.65)	74-0385-83C
Paeoniflorigenone	MeOH	220(3.94),258s(2.84), 263(2.90),270(2.82)	94-0577-83
Spiro[1,3-dioxolane-2,3'-[3H]oxireno-[8,8a]naphtho[2,3-b]furan]-5',9'-dione, 1'a,2',4',4'a-tetrahydro-4',4'a,6'-trimethyl-, (1'aα,4'α-4'aα,9'aS*)-(±)-	EtOH	306.5(3.98)	94-3544-83
$C_{17}H_{18}O_7$			
4aH-Xanthene-4a-carboxylic acid, 2,3,4,9-tetrahydro-1,4-dihydroxy-8-methoxy-6-methyl-9-oxo-, methyl ester	EtOH	233(4.35),254(4.05), 314(3.63)	39-1365-83C
$C_{17}H_{18}O_8$			
2,4-Hexadienoic acid, [6-(2-carboxy-ethenyl)-5-(hydroxymethyl)-4-methoxy-2-oxo-2H-pyran-3-yl]methyl ester, (Z,E,E)- (islandic acid)	MeOH	234(4.51),260(4.36), 335(4.16)	158-83-7
$C_{17}H_{19}BF_4$			
Cycloheptatrienylium, [4-(1,1-dimethyl-ethyl)phenyl]-, tetrafluoroborate	10% HCl	275.5(4.13),394(4.33)	78-4011-83
	MeCN	273.5(4.12),392.5(4.32)	78-4011-83
$C_{17}H_{19}Br_2N_3$			
Pyrido[2',1':3,4][1,4]diazepino[1,2-a]-benzimidazole-5,9-diium, 6,7,8,14-tetrahydro-2,4-dimethyl-, dibromide	EtOH	319(4.20)	4-0029-83
$C_{17}H_{19}FN_2O_{11}$			
β-D-Glucopyranuronic acid, 1-deoxy-1-(5-fluoro-3,4-dihydro-2,4-dioxo-1(2H)-pyrimidinyl)-, methyl ester, 2,3,4-triacetate	MeOH	262(3.97)	65-1483-83
β-D-Glucopyranuronic acid, 1-deoxy-1-(5-fluoro-3,6-dihydro-2,6-dioxo-1(2H)-pyrimidinyl)-, methyl ester, 2,3,4-triacetate	MeOH	272(3.78)	65-1483-83
$C_{17}H_{19}F_3N_2O$			
Pyrrolo[2,3-b]indole, 1,2,3,3a,8,8a-hexahydro-3a-(3-methyl-2-butenyl)-1-(trifluoroacetyl)-, cis-(±)-	EtOH	212(4.11),236(3.95), 296(3.38)	1-0803-83
	EtOH-HCl	212(4.11),236(3.96), 296(3.40)	1-0803-83
$C_{17}H_{19}IN_4O_7$			
9H-Purine, 6-(iodomethyl)-9-(2,3,5-tri-O-acetyl-β-D-ribofuranosyl)-	MeOH	278(3.96)	94-3104-83
$C_{17}H_{19}N$			
Benzenemethanamine, N,N-dimethyl-2-(2-phenylethenyl)-, hydrochloride, (E)-	EtOH	295(4.38)	39-1503-83B
	HOAc	295(4.34)	39-1503-83B

Compound	Solvent	λ_{max}(log ϵ)	Ref.
$C_{17}H_{19}NO$			
Benzeneacetamide, α-ethyl-N-methyl-N-phenyl-	EtOH	256s(3.01),268s(2.82)	33-0262-83
Benzoxazole, 2-tricyclo[3.3.1.1^{3,7}]dec-1-yl-	C_6H_{12}	263s(--),266(--),269(--),272(--),278(3.82)	41-0595-83
1-Naphthalenol, 4-[3-[(1,1-dimethylethyl)imino]-1-propenyl]-	CCl_4	325(3.91),463(3.71)	104-1864-83
	DMSO	335(3.94),520(4.26)	104-1864-83
	DMSO-NaOMe	370(3.92),440(4.08)	104-1864-83
Oxaziridine, 2-(1,1-dimethylethyl)-3,3-diphenyl-, (S)-	isooctane	255(3.390)	32-0799-83
$C_{17}H_{19}NOS$			
Cyclohexanone, 2-[(3-ethyl-2(3H)-benzothiazolylidene)ethylidene]-	MeOH	462(4.69)	104-1854-83
	benzene	420(--),435(--)	104-1854-83
	DMF	438(--)	104-1854-83
$C_{17}H_{19}NO_2$			
Ethanone, 1,1'-(1,4-dihydro-2,6-dimethyl-4-phenyl-3,5-pyridinediyl)bis-	EtOH	205(4.23),258(4.13),267(4.05),387s(3.85)	103-0415-83
2,4,6-Heptatrien-1-one, 1-(4-acetylphenyl)-7-(dimethylamino)-	EtOH	266s(4.05),516(4.55)	70-1882-83
$C_{17}H_{19}NO_2S$			
Benzo[5,6]cycloocta[1,2-b]thiophen-4,10-imine, 4,5,10,11-tetrahydro-7,8-dimethoxy-12-methyl-	MeOH	230(4.12),280(3.55)	83-0353-83
$C_{17}H_{19}NO_3$			
7,8-Isoquinolinediol, 1,2,3,4-tetrahydro-1-[(4-hydroxyphenyl)methyl]-2-methyl-	MeOH	218(3.80),239(3.90),284(3.54)	100-0342-83
Piperidine, 1-[5-(1,3-benzodioxol-5-yl)-1-oxo-2,4-pentadienyl]- (chavicine)	MeOH	318(4.21)	162-0284-83
$C_{17}H_{19}NO_4$			
7-Azabicyclo[4.1.0]hept-4-ene-1,4-dicarboxylic acid, 7-(phenylmethyl)-, dimethyl ester	MeCN	257(3.67)	24-0563-83
1H-Azepine-2,5-dicarboxylic acid, 2,3-dihydro-1-(phenylmethyl)-, dimethyl ester	MeCN	335(3.70)	24-0563-83
Buphanamine	n.s.g.	287(3.18)	162-0206-83
3H-Cyclobut[b]indole-1,2-dicarboxylic acid, 2a,7b-dihydro-2a,3,7b-trimethyl-, dimethyl ester	EtOH	349(3.97),429(2.99)	4-1263-83
Pyrrole-2-carboxylic acid, 3-(2-acetoxyethyl)-4-methyl-, phenylmethyl ester	$CHCl_3$	280(4.23)	78-1849-83
$C_{17}H_{19}NO_4S_2$			
2H-Isoindole-2-acetic acid, 1,3-dihydro-1,3-dioxo-4-(1,3-dithiolan-2-yl)butyl ester	$CHCl_3$	242(4.01),295(3.30)	78-2691-83
Spiro[1,3-dithiolane-2,8'(8H)[1,4]oxazecino[5,4-a]isoindole]-2',13'-(1'H,8'aH)-dione, 4',5',6',7'-tetrahydro-8'a-hydroxy-	MeOH	235s(3.77)	78-2691-83

Compound	Solvent	λ_{max}(log ϵ)	Ref.
$C_{17}H_{19}NO_5$			
2-Oxabicyclo[2.2.1]heptane-1-carboxylic acid, 4,7,7-trimethyl-3-oxo-, 2-cyano-7-oxabicyclo[2.2.1]hept-5-en-2-yl ester	dioxan	212(3.10)(end abs.)	33-1865-83
$C_{17}H_{19}NO_7$			
α-D-Xylofuranose, 5-C-(carboxyphenyl-amino)-1,2-O-(1-methylethylidene)-, intramol. 5,3-ester, 5-acetate	MeOH	208s(3.98),218s(3.76)	65-1917-83
$C_{17}H_{19}NS$			
Benzeneethanethioamide, α-ethyl-N-meth-yl-N-phenyl-	EtOH	280(4.14)	33-0262-83
$C_{17}H_{19}N_2O$			
Xanthylium, 3,6-bis(dimethylamino)-, tetrafluoroborate (pyronin)	H_2O	547(4.99)	35-0279-83
$C_{17}H_{19}N_2S$			
Thioxanthylium, 3,6-bis(dimethylamino)- (thiopyronin)	H_2O	563(4.95)	35-0279-83
$C_{17}H_{19}N_3$			
3,6-Acridinediamine, N,N,N',N'-tetra-methyl- (acridine orange)	pH 7.0	492(4.73)	156-0177-83B
Pyrido[2',1':3,4][1,4]diazepino[1,2-a]-benzimidazole-5,9-diium, 6,7,8,14-tetrahydro-2,14-dimethyl-, dibromide	EtOH	319(4.20)	4-0029-83
$C_{17}H_{19}N_3O_2$			
Propanedinitrile, ethyl[4-methyl-1-(4-nitrophenyl)-2-pentenyl]-, (E)-	EtOH	266(4.05)	12-0527-83
$C_{17}H_{19}N_3O_3$			
1H-Benz[de]isoquinoline-1,3(2H)-dione, 6-methoxy-2,7-dimethyl-4,9-bis(meth-ylamino)-	EtOH	246(4.59),265s(4.48), 288s(4.08),312(3.90), 370s(4.23),390s(4.42), 403(4.62)	78-2283-83
$C_{17}H_{19}N_3O_4S$			
1,2-Diazabicyclo[5.2.0]nona-3,5-diene-2-carboxylic acid, 5-methyl-9-oxo-8-[(2-thienylacetyl)amino]-, ethyl ester, cis-(±)-	MeOH	274(3.96)	5-1374-83
$C_{17}H_{19}N_3O_6$			
5-Pyrimidinepentanoic acid, 2-amino-5-[(4-carboxyphenyl)methyl]-1,4,5,6-tetrahydro-4,6-dioxo-	50%MeOH-HCl	201(4.53),234(4.21)	73-0299-83
	+ NaOH	226(4.46),263(3.99)	73-0299-83
$C_{17}H_{19}N_3S$			
1,3-Benzenediamine, N'-2-benzothiazolyl-N,N-diethyl-	64% MeOH	209s(4.35),223s(4.35), 273s(4.28),288s(4.28), 302(4.35)	33-2165-83
$C_{17}H_{19}N_5OS$			
Acetamide, 2-cyano-N-[(dimethylamino)-methylene]-2-[2-[(dimethylamino)meth-ylene]-2H-1,4-benzothiazin-3(4H)-yli-dene]-	MeOH	222(4.37),272(4.39), 379(4.51)	87-0845-83

Compound	Solvent	λ_{max}(log ε)	Ref.
$C_{17}H_{20}NO_2S$			
Thieno[2,3-c]pyridinium, 5-[(3,4-dimethoxyphenyl)methyl]-4,5-dihydro-6-methyl-, perchlorate	MeOH	225(4.07),278(3.73), 316(4.09)	83-0912-83
Thieno[3,2-c]pyridinium, 6-[(3,4-dimethoxyphenyl)methyl]-6,7-dihydro-5-methyl-, perchlorate	MeOH	233(4.18),281(3.97), 298(3.90),340s(3.46)	83-0244-83
$C_{17}H_{20}N_2$			
20,21-Dinoraspidospermidine, 1,2-didehydro-, (±)-	EtOH	222(4.18),265(3.73)	78-3719-83
$C_{17}H_{20}N_2O$			
20,21-Dinoraspidospermidin-15-ol, 1,2-didehydro-, (5α)-(±)-	EtOH	230(3.80),282(3.58)	78-3719-83
	EtOH-NaOH	236(3.80),307(3.77)	78-3719-83
20,21-Dinoraspidospermidin-4-one, (±)-	EtOH	246(3.85),299(3.40)	78-3719-83
1-Propanone, 2-methyl-1-[4-(1-methylethyl)-2-phenyl-5-pyrimidinyl]-	EtOH	277.5(4.35)	4-0649-83
$C_{17}H_{20}N_2OS$			
Promazine sulfoxide	pH 1	224(3.18),238(3.24), 270.5s(2.88)	36-0050-83
	pH 3.5	225(3.10),236(3.14), 271(2.78)	36-0050-83
	H_2O	224(3.03),235(3.06), 270(2.48)	36-0050-83
	2N H_2SO_4	222s(3.09),235s(3.14), 270(2.91)	36-0050-83
$C_{17}H_{20}N_2O_2$			
Acetamide, N-[2-[2-(1-methyl-3-oxo-1-butenyl)-1H-indol-3-yl]ethyl]-	EtOH	217(4.39),255(3.94), 351(4.16)	4-0267-83
Cinnolinium, 5,6,7,8-tetrahydro-4-hydroxy-3-[(4-methoxyphenyl)methyl]-2-methyl-, hydroxide, inner salt	EtOH	268(3.76),319(3.94)	4-0543-83
Ethanone, 2-(4-methoxyphenyl)-1-(4,5-6,7-tetrahydro-1-methyl-1H-indazol-3-yl)-	EtOH	224(4.02),246(4.05)	4-0543-83
2,5-Heptadienamide, 6-methyl-2-[1-(methylimino)ethyl]-4-oxo-N-phenyl-	MeOH	249(4.16),348(3.70)	83-1000-83
2,4-Heptadienenitrile, 2-ethyl-6,6-dimethyl-3-(4-nitrophenyl)-, (E,E)-	EtOH	260(4.47)	12-0527-83
4-Oxazolecarbonitrile, 4-(1,1-dimethylethyl)-4,5-dihydro-5-oxo-2-(2,4,6-trimethylphenyl)-	EtOH	251(3.96)	39-0979-83C
1H-Pyrido[3,4-b]indole, 2-acetyl-2,3,4,9-tetrahydro-1-methyl-1-(2-oxopropyl)-	EtOH	222(4.59),276(3.89), 288s(3.78)	4-0267-83
1H-Pyrido[3,4-b]indol-1-one, 2,3,4,9-tetrahydro-2-(4-oxohexyl)-	EtOH	232(4.3),242(4.0), 303(4.01)	2-0531-83
$C_{17}H_{20}N_2O_3$			
4(1H)-Cinnolinone, 3-[(3,4-dimethoxyphenyl)methyl]-5,6,7,8-tetrahydro-	EtOH	280(4.17)	4-0543-83
1H,4H-Indolizino[1',8':3,4,5]pyrano-[2,3-b]indol-12(11H)-one, 2,3,3a-5a,6,13a-hexahydro-4-hydroxy-4-methyl-	EtOH	244(3.90),297(3.50)	78-3719-83
1H-Pyrido[3,4-b]indol-1-ol, 2-acetyl-1-ethyl-2,3,4,9-tetrahydro-, acetate	MeOH	310(4.26)	4-0183-83

Compound	Solvent	λ_{max}(log ϵ)	Ref.
$C_{17}H_{20}N_2O_3$			
1H-Pyrrolo[2,3-d]carbazol-2(3H)-one, 7-acetyl-3a,4,5,6,6a,7-hexahydro-5-hydroxy-3a-methyl-	MeOH	253(4.11),278s(3.43), 287(3.28)	5-1744-83
Spiro[1,3-dioxolane-2,5'(4'H)-[1H]-pyrrolo[2,3-d]carbazol-2'(3'H)-one, 3'a,6',6'a,7'-tetrahydro-3'a-methyl-, (3'aα,6'aβ,11'bR*)-(±)-	MeOH	248(3.93),299(3.50)	5-1744-83
$C_{17}H_{20}N_2O_4$			
2H-Pyrrole-3,4-dicarboxylic acid, 2,2-dimethyl-5-(methylphenylamino)-, dimethyl ester	EtOH	247(4.18),323(3.43)	33-0262-83
$C_{17}H_{20}N_2O_6S$			
4-Thia-1-azabicyclo[3.2.0]heptane-2-carboxylic acid, 3,3-dimethyl-7-oxo-6-[(phenoxyacetyl)amino]-, methyl ester, 4-oxide	EtOH	211(3.89)	39-2241-83C
$C_{17}H_{20}N_2S$			
Promazine	pH 1	251(3.32),298s(2.79)	36-0050-83
	pH 3.5	250.5(3.21),298s(2.71)	36-0050-83
	H_2O	250.5(3.18),298s(2.54)	36-0050-83
	2N H_2SO_4	250(3.21),298s(2.72)	36-0050-83
radical cation	9N H_2SO_4	212(3.25),265(3.48), 271(3.51)	36-0050-83
$C_{17}H_{20}N_4O$			
3H-Pyrrole-3,3-dicarbonitrile, 4-ethoxy-2,4-dihydro-2,2-dimethyl-5-(methylphenylamino)-	EtOH	250(4.02)	33-0262-83
	EtOH	249.5(4.02),300s(2.96)	33-0262-83
7H-1,2,3-Triazolo[4,5,1-ij]quinolin-7-one, 6a-ethyl-4,5,6,6a,9a,9b-hexahydro-9-(phenylamino)-, (6aα,9aα,9bα)-	MeOH	205(4.11),225(4.14), 236(4.12),316(4.23)	35-3273-83
$C_{17}H_{20}N_4O_7$			
9H-Purine, 6-methyl-9-(2,3,5-tri-O-acetyl-β-D-ribofuranosyl)-	MeOH	261(3.88)	94-3104-83
$C_{17}H_{20}N_4O_8$			
9H-Purine-6-methanol, 9-(2,3,5-tri-O-acetyl-β-D-ribofuranosyl)-	MeOH	263(3.93)	94-3104-83
$C_{17}H_{20}O_3$			
9(1H)-Phenanthrenone, 2,3,4,4a-tetrahydro-10-hydroxy-6-methoxy-1,4a-dimethyl-, (1S-cis)-	EtOH	245(3.82),272(4.00), 323(3.90)	23-2461-83
$C_{17}H_{20}O_3S$			
1,4-Dioxaspiro[4.5]dec-7-ene, 8-ethenyl-7-[(phenylsulfinyl)methyl]-	EtOH	243(4.20)	78-1123-83
$C_{17}H_{20}O_4$			
2H-1-Benzopyran-2-one, 6-(3-ethoxy-3-methyl-1-butenyl)-7-methoxy-	MeOH	257(4.52),295(3.98), 306(3.98),343(4.15)	94-0901-83
2,4,6-Heptatrienoic acid, 7-(3,5-dimethoxyphenyl)-, ethyl ester, (E,E,E)-	EtOH	255(3.85),342.5(4.65)	39-2211-83C
(E,E,Z)-	EtOH	262(3.86),342(4.51)	39-2211-83C

Compound	Solvent	λ_{max}(log ε)	Ref.
$C_{17}H_{20}O_4S$			
1,4-Dioxaspiro[4.5]decane, 7-methylene-8-[2-(phenylsulfinyl)ethylidene]-	EtOH	237(3.93)	78-1123-83
$C_{17}H_{20}O_5$			
Achelenalin	n.s.g.	221(4.10),316(1.79)	162-0667-83
Spiro[1,3-dioxolane-2,4'-[4H]oxireno-[1,8a]naphtho[2,3-b]furan]-6'(9'bH)-one, 2',3',5',5'a-tetrahydro-5',5'a-7'-trimethyl-, (1'aS*,5'α,5'aα-9'bβ)-(±)-	EtOH	278.5(3.53)	94-3544-83
$C_{17}H_{20}O_6$			
1-Naphthaleneacetic acid, 1,4,4a,7,8,8a-hexahydro-8a-(methoxycarbonyl)-5-methyl-7,8-dioxo-, ethyl ester, (1α,4aβ,8aβ)-	MeOH	264(3.66)	44-4873-83
1-Naphthaleneacetic acid, 1,4,8,8a-tetrahydro-7-hydroxy-8a-(methoxy-carbonyl)-5-methyl-8-oxo-, ethyl ester, trans	MeOH	365(3.51)	44-4873-83
	MeOH-NaOH	410(3.51)	44-4873-83
Spiro[1,3-dioxolane-2,3'-[3H]oxireno-[8,8a]naphtho[2,3-b]furan]-5'(9'H)-one, 1'a,2',4',4'a-tetrahydro-9'-hydroxy-4',4'a,6' -trimethyl-	EtOH	266.5(3.52)	94-3544-83
$C_{17}H_{20}O_7S$			
α-D-xylo-Hex-5-enofuranose, 3-O-methyl-5-O,6-S-methylene-1,2-O-(1-methylethylidene)-6-C-phenyl-6-thio-, S,S-dioxide	EtOH	210(3.76),239(3.66)	159-0139-83
$C_{17}H_{20}O_{10}$			
2H-1-Benzopyran-2-one, 7-(β-D-glucopyranosyloxy)-6,8-dimethoxy-(eleutheroside B_1)	MeOH	209(4.55),229.5(4.32), 290.5(4.07),339(3.86)	94-0064-83
$C_{17}H_{21}BrO_2$			
Benzoic acid, 4-bromo-, 1-ethenyl-1,5-dimethyl-4-hexenyl ester, (R)-	EtOH	206.5(4.16),244.5(4.21)	18-1125-83
$C_{17}H_{21}ClO_5$			
Propanedioic acid, [4-(4-chloro-2-hydroxyphenyl)-2-butenyl]-, diethyl ester, (E)-	EtOH	218s(3.97),277(3.73)	78-1761-83
$C_{17}H_{21}F_3N_2O_3$			
4-Oxazolamine, 2,5-dihydro-5-methoxy-N,2,2-trimethyl-N-[2-(2-propenyl-oxy)phenyl]-5-(trifluoromethyl)-	EtOH	220(4.28),236s(3.87), 274(3.46),278s(3.44)	33-0262-83
$C_{17}H_{21}N$			
4H-Cycloocta[4,5]pyrrolo[3,2,1-ij]quinoline, 5,6,8,9,10,11,12,13-octahydro-	MeOH	210s(4.34),232(4.49), 286(3.93)	39-0515-83C
$C_{17}H_{21}NO$			
8-Azaspiro[5.6]dodec-10-en-9-one, 11-phenyl-	EtOH	210(4.07),268(4.16)	2-0710-83
Butanamide, N-[2-(1-azulenyl)propyl]-	EtOH	238(4.03),258s(4.00), 263s(4.18),268s(4.35), 274s(4.49),278(4.55),	18-2059-83

Compound	Solvent	λ max (log ε)	Ref.
(cont.)		282s(4.51),288s(4.34), 296s(3.55),320s(3.12), 333s(3.33),343(3.51), 359(3.23),517s(1.91), 534s(2.06),557s(2.21), 579s(2.28),597(2.34), 621s(2.28),686s(1.91), 722(1.78)	18-2059-83
1H-Carbazole, 9-(2,2-dimethyl-1-oxo-propyl)-2,3,4,9-tetrahydro-	EtOH	208(4.40),219(4.30), 263(4.00),302(3.50)	78-3767-83
3-Cyclohexen-1-one, 3,4-dimethyl-2-[[(2-phenylethyl)amino]methylene]-	EtOH	249(4.01),368(4.23)	104-0086-83
1-Propanol, 2-[(1,2-diphenylethyl)-amino]-	EtOH	205(4.18),248s(2.30), 252(2.45),258(2.57), 264(2.49),268(2.30)	94-0031-83
$C_{17}H_{21}NOS_2$			
1-Butanone, 4-(dimethylamino)-1-[2-[(2-thienylthio)methyl]phenyl]-, hydrochloride	MeOH	239(4.15),280s(3.57)	73-0623-83
$C_{17}H_{21}NO_2$			
Phenol, 4,4'-[1-(aminomethyl)-2-ethyl-1,2-ethanediyl]bis-	MeOH	208(4.04),223(4.26), 275(3.48),283(3.38)	83-0271-83
$C_{17}H_{21}NO_2S$			
Thieno[2,3-c]pyridine, 7-[(3,4-dimeth-oxyphenyl)methyl]-4,5,6,7-tetra-hydro-6-methyl-	MeOH	234(4.16),280(3.46)	83-0912-83
$C_{17}H_{21}NO_2S_2$			
1H-Isoindole-1,3(2H)-dione, 2-[5-(1,3-dithian-2-yl)pentyl]-	$CHCl_3$	244(4.18),295(3.42)	78-2691-83
$C_{17}H_{21}NO_3$			
Phenol, 4-[2-[[(3,4-dimethoxyphenyl)-methyl]amino]ethyl]- (latisodine)	MeOH	220(4.02),277(3.71), 288s(3.65)	150-0238-83S
$C_{17}H_{21}NO_5$			
Spiro[bicyclo[3.1.0]hex-3-ene-6,7'-[4,6]dioxa[2]azabicyclo[3.2.0]-heptane]-2,3'-dione, 2'-acetyl-1,1',3,4,5,5'-hexamethyl-	MeOH	205(2.90),248(2.98), 285(2.30)	73-2812-83
$C_{17}H_{21}N_2$			
Pyridinium, 2-[2-[4-(dimethylamino)-phenyl]ethenyl]-1-ethyl-, iodide	H_2O	433(4.35)	62-0957-83
	20% dioxan	452(4.42)	62-0957-83
	30% dioxan	456(4.43)	62-0957-83
	40% dioxan	458(4.44)	62-0957-83
	50% dioxan	459(4.46)	62-0957-83
	60% dioxan	460(4.48)	62-0957-83
	70% dioxan	462(4.55)	62-0957-83
	80% dioxan	459(4.47)	62-0957-83
	90% dioxan	454(4.46)	62-0957-83
	dioxan	442(4.40)	62-0957-83
$C_{17}H_{21}N_3O$			
Benzenamine, N,N-diethyl-4-[(4-methoxy-phenyl)azo]-	n.s.g.	317(3.9),400(4.3)	151-0249-83C

Compound	Solvent	λ_{max}(log ϵ)	Ref.
$C_{17}H_{21}N_3O_4$			
5-Pyrimidinepentanoic acid, 1,4-dihydro-6-hydroxy-4-oxo-2-[(phenylmethyl)amino]-, methyl ester	50% MeOH	241(3.889),273(4.174)	73-0304-83
	+ HCl	235(3.978),272(4.102)	73-0304-83
	+ NaOH	242(3.884),276(4.147)	73-0304-83
$C_{17}H_{21}N_5$			
Pyrazolo[5,1-c][1,2,4]triazin-7-amine, N,N-diethyl-5,6-dimethyl-2-phenyl-	EtOH	268(4.47),400(3.79)	44-2330-83
$C_{17}H_{21}N_6O_7P$			
5'-Adenylic acid, N-[(4-aminophenyl)-methyl]-	H_2O	264(4.32)	130-0045-83
$C_{17}H_{22}N_2$			
20,21-Dinoraspidospermidine, (±)-	EtOH	245(3.82),297(3.45)	78-3719-83
$C_{17}H_{22}N_2O$			
Pyrrolo[2,3-b]indole, 1-acetyl-1,2,3-3a,8,8a-hexahydro-3a-(3-methyl-2-butenyl)-	EtOH	213(3.91),244(3.32), 294(2.88)	1-0803-83
	EtOH-HCl	213(3.74),244(3.04), 294(2.57)	1-0803-83
$C_{17}H_{22}N_2O_2$			
Acetamide, N-[2-[2-(1-methyl-3-oxobutyl)-1H-indol-3-yl]ethyl]-	EtOH	225(4.57),275s(3.88), 282(3.92),290(3.86)	4-0267-83
2H-Indeno[5,4-f]quinazoline-2,7(4aH)-dione, 4b,5,6,6a,8,9,9a,9b,10,11-decahydro-4,6a-dimethyl-, [4aS-(4aα,4bβ,6aα,9aβ,9bα)]-	EtOH	204(4.83),223(4.87), 312(4.52)	39-2353-83C
$C_{17}H_{22}N_2O_2S$			
Carbamic acid, (3-cyclohexyl-4-phenyl-1,3-thiazetidin-2-ylidene)-, ethyl ester	dioxan	230(4.31),c.330(3.0)	24-1297-83
$C_{17}H_{22}N_2O_3$			
2-Butenoic acid, 4-[2-[(2,2-dimethyl-2H-azirin-3-yl)methylamino]phenoxy]-, ethyl ester, (E)-	EtOH	237s(3.81),278(3.41)	33-0262-83
(Z)-	EtOH	253.5(4.22),274(4.04)	33-0262-83
2-Pentanone, 1-methoxy-4-[[2-(5-methoxy-1H-indol-3-yl)ethyl]imino]-	EtOH	222(4.44),281s(4.03), 311(4.35)	4-0267-83
2-Pentanone, 5-methoxy-4-[[2-(5-methoxy-1H-indol-3-yl)ethyl]imino]-	EtOH	223(4.42),309(4.31)	4-0267-83
2-Propanone, 1-[2,3,4,9-tetrahydro-6-methoxy-1-(methoxymethyl)-1H-pyrido[3,4-b]indol-1-yl]-, monohydrochloride	EtOH	220(4.45),272(3.95), 295s(3.70),307s(3.59)	4-0267-83
$C_{17}H_{22}N_2O_3Se$			
2,4,6(1H,3H,5H)-Pyrimidinetrione, 5-methyl-5-[6-(phenylseleno)hexyl]-	MeOH	246.1(3.63),268.3(3.58)	101-0171-83S
$C_{17}H_{22}N_2O_5$			
2H-1-Benzopyran-3-carboxylic acid, 4-(acetylhydrazono)-3,4-dihydro-7-methoxy-2,2-dimethyl-, ethyl ester, (±)-	EtOH	210(4.18),239(4.09), 287(4.18),291s(4.22), 323(4.36)	39-0197-83C
$C_{17}H_{22}N_4O_2$			
2H-Pyrrole-3-carboxamide, 3-cyano-4-	EtOH	216s(3.96),254(4.04)	33-0262-83

Compound	Solvent	λ_{max}(log ε)	Ref.
ethoxy-3,4-dihydro-2,2-dimethyl-5-(methylphenylamino)- (cont.)			33-0262-83
$C_{17}H_{22}N_4O_3$			
3-Pyridinecarboxamide, 1,6-dihydro-N-methyl-1-[3-[5-[(methylamino)carbonyl]-1(2H)-pyridinyl]propyl]-6-oxo-	CH_2Cl_2	260(4.1),306(3.7), 347(3.6)	64-0878-83B
$C_{17}H_{22}N_4O_8$			
1H-Purine-6-methanol, 6,9-dihydro-9-(2,3,5-tri-O-acetyl-β-D-ribofuranosyl)-, (R)-	MeOH	297.5(3.62)	94-3104-83
(S)-	MeOH	299(3.68)	94-3104-83
$C_{17}H_{22}O_2$			
2,6α-Cyclo-A-nor-3,5-secoestr-1-ene-5,17-dione	EtOH	223(3.85)	39-2337-83C
2,8,10-Heptadecatriene-4,6-diyne-1,14-diol, (E,E,E)- (enanthotoxin)	n.s.g.	213(4.24),252(4.52), 267(4.46),281(4.24), 296(4.48),315.5(4.60), 337.5(4.46)	162-0515-83
8,10,12-Heptadecatriene-4,6-diyne-1,14-diol, (E,E,E)-(±)- (cicutoxin)	EtOH	242(4.16),252(4.34), 318.5(4.70),335.5(4.78)	162-0323-83
A-Norestr-3(5)-ene-2,17-dione	EtOH	232(4.20)	39-2337-83C
11-Oxatetracyclo[8.4.0.$0^{2,7}$.$0^{4,6}$]-tetradeca-1(10),2-dien-12-one, 4a-(1-methylethyl)-7β-methyl-	MeCN	268.5(4.02)	33-1806-83
$C_{17}H_{22}O_3$			
5H-Benzocyclohepten-5-one, 1,2,3,4-tetrahydro-6,7-dimethoxy-9-methyl-2-(1-methylethenyl)-, (R)-	EtOH	257(4.31),333(3.76)	78-2647-83
4H-1-Benzopyran-4-one, 2,3-dihydro-5-hydroxy-3-(1-methylethylidene)-7-pentyl-	EtOH	227(4.25),251(4.37), 331(3.54)	161-0775-83
$C_{17}H_{22}O_3S_2$			
Spiro[1,3-dithiolane-2,6'(4'H)-naphtho-[2,3-b]furan]-4'-one, 4'a,5',7',8'-8'a,9'-hexahydro-8'a-hydroxy-3',4'a,5'-trimethyl-, (4'aα,5'α,8'aα)-	EtOH	267(3.59)	94-3544-83
$C_{17}H_{22}O_4$			
Azuleno[5,6-c]furan-3(1H)-one, 4-acetoxy-4,4a,5,6,7,7a-hexahydro-6,6,8-trimethyl-, [4S-(4α,4aα,7aα)]-	EtOH	283(3.92)	56-0483-83
$C_{17}H_{22}O_5$			
Chrysanolide, dihydro-	MeOH	205(4.23)	100-0923-83
Cyclobuta[6,7]cycloocta[1,2-b]furan-2(3H)-one, 4-acetoxydecahydro-9-hydroxy-8a-methyl-3,6-bis(methylene)-	MeOH	220(4.21)(end abs.)	44-4251-83
Cyclobuta[6,7]cycloocta[1,2-b]furan-2(3H)-one, 4-acetoxy-3a,4,5,7,8,8a-9,9a-octahydro-9-hydroxy-6,8a-dimethyl-3-methylene-, [3aR-(3aα,4β-8aβ,9β,9aβ)]-	MeOH	220(4.14)(end abs.)	44-4251-83
Cyclodeca[b]furan-2(3H)-one, 4-acetoxy-3a,4,5,8,11,11a-hexahydro-11-hydroxy-6,10-dimethyl-3-methylene-	MeOH	210(4.28)(end abs.)	44-4251-83
α-Cyclolipiferolide	MeOH	220(3.66)(end abs.)	44-4251-83

Compound	Solvent	$\lambda_{max}(\log \epsilon)$	Ref.
β-Cyclolipiferolide	MeOH	210(3.97)(end abs.)	100-0923-83
Propanedioic acid, [4-(2-hydroxyphenyl)-2-butenyl]-, diethyl ester, (E)-	EtOH	213s(3.91),274(3.71)	78-1761-83
Xanthanolide	MeOH	220(4.68)(end abs.)	44-4251-83
$C_{17}H_{22}O_{12}$			
Mollugoside	n.s.g.	230(3.5)	102-0175-83
$C_{17}H_{23}BrO_3$			
3-Furanol, 5-(1-bromo-3-hexenyl)tetrahydro-2-(2-penten-4-ynyl)-, acetate (trans-kumausyne)	EtOH	216s(4.04),223(4.09), 230s(3.97)	138-1643-83
$C_{17}H_{23}ClO_5$			
Azuleno[4,5-b]furan-2(3H)-one, 4-acetoxy-6-chlorodecahydro-9-hydroxy-6,9-dimethyl-3-methylene-, [3aR-(3aα,4α,6α,6aα,9α,9aα,9bβ)]-	MeOH	212(3.86)	44-4251-83
Cyclobuta[6,7]cycloocta[1,2-b]furan-2(3H)-one, 4-acetoxy-6-chlorodecahydro-9-hydroxy-6,8a-dimethyl-3-methylene-	MeOH	220(4.13)(end abs.)	44-4251-83
Cyclodeca[b]furan-2(3H)-one, 4-acetoxy-10-chloro-3a,4,5,8,9,10,11,11a-octahydro-11-hydroxy-6,10-dimethyl-3-methylene-	MeOH	210(4.18)	44-4251-83
$C_{17}H_{23}FO_5$			
Cyclodeca[b]furan-2(3H)-one, 4-acetoxy-10-fluoro-3a,4,5,8,9,10,11,11a-octahydro-11-hydroxy-6,10-dimethyl-3-methylene-	MeOH	220(4.72)(end abs.)	44-4251-83
$C_{17}H_{23}NO$			
8-Azaspiro[5.6]dodecan-9-one, 11-phenyl-	EtOH	210(4.14)	2-0710-83
1H-Isoindole, 2,3-dihydro-1,1,3,3-tetramethyl-2-[(2-methylene-3-butenyl)oxy]-	EtOH	215(3.48),257(2.91), 263(3.01),269(3.00)	12-1957-83
$C_{17}H_{23}NO_2$			
4aH-Carbazol-4a-ol, 9-(2,2-dimethyl-1-oxopropyl)-1,2,3,4,9,9a-hexahydro-, cis	EtOH	208(4.30),252(4.10)	78-3767-83
Ethanone, 1-(4-cyclopentylphenyl)-2-morpholino-	MeOH	256(4.36)	73-0642-83
$C_{17}H_{23}NO_3$			
3-Buten-2-one, 4-(1,2,3,4-tetrahydro-5,6-dimethoxy-2,3-dimethyl-8-isoquinolinyl)-	MeOH	221s(4.19),250(3.95), 303(3.99)	83-0801-83
9H-Carbazole-9-carboxylic acid, 1,2,3,4,4a,9a-hexahydro-4a-hydroxy-, 1,1-dimethylethyl ester, cis	EtOH	205(4.30),234(4.10)	78-3767-83
2H-Isoindole-2-acetic acid, 1-ethyl-1,3-dihydro-3-oxo-, pentyl ester	$CHCl_3$	243(3.80),270s(3.30), 281(3.19)	78-2691-83
$C_{17}H_{23}NO_3S$			
1,2-Benzoxathiin-4-amine, 3,4,5,6,7,8-hexahydro-N-(1-methylethyl)-3-phenyl-, 2,2-dioxide, cis	EtOH	327(3.41)	4-0839-83

Compound	Solvent	λ_{max}(log ϵ)	Ref.
$C_{17}H_{23}NO_4$			
Lunacridine, (±)-	MeOH	239(4.38),258(4.42), 285(3.92),292(3.89), 333(3.49)	162-0800-83
$C_{17}H_{23}N_3$			
1H-Indole-3-ethanamine, 1-ethyl-N-(1-methyl-2-pyrrolidinylidene)-, (E)-2-butenedioate (1:1)	MeOH	282(3.79),286(3.79), 297s(3.68)	87-0230-83
Spiro[cyclohexane-1,3'-[3H]indazole], 2',3'a,4',5',6',7'-hexahydro-2'-(2-pyridinyl)-	EtOH	290(4.29)	25-0202-83
	EtOH-HCl	270(4.22),350(4.03)	25-0202-83
$C_{17}H_{24}N_2$			
4-Azuleneethanamine, 1-[(dimethylamino)methyl]-N,N-dimethyl-	EtOH	242(4.28),278s(4.49), 282(4.54),286(4.53), 290s(4.49),300s(3.98), 336(3.55),345(3.64), 359(3.38),543s(2.48), 575(2.76),617(2.51), 677s(2.09)	18-2311-83
$C_{17}H_{24}N_2O$			
1,3,5,8,10,12-Tridecahexaen-7-one, 1,13-bis(dimethylamino)-	EtOH	310(4.35),550(4.94)	70-1882-83
$C_{17}H_{24}N_2O_2S$			
Benzenesulfonic acid, 4-methyl-, (octahydro-4(1H)-azulenylidene)hydrazide, cis	EtOH	229(4.00),275(3.10)	44-1661-83
trans	EtOH	234(4.03),275(3.02)	44-1661-83
Benzenesulfonic acid, 4-methyl-, (octahydro-1(2H)-naphthalenylidene)hydrazide, trans	EtOH	232(4.02),277(2.95)	44-1661-83
$C_{17}H_{24}N_2O_3$			
1H-Pyrrole-3-propanoic acid, 5-[(3-ethyl-4,5-dihydro-4-methyl-5-oxo-1H-pyrrol-2-yl)methyl]-4-methyl-, methyl ester	MeOH	246(3.81)	5-0585-83
1H-Pyrrole-3-propanoic acid, 5-[(4-ethyl-4,5-dihydro-3-methyl-5-oxo-1H-pyrrol-2-yl)methyl]-4-methyl-, methyl ester	MeOH	212(3.75)	5-0585-83
2H-Pyrrol-2-one, 4-(dimethylamino)-1,5-dihydro-5,5-dimethoxy-1-(2,4,6-trimethylphenyl)-	heptane	200(4.46),215s(4.28), 276(4.11)	5-1504-83
$C_{17}H_{24}N_2O_4S$			
Glycine, N-[N-[(phenylmethoxy)carbonyl]thio-L-isoleucyl]-, methyl ester	$CHCl_3$	269(4.05)	78-1075-83
$C_{17}H_{24}N_2O_4Si$			
1H-Pyrazole-3,4-dicarboxylic acid, 4,5-dihydro-1-phenyl-4-(trimethylsilyl)-, 3-ethyl 4-methyl ester	EtOH	236(3.99),298(3.60), 349(4.19)	44-3189-83
$C_{17}H_{24}N_3O_2$			
1(2H)-Pyridinyloxy, 3,6-dihydro-2,2,6,6-tetramethyl-4-[[[(phenylamino)carbonyl]amino]methyl]-	EtOH	241(4.45),269s(3.45), 275s(3.38),284s(3.18), 432(0.90)	70-1659-83

Compound	Solvent	$\lambda_{max}(\log \epsilon)$	Ref.
$C_{17}H_{24}N_4O$			
2H-Pyrrole-3-carbonitrile, 3-(aminomethyl)-4-ethoxy-3,4-dihydro-2,2-dimethyl-5-(methylphenylamino)-	EtOH	252(3.97)	33-0262-83
$C_{17}H_{24}N_4O_4$			
Cycloheptanone, 3-(1,1-dimethylethyl)-, 2,4-dinitrophenylhydrazone, cis?	EtOH	230(4.07),369(4.12)	44-1670-83
trans?	EtOH	231(4.06),370(4.12)	44-1670-83
1H-1,2,4-Triazole-1,2(3H)-dicarboxylic acid, 3,3-dimethyl-5-(methylphenylamino)-, diethyl ester	EtOH	252(3.95)	33-0262-83
$C_{17}H_{24}N_4O_4S$			
Compd., m. 183-188°	MeOH	218(3.93),244(3.85), 300(3.57),337(4.02)	36-0372-83
Propanoic acid, 2,2-dimethyl-, [[3-[(2,2-dimethyl-1-oxopropoxy)methyl]-3H-purin-6-yl]thio]methyl ester	MeOH	227(3.96),326(4.23)	36-0372-83
Propanoic acid, 2,2-dimethyl-, [[7-[(2,2-dimethyl-1-oxopropoxy)methyl]-7H-purin-6-yl]thio]methyl ester	MeOH	218(3.99),256(3.56), 279s(3.96),287(4.03), 295s(3.93)	36-0372-83
Propanoic acid, 2,2-dimethyl-, [[9-[(2,2-dimethyl-1-oxopropoxy)methyl]-9H-purin-6-yl]thio]methyl ester	MeOH	218(4.17),226s(4.13), 278(4.31),283(4.27)	36-0372-83
$C_{17}H_{24}N_4S$			
1H-Imidazol-2-amine, 4,5-dihydro-1-methyl-N-[2-[(2-propyl-1H-indol-3-yl)thio]ethyl]-, (E)-2-butenedioate (1:1)	MeOH	273s(3.92),278(3.94), 282(3.94),288(3.90)	87-0230-83
$C_{17}H_{24}N_6O_8$			
1,2-Hydrazinedicarboxylic acid, 1,1'-[(dimethylamino)-4-pyridinylethenylidene]bis-, tetramethyl ester	EtOH	230(4.15),320(3.63)	104-0110-83
$C_{17}H_{24}O$			
Cyclohexanone, 4-(1,1-dimethylethyl)-2-(4-methylphenyl)-, cis	MeOH	259(2.56),265(2.61), 273(2.51)	12-0789-83
$C_{17}H_{24}OS_2$			
2,5-Cyclohexadien-1-one, 2,6-bis(1,1-dimethylethyl)-4-(1,3-dithiolan-2-ylidene)-	isooctane	395(4.59)	73-2376-83
$C_{17}H_{24}O_2$			
Farcarindiol	MeOH	232(3.24),245(3.22), 257.5(3.08),266(2.89)	94-2710-83
11-Oxatetracyclo[8.4.0.0^{2,7}.0^{4,6}]tetradec-1(10)-en-12-one, 4a-(1-methylethyl)-7β-methyl-	EtOH	222(3.54)	33-1806-83
$C_{17}H_{24}O_2S$			
Cyclohexanone, 5-(1-methoxy-1-methylethyl)-2-methyl-3-(phenylthio)-	MeOH	214s(3.99),259(3.81), 308(2.00)	49-0195-83
$C_{17}H_{24}O_3$			
1(2H)-Naphthalenone, 4-acetoxy-3,4,4a-5,6,8a-hexahydro-4a,8-dimethyl-2-(1-methylethylidene)-, (4α,4aα,8aα)-(±)-	EtOH	254(3.89)	78-3397-83

Compound	Solvent	λ_{max}(log ϵ)	Ref.
Tricyclo[4.4.0.0^{2,4}]dec-6-ene-7-propanoic acid, 4a-(1-methylethyl)-1-methyl-8-oxo-	EtOH	245(4.01)	33-1806-83
$C_{17}H_{24}O_4$			
1,1-Cyclobutanedicarboxylic acid, 3-(dicyclopropylmethylene)-, diethyl ester	hexane	212(4.18)	44-1500-83
2-Cyclohexen-1-one, 2-(5,5-dimethyl-3-oxo-1-cyclohexenyloxy)-3-methoxy-5,5-dimethyl-	$CHCl_3$	253(4.41)	142-1959-83
1H-Indene-1,5(4H)-dione, hexahydro-4-(5-methoxy-3-oxo-4-hexenyl)-7a-methyl-, [3aS-(3aα,4α,7aβ)]-	EtOH	256(4.15)	39-2353-83C
$C_{17}H_{24}O_5$			
Cyclodeca[b]furan-2(3H)-one, 9-acetoxy-3a,4,5,8,9,11a-hexahydro-4-hydroxy-3,6,10-trimethyl-	EtOH	218(3.37)	102-0197-83
Ethanone, 1-(3,4-diethoxy-3,4-dihydro-7-hydroxy-2,2-dimethyl-2H-1-benzopyran-6-yl)-, cis	EtOH	286(4.11),320(3.95)	39-0883-83C
Lactarolide B, 5-deoxy-, 8-acetate	EtOH	215(3.00)	56-0483-83
$C_{17}H_{24}O_6$			
Benzeneacetic acid, 2-(2-ethoxy-1,1-dimethyl-2-oxoethoxy)-4-methoxy-, ethyl ester	EtOH	212(4.18),222s(4.04), 279(3.53),286(3.51)	39-1431-83C
2,4-Cyclohexadien-1-one, 6-acetoxy-3,5-dimethoxy-2,6-dimethyl-4-(2-methyl-1-oxobutyl)-	EtOH	324(3.61),327s(3.79), 334(3.61),348s(3.60)	88-2531-83
Lactarolide A, 8-acetate	EtOH	209(3.69)	56-0483-83
Oxireno[5,6]cyclodeca[1,2-b]furan-7(2H)-one, 3-acetoxy-1a,3,5a,8,8a-9,10,10a-octahydro-9-hydroxy-4,8,10a-trimethyl-	EtOH	216(2.66)	102-0197-83
Shellolic acid, dimethyl ester	EtOH	231(3.79)	162-1218-83
$C_{17}H_{24}O_{11}$			
Asperulosidic acid, deacetyl-, methyl ester	H_2O	238(4.00)	100-0532-83
Scandoside, methyl ester	H_2O	239(3.87)	100-0532-83
$C_{17}H_{25}Cl_2N_3O_7$			
Alanylbactobolin	H_2O	282(4.00)	158-83-95
	pH 1	260(4.11)	158-83-95
	pH 13	286(4.28)	158-83-95
$C_{17}H_{25}NO$			
1H-Isoindole, 2,3-dihydro-1,1,3,3-tetramethyl-2-[(3-methyl-2-butenyl)oxy]-	EtOH	215(3.36),257(2.88), 263(2.98),270(3.00)	12-1957-83
1H-Isoindole, 2-[(1,1-dimethyl-2-propenyl)oxy]-2,3-dihydro-1,1,3,3-tetramethyl-	EtOH	215(3.39),258(2.91), 263(3.02),270(2.99)	12-1957-83
$C_{17}H_{25}NO_3$			
2-Butanone, 4-(1,2,3,4-tetrahydro-5,6-dimethoxy-2,3-dimethyl-8-isoquinolinyl)-, hydrochloride	MeOH	222s(3.94),282(3.29)	83-0801-83

Compound	Solvent	λ_{max}(log ϵ)	Ref.
3-Buten-2-ol, 4-(1,2,3,4-tetrahydro-5,6-dimethoxy-2,3-dimethyl-8-isoquinolinyl)-	MeOH	212(4.28),259(3.92), 300(3.35)	83-0801-83
Cyclizidine	EtOH	251(4.45)	158-83-162
$C_{17}H_{25}NO_6$			
1,3,5(4H)-Pyridinetricarboxylic acid, 2,4,6-trimethyl-, triethyl ester	EtOH	221(4.14),261(4.10)	73-0608-83
$C_{17}H_{25}N_5O_5$			
L-Lysine, N^{α}-acetyl-N^{t}-carboxyethyl-thymin-1-yl-, methylamide	0.5M NaOH	270(3.87)	47-2813-83
	6M HCl	272.5(c.3.9)	47-2813-83
$C_{17}H_{26}$			
Bicyclo[3.1.1]heptane, 2-(2-ethenyl-4-methyl-2,3-pentadienyl)-6,6-dimethyl-	EtOH	225(4.19)	70-0569-83
$C_{17}H_{26}N_2O_2$			
Ethanone, 2-[1-(3-aminopropyl)-2-pyrrolidinyl]-1-(2-methoxy-6-methylphenyl)-, (±)-	MeOH	225(3.16),245(2.71), 280(2.47)	100-0235-83
$C_{17}H_{26}N_2O_5S$			
4(1H)-Pyrimidinone, 2,3-dihydro-5-[2,3-O-(1-methylethylidene)-1-C-pentyl-β-DL-ribofuranosyl]-2-thioxo-	pH 13	232(4.13),261(4.23), 303(3.93)	18-2680-83
	MeOH	214(3.97),276(4.14), 289(4.10)	18-2680-83
4(1H)-Pyrimidinone, 2,3-dihydro-5-[2,3-O-(1-methylethylidene)-4-C-pentyl-β-DL-ribofuranosyl]-2-thioxo-	pH 13	220(4.24),264(4.13), 282s(4.04)	18-2680-83
	MeOH	213(4.14),276(4.24), 290s(4.20)	18-2680-83
4(1H)-Pyrimidinone, 2,3-dihydro-5-[tetrahydro-6-(1-hydroxyhexyl)-2,2-dimethylfuro[3,4-d]-1,3-dioxol-4-yl]-2-thioxo-	pH 13	222(4.23),264(4.13), 283(4.04)	18-2680-83
	MeOH	214(4.11),276(4.20), 291(4.17)	18-2680-83
$C_{17}H_{26}N_2O_6$			
2,4 (1H,3H)-Pyrimidinedione, 5-[2,3-O-(1-methylethylidene)-1-C-pentyl-β-DL-ribofuranosyl]-	pH 13	289(3.75)	18-2680-83
	MeOH	264(3.90)	18-2680-83
2,4(1H,3H)-Pyrimidinedione, 5-[2,3-O-(1-methylethylidene)-4-C-pentyl-β-DL-ribofuranosyl]-	pH 13	285(3.90)	18-2680-83
	MeOH	264(3.89)	18-2680-83
2,4(1H,3H)-Pyrimidinedione, 5-[tetrahydro-6-(1-hydroxyhexyl)-2,2-dimethylfuro[3,4-d]-1,3-dioxol-4-yl]-, (3aα,4α,6α,6aα)-(±)-	MeOH	263(3.83)	18-2680-83
	MeOH-NaOH	286(3.87)	18-2680-83
$C_{17}H_{26}N_3O_2$			
1-Piperidinyloxy, 2,2,6,6-tetramethyl-4-[[[(phenylamino)carbonyl]amino]-	EtOH	241(4.54),267s(3.56), 275s(3.46),284s(3.26), 449(1.02)	70-1659-83
$C_{17}H_{26}N_3O_4P$			
Sydnone imine, N-(dipropoxyphosphinyl)-3-(1-methyl-2-phenylethyl)-	EtOH	316(3.78)	65-1263-83
$C_{17}H_{26}O$			
2,5-Cyclohexadien-1-one, 2,6-bis(1,1-dimethylethyl)-4-(1-methylethylidene)-	isooctane	316(4.45)	73-2376-83

Compound	Solvent	λ_{max}(log ε)	Ref.
12-Oxatetracyclo[6.4.1.0^{2,10}.0^{4,8}]tridecane, 2,3,3,7-tetramethyl-11-methylene-	C_6H_{12}	209.5(3.95),224.5(3.48)	39-1373-83C
$C_{17}H_{26}OSi_2$			
Silane, [[1,1a-dihydro-1-(trimethylsilyl)-7bH-cyclobut[cd]azulen-7b-yl]oxy]trimethyl-, (1α,1aα,7bα)-	C_6H_{12}	240(4.26),245s(4.25), 306(3.83)	88-4261-83
$C_{17}H_{26}O_2$			
Germacra-4(15),5(E),10(14)-trien-1β-ol, acetate	EtOH	238(4.13)	94-1743-83
$C_{17}H_{26}O_2S$			
Cyclohexene, 4-methyl-1-(1-methylethyl)-2-[1-[1-(methylsulfonyl)ethylidene]-2-butynyl]-	MeCN	190(4.10),247(4.00)	4-0729-83
$C_{17}H_{26}O_3$			
Cyclohexanol, 3-(cyclohexylideneethylidene)-5-(methoxymethoxy)-4-methylene-, (1α,3Z,5α)-(±)-	ether	265(4.12)	88-4257-83
1-Naphthaleneacetic acid, 3,4,4a,5,6-7,8,8a-octahydro-2,5,5,8a-tetramethyl-3-oxo-, methyl ester, (4aS-trans)-	EtOH	248(3.75)	105-0139-83
$C_{17}H_{26}O_4$			
Acetic acid, [[3,5-bis(1,1-dimethylethyl)-1-methyl-4-oxo-2,5-cyclohexadien-1-yl]oxy]-	EtOH	235(4.53)	44-3696-83
Benzeneheptanoic acid, 3,5-dimethoxy-, ethyl ester	EtOH	221.5(3.89),273.5(3.18), 280(3.18)	39-2211-83C
$C_{17}H_{26}O_6$			
Acetic acid, (2-methoxyphenoxy)-, 2-(2-butoxyethoxy)ethyl ester	MeOH	272(3.41)	49-0359-83
$C_{17}H_{26}O_{11}$			
Shanzhiside, methyl ester	MeOH	236(3.9)	94-0780-83
$C_{17}H_{26}O_{12}$			
Ipolamiide, 6β-hydroxy-	MeOH	231(4.0)	102-1185-83
$C_{17}H_{27}NO_2$			
2-Buten-1-one, 3-(2-hydroxyethylamino)-3-methyl-1-(1,2,4a,5,6,7,8,8a-octahydro-2-methyl-1-naphthalenyl)-	EtOH	309(3.62)	88-5373-83
$C_{17}H_{27}NO_3$			
8-Isoquinolinepropanol, 1,2,3,4-tetrahydro-5,6-dimethoxy-α,2,3-trimethyl-	MeOH	221s(4.01),275(3.44)	83-0801-83
Nonanamide, N-[(4-hydroxy-3-methoxyphenyl)methyl]-	EtOH	229(3.89),281(3.51)	98-1326-83
Norcapsaicin, dihydro-	EtOH	228(3.87),280(3.49)	98-1326-83
$C_{17}H_{27}N_2O_2$			
1-Piperidinyloxy, 4-[(2-hydroxy-2-phenylethyl)amino]-2,2,6,6-tetramethyl-	MeOH	270(3)	28-0043-83B

Compound	Solvent	λ_{max}(log ε)	Ref.
$C_{17}H_{27}N_3O_5$			
4(1H)-Pyrimidinone, 2-amino-5-[2,3-O-(1-methylethylidene)-1-C-pentyl-β-DL-ribofuranosyl]-	pH 13	230(3.83),278(3.68)	18-2680-83
	MeOH	225(3.76),292(3.57)	18-2680-83
4(1H)-Pyrimidinone, 2-amino-5-[2,3-O-(1-methylethylidene)-4-C-pentyl-β-DL-ribofuranosyl]-	pH 13	216(3.96),233(3.99), 276(3.89)	18-2680-83
	MeOH	227(3.98),290(3.98)	18-2680-83
4(1H)-Pyrimidinone, 2-amino-5-[tetrahydro-6-(1-hydroxyhexyl)-2,2-dimethylfuro[3,4-d]-1,3-dioxol-4-yl]-, (3aα,4α,6α,6aα)-(±)-	pH 13	233(4.06),277(3.97)	18-2680-83
	MeOH	227(4.03),290(4.03)	18-2680-83
$C_{17}H_{27}N_3O_6$			
Glycine, N-[5-(5-butylhexahydro-2,4,6-trioxo-5-pyrimidinyl)-1-oxopentyl]-, ethyl ester	MeOH	212.5(3.98)	73-0137-83
$C_{17}H_{28}N_2O_2$			
2(3H)-Naphthalenone, 4,4a,5,6,7,8-hexahydro-1-[4-(2-hydroxyethyl)-1-piperazinyl]-4a-methyl-, (S)-	EtOH	242(4.00)	107-1013-83
2(3H)-Naphthalenone, 4,4a,5,6,7,8-hexahydro-8-[4-(2-hydroxyethyl)-1-piperazinyl]-4a-methyl-, (4aS-cis)-	EtOH	238(4.00)	107-1013-83
$C_{17}H_{28}N_4$			
2-Heptenenitrile, 3,3'-(1,3-propanediyldiimino)bis-	EtOH	263(4.49)	39-1223-83C
2-Hexenenitrile, 3,3'-(1,3-propanediyldiimino)bis[4-methyl-	EtOH	263(4.48)	39-1223-83C
$C_{17}H_{28}O_3$			
1-Cyclopentene-1-undecanoic acid, 5-oxo-, methyl ester	EtOH	228(3.87)	150-2701-83M
$C_{17}H_{28}O_3Si$			
Butanoic acid, 2-phenoxy-, (triethylsilyl)methyl ester	heptane	220(3.94),273(3.18)	65-1039-83
$C_{17}H_{30}$			
4-Hexadecyne, 6-methylene-	EtOH	225(4.04)	104-1621-83
$C_{17}H_{34}O_2Se$			
Octanoic acid, 8-(octylseleno)-, methyl ester	MeOH	216.0(3.45)	101-0031-83M
$C_{17}H_{34}O_2Te$			
Decanoic acid, 10-(hexyltelluro)-, methyl ester	MeOH	235.0(3.72)	101-0031-83M
Octanoic acid, 8-(octyltelluro)-, methyl ester	MeOH	233.0(3.64)	101-0031-83M
$C_{17}H_{35}O_3PS$			
Phosphonium, tributyl(3-methyl-4-sulfo-2-butenyl)-, hydroxide, inner salt, (Z)-	EtOH	235s(2.30),294(2.00)	65-1125-83
$C_{17}H_{39}CoN_5$			
Cobalt(2+), (N,N,4,8,11-pentamethyl-1,4,8,11-tetraazacyclotetradecan-1-ethanamine-N^1,N^4,N^8,N^{11})-, perchlorate	H_2O	488(1.34),542(1.36), 730(1.00)	33-2086-83

Compound	Solvent	λ_{max}(log ε)	Ref.
$C_{17}H_{39}CuN_5$			
Copper(2+), (N,N,4,8,11-pentamethyl-1,4,8,11-tetraazacyclotetradecan-1-ethanamine-N^1,N^4,N^8,N^{11})-, diperchlorate	H_2O	642(2.38)	33-2086-83
$C_{17}H_{39}N_5Ni$			
Nickel(2+), (N,N,4,8,11-pentamethyl-1,4,8,11-tetraazacyclotetradecan-1-ethanamine-N^1,N^4,N^8,N^{11})-, diperchlorate	H_2O	395(1.88),517(1.79), 657(1.38)	33-2086-83
$C_{17}H_{40}Si_4$			
Pentadiene, 1,1,3,5-tetrakis(trimethylsilyl)-	THF	262(4.45)	157-0021-83

Compound	Solvent	λ_{max}(log ϵ)	Ref.
$C_{18}H_6Cl_4N_2O_2S_2$			
6,13(7H,14H)-Triphenodithiazinedione, 1,3,8,10-tetrachloro-	H_2SO_4	348(4.83),748(4.83)	18-1482-83
$C_{18}H_8Br_2N_2O_2S_2$			
6,13(7H,14H)-Triphenodithiazinedione, 3,10-dibromo-	H_2SO_4	340(4.72),749(4.75)	18-1482-83
$C_{18}H_8Cl_2N_2O_2S_2$			
6,13(7H,14H)-Triphenodithiazinedione, 3,10-dichloro-	H_2SO_4	337(4.80),733(4.83)	18-1482-83
$C_{18}H_8N_2$			
Benzo[b]biphenylene-5,10-dicarbonitrile	EtOH	252(4.29),259(4.39), 283(4.45),300(4.31), 373(3.62),393(3.92), 417(4.07)	39-1443-83C
$C_{18}H_8N_6O_4S_3$			
Benzenesulfenamide, N,N'-(3,4-dicyano-2,5-thiophenediylidene)bis[4-nitro-, (Z,Z)-	$CHCl_3$	527(4.30)	150-0028-83S +150-0480-83M
$C_{18}H_8O_4$			
1H-Indene-1,3(2H)-dione, 2-(1,3-dihydro-1,3-dioxo-2H-inden-2-ylidene)-	MeCN	278(4.79),344(4.15)	32-0507-83
$C_{18}H_9ClN_2O_3S$			
7H-Benzimidazo[2,1-a]benz[de]isoquinoline-10-sulfonyl chloride, 7-oxo-	benzene	387(4.31)	73-2249-83
	EtOH	382(4.24)	73-2249-83
$C_{18}H_9N_3$			
5,10-Methanocyclopentacycloundecene-1,2,3-tricarbonitrile	MeCN	614(3.43)	88-2151-83
$C_{18}H_{10}$			
Benzo[3,4]cyclobuta[1,2-b]biphenylene	C_6H_{12}	276(4.54),285(4.69), 409(3.83),435(4.12)	88-0299-83
	THF	274(4.75),285(4.90), 406(4.10),432(4.31)	77-0502-83
Cyclopenta[cd]pyrene	MeOH	221(4.83),237(4.70), 276(4.29),286(4.36), 339(4.15),352(4.27), 365(4.06),372(4.15), 384(3.73)	54-0014-83
$C_{18}H_{10}ClNO_2S$			
2H-1-Benzopyran-2-one, 3-[2-(4-chlorophenyl)-4-thiazolyl]-	MeCN	348(4.34)	48-0551-83
$C_{18}H_{10}ClO_3S$			
1,3-Oxathiol-1-ium, 2-(4-chlorophenyl)-5-(2-oxo-2H-1-benzopyran-3-yl)-, perchlorate	MeCN	375(4.16)	48-0551-83
$C_{18}H_{10}Cl_4N_2S_2$			
Benzenesulfenamide, N,N'-(2,3,5,6-tetrachloro-2,5-cyclohexadiene-1,4-diylidene)bis-	$CHCl_3$	510(4.77)	150-0028-83S +150-0480-83M

Compound	Solvent	$\lambda_{max}(\log \epsilon)$	Ref.
$C_{18}H_{10}N_2O$			
7H-Benzimidazo[2,1-a]benz[de]isoquino-lin-7-one	benzene	385(4.08)	73-2249-83
	EtOH	382(4.13)	73-2249-83
$C_{18}H_{10}N_2O_2S_2$			
6,13(7H,14H)-Triphenodithiazinedione	H_2SO_4	334(4.79),723(4.75)	18-1482-83
$C_{18}H_{10}N_4O_2$			
Pyrido[2,1-b]pyrido[1',2':1,2]pyrimido-[4,5-g]quinazoline-7,15-dione	DMSO	458(3.60),487(3.90), 519(3.78)	18-1775-83
	H_2SO_4	420(3.58),437(3.88), 469(3.88)	18-1775-83
$C_{18}H_{10}N_4S_3$			
Benzenesulfenamide, N,N'-(3,4-dicyano-2,5-thiophenediylidene)bis-, (Z,Z)-	$CHCl_3$	542(4.45)	150-0028-83S +150-0480-83M
$C_{18}H_{10}O$			
Cyclopenta[cd]pyran-3(4H)-one	MeOH	228(4.66),248(4.71), 254(4.72),273(4.53), 285(4.56),356(4.37), 397(4.24)	54-0014-83
$C_{18}H_{10}O_2$			
Naphth[2,3-a]azulene-5,12-dione	MeCN	248(4.1),308(4.3), 415(3.6),585(2.8), 610(2.9),655(2.7)	24-2408-83
$C_{18}H_{10}O_3S$			
3H-Naphtho[2,1-b]pyran-3-one, 2-(2-thienylcarbonyl)-	benzene	376(4.11)	151-0325-83A
$C_{18}H_{10}O_4S$			
2H-1-Benzopyran-2-one, 3-[(2-oxo-2H-1-benzopyran-4-yl)thio]-	dioxan	235(4.04),270(4.022), 310(3.932)	117-0321-83
$C_{18}H_{10}O_5$			
4H-Furo[3,2-c][1]benzopyran-4-one, 2-(2-hydroxybenzoyl)-	MeOH	274(4.16),335(4.32)	39-0841-83C
$C_{18}H_{10}O_6$			
Anthra[1,2-b]furan-3-carboxylic acid, 6,11-dihydro-5-hydroxy-2-methyl-6,11-dioxo-	dioxan	275(4.11),422(3.81)	104-0533-83
$C_{18}H_{10}O_7$			
Anthra[2,3-b]furan-3-carboxylic acid, 4,11-dihydroxy-2-methyl-5,10-dioxo-	dioxan	267(4.51),490(4.01)	104-1900-83
$C_{18}H_{11}BrN_2S$			
3-Pyridinecarbonitrile, 4-(4-bromophenyl)-1,2-dihydro-6-phenyl-2-thioxo-	EtOH	246(4.11),286s(4.30), 312(4.33),420(3.59)	103-1202-83
$C_{18}H_{11}ClO_2$			
Azuleno[2,1,8-ija]azulene-5-carboxylic acid, 10-chloro-, methyl ester	MeOH	289(5.15),315s(4.05), 368(3.72),387(3.92), 409(3.98),451s(3.26), 480(3.75),640(2.58)	88-0781-83
$C_{18}H_{11}FN_2S$			
3-Pyridinecarbonitrile, 4-(4-fluoro-	EtOH	244(4.12),290(4.32),	103-1202-83

Compound	Solvent	$\lambda_{max}(\log \epsilon)$	Ref.
phenyl)-1,2-dihydro-6-phenyl-2-thioxo- (cont.)		312s(4.22),418(3.51)	103-1202-83
$C_{18}H_{11}NO_2$			
Chrysene, 4-nitro-	MeCN	251(4.37),275s(4.66?), 280(4.64),286(4.62)	1-0833-83
Chrysene, 5-nitro-	MeCN	277s(4.80?),280(4.78), 289(4.75)	1-0833-83
Chrysene, 6-nitro-	MeCN	260(5.03),370(--)	1-0833-83
Triphenylene, 1-nitro-	MeCN	258(4.88)	1-0833-83
Triphenylene, 2-nitro-	MeCN	244(4.62),252(4.59), 275(4.44),312(3.93), 335(3.94)	1-0833-83
$C_{18}H_{11}NO_2S$			
2H-1-Benzopyran-2-one, 3-(2-phenyl-4-thiazolyl)-	MeCN	350(4.32)	48-0551-83
$C_{18}H_{11}NO_3$			
Isoquinolinium, 2-(4-hydroxy-2-oxo-2H-1-benzopyran-3-yl)-, hydroxide, inner salt	EtOH	230(4.42),280(4.11), 380(3.68)	104-0283-83
	50% EtOH	377(3.62)	104-0283-83
	$CHCl_3$	408(3.83)	104-0283-83
$C_{18}H_{11}NO_5$			
Anthra[1,2-b]furan-3-carboxamide, 6,11-dihydro-5-hydroxy-2-methyl-6,11-dioxo-	dioxan	278(4.23),423(3.83)	104-0533-83
Oxoisocalycinine	EtOH	252(4.07),280(3.96), 320s(3.18)	100-0761-83
	EtOH-HCl	264(4.32),294(4.26), 360(3.66),390(3.66)	100-0761-83
$C_{18}H_{11}N_3$			
Cyclohepta[4,5]pyrrolo[1,2-a]imidazole-10-carbonitrile, 2-phenyl-	$CHCl_3$	247(4.43),280s(4.18), 342(4.45),355s(4.42), 435(3.62),457(3.63), 507s(2.95),546s(2.83), 595s(2.65),650s(2.29)	18-3703-83
2,3,9b-Triazaindeno[6,7,1-ija]azulene, 1-phenyl-	EtOH	234(4.35),261(4.38), 282(4.43),291(4.40), 322(3.93),335(4.01), 364(4.23),380s(4.21), 438(4.08),462(4.15)	18-3703-83
$C_{18}H_{11}N_3O_2S$			
3-Pyridinecarbonitrile, 1,2-dihydro-4-(3-nitrophenyl)-6-phenyl-2-thioxo-	EtOH	245s(4.38),270(4.43), 321(4.21),421(3.58)	103-1202-83
$C_{18}H_{11}N_3O_4$			
Benzonitrile, 2-[1,4-dihydro-3-(methylamino)-1,4-dioxo-2-naphthalenyl]-5-nitro-	toluene	443(3.49)	150-0168-83S
	EtOH	461(--)	150-0168-83S
$C_{18}H_{11}O_3$			
1-Benzopyrylium, 2-(2-oxo-2H-1-benzopyran-3-yl)-, perchlorate	MeCN	455(4.50)	48-0505-83
$C_{18}H_{11}O_3S$			
1,3-Oxathiol-1-ium, 5-(2-oxo-2H-1-benzopyran-3-yl)-2-phenyl-, perchlorate	MeCN	396(4.30)	48-0551-83

Compound	Solvent	λ_{max}(log ϵ)	Ref.
$C_{18}H_{11}O_4$			
1-Benzopyrylium, 7-hydroxy-2-(2-oxo-2H-1-benzopyran-3-yl)-, perchlorate	MeCN	471(4.52)	48-0505-83
$C_{18}H_{11}O_5$			
1-Benzopyrylium, 7-hydroxy-2-(7-hydroxy-2-oxo-2H-1-benzopyran-3-yl)-, perchlorate	MeCN	517(4.60)	48-0505-83
$C_{18}H_{12}$			
1H-Cyclobuta[de]naphthalene, 1-(phenylmethyl)-	EtOH	222(4.90),277(4.34), 287(4.41),397(4.36), 321[sic](4.13)	35-6104-83
Cyclopenta[cd]pyrene, 3,4-dihydro-	MeOH	232(4.55),242(4.74), 254(4.08),265(4.35), 276(4.63),313(3.88), 327(4.23),343(4.35), 356(3.49),369(2.77), 377(3.46)	54-0014-83
$C_{18}H_{12}Br_3N$			
Benzenamine, 4-bromo-N,N-bis(4-bromophenyl)-, radical cation	MeCN	361(4.32),702(4.5)	152-0105-83
$C_{18}H_{12}ClF_3N_2O$			
Bicyclo[4.4.1]undeca-3,6,8,10-tetraene-2,5-dione, mono[[4-chloro-2-(trifluoromethyl)phenyl]hydrazone, (±)-	CH_2Cl_2	393(4.42)	33-2369-83
	EtOH-KOH	548(4.68)	33-2369-83
$C_{18}H_{12}ClNO_3$			
2H-Furo[2,3-h]-1-benzopyran-2-one, 3-chloro-4-(methylphenylamino)-	EtOH	245.5(4.47),265s(3.96), 301(3.96),370(3.93)	111-0113-83
5H-Furo[2,3-f]indole-2-carboxylic acid, 8-chloro-5-phenyl-, methyl ester	CH_2Cl_2	228(4.20),272(4.36), 318(4.45),352(4.07)	4-1059-83
$C_{18}H_{12}ClN_3O_3$			
1H-Naphtho[2,3-g]indole-3-carboxamide, 5-amino-4-chloro-6,11-dihydro-2-methyl-6,11-dioxo-	dioxan	281(4.43),455(3.68), 556(3.91)	104-1892-83
$C_{18}H_{12}Cs_2$			
Cesium, μ-[2,2'-bi-1H-indene]-1,1'-diyldi-	n.s.g.	371(4.51),391(4.48)	35-2502-83
$C_{18}H_{12}Fe_2O_7$			
Iron, hexacarbonyl[μ-[η^4:η^4-5,6,7,8-tetrakis(methylene)bicyclo[2.2.2]octanone]]di-	EtOH	265s(3.42)	33-1134-83
$C_{18}H_{12}N_2$			
Benzo[h]quinazoline, 2-phenyl-	EtOH	237(4.47),266(4.48), 312(4.22),344(3.58), 364(3.53)	88-4351-83
1,4-Ethenobenzo[3,4]cyclobuta[1,2]cyclooctene-2,3-dicarbonitrile, 1,4,4a,10b-tetrahydro-	THF	306s(2.76)	89-1371-83S
$C_{18}H_{12}N_2O$			
Benzo[4,5]cyclohepta[1,2-c]pyrazol-4(1H)-one, 1-phenyl-	EtOH	257(4.54),306s(4.06), 318(4.10)	142-1581-83

Compound	Solvent	λ_{max}(log ϵ)	Ref.
Furo[3,4-d]pyridazine, 5,7-diphenyl-	CH_2Cl_2	248(4.37),270(4.21), 295(4.16),408(4.20)	118-1018-83
	CH_2Cl_2-HCl	239(4.48),290(4.53), 469(4.09)	118-1018-83
$C_{18}H_{12}N_2OS$			
Thiazolo[3,2-a]pyrimidin-4-ium, 3-hydroxy-2,7-diphenyl-, hydroxide, inner salt	MeCN	561(4.22)	124-0857-83
$C_{18}H_{12}N_2O_2$			
1H-Isoindole-1,3(2H)-dione, 2-(3H-3-benzazepin-3-yl)-	EtOH	225(4.38),290(3.02), 340(2.27)	44-1122-83
$C_{18}H_{12}N_2O_2S$			
2H-1-Benzopyran-2-one, 3-(2,3-dihydro-1-phenyl-2-thioxo-1H-imidazol-4-yl)-	MeCN	395(4.21)	48-0551-83
1H-Pyrrole-2,5-dione, 1-[4-(6-methyl-2-benzothiazolyl)phenyl]-	MeOH	329(4.54)	49-0599-83
$C_{18}H_{12}N_4$			
Benz[c]phenanthridine-12-carbonitrile, 6,11-diamino-	EtOH	220(4.37),234(4.36), 254(4.51),271(4.55), 293(4.30),303s(4.28), 319(4.24),360(4.15), 394s(3.79)	39-1137-83C
1,4-Ethanonaphthalene-6,6,7,7-tetracarbonitrile, 1,4,5,8-tetrahydro-8,10-bis(methylene)-	MeCN	236(3.96),243s(3.92), 257s(3.56)	33-0019-83
4-Isoquinolinecarbonitrile, 1-amino-3-[(2-cyanophenyl)methyl]-	THF	247(4.30),276(3.56), 284(3.66),315(4.08), 335(3.95)	39-1137-83C
$C_{18}H_{12}N_4O$			
1,4-Ethanonaphthalene-6,6,7,7-tetracarbonitrile, 1,2,3,4,5,8-hexahydro-2,3-bis(methylene)-9-oxo-	EtOH	249(3.82),293(2.73)	33-1134-83
	dioxan	250(3.83),297(2.70)	33-1134-83
$C_{18}H_{12}N_5O_6$			
Hydrazyl, 2,2-diphenyl-1-(2,4,6-trinitrophenyl)-	hexane	513(4.13)	49-1035-83
	C_6H_{12}	515(4.07)	49-1035-83
	benzene	526(4.04)	49-1035-83
	p-xylene	524(3.97)	49-1035-83
	MeOH	523(3.96)	49-1035-83
	EtOH	523(3.92)	49-1035-83
	isoPrOH	520(4.07)	49-1035-83
	acetone	526(3.96)	49-1035-83
	ether	520(4.08)	49-1035-83
	DMF	533(4.02)	49-1035-83
	$MeNO_2$	528(3.87)	49-1035-83
	pyridine	531(4.02)	49-1035-83
	DMSO	533(4.04)	49-1035-83
	CCl_4	523(4.12)	49-1035-83
(also other solvents)	$C_2H_2Cl_4$	523(4.05)	49-1035-83
$C_{18}H_{12}N_6$			
s-Triazine, 2,4,6-tri-2-pyridinyl-, iron complex	H_2O	593(4.35)	162-1393-83
$C_{18}H_{12}N_6O_2$			
1,3-Benzenediol, 2,4-bis(2H-benzotria-	$CHCl_3$	238(4.21),325(4.56)	49-0937-83

Compound	Solvent	$\lambda_{max}(\log \epsilon)$	Ref.
zol-2-yl)- (cont.)			49-0937-83
$C_{18}H_{12}N_6O_3$			
1,3,5-Benzenetriol, 2,4-bis(2H-benzo-triazol-2-yl)-	$CHCl_3$	239(4.13),340(4.60)	49-0937-83
$C_{18}H_{12}O$			
Cyclopenta[cd]pyren-3(4H)-one, 4a,5-dihydro-	MeOH	236(4.38),242(4.39), 260(4.49),290(4.22), 320(4.05),345(3.65), 366(3.54),398(2.32)	54-0014-83
1,11-Methanocyclopenta[a]phenanthren-17-one, 15,16-dihydro-	EtOH	266.5(4.77),277(4.76), 303(3.34),348(3.03), 365.5(2.86)	39-0087-83C
Oxirane, 1-pyrenyl-	EtOH	201(4.34),231(4.47), 240(4.76),262(4.23), 273(4.51),307(3.83), 320(4.28),336(4.47)	44-2930-83
$C_{18}H_{12}O_2$			
4-Pyreneacetic acid	MeOH	232(4.66),240(4.90), 252s(4.12),263(4.39), 274(4.61),307(4.03), 321(4.42),336(4.62)	54-0014-83
$C_{18}H_{12}O_3$			
9,10[3',4']-Furanoanthracene-12,14-dione, 9,10,11,15-tetrahydro-	EtOH	260(2.87),267(3.02), 274(3.09)	73-0112-83
Tanshinone I	MeOH	244(4.60),260s(--), 325(3.66),417(3.68)	94-1670-83
$C_{18}H_{12}O_4$			
2(3H)-Benzofuranone, 7-methyl-3-(7-methyl-2-oxo-3(2H)-benzofuranylidene)-, (E)-	CH_2Cl_2	227(4.06),254s(4.06), 258(4.09),400s(4.19), 411(4.21)	78-2147-83
Cyclobuta[1,2-c:4,3-c']bis[1]benzopyran-6,7-dione, cis	dioxan	271(3.85),279(3.83)	39-1083-83C
trans	dioxan	271(3.61),278(3.58)	39-1083-83C
1,4-Dioxaspiro[2.4]hept-5-en-7-one, 2-benzoyl-5-phenyl-, (Z)-	EtOH	252(4.20),315(4.15)	4-1389-83
4H-Furo[2,3-h]-1-benzopyran-4-one, 5-methoxy-2-phenyl-	MeOH	270(4.20),316(4.46)	102-0800-83
$C_{18}H_{12}O_5$			
4H-Anthra[1,2-b]pyran-7,12-dione, 6,11-dihydroxy-2-methyl-	MeOH	230(4.45),250(4.08), 273(3.82),285(3.76), 392(3.43),483(3.90)	5-1818-83
4H-Furo[3,2-c][1]benzopyran-4-one, 2,3-dihydro-2-(2-hydroxybenzoyl)-	MeOH	250(4.20),285(4.06), 310(4.06)	39-0841-83C
$C_{18}H_{12}O_6$			
9,10-Anthracenedione, 1,2-diacetoxy-	EtOH	331(3.78)	99-0420-83
9,10-Anthracenedione, 1,4-diacetoxy-	EtOH	340(3.77)	99-0420-83
9,10-Anthracenedione, 1,5-diacetoxy-	EtOH	342(3.76)	99-0420-83
9,10-Anthracenedione, 1,8-diacetoxy-	EtOH	338(3.78)	99-0420-83
[1,2'-Bi-1H-indene]-1',2,3'(2'H,3H)-trione, 2',3,3-trihydroxy-	MeOH	250(3.41),279(2.78)	24-1309-83
$C_{18}H_{12}O_7$			
1,3-Dioxolo[7,8][2]benzopyrano[4,3-b]-[1]benzopyran-6(4H)-one (pulcherrimin)	EtOH	272(4.32),377(3.92)	102-2835-83
	EtOH-NaOAc	269(4.31),377(3.98)	102-2835-83

Compound	Solvent	$\lambda_{max}(\log \epsilon)$	Ref.
Pulcherrimin (cont.)	EtOH-NaOMe	266(4.34),298(4.31), 405(3.87)	102-2835-83
	EtOH-AlCl$_3$	285(4.23),369(3.90), 425(3.96)	102-2835-83
	EtOH-AlCl$_3$-HCl	285(4.20),363(3.93), 423(3.90)	102-2835-83
	EtOH-NaOAc-boric acid	269(4.31),377(3.98)	102-2835-83
$C_{18}H_{12}S_2$			
1,3-Benzodithiole, 2-(7H-benzocyclohepten-7-ylidene)-	MeCN	250(4.4),285(4.0), 316(4.3),375(4.3), 388s(4.3)	142-2039-83
$C_{18}H_{13}AsCl_2O_2$			
1H-Arsole-1-acetic acid, α,α-dichloro-2,5-diphenyl-	EtOH	246(4.21),385(4.13)	101-0335-83I
$C_{18}H_{13}BrN_2S$			
3-Pyridinecarbonitrile, 4-(4-bromophenyl)-1,2,3,4-tetrahydro-6-phenyl-2-thioxo-	EtOH	260(4.21),325(3.96)	103-1202-83
compd. with piperidine (1:1)	EtOH	250s(4.38),289(4.31), 350s(3.60)	103-1202-83
$C_{18}H_{13}ClFN_3$			
Midazolam	isoPrOH	220(4.48)	162-0886-83
$C_{18}H_{13}ClN_2O$			
Pyrimido[2,1-a]isoindol-4(1H)-one, 2-(4-chlorophenyl)-1-methyl-	EtOH	233(4.48),254(4.49), 278(4.42),363(3.98), 375(4.03),426(3.47)	124-0755-83
$C_{18}H_{13}ClN_2O_2$			
Pyrazinol, 6-chloro-3,5-diphenyl-, acetate	EtOH	234s(3.94),251(4.00), 340(3.88)	4-0311-83
$C_{18}H_{13}ClN_2O_3$			
1,2-Benzisoxazole-3-acetonitrile, 5-chloro-α-[(2,5-dimethoxyphenyl)-methylene]-, (E)-	EtOH	288(4.03),380(3.83)	94-2023-83
(Z)-	EtOH	315(4.15),390(3.96)	94-2023-83
$C_{18}H_{13}ClN_2S$			
3-Pyridinecarbonitrile, 4-(4-chlorophenyl)-1,2,3,4-tetrahydro-6-phenyl-2-thioxo-	EtOH	260(4.23),330(3.93)	103-1202-83
compd. with piperidine (1:1)	EtOH	252s(4.30),286(4.28), 350s(3.58)	103-1202-83
$C_{18}H_{13}ClN_2S_2$			
Benzenesulfenamide, N,N'-(2-chloro-2,5-cyclohexadiene-1,4-diylidene)bis-	CHCl$_3$	490(4.70)	150-0028-83S +150-0480-83M
$C_{18}H_{13}ClO_3$			
1,4-Naphthalenedione, 2-chloro-3-(2,6-dimethylphenoxy)-	EtOH	246(4.31),252(4.34), 285(4.23),337(3.58)	104-0144-83
$C_{18}H_{13}Cs$			
Cesium, [2,2'-bi-1H-inden]-1-yl-	n.s.g.	345(4.42)	35-2502-83

Compound	Solvent	λ_{max}(log ε)	Ref.
$C_{18}H_{13}FN_2S$			
3-Pyridinecarbonitrile, 4-(4-fluoro-phenyl)-1,2,3,4-tetrahydro-6-phenyl-2-thioxo-	EtOH	259(4.28),332(3.98)	103-1202-83
compd. with piperidine (1:1)	EtOH	252s(4.25),287(4.24), 346s(3.53)	103-1202-83
$C_{18}H_{13}N$			
6-Chrysenamine	C_6H_{12}	228(4.77),242(4.39), 276(5.28),338(4.34)	151-0163-83C
	MeOH	227(4.72),242(4.59), 273(4.96),339(4.29)	151-0163-83C
	EtOH	227(4.71),242(4.58), 273.5(4.96),339(4.28)	151-0163-83C
	MeCN	228(4.71),242.5(4.59), 276(4.96),341(4.29)	151-0163-83C
	H_2O	233(--),273(--), 340(--)	151-0163-83C
$C_{18}H_{13}NO_2$			
9,10[3',4']-Pyrroloanthracene-12,14-dione, 9,10,11,15-tetrahydro-	EtOH	250(2.88),259(2.90), 265(2.96),272.5(2.98)	73-0112-83
$C_{18}H_{13}NO_2S$			
11H-Benzo[b]phenothiazine-6,11(12H)-dione, N-ethyl-	MeCN	280(4.43),507(4.62)	88-3567-83
11H-Benzo[b]phenothiazin-11-one, 6-ethoxy-	MeCN	240(4.35),312(4.27), 630(4.08)	88-3567-83
$C_{18}H_{13}NO_3$			
5H-Benzo[b]carbazole-6,11-dione, 8-methoxy-5-methyl-	EtOH	211(4.39),220s(4.19), 270(4.44),280(4.44), 304s(3.80),353(3.89), 382s(3.67),440s(3.23)	39-2409-83C
$C_{18}H_{13}NO_3S$			
Furo[2,3-g]quinoline, 2-methyl-9-(phenylsulfonyl)-	EtOH	242(4.82),333(4.30)	44-0774-83
$C_{18}H_{13}NO_4$			
7H-Dibenzo[de,h]quinolin-7-one, 6-hydroxy-5,9-dimethoxy-	EtOH	238s(4.38),254(4.62), 292s(3.76),307(3.53), 319(3.44),358s(3.77), 366(3.79),406s(3.51), 430(3.83),455(3.84)	78-3261-83
7H-Dibenzo[de,h]quinolin-7-one, 6-hydroxy-5,10-dimethoxy-	EtOH	219(4.52),252(4.50), 310(3.73),346(3.98), 380(3.87),402(3.97), 422(4.04)	78-3261-83
Isomoschatoline	EtOH	230(4.10),283(4.26), 363(2.96),467(3.53)	100-0761-83
	EtOH-HCl	222(4.28),288(4.25), 545(3.31)	100-0761-83
Peruvianine	EtOH	238(4.42),271(4.43), 289s(4.04),328s(3.63), 369(3.57),432(3.57)	100-0761-83
$C_{18}H_{13}NO_5$			
Oxosarcocapnidine	EtOH	252(4.26),342(3.34), 396(3.59)	88-2303-83
	EtOH-acid	217(4.28),265(4.05),	88-2303-83

Compound	Solvent	λ_{max}(log ε)	Ref.
Oxosarcocapnidine (cont.)		458(3.55)	88-2303-83
	EtOH-base	243(4.26),340(3.34), 400(3.57)	88-2303-83
$C_{18}H_{13}N_3O$			
Benzo[1,2-b:3,4-b']dipyrrole-8-carboxaldehyde, 1,6-dihydro-3-[(phenylimino)methyl]-	EtOH	210s(4.49),222(4.51), 271(4.20),310(4.44)	103-0488-83
Benzonitrile, 4-[(5-oxobicyclo[4.4.1]undeca-3,6,8,10-tetraen-2-ylidene)hydrazino]-	CH_2Cl_2	293(4.32),412(4.25)	33-2369-83
	EtOH-KOH	570(4.74)	33-2369-83
Benzonitrile, 4-[(2-oxo-4a,8a-methanonaphthalen-1(2H)-ylidene)hydrazino]-	CH_2Cl_2	443(3.90)	138-0653-83
Ethanone, 1-(3-diazo-2,5-diphenyl-3H-pyrrol-4-yl)-	EtOH	250(4.18)	142-0255-83
4H-Pyrrolo[3,4-c]pyridazin-4-one, 1,6-dihydro-5,7-diphenyl-	EtOH	217(4.20),228(4.22), 295(4.30)	142-0255-83
Rutecarpine	EtOH	278(3.83),290(3.88), 332(4.49),345(4.54), 364(4.44)	162-1194-83
$C_{18}H_{13}N_3O_2$			
2-Naphthalenecarboxamide, 4-diazo-3,4-dihydro-N-(4-methylphenyl)-3-oxo-	dioxan	405(4.0),428(3.9)	104-1363-83
Phenanthridinium, 5-[(2-cyano-3-methoxy-3-oxo-1-propenyl)amino]-, hydroxide, inner salt	EtOH	251(4.38),289(4.19), 375(3.41),445(3.89)	150-0260-83S
$C_{18}H_{13}N_3O_2S$			
3-Pyridinecarbonitrile, 1,2,3,4-tetrahydro-4-(3-nitrophenyl)-6-phenyl-2-thioxo-	EtOH	260(4.21),328(4.00)	103-1202-83
compd. with piperidine (1:1)	EtOH	252s(4.33),279(4.29), 338s(3.60)	103-1202-83
3-Pyridinecarbonitrile, 1,2,3,4-tetrahydro-4-(4-nitrophenyl)-6-phenyl-2-thioxo-	EtOH	262(4.22),325(3.86)	103-1202-83
compd. with piperidine (1:1)	EtOH	256s(4.35),281(4.42), 340s(3.81)	103-1202-83
$C_{18}H_{13}N_3O_2S_2$			
Benzenesulfenamide, N,N'-(2-nitro-2,5-cyclohexadiene-1,4-diylidene)bis-	$CHCl_3$	513(4.63)	150-0028-83S +150-0480-83M
$C_{18}H_{13}N_3O_3$			
1H-Naphtho[2,3-g]indole-3-carboxamide, 5-amino-6,11-dihydro-2-methyl-6,11-dioxo-	dioxan	279(4.44),446(3.50), 556(3.85)	104-1892-83
1H-Naphtho[2,3-d]triazole-4,9-dione, 1-[(4-methoxyphenyl)methyl]-	EtOH	228(4.64),244(4.72), 266(4.46),330(3.23)	87-0714-83
$C_{18}H_{13}N_3S$			
Quinoxaline, 2-methyl-3-(2-phenyl-4-thiazolyl)-	EtOH	251(3.99),299(3.68), 341s(3.51)	106-0829-83
Quinoxaline, 2-(4-methyl-2-phenyl-5-thiazolyl)-	EtOH	237(3.81),267(3.56), 308(3.77),377(3.90)	106-0829-83
$C_{18}H_{13}N_5$			
Imidazo[1,2-a]pyrimidine, 2-phenyl-3-(phenylazo)-	EtOH	244(4.31),280(4.38), 387(4.51)	4-0639-83

Compound	Solvent	$\lambda_{max}(\log \epsilon)$	Ref.
$C_{18}H_{13}N_5O$			
4H-Pyrimido[1,2-a]-1,3,5-triazin-4-one, 2,3-dihydro-3-phenyl-2-(phenylimino)-	dioxan	262(4.05),295(4.07), 400(2.99)	48-0463-83
$C_{18}H_{13}N_5O_2S$			
Imidazo[2,1-b]thiazole, 5-methyl-6-[(4-nitrophenyl)azo]-3-phenyl-	EtOH	358s(3.88),395(4.30)	4-0639-83
$C_{18}H_{13}O_2S$			
Thiopyrylium, 4-carboxy-2,6-diphenyl-, perchlorate	MeCN	228(4.21),256(4.42), 432(4.23)	22-0115-83
$C_{18}H_{13}O_3$			
Pyrylium, 4-carboxy-2,6-diphenyl-, perchlorate	MeCN	222(4.28),282(4.22), 320(3.73),434(4.37)	22-0115-83
$C_{18}H_{14}$			
Azuleno[2,1,8-ija]azulene, 5-ethyl-	MeOH	290(5.15),315s(3.04), 368(3.77),386(3.94), 408(3.98),448(3.22), 477(3.69),660(2.74)	88-0781-83
Cyclopenta[cd]pyrene, 3,4,4a,5-tetrahydro-	MeOH	218(4.71),259(4.64), 302(4.05)	54-0014-83
$C_{18}H_{14}BrNO$			
2(1H)-Pyridinone, 4-bromo-1-methyl-3,5-diphenyl-	MeOH	208.1(4.48),241.1(4.14), 322.0(3.87)	88-2973-83
4(1H)-Pyridinone, 2-bromo-1-methyl-3,5-diphenyl-	MeOH	205.9(4.49),238.4(4.41), 283.5(4.09)	88-2973-83
$C_{18}H_{14}BrN_3OS$			
2H-Thiazolo[3,2-a]-1,3,5-triazin-6(7H)-one, 7-[(4-bromophenyl)methylene]-3,4-dihydro-3-phenyl-	EtOH	237(4.26),335(4.53)	103-0961-83
$C_{18}H_{14}ClNO$			
2H-Pyrrol-2-one, 3-(4-chlorophenyl)-1,5-dihydro-4-methyl-5-(phenylmethylene)-	MeOH	228(3.98),236(3.96), 344(4.58)	4-0687-83
$C_{18}H_{14}ClNO_2$			
1,4-Naphthalenedione, 2-chloro-3-[4-(dimethylamino)phenyl]-	toluene	527(3.64)	150-0168-83S
	EtOH	558(--)	150-0168-83S
$C_{18}H_{14}ClNO_3$			
2H-Furo[2,3-h]-1-benzopyran-2-one, 3-chloro-5,6-dihydro-4-(methylphenylamino)-	EtOH	238(4.19),270s(3.70), 376(4.26)	111-0113-83
$C_{18}H_{14}ClN_2O$			
Pyrimido[2,1-a]isoindolium, 2-(4-chlorophenyl)-4,6-dihydro-1-methyl-4-oxo-, perchlorate	EtOH	205(4.45),268(4.45), 313(3.01)	124-0755-83
$C_{18}H_{14}ClN_3$			
Pyridinium, 1,1'-(3-chloro-1,2-indolizinediyl)bis-	H_2O	219(4.36),264(3.96), 390(3.42)	44-2629-83
$C_{18}H_{14}ClN_3OS$			
2H-Thiazolo[3,2-a]-1,3,5-triazin-6(7H)-one, 7-[(4-chlorophenyl)-	EtOH	235(4.36),335(4.51)	103-0961-83

Compound	Solvent	λ_{max}(log ε)	Ref.
methylene]-3,4-dihydro-3-phenyl-			103-0961-83
$C_{18}H_{14}Cl_2$			
1,3,5,11,13-Cyclopentadecapentaene-7,9-diyne, 15-(dichloromethylene)-6,11-dimethyl-	CH_2Cl_2	268s(4.21),282s(4.32),308(4.48),378s(3.71)	39-2997-83C
$C_{18}H_{14}Cl_2N_2O$			
4H-Pyrrolo[3,2-c]quinolin-4-one, 1-(3,4-dichlorophenyl)-1,2,3,5-tetrahydro-5-methyl-	MeOH	232(4.57),267s(4.07),340(4.10)	83-0897-83
$C_{18}H_{14}Cl_2N_4S$			
4H-[1,2,4]Triazolo[4,3-a][1,4]benzodiazepine, 8-chloro-6-(2-chlorophenyl)-1-[(methylthio)methyl]-	MeOH	250s(4.11),295(3.19)	73-0123-83
$C_{18}H_{14}FN_3OS$			
2H-Thiazolo[3,2-a]-1,3,5-triazin-6(7H)-one, 7-[(4-fluorophenyl)methylene]-3,4-dihydro-3-phenyl-	EtOH	235(4.24),330(4.43)	103-0961-83
$C_{18}H_{14}Fe_2O_7$			
Iron, hexacarbonyl[μ-[η^4:η^4-5,6,7,8-tetrakis(methylene)bicyclo[2.2.2]octan-2-ol]]di-	EtOH	287s(3.40)	33-1134-83
$C_{18}H_{14}NO_3S$			
Methanaminium, N-methyl-N-[5-(3-oxo-3H-naphtho[2,1-b]pyran-2-yl)-1,3-oxathiol-2-ylidene]-, perchlorate	MeCN	388(4.38)	48-0551-83
$C_{18}H_{14}N_2$			
Cyclohepta[4,5]pyrrolo[1,2-a]imidazole, 2-(4-methylphenyl)-	EtOH	250(4.28),264(4.28),325(4.61),345s(4.48),385(3.60),405(3.64),429(3.55),470(2.75),505(2.79),545(2.77),587s(2.65),650s(2.39)	18-3703-83
Tricyclo[9.3.1.$1^{4,8}$]hexadeca-1(15),4-6,8(16),11,13-hexaene-15,16-dicarbonitrile	EtOH	235(4.20),300(3.26)	35-0040-83
$C_{18}H_{14}N_2O$			
Benzo[4,5]cyclohepta[1,2-c]pyrazol-4(1H)-one, 9,10-dihydro-1-phenyl-	EtOH	280(4.18)	142-1581-83
4-Isoquinolinecarbonitrile, 1-ethoxy-3-phenyl-	EtOH	217(4.54),256(4.52),309(4.20),325s(4.10),339s(3.84)	39-1813-83C
Methanone, (2-methyl-4-phenyl-5-pyrimidinyl)phenyl-	EtOH	256.5(4.26),277s(4.11)	4-0649-83
Pyrimido[2,1-a]isoindol-4(1H)-one, 1-methyl-2-phenyl-	EtOH	235(4.38),250(4.36),365(3.88),376(3.91),428(3.48)	124-0755-83
$C_{18}H_{14}N_2O_2$			
[1]Benzopyrano[2,3-c]pyrazol-4(1H)-one, 3,6-dimethyl-1-phenyl-	EtOH	240(4.46),264(4.23)	118-0214-83
[1]Benzopyrano[2,3-c]pyrazol-4(1H)-one, 3-ethyl-1-phenyl-	EtOH	242(4.43),264(4.20)	118-0214-83

Compound	Solvent	λ_{max}(log ϵ)	Ref.
[1]Benzopyrano[4,3-c]pyrazol-4(1H)-one, 3,8-dimethyl-1-phenyl-	EtOH	260s(4.24),272s(3.94), 308(3.83)	88-0381-83
[1]Benzopyrano[4,3-c]pyrazol-4(1H)-one, 3-ethyl-1-phenyl-	EtOH	261(4.11),272(4.12), 299(3.91),308(3.86)	88-0381-83
[1]Benzopyrano[4,3-c]pyrazol-4(2H)-one, 3,8-dimethyl-2-phenyl-	EtOH	260s(4.24),272s(3.94), 305(3.83)	88-0381-83
[1]Benzopyrano[4,3-c]pyrazol-4(2H)-one, 3-ethyl-2-phenyl-	EtOH	259s(4.20),273s(4.00), 286s(3.94),299(3.84)	88-0381-83
4-Isoquinolinecarbonitrile, 1-methoxy-3-(4-methoxyphenyl)-	EtOH	218(4.51),234s(4.31), 255(4.13),284(4.31), 316(4.27),350s(3.80)	39-1813-83C
Pyrazinol, 3,5-diphenyl-, acetate	EtOH	244(3.81),255(3.81), 283s(3.56),322(3.56)	4-0311-83
2H-Pyrido[4,3-b]carbazole-9,10-dione, 2,5,11-trimethyl-	MeOH	230(4.33),307(4.33), 405(3.82),608(3.56)	87-0574-83
Pyrimido[2,1-a]isoindol-4(6H)-one, 2-(4-methoxyphenyl)-	EtOH	205(4.52),277(4.52), 308s(3.85)	124-0755-83
$C_{18}H_{14}N_2O_2S$			
4H-Pyrano[2',3':4,5]thieno[3,2-c]pyridin-4-one, 2-amino-7,9-dimethyl-3-phenyl-	MeOH	210(4.28),235(4.32), 330(4.00)	103-1167-83
$C_{18}H_{14}N_2O_5$			
Benzoic acid, 2-[(1,3-dihydro-1-methyl-3-oxo-2H-indol-2-ylidene)methyl]-5-nitro-	EtOH	214(4.20),232s(4.15), 266(4.13),320(3.90), 335s(3.87),505(3.57)	39-2413-83C
1,4-Naphthalenedione, 2-(2-methoxy-4-nitrophenyl)-3-(methylamino)-	toluene	443(3.46)	150-0168-83S
	EtOH	461(--)	150-0168-83S
$C_{18}H_{14}N_2O_5S$			
Benzenesulfonic acid, 3-[(4-acetoxy-1-naphthalenyl)azo]-, sodium salt	pH 7.5	374(4.01)	94-0162-83
Benzenesulfonic acid, 4-[(4-acetoxy-1-naphthalenyl)azo]-, sodium salt	pH 7.5	377(4.08)	94-0162-83
$C_{18}H_{14}N_2O_8S_2$			
1-Naphthalenesulfonic acid, 4-acetoxy-3-[(4-sulfophenyl)azo]-, disodium salt	pH 7.5	333(4.21)	94-0162-83
$C_{18}H_{14}N_2S$			
3-Pyridinecarbonitrile, 1,2,3,4-tetrahydro-4,6-diphenyl-2-thioxo-	EtOH	260(4.29),328(4.01)	103-1202-83
compd. with piperidine (1:1)	EtOH	250s(4.31),288(4.28), 348s(3.58)	103-1202-83
$C_{18}H_{14}N_2S_2$			
Benzenesulfenamide, N,N'-2,5-cyclohexadiene-1,4-diylidenebis-	$CHCl_3$	480(4.65)	150-0028-83S +150-0480-83M
$C_{18}H_{14}N_4$			
1,4-Ethanonaphthalene-6,6,7,7-tetracarbonitrile, 1,2,3,4,5,8-hexahydro-2,3-bis(methylene)-	MeCN	230(3.87),241(3.92), 247(3.91),259s(3.64)	33-0019-83
Pyrazolo[5,1-c][1,2,4]triazine, 8-methyl-3,7-diphenyl-	EtOH	286(4.75),324(3.74), 410(3.15)	44-2330-83
1H-Tetrazolium, 5-(1,3-cyclopentadien-1-yl)-1,3-diphenyl-, hydroxide, inner salt	MeCN	258(4.26),335(4.49), 475(3.08)	77-0789-83
	MeOH	464(--)	77-0789-83
	EtOH	476(--)	77-0789-83

Compound	Solvent	λ_{max}(log ε)	Ref.
$C_{18}H_{14}N_4O$			
1,4-Ethanonaphthalene-6,6,7,7-tetracarbonitrile, 1,2,3,4,6,8-hexahydro-9-hydroxy-2,3-bis(methylene)-, (1α,4α,9R*)-(±)-	EtOH	243(3.90)	33-1134-83
	dioxan	244(3.92)	33-1134-83
[1S-(1α,4α,9R*)]-	EtOH	243(3.90)	33-1134-83
	dioxan	244(3.93)	33-1134-83
4-Isoquinolinecarboxamide, 1-amino-3-[(2-cyanophenyl)methyl]-	EtOH	222(4.62),252(4.21), 277s(3.58),285s(3.76), 312(4.10),342s(3.78)	39-1137-83C
$C_{18}H_{14}N_4OS$			
Methanone, (1,5-diphenylformazanyl)-2-thienyl-	EtOH	230(3.90),327(4.03), 448(4.12)	104-0585-83
$C_{18}H_{14}N_4OS_2$			
Thiazolo[5,4-d]pyrimidin-7(6H)-one, 5-(methylthio)-6-phenyl-2-(phenylamino)-	MeOH	224s(--),258(4.08), 290(4.28),302(4.26), 345(4.31)	2-0243-83
$C_{18}H_{14}N_4OSe$			
Methanone, (1,5-diphenyl-3-formazanyl)-selenophen-2-yl-	EtOH	232(4.18),324(4.30), 446(4.40)	104-0585-83
$C_{18}H_{14}N_4O_2$			
Methanone, (1,5-diphenyl-3-formazanyl)-2-furanyl-	EtOH	230(3.87),322(4.03), 442(4.12)	104-0585-83
$C_{18}H_{14}N_4S_2$			
Cyclohepta[b]pyrrol-2-amine, 3,3'-dithiobis-	MeOH	216(4.48),280(4.75), 426(3.92)	18-1247-83
$C_{18}H_{14}O$			
17H-Cyclopenta[a]phenanthren-17-one, 15,16-dihydro-1-methyl-	EtOH	266(4.84),288(4.53), 303(4.30),359(3.77), 375(3.77)	39-0087-83C
Dibenz[b,g]azulen-5-one, 5,6,11,12-tetrahydro-	EtOH	238.5(4.26),280(3.47), 288(3.49),298(3.48)	94-2868-83
1,11-Methanocyclopenta[a]phenanthren-17-one, 15,16-dihydro-	EtOH	267s(--),277(4.52), 287(4.59),310(2.83), 322(3.92),338(3.88), 375(3.92)	39-0087-83C
$C_{18}H_{14}O_2$			
Benzo[c]phenanthrene-3,4-diol, 3,4-dihydro-	MeOH	272(4.57)	88-1349-83
19,20-Dioxatricyclo[14.2.1.1^{7,10}]eicosa-1,3,5,7,9,11,13,15,17-nonaene	hexane	247(3.69),273(3.35), 284(3.44),345(5.66), 393(3.40),413(3.78), 438(4.21),460(4.62), 503(1.71),520(2.01), 527(1.95),538(2.19), 546(2.88),560(2.51), 570(2.20),584(2.10), 600(2.17),607(2.53), 619(2.82),626(2.80), 632(3.10),641(2.73), 659(3.36),671(4.34), 704(1.66),716(0.96)	89-0471-83S
19,20-Dioxatricyclo[14.2.1.1^{6,9}]eicosa-2,4,6,8,10,12,14,16,18-nonaene	hexane	278(4.63),336(4.86), 345(4.85),382(3.53),	88-1045-83

Compound	Solvent	λ_{max}(log ε)	Ref.
(cont.)		403(3.73),423(4.11), 442(3.35),480(2.14), 489(2.10),498(2.01), 522(2.32),533(2.18), 544(2.05),564s(2.10), 573(2.46),587(2.15), 602(1.95),632(2.53)	88-1045-83
4-Pyreneacetic acid, 4,5-dihydro-	MeOH	258(4.62),280(4.08), 286(4.03),298(4.05)	54-0014-83
$C_{18}H_{14}O_3$			
2,4,6-Cycloheptatrien-1-one, 2-hydroxy-4-(1-oxo-5-phenyl-2,4-pentadienyl)-, (E,?)-	MeOH	258(4.28),345(4.57)	18-3099-83
	MeOH-NaOH	280(4.25),342(4.63), 450(3.78)	18-3099-83
Tanshinone I, 1,2-dihydro-	MeOH	240(4.16),266s(--), 290(3.96),330(2.92), 410(3.27)	94-1670-83
$C_{18}H_{14}O_4$			
2(5H)-Furanone, 4-hydroxy-5-methyl-3-[3-(1-naphthalenyl)-1-oxo-2-propenyl]-, (E)-	MeOH	219(4.57),402(4.17)	83-0115-83
2(5H)-Furanone, 4-hydroxy-5-methyl-3-[3-(2-naphthalenyl)-1-oxo-2-propenyl]-, (E)-	MeOH	218(4.06),269(4.06), 279(4.10),294(4.04), 372(4.39)	83-0115-83
$C_{18}H_{14}O_5$			
2H-1-Benzopyran-2-one, 5-[(benzoyloxy)methyl]-7-methoxy-	EtOH	321.5(4.10)	94-3014-83
$C_{18}H_{14}O_6$			
2-Anthraceneacetic acid, 9,10-dihydro-1,4-dihydroxy-9,10-dioxo-, ethyl ester	dioxan	284(3.76),480(3.76)	104-0533-83
9,10-Anthracenedione, 1,4,8-trihydroxy-2-(3-oxobutyl)-	MeOH	226(4.35),242(4.12), 288(3.51),400(3.31), 458(3.87),487(4.05), 508(3.89),519(3.82)	5-1818-83
2-Hexenedioic acid, 3-hydroxy-4-oxo-2,5-diphenyl-	pH 3.5	274(4.01)	102-0371-83
	pH 7.0	263(4.14),282(4.09)	102-0371-83
	EtOH	270(4.04)	102-0371-83
2-Propen-1-one, 1-(1,3-benzodioxol-5-yl)-3-(6-methoxy-1,3-benzodioxol-4-yl)-	EtOH	248(4.2),340(4.3)	94-3024-83
2-Propen-1-one, 1-(1,3-benzodioxol-5-yl)-3-(6-methoxy-1,3-benzodioxol-5-yl)-, (E)-	EtOH	246(4.12),310(4.01), 399(4.28)	94-3024-83
$C_{18}H_{14}O_7$			
2-Anthracenecarboxylic acid, 9,10-dihydro-3,8-dihydroxy-4-methoxy-1-methyl-9,10-dioxo-, methyl ester	MeOH	223(--),247s(4.04), 268(4.20),286s(4.10), 409(3.75)	94-4206-83
2-Anthracenecarboxylic acid, 9,10-dihydro-4,8-dihydroxy-3-methoxy-1-methyl-9,10-dioxo-, methyl ester	MeOH	229(4.22),254(4.05), 285s(3.71),290(3.71), 426s(3.70),442(3.73)	94-4206-83
9H-Xanthen-9-one, 1,7-diacetoxy-3-methoxy-	MeOH	240(4.58),270(4.05), 300(4.10)	162-0628-83
$C_{18}H_{14}O_8$			
5,12-Naphthacenedione, 7,8,9,10-tetrahydro-6,7,8,9,10,11-hexahydroxy-	EtOH	227s(4.03),235s(4.06), 252(4.32),256s(4.22),	39-0613-83C

Compound	Solvent	λ_{max}(log ε)	Ref.
[7R-(7α,8α,9β,10α)]- (cont.)		285(3.71),325s(3.20), 460(3.55),485(3.62), 519(3.38)	39-0613-83C
$C_{18}H_{14}S$			
4,8:11,15-Dimethenocyclotetradeca[b]-thiophene, 9,10-dihydro-	n.s.g.	225(4.5),275(4.4), 312(4.0)	24-3112-83
4,8:11,15-Dimethenocyclotetradeca[c]-thiophene, 9,10-dihydro-	n.s.g.	221(4.5),239s(4.5), 275(4.0)	24-3112-83
$C_{18}H_{14}S_3$			
6H-Cyclohepta[c]thiophene, 6-(1,3-benzodithiol-2-ylidene)-1,3-dimethyl-	MeCN	260s(4.6),266(4.6), 310(4.2),410(4.6)	142-2039-83
$C_{18}H_{15}B$			
Borane, triphenyl-, TCNE complex	CH_2Cl_2	385(2.34)	35-2175-83
$C_{18}H_{15}B_3O_3$			
Boroxin, triphenyl-	C_6H_{12}	236f(4.88)	32-0515-83
$C_{18}H_{15}ClN_2OS$			
3-Piperidinecarbonitrile, 4-(4-chlorophenyl)-6-hydroxy-6-phenyl-2-thioxo-, compd. with piperidine (1:1)	EtOH	240(4.13),285(4.13)	103-1202-83
$C_{18}H_{15}ClN_4S$			
4H-[1,2,4]Triazolo[4,3-a][1,4]benzodiazepine, 8-chloro-1-[(methylthio)-methyl]-6-phenyl-	MeOH	249(4.19)	73-0123-83
$C_{18}H_{15}ClO_6$			
Isopannarin	EtOH	216(4.23),260(4.19), 295s(4.12),316(4.20), 360(3.84)	12-1057-83
$C_{18}H_{15}Cl_2NO_2S_2$			
Morpholine, 4-[2-[5-chloro-2-[(4-chlorophenyl)thio]phenyl]-2-oxo-1-thioxo-ethyl]-	MeOH	272(4.34),354(3.86)	73-1173-83
$C_{18}H_{15}Cl_2NO_3$			
Acetamide, N-[3-(2,4-dichlorophenyl)-1-(6-hydroxy-5-oxo-1,3,6-cycloheptatrien-1-yl)-2-propenyl]-, (E)-	MeOH	248(4.51),309(3.85), 335(3.97),371(3.77), 397(3.67)	18-3099-83
$C_{18}H_{15}FN_2OS$			
3-Piperidinecarbonitrile, 4-(4-fluorophenyl)-6-hydroxy-6-phenyl-2-thioxo-, compd. with piperidine (1:1)	EtOH	241(4.00),285(4.00)	103-1202-83
$C_{18}H_{15}FeNS$			
Ferrocene, (2-methyl-5-benzothiazolyl)-	EtOH	240(4.68),285(4.29), 450(2.95)	104-0950-83
Ferrocene, (2-methyl-6-benzothiazolyl)-	EtOH	240(4.29),350(3.56), 450(2.99)	104-0950-83
$C_{18}H_{15}N$			
Benzenamine, N,N-diphenyl-	CCl_4	300(4.4)	65-1687-83
	CF_3COOH	255f(3.6)	65-1687-83
Pyridine, 2-methyl-4,6-diphenyl-	neutral	295(3.90)	39-0045-83B
	protonated	298(4.34)	39-0045-83B

Compound	Solvent	λ_{max}(log ε)	Ref.
$C_{18}H_{15}NO$			
2H-Benz[g]indol-2-one, 1,3,3a,9b-tetrahydro-4-phenyl-	EtOH	231(4.09),237(4.08), 303(4.25),313s(4.18), 330s(3.79)	44-1451-83
2H-Benz[g]indol-2-one, 1,3,5,9b-tetrahydro-4-phenyl-	EtOH	247(4.04)	44-1451-83
3-Butynamide, N-bicyclo[4.2.0]octa-1,3,5-trien-7-yl-4-phenyl-	EtOH	238.8(4.32),249.2(4.31), 264.2(3.50),270.8(3.46), 277.6(2.86),287.5(2.32)	44-1451-83
Phenanthro[9,10-d]oxazole, 2-propyl-	CH_2Cl_2	253s(4.83),257(4.89), 277(4.45),293(4.32), 306(4.38)	4-1019-83
2H-Pyrrol-2-one, 1,3-dihydro-4-methyl-3-phenyl-5-(phenylmethylene)-, (E)-	MeOH	226(3.93),322(4.35)	4-0687-83
(Z)-	MeOH	226(4.11),341(4.54)	4-0687-83
$C_{18}H_{15}NO_2$			
5H-Benzo[g]-1,3-benzodioxolo[6,5,4-de]-quinoline, 6,7-dihydro-7-methyl-	EtOH	256s(4.64),263(4.67), 333(4.07),388(3.60), 395s(3.57)	12-1061-83
[1]Benzopyrano[4,3-b]pyrrol-4(1H)-one, 2,3-dihydro-1-(phenylmethyl)-	MeOH	240(4.17),254s(4.06), 312(3.75),336s(3.92), 348(3.96)	5-0165-83
2-Naphthalenecarboxamide, N-hydroxy-N-(phenylmethyl)-	EtOH	226(4.75),270s(3.92), 280(3.91)	34-0433-83
1,4-Naphthalenedione, 2-[4-(dimethyl-amino)phenyl]-	C_6H_{12}	498(3.93)	150-0168-83S
	EtOH	544(--)	150-0168-83S
$C_{18}H_{15}NO_2S_2$			
2H-1,3-Thiazine-2-thione, 4,6-bis(4-methoxyphenyl)-	MeCN	255(3.94),384(4.30), 480(3.56)	118-0827-83
$C_{18}H_{15}NO_3$			
Anonaine, N-formyl-	EtOH	233s(--),274(4.23), 292s(--),315(3.73)	100-0761-83
6H-Anthra[1,9-cd]isoxazol-6-one, 5-butoxy-	EtOH	243(4.99),249(5.08), 446(4.17),464(4.17)	103-1057-83
6H-Anthra[1,9-cd]isoxazol-6-one, 5-(2-methylpropoxy)-	EtOH	244(4.98),249(5.07), 446(4.13),465(4.13)	103-1057-83
1H-Indene-3-carboxylic acid, 2-[[(4-methylphenyl)amino]carbonyl]-	dioxan	303(4.2)	104-1363-83
1,4-Naphthalenedione, 2-(4-methoxy-phenyl)-3-(methylamino)-	toluene	465(3.38)	150-0168-83S
	EtOH	480(--)	150-0168-83S
4H-Quinolizin-4-one, 1-acetyl-2-meth-oxy-3-phenyl-	MeOH	232(4.17),270(4.14), 381(4.15)	49-0349-83
$C_{18}H_{15}NO_3S$			
4H-Furo[3,4-b]indole, 1,3-dimethyl-4-(phenylsulfonyl)-	EtOH	240s(4.21),266s(3.82), 305(3.81)	88-5435-83
$C_{18}H_{15}NO_4$			
9,10-Anthracenedione, 1-acetoxy-2-(dimethylamino)-	EtOH?	463(3.70)	99-0420-83
Guattescidine, (-)-	EtOH	236(4.16),267(4.40), 302(3.93),323(3.83), 344(3.75),358(3.72)	100-0335-83 +100-0761-83
	EtOH-HCl	276(4.46),367(3.92), 420(3.65)	100-0335-83
1H-Indole-3-acetic acid, 1-benzoyl-2,3-dihydro-2-oxo-, methyl ester	MeOH	249(3.97),286(3.46)	142-0421-83

Compound	Solvent	λ_{max}(log ε)	Ref.
Norrufescine	EtOH	225s(3.56),248(3.83), 303(3.68),315s(3.36), 374(2.87)	100-0761-83
$C_{18}H_{15}N_2O$			
Pyrimido[2,1-a]isoindolium, 4,6-dihydro-1-methyl-4-oxo-2-phenyl-, perchlorate	EtOH	204(4.37),267(4.28), 313s(3.02)	124-0755-83
$C_{18}H_{15}N_2O_2$			
6H-Pyrido[4,3-b]carbazolium, 9,10-dihydro-2,5,11-trimethyl-9,10-dioxo-,	H_2O	220(4.33),288(4.38), 420(3.78)	87-0574-83
acetate	MeOH	220(4.31),295(4.36), 400s(3.85)	87-0574-83
$C_{18}H_{15}N_3OS_2$			
5H-Thiazolo[3,4-b][1,2,4]triazol-4-ium, 6-(4-methoxyphenyl)-3-(methylthio)-1-phenyl-, hydroxide, inner salt	MeCN	256(4.30),295(4.41), 412(4.27)	103-0749-83
$C_{18}H_{15}N_3O_2S$			
Anthra[1,2-c][1,2,5]thiadiazole-6,11-dione, 4-(diethylamino)-	toluene	548(4.05)	135-0283-83
	MeOH	560(--)	135-0283-83
	dioxan	545(4.06)	135-0283-83
	HOAc	560(4.15)	135-0283-83
	$CHCl_3$	565(4.16)	135-0283-83
5H-Thiazolo[2,3-b]quinazoline-2-acetamide, 2,3-dihydro-3-oxo-N-phenyl-	MeOH	224(4.56),239(4.53), 293(4.38)	83-0569-83
$C_{18}H_{15}N_3O_2Se$			
Anthra[1,2-c][1,2,5]selenadiazole-6,11-dione, 4-(diethylamino)-	toluene	585(4.02)	135-0283-83
	MeOH	595(--)	135-0283-83
	dioxan	580(3.94)	135-0283-83
	HOAc	600(4.17)	135-0283-83
	$CHCl_3$	600(4.14)	135-0283-83
$C_{18}H_{15}N_3O_3$			
Anthra[1,2-c][1,2,5]oxadiazole-6,11-dione, 4-(butylamino)-	toluene	515(4.00)	135-0283-83
Anthra[1,2-c][1,2,5]oxadiazole-6,11-dione, 4-(diethylamino)-	toluene	555(4.32)	135-0283-83
	MeOH	560(--)	135-0283-83
	dioxan	550(4.16)	135-0283-83
	HOAc	560(4.23)	135-0283-83
	$CHCl_3$	565(4.20)	135-0283-83
Bicyclo[4.4.1]undeca-3,6,8,10-tetraene-2,5-dione, 2-[(2-methyl-5-nitrophenyl)hydrazone]	CH_2Cl_2	423(4.61)	33-2369-83
	EtOH-KOH	630(4.74)	33-2369-83
Bicyclo[4.4.1]undeca-3,6,8,10-tetraene-2,5-dione, 2-[(4-methyl-2-nitrophenyl)hydrazone]-, (Z)-	CH_2Cl_2	382(4.19),455(4.23)	24-2881-83
$C_{18}H_{15}N_3O_3S$			
3-Piperidinecarbonitrile, 6-hydroxy-4-(4-nitrophenyl)-6-phenyl-2-thioxo-, compd. with piperidine (1:1)	EtOH	240(4.11),290(4.28)	103-1202-83
Tropeolin OO	pH 10	443(5.4)	60-0155-83
plus $CaCl_2$	pH 10	350(4.9)	60-0155-83
$C_{18}H_{15}N_3O_4$			
Benzoic acid, 4-[(2-amino-1,4,5,6-	50% MeOH-HCl	235(4.35)	73-0299-83

Compound	Solvent	λ_{max}(log ε)	Ref.
tetrahydro-4,6-dioxo-5-phenyl-5-pyrimidinyl)methyl]- (cont.)	+ NaOH	228(4.52),265(3.95)	73-0299-83
Bicyclo[4.4.1]undeca-3,6,8,10-tetraene-2,5-dione, mono[(4-methoxy-2-nitrophenyl)hydrazone]	CH_2Cl_2	480(4.18)	138-0653-83
1,2-Diazabicyclo[5.2.0]nona-3,5-dien-9-one, 2-benzoyl-8-(2,5-dioxo-1-pyrrolidinyl)-, cis-(±)-	MeOH	286(3.99)	5-1374-83
7H-Pyrazolo[1,5-a]perimidine-8,9-dicarboxylic acid, 7-methyl-, dimethyl ester	EtOH	221(4.51),253(4.16), 316(3.90),354(3.91)	94-1378-83
$C_{18}H_{15}N_3O_5$			
1,2-Diazabicyclo[5.2.0]nona-3,5-diene-2-carboxylic acid, 8-(1,3-dihydro-1,3-dioxo-2H-isoindol-2-yl)-9-oxo-, ethyl ester, cis-(±)-	MeOH	278(3.96)	5-1374-83
1,2-Diazabicyclo[5.2.0]nona-3,5-diene-2-carboxylic acid, 9-oxo-8-[(3-oxo-1(3H)-isobenzofuranylidene)amino]-, ethyl ester, cis-(±)-	MeOH	280(4.02)	5-1374-83
$C_{18}H_{15}N_3O_8$			
Propanedioic acid, [2-[[(2,4-dinitrophenyl)methylene]amino]phenyl]-, dimethyl ester	$CHCl_3$	355(4.12)	44-2468-83
$C_{18}H_{15}N_5$			
3H-Imidazo[1,2-b]pyrazol-3-one, 2-methyl-6-phenyl-, phenylhydrazone	EtOH	255(4.55),370(4.42)	4-0639-83
$C_{18}H_{15}N_7$			
3-Pyridinecarbonitrile, 4,6-diamino-2-(phenylamino)-5-(phenylazo)-	EtOH	224(4.44),252(4.26), 294(4.27),401(4.49)	49-0973-83
$C_{18}H_{15}N_7O_6$			
1H-Pyrazole-4-carboxylic acid, 5-amino-3-[nitro[(4-nitrophenyl)azo]methyl]-1-phenyl-, methyl ester	MeCN	390(4.46)	48-0041-83
$C_{18}H_{15}O$			
Pyrylium, 2-methyl-4,6-diphenyl-, perchlorate	MeCN	252(4.09),335(4.35), 360(4.43),410(3.91)	22-0115-83
Pyrylium, 4-methyl-2,6-diphenyl-, perchlorate	MeCN	236(4.23),274(4.29), 387(4.44)	22-0115-83
$C_{18}H_{15}P$			
Phosphine, triphenyl-	MTsG	260(4.1)	135-0658-83
$C_{18}H_{16}BrCl_3N_2O_5S$			
5-Thia-1-azabicyclo[4.2.0]oct-2-ene-2-carboxylic acid, 3-(bromomethyl)-8-oxo-7-[(phenylacetyl)amino]-, 2,2,2-trichloroethyl ester, 5-oxide	EtOH	284(3.98)	78-0337-83
$C_{18}H_{16}BrCl_3N_2O_6S$			
5-Thia-1-azabicyclo[4.2.0]oct-2-ene-2-carboxylic acid, 3-(bromomethyl)-8-oxo-7-[(phenoxyacetyl)amino]-, 2,2,2-trichloroethyl ester, 5-oxide, [5R-(5α,6α,7β)]-	EtOH	263s(3.77),268(3.83), 275(3.85),282s(3.81)	78-0337-83

Compound	Solvent	λ_{max}(log ε)	Ref.
[5S-(5α,6β,7α)]- (cont.)	EtOH	271s(3.94),276(3.99), 282s(3.98)	78-0337-83
$C_{18}H_{16}BrNO_3$			
Isoquinoline, 1-(2-bromo-4-methoxyphenyl)-6,7-dimethoxy-	EtOH	243(4.69),281s(3.77), 287s(3.70),316s(3.49), 330(3.61)	78-3261-83
Isoquinoline, 1-(2-bromo-5-methoxyphenyl)-6,7-dimethoxy-	EtOH	242(4.75),287(3.74), 317.5(3.60),330.5(3.62)	78-3261-83
$C_{18}H_{16}ClFN_4O$			
3H-1,4-Benzodiazepine-2-carboximidamide, 7-chloro-5-(2-fluorophenyl)-N'-hydroxy-N,N-dimethyl-	isoPrOH	221(4.58),320(4.86)	4-0551-83
$C_{18}H_{16}ClIO_4S$			
Iodonium, (4-chlorophenyl)(5-methyl-2-furanyl)-, 4-methylbenzenesulfonate	MeOH	202.7(4.41),221.6(4.38), 249(4.03)	44-2534-83
$C_{18}H_{16}ClNO$			
Isoquinoline, 7-chloro-3-ethoxy-4-methyl-1-phenyl-	EtOH	239(4.72),279(3.89), 289(3.96),299(3.92), 364(3.98)	44-1275-83
$C_{18}H_{16}ClNOS$			
[1]Benzothieno[3,2-b]azete, 6-chloro-2-ethoxy-2a,7b-dihydro-2a-methyl-7b-phenyl-	EtOH	256(4.09),300(3.27), 310(4.22)	44-1275-83
$C_{18}H_{16}ClNO_3$			
Acetamide, N-[3-(4-chlorophenyl)-1-(6-hydroxy-5-oxo-1,3,6-cycloheptatrien-1-yl)-2-propenyl]-, (E)-	MeOH	246(4.56),335(3.96), 370(3.75),394(3.65)	18-3099-83
$C_{18}H_{16}Cl_2$			
Pentadeca-3,5,7,10,12-pentaene-1,14-diyne, 9-(dichloromethylene)-3,13-dimethyl-, (E,E,Z,Z,E)-	CH_2Cl_2	263s(4.30),273(4.40), 282s(4.36),328(4.69), 343(4.71),362(4.59)	39-2997-83C
$C_{18}H_{16}Cl_2N_2O_2$			
2-Pyrrolidinone, 1-(3,4-dichlorophenyl)-3-[2-(methylamino)benzoyl]-	MeOH	216(4.59),232s(4.43)	83-0897-83
$C_{18}H_{16}Cl_2O_2$			
9,10-Ethanoanthracene, 11,12-dichloro-9,10-dihydro-2,3-dimethoxy-, anti-cis	MeCN	247(3.16),290(3.71)	35-7337-83
syn-cis	MeCN	260s(3.32),286.5(3.72)	35-7337-83
trans	MeCN	250s(3.34),287(3.77)	35-7337-83
$C_{18}H_{16}NS_2$			
1,3-Thiazin-1-ium, 2-(ethylthio)-4,6-diphenyl-, perchlorate	MeCN	264(4.08),370(4.35), 424(4.10)	118-0827-83
$C_{18}H_{16}N_2$			
6,7-Diazatricyclo[3.3.0.0^{2,4}]oct-6-ene, 8,8-diphenyl-, (1α,2β,4β,5α)-	EtOH	257s(4.1),330(2.5)	88-5727-83
$C_{18}H_{16}N_2O$			
10H-Pyrido[3,4-b]carbazole, 5,11-dimethyl-7-methoxy-	EtOH	234(4.26),280s(4.63), 287(4.63),306s(4.08),	44-2690-83

Compound	Solvent	$\lambda_{max}(\log \epsilon)$	Ref.
(cont.)		324s(3.84),338s(3.44), 430s(3.54),440(3.56)	44-2690-83
2(1H)-Pyrimidinone, 6-methyl-4-(4-methylphenyl)-1-phenyl-	MeOH	211(4.32),221(4.29), 285(4.34),325(4.14)	39-1773-83C
$C_{18}H_{16}N_2OS$			
3-Piperidinecarbonitrile, 6-hydroxy-4,6-diphenyl-2-thioxo-, compd. with piperidine (1:1)	EtOH	238(4.15),286(4.13)	103-1202-83
$C_{18}H_{16}N_2OS_2$			
Diazene, [2-furanyl(4,5,6,7-tetrahydro-1,3-benzodithiol-2-ylidene)methyl]-phenyl-	MeCN	517(4.12)	118-0840-83
$C_{18}H_{16}N_2O_2$			
Benz[cd]indol-5(1H)-one, 1-benzoyl-2,2a,3,4-tetrahydro-, 5-oxime	EtOH	237s(4.41),252(4.47), 309(3.81)	87-0522-83
	EtOH-NaOH	271(4.48),310s(4.04), 321s(3.91)	87-0522-83
1-Phenazinecarboxylic acid, 6-(3-methyl-2-butenyl)-	MeOH	255(4.81),355s(4.02), 367(4.15)	73-0527-83
$C_{18}H_{16}N_2O_3$			
Anonaine, N-carbamoyl-	EtOH	232(4.22),275(4.15), 313(3.53)	100-0761-83
4H-Furo[3,2-b]pyrrole-5-carboxylic acid, 4-(2-cyanoethyl)-2-phenyl-, ethyl ester	MeOH	233(3.88),335(4.82)	73-0772-83
$C_{18}H_{16}N_2O_4$			
9H-Pyrido[1,2,3-lm]pyrrolo[2,3-d]-carbazole-2,9,12(1H,3H)-trione, 11a,11b,13,13a-tetrahydro-11-hydroxy-13a-methyl-, (3aR*,11aα-11bβ,13aα)-(±)-	MeOH	216(4.07),253(4.13), 280s(3.78),290(3.72)	5-1744-83
$C_{18}H_{16}N_2O_4S$			
Spiro[2H-1-benzopyran-2,2'(3'H)-benzothiazole], 8-methoxy-3,3'-dimethyl-6-nitro- (photochromic form)	toluene	620(4.30)	33-0342-83
$C_{18}H_{16}N_2O_6$			
3H-Indole-3,3-dicarboxylic acid, 1,2-dihydro-2-(4-nitrophenyl)-, dimethyl ester	$CHCl_3$	255(--)	44-2468-83
Propanedioic acid, [2-[[(4-nitrophenyl)methylene]amino]phenyl]-, dimethyl ester	$CHCl_3$	337(4.00)	44-2468-83
Spiro[2H-1-benzopyran-2,2'(3'H)-benzoxazole], 3,8-dimethoxy-3'-methyl-6-nitro- (photochromic form)	toluene	610(4.20)	33-0342-83
$C_{18}H_{16}N_2S$			
4H-Pyrrolo[3,2-c]quinoline-4-thione, 1,2,3,5-tetrahydro-5-methyl-1-phenyl-	MeOH	270(4.34),382(4.22)	83-0897-83
$C_{18}H_{16}N_2S_3$			
Diazene, phenyl[(4,5,6,7-tetrahydro-1,3-benzodithiol-2-ylidene)-2-thienylmethyl]-	MeCN	513(4.29)	118-0840-83

Compound	Solvent	λ max(log ε)	Ref.
$C_{18}H_{16}N_3OS_2$			
5H-Thiazolo[3,4-b][1,2,4]triazol-4-ium, 6-(4-methoxyphenyl)-3-(methylthio)-1-phenyl-, perchlorate	HOAc	262(4.14),297(4.22), 390(4.05)	103-0749-83
$C_{18}H_{16}N_4$			
Acetaldehyde, (4,6-diphenyl-2-pyrimidinyl)hydrazone	EtOH	270(4.82),315(4.08)	65-0153-83
Isoquinoline, 1,1'-azobis[3,4-dihydro-	pH 1	297(4.04),307s(4.05), 325s(3.86)	4-0121-83
	MeOH	263(4.51),304s(3.66), 374(2.56)	4-0121-83
$C_{18}H_{16}N_4O$			
Propanedinitrile, (2,3,3a,4,6a,7-hexahydro-3a-methyl-2-oxo-1H-pyrrolo-[2,3-d]carbazol-5(6H)-ylidene)-	MeOH	245(4.22),298(3.42)	5-1744-83
$C_{18}H_{16}N_4O_2S$			
Isothiazolo[5,1-e][1,2,3]thiadiazole-7-S^{IV}, 1,6-dihydro-3,4-dimethyl-6-(4-nitrophenyl)-1-phenyl-	C_6H_{12}	203(4.34),284(4.20), 361(4.05),513(4.41)	39-0777-83C
$C_{18}H_{16}N_4O_4$			
1,14-Epoxy-3H-dibenzo[f,i][1,4,8]triazacycloundecine-3,5,10-trione, 4-amino-6,7,8,9-tetrahydro-2,13-dimethyl-	DMF	265(4.12),425(4.18), 446(4.19)	95-0049-83
$C_{18}H_{16}O$			
17H-Cyclopenta[a]phenanthren-17-one, 11,12,15,16-tetrahydro-1-methyl-	EtOH	276(4.55),286(4.62), 330(4.09),343(4.11), 376(3.87)	39-0087-83C
$C_{18}H_{16}O_2$			
Benzeneacetic acid, 2-ethenyl-α-(phenylmethylene)-, methyl ester	MeOH	202(4.21),252(3.88), 260(3.89),276(3.89)	39-1015-83B
2,4,6,12,14-Hexadecapentaene-8,10-diynedial, 7,12-dimethyl-	THF	257s(4.00),268(4.03), 318(4.38),340s(4.38), 375(4.44),405s(4.30)	18-1467-83
5,10-Methanocyclopentacycloundecene-3-carboxylic acid, ethyl ester	EtOH	625(2.90)	88-2151-83
3,4-Phenanthrenedione, 8-methyl-2-(1-methylethyl)-	MeOH	222s(4.41),237(4.54), 290(4.06),425(3.69)	94-1670-83
2-Propenoic acid, 3-phenyl-, 3-phenyl-2-propenyl ester, trans-trans	EtOH	216(3.45),223(3.25)	162-0327-83
Tricyclo[4.4.0.$0^{2,4}$]deca-1(10),6,8-triene-2-carboxylic acid, 3-phenyl-, methyl ester, endo	MeOH	254(2.48),259(3.55), 268(3.59),275(3.42), 304(3.07)	39-1015-83B
Tricyclo[10.2.2.$2^{5,8}$]octadeca-5,7,12,14,15,17-hexaene-3,10-dione	EtOH	220(4.02),271s(2.76), 282s(2.60),299(2.54), 312s(2.40)	94-2868-83
Tricyclo[11.3.1.$1^{5,9}$]octadeca-1(17),5,7,9(18),13,15-hexaene-3,11-dione, anti	EtOH	221s(4.27),276(3.13), 294s(2.96),303s(2.90), 313s(2.73)	94-2868-83
$C_{18}H_{16}O_3$			
1-Benzoxepin, 3,5-dimethoxy-4-phenyl-	heptane	236(4.14),305s(3.81)	64-0895-83B
Cyclobuta[b]benzofuran, 2a,7b-dihydro-2,7b-dimethoxy-1-phenyl-	heptane	272(4.23)	64-0895-83B

Compound	Solvent	λ_{max}(log ε)	Ref
$C_{18}H_{16}O_4$			
4H-1-Benzopyran-4-one, 7-methoxy-2-(4-methoxyphenyl)-3-methyl-	MeOH	210(4.43),245(4.25), 304(4.29)	2-0759-83
4H-1-Benzopyran-4-one, 7-methoxy-3-[(4-methoxyphenyl)methyl]-	EtOH	217(4.35),241(4.24), 285(3.94),295(3.96), 305(3.93)	2-0759-83
$C_{18}H_{16}O_5$			
3(2H)-Benzofuranone, 2-[(4-methoxyphenyl)methylene]-4,6-dimethoxy-	MeOH	340(4.21),392(4.45)	102-0794-83
2H-1-Benzopyran-2-one, 5,7-dimethoxy-4-(4-methoxyphenyl)-	MeOH	250(4.07),325(4.29)	102-1657-83
4H-1-Benzopyran-4-one, 2,3-dihydro-3-[(4-hydroxyphenyl)methylene]-7,8-dimethoxy-	EtOH	328(4.20),358(4.20)	102-2835-83
	EtOH-NaOMe	271(3.96),310(4.04), 400(4.32)	102-2835-83
	EtOH-NaOAc	328(4.23),378(4.23)	102-2835-83
	+ boric acid	328(4.17),358(4.20)	102-2835-83
	EtOH-$AlCl_3$	328(4.19),358(4.20)	102-2835-83
	+ HCl	328(4.19),358(4.20)	102-2835-83
$C_{18}H_{16}O_5$(cont.)			
4H-1-Benzopyran-4-one, 5,7-dimethoxy-2-(4-methoxyphenyl)-	EtOH	264(4.33),324(4.40)	102-0625-83
4H-1-Benzopyran-4-one, 5-hydroxy-2-(4-hydroxyphenyl)-7-methoxy-6,8-dimethyl- (relative absorbance given)	MeOH	280(0.87),294s(0.84), 328(1.00)	102-2107-83
	MeOH-$AlCl_3$	283(0.67),310(0.76), 350(1.00),410s(0.50)	102-2107-83
	MeOH-$AlCl_3$-HCl	284(0.70),309(0.82), 348(1.00),406s(0.47)	102-2107-83
4H-1-Benzopyran-4-one, 5-hydroxy-7-methoxy-2-(4-methoxyphenyl)-6-methyl- (relative absorbance given)	MeOH	270(0.88),328(1.00)	102-2107-83
	MeOH-$AlCl_3$	262s(0.60),279(0.82), 292s(0.83),302(0.87), 346(1.00),388s(0.69)	102-2107-83
	MeOH-$AlCl_3$-HCl	262s(0.61),278(0.85), 292s(0.86),301(0.91), 342(1.00),387s(0.64)	102-2107-83
2,4,6-Cycloheptatrien-1-one, 4-[3-(3,4-dimethoxyphenyl)-1-oxo-2-propenyl]-2-hydroxy-, (E)-	MeOH	252(4.45),368(4.30)	18-3099-83
	MeOH-NaOH	263(4.34),350(4.47), 430(3.86)	18-3099-83
$C_{18}H_{16}O_6$			
2H-1-Benzopyran-2-one, 4-(3-hydroxy-4-methoxyphenyl)-5,7-dimethoxy-	MeOH	252(4.13),329(4.32)	102-1657-83
	MeOH-NaOMe	250(--),288s(--), 329(--),400s(--)	102-1657-83
2H-1-Benzopyran-2-one, 4-(4-hydroxy-3-methoxyphenyl)-5,7-dimethoxy-	MeOH	251(4.07),329(4.19)	102-1657-83
	MeOH-NaOMe	250(--),331(--), 395(--)	102-1657-83
4H-1-Benzopyran-4-one, 5,7-dihydroxy-2-(4-hydroxyphenyl)-3-methoxy-6,8-dimethyl- (relative absorbance given)	MeOH	254s(0.68),275(1.00), 300s(0.76),332(0.70)	102-2107-83
	MeOH-$AlCl_3$	264s(0.77),284(1.00), 310(0.92),360(0.91), 416s(0.56)	102-2107-83
	MeOH-$AlCl_3$-HCl	260s(0.79),284(1.00), 308s(0.91),356(0.88), 414s(0.50)	102-2107-83
4H-1-Benzopyran-4-one, 5-hydroxy-6,7-dimethoxy-2-(4-methoxyphenyl)-(salvigenin)(relative absorbance given)	MeOH	276(0.81),329(1.00)	102-2107-83
	MeOH-$AlCl_3$	264s(0.50),293s(0.74), 301(0.79),356(1.00)	102-2107-83
	MeOH-$AlCl_3$-HCl	262s(0.54),291s(0.82), 301(0.87),348(1.00)	102-2107-83

Compound	Solvent	$\lambda_{max}(\log \epsilon)$	Ref.
4H-1-Benzopyran-4-one, 5-hydroxy-7,8-dimethoxy-2-(4-methoxyphenyl)- (relative absorbance given)	MeOH	273(1.00),299(0.86), 319(0.88),356s(0.51)	102-2107-83
	MeOH-$AlCl_3$	272s(0.71),283(0.83), 310(0.93),343(1.00), 404(0.37)	102-2107-83
	MeOH-$AlCl_3$-HCl	272s(0.79),285(0.96), 311(1.00),342(0.98), 404(0.35)	102-2107-83
4H-1-Benzopyran-4-one, 7-hydroxy-5,8-dimethoxy-2-(4-methoxyphenyl)-	EtOH	272(4.40),300s(4.31), 335s(4.20)	95-0675-83
	EtOH-NaOAc	281(4.56),300s(4.38), 370(4.05)	95-0675-83
1,2-Cyclobutanedicarboxylic acid, 3,4-bis(2-hydroxyphenyl)-	H_2O	224(4.03),271(3.65), 279(3.62)	39-1083-83C
4H,6H-Furo[3',2':3,4]naphtho[1,8-cd]-pyran-4,6-dione, 8,9-dihydro-3,7-dihydroxy-1,8,8,9-tetramethyl-, (R)-	EtOH	256(4.32),290s(3.83), 360(4.08),322(4.11), 376s(3.98),392(4.15)	78-2283-83
	EtOH-NaOH	290s(4.23),306(4.08), 322(4.11),376s(3.08), 392(4.15)	78-2283-83
1H,3H-Naphtho[1,8-cd]pyran-1,3-dione, 4,9-dihydroxy-6-methyl-7-[(3-methyl-2-butenyl)oxy]-	EtOH	249(4.04),290(3.61), 309(3.49),348(3.87), 366s(3.79)	78-2283-83
	EtOH-NaOH	248s(4.11),302(3.92), 317(4.04),352s(3.69), 384(3.96)	78-2283-83
$C_{18}H_{16}O_6S$			
2H-1-Benzopyran-2-one, 4-(hydroxymethyl)-7-methoxy-, 4-methylbenzenesulfonate	EtOH	323(4.16)	94-3014-83
$C_{18}H_{16}O_7$			
4H-1-Benzopyran-4-one, 5,7-dihydroxy-3,6-dimethoxy-2-(4-methoxyphenyl)- (relative absorbance given)	MeOH	272(0.92),338(1.00)	102-2107-83
	MeOH-$AlCl_3$	262s(0.53),282(0.75), 297s(0.67),305s(0.65), 360(1.00),404s(0.54)	102-2107-83
	MeOH-$AlCl_3$-HCl	262(0.53),282(0.80), 296s(0.73),306s(0.67), 360(1.00),404s(0.51)	102-2107-83
4H-1-Benzopyran-4-one, 5,7-dihydroxy-3,8-dimethoxy-2-(4-methoxyphenyl)- (relative absorbance given)	MeOH	273(1.00),300s(0.64), 322(0.65),356s(0.55)	102-2107-83
	MeOH-$AlCl_3$	283(1.00),312(0.76), 350(0.84),416(0.49)	102-2107-83
	MeOH-$AlCl_3$ HCl	284(1.00),312(0.77), 348(0.83),418(0.47)	102-2107-83
4H-1-Benzopyran-4-one, 5,7-dihydroxy-6,8-dimethoxy-2-(4-methoxyphenyl)- (relative absorbance given)	MeOH	284(1.00),329(0.75),	102-2107-83
	MeOH-$AlCl_3$	265s(0.38),290s(0.76), 310(1.00),356(0.97), 413s(0.32)	102-2107-83
	MeOH-$AlCl_3$-HCl	262s(0.37),289s(0.75), 309(1.00),351(0.93), 404s(0.25)	102-2107-83
4H-1-Benzopyran-4-one, 5,7-dihydroxy-2-(3,4,5-trimethoxyphenyl)- (relative absorbance given)	MeOH	270(0.95),310s(0.90), 331(1.00)	102-2107-83
	MeOH-$AlCl_3$	253s(0.66),278(0.97), 300(0.76),348(1.00), 385s(0.72)	102-2107-83
	MeOH-$AlCl_3$-HCl	280(1.00),298s(0.75), 340(0.96),382s(0.47)	102-2107-83

Compound	Solvent	λ_{max}(log ε)	Ref.
4H-1-Benzopyran-4-one, 2-(3,4-dimethoxyphenyl)-5,7-dihydroxy-3-methoxy- (relative absorbance given)	MeOH	253(1.00),267(0.88), 291s(0.50),353(0.94)	102-2107-83
	MeOH-$AlCl_3$	266(1.00),276s(0.96), 300(0.47),362(0.75), 402(0.81)	102-2107-83
	MeOH-$AlCl_3$-HCl	265s(0.99),277(1.00), 299s(0.51),355(0.81), 401(0.73)	102-2107-83
4H-1-Benzopyran-4-one, 2-(3,4-dimethoxyphenyl)-5,8-dihydroxy-7-methoxy-	EtOH	281(4.32),335(4.26)	18-3773-83
	EtOH-NaOAc	283(4.30),335(4.28)	18-3773-83
	EtOH-$AlCl_3$	288(4.25),330(4.27), 350(4.29),415(3.82)	18-3773-83
4H-1-Benzopyran-4-one, 5-hydroxy-2-(3-hydroxy-4-methoxyphenyl)-3,7-dimethoxy- (relative absorbance given)	MeOH	256(1.00),265s(0.81), 293s(0.40),353(0.86)	102-2107-83
	MeOH-$AlCl_3$	270(1.00),275s(0.90), 300(0.45),356(0.68), 402(0.79)	102-2107-83
	MeOH-$AlCl_3$-HCl	268(1.00),274s(0.96), 298s(0.43),360(0.64), 401(0.70)	102-2107-83
4H-1-Benzopyran-4-one, 5-hydroxy-2-(3-hydroxy-4-methoxyphenyl)-6,7-dimethoxy- (eupatorin) (relative absorbance given)	MeOH	244(0.70),252s(0.68), 276(0.70),342(1.00)	102-2107-83
	MeOH-$AlCl_3$	262(0.72),287(0.81), 297s(0.78),370(1.00)	102-2107-83
	MeOH-$AlCl_3$-HCl	261(0.63),284(0.74), 292(0.75),363(1.00)	102-2107-83
	EtOH	243(4.24),254(4.29), 274(4.30),342(4.44)	162-0563-83
4H-1-Benzopyran-4-one, 5-hydroxy-2-(4-hydroxy-3-methoxyphenyl)-3,7-dimethoxy- (pachypodol) (relative absorbance given)	MeOH	254(1.00),262s(0.83), 354(0.99)	102-2107-83
	MeOH-$AlCl_3$	269(1.00),280(0.87), 300(0.45),368(0.75), 404(0.92)	102-2107-83
	MeOH-$AlCl_3$-HCl	268(1.00),278(0.94), 300(0.48),362(0.74), 402(0.79)	102-2107-83
4H-1-Benzopyran-4-one, 5-hydroxy-2-(4-hydroxy-3-methoxyphenyl)-6,7-dimethoxy- (cirsilineol) (relative absorbance given)	MeOH	240s(0.72),254s(0.63), 276(0.67),344(1.00)	102-2107-83
	MeOH-$AlCl_3$	260(0.62),284(0.69), 292s(0.61),374(1.00)	102-2107-83
	MeOH-$AlCl_3$-HCl	260(0.62),286(0.69), 294s(--),368(1.00)	102-2107-83
4H-1-Benzopyran-4-one, 5-hydroxy-2-(4-hydroxyphenyl)-3,6,7-trimethoxy- (penduletin)(relative absorbance given)	MeOH	271(0.92),340(1.00)	102-2107-83
	MeOH-$AlCl_3$	264s(0.58),282(0.71), 302s(0.57),367(1.00), 404s(0.65)	102-2107-83
	MeOH-$AlCl_3$-HCl	262s(0.58),282(0.78), 302s(0.63),361(1.00), 402s(0.55)	102-2107-83
4H-1-Benzopyran-4-one, 5-hydroxy-2-(4-hydroxyphenyl)-3,7,8-trimethoxy- (relative absorbance given)	MeOH	273(1.00),300s(0.58), 328(0.68),362(0.65)	102-2107-83
	MeOH-$AlCl_3$	282(1.00),312(0.64), 354(0.90),423(0.56)	102-2107-83
	MeOH-$AlCl_3$-HCl	283(1.00),311(0.69), 350(0.89),420(0.52)	102-2107-83
4H-1-Benzopyran-4-one, 5-hydroxy-2-(4-hydroxyphenyl)-6,7,8-trimethoxy- (xanthomicrol)	EtOH	281(4.26),293s(4.25), 335(4.38)	40-0161-83
	EtOH-NaOAc	277(4.20),402(4.42)	40-0161-83
	EtOH-$AlCl_3$	288(4.18),311(4.31), 352(4.42)	40-0161-83

Compound	Solvent	λ_{max}(log ε)	Ref.
Xanthomicrol (relative absorbance given) (cont.)	MeOH	281(0.90),294s(0.86), 332(1.00)	102-2107-83
	MeOH-$AlCl_3$	267s(0.35),288(0.45), 311(0.68),361(1.00), 407s(0.32)	102-2107-83
	MeOH-$AlCl_3$-HCl	265s(0.31),290(0.55), 311(0.72),354(1.00), 408s(0.25)	102-2107-83
$C_{18}H_{16}O_8$			
4H-1-Benzopyran-4-one, 5,7-dihydroxy-2-(3-hydroxy-4,5-dimethoxyphenyl)-3-methoxy- (relative absorbance given)	MeOH	250s(--),264(1.00), 304s(--),345(0.83)	102-2107-83
	MeOH-$AlCl_3$	276(1.00),305(0.15), 348(0.66),400(0.50)	102-2107-83
	MeOH-$AlCl_3$-HCl	278(1.00),304(0.26), 345(0.43),398(0.55)	102-2107-83
4H-1-Benzopyran-4-one, 5,7-dihydroxy-2-(3-hydroxy-4,5-dimethoxyphenyl)-6-methoxy- (relative absorbance given)	MeOH	274(0.90),332(1.00)	102-2107-83
	MeOH-$AlCl_3$	284(0.70),298s(0.60), 358(1.00)	102-2107-83
	MeOH-$AlCl_3$-HCl	289(0.80),353(1.00)	102-2107-83
4H-1-Benzopyran-4-one, 5,7-dihydroxy-2-(4-hydroxy-3,5-dimethoxyphenyl)-6-methoxy-	MeOH	242s(0.80),275(0.66), 350(1.00)	102-2107-83
	MeOH-$AlCl_3$	258s(0.62),282(0.67), 310s(0.57),382(1.00)	102-2107-83
	MeOH-$AlCl_3$-HCl	258s(0.67),283(0.72), 308s(0.67),372(1.00)	102-2107-83
4H-1-Benzopyran-4-one, 5,8-dihydroxy-2-(4-hydroxy-3,5-dimethoxyphenyl)-7-methoxy-	EtOH	279(4.23),318s(4.14), 346(4.28)	18-3773-83
	EtOH-NaOAc	268(4.21),313(4.01), 345(4.08),430(4.20)	18-3773-83
	EtOH-$AlCl_3$	288(4.20),332s(4.17), 361(4.31),415s(3.90)	18-3773-83
4H-1-Benzopyran-4-one, 5,6-dihydroxy-2-(4-hydroxy-3-methoxyphenyl)-3,7-dimethoxy- (chrysosplenol C) (relative absorbance given)	MeOH	256s(0.67),281(0.74), 352(1.00)	102-2107-83
	MeOH-$AlCl_3$	243s(0.66),266(0.60), 296(0.73),378(1.00)	102-2107-83
	MeOH-$AlCl_3$-HCl	242s(0.71),263(0.63), 296(0.73),378(1.00)	102-2107-83
4H-1-Benzopyran-4-one, 5,6-dihydroxy-2-(4-hydroxy-3-methoxyphenyl)-7,8-dimethoxy- (thymonin) (relative absorbance given)	MeOH	250s(0.53),292(0.80), 343(1.00)	102-2107-83
	MeOH-$AlCl_3$	241s(0.71),261(0.50), 306(0.69),316(0.64), 380(1.00)	102-2107-83
	MeOH-$AlCl_3$-HCl	240s(0.66),259s(0.47), 305(0.64),316(0.58), 369(1.00)	102-2107-83
4H-1-Benzopyran-4-one, 5,7-dihydroxy-2-(3-hydroxy-4-methoxyphenyl)-3,6-dimethoxy- (relative absorbance given)	MeOH	258(0.87),269(0.77), 346(1.00)	102-2107-83
	MeOH-$AlCl_3$	267(0.92),279s(0.83), 300s(0.51),375(1.00), 406s(0.88)	102-2107-83
	MeOH-$AlCl_3$-HCl	264(0.88),280s(0.83), 300s(0.56),364(1.00), 400s(0.75)	102-2107-83
4H-1-Benzopyran-4-one, 5,7-dihydroxy-2-(3-hydroxy-4-methoxyphenyl)-6,8-dimethoxy- (acerosin) (relative absorbance given)	MeOH	254s(0.62),281(0.79)	102-2107-83
	MeOH-$AlCl_3$	263(0.59),284(0.71), 304s(0.65),368(1.00), 412s(0.53)	102-2107-83
	MeOH-$AlCl_3$-HCl	261(0.61),292(0.75), 304s(0.71),363(1.00),	102-2107-83

Compound	Solvent	λ max (log ε)	Ref.
Acerosin (cont.)		414s(0.42)	102-2107-83
4H-1-Benzopyran-4-one, 5,7-dihydroxy-2-(4-hydroxy-3-methoxyphenyl)-3,6-dimethoxy- (jaceidin) (relative absorbance given)	MeOH	254(0.81),270(0.75), 350(1.00)	102-2107-83
	MeOH-AlCl3	267(0.86),279s(0.78), 302s(0.47),380(1.00), 408s(0.88)	102-2107-83
	MeOH-AlCl3-HCl	264(0.83),280(0.79), 301s(0.49),372(1.00), 408s(0.78)	102-2107-83
4H-1-Benzopyran-4-one, 5,7-dihydroxy-2-(4-hydroxy-3-methoxyphenyl)-3,8-dimethoxy- (relative absorbance given)	MeOH	256(0.93),273(1.00), 334s(0.75),360(0.86)	102-2107-83
	MeOH-AlCl3	266s(0.90),283(1.00), 308(0.64),362(0.84), 418(0.67)	102-2107-83
	MeOH-AlCl3-HCl	262s(0.86),282(1.00), 306(0.55),360(0.86), 418(0.65)	102-2107-83
4H-1-Benzopyran-4-one, 5,7-dihydroxy-2-(4-hydroxy-3-methoxyphenyl)-6,8-dimethoxy- (sudachitin) (relative absorbance given)	MeOH	256s(0.66),281(0.80), 344(1.00)	102-2107-83
	MeOH-AlCl3	262(0.61),290(0.73), 303s(0.69),372(1.00), 412s(0.41)	102-2107-83
	MeOH-AlCl3-HCl	260(0.64),292(0.77), 302s(0.81),366(1.00), 412s(0.41)	102-2107-83
4H-1-Benzopyran-4-one, 2-(3,5-dihydroxy-4-methoxyphenyl)-5,7-dihydroxy-3-methoxy-6-methyl- (relative absorbance given)	MeOH	271(1.00),304s(0.76), 337(0.94)	102-2107-83
	MeOH-AlCl3	252s(0.76),277(1.00), 308s(0.73),356(0.99), 406s(0.64)	102-2107-83
	MeOH-AlCl3-HCl	252s(0.78),280(1.00), 308s(0.73),352(1.00), 404s(0.69)	102-2107-83
4H-1-Benzopyran-4-one, 2-(3,5-dihydroxy-4-methoxyphenyl)-5-hydroxy-3,7-dimethoxy- (relative absorbance given)	MeOH	267(1.00),302s(0.65), 344(0.83)	102-2107-83
	MeOH-AlCl3	275(1.00),308(0.52), 349(0.69),399(0.65)	102-2107-83
	MeOH-AlCl3-HCl	276(1.00),306(0.52), 344(0.66),398(0.55)	102-2107-83
4H-1-Benzopyran-4-one, 2-(3,4-dihydroxyphenyl)-5-hydroxy-3,6,7-trimethoxy- (chrysosplenol D) (relative absorbance given)	MeOH	259(1.00),273(0.87), 354(1.00)	102-2107-83
	MeOH-AlCl3	274(1.00),306(0.38), 342s(0.34),434(0.89)	102-2107-83
	MeOH-AlCl3-HCl	269(1.00),286s(0.87), 300s(0.53),373(0.91), 408s(0.68)	102-2107-83
4H-1-Benzopyran-4-one, 2-(3,4-dihydroxyphenyl)-5-hydroxy-3,7,8-trimethoxy- (relative absorbance given)	MeOH	260(1.00),272s(0.94), 298s(0.44),334s(0.65), 367(0.81)	102-2107-83
	MeOH-AlCl3	281(1.00),312s(0.33), 344s(0.24),454(0.98)	102-2107-83
	MeOH-AlCl3-HCl	273(1.00),282s(0.99), 305(0.50),362(0.77), 422(0.67)	102-2107-83
4H-1-Benzopyran-4-one, 2-(3,4-dihydroxyphenyl)-5-hydroxy-6,7,8-trimethoxy-	EtOH	261(4.21),278(4.24), 353(4.33)	40-0161-83
	EtOH-NaOAc	269(4.29),417(4.36)	40-0161-83
	EtOH-AlCl3	285(4.22),304(4.18), 363(4.30)	40-0161-83
	EtOH-AlCl3-HCl	265s(4.16),286(4.23), 303(4.20),365(4.35)	40-0161-83

Compound	Solvent	λ max(log ε)	Ref.
4H-1-Benzopyran-4-one, 2-(3,4-dihydroxyphenyl)-5-hydroxy-6,7,8-trimethoxy- (relative absorbance given) (cont.)	MeOH	258(0.80),280(0.87), 349(1.00)	102-2107-83
	MeOH-AlCl3	279(0.71),306s(0.54), 342s(0.32),438(1.00)	102-2107-83
	MeOH-AlCl3-HCl	264s(0.68),288(0.81), 304s(0.77),368(1.00), 420s(0.51)	102-2107-83
4H-1-Benzopyran-4-one, 5,7,8-trihydroxy-2-(3,4,5-trimethoxyphenyl)-	EtOH	300(4.35)	18-3773-83
	EtOH-NaOAc	300(4.29)	18-3773-83
	EtOH-AlCl3	298s(4.28),317(4.35), 350s(4.15),405(3.69)	18-3773-83
$C_{18}H_{16}O_9$			
4H-1-Benzopyran-4-one, 2-(3,4-dihydroxy-5-methoxyphenyl)-5,7-dihydroxy-3,6-dimethoxy- (relative absorbance given)	MeOH	255(0.80),268s(0.75), 300s(0.56),357(1.00)	102-2107-83
	MeOH-AlCl3	275(0.93),316s(0.41), 446(1.00)	102-2107-83
	MeOH-AlCl3-HCl	262s(0.81),275(0.87), 310s(0.62),375(1.00), 414s(0.91)	102-2107-83
4H-1-Benzopyran-4-one, 2-(3,5-dihydroxy-4-methoxyphenyl)-5,7-dihydroxy-3,6-dimethoxy- (relative absorbance given)	MeOH	254s(0.77),273(0.89), 341(1.00)	102-2107-83
	MeOH-AlCl3	254s(0.79),278(0.93), 311s(0.61),362(1.00), 407s(0.61)	102-2107-83
	MeOH-AlCl3-HCl	259s(0.75),283(0.96), 308s(0.61),360(1.00), 406s(0.53)	102-2107-83
$C_{18}H_{17}As$			
1H-Arsole, 1-ethyl-2,5-diphenyl-	C_6H_{12}	228(4.22),367(4.27)	101-0335-83I
$C_{18}H_{17}Br$			
Benzene, 1,1'-(2-bromo-4-methyl-1,3-pentadienylidene)bis-	hexane	279(4.02)	104-1438-83
1H-Indene, 2-bromo-1-(1-methylethyl)-3-phenyl-	hexane	270(4.03)	104-1438-83
$C_{18}H_{17}Cl$			
Benzene, 1,1'-(2-chloro-4-methyl-1,3-pentadienylidene)bis-	hexane	281(4.09)	104-1438-83
1H-Indene, 2-chloro-1-(1-methylethyl)-3-phenyl-	hexane	269(3.85)	104-1438-83
1H-Indene, 2-chloro-3-phenyl-1-propyl-	hexane	269(3.97),290(3.56)	104-1438-83
$C_{18}H_{17}ClO_4$			
Benzoic acid, 3-chloro-, 3,4-dihydro-3-hydroxy-2,2-dimethyl-2H-1-benzopyran-4-yl ester	EtOH	211(4.34),224(4.23), 279(3.56),286(3.58)	39-1431-83C
2H-Pyran-2-one, 6-butyl-3-[3-(4-chlorophenyl)-1-oxo-2-propenyl]-4-hydroxy-	MeOH	203(4.5),350(4.4)	83-0845-83
$C_{18}H_{17}Cl_2NOS_2$			
Morpholine, 4-[2-[5-chloro-2-[(4-chlorophenyl)thio]phenyl]-1-thioxoethyl]-	MeOH	256.5(4.31),280(4.35)	73-1173-83
$C_{18}H_{17}Cl_2N_5O$			
Carbamimidic acid, N-[6-(2,6-dichlorophenyl)-2-methylpyrido[2,3-d]pyrimidin-7-yl]-N'-methyl-	MeOH	219(4.58),282(3.82), 348(4.28),363(4.27)	87-0403-83

Compound	Solvent	λ_{max}(log ε)	Ref.
$C_{18}H_{17}Cl_3N_2O_5S$			
5-Thia-1-azabicyclo[4.2.0]oct-2-ene-2-carboxylic acid, 3-methyl-8-oxo-7-[(phenylacetyl)amino]-, 2,2,2-trichloroethyl ester, 5-oxide,[5R-(5α,6β,7α)]-	EtOH	269(3.69)	78-0337-83
[5S-(5α,6β,7α)]-	EtOH	269(3.87)	78-0337-83
$C_{18}H_{17}Cl_3N_2O_7S_2$			
5-Thia-1-azabicyclo[4.2.0]oct-2-ene-2-carboxylic acid, 3-(acetoxymethyl)-8-oxo-7-[(2-thienylacetyl)amino]-, 2,2,2-trichloroethyl ester, 5-oxide, [5S-(5α,6β,7α)]-	EtOH	236(4.05),269(3.88)	78-0461-83
$C_{18}H_{17}FeNO_3$			
Ferrocene, [[5-(carboxymethyl)-1-methyl-1H-pyrrol-2-yl]carbonyl]-, calcium salt	MeOH	237.5(3.96),271(3.85), 325(4.17),472(3.00)	87-0226-83
Ferrocene, [[5-(carboxymethyl)-1-methyl-1H-pyrrol-3-yl]carbonyl]-, calcium salt	MeOH	232s(4.05),272(3.94), 322(4.07),469(3.03)	87-0226-83
$C_{18}H_{17}FeO$			
Methylium, ferrocenyl(4-methoxyphenyl)-	50% H_2SO_4	258(4.06),290s(--), 363(4.16)	32-0721-83
$C_{18}H_{17}N$			
Benzonitrile, 4-(3-methyl-3-phenyl-1-butenyl)-, (E)-	n.s.g.	213.5(4.30),275.5(4.51), 301(3.73)	12-0565-83
$C_{18}H_{17}NO$			
2-Azetidinone, 3-methyl-1-phenyl-4-(2-phenylethenyl)-, erythro-	MeOH	250(4.56),282(3.51), 293(3.20)	5-1393-83
threo-	MeOH	250(4.56),282(3.52), 293(3.23)	5-1393-83
2H-Benz[g]indol-2-one, 1,3,3a,4,5,9b-hexahydro-4-phenyl-	EtOH	247(1.94),252(2.07), 257(2.19),263(2.22), 267s(2.11),271(2.11)	44-1451-83
2H-Pyrrol-2-one, 1,5-dihydro-4-methyl-3-phenyl-5-(phenylmethyl)-	MeOH	235(4.10)	4-0687-83
$C_{18}H_{17}NOS$			
[1]Benzothieno[3,2-b]azete, 2-ethoxy-2a,7b-dihydro-2a-methyl-7b-phenyl-	EtOH	249(3.98),290(3.17), 299(3.11)	44-1275-83
$C_{18}H_{17}NO_2$			
Aporeine	n.s.g.	262(4.3),315(3.7)	162-0110-83
1H-Isoindole-4,7-dione, 1,1,5,6-tetramethyl-3-phenyl-	EtOH	229(4.16),261(4.17), 345(2.73)	33-2252-83
2-Propenamide, N-[2-(4-methoxyphenyl)-ethenyl]-3-phenyl-, (E,E)-	EtOH	290(4.30),305(4.25), 345(4.35)	102-0755-83
$C_{18}H_{17}NO_3$			
Acetamide, N-[1-(6-hydroxy-5-oxo-1,3,5-cycloheptatrien-1-yl)-3-phenyl-2-propenyl]-, (E)-	MeOH	243(4.56),328(3.96), 370s(3.76)	18-3099-83
2,4,6-Cycloheptatrien-1-one, 4-[3-[4-(dimethylamino)phenyl]-1-oxo-2-propenyl]-2-hydroxy-, (E)-	MeOH	252(4.48),312(4.37), 393(4.47)	18-3099-83
	MeOH-NaOH	273(4.35),345(4.18), 430(4.57)	18-3099-83

Compound	Solvent	$\lambda_{max}(\log \epsilon)$	Ref.
Norstephalagine	EtOH	241(4.45),278(4.49)	100-0761-83
Roemeroline	MeOH or EtOH	240s(3.78),281(3.95), 320s(3.31)	100-0761-83
$C_{18}H_{17}NO_4$			
Calycinine	EtOH	222(4.43),268s(4.12), 278(4.23),299(4.08)	100-0862-83
	EtOH-base	233(4.24),286(3.85), 327(3.79)	100-0862-83
Fissoldine, hydrobromide	EtOH	216(4.55),273(4.03), 295(4.15)	142-0813-83
1H-Indole-3-carboxylic acid, 2-(3,4-dimethoxyphenyl)-, methyl ester	EtOH	220(4.69),243(4.53), 310(4.32)	12-1419-83
Isocalycinine	EtOH	218(4.42),282(4.26), 302(4.14)	100-0761-83
Laetine	EtOH	270(4.06),307(3.67)	100-0761-83
Norannuradhapurine, (-)-	EtOH	218s(4.27),281(3.87), 298s(3.69),317s(3.46)	100-0761-83
$C_{18}H_{17}NO_5$			
Benz[f]isoquinoline-1,2-dicarboxylic acid, 5,6-dihydro-8-methoxy-, dimethyl ester	MeOH	208(4.3),231(4.2), 254(4.1),312(4.2)	5-1476-83
Benz[h]isoquinoline-3,4-dicarboxylic acid, 5,6-dihydro-8-methoxy-, dimethyl ester	MeOH	229(4.1),326(4.3)	5-1476-83
2,4,6-Cycloheptatrien-1-one, 4-[3-(3,4-dimethoxyphenyl)-1-(hydroxyimino)-2-propenyl]-2-hydroxy-, higher melting	MeOH	245(4.51),328(4.48), 395(3.90)	18-3099-83
lower melting	MeOH	243(4.49),328(4.46), 390s(3.84)	18-3099-83
6-Oxa-2-azaspiro[4.5]deca-3,7-diene-4-carboxylic acid, 2-methyl-1,9-dioxo-3-phenyl-, ethyl ester	EtOH	247(3.98),303(3.81)	94-0356-83
$C_{18}H_{17}N_2O$			
2H-Pyrrolo[3,2-c]quinolinium, 3,5-dihydro-4-hydroxy-5-methyl-1-phenyl-, perchlorate	MeOH	232(4.60),260s(4.11), 338(4.15)	83-0897-83
$C_{18}H_{17}N_2OS$			
Thiazolium, 4-(acetylamino)-2-methyl-3,5-diphenyl-, perchlorate	MeCN	230s(4.01),283(3.86)	118-0582-83
Thiazolium, 4-(acetylamino)-3-methyl-2,5-diphenyl-, perchlorate	MeCN	248s(3.85),310(4.17)	118-0582-83
$C_{18}H_{17}N_2S$			
Methanaminium, N-(4,6-diphenyl-2H-1,3-thiazin-2-ylidene)-N-methyl-, perchlorate	MeCN	267s(3.78),340(4.19), 380(3.74)	118-0827-83
$C_{18}H_{17}N_3O$			
Benzenepropanenitrile, α-[[4-(dimethylamino)phenyl]imino]-4-methyl- -oxo-	$CHCl_3$	484(4.58)	64-0930-83B
2(1H)-Quinoxalinone, 3-[2-[4-(dimethylamino)phenyl]ethenyl]-	EtOH	450(4.78)	18-0326-83
$C_{18}H_{17}N_3O_2$			
1H-Imidazole-4-carboxamide, 5-(phenylmethoxy)-1-(phenylmethyl)-	H_2O	247(4.03)	4-0875-83
	M NaOH	247(4.03)	4-0875-83
	M HCl	230s(--)	4-0875-83

Compound	Solvent	λ_{max}(log ε)	Ref.
1H-Imidazolium, 4-(aminocarbonyl)-5-hydroxy-1,3-bis(phenylmethyl)-, hydroxide, inner salt	M HCl	284(4.04)	4-0875-83
	H_2O	242(3.70),283(4.08)	4-0875-83
	M NaOH	286(4.10)	4-0875-83
$C_{18}H_{17}N_3O_3$			
Benzoic acid, 4-[5-[4-(dimethylamino)-phenyl]-1,3,4-oxadiazol-2-yl]-, methyl ester	toluene	297(4.12),355(4.36)	103-0022-83
Phenol, 2-[2-amino-5-(3,4-dimethoxy-phenyl)-4-pyrimidinyl]-	MeOH	204(4.89),228(4.73), 264(4.55),340(4.09)	4-1111-83
Phenol, 2-[2-amino-5-(4-methoxyphenyl)-4-pyrimidinyl]-5-methoxy-	MeOH	206(4.59),246(4.44), 272(4.32),346(4.18)	4-1111-83
$C_{18}H_{17}N_3O_4$			
1H-Pyrazole-4,5-dicarboxylic acid, 1-[8-(methylamino)-1-naphthalenyl]-, dimethyl ester	EtOH	212(4.71),252(4.45), 343(3.84)	94-1378-83
$C_{18}H_{17}N_3O_5$			
Butanoic acid, 4-[(2-benzoyl-9-oxo-1,2-diazabicyclo[5.2.0]nona-3,5-dien-8-yl)amino]-4-oxo-, cis-(±)-	MeOH	282(3.84)	5-1374-83
$C_{18}H_{17}N_3O_6$			
1,2-Diazabicyclo[5.2.0]nona-3,5-diene-2-carboxylic acid, 8-[(2-carboxy-benzoyl)amino]-9-oxo-, 2-ethyl ester, cis-(±)-	MeOH	278(3.97)	5-1374-83
$C_{18}H_{17}N_3S$			
Isothiazolo[5,1-e][1,2,3]thiadiazole-7-S^{IV}, 1,6-dihydro-3,4-dimethyl-1,6-diphenyl-	C_6H_{12}	203(4.36),251s(4.14), 292(4.18),476(4.34)	39-0777-83C
$C_{18}H_{18}$			
[2_2](1,6)Cyclooctatetraenyl(1,4)cyclo-phane	n.s.g.	280s(2.70),290s(2.61), 325s(2.19)	35-7384-83
[2.2.2](1,2,4)Cyclophane, TCNE complex (also other complexes)	CH_2Cl_2	540(1.57)	59-0289-83
1,3-Pentadiene, 4-methyl-1,3-diphenyl-, (E)-	EtOH	232(4.43),290(4.42)	44-1834-83
1,3-Pentadiene, 4-methyl-2,3-diphenyl-	EtOH	250(4.52)	44-1834-83
$C_{18}H_{18}ClNOS$			
2-Propen-1-one, 1-(4-chlorophenyl)-3-[1-(1,1-dimethylethyl)-1,2-dihydro-2-thioxo-3-pyridinyl]-, (E)-	EtOH	277(4.30),350(3.94), 417(3.74)	4-1651-83
$C_{18}H_{18}ClNO_2S$			
Benzo[b]thiophene-2-carboxylic acid, 3-amino-5-chloro-2,3-dihydro-2-methyl-3-phenyl-, ethyl ester	EtOH	258(4.01),296(3.70)	44-1275-83
$C_{18}H_{18}ClNO_3$			
2H-Pyrano[3,2-d][1]benzoxepin-2-one, 3-chloro-5,6-dihydro-4-piperidino-	EtOH	212(4.22),236s(3.82), 266(4.07),336(4.11)	4-0539-83
$C_{18}H_{18}ClNS$			
Benzo[b]thiophen-2-amine, 5-chloro-N,N-diethyl-3-phenyl-	EtOH	223(4.37),240(4.44), 258(4.17),312(4.11)	44-1275-83

Compound	Solvent	λ_{max}(log ε)	Ref.
$C_{18}H_{18}Cl_2$			
Benzene, 1,1'-(2,3-dichloro-4-phenyl-1-pentenylidene)bis-	hexane	253(4.08)	104-1438-83
$C_{18}H_{18}CuN_6$			
Copper, [N,N'-bis[2-(phenylazo)ethylidene]-1,2-ethanediaminato(2-)]-, (SP-4-2)-	$CHCl_3$	330s(4.08),360s(4.17), 411(4.40),550s(3.08), 670s(2.45),880(1.93)	65-1214-83
$C_{18}H_{18}FeO_2$			
Ferrocene, [hydroxy(4-methoxyphenyl)-methyl]-	EtOH	227(2.50),265s(--), 437(2.06)	32-0721-83
$C_{18}H_{18}N_2$			
1,2-Diazabicyclo[3.1.0]hex-2-ene, 6,6-dimethyl-3,5-diphenyl-	EtOH	250(4.05)	44-1834-83
2,3-Diazabicyclo[3.1.0]hex-2-ene, 4,4-dimethyl-1,5-diphenyl-	EtOH	330(1.42)	44-1834-83
2,3-Diazabicyclo[2.2.2]oct-2-ene, 1,4-diphenyl-	benzene	386(2.12)	35-7102-83
Diazene, (2-methyl-1-phenyl-1-propenyl)(1-phenylethenyl)-	EtOH	245(4.19)	44-1834-83
2,4,6-Nonatrien-8-ynal, 7-methyl-, (5-methyl-2,4-heptadien-6-ynylidene)hydrazone	THF	246(4.18),253s(4.16), 262s(4.08),298s(4.14), 310s(4.20),362s(4.80), 377(4.91),394(4.93), 414(4.77)	18-1467-83
2,4,6,8-Undecatetraen-10-ynal, 9-methyl-, (3-methyl-2-penten-4-ynylidene)-hydrazone	THF	248(4.04),293(3.84), 305(3.89),359s(4.58), 381s(4.77),397(4.82), 417s(4.71)	18-1467-83
$C_{18}H_{18}N_2O$			
Benzonitrile, 4-[3-[(2-methylphenyl)-methyl]-3-isoxazolidinyl]-	EtOH	231(4.22),266s(3.31), 272(3.34),279(3.29)	88-2279-83
Benzonitrile, 4-[3-[(4-methylphenyl)-methyl]-3-isoxazolidinyl]-	EtOH	224(4.23),232(4.25), 265(3.08),274s(3.04), 280s(2.88)	88-2279-83
$C_{18}H_{18}N_2O_2$			
Benz[cd]indol-5-ol, 4-amino-1-benzoyl-1,2,2a,3,4,5-hexahydro-, (2aα,4β,5α)-(±)-	EtOH	266(4.08),294(3.94)	87-0522-83
1-Phenazinecarboxylic acid, 6-(3-methylbutyl)-	MeOH	255(4.81),355s(4.02), 367(4.15)	73-0527-83
Pyrano[4,3-c]pyrazol-3(2H)-one, 6,7-dihydro-6-methyl-4-(1,3-pentadienyl)-2-phenyl-, (E,E)-	EtOH	254(4.23),350(4.54)	118-0948-83
2-Pyrrolidinone, 3-[2-(methylamino)-benzoyl]-1-phenyl-	MeOH	236(4.45),390(3.86)	83-0897-83
2(1H)-Quinolinone, 4-hydroxy-1-methyl-3-[2-(phenylamino)ethyl]-	MeOH	230(4.69),284(3.92), 318(3.90)	83-0897-83
$C_{18}H_{18}N_2O_3$			
Asimilobine, N-carbamoyl-	EtOH	273(4.08),305(3.43)	100-0761-83
2,4,6-Cycloheptatrien-1-one, 4-[5-[4-(dimethylamino)phenyl]-4,5-dihydro-3-isoxazolyl]-2-hydroxy-	MeOH	266(4.49),380(3.83), 400s(3.78)	18-3099-83
$C_{18}H_{18}N_2O_4$			
2H-Indol-2-one, 1,3-dihydro-4-[[(3,4,5-	$CHCl_3$	314(4.15)	44-2468-83

Compound	Solvent	λ_{max}(log ε)	Ref.
trimethoxyphenyl)methylene]amino]- (cont.)			44-2468-83
$C_{18}H_{18}N_2O_4S$			
Glycine, N-[N-[(phenylmethoxy)carbonyl]thioglycyl]-, phenyl ester	$CHCl_3$	265(4.20)	78-1075-83
$C_{18}H_{18}N_2O_6S$			
1-Azabicyclo[3.2.0]hept-2-ene-2-carboxylic acid, 3-(acetylthio)-6-ethyl-7-oxo-, (4-nitrophenyl)methyl ester, (5R-cis)-	THF	269(4.24),310s(4.06)	158-0407-83
$C_{18}H_{18}N_2O_7$			
2-Butenedioic acid, 2-[[(1,3-dihydro-1,3-dioxo-2H-isoindol-2-yl)acetyl]amino]-, diethyl ester, (E)-	MeOH	260(3.89)	118-0539-83
(Z)-	MeOH	274(4.27)	118-0539-83
$C_{18}H_{18}N_4O_3$			
Carbamic acid, (11-amino-6,6a-dihydro-5H-naphtho[2,1-b]pyrido[3,4-e][1,4]oxazin-9-yl)-, ethyl ester	pH 1	302s(3.85),309(3.86), 367s(4.35),380(4.41) (changing)	87-1614-83
$C_{18}H_{18}N_4O_5$			
1,2-Diazabicyclo[5.2.0]nona-3,5-diene-2-carboxylic acid, 5-methyl-8-[[(4-nitrophenyl)methylene]amino]-9-oxo-, ethyl ester, cis-(±)-	MeOH	277(3.96)	5-1374-83
$C_{18}H_{18}N_4O_{13}$			
1,4:13,16-Diepoxy-5H-8-oxa-4a,6,11,12a-tetraazabenzo[1,7]cyclohepta[1,2-b]-cyclohepta[de]naphthalene-5,7,10,12-(6H,9H,11H)-tetrone, 1,2,3,4,13,14-15,16,16a,16b-decahydro-2,3,9,9,14-15-hexahydroxy-, [1S-(1α,2α,3α,4α-8αS*,13β,14β,15β,16β,16aα,16bβ)]-	H_2O	284(3.89)	4-0753-83
$C_{18}H_{18}N_4S$			
Benzenamine, 2-(3,4-dimethyl-1-phenyl-isothiazolo[5,1-e][1,2,3]thiadiazol-7-S^{IV}-6(1H)-yl)-	C_6H_{12}	210(4.41),235s(4.17), 288(4.10),476(4.20)	39-0777-83C
$C_{18}H_{18}N_6O_5S$			
1-Azabicyclo[3.2.0]hept-2-ene-2-carboxylic acid, 6-ethyl-3-[(1-methyl-1H-tetrazol-5-yl)thio]-7-oxo-, (4-nitrophenyl)methyl ester, (5R-cis)-	THF	269.5(4.04),307(4.09)	158-0407-83
$C_{18}H_{18}N_6O_5S_2$			
Cefamandole (sodium salt)	H_2O	269(4.03)	162-0268-83
5-Thia-1-azabicyclo[4.2.0]oct-2-ene-2-carboxylic acid, 7-[[(4-hydroxyphenyl)acetyl]amino]-3-[[(1-methyl-1H-tetrazol-5-yl)thio]methyl]-8-oxo-, sodium salt, (6R-trans)-	MeOH	274(4.00)	87-1577-83
$C_{18}H_{18}N_6O_6S$			
5-Oxa-1-azabicyclo[4.2.0]oct-2-ene-2-carboxylic acid, 7-[[(4-hydroxyphen-	MeOH	267(3.80)	87-1577-83

Compound	Solvent	λ_{max}(log ϵ)	Ref.
yl)acetyl]amino]-3-[[(1-methyl-1H-tetrazol-5-yl)thio]methyl]-8-oxo-, monosodium salt (cont.)			87-1577-83
$C_{18}H_{18}N_6S_2Zn$			
Zinc, [[2,2'-(1-[1,1'-biphenyl]-4-yl-1,2-ethanediylidene)bis[N-methylhydrazinecarbothioamidato]](2-)-N^2,$N^{2'}$,S,S']-, (T-4)-	EtOH	307(4.45),455(4.07)	161-0143-83
$C_{18}H_{18}N_6S_4$			
Propanedinitrile, 2,2'-[3,6-bis-[bis(methylthio)methylene]-1,4-dimethyl-2,5-piperazinediylidene]bis-	MeCN	225(4.10),274(4.30), 360(4.40)	89-0717-83
$C_{18}H_{18}N_8O_4S_2$			
2,4(1H,3H)-Pteridinedione, 7,7'-dithiobis[1,3,6-trimethyl-	CH_2Cl_2	252(4.42),352(4.43), 369(4.48)	88-5047-83
$C_{18}H_{18}N_8O_7S_3$			
Ceftriaxone, sodium salt	H_2O	242(4.51),272(4.47)	162-0272-83
$C_{18}H_{18}O_2$			
2-Butenoic acid, 4-bicyclo[5.4.1]dodeca-2,5,7,9,11-pentaen-4-ylidene-, ethyl ester	EtOH	405(4.29)	88-2151-83
$C_{18}H_{18}O_3$			
Furan, 2,2'-[(4-methoxyphenyl)methylene]bis[5-methyl-	EtOH	230(4.40)	103-0478-83
$C_{18}H_{18}O_4$			
Benzoic acid, 2-(4-hydroxy-3,5-dimethylbenzoyl)-, ethyl ester	EtOH	299(4.19)	24-0970-83
6H-Dibenzo[b,d]pyran-6-one, 4-acetyl-7,8,9,10-tetrahydro-3-(2-propenyloxy)-	MeOH	318(4.20)	78-1265-83
2(3H)-Furanone, dihydro-3,4-bis[(3-hydroxyphenyl)methyl]-	EtOH	227(4.66),281(4.64)	39-0643-83C
trans	EtOH	272.5(3.31)	162-0687-83
3,9-Phenanthrenedione, 4,4a-dihydro-6,10-dimethoxy-1,4a-dimethyl-, (R)-	EtOH	260(3.95),310(3.99)	23-2461-83
2-Propenoic acid, 3-(4-hydroxy-3-methoxyphenyl)-, 2-phenylethyl ester	MeOH	210.5(4.14),235.5(3.98), 296.5(4.04),325.5(4.2)	94-2710-83
$C_{18}H_{18}O_4S$			
1H-Indene-1,5-dione, 5,6,7,7a-tetrahydro-2,7a-dimethyl-5-[(phenylsulfonyl)methyl]-	EtOH	216.5(4.07),302(4.24)	39-2337-83C
$C_{18}H_{18}O_5$			
1H-Naphtho[2,1-b]pyran-1,2-diol, 2,3-dihydro-3-methyl-, diacetate	EtOH	229(4.63),276(4.55), 288(4.54)	39-0883-83C
$C_{18}H_{18}O_6$			
Acetic acid, [(2-acetyl-7,8,9,10-tetrahydro-4-methyl-6-oxo-6H-dibenzo[b,d]-pyran-3-yl)oxy]-	MeOH	255(4.42),314(3.92)	78-1265-83
3,5-Cyclohexadiene-1,2-diol, 3-[(benzoyloxy)methyl]-, diacetate	EtOH	264(3.81)	88-2019-83

Compound	Solvent	λ_{max}(log ϵ)	Ref.
1,4-Naphthalenedione, 2-(1-acetoxy-4-methyl-3-pentenyl)-5,8-dihydroxy-(shikonin acetate)	MeOH	276(3.95),330(3.02), 488(3.85),520(3.90), 560(3.67)	105-0532-83
$C_{18}H_{18}O_7$			
Frenolicin	MeOH	234(4.26),362(3.72)	162-0608-83
7-Oxabicyclo[4.1.0]hept-4-ene-2,3-diol, 1-[(benzoyloxy)methyl]-, diacetate (β-senepoxide)	EtOH	273(3.06),280(2.97)	88-2019-83
$C_{18}H_{18}S$			
10,9-(Epithiomethano)anthracene, 9,10-dihydro-9,10,12-trimethyl-	MeCN	278(3.36)	78-1487-83
$C_{18}H_{18}S_2$			
Benzene, 1,1'-[1,3-hexadienylidene-bis(thio)]bis-, (E)-	EtOH	255(4.38)	118-0383-83
Benzene, 1,1'-[(4-methyl-1,3-pentadien-ylidene)bis(thio)]bis-	EtOH	260(4.43),304(4.39)	118-0383-83
$C_{18}H_{19}BN_4O_4$			
Formazan, 5-(diacetoxyboryl)-3-methyl-1,5-diphenyl- (or cyclic isomer)	hexane	510(4.34)	65-2061-83
	EtOH	510(4.32)	65-2061-83
$C_{18}H_{19}BrNO_2$			
Pyridinium, 3,5-diacetyl-4-(4-bromo-phenyl)-1,2,6-trimethyl-, perchlorate	EtOH	209(4.46),290(4.13)	103-0415-83
$C_{18}H_{19}BrO_3$			
Furan, 2,2',2"-(2-bromopropylidyne)-tris[5-methyl-	EtOH	227(4.37)	103-0474-83
$C_{18}H_{19}Br_3O_2$			
Estra-1,3,5(10)-trien-17-one, 2,4,16-tribromo-3-hydroxy-, (16α)-	EtOH	283(3.43),290(3.45)	39-0121-83C
(16β)-	EtOH	283(3.43),290(3.45)	39-0121-83C
$C_{18}H_{19}ClN_2O_3S$			
2H-Thiopyran-3-carboxylic acid, 5-(4-chlorophenyl)-2-imino-6-morpholino-, ethyl ester, monoperchlorate	MeCN	238(4.34),314(4.10), 439(4.24)	97-0403-83
$C_{18}H_{19}ClN_2O_5S_2$			
6-Benzofuransulfonamide, 3-chloro-2-[4-[(dimethylamino)sulfonyl]phenyl]-N,N-dimethyl-	n.s.g.	320(4.3),330(4.0)	39-2181-83C
$C_{18}H_{19}ClN_4$			
Clozapine	EtOH	215(4.44),230(4.41), 261(4.23),297(4.02)	162-0344-83
$C_{18}H_{19}ClO_7$			
1,8-Naphthalenedicarboxylic acid, 3-chloro-2,4,7-trimethoxy-5-methyl-, dimethyl ester	EtOH	246(4.76),285s(3.66), 298(3.76),311(3.76), 344(3.49)	78-2283-83
$C_{18}H_{19}CoN_4$			
Cobalt(1+), (6,7,8,9,10,11-hexahydro-dibenzo[i,l][1,4,7,11]tetraazacyclo-tetradecinato-N^6,N^9,N^{12},N^{18})-, (SP-4-3)-, perchlorate	acetone	400(3.16),490(2.89)	12-2395-83

Compound	Solvent	$\lambda_{max}(\log \epsilon)$	Ref.
$C_{18}H_{19}CuN_4$			
Copper(1+), (6,7,8,9,10,11-hexahydrodibenzo[i,l][1,4,7,11]tetraazacyclotetradecinato-N^6,N^9,N^{12},N^{18})-, perchlorate	EtOH-DMF	347(3.65),364(3.84), 396s(3.60),410(3.63), 430s(3.51)	12-2395-83
$C_{18}H_{19}N$			
1H-Indole, 3-(1-methyl-1-phenylpropyl)-	n.s.g.	220(4.7),270(3.8)	103-0638-83
$C_{18}H_{19}NO$			
1H-Pyrrole, 2,5-dihydro-2-[(4-methoxyphenyl)phenylmethyl]-	MeOH	229(4.16),261(3.08), 268(3.16),277(3.25), 283(3.18)	87-0469-83
$C_{18}H_{19}NOS$			
2-Propen-1-one, 3-[1-(1,1-dimethylethyl)-1,2-dihydro-2-thioxo-3-pyridinyl]-1-phenyl-, (E)-	EtOH	274(4.41),349(4.15), 415(3.83)	4-1651-83
$C_{18}H_{19}NOS_2$			
Piperidine, 1-methyl-4-thieno[2,3-c][2]benzothiepin-4(9H)-ylidene-, N-oxide	MeOH	284(3.81),304s(3.66)	73-0623-83
Thieno[2,3-c][2]benzothiepin-4(9H)-one, 7-(1-methyl-4-piperidinyl)-	MeOH	265(4.24),356(3.62)	73-0623-83
$C_{18}H_{19}NO_2$			
Bisnoratherosperminine	HCl	212(4.24),233(4.32), 249(4.65),256(4.78), 274(4.06),304(4.06), 311(4.06)	100-0761-83
6H-Dibenzo[a,g]quinolizin-2-ol, 5,8,13,13a-tetrahydro-3-methoxy-	pH 12	208(4.85),295(3.91), 310(2.71)	88-0291-83
(bharatamine)	EtOH	206(5.08),290(2.77)	88-0291-83
Ethanone, 1,1'-[2,6-dimethyl-4-(4-methylphenyl)-3,5-pyridinediyl]bis-	EtOH	207(4.36),241s(4.18), 265s(4.01)	103-0415-83
1H-Indole-2-carboxaldehyde, 4,5,6,7-tetrahydro-6,6-dimethyl-4-oxo-1-(phenylmethyl)-	EtOH	235(4.25),298(4.18)	136-0255-83E
1H-Isoindole-4,7-diol, 1,1,5,6-tetramethyl-3-phenyl-	EtOH	241(4.12),267s(3.88), 332(3.57)	33-2252-83
6-Oxa-3-azabicyclo[3.1.0]hexane, 2-[(4-methoxyphenyl)phenylmethyl]-	MeOH	229(4.11),262(3.06), 268(3.16),276(3.22), 273[sic](3.16)	87-0469-83
1-Oxa-3-azaspiro[4.5]deca-3,6,9-trien-8-one, 2,2,6,7-tetramethyl-4-phenyl-	EtOH	245(4.27)	33-2252-83
$C_{18}H_{19}NO_2S$			
Benzo[b]thiophene-2-carboxylic acid, 3-amino-2,3-dihydro-2-methyl-3-phenyl-, ethyl ester	EtOH	250(3.67),284(3.29)	44-1275-83
$C_{18}H_{19}NO_2S_2$			
Piperidine, 1-methyl-4-thieno[2,3-c][2]benzothiepin-4(9H)-ylidene-, S,1-dioxide	MeOH	269(4.14)	73-0623-83
$C_{18}H_{19}NO_3$			
Acetamide, N-[1-(6-hydroxy-5-oxo-1,3,6-cycloheptatrien-1-yl)-3-phenylpropyl]-	MeOH	244(4.45),327(3.91), 359s(3.76)	18-3099-83

Compound	Solvent	$\lambda_{max}(\log \epsilon)$	Ref.
Aporphine, 1,9-dihydroxy-2-methoxy-	EtOH	222(4.25),278(4.22), 302(3.97),314(3.77)	100-0761-83
Ethanone, 1,1'-[4-(4-methoxyphenyl)-2,6-dimethyl-3,5-pyridinediyl]bis-	EtOH	206(4.40),246(4.06), 277(4.11)	103-0415-83
Glaufine	EtOH	217(4.60),274(4.21), 308(3.84)	100-0761-83
Norliridinine	EtOH	230s(4.05),274(3.95), 282s(3.89),304s(3.38)	100-0761-83
Nornuciferine, 3-hydroxy-	EtOH	219(4.39),240s(3.90), 280(4.14),292s(4.04)	100-0761-83
1,9,12-Trioxa-3-azaspiro[4.2.4.2]tetradeca-3,6,13-triene, 2,2-dimethyl-4-phenyl-	EtOH	250(4.09)	33-2252-83
$C_{18}H_{19}NO_4$			
Culacorine	MeOH	209(4.50),225s(4.19), 285(3.80),296s(3.64)	100-0881-83
2,4,6-Cycloheptatrien-1-one, 4-[1-amino-3-(3,4-dimethoxyphenyl)-2-propenyl]-2-hydroxy-, (E)-	MeOH	245(4.47),307(4.13), 393(3.70)	18-3099-83
	MeOH-HCl	228(4.52),317(4.20), 350s(3.85)	18-3099-83
Daviculine	EtOH	218(4.59),276(4.10)	88-2303-83
	EtOH-base	240(5.66),292(4.55)	88-2303-83
Laetanine	EtOH	284(4.07),304(4.09)	100-0761-83
Oureguattidine	EtOH	222(4.34),282(4.25), 300s(4.08),316s(4.01)	100-0761-83
1-Oxa-3-azaspiro[4.5]deca-3,6,9-trien-8-one, 6,10-dimethoxy-2,2-dimethyl-4-phenyl-	EtOH	246(4.43),282s(3.62)	33-2252-83
2-Propenamide, 3-(4-hydroxy-3-methoxyphenyl)-N-[2-(4-hydroxyphenyl)ethyl]-, (E)-	EtOH	285(4.08),293(4.08), 317(4.12)	94-0156-83
$C_{18}H_{19}NO_4S$			
2H-Indol-2-one, 1,3-dihydro-3-methyl-3-[2-[[(4-methylphenyl)sulfonyl]oxy]ethyl]-	EtOH	227(4.10),250(3.88)	94-2986-83
$C_{18}H_{19}NO_6$			
2-Butenedioic acid, 2-[[(3,4-dihydro-6-methoxy-2-oxo-1(2H)-naphthalenylidene)methyl]amino]-, dimethyl ester, (Z,E)-	dioxan	257(4.2),288(3.9), 293(3.9),407(4.3)	5-1476-83
6-Oxa-2-azaspiro[4.5]dec-3-ene-4-carboxylic acid, 7-methoxy-1,9-dioxo-3-phenyl-, ethyl ester	EtOH	242(4.03),312(3.88)	94-0356-83
1H-Pyrrole-2,5-dicarboxylic acid, 3-(2-acetoxyethyl)-4-methyl-, 2-(phenylmethyl) ester	$CHCl_3$	280(4.14)	78-1849-83
$C_{18}H_{19}NO_7$			
9(10H)-Acridinone, 1,6-dihydroxy-2,3,4,5-tetramethoxy-10-methyl-(glyfoline)	MeOH	223s(4.17),261s(4.50), 272(4.56),333(4.12), 409(3.83)	39-1681-83C
$C_{18}H_{19}NO_9$			
β-D-Xylofuranose, 5-C-(carboxyphenylamino)-, intramol. 5,3-ester, 1,2,5-triacetate	MeOH	208(3.81),217s(3.71), 259(2.60)	65-1917-83

Compound	Solvent	$\lambda_{max}(\log \epsilon)$	Ref.
$C_{18}H_{19}NS_2$			
Piperidine, 1-methyl-4-thieno[2,3-c]-[2]benzothiepin-4(9H)-ylidene-	MeOH	232(4.27),284(3.80), 306s(3.62)	73-0623-83
$C_{18}H_{19}N_3O_2$			
Benzenamine, N,N-dimethyl-4-(3,3-dimethyl-5-nitro-3H-indol-2-yl)-	$C_6H_{11}Me$	405(4.592)	151-0251-83A
3H-Indol-5-amine, N,N,3,3-tetramethyl-2-(4-nitrophenyl)-	$C_6H_{11}Me$	440(4.370)	151-0251-83A
2-Propenoic acid, 2-cyano-3-(1,2,3,4-tetrahydro-3-methylpyrimido[1,2-a]-indol-10-yl)-, ethyl ester	MeOH	293(4.23),417(4.75)	103-0507-83
$C_{18}H_{19}N_3O_4$			
2-Pyrimidineacetamide, α-acetoxy-N-acetyl-4,6-dimethyl-α-phenyl-	MeCN	248(3.62)	44-0575-83
$C_{18}H_{19}N_3O_6S$			
1,3,4-Thiadiazole, 2-(4,5-dimethoxy-2-nitrophenyl)-5-(3,4-dimethoxyphenyl)-2,3-dihydro-	EtOH	216(4.45),246s(4.29), 312(4.23)	39-2011-83C
$C_{18}H_{19}N_4Ni$			
Nickel(1+), (6,7,8,9,10,11-hexahydrodibenzo[i,l][1,4,7,11]tetraazacyclotetradecinato-N^6,N^9,N^{12},N^{18})-, (SP-4-3)-, perchlorate	DMF	420(3.81),462(3.90), 544(3.40)	12-2395-83
$C_{18}H_{19}N_5O_4$			
Adenosine, 3'-deoxy-N-(4-methylbenzoyl)-	EtOH	257s(4.19),282(4.43)	118-0304-83
$C_{18}H_{19}OS$			
Thiophenium, tetrahydro-1-(2-oxo-1,2-diphenylethyl)-, tetrafluoroborate	n.s.g.	255(4.06),315(3.31)	116-0864-83
$C_{18}H_{19}O_5P$			
Methanone, [4-[(5,5-dimethyl-1,3,2-dioxaphosphorinan-2-yl)oxy]phenyl]-phenyl-, P-oxide	EtOH	260(4.4)	65-0243-83
$C_{18}H_{19}O_6P$			
Methanone, [4-[(5,5-dimethyl-1,3,2-dioxaphosphorinan-2-yl)oxy]-2-hydroxyphenyl]phenyl-, P-oxide	EtOH	270(1.6),335(0.7)	65-0653-83
$C_{18}H_{19}O_8P$			
Methanone, (2,4-dihydroxyphenyl)[4-[(5,5-dimethyl-1,3,2-dioxaphosphorinan-2-yl)oxy]-2-hydroxyphenyl]-, P-oxide	EtOH	295(1.4),330(1.4)	65-0653-83
$C_{18}H_{20}$			
1-Butene, 3-methyl-1-(4-methylphenyl)-3-phenyl-, (E)-	n.s.g.	217(4.17),253.5(4.30), 286.5(3.37),297(3.16)	12-0565-83
[3.3]Metacyclophane	EtOH	262s(2.49),266s(2.41), 275s(2.14)	94-2868-83
$C_{18}H_{20}BrNO_2$			
Ethanone, 1,1'-[4-(4-bromophenyl)-1,2-dihydro-1,2,6-trimethyl-3,5-pyridine-	EtOH	204(4.36),224s(4.17), 277(4.16),337(4.08),	103-0415-83

Compound	Solvent	λ_{max}(log ε)	Ref.
diyl]bis- (cont.)		419(3.95)	103-0415-83
Ethanone, 1,1'-[4-(4-bromophenyl)-1,4-dihydro-1,2,6-trimethyl-3,5-pyridinediyl]bis-	EtOH	205(4.22),267(4.15), 374(3.85)	103-0415-83
$C_{18}H_{20}BrNO_4$			
Estra-1,3,5(10)-trien-17-one, 4-bromo-3-hydroxy-2-nitro-	MeOH	298(3.80)	13-0675-83A
$C_{18}H_{20}BrNO_5$			
Estra-1,3,5(10)-trien-17-one, 4-bromo-3,16-dihydroxy-2-nitro-, (16α)-	MeOH	297(3.76)	13-0675-83A
$C_{18}H_{20}BrN_3OS$			
4(1H)-Quinazolinethione, 1-(4-bromophenyl)-5,6,7,8-tetrahydro-2-morpholino-	MeOH	227(4.25),342(4.39)	73-3575-83
$C_{18}H_{20}Br_2O_2$			
Estra-1,3,5(10)-trien-17-one, 4,16α-dibromo-3-hydroxy-	EtOH	281(3.30),289(3.28)	39-0121-83C
$C_{18}H_{20}Br_2O_3$			
Estra-1,3,5(10)-trien-16-one, 2,4-dibromo-3,17β-dihydroxy-	EtOH	285(3.40),290(3.43)	39-0121-83C
Estra-1,3,5(10)-trien-17-one, 2,4-dibromo-3,16α-dihydroxy-	EtOH	285(3.42),292(3.38)	39-0121-83C
$C_{18}H_{20}ClN_3OS$			
4(1H)-Quinazolinethione, 1-(4-chlorophenyl)-5,6,7,8-tetrahydro-2-morpholino-	MeOH	222(4.24),340(4.45)	73-3575-83
$C_{18}H_{20}ClN_3S$			
1-Piperazinamine, 4-(2-chloro-10,11-dihydrodibenzo[b,f]thiepin-10-yl)-	MeOH	282(4.47),310s(4.38)	73-1173-83
$C_{18}H_{20}FeN_2O_7$			
Iron, tricarbonyl[dimethyl 1-(2,3-diazabicyclo[2.2.1]hept-2-yl)-4-cyclohexene-1,2-dicarboxylato(2-)]-, (SP-5-22)-	benzene	390(3.71),431(3.43), 505s(3.04)	64-0648-83B
$C_{18}H_{20}FeO_6$			
Iron, tricarbonyl[1,11,12,13-η-(12-hydroxy-5,9,13-trimethyl-3-oxatricyclo[7.4.0.0^{2,6}]trideca-11,13-dien-4-one)]-	dioxan	227(4.2),288s(3.4)	24-0243-83
$C_{18}H_{20}NO_2$			
Pyridinium, 3,5-diacetyl-1,2,6-trimethyl-4-phenyl-, perchlorate	EtOH	208(4.41),288(4.08)	103-0415-83
$C_{18}H_{20}N_2$			
Benzenamine, 4-(3,3-dimethyl-3H-indol-2-yl)-N,N-dimethyl-	$C_6H_{11}Me$	348(4.554)	151-0251-83A
3H-Indol-5-amine, N,N,3,3-tetramethyl-2-phenyl-	$C_6H_{11}Me$	364(4.201)	151-0251-83A
$C_{18}H_{20}N_2O$			
Benz[cd]indol-5-ol, 4-amino-1,2,2a,3,4-	EtOH	259(4.02),299(3.42)	87-0522-83

Compound	Solvent	$\lambda_{max}(\log \epsilon)$	Ref.
5-hexahydro-1-(phenylmethyl)-, (2aα,4β,5α)-(±)- (cont.)	EtOH-HCl	259(3.86),299(3.23)	87-0522-83
$C_{18}H_{20}N_2O_2$			
Benzenemethanamine, N,N-diethyl-α-(nitrophenylmethylene)-, (Z)-	EtOH	418(4.26)	22-0339-83
Indolo[4,3-fg]isoquinolin-9(4H)-one, 4-acetyl-5,5a,6,6a,7,8-hexahydro-8,10-dimethyl-	MeOH	253(4.53),295(4.24)	39-1545-83C
Indolo[4,3-fg]isoquinolin-9(4H)-one, 4-acetyl-5,5a,6,8,10,10a-hexahydro-8,10-dimethyl-	MeOH	213(4.49),262(4.30)	39-1545-83C
Piperazinomycin	MeOH	214(4.05),280(3.31), 290s(3.23)	158-83-117
$C_{18}H_{20}N_2O_2S$			
Benzenesulfonic acid, 4-methyl-, (1-phenyl-3-pentenylidene)hydrazide, (E)-	EtOH	265(4.12)	35-0933-83
(Z)-	EtOH	265(4.06)	35-0933-83
$C_{18}H_{20}N_2O_3$			
Erymelanthine	EtOH	230(3.90),270(3.48)	88-5067-83
$C_{18}H_{20}N_2O_3S$			
2H-Thiopyran-3-carboxylic acid, 2-imino-6-morpholino-5-phenyl-, ethyl ester, monoperchlorate	MeCN	235(4.29),314(4.04), 439(4.22)	97-0403-83
$C_{18}H_{20}N_2O_4$			
Ethanone, 1,1'-[1,4-dihydro-1,2,6-trimethyl-4-(4-nitrophenyl)-3,5-pyridinediyl]bis-	EtOH	206(4.23),217s(4.09),* 262(4.31),277s(4.23), 375(3.75)	103-0415-83
$C_{18}H_{20}N_2O_4S$			
Benzenesulfonic acid, 4-methyl-, (3,4-dihydro-8-hydroxy-5-methoxy-1(2H)-naphthalenylidene)hydrazide	EtOH	224(4.38),246s(4.15), 281(4.18),310s(3.81), 342(3.80)	94-2662-83
$C_{18}H_{20}N_2O_5$			
2-Butenoic acid, 2-[[(1,3-dihydro-1,3-dioxo-2H-isoindol-2-yl)acetyl]amino]-, 1,1-dimethylethyl ester, (E)-	MeOH	240.5(3.71)	118-0539-83
(Z)-	MeOH	241.5(3.80)	118-0539-83
$C_{18}H_{20}N_2O_5S$			
4(1H)-Pyrimidinone, 2,3-dihydro-5-[2,3-O-(1-methylethylidene)-4-C-phenyl-β-DL-ribofuranosyl]-2-thioxo-	pH 13	264(4.04),280(3.93)	18-2680-83
	MeOH	276(3.90),289(3.86)	18-2680-83
4(1H)-Pyrimidinone, 2,3-dihydro-5-[2,3-O-(1-methylethylidene)-5-C-phenyl-β-DL-ribofuranosyl]-2-thioxo-, (R*)-	pH 13	264(4.06),281s(4.00)	18-2680-83
	MeOH	275(4.11),291s(4.05)	18-2680-83
4-Thia-1-azabicyclo[3.2.0]heptane-2-carboxylic acid, 3,7-dioxo-6-[(phenylacetyl)amino]-, 1,1-dimethylethyl ester	dioxan	243(3.46)	78-2493-83
$C_{18}H_{20}N_2O_6$			
2,4(1H,3H)-Pyrimidinedione, 5-[(R)-1,2-O-(1-methylethylidene)-5-C-phenyl-β-DL-ribofuranosyl]-	pH 13	285(3.88)	18-2680-83
	MeOH	263(3.90)	18-2680-83

Compound	Solvent	λ max (log ε)	Ref.
$C_{18}H_{20}N_2O_6S$			
1-Azabicyclo[3.2.0]hept-2-ene-2-carboxylic acid, 6-ethyl-3-[(2-hydroxyethyl)thio]-7-oxo-, (4-nitrophenyl)methyl ester	THF	270(4.03),320(4.02)	158-0407-83
$C_{18}H_{20}N_2S$			
Azuleno[2,1-d]thiazole, 2-methyl-4-(1-piperidinylmethyl)-	MeOH	220(4.05),257(3.97), 301(4.84),306(4.80), 352(3.71),368(3.83), 378s(3.36),610(2.59), 663s(2.50),733s(2.12)	142-1263-83
$C_{18}H_{20}N_4O$			
1,3,4-Oxadiazole, 2,5-bis[4-(dimethylamino)phenyl]-	toluene	350(4.72)	103-0022-83
$C_{18}H_{20}N_4O_2S$			
Carbamic acid, (5-amino-2,2-dimethyl-3-phenyl-2H-pyrido[4,3-b]-1,4-thiazin-7-yl)-, ethyl ester	pH 1	251(4.42),373(4.22)	87-1614-83
$C_{18}H_{20}N_4O_9$			
Hypoglycine B, 2,4-dinitrophenylhydrazone	n.s.g.	359(4.22)	162-0711-83
$C_{18}H_{20}N_6$			
Acetaldehyde, 2,2'-(1,2-ethanediyldinitrilo)bis-, bis(phenylhydrazone), (Z,Z,E,E)- (approximate spectral data)	$CHCl_3$	300(4.12),340(4.54)	65-1214-83
$C_{18}H_{20}O$			
Benzene, 1-methoxy-4-(3-methyl-3-phenyl-1-butenyl)-, (E)-	n.s.g.	213.5(4.17),260(4.26), 263(4.26),269(4.20), 274(3.42),307(3.16)	12-0565-83
$C_{18}H_{20}O_2$			
Estra-1,3,5(10),9(11)-tetraen-17-one, 3-hydroxy-	EtOH	263.5(4.26),298(3.50)	12-0339-83
2-Naphthalenol, 5,6,7,8-tetrahydro-8-(4-hydroxyphenyl)-6,7-dimethyl-, (6α,7β,8α)-	MeOH	225(4.26),290(3.63)	102-2281-83
$C_{18}H_{20}O_3S$			
Benzenemethanol, 2-methoxy-α-[1-[(4-methylphenyl)sulfinyl]-2-propenyl]-	n.s.g.	220s(4.27),252(3.81), 271(3.76),277s(3.67)	12-1049-83
Benzenemethanol, 2-methoxy-α-[3-[(4-methylphenyl)sulfinyl]-2-propenyl]-, (E)-	n.s.g.	222s(4.28),269s(3.58), 275s(3.49)	12-1049-83
Benzenemethanol, 4-methoxy-α-[1-[(4-methylphenyl)sulfinyl]-2-propenyl]-	n.s.g.	227(4.24),251s(3.96), 280s(3.32)	12-1049-83
Benzenemethanol, 4-methoxy-α-[3-[(4-methylphenyl)sulfinyl]-2-propenyl]-, (E)-	n.s.g.	228(4.09),279s(3.00)	12-1049-83
$C_{18}H_{20}O_4$			
Benzene, 1,2,4-trimethoxy-5-[2-(3-methoxyphenyl)ethenyl]-, trans	EtOH	240s(4.18),289(4.14), 338(4.24)	2-0101-83
Benzoic acid, 6-hydroxy-2,3-dimethyl-4-(phenylmethoxy)-, ethyl ester	EtOH	267(3.95),310(3.77)	39-0667-83C

Compound	Solvent	λ_{max}(log ε)	Ref.
2H-Pyran-5-carboxylic acid, 3,4-dihydro-2,2-dimethyl-4-oxo-6-(2-phenylethenyl)-, ethyl ester	EtOH	340(4.41)	118-0948-83
$C_{18}H_{20}O_4S$			
1H-Indene-1,5(6H)-dione, 7a-ethyl-2,3,7,7a-tetrahydro-4-[(phenylsulfonyl)methyl]-, (S)-	EtOH	218(4.08),252(3.98)	39-2337-83C
1,6(2H,7H)-Naphthalenedione, 3,4,8,8a-tetrahydro-8a-methyl-5-[(phenylsulfonyl)methyl]-, (8aS)-(+)-	EtOH	219(4.08),253(4.05)	39-2349-83C
$C_{18}H_{20}O_5$			
1-Phenanthrenecarboxylic acid, 1,2,3,4-4a,9-hexahydro-10-hydroxy-6-methoxy-1,4a-dimethyl-9-oxo-, (1S-cis)-	EtOH	238(3.85),288(3.98),323(3.97)	23-2461-83
$C_{18}H_{20}O_6$			
2,5-Methano-1,3-benzodioxol-6(3aH)-one, 8-[(benzoyloxy)methyl]tetrahydro-3a-methoxy-7a-methyl-, [2S-(2α,3aβ,5α-7aβ,8R*)]-	MeOH	228(4.27),268s(3.21),275(3.22),280s(3.20)	94-0577-83
$C_{18}H_{20}O_7$			
7H-Furo[3,2-g][1]benzopyran-7-one, 9-[2-[2-(2-methoxyethoxy)ethoxy]-ethoxy]-	EtOH	218(4.39),248(4.36),263(4.12),300(4.08)	39-1807-83B
D-erythro-Hex-2-enitol, 1,5-anhydro-2-deoxy-1-C-phenyl-, triacetate, (S)-	EtOH	257(2.46)	23-0533-83
threo-	EtOH	257(2.46)	23-0533-83
$C_{18}H_{21}ClN_2OS_2$			
1-Piperazineethanol, 4-(7-chloro-4,5-dihydrothieno[2,3-b][1]benzothiepin-4-yl)-	MeOH	265s(3.90)	73-2970-83
$C_{18}H_{21}FN_4$			
4H-Pyrrolo[2,3-d]pyrimidin-4-imine, 3-(4-fluorophenyl)-3,7-dihydro-2,5-dimethyl-6-(2-methylpropyl)-	EtOH	245(4.38),285(4.30)	5-2066-83
$C_{18}H_{21}N$			
1H-Indole, 2,3-dihydro-3-(1-methyl-1-phenylpropyl)-	n.s.g.	246(3.9),303(3.3)	103-0638-83
$C_{18}H_{21}NO$			
Benzenemethanamine, 2-[2-(4-methoxyphenyl)ethenyl]-N,N-dimethyl-, hydrochloride	HOAc	310(4.37)	39-1503-83B
Cyclopentanone, 2-[(1,3-dihydro-1,3,3-trimethyl-2H-indol-2-ylidene)ethylidene]-	MeOH	435(4.70)	104-1854-83
	benzene	405(--),418(--)	104-1854-83
	DMF	420(--)	104-1854-83
$C_{18}H_{21}NO_2$			
Ethanone, 1,1'-[1,4-dihydro-2,6-dimethyl-4-(4-methylphenyl)-3,5-pyridinediyl]bis-	EtOH	204(4.24),217(4.11),257(4.15),265s(4.12),385(3.90)	103-0415-83
Ethanone, 1,1'-(1,2-dihydro-1,2,6-trimethyl-4-phenyl-3,5-pyridinediyl)bis-	EtOH	204(4.23),238s(4.04),269(4.07),283s(4.03),336(4.07),419(3.95)	103-0415-83

Compound	Solvent	$\lambda_{max}(\log \epsilon)$	Ref.
Ethanone, 1,1'-(1,4-dihydro-1,2,6-trimethyl-4-phenyl-3,5-pyridinediyl)bis-	EtOH	205(4.19),264(4.11), 282s(3.90),376(3.87)	103-0415-83
$C_{18}H_{21}NO_2S$			
Benzene, 1-[2,2-dimethyl-1-[(4-methylphenyl)thio]propyl]-3-nitro-	EtOH	253(3.91)	12-0081-83
$C_{18}H_{21}NO_3$			
Ethanone, 1,1'-[1,4-dihydro-4-(4-methoxyphenyl)-2,6-dimethyl-3,5-pyridinediyl]bis-	EtOH	204(4.32),226(4.19), 258(4.11),269s(4.06), 385(3.91)	103-0415-83
7-Isoquinolinol, 1,2,3,4-tetrahydro-1-[(4-hydroxyphenyl)methyl]-8-methoxy-2-methyl-, hydrochloride	MeOH-NaOH	232(4.18),302(3.53), 336(3.40)	100-0342-83
8-Isoquinolinol, 1,2,3,4-tetrahydro-1-[(4-hydroxyphenyl)methyl]-7-methoxy-2-methyl- (juziphine)	MeOH	232(4.24),281(3.66)	100-0342-83
$C_{18}H_{21}NO_4$			
5-Oxa-1-azabicyclo[4.2.0]oct-2-ene-2-carboxylic acid, 8-oxo-3-(phenylmethyl)-, 1,1-dimethylethyl ester	EtOH	258(4.04)	44-1841-83
Phenol, 2-(1,1-dimethylethyl)-6-[(2-hydroxy-5-methylphenyl)methyl]-4-nitro-	neutral	335(3.94)	126-2361-83
	anion	432(4.31)	126-2361-83
	dianion	449(4.43)	126-2361-83
Phenol, 2-(1,1-dimethylethyl)-6-[(2-hydroxy-5-nitrophenyl)methyl]-4-methyl-	neutral	320(3.95)	126-2361-83
	anion	405(4.24)	126-2361-83
	dianion	407(4.25)	126-2361-83
1H-Pyrrole-2-carboxylic acid, 3-(2-acetoxyethyl)-4,5-dimethyl-, phenylmethyl ester	CHCl	284(4.26)	78-1849-83
$C_{18}H_{21}NO_4S$			
Benzene, 1-[2,2-dimethyl-1-[(4-methylphenyl)sulfonyl]propyl]-3-nitro-	EtOH	265(3.93)	12-0081-83
$C_{18}H_{21}NO_5$			
Butanedioic acid, (2,3,4,5-tetrahydro-1-methyl-2-oxo-1-benzazocin-6(1H)-ylidene)-, dimethyl ester	MeOH	212(4.18),220(4.15)	39-0505-83C
$C_{18}H_{21}N_3O$			
Dibenzepin, hydrochloride	pH 1	204(4.530),220(4.458)	162-0436-83
$C_{18}H_{21}N_3OS$			
4(1H)-Quinazolinethione, 5,6,7,8-tetrahydro-2-morpholino-1-phenyl-	MeOH	229(4.20),340(4.39)	73-3575-83
$C_{18}H_{21}N_3O_2$			
Propanedinitrile, [4,4-dimethyl-1-(4-nitrophenyl)-2-pentenyl]ethyl-, (E)-	EtOH	265(4.00)	12-0527-83
$C_{18}H_{21}N_3O_3S_3$			
1,3,5-Benzenetriethanethioamide, N,N,N',N',N",N"-hexamethyl-α,α',α"-trioxo-	MeCN	232(4.70),252(4.69), 328s(3.75),391(3.69)	5-1116-83
$C_{18}H_{21}N_3O_5$			
4(1H)-Pyrimidinone, 2-amino-5-[2,3-O-(1-methylethylidene)-4-C-phenyl--DL-ribofuranosyl]-	pH 13	233(3.74),277(3.64)	18-2680-83
	MeOH	227(4.09),290(4.12)	18-2680-83

Compound	Solvent	λ_{max}(log ε)	Ref.
4(1H)-Pyrimidinone, 2-amino-5-[2,3-O-(1-methylethylidene)-5-C-phenyl-β-DL-ribofuranosyl]-, (R*)-	pH 13	211(4.18),234(4.08), 276(3.96)	18-2680-83
	MeOH	226(4.09),239(4.02)	18-2680-83
$C_{18}H_{21}N_3O_5S_2$			
1,2,4-Triazin-3(2H)-one, 4,5-dihydro-2-[2,3-O-(1-methylethylidene)-β-D-ribofuranosyl]-6-[(phenylmethyl)-thio]-5-thioxo-	pH 1	324(3.90)	80-0989-83
	pH 13	336(4.0)	80-0989-83
	EtOH	324(3.89)	80-0989-83
$C_{18}H_{21}N_3O_6$			
1-Deazariboflavin	1.2M $HClO_4$	231(4.70),265s(4.42), 304(4.79),372(3.72), 520(3.97)	125-0655-83
$C_{18}H_{21}N_5O_3$			
9H-Purin-6-amine, 9-[(2,2-dimethyl-1,3-dioxolan-4-yl)methyl]-8-(phenylmethoxy)-	pH 2	272(3.96),288(3.95)	73-1910-83
	pH 7 and 12	276(4.05)	73-1910-83
$C_{18}H_{21}N_8O$			
5H-Dipyrimido[1,6-a:4',5'-d]pyrimidinium, 8-[(1,4-dihydro-4-imino-1,2-dimethyl-5-pyrimidinyl)methyl]-7,8-dihydro-2,3-dimethyl-7-oxo-, perchlorate, monoperchlorate	pH 1	222(4.23),250(4.15), 347(4.33),385s(3.59)	44-2481-83
	pH 6.86	223(4.14),239(4.23), 247(4.22),385(4.43)	44-2481-83
$C_{18}H_{22}BrNO_2$			
Estra-1,3,5(10)-trien-17-one, 2-amino-4-bromo-3-hydroxy-	MeOH	298(3.54)	13-0675-83A
$C_{18}H_{22}BrNO_3$			
Estra-1,3,5(10)-trien-16-one, 2-amino-4-bromo-3,17β-dihydroxy-	MeOH	297(3.51)	13-0675-83A
$C_{18}H_{22}BrNO_4$			
Estra-1,3,5(10)-triene-3,17β-diol, 4-bromo-2-nitro-	MeOH	298(3.76)	13-0675-83A
$C_{18}H_{22}BrNO_5$			
Estra-1,3,5(10)-triene-3,16α,17β-triol, 4-bromo-2-nitro-	MeOH	298(3.74)	13-0675-83A
$C_{18}H_{22}ClNO$			
1-Butanol, 2-[[(4-chlorophenyl)phenylmethyl]amino]-3-methyl-, hydrochloride, (R,S)-	EtOH	227(4.20),255(3.24), 260(3.24),268s(3.13), 275s(2.92)	94-2183-83
(S,S)-	EtOH	226(4.40),255s(3.09), 258(3.14),263(3.12), 269s(3.06),276s(2.76)	94-2183-83
$C_{18}H_{22}NO_3$			
Isoquinolinium, 1,2,3,4-tetrahydro-7,8-dihydroxy-1-[(4-hydroxyphenyl)methyl]-2,2-dimethyl-, iodide, (±)-	MeOH	232(4.08),283(3.5)	100-0342-83
$C_{18}H_{22}N_2$			
Benzenamine, 4,4'-(1,2-ethenediyl)-bis[N,N-dimethyl-	MeCN	358(4.48+)	94-0373-83
Desipramine	n.s.g.	213(4.39),252(3.93)	162-0421-83
Piperazine, 1-(diphenylmethyl)-4-methyl- (cyclizine)	pH 1	225(4.05),258(2.84), 263(2.87),269(2.73)	162-0388-83

Compound	Solvent	λ_{max}(log ε)	Ref.
Uleine	MeOH	209(4.35),310(4.28), 319(4.26)	100-0200-83
$C_{18}H_{22}N_2O$			
Indolo[2,3-a]quinolizine-1-carboxaldehyde, 1-ethyl-1,2,3,4,5,6,7,12b-octahydro-, (1RS,12bRS)-	EtOH	224(4.44),282(3.91), 290(3.83)	2-0531-83
$C_{18}H_{22}N_2O_2$			
Indolo[4,3-fg]isoquinolin-9(4H)-one, 4-acetyl-5,5a,6,6a,7,8,10,10a-octahydro-8,10-dimethyl-, (6aR*,10R*,10aR*)	MeOH	218(4.42),257(4.19)	39-1545-83C
(6aR*,10S*,10aS*)-	MeOH	214(4.49),255(4.17), 280(3.68),290(3.63)	39-1545-83C
$C_{18}H_{22}N_2O_3$			
Acetamide, N-[2-[2-[(1-methoxymethylene)-3-oxobutyl]-1H-indol-3-yl]-ethyl]-	EtOH	216(4.39),256(3.94), 352(4.16)	4-0267-83
Acetamide, N-[2-[2-[(1-methoxymethyl)-3-oxo-1-butenyl]-1H-indol-3-yl]-ethyl]-	EtOH	255(3.94),352(4.16)	4-0267-83
Cinnolinium, 3-[(3,4-dimethoxyphenyl)-methyl]-5,6,7,8-tetrahydro-4-hydroxy-2-methyl-, hydroxide, inner salt	EtOH	272(3.75),320(3.92)	4-0543-83
4(1H)-Cinnolinone, 3-[(3,4-dimethoxy-phenyl)methyl]-5,6,7,8-tetrahydro-1-methyl-	EtOH	284(4.19)	4-0543-83
Ethanone, 2-(3,4-dimethoxyphenyl)-1-(4,5,6,7-tetrahydro-1-methyl-1H-indazol-3-yl)-	EtOH	234(4.12),277(3.84)	4-0543-83
1H-Pyrido[3,4-b]indole, 2-acetyl-2,3,4,9-tetrahydro-1-(methoxymethyl)-1-(2-oxopropyl)-	EtOH	223(4.59),274s(3.91), 282(3.92),291(3.83)	4-0267-83
1H-Pyrido[3,4-b]indole, 2-acetyl-2,3,4,9-tetrahydro-6-methoxy-1-methyl-1-(2-oxopropyl)-	EtOH	221(4.59),276(3.89), 288s(3.78)	4-0267-83
$C_{18}H_{22}N_2O_3Si$			
1H-1,2-Diazepine-4-carboxylic acid, 1-benzoyl-, 2-(trimethylsilyl)ethyl ester	MeOH	231(4.27),265s(4.05), 382(3.15)	5-1374-83
1H-1,2-Diazepine-6-carboxylic acid, 1-benzoyl-, 2-(trimethylsilyl)ethyl ester	MeOH	230(4.20),290(3.87)	5-1374-83
$C_{18}H_{22}N_2O_9S$			
Benzenesulfonamide, N-(3,3-dimethoxy-4-methyl-6-oxo-1,4-cyclohexadien-1-yl)-2-(2-methoxyethoxy)-5-nitro-	acetone	297(4.10)	44-0177-83
$C_{18}H_{22}N_4$			
4H-Pyrrolo[2,3-d]pyrimidin-4-imine, 3-(4-butylphenyl)-3,7-dihydro-2,5-dimethyl-	EtOH	238(4.35),280(3.94)	5-2066-83
4H-Pyrrolo[2,3-d]pyrimidin-4-imine, 3,7-dihydro-2,5-dimethyl-6-(2-methylpropyl)-	EtOH	236(4.19),287(3.79)	5-2066-83
$C_{18}H_{22}N_4O_6S$			
Azirino[2',3':3,4]pyrrolo[1,2-a]indole-	MeOH	253(4.40),259s(4.11),	44-5026-83

Compound	Solvent	λ_{max}(log ε)	Ref.
1(7H)-carbothioic acid, 6-amino-8-[[(aminocarbonyl)oxy]methyl]-1a,2,4,8,8a,8b-hexahydro-8a-methoxy-5-methyl-4,7-dioxo-, O-ethyl ester, [1aS-(1aα,8β,8aα,8bα)]- (cont.)		356(4.44)	44-5026-83
$C_{18}H_{22}N_4O_7$			
Galactoflavin	n.s.g.	223(3.44),267(4.45), 370(3.96),445(4.03)	162-0619-83
$C_{18}H_{22}N_4O_{10}S$			
9H-Purine-6-methanol, 9-(2,3,5-tri-O-acetyl-β-D-ribofuranosyl)-, methanesulfonate	MeOH	267(3.91)	94-3104-83
$C_{18}H_{22}O_2$			
Estra-4,9,11-trien-3-one, 17β-hydroxy-(trenbolone)	n.s.g.	239(3.73),340.5(4.45)	162-1371-83
Phenol, 4-(1-methylethyl)-, dimer	5% MeCN	244(3.52),289(3.38)	44-2133-83
Spiro[2.4]hept-1-ene-1-carboxylic acid, 2-phenyl-, 1,1-dimethylethyl ester	heptane	293(3.81)	24-3097-83
$C_{18}H_{22}O_3$			
Estra-1,4-diene-3,17-dione, 10-hydroxy-	EtOH	236(4.07)	89-1025-83S
epimer	EtOH	233(4.08)	89-1025-83S
Estra-1,3,5(10)-trien-6-one, 3,17-dihydroxy-	MeOH	220(4.36),253(3.96), 325(3.54)	145-0347-83
9(1H)-Phenanthrenone, 2,3,4,4a-tetrahydro-6,10-dimethoxy-1,4a-dimethyl-	EtOH	239(4.02),265(4.03), 303(4.10)	23-2461-83
$C_{18}H_{22}O_3S$			
1,4-Dioxaspiro[4.5]dec-7-ene, 7-[(phenylsulfinyl)methyl]-8-(1-propenyl)-	EtOH	253(4.09)	78-1123-83
$C_{18}H_{22}O_4$			
2H-1-Benzopyran-6-carboxaldehyde, 5-hydroxy-7-methoxy-2-methyl-2-(4-methyl-3-pentenyl)-	EtOH	225(3.95),232s(3.90), 269s(4.42),276(4.50), 298(4.09),310(4.09), 361(3.44)	39-1411-83C
Estra-1,4-diene-3,17-dione, 10-hydroperoxy-, (10β)-	EtOH	237(4.08)	89-1025-83S
1,4-Naphthalenedione, 5-ethenyl-5,6,7,8-tetrahydro-2,3-dimethoxy-5-methyl-8-(1-methylethenyl)-(arnebinone)	EtOH	207(4.13),277(4.11), 410(2.76)	88-3247-83
1H-4-Oxabenzo[f]cyclobut[cd]indene-7-carboxaldehyde, 1a,2,3,3a,8b,8c-hexahydro-8-hydroxy-6-methoxy-1,1,3a-trimethyl-, (±)-	EtOH	219(4.20),223s(4.18), 233s(4.04),301(4.31), 339s(3.48)	39-1411-83C
2H-Pyran-3-carboxaldehyde, 4-methoxy-6-(1,2,4a,5,6,7,8,8a-octahydro-2-methyl-1-naphthalenyl)-2-oxo-(solanopyrone A)	EtOH	232(3.96),327(3.97)	88-5373-83
$C_{18}H_{22}O_4S$			
1,4-Dioxaspiro[4.5]decane, 7-methylene-8-[2-[(4-methylphenyl)sulfonyl]ethylidene]-	EtOH	238(3.94)	78-1123-83
1,4-Dioxaspiro[4.5]decane, 7-methylene-8-[2-(phenylsulfonyl)propylidene]-	EtOH	240(3.97)	78-1123-83

Compound	Solvent	λ_{max}(log ε)	Ref.
$C_{18}H_{22}O_5$			
Zearalenone	MeOH	236(4.48),274(4.15), 316(3.78)	162-1454-83
$C_{18}H_{22}O_6$			
2,5-Methano-1,3-benzodioxol-6-ol, 8-[(benzoyloxy)methyl]hexahydro-3a-methoxy-7a-methyl-, [2S-(2α,3aβ,5α,6α,7aβ,8R*)]-	MeOH	228(4.18),267s(3.08), 272(3.11),279(3.02)	94-0577-83
1(2H)-Naphthalenone, 3-[2-(2,2-dimethyl-1,3-dioxolan-4-yl)-1-methoxy-2-oxoethyl]-3,4-dihydro-2-hydroxy-, (R)-	EtOH	247(4.08),287(3.32)	44-3269-83
(S)-	EtOH	255(3.96),290(3.34)	44-3269-83
$C_{18}H_{22}O_7$			
Propanedioic acid, [4-(6-hydroxy-1,3-benzodioxol-5-yl)-2-butenyl]-, diethyl ester	EtOH	236(3.66),301(3.80)	78-1761-83
$C_{18}H_{22}O_{10}$			
Bicyclo[2.2.2]oct-7-ene-2,3,5,6,7-pentacarboxylic acid, pentamethyl ester, (1α,2α,3α,4α,5α,6α)-	MeCN	215(3.98)	33-0019-83
$C_{18}H_{23}ClO_4$			
2H-1-Benzopyran-6-carboxaldehyde, 2-(4-chloro-4-methylpentyl)-5-hydroxy-7-methoxy-2-methyl-, (±)-	EtOH	225(3.96),232s(3.90), 269s(4.42),275(4.48), 298(4.10),310(4.10), 359(3.57)	39-1411-83C
$C_{18}H_{23}ClO_5$			
5-Hexenoic acid, 6-(4-chloro-3-methoxyphenyl)-4-methoxy-2-(methoxymethylene)-3-methyl-, methyl ester (oudemansin B)	dioxan	203s(4.19),219(4.34), 248(4.25),261(4.22), 270s(4.11),299(3.67), 312(3.49)	158-0661-83
$C_{18}H_{23}NO$			
1-Butanol, 2-[(diphenylmethyl)amino]-3-methyl-, hydrochloride, (S)-	EtOH	220(4.27),253s(2.97), 258(3.04),263(3.03), 269(2.92)	94-2183-83
$C_{18}H_{23}NO_2$			
4H-Cyclopenta[5,6]naphtho[2,1-c]isoxazol-6-ol, 6-ethynyl-3b,5,5a,6,7,8-8a,8b,9,10-decahydro-3,5a-dimethyl-, [3bS-(3bα,5aβ,6β,8aα,8bβ)]-	EtOH	205(3.17),227(3.71)	39-2353-83C
Estra-1,3,5(10)-trien-17-one, 3-hydroxy-, oxime, (8α)-	MeOH	216(3.92),228(3.74), 281(3.29),286(3.25)	13-0001-83B
	MeOH-NaOH	241(3.98),298(3.40)	13-0001-83B
1-Propanol, 2-[[1-(4-methoxyphenyl)-2-phenylethyl]amino]-, hydrochloride	EtOH	228(4.07),251(2.65), 260s(2.86),266s(2.98), 270s(3.03),275(3.06)	94-0031-83
$C_{18}H_{23}NO_3$			
Dobutamine	MeOH	223(4.16),281(3.68)	162-0495-83
$C_{18}H_{23}NO_4S$			
Morpholine, 4-(3,4,5,6,7,8-hexahydro-3-phenyl-1,2-benzoxathiin-4-yl)-, S,S-dioxide, cis	EtOH	230s(3.55),320(2.20)	4-0839-83

Compound	Solvent	$\lambda_{max}(\log \epsilon)$	Ref.
$C_{18}H_{23}NO_6$			
Benzenamine, 2,4,6-trimethoxy-N-(2,4,6-trimethoxyphenyl)-, radical cation	MeCN	372(3.87),789(4.10)	152-0105-83
$C_{18}H_{23}NO_{11}$			
2,5-Pyrrolidinedione, 1-(2,3,4,6-tetra-O-acetyl-β-D-glucopyranosyl)-	$CHCl_3$	198(4.10),235(3.15)	102-1278-83
$C_{18}H_{23}N_2O_5$			
1-Piperidinyloxy, 2,2,6,6-tetramethyl-4-[[3-(4-nitrophenyl)-1-oxo-2-propenyl]oxy]-	MeOH	277(3)	28-0043-83B
$C_{18}H_{23}N_3O_2$			
4-Imidazolidinone, 2-(1H-indol-3-ylcarbonyl)-1-methyl-5-(2-methylbutyl)- (martensine B)	MeOH	215(4.00),244(4.04), 261(3.92),303(4.00)	88-2087-83
	MeOH-NaOH	267(4.00),335(4.00)	88-2087-83
$C_{18}H_{23}N_3O_4$			
1H-Pyrrolo[1,2-a][1,3]diazepine-5-carboxylic acid, 3-acetyl-9-cyano-5-ethoxy-4,5-dihydro-7,8-dimethyl-, ethyl ester	EtOH	364(4.11)	4-0081-83
Pyrrolo[1,2-a]pyrimidine-4-carboxylic acid, 3-acetyl-8-cyano-1,4-dihydro-6,7-dimethyl-4-ethoxymethyl-, ethyl ester	EtOH	353(3.95)	4-0081-83
$C_{18}H_{23}N_3O_6$			
1-Deazariboflavin, dihydro-	1.2M $HClO_4$	238s(4.53),276(4.43), 341(4.23)	125-0655-83
$C_{18}H_{23}N_3O_9$			
Acetamide, N-[1,2-dihydro-2-oxo-1-(2,3,6-tri-O-acetyl-5-deoxy-β-D-arabinohexofuranosyl)-4-pyrimidinyl]-	EtOH	249(4.24),299(3.91)	136-0099-83J
$C_{18}H_{23}N_6O_7P$			
5'-Adenylic acid, N-[2-(4-aminophenyl)-ethyl]-	H_2O	267(4.29)	130-0045-83
$C_{18}H_{23}O_3P$			
Phosphorin, 1-(1,1-dimethylethyl)-1,1-dihydro-1-methoxy-2-(methoxycarbonyl)-4-phenyl-	EtOH	235(4.27),290(4.19), 348(3.89)	88-5051-83
Phosphorin, 4-(1,1-dimethylethyl)-1,1-dihydro-1-methoxy-2-(methoxycarbonyl)-1-phenyl-	EtOH	212(4.50),320s(3.45), 380(3.92)	88-5051-83
$C_{18}H_{24}$			
Cyclobutane, 1,2-bis(dicyclopropylmethylene)-	hexane	278(4.14)	44-1500-83
Cyclobutane, 1,1-dicyclopropyl-3-(dicyclopropylmethylene)-2-methylene-	hexane	268(4.19)	44-1500-83
$C_{18}H_{24}BrNO_2$			
Estra-1,3,5(10)-triene-3,17β-diol, 2-amino-4-bromo-	MeOH	298(3.50)	13-0675-83A
$C_{18}H_{24}BrNO_3$			
Estra-1,3,5(10)-triene-3,16α,17β-triol,	MeOH	298(3.50)	13-0675-83A

Compound	Solvent	λ_{max}(log ϵ)	Ref.
2-amino-4-bromo- (cont.)			13-0675-83A
$C_{18}H_{24}CuN_{10}$			
Copper, [6,7,8,9,16,17-hexahydro-N,N,N',N'-tetramethyldipyrimido-[4,5-e:5',4'-m][1,4,8,11]tetraaza-cyclotetradecine-4,11-diaminato(2-)-N^6,N^9,N^{15},N^{18}]-, (SP-4-2)-	DMF	328(3.98),372(3.73)	12-2413-83
$C_{18}H_{24}FeO_5$			
Ferrocene, 1,1'-[oxybis(2,1-ethanediyl-oxy-2,1-ethanediyloxy)]-	MeOH	438(2.10)	18-0537-83
LiSCN complex	MeOH	438(2.10)	18-0537-83
NaSCN complex	MeOH	434(2.07)	18-0537-83
KSCN complex	MeOH	435(2.09)	18-0537-83
$C_{18}H_{24}N_2O$			
4-Nor-2,3-diazaestra-1,5(10)-dien-17β-ol, 17α-ethynyl-1-methyl-	EtOH	203(3.52),226(3.53)	39-2353-83C
$C_{18}H_{24}N_2O_3$			
Acetamide, N-[2-[2-[1-(methoxymethyl)-3-oxobutyl]-1H-indol-3-yl]ethyl]-	EtOH	225(4.58),275s(3.89), 283(3.92),291(3.86)	4-0267-83
Julocrotine	EtOH	252(2.43),258(2.44), 264(2.30),268(2.18)	162-0756-83
$C_{18}H_{24}N_2O_4$			
β-D-Ribofuranosylamine, N-[2-(1H-indol-3-yl)ethyl]-2,3-O-(1-methylethylidene)-	EtOH	221(4.54),280(3.78), 290(3.71)	136-0001-83B
$C_{18}H_{24}N_2O_6$			
[3,3'-Bi-3H-pyrrole]-5,5'-dicarboxylic acid, 4,4'-dihydroxy-2,2',3,3'-tetramethyl-, diethyl ester	MeCN	309(4.08)	33-1902-83
$C_{18}H_{24}N_4O_3Si$			
1H-Imidazole-4-acetonitrile, α-[[(1,1-dimethylethyl)dimethylsilyl]oxy]-5-nitro-1-(phenylmethyl)-	MeOH	300(3.82)	4-0629-83
$C_{18}H_{24}N_4O_4$			
1,8-Naphthalenediamine, N,N,N',N'-tetramethyl-2,4-dinitro-	pH 1	338(3.72)	104-0520-83
	MeOH	282(4.07),460(4.13)	104-0520-83
	HOAc	325(3.76)	104-0520-83
	DMSO	284(4.23),340(3.84), 352(3.86),467(4.26)	104-0520-83
1,8-Naphthalenediamine, N,N,N',N'-tetramethyl-4,5-dinitro-	pH 1	305(--),330(--)	104-0520-83
	MeOH	278(3.97),475(4.03)	104-0520-83
	HOAc	300(3.97),350(3.96), 439(3.46)	104-0520-83
	DMSO	276(4.09),360s(3.96), 415s(4.13),470(4.22)	104-0520-83
$C_{18}H_{24}N_4O_5$			
3-Pyridinecarboxamide, 1,6-dihydro-N-methyl-1-[[2-[[3-[(methylamino)-carbonyl]-1(4H)-pyridinyl]methoxy]-ethoxy]methyl]-6-oxo-	CH_2Cl_2	260(3.9),307(3.7), 345(3.5)	64-0878-83B

Compound	Solvent	λ_{max}(log ε)	Ref.
$C_{18}H_{24}N_6$			
Ethanone, 1-(4,5,6,7-tetrahydro-1H-indazol-3-yl)-, [1-(4,5,6,7-tetrahydro-1H-indazol-3-yl)ethylidene]-hydrazone	EtOH	256(4.04),300(3.93)	142-0061-83
$C_{18}H_{24}N_6O_6$			
1(2H)-Pyrimidinepropanamide, N,N'-1,4-butanediylbis[3,4-dihydro-2,4-dioxo-	H_2O	266(4.15)	103-1228-83
$C_{18}H_{24}O$			
Tetracyclo[8.4.0.0^{2,7}.0^{4,6}]tetradeca-1,10-dien-12-one, 7β-methyl-4a-(1-methylethyl)-	EtOH	305(4.18)	33-1806-83
$C_{18}H_{24}O_2$			
D-Homo-4-norestr-3(5)-ene-2,17-dione, (-)-	EtOH	232(4.16)	39-2349-83C
18-Homo-4-norestr-3(5)-ene-2,17-dione (-)-	EtOH	232(4.32)	39-2337-83C
6,9-Methanocyclohept[e]indene-3,11(2H)-dione, 3a-ethyl-1,3a,4,5,5a,6,9,10-10a,10b-decahydro-8-methyl-, [3aS-(3aα,5aβ,6α,9α10aβ,10bβ)]-	EtOH	223(3.90)	39-2337-83C
Nimbiol	EtOH	234(4.13),283(4.10)	162-0939-83
2,7-Nonadienoic acid, 8-(2-methylphenyl)-, ethyl ester	heptane	<240(3.98)	44-0090-83
$C_{18}H_{24}O_3$			
2H-1-Benzopyran, 5,7-dimethoxy-2-methyl-2-(4-methyl-3-pentenyl)-, (±)-	EtOH	228s(4.24),234(4.26), 240s(4.15),289(3.85), 304s(3.65)	39-1411-83C
2H-1-Benzopyran-2-one, 4-methyl-7-(octyloxy)-	$CHCl_3$	323(4.23)	44-1872-83
H-T dimer	$CHCl_3$	244(3.90),278(3.61)	44-1872-83
10aH-Dibenzo[b,d]pyran-1,7-dione, 1,2,3,4,7,8,9,10-octahydro-3,3,6,9,9-pentamethyl-	MeOH	206(3.60),231(3.81), 247(3.70),311(3.45)	104-2027-83
6-Octynoic acid, 8-[4-oxo-5-(2-pentenyl)-2-cyclopenten-1-yl]- (dicranenone A)	MeOH	217(3.93)	88-3337-83
$C_{18}H_{24}O_4$			
Estra-1,4-dien-3-one, 10-hydroperoxy-17β-hydroxy-	EtOH	239(4.05)	89-1025-83S
2H-Pyran-2-one, 3-(hydroxymethyl)-4-methoxy-6-(1,2,4a,5,6,7,8,8a-octahydro-2-methyl-1-naphthalenyl)-(solanapyrone B)	EtOH	303(3.93)	88-5373-83
$C_{18}H_{24}O_4Si$			
1-Benzoxepin-3-carboxaldehyde, 7-[[[(1,1-dimethylethyl)dimethylsilyl]oxy]methyl]-2,5-dihydro-5-oxo-	EtOH	201(4.35),220(4.27), 257.5(3.99),332.5(3.26)	39-1718-83C
$C_{18}H_{24}O_5$			
Curvularin dimethyl ether	EtOH	223(4.02),267.5(3.71)	162-0383-83
Propanedioic acid, [4-(2-hydroxy-3-methylphenyl)-2-butenyl]-, diethyl ester, (E)-	EtOH	225(3.88),275(3.76)	78-1761-83
Propanedioic acid, [4-(2-hydroxy-5-	EtOH	217s(3.00),280(3.56)	78-1761-83

Compound	Solvent	$\lambda_{max}(\log \epsilon)$	Ref.
methylphenyl)-2-butenyl]-, diethyl ester, (E)- (cont.)			78-1761-83
Propanedioic acid, [4-(2-hydroxyphenyl)-3-methyl-2-butenyl]-, diethyl ester, (E)-	EtOH	226(3.98),277(3.43), 284(3.38)	78-1761-83
$C_{18}H_{24}O_5S$			
α-D-gluco-Heptofuranos-5-ulose, 7-deoxy-3-O-methyl-6-S-methyl-1,2-O-(1-methylethylidene)-7-phenyl-6-thio-, d-	EtOH	209(3.73)	159-0139-83
1-	EtOH	209(3.66)	159-0139-83
α-D-xylo-Hexofuranos-5-ulose, 6-deoxy-3-O-methyl-1,2-O-(1-methylethylidene)-6-[methyl(phenylmethyl)sulfonio]-ylide	EtOH	210(3.70),274(3.87)	159-0139-83
$C_{18}H_{24}O_6$			
2,3-Naphthalenedicarboxylic acid, 1-ethoxy-1,4,5,6,7,8-hexahydro-7,7-dimethyl-5-oxo-, dimethyl ester	EtOH	210(3.77),215(3.80), 224(3.72)	39-1919-83C
Propanedioic acid, [4-(2-hydroxy-5-methoxyphenyl)-2-butenyl]-, diethyl ester, (E)-	EtOH	220s(3.84),294(3.55)	78-1761-83
$C_{18}H_{25}FO_8S_2$			
α-D-xylo-Hexofuranos-5-ulose, 6-deoxy-3-O-methyl-1,2-O-(1-methylethylidene)-6-[methyl(phenylmethyl)sulfonio]-ylide, fluorosulfate	EtOH	217(3.83)	159-0139-83
$C_{18}H_{25}N$			
1-Azuleneethanamine, α,3,8-trimethyl-5-(1-methylethyl)-	EtOH	217(4.07),247(4.22), 287(4.47),306(4.05), 341s(3.48),353(3.62), 369(3.55),581s(2.46), 622(2.52),673s(2.46), 751s(2.05)	18-2059-83
4-Azuleneethanamine, N,N,1-trimethyl-7-(1-methylethyl)-	EtOH	217(4.01),245(4.35), 280s(4.52),284(4.57), 288(4.56),302(3.98), 327s(3.23),338s(3.43), 350(3.60),367(3.45), 563(2.53),585s(2.60), 607(2.65),632s(2.61), 654(2.58),725(2.16)	18-2311-83
hydrochloride	EtOH	218s(4.05),245(4.29), 280s(4.46),285(4.52), 289(4.51),303(3.95), 350(3.58),367(3.41), 565s(2.51),588s(2.57), 605(2.61),633s(2.56), 658(2.53),730(2.09)	18-2311-83
4-Azuleneethanamine, N,α,1-trimethyl-7-(1-methylethyl)-	EtOH	219(4.02),246(4.36), 285(4.58),289(4.58), 303(3.99),349(3.62), 367(3.49),565s(2.55), 607(2.66),656(2.59), 725s(2.15)	18-2311-83
1-Azulenemethanamine, N,N,3,8-tetramethyl-5-(1-methylethyl)-	C_6H_{12}	215(4.02),247(4.28), 282s(4.49),288(4.56),	18-2311-83

Compound	Solvent	$\lambda_{max}(\log \epsilon)$	Ref.
(cont.)		292(4.58),305(4.14), 339s(3.54),347s(3.60), 353(3.73),371(3.73), 568s(2.50),595s(2.58), 613(2.61),640s(2.59), 667(2.53),708s(2.22), 741(2.08)	18-2311-83
$C_{18}H_{25}NO$			
4(1H)-Quinolinone, octahydro-1-(1-methyl-2-phenylethyl)-, trans	heptane	253s(2.67),307(1.83)	103-0425-83
$C_{18}H_{25}NOS$			
1-Pentanol, 4-methyl-2-[[2-phenyl-1-(2-thienyl)ethyl]amino]-, hydrochloride	EtOH	207(4.04),234(3.9), 263s(2.20),265s(1.95)	94-0031-83
$C_{18}H_{25}NO_2$			
Estra-1,3,5(10)-trien-3-ol, 17β-(hydroxyamino)-	MeOH	217(3.88),222(3.86), 229(3.73),281(3.30), 287(3.26)	13-0001-83B
	MeOH-NaOH	241(3.97),298(3.40)	13-0001-83B
hydrochloride	MeOH	217(3.93),222(3.91), 279(3.80),281(3.32), 286(3.28)	13-0001-83B
	MeOH-NaOH	220(4.04),241(4.05), 298(3.43)	13-0001-83B
1,7(2H,5H)-Phenanthridinedione, 3,4,8,9,10,10a-hexahydro-3,3,6,9,9-pentamethyl-	EtOH	199(4.09),248(4.15), 254(4.18),274(3.85), 407(3.95)	104-2027-83
$C_{18}H_{25}NO_3$			
2-Pyrrolidinone, 5-(2,4-dihydroxy-6-pentylphenyl)-1-(2-propenyl)-	EtOH	284(3.24)	150-1156-83M
2-Pyrrolidinone, 5-(2,6-dihydroxy-4-pentylphenyl)-1-(2-propenyl)-	EtOH	284(3.24)	150-1156-83M
$C_{18}H_{25}NO_3S$			
1,2-Benzoxathiin-4-amine, N,N-diethyl-3,4,5,6,7,8-hexahydro-3-phenyl-, 2,2-dioxide, cis	EtOH	237s(3.38),308(2.86), 321s(2.78)	4-0839-83
trans	EtOH	230s(3.36)	4-0839-83
$C_{18}H_{25}NO_5Si$			
5-Isoxazolepropanoic acid, 4,5-dihydro-5-(methoxycarbonyl)-3-phenyl-α-(trimethylsilyl)-, methyl ester	C_6H_{12}	273(4.08)	44-3189-83
$C_{18}H_{25}NO_6$			
Retrorsine	H_2O	217(3.85)	162-1178-83
$C_{18}H_{25}NO_8$			
7-Azabicyclo[2.2.1]hept-2-ene-1,2,3,4-tetracarboxylic acid, 1,4-dimethyl 2,3-bis(1-methylethyl) ester	CCl_4	262(3.32)	4-0001-83
7-Azabicyclo[2.2.1]hept-2-ene-1,2,3,4-tetracarboxylic acid, 1,4-dimethyl 2,3-dipropyl ester	CCl_4	263(3.46)	4-0001-83
1H-Pyrrole-2,3,4,5-tetracarboxylic acid, 3,4-bis(1,1-dimethylethyl) 2,5-dimethyl ester	EtOH	223(4.22),269(4.14)	4-0001-83

Compound	Solvent	λmax(log ε)	Ref.
1H-Pyrrole-2,3,4,5-tetracarboxylic acid, 3,4-dibutyl 2,5-dimethyl ester	EtOH	223(4.16),267(4.02)	4-0001-83
1H-Pyrrole-2,3,4,5-tetracarboxylic acid, 2,5-dimethyl 3,4-bis(1-methylpropyl) ester	EtOH	223(4.19),267(4.06)	4-0001-83
$C_{18}H_{25}N_2$			
Pyridinium, 1-methyl-2-[[3-(1-piperidinyl)-2-cyclohexen-1-ylidene]methyl]-, perchlorate	MeOH	498(4.69)	104-2089-83
Pyridinium, 1-methyl-4-[[3-(1-piperidinyl)-2-cyclohexen-1-ylidene]methyl]-, perchlorate	MeOH	515(4.98)	104-2089-83
$C_{18}H_{25}N_3O$			
2H-Pyrrol-2-one, 5-[(1,1-dimethylethyl)imino]-1,5-dihydro-4-[(1-methylethyl)amino]-1-(4-methylphenyl)-	heptane	200(4.32),244(4.21), 330(3.57)	5-1504-83
$C_{18}H_{25}N_3O_2$			
Martensine A	EtOH	223(4.32),257s(3.72), 268s(3.74),280(3.76), 288(3.72)	88-2087-83
1,8-Naphthalenediamine, N,N,N',N'-tetraethyl-4-nitro-	pH 1	240s(--),342(3.68)	104-0520-83
	MeOH	277(4.42),357(3.68)	104-0520-83
	acetone	355(3.77),454(4.11)	104-0520-83
	HOAc	325(3.80)	104-0520-83
	DMSO	284(4.29),358(3.71), 470(4.12)	104-0520-83
1(4H)-Pyridinepentanoic acid, 3,5-dicyano-2,4,4,6-tetramethyl-, ethyl ester	EtOH	222(4.48),341(3.86)	73-0608-83
$C_{18}H_{25}N_5O_8$			
1,2-Hydrazinedicarboxylic acid, 1,1'-[(dimethylamino)phenylethenylidene]-bis-, tetramethyl ester	EtOH	223(4.32),292(3.75)	104-0110-83
$C_{18}H_{26}ClN_3O$			
2H-Pyrrol-2-one, 3-chloro-5-[(1,1-dimethylethyl)amino]-1,5-dihydro-4-[(1-methylethyl)amino]-1-(4-methylphenyl)-	heptane	195(4.30),243(4.12), 283(3.88)	5-1504-83
$C_{18}H_{26}N_2O_2$			
α-Obscurine, N-acetyl-N-demethyl-, (±)-	MeOH	255(3.38)	5-0220-83
2H-Pyrrol-2-one, 3,4-diethyl-5-[(4-ethyl-2-methoxy-3,5-dimethyl-2H-pyrrol-2-yl)methylene]-1,5-dihydro-	MeOH	312(4.20)	78-1893-83
$C_{18}H_{26}N_2O_4$			
Pyrrolo[3,2-b]pyrrole-1,4-dicarboxylic acid, 3,6-bis(1,1-dimethylethyl)-, dimethyl ester	C_6H_{12}	254(4.23),262(4.26), 283(3.95)	138-0743-83
$C_{18}H_{26}N_4O_3$			
3H,4H-2,6a-Diaza-3a-azoniaphenalene, 1-(dimethylamino)-5-formyl-2,7,8,9-tetrahydro-6-hydroxy-2,7-dimethyl-3-(1-methylethyl)-4-oxo-, hydroxide, inner salt, trans	EtOH	220(4.14),272(3.75), 368(4.48)	142-1891-83

Compound	Solvent	λ_{max}(log ε)	Ref.
$C_{18}H_{26}N_4O_8S$			
1H-Imidazole-4-carboxylic acid, 5-[[(ethoxycarbonyl)amino]thioxomethyl]amino]-1-[2,3-O-1-methylethylidene-β-D-ribofuranosyl]-, ethyl ester	MeOH	212(4.67),248(4.62), 289(4.24)	88-0931-83
Pyrazin-2-onium, 1,2-diamino-4-β-D-ribofuranosyl-, mesitylenesulfonate	pH 3	223(4.26),315(3.93), 325(3.92)	87-0283-83
unstable in base	pH 11	225(4.18),246(3.99), 323(3.86),335s(3.82)	87-0283-83
$C_{18}H_{26}O_2$			
Nandrolone	EtOH	241(4.23)	162-0912-83
$C_{18}H_{26}O_2S$			
Benzene, [[1-(1-butenyl)-1-octenyl]-sulfonyl]-, (E,E)-	MeOH	224(4.24)	88-4315-83
(E,Z)-	MeOH	222(4.26)	88-4315-83
$C_{18}H_{26}O_3$			
Estr-4-en-3-one, 4,17β-dihydroxy-	n.s.g.	278(4.06)	162-0705-83
$C_{18}H_{26}O_4$			
[1,1'-Bicyclohexyl]-2,3',6-trione, 2'-acetyl-4,4,5',5'-tetramethyl-	MeOH	265(4.06)	104-2027-83
$C_{18}H_{26}O_5$			
Benzeneheptanoic acid, 2-formyl-3,5-dimethoxy-, ethyl ester	EtOH	232(4.18),233.5(4.19), 278(4.14),313(4.26)	39-2211-83C
Benzeneheptanoic acid, 4-formyl-3,5-dimethoxy-, ethyl ester	EtOH	279(4.18),323.5(3.59)	39-2211-83C
$C_{18}H_{26}O_6$			
Acetic acid, 2,2'-[1,2-phenylenebis(oxy)]bis-, dibutyl ester	MeOH	271(3.28)	49-0359-83
$C_{18}H_{26}O_9$			
1,1-Cyclopropanedicarboxylic acid, 2-[3-O-methyl-1,2-O-(1-methylethylidene)-α-D-xylofuranuronoyl]-, diethyl ester	EtOH	210(3.67)	159-0139-83
$C_{18}H_{26}O_{10}$			
Tinotuberide	MeOH	221(4.53),265(4.22)	94-0156-83
$C_{18}H_{27}NOS$			
Benzeneethanethioamide, 3,5-bis(1,1-dimethylethyl)-N,N-dimethyl-α-oxo-	MeOH	261(4.34),375(3.02)	5-1116-83
$C_{18}H_{27}NO_3$			
Capsaicin	EtOH	230(3.91),280.5(3.43)	98-1326-83
	n.s.g.	227(3.85),281(3.40)	162-0243-83
2-Pyrrolidinone, 5-(2,4-dihydroxy-6-pentylphenyl)-1-(1-methylethyl)-	EtOH	286(3.40)	150-1156-83M
2-Pyrrolidinone, 5-(2,6-dihydroxy-4-pentylphenyl)-1-(1-methylethyl)-	EtOH	283.5(3.18)	150-1156-83M
$C_{18}H_{27}NO_6$			
3,5-Pyridinedicarboxylic acid, 1-(2-ethoxy-2-oxoethyl)-1,4-dihydro-2,4,6-trimethyl-, diethyl ester	EtOH	231(4.19),258(4.03), 340(3.87)	73-0608-83

Compound	Solvent	$\lambda_{max}(\log \epsilon)$	Ref.
$C_{18}H_{27}NS_2$			
Benzeneethanethioamide, 3,5-bis(1,1-dimethylethyl)-N,N-dimethyl-α-thioxo-	MeOH	242s(4.00),271(4.14), 331(4.15),581(2.26)	5-1116-83
$C_{18}H_{27}N_3O_4S$			
1,2,3-Triazolidine-1,2-dicarboxylic acid, 3-(4-methylphenyl)-5-(methylthio)-, bis(1-methylethyl) ester	EtOH	215s(4.20),280s(3.08)	33-1608-83
$C_{18}H_{28}$			
Ethene, tetrakis(1-methylcyclopropyl)-	C_6H_{12}	217(4.14)	88-5861-83
$C_{18}H_{28}N_2O_4$			
1H-Pyrrole-1-carboxylic acid, 4-(1,1-dimethylethyl)-2-[1-[[(methoxycarbonyl)imino]methyl]-2,2-dimethylpropyl]-, methyl ester	C_6H_{12}	225(4.29),250s(3.75)	138-0743-83
Pyrrolo[3,2-b]pyrrole-1,4-dicarboxylic acid, 3,6-bis(1,1-dimethylethyl)-3a,6a-dihydro-, dimethyl ester	C_6H_{12}	223(4.41)	88-2275-83
$C_{18}H_{28}N_2O_8$			
Propanedioic acid, 2,2'-[1,2-ethanediylbis(iminomethylidyne)]bis-, tetraethyl ester	MeCN	222(4.36),272(4.44), 292(4.43)	97-0064-83
$C_{18}H_{28}O_4$			
Propanoic acid, 2-[[3,5-bis(1,1-dimethylethyl)-1-methyl-4-oxo-2,5-cyclohexadien-1-yl]oxy]-	EtOH	235(4.52)	44-3696-83
2-Propanone, 1-[hexahydro-5-hydroxy-4-(3-hydroxy-1-octenyl)-2H-cyclopenta[b]furan-2-ylidene]-, (E)-	EtOH	267(4.413)	88-1281-83
(Z)-	EtOH	262(4.292)	88-1281-83
$C_{18}H_{28}O_5$			
1-Cyclopenteneheptanoic acid, 5-oxo-4-[(tetrahydro-2H-pyran-2-yl)oxy]-, methyl ester	EtOH	230(3.86)	39-0319-83C
$C_{18}H_{29}NO_2$			
1,3-Benzenediol, 2-[1-(1-methylethyl)-2-pyrrolidinyl]-5-pentyl-	EtOH	283.5(3.24)	150-1156-83M
1,3-Benzenediol, 4-[1-(1-methylethyl)-2-pyrrolidinyl]-5-pentyl-	EtOH	284(3.26)	150-1156-83M
$C_{18}H_{29}NO_3$			
Capsaicin, dihydro-	EtOH	228(3.81),281(3.42)	98-1326-83
$C_{18}H_{29}NO_5$			
4,9-Epoxy-7H-1,3-dioxolo[4,5-d]oxocin-7-one, 8-[(dimethylamino)methylene]-hexahydro-2,2-dimethyl-5-pentyl-	MeOH	297(4.27)	18-2680-83
isomer	MeOH	298(4.38)	18-2680-83
$C_{18}H_{29}NO_6S_2Si$			
4,5-Dithia-1-azabicyclo[4.2.0]oct-2-ene-2-carboxylic acid, 3-(acetoxymethyl)-7-[1-[[(1,1-dimethylethyl)dimethylsilyl]oxy]ethyl]-8-oxo-, methyl ester	$CHCl_3$	279(3.78),332(3.46)	88-3283-83

Compound	Solvent	$\lambda_{max}(\log \epsilon)$	Ref.
$C_{18}H_{29}N_3S$			
2-Hexenenitrile, 3-[[2-[[1-(cyanomethylene)-2-ethyl-1-butenyl]thio]ethyl]amino]-4-ethyl-	EtOH	206(4.03),260(4.25)	39-1223-83C
$C_{18}H_{29}N_4O_6$			
1-Piperidinyloxy, 4-[[1-(2-deoxy-β-D-erythro-pentofuranosyl)-1,2,3,4-tetrahydro-2,4-dioxo-5-pyrimidinyl]amino]-2,2,6,6-tetramethyl-	pH 7	237(3.94),299(3.73)	33-0534-83
$C_{18}H_{29}N_4O_7$			
1-Piperidinyloxy, 2,2,6,6-tetramethyl-4-[(1,2,3,4-tetrahydro-2,4-dioxo-1-β-D-ribofuranosyl-5-pyrimidinyl)amino]-	pH 7	238(3.94),300(3.73)	33-0534-83
$C_{18}H_{29}N_5O_4S_3Si$			
4,5-Dithia-1-azabicyclo[4.2.0]oct-2-ene-2-carboxylic acid, 7-[1-[[(1,1-dimethylethyl)dimethylsilyl]oxy]ethyl]-3-[[(1-methyl-1H-tetrazol-5-yl)thio]methyl]-8-oxo-, methyl ester, [6α,7α(R*)]-	$CHCl_3$	281(3.77),333(3.49)	88-3283-83
$C_{18}H_{30}N_2O$			
Acetamide, N-[4-(decylamino)phenyl]-	MeOH	247(4.21)	5-0802-83
$C_{18}H_{30}N_2O_2$			
Benzenamine, N-dodecyl-4-nitro-	MeOH	231(3.91),389(4.32)	5-0802-83
$C_{18}H_{30}N_4$			
2-Hexenenitrile, 3,3'-(1,2-ethanediyldiimino)bis[4-ethyl-	EtOH	263(4.59)	39-1223-83C
$C_{18}H_{30}O_3Si$			
Butanoic acid, 2-(2-methylphenoxy)-, (triethylsilyl)methyl ester	heptane	217(3.89),274(3.18)	65-1039-83
Butanoic acid, 2-(3-methylphenoxy)-, (triethylsilyl)methyl ester	heptane	218(3.83),273(3.13)	65-1039-83
Butanoic acid, 2-(4-methylphenoxy)-, (triethylsilyl)methyl ester	heptane	223(3.97),280(3.22)	65-1039-83
$C_{18}H_{31}DO_5$			
2,6-Octadienal-1-d, 7-(diethoxymethyl)-8,8-diethoxy-3-methyl-, (E)-	EtOH	237(4.07)	23-1053-83
(Z)-	EtOH	238(4.06)	23-1053-83
$C_{18}H_{32}N_2O_3$			
Cyclopentaneheptanoic acid, 2-[(1-oxohexyl)hydrazono]-	EtOH	215s(3.92),234(4.04)	87-1056-83
$C_{18}H_{34}N_2O_2$			
Cyclopentaneheptanoic acid, 2-(hexylhydrazono)-	EtOH	206(3.56),233(3.59)	87-1056-83
$C_{18}H_{36}O_2Te$			
Hexanoic acid, 6-(undecyltelluro)-, methyl ester	MeOH	233.0(3.71)	101-0031-83M
$C_{18}H_{37}N_5O_5$			
Istamycin C	H_2O	end absorption	158-83-42

Compound	Solvent	λ_{max}(log ε)	Ref.
$C_{19}H_{10}ClN_3O_2$			
5H-Pyrido[1',2':1,2]pyrimido[4,5-b]-acridine-7,15-dione, 2-chloro-	DMSO	464(3.68),491(3.70), 524(3.92)	18-1775-83
	H_2SO_4	460(3.40),488(3.63), 521(3.60)	18-1775-83
5H-Pyrido[1',2':1,2]pyrimido[4,5-b]-acridine-7,15-dione, 4-chloro-	DMSO	456(3.70),483(3.92), 515(3.92)	18-1775-83
	H_2SO_4	463(3.65),490(3.86), 523(3.84)	18-1775-83
$C_{19}H_{10}Cl_2O_5$			
Anthra[1,2-b]furan-6,11-dione, 3-acetyl-7,10-dichloro-5-hydroxy-2-methyl-	dioxan	257(4.50),425(3.97)	104-0533-83
$C_{19}H_{10}F_4N_4$			
1H-Indazole, 4,5,6,7-tetrafluoro-1-phenyl-3-(phenylazo)-	EtOH	357(4.3)	104-0212-83
$C_{19}H_{10}N_2O_2$			
6H-Indolo[2,3-b]acridine-6,12(11H)-dione	EtOH	228s(4.63),270s(4.24), 285(4.26),288s(4.25), 323s(4.11),327s(4.06), 344s(3.82),350(3.78), 390(3.61)	39-2409-83C
$C_{19}H_{10}O$			
6H-Benzo[cd]pyren-6-one	C_6H_{12}	244(4.55),256(4.39), 291(4.47),296(4.49), 307(4.42),376(3.94), 390(4.02),397(4.00), 413(4.02)	54-0220-83
$C_{19}H_{11}BrO_2$			
1H-Phenalen-1-one, 9-(4-bromophenyl)-6-hydroxy-	pyridine	318(3.96),360(3.70), 446(4.04)	39-1267-83C
$C_{19}H_{11}BrO_5$			
Anthra[1,2-b]furan-6,11-dione, 3-acetyl-4-bromo-5-hydroxy-2-methyl-	dioxan	282(4.35),427(3.99)	104-1900-83
$C_{19}H_{11}ClO_5$			
Anthra[1,2-b]furan-6,11-dione, 3-acetyl-4-chloro-5-hydroxy-2-methyl-	dioxan	265(4.65),429(3.99)	104-1900-83
$C_{19}H_{11}Cl_2NO_6$			
Anthra[1,2-b]furan-3-carboxylic acid, 2-amino-7,10-dichloro-6,11-dihydro-5-hydroxy-6,11-dioxo-, ethyl ester	dioxan	330(3.90),476(3.87)	104-0533-83
$C_{19}H_{11}Cs$			
Cesium, 7bH-indeno[1,2,3-jk]fluoren-7b-yl)-	$(MeOCH_2)_2$	363(4.36),376(4.40), 519(3.76),556(3.90)	104-1592-83
$C_{19}H_{11}Li$			
Lithium, 7bH-indeno[1,2,3-jk]fluoren-7b-yl)-	$(MeOCH_2)_2$	386(4.36),526(3.56), 568(3.69)	104-1592-83
$C_{19}H_{11}N_3O_2$			
8H-Benzo[b]pyrrolo[4,3,2-de][1,8]phenanthroline-8,11(10H)-dione, 10-methyl- (amphimedine)	EtOH	210(4.29),233(4.60), 281(3.96),341(3.78)	35-4835-83
	EtOH-NaBH	235(4.11),280(3.96)	35-4835-83

Compound	Solvent	λ_{max}(log ε)	Ref.
5H-Pyrido[1',2':1,2]pyrimido[4,5-b]-acridine-7,15-dione	DMSO	460(3.70),487(3.96), 519(3.99)	18-1775-83
	H_2SO_4	450(3.74),483(3.95), 516(3.92)	18-1775-83
$C_{19}H_{11}N_3O_3$			
5H-Pyrido[1',2':1,2]pyrimido[4,5-b]-acridine-7,15-dione, 2-hydroxy-	DMSO	464(3.51),494(3.76), 528(3.79)	18-1775-83
	H_2SO_4	452(3.60),481(3.83), 513(3.81)	18-1775-83
$C_{19}H_{12}ClNO_6$			
Anthra[1,2-b]furan-3-carboxylic acid, 2-amino-4-chloro-6,11-dihydro-5-hydroxy-6,11-dioxo-, ethyl ester	dioxan	314(4.12),468(3.94)	104-1900-83
$C_{19}H_{12}ClNO_6S$			
2-Propen-1-one, 3-(4-chlorophenyl)-1-(5-nitro-2-furanyl)-2-(phenylsulfonyl)-, (Z)-	dioxan	212(4.52),220(4.37), 288(4.45)	73-1057-83
$C_{19}H_{12}ClN_3O_2S$			
Acetamide, N-[4-(2-benzoyl-4-chlorophenyl)-5-cyano-2-thiazolyl]-	EtOH	237(4.47),254(4.50), 285s(4.06),315s(3.83)	87-0100-83
$C_{19}H_{12}Cl_2N_4O_2$			
2-Furancarboxamide, N-[6-(2,6-dichlorophenyl)-2-methylpyrido[2,3-d]pyrimidin-7-yl]-	pH 1	259(4.37),299(4.32) (changing)	87-0403-83
	MeOH-pH 7	242(4.44),323(4.28), 368(3.09),385(2.99)	87-0403-83
	MeOH-NaOH	360(4.14)	87-0403-83
$C_{19}H_{12}FNO_2S$			
10H-Phenothiazin-3-ol, 7-fluoro-, benzoate	benzene	205(4.30),256(4.46), 206(3.54)	4-0803-83
$C_{19}H_{12}FNO_6S$			
2-Propen-1-one, 3-(4-fluorophenyl)-1-(5-nitro-2-furanyl)-2-(phenylsulfonyl)-, (Z)-	dioxan	211(4.42),219(4.34), 295(4.31)	73-1057-83
$C_{19}H_{12}N_2$			
Cyclohept[1,2,3-hi]imidazo[2,1,5-cd]-indolizine, 2-phenyl-	EtOH	239(4.36),260s(4.34), 270s(4.40),281s(4.51), 288(4.54),295s(4.47), 320(3.75),357(4.26), 374(4.36),417(3.95), 440(4.07)	18-3703-83
$C_{19}H_{12}N_2O_2$			
14H-[1,4]Benzoxazino[3',2':3,4]cyclohepta[1,2-b][1,4]benzoxazine	MeOH	207(4.10),254(3.99), 360(3.43),500(3.68)	18-2756-83
	MeOH-HCl	207(4.08),223(4.02), 275(4.00),325s(3.57), 410(3.68),535(3.54)	18-2756-83
Cyclohepta[2,1-b:2,3-b']bis[1,4]benzoxazine	MeOH	208(4.45),235(4.29), 287(4.30),378(3.86)	18-2756-83
6H-Indolo[2',3':6,7]oxepino[3,4-b]quinolin-13(12H)-one	EtOH	255(4.31),344(4.13), 418s(3.52)	39-2409-83C

Compound	Solvent	λmax(log ε)	Ref.
$C_{19}H_{12}N_2O_8S$			
2-Propen-1-one, 1-(5-nitro-2-furanyl)-3-(4-nitrophenyl)-2-(phenylsulfonyl)-, (Z)-	dioxan	212(4.49),218(4.35), 299(4.45)	73-1057-83
$C_{19}H_{12}N_4O_3$			
1H-Benz[de]isoquinoline-1,3(2H)-dione, 2-(1H-benzotriazol-1-yl)-5-methoxy-	dioxan	249(4.34),288(3.70), 335(4.06),378(3.94)	56-0817-83
1H-Benz[de]isoquinoline-1,3(2H)-dione, 2-(1H-benzotriazol-1-yl)-6-methoxy-	dioxan	254(4.37),289(3.60), 364(4.16)	56-0817-83
1H-Benz[de]isoquinoline-1,3(2H)-dione, 2-(2H-benzotriazol-2-yl)-5-methoxy-	dioxan	243(4.20),282(4.14), 335(4.10),378(3.96)	56-0817-83
1H-Benz[de]isoquinoline-1,3(2H)-dione, 2-(2H-benzotriazol-2-yl)-6-methoxy-	dioxan	257(4.30),282(4.16), 363(4.19)	56-0817-83
$C_{19}H_{12}OS$			
4H-Cyclopenta[b]thiophen-4-one, 5,6-diphenyl-	CH_2Cl_2	267(4.67),448(3.44)	70-1049-83
6H-Cyclopenta[b]thiophen-6-one, 4,5-diphenyl-	CH_2Cl_2	267(4.40),510(3.10)	70-1049-83
2(5H)-Thiophenone, 5-(2,3-diphenyl-2-cyclopropen-1-ylidene)-	C_6H_{12}	275(4.35),352(3.89), 379(4.41),403(4.06), 474(3.58)	88-0205-83
	benzene	378(4.24),400(4.28), 460(3.58)	88-0205-83
	acetone	370(4.13),395(4.40), 420(3.70)	88-0205-83
	MeCN	370(4.01),395(4.35), 414(3.84)	88-0205-83
	CH_2Cl_2	372(4.12),399(4.07), 430(3.44)	88-0205-83
$C_{19}H_{12}O_2$			
1,4-Methanonaphth[2,3-a]azulene-5,12-dione, 1,4-dihydro-, (1R)-	MeCN	255(4.5),303(4.7), 378(4.0),448(3.7), 476(3.6),637(3.4), 677(3.3)	24-2408-83
$C_{19}H_{12}O_3$			
1H-Indene-1,3(2H)-dione, 2-(3,4-dihydro-2-oxo-1(2H)-naphthalenylidene)-	MeCN	326(4.12),342(4.09)	32-0507-83
Naphth[2,3-a]azulene-5,12-dione, 11-methoxy-	MeCN	253(3.9),312(4.0), 454(3.5),678(2.6)	24-2408-83
1H-Phenalen-1-one, 2,6-dihydroxy-9-phenyl- (lachnanthocarpon)	pH 2	491(3.79)	74-0385-83C
	pH 10	592(4.02)	74-0385-83C
	MeOH	488(3.76)	74-0385-83C
$C_{19}H_{12}O_4S$			
2H-1-Benzopyran-2-one, 3-[[(2-oxo-2H-1-benzopyran-4-yl)thio]methyl]-	dioxan	235(3.97),295(4.296)	117-0321-83
$C_{19}H_{12}O_5$			
Anthra[1,2-b]furan-6,11-dione, 3-acetyl-5-hydroxy-2-methyl-	dioxan	254(4.53),428(3.93)	104-0533-83
$C_{19}H_{12}O_6$			
Anthra[2,3-b]furan-5,10-dione, 3-acetyl-4,11-dihydroxy-2-methyl-	dioxan	260(4.70),481(4.12)	104-1900-83
$C_{19}H_{13}BrF_3NO$			
2(1H)-Pyridinone, 4-bromo-1-methyl-3-	MeOH	206.5(4.44),250(4.00),	88-2973-83

Compound	Solvent	$\lambda_{max}(\log \epsilon)$	Ref.
phenyl-5-[3-(trifluoromethyl)phenyl]-		318.5(3.84)	88-2973-83
2(1H)-Pyridinone, 4-bromo-1-methyl-5-phenyl-3-[3-(trifluoromethyl)phenyl]-	MeOH	209.5(4.52),233(4.25), 323(3.89)	88-2973-83
4(1H)-Pyridinone, 2-bromo-1-methyl-5-phenyl-3-[3-(trifluoromethyl)phenyl]-	MeOH	205.3(4.52),238.7(4.43), 285.5(4.06)	88-2973-83
$C_{19}H_{13}BrO_6$			
9,10-Anthracenedione, 2-(1-acetyl-2-oxopropyl)-3-bromo-1,4-dihydroxy-	dioxan	472(4.06)	104-1900-83
$C_{19}H_{13}ClN_2$			
9H-Carbazol-3-amine, N-[(3-chlorophenyl)methylene]-	MeOH	282(4.3),370(4.2)	103-1310-83
$C_{19}H_{13}ClN_2O_2$			
Pyrano[2,3-c]pyrazol-4(1H)-one, 5-chloro-3-methyl-1,6-diphenyl-	EtOH	251(4.42)	44-4078-83
Pyrano[2,3-c]pyrazol-4(1H)-one, 5-chloro-6-methyl-1,3-diphenyl-	EtOH	248(4.41)	44-4078-83
$C_{19}H_{13}ClN_2O_2S$			
1H-Indole, 3-chloro-1-(phenylsulfonyl)-2-(2-pyridinyl)-	EtOH	277s(4.12),235s(4.03), 306s(3.95),325(4.04)	39-2417-83C
$C_{19}H_{13}ClN_2O_3$			
1H-Naphth[2,3-g]indole-6,11-dione, 3-acetyl-5-amino-4-chloro-2-methyl-	dioxan	280(4.50),450(3.68), 550(3.97)	104-1892-83
$C_{19}H_{13}ClO_6$			
9,10-Anthracenedione, 2-(1-acetyl-2-oxopropyl)-3-chloro-1,4-dihydroxy-	dioxan	474(4.53)	104-1900-83
$C_{19}H_{13}Cl_2N_3O_3$			
Pyrazino[1,2-a][1,4]benzodiazepine-4-carboxylic acid, 9-chloro-7-(2-chlorophenyl)-1,2,3,5-tetrahydro-2-oxo-	isoPrOH	217(4.51),250s(4.13), 337(4.26)	4-0791-83
$C_{19}H_{13}Cs$			
Cesium, (9-phenyl-9H-fluoren-9-yl)-	$(MeOCH_2)_2$	395(4.34),480(3.35)	104-1592-83
$C_{19}H_{13}Li$			
Lithium, (9-phenyl-9H-fluoren-9-yl)-	$(MeOCH_2)_2$	409(4.36),489(3.38), 525(3.26)	104-1592-83
$C_{19}H_{13}NO$			
Benzoxazole, 2-[1,1'-biphenyl]-4-yl-	C_6H_{12}	312(4.61)	41-0595-83
$C_{19}H_{13}NO_2S$			
2H-1-Benzopyran-2-one, 3-[2-(4-methylphenyl)-4-thiazolyl]-	MeCN	353(4.32)	48-0551-83
$C_{19}H_{13}NO_4$			
1H-Naphth[2,3-g]indole-6,11-dione, 3-acetyl-5-hydroxy-2-methyl-	dioxan	298(4.46),450(3.97)	104-1892-83
$C_{19}H_{13}NO_5$			
Anthra[1,2-b]furan-3-carboxamide, 6,11-dihydro-5-hydroxy-N,2-dimethyl-6,11-dioxo-	dioxan	278(4.07),423(3.82)	104-0533-83

Compound	Solvent	λ_{max}(log ε)	Ref.
Oxobuxifoline	EtOH	214(4.27),249(4.25), 271s(4.23),282(4.36), 330(3.39)	100-0862-83 +100-0761-83
	EtOH-acid	212(4.37),224s(4.29), 269(4.37),284(4.27), 297(4.18),362(3.12)	100-0862-83
Oxocrebanine	EtOH	249(4.03),273(3.93), 440(3.26)	100-0761-83
	EtOH-HCl	260(4.00),285(3.87), 385(3.26)	100-0761-83
Oxonantenine	EtOH	243(4.44),264s(4.38), 272(4.46),288s(4.22), 318(3.90),357(3.96), 378s(3.90),426(3.56)	100-0761-83
Spiro[4,6-dioxa-2-azabicyclo[3.2.0]-heptane-7,9'(10'H)-phenanthrene]-3,10'-dione, 2-acetyl-	MeOH	208(3.36),248(3.47), 253(3.47),272(2.74), 283(2.70),294(2.48), 310(2.00),340(1.70)	73-2812-83
$C_{19}H_{13}NO_6$			
Anthra[1,2-b]furan-3-carboxylic acid, 2-amino-6,11-dihydro-5-hydroxy-6,11-dioxo-, ethyl ester	dioxan	312(4.19),472(3.86)	104-0533-83
$C_{19}H_{13}NO_6S$			
2-Propen-1-one, 1-(5-nitro-2-furanyl)-3-phenyl-2-(phenylsulfonyl)-	dioxan	221(4.35),308(4.27)	73-1057-83
$C_{19}H_{13}N_3$			
2,3,9b-Triazaindeno[6,7,1-ija]azulene, 1-(3-methylphenyl)-	EtOH	237(4.36),262(4.39), 285(4.36),294(4.37), 327s(3.98),340s(4.10), 367(4.33),377s(4.32), 440(4.07),460(4.20)	18-3703-83
$C_{19}H_{13}N_3OS$			
4H-Pyrido[3,2-e]-1,3-thiazin-4-one, 2-(diphenylamino)-	MeOH	255(4.13)	73-3315-83
$C_{19}H_{13}N_3O_3S$			
7H-Benzimidazo[2,1-a]benz[de]isoquinoline-10-sulfonamide, N-methyl-7-oxo-	benzene	382(3.96)	73-2249-83
	EtOH	382(4.10)	73-2249-83
$C_{19}H_{13}N_3S_2$			
Benzenesulfenamide, N,N'-(2-cyano-2,5-cyclohexadiene-1,4-diylidene)bis-	$CHCl_3$	517(4.67)	150-0028-83S +150-0480-83M
$C_{19}H_{13}N_3S_3$			
Benzenesulfenamide, N,N'-2,6-benzothiazolediylidenebis-, (E)-	$CHCl_3$	576(4.54)	150-0028-83S +150-0480-83M
$C_{19}H_{13}N_5$			
Ethenetricarbonitrile, [4-[methyl(phenylmethylene)hydrazino]phenyl]-	benzene	796.9(3.48)	32-0161-83
	acetone	715.0(2.19)	32-0161-83
	dioxan	753.7(2.43)	32-0161-83
	HOAc	749.3(2.57)	32-0161-83
	CH_2Cl_2	806.0(3.51)	32-0161-83
	CCl_4	806.4(3.47)	32-0161-83
	DMF	703.0(--)	32-0161-83

Compound	Solvent	$\lambda_{max}(\log \epsilon)$	Ref.
$C_{19}H_{13}N_5OSe$			
4H-Pyrido[3,2-e]-1,3-selenazin-4-one, 2-[[4-(phenylazo)phenyl]amino]-	MeOH	213(3.33),340(4.11)	73-3567-83
$C_{19}H_{13}N_9$			
1H-Benzotriazole, 1,1'-(2H-benzotriazol-2-ylmethylene)bis-	EtOH	258(4.26),282(4.30), 288(4.27)	142-1787-83
1H-Benzotriazole, 1-[bis(2H-benzotriazol-2-yl)methyl]-	EtOH	212(4.69),282(4.46), 290(4.44)	142-1787-83
1H-Benzotriazole, 1,1',1"-methylidynetris-	EtOH	253(4.33),284(4.02)	142-1787-83
$C_{19}H_{13}O_4S$			
1,3-Oxathiol-1-ium, 2-(4-methoxyphenyl)-5-(2-oxo-2H-1-benzopyran-3-yl)-, perchlorate	MeCN	385(4.28)	48-0551-83
$C_{19}H_{14}BrNO$			
Methanone, (4-bromophenyl)[4-(phenylamino)phenyl]-	MeOH	255(4.28),364(4.42)	12-0409-83
1-Naphthalenol, 4-[3-[(4-bromophenyl)imino]-1-propenyl]-, (E,?)-	EtOH	390(3.70),535(3.04)	104-1864-83
	acetone	370(4.08),500(3.11)	104-1864-83
	dioxan	370(4.24)	104-1864-83
	DMSO	390(3.70),535(3.04)	104-1864-83
	+ NaOMe	510(4.20)	104-1864-83
$C_{19}H_{14}ClFN_2O_3$			
1H-Pyrazole-3-carboxylic acid, 4-(4-chlorobenzoyl)-1-(4-fluorophenyl)-, ethyl ester	MeOH	262(3.5)	83-0608-83
$C_{19}H_{14}ClN_3OS$			
Acetamide, N-[4-[4-chloro-2-(phenylmethyl)phenyl]-5-cyano-2-thiazolyl]-	EtOH	215s(4.38),252(4.30), 291(3.97)	87-0100-83
Acetamide, N-(8-chloro-6-phenyl-4H-thiazolo[5,4-d][2]benzazepin-2-yl)-	EtOH	228(4.76),277(4.40), 340(3.57)	87-0100-83
$C_{19}H_{14}ClN_3O_3$			
1H-Naphth[2,3-g]indole-3-carboxamide, 5-amino-4-chloro-6,11-dihydro-N,2-dimethyl-6,11-dioxo-	dioxan	283(4.37),455(3.65), 550(3.85)	104-1892-83
$C_{19}H_{14}ClN_5OSe$			
3-Pyridinecarboxamide, 2-chloro-N-[[[4-(phenylazo)phenyl]amino]selenoxomethyl]-	MeOH	210(4.38),355(5.37)	73-3567-83
$C_{19}H_{14}FNOS$			
10H-Phenothiazine, 3-fluoro-7-(phenylmethoxy)-	benzene	207(4.38),255(4.51), 307(3.61)	4-0803-83
$C_{19}H_{14}FNO_2S$			
10H-Phenothiazine, 3-fluoro-7-(phenylmethoxy)-, 5-oxide	EtOH	239(4.56),274(4.38), 299(3.90),342(3.77)	4-0803-83
$C_{19}H_{14}NO_3$			
Oxazolium, 5-(1,2-dihydro-2-oxobenzopyran-3-yl)-2-methyl-3-phenyl-, perchlorate	MeCN	319(4.32)	48-0551-83

Compound	Solvent	$\lambda_{max}(\log \epsilon)$	Ref.
$C_{19}H_{14}NO_4$			
Thailandine (iodide)	EtOH	216(4.34),257(4.35), 288(4.18),325s(3.63), 376(3.83),464(3.73)	100-0761-83
$C_{19}H_{14}N_2O$			
Benzo[4,5]cyclohepta[1,2-c]pyrazol-4(1H)-one, 3-methyl-1-phenyl-	EtOH	259(4.42),319(4.08)	142-1581-83
Benzo[4,5]cyclohepta[1,2-c]pyrazol-4(2H)-one, 2-(phenylmethyl)-	EtOH	254(4.59),298(3.92), 312(3.91)	142-1581-83
$C_{19}H_{14}N_2OS$			
Thiazolo[3,2-a]pyrimidin-4-ium, 3-hydroxy-5-methyl-2,7-diphenyl-, hydroxide, inner salt	MeCN	568(4.15)	124-0857-83
$C_{19}H_{14}N_2O_2$			
Cyclohepta[b][1,4]benzoxazine, 6-(2-hydroxyanilino)-, hydrobromide	MeOH	265(4.32),309(4.08), 435(4.05)	18-2756-83
	MeOH-NaOH	271(4.33),463(4.12), 485(4.07)	18-2756-83
2-Naphthalenecarbonitrile, 3-[4-(dimethylamino)phenyl]-1,4-dihydro-1,4-dioxo-	C_6H_{12}	583(3.91)	150-0168-83S
	EtOH	620(--)	150-0168-83S
Pyrano[2,3-c]pyrazol-4(1H)-one, 3-methyl-1,6-diphenyl-	EtOH	253(4.40)	44-4078-83
Pyrano[2,3-c]pyrazol-4(1H)-one, 6-methyl-1,3-diphenyl-	EtOH	248(4.39)	44-4078-83
$C_{19}H_{14}N_2O_2S$			
1H-Indole, 1-(phenylsulfonyl)-2-(2-pyridinyl)-	EtOH	215s(4.18),268(4.86), 274(3.86),290(3.86)	39-2417-83C
$C_{19}H_{14}N_2O_2S_2$			
1,4-Cyclohexadiene-1-carboxylic acid, 3,6-bis[(phenylthio)imino]-	$CHCl_3$	516(4.60)	150-0028-83S +150-0480-83M
$C_{19}H_{14}N_2O_3$			
Acetamide, N-(6,11-dihydro-5-methyl-6,11-dioxobenzo[b]carbazol-8-yl)-	EtOH	216(3.97),280s(4.09), 286(4.15),296s(3.95), 307s(3.70),354(3.56), 424s(3.25)	39-2409-83C
1-Naphthalenol, 4-[3-[(4-nitrophenyl)-imino]-1-propenyl]-, (E,?)-	toluene	350(4.24)	104-1864-83
	acetone	387(4.33),490(4.08)	104-1864-83
	DMSO	403(3.81),540(3.77), 640(3.64)	104-1864-83
	+ NaOMe	358(3.57),470(3.81), 640(4.16)	104-1864-83
1H-Naphth[2,3-g]indole-6,11-dione, 3-acetyl-5-amino-2-methyl-	dioxan	288(4.48),434(3.53), 550(3.96)	104-1892-83
$C_{19}H_{14}N_2O_4$			
1H-Naphth[2,3-g]indole-3-carboxamide, 6,11-dihydro-5-hydroxy-N,2-dimethyl-6,11-dioxo-	dioxan	299(4.32),450(3.96)	104-1892-83
$C_{19}H_{14}N_4O$			
4H-Pyrido[1,2-a]-1,3,5-triazin-4-one, 2,3-dihydro-3-phenyl-2-(phenylimino)-	dioxan	289(4.23),394(3.62)	48-0463-83

Compound	Solvent	$\lambda_{max}(\log \epsilon)$	Ref.
$C_{19}H_{14}N_4O_4$			
2-Propenal, 3-[(3-benzoyl-1-phenyl-1H-pyrazol-4-yl)amino]-2-nitro-	EtOH	260(4.28),288(4.29), 381(4.37)	103-1163-83
$C_{19}H_{14}N_4O_7$			
Pyridine, 1-benzoyl-1,4-dihydro-4-[(2,4,6-trinitrophenyl)methyl]-	$CHCl_3$	255(3.94),300(3.43), 383(2.60)	56-0829-83
$C_{19}H_{14}N_4O_{10}S_4$			
2,7-Naphthalenedisulfonic acid, 4,5-dihydroxy-3-[[4-[(2-thiazolylamino)sulfonyl]phenyl]azo]-	n.s.g.	520(4.15)	7-0265-83
$C_{19}H_{14}N_4S$			
4H-Pyrido[1,2-a]-1,3,5-triazine-4-thione, 2,3-dihydro-3-phenyl-2-(phenylimino)-	dioxan	256(4.23),318(4.21), 424(3.21)	48-0463-83
$C_{19}H_{14}O$			
3H-Benzo[cd]pyren-3-one, 4,5,5a,6-tetrahydro-	MeOH	236(3.81),244(4.42), 256(4.48),260(4.49), 289(4.01),322(4.08), 366(3.23)	54-0220-83
2,5-Cyclohexadien-1-one, 4-(diphenylmethylene)-	hexane	210(4.15),260(4.20), 267(4.19),359(4.50)	73-2825-83
$C_{19}H_{14}OS$			
6H-Dibenzo[b,d]thiopyran-6-ol, 6-phenyl-	MeOH	256(4.32),277(3.92), 317(2.30)	12-0795-83
$C_{19}H_{14}O_2$			
Azuleno[2,1,8-ija]azulene-5-carboxylic acid, 10-methyl-, methyl ester	MeOH	290(5.12),302s(4.51), 316s(4.07),350s(3.54), 367(3.76),386(3.92), 407(3.95),447(3.17), 477(3.65),640(2.64)	88-0781-83
Phenol, 3-(9H-xanthen-9-yl)-	EtOH	220(4.15),245(3.86), 282(3.58)	142-0481-83
$C_{19}H_{14}O_4$			
4H-1,3-Benzodioxin-4-one, 2-methyl-2-(2-naphthalenyloxy)-	pH 7.4	300(3.72)	1-0351-83
4,11-Ethenoanthra[2,3-c]furan-1,3,5-(3aH)-trione, 4,4a,11,11a-tetrahydro-4a-methyl-	EtOH	241(4.63),247s(4.58), 271(3.65),278(3.66), 288s(3.49),318(3.26), 327s(3.22),370(3.19), 388s(3.07)	35-3234-83
$C_{19}H_{14}O_6$			
9,10-Anthracenedione, 2-(1-acetyl-2-oxopropyl)-1,4-dihydroxy-	dioxan	463(3.99)	104-1900-83
2H-Anthra[1,2-b]pyran-2-carboxaldehyde, 3,4,7,12-tetrahydro-6,11-dihydroxy-2-methyl-7,12-dioxo-, (±)-	MeOH	231(4.59),246(4.20), 288(3.93),417(3.64), 480(4.04)	5-1818-83
2H-1-Benzopyran-2-one, 7-acetoxy-4-[(benzoyloxy)methyl]-	EtOH	275(4.01),312(3.94)	94-3014-83
2H-1-Benzopyran-2-one, 3-[[2,3-dihydro-2-(hydroxymethyl)-3-oxo-2-benzofuranyl]methyl]-4-hydroxy-	MeOH	246(4.05),281(3.90), 308(3.98)	39-0841-83C
4H-Furo[3,2-c][1]benzopyran-4-one, 2,3-dihydro-2-(2-hydroxybenzoyl)-2-(hy-	MeOH	263(4.00),290(3.91), 310(3.96)	39-0841-83C

Compound	Solvent	$\lambda_{max}(\log \epsilon)$	Ref
droxymethyl)- (cont.)			39-0841-83C
$C_{19}H_{14}O_8$			
1,3-Dioxolo[7,8][2]benzopyrano[4,3-b]-[1]benzopyran-6(4H)-one, 7-hydroxy-6,9-dimethoxy- (6-methoxypulcherrimin)	EtOH	267(3.69),365(3.68)	102-2835-83
	EtOH-NaOAc	265(3.75),365(3.73)	102-2835-83
	EtOH-NaOAc-boric acid	265(3.76),365(3.81)	102-2835-83
	EtOH-$AlCl_3$	280(3.77),382(3.77), 427(3.64)	102-2835-83
	+ HCl	279(3.74),380(3.75), 425(3.59)	102-2835-83
$C_{19}H_{14}O_9$			
1H,3H-Naphtho[1,8-cd]pyran-1,3-dione, 4,6,9-triacetoxy-7-methyl-	EtOH	248(3.76),341(3.32)	78-2283-83
$C_{19}H_{14}S$			
6H-Dibenzo[b,d]thiopyran, 6-phenyl-	MeOH	258(4.31),280(3.84), 321(3.43)	12-0795-83
Methanethione, [1,1'-biphenyl]-4-yl-phenyl-	benzene	353(4.23),604(2.38)	44-0214-83
$C_{19}H_{14}S_2$			
1,3-Benzodithiole, 2-(6-methyl-7H-benzocyclohepten-7-ylidene)-	MeCN	238(4.5),246(4.5), 304(4.3),360s(4.1)	142-2039-83
$C_{19}H_{15}ClO_6$			
9,10-Anthracenedione, 2-[2-(3-chloro-2-methyloxiranyl)ethyl]-1,4,8-trihydroxy-	MeOH	226(4.35),242(4.12), 288(3.51),400(3.31), 458(3.87),487(4.05), 508(3.89),519(3.82)	5-1818-83
$C_{19}H_{15}Cl_2NO$			
Methanone, [1-[(2,6-dichlorophenyl)-methyl]-1,4-dihydro-3-pyridinyl]-phenyl-	CH_2Cl_2	376(3.8)	64-0878-83B
$C_{19}H_{15}Cl_3O_5$			
Nornidulin	n.s.g.	266(3.91)	162-0963-83
$C_{19}H_{15}Cl_4NO$			
1(4H)-Naphthalenone, 2-(diethylamino)-4-(2,3,4,5-tetrachloro-2,4-cyclopentadien-1-ylidene)-	$CHCl_3$	267(4.06),355(3.92), 460(3.75),670(4.29)	88-3567-83
$C_{19}H_{15}FN_4$			
4H-Pyrrolo[2,3-d]pyrimidin-4-imine, 3-(4-fluorophenyl)-3,7-dihydro-2-methyl-5-phenyl-	EtOH	247(4.41),285(4.36)	5-2066-83
$C_{19}H_{15}NO$			
Methanone, [4-(phenylamino)phenyl]-phenyl-	MeOH	245(4.25),360(4.46)	12-0409-83
1-Naphthalenol, 4-[3-(phenylimino)-1-propenyl]-	acetone	370(4.42),510(3.34)	104-1864-83
	dioxan	370(3.78)	104-1864-83
	DMSO	385(4.08),530(3.60)	104-1864-83
	+ NaOMe	385(4.05),520(3.61)	104-1864-83
Pyridine, 1-benzoyl-1,4-dihydro-4-(phenylmethylene)-	MeOH	226(4.11),358(4.37)	24-3192-83

Compound	Solvent	λ_{max}(log ε)	Ref.
$C_{19}H_{15}NO_2$			
Methanone, (6-methyl-2-phenoxy-3-pyridinyl)phenyl-	EtOH	250(4.2),290s(4.0)	104-1190-83
	15% H_2SO_4	293(4.2)	104-1190-83
$C_{19}H_{15}NO_3$			
Duguenaine	EtOH	210(4.26),222(4.34), 256(4.63),264(4.69), 334(4.06)	100-0761-83
$C_{19}H_{15}NO_4$			
5H-Benzo[b]carbazole-6,11-dione, 8-methoxy-5-(methoxymethyl)-	EtOH	212(4.28),221s(4.03), 272s(4.43),276(4.45), 304s(3.62),355(3.85), 382s(3.51),440s(2.68)	39-2409-83C
6H-Dibenzo[de,h]quinolin-6-one, 5,7,9-trimethoxy-	EtOH	241(4.60),281s(4.12), 287(4.13),362(3.93), 398s(3.88),448s(3.65)	78-3261-83
	EtOH-HCl	245(4.54),289s(4.08), 299(4.10),317s(3.97), 374s(3.94),392(3.97), 430(3.83),450(3.83)	78-3261-83
7H-Dibenzo[de,g]quinolin-7-one, 1,2,9-trimethoxy-	EtOH	244(4.46),271(4.44), 292s(4.16),377(3.68), 444(3.62)	78-3261-83
7H-Dibenzo[de,g]quinolin-7-one, 1,2,10-trimethoxy-	EtOH	242(4.45),272(4.44), 284s(4.19),312(3.84), 351(4.07),387(4.00)	100-0761-83
7H-Dibenzo[de,h]quinolin-7-one, 5,6,9-trimethoxy- (menisporphine)	EtOH	254(4.45),290s(4.44), 320(3.75),368(3.68), 420(3.78)	78-3261-83
	EtOH	254(4.72),288s(4.13), 310s(3.96),319(3.97), 334s(3.94),368(3.91), 420(3.97)	78-3261-83
	EtOH-HCl	270(4.43),300s(3.81), 320s(3.69),332(3.69), 428(3.70)	78-3261-83
7H-Dibenzo[de,h]quinolin-7-one, 5,6,10-trimethoxy-	EtOH	217(4.66),254(4.46), 262(4.46),270s(4.43), 280s(4.25),302s(3.85), 315(3.93),346(4.13), 376(4.05)	78-3261-83
3,4-Quinolinedicarboxylic acid, 2-phenyl-, dimethyl ester	EtOH	227(4.32),250(4.45), 330(3.63)	12-1419-83
Splendidine	EtOH	237(4.36),270(4.34), 290s(4.15),415(4.05)	100-0761-83
$C_{19}H_{15}NO_5$			
Arosinine	MeOH	244(4.58),317(4.53), 414(4.01),590(3.76)	100-0761-83
	HCl	246(4.65),292s(4.23), 316s(4.35),391(3.99), 480(3.62)	100-0761-83
3,1-Benzoxazepine-4,5-dicarboxylic acid, 2-phenyl-, dimethyl ester	EtOH	244(4.45),302(4.18)	12-1419-83
Furo[3,4-c]quinoline-3a(3H)-carboxylic acid, 1,4,5,9b-tetrahydro-1,4-dioxo-5-phenyl-, methyl ester	EtOH	257(3.62),285(3.11)	12-1419-83
Glaunine	MeOH	250(4.40),272(4.22), 310s(3.97),348(3.87), 406(2.75),600(2.68)	100-0761-83

Compound	Solvent	λ_{max}(log ϵ)	Ref.
Glaunine (cont.)	HCl	248(4.48),263s(4.41), 285(4.32),320s(3.88), 375(2.94),470s(2.60)	100-0761-83
Grandirubrine	EtOH	232(4.96),254(4.79), 274s(4.66),296(4.58), 312s(4.46),343s(4.51), 363(4.72),384(4.41), 400(4.19),480(3.90)	100-0761-83
Norchelidonine, dehydro-, (-)-	EtOH	227(4.41),234s(4.38), 270(3.99),288s(3.89), 338(3.61)	142-1895-83
	EtOH-HCl	236(4.26),296(4.26), 398(3.52)	142-1895-83
Oxocularine	MeOH	214(4.53),254(4.40), 302s(3.62),402(3.71)	100-0881-83
	MeOH-acid	224(4.47),267(4.37), 331s(3.70),345s(3.60), 486(3.61)	100-0881-83
Oxolirioferine	EtOH	244(4.35),274(4.32), 294s(4.12),359(3.82), 394s(3.73)	100-0761-83
6H-1-Pyrindine-6-acetic acid, 6-(methoxycarbonyl)-α-oxo-5-phenyl-, methyl ester	EtOH	232s(4.23),312(3.94)	95-0631-83
3,4-Quinolinedicarboxylic acid, 1,2-dihydro-2-oxo-1-phenyl-, dimethyl ester	EtOH	238(4.42),293(3.92), 351(3.72)	12-1419-83
3,4-Quinolinedicarboxylic acid, 2-phenyl-, dimethyl ester, 1-oxide	EtOH	275(4.26),347(3.96)	12-1419-83
$C_{19}H_{15}NO_6$			
Rugosinone	MeOH	234(4.49),299(4.06), 336s(3.90)	102-2607-83
$C_{19}H_{15}N_2S$			
Thiazolo[3,4-a]pyrimidin-5-ium, 6-methyl-3,8-diphenyl-, perchlorate	MeCN	238(3.19),283(3.44), 392(2.73)	103-0037-83
$C_{19}H_{15}N_3O_3$			
1H-Naphth[2,3-g]indole-3-carboxamide, 5-amino-6,11-dihydro-N,2-dimethyl-6,11-dioxo-	dioxan	279(4.42),447(3.52), 556(3.83)	104-1892-83
$C_{19}H_{15}N_3O_4$			
1H-Naphth[2,3-g]indole-3-carboxylic acid, 2,5-diamino-6,11-dihydro-6,11-dioxo-, ethyl ester	dioxan	286(4.44),313(4.38), 526(4.11)	104-1892-83
$C_{19}H_{15}N_5O_2$			
Formazan, 5-(4-nitrophenyl)-1,3-diphenyl-	EtOH	295(4.18),480(4.17)	104-2104-83
	MeCN	302(4.21),350s(4.21), 484(4.24)	104-2104-83
	CCl_4	310(4.24),340s(4.16), 490(4.24)	104-2104-83
	isooctane	300(--),332(--), 480(--),530s(--)	104-2104-83
$C_{19}H_{15}N_5O_6S_3$			
2-Naphthalenesulfonic acid, 6-amino-4-hydroxy-3-[[4-[(2-thiazolylamino)-sulfonyl]phenyl]azo]-	n.s.g.	513(4.01)	7-0265-83

Compound	Solvent	λ_{max}(log ϵ)	Ref.
$C_{19}H_{15}N_5O_9S_4$			
2,7-Naphthalenedisulfonic acid, 5-amino-4-hydroxy-3-[[4-[(2-thiazolylamino)-sulfonyl]phenyl]azo]-	n.s.g.	539(4.08)	7-0265-83
$C_{19}H_{16}BrNO_6$			
9H-Carbazole-1,2,4-tricarboxylic acid, 6-bromo-9-methyl-, trimethyl ester	MeOH	218(4.24),231(4.38), 277(4.24),328(4.02), 387(3.46)	39-0515-83C
$C_{19}H_{16}ClFN_4$			
4H-Imidazo[1,5-a][1,4]benzodiazepin-3-amine, 8-chloro-6-(2-fluorophenyl)-N,N-dimethyl-	isoPrOH	217(4.67),244s(4.30), 272s(3.83)	4-0551-83
$C_{19}H_{16}ClFN_4O$			
4H-Imidazo[1,5-a][1,4]benzodiazepin-3-amine, 8-chloro-6-(2-fluorophenyl)-N,N-dimethyl-, 5-oxide	isoPrOH	226(4.52),253s(4.28), 298(4.22),400s(2.78)	4-0551-83
$C_{19}H_{16}ClNO_4$			
1H-Indole-3-acetic acid, 1-(4-chlorobenzoyl)-5-methoxy-2-methyl- (indomethacin)	EtOH	230(4.32),260(4.21), 319(3.80)	162-0721-83
$C_{19}H_{16}ClN_3O_6$			
Benzoic acid, 4-[3-chloro-2,5-dihydro-2,5-dioxo-4-(1,2,3,4-tetrahydro-1,3-dimethyl-2,4-dioxo-5-pyrimidinyl)-1H-pyrrol-1-yl]-, ethyl ester	MeOH	254(3.96),358(3.13)	89-0120-83S
$C_{19}H_{16}Cl_2O_5$			
9,10-Anthracenedione, 2-(4,4-dichloro-3-hydroxy-3-methylbutyl)-1,4-dihydroxy-	MeOH	226(4.39),246(4.65), 279(4.10),318(3.71), 456(4.04),478(4.07), 508(3.90)	5-1818-83
$C_{19}H_{16}Cl_2O_6$			
9,10-Anthracenedione, 2-(4,4-dichloro-3-hydroxy-3-methylbutyl)-1,4,8-trihydroxy-	MeOH	229(4.54),245(4.23), 285(3.81),460(3.94), 487(4.05),508(3.93), 520(3.86)	5-1818-83
$C_{19}H_{16}NO_5$			
Thalidastine (chloride)	MeOH	231(4.15),273(4.09), 350(4.02),425(3.36)	102-2607-83
$C_{19}H_{16}N_2$			
Sempervirine	EtOH	243(4.58),249(4.57), 297(4.20),345(4.26), 387(4.24)	162-1214-83
$C_{19}H_{16}N_2O$			
Benzo[4,5]cyclohepta[1,2-c]pyrazol-4(1H)-one, 9,10-dihydro-3-methyl-1-phenyl-	EtOH	282(4.22)	142-1581-83
Dipyrrolo[1,2,3-de:3',2',1'-ij]quinoxalin-2(1H)-one, 4,5-dihydro-7-methyl-8-phenyl-	EtOH	208(4.52),234(4.53), 280(4.06),300(4.11)	103-1340-83
1H-Indeno[1,2-d]pyridazin-1-one, 2,9-dihydro-9,9-dimethyl-2-phenyl-	MeOH	244(4.46),247s(--), 301(4.18)	2-0230-83

Compound	Solvent	λ_{max}(log ε)	Ref.
$C_{19}H_{16}N_2O$			
4-Isoquinolinecarbonitrile, 3-phenyl-1-propoxy-	EtOH	217(4.51),256(4.49), 309(4.16),325s(4.07), 339s(3.79)	39-1813-83C
Methanone, (4-aminophenyl)[4-(phenylamino)phenyl]-	MeOH	236(4.25),360(4.55)	12-0409-83
$C_{19}H_{16}N_2OS$			
3-Pyridinecarbonitrile, 1,2,3,4-tetrahydro-4-(4-methoxyphenyl)-6-phenyl-2-thioxo-	EtOH	258(4.23),326(3.98)	103-1202-83
compd. with piperidine (1:1)	EtOH	252s(4.25),288(4.25), 348s(3.64)	103-1202-83
$C_{19}H_{16}N_2OS_2$			
Benzenesulfenamide, N,N'-(2-methoxy-2,5-cyclohexadiene-1,4-diylidene)-bis-	$CHCl_3$	474(4.63)	150-0028-83S +150-0480-83M
$C_{19}H_{16}N_2O_2$			
[1]Benzopyrano[2,3-c]pyrazol-4(1H)-one, 1-phenyl-3-propyl-	EtOH	242(4.45),264(4.23)	118-0214-83
1H-Indole-3-acetonitrile, α-[(2,5-dimethoxyphenyl)methylene]-, (E)-	EtOH	275(4.12),370(4.02)	94-2023-83
(Z)-	EtOH	280(3.98),375(4.26)	94-2023-83
Pyrano[2,3-c]pyrazol-4(1H)-one, 5,6-dihydro-5-methyl-1,3-diphenyl-	EtOH	254(4.48)	118-0844-83
Pyrano[2,3-c]pyrazol-4(1H)-one, 5,6-dihydro-6-methyl-1,3-diphenyl-	EtOH	254(4.50)	118-0844-83
3H-Pyrazol-3-one, 4-benzoyl-2,4-dihydro-2-phenyl-5-(1-propenyl)-	EtOH	250(4.31)	142-1801-83
Pyrimido[2,1-a]isoindol-4(1H)-one, 2-(4-methoxyphenyl)-1-methyl-	EtOH	227(4.70),283(4.64), 364(4.16),375(4.21), 426(3.43)	124-0755-83
$C_{19}H_{16}N_2O_3$			
Benzonitrile, 2-(6,7-dimethoxy-1-isoquinolinyl)-4-methoxy-	EtOH	210(4.28),243(4.66), 319s(3.55),332(3.61)	78-3261-83
Benzonitrile, 2-(6,7-dimethoxy-1-isoquinolinyl)-5-methoxy-	EtOH	241(4.73),332(3.85)	78-3261-83
1,3-Butanedione, 1-(2,5-dihydro-5-oxo-1,3-diphenyl-1H-pyrazol-4-yl)-	EtOH	252(4.37),316(4.00)	44-4078-83
1,3-Propanedione, 1-(2,5-dihydro-3-methyl-5-oxo-1-phenyl-1H-pyrazol-4-yl)-3-phenyl-	EtOH	248(4.22),360(4.30)	44-4078-83
2H-Pyran-2-one, 4-hydroxy-6-methyl-3-[phenyl(phenylhydrazono)methyl]-	EtOH	288(4.08),370(4.05)	44-4078-83
2H-Pyran-2-one, 4-hydroxy-6-phenyl-3-[1-(phenylhydrazono)ethyl]-	EtOH	286(4.07),350(4.08)	44-4078-83
$C_{19}H_{16}N_2O_4$			
Benzoic acid, 4-methoxy-, 6-(4-methoxyphenyl)-3-pyridazinyl ester	MeOH	272(4.52)	97-0296-83
$C_{19}H_{16}N_4$			
Formazan, 1,3,5-triphenyl-	EtOH	299(4.36),490(4.21), 530s(3.98)	104-2104-83
	MeCN	298(4.33),492(4.14), 540s(3.98)	104-2104-83
	CCl_4	302(4.33),502(4.16), 550s(4.02)	104-2104-83

Compound	Solvent	$\lambda_{max}(\log \epsilon)$	Ref.
$C_{19}H_{16}N_4O_2$			
Methanone, (1,5-diphenylformazanyl)(5-methyl-2-furanyl)-	EtOH	230(4.02),326(4.27), 445(4.31)	104-0585-83
$C_{19}H_{16}N_4O_2S$			
2H-Isothiazolo[4,5,1-hi][1,2,3]benzothiadiazole-3-S^{IV}, 4,6,7,8-tetrahydro-4-(4-nitrophenyl)-2-phenyl-	C_6H_{12}	204(4.37),283s(4.16), 361(4.07),530(4.43)	39-0777-83S
$C_{19}H_{16}N_4O_7S_2$			
1-Azabicyclo[3.2.0]hept-2-ene-2-carboxylic acid, 6-ethyl-3-[(5-nitro-2-thiazolyl)thio]-7-oxo-, (4-nitrophenyl)-methyl ester, (5R-cis)-	THF	270(4.18),301(4.17), 373.5(3.92)	158-0407-83
$C_{19}H_{16}O$			
17H-Cyclopenta[a]phenanthren-17-one, 15,16-dihydro-7,11-dimethyl-	EtOH	275(4.76),293(4.48), 307(4.29),368(3.46), 389(3.51)	39-0087-83C
1-Propanone, 1-(2-naphthalenyl)-3-phenyl-	dioxan	327.5(3.20),341(3.20)	126-1143-83
$C_{19}H_{16}O_2$			
4-Pyrenepropanoic acid, 4,5-dihydro-	MeOH	258(4.62),280(4.08), 286(4.03),298(4.05)	54-0220-83
$C_{19}H_{16}O_3$			
Methanone, (2,7-dimethoxy-1-naphthalenyl)phenyl-	EtOH	275(3.37),308(3.40)	104-0183-83
Methanone, (3,6-dimethoxy-2-naphthalenyl)phenyl-	EtOH	282(3.49),328(3.48)	104-0183-83
$C_{19}H_{16}O_4$			
1-Benzoxepin-3-ol, 5-methoxy-4-phenyl-, acetate	heptane	234(4.12),305(3.78)	64-0895-83B
1-Benzoxepin-5-ol, 3-methoxy-4-phenyl-, acetate	heptane	232(4.12),307(3.67)	64-0895-83B
Naphtho[2,3-d]-1,3-dioxole, 6-(3,5-dimethoxyphenyl)-	EtOH	226(4.57),254(4.67), 286(4.18),335(3.61)	94-3056-83
Warfarin	pH 10	306(4.14)	162-1441-83
$C_{19}H_{16}O_5$			
5,12-Naphthacenedione, 7,8,9,10-tetrahydro-1,6,11-trihydroxy-8-methyl-	MeOH	232(4.49),252(4.49), 289(3.93),433s(--), 459(3.95),476(4.01), 487(4.08),510(3.93), 522(3.95)	5-1818-83
$C_{19}H_{16}O_6$			
5,12-Naphthacenedione, 7,8,9,10-tetrahydro-1,6,8,11-tetrahydroxy-8-methyl-	MeOH	232(4.35),253(4.36), 291(3.78),433s(--), 459(3.91),475(3.95), 488(4.03),510(3.86), 522(3.89)	5-1818-83
$C_{19}H_{16}O_7$			
2-Anthracenecarboxylic acid, 9,10-dihydro-8-hydroxy-3,4-dimethoxy-1-methyl-9,10-dioxo-, methyl ester	MeOH	224(4.26),259(4.17), 286s(3.78),400(3.64)	94-4206-83
4H-1-Benzopyran-4-one, 5,7-dimethoxy-2-(7-methoxy-1,3-benzodioxol-5-yl)-	MeOH	272s(2.13),340(2.25), 376s(1.70)	95-0994-83

Compound	Solvent	λ_{max}(log ϵ)	Ref.
5,12-Naphthacenedione, 7,8,9,10-tetrahydro-1,6,7,8,11-pentahydroxy-8-methyl-, trans	MeOH	232(4.32),247(4.21), 287(3.64),464(3.87), 478(3.92),488(3.99), 511(3.85),522(3.85)	5-1818-83
5,12-Naphthacenedione, 7,8,9,10-tetrahydro-6,r-7,t-8,c-10,11-pentahydroxy-8-methyl-	MeOH	228(3.89),250(4.11), 282(3.64),317(3.23), 459(3.44),482(3.49), 502(3.39),512(3.34), 550(2.78)	5-1818-83
t-10-	MeOH	223(4.04),245(4.37), 278(3.64),319(3.27), 448(3.68),478(3.82), 513(3.59)	5-1818-83
$C_{19}H_{16}S_3$			
6H-Cyclohepta[c]thiophene, 6-(1,3-benzodithiol-2-ylidene)-1,3,5-trimethyl-	MeCN	257(4.6),302(4.2), 370(4.2)	142-2039-83
$C_{19}H_{17}NOS$			
Benzo[b]thiophen-3(2H)-one, 2-[3-(dimethylamino)-2-phenyl-2-propenylidene]-	hexane	350(4.4),465(4.4), 490(4.5)	30-0331-83
Methanone, phenyl[1-[2-(phenylthio)-ethyl]-1H-pyrrol-2-yl]-	MeOH	213(4.24),253(4.19), 306(4.14)	23-1697-83
$C_{19}H_{17}NO_2$			
3-Butenenitrile, 4-(4-methoxyphenyl)-2-[[(4-methoxyphenyl)methylene]-	MeCN	362(4.67)	118-0917-83
1,4-Naphthalenedione, 2-[4-(dimethylamino)-2-methylphenyl]-	C_6H_{12}	506(3.61)	150-0168-83S
	EtOH	539(--)	150-0168-83S
2H-Pyrrol-2-one, 1,5-dihydro-3-(4-methoxyphenyl)-4-methyl-5-(phenylmethylene)-	MeOH	229(4.15),245(3.94), 346(4.57)	4-0687-83
$C_{19}H_{17}NO_2S$			
Methanone, phenyl[1-[2-(phenylsulfinyl)ethyl]-1H-pyrrol-2-yl]-	MeOH	212(4.10),248(3.98), 305(4.03)	23-1697-83
$C_{19}H_{17}NO_3$			
Dehydrostephanine	EtOH	224(4.43),253s(4.55), 262(4.59),336(4.16), 400s(3.58)	100-0761-83
Guadiscidine	EtOH	234(4.12),270(4.38), 306(4.13),322s(3.92), 348(3.78)	100-0761-83
	EtOH-HCl	214(4.35),278(4.43), 368(3.96),420(3.68)	100-0761-83
1H-Indene-3-carboxylic acid, 2-[[[(4-methylphenyl)amino]carbonyl]-, methyl ester	EtOH	299(--)	104-1363-83
	EtOH-KOH	293(--),340(--), 380(--)	104-1363-83
11,14a-Methano-4H,14aH-benzo[ij]cycloocta[b]quinolizine-8,10,15-trione, 5,6,11,12,13,14-hexahydro-	MeOH	236.5(4.03),251.5(3.75), 281(3.34),290(3.32), 334(3.38)	39-0505-83C
Noracronycine	EtOH	227(4.22),256(4.41), 284(4.68),295s(4.63), 312(4.37),342s(3.68), 410(3.69)	100-0391-83
$C_{19}H_{17}NO_3S$			
1H-Indole, 1-[(4-methylphenyl)sulfon-	MeOH	232(4.31),317(4.21)	5-2135-83

Compound	Solvent	λ_{max}(log ε)	Ref.
yl]-4-(3-oxo-1-butenyl)-, (E)- (cont.)			5-2135-83
1H-Indole, 4-(3-oxo-1-butenyl)-1-(phenylmethyl)sulfonyl]-, (E)-	MeOH	230(4.18),314(4.19)	5-2135-83
Methanone, phenyl[1-[2-(phenylsulfonyl)-ethyl]-1H-pyrrol-2-yl]-	MeOH	221(4.10),251(3.93), 265s(3.81),273(3.79), 305(4.14)	23-1697-83
$C_{19}H_{17}NO_4$			
Dehydrophanostenine	EtOH	261(4.75),302(3.93), 337(4.08),385(3.67)	100-0761-83
Dehydrostesakine	EtOH	247s(4.32),271(4.73), 296s(4.12),338(4.10)	100-0761-83
Eschscholtzine, (-)-	EtOH	235s(3.80),275s(3.93), 281(4.03),296(4.04), 307s(3.96)	100-0293-83
Furo[2,3-c]acridin-6(2H)-one, 1,11-dihydro-5-hydroxy-2-(1-hydroxy-1-methylethyl)-11-methyl- (furofoline II)	MeOH	226(4.29),247(4.43), 268s(4.11),276(4.77), 312s(4.03),323(4.07), 409(3.55)	39-1681-83C
Guattescine	EtOH	236(4.14),265(4.37), 278s(4.21),302(3.96), 324(3.88),344(3.77), 358(3.72)	100-0761-83 +100-0335-83
	EtOH-HCl	276(4.45),368(3.92), 420(3.65)	100-0761-83
1H-Indole-2-acetic acid, 1-benzoyl-2,3-dihydro-3-oxo-, methyl ester	MeOH	233(4.33),258(3.81), 400(3.49)	142-0421-83
9-Phenanthrenecarbonitrile, 2,3,6,7-tetramethoxy-	$CHCl_3$	250s(--),266(5.02), 279(4.68),290(4.80), 303s(--),319(4.35), 332(4.46),354(3.97), 372(4.03)	44-4222-83
Puterine, N-formyl-	EtOH	218(4.49),266(4.07), 274(4.09),301(3.93)	100-0761-83
	EtOH	218(4.49),266s(4.07), 274(4.09),301(3.93)	100-0862-83
Reframidine, (-)-	MeOH	235s(4.00),248(3.72), 294(3.93)	100-0293-83
Rufescine	EtOH	247(4.52),285s(4.31), 295(4.34),304(4.29), 315s(3.84),356(3.65), 373(3.78),400s(3.32)	100-0761-83
Xylopine, N-formyl-	EtOH	220(4.39),240s(4.07), 284(4.19)	100-0761-83 +100-0862-83
$C_{19}H_{17}NO_5$			
Acetamide, N-[3-(1,3-benzodioxol-5-yl)-1-(6-hydroxy-5-oxo-1,3,6-cycloheptatrien-1-yl)-2-propenyl]-, (E)-	MeOH	246(4.49),315(4.19), 372(3.75)	18-3099-83
Benzoic acid, 2-(6,7-dimethoxy-1-isoquinolinyl)-4-methoxy-	EtOH	239(4.36),315(3.37), 328(3.36)	78-3261-83
Benzoic acid, 2-(6,7-dimethoxy-1-isoquinolinyl)-5-methoxy-	EtOH	242(4.81),318s(3.92), 330(3.97)	78-3261-83
Guacolidine	EtOH	224(4.19),274(4.12), 328(3.54),361(3.57)	100-0761-83
	EtOH-HCl	228(4.12),278(4.45), 364(3.74),428(3.44)	100-0761-83
Norchelidonine, (±)-	EtOH	236(3.92),288.5(3.89)	33-1119-83
3,4-Quinolinedicarboxylic acid, 1,2,3-4-tetrahydro-2-oxo-1-phenyl-,dimethyl ester	EtOH	215(4.32),253(3.96), 280(3.38)	12-1419-83

Compound	Solvent	λ_{max}(log ϵ)	Ref.
$C_{19}H_{17}NO_6$			
3H-Benz[e]indole-1,8-dicarboxylic acid, 4,5-dihydro-2-methyl-4,5-dioxo-, diet hyl ester	EtOH	212(4.53),247(4.34), 274(4.48),350s(3.67), 384(3.81),460s(3.19)	103-0876-83
$C_{19}H_{17}N_2$			
Pyridinium, 1-phenyl-4-[2-(phenyl-amino)ethenyl]-, tetrafluoroborate	MeOH	444(4.84)	24-1982-83
$C_{19}H_{17}N_2O_2$			
Pyrimido[2,1-a]isoindolium, 4,6-dihy-dro-2-(4-methoxyphenyl)-1-methyl-4-oxo-, methyl sulfate	EtOH	206(4.63),287(4.58)	124-0755-83
$C_{19}H_{17}N_3O$			
Evodiamine	MeCN	272(4.06),280(4.02), 291(3.90),335(3.30)	162-0564-83
$C_{19}H_{17}N_3O_3$			
Bicyclo[4.4.1]undeca-3,6,8,10-tetraene-2,5-dione, mono[(2,5-dimethyl-4-nitro-phenyl)hydrazone], (±)-	CH_2Cl_2	424(4.60)	33-2369-83
	EtOH-KOH	620(4.71)	33-2369-83
$C_{19}H_{17}N_3O_3S$			
Benzyl Orange	pH 10	435(5.4)	60-0155-83
	+ $CaCl_2$	350(4.9)	60-0155-83
Pyrimido[4,5-b][1,4]oxazepine-2,4(1H-3H)-dione, 7,8-dihydro-1,3-dimethyl-8-phenyl-6-(2-thienyl)-	MeOH	230(4.14),259(4.06), 289(4.03),350(3.40)	103-0544-83
$C_{19}H_{17}N_3O_4$			
1,2-Diazabicyclo[5.2.0]nona-3,5-dien-9-one, 2-benzoyl-8-(2,5-dioxo-1-pyrrolidinyl)-5-methyl-, cis-(±)-	MeOH	284(3.88)	5-1374-83
$C_{19}H_{17}N_3O_5$			
1,2-Diazabicyclo[5.2.0]nona-3,5-diene-2-carboxylic acid, 8-(1,3-dihydro-1,3-dioxo-2H-isoindol-2-yl)-5-methyl-9-oxo-, ethyl ester, cis-(±)-	MeOH	277(4.00)	5-1374-83
1,2-Diazabicyclo[5.2.0]nona-3,5-diene-2-carboxylic acid, 5-methyl-9-oxo-8-[(3-oxo-1(3H)-isobenzofuranylidene)-amino]-, ethyl ester, cis-(±)-	MeOH	276(4.10)	5-1374-83
$C_{19}H_{17}N_3O_7$			
2-Furancarboxylic acid, 4-acetoxy-5-[4-(benzoylamino)-2-oxo-1(2H)-pyri-midinyl]-4,5-dihydro-, methyl ester, (2R-trans)-	MeOH	232(4.38),250s(--)	65-1712-83
$C_{19}H_{17}N_3O_8$			
Benzoic acid, 2-[bis[(5-nitro-2-furan-yl)methyl]amino]-, ethyl ester	MeOH	221(4.22),229s(2.60), 293(7.87)[sic]	73-2682-83
Benzoic acid, 4-[bis[(5-nitro-2-furan-yl)methyl]amino]-, ethyl ester	MeOH	221(2.73),294(5.84)	73-2682-83
$C_{19}H_{17}N_3S$			
2H-Isothiazolo[4,5,1-hi][1,2,3]benzo-thiadiazole-3-S^{IV}, 4,6,7,8-tetrahy-dro-2,4-diphenyl-	C_6H_{12}	205(4.41),250s(4.18), 295(4.20),488(4.35)	39-0777-83C

Compound	Solvent	λ_{max}(log ϵ)	Ref.
$C_{19}H_{17}N_3S_2$			
Pyridine, 2-[[phenyl(4,5,6,7-tetrahydro-1,3-benzodithiol-2-ylidene)methyl]azo]-	MeCN	506(3.89)	118-0840-83
$C_{19}H_{17}N_5O_2S$			
4H-Isothiazolo[4,5,1-hi][1,2,3]benzothiadiazol-3-S^{IV}-4-amine, 2,6,7,8-tetrahydro-N-(4-nitrophenyl)-2-phenyl-	C_6H_{12}	200(4.49),225(4.20), 329(4.21),376(4.24), 499(4.34),527(4.31)	39-0777-83C
$C_{19}H_{18}$			
Phenanthrene, 9-(1,2-dimethylcyclopropyl)-	EtOH	210(4.05),252(4.30), 300(3.60)	23-0866-83
Phenanthrene, 9-(1,1-dimethyl-2-propenyl)-	EtOH	214(4.48),252(4.75), 296(4.00)	23-0866-83
Phenanthrene, 9-(3-methyl-3-butenyl)-	EtOH	209(4.60),249(4.75), 291(3.90),322(3.35)	23-0866-83
$C_{19}H_{18}BrN_2OS$			
Isothiazolium, 2-(4-bromophenyl)-5-morpholino-4-phenyl-, perchlorate	MeCN	236s(4.14),354(4.29)	48-0689-83
$C_{19}H_{18}ClN_3O_2$			
2H-Pyrrolo[2,3-b]quinoxalin-2-one, 3-acetyl-1-(4-chlorophenyl)-1,3,3a,4,9,9a-hexahydro-4-methyl-	EtOH	221(4.52),304(3.65)	103-0901-83
$C_{19}H_{18}IP$			
Phosphonium, methyltriphenyl-, iodide	"MTsG"	265(3.4),270(3.4)	135-0658-83
$C_{19}H_{18}NOS$			
Thiazolium, 3-(2-[1,1'-biphenyl]-4-yl-2-oxoethyl)-2,4-dimethyl-, perchlorate	EtOH	294(4.47)	65-2424-83
$C_{19}H_{18}NO_2S_2$			
1,3-Thiazin-1-ium, 4,6-bis(4-methoxyphenyl)-2-(methylthio)-, perchlorate	MeCN	276(4.04),418(4.32), 480(4.47)	118-0827-83
$C_{19}H_{18}NO_4$			
Dehydrodiscretamine (chloride)	MeOH	225(4.28),237s(4.22), 263s(4.25),273(4.26), 346(4.29),434(3.60)	102-2607-83
Dibenzo[a,g]quinolizinium, 5,6-dihydro-2,3-dihydroxy-9,10-dimethoxy-, chloride	MeOH	235(4.63),270(4.52), 325(4.49)	100-0454-83
	MeOH-HCl	225(4.68),263(4.56), 340(4.51)	100-0454-83
	MeOH-NaOH	238(4.61),274(4.48), 330(4.54)	100-0454-83
$C_{19}H_{18}N_2O_2$			
1,4-Naphthalenedione, 2-[4-(dimethylamino)phenyl]-3-(methylamino)-	C_6H_{12}	410s(--),501(3.33)	150-0168-83S
	EtOH	488(--)	150-0168-83S
2-Propenoic acid, 3-(1H-indol-3-ylphenylamino)-, ethyl ester, (E)-	EtOH	207(4.50),289.8(4.21), 312(4.14)	104-1367-83
$C_{19}H_{18}N_2O_3$			
4H-Furo[3,2-b]pyrrole-5-carboxylic acid, 4-(2-cyanoethyl)-2-(4-methylphenyl)-, ethyl ester	MeOH	205(3.78),337(4.58)	73-0772-83

Compound	Solvent	$\lambda_{max}(\log \epsilon)$	Ref.
$C_{19}H_{18}N_2O_4$			
20,21-Dinoraspidospermidine-1-carboxylic acid, 2,3-didehydro-4,10-dioxo-, methyl ester, (±)-	EtOH	241(4.17),280(3.67), 312(3.75)	78-3719-83
20,21-Dinoraspidospermidine-3-carboxylic acid, 2,3-didehydro-4,10-dioxo-, methyl ester, (±)-	EtOH	243(4.09),286(3.54), 296(3.63),345(4.08)	78-3719-83
$C_{19}H_{18}N_2O_5$			
Pseudoindolo[2,3-a]quinolizine(15,20)-muconic acid, 3,5,6,7,14,21-hexahydro-7-hydroxy-	EtOH-HCl	254(4.11),308(3.97)	44-0044-83
$C_{19}H_{18}N_2O_7$			
Sarubicin A O-benzoyl deriv.	EtOH	214(4.35),230(4.27), 263(4.16),272s(4.09), 294s(3.76),478(3.20)	158-0976-83
$C_{19}H_{18}N_3O_3S$			
Isothiazolium, 5-morpholino-2-(3-nitrophenyl)-4-phenyl-, perchlorate	MeCN	230(4.28),249(4.26), 353(4.32)	48-0689-83
$C_{19}H_{18}N_4$			
1H-Benzimidazole-2-acetonitrile, α-[[4-(dimethylamino)phenyl]methylene]-1-methyl-	EtOH	405(4.48)	124-0297-83
$C_{19}H_{18}N_4O_2$			
Benzenamine, 4-nitro-N-[(1,2,3,4-tetrahydro-3-methylpyrimido[1,2-a]indol-10-yl)methylene]-	MeOH	467(4.48)	103-0507-83
1H-Imidazole-5-carboxamide, 1-(2-oxo-2-phenylethyl)-4-[(phenylmethyl)-amino]-	EtOH	244(4.33),303(4.10)	2-0030-83
$C_{19}H_{18}N_4O_3$			
Morpholine, 4-[1-(4-nitrophenyl)-4-phenyl-1H-pyrazol-5-yl]-	MeCN	231(4.25),322(4.16)	48-0689-83
$C_{19}H_{18}N_4O_7$			
Isoquinolinium, 2-methyl-, 1-acetonyl-2,4,6-trinitrocyclohexadienate, cation	MeCN	235(4.56)	103-0773-83
anion	MeCN	465(4.01),568(3.96)	103-0773-83
Quinolinium, 1-methyl-, 1-acetonyl-2,4,6-trinitrocyclohexadienate, cation	MeCN	238(4.59)	103-0773-83
anion	MeCN	463(4.06),568(3.96)	103-0773-83
$C_{19}H_{18}N_4S$			
2H-Isothiazolo[4,5,1-hi][1,2,3]benzothiadiazole-3-S^{IV}, 4,6,7,8-tetrahydro-4-(4-methyl-2-pyridinyl)-2-phenyl-	C_6H_{12}	210(4.46),253s(4.17), 289(4.14),499(4.30)	39-0777-83C
$C_{19}H_{18}N_6O_7S_2$			
5-Thia-1-azabicyclo[4.2.0]oct-2-ene-2-carboxylic acid, 7-[[carboxy(4-hydroxyphenyl)acetyl]amino]-3-[[(1-methyl-1H-tetrazol-5-yl)thio]methyl]-8-oxo-, [6R-(6α,7β)]-	MeOH	271(3.90)	87-1577-83
disodium salt	H_2O	272(4.02)	87-1577-83

Compound	Solvent	λ_{max}(log ε)	Ref.
$C_{19}H_{18}N_6O_8S$			
5-Oxa-1-azabicyclo[4.2.0]oct-2-ene-2-carboxylic acid, 7-[[carboxy(4-hydroxyphenyl)acetyl]amino]-3-[[(1-methyl-1H-tetrazol-5-yl)thio]methyl]-8-oxo-, disodium salt	H_2O	267.5(4.06)	87-1577-83
$C_{19}H_{18}O$			
17H-Cyclopenta[a]phenanthren-17-one, 11,12,15,16-tetrahydro-7,11-dimethyl-	EtOH	265s(4.47),274(4.75), 285(4.84),335(4.39)	39-0087-83C
4,6-Heptadien-3-one, 1,7-diphenyl-, (E,E)-	MeOH	233(3.90),322(4.54)	94-1544-83
$C_{19}H_{18}O_2$			
4,6-Heptadien-3-one, 5-hydroxy-1,7-diphenyl-, (E,E)-	MeOH	341(4.50),358s(--), 377s(--)	94-1544-83
$C_{19}H_{18}O_3$			
2-Cyclohexen-1-one, 3-(2-hydroxy-4-methoxyphenyl)-5-phenyl-	MeOH	202(4.41),330(4.04)	142-2369-83
Phenanthro[1,2-b]furan-10,11-dione, 6,7,8,9-tetrahydro-1,6,6-trimethyl- (tanshinone II)	MeOH	225(4.31),251(4.30), 268(4.41),350(3.22), 460(3.47)	94-1670-83
$C_{19}H_{18}O_5$			
9(2H)-Anthracenone, 2,10-diacetoxy-1,9a-dihydro-9a-methyl-	MeOH	252(4.79),262s(4.39), 273(4.28),328(3.78)	44-2412-83
4H-1-Benzopyran-4-one, 3-[(3,4-dimethoxyphenyl)methylene]-2,3-dihydro-7-methoxy-	MeOH	215(4.37),244(4.16), 304(4.10),358(4.26)	2-1119-83
4H-1-Benzopyran-4-one, 3-[(3,4-dimethoxyphenyl)methyl]-7-methoxy-	MeOH	220(4.64),236s(4.59), 245s(4.50),282(4.35), 292s(4.33),303s(4.28)	2-1119-83
4H-1-Benzopyran-4-one, 5-hydroxy-7-methoxy-2-(4-methoxyphenyl)-6,8-dimethyl- (relative absorbance given)	MeOH	282(0.91),292s(0.90), 326(1.00)	102-2107-83
	MeOH-$AlCl_3$	284(0.91),312(0.93), 342(1.00),410s(0.26)	102-2107-83
	MeOH-$AlCl_3$-HCl	286(0.75),309(0.90), 342(1.00),408s(0.21)	102-2107-83
Ethanone, 1-[3-acetoxy-3,4-dihydro-7-hydroxy-2-phenyl-2H-1-benzopyran-6-yl]-, trans	MeOH	277(4.68),376(4.21)	2-1116-83
9-Phenanthrenecarboxaldehyde, 2,3,6,7-tetramethoxy-	$CHCl_3$	262(4.60),271(4.70), 286(4.46),301(4.44), 345(4.07)	44-4222-83
$C_{19}H_{18}O_6$			
2H-1-Benzopyran-2-one, 4-(3,4-dimethoxyphenyl)-5,7-dimethoxy-	MeOH	248(4.13),328(4.26)	102-1657-83
2,4,6-Cycloheptatrien-1-one, 2-hydroxy-4-[1-oxo-3-(3,4,5-trimethoxyphenyl)-2-propenyl]-, (E)-	MeOH	255(4.41),342(4.31)	18-3099-83
	MeOH-NaOH	260(4.20),347(4.41), 441(3.73)	18-3099-83
5H-Phenanthro[8a,9-c]furan-2,7-dione, 7a,8-dihydro-3,10,11-trimethoxy-	EtOH	265(3.82),290(3.62), 360(3.68)	39-2053-83C
Spiro[furan-3(2H),6'-[6H]naphtho[1,8-bc]furan]-2,2'(4'H)-dione, 5-(3-furanyl)-3',4,5,5',5'a,7'-hexahydro-4'-hydroxy-7'-methyl- (2-hydroxyteucorolide)	EtOH	281(4.18)	102-0727-83

Compound	Solvent	$\lambda_{max}(\log \epsilon)$	Ref.
$C_{19}H_{18}O_7$			
4H-1-Benzopyran-4-one, 2-(3,4-dimethoxyphenyl)-5-hydroxy-3,7-dimethoxy- (relative absorbance given)	MeOH	255(1.00),268(0.84), 299s(0.46),351(0.88)	102-2107-83
	MeOH-AlCl$_3$	271(1.00),277s(0.95), 299s(0.44),364(0.72), 402(0.77)	102-2107-83
	MeOH-AlCl$_3$-HCl	270(1.00),276(0.98), 298s(0.50),356(0.75), 401(0.74)	102-2107-83
4H-1-Benzopyran-4-one, 2-(3,4-dimethoxyphenyl)-5-hydroxy-6,7-dimethoxy- (relative absorbance given)	MeOH	240(0.80),252s(0.69), 276(0.72),340(1.00)	102-2107-83
	MeOH-AlCl$_3$	260(0.65),287(0.72), 297s(0.67),370(1.00)	102-2107-83
	MeOH-AlCl$_3$-HCl	258(0.66),296(0.76), 362(1.00)	102-2107-83
4H-1-Benzopyran-4-one, 2-(3,4-dimethoxyphenyl)-5-hydroxy-7,8-dimethoxy- (relative absorbance given)	MeOH	254(0.88),275(1.00), 298s(0.66),340(0.95)	102-2107-83
	MeOH-AlCl$_3$	263s(0.80),282(0.98), 302(0.72),357(1.00), 406(0.62)	102-2107-83
	MeOH-AlCl$_3$-HCl	262s(0.77),283(0.98), 302(0.75),353(1.00), 406(0.57)	102-2107-83
4H-1-Benzopyran-4-one, 5-hydroxy-3,7,8-trimethoxy-2-(4-methoxyphenyl)- (relative absorbance given)	MeOH	273(1.00),300s(0.70), 322(0.77),361(0.69)	102-2107-83
	MeOH-AlCl$_3$	284(1.00),313(0.80), 350(0.98),422(0.64)	102-2107-83
	MeOH-AlCl$_3$-HCl	283(1.00),312(0.82), 346(0.94),422(0.61)	102-2107-83
4H-1-Benzopyran-4-one, 5-hydroxy-6,7,8-trimethoxy-2-(4-methoxyphenyl)- (gardenin B)(relative absorbance given)	MeOH	282s(0.90),293(0.92), 328(1.00)	102-2107-83
	MeOH-AlCl$_3$	263s(0.28),287s(0.55), 311(0.78),354(1.00), 411s(0.25)	102-2107-83
	MeOH-AlCl$_3$-HCl	264s(0.29),286(0.62), 311(0.89),350(1.00), 408s(0.23)	102-2107-83
	EtOH	292(4.40),330(4.35)	162-0623-83
$C_{19}H_{18}O_8$			
4H-1-Benzopyran-4-one, 2-(3,5-dihydroxy-4-methoxyphenyl)-5,7-dihydroxy-3-methoxy-6,8-dimethyl- (relative absorbance given)	MeOH	258s(0.58),279(1.00), 334(0.75)	102-2107-83
	MeOH-AlCl$_3$	258s(0.60),281(1.00), 311(0.70),352(0.80), 438s(0.19)	102-2107-83
	MeOH-AlCl$_3$-HCl	258s(0.65),284(1.00), 291s(0.97),312s(0.82), 356(0.92),430s(0.23)	102-2107-83
4H-1-Benzopyran-4-one, 5,7-dihydroxy-3-methoxy-2-(3,4,5-trimethoxyphenyl)- (relative absorbance given)	EtOH	268(1.00),305s(0.69), 346(0.92)	102-2107-83
	EtOH-AlCl$_3$-HCl	256s(0.57),280(1.00), 305s(0.66),343(0.83), 399(0.57)	102-2107-83
4H-1-Benzopyran-4-one, 5,7-dihydroxy-3,6,8-trimethoxy-2-(4-methoxyphenyl)- (relative absorbance given)	MeOH	258s(0.52),278(1.00), 335(0.88)	102-2107-83
	MeOH-AlCl$_3$	264s(0.56),288(0.87), 312(0.80),361(1.00), 413s(0.38)	102-2107-83
	MeOH-AlCl$_3$-HCl	262s(0.57),289(0.96), 310(0.92),356(1.00), 412s(0.32)	102-2107-83

Compound	Solvent	λ_{max}(log ϵ)	Ref.
4H-1-Benzopyran-4-one, 5,8-dihydroxy-7-methoxy-2-(3,4,5-trimethoxyphenyl)-	EtOH	283(4.31),307(4.29), 330s(4.19)	18-3773-83
	EtOH-NaOAc	285(4.30),307(4.31), 330s(4.22)	18-3773-83
	EtOH-AlCl$_3$	292(4.26),319(4.32), 345s(4.19),416(3.74)	18-3773-83
4H-1-Benzopyran-4-one, 2-(3,4-dimethoxyphenyl)-5,7-dihydroxy-3,6-dimethoxy- (bonanzin)(relative absorbance given)	MeOH	255(0.88),272(0.78), 348(1.00)	102-2107-83
	MeOH-AlCl$_3$	265(0.99),280s(0.89), 302s(0.56),376(1.00), 408s(0.75)	102-2107-83
	MeOH-AlCl$_3$-HCl	260(0.96),281(0.91), 302s(0.60),364(1.00), 406s(0.58)	102-2107-83
4H-1-Benzopyran-4-one, 2-(3,4-dimethoxyphenyl)-5,7-dihydroxy-6,8-dimethoxy- (hymenoxin)(relative absorbance given)	MeOH	255(0.66),282(0.89), 339(1.00)	102-2107-83
	MeOH-AlCl$_3$	260(0.76),290(0.78), 302s(0.75),362(1.00), 415s(0.42)	102-2107-83
	MeOH-AlCl$_3$-HCl	260(0.62),293(0.81), 361(1.00),412s(0.41)	102-2107-83
4H-1-Benzopyran-4-one, 5-hydroxy-2-(4-hydroxy-3,5-dimethoxyphenyl)-6,7-dimethoxy- (relative absorbance given)	MeOH	240s(0.96),274(0.61), 348(1.00)	102-2107-83
	MeOH-AlCl$_3$	256s(0.62),283(0.62), 310s(0.37),386(1.00)	102-2107-83
	MeOH-AlCl$_3$-HCl	256s(0.64),286(0.64), 308(0.44),374(1.00)	102-2107-83
4H-1-Benzopyran-4-one, 5-hydroxy-2-(3-hydroxy-4-methoxyphenyl)-3,6,7-trimethoxy- (relative absorbance given)	MeOH	257(0.92),270(0.84), 348(1.00)	102-2107-83
	MeOH-AlCl$_3$	267(0.95),280s(0.88), 298s(0.68),380(1.00), 406s(0.84)	102-2107-83
	MeOH-AlCl$_3$-HCl	266(0.94),282(0.92), 300s(0.71),370(1.00), 406s(0.72)	102-2107-83
4H-1-Benzopyran-4-one, 5-hydroxy-2-(3-hydroxy-4-methoxyphenyl)-3,7,8-trimethoxy- (relative absorbance given)	MeOH	299(1.00),273(0.93), 340s(0.64),364(0.73)	102-2107-83
	MeOH-AlCl$_3$	274s(0.97),282(1.00), 307(0.51),365(0.70), 424(0.61)	102-2107-83
	MeOH-AlCl$_3$-HCl	274s(0.97),282(1.00), 307(0.53),361(0.69), 423(0.51)	102-2107-83
4H-1-Benzopyran-4-one, 5-hydroxy-2-(3-hydroxy-4-methoxyphenyl)-6,7,8-trimethoxy- (gardenin D)(relative absorbance given)	MeOH	255(0.85),283(0.96), 344(1.00)	102-2107-83
	MeOH-AlCl$_3$	265(0.71),280(0.81), 304(0.80),370(1.00), 420s(0.36)	102-2107-83
	MeOH-AlCl$_3$-HCl	262(0.63),290(0.82), 303(0.83),363(1.00), 420s(0.32)	102-2107-83
4H-1-Benzopyran-4-one, 5-hydroxy-2-(4-hydroxy-3-methoxyphenyl)-3,6,7-trimethoxy- (relative absorbance given)	MeOH	256(0.86),270(0.77), 350(1.00)	102-2107-83
	MeOH-AlCl$_3$	268(0.85),280s(0.85), 300s(0.58),386(1.00), 408s(0.87)	102-2107-83
	MeOH-AlCl$_3$-HCl	265(0.81),281(0.86), 300s(0.52),382(1.00), 406s(0.80)	102-2107-83
4H-1-Benzopyran-4-one, 5-hydroxy-2-(4-hydroxy-3-methoxyphenyl)-3,7,8-tri-	MeOH	256(1.00),273(0.99), 322s(0.82),366(0.91)	102-2107-83

Compound	Solvent	λ_{max}(log ε)	Ref.
methoxy- (ternatin)(relative absorbance given)	MeOH-AlCl$_3$	272s(0.97),282(1.00), 306(0.57),366(0.85), 424(0.77)	102-2107-83
	MeOH-AlCl$_3$-HCl	268s(0.92),282(1.00), 306(0.63),361(0.88), 422(0.75)	102-2107-83
4H-1-Benzopyran-4-one, 5-hydroxy-2-(4-hydroxy-3-methoxyphenyl)-6,7,8-tri-methoxy- (relative absorbance given)	MeOH	255(0.76),281(0.81), 346(1.00)	102-2107-83
	MeOH-AlCl$_3$	265s(0.62),285(0.72), 304s(0.60),374(1.00), 418s(0.48)	102-2107-83
	MeOH-AlCl$_3$-HCl	264s(0.63),288(0.72), 304s(0.66),366(1.00), 418s(0.33)	102-2107-83
Javanicin, diacetyl-	EtOH	221(4.57),290(4.17), 426(3.79)	162-0755-83
Sakyomicin D	EtOH	213(4.57),249(3.92), 273(3.94),425(3.56)	77-0174-83
$C_{19}H_{18}O_9$			
D-Arabinitol, 1-C-(9,10-dihydro-1,4-dihydroxy-9,10-dioxo-2-anthracenyl)-	EtOH	227s(3.80),250(4.26), 256s(4.18),287(3.63), 322(3.19),480(3.82)	39-0613-83C
4H-1-Benzopyran-4-one, 5,6-dihydroxy-2-(3-hydroxy-4,5-dimethoxyphenyl)-3,7-dimethoxy- (relative absorbance given)	MeOH	240s(0.77),281(0.85), 339(1.00)	102-2107-83
	MeOH-AlCl$_3$	246s(0.75),262s(0.56), 299(0.80),371(1.00)	102-2107-83
	MeOH-AlCl$_3$-HCl	242s(0.75),260s(0.58), 296(0.81),365(1.00)	102-2107-83
4H-1-Benzopyran-4-one, 5,7-dihydroxy-2-(3-hydroxy-4,5-dimethoxyphenyl)-3,6-dimethoxy- (relative absorbance given)	MeOH	255s(0.71),272(0.85), 338(1.00)	102-2107-83
	MeOH-AlCl$_3$	254s(0.65),261s(0.69), 282(0.82),309s(0.54), 363(1.00),406s(0.59)	102-2107-83
	MeOH-AlCl$_3$-HCl	255s(0.71),285(0.97), 308s(0.69),357(1.00), 406s(0.51)	102-2107-83
4H-1-Benzopyran-4-one, 5,7-dihydroxy-2-(3-hydroxy-4,5-dimethoxyphenyl)-3,8-dimethoxy- (relative absorbance given)	MeOH	277(1.00),304s(0.72), 326(0.74),362s(0.67)	102-2107-83
	MeOH-AlCl$_3$	286(1.00),314(0.71), 354(0.82),420(0.59)	102-2107-83
	MeOH-AlCl$_3$-HCl	286(1.00),313(0.77), 350(0.83),420(0.59)	102-2107-83
4H-1-Benzopyran-4-one, 2-(3,5-dihy-droxy-4-methoxyphenyl)-5-hydroxy-3,6,7-trimethoxy- (relative absorbance given)	MeOH	274(0.91),339(1.00)	102-2107-83
	MeOH-AlCl$_3$	258s(0.79),282(0.90), 310(0.68),362(1.00), 406s(0.70)	102-2107-83
	MeOH-AlCl$_3$-HCl	260s(0.78),281(0.93), 308s(0.76),360(1.00), 406s(0.78)	102-2107-83
4H-1-Benzopyran-4-one, 2-(3,5-dihy-droxy-4-methoxyphenyl)-5-hydroxy-3,7,8-trimethoxy- (relative absorbance given)	MeOH	275(1.00),304s(0.62), 326(0.65),360(0.64)	102-2107-83
	MeOH-AlCl$_3$	282(1.00),312(0.57), 348(0.70),424(0.49)	102-2107-83
	MeOH-AlCl$_3$-HCl	282(1.00),312(0.59), 348(0.70),424(0.47)	102-2107-83
4H-1-Benzopyran-4-one, 2-(3,5-dihy-droxy-4-methoxyphenyl)-5-hydroxy-6,7,8-trimethoxy- (gardenin E) (relative absorbance given)	MeOH	285(1.00),296s(0.95), 332(0.97)	102-2107-83
	MeOH-AlCl$_3$	291(0.91),313(0.93), 358(1.00),420s(0.49)	102-2107-83

Compound	Solvent	λ_{max}(log ε)	Ref.
Gardenin E (cont.)	MeOH-$AlCl_3$-HCl	294s(0.87),313(0.98), 352(1.00),420s(0.45)	102-2107-83
$C_{19}H_{18}P$			
Phosphonium, methyltriphenyl-, iodide	"MTsG"	265(3.4),270(3.4)	135-0658-83
$C_{19}H_{19}BN_4O_5$			
Ethanone, 1-[5-(diacetoxyboryl)-1,5-diphenylformazanyl]-	hexane	500(4.02)	65-2061-83
	EtOH	498(4.02)	65-2061-83
$C_{19}H_{19}BrN_2OS$			
Morpholine, 4-[3-[(4-bromophenyl)amino]-2-phenyl-1-thioxo-2-propenyl]-	MeCN	255(4.07),304(4.34), 334s(4.32)	48-0689-83
$C_{19}H_{19}BrN_4O$			
1H-Imidazolium, 4-amino-5-(aminocarbonyl)-1-(2-oxo-2-phenylethyl)-3-(phenylmethyl)-, bromide	EtOH	244(4.37),300(4.18)	2-0030-83
$C_{19}H_{19}BrN_4O_4S$			
1H-Pyrazolo[3,4-d]pyrimidine, 4-[[3-(3-bromophenyl)-2-propenyl]thio]-1-β-D-ribofuranosyl-, (E)-	pH 1	255(4.27),293(4.30), 303s(4.23)	87-1489-83
$C_{19}H_{19}ClO_6$			
1H,3H-Naphtho[1,8-cd]pyran-1,3-dione, 5-chloro-4,6,9-triethoxy-7-methyl-	EtOH	257(4.15),323s(3.49), 339(3.60),366(3.52), 384(3.48)	78-2283-83
	EtOH-NaOH	249(4.23),308(3.49), 345s(3.28)	78-2283-83
$C_{19}H_{19}Cl_2N_5O$			
Urea, N-[6-(2,6-dichlorophenyl)-2-methylpyrido[2,3-d]pyrimidin-7-yl]-N'-(1,1-dimethylethyl)-	MeOH	224(4.75),319(4.24), 330(3.24)	87-0403-83
$C_{19}H_{19}Cl_3N_2O_5S_2$			
5-Thia-1-azabicyclo[4.2.0]oct-2-ene-2-carboxylic acid, 3-[(methylthio)methyl]-8-oxo-7-[(phenoxyacetyl)amino]-, 2,2,2-trichloroethyl ester. (6R-trans)	EtOH	264s(3.88),269(3.92), 275(3.91)	78-0337-83
$C_{19}H_{19}Cl_3N_2O_6S_2$			
5-Thia-1-azabicyclo[4.2.0]oct-2-ene-2-carboxylic acid, 3-[(methylthio)methyl]-8-oxo-7-[(phenoxyacetyl)amino]-, 2,2,2-trichloroethyl ester, 5-oxide, [5S-(5α,6β,7α)]-	EtOH	264s(3.92),269(3.97), 275(3.97)	78-0337-83
$C_{19}H_{19}FeNO_3$			
Ferrocene, [[5-(2-methoxy-2-oxoethyl)-1-methyl-1H-pyrrol-2-yl]carbonyl]-	MeOH	235(4.01),273(3.86), 314.5(4.18),474(3.02)	87-0226-83
$C_{19}H_{19}NO_3$			
Acetamide, N-[1-(6-hydroxy-5-oxo-1,3,6-cycloheptatrien-1-yl)-3-(4-methylphenyl)-2-propenyl]-, (E)-	MeOH	245(4.56),333(3.96), 370(3.75)	18-3099-83
9(10H)-Acridinone, 1,3-dihydroxy-10-methyl-4-(3-methyl-2-butenyl)-(glycocitrine II)	MeOH	226(4.27),251(4.48), 268s(4.54),275(4.74), 304(4.12),334(3.95), 405(3.77)	39-1681-83C

Compound	Solvent	λ_{max}(log ϵ)	Ref.
7H-Benzo[h]cyclopent[c]isoquinoline-5-carboxylic acid, 8,9,10,11-tetrahydro-2-methoxy-, methyl ester	C_6H_{12}	223(4.3),289(4.3)	5-1476-83
1H-Carbazole-1,4(9H)-dione, 7-methoxy-3-methyl-8-(3-methyl-2-butenyl)-(murrayaquinone B)	MeOH	210s(4.28),231(4.58), 264(4.44),310s(3.21), 404(3.66)	142-1267-83
Noracronycine, 1,2-dihydro-	EtOH	228(4.25),252(4.48), 266s(4.55),276(4.69), 300(4.13),332(4.09), 397(3.87)	100-0391-83
Nuciferine, 6a,7-dehydro-3-hydroxy-	EtOH	262(4.73),292s(4.11)	100-0761-83
$C_{19}H_{19}NO_3S$			
1-Naphthalenesulfonic acid, 8-[methyl(3,5-dimethylphenyl)amino]-	dioxan	252(4.15),345s(--), 365(3.29)	35-6236-83
Propanoic acid, 3-[bis(phenylmethyl)-amino]-2-oxo-3-thioxo-, ethyl ester	EtOH	259(4.29),300(3.60), 384(3.09)	5-1694-83
$C_{19}H_{19}NO_4$			
Acetamide, N-[1-(6-hydroxy-5-oxo-1,3,6-cycloheptatrien-1-yl)-3-(4-methoxyphenyl)-2-propenyl]-, (E)-	MeOH	244(4.45),305(3.99), 352(3.78),368s(3.74)	18-3099-83
Acetic acid, (5,6-dihydro-9-oxo-7a,10a-propano-4H-furo[3',2':4,5]pyrrolo-[3,2,1-ij]quinolin-10(9H)-ylidene)-, methyl ester, (E)-	MeOH	235(4.13),297(3.46)	39-0505-83C
(Z)-	MeOH	250(3.94),301(3.20)	39-0505-83C
Acetic acid, (5,6,7a,9,10,10a-hexahydro-9-oxo-4H-7a,10a-butanofuro[3',2'-4,5]pyrrolo[3,2,1-ij]quinolin-10-ylidene)-, (Z)-	MeOH	241(4.01),298(2.90)	39-0505-83C
Amurensine, (-)-	EtOH	230(4.07),250s(3.67), 294(3.95)	100-0293-83
Ayuthianine	EtOH	218(4.40),270(4.05), 302(3.76)	100-0761-83
Benzeneacetonitrile, α-[(3,4-dimethoxyphenyl)methylene]-3,4-dimethoxy-, (Z)-	$CHCl_3$	241(4.25),290s(--), 348(4.33),483(2.99)	44-4222-83
2-Butenedioic acid, 2-[1-methyl-5-(2-phenylethenyl)-1H-pyrrol-2-yl]-, dimethyl ester, (E,E)-	MeOH	393(4.83)	44-2488-83
Calycinine, N-methyl-	EtOH	223(4.49),270s(4.19), 278(4.28),301(4.13)	100-0761-83
	EtOH-base	236(4.49),275s(4.02), 283(4.10),317s(4.03), 332(4.08)	100-0862-83
Calycinine, O-methyl- (discoguattine)	EtOH	218(4.62),279(4.39), 296(4.30)	100-0862-83
Caryachine, (+)-	EtOH	291.5(3.97)	100-0293-83
Caryachine, (-)-	EtOH	291.5(4.02)	100-0293-83
	hexane	278(--),285(--), 290(--),295(--), 303(--)	100-0293-83
Domesticine	EtOH	221(4.56),283(4.01), 310(4.17)	162-0497-83
Epioliveridine, (±)-	EtOH	206(4.64),279(4.16), 320(3.63)	2-0321-83
Guattescine, dihydro-, (+)-	EtOH and EtOH-acid	217(4.45),239(4.21), 283(4.31),293(4.29), 326s(3.92)	100-0335-83
Laetine, O-methyl-	EtOH	269(3.55),303(3.24)	100-0761-83

Compound	Solvent	λ_{max}(log ε)	Ref.
Oliveridine, (±)-	EtOH	208(4.45),222(4.43), 238(4.11),284(4.20), 318(3.63)	2-0321-83
Reframoline, (-)-	MeOH	230s(4.00),248(2.43), 290(3.92)	100-0293-83
Stesakine	EtOH	218(4.47),240s(4.10), 281(4.23),320s(3.69)	100-0761-83
$C_{19}H_{19}NO_5$			
Acetamide, N-[3-(1,3-benzodioxol-5-yl)-1-(6-hydroxy-5-oxo-1,3,6-cyclohepta-trien-1-yl)propyl]-	MeOH	240(4.46),330(3.89), 369s(3.71)	18-3099-83
Bulbocapnine, 4-hydroxy-	EtOH	270(4.10),280s(4.05), 303(3.71)	100-0761-83
Duguevanine	EtOH	222(4.48),269s(4.17), 279(4.28),297(4.09), 305s(4.05)	100-0761-83
	EtOH-base	235(4.46),272s(3.78), 281(4.02),331(4.05)	100-0862-83
Guattouregidine	EtOH	238s(4.17),264(4.23), 296s(3.86),361(3.71)	100-0761-83
	EtOH-HCl	272(4.41),352(3.74), 410(3.59)	100-0761-83
6-Oxa-2-azaspiro[4.5]deca-3,7-diene-4-carboxylic acid, 2-ethyl-1,9-di-oxo-3-phenyl-, ethyl ester	EtOH	250(4.00),300(3.84)	94-0356-83
$C_{19}H_{19}NO_6$			
1,3-Benzodioxole-5-acetamide, N-[2-(1,3-benzodioxol-5-yl)ethyl]-4-(hydroxy-methyl)-	MeOH	232(4.85),283(4.84)	78-1975-83
2,4,6-Cycloheptatrien-1-one, 2-hydroxy-4-(1-hydroxyimino)-3-(3,4,5-trimeth-oxyphenyl)-2-propenyl]-, (?,E)-	MeOH	238(4.58),315(4.48), 375(3.82),395(3.68)	18-3099-83
$C_{19}H_{19}N_2OS$			
Isothiazolium, 5-morpholino-2,4-di-phenyl-, perchlorate	MeCN	220s(4.37),247s(4.00), 351(4.34)	48-0689-83
$C_{19}H_{19}N_3O$			
Suaveoline, N_a-demethyl-19-hydroxy-N_b-methyl-	MeOH	225(4.32),273(3.77), 283(3.75),291(3.64)	102-2297-83
$C_{19}H_{19}N_3OS$			
Isothiazolo[5,1-e][1,2,3]thiadiazole-7-S^{IV}, 1,6-dihydro-6-(4-methoxy-phenyl)-3,4-dimethyl-1-phenyl-	C_6H_{12}	203(4.36),228s(4.14), 292(4.22),475(4.32)	39-0777-83C
$C_{19}H_{19}N_3O_2$			
4H-Pyrido[1,2-a]pyrimidine-7-carbox-amide, 3-ethyl-2-methyl-4-oxo-N-(phenylmethyl)-	EtOH	234(4.39),347(4.37)	106-0218-83
2H-Pyrrolo[2,3-b]quinoxalin-2-one, 3-acetyl-1,3,3a,4,9,9a-hexahydro-4-methyl-1-phenyl-	EtOH	222(4.51),304(3.63)	103-0901-83
$C_{19}H_{19}N_3O_3S$			
Morpholine, 4-[3-[(3-nitrophenyl)ami-no]-2-phenyl-1-thioxo-2-propenyl]-	MeCN	249s(4.42),260(4.44), 299s(4.24),319(4.26), 392(3.39)	48-0689-83

Compound	Solvent	λ_{max}(log ϵ)	Ref.
$C_{19}H_{19}N_3O_4$			
L-Tryptophan, N-[[3-hydroxy-5-(hydroxy-methyl)-2-methyl-4-pyridinyl)methyl-ene]-	MeOH	255(4.06),278s(4.00), 290(3.93),340(3.57), 426(3.56)	35-0803-83
$C_{19}H_{19}N_3O_5$			
Butanoic acid, 4-[(2-benzoyl-9-oxo-1,2-diazabicyclo[5.2.0]nona-3,5-dien-8-yl)amino]-4-oxo-, methyl ester, cis-(±)-	MeOH	287(4.08)	5-1374-83
Quinolinium, 1-methyl-, 1-acetonyl-2,4-dinitrocyclohexadienide, cation	MeCN	245(4.53)	103-0773-83
anion	MeCN	515(3.36)	103-0773-83
$C_{19}H_{19}N_3O_6$			
1,2-Diazabicyclo[5.2.0]nona-3,5-diene-2-carboxylic acid, 8-[(2-carboxy-benzoyl)amino]-5-methyl-9-oxo-, 2-ethyl ester, cis-(±)-	MeOH	275(4.00)	5-1374-83
$C_{19}H_{19}N_4O_2$			
1H-Imidazolium, 4-amino-5-(aminocarbo-nyl)-1-(2-oxo-2-phenylethyl)-3-(phen-ylmethyl)-, bromide	EtOH	244(4.37),300(4.18)	2-0030-83
$C_{19}H_{19}N_7O_6$			
Folic acid	pH 13	256(4.43),283(4.40), 368(3.96)	162-0603-83
$C_{19}H_{20}BrNO_3$			
Thebaine, 1-bromo-, (-)-	MeOH	287.5(3.83)	78-2393-83
$C_{19}H_{20}Br_2O_2$			
3-Heptanone, 1,7-bis(4-bromophenyl)-5-hydroxy-	EtOH	255(2.78),262(2.87), 269(2.94),277(2.85)	18-3353-83
$C_{19}H_{20}Br_2O_4$			
6H-Dibenzo[b,d]pyran-6-one, 3-acetoxy-2-(2,3-dibromopropyl)-7,8,9,10-tetra-hydro-4-methyl-	MeOH	205(4.63),276(4.08), 313(3.91)	78-1265-83
$C_{19}H_{20}CuN_6$			
Copper, [N,N'-bis[2-(phenylazo)ethyli-dene]-1,2-propanediaminato(2-)]-, (SP-4-4)-	$CHCl_3$	330s(4.05),360s(4.14), 412(4.35),540s(3.13), 640s(2.54),860(1.82)	65-1214-83
Copper, [N,N'-bis[2-(phenylazo)ethyli-dene]-1,3-propanediaminato(2-)]-, (T-4)-	$CHCl_3$	340s(4.13),366(4.15), 439(4.40),570s(3.16), 650s(2.83),960(2.06)	65-1214-83
$C_{19}H_{20}N_2$			
1,2-Diazabicyclo[3.1.0]hex-2-ene, 4,6,6-trimethyl-3,5-diphenyl-, exo	EtOH	250(4.00)	44-1834-83
2,3-Diazabicyclo[3.1.0]hex-2-ene, 4,4,6-trimethyl-1,5-diphenyl-, endo	EtOH	220(4.10),260(3.11), 330(2.57)	44-1834-83
exo	EtOH	330(2.40)	44-1834-83
1H-1,4-Diazepine, 2,3-dihydro-2,3-bis(4-methylphenyl)-, cis	MeCN	307(3.826)	5-1207-83
$C_{19}H_{20}N_2O$			
Vellosimine	EtOH	280(3.90),289(3.81)	162-1420-83

Compound	Solvent	$\lambda_{max}(\log \epsilon)$	Ref.
$C_{19}H_{20}N_2OS$			
Morpholine, 4-[2-phenyl-3-(phenylamino)-1-thioxo-2-propenyl]-	MeCN	248(4.12),297(4.38), 304s(4.31),391s(3.42)	48-0689-83
$C_{19}H_{20}N_2OS_2$			
Propanedithioamide, N,N'-dimethyl-2-oxo-N,N'-bis(phenylmethyl)-	MeCN	272(4.29),307(4.15), 375s(3.25)	5-1116-83
$C_{19}H_{20}N_2O_2$			
Cleavamine, 5,16-dioxo-, (±)-	MeOH	244.5(4.12),321.5(4.19)	94-1183-83
1-Phenazinecarboxylic acid, 6-(3-methylbutyl)-, methyl ester	MeOH	254(4.82),349(4.03), 365(4.15)	73-0527-83
Phenylbutazone	MeOH-acid	239.5(4.19)	162-1049-83
Pyrano[4,3-c]pyrazol-3(2H)-one, 6,7-dihydro-6,6-dimethyl-4-(1,3-pentadienyl)-2-phenyl-, (E,E)-	EtOH	254(4.19),350(4.49)	118-0948-83 +142-1801-83
3H-Pyrazol-3-one, 2,4-dihydro-5-(2-methyl-1-propenyl)-4-(1-oxo-2,4-hexadienyl)-2-phenyl-, (E,E)-	EtOH	264(4.23),272(4.25), 340(4.33)	142-1801-83
2-Pyrrolidinecarboxamide, N-(2-formylphenyl)-1-(phenylmethyl)-, (S)-	$CHCl_3$	267(3.953),274(3.940), 336(3.720)	35-2010-83
Quebrachamine, 14,15-dehydro-5,16-dioxo-, (±)-	MeOH	243(4.15),319.5(4.21)	94-1183-83
$C_{19}H_{20}N_2O_2S$			
Carbamic acid, [3-(2,6-dimethylphenyl)-4-phenyl-1,3-thiazetidin-2-ylidene]-, ethyl ester	dioxan	232(4.28),c.325(3.0)	24-1297-83
$C_{19}H_{20}N_2O_3$			
Acetic acid, cyano[3-(diethylamino)-4-oxo-1(4H)-naphthalenylidene]-, ethyl ester	$CHCl_3$	255(4.33),320(4.05), 380(3.73),553(4.02)	88-3567-83
Acetic acid, cyano[4-(diethylamino)-1-oxo-2(1H)-naphthalenylidene]-, ethyl ester	CH_2Cl_2	250(3.97),305(4.20), 575(4.05)	88-3567-83
20,21-Dinoraspidospermidine-1-carboxylic acid, 2,3-didehydro-4-oxo-, methyl ester	EtOH	241(4.21),278(3.85), 305(3.81)	78-3719-83
20,21-Dinoraspidospermidine-3-carboxylic acid, 2,3-didehydro-4-oxo-, methyl ester	EtOH	241(4.05),287(3.51), 296(3.57),345(4.06)	78-3719-83
20,21-Dinoraspidospermidine-3-carboxylic acid, 2,3-didehydro-10-oxo-, methyl ester, (±)-	EtOH	230(4.12),302(4.17), 333(4.28)	44-5006-83
20,21-Dinoraspidospermidine-3-carboxylic acid, 3,4-didehydro-10-oxo-, methyl ester	EtOH	245(3.80),303(3.42)	44-5006-83
$C_{19}H_{20}N_2O_4$			
20,21-Dinoraspidospermidine-3-carboxylic acid, 4,10-dioxo-, methyl ester, (3)-(±)-	EtOH	248(3.88),298(3.75)	44-5006-83
$C_{19}H_{20}N_4O$			
3-Penten-2-one, 4-[(5,7-dimethyl-3-phenylpyrazolo[1,5-a]pyrimidin-2-yl)amino]-	MeCN	278(4.34),330(4.36)	39-0011-83C
$C_{19}H_{20}N_4O_3S$			
Morpholine, 4-[3-[2-(4-nitrophenyl)-	MeCN	224s(4.16),279(4.21),	48-0689-83

Compound	Solvent	$\lambda_{max}(\log \epsilon)$	Ref.
hydrazino]-2-phenyl-1-thioxo-2-propenyl]- (cont.)		389(4.34)	48-0689-83
$C_{19}H_{20}N_4O_4S$			
1H-Purine-2,6-dione, 7-(3,6-dideoxy-2-S-phenyl-2-thio-β-L-erythro-hexopyranos-4-ulos-1-yl)-3,7-dihydro-1,3-dimethyl-	MeOH	275(3.87)	136-0301-83A
threo-	MeOH	276(3.89)	136-0301-83A
1H-Pyrazolo[3,4-d]pyrimidine, 4-[(3-phenyl-2-propenyl)thio]-1-β-D-ribofuranosyl-, (E)-	pH 1	263(4.31),285s(4.25), 295(4.31)	87-1489-83
	pH 13	253(4.32),285s(4.26), 294(4.32)	87-1489-83
(Z)-	pH 1	245s(4.16),297(4.27)	87-1489-83
	pH 13	245s(4.17),295(4.24)	87-1489-83
$C_{19}H_{20}N_4O_5S$			
Phenol, 3-[3-[(1-β-D-ribofuranosyl-1H-pyrazolo[3,4-d]pyrimidin-4-yl)thio]-1-propenyl]-, (E)-	pH 1	257(4.25),297(4.34)	87-1489-83
$C_{19}H_{20}N_4O_6$			
1,2-Diazabicyclo[5.2.0]nona-3,5-diene-2-carboxylic acid, 5-methyl-8-[[(4-nitrophenyl)acetyl]amino]-9-oxo-, ethyl ester, cis-(±)-	MeOH	272(4.21)	5-1374-83
$C_{19}H_{20}N_6O_6S_2$			
5-Thia-1-azabicyclo[4.2.0]oct-2-ene-2-carboxylic acid, 7-[[(4-hydroxyphenyl)acetyl]amino]-7-methoxy-3-[[(1-methyl-1H-tetrazol-5-yl)thio]methyl]-8-oxo-, (6R-cis)-	MeOH	276(3.98)	87-1577-83
sodium salt	MeOH	275(4.02)	87-1577-83
$C_{19}H_{20}N_6O_7S$			
5-Oxa-1-azabicyclo[4.2.0]oct-2-ene-2-carboxylic acid, 7-[[(4-hydroxyphenyl)acetyl]amino]-7-methoxy-3-[[(1-methyl-1H-tetrazol-5-yl)thio]methyl]-8-oxo-, (6R-cis)-	MeOH	274(4.05)	87-1577-83
sodium salt	MeOH	273(4.03)	87-1577-83
$C_{19}H_{20}O_2$			
Cyclohexanone, 2-(4-methoxyphenyl)-2-phenyl-	MeOH	222s(4.08),271s(3.18), 277(3.26),283(3.20)	12-0789-83
14α-Estra-1,3,5(10),6,8-pentaen-17-one, 3-methoxy-	EtOH	232(4.74),257s(3.61), 268(3.67),278.5(3.69), 289.5(3.53),310s(3.05), 322.5(3.28),332s(3.25), 337(3.74)	65-1724-83
6-Hepten-3-one, 5-hydroxy-1,7-diphenyl-, [R-(E)]-	MeOH	251(4.31),283(3.61), 292(3.43)	94-1544-83
$C_{19}H_{20}O_3$			
Acerogenin C	EtOH	275(3.838)	94-1917-83
	EtOH-NaOH	297(--)	94-1917-83
Phenanthro[1,2-b]furan-10,11-dione, 1,2,6,7,8,9-hexahydro-1,6,6-trimethyl- (cryptotanshinone)	MeOH	263(4.46),271(4.41), 292(3.95),355(3.41), 477(3.48)	94-1670-83

Compound	Solvent	λ_{max}(log ε)	Ref.
$C_{19}H_{20}O_4$			
Benzoic acid, 2-(4-hydroxy-3,5-dimethylbenzoyl)-, propyl ester	EtOH	300(4.01)	24-0970-83
6H-Dibenzo[b,d]pyran-6-one, 3-acetoxy-7,8,9,10-tetrahydro-4-methyl-2-(2-propenyl)-	MeOH	206(4.62),278(4.06), 315(3.92)	78-1265-83
1H,7H-4,8-Dioxacyclobut[1,7]indeno[5,6-a]naphthalen-7-one, 1a,2,3,3a,10b,10c-hexahydro-10-hydroxy-1,1,3a-trimethyl-, (±)-	EtOH	234s(4.16),252(3.82), 261(3.77),339(4.23)	39-1411-83C
1H,9H-4,10-Dioxacyclobut[3,4]indeno-[5,6-a]naphthalen-9-one, 1a,2,3,3a-10c,10d-hexahydro-6-hydroxy-1,1,3a-trimethyl-	EtOH	227s(4.05),260s(3.90), 266(3.95),332(4.08)	39-1411-83C
$C_{19}H_{20}O_4S$			
1H-Indene-1,5(6H)-dione, 7a-ethyl-7,7a-dihydro-2-methyl-4-[(phenylsulfonyl)methyl]-, (S)-	EtOH	216.5(4.08),302(4.28)	39-2337-83C
$C_{19}H_{20}O_5$			
Anthra[1,9-de]-1,3-dioxin-7(7aH)-one, 11-acetoxy-8,11,11a,11b-tetrahydro-2,2-dimethyl-, (7aα,11β,11aα,11α)-(±)-	EtOH	227(4.08),262(3.81), 315(3.34)	44-3252-83
$C_{19}H_{20}O_6$			
Benzeneacetic acid, α-[(2,6-dimethoxyphenyl)methylene]-3,5-dimethoxy-, trans	EtOH	297(4.06)	2-0101-83
Benzeneacetic acid, 3-methoxy-α-[(2,4,5-trimethoxyphenyl)methylene]-, trans	EtOH	289(4.09),342(4.23)	2-0101-83
3(2H)-Benzofuranone, 2,4,6-trimethoxy-2-[(4-methoxyphenyl)methyl]-	MeOH	292(4.35)	102-0794-83
$C_{19}H_{20}O_7$			
Murrangatin diacetate	EtOH	248(3.64),258(3.64), 321(4.17)	102-0792-83
Paeoniflorigenone monoacetate	MeOH	229(4.14),268s(2.94), 272(2.99),280(2.89)	94-0577-83
$C_{19}H_{20}O_9$			
Cervicarcin	n.s.g.	227(4.15),264(3.89), 323(3.42)	162-0279-83
$C_{19}H_{20}S_2$			
Benzene, 1,1'-[1,3-heptadienylidenebis(thio)]bis-	dioxan	257(4.38)	118-0383-83
$C_{19}H_{21}FO_8$			
D-threo-Hex-2-enitol, 1,5-anhydro-2-deoxy-1-C-(2-fluoro-5-methoxyphenyl)-, triacetate, (S)-	EtOH	282.5(3.57)	23-0533-83
$C_{19}H_{21}F_3N_2S$			
Trifluomeprazine	EtOH	258(4.55),308(3.60)	162-1383-83
$C_{19}H_{21}N$			
Acetexa (nortriptyline hydrochloride)	MeOH	240(4.14)	162-0964-83
Protriptyline hydrochloride (MK-240)	n.s.g.	290(4.12)	162-1139-83

Compound	Solvent	λ_{max}(log ϵ)	Ref.
$C_{19}H_{21}NO_2$			
Benzenepropanenitrile, β-ethyl-4-methoxy-α-(4-methoxyphenyl)-	MeOH	205(4.02),222(4.40), 273(3.63),280(3.42)	83-0271-83
1-Oxa-3-azaspiro[4.5]deca-3,6,9-trien-8-one, 2,2,6,7,10-pentamethyl-4-phenyl-	EtOH	245(4.26)	33-2252-83
$C_{19}H_{21}NO_3$			
Acetamide, N-[1-(6-hydroxy-5-oxo-1,3,6-cycloheptatrien-1-yl)-3-(4-methylphenyl)propyl]-	MeOH	245(4.43),327(3.87), 368s(3.70)	18-3099-83
	MeOH-NaOH	251(4.43),337(4.05), 400(4.09)	18-3099-83
Aporphine, 1-hydroxy-2,9-dimethoxy-	EtOH	235(3.99),270(4.04), 283s(4.02)	100-0761-83
$C_{19}H_{21}NO_4$			
Bisnorargemonine, (-)- (rotundine)	EtOH	230s(4.07),279s(3.83), 285(3.92),288(3.93), 294(3.89)	100-0293-83
Hernagine	EtOH	223(4.55),272(4.15), 304(3.71)	100-0761-83
Isocorytuberine	EtOH	225(4.39),275(3.87), 313(3.67)	100-0761-83
Isoquinoline, 3-(3,4-dimethoxyphenyl)-3,4-dihydro-6,7-dimethoxy-, hydrochloride	EtOH	237(4.33),282(3.80), 313(3.88),370(3.62)	142-1247-83
Munitagine, (-)-	EtOH	283(3.63)	100-0293-83
Norlirioferine	EtOH	220(4.45),237s(3.95), 280(4.02),305(3.98), 316s(3.91)	100-0761-83
Oureguattidine, N-methyl-	EtOH	213(4.34),276(4.09), 307(3.88)	100-0761-83
Salutaridine	MeOH	236(4.23),279(3.76)	162-1201-83
Sarcocapnidine	EtOH	238(3.8),281(3.5)	88-2303-83
	EtOH-base	250(3.73),294(3.62)	88-2303-83
Thalidicine, (-)-	EtOH	221(4.38),290(3.96)	100-0293-83
Thalidine, (-)-	EtOH	250s(4.08),291(4.30)	100-0293-83
$C_{19}H_{21}NO_4S$			
2H-Indol-2-one, 3-ethyl-1,3-dihydro-3-[2-[[(4-methylphenyl)sulfonyl]-oxy]ethyl]-	EtOH	226(3.98),251(3.76)	94-2986-83
$C_{19}H_{21}NO_5$			
Alkaloid, m. 225-6°	EtOH	273.5(3.86)	94-2574-83
2,4,6-Cycloheptatrien-1-one, 4-[1-amino-3-(3,4,5-trimethoxyphenyl)-2-propenyl]-2-hydroxy-, (E)-	MeOH	245(4.44),335(3.96), 395s(3.52)	18-3099-83
Guattouregidine, dihydro-	EtOH	216(4.31),230s(4.11), 281(4.07),306(3.85)	100-0761-83
$C_{19}H_{21}NO_6$			
6-Oxa-2-azaspiro[4.5]dec-3-ene-4-carboxylic acid, 7-methoxy-2-methyl-1,9-dioxo-3-phenyl-, ethyl ester	EtOH	238s(3.83),309(3.84)	94-0356-83
1H-Pyrrole-3-propanoic acid, 4-(2-methoxy-2-oxoethyl)-2-[(phenylmethoxy)-carbonyl]-, methyl ester	$CHCl_3$	272(4.28)	78-1849-83
$C_{19}H_{21}NO_{10}$			
D-threo-Hex-2-enitol, 1,5-anhydro-2-de-	EtOH	305(4.05)	23-0533-83

Compound	Solvent	λ_{max}(log ε)	Ref.
oxy-1-C-(2-methoxy-5-nitrophenyl)-, 3,4,6-triacetate, (S)- (cont.)			23-0533-83
$C_{19}H_{21}NS$			
1-Propanamine, 3-dibenzo[b,e]thiepin-11(6H)-ylidene-N,N-dimethyl-, hydrochloride	MeOH	232(4.41),260(3.97), 309(3.53)	162-0498-83
$C_{19}H_{21}N_3OS$			
Morpholine, 4-[2-phenyl-3-(2-phenylhydrazino)-1-thioxo-2-propenyl]-	MeCN	282(4.22)	48-0689-83
$C_{19}H_{21}N_3O_4$			
1,2-Diazabicyclo[5.2.0]nona-3,5-diene-2-carboxylic acid, 5-methyl-9-oxo-8-[(phenylacetyl)amino]-, ethyl ester, cis-(±)-	MeOH	279(3.90)	5-1374-83
$C_{19}H_{21}N_3O_4S$			
Hypoglycine B, phenylhydantoin deriv.	n.s.g.	269(4.20)	162-0711-83
$C_{19}H_{21}N_3O_5$			
1,2-Diazabicyclo[5.2.0]nona-3,5-diene-2-carboxylic acid, 5-methyl-9-oxo-8-[[(phenylmethoxy)carbonyl]amino]-, ethyl ester, cis	MeOH	269(4.01),275(4.02)	5-1374-83
$C_{19}H_{21}N_5O_2$			
Pyrimidine, 4,4'-(2,6-pyridinediyl)-bis[6-ethoxy-2-methyl-	C_6H_{12}	220(4.48),276(4.44)	44-4841-83
$C_{19}H_{21}N_5O_2S_2$			
Pyrimidine, 4,4'-(2,6-pyridinediyl)-bis[6-ethoxy-2-(methylthio)-	C_6H_{12}	228(4.51),260(4.59), 302(4.06)	44-4841-83
$C_{19}H_{21}N_5O_3Se$			
Adenosine, 2',3'-O-(1-methylethylidene)-5'-Se-phenyl-5'-seleno-	$CHCl_3$	260(4.18)	78-0759-83
$C_{19}H_{21}N_7O_4$			
1H-1,2,4-Triazol-5-amine, 3-[4-(2-methoxyphenyl)-1-piperazinyl]-1-methyl-N-[(5-nitro-2-furanyl)methylene]-	$CHCl_3$	330(4.18),438(3.92)	24-1547-83
$C_{19}H_{21}N_9O_{13}P_2$			
Adenosine 5'-(trihydrogen diphosphate), 3'-(4-azido-2-nitrobenzenepropanoate)	pH 7.0	249(4.43)	64-0049-83C
$C_{19}H_{22}BrN$			
Benzenemethanamine, 2-[2-(4-bromophenyl)ethenyl]-N,N-diethyl-, hydrochloride, (E)-	HOAc	300(4.32)	39-1503-83B
$C_{19}H_{22}BrNO_5$			
Benzamide, 2-bromo-N-[2-(3,4-dimethoxyphenyl)-2-methoxyethyl]-4-methoxy-	EtOH	232(4.23),279(3.64), 289(3.53)	78-3261-83
Benzamide, 2-bromo-N-[2-(3,4-dimethoxyphenyl)-2-methoxyethyl]-5-methoxy-	EtOH	211(4.56),232(4.33), 284(3.69)	78-3261-83

Compound	Solvent	$\lambda_{max}(\log \epsilon)$	Ref.
$C_{19}H_{22}BrN_6O_3P$			
Phosphonic acid, [2-(1H-benzimidazol-1-yl)-1-bromoethyl]-, bis[2-(1H-imidazol-1-yl)ethyl] ester, dihydrochloride	EtOH	244(3.25),273(3.21), 279(3.13)	65-1555-83
$C_{19}H_{22}ClN$			
Benzenemethanamine, 2-[2-(4-chlorophenyl)ethenyl]-N,N-diethyl-, hydrochloride, (E)-	HOAc	295(4.25)	39-1503-83B
$C_{19}H_{22}NO_2$			
Pyridinium, 3,5-diacetyl-1,2,6-trimethyl-4-(4-methylphenyl)-, perchlorate	EtOH	207(4.40),247(4.06), 275(4.11)	103-0415-83
$C_{19}H_{22}N_2$			
Triprolidine	EtOH	236(4.19),285(3.83)	162-1392-83
hydrochloride hydrate	EtOH	235(4.18),283(3.87)	162-1392-83
oxalate	EtOH	233(4.21),283(3.91)	162-1392-83
$C_{19}H_{22}N_2O$			
Δ^3,Δ^5-Corynantheol, 18,19-dihydro-	EtOH and EtOH-acid	219(3.38),255(4.54), 308(4.36),366(3.59)	100-0694-83
	EtOH-NaOH	219(4.62),285(4.7), 324(3.8),416(3.39)	100-0694-83
Eburnamonine, (+)-	n.s.g.	241(4.30),268(4.01), 296(3.68),302(3.68)	162-0505-83
(-)-	n.s.g.	205(4.28),240(4.16), 265(3.90),290(3.59), 300(3.57)	162-0505-83
(±)-	n.s.g.	227(4.49),287(3.89), 294(3.87)	162-0505-83
1-Propanone, 1-[2-(phenylazo)-4-(1,1-dimethylethyl)phenyl]-	EtOH	234(3.11),328(3.20), 453(2.71)	103-0559-83
Rhazinol	MeOH	225(4.13),266(3.74)	102-1017-83
Tubotaiwinal, (+)-	MeOH	210(4.0),245(3.89), 254(3.85),300(3.55), 368(4.14)	78-3645-83
$C_{19}H_{22}N_2OS$			
Cyclopentanone, 2-[(dimethylamino)methylene]-5-[(3-ethyl-2(3H)-benzothiazolylidene)ethylidene]-	MeOH	498(4.86)	104-1854-83
	DMF	446(--),469(--)	104-1854-83
$C_{19}H_{22}N_2O_2$			
1,3-Cyclopentadiene-1-carboxylic acid, 2-amino-3-cyano-5,5-dimethyl-4-phenyl-, 1,1-dimethylethyl ester	hexane	358(3.54)	24-3097-83
20,21-Dinoraspidospermidine-1-carboxylic acid, 2,3-didehydro-, methyl ester, (±)-	EtOH	249(3.97),282(3.06), 291(3.01)	78-3719-83
20,21-Dinoraspidospermidine-3-carboxylic acid, 2,3-didehydro-, methyl ester, (±)-	EtOH	226(4.11),297(4.05), 328(4.19)	78-3719-83
1-Propanone, 1-[4-(1,1-dimethylethyl)-2-(phenyl-ONN-azoxy)phenyl]-	EtOH	256(2.86),328(3.96)	103-0559-83
Sarpagine	EtOH	230(4.30),278(3.92)	162-1205-83
Wieland-Gumlich aldehyde, hydrochloride	H_2O	240(3.80),290(3.40)	162-1442-83
$C_{19}H_{22}N_2O_2S$			
3-Penten-1-one, 4-methyl-1-phenyl-, N-	EtOH	265(4.13)	35-0933-83

Compound	Solvent	λ_{max}(log ϵ)	Ref.
tosylhydrazone (cont.)			35-0933-83
$C_{19}H_{22}N_2O_3$			
Bumadizon	pH 13	234(4.21),264(3.57)	162-0205-83
20,21-Dinoraspidospermidine-3-carboxylic acid, 4-oxo-, methyl ester, (3β)-(±)-	EtOH	246(3.85),300(3.45)	44-5006-83
20,21-Dinoraspidospermidine-3-carboxylic acid, 10-oxo-, methyl ester, (3α)-(±)-	EtOH	249(3.89),302(3.51)	44-5006-83
1H-Pyrrolo[2,3-d]carbazol-5-ylideneacetic acid, 2,3,3a,4,6a,7-hexahydro-3a-methyl-2-oxo-, ethyl ester	MeOH	220s(4.22),247(3.90), 299(3.38)	5-1744-83
$C_{19}H_{22}N_2O_4$			
20,21-Dinoraspidospermidine-3-carboxylic acid, 4-hydroxy-10-oxo-, methyl ester, (3α,4α)-(±)-	EtOH	251(3.83),302(3.44)	44-5006-83
$C_{19}H_{22}N_2O_5$			
1H-Pyrrole-3-propanoic acid, 5-[(3-ethenyl-1,5-dihydro-4-methyl-5-oxo-2H-pyrrol-2-ylidene)methyl]-2-(methoxycarbonyl)-4-methyl-, methyl ester	$CHCl_3$	253(4.22),403(4.44)	5-0585-83
1H-Pyrrole-3-propanoic acid, 5-[(4-ethenyl-1,5-dihydro-3-methyl-5-oxo-2H-pyrrol-2-ylidene)methyl]-2-(methoxycarbonyl)-4-methyl-, methyl ester	$CHCl_3$	268(4.19),405(4.40), 424(4.38)	5-0585-83
$C_{19}H_{22}N_2O_5S$			
4-Thia-1-azabicyclo[3.2.0]hept-2-ene-2-carboxylic acid, 3-methoxy-7-oxo-6-[(phenylacetyl)amino]-, 1,1-dimethylethyl ester, (5R-trans)-	dioxan	268(3.68),304(3.71)	78-2943-83
$C_{19}H_{22}N_2O_8S$			
Benzenesulfonamide, N-[4-(1,1-dimethylethyl)-3,6-dioxo-1,4-cyclohexadien-1-yl]-2-(2-methoxyethoxy)-5-nitro-	MeOH	271(4.09)	44-0177-83
3(2H)-Pyridazinone, 2-[2,3-O-(1-methylethylidene)-β-D-ribofuranosyl]-5-[[(4-methylphenyl)sulfonyl]oxy]-	pH 1	275(3.71),288(3.69)	44-3765-83
	pH 11	265(3.80),292s(3.55)	44-3765-83
	MeOH	275(3.58),293(3.61)	44-3765-83
$C_{19}H_{22}N_4O_2$			
2-Piperidinone, 6-hydroxy-1-(phenylamino)-6-[1-(phenylhydrazono)ethyl]-	MeOH	355(4.37)	24-1309-83
$C_{19}H_{22}N_4O_3$			
Carbamic acid, [4-(3-azidopropyl)-4-ethyl-3-oxo-1,5-cyclohexadien-1-yl]phenyl-, methyl ester	MeOH	204(4.33),222(4.26), 261(3.88),324(3.96)	35-3273-83
Carbamic acid, (6a-ethyl-5,6,6a,7,9a,9b-hexahydro-7-oxo-4H-1,2,3-triazolo[4,5,1-ij]quinolin-9-yl)phenyl-, methyl ester, (6aα,9aα,9bα)-	MeOH	242(4.11),268(4.05), 286(4.07)	35-3273-83
[1,2,3]Triazolo[1,5-a]pyridine-3-carboxamide, N,N-diethyl-7-[hydroxy(4-methoxyphenyl)methyl]-	EtOH	245s(--),280s(--), 290(4.08),310(4.08)	150-1341-83M
$C_{19}H_{22}N_4O_5$			
Carbamic acid, [5-amino-3-(3,5-dimeth-	pH 1	302s(3.89),307(3.90),	87-1614-83

Compound	Solvent	λ_{max}(log ε)	Ref.
oxyphenyl)-2-methyl-2H-pyrido[4,3-b]-1,4-oxazin-7-yl]-, ethyl ester (cont.)		359s(4.34),373(4.37)	87-1614-83
$C_{19}H_{22}N_4O_6$			
Carbamic acid, [5-amino-3-(3,4,5-trimethoxyphenyl)-2H-pyrido[4,3-b]-1,4-oxazin-7-yl]-, ethyl ester	pH 1	302s(3.89),308(3.90), 367s(4.40),379(4.43) (changing)	87-1614-83
$C_{19}H_{22}N_6$			
Acetaldehyde, 2,2'-[(1-methyl-1,2-ethanediyl)dinitrilo]bis-, bis(phenylhydrazone), (Z,Z,E,E)-	$CHCl_3$	300(4.12),340(4.54)	65-1214-83
Acetaldehyde, 2,2'-(1,3-propanediyldinitrilo)bis-, bis(phenylhydrazone), (Z,Z,E,E)-	$CHCl_3$	300(4.12),340(4.54)	65-1214-83
$C_{19}H_{22}O$			
Benzenepropanol, α-(4-phenyl-3-butenyl)-, (E)-	MeOH	252(4.20),283(3.49), 292(3.33)	94-1544-83
Dibenzo[b,d]pyran, 2,3,4,6,6a,7,8,9-octahydro-6-phenyl-	hexane	257(4.17)	104-2185-83
14β-Estra-1,3,5,7,9-pentaene, 3-methoxy-	EtOH	229(4.82),255.5(3.65), 265(3.73),275(3.73), 284(3.55),308(3.05), 321(3.29),329.5(3.25)	65-1724-83
$C_{19}H_{22}O_2$			
1,3-Benzenediol, 2-(3-methyl-2-butenyl)-5-(2-phenylethyl)-	MeOH	269(3.08),275(3.10), 282(3.11),293(2.80)	102-0573-83
	MeOH-HCl	269(3.59),274(3.62), 282(3.66),295(3.38)	102-0573-83
	MeOH-NaOH	284(3.94),291(3.96), 330(3.68)	102-0573-83
2-Butanone, 1-ethoxy-3-methyl-1,3-diphenyl-	n.s.g.	217(4.44),253.5(3.39), 259.5(3.41),265.5(3.35)	12-0097-83
14α-Estra-1,3,5,7,9-pentaen-17β-ol, 3-methoxy-	EtOH	230(4.84),258(3.57), 268(3.68),278(3.71), 289(3.55),311s(3.01), 323(3.30),332s(3.28), 338(3.41)	65-1724-83
Estra-1,3,5(10),9(11)-tetraen-17-one, 3-methoxy-	EtOH	263(4.28),273s(4.18), 298(3.50)	12-0339-83
3-Heptanone, 5-hydroxy-1,7-diphenyl-	EtOH	261(2.85),264(2.80), 268(2.80),287(2.64)	18-3353-83
1-Heptene-3,5-diol, 1,7-diphenyl-, [S-[R*,R*-(E)]]-	MeOH	252(4.30),283(3.56), 292(3.42)	94-1544-83
2-Naphthalenecarboxylic acid, 2,3,5,6,7,8-hexahydro-3-phenyl-, ethyl ester	C_6H_{12}	262(3.53),273s(3.48)	44-4272-83
2-Propenoic acid, 3-[2-(2-phenylethenyl)-1-cyclohexen-1-yl]-, ethyl ester, (E,E)-	C_6H_{12}	258(4.51),343(4.73)	44-4272-83
$C_{19}H_{22}O_3$			
3b,8-Etheno-3bH-indeno[1,2-c]furan-1,3-dione, 3a,7a,8,8a-tetrahydro-4,5,6,7-7a,8-hexamethyl-, (3aα,3bβ,7aβ,8β,8aα)-	MeCN	277(3.57)	44-0309-83
isomer	MeCN	217(4.01),307(3.91)	44-0309-83
$C_{19}H_{22}O_4$			
Benzenepropanoic acid, β-ethyl-4-meth-	MeOH	207(4.08),227(4.41),	83-0271-83

Compound	Solvent	λ_{max}(log ε)	Ref.
oxy-α-(4-methoxyphenyl)- (cont.)		273(3.56),280(3.44)	83-0271-83
Benzoic acid, 2-hydroxy-3,5,6-trimethyl-4-(phenylmethoxy)-, ethyl ester	EtOH	216(4.60),258(4.15), 318(3.66)	39-0667-83C
4H-1-Benzopyran-4-one, 8-(3,7-dimethyl-2,6-octadienyl)-5,7-dihydroxy-, (E)-	EtOH	218s(4.06),264(3.97), 305(2.89)	94-2859-83
(sophorachromone A)	EtOH-NaOEt	274(4.20),336(3.54)	94-2859-83
	EtOH-NaOAc	267(3.84),275s(3.68)	94-2859-83
	EtOH-$AlCl_3$	274(3.98),316(2.93)	94-2859-83
1,4-Naphthalenediol, 2,5-dimethyl-8-(1-methylethyl)-	EtOH	285(3.81),290(3.85), 298(3.91),318(3.61), 355(3.26)	22-0112-83
$C_{19}H_{22}O_5$			
Anthra[1,9-de]-1,3-dioxin-7,11-diol, 7,7a,8,11,11a,11b-hexahydro-2,2-dimethyl-, 11-acetate, (7α,7aβ-11α,11aβ,11bβ)-	EtOH	221(3.62),277(3.08), 286(3.00)	44-3252-83
5H-Benzocyclohepten-5-one, 6,7-diacetoxy-1,2,3,4-tetrahydro-9-methyl-2-(1-methylethenyl)-, (R)- (manicol diacetate)	EtOH	244(4.44),330(3.84)	78-2647-83
Gibba-1,3,4a(10a)-triene-9,10-dione, 2,7-dimethoxy-, cyclic 10-(1,2-ethanediyl acetal), (4bβ,7α,9aα)-	EtOH	219(3.88),290(3.38)	44-1643-83
1H-4-Oxabenzo[f]cyclobut[cd]indene-7-carboxaldehyde, 8-acetoxy-1a,2-3,3a,8b,8c-hexahydro-6-hydroxy-1,1,3a-trimethyl-, (±)-	EtOH	220(4.23),224(4.23), 236(4.07),289(4.16), 329s(3.71)	39-1411-83C
$C_{19}H_{22}O_6$			
Strigol	n.s.g.	234(4.25)	162-1266-83
$C_{19}H_{22}O_7$			
Gibbane-1,10-dicarboxylic acid, 2,4a,7-trihydroxy-1-methyl-8-methylene-4-oxo-, 1,4a-lactone	MeOH	280(2.06)	78-0449-83
1,8-Naphthalenedicarboxylic acid, 2,4,7-trimethoxy-5-methyl-, 8-ethyl 1-methyl diester	EtOH	254(4.38),320(3.70), 341s(3.57)	78-2283-83
$C_{19}H_{23}Cl_2N_5O$			
Trazodone hydrochloride (AF 1161)	H_2O	211(4.70),246(4.07), 274(3.59),312(3.59)	162-1370-83
$C_{19}H_{23}N$			
1H-Isoindole, 2,3-dihydro-1,1,3,3-tetramethyl-2-(phenylmethyl)-	EtOH	262(3.08),270(2.99)	12-0397-83
$C_{19}H_{23}NO$			
2-Aza-6,8-dimethyl-10-tert-butyl-3,4-benzotricyclo[4.4.0.0^{2,8}]dec-8-en-7-one	MeOH	208(4.13),232(4.10), 322(3.09)	35-3273-83
2-Aza-6,8-dimethyl-10-tert-butyl-3,4-benzotricyclo[4.4.0.0^{2,8}]dec-9-en-7-one	MeOH	211(4.08),324(2.60)	35-3273-83
Cyclohexanone, 2-[(1,3-dihydro-1,3,3-trimethyl-2H-indol-2-ylidene)ethylidene]-	MeOH	434(4.65)	104-1854-83
	benzene	410(--)	104-1854-83
	DMF	416(--)	104-1854-83
$C_{19}H_{23}NO_2$			
Ethanone, 1,1'-[1,2-dihydro-1,2,6-tri-	EtOH	204(4.32),234(4.09),	103-0415-83

Compound	Solvent	$\lambda_{max}(\log \epsilon)$	Ref.
methyl-4-(4-methylphenyl)-3,5-pyridinediyl]bis- (cont.)		291(4.09),335(4.09), 417(3.95)	103-0415-83
Ethanone, 1,1'-[1,4-dihydro-1,2,6-trimethyl-4-(4-methylphenyl)-3,5-pyridinediyl]bis-	EtOH	204(4.24),217s(4.14), 263(4.10),274s(4.05), 375(3.88)	103-0415-83
Ethanone, 1,1'-(1-ethyl-1,4-dihydro-2,6-dimethyl-4-phenyl-3,5-pyridinediyl)bis-	EtOH	204(4.37),221(4.22), 262(4.21),374(3.88)	103-0415-83
$C_{19}H_{23}NO_3$			
Ethanone, 1,1'-[1,4-dihydro-4-(4-methoxyphenyl)-1,2,6-trimethyl-3,5-pyridinediyl]bis-	EtOH	204(4.17),225(4.07), 263(3.94),277s(3.92), 372(3.73)	103-0415-83
7-Isoquinolinol, 1,2,3,4-tetrahydro-8-methoxy-2-methyl-1-[(4-methoxyphenyl)methyl]-, hydrochloride	MeOH	230(4.20),280(3.7)	100-0342-83
8-Isoquinolinol, 1,2,3,4-tetrahydro-7-methoxy-1-[(4-methoxyphenyl)-methyl]-2-methyl- (gortschakoine), hydrochloride	MeOH	232(4.19),282(3.65)	100-0342-83
Phenol, 4-[(1,2,3,4-tetrahydro-7,8-dimethoxy-2-methyl-1-isoquinolinyl)-methyl]-, hydrochloride, (±)-	MeOH	232(4.13),282(3.68)	100-0342-83
2H-Pyrano[3,2-d][1]benzoxepin-2-one, 4-[bis(1-methylethyl)amino]-5,6-dihydro-	EtOH	211.5(4.06),232.5(3.96), 306(3.93),319(3.91), 390(4.26)	4-0539-83
$C_{19}H_{23}NO_4$			
Crassifoline, perchlorate	EtOH	214(4.23),232s(4.19), 282(3.82)	88-2303-83
	EtOH-base	250(3.73),294(3.62)	88-2303-83
Ethanone, 1,1'-[4-(2,3-dimethoxyphenyl)-1,4-dihydro-2,6-dimethyl-3,5-pyridinediyl]bis-	EtOH	206(4.38),257(3.95), 376(3.70)	103-0415-83
1-Naphthalenecarbonitrile, 3,4-dihydro-3-(1-methoxy-2,4-O-isopropylidene-3-hydroxybutyl)-, (R)-	EtOH	224(4.32),273(3.99)	44-3269-83
(S)-	EtOH	224(4.31),273(3.96)	44-3269-83
Reticuline	n.s.g.	284(3.85)	162-1178-83
$C_{19}H_{23}NO_5$			
3,9-Dioxabicyclo[4.2.1]nonan-4-one, 5-[(dimethylamino)methylene]-7,8-isopropylidenedioxy-1-phenyl-, (1R*,6R*,7S*,8S)-	MeOH	298(4.26)	18-2680-83
3,9-Dioxabicyclo[4.2.1]nonan-4-one, 5-[(dimethylamino)methylene]-7,8-isopropylidenedioxy-2-phenyl-, (1S*,2S*.6R*,7R*,8R*)-	MeOH	299(4.14)	18-2680-83
$C_{19}H_{23}NO_{10}$			
Bicyclo[2.2.2]octane-2,3,5,6,7-pentacarboxylic acid, 8-cyano-, pentamethyl ester, (2endo,3exo,5endo,6exo-7syn)-	MeCN	215(2.54)	33-0019-83
$C_{19}H_{23}N_3O$			
Pyrazolidinium, 5-hexyl-3-oxo-1-(2-quinolinylmethylene)-, hydroxide, inner salt	toluene	390(4.52)	48-0205-83
	MeCN	378(4.47)	48-0205-83
Pyrazolidinium, 5-hexyl-3-oxo-1-(4-	toluene	393(4.50)	48-0205-83

Compound	Solvent	$\lambda_{max}(\log \epsilon)$	Ref.
quinolinylmethylene)-, hydroxide, inner salt (cont.)	MeCN	384(4.46)	48-0205-83
5H-1,2,3-Triazolo[4,5,1-de]acridin-5-one, 2a-(1,1-dimethylethyl)-2a,5a,6,11a-tetrahydro-4,5a-dimethyl-, (2aα,5aα,11aα)-	MeOH	207(4.33),234(4.20), 270(3.76),303(3.62), 335(3.64)	35-3273-83
$C_{19}H_{23}N_3O_4S_4$			
Carbonodithioic acid, S-[[7-amino-2-[(ethoxythioxomethyl)amino]-2,3,5,8-tetrahydro-1-mercapto-6-methyl-5,8-dioxo-1H-pyrrolo[1,2-a]indol-9-yl]methyl] O-ethyl ester, (1R-cis)	MeOH	252(4.45),271(4.47), 310s(4.08),355(3.74)	44-5026-83
Carbonodithioic acid, S-[2,7-diamino-9-[[(ethoxythioxomethyl)thio]methyl]-6-methyl-5,8-dioxo-1H-pyrrolo[1,2-a]indol-1-yl] O-ethyl ester, (1S-trans)-	MeOH	250(4.49),270(4.30), 280s(4.29),310s(4.12), 355(3.70)	44-5026-83
$C_{19}H_{23}N_3O_6S_2$			
Carbonothioic acid, S-[[7-amino-2-[(ethoxycarbonyl)amino]-2,3,5,8-tetrahydro-1-mercapto-6-methyl-5,8-dioxo-1H-pyrrolo[1,2-a]indol-9-yl]methyl] O-ethyl ester, (1R-cis)-	MeOH	206(4.00),255(3.92), 310(3.68),351(3.20)	44-5033-83
Carbonothioic acid, S-[2,7-diamino-9-[[(ethoxycarbonyl)thio]methyl]-2,3,5,8-tetrahydro-6-methyl-5,8-dioxo-1H-pyrrolo[1,2-a]indol-1-yl] O-ethyl ester, (1S-trans)-	MeOH	205(4.12),255(4.08), 309(3.82),349(3.36)	44-5033-83
$C_{19}H_{23}N_5O_4$			
1H-Pyrazolo[3,4-d]pyrimidin-4-amine, N-(1-methyl-2-phenylethyl)-1-β-D-ribofuranosyl-, (R)-	MeOH	265(4.94),284(5.13)	87-1601-83
$C_{19}H_{23}N_6O_3P$			
Phosphonic acid, [2-(1H-benzimidazol-1-yl)ethyl]-, bis[2-(1H-imidazol-1-yl)ethyl] ester, dihydrochloride	EtOH	244(3.68),273(3.65), 279(3.63)	65-1555-83
$C_{19}H_{23}N_{11}O$			
6H-Purin-6-one, 2-amino-1,7-bis[(1,4-dihydro-4-imino-1,2-dimethyl-5-pyrimidinyl)methyl]-1,7-dihydro-, diperchlorate	pH 6	219(4.48),249(4.49)	44-2481-83
$C_{19}H_{24}NO_3$			
Isoquinolinium, 1,2,3,4-tetrahydro-7-hydroxy-1-[(4-hydroxyphenyl)methyl]-8-methoxy-2,2-dimethyl-, iodide	MeOH	236(4.02),282(3.73)	100-0342-83
Oblongine (iodide)	MeOH	213(4.17),227(4.09), 283(3.50)	100-0342-83
	MeOH-KOH	218(4.25),250(3.57), 300(3.63)	100-0342-83
$C_{19}H_{24}N_2$			
Cleavamine, (±)-	MeOH	229.5(4.40),285.5(3.76), 292.5(3.72)	94-1183-83
$C_{19}H_{24}N_2O$			
Alloibogamine, 16-hydroxy-, (±)-	MeOH	228(4.43),285(3.85),	77-1018-83

Compound	Solvent	$\lambda_{max}(\log \epsilon)$	Ref.
(cont.)		292(3.81)	77-1018-83
Cinchonamine	MeOH	223(4.60),292(3.88)	162-0324-83
1-Propanone, 1-[4-(1,1-dimethylethyl)-2-phenyl-5-pyrimidinyl]-2,2-dimethyl-	EtOH	268(4.33)	4-0649-83
$C_{19}H_{24}N_2O_2$			
Quinamine	MeOH	242(3.93),301(3.39)	162-1162-83
Spiro[cyclopentane-1,4'(1'H)-pyridazine]-3'-carboxylic acid, 5'-phenyl-, 1,1-dimethylethyl ester	$CHCl_3$	243(3.77)	24-3097-83
Spiro[cyclopentane-1,4'(1'H)-pyridazine]-5'-carboxylic acid, 3'-phenyl-, 1,1-dimethylethyl ester	$CHCl_3$	322(3.68)	24-3097-83
$C_{19}H_{24}N_2O_2S$			
Carbamic acid, (3-cyclohexyl-3,6-dihydro-6-phenyl-2H-1,3-thiazin-2-ylidene)-, ethyl ester	dioxan	271(3.83),293(3.89)	24-1297-83
$C_{19}H_{24}N_2O_4$			
1H-Pyrido[3,4-b]indole, 2-acetyl-2,3,4,9-tetrahydro-6-methoxy-1-(methoxymethyl)-1-(2-oxopropyl)-	EtOH	225(4.46),277(3.96), 296s(3.85),308s(3.60)	4-0267-83
$C_{19}H_{24}N_2O_8$			
α-D-Mannofuranosylamine, 2,3:5,6-bis-O-(1-methylethylidene)-N-[(4-nitrophenyl)methyl]-, N-oxide, (Z)-	CH_2Cl_2	249(3.96),347(4.23)	33-0789-83
$C_{19}H_{24}N_4$			
1(4H)-Pyridinepropanenitrile, 4-[[1-(cyanomethyl)-2-methyl-1-butenyl]imino]-β-(1-methylpropylidene)-	EtOH	207(4.26),265(4.16)	39-1223-83C
$C_{19}H_{24}O$			
Androsta-1,4,6-trien-3-one	EtOH	223(4.05),254(3.96), 298(4.11)	95-1046-83
$C_{19}H_{24}O_2$			
Androsta-4,6-diene-3,17-dione	MeCN	280(4.36)	13-0707-83B
A,19-Dinorpregn-3(5)-en-20-yn-2-one, 17-hydroxy-, (17α)-	EtOH	232.5(4.18)	39-2337-83C
3,5-Heptanediol, 1,7-diphenyl-, (3S,5R)-	MeOH	247(2.65),252(2.70), 258(2.75),262(2.77), 264(2.70),268(2.74)	94-1544-83
Retinal, 13-demethyl-11,14-epoxy-, 9-cis	hexane	242(4.00),290(4.03), 355(4.20)	54-0046-83
$C_{19}H_{24}O_2S$			
2-Thiophenecarboxylic acid, 5-[2-methyl-4-(2,6,6-trimethyl-1-cyclohexen-1-yl)-1,3-butadienyl]-, (E,E)-	EtOH	233(3.97),347(4.51)	87-1282-83
$C_{19}H_{24}O_3$			
Androsta-1,4-diene-3,11-dione, 17β-hydroxy-	MeOH	238(4.18)	87-0078-83
Androst-4-ene-3,17-dione, 6α,7α-epoxy-	MeCN	239(4.06)	13-0707-83B
6H-Dibenzo[b,d]pyran-2-carboxaldehyde, 7,8,9,10-tetrahydro-1-methoxy-3,6,6,9-tetramethyl-	$CHCl_3$	270(4.64),295(3.99)	83-0326-83
2-Furancarboxylic acid, 5-[2-methyl-	EtOH	335(4.56)	87-1282-83

Compound	Solvent	λ_{max}(log ε)	Ref.
4-(2,6,6-trimethyl-1-cyclohexen-1-yl)-1,3-butadienyl]-, (E,E)- (cont.)			87-1282-83
Testolactone	EtOH	242(4.20)	162-1312-83
$C_{19}H_{24}O_4$			
Capillartemisin A	MeOH	219(4.30),234(4.31), 303(4.32)	94-0352-83
Capillartemisin B	MeOH	218(4.35),234(4.32), 303(4.33)	94-0352-83
$C_{19}H_{24}O_5$			
Trichothecin	hexane	217(4.26)	162-1380-83
	MeOH	215(4.28)	162-1380-83
$C_{19}H_{24}O_6$			
Illudin S, diacetate	EtOH	227(4.11),313(3.53)	162-0714-83
$C_{19}H_{25}$			
1H-Tribenzo[a,c,e]cycloheptenylium, 2,3,4,5,6,7,8,10,11,12,13,?-dodecahydro-, (OC-6-11)-, hexafluoroantimonate	MeCN	258(4.20),315(3.65)	44-0596-83
$C_{19}H_{25}N$			
Pyridine, 2,6-bis(1,1-dimethylethyl)-4-phenyl-	neutral	277(4.18)	39-0045-83B
	acid	282(4.42)	39-0045-83B
$C_{19}H_{25}NO$			
1-Butanol, 3-methyl-2-[[(4-methylphenyl)phenylmethyl]amino]-, hydrochloride, (R,S)-	EtOH	224(4.09),254(2.85), 260(2.85),269s(2.69), 272s(2.49)	94-2183-83
(S,S)-	EtOH	224(4.16),254s(2.79), 259s(2.85),263(2.88), 268s(2.77),273s(2.58)	94-2183-83
$C_{19}H_{25}NO_2$			
1-Butanol, 2-[[(4-methoxyphenyl)phenylmethyl]amino]-3-methyl-, hydrochloride, (R,S)-	EtOH	233(4.36),263(3.36), 279(3.39),274s(3.38), 281(3.31)	94-2183-83
(S,S)-	EtOH	237(4.34),263s(3.30), 268(3.33),275s(3.34), 281(3.27)	94-2183-83
Pentanamine, 2,3-bis(4-methoxyphenyl)-, hydrochloride	MeOH	208(3.02),226(4.39), 273(3.49),282(3.44)	83-0271-83
$C_{19}H_{25}NO_3S$			
Piperidine, 1-(3,4,5,6,7,8-hexahydro-3-phenyl-1,2-benzoxathiin-4-yl)-, S,S-dioxide, cis	EtOH	230s(3.43),332(3.35)	4-0839-83
trans	EtOH	230s(3.48)	4-0839-83
$C_{19}H_{25}NO_4$			
2H-Pyran-3-carboxaldehyde, 4-[(2-hydroxyethyl)amino]-6-(1,2,4a,5,6,7-8,8a-octahydro-2-methyl-1-naphthalenyl)-2-oxo- (solanopyrone C)	EtOH	238(4.30),282(3.84), 320(3.86)	88-5373-83
$C_{19}H_{25}NO_6S$			
Benzenesulfonamide, N-(2,2-dimethoxyethyl)-N-[(3-hydroxy-4-methoxyphenyl)methyl]-4-methyl-	MeOH	231(4.21),280(3.50)	83-0520-83

Compound	Solvent	$\lambda_{max}(\log \epsilon)$	Ref.
$C_{19}H_{25}N_3O_4$			
Benzoic acid, 4-[(2-amino-1,4-dihydro-6-hydroxy-4-oxo-5-pyrimidinyl)methyl]-, heptyl ester	MeOH	247(4.26),270(4.28)	73-0299-83
5-Pyrimidinepentanoic acid, 1,4-dihydro-6-hydroxy-4-oxo-2-[(phenylmethyl)-amino]-, propyl ester	50% MeOH	240s(3.87),273(4.19)	73-0304-83
	+ HCl	235(3.808),276(4.126)	73-0304-83
	+ NaOH	245s(3.837),272(4.072)	73-0304-83
$C_{19}H_{26}CuN_{10}$			
Copper, [7,8,9,16,17,18-hexahydro-N,N,N',N'-tetramethyl-6H-dipyrimido-[4,5-f:5',4'-m][1,4,8,12]tetraaza-cyclopentadecine-4,11-diaminato(2-)-N^6,N^9,N^{15},N^{19}]-, (SP-4-2)-	DMF	332(3.80),375s(3.49)	12-2413-83
$C_{19}H_{26}N_2$			
Melonine, (+)-	n.s.g.	248(3.80),298(3.45)	88-0761-83
Quebrachamine	MeOH	230(4.55),287(3.85), 293(3.84)	162-1159-83
$C_{19}H_{26}N_2O$			
Cyclopentanone, 2,5-bis[5-(dimethyl-amino)-2,4-pentadienylidene]-	EtOH	300(3.88),585(4.77)	70-1882-83
	$CHCl_3$	550(--)	70-1882-83
Ethanone, 1-[1-[2-(2-ethyl-1H-indol-3-yl)ethyl]-3-piperidinyl]- (same in acid or base)	EtOH	226(4.76),274(3.81), 283(3.88),291(3.65)	100-0694-83
Geissoschizoline	EtOH	245(3.93),300(3.47)	162-0625-83
Isogeissoschizol (same in acid or base)	EtOH	227(4.67),273(3.72), 284(3.81),290(3.57)	100-0694-83
$C_{19}H_{26}N_2O_2$			
Formic acid, 2-[(17β)-3-hydroxyestra-1,3,5(10)-trien-17-yl]hydrazide	MeOH	224(4.07),230(4.06), 241(3.98),281(3.30), 287(3.07)	13-0001-83B
	MeOH-NaOH	242(4.24),298(3.38)	13-0001-83B
$C_{19}H_{26}N_2O_2S$			
Benzenesulfonic acid, [1-methyl-3-(2,6,6-trimethyl-1-cyclohexen-1-yl)-2-propenylidene]hydrazide, (E,E)-	EtOH	269s(--),274(4.05)	33-1148-83
Benzenesulfonic acid, 4-methyl-, (octa-hydro-3a,7a-dimethyl-4,7-methano-1H-inden-1-ylidene)hydrazide, (3aα,4β-7β,7aα)-(±)-	EtOH	226(4.08)	33-0627-83
$C_{19}H_{26}N_2O_3S$			
Piperazine, 1-(3,4,5,6,7,8-hexahydro-3-phenyl-1,2-benzoxathiin-4-yl)-4-methyl-, S,S-dioxide, cis	EtOH	230s(3.54),305s(2.93), 328s(2.71)	4-0839-83
trans	EtOH	230s(3.06)	4-0839-83
$C_{19}H_{26}N_6O_6$			
1(2H)-Pyrimidinepropanamide, N,N'-(1,5-pentanediylbis[3,4-dihydro-2,4-dioxo-	H_2O	266(4.18)	103-1228-83
$C_{19}H_{26}O_2$			
Nimbiol methyl ether, (+)-	EtOH	207(4.18),232(4.15), 279(4.12)	162-0940-83
(-)-	EtOH	207(4.13),232(4.13), 279(4.11)	162-0940-83

Compound	Solvent	$\lambda_{max}(\log \epsilon)$	Ref.
$C_{19}H_{26}O_2S$			
Cyclopentanone, 3-[3-(phenylsulfinyl)-2-octenyl]-, (E)-	EtOH	222.5(4.10)	104-0835-83
$C_{19}H_{26}O_3S$			
Cyclopentanone, 3-[3-(phenylsulfonyl)-2-octenyl]-, (E)-	EtOH	260.5(2.73),267(2.81), 273.5(2.74)	104-0835-83
S-Petasin	EtOH	237(4.10),290(4.09)	102-1619-83
neo-	EtOH	253(4.12),290(4.12)	102-1619-83
$C_{19}H_{26}O_4$			
1H-Benz[e]indene-3,7(2H,3aH)-dione, 4,5,8,9,9a,9b-hexahydro-3a-methyl-6-[(2-methyl-1,3-dioxolan-2-yl)-methyl]-, [3aS-(3aα,9aα,9bβ)]-	EtOH	248(4.11)	39-2337-83C
4H-1,3-Benzodioxin-4-one, 2-methyl-2-[[5-methyl-2-(1-methylethyl)-cyclohexyl]oxy]-	pH 7.4	304(3.57)	1-0351-83
$C_{19}H_{26}O_5$			
1H-Benz[e]indene-3,7(2H,3aH)-dione, 4,5,8,9,9a,9b-hexahydro-5-hydroxy-3a-methyl-6-[(2-methyl-1,3-dioxolan-2-yl)methyl]-, [3aS-(3aα,5α,9aβ,9bβ)]-	EtOH	245(4.10)	39-2337-83C
Furanoeremophilan-9-one, 10α-hydroxy-6β-isobutyryloxy-	EtOH	281.5(4.15)	94-3544-83
10β-	EtOH	281(4.21)	94-3544-83
Propanedioic acid, [4-(2-hydroxy-3,5-dimethylphenyl)-2-butenyl]-, diethyl ester, (E)-	EtOH	218s(3.98),277(3.63)	78-1761-83
Propanedioic acid, [4-(2-hydroxy-5-methylphenyl)-3-methyl-2-butenyl]-, diethyl ester, (E)-	EtOH	220s(3.97),286(3.39)	78-1761-83
$C_{19}H_{26}O_5S$			
Dibenzo[d,g][1,6,3]dioxaselenocin-1(2H)-one, 8-acetoxy-3,4,8,9,10,11-hexahydro-2,2,10,10-tetramethyl-	$CHCl_3$	242(4.08),255s(4.06)	142-1959-83
$C_{19}H_{26}O_5Si$			
1-Benzoxepin-3-carboxylic acid, 7-[[[(1,1-dimethylethyl)dimethylsilyl]-oxy]methyl]-5-oxo-, methyl ester	EtOH	203.5(4.24),220(4.10), 251.5(3.80),280(3.65), 345(3.15)	39-1718-83C
$C_{19}H_{26}O_6$			
Cyclodeca[b]furan-2(3H)-one, 4,9-di-acetoxy-3a,4,5,8,9,11a-hexahydro-3,6,10-trimethyl-	EtOH	217(3.08)	102-0197-83
$C_{19}H_{27}ClO_2$			
Androst-4-en-3-one, 4-chloro-17β-hy-droxy- (clostebol)	EtOH	256(4.13)	162-0343-83
$C_{19}H_{27}N$			
1-Azuleneethanamine, N,N-diethyl-4,6,8-trimethyl-	EtOH	246(4.42),278s(4.40), 284s(4.57),288(4.66), 293(4.65),298s(4.61), 327s(3.46),338(3.58), 352(3.68),367s(3.18), 555s(2.57),564(2.63), 607s(2.53),667s(2.11)	18-2059-83

Compound	Solvent	λ_{max}(log ε)	Ref.
1-Azuleneethanamine, N,N-diethyl-4,6,8-trimethyl-, hydrochloride	EtOH	247(4.44),277s(4.38), 283s(4.55),288(4.63), 293(4.63),297s(4.60), 336(3.56),345s(3.57), 351(3.61),552(2.52), 592s(2.44),647s(2.03)	18-2059-83
5H-Cyclododec[b]indole, 6,7,8,9,10,11-12,13,14,15-decahydro-5-methyl-	MeOH	211s(4.08),231(4.42), 285(3.77)	39-0501-83C
$C_{19}H_{27}NO_2$			
4-A za-A-homoandrosta-1,5-dien-3-one, 17β-hydroxy-	EtOH	210(4.06)	111-0041-83
$C_{19}H_{27}NO_3$			
Tetrabenazine, hydrochloride	EtOH	230(3.89),284(3.58)	162-1314-83
$C_{19}H_{27}NO_6$			
Propanedioic acid, [2,2-dimethyl-1-(3-nitrophenyl)propyl]methyl-, diethyl ester	EtOH	266(3.91)	12-0081-83
$C_{19}H_{27}N_5O_9$			
1,2-Hydrazinedicarboxylic acid, 1,1'-[(dimethylamino)(3-methoxyphenyl)-ethenylidene]bis-, tetramethyl ester	EtOH	204(4.57),211(4.52), 282(3.89)	104-0110-83
$C_{19}H_{28}N$			
Pyrrolidinium, 1-(3,7,11-trimethyl-2,4,6,8,10-dodecapentaenylidene)-, (all-E)-, perchlorate	$CHCl_3$	500(4.62)	35-4033-83
$C_{19}H_{28}NO_2$			
Pyrrolidinium, 1-[7-(2-carboxycyclo-pentyl)-3-methyl-2,4,6-cycloocta-trienylidene]-, [1α(2E,4E,6E),2β]-, perchlorate	$CHCl_3$	410(4.52)	35-4033-83
	$CHCl_3$-Et_3N	420(--)	35-4033-83
$C_{19}H_{28}N_2O_2$			
Ethanone, 1-(4-cyclopentylphenyl)-2-[4-(2-hydroxyethyl)-1-pipera-zinyl]-	MeOH	257(4.23)	73-0642-83
$C_{19}H_{28}N_2O_3$			
2,2'-Pyrromethene-5'-carboxylic acid, 3,4-dihydro-5-methoxy-3,3,3',4'-tetramethyl-, 1,1-dimethylethyl ester	$CHCl_3$	261(4.00),273s(3.93), 292s(3.88),305s(3.99), 342(4.36)	49-0753-83
2,2'-Pyrromethene-5'-carboxylic acid, 1,3,4,5-tetrahydro-1,3,3,3',4'-penta-methyl-5-oxo-, 1,1-dimethylethyl ester, (E)-	$CHCl_3$	283(4.32),330s(3.81)	49-0753-83
(Z)-	$CHCl_3$	292(4.27)	49-0753-83
$C_{19}H_{28}N_2O_4$			
1-Pyrroline, 4,4-dimethyl-5-methoxy-3-hydroxy-2-(3,4-dimethyl-5-tert-but-oxycarbonyl-2-pyrrolylmethylene)-	$CHCl_3$	254(3.95),342(4.32)	49-0983-83
1-Pyrroline, 4,4-dimethyl-5-methoxy-3-hydroxy-2-(3,4-dimethyl-5-tert-but-oxycarbonyl-2-pyrrolylmethyl)-	$CHCl_3$	278(4.22)	49-0983-83
1-Pyrrolin-3-one, 4,4-dimethyl-5-meth-	$CHCl_3$	246(4.23),280s(3.77),	49-0983-83

Compound	Solvent	λ_{max}(log ε)	Ref.
oxy-2-(3,4-dimethyl-5-tert-butoxy-carbonyl-2-pyrrolylmethylene)- (cont.)		384s(4.37),402(4.68)	49-0983-83
$C_{19}H_{28}N_2O_4S$			
Benzene, 4-[(dodecenylthio)methyl]-1,2-dinitro-	MeOH	247(4.29),374(2.62)	44-0835-83
$C_{19}H_{28}N_4O_8S$			
1H-Imidazole-4-carboxylic acid, 5-[[[(ethoxycarbonyl)imino](methyl-thio)methyl]amino]-1-[2,3-O-(1-meth-ylethylidene)-β-D-ribofuranosyl]-, ethyl ester	MeOH	212(4.41),253(4.26), 267(4.27)	88-0931-83
$C_{19}H_{28}O_2$			
Cyclopropaneacetic acid, 2-(1,3,5,8-tetradecatetraenyl)-, [1α,2β(1E,3E-5Z,8Z)]-	EtOH	273(4.62),283(4.70), 294(4.60)	87-0072-83
D-Homo-18-nor-5α-androst-13(17a)-en-17-one, 3β-hydroxy-	EtOH	240(4.25)	44-1954-83
Normethandrone	EtOH	240(4.23)	162-0962-83
Testosterone	EtOH	238(4.20)	44-4766-83
$C_{19}H_{28}O_2Si$			
Cyclopentanone, 2,2,3-trimethyl-5-(phenylmethylene)-3-[[(trimethyl-silyl)oxy]methyl]-, (R)-	EtOH	295.5(4.38)	39-1465-83C
$C_{19}H_{28}O_3$			
Butanoic acid, 2-methyl-, 1,2,3,7,8,8a-hexahydro-7-methyl-8-(3-oxopropyl)-1-naphthalenyl ester, [1S-[1α(R*),7β,8β,8aβ]]-	MeOH	229(4.04),237(4.10), 245(3.91)	35-0593-83
$C_{19}H_{28}O_4Te$			
Propanedioic acid, ethyl[4-(phenyl-telluro)butyl]-, diethyl ester	MeOH	223(4.20),268.2(3.64), 323.7(2.74)	101-0171-83S
$C_{19}H_{28}O_5$			
Ethanone, 1-[3,4-dihydro-7-hydroxy-2,2-dimethyl-3,4-bis(1-methylethoxy)-2H-1-benzopyran-6-yl]-, cis	EtOH	287(4.17),329(4.02)	39-0883-83C
$C_{19}H_{29}NO$			
Procyclidine	EtOH	258.5(2.37)	162-1118-83
$C_{19}H_{29}NO_2$			
3-Aza-A-homoandrost-4a-en-3-one, 17β-hydroxy-	EtOH	222(4.22)	111-0041-83
4-Aza-A-homoandrost-1-en-3-one, 17β-hydroxy-, (5α)-	EtOH	214(4.06)	111-0041-83
$C_{19}H_{29}NO_2S$			
Benzene, 1-[(dodecenylthio)methyl]-4-nitro-	MeOH	271(4.04)	44-0835-83
$C_{19}H_{29}NO_3$			
Homocapsaicin	EtOH	231(3.89),282(3.50)	98-1326-83
$C_{19}H_{29}N_2$			
Piperidinium, 1-(9-piperidino-2,4,6,8-	MeOH	616(5.42)	35-0646-83

Compound	Solvent	$\lambda_{max}(\log \epsilon)$	Ref.
nonatetraenylidene)-, (all-E)-, perchlorate (cont.)			35-0646-83
$C_{19}H_{30}N_2O_4$			
1-Pyrrolin-3-one, 4,4-dimethyl-5-methoxy-2-(3,4-dimethyl-5-tert-butoxycarbonyl-2-pyrrolylmethyl)-	$CHCl_3$	244s(3.80),280(4.24)	49-0983-83
Pyrrolo[3,2-b]pyrrole-1,4-dicarboxylic acid, 3,6-bis(1,1-dimethylethyl)-3a,6a-dihydro-, ethyl methyl ester, cis	C_6H_{12}	224.5(4.36)	88-2275-83
$C_{19}H_{30}N_2O_6Si$			
4,11-Epoxy-8H-1,3-dioxolo[5,6]oxocino-[2,3-d]pyrimidin-8-one, 11a-[[[(1,1-dimethylethyl)dimethylsilyl]oxy]methyl]-3a,4,5,9,11,11a-hexahydro-2,2-dimethyl-, (3aα,4β,11β,11aα)-(±)-	MeOH	295(3.57)	18-2680-83
$C_{19}H_{30}O_3$			
Butanoic acid, 2-methyl-, 1,2,3,7,8,8a-hexahydro-8-(3-hydroxypropyl)-7-methyl-1-naphthalenyl ester	MeOH	229(4.35),236(4.40), 244(4.26)	35-0593-83
$C_{19}H_{31}N$			
Piperidine, 1-[[2,4-bis(1,1-dimethylethyl)-2,4-cyclopentadien-1-ylidene]methyl]-	hexane	220s(3.82),327(4.47)	89-0490-83
Piperidine, 1-[3,5-bis(1,1-dimethylethyl)phenyl]-	hexane	217(4.36),254(4.03), 285s(3.20)	89-0490-83
$C_{19}H_{31}NO_2$			
1H-Isoindole, 2-[2-(1,1-dimethylethoxy)-1-methylethoxy]-2,3-dihydro-1,1,3,3-tetramethyl-	EtOH	216(3.40),256(2.89), 263(2.99),270(3.00)	12-1957-83
1H-Isoindole, 2-[2-(1,1-dimethylethoxy)propoxy]-2,3-dihydro-1,1,3,3-tetramethyl-	EtOH	215(3.41),256(2.90), 263(3.02),270(3.03)	12-1957-83
$C_{19}H_{31}NO_6$			
Hexanoic acid, (5-methyl-6-oxo-2H-1,4-oxazin-3(6H)-ylidene)bis(methylene) ester	isoPrOH	320(2.08)	44-2989-83
$C_{19}H_{32}N_2O_5S$			
4(1H)-Pyrimidinone, 2,3-dihydro-5-[tetrahydro-3,4-dihydroxy-5-(hydroxymethyl)-1,5-dipentyl-2-furanyl]-2-thioxo-, (2α,3β,4β,5α)-(±)-	pH 1	217(3.80),275(4.00), 289(3.98)	18-2680-83
	pH 13	225(4.01),262(4.03)	18-2680-83
	MeOH	215(3.97),277(4.12), 290s(4.07)	18-2680-83
$C_{19}H_{32}N_2O_6$			
2,4(1H,3H)-Pyrimidinedione, 1-[tetrahydro-3,4-dihydroxy-5-(hydroxymethyl)-1,5-dipentyl-2-furanyl]-	pH 1	265(3.36)	18-2680-83
	pH 13	290(3.86)	18-2680-83
	MeOH	265(3.86)	18-2680-83
$C_{19}H_{32}N_2O_6SSi$			
4(1H)-Pyrimidinone, 5-[2-C-[[[(1,1-dimethylethyl)dimethylsilyl]oxy]methyl]-2,3-O-(1-methylethylidene)-β-DL-ribofuranosyl]-2,3-dihydro-2-thioxo-	pH 13	224(4.00),262(3.96), 296(3.83)	18-2680-83
	MeOH	216(3.96),275(4.07), 299(4.05)	18-2680-83

Compound	Solvent	$\lambda_{max}(\log \epsilon)$	Ref.
$C_{19}H_{32}N_2O_7Si$			
2,4(1H,3H)-Pyrimidinedione, 5-[2-C-[[[(1,1-dimethylethyl)dimethylsilyl]-oxy]methyl]-2,3-O-(1-methylethylidene)-β-DL-ribofuranosyl]-	pH 13	289(3.79)	18-2680-83
	MeOH	266(3.85)	18-2680-83
$C_{19}H_{32}N_4$			
2-Hexenenitrile, 3,3'-(1,3-propanediyldiimino)bis[4-ethyl-	EtOH	263(4.44)	39-1223-83C
$C_{19}H_{32}O_3$			
10-Octadecen-12-ynoic acid, 9-hydroxy-, methyl ester	isooctane	228(4.24),238(4.16)	162-0668-83
$C_{19}H_{32}O_4Si$			
1-Butanone, 1-[4-[[(1,1-dimethylethyl)-dimethylsilyl]oxy]-2,6-dihydroxy-3,5-dimethylphenyl]-2-methyl-, (±)-	EtOH	222s(4.13),288(4.19), 346(3.46)	88-2531-83
	EtOH-NaOH	400(3.57)	88-2531-83
$C_{19}H_{33}NO_4S_2Si$			
4,5-Dithia-1-azabicyclo[4.2.0]oct-2-ene-2-carboxylic acid, 7-[1-[[(1,1-dimethylethyl)dimethylsilyl]oxy]-ethyl]-3-methyl-8-oxo-, 1,1-dimethylethyl ester, [6α,7α(R*)]-	$CHCl_3$	281(3.72),335(3.45)	88-3283-83
$C_{19}H_{33}N_3O_5$			
4(1H)-Pyrimidinone, 2-amino-5-[tetrahydro-3,4-dihydroxy-5-(hydroxymethyl)-1,5-dipentyl-2-furanyl]-, monohydrochloride, (2α,3β,4β,5α)-(±)-	pH 1	265(3.57)	18-2680-83
	pH 13	230(3.83),278(3.59)	18-2680-83
	MeOH	263(3.72)	18-2680-83
$C_{19}H_{33}N_3O_6Si$			
4(1H)-Pyrimidinone, 2-amino-5-[2-C-[[[(1,1-dimethylethyl)dimethylsilyl]-oxy]methyl]-2,3-O-(1-methylethylidene)-β-DL-ribofuranosyl]-	pH 13	233(3.71),280(3.59)	18-2680-83
	MeOH	228(3.98),294(3.95)	18-2680-83
$C_{19}H_{34}N_4O_2$			
1-Piperidinyloxy, 4,4'-(methanetetrayldinitrilo)bis[2,2,6,6-tetramethyl-	MeCN	241(3.63),464(1.31)	70-1906-83
$C_{19}H_{36}N_2O_5$			
Neo-enactin A sulfate (2:1)	MeOH and MeOH-acid	211(3.77)	158-1399-83
	MeOH-NaOH	238(3.77)	158-1399-83
$C_{19}H_{36}S_4$			
1,2-Propadiene, 1,1,3,3-tetrakis[(1,1-dimethylethyl)thio]-	MeOH	269(3.81)	18-0171-83

Compound	Solvent	λ_{max}(log ϵ)	Ref.
$C_{20}Cl_{15}O$			
Methyl, bis(pentachlorophenyl)[2,3,5,6-tetrachloro-4-(chlorocarbonyl)phenyl]-	$CHCl_3$	290(3.87),335s(3.79), 370s(4.31),385(4.59), 480(3.11),510(3.10), 565(3.08)	44-3716-83
$C_{20}HCl_{15}O$			
Benzoyl chloride, 4-[bis(pentachlorophenyl)methyl]-2,3,5,6-tetrachloro-	$CHCl_3$	294(3.28),304(3.34)	88-2121-83
$C_{20}H_2Cl_{14}NO$			
Methyl, [4-(aminocarbonyl)-2,3,5,6-tetrachlorophenyl]bis(pentachlorophenyl)-	$CHCl_3$	287(3.81),337s(3.80), 368s(4.29),382(4.59), 480s(3.08),506(3.09), 560(3.06)	44-3716-83
$C_{20}H_3Cl_{14}NO$			
Benzamide, 4-[bis(pentachlorophenyl)methyl]-2,3,5,6-tetrachloro-	$CHCl_3$	293(3.25),303(3.28)	88-2121-83
$C_{20}H_4Cl_4I_4O_5$			
Rose Bengal (dianion)	MeOH	519(4.6),558(5.0)	35-7465-83
$C_{20}H_8Br_2O_6$			
[2,2'-Binaphthalene]-1,1',4,4'-tetrone, 3,3'-dibromo-5,5'-dihydroxy-	EtOH	288(4.33),440(3.91)	5-1020-83
	EtOH-NaOH	290(4.38),567(4.07)	5-1020-83
$C_{20}H_8I_4O_5$			
Erythrosin	pH 0-1	500(4.06)	104-0705-83
	pH 7-12	525(4.98)	104-0705-83
	14M H_2SO_4	462(4.65)	104-0705-83
$C_{20}H_8N_4$			
Propanedinitrile, 2,2'-(1,4-anthracenediylidene)bis-	MeCN	237(4.53),250(4.44), 339(4.26),387(4.35), 489(3.71),668(3.05)	138-1229-83
Propanedinitrile, 2,2'-(9,10-anthracenediylidene)bis-	MeCN	279(4.46),302(4.22), 342(4.40)	138-1229-83
$C_{20}H_9BrO_6$			
[2,2'-Binaphthalene]-1,1',4,4'-tetrone, 3-bromo-5,5'-dihydroxy-	EtOH	285(4.11),435(3.90)	5-1020-83
	EtOH-NaOH	288(4.28),555(4.03)	5-1020-83
[2,2'-Binaphthalene]-1,1',4,4'-tetrone, 3-bromo-5,8'-dihydroxy-	EtOH	287.5(4.13),433(3.92)	5-1020-83
	EtOH-NaOH	287.5(4.31),550(4.02)	5-1020-83
[2,2'-Binaphthalene]-1,1',4,4'-tetrone, 3-bromo-8,8'-dihydroxy-	EtOH	286(4.11),431(3.91)	5-1020-83
	EtOH-NaOH	289(4.30),546(3.97)	5-1020-83
$C_{20}H_9BrO_7$			
[2,2'-Binaphthalene]-1,1',4,4'-tetrone, 3-bromo-3',5,5'-trihydroxy-	MeOH	217(4.51),247s(4.26), 286(4.34),425(3.96)	5-1886-83
[2,2'-Binaphthalene]-1,1',4,4'-tetrone, 3-bromo-3',8,8'-trihydroxy-	MeOH	215(4.56),244(4.34), 286(4.34),422(3.94)	5-1886-83
$C_{20}H_9ClO_6$			
[2,2'-Binaphthalene]-1,1',4,4'-tetrone, 3-chloro-8,8'-dihydroxy-	EtOH	275s(4.10),435(4.84)	5-1020-83
	EtOH-NaOH	290(4.23),552(3.90)	5-1020-83
$C_{20}H_{10}ClNO_3$			
6H-Anthra[1,9-cd]isoxazol-6-one, 5-(4-chlorophenyl)-	EtOH	243(5.06),248(5.19), 442(4.14),459(4.14)	103-1057-83

Compound	Solvent	λ_{max}(log ε)	Ref.
$C_{20}H_{10}Cl_2N_2O_2$			
5H-8b,9-Diazadicyclohepta[b,l,m]fluorene-5,8(10H)-dione, 6,7-dichloro-	MeCN	263(4.43),290s(4.25), 400(3.95),414(3.97), 636(3.90),680(3.92), 750s(3.64)	24-1963-83
$C_{20}H_{10}O_4$			
[2,2'-Binaphthalene]-1,1',4,4'-tetrone	$CHCl_3$	248(4.43),251s(4.42), 269(4.36),342(3.79)	5-0299-83
$C_{20}H_{10}O_6$			
[2,2'-Binaphthalene]-1,1',4,4'-tetrone, 5,5'-dihydroxy-	$CHCl_3$	248(4.28),274(4.27), 339(3.34),440(3.94)	5-0299-83
[2,2'-Binaphthalene]-1,1',4,4'-tetrone, 5,8'-dihydroxy-	$CHCl_3$	250(4.24),275(4.23), 440(3.88)	5-1020-83
[2,2'-Binaphthalene]-1,1',4,4'-tetrone, 8,8'-dihydroxy-	$CHCl_3$	249(4.28),269(4.27), 335(3.42),438(3.92)	5-0299-83
$C_{20}H_{10}O_7$			
[2,2'-Binaphthalene]-1,1',4,4'-tetrone, 3,8,8'-trihydroxy-	MeOH	213(4.48),247(4.41), 284s(4.15),412(3.93)	5-1886-83
$C_{20}H_{10}O_8$			
[2,2'-Binaphthalene]-1,1',4,4'-tetrone, 3,3',5,5'-tetrahydroxy-	MeOH	210(4.47),226(4.43), 285(4.37),418(3.98)	5-1886-83
$C_{20}H_{11}Cl_4N_5O_8$			
Benzenamine, N-(3-chlorophenyl)-4-[2,2-dichloro-1-(2-chloro-3,5-dinitrophenyl)ethyl]-2,6-dinitro-	MeOH	209.7(4.63),390.6(3.72)	56-1357-83
Benzenamine, N-(4-chlorophenyl)-4-[2,2-dichloro-1-(2-chloro-3,5-dinitrophenyl)ethyl]-2,6-dinitro-	MeOH	207.5(4.57),401.3(3.54)	56-1357-83
$C_{20}H_{11}NO_3$			
6H-Anthra[1,9-cd]isoxazol-6-one, 5-phenoxy-	EtOH	243(5.03),248(5.07), 443(4.16),455(4.19)	103-1057-83
$C_{20}H_{11}N_3O_3$			
5H-Indolo[2,3-a]pyrrolo[3,4-c]carbazole-5,7(6H)-dione, 12,13-dihydro-2-hydroxy- (arcyriaflavin B)	MeOH	230(4.41),272(4.01), 282(4.05),324(4.49), 415(3.58)	88-1441-83
$C_{20}H_{12}$			
Benz[e]aceanthrylene	hexane	231(4.59),258(4.69), 265(4.71),292s(4.56), 316(4.14),331(4.10), 360s(4.00),375(4.10), 394(4.05)	44-1632-83
Benz[j]aceanthrylene	hexane	221(4.56),259(4.56), 261(4.56),278(4.43), 302(4.38),313(4.38), 360(3.72),379(3.85), 392(3.93),414(3.96)	44-1632-83
Benz[l]aceanthrylene	hexane	222(4.51),233(4.38), 254(4.43),284(4.51), 296(4.51),309(4.66), 321(4.21),358(3.66), 375(3.90),396(3.92), 411(3.52)	44-1632-83

Compound	Solvent	λ_{max}(log ε)	Ref.
Benz[k]acephenanthrylene	hexane	257(4.70),265s(4.61), 297(4.30),310(4.56), 340(4.07),352(4.00), 395(3.08),416(3.90)	44-1632-83
Dicycloocta[1,2,3,4-def:1',2',3',4'-jkl]biphenylene	C_6H_{12}	208(5.16),249(4.29), 260(4.27),271(4.42), 282(4.47),294s(3.94), 309(3.94),322(3.84), 338s(3.55),358(3.23), 386s(2.86),425(2.80), 456(2.82),484(2.74), 557(2.74),596(2.77), 648s(2.64),702s(2.17)	35-7191-83
$C_{20}H_{12}BrN_3O_4$			
9,10-Anthracenedione, 1-amino-2-bromo-4-[(3-nitrophenyl)amino]-	DMF	578(4.26),612s(4.19)	2-0808-83
$C_{20}H_{12}Br_2O_2$			
Ethanone, 1,1'-(1,8-pyrenediyl)bis[2-bromo-	EtOH	204(4.55),232(4.67), 243(4.61),288(4.44), 377(4.44)	44-2930-83
$C_{20}H_{12}ClNO_2$			
9,10[1',2']-Benzenoanthracene-1,4-dione, 2-amino-3-chloro-9,10-dihydro-	EtOH	241(4.16),256(3.83), 267(3.75),273(3.51), 315(3.56),413(3.45)	48-0353-83
$C_{20}H_{12}ClNO_4S_2$			
4H,5H-Pyrano[3,4-e][1,3]oxazine-4,5-dione, 7-[(4-chlorophenyl)thio]-2-[(phenylmethyl)thio]-	$CHCl_3$	246(4.17),320(4.01), 378(4.20)	24-3725-83
$C_{20}H_{12}Cl_2N_2O$			
1-Anthracenol, 4-[(2,3-dichlorophenyl)-azo]-	EtOH	485(4.44)	42-0408-83
	acetone	470(4.54)	42-0408-83
1-Anthracenol, 4-[(2,4-dichlorophenyl)-azo]-	EtOH	510(4.26)	42-0408-83
	acetone	505(4.30)	42-0408-83
1-Anthracenol, 4-[(2,6-dichlorophenyl)-azo]-	EtOH	495(4.57)	42-0408-83
	acetone	485(4.68)	42-0408-83
$C_{20}H_{12}Cl_2O_6$			
Anthra[1,2-b]furan-3-carboxylic acid, 7,10-dichloro-6,11-dihydro-5-hydroxy-2-methyl-6,11-dioxo-, ethyl ester	dioxan	259(4.50),423(3.91)	104-0533-83
$C_{20}H_{12}Cl_3N_5O_8$			
Benzenamine, 4-[2,2-dichloro-1-(2-chloro-3,5-dinitrophenyl)ethyl]-2,6-dinitro-N-phenyl-	MeOH	208.3(4.55),425.1(3.70)	56-1357-83
$C_{20}H_{12}Cl_3N_5O_9$			
Phenol, 2-[[4-[2,2-dichloro-1-(2-chloro-3,5-dinitrophenyl)ethyl]-2,6-dinitrophenyl]amino]-	MeOH	210.4(4.63),426.6(3.78)	56-1357-83
Phenol, 3-[[4-[2,2-dichloro-1-(2-chloro-3,5-dinitrophenyl)ethyl]-2,6-dinitrophenyl]amino]-	MeOH	210.1(4.63),423(3.70)	56-1357-83
Phenol, 4-[[4-[2,2-dichloro-1-(2-chloro-3,5-dinitrophenyl)ethyl]-2,6-dinitrophenyl]amino]-	MeOH	210.1(4.53),431.8(3.71)	56-1357-83

Compound	Solvent	λ_{max}(log ϵ)	Ref.
$C_{20}H_{12}Cs_2$			
Indeno[1,2-b]fluorene, 6,12-dihydro-, dicesium deriv.	n.s.g.	355(4.91),428(4.20)	35-2502-83
Indeno[2,1-b]fluorene, 10,12-dihydro-, dicesium deriv.	n.s.g.	343(4.45),391(4.56), 413(4.98),467(3.88), 499(3.70)	35-2502-83
$C_{20}H_{12}N_2O_2$			
Benzo[b]biphenylene-5,10-dicarbonitrile, 7,8-dimethoxy-	CH_2Cl_2	275(4.44),296s(4.06), 314s(4.30),321(4.33), 383s(3.65),401(4.05), 424(4.34),450(4.45)	39-1443-83C
Quino[2,3-b]acridine-7,14-dione, 5,12-dihydro-	DMSO	462(3.56),491(4.00), 525(4.20)	18-1775-83
	H_2SO_4	517(3.65),556(3.98), 597(4.06)	18-1775-83
$C_{20}H_{12}N_2O_6S$			
Benzonitrile, 4-[3-(5-nitro-2-furanyl)-3-oxo-2-(phenylsulfonyl)-1-propenyl]-	dioxan	213(4.55),220s(4.34), 286(4.51)	73-1057-83
$C_{20}H_{12}N_2O_7S$			
4H,5H-Pyrano[3,4-e]-1,3-oxazine-4,5-dione, 7-(4-nitrophenoxy)-2-[(phenylmethyl)thio]-	$CHCl_3$	274(4.34),340(4.15)	24-3725-83
$C_{20}H_{12}N_4S_2$			
Benzenesulfenamide, N,N'-(2,5-dicyano-2,5-cyclohexadiene-1,4-diylidene)bis-	$CHCl_3$	547(4.69)	150-0480-83M
Benzenesulfenamide, N,N'-(2,6-dicyano-2,5-cyclohexadiene-1,4-diylidene)bis-	$CHCl_3$	547(4.67)	150-0480-83M
$C_{20}H_{12}O$			
Benz[e]aceanthrylen-6(5H)-one	MeOH	229(4.79),251s(4.71), 268(4.81),272s(4.72), 294(3.89),308(4.29), 322(4.27),345s(3.56), 363(3.88),381(3.93), 411s(3.57)	44-1632-83
Benz[j]aceanthrylen-2(1H)-one	MeOH	212(4.65),234(4.49), 240s(4.49),273(4.55), 294s(4.62),299(4.64), 306s(4.61),385(3.73), 412(3.87)	44-1632-83
Benz[l]aceanthrylen-1(2H)-one	MeOH	228(4.52),250(4.49), 271(4.39),298(4.72), 380(3.80),396(3.82), 422(3.75)	44-1632-83
$C_{20}H_{12}O_2$			
9,10-Anthracenedione, 1-phenyl-	benzene	335(3.66)	104-1336-83
$C_{20}H_{12}O_3$			
3H-Naphtho[2,1-b]pyran-3-one, 2-benzoyl-	benzene	375(4.08)	151-0325-83A
$C_{20}H_{12}O_3S$			
5,10[3',4']-Furanoanthra[2,3-b]thiophene-13,15-dione, 5,10,12,16-tetrahydro-	EtOH	237(4.95),266(4.36), 271(4.37)	44-4419-83

Compound	Solvent	λ_{max}(log ε)	Ref.
$C_{20}H_{12}O_4$			
1,4-Naphthalenedione, 2-(2,3-dihydro-1,3-dioxo-1H-inden-2-yl)-3-methyl-	EtOH	251(4.48),258s(4.42), 313(3.64),325(3.51)	39-1753-83C
$C_{20}H_{12}O_5$			
[1,2]-Binaphthalene]-1',4'-dione, 4,5,5'-trihydroxy-	EtOH	228(4.4),250(3.9), 319(3.7),334(3.7), 418(--)	102-2579-83
Halenaquinone	MeCN	216(4.26),232s(4.22), 253(4.34),260s(4.31), 278(4.20),325s(3.78)	35-6177-83
5,12-Naphthacenedione, 8-acetyl-6,11-dihydroxy-	MeCN	267(3.57),454s(2.72), 478(2.87),510(2.77)	35-1608-83
$C_{20}H_{12}S$			
Anthra[1,2-b]benzo[d]thiophene	hexane	252(3.45),251(3.44), 283(3.66),294(3.91)	4-0861-83
Anthra[2,1-b]benzo[d]thiophene	hexane	221(3.59),249s(3.64), 256(3.73),284(3.76), 294s(3.68)	4-0861-83
Anthra[2,3-b]benzo[d]thiophene	hexane	248(3.54),281s(3.66), 292(3.88)	4-0861-83
Benzo[b]phenanthro[2,1-d]thiophene	hexane	258(3.59),266s(3.52), 276(3.49),283s(3.51), 292(3.60)	4-0861-83
Benzo[b]phenanthro[2,3-d]thiophene	hexane	220(3.78),240(3.82), 264s(3.59),294(3.56)	4-0861-83
Benzo[b]phenanthro[3,2-d]thiophene	hexane	250(3.51),273s(3.52), 284(3.66),294(3.82), 304(3.68)	4-0861-83
Benzo[b]phenanthro[4,3-d]thiophene	hexane	218(3.92),237(3.80), 281(3.64),298(3.69)	4-0861-83
$C_{20}H_{13}ClN_2O$			
1-Anthracenol, 4-[(2-chlorophenyl)azo]-	EtOH	495(4.78)	42-0408-83
	acetone	480(4.78)	42-0408-83
1-Anthracenol, 4-[(3-chlorophenyl)azo]-	EtOH	505(4.68)	42-0408-83
	acetone	490(4.72)	42-0408-83
1-Anthracenol, 4-[(4-chlorophenyl)azo]-	EtOH	515(4.60)	42-0408-83
	acetone	505(4.61)	42-0408-83
$C_{20}H_{13}ClN_2O_2$			
9,10-Anthracenedione, 1-amino-4-[(4-chlorophenyl)amino]-	DMF	611(4.28)	2-0808-83
$C_{20}H_{13}Cl_2N_5O$			
3-Pyridinecarboxamide N-[6-(2,6-dichlorophenyl)-2-methylpyrido[2,3-d]pyrimidin-7-yl]-	pH 1	256(4.17),297(4.20) (changing)	87-0403-83
	MeOH-pH 7	222(4.72),240s(4.51), 322(4.14),360(3.51), 380(3.38)	87-0403-83
	MeOH-NaOH	348(4.20)	87-0403-83
$C_{20}H_{13}Cs$			
Indeno[1,2-b]fluorene, monocesium salt	n.s.g.	370(4.08),392(4.38), 482(2.90),524(3.00), 562(2.90)	35-2502-83
Indeno[2,1-b]fluorene, monocesium salt	n.s.g.	331(4.61),416(4.30), 472(3.34),500(3.45), 533(3.26)	35-2502-83

Compound	Solvent	λ_{max}(log ϵ)	Ref.
$C_{20}H_{13}NO_2$			
9,10-Anthracenedione, 1-(phenylamino)-	C_6H_5Cl	330(4.0),510(3.9)	104-0139-83
$C_{20}H_{13}N_3O$			
5H-Indolo[2,3-a]pyrrolo[3,4-c]carbazol-5-one, 6,7,12,13-tetrahydro-	MeOH	237(3.73),291(4.12), 321(3.49),335(3.58), 346(3.48),361(3.31)	142-0469-83
$C_{20}H_{13}N_3O_2$			
5H-Pyrido[1',2':1,2]pyrimido[4,5-b]-acridine-7,15-dione, 2-methyl-	DMSO	464(3.66),492(3.90), 525(3.93)	18-1775-83
	H_2SO_4	460(3.42),488(3.66), 520(3.66)	18-1775-83
5H-Pyrido[1',2':1,2]pyrimido[4,5-b]-acridine-7,15-dione, 4-methyl-	DMSO	458(3.67),485(3.91), 517(3.91)	18-1775-83
	H_2SO_4	460(3.66),487(3.86), 520(3.83)	18-1775-83
5H-Pyrido[1',2':1,2]pyrimido[4,5-b]-acridine-7,15-dione, 9-methyl-	DMSO	460(3.59),487(3.69), 519(3.64)	18-1775-83
	H_2SO_4	458(3.53),486(3.73), 519(3.71)	18-1775-83
5H-Pyrido[1',2':1,2]pyrimido[4,5-b]-acridine-7,15-dione, 10-methyl-	DMSO	460(3.72),487(3.94), 520(3.94)	18-1775-83
	H_2SO_4	455(3.49),489(3.79), 522(3.79)	18-1775-83
5H-Pyrido[1',2':1,2]pyrimido[4,5-b]-acridine-7,15-dione, 11-methyl-	DMSO	462(3.63),489(3.86), 522(3.87)	18-1775-83
	H_2SO_4	455(3.69),485(3.82), 517(3.79)	18-1775-83
5H-Pyrido[1',2':1,2]pyrimido[4,5-b]-acridine-7,15-dione, 12-methyl-	DMSO	459(3.67),486(3.91), 517(3.93)	18-1775-83
	H_2SO_4	455(3.58),486(3.71), 518(3.68)	18-1775-83
2,3,9b-Triazaindeno[6,7,1-ija]azulene-8-carboxylic acid, 1-phenyl-, methyl ester	EtOH	235(4.27),263(4.23), 292(4.49),319(3.93), 333(3.95),373(4.21), 388(4.23),423(3.97), 446(4.02)	18-3703-83
$C_{20}H_{13}N_3O_3$			
1-Anthracenol, 4-[(2-nitrophenyl)azo]-	EtOH	495(4.10)	42-0408-83
	acetone	490(4.35)	42-0408-83
5H-Pyrido[1',2':1,2]pyrimido[4,5-b]-acridine-7,15-dione, 4-methoxy-	DMSO	460(3.70),488(3.91), 520(3.91)	18-1775-83
	H_2SO_4	455(3.52),482(3.74), 513(3.77)	18-1775-83
$C_{20}H_{13}N_3O_4$			
9,10-Anthracenedione, 1-amino-4-[(3-nitrophenyl)amino]-	DMF	571(4.23),604(4.17)	2-0808-83
$C_{20}H_{13}N_3O_6$			
4-Benzoxazolecarboxylic acid, 2-[3-hydroxy-2-[[(3-hydroxy-2-pyridinyl)-carbonyl]amino]phenyl]-	dioxan and dioxan-acid	250s(4.26),313(4.45), 322s(3.45)	158-83-143
	dioxan-base	254(4.59),310(4.23), 360(4.04),400(4.02)	158-83-143
$C_{20}H_{13}N_5O_3$			
Acetamide, N-[2-(1H-benzotriazol-1-yl)-2,3-dihydro-1,3-dioxo-1H-benz[de]-isoquinolin-5-yl]-	dioxan	256(4.57),340(4.09), 384(3.82)	56-0817-83

Compound	Solvent	λ_{max}(log ϵ)	Ref.
Acetamide, N-[2-(1H-benzotriazol-1-yl)-2,3-dihydro-1,3-dioxo-1H-benz[de]-isoquinolin-6-yl]-	dioxan	246(4.48),372(4.20)	56-0817-83
Acetamide, N-[2-(2H-benzotriazol-2-yl)-2,3-dihydro-1,3-dioxo-1H-benz[de]-isoquinolin-5-yl]-	dioxan	240(4.35),282(4.20), 335(4.16),384(4.68)	56-0817-83
Acetamide, N-[2-(2H-benzotriazol-2-yl)-2,3-dihydro-1,3-dioxo-1H-benz[de]-isoquinolin-6-yl]-	dioxan	241.5(4.35),288(4.18), 372.5(4.21)	56-0817-83
$C_{20}H_{14}BrNO_2S$			
1H-Indole, 3-bromo-2-phenyl-1-(phenyl-sulfonyl)-	EtOH	218s(4.46),264s(4.25), 268(4.28),275(4.29), 280s(4.25)	39-2417-83C
$C_{20}H_{14}ClNOS$			
2-Propen-1-one, 1-(4-chlorophenyl)-3-(1,2-dihydro-1-phenyl-2-thioxo-3-pyridinyl)-, (E)-	EtOH	277(4.36),345(4.00), 432(3.73)	4-1651-83
$C_{20}H_{14}ClNO_2S$			
1H-Indole, 3-chloro-2-phenyl-1-(phenyl-sulfonyl)-	EtOH	219(4.45),250s(4.13), 295s(3.88)	39-2417-83C
$C_{20}H_{14}ClN_3O_2$			
9,10-Anthracenedione, 1-amino-2-[(4-ni-trophenyl)amino]-8-chloro-	DMF	566(4.06)	2-0808-83
$C_{20}H_{14}ClN_3S_2$			
4-Thiazolamine, N-[(2-chloro-3-quino-linyl)methylene]-2-(methylthio)-5-phenyl-	DMF	305(3.37),400(3.06)	103-0166-83
$C_{20}H_{14}Cl_2$			
5,12-Ethanonapthacene, 13,14-dichloro-5,12-dihydro-, anti-cis	MeCN	213(4.53),232(4.88), 262s(3.79),270(3.88), 278(3.81),290(3.60), 307(2.72),317(2.34), 321(2.64)	35-7337-83
syn-cis	MeCN	197(4.69),214(4.68), 232(4.86),257s(3.83), 266s(3.94),273(3.89), 275(3.88),287(3.67), 306(2.59),315(2.30), 319(2.40)	35-7337-83
trans	MeCN	215(4.66),233(4.86), 257s(3.81),266s(3.92), 271(3.86),277(3.86), 288(3.61),307(2.58), 316(2.26),320(2.38)	35-7337-83
$C_{20}H_{14}Cl_2N_2O_2$			
1H,7H-Pyrazolo[1,2-a]pyrazole-1,7-di-one, 3,5-bis(4-chlorophenyl)-2,6-dimethyl-	dioxan	220(4.30),269(4.25), 358(3.74)	151-0171-83A
$C_{20}H_{14}Cl_2O_4S$			
5,10-Methanocycloundeca[3,4]cyclopenta-[1,2-b]thiophene-13-carboxylic acid, 2,3-dichloro-, ethyl ester, 1,1-di-oxide	MeCN	670(3.00)	88-2151-83

Compound	Solvent	λ_{max}(log ε)	Ref.
$C_{20}H_{14}NO_4$			
Sanguinarine	pH 1	273(4.52),325(4.23)	149-0245-83B
	pH 3.5	273(4.52),325(4.23)	149-0245-83B
	pH 4.3	273(4.52),325(4.23)	149-0245-83B
	pH 5.2	273(4.50),325(4.22)	149-0245-83B
	pH 6.1	273(4.48),325(4.21)	149-0245-83B
	pH 7.0	273(4.47),325(4.18)	149-0245-83B
	H_2O	273(4.47),325(4.18)	149-0245-83B
	pH 7.1	273(4.46),325(4.17)	149-0245-83B
	pH 7.4	273(4.42),325(4.10)	149-0245-83B
	pH 7.6	275(4.37),325(4.03)	149-0245-83B
	pH 8.2	281(4.24),325(3.84)	149-0245-83B
	pH 8.4	283(4.24),325(3.83)	149-0245-83B
	pH 8.7	283(4.24),325(3.80)	149-0245-83B
	pH 9.0	283(4.24),325(3.80)	149-0245-83B
	pH 10.8	283(4.24),325(3.80)	149-0245-83B
	pH 13.0	283(4.24),325(3.80)	149-0245-83B
chloride dihydrate	MeOH	234(4.50),283(4.52), 325(4.18)	162-1202-83
$C_{20}H_{14}N_2$			
Benzaldehyde, 2-ethynyl-, [3-(2-ethynylphenyl)-2-propenylidene]hydrazone	THF	236(4.27),246(4.21), 253s(4.20),265s(4.02), 322s(4.43),336(4.48), 348s(4.45),360s(4.28)	18-1467-83
Cyclohept[1,2,3-hi]imidazo[2,1,5-cd]-indolizine, 2-(4-methylphenyl)-	EtOH	242(4.71),260s(4.65), 271s(4.60),280s(4.69), 290(4.77),300(4.77), 323(4.15),359(4.59), 375(4.66),415(4.26), 438(4.36)	18-3703-83
$C_{20}H_{14}N_2O$			
2(1H)-Quinoxalinone, 3-[2-(1-naphthalenyl)ethenyl]-	EtOH	396(4.45)	18-0326-83
$C_{20}H_{14}N_2O_2$			
9,10-Anthracenedione, 1-amino-4-(phenylamino)-	DMF	574(3.99),615(4.15)	2-0808-83
1H-Isoindole-1,3(2H)-dione, 2-[4-(2-pyridinylmethyl)phenyl]-	EtOH	295(3.36),380(1.96)	103-0970-83
Indolo[2,3-a]carbazole-5-carboxylic acid, 11,12-dihydro-, methyl ester	MeOH	208(3.25),218(3.26), 225(3.30),228(3.31), 288(3.12),325(2.59), 340(2.45),357(2.31)	142-0469-83
$C_{20}H_{14}N_2O_2S_2$			
6,13(7H,14H)-Triphenodithiazinedione, 3,10-dimethyl-	H_2SO_4	335(4.79),749(4.81)	18-1482-83
$C_{20}H_{14}N_2O_3$			
Pyrano[2,3-c]pyrazol-4(1H)-one, 5-acetyl-1,3-diphenyl-	EtOH	254(4.37),315(4.13)	44-4078-83
Pyrano[2,3-c]pyrazol-4(1H)-one, 5-benzoyl-3-methyl-1-phenyl-	EtOH	248(4.37),328(3.73)	44-4078-83
$C_{20}H_{14}N_2O_4S$			
1H-Indole, 2-(2-nitrophenyl)-1-(phenylsulfonyl)-	EtOH	212(4.46),254(4.30), 318s(3.32)	39-2417-83C

Compound	Solvent	$\lambda_{max}(\log \epsilon)$	Ref.
$C_{20}H_{14}N_2O_4S_2$			
6,13(7H,14H)-Triphenodithiazinedione, 3,10-dimethoxy-	H_2SO_4	331(4.81),773(4.72)	18-1482-83
$C_{20}H_{14}N_2O_5S$			
1-Naphthalenesulfonic acid, 3-hydroxy-4-[(2-hydroxy-1-naphthalenyl)azo]- (also metal complexes)	dianion	641(4.45)	97-0105B-83
$C_{20}H_{14}N_2O_6$			
Quino[7,8-h]quinoline-2,11-dicarboxylic acid, 1,4,9,12-tetrahydro-4,9-dioxo-, dimethyl ester	MeOH	216(4.57),221s(--), 235(4.31),249(4.29), 271s(--),278s(--), 307s(--),322(4.10), 330s(--),374(4.03), 393(4.12)	18-2338-83
$C_{20}H_{14}N_2O_7S_2$			
1-Naphthalenesulfonic acid, 4-hydroxy-3-[(4-sulfo-1-naphthalenyl)azo]-, disodium salt (Azorubin S)	pH 7.5	507(4.22)	94-0162-83
$C_{20}H_{14}N_4$			
1,4-Ethanonaphthalene-6,6,7,7-tetracarbonitrile, 1,2,3,4,5,8-hexahydro-2,3,9,10-tetrakis(methylene)-	MeCN	218(4.13),226(4.11), 235(4.06),250(3.97), 257(3.95),269s(3.64)	33-0019-83
$C_{20}H_{14}N_4O$			
Acetamide, N-[4-cyano-3-[(2-cyanophenyl)methyl]-1-isoquinolinyl]-	EtOH	240(4.55),279s(3.84), 285(3.91),294(3.94), 306(4.03),335(4.03), 349s(3.88)	39-1137-83C
$C_{20}H_{14}N_4O_2S_3$			
Benzenesulfenamide, N,N'-(3,4-dicyano-2,5-thiophenediylidene)bis[4-methoxy-, (Z,Z)-	$CHCl_3$	555(4.29)	150-0028-83S +150-0480-83M
$C_{20}H_{14}N_6O_3$			
Ethanone, 1-[3,5-bis(2H-benzotriazol-2-yl)-2,4-dihydroxyphenyl]-	$CHCl_3$	252(4.37),273(4.49), 322(4.47),343s(4.37)	121-0309-83B
$C_{20}H_{14}N_6S_2$			
2-Thiazolamine, N-[2,3-dihydro-3-phenyl-2-(phenylimino)-4H-thiazolo[3,2-a]-1,3,5-triazin-4-ylidene]-	dioxan	274(4.39),314s(4.25)	48-0463-83
$C_{20}H_{14}O_2$			
4H-Cyclopenta[b]furan-4-one, 2-methyl-5,6-diphenyl-	CH_2Cl_2	267(4.47),470(3.66)	70-1049-83
Oxirane, 2,2'-(1,6-pyrenediyl)bis-	EtOH	201(4.37),233(4.51), 242(4.77),256s(3.87), 266(4.35),277(4.65), 317(4.00),332(4.45), 349(4.62)	44-2930-83
Oxirane, 2,2'-(1,8-pyrenediyl)bis-	EtOH	202(4.53),234(4.65), 242(4.84),256s(4.14), 266(4.46),277(4.71), 304s(3.79),318(4.11), 333(4.50),349(4.63)	44-2930-83

Compound	Solvent	λ_{max}(log ϵ)	Ref.
$C_{20}H_{14}O_3$			
1,4-Methanonaphth[2,3-a]azulene-5,12-dione, 1,4-dihydro-11-methoxy-, (1R)-	MeCN	264(4.4),284(4.4), 311(4.5),399(3.8), 496(3.7),693(3.3)	24-2408-83
1H-Phenalen-1-one, 6-hydroxy-9-(4-methoxyphenyl)-	$CHCl_3$	277(4.28),445(4.03)	39-1267-83C
$C_{20}H_{14}O_5$			
5,12-Naphthacenedione, 8-acetyl-9,10-dihydro-6,11-dihydroxy-	MeCN	282(4.57),298(4.53), 493(4.05)	44-2820-83
$C_{20}H_{14}O_6$			
Anthra[1,2-b]furan-3-carboxylic acid, 6,11-dihydro-5-hydroxy-2-methyl-6,11-dioxo-, ethyl ester	dioxan	257(4.63),426(3.90)	104-0533-83
$C_{20}H_{14}O_7$			
Anthra[2,3-b]furan-3-carboxylic acid, 5,10-dihydro-4,11-dihydroxy-2-methyl-5,10-dioxo-, ethyl ester	dioxan	260(4.66),480(4.07)	104-1900-83
$C_{20}H_{15}BrO_5$			
5,12-Naphthacenedione, 8-acetyl-8-bromo-7,8,9,10-tetrahydro-6,11-dihydroxy-	MeCN	249(4.55),287(3.96), 320(3.36),449(4.01), 471(4.04),500(3.85)	44-2820-83
$C_{20}H_{15}ClN_2O_2$			
Benzenemethanamine, N-(4-chlorophenyl)-α-(nitrophenylmethylene)-, (Z)-	EtOH	388(4.18)	22-0339-83
$C_{20}H_{15}ClN_2O_2S$			
1H-Indole, 2-(2-aminophenyl)-3-chloro-1-(phenylsulfonyl)-	EtOH	219(4.48),234s(4.34), 256s(4.19),290s(3.89)	39-2417-83C
$C_{20}H_{15}ClN_2O_4$			
1H-Naphth[2,3-g]indole-3-carboxylic acid, 5-amino-4-chloro-6,11-dihydro-2-methyl-6,11-dioxo-, ethyl ester	dioxan	278(4.54),446(3.68), 550(3.98)	104-1892-83
$C_{20}H_{15}ClN_2O_5S$			
Benzenesulfonic acid, 4-chloro-, 2-benzoyl-9-oxo-1,2-diazabicyclo-[5.2.0]nona-3,5-dien-8-yl ester, cis	MeOH	289(3.88)	5-1361-83
$C_{20}H_{15}Cl_2NO_4$			
3(2H)-Dibenzofuranone, 7,9-dichloro-1,4-dihydro-8-hydroxy-4-[(4-methoxyphenylamino)methylene]-, (Z)-	CH_2Cl_2	406(4.35)	24-0152-83
$C_{20}H_{15}Cl_2N_3O_3$			
Pyrazino[1,2-a][1,4]benzodiazepine-4-carboxylic acid, 9-chloro-7-(2-chlorophenyl)-1,2,3,5-tetrahydro-2-oxo-, methyl ester	isoPrOH	217(4.53),250s(4.16), 341(4.35)	4-0791-83
$C_{20}H_{15}Cs$			
Cesium, [9-(phenylmethyl)-9H-fluoren-9-yl]-	$(MeOCH_2)_2$	372.5(4.20),471(3.18), 500(3.28),534(3.10)	104-1592-83
$C_{20}H_{15}I$			
1,1'-Biphenyl, 4-iodo-4'-(2-phenyleth-	dioxan	328(4.77)	104-0683-83

Compound	Solvent	λ_{max}(log ε)	Ref.
enyl)- (cont.)			104-0683-83
$C_{20}H_{15}Li$			
Lithium, [9-(phenylmethyl)-9H-fluoren-9-yl]-	$(MeOCH_2)_2$	379(4.15),506(3.21), 544(3.10)	104-1592-83
$C_{20}H_{15}NO$			
3H-Indol-3-ol, 2,3-diphenyl-	CH_2Cl_2	245(4.18),320(4.08)	24-2115-83
$C_{20}H_{15}NOS$			
2-Propen-1-one, 3-(1,2-dihydro-1-phenyl-2-thioxo-3-pyridinyl)-1-phenyl-, (E)-	EtOH	246(4.25),274(4.32), 343(3.96),427(3.71)	4-1651-83
$C_{20}H_{15}NO_2$			
1H-Benz[f]isoindole-4,9-dione, 1,1-dimethyl-3-phenyl-	EtOH	231(4.31),249(4.30), 255s(4.28),267(4.19), 273s(4.15),340(3.49)	33-2252-83
2-Propen-1-one, 3-(1,2-dihydro-2-oxo-1-phenyl-3-pyridinyl)-1-phenyl-, (E)-	EtOH	224(4.05),267(3.90), 372(4.02)	4-1651-83
Pyridine, 1-benzoyl-1,4-dihydro-4-(2-oxo-2-phenylethylidene)-	MeOH	235(4.20),420(4.33), 439(4.32)	24-1506-83 +24-3192-83
Pyridinium, 1-benzoyl-2-oxo-2-phenylethylide	H_2O	375(--)	65-2073-83
	85% EtOH	388(4.25)	65-2073-83
	$CHCl_3$	425(3.3)	65-2073-83
$C_{20}H_{15}NO_4S$			
2H-1-Benzopyran-2-one, 3-[2-(3,4-dimethoxyphenyl)-4-thiazolyl]-	MeCN	352(4.32)	48-0551-83
$C_{20}H_{15}NO_5$			
1H-Naphth[2,3-g]indole-3-carboxylic acid, 6,11-dihydro-5-hydroxy-2-methyl-6,11-dioxo-, ethyl ester	dioxan	294(4.38),455(3.94)	104-1892-83
2(3H)-Oxazolone, 3-acetyl-5-[(10-methoxy-9-phenanthrenyl)oxy]-	MeOH	211(3.49),226(3.52), 251(3.67),257(3.73), 274(3.10),280(2.98), 293(2.93),305(2.93)	73-2812-83
$C_{20}H_{15}NO_6$			
Corydione	EtOH	223s(4.48),241(4.59), 246(4.59),286s(4.04), 301(4.14),314(4.30), 326(4.42),466(4.04)	100-0761-83
Crebanine, dehydro-4,5-dioxo-	EtOH	220(4.54),244.5(4.62), 308(4.21),321(4.25), 435(4.21)	100-0761-83
Nor-α-hydrastine, dehydro-	MeOH	242(4.26),246(4.24), 293(3.70),318(3.79), 331(3.78)	44-4879-83
$C_{20}H_{15}NO_6S$			
2-Propen-1-one, 3-(4-methylphenyl)-1-(5-nitro-2-furanyl)-2-(phenylsulfonyl)-, (Z)-	dioxan	212(4.43),224(4.37), 294(4.50)	73-1057-83
$C_{20}H_{15}NO_7S$			
2-Propen-1-one, 3-(4-methoxyphenyl)-1-(5-nitro-2-furanyl)-2-(phenylsulfonyl)-, (Z)-	dioxan	210(4.41),228(4.37), 309(4.55)	73-1057-83

Compound	Solvent	λ_{max}(log ε)	Ref.
$C_{20}H_{15}N_3$			
9-Acridinecarboxaldehyde, phenylhydrazone	hexane	440(4.28)	140-0215-83
	EtOH	455(--)	140-0215-83
trifluoroacetate	EtOH	562(4.46)	140-0215-83
$C_{20}H_{15}N_3O$			
2H-Pyrrol-2-one, 1,5-dihydro-3,4-di-1H-indol-3-yl-	MeOH	223(3.97),277(3.63), 286(3.61),338(3.51), 353(3.41)	142-0469-83
$C_{20}H_{15}N_3O_2$			
9,10-Anthracenedione, 1-amino-4-[(4-aminophenyl)amino]-	DMF	580s(4.11),617(4.19)	2-0808-83
$C_{20}H_{15}N_3O_3$			
1H-Indole-3-acetamide, N-[2-(1H-indol-3-yl)-2-oxoethyl]-α-oxo-	MeOH	214(4.40),243(4.20), 258(4.20),275(4.14), 303(4.19)	142-0469-83
$C_{20}H_{15}N_3O_3S$			
7H-Benzimidazo[2,1-a]benz[de]isoquinoline-10-sulfonamide, N-ethyl-7-oxo-	benzene	385(4.12)	73-2249-83
	EtOH	382(4.18)	73-2249-83
$C_{20}H_{15}N_3O_8S_2$			
2-Naphthalenesulfonic acid, 6-amino-4-hydroxy-5-[(8-hydroxy-6-sulfo-2-naphthalenyl)azo]-	pH 7.00	500(4.34)	33-2002-83
$C_{20}H_{15}N_5O_3$			
Benzoic acid [[(4-nitrophenyl)azo]phenylmethylene]hydrazide	isooctane	243(4.25),282s(4.33), 426(3.78)	104-2104-83
	EtOH	240(4.22),280s(4.26), 303(4.31),410(3.61)	104-2104-83
	dioxan	286s(4.36),303(4.30), 418(3.93)	104-2104-83
	MeCN	280(4.29),302(4.32), 417(3.70)	104-2104-83
	CCl_4	313(4.33),446(3.66)	104-2104-83
Benzoic acid, 4-nitro-, [phenyl(phenylazo)methylene]hydrazide	isooctane	258(4.30),312(4.15), 415(3.86)	104-2104-83
	EtOH	262(4.35),302(4.27), 390(3.95)	104-2104-83
	dioxan	262(4.31),306(4.16), 395(3.91)	104-2104-83
	MeCN	262(4.31),304(4.24), 390(4.00)	104-2104-83
	CCl_4	313(4.20),403(3.91)	104-2104-83
$C_{20}H_{15}N_5S$			
2(3H)-Benzothiazolone, [phenyl(phenylazo)methylene]hydrazone	benzene	460(4.04)	65-2332-83
	EtOH	460(3.95)	65-2332-83
$C_{20}H_{16}$			
Benzo[c]phenanthrene, 3,10-dimethyl-	$CHCl_3$	268s(4.42),276(4.74), 286(4.91),306(4.02), 318(4.00),331s(3.61)	5-2262-83
1,3,9,11-Cyclotridecatetraene-5,7-diyne, 13-(2,4-cyclopentadien-1-ylidene)-4,9-dimethyl-, (E,E,Z,Z)-	THF	260(4.3),270(4.2), 310(4.2),405(4.4)	88-5273-83
Phenylstilbene	C_6H_{12}	325(4.51)	104-0683-83

Compound	Solvent	$\lambda_{max}(\log \epsilon)$	Ref.
$C_{20}H_{16}BrNO$			
Propenal, 3-(1-hydroxy-4-methyl-2-naphthalenyl)-, N-(4-bromophenyl)imine	acetone	333(3.76),340(3.72), 358(3.65),385(3.08), 505(3.04)	103-0825-83
	dioxan	290(4.03),325(3.81), 340(3.75),358(3.76), 380(3.23),500(2.70)	103-0825-83
	MeCN	325(3.78),340(3.72), 357(3.65),375(3.15), 500(3.08)	103-0825-83
	CCl_4	313(3.80),327(3.70), 342(3.65),357(3.61)	103-0825-83
	DMSO	335(3.92),357(3.82), 387(3.71),530(3.38), 555(3.48),580(3.36)	103-0825-83
	+ NaOMe	515(4.44)	103-0825-83
$C_{20}H_{16}Br_2Cl_2$			
2,4,6-Octatriene, 4,5-dibromo-3,6-bis(4-chlorophenyl)-	C_6H_{12}	247.9(4.33),297s(3.43)	19-0233-83
$C_{20}H_{16}ClNO_3$			
2H-Pyrano[3,2-d]benzoxepin-2-one, 3-chloro-5,6-dihydro-4-(methylphenylamino)-	EtOH	215(4.30),243(4.19), 270s(3.82),356(4.16)	4-0539-83
$C_{20}H_{16}Cl_2$			
1,2[1',2']-Benzenoanthracene, 9,10-dichloro-1,2,11,12,13,14-hexahydro-	MeCN	260(4.60),304(4.25), 317(4.18),344(2.72), 362(2.58)	35-0545-83
1,4[1',4']-Benzenoanthracene, 9,10-dichloro-1,4,11,12,15,16-hexahydro-	C_6H_{12}	292(3.82),301(3.71), 326(2.59)	35-0545-83
9,10[1',2']-Benzenoanthracene, 9,10-dichloro-1,2,4a,9,9a,10-hexahydro-	C_6H_{12}	261(2.78),270(2.65)	35-0545-83
9,10[1',4']-Benzenoanthracene, 9,10-dichloro-9,10,11,12,13,14-hexahydro-	C_6H_{12}	220s(4.29),270(2.72), 278(2.79)	35-0545-83
1,3,5,11,13,15-Cycloheptadecahexaene-7,9-diyne, 17-(dichloromethylene)-6,11-dimethyl-, (E,E,Z,Z,E,E)-	CH_2Cl_2	291s(4.47),315(4.59), 390(4.04)	39-2997-83C
2,6-Octadien-4-yne, 3,6-bis(4-chlorophenyl)-	C_6H_{12}	213s(4.43),221s(4.34), 231.2(4.25),248.2(4.47), 259s(4.41),277s(4.17), 288s(4.00),373.3(2.55), 386s(2.50)	19-0233-83
$C_{20}H_{16}HgN_2O_3$			
Mercury, [2-[(2-hydroxy-5-methylphenyl)azo]benzoato]phenyl-	MeOH	250(4.16),325(4.33), 390(3.97)	90-0493-83
$C_{20}H_{16}NO_3S$			
1,3-Oxathiol-1-ium, 2-[4-(dimethylamino)phenyl]-5-(2-oxo-2H-1-benzopyran-3-yl)-, bromide	MeCN	497(4.84)	48-0551-83
$C_{20}H_{16}NO_5$			
1-Benzopyrylium, 2-(7,8-dihydroxy-2-oxo-2H-1-benzopyran-3-yl)-7-(dimethylamino)-, perchlorate	MeCN	569(4.52)	48-0505-83
Uthongine (iodide)	EtOH	217(4.04),229(4.01), 263(3.96),284(3.87), 385(3.12),500(2.94)	100-0761-83

Compound	Solvent	$\lambda_{max}(\log \epsilon)$	Ref.
$C_{20}H_{16}N_2$			
Benzonitrile, 4-[2-[4-(4-cyano-3-butynyl)phenyl]ethyl]-	n.s.g.	258s(3.02),266s(3.02), 272s(2.93),278s(2.66)	35-7384-83
Indolo[2,3-c]carbazole, N,N'-dimethyl-	$C_6H_{11}Me$	223(4.61),262(4.50), 272(4.35),286(4.26), 308(3.59),332(4.29), 347(4.49),379(3.73), 393(3.67),401(3.93)	35-6268-83
Phthalazine, 1,2-dihydro-2,4-diphenyl-	EtOH	247(4.34),373(4.09)	4-0225-83
Tricyclo[8.4.2.2^{4,7}]octadeca-4,6,10,12-14,15,17-heptaene-11,12-dicarbonitrile	n.s.g.	320s(2.61)	35-7384-83
$C_{20}H_{16}N_2O$			
Benzo[4,5]cyclohepta[1,2-c]pyrazol-4(1H)-one, 3-methyl-1-(phenylmethyl)-	EtOH	255(4.49),304s(3.92), 318(4.03)	142-1581-83
Benzo[4,5-cyclohepta[1,2-c]pyrazol-4(2H)-one, 3-methyl-2-(phenylmethyl)-	EtOH	254(4.56),297s(4.00), 311(3.95)	142-1581-83
5(6H)-Quinazolinone, 7,8-dihydro-2,7-diphenyl-	EtOH	294(4.41)	4-0649-83
$C_{20}H_{16}N_2OS_2$			
4(1H)-Quinolinone, 1-methyl-3-[[3-(methylthio)-4-quinolinyl]thio]-	EtOH	337.5(4.29),370(4.07)	49-0281-83
$C_{20}H_{16}N_2O_2$			
Benzenemethanamine, α-(nitrophenylmethylene)-N-phenyl-, (Z)-	EtOH	386(4.20)	22-0339-83
	n.s.g.	390(4.20)	22-0339-83
1H-Benz[de]isoquinoline-1,3(2H)-dione, 6-(dimethylamino)-2-phenyl-	MeOH	430(4.08)	104-2273-83
Cyclohepta[4,5]pyrrolo[1,2-a]imidazole-10-carboxylic acid, 2-phenyl-, ethyl ester	EtOH	253(4.41),280s(4.09), 340(4.36),355s(4.33), 442(3.60),455s(3.59), 530s(2.76),575s(2.56), 640s(1.98)	18-3703-83
7H-Dibenz[f,ij]isoquinolin-7-one, 2-morpholino-	DMF	438(3.57)	2-1197-83
Pyrano[4,3-c]pyrazol-3(2H)-one, 6,7-dihydro-2-phenyl-4-(2-phenylethenyl)-, (E)-	EtOH	246(4.30),360(4.52)	118-0948-83
$C_{20}H_{16}N_2O_2S$			
2H-1-Benzopyran-2-one, 3-[2-[4-(dimethylamino)phenyl]-4-thiazolyl]-	MeCN	354(4.30)	48-0551-83
1H-Indole, 2-(2-aminophenyl)-1-(phenylsulfonyl)-	EtOH	208(4.49),242(4.27), 286s(3.84)	39-2417-83C
Thiazolo[3,2-a]pyrimidin-4-ium, 3-hydroxy-7-(4-methoxyphenyl)-5-methyl-2-phenyl-, hydroxide, inner salt	MeCN	562(4.15)	124-0857-83
$C_{20}H_{16}N_2O_2S_2$			
1,4-Cyclohexadiene-1-carboxylic acid, 3,6-bis[(phenylthio)imino]-, methyl ester	$CHCl_3$	499(4.69)	150-0480-83M
$C_{20}H_{16}N_2O_3$			
Benzo[c][2,7]naphthyridin-4(3H)-one, 5-(3,4-dimethoxyphenyl)-	n.s.g.	240(4.44),254(4.34), 268(4.19),280(4.00), 326(3.98)	12-1431-83
Propenal, 3-(1-hydroxy-4-methyl-2-naphthalenyl)-, (4-nitrophenyl)imine	EtOH	357(4.11)	103-0825-83
	acetone	358(4.22)	103-0825-83

Compound	Solvent	λ_{max}(log ε)	Ref.
(cont.)	dioxan	353(3.86)	103-0825-83
	MeCN	355(4.24)	103-0825-83
	DMSO	370(4.15),380(4.15), 550(3.00)	103-0825-83
	+ NaOMe	380(4.12),550(4.18)	103-0825-83
$C_{20}H_{16}N_2O_4$			
9,10-Anthracenedicarbonitrile, 2,3,6,7-tetramethoxy-	$CHCl_3$	252(4.71),283(5.05), 407(4.23),425(4.40)	39-1193-83C
Benzo[c][2,7]naphthyridin-4(3H)-one, 5-(3,4-dimethoxyphenyl)-, 6-oxide	n.s.g.	242(4.43),280(3.97), 288(3.90),310(3.65), 364(3.82),404(3.62)	12-1431-83
[8,8'-Biquinoline]-5,5'(1H,1'H)-dione, 6,6'-dimethoxy-	CH_2Cl_2	505(4.71)	94-2718-83
Camptothecin	n.s.g.	220(4.57),254(4.47), 290(3.70),370(4.30)	162-0239-83
1H-Naphth[2,3-g]indole-3-carboxylic acid, 5-amino-6,11-dihydro-2-methyl-6,11-dioxo-, ethyl ester	dioxan	279(4.49),434(3.49), 546(3.86)	104-1892-83
2-Oxazolin-5-one, 2-methyl-4-phenyl-, head-head dimer	C_6H_{12}	273(4.21)	44-0695-83
dimer m. 129-30°	C_6H_{12}	273(4.23)	44-0695-83
Perloline, dehydro-	n.s.g.	238(4.46),255(4.28), 275(4.04),340(3.90), 350(3.94),370(3.78)	12-1431-83
9,10-Phenanthrenedicarbonitrile, 2,3,6,7-tetramethoxy-	$CHCl_3$	262(4.54),276(4.65), 287(4.61),298(4.59), 368(4.23),416(3.74)	39-1193-83C
$C_{20}H_{16}N_2O_5$			
1-Azabicyclo[3.2.0]hept-2-ene-2-carboxylic acid, 7-oxo-3-phenyl-, (2-nitrophenyl)methyl ester, (R)-	CH_2Cl_2	268(--),300(3.70)	78-2531-83
1-Azabicyclo[3.2.0]hept-2-ene-2-carboxylic acid, 7-oxo-3-phenyl-, (4-nitrophenyl)methyl ester, (R)-	dioxan	274(4.11),315s(--)	78-2531-83
	+ base	313(3.90)	78-2531-83
$C_{20}H_{16}N_4$			
1,4-Phthalazinediamine, N,N'-diphenyl-	MeOH	210(4.78),266(4.40), 372(4.05)	103-0666-83
	MeCN	210(4.78),267(4.45), 370(4.08)	103-0666-83
$C_{20}H_{16}N_4O$			
Formazan, 1-benzoyl-3,5-diphenyl-	EtOH	275(4.26),330(4.09), 397(3.88)	104-2104-83
	dioxan	278(4.31),325(4.10), 397(3.91)	104-2104-83
	MeCN	275(4.25),327(4.09), 396(3.92)	104-2104-83
	CCl_4	282(4.29),320s(4.11), 420(3.89)	104-2104-83
	isooctane	282(--),315s(--), 426(--)	104-2104-83
Formazan, 3-benzoyl-1,5-diphenyl-	EtOH	250(4.14),320(4.13), 448(4.26)	104-0585-83
$C_{20}H_{16}N_4O_3$			
Pyrazolo[1,5-d][1,2,4]triazine-3-carboxylic acid, 6,7-dihydro-4-methyl-7-oxo-2,6-diphenyl-, methyl ester	EtOH	240(4.47),294(4.17)	44-1069-83

Compound	Solvent	$\lambda_{max}(\log \epsilon)$	Ref.
Pyrazolo[1,5-d][1,2,4]triazine-3-carboxylic acid, 6,7-dihydro-6-methyl-7-oxo-2,4-diphenyl-, methyl ester	EtOH	248(4.53),291(4.06)	44-1069-83
$C_{20}H_{16}N_5O_2$			
1,2,4,5-Tetrazin-1(2H)-yl, 2-(4-nitrophenyl)-4,6-diphenyl-	MeCN	730(3.521)	104-2191-83
$C_{20}H_{16}N_6$			
3,4,7-Metheno-3H-1,2,5,6-tetraazacyclobuta[def]fluorene, 3a,3b,4,6a,7,7a-7b,7c-octahydro-3,7-di-2-pyridinyl-	MeOH	261(3.94),357(2.64), 384(2.32)	24-2366-83
$C_{20}H_{16}N_8O_9$			
3-Furancarboxaldehyde, 2-[2-[(2,4-dinitrophenyl)hydrazono]propyl]-, 2,4-dinitrophenylhydrazone	MeCN	223(4.81),260s(4.67), 295s(4.28),373(4.97)	24-3366-83
$C_{20}H_{16}O$			
2,5-Cyclohexadien-1-one, 4-(diphenylmethylene)-2-methyl-	isooctane	361(4.46)	73-2825-83
2,5-Cyclohexadien-1-one, 4-(diphenylmethylene)-3-methyl-	isooctane	362(4.41)	73-2825-83
4H-Cyclopenta[b]furan, 6-methyl-4,5-diphenyl-	EtOH	232(4.00),330(4.16)	39-0915-83C
6H-Cyclopenta[b]furan, 6-methyl-4,5-diphenyl-	EtOH	254(4.33),320(3.75)	39-0915-83C
Furan, 3-(1-methyl-2,3-diphenyl-2-cyclopropen-1-yl)-	EtOH	238(4.33),298s(4.30), 312(4.40)	39-0915-83C
Furan, 3-(2-methyl-1,3-diphenyl-2-cyclopropen-1-yl)-	EtOH	257(4.15)	39-0915-83C
$C_{20}H_{16}O_2$			
Azuleno[2,1,8-ija]azulene-5-carboxylic acid, 10-ethyl-, methyl ester	MeOH	283(5.22),296s(4.56), 310(4.45),355s(3.68), 370(3.94),377s(3.73), 386s(3.92),391(3.14), 432s(2.84),461s(3.18), 490(3.75),620(1.95)	88-0781-83
Furan, 2-[(9-anthracenylmethoxy)methyl]-	$C_6H_{11}Me$	249(4.94),256(5.19), 330(3.45),346(3.74), 365(3.94),385(3.92)	39-0109-83B
cyclic photoisomer	$C_6H_{11}Me$	214(4.50),265(2.93), 273(3.10),281(3.27)	39-0109-83B
9H-Xanthene, 9-(3-methoxyphenyl)-	EtOH	225(4.22),225[sic](4.00), 284(3.72)	142-0481-83
$C_{20}H_{16}O_4$			
2H,6H-Benzo[1,2-b:5,4-b']dipyran-6-one, 5-hydroxy-2,2-dimethyl-8-phenyl-	MeOH	228(4.32),292(4.48), 336(4.13)	106-0876-83
4H,8H-Benzo[1,2-b:3,4-b']dipyran-4-one, 5-hydroxy-8,8-dimethyl-2-phenyl-	MeOH	233(4.45),279(4.56), 372(3.77)	106-0876-83
2,5-Cyclohexadiene-1,4-dione, 2,6-bis(4-methoxyphenyl)-	EtOH	249(4.549)	150-0098-83S
2(5H)-Furanone, 3-(3-[1,1'-biphenyl]-4-yl-1-oxo-2-propenyl)-4-hydroxy-5-methyl-, (E)-	MeOH	203(4.55),233(4.17), 388(4.35)	83-0115-83
Naphtho[2,3-c]furan-1(3H)-one, 5-methoxy-4-(3-methoxyphenyl)-	$CHCl_3$	283(4.42),370(3.74)	39-0643-83C
Naphtho[2,3-c]furan-1(3H)-one, 7-methoxy-4-(3-methoxyphenyl)-	$CHCl_3$	293(3.38),348(2.93), 361(2.89)	39-0643-83C

Compound	Solvent	λ_{max}(log ϵ)	Ref.
$C_{20}H_{16}O_5$			
2-Cyclopenten-1-one, 4-(benzoyloxy)-2-[(benzoyloxy)methyl]-, (R)-	EtOH	210s(3.88),228(4.37)	39-2441-83C
5,12-Naphthacenedione, 9,10-dihydro-6,11-dihydroxy-8-(1-hydroxyethyl)-	MeCN	255(4.17),269(4.34), 491(3.93),522(3.87)	44-2820-83
$C_{20}H_{16}O_6$			
2-Anthracenebutanal, β-acetyl-9,10-dihydro-4,5-dihydroxy-9,10-dioxo-, (±)-	MeOH	226(4.57),255(4.33), 276(3.98),298(3.98), 407s(--),429(3.97), 440s(--)	89-1267-83S
4H-Furo[3,2-c][1]benzopyran-4-one, 2,3-dihydro-2-(hydroxymethyl)-2-(2-methoxybenzoyl)-	MeOH	290(4.01),310(4.01)	39-0841-83C
5,12-Naphthacenedione, 8-acetyl-7,8,9,10-tetrahydro-1,10,11-trihydroxy-, (±)-	MeOH	227(4.55),259(4.41), 275s(--),288(3.99), 412s(--),431(4.09), 440s(--)	89-1267-83S
Viridin	n.s.g.	242(4.49),300(4.22)	162-1433-83
$C_{20}H_{16}O_7$			
9,10-Anthracenedione, 1,6-diacetoxy-8-methoxy-3-methyl- (questin diacetate)	EtOH	258(4.53),375(3.78)	94-4543-83
1,6,11(2H)-Naphthacenetrione, 3,4-dihydro-2,5,7-trihydroxy-9-methoxy-2-methyl-, (±)-	EtOH	240(4.46),285(4.43), 440(4.23)	88-2761-83
$C_{20}H_{16}O_8$			
1,6,11(2H)-Naphthacenetrione, 3,4-dihydro-2,4,5,7-tetrahydroxy-9-methoxy-2-methyl-, cis-(±)-	EtOH	240(4.38),283(4.21), 455(3.95)	88-2761-83
$C_{20}H_{16}S$			
4H-Cyclopenta[b]thiophene, 4-methyl-5,6-diphenyl-	EtOH	233(3.90),325(4.32)	39-0915-83C
4H-Cyclopenta[b]thiophene, 6-methyl-4,5-diphenyl-	EtOH	238(4.00),320(4.16)	39-0915-83C
6H-Cyclopenta[b]thiophene, 4-methyl-5,6-diphenyl-	EtOH	235(4.02),320(4.14)	39-0915-83C
6H-Cyclopenta[b]thiophene, 6-methyl-4,5-diphenyl-	EtOH	238(4.32),325(3.87)	39-0915-83C
Thiophene, 2-(1-methyl-2,3-diphenyl-2-cyclopropen-1-yl)-	EtOH	228(4.38),300s(4.38), 311(4.47),330(4.36)	39-0915-83C
Thiophene, 3-(1-methyl-2,3-diphenyl-2-cyclopropen-1-yl)-	EtOH	238(4.18),280(4.21), 312(4.26),328(4.16)	39-0915-83C
Thiophene, 3-(2-methyl-1,3-diphenyl-2-cyclopropen-1-yl)-	EtOH	254(4.15)	39-0915-83C
$C_{20}H_{16}S_2$			
1,3-Benzodithiole, 2-(6,8-dimethyl-7H-benzocyclohepten-7-ylidene)-	MeCN	242(4.7),292(4.2)	142-2039-83
$C_{20}H_{17}BrF_3NO_5$			
Morphinan-7-one, 1-bromo-5,6,8,14-tetradehydro-4-hydroxy-3,6-dimethoxy-17-(trifluoroacetyl)-	EtOH	239.5(4.34),280(3.86)	78-2393-83
$C_{20}H_{17}BrN_2O$			
4H-Indol-4-one, 1-(4-bromophenyl)-1,5,6,7-tetrahydro-2-phenyl-, oxime	EtOH	253(4.50),287(4.20)	4-0989-83

Compound	Solvent	$\lambda_{max}(\log \epsilon)$	Ref.
Pyrrolo[3,2-c]azepin-4(1H)-one, 1-(4-bromophenyl)-5,6,7,8-tetrahydro-2-phenyl-	EtOH	246(4.50),282(4.30)	4-0989-83
$C_{20}H_{17}BrO_6$			
3-Benzofurancarboxylic acid, 6-bromo-4,5-dihydroxy-2-methyl-7-(2-oxo-2-phenylethyl)-, ethyl ester	EtOH	214(4.67),244(4.48)	103-0945-83
3-Benzofurancarboxylic acid, 6-bromo-4,5-dihydroxy-7-(2-oxopropyl)-2-phenyl-, ethyl ester	EtOH	217(4.61),318(4.12)	103-0945-83
$C_{20}H_{17}ClN_2O_3$			
Ketazolam	EtOH	202(4.61),241(4.27)	162-0761-83
$C_{20}H_{17}ClN_2O_4$			
1,4-Naphthalenedione, 2-chloro-3-[4-(diethylamino)phenyl]-5-nitro-	C_6H_{12}	592(3.87)	150-0168-83S
	EtOH	608(--)	150-0168-83S
$C_{20}H_{17}ClN_2S_2$			
Diazene, [(4-chlorophenyl)(4,5,6,7-tetrahydro-1,3-benzodithiol-2-ylidene)methyl]phenyl-	MeCN	499(4.34)	118-0840-83
$C_{20}H_{17}ClO_6$			
3-Benzofurancarboxylic acid, 6-chloro-4,5-dihydroxy-2-methyl-7-(2-oxo-2-phenylethyl)-, ethyl ester	EtOH	220(4.56),255(4.38)	103-0945-83
$C_{20}H_{17}FeN$			
Ferrocene, (2-methyl-5-quinolinyl)-	EtOH	235(4.57),275(4.00), 450(2.60)	104-0950-83
$C_{20}H_{17}N$			
5H-1-Pyrindine, 6,7-dihydro-2,4-diphenyl-	neutral	301(3.90)	39-0045-83B
	protonated	292(4.20)	39-0045-83B
$C_{20}H_{17}NO$			
1-Naphthalenol, 4-[3-[(4-methylphenyl)-imino]-1-propenyl]-	DMSO	385(4.13),525(3.61)	104-1864-83
	+ NaOMe	370(3.34),495(3.96)	104-1864-83
1-Naphthalenol, 4-[3-(phenylmethyl)-imino]-1-propenyl]-	CCl_4	335(3.48),420(4.02)	104-1864-83
	DMSO	360(4.05),515(4.18)	104-1864-83
	+ NaOMe	360(4.00),515(4.20)	104-1864-83
Propenal, 3-(1-hydroxy-4-methyl-2-naphthalenyl)-, N-phenylimine	benzene	315(3.71),325(3.69), 340(3.65),358(3.60)	103-0825-83
	acetone	330(4.03),340(3.98), 357(3.81),375(3.48), 505(3.36),535(3.40)	103-0825-83
	dioxan	330(3.96),340(3.89), 358(3.85),370(3.20), 503(2.85),530(2.85)	103-0825-83
	MeCN	315(3.76),340(3.76), 357(3.63),376(3.04), 520(3.36),540(3.43)	103-0825-83
	DMSO	335(4.00),355(3.88), 380(3.70),530(4.13)	103-0825-83
	+ NaOMe	335(4.02),380(3.36), 520(4.28)	103-0825-83
Pyridine, 1,4-dihydro-1-(4-methylbenzoyl)-4-(phenylmethylene)-	MeOH	236(4.11),363(4.42)	24-3192-83

Compound	Solvent	λ_{max}(log ε)	Ref.
Tricyclo[8.4.2.2^{4,7}]octadeca-4,6,10,12,14,15,17-heptaene-11-carbonitrile, 12-formyl-	n.s.g.	368s(2.76)	35-7384-83
$C_{20}H_{17}NO_2$			
2H-Benz[g]indol-2-one, 1-acetyl-1,3,3a,9b-tetrahydro-4-phenyl-	EtOH	227(3.85),233s(3.82), 304(3.86),316(3.85), 330s(3.58)	44-1451-83
1-Naphthalenol, 4-[3-[(4-methoxyphenyl)imino]-1-propenyl]-, (E,?)-	acetone	375(4.23),518(3.08)	104-1864-83
	MeCN	375(4.04),520(3.23)	104-1864-83
	CCl_4	335(3.48)	104-1864-83
	DMSO	380(3.71),497(3.38)	104-1864-83
	+ NaOMe	368(3.30),500(4.03)	104-1864-83
Oxiranecarbonitrile, 3-[2-(2-cyclopenten-1-yloxy)phenyl]-2-phenyl-, trans	EtOH	265s(3.59),277s(3.59), 283(3.62)	24-2383-83
Pyridine, 1,4-dihydro-1-(4-methoxybenzoyl)-4-(phenylmethylene)-	MeOH	245(4.10),273(4.01), 363(4.42)	24-3192-83
$C_{20}H_{17}NO_2S$			
Benzeneacetaldehyde, α-[6-(4-methoxyphenyl)-2H-thiopyran-2-ylidene]-, oxime	$CHCl_3$	460(3.84)	97-0147-83
$C_{20}H_{17}NO_2Se$			
Benzeneacetaldehyde, α-[6-(4-methoxyphenyl)-2H-selenin-2-ylidene]-, oxime	$CHCl_3$	430(3.65)	97-0147-83
$C_{20}H_{17}NO_4$			
7H-Benzo[g]-1,3-benzodioxolo[6,5,4-de]quinoline-7-carboxylic acid, 5,6-dihydro-, ethyl ester	EtOH	257(4.78),278(4.10), 289(4.01),322(4.06), 332(4.07),358(3.55), 377(3.58)	12-1061-83
1,3-Dioxolo[4,5-i][1,3]dioxolo[5,6]indeno[2,1-c][2]benzazepine, 12,12a,13,14-tetrahydro-13-methyl-	EtOH	216(4.48),230s(4.36), 296(4.11),354(4.43)	88-4481-83
Duguecalyne	EtOH	220(4.05),266(4.38), 334(3.74)	100-0761-83
$C_{20}H_{17}NO_5$			
7H-Dibenzo[de,h]quinolin-7-one, 4,5,6,9-tetramethoxy- (dauriporphine)	EtOH	226s(4.58),258(4.75), 285s(4.15),310s(3.91), 344s(3.95),414(4.20)	94-3091-83
	EtOH-HCl	226s(4.58),257(4.74), 285s(4.19),310s(3.91), 344s(3.95),414(4.20)	94-3091-83
	EtOH-NaOH	226s(4.58),257(4.74), 285s(4.14),310s(3.90), 344s(3.95),414(4.20)	94-3091-83
Glaunidine	MeOH	238(4.47),314(4.40), 410(3.78),610(3.60)	100-0761-83
	MeOH-HCl	221(4.50),253(4.57), 282s(4.41),383(3.95), 430s(3.60)	100-0761-83
Imerubrine	EtOH	255(4.48),267(4.52), 295(4.40),350(4.35), 372s(4.20),394(4.11), 450(3.93)	100-0761-83
Isodidehydrochelidonine, (-)-	EtOH	211(4.33),238(3.89), 290(3.86)	142-1895-83
Limogine	MeOH	209(4.45),238s(3.98), 293(3.93)	88-2445-83

Compound	Solvent	λ_{max}(log ε)	Ref.
Limogine (cont.)	MeOH-acid	210(4.49),240(3.96), 297(3.96)	88-2445-83
3,4-Quinolinedicarboxylic acid, 2-(4-methoxyphenyl)-, dimethyl ester	EtOH	230s(4.40),248(4.45), 280s(4.33),343(3.79)	12-1419-83
Ribasine (same spectra in acid or base)	EtOH	213(4.37),240s(4.01), 292(4.07)	88-2029-83
$C_{20}H_{17}NO_6$			
3,1-Benzoxazepine-4,5-dicarboxylic acid, 2-(4-methoxyphenyl)-, dimethyl ester	EtOH	285(4.46)	12-1419-83
Bicuculline	EtOH-acid	225(4.56),296(3.80), 324(3.77)	162-0171-83
Himalayamine	MeOH	210(4.49),235s(3.98), 293(4.02)	88-2445-83
	MeOH-acid	210(4.53),238(3.96), 295(4.04)	88-2445-83
3,4-Quinolinedicarboxylic acid, 1,2-dihydro-1-(4-methoxyphenyl)-2-oxo-, dimethyl ester	EtOH	236(4.53),292(3.95), 353(3.76)	12-1419-83
3,4-Quinolinedicarboxylic acid, 2-(4-methoxyphenyl)-, dimethyl ester, 1-oxide	EtOH	245(4.15),298(3.77)	12-1419-83
Ribasidine	EtOH	211(4.48),235s(3.98), 291(3.98)	88-4481-83
	EtOH-HCl	242(3.93)	88-4481-83
$C_{20}H_{17}N_2S$			
Thiazolo[3,4-a]pyrimidin-5-ium, 4,6-dimethyl-2,8-diphenyl-, perchlorate	MeCN	246(3.28),296(3.44), 398(2.91)	103-0037-83
$C_{20}H_{17}N_3O_2S$			
Anthra[1,2-c][1,2,5]thiadiazole-6,11-dione, 4-(cyclohexylamino)-	toluene	520(3.96)	135-0283-83
	MeOH	532(--)	135-0283-83
	dioxan	515(3.99)	135-0283-83
	HOAc	530(4.08)	135-0283-83
	$CHCl_3$	530(4.09)	135-0283-83
$C_{20}H_{17}N_3O_2S_2$			
Diazene, (4-nitrophenyl)[phenyl(4,5,6-7-tetrahydro-1,3-benzodithiol-2-ylidene)methyl]-	MeCN	559(4.42)	118-0840-83
Diazene, [(4-nitrophenyl)(4,5,6,7-tetrahydro-1,3-benzodithiol-2-ylidene)methyl]phenyl-	MeCN	493(4.36)	118-0840-83
$C_{20}H_{17}N_3O_2Se$			
Anthra[1,2-c][1,2,5]selenadiazole-6,11-dione, 4-(cyclohexylamino)-	toluene	560(3.98)	135-0283-83
	MeOH	570(--)	135-0283-83
	dioxan	550(3.92)	135-0283-83
	HOAc	575(4.11)	135-0283-83
	$CHCl_3$	570(4.00)	135-0283-83
$C_{20}H_{17}N_3O_3$			
Anthra[1,2-c][1,2,5]oxadiazole-6,11-dione, 4-(cyclohexylamino)-	toluene	520(4.10)	135-0283-83
	MeOH	540(--)	135-0283-83
	dioxan	525(4.08)	135-0283-83
	HOAc	535(4.24)	135-0283-83
	$CHCl_3$	530(4.18)	135-0283-83
1H-Indole-3-acetamide, α-hydroxy-N-[2-(1H-indol-3-yl)-2-oxoethyl]-	MeOH	224(4.47),243(4.11), 265(4.06),284(4.12),	142-0469-83

Compound	Solvent	λ_{max}(log ε)	Ref.
(cont.)		292(4.15),303(4.07)	142-0469-83
1H-Naphtho[2,3-d]triazole-4,9-dione, 1-[(4-methoxyphenyl)methyl]-6,7-dimethyl-	EtOH	226(4.41),257(4.61), 279s(4.60),342(3.08)	87-0714-83
$C_{20}H_{17}N_3S$			
2-Propenenitrile, 3-[[2-[(2-cyano-1-phenylethenyl)amino]ethyl]thio]-3-phenyl-	EtOH	205(4.53),225(4.37), 280(4.21)	39-1223-83C
$C_{20}H_{17}N_4$			
Verdazyl, 1,3,5-triphenyl-	MeCN	720(3.633)	104-2191-83
Verdazylium, 1,3,5-triphenyl-, iodide	MeCN	540(4.104)	104-2191-83
$C_{20}H_{17}O_3$			
Pyrylium, 4-(ethoxycarbonyl)-2,6-diphenyl-, perchlorate	MeCN	230(4.25),284(4.16), 325(3.63),436(4.37)	22-0115-83
$C_{20}H_{18}BrNO_4$			
1,3-Dioxolo[4,5-g]isoquinoline-6(5H)-carboxylic acid, 5-[(2-bromophenyl)-methylene]-7,8-dihydro-, ethyl ester, (Z)-	EtOH	295(4.10),322s(4.03)	12-1061-83
$C_{20}H_{18}BrNO_6$			
3-Benzofurancarboxylic acid, 6-bromo-4,5-dihydroxy-7-[2-(hydroxyimino)-2-phenylethyl]-2-methyl-, ethyl ester	EtOH	213(4.57),245(4.25), 300s(3.51)	103-0945-83
3-Benzofurancarboxylic acid, 6-bromo-4,5-dihydroxy-7-[2-(hydroxyimino)-propyl]-2-phenyl-, ethyl ester	EtOH	218(4.60),278(4.11), 300(4.08)	103-0945-83
$C_{20}H_{18}ClNO_2$			
1,4-Naphthalenedione, 2-chloro-3-[4-(diethylamino)phenyl]-	C_6H_{12}	551(3.86)	150-0168-83S
	EtOH	582(--)	150-0168-83S
1,4-Naphthalenedione, 2-[2-chloro-4-(diethylamino)phenyl]-	C_6H_{12}	486(3.63)	150-0168-83S
	EtOH	524(--)	150-0168-83S
$C_{20}H_{18}ClNO_6$			
3-Benzofurancarboxylic acid, 6-chloro-4,5-dihydroxy-7-[2-(hydroxyimino)-2-phenylethyl]-2-methyl-, ethyl ester	EtOH	212(4.56),245(4.33), 300s(3.56)	103-0945-83
$C_{20}H_{18}Cl_2$			
3,5,7,10,12,14-Heptadecahexaene-1,16-diyne, 9-(dichloromethylene)-3,15-dimethyl-, (E,E,Z,Z,E,E)-	CH_2Cl_2	273s(4.18),284(4.40), 296(4.48),329(4.65), 344(4.78),362(4.70), 425s(3.20)	39-2997-83C
$C_{20}H_{18}Cl_2O_5$			
9,10-Anthracenedione, 2-[3-(dichloromethyl)-3-hydroxypentyl]-1,4-dihydroxy-	MeOH	225(4.27),246(4.51), 281(3.97),317(3.43), 456(3.93),477(3.97), 510(3.77)	5-1818-83
$C_{20}H_{18}Cl_2O_6$			
9,10-Anthracenedione, 2-[3-(dichloromethyl)-3-hydroxypentyl]-1,4,8-trihydroxy-, (±)-	MeOH	228(4.58),245(4.34), 284(3.97),461(3.99), 487(4.07),510(3.96), 520(3.91)	5-1818-83

Compound	Solvent	λ_{max}(log ε)	Ref.
$C_{20}H_{18}NO_2S$			
11H-Benzo[b]phenothiazinium, 6-ethoxy-12-ethyl-11-oxo-, tetrafluoroborate	MeCN	260(4.26),460(4.64)	88-3567-83
$C_{20}H_{18}N_2$			
9H-Carbazol-3-amine, N,9-dimethyl-N-phenyl-	$C_6H_{11}Me$	237(4.41),266(4.27), 295(4.32),348(3.49)	35-6268-83
$C_{20}H_{18}N_2O$			
Benzo[4,5]cyclohepta[1,2-c]pyrazol-4(1H)-one, 9,10-dihydro-3-methyl-1-(phenylmethyl)-	EtOH	278(4.13)	142-1581-83
Ethanone, 2-(2,2-diphenylhydrazino)-1-phenyl-	C_6H_{12}	240(3.42),280s(3.04)	4-1739-83
4-Isoquinolinecarbonitrile, 1-butoxy-3-phenyl-	EtOH	217(4.50),256(4.48), 309(4.16),325s(4.06), 340s(3.79)	39-1813-83C
$C_{20}H_{18}N_2O_2$			
1H-Indole-3-ethanol, β-1H-indol-3-yl-, acetate (streptindole)	n.s.g.	273(3.79),283(3.81), 290(3.76)	88-4719-83
1H-Indole-3-propanoic acid, α-1H-indol-3-yl-, methyl ester	MeOH	223(4.34),270(3.95), 279(3.97),288(3.94)	142-0469-83
Pyrano[2,3-c]pyrazol-4(1H)-one, 5,6-dihydro-6,6-dimethyl-1,3-diphenyl-	EtOH	255(4.51)	118-0844-83
Pyrano[4,3-c]pyrazol-3(2H)-one, 6,7-dihydro-6,6-dimethyl-2,4-diphenyl-	EtOH	256(4.28),305(4.27)	142-1801-83
3H-Pyrazol-3-one, 4-benzoyl-2,4-dihydro-5-(1-methyl-1-propenyl)-2-phenyl-	EtOH	250(4.34)	142-1801-83
3H-Pyrazol-3-one, 4-benzoyl-2,4-dihydro-5-(2-methyl-1-propenyl)-2-phenyl-	EtOH	252(4.30)	142-1801-83
$C_{20}H_{18}N_2O_2S_2$			
Benzenesulfenamide, N,N'-2,5-cyclohexadiene-1,4-diylidenebis[4-methoxy-	$CHCl_3$	485(4.56)	150-0480-83M
$C_{20}H_{18}N_2O_3$			
Benzoic acid, 4-[(5-oxo-bicyclo[4.4.1]-undeca-3,6,8,10-tetraen-2-ylidene)hydrazino]-, ethyl ester	EtOH-KOH	569(4.52)	33-2626-83
Benzoic acid, 4-[(4-oxo-4a,8a-methanonaphthalen-1(4H)-ylidene)hydrazino]-, ethyl ester	EtOH-KOH	554(4.65)	33-2626-83
$C_{20}H_{18}N_2O_3S$			
Morpholine, 4-[5-(4-nitrophenyl)-3-phenyl-2-thienyl]-	MeCN	244(4.38),271(4.13), 411(4.54)	48-0168-83
$C_{20}H_{18}N_2O_4$			
2-Butenedicarbonitrile, 2,3-bis(3,4-dimethoxyphenyl)-, (E)-	MeOH	237(4.16),267(4.05), 402(4.20)	39-1193-83C
(Z)-	MeOH	260(4.16),392(4.08)	39-1193-83C
1,4-Naphthalenedione, 2-[4-(diethylamino)phenyl]-5-nitro-	C_6H_{12}	559(3.71)	150-0168-83S
	EtOH	594(--)	150-0168-83S
1,4-Naphthalenedione, 2-[4-(diethylamino)phenyl]-8-nitro-	C_6H_{12}	550(3.98)	150-0168-83S
	EtOH	586(--)	150-0168-83S
Propanedioic acid, [(1H-indol-3-yl)-phenylamino)methylene]-, monoethyl ester	EtOH	210(4.66),298.5(4.32)	104-1367-83

Compound	Solvent	$\lambda_{max}(\log \epsilon)$	Ref.
Pyrimido[2,1-a]isoindol-4(6H)-one, 2-(3,4,5-trimethoxyphenyl)-	EtOH	203(3.20),287(3.32), 308s(3.83)	124-0755-83
$C_{20}H_{18}N_2O_5S$			
Butanoic acid, 4-[(3-sulfophenyl)azo]-1-naphthalenyl ester, sodium salt	pH 7.5	374(4.04)	94-0162-83
Butanoic acid, 4-[(4-sulfophenyl)azo]-1-naphthalenyl ester, sodium salt	pH 7.5	378(4.12)	94-0162-83
$C_{20}H_{18}N_2O_8S_2$			
Butanoic acid, 4-sulfo-2-[(4-sulfophenyl)azo]-1-naphthalenyl ester, disodium salt	pH 7.5	333(4.31)	94-0162-83
$C_{20}H_{18}N_2S_2$			
Diazene, phenyl[phenyl(4,5,6,7-tetrahydro-1,3-benzodithiol-2-ylidene)-methyl]-	MeCN	493(4.06)	118-0840-83
$C_{20}H_{18}N_4$			
2-Propenenitrile, 3,3'-(1,2-ethanediyldiimino)bis[3-phenyl-	EtOH	205(4.50),226(4.40), 282(4.23)	39-1223-83C
$C_{20}H_{18}N_4O_2$			
Propanedinitrile, (7-acetyl-2,3,3a,4-6a,7-hexahydro-3a-methyl-2-oxo-1H-pyrrolo[2,3-d]carbazol-5(6H)-ylidene)-	MeOH	251(4.15),291(3.82), 319s(3.62)	5-1744-83
$C_{20}H_{18}N_4O_2S$			
Carbamic acid, [5-amino-3-(2-naphthalenyl)-2H-pyrido[4,3-b]-1,4-thiazin-7-yl]-, ethyl ester	pH 1	254(4.58),289(4.23), 297(4.22),401(4.34)	87-1614-83
$C_{20}H_{18}N_4O_5S$			
1-Azabicyclo[3.2.0]hept-2-ene-2-carboxylic acid, 6-ethyl-7-oxo-3-(2-pyrimidinylthio)-, (4-nitrophenyl)-methyl ester, (5R-cis)-	THF	268(4.01),322(3.99)	158-0407-83
$C_{20}H_{18}N_6O$			
1,2,4-Triazolo[4,3-b]pyridazinium, 8-hydroxy-6-methyl-7-[(4-methylphenyl)-azo]-1-(phenylmethyl)-, hydroxide, inner salt, (E)-	benzene	400(4.28),487(3.30)	48-1016-83
	MeOH	393(4.33),488(3.30)	48-1016-83
	MeCN	398(4.31),489(3.31)	48-1016-83
	CH_2Cl_2	397(4.31),489(3.31)	48-1016-83
(Z)-	benzene	336(--),481(--)	48-1016-83
	MeOH	332(--),477(--)	48-1016-83
	MeCN	334(--),483(--)	48-1016-83
	CH_2Cl_2	334(--),483(--)	48-1016-83
$C_{20}H_{18}O_2$			
2,4,6,12,14,16-Octadecahexaene-8,10-diynedial, 7,12-dimethyl-, (E,E,Z,Z,E,E)-	THF	239(4.26),283(4.25), 321s(4.50),334(4.55), 377(4.51),391s(4.50), 423(4.34)	18-1467-83
Tricyclo[8.4.2.2^{4,7}]octadeca-4,6,10-12,14,15,17-heptaene-11,12-dicarbaldehyde	n.s.g.	360s(2.79)	35-7384-83
$C_{20}H_{18}O_4$			
9,10-Anthracenedipropanoic acid	pH 9	400(4.08)	149-0271-83A

Compound	Solvent	λ_{max}(log ε)	Ref.
Phaseolin	EtOH	281(4.04)	149-0323-83B
$C_{20}H_{18}O_5$			
2H-Benzo[b]cyclobuta[d]pyran-2-carboxylic acid, 1,2a,3,8b-tetrahydro-1-(2-hydroxyphenyl)-3-oxo-, ethyl ester, (1α,2α,2aα,8bα)-	dioxan	272(3.60),278(3.63)	39-1083-83C
4H-1-Benzopyran-4-one, 3,5,7-trihydroxy-8-(3-methyl-2-butenyl)-2-phenyl- (glepidotin A)	MeOH	216(4.49),239s(4.23), 272(4.32),324(4.11), 360(4.06)	102-0573-83
	MeOH-HCl	216(4.53),240s(4.23), 271(4.34),324(4.14), 360(4.10)	102-0573-83
	MeOH-NaOH	235(4.43),287(4.24), 417(4.15)	102-0573-83
	MeOH-$AlCl_3$	228(4.32),253(4.25), 276(4.30),358(4.07), 420(4.21)	102-0573-83
	MeOH-$AlCl_3$-HCl	229(4.31),252(4.25), 276(4.26),353(4.04), 419(4.20)	102-0573-83
Glyceollin I	EtOH	286(4.01)	149-0323-83B
5,12-Naphthacenedione, 8-ethyl-7,8,9-10-tetrahydro-6,8,11-trihydroxy-	MeOH	253(4.39),282(3.71), 326(3.20),451(3.66), 477(3.75),509(3.59), 553(2.77)	5-1818-83
Tuberosin	EtOH	287(3.82)	149-0323-83B
$C_{20}H_{18}O_6$			
9,10-Anthracenedione, 1,8-dihydroxy-3-[2-(2-hydroxyethyl)-3-oxobutyl]-	MeOH	226(4.56),255(4.32), 277(3.98),413s(--), 430(4.03),440s(--)	89-1267-83S
9,10-Anthracenedione, 1-hydroxy-4,8-dimethoxy-2-(3-oxobutyl)-	MeOH	225(4.57),246(4.19), 282(3.82),410(3.50), 440(3.91),465(3.93), 495(3.79)	5-1818-83
9,10-Anthracenedione, 1,4,8-trihydroxy-2-(3-oxohexyl)-	MeOH	228(4.55),246(4.26), 286(3.76),384(3.22), 462(3.98),476(4.04), 488(4.09),508(3.96), 520(3.89)	5-1818-83
[2,6'-Bi-2H-1-benzopyran]-4(3H)-one, 5,5',7-trihydroxy-2',2'-dimethyl- (sanggenon F)	EtOH	228(4.62),288(4.33), 319s(3.77)	142-0661-83
	EtOH-$AlCl_3$	227(4.68),310(4.32), 370(3.42)	142-0661-83
[2,8'-Bi-2H-1-benzopyran]-4(3H)-one, 5,5',7-trihydroxy-2',2'-dimethyl-, (S)- (sanggenon H)	EtOH	218s(4.51),227(4.54), 289(4.31),318s(3.80)	142-1071-83
	EtOH-$AlCl_3$	223(4.62),282s(3.99), 309(4.36),374(3.52)	142-1071-83
2(3H)-Furanone, 3,4-bis(1,3-benzodioxol-5-ylmethyl)dihydro-, (3R-trans)- (hinokinin)	MeOH	232(4.13),284(4.03)	102-1516-83
1(3H)-Isobenzofuranone, 3-acetoxy-3-(4-acetoxy-3,5-dimethylphenyl)-	EtOH	277(3.15)	24-0970-83
5,12-Naphthacenedione, 8-ethyl-7,8,9,10-tetrahydro-6,7,8,11-tetrahydroxy-, cis	MeOH	248(4.46),285(3.84), 457(3.83),477(3.90), 509(3.71),562(2.93)	5-1818-83
trans	MeOH	249(4.45),284(3.76), 456(3.79),480(3.86), 514(3.70)	5-1818-83

Compound	Solvent	λ_{max}(log ε)	Ref.
5,12-Naphthacenedione, 8-ethyl-7,8,9,10-tetrahydro-6,8,10,11-tetrahydroxy-, cis	MeOH	222(4.29),250(4.62), 279(4.06),323(3.69), 460(3.99),478(4.03), 512(3.86)	5-1818-83
$C_{20}H_{18}O_7$			
2-Anthracenecarboxylic acid, 9,10-dihydro-3,4,8-trimethoxy-1-methyl-9,10-dioxo-, methyl ester	MeOH	220(4.39),257(4.31), 373(3.69)	94-4206-83
5,12-Naphthacenedione, 8-ethyl-7,8,9,10-tetrahydro-1,6,7,8,10-pentahydroxy- (α-citromycinone)	C_6H_{12}	418(4.04),436(4.04)	88-1329-83
synthetic	C_6H_{12}	417(3.92),435(3.92)	88-1329-83
5,12-Naphthacenedione, 8-ethyl-7,8,9,10-tetrahydro-1,6,t-8,t-10,11-pentahydroxy-	MeOH	233(4.46),253(4.38), 295(3.87),433(3.66), 463(3.98),480(4.04), 490(4.10),511(3.96), 526(3.96)	5-1818-83
$C_{20}H_{18}O_8$			
Benzeneacetic acid, α-[2-(1,3-benzodioxol-5-yl)-2-oxoethylidene]-3,4,5-trimethoxy-	EtOH	236s(4.25),286s(3.88), 324(4.00)	94-3039-83
4H-1-Benzopyran-4-one, 5,6,7-trimethoxy-2-(7-methoxy-1,3-benzodioxol-5-yl)-	MeOH	276s(2.75),338(3.10), 375s(2.57)	95-0994-83
Daphneticin	MeOH	242(3.96),260(3.95), 317(4.05)	102-0617-83
1H-Naphtho[2,3-c]pyran-1-one, 7,8,10-triacetoxy-3,4-dihydro-3-methyl-	n.s.g.	246(4.68),296(3.75), 345(3.36)	119-0082-83
$C_{20}H_{18}S_3$			
6H-Cyclohepta[c]thiophene, 6-(1,3-benzodithiol-2-ylidene)-1,3,5,7-tetramethyl-	MeCN	247(4.6),296(4.2)	142-2039-83
$C_{20}H_{19}BrN_2$			
4,7-Methano-1H-indazole, 1-(4-bromophenyl)-3a,4,5,6,7,7a-hexahydro-3-phenyl-	C_6H_{12}	369(4.43)	12-1649-83
	MeOH	365(4.42)	12-1649-83
$C_{20}H_{19}BrO_7$			
Herqueinone, 12-bromo-	C_6H_{12}	230(5.65),255(5.42), 307(5.16),450(4.09)	18-3661-83
$C_{20}H_{19}ClN_2$			
Sempervirine methochloride	n.s.g.	241(4.56),292(4.20), 330(4.28),395(4.22)	162-1214-83
$C_{20}H_{19}ClN_2O_2$			
1,4-Naphthalenedione, 5(and 8)-amino-3-chloro-2-[4-(diethylamino)phenyl]-	C_6H_{12}	498(3.93)	150-0168-83S
	EtOH	535(--)	150-0168-83S
$C_{20}H_{19}ClN_{10}O_4S_2$			
Acetamide, N,N',N",N"'-[(5-chloro-2,3-pyrazinediyl)bis(thio-5,4,6-pyrimidinetriyl)]tetrakis-	MeOH	235s(4.55),275(3.77), 372(4.07)	4-1047-83
$C_{20}H_{19}NO$			
1(4H)-Naphthalenone, 4-[[4-(1,1-dimethylethyl)phenyl]imino]-	n.s.g.	330(3.80)	39-1759-83C

Compound	Solvent	λ_{max}(log ϵ)	Ref.
$C_{20}H_{19}NO_2$			
2,5-Cyclohexadien-1-one, 4-[8-(1,1-dimethylethyl)-2-naphthalenyl]imino-, N-oxide	n.s.g.	381(4.24)	39-1759-83C
1,4-Naphthalenedione, 2-[4-(diethylamino)phenyl]-	C_6H_{12}	516(4.03)	150-0168-83S
	EtOH	561(--)	150-0168-83S
1,4-Naphthalenedione, 2-[4-(dimethylamino)-2,6-dimethylphenyl]-	C_6H_{12}	518(3.13)	150-0168-83S
	EtOH	545(--)	150-0168-83S
1(2H)-Naphthalenone, 8-(1,1-dimethylethyl)-2-(phenylimino)-, N-oxide	n.s.g.	502(3.75)	39-1759-83C
1(4H)-Naphthalenone, 3-(1,1-dimethylethoxy)-4-(phenylimino)-	n.s.g.	211(3.32),251s(3.03), 292(3.18),340(2.78), 443(2.43)	39-2711-83C
1(4H)-Naphthalenone, 4-[[4-(1,1-dimethylethyl)phenyl]imino]-, N-oxide	n.s.g.	382(4.37)	39-1759-83C
$C_{20}H_{19}NO_3$			
Acronycine	EtOH	224(4.39),260s(4.60), 281(4.69),293(4.64), 308(4.41),380(4.01)	100-0391-83
Guadiscine	EtOH	232s(3.95),265(4.33), 310(3.97),316s(3.84), 342(3.68),355s(3.66)	100-0761-83
	EtOH-HCl	274(4.38),364(3.88), 408(3.62)	100-0761-83
$C_{20}H_{19}NO_4$			
Dehydrocrebanine	EtOH	248s(4.36),272(4.77), 296s(4.17),337(4.15), 385(3.49)	100-0761-83
Melosmine	EtOH	218s(4.34),240s(4.41), 252(4.44),309(3.76), 322(3.78),377(3.96)	100-0761-83
	EtOH-HCl	223(4.48),240s(4.47), 279(4.56),331s(3.55), 450(3.83)	100-0761-83
1H-Pyrrole-2,5-dione, 1-[2-hydroxy-1-(hydroxymethyl)-1-methylethyl]-3,4-diphenyl-	EtOH	272(4.0),360(3.6)	33-1078-83
$C_{20}H_{19}NO_5$			
Benzoic acid, 2-(6,7-dimethoxy-1-isoquinolinyl)-4-methoxy-, methyl ester	EtOH	239(4.36),315(3.37), 328(3.36)	78-3261-83
Benzoic acid, 2-(6,7-dimethoxy-1-isoquinolinyl)-5-methoxy-, methyl ester	EtOH	240(4.69),330(3.52)	78-3261-83
Buxifoline, N-formyl-	EtOH	222(4.43),246(4.22), 288(4.30)	100-0761-83
Chelidonine, (±)-	EtOH	236(3.92),288.5(3.89)	33-1119-83
Guacoline	EtOH	222(4.32),266(4.26), 324(3.85),353s(3.63)	100-0761-83
	EtOH-HCl	225(4.25),273(4.16), 372(3.55)	100-0761-83
Hypecorine, (±)-	MeOH and MeOH-acid	208(4.55),248(4.15), 294(3.85),369(3.89)	39-2431-83C
	MeOH-base	212(4.67),236s(3.96), 288(3.88)	39-2431-83C
Imeluteine	EtOH	233(4.48),253(4.49), 288(4.43),317(3.75), 365s(3.72),380(3.85), 400s(3.72)	100-0761-83
Ocominarine	EtOH	221(4.69),292(4.33)	100-0761-83

Compound	Solvent	λ_{max}(log ε)	Ref.
6-Oxa-2-azaspiro[4.5]deca-3,7-diene-4-carboxylic acid, 1,9-dioxo-3-phenyl-2-(2-propenyl)-, ethyl ester	EtOH	254(4.03),300(3.86)	94-0356-83
Protopine	EtOH	293(3.93)	162-1138-83
6H-Pyrano[3,2-b]acridin-6-one, 2,11-dihydro-5,10-dihydroxy-12-methoxy-2,2,11-trimethyl- (pyranofoline)	MeOH	230(4.26),284s(4.53), 306(4.74),332s(4.23), 427(3.64)	39-1681-83C
7H-Pyrano[2,3-c]acridin-7-one, 3,12-dihydro-6,10-dihydroxy-11-methoxy-3,3,12-trimethyl- (citracridone I)	MeOH	209(4.28),270(4.67), 282s(4.63),295s(4.58), 340(4.05),400(3.57)	94-0895-83
	MeOH-AlCl$_3$	209(--),215s(--), 272(--),290(--), 307(--),363(--), 452(--)	94-0895-83
Ribasine, dihydro-	EtOH	210(4.43),238(3.90), 296(4.06)	88-4481-83
$C_{20}H_{19}NO_6$			
3H-Benz[e]indole-1,8-dicarboxylic acid, 4,5-dihydro-2,3-dimethyl-4,5-dioxo-, diethyl ester	EtOH	214(4.49),249(4.35), 274(4.47),340s(3.60), 384(3.83)	103-0876-83
Bis[1,3]benzodioxolo[5,6-e:5',4'-i]-[1,2]oxaazacycloundecin-14(6H)-one, 4,7,8,15-tetrahydro-6-methyl-	MeOH	208(4.48),223s(4.29), 297(3.87)	39-2431-83C
Duguevanine, N-formyl-, (-)-	MeOH	226(4.54),274s(4.26), 282(4.34),301(4.14), 308s(4.11)	100-0862-83
	MeOH-base	224(4.60),275s(4.18), 284(4.23),314s(4.01), 332(4.07)	100-0862-83
Egenine, (+)-	MeOH	212(4.11),233s(3.75), 290(3.71)	78-0577-83
Ethanone, 1-(6-ethenyl-1,3-benzodioxol-5-yl)-2-[4-[(hydroxymethylamino)methyl]-1,3-benzodioxol-5-yl]-	MeOH	208(4.56),239(4.41), 292(4.00),322s(3.75)	39-2431-83C
Ochratoxin B	n.s.g.	218(4.57),318(3.84)	162-0969-83
Papaverrubine A	MeOH	240(3.91),289(4.03)	100-0441-83
Protopine N-oxide	MeOH	209(4.56),228(4.42), 283s(3.85),302(3.97)	39-2431-83C
	EtOH	230(4.36),282s(3.80), 303(3.96)	102-0627-83
Rhoeagenine	MeOH	243(3.97),290(3.96)	100-0441-83
Taspine	HCl	246(4.79),285(3.94), 297s(3.88),330(3.86), 346(3.94)	100-0761-83
$C_{20}H_{19}NO_7$			
Benz[h]isoquinoline-1,3,4-tricarboxylic acid, 5,6-dihydro-8-methoxy-, trimethyl ester	MeOH	235(4.1),329(4.3)	5-1476-83
Spiro[2H-pyran-2,2'(3'H)-pyrrolo[2,1-a]isoquinoline]-1'-carboxylic acid, 3,4,5',6'-tetrahydro-8',9'-dimethoxy-3',4-dioxo-, methyl ester	EtOH	245(4.05),289(3.85), 360(3.95)	94-0356-83
$C_{20}H_{19}N_2OS$			
Morpholinium, 4-(4,6-diphenyl-2H-1,3-thiazin-2-ylidene)-, perchlorate	MeCN	270(4.00),336(4.33), 410(3.82)	118-0827-83
$C_{20}H_{19}N_3O$			
1H-Indole-3-acetamide, N-[2-(1H-indol-	MeOH	273(4.02),280(4.05),	142-0469-83

Compound	Solvent	$\lambda_{max}(\log \epsilon)$	Ref.
3-yl)ethyl]- (cont.)		290(3.99)	142-0469-83
$C_{20}H_{19}N_3OS$			
Benzeneacetamide, α-[(4,6-dimethyl-2-pyrimidinyl)thio]-N-phenyl-	MeOH	248(4.48)	103-0492-83
2H-Isothiazolo[4,5,1-hi][1,2,3]benzothiadiazole-3-S^{IV}, 4,6,7,8-tetrahydro-4-(4-methoxyphenyl)-2-phenyl-	C_6H_{12}	203(4.43),244s(4.16), 295(4.23),488(4.34)	39-0777-83C
$C_{20}H_{19}N_3O_6S$			
1-Azabicyclo[3.2.0]hept-3-ene-2-carboxylic acid, 3-[[2-(acetylamino)ethenyl]thio]-6-ethylidene-7-oxo-, (4-nitrophenyl)methyl ester	EtOH	217(4.37),263(4.19), 310(4.07)	78-2551-83
$C_{20}H_{19}N_3S$			
3-Pyridinecarbonitrile, 4-[4-(dimethylamino)phenyl]-1,2,3,4-tetrahydro-6-phenyl-2-thioxo-, compd. with piperidine (1:1)	EtOH	256(4.33),288s(4.17), 348s(2.90)	103-1202-83
$C_{20}H_{19}N_3S$			
2H-1,3-Thiazin-2-one, 4,6-diphenyl-, (1-methylpropylidene)hydrazone, monoperchlorate	MeCN	289(4.22),356s(3.41), 427(4.07)	118-0827-83
$C_{20}H_{19}N_3S_2$			
Benzenesulfenamide, N,N'-[2-(dimethylamino)-2,5-cyclohexadiene-1,4-diylidene]bis-	$CHCl_3$	480(4.51)	150-0028-83S +150-0480-83M
$C_{20}H_{19}N_4O$			
1H-Benzimidazolium, 2-[2-(acetylphenylamino)-1-cyanoethenyl]-1,3-dimethyl-, iodide	EtOH	326(4.35)	124-0297-83
$C_{20}H_{19}N_5O_5$			
1,2-Naphthalenedione, 4-(diethylamino)-, 2-(2,4-dinitrophenylhydrazone)	CH_2Cl_2	295(4.08),378(4.31), 560(4.27)	88-3567-83
$C_{20}H_{20}BrN_2O_2S$			
Isothiazolium, 2-(4-bromophenyl)-4-(4-methoxyphenyl)-5-morpholino-, perchlorate	MeCN	235(4.32),251s(4.21), 288s(3.92),356(4.29)	48-0689-83
$C_{20}H_{20}ClFN_4O$			
Morpholine, 4-[[7-chloro-5-(2-fluorophenyl)-2,3-dihydro-1H-1,4-benzodiazepin-2-yl]iminomethyl]-	isoPrOH	227(4.48),265s(3.90), 360(3.44)	4-0551-83
$C_{20}H_{20}ClNOS$			
2-Propen-1-one, 1-(4-chlorophenyl)-3-[1-cyclohexyl-1,2-dihydro-2-thioxo-3-pyridinyl)-, (E)-	EtOH	276(4.38),341(3.98), 424(3.75)	4-1651-83
$C_{20}H_{20}ClN_3O_2$			
2H-Pyrrolo[2,3-b]quinoxalin-2-one, 3-acetyl-1-(4-chlorophenyl)-4-ethyl-1,3,3a,4,9,9a-hexahydro-	EtOH	222(4.51),306(3.63)	103-0901-83

Compound	Solvent	$\lambda_{max}(\log \epsilon)$	Ref.
$C_{20}H_{20}Cl_2N_4O_2$			
Hydrazinecarboxylic acid, 2-[7-chloro-5-(2-chlorophenyl)-3H-1,4-benzodiazepin-2-yl]-, butyl ester	MeOH	240(4.42),255s(4.33), 275s(4.22),344(3.29)	73-2395-83
$C_{20}H_{20}Cl_2N_8O_5$			
Dichloromethotrexate	pH 1	240(4.36),330(4.08)	162-0446-83
	pH 13	258(4.41),370(3.88)	162-0446-83
$C_{20}H_{20}Fe$			
[2]Metacyclo[2](1,3)ferrocenophane	n.s.g.	451(2.17)	88-5757-83
[2]Paracyclo[2](1,3)ferrocenophane	n.s.g.	447(2.15)	88-5757-83
$C_{20}H_{20}NO_4$			
Californidine (iodide)	MeOH	292(4.02)	100-0293-83
Dehydrodiscretine (iodide)	MeOH and MeOH-acid	242(4.24),265(4.13), 289(4.32),315s(4.15), 342(3.97),381(3.97)	102-0321-83
	MeOH-base	254(4.46),309(4.24), 379(4.31)	102-0321-83
NaBH$_4$ reduction product	MeOH	224s(3.86),286(3.41)	102-0321-83
Jatrorrhizine (iodide)	EtOH	225(4.6),265(4.5), 370(4.6),410-434(3.9)	102-1671-83
	EtOH-NaOH	220(4.6),248(4.5), 400(4.6)	102-1671-83
$C_{20}H_{20}NS$			
1,3-Thiazin-1-ium, 2-(1,1-dimethylethyl)-4,6-diphenyl-, perchlorate	MeCN	275(4.03),344(3.87), 395(4.27)	118-0827-83
$C_{20}H_{20}N_2$			
1,4-Benzenediamine, N,N'-bis(methylphenyl)-	$C_6H_{11}Me$	248(4.18),304(4.27)	35-6268-83
4,7-Methano-1H-indazole, 3a,4,5,6,7,7a-hexahydro-1,3-diphenyl-	C_6H_{12}	368(4.34)	12-1649-83
	MeOH	363(4.30)	12-1649-83
1,3-Pentadien-1-amine, N,N-dimethyl-5-(3-methyl-9H-indeno[2,1-c]pyridin-9-ylidene)-	EtOH	255(4.66),490(4.76)	70-0780-83
$C_{20}H_{20}N_2O_2$			
1H-Indene-2-carboxylic acid, 3-formyl-1,1-dimethyl-, methyl ester, phenylhydrazone	MeOH	213(4.12),243(4.18), 279(4.10),303s(--), 323s(--),390(4.41)	2-0230-83
7H-Indolo[3,4-gh][1,4]benzoxazin-8-one, 4,5,5a,6,6a,7,9,10a-octahydro-4-(phenylmethyl)-, [5aRS(5aα,6aα,10aα)]-	EtOH-HCl	259(3.88),300(3.28)	87-0522-83
1,4-Naphthalenedione, 5-amino-2-[4-(diethylamino)phenyl]-	C_6H_{12}	527(3.85)	150-0168-83S
	EtOH	564(--)	150-0168-83S
1,4-Naphthalenedione, 8-amino-2-[4-(diethylamino)phenyl]-	C_6H_{12}	494(4.05)	150-0168-83S
2-Propenoic acid, 3-[1H-indol-3-yl-(4-methylphenyl)amino]-, ethyl ester	EtOH	210(4.45),290(4.15), 312.5(4.06)	104-1367-83
2-Propenoic acid, 3-(1H-indol-3-yl-phenylamino)-2-methyl-, ethyl ester	EtOH	210(4.52),322(3.37)	104-1367-83
2-Propenoic acid, 3-[(2-methyl-1H-indol-3-yl)phenylamino]-, ethyl ester	EtOH	220(4.23),288(4.16), 312(4.20)	104-1367-83
3,5-Pyrazolidinedione, 4-(3-methyl-2-butenyl)-1,2-diphenyl-	EtOH	246(4.19)	162-0577-83
	pH 9-12	264(4.32)	162-0577-83

Compound	Solvent	$\lambda_{max}(\log \epsilon)$	Ref.
$C_{20}H_{20}N_2O_3$			
3,16b:7,11-Dimethano-4H-[1,7]oxaaza-cyclotridecino[11,10-b]indole-10,17(12H)-dione, 1,2,7,8-tetrahydro-7-methyl-, [7R-(5Z,7R*,11Z,16bS*)]-	EtOH	232(3.96),306(3.97), 355(4.12)	78-3639-83
$C_{20}H_{20}N_2O_4$			
2,4,6-Cycloheptatrien-1-one, 2,2'-(1,2-ethanediyldimino)bis[7-acetyl-	MeOH	248(4.44),354(4.03), 430(4.22)	18-3358-83
α-D-threo-Hex-2-enopyranoside, methyl 2,3-dideoxy-3-(phenylazo)-4,6-O-(phenylmethylene)-	EtOH	302(4.31)	39-0257-83C
Spiro[2H-1-benzopyran-2,2'-[2H]indole], 1',3'-dihydro-8-methoxy-1',3',3'-tri-methyl-6-nitro- (colored form)	toluene	610(4.54)	33-0342-83
$C_{20}H_{20}N_2O_5$			
Acetamide, N-[2-[(9,10-dihydro-4-hy-droxy-9,10-dioxo-1-anthracenyl)-amino]ethyl]-N-(2-hydroxyethyl)-	$CHCl_3$	253(4.56),287(3.89), 555(4.09),593(4.04)	18-1435-83
$C_{20}H_{20}N_2O_8$			
4,4'-Bipyridinium, bis[2-methoxy-1-(methoxycarbonyl)-2-oxoethylide	H_2O	456(3.60)	65-2073-83
	EtOH	510(3.92)	65-2073-83
	85% EtOH	505(3.68)	65-2073-83
	$CHCl_3$	560(4.08)	65-2073-83
$C_{20}H_{20}N_2O_8S_2$			
Gliovirin	EtOH	205(4.61),280s(--)	158-83-135
	EtOH-NaOH	301(3.73)	158-83-135
$C_{20}H_{20}N_2O_{11}$			
Acetic acid, (2-nitrophenoxy)-, oxydi-2,1-ethanediyl ester	MeOH	252(3.80),310(3.59)	49-0359-83
$C_{20}H_{20}N_3O_{10}S_2$			
1-Azabicyclo[3.2.0]hept-2-ene-2-carbox-ylic acid, 3-[[2-(acetylamino)ethen-yl]thio]-7-oxo-6-[1-(sulfooxy)ethyl]-, 2-[(4-nitrophenyl)methyl] ester, ion(1-)	H_2O	218(4.25),264(4.23), 322(4.23)	78-2551-83
$C_{20}H_{20}N_4O_4$			
1,8-Dioxa-2,5,9,12-tetraazacyclotetra-deca-2,9-diene-6,13-dione, 3,10-di-phenyl-, (Z,Z)-	EtOH	255(4.4)	5-2038-83
1,16-Epoxy-3H-dibenzo[c,f][1,5,9]tri-azacyclotridecine-3,5,12-trione, 4-amino-6,7,8,9,10,11-hexahydro-2,15-dimethyl-	DMF	265(4.27),425(4.32), 446(4.33)	95-0049-83
$C_{20}H_{20}N_4O_5S$			
1-Azabicyclo[3.2.0]hept-2-ene-2-carbox-ylic acid, 6-ethyl-3-[(1H-imidazol-4-ylmethyl)thio]-7-oxo-, (4-nitrophen-yl)methyl ester	THF	269.2(4.08),323.5(4.18)	158-0407-83
$C_{20}H_{20}N_4O_7$			
Isoquinolinium, 2-ethyl-, 1-acetonyl-2,4,6-trinitrocyclohexadienide	cation	232(4.70)	103-0773-83
	anion	454(4.03),568(3.99)	103-0773-83
Quinolinium, 1-ethyl-, 1-acetoxy-2,4,6-	cation	239(4.51)	103-0773-83

Compound	Solvent	λ_{max}(log ε)	Ref.
cyclohexadienide (cont.)(in MeCN)	anion	459(4.08),588(4.02)	103-0773-83
$C_{20}H_{20}N_6O_6$			
5-Deazafolic acid, L-	pH 1	212(4.61),278(4.36), 295s(4.30),350(3.83)	44-4852-83
	pH 13	240(4.37),275(4.38), 292s(4.34),340s(--)	44-4852-83
$C_{20}H_{20}N_6O_8S$			
5-Oxa-1-azabicyclo[4.2.0]oct-2-ene-2-carboxylic acid, 7-[(carboxy(4-hydroxyphenyl)acetyl]amino]-7-methyl-3-[[(1-methyl-1H-tetrazol-5-yl)thio]-methyl]-8-oxo-, disodium salt	H_2O	267(4.05)	87-1577-83
$C_{20}H_{20}N_6O_8S_2$			
5-Thia-1-azabicyclo[4.2.0]oct-2-ene-2-carboxylic acid, 7-[[carboxy(4-hydroxyphenyl)acetyl]amino]-7-methoxy-3-[[(1-methyl-1H-tetrazol-5-yl)thio]-methyl]-8-oxo-	MeOH	276.5(3.99)	87-1577-83
disodium salt	H_2O	273.5(4.05)	87-1577-83
$C_{20}H_{20}N_6O_9S$			
Moxalactam	MeOH	276(4.01)	162-0900-83
disodium salt	H_2O	270(4.08)	162-0900-83
$C_{20}H_{20}O_3$			
2-Cyclopenten-1-one, 4-(phenylmethoxy)-2-[(phenylmethoxy)methyl]-, (R)-	EtOH	217(4.20)	39-2441-83C
$C_{20}H_{20}O_4$			
Carinatidin	$CHCl_3$	243(4.33),307(4.59)	102-2227-83
2-Cyclohexen-1-one, 3-(2-hydroxy-4,6-dimethoxyphenyl)-5-phenyl-	MeOH	204(4.71),306(3.39)	142-2369-83
2-Cyclohexen-1-one, 3-(2-hydroxy-4-methoxyphenyl)-5-(4-methoxyphenyl)-	MeOH	202(4.53),330(4.20)	142-2369-83
Glabranin	MeOH	209(4.17),294(4.24), 336(3.59)	102-0573-83
	MeOH-HCl	294(4.25),337(3.57)	102-0573-83
	MeOH-NaOH	248(3.91),332(4.46)	102-0573-83
	MeOH-$AlCl_3$	317(4.43),392(3.60)	102-0573-83
	MeOH-$AlCl_3$-HCl	315(4.40),393(3.61)	102-0573-83
Tricyclo[6.2.2.0^{2,7}]dodeca-2(7),4-diene-4,5-dicarboxylic acid, 9,10,11,12-tetramethylene-, dimethyl ester	MeCN	215(4.20),220(4.19), 235(3.93),251(3.88), 257(3.87),270s(--)	33-0019-83
$C_{20}H_{20}O_5$			
2H-1-Benzopyran-2-one, 3,4-dihydro-5,7-dihydroxy-6-(2-methyl-1-oxobutyl)-4-phenyl-	EtOH	232(4.05),285(4.09), 350(3.38)	78-3923-83
4H-1-Benzopyran-4-one, 2,3-dihydro-3,5,7-trihydroxy-8-(3-methyl-2-butenyl)-2-phenyl-, (2R-trans)-(glepidotin B)	MeOH	214s(4.48),233s(4.23), 297(4.26),334(3.67)	102-0573-83
	MeOH-HCl	215(4.48),233(4.24), 297(4.26),334(3.49)	102-0573-83
	MeOH-NaOH	250(3.96),333(4.43)	102-0573-83
	MeOH-$AlCl_3$	315(4.22)	102-0573-83
	+ HCl	320(4.32)	102-0573-83
Naphtho[2,3-d]-1,3-dioxole, 5,6-dihydro-7-(3,4,5-trimethoxyphenyl)-	EtOH	225(4.35),338(4.38)	94-3056-83

Compound	Solvent	λ_{max}(log ϵ)	Ref.
Salvifaricin	EtOH	207(3.83),297.5(3.67)	102-0784-83
$C_{20}H_{20}O_6$			
Bahifolin	n.s.g.	210(4.00)	102-2755-83
4H-Phenaleno[1,2-b]furan-4-one, 8,9-dihydro-3,6,7-trihydroxy-5-methoxy-1,8,8,9-tetramethyl-, (R)-	EtOH	370(4.15),407(4.24)	18-3661-83
1,5-Phenanthrenediol, 9,10-dihydro-2,7-dimethoxy-, diacetate (eulophiol diacetate)	MeOH	210(4.53),280(4.34)	102-0747-83
Salvifarin	EtOH	208(3.70),236.5(3.45)	102-0784-83
Sesamin, dihydro-, (-)-	MeOH	235(3.96),286(3.97)	102-0265-83
Sigmoidin B	MeOH	288(4.11)	88-4127-83
	MeOH-NaOMe	325(4.34)	88-4127-83
	MeOH-NaOAc	323(4.29)	88-4127-83
	MeOH-$AlCl_3$	309(4.20)	88-4127-83
	+ HCl	309(4.20)	88-4127-83
$C_{20}H_{20}O_7$			
Herqueinone	C_6H_{12}	223(4.34),254(4.08), 307(4.60),450(3.45)	18-3661-83
	EtOH	220(4.29),250(4.09), 314(4.47),416(3.66)	162-0675-83
$C_{20}H_{20}O_8$			
4H-1-Benzopyran-4-one, 2-(3,4-dimethoxyphenyl)-5-hydroxy-3,6,7-trimethoxy- (artemetin)(relative absorbance given)	MeOH	256(0.85),271(0.78), 346(1.00)	102-2107-83
	MeOH-$AlCl_3$	266(0.90),280s(0.84), 299s(0.55),377(1.00), 408s(0.75)	102-2107-83
	MeOH-$AlCl_3$-HCl	263(0.85),283(0.85), 300s(0.63),366(1.00), 406s(0.58)	102-2107-83
4H-1-Benzopyran-4-one, 2-(3,4-dimethoxyphenyl)-5-hydroxy-6,7,8-trimethoxy- (relative absorbance given)	MeOH	254(0.67),284(0.85), 342(1.00)	102-2107-83
	MeOH-$AlCl_3$	264(0.59),293(0.71), 302s(0.70),362(1.00), 420s(--)	102-2107-83
	MeOH-$AlCl_3$-HCl	261(0.61),294s(0.74), 303(0.75),362(1.00), 420s(0.31)	102-2107-83
4H-1-Benzopyran-4-one, 5-hydroxy-3,7-dimethoxy-2-(3,4,5-trimethoxyphenyl)-(combretol)(relative absorbance given)	MeOH	252s(0.85),266(1.00), 300(0.72),346(0.94)	102-2107-83
	MeOH-$AlCl_3$	252s(0.71),276(1.00), 306(0.63),355(0.88), 399(0.72)	102-2107-83
	MeOH-$AlCl_3$-HCl	253s(0.66),278(1.00), 304(0.62),345(0.79), 398(0.60)	102-2107-83
$C_{20}H_{20}O_9$			
4H-1-Benzopyran-4-one, 5-hydroxy-2-(4-hydroxy-3,5-dimethoxyphenyl)-3,6,7-trimethoxy- (murrayanol)(relative absorbance given)	MeOH	255(0.95),267(0.91), 344s(0.97),360(1.00)	102-2107-83
	MeOH-$AlCl_3$	263(1.00),275(0.99), 310(0.34),360(0.70), 420(0.98)	102-2107-83
	MeOH-$AlCl_3$-HCl	263(1.00),275(0.99), 310(0.35),367(0.74), 420(0.97)	102-2107-83
Gardenin C	MeOH	304(4.18),323(4.20)	162-0623-83

Compound	Solvent	λ_{max}(log ε)	Ref.
$C_{20}H_{21}BN_4O_6$			
3-Formazancarboxylic acid, 5-(diacet-oxyboryl)-1,5-diphenyl-, ethyl ester	hexane	495(4.23)	65-2061-83
	EtOH	495(4.16)	65-2061-83
$C_{20}H_{21}BrN_2O_2S$			
Morpholine, 4-[3-[(4-bromophenyl)amino]-2-(4-methoxyphenyl)-1-thioxo-2-propenyl]-	MeCN	251(4.06),301(4.41), 341s(4.22),392s(3.46)	48-0689-83
$C_{20}H_{21}ClN_2$			
4-Isoquinolinamine, 7-chloro-N,N-di-ethyl-3-methyl-1-phenyl-	EtOH	230(4.61),282(3.79), 294(3.84),342(3.73)	44-1275-83
$C_{20}H_{21}ClN_2O_2$			
Acetamide, 2-chloro-N-[1,2,2a,3,4,5-hexahydro-5-hydroxy-1-(phenylmethyl)-benz[cd]indol-4-yl]-	EtOH	257(4.03),296(3.45)	87-0522-83
	EtOH-HCl	257(3.69),296(3.08)	87-0522-83
$C_{20}H_{21}ClN_2S$			
[1]Benzothieno[3,2-b]azet-2a(7bH)-amine, 6-chloro-N,N-diethyl-2-methyl-7b-phenyl-	EtOH	258(4.11),302(3.43), 314(3.34)	44-1275-83
$C_{20}H_{21}N$			
Cyclobenzaprine	n.s.g.	224(4.57),289(4.02)	162-0388-83
hydrochloride	n.s.g.	226(4.72),295(4.08)	162-0388-83
$C_{20}H_{21}NO_2S_2$			
1-Piperidinecarboxylic acid, 4-thieno-[2,3-c][2]benzothiepin-4(9H)-ylidene-, ethyl ester	MeOH	283(3.86),305s(3.69)	73-0623-83
$C_{20}H_{21}NO_3$			
Acronycine, 1,2-dihydro-	EtOH	226(4.17),253s(4.34), 264s(4.53),272(4.60), 298(4.51),315(4.01), 380(3.89)	100-0391-83
5H-1,3-Benzodioxolo[5,6-b]quinolizin-9-ol, 7,8,9,10,10a,11-hexahydro-9-phenyl-	n.s.g.	237(4.23),285(3.90)	12-0149-83
Dehydrodomesticine	EtOH	252(4.66),264s(4.57), 297(3.95),322(4.03), 396(3.40)	100-0761-83
Guadiscine, 6,6a-dihydro-	EtOH	218(4.47),238s(4.20), 280(4.36),290s(4.31), 320s(3.74)	100-0761-83
Mukonicine	EtOH	226(4.70),240(4.67), 300(4.59),342(4.26)	102-2328-83
$C_{20}H_{21}NO_3S$			
1-Naphthalenesulfonic acid, 8-[(3,5-dimethylphenyl)methyl]amino-, methyl ester	dioxan	255(4.18),285(4.03), 385(3.11)	35-6236-83
$C_{20}H_{21}NO_4$			
Acetic acid, (5,6,8,9,10,11-hexahydro-13-oxo-7a,11a-(epoxyethano)-4H-pyrido[3,2,1-jk]carbazol-12-ylidene)-, methyl ester, (E)-	MeOH	238(4.18),280(3.40)	39-0505-83C
(Z)-	MeOH	241(3.98),298(3.08)	39-0505-83C

Compound	Solvent	λ$_{max}$(log ε)	Ref.
9(10H)-Acridinone, 1,5-dihydroxy-3-methoxy-10-methyl-4-(3-methyl-2-butenyl)- (glycocitrine I)	MeOH	228(4.17),268(4.57), 322s(4.01),337(4.07), 415(3.58)	39-1681-83C
Amurensinine, (-)-	EtOH	230(4.07),250s(3.67), 294(3.95)	100-0293-83
7a,11a-Butano-4H,9H-pyrano[3',2':4,5]-pyrrolo[3,2,1-ij]quinoline-11-carboxylic acid, 5,6-dihydro-9-oxo-, methyl ester	MeOH	220(4.04),285(3.60)	39-0505-39C
Buxifoline, N-methyl-	EtOH	217(4.24),247(4.11), 272(4.25)	100-0761-83
Canadine	MeOH	225s(4.25),284(3.88)	78-1975-83
	EtOH	209(4.45),284(3.72)	162-0240-83
Crebanine	MeOH or EtOH	218(4.48),245s(4.02), 280(4.29),290s(4.25), 320s(3.63)	100-0761-83
Dehydrocorydine	EtOH	220(4.33),310(4.27), 340(4.10)	100-0761-83
Dehydroliriofferine	EtOH	262(4.50),333(3.87)	100-0761-83
Eschscholtzidine, (-)-	EtOH	235s(4.01),282s(3.88), 290(3.97),303(3.72)	100-0293-83
Melosmine, dihydro-	EtOH	234s(4.05),262(4.33), 299s(3.77),357(3.58)	100-0761-83
	EtOH-HCl	272(4.33),356(3.62), 438(3.50)	100-0761-83
Reframine, (-)-	MeOH	235s(4.00),248s(3.75), 292(3.83)	100-0293-83
Sinactine	MeOH	230(3.01),282(2.89)	78-1975-83
$C_{20}H_{21}NO_5$			
Acetamide, N-[3-(3,4-dimethoxyphenyl)-1-(6-hydroxy-5-oxo-1,3,6-cycloheptatrien-1-yl)-2-propenyl]-, (E)-	MeOH	245(4.53),307(4.15), 370(3.97)	18-3099-83
	MeOH-HCl	243(4.52),313(4.17), 350(3.81)	18-3099-83
	MeOH-NaOH	248(4.52),337(4.12), 395(4.14)	18-3099-83
9(10H)-Acridinone, 1,3,6-trihydroxy-5-methoxy-10-methyl-4-(3-methyl-2-butenyl)- (prenylcitpressine)	MeOH	222(4.29),262(4.51), 269(4.49),334(4.18), 387(3.76)	94-0895-83
	MeOH-AlCl$_3$	234(--),250s(--), 265s(--),279(--), 368(--),433(--)	94-0895-83
Corydalisol, (±)-	MeOH	210(4.51),239s(3.87), 293(3.91)	39-2431-83C
(-)-	MeOH	210(4.45),235s(3.83), 293(3.87)	100-0414-83
Crebanine, 4-hydroxy-	EtOH	218(4.53),245s(4.17), 281(4.34),320s(3.58)	100-0761-83
Dicentrine, 4-hydroxy-	EtOH	282(3.96),304(3.98)	100-0761-83
2H-[1,3]Dioxolo[4,5-g]furo[2,3,4-ij]-isoquinoline, 2-(3,4-dimethoxyphenyl)-2a,3,4,5-tetrahydro-3-methyl-	MeOH	217(4.31),280(3.51)	83-0737-83
Duguevanine, N-methyl-	EtOH	225(4.51),271s(4.20), 280(4.28),300(4.09), 309s(4.04)	100-0761-83
	EtOH-base	233(4.49),273s(4.04), 282(4.10),331(4.04)	100-0862-83
Guattouregine	EtOH	236s(4.15),262(4.22), 352(3.63)	100-0761-83
	EtOH-HCl	272(4.29),350(3.66), 438(3.51)	100-0761-83

Compound	Solvent	λ_{max}(log ϵ)	Ref.
2(1H)-Pyridinone, 3-[(5,7-dimethoxy-2,2-dimethyl-2H-1-benzopyran-6-yl)-carbonyl]-1-methyl-	EtOH	210(4.58),280s(3.88), 321s(3.91),340(3.99)	39-0219-83C
Sukhodianine	EtOH	216(4.42),280(4.18)	100-0761-83
$C_{20}H_{21}NO_6$			
Acetic acid, [6-[2-(dimethylamino)ethyl]-9-hydroxy-8-methoxy-1-oxonaphtho[1,2-c]furan-3(1H)-ylidene]-, methyl ester (chiloenine)	MeOH	204(4.19),230s(4.02), 290(3.49),377(3.14)	77-0799-83
2-Azabicyclo[3.2.2]nona-3,6-diene-3,4,5-tricarboxylic acid, 2-phenyl-, trimethyl ester	MeCN	242(3.69),360(3.78)	24-0563-83
Bicucullinediol, (-)-	MeOH	210(4.40),237s(3.83), 291(3.81)	78-0577-83
Bis[1,3]benzodioxolo[5,6-e:5',4'-i]-[1,2]oxaazacycloundecin-14-ol, 4,6,7,8,14,15-hexahydro-6-methyl-	MeOH	208(4.64),237s(3.86), 293(3.88)	39-2431-83C
Epiporphyroxine, N-methyl-14-O-de-methyl-	EtOH	238(--),288(--)	100-0441-83
Papaverrubine C	MeOH	232(3.97),285(3.85)	100-0441-83
Papaverrubine D	MeOH	232(3.90),287(3.83)	100-0441-83
$C_{20}H_{21}N_2OS$			
Isothiazolium, 5-morpholino-4-phenyl-2-(phenylmethyl)-, perchlorate	MeCN	247(4.10),331(4.21)	48-0689-83
$C_{20}H_{21}N_2O_2S$			
Isothiazolium, 2-(2-methoxyphenyl)-5-morpholino-4-phenyl-, perchlorate	MeCN	237s(4.02),249(4.05), 347(4.21)	48-0689-83
$C_{20}H_{21}N_3$			
Benzenamine, 4-methyl-N-[(1,2,3,4-tetrahydro-3-methylpyrimido[1,2-a]indol-10-yl)methylene]-	MeOH	374(4.29)	103-0507-83
$C_{20}H_{21}N_3O_2$			
2H-Pyrrolo[2,3-b]quinoxalin-2-one, 3-acetyl-4-ethyl-1,3,3a,4,9,9a-hexahydro-1-phenyl-	EtOH	223(4.49),306(3.59)	103-0901-83
2H-Pyrrolo[2,3-b]quinoxalin-2-one, 3-acetyl-1,3,3a,4,9,9a-hexahydro-4-methyl-1-(2-methylphenyl)-	EtOH	220(4.46),255(3.69), 304(3.50)	103-0901-83
2H-Pyrrolo[2,3-b]quinoxalin-2-one, 3-acetyl-1,3,3a,4,9,9a-hexahydro-4-methyl-1-(4-methylphenyl)-	EtOH	221(4.55),304(3.63)	103-0901-83
$C_{20}H_{21}N_3O_3$			
2H-Pyrrolo[2,3-b]quinoxalin-2-one, 3-acetyl-1,3,3a,4,9,9a-hexahydro-1-(2-methoxyphenyl)-4-methyl-, (3aα,9aα)-	EtOH	220(4.55),258(3.75), 303(3.53)	103-0901-83
$C_{20}H_{21}N_3O_6S$			
1-Azabicyclo[3.2.0]hept-2-ene-2-carboxylic acid, 3-[[2-(acetylamino)ethyl]-thio]-6-ethylidene-7-oxo-, (4-nitrophenyl)methyl ester, [R-(E)]-	EtOH	267(4.05),299(4.04), 337s(3.87)	78-2551-83
$C_{20}H_{21}N_3O_7S$			
1,3,4-Thiadiazole, 3-acetyl-2-(4,5-di-	EtOH	226(4.48),267s(4.24),	39-2011-83C

Compound	Solvent	$\lambda_{max}(\log \epsilon)$	Ref.
methoxy-2-nitrophenyl)-5-(3,4-dimethoxyphenyl)- (cont.)		316(4.34)	39-2011-83C
$C_{20}H_{21}N_4$			
1H-Benzimidazolium, 2-[1-cyano-2-[4-(dimethylamino)phenyl]ethenyl]-1,3-dimethyl-, perchlorate	EtOH	454(4.56)	124-0297-83
$C_{20}H_{21}N_5O_4$			
4,13-Epoxy-1,3-dioxolo[6,7][1,3]diazocino[1,2-e]purin-9(6H)-one, 3a,4,11-12,13,13a-hexahydro-2,2-dimethyl-11-(phenylmethyl)-, [3aR-(3aα,4β-13β,13aα)]-	MeOH	264(4.23),289s(3.90)	138-1017-83
$C_{20}H_{21}N_5O_6$			
Pyrrolidine, 1,1'-(2',4',6'-trinitro-[1,1'-biphenyl]-2,4-diyl)bis-	$CHCl_3$	250s(3.7),568(3.86)	44-4649-83
$C_{20}H_{21}N_5O_7S$			
α-D-Xylofuranuronamide, 1,2-O-(1-methylethylidene)-N-1H-purin-6-yl-, 3-(4-methylbenzenesulfonate)	EtOH	212(4.40),222s(4.33), 257s(3.91),277(4.11), 283(4.11),294s(3.94)	65-1480-83
$C_{20}H_{21}N_7O_5$			
5-Deazaaminopterin, L-	pH 13	247(4.33),278(4.38), 290s(4.35),342s(3.88)	44-4852-83
$C_{20}H_{22}CuN_6$			
Copper, [N,N'-bis[2-(phenylazo)ethylidene]-1,4-butanediaminato(2-)]-, (T-4)-	$CHCl_3$	350s(4.18),375(4.19), 449(4.33),610(3.28), 1160(2.0)	65-1214-83
$C_{20}H_{22}NO_4$			
Bulbocapnine, N-methyl- (iodide)	EtOH	225(4.75),272(4.25), 310(4.02)	100-0761-83
Caryachinium, N-methyl-, (-)-, perchlorate	MeOH	225s(4.07),291.5(3.95)	100-0293-83
$C_{20}H_{22}N_2$			
4-Isoquinolinamine, N,N-diethyl-3-methyl-1-phenyl-	EtOH	230(4.37),280(3.82), 289(3.86),331(3.79)	44-1275-83
$C_{20}H_{22}N_2O$			
1,6-Naphthalenediamine, 7-methoxy-N,N',3-trimethyl-2-phenyl-	MeOH	212(4.30),233(4.51), 261(4.59),328(3.53), 342(3.53)	103-0066-83
9-Oxaergoline, 2,3-dihydro-1-(phenylmethyl)-, (3)-(±)-	EtOH	259(4.01),299(3.42)	87-0522-83
	EtOH-HCl	260(3.94),301(3.40)	87-0522-83
$C_{20}H_{22}N_2OS$			
Morpholine, 4-[3-(methylphenyl amino]-2-phenyl-1-thioxo-2-propenyl]-	MeCN	251(4.12),300(4.32), 332s(4.20),394s(3.44)	48-0689-83
Morpholine, 4-[2-phenyl-3-[(phenylmethyl)amino]-1-thioxo-2-propenyl]-	MeCN	241(4.01),295(4.41), 399(3.23)	48-0689-83
$C_{20}H_{22}N_2O_2$			
Akuammicine	EtOH	227(4.09),300(4.07), 330(4.24)	162-0031-83
Gelsimine	MeOH	210(4.50),252(3.87), 280(3.15)	162-0625-83

Compound	Solvent	$\lambda_{max}(\log \epsilon)$	Ref.
Limatinine	MeOH	217(4.34),258(3.83), 289(3.45)	100-0200-83
2-Piperidinone, 3-[2-(methylamino)-benzoyl]-1-(phenylmethyl)-	MeOH	237(4.41),270(3.87), 396(3.86)	83-0897-83
1H-Pyrazole-4-carboxylic acid, 5-(1,3-pentadienyl)-1-phenyl-3-(1-propen-yl)-, ethyl ester, (E,E,E)-	EtOH	250(4.41),290(4.26)	118-0948-83
2-Pyrazolidinecarboxamide, N-(2-acetyl-phenyl)-1-(phenylmethyl)-, (S)-	$CHCl_3$	261(4.00),270s(3.89), 300(3.614)	35-2010-83
$C_{20}H_{22}N_2O_2S$			
Morpholine, 4-[3-[(2-methoxyphenyl)-amino]-2-phenyl-1-thioxo-2-propenyl]-	MeCN	252(4.15),293(4.30), 342(4.31),399s(3.61)	48-0689-83
Morpholine, 4-[3-[(4-methoxyphenyl)-amino]-2-phenyl-1-thioxo-2-propenyl]-	MeCN	247(4.18),302(4.39), 338s(4.23),398(3.51)	48-0689-83
$C_{20}H_{22}N_2O_3$			
Indolo[2,3-a]quinolizine-3-carboxylic acid, 6,7,12,12b-tetrahydro-1-(1-methoxyethyl)-, methyl ester	EtOH	227(4.3),293(4.1)	142-0001-83
$C_{20}H_{22}N_2O_4$			
Pyrido[3,2-g]quinoline-2,6,8,10(1H,9H)-tetrone, 3,7-dimethyl-4,6-dipropyl-	MeOH	250s(4.07),260s(4.13), 278s(4.30),286(4.34), 309(3.99),321(3.95), 367(3.61),490(3.06)	88-3643-83
$C_{20}H_{22}N_2O_4S$			
Glycine, N-[N-[(phenylmethoxy)carbo-nyl)thio-L-phenylalanyl]-, methyl ester	$CHCl_3$	269(4.09)	78-1075-83
$C_{20}H_{22}N_2O_8S$			
Uridine, 2'-deoxy-5-[4-[[(4-methyl-phenyl)sulfonyl]oxy]-1-butynyl]-	pH 6	228(4.39),292(4.13)	44-1854-83
	pH 13	228(4.42),287(4.00)	44-1854-83
$C_{20}H_{22}N_2S$			
[1]Benzothieno[3,2-b]azet-2a(7bH)-amine, N,N-diethyl-2-methyl-7b-phenyl-	EtOH	244(4.12),290(3.54), 302(3.45)	44-1275-83
$C_{20}H_{22}N_4O$			
1,2,3-Cyclohexanetrione, 4,6-dimethyl-, 1,3-bis(phenylhydrazone)	MeOH	250(4.14),290s(3.75), 395(4.37)	24-1309-83
$C_{20}H_{22}N_4OS_2$			
1H-Benzimidazole, 2,2'-[oxybis(2,1-ethanediylthiomethylene)]bis-	MeOH	250(4.14),275(4.21), 285s(4.15)	4-1481-83
$C_{20}H_{22}N_4O_9$			
3,6,9,12-Tetraoxabicyclo[12.3.1]octa-deca-1(18),14,16-trien-18-ol, 16-[(2,4-dinitrophenyl)azo]-, cesium complex	$CHCl_3$-MeOH-TEA	580(4.67)	138-1415-83
lithium complex	"	547(4.53)	138-1415-83
potassium complex	"	573(4.62)	138-1415-83
rubidium complex	"	575(4.63)	138-1415-83
sodium complex	"	562(4.63)	138-1415-83
$C_{20}H_{22}N_4S_3$			
1H-Benzimidazole, 2,2'-[thiobis(2,1-	MeOH	250(4.15),275(4.24),	4-1481-83

Compound	Solvent	$\lambda_{max}(\log \epsilon)$	Ref.
ethanediylthiomethylene)]bis- (cont.)		285s(4.18)	4-1481-83
$C_{20}H_{22}N_6$			
1,2,4-Triazolo[5,1-c][1,2,4]triazin-4-amine, N,N-diethyl-1,4-dihydro-3,7-diphenyl-	EtOH	235(4.32),302(4.31)	44-2330-83
$C_{20}H_{22}N_6O_3$			
4,13-Epoxy-1,3-dioxolo[6,7][1,3]diazocino[1,2-e]purin-9-amine, 3a,4,11,12-13,13a-hexahydro-2,2-dimethyl-11-(phenylmethyl)-, [3aR-(3aα-4β,13β,13aα)]-	MeOH	275.5(4.49)	138-1017-83
$C_{20}H_{22}O_2$			
Estra-1,3,5,7,9-pentaen-17-one, 3-methoxy-14-methyl-, (±)-	EtOH	229(4.80)	39-2723-83C
$C_{20}H_{22}O_4$			
Benzoic acid, 2-(4-hydroxy-3,5-dimethylbenzoyl)-, butyl ester	EtOH	298.5(4.14)	24-0970-83
1H,7H-4,8-Dioxacyclobut[1,7]indeno-[5,6-a]naphthalen-7-one, 1a,2,3,3a-10b,10c-hexahydro-10-methoxy-1,1,3a-trimethyl-, (±)-	EtOH	238(4.16),257(3.79), 329(4.21)	39-1411-83C
2(3H)-Furanone, dihydro-3,4-bis[(3-methoxyphenyl)methyl]-	$CHCl_3$	275(3.49),282(3.47)	39-0643-83C
2-Naphthalenol, 7,8-dihydro-8-(4-hydroxy-3-methoxyphenyl)-3-methoxy-6,7-dimethyl-, trans-(±)-	n.s.g.	281(4.50)	31-0991-83
Phenol, 4-[2,3-dihydro-7-methoxy-3-methyl-5-(2-propenyl)-2-benzofuran-yl]-2-methoxy-, (2S-trans)-	$CHCl_3$	244.5(3.93),282(3.83), 285(3.82)	102-2227-83
$C_{20}H_{22}O_5$			
Liliflodione	MeOH	208(4.13),226(4.04), 276(3.59)	102-0763-83
$C_{20}H_{22}O_6$			
Acetic acid, [(2-acetyl-7,8,9,10-tetrahydro-4-methyl-6-oxo-6H-dibenzo[b,d]-pyran-3-yl)oxy]-, ethyl ester	MeOH	254(4.41),281(4.06), 314(3.92)	78-1265-83
Bicyclo[3.2.1]oct-3-en-2-one, 7-(1,3-benzodioxol-5-yl)-3,8-dihydroxy-1-methoxy-6-methyl-5-(2-propenyl)-, (6-endo,7-exo,8-anti)-	MeOH	233(3.68),272(3.65)	102-0561-83
Columbin	EtOH	209(3.78)	162-0354-83
Isocolumbin	EtOH	209(3.80)	162-0354-83
Teuscorodonin	EtOH	216s(3.95),220(4.00)	102-0727-83
$C_{20}H_{22}O_7$			
Acetic acid, phenoxy-, oxydi-2,1-ethanediyl ester	MeOH	288(3.46),295(3.39)	49-0359-83
Shinjulactone C	MeOH	248(4.03)	18-3683-83
$C_{20}H_{22}S_2$			
Benzene, 1,1'-[1,3-octadienylidenebis(thio)]bis-, (E)-	EtOH	254(4.44)	118-0383-83
$C_{20}H_{23}ClN_2O$			
Dibenzo[b,f][1,4]oxazepine, 8-chloro-	MeOH	210(4.42),256(3.86),	118-0288-83

Compound	Solvent	$\lambda_{max}(\log \epsilon)$	Ref.
10,11-dihydro-10-methyl-11-(1-methyl-4-piperidinyl)- (cont.)		307(3.51)	118-0288-83
$C_{20}H_{23}CoN_4$			
Cobalt(1+), 6,7,8,9,10,11,12,13-octahydrodibenzo[b,o][1,5,9,13]tetraazacyclohexadecinato-N^6,N^{10},N^{14},N^{20}]-, (SP-4-3)-, perchlorate	DMF	394(3.70),492(3.43)	12-2395-83
$C_{20}H_{23}CuN_4$			
Copper(1+), 6,7,8,9,10,11,12,13-octahydrodibenzo[b,o][1,5,9,13]tetraazacyclohexadecinato-N^6,N^{10},N^{14},N^{20}]-, (SP-4-3)-, perchlorate	DMF	366(3.71),400s(3.69), 506(3.76)	12-2395-83
$C_{20}H_{23}N$			
Amitriptyline, hydrochloride	MeOH	240(4.14)	162-0073-83
$C_{20}H_{23}NO$			
Cyclohexanone, 2-[phenyl(2-phenylethyl)imino]-	MeOH	254(4.21),291(3.31)	39-0573-83C
$C_{20}H_{23}NOS$			
Butanethioamide, 3,3-dimethyl-2-oxo-N,N-bis(phenylmethyl)-	MeOH	251(4.10),274(4.10), 308s(3.47),322s(3.35), 369(2.57)	5-1116-83
$C_{20}H_{23}NO_2$			
1H-Isoindole-4,7-dione, 3a,7a-dihydro-1,1,3a,5,6,7a-hexamethyl-3-phenyl-	EtOH	252(3.85),280s(3.71), 328(3.06)	33-2252-83
1-Oxa-3-azaspiro[4.5]deca-3,6,9-trien-8-one, 2,2,6,7,9,10-hexamethyl-4-phenyl-	EtOH	244(4.24)	33-2252-83
$C_{20}H_{23}NO_3$			
Aporphine, 1,2,10-trimethoxy-	EtOH	216(4.56),274(4.01), 298(4.13)	100-0761-83
$C_{20}H_{23}NO_4$			
5H,6H-5a,10a-Ethenocyclohept[b]indole-11,12-dicarboxylic acid, 7,8,9,10-tetrahydro-5-methyl-	MeOH	211(4.35),240.5(4.28), 307(3.51),440(3.20)	39-0515-83C
Hernagine, N-methyl-	EtOH	221(4.39),269(4.04), 307(3.73)	100-0761-83
Isocorydine, N-oxide	MeOH	223(4.39),271(3.95), 306(3.96)	100-0761-83
Isonorargemonine, (-)-	MeOH	287(3.96)	100-0293-83
Isoquinoline, 3-(3,4-dimethoxyphenyl)-3,4-dihydro-6,7-dimethoxy-1-methyl-, hydrochloride	EtOH	236(4.30),285(3.79), 307(3.93),358(3.75)	142-1247-83
Melosmine, tetrahydro-	EtOH	218(3.90),235s(3.69), 272s(3.71),282(3.75), 300s(3.59),312s(3.51)	100-0761-83
	EtOH-HCl	218(3.90),235s(3/69), 272s(3.71),283(3.75), 302s(3.58),316(3.57)	100-0761-83
Norargemonine, (-)-	EtOH	205s(4.94),230s(4.12), 278s(3.81),283(3.92), 287(3.93),293(3.92)	100-0293-83
3,4-Pyrrolidinediol, 2-[(4-methoxyphen-	MeOH	230(4.08),261(3.11),	87-0469-83

Compound	Solvent	$\lambda_{max}(\log \epsilon)$	Ref.
yl)phenylmethyl]-, 3-acetate (cont.)		268(3.17),276(4.18), 283(3.11)	87-0469-83
3,4-Pyrrolidinediol, 2-[(4-methoxyphenyl)phenylmethyl]-, 4-acetate	MeOH	230(4.14),261(3.19), 268(3.25),276(3.27), 283(3.17)	87-0469-83
Thalisopavine, (-)-	EtOH	289(4.06)	100-0293-83
Vochysine	EtOH	215(4.45),275(3.34)	100-0681-83
	EtOH-NaOH	228(4.46),245(4.42), 288(3.57)	100-0681-83
$C_{20}H_{23}NO_4S$			
Benzene, 1-[4,4-dimethyl-1-[(4-methylphenyl)sulfonyl]-2-butenyl]-4-nitro-, (E)-	EtOH	268(4.15),273(4.14)	12-0527-83
$C_{20}H_{23}NO_5$			
Acetamide, N-[3-(3,4-dimethoxyphenyl)-1-(6-hydroxy-5-oxo-1,3,6-cycloheptatrien-1-yl)propyl]-	MeOH	233(4.45),325(3.80), 370(3.67)	18-3099-83
	MeOH-NaOH	235(4.32),338(3.90), 399(3.85)	18-3099-83
Aporphine, 1,4-dihydroxy-2,10,11-trimethoxy-	EtOH	224(4.23),270(3.72), 305(3.37)	105-0464-83
Butanedioic acid, 2,3,5,6,7,8-hexahydro-5-oxo-1H,9H-pyrido[3,2,1-kl]-benzazocin-9-ylidene)-, dimethyl ester	MeOH	216(4.35),235(4.16)	39-0505-83C
Corydine N-oxide	EtOH	225(4.43),270(3.89), 313(3.70)	100-0761-83
Erythroculine, 3-demethoxy-2α,3α-(methylenedioxy)-	EtOH	215(4.55),238s(4.07), 305(3.54)	102-2603-83
Glaufidine	EtOH	223(4.51),269(4.03), 305(3.69)	100-0761-83
$C_{20}H_{23}NO_6$			
Benzeneacetamide, N-[2-(1,3-benzodioxol-5-yl)ethyl]-2-(hydroxymethyl)-3,4-dimethoxy-	MeOH	225(4.98),280(4.69)	78-1975-83
1,3-Benzodioxole-5-acetamide, N-[2-(3,4-dimethoxyphenyl)ethyl]-4-(hydroxymethyl)-	MeOH	220(5.07),278(4.73)	78-1975-83
Chiloenamine	MeOH	215(4.49),269(4.32), 327(3.22),341(3.25), 387(3.45)	77-0799-83
6-Oxa-2-azaspiro[4.5]dec-3-ene-4-carboxylic acid, 2-ethyl-7-methoxy-1,9-dioxo-3-phenyl-, ethyl ester	EtOH	308(3.84)	94-0356-83
$C_{20}H_{23}N_2O$			
C-Curarine III (chloride)	MeOH	240(4.00),300(3.60), 360(4.23)	162-0382-83
$C_{20}H_{23}N_3O_2S$			
2H-1-Benzopyran-2-one, 7-(diethylamino)-3-(2-pyrrolidino-4-thiazolyl)-	MeCN	417(4.51)	48-0551-83
Carbamic acid, (tetrahydro-3,5-dimethyl-4,6-diphenyl-2H-1,3,5-thiadiazin-2-ylidene)-, ethyl ester	dioxan	271(3.83),293(3.89)	24-1297-83
$C_{20}H_{23}N_3O_3S$			
3,5-Pyridinedicarbonitrile, 1,4-dihydro-2,4,4,6-tetramethyl-1-[2-[[(4-methylphenyl)sulfonyl]oxy]ethyl]-	EtOH	222(4.57),340(3.80)	73-0617-83

Compound	Solvent	λ_{max}(log ϵ)	Ref.
$C_{20}H_{23}N_3O_4S$			
2H-Thiopyran-4-amine, N-(2,4-dinitrophenyl)tetrahydro-N,2,2-trimethyl-6-phenyl-	EtOH	366(4.17)	2-0410-83
isomer	EtOH	366(4.29)	2-0410-83
$C_{20}H_{23}N_3O_7S$			
1-Azabicyclo[3.2.0]hept-2-ene-2-carboxylic acid, 3-[[2-(acetylamino)ethyl]-sulfinyl]-6-ethyl-7-oxo-, (4-nitrophenyl)methyl ester	$CHCl_3$	267(4.12),310(3.89)	158-0407-83
$C_{20}H_{23}N_3O_{10}S_2$			
1-Azabicyclo[3.2.0]hept-2-ene-2-carboxylic acid, 3-[[2-(acetylamino)ethyl]-thio]-7-oxo-6-[1-(sulfooxy)ethyl]-, 2-[(4-nitrophenyl)methyl] ester, monosodium salt	H_2O	267(4.08),318(4.16)	78-2551-83
$C_{20}H_{23}N_4Ni$			
Nickel(1+), (6,7,8,9,10,11,12,13-octahydrodibenzo[b,o][1,5,9,13]tetraazacyclohexadecinato-N^6,N^{10},N^{14},N^{20})-, (SP-4-3)-, perchlorate	DMF	416(3.56),484(3.91), 610(3.23)	12-2395-83
$C_{20}H_{23}N_9O_{13}P_2$			
Adenosine 5'-(trihydrogen diphosphate), 3'-(4-azido-2-nitrobenzenebutanoate)	pH 7.0	249(4.43)	64-0049-83C
$C_{20}H_{23}O_8P$			
α-D-Ribofuranose, 1,2-O-(1-methylethylidene)-, 3-(diphenyl phosphate)	EtOH	209(3.81)	23-1387-83
$C_{20}H_{24}$			
Dicyclopropa[a,c]naphthalene, 1,1a,1b-2,2a,6b-hexahydro-1a,1b,3,4,5,6-hexamethyl-1,2-bis(methylene)-, (1aα,1bβ,2aβ,6bα)-	C_6H_{12}	220(4.34),280(3.60)	35-4400-83
$C_{20}H_{24}BrN$			
Benzenemethanamine, 2-[2-(4-bromophenyl)ethenyl]-N,N-diethyl-4-methyl-, hydrochloride, (E)-	EtOH	303(4.37)	39-1503-83B
$C_{20}H_{24}BrN_3O$			
Bromolysergide	n.s.g.	240(4.28),301(3.95)	162-0196-83
$C_{20}H_{24}ClN_3OS$			
Prochlorperazine sulfoxide	H_2O	240(3.05)	36-0050-83
	2N H_2SO_4	240s(3.13),275s(2.60)	36-0050-83
$C_{20}H_{24}ClN_3S$			
Prochlorperazine	H_2O	250(3.15),300s(2.54)	36-0050-83
	2N H_2SO_4	250(3.17),300(2.53)	36-0050-83
radical cation	9N H_2SO_4	216(3.20),268(3.37), 275(3.39)	36-0050-83
$C_{20}H_{24}Cl_2$			
Benzocyclooctene, 5,9-dichloro-1,2,3,4,6,7,8,10-octamethyl-	C_6H_{12}	220(4.50),290(2.70)	35-4400-83

Compound	Solvent	$\lambda_{max}(\log \epsilon)$	Ref.
$C_{20}H_{24}NO_4$			
Magnoflorine	MeOH	226(4.7),269(4.1), 308(3.9)	102-1671-83
	n.s.g.	270(3.75),310(3.59)	162-0813-83
$C_{20}H_{24}N_2OS$			
Cyclohexanone, 2-[(dimethylamino)methylene]-6-[(3-ethyl-2(3H)-benzothiazolylidene)ethylidene]-	benzene	437(--),457(4.72)	104-1854-83
	MeOH	479(4.81)	104-1854-83
	DMF	442(--),462(--)	104-1854-83
$C_{20}H_{24}N_2O_2S$			
Hycanthone	EtOH	233(4.29),258(4.57), 329(3.99),438(3.82)	162-0689-83
$C_{20}H_{24}N_2O_3$			
Echitamidine	EtOH	237(4.35),300(4.26), 334(4.47)	32-0533-83
Lochneridine	EtOH	230(4.04),293(3.94), 328(4.07)	162-0795-83
2H-Pyran-5-carboxylic acid, 3,4-dihydro-2-methyl-6-(1,3-pentadienyl)-4-(phenylhydrazono)-, ethyl ester, (E,E)-	EtOH	300(4.29),384(4.32)	118-0948-83
$C_{20}H_{24}N_2O_4$			
Cyclohexanone, 2-[(2-aminoethyl)amino]-3,4,5-trihydroxy-3,5-diphenyl-	MeOH	400(3.28)	70-0088-83
Pyrido[3,2-g]quinoline-2,8(1H,3H)-dione, 5,10-dihydro-3,7-dimethyl-4,6-dipropyl- (diazaquinomycin B)	MeOH	277(4.30),310s(4.17), 325s(4.07),356(3.79), 373(3.78)	88-3643-83
$C_{20}H_{24}N_2O_5S$			
1-Azabicyclo[3.2.0]hept-2-ene-2-carboxylic acid, 3-(butylthio)-6-ethyl-7-oxo-, (4-nitrophenyl)methyl ester	THF	270(4.00),322.5(4.05)	158-0407-83
$C_{20}H_{24}N_2O_7$			
Uridine, 5,6-dihydro-5-methyl-2',3'-O-(1-methylethylidene)-, 5'-benzoate, (R)-	n.s.g.	231(4.11),275(2.88), 282(2.83)	33-0687-83
(S)-	n.s.g.	227(4.10),273(3.12), 281s(3.04)	33-0687-83
$C_{20}H_{24}N_2O_7S$			
1-Azetidineacetic acid, 2-(butylthio)-α-(1-methoxy-3-oxopropylidene)-, (4-nitrophenyl)methyl ester	EtOH	260(4.25)	39-0115-83C
$C_{20}H_{24}N_2O_8$			
1H-Indole-1,4,5-tricarboxylic acid, 3-(1,1-dimethylethyl)-6-[(methoxycarbonyl)amino]-, trimethyl ester	EtOH	249(4.31),284(3.78), 315s(--)	138-0743-83
$C_{20}H_{24}N_6$			
Acetaldehyde, 2,2'-(1,4-butanediyldinitrilo)bis-, bis(phenylhydrazone), (Z,Z,E,E)-	$CHCl_3$	300(4.12),340(4.54)	65-1214-83
$C_{20}H_{24}O_2$			
Anthra[2,3-b]furan-5(5aH)-one, 6,7,8,9-9a,10-hexahydro-5a,9,9,11-tetramethyl-	MeOH	236.5(4.52),258(3.76), 272(3.62),310(3.06)	102-0527-83

Compound	Solvent	$\lambda_{max}(\log \epsilon)$	Ref
Estra-1,3,5(10),8-tetraen-17-one, 3-methoxy-14-methyl-, (±)-	EtOH	275(4.24)	39-2723-83C
Estra-1,3,5(10),9(11)-tetraen-17-one, 3-methoxy-14-methyl-, (±)-	EtOH	264(4.29)	39-2723-83C
$C_{20}H_{24}O_3$			
3,6-Ethanodibenz[b,f]oxepin-2(3H)-one, 8-(1,1-dimethylethyl)-4,4a,10,11-tetrahydro-4a-hydroxy-	$CHCl_3$	243(3.58)	35-6650-83
2-Naphthalenecarboxylic acid, 2,3,5,6-7,8-hexahydro-3-(4-methoxyphenyl)-, ethyl ester	C_6H_{12}	276(3.30),283(3.23)	44-4272-83
Phenanthro[10,1-bc]pyran-5,9(4H,7H)-dione, 7a,8,10,11-tetrahydro-3,8,8,11a-tetramethyl-	EtOH	209(4.09),269(2.72), 278(2.44)	102-2011-83
2-Propenoic acid, 3-[2-[2-(4-methoxyphenyl)ethenyl]-1-cyclohexen-1-yl]-, ethyl ester, (E,E)-	C_6H_{12}	356(4.52)	44-4272-83
$C_{20}H_{24}O_4$			
[1,3'-Bi-1H-indene]-1',3,5',6(2H,3aH-4'H)-tetrone, 3'a,4,5,6',7,7',7a,7'a-octahydro-3a,7'a-dimethyl-	EtOH	233(4.07)	12-0117-83
Estra-1,3,5(10)-trien-6-one, 17-acetoxy-3-hydroxy-, (17β)-	MeOH	215(4.19),253(3.84), 325(3.40)	145-0347-83
5,6-Isojatrophone, 2β-hydroxy-	MeOH	280(4.08)	35-3177-83
Jatrophone, 2α-hydroxy-	MeOH	225s(4.17),283(4.16)	35-3177-83
Jatrophone, 2β-hydroxy-	MeOH	225s(4.17),283(4.16)	35-3177-83
$C_{20}H_{24}O_5$			
2H-1-Benzopyran-6-carboxaldehyde, 5-acetoxy-7-methoxy-2-methyl-2-(4-methyl-3-pentenyl)-	EtOH	223(4.03),256s(4.39), 262(4.41),291s(3.80), 304s(3.62),343(3.77)	39-1411-83C
1H-4-Oxabenzo[f]cyclobut[cd]indene-7-carboxaldehyde, 8-acetoxy-1a,2-3,3a,8b,8c-hexahydro-6-methoxy-1,1,3a-trimethyl-, (±)-	EtOH	216(4.23),220(4.23), 237(4.22),281(4.13), 322(3.89)	39-1411-83C
$C_{20}H_{24}O_6$			
3-Furanmethanol, tetrahydro-2-(4-hydroxy-3-methoxyphenyl)-4-[(4-hydroxy-3-methoxyphenyl)methyl]- (lariciresinol)	MeOH	232(4.17),283(4.02)	36-1285-83
Teuscorodin	EtOH	214(3.84),292(2.20)	102-0727-83
$C_{20}H_{24}O_6Se$			
Spiro[1,3-benzoxaselenole-2,1'-cyclohexane]-2',4,4',6,6'(5H,7H)-pentone, 3',3',5,5,5',5',7,7-octamethyl-	$CHCl_3$	300(3.61),402(2.34)	39-0333-83C
$C_{20}H_{24}O_7$			
2,3-Furandiol, tetrahydro-3,4-bis[(4-hydroxy-3-methoxyphenyl)methyl]- (carissanol)	EtOH	228(4.05),282(3.72)	102-0749-83
Gibbane-1,10-dicarboxylic acid, 2,4a,7-trihydroxy-1-methyl-8-methylene-4-oxo-, 1,4a-lactone, 10-methyl ester, (1α,2β,4aα,4bβ,10β)-	MeOH	280(1.94)	78-0449-83
Hemiacetal isomerization product from ailanthone	EtOH	238(3.91)	18-3683-83

Compound	Solvent	λ_{max}(log ε)	Ref.
1(2H)-Naphthalenone, 2-acetoxy-3-[2-(2,2-dimethyl-1,3-dioxolan-4-yl)-1-methoxy-2-oxoethyl]-3,4-dihydro-	EtOH	252(4.58),295(2.78)	44-3269-83
Paeoniflorigenone monomethyl ether acetate	MeOH	228(4.10),268s(3.04), 273(3.06),280(3.00)	94-0577-83
Shinjudilactone	EtOH	238(4.03)	18-3683-83
13-epi-	EtOH	239(4.01)	18-3683-83
$C_{20}H_{24}O_8$			
Glaucarubolone	MeOH	240(4.02)	100-0359-83
Propanedioic acid, 2,2'-(2,5-cyclohexadiene-1,4-diylidene)bis-, tetraethyl ester	$CHCl_3$	371(4.70)	116-1817-83
$C_{20}H_{24}O_9$			
7H-Furo[3,2-g][1]benzopyran-7-one, 2-[1-(β-D-glucopyranosyloxy)-1-methylethyl]-2,3-dihydro-, (R)- (marmesinin)	MeOH	226(4.01),249(3.58), 259(3.49),300s(3.75), 335(4.20)	102-0553-83
α-D-erythro-Hex-2-enitol, 1,5-anhydro-2-deoxy-1-C-2,4-dimethoxyphenyl-, triacetate	EtOH	277(3.44),282(3.39)	23-0533-83
D-threo-Hex-2-enitol, 1,5-anhydro-2-deoxy-1-C-(2,4-dimethoxyphenyl)-, triacetate, (S)-	EtOH	278(3.44),282(3.39)	23-0533-83
D-threo-Hex-2-enitol, 1,5-anhydro-2-deoxy-1-C-(2,5-dimethoxyphenyl)-, triacetate, (S)-	EtOH	293(3.70)	23-0533-83
D-threo-Hex-2-enitol, 1,5-anhydro-2-deoxy-1-C-(2,6-dimethoxyphenyl)-, triacetate, (S)-	EtOH	278(3.34)	23-0533-83
$C_{20}H_{24}O_{10}$			
7H-Furo[3,2-g][1]benzopyran-7-one, 2-[1-(-D-glucopyranosyloxy)-1-methylethyl]-2,3-dihydro-3-hydroxy-, (2R-trans)-	MeOH	225(4.13),249(3.55), 259(3.49),300s(3.97), 327(4.21)	102-0553-83
Marmesin, 2'-β-D-glucopyranosyloxy-	MeOH	226(4.10),248.5(3.76), 258.5(3.66),300s(3.88), 334(4.16)	102-0553-83
$C_{20}H_{24}S_8$			
Bis(tetramethyltetrathiafulvalene), radical cation tetrafluoroborate	MeCN	400s(--),420s(--), 462(4.17),530(3.53), 658(3.82)	104-1092-83
$C_{20}H_{25}N$			
Benzenemethanamine, N,N-diethyl-2-[2-(4-methylphenyl)ethenyl]-, hydrochloride, (E)-	HOAc	300(4.29)	39-1503-83B
Benzenemethanamine, N,N-diethyl-4-methyl-2-(2-phenylethenyl)-, hydrochloride, (E)-	HOAc	295(4.42)	39-1503-83B
$C_{20}H_{25}NO_2$			
Cyclopentanecarboxylic acid, 2-(10-cyano-1,5,9-trimethyl-1,3,5,7,9-decapentaenyl)-	EtOH	372(4.60)	35-4033-83
Fomocaine	EtOH	220(4.20),269(3.14)	162-0604-83

Compound	Solvent	λ_{max}(log ϵ)	Ref.
$C_{20}H_{25}NO_3$			
Fortuneine	MeOH	233(4.05),282s(3.37)	102-0251-83
$C_{20}H_{25}NO_4$			
Erythroculine	EtOH	214(4.38),238s(3.93), 303(3.49)	102-2603-83
Estra-1,3,5(10)-trien-6-one, 17-acetoxy-3-hydroxy-, oxime, (17β)-	MeOH	212(4.29),255(3.96), 306(3.50)	145-0347-83
Ethanone, 1,1'-[4-(2,3-dimethoxyphenyl)-1,4-dihydro-1,2,6-trimethyl-3,5-pyridinediyl]bis-	EtOH	206(4.40),261(3.88), 279s(3.73),356(3.78)	103-0415-83
Isoquinoline, 4-[(3,4-dimethoxyphenyl)-methyl]-1,2,3,4-tetrahydro-6,7-dimethoxy-, hydrochloride	EtOH	233(4.19),282(3.80)	39-2053-83C
7-Isoquinolinol, 1-[(2,4-dimethoxyphenyl)methyl]-1,2,3,4-tetrahydro-6-methoxy-2-methyl-, (±)-	MeOH	205(4.52),231s(4.04), 283(4.47)	100-0908-83
Laudanine	n.s.g.	284(3.78)	162-0774-83
Platycerine	MeOH	282(3.85),315s(3.38)	100-0293-83
$C_{20}H_{25}NO_5S$			
1-Azetidineacetic acid, 2-(butylthio)-α-(1-methoxy-3-oxopropylidene)-4-oxo-, phenylmethyl ester	EtOH	253(4.00)	39-0115-83C
$C_{20}H_{25}NO_6$			
Benzenepropanamide, α-[2-(hydroxymethyl)-4,5-dimethoxyphenyl]-3,4-dimethoxy-	EtOH	232(4.19),281.5(3.76)	39-2053-83C
$C_{20}H_{25}N_2O$			
C-Alkaloid-O , chloride (same spectra in acid or base	EtOH	206s(4.38),223(4.59), 267(3.97),281(3.78), 291(3.74)	33-0405-83
4,17-Cyclocorynanium, 19,20-didehydro-9-methoxy-, chloride, (19E)-	EtOH	206s(4.47),215(4.52), 273(3.98),296(3.73), 308(3.63)	33-0405-83
Macusine B	H_2O	222(4.61),273(3.84), 280(3.82),291(3.74)	162-0808-83
$C_{20}H_{25}N_2OS$			
Benzothiazolium, 2-[2-[3-[(dimethylamino)methylene]-2-methoxy-1-cyclopenten-1-yl]ethenyl]-3-ethyl-, perchlorate	MeOH	606(5.14)	104-1854-83
$C_{20}H_{25}N_3O$			
Perazine sulfoxide	pH 1	234(3.16),271s(2.92)	36-0050-83
	pH 3.5	234(3.05),271(2.88)	36-0050-83
	H_2O	235(3.03),271s(2.85)	36-0050-83
	2N H_2SO_4	234(3.15),270(2.95)	36-0050-83
$C_{20}H_{25}N_3O_5S$			
1-Azabicyclo[3.2.0]hept-2-ene-2-carboxylic acid, 3-[[2-(dimethylamino)-ethyl]thio]-6-ethyl-7-oxo-, (4-nitrophenyl)methyl ester	THF	270(4.03),322(4.08)	158-0407-83
$C_{20}H_{25}N_3S$			
Perazine	pH 1	251s(3.09),297s(2.62)	36-0050-83
	pH 3.5	251(3.06),297(2.48)	36-0050-83

Compound	Solvent	$\lambda_{max}(\log \epsilon)$	Ref.
Perazine (cont.)	H_2O	251(3.12),297(2.52)	36-0050-83
	2N H_2SO_4	251s(3.04),295s(2.48)	36-0050-83
radical cation	9N H_2SO_4	212(3.25),265(3.48), 271(3.50)	36-0050-83
$C_{20}H_{26}$			
2H-Benzocycloheptene, 5-ethylidene-1,2,3,4,6,7,9-heptamethyl-	C_6H_{12}	213(4.34),223(4.32), 278(3.67)	35-4400-83
Benzocyclooctene, 1,2,3,4,5,7,8,9-octamethyl-	C_6H_{12}	233(4.30),330(1.46)	35-4400-83
$C_{20}H_{26}NO_5$			
Isoquinolinium, 1,2,3,4-tetrahydro-7-hydroxy-8-methoxy-1-[(4-methoxyphenyl)methyl]-2,2-dimethyl-, iodide	MeOH	229(4.0),282(3.36)	100-0342-83
Isoquinolinium, 1,2,3,4-tetrahydro-1-[(4-hydroxyphenyl)methyl]-7,8-dimethoxy-2,2-dimethyl-, iodide	MeOH	230(3.07),282(2.56)	100-0342-83
Petaline (iodide)	MeOH	230(3.98),283(3.24)	100-0342-83
$C_{20}H_{26}NO_4$			
Tembetarine (chloride)	MeOH	236(3.85),287(3.59)	102-2607-83
	MeOH-KOH	255(3.75),300(3.60)	102-2607-83
$C_{20}H_{26}N_2O$			
Ibogaine	MeOH	226(4.39),298(3.93)	162-0712-83
Mehranine	n.s.g.	209(4.03),257(3.85), 305(3.47)	64-1700-83B
$C_{20}H_{26}N_2O_2$			
Isogeissoschizol, 10-methoxy-	EtOH and EtOH-base	229(4.72),284(3.84), 294(3.78),308(3.54)	100-0694-83
	EtOH-HCl	224(4.89),276(3.92), 284(3.71),295(3.64), 308(3.56)	100-0694-83
Spiro[cyclohexane-1,4'(1'H)-pyridazine]-3'-carboxylic acid, 5'-phenyl-, 1,1-dimethylethyl ester	$CHCl_3$	250(3.69)	24-3097-83
$C_{20}H_{26}N_2O_3$			
Tetraphyllinol	EtOH	229(4.38),250s(3.77), 268(3.70),297(3.70), 380(3.08)	33-2059-83
$C_{20}H_{26}N_2O_7S$			
1-Azetidineacetic acid, 2-(butylthio)-α-(3-hydroxy-1-methoxypropylidene)-4-oxo-, (4-nitrophenyl)methyl ester	EtOH	266(4.27)	39-0115-83C
$C_{20}H_{26}N_2O_8S$			
Uridine, 5,6-dihydro-5-methyl-2',3'-O-(1-methylethylidene)-, 5'-(4-methylbenzenesulfonate)	n.s.g.	223(4.03),263(2.88), 274(2.81)	33-0687-83
$C_{20}H_{26}N_3O_4S$			
Phenothiazin-5-ium, 3,7-bis[bis(2-hydroxyethyl)amino]-, bromide	pH 7	662(4.97)	77-1521-83
$C_{20}H_{26}O$			
3'H-Cycloprop[1,2]androsta-1,4,6-trien-3-one, 1,2-dihydro-, (1β,2β)-	EtOH	282(4.32)	95-1046-83

Compound	Solvent	$\lambda_{max}(\log \epsilon)$	Ref.
9(1H)-Phenanthrenone, 8-ethenyl-2,3,4,4a,10,10a-hexahydro-1,1,4a,7-tetramethyl-, (4aS-trans)-	MeOH	229(3.60),262(3.49)	78-3351-83
$C_{20}H_{26}O_2$			
1H-Dibenzo[a,d]cycloheptene-6,7-dione, 2,3,5,10,11,11a-hexahydro-1,1-dimethyl-8-(1-methylethyl)-, (S)-	MeOH	227(3.92),270(3.61), 438(3.00)	78-3603-83
2(1H)-Phenanthrenone, 4a,9,10,10a-tetrahydro-6-hydroxy-1,1,4a-trimethyl-7-(1-methylethyl)- (1,2-dehydrohinokione)	n.s.g.	223s(--),284(3.52)	102-1771-83
$C_{20}H_{26}O_3$			
Androst-4-ene-17-carboxylic acid, 18-hydroxy-3-oxo-, γ-lactone	MeOH	240(4.20)	44-2696-83
Cyclopentanecarboxylic acid, 2-(1,5,9-trimethyl-11-oxo-1,3,5,7,9-undecapentaenyl)-, trans	EtOH	378(4.60)	35-4033-83
1,2-Epoxyhinokione	EtOH	220s(--),284(3.52)	102-1771-83
Estra-3,5-dien-17-one, 3-acetoxy-	MeOH	235(4.28)	78-3609-83
1,4-Naphthalenedione, 2-ethyl-3-(3-furanylmethyl)-4a,5,6,7,8,8a-hexahydro-4a,8,8-trimethyl-, (4aS-trans)-	MeOH	215.5(3.77),253.5(3.98)	102-0527-83
1(4H)-Naphthalenone, 3-acetyl-2-(3-furanylmethyl)-4a,5,6,7,8,8a-hexahydro-5,5,8a-trimethyl-, (4aS-trans)-	MeOH	214.5(3.92),242(3.88)	102-0527-83
$C_{20}H_{26}O_4$			
Cyclopentanemethanol, 3-acetoxy-1,2,2-trimethyl-4-(phenylmethylene)-, acetate	EtOH	258(4.20)	39-1465-83C
9(1H)-Phenanthrenone, 2,3,4,4a-tetrahydro-5,6,10-trihydroxy-1,1,4a-trimethyl-7-(1-methylethyl)-, (R)- (14-deoxycoleon U)	EtOH	215(4.10),251.5(3.99), 288(3.91),343(3.95)	102-2005-83
	EtOH-NaOMe	221(4.21),270(4.16), 295(3.76),412(4.09)	102-2005-83
	EtOH-$AlCl_3$	224(4.22),270(4.12), 280(4.20),329(4.00), 380(3.70),464(4.09)	102-2005-83
	EtOH-$AlCl_3$-HCl	221(4.19),256(4.04), 314(4.00),409(4.08)	102-2005-83
	EtOH-NaOAc-boric acid	219(4.18),266(4.22), 296(3.85),340s(3.70), 390(3.94)	102-2005-83
2-Propenoic acid, 3-[3,4-dihydro-8-(4-hydroxy-3-methyl-2-butenyl)-2,2-dimethyl-2H-1-benzopyran-6-yl]-, methyl ester	MeOH	218(4.25),240(4.20), 319(4.34)	94-0352-83
Pteroatisen P_3	MeOH	225(3.88)	94-1502-83
$C_{20}H_{26}O_4S_2$			
3(2H)-Thiophenone, 2-[(4-methoxyphenyl)methylene]-4-(octylthio)-, 1,1-dioxide, (Z)-	EtOH	215(3.71),255(3.70), 295(3.82),380(4.11)	104-1447-83
$C_{20}H_{26}O_5$			
Allethrin II	EtOH	232(4.36)	162-0040-83
$C_{20}H_{26}O_6$			
1,4-Butanediol, 2,3-bis[(4-hydroxy-3-methoxyphenyl)methyl]-	EtOH	215(4.26),230(4.07), 285(3.72)	95-0279-83

Compound	Solvent	λ_{max}(log ε)	Ref.
(cont.)	EtOH-NaOH	216(4.31),236s(4.04), 288(3.69)	95-0279-83
Murrayatin	EtOH	248(3.54),258(3.58), 323(4.14)	102-2273-83
$C_{20}H_{26}O_7$			
Angelol I	MeOH	222(4.36),251(3.84), 300(3.94),324(4.11)	94-0064-83
Cnicin	n.s.g.	220(4.34)	162-0344-83
Picras-3-ene-2,16-dione, 11,20-epoxy-1,11,12-trihydroxy-	EtOH	240(4.06)	18-3683-83
Teucroxide	EtOH	210(3.60)	44-5123-83
$C_{20}H_{26}O_{12}$			
Bicyclo[2.2.2]octane-2,3,5,6,7,8-hexacarboxylic acid, hexamethyl ester	MeCN	212(2.64)	33-0019-83
$C_{20}H_{26}S$			
Benzene, 1,3-bis(1,1-dimethylethyl)-5-(phenylthio)-	hexane	235(3.96),252(4.08), 276(3.76)	89-0490-83
$C_{20}H_{27}N$			
Benzenamine, 4-(1,1-dimethylethyl)-N-[4-(1,1-dimethylethyl)phenyl]-, cation	MeCN	710(4.45)	70-0251-83
radical	MeCN	760(3.56)	70-0251-83
$C_{20}H_{27}NO$			
1-Pentanol, 2-[(1,2-diphenylethyl)amino]-4-methyl-, hydrochloride, [S-(R*,R*)]-	EtOH	205(4.19),248s(2.28), 253(2.43),258(2.56), 265(2.46),268(2.28)	94-0031-83
$C_{20}H_{27}NO_2$			
Estra-1,3,5(10)-trien-3-ol, 17β-(ethylideneamino)-, N-oxide	MeOH	224(4.13),230(4.12), 281(3.19),287(3.14)	13-0001-83B
$C_{20}H_{27}NO_3$			
Estra-1,3,5(10)-triene-3,17-diol, 6-amino-, 17-acetate	MeOH	210(4.04),220(3.80), 282(3.28),286(3.25)	145-0347-83
$C_{20}H_{27}NO_5S$			
1-Azetidineacetic acid, 2-(butylthio)-α-(3-hydroxy-1-methoxypropylidene)-4-oxo-, phenylmethyl ester, (E)-	EtOH	260(3.98)	39-0115-83C
(Z)-	EtOH	264(3.97)	39-0115-83C
$C_{20}H_{27}N_3O_4$			
1,2-Cyclopentatriazoledicarboxylic acid, 3,3a,4,6a-tetrahydro-3-(4-methylphenyl)-, bis(1-methylethyl) ester, cis	EtOH	232.5(4.11),279.5(3.04)	33-1608-83
$C_{20}H_{27}N_3O_7$			
Glycine, N-[1-[(1,1-dimethylethoxy)-carbonyl]-L-prolyl]-N-[(2-nitrophenyl)methyl]-, methyl ester	MeOH	257(3.8)	5-0231-83
$C_{20}H_{28}FeO_6$			
Ferrocene, 1,1'-(2,5,8,11,14-pentaoxapentadecane-1,15-diyl)-	MeOH	438(2.10)	18-0537-83
LiSCN complex	MeOH	438(2.14)	18-0537-83

Compound	Solvent	λ max(log ε)	Ref.
NaSCN complex (cont.)	MeOH	435(2.11)	18-0537-83
KSCN complex	MeOH	434(2.08)	18-0537-83
$C_{20}H_{28}NO_2$			
Pyrrolidinium, 2-carboxy-1-(3,7,11-trimethyl-2,4,6,8,10-dodecapentaenylidene)-, [S-(E,E,E,E,?)]-, perchlorate	$CHCl_3$	516(4.60)	35-4033-83
	+ Et_3N	496(--)	35-4033-83
$C_{20}H_{28}NO_4$			
Pyrrolidinium, 2-carboxy-1-[7-(2-carboxycyclopentyl)-3-methyl-2,4,6-octatrienylidene]-, perchlorate	$CHCl_3$	413(4.51)	35-4033-83
$C_{20}H_{28}N_2$			
Pyrrolo[2,3-b]indole, 1,2,3,3a,8,8a-hexahydro-3a,8-bis(3-methyl-2-butenyl)-, cis-(±)-	EtOH	212(4.04),258(3.83), 310(3.15)	1-0803-83
	EtOH-HCl	212(4.04),248(3.73), 304(3.11)	1-0803-83
$C_{20}H_{28}N_2O$			
Cyclohexanone, 2,6-bis[5-(dimethylamino)-2,4-pentadienylidene]-	$CHCl_3$	310(4.20),525(4.90)	70-1882-83
	EtOH	550(--)	70-1882-83
$C_{20}H_{28}N_2O_3$			
α-Obscurine, N,N-diacetyl-N-demethyl-	MeOH	254(3.38)	5-0220-83
$C_{20}H_{28}N_2O_5$			
1H-Pyrrole-2-carboxylic acid, 5-[(hexahydro-4,4-dimethyl-2,5-dioxo-6aH-furo[2,3-b]pyrrol-6a-yl)methyl]-3,4-dimethyl-, 1,1-dimethylethyl ester	$CHCl_3$	265(4.42)	49-0983-83
$C_{20}H_{28}N_2O_7$			
Amicoumacin C	MeOH	207(4.53),247(3.81), 315(3.62)	158-83-059
$C_{20}H_{28}N_2O_8$			
Amicoumacin B (hydrate)	MeOH	213(4.34),247(3.79), 315(3.64)	158-83-059
$C_{20}H_{28}N_6O_6$			
1(2H)-Pyrimidinepropanamide, N,N'-1,6-hexanediylbis[3,4-dihydro-2,4-dioxo-	H_2O	266(4.15)	103-1228-83
$C_{20}H_{28}O$			
Androst-4-en-3-one, 6-methylene-	EtOH	258(4.04)	95-1046-83
Cleistantha-8,11,13-trien-7-one	MeOH	222(3.78),257(3.84), 305(3.29)	78-3351-83
9(1H)-Phenanthrenone, 7-ethenyl-2,3,4,4a,7,8,10,10a-octahydro-1,1,4a,7-tetramethyl-	EtOH	309(3.85)	102-2011-83
Retinal, 11-cis	EPA	250(4.3),280(4.1), 368(4.4)	35-1626-83
11,13-di-cis	EPA	222(4.1),300(4.3), 360s(4.1)	35-1626-83
8,11,13-tri-cis	EPA	220(4.1),300(4.2)	35-1626-83
$C_{20}H_{28}O_2$			
1H-Cyclopropa[3,4]cycloocta[1,2-c]furan-1-one, 4-(1,5-dimethyl-4-hexenyl)-	MeOH	321(3.86)	44-1903-83

Compound	Solvent	λ_{max}(log ε)	Ref.
3,4,7,7a,8,8a-hexahydro-7-methyl-			44-1903-83
1H-Cyclopropa[3,4]cycloocta[1,2-c]-furan-1-one, 4-(1,5-dimethyl-4-hexenyl)-3,6,7,7a,8,8a-hexahydro-7-methyl-	MeOH	231(3.66),274(3.73)	44-1903-83
1H-Dibenzo[a,d]cycloheptene-6,7-diol, 2,3,5,10,11,11a-hexahydro-1,1-di-	MeOH	210(4.26),278(3.44), 310(3.15)	78-3603-83
methyl-8-(1-methylethyl)-, (S)-	MeOH-base	234(3.61),246(3.64), 282(3.20),335(3.00)	78-3603-83
Methandrostenolone	n.s.g.	245(4.19)	162-0852-83
3,4-Phenanthrenedione, 4b,5,6,7,8,8a-9,10-octahydro-4b,8,8-trimethyl-2-(1-methylethyl)-, (4bS-trans)-	EtOH	264(3.79),424(3.24)	107-0201-83
3(4bH)-Phenanthrenone, 5,6,7,8,8a,9-hexahydro-4-hydroxy-4b,8,8-trimethyl-2-(1-methylethyl)-, (4bS-trans)-	MeOH	318(3.85),370(3.20)	107-0201-83
Pimara-8(9),15-diene, 7,11-dioxo-	EtOH	264(3.93)	102-2011-83
$C_{20}H_{28}O_3$			
Cafestol	n.s.g.	222(3.78)	162-0225-83
Cyclopropanecarboxylic acid, 2,2-dimethyl-3-(2-methyl-1-propenyl)-, 3-(2-butenyl-2-methyl-4-oxo-2-cyclopenten-1-yl ester (cinerin I)	n.s.g.	222(4.33)	162-0326-83
Galeopsinolone	MeOH	217(3.72),238(3.74)	102-0527-83
isomer	MeOH	218(3.81),242(3.95)	102-0527-83
1-Naphthalenecarboxylic acid, 5-[2-(3-furanyl)ethyl]-1,2,3,4,4a,5,8,8a-octahydro-1,4a,6-trimethyl-	$CHCl_3$	220(3.92)	102-1512-83
PGA_2 1,15-lactone	EtOH	215(3.97)	87-1089-83
	EtOH-base	240(3.98),255s(3.92), 267s(3.88),325(4.30)	87-1089-83
PGB_2 1,15-lactone	EtOH	277(4.23)	87-1089-83
9(1H)-Phenanthrenone, 2,3,4,4a,10,10a-hexahydro-5,6-dihydroxy-1,1,4a-tri-	EtOH	216.5(4.10),235(3.96), 290(3.90),364s(3.36)	102-2005-83
methyl-7-(1-methylethyl)-, (4aS-trans) (demethylcryptojaponol)	EtOH-NaOMe	218.5(4.29),259(4.00), 364(4.21)	102-2005-83
	EtOH-$AlCl_3$	218(4.22),257(4.18), 280s(3.64),318s(3.85), 357(4.02),420s(3.65)	102-2005-83
	EtOH-$AlCl_3$-HCl	217(4.22),235(4.15), 290(4.02),362s(3.23)	102-2005-83
	EtOH-NaOAc-boric acid	217(4.22),252(4.16), 315(3.88),349(3.94)	102-2005-83
Pimara-8(9),15-diene, 12β-hydroxy-7,11-dioxo-	EtOH	264(3.94)	102-2011-83
Sarcophytolide	EtOH	238(4.32)	138-0613-83
$C_{20}H_{28}O_4$			
2-Butenoic acid, 2-methyl-, 6-ethenyl-octahydro-3,6-dimethyl-7-(1-methyl-ethenyl)-2-oxo-4-benzofuranyl ester, [3S-(3α,3aα,4a(Z),6β,7α,7aα]]-	EtOH	215(3.4)	39-2017-83C
Foetidin	EtOH	235(4.11)	102-2775-83
Galeolone	EtOH	211(3.79),265(4.01)	102-0527-83
Galeopsitrione	MeOH	221.5(3.70),245(3.23)	102-0527-83
	MeOH-NaOMe	220(3.74),265(3.64), 300(3.34)	102-0527-83
D-Homo-4-nor-3,5-secoestr-9-ene-2,5,17-trione, 2-ethylene acetal	EtOH	249.5(4.23)	39-2349-83C
14β-isomer	EtOH	249.5(4.05)	39-2349-83C

Compound	Solvent	λ max (log ε)	Ref.
18-Homo-4-nor-3,5-secoestr-9-ene-2,5,17-trione, 3-ethylene acetal, (+)-	EtOH	248(3.11)	39-2337-83C
Marrubiin	n.s.g.	208(3.75),212(3.75), 216(3.70)	162-0820-83
Rabdolatifolin	MeOH	244(3.89)	102-2531-83
Salviphlomone	EtOH	217(3.99),256s(3.52), 422(3.48),582(1.90)	102-2005-83
	EtOH-NaOMe	225(4.18),257(4.02), 364(4.12)	102-2005-83
$C_{20}H_{28}O_4Se$			
Spiro[1,3-benzoxaselenole-2,1'-cyclohexane]-2',3',7(4H)-trione, 5,6-dihydro-4,4,4',4',6,6,6',6'-octamethyl-	$CHCl_3$	300(3.88),351(3.38), 445(2.97)	39-0333-83C
$C_{20}H_{28}O_{10}$			
Glycoside A	MeOH	257(4.13),290s(3.26), 301s(3.04)	95-0679-83
$C_{20}H_{29}FO_3$			
Androst-4-en-3-one, 9-fluoro-11β,17β-dihydroxy-17-methyl-	EtOH	240(4.22)	162-0599-83
$C_{20}H_{29}N$			
4-Azuleneethanamine, N,N,α,α,1-pentamethyl-7-(1-methylethyl)-	EtOH	247(4.29),282s(4.51), 285(4.54),289(4.53), 305s(3.94),340s(3.55), 351(3.66),368(3.52), 566s(2.53),588s(2.60), 609(2.66),634s(2.61), 659(2.58),728(2.14)	18-2311-83
hydrochloride	EtOH	215(4.64),246(5.00), 285(5.28),290(5.27), 303s(4.65),330s(2.52), 351(3.19),368(2.90), 440s(2.43),565s(2.35), 614(2.45),666(2.36), 740(1.89)	18-2311-83
$C_{20}H_{29}NO_3$			
Androst-5-en-3β-ol, 17-(nitromethylene)-	EtOH	269(4.62)	22-0061-83
$C_{20}H_{29}NO_3S$			
Spiro[estr-4-ene-17,5'-[1,2,3]oxathiazolidin]-3-one, 3'-methyl-, 2'-oxide	EtOH	238(4.26)	5-1001-83
$C_{20}H_{29}NO_5S$			
1,2-Benzoxathiin-4-amine, 3,4,5,6,7,8-hexahydro-N,N-bis(2-methoxyethyl)-3-phenyl-, 2,2-dioxide, cis	EtOH	330(2.12)	4-0839-83
Latrunculin B	MeOH	212(4.24),269s(--)	44-3512-83
	MeOH	212(4.24)	162-0774-83
$C_{20}H_{29}NO_8$			
7-Azabicyclo[2.2.1]hept-2-ene-1,2,3,4-tetracarboxylic acid, 2,3-bis(1,1-dimethylethyl) 1,4-dimethyl ester	CCl_4	262(3.27)	4-0001-83
7-Azabicyclo[2.2.1]hept-2-ene-1,2,3,4-tetracarboxylic acid, 2,3-dibutyl 1,4-dimethyl ester	CCl_4	262(3.28)	4-0001-83

Compound	Solvent	$\lambda_{max}(\log \epsilon)$	Ref.
7-Azabicyclo[2.2.1]hept-2-ene-1,2,3,4-tetracarboxylic acid, 1,4-dimethyl 2,3-bis(1-methylpropyl) ester	CCl_4	262(3.32)	4-0001-83
1H-Pyrrole-2,3,4,5-tetracarboxylic acid, 2,5-dimethyl 3,4-dipentyl ester	EtOH	225(4.16),267(4.07)	4-0001-83
$C_{20}H_{29}N_2$			
Pyridinium, 4-[(5,5-dimethyl-3-piperidino-2-cyclohexen-1-ylidene)methyl]-1-methyl-, perchlorate	MeOH	516(4.93)	104-2089-83
$C_{20}H_{29}N_3O_7$			
Al-77-A	MeOH	246(3.76),314(3.61)	158-83-91
$C_{20}H_{29}N_5O_3$			
Urapidil	MeOH	237(4.04),268(4.43)	162-1410-83
$C_{20}H_{29}N_5O_8$			
1,2-Hydrazinedicarboxylic acid, 1,1'-[(diethylamino)phenylethenylidene]-bis-, tetramethyl ester	EtOH	225(4.17),294(3.79)	104-0110-83
$C_{20}H_{30}$			
Bicyclo[11.1.0]tetradeca-2,5,10-triene, 2,6-dimethyl-12-methylene-9-(1-methylethyl)-	heptane	239(4.36)	105-0141-83
Cleistantha-8,11,13-triene	MeOH	226(3.30),270(2.16)	78-3351-83
Phenanthrene, 1,2,3,4,4a,9,10,10a-octahydro-1,1,4a-trimethyl-7-(1-methylethyl)-, (4aS-trans)- (abietatriene)	MeOH	261s(2.84),268(2.98), 276(3.01)	33-0780-83
$C_{20}H_{30}N_2O_2$			
Furazabol	EtOH	217(3.63)	162-0613-83
$C_{20}H_{30}N_2O_5SSi$			
Pyridinium, 1-amino-3-(2-trimethylsilylethoxycarbonyl)-, mesitylenesulfonate	MeOH	260(3.82)	5-1374-83
$C_{20}H_{30}N_2O_8$			
Propanedioic acid, 2,2'-(1,4-piperazinediyldimethylidyne)bis-, tetraethyl ester	MeCN	314(4.52)	97-0064-83
$C_{20}H_{30}N_{10}O_8$			
Miharamycin B, hydrochloride	H_2O	218(4.33),244(3.79), 307(3.80)	88-1805-83
$C_{20}H_{30}N_{10}O_9$			
Miharamycin A, dihydrochloride	H_2O	217.5(4.36),244(3.82), 307(3.83)	88-1805-83
$C_{20}H_{30}O$			
Cembrene, 10-oxo-	EtOH	235(4.17)	138-1719-83
4,8,10-Cyclotetradecatrien-1-one, 4,8-dimethyl-14-methylene-11-(1-methylethyl)-, (E,E,E)-	EtOH	240(4.10),248(4.12)	78-1643-83
Estr-4-en-3-one, 16β-ethyl-	EtOH	239(4.21)	95-1046-83
Retinol, 11-cis	EPA	235(4.2),315(4.5)	35-1626-83
11,13-di-cis	EPA	229(4.0),310(4.4)	35-1626-83
9,11,13-tri-cis	EPA	302(4.4)	35-1626-83

Compound	Solvent	λ_{max}(log ϵ)	Ref.
$C_{20}H_{30}O_2$			
Abietic acid	n.s.g.	235(4.29),241.5(4.34), 250(4.16)	162-0001-83
Bicyclo[4.3.1]deca-2,7-diene-7-carboxaldehyde, 5-(1,5-dimethyl-4-hexenyl)-10-hydroxy-2-methyl-	MeOH	229(4.30)	138-0999-83
α-Crenulal	EtOH	234(3.78)	102-2527-83
β-Crenulal (sanadaol)	EtOH	234(3.85)	102-2527-83
2-Cyclohexen-1-one, 2,2'-(1,2-ethanediyl)bis[3,5,5-trimethyl-	EtOH	244.5(4.22)	23-0078-83
Estr-4-en-3-one, 16β-ethyl-17β-hydroxy-	EtOH	240(4.24)	95-1042-83
2-Naphthalenol, 5-[2-(3-furanyl)ethyl]-1,2,3,4,4a,7,8,8a-octahydro-1,1,4a,6-tetramethyl- (baiyunol)	MeOH	212(3.9)	94-0780-83
18-Nor-5α,13β-pregn-16-en-20-one, 3β-hydroxy-	EtOH	239(3.94)	44-1954-83
2-Oxatricyclo[5.4.0.0^{3,11}]undec-8-ene-8-carboxaldehyde, 6-(1,5-dimethyl-4-hexenyl)-3-methyl-	MeOH	232(4.21)	138-0999-83
Oxendolone	EtOH	240(4.20)	162-0994-83
Pachylactone	n.s.g.	228(4.07)	88-5117-83
Pimara-8(9),15-diene, 11β-hydroxy-7-oxo-	EtOH	252(4.12)	102-2011-83
Pimara-8(9),15-diene, 12β-hydroxy-7-oxo-	EtOH	252(4.15)	102-2011-83
Stenbolone	EtOH	241(3.99)	162-1259-83
$C_{20}H_{30}O_3$			
Crenulide, 9-hydroxy-	EtOH	209(3.63),231(3.76)	44-1906-83
Dictyodial A, 4β-hydroxy-	EtOH	232(3.60)	102-2539-83
2,5,10,14-Hexadecatetraen-13-one, 7,8-epoxy-1-hydroxy-3,7,11,15-tetramethyl-, (E,E,E)-	CHCl	248(4.08)	102-1767-83
Leukotriene A_4	MeOH	270(4.64),278(4.75), 290(4.63)	162-0784-83
3(4bH)-Phenanthrenone, 5,6,7,8,8a,9,10-10a-octahydro-10a-hydroxy-4b-(hydroxymethyl)-8,8-dimethyl-2-(1-methylethyl)-	EtOH	242(4.10)	100-0135-83
$C_{20}H_{30}O_4$			
2-Butenoic acid, 2-methyl-, 6-ethenyloctahydro-2-hydroxy-3,6-dimethyl-7-(1-methylethenyl)-4-benzofuranyl ester	EtOH	212(3.8)	39-2017-83C
Crenulide, 1,9-dihydroxy-	EtOH	208(3.75),231(3.73)	44-1906-83
	EtOH-base	268(3.73)	44-1906-83
Fumotoshidin C	EtOH	263(2.89)	94-4409-83
Hallerin	EtOH	215(3.9)	39-2017-83C
Isolathyrol	MeOH	195(4.08),281(4.05)	102-1791-83
Jolkinol 5β,6β-oxide	MeOH	192(3.74),270(4.18)	102-1791-83
PGD_2 1,15-lactone	EtOH-base	243(3.98),250(3.97), 260(3.90),310(4.20)	87-0790-83
Δ^9-PGD_2, 9-deoxy-	EtOH	216(4.00),305(3.08)	87-0790-83
$\Delta^{12,14}$-PGD_2, 15-deoxy-	EtOH	297(4.23)	87-0790-83
PGE_2, 1,15-lactone	EtOH-base	247(3.99),326(4.30)	87-1089-83
$C_{20}H_{30}O_4Se$			
2-Cyclohexen-1-one, 3,3'-selenobis[2-2-hydroxy-4,4,6,6-tetramethyl-	$CHCl_3$	289(4.08),355(4.00)	39-0333-83C
Propanedioic acid, methyl[6-(phenylseleno)hexyl]-, diethyl ester	MeOH	246(3.56),269(3.51)	101-0171-83S

Compound	Solvent	$\lambda_{max}(\log \epsilon)$	Ref.
$C_{20}H_{30}O_5$			
Andrographolide	n.s.g.	223(4.09)	162-0092-83
$C_{20}H_{30}O_{11}$			
Glycoside B	MeOH	221(4.19)	95-0679-83
$C_{20}H_{31}Br$			
1,3,6,10-Cyclotetradecatetraene, 3-(bromomethyl)-7,11-dimethyl-14-(1-methylethyl)-, [S-(all-E)]-	EtOH	243(3.84)	105-0141-83
1,3,6,10-Cyclotetradecatetraene, 4-bromo-3,7,11-trimethyl-14-(1-methylethyl)-, [S-(all-E)]-	EtOH	254(4.23)	105-0141-83
$C_{20}H_{31}NO_2$			
Estr-4-en-3-one, 17-hydroxy-17-[(methylamino)methyl]-, (17β)-	EtOH	239(4.17)	5-1001-83
1H-Isoindole, 2-[[4-(1,1-dimethylethoxy)-2-butenyl]oxy]-2,3-dihydro-1,1,3,3-tetramethyl-, (E)-	EtOH	216(3.40),256(2.89), 263(2.99),270(3.00)	12-1957-83
1H-Isoindole, 2-[[1-(1,1-dimethylethoxy)phenyl]-2-propenyl]oxy]-2,3-dihydro-1,1,3,3-tetramethyl-	EtOH	215(3.41),257(2.85), 263(3.00),270(3.00)	12-1957-83
$C_{20}H_{31}NO_2S_2$			
18,5-(Nitrilometheno)-4H-[1,8]dithiacyclohexadecino[4,5-d]-1,3-dioxin, 6,8,9,10,11,12,13,14,15,17-decahydro-2,2,20-trimethyl-	EtOH	228(3.93)	44-1282-83
$C_{20}H_{31}NO_7$			
Imidodicarbonic acid, [2-[3-(1,1-dimethylethoxy)-5-oxo-2(5H)furanylidene]ethyl]-, bis(1,1-dimethylethyl) ester	EtOH	259(4.26)	39-2983-83C
$C_{20}H_{31}N_2$			
Piperidinium, 1-(5-methyl-9-piperidino-2,4,6,8-nonatetraenylidene)-, (all-E)-, perchlorate	MeOH	637(5.40)	35-0646-83
$C_{20}H_{31}N_3O_7S$			
OA-6129A, sodium salt	H_2O	300(3.75)	158-1473-83
$C_{20}H_{31}N_3O_8S$			
OA-6129B, sodium salt, (R,R)-	H_2O	300(3.81)	158-1473-83
(R,S)-	H_2O	300(3.73)	158-1473-83
$C_{20}H_{31}N_3O_{11}S_2$			
OA-6129C, sodium salt	H_2O	300.5(3.88)	158-1473-83
$C_{20}H_{32}$			
Isoatiserene	n.s.g.	215(2.81),220(2.36), 225(1.70)(end abs.)	2-0989-83
$C_{20}H_{32}ClNO_2$			
Tetradecanamide, N-(2-chlorophenyl)-N-hydroxy-	n.s.g.	257(4.18)	42-0686-83
Tetradecanamide, N-(3-chlorophenyl)-N-hydroxy-	n.s.g.	257(4.15)	42-0686-83

Compound	Solvent	$\lambda_{max}(\log \epsilon)$	Ref.
$C_{20}H_{32}N_2O_8$			
Propanedioic acid, 2,2'-[1,2-ethanediylbis[(methylimino)methylidyne]]bis-, tetraethyl ester	MeCN	278(4.45),291(4.43)	97-0064-83
$C_{20}H_{32}O$			
Abieta-7,9(11)-dien-13β-ol	MeOH	242(4.15)	33-0780-83
Alcyonol A	EtOH	248(4.00)	78-1643-83
Alcyonol B	EtOH	252(4.13)	78-1643-83
Alcyonol C	EtOH	248(4.04)	78-1643-83
Ethanone, 1-[7,11-dimethyl-4-(1-methylethyl)-1,7,11-cyclotridecatrien-1-yl]-	heptane	228.5(3.99)	105-0141-83
Labda-7,13E-dien-15-al	EtOH	237(4.15)	102-1294-83
$C_{20}H_{32}O_2$			
Anadensin	n.s.g.	227(4.02)	88-3787-83
11,12-Epoxycembra-1,3,7-trien-14-ol, (14S,1E,3E,7E)-	n.s.g.	250(4.31)	12-2289-83
$C_{20}H_{32}O_3$			
6-Oxaheneicosa-8,10,12,15-tetraenoic acid, (E,E,Z,Z)-	n.s.g.	272(4.68),283(4.57)	44-4413-83
$C_{20}H_{32}O_4$			
Leukotriene B_4	MeOH	260(4.58),270.5(4.70), 281(4.59)	162-0784-83
1(4H)-Naphthalenone, 4a,5,6,7,8,8a-hexahydro-2-hydroxy-3,4a,8,8-tetramethyl-4-[[(tetrahydro-2H-pyran-2-yl)oxy]methyl]-	MeOH	272(3.86)	39-2739-83C
Δ^{12}-PGD_2	EtOH	244(3.88)	87-0790-83
PGE_1, 1,15-lactone	EtOH-base	242(3.96),270(3.91), 324(4.30)	87-1089-83
$C_{20}H_{32}O_5$			
Prost-13-en-1-oic acid, 10-hydroxy-9,15-dioxo-	EtOH	225(4.12)	39-0319-83C
$C_{20}H_{32}O_7$			
Propanedioic acid, [5-[1-(ethoxycarbonyl)-2-oxocyclopentyl]pentyl]-, diethyl ester	EtOH	253(2.96)	104-0261-83
$C_{20}H_{32}S$			
Cyclohexene, 2-[1-[1-(butylthio)ethylidene]-2-butynyl]-4-methyl-1-(1-methylethyl)-, (E)-(±)-	MeCN	280(3.98)	4-0729-83
$C_{20}H_{33}BrO_4$			
2,3,17,18-Nonadecanetetrone, 10-(bromomethyl)-	EtOH	264(2.84),468(1.65)	5-0181-83
$C_{20}H_{33}NO_2$			
1H-Isoindole, 2-[2-(1,1-dimethylethoxy)butoxy]-2,3-dihydro-1,1,3,3-tetramethyl-	EtOH	215(3.30),257(2.90), 263(2.99),269(3.03)	12-1957-83
1H-Isoindole, 2-[2-(1,1-dimethylethoxy)-1,1-dimethylethoxy]-2,3-dihydro-1,1,3,3-tetramethyl-	EtOH	215(3.34),258(2.85), 263(2.99),270(2.96)	12-1957-83
1H-Isoindole, 2-[1-[(1,1-dimethylethoxy)methyl]propoxy]-2,3-dihydro-	EtOH	214(3.34),257(2.85), 263(2.98),269(3.03)	12-1957-83

Compound	Solvent	λmax(log ε)	Ref.
1,1,3,3-tetramethyl- (cont.)			12-1957-83
1H-Isoindole, 2-[2-(1,1-dimethylethoxy)-1-methylpropoxy]-2,3-dihydro-1,1,3,3-tetramethyl-	EtOH	216(3.32),256(2.91), 263(3.01),270(3.04)	12-1957-83
$C_{20}H_{33}NO_7$			
Imidodicarbonic acid, [2-[3-(1,1-dimethylethoxy)-2,5-dihydro-5-oxo-2-furanyl]ethyl]-, bis(1,1-dimethylethyl) ester	EtOH	223(4.13)	39-2983-83C
$C_{20}H_{33}N_3O_4$			
Urea, N'-[3-acetyl-4-[3-[(1,1-dimethylethyl)amino]-2-hydroxypropoxy]phenyl]-N,N-diethyl- (celiprolol)	pH 2	232(4.44),323(3.38)	145-0002-83
$C_{20}H_{33}N_5O_6Si$			
Guanosine, 5'-O-[(1,1-dimethylethyl)dimethylsilyl]-N-(2-methyl-1-oxopropyl)-	MeOH	255(4.22),259(4.22), 280(4.08)	33-2069-83
$C_{20}H_{34}N_2O_9SSi$			
2,4(1H,3H)-Pyrimidinedione, 5-[2-C-[[[(1,1-dimethylethyl)dimethylsilyl]oxy]methyl]-2,3-O-(1-methylethylidene)-5-O-(methylsulfonyl)-β-DL-ribofuranosyl]-	pH 13	289(3.79)	18-2680-83
	MeOH	266(3.85)	18-2680-83
$C_{20}H_{34}N_4$			
2-Hexenenitrile, 3,3'-(hexanediyldiimino)bis[4-methyl-	EtOH	262(4.51)	39-1223-83C
$C_{20}H_{34}O$			
Acetaldehyde, [1-(1,5-dimethylhexyl)-octahydro-7a-methyl-4H-inden-4-ylidene]-	C_6H_{12}	236(4.30),319(1.81), 332(1.83),346(1.83), 361(1.71),380(1.42), 389(1.00),398(0.30)	35-3270-83
2-Naphthalenol, decahydro-2,5,5,8a-tetramethyl-1-(3-methylene-4-pentenyl)-	hexane	226(4.04)	102-2779-83
$C_{20}H_{34}O_3$			
Benzeneethanol, 3-hydroxy-5-methoxy-α-undecyl-, (R)-	EtOH	274(3.89),281(3.86)	102-2031-83
$C_{20}H_{36}N_4O_8$			
Deferriferrioxamine H	H_2O	455(3.34)	158-83-114
$C_{20}H_{38}N_2O_4$			
Neoenactin B_1, sulfate(2:1)	MeOH and MeOH-HCl	211(3.70)	158-1399-83
	MeOH-NaOH	238(3.76)	158-1399-83
$C_{20}H_{38}N_2O_5$			
Neoenactin B_2, sulfate(2:1)	MeOH and MeOH-HCl	211(3.73)	158-1399-83
	MeOH-NaOH	238(3.78)	158-1399-83
$C_{20}H_{50}Si_5$			
Cyclopentasilane, decaethyl-	isooctane	210s(4.48),225s(4.04), 266f(3.04)	157-1792-83

Compound	Solvent	$\lambda_{max}(\log \epsilon)$	Ref.
$C_{21}H_5Cl_{14}N_2O$			
Methylium, [4-[(aminoacetyl)amino]-2,3,5,6-tetrachlorophenyl]bis(pentachlorophenyl)-, monohydrobromide	dioxan	232(4.76),284s(3.82),365s(4.25),383(4.51),510(2.98),563(3.01)	44-3716-83
$C_{21}H_{10}BrNO_3$			
6H-Anthra[1,9-cd]isoxazol-6-one, 3-(4-bromobenzoyl)-	toluene	325(4.19),468(3.79)	103-0019-83
$C_{21}H_{10}BrN_3O_3$			
9,10-Anthracenedione, 1-azido-2-(4-bromobenzoyl)-	toluene	335(3.64)	103-0019-83
$C_{21}H_{10}ClNO_3$			
6H-Anthra[1,9-cd]isoxazol-6-one, 3-(4-chlorobenzoyl)-	toluene	325(4.22),468(3.84)	103-0019-83
$C_{21}H_{10}ClN_3O_3$			
9,10-Anthracenedione, 1-azido-2-(4-chlorobenzoyl)-	toluene	335(3.65)	103-0019-83
$C_{21}H_{11}BrClN_3O_2$			
7H-Benzo[e]perimidin-7-one, 4-bromo-6-[(4-chlorophenyl)amino]-, 1-oxide	n.s.g.	495(4.23)	2-0812-83
$C_{21}H_{11}BrN_2O_3$			
6H-Anthra[1,9-cd]isoxazole-3-carboxamide, N-(3-bromophenyl)-6-oxo-	toluene	460(4.22)	103-1279-83
6H-Anthra[1,9-cd]isoxazole-3-carboxamide, N-(4-bromophenyl)-6-oxo-	toluene	460(4.22)	103-1279-83
1H-Anthra[1,2-c]pyrazole-3,6,11(2H)-trione, 2-(3-bromophenyl)-	DMF	462(3.47)	103-1279-83
1H-Anthra[1,2-c]pyrazole-3,6,11(2H)-trione, 2-(4-bromophenyl)-	DMF	462(3.39)	103-1279-83
$C_{21}H_{11}BrN_2O_5$			
2-Anthracenecarboxamide, N-(3-bromophenyl)-9,10-dihydro-1-nitro-9,10-dioxo-	dioxan	363(3.73)	103-1279-83
$C_{21}H_{11}BrN_4O_3$			
2-Anthracenecarboxamide, 1-azido-N-(3-bromophenyl)-9,10-dihydro-9,10-dioxo-	dioxan	327(3.81)	103-1279-83
$C_{21}H_{11}ClN_2O_3$			
6H-Anthra[1,9-cd]isoxazole-3-carboxamide, N-(2-chlorophenyl)-6-oxo-	toluene	461(4.18)	103-1279-83
1H-Anthra[1,2-c]pyrazole-3,6,11(2H)-trione, 2-(2-chlorophenyl)-	DMF	475(3.56)	103-1279-83
$C_{21}H_{11}NO_3$			
6H-Anthra[1,9-cd]isoxazol-6-one, 3-benzoyl-	toluene	323(4.18),470(3.79)	103-0019-83
$C_{21}H_{11}N_3O_3$			
9,10-Anthracenedione, 1-azido-2-benzoyl-	toluene	335(3.62)	103-0019-83
$C_{21}H_{11}N_5S_2$			
Benzenesulfenamide, N,N'-(2,3,5-tricyano-2,5-cyclohexadiene-1,4-diyl-	$CHCl_3$	587(4.69)	150-0480-83M

Compound	Solvent	λ_{max}(log ε)	Ref.
idene)bis- (cont.)			150-0480-83M
$C_{21}H_{12}N_2O_2$			
Indolo[1,2-b]-2,6-naphthyridine-5,11-dione, 12-phenyl- (9λ,10ε)	EtOH	230(4.47),250s(4.44), 255(4.47),268s(4.43), 277s(4.13),322s(4.09), 328(4.13),337s(4.09), 403(4.07),?(4.15)	39-2413-83C
$C_{21}H_{12}N_2O_3$			
6H-Anthra[1,9-cd]isoxazole-3-carboxamide, 6-oxo-N-phenyl-	toluene	459(4.20)	103-1279-83
1H-Anthra[1,2-c]pyrazole-3,6,11(2H)-trione, 2-phenyl-	DMF	462(3.48)	103-1279-83
$C_{21}H_{12}N_2O_3S$			
2H-1-Benzopyran-2-one, 3-[2-(2-imino-2H-1-benzopyran-3-yl)-4-thiazolyl]-	MeCN	383(4.27)	48-0551-83
3-Thiophenecarbonitrile, 4-amino-5-[(2-oxo-2H-1-benzopyran-3-yl)carbonyl]-2-phenyl-	MeCN	327(4.18)	48-0551-83
$C_{21}H_{12}O_4$			
[2,2'-Binaphthalene]-1,1',4,4'-tetrone, 3-methyl-	$CHCl_3$	249s(4.52),252(4.53), 262s(4.38),342(3.80)	5-0299-83
$C_{21}H_{13}ClO_2$			
4H-1-Benzopyran-4-one, 2-(3-chlorophenyl)-6-phenyl-	MeOH	273(4.67),306s(4.29)	18-2037-83
4H-1-Benzopyran-4-one, 2-(4-chlorophenyl)-6-phenyl-	MeOH	276(4.62),310(4.39)	18-2037-83
$C_{21}H_{13}NO_3$			
6H-Anthra[1,9-cd]isoxazol-6-one, 5-(4-methylphenoxy)-	EtOH	243(5.03),249(5.08), 444(4.14),456(4.14)	103-1057-83
$C_{21}H_{13}N_3O_3$			
2-Anthracenecarbonitrile, 4-amino-9,10-dihydro-3-hydroxy-9,10-dioxo-1-(phenylamino)-	DMF	570(4.11)	33-0411-83
5H-Pyrido[1',2':1,2]pyrimido[4,5-b]-acridine-7,15-dione, 2-acetyl-	DMSO	452(3.73),480(3.90), 511(3.88)	18-1775-83
	H_2SO_4	464(3.62),497(3.65), 530(3.59)	18-1775-83
$C_{21}H_{14}BrNO$			
Benzo[f]quinolinium, 2-(4-bromophenyl)-2-oxoethylide	DMF	515(3.70)	104-0568-83
$C_{21}H_{14}BrNO_4$			
3H-Benz[e]indole-1-carboxylic acid, 8-bromo-4,5-dihydro-4,5-dioxo-2-phenyl-, ethyl ester	EtOH	205(4.47),282(4.59), 355(3.92)	103-1086-83
$C_{21}H_{14}ClN_7O_2$			
1(2H)-Phthalazinone, 4-chloro-, [[(4-nitrophenyl)azo]phenylmethylene]hydrazone	MeOH	208(4.55),240s(4.11), 295(4.23),480(4.18)	104-0953-83
	dioxan	240(4.15),295(4.22), 480(4.28)	104-0953-83
	MeCN	208(4.61),240s(4.11), 290(4.30),395s(4.04),	104-0953-83

Compound	Solvent	λ_{max}(log ε)	Ref.
(cont.)		475(4.13)	104-0953-83
$C_{21}H_{14}Cl_2$			
7H-Benzocyclopentadecene, 14,15,16,17-tetradehydro-7-(dichloromethylene)-13-methyl-, (E,E,Z,E)-	CH_2Cl_2	231(4.31),293(4.57), 318s(4.48),371s(4.14)	39-2997-83C
9H-Benzocyclopentadecene, 14,15,16,17-tetradehydro-9-(dichloromethylene)-13-methyl-, (E,E,E,Z)-	CH_2Cl_2	232(4.48),273s(4.54), 285(4.62),305(4.55), 368s(4.15),394s(4.02)	39-2997-83C
$C_{21}H_{14}Cl_2N_2S$			
Benzenamine, 4-chloro-N-[3-(4-chlorophenyl)-4-phenyl-2(3H)-thiazolylidene]-	EtOH	308(4.25)	56-0875-83
$C_{21}H_{14}Cl_3N_3O_2$			
9,10-Anthracenedione, 1,5-diamino-2,4,6-trichloro-8-[(4-methylphenyl)amino]-	DMF	616s(3.91),650(4.30)	2-0808-83
$C_{21}H_{14}Cl_3N_3O_3$			
9,10-Anthracenedione, 1,5-diamino-2,4,6-trichloro-8-[(4-methoxyphenyl)-amino]-	DMF	465(4.27),608s(4.22)	2-0808-83
$C_{21}H_{14}Cl_3N_5O_8$			
Benzenamine, 4-[2,2-dichloro-1-(2-chloro-3,5-dinitrophenyl)ethyl]-N-(4-methylphenyl)-2,6-dinitro-	MeOH	209.5(4.57),428.8(3.73)	56-1357-83
$C_{21}H_{14}Cl_3N_5O_9$			
Benzenamine, 4-[2,2-dichloro-1-(2-chloro-3,5-dinitrophenyl)ethyl]-N-(2-methoxyphenyl)-2,6-dinitro-	MeOH	210.6(4.55),426.6(3.70)	56-1357-83
$C_{21}H_{14}Cl_4O_3$			
Spiro[bicyclo[2.2.2]oct-5-ene-2,4'-[3]-oxatricyclo[4.2.1.0^{2,5}]non[7]ene]-3,9'-dione, 1,4,5,6-tetrachloro-8-phenyl-	$CHCl_3$	249(3.43),288s(2.36), 297(2.39),327s(2.21), 338(2.03)	142-1017-83
$C_{21}H_{14}N_2O_2$			
Cyclohept[1,2,3-hi]imidazo[2,1,5-cd]-indolizine-4-carboxylic acid, 2-phenyl-, methyl ester	EtOH	247(4.33),278s(4.39), 295(4.54),310s(4.37), 358s(4.35),374(4.43), 406(4.08),428(4.08)	18-3703-83
6H-Pyrido[4',3':5,6]oxepino[3,2-b]in-dol-5(12H)-one, 12-phenyl-	EtOH	241(4.32),260s(4.10), 345s(3.60),414(3.48)	39-2413-83C
$C_{21}H_{14}N_2O_2S_3$			
Benzenesulfenamide, N-(4,5-diphenyl-1,3-dithiol-2-ylidene)-2-nitro-	1,2-$C_2H_4Cl_2$	315(4.17),422(3.72)	118-0050-83
$C_{21}H_{14}N_2O_3$			
Benzo[f]quinolinium 2-(4-nitrophenyl)-2-oxoethylide	DMF	523(4.30)	104-0568-83
$C_{21}H_{14}O_2$			
4H-1-Benzopyran-4-one, 2,6-diphenyl-	MeOH	273(4.64),304(4.32)	18-2037-83
4H-1-Benzopyran-4-one, 2,8-diphenyl-	MeOH	253(4.36),302(4.29)	18-2037-83

Compound	Solvent	λ_{max}(log ε)	Ref.
$C_{21}H_{14}O_3$			
4H-1-Benzopyran-4-one, 2-(2-hydroxy-phenyl)-6-phenyl-	MeOH	258(4.31),318(4.12)	18-2037-83
$C_{21}H_{14}O_4$			
3H-Naphtho[2,1-b]pyran-3-one, 2-(4-methoxybenzoyl)-	benzene	373(4.14)	151-0325-83A
$C_{21}H_{14}O_7$			
Benzoic acid, 2-[(2-hydroxybenzoyl)-oxy]-, 2-carboxyphenyl ester	EtOH	228(4.44),281(3.64), 310(3.69)	36-0322-83
$C_{21}H_{15}BrNO$			
Benzo[f]quinolinium, 4-[2-(4-bromo-phenyl)-2-oxoethyl]-, iodide	EtOH	218(4.82),276(4.44), 371(3.93)	104-0568-83
$C_{21}H_{15}BrN_2O_3$			
9,10-Anthracenedione, 1-amino-2-bromo-4-[(4-methoxyphenyl)amino]-	DMF	590(4.08),615(4.08)	2-0808-83
$C_{21}H_{15}BrO_2$			
2-Propen-1-one, 3-(4-bromophenyl)-1-(4-phenoxyphenyl)-	isooctane	282(4.17),314(4.61)	124-1211-83
	dioxan	318(4.57)	124-1211-83
$C_{21}H_{15}ClN_2O_2$			
9,10-Anthracenedione, 1-amino-8-chloro-2-[(4-methylphenyl)amino]-	DMF	543(4.17)	2-0808-83
$C_{21}H_{15}ClN_2O_3$			
9,10-Anthracenedione, 1-amino-8-chloro-2-[(4-methoxyphenyl)amino]-	DMF	552(4.07)	2-0808-83
$C_{21}H_{15}ClN_6$			
1(2H)-Phthalazinone, 4-chloro-, [phen-yl(phenylazo)methylene]hydrazone	MeOH	205(4.60),292(4.24), 410(4.11),450(4.11)	104-0953-83
	dioxan	292(4.20),400s(4.15), 445(4.20)	104-0953-83
	MeCN	205(4.61),290(4.28), 400(4.13),445s(4.00)	104-0953-83
$C_{21}H_{15}ClO_2$			
4H-1-Benzopyran-4-one, 2-(3-chloro-phenyl)-2,3-dihydro-6-phenyl-	MeOH	249(4.46),340(3.28)	18-2037-83
4H-1-Benzopyran-4-one, 2-(4-chloro-phenyl)-2,3-dihydro-6-phenyl-	MeOH	247(4.57),346(3.23)	18-2037-83
2-Propen-1-one, 3-(3-chlorophenyl)-1-(4-hydroxy[1,1'-biphenyl]-3-yl)-	MeOH	258(4.44),313(4.37)	18-2037-83
2-Propen-1-one, 3-(4-chlorophenyl)-1-(4-hydroxy[1,1'-biphenyl]-3-yl)-	MeOH	258(4.44),326(4.42)	18-2037-83
2-Propen-1-one, 3-(4-chlorophenyl)-1-(4-phenoxyphenyl)-	isooctane	282(4.20),314(4.60)	124-1211-83
	dioxan	317(4.58)	124-1211-83
$C_{21}H_{15}Cl_2N_3$			
3-Pyridinecarbonitrile, 1-[(2,6-di-chlorophenyl)methyl]-1,4-dihydro-4-(1H-indol-1-yl)-	CH_2Cl_2	282(3.6),330(3.8)	64-0878-83B
$C_{21}H_{15}Cl_3N_2$			
2H-Pyrazole, 1,3,5-tris(4-chloro-phenyl)-4,5-dihydro-	C_6H_{12}	366(4.38)	12-1649-83
	MeOH	364(4.38)	12-1649-83

Compound	Solvent	$\lambda_{max}(\log \epsilon)$	Ref
$C_{21}H_{15}I$			
9H-Fluorene, 2-[2-(4-iodophenyl)ethen-yl]-	dioxan	344(4.54)	104-0683-83
$C_{21}H_{15}I_4NO_5$			
DL-Tyrosine, O-(4'-hydroxy-3',5'-di-iodo[1,1'-biphenyl]-4-yl)-3,5-diiodo-	MeOH-NaOH	225(4.79),240s(4.62), 301(4.00)	117-0137-83
$C_{21}H_{15}N$			
Indeno[1,2,3-kl]acridine, 1,3-dimethyl-	EtOH	358(3.45),377(3.70), 419(3.50)	35-3723-83
Pyridine, 2-[2-(9-phenanthrenyl)ethen-yl]-, cis	pH 1	348(3.88)	62-0199-83B
	pH 8	298(4.06)	62-0199-83B
trans	pH 1	369(4.08)	62-0199-83B
	pH 8	333(4.34)	62-0199-83B
Pyridine, 3-[2-(9-phenanthrenyl)ethen-yl]-, cis	pH 1	none above 300 nm	62-0199-83B
	pH 8	299(4.03)	62-0199-83B
trans	pH 1	341(4.24)	62-0199-83B
	pH 8	329(4.32)	62-0199-83B
Pyridine, 4-[2-(9-phenanthrenyl)ethen-yl]-, cis	pH 1	361(3.76)	62-0199-83B
	pH 8	294(4.00)	62-0199-83B
trans	pH 1	379(4.33)	62-0199-83B
	pH 8	333(4.33)	62-0199-83B
$C_{21}H_{15}NO_2$			
9,10-Anthracenedione, 1-(methylphenyl-amino)-	C_6H_5Cl	375(3.7),540(3.5)	104-0139-83
9,10-Anthracenedione, 2-methyl-1-(phen-ylamino)-	C_6H_5Cl	335(3.9),505(3.6)	104-0139-83
$C_{21}H_{15}NO_4$			
2-Propen-1-one, 3-(4-nitrophenyl)-1-(4-phenoxyphenyl)-	isooctane	290(4.22),318(4.40)	124-1211-83
	dioxan	321(4.37)	124-1211-83
$C_{21}H_{15}N_2O_3$			
Benzo[f]quinolinium, 4-[2-(4-nitro-phenyl)-2-oxoethyl]-, iodide	EtOH	231(4.82),274(4.57), 368(4.06)	104-0568-83
$C_{21}H_{15}N_3O$			
9H,13cH,14H-4b,9a,13b-Triazadibenzo-[a,e]acephenanthrylen-9-one	EtOH	211(4.46),235(4.30), 333(3.81)	142-0617-83
	50% $HClO_4$	207(4.46),230(4.31), 256(4.27),301(4.01)	142-0617-83
$C_{21}H_{15}N_3O_2$			
7H-Dibenz[f,ij]isoquinoline-1-carbo-nitrile, 2-morpholino-7-oxo-	DMF	448(3.47)	2-1197-83
2,3,9b-Triazaindeno[6,7,1-ija]azulene-8-carboxylic acid, 1-(4-methylphen-yl)-, methyl ester	EtOH	241(4.31),263(4.26), 292(4.45),321(4.01), 336(4.07),370s(4.33), 381(4.35),421(4.16), 443(4.19)	18-3703-83
$C_{21}H_{15}N_3O_3$			
Bicyclo[4.4.1]undeca-3,6,8,10-tetra-ene-2,5-dione, mono[(1-nitro-2-naphthalenyl)hydrazone], (±)-	CH_2Cl_2	367(4.36),454(4.25)	33-2369-83
	EtOH-KOH	567(4.73)	
4a,8a-Methanonaphthalene-1,4-dione, mono[(4-nitro-1-naphthalenyl)hy-drazone]	CH_2Cl_2	450(4.48)	33-2626-83
	EtOH-KOH	626(4.44)	33-2626-83

Compound	Solvent	λ_{max}(log ε)	Ref.
$C_{21}H_{16}$			
9H-Fluorene, 2-(2-phenylethenyl)-	dioxan	339(4.76)	104-0683-83
$C_{21}H_{16}BrNO$			
2-Propen-1-one, 3-(4-bromophenyl)-1-[4-(phenylamino)phenyl]-	EtOH	241(4.06),278(3.83),308(4.30),397(4.41)	124-1211-83
	dioxan	280(3.81),308(4.32),379(4.45)	124-1211-83
$C_{21}H_{16}BrNS$			
1,5-Benzothiazepine, 2-(4-bromophenyl)-2,3-dihydro-4-phenyl-	EtOH	256(4.36),335(3.65)	103-1293-83
1,5-Benzothiazepine, 4-(4-bromophenyl)-2,3-dihydro-2-phenyl-	EtOH	262(4.37),341(3.68)	103-1293-83
$C_{21}H_{16}Br_2N_2$			
1H-Pyrazole, 3,5-bis(4-bromophenyl)-4,5-dihydro-1-phenyl-	C_6H_{12}	365(4.30)	12-1649-83
	MeOH	364(4.32)	12-1649-83
$C_{21}H_{16}ClNO$			
2-Propen-1-one, 3-(4-chlorophenyl)-1-[4-(phenylamino)phenyl]-	EtOH	241(4.00),278(3.86),308(4.28),397(4.40)	124-1211-83
	dioxan	280(3.83),306(4.28),379(4.43)	124-1211-83
$C_{21}H_{16}ClNO_2S$			
1H-Indole, 3-chloro-1-[(4-methylphenyl)sulfonyl]-2-phenyl-	EtOH	224(4.45),241s(4.37),278s(4.24),288(4.27)	39-2417-83C
2-Propen-1-one, 3-[1-(4-chlorophenyl)-1,2-dihydro-2-thioxo-3-pyridinyl]-1-(4-methoxyphenyl)-, (E)-	EtOH	241(4.26),272(4.25),309(4.15),337(4.15),427(3.73)	4-1651-83
$C_{21}H_{16}ClNS$			
1,5-Benzothiazepine, 2-(4-chlorophenyl)-2,3-dihydro-4-phenyl-	EtOH	257(4.36),334(3.64)	103-1293-83
1,5-Benzothiazepine, 4-(4-chlorophenyl)-2,3-dihydro-2-phenyl-	EtOH	261(4.28),340(3.61)	103-1293-83
$C_{21}H_{16}ClN_3OS_2$			
4-Thiazolamine, N-[(2-chloro-6-methoxy-3-quinolinyl)methylene]-2-(methylthio)-5-phenyl-	DMF	305(3.42),400(3.12)	103-0166-83
$C_{21}H_{16}ClN_3S_2$			
4-Thiazolamine, N-[(2-chloro-6-methyl-3-quinolinyl)methylene]-2-(methylthio)-5-phenyl-	DMF	305(3.44),400(3.10)	103-0166-83
$C_{21}H_{16}ClN_7O_2$			
Benzenecarbohydrazonic acid, N-(4-chloro-1-phthalazinyl)-, 2-(4-nitrophenyl)hydrazide	MeOH	210(4.69),289(4.22),375(4.43)	104-0953-83
	dioxan	290(4.26),377(4.48)	104-0953-83
	MeCN	208(4.73),290(4.24),377(4.49)	104-0953-83
$C_{21}H_{16}Cl_2$			
Undeca-3,5,7,10-tetraen-1-yne, 9-(dichloromethylene)-11-(2-ethynylphenyl)-3-methyl-	CH_2Cl_2	237(4.43),264(4.35),312s(4.53),327(4.67),343(4.73),362(4.64),396(3.60),421(3.49)	39-2997-83C

Compound	Solvent	λ_{max}(log ϵ)	Ref.
Undeca-3,5,8,10-tetraen-1-yne, 7-(dichloromethylene)-11-(2-ethynylphenyl)-3-methyl-	CH_2Cl_2	235(4.22),262(4.34), 271(4.34),317s(4.58), 329(4.61),343s(4.54), 372s(4.30)	39-2997-83C
$C_{21}H_{16}Cl_2N_2$			
1H-Pyrazole, 1,3-bis(4-chlorophenyl)-4,5-dihydro-5-phenyl-	C_6H_{12}	367(4.36)	12-1649-83
	MeOH	365(4.36)	12-1649-83
1H-Pyrazole, 1,5-bis(4-chlorophenyl)-4,5-dihydro-3-phenyl-	C_6H_{12}	355(4.30)	12-1649-83
	MeOH	354(4.34)	12-1649-83
1H-Pyrazole, 3,5-bis(4-chlorophenyl)-4,5-dihydro-1-phenyl-	C_6H_{12}	363(4.32)	12-1649-83
	MeOH	362(4.36)	12-1649-83
$C_{21}H_{16}Cl_4O_3$			
Spiro[bicyclo[2.2.2]oct-5-ene-2,4'-[3]oxatricyclo[4.2.1.0^{2,5}]non[7]-en]-3-one, 1,3,5,6-tetrachloro-9'-hydroxy-8-phenyl-	$CHCl_3$	248(3.44),263s(2.93), 271(2.63),325(2.28)	142-1017-83
$C_{21}H_{16}F_5N_2S_2$			
Benzothiazolium, 3-methyl-2-(3,3,4,4,4-pentafluoro-2-[(3-methyl-2(3H)-benzothiazolylidene)methyl]-1-butenyl]-, perchlorate	EtOH	590(4.23)	104-1937-83
$C_{21}H_{16}FeO_4$			
Iron, tricarbonyl[(1,2,5,6-η)-4,4-diphenyl-1,5-cyclohexadien-1-ol]-	dioxan	234s(4.4),275s(3.7), 284s(3.6)	24-0243-83
$C_{21}H_{16}NO$			
Benzo[f]quinolinium, 4-(2-oxo-2-phenylethyl)-, iodide	EtOH	230(4.82),282(4.45), 372(4.03)	104-0568-83
$C_{21}H_{16}N_2$			
9b,10-Diazacyclohepta[a]cyclohepta-[4,5]benz[1,2,3-cd]azulene, 5,11-dihydro-	isooctane	228s(4.42),263(4.24), 331(4.22),422(4.24), 440(4.08),510(2.74), 556(2.90),608(3.04), 678(3.04),762(3.81)	24-1963-83
1H-Pyrazole, 1,3,5-triphenyl-	n.s.g.	252(4.50)	44-0542-83
$C_{21}H_{16}N_2O_2$			
9,10-Anthracenedione, 1-amino-4-[(4-methylphenyl)amino]-	DMF	580(4.24),618(4.25)	2-0808-83
1-Anthracenol, 4-[(4-methoxyphenyl)-azo]-	EtOH	530(4.47)	42-0408-83
	acetone	525(4.51)	42-0408-83
Indolo[1,2-b][2,6]naphthyridin-5(12H)-one, 11,11a-dihydro-11-hydroxy-12-phenyl-	EtOH	230s(3.70),285(3.65), 322(3.65)	39-2413-83C
2-Naphthalenecarbonitrile, 3-[2-(ethylphenylamino)ethenyl]-1,4-dihydro-1,4-dioxo-	C_6H_{12}	547(4.30)	150-0168-83S
	EtOH	572(--)	150-0168-83S
2(1H)-Quinolinone, 4-hydroxy-3-(phenylmethyl)-1-(2-pyridinyl)-	pH 1	213(4.829),228(4.831), 279(4.210),287(4.207), 311(4.107),318(4.143), 329(4.075)	49-0227-83
	pH 10	219(4.743),230s(4.499), 256s(4.272),310(4.299), 322(4.220)	49-0227-83

Compound	Solvent	λ_{max}(log ε)	Ref.
$C_{21}H_{16}N_2O_2S_3$			
Ethane(dithioic) acid, (diphenylamino)-thioxo-, (4-nitrophenyl)methyl ester	CH_2Cl_2	279(4.32),345s(3.87), 435s(3.08)	97-0247-83
$C_{21}H_{16}N_2O_3$			
9,10-Anthracenedione, 1-amino-4-[(4-methoxyphenyl)amino]-	DMF	580(4.27),615(4.31)	2-0808-83
2-Propen-1-one, 3-(4-nitrophenyl)-1-[4-(phenylamino)phenyl]-	EtOH	263(3.78),286(3.83), 321(4.49),417(4.33)	124-1211-83
	dioxan	260(3.86),286(3.68), 315(4.49),395(4.32)	124-1211-83
$C_{21}H_{16}N_2O_5S_2$			
1H-Pyrrole, 3,3'-carbonylbis[1-(phenylsulfonyl)-	MeOH	220(4.32),245(4.30), 266s(4.04),275(4.04), 292(4.08)	44-3214-83
$C_{21}H_{16}N_2S$			
Benzenamine, N-(3,4-diphenyl-2(3H)-thiazolylidene)-	EtOH	303(4.26)	56-0875-83
$C_{21}H_{16}N_4O$			
1H-Indene-1,2,3-trione, 1,3-bis(phenylhydrazone)	MeOH	260(3.92),295(3.88), 355(4.11),480(4.05)	24-1309-83
$C_{21}H_{16}N_4O_2$			
7H-Dibenz[f,ij]isoquinoline-1-carbonitrile, 6-amino-2-morpholino-7-oxo-	DMF	514(4.03)	2-1197-83
7H-Dibenz[f,ij]isoquinoline-1-carbonitrile, 8-amino-2-morpholino-7-oxo-	DMF	505(3.98)	2-1197-83
$C_{21}H_{16}N_6O_3S$			
Formazan, 5-(2-benzothiazolyl)-3-(2-methoxyphenyl)-1-(4-nitrophenyl)-	EtOH	468(4.46)	135-0334-83
	EtOH-HCl	468(4.46)	135-0334-83
	EtOH-KOH	605(4.69)	135-0334-83
$C_{21}H_{16}N_7$			
Phenanthridinium, 3,8-diazido-5-ethyl-6-phenyl-, chloride	pH 3	284(4.72),294(4.78), 432(3.77)	4-0759-83
$C_{21}H_{16}O$			
2,5-Cyclohexadien-1-one, 4-(9H-fluoren-9-ylidene)-2,6-dimethyl-	isooctane	424(4.57)	73-2825-83
2-Propen-1-one, 1-[1,1'-biphenyl]-4-yl-3-phenyl-	dioxan	228.5(4.32),317(4.59)	126-1143-83
$C_{21}H_{16}O_2$			
4H-1-Benzopyran-4-one, 2,3-dihydro-2,6-diphenyl-	MeOH	247(4.55),345(3.34)	18-2037-83
4H-1-Benzopyran-4-one, 2,3-dihydro-2,8-diphenyl-	MeOH	240(4.42),334(4.66)	18-2037-83
2-Propen-1-one, 1-(2-hydroxy[1,1'-biphenyl]-3-yl)-3-phenyl-	MeOH	255(4.21),318(4.35)	18-2037-83
2-Propen-1-one, 1-(4-hydroxy[1,1'-biphenyl]-3-yl)-3-phenyl-	MeOH	240(4.26),322(4.45)	18-2037-83
2-Propen-1-one, 1-(4-phenoxyphenyl)-3-phenyl-	isooctane	227(4.20),278(4.08), 310(4.53)	124-1211-83
	dioxan	313(4.53)	124-1211-83
Spiro[acenaphthylene-1(2H),2'-oxetan]-2-one, 3'-methyl-3'-phenyl-	hexane	223(4.43),252(3.82), 260s(3.76),284s(3.51), 300(3.55),312(3.62),	18-3464-83

Compound	Solvent	λmax(log ε)	Ref.
(cont.)		326s(3.47),337(3.48)	18-3464-83
$C_{21}H_{16}O_3$			
4H-1-Benzopyran-4-one, 2,3-dihydro-2-(2-hydroxyphenyl)-6-phenyl-	MeOH	249(4.39),347(3.32)	18-2037-83
2-Propen-1-one, 1-(4-hydroxy[1,1'-biphenyl]-3-yl)-3-(2-hydroxyphenyl)-	MeOH	258(4.41),314(4.00), 378(4.07)	18-2037-83
$C_{21}H_{16}O_4$			
2(5H)-Furanone, 3-[3-(9H-fluoren-2-yl)-1-oxo-2-propenyl]-4-hydroxy-5-methyl-, (E)-	MeOH	204(4.61),269(3.95), 364(4.36),411(4.33)	83-0115-83
$C_{21}H_{16}O_5$			
6H-Benzo[d]naphtho[1,2-b]pyran-6-one, 8-ethenyl-1-hydroxy-10,12-dimethoxy-	MeOH	243(4.66),263(4.50), 273(4.55),283(4.54), 307(4.26),386(4.21)	23-0323-83
$C_{21}H_{16}O_6$			
Justicidin B	$CHCl_3$	260(4.52),295(4.13), 310(4.13),350(3.41)	162-0756-83
Retrojusticidin B	$CHCl_3$	267(4.08),318(3.76)	39-0643-83C
Viridin, demethoxy-, acetate	n.s.g.	238(4.51),310(4.08)	39-0867-83C
$C_{21}H_{16}S$			
10,9-(Epithiomethano)anthracene, 9,10-dihydro-12-phenyl-	MeCN	212(4.56),279(3.48)	78-1487-83
$C_{21}H_{17}BrN_2$			
1H-Pyrazole, 5-(2-bromophenyl)-4,5-dihydro-1,3-diphenyl-	C_6H_{12}	236(4.17),358(4.31)	39-1679-83B
$C_{21}H_{17}ClN_2$			
1H-Pyrazole, 5-(2-chlorophenyl)-4,5-dihydro-1,3-diphenyl-	C_6H_{12}	236(4.16),357(4.31)	39-1679-83B
$C_{21}H_{17}ClN_6$			
Benzenecarbohydrazonic acid, N-(4-chloro-1-phthalazinyl)-, 2-phenylhydrazide	MeOH	205(4.82),240(4.36), 290(4.26),370(4.20)	104-0953-83
	dioxan	243(4.36),290(4.23), 380(4.22)	104-0953-83
	MeCN	205(4.79),240(4.36), 290(4.24),373(4.26)	104-0953-83
$C_{21}H_{17}Cl_2N_3O_3$			
Proline, 2-[7-chloro-5-(2-chlorophenyl)-3H-1,4-benzodiazepin-2-yl]-5-oxo-, methyl ester	isoPrOH	219(4.62),240s(4.34), 295(3.67),317s(3.60)	4-0791-83
Pyrazino[1,2-a][1,4]benzodiazepine-4-carboxylic acid, 9-chloro-7-(2-chlorophenyl)-3-methyl-2-oxo-1,2,3,5-tetrahydro-, methyl ester	isoPrOH	216(4.57),250s(4.13), 327(4.29)	4-0791-83
$C_{21}H_{17}CoN_5$			
Cobalt(2+), (6,12-dimethyl-11,7-nitrilo-7H-dibenzo[c,n][1,2,5,13]-tetraazacyclopentadecine-N^5,N^{13}-N^{18},N^{20})-, diperchlorate	EtOH-DMF	275(4.36),330(4.20)	12-2387-83
$C_{21}H_{17}CuN_4$			
Copper(1+), (6,12-dimethyl-18H-7,11-	EtOH-DMF	275(4.20),350(3.98)	12-2387-83

Compound	Solvent	λ_{max}(log ϵ)	Ref.
nitrilodibenzo[b,e][1,4,7]triazacyclotetradecinato-N^5,N^{13},N^{18},N^{19})-, (SP-4-1)-, perchlorate (cont.)			12-2387-83
$C_{21}H_{17}CuN_5$			
Copper(2+), (6,12-dimethyl-11,7-nitrilo-7H-dibenzo[c,n][1,2,5,13]tetraazacyclopentadecine-N^5,N^{13},N^{18},N^{20})-, diperchlorate	EtOH-DMF	285(4.56),320(4.59), 340(4.30)	12-2387-83
$C_{21}H_{17}IN_2$			
1H-Pyrazole, 4,5-dihydro-5-(2-iodophenyl)-1,3-diphenyl-	C_6H_{12}	243(4.16),358(4.27)	39-1679-83B
$C_{21}H_{17}N$			
3,5-Hexadienenitrile, 6-phenyl-2-(3-phenyl-2-propenylidene)-	MeCN	382(5.82)	118-0917-83
$C_{21}H_{17}NO$			
2-Propen-1-one, 3-phenyl-1-[4-(phenylamino)phenyl]-, (E)-	EtOH	241(3.96),272(3.91), 303(4.27),392(4.37)	124-1211-83
	dioxan	263(3.81),299(4.28), 370(4.45)	124-1211-83
Pyridine, 1,4-dihydro-1-(1-oxo-3-phenyl-2-propenyl)-4-(phenylmethylene)-	MeOH	221(4.10),327(4.37), 411(4.20)	24-3192-83
$C_{21}H_{17}NO_2$			
4(5H)-Benzofuranone, 5-[(diphenylamino)methylene]-6,7-dihydro-	EtOH	243s(4.07),258s(4.08), 284(4.22),367(4.42)	111-0113-83
3H-Indol-3-ol, 2-(4-methoxyphenyl)-3-phenyl-	MeOH	333(4.28)	24-2115-83
$C_{21}H_{17}NO_3S$			
2-Propenamide, N,3-diphenyl-2-(phenylsulfonyl)-	EtOH	222(4.27),260(4.40), 275(4.35)	104-1314-83
$C_{21}H_{17}NO_4$			
3-Butenenitrile, 4-(2,3-dihydro-1,4-benzodioxin-6-yl)-2-[(2,3-dihydro-1,4-benzodioxin-6-yl)methylene]-	MeCN	378(4.23)	118-0917-83
1,3-Dioxolane, 4-(2-nitrophenyl)-2,2-diphenyl-	EtOH	202.5(4.39),261(4.53)	23-0400-83
$C_{21}H_{17}NO_5$			
Phenanthro[9',10':5,6][1,4]dioxino-[2,3-d]oxazol-11(9aH)-one, 12-acetyl-12,12a-dihydro-9a,12a-dimethyl-	MeOH	212(3.82),222(3.66), 250(3.68),257(3.78), 273(3.20),284(2.90), 295(2.96),309(3.04)	73-2812-83
$C_{21}H_{17}NO_7$			
Benzoic acid, 6-(1,3-dioxolo[4,5-g]isoquinolin-5-ylcarbonyl)-2,3-dimethoxy-, methyl ester	MeOH	240(4.99),282(4.77), 303(4.45),328(4.33)	44-4879-83
Ocominarone	EtOH	218(4.57),254(4.44), 282(4.56),350(3.98), 444(3.91)	100-0761-83
$C_{21}H_{17}NS$			
1,5-Benzothiazepine, 2,3-dihydro-2,4-diphenyl-	EtOH	259(4.23),335(3.5)	103-1293-83

Compound	Solvent	λ_{max}(log ε)	Ref.
$C_{21}H_{17}N_3$			
2H-Isoindole, 1-[(4-methylphenyl)azo]-3-phenyl-	EtOH	247s(4.31),254s(4.34), 261(4.35),274s(4.26), 283(4.26),467(4.57)	78-1401-83
	EtOH-HCl	243(4.18),249s(4.16), 270(4.26),276s(4.23), 303s(4.11),324(4.21), 547(4.56)	78-1401-83
$C_{21}H_{17}N_3OS$			
4H-Pyrido[3,4-e]-1,3-thiazin-4-one, 2-(diphenylamino)-5,7-dimethyl-	MeOH	250(4.16)	73-3426-83
$C_{21}H_{17}N_3O_3S$			
7H-Benzimidazo[2,1-a]benz[de]isoquinoline-10-sulfonamide, N-(1-methylethyl)-7-oxo-	benzene	382(4.17)	73-2249-83
	EtOH	385(4.25)	73-2249-83
$C_{21}H_{17}N_3O_7S$			
4-Thia-1-azabicyclo[3.2.0]heptane-2-carboxylic acid, 3,7-dioxo-6-[(phenylacetyl)amino]-, (4-nitrophenyl)-methyl ester, [2S-(2α,5α,6β)]-	dioxan	260(4.03)	78-2493-83
$C_{21}H_{17}N_3S_3$			
Benzothiazole, 2-[[phenyl(4,5,6,7-tetrahydro-1,3-benzodithiol-2-ylidene)methyl]azo]-	MeCN	566(4.42)	118-0840-83
$C_{21}H_{17}N_4$			
Phenanthridinium, 3-azido-5-ethyl-6-phenyl-, chloride	pH 3	245(4.40),270(4.70), 400(3.81)	4-0759-83
Phenanthridinium, 8-azido-5-ethyl-6-phenyl-, chloride	pH 3	207(4.55),270(4.66), 406(3.63)	4-0759-83
$C_{21}H_{17}N_4Ni$			
Nickel(1+), (6,12-dimethyl-18H-7,11-nitrilodibenzo[b,e][1,4,7]triazacyclotetradecinato-N^5,N^{13},N^{18},N^{19})-, (SP-4-1)-, perchlorate	acetone	300(4.11),340(3.99), 396(3.70)	12-2387-83
$C_{21}H_{17}N_5Ni$			
Nickel(2+), (6,12-dimethyl-11,7-nitrilo-7H-dibenzo[c,n][1,2,5,13]-tetraazacyclopentadecine-N^5,N^{13}-N^{18},N^{20})-, diperchlorate	EtOH-DMF	280(4.45),330(4.16), 390(3.99),540(3.98)	12-2387-83
$C_{21}H_{17}N_5OS$			
Formazan, 1-(2-benzothiazolyl)-3-(2-methoxyphenyl)-5-phenyl-	PrOH	395(4.26)	135-0334-83
	PrOH-HCl	395(4.26)	135-0334-83
	PrOH-KOH	510(4.72)	135-0334-83
$C_{21}H_{18}BrN_3O_3$			
Pyrimido[4,5-b][1,4]oxazepine-2,4(1H-3H)-dione, 6-(4-bromophenyl)-7,8-dihydro-1,3-dimethyl-8-phenyl-	MeOH	262(4.42),318(3.56)	103-0544-83
Pyrimido[4,5-b][1,4]oxazepine-2,4(1H-3H)-dione, 8-(4-bromophenyl)-7,8-dihydro-1,3-dimethyl-6-phenyl-	MeOH	234(4.32),242(4.32), 285s(--),306(3.36)	103-0544-83

Compound	Solvent	λ max (log ε)	Ref.
$C_{21}H_{18}ClN_3O_3$			
Pyrimido[4,5-b]-1,5-oxazepine-2,4(1H-3H)-dione, 6-(4-chlorophenyl)-7,8-dihydro-1,3-dimethyl-8-phenyl-	MeOH	255(4.40),317(3.56)	103-0544-83
Pyrimido[4,5-b]-1,5-oxazepine-2,4(1H-3H)-dione, 8-(4-chlorophenyl)-7,8-dihydro-1,3-dimethyl-6-phenyl-	MeOH	235(4.24),243(4.28), 285s(--),307(3.08)	103-0544-83
$C_{21}H_{18}F_3NO_6$			
Duguevanine, N-(trifluoroacetyl)-	EtOH	224(4.53),278(4.29), 297(4.14),308s(4.07)	100-0862-83
	EtOH-base	217(4.89),233s(4.65), 280(4.19),331(4.09)	100-0862-83
$C_{21}H_{18}HgN_2O_3$			
Mercury, [2-[(2-hydroxy-5-methylphen-yl)azo]benzoato](4-methylphenyl)-, (T-4)-	MeOH	325(4.25),390(3.89)	90-0493-83
$C_{21}H_{18}N$			
Phenanthridinium, 3-ethyl-6-phenyl-, chloride	pH 3	251(4.63),321(3.89), 363(3.69),377(3.68)	4-0759-83
$C_{21}H_{18}NO$			
3H-Indolium, 3-hydroxy-1-methyl-2,3-di-phenyl-, fluorosulfate	CH_2Cl_2	240(4.02),320(4.01)	24-2115-83
tetrafluoroborate	CH_2Cl_2	322(4.0)	24-2115-83
$C_{21}H_{18}NO_4$			
1-Benzopyrylium, 2-[7-(dimethylamino)-2-oxo-2H-1-benzopyran-3-yl]-8-meth-oxy-, perchlorate	MeCN	607(4.83)	48-0505-83
$C_{21}H_{18}N_2$			
1H-1,5-Benzodiazepine, 2,3-dihydro-2,4-diphenyl-	EtOH	256(4.39),367(3.79)	103-1293-83
1H-Indole-3-methanamine, N,α-diphenyl-	EtOH	209(4.56),222(4.52), 283(3.82),291(3.77)	2-0027-83
	EtOH-$HClO_4$	218(4.60),272(3.91), 281(3.90)	2-0027-83
1H-Pyrazole, 4,5-dihydro-1,3,5-tri-phenyl-	C_6H_{12}	356(4.30)	12-1649-83
	C_6H_{12}	238(4.13),360(4.29)	39-1679-83B
	MeOH	355(4.29)	12-1649-83
	n.s.g.	241(4.15),355(4.25)	44-0542-83
$C_{21}H_{18}N_2O$			
Benzenemethanol, α-[bis(phenylimino)-ethyl]-	MeOH	247(4.32)	73-1854-83
7H-Dibenz[f,ij]isoquinolin-7-one, 2-piperidino-	DMF	455(3.66)	2-1197-83
$C_{21}H_{18}N_2OS_2$			
Diazene, benzoyl[phenyl(4,5,6,7-tetra-hydro-1,3-benzodithiol-2-ylidene)-methyl]-	MeCN	515(3.87)	118-0840-83
$C_{21}H_{18}N_2O_2$			
Benzenemethanamine, N-methyl-α-(nitro-phenylmethylene)-N-phenyl-, (Z)-	EtOH	420(4.19)	22-0339-83
Benzenemethanamine, N-(4-methylphenyl)-α-(nitrophenylmethylene)-, (Z)-	EtOH	390(4.21)	22-0339-83

Compound	Solvent	λ_{max}(log ϵ)	Ref.
Benzenemethanamine, α-(nitrophenyl-methylene)-N-(phenylmethyl)-, (Z)-	EtOH	370(4.27)	22-0339-83
Cyclohepta[4,5]pyrrolo[1,2-a]imidazole-10-carboxylic acid, 2-(4-methylphenyl)-, ethyl ester	EtOH	254(4.42),280s(4.11), 347(4.47),360s(4.45), 445(3.68),455s(3.68), 530s(2.85)	18-3703-83
Pyrano[4,3-c]pyrazol-3(2H)-one, 6,7-dihydro-6-methyl-2-phenyl-4-(2-phenylethenyl)-	EtOH	246(4.31),360(4.52)	118-0948-83
$C_{21}H_{18}N_2O_3$			
Benzenemethanamine, N-(4-methoxyphenyl)-α-(nitrophenylmethylene)-, (Z)-	EtOH	393(4.18)	22-0339-83
$C_{21}H_{18}N_2O_4$			
4,9[1',2']-Benzeno-3H-benz[f]indazole-3a,9a-dicarboxylic acid, 4,9-dihydro-, dimethyl ester	EtOH	251(3.03),268(2.88), 275(2.90),321(2.24)	78-1151-83
4,9-Imino-1H-benz[f]isoindole-10-carboxylic acid, 2,3,3a,4,8,9a-hexahydro-1,3-dioxo-2-phenyl-, ethyl ester, endo	EtOH	256s(2.87),263(2.82), 270(2.60)	78-1401-83
exo	EtOH	256s(3.02),263(2.97), 271(2.76)	78-1401-83
Isoquinoline, 6,7-dimethoxy-8-[(6-methoxy-7-isoquinolinyl)oxy]-	MeOH	235(4.92),284(3.81)	83-0694-83
$C_{21}H_{18}N_2O_6$			
1H-Isoindole-1,3(2H)-dione, 2,2'-(1,3-propanediyl)bis[5-methoxy-	THF	245(4.56),254s(4.45), 288(3.70),322(3.62)	47-3425-83
$C_{21}H_{18}N_2O_6S$			
2-Propen-1-one, 3-[4-(dimethylamino)-phenyl]-1-(5-nitro-2-furanyl)-2-(phenylsulfonyl)-, (Z)-	dioxan	210(4.45),220(4.20), 310(4.52)	73-1057-83
$C_{21}H_{18}N_2O_6S_2$			
1-Azabicyclo[3.2.0]hept-2-ene-2-carboxylic acid, 6-ethyl-7-oxo-3-[(2-thienylcarbonyl)thio]-, (4-nitrophenyl)-methyl ester, (5R-cis)-	THF	266.5(4.30),315s(4.16)	158-0407-83
$C_{21}H_{18}N_4$			
1H-Pyrazole-4,5-diamine, N ,1,3-tri-phenyl-	EtOH	242(3.15),293(2.98)	4-1501-83
$C_{21}H_{18}N_4O_7S$			
1-Azabicyclo[3.2.0]hept-2-ene-2-carboxylic acid, 6-ethyl-3-[(5-nitro-2-pyridinyl)thio]-7-oxo-, (4-nitrophenyl)-methyl ester, (5R-cis)-	THF	269(4.21),326(4.18)	158-0407-83
$C_{21}H_{18}N_5$			
Phenanthridinium, 3-amino-8-azido-5-ethyl-6-phenyl-, chloride	pH 3	214(4.56),288(4.73), 462(3.72)	4-0759-83
Phenanthridinium, 8-amino-3-azido-5-ethyl-6-phenyl-, chloride	pH 3	214(4.55),287(4.74), 454(3.71)	4-0759-83
$C_{21}H_{18}O$			
2,5-Cyclohexadien-1-one, 4-(diphenyl-methylene)-2,6-dimethyl-	isooctane	363(4.49)	73-2825-83

Compound	Solvent	λ_{max}(log ε)	Ref.
2,5-Cyclohexadien-1-one, 4-(diphenylmethylene)-3,5-dimethyl-	isooctane	385(4.21)	73-2825-83
1-Propanone, 1-[1,1'-biphenyl]-4-yl-3-phenyl-	dioxan	280.5(4.41)	126-1143-83
$C_{21}H_{18}O_4$			
2H,6H-Benzo[1,2-b:5,4-b']dipyran-6-one, 5-methoxy-2,2-dimethyl-8-phenyl-	MeOH	234(4.52),278(4.63), 320(4.30)	106-0876-83
4H,8H-Benzo[1,2-b:3,4-b']dipyran-4-one, 5-methoxy-8,8-dimethyl-2-phenyl-	MeOH	240(4.30),272(4.49), 344(3.85)	106-0876-83
11H-9,10-endo-Cyclopropanthracene-11,12(13H)-dicarboxylic acid, 9,10-dihydro-, dimethyl ester	EtOH	251(2.79),259(2.79), 268(2.97),275(3.08)	78-1151-83
6H-Dibenzo[b,d]pyran-6-one, 3-(benzoyloxy)-7,8,9,10-tetrahydro-4-methyl-	MeOH	229(4.38),280(4.19)	78-1265-83
6H-Dibenzo[b,d]pyran-6-one, 2-benzoyl-7,8,9,10-tetrahydro-3-hydroxy-4-methyl-	MeOH	274(4.48),351(3.91)	78-1265-83
10,9-Propenoanthracene-11,13-dicarboxylic acid, 9,10-dihydro-, dimethyl ester	C_6H_{12}	266.5(3.19),273.5(3.17)	78-1151-83
5a,1-Propeno-5aH-cyclopropa[k]fluorene-1,11(9bH)-dicarboxylic acid, dimethyl ester	C_6H_{12}	274(3.44),282(3.44)	78-1151-83
$C_{21}H_{18}O_5$			
4H,8H-Benzo[1,2-b:5,4-b']dipyran-4-one, 3-hydroxy-2-(4-methoxyphenyl)-8,8-dimethyl-	MeOH	228(4.15),272(3.99), 366-381(4.04)	18-1267-83
4H,8H-Benzo[1,2-b:5,6-b']dipyran-4-one, 3-hydroxy-2-(4-methoxyphenyl)-8,8-dimethyl-	MeOH	230(4.12),277(3.87), 364(4.25)	18-1267-83
2-Cyclopentene-1,3-dicarboxylic acid, 2-oxo-4,5-diphenyl-, dimethyl ester	MeOH	287(4.08)	18-0175-83
5,12-Naphthacenedione, 9,10-dihydro-1,6,11-trimethoxy-	EtOH	225(4.38),240(4.37), 257(4.39),280s(4.41), 287(4.42),394(3.97)	94-2662-83
$C_{21}H_{18}O_7$			
2-Anthracenecarboxylic acid, 9,10-dihydro-4-hydroxy-5-methoxy-9,10-dioxo-3-(3-oxopentyl)-	MeOH	227(4.44),261(4.37), 396s(--),418(4.03), 429s(--)	5-2151-83
$C_{21}H_{19}Br_2ClO$			
Benzene, 1-[5-bromo-1-(1-bromoethyl)-4-(4-chlorophenyl)-1,2,3-hexatrienyl]-4-methoxy-	C_6H_{12}	228.5(4.40),251s(4.31), 300s(4.01),419.5(3.89)	19-0233-83
Benzene, 1-chloro-4-[2,3-dibromo-2-ethylidene-4-(4-methoxyphenyl)-2,4-hexadienyl]-	C_6H_{12}	248.5(4.44),298s(3.47)	19-0233-83
	CH_2Cl_2	251.0(4.45)	19-0233-83
$C_{21}H_{19}ClN_2$			
Benzenamine, N-[[2-chloro-3-[3-(phenylamino)-2-propenylidene]-1-cyclopenten-1-yl]methylene]-, monoperchlorate	MeOH	647(5.15)	104-1854-83
$C_{21}H_{19}ClO$			
1-Acenaphthylenol, 1-[(4-chlorophenyl)-methyl]-1,2-dihydro-2,2-dimethyl-	hexane	271s(--),280(3.70), 290(3.75),302(3.57), 307(3.38),317(2.84), 322(2.66)	104-0520-83

Compound	Solvent	$\lambda_{max}(\log \epsilon)$	Ref.
Benzene, 1-chloro-4-[1-ethylidene-4-(4-methoxyphenyl)-4-hexen-2-ynyl]-	C_6H_{12}	218s(4.46),229s(4.31), 251.2(4.17),259s(4.48), 295s(3.87)	19-0233-83
$C_{21}H_{19}Cl_4N_3O_5S_2$			
Pyridinium, 1-[[8-oxo-7-[(2-thienylacetyl)amino]-2-[(2,2,2-trichloroethoxy)carbonyl]-5-thia-1-azabicyclo[4.2.0]oct-2-en-3-yl]methyl]-, chloride, S-oxide, [5S-(5α,6β,7α)]-	EtOH	237(4.01),263(3.96), 275s(3.83)	78-0461-83
$C_{21}H_{19}FN_4$			
4H-Pyrrolo[2,3-d]pyrimidin-4-imine, 3-(4-fluorophenyl)-3,7-dihydro-2,5-dimethyl-6-(phenylmethyl)-	EtOH	237(4.25),283(3.89)	5-2066-83
$C_{21}H_{19}N$			
Cyclopenta[b]pyrrole, 1,6-dihydro-1,6-dimethyl-4,5-diphenyl-	EtOH	238(4.22),290(4.08), 322(4.00)	39-0915-83C
1H-Pyrrole, 1-methyl-2-(1-methyl-2,3-diphenyl-2-cyclopropen-1-yl)-	EtOH	230(4.51),260(3.89), 301s(4.42),314(4.48)	39-0915-83C
Quinoline, 5,6,7,8-tetrahydro-2,4-diphenyl-	neutral	286(4.01)	39-0045-83B
	protonated	308(4.22)	39-0045-83B
$C_{21}H_{19}NO$			
1(2H)-Naphthalenone, 4-methyl-2-[3-(phenylmethyl)amino]-2-propenylidene]-	acetone	338(3.80),355(3.72), 490(3.63),510(3.72), 540(3.60)	103-0825-83
	dioxan	292(3.88),330(3.72), 357(3.58),373(3.20), 510(3.18),540(3.60)	103-0825-83
	MeCN	335(3.80),355(3.75), 510(3.72),560(3.76)	103-0825-83
	DMSO	290(3.65),350(3.60), 520(3.80)	103-0825-83
	DMSO-NaOMe	313(3.75),480(3.76)	103-0825-83
Oxaziridine, 3,3-diphenyl-2-(1-phenylethyl)-, [S-(R*,R*)]-	isooctane	215s(4.53),253(2.85), 258(2.89),265(2.82), 272(2.46)	32-0799-83
[S-(R*,S*)]-	isooctane	215(4.48),252(2.61), 258(2.66),265(2.58), 272(2.19)	32-0799-83
$C_{21}H_{19}NO_2$			
6,7-Benzo-1-oxa-3-azaspiro[4.5]deca-3,6,9-trien-8-one, 2,2,9(and 10)-trimethyl-4-phenyl-	EtOH	246(4.34)	33-2252-83
1(2H)-Naphthalenone, 2-[3-[(4-methoxyphenyl)amino]-2-propenylidene]-4-methyl-	EtOH	310(4.12),320(4.08), 340(4.03),355(4.02), 390(3.95),555(3.88), 590(3.86)	103-0825-83
	acetone	333(3.85),357(3.81), 370(3.70),545(3.38)	103-0825-83
	dioxan	338(4.06),357(4.04), 373(3.91),530(3.18)	103-0825-83
	MeCN	308(3.98),340(3.85), 357(3.80),373(3.58), 515(3.53),550(3.56)	103-0825-83
	DMSO	330(3.89),357(3.80), 388(3.77),520(3.99), 585(3.60)	103-0825-83

Compound	Solvent	λ_{max}(log ε)	Ref.
(cont.)	DMSO-NaOMe	518(4.28)	103-0825-83
Oxiranecarbonitrile, 3-[2-(2-cyclohex-en-1-yloxy)phenyl]-2-phenyl-, trans	EtOH	265s(3.38),277s(3.59), 283(3.61)	24-2383-83
3,4-Propano-4H-furo[3,2-c][1]benzo-pyran-2-carbonitrile, 2,3,3a,9b-tetrahydro-2-phenyl-, (2α,3α,3aα-4α,9bβ)-	EtOH	252s(2.73),258s(2.96), 274(3.32)	24-2383-83
$C_{21}H_{19}NO_2S$			
Methanone, (5-morpholino-4-phenyl-2-thienyl)phenyl-	MeCN	232(4.26),275(4.08), 380(4.18)	48-0168-83
$C_{21}H_{19}NO_3$			
Benz[cd]indol-5(1H)-one, 1-benzoyl-2,2a,3,4-tetrahydro-4-(2-oxopropyl)-	MeOH	237(4.38),326(3.66)	39-1545-83C
$C_{21}H_{19}NO_6$			
Benzo[g]-1,3-benzodioxolo[5,6-a]quino-lizinium, 5,6-dihydro-13-hydroxy-8,9,10-trimethoxy-, hydroxide, inner salt	EtOH	235(4.22),262(4.15), 317(4.13),362(3.92), 377(4.12),464(4.13)	94-0947-83
8H-Benzo[g]-1,3-benzodioxolo[5,6-a]-quinolizin-8-one, 5,6-dihydro-9,10,13-trimethoxy-	MeOH	228.5(4.68),335(4.40), 345.5(4.42)	94-0947-83
4H-Pyrido[3,2,1-jk]carbazole-8,9,11-tricarboxylic acid, 5,6-dihydro-, trimethyl ester	MeOH	218(4.28),233(4.39), 281.5(4.34),333(4.15), 390(3.56)	39-0515-83C
3,4-Quinolinedicarboxylic acid, 2-(3,4-dimethoxyphenyl)-, dimethyl ester	EtOH	245(4.46),276(4.24), 345(3.78)	12-1419-83
$C_{21}H_{19}NO_7$			
3,1-Benzoxazepine-4,5-dicarboxylic acid, 2-(3,4-dimethoxyphenyl)-, dimethyl ester	EtOH	235(4.40),273(4.48), 312(4.19)	12-1419-83
3,4-Quinolinedicarboxylic acid, 2-(3,4-dimethoxyphenyl)-, dimethyl ester, 1-oxide	EtOH	244(4.51),250(4.48), 277(4.19),340(4.03)	12-1419-83
3,4-Quinolinedicarboxylic acid, 1-(3,4-dimethoxyphenyl)-1,2-dihydro-2-oxo-, dimethyl ester	MeCN	238(4.42),285(3.90), 352(3.64)	12-1419-83
$C_{21}H_{19}NO_8$			
Benzoic acid, 2-[(7,8-dihydro-5-oxo-1,3-dioxolo[4,5-g]isoquinolin-6(5H)-yl)carbonyl]-3,4-dimethoxy-, methyl ester	MeOH	227.5(4.25),265.5(4.24), 320(4.05)	94-0947-83
$C_{21}H_{19}N_2$			
Phenanthridinium, 3-amino-5-ethyl-6-phenyl-, chloride	pH 3	242(4.38),275(4.52), 437(3.65)	4-0759-83
Phenanthridinium, 8-amino-5-ethyl-6-phenyl-, chloride	pH 3	270(4.59),430(3.53)	4-0759-83
$C_{21}H_{19}N_3$			
Benzonitrile, 4-(3a,4,5,6,7,7a-hexa-hydro-1-phenyl-4,7-methano-1H-ind-azol-3-yl)-	C_6H_{12}	407(4.45)	12-1649-83
	MeOH	404(4.48)	12-1649-83
Benzonitrile, 4-(3a,4,5,6,7,7a-hexa-hydro-3-phenyl-4,7-methano-1H-ind-azol-1-yl)-	C_6H_{12}	367(4.58)	12-1649-83
	MeOH	365(4.58)	12-1649-83

Compound	Solvent	$\lambda_{max}(\log \epsilon)$	Ref.
$C_{21}H_{19}N_3O_2S$			
Hydrazinecarboxamide, 2-[2-[6-(4-methoxyphenyl)-2H-thiopyran-2-ylidene]-2-(phenylethylidene)-	$CHCl_3$	470(3.86)	97-0147-83
$C_{21}H_{19}N_3O_2Se$			
Hydrazinecarboxamide, 2-[2-[6-(4-methoxyphenyl)-2H-selenin-2-ylidene]-2-(phenylethylidene)-	DMSO	468(3.94)	97-0147-83
$C_{21}H_{19}N_3O_3$			
Pyrimido[4,5-b][1,4]oxazepine-2,4(1H-3H)-dione, 7,8-dihydro-1,3-dimethyl-6,8-diphenyl-	MeOH	245(4.24),308(3.15)	103-0544-83
$C_{21}H_{19}N_3O_5S$			
1-Azabicyclo[3.2.0]hept-2-ene-2-carboxylic acid, 6-ethyl-3-(2-pyridinylthio)-7-oxo-, (4-nitrophenyl)methyl ester	THF	269(4.08),325(4.23)	158-0407-83
1-Azabicyclo[3.2.0]hept-2-ene-2-carboxylic acid, 6-ethyl-3-(4-pyridinylthio)-7-oxo-, (4-nitrophenyl)methyl ester	THF	267(3.94),322(3.90)	158-0407-83
$C_{21}H_{19}N_3S_2$			
Pyridine, 2-[[3-phenyl-3-(4,5,6,7-tetrahydro-1,3-benzodithiol-2-ylidene)-1-propenyl]azo]-	MeCN	538(4.07)	118-0840-83
$C_{21}H_{19}N_4S$			
Benzothiazolium, 2-[3-cyano-3-(1,3-dimethyl-2H-benzimidazol-2-ylidene)-1-propenyl]-3-methyl-, perchlorate	EtOH	486(4.55)	124-0297-83
$C_{21}H_{19}N_5O_5$			
Adenosine, 3'-(1-naphthalenecarboxylate)	MeOH	259(4.22)	64-0049-83C
$C_{21}H_{19}N_5O_9$			
Pyridinium, N-(4-nitrobenzyl)-, 1-acetonyl-2,4,6-trinitrocyclohexadienide, cation	MeCN	260(4.41)	103-0773-83
anion	MeCN	459(4.04),580(3.95)	103-0773-83
$C_{21}H_{20}ClFN_4O_3$			
Spiro[2H-1,4-benzodiazepine-2,5'(2'H)-oxazole], 7-chloro-5-(2-fluorophenyl)-1,3-dihydro-4'-morpholino-, 4-oxide	isoPrOH	232(4.46),260s(4.16), 300(4.03),345s(3.59)	4-0551-83
$C_{21}H_{20}ClFN_4O_4$			
3H-1,4-Benzodiazepine-2-carboxaldehyde, 7-chloro-5-(2-fluorophenyl)-α-nitro-, O-(4-methoxybutyl)oxime	isoPrOH	212s(4.40),247(4.46), 275s(4.26),327s(3.75), 347(3.74)	4-0551-83
$C_{21}H_{20}Cl_2N_6$			
4H-[1,2,4]Triazolo[4,3-a][1,4]benzodiazepine, 8-chloro-6-(2-chlorophenyl)-1-(4-methylpiperazino)-	MeOH	210(4.63),245s(4.17)	73-2395-83

Compound	Solvent	$\lambda_{max}(\log \epsilon)$	Ref.
$C_{21}H_{20}Cl_2O_6$			
9,10-Anthracenedione, 2-[3-(dichloromethyl)-3-hydroxyhexyl]-1,4,8-trihydroxy-	MeOH	232(4.57),250(4.33), 290(3.89),434(3.72), 462(4.01),479(4.07), 489(4.11),511(3.99), 523(3.95)	5-1818-83
$C_{21}H_{20}Cl_3NO_3$			
1H-Pyrrole-1-carboxylic acid, 2,5-dihydro-2-[(4-methoxyphenyl)phenylmethyl]-, 2,2,2-trichloroethyl ester	MeOH	225(4.14),268(3.15), 276(3.22),283(3.14)	87-0469-83
$C_{21}H_{20}Cl_3NO_4$			
6-Oxa-3-azabicyclo[3.1.0]hexane-3-carboxylic acid, 2-[(4-methoxyphenyl)-phenylmethyl]-, 2,2,2-trichloroethyl ester	MeOH	227(4.13),262(3.04), 268(3.13),276(3.19), 283(3.12)	87-0469-83
$C_{21}H_{20}NO_4$			
Dehydrocavidine	MeOH and MeOH-base	215(5.54),270(5.52), 348(5.44),450(4.85)	100-0320-83 +100-0466-83
$C_{21}H_{20}N_2O$			
Benzaldehyde, 4-(3a,4,5,6,7,7a-hexahydro-3-phenyl-4,7-methano-1H-indazol-1-yl)-	C_6H_{12}	377(4.65)	12-1649-83
	MeOH	395(4.63)	12-1649-83
4H-Indol-4-one, 1,5,6,7-tetrahydro-1-(3-methylphenyl)-2-phenyl-, oxime, (Z)-	EtOH	246(4.20),270(4.00)	4-0989-83
4H-Indol-4-one, 1,5,6,7-tetrahydro-1-(4-methylphenyl)-2-phenyl-, oxime, (Z)-	EtOH	250(4.30),270(4.30)	4-0989-83
Pyrrolo[3,2-c]azepin-4(1H)-one, 5,6,7,8-tetrahydro-1-(2-methylphenyl)-2-phenyl-	EtOH	231(4.30),281(4.20)	4-0989-83
Pyrrolo[3,2-c]azepin-4(1H)-one, 5,6,7,8-tetrahydro-1-(3-methylphenyl)-2-phenyl-	EtOH	228(4.40),281(4.30)	4-0989-83
Pyrrolo[3,2-c]azepin-4(1H)-one, 5,6,7,8-tetrahydro-1-(4-methylphenyl)-2-phenyl-	EtOH	220(4.20),281(4.20)	4-0989-83
$C_{21}H_{20}N_2OS_2$			
Diazene, (4-methoxyphenyl)(phenyl-(4,5,6,7-tetrahydro-1,3-benzodithiol-2-ylidene)methyl]-	MeCN	498(4.02)	118-0840-83
$C_{21}H_{20}N_2O_3$			
Alstonine	MeOH	252(4.54),289(4.08), 309(4.36),336(3.39), 369(3.60)	162-0047-83
$C_{21}H_{20}N_2O_4$			
Propanedioic acid, [[1H-indol-3-yl(4-methylphenyl)amino]methylene]-, monoethyl ester	EtOH	213(4.59),286(4.08)	104-1367-83
Propanedioic acid, [(1H-indol-3-ylphenylamino)methylene]-, ethyl methyl ester	EtOH	222(4.29),288(4.39), 314(4.44)	104-1367-83
Propanedioic acid, [[(1-methyl-1H-indol-3-yl)phenylamino]methylene]-, ethyl ester	EtOH	225(4.37),290(3.99), 360(2.7)	104-1367-83

Compound	Solvent	$\lambda_{max}(\log \epsilon)$	Ref.
Propanedioic acid, [[(2-methyl-1H-indol-3-yl)phenylamino]methylene]-, monoethyl ester	EtOH	217(4.45),289.8(4.24), 307.7(4.20)	104-1367-83
Pyrimido[2,1-a]isoindol-4(1H)-one, 1-methyl-2-(3,4,5-trimethoxyphenyl)-	EtOH	208(4.52),229(4.53), 284(4.35),365(3.94), 378(3.98),428(3.45)	124-0755-83
$C_{21}H_{20}N_2O_5$			
Fumaramine, (E)-	MeOH	205(4.59),219s(4.42), 238s(4.21),264(3.99), 357(4.04)	102-2073-83
$C_{21}H_{20}N_2O_6$			
1H-Pyrrole-2,3,4-tricarboxylic acid, 1-[8-(methylamino)-1-naphthalenyl]-, trimethyl ester	EtOH	216(4.72),253(4.45), 344(3.90)	94-1378-83
$C_{21}H_{20}N_3$			
Ethidium (chloride)	pH 3	213(4.65),285(4.75), 478(3.75)	4-0759-83
$C_{21}H_{20}N_4$			
1H-Cyclopent[cd]indene-1,1,2,2(2aH)-tetracarbonitrile, 7a,7b-dihydro-2a,3,5,6,7a,7b-hexamethyl-	MeCN	215(4.00),315(3.84)	44-0309-83
2-Propenenitrile, 3,3'-(1,3-propanediyldiimino)bis[3-phenyl-	EtOH	207(4.44),227(4.41), 283(4.20)	39-1223-83C
4H-Pyrrolo[2,3-d]pyrimidin-4-imine, 3,7-dihydro-2,5-dimethyl-3-phenyl-6-(phenylmethyl)-	EtOH	237(4.36),285(4.00)	5-2066-83
$C_{21}H_{20}N_4O_3$			
Pyrimido[4,5-b]-1,4-oxazepine-2,4(1H-3H)-dione, 6-(4-nitrophenyl)-7,8-dihydro-1,3-dimethyl-8-phenyl-	MeOH	226(4.28),318(4.19)	103-0544-83
$C_{21}H_{20}N_4O_5$			
1,4-Naphthalenedione, 2,5-dimethyl-8-(1-methylethyl)-, 4-(2,4-dinitrophenylhydrazone)	EtOH	218(4.55),262(4.22), 412(4.43),449(4.53)	22-0112-83
$C_{21}H_{20}N_6O_9$			
Methanone, bis(3,5-dinitro-4-pyrrolidinophenyl)-	dioxan	260.4(4.21),342.9(4.47)	56-1357-83
$C_{21}H_{20}O$			
1,2,4-Methenocyclobut[cd]inden-5(1H)-one, octahydro-6-(octahydro-1,2,4-metheno-3H-cyclobuta[cd]pentalen-3-ylidene)- (enone A)	EtOH	263(4.13)	39-1791-83C
enone B	EtOH	264(4.14)	39-1791-83C
enone C	EtOH	262.5(4.14)	39-1791-83C
enone D	EtOH	264(4.15)	39-1791-83C
$C_{21}H_{20}O_2S$			
1-Naphthalenemethanol, α-[3-[(4-methylphenyl)sulfinyl]-2-propenyl]-, (E)-	n.s.g.	223(4.91),269s(3.82), 280s(3.86),290s(3.68)	12-1049-83
1-Naphthalenemethanol, α-[1-[(4-methylphenyl)sulfinyl]-2-propenyl]-	n.s.g.	224(4.88),271(3.97), 282(4.00),293(3.83)	12-1049-83
2-Naphthalenemethanol, α-[1-[(4-methylphenyl)sulfinyl]-2-propenyl]-	n.s.g.	226(4.92),250s(4.09)	12-1049-83

Compound	Solvent	λ_{max}(log ε)	Ref.
2-Naphthalenemethanol, α-[3-[(4-methylphenyl)sulfinyl]-2-propenyl]-, (E)-	n.s.g.	223(4.93),262s(3.74), 272s(3.76),284s(3.56)	12-1049-83
$C_{21}H_{20}O_4$			
4H,8H-Benzo[1,2-b:3,4-b']dipyran-4-one, 2,3-dihydro-5-methoxy-8,8-dimethyl-2-phenyl-, (S)- (mixtecacin)	MeOH	230(3.71),273(3.97)	102-2047-83
Butanedioic acid, (3a,4-dihydro-3H-cyclopenta[def]phenanthren-1-yl)-, 1-ethyl ester	EtOH	251s(--),257(4.60), 266(4.64),295s(--), 310(3.71),318(3.75), 331(3.67),347(3.51)	39-0087-83C
2-Cyclohexen-1-one, 6-acetyl-3-(2-hydroxy-4-methoxyphenyl)-5-phenyl-	MeOH	202(4.49),340(4.26)	142-2369-83
1,3,9,11-Cyclotridecatetraene-5,7-diyne-1-carboxylic acid, 13-(2-ethoxy-2-oxoethylidene)-4,9-dimethyl-, methyl ester, (E)-	C_6H_{12}	270s(4.07),305(4.44), 365s(3.77),430s(3.13)	44-2379-83
(Z)-	C_6H_{12}	270s(4.02),302s(4.50), 309(4.53),420s(3.28)	44-2379-83
Ethanone, 1-[3-(2-hydroxy-4-methoxyphenyl)-2-methyl-4-phenyl-4H-pyran-3-yl]-	MeOH	208(4.43),274(4.21)	142-2369-83
7H-Furo[3,2-g][1]benzopyran-7-one, 4-[(3,7-dimethyl-2,5,7-octatrienyl)oxy]-, (E,E)- (anhydronotoptol)	MeOH	225(4.74),248(4.41), 258(4.32),268(4.29), 307.5(4.25)	94-2710-83
2-Propen-1-one, 1-(7-hydroxy-5-methoxy-2,2-dimethyl-2H-1-benzopyran-6-yl)-3-phenyl-, (E)- (oaxacacin)	MeOH	302(4.57),350(4.60)	102-2047-83
$C_{21}H_{20}O_5$			
1(3H)-Isobenzofuranone, 3-(5,6-dihydro-1,4-dimethoxy-2-naphthalenyl)-7-methoxy-	EtOH	213(4.72),226s(3.93), 229(4.59),273(3.95), 286s(3.91),300s(3.95), 307(3.97),322s(3.65)	94-2662-83
$C_{21}H_{20}O_6$			
9,10-Anthracenedione, 1-hydroxy-3-(hydroxymethyl)-8-methoxy-2-(3-oxopentyl)-	MeOH	257(4.39),280s(--), 288(4.53),404s(--), 419(4.02),436s(--)	5-2151-83
2-Furancarboxylic acid, 5,5'-(phenylmethylene)bis-, diethyl ester	EtOH	263(4.46)	103-0478-83
$C_{21}H_{20}O_7$			
5,12-Naphthacenedione, 7,8,9,10-tetrahydro-1,6,7,8,11-pentahydroxy-8-propyl-, cis	MeOH	218s(--),233(4.43), 251(4.31),291(3.78), 464(3.94),477(3.99), 491(4.06),511(3.92), 525(3.90)	5-1818-83
trans	MeOH	220s(--),235(4.41), 252(4.34),290(3.77), 435s(--),466(3.94), 479(3.98),491(4.06), 510(3.91),525(3.92)	5-1818-83
$C_{21}H_{20}O_9$			
Acacetin, 6-C-α-L-arabinofuranosyl-	MeOH	270(4.29),326(4.32)	102-2051-83
	MeOH-NaOAc	278(--),350(--)	102-2051-83
	MeOH-AlCl$_3$	278(--),301(--), 344(--),377s(--)	102-2051-83
Acacetin, 7-O-L-arabinopyranoside	MeOH	267(4.26),322(4.29)	102-2051-83
	MeOH-NaOAc	267(--),322(--)	102-2051-83

Compound	Solvent	λ_{max}(log ε)	Ref.
Acacetin, 6-C-α-L-arabinopyranosyl-	MeOH	270(4.27),325(4.29)	102-2051-83
	MeOH-NaOAc	278(--),350(--)	102-2051-83
Frangulin A	n.s.g.	225(4.52),264(4.28), 282(4.15),300(3.97), 430(4.05)	162-0608-83
$C_{21}H_{20}O_{10}$			
Genistin	85% EtOH	262.5(1.96)	162-0626-83
$C_{21}H_{21}B$			
Borane, tris(2-methylphenyl)-, TCNE complex	CH_2Cl_2	415(2.13)	35-2175-83
$C_{21}H_{21}Cl_2N_3O_3S$			
Benzamide, N-[1-[(2,6-dichlorophenyl)-methyl]-3-[(dimethylamino)sulfonyl]-1,4-dihydro-4-pyridinyl]-	CH_2Cl_2	274(3.5),282(3.5), 321(3.7)	64-0873-83B
$C_{21}H_{21}Cl_2N_5S$			
Ethanamine, 2-[[[8-chloro-6-(2-chloro-phenyl)-4H-[1,2,4]triazolo[4,3-a]-[1,4]benzodiazepin-1-yl]methyl]-thio]-N,N-dimethyl-	MeOH	220.5(4.55),252s(4.02)	73-0123-83
$C_{21}H_{21}I_2NO_3$			
Isoxazole, 3,5-bis[2-(3-iodo-4-methoxy-phenyl)ethyl]-	EtOH	229(4.63),277(4.55), 285(4.54)	39-2577-83C
$C_{21}H_{21}N$			
Pyridine, 2-(1,1-dimethylethyl)-4,6-diphenyl-	neutral	295(4.03)	39-0045-83B
	protonated	289(4.24)	39-0045-83B
$C_{21}H_{21}NO_3$			
1,4-Naphthalenedione, 2-[4-(diethyl-amino)-2-methoxyphenyl]-	C_6H_{12}	509(3.82)	150-0168-83S
	EtOH	559(--)	150-0168-83S
1,4-Naphthalenedione, 2-[4-(diethyl-amino)phenyl]-3-methoxy-	C_6H_{12}	534(3.79)	150-0168-83S
	EtOH	560(--)	150-0168-83S
10-Oxa-11-azatetracyclo[13.3.1.1^{2,6}-1^{9,12}]heneicosa-1(19),2,4,6(21)-9(20),11,15,17-octaene, 3,18-di-methoxy-	EtOH	235(4.62),294(4.53), 303(4.51)	39-2577-83C
$C_{21}H_{21}NO_4$			
2H-1-Benzopyran-2-one, 4-[(benzoyloxy)-methyl]-7-(diethylamino)-	EtOH	236.5(4.30),379(4.33)	94-3014-83
Guadiscoline	EtOH	221(4.33),268(4.22), 320(3.89),356s(3.63)	100-0761-83
	EtOH-HCl	224(4.27),259s(4.06), 273(4.15),368(3.82), 410s(3.50)	100-0761-83
Melosmidine	EtOH	219s(3.90),242(4.03), 328(3.30),365(3.32)	100-0761-83
	EtOH-HCl	229s(3.98),239(4.05), 277(3.84),438(3.15)	100-0761-83
$C_{21}H_{21}NO_5$			
Aporphine, 3-hydroxy-1,2,9,10-tetra-methoxy-	EtOH	222(4.53),284(4.19), 297(4.15),301(4.15), 314(4.07)	5-0744-83
	EtOH-KOH	214(5.55),320(4.56)	5-0744-83

Compound	Solvent	$\lambda_{max}(\log \epsilon)$	Ref.
Buxifoline, N-acetyl-	EtOH	218(4.49),242(4.26), 286(4.33)	100-0862-83
7H-Pyrano[2,3-c]acridin-7-one, 3,12-dihydro-6-hydroxy-10,11-dimethoxy-3,3,12-trimethyl-	MeOH	209(4.37),268(4.67), 283(4.64),341(4.14), 405(3.69)	94-0895-83
	MeOH-$AlCl_3$	248s(--),272(--), 292(--),305s(--), 365(--),417(--)	94-0895-83
$C_{21}H_{21}NO_6$			
Hydrastine	EtOH	202(4.79),218(4.53), 238(4.15),298(3.86), 316(3.63)	162-0690-83
β-Hydrastine, BH_3 complex	MeOH	213(4.08),238s(3.58), 295(3.49),315s(3.38)	83-0737-83
Ocotominarine	EtOH	228(4.64),294(4.36)	100-0761-83
Rhoeadine	MeOH	205(4.91),240(3.96), 292(3.94)	100-0441-83
$C_{21}H_{21}N_2O_4$			
Pyrimido[2,1-a]isoindolium, 4,6-dihydro-1-methyl-4-oxo-2-(3,4,5-trimethoxyphenyl)-, salt with 4-methylbenzenesulfonic acid	EtOH	208(4.65),284(4.27)	124-0755-83
$C_{21}H_{21}N_2S_2$			
Benzothiazolium, 3-ethyl-2-[3-(3-ethyl-2(3H)-benzothiazolylidene)-1-propenyl]-, tetrafluoroborate	MeOH	556.5(5.15)	89-0876-83 +89-1147-83S
salt with 4-methylbenzenesulfonic acid	MeOH	558(5.20)	104-2114-83
	CH_2Cl_2	562(--)	104-2114-83
$C_{21}H_{21}N_5$			
Benzenamine, N-(1,5-dihydro-1,6-dimethyl-5-(2-methylphenyl)-4H-pyrazolo-[3,4-d]pyrimidin-4-ylidene]-2-methyl-	EtOH	268.5(4.07)	11-0171-83B
$C_{21}H_{21}N_5O_{11}P_2$			
Adenosine 5'-(trihydrogen diphosphate), 3'-(1-naphthalenecarboxylate)	pH 7	259(4.08)	64-0049-83C
$C_{21}H_{22}Br_2O_3$			
3-Heptanone, 5-acetoxy-1,7-bis(4-bromophenyl)-, (S)-	EtOH	255(2.84),262(2.88), 269(2.93),277(2.82)	18-3353-83
$C_{21}H_{22}N_2O_2$			
4,9-Iminobenz[f]isoindole-10-carboxylic acid, 2,3,3a,4,9,9a-hexahydro-2-phenyl-, ethyl ester, (3aα,4α,9α,9aα)-	EtOH	257(4.18),298(3.38)	78-1401-83
1,4-Naphthalenedione, 2-[4-(diethylamino)phenyl]-3-(methylamino)-	C_6H_{12}	415s(--),521(3.26)	150-0168-83S
	EtOH	463(--)	150-0168-83S
Vomilenine, 21-deoxy-	MeOH	220(3.73),227s(--), 257(3.28)	102-2297-83
$C_{21}H_{22}N_2O_3$			
2H-1-Benzopyran-2-one, 7-(diethylamino)-3-[(phenylamino)acetyl]-	MeCN	441(4.69)	48-0551-83
Sitsirikine-16R, 3,4,5,6-tetradehydro-	EtOH and EtOH-acid	223(3.9),254(4.5), 309(4.31),370(3.65)	100-0708-83
	EtOH-NaOH	223(3.87),281(4.21), 330(4.15),419(3.34)	100-0708-83

Compound	Solvent	λ_{max}(log ε)	Ref.
Strictamine, 16-formyl-	MeOH	222(4.15),265(3.56)	102-1017-83
$C_{21}H_{22}N_2O_4$			
Spiro[2H-1-benzopyran-2,2'-[2H]indole], 1',3'-dihydro-8-methoxy-1',3,3',3'-tetramethyl-6-nitro- (colored form)	toluene	610(3.51)	33-0342-83
Trichophylline	EtOH	238(4.10),251s(3.46), 306(4.10),362(4.13)	78-3639-83
$C_{21}H_{22}N_2O_5$			
1H-Pyrrolo[2,3-d]carbazol-2(3H)-one, 5-acetoxy-3,7-diacetyl-3a,4,6a,7-tetrahydro-3a-methylpyrrolo[2,3-d]-carbazol-2(3H)-one, (3aα,6aβ,11bR*)	MeOH	252(4.13),278s(3.46), 288s(3.40)	5-1744-83
$C_{21}H_{22}N_2O_8S_2$			
Antibiotic FA 2097	MeOH	284s(3.51)	158-83-46
	MeOH-NaOH	302(--)	158-83-46
$C_{21}H_{22}O_2$			
Endiandric acid A	EtOH	242(2.19),255(2.36), 261(2.45),268(2.32), 286(1.45)	162-0515-83
$C_{21}H_{22}O_3$			
Estra-1,3,5(10),7,9,14-hexaen-17β-ol, 3-methoxy-, acetate	EtOH	246(4.89),255(4.62), 264(4.63),284(4.06), 293.5(4.19),305(4.16), 335(3.11),350.5(3.00)	65-1724-83
$C_{21}H_{22}O_3Se$			
Naphtho[2,3-b]furan-2,6(3H,4H)-dione, 3a,8a,9,9a-tetrahydro-3,5,8a-tri-methyl-3-(phenylseleno)-, [3R-(3α,3aα,8aβ,9aα)]-	EtOH	232(4.03),263s(--)	94-3397-83
$C_{21}H_{22}O_4$			
6H-Benzo[d]naphtho[1,2-b]pyran-6-one, 8-ethyl-1,2,3,4-tetrahydro-10,12-dimethoxy-	MeOH	212s(4.58),221(4.61), 232s(4.69),238(4.70), 262s(4.29),274(4.29), 350(4.16)	23-0323-83
$C_{21}H_{22}O_5$			
2H,6H-Benzo[1,2-b:5,4-b']dipyran-6-one, 5-methoxy-2,2,8-trimethyl-10-(3-methyl-1-oxo-2-butenyl)-	MeOH	262(4.71),335(3.87)	102-2090-83
Benzoic acid, 2-[(5,6-dihydro-1,4-di-methoxy-2-naphthalenyl)methyl]-6-methoxy-	EtOH	223(4.57),226s(3.97), 273(4.04),280s(3.97), 310(3.50),320s(3.43)	94-2662-83
1H,9H-4,10-Dioxacyclobut[3,4]indeno-[5,6-a]naphthalen-9-one, 6-acetoxy-1a,2,3,3a,10c,10d-hexahydro-1,1,3a-trimethyl-, (±)-	EtOH	226s(4.11),255(3.73), 264(3.77),328(4.11)	39-1411-83C
Eriobrucinol, acetate, (±)-	EtOH	228(4.19),249(3.65), 259(3.51),299s(3.94), 310s(4.03),333(4.16)	39-1411-83C
7H-Furo[3,2-g][1]benzopyran-7-one, 4-[(5-hydroxy-3,7-dimethoxy-2,6-octadienyl)oxy]-, (E)-(±)-	MeOH	223(4.36),250.5(4.22), 259(4.17),268(4.17), 309.5(4.12)	94-2710-83
7H-Furo[3,2-g][1]benzopyran-7-one,	MeOH	221.5(4.42),250.5(4.30),	94-2710-83

Compound	Solvent	λ_{max}(log ε)	Ref.
4-[(7-hydroxy-3,7-dimethyl-2,5-octadienyl)oxy]-, (E,E)- (notoptol)		259(4.24),268(4.22), 309.5(4.18)	94-2710-83
1-Naphthalenol, 2-(3,4-dimethoxyphenyl)-6-ethoxy-7-methoxy-	MeOH	226(4.68),263(4.66), 290s(4.28)	103-0066-83
$C_{21}H_{22}O_6$			
6(2H)-Benzofuranone, 2-(1,3-benzodioxol-5-yl)-3,3a-dihydro-5,7-dimethoxy-3-methyl-3a-(2-propenyl)-	MeOH	267(4.24),285(3.96)	102-0561-83
6(2H)-Benzofuranone, 2-(1,3-benzodioxol-5-yl)-3,5-dihydro-5,7-dimethoxy-3-methyl-5-(2-propenyl)-	MeOH	238(4.22),285(4.05), 325(3.96)	102-0561-83
2(3H)-Furanone, 3-(1,3-benzodioxol-5-ylmethyl)-4-[(3,4-dimethoxyphenyl)-methyl]dihydro-, (2R,3R)-	MeOH	229(4.20),280(3.92)	102-1516-83
(2R,3S)-	MeOH	230(4.34),279(4.04)	102-1516-83
2(3H)-Furanone, 4-(1,3-benzodioxol-5-ylmethyl)-3-[(3,4-dimethoxyphenyl)-methyl]dihydro-	$CHCl_3$	246(3.66),285(3.73)	39-0643-83C
$C_{21}H_{22}O_7$			
6-Epiteucrin A acetate	EtOH	218(3.81),224(3.85)	102-0723-83
$C_{21}H_{23}Br_3N_2O_6$			
Carbamic acid, [3-[2,6-dibromo-4-[2-[[(3-bromo-2,5-dihydroxy-4-methoxyphenyl)methylene]amino]ethyl]phenoxy]propyl]-, methyl ester	MeOH	269(4.20),309(4.04), 440(3.43)	78-0667-83
	MeOH-NaOH	220(3.15),260(2.78), 302(2.85)	78-0667-83
$C_{21}H_{23}ClFNO_2$			
Haloperidol	MeOH-HCl	221(4.18),247(4.12)	162-0662-83
$C_{21}H_{23}ClN_6O$			
1H-1,2,4-Triazol-3-amine, N-[(4-chlorophenyl)methylene]-5-[4-(2-methoxyphenyl)-1-piperazinyl]-1-methyl-	$CHCl_3$	282(4.30),321(4.18)	24-1547-83
1H-1,2,4-Triazol-5-amine, N-[(4-chlorophenyl)methylene]-3-[4-(2-methoxyphenyl)-1-piperazinyl]-1-methyl-	$CHCl_3$	290(4.34),382(3.95)	24-1547-83
$C_{21}H_{23}FeN$			
Ferrocene, (2,3-dihydro-2,3,3-trimethyl-1H-indol-5-yl)-	EtOH	295(4.32),450(3.74)	104-0950-83
$C_{21}H_{23}NO_2S$			
Pentanethioamide, 4,4-dimethyl-2,3-dioxo-N,N-bis(phenylmethyl)-	MeCN	297(3.92),317(3.77), 412(2.92)	5-1694-83
$C_{21}H_{23}NO_3$			
Isoxazole, 3,5-bis[2-(4-methoxyphenyl)-ethyl]-	EtOH	226(4.64),277(4.55), 285(4.54)	39-2577-83C
Pentanamide, 4,4-dimethyl-2,3-dioxo-N,N-bis(phenylmethyl)-	MeCN	215(3.92),250(3.18), 256(3.18),414(1.78)	5-1694-83
$C_{21}H_{23}NO_3S$			
1,2-Benzoxathiin-4-amine, 3,4,5,6,7,8-hexahydro-N-methyl-N,3-diphenyl-, 2,2-dioxide, cis	EtOH	249.5(4.36),295(3.55)	4-0839-83
trans	EtOH	250(3.98),294(3.10)	4-0839-83

Compound	Solvent	$\lambda_{max}(\log \epsilon)$	Ref.
$C_{21}H_{23}NO_4$			
Guadiscoline, 6,6a-dihydro-	EtOH	220(4.45),270s(4.17), 278(4.23),302(4.07)	100-0761-83
Hexanoic acid, 6-[[(9H-fluoren-9-yl-methoxy)carbonyl]amino]-	MeOH	205(4.67),219s(4.20), 227(3.79),255s(4.21), 264(4.29),288(3.68), 299(3.77)	64-1015-83B
1-Naphthalenamine, 2-(3,4-dimethoxy-phenyl)-6,7-dimethoxy-N-methyl-	MeOH	231(4.64),261(4.62)	103-0066-83
$C_{21}H_{23}NO_5$			
Acetamide, N-[6-(3,5-dimethoxyphenyl)-5,6,7,8-tetrahydronaphtho[2,3-d]-1,3-dioxol-5-yl]-, cis	EtOH	282(3.69),294(3.66)	94-3056-83
β-Anhydrohydrastinediol	MeOH	217(4.22),286(3.50)	83-0737-83
borane adduct	MeOH	215(4.23),235s(3.68), 285(3.56)	83-0737-83
Canadaline, (±)-	MeOH	232(4.13),288(3.85)	94-2685-83
Furo[2,3-e]-1,3-benzodioxol-6-amine, 7-(3,4-dimethoxyphenyl)-5-ethenyl-6,7-dihydro-N,N-dimethyl-, (6R-trans)-	MeOH	231(3.90),279(3.58)	83-0737-83
Grandisinine	MeOH	224(4.28),262(4.70), 270(4.72),334(4.33), 394(3.87)	102-1493-83
	MeOH-NaOMe	225(--),266(--), 295(--),378(--)	102-1493-83
	MeOH-$AlCl_3$	237(--),268s(--), 278(--),358(--), 440(--)	102-1493-83
Guattouregidine	EtOH	212(4.34),266(4.49), 352(3.64)	100-0761-83
	EtOH-HCl	212(4.36),276(4.48), 336(3.69),421(3.57)	100-0761-83
Pavinane, 2,3-(methylenedioxy)-4,8,9-trimethoxy-, (-)-	EtOH	287(3.84)	100-0293-83
L-Proline, 1-(3,4-dimethoxybenzoyl)-, phenylmethyl ester	MeOH	254(3.78),284(3.49)	83-0773-83
$C_{21}H_{23}NO_6$			
Acetamide, N-[1-(6-hydroxy-5-oxo-1,3,6-cycloheptatrien-1-yl)-3-(3,4,5-tri-methoxyphenyl)-2-propenyl]-, (E)-	MeOH	242(4.52),335s(3.96), 370s(3.79)	18-3099-83
2-Azabicyclo[3.2.2]nona-3,6-diene-3,4,5-tricarboxylic acid, 2-(phenyl-methyl)-, trimethyl ester	MeCN	233(3.66),267s(3.60), 302(3.91)	24-0563-83
Colchiceine	EtOH	244(4.51),351(4.28)	162-0351-83
Glaucamine	MeOH	238(4.0),286(3.8)	100-0441-83
Norleucoxylonine	EtOH-HCl	226(4.54),284(4.33), 305(4.12)	100-0761-83
Oreogenine	MeOH	237(4.08),286(3.83)	100-0441-83
Papaverrubine F	MeOH	237(4.11),287(3.84)	100-0441-83
$C_{21}H_{23}NO_7$			
Propanedioic acid, [2-[[(3,4,5-trimeth-oxyphenyl)methylene]amino]phenyl]-, dimethyl ester	$CHCl_3$	304(4.79)	44-2468-83
$C_{21}H_{23}N_3$			
3-Butenenitrile, 4-[4-(dimethylamino)-phenyl]-2-[[4-(dimethylamino)phenyl]-methylene]-	MeCN	462(5.77)	118-0917-83

Compound	Solvent	$\lambda_{max}(\log \epsilon)$	Ref.
$C_{21}H_{23}N_3O_3$			
2H-Pyrrolo[2,3-b]quinoxalin-2-one, 3-acetyl-4-ethyl-1,3,3a,4,9,9a-hexahydro-1-(2-methoxyphenyl)-	EtOH	221(4.43),263(3.72), 304(3.47)	103-0901-83
$C_{21}H_{23}N_3O_5Si$			
1,2-Diazabicyclo[5.2.0]nona-2,4-diene-4-carboxylic acid, 8-(1,3-dihydro-1,3-dioxo-2H-isoindol-2-yl)-9-oxo-, 2-(trimethylsilyl)ethyl ester, cis-(±)-	MeOH	302(3.92)	5-1374-83
1,2-Diazabicyclo[5.2.0]nona-2,4-diene-6-carboxylic acid, 8-(1,3-dihydro-1,3-dioxo-2H-isoindol-2-yl)-9-oxo-, 2-(trimethylsilyl)ethyl ester, cis	MeOH	294(4.01),364(3.30)	5-1374-83
$C_{21}H_{23}N_3O_7S$			
1-Azabicyclo[3.2.0]hept-2-ene-2-carboxylic acid, 3-[[2-(acetylamino)ethenyl]thio]-6-(1-hydroxy-1-methylethyl)-7-oxo-, (4-nitrophenyl)methyl ester, [3(E),5α,6α]-(±)-	MeOH	229.5(4.24),262(4.23), 322.5(4.23)	39-0403-83C
$C_{21}H_{23}N_5S$			
Methanimidamide, N'-[2-(3,4-dimethyl-1-phenylisothiazolo[5,1-e][1,2,3]-thiadiazole-7-S^{IV}-6(1H)-yl)phenyl]-N,N-dimethyl-	C_6H_{12}	203(4.38),223s(4.34), 247s(4.30),290s(4.23), 476(4.30)	39-0777-83C
$C_{21}H_{24}$			
[2.2.2](1,2,4)Cyclophane, 5,15,16-trimethyl-, TCNE complex	CH_2Cl_2	575(1.46)	59-0289-83
bromanil complex	CH_2Cl_2	537(1.73)	59-0289-83
chloranil complex	CH_2Cl_2	527(1.61)	59-0289-83
DDQ complex	CH_2Cl_2	645(2.35)	59-0289-83
$C_{21}H_{24}F_3N_3OS$			
Trifluoperazine sulfoxide	pH 1	234(3.21),270s(2.76)	36-0050-83
	pH 2.2	235(3.21),270s(2.72)	36-0050-83
	pH 3.5	235.5(3.17),271s(2.72)	36-0050-83
	H_2O	235s(3.10)	36-0050-83
	2N H_2SO_4	234.5(3.15),271s(2.70)	36-0050-83
$C_{21}H_{24}F_3N_3S$			
Trifluoperazine	pH 1	252.5(3.22),301s(2.68)	36-0050-83
	pH 2.2	251.5(3.22),301s(2.65)	36-0050-83
	pH 3.5	251.5(3.21),300(2.62)	36-0050-83
	H_2O	254(3.36),302s(2.54)	36-0050-83
	2N H_2SO_4	251.5(3.16),300s(2.54)	36-0050-83
radical cation	9N H_2SO_4	213(3.24),271(3.60)	36-0050-83
$C_{21}H_{24}Ge_2$			
Digermane, 1,1,1-trimethyl-2,2,2-triphenyl-	C_6H_{12}	233(4.25)	101-0149-83G
Digermane, 1,1,2-trimethyl-1,2,2-triphenyl-	C_6H_{12}	237s(4.54)	101-0149-83G
$C_{21}H_{24}NO_4$			
Remrefine (chloride), (-)-	MeOH	234(4.11),250s(3.75), 291(3.88)	100-0293-83

Compound	Solvent	λ_{max}(log ε)	Ref.
$C_{21}H_{24}N_2O$			
9-Oxaergoline, 2,3-dihydro-6-methyl-1-(phenylmethyl)-, (3)-(±)-	EtOH	212(4.53),258(4.61), 299(3.41)	87-0522-83
	EtOH-HCl	211(4.49),259(3.93), 300(3.33)	87-0522-83
$C_{21}H_{24}N_2O_2$			
Bicyclo[8.3.2]pentadeca-10,13-diene-14,15-dione, 12-(phenylazo)-	MeOH	215s(4.09),220s(4.05), 250s(4.09),258(4.10)	88-3879-83
Catharanthine	EtOH	226(4.56),284(3.92), 292(3.88)	162-0266-83
$C_{21}H_{24}N_2O_2S$			
Octanamide, N-[4-(2-benzothiazolyl)-3-hydroxyphenyl]-	EtOH	344(4.51)	4-1517-83
$C_{21}H_{24}N_2O_3$			
Ajmalicine	MeOH	227(4.61),292(3.79)	162-0030-83
Akuammiline, deacetyl-	MeOH	229(4.79),264(3.76)	102-1017-83
Benzenepropanamide, N-(4-methoxyphenyl)-α-[(1-methyl-3-oxo-1-butenyl)-amino]-, (S)-	EtOH	273(3.57)	33-0262-83
Diaboline	EtOH	249(4.06)	162-0428-83
Ercininamine	EtOH	222(3.91),279(3.64)	105-0454-83
3,6-Ethano-3H-azecino[5,4-b]indole-4,8(5H,9H)-dione, 14-ethylidene-1,2,6,7-tetrahydro-5-(hydroxymethyl)-9-methyl-, [5R-(5R*,6R*,14E)]-	MeOH	220(4.25),241(3.81), 310(3.88),340s(--)	33-2414-83
Lochnericine	EtOH	227(4.10),299(4.15), 328(4.32)	162-0795-83
1,6-Naphthalenediamine, 2-(3,4-dimethoxyphenyl)-7-methoxy-N,N'-dimethyl-	MeOH	230(4.53),344(3.47), 356(4.59)	103-0066-83
Octanamide, N-[4-(2-benzoxazolyl)-3-hydroxyphenyl]-	EtOH	330(4.56)	4-1517-83
β-Yohimbine, 3,4-dehydro-	EtOH and EtOH-base	228(4.43),298(4.21), 308(4.42),318(4.18)	100-0708-83
	EtOH-HCl	218(4.56),246(3.85), 290(3.2),355(4.78)	100-0708-83
β-Yohimbine, 19,20-dehydro-	EtOH	228(4.83),272(3.98), 284(3.92),291(3.78)	100-0708-83
$C_{21}H_{24}N_2O_3S$			
Morpholine, 4-[2-(4-methoxyphenyl)-3-[(4-methoxyphenyl)amino]-1-thioxo-2-propenyl]-	MeCN	251(4.18),299(4.42), 342(4.14),409(3.51)	48-0689-83
$C_{21}H_{24}N_2O_4$			
Ajmalicine, 10-hydroxy-	EtOH	228(4.30),279(3.93), 295s(--),310s(--)	44-3825-83
Alstonine, tetrahydro-10-hydroxy-	EtOH	228(4.28),280(3.95), 294s(--),310s(--)	44-3825-83
Echitamidine, N_a-formyl-	EtOH	210(4.24),252(3.95), 290s(3.48)	32-0533-83
Ercinamine	EtOH	226(3.95),283(3.66)	105-0454-83
Haplocidine, 18-oxo-	EtOH	222(4.54),260(3.89), 298(3.60)	100-0694-83
$C_{21}H_{24}N_2O_5$			
Acetamide, N-acetyl-N-[2-[1-acetyl-2-(1-acetoxy-1-propenyl)-1H-indol-3-yl]ethyl]-	MeOH	240(4.20),280(4.09), 293s(4.04),303s(3.98)	4-0183-83

Compound	Solvent	λ_{max}(log ϵ)	Ref.
$C_{21}H_{24}N_4O_5$			
Cyanocycline F	EtOH	270(3.84)	39-0335-83B
$C_{21}H_{24}N_6O$			
1H-1,2,4-Triazol-3-amine, 5-[4-(2-methoxyphenyl)-1-piperazinyl]-1-methyl-N-(phenylmethylene)-	$CHCl_3$	260(4.26)	24-1547-83
1H-1,2,4-Triazol-5-amine, 3-[4-(2-methoxyphenyl)-1-piperazinyl]-1-methyl-N-(phenylmethylene)-	$CHCl_3$	292(4.27),374(3.89)	24-1547-83
$C_{21}H_{24}N_8O_5$			
L-Glutamic acid, N-[4-[[1-(2,4-diamino-6-pteridinyl)ethyl]methylamino]-benzoyl]-	pH 1	244(4.34),316(4.15)	162-0023-83
	pH 13	255(4.40),306(4.40), 369(3.95)	162-0023-83
$C_{21}H_{24}O_3$			
14α-Estra-1,3,5,7,9-pentaen-17β-ol, 3-methoxy-, acetate	EtOH	258(3.59),268(3.70), 278(3.73),289(3.58), 311s(3.06),323(3.33), 332s(3.30)	65-1724-83
14β-	EtOH	229(4.82),255.5(3.65), 265(3.73),275(3.74), 285(3.55),308(3.01), 321(3.29),330(3.25), 335(3.40)	65-1724-83
Estra-1,3,5,7,9-pentaen-17-one, 3-methoxy-, cyclic 1,2-ethanediyl acetal	EtOH	231(4.78),256s(3.66), 267(3.70),278(3.70), 289(3.55),310(3.11), 323(3.30),332(3.28), 337(3.39)	65-1724-83
3-Heptanone, 5-acetoxy-1,7-diphenyl-	EtOH	260(2.95),264(2.89), 268(2.88),284(2.69)	18-3353-83
$C_{21}H_{24}O_5$			
1(2H)-Anthracenone, 3,4-dihydro-3,8,9-trihydroxy-6-methoxy-3-methyl-7-(3-methyl-2-butenyl)- (vismione E)	$CHCl_3$	245s(4.40),278(4.71), 326(4.11),406(4.19)	102-0539-83
Bicyclo[3.2.1]oct-3-ene-2,8-dione, 7-(3,4-dimethoxyphenyl)-5-methoxy-6-methyl-3-(2-propenyl)-	MeOH	209(4.19),229(4.10), 278(3.62)	102-0763-83
Propanedioic acid, [4-(2-hydroxy-1-naphthalenyl)-2-butenyl]-, diethyl ester	EtOH	228(4.62),268(3.88), 278(3.92),290(3.80), 335(3.46)	78-1761-83
$C_{21}H_{24}O_7$			
Butanoic acid, 3-hydroxy-3-methyl-, 1-(1,4-dihydro-5,8-dihydroxy-1,4-dioxo-2-naphthalenyl)-4-methyl-3-pentenyl ester	MeOH	276(3.80),330(3.05), 487(3.75),517(3.77), 556(3.58)	105-0532-83
Medioresinol	EtOH	220s(4.58),232(4.13), 280(3.58)	36-1285-83
7H-Phenaleno[1,2-b]furan-7-one, 7a,8,9,10a-tetrahydro-3,6,7a-trihydroxy-4,5-dimethoxy-1,8,8,9-tetramethyl-	EtOH	357(3.74),402(3.90)	18-3661-83
$C_{21}H_{24}O_8$			
Benzenebutanoic acid, 3,4-dimethoxy-γ-oxo-α-(2,4,5-trimethoxyphenyl)-	EtOH	229.5(4.40),277(4.11), 295.5(4.07)	94-3039-83

Compound	Solvent	$\lambda_{max}(\log \epsilon)$	Ref.
$C_{21}H_{25}BrO_3$			
Pregna-1,4,6-triene-3,20-dione, 2-bromo-17α-hydroxy-	MeOH	222(4.18),270(4.07), 308(3.98)	39-2793-83C
$C_{21}H_{25}Br_2N_3O_3$			
Pregna-1,4-diene-3,20-dione, 7-azido-2,6-dibromo-17-hydroxy-, (6β,7α)-	MeOH	255(4.15)	39-2793-83C
$C_{21}H_{25}FO_2$			
Benzoic acid, 3-fluoro-4-[2-methyl-4-(2,6,6-trimethyl-1-cyclohexen-1-yl)-1,3-butadienyl]-, (E,E)-	EtOH	313(4.36)	87-1282-83
Benzoic acid, 4-[1-fluoro-2-methyl-4-(2,6,6-trimethyl-1-cyclohexen-1-yl)-1,3-butadienyl]-, (Z,E)-	EtOH	241.5(4.02),317(4.31)	87-1282-83
$C_{21}H_{25}NO_3$			
2H-Benzo[b]quinolizin-2-ol, 1,3,4,6-11,11a-hexahydro-8,9-dimethoxy-2-phenyl-	n.s.g.	230(4.13),285(3.81)	12-0149-83
$C_{21}H_{25}NO_4$			
Argemonine, (-)-	n.s.g.	205s(4.94),230s(4.15), 276s(3.79),282(3.91), 287(3.93),292(3.89)	100-0293-83
2-Butenedioic acid, 2-(5,7,8,9,10,11-hexahydro-5-methyl-11aH-cyclooct[b]-indol-11a-yl)-, dimethyl ester, (Z)-	MeOH	211(4.23),281(4.37)	39-0515-83C
5a,11a-Etheno-5H-cyclooct[b]indole-12,13-dicarboxylic acid, 6,7,8,9-10,11-hexahydro-5-methyl-, dimethyl ester	MeOH	210(4.34),240.5(4.22), 307(3.42),435(3.08)	39-0515-83C
Melosmidine, tetrahydro-	EtOH	218(4.52),237s(4.24), 273s(4.31),284(4.39), 300s(4.25)	100-0761-83
	EtOH-HCl	218(4.52),237s(4.24), 275s(4.29),287(4.39), 302s(4.23)	100-0761-83
Melosmine, tetrahydro-N-methyl-	EtOH	220(4.43),270s(4.13), 286(4.19),301s(4.15), 311s(4.06)	100-0761-83
	EtOH-HCl	220(4.43),275s(4.22), 282(4.23),304(4.11), 315(4.11)	100-0761-83
L-Proline, 1-[(3,4-dimethoxyphenyl)-methyl]-, phenylmethyl ester	MeOH	227(4.06),278(3.44)	83-0773-83
$C_{21}H_{25}NO_5$			
Demecolcine	EtOH	245(4.55),355(4.24)	162-0416-83
Isoconovine	EtOH	224(4.60),274(4.19), 302(3.96)	100-0761-83
$C_{21}H_{25}NO_6$			
6-Oxa-2-azaspiro[4.5]dec-3-ene-4-carb-oxylic acid, 7-methoxy-2-(1-methyl-ethyl)-1,9-dioxo-3-phenyl-, ethyl ester	EtOH	309(3.83)	94-0356-83
$C_{21}H_{25}N_3O$			
3H-Pyrazol-3-one, 2,4-dihydro-5-methyl-	MeOH	432(4.64)	104-2089-83

Compound	Solvent	λ_{max}(log ϵ)	Ref.
2-phenyl-4-(3-piperidino-2-cyclohex-en-1-ylidene)- (cont.)			104-2089-83
$C_{21}H_{25}N_3O_2$			
Octanamide, N-[4-(1H-benzimidazol-2-yl)-3-hydroxyphenyl]-	EtOH	327(4.57)	4-1517-83
$C_{21}H_{25}N_3O_3$			
Pregna-1,4,6-triene-3,20-dione, 4-azido-17-hydroxy-	MeOH	229(4.08),240(4.08), 266s(3.76),337(3.80)	39-2793-83C
Pregna-1,4,6-triene-3,20-dione, 6-azido-17-hydroxy-	MeOH	249(4.18),311(3.81)	39-2787-83C
$C_{21}H_{25}N_3O_7S$			
1-Azabicyclo[3.2.0]hept-2-ene-2-carboxylic acid, 3-[[2-(acetylamino)ethyl]-thio]-6-(1-hydroxy-1-methylethyl)-7-oxo-, (4-nitrophenyl)methyl ester	MeOH	265(4.26),318(4.21)	158-0943-83
cis-(±)-	EtOH	265(4.06),315(4.09)	39-0403-83C
$C_{21}H_{25}N_5O_2$			
3H,4H-2,6a-Diaza-3a-azoniaphenalene, 1-(dimethylamino)-5-formyl-2,7,8,9-tetrahydro-6-hydroxy-4-imino-2,7-dimethyl-3-phenyl-, hydroxide, inner salt	EtOH	234(4.43),266(4.0), 366(4.52)	142-1891-83
$C_{21}H_{25}N_5O_4$			
4,6-Methano-5H-benz[h]oxazolo[3,2-a]-pyrazino[3,2,1-de][1,5]naphthyridine-7-carbonitrile, 11-amino-1,2,3a,4,4a-6,7,9,10,13,13b,13c-dodecahydro-9-(hydroxymethyl)-5,12-dimethyl-10,13-dioxo-, [3aS-(3aα,4α,4aβ,6α,7β,9α,13bβ-13cβ)]-	MeOH	276(3.86)	158-1767-83
$C_{21}H_{25}N_9O_4$			
1-Nordistamycin A, hydrobromide	EtOH	237(4.46),302(4.59)	87-1042-83
2-Nordistamycin A, hydrobromide	EtOH	235(4.46),308(4.56)	87-1042-83
3-Nordistamycin A, hydrobromide	EtOH	237(4.50),305(4.58)	87-1042-83
$C_{21}H_{26}BrN_3O_3$			
Pregna-1,4-diene-3,20-dione, 7-azido-6-bromo-17-hydroxy-, (6β,7α)-	MeOH	248(4.22)	39-2793-83C
Pregna-4,6-diene-3,20-dione, 1-azido-2-bromo-17-hydroxy-, (1α,2β)-	MeOH	301(4.35)	39-2793-83C
$C_{21}H_{26}Cl_2O_4$			
Dichlorisone	MeOH	237(4.19)	162-0443-83
$C_{21}H_{26}IN_3O_3$			
Pregna-1,4-diene-3,20-dione, 7-azido-17-hydroxy-6-iodo-, (6β,7α)-	MeOH	244(4.31)	39-2793-83C
$C_{21}H_{26}I_2O_4S$			
Benzene, 2-iodo-4-[3-[[4-(3-iodo-4-methoxyphenyl)butyl]sulfonyl]propyl]-1-methoxy-	EtOH	277(4.55),283(4.54)	39-2577-83C
$C_{21}H_{26}NO_4$			
Eschscholtzidinium, N-methyl- (chloride)	MeOH	230s(3.91),257(3.46),	100-0293-83

Compound	Solvent	λ_{max}(log ε)	Ref.
(cont.)		289(3.74)	100-0293-83
6H-5,11-Methanodibenz[b,f]azocinium, 11,12-dihydro-2,3,8,9-tetramethoxy-5-methyl-, perchlorate	EtOH	231(4.03),290(3.70)	39-2053-83C
Platycerinium, N-methyl-, perchlorate	MeOH	227(4.12),234(4.14), 260(3.27),284(3.77)	100-0293-83
Xanthoplanine, iodide	MeOH	222(4.19),285(3.71), 305(3.71)	102-2607-83
$C_{21}H_{26}N_2O$			
Cyclopentanone, 2-[(1,3-dihydro-1,3,3-trimethyl-2H-indol-2-ylidene)ethylidene]-5-[(dimethylamino)methylene]-	MeOH	470(4.85)	104-1854-83
	benzene	424(--),447(4.83)	104-1854-83
	DMF	432(--),453(--)	104-1854-83
$C_{21}H_{26}N_2O_2$			
2-Epi-16-epicathafoline	MeOH	212(4.13),247(3.83), 293(3.37)	33-2414-83
$C_{21}H_{26}N_2O_3$			
Corynanthine	EtOH	283(3.87),290(3.80)	151-0171-83C
	MeOH	226(4.56),283(3.87), 290(3.79)	162-0364-83
Isositsirikine, 16-epi-Z-	MeOH	225(4.10),280(3.90), 290(3.75)	100-0409-83
Yohimbine	EtOH	273(3.93),279(3.93), 289(3.85)	151-0171-83C
17-epi-allo-	EtOH	227(4.78),274(3.7), 283(3.92),291(3.85)	100-0708-83
α-Yohimbine, 19,20-dihydro- (same in acid or base)	EtOH	228(4.82),272(3.81), 284(3.9),291(3.78)	100-0708-83
$C_{21}H_{26}N_2O_4$			
Echitamidine, 12-methoxy-	EtOH	212(4.70),240(4.57), 288(4.11),334(4.58)	32-0533-83
Isositsirikine N-oxide (same in acid or base)	EtOH	224(4.76),273(3.78), 284(3.65),290(3.54)	100-0694-83
16-epi-	EtOH	228(4.52),274(3.9), 284(3.91),291(3.86)	100-0694-83
β-Yohimbine, 7(S)-oxindole	EtOH	216(4.50),255(3.87), 284(3.15)	100-0708-83
β-Yohimbine, pseudoindoxyl of	EtOH	234(4.56),255(3.91), 407(3.67)	100-0708-83
β-Yohimbine N-oxide	EtOH	228(4.56),274(3.89), 284(3.95),291(3.81)	100-0708-83
$C_{21}H_{26}N_2O_4Si$			
1,2-Diazabicyclo[5.2.0]nona-3,5-diene-4-carboxylic acid, 2-benzoyl-8-methyl-9-oxo-, (2-trimethylsilyl)ethyl ester	MeOH	312(4.01)	5-1393-83
1,2-Diazabicyclo[5.2.0]nona-3,5-diene-6-carboxylic acid, 2-benzoyl-8-methyl-9-oxo-, (2-trimethylsilyl)ethyl ester	MeOH	326(4.14)	5-1393-83
$C_{21}H_{26}N_2O_6$			
Acetamide, N-[1-(2-amino-6-hydroxy-5-oxo-1,3,6-cycloheptatrien-1-yl)-3-(3,4,5-trimethoxyphenyl)propyl]-	MeOH	232(4.42),360(4.06), 400(3.96)	18-3106-83

Compound	Solvent	λ_{max}(log ε)	Ref.
$C_{21}H_{26}N_2O_8$			
2,7-Dioxa-9-azabicyclo[4.2.2]decane-1-carboxamide, 8-[2-(benzoyloxy)-1-hydroxy-1-methylethyl]-6-hydroxy-N,9-dimethyl-5-methylene-10-oxo-	EtOH	227.5(4.10)	88-5607-83
$C_{21}H_{26}N_4$			
9H-Carbazole, 9-[(2,2,6,6-tetramethyl-1-piperidinyl)azo]-	CH_2Cl_2	231(4.44),238(4.45), 253(4.47),272(4.25), 283(4.14),294(4.15), 320(4.14),325(4.16), 345(4.05)	24-0819-83
$C_{21}H_{26}N_6O_3$			
Pregna-1,4-diene-3,20-dione, 6,7-di-azido-17-hydroxy-, (6β,7α)-	MeOH	244(4.22)	39-2787-83C
$C_{21}H_{26}O$			
2,5-Cyclohexadien-1-one, 2,6-bis(1,1-dimethylethyl)-4-(phenylmethylene)-	isooctane	346(4.50)	73-2825-83
Furan, 2-(1,1-dimethylethyl)tetra-hydro-2-methyl-5,5-diphenyl-	CH_2Cl_2	249(2.85),256(2.93)	152-0105-83
$C_{21}H_{26}O_3$			
Estra-1,3,5(10),9(11)-tetraen-17-one, 3-methoxy-, cyclic 1,2-ethanediyl acetal	EtOH	263(4.29),273s(4.17), 298(3.50)	12-0339-83
Furan, 2,2'-(2-furanylmethylene)bis[5-butyl-	EtOH	227(4.23)	103-0478-83
Furan, 2,2',2"-hexylidynetris[5-meth-yl- (trifurylalkane)	EtOH	223(4.41)	103-0474-83
$C_{21}H_{26}O_4$			
2-Propenoic acid, 3-(4-acetoxyphenyl)-, 1,1,7-trimethylbicyclo[2.2.1]hept-2-yl ester	EtOH	283(4.39)	105-0149-83
Strobilurin C	EtOH	226(4.92),231s(4.90), 240s(4.86),262s(4.65), 297(4.92)	158-0661-83
$C_{21}H_{26}O_5$			
Benzenemethanol, 3,4-dimethoxy-α-[1-[2-(2-methoxy-4-(1-propenyl)phen-oxy]ethyl]-, erythro-	EtOH	261(4.20),265s(4.19)	150-2625-83M
threo-	EtOH	260(4.12)	150-2625-83M
$C_{21}H_{26}O_6$			
Bicyclo[3.2.1]oct-3-ene-2,8-diol, 7-(1,3-benzodioxol-5-yl)-1,3-dimethoxy-6-methyl-5-(2-propenyl)-	MeOH	231(3.83),284(3.68)	102-0561-83
β-	MeOH	234(3.85),285(3.72)	102-0561-83
Dispiro[cyclohexane-1,2'-[1,3]benzodi-oxole-5'(4'H),2"-[1,3]dioxolan]-7'-ol, tetrahydro-, benzoate, (3aR-cis)-	EtOH	205s(3.65),230(4.21), 275(3.04)	39-2441-83C
Picrasin B	MeOH	254(4.05)	100-0359-83
1,3-Propanediol, 1-(3,4-dimethoxyphen-yl)-2-[2-methoxy-4-(2-propenyl)phen-oxy]- (carinatidiol)	$CHCl_3$	248.5(4.09),285.5(4.04)	102-2227-83
$C_{21}H_{26}O_7$			
Carissanol, O-methyl-, epimer I	EtOH	228(4.01),280(3.73)	102-0749-83

Compound	Solvent	λ max (log ε)	Ref.
Carissanol, O-methyl-, epimer II	EtOH	228(4.05),279(3.73)	102-0749-83
Shinjudilactone, 1-O-methyl-	EtOH	239.5(3.73)	18-3683-83
$C_{21}H_{26}O_{11}$			
5H-Furo[3,2-g][1]benzopyran-5-one, 7-[(β-D-glucopyranosyloxy)methyl]-2,3-dihydro-4-hydroxy-2-(1-hydroxy-1-methylethyl)-	MeOH	215(4.43),232.5(4.22), 251.5(4.14),256(4.13), 298(4.03)	94-0064-83
	MeOH-AlCl$_3$	216.5(--),236(--), 266(--),317.5(--), 350(--),370(--)	94-0064-83
$C_{21}H_{27}Cl_2NO_2$			
Phenol, 4,4'-[1-[[bis(2-chloroethyl)-amino]methyl]-2-ethyl-1,2-ethanediyl]bis-, hydrochloride, (R*,S*)-	MeOH	208(4.15),227(4.31), 275(3.61),285(3.34)	83-0271-83
$C_{21}H_{27}FO_5$			
Isoflupredone	EtOH	240(4.20)	162-0743-83
Pregna-1,4-diene-3,20-dione, 9-fluoro-11,16,17-trihydroxy-	MeOH	238(4.20)	162-0421-83
$C_{21}H_{27}NO_3$			
Piperidine, 1-[9-(1,3-benzodioxol-5-yl)-1-oxo-2,8-nonadienyl]-, (E,E)-	EtOH	259(4.21),305(3.79)	94-3562-83
$C_{21}H_{27}NO_4$			
Isoquinoline, 1-[(2,4-dimethoxyphenyl)-methyl]-1,2,3,4-tetrahydro-6,7-dimethoxy-2-methyl-, (S)-	MeOH	205(4.58),232s(4.05), 282(3.58)	100-0908-83
Protostephanine	EtOH	283(3.92)	102-2603-83
$C_{21}H_{27}NO_5$			
Phenol, 4,5-dimethoxy-2-[(1,2,3,4-tetrahydro-6,7-dimethoxy-2-methyl-1-isoquinolinyl)methyl]-, (S)-	MeOH	210(4.45),231s(4.15), 288(3.89)	44-3957-83
Solanopyrone C monoacetate	EtOH	237(3.46),282(3.88), 321(3.96)	88-5373-83
$C_{21}H_{27}NO_6$			
Benzeneacetamide, N-[2-(3,4-dimethoxyphenyl)ethyl]-2-(hydroxymethyl)-3,4-dimethoxy-	MeOH	223(5.17),275(4.67)	78-1975-83
$C_{21}H_{27}N_2O$			
Benzoxazolium, 3-ethyl-2-[(3-piperidino-2-cyclohexen-1-ylidene)methyl]-, perchlorate	MeOH	462(5.15)	104-2089-83
$C_{21}H_{27}N_2OS$			
Benzothiazolium, 2-[2-[3-(dimethylamino)methylene]-2-methoxy-1-cyclohexen-1-yl]ethenyl]-3-ethyl-, perchlorate	MeOH	604(5.00)	104-1854-83
$C_{21}H_{27}N_2S$			
Benzothiazolium, 3-ethyl-2-[(3-piperidino-2-cyclohexen-1-ylidene)methyl]-, iodide	MeOH	492(5.12)	104-2089-83
$C_{21}H_{27}N_3$			
1H-Benzimidazole, 2-(4-methyl-2-pyridinyl)-1-octyl-	EtOH	302(4.28)	4-0023-83

Compound	Solvent	$\lambda_{max}(\log \epsilon)$	Ref.
$C_{21}H_{27}N_3O_4$			
Pregna-1,4-diene-3,20-dione, 7-azido-6,17-dihydroxy-, (6β,7α)-	MeOH	252(4.05)	39-2793-83C
$C_{21}H_{27}N_3O_6$			
Casimiroedine	EtOH	219(4.26),280(4.30)	162-0263-83
$C_{21}H_{27}N_9O_3$			
L-Lysine, N -[4-[[(2,4-diamino-6-pteridinyl)methyl]methylamino]benzoyl]-	pH 1.8	304(4.43)	87-0111-83
	pH 7.0	258(4.41),302(4.40), 370(3.91)	87-0111-83
	pH 13	258(4.40),302(4.40), 370(3.90)	87-0111-83
$C_{21}H_{27}O_8P$			
α-D-Allofuranose, 3-deoxy-3-(hydroxymethoxyphosphinyl)-3-C-(2-hydroxy-2-phenylethenyl)-1,2:5,6-bis-O-(1-methylethylidene)-, intramol. $3,3^2$-ester	EtOH	207(3.97),211(4.02), 218(3.86),258(4.26)	159-0019-83
$C_{21}H_{28}BrFO_2$			
Pregn-4-ene-3,20-dione, 17-bromo-6α-fluoro-	EtOH	236(4.20)	162-0663-83
$C_{21}H_{28}NO_3$			
Isoquinolinium, 1,2,3,4-tetrahydro-7,8-dimethoxy-1-[(4-methoxyphenyl)-methyl]-2,2-dimethyl-, (iodide)	MeOH	235(3.70),283(3.00)	100-0342-83
$C_{21}H_{28}NO_8P$			
α-Ribofuranose, 2-C-(cyanomethyl)-2-deoxy-2-(diethoxyphosphinyl)-1,2-O-(1-methylethylidene)-, 5-benzoate	EtOH	228(3.73)	159-0019-83
α-Xylofuranose, 2-C-(cyanomethyl)-2-deoxy-2-(diethoxyphosphinyl)-1,2-O-(1-methylethylidene)-, 5-benzoate	EtOH	228(3.94)	159-0019-83
$C_{21}H_{28}N_2O_2$			
Aspidospermine, N-deacetyl-N-formyl-	n.s.g.	211(4.47),250(3.94)	162-0123-83
$C_{21}H_{28}N_2O_3$			
Anomaline, demethoxy-	MeOH	221(4.13),260(3.72), 288(3.47)	102-2301-83
Sitsirikine, 18,19-dihydro-, (16R)-	EtOH	226(4.61),282(3.95), 291(3.87)	100-0694-83
(16S)-	EtOH	226(4.82),274(3.89), 284(3.91),291(3.65)	100-0694-83
$C_{21}H_{28}N_6O_3$			
Pregn-4-ene-3,20-dione, 6,7-diazido-17-hydroxy-, (6β,7α)-	MeOH	235(4.10)	39-2787-83C
$C_{21}H_{28}O_2$			
Demegestone	EtOH	214(3.80),302(4.32)	162-0416-83
Phenol, 2,2'-methylenebis[4-(1,1-dimethylethyl)-	EtOH	280(3.81),286s(3.79)	126-1363-83
Retinoic acid, 12-(hydroxymethyl)-, δ-lactone, 13-cis	EtOH	368(4.45)	44-0222-83
11-cis,13-cis	EtOH	365(4.52)	44-0222-83

Compound	Solvent	λ_{max}(log ε)	Ref.
$C_{21}H_{28}O_2S$			
Benzene, 1-methoxy-4-[3-[[4-(4-methoxyphenyl)butyl]thio]propyl]-	EtOH	277(4.55),284(4.54)	39-2577-83C
2-Thiophenecarboxylic acid, 5-[2-methyl-4-(2,6,6-trimethyl-1-cyclohexen-1-yl)-1,3-butadienyl]-, ethyl ester, (E,E)-	EtOH	235(3.96),353(4.49)	87-1282-83
(Z,E)-	EtOH	239(4.15),272(4.04), 351(4.30)	87-1282-83
$C_{21}H_{28}O_3$			
2-Furancarboxylic acid, 5-[2-methyl-4-(2,6,6-trimethyl-1-cyclohexen-1-yl)-1,3-butadienyl]-, ethyl ester, (E,E)-	EtOH	389(4.50)	87-1282-83
18-Norandrost-4-ene-17-carboxaldehyde, 13β-acetyl-3-oxo-	MeOH	240(4.21)	44-2696-83
Oxiranebutanoic acid, 3-(1,3-tetradecadiene-5,8-diynyl)-, methyl ester, [2α,3β(1E,3E)]-(±)-	EtOH	260(4.43),272(4.55), 285(4.47)	35-3656-83
$C_{21}H_{28}O_4$			
2-Pentenal, 5-(6-acetoxy-3,4-dihydro-2,5,7,8-tetramethyl-2H-1-benzopyran-2-yl)-2-methyl-, (E)-(±)-	hexane	277(3.46),283(3.53), 286(3.53)	94-4341-83
$C_{21}H_{28}O_4S$			
Benzene, 1-methoxy-4-[3-[[4-(4-methoxyphenyl)butyl]sulfonyl]propyl]-	EtOH	277(4.55),284(4.54)	39-2577-83C
$C_{21}H_{28}O_5$			
Aldosterone	n.s.g.	240(4.18)	162-0035-83
hydrate	n.s.g.	240(4.20)	162-0035-83
Cinerin II	n.s.g.	229(4.46)	162-0326-83
Cortisone	n.s.g.	237(4.15)	162-0362-83
Ketoemblide	EtOH	284(4.08)	138-0613-83
$C_{21}H_{28}O_5S_2$			
Propanoic acid, 2-methyl-, 4'a,5',7'-8',8'a,9'-hexahydro-8'a-hydroxy-3',4'a,5'-trimethyl-9'-oxospiro-[1,3-dithiolane-2,6'(4'H)-naphtho-[2,3-b]furan]-4'-yl ester, α-	EtOH	282.5(4.15)	94-3544-83
β-	EtOH	285.5(4.15)	94-3544-83
$C_{21}H_{28}O_6$			
10βH-Furanoeremophilan-9-one, 3,3-(ethylenedioxy)-6β-isobutyryloxy-	EtOH	284(4.15)	94-3544-83
Spiro[bicyclo[2.2.2]octa-5,7-diene-2,2'-oxiran]-5,6-dicarboxylic acid, 4,7-bis(1,1-dimethylethyl)-3-oxo-, dimethyl ester	n.s.g.	250s(3.7),327(2.7)	12-1361-83
$C_{21}H_{28}O_7$			
Furanoeremophilan-9-one, 3,3-(ethylenedioxy)-10β-hydroxy-6β-isobutyryloxy-	EtOH	282(4.19)	94-3544-83
$C_{21}H_{29}ClO_3$			
Androst-4-en-3-one, 17β-acetoxy-4-chloro-	n.s.g.	255(4.13)	162-0343-83

Compound	Solvent	λ_{max}(log ε)	Ref.
$C_{21}H_{29}FO_5$			
Pregn-4-ene-3,20-dione, 9-fluoro-11β,17,21-trihydroxy-	EtOH	239(4.25)	162-0591-83
$C_{21}H_{29}N$			
4H-Cyclododeca[4,5]pyrrolo[3,2,1-ij]-quinoline, 5,6,8,9,10,11,12,13,14-15,16,17-dodecahydro-	MeOH	212s(3.91),231(4.31), 285(3.69)	39-0501-83C
$C_{21}H_{29}NO$			
Butanamide, N-[2-[3,8-dimethyl-5-(1-methylethyl)-1-azulenyl]ethyl]-	EtOH	202(4.13),218(4.06), 247(4.26),287(4.54), 290s(4.54),306(4.08), 338s(3.47),352(3.66), 369(3.63),577s(2.44), 621(2.54),673s(2.45), 744s(1.98)	18-2059-83
Spiro[2.5]oct-1-ene-1-carboxamide, N,N-bis(1-methylethyl)-2-phenyl-	$CHCl_3$	452(2.61)	24-3097-83
$C_{21}H_{29}NO_2$			
Estra-1,3,5(10)-trien-3-ol, 17-[(1-methylethylidene)amino]-, (8α,17β)-, N-oxide	MeOH	224(4.03),230(4.04), 242(3.93),281(3.30), 286(3.27)	13-0001-83B
	MeOH-NaOH	241(4.25),298(3.39)	13-0001-83B
17β-	MeOH	223(4.06),231(4.06), 281(3.30),287(3.25)	13-0001-83B
	MeOH-NaOH	241(4.22),298(3.40)	13-0001-83B
1-Pentanol, 2-[[1-(4-methoxyphenyl)-2-phenylethyl]amino]-4-methyl-, hydrochloride	EtOH	228(4.09),254s(2.69), 259s(2.88),265s(3.00), 270s(3.05),274(3.12), 281(3.05)	94-0031-83
$C_{21}H_{29}NO_3S$			
Spiro[androsta-1,4-diene-17,5'-[1,2,3]-oxathiazolidin]-3-one, 3'-methyl-, 2'-oxide, (2'S,17β)	EtOH	242(4.15)	5-1001-83
Spiro[androsta-4,6-diene-17,5'-[1,2,3]-oxathiazolidin]-3-one, 3'-methyl-, 2'-oxide, (2'S,17β)-	EtOH	282(4.33)	5-1001-83
$C_{21}H_{29}NO_4$			
4H-1-Benzopyran-4-one, 2,3-dihydro-5-hydroxy-2,2-dimethyl-3-(4-morpholinylmethylene)-7-pentyl-	EtOH	250s(3.73),288(3.71), 389(4.32)	161-0775-83
Phenol, 4,4'-[1-[[bis(2-hydroxyethyl)-amino]methyl]-2-ethyl-1,2-ethanediyl]bis-, (R*,S*)-	MeOH	209(4.16),228(4.32), 278(3.54),287(3.36)	83-0271-83
$C_{21}H_{29}NO_4S$			
1H,3H-[1,3]Oxaazacyclopentadecino[4,3-a]isoindole-3,18(4H)-dione, 5,6,7,8-9,10,11,12,13,13a-decahydro-13a-hydroxy-13-(methylthio)-	MeOH	222(3.94),229(3.95), 235(3.85)	78-1273-83
13H,15H,17H-[1,6,3]Oxathiaazacycloheptadecino[4,3-a]isoindole-13,17-dione, 1,3,4,5,6,7,8,9,10,11,12,21b-dodecahydro-21b-hydroxy-	MeOH	220(4.01),228(4.00), 234(3.89)	78-1273-83
Undecenoic acid, 11-(methylthio)-, (1,3-dihydro-1,3-dioxo-2H-isoindol-2-yl)methyl ester	$CHCl_3$	296(3.37),303s(3.32)	78-1273-83

Compound	Solvent	λ max (log ε)	Ref.
$C_{21}H_{29}NO_6$			
Propanedioic acid, [4,4-dimethyl-1-(4-nitrophenyl)-2-pentenyl]methyl-, diethyl ester, (E)-	EtOH	210(3.95),275(4.00)	12-0527-83
$C_{21}H_{29}N_5O_{10}$			
Dapiramicin A	MeOH	227(4.60),289(3.94)	88-0495-83
$C_{21}H_{29}O_2P$			
Phosphorin, 1-(1,1-dimethylethyl)-4-(2,2-dimethyl-1-oxopropyl)-1,1-dihydro-1-methoxy-3-phenyl-	EtOH	248(3.94),325(3.76)	88-5051-83
$C_{21}H_{29}O_8P$			
Cortisone 21-dihydrogen phosphate	H_2O	244(4.20)	162-0362-83
	MeOH	238(4.18)	162-0362-83
$C_{21}H_{30}N_2$			
Pyrrolo[2,3-b]indole, 1,2,3,3a,8,8a-hexahydro-1-methyl-3a,8-bis(3-methyl-2-butenyl)-	EtOH	211(4.04),254(3.30), 306(2.73)	1-0803-83
	EtOH-HCl	211(4.11),246(3.28), 295(3.79)	1-0803-83
$C_{21}H_{30}N_2O_2$			
[1,5]Diazacyclohexadecino[2,1-a]isoindole-4,16-dione, 1,2,3,5,6,7,8,9,10-11,12,13,14,20b-tetradecahydro-	$CHCl_3$	246(3.71),268s(3.42), 279(3.23)	78-2691-83
$C_{21}H_{30}N_2O_8$			
L-ribo-Hexar-1-amic acid, 2,3-dideoxy-N-[1-(3,4-dihydro-8-methoxy-1-oxo-1H-2-benzopyran-3-yl)-3-methylbutyl]-, [S-(R*,R*)]-	MeOH	244(3.75),306(3.60)	87-1370-83
$C_{21}H_{30}O_2$			
6,8,11,13-Abietatetraen-11-ol, 12-methoxy-	EtOH	220(4.27),270(3.86)	102-1771-83
Cannabidiol	EtOH	274(3.12),282(3.10)	162-0241-83
Pregna-5,16-dien-20-one, 3β-hydroxy-	EtOH	267(3.95)	83-0678-83
$C_{21}H_{30}O_2S$			
Androst-4-en-3-one, 7α-(acetylthio)-	EtOH	237(4.28)	95-1046-83
$C_{21}H_{30}O_3$			
Atis-16-en-18-oic acid, 13-oxo-, methyl ester	MeOH	286(2.46)	102-0875-83
2H-1-Benzopyran-2-one, 7-(dodecyloxy)-	$CHCl_3$	322(4.19)	44-1872-83
H-T dimer	$CHCl_3$	244(3.39),279(3.15)	44-1872-83
D-Homo-18-nor-5α-androst-13(17a)-en-17-one, 3β-acetoxy-	EtOH	240(4.28)	44-1954-83
1-Naphthalenecarboxylic acid, 5-[2-(3-furanyl)ethyl]-1,2,3,4,4a,5,8,8a-octahydro-1,4a,6-trimethyl-, methyl ester	EtOH	215(3.67)	102-1512-83
18-Norandrost-4-en-3-one, 13β-acetyl-17β-(hydroxymethyl)-	MeOH	240(4.21)	44-2696-83
Retinol, 12-carboxy-, 11-cis,13-cis	MeOH	327(4.43)	44-0222-83
$C_{21}H_{30}O_4$			
Abieta-8,11,13-trien-3-one, 7β,11-dihydroxy-12-methoxy-	EtOH	225s(--),284(3.24)	102-1771-83

Compound	Solvent	λ_{max}(log ϵ)	Ref.
Algestone	n.s.g.	240(4.22)	162-0037-83
2H-1-Benzopyran-6-ol, 3,4-dihydro-2-(5-hydroxy-4-methyl-3-pentenyl)-2,5,7,8-tetramethyl-, 6-acetate, (E)-(±)-	hexane	277(3.46),284(3.53), 286(3.53)	94-4341-83
Cyclohexaneacetic acid, 2,2,6-trimethyl-6-[4-(1-methylethyl)-5,6-dioxo-1,3-cyclohexadien-1-yl]-, methyl ester, trans	EtOH	274(2.90),418(3.20), 586(2.00)	107-0201-83
18-Norandrost-4-en-3-one, 13-(hydroxyacetyl)-17β-(hydroxymethyl)-	MeOH	240(4.18)	44-2696-83
$C_{21}H_{30}O_5$			
Humulon	EtOH	237(4.14),282(3.92)	162-0688-83
Pregn-4-ene-3,20-dione, 11β,19,21-trihydroxy-	MeOH	243(4.00)	39-2945-83C
$C_{21}H_{30}O_{13}$			
Acetylbarlerin	MeOH	234(4.0)	94-0780-83
Xanthoxylin glycoside	EtOH	225(4.16),273(3.88)	102-0790-83
$C_{21}H_{31}N$			
1-Azuleneethanamine, N,N-diethyl-3,8-dimethyl-5-(1-methylethyl)-	EtOH	218(4.06),246(4.22), 287(4.46),304s(4.04), 339s(3.49),352(3.68), 369(3.56),582s(2.41), 622(2.47),675s(2.37), 751s(1.89)	18-2059-83
1-Azuleneethanamine, N,N,α,α,3,8-hexamethyl-5-(1-methylethyl)-	EtOH	218(4.16),247(4.29), 289(4.48),306s(4.13), 342s(3.61),353(3.69), 371(3.65),575(2.41), 618(2.48),672s(2.37), 755s(1.85)	18-2059-83
1-Azuleneethanamine, N,N,β,β,3,8-hexamethyl-5-(1-methylethyl)-	EtOH	247(4.03),289(4.00), 305s(3.93),590(2.16), 665s(2.04)	18-2059-83
$C_{21}H_{31}NO$			
Retinamide, N-methyl-, all trans	EtOH	347(4.70)	34-0422-83
$C_{21}H_{31}NO_2$			
Androisoxazole	EtOH	226(3.71)	162-0092-83
$C_{21}H_{31}NO_3$			
3-Aza-A-homoandrost-4a-en-4-one, 17β-acetoxy-	EtOH	241(4.33)	4-1093-83
4H-1-Benzopyran-4-one, 3-[(diethylamino)methylene]-2,3-dihydro-5-hydroxy-2,2-dimethyl-7-pentyl-	EtOH	245s(3.75),283(3.65), 386(4.24)	161-0775-83
$C_{21}H_{31}N_7O_6$			
1(2H)-Pyrimidinepropanamide, N-[3-[[4-[[3-(3,4-dihydro-2,4-dioxo-1(2H)-pyrimidinyl)-1-oxopropyl]amino]butyl]amino]propyl]-3,4-dihydro-2,4-dioxo-	H_2O	266(4.12)	103-1228-83
$C_{21}H_{32}N_2$			
4-Azuleneethanamine, 3-[(dimethylamino)methyl]-N,N,1-trimethyl-7-(1-methylethyl)-	EtOH	217s(4.09),247(4.34), 289(4.56),304s(4.12), 342s(3.61),353(3.72),	18-2311-83

Compound	Solvent	λ_{max}(log ϵ)	Ref.
(cont.)		372(3.69),407s(2.09), 573s(2.55),609(2.64), 653s(2.57),727s(2.13)	18-2311-83
$C_{21}H_{32}O$			
Androsta-4,6-diene, 3-ethoxy-	EtOH	240(4.29)	95-1046-83
Estr-4-en-3-one, 16β-(1-methylethyl)-	EtOH	239(4.23)	95-1046-83
$C_{21}H_{32}OSi$			
Silane, (estra-1,3,5(10)-trien-3-yl-oxy)trimethyl-	hexane	276(3.18),282(3.15)	65-0539-83
$C_{21}H_{32}O_2$			
1H-Benz[e]indene-1-acetic acid, 2,3,5-5a,6,7,8,9-octahydro-1,3,3,5a,9-pentamethyl-, methyl ester, [1S-(1α,5aβ,9β)]-	n.s.g.	269(3.74)	12-1001-83
Cyclopropanebutanoic acid, 2-(1,3,5,8-tetradecatetraenyl)-, (E,E,Z,Z)-	MeOH	273(4.62),283(4.70), 294(4.59)	87-0072-83
1(2H)-Naphthalenone, 5-[2-(2-acetyl-1-methylcyclopentyl)ethyl]-3,4,5,6-7,8-hexahydro-5-methyl-	n.s.g.	247(4.02)	77-0123-83
Phenol, 4-[(decahydro-1,2,4a,5-tetra-methyl-1-naphthalenyl)methoxy]-	EtOH	223(3.58),378(3.31), 284[sic](3.23)	105-0664-83
$C_{21}H_{32}O_2Si$			
Estra-5,7,9-triene-3,17-diol, 1-(tri-methylsilyl)-, (1α,3β,17β)-	EtOH	266(2.80),271.5(2.81), 281(2.71)	65-1724-83
(1β,3α,17β)-	EtOH	248s(2.68),254(2.53), 261(2.54),273(2.62), 282(2.55)	65-1724-83
$C_{21}H_{32}O_3$			
8,11,13-Abietatriene-7β,11-diol, 12-methoxy-	EtOH	282(3.25)	102-1771-83
Diemensin A	n.s.g.	247(3.95),307(4.13)	88-1917-83
6,8,11,14-Eicosatetraenoic acid, 5-oxo-, methyl ester, (E,Z,Z)-	MeOH	278.5(4.43)	87-0072-83
Oxiranebutanoic acid, 3-(1,3,5,8-tetra-decatetraenyl)-, methyl ester, (±)-	hexane	268(4.53),279(4.63), 291(4.51)	35-3656-83
cis isomer	EtOH	271(4.51),281(4.60), 292(4.46)	35-3656-83
$C_{21}H_{32}O_4$			
1H-2-Benzopyran-1-one, 3,4-dihydro-8-hydroxy-6-methoxy-3-undecyl-, (R)-	EtOH	214(4.60),266(4.46), 302(4)	102-2031-83
Prosta-5,10,13-trien-1-oic acid, 15-hydroxy-9-oxo-, methyl ester	n.s.g.	218(4.01)	39-1573-83C
Prosta-5,10,13-trien-1-oic acid, 15-hydroxy-15-methyl-11-oxo-	EtOH	216(3.97),319(3.13)	87-0790-83
$C_{21}H_{32}O_5$			
4H-Cyclopenta[b]furan-2-pentanoic acid, 5,6-dihydro-5-hydroxy-4-(3-hydroxy-2-octenyl)-, methyl ester	n.s.g.	223(3.95)	88-0047-83
$C_{21}H_{32}O_6$			
Pentanoic acid, 5-[2,3,3a,4,5,6-hexa-hydro-5-hydroxy-4-(3-oxo-2-octenyl)-cyclopenta[b]furan-2-ylidene]-4-oxo-, methyl ester, (E)-	EtOH	265(4.303)	88-1281-83

Compound	Solvent	λ_{max}(log ε)	Ref.
(Z)- (cont.)	EtOH	261(4.264)	88-1281-83
$C_{21}H_{33}NO_2$			
2-Aziridinebutanoic acid, 3-(1,3,5,8-tetradecatetraenyl)-, methyl ester	MeOH	273(--),282(4.68), 290(--)	88-0331-83
1H-Isoindole, 2-[[4-(1,1-dimethylethoxy)-2-methyl-2-butenyl]oxy]-2,3-dihydro-1,1,3,3-tetramethyl-	EtOH	215(3.45),256(2.80), 263(2.99),270(3.01)	12-1957-83
1H-Isoindole, 2-[[4-(1,1-dimethylethoxy)-3-methyl-2-butenyl]oxy]-2,3-dihydro-1,1,3,3-tetramethyl-	EtOH	214(3.52),257(2.90), 263(3.03),270(3.02)	12-1957-83
1H-Isoindole, 2-[[1-[(1,1-dimethylethoxy)methyl]-1-methyl-2-propenyl]oxy]-2,3-dihydro-1,1,3,3-tetramethyl-	EtOH	216(3.51),256(2.88), 263(3.06),270(3.00)	12-1957-83
1H-Isoindole, 2-[[1-[(1,1-dimethylethoxy)methyl]-2-methyl-2-propenyl]oxy]-2,3-dihydro-1,1,3,3-tetramethyl-	EtOH	216(3.33),257(2.87), 263(3.02),270(3.01)	12-1957-83
$C_{21}H_{33}NO_6$			
3,5-Pyridinedicarboxylic acid, 1-(5-ethoxy-5-oxopentyl)-1,4-dihydro-2,4,6-trimethyl-, diethyl ester	EtOH	233(3.71),263(3.54), 347(3.43)	73-0608-83
$C_{21}H_{34}OSi$			
Estr-4-en-3-one, 1α-(trimethylsilyl)-	EtOH	249(4.04)	65-0539-83
Estr-4-en-3-one, 1β-(trimethylsilyl)-	EtOH	249(3.90)	65-0539-83
$C_{21}H_{34}O_2$			
1-Naphthalenepentanoic acid, 1,4,4a,5-8,8a-hexahydro-β,2,5,5,8a-pentamethyl-, methyl ester	EtOH	203(3.65)	102-1292-83
2-Pentenoic acid, 3-methyl-5-(1,4,4a,5-6,7,8,8a-octahydro-2,5,5,8a-tetramethyl-1-naphthalenyl)-, methyl ester	EtOH	219(4.14)	102-1294-83
$C_{21}H_{34}O_2Si$			
Estr-4-en-3-one, 17-hydroxy-1-(trimethylsilyl)-, (1α,17β)-	EtOH	248.5(4.05)	65-0539-83
$C_{21}H_{34}O_3$			
6,8,11,14-Eicosatetraenoic acid, 5-(hydroxymethyl)-, (E,Z,Z,Z)-	EtOH	236(4.45)	87-0072-83
1-Naphthalenepentanoic acid, 1,2,3,4-5,6,7,8-octahydro-β,1,2,5,5-pentamethyl-6-oxo-, methyl ester	EtOH	280(2.20)	102-2783-83
1-Naphthalenepentanoic acid, 1,2,3,4-4a,7,8,8a-octahydro-β,1,2,4a,5-pentamethyl-7-oxo-, methyl ester	EtOH	234(4.11)	102-2805-83
Pentanoic acid, 5-(2,4,6,9-pentadecatetraenyloxy)-, methyl ester, (E,E,Z,Z)-	n.s.g.	272(4.57),283(4.47)	44-4413-83
2H-Pyran-2-one, 4-hydroxy-3,5-dimethyl-6-(1,3,5,7-tetramethyl-1-decenyl)-	n.s.g.	301(3.70)	88-3055-83
$C_{21}H_{34}O_5$			
Prosta-8,13-dien-1-oic acid, 9,15-dihydroxy-10-oxo-, methyl ester, (13E,15S)-(±)-	EtOH	262(4.09)	39-0319-83C
$C_{21}H_{34}O_{11}$			
Urceolide	MeOH	218(4.15)	102-0255-83

Compound	Solvent	λ_{max}(log ϵ)	Ref.
$C_{21}H_{35}NO_2$			
1H-Isoindole, 2-[1-[(1,1-dimethylethoxy)methyl]-2-methylpropoxy]-2,3-dihydro-1,1,3,3-tetramethyl-	EtOH	215(3.47),257(2.87), 263(2.99),270(3.00)	12-1957-83
$C_{21}H_{36}$			
1,3,6,9-Heneicosatetraene, (Z,Z,Z)-	hexane	230(4.415)	44-2270-83
1H-Indene, 1-(1,5-dimethylhexyl)octahydro-7a-methyl-4-(2-propenylidene)-, (de-A vitamin D_3)	C_6H_{12}	234(4.45),241(4.49), 249(4.30)	35-3270-83
$C_{21}H_{36}N_7O_{16}P_3S$			
Coenzyme A	n.s.g.	259.5(4.23)	162-0351-83
$C_{21}H_{36}O_2$			
1,2-Benzenediol, 3-pentadecyl-	MeOH	230(3.19),278(3.23)	36-0792-83
$C_{21}H_{36}O_3Si$			
Butanoic acid, 2-(3-methylphenoxy)-, (tripropylsilyl)methyl ester	heptane	219(3.78),274(3.08)	65-1039-83
Butanoic acid, 2-(4-methylphenoxy)-, (tripropylsilyl)methyl ester	heptane	223(3.89),280(3.16)	65-1039-83
$C_{21}H_{36}O_4$			
Spiro[4.5]dec-7-ene-8-carboxylic acid, 9-hydroxy-1-(6-hydroxy-1,5-dimethylhexyl)-4-methyl-, methyl ester	n.s.g.	236(3.26)	12-0993-83
$C_{21}H_{36}O_4Si$			
1-Butanone, 1-[4-[[(1,1-dimethylethyl)dimethylsilyl]oxy]-2,6-dimethoxy-3,5-dimethylphenyl]-2-methyl-	EtOH	218s(4.21),253(3.63)	88-2531-83
$C_{21}H_{38}N_2O_7Si_2$			
Uridine, 3',5'-O-[1,1,3,3-tetrakis(1-methylethyl)-1,3-disiloxanediyl]-	MeOH	262(3.98)	35-4059-83
$C_{21}H_{38}O_3$			
9-Octadecenoic acid, 9-acetyl-, methyl ester	EtOH	234(3.93)	2-0319-83
$C_{21}H_{39}OS_5$			
Sulfonium, [2-(ethylthio)-2-oxo-1-[1,2,3-tris[(1,1-dimethylethyl)thio]-2-cyclopropen-1-yl]ethyl]dimethyl-, perchlorate	MeCN	244(3.95),282s(3.85), 306(4.04)	18-0171-83
tetrafluoroborate	MeCN	248(4.18),308(4.23)	18-0171-83
$C_{21}H_{39}O_2S_4$			
Sulfonium, [2-ethoxy-2-oxo-1-[1,2,3-tris[(1,1-dimethylethyl)thio]-2-cyclopropen-1-yl]ethyl]dimethyl-, perchlorate	MeCN	272(4.04),285s(3.96)	18-0171-83
tetrafluoroborate	MeCN	274(4.54),286s(4.11)	18-0171-83

Compound	Solvent	λ_{max}(log ϵ)	Ref.
$C_{22}H_4Cl_{14}NO_3$			
Methyl, [4-[[(carboxymethyl)amino]-carbonyl]-2,3,5,6-tetrachloro-phenyl]bis(pentachlorophenyl)-	$CHCl_3$	288(3.76),338s(3.77), 368s(4.25),382(4.54), 480s(3.02),507(3.03), 560(3.0)	44-3716-83
$C_{22}H_5Cl_{14}NO_3$			
Glycine, N-[4-[bis(pentachlorophenyl)-methyl]-2,3,5,6-tetrachlorobenzoyl]-	$CHCl_3$	293(3.27),303(3.30)	88-2121-83
$C_{22}H_6Cl_{14}NO_2$			
L-Alanine, 4-[bis(pentachlorophenyl)-methylene-2,3,5,6-tetrachlorophenyl ester, hydrobromide (radical)	MeCN	219(4.96),270s(3.72), 336s(3.76),366s(4.20), 381(4.45),510(3.00), 564(3.02)	44-3716-83
$C_{22}H_8Cl_4I_4O_5$			
Rose Bengal, ethyl ester	CH_2Cl_2	407(--),496(--)	35-7465-83
monoanion	MeOH	524(4.6),564(5.0)	35-7465-83
$C_{22}H_{10}N_2$			
Benzo[3,4]cyclobut[1,2-b]anthracene-5,12-dicarbonitrile	CH_2Cl_2	233(4.65),247(4.60), 260(4.49),268s(4.51), 280(4.67),294(4.73), 305(4.83),325(4.62), 337(4.83),372(3.99), 393(4.05),415(4.14), 440(4.08)	39-1443-83C
Benzo[3,4]cyclobuta[1,2-b]phenanthrene-7,12-dicarbonitrile	CH_2Cl_2	236(4.45),276(4.65), 290(4.58),325(4.33), 339(4.40),374s(3.63), 397(3.79),419(4.13), 435(4.31)	39-1443-83C
Dibenzo[b,h]biphenylene-5,12-dicarbonitrile	CH_2Cl_2	233(4.45),247(4.30), 258s(4.37),265(4.51), 283s(4.30),308(4.97), 324(4.74),337s(4.26), 363(4.11),390(3.91), 413(4.30),442(4.57)	39-1443-83C
$C_{22}H_{11}Cl_6NO$			
3H-Pyrrol-3-one, 2-(1,2-diphenylethen-yl)-4,5-bis(trichloroethenyl)-	MeCN	265(3.95),315(3.79), 475(3.19)	88-2977-83
$C_{22}H_{12}Br_2O_6$			
[2,2'-Binaphthalene]-1,1',4,4'-tetrone, 3,3'-dibromo-5,5'-dihydroxy-7,7'-di-methyl-	EtOH	293(4.33),445(3.96)	5-1020-83
	EtOH-NaOH	292(4.40),570(4.08)	5-1020-83
$C_{22}H_{12}ClNO_2$			
7H-Dibenz[f,ij]isoquinolin-7-one, 2-chloro-6-phenoxy-	toluene	384(4.01)	104-1172-83
7H-Dibenz[f,ij]isoquinolin-7-one, 2-chloro-8-phenoxy-	toluene	384(4.07)	104-1172-83
$C_{22}H_{12}N_2Se$			
9-Anthracenecarbonitrile, 10-(1,2-benzisoselenazol-3-yl)-	CH_2Cl_2	315(3.6),350(3.6), 365(3.9),390(4.0), 410(4.0)	78-0835-83

Compound	Solvent	λ_{max}(log ϵ)	Ref.
$C_{22}H_{12}N_4O_2$			
2,3-Anthracenedicarbonitrile, 1-amino-9,10-dihydro-9,10-dioxo-4-(phenylamino)-	DMF	655(3.90)	33-0411-83
$C_{22}H_{12}O_2$			
Azuleno[1,2-b]anthracene-6,13-dione	MeCN	214(4.5),275(4.4), 320(4.6),408(4.2), 426(4.2),606(3.2)	24-2408-83
$C_{22}H_{12}O_4$			
Benzo[3,4]cyclobut[1,2-b]anthracene-5,12-dicarboxylic acid	EtOH	220(4.58),250(4.50), 280s(4.68),293(4.75), 304(4.78),331(4.74), 370(4.04),389(4.02), 412(4.01),438(3.86)	39-1443-83C
$C_{22}H_{13}BrO_6$			
[2,2'-Binaphthalene]-1,1',4,4'-tetrone,	EtOH	292.5(4.11),442.5(3.94)	5-1020-83
3-bromo-5,5'-dihydroxy-7,7'-dimethyl-	EtOH-NaOH	290(4.29),557.5(4.05)	5-1020-83
[2,2'-Binaphthalene]-1,1',4,4'-tetrone,	EtOH	292.5(4.13),440(3.96)	5-1020-83
3-bromo-5,8'-dihydroxy-6',7-dimethyl-	EtOH-NaOH	290(4.34),552.5(4.03)	5-1020-83
[2,2'-Binaphthalene]-1,1',4,4'-tetrone,	EtOH	292.5(4.12),437.5(3.94)	5-1020-83
3-bromo-8,8'-dihydroxy-6,6'-dimethyl-	EtOH-NaOH	292.5(4.32),553(4.00)	5-1020-83
$C_{22}H_{13}ClN_2O_2$			
[1]Benzopyrano[2,3-c]pyrazol-4(1H)-one, 6-chloro-1,3-diphenyl-	EtOH	242(4.47),260(4.40)	118-0214-83
$C_{22}H_{13}Cl_2NO_2$			
Aziridino[b]naphthalene-2,7-dione, 1,1a-bis(4-chlorophenyl)-	benzene	353(2.73)	44-4968-83
$C_{22}H_{13}NO_3$			
6H-Anthra[1,9-cd]isoxazol-6-one, 3-(4-methylbenzoyl)-	toluene	323(4.24),475(3.86)	103-0019-83
3H-Dibenzo[f,ij]isoquinoline-2,6-dione, 6-phenoxy-	EtOH	418(3.73)	104-1172-83
3H-Dibenzo[f,ij]isoquinoline-2,6-dione, 8-phenoxy-	toluene	367(4.04)	104-1172-83
$C_{22}H_{13}N_3O_2$			
Cyclohept[1,2,3-hi]imidazo[2,1,5-cd]-indolizine-4-carboxylic acid, 9-cyano-2-phenyl-, methyl ester	$CHCl_3$	257(4.31),282s(4.37), 292s(4.44),305(4.50), 317(4.43),340(4.02), 382s(4.35),400(4.45), 436(4.00),461(3.98)	18-3703-83
$C_{22}H_{13}N_3O_3$			
9,10-Anthracenedione, 1-azido-2-(4-methylbenzoyl)-	toluene	335(3.63)	103-0019-83
$C_{22}H_{13}O_3S$			
1,3-Oxathiol-1-ium, 5-(3-oxo-3H-naphtho[2,1-b]pyran-2-yl)-2-phenyl-, perchlorate	MeCN	407(4.18)	48-0551-83
$C_{22}H_{14}BrNO_2$			
Aziridino[b]naphthalene-2,7-dione, 1-(4-bromophenyl)-1a-phenyl-	benzene	352(2.77)	44-4968-83

Compound	Solvent	λ_{max}(log ε)	Ref.
$C_{22}H_{14}BrN_3O_2$			
7H-Benz[e]perimidin-7-one, 4-bromo-6-[(4-methoxyphenyl)amino]-	n.s.g.	485(4.21)	2-0812-83
$C_{22}H_{14}ClNO_2$			
Aziridino[b]naphthalene-2,7-dione, 1-(4-chlorophenyl)-1a-phenyl-	benzene	353(2.73)	44-4968-83
$C_{22}H_{14}ClNO_4$			
Acetamide, 2-chloro-N-(9,10-dihydro-9,10-dioxo-4-phenoxy-1-anthracenyl)-	toluene	419(3.85)	104-1172-83
Acetamide, 2-chloro-N-(9,10-dihydro-9,10-dioxo-5-phenoxy-1-anthracenyl)-	toluene	402(3.90)	104-1172-83
$C_{22}H_{14}ClN_3O$			
7H-Benzo[e]perimidin-7-one, 6-[(4-chlorophenyl)amino]-2-methyl-	n.s.g.	506(4.40)	2-0812-83
$C_{22}H_{14}Cl_2N_6$			
Benzo[1,2-b:4,5-b']dipyrrole, 3,7-bis-[(4-chlorophenyl)azo]-1,5-dihydro-	EtOH	204.5(4.59),227(4.61), 246.9s(4.45),309(4.67), 239[sic](4.78)	103-0871-83
$C_{22}H_{14}Cl_3N_5O_9$			
Ethanone, 1-[3-[[4-[2,2-dichloro-1-(2-chloro-3,5-dinitrophenyl)ethyl]-2,6-dinitrophenyl]amino]phenyl]-	MeOH	228.9(4.58),393.1(3.72)	56-1357-83
$C_{22}H_{14}N_2$			
Benzenamine, N-11H-indeno[1,2-b]quinolin-11-ylidene-	EtOH	205(5.0),230(5.0), 288(5.1),380s(4.0)	104-0158-83
$C_{22}H_{14}N_2O_2$			
[1]Benzopyrano[2,3-c]pyrazol-4(1H)-one, 1,3-diphenyl-	EtOH	242(4.51),262(4.40)	118-0214-83
$C_{22}H_{14}N_2O_3$			
6H-Anthra[1,9-cd]isoxazole-3-carboxamide, N-(2-methylphenyl)-6-oxo-	toluene	459(4.19)	103-1279-83
6H-Anthra[1,9-cd]isoxazole-3-carboxamide, N-(3-methylphenyl)-6-oxo-	toluene	459(4.19)	103-1279-83
6H-Anthra[1,9-cd]isoxazole-3-carboxamide, N-(4-methylphenyl)-6-oxo-	toluene	460(4.22)	103-1279-83
1H-Anthra[1,2-c]pyrazole-3,6,11(2H)-trione, 2-(2-methylphenyl)-	DMF	480(3.53)	103-1279-83
1H-Anthra[1,2-d]pyrazole-3,6,11(2H)-trione, 2-(3-methylphenyl)-	DMF	462(3.42)	103-1279-83
1H-Anthra[1,2-d]pyrazole-3,6,11(2H)-trione, 2-(4-methylphenyl)-	DMF	466(3.51)	103-1279-83
$C_{22}H_{14}N_2O_4$			
6H-Anthra[1,9-cd]isoxazole-3-carboxamide, N-(3-methoxyphenyl)-6-oxo-	toluene	459(4.09)	103-1279-83
1H-Anthra[1,2-c]pyrazole-3,6,11(2H)-trione, 2-(3-methoxyphenyl)-	DMF	466(3.48)	103-1279-83
$C_{22}H_{14}N_2O_4S$			
3-Thiophenecarbonitrile, 4-amino-2-(4-methoxyphenyl)-5-[(2-oxo-2H-1-benzopyran-3-yl)carbonyl]-	MeCN	385(4.23)	48-0551-83

Compound	Solvent	λ_{max}(log ε)	Ref.
$C_{22}H_{14}N_2O_4Se$			
2,3-Naphthalenedicarboxylic acid, 1-(1,2-benzoselenazol-3-yl)-4-cyano-, dimethyl ester	CH_2Cl_2	318(4.06)	78-0835-83
$C_{22}H_{14}N_2O_6$			
2-Anthracenecarboxamide, 9,10-dihydro-N-(3-methoxyphenyl)-1-nitro-9,10-dioxo-	dioxan	367(3.72)	103-1279-83
$C_{22}H_{14}N_4$			
1,4,5,8-Anthracenetetracarbonitrile, 2,3,6,7-tetramethyl-	$CHCl_3$	270(5.26),343(3.57), 362(4.01),380(4.34), 399(4.18),422(4.12)	39-1443-83C
9,9'-Bi-9H-pyrido[3,4-b]indole	EtOH	227(4.95),279s(4.46), 282(4.58),331(4.12), 339(4.15)	78-0033-83
1,4,5,8-Phenanthrenetetracarbonitrile, 2,3,6,7-tetramethyl-	EtOH	237(4.87),259(4.57), 275(4.60),318(4.76), 341s(4.31)	39-1443-83C
$C_{22}H_{14}N_4O_3$			
7H-Benzo[e]perimidin-7-one, 2-methyl-6-[(4-nitrophenyl)amino]-	n.s.g.	492(4.32)	2-0812-83
$C_{22}H_{14}N_4O_4$			
2-Anthracenecarboxamide, 1-azido-9,10-dihydro-N-(3-methoxyphenyl)-9,10-dioxo-	dioxan	327(3.89)	103-1279-83
$C_{22}H_{14}O$			
Benzo[ghi]perylene, 5-acetyl-	n.s.g.	277(4.49),284(4.34), 305(4.46),398(4.28)	65-1898-83
$C_{22}H_{14}O_4$			
[2,2'-Binaphthalene]-1,1',4,4'-tetrone, 3,3'-dimethyl-	$CHCl_3$	248s(4.60),252(4.63), 263(4.52),339(3.83)	5-0299-83
$C_{22}H_{14}O_6$			
[2,2'-Binaphthalene]-1,1',4,4'-tetrone, 5,5'-dihydroxy-3,3'-dimethyl-	$CHCl_3$	255s(4.33),262s(4.34), 272(4.36),422s(3.94), 450s(3.90)	5-0299-83
[2,2'-Binaphthalene]-1,1',4,4'-tetrone, 5,5'-dihydroxy-7,7'-dimethyl-	$CHCl_3$	249(4.17),255(4.18), 261(4.16),278(4.11), 350s(3.43),445(3.88)	5-0299-83
[2,2'-Binaphthalene]-1,1',4,4'-tetrone, 8,8'-dihydroxy-3,3'-dimethyl-	$CHCl_3$	256(4.34),272(4.36), 420s(3.92),432(3.94), 450s(3.85)	5-0299-83
[2,2'-Binaphthalene]-1,1',4,4'-tetrone, 8,8'-dihydroxy-6,6'-dimethyl-	$CHCl_3$	252(4.26),276(4.19), 350(3.54),440(3.92)	5-0299-83
[2,2'-Binaphthalene]-1,1',4,4'-tetrone, 5,5'-dimethoxy-	$CHCl_3$	246(4.40),271(4.30), 405(3.91)	5-0299-83
[2,2'-Binaphthalene]-1,1',4,4'-tetrone, 8,8'-dimethoxy-	$CHCl_3$	247(4.39),257s(4.37), 320s(3.53),404(3.86)	5-0299-83
$C_{22}H_{14}O_8$			
[2,2'-Binaphthalene]-1,1',4,4'-tetrone, 3,3'-dihydroxy-5,5'-dimethoxy-	MeOH	210(4.25),229(4.45), 279(4.46),385(3.96), 467s(3.62)	5-1886-83
[2,2'-Binaphthalene]-1,1',4,4'-tetrone, 3,3'-dihydroxy-8,8'-dimethoxy-	MeOH	230(4.26),279(4.23), 381(3.84),455s(3.56)	5-1886-83

Compound	Solvent	λ_{max}(log ε)	Ref.
[2,2'-Binaphthalene]-1,1',4,4'-tetrone, 3,3',5,5'-tetrahydroxy-7,7'-dimethyl-	MeOH	212(4.22),230(4.12), 291(4.31),412(3.98)	5-1886-83
$C_{22}H_{15}AsCl_2$			
1H-Arsole, 2,5-bis(4-chlorophenyl)-1-phenyl-	C_6H_{12}	200(4.57),221(4.26), 242s(4.01),392(4.18)	101-0335-83I
$C_{22}H_{15}ClN_2O_4$			
[1,2'-Binaphthalene]-1',4'-dione, 3'-chloro-4-(dimethylamino)-5'(and 8')-nitro-	toluene	547(3.08)	150-0168-83S
	EtOH	531(--)	150-0168-83S
$C_{22}H_{15}NO_2$			
6H-Anthra[1,9-cd]isoxazol-6-one, 3-[(4-methylphenyl)methyl]-	toluene	433(3.78),458(3.76)	103-0019-83
Aziridino[b]naphthalene-2,7-dione, 1,1a-diphenyl-	benzene	355(2.71)	44-4968-83
$C_{22}H_{15}NS$			
Benzene, 1,1',1''-(1-isothiocyanato-2-cyclopropene-1,2,3-triyl)tris-	MeCN	231.5(5.06),256.5(4.79), 303.5(6.25),318(5.75)	97-0018-83
$C_{22}H_{15}N_3$			
Benzenamine, N-11H-indeno[1,2-b]quinoxalin-11-ylidene-3-methyl-	EtOH	205(4.62),230s(4.44), 242(4.34),288(4.6), 320s(4.14),380(3.52), 440s(3.02)	104-0158-83
Benzenamine, N-(9-methyl-11H-indeno-[1,2-b]quinoxalin-11-ylidene)-	EtOH	206(4.98),226s(4.88), 246(4.72),300(4.99), 370s(4.0)	104-0158-83
$C_{22}H_{15}N_3O_2$			
9,10-Anthracenedione, 1-azido-2-[(4-methylphenyl)methyl]-	toluene	375(3.64)	103-0019-83
$C_{22}H_{15}N_3O_4$			
1,2-Diazabicyclo[5.2.0]nona-3,5-dien-9-one, 2-benzoyl-8-(1,3-dihydro-1,3-dioxo-2H-isoindol-2-yl)-, cis-(±)-	MeOH	290(3.98)	5-1374-83
2,3,9b-Triazaindeno[6,7,1-ija]azulene-8,9-dicarboxylic acid, 1-phenyl-, dimethyl ester	EtOH	237(4.42),270s(4.45), 292(4.57),380(4.39), 418(4.18),438(4.16)	18-3703-83
$C_{22}H_{15}N_3O_4S_2$			
2-Naphthalenesulfonic acid, 6-amino-3-(2-dibenzothienylazo)-4-hydroxy-	H_2O	499(4.04)	7-0155-83
	HCOOH-DMF	547(3.97)	7-0155-83
$C_{22}H_{15}N_3O_7S_3$			
2,7-Naphthalenedisulfonic acid, 5-amino-3-(2-dibenzothienylazo)-4-hydroxy-	H_2O	543(4.22)	7-0155-83
	HCOOH-DMF	535(4.24)	7-0155-83
$C_{22}H_{16}BrN_3$			
Benzonitrile, 4-[3-(4-bromophenyl)-4,5-dihydro-1-phenyl-1H-pyrazol-5-yl]-	C_6H_{12}	367(4.30)	12-1649-83
	MeOH	363(4.30)	12-1649-83
Benzonitrile, 4-[3-(4-bromophenyl)-4,5-dihydro-5-phenyl-1H-pyrazol-1-yl]-	C_6H_{12}	368(4.52)	12-1649-83
	MeOH	367(4.56)	12-1649-83
Benzonitrile, 4-[5-(4-bromophenyl)-4,5-dihydro-1-phenyl-1H-pyrazol-3-yl]-	C_6H_{12}	394(4.40)	12-1649-83
	MeOH	393(4.40)	12-1649-83
$C_{22}H_{16}ClNO_2$			
[1,2'-Binaphthalene]-1',4'-dione,	C_6H_{12}	502(3.06)	150-0168-83S

Compound	Solvent	$\lambda_{max}(\log \epsilon)$	Ref.
(cont.)	EtOH	503(--)	150-0168-83S
$C_{22}H_{16}ClN_3$			
Benzonitrile, 4-[1-(4-chlorophenyl)-4,5-dihydro-5-phenyl-1H-pyrazol-3-yl]-	C_6H_{12}	396(4.46)	12-1649-83
	MeOH	395(4.46)	12-1649-83
$C_{22}H_{16}ClN_7O_2$			
1(2H)-Phthalazinone, 4-chloro-2-methyl-, [[(4-nitrophenyl)azo]phenylmethylene]hydrazone	dioxan	278(4.25),418(4.25), 490(4.28)	104-0953-83
	MeCN	210(4.65),288(4.28), 418(4.18),480(4.32)	104-0953-83
	MeOH	210(--),286(--), 418s(--),480(--)	104-0953-83
$C_{22}H_{16}Cl_2N_4OS$			
4H-[1,2,4]Triazolo[4,3-a][1,4]benzodiazepine, 8-chloro-6-(2-chlorophenyl)-1-[[(2-furanylmethyl)thio]methyl]-	MeOH	222(4.63),252s(4.06)	73-0123-83
$C_{22}H_{16}Cl_3N_5O_9$			
Benzenamine, 4-[2,2-dichloro-1-(2-chloro-3,5-dinitrophenyl)ethyl]-N-(2-ethoxyphenyl)-2,6-dinitro-	MeOH	209.7(4.66),423.7(3.54)	56-1357-83
$C_{22}H_{16}Cl_6N_2O_2$			
Acetamide, 2,2'-(3,3,4,4-tetrachloro-1,2-cyclobutanediylidene)bis[2-chloro-N-(4-methylphenyl)-	MeCN	242(4.28),312(4.15)	5-1496-83
$C_{22}H_{16}I_2$			
Benzene, 1,4-bis[2-(4-iodophenyl)ethenyl]-	DMF	358(4.73)	104-0683-83
$C_{22}H_{16}NS$			
1,3-Thiazin-1-ium, 2,4,6-triphenyl-, perchlorate	MeCN	273(4.24),340(4.07), 420(3.63)	118-0827-83
$C_{22}H_{16}N_2$			
Benzaldehyde, 2-ethynyl-, [5-(2-ethynylphenyl)-2,4-pentadienylidene]hydrazone	THF	235(4.24),254s(4.15), 288(3.80),341s(4.53), 357(4.65),370(4.65), 390s(4.46)	18-1467-83
6,10:13,17-Dimethenocyclotetradeca[b]-quinoxaline, 11,12-dihydro-	n.s.g.	245(4.7),280(4.0)	24-3112-83
$C_{22}H_{16}N_2O_2$			
Cyclohept[1,2,3-hi]imidazo[2,1,5-cd]-indolizine-4-carboxylic acid, 2-(4-methylphenyl)-, methyl ester	EtOH	249(4.45),273s(4.43), 297(4.57),311s(4.45), 360s(4.46),376(4.53), 405(4.22),427(4.17)	18-3703-83
Ethanone, 2-(1H-indol-3-yl)-1-phenyl-2-(phenylimino)-, N-oxide	MeOH	266(4.66),334(4.57)	73-1854-83
Methanone, (3,8-dihydro-2-phenyl-2H-isoxazolo[5,4-b]indol-3-yl)phenyl-	MeOH	281(4.18),289(4.12)	73-1854-83
$C_{22}H_{16}N_2O_3S_3$			
Ethane(dithioic) acid, (diphenylamino)thioxo-, 2-(4-nitrophenyl)-2-oxoethyl ester	CH_2Cl_2	264(4.47),340s(3.90), 437s(3.17)	97-0247-83

Compound	Solvent	λ_{max}(log ε)	Ref.
$C_{22}H_{16}N_2O_4$			
[1,2'-Binaphthalene]-1',4'-dione, 4-(dimethylamino)-5'(and 8')-nitro-	C_6H_{12}	518(3.30)	150-0168-83S
	EtOH	517(--)	150-0168-83S
Unidentified alkaloid I, m. 238-9°	MeOH	246(4.04),287(3.48)	102-2847-83
Unidentified alkaloid II, m. 231-2°	MeOH	247(4.17),261(4.23)	102-2847-83
$C_{22}H_{16}N_2O_6S$			
1H-Indole, 2-[2-(hydroxymethyl)-5-nitrobenzoyl]-1-(phenylsulfonyl)-	EtOH	242(4.20),271s(3.95), 290s(3.89),364s(2.84)	39-2409-83C
$C_{22}H_{16}N_2O_8S_2$			
1-Naphthalenesulfonic acid, 4-acetoxy-3-[(4-sulfo-1-naphthalenyl)azo]-, disodium salt (azorubin S acetate)	pH 7.5	394(4.06)	94-0162-83
$C_{22}H_{16}N_4O_8S$			
Isoquinoline, 1,2-dihydro-2-(phenylsulfonyl)-1-[(2,4,6-trinitrophenyl)-methyl]-	$CHCl_3$	246(4.26),284(3.72), 364(2.80)	56-0829-83
$C_{22}H_{16}N_6$			
Benzo[1,2-b:4,5-b']dipyrrole, 1,5-dihydro-3,7-bis(phenylazo)-	EtOH	204(4.41),241(4.37), 306(4.51),236[sic](4.64)	103-0871-83
$C_{22}H_{16}N_8$			
2-Pyridinamine, N,N'-[3-(2-pyridinyl)-2H-pyrido[1,2-a]-1,3,5-triazine-2,4(3H)-diylidene]bis-	dioxan	270s(4.36),291s(4.32)	48-0463-83
$C_{22}H_{16}N_8O_2$			
[1,2,4]Triazolo[4,3-b][1,2,4]triazolo-[4',3':1,6]pyridazino[4,5-d]pyridazine-4,10(5H,11H)-dione, 5,11-dimethyl-1,7-diphenyl-	DMSO	257(4.38),374(4.157)	161-0842-83
$C_{22}H_{16}O_2$			
4H-1-Benzopyran-4-one, 2-(4-methylphenyl)-6-phenyl-	MeOH	275(4.58),314(4.42)	18-2037-83
4H-1-Benzopyran-4-one, 2-(4-methylphenyl)-8-phenyl-	MeOH	253(4.30),312(4.39)	18-2037-83
$C_{22}H_{16}O_3$			
9,10-Anthracenedione, 1-(2,6-dimethylphenoxy)-	EtOH	209(4.65),255(4.63), 375(3.77)	104-0144-83
4H-1-Benzopyran-4-one, 2-(2-methoxy-[1,1'-biphenyl]-3-yl)-	MeOH	248(4.48),304(3.92)	18-3519-83
4H-1-Benzopyran-4-one, 2-(4-methoxy-[1,1'-biphenyl]-3-yl)-	MeOH	255(4.50),344(3.92)	18-3519-83
4H-1-Benzopyran-4-one, 2-(4-methoxyphenyl)-6-phenyl-	MeOH	249(4.36),278(4.42), 331(4.51)	18-2037-83
4H-1-Benzopyran-4-one, 2-(4-methoxyphenyl)-8-phenyl-	MeOH	225(4.50),263s(4.14), 329(4.46)	18-2037-83
$C_{22}H_{16}O_4$			
4H-1-Benzopyran-4-one, 2-(4-hydroxy-3-methoxyphenyl)-6-phenyl-	MeOH	257(4.52),276s(4.36), 347(4.46)	18-2037-83
2(5H)-Furanone, 3-[3-(9-anthracenyl)-1-oxo-2-propenyl]-4-hydroxy-5-methyl-, (E)-	MeOH	224(4.30),253(4.93), 315(3.90),390(3.91)	83-0115-83
2(5H)-Furanone, 4-hydroxy-5-methyl-1-oxo-3-(9-phenanthrenyl)-2-propenyl]-	MeOH	208(4.54),252(4.65), 295(4.16),315(4.12)	83-0115-83

Compound	Solvent	$\lambda_{max}(\log \epsilon)$	Ref.
$C_{22}H_{16}O_4S$			
5,10-Ethenoanthra[2,3-b]thiophene-12,13-dicarboxylic acid, 5,10-dihydro-, dimethyl ester	EtOH	233(4.62)	44-4419-83
$C_{22}H_{16}O_5$			
2H-1-Benzopyran-2-one, 5,7-dimethoxy-3-(1-naphthalenecarbonyl)-	benzene	358(4.34)	151-0325-83A
$C_{22}H_{16}O_6$			
Anthra[2,3-b]benzofuran-1,7,12(2H)-trione, 3,4-dihydro-6,13-dihydroxy-3,3-dimethyl-	dioxan	260(4.42),485(4.10)	104-1900-83
Resistomycin	MeOH	268(4.40),290.2(4.31), 320.4(4.16),338.9(4.17), 369.3(4.06),430s(4.06), 457(4.22)	74-0385-83C
$C_{22}H_{16}S_4$			
Naphthaceno[5,6-cd:11,12-c'd']bis[1,2]-dithiole, 4-(1,1-dimethylethyl)-, radical cation, chloride	MeCN	370s(--),469(3.99), 547(4.02),588(4.04)	104-1092-83
tetrafluoroborate	MeCN	370s(--),462(3.97), 547(4.02),588(4.23)	104-1092-83
$C_{22}H_{17}As$			
1H-Arsole, 1,2,5-triphenyl-	C_6H_{12}	208(4.41),233(4.23), 220s[sic](4.26), 368(4.13)	101-0335-83I
$C_{22}H_{17}BrO$			
2-Propen-1-one, 3-(4-bromophenyl)-1-[4-(phenylmethyl)phenyl]-	isooctane	270(4.08),310(4.60)	124-1211-83
	dioxan	318(4.45)	124-1211-83
$C_{22}H_{17}Cl$			
Tricyclo[3.3.1.0^{2,8}]nona-3,6-diene, 5-chloro-9-(diphenylmethylene)-	C_6H_{12}	209(4.30),235(4.12), 269s(3.97)	138-0523-83
$C_{22}H_{17}ClN_2O_2$			
[1,2'-Binaphthalene]-1',4'-dione, 8'-amino-3'-chloro-4-(dimethylamino)-	C_6H_{12}	486(3.78)	150-0168-83S
	EtOH	522(--)	150-0168-83S
$C_{22}H_{17}ClN_6$			
1(2H)-Phthalazinone, 4-chloro-2-methyl-, [phenyl(phenylazo)methylene]-hydrazone, form A	MeOH	206(4.65),286(4.23), 380(4.14),440(4.32)	104-0953-83
	dioxan	285(4.23),390s(4.18), 442(4.34)	104-0953-83
	MeCN	205(4.64),285(4.23), 380s(4.15),440(4.31)	104-0953-83
form B	MeOH	215(4.64),282(4.36), 300s(4.30),395(4.20)	104-0953-83
	dioxan	282(4.33),300s(4.23), 400(4.18)	104-0953-83
	MeCN	213(4.59),283(4.40), 302s(4.23),397(4.18)	104-0953-83
$C_{22}H_{17}ClO$			
2-Propen-1-one, 3-(4-chlorophenyl)-1-[4-(phenylmethyl)phenyl]-, (E)-	isooctane	270(4.02),309(4.48)	124-1211-83
	dioxan	318(4.48)	124-1211-83

Compound	Solvent	λ_{max}(log ϵ)	Ref.
$C_{22}H_{17}CoN_4O$			
Cobalt(1+), (6,12-dimethyl-7,11-nitrilo-11H-dibenz[b,f][1,4,8]triazacyclopentadecin-19(18H)-onato-N^5,N^{13},N^{18}-N^{20}]-, (SP-4-2)-, perchlorate	EtOH-DMF	270(4.12),345(3.90)	12-2387-83
$C_{22}H_{17}CuN_4O$			
Copper(1+), (6,12-dimethyl-7,11-nitrilo-11H-dibenz[b,f][1,4,8]triazacyclopentadecin-19(18H)-onato-N^5,N^{13},N^{18}-N^{20}]-, (SP-4-2)-, perchlorate	EtOH-DMF	270(4.18),340(4.05)	12-2387-83
$C_{22}H_{17}F_3O_7$			
Acetic acid, trifluoro-, 2-ethyl-1,2,3-4,6,11-hexahydro-2,5,12-trihydroxy-6,11-dioxo-1-naphthacenyl ester, trans-(±)-	MeOH	247(4.46),282(3.84), 456(3.85),480(3.93), 512(3.77),553(3.05)	5-1818-83
$C_{22}H_{17}NOS_3$			
Ethane(dithioic) acid, (diphenylamino)-thioxo-, 2-oxo-2-phenylethyl ester	CH_2Cl_2	248(4.38),278s(4.17), 344s(3.83),434s(3.10)	97-0247-83
$C_{22}H_{17}NO_2$			
[1,2'-Binaphthalene]-1',4'-dione, 4-(dimethylamino)-	C_6H_{12}	483(3.29)	150-0168-83S
	EtOH	494(--)	150-0168-83S
$C_{22}H_{17}NO_2S_2$			
Benzene, 1-[4,4-bis(phenylthio)-1,3-butadienyl]-4-nitro-, (E)-	EtOH	261(4.53),350(4.73)	118-0383-83
$C_{22}H_{17}NO_3$			
9,10-Anthracenedione, 1-amino-4-(2,6-dimethylphenoxy)-	EtOH	247(4.48),280s(3.98), 516(3.83)	104-0144-83
2H-1,4-Benzoxazin-2-ol, 2,3-diphenyl-, acetate	EtOH	251(4.28),292(3.91)	39-0141-83C
2-Propen-1-one, 3-(4-nitrophenyl)-1-[4-(4-phenylmethyl)phenyl]-	isooctane	307(4.55)	124-1211-83
	dioxan	313(4.40)	124-1211-83
$C_{22}H_{17}N_3O$			
7H-Dibenz[f,ij]isoquinoline-1-carbonitrile, 7-oxo-2-piperidino-	DMF	468(3.56)	2-1197-83
Methanone, phenyl(5-phenyl-3-isoxalyl)-, phenylhydrazone, (E)-	MeOH	255(4.37),330(4.21)	4-0931-83
9H,10H,14cH,15H-4b,9a,14b-Triazanaphtho[1,2,3-fg]naphthacen-15-one	EtOH	211(--),280(--), 356(--),357(--)	142-0617-83
	$HClO_4$	212(--),280(--), 356(--),357(--)	142-0617-83
$C_{22}H_{17}N_3O_5S$			
2H-Pyrrolo[4,3,2-cd]indol-2-one, 1,2a,3,4-tetrahydro-4-[(4-methylphenyl)sulfonyl]-3-(4-nitrophenyl)-, trans	$CHCl_3$	256(4.22)	44-2468-83
$C_{22}H_{17}N_4NiO$			
Nickel(1+), (6,12-dimethyl-7,11-nitrilo-11H-dibenz[b,f][1,4,8]triazacyclopentadecin-19(18H)-onato-N^5,N^{13},N^{18}-N^{20}]-, (SP-4-2)-, perchlorate	EtOH-DMF	313(4.18),340(4.05), 380(3.79)	12-2387-83
	DMF	690(3.30)	12-2387-83

Compound	Solvent	λ_{max}(log ϵ)	Ref.
$C_{22}H_{17}N_5O_5$			
1H-Pyrazole-4-carboxylic acid, 5-[(2-hydroxy-1-naphthalenyl)azo]-3-(nitromethyl)-1-phenyl-, methyl ester	EtOH	471(4.07)	48-0041-83
$C_{22}H_{18}$			
Benzene, 1,4-bis(2-phenylethenyl)-	dioxan	356(4.78)	104-0683-83
Cyclotetradeca[a]naphthalene, (E,E,Z,E,E,Z)-	hexane	228(4.29),323(4.40)	5-0687-83
Cyclotetradeca[a]naphthalene, 11,12,13-14-tetradehydro-9,10,15,17-tetrahydro-, (E,E)-	hexane	258(4.63),298(4.07)	5-0687-83
Cyclotetradeca[b]naphthalene, (E,E,Z,Z,E,E)-	hexane	238(4.45),260(4.44), 305(4.65)	5-0687-83
Cyclotetradeca[b]naphthalene, 10,11,12-13-tetradehydro-8,9,14,15-tetrahydro-, (E,E)-	hexane	260(4.72)	5-0687-83
$C_{22}H_{18}BrNO$			
2-Propen-1-one, 3-(4-bromophenyl)-1-[4-(methylphenylamino)phenyl]-	EtOH	394(4.42)	124-1211-83
	dioxan	250(4.18),272(3.78), 299(4.08),358(4.30)	124-1211-83
$C_{22}H_{18}ClNOSe$			
2-Propen-1-one, 1-(4-chlorophenyl)-3-[1,2-dihydro-1-(1-phenylethyl)-2-selenoxo-3-pyridinyl]-, (E)-(±)-	EtOH	254(4.13),301(3.79), 380(3.56)	4-1651-83
$C_{22}H_{18}ClNO_2$			
Benzeneacetic acid, 2-chloro-α-[phenyl(phenylamino)methylene]-, methyl ester, (Z)-	EtOH	240(4.04),337(4.20)	4-0245-83
Benzeneacetic acid, 4-chloro-α-[phenyl(phenylamino)methylene]-, methyl ester, (Z)-	EtOH	245(4.03),337(4.20)	4-0245-83
$C_{22}H_{18}ClN_5$			
1H-Pyrazol-5-amine, N-(4-chlorophenyl)-3-methyl-1-phenyl-4-(phenylazo)-	EtOH	245(4.16),275(4.03), 330(4.06),378(4.00)	4-1501-83
$C_{22}H_{18}ClN_7O_2$			
Benzenecarbohydrazonic acid, N-(4-chloro-2-methyl-1(2H)-phthalazinylidene)-,2-(4-nitrophenyl)hydrazide	MeOH	213(4.65),295(4.20), 380(4.34)	104-0953-83
	dioxan	295(4.06),388(4.33)	104-0953-83
	MeCN	210(4.64),290(4.10), 383(4.42)	104-0953-83
Spiro[phthalazine-1(2H),3'-[3H-[1,2,4]-triazol]-4'(2'H)-amine, 4-chloro-2-methyl-2'-(4-nitrophenyl)-5'-phenyl-	dioxan	250(4.30),322(4.06), 410(4.30)	104-0953-83
	MeCN	248(4.30),315(4.04), 420(4.30)	104-0953-83
	MeOH	248(--),316(--), 420(--)	104-0953-83
$C_{22}H_{18}Cl_2$			
1,3,5,7,13,15,17-Cyclononadecaheptaene-9,11-diyne, 19-(dichloromethylene)-8,13-dimethyl-, (E,E,Z,Z,E,E,E,E)-	CH_2Cl_2	303s(4.43),338(4.69), 391s(4.07),423s(3.78)	39-2997-83C
$C_{22}H_{18}Cl_2N_2O_3$			
Fenpyrithrin	MeOH	212(4.32),277(3.71)	98-1113-83
	aq MeCN	197(4.74),277(3.76)	98-1113-83

Compound	Solvent	λ_{max}(log ϵ)	Ref.
$C_{22}H_{18}NO$			
Benzo[f]quinolinium, 4-[2-(4-methyl-phenyl)-2-oxoethyl]-, iodide	EtOH	222(4.60),285(4.47), 374(3.80)	104-0568-83
$C_{22}H_{18}NO_2$			
Benzo[f]quinolinium, 4-[2-(4-methoxy-phenyl)-2-oxoethyl]-, iodide	EtOH	230(4.71),268(4.50), 373(3.95)	104-0568-83
$C_{22}H_{18}NO_3$			
1-Benzopyrylium, 2-(2-oxo-2H-1-benzo-pyran-3-yl)-7-pyrrolidino-, per-chlorate	MeCN	563(4.41)	48-0505-83
1-Benzopyrylium, 2-(2-oxo-2H-7-pyrroli-dino-2H-1-benzopyran-3-yl)-, per-chlorate	MeCN	639(4.78)	48-0505-83
$C_{22}H_{18}NO_4$			
1-Benzopyrylium, 2-(7-hydroxy-2-oxo-2H-1-benzopyran-3-yl)-7-pyrroli-dino-, perchlorate	MeCN	570(4.53)	48-0505-83
1-Benzopyrylium, 7-hydroxy-2-(2-oxo-7-pyrrolidino-2H-1-benzopyran-3-yl)-, perchlorate	MeCN	610(4.94)	48-0505-83
$C_{22}H_{18}N_2O$			
Benzenamine, N-[[1-(phenylmethyl)-1H-indol-3-yl]methylene]-, N-oxide	EtOH	215(4.42),256(3.99), 270(3.92),324(4.26)	104-1367-83
$C_{22}H_{18}N_2OS_3$			
Ethane(dithioic) acid, (diphenylamino)-thioxo-, 2-oxo-2-(phenylamino)ethyl ester	CH_2Cl_2	248(4.37),281s(4.15), 348s(3.81),433s(3.17)	97-0247-83
$C_{22}H_{18}N_2O_2$			
[1,2'-Binaphthalene]-1',4'-dione, 5'-amino-4-(dimethylamino)-	C_6H_{12}	493(3.62)	150-0168-83S
	EtOH	525(--)	150-0168-83S
[1,2'-Binaphthalene]-1',4'-dione, 8'-amino-4-(dimethylamino)-	C_6H_{12}	480(3.80)	150-0168-83S
	EtOH	514(--)	150-0168-83S
1(2H)-Phthalazinone, 4-(4-hydroxy-3,5-dimethylphenyl)-2-phenyl-	EtOH	208(4.68),312.5(4.02)	24-0970-83
$C_{22}H_{18}N_2O_2S$			
6,13(7H,14H)-Triphenodithiazinedione, 1,3,8,10-tetramethyl-	H_2SO_4	342(4.83),750(4.83)	18-1482-83
$C_{22}H_{18}N_2O_3$			
2-Propen-1-one, 1-[4-(methylphenyl-amino)phenyl]-3-(4-nitrophenyl)-	EtOH	314(4.48),413(4.30)	124-1211-83
	dioxan	256(3.9),282(3.64), 314(4.40),385(4.13)	124-1211-83
$C_{22}H_{18}N_2O_4$			
2-Butenedioic acid, 2-(9H-fluoren-9-yl)-3-(1H-pyrazol-1-yl)-, dimethyl ester	n.s.g.	270(4.17)	24-0856-83
$C_{22}H_{18}N_4$			
Benzonitrile, 4-[1-(4-aminophenyl)-4,5-dihydro-5-phenyl-1H-pyrazol-3-yl]-	C_6H_{12}	423(4.32)	12-1649-83
	MeOH	417(4.30)	12-1649-83

Compound	Solvent	$\lambda_{max}(\log \epsilon)$	Ref.
$C_{22}H_{18}N_4OS$			
Benzaldehyde, (5,7-dimethyl-4-oxo-4H-pyrido[3,4-e]-1,3-thiazin-2-yl)phenylhydrazone	MeOH	247(3.96)	73-3426-83
$C_{22}H_{18}N_4S$			
1H-Imidazo[1,2-b][1,2,4]triazepine, 7,8-dihydro-2,8-diphenyl-6-(2-thienyl)-	EtOH	280(4.26),300(4.31), 367(4.04)	103-0083-83
$C_{22}H_{18}O$			
Naphthalene, 1,1'-[oxybis(methylene)]-bis-	isooctane	268(3.0)	78-1407-83
2-Propen-1-one, 3-phenyl-1-[4-(phenylmethyl)phenyl]-	isooctane	263(3.95),298(4.46)	124-1211-83
	dioxan	230(4.13),273(3.86), 309(4.42)	124-1211-83
$C_{22}H_{18}O_2$			
4H-1-Benzopyran-4-one, 2,3-dihydro-2-(4-methylphenyl)-6-phenyl-	MeOH	247(4.57),346(3.44)	18-2037-83
4H-1-Benzopyran-4-one, 2,3-dihydro-2-(4-methylphenyl)-8-phenyl-	MeOH	240(4.42),338(3.57)	18-2037-83
2-Propen-1-one, 1-(2-hydroxy[1,1'-biphenyl]-3-yl)-3-(4-methylphenyl)-	MeOH	244(4.29),335(4.45)	18-2037-83
2-Propen-1-one, 1-(4-hydroxy[1,1'-biphenyl]-3-yl)-3-(4-methylphenyl)-	MeOH	255(4.44),330(4.36)	18-2037-83
$C_{22}H_{18}O_3$			
5H-Benzofuro[6,5-c][2]benzopyran-5-one, 1,2,3,4-tetrahydro-7-methyl-10-phenyl-	MeOH	247(4.45),300(4.20)	78-1265-83
4H-1-Benzopyran-4-one, 2,3-dihydro-2-(2-methoxy[1,1'-biphenyl]-3-yl)-, (±)-	MeOH	250(4.35),326(3.54)	18-3519-83
4H-1-Benzopyran-4-one, 2,3-dihydro-2-(4-methoxy[1,1'-biphenyl]-3-yl)-	MeOH	257(4.42),319(3.52)	18-3519-83
4H-1-Benzopyran-4-one, 2,3-dihydro-2-(4-methoxyphenyl)-6-phenyl-	MeOH	246(4.58),340(3.28)	18-2037-83
4H-1-Benzopyran-4-one, 2,3-dihydro-2-(4-methoxyphenyl)-8-phenyl-	MeOH	232(4.49),336(3.57)	18-2037-83
Isobenzofuran, 4,7-dimethoxy-1,3-diphenyl-	CH_2Cl_2	249(4.43),290(3.96), 345s(3.95),400(4.36)	118-1018-83
2-Propen-1-one, 1-(2-hydroxy[1,1'-biphenyl]-3-yl)-3-(4-methoxyphenyl)-	MeOH	248(4.36),366(4.48)	18-2037-83
2-Propen-1-one, 1-(4-hydroxy[1,1'-biphenyl]-3-yl)-3-(4-methoxyphenyl)-	MeOH	254(4.54),361(4.40)	18-2037-83
2-Propen-1-one, 1-(2-hydroxyphenyl)-3-(2-methoxy[1,1'-biphenyl]-3-yl)-	MeOH	328(4.34)	18-3519-83
2-Propen-1-one, 1-(2-hydroxyphenyl)-3-(4-methoxy[1,1'-biphenyl]-3-yl)-	MeOH	270(4.40),315(4.24), 384(4.21)	18-3519-83
2-Propen-1-one, 3-(4-methoxyphenyl)-1-(4-phenoxyphenyl)-	isooctane	231(4.31),286(4.03), 331(4.53)	124-1211-83
	dioxan	336(4.42)	124-1211-83
$C_{22}H_{18}O_4$			
4H-1-Benzopyran-4-one, 2,3-dihydro-2-(4-hydroxy-3-methoxyphenyl)-6-phenyl-	MeOH	246(4.60),345(3.46)	18-2037-83
2-Propen-1-one, 1-(4-hydroxy[1,1'-biphenyl]-3-yl)-3-(4-hydroxy-3-methoxyphenyl)-	MeOH	258(4.52),384(4.40)	18-2037-83

Compound	Solvent	λ_{max}(log ε)	Ref.
$C_{22}H_{18}O_5$			
4H,8H-Benzo[1,2-b:3,4-b']dipyran-4-one, 5-acetoxy-8,8-dimethyl-2-phenyl-	MeOH	225(4.48),269(4.52), 322(4.05)	106-0876-83
4H,8H-Benzo[1,2-b:5,4-b']dipyran-4-one, 5-acetoxy-8,8-dimethyl-2-phenyl-	MeOH	225(4.43),270(4.46), 313(4.21)	106-0876-83
5,12-Naphthacenedione, 9-acetyl-7,8-dihydro-6,11-dimethoxy-	MeCN	252(4.03),291(4.38), 380(3.80)	44-2820-83
$C_{22}H_{18}O_6$			
2H-1-Benzopyran-2-one, 7-methoxy-4-methyl-, anti-H-T dimer	n.s.g.	246(3.59),275(3.30)	44-1872-83
Isopongachromene	EtOH	235(4.45),280(3.00), 328(4.09),340(4.13)	102-0308-83
$C_{22}H_{18}O_7$			
6,9-Epoxybenzocyclooctene-7,8-dicarboxylic acid, 5,6,7,8,9,10-hexahydro-5,10-dioxo-6-phenyl-, dimethyl ester	benzene	297(3.28),328(2.92)	44-4968-83
	MeCN	227(4.36),248(3.94), 295(3.11),331(2.82)	44-4968-83
Justicidin A	$CHCl_3$	265(4.35),295(4.13), 315(4.13),335(3.33)	162-0756-83
$C_{22}H_{18}S_2$			
Benzene, 1,1'-[(4-phenyl-1,3-butadienylidene)bis(thio)]bis-	EtOH	344(4.76)	118-0383-83
$C_{22}H_{19}ClNO$			
3H-Indolium, 3-(4-chlorophenyl)-3-hydroxy-1-methyl-2-(4-methylphenyl)-, tetrafluoroborate	CH_2Cl_2	342(4.08)	24-2115-83
$C_{22}H_{19}ClN_2O$			
Cinnoline, 2-benzoyl-4-(3-chlorophenyl)-1,2,3,4-tetrahydro-1-methyl-	$CHCl_3$	249(4.11)	97-0028-83
Cinnoline, 2-benzoyl-4-(4-chlorophenyl)-1,2,3,4-tetrahydro-1-methyl-	$CHCl_3$	248(4.20)	97-0028-83
$C_{22}H_{19}ClN_2O_2S$			
2H-1-Benzopyran-2-one, 3-[2-(4-chlorophenyl)-4-thiazolyl]-7-(diethylamino)-	MeCN	413(4.60)	48-0551-83
$C_{22}H_{19}ClN_2S_2$			
Diazene, (4-chlorophenyl)[3-phenyl-3-(4,5,6,7-tetrahydro-1,3-benzodithiol-2-ylidene)-1-propenyl]-	MeCN	535(4.29)	118-0840-83
$C_{22}H_{19}ClN_6$			
Benzenecarbohydrazonic acid, N-[4-chloro-2-methyl-1(2H)-phthalazinylidene)-, 2-phenylhydrazide	MeOH	213(4.52),240(4.26), 289(4.02),370(4.01)	104-0953-83
	dioxan	242(4.38),290(4.11), 385(4.18)	104-0953-83
	MeCN	213(4.65),240(4.36), 290(4.13),380(4.20)	104-0953-83
$C_{22}H_{19}Cl_2NO_3$			
Cypermethrin	MeOH	208(4.51),279(3.23)	98-1113-83
	aq MeCN	200(4.75),279(3.40)	98-1113-83
in sodium dodecyl sulfate	n.s.g.	196(4.83),277(3.18)	98-1113-83
$C_{22}H_{19}Cl_3N_2O_4S_2$			
5-Thia-1-azabicyclo[4.2.0]oct-2-ene-	EtOH	235(4.11),259(3.89),	78-0461-83

Compound	Solvent	$\lambda_{max}(\log \epsilon)$	Ref.
2-carboxylic acid, 8-oxo-3-(phenylmethyl)-7-[(2-thienylacetyl)amino]-, 2,2,2-trichloroethyl ester, (6R-trans) (cont.)		266(3.91),271(3.90)	78-0461-83
$C_{22}H_{19}Cl_3N_2O_5S_2$			
5-Thia-1-azabicyclo[4.2.0]oct-2-ene-2-carboxylic acid, 8-oxo-3-(phenylmethyl)-7-[(2-thienylacetyl)amino]-, 2,2,2-trichloroethyl ester, 5-oxide	EtOH	233(4.11),261s(3.92), 267s(3.98),273(3.99)	78-0461-83
$C_{22}H_{19}FeN$			
Ferrocene, (diphenylamino)-	n.s.g.	290(4.2),447(2.7)	101-0227-83A
$C_{22}H_{19}NO$			
2-Propen-1-one, 1-[4-(methylphenylamino)phenyl]-3-phenyl-	EtOH	274(3.90),303(4.26), 389(4.40)	124-1211-83
	dioxan	244(4.03),270(3.51), 294(4.08),357(4.28)	124-1211-83
2-Propen-1-one, 3-(4-methylphenyl)-1-[4-(phenylamino)phenyl]-	EtOH	241(4.03),282(3.94), 313(4.20),391(4.40)	124-1211-83
	dioxan	278(3.86),306(4.24), 369(4.47)	124-1211-83
$C_{22}H_{19}NOS$			
1,5-Benzothiazepine, 2,3-dihydro-2-(4-methoxyphenyl)-4-phenyl-	EtOH	258(4.30),334(3.58)	103-1293-83
1,5-Benzothiazepine, 2,3-dihydro-4-(4-methoxyphenyl)-2-phenyl-	EtOH	258(4.29),328(3.60)	103-1293-83
$C_{22}H_{19}NO_2$			
1-Butene, 1-nitro-1,2,4-triphenyl-	EtOH	369(4.22)	22-0339-83
1H-Indole, 2,3-bis(4-methoxyphenyl)-	n.s.g.	282(4.52),308(4.32)	162-0722-83
2-Propenoic acid, 2,3-diphenyl-3-(phenylamino)-, methyl ester	EtOH	240(4.07),336(4.25)	4-0245-83
2-Propen-1-one, 3-(4-methoxyphenyl)-1-[4-(phenylamino)phenyl]-	EtOH	241(4.16),282(3.97), 333(4.16),397(4.52)	124-1211-83
	dioxan	282(3.78),317(4.00), 373(4.55)	124-1211-83
$C_{22}H_{19}NO_3$			
Spiro[1,3-dioxolane-2,4'-[4H]isoindol]-7'(1'H)-one, 3'a,7'a-dihydro-1',3'-diphenyl-	EtOH	240(4.28)	33-2252-83
$C_{22}H_{19}NO_7$			
8H-Benzo[g]-1,3-benzodioxolo[5,6-a]quinolizin-8-one, 13-acetoxy-5,6-dihydro-9,10-dimethoxy-	MeOH	229.5(4.65),343.5(4.43)	94-0947-83
Himalayamine, O-acetyl-	MeOH	209(4.55),239s(3.91), 293(3.85)	88-2445-83
	MeOH-acid	210(4.58),242(3.85), 294(3.89)	88-2445-83
Ribasidine acetate	EtOH	212(--),233s(--), 290(--)	88-4481-83
	EtOH-acid	212(--),243(--), 292(--)	88-4481-83
$C_{22}H_{19}N_3O_2S_2$			
Diazene, (4-nitrophenyl)[3-phenyl-3-(4,5,6,7-tetrahydro-1,3-benzo-	MeCN	594(4.06)	118-0840-83

Compound	Solvent	λ_{max}(log ϵ)	Ref.
dithiol-2-ylidene)-1-propenyl]- (cont.)			118-0840-83
$C_{22}H_{19}N_3O_3$			
Cinnoline, 2-benzoyl-1,2,3,4-tetrahydro-1-methyl-4-(3-nitrophenyl)-	$CHCl_3$	255(4.18)	97-0028-83
9H,10H,14cH,15H-4b,9a,14b-Triazanaphtho[1,2,3-fg]naphthacene-9,10,15-triol	EtOH	208(4.52),223(4.50), 272(3.91),340(3.62)	142-0617-83
	50% $HClO_4$	203(4.53),231(4.42), 271(3.91),374(3.35)	142-0617-83
$C_{22}H_{19}N_3O_3S$			
7H-Benzimidazo[2,1-a]benz[de]isoquinoline-10-sulfonamide, N-butyl-7-oxo-	benzene	387(4.11)	73-2249-83
	EtOH	385(4.14)	73-2249-83
7H-Benzimidazo[2,1-a]benz[de]isoquinoline-10-sulfonamide, N,N-diethyl-7-oxo-	benzene	385(4.17)	73-2249-83
	EtOH	385(4.15)	73-2249-83
$C_{22}H_{19}N_3O_3S_3$			
4H-Bisthiazolo[3,2-a:4',5'-e]pyrimidine-4,8(7H)-dione, 5a-ethoxy-5,5a-dihydro-2-(methylthio)-5,7-diphenyl-	MeOH	220(4.43),276(3.87), 285(3.88),325(4.26)	2-0243-83
$C_{22}H_{19}N_3O_7S$			
4-Thia-1-azabicyclo[3.2.0]hept-2-ene-2-carboxylic acid, 3-methoxy-7-oxo-6-[(phenylacetyl)amino]-, (4-nitrophenyl)methyl ester, (5R-trans)-	dioxan	279(4.12)	78-2493-83
$C_{22}H_{19}N_3S_2$			
1H-Indole, 3-[(phenylazo)(4,5,6,7-tetrahydro-1,3-benzodithiol-2-ylidene)methyl]-	MeCN	504(3.66)	118-0840-83
$C_{22}H_{19}N_7O_3$			
2H-Benzimidazol-2-one, 1,3-dihydro-1-(phenylmethyl)-, [1-[(2-hydroxy-4-nitrophenyl)azo]ethylidene]hydrazone	EtOH	545(4.20)	65-1033-83
	EtOH-base	635(4.39)	65-1033-83
	acetone	545(4.15)	65-1033-83
$C_{22}H_{20}$			
Tetracyclo[9.8.1.1^{3,9}.1^{13,18}]docosa-2,4,7,9,11,13,15,17,19-nonaene	3-Mepentane	228(4.15),328(5.06), 400s(3.70)	35-6211-83
$C_{22}H_{20}Cl_2$			
3,5,7,9,12,14,16-Nonadecaheptaene-1,18-diyne, 11-(dichloromethylene)-3,17-dimethyl-	CH_2Cl_2	297s(4.41),311(4.50), 330s(4.59),347(4.74), 366(4.77),389(4.58), 423s(3.71),447s(3.45)	39-2997-83C
$C_{22}H_{20}Cl_2N_6O_8$			
Pyrrolidine, 1,1'-[(dichloroethenylidene)bis(2,6-dinitro-4,1-phenylene)]bis-	MeOH	228.7(4.42),436.3(4.04)	56-1357-83
$C_{22}H_{20}NO_3$			
1-Benzopyrylium, 2-[7-(diethylamino)-2-oxo-2H-1-benzopyran-3-yl]-, perchlorate	MeCN	617(4.78)	48-0505-83
1-Benzopyrylium, 7-(diethylamino)-2-(2-oxo-2H-1-benzopyran-3-yl)-, perchlorate	MeCN	562(4.33)	48-0505-83

Compound	Solvent	$\lambda_{max}(\log \epsilon)$	Ref.
$C_{22}H_{20}N_2$			
3,4-Diazabicyclo[4.1.0]hepta-2,4-diene, 7-methyl-7-(1-methyl-2-cyclopropen-1-yl)-2,5-diphenyl-	dioxan	259(4.02),325(4.18)	88-1485-83
1H-Pyrazole, 4,5-dihydro-1-(4-methylphenyl)-3,5-diphenyl-	n.s.g.	245(4.23),254s(--), 313s(--),363(4.23)	103-0229-83
$C_{22}H_{20}N_2O$			
Cinnoline, 2-benzoyl-1,2,3,4-tetrahydro-1-methyl-4-phenyl-	$CHCl_3$	248(4.07)	97-0028-83
1H-Pyrazole, 4,5-dihydro-5-(2-methoxyphenyl)-1,3-diphenyl-	C_6H_{12}	237(4.11),360(4.30)	39-1679-83B
Spiro[2H-indole-2,3'-[3H]naphtho[2,1-b][1,4]oxazine], 1,3-dihydro-1,3,3-trimethyl- (colorless form)	EtOH	297(3.82),317(3.83), 345(3.68)	23-0300-83
photochromic form	EtOH at -75°	277(4.51),343(4.43), 410(4.26),578(4.69), 612(4.91)	23-0300-83
$C_{22}H_{20}N_2O_2$			
Pyrano[4,3-c]pyrazol-3(2H)-one, 6,7-dihydro-6,6-dimethyl-2-phenyl-4-(2-phenylethenyl)-, (E)-	EtOH	246(4.28),360(4.49)	118-0948-83 +142-1801-83
7H-Pyrano[4',3':3,4]pyrazolo[1,5-a]quinolin-7-one, 5,6,9,10-tetrahydro-9,9-dimethyl-5-phenyl-	EtOH	278(4.16)	118-0948-83
1H-Pyrazole-4-carboxylic acid, 3-ethenyl-1-phenyl-5-(2-phenylethenyl)-, ethyl ester, (E)-	EtOH	260(4.21),280(4.19)	118-0948-83
3H-Pyrazol-3-one, 2,4-dihydro-5-(2-methyl-1-propenyl)-4-(1-oxo-3-phenyl-2-propenyl)-2-phenyl-, (E)-	EtOH	252(4.28),296(4.24), 340(4.33)	142-1801-83
Pyrazolo[1,5-a]quinoline-3-carboxylic acid, 2-ethenyl-4,5-dihydro-5-phenyl-, ethyl ester	EtOH	264(4.07),280(4.14)	118-0948-83
$C_{22}H_{20}N_2O_2S$			
2H-1-Benzopyran-2-one, 7-(diethylamino)-3-(2-phenyl-4-thiazolyl)-	MeCN	419(4.51)	48-0551-83
$C_{22}H_{20}N_2O_3$			
1H-Pyrrole-3-carboxylic acid, 1-acetyl-3-cyano-2,3-dihydro-5-methyl-2,4-diphenyl-, methyl ester	MeOH	265(4.25)	22-0195-83
1H-Pyrrolo[2,3-d]carbazole-2,5(3H,4H)-dione, 7-benzoyl-3a,6,6a,7-tetrahydro-3a-methyl-, (3aα,6aβ,11bR*)-(±)-	MeOH	261(3.84)	5-1744-83
$C_{22}H_{20}N_2O_5S$			
1-Azabicyclo[3.2.0]hept-2-ene-2-carboxylic acid, 6-ethyl-7-oxo-3-(phenylthio)-, (4-nitrophenyl)methyl ester, (5R-cis)-	THF	271(3.96),320(4.04)	158-0407-83
$C_{22}H_{20}N_2S_2$			
Diazene, phenyl[3-phenyl-3-(4,5,6,7-tetrahydro-1,3-benzodithiol-2-ylidene)-1-propenyl]-	MeCN	520(3.93)	118-0840-83
$C_{22}H_{20}N_3OS_2$			
Thiazolo[3,4-a]pyrimidin-5-ium, 6-[(3-	DMF	450(4.57)	103-0752-83

Compound	Solvent	$\lambda_{max}(\log \epsilon)$	Ref.
ethyl-2(3H)-benzothiazolylidene)methyl]-1,2,3,4-tetrahydro-2-oxo-8-phenyl-, perchlorate (cont.)			103-0752-83
$C_{22}H_{20}N_4O_3$			
Carbamic acid, (5-amino-2,3-diphenyl-2H-pyrido[4,3-b]-1,4-oxazin-7-yl)-, ethyl ester	pH 1	296(3.85),306s(3.83), 358s(4.32),374(4.38) (changing)	87-1614-83
$C_{22}H_{20}N_4O_6$			
Benzoic acid, 4,4'-[azobis[(2-cyano-2-methyl-2,1-ethenediyl)oxy]]bis-	DMF	350(1.26)	126-0543-83A
$C_{22}H_{20}N_4O_8S_2$			
Cefsulodin, sodium salt	H_2O	263(4.16)	162-0271-83
$C_{22}H_{20}N_6O_3$			
1,2,4-Triazolo[4,3-b]pyridazinium, 7-[[4-(ethoxycarbonyl)phenyl]azo]-8-hydroxy-6-methyl-1-(phenylmethyl)-, hydroxide, inner salt, (E)-	benzene	411(4.37),509(3.36)	48-1016-83
	MeOH	403(4.37),510(3.36)	48-1016-83
	MeCN	408(4.36),509(3.35)	48-1016-83
	CH_2Cl_2	408(4.36),509(3.36)	48-1016-83
(Z)-	benzene	336(--),490(--)	48-1016-83
	MeOH	328(--),481(--)	48-1016-83
	MeCN	333(--),490(--)	48-1016-83
	CH_2Cl_2	334(--),490(--)	48-1016-83
$C_{22}H_{20}N_8O_2$			
Pyridazino[4,5-d]pyridazine-1,4,5,8-tetrone, 2,3,6,7-tetrahydro-2,6-dimethyl-, 4,8-bis[(phenylmethylene)-hydrazone]	DMSO	353(4.71),463(3.59)	161-0842-83
$C_{22}H_{20}O$			
Bicyclo[3.2.0]hept-2-en-6-one, 4-(diphenylmethylene)-7,7-dimethyl-	C_6H_{12}	292(4.395)	64-0504-83B
2-Butanone, 1,3,4-triphenyl-	isooctane	243(2.78),249(2.96), 255(3.06),260(3.01), 298(2.49),313(2.27)	35-1309-83
$C_{22}H_{20}O_3$			
9H-Xanthene, 9-ethoxy-9-(3-methoxyphenyl)-	EtOH	218(4.22),280(3.88)	142-0481-83
$C_{22}H_{20}O_4$			
2(3H)-Benzofuranone, 7-(1-methylethyl)-3-[7-(1-methylethyl)-2-oxo-3(2H)-benzofuranylidene]-, (E)-	CH_2Cl_2	227(3.95),255s(4.06), 258(4.08),421(4.28)	78-2147-83
9,10-Ethenoanthracene-11,12-dicarboxylic acid, 9,10-dihydro-2,3-dimethyl-, dimethyl ester	C_6H_{12}	276(3.39),284(3.44)	78-1151-83
2,5-Phenanthrenediol, 9,10-dihydro-8-[(4-hydroxyphenyl)methyl]-7-methoxy-	MeOH	273s(4.31),282(4.39), 300s(4.20)	102-1011-83
$C_{22}H_{20}O_4S_2$			
1,4-Dithiaspiro[4.4]non-6-ene-6-methanol, 8-(benzoyloxy)-, benzoate, (R)-	EtOH	210s(4.06),230(4.45)	39-2441-83C
$C_{22}H_{20}O_6$			
5,10-Methanocyclopentacycloundecene-1,2,3-tricarboxylic acid, 1-ethyl 2,3-dimethyl ester	EtOH	590(3.18)	88-2151-83

Compound	Solvent	λ_{max}(log ε)	Ref.
$C_{22}H_{20}O_7$			
2-Anthracenecarboxylic acid, 9,10-dihydro-4-hydroxy-5-methoxy-9,10-dioxo-3-(3-oxopentyl)-, methyl ester	MeOH	227(4.44),261(4.37), 396s(--),418(4.03), 429s(--)	5-2151-83
1-Naphthacenecarboxylic acid, 2-ethyl-1,2,3,4,6,11-hexahydro-2,5,7-trihydroxy-6,11-dioxo-, methyl ester, cis-(±)-	MeOH	230(4.61),258(4.34), 402s(--),417(4.05), 435s(--)	5-2151-83
trans	MeOH	226(4.52),259(4.41), 283s(--),398s(--), 416(4.03),434s(--)	5-2151-83
$C_{22}H_{20}O_8$			
1,3-Benzenediol, 2-[2-(3,5-diacetoxyphenyl)ethenyl]-, diacetate, (E)-	EtOH	227s(4.34),288(4.34)	2-0101-83
1-Naphthacenecarboxylic acid, 2-ethyl-1,2,3,4,6,11-hexahydro-2,4,5,7-tetrahydroxy-6,11-dioxo-, methyl ester	MeOH	228(4.60),256(4.41), 288(3.98),409s(--), 431(4.09),440s(--)	5-2151-83
$C_{22}H_{20}O_{13}$			
Carminic acid	H_2O	500(3.83)	162-0256-83
	0.02M HCl	495(3.76)	162-0256-83
	pH 10	540(3.54)	162-0256-83
$C_{22}H_{21}BrN_2O$			
4H-Indol-4-one, 1-(4-bromophenyl)-1,5,6,7-tetrahydro-6,6-dimethyl-2-phenyl-, oxime, (Z)-	EtOH	253(4.60),280(4.80)	4-0989-83
Pyrrolo[3,2-c]azepin-4(1H)-one, 1-(4-bromophenyl)-5,6,7,8-tetrahydro-7,7-dimethyl-2-phenyl-	EtOH	245(4.40),277(4.20)	4-0989-83
$C_{22}H_{21}ClN_2$			
Benzenamine, N-[[2-chloro-3-[3-(phenylamino)-2-propenylidene]-1-cyclohexen-1-yl]methylene]-, perchlorate	MeOH	625(5.13)	104-1854-83
$C_{22}H_{21}ClN_2O_8$			
Meclocycline	MeOH-HCl	245(4.34),347(4.10)	162-0823-83
$C_{22}H_{21}NO$			
1H-Indol-3-ol, 2,3-dihydro-1,2-dimethyl-2,3-diphenyl-, cis	EtOH	252(3.97),311(3.39)	24-2115-83
$C_{22}H_{21}NO_2$			
Oxiranecarbonitrile, 3-[2-(2-cyclohept-en-1-yloxy)phenyl]-2-phenyl-, trans	EtOH	265s(3.38),277s(3.58), 283(3.59)	24-2383-83
Spiro[naphthalene-1(4H),5'(2'H)-oxazol]-4-one, 2,2',2',3-tetramethyl-4'-phenyl-	EtOH	248(4.33),276s(3.94)	33-2252-83
$C_{22}H_{21}NO_3$			
Benz[cd]indol-5(1H)-one, 1-benzoyl-2,2a,3,4-tetrahydro-4-(2-oxobutyl)-	MeOH	237(4.38),326(3.66)	39-1545-83C
$C_{22}H_{21}NO_4$			
9,10-Anthracenedione, 1-acetoxy-4-(cyclohexylamino)-	EtOH?	527(3.90)	99-0420-83
$C_{22}H_{21}NO_5$			
Phenol, 2-[(2-hydroxy-5-methylphenyl)-	neutral	327(3.95)	126-2361-83

Compound	Solvent	λ max(log ε)	Ref.
methyl]-6-[(2-hydroxy-5-nitrophenyl)-methyl]-4-methyl- (cont.)	anion	403(4.23)	126-2361-83
	dianion	419(4.29)	126-2361-83
$C_{22}H_{21}NO_6$			
Calycinine, N,O-diacetyl-	EtOH	219(4.54),276(4.35), 317(3.70)	100-0862-83
$C_{22}H_{21}N_2O_3$			
1-Benzopyrylium, 7-(dimethylamino)-2-[7-(dimethylamino)-2-oxo-2H-1-benzopyran-3-yl]-, perchlorate	MeCN	650(4.86)	48-0505-83
$C_{22}H_{21}N_3$			
Cyclohexanamine, N-(9-methyl-11H-indeno[1,2-b]quinoxalin-11-ylidene)-	EtOH	220(4.92),224(4.9), 268(4.8),294(5.10), 340s(4.50),374(4.30)	104-0158-83
$C_{22}H_{21}N_3O_3$			
4H-Indol-4-one, 1,5,6,7-tetrahydro-6,6-dimethyl-1-(3-nitrophenyl)-2-phenyl-, oxime, (Z)-	EtOH	253(4.60),285(4.40)	4-0989-83
Pyrimido[4,5-b][1,4]oxazepine-2,4(1H-3H)-dione, 7,8-dihydro-1,3-dimethyl-6-(4-methylphenyl)-8-phenyl-	MeOH	229(4.49),246(4.40), 311(4.03)	103-0544-83
Pyrrolo[3,2-c]azepin-4(1H)-one, 5,6,7,8-tetrahydro-7,7-dimethyl-1-(3-nitrophenyl)-2-phenyl-	EtOH	245(4.50),274s(4.40)	4-0989-83
$C_{22}H_{21}N_3O_4$			
Pyrimido[4,5-b][1,4]oxazepine-2,4(1H-3H)-dione, 7,8-dihydro-6-(4-methoxyphenyl)-1,3-dimethyl-8-phenyl-	MeOH	277(4.33),317(3.74)	103-0544-83
Pyrimido[4,5-b][1,4]oxazepine-2,4(1H-3H)-dione, 7,8-dihydro-8-(4-methoxyphenyl)-1,3-dimethyl-6-phenyl-	MeOH	231(4.34),246(4.30), 285s(--),310s(--)	103-0544-83
$C_{22}H_{21}N_3O_6S$			
1-Azabicyclo[3.2.0]hept-2-ene-2-carboxylic acid, 6-ethyl-3-[[(3-hydroxy-2-pyridinyl)methyl]thio]-7-oxo-, (4-nitrophenyl)methyl ester	THF	274(4.16),323.5(3.98)	158-0407-83
$C_{22}H_{21}N_4O_2$			
Verdazyl, 1,3-bis(4-methoxyphenyl)-5-phenyl-	MeCN	730(3.628)	104-2191-83
$C_{22}H_{21}N_5O_3$			
1H-Purine-2,6-dione, 3,7-dihydro-1,3-dimethyl-7-(2-oxo-2-phenylethyl)-8-[(phenylmethyl)amino]-	EtOH	243(4.25),296(4.33)	56-0461-83
$C_{22}H_{22}$			
Benzene, 4,4'-bis[(4-methylphenyl)-methyl]-	C_6H_{12}	225s(4.6),267(3.3), 275(3.2)	88-5277-83
[2_4](1,2,5,6)Cyclooctatetraenyl(1,2,4-5)cyclophane	n.s.g.	282(2.85),291(2.87)	35-7384-83
$C_{22}H_{22}BrN_2OS$			
Thiazolium, 3-[2-(4-bromophenyl)-2-oxo-ethyl)-2-[2-[4-(dimethylamino)phenyl]ethenyl]-4-methyl-, perchlorate	EtOH	496(4.21)	65-2424-83

Compound	Solvent	λ_{max}(log ϵ)	Ref.
$C_{22}H_{22}Br_2O_2$			
Benzene, 1,1'-[1,4-bis(1-bromoethyl)-1,2,3-butatriene-1,4-diyl]bis[4-methoxy-	C_6H_{12}	254(4.25),420(3.89)	19-0233-83
Benzene, 1,1'-(2,3-dibromo-1,4-diethylidene-2-butene-1,4-diyl)bis[4-methoxy-	C_6H_{12}	251.5(4.43)	19-0233-83
$C_{22}H_{22}ClN_2OS$			
Thiazolium, 3-[2-(4-chlorophenyl)-2-oxoethyl]-2-[2-[4-(dimethylamino)-phenyl]ethenyl]-4-methyl-, perchlorate	EtOH	497(4.25)	65-2424-83
$C_{22}H_{22}FN_3O_2$			
2H-Benzimidazol-2-one, 1-[1-[4-(4-fluorophenyl)-4-oxobutyl]-1,2,3,6-tetrahydro-4-pyridinyl]-1,3-dihydro- (droperidol)	MeOH-HCl	245(4.19),280(3.88)	162-0501-83
$C_{22}H_{22}N_2$			
Benzo[f]quinazoline, 3-tricyclo-[3.3.1.1^{3,7}]dec-1-yl-	EtOH	227(4.55),256s(4.33), 263(4.34),295s(3.77), 327(3.30),343(3.29)	88-4351-83
Benzo[h]quinazoline, 2-tricyclo-[3.3.1.1^{3,7}]dec-1-yl-	EtOH	219(4.40),238s(4.42), 255(4.57),294(3.99), 321(3.38),336(3.51), 352(3.51)	88-4351-83
$C_{22}H_{22}N_2O$			
Benzenamine, N-[[2-methoxy-3-[3-(phenylamino)-2-propenylidene]-1-cyclopenten-1-yl]methylene]-, monoperchlorate	MeOH	595(5.15)	104-1854-83
Cyclopentanone, 2-[3-(methylphenylamino)-2-propenylidene]-5-[(phenylamino)methylene]-	MeOH	470(5.16)	104-1854-83
Ethanone, 2,2-[bis(2-methylphenyl)-amino]-1-phenyl-	C_6H_{12}	243(3.35),283s(2.76)	4-1739-83
Ethanone, 2,2-[bis(4-methylphenyl)-amino]-1-phenyl-	C_6H_{12}	245(3.37),290(2.63)	4-1739-83
6,10-Imino-11H-cyclooct[b]indol-11-one, 5,6,7,8,9,10-hexahydro-5-methyl-12-(phenylmethyl)-	EtOH	248(4.30),267(4.25), 307(4.09)	35-0907-83
4H-Indol-4-one, 1,5,6,7-tetrahydro-6,6-dimethyl-1,2-diphenyl-, oxime, (E)-	EtOH	243s(2.90),285(2.80)	4-0989-83
(Z)-	EtOH	253(3.30),290(3.20)	4-0989-83
Pyrrolo[3,2-c]azepin-4(1H)-one, 5,6,7,8-tetrahydro-7,7-dimethyl-1,2-diphenyl-	EtOH	240s(3.30),284(3.20)	4-0989-83
Quinolinium, 1-ethyl-2-[2-(3-ethylbenzoxazolium-2-yl)ethenyl]-, diiodide	EtOH	212(4.88),242(4.86), 297(4.51),510(4.06), 570(4.00)	80-0725-83
$C_{22}H_{22}N_2O_3$			
Ethanone, 2,2-[bis(4-methoxyphenyl)-amino]-1-phenyl-	C_6H_{12}	244(3.31),300(2.75), 336s(2.48)	4-1739-83
2H-Pyran-5-carboxylic acid, 3,4-dihydro-6-(2-phenylethenyl)-4-(phenylhydrazono)-, ethyl ester, (?,E)-	EtOH	302(4.26),400(4.50)	118-0948-83

Compound	Solvent	λ_{max}(log ϵ)	Ref.
$C_{22}H_{22}N_2O_4$			
Propanedioic acid, [(1H-indol-3-ylphenylamino)methylene]-, diethyl ester	EtOH	220(4.53),287(4.27), 320(4.27)	104-1367-83
Propanedioic acid, [[(2-methyl-1H-indol-3-ylphenylamino)methylene]-, ethyl ester	EtOH	208(3.43),222.2(3.55), 286(3.31),333.3(3.20)	104-1367-83
$C_{22}H_{22}N_2O_8$			
2-Butenedioic acid, 2,2'-(1,8-naphthalenediyldiimino)bis-, tetramethyl ester, (Z,Z)-	MeOH	233(4.50),329(4.18), 340s(--)	18-2338-83
Methacycline, hydrochloride	MeOH-HCl	253(4.37),345(4.19)	162-0850-83
$C_{22}H_{22}N_3O_3S$			
Thiazolium, 2-[2-[4-(dimethylamino)-phenyl]ethenyl]-4-methyl-3-[2-(4-nitrophenyl)-2-oxoethyl]-, perchlorate	EtOH	504(4.23)	65-2424-83
$C_{22}H_{22}N_4$			
1H-Cyclopent[cd]indene-1,1,2,2(2aH)-tetracarbonitrile, 7a,7b-dihydro-2a,3,5,6,7,7a,7b-heptamethyl-	MeCN	315(3.89)	44-0309-83
Dibenzo-5,9,14,18-tetraazacyclotetradecene, 6,8,15,17-tetramethyl-	$CHCl_3$	352(4.67)	74-0381-83A
	ketone	352(3.60)	74-0381-83A
$C_{22}H_{22}N_4O_2$			
3-Pyridinecarbonitrile, 1,1'-[1,4-phenylenebis(methylene)]bis[1,6-dihydro-6-methoxy-	CH_2Cl_2	240(4.2),314(4.1), 362s(3.6)	64-0878-83B
$C_{22}H_{22}O_2$			
2,6-Octadien-4-yne, 3,6-bis(4-methoxyphenyl)-	C_6H_{12}	215.5(4.53),229s(4.31), 252s(4.46),261(4.55), 278(4.40),296s(4.05), 350s(1.96),402.3(1.95)	19-0233-83
$C_{22}H_{22}O_5$			
2-Cyclohexen-1-one, 6-acetyl-3-(2-hydroxy-4-methoxyphenyl)-5-(4-methoxyphenyl)-	MeOH	202(4.82),334(4.20)	142-2369-83
Ethanone, 1-[6-(2-hydroxy-4,6-dimethoxyphenyl)-2-methyl-4-phenyl-4H-pyran-3-yl]-	MeOH	206(4.63),286(4.45)	142-2369-83
Ethanone, 1-[6-(2-hydroxy-4-methoxyphenyl)-4-(4-methoxyphenyl)-2-methyl-4H-pyran-3-yl]-	MeOH	212(4.69),272(4.45)	142-2369-83
$C_{22}H_{22}O_6$			
9,10-Anthracenedione, 1-hydroxy-4,8-dimethoxy-2-(3-oxohexyl)-	MeOH	225(4.56),245(4.22), 282(3.88),382(3.54), 446(3.96),463(3.98), 498(3.76),529(3.11)	5-1818-83
1(3H)-Isobenzofuranone, 3-[5,6-dihydro-4-methoxy-1-(methoxymethoxy)-2-naphthalenyl]-7-methoxy-	EtOH	227(4.58),273(3.93), 284s(3.60),306(4.00), 322s(3.35)	94-2662-83
$C_{22}H_{22}O_9$			
D-Arabinitol, 1-C-(9,10-dihydro-1,4-dihydroxy-9,10-dioxo-2-anthracenyl)-	EtOH	226s(4.21),251(4.57), 256s(4.48),286(3.93),	39-0613-83C

Compound	Solvent	λ_{max}(log ε)	Ref.
2,3-O-(1-methylethylidene)- (cont.)		320(3.41),486(3.86), 518(3.63)	39-0613-83C
$C_{22}H_{22}O_{11}$			
Tamarixetin 3-rhamnoside	MeOH	255(4.36),264s(--), 345(4.10)	102-0621-83
	MeOH-NaOMe	270(--),380(--)	102-0621-83
	MeOH-NaOAc	273(--),365(--)	102-0621-83
	MeOH-$AlCl_3$	265(--),271s(--), 298s(--),360(--), 390(--)	102-0621-83
aglycone	MeOH	256(3.99),366(3.94)	102-0621-83
$C_{22}H_{22}O_{12}$			
Isorhamnetin 3-O-β-D-glucopyranoside	EtOH	255(3.8),355(--)	105-0500-83
	EtOH-NaOAc	275(--),375(--)	105-0500-83
	+ $Zr(NO_3)_4$	255(--),260(--), 400(--)	105-0500-83
$C_{22}H_{23}D_2NO_6$			
1,3-Dioxolo[4,5-g]isoquinoline, 5-(1,3-dihydro-3-d-4,5-dimethoxy-1-isobenzofuranyl-3-d)-5,6,7,8-tetrahydro-4-methoxy-6-methyl-, [S-(R*,S*)]-	MeOH	224(4.38),281(3.52)	83-0737-83
$C_{22}H_{23}NO_2$			
Roquefortine	MeOH	209(4.47),240(4.21), 328(4.43)	98-0655-83
$C_{22}H_{23}NO_3$			
Atherosperminine N-oxide	EtOH	213(4.30),234(4.33), 252(4.60),258(4.63), 279s(4.01),304(4.04), 313(4.04),346(3.21), 364(3.21)	100-0761-83
$C_{22}H_{23}NO_4$			
3-Azabicyclo[3.2.0]heptane-2,4-dione, 3-butyl-6,7-bis(2-hydroxyphenyl)-, (1α,5α,6β,7β)-	MeOH	224(4.15),275(3.70), 282(3.66)	39-1083-83C
2H-Benzo[b]cyclobuta[d]pyran-2-carboxamide, N-butyl-1,2a,3,8b-tetrahydro-1-(2-hydroxyphenyl)-3-oxo-, (1α,2α,2aα,8bα)-	dioxan	271(3.57),278(3.57)	39-1083-83C
Melosmine, O,O-dimethyl-	EtOH	212(4.24),248(4.27), 255(4.23),313s(3.73), 326(3.86),364(3.81)	100-0761-83
	EtOH-HCl	215(4.23),240(4.21), 275(4.37),337s(3.37), 433(3.72)	100-0761-83
$C_{22}H_{23}NO_5$			
1,7-Ethano-4H,7H-furo[4',3',2':1,8]-naphth[1,2-d]azocine-2-carboxylic acid, 1,5,6,12-tetrahydro-9-methoxy-4-methyl-14-oxo-, methyl ester	EtOH	293(4.28)	44-0173-83
Glaucine, 3-methoxy-	EtOH and EtOH-KOH	223(4.45),268s(3.91), 282(4.12),302(4.10), 312s(4.04)	5-0744-83

Compound	Solvent	$\lambda_{max}(\log \epsilon)$	Ref.
$C_{22}H_{23}NO_6$			
Aureothin	EtOH	257(4.39),346(4.27)	162-0127-83
Dubirheine	MeOH	242(3.90),293(3.97)	100-0441-83
Epiophiocarpine, O-acetyl-, (±)-	MeOH	232.5(4.11),283.5(3.65), 291(3.80)	94-0947-83
$C_{22}H_{23}NO_7$			
6H-Benzo[g]-1,3-benzodioxolo[5,6-a]-quinolizin-13(8H)-one, 5,13a-dihydro-8,9,10,13a-tetramethoxy-	EtOH	230(4.16),291(4.12)	94-0947-83
Bis[1,3]benzodioxolo[5,6-e:5',4'-i]-[1,2]oxaazacycloundecin-14-ol, 4,6,7,8,14,15-hexahydro-6-methyl-, acetate	MeOH	210(4.43),238s(3.83), 293(3.83)	39-2431-83C
α-Narcotine, BD_3 adduct	MeOH	226(4.28),289s(3.52), 310(3.57)	83-0737-83
$C_{22}H_{23}NO_8$			
9-Azabicyclo[4.2.1]nona-4,7-diene-1,4,7,8-tetracarboxylic acid, 9-phenyl-, tetramethyl ester	MeCN	243(4.24),275(3.34), 297(2.95)	24-0563-83
$C_{22}H_{23}N_2OS$			
Thiazolium, 2-[2-[4-(dimethylamino)-phenyl]ethenyl]-4-methyl-3-(2-oxo-2-phenylethyl)-, perchlorate	EtOH	494(4.43)	65-2424-83
$C_{22}H_{23}N_2S_2$			
Benzothiazolium, 3-ethyl-2-[3-(3-ethyl-2(3H)-benzothiazolylidene)-2-methyl-1-propenyl]-, tetrafluoroborate	MeOH	542(5.12)	89-0876-83
$C_{22}H_{23}N_3S_2$			
Benzenamine, N,N-dimethyl-4-[(phenyl-azo)(4,5,6,7-tetrahydro-1,3-benzo-dithiol-2-ylidene)methyl]-	MeCN	502(4.17)	118-0840-83
$C_{22}H_{23}N_5O_3$			
1H-Purine-2,6-dione, 3,7-dihydro-7-(2-hydroxy-2-phenylethyl)-1,3-dimethyl-8-[(phenylmethyl)amino]-	EtOH	296(4.29)	56-0461-83
1H-Purine-2,6,8(3H)-trione, 7,9-dihy-dro-1,3-dimethyl-7-[2-phenyl-2-[(phenylmethyl)amino]ethyl]-	EtOH	296(4.08)	56-0461-83
$C_{22}H_{24}CuN_{10}$			
Copper, [6,11,18,19-tetrahydro-N,N,N'-N'-tetramethyldipyrimido[5,4-c:4',5'-i][1,5,8,12]benzotetraazacyclotetra-decine-4,14-diaminato(2-)-N^6,N^{11},N^{17}-N^{20}]-, (SP-4-2)-	DMF	378(4.34),395(4.31), 432(4.24)	12-2413-83
$C_{22}H_{24}F_3NO_5$			
Isoquinoline, 1,2,3,4-tetrahydro-6,7-dimethoxy-4-[(3,4-dimethoxyphenyl)-methyl]-2-(trifluoroacetyl)-	EtOH	230s(--),283(3.82)	39-2053-83C
$C_{22}H_{24}NO_2Te$			
1-Benzotellurinium, 2-[2-[4-(dimethyl-amino)phenyl]ethenyl]-4-ethoxy-7-methoxy-, fluorosulfate	CH_2Cl_2	485(3.86),695(4.98)	35-0883-83

Compound	Solvent	$\lambda_{max}(\log \epsilon)$	Ref.
$C_{22}H_{24}NO_5P$			
α-D-Xylofuranose, 3-C-(cyanomethyl)-3-deoxy-3-(diphenylphosphinyl)-1,2-O-(1-methylethylidene)-	EtOH	206.5(4.14),223.5(4.26)	159-0019-83
$C_{22}H_{24}N_2O_3$			
Strychnidin-10-one, 2-methoxy-	EtOH	262(4.40),297(3.80)	162-0354-83
Strychnidin-10-one, 3-methoxy- (α-colubrine)	EtOH	255(4.03),297(3.77)	162-0354-83
$C_{22}H_{24}N_2O_4$			
Alstonidine	MeOH	238(4.66),291(4.25), 360(3.74)	162-0047-83
Kopsine	EtOH	240(4.08),278(3.37), 286(3.35)	162-0764-83
$C_{22}H_{24}N_2O_8$			
Carabrol 3,5-dinitrobenzoate	MeOH	211(4.51)	102-2773-83
2-Naphthacenecarboxamide, 4-(dimethylamino)-1,4,4a,5,5a,6,11,12a-octahydro-3,5,10,12,12a-pentahydroxy-6-methyl-1,11-dioxo-, hydrochloride	MeOH-HCl	267(4.24),351(4.12)	162-0499-83
$C_{22}H_{24}N_4O_4$			
1,8-Dioxa-2,5,9,12-tetraazacyclotetradeca-2,9-diene-6,13-dione, 3,10-bis(4-methylphenyl)-	EtOH	260(4.4)	5-2038-83
1,18-Epoxy-3H-dibenzo[c,f][1,5,9]triazacyclopentadecine-3,5,14-trione, 4-amino-6,7,8,9,10,11,12,13-octahydro-2,17-dimethyl-	DMF	265(4.28),425(4.30), 446(4.32)	95-0049-83
Morpholine, 4,4'-[azobis(4,1-phenylenecarbonyl)]bis-	o-$C_6H_4Cl_2$	331(4.42)	18-1700-83
$C_{22}H_{24}N_4O_6$			
1,8-Dioxa-2,5,9,12-tetraazacyclotetradeca-2,9-diene-6,13-dione, 3,10-bis(4-methoxyphenyl)-, (Z,Z)-	EtOH	272(4.5)	5-2038-83
$C_{22}H_{24}O_2$			
2-Naphthalenecarboxylic acid, 6-[2-(2,6,6-trimethyl-1-cyclohexen-1-yl)ethenyl]-, (E)-	EtOH	254(4.43),317(4.38)	87-1653-83
$C_{22}H_{24}O_2S_2$			
1,4-Dithiaspiro[4.4]non-6-ene, 8-(phenylmethoxy)-6-[(phenylmethoxy)methyl]-	EtOH	219(3.45)	39-2441-83C
$C_{22}H_{24}O_5$			
7H-Furo[3,2-g][1]benzopyran-7-one, 4-[(5-methoxy-3,7-dimethyl-2,6-octadienyl)oxy]-, (E)-(±)-	MeOH	222(4.25),250.5(4.12), 258(4.07),267.5(4.06), 310(4.02)	94-2710-83
7H-Furo[3,2-g][1]benzopyran-7-one, 4-[(7-methoxy-3,7-dimethyl-2,5-octadienyl)oxy]-, (E,E)-	MeOH	207(4.43),222(4.43), 250.5(4.30),259(4.24), 267(4.22),309(4.17)	94-2710-83
$C_{22}H_{24}O_6$			
1,2-Cyclobutanedicarboxylic acid, 3,4-bis(2-hydroxyphenyl)-, diethyl ester, cis	EtOH	273(3.79),280(3.86)	39-1083-83C
trans	MeOH	274(3.74),280(3.72)	39-1083-83C

Compound	Solvent	λ_{max}(log ε)	Ref.
Naphtho[2,3-b]furan-2,5,6(3H)-trione, 3-(3-acetoxypropyl)-3,4-dimethyl-7-(1-methylethyl)-	EtOH	265(4.38),358(3.63)	18-2985-83
6H-Phenaleno[1,2-b]furan-6-one, 8,9-dihydro-7-hydroxy-3,4,5-trimethoxy-1,8,8,9-tetramethyl-, (R)-	EtOH	375(4.20),402(4.21)	18-3661-83
Vismione C	$CHCl_3$	245s(4.16),278(4.33), 398(3.95)	102-0539-83
$C_{22}H_{24}O_8$			
Shinjulactone C, 20-O-acetyl-	EtOH	245(4.15)	18-3683-83
$C_{22}H_{25}BrN_2O_8S_2$			
1-Azabicyclo[3.2.0]hept-2-ene-2-carboxylic acid, 3-[[2-(acetylamino)ethenyl]thio]-6-[1-[(ethoxysulfonyl)oxy]ethyl]-7-oxo-, (4-bromophenyl)methyl ester, [5R-[3(E),5α,6β(S*)]]-	EtOH	227(4.39),325(4.24)	78-2551-83
$C_{22}H_{25}FeO_3P$			
Ferrocene, [1-(diethoxyphosphinyl)-2-phenylethenyl]-	EtOH	272(4.1),315(3.8), 440(3.1)	65-0514-83
$C_{22}H_{25}NO_2$			
2-Propenamide, N-[2-[4-[(3-methyl-2-butenyl)oxy]phenyl]ethyl]-3-phenyl-, (E)-	EtOH	218(4.48),225(4.48), 277(4.56)	102-0755-83
$C_{22}H_{25}NO_3$			
2-Propenamide, N-[2-hydroxy-2-[4-[(3-methyl-2-butenyl)oxy]phenyl]ethyl]-3-phenyl-, (E)-	EtOH	218.5(4.40),224(4.41), 276(4.43)	102-0755-83
$C_{22}H_{25}NO_4$			
7a,12a-Etheno-4H,8H-cyclohepta[4,5]-pyrrolo[3,2,1-ij]quinoline-13,14-dicarboxylic acid, 5,6,9,10,11,12-hexahydro-, dimethyl ester	MeOH	212(4.34),243(3.42), 313(4.08),460(3.15)	39-0515-83C
1-Naphthalenamine, 2-(3,4-dimethoxyphenyl)-N-ethyl-6,7-dimethoxy-	MeOH	231(4.69),261(4.65)	103-0066-83
$C_{22}H_{25}NO_6$			
α-Anhydronarcotinediol	MeOH	218(4.36),280(3.47)	83-0737-83
BF_3 adduct	MeOH	226(4.19),280(3.47)	83-0737-83
Colchicine	EtOH	243(4.47),350.5(4.22)	162-0352-83
Epiglaudine	MeOH	237(4.18),287(4.06)	100-0441-83
Oreodine	n.s.g.	235(4.19),285(3.94)	100-0441-83
$C_{22}H_{25}NO_7$			
Chiloenamine, O-acetyl-	MeOH	217(4.52),233(4.33), 253(4.46),288(3.76), 307(3.77),320(3.79), 349(3.79)	77-0799-83
$C_{22}H_{25}NO_8$			
6-Oxa-2-azaspiro[4.5]dec-3-ene-2-acetic acid, 4-(ethoxycarbonyl)-7-methoxy-1,9-dioxo-3-phenyl-, ethyl ester	EtOH	302(3.92)	94-0356-83
$C_{22}H_{25}N_3O_{10}S_2$			
1-Azabicyclo[3.2.0]hept-2-ene-2-carbox-	EtOH	219(4.28),265(4.29),	78-2551-83

Compound	Solvent	λ_{max}(log ϵ)	Ref.
ylic acid, 3-[[2-(acetylamino)ethen-yl]thio]-6-[1-(ethoxysulfonyl)oxy]-ethyl]-7-oxo-, (4-nitrophenyl)methyl ester, [5R-(3(E),5α,6β(S*)]]- (cont.)		325(4.28)	78-2551-83
$C_{22}H_{25}N_5O_6$			
1,1'-Biphenyl, 2,4,6-trinitro-2',4'-di-piperidino-	$CHCl_3$	521(3.65)	44-4649-83
$C_{22}H_{26}CuN_6$			
Copper, [N,N'-bis[2-(phenylazo)ethyli-dene]-1,6-hexanediaminato(2-)]-, (T-4)-	$CHCl_3$	294(4.07),308(4.06), 326(4.05),381(4.14), 448(4.22),616(3.33), 1360(2.25)	65-1214-83
$C_{22}H_{26}N_2O$			
1,6-Naphthalenediamine, N,N'-diethyl-7-methoxy-3-methyl-2-phenyl-	MeOH	211(4.38),234(4.60), 262(4.68),328(3.64), 343(3.64)	103-0066-83
$C_{22}H_{26}N_2O_2$			
2H-Pyrrol-2-one, 5-[[3,4-dimethyl-1-[(phenylmethoxy)methyl]-1H-pyrrol-2-yl)methylene]-4-ethyl-1,5-dihydro-3-methyl-, (Z)-	$CHCl_3$	263(4.15),373(4.11)	49-0753-83
Spiro[4.5]deca-1,3-diene-1-carboxylic acid, 2-amino-3-cyano-4-phenyl-, 1,1-dimethylethyl ester	hexane	348(3.60)	24-3097-83
$C_{22}H_{26}N_2O_3$			
Corynantheine	MeOH	227(4.64),280(3.82), 291(3.80)	162-0364-83
$C_{22}H_{26}N_2O_4$			
Akuammine	EtOH	245(3.96),312(3.66)	162-0031-83
Pyrido[3,2-g]quinoline-5,10-dione, 2,8-dimethoxy-3,7-dimethyl-4,6-dipropyl-	MeOH	245s(4.07),254s(4.28), 274(4.74),305(4.25), 380(3.16)	88-3643-83
Pyrido[3,2-g]quinoline-2,5,8,10(1H,9H)-tetrone, 1,3,7,9-tetramethyl-4,6-di-propyl-	MeOH	250s(4.25),260s(4.36), 284(4.51),312(4.31), 323(4.35),352(3.87), 452(3.13)	88-3643-83
Pyrido[3,2-g]quinoline-2,5,10(1H)-tri-one, 8-methoxy-1,3,7-trimethyl-4,6-dipropyl-	MeOH	246s(4.13),252s(4.19), 278(4.52),309(4.20), 318(4.20),340(3.80), 425(3.19)	88-3643-83
$C_{22}H_{26}N_2O_5$			
Echitamidine, N_a-formyl-12-methoxy-	EtOH	216(4.54),246(4.06), 278(3.59),305s(3.12)	32-0533-83
Pseudoindoxyl of aricine (same spectra in acid and base)	EtOH	218(4.76),270(3.61), 405(3.45)	100-0708-83
$C_{22}H_{26}N_2O_5S$			
1-Azabicyclo[3.2.0]hept-2-ene-2-carbox-ylic acid, 3-(cyclohexylthio)-6-eth-yl-7-oxo-, (4-nitrophenyl)methyl ester, (5R-cis)-	THF	270(4.11),325(4.19)	158-0407-83
$C_{22}H_{26}N_2O_6$			
Acetic acid, 2,2'-(8-ethoxy-2,3,8,13b-	EtOH-HCl	260(4.11),290(3.94)	44-0044-83

Compound	Solvent	$\lambda_{max}(\log \epsilon)$	Ref.
tetrahydro-13b-hydroxy-3-methyl-1H-azecino[5,4-b]indole-5,6(4H,7H)-diylidene)bis- (cont.)			44-0044-83
$C_{22}H_{26}N_4$			
Calycanthine	EtOH	250(4.28),309(3.80)	162-0237-83
$C_{22}H_{26}N_4O_5$			
Cyanocycline A	EtOH	268(4.04)	39-0335-83B
Cyanonaphthyridinomycin	MeOH	269(3.98)	158-1767-83
$C_{22}H_{26}O_3$			
2H-1-Benzopyran, 3,4-dihydro-5,7-dimethoxy-8-(3-methyl-2-butenyl)-2-phenyl-, (S)-	MeOH	209(4.76),263s(2.90), 271(2.96),276s(2.50)	102-1305-83
$C_{22}H_{26}O_4$			
1,1'-Biphenyl, 2,2',3,3'-tetramethoxy-5,5'-di-2-propenyl-	EtOH	281(3.51)	102-0609-83
$C_{22}H_{26}O_5$			
Estra-1,3,5(10)-trien-6-one, 3,7-diacetoxy-	MeOH	211(4.25),246(3.96), 296(3.28)	145-0347-83
$C_{22}H_{26}O_6$			
2,5-Cyclohexadien-1-one, 4-[2-[2,6-dimethoxy-4-(1-propenyl)phenoxy]propylidene]-2,6-dimethoxy-, (E)-	$CHCl_3$	269(4.07),328(4.47)	150-2625-83M
2(3H)-Furanone, 3,4-bis[(3,4-dimethoxyphenyl)methyl]dihydro-	MeOH	229(4.24),277(3.85)	102-1516-83
$C_{22}H_{26}O_7$			
Bicyclo[3.2.1]octan-3-one, 2-acetoxy-7-(1,3-benzodioxol-5-yl)-8-hydroxy-1-methoxy-6-methyl-5-(2-propenyl)-	MeOH	234(3.56),286(3.64)	102-0269-83
Ferolide	n.s.g.	255(4.33)	105-0500-83
7H-Phenaleno[1,2-b]furan-7-one, 7a,8,9,10a-tetrahydro-3,7a-dihydroxy-4,5,6-trimethoxy-1,8,8,9-tetramethyl-	EtOH	375(3.58),388(3.55)	18-3661-83
$C_{22}H_{26}O_8$			
Shinjudilactone, 1-acetoxy-	EtOH	240(3.90)	18-3683-83
Syringaresinol, (+)-	EtOH	220s(4.16),239(4.05), 273(3.27)	36-1285-83
$C_{22}H_{26}O_9$			
Acetic acid, (2-methoxyphenoxy)-, oxydi-2,1-ethanediyl ester	MeOH	272(3.59)	49-0359-83
Altersolanol, tetrahydro-, triacetate	EtOH	225(4.15),273(4.12)	23-0378-83
Holacanthone	MeOH	240(4.01)	100-0359-83
$C_{22}H_{26}O_{10}$			
1(2H)-Anthracenone, 8-β-D-glucopyranosyloxy-3,4-dihydro-4,9-dihydroxy-2-methoxy-6-methyl-, cis	MeOH	266(4.22),297(3.80), 306(3.80),320s(--), 395(3.80)	102-1483-83
	$MeOH-AlCl_3$	278(--),315(--), 450(--)	102-1483-83
[1,1'-Biphenyl]-2,2'-dicarboxylic acid, 4,4',5,5',6,6'-hexamethoxy-, dimethyl ester	EtOH	221(4.65),253(4.24), 300(3.74)	39-1765-83C

Compound	Solvent	λ_{max}(log ϵ)	Ref.
$C_{22}H_{26}O_{11}$			
Heraclenol, tert-O-β-D-glucopyranosyl-, (R)-	MeOH and MeOH-NaOH	220(4.40),249(4.37), 263s(4.14),300(4.07)	102-2035-83
$C_{22}H_{26}O_{12}$			
Catalposide	EtOH	260(4.27)	162-0266-83
	NaOH	303(4.35)	162-0266-83
$C_{22}H_{27}BrN_2O_4$			
Phenol, 2-[[[3,5-bis(1,1-dimethyleth-yl)-4-methoxyphenyl]imino]methyl]-4-bromo-6-nitro-	EtOH	210(4.72),255(4.59), 270(4.48),317(4.62), 340(4.56),357(4.59), 370(4.42),425(2.70)	65-0359-83
$C_{22}H_{27}F_3N_2O$			
Pyrrolo[2,3-b]indole, 1,2,3,3a,8,8a-hexahydro-3a,8-bis(3-methyl-2-buten-yl)-1-(trifluoroacetyl)-, cis-(±)-	EtOH	213(4.20),252(3.92), 304(3.18)	1-0803-83
	EtOH-HCl	212(4.18),253(3.91), 310(3.20)	1-0803-83
$C_{22}H_{27}NO_2$			
Danazol	EtOH	286(4.05)	162-0406-83
$C_{22}H_{27}NO_4$			
Benzo[6,7]cyclohept[1,2,3-de]isoquino-line, 4,5,6,6a,7,8-hexahydro-1,2,10-11-tetramethoxy-5-methyl-	EtOH	270(4.06),293(3.95)	39-2053-83C
Isoquinoline, 3,4-dihydro-5,6,7-trimeth-oxy-1-[1-(3-methoxyphenyl)-1-methyl-ethyl]-	MeOH	225(4.03),272(3.61), 310(2.60)	142-0451-83
Melosmidine, tetrahydro-N-methyl-	EtOH	220(4.26),270s(4.02), 288(4.11),301s(4.02)	100-0761-83
	EtOH-HCl	223(4.26),231s(4.11), 290(4.19),303s(4.03)	100-0761-83
Melosmine, tetrahydro-O,O-dimethyl-	EtOH	222(4.34),236s(4.16), 274s(4.20),286(4.28), 301s(4.13)	100-0761-83
	EtOH-HCl	224(4.41),288(4.29), 303s(4.13)	100-0761-83
$C_{22}H_{27}NO_5$			
4H-Indol-4-one, 1,5,6,7-tetrahydro-2-α-D-lyxopyranosyl-6,6-dimethyl-1-(phenylmethyl)-	EtOH	250(3.92),280(3.74)	136-0255-83E
	EtOH	250(3.98),280(3.85)	136-0255-83E
β-	EtOH	252(3.94),280(3.82)	136-0255-83E
$C_{22}H_{27}NO_6$			
Alpinigenin	n.s.g.	230(4.19),284(3.79)	100-0441-83
$C_{22}H_{27}N_2$			
Quinolinium, 1-methyl-4-[[3-(1-piperi-dinyl)-2-cyclohexen-1-ylidene]meth-yl]-, perchlorate	MeOH	572(4.97)	104-2089-83
$C_{22}H_{27}N_2O_3$			
Macusine C	H_2O	222(4.63),272(3.88), 278(3.87),289(3.76)	162-0808-83
Melinonine A	50% EtOH	225(4.68)	162-0829-83
$C_{22}H_{27}N_3O_{10}S_2$			
1-Azabicyclo[3.2.0]hept-2-ene-2-carbox-	EtOH	270(4.08),318(4.13)	78-2551-83

Compound	Solvent	λ_{max}(log ε)	Ref.
ylic acid, 3-[[2-(acetylamino)ethyl]-thio]-6-[1-[(ethoxysulfonyl)oxo]ethyl]-7-oxo-, (4-nitrophenyl)methyl ester, [5R-[5α,6β(S*)]- (cont.)			78-2551-83
$C_{22}H_{27}N_5O_4$			
4,6-Methano-5H-benz[h]oxazolo[3,2-a]-pyrazino[3,2,1-de][1,5]naphthyridine-7-carbonitrile, 1,2,3a,4,4a,6,7,9,10-13,13b,13c-dodecahydro-9-(hydroxymethyl)-5,12-dimethyl-11-(methylamino)-10,13-dioxo-	MeOH	278(3.78)	158-1767-83
$C_{22}H_{27}N_9O_4$			
1H-Pyrrole-2-carboxamide, N-[5-[[(3-amino-3-iminopropyl)amino]carbonyl]-1-methyl-1H-pyrrol-3-yl]-4-[[[4-(formylamino)-1-methyl-1H-pyrrol-2-yl]-carbonyl]amino]-1-methyl-, monohydrobromide	EtOH	237(4.45),303(4.54)	87-1042-83
$C_{22}H_{28}BrNO_2$			
Phenol, 2-[[[3,5-bis(1,1-dimethylethyl)-4-methoxyphenyl]imino]methyl]-4-bromo-	EtOH	208(4.77),235(4.63), 270(4.52),333(4.56), 355(4.60)	65-0359-83
$C_{22}H_{28}ClNO_2$			
Phenol, 2-[[[3,5-bis(1,1-dimethylethyl)-4-methoxyphenyl]imino]methyl]-4-chloro-	EtOH	212(4.65),230(4.60), 265(4.00),330(4.11), 350(3.18),440(2.11)	65-0359-83
$C_{22}H_{28}ClN_3O$			
Dibenz[b,f][1,4]oxazepine, 8-chloro-10,11-dihydro-10-methyl-11-[3-(4-methyl-1-piperazinyl)propyl]-	MeOH	210(4.73),260(3.93), 310(3.58)	118-0288-83
$C_{22}H_{28}Cl_2N_2$			
Diazene, bis[1-(4-chlorophenyl)-2,2-dimethylpropyl]-	dodecane	367(1.30)	24-1787-83
$C_{22}H_{28}F_2O_4$			
Pregna-1,4-diene-3,20-dione, 6,9-difluoro-11,21-dihydroxy-16-methyl-, (6α,11β,16α)-	n.s.g.	237(4.22)	162-0456-83
$C_{22}H_{28}NO_4$			
Argemonine, N-methyl-, chloride	MeOH	286(3.86)	100-0293-83
Argemonine, N-methyl-, iodide	MeOH	230s(4.42),285(3.90)	100-0293-83
Glaucine, N-methyl-	MeOH or EtOH	223(4.61),283(4.12), 304(4.14)	100-0761-83
$C_{22}H_{28}N_2O$			
Cyclohexanone, 2-[(1,3-dihydro-1,3,3-trimethyl-1H-indol-2-ylidene)ethylidene]-6-[(dimethylamino)methylene]-	benzene	420(--),439(4.73)	104-1854-83
	MeOH	457(4.80)	104-1854-83
	DMF	426(--),440(--)	104-1854-83
$C_{22}H_{28}N_2O_2$			
Pyrifoline, 6-demethoxy-	MeOH	216(4.34),259(3.97), 290s(3.60)	102-1526-83

Compound	Solvent	λ_{max}(log ε)	Ref.
$C_{22}H_{28}N_2O_3$			
Retinal, 3-[(diazoacetyl)oxy]-	hexane	245(4.26),360(4.69)	35-5160-83
$C_{22}H_{28}N_2O_4$			
17-Epialloyohimbane, 10-methoxy-	EtOH and EtOH-base	227(4.46),282(3.99), 297(3.78),308(3.68)	100-0708-83
	EtOH-HCl	226(4.67),272(3.98), 295(3.80),308(3.68)	100-0708-83
Isositsirikine, 10-methoxy-	EtOH and EtOH-base	229(4.56),281(3.81), 296(3.60),310(3.35)	100-0708-83
	EtOH-HCl	223(4.6),275(3.82), 296(3.44),310(3.35)	100-0708-83
Phenol, 2-[[[3,5-bis(1,1-dimethyleth-yl)-4-methoxyphenyl]imino]methyl]-6-nitro-	EtOH	206(4.72),242(4.23), 310(4.18),345(4.20), 427(3.30)	65-0359-83
Sitsirikine, 10-methoxy-	EtOH and EtOH-base	228(4.82),280(3.89), 296(3.87),308(3.52)	100-0708-83
	EtOH-HCl	226(4.84),278(3.8), 298(3.77),308(3.52)	100-0708-83
Strychnosplendine, N_a-acetyl-11-methoxy-	EtOH	224(4.32),251(4.03), 290(3.73),296s(3.70)	32-0773-83
α-Yohimbine, 10-methoxy-	EtOH and EtOH-base	226(4.46),282(3.91), 295(3.82),309(3.61)	100-0708-83
	EtOH-HCl	224(4.58),272(3.98), 295(3.8),305(3.67)	100-0708-83
β-Yohimbine, 10-methoxy-	EtOH and EtOH-base	228(4.52),282(3.96), 297(3.85),307(3.21)	100-0708-83
	EtOH-HCl	226(4.67),274(3.94), 297(3.85),307(3.20)	100-0708-83
$C_{22}H_{28}N_2O_5$			
Reserpic acid	EtOH	271(3.77),295(3.87)	151-0171-83C
$C_{22}H_{28}N_2O_6$			
[3a,3'a-Bi-3aH-indole]-2,2'-dicarbox-ylic acid, 4,4',5,5',6,6',7,7'-octa-hydro-3,3'-dihydroxy-	MeCN	309(4.00)	33-1902-83
[3,3'-Bi-3H-pyrrole]-5,5'-dicarboxylic acid, 4,4'-dihydroxy-2,2'-dimethyl-3,3'-di-2-propenyl-, diethyl ester	MeCN	311(3.95)	33-1902-83
$C_{22}H_{28}N_4OS$			
Phenothiazin-5-ium, 3-(4,4-dimethyl-piperazinium-1-yl)-7-morpholino-, dibromide	pH 7	648(4.86)	77-1521-83
$C_{22}H_{28}N_4O_6$			
9,10-Anthracenedione, 1,2-dihydroxy-5,8-bis[[2-[(2-hydroxyethyl)amino]-ethyl]amino]-, dihydrochloride	H_2O	245(4.56),591(4.28), 635(4.27)	18-1812-83
$C_{22}H_{28}N_4O_9S$			
Propanoic acid, 2,2-dimethyl-, [[9-(2,3,5-tri-O-acetyl-β-D-ribofurano-syl)-9H-purin-6-yl]thio]methyl ester	MeOH	276(4.3),282(4.24)	36-0372-83
$C_{22}H_{28}N_6$			
Acetaldehyde, 2,2'-(1,6-hexanediyldi-nitrilo)bis-, bis(phenylhydrazone), (Z,Z,E,E)-	$CHCl_3$	300(4.12),340(4.54)	65-1214-83

Compound	Solvent	λ_{max}(log ε)	Ref.
$C_{22}H_{28}O$			
Bicyclo[4.2.0]octa-1,3,5-trien-7-ol, 8,8-dimethyl-3,5-bis(1-methylethyl)-7-phenyl-	C_6H_{12}	267(3.15)	35-1117-83
Methanone, phenyl[2,4,6-tris(1-methylethyl)phenyl]-	C_6H_{12}	242(4.20),349(1.80)	35-1117-83
4,6,8,10-Undecatetraen-2-yn-1-al, 5,9-dimethyl-11-(2,6,6-trimethylcyclohexenyl)-	hexane	236(3.85),274(4.03), 386(4.60)	33-1148-83
$C_{22}H_{28}O_3$			
Benzoic acid, 3-methoxy-4-[2-methyl-4-(2,6,6-trimethyl-1-cyclohexen-1-yl)-1,3-butadienyl]-, (E,E)-	MeCN	328(4.40)	87-1282-83
Canrenone	n.s.g.	283(4.43)	162-0241-83
A,19-Dinorpregn-4-en-20-yn-3-one, 17-(1-oxopropoxy)-, (17α)-	EtOH	232.5(4.18)	39-2337-83C
Pregna-4,6-diene-3,20-dione, 17-hydroxy-16-methylene-	n.s.g.	283(4.51)	162-0704-83
$C_{22}H_{28}O_4$			
Estra-1,3,5(10)-triene-3,17β-diol, diacetate	MeOH	210(3.92),267(2.82), 274(2.78)	145-0347-83
$C_{22}H_{28}O_4S_2$			
Dispiro[cyclohexane-1,2'-[1,3]benzodioxole-5'(4'H),2"-[1,3]dithian]-7'-ol, tetrahydro-, benzoate, (3'aR-cis)	EtOH	208s(3.83),229(4.15), 272s(3.15),280s(3.04)	39-2441-83C
$C_{22}H_{28}O_5$			
Galdosol, dimethyl deriv.	EtOH	243(4.04),275(4.00)	102-0585-83
9(1H)-Phenanthrenone, 2,3,4,4a-tetrahydro-10-hydroxy-5,6,8-trimethoxy-4a-methyl-1-methylene-7-(1-methylethyl)-, (S)-	EtOH	249(3.99),280(3.91), 336(4.02)	23-2461-83
9(3H)-Phenanthrenone, 4,4a-dihydro-10-hydroxy-5,6,8-trimethoxy-1,4a-dimethyl-7-(1-methylethyl)-, (S)-	EtOH	249(4.09),284(3.83), 358(4.03)	23-2461-83
2-Propenoic acid, 3-(4-acetoxy-3-methoxyphenyl)-, 1,7,7-trimethylbicyclo[2.2.1]hept-2-yl ester, [1S-[1α-2β(E),4α]]-	EtOH	282(4.35),313s(4.08)	105-0149-83
$C_{22}H_{28}O_6$			
3-Furanmethanol, α,5-bis(3,4-dimethoxyphenyl)tetrahydro-4-methyl- (mangostellin A)	MeOH	206(4.34),230(4.22), 278(3.77)	102-0211-83
Gymnocolin	MeOH	210(3.54)	88-0115-83
Isoquassin	n.s.g.	258(4.10)	162-0750-83
Phenol, 4-[2-[2,6-dimethoxy-4-(2-propenyl)phenoxy]-1-methoxypropyl]-2-methoxy-, (R*,R*)-	EtOH	271(3.50)	150-2625-83M
Phenol, 4-[2-[2,6-dimethoxy-4-(1-propenyl)phenoxy]propyl]-2,6-dimethoxy-, (E)-	EtOH	267(4.12)	150-2625-83M
Piperenone	MeOH	222(4.31),266(4.06)	102-0763-83
$C_{22}H_{28}O_8$			
Lanugon J	EtOH	235s(4.00),244(4.02)	33-0429-83
photoproduct 6a	EtOH	234(4.01)	33-0429-83
photoproduct 7a	EtOH	235(4.02)	33-0429-83

Compound	Solvent	λ max (log ε)	Ref.
$C_{22}H_{29}BrO_2$			
Benzoic acid, 4-bromo-, 1-methyl-1-[2-(1,2,6-trimethyl-2-cyclohexen-1-yl)ethyl]-2-propenyl ester	EtOH	206(4.20),244.5(4.26)	18-1125-83
$C_{22}H_{29}ClO_5$			
Alclometasone	MeOH	242(4.19)	162-0034-83
$C_{22}H_{29}FO_4$			
Pregna-1,4-diene-3,20-dione, 6-fluoro-11,21-dihydroxy-16-methyl-	MeOH	242(4.21)	162-0594-83
Pregna-1,4-diene-3,20-dione, 9-fluoro-11,21-dihydroxy-16-methyl-	n.s.g.	238(4.20)	162-0423-83
$C_{22}H_{29}FO_5$			
Pregna-1,4-diene-3,20-dione, 9-fluoro-11,17,21-trihydroxy-16-methyl- (betamethasone)	MeOH	238(4.18)	162-0168-83
21-phosphate, disodium salt	EtOH	239(4.15)	162-0425-83
$C_{22}H_{29}NO$			
2,5-Cyclohexadien-1-one, 4-[[3,5-bis(1,1-dimethylethyl)phenyl]imino]-2,6-dimethyl-	hexane	270(4.30),291(4.22), 451(3.56)	18-1476-83
Tricyclo[4.4.0.0^{2,4}]dec-6-en-8-one, 4a-(1-methylethyl)-1β-methyl-7a-[2-(6-methyl-2-pyridinyl)ethyl]-	MeOH	253(3.94),273(3.69)	33-1820-83
$C_{22}H_{29}NO_2$			
Phenol, 2-[[[3,5-bis(1,1-dimethylethyl)-4-methoxyphenyl]imino]methyl]-	EtOH	212(4.78),240(4.60), 250(4.23),345(4.36), 438(2.60)	65-0359-83
2-Pyridinecarboxylic acid, 5-[2-methyl-4-(2,6,6-trimethyl-1-cyclohexen-1-yl)-1,3-butadienyl]-, ethyl ester, (E,E)-	EtOH	266(4.08),334(4.35)	87-1282-83
$C_{22}H_{29}NO_3S$			
9H-Carbazole-9-decanesulfonic acid, sodium salt	H_2O	331(--),347.5(2.95)	77-0099-83
9H-Carbazole-3-sulfonic acid, 9-decyl-, sodium salt	H_2O	331(--),347.5(3.39)	77-0099-83
$C_{22}H_{29}NO_4$			
Acetamide, N-(17-acetoxy-3-hydroxyestra-1,3,5(10)-trien-6-yl)-	MeOH	210(4.15),218(3.92), 280(3.28)	145-0347-83
Isoquinoline, 1,2,3,4-tetrahydro-5,6,7-trimethoxy-1-[1-(3-methoxyphenyl)-1-methylethyl]-	MeOH	228(4.20),275(3.56), 283(3.54)	142-0451-83
Phenol, 2-tert-butyl-6-(3-tert-butyl-2-hydroxy-5-nitrobenzyl)-4-methyl-	neutral	335(3.94)	126-2361-83
	monoanion	440(4.37)	126-2361-83
	NaOH	447(4.38)	126-2361-83
$C_{22}H_{29}NO_5$			
Benzeneacetamide, 3-methoxy-α,α-dimethyl-N-[2-(2,3,4-trimethoxyphenyl)ethyl]-	MeOH	238(5.25),280(5.30)	142-0451-83
$C_{22}H_{29}NO_6$			
Indol-4-one, 1-benzyl-4,5,6,7-tetrahydro-6,6-dimethyl-2-(D-galacto-penti-	isoPrOH	251(4.03),282(3.87)	136-0255-83E

Compound	Solvent	$\lambda_{max}(\log \epsilon)$	Ref.
itol-1-yl)- (cont.)			136-0255-83E
$C_{22}H_{29}NO_8$			
β-D-Glucopyranosiduronic acid, 1-[[(1-methylethyl)amino]methyl]-2-(1-naphthalenyloxy)ethyl-, sodium salt, (R)-	EtOH	212(3.63),284(2.78)	87-1687-83
(S)-	EtOH	213(3.63),284(2.78)	87-1687-83
1H-Pyrrole-2,3,4,5-tetracarboxylic acid, 3,4-dicyclohexyl 2,5-dimethyl ester	EtOH	224(4.28),267(4.19)	4-0001-83
$C_{22}H_{29}N_2O$			
3H-Indolium, 2-[2-[3-[(dimethylamino)-methylene]-2-methoxy-1-cyclopenten-1-yl]ethenyl]-1,3,3-trimethyl-, perchlorate	MeOH	583(4.99)	104-1854-83
$C_{22}H_{29}N_2O_4$			
Echitamine, chloride	EtOH	235(3.93),295(3.55)	162-0507-83
$C_{22}H_{29}N_3OS_2$			
Thiethylperazine sulfoxide	2N H_2SO_4	224s(3.07),260(3.25)	36-0050-83
	H_2O	223(3.06),260(3.23)	36-0050-83
$C_{22}H_{29}N_3S_2$			
Thiethylperazine	2N H_2SO_4	267(3.25)	36-0050-83
	H_2O	267(3.25)	36-0050-83
radical cation	9N H_2SO_4	205(2.95),241(2.98), 267(3.12),296(3.29)	36-0050-83
$C_{22}H_{29}N_5O_7$			
1H-Pyrazolo[3,4-d]pyrimidin-4-amine, N-cyclohexyl-1-(2,3,5-tri-O-acetyl-β-D-ribofuranosyl)-	MeOH	226s(--),263(4.94), 284(5.12)	87-1601-83
$C_{22}H_{29}N_5O_9S$			
Propanoic acid, 2,2-dimethyl-, [[2-amino-9-(2,3,5-tri-O-acetyl-β-D-ribofuranosyl)-9H-purin-6-yl]thio]-methyl ester	MeOH	312(4.0)	36-0372-83
$C_{22}H_{30}$			
Benzene, [2-[1,4-bis(1,1-dimethylethyl)-2,5-cyclohexadien-1-yl]ethenyl]-	MeOH	250(4.32)	157-1577-83
isomer	MeOH	286(4.38),306s(4.20)	157-1577-83
$C_{22}H_{30}N_2O$			
Cyclohexanone, 2-[7-(dimethylamino)-2,4,6-heptatrienylidene]-6-[5-(dimethylamino)-2,4-pentadienylidene]-	EtOH	295(4.12),575(4.84)	70-1882-83
Pyrrolo[2,3-b]indole, 1-acetyl-1,2,3-3a,8,8a-hexahydro-3a,8-bis(3-methyl-2-butenyl)-	EtOH	216(4.15),258(3.88), 314(3.28)	1-0803-83
	EtOH-HCl	212(4.11),254(3.78), 308(3.15)	1-0803-83
$C_{22}H_{30}N_2O_2$			
Aspidospermine	MeOH	218(4.52),255(4.04), 280-290(3.53-3.40)	162-0122-83
	EtOH	218(4.44),256(3.39), 291s(3.49)	102-1526-83

Compound	Solvent	λ_{max}(log ε)	Ref.
$C_{22}H_{30}N_2O_4$			
Anomaline	MeOH	226(4.27),259(3.85)	102-2301-83
Sitsirikine, 18,19-dihydro-10-methoxy-, (16R)-	EtOH and EtOH-base	227(4.78),283(3.98), 295(3.85),307(3.54)	100-0708-83
	EtOH-HCl	225(4.82),274(3.92), 295(3.81),308(3.57)	100-0708-83
$C_{22}H_{30}N_3$			
1H-Benzimidazolium, 1-methyl-2-(4-methyl-2-pyridinyl)-3-octyl-, iodide	EtOH	282(4.19)	4-0023-83
$C_{22}H_{30}N_4O_2$			
3-Pyridinecarbonitrile, 1,1'-(1,6-hexanediyl)bis[6-ethoxy-1,6-dihydro-	CH_2Cl_2	313(4.1),360(3.5)	64-0878-83B
$C_{22}H_{30}N_4O_2S_2$			
Thioproperazine (flavianate)	benzene	390(3.95)	106-0203-83
	MeOH	390(3.93),440(3.94)	106-0203-83
	$CHCl_3$	365(4.30)	106-0203-83
picrate	benzene	340(4.20),406(3.86)	106-0203-83
$C_{22}H_{30}N_4O_{11}S$			
Uridine, 2'-deoxy-5-ethyl-, 3',3"-sulfite	H_2O	267(4.29)	83-0667-83
$C_{22}H_{30}O$			
3'H-Cycloprop[16,17]estra-4,9,16-trien-3-one, 16,17-dihydro-16-(1-methylethyl)-, (14β,16α,17α)-	EtOH	304(4.25)	33-1806-83
	EtOH	303(3.93)	33-1820-83
$C_{22}H_{30}O_2S$			
3'H-Cycloprop[1,2]androsta-1,4-dien-3-one, 7-(acetylthio)-1,2-dihydro-, (1β,2β,7α)-	EtOH	236(4.25)	95-1046-83
$C_{22}H_{30}O_2Si$			
Silane, [[(17β)-3-methoxyestra-1,3,5-7,9-pentaen-17-yl]oxy]trimethyl-	hexane	230(4.79),258(3.55), 268(3.67),278(3.71), 289(3.55),304s(2.84), 310s(3.01),317(3.14), 323(3.35),332(3.30), 338(3.51)	65-1724-83
$C_{22}H_{30}O_3$			
Androsta-5,16-dieno[17,16-b]furan-5'(2'H)-one, 3-hydroxy-4'-methyl-	EtOH	223(4.09)	13-0055-83A
Androst-5-ene-16-acetic acid, 3,17-dihydroxy-α-methylene-, γ-lactone, (3β,16β,17β)-	MeOH	211(3.99)	13-0055-83A
2,4,6,8-Nonatetraenoic acid, 4-formyl-7-methyl-9-(2,6,6-trimethyl-1-cyclohexen-1-yl)-, ethyl ester, (all-E)-	EtOH	388(4.48)	77-0077-83
$C_{22}H_{30}O_4$			
Cafestol acetate	n.s.g.	222(3.80)	162-0225-83
2H-9,4a-(Epoxymethano)phenanthren-12-one, 1,3,4,9,10,10a-hexahydro-5,6-dimethoxy-1,1-dimethyl-7-(1-methylethyl)-, [4aR-(4aα,9α,10aβ)]-	MeOH	241(3.83),273(2.88)	78-3603-83
9(1H)-Phenanthrenone, 2,3,4,4a-tetrahydro-10-hydroxy-5,6-dimethoxy-	EtOH	243(3.90),279(3.98), 322(4.02)	107-0201-83

Compound	Solvent	λ_{max}(log ϵ)	Ref.
1,1,4a-trimethyl-7-(1-methylethyl)-, (S)- (cont.)			107-0201-83
Retinoic acid, 12-carboxy-, 15-methyl ester, (11-cis,13-cis)-	MeOH	327(4.52)	44-0222-83
$C_{22}H_{30}O_6$			
Royleanone, 20-hydroxy-7α-acetoxy-	MeOH	280(4.08),407(2.78)	102-1296-83
$C_{22}H_{30}O_8$			
Eucannabinolide, 11,13-dihydro-	EtOH	211(3.58)	102-0197-83
$C_{22}H_{30}O_9$			
2-Butenoic acid, 4-hydroxy-2-(hydroxymethyl)-, 3-acetoxy-1a,2,3,5a,7,8,8a-9,10,10a-decahydro-4,8,10a-trimethyl-7-oxooxireno[5,6]cyclodeca[1,2-b]-furan-9-yl ester	EtOH	222(3.85)	102-0197-83
Phlebotrichin	EtOH	285(2.7)	102-0223-83
$C_{22}H_{31}BrO_2$			
Benzoic acid, 4-bromo-, 3,4-dimethyl-2-methylene-3-(3-methylpentyl)cyclohexyl ester	MeOH	205.5(4.23),244(4.24)	18-1125-83
$C_{22}H_{31}Cl_2NO_2$			
Estra-1,3,5(10)-triene-3,17β-diol, 6α-bis(2-chloroethyl)amino-, hydrochloride	MeOH	210(4.23),225(4.04), 278(3.24),285(3.21)	145-0347-83
$C_{22}H_{31}N$			
Pyrrolidine, 1-[1,1-dimethyl-2-[1-methyl-7-(1-methylethyl)-4-azulenyl]-ethyl]-	EtOH	218s(4.00),246(4.31), 275s(4.37),281s(4.47), 285(4.50),289(4.48), 304s(3.92),568s(2.49), 592(2.57),611(2.61), 634s(2.58),661(2.55), 729(2.11)	18-2311-83
hydrochloride	EtOH	216(3.21),245(3.69), 284(4.03),290(4.02), 304s(2.54),338s(3.38), 352(3.46),368(3.38), 608(2.59),660(2.49), 730(2.02)	18-2311-83
$C_{22}H_{31}NO$			
Butanamide, N-[2-(3,8-dimethyl-5-(1-methylethyl)-1-azulenyl]propyl]-	EtOH	202(4.31),218(4.14), 247(4.22),287(4.40), 305(4.04),353(3.55), 369(3.47),574s(2.40), 617(2.45),670s(2.35), 752s(1.88)	18-2059-83
Cycloprop[a]inden-4(1H)-one, octahydro-1b-methyl-6a-(1-methylethyl)-5-[2-(6-methyl-2-pyridinyl)ethyl]-, [1aS-(1aα,1bβ,5α,5aβ,6aα)]-	EtOH	262(3.68),273(3.56)	33-1820-83
$C_{22}H_{31}NO_3$			
4H-1-Benzopyran-4-one, 2,3-dihydro-5-hydroxy-2,2-dimethyl-7-pentyl-3-(1-piperidinylmethylene)-	EtOH	250s(3.68),288(3.78), 317(3.76),391(4.26)	161-0775-83

Compound	Solvent	λ max (log ε)	Ref.
Retinamide, N-(carboxymethyl)-	EtOH	347(4.69)	34-0422-83
$C_{22}H_{31}NO_5S$			
Latrunculin A	MeOH	218(4.37),268s(--)	44-3512-83
	MeOH	218(4.37)	162-0774-83
$C_{22}H_{31}N_{11}O_6$			
DL-Alanine, N-[N^2-[3-(6-amino-9H-purin-9-yl)-DL-alanyl]-L-lysyl]-3-(3,4-dihydro-2,4-dioxo-1(2H)-pyrimidinyl)-, methyl ester, dihydrobromide	pH 1	262(4.27)	103-0328-83
$C_{22}H_{32}N_2O_2$			
2H-Pyrido[3,4-b]indole-2-carboxylic acid, 1,3,4,9-tetrahydro-9-(3-methylbutyl)-1-(2-methylpropyl)-, methyl ester	EtOH	230(4.57),278s(3.84), 285(3.88),294s(3.81)	88-2171-83
$C_{22}H_{32}N_2O_5$			
2,2'-Dipyrrolylmethane-5'-carboxylic acid, 5-methoxy-3-(methoxycarbonylmethyl)-3',4,4,4'-tetramethyl-, 1,1-dimethylethyl ester	$CHCl_3$	280(4.22),342s(3.56)	49-0983-83
$C_{22}H_{32}O$			
3'H-Cycloprop[16,17]estra-4,16-dien-3-one, 16,17-dihydro-16-(1-methylethyl)-, (14β,16α,17α)-	EtOH	239(4.04)	33-1806-83
$C_{22}H_{32}O_2$			
Androsta-3,5-diene-6-carboxaldehyde, 3-ethoxy-	EtOH	220(4.01),323(4.19)	95-1046-83
Tricyclo[4.4.0.0^{2,4}]decan-3-one, 4a-(1-methylethyl)-1β-methyl-7a-[2-(3-oxo-1-cyclohexenyl)ethyl]-	EtOH	235(4.20)	33-1820-83
$C_{22}H_{32}O_2S$			
Androst-4-en-3-one, 7α-(acetylthio)-6α-methyl-	EtOH	238(4.27)	95-1046-83
$C_{22}H_{32}O_3$			
Acetylcoriacenone	EtOH	229(3.81)	44-1937-83
Estr-4-en-3-one, 17β-acetoxy-16β-ethyl-	EtOH	240(4.23)	95-1042-83
Medroxyprogesterone	EtOH	241(4.20)	162-0825-83
Methenolone 17-acetate (primonabol)	MeOH	240(4.13)	162-0856-83
18-Norpregn-4-en-3-one, 13β-acetyl-20β-hydroxy-	MeOH	241(4.16)	44-2696-83
18-Norpregn-16-en-20-one, 3-acetyloxy-, (3β,5α)-	EtOH	239(3.8)	44-1954-83
9(1H)-Phenanthrenone, 2,3,4,4a,10,10a-hexahydro-5,6-dimethoxy-1,1,4a-trimethyl-7-(1-methylethyl)-	EtOH	273(4.37)	107-0201-83
$C_{22}H_{32}O_4$			
Acetic acid, (3β,17β-dihydroxyandrost-5-en-16-ylidene)-, methyl ester	EtOH	226(4.12)	13-0055-83A
Androst-4-en-3-one, 17β-acetoxy-6β,19-epoxy-16β-ethyl-	EtOH	239(4.24)	95-1042-83
Crenulide, acetoxy-	MeOH	227(4.06)	44-1903-83
Crenulide, 9-acetoxy-	EtOH	208(3.71),227(3.79)	44-1906-83

Compound	Solvent	$\lambda_{max}(\log \epsilon)$	Ref.
Petiodial	hexane	247(3.87)	31-1275-83
Pregn-4-ene-3,20-dione, 17-hydroxy-16α-methoxy-	n.s.g.	234(4.19)	162-0037-83
$C_{22}H_{32}O_5$			
Ajugarin V	EtOH	210(4.04)	138-0223-83
Crenulide, 9-acetoxy-1-hydroxy-	EtOH	208(3.82),225(3.81)	44-1906-83
	EtOH-base	265(3.81)	44-1906-83
11H-Cyclodec[e]inden-11-one, 3,9-di-acetoxy-1,2,3,3a,4,5,5a,8,9,10,12-13,13a,13b-tetradecahydro-3a-methyl-, (E)-	MeOH	212(3.45)	78-3609-83
Hypoestoxide	EtOH	235(3.93)	142-2125-83
Isolathyrol, 5-O-acetyl-	MeOH	193(4.09),283(4.05)	102-1791-83
$C_{22}H_{32}O_5S$			
2(1H)-Naphthalenone, octahydro-7-[1-hydroxy-1-methyl-2-[[(4-methylphen-yl)sulfonyl]oxy]ethyl]-1,4a-dimethyl-	MeOH	225(4.13),257(2.76), 262(2.83),267(2.78), 273(2.73),285(1.48)	23-1111-83
$C_{22}H_{32}O_7$			
Ingol, 12-O-acetyl-	MeOH	210(4.08)	102-2795-83
$C_{22}H_{33}NO$			
Retinamide, N-ethyl-	EtOH	347(4.70)	34-0422-83
13-cis	EtOH	242(3.97),349(4.66)	34-0422-83
$C_{22}H_{33}NO_2$			
Retinamide, N-(2-hydroxyethyl)-	EtOH	347(4.71)	34-0422-83
13-cis	EtOH	242(3.98),350(4.65)	34-0422-83
$C_{22}H_{33}NO_6$			
Ophiocarpine, O-acetyl-, (±)-	MeOH	232.5(4.25),283(3.73), 291.5(3.78)	94-0947-83
$C_{22}H_{33}NO_8$			
7-Azabicyclo[2.2.1]hept-2-ene-1,2,3,4-tetracarboxylic acid, 1,4-dimethyl 2,3-dipentyl ester	CCl_4	262(3.42)	4-0001-83
$C_{22}H_{33}O_5P$			
Phosphonic acid, 7,8,9,10-tetrahydro-1-hydroxy-3,6,6,9-tetramethyl-6H-dibenzo[b,d]pyran-2-yl)-, diethyl ester	$CHCl_3$	278(4.19)	83-0326-83
$C_{22}H_{34}N_2O_2$			
Hexanamide, N-[3-[2-[2-(2-hydroxy-6-methykphenyl)ethenyl]-1-pyrrolidin-yl]propyl]-	MeOH	217(4.24),255(3.92), 297(3.38)	100-0235-83
2(1H)-Pyridinone, 1-hexyl-, dimer	EtOH	end absorption	94-2552-83
$C_{22}H_{34}N_2O_3$			
Peripentadenine	MeOH	228(3.96),253(3.36), 283(3.15),309(3.15)	100-0235-83
	MeOH-base	210(4.08),234(3.99)	100-0235-83
$C_{22}H_{34}N_2O_3S$			
Benzenesulfonic acid, 4-methyl-, [octa-hydro-7-(1-hydroxy-1-methylethyl)-1,4a-dimethyl-2(1H)-naphthalen-	MeOH	226(4.15),260(3.35), 273(3.05)	23-1111-83

Compound	Solvent	$\lambda_{max}(\log \epsilon)$	Ref.
ylidene]hydrazide (cont.)			23-1111-83
$C_{22}H_{34}OSi$			
Silane, (3-methoxyestra-1,3,5(10)-trien-2-yl)trimethyl-	EtOH	283(3.49),293(3.37)	65-0539-83
$C_{22}H_{34}O_2$			
D-Homo-18-norandrost-16-en-3-ol, 17-methyl-, acetate, (3β,5α)-(80%)	C_6H_{12}	188(1.84)	44-1954-83
6,7-Methanoheneicosa-8,10,12,15-tetraenoic acid, (8E,10E,12Z,15Z)-trans-	EtOH	273(4.62),283(4.70), 294(4.60)	87-0072-83
5β-Pregn-20-ene-20-carboxaldehyde, 3β-hydroxy-	EtOH	223(3.85)	44-4258-83
$C_{22}H_{34}O_2Si$			
Silane, [[(17β)-3-methoxyestra-1,3,5-(10)-trien-17-yl]oxy]trimethyl-	hexane	278(3.30),281.5(3.29), 287.5(3.34)	65-0539-83
$C_{22}H_{34}O_3$			
Cembra-1,3,7-triene, 14-acetoxy-11,12-epoxy-, (1E,3E,7E,14S)-	n.s.g.	250(4.37)	12-2289-83
$C_{22}H_{34}O_4$			
3,5-Cyclohexadiene-1,4-dione, 2-hydroxy-5-methoxy-3-(10-pentadecenyl)-, (Z)- (maesanin 1)	EtOH	289(4.48),425(2.80)	88-3825-83
1,4-Cyclononadiene-1-carboxaldehyde, 9-(acetoxymethyl)-8-(1,5-dimethyl-4-hexenyl)-7-hydroxy-5-methyl-	EtOH	244(3.40),270s(3.19)	102-2539-83
$C_{22}H_{34}O_6$			
Lapiferin	EtOH	218(3.96)	105-0281-83
Prostacyclin, 5-oxo-, methyl ester, (E)-	EtOH	266(4.210)	88-1281-83
(Z)-	EtOH	262(4.198)	88-1281-83
$C_{22}H_{34}O_{10}S_2$			
L-xylo-DL-erythro-α-D-xylo-Octodialdo-1,4:12,9-difuranose-5,8-diulose, 3,10-di-O-methyl-6,7-di-S-methyl-1,2:11,12-bis-O-(1-methylethylidene)-6,7-dithio-, (R)-	EtOH	211(3.79)	159-0139-83
$C_{22}H_{36}ClNO_2$			
Hexadecanamide, N-(2-chlorophenyl)-N-hydroxy-	n.s.g.	258(4.11)	42-0686-83
Hexadecanamide, N-(3-chlorophenyl)-N-hydroxy-	n.s.g.	260(3.95)	42-0686-83
$C_{22}H_{36}N_2O_3$			
Hexanamide, N-[3-[2-[2-hydroxy-2-(2-hydroxy-6-methylphenyl)ethyl]-1-pyrrolidinyl]propyl]-	MeOH	217(4.57),276(4.17)	100-0235-83
isomer 4B of dihydroperipentadenine (above compound)	MeOH	208(4.22),218s(4.06), 275(3.53)	100-0235-83
$C_{22}H_{36}N_2O_5S$			
4(1H)-Pyrimidinone, 2,3-dihydro-5-[2,3-O-(1-methylethylidene)-1,4-di-C-pentyl-β-DL-ribofuranosyl]-2-thioxo-	pH 13	227(3.98),261(4.03), 305(3.78)	18-2680-83
	MeOH	215(4.11),276(4.22), 290s(4.18)	18-2680-83

Compound	Solvent	λ_{max}(log ϵ)	Ref.
$C_{22}H_{36}N_2O_6$			
2,4(1H,3H)-Pyrimidinedione, 5-[2,3-O-(1-methylethylidene)-1,4-di-C-pentyl-β-DL-ribofuranosyl]-	pH 13	289(3.63)	18-2680-83
	MeOH	265(3.77)	18-2680-83
$C_{22}H_{36}N_2O_8$			
Propanedioic acid, 2,2'-[1,2-ethanediylbis(iminomethylidyne)]bis-, tetrapropyl ester	MeCN	222(4.35),273(4.42), 292(4.41)	97-0064-83
$C_{22}H_{36}O_3$			
Pectinatone methyl ether	n.s.g.	313(3.84)	88-3055-83
isomer	n.s.g.	261(3.88)	88-3055-83
$C_{22}H_{36}O_4$			
Benzenemethanol, 3-hydroxy-5-methoxy-α-undecyl-, α-acetate, (R)-	EtOH	274(4.15),281(4.12)	102-2031-83
$C_{22}H_{36}O_5$			
PGD_2, 16,16-dimethyl-	EtOH	220(3.08),280s(2.48)	87-0790-83
$C_{22}H_{37}N_3O_5$			
4(1H)-Pyrimidinone, 2-amino-5-[2,3-O-(1-methylethylidene)-1,4-di-C-pentyl-β-DL-ribofuranosyl]-	pH 13	231(3.74),279(3.60)	18-2680-83
	MeOH	225(3.88),292(3.76)	18-2680-83
$C_{22}H_{38}N_6O_7Si_2$			
9H-Purin-6-amine, 9-[2-)-nitro-3,5-O-[1,1,3,3-tetrakis(1-methylethyl)-1,3-disiloxanediyl]-β-D-arabinofuranosyl]-	EtOH	259(4.13)	88-3183-83
$C_{22}H_{38}O_4$			
Citrullol	n.s.g.	242(2.85),272(2.68), 282(2.68)	162-0332-83
Homoprostanoic acid	EtOH	244(4.08)	2-0319-83
$C_{22}H_{38}O_4S$			
Cyclohexanone, 3,3'-thiobis[5-(1-methoxy-1-methylethyl)-2-methyl-	MeOH	234(2.68),282(2.28)	49-0195-83
$C_{22}H_{39}N_5O_5Si_2$			
9H-Purin-6-amine, 9-[3,5-O-[1,1,3,3-tetrakis(1-methylethyl)-1,3-disiloxanediyl]-β-D-arabinofuranosyl]-	pH 1	257(4.17)	35-4059-83
	pH 13	259(4.18)	35-4059-83
	EtOH	259(4.18)	88-3183-83
$C_{22}H_{39}N_5O_6Si_2$			
Guanosine, 3',5'-O-[1,1,3,3-tetrakis(1-methylethyl)-1,3-disiloxanediyl]-	MeOH	256(4.16)	35-4059-83
$C_{22}H_{40}N_2$			
1,4-Benzenediamine, N-hexadecyl-	MeOH	248(4.07),314(3.30)	5-0802-83

Compound	Solvent	$\lambda_{max}(\log \epsilon)$	Ref.
$C_{23}H_6Cl_{14}NO_3$			
Methyl, [4-[[(1-carboxyethyl)amino]-carbonyl]-2,3,5,6-tetrachlorophenyl]bis(pentachlorophenyl)-	$CHCl_3$	288(3.82),338s(3.82), 368s(4.31),382(4.60), 482s(3.07),508(3.08), 560(3.07)	44-3716-83
$C_{23}H_{13}NO_2$			
Naphtho[2,1-g]quinoline-7,12-dione, 5-phenyl-	$CHCl_3$	245(4.15),310(4.16), 372(3.12),426(3.32)	142-1031-83
$C_{23}H_{14}ClNO_3$			
2H-Furo[2,3-h]-1-benzopyran-2-one, 3-chloro-4-(diphenylamino)-	EtOH	249.5(4.69),266s(4.50), 280s(4.39),315s(4.20), 382(4.07)	111-0113-83
$C_{23}H_{14}N_6O_{11}$			
Isoquinoline, 2-(3,5-dinitrobenzoyl)-1,2-dihydro-1-[(2,4,6-trinitrophenyl)methyl]-	$CHCl_3$	248(4.90),288(3.82), 340(3.35)	56-0829-83
$C_{23}H_{14}O_2$			
1H-Cyclopenta[b]benzofuran-1-one, 2,3-diphenyl-	CH_2Cl_2	259(4.59),263(4.58), 464(3.88)	70-1049-83
$C_{23}H_{15}BrN_4$			
Benzonitrile, 4,4'-[5-(4-bromophenyl)-4,5-dihydro-1H-pyrazole-1,3-diyl]bis-	C_6H_{12}	385(4.54)	12-1649-83
	MeOH	387(4.57)	12-1649-83
$C_{23}H_{15}BrN_4S$			
Imidazo[2,1-b]thiazole, 5-(4-bromophenyl)-3-phenyl-6-(phenylazo)-	EtOH	266(4.16),365(4.12)	4-0639-83
$C_{23}H_{15}ClN_2O$			
7H-Dibenz[f,ij]isoquinolin-7-one, 6-[(4-chlorophenyl)amino]-2-methyl-	n.s.g.	476(4.33)	2-0812-83
$C_{23}H_{15}Cl_2N$			
Pyridine, 2-(2,5-dichlorophenyl)-4,6-diphenyl-	DMSO	275(4.20),296s(3.88)	150-0301-83M
$C_{23}H_{15}NO_2S$			
3H-Naphtho[2,1-b]pyran-3-one, 2-[2-(4-methylphenyl)-4-thiazolyl]-	MeCN	384(4.41)	48-0551-83
$C_{23}H_{15}NO_3$			
Anthrapyridone, N-methyl-8-phenoxy-	toluene	368(4.00)	104-1172-83
$C_{23}H_{15}NO_3S$			
1,4-Naphthalenedione, 2,3-epoxy-2,3-dihydro-2-phenyl-, benzothiazole photoadduct, endo	benzene	300(3.62),344s(2.94), 387s(2.53)	44-4968-83
	MeOH	227(4.59),297(3.63), 344s(2.94),385s(2.54)	44-4968-83
exo	benzene	298(3.56),336s(2.90)	44-4968-83
	MeOH	228(4.42),299(3.42), 330s(2.92)	44-4968-83
$C_{23}H_{15}N_3O_3$			
7H-Dibenz[f,ij]isoquinolin-7-one, 2-methyl-6-[(3-nitrophenyl)amino]-	n.s.g.	473(4.31)	2-0812-83

Compound	Solvent	λ_{max}(log ϵ)	Ref.
$C_{23}H_{15}N_3O_6$			
1,2-Diazabicyclo[5.2.0]nona-3,5-diene-6-carboxylic acid, 2-benzoyl-8-(1,3-dihydro-1,3-dioxo-2H-isoindol-2-yl)-9-oxo-, cis-(±)-	MeOH	320(4.02),305s(3.99)	5-1374-83
$C_{23}H_{15}N_5O_9$			
Isoquinoline, 1,2-dihydro-2-(4-nitrobenzoyl)-1-[(2,4,6-trinitrophenyl)-methyl]-	$CHCl_3$	248(4.52),284(4.17), 340(3.32)	56-0829-83
$C_{23}H_{16}BrO$			
Pyrylium, 2-(4-bromophenyl)-4,6-diphenyl-, perchlorate	MeCN	242(4.13),282(4.23), 359(4.52),412(4.47)	22-0115-83
$C_{23}H_{16}ClN$			
Pyridine, 4-(2-chlorophenyl)-2,6-diphenyl-	DMSO	298(3.87)	150-0301-83M
$C_{23}H_{16}ClNO_3$			
2H-Furo[2,3-h]-1-benzopyran-2-one, 3-chloro-4-(diphenylamino)-5,6-dihydro-	EtOH	233s(3.82),278(3.92), 380(4.02)	111-0113-83
1H-Naphth[2,3-b]azirine-2,7-dione, 1a-(4-chlorophenyl)-1a,7a-dihydro-1-(4-methoxyphenyl)-	benzene	360(2.85)	44-4968-83
$C_{23}H_{16}Cl_2N_2$			
1H-Pyrazole, 5-(4-chlorophenyl)-3-[2-(4-chlorophenyl)ethenyl]-1-phenyl-	n.s.g.	225(4.32),233(4.30), 297(4.54)	44-0542-83
$C_{23}H_{16}N_2$			
Benzenamine, N-11H-indeno[1,2-b]quinolin-11-ylidene-2-methyl-	EtOH	205(4.72),223(4.66), 230(4.68),246(4.58), 254(4.56),288(4.85), 420s(3.04)	104-0158-83
Benzenamine, N-11H-indeno[1,2-b]quinolin-11-ylidene-3-methyl-	EtOH	205(4.8),230(4.75), 246(4.64),254(4.63), 288(4.9),420s(3.14)	104-0158-83
Benzenamine, N-11H-indeno[1,2-b]quinolin-11-ylidene-4-methyl-	EtOH	203(4.66),213(4.69), 230(4.78),246(4.68), 252s(4.66),290(4.94), 420(3.4)	104-0158-83
Benzenamine, N-(2-methyl-11H-indeno-[1,2-b]quinolin-11-ylidene)-	EtOH	235f(4.7),295(4.8), 370s(3.6)	104-0158-83
Benzenamine, N-(3-methyl-11H-indeno-[1,2-b]quinolin-11-ylidene)-	EtOH	210(4.63),231(4.64), 246(4.6),255(4.58), 278s(4.78),292(4.85), 385s(3.42)	104-0158-83
Benzenamine, N-(6-methyl-11H-indeno-[1,2-b]quinolin-11-ylidene)-	EtOH	208s(4.78),214(4.8), 226(4.81),245(4.78), 294(4.95),340s(4.2), 420s(3.24)	104-0158-83
Benzenamine, N-(7-methyl-11H-indeno-[1,2-b]quinolin-11-ylidene)-	EtOH	204(4.64),214(4.59), 230(4.67),246(4.54), 255(4.52),293(4.78), 360s(3.88)	104-0158-83
$C_{23}H_{16}N_2O$			
Methanone, (2,4-diphenyl-5-pyrimidinyl)phenyl-	EtOH	266(4.53)	4-0649-83

Compound	Solvent	$\lambda_{max}(\log \epsilon)$	Ref.
$C_{23}H_{16}N_2OS$			
Propanedinitrile, [2-[6-(4-methoxyphenyl)-2H-thiopyran-2-ylidene]-2-phenylethylidene]-	$CHCl_3$	560(4.26)	97-0147-83
Pyrrolo[3,2-b][1,4]benzothiazin-2(1H)-one, 3-(4-methylphenyl)-1-phenyl-	MeOH	248(4.04),284(3.64), 360(3.53),417(3.51)	48-0293-83
$C_{23}H_{16}N_2OSe$			
Propanedinitrile, [2-[6-(4-methoxyphenyl)-2H-selenin-2-ylidene]-2-phenylethylidene]-	$CHCl_3$	549(4.26)	97-0147-83
$C_{23}H_{16}N_2O_2$			
[1,2'-Binaphthalene]-3'-carbonitrile, 4-(dimethylamino)-1',4'-dihydro-1',4'-dioxo-	toluene	567(3.46)	150-0168-83S
	MeOH	566(--)	150-0168-83S
$C_{23}H_{16}N_2O_3$			
6H-Anthra[1,9-cd]isoxazole-3-carboxamide, N-(2,5-dimethylphenyl)-6-oxo-	toluene	460(4.23)	103-1279-83
1H-Anthra[1,2-c]pyrazole-3,6,11(2H)-trione, 2-(2,5-dimethylphenyl)-	DMF	482(3.53)	103-1279-83
$C_{23}H_{16}N_2O_4$			
Cyclohept[1,2,3-hi]imidazo[2,1,5-cd]-indolizine-3,4-dicarboxylic acid, 2-phenyl-, dimethyl ester	EtOH	249(4.43),279s(4.52), 291(4.59),310s(4.36), 357s(4.48),373(4.56), 400s(4.14),423(4.10)	18-3703-83
$C_{23}H_{16}N_4$			
Benzonitrile, 4,4'-(4,5-dihydro-1-phenyl-1H-pyrazole-3,5-diyl)bis-	C_6H_{12}	390(4.36)	12-1649-83
	MeOH	393(4.40)	12-1649-83
Benzonitrile, 4,4'-(4,5-dihydro-3-phenyl-1H-pyrazole-1,5-diyl)bis-	C_6H_{12}	356(--)	12-1649-83
	MeOH	356(4.54)	12-1649-83
Benzonitrile, 4,4'-(4,5-dihydro-5-phenyl-1H-pyrazole-1,3-diyl)bis-	C_6H_{12}	387(4.56)	12-1649-83
	MeOH	390(4.57)	12-1649-83
$C_{23}H_{16}N_4OS_3$			
Bisthiazolo[3,4-a:5',4'-e]pyrimidin-9-ium, 8-[(3-ethyl-2(3H)-benzothiazolylidene)methyl]-1,2-dihydro-2-oxo-6-phenyl-, hydroxide, inner salt	DMF	421(4.03),438(3.97), 544(4.45)	103-0754-83
$C_{23}H_{16}N_4O_4$			
2-Naphthalenecarboxamide, 3-hydroxy-4-[(2-nitrophenyl)azo]-N-phenyl-	EtOH	205(--),217(--), 434s(--),516(--)	104-1145-83
	$CHCl_3$	287s(4.441),420s(4.079), 505(4.444),530(4.456)	104-1145-83
	CCl_4	293(4.250),420s(3.959), 500(4.312),520(4.301)	104-1145-83
	DMSO	280s(4.531),420s(4.000), 530(4.398)	104-1145-83
$C_{23}H_{16}N_4O_7$			
Isoquinoline, 2-benzoyl-1,2-dihydro-1-[(2,4,6-trinitrophenyl)methyl]-	$CHCl_3$	246(4.16),300(3.78), 392(2.90)	56-0829-83
Quinoline, 1-benzoyl-1,2-dihydro-2-[(2,4,6-trinitrophenyl)methyl]-	$CHCl_3$	248(4.37),300(3.54), 363(3.04)	56-0829-83
$C_{23}H_{16}N_4S$			
Imidazo[2,1-b]thiazole, 3,5-diphenyl-	EtOH	256s(4.17),361(4.43)	4-0639-83

Compound	Solvent	λ_{max}(log ε)	Ref.
6-(phenylazo)- (cont.)			4-0639-83
$C_{23}H_{16}O_2S_2$			
Methanone, (2-phenyl-1,3-dithiole-4,5-diyl)bis[phenyl-	dioxan	247(4.18),267(4.18), 425(3.37)	104-0750-83
$C_{23}H_{16}O_7$			
[2,2'-Binaphthalene]-1,4,5',8'-tetrone, 1',5-dihydroxy-7'-methoxy-3',7-dimethyl-	MeOH	249(4.37),291(4.19), 428(3.95)	102-1832-83
	MeOH-NaOH	228(--),280(--), 545(--)	102-1832-83
$C_{23}H_{16}O_8$			
Benzoic acid, 2-[(2-acetoxybenzoyl)-oxy]-, 2-carboxyphenyl ester	EtOH	227(4.48),277(3.65)	36-0322-83
$C_{23}H_{17}Cl_2N_5S$			
4H-[1,2,4]Triazolo[4,3-a][1,4]benzodiazepine, 8-chloro-6-(2-chlorophenyl)-1-[[(3-pyridinylmethyl)thio]methyl]-	MeOH	217s(4.57),252s(4.07)	73-0123-83
$C_{23}H_{17}Cl_3N_2O_2S$			
9H-Thioxanthen-9-one, 7-chloro-3-(2,4-dichlorophenyl)-1-(4-morpholinyl-amino)-	MeOH	232(4.44),272(4.56), 343(3.98),443(3.87)	4-1575-83
$C_{23}H_{17}CoN_4O_2$			
Cobalt(1+), (6,12-dimethyl-18H-7,11-nitrilodibenzo[b,g][1,5,9]triaza-cyclohexadecine-18,20(19H)-dionato-N^5,N^{13},N^{19},N^{21})-, (SP-4-1)-, perchlorate	EtOH-DMF	345(3.78)	12-2387-83
$C_{23}H_{17}CuN_4O_2$			
Copper(1+), (6,12-dimethyl-18H-7,11-nitrilodibenzo[b,g][1,5,9]triaza-cyclohexadecine-18,20(19H)-dionato-N^5,N^{13},N^{19},N^{21})-, (SP-4-1)-, perchlorate	EtOH-DMF	340(3.83)	12-2387-83
$C_{23}H_{17}N$			
Pyridine, 2,4,6-triphenyl-	neutral	311(3.90)	39-0045-83B
	protonated	307(4.38)	39-0045-83B
$C_{23}H_{17}NO_3$			
1H-Naphth[2,3-b]azirine-2,7-dione, 1a,7a-dihydro-1-(4-methoxyphenyl)-1a-phenyl-	benzene	295s(3.60),311s(3.48), 360s(2.78)	44-4968-83
	MeCN	233(4.88),290s(3.56), 310s(3.42),360s(2.78)	44-4968-83
	EPA	310s(3.49),356s(2.83)	44-4968-83
$C_{23}H_{17}NO_4$			
Acetamide, N-(9,10-dihydro-9,10-dioxo-5-phenoxy-1-anthracenyl)-N-methyl-	toluene	366(4.01)	104-1172-83
$C_{23}H_{17}N_3O$			
2H-Pyrrolo[2,3-b]quinoxalin-2-one, 1,4-dihydro-3-(4-methylphenyl)-1-phenyl-	MeOH	242(4.53),280(4.38), 334(3.89),417(3.51)	48-0293-83
$C_{23}H_{17}N_3O_2$			
7H-Benzo[e]perimidin-7-one, 6-[(4-meth-	n.s.g.	488(4.29)	2-0812-83

Compound	Solvent	λ_{max}(log ε)	Ref.
oxyphenyl)amino]-2-methyl- (cont.)			2-0812-83
2-Naphthalenecarboxamide, 3-hydroxy-N-phenyl-4-(phenylazo)-	EtOH	328s(3.935),410(3.949),505(4.354),530(4.365)	104-1145-83
	$CHCl_3$	290s(4.435),340s(4.041),410(4.086),510(4.494),530(4.511)	104-1145-83
	CCl_4	294s(4.265),336s(3.903),426s(3.964),496(4.364),521(4.373)	104-1145-83
	DMSO	340s(3.975),425s(3.903),520(4.328)	104-1145-83
$C_{23}H_{17}N_3O_4$			
Benzamide, N-[2,3-dihydro-3-(4-nitrophenyl)methylene]-2-oxo-1H-indol-4-yl]-4-methyl-	$CHCl_3$	375(3.86)	44-2468-83
1,2-Diazabicyclo[5.2.0]nona-3,5-dien-9-one, 2-benzoyl-8-(1,3-dihydro-1,3-dioxo-2H-isoindol-2-yl)-5-methyl-	MeOH	285(4.08)	5-1374-83
1,2-Diazabicyclo[5.2.0]nona-3,5-dien-9-one, 2-benzoyl-5-methyl-8-[(3-oxo-1(3H)-isobenzofuranylidene)amino]-, cis-(±)-	MeOH	282(4.04)	5-1374-83
2,3,9b-Triazaindeno[6,7,1-ija]azulene-8,9-dicarboxylic acid, 1-(4-methylphenyl)-, dimethyl ester	EtOH	240(4.33),270s(4.30),291(4.35),385(4.29),417(4.14),438(4.09)	18-3703-83
$C_{23}H_{17}N_3O_5$			
Isoquinoline, 2-benzoyl-1-[(2,4-dinitrophenyl)methyl]-1,2-dihydro-	$CHCl_3$	246(4.00),300(3.80)	56-0829-83
$C_{23}H_{17}N_3S$			
Benzaldehyde, (4,6-diphenyl-2H-1,3-thiazin-2-ylidene)hydrazone	MeCN	228(4.38),310(4.78),460(3.69)	118-0827-83
monoperchlorate	MeCN	312(4.64),364s(4.08),445(3.99)	118-0827-83
$C_{23}H_{17}N_4NiO_2$			
Nickel(1+), [6,12-dimethyl-18H-7,11-nitrilodibenzo[b,g][1,5,9]triazacyclohexadecine-18,20(19H)-dionato-N^5,N^{13},N^{19},N^{21}[-, (SP-4-1)-, perchlorate	EtOH-DMF	280(4.07),325(3.97),345(3.88)	12-2387-83
$C_{23}H_{17}N_4OS_3$			
Bisthiazolo[3,4-a:5',4'-e]pyrimidin-9-ium, 8-[(3-ethyl-2(3H)-benzothiazolylidene)methyl]-1,2-dihydro-2-oxo-6-phenyl-, perchlorate	HCOOH	426(4.02),442(4.17),546(4.24)	103-0754-83
$C_{23}H_{17}O$			
Pyrylium, 2,4,6-triphenyl-, perchlorate	MeCN	215(4.28),275(4.29),365(4.55),405(4.43)	22-0115-83
$C_{23}H_{17}P$			
Phosphorin, 2,4,6-triphenyl-	EtOH	228(4.32),278(4.61),314(4.10)	24-0445-83
$C_{23}H_{18}$			
Naphthalene, 1-phenyl-4-(phenylmethyl)-	EtOH	231(4.55),299(3.93)	2-0542-83

Compound	Solvent	λ_{max}(log ε)	Ref.
$C_{23}H_{18}AsCl$			
1H-Arsole, 1-(chlorophenylmethyl)-2,5-diphenyl-	C_6H_{12}	213(4.33),230(4.31), 379(4.04)	101-0335-83I
$C_{23}H_{18}Cl_2N_2$			
1H-Pyrazole, 5-(4-chlorophenyl)-3-[2-(4-chlorophenyl)ethenyl]-4,5-dihydro-1-phenyl-	n.s.g.	263(4.28),384(4.55)	44-0542-83
$C_{23}H_{18}N_2$			
1H-Pyrazole, 1,5-diphenyl-3-(2-phenylethenyl)-	n.s.g.	224(4.32),231(4.29), 298(4.54)	44-0542-83
$C_{23}H_{18}N_2O$			
Ethanone, 2-(1-methyl-1H-indol-3-yl)-1-phenyl-2-(phenylimino)-	MeOH	249(4.41),313(4.28)	73-1854-83
2(1H)-Pyrimidinone, 6-(4-methylphenyl)-1,4-diphenyl-	MeOH	212(4.33),282(4.20), 339(4.11)	39-1773-83C
$C_{23}H_{18}N_2O_2$			
Ethanone, 2-(1-methyl-1H-indol-3-yl)-1-phenyl-2-(phenylimino)-, N-oxide	MeOH	265(4.40),338(4.28)	73-1854-83
Methanone, (3,8-dihydro-8-methyl-2-phenyl-2H-isoxazolo[5,4-b]indol-3-yl)phenyl-	MeOH	289(4.20)	73-1854-83
2(1H)-Pyrimidinone, 1-(3-methoxyphenyl)-4,6-diphenyl-	MeOH	211(4.42),277(4.26), 337(4.07)	39-1773-83C
$C_{23}H_{18}N_2O_2S$			
Carbamic acid, [4-(diphenylmethylene)-3-methyl-1,3-thiazetidin-2-ylidene]-, phenyl ester	CH_2Cl_2	237(3.96),316(4.15)	24-1297-83
$C_{23}H_{18}N_2O_5$			
1-Phenazinecarboxylic acid, 6-[1-[(2-hydroxy-6-methylbenzoyl)oxy]ethyl]-	EtOH	207(4.71),253(4.90), 320s(3.62),354s(4.04), 367(4.19),400s(3.68)	158-83-142
	EtOH-HCl	207(4.72),253(4.90), 320s(3.60),354s(4.04), 369(4.21),400s(3.70)	158-83-142
	EtOH-base	254(4.98),300(4.18), 348(4.24),364(4.34), 400s(4.01)	158-83-142
Saphenamycin	MeOH-HCl	252(4.87),369(4.19)	158-83-140
	MeOH-NaOH	255(5.03),365(4.16)	158-83-140
$C_{23}H_{18}N_4$			
Benzaldehyde, (4,6-diphenyl-2-pyrimidinyl)hydrazone	EtOH	270(4.56),315(4.76)	65-0153-83
$C_{23}H_{18}O_2$			
2,4-Pentadien-1-one, 1-(4-phenoxyphenyl)-5-phenyl-	dioxan	234(4.29),253(3.90), 290(4.08),340(4.64)	124-1211-83
$C_{23}H_{19}As$			
1H-Arsole, 2,5-diphenyl-1-(phenylmethyl)-	C_6H_{12}	225(4.39),373(4.23)	101-0335-83I
$C_{23}H_{19}F_3O_8$			
Acetic acid, trifluoro-, 1,2,3,4,6,11-hexahydro-2,5,7,12-tetrahydroxy-6,11-	MeOH	222s(--),233(4.43), 252(4.33),291(3.84),	5-1818-83

Compound	Solvent	λ_{max}(log ε)	Ref.
dioxo-2-propyl-1-naphthacenyl ester, trans-(±)-		464(3.95),479(4.00), 490(4.06),509(3.93), 524(3.91)	5-1818-83
$C_{23}H_{19}Fe$			
Methylium, [1,1'-biphenyl]-4-ylferrocenyl-	50% H_2SO_4	255(4.03),285(3.98), 333(4.05),408(3.60)	32-0721-83
$C_{23}H_{19}NO$			
2,4-Pentadien-1-one, 5-phenyl-1-[4-(phenylamino)phenyl]-, (E,E)-	dioxan	265(3.94),286(4.10), 328(4.43),379(4.58)	124-1211-83
$C_{23}H_{19}NO_2$			
4H-1-Benzopyran-4-one, 2-[4-(dimethylamino)phenyl]-6-phenyl-	MeOH	257(4.58),333(4.05), 400(4.60)	18-2037-83
$C_{23}H_{19}NO_2S$			
Benzenepropanamide, β-oxo-N,N-bis(phenylmethyl)-α-thioxo-	MeCN	267(4.32),323(3.69), 407(2.99)	5-1694-83
$C_{23}H_{19}NO_3$			
Benzenepropanamide, α,β-dioxo-N,N-bis(phenylmethyl)-	MeCN	263(4.09),418(1.81)	5-1694-83
1H-Indene-1,3(2H)-dione, 2-[3-(diethylamino)-4-oxo-1(4H)-naphthalenylidene]-	$CHCl_3$	270(4.38),340(4.10), 400(2.86),630(4.20)	88-3567-83
$C_{23}H_{19}NO_4$			
2H-Pyran-2-one, 3-[3-(9-ethyl-9H-carbazol-3-yl)-1-oxo-2-propenyl]-4-hydroxy-6-methyl-	MeOH	204(4.4),239(4.5), 293(4.3),325(4.2), 425(4.5)	83-0951-83
$C_{23}H_{19}NO_6$			
Spiro[cyclopentane-1,5'(6'H)-[4H]pyrido[3,2,1-jk]carbazole]-1',3'-dicarboxylic acid, 4',6'-dioxo-, dimethyl ester	MeOH	219s(4.47),234.5(4.51), 274(4.27),399(4.06)	39-0515-83C
$C_{23}H_{19}NO_{10}$			
Tuberosinone N-β-D-glucoside	EtOH	227(4.59),240(4.63), 285(4.08),330(4.09), 370(3.72),388(3.72), 477(4.14)	100-0761-83
$C_{23}H_{19}N_2OS$			
Thiazolium, 4-(acetylamino)-2,3,5-triphenyl-, perchlorate	MeCN	245s(4.11),319(4.15)	118-0582-83
$C_{23}H_{19}N_3O$			
Methanone, phenyl(5-phenyl-3-isoxazolyl)-, methylphenylhydrazone, (E)-	MeOH	252(4.47),340(4.06)	4-0931-83
(Z)-	MeOH	256(4.49),350(4.00)	4-0931-83
$C_{23}H_{19}N_3O_2$			
1H-Benzo[a]pyrrolo[3,2-c]phenazine-3-carboxylic acid, 1,2-dimethyl-, ethyl ester	EtOH	210(4.49),264(4.69), 310(4.49),385(3.96), 430(3.90)	103-1086-83
$C_{23}H_{19}N_3O_2S$			
Benzonitrile, 4-[4,5-dihydro-1-[4-(methylsulfonyl)phenyl]-5-phenyl-1H-pyra-	C_6H_{12}	385(--)	12-1649-83
	MeOH	385(4.52)	12-1649-83

Compound	Solvent	$\lambda_{max}(\log \epsilon)$	Ref.
zol-3-yl]- (cont.)			12-1649-83
$C_{23}H_{19}N_3O_2S_2$			
Benzenesulfonic acid, 4-methyl-, (4,6-diphenyl-2H-1,3-thiazin-2-ylidene)-hydrazide, monoperchlorate	MeCN	221(4.54),289(4.46), 410(3.79)	118-0827-83
$C_{23}H_{19}N_3O_5$			
Benzoic acid, 2-[[(2-benzoyl-5-methyl-9-oxo-1,2-diazabicyclo[5.2.0]nona-3,5-dien-8-yl)amino]carbonyl]-, cis-(±)-	MeOH	284(3.98)	5-1374-83
2-Propanone, 1-(9,10-dihydro-10-methyl-9-acridinyl)-1-(2,4-dinitrophenyl)-	MeCN	213(4.57),525(4.70), 260(4.76),356(2.94)	103-0773-83
$C_{23}H_{19}N_3S_3$			
Benzothiazole, 2-[[3-phenyl-3-(4,5,6,7-tetrahydro-1,3-benzodithiol-2-ylidene)-1-propenyl]azo]-	MeCN	595(4.46)	118-0840-83
$C_{23}H_{20}BN_5O_6$			
3,1,2,4,5-Boratetrazine, 3,3-diacetoxy-4-(4-nitrophenyl)-2,6-diphenyl-	hexane	254(4.11),268(3.75), 300(3.97),325(4.17), 382(3.34),560(4.31)	65-2061-83
	EtOH	251(3.86),320(3.92), 386(3.18),567(4.03)	65-2061-83
3,1,2,4,5-Boratetrazine, 3,3-diacetoxy-6-(4-nitrophenyl)-2,4-diphenyl-	hexane	236(4.22),294(3.95), 329(4.20),345(4.23), 385(3.15),533(4.19)	65-2061-83
	EtOH	240(4.24),296(3.58), 349(4.33),536(4.22)	65-2061-83
$C_{23}H_{20}Cl_2O_{10}$			
Benzoic acid, 5-acetoxy-2-(2,6-diacetoxy-3,5-dichloro-4-methylbenzoyl)-3-methoxy-, methyl ester (dihydrogeodin triacetate)	EtOH	250(3.73),309(3.40)	94-4543-83
$C_{23}H_{20}FeO$			
Ferrocene, ([1,1'-biphenyl]-4-yl-hydroxymethyl)-	EtOH	439(2.04)	32-0721-83
$C_{23}H_{20}NO_4$			
1-Benzopyrylium, 2-(7-methoxy-2-oxo-2H-1-benzopyran-3-yl)-7-pyrrolidino-	MeCN	569(4.50)	48-0505-83
1-Benzopyrylium, 2-(8-methoxy-2-oxo-2H-1-benzopyran-3-yl)-7-pyrrolidino-	MeCN	565(4.35)	48-0505-83
1-Benzopyrylium, 7-methoxy-2-(2-oxo-7-pyrrolidino-2H-1-benzopyran-3-yl)-, perchlorate	MeCN	622(4.92)	48-0505-83
1-Benzopyrylium, 8-methoxy-2-(2-oxo-7-pyrrolidino-2H-1-benzopyran-3-yl)-, perchlorate	MeCN	613(4.52)	48-0505-83
$C_{23}H_{20}N_2$			
1H-1,4-Diazepine, 2,3-dihydro-2,3,6-triphenyl-, cis	MeCN	323(3.716)	5-1207-83
1H-Pyrazole, 3,5-bis(4-methylphenyl)-1-phenyl-	n.s.g.	254(4.26)	44-0542-83
1H-Pyrazole, 4,5-dihydro-1,5-diphenyl-3-(2-phenylethenyl)-	n.s.g.	260(4.18),379(4.51)	44-0542-83

Compound	Solvent	$\lambda_{max}(\log \epsilon)$	Ref
$C_{23}H_{20}N_2O$			
Ethanone, 2-(1-methyl-1H-indol-3-yl)-1-phenyl-2-(phenylamino)-	MeOH	248(4.50),308(4.25)	73-1854-83
$C_{23}H_{20}N_2OS_2$			
Diazene, benzoyl[3-phenyl-3-(4,5,6,7-tetrahydro-1,3-benzodithiol-2-ylidene)-1-propenyl]-	MeCN	561(3.58)	118-0840-83
$C_{23}H_{20}N_2O_2$			
Benzenemethanol, 3-ethenyl-α-[(3-ethenylphenyl)azo]-α-phenyl-, acetate	CH_2Cl_2	398(2.41)	126-2285-83
Benzenemethanol, 4-ethenyl-α-phenyl-α-(phenylazo)-, acetate	CH_2Cl_2	400(2.39)	126-2285-83
[1,2'-Binaphthalene]-1',4'-dione,	C_6H_{12}	460(3.40)	150-0168-83S
4-(dimethylamino)-3'-(methylamino)-	EtOH	472(--)	150-0168-83S
1H-Pyrazole, 3,5-bis(4-methoxyphenyl)-1-phenyl-	n.s.g.	264(4.58)	44-0542-83
$C_{23}H_{20}N_2O_2S$			
Benzenamine, 4-methoxy-N-[3-(4-methoxyphenyl)-4-phenyl-2(3H)-thiazolylidene]-	EtOH	290(4.23)	56-0875-83
$C_{23}H_{20}N_2S$			
Benzenamine, 4-methyl-N-[3-(4-methylphenyl)-4-phenyl-2(3H)-thiazolylidene]-	EtOH	300(4.24)	56-0875-83
$C_{23}H_{20}O_2$			
2-Propen-1-one, 3-(4-methoxyphenyl)-1-[4-(phenylmethyl)phenyl]-	isooctane	229(4.19),267(4.02), 330(4.43)	124-1211-83
	dioxan	341(4.38)	124-1211-83
$C_{23}H_{20}O_3$			
Epoxynaphthalenedione deriv. 5e	benzene	298(3.15),342(2.78)	44-4968-83
	MeOH	228(4.30),245s(4.00), 294(3.08),334(2.74)	44-4968-83
	MeCN	228(4.30),245s(4.00), 293(3.08),335(2.74)	44-4968-83
$C_{23}H_{20}O_5$			
10H-Dipyrano[3,2-a:2',3'-i]xanthen-14(3H)-one, 13-hydroxy-3,3,10,10-tetramethyl- (cudraxanthone A)	EtOH	267s(4.56),275(4.63), 283(4.62),336s(3.85), 354s(4.02),360(4.04), 405(3.61)	142-0213-83
	$EtOH-AlCl_3$	271s(4.42),290(4.65), 297(4.66),345(4.02), 385(3.99),465(3.53)	142-0213-83
$C_{23}H_{20}O_6$			
Acetic acid, [(2-benzoyl-7,8,9,10-tetrahydro-4-methyl-6-oxo-6H-dibenzo[b,d]-pyran-3-yl)oxy]-	MeOH	265(4.38)	78-1265-83
3,5-Cyclohexadiene-1,2-diol, 3-[(benzoyloxy)methyl]-, 2-acetate, 1-benzoate, (1R-trans)-	EtOH	264(3.85)	88-2019-83
$C_{23}H_{20}O_7$			
Benzoic acid, 2-(2,3-dihydro-2,6-dihydroxy-7-methoxynaphtho[1,8-bc]-	MeOH	228(4.76),260(4.38), 308(4.07),336(4.01),	23-0323-83

Compound	Solvent	λmax(log ε)	Ref.
pyran-9-yl)-5-ethenyl-3-methoxy- (cont.)		353(4.11)	23-0323-83
7-Oxabicyclo[4.1.0]hept-4-ene-2,3-diol, 1-[(benzoyloxy)methyl]-, 2-acetate 3-benzoate (tingtanoxide)	EtOH	273(3.28),280(3.15)	88-2019-83
$C_{23}H_{20}S$			
10,9-(Epithiomethano)anthracene, 9,10-dihydro-9,10-dimethyl-12-phenyl-	MeCN	280(3.46)	78-1487-83
$C_{23}H_{20}S_2$			
Benzene, 1-[4,4-bis(phenylthio)-1,3-butadienyl]-4-methyl-, (E)-	EtOH	346(4.77)	118-0383-83
Benzene, 1,1'-[(2-methyl-4-phenyl-1,3-butadienylidene)bis(thio)]bis-, (E)-	EtOH	235(4.57),312(4.60), 346(4.74)	118-0383-83
$C_{23}H_{21}BN_4O_4$			
Boron, bis(acetato-O)[1,1'-(phenylmethylene)bis[2-phenyldiazenato](1-)-N^2,N^2]-, (T-4)-	EtOH	253(4.22),310(4.22), 375(3.36),556(4.26)	65-2061-83
	hexane	254(4.22),273(3.87), 304(4.10),350(3.53), 542(4.21)	65-2061-83
$C_{23}H_{21}N$			
Benz[c]acridine, 5,6,8,9,10,11-hexahydro-7-phenyl-	neutral	313(4.31)	39-0045-83B
	protonated	339(4.33)	39-0045-83B
$C_{23}H_{21}NO_2$			
4H-1-Benzopyran-4-one, 2-[4-(dimethylamino)phenyl]-2,3-dihydro-6-phenyl-	MeOH	265(4.64),346s(--)	18-2037-83
2-Propen-1-one, 3-[4-(dimethylamino)-phenyl]-1-(4-hydroxy[1,1'-biphenyl]-3-yl)-	MeOH	260(4.52),437(4.40)	18-2037-83
2-Propen-1-one, 3-[4-(dimethylamino)-phenyl]-1-(4-phenoxyphenyl)-	isooctane	253(4.00),272(3.48), 288(4.08),322(3.56), 385(4.54)	124-1211-83
	dioxan	402(4.41)	124-1211-83
2-Propen-1-one, 3-(4-methoxyphenyl)-1-[4-(methylphenylamino)phenyl]-, (E)-	EtOH	333(4.23),391(4.51)	124-1211-83
	dioxan	244(4.20),274(3.73), 317(4.08),363(4.35)	124-1211-83
$C_{23}H_{21}NO_6$			
4,7-Epoxy-1,2-benzisoxazole-3a,7a-dicarboxylic acid, 2,3,4,7-tetrahydro-2,3-diphenyl-, dimethyl ester, (3α,3aα,4β,7β,7aα)-	MeOH	253(3.73)	73-1048-83
4,7-Epoxy-1,2-benzisoxazole-5,6-dicarboxylic acid, 2,3,3a,4,7,7a-hexahydro-2,3-diphenyl-, dimethyl ester, (3α,3aβ,4α,7α,7aβ)-	MeOH	251(4.17)	73-1048-83
$C_{23}H_{21}N_2$			
Quinolinium, 1-methyl-2-[3-(1-methyl-2(1H)-quinolinylidene)-1-propenyl]-, salt with 4-methylbenzenesulfonic acid	MeOH	604(5.27)	104-2114-83
	CH_2Cl_2	610(--)	104-2114-83
$C_{23}H_{21}N_3O_2$			
6H-Anthra[1,9-cd]isoxazol-6-one, 5-[[5-[(1,1-dimethylethyl)imino]-1,3-pentadienyl]amino]-	dioxan	555(4.28)	103-0377-83

Compound	Solvent	λ_{max}(log ε)	Ref.
$C_{23}H_{21}N_3O_4S$			
Benzenesulfonic acid, 4-methyl-, [1-(4-nitrophenyl)-4-phenyl-3-butenylidene]hydrazide	EtOH	265(4.17)	35-0933-83
$C_{23}H_{21}N_5$			
1H-Pyrazol-5-amine, 3-methyl-N-(4-methylphenyl)-1-phenyl-4-(phenylazo)-	EtOH	240(4.01),275(3.84), 328(3.89),385(3.93)	4-1501-83
1H-Pyrazol-5-amine, 3-methyl-1-phenyl-4-(phenylazo)-N-(phenylmethyl)-	EtOH	260(3.51),280(3.16), 378(3.21)	4-1501-83
$C_{23}H_{21}N_6$			
1H-Benzimidazolium, 2-[1,3-dicyano-2-(1,3-dihydro-1,3-dimethyl-2H-benzimidazol-2-ylidene)-1-propenyl]-1,3-dimethyl-, methyl sulfate	EtOH	406(4.83)	124-0297-83
$C_{23}H_{21}N_7O_3$			
2H-Benzimidazol-2-one, 1,3-dihydro-1-(phenylmethyl)-, [1-[(2-hydroxy-4-nitrophenyl)azo]propylidene]hydrazone	EtOH	545(4.40)	65-1033-83
	EtOH-base	640(4.50)	65-1033-83
	acetone	545(4.66)	65-1033-83
$C_{23}H_{21}O_4$			
Ethylium, 1,1,2-tris(4-methoxyphenyl)-2-oxo-, (OC-6-11)-, hexafluoroantimonate	1,2-$C_2H_4Cl_2$	542(4.93)	77-0007-83
$C_{23}H_{22}$			
[2_5](1,2,3,4,5)Cyclophane, 4-methyl-(superphane)	MeCN	211(4.63)	24-1682-83
$C_{23}H_{22}BrN_2$			
3H-Indolium, 3-[2-bromo-3-(1,2-dimethyl-1H-indol-3-yl)-2-propenylidene]-1,2-dimethyl-, chloride	$MeNO_2$-HOAc	568(5.16)	103-1306-83
$C_{23}H_{22}ClN_2$			
3H-Indolium, 3-[2-chloro-3-(1,2-dimethyl-1H-indol-3-yl)-2-propenylidene]-1,2-dimethyl-, chloride	$MeNO_2$-HOAc	592(5.06)	103-1306-83
$C_{23}H_{22}Cl_2N_4S$			
4H-[1,2,4]Triazolo[4,3-a][1,4]benzodiazepine, 8-chloro-6-(2-chlorophenyl)-1-[(cyclohexylthio)methyl]-	MeOH	225(4.54),251s(4.07)	73-0123-83
$C_{23}H_{22}CoN_4$			
Cobalt, [4,18-dimethyl-11,12-dihydro-10H-dibenzo[h,m]pyrido[2,1,6-cd]-[1,4,7,11]tetraazacyclotetradecinato]-, diperchlorate	EtOH-DMF	265(4.18),318(4.11)	12-2387-83
$C_{23}H_{22}CuN_4$			
Copper, [4,18-dimethyl-11,12-dihydro-10H-dibenzo[h,m]pyrido[2,1,6-cd]-[1,4,7,11]tetraazacyclotetradecinato]-, diperchlorate	EtOH-DMF	320(4.04)	12-2387-83
$C_{23}H_{22}N_2$			
Cyclohexanamine, N-(2-methyl-11H-ind-	EtOH	240(4.7),292(4.8)	104-0158-83

Compound	Solvent	λ_{max}(log ε)	Ref.
eno[1,2-b]quinolin-11-ylidene)- (cont.)			104-0158-83
Cyclohexanamine, N-(3-methyl-11H-ind-eno[1,2-b]quinolin-11-ylidene)-	EtOH	207(4.6),214s(4.62), 231(4.7),246(4.65), 254(4.64),284(4.92), 254(4.64),284(4.92), 340s(4.1),374s(3.73)	104-0158-83
Cyclohexanamine, N-(6-methyl-11H-ind-eno[1,2-b]quinolin-11-ylidene)-	EtOH	214s(4.4),226(4.44), 232(4.44),245(4.44), 253(4.37),290(4.72), 336s(3.7)	104-0158-83
Cyclohexanamine, N-(7-methyl-11H-ind-eno[1,2-b]quinolin-11-ylidene)-	EtOH	210s(4.6),224(4.76), 232(4.79),246(4.66), 254(4.66),288(4.96), 360s(3.96)	104-0158-83
1H-Pyrazole, 4,5-dihydro-3,5-bis(4-methylphenyl)-1-phenyl-	n.s.g.	242(4.24),354(4.29)	44-0542-83
1H-Pyrazole, 1-(4-ethylphenyl)-4,5-di-hydro-3,5-diphenyl-	n.s.g.	245(4.23),254s(--), 313s(--),363(4.23)	103-0229-83
$C_{23}H_{22}N_2O$			
2-Propen-1-one, 3-[4-(dimethylamino)-phenyl]-1-[4-(phenylamino)phenyl]-	EtOH	287(4.04),313(3.56), 370(4.21),431(4.67)	124-1211-83
	dioxan	283(3.90),308(3.60), 351(4.22),403(4.66)	124-1211-83
$C_{23}H_{22}N_2OS$			
Morpholine, 4-[3-(1-naphthalenylamino)-2-phenyl-1-thioxo-2-propenyl]-	MeCN	253s(4.24),293(4.29), 364(4.26),407s(3.78)	48-0689-83
$C_{23}H_{22}N_2O_2$			
Cinnoline, 2-benzoyl-1,2,3,4-tetra-hydro-4-(4-methoxyphenyl)-1-methyl-	$CHCl_3$	248(4.13)	97-0028-83
1H-Pyrazole, 4,5-dihydro-3,5-bis(4-methoxyphenyl)-1-phenyl-	n.s.g.	227(4.28),247(4.24), 352(4.32)	44-0542-83
1H-Pyrazole-4-carboxylic acid, 1-phen-yl-5-(2-phenylethenyl)-3-(1-propen-yl) , ethyl ester, (E)-	EtOH	265(4.21),280(4.18)	118-0948-83
Pyrazolo[1,5-a]quinoline-3-carboxylic acid, 4,5-dihydro-5-phenyl-2-(1-pro-penyl)-, ethyl ester, (E)-	EtOH	270(4.13),290(4.17)	118-0948-83
$C_{23}H_{22}N_2O_2S$			
Benzenesulfonic acid, 4-methyl-, (1,4-diphenyl-3-butenylidene)hydrazide, (E)-	MeOH	255(4.48)	35-0933-83
(Z)-	MeOH	250(4.36)	35-0933-83
2H-1-Benzopyran-2-one, 7-(diethylami-no)-3-[2-(4-methylphenyl)-4-thiazo-lyl]-	MeCN	416(4.37)	48-0551-83
$C_{23}H_{22}N_2O_4S_2$			
1H-Pyrazole, 4,5-dihydro-3,5-bis[(4-methylsulfonyl)phenyl]-1-phenyl-	MeOH	388(4.42)	12-1649-83
$C_{23}H_{22}N_2O_7$			
9H-Pyrido[1,2,3-lm]pyrrolo[2,3-d]carb-azole-1(12H)-propanoic acid, 2,3,11a-11b,13,13a-hexahydro-11-hydroxy-13a-methyl-β,2,9,12-tetraoxo-, ethyl ester	MeOH	252(4.57),280s(4.23), 290(4.17),307(3.86)	5-1744-83

Compound	Solvent	λ_{max}(log ϵ)	Ref
$C_{23}H_{22}N_2S_2$			
Diazene, (4-methylphenyl)[3-phenyl-3-(4,5,6,7-tetrahydro-1,3-benzodithiol-2-ylidene)-1-propenyl]-	MeCN	523(4.26)	118-0840-83
$C_{23}H_{22}N_4Ni$			
Nickel, [4,18-dimethyl-11,12-dihydro-10H-dibenzo[h,m]pyrido[2,1,6-cd]-[1,4,7,11]tetraazacyclotetradecinato]-, diperchlorate	EtOH-DMF	270(4.54),315(4.39), 360(4.03)	12-2387-83
	DMF	610(2.80),680(2.85)	12-2387-83
$C_{23}H_{22}N_4O_2$			
21H-Biline-1,19-dione, 22,24-dihydro-2,3,17,18-tetramethyl-	CH_2Cl_2	359(4.46),601(4.28)	54-0347-83
[4]Metacyclo[3](6,10)isoalloxazinophane	$CHCl_3$	271(4.402),365(3.892), 447(3.892)	88-1925-83
$C_{23}H_{22}N_6$			
Formazan, 3-methyl-1-(4-methylphenyl)-5-[1-(phenylmethyl)-1H-benzimidazol-2-yl]-	benzene	447(4.53)	65-2332-83
anion	EtOH	532(4.67)	65-2332-83
$C_{23}H_{22}O$			
2,5-Cyclohexadien-1-one, 4-(diphenylmethylene)-2,3,5,6-tetramethyl-	isooctane	401(4.10)	73-2825-83
$C_{23}H_{22}O_4$			
11H-9,10-endo-Cyclopropanthracene-11,12(13H)-dicarboxylic acid, 9,10-dihydro-2,3-dimethyl-, dimethyl ester	C_6H_{12}	276(3.46),282(3.52)	78-1151-83
exo	C_6H_{12}	276(3.35),281(3.44)	78-1151-83
10,9-Propenoanthracene-11,13-dicarboxylic acid, 9,10-dihydro-2,3-dimethyl-, dimethyl ester	C_6H_{12}	241(3.43),250(3.56), 263(3.39),270(3.39)	78-1151-83
5a,1-Propeno-5aH-cyclopropa[k]fluorene-1,11(9bH)-dicarboxylic acid, 3,4-dimethyl-, dimethyl ester	C_6H_{12}	274(3.55),281(3.55)	78-1151-83
$C_{23}H_{22}O_4S_2$			
6,10-Dithiaspiro[4.5]dec-1-ene-1-methanol, 3-(benzoyloxy)-, benzoate, (R)-	EtOH	207(4.05),229(4.36), 272(3.32),279(3.20)	39-2441-83C
$C_{23}H_{22}O_6$			
γ-Butyrolactone, α-(2,4:3,5-di-O-benzylidene-D-ribosylidene)-	$CHCl_3$	244(3.41)	87-0030-83
Cudraxanthone B	EtOH	247(4.13),268(4.17), 288s(3.53),328s(4.06), 334(4.07),383(3.51), 392s(3.49)	142-0213-83
	EtOH-$AlCl_3$	223(4.13),248(4.07), 276(4.11),290s(3.79), 352s(4.11),356(4.13), 428(3.53)	142-0213-83
Naphthaceno[1,2-d]-1,3-dioxole-7,12-dione, 3a-ethyl-3a,4,5,13b-tetrahydro-6,13-dihydroxy-2,2-dimethyl-, cis	MeOH	230(4.19),234(4.21), 251(4.17),285(3.65), 292(3.65),317(3.33), 472(3.49),482(3.51), 506(3.37)	5-1818-83

Compound	Solvent	$\lambda_{max}(\log \epsilon)$	Ref.
$C_{23}H_{22}O_7$			
2-Anthraceneacetic acid, 9,10-dihydro-4-hydroxy-5-methoxy-9,10-dioxo-3-(3-oxopentyl)-, methyl ester	MeOH	228(4.56),258(4.40), 281s(--),405s(--), 417(4.04),427s(--)	5-2151-83
Benzoic acid, 2-(2,3-dihydro-2,6-dihydroxy-7-methoxynaphtho[1,8-bc]-pyran-9-yl)-5-ethyl-3-methoxy-	MeOH	227(4.63),248s(4.22), 294(3.82),317(3.86), 336(3.90),351(3.97)	23-0323-83
1-Naphthacenecarboxylic acid, 2-ethyl-1,2,3,4,6,11-hexahydro-2,5-dihydroxy-7-methoxy-6,11-dioxo-, methyl ester	MeOH	230(4.61),258(4.34), 402s(--),417(4.05), 435s(--)	5-2151-83
$C_{23}H_{22}O_8$			
2-Anthracenepropanoic acid, α-acetyl-9,10-dihydro-1-hydroxy-4,8-dimethoxy-9,10-dioxo-, ethyl ester	MeOH	245(4.10),277(3.78), 375(3.38),415(3.78), 440(3.87),464(3.88), 495(3.64),519(3.17)	5-1818-83
$C_{23}H_{22}O_{10}$			
Benzoic acid, 5-acetoxy-2-(2,6-diacetoxy-4-methylbenzoyl)-3-methoxy-, methyl ester (sulochrin triacetate)	EtOH	251(4.13),303(3.72)	94-4543-83
$C_{23}H_{23}AsO_3S$			
Arsonium, (3-methyl-4-sulfo-2-butenyl)-triphenyl-, hydroxide, inner salt, (Z)-	EtOH	250s(3.61),260(3.49), 268(3.48),275(3.34)	65-1125-83
$C_{23}H_{23}ClO_6$			
2(3H)-Furanone, 3-[(7-chloro-1,3-benzodioxol-5-yl)methyl]dihydro-4-hydroxy-5-[(4-methoxyphenyl)methylene]-4-(1-methylethyl)- (cyanobacterin)	MeOH	266(4.08)	44-4035-83
$C_{23}H_{23}N$			
Benzo[h]quinoline, 2-(1,1-dimethylethyl)-5,6-dihydro-4-phenyl-	neutral	312(4.08)	39-0045-83B
	protonated	339(4.25)	39-0045-83B
$C_{23}H_{23}NO_2$			
1,8-Dioxabenzo[g]cyclooct[cd]indene-2-carbonitrile, 2,2a,3,4,5,6,7,7a,12b-12c-decahydro-2-phenyl-	EtOH	252s(2.70),258s(2.91), 269s(3.26),276(3.40), 283(3.34)	24-2383-83
1H-Indol-2-ol, 3-ethoxy-2,3-dihydro-1-methyl-2,3-diphenyl-, trans	MeOH	254(3.97),318(3.45)	24-2115-83
1H-Indol-3-ol, 2-ethoxy-2,3-dihydro-1-methyl-2,3-diphenyl-, trans	EtOH	247(4.01),310(3.42)	24-2115-83
	acid	240(3.93),322(4.03)	24-2115-83
Oxiranecarbonitrile, 3-[2-(2-cyclooct-en-1-yloxy)phenyl]-2-phenyl-, trans	EtOH	265s(3.38),277s(3.63), 283(3.65)	24-2383-83
$C_{23}H_{23}NO_4$			
1H-Pyrrole-3-acetic acid, 4,5-dimethyl-2-[(phenylmethoxy)carbonyl]-, phenylmethyl ester	$CHCl_3$	285(4.23)	78-1849-83
$C_{23}H_{23}NO_5$			
Phenol, 2-[(2-hydroxy-3,5-dimethylphenyl)methyl]-6-[(2-hydroxy-5-nitrophenyl)methyl]-4-methyl-	neutral	325(3.96)	126-2361-83
	anion	400(4.18)	126-2361-83
	dianion	419(4.31)	126-2361-83
Phenol, 2-[(4-hydroxy-3,5-dimethylphenyl)methyl]-6-[(2-hydroxy-5-nitrophenyl)methyl]-4-methyl-	neutral	325(3.96)	126-2361-83
	anion	404(4.24)	126-2361-83
	dianion	415(4.28)	126-2361-83

Compound	Solvent	λ_{max}(log ε)	Ref.
Phenol, 2-[(2-hydroxy-5-nitrophenyl)-methyl]-6-[(2-methoxy-4-methyl-phenyl)methyl]-	neutral	327(3.94)	126-2361-83
	anion	406(4.21)	126-2361-83
	dianion	416(4.26)	126-2361-83
$C_{23}H_{23}NO_6$			
4,7-Epoxy-1,2-benzisoxazole-5,6-dicarb-oxylic acid, octahydro-2,3-diphenyl-, dimethyl ester, (3α,3aβ,4α,7α,7aβ)-	MeOH	247(3.87)	73-1048-83
Glaucine, 3-acetoxy-	EtOH	221(4.41),281(4.07), 312(3.82)	5-0744-83
$C_{23}H_{23}NO_7$			
Duguevanine, N,O-diacetyl-	MeOH	223(4.55),242s(4.29), 282(4.37)	100-0862-83
$C_{23}H_{23}N_2$			
3H-Indolium, 3-[3-(1,2-dimethyl-1H-indol-3-yl)-2-propenylidene]-1,2-dimethyl-, chloride	$MeNO_2$-HOAc	568(5.12)	103-1306-83
$C_{23}H_{23}N_2OS$			
Benzothiazolium, 2-[[3-(diethylamino)-4-oxo-1(4H)-naphthalenylidene]meth-yl]-3-methyl-, perchlorate	$CHCl_3$	260(4.30),355(3.97), 425(3.96),625(4.41)	88-3567-83
Benzothiazolium, 2-[[4-(diethylamino)-1-oxo-2(1H)-naphthalenylidene]meth-yl]-3-methyl-, perchlorate	CH_2Cl_2	250(4.37),284(4.22), 325(4.31),620(4.33)	88-3567-83
$C_{23}H_{23}N_3O_2$			
Aszonalenin	EtOH	210(4.65),233s(4.40), 290(3.70)	158-83-17
$C_{23}H_{23}N_3O_5S$			
1-Azabicyclo[3.2.0]hept-2-ene-2-carbox-ylic acid, 6-ethyl-7-oxo-3-(2-(4-pyr-idinyl)ethylthio]-, (4-nitrophenyl)-methyl ester	THF	266(4.13),322(4.18)	158-0407-83
$C_{23}H_{23}N_3O_6S$			
4-Thia-1-azabicyclo[3.2.0]hept-5-ene-2-carboxylic acid, 3,3-dimethyl-7-oxo-6-[(phenylacetyl)amino]-, (4-nitrophenyl)methyl ester	MeOH	270(4.27)	88-3419-83
$C_{23}H_{23}N_7O_4$			
1H-Pyrazole-4-carboxylic acid, 5-[[4-[(2-cyanoethyl)ethylamino]phenyl]-azo]-3-(nitromethyl)-1-phenyl-, methyl ester	MeCN	526(4.58)	48-0041-83
$C_{23}H_{23}O_3PS$			
Phosphonium, (3-methyl-4-sulfo-2-but-enyl)triphenyl-, hydroxide, inner salt, (Z)-	EtOH	270(3.48)	65-1125-83
$C_{23}H_{24}Cl_2N_6O$			
4H-[1,2,4]Triazolo[4,3-a][1,4]benzodi-azepine, 8-chloro-6-(2-chlorophenyl)-1-[4-(2-methoxyethyl)piperazino]-	MeOH	218(4.61),245s(4.15)	73-2395-83

Compound	Solvent	λ max(log ε)	Ref.
$C_{23}H_{24}Cl_2N_6S$			
4H-[1,2,4]Triazolo[4,3-a][1,4]benzodi-azepine, 8-chloro-6-(2-chlorophenyl)-1-[4-[2-(methylthio)ethyl]pipera-zino]-	MeOH	214(4.65),244s(4.15)	73-2395-83
$C_{23}H_{24}N_2O$			
Cyclopentanone, 2-[(methylphenylamino)-methylene]-5-[3-(methylphenylamino)-2-propenylidene]-	MeOH	458(5.18)	104-1854-83
4H-Indol-4-one, 1,5,6,7-tetrahydro-6,6-dimethyl-1-(2-methylphenyl)-2-phenyl-, oxime, (Z)-	EtOH	245(4.40),278(4.30)	4-0989-83
4H-Indol-4-one, 1,5,6,7-tetrahydro-6,6-dimethyl-1-(3-methylphenyl)-2-phenyl-, oxime, (Z)-	EtOH	250(4.50),277(4.50)	4-0989-83
4H-Indol-4-one, 1,5,6,7-tetrahydro-6,6-dimethyl-1-(4-methylphenyl)-2-phenyl-, oxime, (Z)-	EtOH	250(4.50),270(4.30)	4-0989-83
Pyrrolo[3,2-c]azepin-4(1H)-one, 5,6,7,8-tetrahydro-7,7-dimethyl-1-(2-methyl-phenyl)-2-phenyl-	EtOH	227(4.50),281(4.40)	4-0989-83
Pyrrolo[3,2-c]azepin-4(1H)-one, 5,6,7,8-tetrahydro-7,7-dimethyl-1-(3-methyl-phenyl)-2-phenyl-	EtOH	228(4.40),281(4.30)	4-0989-83
Pyrrolo[3,2-c]azepin-4(1H)-one, 5,6,7,8-tetrahydro-7,7-dimethyl-1-(4-methyl-phenyl)-2-phenyl-	EtOH	227(4.20),281(4.00)	4-0989-83
$C_{23}H_{24}N_2O_2$			
4H-Indol-4-one, 1,5,6,7-tetrahydro-1-(4-methoxyphenyl)-6,6-dimethyl-2-phenyl-, oxime, (Z)-	EtOH	246(4.50),274(4.40)	4-0989-83
6,9-Methano-9H-azecino[5,4-b]indole-3(2H)-carboxylic acid, 1,4,5,6,7,8-hexahydro-, phenyl ester, (±)-	EtOH	232(4.49),282s(3.80), 286(3.82),293s(3.76)	150-1848-83M
Pyrrolo[3,2-c]azepin-4(1H)-one, 5,6,7,8-tetrahydro-1-(4-methoxyphenyl)-7,7-dimethyl-2-phenyl-	EtOH	231(4.50),282(4.30)	4-0989-83
$C_{23}H_{24}N_2O_3$			
2H-Pyran-5-carboxylic acid, 3,4-dihy-dro-2-methyl-6-(2-phenylethenyl)-4-(phenylhydrazono)-, ethyl ester, (?,E)-	EtOH	310(4.23),382(4.35)	118-0948-83
$C_{23}H_{24}N_2S$			
4H-Cyclopentapyrimidine-4-thione, 1,5,6,7-tetrahydro-1-phenyl-2-(4-phenyl-1-piperazinyl)-	MeOH	250(4.26),340(4.36)	73-3575-83
$C_{23}H_{24}N_6O_5$			
Adenosine, 3'-[5-(dimethylamino)-1-naphthalenecarboxylate]	MeOH	256(4.43)	64-0049-83C
$C_{23}H_{24}N_6O_9$			
Methanone, bis[3,5-dinitro-4-(1-piperi-dinyl)phenyl]-	dioxan	268.2(4.25),319.3(4.39), 430.3(4.15)	56-1357-83
$C_{23}H_{24}O$			
2(3H)-Naphthalenone, 4,4a,5,6,7,8-hexa-	EtOH	208(4.31),245(4.02)	2-0542-83

Compound	Solvent	λ_{max}(log ϵ)	Ref.
hydro-4-phenyl-1-(phenylmethyl)-			2-0542-83
$C_{23}H_{24}O_2$			
Endiandric acid B (5λ,4ε)	EtOH	252(3.11),258(3.08), 262(3.07),265(3.01), 269(?)	162-0515-83
Endiandric acid C	EtOH	222(4.10),228(4.05), 236(3.89),280(4.51), 288(4.53)	162-0515-83
$C_{23}H_{24}O_4$			
Benzene, 1,1'-(phenylmethylene)bis[3,4-dimethoxy-	$CHCl_3$	283(3.88),288(3.85)	78-0623-83
Cyclofenil	EtOH	247(4.23)	162-0390-83
$C_{23}H_{24}O_6$			
2-Propen-1-one, 1-(2-hydroxy-4,6-dimethoxyphenyl)-3-(5-methoxy-2,2-dimethyl-2H-1-benzopyran-6-yl)-	EtOH	296(3.32),372(3.62)	142-0661-83
	EtOH-$AlCl_3$	305(3.17),415(3.68)	142-0661-83
$C_{23}H_{24}O_{12}$			
Quercetin, 3,3'-di-O-methyl-, 4-O-β-D-glucopyranoside	EtOH	253(4.23),271(4.31), 350(4.21)	102-2881-83
	EtOH-NaOEt	280(4.48),308s(4.14), 382(4.11)	102-2881-83
	EtOH-NaOAc	279(4.46),311(4.12), 375(4.11)	102-2881-83
	+ H_3BO_3	272(4.34),318(4.13), 352(4.15)	102-2881-83
	EtOH-$AlCl_3$	262(4.21),280(4.28), 298s(4.11),352(4.20), 402(4.09)	102-2881-83
	EtOH-$AlCl_3$-HCl	259(4.21),281(4.27), 350(4.19),402(4.02)	102-2881-83
$C_{23}H_{25}ClN_6O$			
4H-[1,2,4]Triazolo[4,3-a][1,4]benzodiazepine, 8-chloro-1-[4-(2-methoxyethyl)piperazino]-6-phenyl-	MeOH	220(4.58),245(4.23)	73-2395-83
$C_{23}H_{25}NO_5$			
Benzoic acid, 2-methoxy-5-(4,5,6,7-tetrahydro-10-methoxy-5-methyl-3H-furo[4,3,2-fg][3]benzazocin-6-yl)-, methyl ester, (±)-	EtOH	236(4.11),255(3.92), 289(3.77)	44-0173-83
1,7-Etheno-4H,7H-furo[4',3',2':1,8]-naphth[1,2-d]azocine-2-carboxylic acid, 1,5,6,12-tetrahydro-9,14-dimethoxy-4-methyl-, methyl ester	EtOH	296(4.3)	44-0173-83
$C_{23}H_{25}NO_6$			
3-Butenenitrile, 4-(3,4,5-trimethoxyphenyl)-2-[(3,4,5-trimethoxyphenyl)-methylene]-	MeCN	368(4.34)	118-0917-83
Fumarophycine, O-methyl-	MeOH	208(4.79),232s(4.18), 285(3.86)	100-0433-83
$C_{23}H_{25}NO_8$			
9-Azabicyclo[4.2.1]nona-4,7-diene-1,4,7,8-tetracarboxylic acid, 4-(phenylmethyl)-, tetramethyl ester	MeCN	231(4.05)	24-0563-83

Compound	Solvent	λ_{max}(log ϵ)	Ref.
$C_{23}H_{25}N_2OS$			
Thiazolium, 2-[2-[4-(dimethylamino)-phenyl]ethenyl]-4-methyl-3-[2-(4-methylphenyl)-2-oxoethyl]-, perchlorate	EtOH	492(4.45)	65-2424-83
$C_{23}H_{25}N_2O_2S$			
Thiazolium, 2-[2-[4-(dimethylamino)-phenyl]ethenyl]-3-[2-(4-methoxy-phenyl)-2-oxoethyl]-4-methyl-, perchlorate	EtOH	491(4.49)	65-2424-83
$C_{23}H_{25}N_2S_2$			
Benzothiazolium, 3-ethyl-2-[2-[(3-ethyl-2(3H)-benzothiazolylidene)methyl]-1-butenyl]-, tetrafluoroborate	MeOH	547.5(5.12)	89-0876-83 +89-1147-83S
$C_{23}H_{25}N_3$			
Benzaldehyde, 4-(diethylamino)-, diphenylhydrazone	n.s.g.	331(4.33),363(4.52)	44-0542-83
$C_{23}H_{26}CuN_{10}$			
Copper, (12,13,14,21-tetrahydro-N,N-N',N'-tetramethyl-5H-dipyrimido[5,4-c:4',5'-j][1,5,9,13]benzotetraaza-cyclopentadecine-7,19-diaminato(2-)-N^5,N^{11},N^{15},N^{21}]-, (SP-4-2)-	DMF	366(4.07),398(4.09)	12-2413-83
$C_{23}H_{26}N_2O_4$			
Brucine	EtOH	263(4.09),301(3.93)	162-0201-83
Vomilenine, 17-acetyl-19,20-dihydro-	MeOH	220(4.88),227s(4.26), 257(3.85)	102-2297-83
$C_{23}H_{26}N_2O_6$			
Carbamic acid, [2-[(9,10-dihydro-4-hydroxy-9,10-dioxo-1-anthracenyl)amino]ethyl](2-hydroxyethyl)-, 1,1-dimethylethyl ester	$CHCl_3$	253(4.58),291(3.88), 358(4.11),608(4.07)	18-1435-83
$C_{23}H_{26}N_6O_{11}P_2$			
Adenosine, 5'-(trihydrogen diphosphate), 3'-[5-(dimethylamino)-1-naphthalene-carboxylate]	pH 7.0	252(4.14)	64-0049-83C
$C_{23}H_{26}OS_3$			
Spiro[2H-thiete-2,9'-[9H]xanthene], 3,4-bis[(1,1-dimethylethyl)thio]-	CCl_4	299(4.08),340(3.76), 400(3.30),520(1.85)	54-0083-83
$C_{23}H_{26}O_5$			
Benzoic acid, 5-ethyl-2-(2,3,3a,4,5,6-hexahydro-7-methoxynaphtho[1,8-bc]-pyran-9-yl)-3-methoxy-	MeOH	209(4.60),300(3.78)	23-0323-83
2-Propen-1-one, 1-[4-(methoxymethoxy-phenyl)-3-[2-methoxy-4-[(3-phenyl-2-butenyl)oxy]phenyl]-	MeOH	250(3.62),335(3.54)	2-0276-83
$C_{23}H_{26}S_4$			
Ethane(dithioic) acid, [(1,1-dimethyl-ethyl)thio]-9H-thioxanthen-9-ylidene-, 1,1-dimethylethyl ester	CCl_4	303s(4.04),328(3.98), 390s(2.56),400(3.51), 520(2.18)	54-0083-83

Compound	Solvent	λ_{max}(log ε)	Ref.
$C_{23}H_{27}ClO_2$			
2-Buten-1-one, 4-[2,6-bis(1,1-dimethyl-ethyl)-4H-pyran-4-ylidene]-1-(4-chlorophenyl)-	CH_2Cl_2	456(4.66)	5-1807-83
	benzene	448(--)	5-1807-83
	toluene	443(--)	5-1807-83
	MeOH	469(--)	5-1807-83
	EtOH	469(--)	5-1807-83
	ether	440(--)	5-1807-83
	HOAc	480(--)	5-1807-83
	DMF	455(--)	5-1807-83
and other solvents not listed	MeCN	453(--)	5-1807-83
$C_{23}H_{27}ClO_4$			
Delmadinone acetate	EtOH	229(4.00),258(4.00), 297(4.03)	162-0414-83
$C_{23}H_{27}ClO_6$			
Topilan	EtOH	237(4.19)	162-0304-83
$C_{23}H_{27}FO_2$			
2-Buten-1-one, 4-[2,6-bis(1,1-dimethyl-ethyl)-4H-pyran-4-ylidene]-1-(2-fluorophenyl)-	CH_2Cl_2	460(4.28)	5-1807-83
2-Buten-1-one, 4-[2,6-bis(1,1-dimethyl-ethyl)-4H-pyran-4-ylidene]-1-(4-fluorophenyl)-	CH_2Cl_2	460(4.64)	5-1807-83
$C_{23}H_{27}N$			
Benzenamine, 4-[[3,8-dimethyl-5-(1-methylethyl)-1-azulenyl]methyl]-N-methyl-	EtOH	248(4.41),283s(4.54), 290(4.60),307(4.22), 341s(3.58),354(3.73), 371(3.63),584s(2.49), 625(2.58),676s(2.49), 748s(2.05)	18-2311-83
$C_{23}H_{27}NO_4$			
2-Buten-1-one, 4-[2,6-bis(1,1-dimethyl-ethyl)-4H-pyran-4-ylidene]-1-(4-nitrophenyl)-, (E)-	CH_2Cl_2	493(4.59)	5-1807-83
4H-7a,13a-Ethenocycloocta[4,5]pyrrolo-3,2,1-ij]quinoline-14,15-dicarboxylic acid, 5,6,8,9,10,11,12,13-octahydro-, dimethyl ester	MeOH	212(4.38),244(4.16), 310(3.48),465(3.41)	39-0515-83C
$C_{23}H_{27}NO_6$			
Colchiceine ethyl ether	EtOH	243.5(4.45),351(4.21)	162-0351-83
$C_{23}H_{27}NO_{14}$			
Benzoic acid, 3-nitro-4-[[(2,3,4,6-tetra-O-acetyl-β-D-glucopyranosyl)-oxy]methyl]-, methyl ester	$CHCl_3$	244(3.94),260s(3.84)	136-0023-83M
$C_{23}H_{27}N_3O_2$			
Quinazolinoquinazoline-5,8-dione, N,N'-dibutyl-	EtOH	223(4.67),333(3.73)	142-0617-83
	50% $HClO_4$	203(4.63),229(4.65), 272(3.94),284(4.00), 296(4.00)	142-0617-83
$C_{23}H_{27}N_3O_7$			
Pregn-4-ene-3,6,11,20-tetrone, 21-acet-oxy-7α-azido-17-hydroxy-	MeOH	232(4.04)	39-2781-83C

Compound	Solvent	λ_{max}(log ε)	Ref.
$C_{23}H_{27}N_3O_{10}S$			
Acetamide, N-[1-(2,6-di-O-acetyl-5-deoxy-3-O-[(4-methylphenyl)sulfonyl]-β-D-xylofuranosyl]-1,2-dihydro-2-oxo-4-pyrimidinyl]-	EtOH	216(4.05),230(4.01), 249(3.99),298(3.65)	136-0099-83J
$C_{23}H_{28}ClNO_5$			
2H,5H-Pyrano[3,2-c][1]benzopyran-2-one, 3-chloro-10-hydroxy-5,5-dimethyl-4-(4-morpholinyl)-8-pentyl-	EtOH	225s(4.33),269(3.90), 392(4.27)	161-0775-83
$C_{23}H_{28}Cl_2O_5$			
Pregna-1,4-diene-3,20-dione, 21-acetoxy-9,11β-dichloro-17-hydroxy-	MeOH	237(4.18)	162-0443-83
$C_{23}H_{28}I_2O_4$			
Benzenepentanol, 3-iodo-α-[2-(3-iodo-4-methoxyphenyl)ethyl]-4-methoxy-, acetate	EtOH	284(4.54),292(4.53)	39-2577-83C
$C_{23}H_{28}N_2$			
1H-1,4-Diazepine, 2,3-dihydro-2,3-bis(2,4,6-trimethylphenyl)-, cis	MeCN	305(3.820)	5-1207-83
$C_{23}H_{28}N_2O$			
9-Oxaergoline, 2,3-dihydro-1-(phenylmethyl)-6-propyl-, (3β)-(±)-	EtOH	258(3.99),299(3.40)	87-0522-83
	EtOH-HCl	258(3.90),299(3.31)	87-0522-83
	EtOH-NaOH	258(3.97),298(3.40)	87-0522-83
$C_{23}H_{28}N_2O_2$			
2H-Pyrrol-2-one, 5-[[3,4-dimethyl-1-[(phenylmethoxy)methyl]-1H-pyrrol-2-yl]methylene]-4-ethyl-1,5-dihydro-1,3-dimethyl-, (Z)-	$CHCl_3$	266(4.12),375(3.91)	49-0753-83
$C_{23}H_{28}N_2O_3$			
1,6-Naphthalenediamine, 2-(3,4-dimethoxyphenyl)-N,N-diethyl-7-methoxy-	MeOH	232(4.37),257(4.49), 344s(3.44)	103-0066-83
$C_{23}H_{28}N_2O_4$			
D-Ribitol, 1,4-anhydro-1-C-(1,3-diphenyl-2-imidazolidinyl)-2,3-O-(1-methylethylidene)-, (S)-	MeOH	253(4.54),293(3.64)	23-0312-83
$C_{23}H_{28}N_2O_5$			
Reserpiline	EtOH	229(4.57),302(4.03)	162-1175-83
$C_{23}H_{28}N_2O_6$			
Isoreserpilinol	EtOH	229(4.30),270s(3.87), 302s(3.89),305(3.90), 320s(3.75),336(3.73), 400(3.45)	33-2059-83
Reserpilinol	EtOH	284(3.78),301(3.98), 306(3.98)	33-2059-83
$C_{23}H_{28}O_2$			
2-Buten-1-one, 4-[2,6-bis(1,1-dimethylethyl)-4H-pyran-4-ylidene]-1-phenyl-, (E)-	CH_2Cl_2	453(4.66)	5-1807-83
Furan, 2,2'-(phenylmethylene)bis[5-butyl-	EtOH	227(4.38)	103-0478-83

Compound	Solvent	$\lambda_{max}(\log \epsilon)$	Ref.
2,4-Pentadienoic acid, 5-phenyl-2-[2-(2,6,6-trimethyl-1-cyclohexen-1-yl)ethenyl]-, methyl ester, (E,E,E)-	hexane	238(4.12),338(4.34)	33-1148-83
$C_{23}H_{28}O_3S$			
Benzenesulfonic acid, 4-[2-(5,6,7,8-tetrahydro-5,5,8,8-tetramethyl-2-naphthalenyl)-1-propenyl]-, sodium salt	EtOH	287(4.33)	111-0425-83
$C_{23}H_{28}O_4$			
Tricyclo[12.3.1.1^{2,6}]nonadeca-1(18),2-4,6(19),14,16-hexaen-9-ol, 3,17-dimethoxy-, acetate	EtOH	259(3.35),277(3.46), 284(4.54)	39-2577-83C
$C_{23}H_{28}O_6$			
Melleolide	n.s.g.	218(4.49),267(4.22), 305(3.83)	158-83-75
	base	218(4.57),240(4.34), 313(4.38)	158-83-75
3,9-Phenanthrenedione, 4,4a-dihydro-5,6,8,10-tetramethoxy-1,4a-dimethyl-7-(1-methylethyl)-, (R)-	EtOH	213(4.52),283(4.16), 326(4.02)	23-2461-83
2-Propenoic acid, 3-[4-acetoxy-3-(4-acetoxy-3-methyl-2-butenyl)-5-(3-methyl-2-butenyl)phenyl]-	MeOH	222s(4.43),280(4.26)	94-0352-83
$C_{23}H_{28}O_6S$			
4a(2H)-Phenanthrenecarboxylic acid, 1,3,4,4b,5,6,7,8,10,10a-decahydro-4-[[(4-methylphenyl)sulfonyl]oxy]-1-oxo-, methyl ester, (4α,4aα,4bα-10aβ)-(±)-	EtOH	225(3.68)	32-0187-83
$C_{23}H_{28}O_7$			
Phenanthro[10a,1-b]oxete-8,10,11(4bH)-trione, 5,6,7,7a-tetrahydro-1,3,4-trimethoxy-4b,7a-dimethyl-2-(1-methylethyl)-, [4bR-(4bα,7aβ,9aR*)]-	EtOH	215(4.40),280(4.12), 327(3.59)	23-2461-83
$C_{23}H_{28}O_8$			
Bicyclo[3.2.1]octan-3-one, 2-acetoxy-8-hydroxy-1-methoxy-7-(7-methoxy-1,3-benzodioxol-5-yl)-6-methyl-5-(2-propenyl)-, (2-endo,6-exo,7-endo-8-anti)-	MeOH	237(3.80),275(3.30)	102-0269-83
$C_{23}H_{28}O_{10}$			
Diffutin	MeOH and MeOH-NaOAc	220s(4.30),270(3.21)	150-2601-83M
	MeOH-NaOMe	305(--)	150-2601-83M
$C_{23}H_{28}O_{12}$			
Byakangelicin, sec-O-β-D-glucopyranosyl-, (R)-	MeOH and MeOH-NaOMe	228(4.34),242(4.21), 250(4.21),273(4.29), 314(4.11)	102-2035-83
Byakangelicin, tert-O-β-D-glucopyranosyl-, (R)-	MeOH and MeOH-NaOMe	228(4.34),242(4.19), 250(4.19),272(4.30), 315(4.11)	102-2035-83
Isobyakangelicin, tert-O-β-D-glucopyranosyl-, (R)-	MeOH and MeOH-NaOMe	227(4.31),242(4.18), 251(4.15),271(4.28),	102-2035-83

Compound	Solvent	$\lambda_{max}(\log \epsilon)$	Ref.
(cont.)		310(4.05)	102-2035-83
$C_{23}H_{28}O_{13}$			
Dauroside A	EtOH	215(4.18),241s(3.65), 251(3.53),297s(3.94), 321(4.08)	105-0134-83
$C_{23}H_{29}BrO_4$			
Pregna-4,6-diene-3,20-dione, 17-acetoxy-6-bromo-	MeOH	287(4.30)	39-2793-83C
$C_{23}H_{29}ClO_4$			
Pregna-4,6-diene-3,20-dione, 17-acetoxy-6-chloro- (chlormadinone acetate)	n.s.g.	283.5(4.37),286(4.35)	162-0295-83
$C_{23}H_{29}FO_2$			
Benzoic acid, 3-fluoro-4-[2-methyl-4-(2,6,6-trimethyl-1-cyclohexen-1-yl)-1,3-butadienyl]-, ethyl ester, (E,E)-	EtOH	318(4.36)	87-1282-83
Benzoic acid, 4-[1-fluoro-2-methyl-4-(2,6,6-trimethyl-1-cyclohexen-1-yl)-1,3-butadienyl]-, ethyl ester, (E,E)-	EtOH	231(4.20),252(4.16), 323(4.11)	87-1282-83
(Z,E)-	EtOH	244(4.09),322(4.35)	87-1282-83
$C_{23}H_{29}FO_6$			
Isoflupredone 21-acetate	EtOH	240(4.21)	162-0743-83
$C_{23}H_{29}NO_3$			
Piperidine, 1-[11-(1,3-benzodioxol-5-yl)-1-oxo-2,4,10-undecatrienyl]-, (E,E,E)-	EtOH	262(4.53),309(3.84)	94-3562-83
$C_{23}H_{29}NO_4$			
Melosmine, tetrahydro-N-methyl-O,O-dimethyl-	EtOH	220(4.46),273s(4.15), 288(4.29),303s(4.16)	100-0761-83
	EtOH-HCl	221(4.46),235s(4.23), 288(4.38),303(4.21)	100-0761-83
$C_{23}H_{29}NO_6$			
Alpinine	n.s.g.	231(4.19),286(3.80)	100-0441-83
Epialpinine	MeOH	230(4.20),284(3.85)	100-0441-83
$C_{23}H_{29}N_3O_4$			
Pregna-4,6-diene-3,20-dione, 17α-acetoxy-4-azido-	MeOH	206(3.99),243s(3.73), 322(4.27)	39-2793-83C
Pregna-4,6-diene-3,20-dione, 17α-acetoxy-6-azido-	MeOH	251(4.12),298(4.10)	39-2787-83C
$C_{23}H_{29}N_3O_5$			
Pregna-1,4-diene-3,20-dione, 17-acetoxy-7α-azido-6β-hydroxy-	MeOH	243(4.20)	39-2781-83C
$C_{23}H_{29}N_3O_7$			
Pregn-4-ene-3,11,20-trione, 21-acetoxy-7α-azido-6β,17-dihydroxy-	MeOH	230(4.12)	39-2781-83C
$C_{23}H_{29}N_7O_2$			
Guanidine, [3-[3,5-dihydro-6-(1H-indol-3-yl)-5-methoxy-2-(1-methylpropyl)-3-	isoPrOH	248(3.97),254(3.98), 260(3.96),310(3.89),	88-5753-83

Compound	Solvent	λ_{max}(log ε)	Ref.
oxoimidazo[1,2-a]pyrazin-8-yl]-propyl]- (cont.)		472(4.20)	88-5753-83
$C_{23}H_{29}N_9O_4$			
1H-Pyrrole-2-carboxamide, N-[5-[[(3-amino-3-iminopropyl)amino]carbonyl]-1-ethyl-1H-pyrrol-3-yl]-4-[[[4-(formylamino)-1-methyl-1H-pyrrol-2-yl]-carbonyl]amino]-1-methyl-, monohydrobromide	EtOH	238(4.47),303(4.54)	87-1042-83
1H-Pyrrole-2-carboxamide, N-[5-[[(3-amino-3-iminopropyl)amino]carbonyl]-1-methyl-1H-pyrrol-3-yl]-4-[[1-ethyl-4-(formylamino)-1H-pyrrol-2-yl]carbonyl]amino]-1-methyl-, monohydrobromide	EtOH	239(4.44),303(4.52)	87-1042-83
1H-Pyrrole-2-carboxamide, N-[5-[[(3-amino-3-iminopropyl]amino]carbonyl]-1-methyl-1H-pyrrol-3-yl]-4-[[[4-(formylamino)-1-methyl-1H-pyrrol-2-yl]-carbonyl]amino]-1-ethyl-, monohydrobromide	EtOH	238(4.45),303(4.52)	87-1042-83
1H-Pyrrole-2-carboxamide, 4-[[[4-(formylamino)-1-methyl-1H-pyrrol-2-yl]-carbonyl]amino]-N-[5-[[[3-imino-3-(methylamino)propyl]amino]carbonyl]-1-methyl-1H-pyrrol-3-yl]-1-methyl-, monohydrobromide	EtOH	238(4.45),303(4.52)	87-1042-83
$C_{23}H_{30}BrN_3O_4$			
Pregn-4-ene-3,20-dione, 17-acetoxy-7α-azido-6β-bromo-	MeOH	245(4.07)	39-2787-83C
$C_{23}H_{30}ClNO_4$			
2H,5H-Pyrano[3,2-c][1]benzopyran-2-one, 3-chloro-4-(diethylamino)-10-hydroxy-5,5-dimethyl-8-pentyl-	EtOH	225s(4.31),269(3.90), 374(4.28)	161-0775-83
$C_{23}H_{30}ClN_3O_4$			
Pregn-4-ene-3,20-dione, 17-acetoxy-7α-azido-6β-chloro-	MeOH	237(4.16)	39-2787-83C
$C_{23}H_{30}N_2O_2S$			
Thiourea, N-(3-methoxy-17-oxoestra-1,3,5(10)-trien-2-yl)-N'-2-propenyl-	EtOH	247(4.212),291(3.892)	36-1205-83
Thiourea, N-(3-methoxy-17-oxoestra-1,3,5(10)-trien-4-yl)-N'-2-propenyl-	EtOH	251(4.253),282(3.609)	36-1205-83
$C_{23}H_{30}N_2O_3$			
Pyrifoline, (+)-	EtOH	218(4.35),260(3.93), 288s(3.50)	102-1526-83
$C_{23}H_{30}N_2O_6$			
Vindoline, 17-deacetyl-14,15-dihydro-3-oxo-	EtOH	255(3.68),308(3.55)	100-0884-83
$C_{23}H_{30}N_2O_7$			
3,6,8,12,15-Pentaoxabicyclo[15.3.1]heneicosa-1(21),17,19-trien-21-ol, 19-[(4-methoxyphenyl)azo]-	$CHCl_3$	358(4.34)	18-3253-83

Compound	Solvent	$\lambda_{max}(\log \epsilon)$	Ref.
$C_{23}H_{30}N_4O_6$			
9,10-Anthracenedione, 1-hydroxy-5,8-bis[[2-[(2-hydroxyethyl)amino]-ethyl]amino]-2-methoxy-	H_2O-HCl	227s(4.24),243(4.45), 260(4.38),589(4.15), 635(4.17)	18-1812-83
9,10-Anthracenedione, 2-hydroxy-5,8-bis[[2-[(2-hydroxyethyl)amino]-ethyl]amino]-1-methoxy-	H_2O-HCl	217(4.50),239(4.50), 271(4.50),586(4.20), 629(4.20)	18-1812-83
$C_{23}H_{30}N_6O_4$			
Pregn-4-ene-3,20-dione, 17-acetoxy-6β,7α-diazido-	MeOH	234(4.06),286(3.28)	39-2781-83C
$C_{23}H_{30}O_3$			
Furan, 2,2'-(2-furanylmethylene)bis[5-pentyl-	EtOH	227(4.25)	103-0478-83
$C_{23}H_{30}O_4$			
Benzeneacetic acid, α-[[3,5-bis(1,1-dimethylethyl)-1-methyl-4-oxo-2,5-cyclohexadien-1-yl]oxy]-	EtOH	234(4.52)	44-3696-83
Ethanone, 2-[hexahydro-5-hydroxy-4-(3-hydroxy-1-octenyl)-2H-cyclopenta[b]-furan-2-ylidene]-1-phenyl-	EtOH	254(4.109),292(4.292)	88-1281-83
2'H-18-Norandrost-4-eno[13,17-c]furan-3-one, 2'-(acetoxymethylene)-5',17-dihydro-, (2'Z,17α)-	MeOH	237(4.36)	44-2696-83
$C_{23}H_{30}O_5$			
18-Norandrost-4-ene-17-carboxaldehyde, 13-(acetoxyacetyl)-3-oxo-, (17β)-	MeOH	240(4.20)	44-2696-83
19-Norpregna-4,9-dien-3-one, 16-methyl-17,20:20,21-bis[methylenebis(oxy)]-, (16α)-	EtOH	304(4.31)	78-3083-83
9(3H)-Phenanthrenone, 4,4a-dihydro-5,6,8,10-tetramethoxy-1,4a-dimethyl-7-(1-methylethyl)-, (R)-	EtOH	250s(3.94),259s(3.89), 286(3.94),327(4.02)	23-2461-83
$C_{23}H_{30}O_6$			
Apetalic acid	EtOH	228(3.59),267(4.62), 312(4.02),363(3.22)	39-0703-83C
Benzoic acid, 2,4-dihydroxy-6-methyl-, 2,4,4a,5,6,7,7a,7b-octahydro-4-hydroxy-3-(hydroxymethyl)-6,6,7b-trimethyl-1H-cyclobut[e]inden-2-yl ester	MeOH	263(4.16),300(3.77)	158-83-2
Cortisone 21-acetate	n.s.g.	238(4.20)	162-0362-83
Fenprostalene	MeOH	220(3.99),265(3.11), 271(3.23),278(3.16)	162-0575-83
Isoapetalic acid	EtOH	228(4.01),268(4.42), 274(4.63),301(4.02), 315(4.01),371(3.62)	39-0703-83C
$C_{23}H_{30}O_7$			
1-Phenanthrenecarboxylic acid, 1,2,3,4-4a,9-hexahydro-10-hydroxy-5,6,8-trimethoxy-1,4a-dimethyl-7-(1-methylethyl)-9-oxo-	EtOH	247(3.95),278(3.92), 332(3.96)	23-2461-83
$C_{23}H_{30}O_{11}$			
Urceolatoside D	MeOH	236(3.82)	102-1977-83

Compound	Solvent	$\lambda_{max}(\log \epsilon)$	Ref.
$C_{23}H_{31}ClO_5$			
Pregn-4-ene-3,20-dione, 17-acetoxy-6β-chloro-7α-hydroxy-	MeOH	239(4.12)	39-2781-83C
$C_{23}H_{31}FO_5$			
Pregn-4-ene-3,20-dione, 17-acetoxy-9-fluoro-11-hydroxy-	MeOH	238(4.24)	162-0601-83
$C_{23}H_{31}FO_6$			
Pregn-4-ene-3,20-dione, 9-fluoro-11β,17,21-trihydroxy-, 21-acetate	EtOH	238(4.23)	162-0591-83
$C_{23}H_{31}NO_4$			
Isoquinoline, 1,2,3,4-tetrahydro-5,6,7-trimethoxy-1-[1-(3-methoxyphenyl)-1-methylethyl]-2-methyl-	MeOH	228(4.30),262(4.29), 320(2.80)	142-0451-83
$C_{23}H_{31}NO_5$			
Benzenepropanamide, β-ethyl-N,N-bis(2-hydroxyethyl)-4-methoxy-α-(4-methoxyphenyl)-, (R*,S*)-	MeOH	207(4.20),230(4.44), 273(3.77),282(3.68)	83-0271-83
$C_{23}H_{31}NO_6$			
Carbamic acid, [5-[3-(2,5-dimethyl-1-oxo-2,4-octadienyl)-4-hydroxy-2-oxo-2H-pyran-6-yl]-1-hexenyl]-, methyl ester, [R-(E,E,E)- (myxopyronin A)	MeOH	213(4.50),298(4.31)	5-1656-83
	MeOH	215(4.5),300(4.3)	158-1651-83
$C_{23}H_{31}NO_8$			
Latisoline	MeOH	275(3.69),280(3.61)	150-0238-83S
$C_{23}H_{31}N_2S$			
Benzothiazolium, 2-[[5,5-dimethyl-3-(1-piperidinyl)-2-cyclohexen-1-ylidene]methyl]-3-ethyl-, perchlorate	MeOH	492(5.09)	104-2089-83
$C_{23}H_{31}N_3O_5$			
Pregn-4-ene-3,20-dione, 17-acetoxy-7α-azido-6β-hydroxy-	MeOH	234(4.05)	39-2781-83C
$C_{23}H_{31}N_5O_6Si$			
Guanosine, N-benzoyl-2'-O-[(1,1-dimethylethyl)dimethylsilyl]-	MeOH	236(4.18),257(4.13), 264(4.13),294(4.13)	33-2069-83
Guanosine, N-benzoyl-3'-O-[(1,1-dimethylethyl)dimethylsilyl]-	MeOH	236(4.18),257(4.13), 264(4.13),298(4.13)	33-2069-83
Guanosine, N-benzoyl-5'-O-[(1,1-dimethylethyl)dimethylsilyl]-	MeOH	236(4.19),257(4.15), 264(4.15),293(4.14)	33-2069-83
$C_{23}H_{32}N_2O_2$			
Pyrrolo[2,3-b]indole-1(2H)-carboxylic acid, 3,3a,8,8a-tetrahydro-3a,8-bis(3-methyl-2-butenyl)-, ethyl ester	EtOH	212(4.20),253(3.94), 304(3.46)	1-0803-83
	EtOH-HCl	212(4.20),255(3.88), 304(3.38)	1-0803-83
$C_{23}H_{32}N_2O_3S_2$			
2H-Isoindole-2-undecanamide, N-(1,3-dithiolan-2-ylmethyl)-1,3-dihydro-1,3-dioxo-	$CHCl_3$	244(4.06),295(3.30)	78-2691-83
Spiro[[1,5]Diazacyclohexadecino[2,1-a]-isoindole-1(2H),2'-[1,3]dithiolane]-	MeOH	252s(3.68)	78-2691-83

Compound	Solvent	$\lambda_{max}(\log \epsilon)$	Ref.
4,16(3H,20bH)-dione, 5,6,7,8,9,10,11-12,13,14-decahydro-20b-hydroxy- (cont.)			78-2691-83
$C_{23}H_{32}N_2O_4$			
Anomaline, 12-O-methyl-	MeOH	221(4.29),253(3.80), 288(3.49)	102-2301-83
$C_{23}H_{32}O$			
Acetaldehyde, [2-[3-methyl-5-(2,6,6-trimethyl-1-cyclohexen-1-yl)-2,4-pentadienylidene]cyclohexylidene]-, (E,E,Z,E)-	EtOH	235(4.26),251(4.30), 288(4.28),358(4.00)	35-3588-83
(Z,E,Z,E)-	EtOH	236(4.30),300(4.43), 357(3.70)	35-3588-83
(Z,E,Z,Z)-	EtOH	233(4.28),299(4.36), 362(3.48)	35-3588-83
$C_{23}H_{32}O_2$			
4α,5α-Ethenopregnane-3,20-dione	EtOH	290(2.49)	70-0103-83
4β,5β-Ethenopregnane-3,20-dione	EtOH	290(2.03)	70-0103-83
Medrogestone	n.s.g.	288(4.40)	162-0825-83
$C_{23}H_{32}O_3$			
Pregna-5,16-dien-20-one, 3β-acetoxy-	EtOH	269(4.00)	83-0678-83
$C_{23}H_{32}O_4$			
Androsta-3,5-diene-6-carboxaldehyde, 17β-acetoxy-3-methoxy-	EtOH	220(4.04),323(4.21)	95-1046-83
Androst-4-ene-3,19-dione, 17β-acetoxy-16β-ethyl-	EtOH	247(4.24)	95-1042-83
2'H-18-Norandrost-5-eno[13,17-c]furan-3-ol, 2'-(acetoxymethylene)-5',17-dihydro-	MeOH	220(3.88)(end absorption)	44-2696-83
18-Norandrost-4-en-3-one, 13-acetyl-17β-(acetoxymethyl)-	MeOH	240(4.22)	44-2696-83
4a(2H)-Phenanthrenecarboxylic acid, 1,3,4,10a-tetrahydro-5,6-dimethoxy-1,1-dimethyl-7-(1-methylethyl)-, methyl ester, (4aR-trans)-	MeOH	240(4.15),280(3.95)	78-3603-83
Pregn-4-ene-3,20-dione, 17-acetoxy-	n.s.g.	240(4.33)	162-0706-83
Spiro[5.5]undecane-2,4-dione, 3,3'-methylenebis-	MeOH	258(4.17)	24-1309-83
$C_{23}H_{32}O_5$			
18-Norandrost-4-en-3-one, 13-(acetoxyacetyl)-17β-(hydroxymethylene)-	MeOH	240(4.24)	44-2696-83
9(1H)-Phenanthrenone, 2,3,4,4a-tetrahydro-5,6,8,10-tetramethoxy-1,4a-dimethyl-7-(1-methylethyl)-, (1S-cis)-	EtOH	247(3.91),268(3.87), 294(3.82),325(3.65)	23-2461-83
$C_{23}H_{32}O_6$			
Emblide	EtOH	284(4.13)	138-0613-83
$C_{23}H_{32}O_7$			
Trichodermadienediol A, 16-hydroxy-	MeOH	260(4.25)	158-0459-83
Trichodermadienediol B, 16-hydroxy-	MeOH	260(4.27)	158-0459-83
$C_{23}H_{33}NO_4$			
Ethanol, 2,2'-[[2,3-bis(4-methoxyphenyl)pentyl]imino]bis-, (R*,S*)-	MeOH	205(4.14),226(4.34), 275(3.52),282(3.42)	83-0271-83

Compound	Solvent	$\lambda_{max}(\log \epsilon)$	Ref.
$C_{23}H_{33}NO_5S$			
Latrunculin A methyl ether	MeOH	218(4.41),269s(--)	44-3512-83
$C_{23}H_{33}NO_6$			
U-62162	MeOH	235(4.16),276s(3.91)	158-83-68
	MeOH-H_2SO_4	232(4.17),276(3.91)	158-83-68
$C_{23}H_{33}N_3$			
1H-Benzimidazolium, 2-(1,4-dimethyl-pyridinium-2-yl)-1-methyl-3-octyl-, diiodide	EtOH	284(4.04)	4-0023-83
$C_{23}H_{33}N_3O_2$			
24-Norchola-20,22-dien-14-ol, 3-azido-21,23-epoxy-, (3β,5β,14β)-	MeOH	213(3.78)	13-0189-83B
$C_{23}H_{33}N_3O_3$			
24-Norchola-20,22-diene-12,14-diol, 3-azido-21,23-epoxy-, (3α,5β,12β-14β)-	MeOH	214.5(3.71)	13-0171-83B
(3β,5β,12β,14β)-	MeOH	215(3.68)	13-0171-83B
$C_{23}H_{33}N_5O$			
Androstan-14-ol, 3-azido-17-(4-pyridazinyl)-, (3β,5β,14β,17β)-	MeOH	218(3.79),252s(3.41)	13-0189-83B
$C_{23}H_{33}O_8P$			
Cortisone phosphate, dimethyl ester	MeOH	235(4.19)	162-0362-83
$C_{23}H_{34}N_2O_2$			
5β,14β-Androstane-3α,14-diol, 17β-(4-pyridazinyl)-	MeOH	219.5(3.78),251.5s(3.42)	13-0189-83B
$C_{23}H_{34}N_2O_3$			
Acetamide, N-[(3β)-3-acetoxy-17-aza-D-homoandrosta-5,17-dien-17a-yl]-	EtOH	260(3.94)	70-2141-83
Acetamide, N-[(3β)-3-acetoxy-17a-aza-D-homoandrosta-5,17-dien-17-yl]-	EtOH	255(4.04)	70-2141-83
$C_{23}H_{34}N_2O_5$			
1-Pyrroline, 4,4-dimethyl-5-methoxy-3-(ethoxycarbonylmethylene)-2-(3,4-dimethyl-5-tert-butoxycarbonyl-2-pyrrolylmethyl)-	$CHCl_3$	280(4.24)	49-0983-83
2,2'-Pyrromethene-5'-carboxylic acid, 1,3,4,5-tetrahydro-3-(methoxycarbonylmethyl)-1,1',3',4,4,4'-hexamethyl-5-oxo-, 1,1-dimethylethyl ester, (E)-	MeOH	234(4.08),293(4.37)	49-1107-83
$C_{23}H_{34}O$			
3'H-Cycloprop[16,17]androsta-4,16-dien-3-one, 16,17-dihydro-16-(1-methylethyl)-, (14β,16α,17α)-	MeCN	237(4.21)	33-1806-83
$C_{23}H_{34}O_2$			
24-Norchola-5,7-dien-23-al, 3β-hydroxy-	EtOH	261(3.89),270(4.05), 280(4.06),291(3.78)	5-1031-83
$C_{23}H_{34}O_3$			
24-Norchola-20,22-diene-3,14-diol, 21,23-epoxy-, (3α,5β,14β)-	MeOH	214(3.70)	13-0189-83B

Compound	Solvent	λmax(log ε)	Ref.
$C_{23}H_{34}O_4$			
Androst-4-en-3-one, 17β-acetoxy-16β-ethyl-19-hydroxy-	EtOH	242(4.26)	95-1042-83
24-Norchol-20(22)-en-21-oic acid, 3,14,23-trihydroxy-, γ-lactone, (3α,5β,14β)-	MeOH	217(3.83)	13-0189-83B
$C_{23}H_{34}O_5$			
Butanoic acid, 2-methyl-, 1,2,3,7,8,8a-hexahydro-7-methyl-8-[2-(tetrahydro-4-hydroxy-6-oxo-2H-pyran-2-yl)ethyl]-1-naphthalenyl ester	MeOH	229(4.18),236(4.25), 244(4.06)	35-0593-83
isomer 49	MeOH	229(4.29),236(4.34), 244(4.18)	35-0593-83
isomer 51	MeOH	229(4.19),236(4.26), 244(4.07)	35-0593-83
Compactin	MeOH	229(4.26),237(4.32), 246(4.14)	35-0593-83
$C_{23}H_{34}O_6$			
Diginatigenin	EtOH	318(4.18)	162-0457-83
$C_{23}H_{34}O_7$			
Ingol, 12-acetoxy-8-methoxy-	MeOH	208(4.08)	102-2795-83
$C_{23}H_{35}NO$			
Retinamide, N-propyl-	EtOH	347(4.70)	34-0422-83
$C_{23}H_{35}NO_2$			
24-Norchola-20,22-dien-14-ol, 3-amino-21,23-epoxy-, (3β,5β,14β)-	MeOH	213.5(3.76)	12-0189-83B
Retinamide, N-(2-hydroxypropyl)-	EtOH	347(4.71)	34-0422-83
Retinamide, N-(3-hydroxypropyl)-	EtOH	347(4.71)	34-0422-83
$C_{23}H_{35}NO_3$			
2,4,6,8,10,14-Hexadecahexaenamide, 13-hydroxy-N-(2-hydroxy-1-methylethyl)-2,10,12,14-tetramethyl-, (E,E,Z,E,E,E)-	EtOH	203(3.98),264(3.87), 340s(--),355(4.31), 370s(--)	5-1081-83
24-Norchola-20,22-diene-12,14-diol, 3-amino-21,23-epoxy-, (3α,5β,12β,14β)-	MeOH	215(3.67)	13-0171-83B
(3β,5β,12β,14β)-	MeOH	215(3.64)	13-0171-83B
24-Norchol-20(22)-en-21-oic acid, 3-amino-14,23-dihydroxy-, γ-lactone, (3α,5β,14β)-	MeOH	217(3.82)	13-0189-83B
Retinamide, N-(2,3-dihydroxypropyl)-	EtOH	348(4.70)	34-0422-83
$C_{23}H_{35}NO_3S$			
Spiro[androst-4-en-17,5'-[1,2,3]oxathiazolidin]-3-one, 2,2,3'-trimethyl-, 2'-oxide	EtOH	239(4.22)	5-1001-83
$C_{23}H_{35}NO_4$			
Card-20(22)-enolide, 3-amino-12,14-dihydroxy-, (3α,5β,12β)-	MeOH	218.5(4.17)	13-0171-83B
(3β,5β,12β)-	MeOH	219.5(4.17)	13-0171-83B
$C_{23}H_{35}N_3O$			
Androstan-14-ol, 3-amino-17-(4-pyridazinyl)-, (3β,5β,14β,17β)-	MeOH	218(3.80),252s(3.41)	13-0189-83B

Compound	Solvent	$\lambda_{max}(\log \epsilon)$	Ref.
$C_{23}H_{36}N_2O_2$			
Hexanamide, N-[3-[2-[2-(2-methoxy-6-methylphenyl)ethenyl]-1-pyrrolidinyl]propyl]-	MeOH	223(3.90),258(3.62),295(3.23)	100-0235-83
2H-Isoindole-2-undecanamide, N,1-diethyl-1,3-dihydro-3-oxo-	$CHCl_3$	247(3.84),269s(3.57),279(3.39)	78-2691-83
$C_{23}H_{36}N_2O_3$			
2,5-Cyclohexadien-1-one, 2,6-bis(1,1-dimethylethyl)-4-(di-4-morpholinylmethylene)-	isooctane	425(4.55)	73-2376-83
Hexanamide, N-[3-[[3-(3,4-dihydro-5-methyl-4-oxo-2H-1-benzopyran-2-yl)propyl]methylamino]propyl]-	MeOH	222(2.68),258(2.68),320(2.42)	100-0235-83
Peripentadenine, O-methyl-	MeOH	225(4.05),278(3.35),318(2.42)	100-0235-83
$C_{23}H_{36}N_4O_4$			
1,2,3-Triazolidine-1,2-dicarboxylic acid, 4,4-dimethyl-3-(4-methylphenyl)-5-pyrrolidino-, bis(1-methylethyl) ester	EtOH	239(3.93),280s(2.92)	33-1608-83
$C_{23}H_{36}O_2$			
Androst-8-en-17-one, 3-methoxy-4,4,14-trimethyl-, (3β,5α,13α,14β)-	dioxan	283(1.58)	78-2799-83
$C_{23}H_{36}O_3$			
Benzenehexadecanoic acid, 4-methyl-o-oxo-	MeCN	264(3.92)	35-3951-83
	CTAC	256(3.99)	35-3951-83
	SDS	257(3.97)	35-3951-83
$C_{23}H_{36}O_4$			
9-Phenanthrenol, 1,2,3,4,4a,9,10,10a-octahydro-5,6,10-trimethoxy-1,1,4a-trimethyl-7-(1-methylethyl)-	EtOH	270(2.70)	107-0201-83
Pregn-20-ene-21-carboxylic acid, 3,14-dihydroxy-, methyl ester, (3β,5β-14β,20E)-	MeOH	217(4.20)	13-0037-83B
Prostacyclin, 6-oxo-, methyl ester, (E)-	EtOH	265(4.225)	88-1281-83
(Z)-	EtOH	260(4.190)	88-1281-83
$C_{23}H_{36}O_5$			
Benzeneheptanoic acid, 2-(3-hydroxy-1-octenyl)-3,5-dimethoxy-, (E)-	EtOH	262(4.06),290(3.70),301(3.62),330(3.20)	39-2211-83C
$C_{23}H_{38}N_2O_3$			
Hexanamide, N-[3-[2-[2-hydroxy-2-(2-methoxy-6-methylphenyl)ethyl]-1-pyrrolidinyl]propyl]-	MeOH	208(3.78),285(3.00),320(2.79)	100-0235-83
$C_{23}H_{38}O_3$			
3,5-Cyclohexadiene-1,2-dione, 4-(hexadecyloxy)-5-methyl-	MeCN	460(3.80)	44-0177-83
$C_{23}H_{38}O_5$			
Butanoic acid, 2-methyl-, 8-[3-(1-ethoxyethoxy)propyl]-1,2,3,4,6,7,8,8a-octahydro-7-methyl-4-oxo-1-naphthalenyl ester	MeOH	244(3.87)	35-0593-83

Compound	Solvent	λ_{max}(log ε)	Ref.
$C_{23}H_{40}N_4O_5Si_2$			
7H-Pyrrolo[2,3-d]pyrimidin-4-amine, 7-[3,5-O-[1,1,3,3-tetrakis(1-methylethyl)-1,3-disiloxanediyl]-β-D-ribofuranosyl]-	MeOH	227(4.30),270(4.03)	35-4059-83
$C_{23}H_{40}O_2$			
Benzene, 1,2-dimethoxy-3-pentadecyl-	MeOH	207(3.99),213s(3.84), 270(3.01),276(3.00)	36-0792-83
1,2-Benzenediol, 3-heptadecyl-	MeOH	230(3.19),275(3.21)	36-0792-83
$C_{23}H_{41}N_3O_7Si_2$			
Cytidine, N-acetyl-3',5'-O-[1,1,3,3-tetrakis(1-methylethyl)-1,3-disiloxanediyl]-	MeOH	247(4.18),297(3.91)	35-4059-83
$C_{23}H_{42}N_4O_4Si_2$			
7H-Pyrrolo[2,3-d]pyrimidin-4-amine, 7-[2,3-bis-O-[(1,1-dimethylethyl)-dimethylsilyl]-β-D-ribofuranosyl]-	MeOH	272(4.09)	5-1169-83
$C_{23}H_{45}N_4O_6P$			
1-Piperidinyloxy, 4-[(diethoxyphosphinyl)[[(2,2,6,6-tetramethyl-1-oxy-4-piperidinyl)amino]carbonyl]-amino]-2,2,6,6-tetramethyl-	EtOH	243(3.65),453(1.33)	70-1906-83
$C_{23}H_{50}OSi_5$			
Silane, [2-[(trimethylsilyl)[(trimethylsilyl)oxy]phenyl]-1-ethanyl-2-ylidene]tris[trimethyl-	n.s.g.	284(--),292(3.43)	101-0C13-83M

Compound	Solvent	λ_{max}(log ϵ)	Ref.
$C_{24}H_8Cl_{14}NO_3$			
Methyl, bis(pentachlorophenyl)[2,3,5,6-tetrachloro-4-[[(2-ethoxy-2-oxoethyl)-amino]carbonyl]phenyl]-	$CHCl_3$	287(3.77),335s(3.80), 368s(4.30),382(4.60), 483s(3.08),507(3.09), 562(3.07)	44-3716-83
$C_{24}H_{10}Cl_4I_4O_6$			
Benzoic acid, 2-(6-acetoxy-2,4,5,7-tetraiodo-3-oxo-3H-xanthen-9-yl)-3,4,5,6-tetrachloro-, ethyl ester (Rose Bengal ethyl ester acetate)	MeOH	400(4.0),494(4.0)	35-7465-83
	CH_2Cl_2	395(--),494(--)	35-7465-83
$C_{24}H_{10}Cl_{14}NO_2$			
L-Valine, 4-[bis(pentachlorophenyl)-methylene]-2,3,5,6-tetrachlorophenyl ester, hydrobromide (radical)	MeCN	218(5.00),272s(3.74), 336s(3.75),366(4.21), 380(4.47),506(3.00), 556(3.01)	44-3716-83
$C_{24}H_{10}Cl_{14}N_3O_2$			
Glycinamide, L-alanyl-N-[4-[bis(penta-chlorophenyl)methylene]-2,3,5,6-tetrachlorophenyl]-, monohydro-bromide (radical)	dioxan	226(4.81),285s(3.76) 367s(4.27),385(4.54), 511(3.09),566(3.10)	44-3716-83
$C_{24}H_{10}N_2$			
Biphenyleno[2,3-b]biphenylene-5,12-di-carbonitrile	$CHCl_3$	282s(4.55),298(4.81), 305s(4.71),320s(4.35), 350(3.49),418(3.77), 443(4.15),473(4.37)	39-1443-83C
$C_{24}H_{12}N_4O_2S_2$			
Benzenesulfenamide, N,N'-(3,7-dicyano-benzo[1,2-b:4,5-b']difuran-2,6-diyl-idene)bis-, (Z,Z)-	$CHCl_3$	675(4.81)	150-0480-83M
$C_{24}H_{12}N_4S_5$			
2-Thiazoleacetonitrile, α,α'-1,2,4-tri-thiolane-3,5-diylidenebis[4-phenyl-	DMF	293s(4.05),314s(3.87), 405(4.46)	49-0999-83
$C_{24}H_{12}N_8$			
9,10-Ethenoanthracene-2,2,3,3,6,6,7,7-(1H,4H)-octacarbonitrile, 5,8,9,10-tetrahydro-	MeCN	234(3.35)	33-0019-83
$C_{24}H_{12}N_8O$			
9,10-Ethanoanthracene-2,2,3,3,6,6,7,7-octacarbonitrile, 5,8,9,10-tetrahy-dro-11-oxo-	EtOH	293(2.74)	33-1134-83
$C_{24}H_{14}Cl_2$			
7H-Dibenzo[a,g]cyclopentadecene, 16,17,18,19-tetradehydro-7-(di-chloromethylene)-, (E,E,E)-	CH_2Cl_2	231(4.59),303(4.75), 371s(4.14)	39-2997-83C
$C_{24}H_{14}N_2O_2$			
Benzoxazole, 2,2'-(1,2-naphthalene-diyl)bis-	DMF	323(4.37)	5-0931-83
Benzoxazole, 2,2'-(1,3-naphthalene-diyl)bis-	DMF	312(4.67)	5-0931-83
Benzoxazole, 2,2'-(1,5-naphthalene-diyl)bis-	DMF	347(4.48)	5-0931-83

Compound	Solvent	λ_{max}(log ε)	Ref.
Benzoxazole, 2,2'-(1,6-naphthalene-diyl)bis-	DMF	316(4.73)	5-0931-83
Benzoxazole, 2,2'-(1,7-naphthalene-diyl)bis-	DMF	342(4.39)	5-0931-83
Benzoxazole, 2,2'-(2,3-naphthalene-diyl)bis-	DMF	287(4.66)	5-0931-83
Benzoxazole, 2,2'-(2,6-naphthalene-diyl)bis-	DMF	347(4.82)	5-0931-83
Benzoxazole, 2,2'-(2,7-naphthalene-diyl)bis-	DMF	318(4.86)	5-0931-83
$C_{24}H_{14}N_4S_2$			
Quinoxaline, 2,2'-[2,2'-bithiophene]-3,3'-diylbis-	$CHCl_3$	244(4.6),269(4.5), 341(4.2)	24-0479-83
$C_{24}H_{14}N_8$			
9,10-Ethanoanthracene-2,2,3,3,6,6,7,7-(1H,4H)-octacarbonitrile, 5,8,9,10-tetrahydro-	MeCN	215(3.30)(end abs.)	33-0019-83
$C_{24}H_{14}O_2$			
1(4H)-Anthracenone, 4-(4-oxo-1(4H)-naphthalenylidene)-	EtOH	244s(4.16),250(4.28), 254(4.275),263s(4.16), 273s(4.06),337(4.42)	73-0112-83
$C_{24}H_{14}O_8$			
7H-Furo[3,2-g][1]benzopyran-7-one, 9,9'-(1,2-ethanediylbis(oxy)]bis-	EtOH	218(4.65),248(4.60), 263(4.38),300(4.31)	39-1807-83B
$C_{24}H_{15}BrN_2$			
Benzo[f]quinoline, 1-(4-bromophenyl)-3-(2-pyridinyl)-	EtOH	252(4.49),286(4.73), 351(3.74),369(3.74)	103-0422-83
$C_{24}H_{15}ClN_2O_2$			
Pyrano[2,3-c]pyrazol-4(1H)-one, 5-chloro-1,3,6-triphenyl-	EtOH	254(4.49)	44-4078-83
$C_{24}H_{15}N_3O_2$			
Benzo[f]quinoline, 1-(4-nitrophenyl)-3-(2-pyridinyl)-	EtOH	284(4.85),351(3.97)	103-0422-83
$C_{24}H_{15}N_3O_4$			
Cyclohept[1,2,3-hi]imidazo[2,1,5-cd]-indolizine-3,4-dicarboxylic acid, 9-cyano-2-phenyl-, dimethyl ester	$CHCl_3$	255s(4.35),282(4.47), 292(4.47),303s(4.43), 317(4.34),341(3.97), 385s(4.42),399(4.51), 432s(4.16),453(4.09)	18-3703-83
$C_{24}H_{15}N_5$			
Benzonitrile, 4,4',4"-(4,5-dihydro-1H-pyrazole-1,3,5-triyl)tris-	MeOH	384(4.57)	12-1649-83
$C_{24}H_{15}N_9O_3$			
1,3,5-Benzenetriol, 2,4,6-tris(2H-benzotriazol-2-yl)-	$CHCl_3$	265s(--),300s(--), 337(4.62)	49-0937-83
$C_{24}H_{16}Cl_2$			
Benzene, 1,1'-[5-(dichloromethylene)-1,3,6-heptatriene-1,7-diyl]bis[2-ethynyl-, (E,E,E)-	CH_2Cl_2	235(4.60),255(4.53), 262(4.53),270s(4.46), 326(4.75),341(4.74), 359s(4.53)	39-2997-83C

Compound	Solvent	λ_{max}(log ε)	Ref.
$C_{24}H_{16}F_{11}N_2S_2$			
Benzothiazolium, 3-methyl-2-[3,3,4,4,5-5,6,6,7,7,7-undecafluoro-2-[(3-methyl-2(3H)-benzothiazolylidene)methyl]-1-heptenyl]-, perchlorate	$CHCl_3$	596(4.43)	104-1937-83
$C_{24}H_{16}N_2$			
Benzo[f]quinoline, 1-phenyl-3-(2-pyridinyl)-	EtOH	250(4.33),284(4.57), 350(3.63),365(3.63)	103-0422-83
$C_{24}H_{16}N_2OS$			
Thiazolo[3,2-a]pyrimidin-4-ium, 3-hydroxy-2,5,7-triphenyl-, hydroxide, inner salt	MeCN	600(4.00)	124-0857-83
$C_{24}H_{16}N_2O_2$			
9,10-Anthracenedione, 1-amino-4-(2-naphthalenylamino)-	DMF	585(4.01),620(4.01)	2-0808-83
Pyrano[2,3-c]pyrazol-4(1H)-one, 1,3,6-triphenyl-	EtOH	256(4.49)	44-4078-83
$C_{24}H_{16}N_2O_5$			
Isotriphenodioxazine, 3,8,13-triacetyl-	$CHCl_3$	460s(--),495(4.35), 530(4.42)	44-3649-83
Triphenodioxazine, 1,6,8-triacetyl-	$CHCl_3$	460s(--),490(4.30), 530(4.41)	44-3649-83
$C_{24}H_{16}N_2O_8$			
3,8,13-Isotriphenodioxazinetricarboxylic acid, trimethyl ester	$CHCl_3$	454s(--),490(4.37), 528(4.40)	44-3649-83
1,6,8-Triphenodioxazinetricarboxylic acid, trimethyl ester	$CHCl_3$	460s(--),482(4.39), 517(4.45)	44-3649-83
$C_{24}H_{16}N_4OS_3$			
4-Thiazolidinone, 3-ethyl-5-(3-phenyl-1H-thiazolo[4',3':2,3]pyrimido[4,5-b]quinolin-1-ylidene)-2-thioxo-	DMF	498(3.39)	103-0166-83
$C_{24}H_{16}N_4O_8S_3$			
2,7-Naphthalenedisulfonic acid, 4-[[4-[(2-benzothiazolylamino)carbonyl]-phenyl]azo]-3-hydroxy-	n.s.g.	518(4.07)	7-0265-83
$C_{24}H_{16}N_4O_9S_3$			
2,7-Naphthalenedisulfonic acid, 3-[[4-[(2-benzothiazolylamino)carbonyl]-phenyl]azo]-4,5-dihydroxy-	n.s.g.	530(4.18)	7-0265-83
$C_{24}H_{16}O_2$			
Benzo[ghi]perylene, 5,10-diacetyl-	n.s.g.	282(4.51),310(4.48), 320(4.52),428(4.36)	65-1898-83
$C_{24}H_{17}BiN_2O_2$			
Bismuth, phenylbis(8-quinolinolato-N^1,O^8)-	benzene	385(3.68)	101-0317-83M
$C_{24}H_{17}ClN_2O_2$			
7H-Dibenz[f,ij]isoquinolin-7-one, 4-chloro-6-[(4-methoxyphenyl)amino]-2-methyl-	n.s.g.	480(4.30)	2-0812-83

Compound	Solvent	$\lambda_{max}(\log \epsilon)$	Ref.
$C_{24}H_{17}NO_4S$			
3H-Naphtho[2,1-b]pyran-3-one, 2-[2-(3,4-dimethoxyphenyl)-4-thiazolyl]-	MeCN	385(4.28)	48-0551-83
$C_{24}H_{17}N_3S$			
Triazolo[1,5-a]pyridinium, 2,3-dihydro-1,5,7-triphenyl-2-thioxo-, hydroxide, inner salt	EtOH	262(4.31)	88-3523-83
$C_{24}H_{17}N_4O_6P$			
Benzofuroxan, 4,6-dinitro-, triphenylphosphine complex	MeOH	263(4.21),302s(3.97), 350(3.58),460(4.37)	12-1227-83
$C_{24}H_{17}N_5O_5S_2$			
2-Naphthalenesulfonic acid, 6-amino-3-[[4-[(2-benzothiazolylamino)carbonyl]phenyl]azo]-4-hydroxy-	n.s.g.	560(4.02)	7-0265-83
$C_{24}H_{17}N_5O_8S_3$			
2,7-Naphthalenedisulfonic acid, 5-amino-3-[[4-[(2-benzothiazolylamino)-carbonyl]phenyl]azo]-4-hydroxy-	n.s.g.	545(4.15)	7-0265-83
$C_{24}H_{18}$			
[2_3]Metacyclophanetriene	EtOH	253(4.13)	1-0693-83
[2_3]Metametaparacyclophanetriene	EtOH	274(4.41)	1-0693-83
[2_3]Metaparaparacyclophanetriene	EtOH	227s(4.37)	1-0693-83
m-Terphenyl, 2'-phenyl-	C_6H_{12}	186(4.84),196(4.80), 242(4.53)	94-1572-83
$C_{24}H_{18}BrN$			
Pyridine, 2-(4-bromophenyl)-3-methyl-4,6-diphenyl-	MeCN	245(4.63),286s(4.04), 297s(4.00)	48-0729-83
Pyridine, 4-(4-bromophenyl)-3-methyl-2,6-diphenyl-	MeCN	248(4.59),296s(3.93)	48-0729-83
Pyridine, 6-(4-bromophenyl)-3-methyl-2,4-diphenyl-	MeCN	247(4.54),285s(4.22), 303s(4.06)	48-0729-83
$C_{24}H_{18}BrN_3O$			
Pyrazolo[1,5-c]pyrimidin-7(3H)-one, 4-bromo-3a,6-dihydro-2,3a,5-triphenyl-	isoPrOH	282(4.1),305s(4.0)	103-0564-83
$C_{24}H_{18}BrO$			
Pyrylium, 2-(4-bromophenyl)-3-methyl-4,6-diphenyl-, perchlorate	MeCN	238(4.24),276(4.26), 346(4.33),389(4.41)	48-0729-83
Pyrylium, 4-(4-bromophenyl)-3-methyl-2,6-diphenyl-, perchlorate	MeCN	233(4.29),274(4.29), 364s(4.41),380(4.42)	48-0729-83
Pyrylium, 6-(4-bromophenyl)-3-methyl-2,4-diphenyl-, perchlorate	MeCN	233(4.27),277(4.30), 346(4.33),392(4.47)	48-0729-83
$C_{24}H_{18}ClN$			
Pyridine, 2-(4-chlorophenyl)-3-methyl-4,6-diphenyl-	MeCN	244(4.61),285s(4.03), 296s(3.97)	48-0729-83
Pyridine, 4-(4-chlorophenyl)-3-methyl-2,6-diphenyl-	MeCN	247(4.59),296s(3.95)	48-0729-83
Pyridine, 6-(4-chlorophenyl)-3-methyl-2,4-diphenyl-	MeCN	246(4.55),286s(4.18), 302(4.01)	48-0729-83
$C_{24}H_{18}ClO$			
Pyrylium, 2-(4-chlorophenyl)-3-methyl-	MeCN	234(4.26),275(4.28),	48-0729-83

Compound	Solvent	λ max (log ε)	Ref.
4,6-diphenyl- (cont.)		346(4.34),388(4.42)	48-0729-83
Pyrylium, 4-(4-chlorophenyl)-3-methyl-2,6-diphenyl-	MeCN	233(4.26),274(4.28), 358s(4.39),383(4.41)	48-0729-83
Pyrylium, 6-(4-chlorophenyl)-3-methyl-2,4-diphenyl-	MeCN	232(4.26),277(4.30), 346(4.33),390(4.45)	48-0729-83
$C_{24}H_{18}FN$			
Pyridine, 2-(4-fluorophenyl)-3-methyl-4,6-diphenyl-	MeCN	242(4.55),281s(4.04), 296s(3.97)	48-0729-83
Pyridine, 6-(4-fluorophenyl)-3-methyl-2,4-diphenyl-	MeCN	242(4.55),281s(4.04), 295s(3.91)	48-0729-83
$C_{24}H_{18}FO$			
Pyrylium, 2-(4-fluorophenyl)-3-methyl-4,6-diphenyl-, perchlorate	MeCN	231(4.24),274(4.27), 344s(4.31),387(4.40)	48-0729-83
Pyrylium, 6-(4-fluorophenyl)-3-methyl-2,4-diphenyl-, perchlorate	MeCN	230(4.24),274(4.26), 343s(4.29),387(4.39)	48-0729-83
$C_{24}H_{18}GeO$			
10H-Phenoxagermanin, 10,10-diphenyl-	n.s.g.	295(3.67)	101-0283-83A
$C_{24}H_{18}NO_3$			
Pyrylium, 3-methyl-4-(4-nitrophenyl)-2,6-diphenyl-, perchlorate	MeCN	276(4.39),318(4.29), 392(4.35)	48-0729-83
$C_{24}H_{18}N_2$			
Benzaldehyde, 2-ethynyl-, [7-(2-ethynylphenyl)-2,4,6-heptatrienylidene]-hydrazone, (?,?,E,E,E)-	THF	227(4.45),260s(4.17), 267s(4.13),277s(4.02), 292(3.91),307s(3.96), 380s(4.75),392(4.77), 410s(4.66)	18-1467-83
Benzenamine, N-(2,3-dimethyl-11H-indeno[1,2-b]quinolin-11-ylidene)-	EtOH	206(4.5),228(4.52), 235(4.55),245(4.52), 258(4.46),292(4.86), 370s(3.48)	104-0158-83
Benzenamine, N-(2,4-dimethyl-11H-indeno[1,2-b]quinolin-11-ylidene)-	EtOH	203(4.54),217(4.49), 234(4.5),250(4.41), 260(4.48),294(4.78), 334s(3.76),380(3.23)	104-0158-83
Benzenamine, N-(2,6-dimethyl-11H-indeno[1,2-b]quinolin-11-ylidene)-	EtOH	217s(4.54),228s(4.56), 236(4.62),248(4.6), 256(4.53),294(4.86), 330s(3.83)	104-0158-83
Benzenamine, N-(3,6-dimethyl-11H-indeno[1,2-b]quinolin-11-ylidene)-	EtOH	206(4.74),214(4.74), 236(4.72),246(4.72), 256s(4.64),294(4.92), 340s(4.2),410s(3.33)	104-0158-83
Benzenamine, N-11H-indeno[1,2-b]quinolin-11-ylidene-2,5-dimethyl-	EtOH	206(4.92),223(4.82), 229(4.82),245(4.71), 252(4.7),273s(4.88), 288(4.96),410s(3.22)	104-0158-83
2,4-Pentadienal, 5-(2-ethynylphenyl)-, [3-(2-ethynylphenyl)-2-propenylidene]hydrazone, (?,?,E,E,E)-	THF	227(4.45),257(4.24), 296s(3.91),360s(4.71), 372(3.76),387(4.74), 410s(4.72)	18-1467-83
$C_{24}H_{18}N_2O_2$			
7H-Dibenz[f,ij]isoquinolin-7-one, 6-[(4-methoxyphenyl)amino]-2-methyl-	n.s.g.	476(4.27)	2-0812-83
Pyridine, 3-methyl-4-(4-nitrophenyl)-2,6-diphenyl-	MeCN	237(4.42),266(4.48)	48-0729-83

Compound	Solvent	$\lambda_{max}(\log \epsilon)$	Ref.
$C_{24}H_{18}N_2O_2S$			
3H-Naphtho[2,1-b]pyran-3-one, 2-[2-[4-(dimethylamino)phenyl]-4-thiazolyl]-	MeCN	399(4.32)	48-0551-83
$C_{24}H_{18}N_2O_3$			
2H-Pyran-2-one, 4-hydroxy-6-phenyl-3-[phenyl(phenylhydrazono)methyl]-	EtOH	304(4.33),336(4.42)	44-4078-83
3H-Pyrazol-3-one, 1,2-dihydro-4-(1-hydroxy-3-oxo-3-phenyl-1-propenyl)-2,5-diphenyl-	EtOH	248(4.42),356(4.35)	44-4078-83
$C_{24}H_{18}N_2O_4$			
Cyclohept[1,2,3-hi]imidazo[2,1,5-cd]-indolizine-3,4-dicarboxylic acid, 2-(4-methylphenyl)-, dimethyl ester	EtOH	249(4.44),278s(4.43), 290(4.48),310s(4.37), 357s(4.42),375(4.52), 400s(4.23),421(4.09)	18-3703-83
Cyclohept[1,2,3-hi]imidazo[2,1,5-cd]-indolizine-4,9-dicarboxylic acid, 2-phenyl-, 9-ethyl 4-methyl ester	EtOH	245(4.24),285s(4.30), 300(4.53),310s(4.40), 383(4.42),436(4.21)	18-3703-83
$C_{24}H_{18}N_2O_5$			
Isotriphenodioxazine, 3,8,13-triacetyl-7,14-dihydro-	$CHCl_3$	460(3.86)	44-3649-83
Triphenodioxazine, 1,8,13-triacetyl-7,14-dihydro-	$CHCl_3$	510(3.98)	44-3649-83
$C_{24}H_{18}N_2O_6$			
1-Triphenodioxazinecarboxylic acid, 6,8-diacetul-7,14-dihydro-, methyl ester	$CHCl_3$	491(3.92)	44-3649-83
$C_{24}H_{18}N_2O_7$			
1,13-Triphenodioxazinedicarboxylic acid, 8-acetyl-7,14-dihydro-, dimethyl ester	$CHCl_3$	468(3.84)	44-3649-83
$C_{24}H_{18}N_2O_8$			
Isotriphenodioxazine-3,8,13-tricarboxylic acid, 7,14-dihydro-, trimethyl ester	$CHCl_3$	462(3.90)	44-3649-83
Triphenodioxazine-1,6,8-tricarboxylic acid, 7,14-dihydro-, trimethyl ester	$CHCl_3$	462(3.96)	44-3649-83
$C_{24}H_{18}N_2SSe$			
Diazene, phenyl[3-phenyl-3-(5-phenyl-1,3-thiaselenol-2-ylidene)1-propenyl]-	MeCN	528(3.93)	118-0840-83
$C_{24}H_{18}N_2S_2$			
Diazene, phenyl[3-phenyl-3-(4-phenyl-1,3-dithiol-2-ylidene)-1-propenyl]-	MeCN	515(4.27)	118-0840-83
$C_{24}H_{18}N_4$			
3,9'-Bi-9H-pyrido[3,4-b]indole, 1,1'-dimethyl-	EtOH	236(5.00),251s(4.51), 286s(4.19),338(3.84), 350(3.91)	78-0033-83
9,9'-Bi-9H-pyrido[3,4-b]indole, 1,1'-dimethyl-	EtOH	230(4.86),278s(4.28), 281(4.40),327(4.05), 339(4.11)	78-0033-83

Compound	Solvent	λ_{max}(log ϵ)	Ref.
$C_{24}H_{18}N_4O_4$			
2-Naphthalenecarboxamide, 3-hydroxy-N-methyl-4-[(2-nitrophenyl)azo]-N-phenyl-	EtOH	306(4.076),402s(3.949), 500(4.350)	104-1145-83
	$CHCl_3$	312(4.013),412s(3.935), 504(4.297),530(4.322)	104-1145-83
	CCl_4	306(4.033),400s(3.987), 490(4.354),510(4.342)	104-1145-83
	DMSO	312(4.017),412s(3.892), 504-530(4.229)	104-1145-83
$C_{24}H_{18}N_4O_5$			
2-Naphthalenecarboxamide, 3-hydroxy-4-[(2-methoxy-4-nitrophenyl)azo]-N-phenyl-	EtOH	211(--),287s(--), 439s(--),516-545(--)	104-1145-83
	$CHCl_3$	300(4.314),350s(4.017), 433s(4.033),520(4.549), 545(4.571)	104-1145-83
	CCl_4	305(--),343s(--), 432s(--),516-542(--)	104-1145-83
	DMSO	300(4.134),345s(3.857), 433s(3.898),520(4.408), 540(4.415)	104-1145-83
$C_{24}H_{18}N_4S$			
Imidazo[2,1-b]thiazole, 6-[(3-methylphenyl)azo]-3,5-diphenyl-	EtOH	268(3.91),364(4.06)	4-0639-83
Imidazo[2,1-b]thiazole, 6-[(4-methylphenyl)azo]-3,5-diphenyl-	EtOH	258s(3.88),372(4.20)	4-0639-83
Imidazo[2,1-b]thiazole, 5-(4-methylphenyl)-3-phenyl-6-(phenylazo)-	EtOH	258s(3.90),370(4.19)	4-0639-83
$C_{24}H_{18}N_6$			
Benzenamine, N,N'-[3-(2-pyridinyl)-2H-pyrido[1,2-a]-1,3,5-triazine-2,4(3H)-diylidene]bis-	dioxan	284(4.50),394(3.96)	48-0463-83
$C_{24}H_{18}OSi$			
10H-Phenoxasilin, 10,10-diphenyl-	n.s.g.	300(3.95)	101-0283-83R
$C_{24}H_{18}OSn$			
10H-Phenoxastannin, 10,10-diphenyl-	n.s.g.	295(3.67)	101-0283-83R
$C_{24}H_{18}O_6$			
[2,2'-Binaphthalene]-1,1',4,4'-tetrone, 5,5'-dihydroxy-3,3',7,7'-tetramethyl-	$CHCl_3$	255(4.34),260s(4.33), 275(4.30),432(3.92)	5-0299-83
[2,2'-Binaphthalene]-1,1',4,4'-tetrone, 8,8'-dihydroxy-3,3',6,6'-tetramethyl-	$CHCl_3$	255(4.50),260s(4.49), 277(4.46),350s(3.53), 435(4.08)	5-0299-83
[2,2'-Binaphthalene]-1,1',4,4'-tetrone, 5,5'-dimethoxy-3,3'-dimethyl-	$CHCl_3$	254(4.48),267s(4.42), 402(3.93)	5-0299-83
[2,2'-Binaphthalene]-1,1',4,4'-tetrone, 5,5'-dimethoxy-7,7'-dimethyl-	$CHCl_3$	248(4.39),274s(4.22), 412(3.94)	5-0299-83
[2,2'-Binaphthalene]-1,1',4,4'-tetrone, 8,8'-dimethoxy-3,3'-dimethyl-	$CHCl_3$	249(4.45),253s(4.43), 266s(4.35),397(3.85)	5-0299-83
[2,2'-Binaphthalene]-1,1',4,4'-tetrone, 8,8'-dimethoxy-6,6'-dimethyl-	$CHCl_3$	251(4.41),260s(4.39), 323(3.62),410(3.88)	5-0299-83
$C_{24}H_{18}O_7$			
1H-Benzo[6,7]phenanthro[10,1-bc]furan-3,6(2H,12bH)-dione, 8,11-diacetoxy-12b-methyl-	MeCN	220(4.66),260(4.36), 282(4.20),294(4.21), 306(4.20),317(4.24), 355(3.61)	35-6177-83

Compound	Solvent	λ_{max}(log ε)	Ref.
5,12-Naphthacenedione, 9-acetyl-6,11-diacetoxy-7,8-dihydro-	MeCN	253(4.36),285(4.66), 293(4.60),365(3.95)	44-2820-83
$C_{24}H_{18}O_8$			
[2,2'-Binaphthalene]-1,1',4,4'-tetrone, 3,3'-dihydroxy-5,5'-dimethoxy-7,7'-dimethyl-	MeOH	213(4.50),227s(4.40), 285(4.45),386(3.94), 470s(3.62)	5-1886-83
[2,2'-Binaphthalene]-1,1',4,4'-tetrone, 3,3'-dihydroxy-8,8'-dimethoxy-6,6'-dimethyl-	MeOH	211(4.48),237(4.30), 285(4.28),384(3.90), 472s(3.55)	5-1886-83
$C_{24}H_{18}O_9$			
[1,2'-Bi-1H-indene]-1',2,3'-trione, 2',3,3-triacetoxy-2',3-dihydro-	MeOH	235(3.50),255(3.56), 280(2.87)	24-1309-83
$C_{24}H_{18}SSi$			
10H-Phenothiasilin, 10,10-diphenyl-	n.s.g.	290(3.90)	101-0283-83R
$C_{24}H_{19}BrN_4$			
1H-Imidazo[1,2-b][1,2,4]triazepine, 6-(4-bromophenyl)-7,8-dihydro-2,8-diphenyl-	EtOH	284(4.34),361(3.81)	103-0083-83
$C_{24}H_{19}ClN_4$			
1H-Imidazo[1,2-b][1,2,4]triazepine, 6-(4-chlorophenyl)-7,8-dihydro-2,8-diphenyl-	EtOH	284(4.46),361(3.83)	103-0083-83
1H-Imidazo[1,2-b][1,2,4]triazepine, 8-(4-chlorophenyl)-6,8-dihydro-2,6-diphenyl-	EtOH	279(4.46),351(4.02)	103-0083-83
$C_{24}H_{19}N$			
Pyridine, 4-(2-methylphenyl)-2,6-diphenyl-	DMSO	304(3.92)	150-0301-83M
$C_{24}H_{19}NO_3$			
6H-Anthra[1,9-cd]isoxazol-6-one, 5-[4-(1,1-dimethylethyl)phenoxy]-	EtOH	243(5.06),248(5.11), 442(4.17),459(4.16)	103-1057-83
$C_{24}H_{19}NO_4$			
3-Azabicyclo[3.2.0]heptane-2,4-dione, 6,7-bis(2-hydroxyphenyl)-3-phenyl-	MeOH	222(3.57),275(3.79), 281(3.74)	39-1083-83C
Aziridino[b]naphthalene-2,7-dione, 1,1a-bis(4-methoxyphenyl)-	benzene	360(2.88)	44-4968-83
2H-Benzo[b]cyclobuta[d]pyran-2-carboxamide, 1,2a,3,8b-tetrahydro-1-(2-hydroxyphenyl)-3-oxo-N-phenyl-, (1α,2α,2aα,8bα)-	dioxan	244(4.23),272(3.83), 280(3.64)	39-1083-83C
$C_{24}H_{19}N_2O_4S$			
1,3-Thiazin-1-ium, 4,6-bis(4-methoxyphenyl)-2-(4-nitrophenyl)-, perchlorate	MeCN	274(4.05),417(3.85), 493(3.98)	118-0827-83
$C_{24}H_{19}N_3O$			
Pyrazolo[1,5-c]pyrimidin-7(3H)-one, 3a,6-dihydro-2,3a,5-triphenyl-	isoPrOH	275(4.3),305s(4.2)	103-0564-83
$C_{24}H_{19}N_3O_2$			
Benzoic acid, 4-[3-(4-cyanophenyl)-4,5-dihydro-5-phenyl-1H-pyrazol-1-yl]-,	C_6H_{12}	393(4.56)	12-1649-83
	MeOH	393(4.57)	12-1649-83

Compound	Solvent	λ_{max}(log ε)	Ref.
methyl ester (cont.)			12-1649-83
2-Naphthalenecarboxamide, 3-hydroxy-N-methyl-N-phenyl-4-(phenylazo)-	EtOH	205(4.631),233(4.566), 296s(3.778),323(3.903), 436s(4.025),496(4.276), 520(4.260)	104-1145-83
	$CHCl_3$	265s(4.301),300s(3.763), 326(3.903),436s(3.991), 503(4.297),526(4.283)	104-1145-83
	CCl_4	300s(3.881),326(3.944), 436s(4.083),496(4.288), 520(4.262)	104-1145-83
	DMSO	265s(4.265),320(3.898), 436s(4.017),496(4.250), 520(4.228)	104-1145-83
$C_{24}H_{19}N_3O_3$			
1H-Indole, 3,3'-(2,5-dihydro-2-oxo-1H-pyrrole-3,4-diyl)bis[1-acetyl-	MeOH	221(4.07),255(3.87), 291(3.80),299(3.84), 322(3.52)	142-0469-83
$C_{24}H_{19}N_3O_4$			
2,3,9b-Triazaindeno[6,7,1-ija]azulene-8,9-dicarboxylic acid, 1-phenyl-, diethyl ester	EtOH	238(4.14),269s(4.20), 291(4.30),380(4.05), 419(3.82),440(3.82)	18-3703-83
$C_{24}H_{19}N_3O_6$			
Phenanthridinium 2-[(2-cyano-3-methoxy-3-oxo-1-propenyl)imino]-3-methoxy-1-(methoxycarbonyl)-3-oxopropylide	EtOH	250(4.40),329(3.98), 395(4.43)	150-0260-83S
$C_{24}H_{19}N_3S$			
Pyrazolo[1,5-c]pyrimidine-7(3H)-thione, 3a,6-dihydro-2,3a,5-triphenyl-	isoPrOH	232(4.3),250(4.4), 315s(4.2),325(4.3)	103-0564-83
$C_{24}H_{19}N_5O_2$			
1H-Imidazo[1,2-b][1,2,4]triazepine, 7,8-dihydro-6-(4-nitrophenyl)-2,8-diphenyl-	EtOH	299(4.24),403(4.20)	103-0083-83
$C_{24}H_{20}$			
2,3-Benzobicyclo[3.2.0]hepta-2,6-diene, 1-methyl-6,7-diphenyl-	EtOH	277(3.96),291(3.96)	35-4446-83
5H-Benzocycloheptene, 9-methyl-7,8-diphenyl-	EtOH	242(4.45)	35-4446-83
5H-Benzo[b]cyclopropa[lm]fluorene, 4b,4c,9b,9c-tetrahydro-9c-methyl-4c-phenyl-	EtOH	233s(4.23)	35-4446-83
1H-Cyclopropa[a]naphthalene, 1a,7b-dihydro-3-methyl-1a,2-diphenyl-	EtOH	237(4.38),288(4.05)	35-4446-83
Cyclopropene, 3-(2-ethenylphenyl)-2-methyl-1,3-diphenyl-	EtOH	256(4.34)	35-4446-83
Cyclopropene, 3-(2-ethenylphenyl)-3-methyl-1,2-diphenyl-	EtOH	227(4.48),317(4.35)	35-4446-83
1H-Indene, 1-ethenyl-3-methyl-1,2-diphenyl-	EtOH	222s(4.31),288(4.17)	35-4446-83
1H-Indene, 4-ethenyl-3-methyl-1,2-diphenyl-	EtOH	253(4.22),295(4.30)	35-4446-83
1H-Indene, 3-(2-ethenylphenyl)-2-methyl-1-phenyl-	EtOH	250(4.20)	35-4446-83
1,6-Methanocycloprop[a]indene, 1,1a,6-6a-tetrahydro-1-methyl-6,6a-diphenyl-	EtOH	249(4.12)	35-4446-83

Compound	Solvent	λ_{max}(log ε)	Ref.
1,6-Methanocycloprop[a]indene, 1,1a,6,6a-tetrahydro-1-methyl-6,6a-diphenyl-, (1α,1aβ,6α,6aβ)-	EtOH	249(4.12)	35-4446-83
1,6-Methanocycloprop[a]indene, 1,1a,6,6a-tetrahydro-1a-methyl-1,6a-diphenyl-, (1α,1aβ,6α,6aβ)-	EtOH	223s(4.36)	35-4446-83
1,4-Methanonaphthalene, 1,4-dihydro-2-methyl-1,3-diphenyl-	C_6H_{12}	237(4.21),277(3.95)	35-4446-83
1,4-Methanonaphthalene, 1,4-dihydro-3-methyl-1,2-diphenyl-	EtOH	233s(4.14),272(3.72)	35-4446-83
Naphthalene, 6-methyl-4-phenyl-1-(phenylmethyl)-	EtOH	223(4.63),234(4.65), 238(4.65),300(3.92)	2-0542-83
[2.2](2,6)-Naphthalenophane, achiral	C_6H_{12}	220(5.04),250s(3.75), 266(3.74),276(3.90), 288(4.00),301(3.17), 340s(2.64)	24-0827-83
chiral isomer	C_6H_{12}	219(5.20),240(4.25), 253s(3.97),276(3.68), 286s(3.64),301(3.49), 322(3.18),336s(2.78)	24-0827-83
$C_{24}H_{20}Bi_2$			
Dibismuthine, tetraphenyl-	pentane	216(4.31),310(3.44), 362(3.59)	157-1859-83
$C_{24}H_{20}Br_2$			
Benzene, 1,4-bis[2-(4-bromo-2-methylphenyl)ethenyl]-	dioxan	354(4.72)	104-0683-83
$C_{24}H_{20}ClN_3O_2$			
2H-Pyrrolo[2,3-b]quinoxalin-2-one, 3-benzoyl-1-(4-chlorophenyl)-1,3,3a,4,9,9a-hexahydro-4-methyl-	EtOH	222(4.48),251(4.36), 299(3.60)	103-0901-83
$C_{24}H_{20}Cl_2Ge_2$			
Digermane, 1,2-dichloro-1,1,2,2-tetraphenyl-	C_6H_{12}	225(4.58)	101-0149-83G
$C_{24}H_{20}F_2Ge_2$			
Digermane, 1,2-difluoro-1,1,2,2-tetraphenyl-	C_6H_{12}	226(4.66)	101-0149-83G
$C_{24}H_{20}Fe$			
Ferrocene, 1,1'-(1,2-ethanediyldi-2,1-phenylene)-	dioxan	245(4.2),390(4.0), 450(2.6)	18-2023-83
Ferrocene, 1,1'-(1,2-ethanediyldi-4,1-phenylene)-	dioxan	250(4.5),280s(4.4), 360s(3.5),485(2.7)	18-2023-83
$C_{24}H_{20}NO_2P$			
Ethanone, 1-[5-[(triphenylphosphoranylidene)amino]-2-furanyl]-	EtOH	210(4.55),225s(4.42), 375(4.42)	12-0963-83
$C_{24}H_{20}NO_2S$			
1,3-Thiazin-1-ium, 4,6-bis(4-methoxyphenyl)-2-phenyl-, perchlorate	MeCN	275(4.24),417(4.33), 485(4.56)	118-0827-83
$C_{24}H_{20}N_2$			
3,11:4,10-Dimethenocycloocta[3',4']-cyclobuta[1',2':4,5]benzo[1,2]cyclooctene-7,14-dicarbonitrile, 1,2,5,6-6b,8,9,12,13,14a-decahydro-	n.s.g.	342(3.75),354(3.74)	35-7384-83

Compound	Solvent	$\lambda_{max}(\log \epsilon)$	Ref.
3,12:4,11-Dimethenocyclooct[c]octalene-7,16-dicarbonitrile, 1,2,5,6,9,10,13-14-octahydro-	n.s.g.	260(3.62),318(3.01)	35-7384-83
$C_{24}H_{20}N_2O$			
1-Propanone, 3-(2-naphthalenylamino)-1-phenyl-3-(2-pyridinyl)-	EtOH	245(4.40),282(3.87), 294(3.77),345(3.22)	103-0422-83
$C_{24}H_{20}N_2OS$			
1H-Indene-2-carbothioamide, 1-oxo-N-(phenylmethyl)-3-[(phenylmethyl)-amino]-	EtOH	271(4.39),303(4.47), 330(4.23),343(4.30)	95-1243-83
$C_{24}H_{20}N_2O_2$			
1H-Pyrazole-3-carboxylic acid, 1,4,5-triphenyl-, ethyl ester	MeOH	225(3.5),261(3.1), 274(3.0)	83-0608-83
1H-Pyrazole-5-carboxylic acid, 1,3,4-triphenyl-, ethyl ester	MeOH	221(3.5),277(3.1)	83-0608-83
$C_{24}H_{20}N_2O_8S_2$			
Butanoic acid, 4-sulfo-2-[(4-sulfo-1-naphthalenyl)azo]-1-naphthalenyl ester, disodium salt	pH 7.5	393(4.08)	94-0162-83
$C_{24}H_{20}N_3S_2$			
Thiazolo[3,4-a]pyrimidin-5-ium, 6-[3-(3-ethyl-2(3H)-benzothiazolylidene)-1-propenyl]-8-phenyl-, perchlorate	MeCN	496(3.14),520(3.16), 655(3.76)	103-0037-83
$C_{24}H_{20}N_4$			
1H-Imidazo[1,2-b][1,2,4]triazepine, 7,8-dihydro-2,6,8-triphenyl-	EtOH	280(4.37),352(3.89)	103-0083-83
$C_{24}H_{20}N_4O$			
Acetamide, N-[4-[3-(4-cyanophenyl)-4,5-dihydro-1H-pyrazol-1-yl]phenyl]-	C_6H_{12}	414(4.46)	12-1649-83
	MeOH	408(4.45)	12-1649-83
Phenol, 2-(7,8-dihydro-2,6-diphenyl-1H-imidazo[1,2-b][1,2,4]triazepin-8-yl)-	EtOH	279(4.27),345(3.81)	103-0083-83
$C_{24}H_{20}N_6$			
Formazan, 5-(4,6-diphenyl-2-pyrimidinyl)-3-methyl-1-phenyl-	EtOH	370(4.37)	65-0153-83
	acetone	395(4.25)	65-0153-83
Co chelate	EtOH	430(3.86),525(3.96)	65-0153-83
Cu chelate	EtOH	510(4.13)	65-0153-83
Ni chelate	EtOH	570(4.30)	65-0153-83
$C_{24}H_{20}N_{10}O_8PdS_4$			
Palladium, bis[N,N-dimethyl-N'-[5-nitro-4-(4-nitrophenyl)-2-thiazolyl]-thioureato]-, (SP-4-2)-	$CHCl_3$	275s(4.53),310s(4.32), 430(4.40)	74-0075-83B
$C_{24}H_{20}O$			
2,4-Pentadien-1-one, 5-phenyl-1-[4-(phenylmethyl)phenyl]-	dioxan	235(4.06),274(4.00), 338(4.59)	124-1211-83
$C_{24}H_{20}O_4$			
4H-1-Benzopyran-4-one, 2,3-dihydro-3-[(4-methoxyphenyl)methylene]-7-(phenylmethoxy)-	EtOH	218(3.92),238(3.80), 353(3.90)	2-0759-83

Compound	Solvent	$\lambda_{max}(\log \epsilon)$	Ref.
4H-1-Benzopyran-4-one, 2-(4-methoxy-phenyl)-3-methyl-7-(phenylmethoxy)-	MeOH	206(4.52),245(4.25), 304(4.33)	2-0759-83
4H-1-Benzopyran-4-one, 3-[(4-methoxy-phenyl)methyl]-7-(phenylmethoxy)-	EtOH	227(4.35),241(4.27), 285(3.93),296(3.96), 305(3.95)	2-0759-83
$C_{24}H_{20}O_5$			
2,4,6-Cycloheptatrien-1-one, 2-hydroxy-4-[3-[3-methoxy-4-(phenylmethoxy)-phenyl]-1-oxo-2-propenyl]-, (E)-	MeOH	254(4.44),332s(4.19), 375(4.32)	18-3099-83
	MeOH-NaOH	262(4.31),353(4.44), 440s(3.80)	18-3099-83
2,4,6-Cycloheptatrien-1-one, 2-hydroxy-4-[3-[4-methoxy-3-(phenylmethoxy)-phenyl]-1-oxo-2-propenyl]-, (E)-	MeOH	253(4.48),332s(4.15), 376(4.44)	18-3099-83
	MeOH-NaOH	263(4.33),351(4.44), 440s(3.80)	18-3099-83
$C_{24}H_{20}O_7$			
5,12-Naphthacenedione, 6,11-diacetoxy-7,8-dihydro-9-(1-hydroxyethyl)-	MeCN	252(4.10),285(4.36), 377(3.57)	44-2820-83
$C_{24}H_{20}O_8$			
5,12-Naphthacenedione, 6,11-diacetoxy-8-acetyl-7,8,9,10-tetrahydro-8-hy-droxy-	EtOH	257(4.32),277(3.68), 340(3.24)	35-1608-83
Naphthaceno[1,2-b]oxirene-5,10-dione, 4,11-diacetoxy-1a,2,3,11b-tetrahydro-1a-(1-hydroxyethyl)-, [1aα(R*),11bα]-(±)-	MeCN	269(4.34),280(4.09), 330(3.56)	44-2820-83
[1aα(S*),11bα]-(±)-	MeCN	268(4.37),281(4.11), 330(3.59)	44-2820-83
$C_{24}H_{21}As$			
1H-Arsole, 2,5-bis(4-methylphenyl)-1-phenyl-	C_6H_{12}	210(4.51),338(4.34), 377(4.29)	101-0335-83I
$C_{24}H_{21}BrO_7$			
5,12-Naphthacenedione, 9-acetoxy-9-ace-tyl-7-bromo-7,8,9,10-tetrahydro-6,11-dimethoxy-, trans-(±)-	MeCN	255(4.01),370(3.31), 476(3.29),508(3.28)	35-1608-83
$C_{24}H_{21}Cl_3N_2OS$			
9H-Thioxanthen-9-one, 7-chloro-3-(2,4-dichlorophenyl)-1-[[3-(dimethylamino)-propyl]amino]-	MeOH	230(4.49),268(4.60), 346(4.00),450(3.95)	4-1575-83
$C_{24}H_{21}NO_2$			
5(4H)-Oxazolone, 4,4-diphenyl-2-(2,4,6-trimethylphenyl)-	EtOH	238s(4.04)	39-0979-83C
$C_{24}H_{21}NO_5$			
2,4,6-Cycloheptatrien-1-one, 2-hydroxy-4-[1-(hydroxyimino)-3-[3-methoxy-4-(phenylmethoxy)phenyl]-2-propenyl]-, (?,E)-	MeOH	240(4.51),325(4.47), 375s(3.92)	18-3099-83
	MeOH-NaOH	239(4.39),326(4.33), 402(3.69)	18-3099-83
2,4,6-Cycloheptatrien-1-one, 2-hydroxy-4-[1-(hydroxyimino)-3-[4-methoxy-3-(phenylmethoxy)phenyl]-2-propenyl]-, (?,E)-	MeOH	241(4.50),325(4.42), 380(3.77)	18-3099-83
	MeOH-NaOH	240s(4.38),325(4.31), 402(3.69)	18-3099-83
$C_{24}H_{21}NO_6$			
Spiro[cyclohexane-1,5'(6'H)-[4H]pyrido-	MeOH	219s(4.33),236(4.50),	39-0515-83C

Compound	Solvent	λ_{max}(log ε)	Ref.
[3,2,1-jk]carbazole]-1',3'-dicarboxylic acid, 4',6'-dioxo-, dimethyl ester		275(4.01),338(3.94)	39-0515-83C
$C_{24}H_{21}NO_7$			
4,7-Epoxy-1,2-benzisoxazole-3a,7a-dicarboxylic acid, 3-benzoyl-2,3,4,7-tetrahydro-2-phenyl-, dimethyl ester, (3α,3aα,4β,7β,7aα)-	MeOH	250(4.28)	73-1048-83
(3α,3aβ,4α,7α,7aβ)-	MeOH	250(3.65)	73-1048-83
4,7-Epoxy-1,2-benzisoxazole-5,6-dicarboxylic acid, 3-benzoyl-2,3,3a,4,7,7a-hexahydro-2-phenyl-, dimethyl ester, (3α,3aβ,4α,7α,7aβ)-	MeOH	245(4.39)	73-1048-83
$C_{24}H_{21}N_2OP$			
3-Furancarbonitrile, 4,5-dihydro-5-methyl-2-[(triphenylphosphoranylidene)amino]-	$CHCl_3$	267(4.16),273(4.11), 285(3.99)	24-1691-83
$C_{24}H_{21}N_5O_2$			
Benzeneacetic acid, 2-[[3-methyl-1-phenyl-4-(phenylazo)-1H-pyrazol-5-yl]amino]-	EtOH	215(4.32),280(3.91), 329(4.20),372(3.97)	4-1501-83
$C_{24}H_{22}$			
1H-Indene, 1-ethyl-3-methyl-1,2-diphenyl-	EtOH	222s(3.72),290(3.35)	35-4446-83
1H-Indene, 4-ethyl-3-methyl-1,2-diphenyl-	EtOH	288(4.02)	35-4446-83
$C_{24}H_{22}BN_5O_7$			
Boron, bis(acetato-O)[[[(4-methoxyphenyl)azo](4-nitrophenyl)methyl]phenyldiazenato]-, (T-4)-	hexane	242(4.30),308(4.08), 342(4.27),378(4.20), 553(4.37)	65-2061-83
	EtOH	245(4.20),303(3.62), 347(4.22),384(3.74), 561(4.28)	65-2061-83
Boron, bis(acetato-O)[[(4-methoxyphenyl)[(4-nitrophenyl)azo]methyl]phenyldiazenato]-, (T-4)-	EtOH	256(4.27),277(3.88), 317(4.15),380(3.85), 596(4.15)	65-2061-83
$C_{24}H_{22}ClNOS$			
2-Propen-1-one, 3-[1-(4-chlorophenyl)-1,2-dihydro-2-thioxo-3-pyridinyl]-1-[4-(1,1-dimethylethyl)phenyl]-, (E)-	EtOH	245(3.87),276(3.97), 344(3.61),430(3.36)	4-1651-83
$C_{24}H_{22}F_3NO_2$			
1H-Indol-3-ol, 2-ethoxy-2,3-dihydro-1-methyl-3-phenyl-2-[(4-trifluoromethyl)phenyl]-	MeCN	248(4.01),305(3.26)	24-2115-83
$C_{24}H_{22}Fe$			
Ferrocene, 1,1'-bis(4-methylphenyl)-	dioxan	225(4.5),245(4.4), 275(4.3),450(2.7)	18-2023-83
$C_{24}H_{22}NOTe$			
1-Benzo[b]tellurinium, 4-[4-(dimethylamino)phenyl]-7-methoxy-2-phenyl-, perchlorate	CH_2Cl_2	678(4.46)	35-0883-83

Compound	Solvent	λ max(log ε)	Ref.
$C_{24}H_{22}N_2O_2$			
1H-Pyrrolo[2,3-d]carbazole-2,5(3H,6H)-dione, 3a,4,6a,7-tetrahydro-3a-methyl-6-(3-phenyl-2-propenylidene)-, (3aα,6aβ,11bR*)-(±)-	MeOH	238(3.03),332(3.49)	5-1744-83
$C_{24}H_{22}N_2O_7S$			
1-Azabicyclo[3.2.0]hept-2-ene-2-carboxylic acid, 6-(1-acetoxyethyl)-7-oxo-3-(phenylthio)-, (4-nitrophenyl)methyl ester	$CHCl_3$	271(3.99),320(4.05)	158-1473-83
$C_{24}H_{22}N_4$			
21H,23H-Porphine, tetramethyl-	benzene	420(5.55),487(3.45), 521(4.10),556(3.93), 606(3.51),666(3.78)	64-1240-83B
$C_{24}H_{22}N_4O_4$			
Benzoic acid, 2-methyl-4-(phenylmethoxy)-6-[(1-phenyl-1H-tetrazol-5-yl)oxy]-, ethyl ester	EtOH	239(4.43)	39-0667-83C
$C_{24}H_{22}O_4$			
1(3H)-Isobenzofuranone, 3,3-bis(4-hydroxy-3,5-dimethylphenyl)-	EtOH	205(4.95),276(3.66)	24-0970-83
$C_{24}H_{23}BN_4O_4$			
Boron, bis(acetato-O)[[[(2-methylphenyl)azo]phenylmethyl]phenyldiazenato]-, (T-4)-	EtOH	249(4.21),289(3.80), 309(4.15),358(3.63), 571(4.08)	65-2061-83
$C_{24}H_{23}BN_4O_5$			
Boron, bis(acetato-O)[[[(2-methoxyphenyl)azo]phenylmethyl]phenyldiazenato]-, (T-4)-	EtOH	245(4.18),277(4.02), 298(3.63),307(3.97), 370(3.53),516(4.02)	65-2061-83
Boron, bis(acetato-O)[[[(4-methoxyphenyl)azo]phenylmethyl]phenyldiazenato]-, (T-4)-	EtOH	263(4.27),320(3.96), 374(3.50),570(4.31)	65-2061-83
	hexane	263(4.27),299(3.40), 318(4.08),391(3.30), 557(4.30)	65-2061-83
Boron, bis(acetato-O)[1,1'-[(4-methoxyphenyl)methylene]bis[2-phenyldiazenato(1-)-N^2,$N^{2'}$]-	hexane	264(4.25),284(3.65), 307(4.24),374(3.78), 566(4.08)	65-2061-83
	EtOH	264(4.32),308(4.26), 375(3.36),573(4.06)	65-2061-83
$C_{24}H_{23}Cl_3N_2OS$			
9H-Thioxanthen-9-one, 7-chloro-3-(2,4-dichlorophenyl)-1-[[3-(dimethylamino)propyl]amino]-3,4-dihydro-	MeOH	244(4.41),294(4.22), 332(3.75),346(3.73), 443(3.64)	4-1575-83
$C_{24}H_{23}NO$			
2-Propen-1-one, 3-[4-(dimethylamino)-phenyl]-1-[4-(phenylmethyl)phenyl]-	isooctane	261(4.32),387(4.54)	124-1211-83
	dioxan	402(4.53)	124-1211-83
$C_{24}H_{23}NOS$			
2-Propen-1-one, 3-(1,2-dihydro-1-phenyl-2-thioxo-3-pyridinyl)-1-[4-(1,1-dimethylethyl)phenyl]-, (E)-	EtOH	242(4.23),278(4.38), 343(4.04),430(3.79)	4-1651-83

Compound	Solvent	λ_{max}(log ε)	Ref.
$C_{24}H_{23}NO_4$			
2,4,6-Cycloheptatrien-1-one, 4-[1-amino-3-[3-methoxy-4-(phenylmethoxy)-phenyl]-2-propenyl]-2-hydroxy-, (E)-	MeOH	246(4.47),309(4.16), 393s(3.72)	18-3099-83
Spiro[9H-fluorene-9,1'(5'H)-indolizine]-2',3'-dicarboxylic acid, 6',7',8',8'a-tetrahydro-, dimethyl ester	CH_2Cl_2	327(4.04)	24-3915-83
$C_{24}H_{23}NO_5$			
9H-Phenanthro[9,10-b]quinolizine-11,15-dione, 12,13,14,14a-tetrahydro-2,3,6-trimethoxy-, (S)-	$CHCl_3$	251(4.31),272(4.35), 279s(--),288(4.32), 302s(--),344(3.58)	44-4222-83
Spiro[cyclohex-2-ene-1,9'-[9H-1,3]dioxolo[4,5-h][2]benzazepine]-6'(5'H)-carboxylic acid, 7',8'-dihydro-4-oxo-, phenylmethyl ester, (±)-	n.s.g.	238(4.05),294(3.49)	35-7640-83
$C_{24}H_{23}NO_6$			
Dibenzo[f,h]pyrrolo[1,2-b]isoquinoline-11,14(9H,12H)-dione, 13,13a-dihydro-2,3,6,7-tetramethoxy-, (S)-	$CHCl_3$	262(4.54),275(4.52) 2.1[sic](4.47),298s(--), 353(4.05),407s(--)	44-4222-83
$C_{24}H_{23}NO_7$			
4,7-Epoxy-1,2-benzisoxazole-5,6-dicarboxylic acid, 3-benzoyloctahydro-2-phenyl-, dimethyl ester, (3α,3aβ,4α,7α,7aβ)-	MeOH	247(4.35)	73-1048-83
Spiro[cyclopentane-1,10'-[6,8a]ethano-[8aH]carbazole]-5',7',8'-tricarboxylic acid, 6',9'-dihydro-11'-oxo-, trimethyl ester	MeOH	209(4.27),244.5(4.37), 295(3.83),314(3.69), 435(3.95)	39-0515-83C
$C_{24}H_{23}N_2O_2S$			
Morpholinium, 4-[2-(4-methoxyphenyl)-4-(1-naphthalenyl)-5(2H)-isothiazolylidene]-, perchlorate	MeCN	223(4.83),292(4.09), 322(4.08),352(4.24)	48-0689-83
$C_{24}H_{23}N_3O_2$			
3H-Pyrazol-3-one, 4-[3-(diethylamino)-4-oxo-1(4H)-naphthalenylidene]-2,4-dihydro-5-methyl-2-phenyl-	$CHCl_3$	255(4.45),340(3.97), 400(3.75),620(3.97)	88-3567-83
$C_{24}H_{23}N_3O_4S$			
2H-Thiopyran-4-amine, N-(2,4-dinitrophenyl)tetrahydro-N-methyl-2,6-diphenyl-, axial	EtOH	357(3.94)	2-0410-83
equatorial	EtOH	367(4.26)	2-0410-83
$C_{24}H_{23}N_3O_5$			
2H-Pyran-4-amine, N-(2,4-dinitrophenyl)-tetrahydro-N-methyl-2,6-diphenyl-	EtOH	360(4.01)	2-0410-83
epimer	EtOH	365(4.28)	2-0410-83
$C_{24}H_{24}$			
Tricyclo[16.2.2.$2^{8,11}$]tetracosa-2,6,8-10,12,16,18,20,21,23-decaene, (all-E)-	C_6H_{12}	236s(4.29),266s(4.59), 276(4.69),316(4.12)	88-4413-83
$C_{24}H_{24}Cl_2N_6O_8$			
Piperidine, 1,1'-[(dichloroethenylidene)bis(2,6-dinitro-4,1-phenylene)]-bis-	MeOH	226.3(4.38),263.7(4.39), 442.4(4.05)	56-1357-83

Compound	Solvent	λ max(log ε)	Ref.
$C_{24}H_{24}N_2$			
Cyclohexanamine, N-(2,3-dimethyl-11H-indeno[1,2-b]quinolin-11-ylidene)-	EtOH	208(4.45),216(4.47), 226s(4.5),235(4.57), 247(4.54),257(4.5), 290(4.86),370s(3.46)	104-0158-83
Cyclohexanamine, N-(2,4-dimethyl-11H-indeno[1,2-b]quinolin-11-ylidene)-	EtOH	203s(4.4),219(4.48), 235(4.52),248(4.48), 260(4.48),290(4.77), 350s(3.56),380s(2.93)	104-0158-83
Cyclohexanamine, N-(3,6-dimethyl-11H-indeno[1,2-b]quinolin-11-ylidene)-	EtOH	204(4.46),216(4.44), 227(4.47),234(4.5), 247(4.5),254(4.44), 292(4.71),336s(3.86)	104-0158-83
Cyclohexanamine, N-(2,6-dimethyl-11H-indeno[1,2-b]quinolin-11-ylidene)-	EtOH	217s(4.54),228s(4.56), 236(4.62),248(4.6), 256(4.53),294(4.86), 330s(3.83)	104-0158-83
2,4,6,8-Undecatetraen-10-ynal, 9-methyl-, (9-methyl-2,4,6,8-undecatetraen-10-ynylidene)hydrazone, (?,?,E,Z,Z-E,E,E,E)-	THF	293(4.51),303(4.50), 402s(4.78),418(4.87), 437(4.86),468s(4.64)	18-1467-83
$C_{24}H_{24}N_2O$			
2-Propen-1-one, 3-[4-(dimethylamino)-phenyl]-1-[4-(methylphenylamino)-phenyl]-	EtOH	427(4.65)	124-1211-83
	dioxan	279(3.90),308(3.45), 345(4.08),403(4.56)	124-1211-83
$C_{24}H_{24}N_2OS$			
Cyclohexanone, 2-[(3-ethyl-2(3H)-benzothiazolylidene)ethylidene]-6-[(phenylamino)methylene]-	benzene	474(--).498(4.81)	104-1854-83
	MeOH	494(--),516(4.78)	104-1854-83
	DMF	486(--),506(--)	104-1854-83
Cyclopentanone, 2-[(3-ethyl-2(3H)-benzothiazolylidene)ethylidene]-5-[(methylphenylamino)methylene]-	benzene	440(--),464(4.84)	104-1854-83
	MeOH	512(4.89)	104-1854-83
	DMF	487(--)	104-1854-83
$C_{24}H_{24}N_2O_2$			
Benzeneacetamide, α-(benzoylamino)-2-ethyl-N-methyl-N-phenyl-	EtOH	224(4.31),266s(3.33), 276s(3.03)	33-0262-83
1H-Pyrazole-4-carboxylic acid, 3-(2-methyl-1-propenyl)-1-phenyl-5-(2-phenylethenyl)-, ethyl ester, (E)-	EtOH	268(4.22),285(4.20)	118-0948-83
Pyrazolo[1,5-a]quinoline-3-carboxylic acid, 4,5-dihydro-2-(2-methyl-1-propenyl)-5-phenyl-, ethyl ester	EtOH	270(4.19),286(4.19)	118-0948-83
$C_{24}H_{24}N_2O_2S$			
Morpholine, 4-[3-[(4-methoxyphenyl)-amino]-2-(1-naphthalenyl)-1-thioxo-2-propenyl]-	MeCN	224(4.84),272(4.30), 304s(4.25),317(4.26), 342s(4.14),403s(3.90)	48-0689-83
$C_{24}H_{24}N_2O_3S$			
Benzenesulfonic acid, 4-methyl-, [1-(4-methoxyphenyl)-4-phenyl-3-butenylidene]hydrazide, (E)-	EtOH	255(4.20)	35-0933-83
$C_{24}H_{24}N_2O_4S$			
2H-1-Benzopyran-2-one, 7-(diethylamino)-3-[2-(3,4-dimethoxyphenyl)-4-thiazolyl]-	MeCN	420(4.51)	48-0551-83

Compound	Solvent	λ max(log ε)	Ref.
$C_{24}H_{24}O_3$			
Bicyclo[3.2.0]hept-2-en-6-one, 4-[bis(4-methoxyphenyl)methylene]-7,7-dimethyl-	MeCN	257(4.268),302(4.370), 320(4.186)	64-0504-83B
$C_{24}H_{24}O_4$			
2(3H)-Benzofuranone, 7-(1,1-dimethylethyl)-3-[7-(1,1-dimethylethyl)-2-oxo-3(2H)-benzofuranylidene]-, (E)-	CH_2Cl_2	228(4.06),255s(4.08), 260(4.10),401s(4.19), 415(4.22)	78-2147-83
$C_{24}H_{24}O_7$			
Naphthaceno[1,2-d]-1,3-dioxole-7,12-dione, 3a,4,5,13b-tetrahydro-6,8,13-trihydroxy-2,2-dimethyl-3a-propyl-, cis-(±)-	MeOH	233(4.48),250(4.25), 292(3.77),464(3.93), 480(3.99),490(4.02), 512(3.89),524(3.84)	5-1818-83
$C_{24}H_{24}O_8$			
D-ribo-Oct-3-en-2-ulose, 3-acetyl-1,3,4-trideoxy-, 6,8-dibenzoate	MeOH	245(4.41)	87-0030-83
6H-Phenaleno[1,2-b]furan-6-one, 3,4-diacetoxy-8,9-dihydro-7-hydroxy-5-methoxy-1,8,8,9-tetramethyl-, (R)-	EtOH	360(5.09),410(5.34), 433(5.15)	18-3661-83
$C_{24}H_{25}ClN_3OS$			
Morpholinium, 4-[3-(4-chlorophenyl)-6-[[(dimethylamino)methylene]amino]-5-phenyl-2H-thiopyran-2-ylidene]-, perchlorate	MeCN	243s(4.42),258(4.45), 341(4.29),498(4.35)	97-0403-83
Morpholinium, 4-[5-(4-chlorophenyl)-6-[[(dimethylamino)methylene]amino]-3-phenyl-2H-thiopyran-2-ylidene]-, perchlorate	MeCN	240(4.41),256(4.46), 339(4.31),498(4.36)	97-0403-83
$C_{24}H_{25}ClO_3$			
1,4-Naphthalenedione, 3-chloro-2-[3,5-bis(1,1-dimethylethyl)-4-hydroxyphenyl]-	EtOH	211(4.42),270(4.34), 337(3.53),454(3.45)	104-0144-83
$C_{24}H_{25}NO_2$			
1H-Indol-3-ol, 2-ethoxy-2,3-dihydro-1-methyl-2-(4-methylphenyl)-3-phenyl-, trans	MeOH	244(4.01),306(3.40)	24-2115-83
	MeOH-acid	336(4.10)	24-2115-83
	MeCN	248(3.99),310(3.44)	24-2115-83
$C_{24}H_{25}NO_3$			
1H-Indol-3-ol, 2-ethoxy-2,3-dihydro-2-(4-methoxyphenyl)-1-methyl-3-phenyl-, trans	hexane	232(4.19),250s(3.81), 307(3.26)	24-2115-83
	EtOH-acid	245(4.08),377(4.29)	24-2115-83
$C_{24}H_{25}NO_4$			
Glycofoline	MeOH	225(4.13),268s(4.32), 280s(4.39),297(4.59), 308(4.65),335s(3.92), 415(3.53)	39-1681-83C
Phenanthro[9,10-b]quinolizin-11-one, 9,12,13,14,14a,15-hexahydro-2,3,6-trimethoxy-, (S)-	$CHCl_3$	262(4.50),283s(--), 288(4.30),313s(--), 344(3.11)	44-4222-83
$C_{24}H_{25}NO_5$			
Dibenzo[f,h]pyrrolo[1,2-b]isoquinolin-11(9H)-one, 12,13,13a,14-tetrahydro-	$CHCl_3$	259(4.87),285s(--), 290(4.62),305(3.37),	44-4222-83

Compound	Solvent	λ_{max}(log ϵ)	Ref.
2,3,6,7-tetramethoxy-, (S)- (cont.)		324(3.33),340(3.37), 356(3.29),367(2.56)	44-4222-83
1H-Indole-1-carboxylic acid, 4-(3-oxo-1-butenyl)-, 1-(3,5-dimethoxyphenyl)-1-methylethyl ester	MeOH	262(4.42),317(4.37)	5-2135-83
11H-Phenanthro[9,10-b]quinolizin-11-one, 9,12,13,14,14a,15-hexahydro-15-hydroxy-2,3,6-trimethoxy-, (14aS-cis)	$CHCl_3$	259(4.63),278s(--), 283(4.38),309s(--), 396(3.01)	44-4222-83
Phenol, 2-[(2-hydroxy-5-nitrophenyl)-methyl]-6-[(4-methoxy-3,5-dimethyl-phenyl)methyl]-4-methyl-	neutral	326(3.95)	126-2361-83
	anion	404(4.24)	126-2361-83
	dianion	416(4.29)	126-2361-83
$C_{24}H_{25}NO_6$			
Dibenzo[f,h]pyrrolo[1,2-b]isoquinolin-11(9H)-one, 12,13,13a,14-tetrahydro-14-hydroxy-2,3,6,7-tetramethoxy-, (13aS-cis)-	$CHCl_3$	245s(--),253s(--), 261(4.77),284(4.39), 390s[sic](4.45), 304(4.18),326s(--), 341(3.37),358(3.34)	44-4222-83
(13aS-trans)-	$CHCl_3$	262(4.71),285s(--), 290(4.46),304(4.21), 316s(--),325s(--), 342(3.43),358(3.33)	44-4222-83
2-Piperidinecarboxylic acid, 6-oxo-1-[(2,3,6-trimethoxy-9-phenanthrenyl)-methyl]-, (S)-	$CHCl_3$	257(4.55),280(4.34)	44-4222-83
$C_{24}H_{25}NO_7$			
L-Proline, 5-oxo-1-[(2,3,6,7-tetramethoxy-9-phenanthrenyl)methyl]-	$CHCl_3$	261(4.81),284s(--), 290(4.51),304(4.28), 324(3.32),341(3.40), 357(3.29)	44-4222-83
$C_{24}H_{25}N_2$			
3H-Indolium, 3-[3-(1,2-dimethyl-1H-indol-3-yl)-2-methyl-2-propenylidene]-1,2-dimethyl-, chloride	$MeNO_2$-HOAc	580(4.88)	103-1306-83
$C_{24}H_{25}N_3O$			
2H-Benzo[g]indazol-3-amine, N-cyclohexyl-2-(4-methoxyphenyl)-	$CHCl_3$	240(4.34),272(4.41), 335(3.98)	33-2603-83
$C_{24}H_{25}N_3O_2S$			
2H-1-Benzopyran-2-one, 7-(diethylamino)-3-[2-[4-(dimethylamino)phenyl]-4-thiazolyl]-	MeCN	417(4.64)	48-0551-83
$C_{24}H_{25}N_3O_2S_2$			
Benzenesulfonamide, 4-methyl-N-(tetrahydro-3,5-dimethyl-4,6-diphenyl-2H-1,3,5-thiadiazin-2-ylidene)-	dioxan	246(4.31),c.335(2.91)	24-1297-83
$C_{24}H_{25}N_3O_{11}$			
β-D-Glucopyranuronic acid, 1-(4-benzoylamino)-2-oxo-1(2H)-pyrimidinyl]-1-deoxy-, methyl ester, 2,3,4-triacetate	MeOH	230(4.29),272(3.96)	105-0580-83
$C_{24}H_{25}N_4$			
1H-Benzimidazolium, 2-[1-cyano-3-(1,3-dihydro-1,3,3-trimethyl-2H-indol-2-ylidene)-1-propenyl]-1,3-dimethyl-, iodide	EtOH	459(4.39)	124-0297-83

Compound	Solvent	λmax(log ε)	Ref.
$C_{24}H_{25}N_5O_4$			
Acetamide, N-[2-(1,2,3,6,8,9-hexahydro-1,3-dimethyl-2,6,8-trioxo-7H-purin-7-yl)-1-phenylethyl]-N-(phenylmethyl)-	EtOH	295(4.06)	56-0461-83
$C_{24}H_{26}Cl_2N_6O$			
4H-[1,2,4]Triazolo[4,3-a][1,4]benzodiazepine, 8-chloro-6-(2-chlorophenyl)-1-[4-(2-ethoxyethyl)piperazino]-	MeOH	243(4.11)	73-2395-83
4H-[1,2,4]Triazolo[4,3-a][1,4]benzodiazepine, 8-chloro-6-(2-chlorophenyl)-1-[4-(3-methoxypropyl)piperazino]-	MeOH	215(4.64),245s(4.13)	73-2395-83
$C_{24}H_{26}N_2$			
3,4-Diazabicyclo[4.1.0]hepta-2,4-diene, 7-methyl-7-[2-(1-methylcyclopropyl)ethyl]-2,5-diphenyl-, anti	dioxan	260(4.08),329(4.22)	88-1485-83
syn	dioxan	260(4.07),329(4.22)	88-1485-83
1H-Pyrrolo[1,2-a]indole, 2,3-dihydro-1,1,3,9-tetramethyl-3-(3-methyl-1H-indol-2-yl)-	EtOH	227(4.32),285(3.69), 293(3.68)	142-1355-83
	EtOH-$HClO_4$	204(3.97),224(3.96), 269(3.55),290(3.54)	142-1355-83
$C_{24}H_{26}N_2O_3$			
2H-Pyran-5-carboxylic acid, 3,4-dihydro-2,2-dimethyl-6-(2-phenylethenyl)-4-(phenylhydrazono)-, ethyl ester, (?,E)-	EtOH	302(4.25),400(4.49)	118-0948-83
$C_{24}H_{26}N_2O_4$			
Benzoic acid, 4-(hexyloxy)-, 6-(4-methoxyphenyl)-3-pyridazinyl ester	MeOH	201(4.55),265(4.52)	97-0296-83
$C_{24}H_{26}N_3OS$			
Morpholinium, 4-[6-[[(dimethylamino)-methylene]amino]-3,5-diphenyl-2H-thiopyran-2-ylidene]-, perchlorate	MeCN	239s(4.37),253(4.40), 340(4.26),498(4.34)	97-0403-83
$C_{24}H_{26}N_4O_{13}$			
4,18:13,17-Diepoxy-2H,5H,15H-1,3,8,14-16-pentaoxa-4a,6,11,12a-tetraazabenzo[g']naphtho[1,8-fg:2,3-f']diazulene-5,7,10,12(6H,9H,11H)-tetrone, 3a,4,13,13a,16a,17,17a,17b,18,18a-decahydro-9,9-dihydroxy-2,2,15,15-tetramethyl-, [3aR-(3aα,4β,8aR*,13α-13aβ,16aβ,17α,17aβ,17bα,18β,18aα)]-	H_2O	286(3.89)	4-0753-83
$C_{24}H_{26}O$			
2(3H)-Naphthalenone, 4,4a,5,6,7,8-hexahydro-6-methyl-4-phenyl-1-(phenylmethyl)-	EtOH	208(4.27),255(4.02)	2-0542-83
$C_{24}H_{26}O_3$			
Bicyclo[3.2.0]heptan-6-one, 4-[bis(4-methoxyphenyl)methylene]-7,7-dimethyl-, (1α,5α,6α)-	C_6H_{12}	246(4.313),265(4.258), 310s(3.590)	64-0504-83B
1,4-Naphthalenedione, 2-[3,5-bis(1,1-dimethylethyl)-4-hydroxyphenyl]-	EtOH	201(4.72),260(4.39), 333(3.72),451(3.75)	104-0144-83

Compound	Solvent	λ_{max}(log ϵ)	Ref.
$C_{24}H_{26}O_4$			
Bicyclo[5.2.1]decane-4,10-dione, 2,6-bis(2-methoxyphenyl)-	MeOH	217(4.27),275(3.74), 281(3.73)	2-0864-83
$C_{24}H_{26}O_6$			
Mangostin	EtOH	243(4.54),259(4.44), 318(4.38),351(3.86)	162-0818-83
9H-Xanthen-9-one, 5-(1,1-dimethyl-2-propenyl)-3,6,8-trihydroxy-2-methoxy-1-(3-methyl-2-butenyl)- (cudraxanthone C)	EtOH	246(4.35),258(4.27), 317(4.17),354(3.75)	142-0213-83
	EtOH-AlCl$_3$	237(4.30),270(4.23), 278s(4.17),343(4.23), 401(3.64)	142-0213-83
$C_{24}H_{26}O_8$			
Naphtho[2,3-b]furan-2,5,6(3H)-trione, 8-acetoxy-3-(3-acetoxypropyl)-3,4-dimethyl-7-(1-methylethyl)-, (R)-	EtOH	241.5(4.61),290s(3.74), 298s(3.71),302s(3.71), 335.5(3.53)	18-2985-83
Naphtho[2,3-b]furan-2,5,8(3H)-trione, 6-acetoxy-3-(3-acetoxypropyl)-3,4-dimethyl-7-(1-methylethyl)-, (R)-	EtOH	252(4.31),353(3.45)	18-2985-83
$C_{24}H_{26}O_9$			
Shinjulactone C, 12,20-di-O-acetyl-	EtOH	245.5(3.98)	18-3683-83
$C_{24}H_{26}O_{13}$			
Centaurein	MeOH	258(4.30),349(4.31)	162-0274-83
$C_{24}H_{26}Si_2$			
Silane, benzo[3,4]cyclobuta[1,2-b]biphenylene-2,3-diylbis[trimethyl-	isooctane	279(4.85),290(5.06), 320s(3.80),410(4.24), 422s(4.18),436(4.46)	77-0502-83
$C_{24}H_{27}BrO_2$			
Methanone, (4-bromophenyl)(1,2,3,4,4a-9,10,10a-octahydro-6-methoxy-1,4a-dimethyl-1-phenanthrenyl)-, [1S-(1α,4aα,10aβ)]-	EtOH	247(4.06),277(3.52), 286(3.42)	12-1275-83
$C_{24}H_{27}ClO_2$			
Methanone, (4-chlorophenyl)(1,2,3,4,4a-9,10,10a-octahydro-6-methoxy-1,4a-dimethyl-1-phenanthrenyl)-, [1S-(1α,4aα,10aβ)]-	EtOH	243(4.08),278(3.53), 286(3.46)	12-1275-83
$C_{24}H_{27}FN_6O_6$			
Pregna-1,4-diene-3,11-dione, 6β,7α-diazido-9α-fluoro-16α-methyl-17 ,20-20,21-bis[methylenebis(oxy)]-	MeOH	234(4.20)	39-2787-83C
$C_{24}H_{27}NO$			
2,5-Cyclohexadien-1-one, 4-[[3,7-bis(1,1-dimethylethyl)-1-naphthalenyl]imino]-	n.s.g.	232(3.65),280(3.23), 515(2.57)	39-1759-83C
1(4H)-Naphthalenone, 2,6-bis(1,1-dimethylethyl)-4-(phenylimino)-	n.s.g.	340(4.63)	39-1759-83C
1(4H)-Naphthalenone, 4-[[3,5-bis(1,1-dimethylethyl)phenyl]imino]-	n.s.g.	255(4.34),435s(3.49), 454(3.54)	18-1476-83
$C_{24}H_{27}NO_2$			
2,5-Cyclohexadien-1-one, 4-[[3,7-bis(1,1-dimethylethyl)-1-naphtha-	n.s.g.	362(4.0)	39-1759-83C

Compound	Solvent	λ_{max}(log ε)	Ref.
lenyl]imino]-, N-oxide (cont.)			39-1759-83C
$C_{24}H_{27}NO_6$			
1,7-Ethano-4H,7H-furo[4',3',2':1,8]-naphth[1,2-d]azocine-2-carboxylic acid, 14-acetoxy-1,5,6,12-tetrahydro-9-methoxy-4-methyl-, methyl ester, [1S-(1α,6aR*,7α,14R*)]- (compd. 16)	EtOH	296(4.28)	44-0173-83
dihydro 16	EtOH	232(3.14)	44-0173-83
$C_{24}H_{27}NO_7$			
Colchicine, N-deacetyl-N-3-oxobutyryl-	EtOH and EtOH-base	237s(4.48),243(4.50), 352(4.25)	73-2989-83
$C_{24}H_{27}N_5O_3$			
1H-Purine-2,6-dione, 8-[ethyl(phenylmethyl)amino]-3,7-dihydro-7-(2-hydroxy-2-phenylethyl)-1,3-dimethyl-	EtOH	297(4.19)	56-0461-83
$C_{24}H_{27}N_5O_9$			
Adenosine, N-benzoyl-7,8-dihydro-8-(hydroxymethyl)-, 2',3',5'-triacetate	MeOH	282(3.73),330(3.97)	88-0799-83
$C_{24}H_{28}N_2O$			
Spiro[furo[3,2-b]quinoline-2(4H),2'-[2H]indole], 1',3',9,9a-tetrahydro-1',3',3',4,9,9-hexamethyl-	MeOH	250s(3.50),277(3.92)	24-1309-83
$C_{24}H_{28}N_2O_2$			
2,4,6-Heptatrien-1-one, 1,1'-(1,4-phenylene)bis[7-(dimethylamino)-	$CHCl_3$	290(4.24),360(4.20), 500(4.87)	70-1882-83
$C_{24}H_{28}N_2O_9$			
9,10-Anthracenedione, 1-(β-D-glucopyranosyloxy)-4-[[2-[(2-hydroxyethyl)amino]ethyl]amino]-	MeOH	248(4.56),280(3.93), 526(3.87)	18-1435-83
$C_{24}H_{28}N_4O_4$			
1,20-Epoxy-3H-dibenzo[c,f][1,5,9]triazacycloheptadecine-3,5,16-trione, 4-amino-6,7,8,9,10,11,12,13,14,15-decahydro-2,19-dimethyl-	DMF	265(3.89),425(3.87), 446(3.86)	95-0049-83
$C_{24}H_{28}N_6O_6$			
Pyrrolidine, 1,1',1"-(2',4',6'-trinitro[1,1'-biphenyl]-2,4,6-triyl)tris-	$CHCl_3$	400(3.5),588(3.76)	44-4649-83
$C_{24}H_{28}N_9O_{11}P$			
MG-1	pH 1	260(--),275s(--), 312(--)	94-0861-83
	pH 7	255(4.58),277s(--), 310(3.78)	94-0861-83
	pH 10	255(--),270s(--), 312(--)	94-0861-83
MG-2	pH 1	260(--),310(--)	94-0861-83
	pH 7	255(4.58),275s(--), 308(3.78)	94-0861-83
	pH 10	255(--),275s(--), 310(--)	94-0861-83

Compound	Solvent	λ_{max}(log ϵ)	Ref.
$C_{24}H_{28}O_2$			
Methanone, (1,2,3,4,4a,9,10,10a-octahydro-6-methoxy-1,4a-dimethyl-1-phenanthrenyl)phenyl-, [1S-(1α,4aα,10aβ)]-	EtOH	227s(4.10),278(3.52), 287(3.43)	12-1275-83
$C_{24}H_{28}O_4$			
Diligustilide, (Z,Z')-	n.s.g.	270(4.09)	88-4677-83
$C_{24}H_{28}O_5$			
2H-1-Benzopyran-2-one, 7-[(1,4,4a,5,6-7,8,8a-octahydro-6-hydroxy-2,5,5,8a-tetramethyl-4-oxo-1-naphthalenyl)-methoxy]-	EtOH	222(4.38),253(3.24), 296(3.68),326(3.92)	105-0664-83
$C_{24}H_{28}O_6$			
6H,7aH-Bischromeno[3,4-b:3',4'-e][1,4]-dioxin, 6a,13,13a,14a-tetrahydro-3,10-dimethoxy-6,6,13,13-tetramethyl-	EtOH	212(4.40),220s(4.23), 276(3.76),280s(3.72)	39-1431-83C
$C_{24}H_{28}O_7$			
Benzenemethanol, 4-acetoxy-3-methoxy-α-[1-[2-methoxy-4-(1-propenyl)phenoxy]ethyl]-, acetate, erythro-	EtOH	260(4.21),264s(4.20)	150-2625-83M
threo-	EtOH	259(4.22),264s(4.20)	150-2625-83M
Paraherquonin	EtOH	211.5(4.02),265(3.96)	88-3113-83
$C_{24}H_{28}O_8$			
Ethanone, 1-[3-[(3-acetyl-2,4-dihydroxy-6-methoxy-5-methylphenyl)methyl]-2,4,6-trihydroxy-5-(3-methyl-2-butenyl)phenyl]-	EtOH	293(4.37)	102-0323-83
7H-Phenaleno[1,2-b]furan-7-one, 3-acetoxy-7a,8,9,10a-tetrahydro-7a-hydroxy-4,5,6-trimethoxy-1,8,8,9-tetramethyl-	EtOH	334(4.29),348(4.32)	18-3661-83
$C_{24}H_{28}O_9$			
Ethanone, 1-[3-[(3-acetyl-2,4-dihydroxy-6-methoxy-5-methylphenyl)methyl]-2,4,6-trihydroxy-5-(2-hydroxy-3-methyl-3-butenyl)phenyl]-, (±)-	EtOH	232(4.35),289(4.27), 323(4.21)	102-0323-83
$C_{24}H_{28}O_{12}$			
Procumbide, 6'-O-p-coumaroyl-	MeOH	228(4.10),313(4.36)	94-2296-83
$C_{24}H_{28}O_{13}$			
Gentiopicrin tetraacetate	MeOH	272(3.84)	162-0628-83
$C_{24}H_{29}ClO_4$			
Cyproterone acetate	MeOH	281(4.24)	162-0399-83
$C_{24}H_{29}FN_6O_6$			
Pregna-1,4-diene-3,20-dione, 21-acetoxy-6β,7α-diazido-9α-fluoro-11β,17-dihydroxy-16α-methyl-	MeOH	237(4.19)	39-2787-83C
$C_{24}H_{29}FO_6$			
Pregna-1,4-diene-3,20-dione, 21-acetoxy-9-fluoro-11β,17-dihydroxy-16-methylene-	MeOH	238(4.20)	162-0600-83

Compound	Solvent	$\lambda_{max}(\log \epsilon)$	Ref.
$C_{24}H_{29}NO_3$			
4H-1-Benzopyran-4-one, 2,3-dihydro-5-hydroxy-2,2-dimethyl-3-[(methylphenylamino)methylene]-7-pentyl-	EtOH	261(3.98),291(3.76), 397(4.45)	161-0775-83
$C_{24}H_{29}NO_6$			
2-Propenoic acid, 3-[[2-(3,3a-dihydro-3,3,5-trimethoxyphenanthro[4,5-bcd]-furan-9b(2H)-yl)ethyl]methylamino]-, methyl ester, [3aR-[3aα,9bα(E)]]-	EtOH	282(4.49),326(3.81)	44-0173-83
2-Propenoic acid, 3-[[2-(3,8-dihydro-3,3,5-trimethoxyphenanthro[4,5-bcd]-furan-9b(3aH)-yl)ethyl]methylamino]-, methyl ester, [3aR-[3aα,9bα(E)]]-	EtOH	225s(4.26),280(4.49)	44-0173-83
$C_{24}H_{29}N_3O_6S_2$			
Merocyanine 540 (sodium salt)	H_2O	533(4.67)	35-2928-83
	EtOH	559(5.14)	35-2928-83
	benzene	567(4.51)	35-2928-83
	1-pentanol	562(5.11)	35-2928-83
$C_{24}H_{29}N_5O_4$			
Pyrrolidine, 1,1',1"-(2',4'-dinitro-[1,1'-biphenyl]-2,4,6-triyl)tris-	$CHCl_3$	524(3.80)	44-4649-83
$C_{24}H_{30}ClNO_4$			
2H,5H-Pyrano[3,2-c][1]benzopyran-2-one, 3-chloro-10-hydroxy-5,5-dimethyl-8-pentyl-4-piperidino-	EtOH	225s(4.34),269(3.91), 390(4.25)	161-0775-83
$C_{24}H_{30}F_2O_6$			
Fluocinolone acetonide	n.s.g.	238(4.21)	162-0593-83
Pregna-1,4-diene-3,20-dione, 21-acetoxy-6,9-difluoro-11,17-dihydroxy-16-methyl-	n.s.g.	237(4.16)	162-0592-83
$C_{24}H_{30}N_2O_{10}S$			
Spiro[3,9a-epithio-9aH-cyclopenta[4,5]-pyrrolo[1,2-a]pyrazine-7(8H),2'(3'H)-furan]-1,3',4-trione, 8,8a-diacetoxy-3-(acetoxymethyl)octahydro-2,4',4',5'-tetramethyl-, [3R-[3α,5aβ,7α(R*),8β-8aβ,9aα]-	MeOH	215(3.71),262(3.32)	35-5402-83
$C_{24}H_{30}N_4NiS_2Se_2$			
Nickel, bis[N-[(diethylamino)thioxomethyl]benzenecarboselenoamidato-S,Se]-	$CHCl_3$	284(4.48),372s(4.19), 490(3.35)	97-0230B-83
$C_{24}H_{30}N_4O_2$			
Pyrrolidine, 1,1',1"-(4-nitro[1,1'-biphenyl]-2,4,6-triyl)tris-	$CHCl_3$	457(3.82)	44-4649-83
$C_{24}H_{30}N_4S_3$			
1H-Benzimidazole, 2,2'-[thiobis(2,1-ethanediylthio-2,1-ethanediyl)]bis-[1-methyl-	MeOH	250(4.13),275(4.16), 285(4.09)	4-1481-83
$C_{24}H_{30}N_8O_{14}P_2$			
Adenosine 5'-(trihydrogen diphosphate), 5'→5'-ester with N-(2-β-D-ribofuran-	pH 7	240(4.35),263s(4.13), 311(3.56)	138-0805-83

Compound	Solvent	$\lambda_{max}(\log \epsilon)$	Ref.
osyl-2H-indazol-5-yl)acetamide (cont.)			138-0805-83
Adenosine 5'-(trihydrogen diphosphate), 5'→5'-ester with N-(2-β-D-ribofuranosyl-2H-indazol-6-yl)acetamide	pH 7	239(4.31),265s(4.13), 297(3.84)	138-0805-83
$C_{24}H_{30}O$			
Cyclohexanone, 4-(1,1-dimethylethyl)-2,2-bis(4-methylphenyl)-	MeOH	261(2.91),267(2.97), 276(2.91)	12-0789-83
$C_{24}H_{30}O_2$			
2-Buten-1-one, 4-[2,6-bis(1,1-dimethylethyl)-4H-pyran-4-ylidene]-1-(4-methylphenyl)-, (E)-	CH_2Cl_2	455(4.70)	5-1807-83
Ethanone, 1-[3-[7-hydroxy-8,8-dimethyl-3,5-bis(1-methylethyl)bicyclo[4.2.0]-octa-1,3,5-trien-7-yl]phenyl]-	C_6H_{12}	242(4.10),328s(1.70)	35-1117-83
Ethanone, 1-[4-[7-hydroxy-8,8-dimethyl-3,5-bis(1-methylethyl)bicyclo[4.2.0]-octa-1,3,5-trien-7-yl]phenyl]-	C_6H_{12}	253(4.36),328s(1.91)	35-1117-83
$C_{24}H_{30}O_5$			
Microlobin	EtOH	217(4.16),244(3.60), 253(4.42),295(3.85), 327(4.14)	105-0664-83
$C_{24}H_{30}O_6$			
2-Propenoic acid, 3-[4-acetoxy-3-(4-acetoxy-3-methyl-2-butenyl)-5-(3-methyl-2-butenyl)phenyl]-, (E,E)-	MeOH	223s(4.24),281(4.25)	94-0352-83
(E,Z)-	MeOH	223s(4.23),282(4.23)	94-0352-83
$C_{24}H_{30}O_7$			
Athamantin	EtOH	217(4.18),322(4.17)	162-0124-83
$C_{24}H_{30}O_7S$			
4a,1-(Epithiomethano)-7,9a-methanobenz-[a]azulene-10-carboxylic acid, 2,7-diacetoxydodecahydro-1-methyl-8-methylene-13-oxo-, methyl ester, [1S-(1α,2β,4aα,4bβ,7β,9aα,10β,10aβ)]-	MeOH	242(3.48)	88-0351-83
Gibb-4-ene-10-carboxylic acid, 2,7-diacetoxy-1-methyl-8-methylene-1-thiocarboxy-, 10-methyl ester, (1α,2β,4bβ,10β)-	MeOH	245(3.31)	88-0351-83
$C_{24}H_{30}O_8$			
Bicyclo[3.2.1]oct-3-ene-2,8-diol, 1,3-dimethoxy-7-(7-methoxy-1,3-benzodioxol-5-yl)-6-methyl-5-(2-propenyl)-, 2-acetate, (2-endo,6-exo,7-endo-8-anti)-	MeOH	239(3.72),276(3.18)	102-0269-83
Desaspidin	C_6H_{12}	230(4.36),274(4.23)	162-0420-83
$C_{24}H_{30}O_{12}$			
Harpagide, 8-O-p-coumaroyl-	MeOH	224(3.97),313(4.17)	94-2296-83
$C_{24}H_{30}O_{14}$			
1H,3H-Pyrano[3,4-c]pyrimidin-1-one, 4,4a,5,6-tetrahydro-5-oxiranyl-6-[(2,3,4,6-tetra-O-acetyl-β-D-glucopyranosyl)oxy]-	MeOH	241(3.95)	23-0276-83

Compound	Solvent	λ_{max}(log ϵ)	Ref.
$C_{24}H_{31}ClO_9$			
Brianthein X	CH_2Cl_2	231(3.84)	44-5203-83
$C_{24}H_{31}FO_5$			
Pregna-1,4-diene-3,20-dione, 21-acetoxy-6-fluoro-11-hydroxy-16-methyl-, (6α,11β,16α)-	MeOH	240(4.20)	162-0594-83
$C_{24}H_{31}FO_6$			
Fluperolone acetate	n.s.g.	239(4.19)	162-0600-83
Pregna-1,4-diene-3,20-dione, 21-acetoxy-9-fluoro-11,17-dihydroxy-16-methyl-	MeOH	238(4.17)	162-0168-83
	n.s.g.	239(4.17)	162-0425-83
$C_{24}H_{31}F_3O_4$			
Pregn-4-ene-3,20-dione, 17-acetoxy-6α-(trifluoromethyl)-	EtOH	234(4.19)	162-0591-83
$C_{24}H_{31}NO_3$			
Kayawongine	EtOH	226(4.01),276(3.63), 281(3.61)	100-0353-83
$C_{24}H_{31}NO_6$			
Alpinigenin, O-ethyl-	n.s.g.	231(4.19),286(3.80)	100-0441-83
$C_{24}H_{31}N_9O_4$			
1H-Pyrrole-2-carboxamide, N-[5-[[[3-(dimethylamino)-3-iminopropyl]amino]-carbonyl]-1-methyl-1H-pyrrol-3-yl]-4-[[[4-(formylamino)-1-methyl-1H-pyrrol-2-yl]carbonyl]amino]-1-methyl-, monohydrobromide	EtOH	235(4.46),303(4.53)	87-1042-83
$C_{24}H_{32}ClFO_5$			
Halcinonide	MeOH	238(4.21)	162-0662-83
$C_{24}H_{32}N_2O_6$			
[3,3'-Bi-3H-pyrrole]-5,5'-dicarboxylic acid, 3,3'-di-3-butenyl-4,4'-dihydroxy-2,2'-dimethyl-, diethyl ester	MeCN	309(4.04)	33-1902-83
$C_{24}H_{32}N_2O_7$			
8,11,14,17,20,23,26-Heptaoxa-2,3-diazatricyclo[25.2.2.2^{4,7}]tritriaconta-2,4,6,27,29,30,32-heptaene	o-$C_6H_4Cl_2$	363(4.40)	35-1851-83
$C_{24}H_{32}N_4O_6$			
9,10-Anthracenedione, 5,8-bis[[2-[(2-hydroxyethyl)amino]ethyl]amino]-1,2-dimethoxy-, dihydrochloride	H_2O	221(4.39),239(4.41), 272(6.41),584(4.12), 627(4.12)[sic]	18-1812-83
$C_{24}H_{32}O_3$			
Benzoic acid, 3-methoxy-4-[2-methyl-4-(2,6,6-trimethyl-1-cyclohexen-1-yl)-1,3-butadienyl]-, ethyl ester, (E,E)-	EtOH	230s(4.13),330(4.44)	87-1282-83
Norethisterone 2-methylpropanoate	MeOH	241(4.41)	13-0255-83A
$C_{24}H_{32}O_3S$			
[1,1'-Bicyclopentyl]-2,3'-dione,	EtOH	220.0(3.96)	104-0835-83

Compound	Solvent	λmax(log ε)	Ref.
5-[3-(phenylsulfinyl)-2-octenyl]-			104-0835-83
$C_{24}H_{32}O_4$			
1H-Dibenzo[a,d]cycloheptene-6,7-diol, 2,3,5,10,11,11a-hexahydro-1,1-dimethyl-8-(1-methylethyl)-, diacetate, (S)-	C_6H_{12}	228(3.54),268(3.23), 275(3.26),288(3.21), 296(3.14)	78-3603-83
Megestrol acetate	EtOH	287(4.40)	162-0826-83
2'H-18-Norandrost-4-eno[13,17-c]furan-3-one, 2'-(acetoxymethylene)-5',17-dihydro-5'-methyl-, (2'Z,5'α,17α)-	MeOH	237(4.31)	44-2696-83
Pregna-4,6-diene-3,20-dione, 17-acetoxy-6-methyl-	MeOH	289(4.39)	5-0712-83
	MeOH	288(4.39)	5-0712-83
2-Propenoic acid, 3-phenyl-, 1,2,3,4-4a,5,8,8a-octahydro-2,4-dihydroxy-1,4a-dimethyl-7-(1-methylethyl)-1-naphthalenyl ester	EtOH	275(4.26)	102-0979-83
2-Propenoic acid, 3-phenyl-, 1,2,3,4-4a,5,8,8a-octahydro-3,4-dihydroxy-1,4a-dimethyl-7-(1-methylethyl)-1-naphthalenyl ester	EtOH	275(4.26)	102-0979-83
Resibufogenin	EtOH	298(3.74)	162-1175-83
$C_{24}H_{32}O_4S$			
[1,1'-Bicyclopentyl]-2,3'-dione, 5-[3-(phenylsulfonyl)-2-octenyl]-	EtOH	260(2.68),266.5(2.75), 274(2.69)	104-0835-83
$C_{24}H_{32}O_5$			
2H-1-Benzopyran-2-one, 7-[(decahydro-4,6-dihydroxy-1,2,4a,5-tetramethyl-1-naphthalenyl)methoxy]-	EtOH	223(4.58),253(4.06), 237[sic](4.19)	105-0664-83
$C_{24}H_{32}O_6$			
1-Phenanthrenecarboxylic acid, 1,2,3,4-4a,9-hexahydro-5,6,8-trimethoxy-1,4a-dimethyl-7-(1-methylethyl)-9-oxo-, methyl ester, (1S-cis)-	EtOH	216(4.28),248(4.19), 288(3.96),326(3.66)	23-2461-83
$C_{24}H_{32}O_9$			
Longicornin B	MeOH	207(4.32)	102-2759-83
$C_{24}H_{32}O_{10}$			
Longicornin A	MeOH	213(4.26)	102-2759-83
$C_{24}H_{33}ClO_3S$			
Androsta-1,4-diene-3,11-dione, 17-chloro-17-(pentylsulfinyl)-, (17α)-	MeOH	236(4.21)	87-0078-83
$C_{24}H_{33}FO_6$			
Pregn-4-ene-3,20-dione, 6α-fluoro-11β,21-dihydroxy-16α,17-[(1-methylethylidene)bis(oxy)]-	n.s.g.	236(4.17)	162-0601-83
$C_{24}H_{33}NO_3S$			
9H-Carbazole-9-dodecanesulfonic acid, sodium salt	H_2O	331(--),347.5(3.14)	77-0099-83
9H-Carbazole-3-sulfonic acid, 9-dodecyl-, sodium salt	H_2O	331(--),347.5(3.42)	77-0099-83
$C_{24}H_{33}NO_3S_2$			
2H-1-Benzopyran-6-ol, 2-[5-[(4,5-dihydro-2-thiazolyl)thio]-4-methyl-	MeOH	274(3.40),283(3.38)	94-4341-83

Compound	Solvent	$\lambda_{max}(\log \epsilon)$	Ref.
3-pentenyl]-3,4-dihydro-2,5,7,8-tetramethyl-, acetate, (E)-(±)- (cont.)			94-4341-83
$C_{24}H_{33}NO_6$			
Carbamic acid, [5-[3-(2,5-dimethyl-1-oxo-2,4-nonadienyl)-4-hydroxy-2-oxo-2H-pyran-6-yl]-1-hexenyl]-, methyl ester, [R-(E,E,E)]-	MeOH	213(4.48),297(4.31)	5-1656-83
Myxopyronin B	MeOH	216(4.51),301(4.3)	158-1651-83
$C_{24}H_{33}NO_6S$			
1-Azetidineacetic acid, 2-(butylthio)-α-[1-hydroxy-3-[(tetrahydro-2H-pyran-2-yl)oxy]propylidene]-4-oxo-, phenylmethyl ester	EtOH	265(3.82)	39-0115-83C
$C_{24}H_{33}NO_8$			
7-Azabicyclo[2.2.1]hept-2-ene-1,2,3,4-tetracarboxylic acid, 2,3-dicyclohexyl 1,4-dimethyl ester	CCl_4	263(3.45)	4-0001-83
$C_{24}H_{33}N_3$			
Pyrido[2',1':3,4][1,4]diazepino[1,2-a]-benzimidazole-5,9-diium, 6,7,8,14-tetrahydro-2-methyl-14-octyl-, dibromide	EtOH	319(4.20)	4-0029-83
$C_{24}H_{33}N_3O_8S$			
Benzenesulfonamide, 4-[(21-hydroxy-3,6,9,12,15-pentaoxabicyclo[15.3.1]-heneicosa-1(21),17,19-trien-19-yl)-azo]-N,N-dimethyl-	$CHCl_3$	364(4.46)	18-3253-83
$C_{24}H_{34}NO_2$			
Pyrrolidinium, 1-[[11-(2-carboxycyclopentyl)-3,7-dimethyl-2,4,6,8,10-dodecapentaenylidene]-, [1α(2E,4E,6E-8E,10E),2β]-, perchlorate	$CHCl_3$	502(4.62)	35-4033-83
	+ HOAc	504(--)	35-4033-83
	+ Et_3N	510(--)	35-4033-83
$C_{24}H_{34}N_2O_2S$			
Thiourea, N-butyl-N'-(3-methoxy-17-oxo-estra-1,3,5(10)-trien-2-yl)-	EtOH	245(4.273),290(3.968)	36-1205-83
Thiourea, N-butyl-N'-(3-methoxy-17-oxo-estra-1,3,5(10)-trien-4-yl)-	EtOH	245(4.331),281(3.707)	36-1205-83
$C_{24}H_{34}N_5S$			
Phenothiazin-5-ium, 3,7-bis(4,4-dimethylpiperazinium-1-yl)-, tribromide	pH 7	638(4.96)	77-1521-83
$C_{24}H_{34}O$			
Acetaldehyde, [2-[3-methyl-5-(2,6,6-trimethyl-1-cyclohexen-1-yl)-2,4-pentadienylidene]cycloheptylidene]-, "11-cis"	EtOH	232(4.26),256(4.30), 292(4.26),376(4.08)	35-3588-83
"9-cis,11-cis"	EtOH	232(4.18),252(4.11), 294(4.15),370(3.90)	35-3588-83
"11-cis,13-cis"	EtOH	227(4.36),295(4.32), 370(3.85)	35-3588-83
"9-cis,11-cis,13-cis"	EtOH	235(4.26),300(4.32), 365(3.48)	35-3588-83

Compound	Solvent	$\lambda_{max}(\log \epsilon)$	Ref.
$C_{24}H_{34}O_2S$			
Androsta-1,4-diene-3,11-dione, 17α-(pentylthio)-	MeOH	238(4.21)	87-0078-83
$C_{24}H_{34}O_3$			
Tetracyclo[8.4.0.0^{2,7}.0^{4,6}]tetradeca-1,10-dien-12-one, 11-(3,3-ethylene-dioxybutyl)-4a-(1-methylethyl)-7-methyl-	EtOH	310(4.15)	33-1806-83
$C_{24}H_{34}O_4$			
Androsta-3,5-diene-6-carboxaldehyde, 17β-acetoxy-3-ethoxy-	EtOH	220(4.04),323(4.21)	95-1046-83
Bufalin	EtOH	289(3.77)	33-2632-83
Chola-4,6-dien-24-oic acid, 12α-hy-droxy-3-oxo-	EtOH	284(4.48)	13-0475-83A
α'-Isobufalin	$CHCl_3$	306(3.84)	33-2632-83
Medroxyprogesterone 17-acetate	EtOH	240(4.20)	162-0825-83
2'H-18-Norandrost-5-eno[13,17-c]furan-3-ol, 2'-(acetoxymethylene)-5',17-dihydro-5'-methyl-, (2'Z,3β,5'α,17α)-	MeOH	224(3.86)	44-2696-83
18-Norpregn-4-en-3-one, 20β-acetoxy-13β-acetyl-	MeOH	240(4.20)	44-2696-83
$C_{24}H_{34}O_5$			
18-Norpregn-4-en-3-one, 13-(acetoxy-acetyl)-20-hydroxy-, (20R)-	MeOH	240(4.16)	44-2696-83
$C_{24}H_{34}O_6$			
Isolathyrol, 3,5-di-O-acetyl-	MeOH	194(4.16),283(3.97)	102-1791-83
$C_{24}H_{35}NO_2$			
Pyridine, 2-methyl-6-[2-[octahydro-1b-methyl-6a-(1-methylethyl)spiro[cyclo-prop[a]indene-4(1H),2'-[1,3]dioxol-an]-5-yl]ethyl]-, [1aS-(1aα,1bβ-5aα,6aα)]-	EtOH	266(3.71),273(3.59)	33-1820-83
$C_{24}H_{35}NO_3$			
7H,19H-[1,8]Oxaazacyclononadecino[7,8-a]isoindole-7,19-dione, 1,2,3,4,5,8-9,10,11,12,13,14,15,16,17,23b-hexa-decahydro-	$CHCl_3$	248(3.79),270s(3.51), 280(3.32)	78-2691-83
Retinamide, N-(2-ethoxy-2-oxoethyl)-	EtOH	349(4.69)	34-0422-83
$C_{24}H_{35}NO_3S_2$			
Thiazole, 2-[[5-[3,4-dihydro-6-(meth-oxymethoxy)-2,5,7,8-tetramethyl-2H-1-benzopyran-2-yl]-2-methyl-2-pent-enyl]thio]-4,5-dihydro-, (E)-(±)-	MeOH	278(3.40),285(3.38)	94-4341-83
$C_{24}H_{35}O_9P$			
α-D-Allofuranose, 3-deoxy-3-(diethoxy-phosphinyl)-1,2:5,6-bis-O-(1-methyl-ethylidene)-3-C-(2-oxo-2-phenylethyl)-	EtOH	209(3.68),246(3.96)	159-0019-83
$C_{24}H_{36}$			
Cyclohexane, hexakis(1-methylethyli-dene)-	C_6H_{12}	240s(4.307),248(4.324)	77-1058-83
Cyclohexane, (1-methylethenyl)penta-kis(1-methylethylidene)-	C_6H_{12}	247.5(4.207)	77-1058-83

Compound	Solvent	$\lambda_{max}(\log \epsilon)$	Ref.
$C_{24}H_{36}Fe$			
Iron(2+), bis[(1,2,3,4,5,6-η)-hexamethylbenzene]-, bis(hexafluorophosphate)	MeCN	320(3.7),460(1.7)	157-1214-83
$C_{24}H_{36}N_2O_4$			
Peripentadenine, O-acetyl-	MeOH	215(3.06),305(2.89)	100-0235-83
$C_{24}H_{36}O_2$			
1'H-Cyclobut[4,5]androstan-3-one, 17-hydroxy-4'-propyl-, (4α,5α)-	EtOH	287(2.36)	70-0103-83
(4β,5β)-	EtOH	288(2.19)	70-0103-83
$C_{24}H_{36}O_3$			
Tetracyclo[8.4.0.0^{2,7}.0^{4,6}]tetradec-10-en-12-one, 11-(3,3-ethylenedioxybutyl)-4a-(1-methylethyl)-7β-methyl-	EtOH	250(4.16)	33-1806-83
Tricyclo[4.4.0.0^{2,4}]decan-3-one, 4a-(1-methylethyl)-1β-methyl-7a-[2-(3-oxo-1-cyclohexenyl)ethyl]-, 3,3-ethylene acetal	EtOH	236(4.16)	33-1820-83
$C_{24}H_{36}O_4$			
1,3-Butadiene-1,4-diol, 2-[2-(1,4,4a-5,6,7,8,8a-octahydro-2,5,5,8a-tetramethyl-1-naphthalenyl)ethyl]-, diacetate, [1R-[1α(1E,3E),4aβ,8aα]]-	EtOH	251(4.25)	102-1465-83
Chol-4-en-24-oic acid, 12α-hydroxy-3-oxo-	EtOH	241(4.22)	13-0475-83A
Jolkinol, 3-O-isobutyryl-, 5β,6β-oxide	MeOH	192(3.79),269(4.21)	102-1791-83
Spiro[cyclohexane-1,2'(1'H)-naphthalene]-2,5'(3'H)-dione, 1'-ethoxy-6-(ethoxymethylene)-4',6',7',8'-tetrahydro-4,4,7',7'-tetramethyl-	EtOH	249(3.71)	39-1919-83C
Tricyclo[4.4.0.0^{2,4}]dec-6-en-8-one, 7-(7,7-ethylenedioxy-3-oxooctyl)-4a-(1-methylethyl)-1β-methyl-	EtOH	247(3.98)	33-1806-83
$C_{24}H_{36}O_5$			
Chol-4-en-24-oic acid, 7α,12α-dihydroxy-3-oxo-	EtOH	242(4.13)	13-0475-83A
$C_{24}H_{36}O_8$			
Butanoic acid, 2,3-diacetoxy-2-methyl-, decahydro-1-hydroxy-1,4a-dimethyl-7-(1-methylethylidene)-6-oxo-2-naphthalenyl ester	EtOH	258(4.15)	100-0671-83
$C_{24}H_{36}Zr$			
Zirconium, (η^4-1,3-butadiene)bis[(1,2,3,4,5-η)-1,2,3,4,5-pentamethyl-2,4-cyclopentadien-1-yl]-	hexane	354(3.21),450(2.64)	18-3735-83
$C_{24}H_{37}IO_3$			
Cholan-24-oic acid, 7,12-dihydroxy-3-iodo-, ζ-lactone, (3α,5β,7α,12α)-	MeOH	257(3.93)	78-2815-83
$C_{24}H_{37}NO$			
Retinal, 19-(butylamino)-	EtOH	377(4.51)	77-0077-83
Retinamide, N-butyl-	EtOH	347(4.71)	34-0422-83
13-cis	EtOH	243(3.97),350(4.66)	34-0422-83

Compound	Solvent	λ_{max}(log ε)	Ref.
$C_{24}H_{37}NO_2$			
Retinamide, N-(4-hydroxybutyl)-	EtOH	347(4.71)	34-0422-83
13-cis	EtOH	242(3.96),350(4.63)	34-0422-83
$C_{24}H_{37}NO_3$			
2,4,6,8,10,14-Heptadecahexaenamide, 13-hydroxy-N-(2-hydroxy-1-methyl-ethyl)-2,10,12,14-tetramethyl-	EtOH	204(4.12),265(4.06), 340s(--),356(4.54), 370s(--)	5-1081-83
$C_{24}H_{38}N_2$			
Pyrrolidine, 1,1'-(1,4,6,7,8,8a-hexa-hydro-8,8,9,9-tetramethyl-1,4-eth-anonaphthalene-2,5-diyl)bis-, (1α,4α,8aα)-	C_6H_{12}	244(3.10)	33-0735-83
$C_{24}H_{38}N_2O_2$			
Hexanamide, N-[3-[[6-(2-methoxy-6-meth-ylphenyl)-3,5-hexadienyl]methyl-amino]propyl]-	MeOH	218(3.30),308(3.00)	100-0235-83
$C_{24}H_{38}OSi$			
Silane, (1,1-dimethylethyl)(estra-1,3,5(10)-trien-3-yloxy)dimethyl-	EtOH	277(3.17),283(3.14)	65-0539-83
$C_{24}H_{38}O_3Si$			
Estr-4-en-3-one, 17-(1-oxopropoxy)-1-(trimethylsilyl)-, (1α,17β)-	EtOH	248(3.98)	65-0539-83
$C_{24}H_{38}O_4$			
1,3-Cyclohexanedione, 4-hydroxy-2-(1-oxo-7,9-octadienyl)-, (Z,Z)-	hexane	233(4.53),274(3.97)	39-2161-83C
Pregnane-21-carboxylic acid, 3β,14-di-hydroxy-21-methylene-, methyl ester, (3β,5β,14β)-	MeOH	210(3.94)(end abs.)	13-0037-83B
$C_{24}H_{38}O_6$			
Prosta-6,14-diene-20-carboxylic acid, 11,14-epoxy-5,9-dihydroxy-1-methyl-16-oxo-, (E)-	EtOH	266(4.305)	88-1281-83
(Z)-	EtOH	261(4.180)	88-1281-83
$C_{24}H_{39}IO_4$			
Cholan-24-oic acid, 7,12-dihydroxy-3-iodo-, (3α,5β,7α,12α)-	MeOH	257(3.90)	78-2815-83
$C_{24}H_{39}NO_5S$			
Leukotriene E_4, methyl ester	EtOH	269(4.45),280(4.55), 291(4.46)	162-0784-83
$C_{24}H_{39}N_5O_4Si_2$			
7H-Pyrrolo[2,3-d]pyrimidine-5-carboni-trile, 4-amino-7-[2-deoxy-3,5-O-[1,1,3,3-tetrakis(1-methylethyl)-1,3-disiloxanediyl]-β-D-erythro-pentofuranosyl]-	MeOH	230(4.07),279(4.21)	35-4059-83
$C_{24}H_{39}N_5O_5Si_2$			
7H-Pyrrolo[2,3-d]pyrimidine-5-carboni-trile, 4-amino-7-[3,5-O-[1,1,3,3-tetrakis(1-methylethyl)-1,3-disilox-	MeOH	232(4.08),280(4.24)	35-4059-83

Compound	Solvent	λ_{max}(log ε)	Ref.
anediyl]-β-D-ribofuranosyl]- (cont.)			35-4059-83
$C_{24}H_{40}ClNO_2$			
Octadecanamide, N-(2-chlorophenyl)-N-hydroxy-	n.s.g.	257(4.15)	42-0686-83
Octadecanamide, N-(3-chlorophenyl)-N-hydroxy-	n.s.g.	259(4.04)	42-0686-83
$C_{24}H_{40}Cl_2N_2Rh_2$			
Rhodium, dichlorobis[(1,2,5,6-η)-1,5-cyclooctadiene][μ-[N,N'-1,2-ethanediylidenebis[2-propanamine]-N:N']]di-	CH_2Cl_2	296(3.95),345(3.94), 374s(3.81),600(1.40)	101-0301-83F
$C_{24}H_{40}Ge_3$			
Trigermane, 1,1,2,2,3,3-hexaethyl-1,3-diphenyl-	C_6H_{12}	241(4.31)	101-0149-83G
$C_{24}H_{40}O_6$			
Protylonolide, 19-hydroxy-	MeOH	283(4.41)	158-0921-83
Protylonolide, 23-hydroxy-	MeOH	283(4.21)	158-0921-83
$C_{24}H_{44}$			
Betweenanene	heptane	202s(3.76),208(3.83) 212(3.83)	78-2679-83
Bicyclo[11.11.0]tetracos-1(13)-ene, (Z)-	heptane	200(4.00)	78-2679-83
11-Tricosyne, 13-methylene-	EtOH	223(4.16)	104-1621-83
$C_{24}H_{48}Si_6$			
5,6,11,16,21,26-Hexasilahexaspiro-[4.0.4.0.4.0.4.0.4.0.4.0]triacontane	n.s.g.	208s(4.50),252(3.71), 275s(3.08)	157-0453-83
$C_{24}H_{60}Si_6$			
Cyclohexasilane, dodecaethyl-	isooctane	205s(4.60),237(3.90), 259s(3.28)	157-1792-83

Compound	Solvent	λ_{max}(log ϵ)	Ref.
$C_{25}H_{10}Cl_{14}NO_3$			
Methyl, bis(pentachlorophenyl)[2,3,5,6-tetrachloro-4-[[(2-ethoxy-1-methyl-2-oxoethyl)amino]carbonyl]phenyl]-, (S)-	C_6H_{12}	222(4.95),286s(3.79),335s(3.81),365s(4.29),382(4.60),482s(3.07),507(3.09),560(4.07)	44-3716-83
$C_{25}H_{14}Cl_2N_6O_9$			
Methanone, bis[4-[(3-chlorophenyl)amino]-3,5-dinitrophenyl]-	dioxan	291.4(4.17)	56-1357-83
Methanone, bis[4-[(4-chlorophenyl)amino]-3,5-dinitrophenyl]-	dioxan	291.0(4.13)	56-1357-83
$C_{25}H_{14}Cl_2O_6$			
Anthra[1,2-b]furan-3-carboxylic acid, 7,10-dichloro-6,11-dihydro-5-hydroxy-6,11-dioxo-2-phenyl-, ethyl ester	dioxan	298(4.18),459(4.13)	104-0533-83
$C_{25}H_{14}N_2O_2$			
7,12[1',2']-Benzenoanthra[2',3':4,5]-imidazo[1,2-a]pyridine-6,13-dione, 7,12-dihydro-, endo	EtOH	252(4.51),263(4.42),297(4.18),310(3.70),332(3.63),405(3.54)	48-0353-83
$C_{25}H_{15}ClO_4$			
2H-Furo[2,3-h]-1-benzopyran-2-one, 8-(4-chlorobenzoyl)-4-methyl-9-phenyl-	dioxan	252(3.80),280(3.60)	2-0164-83
$C_{25}H_{15}ClO_6$			
Anthra[1,2-b]furan-3-carboxylic acid, 7-chloro-6,11-dihydro-5-hydroxy-6,11-dioxo-2-phenyl-, ethyl ester	dioxan	294(4.80),485(4.38)	104-1900-83
$C_{25}H_{15}NO_6$			
2H-Furo[2,3-h]-1-benzopyran-2-one, 4-methyl-8-(4-nitrobenzoyl)-9-phenyl-	MeOH	272(2.55),302(2.47)	2-0164-83
$C_{25}H_{16}N_2O_3$			
Pyrano[2,3-c]pyrazol-4(1H)-one, 5-benzoyl-1,3-diphenyl-	EtOH	254(4.50),330(3.88)	44-4078-83
$C_{25}H_{16}N_4$			
4-Isoquinolinecarbonitrile, 1-amino-3-[2-(1-cyano-2-phenylethenyl)-phenyl]-	EtOH	220(4.59),255(4.42),305(4.39),344s(4.08),360s(3.81)	39-1813-83C
$C_{25}H_{16}N_6O_3$			
Methanone, [3,5-bis(2H-benzotriazol-2-yl)-2,4-dihydroxyphenyl]phenyl-	$CHCl_3$	246(4.33),285(4.55),327(4.47)	121-0309-83B
$C_{25}H_{16}N_6O_9$			
Methanone, bis[3,5-dinitro-4-(phenylamino)phenyl]-	dioxan	291.0(4.09)	56-1357-83
$C_{25}H_{16}N_6O_{11}$			
Methanone, bis[4-[(2-hydroxyphenyl)-amino]-3,5-dinitrophenyl]-	dioxan	294.8(4.13)	56-1357-83
Methanone, bis[4-[(3-hydroxyphenyl)-amino]-3,5-dinitrophenyl]-	dioxan	293.4(4.18)	56-1357-83
Methanone, bis[4-[(4-hydroxyphenyl)-amino]-3,5-dinitrophenyl]-	dioxan	298.3(4.06)	56-1357-83

Compound	Solvent	$\lambda_{max}(\log \epsilon)$	Ref.
$C_{25}H_{16}O_4$			
2H-Furo[2,3-h]-1-benzopyran-2-one, 8-benzoyl-4-methyl-9-phenyl-	MeOH	266(2.62),280(2.67), 298(2.71)	2-0164-83
$C_{25}H_{16}O_6$			
Anthra[1,2-b]furan-3-carboxylic acid, 6,11-dihydro-5-hydroxy-6,11-dioxo-2-phenyl-, ethyl ester	dioxan	299(4.33),451(4.22)	104-0533-83
$C_{25}H_{16}O_7$			
Anthra[2,3-b]furan-3-carboxylic acid, 5,10-dihydro-4,11-dihydroxy-5,10-dioxo-2-phenyl-, ethyl ester	dioxan	294(4.80),485(4.38)	104-1900-83
$C_{25}H_{17}NO_2$			
Naphtho[2,1-g]quinoline-7,12-dione, 2-methyl-5-(4-methylphenyl)-	$CHCl_3$	243(4.44),253(4.39), 316(4.41),374(3.45), 444(3.62)	142-1031-83
Naphtho[2,1-g]quinoline-7,12-dione, 3-methyl-5-(3-methylphenyl)-	$CHCl_3$	244(4.42),252(4.38), 315(4.51),384(3.62), 427(3.62)	142-1031-83
$C_{25}H_{17}NO_4$			
Naphtho[2,1-g]quinoline-7,12-dione, 2-methoxy-5-(4-methoxyphenyl)-	$CHCl_3$	248(4.71),324(4.43), 390(3.80),477(3.89)	142-1031-83
Naphtho[2,1-g]quinoline-7,12-dione, 3-methoxy-5-(3-methoxyphenyl)-	$CHCl_3$	243(4.61),322(4.59), 400(3.81),440(3.65)	142-1031-83
Naphtho[2,1-g]quinoline-7,12-dione, 4-methoxy-5-(2-methoxyphenyl)-	$CHCl_3$	245(4.45),271(4.24), 324(4.32),360(3.57), 479(3.44)	142-1031-83
$C_{25}H_{17}N_3O$			
4H-Pyran-3,5-dicarbonitrile, 2-amino-6-[1,1'-biphenyl]-4-yl-4-phenyl-	EtOH	206(4.52),243(4.19), 297(4.32)	73-3123-83
$C_{25}H_{18}ClNO_3$			
2H-Pyrano[3,2-d][1]benzoxepin-2-one, 3-chloro-4-(diphenylamino)-5,6-dihydro-	EtOH	213.5(4.39),259s(4.16), 276(4.25),358(4.16)	4-0539-83
$C_{25}H_{18}N_2$			
Benzo[f]quinoline, 1-(4-methylphenyl)-3-(2-pyridinyl)-	EtOH	242(4.38),285(4.64), 350(3.69),368(3.69)	103-0422-83
$C_{25}H_{18}N_2O_2$			
[1(2H),2'-Biquinolin]-2-one, 4-hydroxy-3-(phenylmethyl)-	pH 1	211(4.803),235(4.856), 279(4.199),286(4.212), 309(4.090),317(4.121), 336s(4.076)	49-0227-83
	pH 10	221(4.770),233s(4.638), 254s(4.173),306(4.21), 316(4.193)	49-0227-83
$C_{25}H_{18}N_2O_4S$			
1H-Indole, 2-[[2-(hydroxymethyl)-3-quinolinyl]carbonyl]-1-(phenylsulfonyl)-	EtOH	230s(4.56),250s(4.58), 253(4.59),295(4.33)	39-2409-83C
$C_{25}H_{18}N_2O_4S_2$			
1H-Indole, 1,3-bis(phenylsulfonyl)-2-(2-pyridinyl)- (5λ,4ε)	EtOH	213s(4.41),260(4.25), 267(4.23),274(4.30), 294(?)	39-2417-83C

Compound	Solvent	λ_{max}(log ε)	Ref.
$C_{25}H_{18}N_4OS_2$			
Bisthiazolo[3,4-a:5',4'-e]pyrimidin-9-ium, 8-[(1-ethyl-4(1H)-quinolinylidene)methyl]-1,2-dihydro-2-oxo-6-phenyl-, hydroxide, inner salt	DMF	630(4.55),650(4.54)	103-0754-83
$C_{25}H_{18}N_4O_2S_3$			
4-Thiazolidinone, 3-ethyl-5-(8-methoxy-3-phenyl-1H-thiazolo[4',3':2,3]pyrimido[4,5-b]quinolin-1-ylidene)-2-thioxo-	DMF	496(3.40)	103-0166-83
$C_{25}H_{19}BrO_6$			
3-Benzofurancarboxylic acid, 6-bromo-4,5-dihydroxy-7-(2-oxo-2-phenylethyl)-2-phenyl-, ethyl ester	EtOH	210(4.53),245(4.33)	103-0945-83
$C_{25}H_{19}N$			
Benzo[h]quinoline, 5,6-dihydro-2,4-diphenyl-	neutral	318(4.24)	39-0045-83B
	protonated	352(4.38)	39-0045-83B
$C_{25}H_{19}NO_2$			
1,1':3',1"-Terphenyl, 4'-methyl-2'-nitro-5'-phenyl-	MeCN	231(4.58)	48-0729-83
$C_{25}H_{19}NO_3$			
4,11-Etheno-1H-naphth[2,3-f]isoindole-1,3,5(2H,3aH)-trione, 4,4a,11,11a-tetrahydro-4a-methyl-2-phenyl-, (3aα,4β,4aα,11β,11aα)-	EtOH	238(4.63),246s(4.57), 270(3.63),277(3.64), 288s(3.48),317(3.21), 327s(3.18),370(3.15), 384s(3.05)	35-3234-83
$C_{25}H_{19}N_2S$			
Thiazolo[3,4-a]pyrimidin-5-ium, 6-methyl-2,4,8-triphenyl-, perchlorate	MeCN	242(3.45),302(3.48), 408(3.05)	103-0037-83
$C_{25}H_{19}N_3O_6$			
1,2-Diazabicyclo[5.2.0]nona-3,5-diene-4-carboxylic acid, 2-benzoyl-8-(1,3-dihydro-1,3-dioxo-2H-isoindol-2-yl)-9-oxo-, ethyl ester, cis-(±)-	MeOH	305(3.99)	5-1374-83
1,2-Diazabicyclo[5.2.0]nona-3,5-diene-6-carboxylic acid, 2-benzoyl-8-(1,3-dihydro-1,3-dioxo-2H-isoindol-2-yl)-9-oxo-, ethyl ester	MeOH	326(3.84)	5-1374-83
$C_{25}H_{19}O$			
Pyrylium, 2,4-diphenyl-6-(2-phenylethenyl)-, perchlorate	MeCN	255(4.31),305(4.23), 364(4.36),448(4.35)	22-0115-83
Pyrylium, 2,6-diphenyl-4-(2-phenylethenyl)-, perchlorate	MeCN	240(3.93),275(3.98), 325(3.54),418(4.43)	22-0115-83
$C_{25}H_{20}BrNO_6$			
3-Benzofurancarboxylic acid, 6-bromo-4,5-dihydroxy-7-[2-(hydroxyimino)-2-phenylethyl]-2-phenyl-, ethyl ester	EtOH	212(4.58),320(4.00)	103-0945-83
$C_{25}H_{20}F_5N_2$			
Quinolinium, 1-methyl-2-[3,3,4,4,4-pentafluoro-2-[(1-methyl-2(1H)-	$CHCl_3$	652(4.81)	104-1937-83

Compound	Solvent	$\lambda_{max}(\log \epsilon)$	Ref.
quinolinylidene)methyl]-1-butenyl]-, perchlorate (cont.)			104-1937-83
$C_{25}H_{20}N_2$			
Benzenamine, N-(2,4,6-trimethyl-11H-indeno[1,2-b]quinolin-11-ylidene)-	EtOH	206s(4.66),220(4.74), 236(4.76),248(4.76), 260(4.75),294(4.96), 370(4.16)	104-0158-83
$C_{25}H_{20}N_2O$			
Methanone, (4-aminophenyl)[4-(diphenylamino)phenyl]-	MeOH	291(4.24),366(4.49)	12-0409-83
Methanone, bis[4-(phenylamino)phenyl]-	MeOH	233(4.29),266(4.24), 370(4.63)	12-0409-83
$C_{25}H_{20}N_2O_4$			
Cyclohept[1,2,3-hi]imidazo[2,1,5-cd]-indolizine-4,9-dicarboxylic acid, 2-(4-methylphenyl)-, 9-ethyl 4-methyl ester	EtOH	250(4.55),279s(4.53), 301(4.68),313s(4.61), 385(4.64),431s(4.18)	18-3703-83
$C_{25}H_{20}N_2O_9$			
α-L-threo-Hex-4-enopyranuronic acid, 1,4-dideoxy-1-(3,4-dihydro-2,4-dioxo-1(2H)-pyrimidinyl)-, methyl ester, 2,3-dibenzoate	MeOH	235(4.66),263s(4.13)	65-1486-83
$C_{25}H_{20}N_4NiO_2$			
Nickel, [5,13-dihydro-15,17-dimethoxy-20-methyl-14,12-metheno-12H-tribenzo-[b,e,k][1,4,7,10]tetraazacyclotridec-inato(2-)-N^5,N^{10},N^{13},N^{19}]-, (SP-4-2)-	EtOH-DMF	345(4.23),380(4.12), 420(4.15),575(3.94)	12-2407-83
$C_{25}H_{20}N_4O_4$			
2-Naphthalenecarboxamide, 3-methoxy-N-methyl-4-[(2-nitrophenyl)azo]-N-phenyl-	EtOH	205(4.789),224(4.839), 277s(4.271),394(4.004), 480s(3.276)	104-1145-83
	$CHCl_3$	277s(4.076),394(3.875), 480s(3.072)	104-1145-83
	CCl_4	275s(4.134),395(3.924), 490s(3.146)	104-1145-83
	DMSO	277s(4.033),394(4.179), 480s(3.078)	104-1145-83
$C_{25}H_{20}N_4O_5$			
2-Naphthalenecarboxamide, 3-hydroxy-4-[(2-methoxy-4-nitrophenyl)azo]-N-methyl-N-phenyl-	EtOH	205(4.619),225s(4.537), 335(3.869),430s(3.954), 520(4.412)	104-1145-83
	$CHCl_3$	310(3.944),345(3.944), 430s(4.049),510(4.551), 540(4.561)	104-1145-83
	CCl_4	300s(3.940),340(3.944), 430s(4.025),510(4.453), 533(4.455)	104-1145-83
	DMSO	310(3.964),340(3.924), 430s(4.017),525(4.524)	104-1145-83
$C_{25}H_{20}O_5$			
1(2H)-Naphthalenone, 4,8-bis(benzoyloxy)-3,4-dihydro-3-methyl-, (3R-cis)-	MeOH	226(4.58),282(3.58), 296(3.41)	102-1245-83

Compound	Solvent	λ_{max}(log ϵ)	Ref.
$C_{25}H_{21}BrO_2$			
2H-Pyran, 2-(4-bromophenyl)-2-methoxy-3-methyl-4,6-diphenyl-	dioxan	235(4.36),243s(4.35), 324(4.16)	48-0729-83
2H-Pyran, 4-(4-bromophenyl)-2-methoxy-3-methyl-2,6-diphenyl-	dioxan	238s(4.29),244(4.32), 253s(4.26),324(4.18)	48-0729-83
2H-Pyran, 6-(4-bromophenyl)-2-methoxy-3-methyl-2,4-diphenyl-	dioxan	242s(4.31),247(4.32), 330(4.24)	48-0729-83
$C_{25}H_{21}ClO_2$			
2H-Pyran, 2-(4-chlorophenyl)-2-methoxy-3-methyl-4,6-diphenyl-	dioxan	236(4.34),242s(4.43), 324(4.17)	48-0729-83
2H-Pyran, 4-(4-chlorophenyl)-2-methoxy-3-methyl-2,6-diphenyl-	dioxan	238(4.30),243(4.31), 252s(4.25),322(4.17)	48-0729-83
2H-Pyran, 6-(4-chlorophenyl)-2-methoxy-3-methyl-2,4-diphenyl-	dioxan	240s(4.28),246(4.30), 329(4.22)	48-0729-83
$C_{25}H_{21}FO_2$			
2H-Pyran, 2-(4-fluorophenyl)-2-methoxy-3-methyl-4,6-diphenyl-	dioxan	237(4.28),243(4.28), 324(4.17)	48-0729-83
2H-Pyran, 6-(4-fluorophenyl)-2-methoxy-3-methyl-2,4-diphenyl-	dioxan	236s(4.24),241(4.25), 322(4.14)	48-0729-83
$C_{25}H_{21}N$			
Pyridine, 3-methyl-2-(4-methylphenyl)-4,6-diphenyl-	MeCN	244(4.60),284s(4.04), 295s(3.98)	48-0729-83
Pyridine, 3-methyl-4-(4-methylphenyl)-2,6-diphenyl-	MeCN	250(4.57),294s(3.95)	48-0729-83
Pyridine, 3-methyl-6-(4-methylphenyl)-2,4-diphenyl-	MeCN	242(4.53),283s(4.14), 299s(4.01)	48-0729-83
$C_{25}H_{21}NO$			
Pyridine, 2-(4-methoxyphenyl)-3-methyl-4,6-diphenyl-	MeCN	248(4.59),299s(4.00)	48-0729-83
Pyridine, 6-(4-methoxyphenyl)-3-methyl-2,4-diphenyl-	MeCN	241(4.44),269(4.39), 286s(4.27),311s(3.99)	48-0729-83
$C_{25}H_{21}NO_3$			
1,4-Naphthalenedione, 2-[1-hydroxy-2-(methylphenylamino)ethyl]-3-phenyl-	EtOH	202(4.72),252(4.48), 305(3.70),333(3.58), 523(2.60)	104-1151-83
$C_{25}H_{21}NO_3S$			
2-Butenoic acid, 2-cyano-4-[6-(4-methoxyphenyl)-2H-thiopyran-2-ylidene]-4-phenyl-, ethyl ester	$CHCl_3$	548(4.07)	97-0147-83
$C_{25}H_{21}NO_3Se$			
2-Butenoic acid, 2-cyano-4-[6-(4-methoxyphenyl)-2H-selenin-2-ylidene]-4-phenyl-, ethyl ester	$CHCl_3$	543(4.21)	97-0147-83
$C_{25}H_{21}NO_4$			
2H-Pyran, 2-methoxy-3-methyl-4-(4-nitrophenyl)-2,6-diphenyl-	dioxan	233s(4.20),244s(4.14), 277(4.19),314(4.26)	48-0729-83
$C_{25}H_{21}N_3O$			
Pyrazolo[1,5-c]pyrimidin-7(3H)-one, 3a,6-dihydro-6-methyl-2,3a,5-triphenyl-	isoPrOH	270(4.2),305(4.3)	103-0564-83

Compound	Solvent	λ_{max}(log ϵ)	Ref.
$C_{25}H_{21}N_3O_2$			
2-Naphthalenecarboxamide, 3-methoxy-N-methyl-N-phenyl-4-(phenylazo)-	EtOH	205(4.655),224(4.706), 277s(4.126),360(3.778), 450s(3.176)	104-1145-83
	$CHCl_3$	277s(4.121),350(3.756), 450s(3.176)	104-1145-83
	CCl_4	279s(4.215),360(3.806), 450s(3.159)	104-1145-83
	DMSO	277s(4.246),370(3.813), 470s(3.079)	104-1145-83
$C_{25}H_{21}N_3O_3S$			
2H-1-Benzopyran-2-one, 3-[2-[7-(diethylamino)-2-imino-2H-1-benzopyran-3-yl]-4-thiazolyl]-	MeCN	443(4.57)	48-0551-83
$C_{25}H_{21}N_3O_4$			
2,3,9b-Triazaindeno[6,7,1-ija]azulene-8,9-dicarboxylic acid, 1-(4-methylphenyl)-, diethyl ester	EtOH	241(4.21),270s(4.23), 291(4.32),383(4.26), 416(4.09),437(4.05)	18-3703-83
$C_{25}H_{21}N_3O_5S$			
1-Azabicyclo[3.2.0]hept-2-ene-2-carboxylic acid, 6-ethyl-7-oxo-3-(8-quinolinylthio)-, (4-nitrophenyl)methyl ester	THF	266(4.04),317.5(4.01)	158-0407-83
$C_{25}H_{21}N_3S$			
4H-Cyclopentapyrimidine-4-thione, 2-(diphenylamino)-1,5,6,7-tetrahydro-1-phenyl-	MeOH	282(4.25),343(4.28)	73-3575-83
Pyrazolo[1,5-c]pyrimidine, 3,3a-dihydro-7-(methylthio)-2,3a,5-triphenyl-	isoPrOH	248(4.5),292(4.3), 340(4.1)	103-0564-83
$C_{25}H_{21}N_5O_9$			
Isoquinolinium, 2-[(4-nitrophenyl)methyl]-, 1-acetonyl-2,4,6-trinitrocyclohexadienate	cation	233(4.51),320(4.10)	103-0773-83
	anion	465(4.03),582(3.92)	103-0773-83
Quinolinium, 1-[(4-nitrophenyl)methyl]-, 1-acetonyl-2,4,6-trinitrocyclohexadienate	cation	240(4.59),265(4.33)	103-0773-83
	anion	460(4.04),582(3.99)	103-0773-83
$C_{25}H_{21}O$			
Pyrylium, 3-methyl-2-(4-methylphenyl)-4,6-diphenyl-, perchlorate	MeCN	233(4.21),277(4.27), 342(4.32),394(4.38)	48-0729-83
Pyrylium, 3-methyl-4-(4-methylphenyl)-2,6-diphenyl-, perchlorate	MeCN	232(4.17),273(4.29), 375(4.54)	48-0729-83
Pyrylium, 3-methyl-6-(4-methylphenyl)-2,4-diphenyl-, perchlorate	MeCN	233(4.23),278(4.30), 339(4.27),398(4.43)	48-0729-83
$C_{25}H_{21}O_2$			
Pyrylium, 2-(4-methoxyphenyl)-3-methyl-4,6-diphenyl-, perchlorate	MeCN	249(4.22),293(4.29), 343(4.33)	48-0729-83
Pyrylium, 6-(4-methoxyphenyl)-3-methyl-2,4-diphenyl-, perchlorate	MeCN	227(4.30),292(4.30), 337(4.25),425(4.48)	48-0729-83
$C_{25}H_{22}$			
Cyclopropene, 3-(2-ethenylbenzyl)-2-methyl-1,3-diphenyl-	EtOH	253(4.39)	35-4446-83
Cyclopropene, 3-(2-ethenylbenzyl)-3-methyl-1,2-diphenyl-	EtOH	322(4.51),339(4.28)	35-4446-83

Compound	Solvent	λ_{max}(log ϵ)	Ref.
Cyclopropene, 2-methyl-1,3-diphenyl-3-(2-propenylphenyl)-	EtOH	262(4.38)	35-4446-83
Cyclopropene, 3-methyl-1,2-diphenyl-3-(2-propenylphenyl)-	EtOH	227(4.49),317(4.35)	35-4446-83
1,6-Methanocycloprop[a]indene, 1,1a,6-6a-tetrahydro-1,6-dimethyl-1a,6a-diphenyl-	EtOH	247(4.17)	35-4446-83
1,6-Methanocycloprop[a]indene, 1,1a,6-6a-tetrahydro-1a,6-dimethyl-1,6a-diphenyl-	EtOH	218s(4.41)	35-4446-83
1,7-Methano-2H-cyclopropa[b]naphthalene, 1,1a,7,7a-tetrahydro-1a-methyl-1,7a-diphenyl-, (1α,1aβ,7α,7aβ)-	EtOH	223(4.32)	35-4446-83
$C_{25}H_{22}ClN_3O_2$			
2H-Pyrrolo[2,3-b]quinoxalin-2-one, 3-benzoyl-1-(4-chlorophenyl)-4-ethyl-1,3,3a,4,9,9a-hexahydro-	EtOH	223(4.48),251(4.38), 306(3.56)	103-0901-83
$C_{25}H_{22}Cl_2N_2O_6$			
3(2H)-Pyridazinone, 5,6-dichloro-2-[2-deoxy-3,5-bis-O-(4-methylbenzoyl)-α-D-erythro-pentofuranosyl]-	MeOH	281(3.80),306(3.81)	4-0369-83
β-	MeOH	281(3.51),306(3.51)	4-0369-83
$C_{25}H_{22}NO_3S$			
1,3-Thiazin-1-ium, 2,4,6-tris(4-methoxyphenyl)-, perchlorate	MeCN	233(4.11),302(4.26), 445(4.57),482(4.59)	118-0827-83
$C_{25}H_{22}N_2O_2$			
Pyrazole, 5-(4-methoxyphenyl)-3-(4-methoxystyryl)-1-phenyl-	n.s.g.	231(4.21),297(4.60)	44-0542-83
$C_{25}H_{22}N_2O_4S$			
Benz[cd]indol-5(1H)-one, 1-benzoyl-2,2a,3,4-tetrahydro-, 5-[O-[(4-methylphenyl)sulfonyl]oxime]	EtOH	230(4.53),250(4.50), 320(3.74)	87-0522-83
$C_{25}H_{22}N_4$			
1H-Imidazo[1,2-b][1,2,4]triazepine, 7,8-dihydro-1-(4-methylphenyl)-2,8-diphenyl-	EtOH	283(4.13),347(3.72)	103-0083-83
$C_{25}H_{22}N_4O$			
1H-Imidazo[1,2-b][1,2,4]triazepine, 7,8-dihydro-6-(4-methoxyphenyl)-2,8-diphenyl-	EtOH	288(4.46),343(4.09)	103-0083-83
1H-Imidazo[1,2-b][1,2,4]triazepine, 7,8-dihydro-8-(4-methoxyphenyl)-2,6-diphenyl-	EtOH	279(4.46),349(4.00)	103-0083-83
$C_{25}H_{22}N_4O_2$			
1H-Pyrazole-3,5-diamine, N,N'-bis[(4-methoxyphenyl)methylene]-4-phenyl-	MeCN	250(4.29),298(4.53), 345(4.59)	39-0011-83C
$C_{25}H_{22}O_2$			
2H-Pyran, 2-methoxy-3-methyl-2,4,6-triphenyl-	dioxan	237(4.27),243(4.27), 324(4.17)	48-0729-83
$C_{25}H_{22}O_5$			
Anhydrocneorin $C_{11.B}$	MeOH	207(4.03),284(4.19),	5-1760-83

Compound	Solvent	$\lambda_{max}(\log \epsilon)$	Ref.
(cont.)		329(3.64)	5-1760-83
4H-1-Benzopyran-4-one, 3-[(3,4-dimethoxyphenyl)methyl]-7-(phenylmethoxy)-	MeOH	204(5.21),216(4.99), 286(4.77),294s(4.77), 306s(4.73)	2-1119-83
4H-1-Benzopyran-4-one, 3-[(3,4-dimethoxyphenyl)methylene]-2,3-dihydro-7-(phenylmethoxy)-	MeOH	218(4.42),226(4.37), 270(4.45),310(4.14)	2-1119-83
4H-1-Benzopyran-4-one, 2-(3,4-dimethoxyphenyl)-3-methyl-7-(phenylmethoxy)-	MeOH	218(4.72),234(4.37), 305(4.28)	2-1119-83
$C_{25}H_{23}BrO_7$			
Spiro[furan-2(3H),7'(4'H)-[1H,3H-3a,8b]-methanobenzo[1,2-b:3,4-c']difuran-1'-one, 5'-bromo-4-(2,5-dihydro-2-methyl-5-oxo-2-furanyl)-3'-(3-furanyl)-4,5-dihydro-5,5-dimethyl-	MeOH	210(4.36),267(4.08)	5-1760-83
$C_{25}H_{23}Cl_2N_3OS$			
Phenol, 2-[[[4-(2,8-dichloro-10,11-dihydrodibenzo[b,f]thiepin-10-yl)-1-piperazinyl]imino]methyl]-	MeOH	281(4.38),288s(4.35), 311(4.24),320s(4.20)	73-1173-83
$C_{25}H_{23}Cl_2N_3S$			
1-Piperazinamine, 4-(2,8-dichloro-10,11-dihydro-dibenzo[b,f]thiepin-10-yl)-N-(phenylmethylene)-	MeOH	287(4.40)	73-1173-83
$C_{25}H_{23}N$			
Pyridine, 1,4-dihydro-4-methyl-3,5-diphenyl-1-(phenylmethyl)-	EtOH	240(4.07),290(3.65), 375(3.53)	23-2126-83
$C_{25}H_{23}NO_2PRe$			
Rhenium, acetyl(η^5-2,4-cyclopentadien-1-yl)nitrosyl(triphenylphosphine)-	$CHCl_3$	268(3.58),288s(3.38)	157-1852-83
$C_{25}H_{23}N_2OP$			
3-Furancarbonitrile, 4,5-dihydro-4,5-dimethyl-2-[(triphenylphosphoranylidene)amino]-	$CHCl_3$	267(4.13),273(4.08), 290(3.97)	24-1691-83
3-Furancarbonitrile, 4,5-dihydro-5,5-dimethyl-2-[(triphenylphosphoranylidene)amino]-	$CHCl_3$	267(4.27),273(4.23), 287(4.11)	24-1691-83
$C_{25}H_{23}N_3O_2$			
6H-Anthra[1,9-cd]isoxazol-6-one, 5-[[5-(cyclohexylimino)-1,3-pentadienyl]amino]-	dioxan	555(4.35)	103-0377-83
$C_{25}H_{24}BrNO_7$			
Spiro[cyclopentane-1,10'-[6,8a]ethano-[8aH-carbazole]-5',7,8'-tricarboxylic acid, 3'-bromo-6',9'-dihydro-9'-methyl-11'-oxo-, trimethyl ester	EtOH and EtOH-acid	204(4.23),246(4.35), 461(3.79)	39-2175-83C
$C_{25}H_{24}ClN_3OS$			
Phenol, 2-[[[4-(2-chloro-10,11-dihydrodibenzo[b,f]thiepin-10-yl)-1-piperazinyl]imino]methyl]-	MeOH	281.5(4.35),312(4.21), 288s(4.34)	73-1173-83
Phenol, 2-[[[4-(8-chloro-10,11-dihydrodibenzo[b,f]thiepin-10-yl)-1-pipera-	MeOH	230(4.26),275(4.32), 311(4.18)	73-1173-83

Compound	Solvent	λ_{max}(log ε)	Ref.
zinyl)imino]methyl]- (cont.)			73-1173-83
$C_{25}H_{24}ClN_3S$			
1-Piperazinamine, 4-(2-chloro-10,11-dihydrodibenzo[b,f]thiepin-10-yl)-N-(phenylmethylene)-	MeOH	289(4.38)	73-1173-83
1-Piperazinamine, 4-(8-chloro-10,11-dihydrodibenzo[b,f]thiepin-10-yl)-N-(phenylmethylene)-	MeOH	285(4.35)	73-1173-83
$C_{25}H_{24}NO_2PS$			
3-Thiophenecarboxylic acid, 4,5-dihydro-2-[(triphenylphosphoranylidene)-amino]-, ethyl ester	$CHCl_3$	267(3.98),275(3.98), 320(4.15)	24-1691-83
$C_{25}H_{24}NO_3P$			
3-Furancarboxylic acid, 4,5-dihydro-2-[(triphenylphosphoranylidene)-amino]-, ethyl ester	MeOH	267(3.35),273(3.39), 304(3.47)	24-1691-83
$C_{25}H_{24}N_2O_2$			
1H-Pyrazole, 4,5-dihydro-5-(4-methoxyphenyl)-3-[2-(4-methoxyphenyl)ethenyl]-1-phenyl-	n.s.g.	225(4.26),285(4.05), 292(4.04),377(4.44)	44-0542-83
$C_{25}H_{24}N_2O_6S$			
Benzenesulfonamide, N-[2,3-dihydro-2-oxo-3-[(3,4,5-trimethoxyphenyl)methylene]-1H-indol-4-yl]-4-methyl-	$CHCl_3$	636(4.02)	44-2468-83
4-Thia-1-azabicyclo[3.2.0]hept-2-ene-2-carboxylic acid, 3-(benzoyloxy)-7-oxo-6-[(phenylacetyl)amino]-, 1,1-dimethylethyl ester, (5R-trans)-	dioxan	240(4.29),313(3.76)	78-2493-83
$C_{25}H_{24}N_4O_4$			
Benzoic acid, 2,3-dimethyl-4-(phenylmethoxy)-6-[(1-phenyl-1H-tetrazol-5-yl)oxy]-, ethyl ester	EtOH	248s(4.19)	39-0667-83C
$C_{25}H_{24}O_4$			
1,3-Cyclohexanedione, 2,2'-methylenebis[5-phenyl-	MeOH	255(4.34)	24-1309-83
$C_{25}H_{24}O_6$			
Acetic acid, [(2-benzoyl-7,8,9,10-tetrahydro-4-methyl-6-oxo-6H-dibenzo-[b,d]pyran-3-yl)oxy]-, ethyl ester	MeOH	265(4.43)	78-1265-83
Anhydrocneorin $C_{11.1}$	MeOH	212(3.95)	5-1760-83
Cneorin $C_{11.B}$	MeOH	208(4.01),284(4.22)	5-1760-83
6H,8H-5b,8a-Methanocyclopent[cd]isobenzofuro[4,5-g]isobenzofuran-1,6-(2aH)-dione, 8-(3-furanyl)-3,9,10b-10c-tetrahydro-3-(1-hydroxy-1-methylethyl)-2a-methyl-	MeOH	248(4.28)	5-1760-83
$C_{25}H_{25}NO_2$			
Propenoic acid, 2,3-diphenyl-3-(phenylamino)-, 1,1-dimethylethyl ester	EtOH	247(4.05),337(4.24)	4-0245-83
$C_{25}H_{25}NO_4$			
Isoquinoline, 3-(3,4-dimethoxyphenyl)-	EtOH	236(4.33),261(4.01),	142-1247-83

Compound	Solvent	$\lambda_{max}(\log \epsilon)$	Ref.
3,4-dihydro-6,7-dimethoxy-1-phenyl-, hydrochloride (cont.)		316(3.94),375(3.70)	142-1247-83
$C_{25}H_{25}NO_7$			
Spiro[cyclohexane-1,10'-[6,8a]ethano-[8aH]carbazole]-5',7',8'-tricarboxylic acid, 6',9'-dihydro-11'-oxo-, trimethyl ester	MeOH	208(4.27),244(4.38), 296(3.80),315(3.58), 433(3.82)	39-0515-83C
Spiro[cyclopentane-1,10'-[6,8a]ethano-[8aH]carbazole]-5',7',8'-tricarboxylic acid, 6',9'-dihydro-9'-methyl-11'-oxo-, trimethyl ester	MeOH	246(4.35),296(3.79), 315(3.64),455(3.89)	39-0515-83C
$C_{25}H_{25}N_3OS$			
Phenol, 2-[[[4-(10,11-dihydrodibenzo-[b,f]thiepin-10-yl)-1-piperazinyl]-imino]methyl]-	MeOH	281(4.33),288(4.32), 312(4.22)	73-1173-83
$C_{25}H_{25}N_3S$			
1-Piperazinamine, 4-(10,11-dihydro-dibenzo[b,f]thiepin-10-yl)-N-(phenylmethylene)-	MeOH	290(4.32)	73-1173-83
$C_{25}H_{26}BrNO_8$			
2H-Furo[2',3':1,7]cyclohept(1,2-b]indole-3,7a(12H)-diacetic acid, 9-bromo-4,5,6,7-tetrahydro-α^{7a}-(2-methoxy-2-oxoethylidene)-12-methyl-2-oxo-, dimethyl ester, [7aR*(Z)-12aR*]-	EtOH and EtOH-$HClO_4$	256(4.26),321(3.60)	39-2175-83C
$C_{25}H_{26}N_2$			
1H-Pyrazole, 4,5-dihydro-1-(4-butyl-phenyl)-3,5-diphenyl-	n.s.g.	245(4.23),254s(--), 313s(--),363(4.23)	103-0229-83
$C_{25}H_{26}N_4O_7$			
Senacarcin A	EtOH	210(3.60),263(3.62), 350s(--),366(2.85)	158-83-78
$C_{25}H_{26}N_4O_7S$			
1H-Pyrazolo[3,4-d]pyrimidine, 4-[(3-phenyl-2-propenyl)thio]-1-(2,3,5-tri-O-acetyl-β-D-ribofuranosyl)-, (E)-	pH 13	257(4.32),287s(4.23), 294(4.25)	87-1489-83
$C_{25}H_{26}N_4O_{13}S_2$			
Benzenesulfonamide, 2-(2-methoxyethoxy)-N-[5-[[[2-(2-methoxyethoxy)-5-nitro-phenyl]sulfonyl]amino]-2-methyl-4-oxo-2,5-cyclohexadien-1-ylidene]-5-nitro-	MeCN	234(4.46),302(4.51)	44-0177-83
$C_{25}H_{26}O_2$			
[2,2'-Binaphthalene]-6-carboxylic acid, 5',6',7',8'-tetrahydro-5',5',8',8'-tetramethyl-	EtOH	227(4.61),260(4.74), 307(4.32)	87-1653-83
$C_{25}H_{26}O_6$			
Sanggenon I	EtOH	214(4.40),228(4.43), 289(4.19),319s(3.71)	142-1071-83
	EtOH-$AlCl_3$	205(4.41),225(4.49), 282s(3.86),309(4.24)	142-1071-83

Compound	Solvent	λ_{max}(log ε)	Ref.
$C_{25}H_{26}O_7$			
Cneorin B	MeOH	210(4.37)	5-1760-83
Cneorin B_1	MeOH	209(4.35)	5-1760-83
Cneorin $B_{11.A}$	MeOH	214(4.04)	5-1760-83
Cneorin B_{111}	MeOH	213(4.02),230s(3.80)	5-1760-83
Cneorin C	MeOH	209(4.39)	5-1760-83
Cneorin C_1	MeOH	209(4.40)	5-1760-83
Cneorin C_{111}	MeOH	215s(4.01),230(3.80)	5-1760-83
Tricoccin R_2	MeOH	207(4.34)	5-1798-83
Tricoccin R_5	MeOH	209(4.30)	5-1798-83
$C_{25}H_{26}O_8$			
2-Anthracenepropanoic acid, 9,10-dihydro-1-hydroxy-4,8-dimethoxy-9,10-dioxo-α-(1-oxobutyl)-, ethyl ester, (±)-	MeOH	226(4.55),250(4.19), 282(3.93),366(3.41), 422(3.77),447(3.92), 468(3.93),504(3.62), 532(3.13)	5-1818-83
$C_{25}H_{26}O_9$			
D-Arabinitol, 1-C-(9,10-dihydro-1,4-dihydroxy-9,10-dioxo-2-anthracenyl)-2,3:4,5-bis-O-(1-methylethylidene)-, (S)-	EtOH	227s(4.25),250(4.56), 256s(4.47),284(3.94), 320(3.45),482(3.99), 518(3.82)	39-0613-83C
Sakyomicin C	EtOH	216(4.30),240(4.06), 310(3.54),415(3.64)	77-0174-83
$C_{25}H_{26}O_{10}$			
D-Arabinitol, 1-C-(9,10-dihydro-1,4,5-trihydroxy-9,10-dioxo-2-anthracenyl)-2,3:4,5-bis-O-(1-methylethylidene)-	MeOH	219s(4.0),233(4.39), 252(4.17),274(3.61), 292(3.72),462(3.67)	39-0613-83C
Sakyomicin A	EtOH	216(4.46),238(4.20), 310(3.72),415(3.65)	77-0174-83
$C_{25}H_{27}ClO_3S$			
Androsta-1,4-diene-3,11-dione, 17-chloro-17-(phenylsulfinyl)-, (17α,17(R)]-	MeOH	242(4.31)	87-0078-83
$C_{25}H_{27}ClO_7$			
6H,8H-5b,8a-Methanocyclopent[cd]isobenzofuro[4,5-g]isobenzofuran-1,6-(2aH)-dione, 10a-chloro-8-(3-furanyl)-3,4,4a,9,10,10a,10b,10c-octahydro-4a-hydroxy-3-(1-hydroxy-1-methylethyl)-2a-methyl- (cneorin $C_{11.A}$)	MeOH	214(4.10)	5-1760-83
$C_{25}H_{27}NO_5$			
Karakoramine, (+)-	MeOH	208(4.56),226s(4.30), 283(3.75)	142-0425-83
$C_{25}H_{27}NO_8$			
2-Butenedioic acid, 2-[4,5,6,7-tetrahydro-3-(2-methoxy-2-oxoethyl)-12-methyl-2-oxo-2H-furo[2',3':1,7]cyclohept[1,2-b]indol-7a(12H)-yl]-, dimethyl ester, [7aR*(Z),12aS*]-	MeOH	217.5(4.36),245(4.17), 315(3.52)	39-0515-83C
$C_{25}H_{27}N_4O_2$			
21H-Bilin-10-ylium, 1,10,19,22,23,24-hexahydro-2,3,17,18,22,23-hexamethyl-	CH_2Cl_2	390(4.46),602(4.18)	54-0347-83

Compound	Solvent	λ_{max}(log ϵ)	Ref.
1,19-dioxo-, salt with trifluoro-acetic acid (cont.)			54-0347-83
$C_{25}H_{27}N_5O_5$			
Acetamide, N-[6,7-dihydro-6-oxo-7-[[2-(phenylmethoxy)-1-[(phenylmethoxy)methyl]ethoxy]methyl]-1H-purin-2-yl]-	MeOH	263(4.14),280(4.01)	87-0759-83
Acetamide, N-[6,9-dihydro-6-oxo-9-[[2-(phenylmethoxy)-1-[(phenylmethoxy)methyl]ethoxy]methyl]-1H-purin-2-yl]-	MeOH	257(4.22),282(4.07)	87-0759-83
$C_{25}H_{28}N_2O_3$			
3-Pyridinecarboxamide, N-(2,2-dimethyl-4-phenyl-1,3-dioxan-5-yl)-1,4-dihydro-1-(phenylmethyl)-	EtOH	350(3.68)	130-0299-83
$C_{25}H_{28}N_3O_2S$			
Morpholinium, 4-[6-[[(dimethylamino)-methylene]amino]-3-(4-methoxyphenyl)-5-phenyl-2H-thiopyran-2-ylidene]-, perchlorate	MeCN	243(4.38),261(4.38), 340(4.33),511(4.36)	97-0403-83
$C_{25}H_{28}N_6O_4$			
Estra-1,3,5(10)-trien-17-one, 3-[2-(6-amino-9H-purin-9-yl)ethoxy]-4-nitro-	EtOH	262(4.20)	87-0162-83
$C_{25}H_{28}O_2S$			
Androsta-1,4-diene-3,11-dione, 17α-(phenylthio)-	MeOH	249(4.29)	87-0078-83
$C_{25}H_{28}O_3$			
Estra-1,3,5(10)-trien-6-one, 17-hydroxy-3-(phenylmethoxy)-, (17β)-	MeOH	222(4.36),254(3.86), 322(3.40)	145-0347-83
Estra-1,3,5(10)-trien-17-one, 3-hydroxy-4-(phenylmethoxy)-	$CHCl_3$	281(3.32)	94-3309-83
Estra-1,3,5(10)-trien-17-one, 4-hydroxy-3-(phenylmethoxy)-	$CHCl_3$	281(3.33)	94-3309-83
2(3H)-Naphthalenone, 4,4a,5,6,7,8-hexahydro-4-(4-methoxyphenyl)-1-[(4-methoxyphenyl)methyl]-	EtOH	205(4.36),227(4.41), 250(4.14)	2-0542-83
$C_{25}H_{28}O_4$			
Ammothamnidin	EtOH	232s(4.18),261s(3.99), 321s(4.01),390(4.46)	105-0417-83
Glabrol	MeOH	280(4.36),312(4.00)	2-1061-83
$C_{25}H_{28}O_4S_2$			
3(2H)-Thiophenone, 4-(octylthio)-2-[(3-phenoxyphenyl)methylene]-, 1,1-dioxide, (Z)-	EtOH	205(3.40),290(4.30), 320(4.18),350(2.85)	104-1447-83
$C_{25}H_{28}O_5$			
2H,8H-Benzo[1,2-b:3,4-b']dipyran-2-one, 3,4,9,10-tetrahydro-5-hydroxy-8,8-dimethyl-6-(2-methyl-1-oxobutyl)-4-phenyl-	EtOH	220(4.50),290(4.31), 350(3.38)	78-3923-83
	EtOH-KOH	230(4.24),254(3.99), 344(4.60)	78-3923-83
Calozeylanolactone	n.s.g.	238(3.73),290(3.64), 324(3.05),392(2.49)	39-0703-83C

Compound	Solvent	λ_{max}(log ϵ)	Ref.
Sophoraflavanone A	EtOH	223s(4.10),297(3.60), 337(3.13)	94-2859-83
	EtOH-NaOEt	250s(4.31),338(4.08)	94-2859-83
	EtOH-NaOAc	297(3.89),340(3.89)	94-2859-83
	EtOH-$AlCl_3$	226s(4.16),314(3.73)	94-2859-83
$C_{25}H_{28}O_6$			
Sigmoidin A	MeOH	288(4.08)	88-4127-83
	MeOH-NaOMe	323(4.32)	88-4127-83
	MeOH-NaOAc	325(4.27)	88-4127-83
	MeOH-$AlCl_3$	305(4.19)	88-4127-83
	+ HCl	305(4.19)	88-4127-83
$C_{25}H_{28}O_8$			
6H,8H-5b,8a-Methanocyclopent[cd]isobenzofuro[4,5-g]isobenzofuran-1,6-(2aH)-dione, 8-(3-furanyl)-3,4,4a-9,10,10a,10b,10c-octahydro-4a,10a-dihydroxy-3-(1-hydroxy-1-methylethyl)-2a-methyl-	MeOH	205(4.20)	5-1760-83
isomer	MeOH	206(4.18)	5-1760-83
$C_{25}H_{28}O_9$			
7H-Phenaleno[1,2-b]furan-7-one, 3,6-diacetoxy-7a,8,9,10a-tetrahydro-7a-hydroxy-4,5-dimethoxy-1,8,8,9-tetramethyl-	EtOH	330(3.67),355(3.80)	18-3661-83
Spiro[furan-2(3H),1'(4'H)-[1H,3H-3a,8b]-methanobenzo[1,2-b:3,4-c']difuran]-1'-one, 4-(2,5-dihydro-2-methyl-5-oxo-2-furanyl)-3'-(3-furanyl)hexahydro-8',8'a-dihydroxy-5,5-dimethyl-	MeOH	214(3.91)	5-1760-83
isomer	MeOH	209(4.19)	5-1760-83
cis-diol C	MeOH	209(4.24)	5-1760-83
trans-diol C_1	MeOH	210(4.11)	5-1760-83
cis-diol R_2	MeOH	210(4.02)	5-1760-83
cis-diol R_3	MeOH	208(3.98)	5-1760-83
$C_{25}H_{29}BrF_2O_7$			
Halopredone acetate	MeOH	246(4.10)	162-0663-83
$C_{25}H_{29}Cl_4NO_6$			
Acetic acid, dichloro-, 3,3-dichloro-3,4-dihydro-5,5-dimethyl-2-oxo-8-pentyl-2H,5H-pyrano[3,2-c][1]benzopyran-10-yl ester	EtOH	235s(4.06),270(3.83), 319(3.69)	161-0775-83
$C_{25}H_{29}NO_3$			
Estra-1,3,5(10)-trien-6-one, 17-hydroxy-3-(phenylmethoxy)-, oxime, (17β)-	MeOH	221(4.41),253(3.93), 320(3.40)	145-0347-83
$C_{25}H_{29}N_2$			
3H-Indolium, 2-[3-(1,3-dihydro-1,3,3-trimethyl-2H-indol-2-ylidene)-1-propenyl]-1,3,3-trimethyl-, tetrafluoroborate	MeOH	544(5.12)	89-0876-83
	MeOH	545(5.11)	104-2114-83
	CH_2Cl_2	550(5.13)	104-2114-83
$C_{25}H_{29}N_2S$			
Naphtho[2,1-d]thiazolium, 3-ethyl-2-	MeOH	509(5.06)	104-2089-83

Compound	Solvent	$\lambda_{max}(\log \epsilon)$	Ref.
[[3-(1-piperidinyl)-2-cyclohexen-1-ylidene]methyl]-, perchlorate (cont.)			104-2089-83
$C_{25}H_{29}N_2S_2$			
Benzothiazolium, 3-ethyl-2-[2-[(3-ethyl-2(3H)-benzothiazolylidene)methyl]-3,3-dimethyl-1-butenyl]-, tetrafluoroborate	MeOH	593(4.32)	89-0876-83
$C_{25}H_{29}N_5O_2$			
[1,1'-Biphenyl]-2-carbonitrile, 4-nitro-2',4',6'-tri-1-pyrrolidinyl-	$CHCl_3$	488(3.88)	44-4649-83
$C_{25}H_{29}N_5O_7$			
1H-Pyrazolo[3,4-d]pyrimidin-4-amine, N-(1-methyl-2-phenylethyl)-1-(2,3,5-tri-O-acetyl-β-D-ribofuranosyl)-, (R)-	MeOH	263(4.98),283(5.17)	87-1601-83
$C_{25}H_{30}$			
1,4,7,10,13-Cyclopentadecapentayne, 3,3,6,6,9,9,12,12,15,15-decamethyl-	pentane	230s(1.48)	35-7760-83
$C_{25}H_{30}N_4O_8$			
Propanedioic acid, [ethoxy[[6-(ethoxycarbonyl)-4,7-dihydro-7-oxo-3-phenylpyrazolo[1,5-a]pyrimidin-2-yl]amino]methyl]-, diethyl ester	MeCN	262(4.04),318(4.34)	39-0011-83C
$C_{25}H_{30}O_2$			
Estra-1,3,5(10)-trien-17β-ol, 3-(phenylmethoxy)-	MeOH	216(4.09),278(3.29), 286(3.24)	145-0347-83
$C_{25}H_{30}O_3$			
Norethisterone 2,4-pentadienoate, (E)-	EtOH	245(4.61)	13-0277-83A
$C_{25}H_{30}O_7$			
Pregna-1,4,9(11)-triene-3,12,20-trione, 21-[(ethoxycarbonyl)oxy]-17-hydroxy-16-methyl-, (16α)-	MeOH	238(4.41)	39-2781-83C
$C_{25}H_{30}O_8$			
2-Oxatricyclo[13.2.2.1^{3,7}]eicosa-3,5,7(20),15,17,18-hexaen-12-one, 4- (β-D-glucopyranosyloxy)- (aceroside IV)	EtOH and EtOH-NaOH	273(3.99)	94-1917-83
$C_{25}H_{31}Cl_4NO_5$			
Acetic acid, dichloro-, 3,3-dichloro-4-(diethylamino)-3,4-dihydro-5,5-dimethyl-2-oxo-8-pentyl-2H,5H-pyrano-[3,2-c][1]benzopyran-10-yl ester	EtOH	235s(4.15),372(3.92), 318(3.77)	161-0775-83
$C_{25}H_{31}FO_7$			
Pregna-4,6-dien-3-one, 9α-fluoro-11β-hydroxy-2-(hydroxymethylene)-16α-methyl-17α,20:20,21-bis[methylenebis(oxy)]-	MeOH	288(4.23)	39-2793-83C
$C_{25}H_{31}NO_2$			
Estra-1,3,5(10)-trien-17-ol, 6-amino-3-(phenylmethoxy)-, (6α,17β)-	MeOH	212(4.20),225(4.01), 279(3.24),286(3.19)	145-0347-83

Compound	Solvent	λ_{max}(log ϵ)	Ref.
$C_{25}H_{31}N_3O_2$			
Hexanamide, 2-[[(4-hydroxyphenyl)methyl]methylamino]-N-[2-(1H-indol-3-yl)ethenyl]-4-methyl- (fragilamide A)	EtOH	229(4.19),280(4.01)	88-2087-83
$C_{25}H_{31}N_3O_3$			
Caesalpinine A	EtOH	222(4.24),256(3.59), 265(3.64),275(3.62)	35-4441-83
$C_{25}H_{31}N_3O_8$			
Pregn-4-ene-3,11,20-trione, 6,21-diacetoxy-7-azido-17-hydroxy-, (6β,7α)-	MeOH	230(4.11)	39-2781-83C
$C_{25}H_{32}ClFO_5$			
Clobetasol 17-propionate	EtOH	237(4.18)	162-0336-83
$C_{25}H_{32}N_2O_2S$			
Dodecanamide, N-[4-(2-benzothiazolyl)-3-hydroxyphenyl]-	EtOH	344(4.51)	4-1517-83
$C_{25}H_{32}N_2O_3$			
Dodecanamide, N-[4-(2-benzoxazolyl)-3-hydroxyphenyl]-	EtOH	330(4.56)	4-1517-83
$C_{25}H_{32}N_2O_6$			
$\Delta^{3(14)}$-Vindoline, 14-acetyl-17-deacetyl-14,15-dihydro-	EtOH	214(4.39),251(3.74), 320(4.31)	100-0884-83
Vindoline, 3-hydroxy-14,15-dihydro-	EtOH	215(4.30),254(3.66), 308(3.57)	100-0884-83
$C_{25}H_{32}N_2O_7$			
Vindoline, 3-oxo-14,15-dihydro-	EtOH	214(4.38),253(4.05), 305(3.96)	100-0884-83
$C_{25}H_{32}OS$			
Naphthalene, 6-[2-[4-(ethylsulfinyl)phenyl]-1-methylethenyl]-1,2,3,4-tetrahydro-1,1,4,4-tetramethyl-, (E)-	EtOH	296(4.42)	111-0425-83
$C_{25}H_{32}O_2$			
Furan, 2,2'-(phenylmethylene)bis[5-pentyl-	EtOH	227(4.37)	103-0478-83
$C_{25}H_{32}O_2S$			
Benzenesulfinic acid, 4-[2-(5,6,7,8-tetrahydro-5,5,8,8-tetramethyl-2-naphthalenyl)-1-propenyl]-, ethyl ester, (E)-	EtOH	228(4.14),299(4.43)	111-0425-83
Naphthalene, 6-[2-[4-(ethylsulfonyl)phenyl]-1-methylethenyl]-1,2,3,4-tetrahydro-1,1,4,4-tetramethyl-, (E)-	EtOH	232(4.13),302(4.39)	111-0425-83
(Z)-	EtOH	236(4.24),292(4.10)	111-0425-83
$C_{25}H_{32}O_3$			
19-Norpregn-4-en-20-yn-3-one, 17-[(cyclobutylcarbonyl)oxy]-, (17α)-	MeOH	240(4.20)	13-0291-83A
$C_{25}H_{32}O_3S$			
Benzenesulfonic acid, 4-[2-(5,6,7,8-tetrahydro-5,5,8,8-tetramethyl-2-naphthalenyl)-1-propenyl]-, ethyl	EtOH	232(4.15),302(4.39)	111-0425-83

Compound	Solvent	λmax(log ε)	Ref.
ester, (E)- (cont.)			111-0425-83
$C_{25}H_{32}O_4$			
Melengestrol acetate	EtOH	287(4.35)	162-0828-83
$C_{25}H_{32}O_5$			
17,19-Dinoratis-15-ene-4,13,14-tricarboxylic acid, 16-(1-methylethenyl)-, cyclic 13,14-anhydride 4-methyl ester, (4α,8α,12α,13S,14R)-	EtOH	234s(--),240(4.46)	49-1259-83
$C_{25}H_{32}O_8$			
Aspidin	C_6H_{12}	230(4.41),290(4.33)	162-0122-83
Bicyclo[3.2.1]oct-2-en-8-one, 4-acetoxy-3,5-dimethoxy-7-methyl-1-(2-propenyl)-6-(3,4,5-trimethoxyphenyl)-, (4-endo,6-endo,7-exo)-	MeOH	280(3.15)	102-0269-83
α-Kosin	n.s.g.	227(4.49),290(4.39)	162-0765-83
β-Kosin	n.s.g.	228(4.48),292(4.33)	162-0765-83
Phenol, 4-[2-[2,6-dimethoxy-4-(1-propenyl)phenoxy]-1-methoxypropyl]-2,6-dimethoxy-, acetate, erythro	EtOH	267(4.31)	150-2625-83M
threo	EtOH	268(4.14)	150-2625-83M
$C_{25}H_{32}S$			
Naphthalene, 6-[2-[4-(ethylthio)phenyl]-1-methylethenyl]-1,2,3,4-tetrahydro-1,1,4,4-tetramethyl-, (E)-	EtOH	299(4.48)	111-0425-83
$C_{25}H_{33}N_3O_2$			
Dodecanamide, N-[4-(1H-benzimidazol-2-yl)-3-hydroxyphenyl]-	EtOH	327(4.58)	4-1517-83
$C_{25}H_{33}N_3O_6$			
Pregn-4-ene-3,20-dione, 6,17-diacetoxy-7-azido-, (6β,7α)-	MeOH	233(4.12)	39-2781-83C
$C_{25}H_{33}N_9O_4$			
1H-Pyrrole-2-carboxamide, N-[5-[[(3-amino-3-iminopropyl)amino]carbonyl]-1-ethyl-1H-pyrrol-3-yl]-1-ethyl-4-[[[1-ethyl-4-(formylamino)-1H-pyrrol-2-yl]carbonyl]amino]-, monohydrobromide	EtOH	238(4.46),303(4.54)	87-1042-83
$C_{25}H_{34}N_4O_8$			
Glycine, N-[1-[1-[(1,1-dimethylethoxy)carbonyl]-L-prolyl]-L-prolyl]-N-[(2-nitrophenyl)methyl]-, methyl ester	MeOH	223(3.8),256(3.8)	5-0231-83
$C_{25}H_{34}O_3$			
18,19-Dinorpregn-4-en-20-yn-3-one, 13-ethyl-17-(2-methyl-1-oxopropoxy)-, (17α)-	MeOH	242(4.22)	13-0349-83A
18,19-Dinorpregn-4-en-20-yn-3-one, 13-ethyl-17-(1-oxobutoxy)-, (17α)-	MeOH	240(4.24)	13-0349-83A
Furan, 2,2'-(2-furanylmethylene)bis[5-hexyl-	EtOH	227(4.29)	103-0478-83
Furan, 2,2',2"-methylidynetris[5-butyl-	hexane	220(3.29)	103-0360-83

Compound	Solvent	λ_{max}(log ε)	Ref.
$C_{25}H_{34}O_5$			
Chola-4,6-dien-24-oic acid, 12α-(formyloxy)-3-oxo-	EtOH	282(4.49)	13-0475-83A
$C_{25}H_{34}O_6$			
Card-20(22)-enolide, 3-acetoxy-14-hydroxy-21-oxo-, (3β,5β)-	MeOH	214(3.92),227(3.85)	13-0189-83B
$C_{25}H_{34}O_7$			
Chapparin, 6α-senecioyloxy-	MeOH	219(4.24)	100-0218-83
$C_{25}H_{34}O_9$			
Longicornin D	MeOH	215(4.40)	102-2759-83
$C_{25}H_{34}O_{10}$			
Longicornin C	MeOH	208(4.26)	102-2759-83
$C_{25}H_{35}NO_6S$			
1-Azetidineacetic acid, 2-(butylthio)-α-[1-methoxy-3-[(tetrahydro-2H-pyran-2-yl)oxy]propylidene]-4-oxo-, phenylmethyl ester	EtOH	263(3.94)	39-0115-83C
$C_{25}H_{35}N_3$			
1H-Benzimidazole, 1-dodecyl-2-(4-methyl-2-pyridinyl)-	EtOH	302(4.28)	4-0023-83
$C_{25}H_{35}N_3O_4$			
24-Norchola-20,22-diene-12,14-diol, 3-azido-21,23-epoxy-, 12-acetate, (3α)-	MeOH	214.5(3.70)	13-0171-83B
(3β)-	MeOH	214(3.69)	13-0171-83B
$C_{25}H_{35}N_5O_3$			
5β,14β-Androstane-12β,14-diol, 3β-azido-17β-(4-pyridazinyl)-, 12-acetate	MeOH	218.5(3.84),251.5(3.41)	13-0171-83B
$C_{25}H_{35}NaO_6$			
Hydroxydione sodium	n.s.g.	280(1.97)	162-0702-83
$C_{25}H_{36}N_2O_3$			
5β,14β-Androstane-3,14-diol, 17β-(4-pyridazinyl)-, 3-acetate, (3α)-	MeOH	218.5(3.79),252s(3.40)	13-0189-83B
(3β)-	MeOH	219(3.79),252.5s(3.40)	13-0189-83B
$C_{25}H_{36}N_2O_4$			
Acetamide, N-acetyl-N-[(3β)-3-acetoxy-17-aza-D-homoandrosta-5,17-dien-17a-yl]-	EtOH	238(3.67)	70-2141-83
Acetamide, N-acetyl-N-[(3β)-3-acetoxy-17a-aza-D-homoandrosta-5,16-dien-17-yl]-	EtOH	232(3.9),330(4.19)	70-2141-83
5β,14β-Androstane-3,14-diol, 17β-(4-pyridazinyl)-, 3-acetate, N^1-oxide	MeOH	204.5(3.90),270.5(4.23), 313(3.75)	13-0189-83B
N^2-oxide	MeOH	217(4.07),265(4.05), 307(3.55)	13-0189-83B
$C_{25}H_{36}O_4$			
5β,14α-Bufadienolide, 3β-acetoxy-	EtOH	223(3.71),298(3.57)	44-4248-83
Cephalonic acid	n.s.g.	259(4.07)	162-0276-83
24-Norchola-20,22-diene-3,14-diol, 21,23-epoxy-, 3-acetate, (3α,5β,14β)-	MeOH	214.5(3.65)	13-0189-83B

Compound	Solvent	$\lambda_{max}(\log \epsilon)$	Ref.
$C_{25}H_{36}O_5$			
Chol-4-en-24-oic acid, 12α-(formyloxy)-3-oxo-	EtOH	239(4.23)	13-0475-83A
24-Norchola-20,22-diene-3,12,14-triol, 21,23-epoxy-, 12-acetate, (3β,5β,12β,14β)-	MeOH	213.5(3.66)	13-0171-83B
5β-Pregn-16-en-20-one, 3α,12α-diacetoxy-	EtOH	237(4.05)	44-1954-83
$C_{25}H_{36}O_5Si$			
Anthra[1,9-de]-1,3-dioxin-11-ol, 7-[[(1,1-dimethylethyl)dimethylsilyl]oxy]-7,7a,8,11,11a,11b-hexahydro-2,2-dimethyl-, acetate, (7α,7aβ,11α,11aβ-11bβ)-	EtOH	220(3.85),273(3.28), 278(3.28)	44-3252-83
$C_{25}H_{36}O_6$			
Card-20(22)-enolide, 3-acetoxy-14,21-dihydroxy-	MeOH	220(3.99)	13-0189-83B
Card-20(22)-enolide, 3-acetoxy-14,22-dihydroxy-	MeOH	218(3.83)	13-0189-83B
$C_{25}H_{37}NO_4$			
Androst-4-en-3-one, 17-acetoxy-4-morpholino-	EtOH	245(4.03)	107-1013-83
Androst-4-en-3-one, 17-acetoxy-6-morpholino-	EtOH	239(4.03)	107-1013-83
24-Norchola-20,22-diene-12,14-diol, 3-amino-21,23-epoxy-, 12-acetate, (3α,5β,12β,14β)-	MeOH	214(3.68)	13-0171-83B
(3β,5β,12β,14β)-	MeOH	213(3.74)	13-0171-83B
$C_{25}H_{37}NO_5$			
5β,14β-Card-20(22)-enolide, 12β-acetoxy-3α-amino-14-hydroxy-	MeOH	216(3.87)	13-0171-83B
3β-	MeOH	215(3.91)	13-0171-83B
24-Norchol-20(22)-en-21-oic acid, 12-acetoxy-3-amino-14,23-dihydroxy-, γ-lactone, (3α)-	MeOH	218(4.17)	13-0171-83B
(3β)-	MeOH	218(4.19)	13-0171-83B
$C_{25}H_{37}N_3O_3$			
5β,14β-Androstane-12β,14-diol, 3β-amino-17β-(4-pyridazinyl)-, 12-acetate	MeOH	219(3.78),251.5(3.32)	13-0171-83B
$C_{25}H_{38}N_2O_3$			
Androst-4-en-3-one, 17-acetoxy-4-piperazino-	EtOH	243(3.92)	107-1013-83
Androst-4-en-3-one, 17-acetoxy-6-piperazino-	EtOH	238(4.11)	107-1013-83
$C_{25}H_{38}N_2O_8$			
Benzenemethanamine, N-(2,2-dimethoxyethyl)-2-[5-[[(2,2-dimethoxyethyl)amino]methyl]-2-methoxyphenoxy]-3,4-dimethoxy-	MeOH	279(3.63)	83-0694-83
$C_{25}H_{38}O_5$			
Benzeneheptanoic acid, 3,5-dimethoxy-2-(3-oxo-1-octenyl)-, ethyl ester, (E)-	EtOH	247.5(3.90),315(3.95), 337(4.01)	39-2211-83C

Compound	Solvent	λ_{max}(log ε)	Ref.
$C_{25}H_{38}O_7$			
Prosta-5,7,14-trien-1-oic acid, 4,12-diacetoxy-9-hydroxy-, methyl ester	EtOH	248(4.30)	88-1549-83
$C_{25}H_{38}Zr$			
Zirconium, [(1,2,3,4-η)-2-methyl-1,3-butadiene]bis[(1,2,3,4,5-η)-1,2,3,4-5-pentamethyl-2,4-cyclopentadien-1-yl]-	hexane	332(3.41),395(3.05)	18-3735-83
$C_{25}H_{39}NO_3$			
2,4,6,8,10,14-Heptadecahexaenamide, 13-hydroxy-N-(2-hydroxy-1-methylethyl)-2,10,12,14,16-pentamethyl-, all-E-	EtOH	206(4.29),262(3.87), 340s(--),357(4.73), 370s(--)	5-1081-83
(2E,4E,6Z,8E,10E,14E)- (myxalamide B)	EtOH	207(4.38),263(4.04), 340s(--),357(4.67), 370s(--)	5-1081-83
$C_{25}H_{39}NO_5$			
Cassamine	n.s.g.	225(4.2)	162-0264-83
$C_{25}H_{39}NO_6$			
Erythrophlamine	n.s.g.	222(4.2)	162-0532-83
$C_{25}H_{39}N_3O_7S$			
Glycine, N-[S-[[3,5-bis(1,1-dimethylethyl)-4-hydroxyphenyl]methyl]-N-L-γ-glutamyl-L-cysteinyl]-	EtOH	275(3.08)	94-3671-83
$C_{25}H_{40}CoN_5O_4$			
Cobalt, [(10-methyl-2,3,17,18-nonadecanetetrone tetraoximato)(3-)-(pyridine)]-, (OC-6-24)-	EtOH	232(3.62),294(3.72), 384s(3.27),461(3.08)	5-0181-83
$C_{25}H_{40}N_2O_6S$			
Leukotriene D_4	MeOH	270(4.49),280(4.60), 290(4.49)	162-0784-83
$C_{25}H_{40}O_2$			
4H-Inden-4-one, 1-(1,5-dimethylhexyl)-octahydro-7a-methyl-5-(1-methyl-2-oxo-3-cyclohexen-1-yl)-	dioxan	224.5(4.01)	56-0403-83
$C_{25}H_{40}O_3Si_2$			
19-Norpregn-4-en-20-yn-3-one, 17-[(pentamethyldisiloxanyl)oxy]-, (17α)-	MeOH	243(4.95)	13-0255-83A
$C_{25}H_{40}O_4$			
1,2-Benzenediol, 3-pentadecyl-, diacetate	MeOH	210(4.11),258(2.45), 278s(1.65)	36-0792-83
$C_{25}H_{40}O_5$			
Cativic acid, 3α-angeloyloxy-2α-hydroxy-	EtOH	215(3.87)	102-1292-83
$C_{25}H_{41}NO_5S$			
7,9,11,14-Eicosatetraenoic acid, 6-[(2-amino-3-methoxy-3-oxopropyl)-thio]-5-hydroxy-, methyl ester, (5R,6S)-	EtOH	276(4.47),282(4.57), 292(4.46)	35-3656-83

Compound	Solvent	λ_{max}(log ϵ)	Ref.
(5S,6R)- (cont.)	EtOH	271(4.41),281(4.56), 292(4.46)	35-3656-83
(5S,6S)-	EtOH	273(4.47),282(4.55), 291(4.45)	35-3656-83
$C_{25}H_{42}OSi_2$			
Silane, (3-methoxyestra-1,3,5(10)-triene-2,4-diyl)bis[trimethyl-	EtOH	242.5(3.66),300.5(3.35)	65-0539-83
$C_{25}H_{43}NO_2$			
Octadecanamide, N-hydroxy-N-(phenylmethyl)-	EtOH	209(4.16)	34-0433-83
$C_{25}H_{44}N_{14}O_7$			
Capreomycin 1B, tetrahydrochloride	pH 1	268(4.36)	162-0242-83
	H_2O	268(4.35)	162-0242-83
	pH 13	290(4.16)	162-0242-83
$C_{25}H_{44}N_{14}O_8$			
Capreomycin 1A, tetrahydrochloride	pH 1	269(4.38)	162-0242-83
	H_2O	268(4.38)	162-0242-83
	pH 13	289(4.20)	162-0242-83
$C_{25}H_{44}O_2$			
Benzene, 1-heptadecyl-2,3-dimethoxy-	MeOH	231(3.16),271(3.03), 276(3.02)	36-0792-83
$C_{25}H_{46}OSi$			
Silane, (1,1-dimethylethyl)dimethyl-(3,4,7,9-nonadecatetraenyloxy), (E,Z)-	n.s.g.	236(4.40)	44-2572-83
$C_{25}H_{48}AsPSi_2$			
Phosphine, [bis(trimethylsilyl)methyl]-[[2,4,6-tris(1,1-dimethylethyl)phenyl]arsinidene]-	n.s.g.	254(4.02),354(3.92), 431(2.34)	35-5506-83
$C_{25}H_{48}As_2Si_2$			
Diarsene, [bis(trimethylsilyl)methyl]-[2,4,6-tris(1,1-dimethylethyl)phenyl]-, (E)-	n.s.g.	255(4.10),368(3.84), 449(2.26)	35-5506-83

Compound	Solvent	$\lambda_{max}(\log \epsilon)$	Ref.
$C_{26}H_{12}Br_2$			
Benzo[no]naphtho[2,1,8,7-ghij]pleiadene, 7,8-dibromo-	C_6H_{12}	245(4.80),270s(4.72), 277(4.73),312(4.48), 326(4.40)	35-7171-83
$C_{26}H_{12}F_6N_2O_2$			
Benzoxazole, 2,2'-(2,6-naphthalenediyl)bis[5-(trifluoromethyl)-	DMF	344(4.80)	5-0931-83
$C_{26}H_{12}N_2$			
Naphtho[2',3':3,4]cyclobut[1,2-b]anthracene-6,13-dicarbonitrile	CH_2Cl_2	232(4.66),249(4.61), 260s(4.54),266s(4.23), 284(4.52),315s(4.86), 327(5.13),348(4.82), 380(3.82),404(3.07), 428(3.38),458(3.56)	39-1443-83C
Naphtho[2',3':3,4]cyclobut[1,2-b]phenanthrene-7,14-dicarbonitrile	CH_2Cl_2	234(4.75),258(3.50), 268(3.50),286(4.61), 303s(4.32),328(4.82), 366(4.12),409s(3.70), 434(4.25),464(4.58)	39-1443-83C
Naphtho[2',3':3,4]cyclobuta[1,2-l]phenanthrene-9,14-dicarbonitrile	CH_2Cl_2	248(4.45),256(4.49), 266s(4.32),280s(4.23), 290s(4.13),318(4.25), 329(4.53),343(4.64), 354s(4.02),375(4.06), 396(3.78),417(4.09), 442(4.19),464s(3.41)	39-1443-83C
$C_{26}H_{14}N_2O_2S_2$			
Dibenzo[c,n]triphenodithiazine-8,17(7H,16H)-dione	H_2SO_4	362(4.82),860(4.76)	18-1482-83
$C_{26}H_{14}N_2O_5$			
1H-2-Benzopyrano[6',5',4':10,5,6]anthra[2,1,9-def]isoquinoline-1,3,8,10(9H)-tetrone, 9-(2-aminoethyl)-, monohydrochloride	HCOOH	460(4.2),488(4.6), 523(4.8)	24-3524-83
$C_{26}H_{14}N_8$			
9,10-Ethanoanthracene, 2,2,3,3,6,6,7,7-(1H,4H)-octacarbonitrile, 5,8,9,10-tetrahydro-11,12-bis(methylene)-	MeCN	231(3.99),241s(3.92), 252s(3.65)	33-0019-83
$C_{26}H_{15}Cl_5N_6O_8$			
Benzenamine, 4,4'-(2,2,2-trichloroethylidene)bis[N-(3-chlorophenyl)-2,6-dinitro-	MeOH	207.8(4.77),392.2(4.04)	56-1357-83
Benzenamine, 4,4'-(2,2,2-trichloroethylidene)bis[N-(4-chlorophenyl)-2,6-dinitro-	MeOH	204.1(4.66),403.9(4.06)	56-1357-83
$C_{26}H_{15}NO_2$			
Anthra[9,1-bc]pyrido[1,2,3-lm]carbazolediol	EtOH	228(4.635),262(4.30), 273s(4.23),405(3.57)	73-0112-83
$C_{26}H_{16}$			
Benzo[a]naphtho[2,1-j]anthracene	dioxan	249(4.60),264(4.40), 289(4.40),303(4.72), 317(5.05),334(4.02), 344(3.96),359(3.96),	20-0893-83

Compound	Solvent	$\lambda_{max}(\log \epsilon)$	Ref.
(cont.)		378(3.68),408(2.46)	20-0893-83
Benzo[a]picene	dioxan	249(4.62),290(4.77), 303(4.88),332(4.45), 338s(4.06),344s(3.90), 359s(2.72),373(2.73), 393(2.53)	20-0893-83
Naphtho[2,1-c]chrysene	dioxan	245(4.74),275(4.52), 306(4.74),337(4.31)	20-0893-83
$C_{26}H_{16}Cs_2$			
Cesium, μ-[9,9'-bi-9H-fluorene]-9,9'-diyldi-	n.s.g.	413(4.20)	35-2502-83
$C_{26}H_{16}N_2O_2$			
7,12[1',2']-Benzenoanthra[2',3':4,5]-imidazo[1,2-a]pyridine-6,13-dione, 7,12-dihydro-3-methyl-	EtOH	253(4.49),273(4.44), 300(4.20),314(3.67) 332(3.60),405(3.51)	48-0353-83
6,6'-Bibenzoxazole, 2,2'-diphenyl-	C_6H_{12}	330(4.18)	41-0595-83
$C_{26}H_{16}O_6$			
2,2'-(3,6-Dimethyl-2,5-dioxo-p-phenylene)bisindan-1,3-dione	$CHCl_3$	254(4.65)	39-1753-83C
$C_{26}H_{17}ClN_4O_2$			
Benzo[f]quinoline, 4-(4-chlorophenyl)-3,4-dihydro-3-[[(4-nitrophenyl)azo]-methylene]-	neutral	568(4.57)	124-0068-83
	protonated	478(4.58)	124-0068-83
Quinoline, 6-chloro-1,2-dihydro-1-(1-naphthalenyl)-2-[[(4-nitrophenyl)-azo]methylene]-	neutral	546(4.35)	124-0068-83
	protonated	468(4.31)	124-0068-83
$C_{26}H_{17}Cl_3N_6O_8$			
Benzenamine, 4,4'-(2,2,2-trichloroethylidene)bis[2,6-dinitro-N-phenyl-	MeOH	207.1(4.52),426.5(3.99)	56-1357-83
$C_{26}H_{17}Cl_3N_6O_{10}$			
Phenol, 2,2'-[(2,2,2-trichloroethylidene)bis[2,6-dinitro-4,1-phenylene)-imino]bis-	MeOH	205.6(4.70),427.4(4.03)	56-1357-83
Phenol, 3,3'-[(2,2,2-trichloroethylidene)bis[2,6-dinitro-4,1-phenylene)-imino]bis-	MeOH	207.3(4.77),421.6(4.02)	56-1357-83
Phenol, 4,4'-[(2,2,2-trichloroethylidene)bis[2,6-dinitro-4,1-phenylene)-imino]bis-	MeOH	201.9(4.70),420.7(4.09)	56-1357-83
$C_{26}H_{17}Cl_3O_2$			
1-Isobenzofuranol, 1,3,3-tris(4-chlorophenyl)-1,3-dihydro-	hexane	256s(3.07),263s(3.12), 266(3.13),276(2.95)	104-0553-83
$C_{26}H_{17}Cs$			
Cesium, μ-[9,9'-bi-9H-fluoren]-9-yl-	n.s.g.	408(3.79)	35-2502-83
$C_{26}H_{17}N$			
9H-Carbazole, 9-(9-anthracenyl)-	EtOH	249(4.98),254(5.00), 291(4.08),325s(3.68), 337(3.79),347(3.71), 366(3.72)	73-0112-83
$C_{26}H_{17}NO_2$			
1,2-Naphthalenedione, 4-(2-naphtha-	n.s.g.	253(4.60),324s(3.84),	39-1759-83C

Compound	Solvent	λ_{max}(log ϵ)	Ref.
lenylphenylamino)- (cont.)		396(3.57),518(3.83)	39-1759-83C
$C_{26}H_{17}N_2O_2$			
7,12[1',2']-Benzenoanthra[2',3':4,5]-imidazo[1,2-a]pyridinium, 6,7,12,13-tetrahydro-5-methyl-6,13-dioxo-, iodide	EtOH	246(4.30),259(4.25), 290(4.10),304(3.51)	48-0353-83
$C_{26}H_{17}N_5O_4$			
Benzo[f]quinoline, 3,4-dihydro-4-(4-nitrophenyl)-3-[[(4-nitrophenyl)-azo]methylene]-	neutral	582(4.29)	124-0068-83
	protonated	456(3.68)	124-0068-83
$C_{26}H_{18}ClN_3$			
3,5-Pyridinedicarbonitrile, 2-[1,1'-bi-phenyl]-4-yl-4-(4-chlorophenyl)-1,4-dihydro-6-methyl-	EtOH	209(4.55),284(4.40), 352(3.64)	73-3123-83
$C_{26}H_{18}Cl_2O_2$			
1-Isobenzofuranol, 3,3-bis(4-chloro-phenyl)-1,3-dihydro-1-phenyl-	hexane	259s(3.11),263(3.13), 268(3.10),277(2.87)	104-0553-83
$C_{26}H_{18}CoN_4$			
Cobalt, [5,16-dihydrotetrabenzo[b,e,i-l][1,4,7,11]tetraazacyclotetradecin-ato(2-)-N^5,N^{10},N^{16},N^{22}]-, (SP-4-1)-	DMF	496(4.10)	12-2395-83
$C_{26}H_{18}CoN_5$			
Cobalt(1+), (5H tetrabenzo[c,g,j,n]-[1,2,5,9,13]pentaazacyclopentadec-inato-N^5,N^{11},N^{17},N^{23}), perchlorate, (SP-4-2)-	EtOH-DMF	293(4.53),456(4.04)	12-2395-83
$C_{26}H_{18}CuN_4$			
Copper, [5,16-dihydrotetrabenzo[b,e,i-l][1,4,7,11]tetraazacyclotetradecin-ato(2-)-N^5,N^{10},N^{16},N^{22}]-, (SP-4-1)-	EtOH-DMF	304(4.48),337(4.54), 497(4.16)	12-2395-83
$C_{26}H_{18}CuN_4O_8$			
Myxin copper complex	DMSO	287(4.84),300(4.80), 356(4.00),408(4.02), 610(3.98)	162-0908-83
$C_{26}H_{18}CuN_5$			
Copper(1+), (5H-tetrabenzo[c,g,j,n]-[1,2,5,9,13]pentaazacyclopentadec-inato-N^5,N^{11},N^{17},N^{23})-, perchlorate, (SP-4-2)-	EtOH-DMF	300(4.77),420(4.26)	12-2395-83
$C_{26}H_{18}N_2O_2$			
9,10-Anthracenedione, 1-amino-4-([1,1'-biphenyl]-4-ylamino)-	DMF	586(3.96),620(3.96)	2-0808-83
Benzoxazole, 2,2'-(2,3-dimethyl-1,5-naphthalenediyl)bis-	DMF	327(4.37)	5-0931-83
Benzoxazole, 2,2'-(3,7-dimethyl-1,5-naphthalenediyl)bis-	DMF	360(4.41)	5-0931-83
Benzoxazole, 2,2'-(4,7-dimethyl-1,3-naphthalenediyl)bis-	DMF	300(4.62)	5-0931-83
Benzoxazole, 2,2'-(1,7-naphthalene-diyl)bis[4-methyl-	DMF	342(4.36)	5-0931-83

Compound	Solvent	λ_{max}(log ϵ)	Ref.
Benzoxazole, 2,2'-(1,7-naphthalene-diyl)bis[5-methyl-	DMF	354(4.40)	5-0931-83
Benzoxazole, 2,2'-(1,7-naphthalene-diyl)bis[6-methyl-	DMF	355(4.40)	5-0931-83
Benzoxazole, 2,2'-(1,7-naphthalene-diyl)bis[7-methyl-	DMF	340(4.36)	5-0931-83
Benzoxazole, 2,2'-(2,6-naphthalene-diyl)bis[4-methyl-	DMF	349(4.80)	5-0931-83
Benzoxazole, 2,2'-(2,6-naphthalene-diyl)bis[5-methyl-	DMF	351(4.81)	5-0931-83
Benzoxazole, 2,2'-(2,6-naphthalene-diyl)bis[6-methyl-	DMF	353(--)	5-0931-83
Benzoxazole, 2,2'-(2,6-naphthalene-diyl)bis[7-methyl-	DMF	348(4.81)	5-0931-83
$C_{26}H_{18}N_2O_4$			
Benzoxazole, 2,2'-(1,5-naphthalene-diyl)bis[5-methoxy-	DMF	359(4.48)	5-0931-83
Benzoxazole, 2,2'-(1,6-naphthalene-diyl)bis[5-methoxy-	DMF	331(4.61)	5-0931-83
$C_{26}H_{18}N_2S_2$			
Diazene, (3-naphtho[1,2-d]-1,3-dithiol-2-ylidene-3-phenyl-1-propenyl)-phenyl]-	MeCN	510(3.66)	118-0840-83
$C_{26}H_{18}N_4Ni$			
Nickel, [5,16-dihydrotetrabenzo[b,e,i-l][1,4,7,11]tetraazacyclotetradecin-ato(2-)-N^5,N^{10},N^{16},N^{22}]-, (SP-4-1)-	EtOH-DMF	295(4.40),333(4.43), 360s(4.15),476(3.94), 570(4.00)	12-2395-83
$C_{26}H_{18}N_4O$			
3-Isoquinolineacetonitrile, 1-amino-4-cyano-α-[(4-methoxyphenyl)phenyl-methylene]-	EtOH	219(4.65),232s(4.53), 256s(4.38),325(4.51)	39-1813-83C
$C_{26}H_{18}N_4O_2$			
Benzo[f]quinoline, 3,4-dihydro-3-[[(4-nitrophenyl)azo]methylene]-4-phenyl-	neutral	590(5.08)	124-0068-83
	protonated	472(4.76)	124-0068-83
3,5-Pyridinedicarbonitrile, 2-[1,1'-bi-phenyl]-4-yl-1,4-dihydro-6-methyl-4-(4-nitrophenyl)-	EtOH	207(4.59),247(4.24), 318(4.42)	73-3123-83
Quinoline, 1,2-dihydro-1-(1-naphthal-enyl)-2-[[(4-nitrophenyl)azo]meth-ylene]-	neutral	558(4.64)	124-0068-83
	protonated	470(4.69)	124-0068-83
$C_{26}H_{18}N_4O_3$			
Benzoic acid, 2-(nitrosophenylamino)-, 1-phenyl-1H-indazol-3-yl ester	heptane	249(4.45),304(4.11)	88-3651-83
$C_{26}H_{18}N_5Ni$			
Nickel(1+), (5H-tetrabenzo[c,g,j,n]-[1,2,5,9,13]pentaazacyclopentadec-inato-N^5,N^{11},N^{17},N^{23})-, perchlorate, (SP-4-2)-	EtOH-DMF	294(4.46),420(3.94)	12-2395-83
$C_{26}H_{18}O$			
Bicyclo[4.4.1]undeca-1,3,5,7,9-penta-ene-2-carboxaldehyde, 7-(9H-fluoren-9-ylidenemethyl)-, (±)-	CH_2Cl_2	240(4.50),250(4.56), 258(4.67),269s(4.42), 289s(4.22),374(3.85)	83-0240-83

Compound	Solvent	$\lambda_{max}(\log \epsilon)$	Ref.
(cont.)	MeCN	226(4.57),238(4.55), 246(4.60),254(4.69), 266s(4.49),380(3.98)	83-0240-83
$C_{26}H_{18}OS_2$			
6H-Dibenzo[b,d]thiopyran, 6,6'-oxybis-	MeOH	250(4.57),272(4.30), 308(3.67)	12-0795-83
$C_{26}H_{18}O_2$			
Spiro[acenaphthylene-1(2H),2'-oxetan]-2-one, 3',3'-diphenyl-	hexane	223(4.20),254s(3.83), 284s(3.50),300s(4.46), 313(3.56),325s(3.44), 342(3.48)	18-3464-83
$C_{26}H_{18}O_5$			
2H-Furo[2,3-h]-1-benzopyran-2-one, 8-(4-methoxybenzoyl)-4-methyl-9-phenyl-	MeOH	280(2.05)	2-0164-83
$C_{26}H_{18}O_8$			
[2,2'-Binaphthalene]-1,1',4,4'-tetrone, 7,7'-diacetyl-8,8'-dihydroxy-6,6'-dimethyl-	dioxan	223(4.65),275(4.27), 430(3.95)	5-0433-83
	$CHCl_3$	243(4.35),275s(4.24), 440(3.95)	5-0433-83
$C_{26}H_{18}O_9$			
7H-Furo[3,2-g][1]benzopyran-7-one, 9,9'-[oxybis(2,1-ethanediyloxy)]bis-	EtOH	218(4.67),248(4.61), 300(4.31)	39-1807-83B
$C_{26}H_{19}N$			
[1,1':3',1"-Terphenyl]-2'-carbonitrile, 4'-methyl-5'-phenyl-	MeCN	243(4.57),296(3.67)	48-0729-83
$C_{26}H_{19}NO_4S$			
Spiro[9H-fluorene-9,3'(3'aH)-pyrrolo-[2,1-b]benzothiazole]-1',2'-dicarboxylic acid, dimethyl ester	n.s.g.	303(3.968)	24-0856-83
$C_{26}H_{19}NO_{12}$			
Laccaic acid A	H_2SO_4	302(4.33),361(4.11), 518(4.32),558(4.37)	162-0767-83
$C_{26}H_{19}NS$			
1H-2,3-Benzothiazine, 1,1,4-triphenyl-, perchlorate	HOAc	36?(3.72)	104-1744-83
$C_{26}H_{19}N_3$			
3,5-Pyridinedicarbonitrile, 2-[1,1'-biphenyl]-4-yl-1,4-dihydro-6-methyl-4-phenyl-	EtOH	209(4.38),283(4.37), 352(3.65)	73-3123-83
$C_{26}H_{19}N_3O_2$			
9,10-Anthracenedione, 1-amino-4-[[4-(phenylamino)phenyl]amino]-	DMF	576s(4.22),617(4.32)	2-0808-83
6H-Anthra[1,9-cd]isoxazol-6-one, 5-[[5-[(phenylmethyl)imino]-1,3-pentadienyl]amino]-	dioxan	555(4.17)	103-0377-83
$C_{26}H_{19}N_5O_8S_2$			
2-Naphthalenesulfonic acid, 6-[(2-ami-	pH 7.00	549(4.36)	33-2002-83

Compound	Solvent	λ_{max}(log ε)	Ref.
no-8-hydroxy-6-sulfo-1-naphthalenyl)-azo]-4-hydroxy-3-(phenylazo)- (cont.)			33-2002-83
$C_{26}H_{20}BrClN_4$			
Benzenecarbohydrazonic acid, N-(4-bromophenyl)-, [(4-chlorophenyl)methylene]phenylhydrazide	EtOH	224(4.40),263(4.18), 351(4.34)	70-1439-83
	dioxan	224(4.44),265(4.20), 351(4.34)	70-1439-83
	MeCN	222(4.41),265(4.22), 351(4.37)	70-1439-83
$C_{26}H_{20}Br_2N_4$			
Benzenecarbohydrazonic acid, N-(4-bromophenyl)-, [(4-bromophenyl)methylene]phenylhydrazide	EtOH	225(4.42),263(4.21), 350(4.39)	70-1439-83
	dioxan	225(4.43),255(4.20), 351(4.38)	70-1439-83
	MeCN	222(4.43),267(4.22), 350(4.38)	70-1439-83
$C_{26}H_{20}FeN_2S_2$			
Ferrocene, 1,1'-bis(2-methyl-5-benzothiazolyl)-	EtOH	240(4.78),450(2.38)	104-0950-83
Ferrocene, 1,1'-bis(2-methyl-6-benzothiazolyl)-	EtOH	279(4.41),455(3.08)	104-0950-83
$C_{26}H_{20}NO_3$			
1-Benzopyrylium, 2-(3-oxo-3H-naphtho-[2,1-b]pyran-2-yl)-7-pyrrolidino-, perchlorate	MeCN	582(4.39)	48-0505-83
Naphtho[2,1-b]pyrylium, 3-(2-oxo-7-pyrrolidino-2H-1-benzopyran-3-yl)-, perchlorate	MeCN	634(4.91)	48-0505-83
$C_{26}H_{20}N_2O_3$			
4H-Pyran-3-carboxylic acid, 2-amino-6-[1,1'-biphenyl]-4-yl-5-cyano-4-phenyl-, methyl ester	EtOH	208(4.13),265(4.20), 298(4.39)	73-3123-83
$C_{26}H_{20}N_2O_3S$			
Benzenamine, N-[2-[6-(4-methoxyphenyl)-2H-thiopyran-2-ylidene]-2-phenylethylidene]-4-nitro-, monohydrobromide	$CHCl_3$	552(3.96)	97-0147-83
$C_{26}H_{20}N_2O_3Se$			
Benzenamine, N-[2-[6-(4-methoxyphenyl)-2H-selenin-2-ylidene]-2-phenylethylidene]-4-nitro-, monohydrobromide	$CHCl_3$	580(--)	97-0147-83
$C_{26}H_{20}N_2O_6$			
Cyclohept[1,2,3-hi]imidazo[2,1,5-cd]-indolizine-3,4,9-tricarboxylic acid, 2-phenyl-, 9-ethyl 3,4-dimethyl ester	EtOH	247(4.37),280s(4.45), 292(4.49),310s(4.33), 381(4.43),427s(4.01)	18-3703-83
$C_{26}H_{20}N_2S_2$			
Diazene, [3-(4,5-dihydronaphtho[1,2-d]-1,3-dithiol-2-ylidene)-3-phenyl-1-propenyl]phenyl-	MeCN	531(3.62)	118-0840-83
$C_{26}H_{20}N_4NiO_3$			
Nickel, [5,14-dihydro-1,3-dimethoxy-	EtOH-DMF	290(4.08),354(4.12),	12-2407-83

Compound	Solvent	$\lambda_{max}(\log \epsilon)$	Ref.
21-methyl-13H-4,6-methenotribenzo-[b,f,l][1,4,8,11]tetraazacyclotetra-decin-13-onato(2-)-N^5,N^8,N^{14},N^{19}]-, (SP-4-2)- (cont.)		380(4.01),470(4.05), 498(4.06)	12-2407-83
$C_{26}H_{20}N_4O_2$			
Benzamide, 4,4'-azobis[N-phenyl-	o-$C_6H_4Cl_2$	345(4.48)	18-1700-83
$C_{26}H_{20}O$			
Anthracene, 9-[(1-naphthalenylmethoxy)-methyl]-	$C_6H_{11}Me$	223.5(4.95),255.5(5.19), 281.5(3.82),292.5(3.68), 313.5(3.65),330.5(3.48), 347(3.80),365(3.99), 385(3.98)	39-0109-83B
photocyclomer	$C_6H_{11}Me$	203.5(4.59),220(4.48), 255(3.23),272.5(3.10), 281.5(3.08),285(3.06)	39-0109-83B
$C_{26}H_{21}BiN_2O_2$			
Bismuth, bis(2-methyl-8-quinolinolato-N^1,O^8)phenyl-	benzene	370(3.70)	101-0317-83M
$C_{26}H_{21}BrN_4$			
Benzenecarbohydrazonic acid, N-(4-bromophenyl)-, phenyl(phenylmethylene)-hydrazide	EtOH	226(4.27),266(4.19), 352(4.41)	70-1439-83
	dioxan	227(4.31),266(4.21), 354(4.40)	70-1439-83
	MeCN	223(4.31),267(4.19), 354(4.38)	70-1439-83
Benzenecarbohydrazonic acid, N-phenyl-, [(4-bromophenyl)methylene]phenyl-hydrazide	EtOH	224(4.44),255(4.23), 346(4.35)	70-1439-83
	dioxan	224(4.45),258(4.19), 350(4.32)	70-1439-83
	MeCN	223(4.42),257(4.20), 347(4.33)	70-1439-83
$C_{26}H_{21}BrO_9$			
5,12-Naphthacenedione, 9-acetyl-6,9,11-triacetoxy-7-bromo-7,8,9,10-tetra-hydro-	EtOH	260(4.30),275(3.96), 344(3.30)	35-1608-83
$C_{26}H_{21}Br_2NSn$			
Phenazastannin, 2,8-dibromo-5-ethyl-5,10-dihydro-10,10-diphenyl-	n.s.g.	320(3.20)	101-0283-83R
$C_{26}H_{21}ClN_4$			
Benzenecarbohydrazonic acid, N-phenyl-, [(4-chlorophenyl)methylene]phenyl-hydrazide	EtOH	225(4.46),246s(4.29), 253(4.27),346(4.42)	70-1439-83
	dioxan	224(4.39),258(4.16), 351(4.30)	70-1439-83
	MeCN	223(4.40),255(4.18), 346(4.33)	70-1439-83
$C_{26}H_{21}N$			
5H-Benzo[6,7]cyclohepta[1,2-b]pyridine, 6,7-dihydro-2,4-diphenyl-	neutral	321(4.14)	39-0045-83B
	protonated	353(4.33)	39-0045-83B
$C_{26}H_{21}NOS$			
Benzenamine, N-[2-[6-(4-methoxyphenyl)-2H-thiopyran-2-ylidene]-2-phenyleth-ylidene]-, perchlorate	$CHCl_3$	252(4.52)	97-0147-83

Compound	Solvent	$\lambda_{max}(\log \epsilon)$	Ref.
$C_{26}H_{21}NOSe$			
Benzenamine, N-[2-[6-(4-methoxyphenyl)-2H-selenin-2-ylidene]-2-phenylethylidene]-, perchlorate	$CHCl_3$	264(--)	97-0147-83
$C_{26}H_{21}NO_2$			
5(2H)-Oxazolone, 2-methyl-2-(1-methyl-2,3-diphenyl-2-cyclopropen-1-yl)-4-phenyl-	EtOH	228(4.25),320(4.44), 338(4.33)	44-0695-83
5(4H)-Oxazolone, 2-methyl-4-(1-methyl-2,3-diphenyl-2-cyclopropen-1-yl)-4-phenyl-	EtOH	228(4.27),320(4.46), 338(4.33?)	44-0695-83
5(4H)-Oxazolone, 4-methyl-4-(1-methyl-2,3-diphenyl-2-cyclopropen-1-yl)-2-phenyl-	C_6H_{12}	232(4.45),237(4.48), 300(4.31),311(4.34), 316(4.45),334(4.34)	44-0695-83
5(4H)-Oxazolone, 4-methyl-4-(2-methyl-1,3-diphenyl-2-cyclopropen-1-yl)-2-phenyl-	EtOH	263(4.21)	44-0695-83
$C_{26}H_{21}N_3O$			
3,5-Pyridinedicarbonitrile, 2-[1,1'-biphenyl]-4-yl-1,2,3,4-tetrahydro-2-hydroxy-6-methyl-	EtOH	208(4.21),260(4.49), 265(4.48),269s(4.42)	73-3123-83
$C_{26}H_{22}ClN_3O$			
7H-Benzo[e]perimidin-7-one, 6-chloro-4-[4-(diethylamino)phenyl]-2-methyl-	n.s.g.	546(4.35)	2-0805-83
$C_{26}H_{22}Cl_2N$			
Pyridinium, 2-(2,5-dichlorophenyl)-1-(1-methylethyl)-4,6-diphenyl-, salt with trifluoromethanesulfonic acid	DMSO	306(4.42)	150-0301-83M
$C_{26}H_{22}D_2F_{12}N_6P_2Ru$			
Ruthenium(II), bis(2,2'-bipyridine)-[2-(1,1-dideuterioaminomethyl)pyridine]-, bis(hexafluorophosphate)	H_2O	341(4.04),471(4.01)	35-7075-83
$C_{26}H_{22}NO_3$			
1-Benzopyrylium, 7-(diethylamino)-2-(2-oxo-2H-naphtho[1,2-b]pyran-3-yl)-, perchlorate	MeCN	609(4.34)	48-0505-83
$C_{26}H_{22}N_2O_3$			
7H-Oxazolo[3,2-a]pyrimidin-7-one, 2,3-dihydro-2-[(triphenylmethoxy)methyl]-	EtOH	221s(4.24),258(3.84)	128-0125-83
$C_{26}H_{22}N_2O_5$			
1H-Pyrrole-3,4-dicarboxylic acid, 2-benzoyl-1-[8-(methylamino)-1-naphthalenyl]-, dimethyl ester	EtOH	213(4.54),251(4.29), 340(3.72)	94-1378-83
$C_{26}H_{22}N_4$			
Benzenecarbohydrazonic acid, N-phenyl-, phenyl(phenylmethylene)hydrazide	EtOH	242(4.49),297(4.26), 339(4.55)	70-1439-83
	dioxan	243(4.52),296(4.27), 337(4.57)	70-1439-83
	MeCN	242(4.49),297(4.26), 339(4.55)	70-1439-83

Compound	Solvent	λ_{max}(log ϵ)	Ref.
$C_{26}H_{22}N_4OS$			
2H-Thiopyran-2-imine, 3-(1H-benzimidazol-2-yl)-6-morpholino-5-(1-naphthalenyl)-, monoperchlorate	MeCN	275(4.68),304(4.79), 309s(4.79),331(4.81), 468(4.88)	97-0403-83
$C_{26}H_{22}N_4O_2$			
3,9-Bi-9H-pyrido[3,4-b]indole, 7,7'-dimethoxy-1,1'-dimethyl-	EtOH	247(4.98),304(4.61), 328s(4.23),334(4.23)	78-0033-83
Propanedinitrile, 2,2'-(13,15-dimethoxytricyclo[10.2.2.2^{5,8}]octadeca-5(18),7,12,14,15-pentaene-6,17-diylidene)bis-, pseudogem.	$CHCl_3$	440(4.5),710(3.6)	24-2785-83
pseudoortho	$CHCl_3$	425(4.6),665(2.2)	24-2785-83
$C_{26}H_{22}O_4$			
1,4-Naphthalenedione, 2,3-bis(2,6-dimethylphenoxy)-	EtOH	254(4.35),275s(4.32), 337(3.55),402(3.27)	104-0144-83
$C_{26}H_{22}O_5S$			
1,3-Dioxane-4,6-dione, 5-[2-[6-(4-methoxyphenyl)-2H-thiopyran-2-ylidene]-2-phenylethylidene]-2,2-dimethyl-	$CHCl_3$	595(4.26)	97-0147-83
$C_{26}H_{22}O_5Se$			
1,3-Dioxane-4,6-dione, 5-[2-[6-(4-methoxyphenyl)-2H-selenin-2-ylidene]-2-phenylethylidene]-2,2-dimethyl-	$CHCl_3$	625(--)	97-0147-83
$C_{26}H_{22}O_6$			
[Bi-3-cyclopenten-1-yl]-2,2'-dicarboxylic acid, 5,5'-dioxo-3,3'-diphenyl-, dimethyl ester, [1α(1'S*,2'R*),2α]-	MeOH	286(4.44)	94-2526-83
[2,2'-Binaphthalene]-1,1',4,4'-tetrone, 5,5'-dimethoxy-3,3',7,7'-tetramethyl-	$CHCl_3$	257(4.48),275s(4.38), 406(3.94)	5-0299-83
[2,2'-Binaphthalene]-1,1',4,4'-tetrone, 8,8'-dimethoxy-3,3',6,6'-tetramethyl-	$CHCl_3$	257(4.50),270s(4.42), 405(3.94)	5-0299-83
$C_{26}H_{22}O_8$			
5,12-Naphthacenedione, 6,11-diacetoxy-8-(1-acetoxyethylidene)-7,8,9,10-tetrahydro-	EtOH	258(4.47),276(4.16), 339(3.36)	35-1608-83
$C_{26}H_{22}O_9$			
5,12-Naphthacenedione, 8-acetyl-6,8,11-triacetoxy-7,8,9,10-tetrahydro-	EtOH	257(4.11),276(3.61), 336(3.22)	35-1608-83
Spiro[naphthacen-2(1H),2'-oxiran]-6,11-dione, 3',5,12-triacetoxy-3,4-dihydro-3'-methyl-	EtOH	258(4.71),275(4.22), 338(3.83)	35-1608-83
$C_{26}H_{22}O_{10}$			
5,12-Naphthacenedione, 9-acetyl-7,9,11-triacetoxy-7,8,9,10-tetrahydro-6-hydroxy-	EtOH	253(3.88),279(3.53), 401(3.30)	35-1608-83
$C_{26}H_{23}ClN$			
Pyridinium, 4-(2-chlorophenyl)-1-(1-methylethyl)-2,6-diphenyl-, tetrafluoroborate	DMSO	308(4.44)	150-0301-83M
$C_{26}H_{23}GeN$			
Phenazagermine, 5-ethyl-5,10-dihydro-	n.s.g.	320(3.83)	101-0283-83R

Compound	Solvent	$\lambda_{max}(\log \epsilon)$	Ref.
10,10-diphenyl- (cont.)			101-0283-83R
$C_{26}H_{23}NO_2$			
1H-Pyrrole, 1-acetyl-3-benzoyl-2,3-dihydro-5-methyl-2,4-diphenyl-	MeOH	255(4.36)	22-0195-83
$C_{26}H_{23}NO_4$			
Acetamide, N-(4-butylphenyl)-N-(9,10-dihydro-8-hydroxy-9,10-dioxo-1-anthracenyl)-	benzene	415(3.79)	104-0745-83
9,10-Anthracenedione, 1-acetoxy-8-[(4-butylphenyl)amino]-	benzene	517(3.85)	104-0745-83
$C_{26}H_{23}NSi$			
Phenazasiline, 5-ethyl-5,10-dihydro-10,10-diphenyl-	n.s.g.	327(3.80)	101-0283-83R
$C_{26}H_{23}N_3O_2$			
1,4-Naphthalenedione, 2-[4-[[4-(diethylamino)phenyl]azo]phenyl]-	C_6H_{12}	406(4.39),479(4.33)	150-0168-83S
	EtOH	425(--),491(--)	150-0168-83S
$C_{26}H_{23}N_3S_3$			
2-Benzothiazolamine, N-[3-(diethylamino)-6-(5-phenyl-3H-1,2-dithiol-3-ylidene)-2,4-cyclohexadien-1-ylidene]-	EtOH	229(4.76),308(4.60), 525(4.57)	33-2165-83
$C_{26}H_{23}N_4$			
Pyridinium, 1-methyl-4-[3-(1-methyl-4(1H)-pyridinylidene)-2,4-di-4-pyridinyl-1-cyclobuten-1-yl]-, iodide	MeOH	632(5.12)	5-0642-83
$C_{26}H_{23}N_5O_3$			
2,4(1H,3H)-Pyrimidinedione, 1-[2-azido-3-(triphenylmethoxy)propyl]-	EtOH	233s(3.87),264(3.92)	128-0125-83
$C_{26}H_{24}$			
Tetracyclo[16.2.2.2^{3,6}.2^{8,11}]hexacosa-3,5,8,10,12,16,18,20,21,23,25-undecaene	C_6H_{12}	232(4.7),257(4.0)	88-5277-83
$C_{26}H_{24}I_2$			
Benzene, 1,4-bis[2-(4-iodo-2,5-dimethylphenyl)ethenyl]-	dioxan	358(4.74)	104-0683-83
$C_{26}H_{24}N_2O$			
Ethanone, 2-(1-butyl-1H-indol-3-yl)-1-phenyl-2-(phenylimino)-	MeOH	257(4.40),313(4.38)	73-1854-83
$C_{26}H_{24}N_2O_2$			
Ethanone, 2-(1-butyl-1H-indol-3-yl)-1-phenyl-2-(phenylimino)-, N-oxide	MeOH	263(4.34),338(4.25)	73-1854-83
Methanone, (8-butyl-3,8-dihydro-2-phenyl-2H-isoxazolo[5,4-b]indol-3-yl)phenyl-	MeOH	287(4.15)	73-1854-83
$C_{26}H_{24}N_2O_4$			
2,4(1H,3H)-Pyrimidinedione, 1-[2-hydroxy-3-(triphenylmethoxy)propyl]-	EtOH	233s(3.97),265(4.02)	128-0125-83

Compound	Solvent	λ_{max}(log ϵ)	Ref.
$C_{26}H_{24}N_2O_5$			
Benzamide, N-[2,3-dihydro-2-oxo-3-[(3,4,5-trimethoxyphenyl)methylene]-1H-indol-4-yl]-4-methyl-	$CHCl_3$	374(3.44)	44-2468-83
$C_{26}H_{24}N_3S_2$			
Thiazolo[3,4-a]pyrimidin-5-ium, 6-[3-(3-ethyl-2(3H)-benzothiazolylidene)-1-propenyl]-2,4-dimethyl-8-phenyl-, perchlorate	MeCN	426(3.05),670(3.80)	103-0037-83
$C_{26}H_{24}N_3S_3$			
1,2-Dithiol-1-ium, 3-[2-(2-benzothiazolylamino)-4-(diethylamino)phenyl]-5-phenyl-, chloride	EtOH	221(4.66),331(4.50), 551(4.70)	33-2165-83
$C_{26}H_{24}N_6O_4$			
[5,5'-Bipyrimido[4,5-b]quinoline]-2,2',4,4'(1H,1'H,3H,3'H)-tetrone, 5,5',10,10'-tetrahydro-3,3',10,10'-tetramethyl-	pH 5.0	314(4.23)	12-1873-83
	pH 8.7	310(--)	12-1873-83
$C_{26}H_{24}O_2$			
2H-Pyran, 2-ethoxy-3-methyl-2,4,6-triphenyl-	dioxan	238(4.29),243(4.29), 325(4.16)	48-0729-83
2H-Pyran, 2-methoxy-3,5-dimethyl-2,4,6-triphenyl-	dioxan	310(4.05)	48-0729-83
2H-Pyran, 2-methoxy-3-methyl-2-(4-methylphenyl)-4,6-diphenyl-	dioxan	238s(4.30),244(4.31), 324(4.16)	48-0729-83
2H-Pyran, 2-methoxy-3-methyl-4-(4-methylphenyl)-2,6-diphenyl-	dioxan	238(4.26),244(4.28), 252s(4.22),324(4.17)	48-0729-83
2H-Pyran, 2-methoxy-3-methyl-6-(4-methylphenyl)-2,4-diphenyl-	dioxan	237s(4.26),244(4.27), 325(4.19)	48-0729-83
$C_{26}H_{24}O_3$			
2H-Pyran, 2-methoxy-2-(4-methoxyphenyl)-3-methyl-4,6-diphenyl-	dioxan	236s(4.30),244(4.30), 325(4.19)	48-0729-83
2H-Pyran, 2-methoxy-6-(4-methoxyphenyl)-3-methyl-2,4-diphenyl-	dioxan	236s(4.18),242(4.19), 255(4.19),327(4.25)	48-0729-83
1H-Xanthene-1,8(2H)-dione, 3,4,5,6,7,9-hexahydro-9-methyl-3,6-diphenyl-	MeOH	230(3.83),290(3.35)	24-1309-83
$C_{26}H_{24}O_8$			
4H-1-Benzopyran-4-one, 3,5,7-triacetoxy-8-(3-methyl-2-butenyl)-2-phenyl- (glepidotin A triacetate)	MeOH	253.5(4.33),296(4.20)	102-0573-83
$C_{26}H_{25}NO_3S$			
1,2-Benzoxathiin-4-amine, 3,4,5,6,7,8-hexahydro-N,N,3-triphenyl-, 2,2-dioxide, trans	EtOH	220s(3.98)	4-0839-83
$C_{26}H_{25}NO_5$			
Acetamide, N-[1-(6-hydroxy-5-oxo-1,3,6-cycloheptatrien-1-yl)-3-[3-methoxy-4-(phenylmethoxy)phenyl]-2-propenyl]-, (E)-	MeOH	247(4.49),309(4.17), 371(3.73)	18-3099-83
	MeOH-NaOH	251(4.48),339(4.11), 402(4.09)	18-3099-83
Acetamide, N-[1-(6-hydroxy-5-oxo-1,3,6-cycloheptatrien-1-yl)-3-[4-methoxy-3-(phenylmethoxy)phenyl]-2-propenyl]-, (E)-	MeOH	250(4.40),305(4.10), 370(3.60)	18-3099-83
	MeOH-NaOH	265s(4.26),340(3.93), 401(3.75)	18-3099-83

Compound	Solvent	λ_{max}(log ϵ)	Ref.
$C_{26}H_{25}N_2O_3$			
1-Benzopyrylium, 2-(2-oxo-7-pyrrolidino-2H-1-benzopyran-3-yl)-7-pyrrolidino-, perchlorate	MeCN	659(4.89)	48-0505-83
$C_{26}H_{25}N_3O_9S$			
1-Azabicyclo[3.2.0]hept-2-ene-2-carboxylic acid, 6-ethyl-3-[[3-[(4-nitrophenyl)methoxy]-3-oxopropyl]thio]-7-oxo-, (4-nitrophenyl)methyl ester	THF	268(4.35),319(4.14)	158-0407-83
$C_{26}H_{26}CoN_4O_4$			
Cobalt, [diethyl 7,16-dihydro-6,15-dimethyldibenzo[b,i][1,4,8,11]tetraazacyclotetradecine-7,16-dicarboxylato(2-)-N^5,N^9,N^{14},N^{18}]-, (SP-4-1)-	n.s.g.	525(3.63)	97-0261-83
$C_{26}H_{26}CuN_4O_4$			
Copper, [diethyl 7,16-dihydro-6,15-dimethyldibenzo[b,i][1,4,8,11]tetraazacyclotetradecine-7,16-dicarboxylato(2-)-N^5,N^9,N^{14},N^{18}]-, (SP-4-1)-	n.s.g.	580(3.06)	97-0261-83
$C_{26}H_{26}Ge_2$			
Digermane, 1,1-dimethyl-1,2,2,2-tetraphenyl-	C_6H_{12}	234(4.41)	101-0149-83G
$C_{26}H_{26}NO_2PS$			
2H-Thiopyran-5-carboxylic acid, 3,4-dihydro-6-[(triphenylphosphoranylidene)amino]-, ethyl ester	$CHCl_3$	267(4.02),275(4.04), 312(4.15)	24-1691-83
$C_{26}H_{26}NO_3P$			
3-Furancarboxylic acid, 4,5-dihydro-5-methyl-2-[(triphenylphosphoranylidene)amino]-, ethyl ester	MeOH	268(3.31),274(3.41), 305(3.52)	24-1691-83
$C_{26}H_{26}N_2O$			
Ethanone, 2-(1-butyl-1H-indol-3-yl)-1-phenyl-2-(phenylamino)-	MeOH	248(4.49),309(4.26)	73-1854-83
$C_{26}H_{26}N_4NiO_4$			
Nickel, [diethyl 7,16-dihydro-6,15-dimethyldibenzo[b,i][1,4,8,11]tetraazacyclotetradecine-7,16-dicarboxylato(2-)-N^5,N^9,N^{14},N^{18}]-, (SP-4-1)-	n.s.g.	550(3.71)	97-0261-83
$C_{26}H_{26}N_4O_4$			
Benzoic-carboxy,2-$^{14}C_2$ acid, 3,5,6-trimethyl-4-(phenylmethoxy)-2-[(1-phenyl-1H-tetrazol-5-yl)oxy]-, ethyl ester	EtOH	256(3.28)	39-0667-83C
$C_{26}H_{26}O$			
Cyclohexanone, 2-(phenylmethylene)-6-[2-(phenylmethylene)cyclohexylidene]-	hexane	286(4.56)	104-2185-83
$C_{26}H_{26}O_4$			
1(3H)-Isobenzofuranone, 3-(4-hydroxy-3,5-dimethylphenyl)-3-[4-hydroxy-3-	EtOH	205(4.97),279(3.68)	24-0970-83

Compound	Solvent	λ_{max}(log ε)	Ref.
methyl-5-(1-methylethyl)phenyl]-			24-0970-83
$C_{26}H_{26}O_8$			
9,10-Ethanoanthracene-2,3,6,7-tetracarboxylic acid, 1,4,5,8,9,10-hexahydro-11,12-bis(methylene)-, tetramethyl ester	MeCN	235(4.04),245(4.00), 255s(3.85)	33-0019-83
$C_{26}H_{26}O_{10}$			
Chrysomycin	MeOH	277.2(4.71),288.3(4.71), 394.4(4.30)	74-0385-83C
$C_{26}H_{27}Cl_3O_3S$			
Androsta-1,4-diene-3,11-dione, 17-chloro-17-[[(2,4-dichlorophenyl)methyl]sulfinyl]-, (17α)-	MeOH	234(4.47)	87-0078-83
$C_{26}H_{27}NO_4$			
Isoquinoline, 3-(3,4-dimethoxyphenyl)-3,4-dihydro-6,7-dimethoxy-1-(phenylmethyl)-, hydrochloride	EtOH	232(4.37),278(4.02), 313(3.96),359(3.62)	142-1247-83
$C_{26}H_{27}NO_7$			
Spiro[cyclohexane-1,10'-[6,8a]ethano-[8aH]carbazole]-5',7',8'-tricarboxylic acid, 6',9'-dihydro-9'-methyl-11'-oxo-, trimethyl ester	MeOH	210(4.18),245(4.36), 296(3.76),318(3.63), 455(3.81)	39-0515-83C
$C_{26}H_{27}NO_8$			
2,5a-Propano-2H,5aH-furo[3',2':1,7a]-pyrrolizino[2,3,4,5-ija]quinoline-1-acetic acid, 1,4,5,9,10,10a-hexahydro-1-(methoxycarbonyl)-5-(2-methoxy-2-oxoethylidene)-4-oxo-, methyl ester, (E)-	MeOH	237(4.08),294(3.42)	39-0505-83C
(Z)-	MeOH	243(3.99),291.5(3.36)	39-0505-83C
isomer	MeOH	236(4.11),288(3.36)	39-0505-83C
$C_{26}H_{27}N_3O_3S$			
7H-Benzimidazo[2,1-a]benz[de]isoquinoline-10-sulfonamide, N-octyl-7-oxo-	benzene	387(3.90)	73-2249-83
	EtOH	385(3.96)	73-2249-83
$C_{26}H_{27}N_4S_2$			
Benzothiazolium, 2-[2-(2-benzothiazolylamino)-4-(diethylamino)phenyl]-3-ethyl-, tetrafluoroborate	64% MeOH-pH 4.25	275(4.47),428(4.34)	33-2165-83
$C_{26}H_{27}N_5O_5$			
Acetamide, N-[7-(2-acetoxy-2-phenylethyl)-2,3,6,7-tetrahydro-1,3-dimethyl-2,6-dioxo-1H-purin-8-yl]-N-(phenylmethyl)-	EtOH	280(4.06)	56-0461-83
$C_{26}H_{27}OP$			
Ethanone, 1-(1-methylcyclopentyl)-2-(triphenylphosphoranylidene)-	EtOH	220s(4.47),265(3.77), 273(3.74),291(3.70)	33-2760-83
$C_{26}H_{28}ClNO_4$			
2H,5H-Pyrano[3,2-c][1]benzopyran-2-one, 3-chloro-10-hydroxy-5,5-dimethyl-4-(methylphenylamino)-8-pentyl-	EtOH	234(4.40),271(4.03), 390(4.31)	161-0775-83

Compound	Solvent	λ$_{max}$(log ε)	Ref.
$C_{26}H_{28}F_3N_2$			
3H-Indolium, 2-[2-[(1,3-dihydro-1,3,3-trimethyl-2H-indol-2-ylidene)methyl]-3,3,3-trifluoro-1-propenyl]-1,3,3-trimethyl-, perchlorate	$CHCl_3$	584(4.66)	104-1937-83
$C_{26}H_{28}N_2O_4$			
4,4'-Bipyridinium bis(4,4-dimethyl-2,6-dioxocyclohexylide)	H_2O	416(3.90)	65-2073-83
	96% EtOH	453(3.95)	65-2073-83
	85% EtOH	446(3.95)	65-2073-83
	$CHCl_3$	520(4.14)	65-2073-83
$C_{26}H_{28}N_2O_9$			
Ellipticinium, N^2-methyl-9-O-β-D-glucuronosyl-	1% HOAc	245(4.21),278(4.19), 301(4.31),359(3.37), 374(3.38),443(3.20)	111-0551-83
$C_{26}H_{28}N_2O_{10}S_2$			
3(2H)-Pyridazinone, 2-[2,3-O-(1-methylethylidene)-5-O-[(4-methylphenyl)sulfonyl]-β-D-ribofuranosyl]-5-[[(4-methylphenyl)sulfonyl]oxy]-	MeOH	266(3.54),273(3.55), 294(3.51)	44-3765-83
$C_{26}H_{28}O_4$			
2(3H)-Benzofuranone, 5-(1,1-dimethylethyl)-3-[5-(1,1-dimethylethyl)-7-methyl-2-oxo-3(2H)-benzofuranylidene]-7-methyl-, (E)-	CH_2Cl_2	227(4.19),258s(4.09), 262(4.09),422(4.26)	78-2147-83
2(3H)-Benzofuranone, 7-(1,1-dimethylethyl)-3-[7-(1,1-dimethylethyl)-5-methyl-2-oxo-3(2H)-benzofuranylidene]-5-methyl-, (E)-	CH_2Cl_2	227(4.06),259s(4.08), 263(4.12),421(4.23)	78-2147-83
$C_{26}H_{28}O_6$			
Fleminone	MeOH	272(4.63),310s(4.08), 361(3.51)	102-2287-83
$C_{26}H_{28}O_{14}$			
Apiin	EtOH	267.5(4.18),342.5(4.30)	162-0107-83
Genistein 7-O-xyloglucoside	EtOH	263(4.56),328s(3.73)	105-0748-83
	EtOH-NaOMe	273(--),355s(--)	105-0748-83
	EtOH-NaOAc	262(--),330s(--)	105-0748-83
	EtOH-$AlCl_3$	273(--),309s(--), 376(--)	105-0748-83
$C_{26}H_{29}BrN_2O_2S$			
Thiourea, N-(4-bromophenyl)-N'-(3-methoxy-17-oxoestra-1,3,5(10)-trien-2-yl)-	EtOH	286(4.279)	36-1205-83
Thiourea, N-(4-bromophenyl)-N'-(3-methoxy-17-oxoestra-1,3,5(10)-trien-4-yl)-	EtOH	244s(4.276),277(4.34)	36-1205-83
$C_{26}H_{29}ClO_3S$			
Androsta-1,4-diene-3,11-dione, 17-[(chlorophenylmethyl)sulfinyl]-, (17α)-	MeOH	230(4.26)	87-0078-83
Androsta-1,4-diene-3,11-dione, 17-chloro-17-[(phenylmethyl)sulfinyl]-, [17α,17(R)]-	MeOH	223(4.33)	87-0078-83
$C_{26}H_{29}ClO_4S$			
Androsta-1,4-diene-3,11-dione, 17-chlo-	MeOH	220(4.26),237(4.19)	87-0078-83

Compound	Solvent	$\lambda_{max}(\log \epsilon)$	Ref.
ro-17-[(phenylmethyl)sulfonyl]-, (17α)- (cont.)			87-0078-83
$C_{26}H_{29}NO_2$			
Estra-4,9-dien-3-one, 17-hydroxy-17-(1-propynyl)-11-(2-pyridinyl)-	EtOH	301(4.28)	150-0294-83S
Estra-4,9-dien-3-one, 17-hydroxy-17-(1-propynyl)-11-(3-pyridinyl)-	EtOH	301(4.26)	150-0294-83S
Estra-4,9-dien-3-one, 17-hydroxy-17-(1-propynyl)-11-(4-pyridinyl)-	EtOH	300(4.27)	150-0294-83S
$C_{26}H_{29}NO_5$			
Phenol, 2-[[3-(1,1-dimethylethyl)-2-hydroxy-5-methylphenyl]methyl-6-[(2-hydroxy-5-nitrophenyl)methyl]-4-methyl-	neutral	322(3.93)	126-2361-83
	anion	399(4.17)	126-2361-83
	dianion	416(4.28)	126-2361-83
Phenol, 2-[[3-(1,1-dimethylethyl)-2-hydroxy-5-nitrophenyl]methyl]-6-[(2-hydroxy-5-methylphenyl)methyl]-4-methyl-	neutral	333(3.96)	126-2361-83
	anion	431(4.25)	126-2361-83
	polyanion	449(4.45)	126-2361-83
$C_{26}H_{29}NO_8$			
2-Butenedioic acid, 2,2'-(7,8,9,10-tetrahydro-5-methylcyclohept[b]indole-6,10a(5H)-diyl)bis-, tetramethyl ester, (Z,Z)-	MeOH	208(4.20),276.4(3.92), 415(3.86)	39-0515-83C
Cyclonon[b]indole-6,7-dicarboxylic acid, 5,9,10,11,12,12a-hexahydro-12a-[3-methoxy-1-(methoxycarbonyl)-3-oxo-1-propenyl]-5-methyl-, dimethyl ester, (Z,Z,Z)-	MeOH	213(4.35),364(4.12)	39-0515-83C
$C_{26}H_{29}NO_{10}$			
Naphtho[2,3-b]furan-2(3H)-one, 5,6-diacetoxy-3-(3-acetoxypropyl)-3,4-dimethyl-7-(1-methylethyl)-9-nitro-, (R)-	EtOH	236(4.86),265s(3.98), 322s(3.47),335(3.52)	18-2985-83
$C_{26}H_{29}NO_{11}$			
Isorhoeagenin α-D-glucoside	MeOH	240(3.90),293(4.00)	100-0441-83
$C_{26}H_{29}N_2O_3$			
1-Benzopyrylium, 7-(diethylamino)-2-[7-(diethylamino)-2-oxo-2H-1-benzopyran-3-yl]-, perchlorate	MeCN	658(4.95)	48-0505-83
$C_{26}H_{29}N_5O_4$			
1H-Purine-2,6-dione, 7-(2-acetoxy-2-phenylethyl)-8-[ethyl(phenylmethyl)-amino]-3,7-dihydro-1,3-dimethyl-	EtOH	297(4.19)	56-0461-83
$C_{26}H_{30}N_2O_2$			
4b,11a:5a,10b-Dibutano-6H,12H-[1,4]-dioxino[2,3-b:5,6-b']diindole, 6,12-dimethyl-	EtOH	252(3.98),305(3.45)	39-0141-83C
$C_{26}H_{30}N_2O_2S$			
Thiourea, N-(3-methoxy-17-oxoestra-1,3,5(10)-trien-2-yl)-N'-phenyl-	EtOH	276(4.285)	36-1205-83

Compound	Solvent	λ_{max}(log ϵ)	Ref.
$C_{26}H_{30}N_2O_8$			
Ellipticinium, N^2-methyl-9-O-β-D-glycosyl-	H_2O	240(4.11),302(4.40), 352(3.34),390(3.36), 425(3.20)	111-0551-83
$C_{26}H_{30}N_4O_2$			
Phenol, 4,4'-[1-[[[[4-(3,3-dimethyl-1-triazenyl)phenyl]methylene]amino]-methyl]-2-ethyl-1,2-ethanediyl]bis-, (R*,S*)-	MeOH	207(4.28),227(4.42), 288(4.12),334(4.39)	83-0271-83
[1,1':3',1"-Terphenyl]-2',5'-diol, 4',6'-di-1-piperazinyl-	MeOH	316(2.73),410(2.98), 620(2.32)	70-0088-83
$C_{26}H_{30}N_6O_4$			
Estra-1,3,5(10)-trien-17-one, 3-[3-(6-amino-9H-purin-9-yl)propoxy]-4-nitro-	EtOH	262(4.23)	87-0162-83
$C_{26}H_{30}O_2S$			
Androsta-1,4-diene-3,11-dione, 17α-[(phenylmethyl)thio]-	MeOH	237(4.23)	87-0078-83
$C_{26}H_{30}O_3$			
Estra-1,3,5(10)-trien-17-one, 3-methoxy-4-(phenylmethoxy)-	$CHCl_3$	278(3.27)	94-3309-83
Estra-1,3,5(10)-trien-17-one, 4-methoxy-3-(phenylmethoxy)-	$CHCl_3$	277(3.14)	94-3309-83
$C_{26}H_{30}O_3S$			
Androsta-1,4-diene-3,11-dione, 17-[(phenylmethyl)sulfinyl]-, [17α(R)]-	MeOH	230(4.28)	87-0078-83
$C_{26}H_{30}O_7$			
6aH-Cyclopropa[e]benzofuran-6a-carboxylic acid, 5a-(3-furanylmethyl)-2-[(hexahydro-2,2,3a-trimethyl-5-oxo-furo[3,2-b]furan-3-yl)methyl]-4,5,5a,6-tetramethyl-, methyl ester, [3S-[3α(5aS*,6aR*),3aα,6aα]]-	MeOH	216(3.94)	5-1760-83
$C_{26}H_{30}O_8$			
Limonin	n.s.g.	207(3.85),285(1.58)	162-0788-83
Naphtho[2,3-b]furan-2(3H)-one, 5,6-di-acetoxy-3-(3-acetoxypropyl)-3,4-di-methyl-7-(1-methylethyl)-, (R)-	EtOH	239(4.94),272(3.78), 283(3.83),294(3.71), 318.5(3.03),332.5(3.11)	18-2985-83
Physodic acid	EtOH	256(4.2)	162-1064-83
$C_{26}H_{30}O_9$			
7H-Phenaleno[1,2-b]furan-7-one, 3,7a-diacetoxy-7a,8,9,10a-tetrahydro-4,5,6-trimethoxy-1,8,8,9-tetramethyl-	EtOH	334(4.02),350(4.07)	18-3661-83
$C_{26}H_{31}ClO_2S$			
Androsta-1,4-dien-3-one, 17-chloro-17-[(phenylmethyl)sulfinyl]-, [17α,17(R)]-	MeOH	220(4.28),239(4.26)	87-0078-83
$C_{26}H_{31}ClO_3S$			
Androsta-1,4-dien-3-one, 17-chloro-11-hydroxy-17-[(phenylmethyl)sulfinyl]-, [11β,17α,17(R)]-	MeOH	220(4.28),240(4.23)	87-0078-83

Compound	Solvent	λ_{max}(log ε)	Ref.
$C_{26}H_{31}Cl_4NO_5$			
Acetic acid, dichloro-, 3,3-dichloro-3,4-dihydro-5,5-dimethyl-2-oxo-8-pentyl-4-piperidino-2H,5H-pyrano-[3,2-c][1]benzopyran-10-yl ester	EtOH	283(3.94),330s(3.65)	161-0775-83
$C_{26}H_{31}NO_8$			
2-Butenedioic acid, 2-[[2-(3,8-dihydro-3,3,5-trimethoxyphenanthro[4,5-bcd]-furan-9b(3aH)-yl)ethyl]methylamino]-, dimethyl ester	EtOH	225(4.09),284(4.27)	44-0173-83
$C_{26}H_{31}N_2$			
3H-Indolium, 2-[3-(1,3-dihydro-1,3,3-trimethyl-2H-indol-2-ylidene)-2-methyl-1-propenyl]-1,3,3-trimethyl-, tetrafluoroborate, (Z,E)-	MeOH	551.5(4.83)	89-0876-83
$C_{26}H_{31}N_3O_2$			
3,5-Pyridinedicarboxamide, 1,4-dihydro-N,N'-bis(1-phenylethyl)-1-propyl-	EtOH	386(3.84)	130-0299-83
$C_{26}H_{31}N_5O_2$			
Estra-1,3,5(10)-trien-17-one, 3-[3-(6-amino-9H-purin-9-yl)propoxy]-	EtOH	264(4.14)	87-0162-83
$C_{26}H_{31}O_9P$			
D-ribo-2-Octulose, 3-acetyl-1,3,4-tri-deoxy-4-(dimethoxyphosphinyl)-5,7-6,8-bis(O-phenylmethylene)-	MeOH	290(3.00)	111-0487-83
$C_{26}H_{32}F_2O_7$			
Fluocinonide	n.s.g.	237(4.18)	162-0593-83
Pregna-1,4-diene-3,20-dione, 17,21-di-acetoxy-6,9-difluoro-11-hydroxy-16-methyl-	EtOH	238(4.24)	162-0456-83
$C_{26}H_{32}F_3N_3O_2S$			
Oxaflumazine (as disuccinate)	MeOH	259(4.53),310(3.58)	162-0991-83
$C_{26}H_{32}N_2O_2$			
Galeopsitrione quinoxaline deriv.	MeOH	237.5(4.23),308s(3.66), 319(3.75),330s(3.61)	102-0527-83
$C_{26}H_{32}N_2O_4$			
Benzene, 1,1'-[1,2-bis(3,3-dimethyl-1-butenyl)-1,2-ethanediyl]bis[4-nitro-, [R*,S*-(E,E)]-	EtOH	285(4.20)	12-0527-83
$C_{26}H_{32}N_2O_6$			
1H-Pyrrole-2-carboxylic acid, 5-[[5-(ethoxycarbonyl)-4-(2-hydroxyethyl)-3-methyl-1H-pyrrol-2-yl]methyl]-4-(2-hydroxyethyl)-3-methyl-, phenylmethyl ester	$CHCl_3$	277(4.35),289(4.39)	78-1849-83
$C_{26}H_{32}N_2O_{10}$			
L-ribo-Hexar-1-amic acid, 4-(acetyl-amino)-N-[1-(8-acetoxy-3,4-dihydro-1-oxo-1H-2-benzopyran-3-yl)-3-meth-	MeOH	236(3.86),288(3.26)	87-1370-83

Compound	Solvent	λ_{max}(log ε)	Ref.
ylbutyl]-4,5-dideoxy-, -lactone, 2-acetate (cont.)			87-1370-83
$C_{26}H_{32}N_3$			
Pyrimido[1,2-a]benzimidazolium, 10-decyl-2-phenyl-, perchlorate	MeCN	267(4.74),333(4.90)	103-0714-83
$C_{26}H_{32}N_4O_4$			
[1,1'-Biphenyl]-2-carboxylic acid, 4-nitro-2',4',6'-tripyrrolidino-, methyl ester	$CHCl_3$	487(3.86)	44-4649-83
1,22-Epoxy-3H-dibenzo[c,f][1,5,9]triazacyclononadecine-3,5,7,8-trione, 4-amino-6,7,8,9,10,11,12,13,14,15-16,17-dodecahydro-2,21-dimethyl-	DMF	265(4.04),425(3.91), 446(3.95)	95-0049-83
$C_{26}H_{32}N_6O_2$			
3H-Pyrazol-3-one, 4,4'-[1,2-ethanediylbis(methylimino)]bis[1,2-dihydro-1,5-dimethyl-2-phenyl-	n.s.g.	243.5(4.23),272(4.20)	80-0375-83
$C_{26}H_{32}OS$			
Androsta-1,4-dien-3-one, 17α-[(phenylmethyl)thio]-	MeOH	242(4.20)	87-0078-83
$C_{26}H_{32}O_2S$			
Androsta-1,4-dien-3-one, 17β-hydroxy-17β-[(phenylmethyl)thio]-	MeOH	242(4.19)	87-0078-83
$C_{26}H_{32}O_3$			
18,19-Dinorpregn-4-en-20-yn-3-one, 13-ethyl-17-[(1-oxo-2,4-pentadienyl)oxy]-, [17α,17(E)]-	EtOH	245(4.66)	13-0339-83A
19-Norpregn-4-en-20-yn-3-one, 17-[(1-oxo-2,4-hexadienyl)oxy]-, [17α-17(2E,4E)]-	EtOH	254(4.68)	13-0277-83A
$C_{26}H_{32}O_4$			
Bixin methyl ester	benzene	475(5.10)	162-0182-83
$C_{26}H_{32}O_{10}$			
Teucroxide triacetate	EtOH	215(3.49)	44-5123-83
$C_{26}H_{33}ClO_{10}$			
Brianthein Z	CH_2Cl_2	230(3.83)	44-5203-83
$C_{26}H_{33}NO_2$			
Retinamide, N-(4-hydroxyphenyl)-	MeOH	229(4.05),362(4.75)	34-0422-83
13-cis	EtOH	240(4.14),366(4.68)	34-0422-83
$C_{26}H_{33}NO_6$			
Mycotrienol I	MeOH	261(4.53),272(4.64), 282(4.53),386(3.15)	158-83-154
$C_{26}H_{34}FN_3O$			
1-Propanone, 3-[4-[2-(4-fluorophenyl)-2-(1-pyrrolidinyl)ethyl]-1-piperazinyl]-2-methyl-1-phenyl-	MeOH	241(4.12)	73-2977-83

Compound	Solvent	λ max(log ε)	Ref.
$C_{26}H_{34}FN_3O_2$			
1-Propanone, 3-[4-[2-(4-fluorophenyl)-2-(1-pyrrolidinyl)ethyl]-1-piperazinyl]-1-(4-hydroxyphenyl)-2-methyl-	MeOH	276(4.12)	73-2977-83
$C_{26}H_{34}FN_3O_3$			
1-Propanone, 3-[4-[2-(4-fluorophenyl)-2-(4-morpholinyl)ethyl]-1-piperazinyl]-1-(4-hydroxyphenyl)-2-methyl-	MeOH	277(4.10)	73-2977-8
$C_{26}H_{34}N_2O_4$			
1,2-Cyclobutanedicarboxamide, N,N'-dibutyl-3,4-bis(2-hydroxyphenyl)-, cis	MeOH	276(3.72),282(3.69)	39-1083-83C
trans	MeOH	274(3.75),280(3.73)	39-1083-83C
	dioxan	273(3.62),279(3.59)	39-1083-83C
$C_{26}H_{34}O_2$			
18-Norandrost-13(17)-en-3-ol, 17-methyl-, benzoate, (3β,5α)-	EtOH	228(4.05)	44-1954-83
$C_{26}H_{34}O_3$			
18,19-Dinorpregn-4-en-20-yn-3-one, 17-[(cyclobutylcarbonyl)oxy]-13-ethyl-, (17α)-	MeOH	241(4.20)	13-0349-83A
19-Norpregn-4-en-20-yn-3-one, 17-[[(3-methylcyclobutyl)carbonyl]oxy]-, [17α,17(E)]-	MeOH	241(4.24)	13-0291-83A
$C_{26}H_{34}O_4$			
2(1H)-Phenanthrenone, 7-(benzoyloxy)-dodecahydro-4b-methyl-1-(3-oxobutyl)-, [1S-(1α,4aβ,4bα,7α,8aβ,10aα)]-	EtOH	228(4.01)	44-1954-83
$C_{26}H_{34}O_5$			
Drimachone	EtOH	208(4.62),227(4.29), 295(4.00),337(3.87)	102-1997-83
Secodrial	EtOH	207(4.63),227(4.30), 298(3.96),340(3.89)	102-1997-83
Tripartol	EtOH	207(4.65),227(4.31), 295(4.00),339(3.89)	102-1997-83
$C_{26}H_{34}O_6$			
2H-1-Benzopyran-2-one, 7-[(4-acetoxy-decahydro-6-hydroxy-1,2,4a,5-tetramethyl-1-naphthalenyl)methoxy]-	EtOH	212(4.08),220(4.03), 253(3.35),294(3.67), 327(3.95)	105-0664-83
2H-1-Benzopyran-2-one, 6,8-dimethoxy-7-[(3,7,11-trimethyl-10-oxo-2,6-dodecadienyl)oxy]-, (E,E)- (oxofarnachrol)	EtOH	206(4.72),227s(4.38), 296(4.00),338(3.86)	100-0510-83
2H-1-Benzopyran-2-one, 6,8-dimethoxy-7-[(1,4,4a,5,6,7,8,8a-octahydro-6-hydroxy-2,5,5,8a-tetramethyl-1-naphthalenyl)methoxy]- (isodrimartol A)	EtOH	207(4.69),227(4.37), 296(4.03),338(3.90)	100-0510-83
Epoxyfarnochrol	EtOH	207(4.72),226s(4.36), 297(4.01),339(3.86)	100-0510-83
Karatavicin	EtOH	218(4.36),243(3.65), 250(3.53),325(4.19)	105-0498-83
$C_{26}H_{34}O_7$			
Cinobufotalin	n.s.g.	295(3.72)	162-0328-83

Compound	Solvent	$\lambda_{max}(\log \epsilon)$	Ref.
$C_{26}H_{34}O_{10}$			
2-Butenoic acid, 4-acetoxy-2-(acetoxy-methyl)-, 9-acetoxy-2,3,3a,4,5,8,9-11a-octahydro-3,6,10-trimethyl-2-oxo-cyclodeca[b]furan-4-yl ester	EtOH	210(4.02)	102-0197-83
$C_{26}H_{34}O_{12}$			
Angeloside A	MeOH	220(4.30),252(3.57), 302(3.75),328(4.03)	94-0064-83
$C_{26}H_{35}FO_5$			
Pregna-1,4-dien-21-oic acid, 6-fluoro-11-hydroxy-16-methyl-3,20-dioxo-, butyl ester	MeOH	242(4.23)	162-0594-83
$C_{26}H_{35}NO_6$			
Mycotrienol II	MeOH	263(4.49),273(4.60), 282(4.49),307(3.52)	158-83-154
$C_{26}H_{35}N_3O_{10}$			
Kristenin, triacetyl deriv.	MeOH	212(4.21),236(3.94), 287(3.36)	158-83-137
unacetylated form	MeOH	210(4.43),247(3.78), 314(3.62)	158-83-137
$C_{26}H_{36}FN_3O_2$			
1-Propanone, 3-[4-[2-(diethylamino)-2-(4-fluorophenyl)ethyl]-1-pipera-zinyl]-1-(4-hydroxyphenyl)-2-methyl-	MeOH	275(4.07)	73-2977-83
$C_{26}H_{36}N_2O_8$			
8,11,14,17,20,23,26,29-Octaoxa-2,3-di-azatricyclo[28.2.2.2^{4,7}]hexatria-conta-2,4,6,30,32,33,35-heptaene	o-$C_6H_4Cl_2$	363(4.38)	35-1851-83
$C_{26}H_{36}N_2O_9$			
Antimycin A_3	MeOH	225(4.52),320(3.86)	162-0104-83
$C_{26}H_{36}N_4$			
Corrin, 1,2,2,7,7,12,12-heptamethyl-, monohydrochloride	MeOH	259(4.42),286s(4.15), 300s(4.43),311(4.58), 346s(3.12),360(3.64), 384(3.42),472s(4.08), 495(4.13)	89-0735-83
$C_{26}H_{36}N_6O_4$			
2-Pyrrolidinecarboxamide, 1,1'-[1,4-butanediylbis(1,3(4H)-pyridinediyl-carbonyl)]bis-, [S-(R*,R*)]-	$CHCl_3$	348(4.04)	44-1370B-83
$C_{26}H_{36}O_3$			
5α-Androstan-17-one, 3β-hydroxy-16α-(4-methoxyphenyl)-	MeOH	229(4.04),279(3.20), 284(3.11)	12-0789-83
18,19-Dinorpregn-4-en-20-yn-3-one, 13-ethyl-17-(2-methyl-1-oxobutoxy)-, (17α)-	MeOH	240(4.18)	13-0349-83A
18,19-Dinorpregn-4-en-20-yn-3-one, 13-ethyl-17-(3-methyl-1-oxobutoxy)-, (17α)-	MeOH	242(4.22)	13-0349-83A
18,19-Dinorpregn-4-en-20-yn-3-one, 13-ethyl-17α-[(1-oxopentyl)oxy]-	MeOH	240(4.24)	13-0349-83A

Compound	Solvent	λ_{max}(log ε)	Ref.
19-Norpregn-4-en-20-yn-3-one, 17-(2,2-dimethyl-1-oxobutoxy)-	EtOH	247(4.46)	13-0255-83A
$C_{26}H_{36}O_5$			
2'H-18-Nor-17α-androst-5-eno[1,17-c]-furan-3β-ol, 2'-(acetoxymethylene)-5',17α-dihydro-5'α-methyl-, acetate, (Z)-	MeOH	224(3.87)	44-2696-83
Secodriol	n.s.g.	207(4.63),226s(4.28), 297(3.95),340(3.88)	102-1997-83
$C_{26}H_{36}O_6$			
Acetic acid, (3β,17β-diacetoxyandrost-5-en-16-ylidene)-, methyl ester	EtOH	221(4.20)	13-0055-83A
Prednisolone 21-trimethylacetate	n.s.g.	244(4.17)	162-1111-83
$C_{26}H_{36}O_7$			
Chol-4-en-24-oic acid, 7α,12α-bis(formyloxy)-3-oxo-	EtOH	236(4.16)	13-0475-83A
$C_{26}H_{36}O_8$			
Lapiferinin	EtOH	211(4.34),264(4.0), 295(3.87)	105-0286-83
$C_{26}H_{36}O_9$			
Atis-16-en-18-oic acid, 9,11-epoxy-15-oxo-, β-D-glucopyranosyl ester, (4α,5β,8α,9β,10α,11β,12α)-	MeOH	223(4.42)	94-1502-83
$C_{26}H_{36}O_{13}$			
Urceolatoside B	MeOH	246(4.04)	102-1977-83
$C_{26}H_{36}O_{15}$			
Diderroside	MeOH	234(3.91)	102-0975-83
$C_{26}H_{37}NO_4$			
5β-Pregn-17(20)-en-21-oic acid, 3β-acetoxy-20ξ-cyano-, ethyl ester	EtOH	240(4.13)	44-4248-83
$C_{26}H_{37}NO_4S_2$			
2H-Isoindole-2-undecanoic acid, 1,3-dihydro-1,3-dioxo-, 1,4-(1,3-dithiolan-2-yl)butyl ester	$CHCl_3$	243(4.09),294(3.33)	78-2691-83
Spiro[1,3-dithiolane-2,1'-[1H,7H][1,8]-oxaazacyclononadecino[7,8-a]isoindole]-7',19'(23'bH)-dione, 2',3',4'-5',8',9',10',11',12',13',14',15'-16',17'-tetradecahydro-23'b-hydroxy-	MeOH	247s(3.74)	78-2691-83
$C_{26}H_{38}$			
Benzene, 1-[2-methyl-1-(1-methylethyl)-1-phenylpropyl]-4-[2-methyl-1-(1-methylethyl)propyl]-	C_6H_{12}	202(4.67)	24-3235-83
$C_{26}H_{38}N_3$			
1H-Benzimidazolium, 1-dodecyl-3-methyl-2-(4-methyl-2-pyridinyl)-, iodide	EtOH	282(4.19)	4-0023-83
$C_{26}H_{38}O_2$			
1'H-Cyclobuta[4,5]pregnane-3,20-dione, 4'-propyl-, (α,α)-	EtOH	288(2.40)	70-0103-83

Compound	Solvent	$\lambda_{max}(\log \epsilon)$	Ref.
(β,β)- (cont.)	EtOH	288(2.18)	70-0103-83
$C_{26}H_{38}O_4$			
Digitoxigenin, 22-allyl-	EtOH	223(4.23)	48-0599-83
19-Norpregn-4-ene-3,20-dione, 17-[(1-oxohexyl)oxy]-	n.s.g.	239(4.25)	162-0630-83
$C_{26}H_{38}O_5$			
5β,14β-Cardanolide, 3β-acetoxy-14-hydroxy-22-methylene-, (20R)-	EtOH	213(3.97)	48-0574-83
(20S)-	EtOH	209.5(3.97)	48-0574-83
Digoxigenin, 22-allyl-	EtOH	227(4.13)	48-0599-83
$C_{26}H_{38}O_6$			
Isolathyrol, 15-O-acetyl-3-O-isobutyryl-	MeOH	196(4.12),278(4.05)	102-1791-83
Isolathyrol, 15-O-acetyl-5-O-isobutyryl-	MeOH	193(4.07),277(4.04)	102-1791-83
Jolkinol, 15-O-acetyl-3-O-isobutyryl-, 5β,6β-epoxide	MeOH	267(4.19)	102-1791-83
$C_{26}H_{39}NO_5$			
5β,14β-Card-20(22)-enolide, 12β-acetoxy-14-hydroxy-3α-(methylamino)-	n.s.g.	217(4.20)	13-0171-83B
3β-	n.s.g.	218.5(4.16)	13-0171-83B
$C_{26}H_{39}NO_6$			
Estra-1,3,5(10)-trien-17-ol, 6-[bis(2-hydroxyethyl)amino]-3-(2-hydroxyethoxy)-, 17-acetate, (6α,17β)-	MeOH	208(4.09),218(4.05), 278(3.30),286(3.26)	145-0347-83
$C_{26}H_{40}N_2O_3$			
Androst-4-en-3-one, 17-acetoxy-4-(4-methyl-1-piperazinyl)-, (17β)-	EtOH	245(3.89)	107-1013-83
Androst-4-en-3-one, 17-acetoxy-6-(4-methyl-1-piperazinyl)-, (6β,17β)-	EtOH	239(3.97)	107-1013-83
$C_{26}H_{40}O_4$			
Digitoxigenin, 22-propyl-	EtOH	223(4.25)	48-0599-83
$C_{26}H_{41}NO_3$			
2,4,6,8,10,14-Octadecahexaenamide, 13-hydroxy-N-(2-hydroxy-1-methylethyl)-2,10,12,14,16-pentamethyl-, (E,E,Z,E,E,E)- (myxalamide A)	EtOH	204(4.21),265(4.06), 340s(--),357(4.59), 370s(--)	5-1081-83
$C_{26}H_{42}O_2$			
19-Norcholecalciferol, 25-hydroxy-10-oxo-, (5E)-	EtOH	307(4.38)	44-3819-83
	EtOH	312(4.43)	69-3636-83
(5Z)-	EtOH	310(4.18)	44-3819-83
$C_{26}H_{42}O_6$			
Prost-13-en-1-oic acid, 9,15-dioxo-10-[(tetrahydro-2H-pyran-2-yl)oxy]-, methyl ester. (10α,13E)-(±)-	EtOH	226(4.08)	39-0319-83C
$C_{26}H_{42}O_9$			
Pseudomonic acid D	EtOH	220(4.19)	39-2655-83C
sodium salt (isomeric mixture)	EtOH	223(4.16)	39-2655-83C

Compound	Solvent	λ_{max}(log ε)	Ref.
$C_{26}H_{42}O_{10}$			
Acetic acid, 2,2'-[1,2-phenylenebis(oxy)]bis-, bis[2-(2-butoxyethoxy)ethyl] ester	MeOH	271(3.11)	49-0359-83
$C_{26}H_{43}NO_5S$			
Homo-LTE$_4$ methyl ester, (6R,7S)-	EtOH	272(4.42),281(4.51), 290(4.42)	35-3656-83
(6S,7R)-	EtOH	271(4.47),281(4.56), 290(4.47)	35-3656-83
$C_{26}H_{44}$			
Diene 9 from vitamin D_3 (no structure given)	C_6H_{12}	236(4.31),243(4.49), 252(4.58),262(4.42)	35-3270-83
$C_{26}H_{44}O_3$			
3,5-Cyclohexadiene-1,2-dione, 4-(1,1-dimethylethyl)-5-(hexadecyloxy)-	MeCN	467(3.82)	44-0177-83
$C_{26}H_{44}Si_2$			
1,4-Etheno-2,3-disilanaphthalene, 2,2,3,3-tetrakis(1,1-dimethylethyl)-1,2,3,4-tetrahydro-	C_6H_{12}	276(2.94),282(2.94)	157-1464-83
$C_{26}H_{47}N_5O_6Si_2$			
Guanosine, 2',3'-bis-O-[(1,1-dimethylethyl)dimethylsilyl]-N-(2-methyl-1-oxopropyl)-	MeOH	256(4.21),260(4.21), 279(4.08)	33-2069-83
Guanosine, 2',5'-bis-O-[(1,1-dimethylethyl)dimethylsilyl]-N-(2-methyl-1-oxopropyl)-	MeOH	255(4.21),259(4.21), 277(4.08)	33-2069-83
Guanosine, 3',5'-bis-O-[(1,1-dimethylethyl)dimethylsilyl]-N-(2-methyl-1-oxopropyl)-	MeOH	255(4.22),259(4.22), 277(4.08)	33-2069-83
$C_{26}H_{54}Cl_2Si_2$			
Silane, (1,2-dichloro-1,2-ethenediyl)bis[tris(1,1-dimethylethyl)-, (E)-	n.s.g.	225(4.23)	64-1062-83B
$C_{26}H_{56}Si_2$			
Silane, 1,2-ethenediylbis[tris(1,1-dimethylethyl)-, (E)-	n.s.g.	218(4.02)	64-1062-83B
$C_{26}H_{60}Si_6$			
Silane, [6-[2,2-bis(trimethylsilyl)ethylidene]-4-cyclohexene-1,2,3,5-tetrayl]tetrakis[trimethyl-	n.s.g.	273(4.41)	101-0C13-83M

Compound	Solvent	λ_{max}(log ϵ)	Ref.
$C_{27}H_{10}Cl_4I_4O_5$			
Rose Bengal, phenylmethyl ester, anion	MeOH	524(4.6),564(5.0)	35-7465-83
$C_{27}H_{14}Cl_{14}NO_3$			
Methyl, bis(pentachlorophenyl)[2,3,5,6-tetrachloro-4-[[[2-(1,1-dimethylethoxy)-1-methyl-2-oxoethyl]amino]carbonyl]phenyl]-, (S)-	$CHCl_3$	285(3.86),335s(3.89), 365s(4.35),382(4.62), 480s(3.09),506(3.10), 560(3.09)	44-3716-83
$C_{27}H_{17}O$			
Acenaphtho[1,2-b]pyrylium, 8,10-diphenyl-, perchlorate	MeCN	234(4.53),350(4.48), 427(4.37)	22-0115-83
$C_{27}H_{18}CoN_4O$			
Cobalt, [5,16-dihydro-17H-tetrabenzo-[b,f,i,m][1,4,8,12]tetraazacyclopentadecin-17-onato(2-)-N^5,N^{11}-N^{16},N^{22}]-, (SP-4-2)-	EtOH-DMF	305(4.42),464(3.68)	12-2395-83
$C_{27}H_{18}CuN_4O$			
Copper, [5,16-dihydro-17H-tetrabenzo-[b,f,i,m][1,4,8,12]tetraazacyclopentadecin-17-onato(2-)-N^5,N^{11}-N^{16},N^{22}]-, (SP-4-2)-	EtOH-DMF	300(4.76),420(4.04), 522(3.91)	12-2395-83
$C_{27}H_{18}N_4NiO$			
Nickel, [5,16-dihydro-17H-tetrabenzo-[b,f,i,m][1,4,8,12]tetraazacyclopentadecin-17-onato(2-)-N^5 ,N^{11}-N^{16},N^{22}]-, (SP-4-2)-	EtOH-DMF	303(4.58),340(4.23), 420(3.91),448(3.90), 478(3.91),660(3.67)	12-2395-83
$C_{27}H_{18}N_4O$			
Acetamide, N-[4-cyano-3-[2-(1-cyano-2-phenylethenyl)phenyl]-1-isoquinolinyl]-	THF	251(4.50),305(4.40), 338s(4.19),352s(3.95)	39-1813-83C
7H-Benzimidazo[2,1-a]benz[de]isoquinolin-7-one, 4-(4,5-dihydro-1-phenyl-1H-pyrazol-3-yl)-	toluene	505(4.32)	103-0214-83
$C_{27}H_{19}ClN_4O_2$			
7H-Dibenz[f,ij]isoquinoline-1-carbonitrile, 6-[(4-chlorophenyl)amino]-2-morpholino-7-oxo-	n.s.g.	536(3.97)	2-0812-83
$C_{27}H_{19}N$			
Pyridine, 2-(1-naphthalenyl)-4,6-diphenyl-	DMSO	273(4.56),295s(4.12)	150-0301-83M
$C_{27}H_{19}NO$			
Benzo[f]quinolinium, 2-[1,1'-biphenyl]-4-yl-2-oxoethylide	DMF	520(3.47)	104-0568-83
$C_{27}H_{20}Br_2N_2$			
1H-Pyrazole, 1,3-bis(4-bromophenyl)-4,5-dihydro-4,5-diphenyl-, trans	C_6H_{12}	373(4.40)	12-1649-83
	MeOH	367(4.43)	12-1649-83
$C_{27}H_{20}ClN_5$			
1H-Pyrazol-5-amine, N-(4-chlorophenyl)-1,3-diphenyl-4-(phenylazo)-	EtOH	248(4.30),335(4.03)	4-1501-83

Compound	Solvent	λ_{max}(log ε)	Ref.
$C_{27}H_{20}Cl_2O_2$			
1-Isobenzofuranol, 3,3-bis(4-chloro-phenyl)-1,3-dihydro-1-(4-methyl-phenyl)-	hexane	257s(3.09),263(3.17), 269(3.15),277(2.93)	104-0553-83
$C_{27}H_{20}CuN_4O_4$			
Copper, [10,12-dimethoxy-22-methyl-9,7-metheno-7H-tribenzo[e,i,n]-[1,4,8,12]tetraazacyclopentadecine-19,21(8H,20H)-dionato(2-)-N^5,N^8-N^{14},N^{20}]-, (SP-4-2)-	EtOH-DMF	440(3.08)	12-2407-83
$C_{27}H_{20}NO$			
Benzo[f]quinolinium, 4-(2-[1,1'-biphen-yl]-4-yl-2-oxoethyl)-, iodide	EtOH	217(4.84),278(4.67), 369(3.98)	104-0568-83
$C_{27}H_{20}N_2O$			
Pyridinium, 2,4,6-triphenyl-, 1-cyano-2-oxopropylide (ethanolate)	toluene	536(3.02)	142-0623-83
	EtOH	446(2.85)	142-0623-83
	CH_2Cl_2	504(3.03)	142-0623-83
$C_{27}H_{20}N_2O_3S$			
Benzamide, N-[2-[1-(phenylsulfonyl)-1H-indol-2-yl]phenyl]-	EtOH	212s(4.52),256(4.35), 280s(4.19)	39-2417-83C
$C_{27}H_{20}N_2O_5S$			
Indolo[2,3-a]carbazole-11(12H)-carbox-ylic acid, 12-[(4-methoxyphenyl)sul-fonyl] , methyl ester	EtOH	206(4.32),232(4.46), 288(4.13),307(4.13), 340(4.21)	78-3725-83
$C_{27}H_{20}N_4NiO_4$			
Nickel, [10,12-dimethoxy-22-methyl-9,7-metheno-7H-tribenzo[e,i,n]-[1,4,8,12]tetraazacyclopentadecine-19,21(8H,20H)-dionato(2-)-N^5,N^8-N^{14},N^{20}]-, (SP-4-2)-	EtOH-DMF	434(3.68),500(3.56)	12-2407-83
$C_{27}H_{20}N_4O_2$			
Benzo[f]quinoline, 3,4-dihydro-4-(4-methylphenyl)-3-[[(4-nitrophenyl)-azo]methylene]-	neutral	592(4.70)	124-0068-83
	protonated	470(4.54)	124-0068-83
Quinoline, 1,2-dihydro-6-methyl-1-(1-naphthalenyl)-2-[[(4-nitrophenyl)-azo]methylene]-	neutral	570(4.48)	124-0068-83
	protonated	476(4.88)	124-0068-83
$C_{27}H_{20}N_4O_3$			
Benzo[f]quinoline, 3,4-dihydro-4-(4-methoxyphenyl)-3-[[(4-nitrophenyl)-azo]methylene]-	neutral	594(5.14)	124-0068-83
	protonated	474(4.82)	124-0068-83
Quinoline, 1,2-dihydro-6-methoxy-1-(1-naphthalenyl)-2-[[(4-nitrophenyl)-azo]methylene]-	neutral	582(4.59)	124-0068-83
	protonated	478(4.79)	124-0068-83
$C_{27}H_{20}N_6O_9$			
Methanone, bis[4-[(4-methylphenyl)ami-no]-3,5-dinitrophenyl]-	dioxan	293.4(4.05)	56-1357-83
$C_{27}H_{20}N_6O_{11}$			
Methanone, bis[4-[(2-methoxyphenyl)ami-no]-3,5-dinitrophenyl]-	dioxan	293.1(4.13)	56-1357-83

Compound	Solvent	$\lambda_{max}(\log \epsilon)$	Ref.
$C_{27}H_{20}N_8$			
Pyrazolo[1,5-a]pyrimidin-2-amine, 3-phenyl-N-[3-[(3-phenylpyrazolo[1,5-a]pyrimidin-2-yl)amino]-2-propenylidene]-	MeCN	275(4.58),377(4.48)	39-0011-83C
$C_{27}H_{20}O_8$			
7H-Furo[3,2-g][1]benzopyran-7-one, 9,9'-[1,5-pentanediylbis(oxy)]bis-	EtOH	218(4.70),248(4.66), 263(4.44),300(4.37)	39-1807-83B
$C_{27}H_{21}BrN_2$			
1H-Pyrazole, 1-(4-bromophenyl)-4,5-dihydro-3,5,5-triphenyl-	C_6H_{12}	364(4.38)	12-1649-83
	MeOH	356(4.38)	12-1649-83
$C_{27}H_{21}ClN_2O$			
Cinnoline, 2-benzoyl-4-(3-chlorophenyl)-1,2,3,4-tetrahydro-1-phenyl-	$CHCl_3$	248(4.03),266(4.08)	97-0028-83
Cinnoline, 2-benzoyl-4-(4-chlorophenyl)-1,2,3,4-tetrahydro-1-phenyl-	$CHCl_3$	248(4.06),265(4.09)	97-0028-83
$C_{27}H_{21}N$			
Dibenz[c,h]acridine, 5,6,8,9-tetrahydro-7-phenyl-	neutral	335(4.19)	39-0045-83B
	protonated	374(4.32)	39-0045-83B
$C_{27}H_{21}N_3O$			
3,5-Pyridinedicarbonitrile, 2-[1,1'-biphenyl]-4-yl-1,4-dihydro-4-(4-methoxyphenyl)-6-methyl-	EtOH	209(4.52),282(4.43), 348(3.63)	73-3123-83
$C_{27}H_{21}N_3O_3$			
Cinnoline, 2-benzoyl-1,2,3,4-tetrahydro-4-(3-nitrophenyl)-1-phenyl-	$CHCl_3$	265(4.26)	97-0028-83
$C_{27}H_{21}N_5$			
1H-Pyrazol-5-amine, N,1,3-triphenyl-4-(phenylazo)-	EtOH	248(3.96),335(3.73)	4-1501-83
$C_{27}H_{22}N_2$			
1H-Pyrazole, 4,5-dihydro-1,3,4,5-tetraphenyl-, trans	C_6H_{12}	362(4.32)	12-1649-83
	MeOH	358(4.28)	12-1649-83
1H-Pyrazole, 4,5-dihydro-1,3,5,5-tetraphenyl-	C_6H_{12}	361(4.32)	12-1649-83
	MeOH	356(4.30)	12-1649-83
$C_{27}H_{22}N_2O$			
Cinnoline, 2-benzoyl-1,2,3,4-tetrahydro-1,4-diphenyl-	$CHCl_3$	248(4.08),266(4.11)	97-0028-83
$C_{27}H_{22}N_2O_6$			
Cyclohept[1,2,3-hi]imidazo[2,1,5-cd]indolizine-3,4,9-tricarboxylic acid, 2-(4-methylphenyl)-, 9-ethyl 3,4-dimethyl ester	EtOH	251(4.43),278s(4.46), 291(4.50),312s(4.35), 384(4.57),427s(4.14)	18-3703-83
$C_{27}H_{22}O_2$			
Bicyclo[3.1.0]hex-3-en-2-one, 6-benzoyl-1,3-dimethyl-4,5-diphenyl-, (1α,5α,6β)-	EtOH	221(4.31),248(4.27), 310(3.83)	138-0489-83
$C_{27}H_{22}O_{18}$			
Strictinin	MeOH	218(4.78),267(4.44)	39-1765-83C

Compound	Solvent	λ_{max}(log ε)	Ref.
$C_{27}H_{23}BrN_4O$			
Benzenecarbohydrazonic acid, N-(4-bromophenyl)-, [(4-methoxyphenyl)methylene]phenylhydrazide	EtOH	225(4.36),263(4.18), 349(4.42)	70-1439-83
	dioxan	225(4.42),268(4.22), 352(4.40)	70-1439-83
	MeCN	226(4.40),245(4.32), 264s(4.20),295(4.26), 341(4.51)	70-1439-83
$C_{27}H_{23}ClN_2O$			
7H-Dibenz[f,ij]isoquinolin-7-one, 6-chloro-1-[4-(diethylamino)phenyl]-2-methyl-	n.s.g.	507(4.27)	2-0805-83
$C_{27}H_{23}N_3O_3$			
Pyrimido[4,5-b][1,4]oxazepine-2,4(1H-3H)-dione, 6-[1,1'-biphenyl]-4-yl-7,8-dihydro-1,3-dimethyl-8-phenyl-	MeOH	230s(--),286(4.42), 370(3.30)	103-0544-83
$C_{27}H_{24}N_2O_9$			
Albofungin	EtOH	228(4.58),254(4.58), 303(4.19),376(4.42)	162-0032-83
$C_{27}H_{24}N_4O$			
Benzenecarbohydrazonic acid, N-phenyl-, [(4-methoxyphenyl)methylene]phenylhydrazide	EtOH	232(4.41),245(4.40), 298(4.28),341(4.53)	70-1439-83
	dioxan	228(4.37),247(4.32), 296(4.17),343(4.46)	70-1439-83
	MeCN	227(4.36),245s(4.28), 286(4.14),295(4.15), 341(4.41)	70-1439-83
$C_{27}H_{24}N_4O_3S$			
Verdazylium, 1,3,5-triphenyl-, p-toluenesulfonate	MeCN	535(4.08)	104-0481-83
$C_{27}H_{24}N_4O_{10}$			
α-D-lyxo-Hexopyranosid-2-ulose, methyl 4,6-O-(phenylmethylene)-, (2,4-dinitrophenylhydrazone), 3-benzoate	EtOH	350(4.67)	39-0257-83C
β-	EtOH	350(4.35)	39-0257-83C
$C_{27}H_{24}O_3$			
1H-Cyclopenta[b]benzofuran-1-one, 3a,8b-dihydro-7-hydroxy-2,5,6,8b-tetramethyl-3,3a-diphenyl-	$CHCl_3$	283(4.10),335s(2.97)	44-2728-83
Tetracyclo[17.2.2.2^{5,8}.2^{12,15}]heptacosa-5,7,12,14,19,21,22,24,26-nonaene-3,10,17-trione	EtOH	226s(4.45),275(3.44), 283s(3.41),300s(3.29), 312(3.07)	94-2868-83
$C_{27}H_{24}O_6S$			
2(5H)-Furanone, 4-[1,3-benzodioxol-5-yl(phenylthio)methyl]-3-[(3,4-dimethoxyphenyl)methyl]-	$CHCl_3$	260(3.76),285(3.83)	39-0643-83C
$C_{27}H_{25}ClN$			
Pyridinium, 1-butyl-4-(2-chlorophenyl)-2,6-diphenyl-, tetrafluoroborate	DMSO	303(4.40)	150-0301-83M
$C_{27}H_{25}NO_5$			
Spiro[bicyclo[3.1.0]hex-3-ene-6,7'-	MeOH	208(3.39),240(2.83),	73-2812-83

Compound	Solvent	$\lambda_{max}(\log \epsilon)$	Ref.
[4,6]dioxa[2]azabicyclo[3.2.0]heptane]-2,3'-dione, 2'-acetyl-1,3,4,5-tetramethyl-1',5'-diphenyl- (cont.)		263(2.56),330(1.30)	73-2812-83
$C_{27}H_{25}N_2S_2$			
Benzothiazolium, 3-ethyl-2-[3-(3-ethyl-2(3H)-benzothiazolylidene)-2-phenyl-1-propenyl]-, tetrafluoroborate	MeOH	560(5.29)	89-0876-83
$C_{27}H_{25}N_3O_2$			
2-Naphthalenecarboxamide, 4-[(4-butylphenyl)azo]-3-hydroxy-N-phenyl-	EtOH	340(3.889),410(3.952), 510(4.398),540(4.408)	104-1145-83
	$CHCl_3$	335s(3.996),414(4.053), 525(4.521),550(4.526)	104-1145-83
	CCl_4	340s(3.863),423(3.968), 502(4.418),535(4.425)	104-1145-83
	DMSO	270(4.663),335s(3.940), 420(3.903),525(4.346)	104-1145-83
$C_{27}H_{25}N_3O_8$			
α-D-lyxo-Hexopyranosid-2-ulose, methyl 4,6-O-(phenylmethylene)-, (4-nitrophenyl)hydrazone, benzoate	EtOH	378(4.42)	39-0257-83C
$C_{27}H_{26}N$			
Pyridinium, 1-(1-methylethyl)-4-(2-methylphenyl)-2,6-diphenyl-, tetrafluoroborate	DMSO	306(4.43)	150-0301-83M
$C_{27}H_{26}N_2O_2$			
3-Pyridazinecarboxylic acid, 1,4-dihydro-4,4,5-triphenyl-, 1,1-dimethylethyl ester	$CHCl_3$	250(3.72)	24-3097-83
$C_{27}H_{26}N_2O_6$			
α-D-lyxo-Hexopyranosid-2-ulose, methyl 4,6-O-(phenylmethylene)-, phenylhydrazone, 3-benzoate	EtOH	278(4.32)	39-0257-83C
$C_{27}H_{26}N_2O_6S$			
2,4(1H,3H)-Pyrimidinedione, 1-[2-[(methylsulfonyl)oxy]-3-(triphenylmethoxy)propyl]-	EtOH	233s(3.99),262(4.06)	128-0125-83
$C_{27}H_{26}N_6Ru$			
Ruthenium(2+), bis(2,2'-bipyridine-N,N')(α-methyl-2-pyridinemethanamine-N^1,N^2)-, (OC-6-31)-, bis(hexafluorophosphate), (R,S)-	H_2O	343(4.04),472(4.01)	35-7075-83
(S,R)-	H_2O	340(4.04),474(3.98)	35-7075-83
$C_{27}H_{26}OP$			
Phosphonium, [(2-methoxy-6-methylphenyl)methyl]triphenyl-, bromide	EtOH	262(3.60),268.5(3.68), 275.5(3.70)	118-1000-83
$C_{27}H_{26}O_2P$			
Phosphonium, [(2,3-dimethoxyphenyl)-methyl]triphenyl-, bromide	MeOH	225(4.46),265s(3.61), 268(3.67),275(3.66), 282s(3.37)	36-0792-83

Compound	Solvent	λ_{max}(log ε)	Ref.
$C_{27}H_{26}O_9$			
Toromycin (isomer A)	MeOH-CH_2Cl_2	245(4.38),265(4.11), 275(4.26),285(4.07), 390(4.00)	78-0599-83
Toromycin (isomer B)	CH_2Cl_2	245(4.52),265(4.27), 275(4.36),395(4.08)	78-0599-83
$C_{27}H_{27}NO_5$			
Hetisan-2,13-dione, 11-(benzoyloxy)-15-hydroxy-	EtOH	303(2.03)	88-3765-83
Hetisan-2,15-dione, 11-(benzoyloxy)-13-hydroxy-	EtOH	230(4.02)	88-3765-83
$C_{27}H_{27}NO_7$			
Spiro[cyclopentane-1,13[10H-7a,10]ethano[4H]pyrido[3,2,1-jk]carbazole]-8',9',11'-tricarboxylic acid, 5',6'-dihydro-12'-oxo-, trimethyl ester	MeOH	250(4.35),310(3.85), 322(3.84),468(4.00)	39-0515-83C
$C_{27}H_{27}N_4S$			
Quinolinium, 2-[2-benzothiazolyl[3-(diethylamino)phenyl]amino]-1-methyl-, tetrafluoroborate	EtOH	222(4.68),250(4.50), 270(4.55),395(4.11)	33-2165-83
$C_{27}H_{27}O_2P$			
Cyclopentanepropanal, 1-methyl-β-oxo-α-(triphenylphosphoranylidene)-	EtOH	219s(4.47),268(4.11)	33-2760-83
$C_{27}H_{28}ClN_3O_2S$			
Phenol, 4-[[[4-(2-chloro-10,11-dihydrodibenzo[b,f]thiepin-10-yl)-1-piperazinyl]imino]methyl]-2-ethoxy-	MeOH	280(4.36),307(4.35)	73-1173-83
Phenol, 4-[[[4-(8-chloro-10,11-dihydrodibenzo[b,f]thiepin-10-yl)-1-piperazinyl]imino]methyl]-2-ethoxy-	MeOH	216(4.59),287(4.33), 307s(4.29)	73-1173-83
1-Piperazinamine, 4-(2-chloro-10,11-dihydrodibenzo[b,f]thiepin-10-yl)-N-[(3,4-dimethoxyphenyl)methylene]-	MeOH	288(4.36),308(4.37)	73-1173-83
1-Piperazinamine, 4-(8-chloro-10,11-dihydrodibenzo[b,f]thiepin-10-yl)-N-[(3,4-dimethoxyphenyl)methylene]-	MeOH	290(4.18),307.5(4.17)	73-1173-83
$C_{27}H_{28}F_3N_2$			
3H-Indolium, 2-[2-[(1,3-dihydro-1,3,3-trimethyl-2H-indol-2-ylidene)methyl]-3,3,4,4,4-pentafluoro-1-butenyl]-1,3,3-trimethyl-, perchlorate	$CHCl_3$	566(4.46)	104-1937-83
$C_{27}H_{28}NO_3P$			
3-Furancarboxylic acid, 4,5-dihydro-4,5-dimethyl-2-[(triphenylphosphoranylidene)amino]-, ethyl ester	MeOH	265(3.27),273(3.31), 310(3.75)	24-1691-83
3-Furancarboxylic acid, 4,5-dihydro-5,5-dimethyl-2-[(triphenylphosphoranylidene)amino]-, ethyl ester	MeOH	265(3.31),272(3.38), 305(3.49)	24-1691-83
$C_{27}H_{28}N_4O_{11}$			
Dysazecine (picrate)	EtOH	230s(4.18),291(3.86)	100-0127-83
Dysoxyline (picrate)	EtOH	230s(4.11),286(3.89)	100-0127-83

Compound	Solvent	λ_{max}(log ε)	Ref.
$C_{27}H_{28}O_9$			
Chrysomycin B	$CDCl_3$	214(3.83),248(4.64), 278(4.52),307(4.64), 326(4.10),339(4.13), 382(4.08)	158-83-113
$C_{27}H_{29}FN_2O_6S_2$			
Xanthylium, 3,6-bis(diethylamino)-9-[2-(fluorosulfonyl)-4-sulfophenyl]-, hydroxide, inner salt	EtOH	242s(4.273),260(4.442), 287(4.285),306s(4.044), 340(3.696),358(3.848), 412(3.477),430s(3.403), 503s(4.052),535(4.600), 571(5.132)	99-0224-83
Xanthylium, 3,6-bis(diethylamino)-9-[4-(fluorosulfonyl)-2-sulfophenyl]-, hydroxide, inner salt	EtOH	243(4.303),260(4.487), 286(4.271),308(4.071), 341s(3.663),357(3.884), 408(3.487),425s(3.417), 493s(4.311),530s(4.575), 562(5.031)	99-0224-83
$C_{27}H_{29}NO_5$			
Hetisan-2-one, 11-(benzoyloxy)-13,15-dihydroxy-, (11α,13R,15β)- (cardiapetamine)	EtOH	299(1.76)	88-3765-83
$C_{27}H_{29}NO_7$			
Stephabenine	$CHCl_3$	260(3.25),294(3.65)	94-2574-83
$C_{27}H_{29}NO_8$			
2-Butenedioic acid, 2-(4,5,6,7,12,13-hexahydro-3-(2-methoxy-2-oxoethyl)-2-oxo-11H-furo[2",3":2',3']cyclohepta[1',2':4,5]pyrrolo[3,2,1-ij]quinolin-7a(2H)-yl]-, dimethyl ester, [7aα(E),14aS*]-	MeOH	212.5(4.42),252(3.94), 313(3.49)	39-0515-83C
[7aα(Z),14aS*]-	MeOH	212(4.45),249(3.96), 310(3.43)	39-0515-83C
$C_{27}H_{29}N_3O_2S$			
1-Piperazinamine, 4-(10,11-dihydrodibenzo[b,f]thiepin-10-yl)-N-(3,4-dimethoxyphenyl)methylene]-	MeOH	290(4.35),308(4.36)	73-1173-83
$C_{27}H_{30}$			
Tetracyclo[17.2.2.2^{5,8}.2^{12,15}]heptacosa-5,7,12,14,19,21,22,24,26-nonaene	EtOH	262(2.55),268(3.11), 275(3.07)	94-2868-83
$C_{27}H_{30}N_2$			
1,1'-Spirobi[1H-pyrrolo[1,2-a]indole, 2,2',3,3'-tetrahydro-3,3,3',3',9,9'-hexamethyl- (isomer mixture)	EtOH	234(4.83),288(4.23), 295(4.21)	142-1355-83
	EtOH-$HClO_4$	207(4.39),288(4.36), 299.5(3.93)	142-1355-83
$C_{27}H_{30}N_2O_7S_2$			
Xanthylium, 3,6-bis(diethylamino)-9-(2,4-disulfophenyl)-, hydroxide, inner salt	EtOH	239(4.302),258(4.473), 282(4.181),305(4.165), 340s(3.623),353(3.89), 402(3.403),416s(3.328), 520s(4.594),551(4.995)	99-0224-83

Compound	Solvent	λ_{max}(log ϵ)	Ref.
$C_{27}H_{30}N_4$			
Benzenamine, 4-[2-[5-[4-(dimethyl-amino)phenyl]-4,5-dihydro-1-phenyl-1H-pyrazol-3-yl]ethenyl]-N,N-dimethyl-	n.s.g.	262(4.32),322(4.26), 398(4.59)	44-0542-83
$C_{27}H_{30}N_4S$			
1-Piperazinamine, 4-(10,11-dihydro-dibenzo[b,f]thiepin-10-yl)-N-[[4-(dimethylamino)phenyl]methylene]-	MeOH	313(4.13),334s(4.08)	73-1173-83
$C_{27}H_{30}O$			
2,5-Cyclohexadien-1-one, 2,6-bis(1,1-dimethylethyl)-4-(diphenylmethylene)-	isooctane	366(4.41)	73-2376-83
	isooctane	367(4.41)	73-2825-83
$C_{27}H_{30}O_2$			
[2,2'-Binaphthalene]-6-carboxylic acid, 5',6',7',8'-tetrahydro-5',5',8',8'-tetramethyl-, ethyl ester	EtOH	231(4.64),264(4.66), 312(4.36)	87-1653-83
$C_{27}H_{30}O_3$			
19-Norpregn-4-en-20-yn-3-one, 17α-(benzoyloxy)-	$CHCl_3$	243(4.39)	13-0309-83A
$C_{27}H_{30}O_{16}$			
Azamicroside	EtOH	260(4.19),300s(4.14), 358(4.18)	105-0275-83
	EtOH-NaOMe	278(--),400(--)	105-0275-83
	EtOH-NaOAc	267(--),300(--), 358(--)	105-0275-83
	EtOH-$AlCl_3$	260(--),300(--), 370(--)	105-0275-83
	+ HCl	260(--),300(--), 364(--)	105-0275-83
$C_{27}H_{31}ClO_3S$			
Androsta-1,4-diene-3,11-dione, 17-chloro-17-[[(2-methylphenyl)methyl]-sulfinyl]-, (17α)-	MeOH	228(4.34)	87-0078-83
Androsta-1,4-diene-3,11-dione, 17-chloro-17-[(2-phenylethyl)sulfin-yl]-, (17α)-	MeOH	236(4.23)	87-0078-83
$C_{27}H_{31}NO_4$			
Isoquinoline, 4-[(3,4-dimethoxyphenyl)-methyl]-1,2,3,4-tetrahydro-6,7-di-methoxy-2-(phenylmethyl)-	EtOH	239(3.92),283(3.83)	39-2053-83C
Ryosenamine	EtOH	230(4.10),273.5(2.97), 281(2.88)	94-3338-83
$C_{27}H_{31}NO_8$			
2-Butenedioic acid, 2,2'-(5,7,8,9,10,11-hexahydro-5-methyl-11aH-cyclooct[b]-indole-6,11a-diyl)bis-, tetramethyl ester, (?,E,Z)-	MeOH	211(4.25),271(4.07), 314(4.05)	39-0515-83C
(?,Z,Z)-	MeOH	211(4.20),231(4.11), 271(4.00),408(3.95)	39-0515-83C
5H-Cyclodeca[b]indole-6,7-dicarboxylic acid, 9,10,11,12,13,13a-hexahydro-13a-[3-methoxy-1-(methoxycarbonyl)-3-oxo-1-propenyl]-5-methyl-, dimethyl ester, (Z,Z,Z)-	MeOH	213(4.34),355(4.11)	39-0515-83C

Compound	Solvent	$\lambda_{max}(\log \epsilon)$	Ref.
$C_{27}H_{32}N_2O_2S$			
Thiourea, N-(3-methoxy-17-oxoestra-1,3,5(10)-trien-2-yl)-N'-(4-methylphenyl)-	EtOH	279(4.324)	36-1205-83
Thiourea, N-(3-methoxy-17-oxoestra-1,3,5(10)-trien-2-yl)-N'-(phenylmethyl)-	EtOH	253(4.233),295(3.893)	36-1205-83
Thiourea, N-(3-methoxy-17-oxoestra-1,3,5(10)-trien-4-yl)-N'-(phenylmethyl)-	EtOH	244(4.233),280(3.482)	36-1205-83
$C_{27}H_{32}N_2O_7$			
Antibiotic X-14885A, sodium salt	EtOH	204(4.49),257(4.18), 306(4.34)	158-1275-83
$C_{27}H_{32}N_2O_{10}$			
Cadambine	MeOH	230(5.25),240(4.69), 362(4.38)	100-0325-83
	MeOH-NaOH	230(5.25),285(4.17), 293(4.01)	100-0325-83
$C_{27}H_{32}N_6O_4$			
Estra-1,3,5(10)-trien-17-one, 3-[4-(6-amino-9H-purin-9-yl)butoxy]-4-nitro-	EtOH	262(4.19)	87-0162-83
$C_{27}H_{32}O_2S$			
Androsta-1,4-diene-3,11-dione, 17α-[(2-phenylethyl)thio]-	MeOH	236(4.24)	87-0078-83
$C_{27}H_{32}O_4$			
19-Norpregn-4-en-20-yn-3-one, 17α-[(5-methyl-2-furanyl)acetoxy]-	$CHCl_3$	241(4.32)	13-0321-83A
$C_{27}H_{33}B$			
Boron, tris(2,4,6-trimethylphenyl)-, TCNE complex	CH_2Cl_2	475(1.95)	35-2175-83
$C_{27}H_{33}N$			
Pyrrolidine, 1-[1-methyl-2-[1-methyl-7-(1-methylethyl)-4-azulenyl]-1-phenylethyl]-	EtOH	247(4.32),286(4.51), 290s(4.49),304s(3.93), 327s(3.40),341s(3.51), 351(3.61),368(3.50), 415(2.22),435(2.26), 566s(2.49),529s(2.62), 633s(2.58),658(2.55), 730(2.13)	18-2311-83
hydrochloride	EtOH	246(4.03),286(4.34), 291(4.31),303(3.52), 329s(3.46),335s(3.54), 351(3.66),368(3.54), 416(2.90),436(2.89), 560s(2.50),602s(2.59), 611(2.60),660(2.52), 730(2.08)	18-2311-83
$C_{27}H_{33}NO_3$			
Acetamide, N-[(6α,17β)-17-hydroxy-3-(phenylmethoxy)estra-1,3,5(10)-trien-6-yl]-	MeOH	278(3.28),285(3.22)	145-0347-83

Compound	Solvent	$\lambda_{max}(\log \epsilon)$	Ref.
$C_{27}H_{33}NO_5$			
6H-Dibenzo[b,d]pyran-1-ol, 7,8,9,10-tetrahydro-3,6,6,9-tetramethyl-2-[[(3,4,5-trimethoxyphenyl)imino]-methyl]-	$CHCl_3$	262(4.32),272(4.32), 361(4.39)	83-0326-83
$C_{27}H_{33}N_2$			
3H-Indolium, 2-[2-[(1,3-dihydro-1,3,3-trimethyl-2H-indol-2-ylidene)methyl]-1-butenyl]-1,3,3-trimethyl-, (Z,E)-, tetrafluoroborate	MeOH	553.5(4.72)	89-0876-83
$C_{27}H_{34}N_2O_8$			
Vindoline, 16-acetoxy-14,15-dihydro-3-oxo-	EtOH	213(4.42),252(3.76), 302(3.60)	100-0884-83
Vindoline, 16-acetoxy-14,15-dihydro-5-oxo-	EtOH	212(4.49),247(3.79), 300(3.66)	100-0884-83
$C_{27}H_{34}N_6O_6$			
Piperidine, 1,1',1"-(2',4',6'-trinitro-[1,1'-biphenyl]-2,4,6-triyl)tris-	$CHCl_3$	548(3.64)	44-4649-83
$C_{27}H_{34}O$			
10'-Apo-ε,ψ-carotenal, 15,15'-didehydro-, (6R)-	EPA	245(4.21),306(4.37), 407(4.69)	33-1148-83
$C_{27}H_{34}O_3$			
19-Norpregn-4-en-20-yn-3-one, 17α-[(2-methyl-1-oxo-2,4-hexadienyl)oxy]-, (E,E)-	EtOH	255(4.49)	13-0277-83A
19-Norpregn-4-en-20-yn-3-one, 17α-[(5-methyl-1-oxo-2,4-hexadienyl)oxy]-, (E,E)-	EtOH	246(4.31)	13-0277-83A
$C_{27}H_{34}O_{10}$			
Pentanoic acid, 2-(methoxycarbonyl)-3-(4-hydroxy-3,5-dimethoxyphenyl)-4-[2,6-dimethoxy-4-(1-propenyl)-phenoxy]-, methyl ester	EtOH	267(4.40)	150-2625-83M
$C_{27}H_{34}O_{15}$			
D-Glucitol, 1-C-[2-carboxy-2',3',4,4'-5,6-hexamethoxy-6'-(methoxycarbonyl)-[1,1'-biphenyl]-3-yl]-, intramol. 1,2-ester	MeOH	221(4.67),253(4.20), 301(3.75)	39-1765-83C
$C_{27}H_{36}FN_3O$			
1-Propanone, 3-[4-[2-(4-fluorophenyl)-2-(1-piperidinyl)ethyl]-1-piperazin-yl]-2-methyl-1-phenyl-	MeOH	241(4.09)	73-1173-83
$C_{27}H_{36}N_2O_5Se$			
Cyclohexanemethanol, α,α,4-trimethyl-3-[1-methyl-2-[(2-nitrophenyl)seleno]ethyl]-4-[2-[(2-nitrophenyl)-seleno]ethyl]-	$CHCl_3$	256(4.47),272(4.14)	23-1111-83
$C_{27}H_{36}N_2O_7$			
Vindoline, 16-acetoxy-14,15-dihydro-	EtOH	211(4.59),247(3.85), 301(3.75)	100-0884-83

Compound	Solvent	$\lambda_{max}(\log \epsilon)$	Ref.
$C_{27}H_{36}N_4O_9S$			
OA-6129-A, (4-nitrophenyl)methyl ester	CH_2Cl_2	270(4.02),319(3.92)	158-1473-83
$C_{27}H_{36}N_4O_{10}S$			
OA-6129-B_2, (4-nitrophenyl)methyl ester	CH_2Cl_2	271(4.02),320(4.02)	158-1473-83
$C_{27}H_{36}O_2$			
Furan, 2,2'-(phenylmethylene)bis[5-hexyl-	EtOH	227(4.36)	103-0478-83
D-Homo-18-nor-5α,13β-androstan-3β-ol, 17-methylene-, benzoate	EtOH	228(4.1)	44-1954-83
$C_{27}H_{36}O_3$			
18,19-Dinorpregn-4-en-20-yn-3-one, 17α-[(cyclopentylcarbonyl)oxy]-13-ethyl-	MeOH	245(4.22)	13-0349-83A
18,19-Dinorpregn-4-en-20-yn-3-one, 13-ethyl-17α-[[(3-methylcyclobutyl)-carbonyl]oxy]-	MeOH	242(4.22)	13-0349-83A
19-Norpregn-4-en-20-yn-3-one, 17α-(3-cyclobutyl-1-oxopropoxy)-	MeOH	241(4.25)	13-0291-83A
19-Norpregn-4-en-20-yn-3-one, 17α-[(2,2-dimethyl-1-oxo-4-pentenyl)oxy]-	MeOH	242(4.29)	13-0255-83A
19-Norpregn-4-en-20-yn-3-one, 17α-[[(3-ethylcyclobutyl)carbonyl]oxy]-	MeOH	240(4.20)	13-0291-83A
$C_{27}H_{36}O_5$			
Androstane-3,17-dipropanoic acid, 3,14,17-trihydroxy-α,α'-bis(meth-ylene)-, di-γ-lactone, (5β,14β)-	n.s.g.	210.5(4.42)	48-0587-83
$C_{27}H_{36}O_6$			
Spiro[5.5]undec-3-en-2-one, 3,3'-meth-ylenebis[4-acetoxy-	MeOH	240(4.17)	24-1309-83
$C_{27}H_{36}O_7$			
Ingenol, 3-O-[(Z)-2-methyl-2-butenoyl]-16-O-acetyl-20-deoxy-16-hydroxy-	MeOH	215(4.19)	100-0723-83
$C_{27}H_{36}O_9$			
10,13,16,19,22-Pentaoxatricyclo[22.2-2.$2^{5,8}$]triaconta-5(30),7,24,26,27-pentaene-6,29-dione, 25,27-dimethoxy-	$CHCl_3$	462(2.51)	89-0334-83
	$+Na^+$	478(2.94)	89-0334-83
Prosta-5,7,10,14-tetraen-1-oic acid, 4,12,20-triacetoxy-9-oxo-, methyl ester	EtOH	230(4.04),288(4.14)	88-4433-83
isomer	EtOH	230(4.15),292(4.27)	88-4433-83
isomer 3	EtOH	230(4.09),295(4.08)	88-4433-83
$C_{27}H_{37}BrO_6$			
17,19-Dinoratis-15-ene-4,13,14-tricarb-oxylic acid, 16-[1-(bromomethyl)eth-enyl]-, trimethyl ester, (4α,8α,12α)-	EtOH	234(3.85)	49-1259-83
17,19-Dinoratis-15-ene-4,13,14-tricarb-oxylic acid, 16-(2-bromo-1-methyl-ethenyl)-, trimethyl ester, (4α,8α,12α,13S,14R)-	EtOH	235s(--),245s(--), 252(3.23),262s(--)	49-1259-83
$C_{27}H_{37}FO_6$			
Pregna-1,4-diene-3,20-dione, 9-fluoro-11,21-dihydroxy-16-methyl-, 17-[(1-	n.s.g.	239(4.20)	162-0168-83

Compound	Solvent	$\lambda_{max}(\log \epsilon)$	Ref.
oxopentyl)oxy]- (cont.)			162-0168-83
$C_{27}H_{37}NO$			
Spiro[2.5]oct-1-ene-1-carboxamide, N,N-dicyclohexyl-2-phenyl-	$CHCl_3$	386(3.86)	24-3097-83
$C_{27}H_{37}N_3O_4$			
2-(Diethylamino)ethyl tetraphyllinate	MeOH	228(4.63),299(3.73)	20-0263-83
$C_{27}H_{37}N_3O_7S$			
OA-6129A, phenylmethyl ester	CH_2Cl_2	318(3.87)	158-1473-83
$C_{27}H_{38}N_6O_4$			
2-Pyrrolidinecarboxamide, 1,1'-[1,5-pentanediylbis(1,3(4H)-pyridinedi-ylcarbonyl)]bis-, [S-(R*,R*)]-	$CHCl_3$	349(4.08)	44-1370B-83
$C_{27}H_{38}O_3$			
18,19-Dinorpregn-4-en-20-yn-3-one, 13-ethyl-17α-(2-ethyl-1-oxobutoxy)-	MeOH	243(4.34)	13-0349-83A
18,19-Dinorpregn-4-en-20-yn-3-one, 13-ethyl-17α-[(4-methyl-1-oxo-pentyl)oxy]-	MeOH	240(4.25)	13-0349-83A
18,19-Dinorpregn-4-en-20-yn-3-one, 13-ethyl-17α-[(1-oxohexyl)oxy]-	MeOH	240(4.23)	13-0349-83A
Furan, 2,2'-(2-furanylmethylene)bis[5-heptyl-	EtOH	227(4.34)	103-0478-83
19-Norpregn-4-en-20-yn-3-one, 17α-[(2,2-dimethyl-1-oxopentyl)oxy]-	EtOH	240(4.49)	13-0255-83A
$C_{27}H_{38}O_4$			
Androst-5-eno[17,16-b]furan-5'-one, 5',16α-dihydro-3β-hydroxy-4'-methyl-, 3-tetrahydropyranyl ether	EtOH	222(4.05)	13-0055-83A
Androst-5-eno[17,16-b]furan-5'-one, 4',5',16α,17α-tetrahydro-3β-hydroxy-4'-methylene-, 3-tetrahydropyranyl ether	MeOH	211(4.00)	13-0055-83A
Pregn-4-ene-3,20-dione, 17-(2,2-dimeth-yl-1-oxopropoxy)-6-methylene-	MeOH	260(3.98)	5-0705-83
3,7,11-Tridecatriene-5,6-diol, 1-(6-hy-droxy-2,8-dimethyl-2H-1-benzopyran-2-yl)-4,8,12-trimethyl-, [2R-[2R*(3E,5R*,6R*,7E)]]-	EtOH	232(4.39),266(3.72), 275(3.66),335(3.41)	94-0106-83
	EtOH-KOH	355(--)	94-0106-83
$C_{27}H_{38}O_5$			
24-Norchola-16,20(22)-dien-23-oic acid, 3-acetoxy-5,6-epoxy-, ethyl ester, (3β,5α,6α)-	n.s.g.	272(4.02)	105-0179-83
19-Norchol-5-en-24-oic acid, 4,4,9,14-tetramethyl-3,7,11-trioxo-, (9β,10α)-	n.s.g.	203(3.65),246.5(3.97)	95-1155-83
$C_{27}H_{38}O_6$			
2-Butenoic acid, 2-methyl-, 4a-acetoxy-1a,1b,3,4,4a,5,7a,8,8a,9,10,10a-do-decahydro-3,6,8,8,10a-pentamethyl-5-oxo-2H-cyclopenta[3,4]cyclopropa-[8,9]cycloundeca[1,2-b]oxiren-4-yl ester	MeOH	215(4.09),268(4.19)	102-1791-83
17,19-Dinoratis-15-ene-4,13,14-tricarb-oxylic acid, 16-(1-methylethenyl)-,	EtOH	235s(--),242(4.20), 250s(--)	49-1259-83

Compound	Solvent	λ max (log ε)	Ref.
trimethyl ester, (4α,8α,12α,13S,14R)-			49-1259-83
24-Norchol-20(22)-en-23-oic acid, 3-acetoxy-5,6α:16,17α-diepoxy-, ethyl ester	n.s.g.	217(4.15)	105-0179-83
$C_{27}H_{38}O_{11}$			
Methyl [2,3,4-tri-O-acetyl-(3,5-di-tert-butyl-4-hydroxyphenyl)-β-D-glucopyranosid]uronate [sic]	EtOH	281(3.40)	94-3671-83
$C_{27}H_{38}O_{16}$			
2,7-Dioxabicyclo[4.1.0]heptane-6-carboxylic acid, 5-(2,2-dimethoxyethyl)-4-ethenyl-3-[(2,3,4,6-tetra-O-acetyl-β-D-glucopyranosyl)oxy]-, methyl ester, [1R-(1α,3α,4β,5β,6α)]-	MeOH	215(2.44)(end abs.)	23-0276-83
2H-Pyran-5-carboxylic acid, 4-(2,2-dimethoxyethyl)-3,4-dihydro-3-oxiranyl-2-[(2,3,4,6-tetra-O-acetyl-β-D-glucopyranosyl)oxy]-, methyl ester, [2S-[2α,3β(R*),4β]]-	MeOH	230(3.91)	23-0276-83
$C_{27}H_{38}O_{17}$			
2,7-Dioxabicyclo[4.1.0]heptane-6-carboxylic acid, 5-(2,2-dimethoxyethyl)-4-oxiranyl-3-[(2,3,4,6-tetra-O-acetyl-β-D-glucopyranosyl)oxy]-, methyl ester	MeOH	215(2.19)(end abs.)	23-0276-83
$C_{27}H_{39}NO_3$			
Jervine	n.s.g.	250(4.18),360(1.78)	162-0755-83
$C_{27}H_{40}O_2$			
2H-1-Benzopyran-6-ol, 3,4-dihydro-2,8-dimethyl-2-(4,8,12-trimethyl-3,7,11-tridecatrienyl)-, [R-(E,E)]-	MeOH	296(3.58)	102-2281-83
Cholest-5-en-22-yn-24-one, 3β-hydroxy-	EtOH	225(3.95)	22-0189-83
$C_{27}H_{40}O_5$			
5β,14β-Cholan-24,20ξ-lactone, 3β-acetoxy-14-hydroxy-23-methylene-	n.s.g.	212(4.08)	48-0587-83
Hexadecanoic acid, (7-methoxy-2-oxo-2H-1-benzopyran-4-yl)methyl ester	EtOH	321.5(4.18)	94-3014-83
Propanoic acid, 2,2-dimethyl-, 4-[2-[5-acetoxy-2-(methoxymethyl)-1-cyclohexen-1-yl]ethenyl]-2,3,3a,6,7,7a-hexahydro-7a-methyl-1H-inden-1-yl ester	ether	278.0(4.48)	56-1267-83
Propanoic acid, 2,2-dimethyl-, 4-[[5-acetoxy-2-(methoxymethylene)cyclohexylidene]ethylidene]octahydro-7a-methyl-1H-inden-1-yl ester	ether	272.5(4.24)	56-1267-83
Spirosta-3,5-diene-3β,12β,15α-triol	MeOH	237(4.33)	102-0787-83
$C_{27}H_{40}O_7S$			
Pregn-4-ene-3,20-dione, 17-(2,2-dimethyl-1-oxopropoxy)-21-[(methylsulfonyl)oxy]-	MeOH	241(4.23)	5-0705-83
$C_{27}H_{40}P_2$			
Diphosphene, (2,4,6-trimethylphenyl)-[2,4,6-tris(1,1-dimethylethyl)phenyl]-	CH_2Cl_2	273(3.99),326(3.40), 456(2.34)	35-2495-83

Compound	Solvent	λ_{max}(log ε)	Ref.
$C_{27}H_{41}NO_5$			
5β,14β-Card-20(22)-enolide, 17β-acetoxy-3α-(dimethylamino)-14-hydroxy-	MeOH	217.5(4.20)	13-0171-83B
3β-	MeOH	218.5(4.15)	13-0171-83B
$C_{27}H_{41}N_3$			
1H-Benzimidazolium, 2-(1,4-dimethylpyridinium-2-yl)-1-dodecyl-3-methyl-, diiodide	EtOH	284(4.04)	4-0023-83
$C_{27}H_{42}N_2O_4$			
Androst-4-en-3-one, 17-acetoxy-4-[4-(2-hydroxyethyl)-1-piperazinyl]-	EtOH	245(4.01)	107-1013-83
Androst-4-en-3-one, 17-acetoxy-6-[4-(2-hydroxyethyl)-1-piperazinyl]-	EtOH	239.5(4.18)	107-1013-83
$C_{27}H_{42}O$			
Cholesta-5,7,24-trien-3β-ol	EtOH	262(3.89),272(4.05), 282(4.06),294(3.83)	5-1031-83
18-Norcholesta-8,11,13-trien-3-ol, 12-methyl-	C_6H_{12}	225(4.07)	39-0587-83C
$C_{27}H_{42}O_2$			
Cholesta-5,22-dien-24-one, 3-hydroxy-, (3β,22E)-	EtOH	230(4.12)	22-0189-83
9,10-Secocholesta-5,7,10(19)-trien-3-ol, 6,19-epoxy-, (3β,7E)-	EtOH	273.5(4.20),283(4.26), 295.5(4.12)	44-3477-83
$C_{27}H_{43}O_4P$			
Cholesta-5,7,9-trien-3β-ol, dihydrogen phosphate	n.s.g.	267s(3.65),277(3.76), 288(3.81),300(3.79), 317(3.81),332(3.87), 346(3.69)	44-1417-83
$C_{27}H_{44}O$			
Cholest-4-en-3-one	EtOH	242(4.23)	44-4766-83
Vitamin D_3, 5(E)-	hexane	270(4.46)	78-1123-83
$C_{27}H_{44}O_2$			
Calcifediol	EtOH	265(4.26)	162-0226-83
Cholecalciferol, 1α-hydroxy-	ether	264(4.31)	162-0701-83
Cholest-5-en-7-one, 3-hydroxy-	EtOH	238(4.10)	44-4766-83
Cholest-8-en-7-one, 3-hydroxy-	EtOH	262(3.93)	44-4766-83
$C_{27}H_{44}O_3$			
Cholest-4-en-6-one, 2α,3α-dihydroxy-	MeOH	233(3.93)	104-1116-83
9,10-Secocholesta-5,7,10(19)-triene-1,3,25-triol (calcitriol)	EtOH	264(4.28)	162-0226-83
$C_{27}H_{44}O_4$			
1,2-Benzenediol, 3-heptadecyl-, diacetate	MeOH	210(4.08),257(2.47), 276s(1.64)	36-0792-83
Cholest-7-en-6-one, 3α,5β,14α-trihydroxy-	EtOH	245(3.95)	78-2779-83
$C_{27}H_{44}O_5$			
2-Deoxyecdysone	MeOH	242(4.16)	152-0587-83
$C_{27}H_{44}O_6$			
α-Ecdysone	n.s.g.	243(4.07)	162-0505-83
Panasterone A	MeOH	244(4.09)	35-0130-83

Compound	Solvent	$\lambda_{max}(\log \epsilon)$	Ref.
$C_{27}H_{44}O_7$			
Ajugasterone C	MeOH	243(4.01)	35-0130-83
β-Ecdysone	n.s.g.	243(4.01)	162-0505-83
$C_{27}H_{44}O_9$			
Pseudomonic acid D, methyl ester	EtOH	219(4.27)	39-2655-83C
$C_{27}H_{45}N_5O_8S$			
L-Cysteine, N-[(1,1-dimethylethoxy)-carbonyl]-D-cysteinyl-L-isoleucyl-2,3-didehydroalanyl-L-leucyl-, methyl ester, cyclic (1→5)-sulfide	MeOH	250(3.18)	18-2044-83
$C_{27}H_{45}O_4P$			
Cholesta-5,7-dien-3β-ol, dihydrogen phosphate	n.s.g.	267s(3.77),277(3.90), 287(3.92),279[sic](3.7)	44-1417-83
$C_{27}H_{46}N_2O_4$			
L-Alanine, 3-pentadecyl-1,2-phenylene ester, dihydrochloride	MeOH	219(3.40),258(2.47), 280s(2.02)	36-0792-83
$C_{27}H_{46}O_2$			
2H-1-Benzopyran-6-ol, 3,4-dihydro-2,8-dimethyl-2-(4,8,12-trimethyl-tridecyl)-	EtOH	299(3.52)	94-0106-83

Compound	Solvent	λ_{max}(log ε)	Ref.
$C_{28}H_{10}Cl_{14}NO_2$			
L-Phenylalanine, 4-[bis(pentachlorophenyl)methylene]-2,3,5,6-tetrachlorophenyl ester, hydrobromide (radical)	MeCN	218(5.02),273s(3.77), 336s(3.77),360s(4.19), 382(4.44),514(3.00), 564(3.02)	44-3716-83
$C_{28}H_{14}$			
[7]Circulene	C_6H_{12}	236s(4.44),266s(4.86), 275(5.14),296(4.46), 331(3.91),388s(2.90), 403(2.83)	35-7171-83
Pentaleno[1,2,3-cd:4,5,6-c'd']diphenalene	THF	230s(4.40),282(4.48), 309(4.49),324(4.61), 346(4.70),400(4.04), 442(3.74),468(3.52), 544s(3.69),610(4.24), 636(4.41),666(5.13)	35-5136-83
$C_{28}H_{14}Br_2O_3$			
6H-Phenaleno[1,9-bc]pyran-6-one, 4-(4-bromobenzoyl)-2-(4-bromophenyl)-	pyridine	325(4.04),358(4.10), 419(3.85),484s(4.24), 510(4.33),543s(4.16)	39-1267-83C
$C_{28}H_{14}O_2$			
Heterocoerdianthrone	CH_2Cl_2	255(4.6),290(4.6), 420(3.6),530(4.3), 572(4.5)	149-0587-83A
$C_{28}H_{14}O_4$			
Heterocoerdianthrone endoperoxide	CH_2Cl_2	270(4.5),320s(3.5)	149-0587-83A
$C_{28}H_{16}$			
Cycloocta[1,2,3,4-def:5,6,7,8-d'e'f']-diphenanthrene	C_6H_{12}	253(5.00)	1-0297-83
$C_{28}H_{16}N_2O_6$			
6H-Phenaleno[1,9-bc]pyran-6-one, 2-(2-nitrophenyl)-4-[(2-nitrophenyl)methyl]-	n.s.g.	265(4.57),362.5(3.94), 485(4.18),510(4.22), 543(3.99)	39-1267-83C
$C_{28}H_{16}N_2O_{10}S_2$			
[1,1'-Bianthracene]-3,3'-disulfonic acid, 4,4'-diamino-9,9',10,10'-tetrahydro-9,9',10,10'-tetraoxo-, disodium salt	DMF	505(4.17)	104-0382-83
$C_{28}H_{16}O_2$			
1(4H)-Anthracenone, 4-(4-oxo-1(4H)-anthracenylidene)-	EtOH	234(4.96),276(4.28), 290(4.37),301(4.42), 412(3.70)	73-0112-83
1(4H)-Anthracenone, 4-(10-oxo-9(10H)-anthracenylidene)-	EtOH	252(4.26),275(4.19), 292(4.125),370(3.38)	73-0112-83
$C_{28}H_{16}O_3$			
6H-Phenaleno[1,9-bc]pyran-6-one, 4-benzoyl-2-phenyl-	$CHCl_3$	276(4.57),320(3.95), 364(3.97),482s(4.18), 510(4.34),544(4.19)	39-1267-83C
$C_{28}H_{16}S_8$			
Bis(dibenzotetrathiafulvalene) chloride (radical cation)	MeCN	400s(--),425s(--), 444(4.20),485s(--),	104-1092-83

Compound	Solvent	λ max (log ε)	Ref.
(cont.)		600s(--),640(3.99)	104-1092-83
perchlorate	MeCN	400s(--),425s(--), 444(4.20),485s(--), 600s(--),640(3.99)	104-1092-83
$C_{28}H_{17}NS$			
Thieno[3',4':1,2]indolizino[3,4,5-ab]-isoindole, 4,6-diphenyl-	$CHCl_3$	251(4.64),314(4.43), 365s(--),385(3.86), 446(4.04)	142-1525-83
$C_{28}H_{18}Cl_6O_8$			
1,4-Naphthalenediol, 2,6,7-trichloro-, diacetate, complex with itself	CCl_4	460(3.08)	78-0199-83
$C_{28}H_{18}CoN_4O_2$			
Cobalt, [tetrabenzo[b,f,k,o][1,5,9,13]-tetraazacyclohexadecine-11,13(6H,12H)-dionato(2-)-N^6,N^{12},N^{18},N^{24}]-, (SP-4-3)-	EtOH-DMF	293(4.16),400(3.74), 690(3.15)	12-2395-83
$C_{28}H_{18}CuN_4O_2$			
Copper, [tetrabenzo[b,f,k,o][1,5,9,13]-tetraazacyclohexadecine-11,13(6H,12H)-dionato(2-)-N^6,N^{12},N^{18},N^{24}]-, (SP-4-3)-	EtOH-DMF	302(4.43),516(3.91)	12-2395-83
$C_{28}H_{18}Fe$			
[2]Paracyclo[2]paracyclo[2](1,1'-ferrocenophan-4-ene-1,17-diyne)	dioxan	298f(4.4),365(3.6), 485(2.8)	18-1143-83
$C_{28}H_{18}N_2O_6$			
5-Benzoxazolecarboxylic acid, 2,2'-(1,5-naphthalenediyl)bis-, dimethyl ester	DMF	350(4.52)	5-0931-83
$C_{28}H_{18}N_4NiO_2$			
Nickel, [tetrabenzo[b,f,k,o][1,5,9,13]-tetrazacyclohexadecine-11,13(6H,12H)-dionato(2-)-N^6,N^{12},N^{18},N^{24}]-, (SP-4-3)-	EtOH-DMF	296(4.00),500(3.54)	12-2395-83
$C_{28}H_{18}O_2$			
[9,9'-Bianthracene]-10,10'(9H,9'H)-dione	EtOH	214(4.22),267(4.06)	2-1191-83
	96% H_2SO_4	215(4.64),288(4.15), 298(4.15),348(4.45)	23-1965-83
6H-Phenaleno[1,9-bc]pyran-6-one, 2-phenyl-4-(phenylmethyl)-	$CHCl_3$	263(4.34),344(3.65), 378(3.51),484(3.98), 509(4.12),542(3.98)	39-1267-83C
$C_{28}H_{18}O_3$			
6H-Phenaleno[1,9-bc]pyran-6-one, 4-(hydroxyphenylmethyl)-2-phenyl-	$CHCl_3$-10% NaOH	276(4.55),308(3.96), 345(3.87),374(3.83), 486(4.18),511(4.38), 544(4.32)	39-1267-83C
$C_{28}H_{19}N$			
9-Phenanthrenamine, N-(9-phenanthrenyl)-	CH_2Cl_2	254(5.01),325(4.18)	4-1019-83
$C_{28}H_{20}$			
9H-Fluorene, 9-(5,10-dimethyl-2,4,10,12-cyclotridecatetraene-6,8-diyn-1-ylidene)-, (E,E,Z,Z)-	EtOH	260s(4.6),270(4.6), 305(4.0),440(4.5)	138-1887-83
[2.0.2.0]Metacyclophanediene	C_6H_{12}	282(4.29)	1-0297-83

Compound	Solvent	λ_{max}(log ε)	Ref.
$C_{28}H_{20}I_2$			
1,1'-Biphenyl, 4,4'-bis[2-(4-iodophenyl)ethenyl]-	tetralin	355(4.75)	104-0683-83
$C_{28}H_{20}N_2$			
9-Anthracenecarboxaldehyde, (diphenylmethylene)hydrazone, (E)-	n.s.g.	260(4.9),300s(4.0), 410(4.0)	131-0239-83N
(Z)-	n.s.g.	255(5.1),380f(4.0)	131-0239-83N
Butanedinitrile, bis(bicyclo[5.4.1]-dodeca-2,5,7,9,11-pentaen-4-ylidene)-	EtOH	230(4.8),300(4.7), 408(4.4)	88-2151-83
isomer	EtOH	230(4.7),300(4.5), 310s(4.4),468(4.4)	88-2151-83
$C_{28}H_{20}N_2O_3S$			
Benzeneacetonitrile, α-[2-[6-(4-methoxyphenyl)-2H-thiopyran-2-ylidene]-2-phenylethylidene]-4-nitro-	$CHCl_3$	568(4.25)	97-0147-83
$C_{28}H_{20}N_2O_3Se$			
Benzeneacetonitrile, α-[2-[6-(4-methoxyphenyl)-2H-selenin-2-ylidene]-2-phenylethylidene]-4-nitro-	$CHCl_3$	559(4.17)	97-0147-83
$C_{28}H_{20}N_4$			
1(4H)-Pyridinecarbonitrile, 4,4'-(2,4-diphenyl-1,3-cyclobutanediylidene)-bis-, trans	CH_2Cl_2	307(4.32),321s(--)	5-0069-83
$C_{28}H_{20}N_4O_3$			
Benzamide, N-(1-cyano-2-morpholino-7-oxo-7H-dibenz[f,ij]isoquinolin-6-yl)-	DMF	496(3.95)	2-1197-83
Benzamide, N-(1-cyano-2-morpholino-7-oxo-7H-dibenz[f,ij]isoquinolin-8-yl)-	DMF	442(4.06)	2-1197-83
$C_{28}H_{20}N_4O_9$			
Oxepino[4,5-c]pyridine-3,4-dicarboxylic acid, 1-cyano-, dimethyl ester, 1:1 compd. with 1-cyano-3,4-isoquinoline-dicarboxylic acid dimethyl ester	CD_2Cl_2	250(4.14),271(4.15), 318(3.25),324(3.28)	24-0097-83
$C_{28}H_{20}O_3$			
4H-1-Benzopyran-4-one, 2-(2-methoxy-[1,1'-biphenyl]-3-yl)-6-phenyl-	MeOH	266(4.64),308s(4.23)	18-3519-83
4H-1-Benzopyran-4-one, 2-(2-methoxy-[1,1'-biphenyl]-3-yl)-8-phenyl-	MeOH	248(4.54),316(4.23)	18-3519-83
4H-1-Benzopyran-4-one, 2-(4-methoxy-[1,1'-biphenyl]-3-yl)-6-phenyl-	MeOH	267(4.31),338(4.03)	18-3519-83
4H-1-Benzopyran-4-one, 2-(4-methoxy-[1,1'-biphenyl]-3-yl)-8-phenyl-	MeOH	254(4.52),295(4.21)	18-3519-83
$C_{28}H_{20}O_{10}$			
[2,2'-Binaphthalene]-1,4-dione, 1',4',8,8'-tetraacetoxy-	$CHCl_3$	246(4.55),295(4.09), 337(3.68)	5-0433-83
$C_{28}H_{20}S_2$			
6H-Dibenzo[b,d]thiopyran, 6-(6H-dibenzo[b,d]thiopyran-6-ylidenemethyl)-6-methyl-	MeOH	239s(4.58),318s(3.63)	12-0795-83

Compound	Solvent	$\lambda_{max}(\log \epsilon)$	Ref.
$C_{28}H_{21}Cl_3N_6O_8$			
Benzenamine, 4,4'-(2,2,2-trichloroethylidene)bis[N-(4-methylphenyl)-2,6-dinitro-	MeOH	205.4(4.65),425.5(4.04)	56-1357-83
$C_{28}H_{21}Cl_3N_6O_{10}$			
Benzenamine, 4,4'-(2,2,2-trichloroethylidene)bis[N-(2-methoxyphenyl)-2,6-dinitro-	MeOH	208.3(4.64),425.8(4.03)	56-1357-83
$C_{28}H_{21}NOS_3$			
Ethane(dithioic) acid, (diphenylamino)-thioxo-, 2-[1,1'-biphenyl]-4-yl-2-oxoethyl ester	CH_2Cl_2	280(4.55),344s(3.87), 437s(3.13)	97-0247-83
$C_{28}H_{21}N_3$			
Benzonitrile, 4-(4,5-dihydro-1,4,5-triphenyl-1H-pyrazol-3-yl)-, trans	C_6H_{12}	396(4.43)	12-1649-83
	MeOH	395(4.38)	12-1649-83
Benzonitrile, 4-(4,5-dihydro-1,5,5-triphenyl-1H-pyrazol-3-yl)-	C_6H_{12}	396(4.45)	12-1649-83
	MeOH	395(4.46)	12-1649-83
Benzonitrile, 4-(4,5-dihydro-3,5,5-triphenyl-1H-pyrazol-3-yl)-	C_6H_{12}	359(4.54)	12-1649-83
	MeOH	356(4.59)	12-1649-83
$C_{28}H_{22}$			
1,1'-Biphenyl, 4,4'-bis(2-phenylethenyl)-	dioxan	352(4.87)	104-0683-83
1,3,9,11-Cyclotridecatetraene-5,7-diyne, 13-(diphenylmethylene)-4,9-dimethyl-	CH_2Cl_2	267(4.50),308(4.34), 405(4.37)	39-2987-83C
Naphthalene, 1,2-dihydro-1,1,4-triphenyl-	CH_2Cl_2	247(3.78),344(2.90)	152-0105-83
$C_{28}H_{22}ClNO_6$			
Benzocycloocten-6,9-imine-7,8-dicarboxylic acid, 11-(4-chlorophenyl)-5,6,7,8,9,10-hexahydro-5,10-dioxo-6-phenyl-, dimethyl ester	benzene	303(3.65),352(2.64), 419(2.20)	44-4968-83
	MeCN	232(4.43),254(4.36), 301(3.57),349(2.54), 412(2.18)	44-4968-83
$C_{28}H_{22}Fe$			
Ferrocene, 1,1'-bis[(4-methylphenyl)-ethynyl]-	dioxan	255(4.2),298(4.3), 450(2.9)	18-1143-83
Ferrocene, 1,1'-[1,2-ethenediylbis(4,1-phenylene-2,1-ethenediyl)]-, (Z,Z,Z)-	EtOH	230(4.56),285s(--), 475(2.70)	1-0693-83
$C_{28}H_{22}N_2O$			
Benzaldehyde, 4-(4,5-dihydro-3,5,5-triphenyl-1H-pyrazol-1-yl)-	C_6H_{12}	370(4.65)	12-1649-83
	MeOH	385(4.66)	12-1649-83
$C_{28}H_{22}N_2O_2$			
Benzoxazole, 2,2'-(1,2-naphthalenediyl)bis[5,6-dimethyl-	DMF	335(4.39)	5-0931-83
Benzoxazole, 2,2'-(1,3-naphthalenediyl)bis[5,6-dimethyl-	DMF	323(4.65)	5-0931-83
Benzoxazole, 2,2'-(1,5-naphthalenediyl)bis[5,6-dimethyl-	DMF	357(4.50)	5-0931-83
Benzoxazole, 2,2'-(1,6-naphthalenediyl)bis[5,6-dimethyl-	DMF	327(4.73)	5-0931-83
Benzoxazole, 2,2'-(1,7-naphthalenediyl)bis[5,6-dimethyl-	DMF	358(4.41)	5-0931-83

Compound	Solvent	$\lambda_{max}(\log \epsilon)$	Ref.
Benzoxazole, 2,2'-(2,3-naphthalenediyl)bis[5,6-dimethyl-	DMF	294(4.62)	5-0931-83
Benzoxazole, 2,2'-(2,7-naphthalenediyl)bis[5,6-dimethyl-	DMF	328(4.83)	5-0931-83
Pyridinium, 2,4,6-triphenyl-, 1-cyano-2-ethoxy-2-oxoethylide	toluene	568(3.15)	142-0623-83
	EtOH	490(2.96)	142-0623-83
	CH_2Cl_2	544(3.38)	142-0623-83
$C_{28}H_{22}N_2O_5S$			
Indolo[2,3-a]carbazole-11(12H)-carboxylic acid, 12-[(4-methoxyphenyl)sulfonyl]-5-methyl-, methyl ester	EtOH	214(4.44),246(4.42), 274(4.13),284(4.11), 306(4.16),318(4.13), 340(3.53)	78-3725-83
$C_{28}H_{22}N_2O_6S_2$			
Benzo[1,2-b:4,5-b']difuran-3,7-dicarboxylic acid, 2,6-dihydro-2,6-bis-[(phenylthio)imino]-, diethyl ester, (Z,Z)-	$CHCl_3$	659(4.84)	150-0480-83M
$C_{28}H_{22}N_4$			
1H-Pyrazole-4,5-diamine, N^5,1,3-triphenyl-N^4-(phenylmethylene)-	EtOH	242(3.57),293(3.59)	4-1501-83
$C_{28}H_{22}N_4O_3$			
Benzo[f]quinoline, 4-(4-ethoxyphenyl)-3,4-dihydro-3-[[(4-nitrophenyl)azo]-methylene]-	neutral	596(4.42)	124-0068-83
	protonated	473(4.15)	124-0068-83
Quinoline, 6-ethoxy-1,2-dihydro-1-(1-naphthalenyl)-2-[[(4-nitrophenyl)-azo]methylene]-	neutral	584(4.30)	124-0068-83
	protonated	476(4.25)	124-0068-83
$C_{28}H_{22}N_6O_2$			
Benzo[1,2,3-de:4,5,6-d',e']diquinoline-1,7-dicarbonitrile, 2,8-dimorpholino-	DMF	415(4.13),485(4.13)	2-1197-83
$C_{28}H_{22}O_3$			
4H-1-Benzopyran-4-one, 2,3-dihydro-2-(2-methoxy[1,1'-biphenyl]-3-yl)-6-phenyl-	MeOH	248(4.69),346(3.35)	18-3519-83
4H-1-Benzopyran-4-one, 2,3-dihydro-2-(2-methoxy[1,1'-biphenyl]-3-yl)-8-phenyl-	MeOH	241(4.57),336(3.65)	18-3519-83
4H-1-Benzopyran-4-one, 2,3-dihydro-2-(4-methoxy[1,1'-biphenyl]-3-yl)-6-phenyl-	MeOH	251(4.69),343(3.36)	18-3519-83
4H-1-Benzopyran-4-one, 2,3-dihydro-2-(4-methoxy[1,1'-biphenyl]-3-yl)-8-phenyl-	MeOH	242(4.50),336(3.63)	18-3519-83
2-Propen-1-one, 1-(2-hydroxy[1,1'-biphenyl]-3-yl)-3-(2-methoxy[1,1'-biphenyl]-3-yl)-	MeOH	249(4.48),334(4.43)	18-3519-83
2-Propen-1-one, 1-(2-hydroxy[1,1'-biphenyl]-3-yl)-3-(4-methoxy[1,1'-biphenyl]-3-yl)-	MeOH	255(4.48),318(4.33), 392(4.26)	18-3519-83
2-Propen-1-one, 1-(4-hydroxy[1,1'-biphenyl]-3-yl)-3-(2-methoxy[1,1'-biphenyl]-3-yl)-	MeOH	256(4.58),328(4.35)	18-3519-83
2-Propen-1-one, 1-(4-hydroxy[1,1'-biphenyl]-3-yl)-3-(4-methoxy[1,1'-biphenyl]-3-yl)-	MeOH	255(4.72),314s(4.08), 394(3.98)	18-3519-83

Compound	Solvent	λ_{max}(log ϵ)	Ref.
$C_{28}H_{22}O_{10}$			
7H-Furo[3,2-g][1]benzopyran-7-one, 9,9'-[1,2-ethanediylbis(oxy-2,1-ethanediyloxy)]bis-	EtOH	218(4.70),248(4.65), 263(4.41),300(4.36)	39-1807-83B
$C_{28}H_{22}S_2$			
Benzene, 1,1'-[4,4-bis(phenylthio)-1,3-butadienylidene]bis-	EtOH	249(4.59),351(4.77)	118-0383-83
Benzene, 1,1'-[3-[bis(phenylthio)methylene]-1-propene-1,3-diyl]bis-, (E)-	EtOH	277(4.32),287(4.29), 353(4.70)	118-0383-83
$C_{28}H_{23}BiBrNO$			
Bismuth, bromo(2-methyl-8-quinolinolato-N^1,O^8)triphenyl-, (OC-6-24)-	benzene	355s(3.53)	101-0317-83M
$C_{28}H_{23}BiClNO$			
Bismuth, chloro(2-methyl-8-quinolinolato-N^1,O^8)triphenyl-, (OC-6-24)-	benzene	355(3.36)	101-0317-83M
$C_{28}H_{23}ClO_2$			
1-Isobenzofuranol, 1-(4-chlorophenyl)-1,3-dihydro-3,3-bis(4-methylphenyl)-	hexane	254(2.84),259(2.92), 266(2.97),271s(2.90), 273s(2.87)	104-0553-83
$C_{28}H_{23}CoN_4$			
Cobalt(1+), (6,11,12,13-tetrahydrotetrabenzo[b,f,k,o][1,5,9,13]tetraazacyclohexadecinato-N^6,N^{12},N^{18},N^{24})-, (SP-4-3)-, perchlorate	EtOH-DMF	288(4.26),380s(3.86)	12-2395-83
$C_{28}H_{23}CuN_4$			
Copper(1+), (6,11,12,13-tetrahydrotetrabenzo[b,f,k,o][1,5,9,13]tetraazacyclohexadecinato-N^6,N^{12},N^{18},N^{24})-, (SP-4-3)-, perchlorate	EtOH-DMF	287(4.45),527(3.76)	12-2395-83
$C_{28}H_{23}F_3N_2O_2$			
2H-Carbazol-2-one, 3,4,4a,9-tetrahydro-9-methyl-4a-[2,2,2-trifluoro-1-(2-hydroxy-9-methyl-9H-carbazol-3-yl)ethyl]-	EtOH	214(4.39),238(4.66), 264(4.35),300(4.22), 326(4.20),334(4.22)	39-2425-83C
$C_{28}H_{23}NO_3$			
1,4-Epoxynaphthalene-2-carbonitrile, 1,2,3,4-tetrahydro-3-(3-methyl-1,2-dioxobutyl)-1,4-diphenyl-, (1α,2α,3α,4α)-	MeOH	253(3.93)	12-0963-83
(1α,2β,3α,4α)-	MeOH	248(3.98)	12-0963-83
$C_{28}H_{23}NO_4$			
Spiro[9H-fluorene-9,1'(10'bH)-pyrrolo-[2,1-a]isoquinoline]-2',3'-dicarboxylic acid, 2',3'-dihydro-, dimethyl ester, cis	CH_2Cl_2?	228(4.41),265(4.29), 311(4.08)	24-3915-83
	betaine	442(--),587(--)	24-3915-83
$C_{28}H_{23}N_3O_4$			
9,10-Anthracenedione, 1-amino-4-[(4'-amino-3,3'-dimethoxy[1,1'-biphenyl]-4-yl)amino]-	DMF	572s(4.18),617(4.28)	2-0808-83

Compound	Solvent	λ_{max}(log ϵ)	Ref.
$C_{28}H_{23}N_4Ni$			
Nickel(1+), (6,11,12,13-tetrahydro-tetrabenzo[b,f,k,o][1,5,9,13]tetra-azacyclohexadecinato-N^6,N^{12},N^{18},N^{24})-, (SP-4-3)-, perchlorate	EtOH-DMF	296(4.24),354(3.98), 442(3.52),524(3.54)	12-2395-83
$C_{28}H_{23}N_5$			
1H-Pyrazol-5-amine, 1,3-diphenyl-4-(phenylazo)-N-(phenylmethyl)-	EtOH	265(3.30),280(3.48), 372(4.43)	4-1501-83
1H-Pyrazol-5-amine, N-(4-methylphenyl)-1,3-diphenyl-4-(phenylazo)-	EtOH	245(4.39),385(4.18)	4-1501-83
$C_{28}H_{23}N_5Zn$			
Zinc, [11,12,13,14-tetrahydro-6H-tetra-benzo[b,h,l,o][1,4,7,10,14]pentaaza-cycloheptadecinato(2-)-N^6,N^{11},N^{14}-N^{19},N^{25}]-	DMF	380(3.67)	12-2395-83
$C_{28}H_{23}O_8P$			
Phosphoric acid, bis(4-benzoyl-3-hy-droxyphenyl) ethyl ester	EtOH	270(3.0),320(1.3)	65-0653-83
$C_{28}H_{24}$			
1-Butene, 1,1,3,3-tetraphenyl-	CH_2Cl_2	253(4.23)	152-0105-83
[1.1.1.1]Paracyclophane	C_6H_{12}	230(4.7),270(3.3), 279(3.2)	88-5277-83
Trideca-3,5,8,10-tetraene-1,12-diyne, 3,11-dimethyl-7-(diphenylmethylene)	CH_2Cl_2	259(4.63),272s(4.54), 333(4.59),365(4.66)	39-2987-83C
$C_{28}H_{24}BrO$			
2H-Furylium, 3-[4-(4-bromophenyl)-2-phenyl-1,3-butadienyl]-2,2-dimethyl-5-phenyl-, perchlorate	MeCN HOAc	520(4.79) 533(--)	56-0579-83 56-0579-83
$C_{28}H_{24}ClN_3O$			
Quinolinium, 2-[5-chloro-3-(3-ethyl-benzoxazolium-2-yl)-1H-indol-2-yl]-1-ethyl-, diiodide	EtOH	207(5.20),238(5.09), 295(4.81),460(4.00)	80-0725-83
$C_{28}H_{24}ClO$			
2H-Furylium, 3-[4-(2-chlorophenyl)-2-phenyl-1,3-butadienyl]-2,2-dimethyl-5-phenyl-, perchlorate	MeCN HOAc	504(4.66) 517(--)	56-0579-83 56-0579-83
2H-Furylium, 3-[4-(3-chlorophenyl)-2-phenyl-1,3-butadienyl]-2,2-dimethyl-5-phenyl-, perchlorate	MeCN HOAc	505(4.56) 517(--)	56-0579-83 56-0579-83
2H-Furylium, 3-[4-(4-chlorophenyl)-2-phenyl-1,3-butadienyl]-2,2-dimethyl-5-phenyl-, perchlorate	MeCN HOAc	521(4.74) 536(--)	56-0579-83 56-0579-83
$C_{28}H_{24}CuN_5$			
Copper(1+), (11,12,13,14-tetrahydro-6H-tetrabenzo[b,h,l,o][1,4,7,10,14]-pentaazacycloheptadecinato-N^6,N^{11}-N^{14},N^{19},N^{25})-, perchlorate	DMF	440(3.54)	12-2395-83
$C_{28}H_{24}FO$			
2H-Furylium, 3-[4-(2-fluorophenyl)-2-phenyl-1,3-butadienyl]-2,2-dimethyl-5-phenyl-	MeCN HOAc	506(4.74) 520(--)	56-0579-83 56-0579-83

Compound	Solvent	$\lambda_{max}(\log \epsilon)$	Ref.
2H-Furylium, 3-[4-(3-fluorophenyl)-2-phenyl-1,3-butadienyl]-2,2-dimethyl-5-phenyl-, perchlorate	MeCN	505(4.74)	56-0579-83
	HOAc	517(--)	56-0579-83
2H-Furylium, 3-[4-(4-fluorophenyl)-2-phenyl-1,3-butadienyl]-2,2-dimethyl-5-phenyl-, perchlorate	MeCN	522(4.70)	56-0579-83
	HOAc	533(--)	56-0579-83
$C_{28}H_{24}N$			
Benzo[h]quinolinium, 1-cyclopropyl-5,6-dihydro-2,4-diphenyl-, tetrafluoroborate	EtOH	353(4.28)	39-1435-83B
$C_{28}H_{24}N_2O_2$			
Benzenemethanamine, α-(nitrophenylmethylene)-N,N-bis(phenylmethyl)-, (Z)-	EtOH	410(4.27)	22-0339-83
9H-Carbazole, 1-methoxy-9-[(1-methoxy-9H-carbazol-3-yl)methyl]-3-methyl- (bismurrayafoline A)	MeOH	228(4.81),244(4.95), 253s(4.89),284s(4.15), 293(4.30),340(3.92)	94-4202-83
Cinnoline, 2-benzoyl-1,2,3,4-tetrahydro-4-(4-methoxyphenyl)-1-phenyl-	$CHCl_3$	248(4.12),268(4.13)	97-0028-83
3-Pyridinecarboxylic acid, 6-[1,1'-biphenyl]-4-yl-5-cyano-1,4-dihydro-2-methyl-4-phenyl-, ethyl ester	EtOH	209(4.26),285(4.42), 360(3.73)	73-3123-83
$C_{28}H_{24}N_2O_6$			
Cyclohept[1,2,3-hi]imidazo[2,1,5-cd]-indolizine-3,4,9-tricarboxylic acid, 2-phenyl-, triethyl ester	EtOH	248(4.34),280s(4.46), 292(4.50),310s(4.35), 381(4.43),427s(4.00)	18-3703-83
$C_{28}H_{24}N_4OS$			
2H-Thiopyran-2-imine, 3-(1H-benzimidazol-2-yl)-5-[1,1'-biphenyl]-4-yl-6-morpholino-, monoperchlorate	MeCN	234s(4.37),267(4.32), 295(4.37),328(4.27), 478(4.32)	97-0403-83
$C_{28}H_{24}N_4O_3$			
Quinolinium, 1-ethyl-2-[3-(3-ethylbenzoxazolium-2-yl)-5-nitro-1H-indol-2-yl]-, diiodide	EtOH	208(5.50),238(5.58), 291(5.18),465(4.95), 486(4.65)	80-0725-83
$C_{28}H_{24}N_6$			
Imidazo[1,2-a]pyridinium, 1,1'-azobis-[3-methyl-2-phenyl-, dibromide	H_2O	285(4.04),297(4.34)	162-0568-83
$C_{28}H_{24}O$			
Furan, tetrahydro-2,2,5,5-tetraphenyl-	EtOH	252(2.88),258(2.94)	152-0105-83
$C_{28}H_{24}O_2$			
Bicyclo[3.2.0]hept-2-en-6-one, 4-[1-(4-methoxyphenyl)ethylidene]-7,7-diphenyl-, (E)-	C_6H_{12}	282(4.152)	64-0504-83B
	MeCN	280(4.317)	64-0504-83B
(Z)-	C_6H_{12}	285(4.274)	64-0504-83B
	MeCN	285(4.579)	64-0504-83B
1-Isobenzofuranol, 1,3-dihydro-3,3-bis(4-methylphenyl)-1-phenyl-	hexane	254s(3.01),258(3.09), 266(3.12),271s(2.97), 275s(2.93)	104-0553-83
$C_{28}H_{24}O_4$			
Riccardin B	EtOH	233(3.74),280(3.64)	44-2164-83
$C_{28}H_{24}O_6$			
2-Oxabicyclo[4.2.0]oct-4-en-3-one, 5-	MeOH	251(4.52),283s(--)	64-0658-83B

Compound	Solvent	λ_{max}(log ϵ)	Ref.
methoxy-8-(4-methoxy-2-oxo-2H-pyran-6-yl)-7-phenyl-1-(2-phenylethenyl)-, [1α(E),6α,7β,8α]-(±)- (cont.)			64-0658-83B
$C_{28}H_{24}O_9$			
β-D-Ribofuranose, 1-acetate 2,3,5-tribenzoate	EtOH	274(3.48),281(3.38)	94-3074-83
$C_{28}H_{24}O_{11}$			
5,12-Naphthacenedione, 9-acetyl-6,7,9-11-tetraacetoxy-7,8,9,10-tetrahydro-	EtOH	256(4.32),275(4.03), 339(3.29)	35-1608-83
$C_{28}H_{25}BrN_4O$			
Benzenecarbohydrazonic acid, N-(4-bromophenyl)-, [(4-ethoxyphenyl)-methylene]phenylhydrazide	EtOH	225(4.39),267(4.21), 349(4.44)	70-1439-83
	dioxan	225(4.44),267(4.24), 352(4.41)	70-1439-83
	MeCN	226(4.38),245(4.31), 297(4.26),341(4.50)	70-1439-83
$C_{28}H_{25}NO_6$			
Acetamide, N-(5-acetoxy-9,10-dihydro-8-hydroxy-9,10-dioxo-1-anthracenyl)-N-(4-butylphenyl)-	benzene	421(3.79)	104-0745-83
9,10-Anthracenedione, 1,4-diacetoxy-5-[(4-butylphenyl)amino]-	benzene	521(3.83)	104-0745-83
$C_{28}H_{25}N_2O_3P$			
3-Oxepincarboxylic acid, 5-cyano-6,7-dihydro-7-methyl-2-[(triphenylphosphoranylidene)amino]-, methyl ester	$CHCl_3$	267(4.08),275(4.09), 346(4.43)	24-1691-83
$C_{28}H_{25}N_3O$			
Quinolinium, 1-ethyl-2-[3-(3-ethylbenzoxazolium-2-yl)-1H-indol-2-yl]-, diiodide	EtOH	210(5.11),240(5.00), 300(4.60),475(3.75)	80-0725-83
$C_{28}H_{25}N_5O_4$			
Uridine, 3'-azido-2',3'-dideoxy-5'-O-(triphenylmethyl)-	MeOH	260(4.00)	87-0544-83
$C_{28}H_{25}O$			
2H-Furylium, 3-(2,4-diphenyl-1,3-butadienyl)-2,2-dimethyl-5-phenyl-, perchlorate	MeCN	520(4.72)	56-0579-83
$C_{28}H_{25}O_2$			
2H-Furylium, 3-[4-(2-hydroxyphenyl)-1,3-butadienyl]-2,2-dimethyl-5-phenyl-, perchlorate	MeCN	560(4.70)	56-0579-83
	HOAc	582(--)	56-0579-83
2H-Furylium, 3-[4-(3-hydroxyphenyl)-1,3-butadienyl]-2,2-dimethyl-5-phenyl-, perchlorate	MeCN	529(4.70)	56-0579-83
	HOAc	543(--)	56-0579-83
2H-Furylium, 3-[4-(4-hydroxyphenyl)-1,3-butadienyl]-2,2-dimethyl-5-phenyl-, perchlorate	MeCN	588(4.73)	56-0579-83
	HOAc	619(--)	56-0579-83
$C_{28}H_{26}Cl_2N_6S$			
4H-[1,2,4]Triazolo[4,3-a][1,4]benzodiazepine, 8-chloro-6-(2-chlorophenyl)-1-[4-[2-(phenylthio)ethyl]piperazino]-	MeOH	245(4.26),302s(3.16)	73-2395-83

Compound	Solvent	λ_{max}(log ϵ)	Ref.
$C_{28}H_{26}FeN_8O_6$			
Iron, bis[4-pyridinecarboxylic acid [[3-hydroxy-5-(hydroxymethyl)-2-methyl-4-pyridinyl]methylene]hydrazidato]-	EtOH	212(4.12),308(4.53), 368(4.35),465(3.79)	87-0298-83
$C_{28}H_{26}N$			
Benzo[h]quinolinium, 5,6-dihydro-1-(1-methylethyl)-2,4-diphenyl-, tetrafluoroborate	pentanol	350(4.08)	39-1443-83B
	HOAc	350(4.14)	39-1443-83B
	C_6H_5Cl	350(4.23)	39-1443-83B
	CF_3COOH	350(4.28)	39-1443-83B
$C_{28}H_{26}N_2$			
Pyridine, 4,4'-(2,4-diphenyl-1,3-cyclobutanediylidene)bis[1,4-dihydro-1-methyl-	MeCN-EtOH	333(4.57),346(4.54)	5-0069-83
	protonated	313(4.24),326(4.17)	5-0069-83
diprotonated	MeOH	220(4.66),257(4.08)	5-0069-83
	MeCN-EtOH	257(4.09)	5-0069-83
Pyridinium, 4,4'-[1,2-bis(phenylmethylene)-1,2-ethanediyl]bis[1-methyl-, perchlorate, (E,E)-	MeCN	243(4.27),340(4.38)	5-0069-83
(Z,Z)-	MeCN	260(4.31),298(4.32)	5-0069-83
Pyridinium, 4,4'-(2,4-diphenylbicyclo[1.1.0]butane-1,3-diyl)bis[1-methyl-, trans, diperchlorate	MeCN	274(4.21),322(4.17)	5-0069-83
$C_{28}H_{26}N_2O_4S_4$			
3-Thiophenecarboxylic acid, 5-[4-(ethoxycarbonyl)-3-methyl-5-[(phenylthio)imino]-2(5H)-thienylidene]-2,5-dihydro-4-methyl-2-[(phenylthio)imino]-, ethyl ester, (E,Z,Z)-	$CHCl_3$	544(4.61)	150-0028-83S +150-0480-83M
$C_{28}H_{26}N_2O_5$			
2,4(1H,3H)-Pyrimidinedione, 1-[2-deoxy-5-O-(triphenylmethyl)-β-D-threo-pentofuranosyl]-	MeOH	260(4.00)	87-0544-83
$C_{28}H_{26}N_2O_7$			
Uridine, 2'-deoxy-5-(1-propynyl)-, 3',5'-bis(4-methylbenzoate)	pH 7.0	237(4.59),283(4.07)	87-0661-83
$C_{28}H_{26}N_4O$			
Benzenecarbohydrazonic acid, N-phenyl-, [(4-ethoxyphenyl)methylene]phenylhydrazide	EtOH	227(4.36),246(4.30), 294(4.14),341(4.41)	70-1439-83
	dioxan	227(4.36),253(4.18), 349(4.36)	70-1439-83
	MeCN	227(4.44),261(4.26), 351(4.39)	70-1439-83
$C_{28}H_{26}N_6O_3S_3$			
Benzeneacetic acid, α-[[5-methyl-4-oxo-2-(phenylamino)pyrimido[4,5-d]thiazol-6-yl]thio]-, S-benzylisothiouronium salt	MeOH	264s(--),290(4.18), 300(4.18),343(4.33)	2-0243-83
$C_{28}H_{26}N_6O_{10}S_2$			
Pyridinium, 1-[[2-carboxy-7-[[[[(5-carboxy-1H-imidazol-4-yl)carbonyl]amino]phenylacetyl]amino]-8-oxo-5-thia-1-azabicyclo[4.2.0]oct-2-en-3-	H_2O	257(4.35)	158-0242-83

Compound	Solvent	λmax(log ε)	Ref.
yl]methyl]-4-(2-sulfoethyl)-, hydroxide, inner salt, sodium salt, [6R-(6α,7β)]- (cont.)			158-0242-83
$C_{28}H_{26}O_{16}$			
Taxillusin	EtOH	292(4.37)	18-0542-83
$C_{28}H_{27}ClO_3$			
9,10-Anthracenedione, 1-[3,5-bis(1,1-dimethylethyl)-4-hydroxyphenyl]-4-chloro-	EtOH	202(4.93),254(4.82), 325(4.08),445s(3.17)	104-0144-83
$C_{28}H_{27}N_2OS$			
Thiazolium, 3-(2-[1,1'-biphenyl]-4-yl-2-oxoethyl)-2-[2-[4-(dimethylamino)-phenyl]ethenyl]-4-methyl-, perchlorate	EtOH	495(4.05)	65-2424-83
$C_{28}H_{27}N_3O_2$			
2-Naphthalenecarboxamide, 4-[(4-butyl-phenyl)azo]-3-hydroxy-N-methyl-N-phenyl-	EtOH	235(4.542),330(3.881), 425(4.064),510(4.27), 526(4.235)	104-1145-83
	CCl_4	330(3.854),420(4.072), 495(4.236),520(4.20)	104-1145-83
$C_{28}H_{27}N_3O_6Si$			
1,2-Diazabicyclo[5.2.0]nona-3,5-diene-4-carboxylic acid, 2-benzoyl-8-(1,3-dihydro-1,3-dioxo-2H-isoindol-2-yl)-9-oxo-, 2-(trimethylsilyl)ethyl ester, cis-(±)-	MeOH	305(4.00)	5-1374-83
1,2-Diazabicyclo[5.2.0]nona-3,5-diene-6-carboxylic acid, 2-benzoyl-8-(1,3-dihydro-1,3-dioxo-2H-isoindol-2-yl)-9-oxo-, 2-(trimethylsilyl)ethyl ester, cis-(±)-	MeOH	325(4.08)	5-1374-83
$C_{28}H_{27}N_3O_7$			
1H-Imidazole-4-carboxylic acid, 5-(cyanomethyl)-1-[2-deoxy-3,5-bis-O-(4-methylbenzoyl)-α-D-erythro-pentofuranosyl]-, methyl ester	$CHCl_3$	245(4.53)	87-0286-83
β-	$CHCl_3$	244(4.57)	87-0286-83
1H-Imidazole-5-carboxylic acid, 4-(cyanomethyl)-1-[2-deoxy-3,5-bis-O-(4-methylbenzoyl)-β-D-erythro-pentofuranosyl]-, methyl ester	$CHCl_3$	247(4.58)	87-0286-83
$C_{28}H_{28}N$			
Pyridinium, 1-butyl-4-(2-methylphenyl)-2,6-diphenyl-, tetrafluoroborate	DMSO	303(4.40)	150-0301-83M
Pyridinium, 2-(2,5-dimethylphenyl)-1-(1-methylethyl)-4,6-diphenyl-, tetrafluoroborate	DMSO	300(4.27)	150-0301-83M
$C_{28}H_{28}N_4$			
Pyridinium, 4,4'-[3,4-bis(1-methyl-4(1H)-pyridinylidene)-1-cyclobutene-1,2-diyl]bis[1-methyl-, diiodide	EtOH-NaOEt	575s(--),632(4.68)	5-0642-83

Compound	Solvent	λ_{max}(log ϵ)	Ref.
$C_{28}H_{28}N_4O_5S$			
7H-Pyrrolo[2,3-d]pyrimidin-4-amine, 7-[2-deoxy-3,5-bis-O-(4-methylbenzoyl)-β-D-erythro-pentofuranosyl]-2-(methylthio)-	MeOH	237(4.73),281(4.20)	5-0876-83
$C_{28}H_{28}N_4O_6$			
Imidazo[4,5-d][1,3]diazepin-8(3H)-one, 3-[2-deoxy-3,5-bis-O-(4-methylbenzoyl)-β-D-erythro-pentofuranosyl]-6,7-dihydro-5-methyl-	MeOH	234(4.70),299(3.53), 346(3.66)	87-1478-83
$C_{28}H_{28}N_4S$			
Spiro[12H-benzothiazolo[2,3-b]quinazoline-12,2'-quinolin]-3-amine, N,N,1'-triethyl-, tetrafluoroborate	64% MeOH-pH 6.23	271(4.44),293(4.41), 471(4.00)	33-2165-83
$C_{28}H_{28}N_6$			
Propanedinitrile, 2,2'-[13,15-bis(dimethylamino)tricyclo[10.2.2.2^{5,8}]octadeca-5(18),7,12,14,15-pentaene-5,17-diylidene]bis-	hexane	827(3.44)	24-2835-83
	C_6H_{12}	830(3.49)	24-2835-83
	benzene	939(3.54)	24-2835-83
	MeOH	979(3.44)	24-2835-83
	acetone	975(3.53)	24-2835-83
	MeCN	1033(3.55)	24-2835-83
	CH_2Cl_2	360(4.3),450s(4.1), 1022(3.53)	24-2835-83
	CCl_4	886(3.53)	24-2835-83
	DMSO	1094(3.55)	24-2835-83
	THF	916(3.50)	24-2835-83
$C_{28}H_{28}O_3$			
9,10-Anthracenedione, 1-[3,5-bis(1,1-dimethylethyl)-4-hydroxyphenyl]-	EtOH	205(4.72),255(4.78), 329(3.86),458s(2.78)	104-0144-83
9,10-Anthracenedione, 2-[3,5-bis(1,1-dimethylethyl)-4-hydroxyphenyl]-	EtOH	208(4.64),250(4.61), 301(4.34),408(3.81)	104-0144-83
$C_{28}H_{28}O_9$			
Chrysomycin A	$CDCl_3$	217(3.97),251(4.47), 265(4.36),307(4.53), 325(4.16),337(4.06), 392(4.10)	158-83-113
$C_{28}H_{28}O_{10}$			
D-ribo-Oct-3-en-2-ulose, 3-acetyl-1,3,4-trideoxy-, 5,7-diacetate 6,8-dibenzoate	MeOH	245(4.23)	87-0030-83
$C_{28}H_{29}Cl_4NO_5$			
Acetic acid, dichloro-, 3,3-dichloro-3,4-dihydro-5,5-dimethyl-4-(methylphenylamino)-2-oxo-8-pentyl-2H,5H-pyrano[3,2-c][1]benzopyran-10-yl ester	EtOH	242(4.32),281(4.03), 320(3.76)	161-0775-83
$C_{28}H_{29}NO_3$			
9,10-Anthracenedione, 1-amino-4-[3,5-bis(1,1-dimethylethyl)-4-hydroxyphenyl]-	EtOH	206(4.63),250(4.65), 316(3.75),506(3.76)	104-0144-83
$C_{28}H_{29}NO_7$			
Spiro[cyclohexane-1,13'-[10H-7a,10]eth-	MeOH	240(4.34),305(3.81),	39-0515-83C

Compound	Solvent	λ_{max}(log ε)	Ref.
ano[4H]pyrido[3,2,1-jk]carbazole]-8',9',11'-tricarboxylic acid, 5',6'-dihydro-12'-oxo-, trimethyl ester		325(3.73),470(3.99)	39-0515-83C
$C_{28}H_{29}N_4S$			
Quinolinium, 4-[2-(2-benzothiazolyl-amino)-4-(diethylamino)phenyl]-1-ethyl-, tetrafluoroborate	64% MeOH-pH 8.72	275(4.49),493(4.09)	33-2165-83
$C_{28}H_{30}N_4$			
21H,23H-Porphine, 5,10,15,20-tetra-ethyl-	benzene	419(5.46),484(3.32), 521(4.03),553(3.80), 603(3.46),661(3.67)	64-1240-83B
$C_{28}H_{30}N_4O_4$			
1H-Pyrimido[5,4-b][1,4]benzodiazepine-2,4(3H,6H)-dione, 7,8,9,9a,10,11-hexahydro-10-(4-methoxyphenyl)-6-[(4-methoxyphenyl)methylene]-1,3-dimethyl-	EtOH	261(4.52),267s(--), 363(3.98)	103-0544-83
$C_{28}H_{30}O_8$			
Alectorone	EtOH	248(4.64),262(3.57), 282(3.28),298(2.38), 334(3.19)	105-0011-83
Propanoic acid, 2,2-dimethyl-, 3-(acet-oxymethyl)-9,10-dihydro-8-methoxy-9,10-dioxo-2-(3-oxopentyl)-1-an-thracenyl ester	MeOH	218(4.49),259(4.54), 340s(--),379(3.75)	5-2151-83
$C_{28}H_{31}NO_8$			
4H,11H-Cyclonona[4,5]pyrrolo[3,2,1-ij]-quinoline-8,9-dicarboxylic acid, 5,6-12,13,14,14a-hexahydro-14a-[3-meth-oxy-1-(methoxycarbonyl)-3-oxo-1-propenyl]-, dimethyl ester, (Z,Z,Z)-	MeOH	207(4.29),368(4.15)	39-0515-83C
$C_{28}H_{31}N_3O_3S$			
7H-Benzimidazo[2,1-a]benz[de]isoquino-line-10-sulfonamide, N-decyl-7-oxo-	benzene	387(4.12)	73-2249-83
	EtOH	385(4.16)	73-2249-83
$C_{28}H_{31}N_3O_6$			
Acetamide, N-[1-[6-hydroxy-2-[(4-meth-ylphenyl)azo]-5-oxo-1,3,6-cyclohepta-trien-1-yl]-3-(3,4,5-trimethoxyphen-yl)propyl]-	MeOH	230(4.48),285(4.06), 430(4.43)	18-3106-83
$C_{28}H_{31}N_4S_2$			
Benzothiazolium, 2-[4-(diethylamino)-2-[(3-ethyl-2(3H)-benzothiazolyli-dene)amino]phenyl]-3-ethyl-, tetra-fluoroborate	64% MeOH-pH 5.40	280(4.38),300s(--), 434(4.40)	33-2165-83
$C_{28}H_{32}N_2O_4$			
Acetamide, N-cyclohexyl-N-[4-(cyclo-hexylamino)-9,10-dihydro-8-hydroxy-9,10-dioxo-1-anthracenyl]-	benzene	408(3.00),534(3.25)	104-0745-83
9,10-Anthracenedione, 5-acetoxy-1,4-bis(cyclohexylamino)-	benzene	573s(--),610(3.80), 661(3.89)	104-0745-83

Compound	Solvent	$\lambda_{max}(\log \epsilon)$	Ref.
$C_{28}H_{32}N_2O_{11}$			
Acetamide, N-[2-[acetyl[4-(β-D-glucopyranosyloxy)-9,10-dihydro-9,10-dioxo-1-anthracenyl]amino]ethyl]-N-(2-hydroxyethyl)-	H_2O	225(4.65),314(4.01), 352(4.07),427(3.88)	18-1435-83
$C_{28}H_{32}N_4O_3$			
21H-Biline-1,3,19(2H)-trione, 17-ethyl-23,24-dihydro-2,2,7,8,12,13,18-heptamethyl-, (4E)-	$CHCl_3$	298(4.34),375(4.44), 610s(4.10),637(4.20)	49-0753-83
(4Z)-	$CHCl_3$	293(4.32),368(4.73), 614s(4.23),646(4.30)	49-0753-83
$C_{28}H_{32}N_4O_{13}S_2$			
Benzenesulfonamide, N-[2-(1,1-dimethylethyl)-5-[[[2-(2-methoxyethoxy)-5-nitrophenyl]sulfonyl]amino]-4-oxo-2,5-cyclohexadien-1-ylidene]-2-(2-methoxyethoxy)-5-nitro-	MeCN	298(4.57)	44-0177-83
$C_{28}H_{32}O_3$			
19-Norpregn-4-en-20-yn-3-one, 17α-[(phenylacetyl)oxy]-	$CHCl_3$	242(4.20)	13-0309-83A
$C_{28}H_{32}O_7$			
Phenol, 4-[2-[2,6-dimethoxy-4-(1-propenyl)phenoxy]-1-phenoxypropyl]-2,6-dimethoxy-	EtOH	270(4.32)	150-2625-83M
$C_{28}H_{32}O_9$			
Benzoic acid, 4,6-dihydroxy-3-[(6-hydroxy 1-oxo-3-pentyl-1H-2-benzopyran-8-yl)oxy]-2-(2-oxoheptyl)-	MeOH	212(4.44),238s(--), 245(4.64),278s(--), 290(3.83),314(3.89)	105-0011-83
$C_{28}H_{32}O_{10}$			
Naphtho[2,3-b]furan-2(3H)-one, 5,6,8-triacetoxy-3-(3-acetoxypropyl)-3,4-dimethyl-7-(1-methylethyl)-	EtOH	241.5(5.00),274(3.81), 285.5(3.86),296.5(3.77), 320.5(3.17),335(3.25)	18-2985-83
$C_{28}H_{32}O_{12}$			
Aloesaponol IV 8-O-β-D-glucopyranoside triacetate?	MeOH	265(4.10),300(3.70), 310(3.70),385(3.50)	102-1483-83
	$MeOH-AlCl_3$	270(--),385(--), 450(--)	102-1483-83
Holacanthone triacetate	MeOH	240(4.05)	100-0359-83
Picrasa-3,12-diene-2,11,16-trione, 1,6,12,20-tetraacetoxy-	EtOH	243(4.08)	94-2179-83
$C_{28}H_{33}NO_8$			
4H-Indol-4-one, 1,5,6,7-tetrahydro-6,6-dimethyl-1-(phenylmethyl)-2-(2,3,5-tri-O-acetyl-α-D-lyxofuranosyl)-	EtOH	248(4.02),275(3.91)	136-0255-83E
4H-Indol-4-one, 1,5,6,7-tetrahydro-6,6-dimethyl-1-(phenylmethyl)-2-(2,3,4-tri-O-acetyl-α-D-lyxopyranosyl)-	EtOH	248(3.93),275(3.88)	136-0255-83E
β-	EtOH	248(3.93),276(3.86)	136-0255-83E
$C_{28}H_{33}N_3O_8$			
25-Dihydrosaframycin B	MeOH	269(4.12)	158-83-22

Compound	Solvent	λ_{max}(log ϵ)	Ref.
$C_{28}H_{34}N_2O_6$			
4,5-Isoquinolinedicarboxylic acid, 1,2,4a,7,8,8a-hexahydro-2-[2-(1H-indol-3-yl)ethyl]-7-(2-methyl-1,3-dioxolan-2-yl)-, 4-ethyl 5-methyl ester	EtOH	285(4.24)	44-4262-83
$C_{28}H_{34}N_4O_2$			
21H-Biline-1,19-dione, 17-ethyl-2,3,23,24-tetrahydro-2,2,7,8,12,13,18-heptamethyl- (isomer mixt.)	$CHCl_3$	272(4.29),345(4.54), 580(4.16)	49-0753-83
(4Z,9Z,15E) + (4E,9Z,15E)-	$CHCl_3$	272(4.17),345(4.26), 536(4.18)	49-0753-83
21H-Biline-1,19-dione, 17-ethyl-2,3,23,24-tetrahydro-3,3,7,8,12,13,18-heptamethyl-, (Z,Z,Z)-	$CHCl_3$	273(4.36),346(4.53), 586(4.18)	49-1107-83
	30% DMSO	592(4.13)	49-1107-83
	+ base	770(4.27)	49-1107-83
$C_{28}H_{34}O_4$			
Levonorgestrel 5-methyl-2-furanylethanoate	$CHCl_3$	241(4.26)	13-0339-83A
19-Norpregn-4-en-20-yn-3-one, 17α-[(5-ethyl-2-furanyl)acetyl]oxy]-	$CHCl_3$	241(4.25)	13-0321-83A
19-Norpregn-4-en-20-yn-3-one, 17α-[3-(5-methyl-2-furanyl)-1-oxopropoxy]-	$CHCl_3$	241(4.24)	13-0321-83A
$C_{28}H_{34}O_8$			
Bisvertinoquinol	MeOH	250s(3.95),301(4.26), 370(4.13),380s(4.11), 400s(3.90)	78-4243-83
	MeOH-NaOH	308(4.35),367(4.06), 400s(3.85)	78-4243-83
$C_{28}H_{35}NO$			
1(4H)-Naphthalenone, 2,6-bis(1,1-dimethylethyl)-4-[[4-(1,1-dimethylethyl)phenyl]imino]-	n.s.g.	330(3.78)	39-1759-83C
$C_{28}H_{35}NO_2$			
1(4H)-Naphthalenone, 2,6-bis(1,1-dimethylethyl)-4-[[4-(1,1-dimethylethyl)phenyl]imino]-, N-oxide	n.s.g.	390(4.59)	39-1759-83C
$C_{28}H_{35}NO_6$			
3-Pyridinecarboxylic acid, 4a-acetoxy-1a,1b,3,4,4a,5,7a,8,8a,9,10,10a-dodecahydro-3,6,8,8,10a-pentamethyl-5-oxo-2H-cyclopenta[3,4]cyclopropa[8,9]-cycloundec[1,12-b]oxiren-4-yl ester	MeOH	219(4.03),265(4.15)	102-1791-83
$C_{28}H_{35}N_3O_6$			
3,5-Pyridinedicarboxamide, 1,4-dihydro-N,N'-[bis[2-hydroxy-1-(hydroxymethyl)-2-phenylethyl]-1-propyl-	EtOH	385(3.88)	130-0299-83
$C_{28}H_{35}N_5O_9S_2$			
Benzenesulfonic acid, N-[3-[[4-[ethyl-[2-[(methylsulfonyl)amino]ethyl]-amino]-2-methylphenyl]imino]-4-methyl-6-oxo-1,4-cyclohexadien-1-yl]-2-(2-methoxyethoxy)-5-nitro-	EtOAc	634(4.33)	44-0177-83

Compound	Solvent	$\lambda_{max}(\log \epsilon)$	Ref.
$C_{28}H_{36}N_2O_5$			
Alangicine, (±)-	pH 1	244.5(4.21),285s(3.63), 306.5(3.98),354.5(3.99)	94-2583-83
	pH 13	239(4.27),293.5(3.97), 327(4.22)	94-2583-83
	EtOH	272s(4.00),276(4.02), 312.5(3.65),408(4.34)	94-2583-83
Alangicine, (+)-	pH 13	239(4.18),293(3.86), 326(4.10)	94-2583-83
	EtOH	271s(3.99),276(4.02), 312(3.65),407(4.36)	94-2583-83
$C_{28}H_{36}N_4O_4$			
1,24-Epoxy-3H-dibenzo[c,f][1,5,9]triazacycloheneicosine-3,5,20-trione, 4-amino-6,7,8,9,10,11,12,13,14,15,16-17,18,19-tetradecahydro-2,23-dimethyl-	DMF	265(3.79),425(3.93), 446(3.94)	95-0049-83
$C_{28}H_{36}N_4O_6$			
Safracin A	MeOH	271(3.93)	158-1284-83
$C_{28}H_{36}N_4O_7$			
Safracin B	MeOH	270(3.89)	158-1284-83
$C_{28}H_{36}O_3$			
18,19-Dinorpregn-4-en-20-yn-3-one, 17α-[(cyclopenten-1-ylacetyl)oxy]-13-ethyl-	MeOH	243(4.21)	13-0339-83A
18,19-Dinorpregn-4-en-20-yn-3-one, 13-ethyl-17α-[(2-methyl-1-oxo-2,4-hexadienyl)oxy]-, (2E,4E)-	EtOH	254(4.48)	13-0339-83A
19-Norpregn-4-en-20-yn-3-one, 17α-[(1-oxo-2-octynyl)oxy]-	MeOH	239(4.20)	13-0267-83A
$C_{28}H_{36}O_6$			
Pregna-1,4,9(11)-triene-3,20-dione, 21-acetoxy-17-(2,2-dimethyl-1-oxopropoxy)-	MeOH	239(4.20)	5-0705-83
$C_{28}H_{36}O_7$			
Acetylisodrimartol A	EtOH	207(4.66),227(4.30), 297(4.01),339(3.88)	100-0510-83
$C_{28}H_{36}O_{11}$			
Bruceantin	EtOH	221(4.26),280(3.94)	162-0200-83
$C_{28}H_{37}ClO_7$			
Pregna-1,4-diene-3,20-dione, 21-acetoxy-9-chloro-17-(2,2-dimethyl-1-oxopropoxy)-11β-hydroxy-	MeOH	239(4.19)	5-0705-83
$C_{28}H_{37}ClO_{10}$			
Brianthein Y	CH_2Cl_2	229(3.81)	44-5203-83
$C_{28}H_{37}FO_7$			
Pregna-1,4-diene-3,20-dione, 9-fluoro-11-hydroxy-16-methyl-17,21-bis(1-oxopropoxy)-	MeOH	238(4.20)	162-0168-83
$C_{28}H_{38}O_3$			
18,19-Dinorpregn-4-en-20-yn-3-one, 17α-(3-cyclobutyl-1-oxopropoxy)-13-ethyl-	MeOH	242(4.24)	13-0349-83A

Compound	Solvent	$\lambda_{max}(\log \epsilon)$	Ref.
18,19-Dinorpregn-4-en-20-yn-3-one, 17α-[(cyclohexylcarbonyl)oxy]-13-ethyl-	MeOH	246(4.23)	13-0349-83A
18,19-Dinorpregn-4-en-20-yn-3-one, 13-ethyl-17α-[[(3-ethylcyclobutyl)-carbonyl]oxy]-	MeOH	242(4.23)	13-0349-83A
$C_{28}H_{38}O_4$			
18-Nor-5α,13β-pregnan-20-one, 3β-(benzoyloxy)-16ξ-methoxy-	EtOH	228(3.95)	44-1954-83
$C_{28}H_{38}O_5$			
19-Norpregn-4-en-20-yn-3-one, 17α-[(5-methoxy-2,2-dimethyl-1,5-dioxopent-yl)oxy]-	EtOH	245(4.40)	13-0255-83A
$C_{28}H_{38}O_6$			
26,27-Dinor-5β,14β-cholesta-20(22),23-diene-22-carboxylic acid, 3β-acetoxy-14,21-dihydroxy-25-oxo-, γ-lactone	EtOH	277.5(4.20)	48-0607-83
$C_{28}H_{38}O_7$			
Docosa-2,4,8,12,14,18,20-heptaenedioic acid, 3-(carboxymethyl)-17-methoxy-(bongkrekic acid)	MeOH	237(4.51),267(4.56)	162-0185-83
$C_{28}H_{39}FO_8$			
Pregna-1,4-diene-3,20-dione, 6-fluoro-11,21-dihydroxy-16-methyl-, 21-hexanoate	MeOH	242(4.21)	162-0594-83
$C_{28}H_{40}N_2O_9$			
Antimycin A_1	EtOH	226(4.54),320(3.68)	162-0104-83
8,11,14,17,20,23,26,29,32-Nonaoxa-2,3-diazatricyclo[31.2.2.$2^{4,7}$]nonatriaconta-2,4,6,33,35,36,38-heptaene	$o-C_6H_4Cl_2$	361(4.44)	35-1851-83
$C_{28}H_{40}N_6O_4$			
2-Pyrrolidinecarboxamide, 1,1'-[1,6-hexanediylbis(1,3(4H)-pyridinediyl-carbonyl)]bis-, [S-(R*,R*)]-	$CHCl_3$	350(4.01)	44-1370B-83
$C_{28}H_{40}O_3$			
Furan, 2,2',2"-methylidynetris[5-pentyl-	hexane	221(3.17)	103-0360-83
18,19-Dinorpregn-4-en-20-yn-3-one, 13-ethyl-17α-[(4-methyl-1-oxo-hexyl)oxy]-	MeOH	239(4.25)	13-0349-83A
18,19-Dinorpregn-4-en-20-yn-3-one, 13-ethyl-17α-[(5-methyl-1-oxo-hexyl)oxy]-	MeOH	240(4.25)	13-0349-83A
18,19-Dinorpregn-4-en-20-yn-3-one, 13-ethyl-17α-[(1-oxoheptyl)oxy]-	MeOH	240(4.26)	13-0349-83A
19-Norpregn-4-en-20-yn-3-one, 17α-[(2,2-dimethyl-1-oxohexyl)oxy]-	EtOH	240(4.55)	13-0255-83A
$C_{28}H_{40}O_4$			
19-Norpregn-4-en-20-yn-3-one, 17α-[[[(2,2-dimethylpentyl)oxy]-carbonyl]oxy]-	EtOH	247(3.92)	13-0333-83A
19-Norpregn-4-en-20-yn-3-one, 10-hydroxy-7-methyl-17-[(1-oxoheptyl)-	MeOH	235.5(4.18)	13-0401-83B

Compound	Solvent	λ max (log ε)	Ref.
oxy]-, (7α,17α)- (cont.)			13-0401-83B
2,10-Tridecadien-5-one, 12-hydroxy-13-(6-methoxy-2,8-dimethyl-2H-1-benzopyran-2-yl)-2,6,10-trimethyl-	MeOH	213(4.00)	78-0629-83
$C_{28}H_{40}O_5$			
Azzonapyrone A	EtOH	210(4.23),290(3.90)	158-83-83
19-Norpregn-4-en-20-yn-3-one, 10-hydroperoxy-7-methoxy-17-[(1-oxoheptyl)-oxy]-, (7α,17α)-	MeOH	236(4.22)	13-0401-83B
$C_{28}H_{40}O_5S$			
Androstane-3,17-diol, 3-acetate 17-(4-methylbenzenesulfonate), (3β,5α,17β)-	EtOH	225(4.07)	44-1954-83
$C_{28}H_{40}O_6$			
Hexadecanoic acid, (7-acetoxy-2-oxo-2H-1-benzopyran-4-yl)methyl ester	EtOH	277(3.99),312(3.94)	94-3014-83
Propanoic acid, 2,2-dimethyl-, 4-[2-[5-acetoxy-2-(acetoxymethyl)-2,3,3a,6,7-7a-hexahydro-7a-methyl-1H-inden-1-yl ester, [1S-[1α,3aβ,4[Z(R*),7aα]]-	ether	360(3.98)	56-1267-83
Propanoic acid, 2,2-dimethyl-, 4-[[5-acetoxy-2-(acetoxymethylene)cyclohexylidene]ethylidene]octahydro-7a-methyl-1H-inden-1-yl ester	ether	266.0(4.27)	56-1267-83
$C_{28}H_{40}O_7$			
Diginin	EtOH	309(1.94)	162-0457-83
Ergosta-5,24-dien-26-oic acid, 3,14,17,20,22-pentahydroxy-1-oxo-, δ-lactone	EtOH	224(4.03)	102-2253-83
Isolathyrol, 5?,15-di-O-acetyl-3-O-isobutyryl-	MeOH	197(4.09),278(4.05)	102-1791-83
Phorbol, 4α-deoxy-12-O-hexanoyl-, 13-acetate	EtOH	208(3.94),263(3.47)	102-1231-83
$C_{28}H_{40}O_8$			
Ingol, 12-O-acetyl-8-methoxy-7-O-tigloyl-	MeOH	218(4.40)	102-2795-83
$C_{28}H_{40}O_9$			
Withaperuvin D	n.s.g.	230(3.85),285(2.11)	150-0152-83S
$C_{28}H_{41}NO$			
2,5-Cyclohexadien-1-one, 4-[[3,5-bis(1,1-dimethylethyl)phenyl]-imino]-2,6-bis(1,1-dimethylethyl)-	hexane	272(4.24),295(4.23), 445(3.61)	18-1476-83
$C_{28}H_{41}NO_8$			
Phorbol, 12-O-leucyl-, 13-acetate	MeOH	230(3.65),252s(3.53)	64-1015-83B
$C_{28}H_{41}N_3$			
Pyrido[2',1':3,4][1,4]diazepino[1,2-a]-benzimidazole-5,9-diium, 14-dodecyl-6,7,8,14-tetrahydro-2-methyl-, dibromide	EtOH	319(4.20)	4-0029-83
$C_{28}H_{42}NO_4$			
1-Pyrrolidinyloxy, 2,2,5,5-tetramethyl-3-[[[(17β)-3-oxoandrost-4-en-17-yl]-	EtOH	238(3.08)	13-0241-83B

Compound	Solvent	λ_{max}(log ε)	Ref.
oxy]carbonyl]- (cont.)			13-0241-83B
$C_{28}H_{42}O_4$			
5β-Chola-20,22-dien-24-oic acid, 3β-acetoxy-, ethyl ester	EtOH	216(3.95)	44-4248-83
2,6,10,14-Hexadecatetraene-5,12-diol, 1-(2-hydroxy-5-methoxy-3-methylphenyl)-3,7,11,15-tetramethyl-, (all-E)-	MeOH	214(3.94),289(3.38)	78-0629-83
$C_{28}H_{42}O_5$			
5β-Chol-20(22)-en-24-oic acid, 3β-acetoxy-, ethyl ester	EtOH	233(3.83),300(3.00)	44-4248-83
2,10,14-Hexadecatrien-4-one, 5,12-dihydroxy-16-(2-hydroxy-5-methoxy-3-methylphenyl)-2,6,10,14-tetramethyl-, (all-E)-	MeOH	225(4.09)	102-2865-83
2,10,14-Hexadecatrien-5-one, 4,12-dihydroxy-16-(2-hydroxy-5-methoxy-3-methylphenyl)-2,6,10,14-tetramethyl-, (all-E)-	MeOH	214(4.06),288(3.49)	78-0629-83
$C_{28}H_{42}O_6$			
Cynanchogenin	EtOH	218(4)	162-0398-83
$C_{28}H_{43}N$			
Benzenamine, 4-(1,1,3,3-tetramethylbutyl)-N-[4-(1,1,3,3-tetramethylbutyl)phenyl]-, cation	MeCN	730(4.52)	70-0251-83
radical	MeCN	760(3.56)	70-0251-83
$C_{28}H_{43}N_7O_7$			
Alanine, β-(9-adeninyl)-α-alanyl-L-lysyl-L-lysyl-DL-β-(1-uracilyl)-, methyl ester, DL-, trihydrobromide	pH 1	264(4.18)	103-0328-83
	pH 7	262(4.13)	103-0328-83
$C_{28}H_{44}FeN_9O_{13}$			
Ferricrocin	H_2O	434(3.39)	35-0810-83
$C_{28}H_{44}N_2O_2S$			
Benzenesulfonic acid, 2,4,6-tris(1-methylethyl)-, [1-methyl-3-(2,6,6-trimethyl-1-cyclohexen-1-yl)-2-propenylidene]hydrazide, (E,E)-	hexane	278.5(4.29)	33-1148-83
(Z,E)-	hexane	284.5(4.20)	33-1148-83
$C_{28}H_{44}O_4$			
Ergosta-7,22-dien-6-one, 3β,5α,9α-trihydroxy-	EtOH	237(3.00)	78-2779-83
Ergosta-7,22-dien-6-one, 3β,5α,14α-trihydroxy-	EtOH	242(4.16)	78-2779-83
$C_{28}H_{44}O_4Si_2$			
Anthra[1,9-de]-1,3,2-dioxasilin-11(7H)-one, 2,2-bis(1,1-dimethylethyl)-7-[[(1,1-dimethylethyl)dimethylsilyl]oxy]-7a,8,11a,11b-tetrahydro-, (7α,7aβ,11aβ,11bβ)-	EtOH	218(3.26),272(3.26), 281(4.18)	44-3252-83
$C_{28}H_{44}O_5$			
Chol-22-en-24-oic acid, 3-acetoxy-21-	EtOH	216(4.08)	44-4248-83

Compound	Solvent	λ_{max}(log ε)	Ref.
hydroxy-, ethyl ester, (3β,5β,20ξ-22E)- (cont.)	EtOH		44-4248-83
$C_{28}H_{45}O_4P$			
Ergosterol dihydrogen phosphate	n.s.g.	267s(3.76),277(3.92), 283(3.94),298(3.71)	44-1417-83
$C_{28}H_{46}O$			
Ergosterol, 22,23-dihydro-	n.s.g.	262(3.90),272(4.05), 282(4.07),294(3.83)	162-0528-83
$C_{28}H_{46}O_2$			
Ergosta-5,7-diene-3β,23ξ-diol	EtOH	271(4.14),282(4.18), 293(3.94)	88-4729-83
Ergosta-5,25-dienolide, 3β,24ξ-dihydroxy-	EtOH	210s(3.88)	102-2253-83
$C_{28}H_{46}O_2Si$			
Pregna-5,7-diene-20-carboxaldehyde, 3β-[[(1,1-dimethylethyl)dimethylsilyl]oxy]-	EtOH	263(3.89),272(4.01), 280(4.08),292(3.83)	5-1031-83
$C_{28}H_{46}O_3Si$			
2,4,6,8-Nonatetraenoic acid, 7-[[[(1,1-dimethylethyl)dimethylsilyl]oxy]methyl]-3-methyl-9-(2,6,6-trimethyl-1-cyclohexen-1-yl)-, ethyl ester, (all-E)-	EtOH	359(4.58)	77-0077-83
$C_{28}H_{47}ClOSi$			
Silane, [[(3β,20S)-21-chloro-20-methylpregna-5,7-dien-3-yl]oxy](1,1-dimethylethyl)dimethyl-	EtOH	262(3.92),271(4.08), 283(4.10),293(3.88)	5-1031-83
$C_{28}H_{47}NO_2$			
Cholest-5-en-3β-ol, 3α-methyl-, nitrite	THF	342(1.43),355(1.66), 367(1.83),384(1.88), 400(1.69)	18-3349-83
Cholest-5-en-3β-ol, 6-methyl-, nitrite	benzene	337(1.65),348(1.78), 362(1.88),374(1.97), 398(1.65)	18-3349-83
$C_{28}H_{48}O_2Si$			
Pregna-5,7-diene-20-methanol, 3-[[(1,1-dimethylethyl)dimethylsilyl]oxy]-, (3β,20S)-	EtOH	263(3.88),271(4.04), 280(4.03),292(3.79)	5-1031-83
$C_{28}H_{48}O_9$			
8,9-Anhydroerythronolide B, 3-O-mycarosyl-, 6,9-hemiketal	MeOH	218(3.84)	88-5527-83
$C_{28}H_{49}FO_{10}$			
Erythronolide B, 8-fluoro-3-O-mycarosyl-, (8S)-	MeOH	287(1.53)	88-5527-83
$C_{28}H_{50}Ge_4$			
Tetragermane, 1,1,2,2,3,3,4,4-octaethyl-1,4-diphenyl-	C_6H_{12}	248(4.43)	101-0149-83G
$C_{28}H_{55}N_5O_5Si_3$			
Guanosine, 2',3',5'-tris-O-[(1,1-di-	MeOH	254(4.19),272s(3.94)	33-2069-83

Compound	Solvent	λ_{max}(log ε)	Ref.
methylethyl)dimethylsilyl]- (cont.)			33-2069-83
$C_{28}H_{56}Si_7$			
Heptasilaheptaspiro[4.0.4.0.4.0.4.0-4.0.4.0.4.0]pentatriacontane	isooctane	209s(4.65),245s(4.02), 270s(3.66)	157-0453-83
$C_{28}H_{58}P_2$			
Phosphonium, 1,3-butadiene-1,4-diyl-bis[tributyl-, dibromide, (E,E)-	EtOH	251(4.28)	118-0374-83
$C_{28}H_{70}Si_7$			
Cycloheptasilane, tetradecaethyl-	isooctane	200s(4.78),227s(4.26), 244(4.00),257s(3.64)	157-1792-83

Compound	Solvent	$\lambda_{max}(\log \epsilon)$	Ref.
$C_{29}H_{10}Cl_{14}NO_3$			
Methyl, [4-[[(1-carboxy-2-phenylethyl)-amino]carbonyl]-2,3,5,6-tetrachloro-phenyl]bis(pentachlorophenyl)-, (S)-	$CHCl_3$	285(3.79),335s(3.80), 365s(4.29),382(4.57), 485s(3.04),507(3.06), 560(3.05)	44-3716-83
$C_{29}H_{11}Cl_{14}N_2O_3$			
Methylium, bis(pentachlorophenyl)-[2,3,5,6-tetrachloro-4-[[[[(phenyl-methoxy)carbonyl]amino]acetyl]-amino]phenyl]-	C_6H_{12}	229(4.92),286(3.91), 336s(3.81),364s(4.25), 383(4.58),480s(3.07), 510(3.11),564(3.13)	44-3716-83
$C_{29}H_{16}N_2O_2$			
9,14[1',2']-Benzenoanthra[2',3':4,5]-imidazo[2,1-a]isoquinoline-8,15-di-one, 9,14-dihydro-	EtOH	259(4.48),262(4.40), 280(4.18),305(3.51), 402(3.42)	48-0353-83
9,14[1',2']-Benzenoanthra[2',3':4,5]-imidazo[1,2-a]quinoline-8,15-dione, 9,14-dihydro-	EtOH	257(4.46),268(4.38), 284(4.25),303(3.55), 398(3.47)	48-0353-83
$C_{29}H_{16}O_5$			
Anthra[1,2-b]furan-6,11-dione, 3-benz-oyl-5-hydroxy-2-phenyl-	dioxan	275(4.29),426(4.09)	104-0533-83
$C_{29}H_{16}O_6$			
Anthra[2,3-b]furan-5,10-dione, 3-benz-oyl-4,11-dihydro-2-phenyl-	dioxan	292(4.59),486(4.18)	104-1900-83
$C_{29}H_{17}ClN_2O_3$			
1H-Naphth[2,3-g]indole-6,11-dione, 5-amino-3-benzoyl-4-chloro-2-phenyl-	dioxan	286(4.60),459(3.85), 560(3.96)	104-1892-83
$C_{29}H_{17}NO_4$			
1H-Naphth[2,3-g]indole-6,11-dione, 3-benzoyl-5-hydroxy-2-phenyl-	dioxan	279(4.63),459(4.21)	104-1892-83
$C_{29}H_{18}$			
1a,6[1',2']:7,11b[1",2"]-Dibenzeno-1H-dibenzo[a,e]cyclopropa[c]cyclooctene	CCl_4	255(3.59),275(3.51), 285(3.36)	44-5120-83
$C_{29}H_{18}Cl_3NO$			
Pyridinium, 2,4,6-tris(4-chlorophenyl)-1-(4-hydroxyphenyl)-, hydroxide, inner salt	MeOH	314(4.58),448(3.39)	5-0860-83
$C_{29}H_{18}I_2$			
Anthracene, 9,9'-methylenebis[10-iodo-	CCl_4	255(4.80),365(3.78), 385(3.88),410(3.90)	44-5120-83
$C_{29}H_{18}N_2O_3$			
1H-Naphth[2,3-g]indole-6,11-dione, 5-amino-3-benzoyl-2-phenyl-	dioxan	281(4.53),440(3.69), 560(3.95)	104-1892-83
$C_{29}H_{18}N_2O_4$			
Anthra[2,3-b]furan-5,10-dione, 4,11-di-amino-3-benzoyl-2-phenyl-	dioxan	266(4.66),532(4.21), 565(4.18)	104-1892-83
2-Benzopyrano[6',5',4':10,5,6]anthra-[2,1,9-def]pyrimido[2,1-a]isoquino-line-1,3,8(10H)-trione, 11,12-dihy-dro-11,11-dimethyl-, monohydro-chloride	HCOOH	474(4.25),503(4.6), 539(4.8)	24-3524-83

Compound	Solvent	λ_{max}(log ϵ)	Ref.
$C_{29}H_{18}O_2$			
2a,6b[1',2']:7a,11b[1",2"]-Dibenzeno-7H-dibenzo[3,4:7,8]cyclopropa[5,6]-cycloocta[1,2-c]-1,2-dioxete	CCl_4	255(3.65),275(3.40), 285(3.18)	44-5120-83
Dispiro[anthracene-9(10H),1'-cycloprop-ane-2',9"(10"H)-anthracene]-10,10"-dione	$CHCl_3$	255(4.21),280(4.22)	44-5120-83
$C_{29}H_{19}Cl_2NO$			
Pyridinium, 4-(2,6-dichlorophenyl)-1-(4-hydroxyphenyl)-2,6-diphenyl-, hydroxide, inner salt	MeOH	290(4.13),455(3.27)	5-0860-83
Pyridinium, 4-(3,5-dichlorophenyl)-1-(4-hydroxyphenyl)-2,6-diphenyl-, hydroxide, inner salt	MeOH	289(4.15),450(3.32)	5-0860-83
$C_{29}H_{19}N_3$			
1H,8H-3a,7b[1',2']:8a,12b[1",2"]-Di-benzenodibenzo[3,4:7,8]cyclopropa-[5,6]cycloocta[1,2-d]triazole	CCl_4	255(3.57),262(3.49), 285s(2.74)	44-5120-83
$C_{29}H_{20}I_2$			
9H-Fluorene, 2,7-bis[2-(4-iodophenyl)-ethenyl]-	C_6H_5Cl	380(4.57)	104-0683-83
$C_{29}H_{20}NO_3$			
Pyrylium, 3-(2-nitrophenyl)-2,4,6-tri-phenyl-, perchlorate	MeCN	278(4.30),358(4.45), 401(4.45)	22-0115-83
$C_{29}H_{20}N_2$			
1H-Isoindole, 3-phenyl-1-[(3-phenyl-2H-isoindol-1-yl)methylene]-	$CHCl_3$	289(4.38),304s(4.35), 338s(4.18),390s(3.59), 588(4.60)	78-1401-83
	+ HCl	298(4.47),335(4.15), 350(4.14),580s(4.39), 625.5(5.03)	78-1401-83
$C_{29}H_{20}N_2O_4$			
Benz[g]imidazo[5,1,2-cd]indolizine-1,2-dicarboxylic acid, 3,5-diphenyl-, dimethyl ester	MeCN	294(4.27),298(4.27), 380(4.20),428(3.71)	138-0763-83
$C_{29}H_{20}N_4$			
Benzonitrile, 4,4'-(4,5-dihydro-4,5-diphenyl-1H-pyrazole-1,3-diyl)bis-, trans	C_6H_{12}	387(4.54)	12-1649-83
	MeOH	386(4.56)	12-1649-83
1H,8H-3a,7b[1',2']:8a,12b[1",2"]-Di-benzenodibenzo[3,4:7,8]cyclopropa-[5,6]cycloocta[1,2-d]triazol-1-amine	CCl_4	257(3.79),263(3.76)	44-5120-83
$C_{29}H_{20}N_6O_{11}$			
Ethanone, 1,1'-[carbonylbis[(2,6-di-nitro-4,1-phenylene)imino-3,1-phen-ylene]]bis-	dioxan	292.4(4.16)	56-1357-83
$C_{29}H_{21}BrN_6$			
2(1H)-Pyrimidinone, 4,6-diphenyl-, [[(4-bromophenyl)azo]phenylmeth-ylene]hydrazone	EtOH	440(--)	65-0153-83
	acetone	450(4.19)	65-0153-83
copper chelate	acetone	590(4.26)	65-0153-83
nickel chelate (also others)	acetone	605(4.40)	65-0153-83

Compound	Solvent	$\lambda_{max}(\log \epsilon)$	Ref.
$C_{29}H_{21}ClN_6$			
2(1H)-Pyrimidinone, 4,6-diphenyl-, [[(4-chlorophenyl)azo]phenylmethylene]hydrazone	EtOH	450(4.20)	65-0153-83
	acetone	450(4.04)	65-0153-83
copper chelate	EtOH	585(4.13)	65-0153-83
nickel chelate (also other chelates)	EtOH	600(4.40)	65-0153-83
$C_{29}H_{21}N$			
3-Butenenitrile, 2-(diphenylmethylene)-4,4-diphenyl-	MeCN	225(4.40),344(4.48)	118-0917-83
Pyridine, 2,3,5,6-tetraphenyl-	neutral	302(4.30)	39-0045-83B
	protonated	329(4.11)	39-0045-83B
$C_{29}H_{21}NO$			
3(2H)-Pyridinone, 2,4,4,6-tetraphenyl-	EtOH	220(4.26),310(3.76), 390(2.73)	39-0837-83C
$C_{29}H_{21}N_4S_2$			
Thiazolo[4',3':2,3]pyrimido[4,5-b]quinolin-12-ium, 1-[(3-ethyl-2(3H)-benzothiazolylidene)methyl]-3-phenyl-, perchlorate	DMF	514(3.62)	103-0166-83
$C_{29}H_{21}O$			
Pyrylium, 2,3,4,6-tetraphenyl-, perchlorate	MeCN	290(4.27),372(4.31), 402(4.32)	22-0115-83
$C_{29}H_{22}$			
1H-Benz[f]indene, 1-(1,3-butadienyl)-4,9-diphenyl-	hexane	243(4.61),255(4.49), 273(4.31),300(4.16), 310(4.17)	33-0718-83
1H-Benz[f]indene, 1-(2-butenylidene)-4,9-diphenyl-, (Z)-	hexane	235(4.54),305(4.81), 326(4.34),400(4.01)	33-0718-83
9H-Fluorene, 2,7-bis(2-phenylethenyl)-, (E,E)-	dioxan	370(4.98)	104-0683-83
$C_{29}H_{22}ClNO_5$			
Benzoic acid, 3-chloro-, 4-acetyl-3,4-dihydro-3-hydroxy-2,3-diphenyl-2H-1,4-benzoxazin-2-yl ester	EtOH	243(4.52),287s(3.67)	39-0141-83C
$C_{29}H_{22}N_2O_3$			
Pyridinium, 4-(ethoxycarbonyl)-2,6-diphenyl-, 1-cyano-2-oxo-2-phenylethylide	CH_2Cl_2	562(3.05)	142-0623-83
$C_{29}H_{22}N_2O_6$			
D-ribo-Hept-2-enononitrile, 2-cyano-2,3-dideoxy-4,6-O-(phenylmethylene)-, 5,7-dibenzoate, (S)-	$CHCl_3$	240(4.00)	87-0030-83
$C_{29}H_{22}N_6$			
2(1H)-Pyrimidinone, 4,6-diphenyl-, [phenyl(phenylazo)methylene]hydrazone	EtOH	435(--)	65-0153-83
	acetone	440(4.15)	65-0153-83
copper chelate	acetone	585(4.04)	65-0153-83
nickel chelate	acetone	590(4.25)	65-0153-83
$C_{29}H_{23}N$			
Pyridine, 1,4-dihydro-2,4,4,6-tetraphenyl-	EtOH	210(4.65),238(4.46), 290(3.76),330(2.89)	39-0837-83C

Compound	Solvent	λ_{max}(log ε)	Ref.
$C_{29}H_{23}N_3O_2$			
Quinazolino[1,2-a]quinazoline-5,8-dione, N,N'-bis(phenylmethyl)-	EtOH	211(4.45),340(3.05)	142-0617-83
	50% $HClO_4$	207(4.48),285(3.37), 295(3.36)	142-0617-83
$C_{29}H_{23}N_3O_9$			
1H-Imidazole, 2-nitro-1-(2,3,5-tri-O-benzoyl-α-D-arabino-furanosyl)-	EtOH	238(4.39),276(3.77), 284(3.78),315(3.79)	87-0020-83
β-	EtOH	237(4.41),275(3.89), 284(3.89),314(3.84)	87-0020-83
$C_{29}H_{23}N_3O_9S$			
1,2,4-Triazine-3,5(2H,4H)-dione, 6-[(2,3,5-tri-O-benzoyl-α-D-ribo-furanosyl)thio]-	EtOH	220(4.67),305(3.83)	80-0733-83
β-	EtOH	220(4.68),305(3.85)	80-0733-83
$C_{29}H_{23}N_5O_2$			
Benzeneacetic acid, 2-[[1,3-diphenyl-4-(phenylazo)-1H-pyrazol-5-yl]amino]-	EtOH	215(3.92),245(4.75), 275(3.70),330(3.47)	4-1501-83
$C_{29}H_{24}N_2O_2$			
1H-Pyrrolo[2,3-d]carbazole-2,5(3H,6H)-dione, 3a,4,6a,7-tetrahydro-3a-methyl-4,6-bis(phenylmethylene)-, (3aα,6aβ,11bR*)-(±)-	MeOH	264(4.09),288(4.16), 334s(3.58)	5-1744-83
$C_{29}H_{24}N_4O$			
1H-Pyrazole-4,5-diamine, N^4-[(4-methoxyphenyl)methylene]-N^5,1,3-triphenyl-	EtOH	240(3.35),282(3.47)	4-1501-83
$C_{29}H_{24}N_6O_{11}$			
Methanone, bis[4-[(2-ethoxyphenyl)-amino]-3,5-dinitrophenyl]-	dioxan	295.5(4.16)	56-1357-83
$C_{29}H_{24}N_8$			
2H-Benzimidazol-2-one, 1,3-dihydro-1-(phenylmethyl)-, [[[1-(phenylmethyl)-1H-benzimidazol-2-yl]azo]methylene]hydrazone	toluene	509(4.82),540(4.82)	65-2332-83
anion	EtOH	600(4.90),614(4.84)	65-2332-83
$C_{29}H_{25}NO_4$			
Spiro[9H-fluorene-9,1'(5'H)-pyrrolo-[2,1-a]isoquinoline-2',3'-dicarboxylic acid, 6',10'b-dihydro-10'b-methyl-, dimethyl ester	$CHCl_3$	314(4.00)	24-3915-83
betaine	$CHCl_3$	590(--)	24-3915-83
$C_{29}H_{25}NO_5$			
2H-Furo[2,3-h]-1-benzopyran-2-one, 8-[4-[2-(dimethylamino)ethoxy]-benzoyl-4-methyl-9-phenyl-	dioxan	250(4.56),280(4.58)	2-0164-83
$C_{29}H_{25}NO_7$			
Benzocycloocten-6,9-imine-7,8-dicarboxylic acid, 5,6,7,8,9,10-hexahydro-11-(4-methoxyphenyl)-5,10-dioxo-6-phenyl-, dimethyl ester	benzene	306(3.68),356s(2.49), 450(2.18)	44-4968-83
	MeOH	301(3.57),352s(2.46), 440(2.00)	44-4968-83
	MeCN	230(4.43),303(3.60),	44-4968-83

Compound	Solvent	$\lambda_{max}(\log \epsilon)$	Ref.
(cont.)		356s(2.49),441(2.18)	44-4968-83
$C_{29}H_{25}N_3O_3$			
Quinolinium, 2-[5-carboxy-3-(3-ethyl-benzoxazolium-2-yl)-1H-indol-2-yl]-1-ethyl-, diiodide	EtOH	212(5.07),240(5.03), 270(4.81),290(4.66), 565(3.93)	80-0725-83
Quinolinium, 2-[7-carboxy-3-(3-ethyl-benzoxazolium-2-yl)-1H-indol-2-yl]-1-ethyl-, diiodide	EtOH	210(4.11),242(5.03), 270(4.81),325(4.48), 580(3.81)	80-0725-83
$C_{29}H_{25}N_5O_2$			
Pyrimidine, 4,4'-(2,6-pyridinediyl)-bis[6-ethoxy-2-phenyl-	CH_2Cl_2	262(4.58)	44-4841-83
$C_{29}H_{25}O_3$			
2H-Furylium, 3-[4-(1,3-benzodioxol-5-yl)-2-phenyl-1,3-butadienyl]-2,2-dimethyl-5-phenyl-, perchlorate	MeCN	588(4.71)	56-0579-83
	HOAc	610(--)	56-0579-83
$C_{29}H_{26}N$			
Benzo[h]quinolinium, 1-cyclobutyl-5,6-dihydro-2,4-diphenyl-, tetrafluoro-borate	EtOH	326(4.28)	39-1435-83B
$C_{29}H_{26}NO$			
3H-Indolium, 1,3,3-trimethyl-2-[3-(2-phenyl-4H-1-benzopyran-4-ylidene)-1-propenyl]-, perchlorate	MeCN	582(4.65),626(4.52)	103-0948-83
$C_{29}H_{26}N_2O_2$			
1,3-Cyclopentadiene-1-carboxylic acid, 2-amino-3-cyano-4,5,5-triphenyl-, 1,1-dimethylethyl ester	hexane	380(3.46)	24-3097-83
$C_{29}H_{26}N_2O_6$			
Cyclohept[1,2,3-hi]imidazo[2,1,5-cd]-indolizine-3,4,9-tricarboxylic acid, 2-(4-methylphenyl)-, triethyl ester	EtOH	251(4.39),280s(4.43), 292(4.45),312s(4.32), 384(4.46),425s(4.13)	18-3703-83
$C_{29}H_{26}N_4O_4$			
1-Butanone, 4-(4,8-dimethoxy-9H-pyrido-[3,4-b]indol-1-yl)-2-methoxy-1-(9H-pyrido[3,4-b]indol-1-yl)-	EtOH	221(4.85),244(4.93), 270(4.20),287(4.45), 330(3.99),385(3.91)	94-3198-83
$C_{29}H_{26}O_2$			
Isobenzofuran, 1,3-dihydro-1-methoxy-3,3-bis(4-methylphenyl)-1-phenyl-	hexane	255s(2.87),258(2.97), 266(3.02),270s(2.83), 274s(2.78)	104-0553-83
1-Isobenzofuranol, 1,3-dihydro-1,3,3-tris(4-methylphenyl)-	hexane	259s(3.04),265(3.10), 270s(2.95),273(2.93)	104-0553-83
$C_{29}H_{26}O_4$			
Riccardin A	EtOH	213(4.59),283(3.90)	44-2164-83
$C_{29}H_{26}O_{10}$			
Cercosporin	MeOH	330(--),469(4.36), 530s(--),563(3.91)	74-0385-83C
	MeOH-NaOH	477(4.44),598(4.04), 642(4.17)	74-0385-83C
	$CHCl_3$	472(4.41),530s(4.04), 565(3.95)	74-0385-83C

Compound	Solvent	λ_{max}(log ϵ)	Ref.
Isocercosporin	MeOH	474(4.33),564(3.87)	74-0385-83C
	MeOH-NaOH	478(4.37),596s(--), 639(4.11)	74-0385-83C
	$CHCl_3$	476(4.27),530s(3.95), 566(3.81)	74-0385-83C
$C_{29}H_{27}N_2$			
3H-Indolium, 3-[3-(1,2-dimethyl-1H-indol-3-yl)-1-phenyl-2-propenylidene]-1,2-dimethyl-, iodide	$MeNO_2$-HOAc	615(4.69)	103-1306-83
$C_{29}H_{27}N_2O_3P$			
3-Oxepincarboxylic acid, 5-cyano-6,7-dihydro-6,7-dimethyl-2-[(triphenylphosphoranylidene)amino]-, methyl ester	$CHCl_3$	270(4.06),275(4.09), 362(4.40)	24-1691-83
3-Oxepincarboxylic acid, 5-cyano-6,7-dihydro-7,7-dimethyl-2-[(triphenylphosphoranylidene)amino]-, methyl ester	$CHCl_3$	268(4.02),274(4.05), 364(4.42)	24-1691-83
$C_{29}H_{27}N_3O$			
Quinolinium, 1-ethyl-2-[3-(3-ethylbenzoxazolium-2-yl)-5-methyl-1H-indol-2-yl]-, diiodide	EtOH	210(5.26),245(5.14), 300(4.92),580(3.95)	80-0725-83
$C_{29}H_{27}N_3O_2$			
Quinolinium, 1-ethyl-2-[3-(3-ethylbenzoxazolium-2-yl)-5-methoxy-1H-indol-2-yl]-, diiodide	EtOH	220(5.07),250(4.83), 300(4.57),510(3.81), 575(3.76)	80-0725-83
$C_{29}H_{27}O$			
2H-Furylium, 2,2-dimethyl-3-[4-(2-methylphenyl)-2-phenyl-1,3-butadienyl]-5-phenyl-, perchlorate	MeCN	526(4.70)	56-0579-83
	HOAc	538(--)	56-0579-83
2H-Furylium, 2,2-dimethyl-3-[4-(3-methylphenyl)-2-phenyl-1,3-butadienyl]-5-phenyl-, perchlorate	MeCN	537(4.75)	56-0579-83
	HOAc	538(--)	56-0579-83
2H-Furylium, 2,2-dimethyl-3-[4-(4-methylphenyl)-2-phenyl-1,3-butadienyl]-5-phenyl-, perchlorate	MeCN	541(4.72)	56-0579-83
	HOAc	554(--)	56-0579-83
$C_{29}H_{27}O_2$			
2H-Furylium, 3-[4-(3-methoxyphenyl)-2-phenyl-1,3-butadienyl]-2,2-dimethyl-5-phenyl-, perchlorate	MeCN	529(4.75)	56-0579-83
	HOAc	547(--)	56-0579-83
2H-Furylium, 3-[4-(4-methoxyphenyl)-2-phenyl-1,3-butadienyl]-2,2-dimethyl-5-phenyl-, perchlorate	MeCN	582(4.78)	56-0579-83
	HOAc	595(--)	56-0579-83
$C_{29}H_{27}O_3$			
2H-Furylium, 3-[4-(4-hydroxy-3-methoxyphenyl)-2-phenyl-1,3-butadienyl]-2,2-dimethyl-5-phenyl-, perchlorate	MeCN	609(4.80)	56-0579-83
$C_{29}H_{28}Cl_2N_6O$			
4H-[1,2,4]Triazolo[4,3-a][1,4]benzodiazepine, 8-chloro-6-(2-chlorophenyl)-1-[[4-(2-phenoxyethyl)-1-piperazinyl]methyl]-	MeOH	221(4.67),251s(4.00), 270s(3.52)	73-3433-83

Compound	Solvent	$\lambda_{max}(\log \epsilon)$	Ref.
$C_{29}H_{28}N$			
Benzo[h]quinolinium, 5,6-dihydro-1-(1-methylpropyl)-2,4-diphenyl-, tetrafluoroborate	pentanol	350(4.00)	39-1443-83B
	HOAc	350(4.00)	39-1443-83B
	C_6H_5Cl	350(4.10)	39-1443-83B
	CF_3COOH	350(4.20)	39-1443-83B
$C_{29}H_{28}NO_4PS$			
3,5-Thiepindicarboxylic acid, 6,7-dihydro-2-[(triphenylphosphoranylidene)amino]-, 5-ethyl 3-methyl ester	$CHCl_3$	267(3.99),275(4.02), 383(4.29)	24-1691-83
$C_{29}H_{28}NO_5P$			
3,5-Oxepindicarboxylic acid, 6,7-dihydro-2-[(triphenylphosphoranylidene)amino]-, 5-ethyl 3-methyl ester	MeOH	272(4.18),280(4.16), 363(4.39)	24-1691-83
$C_{29}H_{28}N_2O_4$			
Acetamide, N-[4-(cyclohexylamino)-9,10-dihydro-8-hydroxy-9,10-dioxo-1-anthracenyl]-N-(4-methylphenyl)-	benzene	413(3.46),546(3.72)	104-0745-83
9,10-Anthracenedione, 5-acetoxy-1-(cyclohexylamino)-4-[(4-methylphenyl)-amino]-	benzene	573s(--),613(4.01), 655(4.03)	104-0745-83
$C_{29}H_{28}N_2O_7$			
Uridine, 5-(1-butynyl)-2'-deoxy-, 3',5'-bis(4-methylbenzoate)	MeOH	239(4.61),284(4.06)	87-0661-83
$C_{29}H_{28}N_2O_7S$			
2,4(1H,3H)-Pyrimidinedione, 1-[2-deoxy-3-O-(methylsulfonyl)-5-O-(triphenylmethyl)-β-D-threo-pentofuranosyl]-	MeOH	260(3.96)	87-0544-83
$C_{29}H_{28}N_2O_8$			
Uridine, 2'-deoxy-5-(3-methoxy-1-propynyl)-, 3',5'-bis(4-methylbenzoate)	MeOH	238(4.55),283(4.08)	87-0661-83
$C_{29}H_{28}O_5$			
Phenol, 3-[2-(3-hydroxyphenyl)ethyl]-2,4-bis[(4-hydroxyphenyl)methyl]-5-methoxy-	MeOH	273s(3.89),281(3.99), 289s(3.88)	102-1011-83
$C_{29}H_{29}NO_6$			
Hetisan-2,13-dione, 15-acetoxy-11-(benzoyloxy)-, (11α,15β)-	EtOH	303(2.20)	88-3765-83
$C_{29}H_{29}N_5O_3$			
1H-Purine-2,6-dione, 8-[bis(phenylmethyl)amino]-3,7-dihydro-7-(2-hydroxy-2-phenylethyl)-1,3-dimethyl-	EtOH	297(4.22)	56-0461-83
$C_{29}H_{30}N_2O_6S$			
L-Phenylalanine, N-[N-[(phenylmethoxy)-carbonyl]-1-thio-L-α-aspartyl]-, 1-methyl 4-(phenylmethyl) ester	EtOH	267(4.07)	78-4121-83
$C_{29}H_{30}N_4O_6$			
Imidazo[4,5-d][1,3]diazepin-8(3H)-one, 3-[2-deoxy-3,5-bis-O-(4-methylbenzoyl)-α-D-erythro-pentofuranosyl]-5-ethyl-6,7-dihydro- (same for β form)	MeOH	235(4.70),347(3.67)	87-1478-83

Compound	Solvent	$\lambda_{max}(\log \epsilon)$	Ref.
$C_{29}H_{30}O_{13}$			
Amarogentin	n.s.g.	230(4.46),266(4.07), 306(3.68)	162-0055-83
$C_{29}H_{30}O_{15}$			
Naphtho[2,3-c]furan-1(3H)-one, 9-acetoxy-6-methoxy-8-[(2,3,4,6-tetra-O-acetyl-β-D-glucopyranosyl)oxy]-	MeOH	257(4.23),330(3.37)	83-0409-83
$C_{29}H_{31}NO_8$			
Ravidomycin, deacetyl-	MeOH	246(4.54),279(4.47), 286(4.56),307(4.21), 319(4.09),334(4.06), 349(3.96)	158-1490-83
$C_{29}H_{32}N_2O_7$			
Antibiotic MF 722-02	99% MeOH	264s(4.51),278(4.56), 318(4.44)	158-83-159
Antibiotic U-56,407	MeOH	265s(4.55),315(4.78)	158-0950-83
	MeOH-HCl	265s(4.55),315(4.78)	158-0950-83
	MeOH-NaOH	265s(4.64),307(4.78)	158-0950-83
$C_{29}H_{32}N_4O$			
1H-Pyrido[3,4-b]indol-6-ol, 1-[[16(R)-19E]-19,20-didehydro-17-norcorynan-16-yl]-2,3,4,9-tetrahydro-	MeOH	227(4.58),284(4.10), 292(4.10)	78-3645-83
16S-	MeOH	225(4.36),282(3.84), 291(3.75)	78-3645-83
$C_{29}H_{32}N_4O_8$			
25-Dihydrosaframycin A	MeOH	268(4.01)	158-83-22
$C_{29}H_{32}O_2$			
Methanone, [3-[7-hydroxy-8,8-dimethyl-3,5-bis(1-methylethyl)bicyclo[4.2.0]-octa-1,3,5-trien-7-yl]phenyl]phenyl-	C_6H_{12}	248(4.30),345(2.15)	35-1117-83
Methanone, [4-[7-hydroxy-8,8-dimethyl-3,5-bis(1-methylethyl)bicyclo[4.2.0]-octa-1,3,5-trien-7-yl]phenyl]phenyl-	C_6H_{12}	261(4.39),348(2.26)	35-1117-83
$C_{29}H_{32}O_8$			
Collatolone	EtOH	240(4.26),248(4.68), 262(3.57),282(3.27), 298(2.34),334(3.17)	105-0011-83
$C_{29}H_{33}ClN_4OS$			
1-Piperazinamine, 4-(8-chloro-10,11-dihydrodibenzo[b,f]thiepin-10-yl)-N-[[2-[2-(dimethylamino)ethoxy]phenyl]methylene]-	MeOH	312.5(4.28),397(4.15)	73-1173-83
1-Piperazinamine, 4-(8-chloro-10,11-dihydrodibenzo[b,f]thiepin-10-yl)-N-[[3-[2-(dimethylamino)ethoxy]phenyl]methylene]-	MeOH	294(4.38)	73-1173-83
$C_{29}H_{33}FN_2O_4$			
2'H-Pregna-2,4,6-trieno[3,2-c]pyrazol-20-one, 9-fluoro-11,17,21-trihydroxy-16-methyl-2'-phenyl-, (11β,16α)-	MeOH	220(4.12),285(4.25), 313(4.29)	39-2793-83C

Compound	Solvent	$\lambda_{max}(\log \epsilon)$	Ref.
$C_{29}H_{33}NO_4$			
Ethanone, 1-[(6β,7α,14β)-3,6-dimethoxy-17-methyl-4-phenoxy-6,14-ethenomorphinan-7-yl]-	EtOH	222(4.23),266s(3.26), 272(3.40),278(3.48), 287s(3.34)	44-4137-83
$C_{29}H_{33}NO_8$			
2-Butenedioic acid, 2,2'-(5,6,10,11,12-13-hexahydro-4H-cycloocta[4,5]pyrrolo[3,2,1-ij]quinoline-8,13a(9H)-diyl)-bis-, tetramethyl ester, (?,Z,Z)-	MeOH	209(4.20),243(3.99), 279(3.93),418(3.82)	39-0515-83C
4H-Cyclodeca[4,5]pyrrolo[3,2,1-ij]quinoline-8,9-dicarboxylic acid, 5,6,11-12,13,14,15,15a-octahydro-15a-[3-methoxy-1-(methoxycarbonyl)-3-oxo-1-propenyl]-, dimethyl ester, (Z,Z,Z)-	MeOH	209(4.26),366(4.19)	39-0515-83C
Ravidomycin, deacetyldihydro-	MeOH	244(4.73),267(4.47), 275(4.59),305(4.10), 318(4.09),326(4.07)	158-1490-83
$C_{29}H_{33}N_2$			
3H-Indolium, 2-[7-(1,3-dihydro-1,3,3-trimethyl-2H-indol-2-ylidene)-1,3,5-heptatrienyl]-1,3,3-trimethyl-, perchlorate	C_6H_{12}	748(5.35)	99-0150-83
	MeOH	740(5.37)	99-0150-83
	EtOH	742(5.36)	99-0150-83
	PrOH	747(5.37)	99-0150-83
	isoPrOH	744(5.37)	99-0150-83
	$PhCH_2OH$	762(5.38)	99-0150-83
	$HOCH_2CH_2OH$	750(5.37)	99-0150-83
	acetone	742(5.33)	99-0150-83
	Ac_2O	744(5.37)	99-0150-83
	Mecellosolve	745(5.36)	99-0150-83
	EtOOCCOOEt	750(5.38)	99-0150-83
	$HCONH_2$	750(5.37)	99-0150-83
	DMF	746(5.30)	99-0150-83
	$MeNO_2$	740(5.37)	99-0150-83
	MeCN	740(5.35)	99-0150-83
	$C_6H_5NO_2$	760(5.39)	99-0150-83
	pyridine	757(5.37)	99-0150-83
	quinoline	770(5.38)	99-0150-83
	CH_2Cl_2	758(5.49)	99-0150-83
	$ClCH_2CH_2Cl$	756(5.46)	99-0150-83
	$o-C_6H_4Cl_2$	767(5.45)	99-0150-83
	DMSO	750(5.28)	99-0150-83
	sulfolane	750(5.34)	99-0150-83
(and other solvents not listed)	o-nitroanisole	760(5.37)	99-0150-83
$C_{29}H_{34}N_2$			
1H-Pyrazole, 4,5-dihydro-1-(4-octylphenyl)-3,5-diphenyl-	n.s.g.	245(4.23),254s(--), 313s(--),363(4.23)	103-0229-83
$C_{29}H_{34}N_2O_8$			
1H-Pyrrole-3-acetic acid, 2-[[5-(ethoxycarbonyl)-4-(2-ethoxy-2-oxoethyl)-3-methyl-1H-pyrrol-2-yl]methyl]-4-methyl-5-[(phenylmethoxy)carbonyl]-, methyl ester	$CHCl_3$	274(4.49),285(4.51)	78-1849-83
$C_{29}H_{34}N_4O$			
Usambarensine, 4',5',6',17-tetrahydro-, N-oxide	EtOH	224(4.58),274(4.02), 284(4.07),290(4.0)	100-0694-83

Compound	Solvent	λ_{max}(log ϵ)	Ref.
$C_{29}H_{34}N_4OS$			
1-Piperazinamine, 4-(10,11-dihydrodibenzo[b,f]thiepin-10-yl)-N-[[2-[2-(dimethylamino)ethyl]phenyl]methylene]-	MeOH	235(4.30),280(4.25), 290s(4.24),313(4.25)	73-1173-83
$C_{29}H_{34}N_4O_2$			
21H-Biline-1,19-dione, 3,17-diethyl-22,24-dihydro-2,7,8,12,13,18-hexamethyl-, (E,Z,Z)-	$CHCl_3$	375(4.16),572(4.23)	108-0187-83
(Z,Z,Z)-	$CHCl_3$	365(4.71),628(4.19)	108-0187-83
$C_{29}H_{34}O_3$			
19-Norpregn-4-en-20-yn-3-one, 17-(1-oxo-3-phenylpropoxy)-, (17α)-	$CHCl_3$	241(4.26)	13-0309-83A
$C_{29}H_{34}O_5$			
Cyclopentanone, 2,2'-[1,5-bis(2-methoxyphenyl)-3-oxo-1,5-pentanediyl]bis-	MeOH	214(4.24),275(3.66), 280(3.65)	2-0864-83
$C_{29}H_{34}O_8$			
24-Nor-12,13-secochola-1,13,20,22-tetraen-12-oic acid, 6-acetoxy-7,15:21,23-diepoxy-4-formyl-4,8-dimethyl-3-oxo-, methyl ester, (4α,5α,6α,7α,15β,17α)-	EtOH	220(3.99)	18-1139-83
$C_{29}H_{34}O_9$			
α-Collatolic acid (2λ,3ε)	EtOH	210(4.45),316(3.99), ?(3.46)	105-0011-83
β-Collatolic acid	EtOH	248(4.42),276(3.70), 290(3.57),320(3.62)	105-0011-83
$C_{29}H_{35}NO_4$			
Acetamide, N-[(6α,17β)-17-acetoxy-3-(phenylmethoxy)estra-1,3,5(10)-trien-6-yl]-	MeOH	211(4.33),279(3.27), 285(3.20)	145-0347-83
$C_{29}H_{36}N_2O_4$			
Eme t amine	EtOH	236(4.85),283(3.86)	162-0513-83
$C_{29}H_{36}N_4O_2$			
21H-Biline-1,19-dione, 17-ethyl-2,3,22,24-tetrahydro-3,3,7,8,12-13,18,21-octamethyl- (isomer mixture)	$CHCl_3$	291(4.28),332(4.61), 588(4.37),668s(4.03)	49-0753-83
21H-Biline-1,19-dione, 17-ethyl-2,3,23,24-tetrahydro-3,3,7,8,12-13,18,21-octamethyl- (isomer mixture)	$CHCl_3$	296(4.42),303(4.46), 344(4.48),528(4.52)	49-0753-83
$C_{29}H_{36}N_4O_3$			
Verbaskine	EtOH	219(4.16),225(4.05), 284(4.20)	88-4381-83
$C_{29}H_{36}O_3$			
19-Norpregn-4-en-20-yn-3-one, 17-[(1-oxo-2-nonen-4-ynyl)oxy]-, [17α,17(E)]-	EtOH	242(4.29)	13-0277-83A
19-Norpregn-4-en-20-yn-3-one, 17-[(1-oxo-4-nonen-6-ynyl)oxy]-, (17α,17(E)]-	EtOH	238(4.47)	13-0277-83A
$C_{29}H_{36}O_4$			
18,19-Dinorpregn-4-en-20-yn-3-one, 13-	$CHCl_3$	241(4.26)	13-0339-83A

Compound	Solvent	$\lambda_{max}(\log \epsilon)$	Ref.
ethyl-17-[3-(5-methyl-2-furanyl)-1-oxopropoxy]- (cont.)			13-0339-83A
19-Norpregn-4-en-20-yn-3-one, 17α-[3-(5- e thyl-2-furanyl)-1-oxopropoxy]-	$CHCl_3$	241(4.42)	13-0321-83A
19-Norpregn-4-en-20-yn-3-one, 17α-[4-(5-methyl-2-furanyl)-1-oxobutoxy]-	$CHCl_3$	241(4.42)	13-0321-83A
$C_{29}H_{36}O_5$			
6H-Dibenzo[b,d]pyran, 7,8,9,10-tetrahydro-1-methoxy-3,6,6,9-tetramethyl-2-[2-(3,4,5-trimethoxyphenyl)ethenyl]-	$CHCl_3$	288(4.46),324(4.45)	83-0326-83
$C_{29}H_{36}O_6$			
4H-1-Benzopyran-4-one, 2,3-dihydro-7-(methoxymethoxy)-2-[4-(methoxymethoxy)-3-(3-methyl-2-butenyl)-phenyl]-8-(3-methyl-2-butenyl)-	MeOH	273(4.21),311(3.98)	2-1061-83
5H-Cyclopenta[3,4]cyclopropa[8,9]cycloundec[1,2-b]oxiren-5-one, 4a-acetoxy-2-(benzoyloxy)-1a,1b,2,3,4,4a,7a,8-8a,9,10,10a-dodecahydro-3,6,8,8,10a-pentamethyl-	MeOH	228(4.08),270(3.96)	102-1791-83
2-Propen-1-one, 1-[2-hydroxy-4-(methoxymethoxy)-3-(3-methyl-2-butenyl)-phenyl]-3-[4-(methoxymethoxy)-3-(3-methyl-2-butenyl)phenyl]-	MeOH	252(3.61),352(3.79)	2-1061-83
$C_{29}H_{36}O_7$			
Zeylasterone, demethyl-	EtOH	210(4.15),220(4.00), 250(3.99),302(3.67), 340(3.64)	88-2025-83
$C_{29}H_{36}O_8$			
1,4-Ethanonaphthalene-2,8,10-trione, 1,3,4,4a,5,8a-hexahydro-3,5-dihydroxy-9-(1-hydroxy-2,4-hexadienyl)-6-methoxy-1,3,5,7-tetramethyl-8a-(1-	MeOH	259(4.19),300s(4.07), 308(4.03),370(4.36), 380(4.37),383(4.37), 405s(4.18)	78-4243-83
oxo-4-hexenyl)-, [1α,3β,4α,4aβ,5α-8aβ(E),9Z(2E,4E)]-(+)-	MeOH-NaOH	226(4.11),274(4.32), 393(4.22)	78-4243-83
Ingol, 12-O-acetyl-8-O-benzoyl-	MeOH	209(4.32),231(4.39), 274(3.26),282(3.22)	102-2795-83
$C_{29}H_{37}Cl_2NO_2$			
Estra-1,3,5(10)-trien-17-ol, 6-[bis(2-chloroethyl)amino]-3-(phenylmethoxy)-, (6α,17β)-	MeOH	210(4.25),225(4.03), 278(3.25),286(3.22)	145-0347-83
$C_{29}H_{37}N_2$			
3H-Indolium, 2-[2-[(1,3-dihydro-1,3,3-trimethyl-2H-indol-2-ylidene)methyl]-3,3-dimethyl-1-butenyl]-1,3,3-trimethyl-, tetrafluoroborate, all-cis	MeOH	529(4.33)	89-0876-83
$C_{29}H_{38}ClFO_8$			
Formocortal	EtOH	216(4.08),324(4.23)	162-0605-83
$C_{29}H_{38}N_2O_4$			
Emetine, 2,3-dehydro-, (+)-, dihydrobromide	EtOH	282(3.87)	162-0414-83
(-)-	EtOH	282(3.86)	162-0414-83

Compound	Solvent	λ_{max}(log ε)	Ref.
Isoemetine, 2,3-dehydro-, (+)-, dihydrobromide	EtOH	285(3.87)	162-0414-83
(-)-	EtOH	285(3.87)	162-0414-83
$C_{29}H_{38}O_3$			
18,19-Dinorpregn-4-en-20-yn-3-one, 17α-[(1-cyclohexen-1-ylacetyl)oxy]-13-ethyl-	MeOH	240(4.21)	13-0339-83A
19-Norpregn-4-en-20-yn-3-one, 17α-[(1-oxo-2,3-nonadienyl)oxy]-	MeOH	239(4.25)	13-0277-83A
19-Norpregn-4-en-20-yn-3-one, 17α-[(1-oxo-4-nonynyl)oxy]-	EtOH	239.5(4.18)	13-0267-83A
19-Norpregn-4-en-20-yn-3-one, 17α-[(1-oxo-5-nonynyl)oxy]-	EtOH	240(4.22)	13-0267-83A
19-Norpregn-4-en-20-yn-3-one, 17α-[(1-oxo-6-nonynyl)oxy]-	EtOH	239.5(4.25)	13-0267-83A
$C_{29}H_{38}O_{10}$			
Ajugamarin, 12-oxo-	EtOH	290(3.56)	94-2192-83
Roridin L-2, 16-hydroxy-	MeOH	261(3.90)	88-3539-83
$C_{29}H_{39}NO_4$			
Estra-1,3,5(10)-trien-17-ol, 6-[bis(2-hydroxyethyl)amino]-3-(phenylmethoxy)-, (6α,17β)-	MeOH	210(4.22),225(4.07), 278(3.24),287(3.19)	145-0347-83
$C_{29}H_{39}N_3O_2$			
Echinuline	EtOH	230(4.60),279(3.98), 286(3.96)	162-0507-83
$C_{29}H_{40}$			
Cyclopentadecane, 1-(diphenylmethylene)-3-methyl-	EtOH	245(4.04)	33-2608-83
$C_{29}H_{40}N_2O_2S$			
Hexadecanamide, N-[4-(2-benzothiazolyl)-3-hydroxyphenyl]-	EtOH	343(4.51)	4-1517-83
$C_{29}H_{40}N_2O_3$			
Hexadecanamide, N-[4-(2-benzoxazolyl)-3-hydroxyphenyl]-	EtOH	330(4.56)	4-1517-83
$C_{29}H_{40}O_2$			
Furan, 2,2'-(phenylmethylene)bis[5-heptyl-	EtOH	227(4.35)	103-0478-83
$C_{29}H_{40}O_3$			
18,19-Dinorpregn-4-en-20-yn-3-one, 17α-[(cyclohexylacetyl)oxy]-13-ethyl-	MeOH	244(4.23)	13-0349-83A
18,19-Dinorpregn-4-en-20-yn-3-one, 13-ethyl-17-[[(4-methylcyclohexyl)-carbonyl]oxy]-, [17α,17(trans)]-	MeOH	242(4.25)	13-0349-83A
19-Norpregn-4-en-20-yn-3-one, 17α-[[(3-butylcyclobutyl)carbonyl]oxy]-	MeOH	240(4.18)	13-0291-83A
19-Norpregn-4-en-20-yn-3-one, 17α-[(5-cyclobutyl-1-oxopentyl)oxy]-	MeOH	241(4.24)	13-0291-83A
19-Norpregn-4-en-20-yn-3-one, 17-[(1-oxo-2-nonenyl)oxy]-, [17α,17(E)]-	EtOH	215(4.29),239(4.26)	13-0267-83A
19-Norpregn-4-en-20-yn-3-one, 17α-[(1-oxo-4-nonenyl)oxy]-	EtOH	240(4.25)	13-0267-83A

Compound	Solvent	$\lambda_{max}(\log \epsilon)$	Ref.
19-Norpregn-4-en-20-yn-3-one, 17-[(1-oxo-5-nonenyl)oxy]-, [17α,17(E)]-	MeOH	239(4.23)	13-0267-83A
(Z)-	EtOH	239.5(4.22)	13-0267-83A
19-Norpregn-4-en-20-yn-3-one, 17-[(1-oxo-6-nonenyl)oxy]-, [17α,17(Z)]-	EtOH	239.5(4.24)	13-0267-83A
$C_{29}H_{40}O_6$			
Cortisone 21β-cyclopentanepropanoate	EtOH	239(4.21)	162-0362-83
$C_{29}H_{40}O_9$			
Affinoside VIII	MeOH	264(4.17)	94-1199-83
Calotropin	n.s.g.	217(4.21),310(1.49)	162-0236-83
Ingol, 3,12-di-O-acetyl-7-O-tigloyl-	MeOH	209(4.13)	102-2795-83
$C_{29}H_{40}O_{10}$			
Affinoside S-III	MeOH	264(4.30)	94-1199-83
Ajugamarin	EtOH	212(4.30)	94-2192-83
$C_{29}H_{40}O_{11}$			
Affinoside VI	MeOH	264(4.04)	94-1199-83
Affinoside S-V	MeOH	264(4.19)	94-1199-83
$C_{29}H_{41}N_3O_2$			
Hexadecanamide, N-[4-(1H-benzimidazol-2-yl)-3-hydroxyphenyl]-	EtOH	327(4.56)	4-1517-83
$C_{29}H_{42}N_6O_4$			
2-Pyrrolidinecarboxamide, 1,1'-[1,7-heptanediylbis(1,3(4H)-pyridinediylcarbonyl)]bis-, [S-(R*,R*)]-	$CHCl_3$	350(4.09)	44-1370B-83
$C_{29}H_{42}O_3$			
18,19-Dinorpregn-4-en-20-yn-3-one, 13-ethyl-17α-[(2-ethyl-1-oxohexyl)oxy]-	MeOH	244(4.20)	13-0349-83A
18,19-Dinorpregn-4-en-20-yn-3-one, 13-ethyl-17α-[(1-oxooctyl)oxy]-	MeOH	240(4.24)	13-0349-83A
Estradiol 17-undecenoate	n.s.g.	281(3.30)	162-0536-83
Furan, 2,2',2"-ethylidynetris[5-pentyl-	EtOH	228(4.35)	103-0474-83
$C_{29}H_{42}O_5$			
Antheridiol	EtOH	220(4.23)	162-0099-83
$C_{29}H_{42}O_7S_2$			
1,2-Cyclohexanediethanol, 4-(1-hydroxy-1-methylethyl)-β^2,1-dimethyl-, α,α'-bis(4-methylbenzenesulfonate), [1S-[1α,2β(R*),4β]]-	MeOH	225(4.39),257(3.98), 262(3.08),267(3.05), 273(3.02)	23-1111-83
$C_{29}H_{42}O_8S$			
Prost-13-en-1-oic acid, 6,9-epoxy-11,15-dihydroxy-5-[(4-methylphenyl)-sulfonyl]-4-oxo-, ethyl ester	EtOH	230(1.110)	88-0315-83
$C_{29}H_{42}O_9$			
Helveticoside	EtOH	217(4.27),304(1.55)	162-0669-83
Ingol, 7,12-di-O-acetyl-8-O-tigloyl-	MeOH	208(4.22)	102-2795-83
$C_{29}H_{42}O_{11}$			
α-Antiarin	n.s.g.	217(4.08),305(1.8)	162-0101-83

Compound	Solvent	$\lambda_{max}(\log \epsilon)$	Ref.
$C_{29}H_{43}NO_5$			
2,4,6,8,10,14-Heptadecahexaenamide, 13-acetoxy-N-(2-acetoxy-1-methylethyl)-2,10,12,14,16-pentamethyl-	hexane	198(3.94),255(3.77), 261(3.80),339(4.45), 353(4.54),372(4.42)	5-1081-83
$C_{29}H_{43}N_3$			
1H-Benzimidazole, 1-hexadecyl-2-(4-methyl-2-pyridinyl)-	EtOH	302(4.28)	4-0023-83
$C_{29}H_{43}N_3O_4$			
Oxayohimban-16-carboxylic acid, 16,17-didehydro-2,7-dihydro-11-methoxy-1,10,19-trimethyl-, 2-(diethylamino)ethyl ester, (2α,7α,19α)-	MeOH	243(4.25),302(3.69)	20-0263-83
$C_{29}H_{43}N_5O_7SSi_2$			
Guanosine, 3',5'-O-[1,1,3,3-tetrakis(1-methylethyl)-1,3-disiloxanediyl]-, 2'-(O-phenyl carbonothioate)	EtOH	250(4.22)	35-4059-83
$C_{29}H_{44}O_3$			
6,19-Epoxy-9,10-secocholesta-5,7,10(19)-trien-3β-ol, acetate	EtOH	237.5(4.22),283(4.29), 295.5(4.17)	44-3477-83
4(5H)-Oxepinone, 5-[3,5-bis(1,1-dimethylethyl)-4-hydroxyphenyl]-3,5-bis-(1,1-dimethylethyl)-7-methyl-	CH_2Cl_2	262(4.20)	44-3700-83
$C_{29}H_{44}O_4$			
Cholest-4-ene-3,6-dione, 2α-acetoxy-	$CHCl_3$	254(3.92)	104-1116-83
5α-Cholest-7-ene-3,6-dione, 2β-acetoxy-	MeOH	246(3.68)	104-1116-83
$C_{29}H_{44}O_5$			
25-Norscalar-17-ene-18,24-carbolactone, 20-acetoxy-12β-hydroxy-20,24-dimethyl-, (20S,24S)-	EtOH	218(3.98)	39-0155-83C
$C_{29}H_{45}IO_3$			
Cholest-4-en-6-one, 3β-acetoxy-2α-iodo-	MeOH	229(3.85)	104-1116-83
$C_{29}H_{45}N_3O_6$			
Digitoxoside, 24-azido-	EtOH	236(4.10)	69-6303-83
$C_{29}H_{45}N_5O_6Si_2$			
Guanosine, N-benzoyl-2',3'-bis-O-[(1,1-dimethylethyl)dimethylsilyl]-	MeOH	234(4.24),257(4.19), 263(4.19),293(4.19)	33-2069-83
Guanosine, N-benzoyl-2',5'-bis-O-[(1,1-dimethylethyl)dimethylsilyl]-	MeOH	234(4.24),257(4.19), 264(4.19),293(4.19)	33-2069-83
Guanosine, N-benzoyl-3',5'-bis-O-[(1,1-dimethylethyl)dimethylsilyl]-	MeOH	236(4.24),258(4.19), 264(4.19),294(4.19)	33-2069-83
$C_{29}H_{46}N_2O_2$			
Cholest-5-en-3β-ol, diazoacetate	n.s.g.	250(4.05)	35-3283-83
$C_{29}H_{46}O_2$			
Stigmast-4-ene-3,6-dione (92.6%)	EtOH	251(4.01)	102-2087-83
Vitamin D_3, 3,3-ethylenedioxy-, (5E)-	hexane	272(3.97)	78-1123-83
$C_{29}H_{46}O_3$			
5α-Cholesta-8,14-dien-7β-ol, 3β-acetoxy-	EtOH	248(4.1)	39-0587-83C
5α-Lanostane-3,11,12-trione	EtOH	283(2.46)	23-1973-83

Compound	Solvent	λ_{max}(log ε)	Ref.
5α-Lanostane-3,11,12-trione (cont.)	EtOH-KOH	330(3.56)	23-1973-83
$C_{29}H_{46}O_4$			
Cholest-4-en-6-one, 2α-acetoxy-3α-hydroxy-	MeOH	228(3.48)	104-1116-83
5α-Cholest-7-en-6-one, 2β-acetoxy-3β-hydroxy-	MeOH	246(3.54)	104-1116-83
Cholest-7-en-6-one, 3α-acetoxy-5β-hydroxy-	EtOH	252(4.16)	78-2779-83
Digitoxigenin, 21ξ,22-dipropyl-	EtOH	226(4.14)	48-0599-83
$C_{29}H_{46}O_5$			
Cholest-7-en-6-one, 3α-acetoxy-5β,14α-dihydroxy-	EtOH	246(3.95)	78-2779-83
$C_{29}H_{46}O_9$			
Pregn-20-ene-21-carboxylic acid, 3-(β-D-galactopyranosyloxy)-14-hydroxy-, methyl ester, (3β,5β,14β,20E)-	MeOH	217(4.23)	13-0037-83B
Pregn-20-ene-21-carboxylic acid, 3-(β-D-glucopyranosyloxy)-14-hydroxy-, methyl ester, (3β,5β,14β,20E)-	MeOH	216(4.27)	13-0037-83B
$C_{29}H_{47}NO_7$			
Mycinamicin VII	MeOH	215(4.29),281(4.32)	158-0175-83
$C_{29}H_{48}O$			
Stigmast-4-en-3-one, (86.2%)	EtOH	241(4.15)	102-2087-83
$C_{29}H_{50}N_2O_4$			
L-Alanine, 3-heptadecyl-1,2-phenylene ester, dihydrochloride	MeOH	210(4.04),264(2.54)	36-0792-83
$C_{29}H_{50}O_4$			
2-Pentenedioic acid, 2-(1-dodecynyl)-3-undecyl-, 1-methyl ester, (Z)-	MeOH	251(3.81)	23-2449-83
$C_{29}H_{52}FNO_9$			
Erythromycin, 3-O-de(2,6-dideoxy-3-C-methyl-3-O-methyl-α-L-ribo-hexopyranosyl)-12-deoxy-8-fluoro-	MeOH	284(1.55)	158-1439-83
$C_{29}H_{52}FNO_{10}$			
Erythromycin, 3-O-de(2,6-dideoxy-3-C-methyl-3-O-methyl-α-L-ribo-hexopyranosyl)-8-fluoro-	MeOH	287(1.35)	158-1439-83
$C_{29}H_{56}N_4O_4Si_3$			
7H-Pyrrolo[2,3-d]pyrimidin-4-amine, 7-[2,3,5-tris-O-[(1,1-dimethylethyl)dimethylsilyl]-β-D-ribofuranosyl]-	MeOH	272(4.09)	5-1169-83

Compound	Solvent	λ_{max}(log ϵ)	Ref.
$C_{30}H_{12}Cl_8N_4S_2$			
6,13-Triphenodithiazinediamine, 1,3,8,10-tetrachloro-N,N'-bis(2,4-dichlorophenyl)-	dioxan	599(4.54)	18-1482-83
	H_2SO_4	345(4.48)	18-1482-83
$C_{30}H_{12}Cl_{14}NO_4$			
L-Alanine, N-[(phenylmethoxy)carbonyl]-4-[bis(pentachlorophenyl)methylene]-2,3,5,6-tetrachlorophenyl ester (radical)	C_6H_{12}	221(4.94),272s(3.76), 336s(3.83),365s(4.29), 381(4.53),482s(3.03), 500(3.01),554(3.00)	44-3716-83
$C_{30}H_{12}O_9$			
5,6,11,12,17,18-Trinaphthylenehexone, 1,7,13-trihydroxy-	$CHCl_3$	233(4.36),277(4.55), 287(4.54),432(4.14)	5-0433-83
$C_{30}H_{14}N_2$			
Phenanthro[9',10':3,4]cyclobut[1,2-b]-anthracene-9,16-dicarbonitrile	CH_2Cl_2	251(4.61),258(4.62), 301s(4.23),305(4.28), 318(4.45),332s(4.31), 345s(4.67),357(4.84), 370(4.78),384(4.54), 400s(4.24),423(4.10), 440s(3.94),465s(3.63), 480s(3.30)	39-1443-83C
$C_{30}H_{16}Br_4N_4S_2$			
6,13-Triphenodithiazinediamine, 3,10-dibromo-N,N'-bis(4-bromophenyl)-	dioxan	586(4.57)	18-1482-83
	H_2SO_4	336(4.51)	18-1482-83
$C_{30}H_{16}N_8S_4$			
2-Benzothiazolamine, N,N'-[3-(2-benzothiazolyl)-2H-1,3,5-triazino[2,1-b]-benzothiazole-2,4(3H)-diylidene]bis-	dioxan	240(4.33),322(3.96), 385(3.82)	48-0463-83
$C_{30}H_{18}$			
5H-5,14a[1',2']-Benzenobenzo[rst]pentaphene	C_6H_{12}	271(4.82),314s(3.51), 371(3.28),391(3.69), 414(4.02),440(4.11)	89-0495-83
$C_{30}H_{18}Br_2N_4S_2$			
6,13-Triphenodithiazinediamine, N,N'-bis(4-bromophenyl)-	dioxan	555(4.64)	18-1482-83
	H_2SO_4	323(4.68)	18-1482-83
6,13-Triphenodithiazinediamine, 3,10-dibromo-N,N'-diphenyl-	dioxan	591(4.56)	18-1482-83
	H_2SO_4	339(4.54)	18-1482-83
$C_{30}H_{18}Cl_2N_4S_2$			
6,13-Triphenodithiazinediamine, N,N'-bis(4-chlorophenyl)-	dioxan	555(4.63)	18-1482-83
	H_2SO_4	323(4.64)	18-1482-83
6,13-Triphenodithiazinediamine, 3,10-dichloro-N,N'-diphenyl-	dioxan	579(4.62)	18-1482-83
	H_2SO_4	357(4.62)	18-1482-83
$C_{30}H_{18}N_4O_6$			
1H,7H-Pyrazolo[1,2-a]pyrazole-1,7-dione, 3,5-bis(4-nitrophenyl)-2,6-diphenyl-	dioxan	255(4.47),297(4.14), 415(4.04)	151-0171-83A
$C_{30}H_{18}O_8$			
9,10-Anthracenedione, 1,1'-[1,2-ethanediylbis(oxy)]bis[8-hydroxy-	CH_2Cl_2	411(4.04)	18-3185-83

Compound	Solvent	$\lambda_{max}(\log \epsilon)$	Ref.
$C_{30}H_{18}O_{10}$			
Amentoflavone	EtOH	227(4.18),267(4.18), 290(4.18),337(4.16)	94-0919-83
$C_{30}H_{18}O_{11}$			
4H-1-Benzopyran-4-one, 8-[5-(5,7-dihydroxy-4-oxo-4H-1-benzopyran-2-yl)-2-hydroxyphenyl]-2-(3,4-dihydroxyphenyl)-5,7-dihydroxy- (5"'-hydroxyamentoflavone)	EtOH	227(4.15),267(4.14), 290(4.11),339(4.10)	94-0919-83
$C_{30}H_{19}BrO$			
Cyclopent[a]inden-1(3aH)-one, 2-bromo-3,3a,8-triphenyl-	$CHCl_3$	244(4.37),289(4.15), 364(3.95)	18-3314-83
Spiro[1H-indene-1,1'(4'H)-naphthalen]-4'-one, 2-bromo-2',3-diphenyl-	$CHCl_3$	243(4.37),270s(4.19), 283(4.21)	18-3314-83
$C_{30}H_{19}Br_3$			
Cyclobuta[b]naphthalene, 1,1,2-tribromo-1,3-dihydro-3,3,8-triphenyl-	$CHCl_3$	252(4.33),326(4.10)	18-3314-83
$C_{30}H_{19}ClO$			
Cyclopent[a]inden-1(3aH)-one, 2-chloro-3,3a,8-triphenyl-	$CHCl_3$	240(4.31),295(4.12), 360(3.96)	18-3314-83
$C_{30}H_{20}$			
5H-5,14a[1',2']-Benzenobenzo[rst]pentaphene, 5a,12d-dihydro-	C_6H_{12}	250(4.5),277(4.5), 290(4.5),320f(4.0)	89-0495-83
$C_{30}H_{20}Cl_4N_6$			
2H-1,4-Benzodiazepin-2-one, 7-chloro-5-(2-chlorophenyl)-1,3-dihydro-, [7-chloro-5-(2-chlorophenyl)-3H-1,4-benzodiazepin-2-yl]hydrazone	MeOH	305(4.53)	73-0123-83
$C_{30}H_{20}N_4O_2$			
Benzo[f]quinoline, 3,4-dihydro-4-(2-naphthalenyl)-3-[[(4-nitrophenyl)-azo]methylene]-	neutral	572(4.79)	124-0068-83
	protonated	474(4.06)	124-0068-83
$C_{30}H_{20}N_4S_2$			
6,13-Triphenodithiazinediamine, N,N'-diphenyl-	dioxan	555(4.70)	18-1482-83
	H_2SO_4	328(4.77)	18-1482-83
$C_{30}H_{20}O_5$			
6H-Phenaleno[1,9-bc]pyran-6-one, 4-(4-methoxybenzoyl)-2-(4-methoxyphenyl)-	$CHCl_3$	283(4.56),386(4.01), 482s(4.20),511(4.41), 546(4.32)	39-1267-83C
$C_{30}H_{20}O_{10}$			
4H-1-Benzopyran-4-one, 2-[3-[5,7-dihydroxy-2-(4-hydroxyphenyl)-4-oxo-4H-1-benzopyran-8-yl]-4-hydroxyphenyl]-2,3-dihydro-5,7-dihydroxy- (2,3-dihydroamentoflavone)	EtOH	223(4.20),288(4.17), 339(4.05)	94-0919-83
$C_{30}H_{21}As$			
1H-Arsole, 2,5-di-2-naphthalenyl-1-phenyl-	C_6H_{12}	217(4.34),240(4.24), 283(3.59),293(3.61), 303(3.54),398(4.18)	101-0335-83I

Compound	Solvent	λ_{max}(log ϵ)	Ref.
$C_{30}H_{21}Cl_2N_3O_{10}$			
3(2H)-Pyridazinone, 4,5-dichloro-6-nitro-2-(2,3,5-tri-O-benzoyl-β-D-ribofuranosyl	MeOH	275(3.77),281(3.77), 302(3.65)	4-0369-83
$C_{30}H_{21}Cl_3N_6O_{10}$			
Ethanone, 1,1'-[(2,2,2-trichloroethylidene)bis[(2,6-dinitro-4,1-phenylene)-imino-3,1-phenylene]]bis-	MeOH	231.2(4.72),394.3(4.03)	56-1357-83
$C_{30}H_{21}N_3O_3$			
2,5-Cyclohexadiene-1-carbonitrile, 1,2,6-triphenyl-4-(tetrahydro-1-methyl-2,4,6-trioxo-5(2H)-pyrimidinylidene)-	$CHCl_3$	385(4.44)	83-0732-83
$C_{30}H_{21}O_2$			
Pyrylium, 3-benzoyl-2,4,6-triphenyl-, perchlorate	MeCN	275(4.32),355(4.58), 404(4.48)	22-0115-83
$C_{30}H_{22}$			
1a,6[1',2']:7,11b[1",2"]-Dibenzeno-1H-dibenzo[a,e]cyclopropa[c]cyclooctene, 6,7-dihydro-6-methyl-	CCl_4	255(3.85),266(3.64), 275(3.54),285(3.40)	44-5120-83
o-Quaterphenyl, 3'-phenyl-	C_6H_{12}	187(4.89),198(4.91), 245(4.61)	94-1572-83
m-Terphenyl, 2,6-diphenyl-	C_6H_{12}	199(4.89),201s(4.88), 237s(4.68),243(4.71)	94-1572-83
p-Terphenyl, 2,6-diphenyl-	C_6H_{12}	205(4.89),251(4.60), 281(4.37)	94-1572-83
p-Terphenyl, 3,5-diphenyl-	C_6H_{12}	202(4.86),263(4.72), 280s(4.61)	94-2313-83
$C_{30}H_{22}ClFN_2O_8$			
3(2H)-Pyridazinone, 6-chloro-5-fluoro-2-(2,3,5-tri-O-benzoyl-β-D-ribofuranosyl)-	MeOH	274(3.74),282(3.73), 300s(3.49)	4-0369-83
$C_{30}H_{22}Cl_2N$			
Pyridinium, 2-(2,5-dichlorophenyl)-4,6-diphenyl-1-(phenylmethyl)-, salt with trifluoromethanesulfonic acid	DMSO	314(4.44)	150-0301-83M
$C_{30}H_{22}Cl_2N_2O_8$			
3(2H)-Pyridazinone, 5,6-dichloro-2-(2,3,5-tri-O-benzoyl-β-D-ribofuranosyl)-	MeOH	274(3.70),282(3.68), 305(3.60)	4-0369-83
$C_{30}H_{22}F_2O_8$			
Cyclobutane, r-1,t-2-bis[(4-hydroxy-6-methyl-2-oxo-2H-pyran-3-yl)oxomethyl]-c-3,t-4-bis(3-fluorophenyl)-	MeOH	202(4.6),311(4.4), 350(4.4)	83-0951-83
$C_{30}H_{22}N_2$			
2-Butene-2,3-diamine, N,N'-di-9H-fluoren-9-ylidene-	MeCN	470(3.0)	35-6833-83
$C_{30}H_{22}N_2O_6$			
5-Benzoxazolecarboxylic acid, 2,2'-(2,3-dimethyl-1,5-naphthalenediyl)bis-, dimethyl ester	DMF	327(4.42)	5-0931-83

Compound	Solvent	λ_{max}(log ϵ)	Ref.
$C_{30}H_{22}N_6O_2$			
Benzoic acid, 4-[5-(4,6-diphenyl-2-pyrimidinyl)-3-phenyl-1-formazano]-	EtOH	450(4.17)	65-0153-83
	acetone	460(4.08)	65-0153-83
cobalt chelate	EtOH	550(3.93),610s(--)	65-0153-83
copper chelate	EtOH	605(4.13)	65-0153-83
nickel chelate	EtOH	605(4.45)	65-0153-83
$C_{30}H_{22}N_8$			
Diazene, 1,1'-(1,4-phenylene)bis[2-[4-(phenylazo)phenyl]-	n.s.g.	290(4.96),522(2.62)	35-3722-83
$C_{30}H_{22}O_5$			
6H-Phenaleno[1,9-bc]pyran-6-one, 2-[hydroxy(4-methoxyphenyl)methyl]-2-(4-methoxyphenyl)-	$CHCl_3$-10% MeOH	282(4.63),310s(4.07), 486s(4.15),514(4.40), 549(4.39)	39-1267-83C
$C_{30}H_{22}O_{10}$			
[3,8'-Bi-4H-1-benzopyran]-4,4'-dione, 2,2',3,3'-tetrahydro-5,5',7,7'-tetrahydroxy-2,2'-bis(4-hydroxyphenyl)-	EtOH	293(4.4),330(3.5)	102-0233-83
$C_{30}H_{23}ClN$			
Pyridinium, 4-(2-chlorophenyl)-2,6-diphenyl-1-(phenylmethyl)-, tetrafluoroborate	DMSO	307(4.39)	150-0301-83M
$C_{30}H_{23}N_2O_2$			
Pyridinium, 2-(4-nitrophenyl)-4,6-diphenyl-1-(phenylmethyl)-, salt with trifluoromethanesulfonic acid	DMSO	310(4.42)	150-0301-83M
$C_{30}H_{23}N_4OS_2$			
Thiazolo[4',3':2,3]pyrimido[4,5-b]quinolin-12-ium, 1-[(3-ethyl-2(3H)-benzothiazolylidene)methyl]-8-methoxy-3-phenyl-, perchlorate	DMF	514(3.57)	103-0166-83
$C_{30}H_{24}$			
1,3,5,11,13-Cyclopentadecapentaene-7,9-diyne, 15-(diphenylmethylene)-6,11-dimethyl-	CH_2Cl_2	275(4.39),332(4.40), 410s(4.06)	39-2987-83C
[2.2.2.0]Metaparacyclophanediene	C_6H_{12}	262(4.38)	1-0297-83
$C_{30}H_{24}N_2$			
1,4-Benzenediamine, N,N,N',N'-tetraphenyl-, radical cation	MeCN	405(4.22),815(4.14)	152-0105-83
2,4,6-Heptatrienal, 7-(2-ethynylphenyl)-, [7-(2-ethynylphenyl)-2,4,6-heptatrienylidene]hydrazone	THF	230(4.22),268(4.07), 295s(3.96),357s(4.25), 418(4.80),437(4.79), 459s(4.62)	18-1467-83
$C_{30}H_{24}N_3S_2$			
Thiazolo[3,4-a]pyrimidin-5-ium, 6-[3-(3-ethyl-2(3H)-benzothiazolylidene)-1-propenyl]-3,8-diphenyl-, perchlorate	MeCN	496(3.19),530(3.19), 572(3.75)	103-0037-83
$C_{30}H_{24}N_4$			
1H-Pyrazole-4,5-diamine, N^5,1,3-triphenyl-N^4-(3-phenyl-2-propenylidene)-	EtOH	238(3.46),287(3.47)	4-1501-83

Compound	Solvent	λ_{max}(log ε)	Ref.
$C_{30}H_{24}N_6$			
Pyrido[2,3-b]pyrazine, 3,3'-(1,2-ethanediylidene)bis[3,4-dihydro-2-methyl-4-phenyl-	$CHCl_3$-CF_3COOH	267(4.26),381(4.05), 559(4.39),596(4.37)	33-0379-83
2(1H)-Pyrimidinone, 4,6-diphenyl-, [[(4-methylphenyl)azo]phenylmethylene]hydrazone	EtOH	435(4.18)	65-0153-83
	acetone	445(4.19)	65-0153-83
cobalt chelate	EtOH	540(3.78),600(3.75)	65-0153-83
copper chelate	EtOH	600(3.92)	65-0153-83
nickel chelate	EtOH	595(4.16)	65-0153-83
$C_{30}H_{24}N_6O$			
2(1H)-Pyrimidinone, 4,6-diphenyl-, [[(2-methoxyphenyl)azo]phenylmethylene]hydrazone	EtOH	470(4.31)	65-0153-83
	acetone	470(4.24)	65-0153-83
cobalt chelate	EtOH	570(3.75),620s(--)	65-0153-83
copper chelate	EtOH	520(3.90)	65-0153-83
nickel chelate	EtOH	590(4.38)	65-0153-83
2(1H)-Pyrimidinone, 4,6-diphenyl-, [[(4-methoxyphenyl)azo]phenylmethylene]hydrazone	EtOH	435(4.21)	65-0153-83
	acetone	450(4.16)	65-0153-83
cobalt chelate	EtOH	550s(--),600(3.56)	65-0153-83
copper chelate	EtOH	610(3.76)	65-0153-83
nickel chelate	EtOH	600(3.38)	65-0153-83
$C_{30}H_{24}O_2$			
Anthracenone dimer, m. 177-9°	EtOH	237(4.70),254(4.80), 262s(4.74),290(4.23), 300(4.23),310s(4.12), 370(3.74),388s(3.59)	35-3234-83
2H-Pyran, 2-methoxy-2,3,4,6-tetraphenyl-	dioxan	237(4.30),243s(4.30), 334(4.20)	48-0729-83
$C_{30}H_{24}O_4$			
5H-Naphtho[2,3-b]pyran-5-one, 10-(2,10-dihydro-2,2-dimethyl-10-oxo-5H-naphtho[2,3-b]pyran-5-ylidene)-2,10-dihydro-2,2-dimethyl-	EtOH	215(4.54),268(4.36)	2-0886-83
$C_{30}H_{24}O_{10}$			
Elsinochrome A	MeOH	457(4.11),527(3.87), 567(3.94)	74-0385-83C
	MeOH-NaOH	461(4.29),570(3.89), 614(4.11)	74-0385-83
	$CHCl_3$	458(4.31),528(4.00), 571(4.07)	74-0385-83C
$C_{30}H_{25}Cl_3N_6O_{10}$			
Benzenamine, 4,4'-(2,2,2-trichloroethylidene)bis[N-(2-ethoxyphenyl)-2,6-dinitro-	MeOH	207.4(4.78),420.9(4.05)	56-1357-83
$C_{30}H_{25}NO_2PRe$			
Rhenium, benzoyl(η^5-2,4-cyclopentadien-1-yl)nitrosyl(triphenylphosphine)-	$CHCl_3$	265(3.63),295s(3.34)	157-1852-83
$C_{30}H_{26}$			
Pentadeca-3,5,8,10,12-pentaene-1,14-diyne, 3,13-dimethyl-7-(diphenylmethylene)-	CH_2Cl_2	271(4.56),289s(4.41), 334s(4.56),350(4.65), 380(4.70)	39-2987-83C

Compound	Solvent	$\lambda_{max}(\log \epsilon)$	Ref.
$C_{30}H_{26}N$			
Pyridinium, 1-(1-methylethyl)-2-(1-naphthalenyl)-4,6-diphenyl-, tetrafluoroborate	DMSO	305(4.37)	150-0301-83M
$C_{30}H_{26}N_2O_4$			
1,2-Cyclobutanedicarboxamide, 3,4-bis-(2-hydroxyphenyl)-N,N'-diphenyl-	MeOH	245(4.43),273(3.96), 281(3.76)	39-1083-83C
$C_{30}H_{26}N_2O_6S_2$			
Benzo[1,2-b:4,5-b']difuran-3,7-dicarboxylic acid, 2,6-dihydro-2,6-bis-[(phenylthio)imino]-, bis(1-methylethyl) ester, (Z,Z)-	$CHCl_3$	658(4.86)	150-0480-83M +150-0028-83S
$C_{30}H_{26}N_2O_8S_2$			
Benzo[1,2-b:4,5-b']difuran-3,7-dicarboxylic acid, 2,6-dihydro-2,6-bis[[(2-methoxyphenyl)thio]imino]-, diethyl ester	$CHCl_3$	650(4.92)	150-0028-83S
$C_{30}H_{26}N_8$			
2H-Benzimidazol-2-one, 1,3-dihydro-1-(phenylmethyl)-, [1-[[1-(phenylmethyl)-1H-benzimidazol-2-yl]azo]-ethylidene]hydrazone	benzene	530(4.64)	65-2332-83
	EtOH	530(4.64)	65-2332-83
	acetone	530(4.64)	65-2332-83
anion	EtOH	575(--),607(4.82)	65-2332-83
	acetone	575(--),607(4.82)	65-2332-83
$C_{30}H_{26}O_2$			
Butadiene, 1,1-bis(4-methoxyphenyl)-4,4-diphenyl-	CH_2Cl_2	258(4.23),352(4.47)	152-0105-83
$C_{30}H_{26}O_2Si$			
2,4-Dioxa-3-silabicyclo[3.2.0]hept-6-ene, 3,3-dimethyl-1,5,6,7-tetraphenyl-	hexane	225(4.54),294(4.19)	138-0209-83
1,3-Dioxa-2-silacyclohepta-4,6-diene, 2,2-dimethyl-4,5,6,7-tetraphenyl-	hexane	288(4.21),317(4.26)	138-0209-83
$C_{30}H_{26}O_4$			
Anthracene, 2,3,6,7-tetramethoxy-9,10-diphenyl-	$CHCl_3$	278(4.95),366(4.05), 375(4.06)	78-0623-83
$C_{30}H_{26}O_5$			
2H-Naphtho[2,3-b]pyran-5,10-dione, 4-[1,4-dihydro-3-(3-methyl-2-butenyl)-1,4-dioxo-2-naphtalenyl]-3,4-dihydro-2,2-dimethyl-	EtOH	212(4.085),248(4.10), 278(3.664)	2-0886-83
$C_{30}H_{26}O_8$			
7H-Furo[3,2-g][1]benzopyran-7-one, 9,9'-[1,8-octanediylbis(oxy)]bis-	EtOH	218(4.69),248(4.65), 263(4.43),300(4.37)	39-1807-83B
D-ribo-Hept-2-enonic acid, 2,3-dideoxy-2-(2-hydroxyethyl)-4,6-O-(phenylmethylene)-, γ-lactone, 6,8-dibenzoate, (S)-	$CHCl_3$	245(3.85)	87-0030-83
$C_{30}H_{27}Fe_3N$			
Ferrocene, 1,1",1""-nitrilotris-	n.s.g.	317(3.8),454(2.9)	101-0227-83A

Compound	Solvent	λmax(log ε)	Ref.
$C_{30}H_{27}NO_3$			
1,4-Methanodibenzo[a,e]cycloocten-5,12-imine-6,11-dione, 1,2,3,4,4a-5,12,12a-octahydro-13-(4-methoxyphenyl)-5-phenyl-	benzene	302(3.62),356s(2.56), 370s(2.49),440(2.04)	44-4968-83
	MeOH	246(4.08),298(3.34), 355s(2.46),366s(2.43), 440(1.95)	44-4968-83
	MeCN	246(4.30),298(3.56), 355s(2.52),366s(2.51), 440(2.20)	44-4968-83
$C_{30}H_{27}N_2O_5P$			
3,4-Oxepindicarboxylic acid, 5-cyano-6,7-dihydro-7-methyl-2-[(triphenylphosphoranylidene)amino]-, dimethyl ester	$CHCl_3$	275(3.94),298(3.94), 368(4.34)	24-1691-83
$C_{30}H_{27}N_3O_8$			
1H-Indole, 3,3'-(1-acetyl-3,4-diacetoxy-2-oxo-3,4-pyrrolidinediyl)bis-[1-acetyl-	MeOH	236(4.39),262(4.10), 290(4.05),298(4.06)	142-0469-83
$C_{30}H_{27}N_3O_{15}$			
Enterobactin	EtOAc	316(3.97)	162-0518-83
$C_{30}H_{27}O$			
2H-Furylium, 3-(2,6-diphenyl-1,3,5-hexatrienyl)-2,2-dimethyl-5-phenyl-, perchlorate	MeCN	572(4.83)	56-0579-83
	HOAc	588(--)	56-0579-83
$C_{30}H_{28}N$			
Benzo[h]quinolinium, 1-cyclopentyl-5,6-dihydro-2,4-diphenyl-, tetrafluoroborate	EtOH	352(4.23)	39-1435-83B
$C_{30}H_{28}NO_2$			
3H-Indolium, 2-[3-[2-(2-methoxyphenyl)-4H-1-benzopyran-4-ylidene]-1-propenyl]-1,3,3-trimethyl-, perchlorate	MeCN	590(4.78),634(4.69)	103-0948-83
	CH_2Cl_2	602(4.83),648(4.88)	103-0948-83
3H-Indolium, 2-[3-[2-(3-methoxyphenyl)-4H-1-benzopyran-4-ylidene]-1-propenyl]-1,3,3-trimethyl-, perchlorate	MeCN	584(4.80),626(4.66)	103-0948-83
	CH_2Cl_2	596(4.85),640(4.82)	103-0948-83
3H-Indolium, 2-[3-[2-(4-methoxyphenyl)-4H-1-benzopyran-4-ylidene]-1-propenyl]-1,3,3-trimethyl-, perchlorate	MeCN	596(4.82),640(4.79)	103-0948-83
	CH_2Cl_2	606(4.85),650(4.94)	103-0948-83
3H-Indolium, 2-[3-(5-methoxy-2-phenyl-4H-1-benzopyran-4-ylidene)-1-propenyl]-1,3,3-trimethyl-, perchlorate	MeCN	584(4.77),624(4.68)	103-0948-83
	CH_2Cl_2	596(4.84),636(4.87)	103-0948-83
3H-Indolium, 2-[3-(6-methoxy-2-phenyl-4H-1-benzopyran-4-ylidene)-1-propenyl]-1,3,3-trimethyl-, perchlorate	MeCN	592(4.84),636(4.77)	103-0948-83
	CH_2Cl_2	604(4.86),646(4.91)	103-0948-83
3H-Indolium, 2-[3-(7-methoxy-2-phenyl-4H-1-benzopyran-4-ylidene)-1-propenyl]-1,3,3-trimethyl-, perchlorate	MeCN	588(4.83),630(4.81)	103-0948-83
	CH_2Cl_2	598(4.88),638(4.95)	103-0948-83
3H-Indolium, 2-[3-(8-methoxy-2-phenyl-4H-1-benzopyran-4-ylidene)-1-propenyl]-1,3,3-trimethyl-, perchlorate	MeCN	584(4.64),626(4.54)	103-0948-83
	CH_2Cl_2	598(4.76),639(4.76)	103-0948-83
$C_{30}H_{28}N_2O_3$			
6,9-Methano-2H-pyrimido[2,1-b][1,3]oxazepin-2-one, 6,7,8,9-tetrahydro-3-methyl-8-[(triphenylmethoxy)methyl]-	MeOH	257(3.86)	4-0655-83

Compound	Solvent	λ_{max}(log ϵ)	Ref.
$C_{30}H_{28}N_4O_2$			
Benzamide, 4,4'-azobis[N-ethyl-N-phenyl-	o-$C_6H_4Cl_2$	343(4.35)	18-1700-83
$C_{30}H_{28}N_6S_2$			
Isothiazolo[5,1-e][1,2,3]thiadiazole-7-S^{IV}, 6,6'-(1,2-phenylene)bis[1,6-dihydro-3,4-dimethyl-1-phenyl-	C_6H_{12}	205(4.58),245(4.43), 291(4.36),470(4.57)	39-0777-83C
$C_{30}H_{28}O_2$			
Benzene, 1,1'-(1-methyl-3,3-diphenyl-2-propenylidene)bis[4-methoxy-	CH_2Cl_2	258(4.26)	152-0105-83
Isobenzofuran, 1,3-dihydro-1-methoxy-1,3,3-tris(4-methylphenyl)-	hexane	259s(3.06),265(3.12), 269s(2.97),273(2.95)	104-0553-83
$C_{30}H_{28}O_4$			
5H-Naphtho[2,3-b]pyran-5-one, 10-(2,10-dihydro-2,2-dimethyl-10-oxo-5H-naphtho[2,3-b]pyran-5-ylidene)-2,10-dihydro-2,2-dimethyl-, tetrahydro deriv.	EtOH	420(4.25)	2-0886-83
Riccardin B dimethyl ether	EtOH	217(4.41),277(3.79)	44-2164-83
$C_{30}H_{28}O_5$			
2H-Naphtho[2,3-b]pyran-5,10-dione, 4-[1,4-dihydro-3-(3-methyl-2-butenyl)-1,4-dioxo-2-naphthalenyl]-3,4-dihydro-2,2-dimethyl-, dihydro deriv.	EtOH	415(3.46)	2-0886-83
$C_{30}H_{28}O_7$			
Curvularin dibenzoate	EtOH	236(4.57)	162-0383-83
$C_{30}H_{29}NO_6$			
Phenol, 2-[[2-hydroxy-3-[(2-hydroxy-5-methylphenyl)methyl]-5-methylphenyl]-methyl]-6-[(2-hydroxy-5-nitrophenyl)-methyl]-4-methyl-	neutral	325(3.95)	126-2361-83
	anion	403(4.18)	126-2361-83
	dianion	418(4.29)	126-2361-83
$C_{30}H_{29}O_3$			
2H-Furylium, 3-[4-(2,4-dimethoxyphenyl)-2-phenyl-1,3-butadienyl]-2,2-dimethyl-5-phenyl-, perchlorate	MeCN	592(4.67)	56-0579-83
	HOAc	574(--),645(--)	56-0579-83
2H-Furylium, 3-[4-(2,5-dimethoxyphenyl)-2-phenyl-1,3-butadienyl]-2,2-dimethyl-5-phenyl-, perchlorate	MeCN	552(4.58)	56-0579-83
	HOAc	588(--)	56-0579-83
2H-Furylium, 3-[4-(3,4-dimethoxyphenyl)-2-phenyl-1,3-butadienyl]-2,2-dimethyl-5-phenyl-, perchlorate	MeCN	580(4.71)	56-0579-83
	HOAc	631(--)	56-0579-83
$C_{30}H_{30}ClN_2S_2$			
Benzothiazolium, 2-[2-[2-chloro-3-[4-(3-ethyl-2(3H)-benzothiazolylidene)-2-butenylidene]-1-cyclohexen-1-yl]-ethenyl]-3-ethyl-, iodide	MeOH	902(5.16)	104-1854-83
$C_{30}H_{30}N$			
Benzo[h]quinolinium, 5,6-dihydro-1-(1-methylbutyl)-2,4-diphenyl-, tetrafluoroborate	pentanol	350(4.07)	39-1443-83B
	EtOH-HOAc	350(4.24)	39-1443-83B
	EtOH-CF_3COOH	350(4.25)	39-1443-83B
	C_6H_5Cl	350(4.23)	39-1443-83B
Benzo[h]quinolinium, 1-(1,2-dimethylpropyl)-5,6-dihydro-2,4-diphenyl-,	pentanol	350(4.06)	39-1443-83B
	EtOH-HOAc	350(4.27)	39-1443-83B

Compound	Solvent	$\lambda_{max}(\log \epsilon)$	Ref.
tetrafluoroborate (cont.)	EtOH-CF_3COOH	350(4.30)	39-1443-83B
	C_6H_5Cl	350(4.24)	39-1443-83B
Benz[h]quinolinium, 1-(1-ethylpropyl)-5,6-dihydro-2,4-diphenyl-, tetrafluoroborate	pentanol	350(4.10)	39-1443-83B
	EtOH-HOAc	350(4.24)	39-1443-83B
	EtOH-CF_3COOH	350(4.29)	39-1443-83B
	C_6H_5Cl	350(4.26)	39-1443-83B
$C_{30}H_{30}NO$			
2H-Furylium, 3-[4-[4-(dimethylamino)-phenyl]-2-phenyl-1,3-butadienyl]-2,2-dimethyl-5-phenyl-, perchlorate	MeCN	764(5.04)	56-0579-83
	HOAc	774(--)	56-0579-83
$C_{30}H_{30}NO_4PS$			
2H-Thiocin-5,7-dicarboxylic acid, 3,4-dihydro-8-[(triphenylphosphoranylidene)amino]-, 5-ethyl 7-methyl ester	$CHCl_3$	267(3.95),275(3.93), 353(4.20)	24-1691-83
$C_{30}H_{30}NO_5P$			
3,5-Oxepindicarboxylic acid, 6,7-dihydro-7-methyl-2-[(triphenylphosphoranylidene)amino]-, 5-ethyl 3-methyl ester	MeOH	272(4.10),280(4.11), 365(4.46)	24-1691-83
$C_{30}H_{30}N_2$			
Pyridine, 4,4'-(2,4-dimethyl-2,4-diphenyl-1,3-cyclobutanediylidene)-bis[1,4-dihydro-1-methyl-, trans	5:1 MeCN-EtOH	330(4.57),343(4.53)	5-0069-83
	protonated	311(4.27),324(4.18)	5-0069-83
	diprotonated	256(4.24)	5-0069-83
Pyridinium, 4,4'-(2,4-dimethyl-2,4-diphenylbicyclo[1.1.0]butane-1,3-diyl)-bis[1-methyl-, (1α,2α,3α,4β)-, diperchlorate	MeCN	268(4.15),298(4.06)	5-0069-83
$C_{30}H_{30}N_2O_4$			
2,4(1H,3H)-Pyrimidinedione, 1-[3-hydroxy-4-[(triphenylmethoxy)methyl]-cyclopentyl]-5-methyl-, (1α,3β,4α)-(±)-	MeOH	271(3.99)	4-0655-83
$C_{30}H_{30}N_2O_{10}$			
6,15-Dioxa-1,10-diazatricyclo[14.2.0-$0^{7,10}$]octadeca-2,11-diene-2,11-dicarboxylic acid, 3,12-dihydroxy-9,18-dioxo-, bis(phenylmethyl) ester, (7RS,16RS)-	EtOH	254(4.31)	39-0115-83C
(7RS,16SR)-	EtOH	252(4.29)	39-0115-83C
$C_{30}H_{30}O_5$			
Phenol, 2,4-bis[(4-hydroxyphenyl)methyl]-5-methoxy-3-[2-(3-methoxyphenyl)-ethyl]-	MeOH	277s(3.86),283(3.96), 290s(3.89)	102-1011-83
$C_{30}H_{30}O_9$			
Toromycin acetonide	MeOH	245(4.43),265s(--), 275(4.30),285(4.32), 390(4.11)	78-0599-83
$C_{30}H_{30}O_{10}$			
Phleichrome	MeOH	350s(--),470(4.29), 539(3.99),581(3.99)	74-0385-83C

Compound	Solvent	λ_{max}(log ε)	Ref.
Phleichrome (cont.)	MeOH-NaOH	481(3.38),598(3.97), 640(4.20)	74-0385-83C
	$CHCl_3$	472(4.23),542(3.90), 585(3.90)	74-0385-83C
$C_{30}H_{30}O_{19}$			
β-D-Glucopyranosiduronic acid, 2-[4-[[6-O-(carboxyacetyl)-β-D-glucopyranosyl]oxy]phenyl]-5-hydroxy-4-oxo-4H-1-benzopyran-7-yl	MeOH-DMSO	269(4.23),317(4.18)	88-5749-83
$C_{30}H_{32}N_2O$			
1(4H)-Naphthalenone, 2-[[4-(1,1-dimethyl)phenyl]amino]-4-[[4-(1,1-dimethylethyl)phenyl]imino]-	n.s.g.	208(3.46),284(3.27), 304s(2.76),492(2.76)	39-2711-83C
$C_{30}H_{32}O_6$			
Sanggenon J	EtOH	209(4.52),235(4.47), 261(4.35),282s(4.12), 320(3.99)	142-1071-83
	EtOH-$AlCl_3$	209(4.66),220s(4.55), 270(4.47),300s(4.09), 370(4.03)	142-1071-83
Sanggenon K	EtOH	208(4.52),228(4.42), 260(4.31),282(4.05), 320(3.93)	142-1071-83
	EtOH-$AlCl_3$	208(4.61),270(4.40), 286s(4.05),315(3.90), 373(3.90)	142-1071-83
$C_{30}H_{32}O_{14}$			
Dauroside B	EtOH	229s(4.19),294s(4.34), 318(4.45)	105-0134-83
Plumieride coumarate	EtOH	230(4.34),302s(--), 318(4.28)	102-0179-83
$C_{30}H_{34}N_2O_4$			
D-Ribitol, 1,4-anhydro-1-C-(1,3-diphenyl-2-imidazolidinyl)-2,3-O-(1-methylethylidene)-5-O-(phenylmethyl)-, (S)-	MeOH	253(4.50),293s(3.61)	23-0317-83
$C_{30}H_{34}O_3$			
19-Norpregn-4-en-20-yn-3-one, 17α-[[(2-phenylcyclopropyl)carbonyl]oxy]-	MeCN	229(3.94)	13-0291-83A
$C_{30}H_{35}ClO_4$			
19-Norpregn-4-en-20-yn-3-one, 17α-[2-(4-chlorophenoxy)-2-methyl-1-oxopropoxy]-	MeOH	236(4.36)	13-0309-83A
$C_{30}H_{36}N_4O_2$			
21H-Bilin-1(23H)-one, 3,17-diethyl-19-methoxy-2,7,8,12,13,18-hexamethyl-, (Z,Z,Z)-	$CHCl_3$	367(4.49),665(4.00)	108-0187-83
$C_{30}H_{36}O_3$			
19-Norpregn-4-en-20-yn-3-one, 17α-(1-oxo-4-phenylbutoxy)-	$CHCl_3$	242(4.20)	13-0309-83A
$C_{30}H_{36}O_5$			
2H-1-Benzopyran-2,7(3H)-dione, 4,8-di-	EtOH	212(4.19),245(4.11),	78-3923-83

Compound	Solvent	λ_{max}(log ε)	Ref.
hydro-5-hydroxy-8,8-bis(3-methyl-2-butenyl)-6-(2-methyl-1-oxobutyl)-4-phenyl-	EtOH	295(3.80),335(3.62)	78-3923-83
$C_{30}H_{37}NO_5$			
Phenol, 2,6-bis[[3-(1,1-dimethylethyl)-2-hydroxy-5-methylphenyl]methyl]-4-	neutral	333(3.95)	126-2361-83
	anion	428(4.21)	126-2361-83
	polyanion	449(4.45)	126-2361-83
$C_{30}H_{38}N_2O_6$			
4,5-Isoquinolinedicarboxylic acid, 1,2,4a,7,8,8a-hexahydro-2-[2-(1H-indol-3-yl)ethyl]-7-(2-methyl-1,3-dioxolan-2-yl)-, 4-(1,1-dimethyl ethyl) 5-methyl ester, (4aα,7α,8aα)-	EtOH	290(4.25)	44-4262-83
$C_{30}H_{38}O_3$			
Chamaecydinol	EtOH	332(4.24),346(4.29), 396(3.44)	88-1535-83
18,19-Dinorpregn-4-en-20-yn-3-one, 13-ethyl-17-[(1-oxo-2-nonen-4-ynyl)-oxy]-, [17α,17(E)]-	EtOH	242(4.34)	13-0339-83A
$C_{30}H_{38}O_4$			
18,19-Dinorpregn-4-en-20-yn-3-one, 13-ethyl-17α-[3-(5-ethyl-2-furanyl)-1-oxopropoxy]-	$CHCl_3$	240(4.24)	13-0339-83A
18,19-Dinorpregn-4-en-20-yn-3-one, 13-ethyl-17α-[4-(5-methyl-2-furanyl)-1-oxobutoxy]-	$CHCl_3$	241(4.29)	13-0339-83A
$C_{30}H_{38}O_6$			
Zeylasteral	EtOH	211(4.16),225(4.09), 253(4.08),291(3.71), 336(3.67)	88-2025-83
$C_{30}H_{38}O_7$			
D:A-Friedo-24-noroleana-1,3,5(10),7-tetraene-23,29-dioic acid, 2,3-di-hydroxy-6-oxo-, 29-methyl ester (zeylasterone)	EtOH	211(4.19),226(4.05), 255(4.08),295(3.79), 340(3.70)	39-2845-83C
$C_{30}H_{38}O_8$			
Ingol, 12-O-acetyl-7-O-benzoyl-8-meth-oxy-	MeOH	210(4.14),230(4.24), 274(3.18),282(3.07)	102-2795-83
$C_{30}H_{39}N_4Zn$			
Zinc(1+), (7,8,12,13,17,18-hexahydro-2,2,3,5,7,7,12,12,17,17-decamethyl-21H,23H-porphinato-N^{21},N^{22},N^{23},N^{24})-, (SP-4-3)-, perchlorate	EtOH	305(4.48),354(4.51), 516(3.88)	77-1404-83
$C_{30}H_{40}Cl_2N_2O_3$			
2-Azaandrost-4-en-3-one, 17α-[[[4-[bis(2-chloroethyl)amino]phenyl]-acetyl]oxy]-	EtOH	214(4.37),261(4.34), 300(3.34)	111-0041-83
$C_{30}H_{40}N_4$			
21H,23H-Porphine, hexahydro-2,2,7,7-12,13,17,18,18,20-decamethyl-	hexane	284(4.31),364/377(4.4), 493(3.77)	77-1404-83

Compound	Solvent	$\lambda_{max}(\log \epsilon)$	Ref.
$C_{30}H_{40}N_4O_9S$			
1-Azabicyclo[3.2.0]hept-2-ene-2-carboxylic acid, 6-ethyl-7-oxo-3-[[2-[[1-oxo-3-[[(2,2,5,5-tetramethyl-1,3-dioxan-4-yl)carbonyl]amino]propyl]amino]ethyl]thio]-, (4-nitrophenyl)-methyl ester	$CHCl_3$	270(3.99),319(3.79)	158-1473-83
$C_{30}H_{40}N_4O_{10}S$			
1-Azabicyclo[3.2.0]hept-2-ene-2-carboxylic acid, 6-(1-hydroxyethyl)-7-oxo-3-[[2-[[1-oxo-3-[[(2,2,5,5-tetramethyl-1,3-dioxan-4-yl)carbonyl]amino]-propyl]amino]ethyl]thio]-, (4-nitrophenyl)methyl ester	$CHCl_3$	270(4.08),319(3.99)	158-1473-83
$C_{30}H_{40}N_4O_{11}S$			
1-Azabicyclo[3.2.0]hept-2-ene-2-carboxylic acid, 6-(1-hydroxyethyl)-7-oxo-3-[[2-[[1-oxo-3-[[(2,2,5,5-tetramethyl-1,3-dioxan-4-yl)carbonyl]amino]-propyl]amino]ethyl]sulfinyl]-, (4-nitrophenyl)methyl ester	$CHCl_3$	268(4.08),314(3.88)	158-1473-83
$C_{30}H_{40}N_6O_{10}$			
Adenosine, 5'-O-[β-[(17-oxo-4-nitro-estra-1,3,5(10)-trien-3-yl)oxy]ethyl]-	EtOH	260(4.14)	87-0162-83
$C_{30}H_{40}O_3$			
Chamaecydin	EtOH	330(4.35),346(4.42), 396(3.47)	88-1535-83
Isochamaecydin	EtOH	348(4.43),333(4.37)[sic], 392(3.77)	88-1535-83
19-Norpregn-4-en-20-yn-3-one, 17α-[(spiro[3.5]non-1-ylcarbonyl)oxy]-	MeCN	235(4.18)	13-0291-83A
$C_{30}H_{40}O_4$			
Schisanlactone A	MeOH	212(4.11),265(3.89), 334(4.25)	88-2351-83
$C_{30}H_{40}O_5$			
Benzoic acid, 4-benzoyl-, 2-[(1-oxo-tetradecyl)oxy]ethyl ester	EtOH	257(4.38),343(2.25)	24-0761-83
$C_{30}H_{40}O_{12}$			
Aceroside III	EtOH	277(3.89)	94-1923-83
	EtOH-NaOH	297(--)	94-1923-83
Ingol, 3,7,8,12,18-penta-O-acetyl-18-hydroxy-	MeOH	209(4.16)	102-2795-83
$C_{30}H_{42}NO_7$			
Pregn-4-ene-3,11,20-trione, 17-hydroxy-21-[[(2,2,5,5-tetramethyl-1-oxo-3-pyrrolidinyl)carbonyl]oxy]-	EtOH	226(3.20)	13-0241-83B
$C_{30}H_{42}N_2O_5S$			
17-Aza-D-homoandrost-5-en-17a-one, 3β-acetoxy-, 17a-[O-[(2,4,6-trimethyl-phenyl)sulfonyl]oxime]	EtOH	220s(3.87),275(2.95), 284(2.91)	70-2141-83
17a-Aza-D-homoandrost-5-en-17-one, 3β-	EtOH	217s(4.13),275(2.70),	70-2141-83

Compound	Solvent	$\lambda_{max}(\log \epsilon)$	Ref.
acetoxy-, 17-[O-[(2,4,6-trimethyl-phenyl)sulfonyl]oxime] (cont.)		281(2.68)	70-2141-83
$C_{30}H_{42}N_4O_6$			
Porphyrin-2,18-dipropanoic acid, 8,12-bis(1-hydroxyethyl)-3,7,13,17-tetra-methyl-, dimethyl ester	$CHCl_3$	402(5.27),499(4.16), 533(3.92),568(3.83), 621.5(3.63)	12-1639-83
$C_{30}H_{42}O_3$			
Chamaecydin, dihydro-	EtOH	294(3.32)	88-1535-83
18,19-Dinorpregn-4-en-20-yn-3-one, 17α-[(5-cyclobutyl-1-oxopentyl)oxy]-13-ethyl-	MeOH	242(4.38)	13-0349-83A
19-Norpregn-4-en-20-yn-3-one, 17α-[[(2-hexylcyclopropyl)carbonyl]oxy]-	MeCN	234(4.21)	13-0291-83A
$C_{30}H_{42}O_4$			
Schisanlactone B	MeOH	251(4.15)	88-2355-83
$C_{30}H_{42}O_5$			
19-Norpregn-4-en-20-yn-3-one, 17α-[[2,2-dimethyl-5-(1-methylethoxy)-1,5-dioxopentyl]oxy]-	MeOH	247(4.56)	13-0255-83A
$C_{30}H_{42}O_6$			
19-Norpregn-4-en-20-yn-3-one, 10-(acet-yldioxy)-7-methyl-17-[(1-oxoheptyl)-oxy]-, (7α,17α)-	MeOH	235.8(4.09)	13-0401-83B
$C_{30}H_{42}O_7$			
Gitoxigenin, 22-allyl-, 3,16-diacetate	EtOH	222(4.18)	48-0599-83
$C_{30}H_{42}O_9$			
Ingol, 3,12-di-O-acetyl-7-O-angeloyl-8-methoxy-	MeOH	213(4.25)	102-2795-83
Ingol, 3,12-di-O-acetyl-7-O-tigloyl-8-methoxy-	MeOH	210(4.26)	102-2795-83
$C_{30}H_{42}O_{10}$			
Affinoside S-II	MeOH	265(4.22)	94-1199-83
acetate	MeOH	264(4.30)	94-1199-83
$C_{30}H_{43}N_9O_5$			
L-Lysine, N^2-[4-[[(2,4-diamino-6-pteri-dinyl)methyl]methylamino]benzoyl]-N^6-[(1,1-dimethylethoxy)carbonyl]-, 1,1-dimethylethyl ester	CH_2Cl_2	257(4.39),303(4.17), 370(3.85)	87-0111-83
$C_{30}H_{44}NO_7$			
1-Pyrrolidinyloxy, 3-[[[(11β)-11,17-di-hydroxy-3,20-dioxopregn-4-en-21-yl]-oxy]carbonyl]-2,2,5,5-tetramethyl-	EtOH	250(3.19)	13-0241-83
$C_{30}H_{44}N_6O_4$			
2-Pyrrolidinecarboxamide, 1,1'-[1,8-octanediylbis(1,3(4H)-pyridinediyl-carbonyl)]bis-	$CHCl_3$	354(4.01)	44-1370B-83
$C_{30}H_{44}O_3$			
18,19-Dinorpregn-4-en-20-yn-3-one, 13-	MeOH	245(4.17)	13-0349-83A

Compound	Solvent	λ_{max}(log ϵ)	Ref.
ethyl-17α-[(1-oxononyl)oxy]- (cont.)			13-0349-83A
19-Norpregn-4-en-20-yn-3-one, 17α-[(2-methyl-1-oxononyl)oxy]-	EtOH	246(4.15)	13-0255-83A
$C_{30}H_{44}O_6S$			
1-Phenanthrenemethanol, 7-(2,2-dimethyl-1,3-dioxolan-4-yl)-1,2,3,4,4a,4b-5,6,7,9,10,10a-dodecahydro-3-hydroxy-1,4a,7-trimethyl-, α-(4-methylbenzenesulfonate)	EtOH	225(4.62)	94-4409-83
$C_{30}H_{44}O_8$			
Simplexin	MeOH	243(3.92)	100-0563-83
$C_{30}H_{44}O_{10}$			
Affinoside S-I	MeOH	214(4.11)	94-1199-83
acetate	MeOH	215(4.27)	94-1199-83
$C_{30}H_{46}N_2$			
Diazene, bis[1-[4-(1,1-dimethylethyl)-2,2-dimethylpropyl]-	C_6H_{12}	361(1.34)	24-1787-83
$C_{30}H_{46}N_3$			
1H-Benzimidazolium, 1-hexadecyl-3-methyl-2-(4-methyl-2-pyridinyl)-, iodide	n.s.g.	282(4.19)	4-0023-83
$C_{30}H_{46}N_4O_5Si_2$			
Benzamide, N-[7-[2,3-bis-O-[(1,1-dimethylethyl)dimethylsilyl]- -D-ribofuranosyl]-7H-pyrrolo[2,3-d]pyrimidin-4-yl]-	MeOH	301(4.00)	5-1169-83
$C_{30}H_{46}O_2$			
19-Norlanosta-1,5(10)-diene-3,11-dione, 9β-methyl-	EtOH	325(3.68)	23-1973-83
$C_{30}H_{46}O_3$			
Cholesta-4,6-diene-7-propanoic acid, 3-oxo-	EtOH	303(4.38)	35-4033-83
Ebelin lactone	MeOH	268(4.58),278(4.67), 288(4.55)	102-1469-83
19-Norlanost-5-ene-3,7,11-trione, 9-methyl-, (9β,10α)-	n.s.g.	248(3.94)	138-1171-83
$C_{30}H_{46}O_4$			
Ergosta-7,22-dien-6-one, 3β-acetoxy-5α-hydroxy-	$CHCl_3$	255(4.10)	78-2779-83
Ergosta-8(14),22-dien-15-one, 3-acetoxy-9-hydroxy-, (3β,5α,22E)-	$CHCl_3$	255(4.15)	78-2201-83
$C_{30}H_{46}O_7$			
Propanoic acid, 2,2-dimethyl-, 4-[2-[5-acetoxy-2-[[(2-methoxyethoxy)-methoxy]methyl]-1-cyclohexen-1-yl]-ethenyl]-2,3,3a,6,7,7a-hexahydro-7a-methyl-1H-inden-1-yl ester	ether	277.5(4.47)	56-1267-83
Propanoic acid, 2,2-dimethyl-, 4-[[5-acetoxy-2-[[(2-methoxyethoxy)methoxy]methylene]cyclohexylidene]ethylidene]octahydro-7a-methyl-1H-inden-1-yl ester	ether	270.5(4.24)	56-1267-83

Compound	Solvent	$\lambda_{max}(\log \epsilon)$	Ref.
$C_{30}H_{47}NO_3$			
Ergosta-8(14),22-dien-15-one, 3-acetoxy-, 15-oxime, (3β,5α,22E)-	n.s.g.	252(4.08)	78-2201-83
$C_{30}H_{47}NO_4$			
Hexadecanoic acid, [7-(diethylamino)-2-oxo-2H-1-benzopyran-4-yl]methyl ester	EtOH	247(4.15),378.5(4.36)	94-3014-83
$C_{30}H_{47}N_3O_9S$			
Leukotriene C_4	MeOH	270(4.51),280(4.60), 290(4.49)	162-0784-83
$C_{30}H_{48}$			
Ferna-7,9(11)-diene	EtOH	204(4.17),232(4.11), 248(3.99)	102-1801-83
Multiflora-7,9(11)-diene	EtOH	228(4.09),240(4.16), 248(3.96)	102-1801-83
Oleana-11,13(18)-diene	EtOH	242(4.42),250(4.48), 260(4.32)	102-1801-83
$C_{30}H_{48}O$			
D:A-Friedoolean-1-en-3-one	EtOH	233(3.97)	2-0741-83
19-Norlanosta-1(10),2-dien-11-one, 9β-methyl-	EtOH	270(3.81)	23-1973-83
19-Norlanosta-1(10),5-dien-11-one, 9β-methyl-	EtOH	242(4.02)	23-1973-83
9,10-Secocholesta-1,3,5(10)-trien-11-one, 1,9,14-trimethyl-	EtOH	263(2.04),271(1.85)	23-1973-83
$C_{30}H_{48}O_2$			
D:A-Friedoolean-3-en-2-one, 3-hydroxy-	EtOH	278(3.72)	2-0741-83
D(17a)-Homo-C,18-dinorlanost-13(17a)-ene-3,11-dione, 17a-methyl-	EtOH	255(4.19)	23-1973-83
19-Norlanost-5(10)-ene-3,11-dione, 9β-methyl-	EtOH	293(2.18)	23-1973-83
$C_{30}H_{48}O_5$			
9,19-Cyclolanost-17(20)-en-16-one, 3,6,24,25-tetrahydroxy-, (3β,6α,24S)-	MeOH	256(3.92)	94-0689-83
Escigenin	EtOH	275(1.58)	162-0534-83
$C_{30}H_{48}O_9$			
Androstane-17-propanoic acid, 3-(β-D-galactopyranosyloxy)-14-hydroxy-α-methylene-, methyl ester, (3β,5β-14β,17β)-	MeOH	210(4.06)(end abs.)	13-0037-83B
Androstane-17-propanoic acid, 3-(β-D-glucopyranosyloxy)-14-hydroxy-α-methylene-, methyl ester, (3β,5β,14β,17β)-	MeOH	210(3.96)(end abs.)	13-0037-83B
$C_{30}H_{50}O$			
4,5-Secocholest-5-en-3-one, 5-(1-methylethenyl)-	EtOH	203(4.76),274(3.60)	18-3349-83
$C_{30}H_{50}O_2$			
D(17a)-Homo-C,18-dinorlanost-13(17a)-en-11-one, 3β-hydroxy-17a-methyl-	EtOH	255(4.15),356(2.00)	23-1973-83
19-Norlanost-1(10)-en-11-one, 3β-hydroxy-9β-methyl-	EtOH	296(1.94)	23-1973-83

Compound	Solvent	$\lambda_{max}(\log \epsilon)$	Ref.
19-Norlanost-5(10)-en-11-one, 3β-hydroxy-9β-methyl-	EtOH	296(1.94)	23-1973-83
$C_{30}H_{50}O_3$			
Iridogermanal, 10-deoxy-	EtOH	256(4.13)	64-0179-83C
Iridogermanal, 21-deoxy-	EtOH	255(4.05)	64-0179-83C
$C_{30}H_{50}O_3Si$			
Pregna-5,7-diene-20-methanol, 3-[[(1,1-dimethylethyl)dimethylsilyl]oxy]-, acetate, (3β,20S)-	EtOH	262(3.93),270(4.07), 283(4.10),294(3.88)	5-1031-83
$C_{30}H_{50}O_4$			
Isoiridogermanal	EtOH	256(4.24)	64-0179-83C
19-Norlanost-5-en-11-one, 3,24,25-trihydroxy-9-methyl- (11-oxomogrol)	n.s.g.	289(2.12)	95-1155-83
$C_{30}H_{51}NO_2$			
Cholest-5-en-3β-ol, 4,4,6-trimethyl-, nitrite	benzene	334(1.68),345(1.81), 357(1.95),373(1.95), 386(1.70)	18-3349-83
$C_{30}H_{52}N_2O_2$			
9-Hexadecenoic acid, 16-[(4-aminophenyl)octylamino]-, (E)-, hydrochloride	MeOH	253(4.09),295(3.32)	5-0802-83
$C_{30}H_{52}O_3$			
5α-Lanostan-11-one, 3β,12β-dihydroxy-	EtOH	277(1.94)	23-1973-83
$C_{30}H_{52}O_4$			
2-Pentenedioic acid, 2-(1-dodecynyl)-3-undecyl-, dimethyl ester, (Z)-	MeOH	261(3.89)	23-2449-83
$C_{30}H_{66}Si_3$			
Cyclotrisilane, hexakis(2,2-dimethylpropyl)-	C_6H_{12}	240s(3.89),265s(3.32), 310(2.52)	77-0781-83
Cyclotrisilane, hexakis(1-ethylpropyl)-	$C_6H_{11}Me$	304(2.43),328(2.38)	35-6524-83

Compound	Solvent	λmax(log ε)	Ref.
$C_{31}H_{14}Cl_{14}NO_3$			
Methyl, bis(pentachlorophenyl)[2,3,5,6-tetrachloro-4-[[[2-ethoxy-2-oxo-1-(phenylmethyl)ethyl]amino]carbonyl]phenyl]-, (S)-	C_6H_{12}	222(4.95),287(3.82), 336s(3.83),368s(4.30), 382(4.59),472s(3.06), 505(3.08),560(3.05)	44-3716-83
$C_{31}H_{17}O$			
Diacenaphtho[1,2-b:1',2'-e]pyrylium, 14-phenyl-, perchlorate	MeCN	230(4.42),265(4.18), 330(4.35),470(4.32), 496(4.47)	22-0115-83
$C_{31}H_{20}$			
9H-Fluorene, 9-(12,13,14,15-tetradehydro-11-methyl-7H-benzocyclotridecen-7-ylidene)-, (E,E,Z)-	EtOH	250(4.6),275(4.6), 330(4.1),440(4.5)	138-1887-83
$C_{31}H_{20}O_4$			
2H-Furo[2,3-h]-1-benzopyran-2-one, 8-([1,1'-biphenyl]-4-ylcarbonyl)-4-methyl-9-phenyl-	dioxan	252(2.35),292(2.56)	2-0164-83
$C_{31}H_{21}NO_5$			
Phenanthro[9',10':5,6][1,4]dioxino-[2,3-d]oxazol-11(9aH)-one, 12-acetyl-12,12a-dihydro-9a,12a-diphenyl-	MeOH	223(4.43),250(3.56), 257(4.64),270(3.18), 295(2.92),296(3.01), 310(3.09),344(2.00)	73-2812-83
$C_{31}H_{21}N_3$			
3,5-Pyridinedicarbonitrile, 2-[1,1'-biphenyl]-4-yl-1,4-dihydro-4,6-diphenyl-	EtOH	205(4.78),284(4.49), 358(3.64)	73-3123-83
$C_{31}H_{22}GeOS$			
4H-Cyclopenta[b]thiophen-4-one, 6-phenyl-5-(triphenylgermyl)-	CH_2Cl_2	265(4.52),463(3.55)	70-1049-83
$C_{31}H_{22}N_4O_2$			
2-Oxabicyclo[2.2.2]oct-7-ene-5,5,6,6-tetracarbonitrile, 3-methoxy-4-methyl-1,8-diphenyl-	dioxan	244(4.01)	48-0729-83
$C_{31}H_{23}BrO_2$			
Cyclopent[a]inden-1(3aH)-one, 2-bromo-8,8a-dihydro-8-methoxy-3,3a,8-triphenyl-	$CHCl_3$	288(4.27),358(2.95)	18-3314-83
$C_{31}H_{23}NO_2$			
5(2H)-Oxazolone, 2-methyl-4-phenyl-2-(1,2,3-triphenyl-2-cyclopropen-1-yl)-	EtOH	228(4.46),236(4.39), 266(4.23),300(4.33), 312(4.44),334(4.29)	44-0695-83
5(4H)-Oxazolone, 2-methyl-4-phenyl-4-(1,2,3-triphenyl-2-cyclopropen-1-yl)-	C_6H_{12}	238(4.34),299(4.43), 305(4.25),319(4.41), 337(4.31)	44-0695-83
$C_{31}H_{23}N_3O_3$			
2,5-Cyclohexadiene-1-carbonitrile, 1,2,6-triphenyl-4-(tetrahydro-1,3-dimethyl-2,4,6-trioxo-5(2H)-pyrimidinylidene)-	$CHCl_3$	380(4.45)	83-0732-83
$C_{31}H_{24}N_2O$			
Methanone, [4-(diphenylamino)phenyl]-	MeOH	280(4.26),375(4.58)	12-0409-83

Compound	Solvent	$\lambda_{max}(\log \epsilon)$	Ref.
[4-(phenylamino)phenyl]- (cont.)			12-0409-83
$C_{31}H_{24}N_4$			
[1,2,4]Triazolo[1,5-a]pyridinium, 5,7-diphenyl-2-(phenylamino)-1-(phenylmethyl)-, hydroxide, inner salt	EtOH	257(4.75)	88-3523-83
$C_{31}H_{24}N_6O_2$			
Benzoic acid, 4-[5-(4,6-diphenyl-2-pyrimidinyl)-3-phenyl-1-formazano]-, methyl ester	EtOH	450(--)	65-0153-83
	acetone	460(4.06)	65-0153-83
cobalt chelate	acetone	550(3.81),600s(--)	65-0153-83
copper chelate	acetone	595(4.35)	65-0153-83
nickel chelate	acetone	610(4.05)	65-0153-83
$C_{31}H_{26}N$			
Pyridinium, 4-(2-methylphenyl)-2,6-diphenyl-1-(phenylmethyl)-, tetrafluoroborate	DMSO	306(4.38)	150-0301-83M
$C_{31}H_{26}N_3S_2$			
Thiazolo[3,4-a]pyrimidin-5-ium, 6-[3-(3-ethyl-2(3H)-benzothiazolylidene)-1-propenyl]-4-methyl-2,8-diphenyl-, perchlorate	MeCN	560(3.27),678(3.71)	103-0037-83
$C_{31}H_{26}O_2$			
2H-Pyran, 2-methoxy-2,4,6-triphenyl-3-(phenylmethyl)-	dioxan	235(4.23),242(4.22), 323(4.24)	48-0729-83
$C_{31}H_{26}O_5$			
4H-1-Benzopyran-4-one, 2,3-dihydro-3-[[3-methoxy-4-(phenylmethoxy)-phenyl]methylene]-7-(phenylmethoxy)-	MeOH	216(4.53),243(4.41), 300(4.29),360(4.42)	2-1119-83
4H-1-Benzopyran-4-one, 3-[[3-methoxy-4-(phenylmethoxy)phenyl]methyl]-7-(phenylmethoxy)-	MeOH	212(4.98),282(4.53), 292s(4.52),306s(4.49)	2-1119-83
4H-1-Benzopyran-4-one, 3-methyl-7-(phenylmethoxy)-2- [3-methoxy-4-(phenylmethoxy)phenyl]-	MeOH	218(4.56),232(4.53), 304(4.42)	2-1119-83
$C_{31}H_{27}NO_5$			
2H-Furo[2,3-h]-1-benzopyran-2-one, 4-methyl-9-phenyl-8-[4-[2-(1-pyrrolidinyl)ethoxy]benzoyl]-	dioxan	252(4.39),282(4.46), 306(4.54)	2-0164-83
$C_{31}H_{27}NO_6$			
2H-Furo[2,3-h]-1-benzopyran-2-one, 2-methyl-8-[4-[2-(4-morpholinyl)ethoxy]benzoyl]-9-phenyl-	dioxan	250(4.21),280(4.36), 294(4.38)	2-0164-83
$C_{31}H_{28}N_2OS_2$			
Propanamide, 3-[bis(phenylmethyl)amino]-N,N-bis(phenylmethyl)-2,3-dithioxo-	MeCN	272(4.26),308(4.03), 377(2.85),450s(2.90)	5-1694-83
Propanedithioamide, 2-oxo-N,N,N',N'-tetrakis(phenylmethyl)-	MeCN	272(4.30),307(4.15), 370s(3.28)	5-1694-83
$C_{31}H_{28}N_2O_2S$			
Propanamide, 3-[bis(phenylmethyl)amino]-2-oxo-N,N-bis(phenylmethyl)-3-thioxo-	MeCN	269(4.70),370s(2.96)	5-1694-83

Compound	Solvent	$\lambda_{max}(\log \epsilon)$	Ref.
Propanediamide, N,N,N',N'-tetrakis-(phenylmethyl)-2-thioxo-	MeCN	224(3.69),247(3.57), 287(3.63),566(1.67)	5-1694-83
$C_{31}H_{28}N_2O_3$			
Propanediamide, 2-oxo-N,N,N',N'-tetra-kis(phenylmethyl)-	EtOH	218(4.28),257(3.77), 350(2.31)	5-1694-83
$C_{31}H_{28}N_8$			
Pyrazolo[1,5-a]pyrimidin-2-amine, N-[3-[(5,7-dimethyl-3-phenylpyrazolo-[1,5-a]pyrimidin-2-yl)amino]-2-prop-enylidene]-5,7-dimethyl-3-phenyl-	MeCN	276(4.61),370(4.51)	39-0011-83C
$C_{31}H_{29}NO_5$			
2H-Furo[2,3-h]-1-benzopyran-2-one, 8-[4-[2-(diethylamino)ethoxy]benzoyl]-4-methyl-9-phenyl-	dioxan	248(4.58),280(4.70), 292(4.71)	2-0164-83
$C_{31}H_{29}NO_7$			
D-Ribitol, 1-deoxy-1-(3,5-dimethyl-4-isoxazolyl)-2,4-O-(phenylmethylene)-, 3,5-dibenzoate, (S)-	$CHCl_3$	245(4.32)	87-0030-83
$C_{31}H_{29}N_2O_5P$			
3,4-Oxepindicarboxylic acid, 5-cyano-6,7-dihydro-6,7-dimethyl-2-[(tri-phenylphosphoranylidene)amino]-, dimethyl ester	$CHCl_3$	275(3.90),298(3.93), 371(4.39)	24-1691-83
3,4-Oxepindicarboxylic acid, 5-cyano-6,7-dihydro-7,7-dimethyl-2-[(tri-phenylphosphoranylidene)amino]-, dimethyl ester	$CHCl_3$	275(4.29),295(3.87), 372(4.31)	24-1691-83
$C_{31}H_{30}N$			
Benzo[h]quinolinium, 1-cyclohexyl-5,6-dihydro-2,4-diphenyl-, tetrafluoro-borate	C_6H_5Cl	347(4.27)	39-1435-83B
$C_{31}H_{30}NO_6PS$			
3,4,5-Thiepintricarboxylic acid, 6,7-dihydro-2-[(triphenylphosphoranyli-dene)amino]-, 5-ethyl 3,4-dimethyl ester	$CHCl_3$	276(3.90),314(4.07), 355(4.19)	24-1691-83
$C_{31}H_{30}NO_7P$			
3,4,5-Oxepintricarboxylic acid, 6,7-di-hydro-2-[(triphenylphosphoranyli-dene)amino]-, 5-ethyl 3,4-dimethyl ester	MeOH	270(4.00),354(4.23)	24-1691-83
$C_{31}H_{30}N_2O_6$			
D-Ribitol, 1-deoxy-1-(3,5-dimethyl-1H-pyrazol-4-yl)-2,4-O-(phenylmethyl-ene)-, 3,5-dibenzoate, (S)-	$CHCl_3$	245(4.19)	87-0030-83
$C_{31}H_{30}O_4$			
Riccardin A trimethyl ether	EtOH	210(4.77),280(3.96)	44-2164-83
$C_{31}H_{30}O_8$			
D-ribo-2-Octulose, 3-acetyl-1,3,4-tri-deoxy-5,7-O-(phenylmethylene)-, dibenzoate, (S)-	$CHCl_3$	285(4.20)	87-0030-83

Compound	Solvent	$\lambda_{max}(\log \epsilon)$	Ref.
$C_{31}H_{31}Cl_6O_8P$			
α-D-Ribofuranose, 1,2-O-(1-methylethylidene)-5-O-(triphenylmethyl)-, bis(2,2,2-trichloroethyl) phosphate	EtOH	211(4.16)	23-1387-83
$C_{31}H_{31}NO_6$			
Phenol, 2-[(2-hydroxy-3,5-dimethylphenyl)methyl]-6-[[2-hydroxy-3-[(2-hydroxy-5-nitrophenyl)methyl]-5-methylphenyl]methyl]-4-methyl-	neutral	327(3.95)	126-2361-83
	anion	403(4.16)	126-2361-83
	dianion	417(4.27)	126-2361-83
Phenol, 2-[(4-hydroxy-3,5-dimethylphenyl)methyl]-6-[[2-hydroxy-3-[(2-hydroxy-5-nitrophenyl)methyl]-5-methylphenyl]methyl]-4-methyl-	neutral	327(3.95)	126-2361-83
	anion	401(4.18)	126-2361-83
	dianion	419(4.30)	126-2361-83
$C_{31}H_{31}N_5O_4$			
1H-Purine-2,6-dione, 7-(2-acetoxy-2-phenylethyl)-8-[bis(phenylmethyl)amino]-3,7-dihydro-1,3-dimethyl-	EtOH	296(4.15)	56-0461-83
$C_{31}H_{32}NO_5P$			
3,5-Oxepindicarboxylic acid, 6,7-dihydro-6,7-dimethyl-2-[(triphenylphosphoranylidene)amino]-, 5-ethyl 3-methyl ester	MeOH	267(4.10),277(4.07), 367(4.42)	24-1691-83
3,5-Oxepindicarboxylic acid, 6,7-dihydro-7,7-dimethyl-2-[(triphenylphosphoranylidene)amino]-, 5-ethyl 3-methyl ester	MeOH	365(4.28)	24-1691-83
$C_{31}H_{32}N_2O_8$			
D-ribo-2-Octulose, 1,3,4-trideoxy-3-[1-(hydroxyimino)ethyl]-5,7-O-(phenylmethylene)-, oxime, 6,8-dibenzoate, (S)-	$CHCl_3$	245(4.08)	87-0030-83
$C_{31}H_{33}ClN_2O_2S$			
Estra-1,3,5(10)-trien-17-one, 2-[4-(4-chlorophenyl)-2-(2-propenylimino)-3(2H)-thiazolyl]-3-methoxy-	EtOH	296(4.194)	36-1205-83
	EtOH-HCl	264(4.303),288s(4.599)	36-1205-83
Estra-1,3,5(10)-trien-17-one, 4-[4-(4-chlorophenyl)-2-(2-propenylimino)-3(2H)-thiazolyl]-3-methoxy-	EtOH	266s(4.272),290(4.196)	36-1205-83
	EtOH-HCl	226s(4.635),268(4.404)	36-1205-83
$C_{31}H_{33}NO_9$			
Ravidomycin	MeOH	244(4.68),263s(4.54), 277(4.60),285(4.65), 308(4.33),320(4.30), 335(4.20),350(4.08), 392(4.24)	23-0323-83
	$CHCl_3$	246(--),265(--), 278(--),288(--), 308(--),322(--), 334(--),350(--), 395(--)	23-0323-83
$C_{31}H_{33}N_2$			
3H-Indolium, 2-[3-(1,3-dihydro-1,3,3-trimethyl-2H-indol-2-ylidene)-2-phenyl-1-propenyl]-1,3,3-trimethyl-, (Z,E)-, tetrafluoroborate	MeOH	570.5(4.76)	89-0876-83

Compound	Solvent	λ_{max}(log ε)	Ref.
$C_{31}H_{34}FN_5O_5$			
2'H-Pregna-2,4,6-trieno[3,2-c]pyrazol-20-one, 21-acetoxy-6-azido-9α-fluoro-11β,17-dihydroxy-16α-methyl-2'-phenyl-	MeOH	222(4.16),293(4.26), 318(4.32)	39-2793-83C
$C_{31}H_{34}N_2O_2S$			
Estra-1,3,5(10)-trien-17-one, 3-methoxy-4-[4-phenyl-2-(2-propenylimino)-3(2H)-thiazolyl]-	EtOH	288(4.149)	36-1205-83
	EtOH-HCl	266(4.201)	36-1205-83
$C_{31}H_{34}N_4O_{10}$			
Saframycin R	MeOH	270(3.95)	158-83-170
$C_{31}H_{35}Cl_2N_3O_2$			
1H-Pyrrole-2-carboxylic acid, 4-(2-chloroethyl)-5-[[3-(2-chloroethyl)-4-methyl-2-[(3,4,5-trimethyl-1H-pyrrol-2-yl)methylene]-2H-pyrrol-5-yl]methyl]-3-methyl-, phenylmethyl ester, monohydrobromide, (Z)-	CH_2Cl_2	494(4.93)	44-4302-83
$C_{31}H_{35}FN_2O_5$			
2'H-Pregna-2,4,6-trieno[3,2-c]pyrazol-11β-ol, 9α-fluoro-16α-methyl-17α,20-20,21-bis[methylenebis(oxy)]-2'-phenyl-	MeOH	223(4.01),286(4.16), 314(4.24)	39-2793-83C
2'H-Pregna-2,4,6-trieno[3,2-c]pyrazol-20-one, 21-acetoxy-9α-fluoro-11β,17-16α-methyl-2'-phenyl-	MeOH	220(4.16),285(4.28), 314(4.35)	39-2793-83C
$C_{31}H_{35}NO_9$			
6H-Benzo[d]naphtho[1,2-b]pyran-6-one, 4-[4-O-acetyl-3,6-dideoxy-3-(dimethylamino)-α-altropyranosyl]-8-ethyl-1-hydroxy-10,12-dimethoxy- (dihydroravidomycin)	MeOH	245(4.73),266(4.48), 274(4.60),305(4.10), 318(4.10)	158-1490-83
$C_{31}H_{35}O_{19}$			
Succinylcyanin	HCl	277(4.12),290s(3.92), 320(3.38),510(4.31)	88-5749-83
$C_{31}H_{36}Br_2O_{12}$			
β-D-Glucopyranosiduronic acid, (16α)-2,4-dibromo-3-hydroxy-17-oxoestra-1,3,5(10)-trien-16-yl-, methyl ester, 2,3,4-triacetate	EtOH	283(3.36),290(3.40)	39-0121-83C
$C_{31}H_{36}O_3$			
18,19-Dinorpregn-4-en-20-yn-3-one, 13-ethyl-17-[[(2-phenylcyclopropyl)-carbonyl]oxy]-, [17α,17(1R,2R)]-	MeOH	227(4.27)	13-0349-83A
$C_{31}H_{37}ClN_4O_2S$			
1-Piperazinamine, 4-(8-chloro-10,11-dihydrodibenzo[b,f]thiepin-10-yl)-N-[[4-[2-(dimethylamino)ethoxy]-3-ethoxyphenyl]methylene]-	MeOH	280s(4.33),291(4.35), 308(4.35)	73-1173-83
$C_{31}H_{37}NOS$			
Sulfilimine, [2,4,6-tris(1,1-dimethylethyl)phenyl]-9H-xanthen-9-ylidene-	MeOH	470(4.14)	44-4582-83

Compound	Solvent	λ_{max}(log ε)	Ref.
$C_{31}H_{37}NS$			
Sulfilimine, 9H-fluoren-9-ylidene-[2,4,6-tris(1,1-dimethylethyl)phenyl]-	MeOH	430(4.36)	44-4582-83
$C_{31}H_{37}N_5O_4$			
21H-Biline-1,19-dione, 3,8,12,17-tetraethyl-22,24-dihydro-2,7,13,18-tetramethyl-5-nitro-	$CHCl_3$	323(4.58),506s(4.27), 535(4.28)	5-0001-83
$C_{31}H_{38}Br_2O_{11}$			
β-D-Glucopyranosiduronic acid, (17β)-2,4-dibromo-3-hydroxyestra-1,3,5(10)-trien-17-yl, methyl ester, 2,3,4-triacetate	EtOH	284(3.42),291(3.43)	39-0121-83C
$C_{31}H_{38}N_4$			
Benzenamine, 4-[2-[5-[4-(diethylamino)-phenyl]-4,5-dihydro-1-phenyl-1H-pyrazol-3-yl]ethenyl]-N,N-diethyl-	n.s.g.	264(4.37),322(4.08), 396(4.66)	44-0542-83
$C_{31}H_{38}N_4O_2$			
21H-Biline-1,19-dione, 2,7,13,18-tetraethyl-22,24-dihydro-3,8,12,17-tetramethyl-	CH_2Cl_2	366(4.70),654(4.17)	78-1841-83
$C_{31}H_{38}N_4O_4$			
21H-Biline-3-acetic acid, 17-ethyl-1,2,3,19,23,24-hexahydro-2,2,7,8-12,13,18-heptamethyl-1,19-dioxo-, methyl ester, (4Z,10Z,15E)-	$CHCl_3$	272(4.16),348(4.27), 540(4.22)	49-0753-83
(Z,Z,Z)-	$CHCl_3$	275(4.31),347(4.58), 588(4.21)	108-0187-83
$C_{31}H_{38}O_3$			
21-Norchola-14,20(22),23-triene-24-carboxylic acid, 21-hydroxy-3-(phenylmethoxy)-, δ-lactone, (3β,5β)-	$CHCl_3$	304(3.92)	33-2632-83
19-Norpregn-4-en-20-yn-3-one, 17α-[(4-butylbenzoyl)oxy]-	$CHCl_3$	241(4.50)	13-0309-83A
19-Norpregn-4-en-20-yn-3-one, 17α-(1-oxo-5-phenylpentoxy)-	$CHCl_3$	241(4.22)	13-0309-83A
3(2H)-Phenanthrenone, 1,4b,5,6,7,8,8a-9,10,10a-decahydro-4b,8-dimethyl-7-(phenylmethoxy)-8-[(phenylmethoxy)-methyl]-, (4bα,7β,8β,8aβ,10aα)-	EtOH	238(4.17)	33-1922-83
$C_{31}H_{39}N_5O_5$			
Ergocornine	MeOH	311(3.91)	162-0526-83
Ergocorninine	MeOH	240.5(4.31),312.5(3.92)	162-0526-83
$C_{31}H_{40}Cl_2N_2O_3$			
4-Aza-A-homoandrosta-1,5-dien-3-one, 4-[[4-bis(2-chloroethyl)amino]-phenyl]acetyl]-17β-hydroxy-	EtOH	214(4.23),260(4.30), 302(3.32)	111-0041-83
$C_{31}H_{40}N_4O_{11}S$			
1-Azabicyclo[3.2.0]hept-2-ene-2-carboxylic acid, 3-[[2-[[3-[(2,4-diacetoxy-3,3-dimethyl-1-oxobutyl)amino]-1-oxopropyl]amino]ethyl]thio]-6-ethyl-7-oxo-, (4-nitrophenyl)methyl ester	CH_2Cl_2	269(4.03),316.5(3.88)	158-1473-83

Compound	Solvent	λ_{max}(log ε)	Ref.
$C_{31}H_{40}O_4$			
2H-Pyran-2-one, 6-(3β-benzyloxy-14-hydroxy-5β,14β-androstan-17β-yl)-	$CHCl_3$	306(3.93)	33-2632-83
$C_{31}H_{40}O_{10}$			
Ingenol, 3-O-[(Z)-2-methyl-2-butenyl]-5,16,20-O-triacetyl-16-hydroxy-	MeOH	215(4.37)	100-0723-83
$C_{31}H_{41}NO_3$			
Retinamide, N-[4-(2,2-dimethyl-1-oxopropoxy)phenyl]-	n.s.g.	229(4.18),362(4.77)	34-0422-83
$C_{31}H_{42}Cl_2N_2O_3$			
3-Aza-A-homoandrost-4a-en-4-one, 17β-[[[4-[bis(2-chloroethyl)amino]phenyl]acetyl]oxy]-	EtOH	210(4.36),260(4.31), 301(3.34)	111-0041-83
4-Aza-A-homoandrost-1-en-3-one, 17-[[[4-[bis(2-chloroethyl)amino]phenyl]acetyl]oxy]-, (5α,17β)-	EtOH	212(4.36),261(4.32), 300(3.32)	111-0041-83
$C_{31}H_{42}N_2O_6$			
Batrachotoxin	MeOH-HCl	234(3.96),267(3.71)	162-0143-83
$C_{31}H_{42}O_3$			
18,19-Dinorpregn-4-en-20-yn-3-one, 13-ethyl-17α-[(spiro[3.5]non-1-ylcarbonyl)oxy]-	MeOH	235(4.17)	13-0349-83A
$C_{31}H_{42}O_7$			
Milbemycin J	EtOH	240(4.45)	158-0509-83 +158-0980-83
$C_{31}H_{42}O_{10}$			
Ingol, 3,7,12-O-triacetyl-8-O-tigloyl-	MeOH	218(4.32)	102-2795-83
$C_{31}H_{43}N_5O_6SSi_2$			
7H-Pyrrolo[2,3-d]pyrimidine-5-carbonitrile, 4-amino-7-[2-O-(phenoxythioxomethyl)-3,5-O-[1,1,3,3-tetrakis(1-methylethyl)-1,3-disiloxanediyl]-β-D-ribofuranosyl]-	MeOH	230(4.23),278(4.22)	35-4059-83
$C_{31}H_{44}O_3$			
18,19-Dinorpregn-4-en-20-yn-3-one, 13-ethyl-17-[[(2-hexylcyclopropyl)carbonyl]oxy]-, [17α,17(1R,2R)]-	MeOH	234(4.13)	13-0349-83A
19-Norpregn-4-en-20-yn-3-one, 17α-[(7-cyclobutyl-1-oxoheptyl)oxy]-	MeOH	241(4.14)	13-0291-83A
19-Norpregn-4-en-20-yn-3-one, 17α-[[(2-hexylcyclobutyl)carbonyl]oxy]-	MeCN	235(4.09)	13-0291-83A
19-Norpregn-4-en-20-yn-3-one, 17α-[[(3-hexylcyclobutyl)carbonyl]oxy]-	MeOH	240(4.23)	13-0291-83A
$C_{31}H_{44}O_5$			
19-Norpregn-4-en-20-yn-3-one, 17α-[[2,2-dimethyl-1-oxo-6-(1-oxopropoxy)hexyl]oxy]-	EtOH	243(3.97)	13-0255-83A
$C_{31}H_{46}N_2O_2$			
23,24-Dinor-2,3-secolup-3-en-28-oic acid, 2,4-dicyano-, methyl ester	C_6H_{12}	204(3.83)	73-0649-83

Compound	Solvent	λ_{max}(log ε)	Ref.
$C_{31}H_{46}N_6O_4$			
2-Pyrrolidinecarboxamide, 1,1'-[1,9-nonanediylbis(1,3(4H)-pyridinediyl-carbonyl)]bis-, [S-(R*,R*)]-	$CHCl_3$	353(4.07)	44-1370B-83
$C_{31}H_{46}O_3$			
18,19-Dinorpregn-4-en-20-yn-3-one, 13-ethyl-17α-[(3-methyl-1-oxononyl)oxy]-	MeOH	244(4.18)	13-0349-83A
19-Norpregn-4-en-20-yn-3-one, 17α-[(2,3-dimethyl-1-oxononyl)oxy]-	EtOH	244(4.21)	13-0255-83A
19-Norpregn-4-en-20-yn-3-one, 17α-[(2-ethyl-1-oxononyl)oxy]-	EtOH	246(4.13)	13-0255-83A
Spiro[acephenanthrylene-4(5H),2'-bi-cyclo[3.1.0]hexane]-2,5-diol, 5a,6-6a,7,8,9,10,10a-octahydro-1-methoxy-7,7,10a-trimethyl-3,5'-bis(1-methyl-ethyl)-	n.s.g.	288(3.46)	88-1535-83
isomer	n.s.g.	289(3.46)	88-1535-83
$C_{31}H_{46}O_4$			
2(5H)-Furanone, 3-[19-(1,3-benzodioxol-5-yl)-7-nonadecenyl]-5-methyl-, [S-(Z)]- (juruenolide B)	EtOH	236s(3.52),290(3.59)	102-0711-83
$C_{31}H_{46}O_5$			
5α-Cholesta-9(11),20(22)-dien-23-one, 3β,6α-diacetoxy-	MeOH	255(4.05)	5-0056-83
$C_{31}H_{46}O_6$			
Cucurbit-5-ene, 16,24-anhydro-16,24,25-trihydroxy-2β-methoxy-3,11,22-trioxo-	EtOH	234(3.30),297(2.16)	39-2821-83C
Cucurbit-5-ene, 16,24:22,25-dianhydro-16,22,22,24,25-pentahydroxy-2β-meth-oxy-3,11-dioxo-, (22S,24S)-	EtOH	229(2.75),295(2.05)	39-2821-83C
Propanoic acid, 2,2-dimethyl-, 4-[2-[5-acetoxy-2-[[(tetrahydro-2H-pyran-2-yl)oxy]methyl]-1-cyclohexen-1-yl]-ethenyl]-2,3,3a,6,7,7a-hexahydro-7a-methyl-1H-inden-1-yl ester	ether	278.0(4.45)	56-1267-83
Propanoic acid, 2,2-dimethyl-, 4-[[5-acetoxy-2-[[(tetrahydro-2H-pyran-2-yl)oxy]methylene]cyclohexylidene]-ethylidene]octahydro-7a-methyl-1H-inden-1-yl ester	ether	271.0(4.27)	56-1267-83
$C_{31}H_{47}NO_3$			
24-Nor-2,3-secolup-3-en-28-oic acid, 2-cyano-23-oxo-, methyl ester	C_6H_{12}	216(3.86)	73-0649-83
$C_{31}H_{48}O_3$			
α-Tocotrienol methoxymethyl ether	MeOH	276(3.42),289(3.51)	94-4341-83
$C_{31}H_{48}O_4$			
2(5H)-Furanone, 3-[19-(1,3-benzodioxol-5-yl)nonadecyl]-5-methyl-, (S)-	MeOH	230(3.51),290(3.61)	102-0711-83
$C_{31}H_{48}O_5$			
Cholest-4-en-6-one, 2,3-diacetoxy-, (2α,3α)-	MeOH	229(3.92)	104-1116-83
(2β,3β)-	MeOH	232(4.03)	104-1116-83

Compound	Solvent	λ_{max}(log ϵ)	Ref.
$C_{31}H_{48}O_6$			
Fusidic acid	n.s.g.	204(4.00)	162-0616-83
$C_{31}H_{48}O_{11}$			
Baiyunoside	EtOH	213(3.8)	94-0780-83
$C_{31}H_{49}NO_4S$			
2H-Isoindole-2-undecanoic acid, 1,3-dihydro-1,3-dioxo-, 11-(methylthio)undecyl ester	MeOH	293(3.56)	78-1273-83
$C_{31}H_{49}N_3$			
1H-Benzimidazolium, 2-(1,4-dimethylpyridinium-2-yl)-1-hexadecyl-3-methyl-, diiodide	EtOH	284(4.04)	4-0023-83
$C_{31}H_{49}OPSi$			
Phosphine, [[[(1,1-dimethylethyl)dimethylsilyl]oxy]phenylmethylene]-[2,4,6-tris(1,1-dimethylethyl)-phenyl]-, (E)-	CH_2Cl_2	247(4.17),321(4.00)	138-1653-83
(Z)-	CH_2Cl_2	244(4.20),306(4.03)	138-1653-83
$C_{31}H_{50}O_4$			
Iriflorental	EtOH	237.5(4.43)	64-0179-83C
Iripallidal	EtOH	238(4.47)	64-0179-83C
Olean-12-en-28-oic acid, 3β,27-dihydroxy-, methyl ester	EtOH	260(3.24)	150-0130-83S
$C_{31}H_{51}NO_2$			
Cholest-4-en-3-one, 4-morpholino-	EtOH	245(4.13)	107-1013-83
Cholest-4-en-3-one, 6-morpholino-	EtOH	240(4.13)	107-1013-83
$C_{31}H_{52}N_2O$			
Cholest-4-en-3-one, 4-(1-piperazinyl)-	EtOH	245(4.05)	107-1013-83
Cholest-4-en-3-one, 6β-(1-piperazinyl)-	EtOH	240(4.11)	107-1013-83
$C_{31}H_{52}O_3$			
Iriversical	EtOH	256(4.18)	64-0689-83C
$C_{31}H_{53}NO_7$			
Cirramycin A_1, 12,13-deepoxy-12,13-didehydro-20-deoxo-4'-deoxy-	EtOH	283(4.30)	77-1166-83
$C_{31}H_{53}NO_8$			
DODOMT	EtOH	282(4.32)	158-0376-83
Rosaramicin, 20-dihydro-20-deoxy-	EtOH	238(4.16)	77-1166-83
$C_{31}H_{53}NO_9$			
DOOMT	EtOH	282(4.35)	158-0376-83
$C_{31}H_{54}N_2O_2$			
9-Hexadecenoic acid, 16-[(4-aminophenyl)octylamino]-, methyl ester, (E)-	MeOH	266(3.93),316(3.52)	5-0802-83
$C_{31}H_{55}N_3O_6$			
Enocitabine	isoPrOH	216(4.21),248(4.18), 303(3.91)	162-0518-83
$C_{31}H_{62}O_2Te$			
Dodecanoic acid, 12-(octadecyltelluro)-, methyl ester	MeOH	233.0(3.45)	101-0031-83M

Compound	Solvent	λ_{max}(log ε)	Ref.
$C_{32}H_{16}Cl_{14}NO_4$			
L-Valine, N-[(phenylmethoxy)carbonyl]-, 4-[bis(pentachlorophenyl)methylene]-2,3,5,6-tetrachlorophenyl ester (radical)	C_6H_{12}	218(4.95),276s(3.71), 334s(3.81),364s(4.27), 378(4.54),480s(3.04), 500(3.05),554(3.04)	44-3716-83
$C_{32}H_{16}Cl_{14}N_3O_4$			
Glycinamide, N-[(phenylmethoxy)carbonyl]-L-alanyl-N-[4-[bis(pentachlorophenyl)methylene]-2,3,5,6-tetrachlorophenyl]- (radical)	C_6H_{12}	230(4.86),284(3.79), 334s(3.79),365s(4.27), 383(4.56),510(3.11), 563(3.12)	44-3716-83
$C_{32}H_{16}N_4S_2$			
Bis[1]benzothieno[2',3':3,4;2",3":7,8]-cycloocta[1,2-b:5,6-b']diquinoxaline	$CHCl_3$	253(4.81),294(4.46), 342(4.28),380(4.13)	24-0980-83
$C_{32}H_{18}N_2O_2$			
Naphtho[1,2-d]oxazole, 2,2'-(1,6-naphthalenylenediyl)bis-	DMF	364(4.72)	5-0931-83
$C_{32}H_{18}O_{10}$			
[2,2'-Bianthra[1,2-b]furan]-6,6',11,11'-tetrone, 2,2',3,3'-tetrahydro-2,2',5,5'-tetrahydroxy-	dioxan	328(3.75),405(4.06)	104-0533-83
$C_{32}H_{19}N_3O_2$			
Methanone, (1-phenyl-2,3,9b-triazaindeno[6,7,1-ija]azulene-8,9-diyl)bis-[phenyl-	EtOH	254(4.57),306(4.33), 385(4.24),450(3.97)	18-3703-83
$C_{32}H_{20}CdN_4$			
Tetrabenzoporphine cadmium complex	DMF	320(3.7),420(4.0), 445(5.0),500f(3.0), 590(3.5),625(4.5)	109-0295-83
$C_{32}H_{20}N_2O$			
Methanone, (3,5-diphenylbenz[g]imidazo-[5,1,2-cd]indolizin-1-yl)phenyl-	MeCN	245(4.26),320(4.32), 383(4.19),412(3.79), 437(3.73)	138-0763-83
$C_{32}H_{20}N_2O_8$			
[2,2'-Bianthra[1,2-b]furan]-6,6',11,11'-tetrone, 2,2'-diamino-2,2',3,3'-tetrahydro-5,5'-dihydroxy-	dioxan	328(3.80),405(4.11)	104-0533-83
$C_{32}H_{20}N_4O_4S_2$			
Benzoic acid, 2,2'-(6,13-triphenodithiazinediyldiimino)bis-	dioxan	558(4.60)	18-1482-83
	H_2SO_4	330(4.71)	18-1482-83
$C_{32}H_{21}ClN_2O$			
Pyridinium, 2,4,6-triphenyl-, 2-(4-chlorophenyl)-1-cyano-2-oxoethylide	toluene	526(3.03)	142-0623-83
	EtOH	444(3.00)	142-0623-83
	CH_2Cl_2	492(3.14)	142-0623-83
$C_{32}H_{21}N_3$			
Spiro[9H-fluorene-9,1'(10'bH)-pyrrolo-2,1-a]isoquinoline]-2',3'-dicarbonitrile, 5',6'-dihydro-10'b-phenyl-	CH_2Cl_2	376(3.83)	24-3915-83
betaine	CH_2Cl_2	634(4.42)	24-3915-83

Compound	Solvent	λ_{max}(log ε)	Ref.
$C_{32}H_{22}N_2$			
2,6-Naphthyridine, 1,4,5,8-tetraphenyl-	$CHCl_3$	313(3.91),380(4.0)	4-0971-83
2,7-Naphthyridine, 1,4,5,8-tetraphenyl-	$CHCl_3$	342(4.19)	4-0971-83
$C_{32}H_{22}N_2O$			
Pyridinium, 2,4,6-triphenyl-, 1-cyano-2-oxo-2-phenylethylide	toluene	532(2.94)	142-0623-83
	EtOH	448(2.92)	142-0623-83
	CH_2Cl_2	496(3.06)	142-0623-83
$C_{32}H_{22}N_2O_2$			
1,4-Naphthalenedione, 2-(1-naphthalenylphenylamino)-3-(phenylamino)-	n.s.g.	223(4.47),252(4.29), 486(3.56)	39-1759-83C
$C_{32}H_{23}Cs$			
Cesium, [9-(triphenylmethyl)-9H-fluoren-9-yl]-	$MeOCH_2CH_2OMe$	365(4.36),458(3.56), 489(3.66)	104-1592-83
$C_{32}H_{23}Li$			
Lithium, [9-(triphenylmethyl)-9H-fluoren-9-yl]-	$MeOCH_2CH_2OMe$	374(4.36),464(3.64), 496(3.64),532(3.48)	104-1592-83
$C_{32}H_{23}N_3O$			
Pyridinium, 2,4,6-triphenyl-, 1-cyano-2-oxo-2-(phenylamino)ethylide	toluene	580(3.06)	142-0623-83
	EtOH	516(3.13)	142-0623-83
	CH_2Cl_2	544(3.38)	142-0623-83
$C_{32}H_{23}N_3S$			
Pyridinium, 2,4,6-triphenyl-, 1-cyano-2-(phenylamino)-2-thioxoethylide	toluene	596(2.92)	142-0623-83
	EtOH	512(2.78)	142-0623-83
	CH_2Cl_2	564(3.03)	142-0623-83
$C_{32}H_{24}$			
[2_2](2,13)Pentahelicenoparacyclophane	EtOH	235(4.91),263s(--), 271(4.50),293s(--), 295(4.50)	1-0589-83
[2.2](3,6)Phenanthrenophane	C_6H_{12}	247(5.02)	1-0589-83
$C_{32}H_{24}As_2$			
1,1'-Bi-1H-arsole, 2,2',5,5'-tetraphenyl-	$MeOCH_2CH_2OMe$	262(4.23),338(4.38), 414(3.96)	101-0269-83H
$C_{32}H_{24}Br_2N_4$			
Quinoxaline, 2,2'-(1,2-ethanediylidene)bis[6-bromo-1,2-dihydro-3-methyl-1-phenyl-	$CHCl_3$-CF_3COOH	314(4.45),369(4.21), 679s(--),738(4.63)	33-0379-83
$C_{32}H_{24}Cl_2N_4$			
Quinoxaline, 2,2'-(1,2-ethanediylidene)bis[6-chloro-1,2-dihydro-3-methyl-1-phenyl-	$CHCl_3$-CF_3COOH	313(4.37),367(4.18), 680s(--),741(4.52)	33-0379-83
Quinoxaline, 2,2'-(1,2-ethanediylidene)bis[1-(4-chlorophenyl)-1,2-dihydro-3-methyl-	$CHCl_3$-CF_3COOH	310(4.37),360(4.13), 660s(--),715(4.26)	33-0379-83
$C_{32}H_{24}N_2O_{11}$			
Furo[3,4-d]pyrimidine-2,4,7(3H)-trione, 1,5-dihydro-1-(2,3,5-tri-O-benzoyl-β-D-dibofuranosyl)-	EtOH	276(4.06),282(4.07)	94-3074-83
$C_{32}H_{24}N_4O_2S_2$			
6,13-Triphenodithiazinediamine, N,N'-	dioxan	559(4.72)	18-1482-83

Compound	Solvent	$\lambda_{max}(\log \epsilon)$	Ref.
bis(4-methoxyphenyl)- (cont.)	H_2SO_4	324(4.72)	18-1482-83
6,13-Triphenodithiazinediamine, 3,10-dimethoxy-N,N'-diphenyl-	dioxan	578(4.66)	18-1482-83
	H_2SO_4	336(4.63)	18-1482-83
$C_{32}H_{24}N_4S_2$			
6,13-Triphenodithiazinediamine, N,N'-bis(4-methylphenyl)-	dioxan	557(4.60)	18-1482-83
	H_2SO_4	320(4.63)	18-1482-83
6,13-Triphenodithiazinediamine, 3,10-dimethyl-N,N'-diphenyl-	dioxan	591(4.72)	18-1482-83
	H_2SO_4	352(4.71)	18-1482-83
$C_{32}H_{24}N_6O_4$			
Quinoxaline, 2,2'-(1,2-ethanediylidene)bis[1,2-dihydro-3-methyl-6-nitro-1-phenyl-	$CHCl_3$-CF_3COOH	325(4.31),425(4.18), 645s(--),690(4.56)	33-0379-83
$C_{32}H_{24}O$			
Bicyclo[3.2.0]hept-2-en-6-one, 4-(diphenylmethylene)-7,7-diphenyl-	C_6H_{12}	292(4.393)	64-0504-83B
Ethanone, 1-[5-(diphenylmethylene)-1,3-cyclopentadien-1-yl]-2,2-diphenyl-	MeCN	253(4.303),357(4.230)	64-0504-83B
$C_{32}H_{26}$			
1,3,5,11,13,15-Cycloheptadecahexaene-7,9-diyne, 17-(diphenylmethylene)-6,11-dimethyl-, (E,E,Z,Z,E,E)-	CH_2Cl_2	278s(4.40),292(4.44), 343(4.38),437(4.25)	39-2987-83C
$C_{32}H_{26}Cl_2O_8$			
2H-Pyran-2-one, 3,3'-[[2,4-bis(4-chlorophenyl)-1,3-cyclobutanediyl]dicarbonyl]bis[6-ethyl-4-hydroxy-, (1α,2α,3β,4β)-	MeOH	202(4.5),225(4.4), 315(4.2)	83-0845-83
$C_{32}H_{26}N_2O_{10}S_2$			
Benzo[1,2-b:4,5-b']difuran-3,7-dicarboxylic acid, 2,6-dihydro-2,6-bis-[[[2-(methoxycarbonyl)phenyl]thio]-imino]-, diethyl ester, (Z,Z)-	$CHCl_3$	650(4.92)	150-0480-83M
$C_{32}H_{26}N_4$			
Quinoxaline, 2,2'-(1,2-ethanediylidene)bis[1,2-dihydro-3-methyl-1-phenyl-	$CHCl_3$-CF_3COOH	308(4.20),370(4.23), 674s(--),723(4.37)	33-0379-83
1,2,4,5-Tetrazine, 1,2,5,6-tetrahydro-1,3,5,6,6-pentaphenyl-	n.s.g.	293(3.57),365(3.65)	104-1722-83
$C_{32}H_{26}O$			
Bicyclo[3.2.0]hept-2-en-6-ol, 4-(diphenylmethylene)-7,7-diphenyl-, (1α,5α,6α)-	C_6H_{12}	293(4.210)	64-0504-83B
(1α,5α,6β)-	C_6H_{12}	295(4.480)	64-0504-83B
$C_{32}H_{26}O_{12}$			
Kelletinin I	EtOH	259(4.39)	35-7396-83
	EtOH-HCl	257(4.53)	35-7396-83
	EtOH-NaOH	302(4.66)	35-7396-83
$C_{32}H_{27}N_3O$			
Quinolinium, 1-ethyl-2-[3-(3-ethylbenzoxazolium-2-yl)-1H-benz[f]indol-2-yl]-, diiodide	EtOH	220(5.52),245(5.32), 300(5.00),540(4.45)	80-0725-83
Quinolinium, 1-ethyl-2-[3-(3-ethylbenzoxazolium-2-yl)-1H-benz[g]indol-2-	EtOH	218(5.24),245(5.09), 320(4.67),580(3.90)	80-0725-83

Compound	Solvent	λmax(log ε)	Ref.
yl]-, diiodide (cont.)			80-0725-83
$C_{32}H_{27}O$			
2H-Furylium, 2,2-dimethyl-3-[4-(1-naphthalenyl)-2-phenyl-1,3-butadienyl]-5-phenyl-, perchlorate	MeCN	555(4.66)	56-0579-83
	HOAc	582(--)	56-0579-83
2H-Furylium, 2,2-dimethyl-3-[4-(2-naphthalenyl)-2-phenyl-1,3-butadienyl]-5-phenyl-, perchlorate	MeCN	549(4.73)	56-0579-83
	HOAc	550(--)	56-0579-83
$C_{32}H_{28}$			
Heptadeca-3,5,7,10,12,14-hexaene-1,16-diyne, 3,15-dimethyl-9-(diphenylmethylene)-	CH_2Cl_2	278s(4.58),290(4.64), 301(4.61),347s(4.75), 366(4.83),396s(4.78)	39-2987-83C
[2.2.2.2]Paracyclophanediene	EtOH	265(4.37)	1-0297-83
$C_{32}H_{28}N$			
Pyridinium, 2-(2,5-dimethylphenyl)-4,6-diphenyl-1-(phenylmethyl)-, salt with trifluoromethanesulfonic acid	DMSO	310(4.44)	150-0301-83M
$C_{32}H_{28}NO_2$			
Pyridinium, 2,6-bis(4-methoxyphenyl)-4-phenyl-1-(phenylmethyl)-, salt with trifluoromethanesulfonic acid	DMSO	306(4.37)	150-0301-83M
$C_{32}H_{28}N_6S_2$			
2H-Isothiazolo[4,5,1-hi][1,2,3]benzothiadiazole-3-S^{IV}, 4,4'-(1,2-phenylene)bis[4,6,7,8-tetrahydro-2-phenyl-	C_6H_{12}	207(4.59),247(4.41), 294(4.40),482(4.55)	39-0777-83C
$C_{32}H_{28}O_6$			
Riccardin B diacetate	EtOH	215(4.31),273(3.50)	44-2164-83
$C_{32}H_{28}P_2$			
Phosphine, 1,3,5,7-octatetraene-1,8-diylbis[diphenyl-, (E,E,Z,Z)-	$CHCl_3$	266(3.94),273(3.93), 312(4.45),326(4.58), 341(4.49)	88-1955-83
$C_{32}H_{29}NO_5$			
2H-Furo[2,3-h][1]benzopyran-2-one, 4-methyl-9-phenyl-8-[4-[2-(1-piperidinyl)ethoxy]benzoyl]-	MeOH	280(4.35),298(4.42)	2-0164-83
$C_{32}H_{30}$			
Benzene, 1,1',1",1"'-(1,3-butadiene-1,4-diylidene)tetrakis[4-methyl-	CH_2Cl_2	255(4.28),353(4.32)	152-0105-83
Naphthalene, 1,2-dihydro-7-methyl-1,1,4-tris(4-methylphenyl)-	CH_2Cl_2	248(3.81),347(2.90)	152-0105-83
$C_{32}H_{30}Cl_2N_2O_4$			
D-ribo-Hex-1-enitol, 1-chloro-1-(6-chloro-4-pyrimidinyl)-1,2-dideoxy-3,4-O-(1-methylethylidene)-6-O-(triphenylmethyl)-	MeOH	253(3.80)	39-0201-83C
$C_{32}H_{30}FeN_8O_8$			
Iron, bis[4-pyridinecarboxylic acid [[5-(acetoxymethyl)-3-hydroxy-2-methyl-4-pyridinyl]methylene]hydrazidato]-	MeOH	226(4.68),315(4.34), 365(4.16),460(3.51)	87-0298-83

Compound	Solvent	$\lambda_{max}(\log \epsilon)$	Ref.
$C_{32}H_{30}NOPS$			
Thiazolium, 2,4-dimethyl-3-[2-oxo-2-[4-(triphenylphosphonio)methyl]-phenyl]ethyl]-, perchlorate	EtOH	258(4.37)	65-2424-83
$C_{32}H_{30}N_2O_4$			
Cochliodinol	EtOH	280(4.49),471(3.65)	94-2998-83
Isocochliodinol	EtOH	290(4.48),470(3.63)	94-2998-83
Neocochliodinol	EtOH	290(4.49),470(3.60)	94-2998-83
1(4H)-Pyridinecarboxylic acid, 4,4'-(2,4-diphenyl-1,3-cyclobutanediylidene)bis-, diethyl ester, trans	5:1 MeCN-EtOH	329(4.72),343s(--)	5-0069-83
	protonated	316(--)	5-0069-83
$C_{32}H_{30}O_4$			
Benzene, 1,1',1",1"'-(1,3-butadiene-1,4-diylidene)tetrakis[4-methoxy-	CH_2Cl_2	265(4.44),360(4.55)	152-0105-83
$C_{32}H_{30}O_4S_2$			
2(3H)-Furanone, dihydro-4-[(3-methoxyphenyl)bis(phenylthio)methyl]-3-[(3-methoxyphenyl)methyl]-	$CHCl_3$	260s(3.81),275(3.85)	39-0643-83C
$C_{32}H_{30}O_{14}$			
Chrysergonic acid	HOAc	335(4.57)	162-0321-83
$C_{32}H_{31}ClN_2O_4$			
D-ribo-Hex-1-enitol, 1-(6-chloro-4-pyrimidinyl)-1,2-dideoxy-3,4-O-(1-methylethylidene)-6-O-(triphenylmethyl)-, (E)-	MeOH	276(4.09)	39-0201-83C
$C_{32}H_{31}NO_6$			
D-Allitol, 2,5-anhydro-1-deoxy-1-(2,5-dihydro-2,5-dioxo-1H-pyrrol-3-yl)-3,4-O-(1-methylethylidene)-6-O-(triphenylmethyl)-	MeOH	221s(4.13)	18-2700-83
$C_{32}H_{31}N_3O_4$			
7H-Pyrrolo[2,3-d]pyrimidine, 7-[2,3,5-tris-O-(phenylmethyl)-β-D-arabinofuranosyl]-	MeOH	264(3.60),267(3.60)	5-1576-83
$C_{32}H_{32}N$			
Benzo[h]quinolinium, 1-cycloheptyl-5,6-dihydro-2,4-diphenyl-, tetrafluoroborate	EtOH	341(4.28)	39-1435-83B
$C_{32}H_{32}NO_6PS$			
2H-Thiocin-5,6,7-tricarboxylic acid, 3,4-dihydro-8-[(triphenylphosphoranylidene)amino]-, 5-ethyl 6,7-dimethyl ester	$CHCl_3$	276(3.99),309(4.12), 343(4.09)	24-1691-83
$C_{32}H_{32}NO_7P$			
3,4,5-Oxepintricarboxylic acid, 6,7-dihydro-7-methyl-2-[(triphenylphosphoranylidene)amino]-, 5-ethyl 3,4-dimethyl ester	MeOH	260(3.83),270(3.84), 350(4.26)	24-1691-83
$C_{32}H_{32}N_2O_5S$			
D-Allitol, 2,5-anhydro-1-deoxy-3,4-	MeOH	274(4.06),293(4.00)	18-2700-83

Compound	Solvent	$\lambda_{max}(\log \epsilon)$	Ref.
O-(1-methylethylidene)-1-(1,2,3,4-tetrahydro-4-oxo-2-thioxopyrimidinyl)-6-O-(triphenylmethyl)- (cont.)			18-2700-83
$C_{32}H_{32}N_2O_6$			
D-Allitol, 2,5-anhydro-1-deoxy-3,4-O-(1-methylethylidene)-1-(1,2,3,4-tetrahydro-2,4-dioxo-5-pyrimidinyl)-6-O-(triphenylmethyl)-	MeOH	264(3.84)	18-2700-83
$C_{32}H_{32}N_4O_2$			
1(4H)-Pyridinecarboxamide, 4,4'-(2,4-diphenyl-1,3-cyclobutanediylidene)-bis[N,N-dimethyl-	MeOH	316(4.58),351s(--)	5-0069-83
$C_{32}H_{32}O_{11}P_2$			
α-D-Ribofuranose, 1,2-O-(1-methylethylidene)-, bis(diphenyl phosphate)	EtOH	210.5(3.98)	23-1387-83
$C_{32}H_{32}O_{12}$			
9,10[1',2']-Benzenoanthracene-2,3,6,7-14,15-hexacarboxylic acid, 1,4,5,8-9,10,13,16-octahydro-, hexamethyl ester	MeCN	232(4.03)	33-0019-83
$C_{32}H_{33}N_3O_5$			
D-Allitol, 1-(2-amino-1,4-dihydro-4-oxo-5-pyrimidinyl)-2,5-anhydro-1-deoxy-3,4-O-(1-methylethylidene)-6-O-(triphenylmethyl)-	MeOH	293(4.01)	18-2700-83
$C_{32}H_{33}N_3O_6$			
D-Allitol, 1-[5-(aminocarbonyl)-4-hydroxy-1H-pyrazol-3-yl]-2,5-anhydro-1-deoxy-3,4-O-(1-methylethylidene)-6-O-(triphenylmethyl)-	MeOH	227s(4.16),268(3.71)	18-2700-83
$C_{32}H_{34}N$			
Benzo[h]quinolinium, 5,6-dihydro-1-(1-methylhexyl)-2,4-diphenyl-, tetrafluoroborate	pentanol	350(4.06)	39-1443-83B
	EtOH-HOAc	350(4.03)	39-1443-83B
	$EtOH-CF_3COOH$	350(4.26)	39-1443-83B
	C_6H_5Cl	350(4.17)	39-1443-83B
$C_{32}H_{34}N_2O_{10}$			
6,15-Dioxa-1,10-diazatricyclo[14.2.0-$0^{7,10}$]octadeca-2,11-diene-2,11-dicarboxylic acid, 3,12-dimethoxy-9,18-dioxo-, bis(phenylmethyl) ester (7RS,16RS)-	EtOH	250(4.33)	39-0115-83C
(7RS,16SR)-	EtOH	251(4.35)	39-0115-83C
$C_{32}H_{34}N_4O_2$			
21H,23H-Porphine-2,7-dicarboxaldehyde, 3,8,12,18-tetramethyl-13,17-dipropyl-	CH_2Cl_2	435(5.15),524(4.13), 560(3.89),592(3.82), 647(3.56)	39-0371-83B
$C_{32}H_{34}N_4O_5$			
2H,6H-1,4-Oxazino[3,2-b]phenoxazine-5,7-dicarboxamide, N,N,N',N'-tetraethyl-10,12-dimethyl-2-oxo-3-phenyl-	$CHCl_3$	502(4.05)	87-1631-83

Compound	Solvent	λ_{max}(log ϵ)	Ref.
$C_{32}H_{34}O_3$			
Estra-1,3,5(10)-trien-17-one, 3,4-bis(phenylmethoxy)-	$CHCl_3$	277(3.33)	94-3309-83
$C_{32}H_{35}N_5$			
Tricyclo[3.3.1.1^{3,7}]decan-1-amine, N-[1-(3,4-dihydro-2,4,6-triphenyl-1,2,4,5-tetrazin-1(2H)-yl)ethylidene]-	MeCN	332(4.32)	104-0481-83
$C_{32}H_{36}N_2O_4$			
Methanone, [1,4-butanediylbis(6-ethoxy-1,3(6H)-pyridinediyl)]bis[phenyl-	CH_2Cl_2	289(3.9),332(3.9)	64-0878-83B
$C_{32}H_{36}O_2$			
Estra-1,3,5(10)-triene, 3,17-bis(phenylmethoxy)-, (17β)-	MeOH	212(4.13),276(3.35), 282(3.33)	145-0347-83
4aH-Xanthen-4a-ol, 4-(1-cyclohexen-1-yl)-1,2,3,4,4a,5,6,7,8,9a-decahydro-9-phenyl-5-(phenylmethylene)-	hexane	293(4.59)	104-2185-83
$C_{32}H_{37}BrN_2O_2S$			
Estra-1,3,5(10)-trien-17-one, 4-[4-(4-bromophenyl)-2-(butylimino)-3(2H)-thiazolyl]-3-methoxy-	EtOH	231s(4.604),265s(4.201), 290(4.104)	36-1205-83
	EtOH-HCl	231(4.536),268(4.225)	36-1205-83
$C_{32}H_{37}ClN_2O_2S$			
Estra-1,3,5(10)-trien-17-one, 4-[2-(butylimino)-4-(4-chlorophenyl)-3(2H)-thiazolyl]-3-methoxy-	EtOH	262s(4.287),289(4.225)	36-1205-83
	EtOH-HCl	220(4.659),263(4.411)	36-1205-83
$C_{32}H_{38}$			
Tricyclo[3.3.1.1^{3,7}]decane, 2-phenyl-2-(4-tricyclo[3.3.1.1^{3,7}]decylidene-2,5-cyclohexadien-1-yl)-	hexane	268(4.54)	24-3235-83
$C_{32}H_{38}ClN_3O_2$			
1H-Pyrrole-2-carboxylic acid, 5-[[2-[[4-(2-chloroethyl)-3,5-dimethyl-1H-pyrrol-2-yl]methylene]-3-ethyl-4-methyl-2H-pyrrol-5-yl]methyl]-4-ethyl-3-methyl-, phenylmethyl ester, monohydrobromide, (Z)-	CH_2Cl_2	494(4.95)	44-4302-83
$C_{32}H_{38}N_2O_5$			
Cortivazol	MeOH	283(4.20),315(4.28)	162-0362-83
$C_{32}H_{38}N_2O_8$			
Deserpidine	EtOH	218(4.79),272(4.26), 290(4.07)	162-0421-83
$C_{32}H_{38}N_4$			
Etioporphyrin	n.s.g.	246(3.90),269(3.89), 396(5.22),497(4.13), 532(3.99),566(3.81), 620(3.65),645(2.62)	162-0559-83
$C_{32}H_{39}FO_3$			
19-Norpregn-4-en-20-yn-3-one, 17α-[[5-(4-fluorophenyl)-1-oxohexyl)oxy]-	$CHCl_3$	241(4.12)	13-0309-83A

Compound	Solvent	λ_{max}(log ϵ)	Ref.
$C_{32}H_{39}NO_{11}$			
D-Arabinitol, 5-C-[4,5,6,7-tetrahydro-6,6-dimethyl-4-oxo-1-(phenylmethyl)-1H-indol-2-yl]-, 1,2,3,4,5-pentaacetate, (S)-	EtOH	248(3.79),277(3.71)	136-0255-83E
$C_{32}H_{39}N_3O_5$			
Gentiacraline (same spectrum in acid or base)	EtOH?	222(4.37),255(3.85), 295(4.30)	28-0977-83A
$C_{32}H_{40}N_4O_4$			
21H-Biline-3-acetic acid, 17-ethyl-1,2,3,19,22,24-hexahydro-2,2,7,8-12,13,18,22-octamethyl-1,19-dioxo-, methyl ester, (Z,Z,Z)-(±)-	$CHCl_3$	329(4.38),558(4.35), 652s(3.57)	49-1107-83
	20% DMSO	574(4.28)	49-1107-83
	+ base	654(4.36)	49-1107-83
21H-Biline-3-acetic acid, 17-ethyl-1,2,3,19,23,24-hexahydro-2,2,7,8-12,13,18,23-octamethyl-1,19-dioxo-, methyl ester, (Z,Z,Z)-	$CHCl_3$	267(4.27),387(4.22), 533(4.35),606s(4.07)	108-0187-83
21H-Biline-3-acetic acid, 17-ethyl-1,2,3,19,23,24-hexahydro-2,2,7,8-12,13,18,24-octamethyl-1,19-dioxo-, methyl ester, (Z,Z,Z)-(±)-	$CHCl_3$	282(4.83),329(4.78), 384s(4.26),550s(4.14), 590(4.26),640(4.18)	49-1107-83
	30% DMSO	574(4.29)	49-1107-83
	+ base	688(4.37)	49-1107-83
21H-Biline-3-acetic acid, 17-ethyl-1,2,3,23-tetrahydro-19-methoxy-2,2,7,8,12,13,18-heptamethyl-1-oxo-, methyl ester, (Z,Z,Z)-(±)-	$CHCl_3$	270(4.32),350(3.59), 582s(4.19),615(4.23)	49-0983-83
21H-Biline-17-acetic acid, 3-ethyl-1,17,18,23-tetrahydro-19-methoxy-2,7,8,12,13,18,18-heptamethyl-1-oxo-, methyl ester, (Z,Z,Z)-(±)-	$CHCl_3$	277(4.32),349(4.60), 642s(4.23),673(4.32)	49-0983-83
$C_{32}H_{40}N_4O_{10}$			
Acetamide, N,N'-[(9,10-dihydro-5,6-dimethoxy-9,10-dioxo-1,4-anthracenediyl)bis(imino-2,1-ethanediyl)]bis[N-(2-acetoxyethyl)-	EtOH	221(4.08),239(4.06), 590(3.78),635(3.79)	18-1812-83
$C_{32}H_{40}O_4$			
19-Norpregn-4-en-20-yn-3-one, 17α-[[4-(butoxymethyl)benzoyl]oxy]-	$CHCl_3$	241(4.94)	13-0309-83A
19-Norpregn-4-en-20-yn-3-one, 17α-[[(4-butoxyphenyl)acetyl]oxy]-	EtOH	240(4.32)	13-0309-83A
19-Norpregn-4-en-20-yn-3-one, 17α-[[5-(4-methylphenoxy)-1-oxopentyl]oxy]-	$CHCl_3$	240(4.53)	13-0309-83A
$C_{32}H_{40}O_9$			
Ingol, 3,12-di-O-acetyl-7-O-benzoyl-8-methoxy-	MeOH	208(4.32),225(4.23), 273(2.93),282(2.88)	102-2795-83
$C_{32}H_{41}BrO_8$			
Prosta-5,7,14-trien-1-oic acid, 4,12-diacetoxy-9-[(4-bromobenzoyl)oxy]-, methyl ester, cis	EtOH	242(4.41)	88-1549-83
trans	EtOH	243(4.54),248s(4.53)	88-1549-83
$C_{32}H_{41}N_5O_5$			
α-Ergocryptine	MeOH	241(4.31),312.5(3.95)	162-0526-83
β-Ergocryptine	MeOH	312(3.93)	162-0526-83

Compound	Solvent	λ$_{max}$(log ε)	Ref.
α-Ergocryptinine	MeOH	241.5(4.30),312.5(3.94)	162-0526-83
β-Ergocryptinine	MeOH	240.5(4.31),312(3.94)	162-0526-83
$C_{32}H_{42}Cl_2N_2O_3$			
4-Aza-5β-androst-1-en-3-one, 17β-hydroxy-N-methyl-, 4-[N,N-bis(2-chloroethyl)amino]phenylacetate	EtOH	211(4.24),261(4.30), 300(3.32)	111-0041-83
$C_{32}H_{42}N_4$			
1H-Benzimidazolium, 2-[4-[2-[4-(dimethylamino)phenyl]ethenyl]-1-methylpyridinium-2-yl]-1-methyl-3-octyl-, diiodide	EtOH	558(4.58)	4-0023-83
$C_{32}H_{42}N_9O_{12}P$			
Adenosine, thymidylyl-(3'→5')-3'-[[2-amino-3-(4-methoxyphenyl)-1-oxopropyl]amino]-3'-deoxy-N,N-dimethyl-, (S)-, salt with diethylamine	pH 4.5	274(4.44)	118-0443-83
$C_{32}H_{42}O_3$			
Phenol, 4-(1,1-dimethylethyl)-2,6-bis-[[5-(1,1-dimethylethyl)-2-hydroxyphenyl]methyl]-	EtOH	281(3.96),286s(3.96)	126-1363-83
$C_{32}H_{42}O_8$			
Prosta-5,7,14-trien-1-oic acid, 4,12-diacetoxy-9-(benzoyloxy)-, methyl ester	MeOH	229(4.40),245s(4.30)	94-1440-83
isomer	MeOH	242(4.40)	94-1440-83
$C_{32}H_{44}N_2O_4$			
24,30-Dinor-3,4-secolupa-4(23),20(29)-diene-3,28-dioic acid, 4,20-dicyano-, dimethyl ester	C_6H_{12}	210(4.11)	73-0649-83
$C_{32}H_{44}N_4O_{10}$			
Carbamic acid, [9,10-dihydro-5,6-dihydroxy-9,10-dioxo-1,4-anthracenediyl)bis(imino-2,1-ethanediyl)]-bis[(2-hydroxyethyl)-, bis(1,1-dimethylethyl) ester	$CHCl_3$	260s(4.36),274s(4.26), 565s(3.87),602(4.19), 650(4.27)	18-1812-83
$C_{32}H_{44}O_7$			
Milbemycin K	EtOH	240(4.47)	158-0509-83 +158-0980-83
Pristimerol, dimethyl-6-oxo-	EtOH	210(4.13),225(4.00), 247(4.07),285(3.80), 300(3.93)	39-2845-83C
$C_{32}H_{44}O_8$			
Cucurbitacin E	$CHCl_3$	234(4.07),267(3.92)	162-0375-83
Phorbol, 12-O-(2E,4E-decadienoyl)-4-deoxy-16-hydroxy-, 13-acetate	MeOH	204(4.09),262(4.13)	102-1231-83
$C_{32}H_{44}O_9$			
Affinoside VIII acetonide	MeOH	265(4.24)	94-1199-83
$C_{32}H_{44}O_{10}$			
Affinoside S-III acetonide	MeOH	265(4.23)	94-1199-83

Compound	Solvent	$\lambda_{max}(\log \epsilon)$	Ref.
$C_{32}H_{44}O_{11}$			
Affinoside S-VII	MeOH	264(4.28)	94-1199-83
$C_{32}H_{45}NO_{10}$			
Phorbol, 12-O-(N-acetyl-L-leucyl)-, 13,20-diacetate	MeOH	246s(3.61)	64-1015-83B
$C_{32}H_{46}N_2O_6S$			
17-Aza-D-homoandrost-5-en-3β-ol, 17-methylenesulfonyloximino-, 3,17-diacetate	EtOH	225(4.14),275(2.64), 285(2.64)	70-2141-83
17a-Aza-D-homoandrost-5-en-3β-ol, 17-methylenesulfonyloximino-, 3,17a-diacetate	EtOH	225(4.29),273(3.20), 281(3.10)	70-2141-83
$C_{32}H_{46}O_3$			
18,19-Dinorpregn-4-en-20-yn-3-one, 17α-[(7-cyclobutyl-1-oxoheptyl)oxy]-13-ethyl-	MeOH	242(4.11)	13-0349-83A
18,19-Dinorpregn-4-en-20-yn-3-one, 13-ethyl-17α-[[(2-hexylcyclobutyl)carbonyl]oxy]-	MeOH	235(4.06)	13-0349-83A
18,19-Dinorpregn-4-en-20-yn-3-one, 13-ethyl-17α-[[(3-hexylcyclobutyl)carbonyl]oxy]-	MeOH	241(4.10)	13-0349-83A
$C_{32}H_{46}O_6$			
Leptomycin A	EtOH	225(4.28),243s(4.23)	158-0639-83
3,4-Secolupa-4(23),20(29)-diene-3,28-dioic acid, 24,30-dioxo-, dimethyl ester	C_6H_{12}	225(4.10)	73-0649-83
$C_{32}H_{48}N_2O_8S$			
Benzenesulfonamide, N-[3-(hexadecyloxy)-4-methyl-6-oxo-2,4-cyclohexadien-1-ylidene]-2-(2-methoxyethoxy)-5-nitro-	acetone	454(3.64)	44-0177-83
$C_{32}H_{48}O_3$			
Furan, 2,2',2"-ethylidynetris[5-methyl-3-pentyl-	EtOH	226(4.36)	103-0474-83
Oleana-12,15-dien-11-one, 3β-acetoxy-	EtOH	244(4.05)	78-2819-83
$C_{32}H_{48}O_6$			
19-Norlanost-5-ene-3,11-dione, 16,24-22,25-diepoxy-2,22-dimethoxy-, (2β,9β,10α,16α,22S,24S)-	EtOH	224(3.25),295(2.38)	39-2821-83C
$C_{32}H_{49}N_3$			
Pyrido[2',1':3,4][1,4]diazepino[1,2-a]benzimidazole-5,9-diium, 14-hexadecyl-6,7,8,14-tetrahydro-2-methyl-, dibromide	EtOH	319(4.20)	4-0029-83
$C_{32}H_{50}O_2$			
Oleana-11,13(18)-dien-3β-ol, acetate	EtOH	241(4.34),251(4.42), 260(4.29)	102-1801-83
$C_{32}H_{50}O_4$			
Dammar-8-en-7-one, 3-acetoxy-13,17-epoxy-12-methyl-, (3β,12β,13ξ,17ξ)-	n.s.g.	251(4.00)	138-1171-83

Compound	Solvent	λ_{max}(log ε)	Ref.
$C_{32}H_{50}O_5$			
13,17-Secodammar-8-ene-7,13,16-trione, 3-acetoxy-12-methyl-, (3β,12β)-	n.s.g.	248(3.91)	138-1171-83
$C_{32}H_{50}O_8$			
Olean-12-ene-23,28-dioic acid, 2,3,22,27-tetrahydroxy-, dimethyl ester, (2β,3β,4α,22β)-	EtOH	208(3.5)	20-0355-83
$C_{32}H_{54}N_2O$			
Cholest-4-en-3-one, 4-(4-methyl-1-piperazinyl)-	EtOH	245(4.06)	107-1013-83
Cholest-4-en-3-one, 6β-(4-methyl-1-piperazinyl)-	EtOH	240(4.09)	107-1013-83
$C_{32}H_{58}N_2O$			
Acetamide, N-[4-(didodecylamino)phenyl]-	MeOH	280(4.29)	5-0802-83
$C_{32}H_{60}Ge_5$			
Pentagermane, 1,1,2,2,3,3,4,4,5,5-decaethyl-1,5-diphenyl- (impure)	C_6H_{12}	256(4.50)	101-0149-83G
$C_{32}H_{61}N_5O_6Si_3$			
Guanosine, 2',3',5'-tris-O-[(1,1-dimethylethyl)dimethylsilyl]-N-(2-methyl-1-oxopropyl)-	MeOH	255(4.23),258(4.23), 277(4.10)	33-2069-83
$C_{32}H_{80}Si_8$			
Cyclooctasilane, hexadecaethyl-	isooctane	196s(4.91),221s(4.38), 244(4.32),260s(3.79)	157-1792-83

Compound	Solvent	λ_{max}(log ε)	Ref.
$C_{33}H_{16}Cl_{14}N_3O_5$			
L-Phenylalanine, N-[N-[N-[4-[bis(pentachlorophenyl)methylene]-2,3,5,6-tetrachlorobenzoyl]glycyl]glycyl]-, (radical)	$CHCl_3$	288(3.71),336s(3.68), 366s(4.15),384(4.43), 510s(2.97),562(2.91)	44-3716-83
$C_{33}H_{18}O_9$			
5,6,11,12,17,18-Trinaphthylenehexone, 1,7,16-trihydroxy-3,9,14-trimethyl-	dioxan	273(4.49),299(4.52), 423(4.16)	5-0433-83
$C_{33}H_{20}Cl_6NOP$			
3H-Pyrrol-3-one, 2-[phenyl(triphenylphosphoranylidene)methyl]-4,5-bis-(trichloroethenyl)-	MeCN	268(4.13),277(4.11), 308(4.18),620(3.59)	88-2977-83
$C_{33}H_{20}O_9$			
[2,2':7',12"-Ternaphthalene]-1,1',1"-4,4',4"-hexone, 8,8',8"-trihydroxy-6,6',6"-trimethyl-	EtOH	218(4.69),250(4.59), 435(4.16)	5-0433-83
$C_{33}H_{21}N_3O_2$			
Methanone, [1-(4-methylphenyl)-2,3,9b-triazaindeno[6,7,1-ija]azulene-8,9-diyl]bis[phenyl-	EtOH	253(4.48),307(4.22), 388(4.22),445(4.02)	18-3703-83
$C_{33}H_{22}N_4O$			
7H-Benzimidazo[2,1-a]benz[de]isoquinolin-7-one, 4-(4,5-dihydro-1,5-diphenyl-1H-pyrazol-3-yl)-	toluene	505(4.40)	103-0214-83
$C_{33}H_{23}O_2$			
1-Benzopyrylium, 2-phenyl-3-[3-(2-phenyl-4H-1-benzopyran-4-ylidene)-1-propenyl]-, perchlorate	CH_2Cl_2	715(5.31)	103-0243-83
	MeCN	704(5.28)	103-0243-83
$C_{33}H_{24}$			
7H-Benzocyclopentadecene, 14,15,16,17-tetradehydro-7-(diphenylmethylene)-13-methyl-, (E,E,Z,E)-	CH_2Cl_2	229(4.40),270s(4.48), 288(4.52),301s(4.49), 337(4.40),374(4.25)	39-2987-83C
9H-Benzocyclopentadecene, 14,15,16,17-tetradehydro-9-(diphenylmethylene)-13-methyl-, (E,E,E,Z)-	CH_2Cl_2	231(4.37),247s(4.31), 277(4.37),291s(4.36), 329(4.30),383(4.23)	39-2987-83C
$C_{33}H_{24}N_2O$			
Pyridinium, 2,4,6-triphenyl-, 1-cyano-2-(4-methylphenyl)-2-oxoethylide	toluene	536(3.14)	142-0623-83
	MeOH	440(--)	142-0623-83
	EtOH	452(2.98)	142-0623-83
	acetone	506(--)	142-0623-83
	CH_2Cl_2	502(3.16)	142-0623-83
	CCl_4	545(--)	142-0623-83
	MeCN	488(--)	142-0623-83
	DMF	500(--)	142-0623-83
(also other solvents not listed)	DMSO	494(--)	142-0623-83
$C_{33}H_{25}N_3$			
3,5-Pyridinedicarbonitrile, 2,6-bis-([1,1'-biphenyl]-4-yl)-1,4-dihydro-4,4-dimethyl-	EtOH	206(4.86),283(4.68), 359(3.66)	73-3112-83
$C_{33}H_{25}OP$			
Ethanone, 1-(9H-fluoren-2-yl)-2-(tri-	EtOH	220(4.810),274(4.450),	65-1571-83

Compound	Solvent	$\lambda_{max}(\log \epsilon)$	Ref.
phenylphosphoranylidene)- (cont.)		329(4.540)	65-1571-83
$C_{33}H_{26}$			
Undeca-3,5,7,10-tetraen-1-yne, 11-(2-ethynylphenyl)-3-methyl-9-(diphenylmethylene)-	CH_2Cl_2	241(4.34),268(4.41), 375(4.59)	39-2987-83C
Undeca-3,5,8,10-tetraen-1-yne, 11-(2-ethynylphenyl)-3-methyl-7-(diphenylmethylene)-	CH_2Cl_2	239(4.55),267(4.67), 347(4.73),373(4.76)	39-2987-83C
$C_{33}H_{26}As_2$			
1H-Arsole, 1,1'-methylenebis[2,5-diphenyl-	C_6H_{12}	220(4.52),342(4.28), 382(4.42)	101-0335-83I
$C_{33}H_{26}OP$			
Phosphonium, [2-(9H-fluoren-2-yl)-2-oxoethyl)triphenyl-, bromide	EtOH	275(4.371),330(4.446)	65-1571-83
$C_{33}H_{26}O_2$			
1,3-Cyclopentadiene-1-carboxylic acid, 2-(diphenylmethyl)-5-(diphenylmethylene)-, methyl ester	C_6H_{12}	244(4.276),347(4.401)	64-0504-83B
$C_{33}H_{27}N_2$			
3H-Indolium, 1-methyl-3-[3-(1-methyl-2-phenyl-1H-indol-3-yl)-2-propenylidene]-2-phenyl-, iodide	$MeNO_2$-HOAc	592(5.00)	103-1306-83
$C_{33}H_{27}N_3O_2S$			
2,5-Cyclohexadiene-1-carbonitrile, 4-(1,3-diethyltetrahydro-4,6-dioxo-2-thioxo-5(2H)-pyrimidinylidene)-1,2,6-triphenyl-	$CHCl_3$	270(4.07),415(4.60)	83-0732-83
$C_{33}H_{28}N_6O_3S_3$			
Benzeneacetic acid, α-[[4,5-dihydro-4-oxo-5-phenyl-2-(phenylamino)thiazolo-[4,5-d]pyrimidin-6-yl]thio]-, S-benzylisothiouronium salt	MeOH	258s(--),292(4.25), 302(4.25),346(4.33)	2-0243-83
$C_{33}H_{28}N_6O_7$			
β-D-arabino-Heptofuranuronamide, 1-[6-(benzoylamino)-9H-purin-9-yl]-1,5,6-trideoxy-, 2,3-dibenzoate	MeOH	232(4.55),278(4.31)	87-1530-83
$C_{33}H_{28}O_2$			
3-Cyclopentene-1-carboxylic acid, 2-(diphenylmethyl)-5-(diphenylmethylene)-, methyl ester	C_6H_{12}	287(4.365)	64-0504-83B
$C_{33}H_{30}N_4O_2$			
7H-Benzo[e]perimidin-7-one, 4-[4-(diethylamino)phenyl]-6-[(4-methoxyphenyl)amino]-2-methyl-	n.s.g.	515(4.43)	2-0805-83
$C_{33}H_{30}O_6$			
Riccardin A diacetate	EtOH	218(4.62),250(4.38)	44-2164-83
$C_{33}H_{31}NO_3Si_2$			
6H-[1,2,5]Oxadisilolo[3,4-d]azepine-6-carboxylic acid, 1,3,3a,8a-tetrahy-	MeOH	244(4.33),254(4.17), 260(4.11),265(4.03),	18-0175-83

Compound	Solvent	$\lambda_{max}(\log \epsilon)$	Ref.
dro-1,1,3,3-tetraphenyl-, ethyl ester, trans (cont.)		272(3.76)	18-0175-83
$C_{33}H_{31}NO_5$			
2H-Furo[2,3-h][1]benzopyran-2-one, 4-methyl-9-phenyl-8-[4-[3-(1-piperidinyl)propoxy]benzoyl]-	dioxan	250(4.40),280(4.56), 292(4.58)	2-0164-83
$C_{33}H_{32}O_8$			
7H-Furo[3,2-g][1]benzopyran-7-one, 9,9'-[1,11-undecanediylbis(oxy)]bis-	EtOH	218(4.69),248(4.65), 263(4.44),300(4.37)	39-1807-83B
$C_{33}H_{33}NO_4Si_2$			
1H-Azepine-1-carboxylic acid, 4,5-dihydro-4,5-bis(hydroxydiphenylsilyl)-, ethyl ester	MeOH	244(4.37),248(4.39), 254(4.29),260(4.23), 265(4.01),272(3.83)	18-0175-83
$C_{33}H_{33}N_3O_4S$			
1H-Pyrrolo[2,3-d]pyrimidine, 4-(methylthio)-1-[2,3,5-tris-O-(phenylmethyl)-β-D-arabinofuranosyl]-	MeOH	243(4.26),264(3.81), 311(4.16)	5-1576-83
7H-Pyrrolo[2,3-d]pyrimidine, 4-(methylthio)-7-[2,3,5-tris-O-(phenylmethyl)-α-D-arabinofuranosyl]-	MeOH	250(3.78),292(4.08)	5-1576-83
β-	MeOH	250(3.81),292(4.07)	5-1576-83
$C_{33}H_{34}FeO_3P_2$			
Ferrocene, [(diethoxyphosphinyl)(triphenylphosphoranylidene)methyl]-	EtOH	262(4.1),310(3.7), 445(2.9)	65-0514-83
$C_{33}H_{34}NO_7P$			
3,4,5-Oxepintricarboxylic acid, 6,7-dihydro-6,7-dimethyl-2-[(triphenylphosphoranylidene)amino]-, 5-ethyl 3,4-dimethyl ester	MeOH	275(4.03),270(4.08), 385(4.43)[sic]	24-1691-83
3,4,5-Oxepintricarboxylic acid, 6,7-dihydro-7,7-dimethyl-2-[(triphenylphosphoranylidene)amino]-, 5-ethyl 3,4-dimethyl ester	MeOH	260(4.05),275(4.03), 350(4.26)	24-1691-83
$C_{33}H_{34}N_2O_7$			
D-Allitol, 2,5-anhydro-1-deoxy-1-[4-hydroxy-5-(methoxycarbonyl)-1H-pyrazol-3-yl]-3,4-O-(1-methylethylidene)-6-O-(triphenylmethyl)-	MeOH	228s(4.15),268(3.68)	18-2700-83
$C_{33}H_{34}N_2O_7S_2$			
Xanthylium, 3,6-bis(diethylamino)-9-[2-(phenoxysulfonyl)-4-sulfophenyl]-, hydroxide, inner salt	EtOH	243s(4.318),260(4.452), 286(4.281),308s(4.079), 339(3.730),355(3.910), 409(3.634),427s(3.588), 495s(3.943),535s(4.697), 567(5.125)	99-0224-83
Xanthylium, 3,6-bis(diethylamino)-9-[4-(phenoxysulfonyl)-2-sulfophenyl]-, hydroxide, inner salt	EtOH	241(4.270),260(4.439), 285(4.230),307(4.069), 340s(3.642),354(3.882), 407(3.496),425s(3.430), 490s(3.861),528s(4.633), 561(5.057)	99-0224-83

Compound	Solvent	λ_{max}(log ϵ)	Ref.
$C_{33}H_{34}N_4O_2$			
21H,23H-Porphine-2-propanoic acid, 7,12-diethenyl-3,8,13,17,18-pentamethyl-, methyl ester	CH_2Cl_2	404(5.24),504(4.12), 540(4.03),576(3.80), 630(3.66)	44-4302-83
21H,23H-Porphine-2-propanoic acid, 8,13-diethenyl-3,7,12,17,18-pentamethyl-, methyl ester	CH_2Cl_2	404(5.26),504(4.13), 540(4.02),574(3.80), 630(3.71)	44-4302-83
$C_{33}H_{34}N_4O_5$			
7H-Pyrrolo[2,3-d]pyrimidin-2-amine, 4-methoxy-7-[2,3,5-tris-O-(phenylmethyl)-α-D-arabinofuranosyl]-	MeOH	260(4.00),286(3.88)	136-0029-83G
β-	MeOH	259(4.00),286(3.87)	136-0029-83G
$C_{33}H_{34}O_3$			
19-Norpregn-4-en-20-yn-3-one, 17α-[([1,1'-biphenyl]-4-ylcarbonyl)oxy]-	$CHCl_3$	272(4.38)	13-0309-83A
$C_{33}H_{34}O_4$			
19-Norpregn-4-en-20-yn-3-one, 17α-[(4-phenoxybenzoyl)oxy]-	$CHCl_3$	248(4.39)	13-0309-83A
$C_{33}H_{35}ClN_2O_6$			
1H-Pyrrole-2-carboxylic acid, 3-(2-acetoxyethyl)-5-[[3-(2-chloroethyl)-4-methyl-5-[(phenylmethoxy)carbonyl]-1H-pyrrol-2-yl]methyl]-4-methyl-, phenylmethyl ester	$CHCl_3$	272(5.15),285(5.15)	78-1849-83
$C_{33}H_{35}FeO_3P_2$			
Phosphonium, [(diethoxyphosphinyl)-ferrocenylmethyl]triphenyl-, tetrafluoroborate	EtOH	251(4.1),310(3.8), 420(3.2)	65-0514-83
$C_{33}H_{35}NO_3$			
19-Norpregn-4-en-20-yn-3-one, 17α-[[(diphenylamino)carbonyl]oxy]-	EtOH	247(4.63)	13-0333-83A
$C_{33}H_{35}O_{11}P$			
D-ribo-2-Octulose, 3-acetyl-1,3,4-trideoxy-4-(dimethoxyphosphinyl)-5,7-O-(phenylmethylene)-, 6,8-dibenzoate, (4ξ)-	MeOH	275(3.60)	111-0487-83
$C_{33}H_{36}ClN_2$			
3H-Indolium, 2-[2-[2-chloro-3-[4-(1,3-dihydro-1,3,3-trimethyl-2H-indol-2-ylidene)-2-butenylidene]-1-cyclopenten-1-yl]ethenyl]-1,3,3-trimethyl-, perchlorate	MeOH	910(5.24)	104-1854-83
$C_{33}H_{36}Cl_2N_4O_2$			
21H,23H-Porphine-2-propanoic acid, 7,12-bis(2-chloroethyl)-3,8,13,17,18-pentamethyl-, methyl ester	CH_2Cl_2	400(5.26),500(4.18), 532(4.03),568(3.86), 622(3.60)	44-4302-83
21H,23H-Porphine-2-propanoic acid, 8,13-bis(2-chloroethyl)-3,7,12,17,18-pentamethyl-, methyl ester	CH_2Cl_2	400(5.21),498(4.11), 532(3.99),568(3.86), 622(3.59)	44-4302-83
$C_{33}H_{36}N_4O$			
21H,23H-Porphine-2-carboxaldehyde, 17-	CH_2Cl_2	417(5.20),517(4.05),	39-0371-83B

Compound	Solvent	$\lambda_{max}(\log \epsilon)$	Ref.
ethenyl-3,7,13,18-tetramethyl-8,12-dipropyl- (cont.)		557(4.17),583(3.98), 642(3.48)	39-0371-83B
$C_{33}H_{38}NO_7P$			
α-D-Ribofuranose, 3-C-(cyanomethyl)-3-deoxy-3-(diethoxyphosphinyl)-1,2-O-(1-methylethylidene)-5-O-(triphenylmethyl)-	EtOH	208.5(4.34)	159-0019-83
$C_{33}H_{38}N_4O_6$			
Benzeneacetic acid, α-[[1,9-bis[(diethylamino)carbonyl]-4,6-dimethyl-3-oxo-3H-phenoxazin-2-yl]amino]-, methyl ester	$CHCl_3$	424(4.17)	87-1631-83
$C_{33}H_{38}N_6O_2S_2$			
4-Thiazolidinone, 2-[[[4-oxo-5-(phenylmethylene)-2-[(1-piperidinylmethyl)imino]-3-thiazolidinyl]methyl]imino]-5-(phenylmethylene)-3-(1-piperidinylmethyl)-	EtOH	239(4.21),289(4.36), 331(4.60)	103-1076-83
$C_{33}H_{38}N_{10}O_5S$			
L-Lysine, N^2-[4-[[(2,4-diamino-6-pteridinyl)methyl]methylamino]benzoyl]-N^6-[[5-(dimethylamino)-1-naphthalenyl]sulfonyl]-	pH 1.8	296(4.18)	87-0111-83
	pH 7.0	252(4.23),302(4.08), 370(3.58)	87-0111-83
	pH 13	252(4.19),304(4.12), 370(3.56)	87-0111-83
$C_{33}H_{39}S_3$			
Cyclopropenylium, tris[[4-(1,1-dimethylethyl)phenyl]thio]-, tetrafluoroborate	MeOH	250(3.63),275(3.56)	18-0171-83
$C_{33}H_{40}N_2O_9$			
Reserpine	EtOH	269(4.16),297(3.98)	151-0171-83C
	$CHCl_3$	216(4.79),267(4.23), 295(4.01)	162-1175-83
$C_{33}H_{40}N_4O_4$			
21H-Biline-3-propanoic acid, 7,13,18-triethyl-1,19,22,24-tetrahydro-2,8,12,17-tetramethyl-1,19-dioxo-	CH_2Cl_2	372(4.78),654(4.25)	78-1841-83
$C_{33}H_{40}N_4O_{11}$			
Acetamide, N,N'-[(5-acetoxy-9,10-dihydro-6-methoxy-9,10-dioxo-1,4-anthracenediyl)bis(imino-2,1-ethanediyl]-bis[N-(2-acetoxyethyl)-	$CHCl_3$	274(4.48),594(4.14), 640(4.17)	18-1812-83
$C_{33}H_{40}N_4S$			
Benzothiazolium, 3-methyl-2-[3-[1-methyl-2-(1-methyl-3-octyl-1H-benzimidazolium-2-yl)-4(1H)-pyridinylidene]-1-propenyl]-, diiodide	EtOH	605(5.03)	4-0023-83
$C_{33}H_{40}O_3$			
17-Nor-14-aphidicolen-16-one, 3a,18-bis(benzyloxy)-	EtOH	235(4.00)	33-1922-83
19-Norpregn-4-en-20-yn-3-one, 17α-[(4-cyclohexylbenzoyl)oxy]-	$CHCl_3$	241(4.53)	13-0309-83A

Compound	Solvent	$\lambda_{max}(\log \epsilon)$	Ref.
$C_{33}H_{40}O_{10}$			
Ingol, 3,7,12-tri-O-acetyl-8-O-benzoyl-	MeOH	210(4.36),232(4.44), 274(3.33),282(3.26)	102-2795-83
Phenol, 4-[2-[2,6-dimethoxy-4-(1-propenyl)phenoxy]-1-(2,4,6-trimethoxyphenyl)propyl]-2,6-dimethoxy-, acetate	EtOH	271(4.12)	150-2625-83M
$C_{33}H_{40}O_{11}$			
Ingol, 3,7,12-tri-O-acetyl-8-O-benzoyl-18-hydroxy-	MeOH	209(4.32),233(4.43), 275(3.30),282(3.21)	102-2795-83
$C_{33}H_{40}O_{19}$			
Kaempferol 7-O-α-L-rhamnopyranoside 3-O-β-rutinoside	EtOH	260(4.4),350(--)	105-0500-83
	EtOH-NaOEt	270(--),395(--)	105-0500-83
$C_{33}H_{41}N_2O_2$			
3H-Indolium, 2-[7-(1,3-dihydro-1,3,3-trimethyl-2H-indol-2-ylidene)-4-(2,2-dimethoxyethyl)-1,3,5-heptatrienyl]-1,3,3-trimethyl-, tetrafluoroborate	benzene	800(--)	24-1982-83
	MeOH	765(5.37)	24-1982-83
$C_{33}H_{41}O_{21}$			
Bisdeacylplatyconin chloride	MeOH-HCl	282(4.32),300s(3.98), 350(3.71),540(4.59)	88-2181-83
$C_{33}H_{42}N_4O_4$			
21H-Biline-3-acetic acid, 17-ethyl-1,2,3,19,23,24-hexahydro-2,2,7,8-12,13,18,21,24-nonamethyl-1,19-dioxo-, methyl ester, (5E,9Z,15E)-	$CHCl_3$	272(4.17),292(4.19), 342(4.14),523(4.19)	49-0753-83
(5E,9Z,15Z)-	$CHCl_3$	270(4.17),296s(4.15), 344(4.30),535(4.28)	49-0753-83
$C_{33}H_{42}N_4O_{13}S$			
Antibiotic OA-6129B_2, triacetyl-, p-nitrobenzyl ester	$CHCl_3$	270(4.08),320(4.08)	158-1473-83
$C_{33}H_{42}O_3$			
19-Norpregn-4-en-20-yn-3-one, 17α-[3-(4-butylphenyl)-1-oxopropoxy]-	EtOH	239(4.20)	13-0309-83A
19-Norpregn-4-en-20-yn-3-one, 17α-[[5-(4-ethylphenyl)-1-oxopentyl]oxy]-	EtOH	240(4.25)	13-0309-83A
19-Norpregn-4-en-20-yn-3-one, 17α-[(4-hexylbenzoyl)oxy]-	$CHCl_3$	241(4.59)	13-0309-83A
$C_{33}H_{42}O_{14}$			
Urceolatoside A	MeOH	204s(3.82),228(4.26), 240(4.21)	102-1977-83
$C_{33}H_{42}O_{15}$			
Urceolatoside C	MeOH	228(4.04),238s(4.03)	102-1977-83
$C_{33}H_{44}ClN_3O_6$			
1H-Pyrrole-3-propanoic acid, 5-[[4-(2-chloroethyl)-3,5-dimethyl-2H-pyrrol-2-ylidene]methyl]-2-[[5-[(1,1-dimethylethoxy)carbonyl]-3-(3-methoxy-3-oxopropyl)-4-methyl-1H-pyrrol-2-yl]-methyl]-4-methyl-, methyl ester, monohydrobromide, (Z)-	CH_2Cl_2	486(4.90)	4-1383-83

Compound	Solvent	λ max(log ε)	Ref.
$C_{33}H_{44}N_2O_{17}S$			
Paulomycin B	EtOH	236(4.19),276(4.02), 322(3.97)	158-83-44
$C_{33}H_{44}O_4$			
Furan, 2,2',2",2"'-methanetetrayl-tetrakis[5-butyl-	EtOH	227(4.29)	103-0360-83
$C_{33}H_{44}O_7$			
Sargatriol triacetate	EtOH	228(4.54),267(3.74), 277(3.65),320(3.52)	94-0106-83
Zeyleasterone, trimethyl-	EtOH	207(4.00),225(3.88), 245(3.99),287(3.72), 312(3.72)	39-2845-83C +88-2025-83
$C_{33}H_{44}O_8$			
Helvolic acid	EtOH	231(4.24)	162-0669-83
$C_{33}H_{44}O_9$			
Prosta-5,7,14-trien-1-oic acid, 4,12-diacetoxy-9-[(4-methoxybenzoyl)oxy]-, methyl ester, (4R,5E,7Z,9β,12α,14Z)-	MeOH	249(4.56)	94-1440-83
$C_{33}H_{46}O_7$			
Gitoxigenin, 21ξ,22-diallyl-, 3,16-di-acetate	EtOH	223.5(4.17)	48-0599-83
$C_{33}H_{46}O_8$			
Phorbol, 12-O-undecadienoyl-, 13-acet-ate	MeOH	230(4.21)	100-0123-83
$C_{33}H_{46}O_{10}$			
Ostodin	MeOH	230(4.14)	100-0123-83
$C_{33}H_{46}O_{17}P_4$			
Methanone, bis[2,4-bis[(5,5-dimethyl-1,3,2-dioxaphosphorinan-2-yl)oxy]-phenyl]-, P,P',P",P"'-tetraoxide	EtOH	263(1.7)	65-0653-83
$C_{33}H_{48}O_2$			
Chol-8-en-24-one, 3-hydroxy-4,4,14-tri-methyl-24-phenyl-, (3β,5α,13α,14β,17α)-	EtOH	242(4.15)	78-2799-83
$C_{33}H_{48}O_6$			
Leptomycin B	EtOH	225(4.30),240s(4.18)	158-0639-83
Milbemycin H	EtOH	237(4.40)	158-0502-83 +158-0980-83
$C_{33}H_{48}O_7$			
Milbemycin D	EtOH	238s(--),244(4.49), 253s(--)	158-0502-83 +158-0980-83
$C_{33}H_{50}N_2O_2S$			
Benzenesulfonic acid, 2,4,6-tris(1-methylethyl)-, [1,5-dimethyl-7-(2,6,6-trimethyl-2-cyclohexen-1-yl)-2,4,6-heptatrienylidene]hydrazide, (all-E)-	hexane	222s(4.13),308s(4.64), 320(4.78),336(4.72)	33-1148-83
$C_{33}H_{50}O_3$			
Cholesta-4,6-diene-7-propanoic acid, 3-(1-methyl-2-oxoethylidene)-, (3E)-	EtOH	360(4.51)	35-4033-83

Compound	Solvent	λ_{max}(log ϵ)	Ref.
$C_{33}H_{50}O_5$			
Mogrol, 3,7,11-trioxo-, acetonide	n.s.g.	246(4.02)	95-1155-83
$C_{33}H_{50}O_8$			
Cephalosporin P_1	n.s.g.	211(3.96)	162-0276-83
$C_{33}H_{52}N_2O_9S$			
Benzenesulfonamide, N-[3-(hexadecyl-oxy)-3-methoxy-4-methyl-6-oxo-1,4-cyclohexadien-1-yl]-2-(2-methoxy-ethoxy)-5-nitro-	MeOH	297(4.08)	44-0177-83
$C_{33}H_{52}O_6$			
5β,20ξ-Chol-22-en-24-oic acid, 3β-acet-oxy-21-[(tetrahydro-2H-pyranyl)oxy]-, ethyl ester, (22E)-	n.s.g.	212(4.10)	44-4248-83
$C_{33}H_{52}P_2$			
Diphosphene, [2,4,6-tris(1,1-dimethyl-ethyl)phenyl][2,4,6-tris(1-methyl-ethyl)phenyl]-	CH_2Cl_2	279(4.14),330(3.89), 461(2.67)	35-2495-83
$C_{33}H_{54}O_6$			
β-D-Glucopyranoside, 9,10-secocholesta-5,7,10(19)-trien-3-yl, (3β,5Z,7E)-	n.s.g.	265(4.26)	33-2093-83
$C_{33}H_{54}O_7$			
β-D-Glucopyranoside, 1-hydroxy-9,10-secocholesta-5,7,10(19)-trien-3-yl, (1α,3β,5Z,7E)-	n.s.g.	265(4.21)	33-2093-83
β-D-Glucopyranoside, 25-hydroxy-9,10-secocholesta-5,7,10(19)-trien-3-yl, (3β,5Z,7E)-	n.s.g.	265(4.24)	33-2093-83
$C_{33}H_{54}O_8$			
β-D-Glucopyranoside, 1,2,5-trihydroxy-9,10-secocholesta-5,7,10(19)-trien-3-yl, (3β,5Z,7E)-	n.s.g.	265(4.18)	33-2093-83
$C_{33}H_{55}NO_9$			
Antibiotic M-4635G_3, 3-acetoxy-	n.s.g.	282.5(4.07)	158-83-70
$C_{33}H_{56}N_2O_2$			
Cholest-4-en-3-one, 4-[4-(2-hydroxy-ethyl)-1-piperazinyl]-	EtOH	245(4.11)	107-1013-83
Cholest-4-en-3-one, 6β-[4-(2-hydroxy-ethyl)-1-piperazinyl]-	EtOH	240(4.17)	107-1013-83
$C_{33}H_{56}N_2O_3$			
9-Hexadecenoic acid, 16-[[4-(acetyl-amino)phenyl]octylamino]-, methyl ester	MeOH	282(4.30)	5-0802-83
$C_{33}H_{57}NO_9$			
Antibiotic M-4365G_3, 3-acetoxydihydro-	n.s.g.	234(4.30)	158-83-70
$C_{33}H_{59}N_2O_2$			
Benzenaminium, 4-[(15-carboxypentadec-7-enyl)octylamino]-N,N,N-trimethyl-, chloride	MeOH	270(4.36),306(3.42)	5-0802-83

Compound	Solvent	$\lambda_{max}(\log \epsilon)$	Ref.
$C_{33}H_{62}O_8$			
8,16,24-Tritriacontanetrione, 6,14,22,30,32-pentahydroxy-, (PM-toxin B)	MeOH	275(2.14)	88-3803-83
$C_{33}H_{65}N_2O_2$			
1-Hexadecanaminium, N-[[11-(2-isocyano-1-oxopropoxy)undecyl]-N,N-dimethyl-, bromide	H_2O	340(1.65)	35-4507-83

Compound	Solvent	λmax(log ε)	Ref.
$C_{34}H_{20}$			
15H-Dibenzo[a,g]cyclotridecene, 5,6,7,8-tetradehydro-15-(9H-fluoren-9-ylidene)-, (E,E)-	EtOH	230(4.7),280(4.5), 305(4.5),330s(4.0), 415(4.4)	138-1887-83
$C_{34}H_{20}O_2$			
5,14[1',2']:7,12[1",2"]-Dibenzenopentacene-6,13-dione	EtOH	268s(3.84),274(3.86)	73-0112-83
$C_{34}H_{23}N_3OS$			
Propanedinitrile, [1-[[3-(2-oxo-2-phenylethyl)-2(3H)-benzothiazolylidene]-phenylmethyl]-2-phenyl-2-propenylidene]-	EtOH	241(4.35),511(3.87)	18-1688-83
$C_{34}H_{24}F_6N_4$			
Quinoxaline, 2,2'-(1,2-ethanediylidene)bis[1,2-dihydro-3-methyl-1-phenyl-6-(trifluoromethyl)-	$CHCl_3$-CF_3COOH	300(4.27),370(4.15), 660s(--),700(4.33)	33-0379-83
$C_{34}H_{24}I_2$			
1,1'-Biphenyl, 4,4"-(1,4-phenylenedi-2,1-ethanediyl)bis[4'-iodo-	$C_6H_3Cl_3$	313(4.49)	104-0683-83
$C_{34}H_{24}N_4O_2$			
7H-Benzimidazo[2,1-a]benz[de]isoquinolin-7-one, 4-[4,5-dihydro-5-(4-methoxyphenyl)-1-phenyl-1H-pyrazol-3-yl]-	toluene	510(4.50)	103-0214-83
$C_{34}H_{24}N_4O_2S_2$			
Triphenodithiazine, 6,13-bis[4-(acetylamino)phenyl]-	H_2SO_4	328(4.79)	18-1482-83
	dioxan	560(--)	18-1482-83
$C_{34}H_{24}N_6$			
6-Quinoxalinecarbonitrile, 2,2'-(1,2-ethanediylidene)bis[1,2-dihydro-3-methyl-1-phenyl-	$CHCl_3$-CF_3COOH	313(4.32),387(4.28), 650s(--),695(4.49)	33-0379-83
$C_{34}H_{24}O_2S_2$			
Bicyclo[4.2.0]octa-1,3,5-triene-7,8-dione, 3,4-bis[(4-methylphenyl)thio]-2,5-diphenyl-	heptane	265(4.45),340(4.17)	78-0645-83
$C_{34}H_{24}O_9$			
Ismailin trimethyl ether	MeOH	287(4.41)	88-1085-83
$C_{34}H_{24}O_{22}$			
Casuariin	MeOH	213(4.70),233(4.74), 257s(4.57)	39-1765-83C
Pedunculagin	MeOH	232(4.34),256s(4.57)	39-1765-83C
$C_{34}H_{26}BrNO$			
Quinoline, 8-benzoyl-1-(4-bromophenyl)-1,5,6,7-tetrahydro-2,4-diphenyl-	EtOH	380(3.52),510(3.73)	39-2601-83C
Quinoline, 8-[(4-bromophenoxy)phenylmethylene]-5,6,7,8-tetrahydro-2,4-diphenyl-	EtOH	285(4.50),370(4.13)	39-2601-83C
$C_{34}H_{26}ClNO$			
Quinoline, 8-benzoyl-1-(4-chlorophenyl)-1,5,6,7-tetrahydro-2,4-diphenyl-	EtOH	380(3.42),510(3.84)	39-2601-83C

Compound	Solvent	$\lambda_{max}(\log \epsilon)$	Ref.
Quinoline, 8-[(4-chlorophenoxy)phenylmethylene]-5,6,7,8-tetrahydro-2,4-diphenyl-	EtOH	285(4.35),370(3.90)	39-2601-83C
$C_{34}H_{26}N$			
Pyridinium, 2-(1-naphthalenyl)-4,6-diphenyl-1-(phenylmethyl)-, salt with trifluoromethanesulfonic acid	DMSO	310(4.47)	150-0301-83M
Pyridinium, 2-(2-naphthalenyl)-4,6-diphenyl-1-(phenylmethyl)-, salt with trifluoromethanesulfonic acid	DMSO	312(4.50)	150-0301-83M
$C_{34}H_{26}N_3$			
3H-Indolium, 3-[2-cyano-3-(1-methyl-2-phenyl-1H-indol-3-yl)-2-propenylidene]-1-methyl-2-phenyl-, perchlorate	$MeNO_2$-HOAc	586(5.00)	103-1306-83
$C_{34}H_{26}N_4O_4$			
1H-Pyrazolo[1,2-a]pyrazol-4-ium, 3-hydroxy-5,7-bis[[(4-methoxyphenyl)methylene]amino]-1-oxo-2,6-diphenyl-, hydroxide, inner salt	MeCN	275(4.64),322(4.49), 383(4.48)	39-0011-83C
$C_{34}H_{26}O_{22}$			
Praecoxin B	MeOH	210(4.85),297(4.34)	94-0333-83
$C_{34}H_{27}Br_2N_2$			
3H-Indolium, 2-(4-bromophenyl)-3-[3-[2-(4-bromophenyl)-1-methyl-1H-indol-3-yl]-2-methyl-2-propenylidene]-1-methyl-, chloride	$MeNO_2$-HOAc	605(4.83)	103-1306-83
$C_{34}H_{27}NO$			
Quinoline, 8-benzoyl-1,5,6,7-tetrahydro-1,2,4-triphenyl-	EtOH	390(3.58),515(4.03)	39-2601-83C
Quinoline, 5,6,7,8-tetrahydro-8-(phenoxyphenylmethylene)-2,4-diphenyl-	EtOH	290(4.54),375(4.35)	39-2601-83C
$C_{34}H_{27}NO_2P$			
Phosphonium, [2-(9-acetyl-9H-carbazol-2-yl)-2-oxoethyl]triphenyl-, chloride	EtOH	231(4.416),324(4.350)	65-1571-83
$C_{34}H_{27}NO_3$			
3-Pyridinecarboxylic acid, 1,4-dihydro-1-(phenylmethyl)-, 2'-methoxy[1,1'-binaphthalen]-2-yl ester, (R)-	MeCN	361(3.94)	18-3672-83
Spiro[9H-fluorene-9,1'(5'H)-pyrrolo-[2,1-a]isoquinoline]-2'-carboxylic acid, 3'-benzoyl-6',10'b-dihydro-10'b-methyl-, methyl ester	CH_2Cl_2	373(4.02)	24-3915-83
betaine	CH_2Cl_2	604(4.94)	24-3915-83
$C_{34}H_{27}NO_4$			
Spiro[9H-fluorene-9,1'(5'H)-pyrrolo-[2,1-a]isoquinoline]-2',3'-dicarboxylic acid, 6',10'b-dihydro-10'b-phenyl-, dimethyl ester	CH_2Cl_2	320(3.96)	24-3915-83
betaine	CH_2Cl_2	470(--)	24-3915-83

Compound	Solvent	λ max (log ε)	Ref.
$C_{34}H_{27}N_3O_4$			
[1,2':3',1"-Ternaphthalene]-1',4'-dione, 4,4"-bis(dimethylamino)-5'-nitro-	C_6H_{12}	520(3.30)	150-0168-83S
	EtOH	529(--)	150-0168-83S
$C_{34}H_{28}$			
1,3,5,7,13,15,17-Cyclononadecaheptaene-9,11-diyne, 19-(diphenylmethylene)-8,13-dimethyl-, (E,E,Z,Z,E,E,E)-	CH_2Cl_2	305(4.40),363(4.58), 442s(3.99)	39-2987-83C
$C_{34}H_{28}N_2O_2$			
[1,2':3',1"-Ternaphthalene]-1',4'-dione, 4,4"-bis(dimethylamino)-	toluene	488(3.20)	150-0168-83S
	EtOH	489(--)	150-0168-83S
$C_{34}H_{28}N_4O_2$			
Dipyrimido[2,1-a:2',1'-a']phenanthro-[2,1,10-def:7,8,9-d'e'f']diisoquinoline-6,11-dione, 2,3,4,13,14,15-hexahydro-3,3,14,14-tetramethyl-	$CHCl_3$	474(5.4),503(4.8), 541(4.9)	24-3524-83
hydrochloride	$CHCl_3$	487(4.5),517(4.8), 554(4.9)	24-3524-83
$C_{34}H_{28}O_2$			
Bicyclo[3.2.0]hept-2-en-6-ol, 4-(diphenylmethylene)-7,7-diphenyl-, acetate, (1α,5α,6β)-	C_6H_{12}	270(4.207)	64-0504-83B
$C_{34}H_{28}O_3$			
Ethanone, 1-[5-[bis(4-methoxyphenyl)-methylene]-1,3-cyclopentadien-1-yl]-2,2-diphenyl-	MeCN	255(4.234),346(4.068), 395(4.178)	64-0504-83B
$C_{34}H_{28}O_6$			
8,15-Ethenohexaphene-14,16-dione, 5,9-diacetoxy-7a,8,14a,15,15a,15b-hexahydro-14a,15b-dimethyl-	MeOH	244(4.67),259(4.73), 297(4.26),369(3.78), 379s(3.61)	44-2412-83
$C_{34}H_{28}O_9$			
Antibiotic A-39183B	MeOH and MeOH-acid	228(4.76),266(4.68), 325s(3.90),420(4.18)	158-83-105
	MeOH-base	228(4.79),266(4.62), 340(4.15),426(4.29)	158-83-105
$C_{34}H_{28}O_{11}$			
"Anthraquinone 24a"	MeOH	208(4.66),258(4.75), 344(3.76),445(3.89), 472(3.97),490(3.90), 523(3.57)	5-0471-83
	MeOH-NaOH	256(4.71),273s(--), 340(3.91),561(4.05)	5-0471-83
	$CHCl_3$	263(4.50),288s(--), 349(3.72),448s(--), 478(3.79),495(3.73), 529(3.54)	5-0471-83
$C_{34}H_{29}NO_6S$			
Spiro[9H-fluorene-9,1'(5'H)-pyrrolo-[2,1-a]isoquinoline-2',3'-dicarboxylic acid, 6',10'b-dihydro-8',9'-dimethoxy-10'b-(2-thienyl)-, dimethyl ester	CH_2Cl_2	314(3.98)	24-3915-83
	betaine	450(--)	24-3915-83

Compound	Solvent	$\lambda_{max}(\log \epsilon)$	Ref.
$C_{34}H_{29}N_2$			
3H-Indolium, 1-methyl-3-[1-methyl-3-(1-methyl-2-phenyl-1H-indol-3-yl)-2-propenylidene]-2-phenyl-, perchlorate	$MeNO_2$-HOAc	605(4.81)	103-1306-83
3H-Indolium, 1-methyl-3-[2-methyl-3-(1-methyl-2-phenyl-1H-indol-3-yl)-2-propenylidene]-2-phenyl-, chloride	$MeNO_2$-HOAc	612(4.89)	103-1306-83
$C_{34}H_{30}$			
3,5,7,10,12,14,16-Nonadecaheptaene-1,18-diyne, 9-(diphenylmethylene)-3,7-dimethyl-	CH_2Cl_2	279s(4.24),292s(4.36), 304(4.42),317(4.37), 363s(4.63),381(4.69), 406s(4.62)	39-2987-83C
$C_{34}H_{30}N_3OS$			
Pyrylium, 4-[2-(2-benzothiazolylamino)-4-(diethylamino)phenyl]-2,6-diphenyl-, tetrafluoroborate	64% MeOH-pH 4.30	284(4.48),383(4.32), 555(4.72)	33-2165-83
$C_{34}H_{30}N_3S_2$			
Thiopyrylium, 4-[2-(2-benzothiazolyl-amino)-4-(diethylamino)phenyl]-2,6-diphenyl-, tetrafluoroborate	64% MeOH-pH 4.20	283(4.48),389(4.20), 604(4.60)	33-2165-83
$C_{34}H_{30}N_4$			
Quinoxaline, 2,2'-(1,2-ethanediylidene)bis[1,2-dihydro-3,6-dimethyl-1-phenyl-	$CHCl_3$-CF_3COOH	313(4.31),368(4.23), 685s(--),746(4.50)	33-0379-83
$C_{34}H_{30}N_4O_2$			
3-Pyridinecarboxamide, 1,1'-[[1,1'-binaphthalene]-2,2'-diylbis(methylene)]bis[1,4-dihydro-, (S)-	CH_2Cl_2	356(4.08)	18-3672-83
Quinoxaline, 2,2'-(1,2-ethanediylidene)bis[1,2-dihydro-6-methoxy-3-methyl-1-phenyl-	$CHCl_3$-CF_3COOH	327(4.23),390(3.98), 793(4.49)	33-0379-83
$C_{34}H_{30}N_4O_4S_2$			
Quinoxaline, 2,2'-(1,2-ethanediylidene)bis[1,2-dihydro-3-methyl-6-(methylsulfonyl)-1-phenyl-	$CHCl_3$-CF_3COOH	313(4.33),384(4.27), 650s(--),690(4.50)	33-0379-83
$C_{34}H_{30}O_9$			
Kuwanon P	EtOH	218(4.42),284(4.19), 310s(4.12),331(4.16)	94-2936-83
	EtOH-$AlCl_3$	221(4.43),299s(4.25), 305(4.23),331s(4.15)	94-2936-83
$C_{34}H_{30}O_{10}$			
Antibiotic A-39183A	MeOH and MeOH-acid	227(4.72),269(4.84), 300s(3.86),312(3.79), 328(3.83),408(4.21)	158-83-105
	MeOH-base	230(4.76),266(4.68), 340(4.18),423(4.40)	158-83-105
Antibiotic A-39183C	MeOH and MeOH-acid	226(4.65),268(4.75), 300s(3.79),310s(3.72), 327(3.72),410(4.14)	158-83-105
	MeOH-base	229(4.70),266(4.69), 328s(4.04),340(4.08), 340(4.07),420(4.23)	158-83-105

Compound	Solvent	λ_{max}(log ϵ)	Ref.
$C_{34}H_{30}O_{12}$			
ε-Naphthocyclinone	MeOH	222(4.47),264(4.31), 293s(--),353(3.66), 525(3.94)	5-0471-83
	MeOH-NaOH	248(--),293s(--), 360(--),608(--), 651(--)	5-0471-83
	$CHCl_3$	269(4.33),297s(--), 362(3.69),448s(--), 534(3.96),564s(--)	5-0471-83
$C_{34}H_{30}O_{13}$			
Anhydro-seco-α-naphthocyclinone, 7',8'-dehydro-8-methyl-, methyl ether	MeOH	251(4.30),285s(--), 480(3.88)	5-0510-83
methyl ester	MeOH-NaOH	294(--),580(--)	5-0510-83
Seco-γ-naphthocyclinone, 7-methyl-	$CHCl_3$-MeOH	252(4.37),278(4.11), 480(3.95),555s(--)	5-0510-83
	+ base	261(--),291s(--), 575(--)	5-0510-83
$C_{34}H_{31}N_3O_2$			
7H-Dibenz[f,ij]isoquinolin-7-one, 1-[4-(diethylamino)phenyl]-6-[(4-methoxyphenyl)amino]-2-methyl-	n.s.g.	495(4.47)	2-0805-83
$C_{34}H_{31}N_5O_{13}S$			
1-Azabicyclo[3.2.0]hept-2-ene-2-carboxylic acid, 6-ethyl-3-[[3-[(4-nitrophenyl)methoxy]-2-[[[(4-nitrophenyl)-methoxy]carbonyl]amino]-3-oxopropyl]-thio]-7-oxo-, (4-nitrophenyl)methyl ester, [5R-[3(S*),5α,6α]]-	THF	266(4.48),320(4.08)	158-0407-83
$C_{34}H_{32}N_2O_4$			
Methanone, [1,4-phenylenebis[methylene(6-methoxy-1,3(6H)-pyridinedi-yl]]]bis[phenyl-	CH_2Cl_2	240(4.1),314(3.8)	64-0878-83B
$C_{34}H_{32}N_2O_5$			
Apateline	MeOH	283(3.5),305s(--)	100-0001-83
	MeOH-NaOH	297(3.6)	100-0001-83
Norapateline, N-methyl-	MeOH	282(3.72)	100-0001-83
	MeOH-NaOH	302(3.95)	100-0001-83
Thalibrine, O-methyl-	MeOH	280(4.02),285s(4.01)	100-0001-83
$C_{34}H_{32}N_2O_6$			
Pachygonamine	MeOH	234s(4.71),291(4.05)	142-1927-83
$C_{34}H_{32}O_{13}$			
1H-Naphtho[2,3-c]pyran-3-acetic acid, 3,4,6,7,8,9-hexahydro-10-hydroxy-1-methyl-6,9-dioxo-8-[2,3,6,9-tetrahy-	MeOH	235(4.60),282(4.09), 364(3.81),478(3.79), 506(3.79),543(3.48)	5-0510-83
dro-5-hydroxy-4-methoxy-2-(2-methoxy-2-oxoethyl)-8-methyl-6,9-dioxonaphtho[1,2-b]furan-7-yl, methyl ester, [1S-[1α,3β,8β(R*)]]-	MeOH-NaOH	294(--),579(--)	5-0510-83
β-Naphthocyclinone, deacetyl-, methyl ester	MeOH	276(4.12),294s(--), 319s(--),488(3.84), 514(3.88),553(3.63)	5-0471-83
	MeOH-NaOH	225(4.60),282(3.89), 356(3.73),587(4.03),	5-0471-83

Compound	Solvent	λ_{max}(log ϵ)	Ref.
(cont.)		626(4.05)	5-0471-83
$C_{34}H_{34}N_2O_4$			
Neocochliodinol dimethyl ether	EtOH	290(4.48),480(3.58)	94-2998-83
1(4H)-Pyridinecarboxylic acid, 4,4'-(2,4-dimethyl-2,4-diphenyl-1,3-cyclobutanediylidene)bis[1,4-dihydro-	CH_2Cl_2	311s(--),320(4.71)	5-0069-83
$C_{34}H_{34}N_2O_6S_2$			
Benzo[1,2-b:4,5-b']difuran-3,7-dicarboxylic acid, 2,6-dihydro-2,6-bis-[[(2-methylphenyl)thio]imino]-, bis(1,1-dimethylethyl) ester, (Z,Z)-	$CHCl_3$	661(4.84)	150-0028-83S
$C_{34}H_{34}O_{11}$			
Toromycin acetonide diacetate	CH_2Cl_2	245(4.41),265s(--), 275(4.30),285(4.32), 320(4.10),380(4.17)	78-0599-83
$C_{34}H_{34}O_{15}$			
β-Naphthocyclinone, lyxo-4a,10a-dihydro-4a,10a-dihydroxy-deacetyl-, methyl ester	MeOH	222(4.30),248(4.42), 267s(--),352(3.77), 410(3.95),420s(--), 460(3.45)	5-0471-83
	MeOH-NaOH	243(4.41),274(4.22), 338(3.75),480(4.00), 510s(--)	5-0471-83
	$CHCl_3$	251(4.53),267s(--), 401(4.11),413s(--)	5-0471-83
β-Naphthocyclinone, xylo-4a,10a-dihydro-4a,10a-dihydroxy-deacetyl-, methyl ester	MeOH	224(4.28),247(4.40), 267s(--),353(3.77), 411(3.92),420s(--), 462(3.58)	5-0471-83
	MeOH-NaOH	241(4.48),272(4.22), 358(3.71),482(4.00), 513s(--)	5-0471-83
	$CHCl_3$	247(4.39),266s(--), 405(3.97),414s(--)	5-0471-83
$C_{34}H_{36}N_2O_2$			
6,6'-Bibenzoxazole, 2,2'-bis(tricyclo-[3.3.1.1^{3,7}]dec-1-yl)-	C_6H_{12}	275(--),294(4.37)	41-0595-83
$C_{34}H_{36}N_4O_2$			
1(4H)-Pyridinecarboxamide, 4,4'-(2,4-dimethyl-2,4-diphenyl-1,3-cyclobutanediylidene)bis[N,N-dimethyl-, trans	CH_2Cl_2	309(4.61),321s(--)	5-0069-83
$C_{34}H_{36}O_3$			
19-Norpregn-4-en-20-yn-3-one, 17α-[[4-(phenylmethyl)benzoyl]oxy]-	$CHCl_3$	245(4.56)	13-0309-83A
$C_{34}H_{37}NO_6$			
Phenol, 2-[[3-(1,1-dimethylethyl)-2-hydroxy-5-nitrophenyl]methyl]-6-[[2-hydroxy-3-[(2-hydroxy-5-methylphenyl)methyl]-5-methylphenyl]methyl]-4-methyl-	neutral	321(3.95)	126-2361-83
	anion	401(4.15)	126-2361-83
	dianion	415(4.28)	126-2361-83
$C_{34}H_{38}ClN_2$			
3H-Indolium, 2-[2-[2-chloro-3-[4-(1,3-	MeOH	885(5.21)	104-1854-83

Compound	Solvent	λmax(log ε)	Ref.
dihydro-1,3,3-trimethyl-2H-indol-2-ylidene)-2-butenylidene]-1-cyclohexen-1-yl]ethenyl]-1,3,3-trimethyl-, perchlorate (cont.)			104-1854-83
$C_{34}H_{38}N_2O_{13}$			
Acetamide, N-(2-acetoxyethyl)-N-[2-[[9,10-dihydro-9,10-dioxo-4-[(2,3,4-tri-O-acetyl-6-deoxy-β-D-glucopyranosyl)oxy]-1-anthracenyl]amino]ethyl]-	$CHCl_3$	250(4.53),276(4.00), 320(3.72),516(3.85)	18-1435-83
Acetamide, N-(2-acetoxyethyl)-N-[2-[[9,10-dihydro-9,10-dioxo-4-[(2,3,4-tri-O-acetyl-6-deoxy-α-L-mannopyranosyl)oxy]-1-anthracenyl]amino]ethyl]-	$CHCl_3$	249(4.71),276(4.18), 320(3.88),514(4.04)	18-1435-83
$C_{34}H_{38}N_4O_4$			
21H,23H-Porphine-2,12-dicarboxylic acid, 7,17-diethyl-3,8,13,18-tetramethyl-, diethyl ester	CH_2Cl_2	410(5.38),517(3.83), 558(4.27),582(3.99), 636(3.39)	35-2704-83
$C_{34}H_{38}N_4O_6$			
Hematoporphyrin	n.s.g.	397(5.3)(anom.)	64-0083-83C
$C_{34}H_{38}N_4O_{10}S_2$			
21H,23H-Porphine-2,18-dipropanoic acid, 7,12-bis(methoxysulfonyl)-3,8,13,17-tetramethyl-, dimethyl ester	$CHCl_3$	414(5.30),511(4.20), 547(3.85),579(3.86), 634(3.53)	5-0204-83
$C_{34}H_{38}O_{17}$			
Procumbide, 6'-O-p-coumaroyl-, pentaacetate	MeOH	283(4.17)	94-2296-83
$C_{34}H_{39}ClN_4O_2$			
21H,23H-Porphine-2-propanoic acid, 18-(2-chloroethyl)-7,12-diethyl-3,8,13,17-tetramethyl-, methyl ester	CH_2Cl_2	400(5.21),498(4.08), 532(3.98),568(3.83), 622(3.56)	44-4302-83
21H,23H-Porphine-2-propanoic acid, 18-(2-chloroethyl)-8,13-diethyl-3,7,12,17-tetramethyl-, methyl ester	CH_2Cl_2	400(5.22),500(4.13), 532(3.93),568(3.81), 620(3.56)	44-4302-83
$C_{34}H_{39}Cl_2N_3O_4$			
1H-Pyrrole-3-propanoic acid, 5-[[3-(2-chloroethyl)-5-[[3-(2-chloroethyl)-4-methyl-5-[(phenylmethoxy)carbonyl]-1H-pyrrol-2-yl]methyl]-4-methyl-2H-pyrrol-2-ylidene]methyl]-2,4-dimethyl-, methyl ester, (Z)-, hydrobromide	CH_2Cl_2	494(4.85)	44-4302-83
$C_{34}H_{39}N_2O$			
3H-Indolium, 2-[2-[3-[4-(1,3-dihydro-1,3,3-trimethyl-2H-indol-2-ylidene)-2-butenylidene]-2-methoxy-1-cyclopenten-1-yl]ethenyl]-1,3,3-trimethyl-, perchlorate	MeOH	839(5.32)	104-1854-83
$C_{34}H_{40}O_5$			
Benzoic acid, 4-benzoyl-, 2-[(1-oxotetradecyl)oxy]phenyl ester	EtOH	256(4.45),344(2.30)	24-0761-83
$C_{34}H_{40}O_{10}$			
1,2-Dehydroisophorosantonic lactone dimer	n.s.g.	240(3.93)	88-4417-83

Compound	Solvent	$\lambda_{max}(\log \epsilon)$	Ref.
$C_{34}H_{41}N_3O_4$			
1H-Pyrrole-3-propanoic acid, 5-[[3-ethyl-5-[[3-ethyl-4-methyl-5-[(phenylmethoxy)carbonyl]-1H-pyrrol-2-yl]methyl]-4-methyl-2H-pyrrol-2-ylidene]methyl]-2,4-dimethyl-, methyl ester, monohydrobromide, (Z)-	CH_2Cl_2	494(4.83)	44-4302-83
$C_{34}H_{42}Cl_2N_4O_2$			
21H-Biline-2-propanoic acid, 7,12-bis(2-chloroethyl)-10,23-dihydro-1,3,8,13,17,18,19-heptamethyl-, methyl ester, dihydrobromide	CH_2Cl_2	448(4.94),522(5.11)	44-4302-83
21H-Biline-18-propanoic acid, 7,12-bis(2-chloroethyl)-10,23-dihydro-1,2,3,8,13,17,19-heptamethyl-, methyl ester, dihydrobromide	CH_2Cl_2	448(4.75),528(5.05)	44-4302-83
$C_{34}H_{42}O_{11}$			
Meliatoxin B_2	MeOH	207(3.93)	102-0531-83
$C_{34}H_{44}F_3N_3O_{10}S$			
Oxaflumazine disuccinate	MeOH	259(4.53),310(3.58)	162-0991-83
$C_{34}H_{44}N_6O_{12}$			
Adenosine, 5'-O-[2-[(17-oxo-4-nitroestra-1,3,5(10)-trien-3-yl)oxy]ethyl]-, 2',3'-diacetate	EtOH	262(4.15)	87-0162-83
$C_{34}H_{44}O_3$			
19-Norpregn-4-en-20-yn-3-one, 17α-[[1-oxo-5-(4-propylphenyl)pentyl]oxy]-	EtOH	241(4.23)	13-0309-83A
$C_{34}H_{44}O_{10}$			
Ohchinolal	EtOH	215(4.18)	18-1139-83
$C_{34}H_{44}O_{12}$			
Meliatoxin A_2	MeOH	210(3.82)	102-0531-83
$C_{34}H_{46}N_2O_{17}S$			
Paulomycin A	EtOH	236(4.21),276(4.03), 322(3.99)	158-83-44
$C_{34}H_{46}O_5$			
Benzoic acid, 4-benzoyl-, 2-[(1-oxotetradecyl)oxy]cyclohexyl ester, trans	EtOH	255(4.37),341(2.19)	24-0761-83
$C_{34}H_{46}O_7$			
Kijanolide, 32-O-methyl-	CF_3CH_2OH	245(3.85),262(3.90)	39-1497-83C
$C_{34}H_{46}O_{11}$			
Affinoside VIII monomethylate diacetate	MeOH	263(4.13)	94-1199-83
$C_{34}H_{48}N_2O_8$			
Ansatrienin A_2	MeOH	230(4.39),264s(--), 271(4.67),279(4.57), 387(3.26)	158-0187-83
	MeOH-NaOH	261(4.64),269(4.69), 278(4.64),481(3.28)	158-0187-83
same spectra for ansatrienin A_3			

Compound	Solvent	λ_{max}(log ε)	Ref.
$C_{34}H_{48}N_2O_9$			
Ajacine	n.s.g.	223(4.45),252(4.22), 310(3.73)	162-0029-83
$C_{34}H_{48}O_3$			
18,19-Dinorpregn-4-en-20-yn-3-one, 17α-[([1,1'-bicyclohexyl]-4-ylcarbonyl)oxy]-13-ethyl-	MeOH	243(4.18)	13-0349-83A
$C_{34}H_{50}N_2O_4$			
1,2-Cyclobutanedicarboxamide, N,N,N',N'-tetrabutyl-3,4-bis(2-hydroxyphenyl)-, cis	MeOH	277(3.81),284(3.77)	39-1083-83C
trans	MeOH	278(3.80),282(3.78)	39-1083-83C
$C_{34}H_{50}O_2$			
1'H-5α-Cholest-2-eno[2,3-c][2]benzopyran-1'-one, 2α,3α-dihydro-	C_6H_{12}	205(3.54),229(3.45), 240s(3.36),268s(2.74), 277(2.98),286(2.95)	5-2247-83
2β,3α-	C_6H_{12}	206(3.87),230(3.89), 240s(3.65),269s(2.93), 277(3.08),286(3.04)	5-2247-83
	EtOH	224(--),234(--), 276(--),289(--)	5-2247-83
2β,3β-	C_6H_{12}	206(3.93),232(3.85), 240s(3.67),270s(2.85), 278(3.00),288(2.98)	5-2247-83
5α-Cholest-1-en-3-one, 2-(4-methoxyphenyl)-	MeOH	234(4.26),284(3.70), 295s(3.60)	12-0789-83
$C_{34}H_{50}O_3$			
18,19-Dinorpregn-4-en-20-yn-3-one, 13-ethyl-17α-[[(4-hexylcyclohexyl)carbonyl]oxy]-	MeOH	244(4.16)	13-0349-83A
19-Norpregn-4-en-20-yn-3-one, 17α-[(2-cyclopentyl-1-oxononyl)oxy]-	EtOH	244(4.21)	13-0255-83A
$C_{34}H_{50}O_7$			
Milbemycin G	EtOH	238s(--),244(4.48), 253s(--)	158-0502-83 +158-0980-83
$C_{34}H_{50}O_8$			
Tetronomycin	MeOH	256(3.97),290(4.02)	158-83-28
sodium salt	MeOH	252(4.25),301(3.95)	158-83-28
$C_{34}H_{51}NO_3S_2$			
2H-1-Benzopyran, 2-[5-[(4,5-dihydro-2-thiazolyl)thio]-4,8,12-trimethyl-3,7,11-tridecatrienyl]-3,4-dihydro-6-(methoxymethoxy)-2,5,7,8-tetramethyl-	MeOH	278(3.42),286(3.38)	94-4341-83
$C_{34}H_{52}Cl_4N_5O_9PSi_2$			
2'-Guanylic acid, 3',5'-bis-O-[(1,1-dimethylethyl)dimethylsilyl]-N-(2-methyl-1-oxopropyl)-, 2-chlorophenyl 2,2,2-trichloroethyl ester	MeOH	258(4.18),277(4.05)	33-2069-83
3'-Guanylic acid, 2',5'-bis-O-[(1,1-dimethylethyl)dimethylsilyl]-N-(2-methyl-1-oxopropyl)-, 2-chlorophenyl 2,2,2-trichloroethyl ester	MeOH	257(4.23),271(4.10), 283s(4.08)	33-2069-83

Compound	Solvent	$\lambda_{max}(\log \epsilon)$	Ref.
$C_{34}H_{52}O$			
1'H-5α-Cholest-2-eno[2,3-c][2]benzo-pyran, 2α,3α-dihydro-	C_6H_{12}	203(3.78),250s(2.18) 256s(2.40),264(2.70), 271(2.74)	5-2247-83
2β,3α-	C_6H_{12}	212(3.94),252s(2.48), 260s(2.70),265(2.85), 272(2.85)	5-2247-83
2β,3β-	C_6H_{12}	207(3.88),250s(2.30), 258s(2.54),265(2.65), 272(2.70)	5-2247-83
$C_{34}H_{52}O_2$			
5α-Cholestan-3-one, 2α-(4-methoxyphen-yl)-	MeOH	228(4.08),277(3.36), 284(3.30)	12-0789-83
$C_{34}H_{52}O_3$			
Furan, 2,2',2"-methylidynetris[5-heptyl-	hexane	227(3.31)	103-0360-83
$C_{34}H_{52}O_7$			
Milbemycin E	EtOH	241(4.42)	158-0502-83 +158-0980-83
$C_{34}H_{54}O_5$			
19-Norlanost-5-en-7-one, 3,11-diacet-oxy-9-methyl- (cucurbitane)	n.s.g.	244(4.06)	138-1171-83
$C_{34}H_{54}O_6$			
9,10-Secoergosta-5,7,10(19),22-tetra-en-3-ol, β-D-glucopyranoside, (3β,5Z,7E,22E)-	n.s.g.	265(4.27)	33-2093-83
$C_{34}H_{54}O_7$			
Irumanolide I	MeOH	232(4.07)	158-0931-83
$C_{34}H_{54}O_8$			
Lasalocid A	50% isoPrOH	248(3.83),318(3.62)	162-0773-83
sodium salt	50% isoPrOH	308(3.61)	162-0773-83
$C_{34}H_{56}O_6$			
19-Norlanost-5-ene-3,11,24,25-tetrol, 9-methyl-, 3,11-diacetate, (3β,9β-10α,11α,24R)- (mogrol 3,11-diacetate)	n.s.g.	295(4.08)	95-1155-83
$C_{34}H_{60}O_2Si_2$			
Silane, [2-[[(3β,22E)-3-[[(1,1-dimeth-ylethyl)dimethylsilyl]oxy]-24-nor-chola-5,7,22-trien-23-yl]oxy]ethyl]-trimethyl-	EtOH	262(3.93),273(4.03), 284(4.10),294(3.89)	5-1031-83
$C_{34}H_{61}N_2O_2$			
Benzenaminium, 4-[(16-methoxy-16-oxo-7-hexadecenyl)octylamino]-N,N,N-trimethyl-, iodide, (E)-	MeOH	269(4.32),303(3.34)	5-0802-83
$C_{34}H_{62}N_2O_2$			
Benzenamine, 4-nitro-N,N-ditetradecyl-	MeOH	234(3.93),365(4.35)	5-0802-83
$C_{34}H_{64}N_2$			
1,4-Benzenediamine, N,N-ditetradecyl-, monohydrochloride	MeOH	259(4.09),310(3.30)	5-0802-83

Compound	Solvent	$\lambda_{max}(\log \epsilon)$	Ref.
$C_{35}H_{23}NO_4$			
Phenanthro[9',10':5,6][1,4]dioxino-[2,3-d]oxazol-11(9aH)-one, 12,12a-dihydro-9a,12,12a-triphenyl-	MeOH	218(3.70),248(3.60), 258(3.72),270(3.50), 283(3.15),297(3.26), 309(3.30)	73-2812-83
$C_{35}H_{23}N_3O$			
3,5-Pyridinedicarbonitrile, 2,6-bis([1,1'-biphenyl]-4-yl)-4-(2-furanyl)-1,4-dihydro-	EtOH	207(4.90),287(4.69)	73-3112-83
$C_{35}H_{23}N_3O_3$			
2,5-Cyclohexadiene-1-carbonitrile, 1,2,6-triphenyl-4-(tetrahydro-2,4,6-trioxo-1-phenyl-5(2H)-pyrimidinylidene)-	$CHCl_3$	387(4.40)	83-0732-83
$C_{35}H_{25}N$			
Pyridine, pentaphenyl-	neutral	297(3.95)	39-0045-83B
	protonated	325(3.82)	39-0045-83B
$C_{35}H_{25}O$			
Pyrylium, pentaphenyl-, bromide	MeCN	292(4.34),402(4.28)	22-0115-83
$C_{35}H_{27}F_3O_3$			
9-Phenanthrenol, 9,10-dihydro-9,10-bis(4-methoxyphenyl)-10-[4-(trifluoromethyl)phenyl]-	heptane	248s(--),277(4.15), 283s(--)	70-0271-83
$C_{35}H_{27}N_5O$			
7H-Benzimidazo[2,1-a]benz[de]isoquinolin-7-one, 4-[5-[4-(dimethylamino)-phenyl]-4,5-diphydro-1-phenyl-1H-pyrazol-3-yl]-	toluene	515(4.44)	103-0214-83
$C_{35}H_{27}O_4$			
1-Benzopyrylium, 5-methoxy-4-[3-(5-methoxy-2-phenyl-4H-1-benzopyran-4-ylidene)-1-propenyl]-2-phenyl-, perchlorate	CH_2Cl_2	706(5.31)	103-0243-83
	MeCN	696(5.20)	103-0243-83
1-Benzopyrylium, 6-methoxy-4-[3-(6-methoxy-2-phenyl-4H-1-benzopyran-4-ylidene)-1-propenyl]-2-phenyl-, perchlorate	CH_2Cl_2	728(5.37)	103-0243-83
	MeCN	718(5.13)	103-0243-83
1-Benzopyrylium, 7-methoxy-4-[3-(7-methoxy-2-phenyl-4H-1-benzopyran-4-ylidene)-1-propenyl]-2-phenyl-, perchlorate	CH_2Cl_2	702(5.40)	103-0243-83
	MeCN	696(5.30)	103-0243-83
1-Benzopyrylium, 8-methoxy-4-[3-(8-methoxy-2-phenyl-4H-1-benzopyran-4-ylidene)-1-propenyl]-2-phenyl-, perchlorate	CH_2Cl_2	712(5.34)	103-0243-83
	MeCN	702(5.22)	103-0243-83
1-Benzopyrylium, 2-(2-methoxyphenyl)-4-[3-[2-(2-methoxyphenyl)-4H-1-benzopyran-4-ylidene]-1-propenyl]-, perchlorate	CH_2Cl_2	720(5.23)	103-0243-83
	MeCN	710(5.4)	103-0243-83
1-Benzopyrylium, 2-(3-methoxyphenyl)-4-[3-[2-(3-methoxyphenyl)-4H-1-benzopyran-4-ylidene]-1-propenyl]-, perchlorate	CH_2Cl_2	716(5.30)	103-0243-83
	MeCN	706(5.11)	103-0243-83

Compound	Solvent	λ_{max}(log ϵ)	Ref.
1-Benzopyrylium, 2-(4-methoxyphenyl)-4-[3-[2-(4-methoxyphenyl)-4H-1-benzopyran-4-ylidene]-1-propenyl]-, perchlorate	CH_2Cl_2	736(5.30)	103-0243-83
	MeCN	724(5.18)	103-0243-83
$C_{35}H_{28}N_2O_2S$			
2,4-Pentadienoic acid, 2-cyano-4-phenyl-3-[phenyl[3-(phenylmethyl)-2(3H)-benzothiazolylidene]methyl]-, ethyl ester	$CHCl_3$	243(4.47),530(4.09)	18-1688-83
$C_{35}H_{29}N_5O_8$			
β-D-arabino-Hept-5-enofuranuronic acid, 1-[6-(benzoylamino)-9H-purin-9-yl)-1,5,6-trideoxy-, ethyl ester, 2,3-dibenzoate, (E)-	MeOH	232(4.62),278(4.34)	87-1530-83
$C_{35}H_{30}O_4$			
1,3-Cyclopentadiene-1-carboxylic acid, 5-[bis(4-methoxyphenyl)methylene]-2-(diphenylmethyl)-, methyl ester	C_6H_{12}	247(4.386),297s(3.981), 332(4.192),379(4.500)	64-0504-83B
$C_{35}H_{30}O_{14}$			
γ-Isonaphthocyclinone	MeOH-NaOH	217(4.61),276(3.86), 363(3.75),605(4.00), 646(4.04)	5-0471-83
	$CHCl_3$	239(4.33),278(4.21), 329(3.77),500(3.90), 533(3.98),577(3.79)	5-0471-83
$C_{35}H_{31}N_2$			
3H-Indolium, 1-methyl-3-[1-methyl-3-(1-methyl-2-phenyl-1H-indol-3-yl)-2-butenylidene]-2-phenyl-, perchlorate	$MeNO_2$-HOAc	630(4.65)	103-1306-83
$C_{35}H_{31}N_5O_8$			
β-D-arabino-Heptofuranuronic acid, 1-[6-(benzoylamino)-9H-purin-9-yl)-1,5,6-trideoxy-, ethyl ester, 2,3-dibenzoate	MeOH	232(4.58),278(4.33)	87-1530-83
$C_{35}H_{32}N_2O_5$			
Apateline, 1,2-dehydro-	MeOH	288s(--),335(3.46)	100-0001-83
	MeOH-base	292s(--),337(3.56)	100-0001-83
Telobine, 1,2-dehydro-	MeOH and MeOH-base	287s(--),336(3.65)	100-0001-83
$C_{35}H_{32}O_4$			
3-Cyclopentene-1-carboxylic acid, 2-[bis(4-methoxyphenyl)methylene]-5-(diphenylmethyl)-, methyl ester	C_6H_{12}	255(4.314),297(4.433)	64-0504-83B
$C_{35}H_{32}O_{13}$			
[8,8'-Bi-1H-naphtho[2,3-c]pyran]-3,3'-diacetic acid, 3,3',4,4',5,6',9',10-octahydro-6,9,10'-trihydroxy-1,1',7-trimethyl-5,6',9',10-tetraoxo-, dimethyl ester	$CHCl_3$	278(4.06),353(3.61), 458s(--),517(3.85), 558(3.59)	5-0510-83

Compound	Solvent	λ_{max}(log ε)	Ref.
$C_{35}H_{32}O_{14}$			
γ-Naphthocyclinone, dihydro-	MeOH	227(4.68),275(3.95), 283(3.98),293s(--), 472s(--),494(3.94)	5-0471-83
	MeOH-NaOH	297(--),533s(--), 589(--),628(--)	5-0471-83
$C_{35}H_{32}O_{15}$			
β-Naphthocyclinone epoxide	MeOH	206(4.54),212s(--), 252(4.36),264s(--), 352(3.76),423(3.80)	5-0471-83
	MeOH-NaOH	223(4.44),248(4.34), 282s(--),342(3.70), 500(3.80)	5-0471-83
	$CHCl_3$	256(4.39),278s(--), 355(3.70),428(3.98)	5-0471-83
$C_{35}H_{33}BrO_{15}$			
β-Naphthocyclinone bromohydrin	$CHCl_3$	252(4.40),351(3.78), 416(4.03),432s(--)	5-0471-83
$C_{35}H_{33}ClO_{15}$			
β-Naphthocyclinone chlorohydrin	$CHCl_3$	256(4.46),278(4.19), 350(3.73),413(4.01), 432s(--)	5-0471-83
10a-epi-	$CHCl_3$	252(4.28),273(4.09), 360(3.67),410(3.96), 425s(--)	5-0471-83
$C_{35}H_{34}N_2O_5$			
Apateline, N-methyl-	MeOH	280(3.61)	100-0001-83
	MeOH-NaOH	290(3.72)	100-0001-83
$C_{35}H_{34}N_2O_6$			
Gilletine	MeOH	237(4.34),274s(3.33), 290(3.41),301s(3.36)	100-0001-83
Pachygonamine, N-methyl-	MeOH	235s(4.19),291(3.59)	142-1927-83
$C_{35}H_{34}N_2O_7$			
Isogilletine N-oxide	MeOH	223s(4.49),229s(4.54), 240(4.55),289(3.80), 296s(3.75)	100-0001-83
$C_{35}H_{34}O_{13}$			
Seco-β-naphthocyclinone, 7-methyl-, methyl ester	MeOH	237s(--),282(4.03), 350(3.60),480(3.85), 510(3.88),551(3.66)	5-0510-83
	MeOH-NaOH	296(--),585(--), 622(--)	5-0510-83
Toromycin tetraacetate (isomer A)	CH_2Cl_2	245(4.49),268(4.27), 277(4.47),288s(--), 320(4.16),335(4.16), 380(4.17)	78-0599-83
isomer B	n.s.g.	245(4.57),265(4.33), 320(4.17),335(4.16), 385(4.17)	78-0599-83
$C_{35}H_{34}O_{14}$			
β-Naphthocyclinone, dihydro-	MeOH	206(4.50),229(4.62), 283(3.90),300s(--), 483s(--),513(3.91), 552(3.75)	5-0471-83

Compound	Solvent	λ_{max}(log ϵ)	Ref.
(cont.)	$CHCl_3$	241(4.27),286(3.90), 300s(--),483(3.83), 516(3.90),558(3.72)	5-0471-83
$C_{35}H_{34}O_{15}$			
δ-Naphthocyclinone periodate oxidation product	MeOH	218(4.60),267(4.37), 298(4.06),335s(--), 460(3.62)	5-0471-83
	$CHCl_3$	270(4.04),305(3.57), 328(3.49),445(3.42)	5-0471-83
$C_{35}H_{35}BrN_2O_2S$			
Estra-1,3,5(10)-trien-17-one, 4-[4-(4-bromophenyl)-2-[(phenylmethyl)imino]-3(2H)-thiazolyl]-3-methoxy-	EtOH	277s(4.632),289(4.176)	36-1205-83
	EtOH-HCl	265(4.376)	36-1205-83
$C_{35}H_{35}ClN_2O_2S$			
Estra-1,3,5(10)-trien-17-one, 2-[4-(4-chlorophenyl)-2-[(phenylmethyl)imino]-3(2H)-thiazolyl]-3-methoxy-	EtOH	300(4.252)	36-1205-83
	EtOH-HCl	260s(4.280),266(4.307)	36-1205-83
$C_{35}H_{36}N_2O_2S$			
Estra-1,3,5(10)-trien-17-one, 3-methoxy-2-[4-phenyl-2-[(phenylmethyl)-imino]-3(2H)-thiazolyl]-	EtOH	295(4.177)	36-1205-83
	EtOH-HCl	262(4.253),290s(4.156)	36-1205-83
Estra-1,3,5(10)-trien-17-one, 3-methoxy-4-[4-phenyl-2-[(phenylmethyl)-imino]-3(2H)-thiazolyl]-	EtOH	288(4.080)	36-1205-83
	EtOH-HCl	269(4.212)	36-1205-83
$C_{35}H_{36}N_2O_6$			
Daphnoline	MeOH	285(3.9)	162-407-83
$C_{35}H_{36}N_4O_5$			
Deuteroporphyrin IX, 4-ethenyl-2-formyl-, dimethyl ester	CH_2Cl_2	421(5.14),516(4.06), 556(4.18),582(3.97), 642(3.40)	35-6638-83
$C_{35}H_{36}N_6O_{10}$			
Adenosine, 5'-O-[γ-[(17-oxo-4-nitro-estra-1,3,5(10)-trien-3-yl)oxy]-propyl]-	EtOH	260(4.13)	87-0162-83
$C_{35}H_{37}NO_{11}$			
Ravidomycin diacetate	MeOH	245(4.50),278(4.48), 285(4.51),306s(4.17), 323(4.08),335(4.06), 346s(4.01),383(4.13)	23-0323-83
	MeOH	249(4.55),267(4.39), 276(4.51),322(4.01), 336(4.03)	158-1490-83
$C_{35}H_{38}N_4O_4$			
Porphyrin, 6,7-bis(2-methoxycarbonylethyl)-1,2,3,5,8-pentamethyl-4-vinyl-	CH_2Cl_2	400(5.17),502(4.09), 538(4.04),570(3.81), 626(3.54)	39-2329-83C
Porphyrin, 6,7-bis(2-methoxycarbonylethyl)-1,3,4,5,8-pentamethyl-2-vinyl-	CH_2Cl_2	400(5.20),502(4.13), 538(4.08),570(3.86), 626(3.60)	39-2329-83C
$C_{35}H_{39}ClN_4O_4$			
Porphyrin, 2-(2-chloroethyl)-6,7-	CH_2Cl_2	400(5.22),498(4.07),	39-2329-83C

Compound	Solvent	λ_{max}(log ε)	Ref.
bis(2-methoxycarbonylethyl)-1,3,4,5-8-pentamethyl- (cont.)		532(3.90),566(3.72), 622(3.48)	39-2329-83C
Porphyrin, 4-(2-chloroethyl)-6,7-bis(2-methoxycarbonylethyl)-1,2,3,5-8-pentamethyl-	CH_2Cl_2	400(5.22),498(4.08), 532(3.92),566(3.76), 622(3.52)	39-2329-83C
$C_{35}H_{39}NO_{11}$			
Ravidomycin, diacetyldihydro-	MeOH	244(4.64),267(4.46), 276(4.61),300(4.03), 322(4.07),336(4.08)	158-1490-83
$C_{35}H_{40}N_2O_{14}$			
Cadambine tetraacetate	MeOH	230(4.76),240(4.13), 270(3.93),285(3.76), 362(4.17)	100-0325-83
	MeOH-NaOH	230(4.76),270(3.93), 285(3.75)	100-0325-83
$C_{35}H_{40}N_4O_6$			
Isophorcarubin dimethyl ester	MeOH	405s(4.50),435(4.60), 455s(4.55)	108-0233-83
	EtOH	431(--),460s(--)	108-0233-83
Phorcarubin dimethyl ester	MeOH	417(4.72)	108-0233-83
	EtOH	418(--)	108-0233-83
	$CHCl_3$	416(--)	108-0233-83
$C_{35}H_{42}N_2O_9$			
Rescinnamine	EtOH	304(4.42)	151-0171-83C
$C_{35}H_{42}N_4$			
Quinolinium, 1-methyl-4-[3-[1-methyl-2-(1-methyl-3-octyl-1H-benzimidazol-ium-2-yl)-4(1H)-pyridinylidene]-1-propenyl]-, diiodide	EtOH	676(5.09)	4-0023-83
$C_{35}H_{42}N_4O_6$			
21H-Biline-8,12-dipropanoic acid, 3,17-diethyl-1,19,22,24-tetrahydro-2,7,13,18-tetramethyl-1,19-dioxo-, dimethyl ester	CH_2Cl_2	372(4.83),642(4.33)	78-1841-83
21H-Biline-8,12-dipropanoic acid, 18-ethenyl-3-ethyl-1,2,3,19,21,24-hexa-hydro-2,7,13,17-tetramethyl-1,19-di-oxo-, dimethyl ester, trans	MeOH	214(4.78),280(4.26), 354(4.39),592(4.18)	5-0585-83
$C_{35}H_{42}O_{12}$			
Ingol, 3,7,12,18-tetra-O-acetyl-8-O-benzoyl-18-hydroxy-	MeOH	209(4.41),233(4.47), 275(3.49),282(3.42)	102-2795-83
$C_{35}H_{43}NO_{22}$			
Benzoic acid, 3-nitro-4-[[[2,3,6-tri-O-acetyl-4-O-(2,3,4,6-tetra-O-acetyl-β-D-glucopyranosyl)-β-D-glucopyrano-syl]oxy]methyl]-, methyl ester	$CHCl_3$	244(3.99),260s(3.87)	136-0023-83M
$C_{35}H_{44}N_4O_4$			
21H-Biline-2,18-dipropanoic acid, 10,23-dihydro-1,3,7,8,12,13,17,19-octamethyl-, dimethyl ester, dihydrobromide	CH_2Cl_2	372(4.17),454(4.58), 522(5.33)	39-2329-83C

Compound	Solvent	λ_{max}(log ϵ)	Ref.
$C_{35}H_{44}N_4O_6$			
21H-Biline-8,12-dipropanoic acid, 2,17-diethyl-1,2,3,19,22,24-hexahydro-3,7,13,18-tetramethyl-1,19-dioxo-, dimethyl ester, cis	MeOH	280(4.34),346(4.61), 582(4.05)	5-0585-83
trans	MeOH	276(4.57),345(4.83), 588(4.43)	5-0585-83
21H-Biline-8,12-dipropanoic acid, 3,18-diethyl-1,2,3,19,22,24-hexahydro-2,7,13,17-tetramethyl-1,19-dioxo-, dimethyl ester, cis	MeOH	275(4.35),344(4.61), 589(4.35)	5-0585-83
$C_{35}H_{44}O_{11}$			
Meliatoxin B_1	MeOH	207(3.93)	102-0531-83
$C_{35}H_{45}BF_2N_4O_2$			
Boron, difluoro(2,3,7,8,12,13,17,18-octaethyl-22,24-dihydro-21H-biline-1,19-dionato-N^{22},N^{23})-, (T-4)-	$CHCl_3$	330s(4.24),364(4.58), 430s(3.85),629(4.76)	78-1865-83
$C_{35}H_{45}ClN_4O_2$			
21H-Biline-2-propanoic acid, 18-(2-chloroethyl)-7,12-diethyl-10,23-dihydro-1,3,8,13,17,19-hexamethyl-, methyl ester, dihydrobromide	CH_2Cl_2	450(4.67),526(4.38)	44-4302-83
21H-Biline-18-propanoic acid, 2-(2-chloroethyl)-7,12-diethyl-10,23-dihydro-1,3,8,13,17,19-hexamethyl-, methyl ester, dihydrobromide	CH_2Cl_2	450(4.83),534(4.81)	44-4302-83
$C_{35}H_{46}N_4O_6$			
21H-Biline-8,12-dipropanoic acid, 2,17-diethyl-1,2,3,15,16,19,23,24-octahydro-3,7,13,18-tetramethyl-1,19-dioxo-, dimethyl ester, cis	MeOH	215(3.84),297(3.66), 499(3.86),584(3.12)	5-0585-83
21H-Biline-8,12-dipropanoic acid, 3,18-diethyl-1,2,3,15,61,19,23,24-octahydro-2,7,13,17-tetramethyl-1,19-dioxo-, dimethyl ester, cis	MeOH	215(3.84),297(3.66), 499(3.85),584(3.12)	5-0585-83
$C_{35}H_{46}O_6$			
Calozeylanic acid	n.s.g.	282(4.11),362(3.52)	39-0703-83C
$C_{35}H_{46}O_{11}$			
Trichoverritone	MeOH	259(4.57)	88-3539-83
$C_{35}H_{46}O_{12}$			
Affinoside VIII triacetate	MeOH	264(4.24)	94-1199-83
Meliatoxin A_1	MeOH	210(3.82)	102-0531-83
$C_{35}H_{46}O_{14}$			
Urceolatoside A dimethyl ether	MeOH	218(4.00)	102-1977-83
$C_{35}H_{48}N_2O$			
2,5-Cyclohexadien-1-one, 2-[[3,5-bis(1,1-dimethylethyl)phenyl]amino]-4-[[3,5-bis(1,1-dimethylethyl)phenyl]imino]-6-methyl-	hexane	275(4.35),486(3.78)	18-1476-83
$C_{35}H_{48}O_3$			
Ergosta-8(14),22-dien-15-one, 3-hy-	n.s.g.	233(4.23),260(4.23)	78-2201-83

Compound	Solvent	λ_{max}(log ε)	Ref.
droxy-, benzoate (cont.)			78-2201-83
$C_{35}H_{48}O_4$			
Ergosta-8(14),22-dien-15-one, 3β,9-dihydroxy-, 3-benzoate	$CHCl_3$	232(4.32),255(4.23)	78-2201-83
9,10-Secoergosta-5(10),7,22-trien-3-ol, 6,19-epidioxy-, benzoate, (3β,6S,7E,22E)-	EtOH	229(4.18)	44-3477-83
$C_{35}H_{48}O_6$			
Spiro[benzofuran-2(3H),2'-bicyclo-[2.2.2]octa[5,7]diene]-5',6'-dicarboxylic acid, 4',5,7,7'-tetrakis(1,1-dimethylethyl)-3'-oxo-, dimethyl ester	n.s.g.	280(4.6),290s(4.6), 345(3.8)	12-1361-83
$C_{35}H_{48}O_7$			
Kijanolide, 26,32-di-O-methyl-	CF_3CH_2OH	254(3.98)	39-1497-83C
$C_{35}H_{49}NO_{11}$			
Deoxyjesaconitine	EtOH	257(4.23)	94-2884-83
$C_{35}H_{53}FeN_6O_{13}$			
Coprogen	H_2O	434(3.45)	35-0810-83
	EtOH	217(4.46),250(4.22), 440(3.47)	162-0359-83
$C_{35}H_{54}ClN_6O_9PSi_2$			
2'-Guanylic acid, 3',5'-bis-O-[(1,1-dimethylethyl)dimethylsilyl]-N-(2-methyl-1-oxopropyl)-, 2-chlorophenyl 2-cyanoethyl ester	MeOH	257(4.11),271(3.99), 282s(3.95)	33-2069-83
3'-Guanylic acid, 2',5'-bis-O-[(1,1-dimethylethyl)dimethylsilyl]-N-(2-methyl-1-oxopropyl)-, 2-chlorophenyl 2-cyanoethyl ester	MeOH	257(4.15),271(4.01), 282s(3.99)	33-2069-83
$C_{35}H_{54}N_2O_8S$			
Benzenesulfonamide, N-[4-(1,1-dimethylethyl)-3-(hexadecyloxy)-6-oxo-2,4-cyclohexadien-1-ylidene]-2-(2-methoxyethoxy)-5-nitro-	acetone	456(4.66)	44-0177-83
$C_{35}H_{54}O_3$			
Furan, 2,2',2"-ethylidynetris[5-heptyl-	EtOH	230(4.34)	103-0474-83
$C_{35}H_{54}O_4SSi$			
Pregna-5,7-diene-20-methanol, 3-[[(1,1-dimethylethyl)dimethylsilyl]oxy]-, 4-methylbenzenesulfonate, (3β,20S)-	EtOH	262(3.93),271(4.08), 282(4.10),292(3.88)	5-1031-83
$C_{35}H_{56}O_5$			
Olean-12-ene-3,16,22,28-tetrol, 16-(2-methyl-2-butenoate)	MeOH	223(3.82)	20-0473-83
Olean-12-ene-3,16,22,28-tetrol, 16-(3-methyl-2-butenoate)	MeOH	220(3.85)	20-0473-83
Olean-12-ene-3,16,22,28-tetrol, 22-(2-methyl-2-butenoate)	MeOH	222(3.82)	20-0473-83
Olean-12-ene-3,16,22,28-tetrol, 22-(3-methyl-2-butenoate)	MeOH	221(3.62)	20-0473-83

Compound	Solvent	λ_{max}(log ϵ)	Ref.
$C_{35}H_{56}O_6$			
Olean-12-ene-3,15,16,22,28-pentol, 22-(2-methyl-2-butenoate)	MeOH	217(3.82)	20-0473-83
Olean-12-ene-3,15,16,22,28-pentol, 22-(3-methyl-2-butenoate)	MeOH	220(3.84)	20-0473-83
$C_{35}H_{56}O_{14}$			
Chalcomycin	n.s.g.	218(4.36)	162-0283-83
$C_{35}H_{57}NO_{11}$			
Mycinamicin VI	MeOH	215(4.31),281.5(4.33)	158-0175-83
$C_{35}H_{58}O_3$			
Pentanoic acid, 5-[(1α,5Z,7E)-1-hydroxy-19-nor-9,10-secocholesta-5,7-dien-4-ylidene]-, 1,1-dimethylethyl ester	EtOH	264(4.28)	44-3483-83
$C_{35}H_{58}O_9$			
Bafilomycin A_1	MeOH	245(4.40),280(4.08)	88-5193-83
$C_{35}H_{59}N_5O_5Si_3$			
Adenosine, N-benzoyl-2',3',5'-tris-O-[(1,1-dimethylethyl)dimethylsilyl]-	MeOH	230(4.12),280(4.31)	35-5879-83
$C_{35}H_{59}N_5O_6Si_3$			
Guanosine, N-benzoyl-2',3',5'-tris-O-[(1,1-dimethylethyl)dimethylsilyl]-	MeOH	234(4.23),257(4.19), 264(4.19),293(4.19)	33-2069-83
$C_{35}H_{61}NO_8$			
Myxovirescin A	MeOH	283(4.32)	158-83-155
	MeOH	238(4.34),280s(2.48)	158-83-155
$C_{35}H_{66}O_9$			
8,16,24-Pentatriacontanetrione, 6,14,22,30,32,34-hexahydroxy-	MeOH	276(2.11)	88-3803-83

Compound	Solvent	λ_{max}(log ε)	Ref.
$C_{36}H_{16}Cl_{14}NO_{14}$			
L-Phenylalanine, N-[(phenylmethoxy)-carbonyl]-, 4-[bis(pentachlorophenyl)methylene]-2,3,5,6-tetrachlorophenyl ester (radical)	C_6H_{12}	218(4.94),274s(3.77), 334s(3.82),365s(4.27), 380(4.55),480s(3.04), 500(3.05),556(3.06)	44-3716-83
$C_{36}H_{22}N_6$			
3,5:14,16-Dinitrilo-5H,16H-dicyclohepta[e,k]diimidazo[1,5-a:1',5'-g][1,7]-diazacyclodecine, 1,12-diphenyl-	EtOH	238(4.91),250s(4.87), 266s(4.75),277s(4.66), 328(4.70),340s(4.66), 415s(4.53),437(4.70), 525s(3.78),555(3.81), 585s(3.72),640s(3.25)	18-3703-83
$C_{36}H_{24}$			
7H-Dibenzo[a,g]cyclopentadecene, 16,17,18,19-tetradehydro-7-(diphenylmethylene)-	CH_2Cl_2	230(4.50),286s(4.46), 300(4.51),335s(4.38), 368(4.27)	39-2987-83C
$C_{36}H_{24}Fe_2O_8$			
Ferrocene, 1,1":1',1"'-bis[1,4-phenylenebis(oxycarbonyl)]bis-	THF	220(6.7),250s(--), 450(4.8)[sic]	18-2865-83
$C_{36}H_{24}N_2OS$			
Ethanone, 1-[4-[[4-[3-(2-quinolinyl)-benzo[f]quinolin-1-yl]phenyl]thio]-phenyl]-	isoAmOH	256(4.67),286(4.71), 320(4.54),370(3.88)	103-0894-83
Cu(I) complex	isoAmOH	542(3.95)	103-0894-83
$C_{36}H_{24}N_2O_2$			
Ethanone, 1-[4-[4-[3-(2-quinolinyl)-benzo[f]quinolin-1-yl]phenoxy]-phenyl]-	isoAmOH	260(4.72),283(4.86), 315(4.42),368(3.91)	103-0894-83
Cu(I) complex	isoAmOH	539(3.94)	103-0894-83
$C_{36}H_{24}N_2O_4$			
Cyclohept[1,2,3-hi]imidazo[2,1,5-cd]-indolizine-9-carboxylic acid, 3,4-dibenzoyl-2-phenyl-, ethyl ester	$CHCl_3$	257(4.68),285s(4.45), 323(4.38),394(4.46), 445s(4.09)	18-3703-83
$C_{36}H_{24}O_9$			
5,6,11,12,17,18-Trinaphthylenehexone, 1,7,16-trimethoxy-3,9,14-trimethyl-	$CHCl_3$	241(4.45),263(4.54), 280s(4.51),395(4.06)	5-0433-83
$C_{36}H_{25}BrO$			
Cyclopent[a]inden-1(3aH)-one, 2-bromo-8,8a-dihydro-3,3a,8,8-tetraphenyl-	$CHCl_3$	215(4.64),285(3.93)	18-3314-83
$C_{36}H_{26}$			
m-Quaterphenyl, 2',2"-diphenyl-	C_6H_{12}	243(4.66)	94-1572-83
m-Quaterphenyl, 5',5"-diphenyl-	C_6H_{12}	203(4.95),253(4.98)	94-1572-83
o-Quaterphenyl, 3',6"-diphenyl-	C_6H_{12}	188(4.94),254(4.70)	94-1572-83
m-Terphenyl, 2,2",5'-triphenyl-	C_6H_{12}	198(4.97),240(4.78), 254s(4.65)	94-2313-83
m-Terphenyl-, 2,5',6-triphenyl-	C_6H_{12}	201(4.94),204s(4.92), 250(4.79)	94-1572-83
$C_{36}H_{26}N_2O_2$			
1,6-Benzodiazocine-2,5-dione, 3,4-bis(diphenylmethylene)-1,3,4,6-tetrahydro- (or isomer)	$CHCl_3$	298(3.92),356(3.53)	18-3193-83

Compound	Solvent	$\lambda_{max}(\log \epsilon)$	Ref.
$C_{36}H_{27}O_2$			
Pyrylium, 2,6-diphenyl-4-(triphenylmethoxy)-, perchlorate	MeCN	254(4.55),280(4.61), 348(4.47)	22-0115-83
$C_{36}H_{28}N_2$			
[1,1'-Biphenyl]-4,4'-diamine, N,N,N',N'-tetraphenyl-, radical cation	MeCN	482(4.47),680(3.27)	152-0105-83
$C_{36}H_{28}N_2O_2$			
1H-Naphth[1,2-e][1,3]oxazine, 2,3-dihydro-2-[3-methyl-5-(2-phenylethenyl)-4-isoxazolyl]-1,3-diphenyl-	MeOH	233(4.70),268(4.01), 278(4.12),290(4.17), 318(4.21)	2-0425-83
2H-Naphth[2,1-e][1,3]oxazine, 3,4-dihydro-3-[3-methyl-5-(2-phenylethenyl)-4-isoxazolyl]-2,4-diphenyl-	MeOH	237(3.69),274(3.31), 325(3.28)	2-0425-83
$C_{36}H_{28}N_2O_3S$			
2,4-Pentadienoic acid, 2-cyano-3-[3-(2-oxo-2-phenylethyl)-2(3H)-benzothiazolylidene]phenylmethyl]-4-phenyl-, ethyl ester	$CHCl_3$	243(4.31),522(4.00)	18-1688-83
$C_{36}H_{28}N_3S_2$			
Thiazolo[3,4-a]pyrimidin-5-ium, 6-[3-(3-ethyl-2(3H)-benzothiazolylidene)-1-propenyl]-2,4,8-triphenyl-, perchlorate	MeCN	570(3.44),674(3.72)	103-0037-83
$C_{36}H_{28}O_{10}$			
4,6:20,22-Dimethano-5H,21H-tetrabenzo-[h,k,t,w][1,4,7,13,16,19]hexaoxacyclotetracosin-5,21,33,34-tetrone, 11,12,14,15,27,28,30,31-octahydro-	CH_2Cl_2	396(3.99)	18-3185-83
$C_{36}H_{29}N_3O_8S_2$			
1,2,4-Triazin-3(2H)-one, 4,5-dihydro-6-[(phenylmethyl)thio]-5-thioxo-2-(2,3,5-tri-O-benzoyl-ß-D-ribofuranosyl)-	EtOH	230(4.72),326(4.09)	80-0989-83
$C_{36}H_{29}O$			
2H-Furylium, 3-[4-(9-anthracenyl)-2-phenyl-1,3-butadienyl]-2,2-dimethyl-5-phenyl-, perchlorate	MeCN	625(4.34)	56-0579-83
	HOAc	441(--),666(--)	56-0579-83
$C_{36}H_{30}Ge_2$			
Digermane, hexaphenyl-	C_6H_{12}	241(4.49)	101-0149-83G
$C_{36}H_{30}N_3O_6P_3$			
1,3,5,2,4,6-Triazatriphosphorine, 2,2,4,4,6,6-hexahydro-2,2,4,4,6,6-hexaphenoxy-	n.s.g.	206(4.70),259(3.23), 266(3.31),272(3.18)	80-0003-83
$C_{36}H_{30}N_3P_3$			
1,3,5,2,4,6-Triazatriphosphorine, 2,2,4,4,6,6-hexahydro-2,2,4,4,6,6-hexaphenyl-	n.s.g.	266(3.09),273(2.89)	80-0003-83
$C_{36}H_{30}O_{12}$			
"Anthraquinone acetate 24b"	$CHCl_3$	261(4.66),274s(--), 355(3.89),395(3.92),	5-0471-83

Compound	Solvent	$\lambda_{max}(\log \epsilon)$	Ref.
(cont.)		413(3.95),433(3.84)	5-0471-83
$C_{36}H_{32}N_2O_9$			
2,4(1H,3H)-Pyrimidinedione, 5-(1-propynyl)-1-[2,3,5-tris-O-(4-methylbenzoyl)-β-D-arabinofuranosyl]-	MeOH	240(4.73),283(4.03)	87-0661-83
$C_{36}H_{32}N_4$			
1,2,17,18-Tetraazacyclohexatriaconta-2,4,6,12,14,16,18,20,22,24,30,32,34-36-tetradecaene-8,10,26,28-tetrayne, 7,12,25,30-tetramethyl-	THF	252(4.24),367(4.80)	18-1467-83
1,2,19,20-Tetraazacyclohexatriaconta-2,4,6,8,14,16,18,20,22,24,26,32,34-36-tetradecaene-10,12,28,30-tetrayne, 9,14,27,32-tetramethyl-	THF	251(4.20),367(4.77)	18-1467-83
$C_{36}H_{32}N_6O_4$			
1(4H)-Pyridineacetamide, 3,3'-[[1,1'-binaphthalene]-2,2'-diylbis(iminocarbonyl)]bis-, (S)-	$CHCl_3$	362(4.31)	18-3672-83
$C_{36}H_{34}Cl_2O_8$			
2H-Pyran-2-one, 3,3'-[[2,4-bis(4-chlorophenyl)-1,3-cyclobutanediyl]-dicarbonyl]bis[6-butyl-4-hydroxy-, (1α,2α,3β,4β)-	MeOH	203(4.8),225(4.7), 316(4.4),402(3.8)	83-0845-83
$C_{36}H_{34}N_2O_6$			
Isocochliodinol diacetate	EtOH	288(4.48),510(3.85)	94-2998-83
Neocochliodinol diacetate	EtOH	290(4.48),498(3.85)	94-2998-83
$C_{36}H_{34}N_3S_2$			
Thiopyrylium, 4-[4-(diethylamino)-2-[(3-ethyl-2(3H)-benzothiazolylidene)amino]phenyl]-2,6-diphenyl-, tetrafluoroborate	64% EtOH-pH 5.34	286(4.49),397(4.24), 597(4.59)	33-2165-83
$C_{36}H_{34}O_6$			
Phenol, 3-[2-[3-hydroxy-5-[(4-hydroxyphenyl)methyl]phenyl]ethyl]-2,4-bis-[(4-hydroxyphenyl)methyl]-5-methoxy-	MeOH	283(4.09),289s(4.06)	102-1011-83
$C_{36}H_{34}O_{12}$			
Benzoic acid, 4-methoxy-, 1,2,3,4-butanetetrayl ester, (R*,R*)-(±)-	EtOH	257(4.68)	35-7396-83
(R*,S*)-	EtOH	256(4.43)	35-7396-83
Kelletinin I tetramethyl ether	EtOH	258(4.57)	35-7396-83
$C_{36}H_{34}O_{15}$			
β-Naphthocyclinone epoxide, methyl ester	$CHCl_3$	254(4.41),275s(--), 353(3.75),429(4.04), 444s(--),455s(--)	5-0471-83
$C_{36}H_{35}ClO_{15}$			
β-Naphthocyclinone chlorohydrin methyl ester	$CHCl_3$	253(4.47),269s(--), 354(3.83),413(4.12), 429(4.03),443s(--)	5-0471-83
$C_{36}H_{36}N_2O_5$			
Acetamide, N-(4-butylphenyl)-N-[4-[(4-	benzene	584(4.11)	104-0745-83

Compound	Solvent	λ_{max}(log ε)	Ref.
butylphenyl)amino]-9,10-dihydro-5,8-dihydroxy-9,10-dioxo-1-anthracenyl]-			104-0745-83
9,10-Anthracenedione, 1-acetoxy-5,8-bis[(4-butylphenyl)amino]-4-hydroxy-	benzene	420(3.58),551(3.78)	104-0745-83
$C_{36}H_{36}N_2O_6$			
Pachygonamine, N,N-dimethyl-	MeOH	235s(3.95),290(3.26)	142-1927-83
Sciadoferine	MeOH	277(4.12),312(3.86)	100-0001-83
Tiliamosine (same in acid or base)	MeOH	235s(4.70),290(3.98)	142-1927-83
$C_{36}H_{36}N_4O_8$			
1(4H)-Pyridinecarboxylic acid, 4,4',4"-4"'-(1,2,3,4-cyclobutanetetrayli-dene)tetrakis-, tetraethyl ester	$CHCl_3$	309s(--),320(4.39), 374(4.75),385(4.74), 413s(--),478(4.57), 515(4.27),557(4.35)	5-0658-83
	MeCN	320(4.3),380(4.8), 450(4.6),490(4.6), 520(4.3),570(4.3)	5-0658-83
Pyridinium, 4,4'-[3,4-bis[1-(ethoxy-carbonyl)-4(1H)-pyridinylidene]-1-cyclobutene-1,2-diyl]bis[1-(ethoxy-carbonyl)-	$CHCl_3$-MeCN	622s(--),676(5.33+)	5-0658-83
$C_{36}H_{37}N_3O_4$			
1H-Pyrrole-2-carboxylic acid, 5-[[5-[[3,4-dimethyl-5-[(phenylmethoxy)-carbonyl]-1H-pyrrol-2-yl]methyl]-3,4-dimethyl-2H-pyrrol-2-ylidene]-methyl]-3,4-dimethyl-, phenylmethyl ester, monohydrochloride	$CHCl_3$	256(4.37),280(4.35), 499(4.61)	35-6429-83
$C_{36}H_{38}N_2O_6$			
Baluchistine	EtOH	283(3.67)	100-0001-83
	EtOH-base	290(3.80)	100-0001-83
Daphnandrine	MeOH	284(3.92)	162-0407-83
Isothalicberine, 7-O-demethyl-	EtOH	285(3.88)	100-0001-83
$C_{36}H_{38}N_2O_7$			
Jhelumine	MeOH	211(4.72),227s(4.56), 281(4.17),326(3.89)	142-0425-83
	MeOH-base	214(4.78),233s(4.47), 287(4.07),341(4.31)	142-0425-83
$C_{36}H_{38}N_4O_7$			
Norisohaplophytine	EtOH	230(4.18),255(4.04), 290(3.77)	142-1511-83
	EtOH-NaOH	305(3.88)	142-1511-83
$C_{36}H_{40}$			
Pentacyclo[24.2.2.2^{5,8}.2^{12,15}2^{19,22}]-hexatriaconta-7,12,14,19,21,26,28,29-31,33,35-dodecaene	EtOH	261(3.16),266(3.26), 268(3.25),274(3.25)	94-2868-83
Pentacyclo[27.3.1.1^{5,9}.1^{13,17}.1^{21,25}]-hexatriaconta-1(33),5,7,9(36),13,15-17(35),21,23,25(34),29,31-dodecaene	EtOH	258(3.02),265(3.11), 269(2.97),273(3.01)	94-2868-83
$C_{36}H_{40}N_2O_{15}$			
Acetamide, N-(2-acetoxyethyl)-N-[2-[[9,10-dihydro-9,10-dioxo-4-[(2,3,4-6-tetra-O-acetyl-β-D-galactopyrano-syloxy]-1-anthracenyl]amino]ethyl]-	EtOH	206(4.43),247(4.61), 316(3.74),511(3.89)	18-1435-83

Compound	Solvent	$\lambda_{max}(\log \epsilon)$	Ref.
$C_{36}H_{40}N_4O_4S_2$			
21H,23H-Porphine-2,18-dipropanoic acid, 3,8,13,17-tetramethyl-7-(2-methyl-1,3-dithiolan-2-yl)-, dimethyl ester	CH_2Cl_2	403(5.26),500(4.15), 534(3.98),570(3.81), 623(3.56)	44-0500-83
$C_{36}H_{40}N_4O_{10}$			
D-Ribitol, 1,1-bis(5-acetoxy-3-methyl-1-phenyl-1H-pyrazol-4-yl)-1-deoxy-2,3-O-(1-methylethylidene)-, 4,5-diacetate	MeOH	255(4.46)	111-0481-83
$C_{36}H_{40}O_8$			
Naphthalene, 1,2,3,4-tetrahydro-1,2,3-tris(3,4-dimethoxyphenyl)-6,7-dimethoxy-	EtOH	283(4.22)	142-1247-83
$C_{36}H_{41}N_3O_{14}$			
Acetamide, N-(2-acetoxyethyl)-N-[2-[[9,10-dihydro-9,10-dioxo-4-[[2,4,6-tri-O-acetyl-3-(acetylamino)-3-deoxy-β-D-glucopyranosyl]oxy]-1-anthracenyl-amino]ethyl]-	EtOH	205(4.51),247(4.61), 514(3.91)	18-1435-83
Acetamide, N-(2-acetoxyethyl)-N-[2-[[9,10-dihydro-9,10-dioxo-4-[[3,4,6-tri-O-acetyl-2-(acetylamino)-2-deoxy-β-D-glucopyranosyl]oxy]-1-anthracenyl-amino]ethyl]-	CHCl	250(4.70),280(4.17), 319(3.96),514(4.00)	18-1435-83
$C_{36}H_{41}N_7O_6S$			
Ascidiacyclamide	MeOH	232(4.32)	77-0323-83
$C_{36}H_{42}N_4O_2$			
21H,23H-Porphine-2,12-dicarboxaldehyde, 3,8,13,18-tetramethyl-7,17-dipentyl-	CH_2Cl_2	416(5.39),526(3.81), 573(4.37),594(4.24), 646(3.70)	39-0371-83B
$C_{36}H_{42}O_8$			
Ohchinin	EtOH	215(4.34),276(4.28)	18-1139-83
$C_{36}H_{42}O_{18}$			
Harpagide, 8-O-p-coumaroyl-, hexaacetate	MeOH	281(4.38)	94-2296-83
$C_{36}H_{42}O_{19}$			
Plumieride coumarate glucoside	EtOH	226(4.46),300s(--), 309(4.36)	102-0179-83
$C_{36}H_{44}ClGaN_4$			
Gallium, chloro[2,3,7,8,12,13,17,18-octaethyl-21H,23H-porphinato(2-)-$N^{21},N^{22},N^{23},N^{24}$]-, (SP-5-15)-	MeOH	377(4.20),397(5.14), 491(2.85),528(3.64), 567(3.77)	101-0273-83M
$C_{36}H_{45}ClN_4O_4$			
21H-Biline-2,18-dipropanoic acid, 13-(2-chloroethyl)-10,23-dihydro-1,3,7,8,12,17,19-heptamethyl-, dimethyl ester, dihydrobromide	CH_2Cl_2	376(4.09),452(4.50), 526(5.14)	39-2329-83C
isomer	CH_2Cl_2	372(4.17),456(4.39), 524(5.25)	39-2329-83C

Compound	Solvent	λ_{max}(log ϵ)	Ref.
$C_{36}H_{45}ClN_6O_{12}$			
Phomopsin A	MeOH	209(4.72),222s(4.39), 288(4.14)	77-1259-83
$C_{36}H_{46}N_4O$			
21H,23H-Porphine, 2,3,7,8,12,13,17,18-octaethyl-, 22-oxide	$CHCl_3$	395(5.03),528(3.93), 549s(--)	39-0103-83C
$C_{36}H_{46}N_4O_8$			
21H,23H-Porphine-2,3,7,8,12,13,17,18-octaethanol	pyridine	403(5.42),499(4.50), 533(4.31),569(4.17), 622(3.95)	12-1639-83
$C_{36}H_{48}N_2O_8$			
Ansatrienin A	MeOH and MeOH-acid	261(4.55),270(4.66), 281(4.55),388(3.18)	158-83-1
	MeOH-base	261(--),268(--), 278(--),477(3.23)	158-83-1
$C_{36}H_{48}O_6$			
Calozeylanic acid methyl ester	n.s.g.	275(4.40),360(3.45)	39-0703-83C
Kijanolide, 9,17,32-tri-O-methyl-	CF_3CH_2OH	200(4.32),253(3.67)	39-1497-83C
$C_{36}H_{48}O_{12}$			
Affinoside S-III acetonide diacetate	MeOH	263(4.18)	94-1199-83
$C_{36}H_{50}ClN_3O_{11}$			
Colubrinol	EtOH	233(4.23),241(4.15), 253(4.17),280(3.61), 289(3.59)	100-0660-83
$C_{36}H_{50}N_4$			
1H-Benzimidazolium, 2-[4-[2-[4-(dimethylamino)phenyl]ethenyl]-1-methylpyridinium-2-yl]-1-dodecyl-3-methyl-, diiodide	EtOH	558(4.58)	4-0023-83
$C_{36}H_{54}O_8$			
9,19-Cyclolanost-17(20)-en-16-one, 3,6,24-triacetoxy-25-hydroxy-, (3β,6α,17E,24S)-	MeOH	254(3.89)	94-0689-83
(3β,6α,17Z,24S)-	MeOH	254(3.87)	94-0689-83
$C_{36}H_{56}NO_2$			
Pyrrolidinium, 1-[2-[(3E)-7-(2-carboxyethyl)cholesta-4,6-dien-3-ylidene]ethylidene]-, perchlorate	$CHCl_3$	420(4.51)	35-4033-83
$C_{36}H_{56}O_3$			
19-Norpregn-4-en-20-yn-3-one, 17α-[(2-heptyl-1-oxononyl)oxy]-	EtOH	244(4.15)	13-0255-83A
$C_{36}H_{57}N_5O_{11}S$			
Benzenesulfonamide, 4-[[4-hydroxy-3,5-bis(1,4,7,10-tetraoxa-13-azacyclopentadec-13-ylmethyl)phenyl]azo]-N,N-dimethyl-	$CHCl_3$	378(4.42)	18-3253-83
$C_{36}H_{58}P_2$			
Diphosphene, bis[2,4,6-tris(1,1-dimethylethyl)phenyl]-	CH_2Cl_2	284(4.20),340(3.89), 460(3.13)	35-2495-83

Compound	Solvent	$\lambda_{max}(\log \epsilon)$	Ref.
$C_{36}H_{58}P_2S$			
Diphosphene, bis[2,4,6-tris(1,1-dimethylethyl)phenyl]-, 1-sulfide, trans	CH_2Cl_2	267(4.26),384(3.82)	77-0862-83
$C_{36}H_{60}N_4O_5Si_3$			
Benzamide, N-[7-[2,3,5-tris-O-[(1,1-dimethylethyl)dimethylsilyl]-β-D-ribofuranosyl]-7H-pyrrolo[2,3-d]pyrimidin-4-yl]-	MeOH	301(4.01)	5-1169-83
$C_{36}H_{63}NO_{13}$			
Erythronolide B, 3-O-oleandrosyl-5-O-desosaminyl-8,19-epoxy-, (8R)-	MeOH	294(1.52)	158-0365-83
$C_{36}H_{64}FNO_{13}$			
Erythromycin C, 8-fluoro-, (8S)-	MeOH	284(1.37)	158-1439-83
$C_{36}H_{65}NO_{12}$			
Erythronolide B, 3-O-oleandrosyl-5-O-desosaminyl-	MeOH	290(1.78)	158-0365-83
$C_{36}H_{65}NO_{13}$			
Erythronolide B, 3-O-oleandrosyl-5-O-desosaminyl-8-hydroxy-, (8S)-	MeOH	280(1.59)	158-0365-83
Erythronolide B, 3-O-oleandrosyl-5-O-desosaminyl-15-hydroxy-	MeOH	288(1.62)	158-0365-83
$C_{36}H_{66}N_2O$			
Acetamide, N-[4-(ditetradecylamino)-phenyl]-	MeOH	281(4.32)	5-0802-83
$C_{36}H_{66}N_2O_2$			
Benzenamine, N-hexadecyl-4-nitro-N-tetradecyl-	MeOH	236(3.93),400(4.37)	5-0802-83
$C_{36}H_{70}Ge_6$			
Hexagermane, 1,1,2,2,3,3,4,4,5,5,6,6-dodecaethyl-	C_6H_{12}	264(4.63)	101-0149-83G

Compound	Solvent	λ_{max}(log ε)	Ref.
$C_{37}H_{24}BrN_3$			
3,5-Pyridinedicarbonitrile, 2,6-bis([1,1'-biphenyl]-4-yl)-4-(4-bromophenyl)-1,4-dihydro-	EtOH	207(4.74),288(4.54), 356(3.40)	73-3112-83
$C_{37}H_{24}Cl_{14}N_3O_8$			
L-Phenylalanine, N-[N-[N-[4-[bis(pentachlorophenyl)methylene]-2,3,5,6-tetrachlorobenzoyl]glycyl]glycyl]-, 1,1-dimethylethyl ester	$CHCl_3$	287(3.80),336s(3.81), 368s(4.29),384(4.58), 480s(3.06),510(3.07), 560(3.05)	44-3716-83
$C_{37}H_{24}FN_3$			
3,5-Pyridinedicarbonitrile, 2,6-bis([1,1'-biphenyl]-4-yl)-4-(4-fluorophenyl)-1,4-dihydro-	EtOH	206(4.87),286(4.69), 360(3.60)	73-3112-83
$C_{37}H_{24}N_4O$			
7H-Benzimidazo[2,1-a]benz[de]isoquinolin-7-one, 4-[4,5-dihydro-1-(2-naphthalenyl)-5-phenyl-1H-pyrazol-3-yl]-	toluene	520(4.35)	103-0214-83
8H-Benzo[de]naphth[1',2':4,5]imidazo-[2,1-a]isoquinolin-8-one, 11-(4,5-dihydro-1,5-diphenyl-1H-pyrazol-3-yl)-	toluene	525(4.43)	103-0214-83
$C_{37}H_{25}N_3$			
3,5-Pyridinedicarbonitrile, 2,6-bis([1,1'-biphenyl]-4-yl)-1,4-dihydro-4-phenyl-	EtOH	206(4.95),286(4.72), 354(3.65)	73-3112-83
$C_{37}H_{25}N_3O$			
3,5-Pyridinedicarbonitrile, 2,6-bis([1,1'-biphenyl]-4-yl)-1,4-dihydro-4-(4-hydroxyphenyl)-	EtOH	207(4.90),228s(4.45), 286(4.71),360(3.71)	73-3112-83
$C_{37}H_{26}N_2O$			
Ethanone, 1-[4-[[4-[3-(2-quinolinyl)-benzo[f]quinolin-1-yl]phenyl]methyl]-phenyl]-	isoAmOH	258(4.80),280(4.81), 325(4.48),366(3.90)	103-0894-83
Cu(I) complex	isoAmOH	535(3.93)	103-0894-83
$C_{37}H_{26}N_2O_4$			
Cyclohept[1,2,3-hi]imidazo[2,1,5-cd]-indolizine-9-carboxylic acid, 3,4-dibenzoyl-2-(4-methylphenyl)-, ethyl ester	$CHCl_3$	258(4.39),285s(4.36), 325(4.29),397(4.40), 445s(4.09)	18-3703-83
$C_{37}H_{28}Cl_2N_4O$			
Pyrazolo[3',4':4,5]pyrido[2,3-c]carbazole, 4-(3-chlorophenyl)-3-[(3-chlorophenyl)methoxy]-1,2,8,12d-tetrahydro-1,12d-dimethyl-2-phenyl-	MeOH	280(4.6),353(4.2)	103-1310-83
$C_{37}H_{28}N_2O$			
Methanone, bis[4-(diphenylamino)phenyl]-	MeOH	288(4.41),380(4.52)	12-0409-83
$C_{37}H_{30}ClOP_2Pt$			
Platinum(1+), carbonylchlorobis(triphenylphosphine)-, (SP-4-3)-, hexafluorophosphate	CH_2Cl_2	284(4.34),309(4.15)	101-0119-83A

Compound	Solvent	λ max (log ε)	Ref.
$C_{37}H_{30}N_4O$			
Pyrazolo[3',4':4,5]pyrido[2,3-c]carbazole, 1,2,8,12d-tetrahydro-1,12d-dimethyl-2,4-diphenyl-3-(phenylmethoxy)-	MeOH	290(4.6),310s(4.3), 372(4.3)	103-1310-83
$C_{37}H_{32}N_2O_7$			
Daphnine	$CHCl_3$	261(4.72),326(4.27), 444(3.94)	100-0001-83
dihydrochloride	EtOH	257(4.69),324(4.33), 420(3.78)	100-0001-83
hexahydro derivative	EtOH	280(3.76)	100-0001-83
$C_{37}H_{34}N_2O_9$			
2,4(1H,3H)-Pyrimidinedione, 5-(1-butynyl)-1-[2,3,5-tris-O-(4-methylbenzoyl)-β-D-arabinofuranosyl]-	MeOH	240(4.74),283(4.04)	87-0661-83
$C_{37}H_{36}N_2O_9$			
Thalpindione (same in acid or base)	MeOH	275(3.78),283s(3.77)	100-0001-83
$C_{37}H_{36}O_{12}$			
Benzoic acid, 4-methoxy-, 1,2,3,5-pentanetetrayl ester, [S-(R*,S*)]-	EtOH	260(4.71)	35-7396-83
Kelletin II tetramethyl ether	EtOH	258(4.90)	35-7396-83
$C_{37}H_{38}N_2O_2S$			
Estra-1,3,5(10)-trien-17-one, 2-[4-[1,1'-biphenyl]yl]-2-(2-propenylimino)-3(2H)-thiazolyl]-3-methoxy-	EtOH	264(4.614),300(4.374)	36-1205-83
	EtOH-HCl	278(4.633)	36-1205-83
Estra-1,3,5(10)-trien-17-one, 4-[4-[1,1'-biphenyl]yl]-2-(2-propenylimino)-3(2H)-thiazolyl]-3-methoxy-	EtOH	258(4.594),320s(5.034)	36-1205-83
	EtOH-HCl	270(4.622)	36-1205-83
$C_{37}H_{38}N_2O_6$			
Pachygonamine, N,N,O-trimethyl-	MeOH	235s(4.06),288(3.61)	142-1927-83
$C_{37}H_{38}N_4O_8$			
Haplocidiphytine	EtOH	235(4.52),265(4.32), 290s(3.85)	142-1511-83
$C_{37}H_{40}N_2O_6$			
Isothalicberine	EtOH	284(3.98)	100-0001-83
Johnsonine	MeOH	281(3.82)	100-0001-83
	MeOH-NaOH	284(3.85)	100-0001-83
Nor-2'-isotetrandrine	EtOH	282(4.11)	100-0001-83
$C_{37}H_{40}N_2O_7$			
Chenabine	MeOH	209(4.72),227s(4.49), 281(4.08),326(3.87)	142-0425-83
	MeOH-base	211(4.83),283(3.91), 342(4.28)	142-0425-83
Thalrugosidine, N-demethyl- (same in acid or base)	MeOH	278(3.90),283(3.91)	100-0001-83
$C_{37}H_{40}N_4O_7$			
Haplophytine	EtOH	220(4.69),265(4.16), 305(3.65)	162-0665-83
$C_{37}H_{40}O_{20}$			
Sorinin heptaacetate	MeOH	245(4.73),300(3.73), 350(3.76)	83-0399-83

Compound	Solvent	$\lambda_{max}(\log \epsilon)$	Ref.
$C_{37}H_{42}N_2O_6$ Daurisoline	MeOH	284(4.01)	100-0001-83
$C_{37}H_{42}N_2O_7$ Chillanamine	MeOH	209(4.81),226s(4.54), 283(4.02)	100-0908-83
$C_{37}H_{44}ClNO_4$ Penitrem C	MeOH	232(4.58),292(4.10)	39-1847-83C
$C_{37}H_{44}ClNO_5$ Penitrem F	MeOH	232(4.55),292(4.05)	39-1847-83C
$C_{37}H_{44}ClNO_6$ Penitrem A	MeOH	233(4.57),295(4.06)	39-1847-83C
$C_{37}H_{44}N_4O$ 21H,23H-Porphine-2-carboxaldehyde, 12-ethenyl-3,8,13,18-tetramethyl-7,17-dipentyl-	CH_2Cl_2	413(5.23),519(3.90), 562(4.30),580(4.15), 634(3.28)	39-0371-83B
$C_{37}H_{44}N_4ORu$ Ruthenium, carbonyl[2,3,7,8,12,13,17-18-octaethyl-21H,23H-porphinato(2-)-N^{21},N^{22},N^{23},N^{24}]-	CH_2Cl_2	393(5.40),512(4.20), 547(4.63)	23-2389-83
$C_{37}H_{45}NO_4$ Penitrem D	MeOH	220(4.54),286(4.05)	39-1847-83C
$C_{37}H_{45}NO_5$ Penitrem B	MeOH	220(4.51),286(4.08)	39-1847-83C
$C_{37}H_{45}NO_6$ Penitrem E	MeOH	220(4.53),286(4.05)	39-1847-83C
$C_{37}H_{46}Cl_2N_4O_4$ 21H-Biline-2,18-dipropanoic acid, 8,13-bis(2-chloroethyl)-10,23-dihydro-1,3,7,12,17,19-hexamethyl-, dimethyl ester, dihydrobromide	CH_2Cl_2	455(4.67),518(4.94)	4-1383-83
$C_{37}H_{47}FeN_4$ Iron, methyl[2,3,7,8,12,13,17,18-octa-ethyl-21H,23H-porphinato(2-)-N^{21},N^{22}-N^{23},N^{24}]-, (SP-5-31)-	benzene	376(4.89),391(5.05), 405s(--),512(3.97), 551(4.33)	101-0065-83M
$C_{37}H_{47}GaN_4$ Gallium, methyl[2,3,7,8,12,13,17,18-octaethyl-21H,23H-porphinato(2-)-N^{21},N^{22},N^{23},N^{24}]-, (SP-5-31)-	benzene	349(4.62),405s(4.78), 426(5.27),509(3.46), 549(4.19),586(3.97)	101-0273-83M
$C_{37}H_{47}NO_{12}$ Rifamycin S, 16,17-dihydro-	EtOH	235(4.36),275(4.27), 304(4.13),330s(--), 385(3.61),530s(--)	158-83-60
$C_{37}H_{47}NO_{13}$ Rifamycin S, 16,17-dihydro-17-hydroxy-	EtOH	235(4.36),275(4.27), 304(4.13),330s(--), 385(3.61),530s(--)	158-83-60

Compound	Solvent	λ_{max}(log ε)	Ref.
$C_{37}H_{48}N_4O_4$			
21H-Biline-2,18-dipropanoic acid, 7,12-diethyl-10,23-dihydro-1,3,8,13,17,19-hexamethyl-, dimethyl ester, dihydrobromide	CH_2Cl_2	450(4.70),538(4.71)	44-4302-83
$C_{37}H_{48}N_4S$			
Benzothiazolium, 2-[3-[2-(1-dodecyl-3-methyl-1H-benzimidazolium-2-yl)-1-methyl-4(1H)-pyridinylidene]-1-propenyl]-3-methyl-, diiodide	EtOH	605(5.03)	4-0023-83
$C_{37}H_{48}O_{14}$			
Affinoside S-III tetraacetate	MeOH	263(4.14)	94-1199-83
$C_{37}H_{50}Cl_4N_5O_9PSi_2$			
2'-Guanylic acid, N-benzoyl-3',5'-bis-O-[(1,1-dimethylethyl)dimethylsilyl]-, 2-chlorophenyl 2,2,2-trichloroethyl ester	MeOH	235(4.18),255(4.12), 262(4.10),290(4.09)	33-2069-83
3'-Guanylic acid, N-benzoyl-2',5'-bis-O-[(1,1-dimethylethyl)dimethylsilyl]-, 2-chlorophenyl 2,2,2-trichloroethyl ester	MeOH	237(4.19),258(4.18), 263(4.18),292(4.16)	33-2069-83
$C_{37}H_{50}O_9$			
Pimelea factor P_2	MeOH	202(3.96),229(4.07), 272(2.95)	100-0563-83
$C_{37}H_{50}O_{10}$			
Pimelea factor P_2	MeOH	229(3.85),279(3.10)	36-1285-83
$C_{37}H_{52}ClN_3O_{11}$			
10-Epitrewiasine	EtOH	233(4.34),243s(4.24), 255(4.26),282(3.70), 289(3.70)	100-0660-83
$C_{37}H_{52}N_4O_{12}$			
Carbonic acid, 5,8-bis[[2-[[(1,1-dimethylethoxy)carbonyl](2-hydroxyethyl)amino]ethyl]amino]-9,10-dihydro-1-hydroxy-9,10-dioxo-2-anthracenyl 1,1-dimethylethyl ester	$CHCl_3$	260(4.41),381(3.68), 606(4.25),656(4.36)	18-1812-83
$C_{37}H_{52}O_4$			
Furan, 2,2',2",2"'-methanetetrayltetrakis[5-pentyl-	EtOH	229(4.27)	103-0360-83
$C_{37}H_{52}O_5$			
Urs-12-en-29-oic acid, 3β-[(4-hydroxybenzoyl)oxy]-	MeOH	256(4.17)	100-0537-83
$C_{37}H_{55}N_3O_{13}$			
β-D-Glucopyranosiduronic acid, (3α,5β)-21-acetoxy-20-[(aminocarbonyl)hydrazono]pregnan-3-yl methyl ester, 2,3,4-triacetate	MeOH	238(4.06)	13-0349-83B
$C_{37}H_{55}N_3O_{14}$			
β-D-Glucopyranosiduronic acid, (3α,5β,11β)-21-acetoxy-20-[(amino-	MeOH	236(4.11)	13-0349-83B

Compound	Solvent	λ_{max}(log ε)	Ref.
carbonyl)hydrazono]-11-hydroxy-pregnan-3-yl methyl ester, 2,3,4-triacetate (cont.)			13-0349-83B
β-D-Glucopyranosiduronic acid, (3α,5β,11β)-3-acetoxy-20-[(amino-carbonyl)hydrazono]-11-hydroxy-pregnan-21-yl methyl ester, 2,3,4-triacetate	MeOH	236(4.11)	13-0349-83B
$C_{37}H_{56}O_{10}$			
Preatroxigenin, dimethyl ester, tri-acetate	EtOH	217(2.4)	20-0355-83
$C_{37}H_{58}O_3$			
Furan, 2,2',2"-methylidynetris[5-hep-tyl-3-methyl-	hexane	231(3.15)	103-0360-83
$C_{37}H_{58}O_7$			
19-Norlanost-5-en-7-one, 3,11-diacet-oxy-9-methyl-24,25-[(1-methylethyl-idene)bis(oxy)]-, (3β,9β,10α,11α,24R)-	n.s.g.	201(3.64),243.5(4.12)	96-1155-83
$C_{37}H_{61}NO_{14}$			
DOML	EtOH	283(4.33)	158-0376-83
$C_{37}H_{66}FNO_{12}$			
Erythromycin B, 8-fluoro-, (8S)-	MeOH	285(1.47)	158-1439-83
Erythromycin D, 8-fluoro-, (8S)-	MeOH	285(1.49)	158-1439-83
$C_{37}H_{66}FNO_{13}$			
Erythromycin A, 8-fluoro-, (8S)-	MeOH	283(1.25)	158-1439-83

Compound	Solvent	λ_{max}(log ϵ)	Ref.
$C_{38}H_{20}N_8$			
3,5:14,16-Dinitrilo-5H,16H-dicyclohepta[e,k]diimidazo[1,5-a:1',5'-g]-[1,7]diazacyclododecine-10,21-dicarbonitrile, 1,12-diphenyl-	EtOH	239(4.43),254(4.46), 301(4.38),329(4.33), 417s(4.15),440(4.32), 550(3.45)	18-3703-83
$C_{38}H_{24}N_4$			
3,5:14,16-Dimetheno-5H,16H-dicyclohepta[e,k]diimidazo[1,5-a:1',5'-g]-[1,7]diazacyclododecine, 1,12-diphenyl-	EtOH	237(4.77),262(5.01), 285s(4.38),344(4.60), 360(4.51),384s(4.21), 403(4.46),427(4.46), 485s(3.58),515(3.70), 547(3.70),590(3.50), 645s(2.94)	18-3703-83
$C_{38}H_{24}N_4S_2$			
Dibenzo[c,n]triphenodithiazine-8,17-diamine, N,N'-diphenyl-	dioxan	621(4.62)	18-1482-83
	H_2SO_4	358(4.67)	18-1482-83
6,13-Triphenodithiazinediamine, N,N'-di-2-naphthalenyl-	dioxan	557(4.53)	18-1482-83
	H_2SO_4	326(4.54)	18-1482-83
$C_{38}H_{25}NO_2$			
1,7-Indolizinedione, 2,3,5,6,8-pentaphenyl-	MeCN	260(4.40),297(4.30), 344(3.71),478(3.35)	88-2977-83
$C_{38}H_{26}$			
Anthracene, 1,4,9,10-tetraphenyl-	CH_2Cl_2	278(4.8),410(4.1)	151-0379-83C
$C_{38}H_{26}Cl_6NO_4P$			
3,4-Pyridinedicarboxylic acid, 2-[phenyl(triphenylphosphoranylidene)methyl]-5,6-bis(trichloroethenyl)-, dimethyl ester	MeCN	277(4.12),286(4.09), 310(3.99),409(4.11)	88-2977-83
$C_{38}H_{26}N_2$			
2,6-Naphthyridine, 1,3,4,5,8-pentaphenyl-	$CHCl_3$	322(4.2),390(4.1)	4-0971-83
2,7-Naphthyridine, 1,3,4,5,8-pentaphenyl-	$CHCl_3$	350(3.7)	4-0971-83
$C_{38}H_{26}N_4O_2$			
7H-Benzimidazo[2,1-a]benz[de]isoquinolin-7-one, 4-[4,5-dihydro-5-(4-methoxyphenyl)-1-(2-naphthalenyl)-1H-pyrazol-3-yl]-	toluene	530(4.34)	103-0214-83
$C_{38}H_{26}N_6$			
3,5:14,16-Dinitrilo-5H,16H-dicyclohepta[e,k]diimidazo[1,5-a:1',5'-g][1,7]-diazacyclododecine, 1,12-bis(4-methylphenyl)-	EtOH	239(4.86),250s(4.79), 266s(4.66),281(4.42), 335(4.65),343s(4.62), 418s(4.51),440(4.69), 530s(3.78),560(3.82), 595s(3.72),647s(3.26)	18-3703-83
$C_{38}H_{26}O_2$			
9,10-Epidioxyanthracene, 9,10-dihydro-1,4,9,10-tetraphenyl-	CH_2Cl_2	275s(4.0)	151-0379-83C
$C_{38}H_{27}N_3$			
3,5-Pyridinedicarbonitrile, 2,6-bis([1,1'-biphenyl]-4-yl)-1,4-dihydro-	EtOH	207(5.02),287(4.76), 359(3.85)	73-3112-83

Compound	Solvent	λ_{max}(log ε)	Ref.
4-(4-methylphenyl)- (cont.)			73-3112-83
$C_{38}H_{27}N_3O$			
3,5-Pyridinedicarbonitrile, 2,6-bis-([1,1'-biphenyl]-4-yl)-1,4-dihydro-4-(4-methoxyphenyl)-	EtOH	206(4.88),227s(4.48), 286(4.68),358(3.66)	73-3112-83
$C_{38}H_{28}N_2O_{12}$			
[2,2'-Bianthra[1,2-b]furan]-3,3'-dicarboxylic acid, 2,2'-diamino-2,2'-3,3',6,6',11,11'-octahydro-5,5'-dihydroxy-6,6',11,11'-tetraoxo-, diethyl ester	dioxan	328(3.91),405(4.19)	104-0533-83
$C_{38}H_{29}N_3$			
1H-Indole, 2,3-bis(1H-indol-3-ylphenylmethyl)-	EtOH	223(4.72),283(5.07), 291(4.03)	2-0027-83
	EtOH-$HClO_4$	218(4.40),255(4.23), 276s(4.16),333(5.05)	2-0027-83
$C_{38}H_{29}O$			
2H-Furylium, 2,2-dimethyl-5-phenyl-3-[2-phenyl-4-(1-pyrenyl)-1,3-butadienyl]-, perchlorate	MeCN	630(4.69)	56-0579-83
	HOAc	664(--)	56-0579-83
$C_{38}H_{30}NO$			
Pyridinium, 4-[3-(2,6-diphenyl-4H-pyran-4-ylidene)-1-propenyl]-1-methyl-2,6-diphenyl-, perchlorate	C_6H_{12}	556(4.72)	99-0150-83
	MeOH	552(4.74)	99-0150 83
	EtOH	560(4.74)	99-0150-83
	PrOH	564(4.74)	99-0150-83
	isoPrCH	562(4.73)	99-0150-83
	acetone	549(4.72)	99-0150-83
	acetophenone	576(4.73)	99-0150-83
	$(EtOOC)_2$	556(4.74)	99-0150-83
	$HOCH_2CH_2OH$	560(4.73)	99-0150-83
	$HCONH_2$	562(4.73)	99-0150-83
	MeCN	550(4.74)	99-0150-83
	DMF	550(4.73)	99-0150-83
	C_6H_5CN	572(4.75)	99-0150-83
	pyridine	572(4.75)	99-0150-83
	MeNO	554(4.73)	99-0150-83
	quinoline	586(4.74)	99-0150-83
	CH_2Cl_2	578(4.81)	99-0150-83
	$ClCH_2CH_2Cl$	576(4.80)	99-0150-83
	o-$C_6H_4Cl_2$	586(4.80)	99-0150-83
	DMSO	554(4.73)	99-0150-83
	sulfolane	552(4.73)	99-0150-83
$C_{38}H_{32}$			
Triple-layered naphthalenophane	THF	220(5.2),275s(4.0)	88-4851-83
$C_{38}H_{33}ClO_2P_2Pt$			
Platinum, chloro(methoxycarbonyl)-bis(triphenylphosphine)-, (SP-4-3)-	CH_2Cl_2	257(4.36),268(4.20), 275(4.04)	101-0119-83A
$C_{38}H_{33}NO_2P_2$			
Bis(triphenylphosphine)iminium acetate	MeOH	268(3.60),275(3.51)	126-0811-83
$C_{38}H_{34}N_4O_4$			
6-Quinoxalinecarboxylic acid, 2,2'-(1,2-ethanediylidene)bis[1,2-dihydro-3-	$CHCl_3$-CF_3COOH	310(4.31),388(4.29), 650s(--),701(4.47)	33-0379-83

Compound	Solvent	λ_{max}(log ε)	Ref.
methyl-1-phenyl-, diethyl ester (cont.)			33-0379-83
$C_{38}H_{34}N_4O_6$			
Benzamide, N-[7-[5-O-[(4-methoxyphenyl)diphenylmethyl]-β-D-ribofuranosyl]-7H-pyrrolo[2,3-d]pyrimidin-4-yl]-	MeOH	282(4.00),301(3.98)	5-1169-83
$C_{38}H_{34}O_2$			
2-Cyclopenten-1-one, 5-(1,4-dimethyl-5-oxo-2,3-diphenyl-2-cyclopenten-1-yl)-2,5-dimethyl-3,4-diphenyl-	MeOH	275(4.26)	18-0175-83
$C_{38}H_{36}Ge_2Si$			
Silane, dimethylbis(triphenylgermyl)-	C_6H_{12}	244(4.42)	101-0149-83G
$C_{38}H_{36}Ge_3$			
Trigermane, 2,2-dimethyl-1,1,1,3,3,3-hexaphenyl-	C_6H_{12}	245(4.48)	101-0149-83G
$C_{38}H_{36}N_2O_9$			
Thalictrinine	MeOH	205s(4.79),236(4.62), 251s(4.50),285s(4.01), 301s(3.84),330(3.73)	100-0001-83
	MeOH-HCl	282s(4.13),340(3.64)	100-0001-83
$C_{38}H_{38}N_2O_6$			
Acetamide, N-[5-acetoxy-4-[(4-butylphenyl)amino]-9,10-dihydro-8-hydroxy-9,10-dioxo-1-anthracenyl]-N-(4-butylphenyl)-	benzene	420(3.58),551(3.78)	104-0745-83
9,10-Anthracenedione, 1,4-bis[(N-acetyl-N-(4-butylphenyl)amino]-	benzene	506(4.06)	104-0745-83
9,10-Anthracenedione, 1,4-diacetoxy-5,8-bis[(4-butylphenyl)amino]-	benzene	417(3.78),625(4.19), 665(4.20)	104-0745-83
$C_{38}H_{38}N_2O_9$			
Thalibrunimine, oxo-	MeOH	240s(4.10),270s(3.86), 330s(3.40)	100-0001-83
	MeOH-HCl	250s(4.00),284(3.60), 346s(3.31)	100-0001-83
Thalictrinine, dihydro-	MeOH	238(4.81),249s(4.73), 285s(4.05),299s(3.95)	100-0001-83
	MeOH-HCl	210s(4.96),240s(4.69), 252(4.75),303s(4.05), 340s(3.74)	100-0001-83
Thalrugosinone (same in acid or base)	MeOH	274(3.89),283(3.86)	100-0001-83
$C_{38}H_{38}N_2O_9Si$			
2,4(1H,3H)-Pyrimidinedione, 5-[(trimethylsilyl)ethynyl]-1-[2,3,5-tris-O-(4-methylbenzoyl)-β-D-arabinofuranosyl]-	MeOH	240(4.76),283(4.10)	87-0661-83
$C_{38}H_{38}N_{10}O_7$			
1H-Pyrrole-2-carboxylic acid, 4-[[[4-[[[4-[[[4-[[[4-(formylamino)-1-methyl-1H-pyrrol-2-yl]carbonyl]amino-1-methyl-1H-pyrrol-2-yl]carbonyl]amino]-1-methyl-1H-pyrrol-2-yl]carbonyl]amino]-1-methyl-1H-pyrrol-2-yl]carbonyl]-	EtOH	238(4.62),308(4.74)	87-1042-83

Compound	Solvent	λ_{max}(log ε)	Ref.
amino]-1-methyl-, phenylmethyl ester (cont.)			87-1042-83
$C_{38}H_{38}O_{16}$			
δ-Naphthocyclinone	MeOH	248(4.50),268s(--), 360(3.81),405(4.08), 423s(--),432s(--)	5-0471-83
	MeOH-NaOH	223(4.46),244(4.41), 275s(--),344(3.72), 478(4.09)	5-0471-83
	$CHCl_3$	250(4.45),268s(--), 355s(--),405(4.02), 420s(--)	5-0471-83
$C_{38}H_{40}N_2O_2$			
Bismurrayafoline B	MeOH	225s(--),240(--), 265s(--),285s(--), 312(--),333s(--)	94-4202-83
$C_{38}H_{40}N_2O_7$			
Calafatimine	MeOH and MeOH-base	235s(4.85),280(4.40), 292s(3.93)	100-0001-83
$C_{38}H_{40}N_4O$			
Longicaudatine	n.s.g.	223(4.66),270s(--), 284(4.28),290(4.26), 307s(3.97)	44-1869-83
	$HClO_4$	268(--)	44-1869-83
$C_{38}H_{40}N_4O_7$			
21H,23H-Porphine-2,18-dipropanoic acid, 7-acetyl-12-(3-methoxy-3-oxo-1-propenyl)-3,8,13,17-tetramethyl-, dimethyl ester, (E)-	CH_2Cl_2	425(5.17),516(4.19), 550(4.04),585(3.88), 639(3.71)	44-0500-83
21H,23H-Porphine-2,18-dipropanoic acid, 8-acetyl-13-(3-methoxy-3-oxo-1-propenyl)-3,7,12,17-tetramethyl-, dimethyl ester, (E)-	CH_2Cl_2	426(5.19),518(4.17), 553(4.03),587(3.87), 639(3.62)	44-0500-83
$C_{38}H_{42}N_2O_2S$			
Estra-1,3,5(10)-trien-17-one, 2-[4-[1,1'-biphenyl]yl-2-(butylimino)-3(2H)-thiazolyl]-3-methoxy-	EtOH	266(4.570),300s(4.326)	36-1205-83
	EtOH-HCl	278(4.589)	36-1205-83
Estra-1,3,5(10)-trien-17-one, 4-[4-[1,1'-biphenyl]yl-2-(butylimino)-3(2H)-thiazolyl]-3-methoxy-	EtOH	258(4.612),320s(5.027)	36-1205-83
	EtOH-HCl	270(4.632)	36-1205-83
$C_{38}H_{42}N_2O_7$			
Osornine, (-)-	MeOH	207(4.80),230s(4.53), 282(3.75)	100-0908-83
N-Oxy-2'-isotetrandrine	MeOH	282(3.98)	100-0001-83
$C_{38}H_{42}N_2O_8$			
N'-Northalibrunine	EtOH	226(4.41),236(4.56), 284(4.24)	100-0001-83
$C_{38}H_{42}N_2O_{16}$			
Acetamide, N-[2-[acetyl(2-acetoxyethyl)amino]ethyl]-N-[9,10-dihydro-9,10-dioxo-4-[(2,3,4,6-tetra-O-acetyl-β-D-glucopyranosyl)oxy]-1-anthracenyl]-	EtOH	208(4.91),230(4.92), 311(4.31),346(4.34), 421(4.12)	18-1435-83

Compound	Solvent	λ_{max}(log ε)	Ref.
$C_{38}H_{42}N_4$			
21H,23H-Porphine, 2,8,12,18-tetraethyl-3,7,13,17-tetramethyl-5-phenyl-	CH_2Cl_2	402(5.28),501(4.19), 534(3.85),559(3.83), 626(3.40)	44-5388-83
$C_{38}H_{42}N_4O$			
Matopensine	n.s.g.	217(4.46),263(4.39), 313(3.95)	142-2339-83
$C_{38}H_{42}N_4O_7$			
21H,23H-Porphine-2,18-dipropanoic acid, 7-acetyl-12-(2-acetoxyethyl)-3,8,13-17-tetramethyl-, dimethyl ester	CH_2Cl_2	410(5.29),510(4.03), 550(4.12),577(3.91), 635(3.24)	35-6638-83
21H,23H-Porphine-2,7,18-tripropanoic acid, 12-acetyl-3,8,13,17-tetra-methyl-, trimethyl ester	CH_2Cl_2	408(5.25),509(4.01), 548(4.10),575(3.90), 635(3.18)	44-0500-83
21H,23H-Porphine-2,8,12-tripropanoic acid, 17-acetyl-3,7,13,18-tetra-methyl-, trimethyl ester	CH_2Cl_2	409(5.20),510(3.97), 549(4.06),576(3.86), 635(3.11)	44-0500-83
$C_{38}H_{42}N_4O_8$			
Cimilophytine	EtOH	228(4.52),266(4.20), 300(3.60)	44-3015-83
$C_{38}H_{42}O_{21}$			
Sorinin, 6-methoxy-, heptaacetate	MeOH	257(4.92),330(4.03)	83-0399-83
$C_{38}H_{43}N_3O_6$			
3,5-Pyridinedicarboxamide, N,N'-bis(2,2-dimethyl-4-phenyl-1,3-dioxan-5-yl)-1,4-dihydro-1-(phenylmethyl)-	EtOH	380(3.90)	130-0299-83
$C_{38}H_{44}N_2O_6$			
Neothalibrine	MeOH	284(4.10)	100-0001-83
	MeOH-base	285(4.10),310s(3.68)	100-0001-83
$C_{38}H_{44}N_2O_8$			
Vateamine, (+)-	MeOH	212(4.72),230s(4.48), 283(4.07)	44-3957-83
$C_{38}H_{44}N_6$			
21H,23H-Porphine-5,10-dicarbonitrile, 2,3,7,8,12,13,17,18-octaethyl-	CH_2Cl_2	410(5.08),528(3.99), 565(3.81),600(3.71), 658(3.92)	22-0317-83
21H,23H-Porphine-5,15-dicarbonitrile, 2,3,7,8,12,13,17,18-octaethyl-	CH_2Cl_2	400(5.27),524(3.81), 566(4.28),604(3.68), 656(4.41)	22-0317-83
$C_{38}H_{44}O_8$			
Gambogic acid	EtOH	217(4.42),280(4.22), 291(4.23),362(4.17)	162-0622-83
pyridine salt	EtOH	291.5(4.35),359.5(4.26)	162-0622-83
$C_{38}H_{46}O_4$			
1,4-Naphthalenedione, 2,3-bis[3,5-bis(1,1-dimethylethyl)-4-hydroxy-phenyl]-	EtOH	239(4.36),264(4.40), 333(3.72),407(3.45), 502(3.45)	104-0144-83
$C_{38}H_{47}NO_{17}$			
Alatusamine	EtOH	221(3.95),268(3.54)	102-2839-83

Compound	Solvent	λ_{max}(log ϵ)	Ref.
$C_{38}H_{47}NO_{19}$			
Alatusinine	EtOH	223(3.88),268(3.51)	102-2839-83
$C_{38}H_{48}N_2O$			
1(4H)-Naphthalenone, 2-[[3,5-bis(1,1-dimethylethyl)phenyl]amino]-4-[[3,5-bis(1,1-dimethylethyl)phenyl]imino]-	hexane	235s(4.37),248s(4.32), 277(4.41),468(3.83)	18-1476-83
$C_{38}H_{48}N_4O_2$			
21H,23H-Porphin-5-ol, 2,3,7,8,12,13,17-18-octaethyl-, acetate	$CHCl_3$	403(5.21),501(4.14), 533(3.79),570(3.77), 622(3.30)	39-0103-83C
$C_{38}H_{48}N_{10}O_6$			
3H-Phenoxazine-1,9-dicarboxamide, 2-amino-N,N'-bis[5-[[[3-(dimethylamino)propyl]amino]carbonyl]-1-methyl-1H-pyrrol-3-yl]-4,6-dimethyl-3-oxo-, dihydrochloride	H_2O	435(4.32),455(4.30)	104-0787-83
$C_{38}H_{48}O_4$			
6H-Dibenzo[b,d]pyran, 2,2'-(1,2-ethenediyl)bis[7,8,9,10-tetrahydro-1-methoxy-3,6,6,9-tetramethyl-	$CHCl_3$	275(4.67)	83-0326-83
$C_{38}H_{49}GaN_4$			
Gallium, ethyl[2,3,7,8,12,13,17,18-octaethyl-21H,23H-porphinato(2-)-N^{21},N^{22},N^{23},N^{24}]-, (SP-5-31)-	benzene	358(4.80),432(5.17), 515(3.51),552(4.24), 587(3.92)	101-0273-83M
$C_{38}H_{49}N_7O_{10}$			
Virginiamycin S_5	MeOH	304(3.72),358(2.66)	158-83-88
$C_{38}H_{50}O_2$			
Methanone, 3-[7-hydroxy-8,8-dimethyl-3,5-bis(1-methylethyl)bicyclo[4.2.0]-octa-1,3,5-trien-7-yl]phenyl][2,4,6-tris(1-methylethyl)phenyl]-	C_6H_{12}	249(4.11),346(1.89)	35-1117-83
Methanone, 4-[7-hydroxy-8,8-dimethyl-3,5-bis(1-methylethyl)bicyclo[4.2.0]-octa-1,3,5-trien-7-yl]phenyl][2,4,6-tris(1-methylethyl)phenyl]-	C_6H_{12}	259(4.33),348(1.93)	35-1117-83
$C_{38}H_{51}NO_9$			
Milbemycin F	EtOH	238s(--),245(4.38), 253(4.33),266s(--)	158-0502-83
	EtOH	245(4.38),253(4.33)	158-0980-83
$C_{38}H_{52}ClN_6O_9PSi_2$			
2'-Guanylic acid, N-benzoyl-3',5'-bis-O-[(1,1-dimethylethyl)dimethylsilyl]-, 2-chlorophenyl 2-cyanoethyl ester	MeOH	235(4.18),260(4.14), 292(4.13)	33-2069-83
3'-Guanylic acid, N-benzoyl-2',5'-bis-O-[(1,1-dimethylethyl)dimethylsilyl]-, 2-chlorophenyl 2-cyanoethyl ester	MeOH	236(4.21),256(4.20), 264(4.20),292(4.18)	33-2069-83
$C_{38}H_{54}$			
Benzene, [dicyclohexyl[4-(dicyclohexylmethylene)-2,5-cyclohexadien-1-yl]methyl]-	hexane	268(4.07)	24-3235-83

Compound	Solvent	λ_{max}(log ϵ)	Ref.
$C_{38}H_{54}Cl_2N_5O_9PSi_2$			
2'-Guanylic acid, 3',5'-bis-O-[(1,1-dimethylethyl)dimethylsilyl]-N-(2-methyl-1-oxopropyl)-, bis(2-chlorophenyl) ester	MeOH	260(4.25),272(4.12), 283s(4.11)	33-2069-83
3'-Guanylic acid, 2',5'-bis-O-[(1,1-dimethylethyl)dimethylsilyl]-N-(2-methyl-1-oxopropyl)-, bis(2-chlorophenyl) ester	MeOH	255s(4.24),260(4.25), 272(4.12),282s(4.10)	33-2069-83
$C_{38}H_{54}O_{13}$			
Colocynthin	n.s.g.	235(4.11)	162-0353-83
$C_{38}H_{55}NO_2$			
Cholesta-4,6-diene-7-propanoic acid, 3-(5-cyano-1,4-dimethyl-2,4-pentadienylidene)- (cis and trans)	EtOH	392(4.58)	35-4033-83
$C_{38}H_{56}N_5O_9PSi_2$			
2'-Guanylic acid, 3',5'-bis-O-[(1,1-dimethylethyl)dimethylsilyl]-N-(2-methyl-1-oxopropyl)-, diphenyl ester	MeOH	255s(4.20),259(4.21), 277(4.08)	33-2069-83
3'-Guanylic acid, 2',5'-bis-O-[(1,1-dimethylethyl)dimethylsilyl]-N-(2-methyl-1-oxopropyl)-, diphenyl ester	MeOH	257(4.24),277(4.09)	33-2069-83
$C_{38}H_{56}O_3$			
Cholesta-4,6-diene-7-propanoic acid, 3-(1,4-dimethyl-6-oxo-2,4-hexadienylidene)-	EtOH	395(4.58)	35-4033-83
$C_{38}H_{58}O_{11}$			
Hygrolidin	MeOH	246(4.45),277(4.06)	158-83-74
$C_{38}H_{58}O_{15}$			
Digitalin, 16-acetyl-	n.s.g.	217(4.16)	162-0458-83
$C_{38}H_{60}O_3$			
Furan, 2,2',2"-ethylidynetris[3-heptyl-5-methyl-	EtOH	227(4.35)	103-0474-83
$C_{38}H_{63}NO_{12}$			
DMOT	EtOH	283(4.33)	158-0376-83
$C_{38}H_{63}NO_{13}$			
DMT	EtOH	283(4.35)	158-0376-83
$C_{38}H_{65}NO_{11}$			
DODMOT	EtOH	282(4.33)	158-0376-83
$C_{38}H_{65}NO_{12}$			
DODMT	EtOH	283(4.34)	158-0376-83
$C_{38}H_{70}N_2O_2$			
Benzenamine, N,N-dihexadecyl-4-nitro-	MeOH	234(3.93),400(4.33)	5-0802-83
$C_{38}H_{70}N_3$			
Benzenediazonium, 4-(dihexadecylamino)-	MeOH	385(4.60)	5-0802-83
$C_{38}H_{72}N_2$			
1,4-Benzenediamine, N,N-dihexadecyl-	MeOH-HCl	251(4.10),306(3.18)	5-0802-83

Compound	Solvent	λ_{max}(log ε)	Ref.
$C_{39}H_{24}O_{12}$			
5,6,11,12,17,18-Trinaphthylenehexone, 2,8,15-triacetyl-1,7,16-trihydroxy-3,9,14-trimethyl-	dioxan	420(4.15)	5-0433-83
	$CHCl_3$	238(4.56),274(4.54), 307(4.48),435(4.21)	5-0433-83
	EtOH	228(--),280(--), 410(--)	5-0433-83
$C_{39}H_{26}$			
9H-Fluorene, 9,9'-(bicyclo[4.4.1]undeca-1,3,5,7,9-pentaene-2,7-diyldimethylidyne)bis-, (±)-	CH_2Cl_2	233(4.86),248(4.82), 257(4.85),286(4.45), 405(4.29)	83-0240-83
	MeCN	226(3.37),244s(3.32), 280s(2.92),388(2.78)	83-0240-83
$C_{39}H_{28}N_2OS$			
Spiro[benzothiazole-2(3H),6'-[6H]cyclopenta[b]furan]-3'-carbonitrile, 5',6'a-dihydro-2',4',6'a-triphenyl-3-(phenylmethyl)-	$CHCl_3$	265(4.35),330(4.29), 360(4.34)	18-1688-83
$C_{39}H_{28}N_4O$			
Acetamide, N-[4-[2,6-bis([1,1'-biphenyl]-4-yl)-3,5-dicyano-1,4-dihydro-4-pyridinyl]phenyl]-	EtOH	207(4.89),249(4.52), 287(4.74),357(3.76)	73-3112-83
$C_{39}H_{29}Br_2N_2$			
3H-Indolium, 2-(4-bromophenyl)-3-[3-[2-(4-bromophenyl)-1-methyl-1H-indol-3-yl]-1-phenyl-2-propenylidene]-1-methyl-, bromide	$MeNO_2$-HOAc	628(5.00)	103-1306-83
$C_{39}H_{29}N_5O$			
7H-Benzimidazo[2,1-a]benz[de]isoquinolin-7-one, 4-[5-[4-(dimethylamino)-phenyl]-4,5-dihydro-1-(2-naphthalenyl)-1H-pyrazol-3-yl]-	toluene	520(4.34)	103-0204-83
$C_{39}H_{30}As_2$			
1H-Arsole, 1,1'-(phenylmethylene)-bis[2,5-diphenyl-	C_6H_{12}	230(4.51),346s(4.08), 376(4.12)	101-0335-83I
$C_{39}H_{30}N_4$			
3,5-Pyridinedicarbonitrile, 2,6-bis([1,1'-biphenyl]-4-yl)-4-[4-(dimethylamino)phenyl]-1,4-dihydro-	EtOH	206(4.93),286(4.71), 360(4.09)	73-3112-83
$C_{39}H_{31}N_2$			
3H-Indolium, 1-methyl-3-[3-(1-methyl-2-phenyl-1H-indol-3-yl)-1-phenyl-2-propenylidene]-2-phenyl-, salt with 4-methylbenzenesulfonic acid	$MeNO_2$-HOAc	625(5.28)	103-1306-83
$C_{39}H_{32}O_8$			
Albanol B	EtOH	284(4.36),318(4.53), 332(4.7),347(4.82), 365(4.77)	88-3013-83
$C_{39}H_{34}N_2O_5$			
Acetamide, N-[2-[(9,10-dihydro-4-hydroxy-9,10-dioxo-1-anthracenyl)amino]ethyl]-N-[2-(triphenylmethoxy)ethyl]-	$CHCl_3$	556(4.24),594(4.21)	18-1435-83

Compound	Solvent	$\lambda_{max}(\log \epsilon)$	Ref.
$C_{39}H_{35}ClO_2P_2Pt$			
Platinum, chloro(ethoxycarbonyl)bis-(triphenylphosphine)-, (SP-4-3)-	CH_2Cl_2	255(4.36),267(4.20), 275(4.04)	101-0119-83A
$C_{39}H_{36}O_8$			
Albanol A	EtOH	277(3.1),314(4.5)	88-3013-83
$C_{39}H_{37}O_8P$			
α-D-Ribofuranose, 1,2-O-(1-methylethylidene)-5-O-(triphenylmethyl)-, diphenyl phosphate	EtOH	210.5(4.30)	23-1387-83
$C_{39}H_{38}N_2O_9$			
2,4(1H,3H)-Pyrimidinedione, 5-(3,3-dimethyl-1-butynyl)-1-[2,3,5-tris-O-(4-methylbenzoyl)-β-D-arabinofuranosyl]-	MeOH	240(4.75),284(4.05)	87-0661-83
$C_{39}H_{39}O_{22}$			
1-Benzopyrylium, 5-[[6-O-(carboxyacetyl)-β-D-glucopyranosyl]oxy]-7-hydroxy-3-[[6-O-[3-(4-hydroxyphenyl)-1-oxo-2-propenyl]-β-D-glucopyranosyl]oxy]-2-(3,4,5-trihydroxyphenyl)-, chloride, (E)-	aq HCl	277(4.31),295(4.23), 312(4.19),525.5(4.44)	88-4863-83
$C_{39}H_{40}N_6O_{12}$			
Adenosine, 5'-O-[3-[(17-oxo-4-nitroestra-1,3,5(10)-trien-3-yl)oxy]-propyl]-, 2',3'-diacetate	EtOH	262(4.14)	87-0162-83
$C_{39}H_{40}O_{16}$			
δ-Naphthocyclinone methyl ester	$CHCl_3$	251(4.60),268s(--), 355s(--),407(4.18), 426s(--)	5-0471-83
$C_{39}H_{40}O_{17}$			
Hydropiperoside	MeOH	228(4.18),315(4.42)	102-0549-83
$C_{39}H_{42}N_2O_8$			
Thalibrunimine, O-methyl-	EtOH	240s(4.46),282(4.01), 305s(3.92)	100-0001-83
$C_{39}H_{42}N_2O_9$			
Curacautine, (-)-	MeOH	207(4.85),223s(4.74), 271(4.35),282(4.29)	100-0908-83
$C_{39}H_{42}O_{16}$			
δ-Naphthocyclinone, dihydro-, methyl ester	$CHCl_3$	242(4.07),275(4.34), 295s(--),343s(--), 386(3.97)	5-0471-83
$C_{39}H_{43}NO_9$			
Naphthomycin C	MeOH	232(4.51),286s(--), 307(4.45),350s(4.26)	158-0484-83
$C_{39}H_{43}NO_{17}$			
Neoalatamine	EtOH	228(4.23),267(3.61)	102-2839-83
$C_{39}H_{43}N_7$			
21H,23H-Porphine-5,10,15-tricarbonitrile, 2,3,7,8,12,13,17,18-octa-	CH_2Cl_2	418(5.09),542(3.95), 582(3.84),626(3.69),	22-0317-83

Compound	Solvent	λ_{max}(log ϵ)	Ref.
ethyl- (cont.)		685(3.97)	22-0317-83
$C_{39}H_{44}ClNO_9$			
Naphthomycin B	EtOH	235(4.63),307(4.56), 360s(4.05)	158-0484-83
	EtOH-NaOH	235(4.59),297(4.54), 345s(4.14),434(4.17), 570(3.08)	158-0484-83
$C_{39}H_{44}N_2O_7$			
Calafatine	MeOH	258(3.32),281(3.82)	100-0001-83
$C_{39}H_{44}N_2O_8$			
Thalistine	MeOH	278(3.90)	100-0001-83
Thalmirabine	MeOH	280(3.95),314s(3.34)	100-0001-83
$C_{39}H_{46}N_2O_8$			
Malekulatine	MeOH	211(4.79),230s(4.50), 284(4.18)	44-3957-83
Vanuatine, (+)-	MeOH	210(4.83),230s(4.48), 286(4.11)	44-3957-83
$C_{39}H_{46}N_4O_{10}$			
21H-Biline-3,8,12,17-tetrapropanoic acid, 1,19,22,24-tetrahydro-2,7,13-18-tetramethyl-1,19-dioxo-, tetramethyl ester	CH_2Cl_2	378(4.75),646(4.26)	78-1841-83
$C_{39}H_{49}N_4O_2Ru$			
Ruthenium(II), carbonylethanol(octaethylporphinato)-	CH_2Cl_2	395(5.50),515(4.23), 548(4.53)	23-2389-83
$C_{39}H_{50}N_2O_{12}S_2$			
Benzenesulfonamide, N-(2,2-dimethoxyethyl)-N-[[3-[6-[[(2,2-dimethoxyethyl)(4-methylphenyl)sulfonyl]amino]methyl]-2,3-dimethoxyphenoxy]-4-methoxyphenyl]methyl]-4-methyl-	MeOH	231(4.01),275(3.66)	83-0694-83
$C_{39}H_{50}N_4$			
Quinolinium, 4-[3-[2-(1-dodecyl-3-methyl-1H-benzimidazolium-2-yl)-1-methyl-4(1H)-pyridinylidene]-1-propenyl]-1-methyl-, diiodide	EtOH	676(5.09)	4-0023-83
$C_{39}H_{50}O_{16}$			
Affinoside S-IV pentaacetate	MeOH	264(4.26)	94-1199-83
Affinoside S-V pentaacetate	MeOH	264(4.14)	94-1199-83
Affinoside VI pentaacetate	MeOH	264(4.08)	94-1199-83
$C_{39}H_{51}O_2S_4$			
Sulfonium, [2-ethoxy-2-oxo-1-[1,2,3-tris[[4-(1,1-dimethylethyl)phenyl]thio]-2-cyclopropen-1-yl]ethyl]dimethyl-, perchlorate	MeCN	274(4.00)	18-0171-83
$C_{39}H_{52}O_9$			
Kijanolide, 26,32-di-O-methyl-, 9,17-diacetate	CF_3CH_2OH	253(4.01)	39-1497-83C

Compound	Solvent	λ_{max}(log ϵ)	Ref.
$C_{39}H_{52}O_{11}$			
Dircin	MeOH	228(3.91),278(3.10)	36-1285-83
$C_{39}H_{54}O_6$			
Lup-20(29)-en-27-oic acid, 3β-hydroxy-28-[[3-(4-hydroxyphenyl)-1-oxo-2-propenyl]oxy]-	MeOH	227(4.03),291s(4.21),304(4.31)	100-0118-83
	MeOH-KOH	240(--),302s(--),360(--)	100-0118-83
$C_{39}H_{56}O_5$			
Urs-12-en-29-oic acid, 3β-[(4-methoxybenzoyl)oxy]-, methyl ester	MeOH	256(4.24)	100-0537-83
$C_{39}H_{56}O_6$			
Urs-12-en-29-oic acid, 7β-hydroxy-3β-[(4-methoxybenzoyl)oxy]-, methyl ester	MeOH	256(4.32)	100-0537-83
$C_{39}H_{58}CrO_3P_2$			
Chromium, tricarbonyl[2-[(1,2,3,4,5,6-η)-2,4,6-tris(1,1-dimethylethyl)phenyl]-1-[2,4,6-tris(1,1-dimethylethyl)-phenyl]diphosphene 1-sulfide]-	CH_2Cl_2	304(4.09),324(4.11),480(2.88)	88-4855-83
$C_{39}H_{58}N_4O_4S$			
2-Naphthalenecarboxamide, 4-[[4-[ethyl-[2-[(methylsulfonyl)amino]ethyl]amino]-2-methylphenyl]imino]-N-hexadecyl-1,4-dihydro-1-oxo-	EtOAc	654(4.42)	44-0177-83
$C_{39}H_{58}O_{12}$			
Preatoxigenin tetraacetate dimethyl ester	MeOH	218(2.4)	20-0355-83
$C_{39}H_{60}N_2O_{15}$			
3,6,9,12,15-Pentaoxabicyclo[15.3.1]heneicosa-1(21),17,19-trien-21-ol, 19-[[4-[2-[2-[2-(1,4,7,10,13-pentaoxacyclopentadec-2-ylmethoxy)ethoxy]-ethoxy]ethoxy]phenyl]azo]-	$CHCl_3$	358(4.34)	18-3253-83
$C_{39}H_{60}O_7$			
Olean-12-ene-3,16,22,28-tetrol, 3,28-diacetate 22-(2-methyl-2-butenoate), [3β,16α,22α(Z)]-	EtOH	224(3.68)	20-0473-83
Olean-12-ene-3,16,22,28-tetrol, 3,28-diacetate 22-(3-methyl-2-butenoate), (3β,16α,22α)-	EtOH	220(3.93)	20-0473-83
$C_{39}H_{60}O_{12}$			
Bafilomycin C_1	MeOH	245(4.34),280s(4.08),335s(3.68),350s(3.40)	88-5193-83
$C_{39}H_{64}O_{12}$			
β-D-Glucopyranoside, (1α,3β,5Z,7E)-1-hydroxy-9,10-secocholesta-5,7,10(19)-trien-3-yl 4-O-β-D-glucopyranosyl-	n.s.g.	265(4.23)	33-2093-83
$C_{39}H_{66}N_2O_8$			
L-Alanine, N-[(1,1-dimethylethoxy)-	MeOH	220(3.40),257(2.40)	36-0792-83

Compound	Solvent	$\lambda_{max}(\log \epsilon)$	Ref.
carbonyl]-, 3-heptadecyl-1,2-phenylene ester (cont.)			36-0792-83
$C_{39}H_{66}O_5Si_2$			
Prosta-5,7,14-triene-4,12-diol, 1,9-bis[[(1,1-dimethylethyl)dimethylsilyl]oxy]-, 4-benzoate	MeOH	239(4.45),247s(4.44)	94-1440-83
isomer	MeOH	238(4.46),244s(4.46)	94-1440-83
$C_{39}H_{67}NO_{13}$			
Oxamacrocyclic compound 11	EtOH	283(4.29)	77-1166-83

Compound	Solvent	λ_{max}(log ϵ)	Ref.
$C_{40}H_{22}$			
14,21-Ethenodinaphtho[1,2-a:2',1'-o]-pentaphene	$CHCl_3$	264(4.76),277s(4.76), 291s(4.87),300(4.91), 305s(4.87),338(4.43), 361(4.29),380(4.33), 407(4.20),437(2.69), 449(2.64)	5-2262-83
$C_{40}H_{24}$			
Dibenzo[a,o]perylene, 7,16-diphenyl-	toluene	300(4.5),350(3.3), 370(3.3),530(4.2), 585(4.5)	149-0527-83B
14,21-Ethanodinaphtho[1,2-a:2',1'-o]-pentaphene	$CHCl_3$	247s(4.63),262s(4.62), 286(4.88),301s(4.68), 314(4.61),344s(4.02), 361(4.24),372s(4.25), 378(4.29),388(4.14), 400(4.12),419s(2.84), 437(2.54),449(2.44)	5-2262-83
1,26:6,9:12,15:20,23-Tetramethenocyclo-docosa[1,2,3-de:12,13,14-d'e']dinaph-thalene	$CHCl_3$	268s(4.97),278s(5.05), 284(5.17),312s(4.32), 363(4.23)	5-2262-83
$C_{40}H_{24}N_8$			
3,5:14,16-Dinitrilo-5H,16H-dicyclo-hepta[e,k]diimidazo[1,5-a:1',5'-g][1,7]diazacyclododecine-10,21-dicarbonitrile, 1,12-bis(4-methyl-phenyl)-	EtOH	241(4.40),266(4.40), 304(4.37),335(4.28), 423s(4.15),443(4.31), 555(3.49)	18-3703-83
$C_{40}H_{24}O_2$			
7H-7,11b-Epidioxydibenzo[a,o]perylene, 7,16-diphenyl-	n.s.g.	385(4.0),395(4.2), 408(4.2)	149-0527-83B
$C_{40}H_{26}N_6O_4$			
3,5:14,16-Dinitrilo-5H,16H-dicyclo-hepta[e,k]diimidazo[1,5-a:1',5'-g]-[1,7]diazacyclododecine-10,21-dicarb-oxylic acid, 1,12-diphenyl-, dimethyl ester	EtOH	242(4.63),260(4.60), 302(4.56),325(4.55), 417s(4.36),437(4.49), 542(3.65)	18-3703-83
$C_{40}H_{28}$			
[2.2](3,10)Benzo[c]phenanthrenophane	$CHCl_3$	278(5.15),312(4.12), 324(4.15),338(3.97)	5-2262-83
$C_{40}H_{28}N_2O_2S$			
Spiro[benzothiazole-2(3H),6'-[6H]cyclo-penta[b]furan]-3'-carbonitrile, 5',6'a-dihydro-3-(2-oxo-2-phenyl-ethyl)-2',4',6'a-triphenyl-	$CHCl_3$	270(4.28),325(4.21), 360(4.27)	18-1688-83
$C_{40}H_{28}N_2O_4$			
4,4'-Bipyridinium bis(1-benzoyl-2-oxo-2-phenylethylide)	EtOH	466(3.70)	65-2073-83
	$CHCl_3$	546(4.07)	65-2073-83
$C_{40}H_{28}N_4$			
3,5:14,16-Dimetheno-5H,16H-dicyclo-hepta[e,k]diimidazo[1,5-a:1',5'-g]-[1,7]diazacyclododecine, 1,12-bis(4-methylphenyl)-	EtOH	240(4.75),264(4.98), 287s(4.44),345(4.55), 363(4.48),386s(4.27), 406(4.41),428(4.38), 487s(3.57),520(3.62),	18-3703-83

Compound	Solvent	λ_{max}(log ε)	Ref.
(cont.)		553(3.63),592(3.47), 650s(3.03)	18-3703-83
$C_{40}H_{28}O_2$			
Pyrylium, 2,2'-(1,4-phenylene)bis[4,6-diphenyl-, diperchlorate	MeCN	238(4.28),266(4.34), 376(4.72),444(4.66)	22-0115-83
$C_{40}H_{31}NO$			
Quinoline, 8-benzoyl-1-(4-biphenylyl)-1,5,6,7-tetrahydro-2,4-diphenyl-	EtOH	390(3.54),515(3.94)	39-2601-83C
Quinoline, 5,6,7,8-tetrahydro-2,4-diphenyl-8-[[(4-biphenylyl)oxy]phenyl]methylene]-	EtOH	290(4.51),375(4.20)	39-2601-83C
$C_{40}H_{31}NOP$			
Benzo[f]quinolinium, 2-oxo-2-[4-(triphenylphosphonio)methyl]phenyl]ethylide, iodide	DMF	518(4.08)	104-0568-83
$C_{40}H_{31}NO_3$			
3-Pyridinecarboxylic acid, 1,4-dihydro-1-(phenylmethyl)-, 2'-(phenylmethoxy)[1,1'-binaphthalen]-2-yl ester, (S)-	MeCN	361(3.93)	18-3672-83
$C_{40}H_{32}NOP$			
Benzo[f]quinolinium, 4-[2-oxo-2-[4-(triphenylphosphonio)methyl]phenyl]ethyl]-, diiodide	EtOH	225(5.02),275(4.55), 369(3.99)	104-0568-83
$C_{40}H_{33}N_2O$			
3H-Indolium, 3-[1-(4-methoxyphenyl)-3-(1-methyl-2-phenyl-1H-indol-3-yl)-2-propenylidene]-1-methyl-2-phenyl-, bromide	$MeNO_2$-HOAc	623(5.01)	103-1306-83
$C_{40}H_{33}N_8$			
1(2H),3'-Bi-1,2,4,5-tetrazinium, 3,3',4,4'-tetrahydro-2,2',4,4',6,6'-hexaphenyl-, bromide	MeCN	490(4.37)	103-0224-83
oxidation product ($C_{40}H_{33}BrN_8O$)	MeCN	655(4.11)	103-0224-83
free base	MeCN	585(4.08),845(4.14), 925(4.12)	103-0224-83
$C_{40}H_{34}N_2$			
Pyridinium, 4,4'-(2,4-diphenylbicyclo-[1.1.0]butane-1,3-diyl)bis[1-(phenylmethyl)-, perchlorate	MeCN	278(4.30),331(4.23)	5-0069-83
$C_{40}H_{34}O_{16}$			
Kelletinin I tetraacetate	EtOH	233(4.47)	35-7396-83
$C_{40}H_{34}P_2$			
Phosphonium, 1,3-butadiene-1,4-diylbis[triphenyl-, dibromide	EtOH	268(4.44)	118-0374-83
$C_{40}H_{36}O_{12}$			
4,6:23,25-Dimethano-5H,24H-tetrabenzo-[k,n,z,c_1][1,4,7,10,16,19,22,25]octaoxacyclotriaconta-5,24,39,40-tetrone, 11,12,14,15,17,18,30,31,33,34,36,37-	CH_2Cl_2	380(3.93)	18-3185-83

Compound	Solvent	$\lambda_{max}(\log \epsilon)$	Ref.
dodecahydro- (cont.)			18-3185-83
$C_{40}H_{38}O_{11}$			
Sanggenon G	EtOH	226s(4.63),285(4.43), 290s(4.43),318s(4.10)	142-0611-83
	EtOH-$AlCl_3$	219s(4.72),290s(4.41), 298(4.42),333s(3.95)	142-0611-83
$C_{40}H_{40}Ge_3$			
Trigermane, 2,2-diethyl-1,1,1,3,3,3-hexaphenyl-	C_6H_{12}	247s(4.43)	101-0149-83G
$C_{40}H_{40}N_2O_8$			
Tiliamosine, N,O-diacetyl-	MeOH	236s(4.59),291(3.79)	142-1927-83
$C_{40}H_{41}O_{22}$			
Malonylawobanin methyl ester	aq HCl	278(4.29),295(4.21), 312(4.17),527(4.40)	88-4863-83
$C_{40}H_{42}N_4O_8$			
21H,23H-Porphine-2,18-dipropanoic acid, 7,12-bis(3-methoxy-3-oxo-1-propenyl)-3,8,13,17-tetramethyl-, dimethyl ester, (E,E)-	CH_2Cl_2	428(5.16),518(4.18), 554(4.17),586(3.92), 641(3.83)	44-0500-83
$C_{40}H_{42}N_8$			
21H,23H-Porphine-5,10,15,20-tetracarbonitrile, 2,3,7,8,12,13,17,18-octaethyl-	CH_2Cl_2	414(5.13),554(3.94), 600(3.75),644(3.69), 708(3.94)	22-0317-83
$C_{40}H_{44}N_2O_{10}$			
Talcamine	MeOH	208(4.93),225s(4.79), 260(4.41),272s(4.27), 305(3.67)	100-0908-83
$C_{40}H_{44}N_4O$			
C-Curarine I (dichloride)	EtOH	260(4.41),296(4.07)	162-0382-83
$C_{40}H_{47}N_3O_{10}$			
Glycine, N,N-bis[[3-[[(carboxymethyl)-[(2-hydroxy-5-methylphenyl)methyl]-amino]methyl]-2-hydroxy-5-methyl-phenyl]methyl]-	pH 11.03	288(4.0),305s(--)	125-2795-83
	pH 11.65	290(4.0),305(4.0)	125-2795-83
	pH 14.01	305(4.2)	125-2795-83
$C_{40}H_{48}Fe$			
Iron(2+), bis[(4,5,6,7,15,16-η)-5,6,15,16-tetramethyltricyclo-[8.2.2.2^{4,7}]hexadeca-4,6,10,12,13,15-hexaene]-, bis(hexafluorophosphate)	MeCN	295(4.0),355(3.7)	157-1214-83
$C_{40}H_{48}N_2O_6$			
Funiferine dimethiodide	EtOH	229(4.82),286(4.10)	100-0001-83
$C_{40}H_{48}N_4O_2$			
C-Calebassine (dichloride)	H_2O	253(4.37),302(3.77)	162-0235-83
$C_{40}H_{48}N_4O_3$			
Geissospermine	MeOH	251(4.10),285(3.91), 293(3.90)	162-0625-83

Compound	Solvent	λ_{max}(log ϵ)	Ref.
$C_{40}H_{48}N_6O_9$			
Bouvardin, 5-(N-methyl-L-tyrosine)-	EtOH	276(3.58),282(3.52)	94-1424-83
$C_{40}H_{48}O_8$			
Gambogic acid methyl ester monomethyl ether	EtOH	224(4.56),299(4.13)	162-0622-83
$C_{40}H_{53}FeN_4$			
Iron, butyl[2,3,7,8,12,13,17,18-octaethyl-21H,23H-porphinato(2-)-N[21],N[22],N[23],N[24]]-, (SP-5-31)-	benzene	380(4.82),392(4.95), 425s(--),515(3.94), 552(4.22)	101-0065-83M
$C_{40}H_{53}GaN_4$			
Gallium, butyl[2,3,7,8,12,13,17,18-octaethyl-21H,23H-porphinato(2-)-N[21],N[22],N[23],N[24]]-, (SP-5-31)-	benzene	338(4.56),407s(4.51), 431(4.91),511(3.23), 551(4.00),586(3.66)	101-0273-83M
$C_{40}H_{54}O$			
β,β-Caroten-19-al	hexane	275(4.02),305(3.95), 474(4.75)	33-1148-83
β,β-Caroten-10-ol, 9,15,15',19-tetradehydro-9,10-dihydro-	hexane	232(4.25),387(4.72), 406(4.66)	33-1148-83
β,β-Caroten-19-ol, 15,15'-didehydro-, (9E)-	hexane	274(4.22),340(4.38), 424(3.96),448s(4.87)	33-1148-83
β,ε-Caroten-14'-ol, 11,12,13',20'-tetradehydro-13',14'-dihydro-, (±)-	hexane	236(4.37),297s(4.53), 311(4.62),322(4.62), 346s(4.45)	33-1148-83
β,β-Caroten-10-one, 9,19-didehydro-9,10-dihydro-	hexane	255(4.02),322(4.01), 436(4.70)	33-1148-83
Rubixanthin, 15,15'-didehydro-	hexane	283(4.34),350(4.35), 423s(4.92),440(4.99), 468(4.91)	33-0494-83
$C_{40}H_{54}O_{16}$			
α-D-Glucopyranoside, 1,5-diacetoxy-3-(2-acetoxy-1-methylethyl)-2,4,5-6,6a,7,8,11a-octahydro-6,11a-dimethyl-1H-benzo[a]cyclopenta[d]cyclooct-en-4-yl, tetraacetate	MeCN	245(4.30)	32-0717-83
$C_{40}H_{56}$			
β,β-Carotene	hexane	272(4.34),348s(3.90), 429s(4.99),452(5.14), 478(5.09)	33-1148-83
9-cis	hexane	265(4.25),342.5(4.10), 426s(4.98),446.5(5.13), 474(5.07)	33-1148-83
Lycopene, (7Z,7'Z,9Z,9'Z)-	hexane	232(4.42),255(4.34), 295(4.27),362s(4.26), 392s(4.65),417s(4.91), 437(5.03),461s(4.88), 487s(4.41)	39-3011-83C
$C_{40}H_{56}O$			
β,β-Caroten-10-ol, 9,19-didehydro-9,10-dihydro-	hexane	232(4.30),280(4.14), 384s(4.73),401(4.84), 422(4.78)	33-1148-83
β,β-Caroten-19-ol	hexane	278(4.31),332(4.41), 420s(4.86),441(4.98), 466(4.89)	33-1148-83

Compound	Solvent	λ_{max}(log ε)	Ref.
Rubixanthin, (all-E)-	EPA	281(4.44),347(3.94), 435(5.06),460(5.23), 491(5.17)	33-0494-83
(5'Z)-	EPA	281(4.44),347(3.93), 435(5.04),459.5(5.20), 489(5.14)	33-0494-83
$C_{40}H_{56}O_2$			
β,κ-Caroten-6'-one, 3'-hydroxy-, trans-(±)-	benzene	486(5.06),520(4.97)	39-1465-83C
Lutein	EtOH	267(4.46),332(3.92), 423(5.03),447(5.19), 475(5.16)	33-1175-83
$C_{40}H_{56}O_3$			
β,κ-Caroten-6'-one, 3,3'-dihydroxy-	n.s.g.	483(5.08)	162-0244-83
β,ε-Carotene-3,3',19-triol, (3R,3'R-6R',9E)- (relative absorbance given)	EPA	262(0.26),326(0.15), 418(0.74),438(1.00), 465(0.85)	33-1175-83
(9Z)-	EPA	263(0.27),328(0.12), 422(0.73),441(1.00), 468(0.88)	33-1175-83
$C_{40}H_{56}O_4$			
Capsorubin, (3S,3'S,5R,5'R)-	benzene	459s(--),486(5.09), 530(5.01)	39-1465-83C
κ,κ-Carotene-6,6'-dione, 3,3'-dihydroxy-, cis	benzene	457s(--),488(5.08), 522(5.00)	39-1465-83C
trans (isomer mixture)	benzene	455s(--),486(5.08), 519(5.00)	39-1465-83C
$C_{40}H_{57}N_3O_6$			
Acetamide, 2,2'-[[2-(2,5-dihydro-2,5-dioxo-3,4-diphenyl-1H-pyrrol-1-yl)-2-methyl-1,3-propanediyl]bis(oxy)]-bis[N-heptyl-N-methyl-	EtOH	272(3.96),360(3.59)	33-1078-83
	+ Ca^{++}	none	33-1078-83
$C_{40}H_{58}N_4$			
1H-Benzimidazolium, 2-[4-[2-[4-(dimethylamino)phenyl]ethenyl]-1-methylpyridinium-2-yl]-1-hexadecyl-3-methyl-, diiodide	EtOH	558(4.58)	4-0023-83
$C_{40}H_{60}N_4O_{10}$			
Bufotoxin	n.s.g.	295(3.74)	162-0204-83
$C_{40}H_{63}ClO_7Si$			
Milbemycin B, 13-chloro-5-O-demethyl-28-deoxy-5-O-[(1,1-dimethylethyl)-dimethylsilyl]-6,28-epoxy-25-(1-methylpropyl)-	MeOH	246(4.49)	88-5333-83
$C_{40}H_{64}$			
Phytoene, (15Z)-	hexane	265s(4.26),276s(4.49), 286(4.57),296s(4.43)	39-3011-83C
$C_{40}H_{64}O_7Si$			
Milbemycin B, 5-O-demethyl-28-deoxy-5-O-[(1,1-dimethylethyl)dimethylsilyl]-6,28-epoxy-25-(1-methylpropyl)-	MeOH	244(4.45)	88-5333-83

Compound	Solvent	$\lambda_{max}(\log \epsilon)$	Ref.
$C_{40}H_{64}O_8Si$			
Milbemycin B, 5-O-demethyl-28-deoxy-5-O-[(1,1-dimethylethyl)dimethylsilyl]-6,28-epoxy-13-hydroxy-25-(1-methylpropyl)-	MeOH	244(4.46)	88-5333-83
$C_{40}H_{74}N_2O$			
Acetamide, N-[4-(dihexadecylamino)-phenyl]-	MeOH	280(4.30)	5-0802-83
$C_{40}H_{75}ClN_2O$			
Acetamide, N-[4-(dihexadecylamino)-phenyl]-, monohydrochloride	MeOH	253(4.10),289(3.57)	5-0802-83

Compound	Solvent	$\lambda_{max}(\log \epsilon)$	Ref.
$C_{41}H_{23}Cl_{12}NOP$			
1H,5H,7aH-Pyrrolizin-1-one, 6-methyl-2,3,7,7a-tetrakis(1,2,2-trichloroethenyl)-5-(triphenylphosphonium)-phenylmethylene-	MeCN	285(4.46),350(3.87), 475(3.97)	88-2977-83
$C_{41}H_{26}Cl_3NO$			
Pyridinium, 2,4,6-tris(4-chlorophenyl)-1-(2'-hydroxy[1,1':3',1"-terphenyl]-5'-yl)-, hydroxide, inner salt	MeOH	314(4.63),528(3.54)	5-0860-83
$C_{41}H_{26}N_4O$			
8H-Benzo[de]naphth[1',2':4,5]imidazo-[2,1-a]isoquinolin-8-one, 11-[4,5-dihydro-1-(2-naphthalenyl)-5-phenyl-1H-pyrazol-3-yl]-	toluene	530(4.43)	103-0214-83
$C_{41}H_{26}O_{26}$			
Praecoxin D	MeOH	206(4.85),235s(4.77), 293(4.46)	94-0333-83
$C_{41}H_{28}O_{26}$			
Casuarictin	MeOH	220(4.90),260s(4.59)	39-1765-83C
Casuarinin	MeOH	221(4.87),267s(4.51)	39-1765-83C
Stachyurin	MeOH	221(4.88),267s(4.53)	39-1765-83C
$C_{41}H_{30}O_2$			
5H-Inden-5-one, 7-benzoyl-1-(diphenylmethylene)-1,6,7,7a-tetrahydro-4,6-diphenyl-, (6α,7β,7aα)-	MeCN	248(4.41),264s(4.26), 393(4.41)	4-1621-83
11-Oxatricyclo[5.3.1.0^{2,6}]undeca-3,8-dien-10-one, 5-(diphenylmethylene)-1,7,9-triphenyl-, endo	C_6H_{12}	221s(4.53),285(4.23)	4-1621-83
exo	C_6H_{12}	210(4.40),268s(3.92), 288(4.01)	4-1621-83
	MeCN	223s(4.48),288(4.24)	4-1621-83
11-Oxatricyclo[5.3.1.0^{2,6}]undeca-3,9-dien-8-one, 5-(diphenylmethylene)-1,7,9-triphenyl-, endo	MeCN	223s(4.45),268(4.16), 285s(4.10),427s(1.82)	4-1621-83
$C_{41}H_{32}O_2$			
4,8-Epoxyazulen-5-ol, 1-(diphenylmethylene)-1,3a,4,5,8,8a-hexahydro-4,6,8-triphenyl-, (3aα,4β,5α,8β,8aα)-	MeCN	241s(4.14),264s(4.07), 294(4.26)	4-1621-83
isomer, (S)-exo-	MeCN	239(4.38),265s(4.25), 295(4.29)	4-1621-83
$C_{41}H_{34}O_3$			
4,8-Epoxyazulene-5,6-diol, 1-(diphenylmethylene)-1,3a,4,5,6,7,8,8a-octahydro-4,6,8-triphenyl-, (3aα,4β,5α,6α-8β,8aα)-	MeCN	243s(3.97),264s(3.97), 296(4.23)	4-1621-83
$C_{41}H_{35}N_2$			
3H-Indolium, 1-methyl-3-[3-[1-methyl-2-(4-methylphenyl)-1H-indol-3-yl]-1-phenyl-2-propenylidene]-2-(4-methylphenyl)-, bromide	$MeNO_2$-HOAc	628(5.00)	103-1306-83
$C_{41}H_{36}N_2O_6$			
Acetamide, N-[2-[acetyl(9,10-dihydro-	$CHCl_3$	313(4.36),348(4.42),	18-1435-83

Compound	Solvent	λ_{max}(log ϵ)	Ref.
4-hydroxy-9,10-dioxo-1-anthracenyl)-amino]ethyl]-N-[2-(triphenylmethoxy)ethyl]- (cont.)		503(4.23)	18-1435-83
Acetamide, N-[2-[[4-(acetyloxy)-9,10-dihydro-9,10-dioxo-1-anthracenyl]-amino]ethyl]-N-[2-(triphenylmethoxy)ethyl]-	$CHCl_3$	249(4.52),277(3.99), 321(3.77),508(3.83)	18-1435-83
$C_{41}H_{36}O_{16}$			
Benzoic acid, 4-acetoxy-, 1,2,3,5-pentanetetrayl ester, [S-(R*,S*)]-	EtOH	234(4.56)	35-7396-83
$C_{41}H_{36}P_2$			
Phosphonium, (1-methyl-1,3-butadiene-1,4-diyl)bis[triphenyl-, diiodide, (E,E)-	EtOH	269(4.68)	118-0374-83
$C_{41}H_{38}NO_5P$			
α-D-Ribofuranose, 3-C-(cyanomethyl)-3-deoxy-3-(diphenylphosphinyl)-1,2-O-(1-methylethylidene)-5-O-(triphenylmethyl)-	EtOH	207(4.45),220(4.28)	159-0019-83
$C_{41}H_{39}N_2OPS$			
Thiazolium, 2-[2-[4-(dimethylamino)-phenyl]ethenyl]-4-methyl-3-[2-oxo-2-[4-(triphenylphosphonio)methyl]-phenyl]ethyl]-, diperchlorate	EtOH	500(4.28)	65-2424-83
$C_{41}H_{40}N_2O_2S$			
Estra-1,3,5(10)-trien-17-one, 2-[4-[1,1'-biphenyl]yl-2-[(phenylmethyl)-imino]-3(2H)-thiazolyl]-3-methoxy-	EtOH	256(4.596),296(4.374)	36-1205-83
	EtOH-HCl	271(4.619)	36-1205-83
Estra-1,3,5(10)-trien-17-one, 4-[4-[1,1'-biphenyl]yl-2-[(phenylmethyl)-imino]-3(2H)-thiazolyl]-3-methoxy-	EtOH	263(4.100),300s(4.563)	36-1205-83
	EtOH-HCl	282(4.602)	36-1205-83
$C_{41}H_{42}N_2O_2$			
Murrafoline, (±)-	MeOH	218(4.66),243(4.86), 260s(4.66),307(4.45), 332s(3.91)	88-5377-83
$C_{41}H_{45}N_5O_4$			
Pentaphyrin	CH_2Cl_2	458(5.38)	27-0341-83
	CH_2Cl_2	367(3.74),458(5.38), 642(3.78),695(3.55)	77-0275-83
$C_{41}H_{50}N_6O_9$			
Bouvardin, 5-(N-methyl-L-tyrosine)-6-(3-hydroxy-N,O-dimethyl-L-tyrosine)-	EtOH	276(3.58),282(3.51)	94-1424-83
$C_{41}H_{50}N_6O_{10}$			
Bouvardin, 5-(N-methyl-L-tyrosine)-6-[(S)-β,3-dihydroxy-N,O-dimethyl-L-tyrosine]-	EtOH	276(3.42),284(3.30)	94-1424-83
Cyclic (D-alanyl-L-seryl-N,O-dimethyl-L-tyrosyl-L-alanyl-N-methyl-L-tyrosyl-3-hydroxy-N,O-dimethyl-L-tyrosyl), cyclic (5^4→6^3)-ether	EtOH	276(3.36),281(3.26)	94-1424-83

Compound	Solvent	λ_{max}(log ϵ)	Ref.
$C_{41}H_{52}Cl_2N_5O_9PSi_2$			
2'-Guanylic acid, N-benzoyl-3',5'-bis-O-[(1,1-dimethylethyl)dimethylsilyl]-, bis(2-chlorophenyl) ester	MeOH	237(4.15),260(4.11), 290(4.08)	33-2069-83
3'-Guanylic acid, N-benzoyl-2',5'-bis-O-[(1,1-dimethylethyl)dimethylsilyl]-, bis(2-chlorophenyl) ester	MeOH	237(4.17),258(4.19), 259(4.17),292(4.14)	33-2069-83
$C_{41}H_{54}N_2O_{13}$			
Tetrocarcin F-1	MeOH	232s(4.25),268(4.00), 278s(3.95)	158-83-103
$C_{41}H_{54}N_5O_9PSi_2$			
2'-Guanylic acid, N-benzoyl-3',5'-bis-O-[(1,1-dimethylethyl)dimethylsilyl]-, diphenyl ester	MeOH	237(4.15),253(4.11), 290(4.08)	33-2069-83
3'-Guanylic acid, N-benzoyl-2',5'-bis-O-[(1,1-dimethylethyl)dimethylsilyl]-, diphenyl ester	MeOH	237(4.20),257(4.18), 261(4.18),292(4.16)	33-2069-83
$C_{41}H_{56}N_4S$			
Benzothiazolium, 2-[3-[2-(1-hexadecyl)-3-methyl-1H-benzimidazolium-2-yl)-1-methyl-4(1H)-pyridinylidene]-1-propenyl]-3-methyl-, diiodide	EtOH	605(5.03)	4-0023-83
$C_{41}H_{58}N_4O_4$			
21H-Biline-2,19-dicarboxylic acid, 2,7,13,18-tetraethyl-5,15,22,24-tetrahydro-3,8,12,17-tetramethyl-, bis(1,1-dimethylethyl) ester, monohydrobromide	CH_2Cl_2	502(4.82)	78-1841-83
$C_{41}H_{58}O_7$			
Urs-12-en-29-oic acid, 7-acetoxy-3-[(4-methoxybenzoyl)oxy]-, methyl ester, (3β,7β)-	MeOH	356(4.28)	100-0537-83
$C_{41}H_{62}N_2O_{12}$			
1,7-Heptanediamine, N,N'-bis[(2,3,5,6-8,9,11,12,14,15-decahydro-1,4,7,10-13,16-benzohexaoxacyclooctadecin-18-yl)methylene]-	EtOH	288(3.90)	87-0007-83
$C_{41}H_{62}O_8$			
Olean-12-ene-3,16,22,28-tetrol, 3,22,28-triacetate 16-(2-methyl-2-butenoate), [3β,16α(E),22α]-	EtOH	214(3.80)	20-0473-83
Olean-12-ene-3,16,22,28-tetrol, 3,22,28-triacetate 16-(3-methyl-2-butenoate), (3β,16α,22α)-	EtOH	220(3.81)	20-0473-83
$C_{41}H_{62}O_9$			
Olean-12-ene-3,15,16,22,28-pentol, 3,15,28-triacetate 22-(2-methyl-2-butenoate), [3β,15α,16α,22α(Z)]-	EtOH	213(3.69)	20-0473-83
Olean-12-ene-3,15,16,22,28-pentol, 3,15,28-triacetate 22-(3-methyl-2-butenoate), (3β,15α,16α,22α)-	EtOH	220(3.87)	20-0473-83

Compound	Solvent	λ_{max}(log ϵ)	Ref.
$C_{41}H_{62}O_{10}$			
β-D-Glucopyranoside, (3 ,5Z,7E)-9,10-secocholesta-5,7,10(19)-trien-3-yl, tetraacetate	n.s.g.	265(4.28)	33-2093-83
$C_{41}H_{62}O_{11}$			
β-D-Glucopyranoside, (1α,3β)-1-hydroxy-cholesta-5,7-dien-3-yl, 2,3,4,6-tetra-acetate	MeOH	271(4.07),282(4.11), 294(3.87)	33-2093-83
β-D-Glucopyranoside, (1α,3β,5Z,7E)-1-hydroxy-9,10-secocholesta-5,7,10(19)-trien-3-yl, 2,3,4,6-tetraacetate	n.s.g.	264(4.26)	33-2093-83
β-D-Glucopyranoside, (3β,5Z,7E)-3-hy-droxy-9,10-secocholesta-5,7,10(19)-trien-25-yl, 2,3,4,6-tetraacetate	n.s.g.	264(4.16)	33-2093-83
β-D-Glucopyranoside, (3β,5Z,7E)-25-hy-droxy-9,10-secocholesta-5,7,10(19)-trien-3-yl, 2,3,4,6-tetraacetate	n.s.g.	265(4.25)	33-2093-83
$C_{41}H_{62}O_{12}$			
β-D-Glucopyranoside, (1α,3β,5Z,7E)-1,3-dihydroxy-9,10-secocholesta-5,7-10(19)-trien-25-yl, 2,3,4,6-tetraacetate	n.s.g.	265(4.23)	33-2093-83
β-D-Glucopyranoside, (1α,3β,5Z,7E)-1,25-dihydroxy-9,10-secocholesta-5,7,10(19)-trien-3-yl, 2,3,4,6-tetraacetate	n.s.g.	264(4.26)	33-2093-83
β-D-Glucopyranoside, (1α,3β,5Z,7E)-3,25-dihydroxy-9,10-secocholesta-5,7,10(19)-trien-1-yl, 2,3,4,6-tetraacetate	n.s.g.	241(4.12),274(4.11)	33-2093-83
$C_{41}H_{64}O_{14}$			
Digoxin	EtOH	220(4.11)	162-0460-83

Compound	Solvent	λ_{max}(log ϵ)	Ref.
$C_{42}F_{30}S_6$			
Benzene, hexakis[(pentafluorophenyl)-thio]-	hexane	340(2.78),380(3.76)	104-2313-83
$C_{42}H_{23}Cl_9O_{16}$			
Complex of 2,3,6-trichloronaphthazarin diacetate with 2,6,7-trichloro-1,4-diacetoxynaphthalene (2:1)	CCl_4	438(4.60),445(4.72), 465(4.66)	78-0199-83
$C_{42}H_{24}S_{12}$			
Tris(dibenzotetrathiafulvalene), radical cation tetrafluoroborate	MeCN	400s(--),425s(--), 444(4.19),485s(--), 600s(--),640(3.97)	104-1092-83
$C_{42}H_{26}$			
5,18[1',2']:9,14[1",2"]-Dibenzenoheptacene, 5,9,14,18-tetrahydro-	MeCN	243(4.37),260(4.79), 266(4.78),272(4.91), 284(5.33),306(3.31), 320(3.63),336(3.86), 353(4.95),371(3.82)	44-4357-83
$C_{42}H_{28}Cl_2N_4$			
Quinoxaline, 2,2'-(1,2-ethanediylidene)bis[1-(4-chlorophenyl)-1,2-dihydro-3-phenyl-	$CHCl_3$-CF_3COOH	309(4.26),387(3.98), 605s(--),773(4.26)	33-0379-83
$C_{42}H_{28}N_4O_2$			
8H-Benzo[de]naphth[1',2':4,5]imidazo-[2,1-a]isoquinolin-8-one, 11-[4,5-dihydro-5-(4-methoxyphenyl)-1-(2-naphthalenyl)-1H-pyrazol-3-yl]-	toluene	530(4.44)	103-0214-83
$C_{42}H_{28}N_4O_4$			
3,5:14,16-Dimetheno-5H,16H-dicyclohepta[e,k]diimidazo[1,5-a:1',5'-g][1,7]-diazacyclododecine-10,21-dicarboxylic acid, 1,12-diphenyl-, dimethyl ester	EtOH	222(4.79),269(4.91), 312(4.76),322(4.76), 353s(4.50),414(4.18), 436(4.19),508(3.64), 538(3.64),580s(3.46), 630s(2.96)	18-3703-83
$C_{42}H_{30}$			
m-Quinquephenyl, 5"-(3-biphenylyl)-	C_6H_{12}	204(4.90),252(5.08)	94-2313-83
o-Terphenyl, 3,5-di-2-biphenylyl-	C_6H_{12}	196(5.01),207s(4.94), 235(4.86),254s(4.62)	94-2313-83
p-Terphenyl, 3,5-di-4-biphenylyl-	C_6H_{12}	204(4.95),285(4.95)	94-2313-83
$C_{42}H_{30}N_6O_4$			
3,5:14,16-Dinitrilo-5H,16H-dicyclohepta[e,k]diimidazo[1,5-a:1',5'-g][1,7]-diazacyclododecine-10,21-dicarboxylic acid, 1,12-bis(4-methylphenyl)-, dimethyl ester	EtOH	243(4.50),261(4.45), 302(4.41),327(4.35), 423s(4.19),441(4.30), 546(3.52)	18-3703-83
3,5:14,16-Dinitrilo-5H,16H-dicyclohepta[e,k]diimidazo[1,5-a:1',5'-g][1,7]-diazacyclododecine-10,21-dicarboxylic acid, 1,12-diphenyl-, diethyl ester	EtOH	242(4.67),260(4.65), 302(4.59),325(4.59), 417s(4.39),437(4.52), 543(3.72)	18-3703-83
$C_{42}H_{30}N_{12}O_6S_2Zn$			
Zinc, bis[2-[[(2-methoxyphenyl)[(4-nitrophenyl)azo]methyl]azo]benzothiazolato]-, (T-4)-	PrOH	648(4.59)	135-0334-83
	+ HCl	468(4.46)	135-0334-83
	+ KOH	605(--)	135-0334-83

Compound	Solvent	λ_{max}(log ε)	Ref.
$C_{42}H_{32}CdN_{10}O_2S_2$			
Cadmium, bis[2-[[(2-methoxyphenyl)-(phenylazo)methyl]azo]benzothiazolato]-, (T-4)-	PrOH	615(4.59)	135-0334-83
	+ HCl	395(4.26)	135-0334-83
	+ KOH	510(--)	135-0334-83
$C_{42}H_{32}N_8$			
1H-Pyrazole-4,5-diamine, N^4-[1,3-diphenyl-4-(phenylazo)-1H-pyrazol-5-yl]-N^5,1,3-triphenyl-	EtOH	246(3.11)	4-1501-83
$C_{42}H_{32}N_{10}O_2S_2Zn$			
Zinc, bis[2-[[(2-methoxyphenyl)(phenylazo)methyl]azo]benzothiazolato]-, (T-4)-	PrOH	600(4.54)	135-0334-83
	+ HCl	395(4.26)	135-0334-83
	+ KOH	510(--)	135-0334-83
$C_{42}H_{32}O_9$			
9H-Benzo[4,5]cyclohepta[1,2-1]phenanthrene-3,5,7,10,12-pentol, 15-(3,5-dihydroxyphenyl)-8b,14,15,15a-tetrahydro-9-(4-hydroxyphenyl)-14-[(4-hydroxyphenyl)methylene]-, (8bα,9α-15β,15aα)-(-)- (stemonoporol)	MeOH	283(3.41)	39-0699-83C
	MeOH-NaOH	296(3.63)	39-0699-83C
Copalliferol A	EtOH	282(2.94)	39-0699-83C
	EtOH-NaOH	288(3.02)	39-0699-83C
	EtOH-NaOMe	291(3.08)	39-0699-83C
$C_{42}H_{32}S_2$			
4H-Thiopyran, 4,4'-(2,4,6-octatriene-1,8-diylidene)bis[2,6-diphenyl-	CH_2Cl_2	578(4.97)	44-2757-83
dication	CH_2Cl_2	615(4.91)	44-2757-83
radical cation	CH_2Cl_2	1640(4.36)	44-2757-83
$C_{42}H_{38}N_2$			
Pyridinium, 4,4'-(2,4-dimethyl-2,4-diphenylbicyclo[1.1.0]butane-1,3-diyl)-bis[1-(phenylmethyl)-, (1α,2α,3α,4β)-, diperchlorate	MeCN	274(4.21),291(4.23)	5-0069-83
$C_{42}H_{38}N_2O_3$			
[1,1':3',1"-Terphenyl]-2',5'-diol, 4'-[[6-[(2'-hydroxy[1,1':3',1"-terphenyl]-5'-yl)imino]hexyl]amino]-	CH_2Cl_2	315(4.10),540s(2.64)	70-0088-83
$C_{42}H_{40}N_2O_6$			
Carbamic acid, [2-[(9,10-dihydro-4-hydroxy-9,10-dioxo-1-anthracenyl)amino]ethyl][2-(triphenylmethoxy)ethyl]-, 1,1-dimethylethyl ester	$CHCl_3$	254(4.45),294(3.65), 559(3.68),601(3.64)	18-1435-83
$C_{42}H_{42}O_{17}$			
δ-Naphthocyclinone methyl ester 6-monoacetate	$CHCl_3$	247(4.54),267s(--), 363(4.03)	5-0471-83
δ-Naphthocyclinone methyl ester 9-monoacetate	$CHCl_3$	247(4.34),272s(--), 336s(--),351(3.82), 361(3.83),378s(--)	5-0471-83
$C_{42}H_{45}N_5O_5Os$			
Osmium, carbonyl[dimethyl 7,12-diethyl-3,8,13,17-tetramethyl-21H,23H-porphine-2,18-dipropanoato(2-)-N^{21},N^{22}-N^{23},N^{24}](pyridine)-, (OC-6-26)-	CH_2Cl_2	393(5.39),509(4.08), 538(4.26)	5-2164-83

Compound	Solvent	λ_{max}(log ϵ)	Ref.
$C_{42}H_{46}N_4O_4$			
Voafrine B	MeOH	223(4.27),298(4.27), 328(4.38)	33-2525-83
$C_{42}H_{46}O_9$			
Kuwanon P, octamethyl ether	EtOH	218s(4.55),283s(4.23), 297s(4.28),304(4.29), 330(4.28)	94-2936-83
	EtOH-$AlCl_3$	220s(4.55),283s(4.23), 297s(4.29),304(4.30), 330(4.29)	94-2936-83
$C_{42}H_{49}FeN_4$			
Iron, [2,3,7,8,12,13,17,18-octaethyl-21H,23H-porphinato(2-)-N^{21},N^{22},N^{23}-N^{24}]phenyl-, (SP-5-31)-	benzene	368(4.78),392(5.11), 407s(--),515(4.10), 555(4.32)	101-0065-83M
$C_{42}H_{49}GaN_4$			
Gallium, [2,3,7,8,12,13,17,18-octaethyl-21H,23H-porphinato(2-)-N^{21},N^{22}-N^{23},N^{24}]phenyl-, (SP-5-31)-	benzene	348(4.13),404s(4.18), 423(4.81),507(3.15), 547(3.69),587(3.56)	101-0273-83M
$C_{42}H_{50}N_4O_{11}$			
20,23,26,29,32,35,38,41,44-Nonaoxa-2,8,9,15-tetraazapentacyclo[43.2-2.$2^{4,7}$.$2^{10,13}$.$2^{16,19}$]pentapentaconta-4,6,8,10,12,16,18,45,47,48,50,52,54-tridecaene-3,14-dione	o-$C_6H_4Cl_2$	340(4.42)	18-1700-83
$C_{42}H_{50}N_6O_{10}$			
Bouvardin, 5-(N-methyl-L-tyrosine)-, acetate	EtOH	274(3.53),278(3.53), 282(3.40)	94-1424-83
$C_{42}H_{52}O_{11}$			
Manassantin A	EtOH	235(4.62),280(4.16)	88-4947-83
$C_{42}H_{54}O_6$			
Bispuupehenone	EtOH	230(4.45),284(3.98), 295(4.03),341(3.85), 353s(3.77)	33-1672-83
$C_{42}H_{56}O_2$			
β,β-Caroten-10-ol, 9,15,15',19-tetradehydro-9,10-dihydro-, acetate	hexane	232(4.19),387(4.67), 406(4.60)	33-1148-83
$C_{42}H_{58}Cr_2O_6P_2$			
Chromium, [μ-[bis[(1,2,3,4,5,6-η)-2,4,6-tris(1,1-dimethylethyl)phenyl]diphosphene]hexacarbonyldi-	CH_2Cl_2	328(4.48),416(3.54), 520(2.96)	88-4855-83
$C_{42}H_{58}N_2O_{12}$			
Kijanolide, 17-O-[2,3,4,6-tetradeoxy-4-[(methoxycarbonyl)amino]-3-C-methyl-3-nitro-β-D-xylo-hexopyranosyl]-	MeOH	204(4.44),240(4.02), 266(3.91),276(3.88)	39-1497-83C
$C_{42}H_{58}O_2$			
β,β-Caroten-10-ol, 9,19-didehydro-9,10-dihydro-, acetate	hexane	232(4.04),295(3.92), 385s(4.74),402(4.85), 422(4.80)	33-1148-83
β,β-Caroten-19-ol, acetate (9E/Z)-	hexane	267(4.25),338(4.29), 443(4.88),465s(4.79)	33-1148-83

Compound	Solvent	λ_{max}(log ε)	Ref.
$C_{42}H_{58}O_7$			
Urs-12-en-28-oic acid, 2-acetoxy-3-[[3-(4-hydroxyphenyl)-1-oxo-2-propenyl]oxy]-, methyl ester, [2α,3β(E)]-	EtOH	239(4.05),316(4.37)	102-2559-83
	EtOH-base	245(3.90),312(3.80), 372(4.49)	102-2559-83
$C_{42}H_{59}ClO_7$			
19-Norlanost-5-en-7-one, 11-acetoxy-3-[(4-chlorobenzoyl)oxy]-9-methyl-24,25-[(1-methylethylidene)bis(oxy)]-, (3β,9β,10α,11α,24R)-	MeOH	206(4.23),240(4.34)	95-1155-83
$C_{42}H_{61}N_3O_6$			
Propanamide, 3,3'-[[2-(2,5-dihydro-2,5-dioxo-3,4-diphenyl-1H-pyrrol-1-yl)-2-methyl-1,3-propanediyl]bis(oxy)]-bis[N-heptyl-N-methyl-	EtOH	272(3.95),360(3.61)	33-1078-83
$C_{42}H_{62}O_{10}$			
β-D-Glucopyranoside, (3 ,5Z,7E)-9,10-secoergosta-5,7,10(19),22-tetraen-3-yl, 2,3,4,6-tetraacetate	n.s.g.	265(4.27)	33-2093-83
$C_{42}H_{62}O_{16}$			
Glycyrrhizic acid	n.s.g.	248(4.06)	162-0647-83
$C_{42}H_{64}Co_2N_{10}O_8$			
Cobalt, bis[μ-[8-methyl-2,3,13,14-pentadecanetetronetetraoximato(3-)]]-bis(pyridine)di-	EtOH	456(3.11)	5-0181-83
isomer	EtOH	460(3.12)	5-0181-83
$C_{42}H_{64}NO_2$			
Pyrrolidinium, 1-[6-[7-(2-carboxyethyl)cholesta-4,6-dien-3-ylidene]-3-methyl-2,4-heptadienylidene]-, (E,E,E)-, perchlorate	$CHCl_3$	541(4.64)	35-4033-83
	$+Et_3N$	545(--)	35-4033-83
$C_{42}H_{64}O_8$			
15-Dehydro-PGB_1 dimer I	MeOH	238(4.29),299(4.28)	94-0557-83
15-Dehydro-PGB_1 dimer II	MeOH	238(4.42)	94-0557-83
$C_{42}H_{65}N_7O_9$			
Rhizonin A	MeOH	206(4.52)	77-0047-83
$C_{42}H_{68}O_{14}$			
Gratioside	MeOH	295(1.84)	162-0652-83
$C_{42}H_{72}O_{16}$			
Lankamycin	n.s.g.	289(1.50)	162-0770-83

Compound	Solvent	$\lambda_{max}(\log \epsilon)$	Ref.
$C_{43}H_{31}N_5O$			
8H-Benzo[de]naphth[1',2':4,5]imidazo-[2,1-a]isoquinolin-8-one, 11-[5-[4-(dimethylamino)phenyl]-4,5-dihydro-1-(2-naphthalenyl)-1H-pyrazol-3-yl]-	toluene	525(4.44)	103-0214-83
$C_{43}H_{34}O_3$			
4,8-Epoxyazulen-5-ol, 1-(diphenylmethylene)-1,3a,4,5,8,8a-hexahydro-4,6,8-triphenyl-, acetate, (3aα,4β,5α,8β,8aα)-	MeCN	239(4.28),264(4.18), 271(4.16),296(4.28)	4-1621-83
$C_{43}H_{38}Ge_3$			
Trigermane, 2-methyl-1,1,1,2,3,3,3-heptaphenyl-	C_6H_{12}	250(4.56)	101-0149-83G
$C_{43}H_{38}N_2O_7$			
Acetamide, N-[2-[acetyl(4-acetoxy-9,10-dihydro-9,10-dioxo-1-anthracenyl)amino]ethyl]-N-[2-(triphenylmethoxy)ethyl]-	$CHCl_3$	320(3.97),348(4.04), 408(3.73)	18-1435-83
$C_{43}H_{44}N_4O_{12}$			
D-Glucitol, 1,1-bis(5-acetoxy-3-methyl-1-phenyl-1H-pyrazol-4-yl)-1-deoxy-4,6-O-(phenylmethylene)-, 2,3,5-triacetate	MeOH	255(4.34)	111-0481-83
$C_{43}H_{46}N_4O_5$			
Robtusamine, 2',16,16',17-tetradehydro-2',16-dideoxy-	MeOH	228(4.45),275(4.20), 314(3.92)	44-0381-83
$C_{43}H_{46}O_{22}$			
Glucofrangulin A, octaacetate	n.s.g.	212(4.57),264(4.56), 360(4.26)	162-0637-83
$C_{43}H_{44}N_3$			
3H-Indolium, 2-[7-(1,3-dihydro-1,3,3-trimethyl-2H-indol-2-ylidene)-4-[3-(1,3-dihydro-1,3,3-trimethyl-2H-indol-2-ylidene)-1-propenyl]-1,3,5-heptatrienyl]-1,3,3-trimethyl-, tetrafluoroborate	benzene	700(4.88),745(4.83), 820(5.16)	24-1982-83
	MeOH	755(5.20)	24-1982-83
$C_{43}H_{48}N_4O_6$			
Robtusamine, 16,17-didehydro-16-deoxy-	MeOH	228(4.52),263(4.24), 312(3.98)	44-0381-83
$C_{43}H_{50}N_4O_7$			
Vobtusamine	MeOH	228(4.48),263(4.14), 280(3.93),290(3.85), 310(3.46)	44-0381-83
16-iso-	MeOH	228(4.45),263(4.08), 280(3.93),289(3.84), 308(3.48)	44-0381-83
Vobtusine, 16-hydroxy-	MeOH	222(4.46),263(4.13), 290(3.75)	44-0381-83
$C_{43}H_{51}FeN_4$			
Iron, (4-methylphenyl)[2,3,7,8,12,13-	benzene	365(4.75),391(5.06),	101-0065-83M

Compound	Solvent	λ_{max}(log ϵ)	Ref.
17,18-octaethyl-21H,23H-porphinato-(2-)-N^{21},N^{22},N^{23},N^{24}]- (cont.)		406s(--),514(4.06), 552(4.32)	101-0065-83M
$C_{43}H_{51}GaN_4$			
Gallium, (4-methylphenyl)[2,3,7,8,12-13,17,18-octaethyl-21H,23H-porphinato(2-)-N^{21},N^{22},N^{23},N^{24}]-	benzene	350(3.93),403s(4.03), 424(4.64),505(2.78), 546(3.51),582(3.32)	101-0273-83M
$C_{43}H_{51}NO_{10}$			
Phorbol, 12-O-[N-(9-fluorenylmethyloxycarbonyl)-L-leucyl]-, 13-acetate	MeOH	220s(4.32),228(4.04), 255s(4.29),261s(4.32), 265(4.33),275s(4.13), 288(3.70),300(3.79)	64-1015-83B
$C_{43}H_{56}O_4$			
Phenol, 2,2'-methylenebis[4-(1,1-dimethylethyl)-6-[[5-(1,1-dimethylethyl)-2-hydroxyphenyl]methyl]-	EtOH	281s(4.01),286(4.01)	126-1363-83
$C_{43}H_{58}N_4$			
Quinolinium, 4-[3-[2-(1-hexadecyl-3-methyl-1H-benzimidazolium-2-yl)-1-methyl-4(1H)-pyridinylidene]-1-propenyl]-1-methyl-, diiodide	EtOH	676(5.09)	4-0023-83
$C_{43}H_{60}N_2O_{12}$			
Kijanolide, 26-O-methyl-17-O-[2,3,4,6-tetradeoxy-4-[(methoxycarbonyl)amino]-3-C-methyl-3-nitro-β-D-xylo-hexopyranosyl]-	CF_3CH_2OH	199(4.61),254(3.97)	39-1497-83C
$C_{43}H_{60}N_4O_6$			
21H-Biline-1,19-dicarboxylic acid, 7,13,18-triethyl-5,15,22,24-tetrahydro-3-(3-methoxy-3-oxopropyl)-2,8,12,17-tetramethyl-, bis(1,1-dimethylethyl) ester, monohydrobromide	CH_2Cl_2	504(4.81)	78-1841-83
$C_{43}H_{60}N_8O_{11}$			
Viridogrisein II	H_2O	305(3.56),349(3.77)	158-83-88
$C_{43}H_{64}NO_4$			
Pyrrolidinium, 2-carboxy-1-[6-[7-(2-carboxyethyl)cholesta-4,6-dien-3-ylidene]-3-methyl-2,4-heptadienylidene]-, [S-(all-E)]-, perchlorate	$CHCl_3$	565(4.61)	35-4033-83
	+ Et_3N	542(--)	35-4033-83

Compound	Solvent	λ max (log ε)	Ref.
$C_{44}H_{10}Cl_{20}N_4$			
21H,23H-Porphine, 5,10,15,20-tetrakis(pentachlorophenyl)-	$CHCl_3$	421(5.55),514(4.30), 543s(3.39),591(3.81), 651(2.82)	112-0275-83
$C_{44}H_{10}F_{20}N_4$			
21H,23H-Porphine, 5,10,15,20-tetrakis(pentafluorophenyl)-	toluene	417(4.44),509(4.32), 537(3.41),587(3.81), 640(2.95)	112-0275-83
$C_{44}H_{22}Cl_8N_4$			
21H,23H-Porphine, 5,10,15,20-tetrakis(2,6-dichlorophenyl)-	$CHCl_3$	419(5.56),513(4.29), 543(3.36),589(3.79), 643(2.60)	112-0275-83
$C_{44}H_{26}N_2O_2S_2$			
Triphenodithiazine, 6,13-bis(4-benzoylphenyl)-	H_2SO_4	326(4.56)	18-1482-83
	dioxan	560(--)	18-1482-83
$C_{44}H_{26}N_6O_4$			
3,5:14,16-Dimetheno-5H,16H-dicyclohepta[e,k]diimidazo[1,5-a:1',5'-g][1,7]-diazacyclododecine-10,21-dicarboxylic acid, 23,24-dicyano-1,12-diphenyl-, dimethyl ester	$CHCl_3$	271(4.85),319(4.72), 342(4.70),365s(4.59), 422(4.38),448(4.52), 520s(3.79),543(3.83), 580s(3.70),635s(3.20)	18-3703-83
$C_{44}H_{26}O_{12}$			
[2,7':2',2":7",2"'-Quaternaphthalene]-1,1',1",1"',4,4',4",4"'-octone, 8,8',8",8"'-tetrahydroxy-6,6',6",6"'-tetramethyl-	EtOH	219(4.83),245s(4.54), 435(4.13)	5-0433-83
	dioxan	433(4.20)	5-0433-83
$C_{44}H_{28}CdN_4$			
Cadmium, [5,10,15,20-tetraphenyl-21H,23H-porphinato(2-)-N^{21},N^{22}-N^{23},N^{24}]-, (SP-4-1)-	CH_2Cl_2	430.5(5.18),564(4.05), 604.5(3.77)	94-2110-83
THF adduct	CH_2Cl_2	433(5.38),569(3.97), 610(3.83)	94-2110-83
bromide	DMSO	439(5.40),577(4.00), 621(3.95)	94-2110-83
chloride	DMSO	440.5(5.47),581(4.03), 624(4.02)	94-2110-83
MeCN adduct	CH_2Cl_2	433(5.39),570(3.95), 612(3.78)	94-2110-83
pyridine adduct	CH_2Cl_2	436(5.44),575(3.95), 677(3.83)	94-2110-83
$C_{44}H_{28}ClGaN_4$			
Gallium, chloro[5,10,15,20-tetraphenyl-21H,23H-porphinato(2-)-N^{21},N^{22},N^{23}-N^{24}]-, (SP-5-12)-	MeOH	395(3.65),414(4.80), 548(3.36),587(2.90), 621(2.48)	101-0273-83M
$C_{44}H_{28}Cl_2N_6O_2S_2$			
Benzamide, N,N'-[(3,10-dichloro-6,13-triphenodithiazinediyl)bis(imino-4,1-phenylene)]bis-	dioxan	590(4.59)	18-1482-83
	H_2SO_4	337(4.63)	18-1482-83
$C_{44}H_{28}Cl_3N_4P$			
Phosphorus, dichloro[5,10,15,20-tetraphenyl-21H,23H-porphinato(2-)-N^{21}-N^{22},N^{23},N^{24}]-, chloride	MeCN	437(5.38),567(4.02), 610(3.68)	125-1858-83

Compound	Solvent	λ_{max}(log ϵ)	Ref.
$C_{44}H_{28}N_4O_2$			
14H-Acenaphth[4',3':4,5]imidazo[2,1-a]-benz[de]isoquinolin-14-one, 11-[4,5-dihydro-5-(4-methoxyphenyl)-1-(2-naphthalenyl)-1H-pyrazol-3-yl]-	toluene	530(4.45)	103-0214-83
$C_{44}H_{30}N_2$			
2,6-Naphthyridine, 1,3,4,5,7,8-hexa-phenyl-	$CHCl_3$	370(4.4)	4-0971-83
2,7-Naphthyridine, 1,3,4,5,6,8-hexa-phenyl-	$CHCl_3$	358(4.1)	4-0971-83
$C_{44}H_{30}N_4$			
21H,23H-Porphine, 5,10,15,20-tetra-phenyl-	toluene	419(5.67),515(4.29), 549(3.91),591(3.74), 647(3.56)	112-0275-83
$C_{44}H_{30}N_4O_4$			
Phenol, 2,2',2",2"'-(21H,23H-porphine-5,10,15,20-tetrayl)tetrakis-	pyridine	423(5.39),516(4.04), 550(3.63),594(3.53), 651(3.36)	103-1082-83
Phenol, 3,3',3",3"'-(21H,23H-porphine-5,10,15,20-tetrayl)tetrakis-	pyridine	423(5.65),517(4.26), 552(3.87),593(3.72), 649(3.57)	103-1082-83
Phenol, 4,4',4",4"'-(21H,23H-porphine-5,10,15,20-tetrayl)tetrakis-	pyridine	426(5.64),521(4.16), 561(4.11),598(3.67), 656(3.81)	103-1082-83
$C_{44}H_{30}N_4O_8$			
1,2-Benzenediol, 3,3',3",3"'-(21H,23H-porphine-5,10,15,20-tetrayl)tetrakis-	pyridine	426(5.60),518(4.42), 553(4.09),593(3.99), 654(3.91)	103-1082-83
1,2-Benzenediol, 4,4',4",4"'-(21H,23H-porphine-5,10,15,20-tetrayl)tetrakis-	pyridine	431(5.42),523(4.31), 564(4.23),595(3.99), 657(3.92)	103-1082-83
$C_{44}H_{30}N_6O_8$			
3,5:14,16-Dinitrilo-5H,16H-dicyclohep-ta[e,k]diimidazo[1,5-a:1',5'-g][1,7]-diazacyclododecine-10,11,21,22-tetracarboxylic acid, 1,12-di-phenyl-, tetramethyl ester	EtOH	258(4.73),293(4.48), 320(4.48),410s(4.15), 424(4.28),427s(4.27), 550(3.60)	18-3703-83
$C_{44}H_{32}N_4O_2$			
Perylenodyestuff 1	DMF	562(4.30),605(4.36)	24-3524-83
$C_{44}H_{32}N_4O_4$			
3,5:14,16-Dimetheno-5H,16H-dicyclohep-ta[e,k]diimidazo[1,5-a:1',5'-g][1,7]-diazacyclododecine-10,21-dicarboxylic acid, 1,12-bis(4-methylphenyl)-, dimethyl ester	EtOH	224(4.83),270(4.93), 314(4.79),324(4.78), 355s(4.54),415(4.25), 438(4.25),511(3.69), 542(3.68),583s(3.50), 635s(3.14)	18-3703-83
$C_{44}H_{34}N_6O_4$			
3,5:14,16-Dinitrilo-5H,16H-dicyclohep-ta[e,k]diimidazo[1,5-a:1',5'-g][1,7]-diazacyclododecine-10,21-dicarboxylic acid, 1,12-bis(4-methylphenyl)-, diethyl ester	EtOH	243(4.69),262(4.64), 303(4.63),328(4.61), 422s(4.48),441(4.60), 547(3.75)	18-3703-83

Compound	Solvent	λ_{max}(log ε)	Ref.
$C_{44}H_{37}ClO_2P_2Pt$			
Platinum, chloro[(phenylmethoxy)carbonyl]bis(triphenylphosphine)-, (SP-4-3)-	CH_2Cl_2	257(4.36),268(4.23), 275(4.08)	101-0119-83A
$C_{44}H_{39}F_3N_2O_7$			
Carbamic acid, [2-[(9,10-dihydro-4-hydroxy-9,10-dioxo-1-anthracenyl)(trifluoroacetyl)amino]ethyl][2-(triphenylmethoxy)ethyl]-, 1,1-dimethylethyl ester	$CHCl_3$	257(4.59),338(3.60), 402(3.82)	18-1435-83
$C_{44}H_{42}N_2O_7$			
Carbamic acid, [2-[(4-acetoxy-9,10-dihydro-9,10-dioxo-1-anthracenyl)amino]ethyl][2-(triphenylmethoxy)ethyl]-, 1,1-dimethylethyl ester	$CHCl_3$	250(4.66),278(4.19), 319(4.03),511(4.01)	18-1435-83
$C_{44}H_{48}N_4O_6Si$			
Benzamide, N-[7-[2-O-[(1,1-dimethylethyl)dimethylsilyl]-5-O-[(4-methoxyphenyl)diphenylmethyl]-β-D-ribofuranosyl-7H-pyrrolo[2,3-d]pyrimidin-4-yl]-	MeOH	277(3.99),301(3.97)	5-1169-83
Benzamide, N-[7-[3-O-[(1,1-dimethyl ethyl)dimethylsilyl]-5-O-[(4-methoxyphenyl)diphenylmethyl]-β-D-ribofuranosyl-7H-pyrrolo[2,3-d]pyrimidin-4-yl]-	MeOH	277(3.99),301(3.97)	5-1169-83
$C_{48}H_{50}Cl_2N_4O_2$			
Alcuronium dichloride	MeOH	292(4.63)	162-0035-83
$C_{44}H_{50}Ge_4$			
Tetragermane, 2,2,3,3-tetraethyl-1,1,1,4,4,4-hexaphenyl-	C_6H_{12}	256s(4.60)	101-0149-83G
$C_{44}H_{50}I_2N_4O_2$			
Alcuronium diiodide	MeOH	291(4.60)	162-0035-83
$C_{44}H_{54}N_{12}O_8$			
21H,23H-Porphine-2,3,7,8,12,13,17,18-octacarboxamide, N,N,N',N',N",N"-...N"""',N"""'-hexadecamethyl-	H_2O	416(5.42),513(4.24), 547(3.83),585(3.87), 636(3.46)	44-4399-83
	KOH	252s(3.83),434(5.33), 567(4.15),603s(3.79)	44-4399-83
	$CHCl_3$	418(5.38),510(4.23), 543(3.78),584(3.78), 636(3.30)	44-4399-83
$C_{44}H_{56}O_8$			
Maprounic acid, 2α-hydroxy-, 2,3-bis(4-hydroxybenzoate)	MeOH	256(4.46)	100-0537-83
$C_{44}H_{58}N_{12}O_8$			
3H-Phenoxazine-1,9-dicarboxamide, 2-amino-N,N'-bis[3-[[5-[[[3-(dimethylamino)propyl]amino]carbonyl]-1-methyl-1H-pyrrol-3-yl]amino]-3-oxopropyl]-4,6-dimethyl-3-oxo-, dihydrochloride	pH 6	235(4.67),275s(4.46), 442s(4.14)	104-1380-83

Compound	Solvent	λ_{max}(log ε)	Ref.
$C_{44}H_{60}O_5$			
Pregn-5-ene-20-carboxylic acid, 3-acetoxy-, (17α)-3-oxo-19-norpregn-4-en-20-yn-17-yl ester, (3β,20S)-	C_6H_{12}	233(4.25)	13-0327-83A
$C_{44}H_{60}O_8$			
Urs-12-en-28-oic acid, 2-acetoxy-3-[[3-(4-acetoxyphenyl)-1-oxo-2-propenyl]oxy]-, methyl ester, [2α,3β(E)]-	EtOH	225(4.38),281(4.35)	102-2559-83
$C_{44}H_{62}N_2O_{12}$			
Kijanolide, 32-O-methyl-17-O-[2,3,4,6-tetradeoxy-4-[(methoxycarbonyl)methylamino]-3-C-methyl-3-nitro-β-D-xylo-hexopyranosyl]-	CF_3CH_2OH	245s(4.01),262(4.04)	39-1497-83C
$C_{44}H_{62}N_4$			
21H,23H-Porphine, 2,3,7,8,12,13,17,18-octakis(1-methylethyl)-	CH_2Cl_2	402(5.16),500(4.16), 533(4.00),568(3.86), 622(3.59)	22-0317-83
$C_{44}H_{62}N_4O_8$			
21H,23H-Porphine-2,3,7,8,12,13,17,18-octapropanol	pyridine	402(5.70),499(4.46), 532(4.32),569(4.20), 621(3.98)	12-1639-83
$C_{44}H_{62}Si_4$			
Disilacyclopropane, 3-[bis(trimethylsilyl)methylene]-1,1,2,2-tetrakis-(2,4,6-trimethylphenyl)-	n.s.g.	288(4.35),364(3.95), 415(3.75)	157-0174-83
$C_{44}H_{65}NO_{13}$			
Bafilomycin B_1	MeOH	248(4.54),285s(4.27), 355(4.10)	88-5193-83
$C_{44}H_{73}NO_{17}$			
DOMM	EtOH	283(4.35)	158-0376-83
$C_{44}H_{74}N_2O$			
2,5-Cyclohexadien-1-one, 4-[[4-(dihexadecylamino)phenyl]imino]-	C_6H_{12}	275(4.24),355(3.76), 567(4.26)	5-0802-83

Compound	Solvent	λ_{max}(log ϵ)	Ref.
$C_{45}H_{31}FeN_4$			
Iron, methyl[5,10,15,20-tetraphenyl-21H,23H-porphinato(2-)-N^{21},N^{22},N^{23}-N^{24}]-, (SP-5-31)-	benzene	390(4.82),413(5.16), 425s(--),518(4.08), 546(3.86)	101-0065-83M
$C_{45}H_{31}GaN_4$			
Gallium, methyl[5,10,15,20-tetraphenyl-21H,23H-porphinato(2-)-N^{21},N^{22},N^{23}-N^{24}]-	benzene	337(4.55),419s(4.56), 436(5.51),531(3.64), 570(3.69),612(3.86)	101-0273-83M
$C_{45}H_{31}N_5O$			
14H-Acenaphth[4',3':4,5]imidazo[2,1-a]-benz[de]isoquinolin-14-one, 11-[5-[4-(dimethylamino)phenyl]-4,5-di-hydro-1-(2-naphthalenyl)-1H-pyra-zol-3-yl]-	toluene	525(4.47)	103-0214-83
$C_{45}H_{32}N_4SZn$			
Zinc, mercapto(21-methyl-5,10,15,20-tetraphenyl-21H,23H-porphinato-N^{21},N^{22},N^{23},N^{24})-, (SP-5-12)-	CH_2Cl_2	438(5.36),450s(--), 527s(--),565(3.89), 616(4.08),661(3.84)	18-2055-83
$C_{45}H_{45}BO_{16}$			
δ-Naphthocyclinone dimethyl ester, phenylboric acid ester	MeOH	215(4.49),240(4.13), 272(4.22),282s(--), 347(3.72),385(3.79)	5-0471-83
	$CHCl_3$	274(4.27),286s(--), 355(3.82),374(3.81)	5-0471-83
$C_{45}H_{53}BrN_4O_8$			
C'-Norvincaleukoblastine, 17-bromo-3',4'-didehydro-4'-deoxy-	EtOH	222(4.58),275(4.10), 285(4.04),295(3.93), 319(3.46)	111-0419-83
$C_{45}H_{53}FeN_6O$			
Iron(1+), (2,3,7,8,12,13,17,18-octa-ethyl-21H-5-oxaporphinato-N^{21},N^{22}-N^{23},N^{24})bis(pyridine)-, chloride, (OC-6-35)-	THF	330(4.48),375s(4.53), 394(4.69),516(3.85), 549(3.83),620s(4.08), 665(4.72)	88-0995-83
	pyridine	323(4.33),386(4.76), 425s(4.30),492(3.95), 526(4.18),608s(4.16), 652(4.66)	88-0995-83
$C_{45}H_{53}NO_{11}$			
Phorbol, 12-O-[N-(9-fluorenylmethyl-oxycarbonyl)-L-leucyl]-, 13,20-di-acetate	MeOH	220s(4.30),227(4.00), 256s(4.27),261s(4.30), 264(4.31),276(4.08), 288(3.65),299(3.73)	64-1015-83B
$C_{45}H_{54}O_{20}$			
Urceolatoside A hexaacetate	MeOH	228(4.44)	102-1977-83
$C_{45}H_{60}N_4O_8$			
22H-Biline-8,12-dipropanoic acid, 1,19-bis[(1,1-dimethylethoxy)carbo-nyl]-3,17-diethyl-10,23-dihydro-2,7,13,18-tetramethyl-, dimethyl ester, hydrobromide	CH_2Cl_2	460(4.79),530(4.63)	78-1841-83

Compound	Solvent	λ_{max}(log ε)	Ref.
$C_{45}H_{64}O_5$			
5β-Cholan-24-oic acid, 3α-(formyloxy)-, (17α)-3-oxo-19-norpregn-4-en-20-yn-17-yl ester	EtOH	242(4.68)	13-0327-83A
$C_{45}H_{66}O_{16}$			
Bryostatin 2	MeOH	230(4.56),261(4.55)	100-0528-83
$C_{45}H_{68}O_4$			
Furan, 2,2',2",2"'-methanetetrayltetrakis[5-heptyl-	EtOH	230(4.22)	103-0360-83
$C_{45}H_{79}NO_7Si_3$			
Prosta-5,7,14-trien-12-ol, 1,4,9-tris-[[(1,1-dimethylethyl)dimethylsilyl]-oxy]-, 4-nitrobenzoate, (4R,5E,7Z-9β,12α,14Z)-	MeOH	247(4.57)	94-1440-83
$C_{46}H_{32}N_4O_8$			
3,5:14,16-Dimetheno-5H,16H-dicyclohepta[e,k]diimidazo[1,5-a:1',5'-g][1,7]-diazacyclododecine-10,11,21,22-tetracarboxylic acid, 1,12-diphenyl-, tetramethyl ester	EtOH	268(4.90),295(4.65), 310s(4.58),360(4.53), 404(4.17),423(4.13), 523s(3.38),550(3.42), 590s(3.31),640s(2.94)	18-3703-83
$C_{46}H_{33}GaN_4$			
Gallium, ethyl[5,10,15,20-tetraphenyl-21H,23H-porphinato(2-)-N^{21},N^{22},N^{23}-N^{24}]-, (SP-5-31)-	benzene	346(4.50),423s(4.66), 441(5.32),535(3.52), 573(3.98),618(3.92)	101-0273-83M
$C_{46}H_{34}N_6O_8$			
3,5:14,16-Dinitrilo-5H,16H-dicyclohepta[e,k]diimidazo[1,5-a:1',5'-g][1,7]-diazacyclododecine-10,11,21,22-tetracarboxylic acid, 1,12-bis(4-methylphenyl)-, tetramethyl ester	EtOH	257(4.73),298(4.54), 320(4.49),410s(4.19), 424s(4.31),430(4.32), 552(3.63)	18-3703-83
$C_{46}H_{36}N_2O_4$			
3-Pyridinecarboxylic acid, 1,4-dihydro-1-(phenylmethyl)-, [1,1'-binaphthalene]-2,2'-diyl ester, (S)-	EtOH	362(4.24)	18-3672-83
$C_{46}H_{38}N_4O_2$			
3-Pyridinecarboxamide, N,N'-[1,1'-binaphthalene]-2,2'-diylbis[1,4-dihydro-1-(phenylmethyl)-, (S)-	MeCN	363(4.32)	18-3672-83
$C_{46}H_{46}O_6P_2$			
Phosphonium, 1,3-butadiene-1,4-diylbis[tris(4-methoxyphenyl)-, dibromide, (E,E)-	EtOH	255(4.93)	118-0374-83
$C_{46}H_{46}P_2$			
Phosphonium, 1,3-butadiene-1,4-diylbis[tris(4-methylphenyl)-, dibromide, (E,E)-	EtOH	265(4.51)	118-0374-83
$C_{46}H_{58}N_4O_{14}$			
3,6,9,12-Tetraoxabicyclo[12.3.1]octadeca-1(18),14,16-trien-18-ol, 16,16'-[1,2-ethanediylbis(oxy-2,1-ethanedi-	$CHCl_3$	358(4.72)	18-3253-83

Compound	Solvent	λ_{max}(log ϵ)	Ref.
yloxy-4,1-phenyleneazo)]bis- (cont.)			18-3253-83
$C_{46}H_{62}N_2O_{14}$			
Kijanolide, 17-O-[2,3,4,6-tetradeoxy-4-[(methoxycarbonyl)amino]-3-C-methyl-3-nitro-β-D-xylo-hexopyranosyl]-, 9,32-diacetate	CF_3CH_2OH	199(4.61),242s(3.84), 262(3.90),276s(3.85)	39-1497-83C
$C_{46}H_{64}O_7$			
5β-Cholan-24-oic acid, 3,12-bis(formyloxy)-, (17α)-3-oxo-19-norpregn-4-en-20-yn-17-yl ester, (12α)-	C_6H_{12}	236(4.21)	13-0327-83A
$C_{46}H_{64}O_{17}$			
Bryostatin 3	MeOH	230(4.56),261(4.54)	44-5354-83
$C_{46}H_{77}NO_{17}$			
Tylosin	EtOH	282(4.46)	158-0376-83
$C_{46}H_{79}NO_{16}$			
Tylosin, 20-deoxo-	EtOH	282(4.34)	77-1166-83
$C_{47}H_{31}Cl_4FeN_5O$			
Iron, (2-nitrosopropane-N)[5,10,15,20-tetrakis(4-chlorophenyl)-21H,23H-porphinato(2-)-N^{21},N^{22},N^{23},N^{24}]-, (SP-5-12)-	benzene	424(5.37),535(4.04)	35-0455-83
$C_{47}H_{35}FeN_5O$			
Iron, (2-nitrosopropane-N)[5,10,15,20-tetraphenyl-21H,23H-porphinato(2-)-N^{21},N^{22},N^{23},N^{24}]-, (SP-5-12)-	benzene	413(5.34),525(4.04)	35-0455-83
$C_{47}H_{50}O_{22}$			
Trideca-O-methyl-α-pedunculagin	MeOH	222(4.84),247s(4.63), 288s(4.01)	39-1765-83C
$C_{47}H_{56}O_{25}$			
Hydropiperoside octaacetate	MeOH	214(4.07),278(4.37)	102-0549-83
$C_{47}H_{58}N_4O_9$			
Roseadine	MeOH	214(4.69),225(4.68), 258(4.08),286(4.08), 294(4.09),326(3.93)	100-0517-83
$C_{47}H_{58}Si_3$			
Disilacyclopropane, 3-[phenyl(trimethylsilyl)methylene]-1,1,2,2-tetrakis(2,4,6-trimethylphenyl)-	n.s.g.	284(4.26),316(4.13), 382(3.71)	157-0174-83
$C_{47}H_{59}N_4O_8$			
C-Norvincaleukoblastinium, 3',4'-didehydro-4'-deoxy-6'-ethyl-, iodide	EtOH	218(--),270(--), 284(--),291(--), 315(--)	111-0419-83
$C_{47}H_{60}O_6$			
19-Norpregn-4-en-20-yn-3-one, 17,17'-[(2,2-dimethyl-1,5-dioxo-1,5-pentanediyl)bis(oxy)]bis-, (17α)-(17'α)-	MeOH	248(4.81)	13-0255-83A

Compound	Solvent	λ_{max}(log ε)	Ref.
$C_{47}H_{62}O_8$			
Urs-12-en-29-oic acid, 2,3-bis[(4-methoxybenzoyl)oxy]-, methyl ester, (2α,3β)-	MeOH	256(4.51)	100-0537-83
$C_{47}H_{64}N_2O_{14}$			
Kijanolide, 26-O-methyl-17-O-[2,3,4,6-tetradeoxy-4-[(methoxycarbonyl)amino]-3-C-methyl-3-nitro-β-D-xylo-hexopyranosyl]-, 9,32-diacetate	CF_3CH_2OH	200(4.62),254(3.99)	39-1497-83C
$C_{47}H_{64}O_9$			
5β-Cholan-24-oic acid, 3,7,12-triacetoxy-, (17α)-3-oxo-19-norpregn-4-en-20-yn-17-yl ester, (3α,5β,7α,12α)-	MeOH	242(4.24)	13-0327-83A
$C_{47}H_{68}N_2O_9$			
A Jeffamine derivative	EtOH	271(3.75),278(3.71)	87-0007-83
$C_{47}H_{76}O_{17}$			
Zizynummin	MeOH	none above 200 nm	102-1469-83
$C_{47}H_{78}N_2O$			
2,5-Cyclohexadien-1-one, 4-[[4-(dihexadecylamino)phenyl]imino]-2-(2-propenyl)-	C_6H_{12}	275(4.24),355(3.76), 567(4.26)	5-0802-83
$C_{47}H_{82}N_2O_{16}$			
Tylosin, 20-deoxo-20-(methylamino)-	MeOH	284(4.28)	158-1713-83
$C_{47}H_{82}N_2O_{17}$			
Tylosin, 20-deoxo-20-(methoxyamino)-	MeOH	278(4.16)	158-1713-83
$C_{48}H_{24}$			
Kekulene	1,2,4-$C_6H_3Cl_3$	325(5.2),395(4.4)	24-3504-83
$C_{48}H_{30}$			
Heptiptycene	MeCN	272(3.97),282(4.15), 295(4.21)	44-4357-83
$C_{48}H_{30}N_6O_8$			
3,5:14,16-Dimetheno-5H,16H-dicyclohepta[e,k]diimidazo[1,5-a:1',5'-g][1,7]-diazacyclododecine-10,11,21,22-tetracarboxylic acid, 23,24-dicyano-1,12-diphenyl-, tetramethyl ester	$CHCl_3$	267(5.03),309(4.49), 329(4.51),350(4.49), 365s(4.48),414(4.21), 436(4.30),525s(3.77), 550(3.82),585s(3.76), 635s(3.39)	18-3703-83
$C_{48}H_{30}O_8$			
Praecoxin E	MeOH	205(4.88),276(4.34)	94-0333-83
$C_{48}H_{30}O_{30}$			
Praecoxin C	MeOH	209(4.93),302(4.47)	94-0333-83
$C_{48}H_{32}N_4$			
Tetrabenzo[f,l,x,d][1,2,17,18-tetraazacyclohexatriacontine, 5,6,7,8-27,28,29,30-octadehydro-, (?,?,?,?,E,E,E,E,E,E)-	THF	267s(4.59),282(4.70), 355(4.96),410s(4.49), 455s(4.23)	18-1467-83

Compound	Solvent	λ_{max}(log ϵ)	Ref.
$C_{48}H_{36}$			
[2]Paracyclophanehexaene, (Z,E,Z,E,Z,E)-	EtOH	243s(4.29),290s(4.37), 332(4.80)	88-5411-83
(Z,Z,E,Z,Z,E)-	EtOH	342(4.99)	88-5411-83
(Z,Z,Z,Z,Z,Z)-	EtOH	242(4.49),319(4.41)	88-5411-83
$C_{48}H_{36}N_4O_8$			
3,5:14,16-Dimetheno-5H,16H-dicyclohepta[e,k]diimidazo[1,5-a:1',5'-g][1,7]-diazacyclododecine-10,11,21,22-tetracarboxylic acid, 1,12-bis(4-methylphenyl)-, tetramethyl ester	EtOH	269(5.05),305(4.68), 315(4.66),345s(4.55), 407(4.18),423(4.13), 523s(3.63),550(3.66), 590s(3.56),640s(3.19)	18-3703-83
3,5:14,16-Dimetheno-5H,16H-dicyclohepta[e,k]diimidazo[1,5-a:1',5'-g][1,7]-diazacyclododecine-10,11,21,22-tetracarboxylic acid, 1,12-diphenyl-, 23,24-diethyl 10,21-dimethyl ester	$CHCl_3$	271(4.87),320(4.73), 347(4.73),366s(4.60), 425(4.33),450(4.53), 515(3.80),539(3.84), 574s(3.73),625s(3.26)	18-3703-83
$C_{48}H_{37}FeN_4$			
Iron, butyl[5,10,15,20-tetraphenyl-21H,23H-porphinato(2-)-N^{21},$N^{22}N^{23}$-N^{24}]-, (SP-5-31)-	benzene	392(4.88),412(5.14), 434s(--),518(4.03), 552(3.75)	101-0065-83M
$C_{48}H_{37}GaN_4$			
Gallium, butyl[5,10,15,20-tetraphenyl-21H,23H-porphinato(2-)-N^{21},N^{22},N^{23}-N^{24}]-	benzene	347(3.59),424s(3.83), 441(4.36),531(2.70), 573(3.08),617(3.00)	101-0273-83M
$C_{48}H_{38}N_4SZn$			
Zinc, (21-methyl-5,10,15,20-tetraphenyl-21H,23H-porphinato-N^{21},N^{22},$N^{23}N^{24}$)-(1-propanethiolato)-, (SP-5-12)-	CH_2Cl_2	438(5.36),450s(--), 530s(--),567(3.83), 619(4.08),661(3.86)	18-2055-83
$C_{48}H_{38}N_6O_8$			
3,5:14,16-Dinitrilo-5H,16H-dicyclohepta[e,k]diimidazo[1,5-a:1',5'-g][1,7]-diazacyclododecine-10,11,21,22-tetracarboxylic acid, 1,12-diphenyl-, tetraethyl ester	EtOH	258(4.86),293(4.56), 322(4.58),410s(4.35), 425(4.50),428s(4.49), 550(3.75)	18-3703-83
$C_{48}H_{39}FeN_5O_2$			
Iron, (methanol)(2-nitrosopropane-N)-[5,10,15,20-tetraphenyl-21H,23H-porphinato(2-)-N^{21},N^{22},N^{23},N^{24}]-, (OC-6-23)-	benzene	420(5.42),536(4.08)	35-0455-83
$C_{48}H_{40}Ge_3$			
Trigermane, octaphenyl-	C_6H_{12}	250(4.42)	101-0149-83G
$C_{48}H_{40}N_2O_4$			
3-Pyridinecarboxylic acid, 1,4-dihydro-1-(phenylmethyl)-, [1,1'-binaphthalene]-2,2'-diylbis(methylene) ester, (R)-	MeCN	357(4.13)	18-3672-83
$C_{48}H_{41}S_2$			
Thiopyrylium, 2-[3-[3-[3-(4,6-diphenyl-2H-thiopyran-2-ylidene)-1-propenyl]-5,5-dimethyl-2-cyclohexen-1-ylidene]-1-propenyl]-4,6-diphenyl-, perchlorate	CH_2Cl_2	1076(4.96),1223(5.12)	103-1241-83

Compound	Solvent	λ_{max}(log ε)	Ref.
$C_{48}H_{44}N_4$			
1,2,25,26-Tetraazacyclooctatetraconta-2,4,6,8,10,16,18,20,22,24,26,28,30-32,34,40,42,44,46,48-eicosaene-12,14,36,38-tetrayne, 11,16,35,40-tetramethyl-	THF	300s(4.01),420(4.87)	18-1467-83
$C_{48}H_{46}O_{26}$			
Taxillusin decaacetate	EtOH	241s(4.50),263(4.25), 312(3.81)	18-0542-83
$C_{48}H_{50}N_4O_6$			
21H,23H-Porphine-2,18-dipropanoic acid, 8,12-bis(1-hydroxyethyl)-3,7,13,17-tetramethyl-, bis(phenylmethyl) ester	$CHCl_3$	402(5.22),500(4.13), 535(3.94),570(3.77), 624(3.57)	12-1639-83
$C_{48}H_{56}N_2O_{23}$			
Acetamide, N-(2-acetoxyethyl)-N-[2-[[9,10-dihydro-9,10-dioxo-4-[[2,3,6-tri-O-acetyl-4-O-(2,3,4,6-tetra-O-acetyl-β-D-glucopyranosyl)-β-D-gluco-pyranosyl]oxy]-1-anthracenyl]amino]-ethyl]-	EtOH	247(4.64),271(3.99), 316(3.72),509(3.95)	18-1435-83
$C_{48}H_{60}Ge_5$			
Pentagermane, 2,2,3,3,4,4-hexaethyl-1,1,1,3,3,3-hexaphenyl-	C_6H_{12}	269s(4.59)	101-0149-83G
$C_{48}H_{64}Fe$			
Iron(2+), bis[(4,5,6,7,15,16-η)-5,6,11,12,13,14,15,16-octamethyl-tricyclo[8.2.2.2^{4,7}]hexadeca-4,6,10,12,13,15-hexaene]-, bis(hexafluorophosphate)	MeCN	300s(3.9),475(3.8)	157-1214-83
$C_{48}H_{67}N_3O$			
2,5-Cyclohexadien-1-one, 2,6-bis[[3,5-bis(1,1-dimethylethyl)phenyl]amino]-4-[[3,5-bis(1,1-dimethylethyl)phenyl]-imino]-	hexane	271(4.57),414(3.86), 495(3.92)	18-1476-83
$C_{48}H_{72}O_{14}$			
Avermectin B_{1a}	MeOH	237(4.46),243(4.50), 252(4.31)	162-0128-83
$C_{48}H_{74}O_{15}$			
Avermectin B_{2a}	MeOH	237(4.44),243(4.49), 252(4.30)	162-0128-83
$C_{48}H_{78}N_4O_4S$			
Benzenesulfonic acid, 4-[4-[[4-(dihexa-decylamino)phenyl]imino]-4,5-dihydro-2-methyl-5-oxo-1H-imidazol-1-yl]-, sodium salt	EtOH	258(4.28),440(3.94), 526(4.43)	5-0802-83
$C_{48}H_{84}N_2O_{16}$			
Tylosin, 20-deoxo-20-(dimethylamino)-	MeOH	283(4.10)	158-1713-83
Tylosin, 20-deoxo-20-(ethylamino)-	MeOH	284(4.29)	158-1713-83
$C_{49}H_{39}FeN_4$			
Iron, methyl[5,10,15,20-tetrakis(3-	benzene	390(4.88),415(5.17),	101-0065-83M

Compound	Solvent	λ max(log ε)	Ref.
methylphenyl)-21H,23H-porphinato(2-)-N[21],N[22],N[23],N[24]]-, (SP-5-31)- (cont.)		426s(--),518(4.08), 546(3.88)	101-0065-83M
Iron, methyl[5,10,15,20-tetrakis(4-methylphenyl)-21H,23H-porphinato(2-)-N[21],N[22],N[23],N[24]]-, (SP-5-31)-	benzene	390(4.86),414(5.14), 426s(--),518(4.03), 546(3.75)	101-0065-83M
$C_{49}H_{40}N_4SZn$			
Zinc, (2-methyl-2-propanethiolato)(21-methyl-5,10,15,20-tetraphenyl-21H-23H-porphinato-N[21],N[22],N[23],N[24])-, (SP-5-12)-	CH_2Cl_2	438(5.28),450s(--), 528s(--),566(3.80), 615(3.96),661(3.74)	18-2055-83
$C_{49}H_{61}IN_2O_{13}$			
Kijanolide, 17-O-[2,3,4,6-tetradeoxy-4-[(methoxycarbonyl)amino]-3-C-methyl-3-nitro-β-D-xylo-hexopyranosyl]-, 32-(4-iodobenzoate)	CF_3CH_2OH	257(4.40)	39-1497-83C
$C_{49}H_{64}N_4O_{12}$			
22H-Biline-3,8,12,17-tetrapropanoic acid, 1,19-bis[(1,1-dimethylethoxy)-carbonyl]-10,23-dihydro-2,7,13,18-tetramethyl-, tetramethyl ester, dihydrobromide	CH_2Cl_2	460(4.80),530(4.64)	78-1841-83
$C_{49}H_{66}N_2O_{17}$			
Tetracarcin E_1	MeOH	232s(4.23),268(4.01), 278s(3.96)	158-83-103
Tetracarcin E_2	MeOH	232s(4.23),268(4.01), 278s(3.96)	158-83-103
$C_{49}H_{69}ClO_{14}$			
Brevetoxin C	MeOH	208(4.05)	162-0190-83
$C_{49}H_{74}O_{14}$			
Avermectin A_{1a}	MeOH	237(4.46),243(4.50), 252(4.31)	162-0128-83
$C_{49}H_{76}O_4$			
Furan, 2,2',2",2"'-methanetetrayltetrakis[5-heptyl-3-methyl-	EtOH	231(4.33)	103-0360-83
$C_{49}H_{76}O_{15}$			
Avermectin A_{2a}	MeOH	237(4.46),243(4.50), 245(4.31)	162-0128-83
$C_{49}H_{76}O_{21}$			
Lanatoside D	n.s.g.	220(4.16)	162-0770-83
$C_{50}H_{33}FeN_4$			
Iron, phenyl[5,10,15,20-tetraphenyl-21H,23H-porphinato(2-)-N[21],N[22],N[23]-N[24]]-, (SP-5-31)-	benzene	389(4.28),408(4.97), 423s(--),518(3.87), 548(3.74)	101-0065-83M
$C_{50}H_{33}GaN_4$			
Gallium, phenyl[5,10,15,20-tetraphenyl-21H,23H-porphinato(2-)-N[21],N[22],N[23]-N[24]]-, (SP-5-31)-	benzene	334(3.58),416s(3.62), 436(4.55),524(2.70), 567(3.18),608(3.00)	101-0273-83M
$C_{50}H_{38}O_6$			
Benz[e]-as-indacen-3(6H)-one, 4-hydroxy-	$CHCl_3$	605(3.47)	150-2650-83M

Compound	Solvent	$\lambda_{max}(\log \epsilon)$	Ref.
8-methoxy-1,6,6-tris(4-methoxyphenyl)-2,5-diphenyl- (cont.)			150-2650-83M
$C_{50}H_{40}N_4O_8$			
3,5:14,16-Dimetheno-5H,16H-dicyclohepta[e,k]diimidazo[1,5-a:1',5'-g][1,7]-diazacyclododecine-10,21,23,24-tetracarboxylic acid, 1,12-bis(4-methylphenyl)-, 23,24-diethyl 10,21-dimethyl ester	EtOH	233(4.81),271(4.85), 319(4.69),345(4.68), 366s(4.69),425(4.33), 448(4.50),512s(2.77), 542(3.83),578s(3.72), 625s(3.29)	18-3703-83
3,5:14,16-Dimetheno-5H,16H-dicyclohepta[e,k]diimidazo[1,5-a:1',5'-g][1,7]-diazacyclododecine-10,21,23,24-tetracarboxylic acid, 1,12-diphenyl-, tetraethyl ester	EtOH	232(4.93),268(4.99), 317(4.79),343(4.80), 364s(4.67),421(4.42), 445(4.60),515s(3.91), 537(3.95),572s(3.84), 625s(3.38)	18-3703-83
$C_{50}H_{42}N_6O_8$			
3,5:14,16-Dinitrilo-5H,16H-dicyclohepta[e,k]diimidazo[1,5-a:1',5'-g][1,7]-diazacyclododecine-10,11,21,22-tetracarboxylic acid, 1,12-bis(4-methylphenyl)-, tetraethyl ester	EtOH	255(4.73),298(4.54), 323(4.52),410s(4.27), 425s(4.42),429(4.43), 553(3.73)	18-3703-83
$C_{50}H_{44}FeN_6O$			
Iron, (2-nitrosopropane-N)(2-propanamine)[5,10,15,20-tetraphenyl-21H,23H-porphinato(2-)-N^{21},N^{22},N^{23},N^{24}]-, (OC-6-42)-	benzene	425(5.41),539(4.08)	35-0455-83
$C_{50}H_{44}FeN_6O_2$			
Iron, (N-hydroxy-2-propanamine-N)(2-nitrosopropane-N)[5,10,15,20-tetraphenyl-21H,23H-porphinato(2-)-N^{21},N^{22},N^{23},N^{24}]-, (OC-6-23)-	benzene	422(5.43),535(4.08)	35-0455-83
$C_{50}H_{44}N_2O_6$			
3-Pyridinecarboxylic acid, 1,4-dihydro-1-(phenylmethyl)-, [1,1'-binaphthalene]-2,2'-diylbis(oxy-2,1-ethanediyl) ester, (R)-	MeCN	355(4.10)	18-3672-83
$C_{50}H_{44}O_2$			
2H-Furylium, 3,3'-[1,4-phenylenebis(2-phenyl-1,3-butadiene-4,1-diyl)]bis-[2,2-dimethyl-5-phenyl-, diperchlorate	MeCN	597(4.93)	56-0579-83
$C_{50}H_{48}O_{12}$			
Kuwanon M	EtOH	208(4.99),265(4.84), 310s(4.39),326(4.40)	142-0585-83
	EtOH-$AlCl_3$	210(5.02),275(4.87), 320s(4.34),332(4.36), 380(4.27)	142-0585-83
$C_{50}H_{63}ClO_{16}$			
Chlorothricin	EtOH	222(4.20),260(3.81)	162-0306-83
	EtOH-KOH	221(4.09),259(3.95)	162-0306-83
$C_{50}H_{63}IN_2O_{13}$			
Kijanolide, 26-O-methyl-17-O-[2,3,4,6-	CF_3CH_2OH	256(4.43)	39-1497-83C

Compound	Solvent	λ_{max}(log ε)	Ref.
tetradeoxy-4-[(methoxycarbonyl)amino]-3-C-methyl-3-nitro-β-D-xylo-hexopyranosyl]-, 32-(4-iodobenzoate)			39-1497-83C
$C_{50}H_{66}N_4O_{16}$			
3,6,9,12,15-Pentaoxabicyclo[15.3.1]heneicosa-1(21),17,19-trien-21-ol, 19,19'-[1,2-ethanediylbis(oxy-2,1-ethanediyloxy-4,1-phenyleneazo)]bis-	$CHCl_3$	358(4.73)	18-3253-83
$C_{50}H_{70}N_{12}O_8$			
3H-Phenoxazine-1,9-dicarboxamide, 2-amino-N,N'-bis[6-[[5-[[[3-(dimethylamino)propyl]amino]carbonyl]-1-methyl-1H-pyrrol-3-yl]amino]-6-oxohexyl]-4,6-dimethyl-3-oxo-, bis(trifluoroacetate)	pH 6	235(4.73),275s(4.46), 455(4.23)	104-1380-83
$C_{50}H_{70}O_{14}$			
Brevetoxin B	MeOH	208(4.20)	162-0190-83
$C_{50}H_{85}NO_{13}$			
Algacidin B	90% MeOH	212(4.16),256(4.24)	158-1777-83
$C_{50}H_{86}N_2O_{17}$			
Tylosin, 20-deoxo-20-morpholino-	MeOH	284(4.29)	158-1713-83
$C_{50}H_{87}NO_{14}$			
Algacidin A	90% MeOH	210(4.14)	158-1777-83
	+ acid	256(4.31)	158-1777-83
	+ base	210(4.14)	158-1777-83
$C_{51}H_{35}FeN_4$			
Iron, (4-methylphenyl)[5,10,15,20-tetraphenyl-21H,23H-porphinato(2-)-N^{21},N^{22},N^{23},N^{24}]-, (SP-5-31)-	hexane	390(4.84),409(5.04), 426s(--),518(3.96), 548(3.81)	101-0065-83M
$C_{51}H_{35}GaN_4$			
Gallium, (4-methylphenyl)[5,10,15,20-tetraphenyl-21H,23H-porphinato(2-)-N^{21},N^{22},N^{23},N^{24}]-, (SP-5-31)-	benzene	336(4.36),415s(4.51), 435(5.45),528(3.40), 568(4.03),610(3.84)	101-0273-83M
$C_{51}H_{36}N_4SZn$			
Zinc, (benzenethiolato)(21-methyl-5,10,15,20-tetraphenyl-21H,23H-porphinato-N^{21},N^{22},N^{23},N^{24})-, (SP-5-12)-	CH_2Cl_2	438(5.26),450s(--), 530s(--),566(3.68), 619(3.92),661(3.71)	18-2055-83
$C_{51}H_{41}FeN_7O$			
Iron, (1-methyl-1H-imidazole-N^3)(2-nitrosopropane-N)[5,10,15,20-tetraphenyl-21H,23H-porphinato(2-)-N^{21},N^{22},N^{23},N^{24}]-, (OC-6-23)-	benzene	426(5.42),540(4.08)	35-0455-83
$C_{51}H_{66}N_4O_{12}$			
Vindolicine	MeOH	219(4.39),260(4.37), 310(4.24)	100-0517-83
$C_{51}H_{67}CoN_6O_{14}$			
Hexamethyl Coα,Coβ-dicyano-5,6-dioxo-7-de(carboxymethyl)-7,8-didehydro-	MeOH-HCl	300s(4.06),309(4.09), 366s(3.62),397s(3.40),	33-0044-83

Compound	Solvent	λ_{max}(log ε)	Ref.
5,6-secobyrinate (cont.)		561(3.83),615s(3.57)	33-0044-83
$C_{51}H_{86}N_7O_{13}PSi_4$			
Adenosine, 2',5'-bis-O-[(1,1-dimethylethyl)dimethylsilyl]-O-(P-methyluridylyl)(3'→5')-N^6-benzoyl-2',3'-bis-O-[(1,1-dimethylethyl)dimethylsilyl]-, (Rp)-	MeOH	271(4.44)	35-5879-83
(Sp)-	MeOH	271(4.42)	35-5879-83
$C_{51}H_{90}N_4O_{16}$			
Tylosin, 20-deoxo-20-[(4-methyl-1-piperazinyl)amino]-	MeOH	283(4.25)	158-1713-83
$C_{52}H_{40}FeN_6O$			
Iron, (2-nitrosopropane-N)(pyridine)-[5,10,15,20-tetraphenyl-21H,23H-porphinato(2-)-N^{21},N^{22},N^{23},N^{24}]-, (OC-6-32)-	benzene	424(5.41),536(4.04)	35-0455-83
$C_{52}H_{40}N_4O_{12}$			
3,5:14,16-Dimetheno-5H,16H-dicyclohepta[e,k]diimidazo[1,5-a:1',5'-g][1,7]-diazacyclododecine-10,11,21,22,23,24-hexacarboxylic acid, 1,12-diphenyl-, 23,24-diethyl 10,11,21,22-tetramethyl ester	EtOH	265(4.88),310(4.40), 329(4.45),348(4.44), 366(4.45),413(4.11), 436(4.26),515s(3.63), 547(3.73),580s(3.69), 625s(3.37)	18-3703-83
$C_{52}H_{45}O_2$			
5H-Cyclopenta[b]pyrylium, 7-[[3-[2-(5,6-dihydro-2,4-diphenylcyclopenta[b]-pyran-7-yl)ethenyl]-5,5-dimethyl-2-cyclohexen-1-ylidene]ethylidene]-6,7-dihydro-2,4-diphenyl-, perchlorate	CH_2Cl_2	1040(4.97),1180(5.32)	103-1241-83
$C_{52}H_{45}S_2$			
5H-Cyclopenta[b]thiopyrylium, 7-[[3-[2-(5,6-dihydro-2,4-diphenylcyclopenta[b]thiopyran-7-yl)ethenyl]-5,5-dimethyl-2-cyclohexen-1-ylidene]-6,7-dihydro-2,4-diphenyl-, perchlorate	CH_2Cl_2	1270(5.13)	103-1241-83
$C_{52}H_{45}Se_2$			
5H-Cyclopenta[b]seleninium, 7-[[3-[2-(5,6-dihydro-2,4-diphenylcyclopenta[b]selenin-7-yl)-5,5-dimethyl-2-cyclohexen-1-ylidene]ethylidene]-6,7-dihydro-2,4-diphenyl-, perchlorate	CH_2Cl_2	1280(5.23)	103-1241-83
$C_{52}H_{54}N_4O_8$			
21H,23H-Porphine-2,18-dipropanoic acid, 8,12-bis(1-acetoxyethyl)-3,7,13,17-tetramethyl-, bis(phenylmethyl) ester	$CHCl_3$	402(5.29),499(4.17), 534(3.95),570(3.81), 623(3.65)	12-1639-83
21H,23H-Porphine-2,18-dipropanoic acid, 8,12-bis(2-acetoxyethyl)-3,7,13,17-tetramethyl-, bis(phenylmethyl) ester	$CHCl_3$	401(5.26),498(4.17), 533(3.99),568(3.83), 622(3.69)	12-1639-83
$C_{52}H_{62}N_4O_{16}$			
21H,23H-Porphine-2,3,7,8,12,13,17,18-octaethanol, octaacetate	pyridine	404(5.34),500(4.30), 533(4.04),569(3.94),	12-1639-83

Compound	Solvent	λ_{max}(log ϵ)	Ref.
(cont.)		622(3.69)	12-1639-83
21H,23H-Porphine-2,3,7,8,12,13,17,18-octapropanoic acid, octamethyl ester	$CHCl_3$	403.5(5.30),500(4.18), 534(3.99),569.5(3.85), 623(3.68)	12-1639-83
$C_{52}H_{64}N_8O_{12}$			
17,20,39,42,47,50,59,62-Octaoxa-1,7,8-14,23,29,30,36-octaazaheptacyclo-[34.8.8.8^{14,23}.2^{3,6}.2^{9,12}.2^{25,28}-2^{31,34}]octahexaconta-3,5,7,9,11-25,27,29,31,33,53,55,65,67-tetradecaene-2,13,24,35-tetrone	o-$C_6H_4Cl_2$	332(4.73)	18-1700-83
$C_{52}H_{70}Ge_6$			
Hexagermane, 2,2,3,3,4,4,5,5-octaethyl-1,1,1,6,6,6-hexaphenyl-	C_6H_{12}	278s(4.78)	101-0149-83G
$C_{52}H_{76}O_{24}$			
Variamycin B	EtOH	230(4.27),279(4.6), 317(3.83),330(3.75), 415(3.95)	158-83-29
$C_{52}H_{84}N_2O_{16}$			
Tylosin, 20-deoxo-20-(phenylamino)-	MeOH	250(4.19),282(4.33)	158-1713-83
$C_{52}H_{90}N_2O_{16}$			
Tylosin, 20-(cyclohexylamino)-20-deoxo-	MeOH	284(4.30)	158-1713-83
$C_{53}H_{42}O_2P_2$			
Phosphonium, [9H-fluorene-2,5-diyl-bis(2-oxo-2,1-ethanediyl)]bis[triphenyl-, dibromide	EtOH	225(4.61),290(4.45), 350(4.796)	65-1571-83
$C_{53}H_{52}N_4O_9$			
D-Ribitol, 1,1-bis(5-acetoxy-3-methyl-1-phenyl-1H-pyrazol-4-yl)-1-deoxy-2,3-O-(1-methylethylidene)-5-O-(triphenylmethyl)-, 4-acetate	MeOH	255(4.39)	111-0481-83
$C_{53}H_{78}O_{19}$			
β-D-Glucopyranoside, (1α,3β)-1-hydroxy-cholesta-5,7-dien-3-yl, 4-O-(2,3,4,6-tetra-O-acetyl-β-D-glucopyranosyl)-, 2,3,6-triacetate	n.s.g.	271(4.05),282(4.08), 293(3.85)	33-2093-83
β-D-Glucopyranoside, (1α,3β,5Z,7E)-1-hydroxy-9,10-secocholesta-5,7,10(19)-trien-3-yl, 4-O-(2,3,4,6-tetra-O-acetyl-β-D-glucopyranosyl)-, 2,3,6-triacetate	n.s.g.	264(4.25)	33-2093-83
$C_{53}H_{81}N_3O_2$			
Benzenepropanamide, α-[[4-(dihexadecyl-amino)phenyl]imino]-	MeOH	253(4.32),434(4.22)	5-0802-83
$C_{53}H_{83}N_7O_{16}$			
Antibiotic A-30912H	MeOH and MeOH-acid	223(4.12),275(3.32)	158-83-31
	MeOH-base	245(4.17),290(3.54)	158-83-31
$C_{53}H_{86}N_2O_{16}$			
Tylosin, 20-deoxy-20-(methylphenylamino)-	MeOH	260(4.30),282(4.32)	158-1713-83

Compound	Solvent	$\lambda_{max}(\log \epsilon)$	Ref.
Tylosin, 20-deoxo-20-[(phenylmethyl)-amino]-	MeOH	283(4.36)	158-1713-83
$C_{54}H_{41}FeN_4$			
Iron, phenyl[5,10,15,20-tetrakis(3-methylphenyl)-21H,23H-porphinato-(2-)-N^{21},N^{22},N^{23},N^{24}]-, (SP-5-31)-	benzene	392(4.32),408(4.98), 424s(--),518(3.84), 550(3.66)	101-0065-83M
Iron, phenyl[5,10,15,20-tetrakis(4-methylphenyl)-21H,23H-porphinato-(2-)-N^{21},N^{22},N^{23},N^{24}]-, (SP-5-31)-	benzene	391(4.28),409(4.97), 424s(--),518(3.96), 546(3.81)	101-0065-83M
$C_{54}H_{44}N_4O_{12}$			
3,5:14,16-Dimetheno-5H,16H-dicyclohepta[e,k]diimidazo[1,5-a:1',5'-g][1,7]-diazacyclododecine-10,11,21,22,23,24-hexacarboxylic acid, 1,12-bis(4-methylphenyl)-, 23,24-diethyl 10,11,21,22-tetramethyl ester	EtOH	268(4.90),311(4.53), 330(4.53),350(4.51), 366(4.51),416(4.15), 439(4.28),520s(3.70), 550(3.78),580s(3.74), 625s(3.47)	18-3703-83
$C_{54}H_{49}S_2$			
1-Benzothiopyrylium, 8-[[3-[2-(6,7-dihydro-2,4-diphenyl-5H-1-benzothiopyran-8-yl)ethenyl]-5,5-dimethyl-2-cyclohexen-1-ylidene]ethylidene]-5,6,7,8-tetrahydro-2,4-diphenyl-, perchlorate	CH_2Cl_2	1220(4.59)	103-1241-83
$C_{54}H_{70}O_5$			
Phenol, 4-(1,1-dimethylethyl)-2,6-bis-[[5-(1,1-dimethylethyl)-3-[[5-(1,1-dimethylethyl)-2-hydroxyphenyl]methyl]-2-hydroxyphenyl]methyl]-	EtOH	281s(4.07),286(4.11)	126-1363-83
$C_{54}H_{78}O_4$			
2H-1-Benzopyran-6-ol, 5-[[3,4-dihydro-2,8-dimethyl-2-(4,8,12-trimethyl-3,7,11-tridecatrienyl)-2H-1-benzopyran-6-yl]oxy]-3,4-dihydro-2,8-dimethyl-2-(4,8,12-trimethyl-3,7,11-tridecatrienyl)-	MeOH MeOH-NaOH	250(3.92),290(3.60) 250(3.92),300(3.61)	102-2281-83 102-2281-83
$C_{55}H_{54}N_2O_{15}$			
Acetamide, N-[2-[acetyl[9,10-dihydro-9,10-dioxo-4-[(2,3,4,6-tetra-O-acetyl-β-D-glucopyranosyl)oxy]-1-anthracenyl]amino]ethyl]-N-[2-(triphenylmethoxy)ethyl]-	EtOH	325(4.24),348(4.28), 430(4.04)	18-1435-83
$C_{55}H_{70}MgN_4O_6$			
Chlorophyll b, (10R)-	ether	234s(4.34),252(4.43), 288s(4.32),308(4.42), 332(4.46),357s(4.37), 376s(4.28),428(4.74), 453(5.20),545(3.77), 567(3.83),593(4.03), 611(3.89),642.0(4.76)	118-0705-83
Chlorophyll b, (10S)-	ether	232s(4.33),252(4.42), 288s(4.31),307(4.41), 331(4.44),356s(4.36), 375s(4.27),428(4.72), 452.5(5.19),546(3.76),	118-0705-83

Compound	Solvent	λ max (log ε)	Ref.
(cont.)		566(3.81),593(4.00), 610(3.88),642.0(4.74)	118-0705-83
$C_{55}H_{72}MgN_4O_5$			
Chlorophyll a, (10R)-	ether	245(4.36),280s(4.21), 294s(4.26),310s(4.31), 324(4.37),380(4.64), 409(4.86),428(5.06), 495(3.20),530(3.54), 575(3.86),614(4.13), 660.0(4.96)	118-0705-83
Chlorophyll a, (10S)-	ether	245(4.34),280s(4.19), 295s(4.24),310s(4.29), 325(4.35),381(4.63), 410(4.84),428.7(5.03), 496(3.13),531(3.53), 576(3.86),614(4.11), 660.6(4.94)	118-0705-83
$C_{55}H_{72}N_4O_6$			
Pheophytin b, (10R)-	THF	325(4.54),369(4.54), 414(4.94),435(5.33), 490(3.82),525(4.19), 554(3.99),599(4.02), 654.0(4.64)	118-0708-83
Pheophytin b, (10S)-	THF	325(4.56),371(4.54), 415(4.94),434.8(5.35), 489(3.82),525(4.19), 556(3.98),600(4.03), 655.5(4.64)	118-0708-83
$C_{55}H_{74}N_4O_5$			
Pheophytin a, (10R)-	THF	277(4.13),322(4.37), 370s(4.78),395s(5.00), 411.4(5.08),470(3.64), 505(4.12),534(4.05), 560(3.58),609(3.98), 668.0(4.75)	118-0708-83
Pheophytin a, (10S)-	THF	321(4.38),370s(4.78), 395s(5.00),411.4(5.08), 470(3.70),505(4.11), 535(4.08),560(3.55), 610(3.98),668.3(4.75)	118-0708-83
$C_{55}H_{78}O_{19}$			
Wedelin	EtOH	234(3.70)	100-0836-83
$C_{56}H_{40}NO_2P$			
Phosphonium, [(7,7a-dihydro-2-hydroxy-7-oxo-1,5,6,7a-tetraphenyl-3H-pyrrolizin-3-ylidene)phenylmethyl]triphenyl-, hydroxide, inner salt	MeCN	245(4.43),350(4.19), 400(4.02)	88-2977-83
$C_{56}H_{48}N_4O_{12}$			
3,5:14,16-Dimetheno-5H,16H-dicyclohepta[e,k]diimidazo[1,5-a:1',5'-g][1,7]-diazacyclododecine-10,11,21,22,23,24-hexacarboxylic acid, 1,12-diphenyl-, hexaethyl ester	EtOH	265(5.02),308(4.57), 328(4.62),348(4.61), 365(4.62),413(4.24), 437(4.38),517s(3.78), 547(3.73),580s(3.81), 625s(3.50)	18-3703-83

Compound	Solvent	$\lambda_{max}(\log \epsilon)$	Ref.
$C_{56}H_{52}ClFeN_4$			
Iron, chloro[5,10,15,20-tetrakis(2,4,6-trimethylphenyl)-21H,23H-porphinato-(2-)-N^{21},N^{22},N^{23},N^{24}]-, (SP-5-12)	CH_2Cl_2	375(4.76),417(5.64), 509(4.21),576(3.59), 664(3.48),694(3.51)	35-6243-83
$C_{56}H_{55}F_3N_2O_{14}$			
Carbamic acid, [2-[[9,10-dihydro-9,10-dioxo-4-[(2,3,4-tri-O-acetyl-6-deoxy-β-D-glucopyranosyl)oxy]-1-anthracen-yl](trifluoroacetyl)amino]ethyl]-[2-(triphenylmethoxy)ethyl]-, 1,1-dimethylethyl ester	$CHCl_3$	256(4.62),350(3.71)	18-1435-83
Carbamic acid, [2-[[9,10-dihydro-9,10-dioxo-4-[(2,3,4-tri-O-acetyl-6-deoxy-α-L-mannopyranosyl)oxy]-1-anthracen-yl](trifluoroacetyl)amino]ethyl]-[2-(triphenylmethoxy)ethyl]-, 1,1-dimethylethyl ester	$CHCl_3$	256(4.67),354(3.75)	18-1435-83
$C_{56}H_{58}O_{26}$			
Casuarinin, pentadeca-O-methyl-	MeOH	219(4.94),258s(4.54)	39-1765-83C
Stachyurin, pentadeca-O-methyl-	MeOH	216(4.96),257s(4.53)	39-1765-83C
$C_{56}H_{64}I_2N_2O_{14}$			
Kijanolide, 17-O-[2,3,4,6-tetradeoxy-4-[(methoxycarbonyl)amino]-3-C-methyl-3-nitro-β-D-xylo-hexopyranosyl]-, 9,32-bis(4-iodobenzoate)	CF_3CH_2OH	257(4.64)	39-1497-83C
$C_{56}H_{70}N_{16}O_{10}$			
3H-Phenoxazine-1,9-dicarboxamide, 2-amino-N,N'-bis[3-[[5-[[[5-[[[3-(dimethylamino)propyl]amino]carbonyl]-1-methyl-1H-pyrrol-3-yl]amino]carbonyl]-1-methyl-1H-pyrrol-3-yl]amino]-3-oxo-propyl]-4,6-dimethyl-3-oxo-, bis(trifluoroacetate)	pH 6	235(4.79),298(4.60), 448s(4.17)	104-1380-83
$C_{56}H_{72}N_8O_{16}$			
4,15,21,24,30,41,47,50,55,58,67,70-Dodecaoxa-1,9,10,18,27,35,36,44-octa-azaheptacyclo[42.8.8.$8^{18,27}$.$2^{5,8}$-$2^{11,14}$.$2^{31,34}$.$2^{37,40}$]hexaheptaconta-5,7,9,11,13,31,33,35,37,39,61,63,73-75-tetradecaene-2,17,28,43-tetrone	o-$C_6H_4Cl_2$	356(4.63)	18-1700-83
$C_{56}H_{94}N_2O_4$			
4,5-Secocholest-5-en-3-one, 5,5'-[azo-bis(methylene)]bis-, N,N'-dioxide	dioxan	294(3.92)	18-3349-83
$C_{56}H_{98}N_8O_{13}PSi_4$			
Adenosine, 2',5'-bis-O-[(1,1-dimethyl-ethyl)dimethylsilyl]uridylyl(3'→5')-N^6-benzoyl-2',3'-bis-O-[(1,1-dimethylethyl)dimethylsilyl]-, triethyl-amine salt	MeOH	271(4.40)	35-5879-83
$C_{57}H_{49}NO_{11}$			
D-Arabinitol, 5-C-[4,5,6,7-tetrahydro-6,6-dimethyl-4-oxo-1-(phenylmethyl)-1H-indol-2-yl]-, 1,2,3,4,5-penta-	EtOH	233(4.47),275(3.59)	136-0255-83E

Compound	Solvent	λ_{max}(log ϵ)	Ref.
benzoate, (S)- (cont.)			136-0255-83E
$C_{57}H_{62}O_{12}$			
Kuwanon M heptamethyl ether	EtOH	205(5.16),222a(4.93), 260(4.89),296(4.53), 316(4.54)	142-0585-83
	EtOH-$AlCl_3$	206(5.16),222s(4.95), 260(4.90),296(4.53), 316(4.56)	142-0585-83
$C_{57}H_{66}I_2N_2O_{14}$			
Kijanolide, 26-O-methyl-17-O-[2,3,4,6-tetradeoxy-4-[(methoxycarbonyl)amino]-3-C-methyl-3-nitro-β-D-xylo-hexopyranosyl]-, 4,32-bis(4-iodobenzoate)	CF_3CH_2OH	257(4.70)	39-1497-83C
$C_{57}H_{82}O_{26}$			
Toyomycin	EtOH	230(4.39),281(4.72), 304(3.85),318(3.92), 330(3.84),412(4.07)	162-0318-83
$C_{58}H_{36}Cl_2N_4Ni$			
Nickel, [21-[2,2-bis(4-chlorophenyl)-ethenyl]-5,10,15,20-tetraphenyl-21H,23H-porphinato(2-)]-, (SP-4-3)	n.s.g.	426(4.97),554(3.87)	157-1888-83
$C_{58}H_{36}Cl_2N_4Ru$			
Ruthenium, [bis(4-chlorophenyl)ethenylidene][5,10,15,20-tetraphenyl-21H,23H-porphinato(2-)-N^{21},N^{22},N^{23},N^{24}]-, (SP-5-31)-	n.s.g.	416(4.86),522(4.18), 542s(--)	157-1888-83
$C_{58}H_{44}N_2O_{18}$			
Furo[3,4-d]pyrimidine-2,4,7(3H)-trione, 1,5-dihydro-1,3-bis(2,3,5-tri-O-benzoyl-β-D-ribofuranosyl)-	EtOH	276(4.17),282(4.17)	94-3074-83
$C_{58}H_{52}N_4O_{12}$			
3,5:14,16-Dimetheno-5H,16H-dicyclohepta[e,k]diimidazo[1,5-a:1',5'-g][1,7]-diazacyclododecine-10,11,21,22,23,24-hexacarboxylic acid, 1,12-bis(4-methylphenyl)-, hexaethyl ester	EtOH	267(4.98),310(4.61), 331(4.61),350(4.60), 367(4.61),416(4.21), 439(4.34),520s(3.76), 552(3.83),583s(3.79), 630s(3.51)	18-3703-83
$C_{58}H_{57}F_3N_2O_{16}$			
Carbamic acid, [2-[[9,10-dihydro-9,10-dioxo-4-[(2,3,4,6-tetra-O-acetyl-β-D-galactopyranosyl)oxy]-1-anthracenyl](trifluoroacetyl)amino]ethyl]-[2-(triphenylmethoxy)ethyl]-, 1,1-dimethylethyl ester	EtOH	225(4.83),258(4.75), 350(3.80)	18-1435-83
Carbamic acid, [2-[[9,10-dihydro-9,10-dioxo-4-[(2,3,4,6-tetra-O-acetyl-β-D-glucopyranosyl)oxy]-1-anthracenyl](trifluoroacetyl)amino]ethyl]-[2-(triphenylmethoxy)ethyl]-, 1,1-dimethylethyl ester	EtOH	255(4.55),347(3.68)	18-1435-83

Compound	Solvent	λ_{max}(log ϵ)	Ref.
$C_{58}H_{58}F_3N_3O_{15}$			
Carbamic acid, [2-[[9,10-dihydro-9,10-dioxo-4-[[2,4,6-tri-O-acetyl-3-(acetylamino)-3-deoxy-β-D-glucopyranosyl]oxy]-1-anthracenyl](trifluoroacetyl)amino]ethyl][2-(triphenylmethoxy)ethyl]-, 1,1-dimethylethyl ester	EtOH	211(4.61),254(4.52), 346(3.63)	18-1435-83
Carbamic acid, [2-[[9,10-dihydro-9,10-dioxo-4-[[3,4,6-tri-O-acetyl-2-(acetylamino)-2-deoxy-β-D-glucopyranosyl]oxy]-1-anthracenyl](trifluoroacetyl)amino]ethyl][2-(triphenylmethoxy)ethyl]-, 1,1-dimethylethyl ester	$CHCl_3$	256(4.64),345(3.71)	18-1435-83
$C_{58}H_{60}O_{27}$			
Casuarinin, pentadeca-O-methyl-, monoacetate	MeOH	216(5.04),256s(4.64)	39-1765-83C
Stachyurin, pentadeca-O-methyl-, monoacetate	MeOH	214(5.01),256s(4.58)	39-1765-83C
$C_{58}H_{89}N_{19}O_{21}S_2$			
Cleomycin, 3-[N-methyl-N-(3-aminopropyl)aminopropylamino]-, trihydrochloride	pH 1	240s(2.19),289(1.69)	158-83-36
$C_{59}H_{86}O_{22}$			
Dregeoside C_{11}	EtOH	218(4.09),224(4.03), 281(4.17)	94-3971-83
$C_{59}H_{103}N_3O_{18}$			
Niphimycin I	EtOH	228s(4.48),233(4.50), 240s(4.35)	33-0092-83
Niphimycin II	EtOH	228s(4.23),234(4.25), 240s(4.14)	33-0092-83
$C_{60}H_{36}N_4Zn$			
Zinc, [6,13,20,27 tetraphenyl-29H,31H-tetrabenzo[b,g,l,q]porphinato(2-)-N^{29},N^{30},N^{31},N^{32}]-, (SP-4-1)-	pyridine	446(5.26),472(5.34), 595(4.40),635(4.80), 660(4.58)	88-1451-83
$C_{60}H_{42}$			
m-Quaterphenyl, 5',5"-di-2-biphenylyl-2,2"'-diphenyl-	C_6H_{12}	197(5.15),237(5.00), 254s(4.84)	94-2313-83
$C_{60}H_{44}N_4$			
Tetrabenzo[48]annulene	THF	290(4.44),416(4.97), 470s(4.65)	18-1467-83
$C_{60}H_{78}N_4O_{16}$			
21H,23H-Porphine-2,3,7,8,12,13,17,18-octapropanol, octaacetate	$CHCl_3$	401.5(5.28),500(4.17), 534(4.02),568(3.85), 622(3.69)	12-1639-83
$C_{60}H_{82}N_4O_{26}$			
Decilorubicin	MeOH	220(4.45),235(4.54), 254(4.44),290(3.83), 380(3.51),476(3.97), 496(4.01),535(3.88), 586(3.63)	158-0451-83
	MeOH-base	253(4.55),295s(3.81), 360(3.70),560(4.13),	158-0451-83

Compound	Solvent	λ max(log ε)	Ref.
(cont.)		597(4.13)	158-0451-83
$C_{60}H_{82}N_{11}O_{14}PSi_3$			
Adenosine, N-benzoyl-2'-O-[(1,1-dimethylethyl)dimethylsilyl]-P-[2-(4-nitrophenyl)ethyl]adenylyl-(3'→5')-N-benzoyl-2',3'-bis-O-[(1,1-dimethylethyl)-dimethylsilyl]-	MeOH	278(4.71)	33-2018-83
$C_{60}H_{86}N_{12}O_8$			
21H,23H-Porphine-2,3,7,8,12,13,17,18-octacarboxamide, hexadecaethyl-	$CHCl_3$	416(5.40),508(4.30), 540(3.89),581(3.56), 634(3.46)	44-4399-83
$C_{60}H_{90}O_4Se_2$			
19-Norlanosta-2,5(10)-diene-3,11-dione, 2,2'-diselenobis[9-methyl-, (9β)-(9'β)-	EtOH	370(4.08)	23-1973-83
$C_{60}H_{98}N_4P_2Ru$			
Ruthenium, [2,3,7,8,12,13,17,18-octaethyl-21H,23H-porphinato(2-)-N^{21}-N^{22},N^{23},N^{24}]bis(tributylphosphine)-, (OC-6-12)-	CH_2Cl_2	358(4.65),428(5.28), 511(4.20),535(4.04)	23-2389-83
$C_{61}H_{46}FeN_8O_4$			
Iron, [2,2-dimethyl-N-[2-(26,27,34,35-tetrahydro-26,34,38-trioxo-33H,47H-10,30-([1,2]benzeniminoethano)-11,14-imino-28,32-metheno-6,9:19,16-dinitrilo-5,20[2',5']-endo-pyrrolo-25H-dibenzo[1,d_1][1,11]diazacyclohentriacontin-15-yl)phenyl]propanamidato(2-)-N^{47},N^{51},N^{52},N^{53}]-	toluene	417(2.16),443(2.11), 538(1.32)	35-3038-83
$C_{61}H_{48}N_8O_4$			
Propanamide, 2,2-dimethyl-N-[2-(26,27-34,35-tetrahydro-26,34,38-trioxo-33H,47H-10,30-([1,2]benzeniminoethano)-11,14-imino-28,32-metheno-6,9-19,16-dinitrilo-5,20[2',5']-endo-pyrrolo-25H-dibenzo[1,d_1][1,11]diazacyclohentriacontin-15-yl)phenyl]-	$CHCl_3$	421(4.42),516(4.11), 547(3.45),588(3.62), 644(3.09)	35-3038-83
$C_{61}H_{69}NO$			
Pyridinium, 1-[4,4"-bis(1,1-dimethylethyl)-2'-hydroxy[1,1':3',1"-terphenyl]-5'-yl]-2,4,6-tris[4-(1,1-dimethylethyl)phenyl]-, hydroxide, inner salt	C_6H_{12}	216(4.84),238(4.81), 311(4.68),406.5(4.23), 849(4.07)	5-0721-83
	MeOH	236(4.65),323(4.60), 525(2.53)	5-0721-83
$C_{61}H_{71}Cl_2N_{11}O_{15}P_2Si_2$			
3'-Adenylic acid, N-benzoyl-P-(2-chlorophenyl)-2'-O-[(1,1-dimethylethyl)-dimethylsilyl]adenylyl-(3'→5')-N-benzoyl-2'-O-[(1,1-dimethylethyl)dimethylsilyl]-, 2-chlorophenyl 2-cyanoethyl ester	MeOH	234(4.43),280(4.63)	33-2018-83
$C_{61}H_{84}N_4O_{26}$			
Decilorubicin, methyl ester	MeOH	220(4.46),235(4.57),	158-0451-83

Compound	Solvent	λ_{max}(log ε)	Ref.
Decilorubicin, methyl ester (cont.)		254(4.43),290(3.89), 380(3.56),476(4.05), 495(4.06),533(3.85), 580(2.78)	158-0451-83
	+ NaOH	253(4.53),295s(3.83), 360(3.72),561(4.13), 598(4.12)	158-0451-83
$C_{61}H_{90}N_2O_{21}$			
Kijanimicin, 3B-O-de(2,6-dideoxy-α-L-ribo-hexopyranosyl)-	CF_3CH_2OH	200(4.60),236(4.11), 265s(4.01),275(4.05)	49-1497-83C
$C_{61}H_{109}N_{11}O_{13}$			
Leucinostatin B	EtOH	204(4.38),213s(4.27)	158-1084-83
hydrochloride	EtOH	204(4.38),213(4.27)	158-1606-83
$C_{62}H_{82}N_{16}O_{10}$			
3H-Phenoxazine-1,9-dicarboxamide, 2-amino-N,N'-bis[6-[[5-[[[5-[[[3-(dimethylamino)propyl]amino]carbonyl]-1-methyl-1H-pyrrol-3-yl]amino]amino]-carbonyl]-1-methyl-1H-pyrrol-3-yl]-amino]-6-oxohexyl]-4,6-dimethyl-3-oxo-, bis(trifluoroacetate)	pH 6	235(4.80),300(4.64), 445s(4.02)	104-1380-83
$C_{62}H_{86}N_{12}O_{16}$			
Actinomycin D	pH 7.0	442(4.32)	87-1631-83
	$CHCl_3$	424(4.32),442(4.36) (anom.)	87-1631-83
$C_{62}H_{86}N_{12}O_{17}$			
Actinomycin D, 7-hydroxy-	pH 7.0	550(4.08)	87-1631-83
$C_{62}H_{88}N_{18}O_{21}S_2$			
Cleomycin, 3-[(S)-1'-phenylethylamino]-propylamino-	pH 1	240s(2.09),290(1.62)	158-83-36
	+ Cu	243(2.32),292(1.90)	158-83-36
$C_{62}H_{111}N_{11}O_{13}$			
Leucinostatin A, hydrochloride	EtOH	202(4.05),220s(3.83)	158-1606-83
$C_{63}H_{73}O_{37}$			
Platyconin (chloride)	MeOH-HCl	237(4.60),286(4.63), 303s(4.51),323s(4.43), 549(4.54)	88-2181-83
$C_{63}H_{74}N_2O_{11}$			
A Jeffamine derivative	EtOH	293(4.31),320(3.86)	87-0007-83
$C_{63}H_{88}N_{12}O_{16}$			
Actinomycin C	MeOH	443(4.41)	162-0222-83
$C_{64}H_{32}ErN_{16}$			
Diphthalocyanine, erbium chelate	$o-C_6H_4Cl_2$	312(4.92),452(4.37), 594(4.30),630s(4.24), 662(5.03)	65-2339-83
one electron reduction	$o-C_6H_4Cl_2$	330(4.91),620(4.87), 690(4.56)	65-2339-83
one electron oxidation (also other rare earths, not listed)	$o-C_6H_4Cl_2$	312(4.80),344(4.71), 480(4.52),696(4.50)	65-2339-83

Compound	Solvent	λmax(log ε)	Ref.
$C_{64}H_{38}N_6O_4$			
Methanone, (1,12-diphenyl-3,5:14,16-dinitrilo-5H,16H-dicyclohepta[e,k]diimazo[1,5-a:1',5'-g][1,7]diazacyclododecine-10,11,21,22-tetrayl)tetrakis-[phenyl-	EtOH	255(4.96),328(4.60), 415s(4.31),433(4.41), 558(3.73)	18-3703-83
$C_{64}H_{62}N_4O_8$			
21H,23H-Porphine-2,8,12,18-tetrapropanoic acid, 3,7,13,17-tetramethyl-, tetrakis(phenylmethyl) ester	$CHCl_3$	401(5.26),499(4.16), 533(3.99),568(3.83), 622(3.69)	12-1639-83
$C_{64}H_{66}N_4O_4$			
6,9:32,35-Diimino-5,31:10,36-di[2]pyrrolyl[5]ylidenetetrabenzo[b,j,x,f_1]-[1,12,23,34]tetraoxacyclotetratetracontin, 16,17,18,19,20,21,22,23,24-25,42,43,44,45,46,47,48,49,50,51-eicosahydro-	toluene	419(5.59),513(4.30), 546(3.79),591(3.78), 647(3.45)	39-0189-83C
6,9:32,35-Diimino-5,36:10,31-di[2]pyrrolyl[5]ylidenetetrabenzo[b,j,x,f_1]-[1,12,23,34]tetraoxacyclotetratetracontin, 16,17,18,19,20,21,22,23,24-25,42,43,44,45,46,47,48,49,50,51-eicosahydro-, cis-linked	toluene	419(5.56),513(4.28), 546.5(3.83),592(3.75), 649(4.48)	39-0189-83C
trans-linked	toluene	419(5.60),513(4.31), 546.5(3.81),591(3.79), 548(3.46)	39-0189-83C
$C_{64}H_{86}N_{12}O_{17}$			
L-Valine, N,N'-[(2,9,11-trimethyl-8-oxo-8H-oxazolo[4,5-b]phenoxazine-4,6-diyl)dicarbonyl]bis[1-threonyl-D-valyl-L-prolyl-N-methylglycyl-N-methyl-, di-ξ-lactone	pH 7.0	360(4.33),515(4.20)	87-1631-83
	$CHCl_3$	383(4.24),510(4.07)	87-1631-83
$C_{64}H_{88}N_{12}O_{16}$			
L-Valine, N,N'-[(2,9,11-trimethyl-5H-oxazolo[4,5-b]phenoxazine-4,6-diyl)-dicarbonyl]bis[1-threonyl-D-valyl-L-prolyl-N-methylglycyl-N-methyl-, di-ξ-lactone	pH 7.0	415(4.02)	87-1631-83
	$CHCl_3$	383(4.24),510(4.07)	87-1631-83
$C_{64}H_{90}N_{12}O_{16}$			
Actinomycin C_3	MeOH	443(4.38)	162-0222-83
$C_{65}H_{79}Cl_2N_5O_{18}P_2Si_2$			
3'-Uridylic acid, P-(2-chlorophenyl)-2'-O-[(1,1-dimethylethyl)dimethylsilyl]-5,6-dihydro-5'-O-[(4-methoxyphenyl)diphenylmethyl]uridylyl-(3'→5')-2'-O-[(1,1-dimethylethyl)dimethylsilyl]-, 2-chlorophenyl 2-cyanoethyl ester	MeOH	230s(4.24),261(4.07)	33-2641-83
$C_{65}H_{81}Cl_2N_5O_{18}P_2Si_2$			
3'-Uridylic acid, P-(2-chlorophenyl)-2'-O-[(1,1-dimethylethyl)dimethylsilyl]-5,6-dihydro-5'-O-[(4-methoxyphenyl)diphenylmethyl]uridylyl-(3'→5')-2'-O-[(1,1-dimethylethyl)dimethylsil-	MeOH	231s(4.23),265(3.44), 273(3.38),281s(3.15)	33-2641-83

Compound	Solvent	λ max(log ε)	Ref.
yl]-5,6-dihydro-, 2-chlorophenyl 2-cyanoethyl ester (cont.)			33-2641-83
$C_{65}H_{84}O_6$			
Phenol, 2,2'-methylenebis[4-(1,1-dimethylethyl)-6-[[5-(1,1-dimethylethyl)-3-[[5-(1,1-dimethylethyl)-2-hydroxyphenyl]methyl]-2-hydroxyphenyl]methyl]-	EtOH	281s(4.20),286(4.22)	126-1363-83
$C_{66}H_{42}N_6O_4$			
Methanone, [1,12-bis(4-methylphenyl)-3,5:14,16-dinitrilo-5H,16H-dicyclohepta[e,k]diimidazo[1,5-a:1',5'-g]-[1,7]diazacyclododecine-10,11,21,22-tetrayl]tetrakis[phenyl-	EtOH	252(4.73),334(4.32), 420s(4.10),435(4.18), 560(3.44)	18-3703-83
$C_{66}H_{70}N_4O_4$			
6,9:33,36-Diimino-5,32:10,37-di[2]pyrrolyl[5]ylidene-16H,43H-tetrabenzo-[b,j,y,g1][1,12,24,35]tetraoxacyclohexatetratetracontin, 17,18,19,20,21-22,23,24,25,26,44,45,46,47,48,49,50-51,52,53-eicosahydro-	toluene	419(5.60),513(4.30), 546(3.81),591(3.81), 645.5(3.40)	39-0189-83C
6,9:33,36-Diimino-5,37:10,32-di[2]pyrrolyl[5]ylidene-16H,43H-tetrabenzo-[b,j,y,g1][1,12,24,35]tetraoxacyclohexatetratetracontin, 17,18,19,20,21-22,23,24,25,26,44,45,46,47,48,49,50-51,52,53-eicosahydro-, cis	toluene	419(5.60),513(4.30), 546(3.83),591(3.80), 646(3.48)	39-0189-83C
trans	toluene	419(5.61),513(4.31), 546(3.83),591(3.80), 645.5(3.38)	39-0189-83C
$C_{67}H_{100}N_2O_{24}$			
Kijanimicin	MeOH-HCl	205(4.58),258(3.98)	39-1497-83C
	MeOH-NaOH	236(4.17),266s(4.08), 276(4.08)	39-1497-83C
	CF_3CH_2OH	200(4.63),241(3.95), 264s(3.99),274(3.98)	39-1497-83C
potassium salt	CF_3CH_2OH	201(4.50),239(3.91), 264s(3.91),273(3.90)	39-1497-83C
$C_{68}H_{44}ClFeN_4$			
Iron, chloro[5,10,15,20-tetrakis([1,1'-biphenyl]-2-yl)-21H,23H-porphinato-(2-)-N^{21},N^{22},N^{23},N^{24}]-, (SP-5-12)-	CH_2Cl_2	327(4.84),352s(4.60), 379(4.67),430(5.04), 514(4.14),580(3.61), 668(3.51),697(3.55)	35-5786-83
$C_{68}H_{58}N_4O_4$			
19,22"45,48-Dietheno-6,9:32,35-diimino-5,31:10,36-di[2]pyrrolyl[5]ylidene-tetrabenzo[b,j,x,f1][1,12,23,34]-tetraoxacyclotetratetracontin, 16,17-18,23,24,25,42,43,44,49,50,51-dodecahydro-	n.s.g.	420(5.59),513.5(4.27), 547(3.80),590(3.76), 647.5(3.45)	39-0189-83C
$C_{68}H_{74}N_4O_4$			
6,9:34,37-Diimino-5,33:10,38-di[2]pyrrolyl[5]ylidenetetrabenzo[b,j,z,h]-[1,12,25,36]tetraoxacyclooctatetra-	n.s.g.	419(5.61),513(4.30), 546(3.79),592(3.80), 647(3.34)	39-0189-83C

Compound	Solvent	λ_{max}(log ε)	Ref.
contin, 16,17,18,19,20,21,22,23,24-25,26,27,44,45,46,47,48,49,50,51,52-53,54,55-tetracosahydro- (cont.)			39-0189-83C
$C_{68}H_{102}N_2O_{24}$			
Kijanimicin, 26-O-methyl-	CF_3CH_2OH	200(4.63),254(3.99)	39-1497-83C
$C_{69}H_{83}F_5N_{12}O_{17}$			
L-Valine, N,N'-[[9,11-dimethyl-8-oxo-2-(pentafluorophenyl)-8H-oxazolo-[4,5-b]phenoxazine-4,6-diyl]dicarbonyl]bis[L-threonyl-D-valyl-L-prolyl-N-methylglycyl-N-methyl-, di-ξ-lactone	$CHCl_3$	380(4.25),409(4.09)	87-1631-83
$C_{69}H_{85}F_5N_{12}O_{16}$			
L-Valine, N,N'-[[9,11-dimethyl-2-(pentafluorophenyl)-5H-oxazolo[4,5-b]-phenoxazine-4,6-diyl]dicarbonyl]bis-[L-threonyl-D-valyl-L-prolyl-N-methylglycyl-N-methyl-, di-ξ-lactone	$CHCl_3$	379(3.91)	87-1631-83
$C_{69}H_{88}N_{12}O_{17}$			
L-Valine, N,N'-[(9,11-dimethyl-8-oxo-2-phenyl-8H-oxazolo[4,5-b]phenoxazine-4,6-diyl)dicarbonyl]bis[L-threonyl-D-valyl-L-prolyl-N-methylglycyl-N-methyl-, di-ξ-lactone	$CHCl_3$	387(4.30),498(4.14)	87-1631-83
$C_{69}H_{90}N_{12}O_{16}$			
L-Valine, N,N'-[(9,11-dimethyl-2-phenyl-5H-oxazolo[4,5-b]phenoxazine-4,6-diyl)dicarbonyl]bis[L-threonyl-D-valyl-L-prolyl-N-methylglycyl-N-methyl-, di-ξ-lactone	$CHCl_3$	387(4.04)	87-1631-83
$C_{70}H_{72}N_4O_{10}$			
Carbamic acid, [(9,10-dihydro-5,6-dihydroxy-9,10-dioxo-1,4-anthracenediyl)bis(imino-2,1-ethanediyl)]bis-[[2-(triphenylmethoxy)ethyl]-, bis(1,1-dimethylethyl) ester	$CHCl_3$	261s(4.56),560s(4.02), 602(4.36),652(4.46)	18-1812-83
$C_{70}H_{73}F_3N_2O_{24}$			
Carbamic acid, [2-[(9,10-dihydro-9,10-dioxo-4-[[2,3,6-tri-O-acetyl-4-O-(2,3,4,6-tetra-O-acetyl-β-D-glucopyranosyl]oxy]-1-anthracenyl](trifluoroacetyl)amino]ethyl][2-(triphenylmethoxy)ethyl]-, 1,1-dimethylethyl ester	EtOH	214(4.61),254(4.51), 341(3.70)	18-1435-83
$C_{70}H_{85}F_5N_{12}O_{17}$			
L-Valine, N,N'-[[10,12-dimethyl-2-oxo-3-(pentafluorophenyl)-2H,6H-1,4-oxazino[3,2-b]phenoxazine-5,7-diyl]dicarbonyl]bis[L-threonyl-D-valyl-L-prolyl-N-methylglycyl-N-methyl-, di-ξ-lactone	$CHCl_3$	350(3.91),526(3.62) (anom.)	87-1631-83
$C_{70}H_{88}Cl_2N_{12}O_{17}$			
L-Valine, N,N'-[[3-(2,4-dichlorophen-	$CHCl_3$	350(3.93),521(3.68)	87-1631-83

Compound	Solvent	λ max(log ε)	Ref.
yl)-10,12-dimethyl-2-oxo-2H,6H-1,4-oxazino[3,2-b]phenoxazine-5,7-diyl]dicarbonyl]bis[L-threonyl-D-valyl-L-prolyl-N-methylglycyl-N-methyl-, di-ξ-lactone (cont.)		(anom.)	87-1631-83
$C_{70}H_{89}ClN_{12}O_{17}$			
L-Valine, N,N'-[[3-(2-chlorophenyl)-10,12-dimethyl-2H,6H-1,4-oxazino-[3,2-b]phenoxazine-5,7-diyl]dicarbonyl]bis[L-threonyl-D-valyl-L-prolyl-N-methylglycyl-N-methyl-, di-ξ-lactone	$CHCl_3$	350(4.01),523(3.76) (anom.)	87-1631-83
3-chlorophenyl isomer	$CHCl_3$	350(3.99),524(3.74)	87-1631-83
4-chlorophenyl isomer	$CHCl_3$	351(4.05),523(3.80)	87-1631-83
$C_{70}H_{92}N_{12}O_{17}$			
L-Valine, N,N'-[(8-methoxy-9,11-dimethyl-2-phenyl-5H-oxazolo[4,5-b]phenoxazine-4,6-diyl)dicarbonyl]bis[L-threonyl-D-valyl-L-prolyl-N-methylglycyl-N-methyl-, di-ξ-lactone	$CHCl_3$	394(4.08)	87-1631-83
$C_{70}H_{98}N_{12}O_{17}$			
L-Valine, N,N'-[(3-hexyl-10,12-dimethyl-2-oxo-2H,6H-1,4-oxazino[3,2-b]-phenoxazine-5,7-diyl)dicarbonyl]bis-[L-threonyl-D-valyl-L-prolyl-N-methylglycyl-N-methyl-, di-ξ-lactone	$CHCl_3$	320(3.91),401(4.98) (anom.)	87-1631-83
$C_{71}H_{74}N_4O_{10}$			
Carbamic acid, [(9,10-dihydro-5-hydroxy-6-methoxy-9,10-dioxo-1,4-anthracenediyl)bis(imino-2,1-ethanediyl)]-bis[[2-(triphenylmethoxy)ethyl]-, bis(1,1-dimethylethyl) ester	$CHCl_3$	262s(4.52),278(4.38), 564s(3.99),520(4.33), 652(4.43)	18-1812-83
$C_{71}H_{92}N_{12}O_{18}$			
L-Valine, N,N'-[(9-methoxy-10,12-dimethyl-2-oxo-3-phenyl-2H,6H-1,4-oxazino[3,2-b]phenoxazine-5,7-diyl)dicarbonyl]bis[L-threonyl-D-valyl-L-prolyl-N-methylglycyl-N-methyl-, di-ξ-lactone	$CHCl_3$	350(4.08),526(3.83) (anom.)	87-1631-83
$C_{71}H_{94}N_{12}O_{18}$			
Actinomycin D, N-(2-methoxy-2-oxo-1-phenylethyl)-	$CHCl_3$	420(4.11)	87-1631-83
$C_{72}H_{48}N_4O_8$			
3,5:14,16-Dimetheno-5H,16H-dicyclohepta[e,k]diimidazo[1,5-a:1',5'-g][1,7]-diazacyclododecine-23,24-dicarboxylic acid, 10,11,21,22-tetrabenzoyl-1,12-diphenyl-, diethyl ester	EtOH	258(5.07),351(4.59), 366(4.59),417(4.18), 441(4.32),520s(3.69), 551(3.77),587s(3.71), 635s(3.36)	18-3703-83
$C_{72}H_{66}N_4O_4$			
Porphyrin, α(5,15):β(10,20)-bis[2,2'-[(4,4'-p-phenylene)dibutoxy]diphenyl]-	n.s.g.	419.5(5.61),512(4.30), 546(3.81),591(3.80), 646(3.46)	39-0189-83C

Compound	Solvent	λ_{max}(log ε)	Ref.
$C_{72}H_{76}N_4O_{10}$			
9,10-Anthracenedione, 5,8-bis[2-[N-(t-butoxycarbonyl)-2-(trityloxy)ethylamino]ethylamino]-1,2-dimethoxy-	EtOH	212(4.83),271(4.49), 593(4.18),640(4.21)	18-1812-83
$C_{72}H_{84}Cl_2N_6O_{18}P_2Si_2$			
3'-Cytidylic acid, P-(2-chlorophenyl)-2'-O-[(1,1-dimethylethyl)dimethylsilyl]-5,6-dihydro-5'-O-[(4-methoxyphenyl)diphenylmethyl]uridylyl-(3'-5')-N-benzoyl-2'-O-[(1,1-dimethylethyl)dimethylsilyl]-, 2-chlorophenyl 2-cyanoethyl ester	MeOH	233(4.39),261(4.41), 305(4.00)	33-2641-83
3'-Uridylic acid, N-benzoyl-P-(2-chlorophenyl)-2'-O-[(1,1-dimethylethyl)dimethylsilyl]-5'-O-[(4-methoxyphenyl)diphenylmethyl]-cytidylyl-(3'→5')-2'-O-[(1,1-dimethylethyl)dimethylsilyl]-5,6-dihydro-, 2-chlorophenyl 2-cyanoethyl ester	MeOH	232s(4.41),261(4.43), 304(3.99)	33-2641-83
$C_{72}H_{98}ClN_{10}O_{15}PSi_3$			
Guanosine, 2'-O-(t-butyldimethylsilyl)-N^2-isobutyryl-5'-O-(4-methoxytrityl)-guanosyl-[3'-[O^P-(2-chlorophenyl)]-5']-2',3'-bis-O-(t-butyldimethylsilyl)-N^2-isobutyryl-	MeOH	237s(4.40),257(4.54), 259(4.54),274(4.40)	33-2069-83
$C_{73}H_{84}Cl_2N_8O_{17}P_2Si_2$			
3'-Adenylic acid, P-(2-chlorophenyl)-2'-O-[(1,1-dimethylethyl)dimethylsilyl]-5,6-dihydro-5'-O-[(4-methoxyphenyl)diphenylmethyl]uridylyl(3'-5')-N-benzoyl-2'-O-[(1,1-dimethylethyl)dimethylsilyl]-, 2-chlorophenyl 2-cyanoethyl ester	MeOH	230s(4.48),279(4.36)	33-2641-83
$C_{74}H_{52}N_4O_8$			
3,5:14,16-Dimetheno-5H,16H-dicyclohepta[e,k]diimidazo[1,5-a:1',5'-g][1,7]-diazacyclododecine-23,24-dicarboxylic acid, 10,11,21,22-tetrabenzoyl-1,12-bis(4-methylphenyl)-, diethyl ester	EtOH	257(5.03),353(4.55), 366s(4.54),421(4.16), 443(4.29),525s(3.65), 555(3.73),590s(3.68), 635s(3.29)	18-3703-83
$C_{74}H_{92}N_{12}O_{17}$			
L-Valine, N,N'-[[10,12-dimethyl-3-(2-naphthalenyl)-2-oxo-2H,6H-1,4-oxazino[3,2-b]phenoxazine-5,7-diyl]dicarbonyl]bis[L-threonyl-D-valyl-L-prolyl-N-methylglycyl-N-methyl-, di-ξ-lactone	$CHCl_3$	320(3.91),369(3.79), 498(3.79)(anom.)	87-1631-83
$C_{74}H_{103}IN_2O_{25}$			
Kijanimicin, 32-(4-iodobenzoate)	MeOH	196(4.80),256(4.42)	39-1497-83C
$C_{75}H_{80}N_4O_{12}$			
9,10-Anthracenedione, 2-(t-butoxycarbonyl)-5,8-bis[2-[N-(t-butoxycarbonyl)-2-(trityloxy)ethylamino]ethylamino]-1-hydroxy-	EtOH	210(4.93),368(3.77), 564(3.96),603(4.33), 652(4.45)	18-1812-83

Compound	Solvent	$\lambda_{max}(\log \epsilon)$	Ref.
$C_{75}H_{91}Cl_2N_{11}O_{18}P_2Si_2$			
3'-Guanylic acid, P-(2-chlorophenyl)-2'-O-[(1,1-dimethylethyl)dimethylsilyl]-5'-O-[(4-methoxyphenyl)diphenylmethyl]-N-(2-methyl-1-oxopropyl)guanylyl-(3'→5')-2'-O-[(1,1-dimethylethyl)dimethylsilyl]-N-(2-methyl-1-oxopropyl)-, 2-chlorophenyl 2-cyanoethyl ester	MeOH	237(4.42),254(4.51), 259(4.51),276(4.38)	33-2069-83
$C_{75}H_{96}ClN_{10}O_{14}PSi_3$			
Guanosine, N-benzoyl-P-(2-chlorophenyl)-2'-O-[(1,1-dimethylethyl)dimethylsilyl]-5'-O-[(4-methoxyphenyl)diphenylmethyl]adenylyl-(3'→5')-2',3'-bis-O-[(1,1-dimethylethyl)dimethylsilyl]-N-(2-methyl-1-oxopropyl)-	MeOH	232(4.51),256s(4.45), 261(4.47),279(4.53)	33-2069-83
$C_{76}H_{82}N_4O_{12}$			
9,10-Anthracenedione, 1-(t-butoxycarbonyloxy)-5,8-bis[2-[N-(t-butoxycarbonyl)-2-(trityloxy)ethylamino]ethylamino]-2-methoxy-	$CHCl_3$	272(4.63),599(4.32), 646(4.37)	18-1812-83
$C_{76}H_{92}N_4O_4$			
2,5-Cyclohexadien-1-one, 4,4'-[10,20-bis[3,5-bis(1,1-dimethylethyl)-4-hydroxyphenyl]-21H,23H-porphine-5,15-diylidene]bis[2,6-bis(1,1-dimethylethyl)-	CH_2Cl_2	406(1.36),508(2.04), 578s(1.80)	78-3895-83
	+ Et_3N	367(1.23),530(1.93), 590s(1.86)	78-3895-83
	+ TFA	347(1.23),476(1.83), 608s(1.72)	78-3895-83
Phenol, 4,4',4",4"'-(21H,23H-porphine-5,10,15,20-tetrayl)tetrakis[2,6-bis(1,1-dimethylethyl)-, ion(2-)-	CH_2Cl_2	426(2.52),522(1.15), 560(1.11),597(0.7), 654(0.85)	78-3895-83
$C_{76}H_{98}O_7$			
Phenol, 4-(1,1-dimethylethyl)-2,6-bis-[[5-(1,1-dimethylethyl)-3-[[5-(1,1-dimethylethyl)-3-[[5-(1,1-dimethylethyl)-2-hydroxyphenyl]methyl]-2-hydroxyphenyl]methyl]-2-hydroxyphenyl]methyl]-	EtOH	281s(4.23),286(4.26)	126-1363-83
$C_{76}H_{110}Cl_2N_{12}O_{21}P_2Si_4$			
Guanosine, P-(2-chlorophenyl)-2'-O-[(1,1-dimethylethyl)dimethylsilyl]-5,6-dihydrouridylyl-(3'→5')-N-benzoyl-P-(2-chlorophenyl)-2'-O-[(1,1-dimethylethyl)dimethylsilyl]adenylyl-(3'→5')-2',3'-bis-O-[(1,1-dimethylethyl)dimethylsilyl]-N-(2-methyl-1-oxopropyl)-	MeOH	256s(4.45),261(4.46), 280(4.51)	33-2641-83
$C_{77}H_{89}Cl_2N_9O_{18}P_2Si_2$			
3'-Cytidylic acid, P-(2-chlorophenyl)-2'-O-[(1,1-dimethylethyl)dimethylsilyl]-5'-O-[(4-methoxyphenyl)diphenylmethyl]-N-(2-methyl-1-oxopropyl)guanylyl-(3'→5')-N-benzoyl-2'-O-[(1,1-dimethylethyl)dimethylsilyl]-, 2-chlorophenyl 2-cyanoethyl ester	MeOH	237(4.48),260(4.61), 275s(4.45),305s(4.08)	33-2069-83

Compound	Solvent	λmax(log ε)	Ref.
3'-Guanylic acid, N-benzoyl-P-(2-chlorophenyl)-2'-O-[(1,1-dimethylethyl)dimethylsilyl]-5'-O-[(4-methoxyphenyl)diphenylmethyl]cytidylyl-(3'→5')-2'-O-[(1,1-dimethylethyl)dimethylsilyl]-N-(2-methyl-1-oxopropyl)-, 2-chlorophenyl 2-cyanoethyl ester	MeOH	231(4.45),260(4.59), 274s(4.41),302s(4.17)	33-2069-83
$C_{78}H_{80}N_5O_{11}$			
9,10-Anthracenedione, 5,8-bis[2-[N-(t-butoxycarbonyl)-2-(trityloxy)ethylamino]ethylamino]-1-hydroxy-2-(4-methoxybenzyloxy)-	$CHCl_3$	260s(4.51),280s(4.34), 368s(3.95),602(4.28), 654(4.38)	18-1812-83
$C_{78}H_{82}N_8$			
21H,23H-Porphine, 5,5'-(1,8-anthracenediyl)bis[2,8,13,17-tetraethyl-3,7,12,18-tetramethyl-	CH_2Cl_2	394(5.37),503(4.34), 537(4.00),572(3.95), 625(3.60)	44-5388-83
$C_{78}H_{89}Cl_2N_{11}O_{17}P_2Si_2$			
3'-Adenylic acid, P-(2-chlorophenyl)-2'-O-[(1,1-dimethylethyl)dimethylsilyl]-5'-O-[(4-methoxyphenyl)diphenylmethyl]-N-(2-methyl-1-oxopropyl)-guanylyl-(3'→5')-N-benzoyl-2'-O-[(1,1-dimethylethyl)dimethylsilyl]-, 2-chlorophenyl 2-cyanoethyl ester	MeOH	235(4.48),256(4.47), 261(4.49),280(4.52)	33-2069-83
3'-Guanylic acid, N-benzoyl-P-(2-chlorophenyl)-2'-O-[(1,1-dimethylethyl)dimethylsilyl]-5'-O-[(4-methoxyphenyl)diphenylmethyl]adenylyl-(3'→5')-2'-O-[(1,1-dimethylethyl)dimethylsilyl]-N-(2-methyl-1-oxopropyl)-, 2-chlorophenyl 2-cyanoethyl ester	MeOH	233(4.52),255s(4.46), 261(4.48),280(4.53)	33-2069-83
$C_{78}H_{94}ClN_{10}O_{13}PSi_3$			
Adenosine, N-benzoyl-P-(2-chlorophenyl)-2',3'-bis-O-[(1,1-dimethylethyl)dimethylsilyl]adenylyl-(5'→3')-N-benzoyl-2'-O-[(1,1-dimethylethyl)dimethylsilyl]-5'-O-[(4-methoxyphenyl)diphenylmethyl]-	MeOH	231(4.62),280(4.66)	33-2018-83
$C_{79}H_{82}N_4O_{11}$			
9,10-Anthracenedione, 5,8-bis[2-[N-(t-butoxycarbonyl)-2-(trityloxy)ethylamino]ethylamino]-1-methoxy-2-(4-methoxybenzyloxy)-	$CHCl_3$	273(4.56),597(4.23), 645(4.26)	18-1812-83
$C_{79}H_{108}Cl_2N_{12}O_{20}P_2Si_4$			
Adenosine, P-(2-chlorophenyl)-2'-O-[(1,1-dimethylethyl)dimethylsilyl]-5,6-dihydrouridylyl-(3'→5')-N-benzoyl-P-(2-chlorophenyl)-2'-O-[(1,1-dimethylethyl)dimethylsilyl]adenylyl-(3'→5')-N-benzoyl-2',3'-bis-O-[(1,1-dimethylethyl)dimethylsilyl]-	MeOH	232s(4.46),279(4.63)	33-2641-83
$C_{80}H_{64}ClFeN_8O_4$			
Iron(1+), [[N,N',N'',N'''-(21H,23H-por-	CH_2Cl_2	379(4.66),419(4.95),	35-5791-83

Compound	Solvent	λ_{max}(log ε)	Ref.
phine-5,10,15,20-tetrayltetra-2,1-phenylene)tetrakis[α-methylbenzene-acetamidato]](2-)-N[21],N[22],N[23],N[24]]-, chloride		509(4.09),579(3.65), 649(3.58),672(3.52)	35-5791-83
isomer	CH_2Cl_2	379(4.69),419(5.00), 509(4.11),579(3.60), 649(4.57),672(3.49)	35-5791-83
$C_{80}H_{66}N_8O_4$			
Benzeneacetamide, N,N',N",N"'-(21H,23H-porphine-5,10,15,20-tetrayltetra-2,1-phenylene)tetrakis[α-methyl-	CH_2Cl_2	421(5.51),482(3.52), 513(4.29),546(3.63), 588(3.79),644(3.24)	35-5791-83
$C_{80}H_{98}N_{11}O_{15}PSi_3$			
Adenosine, N-benzoyl-2',3'-bis-O-[(1,1-dimethylethyl)dimethylsilyl]-P-[2-(4-nitrophenyl)ethyl]adenylyl-(5'→3')-N-benzoyl-2'-O-[(1,1-dimethylethyl)dimethylsilyl]-5'-O-[(4-methoxyphenyl)-diphenylmethyl]-	MeOH	278(4.71)	33-2018-83
$C_{81}H_{87}Cl_2N_{11}O_{16}P_2Si_2$			
3'-Adenylic acid, N-benzoyl-P-(2-chlorophenyl)-2'-O-[(1,1-dimethylethyl)-dimethylsilyl]-5'-O-[(4-methoxyphenyl)diphenylmethyl]adenylyl-(3'→5')-N-benzoyl-2'-O-[(1,1-dimethylethyl)-dimethylsilyl]-, 2-chlorophenyl 2-cyanoethyl ester	MeOH	230(4.63),280(4.66)	33-2018-83
$C_{87}H_{111}Cl_2N_{15}O_{19}P_2Si_4$			
Adenosine, N-benzoyl-P-(2-chlorophenyl)-2'-O-[(1,1-dimethylethyl)dimethylsilyl]adenylyl-(3'→5')-N-benzoyl-P-(2-chlorophenyl)-2'-O-[(1,1-dimethylsilyl]adenylyl-(3'→5')-N-benzoyl-2',3'-bis-O-[(1,1-dimethylethyl)dimethylsilyl]-	MeOH	230(4.73),280(4.80)	33-2018-83
$C_{87}H_{112}O_8$			
Phenol, 2,2'-methylenebis[4-(1,1-dimethylethyl)-6-[[5-(1,1-dimethylethyl)-3-[[5-(1,1-dimethylethyl)-3-[[5-(1,1-dimethylethyl)-2-hydroxyphenyl]methyl]-2-hydroxyphenyl]-methyl]-2-hydroxyphenyl]methyl]-	EtOH	281s(4.29),286(4.31)	126-1363-83
$C_{88}H_{48}CuN_8$			
Tetra-2,3-tryptycenoporphyrazine, copper chelate	$CHCl_3$	340(5.04),377(4.70), 606(4.60),628(4.75), 678(5.44)	65-2346-83
$C_{88}H_{50}N_8$			
Tetra-2,3-tryptycenoporphyrazine	$CHCl_3$	348(4.85),362(4.57), 412(4.48),548(3.55), 600(4.45),636(4.75), 642(4.77),662(5.11), 698(5.18)	65-2346-83

Compound	Solvent	$\lambda_{max}(\log \epsilon)$	Ref.
$C_{88}H_{94}Cl_2N_{12}O_{20}P_2Si_2$			
3'-Adenylic acid, N-benzoyl-2'-O-[(1,1-dimethylethyl)dimethylsilyl]-5'-O-[(4-methoxyphenyl)diphenylmethyl]-P-[2-(4-nitrophenyl)ethyl]adenylyl-(3'→5')-N-benzoyl-2'-O-[(1,1-dimethylethyl)dimethylsilyl]-, 2,5-dichlorophenyl 2-(4-nitrophenyl)ethyl ester	MeOH	278(4.78)	33-2018-83
$C_{88}H_{112}N_4O_{16}S_4Zn$			
Zincate(4-), [[11,11',11",11"'-[21H-23H-porphine-5,10,15,20-tetrayltetrakis(4,1-phenyleneoxy)]tetrakis[1-undecanesulfonato]](6-)-N^{21},N^{22},N^{23}-N^{24}]-, tetrasodium, (SP-4-1)-	H_2O	410s(5.16),428(5.19), 558(3.98),601(3.79)	39-2535-83C
	MeOH	426(5.76),559(4.21), 600(4.01)	39-2535-83C
$C_{88}H_{118}N_4O_8$			
1-Undecanol, 11,11',11",11"'-[21H,23H-porphine-5,10,15,20-tetrayltetrakis-(4,1-phenyleneoxy)]tetrakis-	benzene	426(5.35),518(3.85), 547(3.92),582(3.48), 652(3.36)	39-2535-83C
$C_{89}H_{99}ClN_5O_{15}PSi_3$			
Uridine, N-benzoyl-P-(2-chlorophenyl)-2'-O-[(1,1-dimethylethyl)dimethylsilyl]-5'-O-[(4-methoxyphenyl)diphenylmethyl]cytidylyl-(3'→5')-2',3'-bis-O-[(1,1-dimethylethyl)diphenylsilyl]-5,6-dihydro-	MeOH	262(4.43),304(4.03)	33-2641-83
$C_{96}H_{72}CeN_8$			
Cerium, bis[5,10,15,20-tetrakis(4-methylphenyl)-21H,23H-porphinato(2-)-N^{21},N^{22},N^{23},N^{24}]-	$CHCl_3$	627(3.15),542(3.85), 485(4.07),398(5.32)	64-1339-83B
$C_{96}H_{72}N_8Pr$			
Praseodymate(1-), bis[5,10,15,20-tetrakis(4-methylphenyl)-21H,23H-porphinato(2-)-N^{21},N^{22},N^{23},N^{24}]-	$CHCl_3$	413(5.60),510(3.94), 556(4.03)	64-1339-83B
$C_{96}H_{96}ErN_{16}$			
Erbium, bis[2,9,16,23-tetrakis(1,1-dimethylethyl)-29H,31H-phthalocyaninato(2-)-N^{29},N^{30},N^{31},N^{32}]-	o-$C_6H_4Cl_2$	318(4.87),342(4.82), 464(4.54),600(4.35), 638s(4.26),666(5.05)	65-2339-83
one electron reduction	o-$C_6H_4Cl_2$	333(4.96),626(4.94), 696(4.58)	65-2339-83
one electron oxidation (also other rare earth chelates)	o-$C_6H_4Cl_2$	318(4.87),352(4.78), 486(4.54),702(4.47)	65-2339-83
$C_{96}H_{126}Cl_2N_{12}O_{22}P_2Si_4$			
Guanosine, P-(2-chlorophenyl)-2'-O-[(1,1-dimethylethyl)dimethylsilyl]-5,6-dihydro-5'-O-[(4-methoxyphenyl)-diphenylmethyl]uridylyl-(3'→5')-N-benzoyl-P-(2-chlorophenyl)-2'-O-[(1,1-dimethylethyl)dimethylsilyl]-adenylyl-(3'→5')-2',3'-bis-O-[(1,1-dimethylethyl)dimethylsilyl]-N-(2-methyl-1-oxopropyl)-	MeOH	232(4.52),255s(4.46), 261(4.48),280(4.53)	33-2641-83
$C_{98}H_{126}O_9$			
Phenol, 4-(1,1-dimethylethyl)-2,6-bis-	EtOH	281s(4.33),286(4.36)	126-1363-83

Compound	Solvent	λ_{max}(log ε)	Ref.
[[5-(1,1-dimethylethyl)-3-[[5-(1,1-dimethylethyl)-3-[[5-(1,1-dimethylethyl)-3-[[5-(1,1-dimethylethyl)-3-[[5-(1,1-dimethylethyl)-2-hydroxyphenyl]methyl]-2-hydroxyphenyl]methyl]-2-hydroxyphenyl]methyl]-2-hydroxyphenyl]methyl]-			126-1363-83
$C_{99}H_{119}Cl_3N_{13}O_{25}P_3Si_3$			
3'-Guanylic acid, P-(2-chlorophenyl)-2'-O-[(1,1-dimethylethyl)dimethylsilyl]-5,6-dihydro-5'-O-[(4-methoxyphenyl)diphenylmethyl]uridylyl-(3'-5')-N-benzoyl-P-(2-chlorophenyl)-2'-O-[(1,1-dimethylethyl)dimethylsilyl]adenylyl-3'→5')-2'-O-[(1,1-dimethylethyl)dimethylsilyl]-N-(2-methyl-1-oxopropyl)-, 2-chlorophenyl 2-cyanoethyl ester	MeOH	232(4.54),258(4.47), 261(4.49),280(4.54)	33-2641-83
$C_{99}H_{124}Cl_2N_{12}O_{21}P_2Si_4$			
Adenosine, P-(2-chlorophenyl)-2'-O-[(1,1-dimethylethyl)dimethylsilyl]-5,6-dihydro-5'-O-[(4-methoxyphenyl)diphenylmethyl]uridylyl-(3'→5')-N-benzoyl-P-(2-chlorophenyl)-2'-O-[(1,1-dimethylethyl)dimethylsilyl]adenylyl-(3'→5')-N-benzoyl-2',3'-bis-O-[(1,1-dimethylethyl)dimethylsilyl]-	MeOH	228s(4.70),280(4.62)	33-2641-83
$C_{100}H_{94}N_{12}O_4$			
4,4'-Bipyridinium, 1,1",1‴',1‴''-[21H,23H-porphine-5,10,15,20-tetrayltetrakis(4,1-phenyleneoxy-3,1-propanediyl)]tetrakis[4'-methyl-, tetrachloride tetraiodide	H_2O	423(5.49),521(3.98), 560(4.09),598(3.85), 645(3.60)	39-2535-83C
$C_{100}H_{156}N_{34}O_{22}S$			
Giractide	pH 13	281(3.83),288(3.81)	162-0632-83
$C_{107}H_{127}Cl_2N_{15}O_{20}P_2Si_4$			
Adenosine, N-benzoyl-P-(2-chlorophenyl)-2',3'-bis-O-[(1,1-dimethylethyl)dimethylsilyl]adenylyl-(5'→3')-N-benzoyl-P-(2-chlorophenyl)-2'-O-[(1,1-dimethylethyl)dimethylsilyl]adenylyl-(5'→3')-N-benzoyl-2'-O-[(1,1-dimethylethyl)dimethylsilyl]-5'-O-[(4-methoxyphenyl)diphenylmethyl]-	MeOH	230(4.73),280(4.80)	33-2018-83
$C_{109}H_{140}O_{10}$			
Phenol, 2,2'-methylenebis[4-(1,1-dimethylethyl)-6-[[5-(1,1-dimethylethyl)-3-[[5-(1,1-dimethylethyl)-3-[[5-(1,1-dimethylethyl)-3-[[5-(1,1-dimethylethyl)-2-hydroxyphenyl]methyl]-2-hydroxyphenyl]methyl]-2-hydroxyphenyl]methyl]-2-hydroxyphenyl]methyl-	EtOH	281s(4.40)	126-1363-83

Compound	Solvent	λ_{max}(log ϵ)	Ref.
$C_{112}H_{108}N_9Pr$			
Praseodymium(III), bis[ms-tetrakis(4-methylphenyl)porphinato]-, tetrabutyl ammonium	CH_2Cl_2	410(5.71),562(4.00), 614(3.92)	64-1339-83B
$C_{120}H_{92}AsN_8Pr$			
Praseodymium(III), bis[ms-tetrakis(4-methylphenyl)porphinato]-, tetraphenylarsonium	CH_2Cl_2	410(5.76),563(3.95), 614(3.89)	64-1339-83B
$C_{136}H_{90}N_8O_{12}$			
[1,1'-Binaphthalene]-2-carboxylic acid, 2',2''',2''''',2'''''''-[21H,23H-porphine-5,10,15,20-tetrayltetrakis(2,1-phenyleniminocarbonyl)]tetrakis-, tetramethyl ester	CH_2Cl_2	423(5.40),484(3.51), 515(4.19),547(3.60), 590(3.71),645(3.21)	35-5791-83
$C_{139}H_{153}Cl_4N_{21}O_{30}P_4Si_4$			
3'-Adenylic acid, N-benzoyl-P-(2-chlorophenyl)-2'-O-[(1,1-dimethylethyl)-dimethylsilyl]-5'-O-[(4-methoxyphenyl)diphenylmethyl]adenylyl-(3'→5')-N-benzoyl-P-(2-chlorophenyl)-2'-O-[(1,1-dimethylethyl)dimethylsilyl]-adenylyl-(3'→5')-N-benzoyl-P-(2-chlorophenyl)-2'-O-[(1,1-dimethylethyl)dimethylsilyl]adenylyl-(3'→5')-N-benzoyl-2'-O-[(1,1-dimethylethyl)-dimethylsilyl]-, 2-chlorophenyl 2-cyanoethyl ester	MeOH	230(4.83),280(4.93)	33-2018-83
$C_{142}H_{172}N_{23}O_{33}P_8Si_5$			
Adenosine, N-benzoyl-2',3'-bis-O-[(1,1-dimethylethyl)dimethylsilyl]-P-[2-(4-nitrophenyl)ethyl]adenylyl-(5'→3')-N-benzoyl-2'-O-[(1,1-dimethylethyl)dimethylsilyl]-P-[2-(4-nitrophenyl)ethyl]adenylyl-(5'→3')-N-benzoyl-2'-O-[(1,1-dimethylethyl)dimethylsilyl]-P-[2-(4-nitrophenyl)ethyl]adenylyl-(5'→3')-N-benzoyl-2'-O-[(1,1-dimethylethyl)dimethylsilyl]-5'-O-[(4-methoxyphenyl)diphenylmethyl]-	MeOH	278(5.03)	33-2018-83
$C_{155}H_{206}Cl_5N_{24}O_{43}P_5Si_7$			
Adenosine, P-(2-chlorophenyl)-2'-O-[(1,1-dimethylethyl)dimethylsilyl]-5,6-dihydrouridylyl-(3'→5')-N-benzoyl-P-(2-chlorophenyl)-2'-O-[(1,1-dimethylethyl)dimethylsilyl]adenylyl-(3'→5')-P-(2-chlorophenyl)-2'-O-[(1,1-dimethylethyl)dimethylsilyl]-N-(2-methyl-1-oxopropyl)guanylyl-(3'→5')-P-(2-chlorophenyl)-2'-O-[(1,1-dimethylethyl)dimethylsilyl]-5,6-dihydrouridylyl-(3'→5')-N-benzoyl-P-(2-chlorophenyl)-2'-O-[(1,1-dimethylethyl)dimethylsilyl]adenylyl-(3'→5')-N-benzoyl-2',3'-bis-O-[(1,1-dimethylethyl)dimethylsilyl]-	MeOH	234s(4.66),253s(4.69), 272(4.74),280(4.85), 326s(3.36)	33-2641-83

1- -83, Acta Chem. Scand., B37 (1983)
0097 M. Begtrup and N.O. Knudsen
0147 C.J. Welch and J. Chattopadhyaya
0297 B. Thulin and O. Wennerström
0351 A.B. Hansen and A. Senning
0585 M. Jacoben et al.
0589 B. Thulin and O. Wennerström
0687 J. Klaveness and K. Undheim
0693 D. Tanner and O. Wennerström
0803 P. Muthusubramanian et al.
0823 P. Kolsaker et al.
0833 H. Svendsen et al.
0917 P.-I. Ohlsson and K.-G. Paul

2- -83, Indian J. Chem.Sect. B, 22 (1983)
0027 J. Banerji et al.
0030 A.K. Sen and D.K. Sengupta
0101 A. Zaman et al.
0164 Y. Geetanjali et al.
0215 H.R. Shitole and U.R. Nayak
0230 C.S. Rao et al.
0243 P.B. Talukdar et al.
0274 A.C. Jain et al.
0276 S.A. Khan and M. Krishnamurti
0286 P. Bhandari and R.P. Rastogi
0290 V.S. Ekkundi et al.
0319 V.K. Singh and S. Dev
0321 S.V. Kessar et al.
0331 R.G. Bhandari and G.V. Bhide
0352 R.S. Mali et al.
0410 P.K. Subramanian et al.
0425 E. Rajanarendur et al.
0531 T.R. Govindachari and S. Rajeswari
0542 L. Anandan et al.
0619 R. Das Gupta et al.
0710 M. Bose et al.
0741 B. Talapatra et al.
0759 A.C. Jain et al.
0805 U.T. Nabar et al.
0808 G. Philip et al.
0812 U.T. Nabar et al.
0815 S.N. Dehuri et al.
0824 T.P. Veluchamy and G.S. KrishnaRao
0864 T.V. Rao et al.
0886 B.K. Rohatgi et al.
0927 K.M. Biswas and H. Mallik
0989 R. Soman et al.
1061 S.A. Khan and M. Krishnamurti
1103 D.K. Banerjee et al.
1108 N.R. Ayyangar and K.V. Srinivasan
1116 A.C. Jain et al.
1119 A.C. Jain et al.
1154 R.G. Bhandari and G.V. Bhide
1191 P.L. Majumder and M. Joardar
1197 D.R. Tatke and S. Seshadri
1217 M.S.A. Abd-El-Motteleb and M.H. Abd-El-Kader
1257 N.S. Narasimhan et al.

3- -83, Anal. Chem., 55 (1983)
1816 A.T. Gowda et al.

4- -83, J. Heterocyclic Chem., 20 (1983)
0001 T. Zaima and Y. Matsunaga
0005 E. Belgodere et al.
0023 E. Barni et al.
0029 E. Barni et al.
0033 T.A. Hamor et al.
0041 L. Bell et al.
0073 T. Saito et al.
0081 T. Kurihara et al.
0105 S. Chimichi et al.
0121 B.F. Powell et al.
0129 S.M.A.D. Zayed and A. Attia
0183 G. Bobowski
0213 L.H. Klemm and D.R. Muchiri
0219 H. Gershon et al.
0225 L. Baiocchi and M. Giannangeli
0245 T. Sheradsky and D. Zbaida
0251 G.D. Hobbs et al.
0267 G. Bobowski
0311 A. Ohta et al.
0369 D.J. Katz et al.
0393 M. Langlois et al.
0407 G.M. Sanders et al.
0539 G. Menozzi et al.
0543 S. Gelin and R. Dolmazon
0551 A. Walser and R.I. Fryer
0629 D.C. Baker et al.
0639 A.O. Abdelhamid et al.
0645 G. Menozzi et al.
0649 L. Mosti et al.
0655 Y.F. Shealy et al.
0687 C. Anselmi et al.
0729 A. Svensson et al.
0753 I.M. Sasson et al.
0759 W.J. Firth, III et al.
0775 J.R. Merchant et al.
0791 A. Walser et al.
0803 S.B. Tambi et al.
0839 A. Bargagna et al.
0861 M.L. Tedjamulia et al.
0875 Y. Tarumi and T. Atsumi
0919 J. Vekemans et al.
0931 N. Vivona et al.
0951 A. Ohta et al.
0971 S. Mataka et al.
0989 C. Dagher et al.
1019 M. Weizberg et al.
1037 G.E. Wright
1047 C.O. Okafor et al.
1059 J.J. Plattner and J.A. Parks
1085 J.A. Charonnat et al.
1093 C. Camoutsis and P. Catsoulacos
1107 E.W. Gill and A.W. Bracher
1111 J.O. Oluwadiya
1169 F. Babin et al.
1263 J.A.R. Rodrigues and L.I. Verardo
1277 T. Watanabe et al.
1287 L. Della Vecchia et al.
1359 W.D. Jones, Jr. et al.
1383 K.M. Smith and R.K. Pandey
1389 M. Pote et al.
1411 D.J. Barker and L.A. Summers
1481 A.W. Addison et al.
1501 E.E. Eid et al.
1517 E. Barni et al.
1549 A. Bargagna et al.
1575 J. Hung et al.
1597 E. Abignente et al.
1601 B. Renger
1609 P. Beltrame et al.

1621 W. Friedrichsen et al.
1629 G. Ege et al.
1651 M.A. Michael et al.
1657 M. Bonanomi and L. Baiocchi
1717 L.H. Klemm and D.R. Muchiri
1739 A.A.H. Saeed and E.K. Ebraheem

5- -83, Ann. Chem. Liebigs (1983)
0001 J.M. Ribo and F. Trull
0013 A. Roedig and W. Ritschel
0056 Y. Itakura et al.
0069 M. Horner and S. Hunig
0112 K. Schank and W. Lorig
0137 F. Seela et al.
0154 H. Junek et al.
0165 F. Eiden et al.
0181 J.A. Robinson et al.
0204 J.-H. Fuhrhop et al.
0220 D. Schumann and A. Naumann
0226 M. Entzeroth et al.
0231 H. Kessler et al.
0299 H. Laatsch
0433 H. Brockmann and H. Laatsch
0471 B. Krone and A. Zeeck
0510 B. Krone and A. Zeeck
0521 W. Flitsch and P. Russkamp
0585 H. Plieninger and I. Preuss
0642 M. Horner and S. Hünig
0658 M. Horner et al.
0687 U.E. Meissner et al.
0695 H.-W. Schmidt et al.
0705 K. Annen et al.
0712 K. Annen et al.
0721 C. Reichardt and E. Harbusch-Görnert
0744 H. Rönsch et al.
0761 T.-L. Ho and S.-H. Liu
0802 J. Führhop and H. Bartsch
0852 A. Perez-Rubalcaba and W. Pfliederer
0860 P. Plieninger and H. Baumgärtel
0876 F. Seela and A. Kehne
0894 S. Sepulveda-Buza and E. Breitmaier
0897 S. Shatzmiller and E. Shalom
0931 H. Frischkorn et al.
1001 S. Solyom et al.
1020 H. Laatsch
1031 K. Schonauer and E. Zbiral
1081 R. Jansen et al.
1107 M. Mittelbach et al.
1116 G. Adiwidjaja et al.
1169 F. Seela et al.
1207 H. Quast et al.
1361 R. Allmann et al.
1374 J. Streith et al.
1393 J. Streith and T. Tschamber
1476 H. Neunhoeffer and H.-J. Metz
1496 A. Roedig and W. Ritschel
1504 A. Roedig and W. Ritschel
1510 M. Nassal
1576 F. Seela and H. Steker
1656 W. Kohl et al.
1694 W.-D. Malmberg et al.
1744 H.-J. Teuber et al.
1760 A. Mondon, B. Epe and H. Callsen
1798 A. Mondon et al.
1807 A.T. Balaban et al.
1818 K. Krohn and B. Behnke
1886 H. Laatsch
2038 H. Gnichtel and C. Grasshoff
2066 N.S. Girgis et al.
2135 A. Nonnenmacher et al.
2141 G. Kiehl et al.
2151 K. Krohn
2164 J.W. Buchler et al.
2247 S. Antus et al.
2262 H.A. Staab et al.

7- -83, Ann. Chim.(Rome), 73 (1983)
0055 G. Alberti et al.
0155 G. Alberti et al.
0265 G. Alberti et al.

11- -83, Chemica Scripta, 22 (1983)
0055 S. Gronowitz et al.
0171 P. Finlander et al.

12- -83, Australian J. Chem., 36 (1983)
0081 S.D. Barker and R.K. Norris
0097 J.W. Blunt et al.
0117 R.F.C. Brown et al.
0149 S.F. Dyke et al.
0165 R. Kazlauskas et al.
0211 B.F. Bowden et al.
0297 W.P. Norris et al.
0311 R.P. Kopinski and J.T. Pinhey
0339 D.J. Collins and J. Sjovall
0361 D.J. Collins et al.
0371 B.F. Bowden et al.
0397 P.G. Griffiths et al.
0409 N.A. Evans
0527 S.D. Barker and R.K. Norris
0565 J.W. Blunt et al.
0581 J.W. Blunt et al.
0789 J.T. Pinhey and B.A. Rowe
0795 D.D. Ridley and M.A. Smal
0839 M.P. Hartshorn et al.
0963 B.J. Barnes et al.
0993 E.L. Ghisalberti et al.
1001 R.E. Corbett et al.
1049 D.D. Ridley and M.A. Smal
1057 P. Djura and M.V. Sargent
1061 S. Nimgirawath and W.C. Taylor
1221 R.D. Allan and J. Fong
1227 R.W. Read et al.
1263 R.F.C. Brown et al.
1275 R.A.J. Smith
1361 H.-D. Becker et al.
1419 R.W. Irvine et al.
1431 T. Duong et al.
1639 D.D. Chan et al.
1649 D.E. Rivett et al.
1873 M. Bliese et al.
1957 M.J. Cuthbertson et al.
2289 B.F. Bowden et al.
2339 M.P. Hartshorn et al.
2387 D. St.C. Black and N.E. Rothnie
2395 D. St.C. Black and N.E. Rothnie
2407 D. St.C. Black et al.
2413 D. St.C. Black and N.E. Rothnie

13- -83A, Steroids, 41 (1983)
0055 M. Kocor et al.

0255 T.G. Watson et al.
0267 C.G. Francisco et al.
0277 C.G. Francisco et al.
0291 A. Shafiee et al.
0309 A.S.C. Wan et al.
0321 A.S.C. Wan et al.
0327 J.E. Herz and J. Sandoval
0333 J.E. Herz et al.
0339 A.S.C. Wan et al.
0349 A. Shafiee et al.
0475 R.A. Leppik
0675 M. Numazawa and K. Kimura

13- 83B, Steroids, 42 (1983)
0001 P. Kaspar and H. Witzel
0037 M. Kihara et al.
0171 D.C. Humber et al.
0189 D.C. Humber et al.
0241 J.R. Dodd and A.E. Mathew
0349 V.R. Mattox et al.
0401 C.M. DiNunno et al.
0493 S.H. Tailor et al.
0707 C.K. Lai, C.Y. Byon and M. Gut

18- -83, Bull. Chem. Soc. Japan, 56 (1983)
0006 T. Moriya
0171 S. Inoue and T. Hori
0175 K. Saito et al.
0326 M.Z.A. Badr et al.
0467 Y. Abe et al.
0481 K. Kusuda
0537 S. Akabori et al.
0542 A. Sakurai and Y. Okumura
0965 M.E. Cracknell et al.
1125 R. Takeda et al.
1139 Y. Fukuyama et al.
1143 A. Kasahara et al.
1192 H. Kidokoro et al.
1206 T. Abe et al.
1247 K. Fujimori et al.
1259 M. Yoshida et al.
1267 A.C. Jain et al.
1362 Y. Ishibashi et al.
1435 K. Nakamura et al.
1450 M. Tada et al.
1467 J. Ojima et al.
1476 Y. Miura et al.
1482 H. Nishi et al.
1514 S. Shiraishi et al.
1688 O. Tsuge et al.
1694 S. Shinkai et al.
1700 S. Shinkai et al.
1775 Y. Yokoyama
1799 H. Seto et al.
1812 K. Nakamura et al.
1879 K. Imafuku et al.
2023 I. Shimizu et al.
2037 H. Matsumura et al.
2044 T. Wakamiya et al.
2055 M. Nukui et al.
2059 S. Kurokawa and A.G. Anderson, Jr.
2173 S. Sekiguchi et al.
2311 S. Kurokawa et al.
2338 K. Honda et al.
2535 H. Yamamoto and A. Nakazawa
2680 T. Sato et al.
2700 T. Sato and R. Noyari
2756 T. Someya et al.
2798 T. Masuda et al.
2865 A. Kasahara et al.
2969 A. Saito and B. Shimizu
2985 T. Matsumoto et al.
3009 H. Matsumoto et al.
3099 T. Nozoe et al.
3106 H. Yamamoto et al.
3185 S. Nakatsuji et al.
3193 Z. Ro, K. Tanaka and F. Toda
3253 S. Kitazawa et al.
3314 F. Toda et al.
3349 H. Suginome et al.
3353 T. Suga et al.
3358 Y. Sudoh et al.
3449 H. Kobayashi et al.
3464 Y. Sawaki et al.
3519 H. Matsumura et al.
3661 T. Suga et al.
3672 M. Amano et al.
3683 M. Ishibashi et al.
3703 N. Abe et al.
3735 Y. Kai et al.
3773 T. Horie et al.
3824 M. Suzuki et al.

19- -83, Bull. Polish Acad. Sci., 31 (1983)
0233 M. Olejnik et al.

20- -83, Bull. soc. chim. Belges, 92 (1983)
0067 A. Maquestian et al.
0263 L. Dupont et al.
0355 Babady-Bila et al.
0451 A. Maquestian et al.
0473 M.Z. Dimbi et al.
0893 J. Brison and R.H. Martin

22- -83, Bull. soc. chim. France, Part II (1983)
0061 D.H.R. Barton et al.
0073 M.M. Al Sabbagh et al.
0096 R. Gruber et al.
0112 L. LeThiNgoc and G. Ourisson
0115 V. Wintgens et al.
0175 J.-P. Morizur and J. Tortajuda
0180 J. Denian et al.
0189 P. Koch et al.
0195 O. Yebdri and F. Texier
0317 H.J. Callot et al.
0339 I. Allade et al.

23- -83, Can. J. Chem., 61 (1983)
0078 P. Yates et al.
0276 G.N. Saunders et al.
0288 E. Piers et al.
0300 N.Y.C. Chu
0312 M.J. Robins and J.M.R. Parker
0317 M.J. Robins and J.M.R. Parker
0323 J.A. Findlay et al.
0334 P. Ribereau and G. Quegainer
0368 S. Fung et al.
0372 A. Stoessl et al.
0378 A. Stoessl et al.
0400 D. Gravel et al.
0454 E. Galeazzi et al.

0533 S. Czernecki and V. Dechavanne
0545 A.B. McKagne et al.
0866 H.E. Zimmerman and G.-S. Wu
0894 A. Safarzadeh-Amiri et al.
0936 P. Yates and P.H. Helferty
0993 H.E. Hunziker et al.
1053 J. Balsevich
1111 J.P. Kutney et al.
1226 E. Piers et al.
1387 J.R. Nesser et al.
1440 A.C. Storer et al.
1465 F.-X. Garneau et al.
1549 K.M. Baines et al.
1562 P.W. Codding et al.
1697 C. Gonzalez et al.
1743 M.Y. Boluk et al.
1771 T.A. Montzka et al.
1890 J. Hoot et al.
1933 S.J. Lo et al.
1965 A.D. Broadbent and J.M. Stewart
1973 O.E. Edwards and Z. Paryzek
2126 M.E. Mekki Attia et al.
2257 Y. Ueda et al.
2285 C.A. Boulet and G.A. Poulton
2389 M. Barley et al.
2449 S.R. Abrams et al.
2461 R.H. Burnell et al.
2790 S. Yamamoto and R.A. Buck

24- -83, Chem. Ber., 116 (1983)
0066 E. Schaumann et al.
0097 R. Dhar et al.
0136 H. tom Dieck et al.
0152 U. Kücklander and H. Töberich
0186 G. Scherowsky and J. Pickardt
0230 K.H. Pannell et al.
0243 P. Eilbracht and R. Jelitte
0264 G. Becher and A. Mannschreck
0299 F.-G. Klärner and F. Adamsky
0445 G. Märkl et al.
0479 T. Kauffmann et al.
0563 W. Eberbach and J.C. Carré
0681 H. Leininger et al.
0761 B. Dors et al.
0819 E. Stöldt and R. Kreher
0827 N.E. Blank and M.W. Haenel
0856 H. Dürr et al.
0882 U.H. Brinker and L. König
0894 U.H. Brinker and L. König
0970 J.K. Ruminski
0980 T. Kauffmann and R. Otter
1154 W. Bauer et al.
1174 W. Flitsch and E.R.F. Gesing
1257 D. Bielefeldt and A. Haas
1297 J. Goerdeler and C. Ho
1309 H.-J. Teuber et al.
1506 E. Anders et al.
1520 W. Sucrow and G. Bredthauer
1525 W. Sucrow et al.
1547 W. Ried et al.
1682 S. El-tamany and H. Hopf
1691 H. Wamhoff et al.
1756 G. Markl and K. Hock
1777 W. Bauer et al.
1787 K.-H. Eichin et al.
1848 W. Adam et al.
1897 D. Kaufmann and A. de Meijere
1963 A. Beck et al.
1982 C. Reichardt et al.
2115 R.R. Schmidt and B. Beitzke
2261 F.A. Neugebauer et al.
2366 R. Askani and W. Schneider
2383 J. Brokatzky-Geiger and W. Eberbach
2408 J. Bindl et al.
2591 H. Leismann et al.
2785 H.A. Staab et al.
2835 H.A. Staab et al.
2881 R. Neidlein et al.
2903 M. Glanzmann and G. Schröeder
2914 M. Glanzmann and G. Schröeder
3097 W. Norden et al.
3112 F. Vögtle et al.
3192 E. Anders et al.
3235 G. Kratt et al.
3325 D.M. Ceacareanu et al.
3366 H. Glombik and W. Tochtermann
3427 H. Quast and U. Nahr
3504 H.A. Staab et al.
3516 G. Pilidis
3524 I. Lukac and H. Langhals
3725 W. Ried, J. Nenninger and J.W. Bats
3751 I. Sellner et al.
3800 K. Sarma et al.
3884 B. Hagenbruch and S. Hünig
3915 H. Dürr et al.

25- -83, Chem. and Ind.(London) (1983)
0202 B. Robinson

27- -83, Chimia, 37 (1983)
0341 A. Gossauer

28- -83A, Compt. rend., 296 (1983)
0977 G. Massiot et al.

28- -83B, Compt. rend., 297 (1983)
0043 C. Blaquiere and M. Massol
0661 M. Ambrioso and C. Feugeas

30- -83, Doklady Akad. Nauk. S.S.S.R., 268-273 (1983)
0289 S.V. Chapyshev and B.G. Kartsev
0316 A.T. Pilipenko et al.
0331 V.A. Bren' et al.

31- -83, Experientia, 39 (1983)
0991 Y.H. Kuo and S.T. Lin
1091 M. D'Ambrosio et al.
1275 E. Fattorusso et al.

32- -83, Gazz. chim. ital., 113 (1983)
0011 P. Beltrame et al.
0069 P. De Maria, A. Fini and F.M. Hall
0161 G. Tosi et al.
0183 F. Orsini et al.
0187 F. Orsini et al.
0427 A.M. Celli et al.
0489 G. Adembri et al.
0507 F. Campagna et al.
0515 L. Santucci
0533 J.U. Oguakwa et al.

0569 V. Carelli et al.
0717 S. Chiosi et al.
0721 C. LoSterzo and G. Ortaggi
0757 F. Orsini et al.
0773 W.A. Chapya et al.
0799 A. Forni et al.
0811 R. Calvino et al.
0845 R.M. Srivastava et al.
0855 M. Alpegiani et al.
0863 L. Mayol et al.

33- -83, Helv. Chim. Acta, 66 (1983)
0019 O. Pilet et al.
0044 B. Kräutler et al.
0092 L. Bassi et al.
0262 K. Dietliker and H. Heimgartner
0342 D. Gaude et al.
0379 D. Schelz
0405 R.P. Borris et al.
0411 J.-M. Adam and T. Winkler
0429 P. Rüedi et al.
0494 E. Märki-Fischer et al.
0534 A.M. Bobst et al.
0586 A. Fredenhagen and U. Sequin
0606 A. Haider and H. Wyler
0620 D.G. Markecs
0627 G.G.G. Manzardo et al.
0687 V. Skaric and J. Matulic-Adamic
0718 A.G. Anastassiou et al.
0735 R. Kilger and P. Margaretha
0780 R. Zelnik et al.
0789 B. Aebischer and A. Vasella
0960 R. Leduc et al.
1061 N. Bischofberger et al.
1078 A. Villiger et al.
1119 W. Oppolzer and C. Robbiani
1134 R. Gabioud and P. Vogel
1148 P.A. Bentikofer and C.H. Eugster
1175 E. Märki-Fischer et al.
1416 N. Egger et al.
1427 U. Stämpfli and M. Neuenschwander
1475 R.O. Duthaler
1599 N. Egger et al.
1608 N. Egger et al.
1631 U. Stämpfli et al.
1638 N. Bischofberger et al.
1672 P. Amade et al.
1806 J.P. Kutney et al.
1820 J.P. Kutney et al.
1835 R. Kaiser and D. Lamparsky
1865 E. Viera and P. Vogel
1876 O. Wallquist et al.
1902 R. Ghaffari-Tabrizi and P. Margaretha
1915 M. Jung et al.
1922 R.M. Bettolo et al.
1939 A. Ruttimann et al.
1961 K. Maeda and E. Fischer
2002 R. Kaminski et al.
2018 D. Flockerzi et al.
2059 E. Seguin et al.
2069 D. Flockerzi et al.
2086 A.K. Basak and T.A. Kaden
2093 A. Furst et al.
2135 H. Balli and M. Zeller
2165 H. Ziegler and H. Balli
2236 G. de Weck et al.
2252 W. Stegmann et al.
2322 V. Bilinski et al.
2369 R. Neidlein and C.M. Radke
2414 G. Massiot et al.
2519 C. Fehr
2525 J. Stockigt et al.
2603 K. Gehrig et al.
2608 R.K. Brunner and H.-J. Borschberg
2626 R. Neidlein and C.M. Radke
2632 K. Wiesner et al.
2641 D. Flockerzi et al.
2760 M. Koller et al.

34- -83, J. Chem. Eng. Data, 28 (1983)
0132 N.R. El-Rayes and A.H. Katrib
0422 W.C. Coburnm Jr. et al.
0433 O. Tandon and B.R. Sahu

35- -83, J. Am. Chem. Soc., 105 (1983)
0040 D.E. Richardson and H. Taube
0079 P.H. Schippers and H.P.J.M. Dekkers
0130 R.J. Stonard et al.
0145 P.H. Schippers and H.P.J.M. Dekkers
0181 S. Chao et al.
0265 J.R. Keefe and W.P. Jencks
0279 C.D. Ritchie et al.
0455 D. Mansuy et al.
0514 W.J. Leigh and R. Srinivasan
0545 W.K. Smothers et al.
0593 C.-T. Hsu
0646 F. Derguini et al.
0681 R.A. Moss et al.
0803 L. Casella and M. Gullotti
0810 G.B. Wong et al.
0875 M.R. Detty et al.
0883 M.R. Detty and B.J. Murray
0902 P.N. Confalone and R.B. Woodward
0907 M. Cain et al.
0933 A. Padwa et al.
0956 I. Saito et al.
0963 I. Saito et al.
1117 Y. Ito et al.
1204 U.C. Yoon et al.
1276 G. Jones, II and W.G. Becker
1309 B.H. Baretz and W.J. Turro
1578 H.E. Smith et al.
1608 D. Dominguez et al.
1626 C.G. Knudsen et al.
1676 K.P.C. Vollhardt and T.W. Weidman
1683 E. Vedejs and D.A. Perry
1738 D. Cohen et al.
1839 M. Zandomeneghi et al.
1851 S. Shinkai et al.
2010 Y.N. Belokon' et al.
2175 J.E. Frey and E.C. Kitchen
2382 R.S. Brown and J.G. Ulan
2495 M. Yoshifuji et al.
2502 A. Streitwieser, Jr. and J.T. Swanson
2704 J.P. Collman et al.
2800 L.A. Paquette et al.
2928 N.S. Dinit and R.A. Mackay
3038 J.P. Collman et al.
3177 M.D. Taylor et al.
3226 S.J. Cristol et al.

3234 B. Miller and A.K. Bhattacharya
3252 M. Duraisamy and H.M. Walborsky
3264 M. Duraisamy and H.M. Walborsky
3270 M. Duraisamy and H.M. Walborsky
3273 A.G. Schultz et al.
3283 S.A. Keilbaugh and E.R. Thornton
3304 E.B. Skibor and T.C. Bruice
3344 F.D. Lewis et al.
3346 F.R. Busch and G.A. Berchtold
3359 H. Sakurai et al.
3375 K.A. Klingensmith et al.
3588 R.A.S. Chandraratna et al.
3656 M. Rosenberger et al.
3661 N. Cohen et al.
3722 Y. Teki et al.
3723 S. Murata et al.
3739 R.G. Powell et al.
3951 J.R. Winkle et al.
4033 M. Sheves and K. Nakanishi
4056 R. Takeda and K. Katoh
4059 M.J. Robins et al.
4337 R. Sinta et al.
4400 A. Oku et al.
4431 T.S. Eckert and J.C. Bruice
4441 S.B. Mahato et al.
4446 A. Padwa et al.
4480 Y. Ohta et al.
4507 M.F.M. Roks et al.
4809 R. Wessiak and T.C. Bruice
4835 F.J. Schmitz et al.
5136 K. Nakasuji et al.
5160 R. Sen et al.
5354 W.K. Appel et al.
5402 J.P. Ferezon et al.
5506 A.H. Cowley et al.
5665 J.W. Wilt et al.
5695 V.H. Houlding and M. Gratzel
5786 J.T. Groves and T.E. Nemo
5791 J.T. Groves and R.S. Myers
5879 F. Seela et al.
6038 X. Yang and C. Kutal
6096 R.J. Bailey, P.J. Card and H. Shechter
6104 P.J. Card et al.
6123 A.R. Browne et al.
6177 D.M. Roll and P.J. Scheuer
6211 W. Frolich et al.
6223 M. Swaminathan and S.K. Dogra
6236 E.M. Kosower and H. Kanety
6243 J.T. Groves and T.E. Nemo
6264 J.H. Hoare et al.
6268 H. Weller and K.-H. Grellmann
6429 V.J. Bauer et al.
6513 D.P. Cox et al.
6524 S. Masamune et al.
6638 K.M. Smith et al.
6650 M. Tashiro et al.
6679 G. Eberlein and T.C. Bruice
6718 Y. Sugihara et al.
6833 P.B. Grasse et al.
6907 P.J. Kropp et al.
6982 E. Vogel et al.
6989 I. Saito et al.
7075 F.R. Keene et al.
7093 P. Mathur and G.C. Dismukes
7102 P.S. Engel et al.
7108 P.S. Engel et al.
7171 K. Yamamoto et al.
7191 C.F. Wilcox, Jr. and E.N. Farley
7241 H. Kunkely et al.
7337 S.J. Cristol et al.
7384 J.E. Garbe and V. Boekelheide
7396 A.A. Tymiak and K.L. Rinehart, Jr.
7465 J.J.M. Lamberts and D.C. Neckers
7469 M. Kira et al.
7473 T. Masuda et al.
7640 I.H. Sanchez et al.
7760 L.T. Scott et al.

36- -83, J. Pharm. Sci., 72 (1983)
0050 A. Ortiz et al.
0322 J.C. Reepmeyer
0372 K.B. Sloan et al.
0792 M.A. Elsohly et al.
1205 E.S.A. Ibrahim et al.
1285 M.M. Badawi et al.

39- -83B, J. Chem. Soc., Perkin II (1983)
0045 A.R. Katritzky et al.
0109 J.-P. Desvergne et al.
0335 T. Hayashi and Y. Nawata
0371 C.K. Chang et al.
0399 P. Salvadori et al.
0859 N.J. Bunce et al.
0983 Y. Inoue et al.
1015 P.M. op den Brouw and W.H. Laarhoven
1053 S.A. Guerrero et al.
1191 P. Helsby and J.H. Ridd
1197 P.J. Atkins et al.
1413 I. Hermecz et al.
1435 A.R. Katritzky et al.
1443 A.R. Katritzky et al.
1503 A.D. Ryabov et al.
1569 G. Saba et al.
1581 H. Inoue and K. Nagaya
1641 M. Swaminathan and S.K. Dogra
1679 J. Grimshaw and A.P. de Silva
1807 A. Castellan et al.

39- -83C, J. Chem. Soc., Perkin I (1983)
0011 G. Zvilichovsky and M. David
0039 G.D. Manners
0083 T.L. Gilchrist et al.
0087 O. Ribeiro et al.
0103 L.E. Andrews et al.
0115 G. Brooks and E. Hunt
0121 M. Numazawa et al.
0141 T. Hino et al.
0155 K.D. Croft et al.
0161 J.N. Herron and A.R. Pinder
0189 M. Momenteau et al.
0197 P. Anastasis and P.E. Brown
0201 N. Katagiri et al.
0219 J. Reisch et al.
0225 J. Moron et al.
0257 N. Dang et al.
0285 J. Tsunetsugu et al.
0305 T. Polonski et al.
0319 M.P.L. Caton et al.
0333 T. Laitalainen et al.
0349 M. Ikeda et al.

0355 F.S. El-Feraly et al.
0369 A. Horvath et al.
0403 H. Natsugari et al.
0413 M.M. Mahandru and A. Tajbakhsh
0501 R.M. Letcher et al.
0505 R.M. Letcher et al.
0515 R.M. Letcher et al.
0573 A.A. El-Hamany et al.
0581 A. Albini et al.
0587 M. Anastasia et al.
0613 D.J. Mincher et al.
0643 A. Pelter et al.
0649 J. Brennan et al.
0667 A.J. Bartlett et al.
0699 S. Sotheeswaran et al.
0703 U. Samaraweera et al.
0751 J.A. Amupitan et al.
0777 J. Czyzewski and D.H. Reid
0795 A.S. Bailey et al.
0837 K. Maeda et al.
0841 K.-Z. Khan et al.
0867 J.R. Hanson et al.
0883 M.J. Begley et al.
0903 T. Terasawa et al.
0915 U. Chiacchio et al.
0979 A.R. Evans and G.A. Taylor
0985 A.B. Turner et al.
1083 N. Yonezawa et al.
1137 I.F. Barnard and J.A. Elvidge
1193 S.S. Bansal et al.
1223 S.R. Landor et al.
1255 G.M. Elgy et al.
1267 J.L. Carey et al.
1315 I. Nakagawa et al.
1365 J. Holker et al.
1373 B.A. McAndrew et al.
1411 L. Crombie et al.
1431 P. Anastasis and P.E. Brown
1443 P.R. Buckland et al.
1465 R.D. Bowden et al.
1497 A.K. Mallams et al.
1545 D.C. Horwell et al.
1573 Z. Cai et al.
1649 G. Casiraghi et al.
1681 T.-S. Wu et al.
1719 M.Y. Jarrah and V. Thaller
1753 M.F. Aldersley et al.
1759 A.R. Forrester et al.
1765 T. Okuda et al.
1773 T. Nishio and Y. Omote
1791 K. Hirao et al.
1799 C. Kashima et al.
1813 I.F. Barnard and J.A. Elvidge
1819 K. Brown et al.
1847 A.E. de Jesus et al.
1885 A.J. Barker and G. Pattenden
1893 A.J. Barker et al.
1901 A.J. Barker and G. Pattenden
1919 A.J. Barker et al.
1937 Y. Tamura et al.
1983 J. Tsunetsugu et al.
2011 M. Lempert-Sreter et al.
2017 G. Appendino and P. Gariboldi
2031 D.A. Young et al.
2053 P. Bird et al.
2161 A. Mudd
2175 R.M. Letcher et al.
2181 R.J. Cremlyn et al.
2185 J.R. Anderson et al.
2211 G. Durrant et al.
2215 A.D. Buss et al.
2241 J. Steele and R.J. Stoodley
2259 J.M.R. Al-Zaidi et al.
2267 M.S. Kemp et al.
2329 K.M. Smith and L.A. Kehres
2337 B. Nassim et al.
2349 Y. Li et al.
2353 Y. Ding et al.
2399 R. McCague et al.
2409 W.R. Ashcroft et al.
2413 G.L. Humphrey et al.
2417 L. Dalton et al.
2425 A.S. Bailey et al.
2431 B. Gozler and M. Shamma
2441 J.D. Elliott et al.
2513 G. Brooks and E. Hunt
2519 G. Sinha et al.
2535 L.R. Milgrom
2541 J.M. Mellor et al.
2545 J.M. Mellor and R. Pathirana
2577 S.E.N. Mohamed and D.A. Whiting
2601 A.R. Katritzky et al.
2645 G. Jones et al.
2655 P.J. O'Hanlon et al.
2659 O. Abou-Teim et al.
2705 G. Appendino et al.
2711 A.R. Forrester et al.
2723 J.R. Bull and K. Bischofberger
2735 S. Fatutta et al.
2739 M. DeBernardi et al.
2781 R.W. Draper
2787 R.W. Draper
2793 F.E. Carlon and R.W. Draper
2821 P.J. Hylands and E.S. Mansour
2845 G.M.K.B. Gunaherath and A.A.L. Gunatilaka
2945 D.N. Kirk and B.L. Yeoh
2983 R.D. Allen et al.
2987 S. Kuroda et al.
2997 J. Ojima et al.
3005 L. Carey et al.
3011 J.M. Clough and G. Pattenden

40- -83, Nippon Kagaku Kaishi (1983)
0088 T. Tokumitsu and T. Hayashi
0161 M. Nakayama et al.
1678 K. Takahashi et al.

41- -83, J. Chim. Phys., 80 (1983)
0363 M. Duquesne et al.
0559 D. Denis-Courtois and B. Vidal
0595 J. Roussilhe and N. Paillons
0603 P. Jardon et al.

42- -83, J. Indian Chem. Soc, 60 (1983)
0137 S.H. Etaiw et al.
0303 P.K. Sen and B. Kundu
0408 C.P. Upasani et al.
0475 S.N. Dehuri et al.
0686 G.D. Mehd and Y.K. Agrawal

44- -83, J. Org. Chem., 48 (1983)

0033 R. Balicki et al.
0044 L. Liu et al.
0090 J.M. Hornback and R.D. Barrows
0127 R.T. Scannell and R. Stevenson
0129 J.A. Hyatt
0136 P. Wan and K. Yates
0162 G. Devincenzis et al.
0173 A. Singh et al.
0177 S. Fujita
0214 N. Ramnath et al.
0222 A.H. Lewin et al.
0268 P. Tantivatana et al.
0283 D. Wenkert and R.B. Woodward
0309 D. Dytnerski et al.
0381 B. Danieli et al.
0395 E. Dilip de Silva et al.
0500 K.M. Smith and K.C. Langry
0537 H. Teeninga and J.B.F.N. Engberts
0542 W. Ando et al.
0551 Y. Tobe et al.
0575 G. Zvilichovsky and M. David
0584 J. Hine and S.OM. Linden
0596 R.P. Thummel and P. Chayangkoon
0670 D. Solas and J. Wolinsky
0695 A. Padwa et al.
0759 O.S. Tee et al.
0774 E. Ghera et al.
0780 M.-I. Lim et al.
0835 M.A. Fox and C.A. Triebel
1069 A. Padwa et al.
1122 K. Narasimhan and P.R. Kumar
1271 C.C. Tzeng et al.
1275 M. Sindler-Kulyk and D.C. Neckers
1282 M. Iwata and H. Kuzuhara
1370 M. Seki et al.
1417 F. Ramirez et al.
1451 L.H. Klemm et al.
1500 Z. Komiya and S. Nishida
1613 N.S. Nudelman and D. Palleros
1628 S. Nishigaki et al.
1632 R. Sangaiah et al.
1643 H.O. House et al.
1654 H.O. House et al.
1661 H.O. House et al.
1670 H.O. House et al.
1708 J.H. Hall
1718 S. Ayral-Kaloustian and W.C. Agosta
1725 D.C. Neckers et al.
1732 Y. Inoue et al.
1834 A. Padwa et al.
1841 M.D. Bachi et al.
1854 M.J. Robins and P.J. Barr
1862 R.R. Sauers et al.
1866 R.K. Okuda et al.
1869 G. Massiot et al.
1872 K. Muthuramu et al.
1903 H.A. Sun et al.
1906 S.L. Midland et al.
1925 P. Margaretha et al.
1937 M. Ishitsuka et al.
1954 C.R. Engel et al.
2029 M.P. Mack et al.
2053 R.M.G. Roberts et al.
2084 P.J. Kropp and N.J. Pienta
2115 R.C. Haddon et al.
2133 A. Pohlman and T. Mill
2164 Y. Asakawa et al.
2202 J.J.M. Lamberts and W.H. Laarhoven
2270 W. Huang et al.
2314 B. Carte and D.J. Faulkner
2318 A.G. Schultz and J.P. Dittami
2330 A. Padwa et al.
2360 J.E. Rice et al.
2379 V.K. Sharma et al.
2412 A.K. Bhattacharya and B. Miller
2432 A.G. Schultz and S.O. Myong
2468 F.J. Goetz et al.
2476 J.A. Zoltewicz et al.
2481 J.A. Zoltewicz and E. Wyrzykiewicz
2488 W.E. Noland et al.
2534 C.S. Carman and G.F. Koser
2572 J.W. Patterson, Jr.
2578 R. Arad-Yellin et al.
2584 D. Becker et al.
2629 K.A. Smith and A. Streitwieser, Jr.
2690 M.G. Saulnier and G.W. Gribble
2696 G.R. Lenz and C.R. Dorn
2709 D.R. Bender et al.
2719 T. Maruyama et al.
2728 K. Harano et al.
2757 C.H. Chen et al.
2820 D. Dominguez and M.P. Cava
2897 C.L. Go and W.H. Waddell
2930 R.G. Harvey et al.
2949 L.I. Rieke et al.
2989 R.J. Himmelsbach et al.
3015 A. Adesomoju et al.
3119 H.-D. Winkeler and F. Seela
3146 T.J. Holmes, Jr. and R.G. Lawton
3189 A. Padwa and J.G. MacDonald
3214 M. Kakushima et al.
3220 V. Girard et al.
3252 B.M. Trost et al.
3269 R.W. Franck and T.V. John
3301 P.C. Preusch and J.W. Suttie
3408 A.G. Schultz et al.
3428 C.H. Heathcock et al.
3458 S. Bank et al.
3477 S. Yamada et al.
3483 S. Yamada et al.
3512 A. Groweiss et al.
3574 R.E. Lyle et al.
3649 A. Bolognese et al.
3696 A. Nishinaga et al.
3700 A. Nishinaga et al.
3716 M. Ballester et al.
3728 C.A. Ogle et al.
3750 L. Moore et al.
3765 D.J. Katz et al.
3819 H.E. Paaren et al.
3825 C. Lesma et al.
3852 J. Baghdadchi and C.A. Panetta
3957 M. Shamma et al.
3994 D.P. Fox et al.
4022 P.L. Wylie et al.
4035 J.J. Pignatello et al.
4038 J.C. Oberti et al.
4078 S. Gelin et al.
4137 W.L. Duax and G.D. Smith
4186 J.F. Outlaw, Jr. et al.
4190 J.R. Cozort et al.

4202 F.I. Flower et al.
4222 T.F. Buckley, III and H. Rapoport
4232 S.K. Das and S.N. Balasubrahmanyam
4241 M. Ikeda et al.
4248 M.M. Kabat et al.
4251 J.H. Wilton and R.W. Doskotch
4262 F. Kungg et al.
4272 E.N. Marvell et al.
4302 K.M. Smith and G.W. Craig
4357 H. Hart et al.
4399 R.W. Kaesler and E. LeGoff
4407 C.A. Loeschorn et al.
4410 M.N. Do and K.L. Erickson
4413 J.W. Patterson, Jr. and D.V. Krishnamurthy
4419 W.A. Lindley et al.
4482 K. Muthuramu et al.
4497 R.H. Kayser et al.
4582 P.A.T.W. Porskamp and B. Zwanenburg
4585 M. Mizutani et al.
4642 W.G. Dauben and R.A. Bunce
4649 F. Effenberger et al.
4766 E.J. Parish and A.D. Scott
4808 J. Toullec et al.
4841 K.T. Potts et al.
4852 E.C. Taylor et al.
4873 D.D. Weller and E.P. Stirchak
4879 V. Elango and M. Shamma
4968 K. Maruyama and T. Ogawa
5006 E. Wenkert et al.
5017 A.G.M. Barrett and H.G. Sheth
5026 U. Hornemann et al.
5033 M. Bean and H. Kohn
5041 R.A. Mustill and A.H. Rees
5120 J.Z. Suits et al.
5123 M.C. Garcia-Alvarez et al.
5130 L.L. Melhado and N.J. Leonard
5149 M.R. Detty and H.R. Luss
5203 S.H. Grode et al.
5318 A. Perales et al.
5348 C.K. Govindan and G. Taylor
5354 G.R. Pettit et al.
5359 M.E. Jung and J.A. Hagenah
5379 E.N. Marvell et al.
5388 C.K. Chang and I. Abdalmuhdi

46- -83, J. Phys. Chem., 87 (1983)
0007 H. Nikl et al.
1960 R.F.C. Claridge and H. Fischer
3024 W.M.D. Wijekoon et al.
4585 D. Cohen et al.
4641 U. Mazur and K.W. Hipps

47- -83, J. Polymer Sci., Polymer Chem. Ed., 21 (1983)
1263 X.-B. Li et al.
2813 Y. Suda et al.
3425 M.D. Shalati and C.G. Overberger

48- -83, J. prakt. Chem., 325 (1983)
0041 H. Schafer and K. Gewald
0168 J. Liebscher et al.
0205 G. Geissler et al.
0293 M. Augustin et al.
0353 M.L. Jain and R.P. Soni
0387 D. Cech et al.
0463 J. Bodeker et al.
0505 P. Czerney and H. Hartmann
0517 S.I. Al-Khaffaf and M. Shanshal
0551 P. Czerney and H. Hartmann
0574 C. Lindig and K.H.R. Repke
0587 C. Lindig
0599 B. Streckenbach et al.
0607 B. Streckenbach and K.R.H. Repke
0689 J. Liebscher et al.
0729 G.W. Fischer et al.
1002 H. Koepernik and R. Borsdorf
1016 H.-J. Timpe and H. Rautschek

49- -83, Monatsh. Chem., 114 (1983)
0101 A. Mannschreck et al.
0195 M.K. Hargreaves and L.F. Rabari
0227 T. Kappe et al.
0281 A. Maslankiewicz and K. Pluta
0349 T. Kappe and Y. Linnau
0359 W.O. Lin et al.
0599 O.S. Wolfbeis and H. Marhold
0753 H. Falk et al.
0915 S. Leistner et al.
0937 S. Li, A. Gupta and O. Vogl
0973 H. Junek et al.
0983 H. Falk and U. Zrunek
0999 K. Gewald and H. Rollig
1035 M.K. Kalinowski and J. Klimkiewicz
1107 H. Falk and U. Zrunek
1259 E. Haslinger et al.

54- -83, Rec. trav. chim., 102 (1983)
0014 C. Tintel et al.
0046 A.D. Brock et al.
0083 A.C. Brouwer et al.
0091 A.C. Brouwer and H.J.T. Bos
0103 A.C. Brouwer et al.
0114 J.A. de Groot et al.
0220 C. Tintel et al.
0299 H.J.A. Lambrechts and H. Cerfontain
0302 C.P. Visser et al.
0307 C.P. Visser and H. Cerfontain
0331 S.A.G.F. Angelino et al.
0347 J.A. de Groot et al.
0364 R.E. van der Stoel et al.
0465 T.A.M. van Schaik and A. van der Gen
0515 M.J. Caus and H. Cerfontain

56- -83, Polish. J. Chem., 57 (1983)
0129 I.M. Dzierkacz and K. Kostka
0403 J. Gumulka et al.
0461 M. Pawloski
0483 W.M. Daniewski et al.
0547 E. Salwinska and J. Suwinski
0579 Z. Wichert-Tur
0767 M. Janczewski et al.
0779 W.J. Krzyzosiak et al.
0817 K. Celnik et al.
0829 J. Urbanski and L. Wrobel
0875 M.J. Korohoda
0971 D. Maciejewska and L. Skulski
1027 M. Szajda and F. Wyrzykiewicz
1219 R. Balicki
1267 W.J. Rodewald and R.R. Sicinski

1357 J. Nowakowski and T. Lesiak
1371 S. Zommer-Urbanska

59- -83, Spectrochim. Acta, 39A (1983)
0289 A.E. Moural et al.
0609 A.K. Mishra and S.K. Dogra
0729 M.R. Mahmoud et al.
0933 A.E. Mourad

60- -83, J. Chem. Soc., Faraday Trans. I (1983)
0155 M. de Vijlder

61- -83, Ber. Bunsen Gesell. Phys. Chem., 87 (1983)
0391 J. Vogel et al.

62- -83, Z. phys. Chem.(Leipzig), 264 (1983)
0950 K. Gustav and M. Bolke
0957 M.S.A. Abd-El-Mottaleb

62- -83B, Z. phys. Chem.(Frankfurt), 138 (1983)
0199 A. Spalletti et al.

64- -83B, Z. Naturforsch., 38B (1983)
0108 J. Schallenberg and E. Meyer
0248 R. Wintersteiger and O.S. Wolfbeis
0392 F. Dallacker et al.
0497 E.L. Michelotti and E.L. Sanchez
0504 W. Friedrichsen et al.
0516 H. Budzikiewicz et al.
0621 U. Bayer and H.A. Brune
0648 R. Battaglia et al.
0658 M. Kaloga and I. Christiansen
0752 F. Dallacker et al.
0866 A. Römer and M. Sammet
0873 W.H. Gündel
0895 H. Hofmann et al.
0930 D. Knittel
1015 A. Marston and E. Hecker
1034 J.E. Rockley and L.A. Summers
1062 M. Weidenbruch et al.
1240 H. Volz and G. Herb
1339 J.W. Buchler et al.
1585 S. Kato et al.
1700 Atta-ur-Rahman et al.

64- -83C, Z. Naturforsch., 38C (1983)
0017 H.W. Rauwald and H. Miething
0049 G. Onur et al.
0083 A. Andreoni et al.
0179 W. Krick et al.
0492 H. Besl et al.
0563 H. Meisch and R. Maus
0689 W. Krick et al.
0701 P. Nielsen et al.

65- -83, Zhur. Obshchei Khim., 53 (1983) (English translation pagination)
0153 G.N. Lipunova et al.
0243 O.I. Chekmacheva et al.
0359 A.A. Medzhidor and V.T. Kasumov
0449 A.M. Panov et al.
0514 V.I. Boev and A.V. Dombrovskii
0519 R.I. Vinokurova et al.
0539 O.N. Minailova et al.
0653 O.I. Chekmacheva
0668 A.A. Tumanov et al.
0814 M.G. Voronkov et al.
1033 T.I. Krivonogova et al.
1039 M.G. Voronkov et al.
1054 G.V. Ratovskii et al.
1072 B.A. Suborov
1125 A.M. Moiseenkov et al.
1214 V.M. Potapov et al.
1263 E.V. Borisov
1294 L.N. Kulinkovich and V.A. Timoshchuk
1315 Y.I. Usatenko et al.
1480 L.N. Kulinkovich and V.A. Timoshchuk
1483 L.N. Kulinkovich and V.A. Timoshchuk
1486 L.N. Kulinkovich and V.A. Timoshchuk
1555 G.L. Matevosyan and P.M. Zarlin
1571 A.S. Antonyuk and A.V. Dombrovskii
1687 A.I. Serebryanskaya et al.
1712 L.N. Kulinkovich and V.A. Timoshchuk
1724 O.N. Minailova et al.
1898 A.A. Ivanenko et al.
1902 B.E. Zaitsev et al.
1917 L.N. Kulinkovich et al.
2021 V.V. Belyaeva et al.
2058 G.A. Tolstikov et al.
2061 G.V. Avramenko et al.
2073 V.E. Kampar et al.
2332 A.V. El'tsov et al.
2339 L.G. Tomilova et al.
2346 M.G. Gal'pern et al.
2351 V.M. Berezovskii et al.
2424 I.N. Chernyuk et al.

69- -83, Biochemistry, 22 (1983)
1342 K. Bose and A.A. Bothner-by
1696 P.J. Barr et al.
3636 J.L. Napoli et al.
6303 T. Deffo et al.

70- -83, Izvest. Akad. Nauk S.S.S.R., 32 (1983)(English trans. pagination)
0088 B.D. Sviridov et al.
0103 E.P. Serebryakov and I.M. Vol'pin
0251 P.P. Levin et al.
0271 V.F. Loktev et al.
0485 R.I. Buzykin et al.
0550 R.D. Malysheva et al.
0569 G.A. Tolstikov et al.
0752 L.A. Myshkina et al.
0780 Z.A. Krasnaya and T.S. Stytsenko
1016 T.I. Cherkasova et al.
1049 L.V. Goncharenko et al.
1320 G.A. Tolstikov and E.E. Shul'ts
1439 B.I. Buzykin et al.
1444 F.M. Stoyanovich et al.
1532 E.K. Trutneva and Y.A. Levin
1655 G.T. Katvalyan and E.A. Mistryukov
1659 V.D. Sen' et al.
1754 G.V. Shustov et al.
1757 I.B. Al'tman et al.

1882 Z.A. Krasnaya et al.
1894 L.M. Korotaeva et al.
1897 T.G. Konstantinova et al.
1906 V.D. Sen et al.
2119 G.T. Katvalyan and E.A. Mistryukov
2123 G.I. Shchukin et al.
2141 A.V. Kamernitskii et al.
2322 R.G. Kostyanovskii and Y.I. El'na-tov
2379 S.M. Lukonina et al.
2437 A.S. Dvornikov et al.
2501 I.A. Grigor'ev et al.

73- -83, Coll. Czech. Chem. Comm., 48 (1983)
0112 I. Chvatal et al.
0123 Z.J. Vejdelek et al.
0137 R. Kotva et al.
0144 K. Sindelar et al.
0228 Z. Janonsek et al.
0292 M. Beran et al.
0299 R. Kotva et al.
0304 J. Krepelka et al.
0471 J. Cacho et al.
0527 J. Tax et al.
0608 J. Palecek et al.
0617 J. Palecek et al.
0623 Z. Polivka et al.
0642 Z. Vejdelek and M. Protiva
0649 I. Valterova et al.
0772 A. Krutosikova et al.
0906 J. Jilek et al.
1048 L. Fisera et al.
1057 A. Jurasek et al.
1062 A. Jurasek et al.
1173 V. Bartl et al.
1187 K. Sindelar et al.
1854 L. Fisera et al.
1878 A. Krutosikova et al.
1891 D. Vegh et al.
1898 K. Sindelar et al.
1910 A. Holy
2249 D. Tocksteinova et al.
2376 L. Pavlickova et al.
2395 Z. Polivka et al.
2676 J. Farkas
2682 R. Mocelo and J. Kovac
2812 S. Sekretar et al.
2825 L. Musil et al.
2970 Z. Polivka et al.
2977 Z. Vejdelek and M. Protiva
2989 F. Santavy et al.
3112 S. Marchalin and J. Kuthan
3123 S. Marchalin and J. Kuthan
3315 D. Koscik et al.
3426 D. Koscik et al.
3433 Z. Polikva et al.
3559 K. Spirkova et al.
3567 P. Kristian et al.
3575 S. Stankovsky

74- -83A, Mikrochim. Acta I (1983)
0371 A. Izquierdo and R. Compano
0381 Y.A. Zolotov et al.

74- -83B, Mikrochim. Acta II (1983)
0075 R. Yoda et al.

74- -83C, Mikrochim. Acta III (1983)
0011 M. Blanto and S. Maspoch
0095 S. Maspoch et al.
0301 J.A. Vinson et al.
0385 O.S. Wolfbeis and E. Furlinger

77- -83, J. Chem. Soc., Chem. Comm. (1983)
0004 M.R. Bryce
0007 K. Takeuchi et al.
0047 P.S. Steyn et al.
0073 H. Koenig and R.T. Oakley
0077 M. Sheves et al.
0099 H. Hidaka et al.
0123 P.M. Bishop et al.
0174 H. Irie et al.
0183 J.A. Zoltewicz and E. Wyrzykiewicz
0239 K.J.H. Kruithof et al.
0275 H. Rexhausen and A. Gossauer
0287 C. Rosini et al.
0295 I. Johannsen et al.
0317 Z. Lidert and C.W. Rees
0323 Y. Hamamoto et al.
0335 M. Yamauchi et al.
0353 J.A. Schneider and K. Nakanishi
0390 R.L. Beddoes et al.
0399 A.J. O'Connell et al.
0425 T.J. King et al.
0502 B.C. Berris et al.
0508 R.J. Stoodley and A. Whiting
0605 C.C. Chang and K. Nakanishi
0781 H. Watanabe et al.
0789 S. Araki and Y. Butsugan
0797 E. Schaumann et al.
0799 M. Shamma et al.
0862 M. Yoshifuji et al.
0995 S. Tsuchiya et al.
1018 G. Massiot et al.
1037 S. Arai et al.
1058 M. Iyoda et al.
1107 H. Nozaki et al.
1120 J. Kervagoret et al.
1166 A.K. Ganguly et al.
1188 K. Kurihara and J.H. Fendler
1216 G. Ferguson et al.
1246 O. Meth-Cohn and C. Moore
1259 C.C.J. Culvenor et al.
1344 J.J.A. Campbell et al.
1404 C. Leumann et al.
1455 Z.T. Fomum et al.
1521 D. Creed et al.

78- -83, Tetrahedron, 39 (1983)
0033 R.E. Balselli and A.R. Frasca
0075 S. Harada et al.
0169 F. Bigi
0187 M. Bucciarelli et al.
0199 J.D. Rodriguez et al.
0337 B.R. Cowley et al.
0427 T.L. Chan et al.
0449 B. Voigt and G. Adam
0461 B.R. Cowley et al.
0577 B. Goezler et al.
0581 M.E. Garst et al.
0599 T.C. Jain et al.

0623 J.M. Aubry et al.
0629 B. Banaigs et al.
0645 A. Roedig et al.
0667 M. Rotem et al.
0733 D.L. Cullen et al.
0759 C. Boullais et al.
0831 D.E. Ames et al.
0835 J.M. Vernon et al.
0883 H. Sakurai et al.
0949 H.J. Reich et al.
1075 D.W. Brown et al.
1109 W.R. Bergmark et al.
1123 H. Nemoto et al.
1151 H. Hemetsberger et al.
1247 M. Franck-Neumann and M. Miesch
1265 P.N. Confalone and D.L. Confalone
1273 M. Wada et al.
1281 T. Ishizu et al.
1401 R. Bonnett et al.
1407 R. Todesco et al.
1487 J.E. Baldwin and R.C. Gerald Lopez
1539 G. Hugel and J. Levy
1551 M. Cariou et al.
1643 Z. Kinamoni et al.
1761 G. Sartori et al.
1841 K.M. Smith and D. Kishore
1849 A.H. Jackson et al.
1865 J.V. Bonfiglio et al.
1893 G.L. Landen et al.
1975 N.S. Narasimhan et al.
2147 F. Bigi et al.
2201 D.H.R. Barton et al.
2255 H.S. Vargha et al.
2277 G.V.P. Chandra Mouli et al.
2283 N.J. McCorkindale et al.
2393 J.D. White et al.
2493 L. Ghosez et al.
2527 D.O. Spry
2531 L.D. Cama et al.
2551 A.G. Brown et al.
2599 F.R. Atherton and R.W. Lambert
2647 J. Polonsky et al.
2679 A. Nickon et al.
2691 M. Wada et al.
2719 K. Muthuramu et al.
2779 J. Valisolalao et al.
2799 M. Audouin and J. Levisalles
2815 P.E. Schulze et al.
2819 B.P. Pradhan et al.
2965 A. Chatterjee et al.
3073 K.J.H. Kruithof et al.
3083 D.F. Crowe et al.
3261 J. Kunitomo et al.
3307 A. Lechevallier et al.
3351 A.C. Pinto et al.
3359 H.I.X. Mager and R. Addink
3397 E. Tsankova et al.
3405 J.V. Greenhill and M.A. Moten
3523 E. Gerlinger et al.
3603 A. Kelecom
3609 L. Lorenc et al.
3639 S. Mukhopadhyay et al.
3645 G. Massiot et al.
3695 W. Oppolzer et al.
3719 E. Wenkert et al.
3725 P.D. Magnus et al.
3767 P.A. Wender and A.W. White
3895 L.R. Milgrom
3923 F. Ramandrasoa et al.
3929 C. Schmidt and T. Breining
4011 K. Okamoto et al.
4121 B. Yde et al.
4221 S. Bhattacharyya et al.
4243 L.S. Trifonov et al.

80- -83, Revue Roumaine Chim., 28 (1983)
0003 C. Mihart et al.
0045 V. Armenau and S. Visan
0065 M. Popescu and A. Danet
0375 I. Havlik et al.
0381 M. Raileanu et al.
0555 M. Ciureanu et al.
0725 Z.H. Khalil and A. Ibrahim
0733 C. Cristescu
0903 I. Havlik et al.
0989 C. Cristescu

83- -83, Arch. Pharm., 316 (1983)
0115 K. Rehse and U. Emisch
0146 H. Griengl et al.
0240 R. Neidlein and H. Zeiner
0244 J. Knabe and J. Lorenz
0264 K. Görlitzer et al.
0271 H. Hamacher and C. Mengold
0283 H.-J. Kallmayer and K. Seyfang
0326 F. Eiden et al.
0353 J. Knabe and J. Lorenz
0379 K.-C. Liu et al.
0399 H.W. Rauwald and H.-D. Just
0409 H.W. Rauwald and H.-D. Just
0445 J. Knabe and W. Weirich
0472 G. Seitz et al.
0476 R.W. Grauert
0520 J. Knabe and W. Weirich
0569 K.-C. Liu et al.
0608 H. Biere et al.
0624 J. Knabe and W. Weirich
0667 K. Keppeler and G. Kiefer
0678 G. Willuhn and U. Köthe
0694 J. Knabe and W. Weirich
0728 K.-C. Liu, H.H. Chen and Y.O. Lin
0730 G. Seitz and H.-S. The
0732 R. Kühn and H.-H. Otto
0737 S. Prior and W. Wiegrebe
0773 F. Knefeli et al.
0801 K.K. Mayer et al.
0831 J. Knabe and J. Lorenz
0845 K. Rehse and W. Schinkel
0862 K.K. Mayer et al.
0889 H.-J. Kallmayer and K. Seyfang
0908 O. Motl et al.
0912 J. Knabe and J. Lorenz
0921 F. Eiden et al.
0951 K. Rehse and W. Schinkel
0971 H. Linde
1000 J. Reisch and M. Abdel-Khalik

86- -83, Talanta, 30 (1983)
0555 F.J. Barragan de la Rosa et al.

87- -83, J. Med. Chem., 26 (1983)
0007 J. Pitha et al.

0020 M. Sakaguchi et al.
0030 E. Breuer et al.
0055 A. Harris et al.
0072 Y. Arai et al.
0078 M.J. Green et al.
0100 L.E. Benjamin, Sr. et al.
0111 A.A. Kumar et al.
0156 Y.F. Shealy et al.
0162 V.K. Iyer et al.
0226 B.E. Maryanoff et al.
0230 M.J. Zelesko et al.
0271 S. Harada et al.
0280 T.L. Chwang et al.
0283 T.-C. Lee et al.
0286 A.M. Mian and T.A. Khwaja
0298 S. Avramovici-Grisaru et al.
0403 C.J. Blankley et al.
0469 S.S. Hall et al.
0522 L. Nedelec et al.
0544 T.-S. Lin and W.R. Mancini
0559 R.N. Henrie et al.
0564 W.S. Saari et al.
0574 J. Bernadou et al.
0598 T.-S. Lin and Y. Gao
0602 L. Colla et al.
0661 E. DeClerq et al.
0667 A. Stuart et al.
0714 D.R. Buckle et al.
0737 C.D. Meyer et al.
0759 J.C. Martin et al.
0786 P.W. Collins
0790 G.L. Bundy et al.
0845 R.J. Chorvat et al.
0891 T.A. Krenitsky
0895 P. Krogsgaard-Larsen et al.
1028 L. Maggiora et al.
1042 L. Grehn et al.
1056 N.I. Ghali et al.
1089 G.L. Bundy et al.
1153 D. Farquhar et al.
1282 M.I. Dawson et al.
1370 Y. Shimojima and H. Hayashi
1478 H.O.H. Showalter et al.
1483 J.A. Montgomery et al,
1489 J.L. Rideout et al.
1527 V. Nelson and H.S. El Khadem
1530 D.C. Baker et al.
1577 M. Narisada et al.
1601 H.W. Hamilton and J.A. Bristol
1614 C. Temple, Jr. et al.
1631 S.K. Sengupta et al.
1653 M.I. Dawson et al.
1687 J.E. Oatis, Jr,
1691 T.-S. Lin, Y.-S. Gao and W.R. Man cini

88- -83, Tetrahedron Letters (1983)
0047 R.C. Nickolson and H. Vorbruggen
0051 M.-L. Bouillant et al.
0069 M. Yasunami et al.
0115 S. Huneck et al.
0205 K. Takahashi et al.
0287 D. Ghiringhelli
0291 S.C. Pakrashi et al.
0299 J.W. Barton and D.J. Rowe
0315 J. Ivanics et al.
0331 R. Zamboni and J. Rokach
0351 G. Adam and A. Schierhorn
0381 B. Chantegrel et al.
0481 A. Sato and W. Fenical
0495 H. Seto et al.
0761 S. Baasou et al.
0781 K. Fujimori et al.
0799 Y. Maki et al.
0847 A.G. Gonzalez et al.
0931 M. Yagisawa et al.
0947 A.S. Horahovats et al.
0995 T. Hirota and H.A. Itano
1045 H. Ogawa et al.
1085 J.A.D. Jeffreys et al.
1281 G. Galambos et al.
1329 D.K. Anderson
1333 C. Che et al.
1349 H. Yagi et al.
1441 I. Hughes and R.A. Raphael
1451 D.E. Remy
1485 H. Schuster et al.
1535 Y. Hirose et al.
1549 H. Kikuchi et al.
1555 R.J. Spear et al.
1567 T.A. Blinka and R. West
1611 M. Franck-Neumann et al.
1631 E. Perrone et al.
1727 Y. Kanao et al.
1749 P.L. Netra and M.D. Sutherland
1775 M. Franck-Neumann et al.
1805 H. Seto et al.
1917 J.E. Hochlowski and D.J. Faulkner
1925 M.F. Zipplies et al.
1955 G. Märkl, B. Alig and E. Eckl
1983 R. Gabioud and P. Vogel
2019 M. Kodpinid et al.
2025 G.M.K.B. Gunaherat-L and A.A.L. Gunatilaka
2029 J.M. Boente et al.
2087 M.P. Kirkup and R.E. Moore
2121 M. Ballester et al.
2147 H. Prinzbach et al.
2151 A. Beck et al.
2171 M. Nakagawa et al.
2181 T. Goto et al.
2209 S.T. Reid et al.
2275 T. Kumagai et al.
2279 T. Kumagai et al.
2303 J.M. Boente et al.
2331 T.R. Kelly et al.
2351 J.-S. Liu et al.
2355 J.-S. Liu et al.
2401 S. Ohnishi et al.
2407 Y. Xin-sheng et al.
2445 D.P. Allais et al.
2505 F. Buzzetti et al.
2527 G. Büchi et al.
2531 G. Büchi et al.
2761 J.P. Gesson et al.
2781 T. Ogino et al.
2973 H.J. Knops and L. Born
2977 T. Eicher et al.
3013 A.V. Rama Rao et al.
3025 H.R. Sonawane et al.
3055 J.E. Biskupiak and C.M. Ireland
3113 E. Okuyama and M. Yamazaki

3183 T.L. Chwang et al.
3247 Y. Xin-sheng et al.
3283 E. Perrone et al.
3337 T. Ichikawa et al.
3361 T. Tsuji and S. Nishida
3419 E.M. Gordon and J. Plusces
3469 Z. Yoshida et al.
3523 P. Molina et al.
3539 B.B. Jarvis et al.
3563 R. Gompper et al.
3567 R. Gompper and N. Sengüler
3577 A.M. Richter and E. Fanghänel
3603 C. Mahaim et al.
3625 J. Escudie et al.
3643 S. Omura et al.
3647 T. Kondo et al.
3651 L. Baiocchi and G. Picconi
3687 N.N. Girotra and N.L. Wendler
3765 A.G. Gonzalez et al.
3787 S. Huneck et al.
3803 Y. Kono et al.
3825 I. Kubo et al.
3879 H. Saito et al.
3897 G. Guella et al.
4117 H. Niwa et al.
4127 Z.T. Fomum et al.
4257 H. Nemoto et al.
4261 Y. Fujise et al.
4315 T. Cuvigny et al.
4351 S.A. Robev
4381 Z. Koblicova et al.
4413 D.T. Glatzhofer and D.T. Longone
4417 J. Beauhaire et al.
4433 K. Iguchi et al.
4481 J.M. Boente et al.
4631 J. Kihlberg et al.
4649 A.E. Wright et al.
4675 M. Kaouadji et al.
4677 M. Kaouadji et al.
4679 S. Deshayes and S. Gelin
4719 T. Osawa and M. Namiki
4729 S. Hata et al.
4747 K. Mackenzie et al.
4755 R.C.F. Jones and G.E. Peterson
4789 O.L. Acevedo et al.
4821 A. Krebs et al.
4843 M. Ishibashi et al.
4851 T. Otsubo et al.
4855 M. Yoshifuji and N. Inamoto
4863 T. Goto et al.
4947 K.V. Rao and F.M. Alvarez
5047 A. Heckel and W. Pfleiderer
5051 G. Märkl and K. Hock
5063 R. Gleiter and G. Jähne
5067 E. Dagne and W. Steglich
5117 M. Ishitsuka et al.
5193 G. Werner et al.
5273 J. Ojima et al.
5277 Y. Miyahara et al.
5333 H. Mrozik et al.
5373 A. Ichihara et al.
5377 A.T. McPhail et al.
5395 A. Małkiewicz et al.
5399 G. Adembri et al.
5411 U. Norinder et al.
5435 M.G. Saulnier and G.W. Gribble
5527 L. Toscano and E. Seghetti
5555 R.S. Phillips and L.A. Cohen
5563 P.N. Confalone et al.
5607 O. Wacker et al.
5641 Y. Hayakawa et al.
5727 W. Adam et al.
5749 H. Tamura et al.
5753 Y. Toya et al.
5757 M. Hisatome et al.
5861 T. Loerzer et al.

89- -83, <u>Angew. Chem., Intl. Ed.</u>, <u>22</u> (1983)
0057 G. Märkl and E. Seidl
0334 H. Bauer et al.
0405 J.F. McGarrity et al.
0490 K. Hafner et al.
0495 H.-D. Becker et al.
0623 G. Wenska et al.
0717 R. Gompper and W. Breitschaft
0735 N.J. Lewis et al.
0876 R. Allman et al.
0879 G. Märkl et al.

89- -83S, <u>Angew. Chem., Intl. Ed.</u>, <u>22S</u> (1983)
0075 G. Markl and E. Seidel
0120 G. Szilagyi and H. Wamhoff
0282 B.C. Becker et al.
0471 H. Ogawa et al.
0480 H. Ogawa et al.
0551 J.F. McGarrity et al.
0571 H.-J. Altenbach et al.
0661 G. Jahne and R. Gleiter
1025 P. Lupon et al.
1147 R. Allmann et al.
1267 K. Krohn and B. Sarstedt
1371 D. Kowko et al.

90- -83, <u>Polyhedron</u>, <u>2</u> (1983)
0309 E.C. Okafor
0493 K.K. Sarkar et al.

93- -83, <u>J. Applied Chem. U.S.S.R.</u>, <u>56</u> (1983)(English translation)
0783 G.V. Sidorov et al.

94- -83, <u>Chem. Pharm. Bull.</u>, <u>31</u> (1983)
0020 A. Ohta and M. Ohta
0031 Y. Suzuki and H. Takahashi
0064 M. Kozawa et al.
0106 T. Kikuchi et al.
0135 S. Kanatomo et al.
0156 N. Fukuda et al.
0162 S. Kamachi et al.
0333 T. Okuda et al.
0352 I. Kitagawa et al.
0356 T. Sano et al.
0373 T. Kubota et al.
0557 N. Hamanaka et al.
0577 M. Shimizu et al.
0689 I. Kitagawa et al.
0723 H. Saikachi et al.
0780 T. Tanaka et al.
0861 Y. Hashimoto et al.
0895 T.-S. Wu et al.

0901 T.-S. Wu and H. Furukawa
0919 T. Ohmoto and O. Yoshida
0925 K. Arai et al.
0947 M. Hanaoka et al.
1183 T. Imanishi et al.
1199 F. Abe and T. Yamauchi
1222 H. Tanaka et al.
1228 H. Ochi and T. Miyasaka
1378 Y. Tamura et al.
1424 S. Itokawa et al.
1440 M. Kobayashi et al.
1474 S. Saeki et al.
1502 T. Murakami et al.
1544 M. Kwoyanagi et al.
1572 S. Ozasa et al.
1670 M. Onitsuka et al.
1733 M. Yamato et al.
1743 H. Itokawa et al.
1746 J. Adachi et al.
1754 T. Mori et al.
1806 T. Hino et al.
1856 M. Taniguchi et al.
1917 M. Kubo et al.
1923 M. Nagai et al.
1991 H. Itokawa et al.
2023 S. Naruto et al.
2110 T. Ozawa and A. Hanaki
2114 Y. Okamoto et al.
2146 A. Numata et al.
2160 N. Kobayashi et al.
2179 M. Ishibashi et al.
2183 H. Takahashi et al.
2192 H. Shimomura et al.
2269 Y. Tagawa et al.
2296 T. Kikuchi et al.
2308 C. Iwata et al.
2313 S. Ozasa et al.
2321 I. Kitagawa et al.
2353 H. Itokawa et al.
2473 M. Furukawa et al.
2526 Y. Okamoto et al.
2552 T. Kato et al.
2574 S. Kondo et al.
2583 T. Fujii et al.
2662 M. Watanabe et al.
2685 M. Hanaoka et al.
2712 M. Kozawa et al.
2718 A. Takada et al.
2859 Y. Shirataki et al.
2868 H. Sasaki and T. Kitagawa
2879 K. Yakushijin et al.
2884 T. Mori et al.
2910 M. Nakamura et al.
2936 Y. Hano et al.
2986 T. Nozoye et al.
2998 S. Sekita
3009 Y. Kurasawa et al.
3014 K. Ito and J. Maruyama
3024 H. Ishii et al.
3039 H. Ishii et al.
3056 H. Ishii et al.
3074 J. Okada et al.
3091 M. Takani et al.
3104 M. Shimazaki et al.
3113 M. Fukuoka et al.
3149 T. Fujii et al.
3198 T. Ohmoto and K. Koike
3302 M. Miyashita et al.
3309 M. Teranishi et al.
3330 H. Ishii et al.
3338 S. Sakai et al.
3397 K. Yamakawa et al.
3454 M. Mano, T. Seo and K. Imai
3496 M. Sako et al.
3521 T. Fujii et al.
3544 K. Yamakawa et al.
3562 W. Tabuneng et al.
3671 K. Tajima et al.
3678 M. Yamamoto et al.
3684 J. Kurita et al.
3728 H. Morita et al.
3781 S. Shimizu et al.
3865 T. Sataka et al.
3971 S. Yoshimura et al.
4202 H. Furukawa, T.Wu and T. Ohta
4206 H. Komura et al.
4270 T. Fujii et al.
4341 S. Urano et al.
4409 T. Kuraishi et al.
4543 Y. Inamori et al.

95- -83, J. Pharm. Soc. Japan, 103 (1983)
0049 N. Motohashi
0193 A. Takadate et al.
0279 H. Ishii et al.
0508 H. Kobayashi and J. Komatsu
0594 T. Higashino et al.
0601 E. Tomitori et al.
0631 E. Oishi et al.
0675 M. Nakayama et al.
0679 T. Kuraishi et al.
0994 M. Iinuma et al.
1042 G. Goto et al.
1046 Y. Masuoka et al.
1129 E. Tomitori et al.
1155 T. Takemoto et al.
1243 Y. Tominaga et al.
1278 A. Takadate et al.

96- -83, The Analyst, 108 (1983)
0380 M.E. El-kommos

97- -83, Z. Chemie, 23 (1983)
0018 J. Gonda et al.
0019 E. Mitzner and J. Liebscher
0020 A. Knoll et al.
0028 W. Jugelt and S. Schwartner
0056 U.-W. Grummt et al.
0064 B. Vieth and W. Jugelt
0096 K. Schulze and F. Richter
0105B H.R. Saguster and G. Roebisch
0147 M. Pulst et al.
0215 S. Leistner et al.
0230B L. Beyer and A. Hantschmann
0247 W. Thiel and R. Mayer
0261 E.-G. Jaeger and D. Seidel
0296 A. Isenberg and H. Zaschke
0341 T. Jira and C. Troeltzsch
0403 J. Liebscher et al.

98- -83, J. Agr. Food Chem., 31 (1983)
0227 W.M. Draper and J.E. Casida

0621 G.B. Quistad and K.M. Mulholland
0625 K.T. Koshy et al.
0655 R.J. Cole et al.
0734 W.M. Draper and D.G. Crosby
0780 K. Kikugawa et al.
1113 L.O. Ruzo
1326 P.G. Hoffman et al.

99- -83, Theor. Exptl. Chem., 19 (1983)
0023 R.A. Loktionova et al.
0034 V.P. Chuer et al.
0150 N.A. Derevyanko et al.
0206 P.A. Krasutskii et al.
0224 Y.S. Ryabokobylko et al.
0420 N.P. Gritsan et al.
0539 B.V. Zaitsev et al.
0612 V.P. Chuev et al.

100- -83, J. Natural Products, 46 (1983)
0001 P.L. Schiff, Jr.
0118 S.P. Gunasekera et al.
0123 S.S. Handa et al.
0127 A.J. Aladesanmi et al.
0135 D.A. Cairnes et al.
0161 G.M. Laekeman et al.
0200 R.P. Borris et al.
0218 M. Arisawa et al.
0222 M. Arisawa et al.
0235 J.A. Lamberton et al.
0248 S.S. Handa et al.
0293 B. Gozler et al.
0310 M. Doe et al.
0314 A. Bianco et al.
0320 D.S. Bhakuni and R. Chaturvedi
0325 S.S. Handa et al.
0335 R. Hocquemiller et al.
0342 D. Dwuma-Badu et al.
0353 E. Saifah et al.
0359 S.S. Handa et al.
0391 S. Funayama et al.
0409 S. Mukhopadhyay et al.
0414 T. Gozler et al.
0424 B. Abegaz et al.
0433 B. Gozler et al.
0441 C.T. Montgomery et al.
0454 S. Badahur and A.K. Shukla
0466 D.S. Bhakuni and R. Chaturvedi
0510 H. Greger et al.
0517 A. El-Sayed et al.
0528 G.R. Pettit et al.
0532 K. Ishiguro et al.
0537 M.C. Wani et al.
0563 G.R. Pettit et al.
0614 P. Esposito et al.
0646 J. Kagan et al.
0660 R.G. Powell et al.
0671 S. Mukhopadhyay et al.
0681 G. Baudouin et al.
0694 G.M.T. Robert et al.
0708 G.M.T. Robert et al.
0723 L.-J. Lin et al.
0732 A.L. Skaltsounis et al.
0761 H. Guinaudeau et al.
0836 M.E.O. Matos and T.C.B. Tomassini
0852 K.-C. Luk et al.
0862 F. Roblot et al.
0881 D.P. Allais and H. Guinaudeau
0884 F.M. Eckenrode and J.P. Rosazza
0908 J.E. Leet et al.
0923 R.W. Doskotch et al.

101- -83A, J. Organomet. Chem., 241 (1983)
0119 C.F. Shibar and W.H. Waddell
0227 M. Herberhold et al.

101- -83F, J. Organomet. Chem., 246 (1983)
0C19 R. McGrindle et al.
0301 H.T. Dieck and J. Klaus

101- -83G, J. Organomet. Chem., 247 (1983)
0149 A. Castel et al.

101- -83H, J. Organomet. Chem., 248 (1983)
0269 G. Märkl and H. Hauptmann

101- -83I, J. Organomet. Chem., 249 (1983)
0335 G. Märkl et al.

101- -83K, J. Organomet. Chem., 251 (1983)
0093 J.E. Sheats et al.
0175 J.-P. Quintard et al.

101- -83L, J. Organomet. Chem., 252 (1983)
0133 E.A. Chernyshev et al.
0143 E.A. Chernyshev et al.
0C73 J. Barrau et al.

101- -83M, J. Organomet. Chem., 253 (1983)
0031 R.A. Grigsby, Jr. et al.
0065 P. Cocolios et al.
0C13 C. Biran et al.
0273 A. Coutsolelos and R. Guilard
0317 G. Faraglia et al.

101- -83P, J. Organomet. Chem., 256 (1983)
0217 D.R. Breitinger and W. Kress

101- -83R, J. Organomet. Chem., 258 (1983)
0257 J. Heinicke et al.
0283 V.K. Belsky et al.

101- -83S, J. Organomet. Chem., 259 (1983)
0171 R.A. Grigsby, Jr. et al.

102- -83, Phytochemistry, 22 (1983)
0175 C. Iavarone et al.
0179 J.J.W. Coppen
0187 C.H. Brieskorn and P. Noble
0197 F. Gomez G. et al.
0211 T. Iida, Y. Noro and K. Ito
0223 M. Takido et al.
0233 A.A.L. Gunatilaka et al.
0251 G. Ma et al.
0255 T. Iwagawa and T. Hase
0265 Z. Lin-gen et al.
0269 M.C.C.P. Gomes et al.
0308 V.P. Pathak et al.
0321 F.K. Duah et al.
0323 N. Shigematsu et al.
0371 H. Iijima et al.
0527 L. Perez-Sirvent et al.
0531 P.B. Oelrichs et al.

0539 B. Botta et al.
0549 Y. Fukuyama et al.
0553 J. Lemmich et al.
0561 M. Haraguchi et al.
0565 U. Samaraweera et al.
0573 L.A. Mitscher et al.
0585 J. Escudero et al.
0592 S. El-Masry et al.
0596 Y. Kiso et al.
0609 M. Suarez et al.
0617 Z. Lin-gen et al.
0621 C.P. Rao et al.
0625 T. Jaipetch et al.
0627 K. Iwasa et al.
0711 P.C. Vieira et al.
0723 F. Fernandez-Gadea et al.
0727 J.L. Marco et al.
0737 K. Inoue et al.
0743 H.H. Sun et al.
0747 S.R. Bhandari and A.H. Kapadi
0749 H. Achenbach et al.
0755 T.R. Govindachari and M.S. Premila
0763 T. Iida and K. Ito
0777 A. Villar et al.
0784 G. Savoni et al.
0787 F. Coll et al.
0790 I. Kouno et al.
0792 B.R. Barik et al.
0794 J. Mathew and A.V. SubbaRao
0800 K.C. Reddy et al.
0875 M.H. Beale et al.
0975 A.O. Adeoye and R.D. Waigh
0979 M. Martinez et al.
0987 M. Pinar et al.
1011 S. Takagi et al.
1017 Y. Ahmad et al.
1026 M.J. McCorkindale et al.
1053 M. Taneyama et al.
1064 S. Mukherjee et al.
1185 C. DeLuca et al.
1207 C.H. Brieskorn and P. Noble
1231 S.E. Taylor et al.
1245 G.M.K.B. Gunaherath
1249 K. Nakano et al.
1277 H. Lehmann et al.
1278 M. Masuko et al.
1292 F. Gomez et al.
1294 H. Suzuki et al.
1296 K.S. Mukherjee et al.
1305 F. Gomez et al.
1465 R.J. Capon et al.
1469 S.C. Sharma and R. Kumar
1483 A. Yagi et al.
1489 J. Jonnenbichler et al.
1493 T.-S. Wu et al.
1507 C.A. Catalan et al.
1512 M. Salmon et al.
1515 A.G. Gonzalez et al.
1516 L.M.X. Lopes et al.
1526 A.A. Craveiro et al.
1591 S.W. Banks and P.M. Dewick
1619 K. Sugama et al.
1657 G.D. Monache et al.
1671 T. Furuya et al.
1771 N. Ozaki et al.
1787 G. Combaut and L. Piovetti
1791 E.H. Seip and E. Hecker
1801 H. Ageta and Y. Arai
1819 S. Nishikawaji et al.
1832 J.A.D. Jeffreys et al.
1977 T. Hase et al.
1985 J. de Pascual-T. et al.
1993 P.S. Kalsi et al.
1997 H. Greger et al.
2005 J.A. Hueso-Rodriguez et al.
2011 A.G. Pinto and C. Borges
2031 A. San Feliciano et al.
2035 O. Thastrup and J. Lemmich
2047 X.A. Dominguez et al.
2051 E. Besson and J. Chopin
2073 B. Sener et al.
2080 S.A. Padwardhan and A.S. Gupta
2082 Y. Tirilly et al.
2087 M.I. Fernandez et al.
2090 M. Diaz et al.
2107 B. Voirin
2193 A. Evidente et al.
2227 A.J. Burbott et al.
2235 J. de Pascual-T. et al.
2253 V.V. Velde et al.
2273 B.R. Barik et al.
2277 K. Kawanishi et al.
2281 P.C. Vieira et al.
2287 K.N. Rao and G. Srimannarayana
2297 A.M.A.G. Nasser and W.E. Court
2301 A.I. Reis Luz et al.
2305 S. Ghosal et al.
2328 M. Mukherjee et al.
2509 T. Furuya et al.
2527 M.P. Kirkup and R.E. Moore
2531 Y. Takeda, T. Fujita and A. Ueno
2539 M.P. Kirkup and R.E. Moore
2559 B. Talapatra et al.
2579 B. Bodo et al.
2583 B.K. Rao et al.
2587 J. de Pascual-T. et al.
2591 S. Ghosal et al.
2603 H. Ripperger et al.
2607 S.K. Chattopadhyay et al.
2625 H.M. Chawla and R.S. Mittal
2753 J. de Pasual Tetresa et al.
2755 P. Nelson and R.O. Asplund
2759 A.J. Malcolm et al.
2767 G. Appendino et al.
2773 M. Maruyama et al.
2775 M. Pinar et al.
2779 E. Cabrera et al.
2783 J. de Pascual Teresa et al.
2795 L.J. Lin and A.D. Kinghorn
2805 J. de Pascual Teresa et al.
2819 M.I. Fernandez et al.
2835 D.D. McPherson et al.
2839 H. Ishiwata et al.
2847 A.A. Leslie Gunatilaka et al.
2865 B. Banaigs et al.
2881 W.S. Woo, J.S. Choi and S.S. Kang

103- -83, <u>Khim. Geterosikl. Soedin.</u>, <u>19</u> (1983)(English translation)
0019 L.M. Gornostaev et al.
0022 N.A. Popova et al.
0029 B.M. Krasovitskii et al.

0037 E.K. Mikitenko and N.N. Romanov
0052 N.N. Bychikhina et al.
0066 V.I. Terenin et al.
0083 V.D. Orlov et al.
0091 V.F. Sedova et al.
0095 A.V. Upadysheva et al.
0146 I.M. Andreeva et al.
0166 E.K. Mikitenko and N.N. Romanov
0184 M.A. Yurovskaya et al.
0214 E.A. Shevchenko et al.
0222 A.F. Shivanyuk et al.
0224 P.V. Tarasenko et al.
0226 R.U. Kochkanyan et al.
0229 V.D. Orlov et al.
0231 B.A. Trofimov et al.
0243 I.M. Gavrilyuk et al.
0273 I.B. Levshin et al.
0286 O.M. Radul et al.
0289 M.A. Yurovskaya et al.
0303 S.K. Kotovskaya et al.
0328 M.Y. Lidak et al.
0360 V.G. Kul'nevich et al.
0377 L.M. Gornostaev et al.
0401 S.A. Yamashkin et al.
0415 V.K. Lusis et al.
0421 N.S. Kozlov et al.
0425 G.V. Grishina et al.
0435 B.M. Khutova et al.
0474 V.G. Kul'nevich et al.
0478 S.V. Zhuravlev and V.G. Kul'nevich
0481 S.M. Ramsh et al.
0488 S.V. Dolidze et al.
0492 K.V. Fedotov et al.
0507 B.I. Gorin et al.
0511 G.P. Sharnin et al.
0518 L.N. Koikov et al.
0528 V.M. Potapov et al.
0544 V.D. Orlov et al.
0553 A.A. Gall' and G.V. Shishkin
0559 S.S. Mochalov et al.
0564 E.A. Zvezdina et al.
0571 N.V. Alekseeva et al.
0593 S.S. Mochalov et al.
0611 S.M. Ramsh et al.
0618 L.G. Levkovskaya et al.
0638 A.N. Kost et al.
0650 V.S. Mokrushin et al.
0657 V.G. Granik and S.I. Kainmanakova
0666 N.N. Bystrykh et al.
0714 E.A. Zvesdina et al.
0745 S.M. Ramsh et al.
0749 E.K. Mikitenko and N.N. Romanov
0752 E.K. Mikitenko and N.N. Romanov
0754 E.K. Mikitenko and N.N. Romanov
0765 A.G. Gorshkov et al.
0773 A.K. Sheinkman et al.
0824 I.M. Andreeva et al.
0868 V.V. Men'shikov and R.S. Sagitullin
0871 D.O. Kadzhrishvili et al.
0876 A.N. Grinev and I.K. Sorokina
0882 S.V. Chapyshev et al.
0886 V.G. Kartser et al.
0894 A.A. Verezubova et al.
0897 S.P. Epshtein et al.
0901 V.N. Charushin et al.
0945 A.N. Grinev and I.K. Sorokina
0948 I.M. Gavrilyuk et al.
0954 V.Y. Denisov et al.
0959 A.N. Grinev et al.
0961 S.Y. Solov'eva et al.
0970 A.N. Kost et al.
1050 N.B. Marchenko and V.G. Granik
1053 V.Y. Denisov and E.P. Fokin
1057 G.I. Zolotareva and L.M. Gornostaev
1076 S.Y. Solov'eva et al.
1082 A.S. Semeikin et al.
1086 A.N. Grinev and I.K. Sorokina
1102 N.S. Prostakov et al.
1104 L.N. Grigor'eva et al.
1120 I.I. Naumenko and G.V. Shishkin
1128 L.B. Volodarskii et al.
1135 E.P. Frolova et al.
1163 H. Schaefer et al.
1167 Y.M. Volovenko et al.
1170 N.K. Rozhkova et al.
1173 K. Sabirov and N.K. Rozhkova
1175 Y.N. Portnov et al.
1202 A.A. Krauze et al.
1228 I.Z. Lulle et al.
1231 V.S. Mokrushin et al.
1235 V.S. Mokrushin et al.
1241 M.A. Kudinova et al.
1279 L.M. Gornostaev and T.I. Lavrikova
1289 V.I. Slutskii and T.E. Bezmenova
1293 V.D. Orlov et al.
1302 E.S. Krichevskii et al.
1306 T.N. Galiullina and P.I. Abramenko
1310 V.I. Letunov et al.
1312 A.N. Grinev et al.
1333 V.N. Charushkin et al.
1340 A.N. Grinev and E.V. Lomanova
1345 L.L. Rodina et al.

104- -83, Zhur. Organ. Khim., 19 (1983) (English translation)
0086 E.N. Rozhkov et al.
0103 G.I. Borodkin et al.
0110 L.N. Koikov et al.
0139 M.V. Kazankov et al.
0144 S.A. Russkikh et al.
0158 N.S. Prostakov et al.
0183 A.M. Gorelik et al.
0189 F.N. Zeiberlikh et al.
0212 I.G. Ryabokon' et al.
0261 M.A. Lapitskaya et al.
0265 G.A. Tolstikov et al.
0283 G.A. Tolstikov et al.
0319 B.Y. Adamsone and O.Y. Neiland
0322 V.E. Mileiko et al.
0376 N.D. Shashurina et al.
0382 Y.V. Vinokurov et al.
0385 S.A. Samsoniya et al.
0388 F.N. Zeiberlikh et al.
0393 A.M. Likhosherstov et al.
0403 E.I. Kvasyuk et al.
0446 B.A. Shainyan et al.
0458 V.K. Daukshas et al.
0469 N.S. Zefirov et al.
0481 E.A. Ponomareva et al.

0520 L.A. Kurasov et al.
0533 M.V. Gorelik and E.V. Mishina
0553 D.A. Oparin et al.
0568 M.I. Shevchuk et al.
0582 E.N. Manukov et al.
0585 N.N. Magdesieva et al.
0586 V.G. Plechakov et al.
0610 V.I. P'yankova et al.
0683 E.A. Andreeshchev et al.
0689 G.V. Bazanova and A.A. Stotskii
0705 N.O. Mchedlov-Petrosyan
0742 V.S. Russkikh and G.G. Abashev
0745 S.I. Popov and V.P. Volosenko
0750 V.N. Drozd et al.
0783 V.A. Buevich and N.N. Lazareva
0787 M.A. Krivtsova et al.
0835 L.L. Vasil'eva et al.
0884 P.N. Dobronravov and V.D. Shteingarts
0899 S.V. Morozov et al.
0950 V.I. Boev and M.S. Lyubich
0953 B.I. Buzykin and N.N. Bystrykh
1092 V.R. Kokars et al.
1116 A.A. Akhrem et al.
1122 I.G. Il'ina et al.
1145 B.E. Zaitsev et al.
1151 A.F. Mishnev et al.
1155 N.I. Golovanova et al.
1172 Y.E. Gerasimenko et al.
1190 Y.M. Petrichenko and M.E. Konshin
1270 V.P. Ivshin et al.
1314 Z.A. Talaikite et al.
1336 L.L. Pushkina et al.
1363 R.D. Erlikh et al.
1367 V.S. Velezheva et al.
1380 N.G. Plekhanova et al.
1402 V.S. Fedenko et al.
1438 R.R. Kostikov et al.
1447 G.A. Tolstikov et al.
1463 V.P. Ivshin and M.S. Komelin
1490 M. Kopacz et al.
1493 I.L. Bagal et al.
1530 N.K. Genkina et al.
1533 O.P. Shvaika et al.
1561 G.A. Tolstikov et al.
1562 B.A. Trofimov et al.
1568 B.M. Berestovitskaya et al.
1592 A.A. Solov'yanov et al.
1621 F.A. Selimov et al.
1722 A.M. Nesterenko et al.
1744 D.A. Oparin et al.
1787 Y.M. Slobodin et al.
1845 G.V. Bazanova and A.A. Stotskii
1854 Y.L. Slominskii et al.
1864 I.M. Andreeva et al.
1880 V.M. Karpov et al.
1892 M.V. Gorelik and E.V. Mishina
1900 M.V. Gorelik and E.V. Mishina
1925 V.N. Drozd and A.S. Vyazgin
1931 A.I. Mel'nikov et al.
1935 T.P. Bochkareva and B.V. Passet
1937 L.M. Yagupol'skii et al.
2027 A.A. Akhrem et al.
2089 Y.L. Slominskii et al.
2094 K.L. Muravich-Aleksandr et al.
2104 B.I. Buzykin et al.
2114 L.A. Lazukina et al.
2148 O.A. Tarasova et al.
2185 T.I. Akimova and M.N. Tilichenko
2191 E.A. Ponomareva et al.
2260 M.A. Kuznetsov and A.A. Suvorov
2270 V.N. Lisitsyn and A.G. Borovkov
2273 V.L. Plakidin and V.N. Vostrova
2313 A.A. Kolomeitsev et al.
2317 I.V. Voznyi et al.

105- -83, Khim. Prirodn. Soedin., 19 (1983)(English translation)
0011 O.E. Krivoshchekova et al.
0029 B.A. Priimenko et al.
0134 D. Batsuren et al.
0139 P.F. Vlad and E.A. Vorob'eva
0141 V.A. Raldugin et al.
0149 V.A. Raldugin and V.A. Pentegova
0179 A.V. Kamernitskii et al.
0202 A.A. Ibragimov et al.
0270 O.E. Krivoshchekova et al.
0275 T.G. Sagareishvili et al.
0281 L.A. Golovina et al.
0286 L.A. Golovina et al.
0376 S. Mukhamedova et al.
0413 A.D. Vdovin et al.
0417 A. Sattikulov et al.
0422 E.N. Manukov and V.A. Chuiko
0440 K.K. Koskoev et al.
0454 M.R. Yagudaev et al.
0464 S.U. Karimova et al.
0478 R.G. Aflyatunova et al.
0498 A.A. Nabiev et al.
0500 M.D. Alaniya et al.
0506 M.R. Nurmukhamedova et al.
0532 A.A. Nabiullin et al.
0580 L.N. Kulinkovich and V.A. Timoshchuk
0610 E.B. Zorin et al.
0664 A.A. Nabiev and V.M. Malikov
0748 A. Sattikulov et al.
0763 I.D. Sham'yanov et al.

106- -83, Die Pharmazie, 38 (1983)
0072 G.R. Nagarajan et al.
0081 P. Richter et al.
0132 B. Proska et al.
0203 M. Tarasiewicz et al.
0218 F. Fülöp et al.
0373 P. Pflegel et al.
0591 J. Oehlke et al.
0829 G. Sarodnick and G. Kempter
0876 C. Vilain

107- -83, Synthetic Comm., 13 (1983)
0201 D. Poirier et al.
0691 P.R.R. Costa et al.
1013 T. Koga et al.

108- -83, Israel. J. Chem., 23 (1983)
0187 H. Falk et al.
0233 W. Kufer et al.

109- -83, Doklady Akad. Nauk S.S.S.R., Phys. Chem. Sect., 268-273 (1983)
0295 A.T. Vartanyan

110- -83, Russian J. Phys. Chem., 57 (1983)
0378 G.K. Glushonok et al.

111- -83, European J. Med. Chem., 18 (1983)
0009 M. Faulques et al.
0015 D. Averbeck et al.
0041 P. Lupon et al.
0113 L. Mosti et al.
0419 F. Gueritte et al.
0425 M. Klaus et al.
0441 G. LeBaut et al.
0447 G. LeBaut et al.
0457 G. LeBaut et al.
0481 E. Breuer et al.
0521 J.F. Ménez
0551 B. Dugué et al.
0555 P. Roveri et al.

112- -83, Spectroscopy Letters, 16 (1983)
0275 J.D. Keegan et al.
0601 F. Campadelli et al.

116- -83, Macromolecules, 16 (1983)
0291 E. Schacht et al.
0864 J.V. Crivello and J.L. Lee
1564 J. Morcellet-Sauvage et al.
1679 A.J.M. van Beijnen et al.
1817 S. Iwatsuki et al.

117- -83, Org. Preps. and Procedures Intl., 15 (1983)
0137 K.G. Boldt and G.A. Brine
0321 J.R. Merchant and N.M. Koshti

118- -83, Synthesis (1983)
0030 P.F. Alewood et al.
0050 E. Fanghaenel et al.
0214 B. Chantegrel et al.
0222 C.W. Thornber et al.
0288 T.C. McKenzie et al.
0304 D.G. Norman and C.B. Reese
0310 C.G. Shanker et al.
0374 H.-J. Cristau, G. Duc and H. Christol
0383 O.G. Kulinkovich et al.
0443 F. Ramirez et al.
0463 R. Neidlein and G, Hartz
0539 T. Kolasa
0566 S. Gelin and C. Deshayes
0577 M. Eckstein and A. Drabczynska
0582 W. Friedrichsen and M. König
0705 P.H. Hynninen and S. Lötjönen
0708 S. Lötjönen and P.H. Hynninen
0727 A. Mandal et al.
0824 F. Lucchesini and V. Bertini
0827 W. Schroth et al.
0835 S.R. Deshpande et al.
0840 A.M. Richter et al.
0842 C.K. Reddy et al.
0844 B. Chantegrel et al.
0917 M. Nakano and Y. Okamoto
0948 B. Chantegrel et al.
1000 S.D. Carter and T.W. Wallace
1010 H.H. Massoudi et al.
1018 L. Lepage and Y. Lepage
1025 F.M. Asaad and J. Becher
1037 A. Ohsawa et al.

119- -83, S. African J. Chem., 36 (1983)
0082 A.E. de Jesus et al.

120- -83, Pakistan J. Sci. Ind. Research, 26 (1983)
0007 N.A. Zaidi et al.
0364 T.U. Qazi
0367 Y. Ahmad et al.

121- -83A, J. Macromol. Sci., A19 (1983)
0017 M. Senga et al.

121- -83B, J. Macromol. Sci., A20 (1983)
0309 S. Li et al.
0433 S. Kondo et al.

124- -83, Ukrain. Khim. Zhur., 49 (1983)
0068 S.V. Shinkorenko et al.
0297 A.Y. Il'chenko et al.
0755 F.S. Babichev et al.
0857 N.N. Romanov et al.
1211 V.M. Nikitchenko et al.

125- -83, Inorg. Chem., 22 (1983)
0655 A.N. Singh et al.
1858 C.A. Marrese and C.J. Carrano
2577 D.T. Sawyer et al.
2795 I. Yoshida et al.

126- -83, Makromol. Chem., 184 (1983)
0811 T. Biela et al.
1143 H. Kammerer and A. Seyed-Mozaffari
1363 G. Casiraghi et al.
2285 O. Nuyken et al.
2361 V. Boehmer et al.

126- -83A, Makromol. Chem., Rapid Comm. 4 (1983)
0543 N.D. Ghatge and S.S. Mohite

128- -83, Croatica Chem. Acta, 56 (1983)
0125 V. Skaric and M. Jokic
0141 M. Kovacevic et al.
0157 J.J. Aaron et al.

130- -83, Bioorg. Chem., 12 (1983-4)
0045 C. Woenckhaus et al.
0299 M. Amano et al.

131- -83H, J. Mol. Structure, 98 (1983)
0277 H. Lumbroso et al.
0309 G.A. Foulds and D.A. Thornton

131- -83K, J. Mol. Structure, 102 (1983)
0001 L. Kania et al.

131- -83N, J. Mol. Structure, 105 (1983)
0055 D. Pitea and G. Moro
0239 K. Gustav and S. Vettermann
0291 D. Pitea et al.

135- -83, J. Applied Spectroscopy S.S.S.R, 38-39 (1983)

0283 D.G. Pereyaslova et al.
0303 V.Z. Kurbako et al.
0334 G.I. Validuda et al.
0658 S.G. Smirnov et al.
1185 G.N. Rodionova et al.
1304 P.M. Zamotaev

136- -83A, Carbohydrate Research, 112 (1983)
0301 J. Herscovici et al.

136- -83B, Carbohydrate Research, 113 (1983)
0001 C. Chavis et al.

136- -83C, Carbohydrate Research, 114 (1983)
0158 J.A. Galbis Perez et al.

136- -83E, Carbohydrate Research, 116 (1983)
0255 E.R. Galan et al.

136- -83G, Carbohydrate Research, 118 (1983)
0029 F. Seela and H.-D. Winkeler
0286 M.V. Fernandez et al.

136- -83J, Carbohydrate Research, 121 (1983)
0099 M. Iwakawa et al.

136- -83M, Carbohydrate Research, 124 (1983)
0023 U. Zehavi et al.
0075 S.N. Mikhailov et al.
0333 M.V. Fernandez et al.

137- -83, Finnish Chem. Letters (1983)
0104 J. Vaskuri and P.O.I. Virtanen

138- -83, Chemistry Letters (1983)
0029 K. Kurata et al.
0145 K. Satake et al.
0209 Y. Nakadaira et al.
0223 I. Kubo et al.
0299 K. Kurata et al.
0347 A. Yasuhara and K. Fuwa
0463 K. Saito
0489 Y. Yamashita et al.
0523 T. Toda et al.
0613 Y. Uchio et al.
0653 R. Neidlein et al.
0715 Y. Tohda et al.
0743 K. Satake et al.
0763 O. Tsuge et al.
0775 Y. Yano
0779 M. Suzuki et al.
0805 S. Tono-Oka et al.
0905 M. Nakatsuka et al.
0923 T. Adachi et al.
0979 I. Kubo et al.
0999 M. Ishitsuka et al.
1017 T. Sasaki et al.
1093 Y. Maki et al.
1171 Z. Paryzek et al.
1229 S. Yamaguchi et al.
1415 K. Nakashima et al.
1445 S. Takahashi et al.
1643 T. Suzuki et al.
1653 M. Yoshifuji et al.
1719 Y. Uchio et al.
1721 K. Fujimori et al.
1887 N. Morita et al.

139- -83B, P and S and Related Elements, 16 (1983)
0059 H. Nakazumi et al.
0361 U. Hildebrand et al.

140- -83, J. Anal. Chem. U.S.S.R., 38 (1983)
0091 V.Y. Veselov et al.
0094 L.K. Maslii et al.
0215 L.K. Maslii et al.
0272 E.A. Biryuk et al.
0666 D.B. Gladilovich and K.P. Stolyarov
0681 V.Y. Veselov et al.

142- -83, Heterocycles, 20 (1983)
0001 M. Lounasmaa and T. Ranta
0009 K. Imafuku and A. Shimazu
0023 W. Friedrichsen et al.
0027 S. Naruto et al.
0039 L. Camarda et al.
0061 S. Gelin and R. Dolmazon
0197 W. Friedrichsen and E. Bueldt
0213 T. Nomura et al.
0255 G. Dattalo et al.
0263 S. Chimichi et al.
0421 M. Ihara et al.
0425 J.E. Leet et al.
0451 H.A. Ammar et al.
0469 B. Sarstedt and E. Winterfeldt
0481 J.L. Hinds et al.
0489 M.R. Bryce et al.
0501 E. Belgodere et al.
0585 T. Nomura et al.
0607 H. Takahashi et al.
0611 T. Fukai et al.
0617 A. Banerji and A. Sahu
0623 A.R. Katritzky et al.
0661 T. Nomura et al.
0771 B. Achari et al.
0797 A. Ohta et al.
0813 S.-T. Lu and Y.-C. Wu
1001 M. Sako et al.
1005 M. Ikeda et al.
1017 W. Friedrichsen and E. Bueldt
1031 K. Maruyama et al.
1067 H. Yamamoto et al.
1071 Y. Hano and T. Nomura
1083 M. Balogh et al.
1117 Y. Sudoh and K. Imafuku
1247 E. Dominguez and E. Lete
1263 K. Yamane et al.
1267 T.-S. Wu et al.
1275 C. Kaneko et al.
1279 F. DeSio et al.
1315 F. Ponticelli and P. Tedeschi
1355 J. Banerji et al.
1511 A.A. Adesomoju et al.
1525 K. Matsumoto et al.
1581 C. Deshayes et al.
1709 H. Takeshita et al.
1745 S. Kohra et al.
1751 E. Buncel and S.-R. Keum
1769 Y.H. Kim and N.J. Lee

1787 L. Avila et al.
1793 S. Fukuda et al.
1801 B. Chantegrel et al.
1891 I. Bitter et al.
1895 L. Castedo et al.
1917 Y. Kurasawa and A. Takada
1927 M.U.S. Sultanbawa et al.
1959 K. Matoba and T. Yamazaki
2019 E. Belgodere et al.
2039 M. Hori et al.
2125 A. Adesomoju et al.
2173 J. Kurita et al.
2211 A.S. Shawali et al.
2339 G. Massiot et al.
2343 T.D. Lash and Y.S. Motta
2351 P. Fernandez-Resa et al.
2369 A.C. Jain et al.
2391 Y. Gelas-Mialhe et al.

145- -83, Arzneimittel. Forsch., 33 (1983)
0002 G. Zölss
0198 A. Marzo et al.
0347 H. Hamacher and E. Christ
0621 J.S. Kaltenbronn et al.

149- -83A, Photochem. Photobiol., 37 (1983)
0271 E. Gandin et al.
0587 H.-D. Brauer and R. Schmidt

149- -83B, Photochem. Photobiol., 38 (1983)
0141 C.V. Kumar et al.
0245 M. Maiti et al.
0323 J. Bakker et al.
0527 A. Acs et al.

150- -83M, J. Chem. Research (microfiche) (1983)
0301 A.R. Katritzky et al.
0358 P. LeMatais and J.M. Carpentier
0480 R.J. Napier et al.
1156 A. Arnoldi et al.
1301 S.C. Roy and U.R. Ghatak
1341 B. Abarca et al.
1556 C. Blackburn and J. Griffiths
1848 M.J. Calverly
2326 G. Paglietti et al.
2601 S. Ghosal et al.
2625 A. Zanarotti et al.
2650 R.S. Atkinson
2701 L.L.F. Gomes et al.

150- -83S, J. Chem. Research (synopses) (1983)
0028 L. Cheng et al.
0098 J. Einhorn et al.
0130 J. Borges-Del-Castillo et al.
0152 M. Sahai et al.
0168 C. Blackburn and J. Griffiths
0236 C. Bertucci et al.
0238 S. Ghosal et al.
0260 R.C. Gupta and R.C. Storr
0294 G. Tentsch and G. Costerousse
0314 M.R. Crampton et al.

151- -83A, J. Photochem., 21 (1983)
0001 R.A. Back and J.M. Parsons
0009 G.B. Fazekas and G.A. Takacs
0067 O.S. Wolfbeis et al.
0149 D. Gegion et al.
0157 R.R. Lembke et al.
0171 E.M. Kosower et al.
0245 M. Swaminathan and S.K. Dogra
0251 M. Belletete and G. Durocher
0325 W.G. Herkstroeter et al.

151- -83B, J. Photochem., 22 (1983)
0061 T. Wolff et al.
0131 N.P. Hacker and N.J. Turro
0245 J. Andersson
0255 J. Andersson
0263 J. Andersson

151- -83C, J. Photochem., 23 (1983)
0061 J. Zechner et al.
0073 I. Naito et al.
0103 B. Simard et al.
0131 T. Wolff and N. Muller
0163 A.K. Mishra and S.K. Dogra
0171 B. Savory and J.H. Turnbull
0183 L.L. Costanzo et al.
0233 W. Vonach and N. Getoff
0249 S. Monti et al.
0379 R. Schmidt et al.

152- -83, Nouveau J. Chim., 7 (1983)
0105 J.-C. Moutet and G. Reverdy
0269 M. Torres et al.
0399 J.-M. Conia and L. Blanco
0413 J.-M. Lehn et al.
0587 C. Hetru et al.
0645 A. Baceiredo et al.

156- -83B, Photobiochem. Photobiophys., 6 (1983)
0177 L.A. Guillo et al.

157- -83, Organometallics, 2 (1983)
0021 H. Yasuda et al.
0174 M. Ishikawa et al.
0332 J. Otera et al.
0453 C.W. Carlson et al.
0903 I.S. Alnaimi and W.P. Weber
1214 J. Elzinga and M. Rosenblum
1464 S. Masamune et al.
1573 A.J. Ashe, III et al.
1577 J.G. Smith et al.
1792 C.W. Carlson and R. West
1852 W.E. Buhro et al.
1859 A.J. Ashe, III et al.
1888 Y.W. Chan et al.

158- -83, J. Antibiotics, 36 (1983)
0001 S. Satoi et al.
0109 S. Omura et al.
0175 M. Hayashi et al.
0187 G. Lazar et al.
0200 Y. Kumada et al.
0242 N. Yasuda et al.
0365 R. Spagnoli et al.
0376 H.A. Kirst et al.
0407 K. Yamamoto et al.
0445 M. Hamada et al.

0448 H. Iinuma et al.
0451 K. Ishii et al.
0459 B.B. Jarvis and V.M. Vrudhula
0484 W. Keller-Schierlein et al.
0502 Y. Takiguchi et al.
0509 M. Ono et al.
0639 T. Hamamoto et al.
0661 T. Anke et al.
0688 N.N. Gerber et al.
0913 Y. Chen et al.
0916 L.A. Dolak and T.M. Castle
0921 N. Sadakane et al.
0931 N. Sadakane et al.
0943 M. Nakayama et al.
0950 T.F. Brodasky et al.
0976 K. Eckardt et al.
0980 H. Mishima et al.
1034 T. Miyadera et al.
1084 Y. Mori et al.
1275 J.W. Westley et al.
1284 Y. Ikeda et al.
1396 S. Kobaru et al.
1399 M. Nishio et al.
1425 L.A. Dolak et al.
1439 L. Toscano et al.
1473 T. Yoshioka et al.
1490 S. Rakhit et al.
1572 H. Umezawa et al.
1576 S. Ohuchi et al.
1606 K. Fukushima et al.
1651 H. Irschik et al.
1713 H. Matsubara et al.
1767 M.J. Zmijewski, Jr. and M.J. Mikolajczak
1777 T. Kihara et al.
1781 S. Omura et al.
Also spectra numbered 83-1, 83-2, etc.

159- -83, J. Carbohydrate Chem., 2 (1983)
0019 J.M.J. Tronchet et al.
0139 J.M.J. Tronchet and H. Eder

160- -83, Anal. Letters, 16 (1983)
1403 K.W. Street and V. Hocson

161- -83, Il Farmaco, 38 (1983)
0024 L. Cecchi et al.
0067 E. Castagnino et al.
0143 E.A. Coats et al.
0330 G. Pirisino et al.
0352 P. Pecorari et al.
0775 S. Cafaggi et al.
0842 G. Auzzi et al.

162- -83, The Merck Index, Tenth Edition (1983)
Page numbers of compounds are given.